WITHDRAWN
WRIGHT STATE UNIVERSITY LIBRARIES

POST-GENOMIC CARDIOLOGY

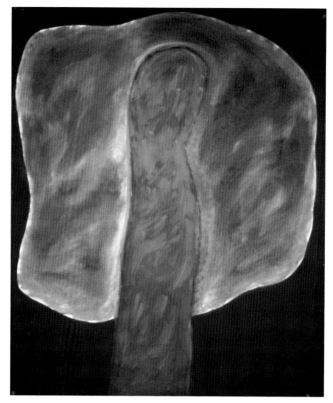

Signal & Pathway #1 *Danièle M. Marin*

Signal & Pathway #2 *Danièle M. Marin*

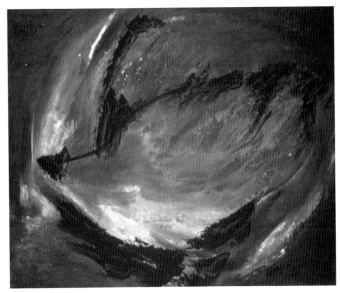

Signal & Pathway #3   *Danièle M. Marin*

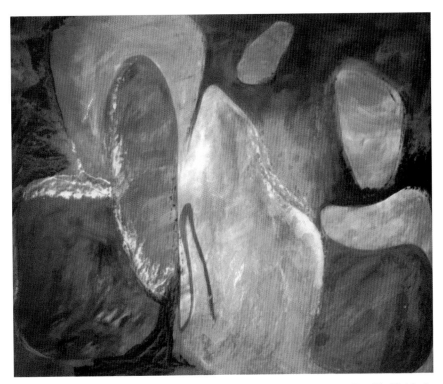

Signal & Pathway #4   *Danièle M. Marin*

The molecular phenomena of signaling pathways are and will be of particular significance to new discoveries in cardiovascular medicine. Many such pathways are described in this book; four are of particular interest to the author. These four paintings, the work of the author's wife, Danièle M. Marin, were inspired by these four signaling pathways.

Artwork #1   Signaling in cardiac growth and development.
Artwork #2   G protein-coupled receptors signaling (including β-adrenergic receptors).
Artwork #3   Signaling at the mitochondria (including energy, ROS and cell death).
Artwork #4   Calcium signaling (including calcineurin, calmodulin, SERCA, etc.).

POST-GENOMIC CARDIOLOGY

By
JOSÉ MARÍN-GARCÍA, M.D.
Director
The Molecular Cardiology and Neuromuscular Institute
Highland Park, New Jersey

With the collaboration of

MICHAEL J. GOLDENTHAL, Ph.D.
Senior Research Scientist
The Molecular Cardiology and Neuromuscular Institute
Highland Park, New Jersey

GORDON W. MOE, M.D.
Associate Professor
University of Toronto
St. Michael Hospital
Toronto, Canada

AMSTERDAM • BOSTON • HEIDELBERG • LONDON
NEW YORK • OXFORD • PARIS • SAN DIEGO
SAN FRANCISCO • SINGAPORE • SYDNEY • TOKYO
Academic Press is an imprint of Elsevier

Cover image entitled *Intercommunication* courtesy of Danièle M. Marin

Academic Press is an imprint of Elsevier
30 Corporate Drive, Suite 400, Burlington, MA 01803, USA
525 B Street, Suite 1900, San Diego, California 92101-4495, USA
84 Theobald's Road, London WC1X 8RR, UK

This book is printed on acid-free paper. ∞

Copyright © 2007, Elsevier Inc. All rights reserved.

No part of this publication may be reproduced or transmitted in any form or by any means, electronic or mechanical, including photocopy, recording, or any information storage and retrieval system, without permission in writing from the publisher.

Permissions may be sought directly from Elsevier's Science & Technology Rights Department in Oxford, UK: phone: (+44) 1865 843830, fax: (+44) 1865 853333, E-mail: permissions@elsevier.com. You may also complete your request on-line via the Elsevier homepage (http://elsevier.com), by selecting "Support & Contact" then "Copyright and Permission" and then "Obtaining Permissions."

Library of Congress Cataloging-in-Publication Data
Marín-García, José, 1936-
 Post-genomic cardiology / by José Marín-García; with the collaboration of Michael J. Goldenthal, Gordon W. Moe.
   p.; cm.
 Includes bibliographical references and index.
 ISBN-13: 978-0-12-373698-7 (alk. paper)
 ISBN-10: 0-12-373698-6 (alk. paper)
 1. Heart--Diseases--Molecular aspects. 2. Heart--Diseases--Genetic aspects.
 I. Goldenthal, Michael J. II. Moe, Gordon W. III. Title.
 [DNLM: 1. Cardiovascular Diseases--genetics. WG 120 M337p 2007]
 RC682.9.M37 2007
 616.1'2042--dc22

                        2006103545

British Library Cataloguing-in-Publication Data
A catalogue record for this book is available from the British Library.

ISBN 13: 978-0-12-373698-7
ISBN 10: 0-12-373698-6

For information on all Academic Press publications
visit our Web site at www.books.elsevier.com

Printed in the United States of America
07 08 09 10 11   9 8 7 6 5 4 3 2 1

Working together to grow
libraries in developing countries

www.elsevier.com | www.bookaid.org | www.sabre.org

ELSEVIER   BOOK AID International   Sabre Foundation

To my wife, Danièle, and daughter, Mèlanie, with love

Contents

Preface vii

Section I
Biochemical, Cellular, and Molecular Functioning of the Heart

1. Introduction to Post-Genomic Cardiology 3
2. Molecular and Biochemical Methodology in the Post-Genomic Era 11
3. Cardiovascular Gene Expression 27
4. Cellular Techniques 51
5. Cardiovascular Signaling Pathways 77

Section II
Pediatric Cardiology in the Post-Genomic Era

6. Cardiac Development: Molecular and Genetic Analysis 117
7. Congenital and Acquired Heart Disease 165

Section III
Post-Genomic Analysis of Coronary Artery Disease, Angiogenesis, and Hypertension

8. Molecular Basis of Lipoprotein Disorders, Atherogenesis, and Thrombosis 211
9. Ischemia and Myocardial Infarction: A Post-Genomic Analysis 261
10. Cellular Pathways and Molecular Events in Cardioprotection 281
11. Cardiac Neovascularization: Angiogenesis, Arteriogenesis, and Vasculogenesis 315
12. Systemic and Pulmonary Hypertension 341

Section IV
Post-Genomic Analysis of the Myocardium

13. Cardiomyopathies 363
14. Heart Response to Inflammation and Infection 415

Section V
The Failing Heart

15. Molecular Analysis of Heart Failure and Remodeling 441

Section VI
Molecular and Genetic Analysis of Metabolic Disorders

16. Fatty Acid and Glucose Metabolism in Cardiac Disease 473

Section VII
Molecular Genetics of Dysrhythmias

17. Dysrhythmias and Sudden Death 513

Section VIII
Genes, Gender, and Environment

18. Gender and Cardiovascular Disease 555

Section IX
Aging and the Cardiovascular System

19. The Aging Heart: A Post-Genomic Appraisal 579

Section X
Looking to the Future

20. Future of Post-Genomic Cardiology 619

Section XI

Glossary 639

Index 657

Preface

Heart disease is an endemic health problem of great magnitude in the world. In spite of considerable clinical and research effort during the last decade and the development of new drugs and surgical modalities of therapy, the mortality and morbidity rates remain very high. Moreover, many fundamental questions regarding the basic underlying mechanisms and pathophysiology of most cardiovascular diseases (CVDs), including congenital and acquired defects, remain unanswered. Breakthroughs in molecular genetic technology have just begun to be applied in studies of cardiovascular disease, allowing chromosomal mapping and the identification of many genes involved in both the primary etiology and also as significant risk factors in the development of these anomalies. Identification of genes responsible for rare familial forms of cardiovascular disease has proved to be informative in the study of nonsyndromic patients with cardiac pathology. Common cardiovascular anomalies (e.g., cardiomyopathy, congenital heart disease, atherosclerosis, hypertension, cardiac arrhythmias) seem to be united by association with distinct subsets of genes. These include genes responsible for subcellular structures (e.g., sarcomere, cytoskeleton, channels), metabolic regulatory enzymes (e.g., renin-angiotensin system, cholesterol metabolic pathway), or intracellular signaling pathways (e.g., calcineurin, CaMK, TNFα).

At present, the following areas of research appear particularly promising:

(1) With the completion of the Human Genome Project the likely identification of novel genes involved in non-syndromic cardiac disease, cardiac organogenesis, and vascular development will serve as an important foundation for our understanding of how specific gene defects generate their cardiovascular phenotypes. Bioinformatic methods can be employed to search existing databases with the routine use of reverse genetics techniques, allowing subsequent cloning of novel genes/cDNAs of interest, followed by the characterization of spatial-temporal patterns of specific gene expression. Moreover, post-genomic analysis including both transcriptome and proteomic methodologies can be used to further delineate the functions of the gene products, defining their precise role in pathogenesis, elucidating their interaction with other proteins in the subcellular pathways, and potentially enabling their application as clinical markers of specific CVDs.

(2) The mechanisms governing the early specification of cardiac chambers in the developing heart tube have not yet been precisely delineated but are thought to involve novel cell-to-cell signaling among migrating cells, as well as the triggering of chamber-specific gene expression programs, mediated by specific transcription factors and growth factors such as bone morphogenetic protein (BMP). Future areas of study will focus on elucidating the role of signaling molecules (e.g., WNT) using conditional gene knock-outs (in a variety of genetic backgrounds) and accessing their interaction with critical transcription factors such as dHAND, NKX2.5, GATA4, and TBX. Similar approaches may also prove informative in probing the origins of the cardiac conduction system, and in deciphering the role of signaling systems as participants in vascular formation in endothelial cells, focusing on the interaction of VEGF, angiopoietin, TGF, and the Notch pathway.

(3) Another critical area of research is the identification of molecular regulators that control cardiomyocyte proliferation. Cardiomyocytes are mitotically active during embryogenesis and generally cease proliferation shortly after birth. Understanding the molecular basis of cardiomyocyte proliferation could greatly impact on our clinical attempts to repair the damaged heart. Mechanism of cell growth regulation is being investigated by careful comparison of comprehensive gene expression profiles of embryonic and post-natal myocytes, as well as by the generation of myocyte cell culture lines with the capacity to respond to proliferative inducers.

(4) Cellular transplantation is an alternative mechanism with which to augment myocyte number in diseased or ischemia damaged hearts. New research efforts will be necessary to further define the optimal conditions necessary for cardiomyocyte differentiation and proliferation and for the fully functional integration of stem cells in the myocardium, as well as to investigate the ability of transplanted stem cells to repair defects in the young and adult heart. It will be critical to learn whether cardiac failure secondary to myocardial ischemia or dilated cardiomyopathy, myocardial infarct in adults, or children with Kawasaki disease and myocardial damage or with ARVD, can be treated with stem-cell transplantation.

With the beginning and eventually rapid progress in cell engineering, we expect to see the end of many of these cardiac abnormalities that weaken human life and bankrupt the health care system. Moreover, insight into the cardiovascular consequences of abnormal gene function and expression should ultimately impact the development of targeted therapeutic strategies and disease management and may replace less effective treatment modalities directed solely at rectifying structural cardiac defects and temporal improvement of function.

As the role of genetic screening in cardiology is strengthened and as research on the multiple signaling pathways involved in cardiac organogenesis and pathology progresses, the time seems appropriate for a book that comprehensively integrates known facts, current developments, and future knowledge. In addition to providing a recount of past discoveries, this book deals with areas that are of emerging interest to cardiologists and researchers in diverse fields, with specific attention paid to pediatric, aging, and gender-based cardiovascular medicine, eyeing new therapeutic modalities that may improve currently available therapies and interventions in the management of human cardiac diseases. Furthermore, we are witnessing the transition from the cardiology of the past to the study of systems biology, the constructive cycle of computational model building, and experimental verification capable to provide the input for exciting new discoveries and hope.

"Tomorrow is here and...there, keep looking up at it."

José Marín-García
Highland Park, 2006

SECTION I

Biochemical, Cellular, and Molecular Functioning of the Heart

CHAPTER 1

Introduction to Post-Genomic Cardiology

OVERVIEW

At the beginning of the 21st century, the application of molecular cellular biology and genetics to clinical cardiology has become increasingly compelling. The great strides that have been made in our understanding of the gene and its expression are now being applied to our improved understanding of normal cardiovascular development, physiology, and aging, as well as to abnormalities that occur during physiological insults and cardiac disease.

The information derived from high-resolution studies of the human genome together with the development of animal models of cardiac disease, including transgenic studies, and from increased bioengineering of the cell (e.g., stem cells) has begun to be used in earnest in the diagnosis and treatment of cardiovascular disease. The complete delineation of the human genome and its approximately 75,000 gene products clearly represents a first and important step in unraveling the complexity of the cardiomyocyte and of cardiovascular disease. Moreover, the next stage of analysis, including the understanding of the gene and its interactions within its cellular environment (termed a post-genomic approach), is now underway. In this book, we will attempt to describe where we have been, where we are, and provide a vision of where we are going on this journey.

The basic concepts and principles of molecular biology and genetics focusing on the gene, its environment within the cell, and the regulation of its expression will be presented in this chapter. It will include a brief review and recapitulation to provide the reader with enough background and terminology to readily understand specific techniques and their applications in clinical cardiology. The information in this introduction includes background material available in a variety of different textbooks dedicated to molecular and cell biology, several of which are cited in the reference section.[1-4] An effort has been made to present the material in an accessible way to readers who may not have had recent coursework in these subjects, reinforcing in this way the basic concepts of molecular and cell biology and highlighting the terminology that is used throughout the text.

GENE STRUCTURE, TRANSCRIPTION, AND TRANSLATION

The *gene* is the fundamental unit of inherited information and can be defined as a segment of DNA involved in producing a polypeptide chain (protein) by virtue of its synthesis of RNA by transcription. Only a small number of the DNA sequences carried on the human chromosomes are organized as genes, whereas other sequences act as regulatory elements; however, no function has yet been found for most of the chromosomal DNA.

The DNA of the chromosomes consists primarily of two long strands that wind around each other in the form of a double *helix*. Each strand is composed of a chain of nucleotides (each containing a phosphate group, a base, and a sugar called deoxyribose); these chains are millions of nucleotides long. The sugar and phosphate groups of each nucleotide and the covalent phosphodiester bond that links them are said to form the backbone of each strand and lie on the outside of the helix; they carry a negative charge from the phosphate groups. The hydrophobic bases are arranged on the inside of the helix structure.

Each strand has a polarity, a 5′ and a 3′ end; the two strands are aligned in an antiparallel fashion so that they have opposite polarities (Fig. 1). There are only four types of nucleotide—each with its own distinct base: adenine, guanine, cytosine, and thymine. The sequence of the covalently joined nucleotides in each DNA strand constitutes the DNA *primary structure*. In double-stranded DNA, the bases in one DNA strand are arranged in such a way as to interact with the bases in the other strand by the formation of hydrogen bonds between complementary bases. These essentially form the rungs on a ladder structure. Adenine (A) always pairs with thymine (T), forming two hydrogen bonds, whereas guanine (G) base pairs with cytosine (C), forming a slightly stronger three-hydrogen bond. The double-stranded structure of DNA, which is stabilized by weak hydrogen bonds and by hydrophobic base-stacking, constitutes the DNA *secondary structure*; this allows the structure to be easily opened during either DNA replication or transcription so that polymerizing enzymes can have access to read the sequence of each strand's DNA as a template for making more DNA or RNA.

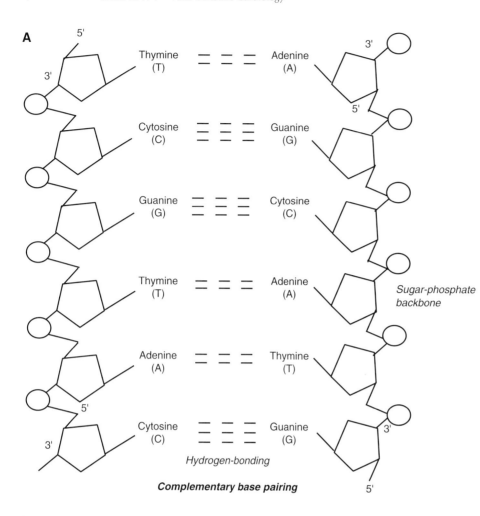

FIGURE 1 Double-strand structure of DNA. Depicted is the complementary base pairing of two strands of DNA located at the center of the double helix with the sugar-phosphate backbone of the two chains oriented on the outside of the helix. Also shown is the antiparallel nature of the strands with respect to their 5′ to 3′ orientation. The structure is largely held together by the hydrogen bonding between guanine (G) and cytosine (C), which involves three bonds and the double hydrogen bonds between adenine (A) and thymine (T) as shown.

In DNA replication, the sequence of each parental strand determines the complementary nucleotide sequence on the nascent strand and, therefore, defines its new partner. The invariant base-pairing rule ensures that an exact complement will be present on the daughter strand during replication and forms the basis for both RNA transcription and DNA repair. RNA produced from the DNA is also a linear polynucleotide molecule containing four kinds of bases, three the same as DNA (A, C, and G), whereas the fourth is uracil (U) instead of thymine; it is generally single stranded and has a different sugar (ribose) in its backbone.

The *central dogma* of molecular biology deals with the detailed residue-by-residue transfer of sequential information, and this information cannot be transferred from protein to either protein or nucleic acid. The flow goes from DNA to RNA to protein. In Fig. 2, we show the flow of genetic information going from DNA to mRNA (*transcription*); the mRNA copy is sent out of the nucleus to specialized structures in the cytoplasm (i.e., *ribosomes*), where the mRNA is used to form proteins from amino acids joined together by peptide bonds (*translation*). In this way, the types of protein a cell makes, which largely confer the structural and functional characteristics specific to each cell type, are genetically controlled. The flow of information from DNA to protein can be reversed from mRNA to DNA only in certain viruses (e.g., retroviruses) that contain specialized enzymes called *reverse transcriptase* and in retrotransposons but never from protein to mRNA. Another unique exception to the normal pathway of information is found in prions (proteins replicating themselves).

The first step in gene expression is transcription, a process that, like DNA replication, involves DNA as a template and takes place accordingly in the nucleus. Transcription is mediated by *RNA polymerase*, which in conjunction with specific transcription factors binds with high affinity to specific sequence elements (*promoter* regions) on the DNA usually located in front (5′) or upstream of genes (Fig. 3A). After binding, the RNA polymerase enzyme initiates, elongates, and terminates the synthesis of a discrete mRNA chain complementary to a portion of one of the two strands (i.e., the *antisense* strand) (Fig. 3B). The RNA made is essentially an exact copy of the untranscribed "sense" strand (except for the substitution of a U for every T). The growing RNA chain is

DNA replication
Nuclear localized, requires a DNA template, oligonucleotide primer, deoxyribonucleotides, DNA polymerase and additional factors

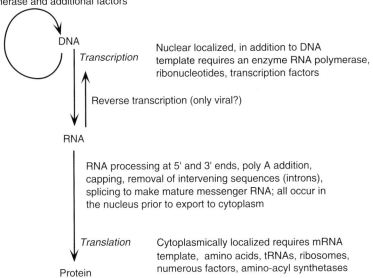

FIGURE 2 Flow of genetic information.

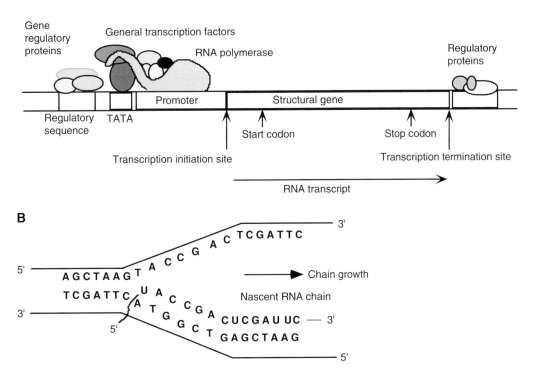

FIGURE 3 Transcription involves both regulatory proteins and DNA sequence. Elements in the formation of new RNA. (A) Involvement of RNA polymerase and transcription factors in binding both TATA and the promoter region upstream of the gene to be transcribed. Other regulatory elements, which bind regulatory proteins and modulate the levels of transcription, are located both further upstream, as well as downstream of the gene coding region. Also shown are the relative positions of the transcription initiation and termination sites and the start and stop codons. (B) The separation of the DNA strands required for transcription to occur and the relative orientation of the growing nascent RNA strand (5' to 3') relative to its DNA template strand.

polymerized in the 5' to 3' direction from the *antisense* strand template (3' to 5'). Discrete regulatory signals are present on the DNA molecule called *cis-acting* elements, which are recognized by proteins (e.g., transcription factors) termed *trans-acting* factors that can either enhance or inhibit the initiation or termination of the RNA transcript. Genes contain multiple *cis-acting* elements responsive to a variety of intracellular and extracellular physiological and pathophysiological stimuli. The interaction between *trans*-acting factors and *cis*-elements constitutes the primary basis for the regulation of transcription that can govern gene activity levels during cell differentiation, development, and in response to physiological and pathophysiological stimuli and is discussed throughout much of this book. After the transcription of a gene is complete, the RNA is separated from the DNA template.

Most nuclear genes also contain intervening sequences, which are noncoding (*introns*), varying in number and size, often larger than the coding region (*exons*); these sequences are not found in the mature mRNA to be translated into proteins. However, introns are initially transcribed in their entirety in a short-lived *primary nuclear transcript* along with the contiguous sequences that are found in the mature message (*exons*) (Fig. 4). Before the transcript is exported from the nucleus, the intron portions of the primary transcripts are rapidly excised, and the exon RNAs are spliced together to form the mature transcript, making the coding region continuous for translation. Although apparently it would seem wasteful to transcribe large stretches of DNA into RNA only to excise major fragments, what is gained is versatility in structure and in regulation that arises from the variable splicing patterns that can produce different mRNAs, depending on the cell type or developmental stage, allowing the production of multiple distinct proteins (*isoforms*) from the same gene. Before the RNA is exported from the nucleus for translation, other processing occurs at both the 5' end ("capping" with 7-methyl guanosine) and at the 3' end, a trimming accompanied by the addition of several adenine bases generating a poly A tail. These modifications to the mRNA increase translation efficiency and contribute to message stability and may play a role in its transport from nucleus to cytoplasm. Modulation of transcript stability and or turnover can be an effective mechanism influencing the control of gene expression.

In the translation process, each consecutive three nucleotides or triplet on the mRNA forms a *codon* read by the ribosome to specify a specific amino acid, which is then inserted into a growing polypeptide chain. This process uses an adaptor transfer RNA (*tRNA*) molecule that carries both a

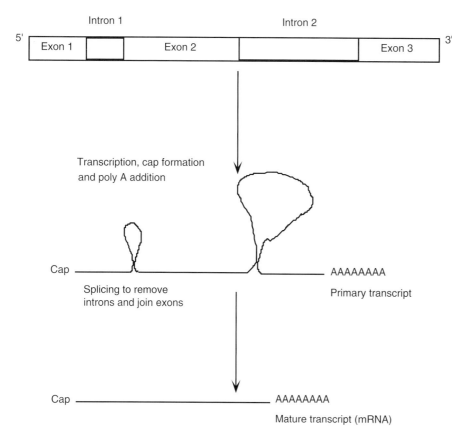

FIGURE 4 Processing of the RNA transcript. The structure of most eucaryotic genes involves the presence of both exons and introns (intervening sequences) whose number and size vary greatly in different genes. The example shown here is with the β-globin gene. After transcription in the nucleus to form a large precursor primary transcript, the RNA is capped at its 5' terminus and polyadenylated at a site near its 3' terminus. A series of RNA splicing events follows to excise the intron sequences and rejoin the exons to form the mature transcript; these processing events are required before the export of the RNA from the nucleus to the cellular location where it will eventually be translated on ribosomes.

specific amino acid (e.g., tRNA^{GLY} carries glycine etc.) and an anticodon region that can recognize specific codons on the mRNA. There are at least 20 types of tRNA molecules; the covalent linkage of each adjoining peptide releases the amino acid from the tRNA, allowing the "free" uncharged tRNA to bind more amino acid for further translation. The codon AUG usually is the initiating codon, which encodes methionine. The first amino acid has a free NH$_2$ group and defines the *amino or N-terminus*, whereas the growing end of the peptide chain has a free COOH group defining the *carboxyl or C-terminus*.

The nearly universal *genetic code*, which is found throughout the entire plant and animal kingdoms as well as in bacteria, specifies which amino acid corresponds to which of the 64 possible 3-nucleotide codons. The code shows degeneracy (i.e., 61 of the possible 64 triplet *codons* represent 20 amino acids), with almost every amino acid represented by several codons. Three codons (UAA, UGA, and UAG) do not represent amino acids and play a role in terminating protein synthesis and are thereby termed *stop codons*. The initiation site on the mRNA determines the *reading frame* of the codons (uninterrupted coding region) and, therefore, determines the amino-acid sequence of the synthesized proteins. Most mature mRNA contains a single reading frame to generate a specific protein. The deletion or insertion of a single base can cause a shift in the reading frame (*frameshift*); mutations with such an insertion/deletion can lead to the formation of a protein with altered sequence (beyond the modified site) and size (the termination sites will likely be different) and can lead to dysfunctional phenotypes and hereditary disorders.

The process of translation can also be subject to regulation. Several key initiating and elongating factors play pivotal roles in the regulation of protein synthesis. Proteins translated on the ribosomes can undergo a variety of cotranslational and post-translational modifications, which can affect their ultimate placement within the cell, as well as their function. For example, specific signal sequences (usually 15–60 amino acids long) at either the N-terminus or C-terminus can target a protein for import into the nucleus or the mitochondria. These signal sequences may or may not be cleaved from the protein on entry to the appropriate cell compartment. Proteins may also undergo modifications such as glycosylation, proteolytic cleavage, and phosphorylation, which can affect their functional activity, stability, or translocation between compartments in the cell.

THE NUCLEAR ENVIRONMENT

Within the nucleus of the animal cells the *chromosomes* reside, each a collection of genes linearly arrayed on a long DNA molecule surrounded by a protein structure or scaffolding structure called *chromatin*. The chromatin is composed primarily of proteins including five basic (positively charged) *histones* (H1, H2A, H2B, H3, and H4) and numerous non-histone proteins. Under specific conditions, the histones wrapped around the nuclear DNA can be arranged in a regularly repeating unit called a *nucleosome*. Modifications to the charged residues of the histone proteins (e.g., by phosphorylation, acetylation, or methylation of the lysine residues) can strikingly alter the assembly or *condensation of chromatin*. This remodeling of chromatin can modulate the global expression of genes, largely by impeding the progression of their transcription into RNA. Recent studies have also identified the presence of small RNA species in chromatin that play a critical role in RNA processing and transport from the nucleus. In contrast, prokaryotes (e.g., bacteria) have neither a separate nuclear compartment nor chromatin and frequently have non-chromosomal DNA molecules called *plasmids*.

CELL CYCLE

The genes on each chromosome are said to be *linked* and are generally inherited as a unit; recombination can occur as a fundamental function of crossing-over during *meiosis*, with increased recombination found between more distally located genes. The *cell cycle* constitutes the period between the release of a cell as one of the progeny of a division and its own subsequent division by *mitosis* into two daughter cells. It is composed of two major phases, an interphase during which there is little visible change, although the cell is active synthetically and bioenergetically, and the *mitotic* stage (M) during which cell division actually occurs (Fig. 5A).

During M, when the chromosomes are segregated before division, the chromatin becomes highly condensed and visible as chromosomes because of this condensation (they take up stain more easily). The chromatin structure enables the efficient packaging and distribution of the nuclear genetic material, which is segregated for distribution within the cell to the new daughter cells at *mitosis*. During the *metaphase* stage of mitosis, a *karyotype*, essentially a snapshot of the entire chromosome complement of the cell, can be made, because the entire complement of chromosomes are highly visible as an ordered array lined up. During *interphase*, the chromatin is reorganized into areas that are either more condensed (and stainable), called *heterochromatin* associated with less active genes, and less condensed regions termed *euchromatin* in which the DNA is more actively expressed (Fig. 5B).

Within the cell cycle interphase, nuclear DNA replication occurs during a discrete period, S phase. In nuclear DNA replication and regulation, all sequences are replicated in a tightly regulated process. Both strands are completely copied—albeit in different regions (or replication units)—on the same chromosome and replicate at distinct times. Highly condensed chromatin replicates late in S phase, whereas genes in active chromatin replicate early. Replication requires a DNA polymerase acting in concert with various enzymes, which are present to unwind the strands for initiation and elongation

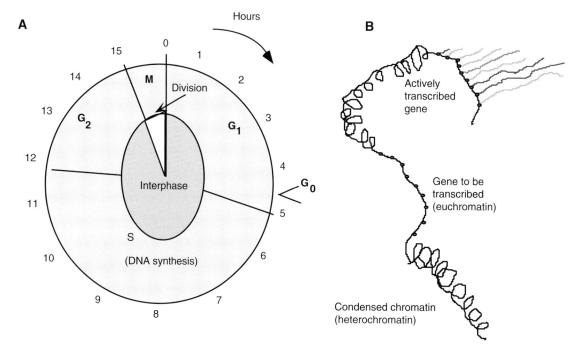

FIGURE 5 Eucaryotic cell cycle. (A) The relative timing of interphase (composed of S, G_1, and G_2 phases and the mitotic (M) events of division. Many quiescent or non-dividing cells such as adult cardiomyocytes are withdrawn from the cell cycle and remain at G_0. (B) Less condensed chromatin with more actively transcribing genes.

to occur and ligate the newly replicated fragments together (e.g., helicases and DNA ligases, respectively). The accurate duplication of DNA is profoundly important, and the polymerizing enzymes such as DNA polymerase also possess proofreading functions. Moreover, the nucleus contains a dedicated group of enzymes to repair DNA damage as it occurs either during replication or post-replication. As we will discuss later, DNA damage is frequently associated with cellular oxidative injury and aging and can be induced by a variety of physical and chemical agents or mutagens such as ultraviolet light, gamma irradiation, and a host of chemicals.

The cell cycle is a central feature of proliferative or dividing cells. A number of critical proteins (e.g., cyclins) have been identified that regulate the progression of events at several discrete stages (or *checkpoints*) of the cycle. Terminally differentiated cells (e.g., cardiomyocytes), which are said to be post-mitotic generally, do not enter mitosis, remain in a perpetual interphase, and are, therefore, said to be withdrawn from the cell cycle. As we will see, this topic with respect to cardiomyocyte replication remains a highly contentious, but important, question to which we will return.

TELOMERES AND mtDNA

At the end of the chromosomes, specialized cap structures called *telomeres* are present. They contain tandem repeats of a short G-rich sequence (GGGTTA in humans) bound to an array of specialized proteins, which constitute this heterochromatin-associated cap structure. To replicate the ends of the linear chromosome during cell division, an enzyme called *telomerase* recognizes the G-rich strand and elongates it using an RNA template that is a component of the enzyme itself. Both the length of telomere repeats and the integrity of the telomere-binding proteins is also important for telomere protection, preventing chromosome ends from being detected as damaged DNA. In cultured cells lacking telomerase, the shortening of telomeres results in both senescence-associated gene expression and inhibition of cell replication. Age-dependent telomere shortening in most somatic cells, including vascular endothelial cells, smooth muscle cells, and cardiomyocytes, is associated with impaired cellular function and viability of the aged organism. In later chapters, we shall review the evidence that telomere dysfunction is implicated in the aging heart and is also a contributory factor in the cardiovascular dysfunction associated with heart failure and hypertension.

In addition to the nuclear DNA organized within the chromosomes (23) present in *haploid* human cells (e.g., germline cells such as sperm and egg) and 46 in human *diploid* cells (e.g., somatic cells), DNA is also present in mitochondria (*mtDNA*). The mtDNA (Fig. 6) encodes a small set of proteins (13 subunits) involved in the mitochondrial respiratory chain and oxidative phosphorylation, as well as components of mitochondrial-specific ribosomes (i.e., 2 rRNAs and 22 tRNA molecules) for the translation of the mtDNA-encoded proteins. Interestingly, its genetic code is slightly different than the nuclear code. Most proteins (estimated at more than 1000 kinds of protein) that define mitochondrial structure and function are encoded by nuclear DNA, translated on cytosolic ribosomes, and imported into the mitochondrial organelle.

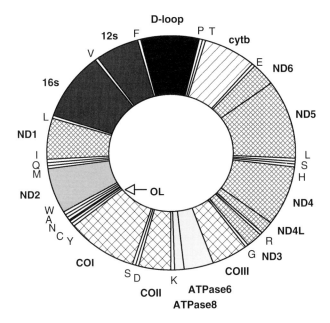

FIGURE 6 Human mitochondrial double-stranded circular DNA. This circular DNA encodes 13 protein components of four of the five enzyme complexes involved in electron transport and OXPHOS, two ribosomal RNAs (12S and 16S), and 22 tRNAs, as shown. The noncoding D-loop region is also shown.

PHENOTYPE/GENOTYPE

Visible or observable properties that define the appearance of a trait or organism constitute what is termed the *phenotype*, whereas the genetic factors, which are largely responsible for creating the phenotype, are called the *genotype*. A gene may exist in alternate forms that determine the expression of particular phenotypes termed *alleles*. Diploid organisms, which carry two identical alleles for a specific gene, are said to be *homozygous* with respect to that gene, whereas the possession of two different alleles is termed *heterozygous*. In some cases, one allele is said to be *dominant* when it determines the phenotype irrespective of the presence of the other allele (*recessive*). Alleles are said to be *codominant* when they contribute equally to the phenotype. *Penetrance* refers to the proportion of individuals with a specific genotype that expresses the related phenotype. A highly *penetrant genotype* is more amenable to mapping by linkage analysis.

MUTATIONS

Differences in alleles are primarily due to *mutation*, a change in the sequence of the gene. Mutations can be deleterious, advantageous, or neutral with respect to the organism. Variations in nucleotide sequence that are frequently present in the population (>1%) are designated *polymorphic alleles*, which can either be harmful or neutral. Mutations can be either *point mutations* involving a single nucleotide change or multiple nucleotide changes or *rearrangements* in which larger areas of sequence are modified either by *insertion, deletion,* or *transposition* of sequences. *Point mutations* can be nucleotide substitutions, single insertions, or deletions. A substitution can cause a change in the codon, resulting in a different amino acid replacement in the protein, or it may cause no change; an insertion or deletion can cause a change in the reading frame, resulting in changed amino acids downstream of the mutation site. Changes in protein caused by substituted amino acid residues are referred to as *missense* mutations; the severity of the change in regard to phenotype depends on the kind of amino acid replaced (e.g., a charged amino acid for a neutral one). A mutation can also change a codon from one specifying an amino-acid residue to one that is a stop codon; this causes a premature termination of the polypeptide chain and is termed a *nonsense* mutation. The effect on the mutated protein depends on its location within the chain; if it is located near the N-terminus, the protein will be drastically shorter with attendant effects on its function and, depending on the role of the protein within the cell, the cellular phenotype. Nonsense mutations residing near the C-terminus of the protein may have limited effects on function and phenotype.

Some traits or phenotypes are determined by a single nuclear gene (*monogenic*). A specific defect in single genes can cause a number of cardiovascular disorders, including familial forms of hypertension, hypertrophic and dilated cardiomyopathy, long QT syndrome, structural anomalies of the heart and large vessels, and atherosclerosis because of defects in lipoprotein metabolism.

Many traits are caused by more complex genotypes involving several genes (*polygenic*). Cardiovascular diseases such as hypertrophic cardiomyopathy (HCM) and hypertension have been shown to have complex genotypes encompassing several genetic loci. In addition, the expression of specific phenotypes can be variable because of other influences including environment. Moreover, a single mutant allele can often have pleiotropic consequences affecting more than just the protein that it encodes. In addition to mutations in protein-encoding genes (i.e., structural gene mutations), mutations can also be found in nonprotein genes such as those involved in coding for the small RNAs (e.g., tRNAs) or larger RNA (e.g., ribosomal RNAs) that can effect protein synthesis or in the noncoding regulatory areas of structural genes such as regulatory mutations, which can impact the levels of specific gene expression.

The *inheritance pattern* can be highly informative as to the chromosomal location of the gene causing dysfunction (Fig. 7). For instance, an *X-linked inheritance* pattern (indicating the presence of the gene of interest on the X chromosome) can be readily distinguished from an *autosomal* inheritance pattern, in which the gene is present on any of the non-sex chromosomes. Either of these inheritance patterns follows the Mendelian rules of transmission. In contrast, a *maternal pattern of inheritance* in which the paternal contribution is almost nil is also readily distinguishable and is a possible indicator of an mtDNA location, because mammalian mitochondria are derived mostly from the mother.

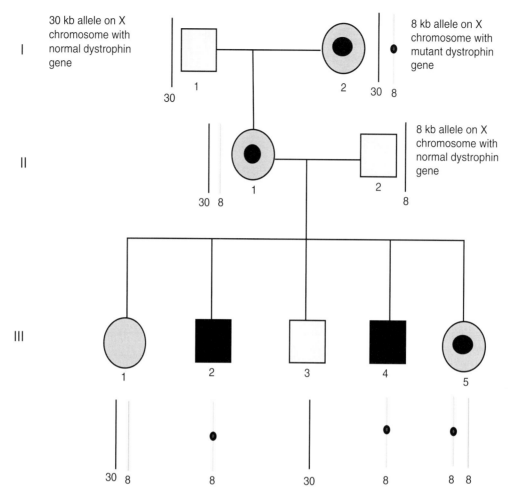

FIGURE 7 Inheritance pattern of an X-linked disease (Duchenne muscular disease or DMD), with a molecular mutation in the X chromosome indicated by a restriction fragment length polymorphism (RFLP). Southern blotting was used to detect a polymorphism of 30 kb and 8 kb fragments; in this family, the mutation causing DMD was located in 8 kb fragment.

The genetic background in which deleterious mutations occur can significantly modulate their phenotypic expression. For instance, a large number of recent studies have reported the presence of modifier genes in the genetic background that influence the phenotypic expression and severity of pathogenic HCM genes. The identification of modifier genes, which will markedly improve the elucidation of genetic risk factors, has been advanced by large-scale genome-wide approaches to detect polymorphic variants correlated with disease severity.

A variety of molecular techniques are currently available for the detection of single nucleotide polymorphisms (*SNPs*). SNP association studies have identified several candidate modifier genes for various cardiac disorders, and a number of specific DNA polymorphisms have been reported in association with myocardial infarction, coronary artery disease, and HCM. With the increased cataloging of SNPs, either alone or within a larger chromosomal region (haplotypes) in available shared databases, these modifier loci can be evaluated for their effects in predisposing to specific cardiac defects and may have an impact on the choice of diagnostic and treatment options (e.g., pharmacogenomics). Furthermore, the intensive effort that is underway to identify SNPs in the human genome is also noteworthy, because these SNPs can be used as specific gene-mapping markers. It is hoped that this genetic information will prove to be clinically useful in the development of highly individualized cardiovascular medicine.

References

1. Alberts, B., Johnson, A., Lewis, J., Raff, M., Roberts, K., and Walter, P. (2002). "Molecular Biology of the Cell." 4th ed. Garland Publishing, New York.
2. Lodish, H., Berk, A., Zipursky, S. L., Matsudaira, P., Baltimore, D., and Darnell, J. E. (1999). "Molecular Cell Biology." 4th ed. W. H. Freeman & Co., New York.
3. Watson, J. D., Baker, T. A., Bell, S. B., Gann, A., Levine, M., and Losick, R. (2004). "Molecular Biology of the Gene." 5th ed. Benjamin Cummings, San Francisco.
4. Lewin B. (2004). "Genes VIII." Pearson Prentice Hall, Upper Saddle River, NJ.

CHAPTER 2

Molecular and Biochemical Methodology in the Post-Genomic Era

OVERVIEW

An overview of standard molecular and biochemical methods, which have been used in the investigation of cardiovascular gene analysis and expression studies, will be presented in this chapter. It will also include a review of techniques, which are currently in development, as well as prospective methods, which are under consideration or have not yet been well characterized.

INTRODUCTION TO STANDARD METHODS OF DNA ANALYSIS

Most DNA sequencing today uses the Sanger technique with DNA polymerase and specific DNA chain terminators called dideoxyribonucleotides.[1] The sources of the DNA analyzed are primarily either fragments produced by recombinant cloning techniques or from polymerase chain reaction (PCR) amplification.

Once the nucleotide sequence is established, the reading frames, coding, and noncoding regions can be recognized. An amino acid sequence of an encoded protein can be determined, and the primary sequence of the gene of interest can be readily compared by computer analysis with other proteins. This may reveal areas involved in either protein secondary structure (e.g., α-helixes or β-pleated sheets); interacting sites, which interface with other proteins, lipids, or DNA; areas of protein involved in their particular cellular translocation (e.g., organelle-specific signal sequences, transmembrane domains); or function (e.g., protein kinase, cell receptor). The search for patterns between proteins is termed *motif homology searching* and can prove highly informative about the structural and functional properties of the encoded protein.

Use of Recombinant DNA

Genetic engineering techniques have been developed taking advantage of the universality of DNA sequences and the ability to shuffle the elements involved in the regulation of gene expression. We can ligate mammalian DNA of interest to a prokaryotic DNA vector, thereby generating hybrid recombinant DNA molecules, which can be introduced into a host cell for replication. By appropriate selection the recombinant DNA can be obtained in large quantities as highly purified DNAs from the bacterial colonies. *Plasmid* vectors are small, double-stranded circular DNA molecules with a bacterial replication origin capable of producing high levels of replication (hundreds of copies can be made per cell) and convenient restriction sites. Other types of vectors include bacterial viruses (e.g., *bacteriophages*) of either single- or double-stranded DNA. This technology has been made possible largely by the discovery in bacteria of *restriction endonuclease enzymes* that cleave DNA at specific sites of defined nucleotide sequences (e.g., most commonly 4 or 6 bp sequences).[2] These restriction enzymes of different specificities are commercially available and have been used to direct the cutting of DNA from any source into discrete fragments that can subsequently be isolated and recombined *in vitro*. The specific nucleotide sites recognized by these restriction enzymes tend to be short, symmetrical sequences called palindromes that are repeated on both DNA strands albeit in opposite orientation. For instance, the *Eco*R1 enzyme recognizes 5′ GAATTC 3′, which is the same sequence on the other strand (Fig. 1). More than 150 different cleavage sites have thus far been identified that are the specific targets of more than 200 restriction enzymes; some sites are targeted by more than one enzyme, and these enzymes with related specificity are called *isoschizomers*. The fragments generated when DNA is cut by restriction enzymes can be separated, thru their size, by their migration during agarose gel electrophoresis. A unique *restriction map* for any DNA molecule can be constructed from the data generated by its digestion with restriction endonucleases and agarose gel electrophoretic analysis. Furthermore, these enzymes can either make a blunt cut at or near the recognition site on the DNA or can make a staggered cut across the two strands of DNA, resulting in one of the two molecules containing a short single-stranded unique overhang. These ends are termed *sticky ends*.

As shown in Fig. 1, *Eco*R1, which cuts between the G and A, leaves an AATT at the 5′ end of each strand that it cuts. The overhangs are capable of complementary base pairing,

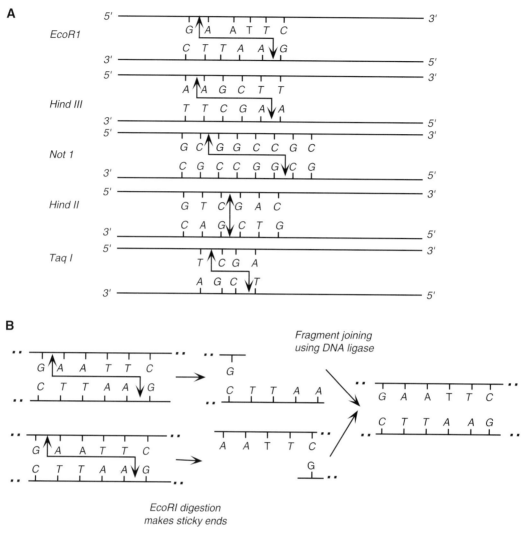

FIGURE 1 Restriction endonuclease digestion. (A) Restriction sites of five commonly used restriction endonucleases. Three of these enzymes (*Hind*II, *Eco*R1, and *Hind*III) recognize 6 base-pair sequences; one targets an 8 base pair sequence (*Not*1) and one a 4 base-pair sequence (*Taq*1). The digestion with four of the enzymes results in an uneven cut, leaving a 5′ overhang or sticky end (*Eco*R1, *Hind*III, *Taq*1, and *Not*1), whereas the digestion with the *Hind*II enzyme is symmetrical, leaving blunt ends. (B) Formation of sticky ended fragments with *Eco*R1 digestion, and the rejoining of *Eco*R1 fragments (regardless of their source) by use of a DNA ligase leading to recombinant DNA.

with the overhangs resulting from cleavage, thereby allowing two different DNA fragments cut with that same enzyme (the source of the DNA does not matter) to be joined together as depicted. The paired or cohesive ends of the two molecules can be covalently ligated by a DNA ligase *in vitro* making a stable recombinant molecule. Conversely, ends with a different sequence generated by other restriction enzymes cannot be easily joined without alternative engineering. In Fig. 1, the recognition site (GTCGAC) of the *Hind*II enzyme is shown. This enzyme cuts the DNA in the center of the restriction site between the C and G leaving *blunt ends*; blunt-ended fragments can also be covalently joined by the action of DNA ligase. Moreover, restriction sites can be added to the DNA of interest by the attachment of linker sequences (commercially available), which are short sequences of synthetic double-stranded DNA containing the sequence of a desired restriction site.

After the inserted DNA of interest is covalently ligated *in vitro* to the vector molecule (e.g., a plasmid or viral DNA), which contains a *replication origin* (Ori) recognizable by the host replication apparatus, these hybrid recombinant DNA molecules can be easily inserted, either by transformation or transfection, into bacteria where they will be replicated *in vivo* (cloned), generating large quantities of the DNA of interest (Fig. 2).[3] Various sequences and genes on the vectors can affect their usefulness. *Antibiotic-resistance* genes on the vector, usually found on plasmid vectors, can allow for antibiotic selection of only those bacterial host cells that contain the plasmid construct. As shown in Table I, a variety of *selectable markers*, operative in different host cells,

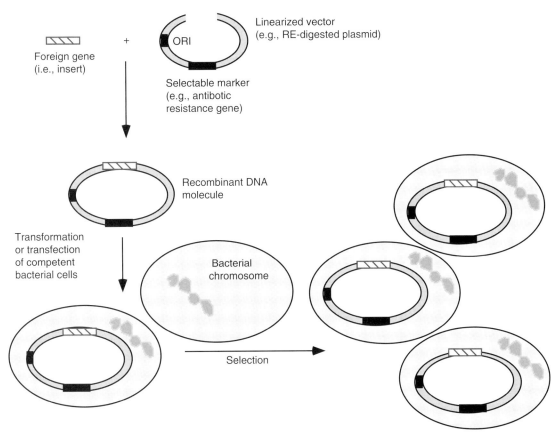

FIGURE 2 Plasmids can incorporate gene fragments and be transferred into bacterial cells. Plasmid vectors can be linearized (by restriction enzyme digestion) and ligated (using DNA ligase) to inserts of DNA from any source (i.e., heterologous or foreign DNA), which contains the appropriate sticky ends (also possible but less efficient using blunt ends). The resultant recombinant DNA molecules can be used to transform or transfect competent bacterial cells (usually *E. coli*). Selection of cells that contain the recombinant molecule/plasmid can be accomplished if the initial plasmid vector contained an antibiotic-selectable gene marker mediating resistance to ampicillin, tetracycline, or chloramphenicol and with subsequent growth of the cells on media containing the appropriate antibiotic.

can be introduced on plasmid vectors, including resistance to drugs targeting cell protein functions such as synthesis (e.g., tetracycline and chloramphenicol), DNA damage (e.g.,

TABLE I Markers of Choice in Different Hosts

Gene	Organism	Selective agent
URA3	Yeast	Prototrophy
LEU2	Yeast	Prototrophy
TRP1	Yeast	Prototrophy
AMP-R	Bacteria	Ampicillin
TET-R	Bacteria	Tetracycline
CAP-R	Bacteria	Chloramphenicol
NEO-R	Yeast/mammalian	Neomycin/G418
DHF-R	Mammalian	Methotrexate/MTX
PURO-R	Mammalian	Puromycin
Hygro-R	Mammalian	Hygromycin
Zeo-R	Mammalian	Zeocin
GFP-Zeo-R	Mammalian	Zeocin

MTX), or supplements to alleviate metabolic defects (e.g., LEU, HIS). *Shuttle vectors,* which can be operative for introduction and replication in several host organisms (e.g., bacteria/yeast, bacteria/mammalian cells), can be designed containing the necessary combinations of selectable markers and applicable replication origins. A hybrid Zeo-R/green fluorescent protein gene has been created and used as a selectable marker on shuttle vectors for the identification and selection of genes introduced in mammalian, insect, and prokaryotic cells.[4] In some cases, *episomal* plasmids, whose extrachromosomal location ensures that their replication and expression are independent from chromosomal DNA, can be transformed into integrating vectors that can be incorporated into the host nuclear genome.[5–7]

The insert capacity of the different vectors available varies markedly as shown in Table II. For instance, the fragment inserted into a plasmid is of a limited size, because plasmids beyond 15 kb (10-kb inserts) tend to become unstable. Phage vectors can incorporate larger DNA inserts than plasmids.[8] A cosmid vector that combines elements of both the plasmid

TABLE II Comparison of Different Recombinant Vectors

Vectors	Size insert	Properties	Limitations
Plasmid	0.1–10 kb	Small circular molecules; can be replicated by transfer into bacteria or yeast; extrachromosomal.	Limited size insert capacity; lower transformation efficiency.
Bacteriophage	8–25 kb	Linear molecule; high transformation efficiency with far more (1000×) clones produced per microgram of DNA. Similarly, far more plaques than *E. coli* colonies can be plated and detected on a single culture plate, simplifying storage and screening.	Phage cloning is more complex than plasmid cloning.
Cosmid	35–50 kb	Extrachromosomal circular molecules with properties of both phage and plasmid; high transformation efficiency.	Requires *in vitro* packaging into phage.
Yeast artificial chromosome (YAC)	100–10,000 kb	Vector with chromosomal elements behaves as chromosomes in yeast.	Very low transformation efficiency; not very stable.
Bacterial artificial chromosome (BAC)	75–300 kb	Similar to plasmid vectors, except they contain the origin and genes encoding the ORI binding proteins required for plasmid replication.	Very low transformation efficiency; electroporation necessary.

and the phage vectors allows the cloning in bacteria of much larger DNA inserts up to 45 kb.[9] Even larger fragments of DNA can be cloned on large episomal vectors called artificial chromosomes in bacteria (BAC),[10] in yeast (YACs), and, most recently, in mammalian cells (MACs),[11] allowing the cloning of very large genomic fragments of DNA of up to several hundreds of kilobases.[12] For instance, YACs have been designed to contain centromeric sequences and telomeric sequences, allowing them to effectively segregate as chromosomes. By examining overlapping fragments between clones, one can essentially "walk" down the chromosome (*chromosome walking*) using a DNA sequence that lies near a gene of interest as a probe to find clones containing that gene and for the analysis of regulatory regions that are either 5′ or 3′ to the gene.[13]

A similar process is involved in the technique called *positional cloning* in which genetic linkage information is used to isolate and clone genes implicated in human disease for which little information (other than their generalized chromosomal location) is available. A marker gene known to be closely genetically linked to the human disease locus of interest is used as a starting point or probe with which to "walk" and isolate the gene of interest. Using positional cloning, genes involved in the pathogenesis of cystic fibrosis,[14] as well as in a number of congenital heart defects,[15–19] have been identified and cloned.

The sources of DNA for cloning can include the chromosomal DNA itself (also called *genomic DNA*) or a DNA copy of an mRNA termed complementary *DNA* or *cDNA*.[8] A genomic DNA can contain noncoding regions of a gene, including the promoter and introns, allowing the analysis of regulatory gene elements. A cDNA can more easily render the coding section of a gene.

cDNA Expression

If the cDNA is full length (most gene-coding sequences can be contained within cloning fragment size), the cDNA can be introduced into an expression vector, and its ability to produce a functional recombinant protein can be assessed. These expression vectors generally feature high-level bacterial promoter sequences and contain important transcription signals to ensure efficient protein production. Some of these vectors contain *regulatable promoters* (e.g., inducible) enabling the switching on or off of foreign, or heterologous, protein production during growth of the culture. This may be important, because foreign proteins can be toxic in a different setting, such as the bacterial cell, particularly with high levels of production (which can exceed 1–10% of the cell protein) or with membrane proteins. Some vectors also include sections of coding sequence from a bacterial protein, which will be fused to the cloned protein of interest. These *fusion proteins* frequently have improved stability and solubility in bacteria compared with entirely heterologous proteins. They can also provide an "*affinity tag*" for subsequent protein identification and isolation. Other features of the fusion protein can include the attachment to a protein of interest of a specific epitope, which can be recognized by an available antiserum to enhance the immunogenicity of the protein or the addition of a signal sequence to redirect the fusion protein to an organelle of interest (e.g., mitochondria, nucleus). Another application of fusion protein technology is *phage display*, a selection technique involving fusing proteins with a bacteriophage coat protein resulting in the display of the fused protein on the exterior surface of the phage, while the DNA encoding the fusion protein resides within. This method permits the selection of proteins (and their physically attached DNAs) for specific binding characteristics (e.g., antibody, DNA, or specific ligand binding) or functional features (e.g., enzymatic assay) by an *in vitro* screening process termed *biopanning*. A relevant application of fusion protein technology to cardiovascular biology used the engineering of highly effective plasminogen activator chimeras with altered pharmacokinetic and functional features for thrombolytic therapy.[20]

High-level *production of heterologous proteins* is now no longer limited to bacteria but possible in yeast, insect, and mammalian cells. The relative advantages and disadvantages in using these different expression systems for the heterologous production of specific gene products are shown in Table III. These hosts have the advantage of containing the necessary machinery for post-translational modification that some proteins require for functioning (e.g., glycosylation), which are not available in bacteria. These techniques can be used to produce large quantities of proteins for further basic studies or for therapeutic applications.

The use of recombinant DNA technology to produce therapeutic proteins in a large-scale manner for treating human cardiovascular disease (CVD) has greatly evolved over the past decade. Many of these recombinant products are currently in their second generation. In addition to generating recombinant proteins with enhanced potency compared with their natural counterparts, the use of recombinant DNA technology, in combination with advanced hybridoma methods, has also spawned antibody engineering resulting in the creation of fusion and chimeric monoclonal antibodies (mABs), some of which have been applied to the treatment of CVD. A representative list of some of the more widely used recombinant proteins, mostly of human origin, and mABs used in clinical cardiovascular medicine is provided in Table IV.

Gene Libraries

Chromosomal DNA digested with restriction endonucleases will yield DNA fragments that after ligation into a vector can be introduced and cloned in bacteria. A collection of cloned fragments propagated in bacteria (from digestion of chromosomal DNA) is called a *genomic library*; it should contain representatives of every sequence in the chromosomal DNA, regardless of function.

In contrast, the production of cDNA involves the isolation of mRNA and the generation of cDNA molecules using the reverse transcriptase enzyme. After ligation of these fragments into an appropriate vector, and introduction into bacteria, the collection of cloned DNA fragments (each present in a single bacteria) is called a *cDNA library* and should be representative of cDNA molecules expressed from every different mRNA. Unlike the genomic library, which should generally be the same for any tissue from which chromosomal DNA is isolated, the cDNA library will be very different, depending on the tissue and developmental expression of the genes.

Screening of these fragments from a specific genomic or cDNA clone can be done using a highly specific probe. A labeled nucleic acid probe (which can specifically hybridize to the gene of interest) can be used to screen among the bacterial colonies for either the genomic or cDNA clone (*colony hybridization*).[21] This probe can either be an isolated DNA fragment from the gene of interest or a synthetic oligonucleotide probe. A sequence deduced from a partial amino-acid sequence of the purified protein can be used in the design of the oligonucleotide. Alternately, in the case of expressed cDNAs, a screen for the produced protein can be used with either specific antibodies to that protein, if available, or an assay for that protein function.

TABLE III Comparison of Expression Systems

Expression system	Advantages	Disadvantages
Bacteria	Short generation; high yield; low cost.	Limited post-translational modification (e.g., glycosylation); foreign proteins are frequently in insoluble form.
Yeast	Most eucaryotic post-translational modifications; low cost; moderate generation time.	Frequent overglycosylation; foreign proteins are often difficult to extract (because of periplasmic entrapment and extensive protease degradation); episomal vectors are unstable in large-scale culture.
Baculovirus (in cultured insect cells)	High yield; some post-translational modification; correct protein folding.	More expensive; long generation; shorter production period; some proteins cannot be properly modified.
Mammalian cells	All post-translational modifications.	Long generation; high cost; lower transfection efficiency causes reduced expression levels.
Transgenic animals	All post-translational modifications; high yield.	Long-term process to develop transgenic animal.

TABLE IV Therapeutic recombinant proteins for cardiovascular disease

Recombinant protein	Description	Targeted condition
Humulin	Insulin	Diabetes
Insulin-like growth factor (rhIGF-1)	IGF-1	CHF
Somatropin (rhGH)	Human growth hormone	Pediatric growth deficiency; DCM
Nesiritide (rhBNP)	B-type natriuretic peptide	CHF
Alteplase	Second-generation recombinant tissue plasminogen-activator (t-PA)	AMI
Reteplase	Mutant t-PA with longer half-life and fibrinolytic potency compared with t-PA	AMI
Tenecteplase (TNK-rt-PA)	Mutant t-PA with enhanced fibrinolytic potency	AMI
Etanercept (Enbrel)	Fusion protein of extracellular ligand domain of TNF receptor and the Fc portion of human IgG	Rheumatoid arthritis; HF
Recombinant mABS		
Abciximab	Human-mouse chimeric antibody; binds to glycoprotein receptors (GP IIb/IIIa) of human platelets.	Inhibits platelet aggregation; used to treat unstable angina and as adjunct to coronary angioplasty.
Daclizumab	Humanized mAB; binds to α-subunit of IL-2 receptor on the surface of activated lymphocytes.	Immunosuppressive; in cardiac transplant, reduced the rate of acute rejection.
Infliximab	Human-mouse chimeric antibody; binds to human TNF-α.	Recent use in Kawasaki disease.

CHF, Congestive heart failure; DCM, dilated cardiomyopathy; HF, heart failure; AMI, acute myocardial infarction.

GENE IDENTIFICATION: EXPRESSED SEQUENCE TAGS

In a major contribution to the human genome project, cDNA libraries were extensively used for gene identification by the expressed sequence tag (EST) approach initiated by Venter and associates.[22,23] Partial gene sequences are selected at random from libraries representing the genes in specific cell types, tissues, or organs; the sequence is compared with an existing database to determine whether it represents a previously known or a novel gene. Sequence information has considerably grown from just examining the expressed sequences in the human genome allowing the definition of gene families, the search for novel genes and their proteins (involved in genetic disease), and in providing information regarding the physical mapping and the expression patterns of these genes in different tissues and disease states. This technique has made possible the evaluation of differentially expressed cardiac genes in normal and diseased hearts (e.g., hypertrophy).[24–27] In combination with data derived from genetic linkage, association mapping, and positional cloning, EST data has allowed the identification of monogenic disorders whose etiology involves defects in single genes. Specific monogenic defects leading to cardiovascular disorders such as familial hypertrophic cardiomyopathy (HCM), dilated cardiomyopathy (DCM), long QT syndrome (LQTS), and Marfan syndrome have been identified (Table V).

On the other hand, CVD such as atherosclerosis, hypertension, and HF involve two or more genes (polygenic) in association with environmental or physiological modulators. In combination with mapping and EST data, new technologies will be necessary to assess these complex diseases.

GENOMICS

The separation and formation of the double-stranded helix underlies the basis of several powerful DNA technologies to characterize the structure and function of genes. Double-stranded DNA can be denatured into single-stranded components by extremes of pH or by heat (i.e., near boiling conditions). The renaturation or reannealing of the complementary strands to form a native double helix can be attained at temperatures close to 65°C. Then the strands are exactly complementary and available in sufficient amount. Nearly complementary strands can be annealed only at lower temperature and less stringent conditions. Renaturation also can occur between DNA and RNA chains containing complementary sequences. These findings underlie the powerful techniques of Southern blotting (e.g., DNA–DNA hybrid-

TABLE V Genes involved in HCM, DCM, LQTS, and Marfan syndrome

Condition	Genes	Associated protein
HCM	*MYH7*	Cardiac β-myosin heavy chain (*β-MHC*)
	MYL3	β-Myosin light chain
	MYBPC3	Cardiac myosin binding protein 3
	TNNT2	Cardiac troponin T
	TPM1	α-Tropomyosin
DCM	*DMD*	Dystrophin
	TAZ / G4.5	Tafazzin
	ACTC	Actin
	DES	Desmin
	TTN	Titin
LQTS	*SCN5A*	Cardiac sodium channel
	HERG	Potassium channel
	KCNQ1	α-Subunit potassium channel
	(KVLQT1)	β-Subunit potassium channel
	KCNE1 (Mink)	
Marfan syndrome	*FBN*	Fibrillin

ization) Northern blotting (e.g., DNA–RNA hybridization), PCR amplification, and the use of renaturation kinetics (e.g., COT curves) to assess the genetic relatedness of different species and the copy number of DNA.

Southern Blots

A widely used technique for the characterization of cloned DNA is the hybridization of restriction enzyme digested fragments with specific probes.[28] Hybridization usually follows electrophoretic size-dependent separation of DNA fragments on agarose gels, denaturation *in situ*, and transfer of the fragments to a membrane filter (Fig. 3). The filter is hybridized with either radiolabeled or nonradioactive labeled probe, which similar to the colony hybridization probe could be a full-length cDNA, a DNA gene fragment, or a synthetic oligonucleotide. The resulting data can be used to provide information about the quantity, size, and location of specific gene sequences and the mapping of gene fragments and flanking regulatory elements. Furthermore, *Southern blot analysis* can be used in the evaluation of gene copy number, the detection of major gene rearrangements, and screening of deletions associated with CVDs.

Hybridization and Gene Structure

The use of DNA hybridization has also permitted a significant glimpse of the complexity of the human genome, because it has allowed the identification of *pseudogenes*, nonfunctional genes that are likely relics of evolution with strong homology to functional gene, and the characterization of sequences as either *single copy, moderately,* or *highly repetitive* according to their kinetics of hybridization. Most protein coding genes are single-copy sequences, although a few cases exist in which several copies of a structural gene have been detected. Although most genes are also relatively well interspersed throughout the chromosomes, gene clusters containing several related genes have been found, including the globin gene family. Although tandem clustering of genes has been well documented for multicopy rRNA genes (non-protein encoding), this type of gene cluster is rare for protein-coding genes; an exception is the histone genes for which extensive repetition may reflect the need of the cell to produce large amounts. In contrast, the genome is replete with families of repeated sequences (i.e., repetitive elements that have been estimated to compose nearly half of the human genome). Human repeats in tandem array include *minisatellite sequences* (ranging from 15–100 bp) that can be present in 20–50 copies; because the number of these repeats (thought to be nontranscribed) is variable in different individuals, they have been applied forensically as the basis for DNA fingerprinting strategies. Another family of repeated sequence present in the human genome is single interspersed elements (SINES), which are not repeated in tandem, often transcribed, and present in thousands to millions of copies per genome.[29]

PCR Amplification

A segment of DNA sequence can be amplified entirely in a test tube without intermediaries of the bacterial cell. Two short synthetic single-stranded DNA fragments or oligonucleotide

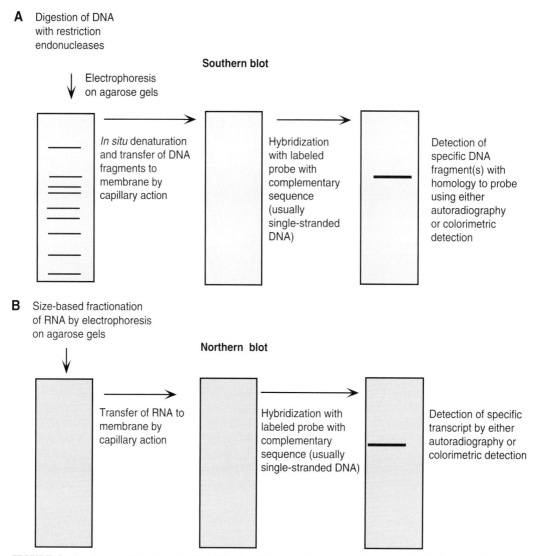

FIGURE 3 Southern and Northern blot hybridization. The technology used to both quantitatively and qualitatively identify specific DNA fragments (Southern blots) and specific RNA transcripts (Northern blots), using hybridization with a complementary labeled DNA probe is shown in (A) and (B), respectively, and is further described in the text.

primers define the ends of the fragment to be synthesized.[30] On the annealing of the primers to a DNA template, DNA polymerase elongates the target DNA; on successive cycles of annealing, elongation, and denaturation (up to 40 cycles), a logarithmic doubling of the fragment number proceeds (Fig. 4). In a few hours, this process can generate millions of copies of a specific DNA fragments. A heat-stable polymerase from thermophilic bacteria (*Taq polymerase*) has proved extremely useful (allowing the enzymatic amplification even after the high temperatures involved in denaturation).[31] Other techniques have emerged that allow the fragment size to be increased up to 20–30 kb. The amount of DNA needed as a template for PCR is extremely small, and PCR can even be used to amplify sequences from a single molecule. The starting material does not have to be purified, and the specificity of the primers (normally approximately 20 nucleotides), if homologous to the template sequence, is adequate for amplification. A greater degree in the fidelity of amplification can be attained by the use of other thermally stable DNA polymerases (e.g., Vent and Pfu) with *proofreading* ability. The amplified PCR fragments can subsequently be treated to many downstream reactions including nucleotide sequence analysis, restriction digestion, cloning, used as hybridization probes, and templates for *in vitro* transcription and *in vitro* mutagenesis.

Reverse transcription of RNA to DNA with the enzyme *reverse transcriptase* can be combined with traditional PCR to allow the amplification and determination of the abundance of specific RNA. After reverse transcription, a second strand of DNA is synthesized through the use of a deoxyoligonucleotide primer and DNA polymerase. Both DNA strands are then exponentially amplified by PCR. *Real-time PCR*, also called quantitative PCR, is a method of simultaneous DNA amplification and quanti-

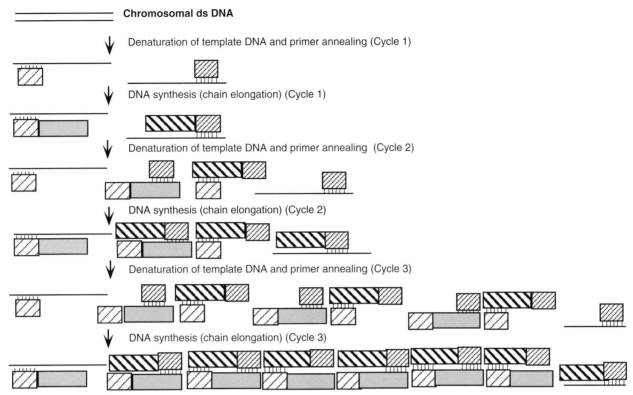

FIGURE 4 The polymerase chain reaction (PCR) leads to amplification of a targeted sequence. The amount of discretely sized targeted DNA is doubled in each cycle of DNA synthesis (normally 30–35 cycles are used). Three steps constitute each cycle, including denaturation (strand separation), annealing of specific oligonucleotide primers (essentially a hybridization reaction), and synthetic elongation of the DNA strand. Each step involves a specific temperature (which can be optimized if characteristics of the DNA sequence are known). In the example shown, three cycles of reaction produce 16 DNA chains.

fication often using fluorescent dyes that intercalate with double-strand DNA and modified DNA oligonucleotide probes that fluoresce when hybridized with a complementary DNA.

DNA–RNA Hybridization: Studying Gene Expression

Northern blotting similar to Southern blotting involves DNA hybridization but focuses on evaluating RNA rather than DNA. RNA separated by gel electrophoresis into discrete sizes is hybridized with a labeled probe. Specific gene expression at the mRNA level can be either quantitatively or qualitatively evaluated from RNA samples prepared from either cells or tissues. This technique also provides information concerning the size and cellular distribution of related mRNA transcripts. Other related techniques to evaluate specific mRNA levels include *nuclease protection* and *primer extension* assays. These techniques can be useful in the differentiation between closely related (and/or similarly sized) transcripts and permit the identification and the sequencing of the 5′ end of the transcript, allowing a fine mapping of the transcriptional start site (which may not be provided from DNA sequence analysis alone).

In situ hybridization using DNA probes can be used to localize specific transcripts within the cell, in conjunction with microscopy. Finally, several techniques, which will be described more comprehensively in Chapter 3, have been developed for the assessment of many transcripts at one time. These *gene-profiling* techniques have been used to identify the abundance of transcripts that are significantly increased or decreased at particular stages of early cardiac development, aging, and in CVDs (e.g., myocardial ischemia, myocardial infarct, heart failure). These methods include subtractive hybridization, differential display, and more recently developed, molecular tagging techniques such as serial analysis of gene expression (SAGE) and microarray (DNA chip) analysis.

Further Analysis of Gene Expression

Recombinant DNA methods have been used to cause directed mutagenesis of cloned gene sequences and reintroduction of the modified genes to the cell (Fig. 5). The assignment of a regulatory function to specific DNA sequences has been convincingly demonstrated on their reintroduction into cultured cardiomyocytes or into animal models. As we shall see in Chapter 3, the use of gene transfer to cardiomyocyte can

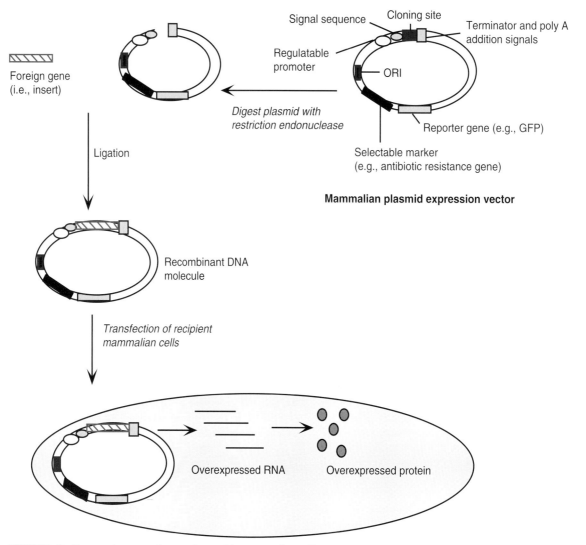

FIGURE 5 Transgenic expression can be used in mammalian cells. Mammalian expression vectors as shown here (with a double-stranded DNA plasmid vector) can contain sequences for selection and replication in *E. coli* (e.g., ORI and antibiotic resistance) and for optimal cell-specific gene expression (e.g., gene regulatory sequences such as promoters and terminators), detection (e.g., GFP), and specific organelle-targeting (e.g., signal sequences), resulting in a overexpression of a specific RNA and protein in an novel environment (e.g., cardiomyocyte).

provide significant information about cardiac gene expression and cell–cell interactions, although it cannot fully recapitulate the events that occur within the myocardium during cardiac development. In these settings, the ability of the modified gene to be transcribed can be determined by assessing specific mRNA levels of that gene. Some of the signals show a defined cell-type or tissue-specific pattern of expression, whereas others are more constitutively expressed. For instance, the promoter elements associated with the myosin light chain 2-gene drive highly specific expression in cardiac

TABLE VI Commonly Used Reporter Genes

Reporter genes	Detection
Lac Z (β-galactosidase)	Activity in lysate; ISH; immunocytochemical
Luciferase	Fluorescence (*in situ* or with lysate); ISH; immunocytochemical
CAT (chloramphenicol acetyltransferase)	Activity in lysate; ISH; immunocytochemical
GFP (green fluorescent protein)	Fluorescence (*in situ* or with lysate); ISH; immunocytochemical

ISH, *In situ* hybridization.

muscle and cardiomyocytes, whereas the creatine kinase promoter drives expression in both skeletal myoblasts and cardiomyocytes.

Another widely used approach has been to assess the activity of a *reporter gene* fused downstream of the modified regulatory element. As shown in Table VI, reporter genes have included chloramphenicol acetyltransferase, β-galactosidase, and more recently luciferase and green fluorescent protein (GFP), proteins that are easily and sensitively measured using either immunocytochemical, colorimetric, or fluorescent methods of detection.

These techniques have allowed the identification of common consensus sequences found near the transcription start site of many genes including the TATA box, the binding site for RNA polymerase, and various transcription factors and other regulatory sequence motifs within the promoter region of the gene. The position and orientation of these elements relative to the gene are highly significant in their efficacy in gene regulation. Also, these techniques have permitted the identification of another class of regulatory DNA element, the *enhancer*, which can dramatically upregulate gene expression, is often located outside the promoter, and is more randomly positioned (either upstream or downstream) relative to the coding sequence and orientation independent.

Understanding that these promoter-associated regulatory sites primarily exert their impact on gene expression as a function of their occupancy by various proteins and cofactors provides a useful strategy for the isolation and characterization of such factors/proteins using *in vitro* DNA-binding methods. The *electrophoretic mobility shift assay*, also called gel shift, uses differential mobility during gel electrophoresis to gauge whether a specific DNA fragment (containing the regulatory motif of interest) is bound by an extract of nuclear proteins. The specific protein–DNA interaction is detected by retardation in the mobility of the DNA fragment, which is successfully bound. This technique can be combined with antibodies to specific transcription factors to assess whether specific "retarded" DNA–protein complexes contain specific factors.

The precise nucleotides with which the regulatory protein(s) interact on the DNA can be determined with a technique called *DNA footprinting*. The full-scale isolation of transcription factors can use specific binding to the DNA sequence using *affinity chromatography* or *magnetic beads* to which the DNA is attached.

The nucleotide sequence of either promoter or coding region can be altered *in vitro* either in a limited fashion at one or a few nucleotides (site-directed mutagenesis) or by the systematic addition or deletion of a block of sequence. Site-directed mutagenesis can be effectively performed using synthetic oligonucleotides.

Mutation Detection

Several methods of mutation detection are currently in use, including single-strand conformation polymorphism (SSCP) analysis, allele-specific oligonucleotides (ASO), restriction fragment length polymorphism (RFLP) analysis, mismatch detection/cleavage, and denaturing gradient gel electrophoresis (DGGE). SSCP is based on the differences in secondary structure of single-strand DNA molecules differing in a single nucleotide and gauged by an alteration of their electrophoretic mobility in nondenaturing gels. This technique has been broadly used in the screening and detection of mutations in patients with cardiac dysrhythmias, including Brugada syndrome and LQTS, as well as DCM and HCM.[32–37]

Single point mutations have been also detected by RFLP. This technique has been used in the identification of missense mutations in familial HCM and their role in the variability of its phenotypic expression.[38] Another approach for mutation detection involves hybridization with radioactively labeled allele-specific oligonucleotides (ASO). This technique is based on the differences in the melting temperature of short DNA fragments differing by a single nucleotide, and it has led to the identification of novel mutations in cardiac troponin and myosin genes involved in familial HCM.[39–41] RFLP and ASO can be used to identify either pathogenic mutations or polymorphic nucleotide variants.

Another technique used in mutations detection is *RNase A mismatch cleavage*, which is based on the ability of ribonuclease to recognize and cleave single base mismatches in RNA:RNA and RNA:DNA heteroduplexes. The technique is performed by hybridization of the target sequence to a labeled, complementary riboprobe, RNase A digestion, and analysis of the digestion resistant products by polyacrylamide gel electrophoresis (PAGE) in denaturing gels. Mutations are detected and localized by the presence and size of the RNA fragments generated by cleavage at the mismatches.

Finally, denaturing gradient gel electrophoresis (DGGE) is a highly sensitive technique based on the differential melting of double-stranded DNA molecules in a gradient with increasing concentration of denaturant (e.g., urea and formamide). The denaturation of the double-stranded DNA in this system results in marked change in conformation and electrophoretic mobility, allowing the resolution of DNA fragments differing by as little as a single nucleotide. The addition of an extra sequence of very high stability to the target sequence of PCR (using one primer that has 40 nucleotides of an artificial GC-rich sequence [GC clamp]) extending at its 5′ end) provides DGGE with the capacity to detect nearly all possible mutations in any given sequences. DGGE detected mutations in the connexin 43 gene, which is involved in gap junctions in children with hypoplastic left heart syndrome.[42]

STANDARD METHODS OF PROTEIN ANALYSIS

The role of specific protein defects in cardiac disease has received wide attention.[43,44] Standard methods of protein analysis have been applied to humans, as well as to animal models, of cardiovascular disease; these methods include the following:

One-Dimensional Gel Electrophoresis (1-DE)

One-dimensional gel electrophoresis, using PAGE, is applied to separate proteins in one dimension, either by charge (isoelectric focusing) or by size (SDS); the latter is often combined with a subsequent transfer of the proteins to a membrane and probing with a specific antibody. This technique is known as *Western immunoblot* analysis.

Two-Dimensional Electrophoresis (2-DE)

Two-dimensional electrophoresis is a powerful technique that combines both of these forms of protein analysis providing a broad survey or snapshot of proteins in specific cells or tissues, either in a developmental or in a physiological stage. Two-dimensional electrophoresis begins with 1-D electrophoresis but then separates the analytes by a second property in a direction 90 degrees from the first. These analytes are spread out across a 2-D surface, rather than along a line. Analytes are more effectively separated in 2-D electrophoresis than in 1-D electrophoresis, because it is less likely that two analytes will be the same in two than in one property. After protein separation, further identification of specific proteins can be achieved by peptide sequencing, peptide mapping, and by mass spectroscopy. The development of broadly available *databases* has greatly assisted in the sharing and interpretation of the enormous volume of data emerging from proteomic studies, as well as in the generation of reference and data standards, to ensure data reproducibility and validity. These databases include a large range of relevant information about the source of the proteins, including organism, tissue, disease/physiological state, age, gender, genetic variation (transgenic or knockout), and cell type. Protein microarrays can similarly furnish a global portrait of the proteins within a tissue under a specific set of conditions (e.g., development, aging).[45,46]

An increasing area of interest in regard to protein function includes the screening of protein–protein interactions. Software is available for the assessment of protein homology to identify highly conserved motifs involved in specific function. Software can also be used in determining the presence of protein secondary structures, including α-helix and β-pleated sheets indicative of patterns of protein folding, as well as the presence of transmembrane domains common to integral membrane proteins, such as receptors and transporters. A similar kind of analysis has been used to study specific regions of proteins involved in cellular localization and import by specific organelles and in targeting regions that are subject to post-translational modification.

Yeast 2-Hybrid

This technique uses cloned human proteins, which are used as a "bait" to fish out interacting proteins. It is effective in the identification of both novel proteins and their encoding genes and provides a fresh perspective to the function of other previously characterized proteins, as well as on the roles they play in specific pathways.[47]

DEVELOPING TECHNOLOGIES

The development of highly sensitive and rapid techniques for high-throughput screening of molecular genotypes will be useful in understanding the genetic components of CVD, as well as in identifying polymorphisms. These techniques, currently in development, are being used for mutation detection.

Molecular Beacons (MBs)

This technique uses highly specific fluorescent DNA probes termed MBs, usually in combination with real-time PCR. It can discriminate between alleles with a single-base mutation. MBs are hairpin-forming oligonucleotides labeled at one end with a quencher and at the other end with a fluorescent reporter dye. In the absence of a target, the fluorescence is quenched while in the presence of a target, and subsequent beacon/target hybridization, the hairpin structure opens resulting in the restoration of fluorescence.[48,49]

Temperature Gradient Capillary Electrophoresis (TGCE)

Temperature gradient capillary electrophoresis is a highly sensitive technique that couples heteroduplex analysis to capillary electrophoresis. It can be used to efficiently scan an entire coding region to identify a wide spectrum of mutations.[50] TGCE was useful in the identification of novel mutations in the fibrillin gene that causes Marfan syndrome[51] and in the detection of the factor V Leiden mutation (G1691A), a clinically important polymorphism that increases the risk of thrombosis.[52]

Other Techniques

HDOA is a technique that has been used in combination with sequencing to construct a MITOCHIP. Maitra and associates[53] have reported that in matched fluid samples (urine and pancreatic juice, respectively) obtained from five patients with bladder cancer and four with pancreatic cancer, the MitoChip detected at least one cancer-associated mitochondrial mutation in six (66%) of nine samples. Other techniques whose development in cardiovascular genotyping is underway include *oligonucleotide-ligation assay*,[54] the use of *peptide-conjugated nucleic acids* (PNA),[55–58] and mass spectroscopy.[59,60] In addition to the mutation detection approaches discussed previously, the preceding techniques only represent a small portion of the numerous molecular techniques that are currently being developed to screen SNPs. For example, genetic engineering (for gene replacement and therapy) encompasses a group of techniques currently under development, and they will be further discussed in Chapter 3.

CONCLUSION

As highlighted in this book, the application of new and exciting technology to cardiovascular medicine has just begun, and in this century it will provide highly effective tools for diagnosis and therapy. Furthermore, these advances will facilitate understanding of the molecular, genetic, and cellular mechanisms underlying normal physiology, cardiovascular development, and pathology.

References

1. Sanger, F., Nicklen, S., and Coulson, A. R. (1977). DNA sequencing with chain-terminating inhibitors. *Proc. Natl. Acad. Sci. USA* **74,** 5463–5467.
2. Roberts, R. J. (1976). Restriction endonucleases. *CRC Crit. Rev. Biochem.* **4,** 123–164.
3. Cohen, S. N., and Chang, A. C. (1975). Replication and expression of constructed plasmid chimeras in transformed *Escherichia coli*—A review. *Basic Life. Sci.* **5A,** 335–344.
4. Bennett, R. P., Cox, C. A., and Hoeffler, J. P. (1998). Fusion of green fluorescent protein with the Zeocin-resistance marker allows visual screening and drug selection of transfected eukaryotic cells. *Biotechniques* **24,** 478–482.
5. Niidome, T., and Huang, L. (2002). Gene therapy progress and prospects: nonviral vectors. *Gene Ther.* **9,** 1647–1652.
6. Struhl, K. (1983). Direct selection for gene replacement events in yeast. *Gene* **26,** 231–241.
7. Valancius, V., and Smithies, O. (1991). Testing an "in-out" targeting procedure for making subtle genomic modifications in mouse embryonic stem cells. *Mol. Cell Biol.* **11,** 1402–1408.
8. Maniatis, T., Hardison, R. C., Lacy, E., Lauer, J., O'Connell, C., Quon, D., Sim, G. K., and Efstratiadis, A. (1978). The isolation of structural genes from libraries of eucaryotic DNA. *Cell* **15,** 687–701.
9. Hohn, B., and Collins, J. (1980). A small cosmid for efficient cloning of large DNA fragments. *Gene* **11,** 291–298.
10. Krzywinski, M., Bosdet, I., Smailus, D., Chiu, R., Mathewson, C., Wye, N., Barber, S., Brown-John, M., Chan, S., Chand, S., Cloutier, A., Girn, N., Lee, D., Masson, A., Mayo, M., Olson, T., Pandoh, P., Prabhu, A. L., Schoenmakers, E., Tsai, M., Albertson, D., Lam, W., Choy, C. O., Osoegawa, K., Zhao, S., de Jong, P. J., Schein, J., Jones, S., and Marra, M. A. (2004). A set of BAC clones spanning the human genome. *Nucleic Acids Res.* **32,** 3651–3660.
11. Lindenbaum, M., Perkins, E., Csonka, E., Fleming, E., Garcia, L., Greene, A., Gung, L., Hadlaczky, G., Lee, E., Leung, J., MacDonald, N., Maxwell, A., Mills, K., Monteith, D., Perez, C. F., Shellard, J., Stewart, S., Stodola, T., Vandenborre, D., Vanderbyl, S., and Ledebur, H. C. Jr. (2004). A mammalian artificial chromosome engineering system (ACE System) applicable to biopharmaceutical protein production, transgenesis and gene-based cell therapy. *Nucleic Acids Res.* **32,** e172.
12. Green, E. D., and Olson, M. V. (1990). Systematic screening of yeast artificial-chromosome libraries by use of the polymerase chain reaction. *Proc. Natl. Acad. Sci. USA* **87,** 1213–1217.
13. Steinmetz, M., Stephan, D., and Fischer Lindahl, K. (1986). Gene organization and recombinational hotspots in the murine major histocompatibility complex. *Cell* **44,** 895–904.
14. McIntosh, I., and Cutting, G. R. (1992). Cystic fibrosis transmembrane conductance regulator and the etiology and pathogenesis of cystic fibrosis. *FASEB J.* **6,** 2775–2782.
15. Andelfinger, G., Wright, K. N., Lee, H. S., Siemens, L. M., and Benson, D. W. (2003). Canine tricuspid valve malformation, a model of human Ebstein anomaly, maps to dog chromosome 9. *J. Med. Genet.* **40,** 320–324.
16. Muncke, N., Jung, C., Rudiger, H., Ulmer, H., Roeth, R., Hubert, A., Goldmuntz, E., Driscoll, D., Goodship, J., Schon, K., and Rappold, G. (2003). Missense mutations and gene interruption in PROSIT240, a novel TRAP240-like gene, in patients with congenital heart defect (transposition of the great arteries). *Circulation* **108,** 2843–2850.
17. Goldmuntz, E. (1999). Recent advances in understanding the genetic etiology of congenital heart disease. *Curr. Opin. Pediatr.* **11,** 437–443.
18. Satoda, M., Pierpont, M. E., Diaz, G. A., Bornemeier, R. A., and Gelb, B. D. (1999). Char syndrome, an inherited disorder with patent ductus arteriosus, maps to chromosome 6p12–p21. *Circulation* **99,** 3036–3042.
19. Mulder, M. P., Wilke, M., Langeveld, A., Wilming, L. G., Hagemeijer, A., van Drunen, E., Zwarthoff, E. C., Riegman, P. H., Deelen, W. H., van den Ouweland, A. M., *et al.* (1995). Positional mapping of loci in the DiGeorge critical region at chromosome 22q11 using a new marker (D22S183). *Hum. Genet.* **96,** 133–141.
20. Lasters, I., Van Herzeele, N., Lijnen, H. R., Collen, D., and Jespers, L. (1997). Enzymatic properties of phage-displayed fragments of human plasminogen. *Eur. J. Biochem.* **244,** 946–952.
21. Grunstein, M., and Hogness, D. S. (1975). Colony hybridization: A method for the isolation of cloned DNAs that contain a specific gene. *Proc. Natl. Acad. Sci. USA* **72,** 3961–3965.
22. Venter, J. C. (1993). Identification of new human receptor and transporter genes by high throughput cDNA (EST) sequencing. *J. Pharm. Pharmacol.* **45,** 355–360.
23. Adams, M. D., Soares, M. B., Kerlavage, A. R., Fields, C., and Venter, J. C. (1993). Rapid cDNA sequencing (expressed sequence tags) from a directionally cloned human infant brain cDNA library. *Nat. Genet.* **4,** 373–380.
24. Hwang, D. M., Dempsey, A. A., Lee, C. Y., and Liew, C. C. (2000). Identification of differentially expressed genes in cardiac hypertrophy by analysis of expressed sequence tags. *Genomics* **66,** 1–14.
25. Megy, K., Audic, S., and Claverie, J. M. (2002), Heart-specific genes revealed by expressed sequence tag (EST) sampling. *Genome Biol.* **3,** 1–11.
26. Chim, S. S., Cheung, S. S., and Tsui, S. K. (2000). Differential gene expression of rat neonatal heart analyzed by suppression

26. subtractive hybridization and expressed sequence tag sequencing. *J. Cell Biochem.* **80,** 24–36.
27. Wang, R., Hwang, D. M., Cukerman, E., and Liew, C. C. (1997). Identification of genes encoding zinc finger motifs in the cardiovascular system. *J. Mol. Cell Cardiol.* **29,** 281–287.
28. Southern, E. M. (1975). Detection of specific sequences among DNA fragments separated by gel electrophoresis. *J. Mol. Biol.* **98,** 503–517.
29. Jasinska, A., and Krzyzosiak, W. J. (2004). Repetitive sequences that shape the human transcriptome. *FEBS Lett.* **567,** 136–141.
30. Mullis, K. B., and Faloona, F. A. (1987). Specific synthesis of DNA in vitro via a polymerase- catalyzed chain reaction. *Methods Enzymol.* **155,** 335–350.
31. Saiki, R. K., Gelfand, D. H., Stoffel, S., Scharf, S. J., Higuchi, R., Horn, G. T., Mullis, K. B., and Erlich, H. A. (1988). Primer-directed enzymatic amplification of DNA with a thermostable DNA polymerase. *Science* **239,** 487–491.
32. Larsen, L. A., Andersen, P. S., Kanters, J., Svendsen, I. H., Jacobsen, J. R., Vuust, J., Wettrell, G., Tranebjaerg, L., Bathen, J., and Christiansen, M. (2001). Screening for mutations and polymorphisms in the genes KCNH2 and KCNE2 encoding the cardiac HERG/MiRP1 ion channel: implications for acquired and congenital long Q-T syndrome. *Clin. Chem.* **47,** 1390–1395.
33. Tsubata, S., Bowles, K. R., Vatta, M., Zintz, C., Titus, J., Muhonen, L., Bowles, N. E., and Towbin, J. A. (2000). Mutations in the human delta-sarcoglycan gene in familial and sporadic dilated cardiomyopathy. *J. Clin. Invest.* **106,** 655–662.
34. Enjuto, M., Francino, A., Navarro-Lopez, F., Viles, D., Pare, J. C., and Ballesta, A. M. (2000). Malignant hypertrophic cardiomyopathy caused by the Arg723Gly mutation in beta-myosin heavy chain gene. *J. Mol. Cell. Cardiol.* **32,** 2307–2313.
35. Yoshida, H., Horie, M., Otani, H., Takano, M., Tsuji, K., Kubota, T., Fukunami, M., and Sasayama, S. (1999). Characterization of a novel missense mutation in the pore of HERG in a patient with long QT syndrome. *J. Cardiovasc. Electrophysiol.* **10,** 1262–1270.
36. Forissier, J. F., Carrier, L., Farza, H., Bonne, G., Bercovici, J., Richard, P., Hainque, B., Townsend, P. J., Yacoub, M. H., Faure, S., Dubourg, O., Millaire, A., Hagege, A. A., Desnos, M., Komajda, M., and Schwartz, K. (1996). Codon 102 of the cardiac troponin T gene is a putative hot spot for mutations in familial hypertrophic cardiomyopathy. *Circulation* **94,** 3069–3073.
37. Saarinen, K., Swan, H., Kainulainen, K., Toivonen, L., Viitasalo, M., and Kontula, K. (1998). Molecular genetics of the long QT syndrome: two novel mutations of the KVLQT1 gene and phenotypic expression of the mutant gene in a large kindred. *Hum. Mutat.* **11,** 158–165.
38. Marian, A. J., Mares, A., Jr., Kelly, D. P., Yu, Q. T., Abchee, A. B., Hill, R., and Roberts, R. (1995). Sudden cardiac death in hypertrophic cardiomyopathy. Variability in phenotypic expression of beta-myosin heavy chain mutations. *Eur. Heart J.* **16,** 368–376.
39. Tardiff, J. C., Hewett, T. E., Palmer, B. M., Olsson, C., Factor, S. M., Moore, R. L., Robbins, J., and Leinwand, L. A. (1999). Cardiac troponin T mutations result in allele-specific phenotypes in a mouse model for hypertrophic cardiomyopathy. *J. Clin. Invest.* **104,** 469–481.
40. Nishi, H., Kimura, A., Harada, H., Toshima, H., and Sasazuki, T. (1992). Novel missense mutation in cardiac beta myosin heavy chain gene found in a Japanese patient with hypertrophic cardiomyopathy. *Biochem. Biophys. Res. Commun.* **188,** 379–387.
41. Nishi, H., Kimura, A., Harada, H., Koga, Y., Adachi, K., Matsuyama, K., Koyanagi, T., Yasunaga, S., Imaizumi, T., Toshima, H., and Sasazuki, T. (1995). A myosin missense mutation, not a null allele, causes familial hypertrophic cardiomyopathy. *Circulation* **91,** 2911–2915.
42. Dasgupta, C., Martinez, A. M., Zuppan, C. W., Shah, M. M., Bailey, L. L., and Fletcher, W. H. (2001). Identification of connexin43 (alpha1) gap junction gene mutations in patients with hypoplastic left heart syndrome by denaturing gradient gel electrophoresis (DGGE). *Mutat. Res.* **479,** 173–186.
43. Van Eyk, J. E. (2001). Proteomics: unraveling the complexity of heart disease and striving to change cardiology. *Curr. Opin. Mol. Ther.* **3,** 546–553.
44. de Hoog, C. L., and Mann, M. (2004). Proteomics. *Annu. Rev. Genomics Hum. Genet.* **5,** 267–293.
45. Mobasheri, A., Airley, R., Foster, C. S., Schulze-Tanzil, G., and Shakibaei, M. (2004). Post-genomic applications of tissue microarrays: basic research, prognostic oncology, clinical genomics and drug discovery. *Histol. Histopathol.* **19,** 325–335.
46. Lal, S. P., Christopherson, R. I., and dos Remedios, C. G. (2002). Antibody arrays: an embryonic but rapidly growing technology. *Drug Discov. Today* **7,** S143–S149.
47. Fields, S., and Song, O. (1989). A novel genetic system to detect protein–protein interactions. *Nature* **340,** 245–246.
48. Tyagi, S., and Kramer, F. R. (1996). Molecular beacons: probes that fluoresce upon hybridization. *Nat. Biotechnol.* **14,** 303–308.
49. Vet, J. A., and Marras, S. A. (2005). Design and optimization of molecular beacon real-time polymerase chain reaction assays. *Methods Mol. Biol.* **288,** 273–290.
50. Gelfi, C., Cremonesi, L., Ferrari, M., and Righetti, P. G. (1996). Temperature-programmed capillary electrophoresis for detection of DNA point mutations. *Biotechniques* **21,** 926–932.
51. Katzke, S., Booms, P., Tiecke, F., Palz, M., Pletschacher, A., Turkmen, S., Neumann, L. M., Pregla, R., Leitner, C., Schramm, C., Lorenz, P., Hagemeier, C., Fuchs, J., Skovby, .F, Rosenberg, T., and Robinson, P. N. (2002). TGGE screening of the entire FBN1 coding sequence in 126 individuals with Marfan syndrome and related fibrillinopathies. *Hum. Mutat.* **20,** 197–208.
52. Murphy, K., Hafez, M., Philips, J., Yarnell, K., Gutshall, K., and Berg, K. (2003). Evaluation of temperature gradient capillary electrophoresis for detection of the Factor V Leiden mutation: Coincident identification of a novel polymorphism in Factor V. *Mol. Diagn.* **7,** 35–40.
53. Maitra, A., Cohen, Y., Gillespie, S. E., Mambo, E., Fukushima, N., Hoque, M. O., Shah, N., Goggins, M., Califano, J., Sidransky, D., and Chakravarti, A. (2004). The Human MitoChip: A high-throughput sequencing microarray for mitochondrial mutation detection. *Genome Res.* **14,** 812–819.
54. Nickerson, D. A., Kaiser, R., Lappin, S., Stewart, J., Hood, L., and Landegren, U. (1990). Automated DNA diagnostics using an ELISA-based oligonucleotide ligation assay. *Proc. Natl. Acad. Sci. USA* **87,** 8923–8927.

55. Gaylord, B. S., Massie, M. R., Feinstein, S. C., and Bazan, G. C. (2005). SNP detection using peptide nucleic acid probes and conjugated polymers: applications in neurodegenerative disease identification. *Proc. Natl. Acad. Sci. USA* **102,** 34–39.

56. Griffin, T. J., Tang, W., and Smith, L. M. (1997). Genetic analysis by peptide nucleic acid affinity MALDI–TOF mass spectrometry. *Nat. Biotechnol.* **15,** 1368–1372.

57. Wang, J., Nielsen, P. E., Jiang, M., Cai, X., Fernandes, J. R., Grant, D. H., Ozsoz, M., Beglieter, A., and Mowat, M. (1997). Mismatch-sensitive hybridization detection by peptide nucleic acids immobilized on a quartz crystal microbalance. *Anal. Chem.* **69,** 5200–5202.

58. Rockenbauer, E., Petersen, K., Vogel, U., Bolund, L., Kolvraa, S., Nielsen, K. V., and Nexo, B. A. (2005). SNP genotyping using microsphere-linked PNA and flow cytometric detection. *Cytometry A* **64**, 80–86.

59. Elso, C., Toohey, B., Reid, G. E., Poetter, K., Simpson, R. J., and Foote, S. J. (2002). Mutation detection using mass spectrometric separation of tiny oligonucleotide fragments. *Genome Res.* **12,** 1428–1433.

60. Tost, J., and Gut, I. G. (2005). Genotyping single nucleotide polymorphisms by MALDI mass spectrometry in clinical applications. *Clin. Biochem.* **38,** 335–350.

CHAPTER 3

Cardiovascular Gene Expression

OVERVIEW

Mammalian cells, both *in vitro* and *in vivo*, have been used to examine how gene expression can be regulated. Gene regulation can be modulated in both cultured cardiomyocytes and in intact animals, and, potentially, this can be applied to the treatment of cardiovascular disease (CVD) in humans. In this chapter, methods of gene delivery to cardiomyocytes and endothelial cells, classical and current strategies of gene targeting, overexpression, and specific inhibition in these cells in tissues will be discussed together with an appraisal of new and developing methods. A number of antisense strategies such as ribozymes, oligonucleotides, and RNAi, which can be used to modulate a specific gene, methods of gene profiling (to assess global patterns of gene expression), and specific issues and technologies involved in cardiovascular gene therapy in animal and human will be reviewed.

INTRODUCTION

The availability of genetically altered cells is an essential prerequisite for many scientific and therapeutic applications, such as functional genomics, drug development, and gene therapy. Somatic gene therapy for the treatment of myocardial diseases relies on efficient gene transfer into cardiac muscle cells. Given the relative difficulties in transducing primary cells *in vivo*, a common strategy has involved *ex vivo* gene transfer using first the removal of cells from a host organism, subsequent gene transfer, and cell growth *in vitro* and, finally, reintroduction of the modified cells in the host. For instance, primary hepatocytes (derived from partial hepatectomy) and skeletal myoblasts could be isolated, transduced *in vitro* with either viral or plasmid-containing genes, and grown and reimplanted by portal vein infusion or intramuscular injection.

Notwithstanding current technological advances, the possibilities for *ex vivo* gene transfer are currently limited, particularly for cardiomyocytes, because they are post-mitotic in nature, have limited growth in culture, and they are difficult to reimplant. In contrast to transfections of most cell lines, which can be successfully performed using a variety of methods, delivering genes into adult cardiomyocytes and endothelial cells, both *in vivo*, as well as in primary cells in culture, is rather difficult. However, these difficulties can be circumvented by a careful selection of gene transfer vectors and delivery techniques (e.g., catheters and stents).

Gene Transfection

VECTORS

Both viral and plasmid DNAs have been used as vectors in cardiovascular gene transfer studies with mammalian cells and tissues. These vectors have been used in a variety of investigative studies involving the transfer of genes of interest to isolated cardiomyocytes, endothelial cells, and to both vascular and cardiac tissues as potential targets for gene therapy.

Plasmid DNA vectors are easy to isolate, manipulate, and produce in large scale and require only a small number of proteins for their replication. However, the use of plasmid vectors poses significant limitations, because they display a markedly lower efficiency of transfection (50–100 fold), particularly with cardiomyocytes compared with using viral vectors. Generally, the plasmid vectors are present as *episomes* and do not integrate into the host chromosomes and, therefore, are more prone to degradation. This can result in prolonged low-level expression *in vivo*. Interestingly, with *in vivo* studies, plasmid DNA has been shown to have good entry and expression in normal and ischemic muscle.[1] Plasmid-mediated gene delivery may be adequate in achieving satisfactory biological effects when the therapeutic transgene is a potent secretory product (e.g., VEGF).

The use of plasmid DNA as vectors has significantly increased in cardiovascular research, mainly because long-term transgene expression from plasmid DNAs can be greatly enhanced by the introduction of *transposon* elements linked to the gene of interest and subsequent directed integration of the plasmid construct into specific host chromosome sequences.[2] Moreover, plasmid vectors, compared with virus vectors, offer significant advantages such as less technical difficulties and side effects. Notwithstanding these advantages and given the markedly lower efficiency of transfection with naked plasmid vectors, gene transfer to cardiomyocytes and myocardium has primarily been carried out by use of viral vectors. The characteristics of the viral vector tend to predetermine the range of host cells that can be transduced, efficiency, level, and duration of transgene expression.

Adenoviral vectors can be produced in high titers, have a broad host range, and effectively transduce both dividing and quiescent, nondividing cells both *in vivo* and *in vitro*. They are particularly efficient in transfecting post-mitotic cells (i.e., cardiomyocytes and to a lesser extent vascular cells) and they have also been the viral vectors of choice in cardiovascular gene transfer in experimental models. Adenovirus propagates primarily as episomes and rarely integrates into the host genome. However, a limitation of the adenoviral vectors is that they provide transient rather than prolonged transgene expression. In addition, adenoviral vectors pose additional safety problems. These vectors, whose replication requires the production and activity of exogenous proteins, can produce severe inflammatory reaction. Moreover, long-term cell-mediated and antibody-mediated immune responses to adenoviral vector proteins limit transgene expression and may also cause myocardial necrosis. To reduce the host immune response, a new generation of adenovirus vectors is being developed to eliminate the adenoviral coding sequences while retaining the regulatory sequences needed for viral replication and packaging.[3]

Retroviruses have proven to be useful vectors, particularly for *ex vivo* approaches. These vectors are relatively nonimmunogenic and can efficiently transduce dividing cells *in vitro*. Because they are primarily effective with proliferative cells, they have generally had limited applicability as vectors for gene transfer to post-mitotic cells, such as cardiomyocytes, in myocardial transfection and gene therapy. Retroviruses integrate into the genome, which leads to long-term gene expression; however, random insertion of the virus into the host genome can trigger neoplastic proliferation, enhancing protooncogene expression, and can promote the insertional mutagenesis of host genes.

Several viruses are emerging as the most promising vectors in regard to cardiovascular gene transfer and therapies, including lentivirus and recombinant adenovirus-associated virus (AAV). Furthermore, they are able to infect both proliferating and nonproliferating cells, have limited inflammatory response, and can induce prolonged, high-level transgene expression. These properties have made both lentivirus and recombinant AAV highly attractive vectors for cardiovascular gene transfer and therapy in both vascular and myocardial cells, into which both vectors can effectively transfect *in vivo* as well as *in vitro*.[4–6] Nevertheless, AAV vectors have some disadvantages, including lower yields of pure virus production, largely as a consequence of their complex packaging requirements, and a limited size capacity for the transgene insert that can be accommodated in the AAV vectors (i.e., generally no more than 4.7 kb). Recently, as a result of AAV virus engineering, its capacity has been increased permitting it to carry the full length of the human dystrophin–coding sequences (greater than 14 kb), with transfection and stable expression in rat cardiomyocytes *in vitro*. This approach was also used to restore dystrophin synthesis *in vivo* in tissues of the mouse model of Duchenne muscular dystrophy (mdx).[7]

Because modification of these viruses, to accommodate their inserts, has been shown to abrogate their ability to integrate at specific sites on the chromosomes, the enhanced possibility of random integration on the host chromosome may result in insertional mutagenesis and neoplastic transformation, and this poses a significant concern. On the other hand, whereas the risk of random integration is low, its occurrence can be largely surmounted by the inclusion of integration cassettes on the AAV vector to direct site-specific integration.[8] However, despite the promising applicability of retroviral-related lentiviral vectors to cardiovascular gene transfer,[9,10] their use in gene therapy has been largely excluded because of their close relationship to HIV.

Besides viral methods of gene delivery, liposomes composed of lipid spheres with a fraction of aqueous fluid in the center have also been used. DNA (either as plasmids or oligonucleotides) mixed with various cationic and neutral lipids can form DNA–liposome complexes, which increase the stability of the DNA. The DNA-containing liposome fuses with cell membranes (without an infectious vector), and DNA within the liposome is transferred into the target cell. Although liposome transfer has been considerably more successful with *in vivo* cell transfection and less effective with *in vivo* animal models, numerous studies have shown that markedly increased transfer efficiency, *in vitro* and *in vivo*, may be achieved by the optimization of known variables that influence liposomal transfection, including the appropriate selection of cationic lipids for the targeted cell type, the DNA/liposome ratio chosen, and the state of proliferation of the targeted cells.[11] Recently, a modification to this approach has been used with only the viral envelope of the Sendai virus (HVJ: hemagglutinating virus of Japan) as a carrier, to deliver plasmid-carrying genes (e.g., luciferase) and oligonucleotides (ODNs) to neonatal cardiomyocytes.[12] A high transfection efficiency and expression of the luciferase plasmid in transfected cardiomyocytes was achieved with no evidence of cytotoxicity. As found with most nonvirally delivered DNA, liposome constructs tend to be limited to transient gene expression.

Transgene Expression

The level and duration of transgene expression is a major variable in cardiovascular gene transfer studies. The transgene product, usually a protein, can be produced at normal cellular levels or can be produced in greater amounts than normal (i.e., overexpressed). Occasionally, overexpression can trigger a toxic reaction in the cell. The addition of appropriate regulatory sequences within the genetic construct to be introduced may modulate transgene expression. These can include the use of constitutive promoters or regulatable promoter elements and enhancers that are inducible in response to a variety of endogenous or exogenous molecular signals (e.g., steroid hormones, cytokines, growth factors).

Regulatable promoters have the advantage of allowing gene expression to be switched on and off.

Targeting of the gene product to the appropriate cellular compartment can also be regulated by the addition of specific peptide presequences to the transgene. In addition, the incorporation of tissue-specific regulatory elements, such as myocyte-specific promoter sequences of *mlc-v* (ventricle-specific myosin light chain-2) and *cTNT* (cardiac troponin T), can enhance cardiac or cardiomyocyte gene expression, as well as prevent unwanted transgene expression in nontarget cells.

The duration of transgene expression varies with the vector type used (Table I) or the specific transgene. The requirements for sustained or long-term gene expression will vary from case to case. For instance, certain cardiovascular disorders and/or clinical settings may require shorter term gene expression, in which transgene expression is required only during a period of defined risk, such as remodeling after myocardial infarct or in preventing myocardial ischemia. In addition, short-term expression (e.g., 2–3 weeks) may be sufficient to promote neovascularization or to inhibit restenosis.[13,14]

Nature of Targeted Gene

A large assortment of genes has thus far been used in cardiovascular gene transfer and myocardial therapy, primarily in animal models; a representative list is presented in Table II. Which genes to use in cardiovascular gene therapy markedly depends on the specific disorder to be treated, as well as which cell type is being targeted. These genes range from SERCA, plasma membrane channel proteins, cytokines, transcription factors, and signaling pathway components. Early studies found the transfer of genes encoding extracellular secreted proteins such as VEGF rather than intracellular proteins was advantageous because a limited number of injections/transfections were needed to produce a large enough quantity of protein. This was particularly evident with the vascular treatments involving therapeutic angiogenesis (e.g., VEGF, FGF). Active secretion can either be mediated by a native or ligated signal sequence.

Gene-Targeting Approaches

In addition to transgene overexpression and activation, gene transfer approaches can be used to negatively modulate gene expression involving the use of antisense strategies (e.g., either ribozymes, antisense oligonucleotides, or RNA interference) as shown in Fig. 1. These approaches regulate the transcription of targeted endogenous genes and selectively inhibit their expression in both cultured cells and in specific animal and human tissues.

Ribozymes

Ribozymes are RNA molecules that catalyze the cleavage of RNA substrates and the formation of covalent bonds in RNA strands at specific sites as shown in Fig. 2. Ribozymes have been shown to be highly specific, efficient, and stable. They can be delivered to cells as preformed ribozymes using primarily lipofection or electroporation or as ribozyme genes. Ribozyme genes can be packaged into viral vectors (e.g., adenovirus or AAV vectors) to enhance transfer into cells and to achieve longer expression compared with naked oligonucleotides. The choice of promoter (i.e., pol II, pol III, viral) can be critical for their expression. The "hammerhead" motif, approximately 30 nucleotides long, is a small endonucleolytic ribozyme. Hammerhead ribozymes can be directed against RNA sequences of interest to selectively target their specific gene expression. Recently, a hammerhead ribozyme, directed against the mannose 6-phosphate/IGF-2 receptor (M6P/IGF2R), was used to probe the receptor's role in regulating cardiac myocyte growth and apoptosis. Downregulation of the expression of M6P/IGF2R in ribozyme-treated neonatal rat cardiac myocytes resulted in a marked increase in cell proliferation and a reduced cell susceptibility to hypoxia-induced and TNF-induced apoptosis.[15]

Antisense Oligonucleotides

The antisense oligonucleotide approach most commonly uses either single-strand RNA or DNA oligonucleotides to target specific gene expression and has been applied with

TABLE I Comparison of Vectors for Gene Transfer and Expression

Vector	Chromosomal integration	*In vivo* transfer efficiency	Target cells	Expression (level/onset/duration)	Disadvantage in gene therapy
Liposome	No	+	Quiescent and dividing	+/rapid/transient	Low transduction efficiency and stability
Plasmid	No	+	Quiescent and dividing	+/moderate/transient	Low transduction efficiency
Adenovirus	No	++++	Quiescent and dividing	++++/rapid/transient	Immunogenic; viral mutation
Lentivirus	Yes	+++	Quiescent and dividing	+++/rapid/long term	Insertional mutagenesis
Retrovirus	Yes	++	Dividing	+++/rapid/long term	Oncogenic potential
AAV	Yes	+++	Quiescent and dividing	+++/slow/long term	Insertional mutagenesis

TABLE II Targeted Gene Expression in CVD

Function	Targeted gene	O or I
Antioxidant	Heme oxygenase (HO-1)	O
	Superoxide dismutase (SOD)	O
	Catalase (CAT)	O
	Glutathione peroxidase (GPx)	O
Pro-apoptotic proteins	p53	I
	Bad	I
	Fas ligand	I
Ca^{++} regulation	β-adrenergic receptor kinase β-ARK)	I
	β-adrenergic receptor (β-AR)	O
	Phospholamban (PLN)	I
	Sarcoplasmic reticulum Ca^{++} ATPase (SERCA)	O
Channel proteins	Cardiac sodium channel (SCN5A)	O
	Hyperpolarization-activated cyclic nucleotide-gated channel (HCN2)	I
Coronary vessel tone	Endothelial nitric oxide synthase (eNOS)	O
	Adenosine receptors	O
Contractile proteins	Sarcomeric proteins	O
	Sarcoglycans (δ-SG)	O
Survival genes	Protein kinase B (Akt)	O
	Bcl-2	O
	Insulin-like growth factor (IGF-1)	O
Pro-angiogenic factors	Vascular endothelial growth factor (VEGF)	O
	Fibroblast growth factor (FGF)	O
	Hypoxia-inducing factor (HIF-1α)	O
Inflammatory cytokines	ICAM	I
	TNF-α	I
	NF-κB	I

O, Overexpression; I, inhibition.

success to modulating the progression of CVD in animal models.[16–18] These synthetic oligonucleotides are usually short, ranging from 10–30 bp. These oligonucleotides can be chemically modified to enhance stability; the substitution of sulfur for one of the oxygens in the phosphate backbone, termed a phosphorothioate modification, renders the oligonucleotide more stable to nuclease degradation.

Phosphorothioate-modified antisense oligonucleotides have been shown to be more stable than natural oligomers to both serum and cellular nucleases. Another type of modified oligonucleotide is the phosphorodiamidate morpholino oligomers, which comprise a novel class of nonionic antisense agents that inhibit gene expression by binding to RNA and blocking its processing or translation (Fig. 3A).[19,20] Although morpholino oligonucleotides have shown improved sequence specificity, biostability, and low toxicity compared with the phosphorothioate oligomers, making them effective antisense modulators of gene function in embryonic and adult tissues, their limited ability to cross cell membranes has limited their use in cell culture.

They have proved to be a highly informative tool for "knocking-down" (inhibiting) specific transcripts in studies of early cardiac development in zebrafish and other model organisms.[21–26]

Another relevant modification of antisense oligonucleotides involves the addition of a conjugated peptide to the oligonucleotide of interest. The use of peptide nucleic acids (PNAs) has been applied in both cells and tissues.[27,28] This approach can be used to better direct the oligonucleotide to the organelle of interest. Recently, it has been shown that PNA-containing oligonucleotides fused with a mitochondrial targeting peptide can be specifically targeted to the mitochondrial organelle for mitochondrial gene repair.[29]

RNA Interference

The RNA interference approach involves the use of a specific double-stranded RNA (dsRNA) construct to post-transcriptionally silence specific gene expression (RNAi). A dsRNA homologous in sequence to the targeted gene is processed into small interfering RNA (siRNA) by an RNAse

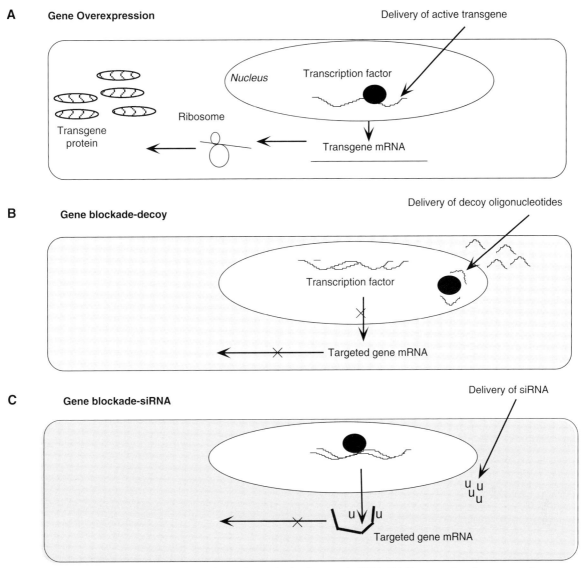

FIGURE 1 Approaches to gene overexpression and targeting by decoy and siRNA. (A) Gene transfer of a transgene by a vector allowing expression of a therapeutic protein in the host. The goal is gain-of-function or enhanced function overcoming a deficiency in the host cell. (B) Gene blockade of a pathogenic pathway by use of a transcriptional decoy (a double-stranded oligonucleotide containing the cis sequence bound by a transcription factor normally involved in the activation of pathogenic gene expression). Transfection of this decoy in molar excess prevents the binding and transactivation of the host pathogenic genes. In this case neither the target gene mRNA or protein product are expressed. (C) Gene blockade strategy using either an antisense oligonucleotide or small interfering RNA (siRNA). A small single-stranded antisense oligonucleotide or siRNA (~21 nt), complementary to the target mRNA, is transfected into the host cell, binds the target mRNA, and prevents it from being translated.

III family member enzyme called Dicer, and the siRNAs are then incorporated into multicomponent RNA-induced silencing ribonucleoprotein complexes (RISC), which find and cleave the target mRNA (Fig. 3B). The siRNA mediating this mRNA cleavage is usually 21–23 nucleotides and can be expressed either as two separate strands or as a single short hairpin RNA.

Expression cassettes encoding engineered siRNA (directed to specific mRNAs) can be efficiently introduced into the host cell by use of viral vector systems (e.g., retroviral, adenoviral, and lentiviral vectors) resulting in long-term silencing of target gene expression, both in cultured cells and in animal models.[30–32] Expression vectors for the induction of siRNA in mammalian cells frequently use polymerase III–dependent promoters (e.g., U6 or H1). They transcribe short hairpin RNAs (shRNA) that, after being directly processed into siRNAs, mediate target mRNA degradation. Although preliminary data have shown that the use of siRNAs allows gene-specific knock-down without induction of the nonspecific interferon response in mammalian cells (which longer dsRNA species promote), recent data have demonstrated an induction of interferon-activated

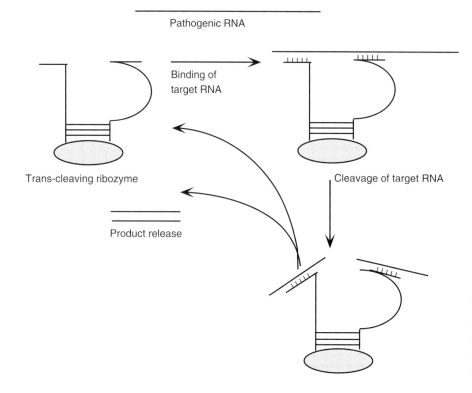

FIGURE 2 Trans-cleaving ribozyme modulates level of specific RNA. Ribozymes can bind specific pathogenic targeted transcripts, cleave the target mRNA, release the cleaved products, and repeat the process with other target mRNAs.

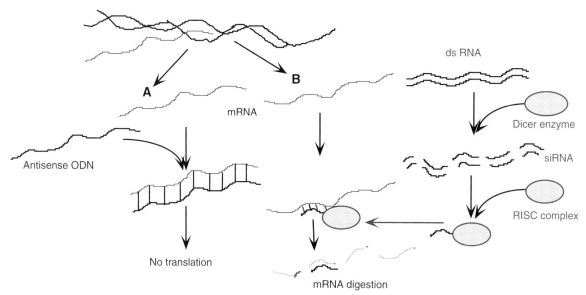

FIGURE 3 Two post-transcriptional approaches to silence gene expression. (A) After production of a specific mRNA, how a short single-strand antisense oligonucleotide, complementary to the mRNA, can attenuate further specific protein translation. (B) The RNA interference (RNAi) approach shows how a double-stranded RNA (dsRNA) with sequence complementary to the mRNA of interest can be cleaved by the Dicer RNAse III enzyme to short dsRNA fragments (termed siRNA), which, on interaction with the RISC complex, can present a short 20–21 bp single-stranded RNA homologous to and accessible for specific binding to the targeted mRNA and enabling its endonucleolytic cleavage.

gene expression in human cells transfected with siRNA. This interferon-mediated response to siRNA can be diminished by reducing the size (fewer than 21 nucleotides) and changing the initiating nucleotide sequence of the siRNA. Exogenous delivery of synthetic siRNA to target cells can also be used to induce specific gene expression knock-down; however, the resulting gene silencing is transient and tends to be less effective than with vectors.

Specific RNA knock-down by siRNA of the sarcoplasmic reticulum Ca^{++} ATPase (SERCA) gene has been achieved in primary myocyte cultures from embryonic chicken and neonatal rats.[33] Although previous studies had displayed a low efficiency in the percentage of cells affected by cardiac myocyte siRNA transfection, the use of adenovirus vectors allowed effective introduction and endogenous production of siRNA with marked reduction of SERCA2 gene expression. Interestingly, even with the pronounced reduction of SERCA2 protein and sarcoplasmic reticulum Ca^{++} uptake and cycling, the cardiomyocytes retained the ability to increase cytosolic Ca^{++} flux in response to stimulation. It has been suggested that the intracellular store deficiency caused by reduced SERCA could be compensated for by Ca^{++} fluxes through the plasma membrane. This was largely achieved by increased transcriptional activation of other Ca^{++} proteins, including the up-regulation of transient receptor potential (TRP) channel proteins (TRPC4 and TRPC5), Na^+/Ca^{++} exchanger, and related transcription factors such as stimulating protein 1 (Sp1), myocyte enhancer factor 2 (MEF-2), and nuclear factor of activated cells 4 (NFATc4), suggesting significant remodeling of the Ca^{++} signaling pathway.

RNAi-mediated gene knock-down has shown a high level of specificity. This approach has been particularly useful in assigning and differentiating the physiological actions of highly similar G-protein–coupled receptors (i.e., AT1a, AT1b, and AT-2) in response to angiotensin II. This has been possible with the development of specific siRNA to AT1a receptor (AT1aR) subtype, which had no significant effect on either AT1bR or AT2R subtypes.[34]

Gene profiling

Recent methodological advances have made it possible to simultaneously assess the entire profile of expressed genes in affected tissues (e.g., myocardium) using only a very limited amount of tissue (endomyocardial biopsy). Gene expression profiling, to comprehensively evaluate which genes are increased and which decreased in expression, has been achieved by several methods. The techniques of differential display and subtractive hybridization have largely been superseded by methods of DNA microarray analysis, which makes use of DNA chips, and the serial analysis of gene expression (SAGE), a method that efficiently quantifies large numbers of mRNA transcripts by sequencing short tags. SAGE and microarray (i.e., Affymetrix GeneChip) approaches have been found to be comparable, indicating that both can lead to similar conclusions regarding transcript levels and differential gene expression[35]; although microarray has been used more extensively in the screening of CVDs. Rigorously performed DNA microarray analysis has proved to be highly informative in documenting gene expression patterns associated with diverse pathologic conditions such as cardiac hypertrophy,[36] myocardial ischemia,[37] myocarditis,[38] DCM,[36] CAD,[39] primary pulmonary hypertension,[40] atrial fibrillation,[41] and HF[42] in both animal models and humans (Table III).

TABLE III Gene Profiling in Human Cardiovascular Disease

Condition	Technique	References
Cardiac hypertrophy	MA	36, 47
Dilated cardiomyopathy	MA	36
	MA	48
	MA	49
	MA	50
	DD	51
	SH, MA	52
Hypertrophic cardiomyopathy	MA	36
	SH	53
Myocardial ischemia	MA	54
	MA/SAGE	55
Heart failure (HF)	MA	56
	MA	57
	MA	58
Pulmonary hypertension	MA	40
		59
Atrial fibrillation	MA	41
HF with LVAD	MA	60, 58
CAD	MA	39
	MA	61
	MA	62
Myocarditis	MA, DD	63
	MA	64
Valvular heart disease	MA	65

MA, Microarray; DD, differential display; SH, subtractive hybridization.

Gene expression profiling (also termed transcriptome analysis) has also been used in models of preconditioning[43] and in studies of early cardiac morphogenesis and development[44] and has also shown applicability in both clinical diagnosis and in the assessment of patients' response to therapy.[45,46]

The limitations inherent in transcriptome analysis should be considered: (1) The correlation between mRNA and protein levels for a particular gene can be highly variable, and because proteins are the primary effectors of most cellular processes, altered gene transcription may not always be related to phenotype changes or to altered protein levels. (2) In myocardial studies there are numerous examples showing that the biological activity of proteins is subject to regulation by post-translational modifications, including their subcellular localization, with subsequent effects on the cardiac phenotype that will not be detected by transcriptional analysis.

Progressive evolution and improvement in microarray techniques requires the normalization of microarray data to remove thru filtering the "noise from signal," as well as the availability of computational software and statistical analysis. Unfortunately, a consensus on standardized approaches for normalization and statistical treatment of microarray data has not yet been reached, making the comparison between experiments difficult. Moreover, further validation of specific gene expression patterns ascertained by microarray data are often necessary by RT-PCR, RNAse protection or Northern blot analysis. Awareness of these limitations underscores the need for complementary approaches (e.g., proteomic analysis) to advance our understanding of complex cardiac disorders.

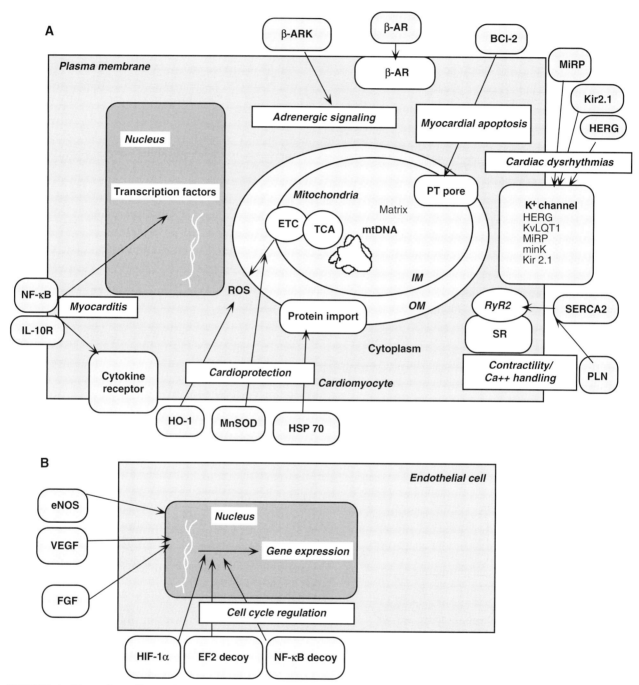

FIGURE 4 Targets for cardiovascular gene therapy. The pathogenesis of CVD with their many shared elements provides multiple therapeutic targets for gene therapy. (A) Specific targets for gene therapy in cardiomyocytes include K+ channels in the treatment of dysrhythmias, SERCA2, and Ca^{++} loading to improve myocardial contractility and cytokine receptors in myocarditis. (B) In endothelial cells VEGF, FGF, and eNOS are some of the important targets to modify nuclear gene expression. The vasculoproliferative characteristics of coronary stenting and vein graft occlusion have been inhibited in pre-clinical studies by overexpression of genes, including isoforms of eNOS, as well as targeting gene expression of cell cycle progression, including the use of decoy and antisense oligonucleotides to knock-down proliferation-activation gene expression in vascular smooth muscle.

Gene Therapy

Advances in the identification of genes affected in CVD have led to improved therapies, either by the use of gene replacement or gene-suppression methods. Preclinical studies in a variety of animal models have suggested that gene therapy may be beneficial in the clinical treatment of HF, hypertension, myocardial hypertrophy, cardiac dysrhythmias, and myocarditis, as well as in disorders of the vascular wall, particularly in cases in which drug therapy has proved to be of limited value. The pathogenesis of CVD with its many shared elements provides multiple therapeutic targets for applied gene therapy (Fig. 4). Gene therapy enables therapeutic concentrations of a gene product to be accumulated and maintained at optimal levels at a localized target site of action and also offers the possibility of minimizing systemic side effects by avoiding high plasma levels of the gene product. Clinical gene therapy trials for CVD have, in many instances, shown promising results. However, the development of improved vectors, methods of delivery, and the acquisition of safety and toxicity data need to be critically improved before these therapies can be routinely used in a clinical setting.

The basic elements of cardiovascular gene therapy are relatively simple as depicted in Fig. 5. As previously mentioned, both viral vectors and naked plasmid DNAs have been used in preclinical and clinical cardiovascular gene therapy studies. These have been used to transfer genes specifying a large variety of structural and regulatory functions to the myocardium and or to the vasculature. In gene therapy, as previously mentioned, a number of antisense strategies such as ribozymes, antisense oligonucleotides, or siRNA have been used to modulate or knock-down the specific transcription of targeted endogenous genes. Their delivery often involves the use of viral vectors for efficient and targeted transduction.

The application of therapeutic ribozymes has recently shown some success in preclinical studies in the treatment of CVD. Ribozymes directed against specific matrix metalloproteinase (MMP-2), which contributes to the degradation or remodeling of the extracellular matrix (ECM) that occurs during cardiac allograft vasculopathy (CAV), have been shown to effect the progression of vascular remodeling in a model of mouse CAV. After gene transfer of MMP-2 ribozymes to donor hearts before transplantation, lumen occlusion was significantly decreased compared with nontreated allografts.[66] This suggests a potential role for ribozyme-mediated anti-MMP-2 expression as a supplemental therapy for CAV. Furthermore, the use of hammerhead ribozymes has been suggested in the treatment of post-myocardial infarction HF in rats using ribozyme-targeted TNF-α, which could lead to substantial improvement in cardiac function, concomitant with a restoration of the hemodynamic status in the animals.[67] Also, gene therapy using ribozymes has been effective in targeting the expression of fibrillin 1 (FBN1), a protein implicated in Marfan syndrome, an autosomal dominant connective tissue disorder whose cardiovascular manifestations includes aortic dissection and rupture. Hammerhead antisense ribozymes directed against FBN1 in cultured fibroblasts can discriminate between a wild-type and a mutant allele, differing just by a single base pair.[68] It is worth noting that FBN1 ribozymes can suppress the expression of the mutant *FBN1* allele.[69] Moreover, hammerhead ribozymes targeted against mRNAs encoding the potent cell-proliferating activators cyclin E, and cyclin E2F1 mediated the down-regulation of smooth muscle cell (SMC) proliferation that plays a critical role in the progression of in-stent restenosis.[70] The administration of both ribozymes (delivered by liposomes) induced a significant reduction in cells in S phase and in BRDU accumulation. Similarly, ribozymes targeting growth factor transcripts have been used to modulate cell proliferation, including that involved in neo-intimal formation after arterial injury.[71,72] The administration of a ribozymes (using adenoviral vectors) directed against PDGF A mRNA markedly reduced the increased expression of PDGF A-chain mRNA and protein in rat carotid arteries after balloon injury, significantly decreased the intima/media ratio of the injured artery, and completely inhibited thrombus formation.[71] Treatment with ribozymes directed against TGF-β mRNA also reduced the growth of vascular SMCs and may prove useful in targeting coronary artery restenosis after percutaneous transluminal coronary angioplasty (PTCA).[72]

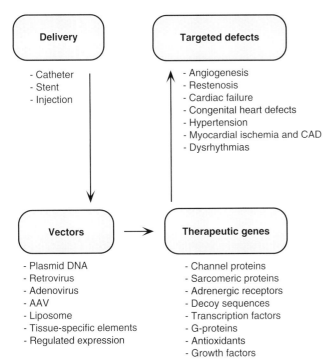

FIGURE 5 Basic elements of cardiovascular gene therapy. The elements of delivery techniques, vectors, therapeutic genes, and the targeted cardiovascular defects are depicted as are representative examples of each.

Double-strand oligonucleotides homologous to the *cis* regulatory sequences of the promoter of a gene of interest can also be similarly used.[17] These oligonucleotides can function as molecular decoys to bind specific transcription factors and, therefore, block the expression of genes requiring those transcription factors.

SPECIFIC CVD TARGETS OF GENE THERAPY

Vascular Disease

RESTENOSIS

One of the first CVDs targeted for gene therapy was restenosis after coronary stenting and vein graft occlusion. The vasculoproliferative characteristics of these disorders have been effectively inhibited in preclinical studies by overexpression of genes, including isoforms of nitric oxide synthase (NOS),[73,74] as well as targeting gene expression of cell cycle progression, including the use of decoy and antisense oligonucleotides to knock-down proliferation-activation gene expression in vascular smooth muscle.[17,75] Treatment of the rabbit jugular veins with antisense oligonucleotides directed against either of two cell cycle–regulatory proteins (i.e., cdc2 kinase [cdc2] and proliferating cell nuclear antigen [PCNA]) before carotid grafting inhibited subsequent atherosclerosis at the graft site.[76] The transfected graft also preserved its endothelial function compared with untreated grafts.[77] This strategy of effectively blocking the proliferative transcription program involved in vascular SMC proliferation was also applied using a double-strand oligonucleotide DNA *cis* recognition site with high affinity for the E2F regulatory factor (illustrated in Figs. 1B and 4), which plays a pivotal role in the coordinated transcriptional activation of numerous cell cycle factors.[78] Introduced *in vivo*, the E2F "decoy" oligonucleotide-bound endogenous E2F blocking the expression of genes mediating cell cycle progression (e.g., c-myc, cdc2, PCNA, and thymidine kinase) in vascular SMCs *in vitro* and stemmed both gene expression and hyperplasia after vascular injury in an *in vivo* model of rat carotid injury. This effect lasted up to 8 weeks after a single transfection.

Therapeutic Angiogenesis

Numerous observations have shown that angiogenic peptide delivery, either in direct intramuscular or intraarterial administration, enhanced blood flow in animal models of ischemia. Similar results were also obtained with gene transfer of an angiogenic peptide, VEGF, a secreted endothelial mitogen that significantly increased capillary density, and blood flow in rabbits with surgically induced hindlimb ischemia.[79] Neovascularization caused by angiogenesis as a result of the gene transfer of VEGF has been demonstrated in several animal models, as well as in a number of human clinical studies.[80–82] Interestingly, gene transfer of HIF-1α, a transcription factor that regulates the expression of hypoxia-inducible genes, also induces angiogenesis at ischemic sites.[83] Besides delivery of VEGF into ischemic limbs, this method has been extended to other angiogenic transgenes, including several VEGF isoforms and fibroblast growth factor (FGF), using both injected plasmid DNA and a variety of adenoviral constructs (Fig. 4).

Hypertension

Gene therapy has been effectively applied in several preclinical models of both systemic and pulmonary hypertension. The principal approaches for gene therapy in systemic hypertension use either the overexpression of vasodilator genes or the reduction of vasoconstriction gene expression, which often is accomplished by antisense inhibition. In addition, several studies have demonstrated that transfer and expression of genes, including an O_2-sensitive voltage-gated potassium channel (Kv1.5),[84] endothelial nitric oxide synthase (eNOS),[85] and prostaglandin synthase (PGS)[86,87] markedly improve pulmonary hypertension.

Because antihypertensive treatments often are not aimed at a specific identifiable cause, traditional pharmacological therapy has focused mainly on pathways, such as the renin–angiotensin system (RAS), that are known to be directly involved in the control of blood pressure (BP). Gene therapy directed at the modulation of RAS regulation represents a potential advance toward managing hypertension and reversing its associated pathophysiology. In spontaneously hypertensive rats (SHRs), administration of antisense constructs to the angiotensin II type 1 receptor (AT1R) successfully prevented BP elevation for up to 3–4 months compared with control rats.[88] These results suggest that antisense gene delivery is useful in the relatively long-term treatment of hypertension in animal models. Furthermore, recent observations have shown that concomitant with a reduction in BP, antisense therapy directed against angiotensin receptor resulted in significant attenuation of cardiac hypertrophy.[89] In addition, angiotensin receptor–targeted gene therapy can differentially target cardiac hypertrophic and hypertensive phenotypes. Interestingly, the intracardiac lentivirus-mediated transfer of angiotensin II type 1 receptor (AT2R) in either the SHR model or in rats with angiotensin II induced–hypertension resulted in a marked reduction in cardiac hypertrophy without lowering BP.[90,91]

Another viable target for gene therapy designed to stem hypertension using antisense technology is the β-adrenergic signaling system. Transfer of antisense oligonucleotides against rat $β_1$-adrenergic receptor ($β_1$-AR) mRNA provided a significant and prolonged reduction in BP in SHRs.[92]

Genes involved in the regulation of vasodilation are also excellent targets for gene therapy of systemic hypertension. Systemic hypertension in adult rats, over a 6- to 12-week period, can be significantly reversed by the transfer and overexpression of genes encoding atrial natriuretic peptide, kallikrein, the hypotensive peptide adrenomedullin, and eNOS.[93–97] In several models of systemic hypertension including SHR and Dahl salt-sensitive rats, delivery of these genes was accomplished with either nonviral (naked plasmid DNA) or adenoviral vectors. Interestingly, reduction in BP, mediated by transfer and overexpression of kallikrein and adrenomedullin, was also accompanied by attenuation of both cardiac hypertrophy and myocardial apoptosis. Furthermore, in adult SHRs long-term and stable reduction of BP and cardiac remodeling was accomplished with a single dose of an AAV-mediated adrenomedullin construct.[98]

Cardiac Dysrhythmias

Treatment of cardiac dysrhythmias using gene therapy may provide a significant advance, because current therapies for the most part are inadequate. A plurality of well-defined genetic loci have been characterized in inherited cardiac dysrhythmias, including gene defects in membrane transporters associated with specific ion channels, and in preclinical studies, these loci have provided suitable targets for gene therapy (Fig. 4).

Adenoviral-mediated transfection of a K+ channel gene has been attempted in cardiomyocytes isolated from failing dog hearts.[99] Interestingly, a moderate level of transgene expression increased the K+ channel current and reversed the channel deficiency mimicking the normal phenotype. However, a robust level of transgene expression had an adverse impact on cardiomyocyte excitation–contraction coupling.

To reverse the K+ channel deficiency or down-regulation without depressing contractility, Ennis and associates have used a gene therapy strategy to express genes involved in both the potassium channel and cardiac contractility.[100] An adenoviral vector was constructed that enabled the coexpression of two genes driven by a single promoter. To boost contractility, genes encoding the K+ channel Kir2.1 and SERCA1 were directly injected into the myocardium of adult guinea pigs. Both genes were amply expressed, and myocytes from transfected hearts exhibited significantly shorter action potentials, as well as larger calcium transients without altering cardiac contractility.

These same investigators also used an alternative gene therapy approach (i.e., viral vector carrying a defective K+ channel protein) to convert quiescent cardiomyocytes into pacemaker cells capable of generating spontaneous, rhythmic electrical activity in the *in vivo* guinea pig heart.[101]

Most cardiomyocytes have a functional channel that maintains the K+ balance and limits the cells from generating their own current; therefore, they are dependent on the pacemaker cells for stimulation. By essentially blocking the complete K+ channel formation in cells treated with defective transporter, a subset of the heart cardiomyocytes began to spontaneously and rhythmically generate current as pacemaker cells.

Qu and associates have reported the reestablishment of pacemaker function after suppression of sinus rhythm by localized overexpression in canine left atrium of the hyperpolarization-activated cyclic nucleotide-gated (HCN2) pacemaker current isoform.[102] This isoform of the gene encoding the wild-type pacemaker channel provides the pacemaker current [I_f] endogenous to the heart, and it is selectively activated during diastole. Interestingly, this adenoviral construct, if implanted in the left bundle branch (LBB) of the canine heart could adequately propagate the physiological activation of the ventricular conduction after vagal stimulation–induced atrioventricular block and clearly provided an escape pacemaker function.[103] However, these observations, using relatively transient transgene expression and pacemaker function, will need to be extended to other vectors to assess long-term expression and pacemaker function and also to carefully avoid the development of adverse effects. Delivery of a biological pacemaker to the heart has also been achieved by transplantation of human mesenchymal stem cells (HMSCs) transfected with plasmid-localized, highly expressed HCN2.[104] Transfected HMSCs have shown to influence myocyte beating rate when combined with neonatal rat ventricular myocytes *in vitro*. Also, HMSCs transfected with the HCN2 gene construct introduced into the canine left ventricular subepicardial wall *in situ* reestablished spontaneous left-sided rhythm after sinus arrest.

Similarly, treatment of atrial dysrhythmias may be achieved by gene transfer of not only HCN2 but also the K+ channel accessory subunit MiRP1, which is associated in cell membranes with HCN2, resulting in increased pacemaker current.[105] Recently, the plasmid-mediated gene transfer of specific mutant alleles of this K+ channel gene associated with long QT syndrome (Q9E-hMiRP1) into human cell lines and also directly into pig atria was reported. Data from both *in vitro* and *in vivo* studies revealed significant levels of transgene expression and altered myocyte electrophysiological phenotype, supporting the feasibility of site-specific gene transfer in the treatment of atrial dysrhythmias.

In a swine model of atrial fibrillation (AF), successful intracoronary adenoviral-mediated gene transfer and overexpression of an inhibitory component ($G_{\alpha i2}$) of the β-adrenergic pathway, directed to the atrioventricular (AV) node to suppress AV node conduction, was reported.[106,107] Calcium channel blockade is highly effective in inhibiting AV nodal conduction, but this therapy also triggers potent vasodilation and hypotension, largely attributed to the blockade of noncardiac channels. Murata and associates have shown that by targeting the AV node with the gene transfer and overexpression of Gem, a signaling protein capable of down-regulating the cardiac Ca^{++} channel, AV nodal conduction can be slowed in AF with subsequent rate control.[108] In a swine model of AF, 7 days after gene transfer, conduction in the AV node slowed. Also, during acute episodes of AF, GEM overexpression in the AV node resulted in significant reduction in ventricular rate, and this effect persisted in the setting of β-adrenergic stimulation, as well as cholinergic inhibition.

Myocarditis and Inflammation

Inflammatory cytokines play a critical role in the pathogenesis of viral myocarditis. In a rat model of experimental autoimmune myocarditis (EAM), introduction to muscle (by electroporation) of plasmid DNA containing the gene encoding murine interleukin-10 (IL-10) significantly affected survival rates, attenuated myocardial lesions, and improved the hemodynamic parameters.[109] Similarly, gene therapy with type 1 interferon transgene, introduced by DNA inoculation, resulted in protection against murine cytomegalovirus infection and myocarditis.[110]

The role of specific interferon 1 subtypes in stemming either the acute or chronic phase of the viral infection has been identified.[110] Recently, Bartlett and associates have shown that interferon β gene therapy significantly reduced CD8 (+) T cells, the principal cell type infiltrating the myocardium affected with inflammation (mainly evident during chronic disease).[111]

The transcription factor NF-κB modulates the expression of TNF-α, inducible nitric oxide synthase (iNOS), as well as adhesion molecule genes (iCAM) and represents a potential target to reduce myocarditis in the EAM model. After infusion of a decoy sequence into the rat coronary artery, knockdown of NF-κB expression, directed to the *cis*-regulatory sequence within the NF-κB promoter, reduced the areas of myocarditis, as well as myocardial gene expression of iNOS, iCAM, and TNF–α.[112] This finding is consistent with earlier observations that *in vivo* transfection of the NF-κB decoy significantly reduced the size of myocardial infarction (MI) after ischemia and reperfusion in a rat model of coronary artery ligation.[113]

Among the most commonly identified infectious agents causing human viral myocarditis is the coxsackievirus B3 (CVB3) to be discussed more extensively in Chapter 14. Recently, it has been shown that the use of an RNA interference strategy to target distinct regions of the viral genome can provide protection against CVB3 infection. Five CVB3-specific siRNAs were tested, and the most effective seemed to be the siRNA-4, which targets the viral protease 2A, resulting in a 92% inhibition of CVB3 replication. Viral replication inhibition persisted even when the administration of siRNA followed viral infection, supporting its therapeutic potential.[114]

Heart Failure and Cardiomyopathy

The failing myocardium is characterized by alterations in calcium handling, adrenergic receptor down-regulation and desensitization, and excessive catecholamine release, leading to decreased contractility. For a number of years, β-adrenergic receptor (β-AR) signaling and calcium regulating/handling pathways have been primary foci in the treatment of HF. A potential approach to gene therapy for rescuing the failing myocardium involves the myocardial delivery and overexpression of genes involved in cardiomyocyte signaling and in the regulation of contractile function, such as sarcoplasmic reticulum calcium ATPase (SERCA2), β-adrenergic receptor, and adenylyl cyclase (AC).

In both human and experimental models of HF, SERCA2 activity is decreased, resulting in reduced calcium uptake by the sarcoplasmic reticulum (SR) and abnormal calcium handling, which are thought to contribute to marked reduction in cardiac contractility. Myocardial SERCA2 Ca^{++} pumping activity is inhibited by phospholamban (PLN), and the ratio of phospholamban to SERCA2a isoform is increased in HF. Intracoronary delivery and overexpression of adenovirus-mediated SERCA2 in a pressure-overload rat HF model restored cardiac function, including improved Ca^{++} cycling and contractility.[115] Similarly, adenovirus-mediated overexpression of SERCA2a significantly enhanced contractile function and calcium transients in both cardiomyocytes derived from rat neonates and from patients with end-stage HF.[116,117] The adenoviral-mediated transfer and overexpression of an antisense PLN construct or a dominant–negative mutant PLN allele enhanced both SERCA 2 activity and contractility in myocytes derived from failing rat and human hearts.[118,119] However, caution must be used in translating these findings to clinical therapy, because all models of HF are not equivalent.[120] The ablation of the PLN allele reversed cardiac dysfunction in some mouse HF models,[121] whereas other models showed no such beneficial effect.[120,122] Moreover, in contrast to the benefits of phospholamban ablation in the mouse, null-PLN alleles resulting in the virtual absence of phospholamban protein have been recently identified in several individuals with lethal DCM.[123]

It is well established that both experimental models and clinical HF exhibit marked abnormalities in β-adrenergic signaling, including the down-regulation of β-adrenergic receptors (β-AR), their uncoupling with second messenger pathways, and modulation by up-regulated β-AR kinase (β-ARK). Transgenic mice that overexpress β-AR when under the control of the cardiac α-MHC promoter exhibit increased heart rates and enhanced contractility in the absence of β-agonists, as well as increased adenylyl cyclase (AC) activity.[124] Similarly, intracoronary adenoviral transfer of an adenovirus encoding a peptide inhibitor of β-ARK reversed cardiac dysfunction in both rabbit and mouse models of HF.[125-127] However, liabilities with these approaches have been raised by the observation of sustained adrenergic stimulation, which can be both cardiotoxic and dysrhythmogenic.[128]

AC is another component of the β–adrenergic signaling pathway that has proven to be a productive target for gene therapy directed at HF. Cardiac-specific overexpression of AC restored β-AR–stimulated cAMP generation, catecholamine responsiveness, and improved ventricular function, increasing long-term survival in mice with cardiomyopathy.[129,130] Intracoronary delivery of an adenovirus containing AC increased left ventricular function, markedly stimulated myocardial cAMP production, and attenuated deleterious ventricular remodeling in pigs with congestive HF.[131]

Myocardial Protection

Preclinical studies have demonstrated that short-term protection of the heart from ischemia can be provided by the gene transfer and overexpression of cardioprotective genes such as superoxide dismutase (SOD) and heme oxygenase (HO-1). Administration of a myocardial protective gene (e.g., HO-1) using a recombinant AAV vector significantly reduced the myocardial infarct size in a rat model of ischemia and reperfusion when introduced into myocardium before coronary artery ligation.[132] In addition, decreased levels of myocardial oxidative stress, inflammation, interstitial fibrosis, and improved ventricular dimension and function were found, indicating that gene therapy prevented long-term remodeling of the infarcted myocardium. Furthermore, cardioprotection against myocardial ischemia has been achieved by introducing and overexpressing genes encoding the antioxidant scavenger MnSOD,[133] eNOS,[133] the heat shock protein HSP70,[134] the anti-apoptotic proteins Bcl-2,[135] and Akt.[136] Additional studies are warranted to see whether these vectors and genes can provide long-term cardioprotection against repeated, chronic forms of ischemic insult. On the other hand, myocardial gene therapy as a method for targeted intervention against acute cardiac insult (e.g., myocardial ischemia) may have limited efficacy, because an extensive period of time is needed for transgene introduction, transcription, translation, and processing of the transgene product.

Prophylactic use of gene therapy to mount a protective response to a potential adverse cardiac event may be more feasible. Recently, a cardioprotective vigilant vector was developed for cardiac-specific expression and hypoxia-regulatable expression of therapeutic transgenes.[137,138] As originally conceived, this vector contained several elements, including a cardiac-specific promoter (MLC-2v), a hypoxia response element (HRE), a therapeutic transgene, and a reporter gene (green fluorescence protein) incorporated into an AAV vector. Because of the low efficiency of the HRE obtained with *in vivo* studies, a double-vector system has been devised that provided an oxygen biosensor (i.e., the oxygen-dependent degradation, ODD, domain derived from HIF-1α) fused to the GAL4 transactivator protein under the control of the MLC-2v promoter located on a sensor plasmid vector. Under hypoxic conditions, this GAL4-ODD protein is produced in myocytes that binds and activates expression of a therapeutic transgene (e.g., HO-1) equipped with a GAL4 upstream activating sequence site located on a second (effector) plasmid.[139] Rapid and robust cardiac-specific HO-1 transgene expression has been reported with this vector system in transfected cardiomyocytes (H9C2 cells) on exposure to low oxygen. In a mouse model of MI, transfection with a similar vigilant vector containing HO-1 resulted in marked HO-1 gene up-regulation in response to ischemia, significantly reduced apoptosis in the infarct area, and improved cardiac function.[140] In summary, gene transfer has shown beneficial effects in the treatment of animal models of CVDs as depicted in Table IV.

Myocardial Transplantation

Another potential application of targeted gene therapy includes protection against myocardial ischemia, inflammation, and rejection, which can follow cardiac transplantation. Initial findings showed that *ex vivo* cardiac transfection with antisense oligonucleotides directed against the intercellular adhesion molecule-1 (ICAM-1), a critical mediator of T-cell adhesion and co-stimulation, prolonged cardiac allograft tolerance and survival rates (over 2 months) when administered before transplantation into the host.[141] Such an approach may prove useful in the preparation of the donor heart for transplantation.

Considerable interest has been generated concerning the potential of *ex vivo* gene transfer for heart grafts, including the use of adenoviral, AAV, and lentiviral vectors, to enhance long-term protection against ischemia and tissue rejection.[142,143] *Ex vivo* gene transfection of the donor organ is also attractive, because targeted genes, and vectors, can be more easily restricted to be organ specific. Adenovirus-mediated gene transfer of HO-1 in an acute cardiac allograft rejection model has been reported.[144] HO-1 overexpression prolonged the survival of vascularized allografts, decreasing the number of graft-infiltrating leukocytes, cytokine expression, and extensive apoptosis in transplanted hearts.

Inherited Cardiomyopathies and Congenital Heart Disease

As previously indicated, there are numerous examples showing that myocardial disease, resulting from single gene mutations, can be corrected by exogenous delivery of the normal gene. Preclinical studies have shown the feasibility of gene therapy for some forms of inherited cardiomyopathies (e.g., caused by δ-sarcoglycan mutation), and defects in transporters involved in dysrhythmias, in particular AF. Although there is strong indication from these studies that supports the proof-in-concept of targeted gene therapy to reverse specific cardiac pathological conditions, the clinical application of this type of gene therapy must await the development of high-efficacy vectors that are safer and have enhanced capacity for long-term expression.

Similarly, there is strong interest and support for the use of gene therapy in the treatment of congenital heart defects (CHD). CHD has frequently been found to be a consequence of specific defects in developmentally regulated genes. Although much remains to be learned about the identification

TABLE IV Gene therapy in animal models of CVDs

CVD	Transgene	Vector	Animal	Ref
Acquired				
CAD/MI	*VEGF*	Plasmid DNA	Pig	82
	Human *FGF4*	Adenovirus	Pig	145
	Human *HGF*	HVJ-liposome	Rat	146
	Human *eNOS*	Adenovirus	Rat	147
	Human *GH*	Adenovirus	Rat	148
Hypertension	*AT1R*-antisense)	Retrovirus	Rat	149
	Human *eNOS*	Plasmid DNA	Rat	96
	PGIS	Naked plasmid	Rat	87
	Human kallikrein (*KLK*)	Adenovirus	Rat	95
	Adrenomedullin (*AM*)	Adenovirus	Rat	97
	Human *ANP*	Adenovirus	Rat	93
Myocarditis	*IFNB*	Naked DNA	Mouse	111
	NF-κB (decoy)	HVJ-liposome	Mouse	112
Restenosis	*eNOS*	Adenovirus	Rat	150
	E2F (decoy)	HVJ-liposome	Rat	78
	Human cyclin G1	Retrovirus	Rat	151
	PDGFA (ribozyme)	Adenovirus	Rat	71
Rescue/cardioprotection				
Heart failure	*SERCA2A*	Adenovirus	Rat	115
	S100A1	Adenovirus	Rat	152
	Human *TIMP-1*	Adenovirus	Rat	153
	Akt	Adenovirus	Rat	136
	β-ARK inhibitor (*BARKct*)	Adenovirus	Rabbit	125–126
Ischemia/reperfusion	Human *HO-1*	AAV	Rat	132
	Extracellular SOD (*SOD3*)	Adenovirus	Rat	154
	Catalase (*CAT*)	Adenovirus	Mouse	155–156
	HSP70	HVJ-liposome	Rat	134
	HSP90	HVJ-liposome	Pig	157
Inherited				
Cardiomyopathy	Human BCL2	Adenovirus	Rabbit	135
	Adenylyl cyclase (ACVI)	Adenovirus	Pig	131
	Sarcoglycan (δSG)	AAV	Hamster	158
Atrial fibrillation	*GEM*	Adenovirus	Guinea pig	108
	Gα	Adenovirus	Pig	107
Other arrhythmias	KIR2.1 (*KCNJ2*)	Adenovirus	Guinea pig	100–101
	HCN2	Adenovirus	Dog	102
	MiRP1	Plasmid	Pig	105
Congenital heart defects	*Nkx2-5*	Adenovirus	Mouse	159
	Human *TBX5*	Retrovirus	Chicken	160
	Fibronectin mRNA binding protein (decoy)	Plasmid/liposome	Lamb	161

of critical genetic loci, including transcription factors such as NKX2.5 and TBX5,[159,160] their mechanism of action, and the temporal sequences in which developmental genes act (see Chapter 6), replacement of defective genes in fetal tissues has initially been successful in preclinical studies. The timing for applying somatic gene therapy (i.e., early delivery) seems to be critical, at least for some CHD (e.g., myotonic dystrophy) in which gene therapy applied before the onset of pathology is more effective than following the often-irreversible damage caused by the underlying genetic defect.[162] Furthermore, early detection of specific DNA defects, their precise location, and timing of gene transfer (which must precede the onset of the developmental programming) is of paramount importance for the developmental defect to be averted. The impact of improved molecular techniques in early mutation detection (e.g., PCR amplification) and the efficient transfer of genes *in utero* (either by intraplacental or embryonic injection) have already been shown in several animal models.[163–166] These studies have demonstrated that both adenoviral and AAV vectors in fetal cardiac gene transfer can be deployed in mice with sustained expression of up to 31 weeks after birth.[167] In addition, fetal delivery of the reporter gene with a lentiviral vector displayed sustained high-level transgene expression in both skeletal and heart tissues for more than 15 months after its administration.[168]

Clinical Studies of Cardiovascular Gene Therapy

Using either pharmaceutical or physical interventions, little progress has been made in the treatment of luminal loss by restenosis occurring in patients after coronary angioplasty. Antiproliferative gene therapy to inhibit vasculoproliferation by targeting cell-cycle activation has been shown to be highly effective in stemming restenosis in several animal models. Therefore, it has been applied in several clinical trials to prevent bypass vein graft failure after coronary artery bypass surgery.[169] Data obtained from the Project in Ex-Vivo Vein Graft Engineering via Transfection (PREVENT I and II), randomized, placebo-controlled studies confirmed the overall safety and feasibility of using a synthetic DNA decoy to sequester the E2F family of transcription factors and arrest cells at the gap period (G_1) checkpoint in the cell cycle. This antiproliferative treatment has been highly effective in preventing intimal hyperplasia in patients (up to 1 year after treatment) and has demonstrated long-term potency in humans.[170]

On the basis of successful gene-mediated therapeutic angiogenesis in the treatment of myocardial ischemia and peripheral vascular disorders in animal studies, several small-scale clinical studies have been conducted with adenovirus-based and plasmid-based delivery of proangiogenic growth factors (e.g., VEGF and FGF) to patients with CAD. The intramyocardial delivery of a plasmid encoding VEGF in patients with untreatable CAD resulted in reduced angina and improved LV function.[171] Similarly, adenovirus-mediated intramyocardial delivery of VEGF in patients with severe CAD undergoing conventional CABG seems to be well tolerated and resulted in improved anginal symptoms, as well as in regional ventricular function and wall motion in the region of vector administration 30 days after therapy.[172] Subsequent studies found that the catheter-based myocardial delivery of plasmid-containing VEGF-2 in patients with chronic myocardial ischemia, as assessed by left ventricular electromechanical mapping, showed reduced areas of myocardial ischemia and improvements in myocardial perfusion.[173] In a randomized placebo-controlled pilot study in patients with chronic myocardial ischemia treated with catheter-based endocardial delivery of this VEGF construct, significant improvement in anginal symptoms occurred after 12 weeks.[174] Although these studies indicate that new collateral vessels are formed and functional, the degree of therapeutic success is difficult to assess in patients. Furthermore, data from larger scale randomized trials using adenoviral-mediated FGF delivery in patients with CAD have also shown evidence of only modestly increased myocardial perfusion, increased treadmill exercise tolerance, and symptomatic improvement in patients with severe disease, but the degree of success remains less conclusive than in the previous studies.[175,176] In another clinical trial, catheter-based intracoronary delivery of adenoviral-VEGF construct was applied during angioplasty (PTCA) in patients with CAD. Although the levels of clinical restenosis were unaffected by VEGF treatment, significant improvement in myocardial perfusion was evidenced in the VEGF-Adv–treated patients after a 6-month follow-up.[177] Importantly, these studies, within their limited timeframe, did not show significant side effects either with gene transfer or overexpression.

Although the promise of effective gene therapy for a large variety of cardiovascular disorders seems closer to being realized with more than 40 clinical trials currently in progress, more information concerning the safety of specific transgenes and vectors, efficacy of delivery, and the long-range consequences of cardiovascular gene therapy is critically needed.

Delivery of DNA

In cardiovascular gene therapy for the effective delivery of genes, several percutaneous catheter systems have been tested in animal models.[178,179] These systems have sought to develop a relatively noninvasive gene delivery, maximizing transgene expression with minimal mortality after infusion of gene constructs to the myocardium and vasculature. The catheter design includes deployment of double balloons, porous, microporous, and hydrogel for intraarterial delivery of plasmid DNAs, because viral vectors may not survive the drying out stage in the preparation of the gel.[180] Major limitations for the use of catheter-based gene delivery include the following: (1) most catheters require prolonged vascular occlusion of the targeted vessel for effective gene delivery; (2) frequent vascular wall damage occurs with subsequent inflammatory response. For gene-mediated treatment of disease with diffused myocardial damage, such as HF and cardiomyopathy, the approach generally used has been intracoronary delivery of the therapeutic gene.

Stents have also been an attractive alternative for localized vascular gene delivery, with either synthetic (e.g., polyurethane, polylactic–polyglycolic acid) or naturally occurring stent coatings (e.g., collagen and phosphorylcholine), which are used as reservoirs for therapeutic agents.[178] These coatings can be impregnated with therapeutic genes, providing a setting for prolonged gene elution and the efficient transduction of arterial walls. Moreover, this gene delivery strategy may decrease the systemic spread of viral vectors and reduce unwanted host immune responses. Furthermore, direct myocardial injection has been frequently used in gene transfer and therapy, particularly for gene delivery to areas of localized, regional myocardial disease. This approach has been successfully used in several studies to deliver angiogenic and cardioprotective genes to the ischemic myocardium. However, the expression of the therapeutic transgene is often restricted to the region surrounding the site of injection and may require multiple injections to adequately cover the entire affected area.

Catheters have been designed to allow subendocardial DNA injections. Such an approach has been successfully used in both animal models and in patients with chronic myocardial ischemia. In the later, introduction of the VEGF gene on plasmid DNA into the left ventricle significantly alleviated anginal symptoms.[180]

Other approaches for effective myocardial gene transfer include coronary artery perfusion and intrapericardial injection. However, most of these techniques, using viral gene transfer, have shown a limited capacity to modify most of the cardiac myocytes rapidly and homogeneously in the *in vivo* heart and, significantly, may result in infection of other organs, including the liver. Retrograde perfusion, using adenoviral constructs, has recently proved to be a highly efficient approach in targeted myocardial infection (greater than 70% infection rate) with limited infection of peripheral organs[181]; however, its use has been rather limited. Finally, another modality of myocardial gene delivery involves the transplantation of genetically engineered cells, usually by direct injection (e.g., adult or embryonic stem cells or skeletal myoblasts).[182] This technique will be further discussed in Chapter 4.

CONCLUSION

The availability of powerful molecular tools for genetic analysis and manipulation of the cardiovascular system has substantially enhanced our understanding of the role that specific gene products play in the normal and diseased heart, as well in cardiovascular development. Moreover, these tools may facilitate new opportunities to develop new and effective therapies. Central to this molecular and cellular analysis is the fact that the cardiomyocyte and the vascular endothelium play paramount roles in the pathogenesis of CVD, and, therefore, they are important targets for interventions in gene expression. Furthermore, a large number of genes critical to growth, signaling, metabolic regulation, structure, and function of these cell types have already been identified. The ability to overexpress specific genes in targeted cell types, as well as the ability to shut down specific gene expression with antisense technologies, including antisense oligonucleotides, ribozymes, and RNA interference, has been well demonstrated in a large variety of preclinical models. Some of these gene-targeting strategies, as well as the introduction of specific genes to enhance cardioprotection and cell growth, have shown potential in the treatment of cardiovascular disorders such as hypertension, atherosclerosis, HF, cardiac dysrhythmias, and myocardial ischemia in animal models, and currently they are under investigation in a number of experimental, multicenter randomized human clinical trials.

Finally, the successful transition of these therapies into mainstream clinical practice is tantalizingly close but, nevertheless, must await substantial improvements and new developments in both vector platforms and delivery systems to provide a documentably safe and efficacious treatment.

SUMMARY

- Viral and plasmid DNAs have been used as vectors in cardiovascular gene transfer into mammalian cells, including cardiomyocytes and endothelial cells, and to both vascular and cardiac tissues.
- Plasmid DNA vectors are easy to isolate, manipulate, and produce in large scale and tend to pose less safety concerns, but display a markedly lower efficiency of transfection (50–100 fold) compared with viral vectors.
- Plasmids can either integrate into chromosomal DNA sequences or be nonintegrating episomes.
- Liposomes can be used effectively as a carrier of genes or of oligonucleotides.
- Adenoviral vectors can readily infect both dividing and nonproliferative cells like cardiomyocytes; however, they provoke unwanted immunogenic responses from the host cell and exhibit transient expression.
- Both adenoviral-associated viruses (AAV) and lentiviruses have a more stable expression and are less immunogenic.
- Specific gene expression can be enhanced (e.g., overexpression), down-regulated, or entirely abolished. Down-regulation can be mediated by a post-transcriptional silencing of gene expression, by the use of antisense oligonucleotides, ribozymes, or RNA interference (RNAi).

- In various animal models, targeted gene therapies have been used to treat a plurality of cardiovascular disorders involving vascular function (e.g., hypertension, restenosis, therapeutic angiogenesis, atherosclerosis), myocardial function (e.g., hypertrophy, HF, cardiac dysrhythmias, and myocardial ischemia), and congenital cardiovascular defects.
- Targeted gene therapy in humans has so far been shown to be less effective than in animal models. A number of large clinical trials are currently underway to assess the impact of specific gene therapy.

References

1. Baumgartner, I., and Isner, J. M. (2001). Somatic gene therapy in the cardiovascular system. *Annu. Rev. Physiol.* **63,** 427–150.
2. Geurts, A. M., Yang, Y., Clark, K. J., Liu, G., Cui, Z., Dupuy, A. J., Bell, J. B., Largaespada, D. A., and Hackett, P. B. (2003). Gene transfer into genomes of human cells by the sleeping beauty transposon system. *Mol. Ther.* **8,** 108–117.
3. Fleury, S., Driscoll, R., Simeoni, E., Dudler, J., von Segesser, L. K., Kappenberger, L., and Vassalli, G. (2004). Helper-dependent adenovirus vectors devoid of all viral genes cause less myocardial inflammation compared with first-generation adenovirus vectors. *Basic Res. Cardiol.* **99,** 247–256.
4. Yoshioka, T., Okada, T., Maeda, Y., Ikeda, U., Shimpo, M., Nomoto, T., Takeuchi, K., Nonaka-Sarukawa, M., Ito, T., Takahashi, M., Matsushita, T., Mizukami, H., Hanazono, Y., Kume, A., Ookawara, S., Kawano, M., Ishibashi, S., Shimada, K., and Ozawa, K. (2004). Adeno-associated virus vector-mediated interleukin-10 gene transfer inhibits atherosclerosis in apolipoprotein E-deficient mice. *Gene Ther.* **11,** 1772–1779.
5. Maeda, Y., Ikeda, U., Shimpo, M., Ueno, S., Ogasawara, Y., Urabe, M., Kume, A., Takizawa, T., Saito, T., Colosi, P., Kurtzman, G., Shimada, K., and , Ozawa K. (1998). Efficient gene transfer into cardiac myocytes using adeno-associated virus (AAV) vectors. *J. Mol. Cell Cardiol.* **30,** 1341–1348.
6. Maeda, Y., Ikeda, U., Ogasawara, Y., Urabe, M., Takizawa, T., Saito, T., Colosi, P., Kurtzman, G., Shimada, K., and Ozawa, K. (1997). Gene transfer into vascular cells using adeno-associated virus (AAV) vectors. *Cardiovasc. Res.* **35,** 514–521.
7. Goncalves, M. A., van Nierop, G. P., Tijssen, M. R., Lefesvre, P., Knaan-Shanzer, S., van der Velde, I., van Bekkum, D. W., Valerio, D., and de Vries, A. A. (2005). Transfer of the full-length dystrophin-coding sequence into muscle cells by a dual high-capacity hybrid viral vector with site-specific integration ability. *J. Virol.* **79,** 3146–3162.
8. Recchia, A., Perani, L., Sartori, D., Olgiati, C., and Mavilio, F. (2004). Site-specific integration of functional transgenes into the human genome by adeno/AAV hybrid vectors. *Mol. Ther.* **10,** 660–670.
9. Fleury, S., Simeoni, E., Zuppinger, C., Deglon, N., von Segesser, L. K., Kappenberger, L., and Vassalli, G. (2003). Multiply attenuated, self-inactivating lentiviral vectors efficiently deliver and express genes for extended periods of time in adult rat cardiomyocytes in vivo. *Circulation* 107, 2375–2382.
10. Bonci, D., Cittadini, A., Latronico, M. V., Borello, U., Aycock, J. K., Drusco, A., Innocenzi, A., Follenzi, A., Lavitrano, M., Monti, M. G., Ross, J. Jr., Naldini, L., Peschle, C., Cossu, G., and Condorelli, G. (2003). 'Advanced' generation lentiviruses as efficient vectors for cardiomyocyte gene transduction in vitro and in vivo. *Gene Ther.* **10,** 630–636.
11. Pelisek, J., Engelmann, M. G., Golda, A., Fuchs, A., Armeanu, S., Shimizu, M., Mekkaoui, C., Rolland, P. H., and Nikol, S. (2002). Optimization of nonviral transfection: variables influencing liposome-mediated gene transfer in proliferating vs. quiescent cells in culture and in vivo using a porcine restenosis model. *J. Mol. Med.* **80,** 724–736.
12. Tashiro, H., Aoki, M., Isobe, M., Hashiya, N., Makino, H., Kaneda, Y., Ogihara, T., and Morishita, R. (2005). Development of novel method of non-viral efficient gene transfer into neonatal cardiac myocytes. *J. Mol. Cell. Cardiol.* **39,** 503–509.
13. Barbato, J. E., Kibbe, M. R., and Tzeng, E. (2003). The emerging role of gene therapy in the treatment of cardiovascular diseases. *Crit. Rev. Clin. Lab. Sci.* **40,** 499–545.
14. Isner, J. M. (2002). Myocardial gene therapy. *Nature* **415,** 234–239.
15. Chen, Z., Ge, Y., and Kang, J. X. (2004). Down-regulation of the M6P/IGF-II receptor increases cell proliferation and reduces apoptosis in neonatal rat cardiac myocytes. *BMC Cell Biol.* **5,** 15.
16. Tomita, N., and Morishita, R. (2004). Antisense oligonucleotides as a powerful molecular strategy for gene therapy in cardiovascular diseases. *Curr. Pharm. Des.* **10,** 797–803.
17. Morishita, R., Aoki, M., and Kaneda, Y. (2001). Decoy oligodeoxynucleotides as novel cardiovascular drugs for cardiovascular disease. *Ann. N. Y. Acad. Sci.* **947,** 294–301.
18. Ehsan, A., and Mann, M. J. (2000). Antisense and gene therapy to prevent restenosis. *Vasc. Med.* **5,** 103–114.
19. Arora, V., Devi, G. R., and Iversen, P. L. (2004). Neutrally charged phosphorodiamidate morpholino antisense oligomers: uptake, efficacy and pharmacokinetics. *Curr. Pharm. Biotechnol.* **5,** 431–439.
20. Summerton, J., and Weller, D. (1997). Morpholino antisense oligomers: design, preparation, and properties. *Antisense Nucleic Acid Drug Dev.* **7,** 187–195.
21. Chen, E., and Ekker, S. C. (2004). Zebrafish as a genomics research model. *Curr. Pharm. Biotechnol.* **5,** 409–413.
22. Heasman, J. (2002). Morpholino oligos: making sense of antisense? *Dev. Biol.* **243,** 209–214.
23. Small, E. M., Warkman, A. S., Wang, D. Z., Sutherland, L. B., Olson, E. N., and Krieg, P. A. (2005). Myocardin is sufficient and necessary for cardiac gene expression in *Xenopus*. *Development* 132, 987–997.
24. Wood, A. W., Schlueter, P. J., and Duan, C. (2005). Targeted knockdown of insulin-like growth factor binding protein-2 disrupts cardiovascular development in zebrafish embryos. *Mol. Endocrinol.* **19,** 1024–1034.
25. Shu, X., Cheng, K., Patel, N., Chen, F., Joseph, E., Tsai, H. J., and Chen, J. N. (2003). Na,K-ATPase is essential for embryonic heart development in the zebrafish. *Development* **130,** 6165–6173.
26. Peterkin, T., Gibson, A., and Patient, R. (2003). GATA-6 maintains BMP-4 and Nkx2 expression during cardiomyocyte precursor maturation. *EMBO J.* **22,** 4260–4273.
27. Chinnery, P. F., Taylor, R. W., Diekert, K., Lill, R., Turnbull, D. M., and Lightowlers, R. N. (1999). Peptide nucleic acid delivery to human mitochondria. *Gene Ther.* **6,** 1919–1928.

28. Tyler, B. M., Jansen, K., McCormick, D. J., Douglas, C. L., Boules, M., Stewart, J. A., Zhao, L., Lacy, B., Cusack, B., Fauq, A., and Richelson, E. (1999). Peptide nucleic acids targeted to the neurotensin receptor and administered i.p. cross the blood–brain barrier and specifically reduce gene expression. *Proc. Natl. Acad. Sci. USA* **96**, 7053–7058.
29. Flierl, A., Jackson, C., Cottrell, B., Murdock, D., Seibel, P., and Wallace, D. C. (2003). Targeted delivery of DNA to the mitochondrial compartment via import sequence-conjugated peptide nucleic acid. *Mol. Ther.* **7**, 550–557.
30. An, D. S., Xie, Y., Mao, S. H., Morizono, K., Kung, S. K., and Chen, I. S. (2003). Efficient lentiviral vectors for short hairpin RNA delivery into human cells. *Hum. Gene Ther.* **14**, 1207–1212.
31. Arts, G. J., Langemeijer, E., Tissingh, R., Ma, L., Pavliska, H., Dokic, K., Dooijes, R., Mesic, E., Clasen, R., Michiels, F., van der Schueren, J., Lambrecht, M., Herman, S., Brys, R., Thys, K., Hoffmann, M., Tomme, P., and van Es, H. (2003). Adenoviral vectors expressing siRNAs for discovery and validation of gene function. *Genome Res.* **13**, 2325–2332.
32. Hurtado, C., Ander, B. P., Maddaford, T. G., Lukas, A., Hryshko, L. V., and Pierce, G. N. (2005). Adenovirally delivered shRNA strongly inhibits Na(+)-Ca(2+) exchanger expression but does not prevent contraction of neonatal cardiomyocytes. *J. Mol. Cell. Cardiol.* **38**, 647–654.
33. Seth, M., Sumbilla, C., Mullen, S. P., Lewis, D., Klein, M. G., Hussain, A., Soboloff, J., Gill, D. L., and Inesi, G. (2004). Sarco(endo)plasmic reticulum Ca2+ ATPase (SERCA) gene silencing and remodeling of the Ca2+ signaling mechanism in cardiac myocytes. *Proc. Natl. Acad. Sci. USA* **101**, 16683–16688.
34. Vazquez, J., Correa de Adjounian, M. F., Sumners, C., Gonzalez, A., Diez-Freire, C., and Raizada, M. K. (2005). Selective silencing of angiotensin receptor subtype 1a (AT1aR) by RNA interference. *Hypertension* **45**, 115–119.
35. Ibrahim, A. F., Hedley, P. E., Cardle, L., Kruger, W., Marshall, D. F., Muehlbauer, G. J., and Waugh, R. (2005). A comparative analysis of transcript abundance using SAGE and Affymetrix arrays. *Funct. Integr. Genomics* **5**, 163–174.
36. Hwang, J. J., Allen, P. D., Tseng, G. C., Lam, C. W., Fananapazir, L., Dzau, V. J., and Liew, C. C. (2002). Microarray gene expression profiles in dilated and hypertrophic cardiomyopathic end-stage heart failure. *Physiol. Genomics* **10**, 31–44.
37. Stanton, L. W., Garrard, L. J., Damm, D., Garrick, B. L., Lam, A., Kapoun, A. M., Zheng, Q., Protter, A. A., Schreiner, G. F., and White, R. T. (2000). Altered patterns of gene expression in response to myocardial infarction. *Circ. Res.* **86**, 939–945.
38. Taylor, L. A., Carthy, C. M., Yang, D., Saad, K., Wong, D., Schreiner, G., Stanton, L. W., and McManus, B. M. (2000). Host gene regulation during coxsackievirus B3 infection in mice: assessment by microarrays. *Circ. Res.* **87**, 328–334.
39. Archacki, S. R., Angheloiu, G., Tian, X. L., Tan, F. L., DiPaola, N., Shen, G. Q., Moravec, C., Ellis, S., Topol, E. J., and Wang, Q. (2003). Identification of new genes differentially expressed in coronary artery disease by expression profiling. *Physiol. Genomics* **15**, 65–74.
40. Bull, T. M., Coldren, C. D., Moore, M., Sotto-Santiago, S. M., Pham, D. V., Nana-Sinkam, S. P., Voelkel, N. F., and Geraci, M. W. (2004). Gene microarray analysis of peripheral blood cells in pulmonary arterial hypertension. *Am. J. Respir. Crit. Care Med.* **170**, 911–919.
41. Kim, Y. H., Lim do, S., Lee, J. H., Shim, W. J., Ro, Y. M., Park, G. H., Becker, K. G., Cho-Chung, Y. S., and Kim, M. K. (2003). Gene expression profiling of oxidative stress on atrial fibrillation in humans. *Exp. Mol. Med.* **35**, 336–349.
42. Ueno, S., Ohki, R., Hashimoto, T., Takizawa, T., Takeuchi, K., Yamashita, Y., Ota, J., Choi, Y. L., Wada, T., Koinuma, K., Yamamoto, K., Ikeda, U., Shimada, K., and Mano, H. (2003). DNA microarray analysis of in vivo progression mechanism of heart failure. *Biochem. Biophys. Res. Commun.* **307**, 771–777.
43. Sergeev, P., da Silva, R., Lucchinetti, E., Zaugg, K., Pasch, T., Schaub, M. C., and Zaugg, M. (2004). Trigger-dependent gene expression profiles in cardiac preconditioning: Evidence for distinct genetic programs in ischemic and anesthetic preconditioning. *Anesthesiology* **100**, 474–488.
44. Masino, A. M., Gallardo, T. D., Wilcox, C. A., Olson, E. N., Williams, R. S., and Garry, D. J. (2004). Transcriptional regulation of cardiac progenitor cell populations. *Circ. Res.* **95**, 389–397.
45. Konstantinov, I. E., Coles, J. G., Boscarino, C., Takahashi, M., Goncalves, J., Ritter, J., and Van Arsdell, G. S. (2004). Gene expression profiles in children undergoing cardiac surgery for right heart obstructive lesions. *J. Thorac. Cardiovasc. Surg.* **127**, 746–754.
46. Blaxall, B. C., Tschannen-Moran, B. M., Milano, C. A., and Koch, W. J. (2003). Differential gene expression and genomic patient stratification following left ventricular assist device support. *J. Am. Coll. Cardiol.* **41**, 1096–1106.
47. Ohki, R., Yamamoto, K., Ueno, S., Mano, H., Misawa, Y., Fuse, K., Ikeda, U., and Shimada, K. (2004). Transcriptional profile of genes induced in human atrial myocardium with pressure overload. *Int. J. Cardiol.* **96**, 381–387.
48. Barrans, J. D., Allen, P. D., Stamatiou, D., Dzau, V. J., and Liew C. C. (2002), Global gene expression profiling of end-stage dilated cardiomyopathy using a human cardiovascular-based cDNA microarray. *Am. J. Pathol.* **160**, 2035–2043.
49. Steenbergen, C., Afshari, C. A., Petranka, J. G., Collins, J., Martin, K., Bennett, L., Haugen, A., Bushel, P., and Murphy, E. (2003). Alterations in apoptotic signaling in human idiopathic cardiomyopathic hearts in failure. *Am. J. Physiol. Heart Circ. Physiol.* **284**, H268–H276.
50. Yung, C. K., Halperin, V. L., Tomaselli, G. F., and Winslow, R. L. (2004). Gene expression profiles in end-stage human idiopathic dilated cardiomyopathy: altered expression of apoptotic and cytoskeletal genes. *Genomics* **83**, 281–297.
51. Tyagi, S. C., Kumar, S., Voelker, D. J., Reddy, H. K, Janicki, J. S., and Curtis, J. J. (1996). Differential gene expression of extracellular matrix components in dilated cardiomyopathy. *J. Cell. Biochem.* **63**, 185–198.
52. Haase, D., Lehmann, M. H., Korner, M. M., Korfer, R., Sigusch, H. H., and Figulla, H. R. (2002). Identification and validation of selective upregulation of ventricular myosin light chain type 2 mRNA in idiopathic dilated cardiomyopathy. *Eur. J. Heart Fail.* **4**, 23–31.
53. Lim, D. S., Roberts, R., and Marian, A. J. (2001). Expression profiling of cardiac genes in human hypertrophic cardiomyopathy: insight into the pathogenesis of phenotypes. *J. Am. Coll. Cardiol.* **38**, 1175–1180.

54. Huang, J., Qi, R., Quackenbush, J., Dauway, E., Lazaridis, E., and Yeatman, T. (2001). Effects of ischemia on gene expression. *J. Surg. Res.* **99,** 222–227.
55. Simkhovich, B. Z., Kloner, R. A., Poizat, C., Marjoram, P., and Kedes, L. H. (2003). Gene expression profiling—a new approach in the study of myocardial ischemia. *Cardiovasc. Pathol.* **12,** 180–185.
56. Kaab, S., Barth, A. S., Margerie, D., Dugas, M., Gebauer, M., Zwermann, L., Merk, S., Pfeufer, A., Steinmeyer, K., Bleich, M., Kreuzer, E., Steinbeck, G., and Nabauer, M. (2004). Global gene expression in human myocardium-oligonucleotide microarray analysis of regional diversity and transcriptional regulation in heart failure. *J. Mol. Med.* **82,** 308–316.
57. Steenman, M., Lamirault, G., Le Meur, N., Le Cunff, M., Escande, D., and Leger, J. J. (2005). Distinct molecular portraits of human failing hearts identified by dedicated cDNA microarrays. *Eur. J. Heart Fail.* **7,** 157–165.
58. Margulies, K. B., Matiwala, S., Cornejo, C., Olsen, H., Craven, W. A., and Bednarik, D. (2005). Mixed messages: transcription patterns in failing and recovering human myocardium. *Circ. Res.* **96,** 592–599
59. Geraci, M. W., Hoshikawa, Y., Yeager, M., Golpon, H., Gesell, T., Tuder, R. M., and Voelkel, N. F. (2002). Gene expression profiles in pulmonary hypertension. *Chest* **121,** 104S–105S.
60. Chen, Y., Park, S., Li, Y., Missov, E., Hou, M., Han, X., Hall, J. L., Miller, L. W., and Bache, R. J. (2003). Alterations of gene expression in failing myocardium following left ventricular assist device support. *Physiol. Genomics* **14,** 251–260.
61. Satterthwaite, G., Francis, S. E., Suvarna, K., Blakemore, S., Ward, C., Wallace, D., Braddock, M., and Crossman, D. (2005). Differential gene expression in coronary arteries from patients presenting with ischemic heart disease: further evidence for the inflammatory basis of atherosclerosis. *Am. Heart J.* **150,** 488–499.
62. Waehre, T., Yndestad, A., Smith, C., Haug, T., Tunheim, S. H., Gullestad, L., Froland, S. S., Semb, A. G., Aukrust, P., and Damas, J. K. (2004). Increased expression of interleukin-1 in coronary artery disease with downregulatory effects of HMG-CoA reductase inhibitors. *Circulation* **109,** 1966–1972.
63. McManus, B. M., Yanagawa, B., Rezai, N., Luo, H., Taylor, L., Zhang, M., Yuan, J., Buckley, J., Triche, T., Schreiner, G., and Yang, D. (2002). Genetic determinants of coxsackievirus B3 pathogenesis. *Ann. N. Y. Acad. Sci.* **975,** 169–179.
64. Luppi, P., Rudert, W., Licata, A., Riboni, S., Betters, D., Cotrufo, M., Frati, G., Condorelli, G., and Trucco, M. (2003). Expansion of specific alphabeta+ T-cell subsets in the myocardium of patients with myocarditis and idiopathic dilated cardiomyopathy associated with Coxsackievirus B infection. *Hum. Immunol.* **64,** 194–210.
65. Gaborit, N., Steenman, M., Lamirault, G., Le Meur, N., Le Bouter, S., Lande, G., Leger, J., Charpentier, F., Christ, T., Dobrev, D., Escande, D., Nattel, S., and Demolombe, S. (2005). Human atrial ion channel and transporter subunit gene-expression remodeling associated with valvular heart disease and atrial fibrillation. *Circulation* **112,** 471–481.
66. Tsukioka, K., Suzuki, J., Fujimori, M., Wada, Y., Yamaura, K., Ito, K., Morishita, R., Kaneda, Y., Isobe, M., and Amano, J. (2002). Expression of matrix metalloproteinases in cardiac allograft vasculopathy and its attenuation by anti MMP-2 ribozyme gene transfection. *Cardiovasc. Res.* **56,** 472–478.
67. Iversen, P. O., and Sioud, M. (2004). Inhibition of gene expression by nucleic acid enzymes in rodent models of human disease. *Methods Mol. Biol.* **252,** 451–456.
68. Phylactou, L. A., Tsipouras, P., and Kilpatrick, M. W. (1998). Hammerhead ribozymes targeted to the FBN1 mRNA can discriminate a single base mismatch between ribozyme and target. *Biochem. Biophys. Res. Commun.* **249,** 804–810.
69. Igondjo-Tchen, S., Pages, N., Bac, P., Godeau, G., and Durlach, J. (2003). Marfan syndrome, magnesium status and medical prevention of cardiovascular complications by hemodynamic treatments and antisense gene therapy. *Magnet. Res.* **16,** 59–64.
70. Grassi, G., Schneider, A., Engel, S., Racchi, G., Kandolf, R., and Kuhn, A. (2005). Hammerhead ribozymes targeted against cyclin E and E2F1 cooperate to down-regulate coronary smooth muscle cell proliferation. *J. Gene Med.* **7,** 1223–1234.
71. Lin, Z. H., Fukuda, N., Suzuki, R., Takagi, H., Ikeda, Y., Saito, S., Matsumoto, K., Kanmatsuse, K., and Mugishima, H. (2004). Adenovirus-encoded hammerhead ribozyme to PDGF A-chain mRNA inhibits neointima formation after arterial injury. *J. Vasc. Res.* **41,** 305–313.
72. Ando, H., Fukuda, N., Kotani, M., Yokoyama, S., Kunimoto, S., Matsumoto, K., Saito, S., Kanmatsuse, K., and Mugishima, H. (2004). Chimeric DNA-RNA hammerhead ribozyme targeting transforming growth factor-beta 1 mRNA inhibits neointima formation in rat carotid artery after balloon injury. *Eur. J. Pharmacol.* **483,** 207–214.
73. Kibbe, M. R., Billiar, T. R., and Tzeng, E. (2000). Gene therapy for restenosis. *Circ. Res.* **86,** 829–833.
74. Shears, L, L, 2nd, Kibbe, M. R., Murdock, A. D., Billiar, T. R., Lizonova, A., Kovesdi, I., Watkins, S. C., and Tzeng, E. (1998). Efficient inhibition of intimal hyperplasia by adenovirus-mediated inducible nitric oxide synthase gene transfer to rats and pigs in vivo. *J. Am. Coll. Surg.* **187,** 295–306.
75. Andres, V. (1998). Control of vascular smooth muscle cell growth and its implication in atherosclerosis and restenosis *Int. J. Mol. Med.* **2,** 81–89.
76. Mann, M. J., Gibbons, G. H., Kernoff, R. S., Diet, F. P., Tsao, P. S., Cooke, J. P., Kaneda, Y., and Dzau, V. J. (1995). Genetic engineering of vein grafts resistant to atherosclerosis. *Proc. Natl. Acad. Sci. USA* **92,** 4502–4506.
77. Mann, M. J., Gibbons, G. H., Tsao, P. S., von der Leyen, H. E., Cooke, J. P., Buitrago, R., Kernoff, R., and Dzau, V. J. (1997). Cell cycle inhibition preserves endothelial function in genetically engineered rabbit vein grafts. *J. Clin. Invest.* **99,** 1295–1301.
78. Morishita, R., Gibbons, G. H., Horiuchi, M., Ellison, K. E., Nakama, M., Zhang, L., Kaneda, Y., Ogihara, T., and Dzau, V. J. (1995). A gene therapy strategy using a transcription factor decoy of the E2F binding site inhibits smooth muscle proliferation in vivo. *Proc. Natl. Acad. Sci. USA* **92,** 5855–5859.
79. Takeshita, S., Weir, L., Chen, D., Zheng, L. P., Riessen, R., Bauters, C., Symes, J. F., Ferrara, N., and Isner, J. M. (1996). Therapeutic angiogenesis following arterial gene transfer of vascular endothelial growth factor in a rabbit model of hindlimb ischemia. *Biochem. Biophys. Res. Commun.* **227,** 628–635.
80. Schwarz, E. R., Speakman, M. T., Patterson, M., Hale, S. S., Isner, J. M., Kedes, L. H., and Kloner, R. A. (2000). Evaluation of the effects of intramyocardial injection of DNA expressing vascular endothelial growth factor (VEGF) in a myocardial

infarction model in the rat—angiogenesis and angioma formation. *J. Am. Coll. Cardiol.* **35,** 1323–1330.
81. Isner, J. M. (1998). Arterial gene transfer of naked DNA for therapeutic angiogenesis: early clinical results. *Adv. Drug Deliv. Rev.* **30,** 185–197.
82. Tio, R. A., Tkebuchava, T., Scheuermann, T. H., Lebherz, C., Magner, M., Kearny, M., Esakof, D. D., Isner, J. M., and Symes, J. F. (1999). Intramyocardial gene therapy with naked DNA encoding vascular endothelial growth factor improves collateral flow to ischemic myocardium. *Hum. Gene Ther.* **10,** 2953–2960.
83. Vincent, K. A., Shyu, K. G., Luo, Y., Magner, M., Tio, R. A., Jiang, C., Goldberg, M. A., Akita, G. Y., Gregory, R. J., and Isner, J. M. (2000). Angiogenesis is induced in a rabbit model of hindlimb ischemia by naked DNA encoding an HIF-1alpha/VP16 hybrid transcription factor. *Circulation* **102,** 2255–2261.
84. Pozeg, Z. I., Michelakis, E. D., McMurtry, M. S., Thebaud, B., Wu, X. C., Dyck, J. R., Hashimoto, K., Wang, S., Moudgil, R., Harry, G., Sultanian, R., Koshal, A., and Archer, S. L. (2003). In vivo gene transfer of the O_2-sensitive potassium channel Kv1.5 reduces pulmonary hypertension and restores hypoxic pulmonary vasoconstriction in chronically hypoxic rats. *Circulation* **107,** 2037–2044.
85. Janssens, S. P., Bloch, K. D., Nong, Z., Gerard, R. D., Zoldhelyi, P., and Collen, D. (1996). Adenoviral-mediated transfer of the human endothelial nitric oxide synthase gene reduces acute hypoxic pulmonary vasoconstriction in rats. *J. Clin. Invest.* **98,** 317–324.
86. Suhara, H., Sawa, Y., Fukushima, N., Kagisaki, K., Yokoyama, C., Tanabe, T., Ohtake, S., and Matsuda, H. (2002). Gene transfer of human prostacyclin synthase into the liver is effective for the treatment of pulmonary hypertension in rats. *J. Thorac. Cardiovasc. Surg.* **123,** 855–861.
87. Tahara, N., Kai, H., Niiyama, H., Mori, T., Sugi, Y., Takayama, N., Yasukawa, H., Numaguchi, Y., Matsui, H., Okumura, K., and Imaizumi, T. (2004). Repeated gene transfer of naked prostacyclin synthase plasmid into skeletal muscles attenuates monocrotaline-induced pulmonary hypertension and prolongs survival in rats. *Hum. Gene Ther.* **15,** 1270–1278.
88. Iyer, S. N., Lu, D., Katovich, M. J., and Raizada, M. K. (1996). Chronic control of high blood pressure in the spontaneously hypertensive rat by delivery of angiotensin type 1 receptor antisense. *Proc. Natl. Acad. Sci. USA* **93,** 9960–9965.
89. Kimura, B., Mohuczy, D., Tang, X., and Phillips, M. I. (2001). Attenuation of hypertension and heart hypertrophy by adeno-associated virus delivering angiotensinogen antisense. *Hypertension* **37,** 376–380.
90. Falcon, B. L., Stewart, J. M., Bourassa, E., Katovich, M. J., Walter, G., Speth, R. C., Sumners, C., and Raizada, M. K. (2004). Angiotensin II type 2 receptor gene transfer elicits cardioprotective effects in an angiotensin II infusion rat model of hypertension. *Physiol. Genomics* **19,** 255–261.
91. Metcalfe, B. L., Huentelman, M. J., Parilak, L. D., Taylor, D. G., Katovich, M. J., Knot, H. J., Sumners, C., and Raizada, M. K. (2004). Prevention of cardiac hypertrophy by angiotensin II type-2 receptor gene transfer. *Hypertension* **43,** 1233–1238.
92. Zhang, Y. C., Bui, J. D., Shen, L, and Phillips, M. I. (2000). Antisense inhibition of beta(1)-adrenergic receptor mRNA in a single dose produces a profound and prolonged reduction in high blood pressure in spontaneously hypertensive rats. *Circulation* **101,** 682–688.
93. Lin, K. F., Chao, J., and Chao, L. (1998). Atrial natriuretic peptide gene delivery attenuates hypertension, cardiac hypertrophy, and renal injury in salt-sensitive rats. *Hum. Gene Ther.* **9,** 1429–1438.
94. Agata, J., Chao, L., and Chao, J. (2002). Kallikrein gene delivery improves cardiac reserve and attenuates remodeling after myocardial infarction. *Hypertension* **40,** 653–659.
95. Chao, J., Zhang, J. J., Lin, K. F., and Chao, L. (1998). Human kallikrein gene delivery attenuates hypertension, cardiac hypertrophy, and renal injury in Dahl salt-sensitive rats. *Hum. Gene Ther.* **9,** 21–31.
96. Lin, K. F., Chao, L., and Chao, J. (1997). Prolonged reduction of high blood pressure with human nitric oxide synthase gene delivery. *Hypertension* **30,** 307–313.
97. Zhang, J. J., Yoshida, H., Chao, L., and Chao, J. (2000). Human adrenomedullin gene delivery protects against cardiac hypertrophy, fibrosis, and renal damage in hypertensive Dahl salt-sensitive rats. *Hum. Gene Ther.* **11,** 1817–1827.
98. Wei, X., Zhao, C., Jiang, J., Li, J., Xiao, X., and Wang, D. W. (2005). Adrenomedullin gene delivery alleviates hypertension and its secondary injuries of cardiovascular system. *Hum. Gene Ther.* **16,** 372–380.
99. Nuss, H. B., Johns, D. C., Kaab, S., Tomaselli, G. F., Kass, D., Lawrence, J. H., and Marban, E. (1996). Reversal of potassium channel deficiency in cells from failing hearts by adenoviral gene transfer: a prototype for gene therapy for disorders of cardiac excitability and contractility. *Gene Ther.* **3,** 900–912.
100. Ennis, I. L., Li, R. A., Murphy, A. M., Marban, E., and Nuss, H. B. (2002). Dual gene therapy with SERCA1 and Kir2.1 abbreviates excitation without suppressing contractility. *J. Clin. Invest.* **109,** 393–400.
101. Miake, J., Marban, E., and Nuss, H. B. (2002). Biological pacemaker created by gene transfer. *Nature* **419,** 132–133.
102. Qu, J., Plotnikov, A. N., Danilo, P. Jr., Shlapakova, I., Cohen, I. S., Robinson, R. B., and Rosen, M. R. (2003). Expression and function of a biological pacemaker in canine heart. *Circulation* **107,** 1106–1109.
103. Plotnikov, A. N., Sosunov, E. A., Qu, J., Shlapakova, I. N., Anyukhovsky, E. P., Liu, L., Janse, M. J., Brink, P. R., Cohen, I. S., Robinson, R. B., Danilo, P., Jr., and Rosen, M. R. (2004). Biological pacemaker implanted in canine left bundle branch provides ventricular escape rhythms that have physiologically acceptable rates. *Circulation* **109,** 506–512.
104. Potapova, I., Plotnikov, A., Lu, Z., Danilo, P., Valiunas, V., Qu, J., Doronin, S., Zuckerman, J., Shlapakova, I. N., Gao, J., Pan, Z., Herron, A. J., Robinson, R. B., Brink, .P, Rosen, M. R., and Cohen, I. S. (2004). Human mesenchymal stem cells as a gene delivery system to create cardiac pacemakers. *Circ. Res.* **94,** 952–959.
105. Burton, D. Y., Song, C., Fishbein, I., Hazelwood, S., Li, Q., DeFelice, S., Connolly, J. M., Perlstein, I., Coulter, D. A., and Levy, R. J. (2003). The incorporation of an ion channel gene mutation associated with the long QT syndrome (Q9E-hMiRP1) in a plasmid vector for site-specific arrhythmia gene therapy: in vitro and in vivo feasibility studies. *Hum. Gene Ther.* **14,** 907–922.
106. Donahue, J. K., Heldman, A. W., Fraser, H., McDonald, A. D., Miller, J. M., Rade, J. J., Eschenhagen, T., and Marban,

E. (2000). Focal modification of electrical conduction in the heart by viral gene transfer. *Nat. Med.* **6,** 1395–1398.

107. Bauer, A., McDonald, A. D., Nasir, K., Peller, L., Rade, J. J., Miller, J. M., Heldman, A. W., and Donahue, J. K. (2004). Inhibitory G protein overexpression provides physiologically relevant heart rate control in persistent atrial fibrillation. *Circulation* **110,** 3115–3120.

108. Murata, M., Cingolani, E., McDonald, A. D., Donahue, J. K., and Marban, E. (2004). Creation of a genetic calcium channel blocker by targeted gem gene transfer in the heart. *Circ. Res.* **95,** 398–405.

109. Watanabe, K., Nakazawa, M., Fuse, K., Hanawa, H., Kodama, M., Aizawa, Y., Ohnuki, T., Gejyo, F., Maruyama, H., and Miyazaki, J. (2001). Protection against autoimmune myocarditis by gene transfer of interleukin-10 by electroporation. *Circulation* **104,** 1098–1100.

110. Cull, V. S., Bartlett, E. J., and James, C. M. (2002). Type I interferon gene therapy protects against cytomegalovirus-induced myocarditis. *Immunology* **106,** 428–437.

111. Bartlett, E. J., Lenzo, J. C., Sivamoorthy, S., Mansfield, J. P., Cull, V. S., and James, C. M. (2004). Type I IFN-beta gene therapy suppresses cardiac CD8+ T-cell infiltration during autoimmune myocarditis. *Immunol. Cell Biol.* **82,** 119–126.

112. Yokoseki, O., Suzuki, J., Kitabayashi, H., Watanabe, N., Wada, Y., Aoki, M., Morishita, R., Kaneda, Y., Ogihara, T., Futamatsu, H., Kobayashi, Y., and Isobe, M. (2001). Element decoy against nuclear factor-kappaB attenuates development of experimental autoimmune myocarditis in rats. *Circ. Res.* **89,** 899–900.

113. Morishita, R., Sugimoto, T., Aoki, M., Kida, I., Tomita, N., Moriguchi, A., Maeda, K., Sawa, Y., Kaneda, Y., Higaki, J., and Ogihara, T. (1997). In vivo transfection of cis element "decoy" against nuclear factor-kappaB binding site prevents myocardial infarction. *Nat. Med.* **3,** 894–900.

114. Yuan, J., Cheung, P. K., Zhang, H. M., Chau, D., and Yang, D. (2005). Inhibition of coxsackie-virus B3 replication by small interfering RNAs requires perfect sequence match in the central region of the viral positive strand. *J. Virol.* **79,** 2151–2159.

115. Miyamoto, M. I., del Monte, F., Schmidt, U., DiSalvo, T. S., Kang, Z. B., Matsui, T., Guerrero, J. L., Gwathmey, J. K., Rosenzweig, A., and Hajjar, R. J. (2000). Adenoviral gene transfer of SERCA2a improves left-ventricular function in aortic-banded rats in transition to heart failure. *Proc. Natl. Acad. Sci. USA* **97,** 793–798.

116. Hajjar, R. J., Kang, J. X., Gwathmey, J. K., and Rosenzweig, A. (1997). Physiological effects of adenoviral gene transfer of sarcoplasmic reticulum ATPase in isolated rat myocytes. *Circulation* **95,** 423–429.

117. del Monte, F., Harding, S. E., Schmidt, U., Matsui, T., Kang, Z. B., Dec, G. W., Gwathmey, J. K., Rosenzweig, A., and Hajjar, R. J. (1999). Restoration of contractile function in isolated cardiomyocytes from failing human hearts by gene transfer of SERCA2a. *Circulation* **100,** 2308–2311.

118. Del Monte, F., Harding, S. E., Dec, W., Gwathmey, J. K., and Hajjar, R. J. (2002). Targeting phospholamban by gene transfer in human heart failure. *Circulation* **105,** 904–907.

119. He, H., Meyer, M., Martin, J. L., McDonough, P. M., Ho, P., Lou, X., Lew, W. Y., Hilal-Dandan, R., and Dillmann, W. H. (1999). Effects of mutant and antisense RNA of phospholamban on SR Ca(2+)-ATPase activity and cardiac myocyte contractility. *Circulation* **100,** 974–980.

120. Dorn, G. W., 2nd, and Molkentin, J. D. (2004). Manipulating cardiac contractility in heart failure: data from mice and men. *Circulation* **109,** 150–158.

121. Minamisawa, S., Hoshijima, M., Chu, G., Ward, C. A., Frank, K., Gu, Y., Martone, M. E., Wang, Y., Ross, J., Jr., Kranias, E. G., Giles, W. R., and Chien, K. R. (1999). Chronic phospholamban-sarcoplasmic reticulum calcium ATPase interaction is the critical calcium cycling defect in dilated cardiomyopathy. *Cell* **99,** 313–322.

122. Song, Q., Schmidt, A. G., Hahn, H. S., Carr, A. N., Frank, B., Pater, L., Gerst, M., Young, K., Hoit, B. D., McConnell, B. K., Haghighi, K., Seidman, C. E., Seidman, J. G., Dorn, G. W., 2nd, and Kranias, E. G. (2003). Rescue of cardiomyocyte dysfunction by phospholamban ablation does not prevent ventricular failure in genetic hypertrophy. *J. Clin. Invest.* **111,** 859–867.

123. Haghighi, K., Kolokathis, F., Pater, L., Lynch, R. A., Asahi, M., Gramolini, A. O., Fan, G. C., Tsiapras, D., Hahn, H. S., Adamopoulos, S., Liggett, S. B., Dorn, G. W., 2nd, MacLennan, D. H., Kremastinos, D. T., and Kranias, E. G. (2003). Human phospholamban null results in lethal dilated cardiomyopathy revealing a critical difference between mouse and human. *J. Clin. Invest.* **111,** 869–876.

124. Milano, C. A., Allen, L. F., Rockman, H. A., Dolber, P. C., McMinn, T. R., Chien, K. R., Johnson, T. D., Bond, R. A., and Lefkowitz, R. J. (1994). Enhanced myocardial function in transgenic mice overexpressing the beta 2-adrenergic receptor. *Science* **264,** 582–586.

125. Shah, A. S., White, D. C., Emani, S., Kypson, A. P., Lilly, R. E., Wilson, K., Glower, D. D., Lefkowitz, R. J., and Koch, W. J. (2001). In vivo ventricular gene delivery of a beta-adrenergic receptor kinase inhibitor to the failing heart reverses cardiac expression. *Circulation* **103,** 1311–1316.

126. Rockman, H. A., Chien, K. R., Choi, D. J., Iaccarino, G., Hunter, J. J., Ross, J., Jr., Lefkowitz, R. J., and Koch, W. J. (1998). Expression of a beta-adrenergic receptor kinase 1 inhibitor prevents the development of myocardial failure in gene-targeted mice. *Proc. Natl. Acad. Sci. USA* **95,** 7000–7005.

127. Williams, M. L., Hata, J. A., Schroder, J., Rampersaud, E., Petrofski, J., Jakoi, A., Milano, C. A., and Koch, W. J. (2004). Targeted beta-adrenergic receptor kinase (betaARK1) inhibition by gene transfer in failing human hearts. *Circulation* **109,** 1590–1593.

128. Hajjar, R. J., del Monte, F., Matsui, T., and Rosenzweig, A. (2000). Prospects for gene therapy for heart failure. *Circulation* **86,** 616–621.

129. Roth, D. M., Gao, M. H., Lai, N. C., Drumm, J., Dalton, N., Zhou, J. Y., Zhu, J., Entrikin, D., and Hammond, H. K. (1999). Cardiac-directed adenylyl cyclase expression improves heart function in murine cardiomyopathy. *Circulation* **99,** 3099–3102.

130. Roth, D. M., Bayat, H., Drumm, J. D., Gao, M. H., Swaney, J. S., Ander, A., and Hammond, H. K. (2002). Adenylyl cyclase increases survival in cardiomyopathy. *Circulation* **105,** 1989–1994.

131. Lai, N. C., Roth, D. M., Gao, M. H., Tang, T., Dalton, N., Lai, Y. Y., Spellman, M., Clopton, P., and Hammond, H.

K. (2004). Intracoronary adenovirus encoding adenylyl cyclase VI increases left ventricular function in heart failure. *Circulation* **110**, 330–336.

132. Melo, L. G., Agrawal, R., Zhang, L., Rezvani, M., Mangi, A. A., Ehsan, A., Griese, D. P., Dell'Acqua, G., Mann, M. J., Oyama, J., Yet, S. F., Layne, M. D., Perrella, M. A., and Dzau, V. J. (2002). Gene therapy strategy for long-term myocardial protection using adeno-associated virus-mediated delivery of heme oxygenase gene. *Circulation* **105**, 602–607.

133. Abunasra, H. J., Smolenski, R. T., Yap, J., Sheppard, M., O'Brien, T., and Yacoub, M. H. (2003). Multigene adenoviral therapy for the attenuation of ischemia-reperfusion injury after preservation for cardiac transplantation. *J. Thorac. Cardiovasc. Surg.* **125**, 998–1006.

134. Jayakumar, J., Suzuki, K., Sammut, I. A., Smolenski, R. T., Khan, M., Latif, N., Abunasra, H., Murtuza, B., Amrani, M., and Yacoub, M. H. (2001). Heat shock protein 70 gene transfection protects mitochondrial and ventricular function against ischemia–reperfusion injury. *Circulation* **104**, I303–I307.

135. Chatterjee, S., Stewart, A. S., Bish, L. T., Jayasankar, V., Kim, E. M., Pirolli, T., Burdick, J., Woo, Y. J., Gardner, T. J., and Sweeney, H. L. (2002). Viral gene transfer of the anti-apoptotic factor Bcl-2 protects against chronic ischemic heart failure. *Circulation* **106**, I212–I217.

136. Matsui, T., Li, L., del Monte, F., Fukui, Y., Franke, T. F., Hajjar, R. J., and Rosenzweig, A. (1999). Adenoviral gene transfer of activated phosphatidylinositol 3-kinase and Akt inhibits apoptosis of hypoxic cardiomyocytes in vitro. *Circulation* **100**, 2373–2379.

137. Tang, Y. L., Qian, K., Zhang, Y. C., Shen, L., and Phillips, M. I. (2005). A vigilant, hypoxia-regulated heme oxygenase-1 gene vector in the heart limits cardiac injury after ischemia-reperfusion in vivo. *J. Cardiovasc. Pharmacol. Ther.* **10**, 251–263.

138. Phillips, M. I., Tang, Y., Schmidt-Ott, K., Qian, K., and Kagiyama, S. (2002). Vigilant vector: heart-specific promoter in an adeno-associated virus vector for cardioprotection. *Hypertension* **39**, 651–655.

139. Tang, Y., Schmitt-Ott, K., Qian, K., Kagiyama, S., and Phillips, M. I. (2002). Vigilant vectors: adeno-associated virus with a biosensor to switch on amplified therapeutic genes in specific tissues in life-threatening diseases. *Methods* **28**, 259–266.

140. Tang, Y. L., Tang, Y., Zhang, Y. C., Qian, K., Shen, L., and Phillips, M. I. (2004). Protection from ischemic heart injury by a vigilant heme oxygenase-1 plasmid system. *Hypertension* **43**, 746–751.

141. Poston, R. S., Mann, M. J., Hoyt, E. G., Ennen, M., Dzau, V. J., and Robbins, R. C. (1999). Antisense oligodeoxynucleotides prevent acute cardiac allograft rejection via a novel, nontoxic, highly efficient transfection method. *Transplantation* **68**, 825–832.

142. Fujishiro, J., Kawana, H., Inoue, S., Shimizu, H., Yoshino, H., Hakamata, Y., Kaneko, T., Murakami, T., Hashizume, K., and Kobayashi, E. (2005). Efficiency of adenovirus-mediated gene transduction in heart grafts in rats. *Transplant. Proc.* **37**, 67–69.

143. Isobe, M., Kosuge, H., Koga, N., Futamatsu, H., and Suzuki, J. (2004). Gene therapy for heart transplantation-associated acute rejection, ischemia/reperfusion injury and coronary arteriosclerosis. *Curr. Gene Ther.* **4**, 145–152.

144. Braudeau, C., Bouchet, D., Tesson, L., Iyer, S., Remy, S., Buelow, R., Anegon, I., and Chauveau, C. (2004). Induction of long-term cardiac allograft survival by heme oxygenase-1 gene transfer. *Gene Ther.* **11**, 701–710.

145. Gao, M. H., Lai, N. C., McKirnan, M. D., Roth, D. A., Rubanyi, G. M., Dalton, N., Roth, D. M., and Hammond, H. K. (2004). Increased regional function and perfusion after intracoronary delivery of adenovirus encoding fibroblast growth factor 4: report of preclinical data. *Hum. Gene Ther.* **15**, 574–187.

146. Ueda, H., Sawa, Y., Matsumoto, K., Kitagawa-Sakakida, S., Kawahira, Y., Nakamura, T., Kaneda, Y., and Matsuda, H. (1999). Gene transfection of hepatocyte growth factor attenuates reperfusion injury in the heart. *Ann. Thorac. Surg.* **67**, 1726–1731.

147. Smith, R. S., Jr., Agata, J., Xia, C. F., Chao, L., and Chao, J. (2005). Human endothelial nitric oxide synthase gene delivery protects against cardiac remodeling and reduces oxidative stress after myocardial infarction. *Life Sci.* **76**, 2457–2471.

148. Jayasankar, V., Pirolli, T. J., Bish, L. T., Berry, M. F., Burdick, J., Grand, T., and Woo, Y. J. (2004). Targeted overexpression of growth hormone by adenoviral gene transfer preserves myocardial function and ventricular geometry in ischemic cardiomyopathy. *J. Mol. Cell. Cardiol.* **36**, 531–538.

149. Katovich, M. J., Gelband, C. H., Reaves, P., Wang, H. W., and Raizada, M. K. (1999). Reversal of hypertension by angiotensin II type 1 receptor antisense gene therapy in the adult SHR. *Am. J. Physiol.* **277**, H1260–H1264.

150. Janssens, S., Flaherty, D., Nong, Z., Varenne, O., van Pelt, N., Haustermans, C., Zoldhelyi, P., Gerard, R., and Collen, D. (1998). Human endothelial nitric oxide synthase gene transfer inhibits vascular smooth muscle cell proliferation and neointima formation after balloon injury in rats. *Circulation* **97**, 1274–1281.

151. Zhu, N. L., Wu, L., Liu, P. X., Gordon, E. M., Anderson, W. F., Starnes, V. A., and Hall, F. L. (1997). Downregulation of cyclin G1 expression by retrovirus-mediated antisense gene transfer inhibits vascular smooth muscle cell proliferation and neointima formation. *Circulation* **96**, 628–635.

152. Most, P., Pleger, S. T., Volkers, M., Heidt, B., Boerries, M., Weichenhan, D., Loffler, E., Janssen, P. M., Eckhart, A. D., Martini, J., Williams, M. L., Katus, H. A., Remppis, A., and Koch, W. J. (2004). Cardiac adenoviral S100A1 gene delivery rescues failing myocardium. *J. Clin. Invest.* **114**, 1550–1563.

153. Jayasankar, V., Woo, Y. J., Bish, L. T., Pirolli, T. J., Berry, M. F., Burdick, J., Bhalla, R. C., Sharma, R. V., Gardner, T. J., and Sweeney, H. L. (2004). Inhibition of matrix metalloproteinase activity by TIMP-1 gene transfer effectively treats ischemic cardiomyopathy. *Circulation* **110**, II180–II186.

154. Agrawal, R. S., Muangman, S., Layne, M. D., Melo, L., Perrella, M. A., Lee, R. T., Zhang, L., Lopez-Ilasaca, M., and Dzau, V. J. (2004). Pre-emptive gene therapy using recombinant adeno-associated virus delivery of extracellular superoxide dismutase protects heart against ischemic reperfusion injury, improves ventricular function and prolongs survival. *Gene Ther.* **11**, 962–969.

155. Woo, Y. J., Zhang, J. C., Vijayasarathy, C., Zwacka, R. M., Englehardt, J. F., Gardner, T. J., and Sweeney, H. L. (1998). Recombinant adenovirus-mediated cardiac gene transfer of superoxide dismutase and catalase attenuates postischemic contractile dysfunction. *Circulation* **98**, II255–II2560.

156. Zhu, H. L., Stewart, A. S., and Taylor, M. D. (2000). Blocking free radical production via adenoviral gene transfer decreases cardiac ischemia-reperfusion injury. *Mol. Ther.* **2,** 470–475.
157. Kupatt, C., Dessy, C., Hinkel, R., Raake, P., Daneau, G., Bouzin, C., Boekstegers, P., and Feron, O. (2004). Heat shock protein 90 transfection reduces ischemia-reperfusion-induced myocardial dysfunction via reciprocal endothelial NO synthase serine 1177 phosphorylation and threonine 495 dephosphorylation. *Arterioscler. Thromb. Vasc. Biol.* **24,** 1435–1441.
158. Kawada, T., Nakazawa, M., Nakauchi, S., Yamazaki, K., Shimamoto, R., Urabe, M., Nakata, J., Hemmi, C., Masui, F., Nakajima, T., Suzuki, J., Monahan, J., Sato, H., Masaki, T., Ozawa, K., and Toyo-Oka, T. (2002). Rescue of hereditary form of dilated cardiomyopathy by rAAV-mediated somatic gene therapy: amelioration of morphological findings, sarcolemmal permeability, cardiac performance and the prognosis of TO-2 hamsters. *Proc. Natl. Acad. Sci. USA* **99,** 901–906.
159. Kasahara, H., Ueyama, T., Wakimoto, H., Liu, M. K., Maguire, C. T., Converso, K. L., Kang, P. M., Manning, W. J., Lawitts, J., Paul, D. L., Berul, C. I., and Izumo, S. (2003). Nkx2.5 homeoprotein regulates expression of gap junction protein connexin 43 and sarcomere organization in postnatal cardiomyocytes. *J. Mol. Cell. Cardiol.* **35,** 243–256.
160. Hatcher, C. J., Kim, M. S., Mah, C. S., Goldstein, M. M., Wong, B., Mikawa, T., and Basson, C. T. (2001). TBX5 transcription factor regulates cell proliferation during cardiogenesis. *Dev. Biol.* **230,** 177–188.
161. Mason, C. A., Bigras, J. L., O'Blenes, S. B., Zhou, B., McIntyre, B., Nakamura, N., Kaneda, Y., and Rabinovitch, M. (1999). Gene transfer in utero biologically engineers a patent ductus arteriosus in lambs by arresting fibronectin-dependent neointimal formation. *Nat. Med.* **5,** 176–182.
162. Waddington, S. N., Kennea, N. L., Buckley, S. M., Gregory, L. G., Themis, M., and Coutelle, C. (2004). Fetal and neonatal gene therapy: benefits and pitfalls. *Gene Ther.* **11,** S92–S97.
163. Woo, Y. J., Raju, G. P., Swain, J. L., Richmond, M. E., Gardner, T. J., and Balice-Gordon, R. J. (1997). In utero cardiac gene transfer via intraplacental delivery of recombinant adenovirus. *Circulation* **96,** 3561–3569.
164. Ryan, K., Russ, A. P., Levy, R. J., Wehr, D. J., You, J., and Easterday, M. C. (2004). Modulation of eomes activity alters the size of the developing heart: implications for in utero cardiac gene therapy. *Hum. Gene Ther.* **15,** 842–855.
165. Christensen, G., Minamisawa, S., Gruber, P. J., Wang, Y., and Chien, K. R. (2000). High-efficiency, long-term cardiac expression of foreign genes in living mouse embryos and neonates. *Circulation* **101,** 178–184.
166. Senoo, M., Matsubara, Y., Fujii, K., Nagasaki, Y., Hiratsuka, M., Kure, S., Uehara, S., Okamura, K., Yajima, A., and Narisawa, K. (2000). Adenovirus-mediated in utero gene transfer in mice and guinea pigs: tissue distribution of recombinant adenovirus determined by quantitative TaqMan-polymerase chain reaction assay. *Mol. Genet. Metab.* **69,** 269–276.
167. Bouchard, S., MacKenzie, T. C., Radu, A. P., Hayashi, S., Peranteau, W. H., Chirmule, N., and Flake, A. W. (2003). Long-term transgene expression in cardiac and skeletal muscle following fetal administration of adenoviral or adeno-associated viral vectors in mice. *J. Gene Med.* **5,** 941–950.
168. Gregory, L. G., Waddington, S. N., Holder, M. V., Mitrophanous, K. A., Buckley, S. M., Mosley, K. L., Bigger, B. W., Ellard, F. M., Walmsley, L. E., Lawrence, L., Al-Allaf, F., Kingsman, S., Coutelle, C., and Themis, M. (2004). Highly efficient EIAV-mediated in utero gene transfer and expression in the major muscle groups affected by Duchenne muscular dystrophy. *Gene Ther.* **11,** 1117–1125.
169. Mann, M. J., Whittemore, A. D., Donaldson, M. C., Belkin, M., Conte, M. S., Polak, J. F., Orav, E. J., Ehsan, A., Dell'Acqua, G., and Dzau, V. J. (1999). Ex-vivo gene therapy of human vascular bypass grafts with E2F decoy: the PREVENT single-centre, randomised, controlled trial. *Lancet* **354,** 1493–1498.
170. Gruchala, M., Roy, H., Bhardwaj, S., and Yla-Herttuala, S. (2004). Gene therapy for cardiovascular diseases. *Curr. Pharm. Des.* **10,** 407–423.
171. Losordo, D. W., Vale, P. R., Symes, J. F., Dunnington, C. H., Esakof, D. D., Maysky, M., Ashare, A. B., Lathi, K., and Isner, J. M. (1998). Gene therapy for myocardial angiogenesis: initial clinical results with direct myocardial injection of phVEGF165 as sole therapy for myocardial ischemia. *Circulation* **98,** 2800–2804.
172. Rosengart, T. K., Lee, L. Y., Patel, S. R., Sanborn, T. A., Parikh, M., Bergman, G. W., Hachamovitch, R., Szulc, M., Kligfield, P. D., Okin, P. M., Hahn, R. T., Devereux, R. B., Post, M. R., Hackett, N. R., Foster, T., Grasso, T. M., Lesser, M. L., Isom, O. W., and Crystal, R. G. (1999). Angiogenesis gene therapy: phase I assessment of direct intramyocardial administration of an adenovirus vector expressing VEGF121 cDNA to individuals with clinically significant severe coronary artery disease. *Circulation* **100,** 468–4674.
173. Vale, P. R., Losordo, D. W., Milliken, C. E., McDonald, M. C., Gravelin, L. M., Curry, C. M., Esakof, D. D., Maysky, M., Symes, J. F., and Isner, J. M. (2001). Randomized, single-blind, placebo-controlled pilot study of catheter-based myocardial gene transfer for therapeutic angiogenesis using left ventricular electromechanical mapping in patients with chronic myocardial ischemia. *Circulation* **103,** 2138–2143.
174. Losordo, D. W., Vale, P. R., Hendel, R. C, Milliken, C. E., Fortuin, F. D., Cummings, N., Schatz, R. A., Asahara, T., Isner, J. M., and Kuntz, R. E. (2002). Phase 1/2 placebo-controlled, double-blind, dose-escalating trial of myocardial vascular endothelial growth factor 2 gene transfer by catheter delivery in patients with chronic myocardial ischemia. *Circulation* **105,** 2012–2018.
175. Grines, C. L., Watkins, M. W., Mahmarian, J. J., Iskandrian, A. E., Rade, J. J., Marrott, P., Pratt, C., and Kleiman, N. (2003). Angiogene GENe Therapy (AGENT-2) Study Group. A randomized, double-blind, placebo-controlled trial of Ad5FGF-4 gene therapy and its effect on myocardial perfusion in patients with stable angina. *J. Am. Coll. Cardiol.* **42,** 1339–1347.
176. Grines, C. L., Watkins, M. W., Helmer, G., Penny, W., Brinker, J., Marmur, J. D., West, A., Rade, J. J., Marrott, P., Hammond, H. K., and Engler, R. L. (2002). Angiogenic Gene Therapy (AGENT) trial in patients with stable angina pectoris. *Circulation* **105,** 1291–1297.
177. Hedman, M., Hartikainen, J., Syvanne, M., Stjernvall, J., Hedman, A., Kivela, A., Vanninen, E., Mussalo, H., Kauppila, E., Simula, S., Narvanen, O., Rantala, A., Peuhkurinen, K., Nieminen, M. S., Laakso, M., and Yla-Herttuala, S. (2003). Safety and feasibility of catheter-based local intracoronary vascular endothelial growth factor gene transfer in the

prevention of postangioplasty and in-stent restenosis and in the treatment of chronic myocardial ischemia: phase II results of the Kuopio Angiogenesis Trial (KAT). *Circulation* **107,** 2677–2683.

178. Sharif, F., Daly, K., Crowley, J., and O'Brien, T. (2004). Current status of catheter- and stent-based gene therapy. *Cardiovasc. Res.* **64,** 208–216.

179. Herttuala, S. Y., and Martin, J. F. (2000). Cardiovascular gene therapy. *Lancet* **355,** 213–222.

180. Sylven, C., Sarkar, N., Insulander, P., Kenneback, G., Blomberg, P., Islam, K., and Drvota, V. (2002). Catheter-based transendocardial myocardial gene transfer. *J. Interv. Cardiol.* **15,** 7–13.

181. O'Donnell, J. M., and Lewandowski, E. D. (2005). Efficient, cardiac-specific adenoviral gene transfer in rat heart by isolated retrograde perfusion in vivo. *Gene Ther.* **12,** 958–964.

182. Melo, L. G., Pachori, A. S., Kong, D., Gnecchi, M., Wang, K., Pratt, R. E., and Dzau, V. J. (2004). Gene and cell-based therapies for heart disease. *FASEB J.* **18,** 648–663.

CHAPTER 4

Cellular Techniques

OVERVIEW

Examination and manipulation of the cells comprising the heart and cardiovascular system have greatly furthered our understanding of the molecular and cellular basis of both their normal and abnormal functioning, as well as providing novel cell therapies for the treatment of cardiovascular disease (CVD). In this chapter, we review advances in techniques used in cell imaging, cell growth, cell transplantation and engineering and speculate about further developments in these areas.

INTRODUCTION

The past decade has brought about a revolution in the techniques available to the cardiologist for both clinical and research endeavors, in particular *in vivo* imaging techniques. The use of radionuclides has allowed the relatively noninvasive assessment of a variety of events, ranging from myocardial perfusion using thallium in the diagnosis of coronary artery disease (CAD) to gauging the increasing apoptosis occurring with cardiac allograft rejection (using technetium-99m-labeled annexin-V).[1,2] Myocardial ischemia can be visualized with nitroimidazole compounds, myocardial necrosis with antimyosin, and myocardial innervation can be evaluated with radiolabeled MIBG. Myocardial metabolism in the normal and ischemic heart can be assessed with PET and single-photon emission computed tomography (SPECT) by using fatty acid and glucose analogs. SPECT is also an excellent method to screen for CAD, whereas PET, because of its enhanced spatial and temporal resolution (compared with SPECT), may be better in the evaluation of an array of CVDs.[3–5] Contrast-enhanced magnetic resonance imaging is another important imaging tool in noninvasive cardiac analysis.[6]

For the analysis of events occurring at the cellular level (usually performed with isolated cells) different imaging options are currently available, and they will be addressed in the first section of this chapter.

CELLULAR IMAGING TECHNIQUES

Metabolic Imaging

The supply of appropriate levels of bioenergetic substrate and the ample production of energy is critical to the normal functioning of the heart. Defects in myocardial energy levels have been implicated in hypertrophic (HCM) and dilated cardiomyopathy (DCM), heart failure (HF), and in the progression of myocardial ischemia. The availability and the selection of specific metabolic substrates (e.g., fatty acids) as an energy source can play a contributory role in the progression of cardiac disease.[7,8] Therefore, imaging to assess metabolic intermediates levels in the myocardium and in the cardiomyocyte may be an important source of information for the clinical cardiologist and researcher.

Several probes can be used to access the important mitochondrial intermediates. For example, nicotinamide adenine dinucleotide (NADH), the primary source of reducing equivalents to the electron transport chain (ETC) and oxidative phosphorylation pathways responsible for the primary production of ATP for the heart, is able to generate an autofluorescence that can be imaged within single cardiomyocytes.[9] This signal can be detected with confocal microscopy in living cells and also can be quantitated *in vitro* with fluorescence spectroscopy. NADH, as a fluorophore, has recently been adapted to measure the dehydrogenase activities associated with mitochondrial NADH generation using enzyme-dependent fluorescence recovery after photobleaching (ED-FRAP).[10] Other aspects of mitochondrial function can also be imaged.[11–13] The mitochondrial membrane potential can be effectively measured by the fluorometric dye, JC-1. Imaging with other fluorometric dyes (e.g., dihydrorhodamine) can also be used to assess the roles of the generation of oxidative stress (OS) by cardiomyocyte mitochondria, the major cellular site for the generation of reactive oxygen species (ROS) and in the progression of apoptosis (e.g., using annexin). The membrane-permeable probes 2′,7′-dichlorodihydrofluorescein diacetate (DCFH-DA) and 2′,7′-dichlorofluorescein (DHF) enter the cardiomyocyte and produce fluorescent signals after intracellular oxidation by ROS, such as hydrogen peroxide and hydroxyl radical, DHF being a better probe in the detection of mitochondrial matrix localized ROS and DCFH-DA being a better sensor of ROS released by mitochondria in the cytosol.[14–17]

Fluorescent dyes can also be used to assess the variable levels of antioxidants present in cardiomyocytes in response to pathological conditions. Levels of free glutathione (GSH), an important antioxidant, whose content changes during cardiac insults, can be measured with an inverted fluorescence microscope in single, isolated cardiomyocytes using CellTracker Blue CMAC, a member of a new family of thiol-sensitive dyes.[18] Moreover, intracellular pH can be assessed in cardiomyocytes with 2′,7′-bis(carboxyethyl)-5,6-carboxyfluorescein.[19] Other specific fluorometric dyes are available that may allow the localized measurement of pH levels within the cardiomyocyte, including seminaphthorhodafluor-1 using confocal imaging.[20] Organelle pH has been also imaged in different subcellular compartments of rat neonatal cardiomyocytes by use of gene transfer.[21] Specific organelles (e.g., Golgi apparatus, mitochondrial matrix, and cytosol) were targeted with target-sequences fused to green fluorescent proteins having defined pH-dependent absorbance and fluorescent emission optima. This allowed the precise measurement of localized pH in these cells. A similar strategy has recently been applied in the introduction of specifically targeted chimeras of the ATP-dependent photoprotein luciferase to fluorometrically assess the specific organelle levels of ATP.[22] However, although these fusion constructs have been evaluated in humans with mitochondrial cytopathies, they have not yet been generated in cardiomyocytes.[23]

Another important element in assessing the cardiomyocyte bioenergetic status involves oxygen content and oxygen consumption. Digital imaging of the oxygenation state within an isolated single rat cardiomyocyte has been made possible by applying three-wavelength microspectrophotometry with enhanced sensitivity augmented by video-enhanced microscopy.[24] This approach has allowed the visualization of gradients of PO_2 within the actively respiring cardiomyocyte, with shallow gradients near the cell center suggesting a site of limited oxygen consumption.[25] The relative hypoxia at the cardiomyocyte core may take on more significance in some pathophysiological conditions such as myocardial ischemia and hypertrophy.

Although PET, as previously noted, has been used in the evaluation of myocardial bioenergetic substrates levels, such as glucose using fluorodeoxyglucose positron emission tomography (FDG-PET), its use in cardiomyocytes has been limited thus far to three-dimensional tissue culture assessment.[26]

Calcium/Ion Flux and Electrophysiological Imaging

It has been well established that calcium flux is pivotally involved in the excitation–contraction cycle in the cardiomyocyte, in bioenergetic activation, in the modulation of gene expression, and in a variety of signaling events in several pathways in cardiomyocyte signal transduction. Understanding of the role of calcium in these diverse aspects of the cardiomyocyte has been greatly expanded by the development and use of sensitive fluorescent dyes (e.g., Fura-2, Fluo-3, and Rhod-2) included in Table I that can be used in both localization and quantitative assessment of calcium flux

TABLE I Fluorescent Dyes Used in Cardiomyocytes

Fluorescent dyes	To evaluate	Reference
2′,7′-Dichlorofluorescin diacetate (DCFH-DA)	ROS	44, 45
Dihydrofluorescein diacetate (DHF)	ROS	16
Dihydroethidium	ROS	14
4,5-Diaminofluorescein-2/diacetate	Intracellular NO	46
di-4-ANEPPS	Action potentials	47
2′,7′-bis(carboxyethyl)-5,6-carboxyfluorescein	Intracellular pH	48
Seminaphthorhodafluor-1	Intracellular pH	49
Mitotracker green	Mitochondrial number	50
Mitotracker red	Membrane potential	51
Chloromethyl-X-rosamine (CMX-ros)	ROS	52
JC-1	Membrane potential	53
Rhodamine 123	Membrane potential	54
Dihydrorhodamine 123	ROS	55
CellTracker Blue CMAC	Glutathione levels	18
Fura-2 AM	Intracellular calcium	27
Fluo-3 AM	Intracellular calcium	27, 56
Rhod-2 AM	Mitochondrial calcium	57
Lucifer yellow	Gap junction	58, 59
Tetramethylrhodamine methyl ester	OS; PT inducer	60
Calcein-AM	Mitochondrial PT pore	61, 62

PT, Permeability transition; NO, nitric oxide; OS, oxidative stress; ROS, reactive oxygen species.

in living cardiomyocytes by use of laser-scanning confocal microscopy.[27] This technique was used to reveal waves of fluorescence throughout the cytosol present in spontaneously contracting cardiac cells containing Fluo-3 that represent localized areas of elevated Ca^{++} initiated by its spontaneous release from the sarcoplasmic reticulum (SR) and propagated through cells.[28] A similar technique to simultaneously measure intracellular calcium transients and contraction in cardiomyocytes *in vivo* was reported using the fluorophores (e.g., Fura-2, Fluo-3) and high-speed digital imaging methods (e.g., using a charge-coupled device [CCD] camera).[29,30] These techniques have shown that calcium transients (and localized gradients) in subcellular regions of cardiomyocytes are affected by α-adrenergic activation, electrical stimulation,[31] and IGF-1.[32] Confocal imaging in combination with patch-clamp analysis has also provided information about the cardiac calcium channels and currents,[33–36] which will be examined in greater detail in a later chapter. Two-dimensional confocal imaging has also been used to study localized calcium release in cardiomyocytes.[34,37]

In addition to the use of fluorophore imaging, novel fluorescent indicators for Ca^{++} termed chameleons have been constructed that are targeted to specific intracellular locations. These indicators are tandem fusions of a blue-emitting green fluorescent protein (GFP), the calcium-binding protein calmodulin, the calmodulin-binding peptide M13, and an enhanced green-emitting or yellow-emitting GFP. Calcium binding makes calmodulin wrap around the M13 domain, increasing the fluorescence resonance energy transfer (FRET) between the flanking GFPs.[38] This technique allowed the visualization and quantitation of small localized changes in calcium levels in the cytosol, nucleus, and endoplasmic reticulum of recipient cells containing cDNAs encoding the chameleons bearing appropriate localization signals.

There is increasing evidence that mitochondrial calcium uptake and release play a pivotal role in influencing cell signaling and in the regulation of mitochondrial bioenergetic function (e.g., Krebs cycle activation).[39] In addition, excessive mitochondrial calcium accumulation has been implicated in the cardiomyocyte dysfunction occurring during reperfusion after ischemia and in combination with increased levels of OS plays a contributory role in the mitochondrial permeability transition (PT) pore opening and progression of cell apoptosis. Therefore, the ability to investigate mitochondrial-localized calcium changes has become of increasing interest. Calcium uptake and levels in cardiomyocyte mitochondria can be quantitatively assessed directly by use of the fluorescent calcium indicator Rhod-2 (Fig. 1). The AM ester of this dye is positively charged and partitions in mitochondria in response to mitochondrial membrane potential, where it is trapped after de-esterification and easily visualized. Its localization is most clearly shown when cells are co-loaded with other mitochondrial-specific staining dyes such as Mitotracker green or after transfection with mitochondrial-targeted GFP.[40,41] Another potential technique for

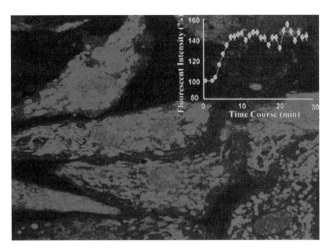

FIGURE 1 Fluorescent imaging of mitochondrial calcium in cultured neonatal rat cardiomyocytes. Cells were loaded with Rhod-2, a fluorescent indicator of mitochondrial calcium level at 4° for 60 min followed by incubation at 37° for 30 min. The graph inside shows the time-lapse increases in mitochondrial-localized Rhod-2 fluorescence in cardiomyocytes subjected to H_2O_2 treatment (see color insert). (Reproduced with permission from Springer Science and Business Media, New York.)

mitochondrial calcium uses the transfection of the recombinant chimeras of the calcium-binding photoprotein, aequorin, that contains a mitochondrial-targeting sequence.[42] This construct has been effectively used for mitochondrial calcium determination in HeLa cells but has not yet been examined in regard to cardiomyocytes.

In addition to (and usually in combination with) calcium imaging, fluorescence imaging has recently been applied to the measurement of electrical activity or transmembrane potentials, V(m) in cardiomyocytes.[43] This technique uses high-power blue-light and green-light emitting diodes (LEDs) to excite cells stained with a V(m)–sensitive dye (di-8-ANEPPS).

Signaling and Real-Time Measurements

The cloning and the heterologous expression of the green fluorescence protein (GFP) have enabled a broad spectrum of investigative applications in which the specific protein location within the cell can be revealed by fluorescence imaging of living cells. Molecularly engineered GFPs can also be exploited not only as a fluorescent reporter but also as a dynamic sensor of intracellular signaling events, such as in the levels of the second messengers such as Ca^{++} and cyclic AMP (cAMP) and in monitoring protein–protein interactions in various cell sub-compartments. This approach allows the study of signaling pathways and messenger fluctuations in living cells with high resolution both in time and in space.[63]

It is widely accepted that cyclic AMP controls several signaling cascades within cells, and changes in the levels of this second messenger have an intrinsic role in many cellular events. As with other second messengers, the intracellular levels of cAMP are responsive to a large number of hormones and neurotransmitters with subsequent effects on cellular behavior. For example, the cardiomyocyte cAMP levels are greatly increased with β-adrenergic stimulation, resulting in downstream stimulation of protein kinase A (PKA) and the phosphorylation of key target proteins, including calcium channel subunits, ryanodine receptors, phospholamban, and troponin I. Although the effector proteins that bind cAMP have been identified, it has largely remained undetermined how this one messenger can effectively regulate the different activities of a multiplicity of cellular proteins. The spatial and temporal nature of cAMP signals remain largely undefined, although there is increasing evidence for the subcellular compartmentation of the myocardial cAMP/G-protein pathway in the myocardium.[64]

A novel method using modified GFP as a biosensor was adapted to monitor the fluctuations and localization of cAMP in living cells. A fusion protein biosensor was constructed using two different GFP fluorophores (i.e., mutant versions) linked to the regulatory and catalytic subunits of the cAMP effector PKA.[65] When intracellular cAMP is low, the two GFP fluorophores are in close proximity and generate FRET between the two fluorescent moieties, whereas increasing levels of cAMP leads to progressive reduction of FRET as the two subunits linked to the GFPs diffuse apart. On selective stimulation of neonatal rat cardiomyocytes with β-adrenergic agonists, a highly localized cAMP response was found defining multiple microdomains in the region of transverse tubule and junctional SR membrane with this biosensor.[66] This approach using real-time imaging of fluorescent biosensors has also shown that the free diffusion of the cAMP is limited and regulated by the activity of specific phosphodiesterases that degrade cAMP,[67] providing a mechanism by which microdomain cAMP levels can be finely modulated, which, in turn, specifically activates a subset of PKA molecules anchored in proximity to the T tubule.[66] On introduction of this biosensor into adult cardiomyocytes using an adenoviral-based vector, cAMP activity was generated by physiologically relevant levels of β-adrenergic receptor activation with normal functional responses. In addition, the complex temporal effects of muscarinic receptor activation on cAMP levels could be explained using the biosensor-containing cells.[68] A nonviral delivery system for the effective cardiomyocyte expression of GFP constructs has also been recently developed.[69]

An alternate approach to optically assess cAMP levels in living cells uses genetically modified cyclic nucleotide-gated (CNG) ion channels to accurately measure plasma membrane–localized cAMP levels in either cell populations or single cells.[70] An olfactory channel A subunit (CNGA2) was modified to improve its sensitivity and selectivity for cAMP and channel activity assessed by measuring Ca^{++} influx using standard fluorometric techniques. In both excitable and nonexcitable cell types, the modified CNG channel sensors also demonstrated the presence of cAMP-rich subcompartments near the plasma membrane, with little evidence of diffusion.

The rapid gating kinetics of these channels allows real-time measurement of cAMP concentrations.

Another method for directly evaluating cAMP levels in living cardiomyocytes involves the use of the fluorescently labeled cAMP-dependent PKA, FlCRhR.[71] A limitation of this technique is that the fluorescent probe must be either delivered by patch-clamp pipette or by microinjection.

Other signaling pathway components can be similarly approached. The analogous use of a fluorescent reporter gene FRET-based approach has also been applied to the detection of protein tyrosine kinase activities[72] and to imaging protein phosphorylation in living cells,[73] although these constructs have not yet been applied to studies with cardiomyocytes to the best of our knowledge. However, the FRET approach has been gainfully used in studies with the cardiomyocyte $Na^+–Ca^{++}$ exchanger, a pivotal plasma membrane regulator of myocyte Ca^{++} levels, Ca^{++} currents, and myocardial contractility.[74] Two fluorophores, a yellow (YFP) and a cyan (CFP) fluorescent protein, were linked to the N- and C-termini of the exchanger Ca^{++} binding domain (CBD) of the $Na^+–Ca^{++}$ exchanger protein to generate a construct (YFP-CBD-CFP) capable of sensitively responding to changes in intracellular Ca^{++} concentrations by FRET detection when expressed in cardiomyocytes. More recently, a similar FRET-based biosensor construct incorporating both YFP and CFP fluorophores sandwiched between the binding domain for the second messenger inositol 1,4,5-trisphosphate (IP3) permitted a detailed spatiotemporal monitoring of intracellular IP3 levels when expressed in either neonatal rat cardiomyocytes or adult cat ventricular myocytes.[75]

Cell imaging technology may also be applied in examining intracellular signaling pathways relevant for cell-death regulation with the potential revelation of strategies to manipulate these pathways for therapeutic effect. Annexin-V binds to externalized phosphatidylserine of apoptotic cells. Real-time imaging of single cell cardiomyocyte apoptosis has been achieved using fluorescent-labeled annexin-V.[76] Interestingly, ^{99m}Tc-annexin scintigraphy has also been used to noninvasively detect cardiomyocyte apoptosis *in vivo* in rats[77] and in rat hearts after myocardial ischemia–reperfusion.[78]

The highly sensitive detection of apoptosis in single cardiomyocytes was recently reported by use of a microfluidic device with electrophoresis on a microdisc, laser-induced ethidium bromide–based fluorescence of cardiomyocyte DNA fragments, monitored by a photomultiplier tube mounted on a confocal microscope.[79]

GROWING CARDIOVASCULAR CELLS: *IN VITRO* CULTURE

Cell Cycle and Cell Sorting Techniques

As previously noted, there are marked alterations in the proliferative capacity of cardiomyocytes during development, although the molecular mechanisms responsible remain largely unidentified. In adult cardiac myocytes, levels of gene expression of an entire constellation of proteins known to be triggers in the cell cycle (e.g., cyclins and cyclin-kinases) are down-regulated, concomitant with the loss of proliferative capacity, and may be involved in the withdrawal of the cardiomyocyte from the cell cycle.[80] The determination of the cardiomyocyte progression through the cell cycle can be assessed by the technique of fluorescence-activated cell sorting (FACS) to evaluate the percentage of S phase myocytes that decreases during development concomitant with a significant increase in the percentage of G_0/G_1 and G_2/M phase cells.[81] This change in S phase myocytes parallels the transition from hyperplasia or hyperplastic growth (i.e., cell proliferation) to hypertrophic myocyte growth, generally occurring shortly after birth.

The use of FACS can also allow the selection and isolation of specific subpopulations of cells from more heterogenous cell mixtures (e.g., pure rod-shaped adult cardiomyocytes, coronary microvascular endothelial cells), enabling targeted studies of gene expression in these subpopulations.[82,83] Recent studies have used FACS (in association with flow cytometry) to directly and quantitatively assess intracellular nitric oxide (NO) in adult cardiomyocytes using the fluorescent NO-specific probe 4,5-diaminofluorescein-2/diacetate (DAF-2/DA).[84] It also has been used to evaluate the levels of cardiomyocyte apoptosis by assessing cells with DNA loss (i.e., with reduced propidium iodine staining).[85] These studies also used FACS of intact cardiomyocytes stained with fluorescein diacetate (FDA) to discriminate between necrosis and apoptosis. FDA is a substrate for cellular esterases whose product, the charged, fluorochrome fluorescein, is entrapped in live cardiomyocytes including apoptotic cells (primarily in the earlier stages), whose membrane remains intact but escapes from necrotic cells whose plasma membrane is damaged.[86]

Cardiomyocyte Cell Culture

A large proportion (estimated at more than 70%) of the cells comprising the ventricular myocardium are not cardiac myocytes. This heterogeneous group of cells consists primarily of fibroblasts but also includes endothelial cells, smooth muscle cells (SMCs), and macrophages. The pronounced heterogeneity of cell types in the heart complicates biochemical and molecular assessment of the intact heart. This has made the study of many phenomena in the heart driven by receptors, kinases, and signaling cascade pathways such as myocardial hypertrophy, and response to growth factors extremely difficult to assess. The development of well-defined cardiac myocyte culture systems can markedly reduce the heterogeneity and complexity involved with *in vivo* myocardial studies. With this cell culture approach, more than 90% of the cells represent cardiomyocytes, whereas contaminating fibroblasts can be reduced to <10% of the cell population by the supplementation to the growth medium of a DNA synthesis inhibitor, bromodeoxyuridine (BRDU).

Generally, serum is not used after the period of growth attachment so as to eliminate undefined growth factors. With serum-free media, nonmyocyte proliferation is reduced, eliminating the need for BRDU. Changes in myocyte size and contractility in response to defined signals can be carefully monitored. Increases in cell size are defined as hypertrophy. These cardiomyocyte cultures can be prepared from either neonatal or adult hearts.

With myocytes derived from neonatal rat heart, insulin stimulated DNA and protein synthesis in cells cultured on a fibronectin-coated surface in serum-free medium.[87] Moreover, IGF-1 stimulated both DNA synthesis and cell proliferation in neonatal myocytes cultured in serum-free medium, without inducing cellular hypertrophy.[88] These findings were somewhat surprising, because it is well known that cardiac myocytes terminate mitotic activity in the neonatal period, and regeneration of cardiac muscle does not occur after myocardial injury in adult hearts. Even embryonic myocytes, which actively proliferate *in vivo*, quickly lose mitotic activity when placed in cell culture.

GROWTH FACTORS IN CARDIOMYOCYTE PROLIFERATION

Several growth factors, including both acidic and basic fibroblast growth factor (FGF), platelet-derived growth factor (PDGF), and transforming growth factor (TGF), have been documented in embryonic hearts, as well as in neonatal cardiomyocytes, and alter myocyte terminal differentiation in culture.[89,90] Although, early studies were unable to demonstrate growth factor–mediated increases in cell division in post-mitotic myocytes,[91] a variety of factors can stimulate proliferative growth of cardiomyocytes derived from either neonatal or embryonic hearts, whereas other factors can inhibit this growth program (see Table II). For example, studies have shown that a homodimeric form of PDGF (i.e., PDGF-BB) can markedly increase rat neonatal cardiomyocyte growth as gauged by increased tritiated leucine incorporation,[92] that the cytokine cardiotrophin-1 can promote proliferation of neonatal cardiomyocytes,[93] as can the increased addition of glucose to the cell culture media.[94]

Moreover, the finding that proliferative pathways in the neonatal cardiomyocyte can be reactivated has been further supported by studies showing a significantly increased proliferation of cardiomyocytes containing overexpressed genes (Table III), such as FGF-2,[95] FGF-receptor,[96] cyclin D2,[97] and cyclin D1.[98]

Similar findings have been recently obtained in adult cardiomyocytes containing overexpression of the cyclin B1-CDC2 (cell division cycle 2 kinase) genes.[99] Previous studies had shown that basic FGF and IGF can modestly

TABLE II Effectors of Proliferation of Primary Cardiomyocyte Culture

Factor	Cell type	Species	Ref
Positive			
FGF-2	Embryonic	Chick	100
	Neonatal	Rat	
IGF	Embryonic	Chick	100
PDGF	Neonatal	Rat	92
CT-1	Embryonic	Mouse	93
EGF	Embryonic	Chick	101
Neuregulin (NRG1)	Neonatal	Rat	102
ANF	Embryonic	Chick	103
Erythropoietin (rHuEpo)	Neonatal	Rat	104
Glucose	Neonatal	Rat	94
Negative			
TGF-β	Neonatal	Rat	100
Rapamycin	Embryonic	Rat	105
Dihydroxyvitamin D3	Neonatal	Rat	106

stimulate DNA synthesis and cell proliferation in adult cardiomyocytes.[100] Studies with transgenic mouse models in which IGF-1 is overexpressed in a cardiac-specific manner leading to increased cardiomyocyte number in adult hearts have also supported the role of the IGF/ IGFR pathway in cardiomyocyte proliferative growth.[107] The findings of reactivation of a proliferative phenotype with adult cardiomyocytes are particularly of interest, because it has been argued that the marked physiological and molecular differences between fetal and neonatal cardiomyocytes compared with terminally differentiated adult cells may obviate the relevancy of studies aimed at cell cycle reactivation in quiescent adult cells.[108]

The administration of a variety of stimuli (i.e., hormones, cytokines, growth factors, vasoactive peptides, and catecholamines) to cultured cardiomyocytes can lead to a hypertrophic phenotype.[112–114] Stimuli such as IGF-1, β-FGF, and triiodothyronine (T3) can lead to increased cell size and protein content in both neonatal and adult cardiomyocytes. These stimuli trigger hypertrophic signaling pathways (which will be discussed in later chapters) with both common elements and distinct features and have provided valuable information in delineating the development of myocardial hypertrophy.

CELLULAR TRANSPLANTATION

Heart disease, including myocardial infarction (MI) and ischemia, is associated with the irreversible loss of cardiomyocytes and vasculature, which occur as a result of either apoptosis or necrosis. The native capacity for renewal and repair of the cardiomyocyte is inadequate as have been the available therapeutic measures to prevent left ventricular remodeling. Until recently, reperfusion of the ischemic myocardium was the primary intervention available to restore the various cellular functions affected by myocardial ischemia, including preventing cell death by necrosis or apoptosis. However, reperfusion often results in extensive myocardial damage, including myocardial stunning, and the functional recovery of the heart may occur only after a period of contractile dysfunction that can last for several hours or days. Moreover, the limited capacity of regeneration and proliferation of human cardiomyocytes can prevent neither the scar formation that follows MI nor the loss of cardiac function occurring in patients with cardiomyopathy and HF. The replacement and regeneration of functional cardiac muscle is an important objective that could be achieved either by the stimulation of autologous resident cardiomyocytes or by

TABLE III Gene Transfer Manipulation of Cardiomyocyte Proliferation *In Vitro*

Gene	Cell type	Modulation	Ref
Positive			
Cyclin D2	Neonatal	Adenoviral overexpression	97
Cyclin B1-CDC2	Adult	Overexpression	99
	Neonatal		
Cyclin D1	Neonatal	Adenoviral nuclear-targeted expression	98
Cyclin B1	Neonatal	Overexpression	99
CDK4	Neonatal	Adenoviral nuclear-targeted expression	98
E2F2	Neonatal	Adenoviral overexpression	109
IGF receptor	Neonatal	Antisense	110
FGF receptor	Neonatal	DNA transfection	96
FGF2	Neonatal	DNA transfection	95
Negative			
p38 MAP kinase	Embryonic	Overexpression	111

the transplantation of stem cells (e.g., embryonic stem cells, bone marrow mesenchymal cells, or skeletal myoblasts) to directly repopulate these tissues, a viable therapeutic approach for repairing the injured myocardium. In this section, we review the cell types used in myocardial transplantation and present information concerning both the preclinical findings and potential clinical application of myocardial cell engineering.

Embryonic Stem Cells

The most primitive of all stem cells are the embryonic stem (ES) cells that develop as the inner cell mass in the human blastocyst at day 5 after fertilization. At this early stage, ES cells exhibit vast developmental potential, because they can give rise to cells of the three embryonic germ layers. When isolated and grown in the appropriate culture media, the pluripotent mouse and human ES cells can undergo proliferation and form embryo-like aggregates (termed embryoid bodies) *in vitro*, some of which can spontaneously contract (Fig. 2). The beating embryoid bodies contain a mixed population of newly differentiated cell types, including cardiomyocytes, based on the expression of cardiac-specific genes encoding proteins such as cardiac-myosin heavy chain, cardiac troponin I and T, atrial natriuretic factor, and cardiac transcription factors GATA-4, Nkx2.5, and MEF-2, their cellular ultrastructure, and extracellular electrical activity.[115–117] These cardiomyocytes can be of the pacemaker-atrium and ventricle-like type distinguishable by their specific patterns of action potential.[118–120]

Although the precise cellular and molecular events comprising the pathway of ES cell-mediated cardiomyocyte-specific differentiation remain largely undetermined, several regulatory factors that can enhance or inhibit the process have been identified, a number of which are listed in Table IV and depicted in Fig. 3.

For instance, early morphogenetic stimuli such as bone morphogenetic protein (BMP), Wnt11, and retinoic acid (RA) can trigger signaling events that modulate cardiomyocyte-specific differentiation from ES cells. Inhibition of BMP signaling by its antagonist Noggin induces cardiomyocyte differentiation from mouse ES cells.[121] Although there is evidence that RA can induce the formation of ventricular-specific cardiomyocytes from ES cells,[122,123] it can also induce neural differentiation as well. NO generated either by NO synthase activity or exogenous origin has also been implicated in the promotion of cardiomyocyte-specific differentiation from mouse ES cells.[124] Moreover, several growth factors have been implicated in promoting cardiomyocyte differentiation from ES cells. TGF-β, FGF-2, PDGF, and IGF-1 promote cardiomyocyte differentiation phenotype and the expression of the cardiomyocyte phenotype in ES cells *in vivo*.[125–129] A newly discovered member of the LIM protein family (hhLIM) plays a role in ES cell–mediated cardiogenesis as a potent transcriptional activator of several cardiac muscle–specific genes; its overexpression

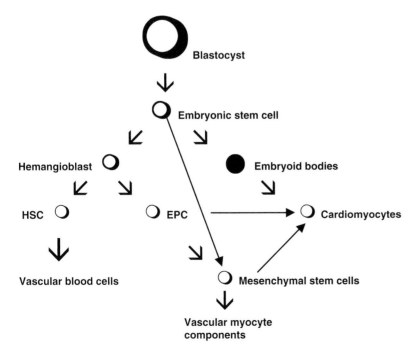

FIGURE 2 Potential development of embryonic stem cells. Pluripotent embryonic stem cells spontaneously differentiate into endothelial progenitor cells (EPC), hemangioblasts, mesenchymal stem cells, and embryoid bodies (embryo-like aggregates). Hemangioblasts further differentiate, generating both hematopoietic stem cells (HSC) and EPC, which give rise to both vascular blood and myocyte components. Under the appropriate conditions (most of which remain to be determined), cardiomyocytes can form from embryoid bodies, as well as from EPC and mesenchymal stem cells.

TABLE IV Regulatory Elements in ES Cell Cardiomyocyte Differentiation

Inducer	References
IGF-1	125
5-aza-2-deoxycytidine	131
Retinoic acid	122
BMP signaling inhibition	121
Oxytocin	132
NO	124
Transforming growth factor-β	129
Icariin	133
GATA-4 acetylation	134
Basic fibroblast growth factor-2	126
Bone morphogenetic protein-2	126
Wnt 11	135
hhLIM	130

results in enhanced expression of cardiac marker genes *Nkx2.5* and *GATA-4* and development of cardiomyocyte-like morphology, whereas its silencing (mediated by antisense targeting) interferes with cardiac muscle gene expression genes and blocks the development of beating cardiomyocytes.[130]

Considerable evidence has implicated epigenetic DNA modifications (e.g., DNA methylation), chromatin modifications (e.g., histone acetylation), and increased gene transcription mediated by the activation and recruitment of specific transcription factors as pivotal regulatory events in promoting cardiomyocyte differentiation from ES cells. Treatment with 5-aza-2′-deoxycytidine, an inhibitor of DNA methylation, promotes cardiomyocyte differentiation of human ES cells,[131] and with the P19 embryonic carcinoma cell-line,[136] a useful model cell for studies on cardiac differentiation. Severe DNA hypomethylation in ES cells markedly abrogates differentiation and induces histone hyperacetylation; restoration of DNA methylation levels rescues these defects.[137]

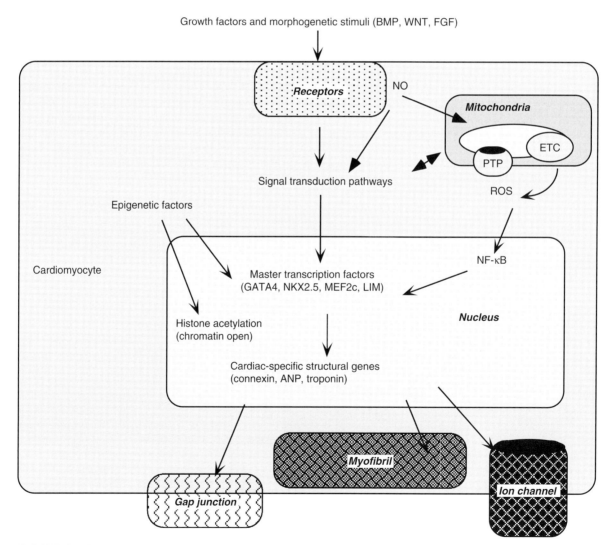

FIGURE 3 Signaling pathways potentially involved in cardiomyocyte differentiation. BMP, bone morphogenetic protein; Wnt, amalgam of wingless (Wg) and int (integration loci); FGF, fibroblast growth factor; ETC, electron transport chain; permeability transition pore (PTP) opening.

Numerous studies have implicated the transcription factors GATA-4 and Nkx2.5 as important regulators of early cardiac morphogenesis and differentiation. Kawamura and associates have recently shown that the post-translational acetylation of GATA-4 is critical for its DNA binding and transcriptional activating function and is involved in the differentiation of ES cells into cardiac myocytes.[134] Histones are also targeted by acetylation during ES-mediated cardiomyocyte differentiation with subsequent remodeling of the chromatin configuration to promote the transcriptional activation of the cardiac-specific gene program.

Interestingly, increased levels of OS seem to modulate the cardiotypic development of embryoid bodies,[138] with recent evidence implicating the cytokine cardiotrophin-1 as mediating both an increase in myocyte ROS levels and increased cardiomyogenesis.[139] Treatment of embryoid bodies with either H_2O_2 or menadione enhanced cardiomyogenesis, whereas incubation with the radical scavengers trolox and N-acetylcysteine exerted inhibitory effects on the differentiation process. Moreover, several redox-sensitive signaling events in the ROS-mediated cardiomyocyte differentiation pathway have been identified by Sauer and associates, including PI3 K activation, the phosphorylation of the Janus kinase signal transducer-2 (Jak-2), the signal transducer and activator of transcription-3 (STAT-3), and the extracellular signal-regulated kinase 1,2 (ERK1/2), as well as the downstream activation of the NF-κB transcription factor.[138,139]

Early studies with fetal cardiomyocyte transplantation demonstrated successful formation of stable grafts and nascent intercalated discs between the grafted and the host myocardial cells.[140,141] A significant feature of ES cell–generated cardiomyocytes is that they maintain myocardial electromechanical properties; human ES cell–derived cardiomyocytes are able to effectively form structural and electromechanical connections with cultured rat cardiomyocytes.[116,117] Transplanted human embryonic cell–derived cardiomyocytes were able to integrate and pace *in vivo* the swine heart with complete atrioventricular block, as demonstrated by several detailed electrophysiological mapping and histopathological studies.[116] Moreover, electrically active, human ES cell–derived cardiomyocytes are capable of both integrating with quiescent, recipient, ventricular cardiomyocytes and inducing rhythmic electrical and contractile activities *in vitro*.[142] In addition, beating embryoid bodies derived from the human ES cells on injection into the LV anterior wall of cryoablated hearts of adult breeder guinea pigs induced pacing in the otherwise electrically silent guinea pig heart.[142]

When fetal rat cardiomyocytes were transplanted into ischemic damaged hearts, a large percentage of cardiomyocytes die after transplantation.[143] This, as well as the finding that no increase in graft size occurred while using increasing number of injected cardiomyocytes, indicates limitations for the clinical application of cardiomyocyte transplantation in the treatment of ischemic heart disease. However, the response of transplanted embryonic stem cells to the various proapoptotic and cytotoxic factors present in the ischemic heart is unknown at present. It is apparent that more research is needed to develop a successful strategy that can maximize grafted cardiomyocyte cell survival and accelerate the differentiation process.

Advantages/Limits to the Use of ES Cell Transplantation

ES cells can be readily and reproducibly obtained from the inner layer of the blastocyst and exhibit an excellent growth phenotype, both *in vitro* and *in vivo*. In addition, the application of ES-like cell lines (e.g., P19) that have been highly informative in the identification and characterization of regulatory factors, transcriptional activators, and signal transduction events involved in cardiomyocyte differentiation[132,136,144,145] may also prove useful in cell transplant therapy.

Preliminary evidence suggests that ES cells may be of particular value in targeting and modifying congenital heart defect phenotypes.[146,147] Once their safety is confirmed, further clinical studies should address the targeted use of ES cell therapy in infants and children with severe cardiac diseases, including cardiomyopathies, congenital heart defects, and dysrhythmias.

ES cells may also be particularly viable substrates for *ex vivo* engineering by means of DNA modifications (e.g., gene therapy, viral transfection, knock-outs, and overexpressed genes). The technology of targeted gene replacement by homologous recombination is particularly effective with mouse embryonic stem cells for the generation of gene knock-outs, as well in studies of cell differentiation, and work is underway to further apply this important technique to human embryonic cell engineering.[148] In addition, the introduction and stable expression of exogenous genes into both undifferentiated and differentiated human embryonic stem cells with a variety of vectors has been achieved by several groups.[149–151] By use of this approach, ES cells transfected with overexpressed therapeutic genes such as adrenergic receptors, growth factors, or ion channel proteins could be transplanted to restore function in defective myocardial cells. However, the safety and efficacy of these methods need to be firmly proven before their use in treating patients.

Significant ethical, moral, and legal concerns about the use of ES cells have been raised, and these concerns have significantly hampered further research efforts needed to provide available cell lines and answers to critical questions regarding the efficacy, integration, long-term stability, function, and even the extent of the negative effects of ES cell transplantation in CVD. A further concern relates to the source of the ES cells, which most often is of heterologous origin, posing the potential problem of generating a host allogenic response such as immunorejection on transplantation.

Recently, it was reported that the differentiation of ES cells transplanted into ischemic myocardium significantly enhanced their immunogenicity.[152] The clinical application of allogenic ES cells may require accompanying immunosuppressive therapy. In addition, the pluripotency and unlimited growth potential of ES cells can have tumorigenic side effects, making the screening for teratoma formation well advised. Moreover, the differentiation of a heterogeneous ES cell population has generally been shown to be inefficient, even with the identification of numerous regulatory factors involved in ES cell–mediated cardiomyocyte-specific differentiation (Table IV). The long-term stability of the ES cell–differentiated cardiomyocyte phenotype is also not well defined, because several studies have demonstrated a loss of ES cell–differentiated cardiomyocytes over time.

Skeletal Myoblasts

Transplanted satellite cells (myoblasts) from skeletal muscle can successfully home and engraft within a damaged myocardium, preventing progressive ventricular dilatation and improving both systolic and diastolic cardiac function.[153–155] Skeletal muscle cells can be delivered into the myocardium by either intramural implantation or arterial delivery[156,157]; a less-invasive catheter approach has been recently deployed.[158] Skeletal muscle satellite cells can proliferate abundantly in culture and can be easily isolated from the patients themselves (i.e., self-derived or autologous), thereby avoiding a potential immune response. Skeletal myoblasts are relatively ischemia resistant, because they can withstand several hours of severe ischemia without becoming irreversibly injured compared with cardiomyocytes that become injured within 20 min.[160] The functional benefits of transplanted skeletal myoblasts in improving the damaged myocardium secondary to ischemia have been well documented in several preclinical models.[160,161] In initial clinical trials, the efficacy of autologous skeletal myoblast transplantation was examined primarily in patients with LV dysfunction.[155] Other preclinical studies have reported that the functional benefits of transplanted skeletal myoblast can be extended to more global forms of nonischemic cardiomyopathy, including DCM, finding that skeletal myoblasts delivered by multiple intramyocardial injections were effective in restoring LV function in the genetically determined Syrian hamster model of DCM.[162] Further studies have recently revealed that graft implantation of a sheet of cultured myoblasts significantly improved cardiac performance and prolonged life expectancy associated with a reduction in myocardial fibrosis and reorganization of the cytoskeletal proteins in DCM hamsters.[163]

Advantages/Limitations of Myoblast Transplantation in Cardiac Therapy

Because myoblasts can be of autologous origin and can be robustly expanded in culture, a large number of cells can be obtained from only a small skeletal muscle sample (derived from the patient) in a relatively short period of time. Compared with transplanted cardiomyocytes, myoblast cells seem to be more resistant to ongoing apoptotic damage prevalent at ischemic sites.

A number of initial studies suggested that a subpopulation of transplanted skeletal muscle cells were capable of transdifferentiation to a cardiomyocyte phenotype with increased expression of cardiac genes.[164,165] However, others have been unable to replicate the transdifferentiation of donor myoblasts to cardiomyocytes,[166] and currently the consensus of most researchers in this field is that grafted myoblast cells primarily remain noncardiomyocytes. On the other hand, there is also evidence suggesting that the cardiac milieu can impact the developmental program of implanted myoblasts, enabling them to better assist cardiac performance. Skeletal myoblasts engrafted to an injured myocardium differentiated to a fatigue-resistant, slow-twitch phenotype better adapted to the chronic workload of the myocardium.[167]

There is also considerable evidence that grafted myoblasts generally display incompatible "wiring" or cell-to-cell connections with resident cardiomyocytes and do not respond in the same way to electrical signaling and stimuli[168]; skeletal muscle does not normally express gap junction proteins (e.g., connexin) needed to form operative electromechanical junctions between cardiomyocytes. Although early preclinical studies did not detect evidence of dysrhythmias, recently it has been shown that a subset of patients receiving skeletal myoblast transplants can experience severe and often life-threatening dysrhythmias.[169] At this time, the precise reason for these dysrhythmias has not been established, but they may be related to the heterogeneous electrical properties and interactions between donor and recipient cells. On the other hand, the dysrhythmias may be promoted by the medium used to introduce the cells rather than by the myoblasts themselves.[170]

It is also possible that some of the functional benefits of myoblast transplantation may not be related to the grafted-myoblasts enhancement of ventricular systolic function but rather to their limitation of adverse post-infarction remodeling or potential paracrine effects on recipient myocardial tissue. Transplants of other cell types, including fibroblasts and smooth muscle cells (neither of which can contract like cardiomyocytes), have also been reported to enhance function of the injured heart.[171,172] This cell-based repair rather than directly involving contractility has been attributed to paracrine effects, whereby transplanted cells produce growth factors, cytokines, and other local signaling molecules that likely promote increasing myocardial perfusion through angiogenesis and arteriogenesis, reducing the infarct tissue such that less ventricular dilation occurs, and enhancement of myocyte survival.

Although preclinical studies with stem cell and myoblast transplantation have shown similar levels of efficacy,[173,174] there is a need for a detailed evaluation of the relative benefits, adverse effects, and efficiency of skeletal myoblast and stem cell transplants in the clinical setting (e.g., HF) *vis á vis* the restoration of myocardial function. New methods to better assess and optimize post-transplanted myoblast recruitment and survival, particularly in the long term, need to be developed, and the repertoire of effective, less-invasive cell delivery technologies needs to be expanded.

Bone Marrow–Derived Cells

The bone marrow contains a wide spectrum of stem cell populations with overlapping phenotypes, including hematopoietic stem cells (HSCs), endothelial progenitor cells (EPCs), mesenchymal stem cells (MSCs), multipotent adult precursor cells (MAPCs), and less abundant stem cell populations such as the side population (SP) cells. These stem cells (most of which are located in the bone marrow stroma) exhibit substantial plasticity with the ability to give rise to a wide range of SMCs.

Interest in the use of bone marrow (BM) stem cells was initially motivated by early findings of their neovascularization properties. These effects on neovascularization and angiogenesis could be enhanced by the presence of specific growth factors and cytokines (e.g., G-CSF). Beneficial effects of implanted BM cells to a damaged vascular system were confirmed and subsequently extended to studies on myocardial damage in mice in which the BM cells were reported to differentiate into myocytes and coronary vessels and thereby ameliorate the function of the injured heart.[175] A less-invasive approach used cytokine treatment, stem cell factor (SCF), and granulocyte colony-stimulating factor (G-CSF) to mobilize endogenous BM cells and direct their integration or homing to the infarcted heart, promoting repair.[176] Mice injected with SCF (200 μg/kg/day) and G-CSF (50 μg/kg/day) exhibited a substantial increase in the number of circulating stem cells from 29 in untreated controls to 7200 in cytokine-treated mice. The BM cells were shown to give rise to new cardiac myocytes and coronary vasculature, and the BM cell–derived myocardial regeneration resulted in improved cardiac function and survival. Bone–marrow derived cardiomyocytes can also arise from either bone marrow stromal cells (MSCs) or from either embryonic or adult endothelial precursor cells (EPCs), as shown in Fig. 2.[177–180] The EPCs can be derived from peripheral blood, as well as bone marrow.[181] Injection of BM stem cells after their expansion in culture can also be used in the rescue of an abnormal cardiac phenotype.[182,183]

Advantages/Concerns in the Use of Adult BM Cell Transplantation

There is evidence that BM cell treatment can ameliorate both myocardial function and vascular damage by increasing angiogenesis. The beneficial effect on vascular growth can significantly impact the recovery of the damaged heart (i.e., by improving oxygen availability). Moreover, autologously derived cells for transplantation are an attractive alternative, because bone marrow and mesenchymal cells can be readily isolated from most subjects, including older adults. In addition, expansion of BM cell number by *in vitro* growth can be readily achieved by vigorous growth of mesenchymal cells in culture. Of significance, this method bypasses much of the ethical and legal maelstrom associated with the use of ES cells.

The mechanism of BM cell–mediated augmentation of cardiomyocyte number and function remains highly controversial. Murry and associates have suggested that the effects of BM cell transplantation on the recipient heart are primarily a result of cell fusion with preexisting cardiomyocytes or occur as a function of paracrine effects of the transfected BM cells.[184–186] Others have presented evidence that under appropriate conditions transdetermination events occur in which BM-derived cells or endothelial progenitor cells differentiate into functionally active cardiomyocytes.[187–191] Although cell fusion between cardiomyocytes and noncardiomyocytes has been documented both *in vivo* and *in vitro*,[186,192] others have excluded its involvement as a factor in their studies.[188,193] Further research is needed to clarify these issues and reconcile these contradictory claims, as well as to provide additional information about the extent of cell fusion and when it occurs. Similarly, a more careful delineation of transdetermination from a well-defined adult stem cell type is warranted. A critical problem in the replication of these experiments and in the determination of the effects of BM cells lies in the overall heterogeneity of the BM cell populations used.

Because most BM cell transplantation studies are performed in mice or rats at present, a critical question that remains unanswered is whether this model is truly applicable to humans. Preliminary clinical studies suggest that BM cell transplantation, cytokine, and growth factor supplementation can increase cell homing into areas of injury and promote neovascularization in those areas where they are needed. Whether enough BM cells can be transplanted to repair the damaged regions in the human heart, which tend to be larger in size than in the mouse heart, remains to be seen. Moreover, it is also undetermined whether adult human BM stem cell therapy is effective in treating apoptotic cell death, HCM, the cardiomyopathy of aging, cardiac conduction/dysrhythmia defects, and cardiac defects in infants and children.

Results of several human clinical trials of BM cell transplantation have shown a modest, but significant, improvement in cardiac function in patients with acute myocardial ischemia and infarct.[194–200] In the BOOST clinical trial that included 60 patients with acute MI, of which 30 randomly assigned subjects received intracoronary injections of unfractionated autologous BM cells an average of 6 days after occlusion, cardiac MRI at 6 month follow-up indicated that the transplant group had a significant increase in ejection fraction compared with controls.[195] In another randomized study, Chen and associates used intracoronary delivery of autologous bone marrow–derived mesenchymal cells in 34 patients 18 days after occlusion and found that the transplant group exhibited significantly improved LV function as shown by improvement in wall movement velocity, reduction in ventricular end-systolic and end-diastolic volumes, and an increase in ejection fraction compared with a saline-infused control group.[199] In a randomized, double-blind, placebo-controlled study in 67 patients, Janssens and associates found that patients receiving intracoronary injections of autologous BM cells within 24 hours of acute MI had significantly decreased infarct size (as gauged by MRI) at 4 month follow-up and an enhanced recovery of regional systolic function.[201] When cell transplantation was applied to patients with chronic myocardial disease or damage secondary to MI, the beneficial results were less definitive.

A limitation of most of the clinical studies with adult noncardiac stem cell transplantation relates to the stability of the differentiated phenotype, because these studies have primarily examined the short-term benefits. However, it is important to underline the absence of adverse events in more than 100 patients studied. This is in contrast with the dysrhythmia-prone myoblast transplantation.[169] With the scarcity of successful techniques to effectively treat HF, there is mounting pressure to expedite the clinical application of stem cell transplantation even before the mechanisms (as well as long-term effects) are fully understood.

Adult Cardiac Stem (ACS) Cells

The discovery of a resident stem cell population in the adult heart emerged recently in several well-established laboratories. Many questions remain concerning the origin, structure, precise location, function, and regulation of these cardiac stem cells.

The existence of a subpopulation of Lin⁻ c-kit⁺ cells in adult rat myocardium that have many of the properties of cardiac stem cells has been described.[202] These cells are self-renewing (can be propagated for months), expandable in culture, and multipotent, giving rise to cardiomyocytes, SMCs, and endothelial cells. When injected into an ischemic heart, these cells contribute to the formation of endothelium and vascular smooth muscle and to the regeneration of myocardium in the region of necrosis, improving its pump function and the ventricular chamber geometry.[202,203]

The isolation and characterization of a small population of adult heart–derived cardiac progenitor cells (derived from post-natal mouse myocardium) expressing the surface marker stem cell antigen-1 (Sca-1⁺) and telomerase activity associated with self-renewal potential was recently reported by Oh and associates.[204,205] These ACS cells that can be selectively isolated by a magnetic cell-sorting system expressed neither cardiac structural genes nor Nkx2.5; they can differentiate *in vitro* forming beating cardiomyocytes, in response to the DNA demethylating agent 5′-azacytidine, in part depending on Bmpr1a, a receptor for bone morphogenetic proteins. Microarray profiling of ACS cells has shown the expression of other cardiogenic transcription factors (GATA-4, MEF-2C, TEF-1), as previously reported in BM stromal cells with cardiogenic potential. Similarly, when treated with oxytocin, Sca-1⁺ cells expressed genes of cardiac transcription factors and contractile proteins and exhibited sarcomeric structure and spontaneous beating.[206] The Sca-1⁺ cardiac stem cell after intravenous delivery can home to myocardium injured by ischemia/reperfusion and can functionally differentiate *in situ*.

Laugwitz and associates recently described a population of cardioblasts in both embryonic and post-natal hearts (mouse, rats, and human) numbering just a few hundred per heart, which were identified on the basis of their expression of a LIM-homeodomain transcription factor, Isl1.[207] The cells were primarily localized in the atria, right ventricle, and outflow tract regions (where Isl1 is most prevalently expressed during cardiac organogenesis). These myocardial-derived stem cells can be isolated, transplanted, survive, and replicate in the damaged heart with evidence of functional improvement.[208]

A distinct population of located SP (side population) cells containing the ABCG2 marker, a member of the ATP binding cassette (ABC) transporters, has been reported in the post-natal and adult heart.[209-211] These rare cardiac progenitor cells are capable of proliferation and differentiation and can be isolated on the basis of their differential staining with the Hoechst 33342 dye and subsequent FACS. Multipotent SP cells are present in many adult tissues and show overlap of cell surface antigens and markers such as Sca-1 and ABCG2 but also display tissue-specific markers.[212] It is unclear whether all tissue-derived SP cells share a common progenitor cell. Although it has been shown that cardiac-derived SP cells can differentiate into cardiomyocytes, evidence that bone marrow SP cells engrafted into an ischemic heart can similarly differentiate[213] has been recently challenged.[214] The use of SP cell transplantation may also be limited in that they represent a heterogenous population and cannot be clonally replicated.

Advantages/Limitations of Cardiac Stem Cell Transplantation

Until recently, there has been limited data concerning the presence of ACS cells. This population of cardiac stem cells seems to be limited in number and difficult to identify and expand in culture, thereby limiting their characterization and use and likely contributing to the difficulty in reproducing experiments concerning their transplantation. In addition, at this time, there is no consensus in the definition of selective markers specific to this cell type (Table V).

Further research remains to be done to fully delineate the relevant cardiac progenitor cell populations and the optimal conditions for their efficient homing, differentiation, and integration into myocardium. Understanding the factors that are responsible for ACS growth (both *in vitro* and *in vivo*), homing, and differentiation may allow enhanced strategies

TABLE V Markers of Stem Cell Types Used for Cardiac Transplantation

Cell type	Differentiation agent	Markers of differentiated cardiomyocyte	Refs
ESC			
Embryonic stem cells	IGF-1, TGF-β	α-Sarcomeric actin, connexin 43, major histocompatibility complex class I, sarcomeric myosin, α-actin	125, 129
P19 embryonal carcinoma line	5′-Azacytidine	Bone morphogenetic protein-2 (BMP-2), BMP-4, Bmpr 1a, Smad1, GATA-4, Nkx2.5, cardiac troponin I, and desmin	136
BMC			
Bone marrow (MSC)	Insulin, ascorbic acid, dexamethasone	α-Skeletal actin, β-myosin heavy chain (MHC), MLC-2v, CaV1.2, cardiac troponin I, sarcomeric tropomyosin, cardiac titin	176, 215
Cardiac stem cell			
c-kit⁺Lin⁻	ND	c-kit⁺	202
isl1⁺	ND	Csx/Nkx-2.5, GATA4	207
Sca-1⁺ c-kit⁻	5′-Azacytidine oxytocin	High telomerase activity, Sca-1⁺ Nkx-2.5, GATA4, MEF-2C, α- and β MHC, MLC-2, MLC-2v, cardiac-α actin	204–206
Cardiosphere	c-kit⁺	Cardiac troponin I, myosin heavy chain, atrial natriuretic peptide	208
SP cells	ND	ATP-binding cassette transporter (ABCG2)	209, 211

ND, not determined.

to optimize their production and functional benefits on transplantation. Moreover, such information may also shed light on the activation of endogenous cardiac stem cells contributing to cardiac repair, as well as to explain their actual physiological role in the heart. Also to be defined are the kinds of cardiac defects, as well as type of injuries, that can be best treated with these cells, including a clear knowledge of the best place in the heart to deliver or direct these cells. For instance, implanting cells within an area of necrosis and/or low oxygen availability may be counterproductive, whereas implanting cells in regions of hibernating myocardium may be successful.

The long-term stability and functionality of transplanted ACS cells remains to be defined. What happens with introduction of genes in the ACS-transplanted cells? Can robust expression of specific genes be directed in such cells? Whether an increased proliferative response in the cardiac progenitor cells can be modulated by the introduction of direct gene transfer of cell-cycle progression genes remains to be seen. ACS transplantation may prove more effective than adult BM cell transplantation, because cardiac stem cells could be better programmed, allowing for greater specificity. The identification, purification, and characterization of the ACS cells, as well a detailed knowledge of the interactions with their milieu or niche, are essential if we are to achieve the major goal of regenerating/transplanting the tissue to treat MI.

Establishment of Stem Cell Identity

From the foregoing discussion, it should be evident that a critical element in the identification of the grafted cell in the heart and in a number of cases even before the transplant is the unequivocal assignment of cell-type identity. In Table V, we provide a list of endogenous molecular markers that have been used to establish a differentiated cardiac phenotype resulting from transplanting different stem cell types, including BM cells, ES cells, and cardiac-derived stem cells.

In addition to the endogenous markers available to establish cell identity, GFP has been extensively used as a reporter to define donor cells. Marking cells with the chromosome stain DAPI has been recently shown to be flawed, because the DAPI stain from dead cells can be readily incorporated by unmarked cells.[216]

Genotype marking has also been shown to be a powerful tool in assessing cell identity. In several studies of cardiovascular self-repair in which female hearts were allografted into human male recipients, the presence of extracardiac recipient cells bearing a Y chromosome were assessed both in the coronary vasculature and in cardiomyocytes, because the Y chromosome can be easily viewed by cytochemical staining or by fluorescence *in situ* hybridization.[217–220] Interestingly, the extent of cardiac chimerism reported in these studies exhibited striking variation ranging from very low level of Y-chromosome containing cardiomyocytes (0.02–01%) to high levels (30%), underscoring the critical need for establishing rigorous criteria by which chimerism is identified.[218–220] The identification of a nucleus with a Y chromosome is in itself not sufficient but should be unequivocally associated with either myocardial vessels or cardiomyocyte structure (i.e., by confocal microscopy). Otherwise, it is possible to attribute the Y chromosome–positive nuclei to host cells involved in immune response and inflammatory infiltration and not to cardiac regeneration. There is also some indication that the use of chromosomal analysis can result in an underestimation of the transfected cells because of the presence of nuclei that may not be counted in the histological section.[177]

The detection of cell phenotype markers by real-time assays, confocal microscopy, and noninvasive detection methods using magnetic resonance imaging (MRI) has just begun to be applied in the assessment of cell transplantation. Real-time visualization can provide identification of regions of MI and precise MRI-guided delivery of therapeutic agents, with injection sites identified by contrast agents. Novel contrast agents can permit the MRI-mediated visualization of gene expression at a cellular resolution and can be used as well to detect apoptotic cells.[221,222] Appropriate labeling of stem cells and their detection by MRI should enable the determination of their *in vivo* distribution and provide a glimpse of their fate over time.[222,223]

Which Stem Cell Type Is Best to Treat Cardiac Disease?

Although preclinical studies with stem cell and myoblast transplantation have shown similar levels of efficacy, there is an critical need for a detailed evaluation of the relative benefits, adverse effects, and efficiency of skeletal myoblast and stem cell transplants in the clinical setting (e.g., HF). A brief comparison of the advantages and limitations of the cell types used in cardiac transplantation at present is featured in Table VI. Although no clear-cut choice has yet emerged as to the best cell type to transplant for myocardial repair, there are increasing reasons to believe that a multiplicity of approaches in the application of cell engineering will be required to advance therapies of different cardiac disorders. The cell-based approach to ameliorating HF may require the transplantation of different cell types (e.g., skeletal myoblasts) than might be used in the targeted treatment of cardiac dysrhythmias, conduction disorders, and congenital defects. There also is the possibility that the long-term repair of a fully functioning myocardium may require the incorporation of more than a single cell type—for instance, the addition of cardiomyocytes, fibroblasts, and endothelial cells—or mixed populations of stem cells[224] for the integration and generation of a stable and responsive cardiac graft.

OTHER DEVELOPING TECHNOLOGIES IN CELL ENGINEERING

The refinement of techniques for nuclear transfer, cybrid, and cell fusion may promote the further engineering of stem cells to provide enhanced cardioprotection or to stimulate antioxidant or antiapoptotic responses in the myocardium. Therapeutic cloning using nuclear transfer techniques has been proposed as a viable strategy for generating an unlimited supply of rejuvenated and histocompatible stem cells carrying the nuclear genome of the patient.[225] Transplanted c-kit$^+$ stem cells derived from cloned mouse embryos constructed from nuclear transfer techniques restored myocardial function in severely infarcted hearts more fully and effectively than transplanted adult bone marrow c-kit$^+$ stem cells.[226] A similar approach to cell-engineering techniques might also allow the specific targeting of mitochondrial-based cytopathies.[227]

The identification of features of the cardiac milieu that contribute to the growth and development of transplanted cells *in vivo* has been advanced by the use of three-dimensional matrices designed as a novel *in vitro* system to mimic aspects of the electrical and biochemical environment of the native myocardium. This approach can provide a finer resolution of electrical and biochemical signals that may be involved in cell proliferation and plasticity. Myoblasts have been grown on three-dimensional polyglycolic acid mesh scaffolds under control conditions in the presence of cardiac-like electrical current fluxes, and in the presence of culture medium that had been conditioned by mature cardiomyocytes.[228] The scaffolds generally use an immunocompatible biodegradable material such as collagen or gelatin that can be degraded shortly after grafting. Such scaffolds containing either fetal or neonatal aggregates of contracting cardiac cells have been used to generate artificial cardiac grafts transplanted into injured myocardium with recuperation of ventricular function and formation of functional gap junctions between the grafted cells and the myocardium.[229–232]

The combination of gene therapy and stem cell engineering is an increasingly attractive approach for treating cardiac disorders. As discussed in Chapter 3, there have been numerous demonstrations that the overexpression (and in some cases the inhibition of gene expression) of specific cardiomyocyte proteins can result in striking changes in cardiomyocyte and cardiac phenotypes. Specific cardiomyocyte functions, including ion channel and cardiac conduction, contractility, and cell proliferation, have been shown to be effected by the gene transfer and expression of specific proteins.[111,233,234] Cell-based therapies for

TABLE VI Myocardial Transplant: Cell Type–specific Advantages and Limitations

Cell type	Source	Advantages	Limitations
Cardiac stem cells	Allogenic fetal, neonatal, or adult heart	1. Recognition of myocardial growth factors and recruitment to myocardium are likely faster and more efficient than other cell-types. 2. *In vivo* electrical coupling of transplanted cells to existing myocardium has been demonstrated.	1. Poor cell growth *in vitro*. 2. Transplanted cells are very sensitive to ischemic insult and apoptosis. 3. Availability from either fetal (F), neonatal (N), or adult sources is low at present; likely immune rejection; F and N cells pose ethical difficulties.
Skeletal myoblast	Autologous skeletal muscle biopsy	1. Cells proliferate *in vitro* (allowing for autologous transplant). 2. Ischemia resistant. 3. Transplanted myoblasts can differentiate into slow-twitch myocytes (similar to cardiomyocytes) enabling cellular cardiomyoplasty. 4. Reduces progressive ventricular dilatation and improves cardiac function. 5. Can use adult cells.	1. Likely do not develop new cardiomyocytes *in vivo*. 2. Electrical coupling to surrounding myocardial cells is unclear (may cause dysrhythmias). 3. Long-term stability of differentiated phenotype is unknown.
Adult bone marrow stem cells	Autologous bone marrow stromal cells (mesenchymal); bone marrow (endothelial progenitor cells)	1. Pluripotent stem cells can develop into cardiomyocytes. 2. Easy to isolate and grow well in culture. 3. Neovascularization can occur at site of myocardial scar reducing ischemia. 4. Transdifferentiation of cells into cardiomyocyte *in vivo* has been shown. 5. Can be derived from autologous source; no immune suppression treatment. 6. Can improve myocardial contractility.	1. New program of cell differentiation is required. 2. Efficiency of the differentiation into adult cardiomyocytes appears limited. 3. Signaling, stability, and regulation of differentiation is unknown.
Embryonic stem cells	Allogenic blastocyst (inner mass)	1. Easy propagation and well-defined cardiomyocyte differentiation process. 2. *In vivo* electrical coupling of transplanted cells to existing myocardial cells. 3. Pluripotent cells.	1. Potential for tumor formation and immune rejection (allogenic). 2. Incomplete response to physiological stimuli. 3. Legal and ethical issues. 4. Donor availability.

injured or dysfunctional hearts can be enhanced by the use of *ex vivo* genetically modified stem cells to deliver therapeutic genes and proteins. For instance, transplanted MSCs were recently shown to be effective delivery platforms for introducing channel proteins involved in pacemaking activity (e.g., channel protein HCN2), resulting in modifying the rhythms of the heart *in vivo*.[235] The introduction of vascular endothelial growth factor (VEGF) and its salutary effect on both angiogenesis and left ventricular function in an animal model of ischemic cardiomyopathy was markedly enhanced in hearts with VEGF-transfected skeletal myoblasts compared with hearts directly injected with the adenoviral–VEGF construct.[236]

CONCLUSION

It is clear that much of the future of clinical cardiology lies in the use and development of molecular and cellular techniques of imaging. The use of isolated cells serves as a useful starting point for the discovery of factors involved in cardiomyocyte differentiation, cell signaling, responsiveness to physiological stimuli, and roles in pathophysiological events, such as myocardial ischemia, infarct, and hypertrophy.

The study of isolated cardiomyocyte growth has also provided a unique window to probe the molecular nature of cardiac growth and repair. The discovery of cardiogenesis in adult animals and humans represents one of the most significant advances in cardiology in the past 25 years. Previously, most cardiologists believed that the birth of new cardiomyocytes was only confined to the fetal and neonatal heart. This dogma collapsed when researchers discovered that the hearts of adult rats, mice, and humans undergo significant cardiac changes as a function of age and that cardiomyocytes have the capacity to proliferate, home, and integrate structurally into injured myocardium so that myocardial function can be restored and new tissue can be produced. These findings have set off a large number of parallel discoveries in rats, mice, and humans, with dramatic implications for how we think about cardiac plasticity and its potential role in rehabilitating individuals with acquired myocardial ischemia/infarct, HF, and different types of cardiomyopathies, including the cardiomyopathy of aging.

Finally, the use of cellular engineering and transplantation with either embryonic or adult stem cells offers a phenomenal opportunity for effective therapeutic repair of myocardial and cardiovascular dysfunction.

SUMMARY

- Metabolic imaging of cardiomyocytes can be performed with fluorophores such as NADH. Fluorometric dyes can be used to assess both quantitatively and qualitatively changes in mitochondrial membrane potential, OS and the generation of ROS, oxidative phosphorylation and electron transport, apoptotic progression and events (e.g., permeability transition pore opening), localized pH and Ca^{++} levels.
- Oxygen content and consumption can be assessed by microspectrophotometry. PET can be used to assess both cellular and cardiac metabolism. Calcium and ion channel currents can be assessed by use of both fluorescence techniques and patch-clamp measurements in combination with confocal imaging. GFP can be used as a sensitive reporter to localize specific events within the subcellular compartments. Signaling molecules such as cAMP can be assessed with specific fluorometric dyes.
- Cardiomyocytes can be cultured *in vitro*. Treatment with specific growth factors and genetic modifications to these cells can be used to modulate both their proliferative capacity, as well as their response to pharmacological stimuli.
- Cell transplantation to repair cardiac function can be provided by use of embryonic stem cells, adult bone marrow cells, endothelial progenitor cells, skeletal myoblasts, and various cardiac progenitor cells. Transplantation of most of these cell types has been shown to provide beneficial effects to injured myocardium (e.g., ischemic) in preclinical studies. Each cell type has both advantages and disadvantages. Legal and ethical issues with the use of embryonic stem cells have greatly hampered their potential development as therapeutic tools.
- Clinical studies (which have only recently been initiated) have thus far displayed moderate benefits (i.e., enhanced contractility and increased perfusion) to patients with myocardial ischemia and infarction with bone marrow and skeletal myoblast transplantation, although the overall long-term benefits have not yet been studied. The mechanism underlying the cardiac-specific benefits of bone marrow cell transplant, which are relatively heterogenous, are not currently well understood; however, clinical treatment with this stem cell type thus far shows no evidence of harmful side effects. Myoblast transplants provide good contractile supplementation to the damaged myocardium, but dysrhythmias may arise.
- Stem cells and cardiomyocytes can be used in combination, grown in sheets, or in three-dimensional arrays for use in remodeling tissue (either cardiac or vascular). These cells can be modified *ex vivo* and can be used to introduce genes of interest or silencing elements to modulate gene expression and cell proliferation. Therapeutic cloning using these cells may be eventually used.

References

1. Bhatnagar, A., and Narula, J. (1999). Radionuclide imaging of cardiac pathology: a mechanistic perspective. *Adv. Drug Deliv. Rev.* **37,** 213–223.
2. Narula, J., Acio, E. R., Narula, N., Samuels, L. E., Fyfe, B., Wood, D., Fitzpatrick, J. M., Raghunath, P. N., Tomaszewski, J. E., Kelly, C., Steinmetz, N., Green, A., Tait, J. F., Leppo, J., Blankenberg, F. G., Jain, D., and Strauss, H. W. (2001). Annexin-V imaging for non-invasive detection of cardiac allograft rejection. *Nat. Med.* **7,** 1347–1352.
3. Segall, G. (2002). Assessment of myocardial viability by positron emission tomography. *Nucl. Med. Commun.* **23,** 323–330.
4. Dobrucki, L. W., and Sinusas, A. J. (2005). Molecular cardiovascular imaging. *Curr. Cardiol. Rep.* **7,** 130–135.
5. Clark, A. N., and Beller, G. A. (2005). The present role of nuclear cardiology in clinical practice. *Q. J. Nucl. Med. Mol. Imaging* **49,** 43–58.
6. Cuocolo, A., Acampa, W., Imbriaco, M., De Luca, N., Iovino, G. L., and Salvatore, M. (2005). The many ways to myocardial perfusion imaging. *Q. J. Nucl. Med. Mol. Imaging* **49,** 4–18.
7. Herrero, P., and Gropler, R. J. (2005). Imaging of myocardial metabolism. *J. Nucl. Cardiol.* **12,** 345–358.
8. Visser, F. C. (2001). Imaging of cardiac metabolism using radiolabelled glucose, fatty acids and acetate. *Coron. Artery Dis.* **12,** S12–S18.

9. Eng, J., Lynch, R. M., and Balaban, R. S. (1989). Nicotinamide adenine dinucleotide fluorescence spectroscopy and imaging of isolated cardiac myocytes. *Biophys. J.* **55**, 621–630.
10. Joubert, F., Fales, H. M., Wen, H., Combs, C. A., and Balaban, R. S. (2004). NADH enzyme-dependent fluorescence recovery after photobleaching (ED-FRAP): applications to enzyme and mitochondrial reaction kinetics, in vitro. *Biophys. J.* **86**, 629–645.
11. Combs, C. A., and Balaban, R. S. (2001). Direct imaging of dehydrogenase activity within living cells using enzyme-dependent fluorescence recovery after photobleaching (ED-FRAP). *Biophys. J.* **80**, 2018–2028.
12. Pelloux, S., Robillard, J., Ferrera, R., Bilbaut, A., Ojeda, C., Saks, V., Ovize, M., and Tourneur, Y. (2006). Non-beating HL-1 cells for confocal microscopy: application to mitochondrial functions during cardiac preconditioning. *Prog. Biophys. Mol. Biol.* **90**, 270–298.
13. Duchen, M. R., Surin, A., and Jacobson, J. (2003). Imaging mitochondrial function in intact cells. *Methods Enzymol.* **361**, 353–389.
14. Vanden Hoek, T., Li, C.-Q., Shao, Z.-H., Schumacker, P., and Becker, L. (1997). Significant levels of oxidants are generated by isolated cardiomyocytes during ischemia prior to reperfusion. *J. Mol. Cell. Cardiol.* **29**, 2571–2583.
15. Sarvazyan, N. (1996). Visualization of doxorubicin-induced oxidative stress in isolated cardiac myocytes. *Am. J. Physiol.* **271**, H2079–H2085.
16. Diaz, G., Liu, S., Isola, R., Diana, A., and Falchi, A. M. (2003). Mitochondrial localization of reactive oxygen species by dihydrofluorescein probes. *Histochem. Cell Biol.* **120**, 319–325.
17. King, N., McGivan, J. D., Griffiths, E. J., Halestrap, A. P., and Suleiman, M. S. (2003). Glutamate loading protects freshly isolated and perfused adult cardiomyocytes against intracellular ROS generation. *J. Mol. Cell. Cardiol.* **35**, 975–984.
18. King, N., Korolchuk, S., McGivan, J. D., and Suleiman, M. S. (2004). A new method of quantifying glutathione levels in freshly isolated single superfused rat cardiomyocytes. *J. Pharmacol. Toxicol. Methods* **50**, 215–222.
19. Bond, J. M., Chacon, E., Herman, B., and Lemasters, J. J. (1993). Intracellular pH and Ca2+ homeostasis in the pH paradox of reperfusion injury to neonatal rat cardiac myocytes. *Am. J. Physiol.* **265**, C129–C137.
20. Spitzer, K. W., Ershler, P. R., Skolnick, R. L., and Vaughan-Jones, R. D. (2000). Generation of intracellular pH gradients in single cardiac myocytes with a microperfusion system. *Am. J. Physiol. Heart Circ. Physiol.* **278**, H1371–H1382.
21. Llopis, J., McCaffery, J. M., Miyawaki, A., Farquhar, M. G., and Tsien, R. Y. (1998). Measurement of cytosolic, mitochondrial, and Golgi pH in single living cells with green fluorescent proteins. *Proc. Natl. Acad. Sci. USA* **95**, 6803–6808.
22. Jouaville, L. S., Pinton, P., Bastianutto, C., Rutter, G. A., and Rizzuto, R. (1999). Regulation of mitochondrial ATP synthesis by calcium: evidence for a long-term metabolic priming. *Proc. Natl. Acad. Sci. USA* **96**, 13807–13812.
23. Gajewski, C. D., Yang, L., Schon, E. A., and Manfredi, G. (2003). New insights into the bioenergetics of mitochondrial disorders using intracellular ATP reporters. *Mol. Biol. Cell* **14**, 3628–3635.
24. Takahashi, E., and Doi, K. (1995). Visualization of oxygen level inside a single cardiac myocyte. *Am. J. Physiol.* **268**, H2561–H2568.
25. Takahashi, E., Sato, K., Endoh, H., Xu, Z. L., and Doi, K. (1998). Direct observation of radial intracellular PO_2 gradients in a single cardiomyocyte of the rat. *Am. J. Physiol.* **275**, H225–H233.
26. Kofidis, T., Lenz, A., Boublik, J., Akhyari, P., Wachsmann, B., Mueller-Stahl, K., Hofmann, M., and Haverich, A. (2003). Pulsatile perfusion and cardiomyocyte viability in a solid three-dimensional matrix. *Biomaterials* **24**, 5009–5014.
27. Williams, D. A. (1990). Quantitative intracellular calcium imaging with laser-scanning confocal microscopy. *Cell. Calcium* **11**, 589–597.
28. Williams, D. A., Delbridge, L. M., Cody, S. H., Harris, P. J., and Morgan, T. O. (1992). Spontaneous and propagated calcium release in isolated cardiac myocytes viewed by confocal microscopy. *Am. J. Physiol.* **262**, C731–C742.
29. O'Rourke, B., Reibel, D. K., and Thomas, A. P. (1990). High-speed digital imaging of cytosolic Ca2+ and contraction in single cardiomyocytes. *Am. J. Physiol.* **259**, H230–H242.
30. Isenberg, G., Etter, E. F., Wendt-Gallitelli, M. F., Schiefer, A., Carrington, W. A., Tuft, R. A., and Fay, F. S. (1996). Intrasarcomere [Ca2+] gradients in ventricular myocytes revealed by high speed digital imaging microscopy. *Proc. Natl. Acad. Sci. USA* **93**, 5413–5418.
31. O'Rourke, B., Reibel, D. K., and Thomas, A. P. (1992). Alpha-adrenergic modification of the Ca2+ transient and contraction in single rat cardiomyocytes. *J. Mol. Cell. Cardiol.* **24**, 809–820.
32. Ibarra, C., Estrada, M., Carrasco, L., Chiong, M., Liberona, J. L., Cardenas, C., Diaz-Araya, G., Jaimovich, E., and Lavandero, S. (2004). Insulin-like growth factor-1 induces an inositol 1,4,5-trisphosphate-dependent increase in nuclear and cytosolic calcium in cultured rat cardiac myocytes. *J. Biol. Chem.* **279**, 7554–7565.
33. Shevchuk, A. I., Gorelik, J., Harding, S. E., Lab, M. J., Klenerman, D., and Korchev, Y. E. (2001). Simultaneous measurement of Ca2+ and cellular dynamics: combined scanning ion conductance and optical microscopy to study contracting cardiac myocytes. *Biophys. J.* **81**, 1759–1764.
34. Woo, S. H., Cleemann, L., and Morad, M. (2002). Ca2+ current-gated focal and local Ca2+ release in rat atrial myocytes: evidence from rapid 2-D confocal imaging. *J. Physiol.* **543**, 439–453.
35. Ianoul, A., Street, M., Grant, D., Pezacki, J., Taylor, R. S., and Johnston, L. J. (2004). Near-field scanning fluorescence microscopy study of ion channel clusters in cardiac myocyte membranes. *Biophys. J.* **87**, 3525–3535.
36. Guatimosim, S., Sobie, E. A., dos Santos Cruz, J., Martin, L. A., and Lederer, W. J. (2001). Molecular identification of a TTX-sensitive Ca(2+) current. *Am. J. Physiol. Cell Physiol.* **280**, C1327–C1339.
37. Cleemann, L., Wang, W., and Morad, M. (1998). Two-dimensional confocal images of organization, density, and gating of focal Ca2+ release sites in rat cardiac myocytes. *Proc. Natl. Acad. Sci. USA* **95**, 10984–10989.
38. Miyawaki, A., Llopis, J., Heim, R., McCaffery, J. M., Adams, J. A., Ikura, M., and Tsien, R. Y. (1997). Fluorescent indicators

for Ca^{++} based on green fluorescent proteins and calmodulin. *Nature* **388**, 882–887.

39. Jacobson, J., and Duchen, M. R. (2004). Interplay between mitochondria and cellular calcium signalling. *Mol. Cell. Biochem.* **256–257**, 209–218.

40. Duchen, M. R. (2000). Mitochondria and calcium: from cell signalling to cell death. *J. Physiol.* **529**, 57–68.

41. Murata, M., Akao, M., O'Rourke, B., and Marban, E. (2001). Mitochondrial ATP-sensitive potassium channels attenuate matrix Ca(2+) overload during simulated ischemia and reperfusion: possible mechanism of cardioprotection. *Circ. Res.* **89**, 891–898.

42. Rizzuto, R., Brini, M., Bastianutto, C., Marsault, R., and Pozzan, T. (1995). Photoprotein-mediated measurement of calcium ion concentration in mitochondria of living cells. *Methods Enzymol.* **260**, 417–428.

43. Entcheva, E., Kostov, Y., Tchernev, E., and Tung, L. (2004). Fluorescence imaging of electrical activity in cardiac cells using an all-solid-state system. *IEEE Trans. Biomed. Eng.* **51**, 333–341.

44. Liu, J. C., Chan, P., Chen, J. J., Lee, H. M., Lee, W. S., Shih, N. L., Chen, Y. L., Hong, H. J., and Cheng, T. H. (2004). The inhibitory effect of trilinolein on norepinephrine-induced beta-myosin heavy chain promoter activity, reactive oxygen species generation, and extracellular signal-regulated kinase phosphorylation in neonatal rat cardiomyocytes. *J. Biomed. Sci.* **11**, 11–18.

45. Swift, L. M., and Sarvazyan, N. (2000). Localization of dichlorofluorescein in cardiac myocytes: implications for assessment of oxidative stress. *Am. J. Physiol. Heart Circ. Physiol.* **278**, H982–H990.

46. Strijdom, H., Muller, C., and Lochner, A. (2004). Direct intracellular nitric oxide detection in isolated adult cardiomyocytes: flow cytometric analysis using the fluorescent probe, diaminofluorescein. *J. Mol. Cell. Cardiol.* **37**, 897–902.

47. Fijnvandraat, A. C., van Ginneken, A. C., Schumacher, C. A., Boheler, K. R., Lekanne Deprez, R. H., Christoffels, V. M., and Moorman, A. F. (2003). Cardiomyocytes purified from differentiated embryonic stem cells exhibit characteristics of early chamber myocardium. *J. Mol. Cell. Cardiol.* **35**, 1461–1472.

48. Loh, S. H., Chen, W. H., Chiang, C. H., Tsai, C. S., Lee, G. C., Jin, J. S., Cheng, T. H., and Chen, J. J. (2002). Intracellular pH regulatory mechanism in human atrial myocardium: functional evidence for Na(+)/H(+) exchanger and Na(+)/HCO3(-) symporter. *J. Biomed. Sci.* **9**, 198–205.

49. Zaniboni, M., Rossini, A., Swietach, P., Banger, N., Spitzer, K. W., and Vaughan-Jones, R. D. (2003). Proton permeation through the myocardial gap junction. *Circ. Res.* **93**, 726–735.

50. Bowser, D. N., Minamikawa, T., Nagley, P., and Williams, D. A. (1998). Role of mitochondria in calcium regulation of spontaneously contracting cardiac muscle cells. *Biophys. J.* **75**, 2004–2014.

51. Kong, J. Y., and Rabkin, S. W. (2003). Mitochondrial effects with ceramide-induced cardiac apoptosis are different from those of palmitate. *Arch. Biochem. Biophys.* **412**, 196–206.

52. Krieg, T., Cui, L., Qin, Q., Cohen, M. V., and Downey, J. M. (2004). Mitochondrial ROS generation following acetylcholine-induced EGF receptor transactivation requires metalloproteinase cleavage of proHB-EGF. *J. Mol. Cell. Cardiol.* **36**, 435–443.

53. Di Lisa, F., Blank, P. S., Colonna, R., Gambassi, G., Silverman, H. S., Stern, M. D., and Hansford, R. G. (1995). Mitochondrial membrane potential in single living adult rat cardiac myocytes exposed to anoxia or metabolic inhibition. *J. Physiol.* **486**, 1–13.

54. Mathur, A., Hong, Y., Kemp, B. K., Barrientos, A. A., and Erusalimsky, J. D. (2000). Evaluation of fluorescent dyes for the detection of mitochondrial membrane potential changes in cultured cardiomyocytes. *Cardiovasc. Res.* **46**, 126–138.

55. Mashimo, K., and Ohno, Y. (2006). Ethanol hyperpolarizes mitochondrial membrane potential and increases mitochondrial fraction in cultured mouse myocardial cells. *Arch. Toxicol.* **11**, 1–8.

56. Lipp, P., and Niggli, E. (1993). Ratiometric confocal Ca(2+)-measurements with visible wavelength indicators in isolated cardiac myocytes. *Cell. Calcium* **14**, 359–372.

57. Hudman, D., Rainbow, R. D., Lawrence, C. L., and Standen, N. B. (2002). The origin of calcium overload in rat cardiac myocytes following metabolic inhibition with 2,4-dinitrophenol. *J. Mol. Cell. Cardiol.* **34**, 859–871.

58. Vink, M. J., Suadicani, S. O., Vieira, D. M., Urban-Maldonado, M., Gao, Y., Fishman, G. I., and Spray, D. C. (2004). Alterations of intercellular communication in neonatal cardiac myocytes from connexin43 null mice. *Cardiovasc. Res.* **62**, 397–406.

59. Camelliti, P., Green, C. R., LeGrice, I., and Kohl, P. (2004). Fibroblast network in rabbit sinoatrial node: structural and functional identification of homogeneous and heterogeneous cell coupling. *Circ. Res.* **94**, 828–835.

60. Hausenloy, D. J., Yellon, D. M., Mani-Babu, S., and Duchen, M. R. (2004). Preconditioning protects by inhibiting the mitochondrial permeability transition. *Am. J. Physiol. Heart Circ. Physiol.* **287**, H841–H849.

61. Hausenloy, D., Wynne, A., Duchen, M., and Yellon, D. (2004). Transient mitochondrial permeability transition pore opening mediates preconditioning-induced protection. *Circulation* **109**, 1714–1717.

62. Katoh, H., Nishigaki, N., and Hayashi, H. (2002). Diazoxide opens the mitochondrial permeability transition pore and alters Ca2+ transients in rat ventricular myocytes. *Circulation* **105**, 2666–2671.

63. Zaccolo, M. (2004). Use of chimeric fluorescent proteins and fluorescence resonance energy transfer to monitor cellular responses. *Circ. Res.* **94**, 866–873.

64. Steinberg, S. F., and Brunton, L. L. (2001). Compartmentation of G protein-coupled signaling pathways in cardiac myocytes. *Annu. Rev. Pharmacol. Toxicol.* **41**, 751–773.

65. Zaccolo, M., De Giorgi, F., Cho, C. Y., Feng, L., Knapp, T., Negulescu, P. A., Taylor, S. S., Tsien, R. Y., and Pozzan, T. (2000). A genetically encoded, fluorescent indicator for cyclic AMP in living cells. *Nat. Cell. Biol.* **2**, 25–29.

66. Zaccolo, M., and Pozzan, T. (2002). Discrete microdomains with high concentration of cAMP in stimulated rat neonatal cardiac myocytes. *Science* **295**, 1711–1715.

67. Mongillo, M., McSorley, T., Evellin, S., Sood, A., Lissandron, V., Terrin, A., Huston, E., Hannawacker, A., Lohse, M. J., Pozzan, T., Houslay, M. D., and Zaccolo, M. (2004). Fluorescence resonance energy transfer-based analysis of cAMP dynamics in live neonatal rat cardiac myocytes reveals distinct functions of compartmentalized phosphodiesterases. *Circ. Res.* **95**, 67–75.

68. Warrier, S., Belevych, A. E., Ruse, M., Eckert, R. L., Zaccolo, M., Pozzan, T., and Harvey, R. D. (2005). Beta-adrenergic and muscarinic receptor induced changes in cAMP activity in adult cardiac myocytes detected using a FRET based biosensor. *Am. J. Physiol. Cell. Physiol.* **289,** C455–C461.
69. Tyner, K. M., Roberson, M. S., Berghorn, K. A., Li, L., Gilmour, R. F., Jr., Batt, C. A., and Giannelis, E. P. (2004). Intercalation, delivery, and expression of the gene encoding green fluorescence protein utilizing nanobiohybrids. *J. Control Release* **100,** 399–409.
70. Rich, T. C., and Karpen, J. W. (2002). Review article: Cyclic AMP sensors in living cells: what signals can they actually measure? *Ann. Biomed. Eng.* **30,** 1088–1099.
71. Goaillard, J. M., Vincent, P. V., and Fischmeister, R. (2001). Simultaneous measurements of intracellular cAMP and L-type Ca2+ current in single frog ventricular myocytes. *J. Physiol.* **530,** 79–91.
72. Ting, A. Y., Kain, K. H., Klemke, R. L., and Tsien, R. Y. (2001). Genetically encoded fluorescent reporters of protein tyrosine kinase activities in living cells. *Proc. Natl. Acad. Sci. USA* **98,** 15003–15008.
73. Sato, M., Ozawa, T., Inukai, K., Asano, T., and Umezawa, Y. (2002). Fluorescent indicators for imaging protein phosphorylation in single living cells. *Nat. Biotechnol.* **20,** 287–294.
74. Ottolia, M., Philipson, K. D., and John, S. (2004). Conformational changes of the Ca(2+) regulatory site of the Na(+)-Ca(2+) exchanger detected by FRET. *Biophys. J.* **87,** 899–906.
75. Remus, T. P., Zima, A. V., Bossuyt, J., Bare, D. J., Martin, J. L., Blatter, L. A., Bers, D. M., and Mignery, G. A. (2006). Biosensors to measure inositol 1,4,5-trisphosphate concentration in living cells with spatiotemporal resolution. *J. Biol. Chem.* **281,** 608–616.
76. Dumont, E. A., Reutelingsperger, C. P., Smits, J. F., Daemen, M. J., Doevendans, P. A., Wellens, H. J., and Hofstra, L. (2001). Real-time imaging of apoptotic cell-membrane changes at the single-cell level in the beating murine heart. *Nat. Med.* **7,** 1352–1355.
77. Bennink, R. J., van den Hoff, M. J., van Hemert, F. J., de Bruin, K. M., Spijkerboer, A. L., Vanderheyden, J. L., Steinmetz, N., and van Eck-Smit, B. L. (2004). Annexin V imaging of acute doxorubicin cardiotoxicity (apoptosis) in rats. *J. Nucl. Med.* **45,** 842–848.
78. Taki, J., Higuchi, T., Kawashima, A., Tait, J. F., Kinuya, S., Muramori, A., Matsunari, I., Nakajima, K., Tonami, N., and Strauss, H. W. (2004). Detection of cardiomyocyte death in a rat model of ischemia and reperfusion using 99mTc-labeled annexin V. *J. Nucl. Med.* **45,** 1536–1541.
79. Kleparnik, K., and Horky, M. (2003). Detection of DNA fragmentation in a single apoptotic cardiomyocyte by electrophoresis on a microfluidic device. *Electrophoresis* **24,** 3778–3783.
80. Bicknell, K. A., Surry, E. L., and Brooks, G. (2003). Targeting the cell cycle machinery for the treatment of cardiovascular disease. *J. Pharm. Pharmacol.* **55,** 571–591.
81. Poolman, R. A., and Brooks, G. (1998). Expressions and activities of cell cycle regulatory molecules during the transition from myocyte hyperplasia to hypertrophy. *J. Mol. Cell. Cardiol.* **30,** 2121–2135.
82. Diez, C., and Simm, A. (1998). Gene expression in rod shaped cardiac myocytes, sorted by flow cytometry. *Cardiovasc. Res.* **40,** 530–537.
83. Li, J. M., Mullen, A. M., and Shah, A. M. (2001). Phenotypic properties and characteristics of superoxide production by mouse coronary microvascular endothelial cells. *J. Mol. Cell. Cardiol.* **33,** 1119–1131.
84. Strijdom, H., Muller, C., and Lochner, A. (2004). Direct intracellular nitric oxide detection in isolated adult cardiomyocytes: flow cytometric analysis using the fluorescent probe, diaminofluorescein. *J. Mol. Cell. Cardiol.* **37,** 897–902.
85. Rabkin, S. W., and Kong, J. Y. (2000). Nitroprusside induces cardiomyocyte death: interaction with hydrogen peroxide. *Am. J. Physiol. Heart Circ. Physiol.* **279,** H3089–H3100.
86. Frey, T. (1997). Correlated flow cytometric analysis of terminal events in apoptosis reveals the absence of some changes in some model systems. *Cytometry* **28,** 253–263.
87. Suzuki, T., Ohta, M., and Hoshi, H. (1989). Serum-free, chemically defined medium to evaluate the direct effects of growth factors and inhibitors on proliferation and function of neonatal rat cardiac muscle cells in culture. *In Vitro Cell Dev. Biol.* **25,** 601–606.
88. Kajstura, J., Cheng, W., Reiss, K., and Anversa, P. (1994). The IGF-1-IGF-1 receptor system modulates myocyte proliferation but not myocyte cellular hypertrophy in vitro. *Exp. Cell Res.* **215,** 273–283.
89. Weiner, H. L., and Swain, J. L. (1989). Acidic fibroblast growth factor mRNA is expressed by cardiac myocytes in culture and the protein is localized to the extracellular matrix. *Proc. Natl. Acad. Sci. USA* **86,** 2683–2687.
90. Long, C. S., Kariya, .K, Karns, L, and Simpson, P. C. (1990). Trophic factors for cardiac myocytes. *J. Hypertens. Suppl.* **8,** S219–S224.
91. Mima, T., Ueno, H., Fischman, D. A., Williams, L. T., and Mikawa, T. (1995). Fibroblast growth factor receptor is required for in vivo cardiac myocyte proliferation at early embryonic stages of heart development. *Proc. Natl. Acad. Sci. USA* **92,** 467–471.
92. Liu, J., Wu, L. L., Li, L., Zhang, L., and Song, Z. E. (2005). Growth-promoting effect of platelet-derived growth factor on rat cardiac myocytes. *Regul. Pept.* **127,** 11–18.
93. Sheng, Z., Pennica, D., Wood, W. I., and Chien, K. R. (1996). Cardiotrophin-1 displays early expression in the murine heart tube and promotes cardiac myocyte survival. *Development* **122,** 419–428.
94. Li, M., Zhang, M., Huang, L., Zhou, J., Zhuang, H., Taylor, J. T., Keyser, B. M., and Whitehurst, R. M., Jr. (2005). T-type Ca2+ channels are involved in high glucose-induced rat neonatal cardiomyocyte proliferation. *Pediatr. Res.* **57,** 550–556.
95. Pasumarthi, K. B., Kardami, E., and Cattini, P. A. (1996). High and low molecular weight fibroblast growth factor-2 increase proliferation of neonatal rat cardiac myocytes but have differential effects on binucleation and nuclear morphology. Evidence for both paracrine and intracrine actions of fibroblast growth factor-2. *Circ. Res.* **78,** 126–136.
96. Sheikh, F., Fandrich, R. R., Kardami, E., and Cattini, P. A. (1999). Overexpression of long or short FGFR-1 results in

FGF-2-mediated proliferation in neonatal cardiac myocyte cultures. *Cardiovasc. Res.* **42,** 696–705.

97. Busk, P. K., Hinrichsen, R., Bartkova, J., Hansen, A. H., Christoffersen, T. E., Bartek, J., and Haunso, S. (2005). Cyclin D2 induces proliferation of cardiac myocytes and represses hypertrophy. *Exp. Cell Res.* **304,** 149–161.

98. Tamamori-Adachi, M., Ito, H., Sumrejkanchanakij, P., Adachi, S., Hiroe, M., Shimizu, M., Kawauchi, J., Sunamori, M., Marumo, F., Kitajima, S., and Ikeda, M. A. (2003). Critical role of cyclin D1 nuclear import in cardiomyocyte proliferation. *Circ. Res.* **92,** e12–9.

99. Bicknell, K. A., Coxon, C. H., and Brooks, G. (2004). Forced expression of the cyclin B1-CDC2 complex induces proliferation in adult rat cardiomyocytes. *Biochem. J.* **382,** 411–416.

100. Kardami, E. (1990). Stimulation and inhibition of cardiac myocyte proliferation in vitro. *Mol. Cell Biochem.* **92,** 129–135.

101. Goldman, B., Mach, A., and Wurzel, J. (1996). Epidermal growth factor promotes a cardiomyoblastic phenotype in human fetal cardiac myocytes. *Exp. Cell Res.* **228,** 237–245.

102. Zhao, Y. Y., Sawyer, D. R., Baliga, R. R., Opel, D. J., Han, X., Marchionni, M. A., and Kelly, R. A. (1998). Neuregulins promote survival and growth of cardiac myocytes: persistence of ErbB2 and ErbB4 expression in neonatal and adult ventricular myocytes. *J. Biol. Chem.* **273,** 10261–10269.

103. Koide, M., Akins, R. E., Harayama, H., Yasui, K., Yokota M., and Tuan, R. S. (1996). Atrial natriuretic peptide accelerates proliferation of chick embryonic cardiomyocytes in vitro. *Differentiation* **61,** 1–11.

104. Wald, M. R., Borda, E. S., and Sterin-Borda, L. (1996). Mitogenic effect of erythropoietin on neonatal rat cardiomyocytes: signal transduction pathways. *J. Cell. Physiol.* **167,** 461–468.

105. Burton, P. B., Yacoub, M. H., and Barton, P. J. (1998). Rapamycin (sirolimus) inhibits heart cell growth in vitro. *Pediatr. Cardiol.* **19,** 468–470.

106. O'Connell, T. D., Berry, J. E., Jarvis, A. K., Somerman, M. J., and Simpson, R. U. (1997). 1,25-Dihydroxyvitamin D3 regulation of cardiac myocyte proliferation and hypertrophy. *Am. J. Physiol.* **272,** H1751–H1758.

107. Reiss, K., Cheng, W., Ferber, A., Kajstura, J., Li, P., Li, B., Olivetti, G., Homcy, C. J., Baserga, R., and Anversa, P. (1996). Overexpression of insulin-like growth factor-1 in the heart is coupled with myocyte proliferation in transgenic mice. *Proc. Natl. Acad. Sci. USA* **93,** 8630–8635.

108. Pasumarthi, K. B. S., and Field, L. J. (2002). Cardiomyocyte cell cycle regulation. *Circ. Res.* **90,** 1044–1054.

109. Ebelt, H., Hufnagel, N., Neuhaus, P., Neuhaus, H., Gajawada, P., Simm, A., Muller-Werdan, U., Werdan, K., and Braun, T. (2005). Divergent siblings: E2F2 and E2F4 but not E2F1 and E2F3 induce DNA synthesis in cardiomyocytes without activation of apoptosis. *Circ. Res.* **96,** 509–517.

110. Kajstura, J., Cheng, W., Reiss, K., and Anversa, P. (1994). The IGF-1-IGF-1 receptor system modulates myocyte proliferation but not myocyte cellular hypertrophy in vitro. *Exp. Cell Res.* **215,** 273–283.

111. Engel, F. B., Schebesta, M., Duong, M. T., Lu, G., Ren, S., Madwed, J. B., Jiang, H., Wang, Y., and Keating, M. T. (2005). p38 MAP kinase inhibition enables proliferation of adult mammalian cardiomyocytes. *Genes Dev.* **19,** 1175–1187.

112. Bell, D., and McDermott, B. J. (2000). Contribution of de novo protein synthesis to the hypertrophic effect of IGF-1 but not of thyroid hormones in adult ventricular cardiomyocytes. *Mol. Cell. Biochem.* **206,** 113–124.

113. Guo, W., Kamiya, K., Hojo, M., Kodama, I., and Toyama, J. (1998). Regulation of Kv4.2 and Kv1.4 K+ channel expression by myocardial hypertrophic factors in cultured newborn rat ventricular cells. *J. Mol. Cell. Cardiol.* **30,** 1449–1455.

114. Schaub, M. C., Hefti, M. A., Harder, B. A., and Eppenberger, H. M. (1997). Various hypertrophic stimuli induce distinct phenotypes in cardiomyocytes. *J. Mol. Med.* **75,** 901–920.

115. Doevendans, P. A., Kubalak, S. W., An, R. H., Becker, D. K., Chien, K. R., and Kass, R. S. (2000). Differentiation of cardiomyocytes in floating embryoid bodies is comparable to fetal cardiomyocytes. *J. Mol. Cell. Cardiol.* **32,** 839–851.

116. Kehat, I., Khimovich, L., Caspi, O., Gepstein, A., Shofti, R., Arbel, G., Huber, I., Satin, J., Itskovitz-Eldor, J., and Gepstein, L. (2004). Electromechanical integration of cardiomyocytes derived from human embryonic stem cells. *Nat. Biotechnol.* **22,** 1237–1238.

117. Doetschman, T., Shull, M., Kier, A., and Coffin, J. D. (1993). Embryonic stem cell model systems for vascular morphogenesis and cardiac disorders. *Hypertension* **22,** 618–629.

118. He, J. Q., Ma, Y., Lee, Y., Thomson, J. A., and Kamp, T. J. (2003). Human embryonic stem cells develop into multiple types of cardiac myocytes: action potential characterization. *Circ. Res.* **93,** 32–39.

119. Muller, M., Fleischmann, B. K., Selbert, S., Ji, G. J., Endl, E., Middeler, G., Muller, O. J., Schlenke, P., Frese, S., Wobus, A. M., Hescheler, J., Katus, H. A., and Franz, W. M. (2000). Selection of ventricular-like cardiomyocytes from ES cells in vitro. *FASEB J.* **14,** 2540–2548.

120. Wobus, A. M., and Boheler, K. R. (2005). Embryonic stem cells: prospects for developmental biology and cell therapy. *Physiol. Rev.* **85,** 635–678.

121. Yuasa, S., Itabashi, Y., Koshimizu, U., Tanaka, T., Sugimura, K., Kinoshita, M., Hattori, F., Fukami, S. I., Shimazaki, T., Okano, H., Ogawa, S., and Fukuda, K. (2005). Transient inhibition of BMP signaling by Noggin induces cardiomyocyte differentiation of mouse embryonic stem cells. *Nat. Biotechnol.* **23,** 607–611.

122. Wobus, A. M., Kaomei, G., Shan, J., Wellner, M. C., Rohwedel, J., Ji Guanju, Fleischmann, B., Katus, H. A., Hescheler, J., and Franz, W. M. (1997). Retinoic acid accelerates embryonic stem cell-derived cardiac differentiation and enhances development of ventricular cardio-myocytes. *J. Mol. Cell. Cardiol.* **29,** 1525–1539.

123. Zandstra, P. W., Bauwens, C., Yin, T., Liu, Q., Schiller, H., Zweigerdt, R., Pasumarthi, K. B., and Field, L. J. (2003). Scalable production of embryonic stem cell-derived cardiomyocytes. *Tissue Eng.* **9,** 767–778.

124. Kanno, S., Kim, P. K., Sallam, K., Lei, J., Billiar, T. R., and Shears, L. L., 2nd. (2004). Nitric oxide facilitates cardiomyogenesis in mouse embryonic stem cells. *Proc. Natl. Acad. Sci. USA* **101,** 12277–12281.

125. Kofidis, T., de Bruin, J. L., Yamane, T., Balsam, L. B., Lebl, D. R., Swijnenburg, R. J., Tanaka, M., Weissman, I. L., and Robbins, R. C. (2004). Insulin-like growth factor promotes engraftment, differentiation, and functional improvement after transfer of embryonic stem cells for myocardial restoration. *Stem Cells* **22,** 1239–1245.

126. Kawai, T., Takahashi, T., Esaki, M., Ushikoshi, H., Nagano, S., Fujiwara, H., and Kosai, K. (2004). Efficient cardiomyogenic differentiation of embryonic stem cell by fibroblast growth factor 2 and bone morphogenetic protein 2. *Circ. J.* **68,** 691–702.
127. Sachinidis, A., Gissel, C., Nierhoff, D., Hippler-Altenburg, R., Sauer, H., Wartenberg, M., and Hescheler, J. (2003). Identification of platelet-derived growth factor-BB as cardiogenesis-inducing factor in mouse embryonic stem cells under serum-free conditions. *Cell Physiol. Biochem.* **13,** 423–429.
128. Sachinidis, A., Kolossov, E., Fleischmann, B. K., and Hescheler, J. (2002). Generation of cardiomyocytes from embryonic stem cells experimental studies. *Herz* **27,** 589–597.
129. Kumar, D., and Sun, B. (2005). Transforming growth factor-beta2 enhances differentiation of cardiac myocytes from embryonic stem cells. *Biochem. Biophys. Res. Commun.* **332,** 135–141.
130. Zheng, B., Wen, J. K., and Han, M. (2006). hhLIM is involved in cardiomyogenesis of embryonic stem cells. *Biochemistry (Mosc)* **71,** S71–S76.
131. Xu, C., Police, S., Rao, N., and Carpenter, M. K. (2002). Characterization and enrichment of cardiomyocytes derived from human embryonic stem cells. *Circ. Res.* **91,** 501–508.
132. Paquin, J., Danalache, B. A., Jankowski, M., McCann, S. M., and Gutkowska, J. (2002). Oxytocin induces differentiation of P19 embryonic stem cells to cardiomyocytes. *Proc. Natl. Acad. Sci. USA* **99,** 9550–9555.
133. Zhu, D. Y., and Lou, Y. J. (2006). Icariin-mediated expression of cardiac genes and modulation of nitric oxide signaling pathway during differentiation of mouse embryonic stem cells into cardiomyocytes in vitro. *Acta Pharmacol. Sin.* **27,** 311–320.
134. Kawamura, T., Ono, K., Morimoto, T., Wada, H., Hirai, M., Hidaka, K., Morisaki, T., Heike, T., Nakahata, T., Kita, T., and Hasegawa, K. (2005). Acetylation of GATA-4 is involved in the differentiation of embryonic stem cells into cardiac myocytes. *J. Biol. Chem.* **280,** 19682–19688.
135. Terami, H., Hidaka, K., Katsumata, T., Iio, A., and Morisaki, T. (2004). Wnt11 facilitates embryonic stem cell differentiation to Nkx2.5-positive cardiomyocytes. *Biochem. Biophys. Res. Commun.* **325,** 968–975.
136. Choi, S. C., Yoon, J., Shim, W. J., Ro, Y. M., and Lim, D. S. (2004). 5-azacytidine induces cardiac differentiation of P19 embryonic stem cells. *Exp. Mol. Med.* **36,** 515–523.
137. Jackson, M., Krassowska, A., Gilbert, N., Chevassut, T., Forrester, L., Ansell, J., and Ramsahoye, B. (2004). Severe global DNA hypomethylation blocks differentiation and induces histone hyperacetylation in embryonic stem cells. *Mol. Cell Biol.* **24,** 8862–8871.
138. Sauer, H., Rahimi, G., Hescheler, J., and Wartenberg, M. (2000). Role of reactive oxygen species and phosphatidylinositol 3-kinase in cardiomyocyte differentiation of embryonic stem cells. *FEBS Lett.* **476,** 218–223.
139. Sauer, H., Neukirchen, W., Rahimi, G., Grunheck, F., Hescheler, J., and Wartenberg, M. (2004). Involvement of reactive oxygen species in cardiotrophin-1-induced proliferation of cardiomyocytes differentiated from murine embryonic stem cells. *Exp. Cell Res.* **294,** 313–324.
140. Klug, M. G., Soonpaa, M. H., Koh, G. Y., and Field, L. J. (1996). Genetically selected cardiomyocytes from differentiating embryonic stem cells form stable intracardiac grafts. *J. Clin. Invest.* **98,** 216–224.
141. Koh, G. Y., Soonpaa, M. H., Klug, M. G., Pride, H. P., Cooper, B. J., Zipes, D. P., and Field, L. J. (1995). Stable fetal cardiomyocyte grafts in the hearts of dystrophic mice and dogs. *J. Clin. Invest.* **96,** 2034–2042.
142. Xue, T., Cho, H. C., Akar, F. G., Tsang, S. Y., Jones, S. P., Marban, E., Tomaselli, G. F., and Li, R. A. (2005). Functional integration of electrically active cardiac derivatives from genetically engineered human embryonic stem cells with quiescent recipient ventricular cardiomyocytes: insights into the development of cell-based pacemakers. *Circulation* **111,** 11–20.
143. Zhang, M., Methot, D., Poppa, V., Fujio, Y., Walsh, K., and Murry, C. E. (2001). Cardiomyocyte grafting for cardiac repair: graft cell death and anti-death strategies. *J. Mol. Cell. Cardiol.* **33,** 907–921.
144. Monge, J. C., Stewart, D. J., and Cernacek, P. (1995). Differentiation of embryonal carcinoma cells to a neural or cardiomyocyte lineage is associated with selective expression of endothelin receptors. *J. Biol. Chem.* **270,** 15385–15390.
145. Wobus, A. M., Kleppisch, T., Maltsev, V., and Hescheler, J. (1994). Cardiomyocyte-like cells differentiated in vitro from embryonic carcinoma cells P19 are characterized by functional expression of adrenoceptors and Ca2+ channels. *In Vitro Cell Dev. Biol. Anim.* **30A,** 425–434.
146. Coburn, B. (2005). Beating congenital heart defects with embryonic stem cells. *Clin. Genet.* **67,** 224–225.
147. Fraidenraich, D., Stillwell, E., Romero, E., Wilkes, D., Manova, K., Basson, C. T., and Benezra, R. (2004). Rescue of cardiac defects in id knockout embryos by injection of embryonic stem cells. *Science* **306,** 247–252.
148. Zwaka, T. P., and Thomson, J. A. (2003). Homologous recombination in human embryonic stem cells. *Nat. Biotechnol.* **21,** 319–321.
149. Xiong, C., Tang, D. Q., Xie, C. Q., Zhang, L., Xu, K. F., Thompson, W. E., Chou, W., Gibbons, G. H., Chang, L. J., Yang, L. J., and Chen, Y. E. (2005). Genetic engineering of human embryonic stem cells with lentiviral vectors. *Stem Cells Dev.* **14,** 367–377.
150. Liu, Y. P., Dovzhenko, O. V., Garthwaite, M. A., Dambaeva, S. V., Durning, M., Pollastrini, L. M., and Golos, T. G. (2004). Maintenance of pluripotency in human embryonic stem cells stably over-expressing enhanced green fluorescent protein. *Stem Cells Dev.* **13,** 636–645.
151. Ren, C., Zhao, M., Yang, X., Li, D., Jiang, X., Wang, L., Shan, W., Yang, H., Zhou, L., Zhou, W., and Zhang, H. (2006). Establishment and applications of Epstein-Barr virus-based episomal vectors in human embryonic stem cells. *Stem Cells* Mar 9; [Epub ahead of print].
152. Swijnenburg, R. J., Tanaka, M., Vogel, H., Baker, J., Kofidis, T., Gunawan, F., Lebl, D. R., Caffarelli, A. D, de Bruin, J. L., Fedoseyeva, E. V., and Robbins, R. C. (2005). Embryonic stem cell immunogenicity increases upon differentiation after transplantation into ischemic myocardium. *Circulation* **112,** I166–I172.
153. Taylor, D. A., Atkins, B. Z., Hungspreugs, P., Jones, T. R., Reedy, M. C., Hutcheson, K. A., Glower, D. D., and Kraus,

W. (1998). Regenerating functional myocardium: improved performance after skeletal myoblast transplantation. *Nat. Med.* **4**, 929–933.
154. Kessler, P. D., and Byrne, B. J. (1999). Myoblast cell grafting into heart muscle: cellular biology and potential applications. *Annu. Rev. Physiol.* **61**, 219–242.
155. Menasche, P. (2003). Skeletal muscle satellite cell transplantation. *Cardiovasc. Res.* **58**, 351–357.
156. Robinson, S. W., Cho, P. W., Levitsky, H. I., Olson, J. L., Hruban, R. H., Acker, M. A., and Kessler, P. D. (1996). Arterial delivery of genetically labelled skeletal myoblasts to the murine heart: long-term survival and phenotypic modification of implanted myoblasts. *Cell Transpl.* **5**, 77–91.
157. Menasche, P. (2005). Skeletal myoblast for cell therapy. *Coron. Artery Dis.* **16**, 105–110.
158. Brasselet, C., Morichetti, M. C., Messas, E., Carrion, C., Bissery, A., Bruneval, P., Vilquin, J. T., Lafont, A., Hagege, A. A., Menasche, P., and Desnos, M. (2005). Skeletal myoblast transplantation through a catheter-based coronary sinus approach: an effective means of improving function of infarcted myocardium. *Eur. Heart J.* **26**, 1551–1556.
159. Jennings, R. B., and Reimer, K. A. (1981). Lethal myocardial ischemic injury. *Am. J. Pathol.* **102**, 241–255.
160. Van Den Bos, E. J., and Taylor, D. A. (2003). Cardiac transplantation of skeletal myoblasts for heart failure. *Minerva Cardioangiol.* **51**, 227–243.
161. Menasche, P. (2002). Cell transplantation for the treatment of heart failure. *Semin. Thorac. Cardiovasc. Surg.* **14**, 157–166.
162. Pouly, J., Hagege, A. A., Vilquin, J. T., Bissery, A., Rouche, A., Bruneval, P., Duboc, D., Desnos, M., Fiszman, M., Fromes, Y., and Menasche, P. (2004). Does the functional efficacy of skeletal myoblast transplantation extend to nonischemic cardiomyopathy? *Circulation* **110**, 1626–1631.
163. Kondoh, H., Sawa, Y., Miyagawa, S., Sakakida-Kitagawa, S., Memon, I. A., Kawaguchi, N., Matsuura, N., Shimizu, T., Okano, T., and Matsuda, H. (2006). Longer preservation of cardiac performance by sheet-shaped myoblast implantation in dilated cardiomyopathic hamsters. *Cardiovasc. Res.* **69**, 466–475.
164. Iijima, Y., Nagai, T., Mizukami, M., Matsuura, K., Ogura, T., Wada, H., Toko, H., Akazawa, H., Takano, H., Nakaya, H., and Komuro, I. (2003). Beating is necessary for trans-differentiation of skeletal muscle-derived cells into cardiomyocytes. *FASEB J.* **17**, 1361–1363.
165. Winitsky, S. O., Gopal, T. V., Hassanzadeh, S., Takahashi, H., Gryder, D., Rogawski, M. A., Takeda, K., Yu, Z. X., Xu, Y. H., and Epstein, N. D. (2005). Adult murine skeletal muscle contains cells that can differentiate into beating cardiomyocytes in vitro. *PLoS Biol.* **3**, e87.
166. Reinecke, H., Poppa, V., and Murry, C. E. (2002). Skeletal muscle stem cells do not transdifferentiate into cardiomyocytes after cardiac grafting. *J. Mol. Cell. Cardiol.* **34**, 241–249.
167. Murry, C. E., Wiseman, R. W., Schwartz, S. M., and Hauschka, S. D. (1996). Skeletal myoblast transplantation for repair of myocardial necrosis. J. Clin. Invest. **98**, 2512–2523.
168. Leobon, B., Garcin, I., Menasche, P., Vilquin, J. T., Audinat, E., and Charpak, S. (2003). Myoblasts transplanted into rat infarcted myocardium are functionally isolated from their host. *Proc. Natl. Acad. Sci. USA* **100**, 7808–7811.
169. Menasche, P., Hagege, A. A., Vilquin, J. T., Desnos, M., Abergel, E., Pouzet, B., Bel, A., Sarateanu, S., Scorsin, M., Schwartz, K., Bruneval, P., Benbunan, M., Marolleau, J. P., and Duboc, D. (2003). Autologous skeletal myoblast transplantation for severe postinfarction left ventricular dysfunction. *J. Am. Coll. Cardiol.* **41**, 1078–1083.
170. Chachques, J. C., Herreros, J., Trainini, J., Juffe, A., Rendal, E., Prosper, F., and Genovese, J. (2004). Autologous human serum for cell culture avoids the implantation of cardioverter-defibrillators in cellular cardiomyoplasty. *Int. J. Cardiol.* **95**, S29–S33.
171. Hutcheson, K. A., Atkins, B. Z., Hueman, M. T., Hopkins, M. B., Glower, D. D., and Taylor, D. A. (2000). Comparison of benefits on myocardial performance of cellular cardiomyoplasty with skeletal myoblasts and fibroblasts. *Cell Transplant.* **9**, 359–368.
172. Fujii, T., Yau, T. M., Weisel, R. D., Ohno, N., Mickle, D. A., Shiono, N., Ozawa, T., Matsubayashi, K., and Li, R. K. (2003). Cell transplantation to prevent heart failure: a comparison of cell types. *Ann. Thorac. Surg.* **76**, 2062–2070.
173. Thompson, R. B., Emani, S. M., Davis, B. H., van den Bos, E. J., Morimoto, Y., Craig, D., Glower, D., and Taylor, D. A. (2003). Comparison of intracardiac cell transplantation: autologous skeletal myoblasts versus bone marrow cells. *Circulation* **108**, II264–I1271.
174. Agbulut, O., Vandervelde, S., Al Attar, N., Larghero, J., Ghostine, S., Leobon, B., Robidel, E., Borsani, P., Le Lorc'h, M., Bissery, A., Chomienne, C., Bruneval, P., Marolleau, J. P., Vilquin, J. T., Hagege, A., Samuel, J. L., and Menasche, P. (2004). Comparison of human skeletal myoblasts and bone marrow-derived CD133+ progenitors for the repair of infarcted myocardium. *J. Am. Coll. Cardiol.* **44**, 458–463.
175. Orlic, D., Kajstura, J., Chimenti, S., Limana, F., Jakoniuk, I., Quaini, F., Nadal-Ginard, B., Bodine, D. M., Leri, A., and Anversa, P. (2001). Mobilized bone marrow cells repair the infarcted heart, improving function and survival. *Proc. Natl. Acad. Sci. USA* **98**, 10344–10349.
176. Orlic, D., Kajstura, J., Chimenti, S., Jakoniuk, I., Anderson, S. M., Li, B., Pickel, J., McKay, R., Nadal-Ginard, B., Bodine, D. M., Leri, A., and Anversa, P. (2001). Bone marrow cells regenerate infarcted myocardium. *Nature* **410**, 701–705.
177. Orlic, D., Hill, J. M., and Arai, A. E. (2002). Stem cells for myocardial regeneration. *Circ. Res.* **91**, 1092–1102.
178. Kudo, M., Wang, Y., Wani, M. A., Xu, M., Ayub, A., and Ashraf, M. (2003). Implantation of bone marrow stem cells reduces the infarction and fibrosis in ischemic mouse heart. *J. Mol. Cell. Cardiol.* **35**, 1113–1119.
179. Yoon, Y. S., Lee, N., and Scadova, H. (2005). Myocardial regeneration with bone-marrow-derived stem cells. *Biol. Cell* **97**, 253–263.
180. Yoon, Y. S., Wecker, A., Heyd, L., Park, J. S., Tkebuchava, T., Kusano, K., Hanley, A., Scadova, H., Qin, G., Cha, D. H., Johnson, K. L., Aikawa, R., Asahara, T., and Losordo, D. W. (2005). Clonally expanded novel multipotent stem cells from human bone marrow regenerate myocardium after myocardial infarction. *J. Clin. Invest.* **115**, 326–338.
181. Badorff, C., Brandes, R. P., Popp, R., Rupp, S., Urbich, C., Aicher, A., Fleming, I., Busse, R., Zeiher, A. M., and Dimmeler, S. (2003). Transdifferentiation of blood-derived human adult endothelial progenitor cells into functionally active cardiomyocytes. *Circulation* **107**, 1024–1032.

182. Tomita, S., Li, R. K., Weisel, R. D., Mickle, D. A., Kim, E. J, Sakai, T., and Jia, Z. Q. (1999). Autologous transplantation of bone marrow cells improves damaged heart function. *Circulation* **100**, II247–I256.

183. Toma, C., Pittenger, M. F., Cahill, K. S., Byrne, B. J., and Kessler, P. D. (2002). Human mesenchymal stem cells differentiate to a cardiomyocyte phenotype in the adult murine heart. *Circulation* **105**, 93–98.

184. Murry, C. E., Soonpaa, M. H., Reinecke, H., Nakajima, H., Nakajima, H. O., Rubart, M., Pasumarthi, K. B., Virag, J. I., Bartelmez, S. H., Poppa, V., Bradford, G., Dowell, J. D., Williams, D. A., and Field, L. J. (2004). Haematopoietic stem cells do not transdifferentiate into cardiac myocytes in myocardial infarcts. *Nature* **428**, 664–668.

185. Murry, C. E., Field, L. J., and Menasche, P. (2005). Cell-based cardiac repair: reflections at the 10-year point. *Circulation* **112**, 3174–3183.

186. Reinecke, H., Minami, E., Poppa, V., and Murry, C. E. (2004). Evidence for fusion between cardiac and skeletal muscle cells. *Circ. Res.* **94**, e56–60.

187. Shim, W. S., Jiang, S., Wong, P., Tan J., Chua, Y. L., Tan, Y. S., Sin, Y. K., Lim, C. H., Chua, T., Teh, M., Liu, T. C., and Sim, E. (2004). Ex vivo differentiation of human adult bone marrow stem cells into cardiomyocyte-like cells. *Biochem. Biophys. Res. Commun.* **324**, 481–488.

188. Eisenberg, C. A., Burch, J. B., and Eisenberg, L. M. (2006). Bone marrow cells transdifferentiate to cardiomyocytes when introduced into the embryonic heart. *Stem Cells* Jan 12; [Epub ahead of print].

189. Xu, M., Wani, M., Dai, Y. S., Wang, J., Yan, M., Ayub, A., and Ashraf, M. (2004). Differentiation of bone marrow stromal cells into the cardiac phenotype requires intercellular communication with myocytes. *Circulation* **110**, 2658–2665.

190. Badorff, C., and Dimmeler, S. (2006). Neovascularization and cardiac repair by bone marrow-derived stem cells. *Handb. Exp. Pharmacol.* **174**, 283–298.

191. Fukuda, K., and Fujita, J. (2005). Mesenchymal, but not hematopoietic, stem cells can be mobilized and differentiate into cardiomyocytes after myocardial infarction in mice. *Kidney Int.* **68**, 1940–1943.

192. Matsuura, K., Wada, H., Nagai, T., Iijima, Y., Minamino, T., Sano, M., Akazawa, H., Molkentin, J. D., Kasanuki, H., and Komuro, I. (2004). Cardiomyocytes fuse with surrounding noncardiomyocytes and reenter the cell cycle. *J. Cell. Biol.* **167**, 351–363.

193. Kajstura, J., Rota, M., Whang, B., Cascapera, S., Hosoda, T., Bearzi, C., Nurzynska, D., Kasahara, H., Zias, E., Bonafe, M., Nadal-Ginard, B., Torella, D., Nascimbene, A., Quaini, F., Urbanek, K., Leri, A., and Anversa, P. (2005). Bone marrow cells differentiate in cardiac cell lineages after infarction independently of cell fusion. *Circ. Res.* **96**, 127–137.

194. Galinanes, M., Loubani, M., Davies, J., Chin, D., Pasi, J., and Bell, P. R. (2004). Autotransplantation of unmanipulated bone marrow into scarred myocardium is safe and enhances cardiac function in humans. *Cell Transplant.* **13**, 7–13.

195. Wollert, K. C., Meyer, G. P., Lotz, J., Ringes-Lichtenberg, S., Lippolt, P., Breidenbach, C., Fichtner, S., Korte, T., Hornig, B., Messinger, D., Arseniev, L., Hertenstein, B., Ganser, A., and Drexler, H. (2004). Intracoronary autologous bone-marrow cell transfer after myocardial infarction: the BOOST randomised controlled clinical trial. *Lancet* **364**, 141–148.

196. Lee, M. S., and Makkar, R. R. (2004). Stem-cell transplantation in myocardial infarction: a status report. *Ann. Intern. Med.* **140**, 729–737.

197. Strauer, B. E., Brehm, M., Zeus, T., Kostering, M., Hernandez, A., Sorg, R. V., Kogler, G., and Wernet, P. (2002). Repair of infarcted myocardium by autologous intracoronary mononuclear bone marrow cell transplantation in humans. *Circulation* **106**, 1913–1918.

198. Assmus, B., Schachinger, V., Teupe, C., Britten, M., Lehmann, R., Dobert, N., Grunwald, F., Aicher, A., Urbich, C., Martin, H., Hoelzer, D., Dimmeler, S., and Zeiher, A. M. (2002). Transplantation of progenitor cells and regeneration enhancement in acute myocardial infarction (TOPCARE-AMI). *Circulation* **106**, 3009–3017.

199. Chen, S. L., Fang, W. W., Ye, F., Liu, Y. H., Qian, J., Shan, S. J., Zhang, J. J., Chunhua, R. Z., Liao, L. M., Lin, S., and Sun, J. P. (2004). Effect on left ventricular function of intracoronary transplantation of autologous bone marrow mesenchymal stem cell in patients with acute myocardial infarction. *Am. J. Cardiol.* **94**, 92–95.

200. Perin, E. C., Dohmann, H. F., Borojevic, R., Silva, S. A., Sousa, A. L., Mesquita, C. T., Rossi, M. I., Carvalho, A. C, Dutra, H. S., Dohmann, H. J., Silva, G. V., Belem, L., Vivacqua, R., Rangel, F. O., Esporcatte, R., Geng, Y. J., Vaughn, W. K., Assad, J. A., Mesquita, E. T., and Willerson, J. T. (2003). Transendocardial, autologous bone marrow cell transplantation for severe, chronic ischemic heart failure. *Circulation* **107**, 2294–2302.

201. Janssens, S., Dubois, C., Bogaert, J., Theunissen, K., Deroose, C., Desmet, W., Kalantzi, M., Herbots, L., Sinnaeve, P., Dens, J., Maertens, J., Rademakers, F., Dymarkowski, S., Gheysens, O., Van Cleemput, J., Bormans, G., Nuyts, J., Belmans, A., Mortelmans, L., Boogaerts, M., and Van de Werf, F. (2006). Autologous bone marrow-derived stem-cell transfer in patients with ST-segment elevation myocardial infarction: double-blind, randomised controlled trial. *Lancet* **367**, 113–121.

202. Beltrami, A. P., Barlucchi, L., Torella D., Baker, M., Limana, F., Chimenti, S., Kasahara, H., Rota, M., Musso, E., Urbanek, K., Leri, A., Kajstura, J., Nadal-Ginard, B., and Anversa, P. (2003). Adult cardiac stem cells are multipotent and support myocardial regeneration. *Cell* **114**, 763–776

203. Dawn, B., Stein, A. B., Urbanek, K., Rota, M., Whang, B., Rastaldo, R., Torella, D., Tang, X. L., Rezazadeh, A., Kajstura, J., Leri, A., Hunt, G., Varma, J., Prabhu, S. D., Anversa, P., and Bolli, R. (2005). Cardiac stem cells delivered intravascularly traverse the vessel barrier, regenerate infarcted myocardium, and improve cardiac function. *Proc. Natl. Acad. Sci. USA* **102**, 3766–3771.

204. Oh, H., Chi, X., Bradfute, S. B., Mishina, Y., Pocius, J., Michael, L. H., Behringer, R. R., Schwartz, R. J., Entman, M. L., and Schneider, M. D. (2004). Cardiac muscle plasticity in adult and embryo by heart-derived progenitor cells. *Ann. N. Y. Acad. Sci.* **1015**, 182–189.

205. Oh, H., Bradfute, S. B., Gallardo, T. D., Nakamura, T., Gaussin, V., Mishina, Y., Pocius, J., Michael, L. H., Behringer, R. R., Garry, D. J., Entman, M. L., and Schneider, M. D. (2003). Cardiac progenitor cells from adult myocardium: homing, differentiation, and fusion after infarction. *Proc. Natl. Acad. Sci. USA* **100**, 12313–12318.

206. Matsuura, K., Nagai, T., Nishigaki, N., Oyama, T., Nishi, J., Wada, H., Sano, M., Toko, H., Akazawa, H., Sato, T., Nakaya, H., Kasanuki, H., and Komuro, I. (2004). Adult cardiac Sca-1-positive cells differentiate into beating cardiomyocytes. *J. Biol. Chem.* **279**, 11384–11391.
207. Laugwitz, K. L., Moretti, A., Lam, J., Gruber, P., Chen, Y., Woodard, S., Lin, L. Z., Cai, C. L., Lu, M. M., Reth, M., Platoshyn, O., Yuan, J. X., Evans, S., and Chien, K. R. (2005). Postnatal isl1+ cardioblasts enter fully differentiated cardiomyocyte lineages. *Nature* **433**, 647–653.
208. Messina, E., De Angelis, L., Frati, G., Morrone, S., Chimenti, S., Fiordaliso, F., Salio, M., Battaglia, M., Latronico, M. V., Coletta, M., Vivarelli, E., Frati, L., Cossu, G., and Giacomello, A. (2004). Isolation and expansion of adult cardiac stem cells from human and murine heart. *Circ. Res.* **95**, 911–921.
209. Martin, C. M., Meeson, A. P., Robertson, S. M., Hawke, T. J., Richardson, J. A., Bates, S., Goetsch, S. C., Gallardo, T. D., and Garry, D. J. (2004). Persistent expression of the ATP-binding cassette transporter, Abcg2, identifies cardiac SP cells in the developing and adult heart. *Dev. Biol.* **265**, 262–275.
210. Hierlihy, A. M., Seale, P., Lobe, C. G., Rudnicki, M. A., and Megeney, L. A. (2002). The post-natal heart contains a myocardial stem cell population. *FEBS Lett.* **530**, 239–243.
211. Garry, D. J., and Martin, C. M. (2004). Cardiac regeneration: self-service at the pump. *Circ. Res.* **95**, 852–854.
212. Liadaki, K., Kho, A. T., Sanoudou, D., Schienda, J., Flint, A., Beggs, A. H., Kohane, I. S., and Kunkel, L. M. (2005). Side population cells isolated from different tissues share transcriptome signatures and express tissue-specific markers. *Exp. Cell Res.* **303**, 360–374.
213. Jackson, K. A., Majka, S. M., Wang, H., Pocius, J., Hartley, C. J., Majesky, M. W., Entman, M. L., Michael, L. H., Hirschi, K. K, and Goodell, M. A. (2001). Regeneration of ischemic cardiac muscle and vascular endothelium by adult stem cells. *J. Clin. Invest.* **107**, 1395–1402.
214. Lapidos, K. A., Chen, Y. E., Earley, J. U., Heydemann, A., Huber, J. M., Chien, M., Ma, A., and McNally, E. M. (2004). Transplanted hematopoietic stem cells demonstrate impaired sarcoglycan expression after engraftment into cardiac and skeletal muscle. *J. Clin. Invest.* **114**, 1577–1585.
215. Hattan, N., Kawaguchi, H., Ando, K., Kuwabara, E., Fujita, J., Murata, M., Suematsu, M., Mori, H., and Fukuda, K. (2005). Purified cardiomyocytes from bone marrow mesenchymal stem cells produce stable intracardiac grafts in mice. *Cardiovasc. Res.* **65**, 334–344.
216. Borenstein, N., Hekmati, M., Bruneval, P., and Montarras, D. (2004). Unambiguous identification of implanted cells after cellular cardiomyoplasty: a critical issue. *Circulation* **109**, e209–e210.
217. Hruban, R. H., Long, P. P., Perlman, E. J., Hutchins, G. M., Baumgartner, W. A., Baughman, K. L., and Griffin, C. A. (1993). Fluorescence in situ hybridization for the Y-chromosome can be used to detect cells of recipient origin in allografted hearts following cardiac transplantation. *Am. J. Pathol.* **142**, 975–980.
218. Taylor, D. A., Hruban, R., Rodriguez, E. R., and Goldschmidt-Clermont, P. J. (2002). Cardiac chimerism as a mechanism for self-repair: does it happen and if so to what degree? *Circulation* 106, 2–4.
219. Laflamme, M. A., Myerson, D., Saffitz, J. E., and Murry, C. E. (2002). Evidence for cardiomyocyte repopulation by extracardiac progenitors in transplanted human hearts. *Circ. Res.* **90**, 634–640.
220. Quaini, F., Urbanek, K., Beltrami, A. P., Finato, N., Beltrami, C. A., Nadal-Ginard, B., Kajstura, J., Leri, A., and Anversa, P. (2002). Chimerism of the transplanted heart. *N. Engl. J. Med.* **346**, 5–15.
221. Louie, A. Y., Huber, M. M., Ahrens, E. T., Rothbacher, U., Moats, R., Jacobs, R. E., Fraser, S. E., and Meade, T. J. (2000). In vivo visualization of gene expression using magnetic resonance imaging. *Nat. Biotechnol.* **18**, 321–325.
222. Zhao, M., Beauregard, D. A., Loizou, L., Davletov, B., and Brindle, K. M. (2001). Non-invasive detection of apoptosis using magnetic resonance imaging and a targeted contrast agent. *Nat. Med.* **7**, 1241–1244.
223. Taylor, D. A. (2004). Cell-based myocardial repair: how should we proceed? *Int. J. Cardiol.* **95**, S8–S12.
224. Thompson, R. B., van den Bos, E. J., Davis, B. H., Morimoto, Y., Craig, D., Sutton, B. S., Glower, D. D., and Taylor, D. A. (2005). Intracardiac transplantation of a mixed population of bone marrow cells improves both regional systolic contractility and diastolic relaxation. *J. Heart Lung Transplant.* **24**, 205–214.
225. Lanza, R. P., Chung, H. Y., Yoo, J. J., Wettstein, P. J., Blackwell, C., Borson, N., Hofmeister, E., Schuch, G., Soker, S., Moraes, C. T., West, M. D., and Atala, A. (2002). Generation of histocompatible tissues using nuclear transplantation. *Nat. Biotechnol.* **20**, 689–696.
226. Lanza, R., Moore, M. A., Wakayama, T., Perry, A. C., Shieh, J. H., Hendrikx, J., Leri, A., Chimenti, S., Monsen, A., Nurzynska, D., West, M. D., Kajstura, J., and Anversa, P. (2004). Regeneration of the infarcted heart with stem cells derived by nuclear transplantation. *Circ. Res.* **94**, 820–827.
227. Zullo, S. J. (2001). Gene therapy of mitochondrial DNA mutations: a brief, biased history of allotopic expression in mammalian cells. *Semin. Neurol.* **21**, 327–335.
228. Pedrotty, D. M., Koh, J., Davis, B. H., Taylor, D. A., Wolf, P., and Niklason, L. E. (2005). Engineering skeletal myoblasts: roles of three-dimensional culture and electrical stimulation. *Am. J. Physiol. Heart Circ. Physiol.* **288**, H1620–H1626.
229. Dar, A., Shachar, M., Leor, J., and Cohen, S. (2002). Optimization of cardiac cell seeding and distribution in 3D porous alginate scaffolds. *Biotechnol. Bioeng.* **80**, 305–312.
230. Leor, J., Amsalem, Y., and Cohen, S. (2005). Cells, scaffolds, and molecules for myocardial tissue engineering. *Pharmacol. Ther.* **105**, 151–163.
231. Shimizu, T,. Yamato, M., Isoi, Y., Akutsu, T., Setomaru, T., Abe, K., Kikuchi, A., Umezu, M., and Okano, T. (2002). Fabrication of pulsatile cardiac tissue grafts using a novel 3-dimensional cell sheet manipulation technique and temperature-responsive cell culture surfaces. *Circ. Res.* **90**, e40.
232. Shimizu, T., Yamato, M., Kikuchi, A., and Okano, T. (2003). Cell sheet engineering for myocardial tissue reconstruction. *Biomaterials* **24**, 2309–2316.
233. Teng, G., Zhao, X., Cross, J. C., Li, P., Lees-Miller, J. P., Guo J., Dyck, J. R., and Duff, H. J. (2004). Prolonged repolarization and triggered activity induced by adenoviral expression of HERG N629D in cardiomyocytes derived from stem cells. *Cardiovasc. Res.* **61**, 268–277.

234. Most, P., Pleger, S. T., Volkers, M., Heidt, B., Boerries, M., Weichenhan, D., Loffler, E., Janssen, P. M., Eckhart, A. D., Martini, J., Williams, M. L., Katus, H. A., Remppis, A., and Koch, W. J. (2004). Cardiac adenoviral S100A1 gene delivery rescues failing myocardium. *J. Clin. Invest.* **114,** 1550–1563.
235. Potapova, I., Plotnikov, A., Lu, Z., Danilo, P., Jr., Valiunas, V, Qu, J., Doronin, S., Zuckerman, J., Shlapakova, I. N., Gao, J., Pan, Z., Herron, A. J., Robinson, R. B., Brink, P. R., Rosen, M. R., and Cohen, I. S. (2004). Human mesenchymal stem cells as a gene delivery system to create cardiac pacemakers. *Circ. Res.* **94,** 952–959.
236. Askari, A., Unzek, S., Goldman, C. K., Ellis, S. G., Thomas, J. D., DiCorleto, P. E., Topol, E. J., and Penn, M. S. (2004). Cellular, but not direct, adenoviral delivery of vascular endothelial growth factor results in improved left ventricular function and neovascularization in dilated ischemic cardiomyopathy. *J. Am. Coll. Cardiol.* **43,** 1908–1914.

CHAPTER 5

Cardiovascular Signaling Pathways

OVERVIEW

A broad analysis of the cardiovascular signaling pathways, focusing primarily on the myocardium, will be presented in this chapter. Major components of these pathways such as receptors, their ligands, G-proteins, and protein kinases, as well as regulators from both a physiological and a postgenomic perspective will be assessed together, with a critical analysis of available information on interorganellar distribution, regulation, transcriptional, and post-transcriptional components of the signaling pathways. Finally, the present and future use of this information, which could advance our understanding of cardiac development and cardiac pathophysiology, as well as its eventual application in clinical cardiology, will be discussed.

INTRODUCTION

At the outset, it is important to understand that the complex field of cardiovascular signaling is in its infancy and that much of the currently available information has come to light as a by product of extensive research effort to understand the mechanisms of hypertrophy, apoptosis, cell death, and subsequent myocardial remodeling. Evidence suggests that the heart signaling pathways act as both transmitter and dynamic receiver of a variety of intracellular and extracellular stimuli, as well as an integrator of numerous interacting transducers, including protein kinases and effectors, the G-proteins, and small G-protein activators that are profoundly influenced by their location in the cell.

Given that the targeting and localization of signaling factors and enzymes to discrete subcellular compartments or substrates are important regulatory mechanisms, ensuring specificity of signaling events in response to local stimuli, these systems merit examination both from a subcellular/organellar and from a functional standpoint under both physiological and pathophysiological conditions. Moreover, cardiovascular signaling includes both a built-in specificity, reversibility and a redundancy of its components, which while making their analysis a very complex undertaking, provides the cardiac cells with great plasticity to respond to insult, as well as to growth stimuli.

Although the focus in this chapter will be mainly on cell signaling in the heart, a number of signaling pathways involved in vascular cells will also be discussed.

CARDIAC SIGNALING IN PHYSIOLOGICAL GROWTH

Several signal transduction pathways regulate cardiomyocyte growth and/or proliferation. These are redundant mechanisms converging on one or several serine/threonine kinases. Several G-protein–coupled receptors (GPCRs) such as AR, β-AR, angiotensin II, and endothelin-1 are able to activate these signaling cascades and induce changes in cell growth and proliferation. A general scheme involves the following: Signals received at the plasma membrane receptors are transmitted by GPCR/G-proteins/second messengers to a wide spectrum of protein kinases and phosphatases, which are in turn activated. These activated protein modifiers may lead to the activation and/or deactivation of specific transcription factors, which modulate specific gene expression affecting a broad spectrum of cellular events, or they can target directly proteins involved in metabolic pathways, ion transport, Ca^{++} regulation, and handling, which affect contractility and excitability, as well as the pathways of cardiomyocyte apoptosis and/or cell survival.

RECEPTORS

Adrenergic Receptors

Adrenoceptors (ARs) are members of the G-protein–coupled receptor superfamily, interfacing between the sympathetic nervous system and the cardiovascular system, with integral roles in the rapid regulation of myocardial function. In heart failure (HF), chronic catecholamine stimulation of adrenoceptors has been linked to pathologic cardiac remodeling, including cardiomyocyte apoptosis and hypertrophy.

Stimulation of β1-adrenergic receptors (the predominant subtype found in both neonatal and adult ventricular myocytes) results in the activation of G_s and adenylyl cyclase (AC) with increased intracellular cAMP and induction of protein kinase A (PKA). This leads to the downstream phosphorylation of key target proteins, including L-type calcium channels, phospholamban, troponin I, and ryanodine receptors, resulting in the modulation of cardiac contractility in both neonatal and adult ventricular myocytes (Fig. 1).[1]

In contrast, β2-adrenergic receptors (which play a paramount role in the inotropic support of the failing, aged, or transplanted heart) are linked to different pathways in neonatal and adult ventricular myocytes. For instance, the pathway linking β2-adrenergic receptors to cAMP-dependent changes in contractile function is only expressed in the neonate.[2]

Among the proteins targeted for PKA phosphorylation are the β-adrenergic receptors. Phosphorylation of the β-adrenergic receptors promotes an uncoupling and desensitization of the receptors. Phosphorylation by the G-protein receptor kinase (β-ARK) also caused uncoupling and reduced the β-adrenergic responsiveness.[3] The uncoupling of the receptor is the prerequisite for receptor internalization, in which the receptor is translocated from the sarcolemmal membrane into cytosolic compartments. Chronic β-AR stimulation causes down-regulation of the receptors. During this process of desensitization, the expression of the β-adrenergic receptor at the mRNA and protein level is reduced.

Members of the α1-adrenoceptors (α1-AR) receptor family each contain a seven-transmembrane–spanning domain and are linked to G-proteins. On stimulation with agonists such as noradrenaline and adrenaline, α1-ARs activate G_q-proteins and subsequently activate phospholipase Cβ (PLC-β), resulting in increased levels of second messengers, inositol trisphosphate (IP3), and diacylglycerol (DAG), which promote an increase in intracellular Ca^{++} levels and protein kinase C (PKC) activation that modulate myocardial contraction.[4] However activation of the α-adrenergic receptors and the G_q/phospholipase C/PKC pathway has limited acute effect on instantaneous myocardial contractility secondary to β–AR regulation but rather serves as a potent stimulus for cardiac hypertrophy. PKC activation of ERK has also been implicated in the α1-adrenoceptor stimulation of cardiomyocyte hypertrophy. The heart contains a relatively small number of α1-ARs; the β/α-AR ratio is 10:1 in human myocardium. In the failing heart, with diminished β-AR levels, there is an increase of the α1-AR to the β-AR ratio, and it has been suggested that α1-AR may, therefore, assume a greater functional role in the failing heart by providing a secondary inotropic system.

The α2-adrenoceptors (α2-ARs) are receptors for endogenous catecholamine agonists (e.g., norepinephrine and epinephrine) that mediate a number of physiological and pharmacological responses such as changes in blood pressure and heart rate. Three distinct subtypes, denoted α2A-, α2B-, and α2C-AR, have been characterized and cloned.[5] Screening of human populations from various ethnic backgrounds has shown that α2-AR genes are polymorphic. Functional changes in G-protein coupling, in agonist-promoted receptor phosphorylation and desensitization, have been found in heterologous systems such as CHO and

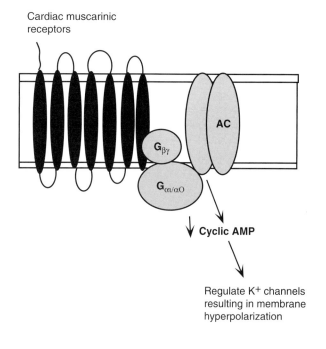

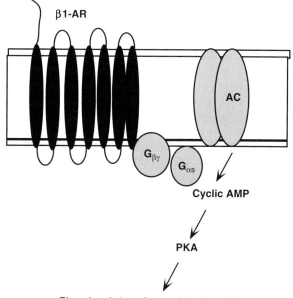

FIGURE 1 Stimulation of β1-adrenergic receptors. The activation of G_s and adenylyl cyclase (AC) with increased intracellular cAMP and induction of protein kinase A is depicted. This pathway leads to the phosphorylation of key target proteins, including L-type calcium channels, phospholamban, troponin I, and ryanodine receptors.

COS-7 cells, which express these genetic polymorphisms compared with wild-type receptors.[6]

Muscarinic Receptors

Muscarinic acetylcholine receptors (mAChR) mediate a variety of cellular responses, including inhibition of AC (Fig. 1), modulation of K$^+$ channels, and increased phosphoinositide breakdown.[4] These diverse effects of mAChR activation elicit both negative and positive inotropic and chronotropic effects in the heart. Positive inotropic effects of cholinergic agonists are present only at high agonist concentration (>10 µmol/L) and tend to be pertussis toxin (PTX)–insensitive in contrast to the negative inotropic effects observed at lower agonist concentrations, which are sensitive to inactivation by PTX. These dual effects of mAChR activation in heart may be a result of the presence of multiple subtypes of mAChRs.[7] Thus far, five mAChR subtypes (M1–M5) have been identified, and each subtype is encoded by a different gene. The mACHR proteins contain seven transmembrane-spanning domains, coupled to G-proteins of the G_I and G_q families to inhibit AC and activate PLC, respectively. Although M2 receptors have been considered to be the only functional mAChRs in the myocardium, new observations reveal that M3 receptors are also present in the hearts of various species.[8]

Stimulation of mAChR results in the activation of an inward rectifier K$^+$ current termed I_{KACh} in cardiac myocytes, primarily mediated by the M2 subtype of mAChR. However, a novel delayed rectifier–like K$^+$ current designated I_{KM3} has been recently identified that is distinct from I_{KACh} and other known K$^+$ currents and that is mediated by the activation of the cardiac M3 receptors.[9] Although I_{KACh} is known to be a G_i-protein–gated K$^+$ channel, I_{KM3} represents the first G_q-protein–coupled K$^+$ channel described in cardiomyocytes. Interestingly, the regulation of these channels is fundamentally different during atrial fibrillation; the atrial levels of I_{KM3} are increased in both animal models and human hearts, whereas the atrial M2 receptor density decreased, indicating down-regulation.[10]

Endothelin

Three endothelin (ET) signaling peptides have been identified: ET-1, ET-2, and ET-3 with well-established effects on the cardiomyocyte, including modulation of contractile function and growth stimulation.[11] ET-1 binds to the ET (A) receptor on the cell surface, coupled to the G_q class of GTP binding proteins, and stimulates hydrolysis of phosphatidylinositol 4′, 5′-bisphosphate to DAG and IP3. DAG remains in the plane of the membrane causing translocation and activation of PKC δ- and ε-isoforms. This is followed by activation of the small G-protein Ras and by an ERK cascade. Over a longer time course, two protein kinase cascades related to the ERK1/2 cascade, the JNK and p38 MAP kinase cascades, also become activated. Downstream activation of nuclear transcription factors (e.g., GATA-4, c-Jun), protein kinases (e.g., 90-kDa ribosomal protein S6 kinase, MAPK-activated protein kinase 2), and ion exchangers/channels (e.g., the Na$^+$/H$^+$ exchanger 1) follows. These changes are responsible for the overall biological effects of ET isopeptides on the cardiomyocyte.[12]

Angiotensin

Angiotensin-converting enzyme (ACE), a central element of the renin–angiotensin system, converts the decapeptide angiotensin I to the potent pressor octapeptide angiotensin II (Ang II), mediating peripheral vascular tone, as well as glomerular filtration in the kidney. In addition to its direct effect on blood flow, which will be discussed in Chapter 12, Ang II directly causes changes in cell phenotype, cell growth, and apoptosis and regulates gene expression of a broad range of bioactive molecules (e.g., vasoactive hormones, growth factors, extracellular matrix components, cytokines). In addition, Ang II activates multiple intracellular signaling cascades, involving numerous transduction components such as MAP kinases, tyrosine kinases, and various transcription factors in cardiomyocytes, fibroblasts, vascular endothelial, smooth muscle cells (SMCs), and renal cells.[13] Ang II also promotes cardiomyocyte enlargement and protein synthesis, as well as hypertrophy-associated alterations in the cardiac gene expression program through specific cellular receptor subtypes AT1 and AT2. AT1 is the more abundant in the adult heart and has been linked to control of both hypertrophy and apoptosis in the cardiomyocyte. Although several studies have found AT2 receptor to act in opposition to AT1, the myocardial AT2 receptor is less well understood. Through the G_q proteins, the AT1 receptor is coupled to a variety of intracellular signals, including the generation of oxygen free radicals, the activation of Ras, and the ERK/MAPK protein kinase family.[14] Ang II activates NF-κB–dependent transcription in other cell types, likely through its effects on the cellular redox state.[15] These actions contribute to the pathophysiology of cardiac hypertrophy and remodeling, HF, vascular thickening, and atherosclerosis.[16]

Growth Factors

The extracellular signal–regulated kinases 1/2 (ERK1/2) are activated in cardiomyocytes by G_q-protein–coupled receptors and are associated with the induction of hypertrophy. In primary cardiomyocyte cultures, platelet-derived growth factor (PDGF), epidermal growth factor (EGF), and fibroblast growth factor (FGF) promoted receptor-coupled ERK1 activation and significantly increased cardiomyocyte size in contrast to insulin, IGF-1, and nerve growth factor (NGF), which had little effect. Peptide growth factors activate phospholipase C γ1 (PLCγ1), and PKC. In cardiomyocytes, only PDGF stimulated tyrosine phosphorylation of PLCγ1 and PKCδ. Furthermore, activation of ERK1/2 by PDGF, but not

EGF, required PKC activity. In contrast, EGF substantially increased Ras-GTP with rapid activation of c-Raf, whereas stimulation of Ras-GTP loading by PDGF was minimal, and activation of c-Raf was delayed, suggesting differential coupling of PDGF and EGF receptors to the ERK1/2 cascade.[17]

Toll-Like Receptors

There is increased expression of innate immune response proteins, including IL-1β, TNF-α, and the cytokine-inducible isoform of nitric oxide synthase (iNOS) in the failing heart of humans and experimental animals, regardless of etiology. Transmembrane signaling proteins of the Toll-like receptor (Toll-R) family constitute key signaling elements in both macrophages and in atherosclerotic lesions (see Chapter 8). The Toll-like receptor 4 (TLR4) is highly expressed in the heart. The expression is strongly up-regulated in mice with cardiac ischemia (relative to controls), as well as in patients with idiopathic DCM[18] and is consistent with the activation of signaling pathways leading to the expression of proinflammatory cytokines, which have been implicated in the etiology of DCM. Moreover, myocardial TLR4 levels were positively correlated with levels of enteroviral replication in DCM.[19] Data suggesting that TLR4 serves a proinflammatory role in ischemia–reperfusion injury have been provided by observations that TLR4-deficient mice sustain smaller infarctions and exhibit less inflammation after myocardial ischemia–reperfusion (I/R) injury.[20] Interestingly, systemic administration of lipopolysaccharide (LPS), a TLR4 agonist, confers a cardioprotective effect against ischemic injury and myocardial infarction (MI).[21] On the other hand, data from chimeric mice have implicated TLR4 effect on leukocytes (not on cardiac myocytes) as an important factor for cardiac myocyte impairment during endotoxemia.[22] Furthermore, the Toll-like receptor 2 (TLR2) is directly involved in mediating the response of cardiomyocytes to oxidative stress (e.g., H_2O_2),[23] and oxidative stress-induced cytotoxicity and apoptosis are enhanced by blocking TLR2.

Protease-Activated Receptors (PARs)

Recently, a novel class of protease-activated receptors (PAR1, PAR2, PAR3, and PAR4), containing seven-transmembrane G-protein–coupled domains has been identified. These receptors are activated by cleavage with serine proteases such as thrombin and trypsin.[24,25] Extracellular proteolytic activation of PARs results in: (1) cleavage of specific sites in the extracellular domain; (2) formation of a new N-terminus (often containing the sequence SFLLRN) that functions as a tethered ligand and binds to an exposed site in the second transmembrane loop triggering G-protein binding; and (3) intracellular signaling (Fig. 2).

In cardiomyocytes expressing PAR-1, a high-affinity receptor for thrombin, agonist binding, and activation of PAR leads to

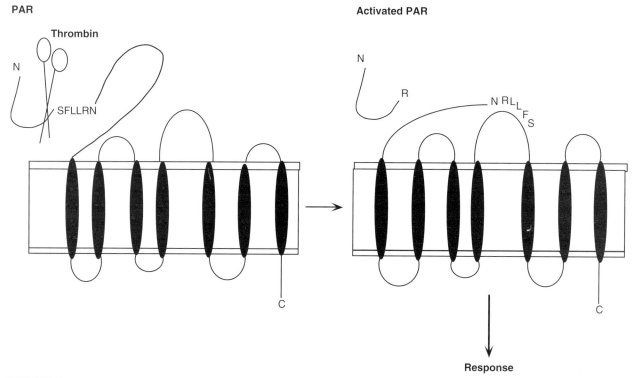

FIGURE 2 Protease activated receptors. PARs are composed of seven transmembrane G-protein–coupled domains, which are activated by serine proteases such as thrombin and trypsin.

IP3 accumulation, stimulation of extracellular signal-regulated (ERK) protein kinase, and modulated contractile function.

Coexpression in cardiomyocytes of PAR-2, activated by trypsin/tryptase but not thrombin, with PAR-1 leads to a more extensive signaling response, including IP3 accumulation, stimulation of MAP kinases (both ERK and p38-MAP kinase), elevated Ca^{++} levels, and contractile function, as well as the activation of JNK and Akt associated with growth and/or survival pathways and induction of both cardiomyocyte hypertrophy and elongation.[25] Furthermore, PAR-1, PAR-2 and PAR-4 have been implicated in vascular development, as well as in a variety of other biological processes, including apoptosis and remodeling.[26]

Receptor Tyrosine Kinases (RTK)

A large family containing more than 20 RTK classes has been identified, all of which share a similar structure that includes a ligand-binding extracellular domain, a single transmembrane domain, and an intracellular tyrosine kinase domain. This large protein family includes the receptors for many growth factors and for insulin. Most of the RTK subfamilies are defined by the extracellular region containing the ligand-binding domains, which exhibit variable length and subdomain composition with highly conserved structural motifs, including domains that are immunoglobulin-binding, cysteine-rich, ephrin-binding, and fibronectin repeats.[27] Except in the case of the insulin receptor, which exists as a dimer in the absence of ligand, ligand binding to the extracellular portion of these receptors results in receptor dimerization, which facilitates the transautophosphorylation of specific tyrosine residues in the highly conserved cytoplasmic portion (Fig. 3). The phosphotyrosine residues enhance the receptor catalytic activity and can provide docking sites for downstream signaling proteins. The creation of docking sites allows the recruitment to the receptor kinase complex of a variety of

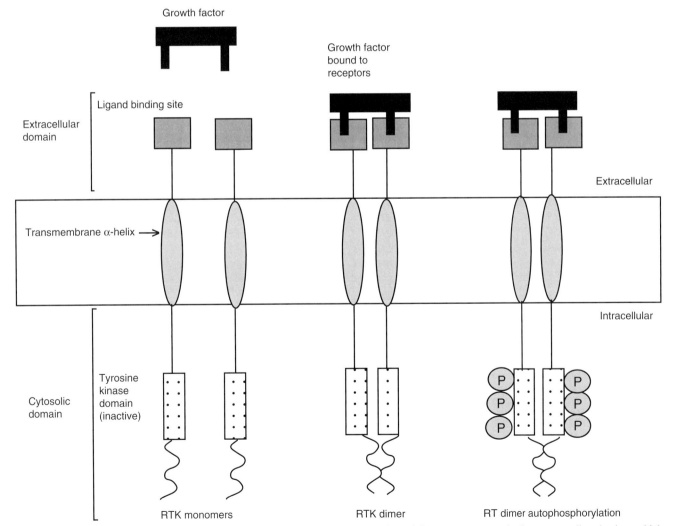

FIGURE 3 Receptor tyrosine kinases. Ligand binding to the extracellular portion of these receptors results in receptor dimerization, which facilitates the trans-autophosphorylation of specific tyrosine residues in the highly conserved cytoplasmic portion. RTKs are phosphorylated in response to stimulation by cytokines, cell adhesion, and stress stimuli.

proteins containing specific binding domains, such as Src homology 2 and 3 or phosphotyrosine binding domains, which can broaden the signaling capacity of the RTKs. For instance, GPCRs that lack an intrinsic kinase activity possessed by RTKs, such as PDGFR or EGFR, have been shown to activate RTKs in response to stimulation by cytokines, cell adhesion, and stress stimuli. In this way, RTKs can function in integrating a large array of stimuli from diverse environmental and intracellular inputs.

Phosphorylation, although necessary, may not be sufficient to fully activate many RTKs. Oligomerization-induced conformational changes may be necessary to modulate the kinetic properties of RTKs and render them fully functional. Because of the critical roles played by RTKs in cellular signaling processes, their catalytic activity is normally under tight control by a variety of intrinsic regulatory mechanisms, as well as by protein tyrosine phosphatases.

A number of the RTKs have been shown to play essential roles in early cardiac development, as well as in the growth, repair, and survival of adult cardiomyocytes as part of a signaling network. For example, the erbB2 RTK is known to have a critical role in cardiac development. In addition, erbB2 participates in an important pathway that involves neuregulins, cell–cell signaling proteins that are ligands for RTKs of the ErbB family and the neuregulin receptor erbB4. Two of the neuregulins (NRG1 and NRG2) and their receptors (erbB2 and erbB4) are essential for normal cardiac development and can mediate hypertrophic growth and enhance the survival of embryonic, post-natal, and adult ventricular cardiomyocytes. Targeting the neuregulin receptors to caveolae microdomains within cardiac myocytes has been shown to be a viable approach to regulating neuregulin signaling in the heart.[28,29]

RTKs also play a pivotal role in the growth responses of vascular cells. RTKs include the VEGF receptors, Eph receptors, Tie1, and Tie2, all of which are expressed on vascular endothelial cells, as well as the PDGF receptors that are expressed on vascular SMCs.[30,31] Although all of these RTKs activate many similar effector molecules, some of the signals initiated seem to be distinct. This could explain, at least in part, how different RTKs expressed in the developing vasculature can direct unique biological functions.

G-Proteins

On activation with the appropriate ligands, GPCRs are converted into the active conformation and are able to complex with and activate heterotrimeric G-proteins.[32] The heterotrimeric G-proteins are composed of three subunits: the α-subunit that carries the guanine-nucleotide binding site and the β- and γ-subunits that form a tightly bound dimer. Inactive G-proteins are heterotrimers composed of a GDP-bound α-subunit associated with the G βγ-dimer, which serves to anchor the heterotrimeric G-protein to the membrane. The activated GPCRs function as GDP/GTP exchange factors and promote the release of GDP and the binding of GTP to the α-subunits, leading to dissociation of the α-subunit and the Gβγ dimer. Both GTP-G_α and $G_{\beta\gamma}$ can interact with a variety of effectors such as AC and PLC to modulate cellular signaling pathways. The deactivation of GPCR signaling occurs at several levels. Importantly, the G_α subunit has an innate GTPase activity that hydrolyses GTP to GDP and promotes reassociation with $G_{\beta\gamma}$ to form the inactive heterotrimer. In addition, ligand dissociation from the GPCRs converts the receptors back to their inactive state (Fig. 4).

Heterotrimeric G-proteins are classified into subclasses according to the α-subunit, with each subfamily designated by its corresponding downstream signaling effect. The $G\alpha_s$ or more simply G_α-proteins are stimulatory regulators of AC linking ligand stimulation (e.g., β-AR) to the accumulation of the second messenger cyclic AMP (cAMP). The G_S subunit is a target of covalent modification by cholera toxin (CTX) that slows GTP hydrolysis, locking G_S in an active GTP-bound form that constitutively stimulates AC. In contrast, the $G_{\beta\alpha i/o}$ or $G_{i/o}$ proteins inhibit AC activity. These proteins are targets for ADP-ribosylation by the pertussis toxin (PTX), which prevents their interaction with receptors and inhibits their downstream signaling. The other two subfamilies, G_q and $G_{12/13}$ proteins, are insensitive to PTX and CTX. The GTP-bound $G_{\alpha q}$-protein activates phosphoinositide phospholipase C-β (PLC-β), leading to generation of IP3 and DAG, accompanied by mobilization of calcium and the activation of PKC, respectively. Dissociated $G_{\beta\gamma}$ can activate small GTP-binding protein Ras and initiate a tyrosine kinase cascade, leading to the activation of MAPK. Furthermore, G_q may activate MAPK independently of $G_{\beta\gamma}$ by a mechanism that is PKC dependent.[33]

Hormones such as angiotensin II, endothelin 1, and norepinephrine, which bind and activate cardiomyocyte membrane receptors coupled to the G_q-proteins, have been implicated in the development and ultimate decompensation of cardiac hypertrophy.[34]

Regulators of G-protein signaling (RGS) proteins are a family of proteins that accelerate intrinsic GTP hydrolysis on α-subunits of trimeric G-proteins.[35] They play crucial roles in the physiological regulation of G-protein–mediated cell signaling. In addition, the small G-proteins are a superfamily of guanine nucleotide–binding proteins with a size ranging from 20–25 kDa, including several subfamilies such as Ras, Rho, Rab, Ran, and ADP ribosylation factor(s). These small G-proteins act as molecular switches to regulate numerous cellular responses, including cardiac myocyte hypertrophy and cell survival associated with cell growth and division, multiple changes in the cytoskeleton, vesicular transport, and myofibrillar apparatus. They share some features with the heterotrimeric G-proteins, including activation by the exchange of GDP to GTP and inactivation by their return to a GDP-bound state, which is enhanced by GTPase-activating proteins. Not surprisingly, there are regions of homology shared between these proteins and the Gα subunit.

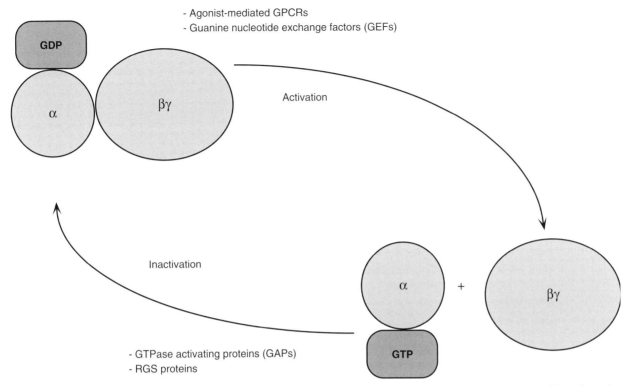

FIGURE 4 G-protein–coupled receptors (GPCRs). On activation with the appropriate ligands, GPCRs are converted into the active conformation and are able to complex with and activate heterotrimeric G-proteins.[32]

Modification of these proteins by isoprenylation promotes their attachment to the membrane. However, the activation of the small G-proteins differs from that of the heterotrimeric G-proteins in one critical respect. With the heterotrimeric G-proteins, ligand binding to a GPCR is the primary stimulus that promotes GDP release from, and GTP binding, to the α-subunit, whereas an association with agonist-occupied receptors is not found with small G-proteins (e.g., Ras and Rho). Instead, activation by the release of GDP from the small G-proteins is primarily mediated by the activation of guanine nucleotide exchange factors (GEFs).

Hearts from transgenic mice expressing activated Ras develop features consistent with myocardial hypertrophy, whereas mice overexpressing RhoA develop lethal HF. In isolated neonatal rat cardiac myocytes, transfection or infection with activated Ras, RhoA, or Rac1 induces features of hypertrophy.

G-proteins and second messenger pathways function differently in cardiac fibroblasts than those in cardiac myocytes. Cardiac fibroblasts are an important cellular component of the myocardial responses to injury and to hypertrophic stimuli. In cardiac fibroblasts, agonists such as bradykinin stimulate inositol phosphate production and increased intracellular Ca^{++} levels, whereas endothelin-1 and norepinephrine do not, in contrast to their action in cardiac myocytes. Cardiac fibroblasts express functional G-protein–linked receptors that couple to G_q and G_s, with little or no coupling to G_i. The expression of receptors and their coupling to G_q, but not to G_i-linked, responses distinguishes the signaling in cardiac fibroblasts from that in myocytes. Furthermore, agonists that activate G_q in fibroblasts also potentiate the stimulation of G_s, an example of signaling cross-talk not previously observed in adult cardiomyocytes.[36]

EFFECTORS

Adenylyl Cyclase (AC)

The amount of AC sets a limit on cardiac β-adrenergic signaling *in vivo*, and increased AC, independent of β-AR number and G-protein content, provides a means to regulate cardiac responsiveness to β–AR stimulation.[37] Overexpressing an effector such as AC does not alter transmembrane signaling, except when receptors are activated, in contrast to receptor/G-protein overexpression, which promotes continuous activation with detrimental consequences.[38,39] These data suggest AC overexpression as a novel target for safely increasing cardiac responsiveness to β–AR stimulation.[37]

Expression of type V AC isoform is restricted to the heart and brain and generates the major AC isoform found in the adult heart. Type V AC is potently activated through PKC-mediated phosphorylation. The degree of this activation is greater than that achieved by forskolin, the most potent AC agonist. Furthermore, the two PKC isoenzymes are additive in their capacity to activate AC. In contrast, PKA-mediated phosphorylation inhibits type V AC. Thus, type V AC is subject to dual regulation by phosphorylation: activation by PKC and inhibition by PKA, mediated by phosphorylation at unique residues within the type V molecule.[40]

PKA-mediated inactivation of AC creates a feedback system within the cAMP-signaling pathway analogous to PKC-mediated inhibition of the phospholipase C pathway. Catecholamine stimulation in the heart activates both the phospholipase C/PKC pathway through α-adrenoreceptors and the AC/PKA pathway through β-adrenergic receptors. Dual regulation of AC by PKC and PKA may play a major role in integrating these two principal signal transduction pathways, modulating neuronal and hormonal input to the heart.

Phospholipase C (PLC)

Diverse and distinct hormonal stimuli engage specific surface receptors of the cardiomyocyte to initiate the hydrolysis of inositol phospholipids mediated by the effector PLC, whereas changes in intracellular levels of IP3 and inositol 1,3,4,5-tetrakisphosphate, DAG, and Ca^{++} result in the specific phosphorylation of cellular proteins by various protein kinases, such as the PKC family, Ca^{++}-calmodulin–dependent kinase and MAPK. Four classes of PLC isozymes are considered to underlie these signaling responses.[41]

A myriad of seven transmembrane-spanning receptors activate isozymes of the PLC-β class through release of α-subunits of the G_q family of heterotrimeric G-proteins. A subset of PLC-β isozymes can also be activated by $G_{\beta\gamma}$. PLC-γ isozymes are activated by protein phosphorylation after the activation of RTKs. A fourth class of PLC isozymes (PLC-ε) has been found to be involved in signaling[42] and exhibits a novel pattern of regulation mediated by the Ras oncoprotein and $G_{\alpha12}$-subunits of heterotrimeric G-proteins.

Caveolae

Caveolae are vesicular organelles (50–100-nm in diameter), which are specialized subdomains of the plasma membrane. They are particularly abundant in cells of the cardiovascular system, including endothelial cells, SMCs, macrophages, cardiac myocytes, and fibroblasts. In these cell types, caveolae function both in protein trafficking and signal transduction. Caveolins (primarily caveolin-2 and caveolin-3 in cardiomyocytes) are structural proteins that are both necessary and sufficient for the formation of caveolae membrane domains. In a number of ways, caveolins serve both to compartmentalize and to concentrate key signaling proteins, thereby regulating cardiomyocyte signaling. Multiple components of signaling cascades, including β-ARs, G-proteins, AC, the Rho family of small GTPases, PKCα and PKCε, and ERK, have been localized to caveolae.[43-45] Colocalization of G-protein pathway signaling molecules may be a contributory factor in both the spatial and temporal regulation of cardiomyocyte signal transduction.

Recent observations using caveolin-deficient mouse models have demonstrated that caveolae and caveolins can promote pathological phenotypes, including atherosclerosis, cardiac hypertrophy, and cardiomyopathy.

KINASES AND PHOSPHATASES

Protein Kinase A (PKA)

This enzyme is structurally organized as a heterotetramer composed of two regulatory (R) subunits that on binding the two catalytic (C) subunits maintain the overall complex in a dormant state. The binding of two cyclic AMP molecules to tandem sites on each R subunit results in the release of the C subunits and the activation of their enzymatic activity. The dissociated C subunits phosphorylate serine or threonine residues on target proteins in the nucleus and cytoplasm, leading to changes in cardiomyocyte metabolism, ion channel function, growth, and gene expression. The catalytic subunits are encoded by three different genes (Cα, Cβ, and Cγ), whereas the regulatory subunits are encoded by four genes (RIα, RIβ, RIIα, and RIIβ). The regulatory subunit contains an N-terminal dimerization domain, an autophosphorylation site that also comprises the primary site for catalytic subunit binding, and two tandem cAMP binding sites.

Compartmentalization of these enzymes can be achieved through association with anchoring or adaptor proteins, which target them to subcellular organelles or tether them directly to target substrates by means of protein–protein interactions. Specific PKA-anchoring proteins (AKAPs) serve as important regulators of PKA function and signaling by directing the subcellular localization of PKA by binding to its regulatory (R) subunits, in effect concentrating PKA at specific intracellular locations. By use of a variety of experimental approaches, including yeast two hybrid screening, proteomic analysis, and interaction cloning, two major anchoring proteins for PKA, MAP2 and AKAP75, and more than 13 different AKAPs have been found in the heart.[46] Targeting of AKAPs to specific sites within the cell is governed by sequences in the AKAP. Interestingly, despite their diverse structure, the AKAPs all contain an amphipathic helical region of 14–18 amino acids that binds to the N-terminus of the RII subunit, underlying their interaction with PKA. Besides PKA, AKAPs also interact with other signaling components, including phosphodiesterase inhibitors, phosphatases, and PKA substrates.[47] Myocyte AKAPs have been identified in association with specific plasma membrane ion channels (e.g., hKCNQ1, the L-type Ca^{++} channel), the β–AR complex, and the sodium-calcium exchanger (NCX1), in association with RyR at both the sarcoplasmic reticulum (SR) and T-tubule junction. In addition, AKAPs have been found on the nuclear membrane and in association with the mitochondrial outer membrane. The precise functional role of a number of the identified AKAPs in the cardiac myocyte has not yet been established.

A novel role for leucine zipper motifs in targeting kinases and phosphatases by means of anchoring proteins has also been identified. Several cardiac ion channels contain a domain to anchor phosphorylation modulatory proteins to the channel, essentially allowing the formation of a scaffolding structure for regulatory proteins. Ion channels such as the ryanodine-sensitive Ca^{++} release channel (RyR), the dihydropyridine receptor (L-type Ca^{++} channel), and the delayed rectifier K^+ channel subunit KCNQ1, all contain a modified leucine zipper termed a LIZ (leucine/isoleucine zipper), which promotes protein–protein interaction and protein oligomerization.[48]

Protein Kinase B (PKB) and Phosphoinositide 3 Kinases (PI3K)

Key signaling pathways in cardiomyocyte proliferative growth and cell survival involve phosphoinositide 3 kinases (PI3K) and protein kinase B (PKB), also known as Akt. The class I PI3Ks phosphorylate PI $(4,5)P_2$ on the 3′ position of the inositol ring to form PI $(3,4,5)P_3$, otherwise termed PIP3, which can be mediated through RTK or GPCR signaling. Class IA PI3Ks are composed of a p110 catalytic subunit (either α, β, or γ) in association with a p85 regulatory protein.[49] Activation of PI3K activity is promoted by the direct interaction of p85 with specific phosphorylated tyrosine residues present on growth factor receptors or by Ras activation. Another PI3K type, class 1B PI3K, contains a p110γ catalytic subunit in association with a p101 adaptor molecule, and its activation proceeds by means of interaction of the γ isoform of the p110 catalytic subunit with heterotrimeric G-protein βγ subunits, resulting from GPCR stimulation.

As a highly potent signaling lipid involved in a multiplicity of cellular processes, the levels of PIP3 are tightly regulated not only at the level of synthesis by PI3K activity but also by rapid dephosphorylation by lipid phosphatases, PTEN and SHP enzymes. Downstream of the PIP3 production at the plasma membrane, PDK-1 (phosphoinositide dependent kinase 1) becomes activated in part by its translocation to the plasma membrane and proximity to its substrates, which include Akt (PKB) (Fig. 5).

Akt represents a family of serine and threonine kinases that include Akt1, Akt2, and Akt3 encoded by three distinct genetic loci with extensive homology (approximately 80%).[50] There are considerable differences in both the expression and function of the Akt isoforms, with only Akt1 and Akt2 being highly expressed in the heart. All three Akt isoforms contain a kinase domain (with structural homology to PKA and PKC), which contains the primary site (Thr^{308}) of phosphorylation by PDK-1. Although Thr^{308} phosphorylation partially activates Akt, subsequent phosphorylation of Akt at a C-terminal site, Ser^{473}, is required for its full activation; under some conditions, this phosphorylation may be produced by PDK-1, another kinase (PDK-2), or by autophosphorylation.

The stimuli that result in Akt activation in the heart are shown in Table I. On activation, myocardial Akt can phosphorylate a number of downstream targets, including cardioprotective factors involved in glucose and mitochondrial metabolism, apoptosis, and regulators of protein synthesis, as discussed later. The regulation of Akt is also achieved by its dephosphorylation effected by the protein phosphatase PP2A. Inhibitors of PP2A activity, including okadaic acid, increase Akt activity, whereas ceramide, which is involved in enhanced apoptotic signaling, stimulates PP2A. In the heart, Akt signaling has a pronounced anti-apoptotic effect, significantly increases cardiomyocyte growth, and enhances function.

Apoptotic progression of hypoxic cardiomyocytes is abrogated with IGF-1 treatment, which activates PI3K and Akt.[53] Overexpression of constitutively active transgenes, either PI3K or Akt, in cultured hypoxic cardiomyocytes reduces apoptosis. Moreover, after transient ischemia *in vivo,* gene transfer of constitutively active Akt to the heart resulted in reduced apoptosis and infarct size.[64] Several mechanisms and/or effectors by which the PI3K/Akt pathway stems apoptosis have been identified, and it seems likely that these effectors might be enhanced when applied in combination. These include the phosphorylation and inactivation of the proapoptotic protein Bad, NF-κB activation, enhanced NO release, eNOS activation, changes in the mitochondrial membrane pores, and membrane potential–suppressing apoptotic progression, as well as cytochrome *c* release induced by several proapoptotic proteins. Recent observations have also shown that Akt's effect on apoptosis and cell survival is mediated by its phosphorylation of the Forkhead transcription factor, FOXO3, which, in turn, reduces the transcription of specific proapoptotic molecules.[65]

Both PI3K and Akt can modulate cardiomyocyte growth with significant effects on both cell and organ size. Transgenic mice with cardiac-specific overexpression of either the constitutively active or dominant-negative alleles of PI3K exhibited an increase or decrease in cardiomyocyte size, respectively.[66] The constitutive activation of PI3K led to an adaptive hypertrophy and did not change into a maladaptive hypertrophy, consistent with a critical role for the PI3K/PDK-1/Akt pathway in regulating normal cardiac growth. In addition, Akt expression confers protection from ischemia-induced cell death and cardiac dysfunction. Furthermore, Akt1 knock-out mice weigh approximately 20% less than wild-type litter mates and have a proportional reduction in the size of all somatic tissues, including the heart. In contrast, Akt2 knock-out mice have only a modest reduction

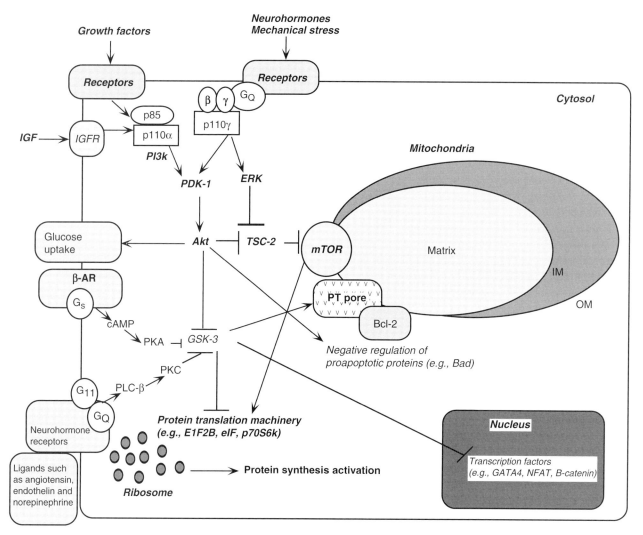

FIGURE 5 Akt signaling. Akt is positioned at a signaling cascade branch point. One branch leads to mammalian target of rapamycin (mTOR) and the activation of the protein synthetic machinery, which is essential for all forms of hypertrophy. Downstream effects of Akt kinase activity include changes in myocardial bioenergetic substrates effected by increasing glucose uptake and may also impact mitochondrial OXPHOS activities, which decline in parallel with increased Akt activity. Both mTOR and Akt modulate cytoplasmic protein synthesis by activation of translation initiation factor (IF) and of ribosomal proteins. Also depicted are peptide growth factors (GH, IGF) and other downstream targets of mTOR.

in organ size. Thus, data from the available Akt knock-out models support a critical role specifically for Akt1 in the normal growth of the heart. On the other hand, Akt1/Akt2 double-knock-out mice have marked growth deficiency and a striking defect in cell proliferation.

Interestingly, activation of Akt in these transgenic models induced cardiac hypertrophy primarily by increasing the size of cardiomyocytes.[67] It is noteworthy that none of the transgenic models in which PI3K or Akt activation were associated with increased heart size demonstrated increased cardiomyocyte proliferation, in contrast to findings in transgenic mice with cardiac specific expression of IGF-1. This suggests that PI3K activation is not sufficient to induce cardiomyocyte proliferation, which likely involves the coordination of other signaling pathways downstream of IGF-1 and upstream of PI3K.

TABLE I Stimuli Activating Myocardial Akt

Stimuli	References
Insulin	51–52
IGF-1	53–54
Cardiotrophin-1	55
LIF	56
β-AR agonists	57
Angiotensin-II	58
Endothelin-1	59
Acetylcholine	60
Adrenomedullin	61
Pressure overload	62
Ischemia (hypoxia)	63

LIF, Leukemia inhibitory factor.

Akt is positioned at a signaling cascade branch point (Fig. 5).[49] One branch leads to mammalian target of rapamycin (mTOR) and the activation of the protein synthetic machinery, which is essential for all forms of hypertrophy. It has been reported that insulin rapidly activates the 70-kDa ribosomal S6 kinase (p70S6k), and this effect is inhibited both by rapamycin and by inhibitors of PI3K.[68] Peptide growth factors (e.g., GH, IGF) are primary activators of mTOR in mammalian cells, and activate mTOR primarily through Akt. Interestingly, downstream activation of p70S6k is mediated by a signaling pathway involving mTOR, a molecule that responds to the nutritional status and amino acid availability and is centrally involved in cell growth and proliferation. One downstream target of mTOR signaling is the 4E binding protein (4E-BP), a translational repressor that directly regulates the activity of the eIF4 translational initiation factor. In addition, activated mTOR is able to phosphorylate p70S6k, which can inactivate eEF2 kinase (regulating translational elongation), as well as phosphorylate the 40S ribosomal protein S6. In several cell types, activation of the TSC1–TSC2 complex (tuberous sclerosis gene) negatively regulates p70S6k; it also inhibits mTOR signaling, reducing cell growth (and insulin signaling), and is inhibited by Akt-dependent phosphorylation.[69,70] The TSC complex mediates its effect on mTOR signaling by targeting Rheb, which is a small Ras-homologous GTPase implicated with mTOR in the activation of p70S6k.[71,72] However, the identification and the contribution of the TSC complex and Rheb to the signaling events occurring in cardiomyocyte growth and cardiac hypertrophy remain to be fully determined.

A second branch of the signaling cascade (Fig. 5) leads to glycogen synthase kinase-3 (GSK-3), which also regulates the general protein translational machinery, as well as specific transcription factors implicated in both normal and pathological cardiac growth. GSK-3β, which was among the first negative regulators of cardiac hypertrophy to be identified, blocks cardiomyocyte hypertrophy in response to ET-1, isoproterenol, and Fas signaling.[73-75] In addition, GSK-3β has been found to be a negative regulator of both normal and pathological stress-induced growth (e.g., pressure overload).[76] GSK-3β plays a key inhibitory role in both insulin signaling and in the Wnt signaling pathway that has been implicated in early cardiomyocyte differentiation, as well as in myocardial hypertrophic growth responses. In unstimulated cells, GSK-3β phosphorylates the N-terminal domain of β-catenin, thereby targeting it for ubiquitylation and proteasomal degradation. GSK-3β is constitutively active unlike most kinases; it is turned "off" by cell stimulation by growth factors and hypertrophic agonists. GSK-3β negatively regulates most of its substrates, including the protein translation initiation factor eIF2B,[77] as well as transcription factors implicated in cardiac growth, including c-Myc, GATA-4, and β-catenin.[78-80] In addition, GSK-3β is a counter-regulator of calcineurin/NFAT signaling phosphorylating NFAT N-terminal residues, which are dephosphorylated by calcineurin, preventing nuclear translocation of the NFATs.[74] Also, Akt phosphorylates GSK-3β at serine-9, inhibiting its activity.

Inhibition of GSK-3β releases a number of transcription factors from tonic inhibition and also releases eIF2B, allowing translational activation. Transgenic mice overexpressing GSK-3β in the heart exhibit significantly defective post-natal cardiomyocyte growth, as well as markedly abnormal cardiac contractile function related to down-regulation of SERCA expression (resulting in abnormal calcium handling) and severe diastolic dysfunction with progressive HF.[81] Moreover, it has been suggested that a family of dimeric phosphoserine–binding molecules, the 14-3-3 proteins (which are implicated in cell-cycle control and the stress response), participate in the regulation of GSK-3β phosphorylation.[82]

It is noteworthy that the activation of protein translation affected by both of these signaling branches can also be regulated by stress-activated mechanisms, which are independent of Akt. For instance, AMP-activated protein kinase (AMPK), a key regulator of cellular energy homeostasis, is involved in modulating the activity of mTOR and can affect the translational response in cardiac hypertrophy.[83] Moreover, hypoxia can rapidly and reversibly trigger mTOR hypophosphorylation and mediate changes in its effectors such as 4E-BP1 and p70S6K, independent of Akt or AMPK signaling.[84] Also, the TSC complex can be phosphorylated and inactivated by stress-mediated stimuli and the ERK pathway independent of Akt signaling. The Akt-independent mechanism of activation of mTOR may be particularly relevant to pathological stress-induced growth.

Inactivation of GSK-3β by S9 phosphorylation may also occur independent of the PI3K/Akt pathway, and this includes involvement of growth factors such as EGF and PDGF, which stimulate the GSK-3β-inactivating p90 ribosomal S6 kinase (RSK) through MAP kinases, activators of cAMP-activated PKA, and PKC activators. Moreover, exposure of cells to Wnt protein ligands leads to inactivation of GSK-3β by an undefined mechanism.[85]

Protein Kinase C (PKC)

The serine/threonine kinase PKC has been implicated as the intracellular mediator of a variety of factors acting through multiple signal transduction pathways. The PKC family of isozymes is increasingly recognized as playing a pivotal role in the cardiac phenotype expressed during post-natal growth and development and in response to pathological stimuli and in the development of cardiac hypertrophy and HF. The expression of multiple PKC isoforms contribute to both a broad spectrum of adaptive and maladaptive cardiac responses with significantly different responses provided by each isoform.[86] Although more than 12 isoforms of PKC have been reported, in the heart, the 4 most functionally significant members of the PKC family are PKC-α and -β (both calcium- and DAG-activated), and PKC-δ and -ε (DAG-activated with no requirement for calcium). These PKC isoforms are activated by membrane receptors coupled to PLC by means of G_q/G_{11} heterotrimeric G-proteins. Activation of PKC phosphorylation is dependent on translocation from the cytosol to the site of action (e.g., to the plasma membrane and to the mitochondria). PKC has an N-terminal regulatory region and a C-terminal catalytic region; protein–lipid interactions are implicated in PKC targeting with the N-terminus, which is required for PKC interaction with the second messenger DAG and Ca^{++}.

Anchoring proteins recruit PKC to specific sites, and receptors for activated C-kinase (RACKs) have been found to be operative in the translocation of PKC-ε in cardiomyocyte growth signaling.[87]

Protein Kinase G

Protein kinase G (PKG) activity modulates several targets involved in muscular contraction. In contrast to the multisubunit nature of other protein kinases, PKG is composed of a single polypeptide sequence containing both regulatory and catalytic information. Compared with protein kinases A, B, and C, the role of cGMP/PKG- mediated signaling in cardiac tissue has been less documented.

Natriuretic peptide binding to type I receptors (NPRA and NPRB) on target cells activates their intrinsic guanylyl cyclase (GC) activity, resulting in a rapid increase in cGMP. Diffusible cGMP acts as a second messenger primarily by stimulating PKG, the primary mediator of cGMP-induced smooth muscle relaxation.[88] Downstream effects that have been directly linked to activated PKG include modulation of the L-type calcium channel and cross-talk with heterologous receptors, such as GPCRs. PKG substrates are membrane-bound, cytosolic, and intranuclear. Evidence indicates that the membrane-bound GC, but not the soluble cyclases that are activated by NO, has potent effects on plasma membrane control of the calcium ATPase pump, suggesting that NO- and natriuretic peptide–mediated effects are compartmentalized in cells. Moreover, using a cytosolic yeast two–hybrid system with PKG as bait, PKG was found to directly interact with NPRA.[89]

NO donors increase heart rate through a GC-dependent stimulation of the pacemaker current I_f, without affecting basal I_{Ca-L}. NO signaling by means of cGMP and cGMP-dependent protein kinase type I (PKG I) has been recognized as a negative regulator of cardiac myocyte hypertrophy. Calcineurin, a Ca^{++}-dependent phosphatase, promotes hypertrophy partly by activating NFAT transcription factors that promote the expression of hypertrophic genes, including BNP. Activation of PKG I by NO/cGMP suppressed NFAT transcriptional activity, BNP induction, and cell enlargement. PKG I inhibits cardiomyocyte hypertrophy by targeting the calcineurin-NFAT signaling pathway and provides a framework for understanding how NO inhibits cardiomyocyte hypertrophy.[90]

The NO/cGMP–dependent protein kinase I (G-kinase I) signal transduction pathway also plays an important role in vascular biology, regulating smooth muscle tone by decreasing Ca^{++} release from intracellular stores and by reducing calcium sensitivity of the contractile apparatus together with SMC proliferation and differentiation. G-kinase I also regulates endothelial cell permeability, motility, and platelet aggregation. Also, insulin-induced relaxation of vascular SMCs is mediated by NO/cGMP/G-kinase inhibition of RhoA.[91] G-kinase regulates gene expression both at the transcriptional and post-transcriptional level, increasing the expression of several genes, including c-fos, heme oxygenase, and MAP kinase phosphatase 1, and decreasing the expression of others, such as thrombospondin, gonadotropin-releasing hormone, and soluble GC.[92-95]

NO can directly influence cardiac contractile function. In the absence of stimulation by extrinsic agonists, both endothelium-derived NO and exogenous NO donors accelerate myocardial relaxation and/or reduce diastolic pressure. NO can also modulate the myocardial inotropic state; however, whether it is positively or negatively modulated may depend on several factors, including the concentration of NO, the rate of NO release, and/or the presence of β-adrenergic stimulation.

Elevation of intracellular cGMP in cardiac myocytes can potentially influence several different pathways, including PKG activation and inhibition or stimulation of phosphodiesterase activity and consequent changes in cAMP levels. It is also feasible that high levels of cGMP could induce changes in contractility by cross-activation of PKA. Rat ventricular myocytes are known to express low levels of PKG (approximately 10-fold lower than in smooth muscle).[96] Despite these low levels, evidence suggests that myocardial PKG mediates the cGMP-induced reduction in the L-type Ca^{++} current after cAMP stimulation.

The intracellular signaling mechanisms responsible for the contractile effects of NO in cardiac myocytes are not known. It is possible that phosphorylation of troponin I by PKG may have comparable effects to PKA-induced phosphorylation. Furthermore, PKG can phosphorylate troponin I *in vitro*, and the contractile effects of NO may be related to troponin I phosphorylation.[97] Earlier *in vitro* observations suggested that PKG phosphorylates cardiac troponin I at the same sites (Ser23/24) as those phosphorylated by PKA. Moreover, reduction in myofilament Ca^{++} responsiveness produced by NO is also mediated by PKG-dependent phosphorylation of troponin I at the same site(s) as those phosphorylated by PKA.

Whereas the signal transduction pathways promoting cardiomyocyte hypertrophy have been well characterized, information concerning signaling pathways that oppose cardiomyocyte hypertrophy is more limited. NO, through activation of soluble GC and cGMP formation, attenuates the hypertrophic response to growth factor stimulation in cardiomyocytes. In addition to its antihypertrophic effect, NO promotes apoptosis in cardiomyocytes in a dose-dependent manner. On the other hand, the role of cGMP in the proapoptotic effects of NO is rather controversial, because cGMP analogs may or may not induce cardiomyocyte apoptosis.

In general, cGMP effectors include cGMP-regulated phosphodiesterases, cGMP-regulated ion channels, and cGMP-dependent protein kinases (PKGs).[98] Two PKG genes have been identified in mammalian cells, encoding for PKG type I (including α- and β- splice variants), and PKG type II. In cardiomyocytes, PKG I has been suggested to mediate negative inotropic effects of NO/cGMP, possibly through regulation of the L-type Ca^{++} channel and troponin I, thereby reducing Ca^{++} influx and myofilament Ca^{++} sensitivity. However, a role for PKG II in controlling cardiomyocyte hypertrophy and/or apoptosis has not been reported.

Potential targets for PKG I in cardiomyocytes include Ca^{++}-dependent signaling pathways, RhoA, and VASP.[99,100] Localization of VASP at intercalated discs in cardiomyocytes suggests that VASP may be involved in PKG I regulation of electrical coupling.

Ca^{++}-dependent signaling pathways, such as calcineurin and Ca^{++}/calmodulin–dependent kinases, are crucial regulators of the hypertrophic response in cardiomyocytes. PKG I regulates intracellular Ca^{++} at multiple levels, including the L-type Ca^{++} channel and the IP3 receptor. PKG I inhibition of the L-type Ca^{++} current may mediate negative inotropic effects of NO/cGMP in cardiomyocytes. Importantly, Ca^{++} influx through the L-type Ca^{++} channel has also been implicated in the regulation of cardiomyocyte hypertrophy. Therefore, antihypertrophic effects of PKG I may be mediated in part through inhibition of Ca^{++}-dependent signaling pathways in cardiomyocytes. The low-molecular-weight GTPase RhoA, which is required for α1–AR signaling in cardiomyocytes, may represent an additional PKG I target. PKG I has recently been shown to phosphorylate RhoA and inhibits its biological activity in vascular SMCs, suggesting that inhibition of RhoA may also contribute to the antihypertrophic effects of PKG I. Notwithstanding the previous observations, further research is needed to identify which of its many molecular targets PKG I uses to inhibit cardiomyocyte hypertrophy.

CA^{++}-MEDIATED KINASE SIGNALING

Calcineurin/Calmodulin

Because Ca^{++} functions as a second messenger activating multiple signaling cascades, it has been suggested that in addition to the effects of altered Ca^{++} handling on cardiac contractile parameters, Ca^{++} may directly affect gene expression in the heart.[101] Many of the actions of Ca^{++} are mediated by means of its interaction with calmodulin (CaM), which is an intracellular Ca^{++} sensor and selectively activates downstream signaling pathways in response to local changes in Ca^{++} (Fig. 6).[102]

Major signaling pathways that are both directly activated by Ca^{++} and sufficient to promote the development of cardiac hypertrophy include Ca^{++} CaM regulated kinase and phosphatase pathways and PKC isozymes.[103]

Major cardiac Ca^{++}/CaM–dependent enzymes include Ca^{++}/CaM–dependent protein kinase (CaMK), myosin light chain kinase (MLCK), and the phosphatase calcineurin.[103] In the heart, the main CaMKII isoform is CaMKIId, which is localized in the nucleus, whereas other isoforms are localized in the SR.[104] Increased intracellular [Ca^{++}] results in autophosphorylation of CaMKII, which switches it to a Ca^{++}-independent state and prolongs its activation. CaMKI, which is ubiquitously expressed, and CaMKIV, mainly expressed in testis and brain, are activated by upstream Ca^{++}/CaM–dependent protein kinases.[103] CaMKII isoforms associated with SR are capable of phosphorylating Ca^{++}-cycling proteins, and hence altering Ca^{++} reuptake and release. In cultured cardiac myocytes, pharmacological inhibitors of CaMKII have been found to attenuate both ET-1 induced hypertrophy[105] and mechanically stretch activated BNP gene transcription.[106] Moreover, overexpression of CaMKIId in the heart of transgenic mice is sufficient to promote hypertrophic growth,[107] as well as cardiac-specific overexpression of either CaMKI or CaMKIV.[108] Furthermore, MLCK is activated by Ca^{++}/CaM, leading to subsequent phosphorylation of its single main substrate, ventricular specific isoform of myosin light chain-2 (MLC-2).

Besides cardiac kinases, Ca^{++}/CaM activates phosphatases, including calcineurin.[109] Activation of calcineurin by the Ca^{++}/CaM complex results in dephosphorylation of its substrates, including nuclear factor of activated T-cells (NFAT) or Ets-like gene-1 (Elk-1). Interaction of Ca^{++}/CaM and calcineurin has been found to increase in the failing human heart, and increased levels of cardiac calcineurin activity have been reported in experimental animal models of cardiac hypertrophy.[110] In addition, calcineurin is sufficient to produce cardiac hypertrophy when overexpressed in the heart of transgenic mice.[109] In several rodent models, pharmacological inhibition of calcineurin for the prevention of cardiac hypertrophy has yielded conflicting results.[111,112] This may be due to the absence of well-tolerated calcineurin inhibitors and also because calcineurin is not a cardiac-restricted enzyme. Compared with normal controls, the human hypertrophied heart exhibits higher calcineurin activity,[113] and those patients maintained with partially inhibited calcineurin activity still can have cardiac hypertrophy develop. This is probably related to the presence of additional pathways making calcineurin signaling not indispensable for hypertrophy.[114,115]

G-Protein–Regulated Kinases (GRKs)

The regulation of myocardial adrenergic receptors, like that of most GPCRs, involves a desensitization mechanism characterized by a rapid loss of receptor responsiveness

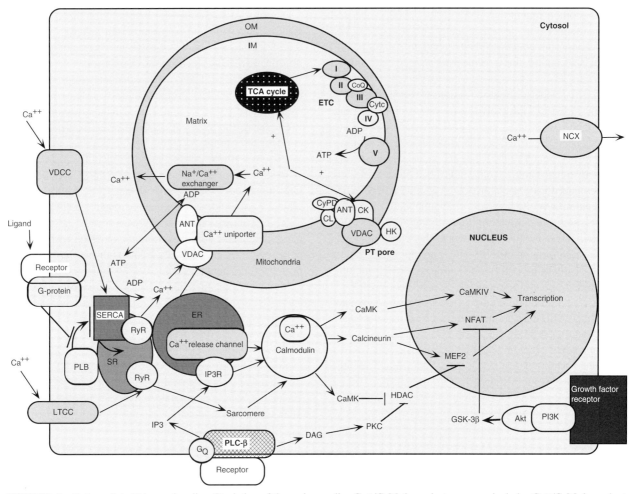

FIGURE 6 Ca^{++}-mediated kinase signaling. Depiction of the major cardiac Ca^{++}/CaM-dependent enzymes includes Ca^{++}/CaM-dependent protein kinase (CaMK), myosin light chain kinase (MLCK), and the phosphatase calcineurin. These signaling pathways are directly activated by Ca^{++}, and they are sufficient to promote the development of cardiac hypertrophy, including Ca^{++} CaM regulated kinase, and phosphatase pathways and PKC isozymes.

despite the continued presence of agonist. The desensitization process has been particularly well characterized using the β2–AR system promoted by a phosphorylation event targeting only the agonist-occupied receptors by a serine/threonine kinase known as β-ARK. β-ARK1 and a highly homologous isozyme, β-ARK2, are two of the most studied members of the GRK family, which currently consists of six members. Desensitization of GPCRs requires not only GRK-mediated phosphorylation but also the binding of a second class of inhibitory proteins, the β-arrestins (β-arrestin-1 and β-arrestin-2), which bind to phosphorylated receptors and sterically restrict the further activation of G-proteins, in part by their displacement from the receptors, resulting in receptor-G–protein uncoupling.[116] The GRKs shown to be expressed in the heart are β-ARK1 (the most abundant), β-ARK2, GRK5, and GRK6. Because β1–AR is the most critical receptor mediating acute changes in myocardial rate and contractility, it is important to realize that three of these GRKs (β-ARK1, β-ARK2, and GRK5) have been shown to phosphorylate and desensitize β1–ARs *in vitro*.[117]

Like most GRKs, β-ARK1 is a cytosolic enzyme that has to be translocated to the plasma membrane to phosphorylate the activated receptor substrate. The mechanism for translocation of β-ARK1 and β-ARK2 involve the physical interaction between the kinase and the membrane-bound β subunits of G-proteins (G$_\beta$). G$_\beta$, anchored to the membrane through a lipid modification on the C- terminus of the subunit (termed prenylation), is available to interact with β-ARK after G-protein activation and dissociation. The region of β-ARK responsible for binding G$_\beta$ has been mapped to a 125-amino acid domain located within the C-terminus of the enzyme. Recently, peptides derived from the G$_\beta$-binding domain of β-ARK have been shown to act as *in vitro* β-ARK inhibitors by competing for G$_\beta$ and preventing translocation.[118]

A pathophysiological role for GRKs can be inferred from recent studies on HF, as well as from the observation that chronic treatment with various agonists or antagonists for

G-protein–coupled receptors results in alterations of GRK expression.[119,120]

MAP Kinases

The MAPK cascade consists of a series of successively acting protein kinases that include three well-characterized branches, the extracellular signal-regulated kinases (ERKs), the c-Jun N-terminal kinases (JNKs), and the p38 MAPKs. Signaling through each of these MAPK branches is initiated by diverse stress and mitogenic stimuli localized to the cell membrane or within the cytoplasm. Activation of ERKs, JNKs, and p38 MAPKs facilitates the phosphorylation of multiple transcriptional regulators such as myocyte enhancer factor-2 (MEF2), activating transcription factor-2 (ATF-2), p53, NFAT, c-Jun, and c-Myc. MAPK-mediated phosphorylation of these and other transcriptional regulators profoundly influences adaptive and inducible gene expression in many cell types. Members of the MAPK signaling cascade are also important regulators of cardiomyocyte hypertrophy, although the downstream transcriptional mechanisms that alter cardiac gene expression have not been characterized.

INTEGRATING RESPONSES: TRANSCRIPTION FACTORS, AND TRANSLATIONAL CONTROL

NF-κB

NF-κB is a pleiotropic family of transcription factors implicated in the regulation of diverse biological phenomena, including apoptosis, cell survival, growth, division and differentiation, innate immunity, and the responses to stress, hypoxia, stretch, and ischemia. In the heart, NF-κB is activated in atherosclerosis, myocarditis, during transplant rejection, after myocardial I/R, in congestive HF, DCM, after ischemic and pharmacological preconditioning, heat shock, and in hypertrophy of isolated cardiomyocytes. In addition to being activated by cytokine-mediated pathways, NF-κB is modulated by many of the signal transduction cascades associated with development of cardiac hypertrophy and response to oxidative stress. Many of these signaling cascades activate NF-κB by activating the IκB kinase (IKK) complex. These signaling interactions primarily involve the MAP kinase/ERK kinases (MEKKs), which are components of MAPK signaling pathways. In addition, other signaling factors directly activate NF-κB by IκB or by direct phosphorylation of NF-κB subunits. Combinatorial interactions have been reported at the level of the promoter between NF-κB, its coactivators, and other transcription factors, several of which are activated by MAPK and cytokine signaling pathways. In addition to being a major mediator of cytokine effects in the heart, NF-κB represents a signaling integrator, functioning as a key regulator of cardiac gene expression programs downstream of multiple signal transduction cascades in a variety of physiological and pathophysiological states. Genetic blockade of NF-κB can reduce the size of infarcts resulting from I/R in the murine heart, consistent with its role as a major determinant of cell death after I/R, and suggests that NF-κB may constitute an important therapeutic target in specific cardiovascular diseases.[121]

PPAR-α, -γ, and Cofactors (RXR and PGC)

Nuclear receptor transcription factors are important regulatory players governing the cardiac metabolic gene program (Table II).[122] This superfamily of receptors was originally described as ligand-dependent transcription factors, which is in fact the case for nearly 50% of those characterized so far.[123] Such ligand-activated receptors include the classical endocrine receptors that respond to steroid or thyroid hormones. Recently, a number of receptors have been identified (without prior insight to their ligands) that respond to dietary-derived lipid intermediates, including long-chain fatty acids (LCFAs) and bile acids. These receptors generally participate in the regulation of pathways involved in the metabolism of the activating ligands. On the other hand, a group of "orphan" nuclear receptors have no identifiable ligands (although modulating ligands may be soon identified for some of these receptors).

Because the heart must adapt to continuously changing energy demands, but has limited capacity for storing fatty acids or glucose, myocardial energy substrate flux must be tightly matched with the demand. Therefore, ligand-activated nuclear receptors, as metabolite sensors, participate in a rapidly activated program of gene expression in response to fluctuating substrate levels. Of particular interest are the peroxisome proliferator–activated receptors (PPARs), fatty acid–activated nuclear receptors, increasingly recognized as key regulators of cardiac fatty acid metabolism. The PPAR-α isoform has been characterized as the central regulator of mitochondrial fatty acid catabolism including fatty acid oxidation (FAO), whereas PPAR-γ primarily regulates lipid storage.[122] In addition, a select group of orphan nuclear receptors have been identified that serve new roles in the regulation of cardiac energy metabolism (Table II).

The nuclear receptors have a conserved modular domain structure.[122–124] These proteins possess an N-terminal region containing a ligand-independent transcriptional activation function (AF-1) domain, a conserved DNA-binding domain (DBD) containing two highly conserved zinc-finger sequences that target the receptor to specific regulatory regions termed hormone responses elements (HREs), and a composite C-terminal region that includes the ligand-binding domain (LBD), a dimerization domain, and a conserved ligand-dependent activation function (AF-2). The nuclear receptors bind to regulatory DNA elements in target genes as homodimers, heterodimers, or in some cases as monomers.

TABLE II Myocardial Nuclear Receptor Transcription Factors and Co-activators

Receptor	Cardiac function	Target genes	Reference
Peroxisome proliferator-activated receptor-α (PPAR-α)	Lipid utilization	FAO-CPT1, MLC-2, PDK4, UCP3	122, 124, 125
Peroxisome proliferator-activated receptor-β (PPAR-β)	Lipid homeostasis	FAO	122
Peroxisome proliferator-activated receptor-γ (PPAR-γ)	Lipid storage Increased glucose oxidation	iNOS	126
Peroxisome proliferator-activated receptor-δ (PPAR-δ)	Lipid homeostasis	Same as PPAR-α	127
Peroxisome proliferator-activated receptor-γ coactivator-1α (PGC-1)	Cardiac metabolism	FAO, ETC, UCP, NRF1, NRF2, mtTFA, Glut4	128–129
Retinoid X receptor (RXR-α)	Cardiac development	GATA-4, FAO ANF, SERCA	130
Thyroid hormone receptor (TR)	Cardiac hypertrophy	BNP, β-AR SERCA	131–132
Vitamin D receptor (VDR)	Cardiac morphogenesis	ANP	133
Estrogen-related receptor (ERR α)	Orphan receptor	TRα, FA uptake, FAO, ETC	125, 134
Estrogen-receptor α. (ER-α.)	I/R Cardioprotection	Nd ANP, eNOS/iNOS	135–137
Estrogen-receptor β (ER-β.)	Mitochondrial localization	Nd	138
Retinoic acid receptor (RAR)	Cardiac morphogenesis	GATA-4	139
Glucocorticoid receptor (GR)	Decreased growth	β-AR	140
Androgen receptor (AR)	Cardiac hypertrophy	Nd	141
Mineralocorticoid receptor (MR)	Cardioprotective Electrical remodelling	Nd Nd	142–143

Nd, Not determined.

Unlike the classic steroid receptors that function as homodimers, a number of the nuclear receptors involved in nutrient sensing and metabolic regulation (e.g., PPARs, TR, RAR) heterodimerize with the retinoid X receptor (RXR). These receptors interact with regulatory DNA elements within the 5′ regulatory region of their target genes, which are composed of variably spaced hexameric half-sites (AGGTCA) arranged as direct, indirect, or everted repeats. Once bound to their specific response element, the receptors recruit coactivator proteins often in concert with the displacement of corepressor proteins. For ligand-activated receptors, it is through ligand binding that the receptor adopts a permissive conformation for the recruitment of coactivator proteins to enable transcriptional activation. More than 100 coactivator proteins have been identified for nuclear receptors, including ATP-dependent chromatin-remodeling complexes, histone acetylases, histone methyltransferases, and RNA polymerase II–recruiting complexes that open up the chromatin structure and facilitate the binding of basal transcription factors and RNA polymerase II. Histone-modifying proteins are often recruited into complexes by adapter proteins, which lack catalytic activity. One such adapter/coactivator, the PPAR gamma coactivator-1 (PGC-1), serves as a key link between physiological cues and metabolic regulation in heart.

TRANSLATION CONTROL

Activation of p70S6k has been identified as another key step in stimulation of protein synthesis under β-adrenoceptor stimulation and downstream of PI3K activation.[144] The p70S6k phosphorylates the S6 protein of the 40S ribosomal subunit, which may modulate overall translational activity. Inhibition of the activation of p70S6k attenuates the growth effect and increased protein synthesis of β1 and α1-adrenoceptor agonists in adult cardiomyocytes. Another factor that contributes to the increase in translational activity is the activation of the peptide chain initiation factor eIF-4E. The phosphorylation and activation of this translational initiation factor also depends on the activation of PKC[145] and on the presence of oxidative stress.[146] Therefore, activation of PKC, PI3K, and p70S6k represent key steps of the intracellular signaling that control protein synthesis in adult cardiomyocytes.

In resting cardiomyocytes, activation of p70S6k can recruit a subset of the available ribosomes to participate in protein synthesis. In active cardiomyocytes, however, virtually all ribosomes seem to be functionally active, and the *de novo* synthesis of ribosomal RNA is required for a significant acceleration of protein synthesis. On both β1- and α1-adrenoceptor stimulation or direct stimulation of PKC, RNA polymerase I is activated, and increased levels of the ribosomal DNA transcription factor UBF are detected, resulting in the increased synthesis of rRNA.[147,148] The stimulation in RNA synthesis is also mediated through PI3K and p70S6k, suggesting that p70S6k has other targets in addition to the S6 protein of ribosomes. These observations also support the premise that increased translational efficiency of existing ribosomes alone is insufficient to account for the hypertrophic growth of cardiomyocytes and that synthesis of new functional ribosomes must occur. Moreover, gene expression in vascular and endothelial cells can also be similarly regulated at the level of translation.[149]

FROM PLASMA MEMBRANE TO MITOCHONDRIA

Signaling at the Plasma Membrane

SARCOLEMMAL K_{ATP} CHANNEL

ATP-sensitive potassium (K_{ATP}) channels link membrane excitability to metabolism.[150] They are regulated by intracellular nucleotides and by other factors, including, membrane phospholipids, protein kinases, and phosphatases. K_{ATP} channels comprise octamers of four Kir6 pore-forming subunits associated with four sulfonylurea receptor (SUR) subunits. K_{ATP} channels are targets for antidiabetic sulfonylurea blockers and for channel-opening drugs that are used as antianginals and antihypertensives. In vascular smooth muscle, K_{ATP} channels are regulated by diverse signaling pathways and cause vasodilation, contributing to both resting blood flow and vasodilator-induced increases in flow. In cardiac muscle, sarcolemmal K_{ATP} channels open to protect cells under stress conditions, such as ischemia and exercise, and seem central to the protection induced by ischemic preconditioning (IPC).

Signaling at the Mitochondria

REACTIVE OXYGEN SPECIES GENERATION AND SIGNALING

A critical byproduct of mitochondrial bioenergetic activity is the generation of reactive oxygen species (ROS), including superoxide, hydroxyl radicals, and hydrogen peroxide (H_2O_2) (Fig. 7). Side reactions of mitochondrial electron transport chain (ETC) enzymes with oxygen directly generate the superoxide anion radical. The primary sites for mitochondrial ROS generation, as a by-product of normal metabolism, are at complex I, II, and III of the respiratory chain; either excessive or diminished electron flux at these sites can stimulate the autooxidation of flavins and quinones (including coenzyme Q), producing superoxide radicals.[151] The superoxide radicals can react with NO to form peroxynitrite, which is a highly reactive and deleterious free radical species or can be converted by superoxide dismutase (SOD) to H_2O_2, which can further react to form hydroxyl radicals. Generation of the hydroxyl radical (the most reactive and deleterious form of ROS) is primarily responsible for the damage to cellular macromolecules such as proteins, DNA, and lipids. The highly reactive hydroxyl radical is generated from reactions involving other ROS species (e.g., the Fenton reaction) in which ubiquitous metal ions, such as Fe (II) or Cu (I), react with H_2O_2. The high reactivity of the hydroxyl radical and its extremely short physiological half-life of 10^{-9} sec restrict its damage to a small radius from its origin, because it is too short-lived to diffuse a considerable distance.[152]

In contrast, the less reactive superoxide radicals produced in mitochondria can be delivered to the cytosol through anion channels (e.g., VDAC) and thereby may impact sites far from its generation, including activation of transcription factors such as NF-κB among other effects.[153] Similarly, the relatively stable H_2O_2 produced by mitochondria can freely diffuse to the cytosol to act as a signaling entity impacting on cytosolic events.

Under normal physiological conditions, the primary source of ROS is the ETC located in the mitochondrial inner membrane, where oxygen can be activated to form superoxide radicals by a nonenzymatic process. Mitochondrial ROS generation can be amplified in cells with abnormal respiratory chain function, as well as under physiological and pathological conditions where oxygen consumption is increased.

NEGATIVE EFFECTS OF ROS

Increased ROS generation resulting from myocardial I/R, inflammation, impaired antioxidant defenses, and aging may cause profound effects on cells, including elevated lipid peroxidation targeting membrane phospholipids and proteins. Protein modifications, such as carboxylation, nitration, and the formation of lipid peroxidation adducts (e.g., 4-hydroxynonenal [HNE]), are products of oxidative damage secondary to ROS.[154] ROS-mediated nitration, carboxylation, and HNE adduct formation reduces the enzymatic activity of myocardial respiratory complexes I–V as shown with *in vitro* studies.[155] Superoxide is also particularly damaging to the Fe-S centers of enzymes, such as complex I, aconitase, and succinic dehydrogenase (SDH), causing inhibition of mitochondrial bioenergetic function. Moreover, the inactivation of mitochondrial aconitase by superoxide, which generates Fe (II) and H_2O_2, also increases hydroxyl radical formation through the Fenton reaction,[156] thereby amplifying the deleterious effects of ROS production. Lipids

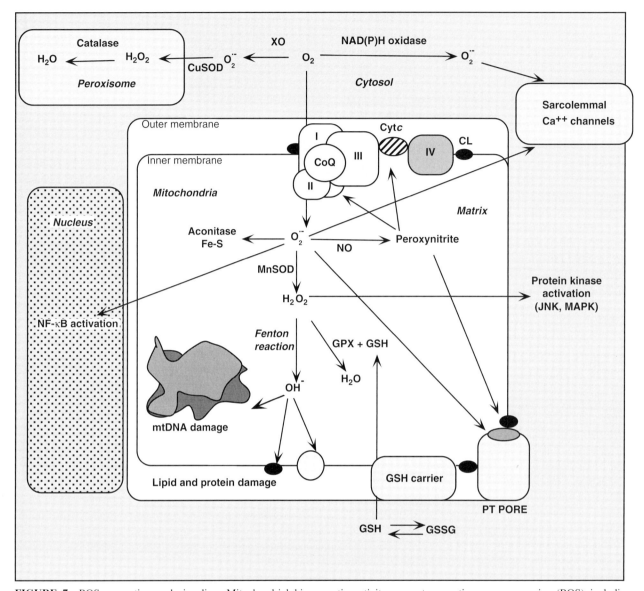

FIGURE 7 ROS generation and signaling. Mitochondrial bioenergetic activity generates reactive oxygen species (ROS) including superoxide, hydroxyl radicals and hydrogen peroxide (H_2O_2). Sites of mitochondrial superoxide $O_2^{\cdot-}$ radical (via respiratory complexes I, II, and III) and cytosolic $O_2^{\cdot-}$ generation (by NADPH oxidase or xanthine oxidase) are depicted. Also shown are reactions of the $O_2^{\cdot-}$ radical with NO to form the highly reactive peroxynitrite, which can target PT pore opening and the inactivation of mitochondrial aconitase by $O_2^{\cdot-}$. Antioxidant enzymes MnSOD (in mitochondria) and CuSOD (in cytosol) that form H_2O_2 are also displayed. The H_2O_2 is then either further neutralized in the mitochondria by glutathione peroxidase (GPx) and glutathione, in the peroxisome by catalase, or in the presence of Fe^{++} by means of the Fenton reaction, which forms the highly reactive OH· radical, which can cause severe lipid peroxidation and extensive oxidative damage to proteins and mtDNA. Superoxide radicals produced in mitochondria can be delivered to the cytosol through anion channels (e.g., VDAC) and may impact sites far from its generation, including activation of transcription factor NF-κB.

and, in particular, the mitochondrial-specific phospholipid cardiolipin serve as a focal target for ROS damage. A large accumulation of superoxide radicals produced *in vitro*, with submitochondrial particles from heart, resulted in extensive cardiolipin peroxidation with a parallel loss of cytochrome *c* oxidase activity.[157,158] Oxidative damage also targets nucleic acids, and in particular mtDNA, by inducing single- and double-strand breaks, base damage and modification (including 8-oxoguanosine formation), resulting in the genera-

tion of point mutations and deletions in mtDNA. Inhibition of mitochondrial respiration by NO can result in further increases in mitochondrial ROS production; interaction with NO enhances the potency of superoxide as an inhibitor of respiration.[159] In addition, the highly reactive peroxynitrite irreversibly impairs mitochondrial respiration,[160] because it inhibits complex I activity,[160] largely by tyrosine nitration of several targeted subunits,[161,162] modifies cytochrome *c* structure and function,[163] affects cytochrome *c* oxidase

(COX) activity,[160] inhibits mitochondrial aconitase,[164] and causes induction of the PT pore.[165] A number of peroxynitrite effects on its mitochondrial targets (e.g., the PT pore) are potentiated by increased calcium levels[166] and can be clearly distinguished from the effects of NO, which often are reversible.[160]

Not surprisingly, mitochondria (a major site of intracellular ROS generation) are also a primary locus of its damaging effects. ROS-induced damage to mtDNA induces abnormalities in the mtDNA-encoded polypeptides of the respiratory complexes located in the inner membrane, with consequent decrease of electron transfer and further production of ROS, thus establishing a vicious circle of oxidative stress, mitochondrial function, and bioenergetic decline.

It is worth noting that ROS produced from other cellular sources, besides mitochondria, can have substantial effects on cardiovascular function. Superoxide radicals are generated from reactions of oxygen with microsomal cytochrome p450, which has an endogenous NAD(P)H oxidase activity, usually in the presence of metal ions. Phagocytic cells (present at sites of active inflammation), vascular endothelial cells, and SMCs have a NAD(P)H oxidase activity that can be induced by certain stimuli such as angiotensin II,[167] TNF-α,[168] and thrombin[169] to generate ROS. NAD(P)H oxidase also produces ROS in response to endothelin-1 in vascular SMCs and cardiac muscle cells. As a result, NAD(P)H oxidases may be a key source of ROS that participate in vascular oxidant–related signaling mechanisms under physiological and pathophysiological conditions. In addition, xanthine oxidase (XO), a primarily cytosolic enzyme involved in purine metabolism, is also a source of the superoxide radical. Notably, XO activity and its superoxide generation are markedly increased in the heart after I/R damage. Its location within the human myocardium is primarily in the endothelial cells of capillaries and smaller vessels.[170] Ischemia and hypoxia promote the accumulation of XO substrates, hypoxanthine and xanthine. Numerous studies have shown that the XO inhibitor allopurinol can provide protection against the cardiac damage resulting from anoxia. Recently, a provocative link was proposed between XO activity and abnormal cardiac energy metabolism in patients with idiopathic DCM, because inhibition of XO with allopurinol significantly improved myocardial function.[171] These toxic metabolic by-products, which are potent cell-damaging oxidants, are normally neutralized by antioxidant enzymes, some of which are mitochondrially located (e.g., Mn-SOD and glutathione peroxidase), whereas others are cytosolic (e.g., Cu-SOD and catalase).

ROLE OF ROS IN CELL SIGNALING

In addition to their cell-damaging effects, ROS generation and oxidative stress play a critical role in cell regulation and signaling. Oxidative species such as H_2O_2 and the superoxide anion can be deployed as potent signals sent from mitochondria to other cellular sites rapidly and reversibly triggering an array of intracellular cascades leading to diverse physiological end points for the cardiomyocyte, some negative (e.g., apoptosis and necrosis) and others positive (e.g., cardioprotection and cell proliferation). Mitochondrial-produced H_2O_2 exported to the cytosol is involved in several signal transduction pathways, including the activation of JNK1 and MAPK activities[172–174] and can impact the regulation of redox-sensitive K^+ channels affecting arteriole constriction.[175] The release of H_2O_2 from mitochondria and its subsequent cellular effects are increased in cardiomyocytes treated with antimycin and high Ca^{++} and further enhanced by treatment with CoQ. CoQ plays a dual role in the mitochondrial generation of intracellular redox signaling by acting both as a prooxidant involved in ROS generation and as an antioxidant.[176] Increased mitochondrial H_2O_2 generation and signaling also occur with NO modulation of the ETC,[177] as well as with the induction of myocardial mitochondrial NO production, resulting from treatment with enalapril.[178] Furthermore, ROS plays a fundamental role in the cardioprotective signaling pathways of IPC, in oxygen sensing, and in the induction of stress responses that promote cell survival.

MITOCHONDRIAL K_{ATP} CHANNEL

ATP-sensitive potassium channels of the inner mitochondrial membrane (mitoK$_{ATP}$) are blocked by ATP and have been implicated as potential mediators of cardioprotective mechanisms such as IPC.[179] This cardioprotective effect is partially mediated by attenuating Ca^{++} overloading in the mitochondrial matrix and by increased ROS generation during preconditioning, further leading to protein kinase activation and decreased ROS levels generated during reperfusion.[180] The mitoK$_{ATP}$ is also regulated by a variety of ligands (e.g., adenosine, opioids, bradykinin, acetylcholine), which bind sarcolemmal G-protein coupled receptors with subsequent activation of calcium flux, tyrosine protein kinases, and the PI3K/Akt pathway.[181,182] In addition, marked changes in mitochondrial matrix volume associated with mitoK$_{ATP}$ channel opening may play a contributory role in the cytoprotection process,[183] although this has been recently challenged.[184] Drugs such as diazoxide and nicorandil specifically activate the mitoK$_{ATP}$ opening and can also inhibit H_2O_2-induced apoptotic progression in cardiomyocytes, suggesting that mitoK$_{ATP}$ channels may also play a significant role in mediating oxidative-stress signals in the mitochondrial apoptotic pathway.[185,186] Another ion channel (i.e., the calcium-activated K^+ channel) has recently been identified on the mitochondrial inner membrane and has been shown to have a cardioprotective function.[187] Nevertheless, the precise temporal order of events in the mitochondrial cardioprotection cascade and the exact molecular nature of the mitoK$_{ATP}$ channel remain to be defined.[188] Further discussion on the relationship of the mitoK$_{ATP}$ channel to cardioprotection (CP) will be presented in Chapter 10.

PT Pore

The opening of another mitochondrial membrane mega-channel, the permeability transition (PT) pore, located at contact sites between the inner and outer membranes, has been suggested to cause a number of important changes in mitochondrial structure and metabolism, including increased mitochondrial matrix volume (leading to mitochondrial swelling), release of matrix Ca^{++}, altered cristae, cessation of ATP production, primarily caused by uncoupling of the ETC, and dissipation of the mitochondrial membrane potential. At the onset of reperfusion after an episode of myocardial ischemia, opening of this nonspecific pore is a critical determinant of myocyte death. Besides its role in the mitochondrial pathway of apoptosis, the opening of the PT pore, if unrestrained, leads to the loss of ionic homeostasis and ultimately to necrotic cell death.[189] The PT pore seems to be composed of several mitochondrial membrane proteins, including the voltage-dependent anion channel (VDAC/porin), peripheral benzodiazepine receptor (PBR), the adenine nucleotide translocator (ANT), cytosolic proteins (e.g., hexokinase II, glycerol kinase), matrix proteins (e.g., cyclophilin D [Cyp-D]), and from proteins of the intermembrane space such as creatine kinase. Fatty acids, high matrix Ca^{++} levels, prooxidants, metabolic uncouplers, NO, and excessive mitochondrial ROS production (primarily from respiratory complex I and III) promote the opening of the PT pore.

The PT pore may also be an important target of cardioprotection. New observations have shown that suppressing PT pore opening at the onset of reoxygenation can protect human myocardium against lethal hypoxia-reoxygenation injury.[190] The inhibition of PT pore opening can be mediated either directly by cyclosporin A (CsA) and Sanglifehrin A (SfA) or indirectly by decreasing calcium loading and ROS levels.

Mitochondrial Kinases

Evidence has been gathered that mitochondria contain multiple phosphoprotein substrates for protein kinases and also that a number of protein kinases are translocated into heart mitochondria. This suggests that protein phosphorylation within the mitochondria is a critical component of the mitochondrial signaling pathways.[191] Protein kinases identified in heart mitochondria include pyruvate dehydrogenase kinase, PKA, PKC-δ and -ε isoforms, and JNK kinase (Table III). Characterization of these proteins has provided new insights into the fundamental mechanisms regulating the mitochondrial response to diverse physiological stimuli and stresses.

In cardiomyocytes, isoforms of PKC (PKCs-δ and -ε) translocate from the cytoplasm to mitochondria for subsequent signal transduction.[195] PKC-ε, after translocation, forms a "signaling module" by complexing with specific MAP kinases (e.g., ERK, p38 and JNK), resulting in phosphorylation of the proapoptotic protein Bad. Also, PKC-ε forms physical interaction with components of the cardiac mitochondrial PT pore, in particular VDAC and ANT.[200] This interaction may inhibit pathological opening of the pore, including Ca^{++}-induced opening and subsequent mitochondrial swelling, contributing to PKC-ε-induced cardioprotection. Activation of PKC in CP likely precedes mitoK$_{ATP}$ channel opening; nevertheless, a direct interaction of these kinases with the mitoK$_{ATP}$ channels has not yet been proved. After diazoxide treatment, PKC-δ is translocated to cardiac mitochondria, which triggers mitoK$_{ATP}$ channel opening, leading to CP;[194] however, other studies have shown that PKC-δ does not play a contributory role in the CP provided by IPC, although the role of PKC-ε translocation has been confirmed.

A mitochondrial cAMP-dependent protein kinase A (mtPKA), as well as its protein substrates, has been localized to the matrix side of the inner mitochondrial membrane.[193] In cardiomyocytes, mtPKA phosphorylates the 18-kDa subunit of complex I (NDUFS4), and increased levels of cAMP promote NDUFS4 phosphorylation, enhancing both complex I activity and NAD-linked mitochondrial respiration.[201] These post-translational changes can be reversed by dephosphorylation mediated by a mitochondrial-localized phosphatase. In addition, PKA phosphorylation of several subunits of cytochrome c oxidase (COXI, III and Vb) at serine residues modulates the activity of this important respiratory enzyme[202] and is considered to be a critical element of respiratory control. This cAMP-dependent phosphorylation occurs with high ATP/ADP ratios, resulting in the allosteric inhibition of COX activity. In the resting state, this regulatory control results in reduced membrane potential and more efficient energy transduction. Conversely, increases in mitochondrial phosphatase (Ca^{++}-induced) reverse the allosteric COX inhibition/respiratory control, resulting in increased membrane potential and ROS formation. Similarly, various stress stimuli leading to increased Ca^{++} flux (activating the phosphatase) result in increased membrane potential and ROS formation.

TABLE III Myocardial Mitochondrial-Located Protein Kinases

Protein kinase	Reference
PDH kinase	192
PKA	193
PKC-δ	194
PKC-ε	195
PKG	196
JNK	197, 198
p38 MAPK	195
Stress-activated protein kinase-3	199
ERK 1/2	195
MAP kinase	198

New techniques of proteomic analysis have led to the identification of mitochondrial phosphoprotein targets for these kinases. Interestingly, a group of proteins constituting a mitochondrial phosphoprotein proteome has been identified using a proteomic approach in bovine heart and characterized as protein targets of kinase-mediated phosphorylation.[203] Most of the identified phosphoproteins were involved in mitochondrial bioenergetic pathways, including the TCA cycle (e.g., aconitase, isocitrate, and pyruvate dehydrogenase) and mitochondrial respiratory complexes, including NDUFA 10 (complex I), the flavoprotein subunit of SDH (complex II), core I and III subunits (complex III), and α- and β- subunits (complex V), whereas others are essential elements for the homeostasis of mitochondrial bioenergetics (e.g., creatine kinase and ANT).

Mitochondrial Translocation

An important subject concerning cell signaling and activation includes the stimuli-dependent translocation to, and incorporation of, specific cytosolic proteins into the cardiomyocyte mitochondria. A growing list of such translocated molecules include several of the proapoptotic proteins (e.g., Bax, Bid), as well as some of the aforementioned protein kinases. Many of these proteins target or interact with specific proteins on the outer mitochondria membrane, and others are imported as preproteins by virtue of recognizing a small set of specific receptors (*translocases*) on the mitochondrial *o*uter *m*embrane (TOM). Mitochondrial protein import is often mediated by heat shock proteins (e.g., HSP60, HSP70), which specifically interact with a complex mitochondrial protein import apparatus, including matrix proteases. In addition, physiological stimuli and stresses, including temperature changes and hormone treatment (e.g., thyroid hormone), impact the regulation of the heart mitochondrial import apparatus.[204,205]

Nuclear transcription factors have been described and characterized in other tissues/cell types as translocating to the mitochondria, including p53, NF-κB, PPAR-α, RXR, and TR3, although they have not yet been detected in heart mitochondria. Furthermore, no specific mitochondrial receptors have yet been found in cardiomyocytes or myocardium that bind TNF-α or the various cytokines known to effect cardiac mitochondrial function.

Mitochondrial retrograde signaling

Mitochondrial retrograde signaling is a pathway of communication from mitochondria to the nucleus that influences many cellular activities under both normal and pathophysiological conditions. In both yeast and animal cells, retrograde signaling is linked to mTOR signaling, but the precise connections in cardiomyocytes have not yet been determined. In mammalian cells, mitochondrial dysfunction sets off signaling cascades through altered Ca^{++} dynamics, including calcineurin activation, which activate several protein kinase pathways (e.g., PKC and MAPK) and transcription factors such as NF-κB, calcineurin-dependent NFAT, CREB, and ATF, leading to stress protein expression (e.g., chaperone proteins) and activities.[206] These can result in both alterations in both cell morphology and phenotype, including proliferative growth, apoptotic signaling, and glucose metabolism.

Endoplasmic reticulum

The endoplasmic reticulum (ER) is a multifunctional signaling organelle that contributes to the regulation of cellular processes such as the entry and release of Ca^{++}, sterol biosynthesis, apoptosis, and the release of arachidonic acid.[207] One of its primary functions is as a source of the Ca^{++} signals that are released through either IP3 or ryanodine receptors (RyRs), which are themselves Ca^{++}-sensitive. The capability of ER in spreading signals throughout the cell mediated by a process of Ca^{++}-induced Ca^{++} release is particularly important in the control of cardiomyocyte function. The role of ER as an internal reservoir of Ca^{++} is coordinated with its role in protein synthesis, because a constant luminal level of Ca^{++} is essential for protein folding. To achieve this regulation, the ER also contains several stress signaling pathways that can activate transcriptional cascades to regulate the luminal content of the Ca^{++}-dependent chaperones responsible for the folding and packaging of secretory proteins. Another significant function of the cardiomyocyte ER is to regulate apoptosis by operating in tandem with mitochondria. Antiapoptotic regulators of apoptosis such as Bcl-2 may act by reducing the ebb and flow of Ca^{++} through ER/mitochondrial cross-talk.

Cell-Cycle Signaling, Cell Proliferation, and Apoptosis

In both animal models and human clinical studies, apoptosis may be causally linked to the myocardial dysfunction stemming from I/R, MI, and HF. Furthermore, cultured cardiomyocytes undergo apoptosis in response to a variety of stimuli, including hypoxia (particularly when followed by reoxygenation), acidosis, increased oxidative stress, serum deprivation, glucose deprivation and metabolic inhibition, β1-adrenergic agonists, TNF-α, Fas ligand, and doxorubicin. The apoptotic death process is mediated by two central pathways, an extrinsic pathway featuring cell surface receptors and an intrinsic pathway involving mitochondria and the ER.[208] The two pathways share a number of components and seem to be substantially intertwined; in both pathways, signaling leads to the activation of a family of cysteine proteases called caspases that trigger cell breakdown and death. We will summarize the pivotal signaling events in both pathways.

Extrinsic signaling is initiated by the binding of a death ligand (usually present as a trimer) to its cognate cell surface receptor (Fig. 8). The death ligand may be an integral

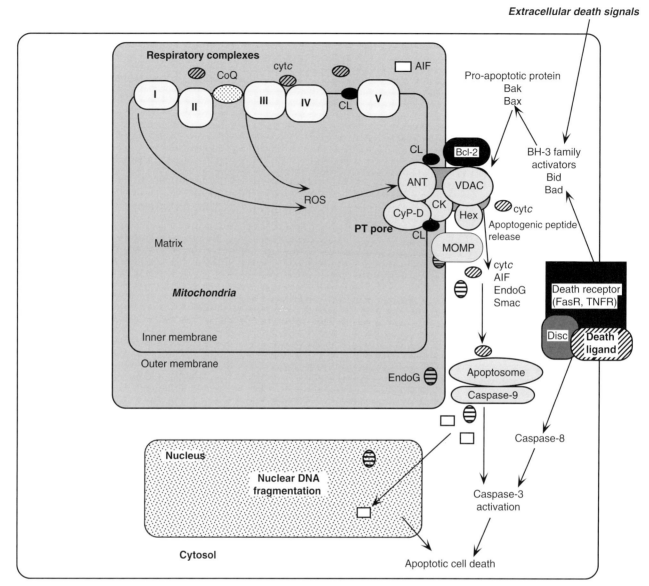

FIGURE 8 Cell-cycle signaling and apoptosis. An extrinsic pathway featuring cell surface receptors and an intrinsic pathway involving mitochondria and the endoplasmic reticulum mediates the apoptotic process. Cellular signals, including ROS and UV-induced DNA damage, trigger the apoptotic pathway regulated by proapoptotic proteins (e.g., Bax, Bid, and Bad) binding to outer mitochondrial membrane leading to mitochondrial outer membrane permeabilization (MOMP) and PT pore opening. Elevated levels of mitochondrial Ca^{++}, as well as ETC-generated ROS, also promote PT pore opening. This is followed by the release of cytochrome c (Cytc), Smac, endonuclease G (Endo G), and apoptosis-inducing factor (AIF) from the mitochondrial intermembrane space to the cytosol and apoptosome formation leading to caspase and endonuclease activation, DNA fragmentation, and cell death. Association of Bax with mitochondria is prevented by antiapoptogenic proteins (e.g., Bcl-2). Ligand binding to the death receptor initiates the formation of a multiprotein complex termed the death-inducing signaling complex (DISC). The formation of this complex signals the recruitment of procaspase-8 into the DISC with subsequent dimerization and activation of procaspase-8, followed by caspase-8 cleavage, and downstream activation of procaspase-3 and Bid (a proapoptotic Bcl-2 protein), which links the extrinsic and intrinsic pathways. Also shown are major proteins that make up the PT pore, including hexokinase (Hex), adenine nucleotide translocator (ANT), creatine kinase (CK), cyclophilin D (CyP-D), and porin (VDAC), as well as the inner membrane phospholipid cardiolipin (CL).

membrane protein on the surface of a neighboring cell (e.g., Fas ligand) or a soluble extracellular protein (e.g., TNF-α). Death receptors are single transmembrane-spanning proteins with domains containing cysteine-rich repeats and cytoplasmic regions, including a death domain sequence of approximately 80 amino acids. Ligand binding to the death receptor initiates the formation of a multiprotein complex termed the death-inducing signaling complex (DISC). On the binding of the death ligand, conformational change and trimerization of the death receptors result, as does the recruitment of an adaptor protein (e.g., FADD [Fas-associated death domain]) through interactions involving death domains in each of the proteins. The formation of this complex signals the recruitment of procaspase-8 into the DISC, resulting in procaspase-8 dimerization and activation. Once activated, caspase-8 cleaves and activates downstream procaspase-3 and Bid (a proapoptotic Bcl-2 related protein), which links the extrinsic and intrinsic pathways.

A variety of extracellular and intracellular signals can contribute to the initiation of the intrinsic pathway. Extracellular stimuli include deficiencies in survival/trophic factors/nutrients, radiation, and chemical (e.g., doxorubicin) and physical stresses, whereas intracellular stimuli include oxidative stress or ROS, DNA damage, protein misfolding, and changes in intracellular Ca^{++} that can be directed in part by the ER. This myriad of signals converges on the mitochondria, leading to pronounced changes in the membrane organization and dysfunction of this organelle, the release of apoptogenic proteins, and the subsequent activation of caspases.

Although in many cases the precise interaction of the heterogenous and diverse apoptotic signaling with the mitochondria remains not well defined, a common element that has emerged involves the Bcl-2 protein family, which has both proapoptotic and antiapoptotic elements. Although these are present in both the ER and in the cytoplasm, the translocation and presence in the mitochondrial outer membrane is a key element governing apoptotic progression.

A change in mitochondrial membrane integrity is regulated by the complex and dynamic interactions of different members of the Bcl-2 family, including Bax, Bid, Bcl-2, and Bcl-X_L. Proteins of the Bcl-2 family share one or several Bcl-2 homology (BH) regions and behave as either pro- or antiapoptotic proteins. The highly conserved BH domains (BH1-4) are essential for homo- and heterocomplex formation, as well as to induce cell death. Proapoptotic homologs can be subdivided into two major subtypes, the multi-domain Bax subfamily (e.g., Bax and Bak) that possesses BH1-3 domains and the BH3-only subfamily (e.g., Bad and Bid). Both proapoptotic subtypes promote cell-death signaling by targeting mitochondrial membranes, albeit by different mechanisms.[209] Proapoptotic membrane–binding proteins (e.g., Bax) on translocation from the cytosol to mitochondria potentiate cytochrome c release, presumably by forming channels in the outer membrane. This is supported by data showing that Bax can form channels and release cytochrome c from artificial membranes or liposomes.[210] The BH3-only proteins (e.g., Bid) act by activating the multidomain proapoptotic proteins or by binding and antagonizing the antiapoptotic proteins. Activation of proapoptotic proteins such as Bax to oligomerize, translocate, and bind to the mitochondria represents a critical control point for apoptosis. This process requires extensive conformational changes in response to a multitude of death signals involving the binding of several factors (e.g., BH3-only proteins) and phosphorylation by several kinases including p38 MAP kinase. Cytoplasmic p53 can directly activate Bax and trigger apoptosis by functioning similarly to the BH3-only proteins.[211]

The antiapoptotic proteins Bcl-2 and Bcl-X_L display conservation in all four BH1-4 domains and act to preserve mitochondrial outer membrane integrity by binding and sequestering proapoptotic activating factors (e.g., Bad or Bid), antagonizing the events of channel formation and cytochrome c release. Bcl-2 prevents the functional association of Bax with the mitochondria and interferes with the release of apoptogenic peptides (e.g., cytochrome c and AIF) from the mitochondria (Fig. 8).

An early event in the mitochondrial apoptotic pathway is the release from the intermembrane space into the cytosol of a group of proteins (e.g., cytochrome c, Smac/diablo, AIF, and endonuclease G).[212] These mitochondrial proteins are involved in triggering the subsequent activation of downstream caspases, initiating cell self-digestion (e.g., cytochrome c and Smac/diablo) and nuclear DNA fragmentation by endonuclease activation (e.g., endonuclease G and AIF), leading to apoptotic cell death.[213] Caspases, which are normally inactive enzymes, require specific proteolytic cleavage for their activation. This is achieved by the formation in the cytosol of large cytosolic complexes termed apoptosomes, which incorporate released cytochrome c, Apaf-1, and recruited caspases activation (e.g., caspase-9). The assembly and function of the apoptosome are regulated by Smac/diablo, intracellular K^+ levels, and a class of proteins termed IAPs. The release of the mitochondrial intermembrane peptides to the cytosol occurs primarily as a result of the disruption of the mitochondrial outer membrane.

Protein release from the mitochondrial cristae, where most cytochrome c is located, may also be associated with the opening of the membrane PT pore.[214] However, the role of the transient opening of the PT pore in the release of cytochrome c is not yet fully understood, because cytochrome c release occurs before any discernible mitochondrial swelling. Nevertheless, PT pore opening is an early requisite feature of apoptosis preceding the activation of caspases. Both Bax and Bcl-2 directly interact with VDAC/porin, both a component of the PT pore as well as a major contributor to the mitochondrial outer membrane permeability.[215] The efflux of cytochrome c is, therefore, coordinated with the activation of a mitochondrial remodeling pathway characterized by changes in inner mitochondrial membrane morphology and organization, ensuring the complete release of cytochrome c and the onset of mitochondrial dysfunction

STRESS, SURVIVAL, AND METABOLIC SIGNALING

The list of extracellular influences and intracellular-generated signals, which impact the cardiomyocyte, is growing. In addition to hormonal and cytokines stimuli (e.g., TH, TNF-α, interleukins), there are also pro/antiapoptotic modulators, nutrient, serum, growth, and mitotic factors, as well as stress and metabolic stimuli, which we describe in more detail in this section.

Survival/Growth Signals

It is well established that cardiac myocytes rapidly proliferate in the embryo but exit the cell cycle irreversibly shortly after birth, with the predominant form of growth shifting from hyperplastic to hypertrophic. Extensive research has focused on identifying the mitogenic stimuli and signaling pathways that mediate these distinct growth processes in isolated cells and with *in vivo* hearts. The molecular mechanisms underlying the proliferative growth of embryonic myocardium *in vivo* and adult cardiac myocyte hypertrophy *in vivo* remain largely undetermined, although considerable progress has been made using post-genomic analysis including studies involving manipulation of the murine genome in concert with mutational analysis of these signaling and growth control pathways *in vivo* and in cardiomyocytes grown *in vitro*, including the use of gene transfer/knock-out. For instance, cell cycle control can be mediated by p38 MAP kinase activity, which regulates the expression of genes required for mitosis in cardiomyocytes including cyclin A and cyclin B. Cardiac-specific p38MAPK knock-out mice show a 92% increase in neonatal cardiomyocyte mitosis. Furthermore, inhibition of p38 MAPK promotes cytokinesis in adult cardiomyocytes.[216] Recent studies have indicated that cyclin D1, a cell-cycle regulator involved in promoting the G_1-to-S phase progression by phosphorylation of the retinoblastoma (Rb) protein, is localized in the nucleus of fetal cardiomyocytes but is primarily cytoplasmic in neonatal and adult cardiomyocytes (concomitant with Rb underphosphorylation). Ectopic expression of a variant of cyclin D1 equipped with nuclear localization signals dramatically promoted neonatal cardiomyocyte proliferation and Rb phosphorylation.[217] Growth factors such as FGF-2 significantly promote neonatal cardiac myocyte proliferation.[218] Overexpression of the FRF-2 receptor (FGF-R1) in neonatal rat cardiomyocytes resulted in marked proliferation.[219] Cardiotrophin (CT-1), an interleukin 6-related cytokine, has been shown to promote both the survival and proliferation of cultured neonatal cardiac myocytes.[220] This likely is mediated by the PI3K/Akt pathway, because CT-1 phosphorylates and activates Akt.[221] These diverse approaches have confirmed the importance of suspected pathways, as well as implicated unexpected pathways leading to new paradigms for the control of cardiac growth.

The PI3K/Akt pathway promotes cell survival in several ways. By intervening in the mitochondrial apoptosis cascade at events before cytochrome *c* release and caspase activation occur, Akt activation inhibits changes in the inner mitochondrial membrane potential that occur in apoptosis (suppressing apoptotic progression and cytochrome *c* release induced by several proapoptotic proteins). Although Akt also contributes to the phosphorylation and inactivation of the proapoptotic protein Bad, it remains unclear whether Bad phosphorylation is the mechanism by which Akt ensures cell survival and mitochondrial integrity, because other mitochondrial targets of Akt remain to be identified. With regard to the heart, PI3K/Akt signaling promotes glucose uptake, growth, and survival of cardiomyocytes and has been implicated in heart growth.[222] Growth factors known to effect cardiomyocyte growth (e.g., IGF-1) signal through the PI3K/Akt pathway.[223] Recently, microarray analysis of cardiomyocytes demonstrated that treatment with IGF-1 results in the differential expression of genes involved in cellular signaling and mitochondrial function and confirmed that this IGF-1–mediated gene regulation required the activation of ERK and PI3K.[224] Transgenic mice with cardiac-specific expression of activated Akt exhibit up-regulation of IGF-binding protein (consistent with its growth signaling/antiapoptotic role) and down-regulation of both PGC-1 and PPAR-α (activators of mitochondrial FAO and mitochondrial biogenesis) presumably shifting cardiomyocytes toward glycolytic metabolism. Deprivation of nutrients (e.g., specific amino acids), glucose serum, and growth factors that can lead to cardiomyocyte apoptosis[225] have been found to signal by means of the mitochondrial associated mTOR protein.[226] Moreover, both the Akt pathway and the downstream mTOR proteins impact cardiomyocyte survival and cell size largely through promoting cytoplasmic protein synthesis mediated by the activation of translational initiation factors and ribosomal proteins.

Finally, Akt also can promote cardioprotection against ischemic injury afforded by exposure of cardiomyocytes to diverse treatments, including cardiotrophin-1, acetylcholine, adenosine, and bradykinin-mediated preconditioning. The precise target of Akt action in cardioprotection remains undetermined, because Akt has not been associated directly with mitoK$_{ATP}$ channels. Recent studies have suggested that cardioprotective signaling is mediated in part by blocking PT pore opening (functioning as an end-effector) regulated by modulation of GSK-3β activity arising from either of several convergent signaling pathways including ROS-activated PKC or by RTKs triggering Akt and mTOR/p70S6k pathways (shown in Fig. 5).[227]

Stress Signals

Stresses in cardiac hypertrophy (e.g., mechanical) and ischemia/hypoxia (e.g., oxidative) elicit a variety of adaptive responses at the tissue, cellular, and molecular levels. A current model displaying the cardiac physiological response to hypoxia suggests the existence of a mitochondrial O_2 sensor coupled to a signal transduction system, which in turn activates a functional response.[228] Myocardial mitochondria may function as O_2 sensors by increasing their generation of ROS during hypoxia and with their abundant heme proteins (e.g., cytochrome *c* oxidase) that reversibly bind oxygen. Similarly, mitochondria in cultured cells from pulmonary artery function as O_2 sensors, which underlies hypoxic pulmonary vasoconstriction.[229] Respiratory inhibitors such as rotenone abolish the hypoxic vasoconstriction; ROS plays a significant role in the mitochondrial signaling in the hypoxic response. Oxidant signals such as ROS act as second messengers initiating signaling cascades and are prominent features in both adaptive responses to hypoxia and the mechanically stressed heart. Down-regulation of cytochrome *c* oxidase (COX) activity contributes to the increased ROS generation and signaling observed in cardiomyocytes during hypoxia.[230] Also, hypoxia stimulates NO synthesis in cardiomyocytes,[231] and NO down-regulates COX activity with subsequent mitochondrial H_2O_2 production. This event has been proposed to provide a mitochondrial-generated signal for further regulating redox-sensitive signaling pathways, including apoptosis, and can proceed even in the absence of marked changes in ATP levels.[231] Interestingly, NOS has been identified in heart mitochondria, although its role in regulating OXPHOS is not clear.[232] Mitochondrial ROS has also been shown to activate p38 MAPK in hypoxic cardiomyocytes.[233] Longer term responses to hypoxia have been shown to include increased gene expression of hypoxia-induced factors (HIF) and the activation of transcription factors such as NF-κB, which have also been implicated in the complex regulation of cardiac hypertrophy and inflammatory cytokines (e.g., TNF-α, IL-1). Although increased ROS has been shown to be an important element in NF-κB gene activation, there is evidence that cardiomyocyte HIF gene activation can also occur in the absence of ROS.[234]

Metabolic Signals

The cardiomyocyte responds to changes in cellular levels of key metabolites such as adenosine, ATP, ADP, oxygen, NADH, as well as numerous substrates and coenzymes. After birth, cardiac FAO becomes critical as a bioenergetic substrate and source of electrons/NADH for the TCA cycle and respiratory chain function.[235] Fatty acids also physically interact with mitochondrial membranes impacting on membrane structure and function, such as transport and excitability. In addition, amphiphilic LCFAs possess detergent-like properties and mediate a variety of toxic effects on the electrophysiological properties of cardiac membranes, including disturbed ion transport and impaired gap junction activity.[236] Increased accumulation of intermediary metabolites of fatty acids, which occurs with defective mitochondrial FAO and transport, is considered responsible for cardiac dysrhythmias and also contributes to HF and sudden death.[237] LCFAs (e.g., palmitate) can modulate the proton conductance of the inner mitochondrial membrane (increased uncoupling) and affect the opening of the PT pore, determining the release of apoptogenic proteins into the cytosol.[238] Major myocardial targets of hormone signaling (thyroid hormone), as well as of LCFAs such as palmitate, include the uncoupling proteins (UCP1-UCP5), carrier proteins located within the inner mitochondrial membrane that function to dissipate the proton gradient across the membrane. These proteins are up-regulated transcriptionally in the presence of palmitate and T3.[239,240] Interestingly, cardiac expression of one of the uncoupling protein genes (UCP3) has been reported to be PPAR-α dependent.[241] In addition, the increased expression of the uncoupling proteins in cardiac muscle results in increased uncoupling of oxidative phosphorylation from respiration, decreased myocardial efficiency, and mitochondrial membrane potential.[240]

Extracellular Signals and Matrix

In addition to the numerous cytokine, neurohumoral, and hormonal stimuli that impact cardiomyocyte signaling processes, the extracellular matrix (ECM) plays an important role in cellular signaling in the heart in addition to providing structural and mechanical support.[242] This signaling becomes most apparent during myocardial remodeling, which occurs as either a physiological or pathological response such as occurs in cardiac hypertrophy and in the progression of HF.

There is an important link between the ECM and the cardiomyocyte cytoskeleton. One class of cell-surface receptor molecules that constitute part of that link are the transmembrane integrins that act as signaling molecules and transducers of mechanical force.[243] Interestingly, integrins are expressed in all cellular components of the cardiovascular system, including the vasculature, blood, cardiac myocytes, and nonmuscle cardiac cells. In response to specific changes in their microenvironment, these receptors become activated and form focal adhesions, areas of attachment of the cells to ECM proteins involving the colocalization of cytoskeletal proteins, intracellular signaling molecules, and growth factor receptors. Growth factor–mediated integrin activation can stimulate cell growth, gene expression, and adhesion in cardiac fibroblasts.[244] In addition, a large number of bioactive signaling molecules and growth factors, proteases, and structural proteins have been identified that influence cell–matrix interactions. ECM proteins can be degraded by matrix metalloproteinases (MMPs), whose activation is associated with ventricular remodeling. The dynamic balance of degradative enzymes, the MMPs, their regulation by multifunctional endogenous inhibitors, the TIMPs, and their highly regulated synthesis largely determines ECM remodeling.

The myocardial fibroblast, the most numerous cell type in the heart, plays a central role in the structure of the myocardial ECM. In response to biochemical and physical stimuli, myocardial fibroblasts can rapidly proliferate with their gene expression program significantly contributing to the remodeling process occurring with cardiac pathologies such as MI. For instance, myocardial ECM collagen is markedly increased with pressure overload hypertrophy because of a combination of increased synthesis and deposition by fibroblasts and decreased collagen degradation. Pro-fibrotic factors such as norepinephrine, angiotensin II, and endothelin-1 stimulate fibrosis by modulating collagen synthesis and

MMP/TIMP activity. In contrast, other bioactive molecules, including the natriuretic peptides (e.g., BNP), and NO demonstrate antifibrotic action by inhibiting collagen synthesis and by stimulating MMP activity. These signal primarily by means of the second messenger cGMP/PKG pathway. The resultant formation and activation of specific MMPs targeted to the cell membrane, as well as released into the local interstitial space, has an amplification effect on MMP activity; the targeting of MMPs to the membrane (mtMMP) can stimulate the release of extracellular signaling molecules such as TNF-α. A variety of signaling systems can be integrated to effect ECM turnover and ventricular remodeling. These findings have provided support for the use of pharmacological intervention aimed at adjusting the level of bioactive molecules as a potential therapeutic strategy for attenuating the adverse ventricular remodeling associated with HF.

SIGNALING IN CARDIAC DEVELOPMENT AND AGING

Studies of the developing heart have begun to unveil a complex molecular circuitry in which gene regulation governs a network of signaling pathways responsive to both physiological stimuli and genetic programming. In this chapter, we review some of the major players in several well-characterized developmental signaling pathways and describe how they have been approached from a post-genomic perspective. In Chapter 6, we will more exhaustively examine the full panoply of the signaling cascades, their salient interactions, and present evidence of their critical involvement in cardiac development.

The GATA family of transcription factors was initially identified as an essential regulator of the two natriuretic peptide genes, ANP and BNP, important markers of cardiomyocyte differentiation, with distinct spatial, developmental, and hormonal regulation.[245] The analysis of the ANP and BNP promoters in association with GATA binding has led to the delineation of combinatorial interactions of multiple factors required for proper regulation of the cardiac genes as key regulators of myocardial development.[246,247]

Interestingly, the role played by the GATA family of transcription factors emerged from using a variety of methodological approaches. The use of standard molecular biology, gene transfections, and transgenic animals proved complementary in their findings.[248–250] GATA-4 is also an intrinsic component of the retinoid-mediated cardiogenic pathway involved in the morphogenesis of the posterior heart tube and the development of the cardiac inflow tract.[251] The importance of GATA as a developmental factor also was established from the evidence of its early expression in a variety of developmental models and was further underscored by clinical studies demonstrating that specific mutations in GATA were associated with severe congenital heart defects[252] and atrial septal defects.[253] Other studies using antisense constructs to block GATA expression further supported the intrinsic role that GATA has in cardiac development. More recently, GATA-4 has emerged as the nuclear effector of several signaling pathways involved in cardiac growth, development, and differentiation whose transcriptional activity function can be modulated by post-translational modifications (e.g., p38K-mediated phosphorylation, p300-mediated acetylation) and protein–protein interactions (e.g., JMJ interaction).[254–259] Bone morphogenetic protein (BMP), an indispensable factor for cardiomyocyte differentiation, acts in part by inducing the expression of two cardiac transcription factors, Csx/Nkx2-5 and GATA-4.[260] Moreover, dynorphin B, a product of the prodynorphin gene, has been found to promote cardiogenesis in embryonic cells by inducing the expression of GATA-4 and Nkx-2.5.[261]

Recent studies have demonstrated the vital nature of calcium/calcineurin/NFAT signaling in cardiac muscle development in vertebrates. As further described in Chapter 6, inhibition, mutation, or forced expression of calcineurin pathway genes result in defects or alterations in cardiomyocyte maturation, heart valve formation, vascular development, skeletal muscle differentiation and fiber-type switching, and cardiac hypertrophy.

Over the past decade, gene disruption in mice and large-scale mutagenesis screens in zebrafish have proved informative in delineating fundamental genetic pathways governing early heart patterning and differentiation. A number of genes have been identified that play important and selective roles in cardiac valve development, in endothelial and cardiac cell proliferation, and in the early development of the cardiac pacemaking and conduction system. The serendipitous discovery of the convergence of several signaling pathways that regulate endothelial proliferation and differentiation in developing and post-natal heart valves has provided insight into the roles played by highly conserved signaling elements including VEGF, NFATc1, Notch, Wnt/β-catenin, BMP/TGF-β, ErbB, and NF1/Ras.

Signaling in Cardiac Aging

The contribution of signaling events both as a mediator and a consequence of cardiac aging is examined in Chapter 19 to which the reader is referred.

CONCLUSION

At the outset, it is important to understand that the complex field of heart signaling is presently in its infancy. The wide assortment of post-genomic approaches presently available have implicated an increasing number of genes and their products as being critical to cardiomyocyte signaling in both physiological and pathological conditions. Although considerable progress has been made in the identification of the downstream targets of these genes and the pathways they constitute, most targets and the complex interrelationships between the signaling pathways involved still remain largely undetermined. In particular, a large number of signaling interactions between cellular organelles remain to be identified. The cross-talk among mitochondria, plasma membrane receptors, exocytotic events, and excitable channels, with ER, Golgi bodies, peroxisomes, and with the nucleus in effecting nuclear import/export, regulatory gene expression, signaling cell cycle progression, as well as the determination of cardiac-specific and developmentally specific factors involved in myocardial signaling, will require an extensive and continuous research effort.

There is increasing recognition that important information about the nature of the signaling process can be gleaned from the study of the multiprotein complexes involved in signaling. Post-genomic analysis has increasingly focused on these macromolecular aggregates, including the delineation of protein–protein interactions, and subproteomic analysis has resulted in the identification of protein components within tyrosine kinase modules, G-protein complexes, ion, and PT pore channels. The description not only of the kinases, ligands, and second messengers but also of the entire panoply of docking proteins, scaffolding proteins that organize the aggregate structure, and numerous modulators that affect signaling processes should eventually allow a three-dimensional architecture of cardiomyocyte signaling to be ascertained. The manipulation of these components within the hearts of transgenic animals and in cardiomyocytes grown *in vitro* is increasingly possible using sophisticated techniques of gene transfer and RNAi to target specific gene expression. Interestingly, the development of techniques to reconstitute functional complexes within artificial membranes has shown less success in signaling studies, underlining the view that the context of molecular interactions that constitute the cardiomyocyte signaling environment can be lost in such reductionist schemes.

A driving force for the assessment of cardiovascular signaling is that it can enhance our understanding of the widening spectra of cardiac abnormalities that are known to result from signaling dysfunction, as well as potentially contributing to treatment modalities. The recognition that cytoprotective signaling plays an essential role in cardioprotection has generated great interest in the pharmacological manipulation of metabolism and signaling. Several important caveats pertain to the application of targeting specific signaling molecules within the clinical setting. Many signaling molecules participate in multiple pathways. For instance, inhibiting ROS production might prove helpful in reducing the negative consequences of myocardial I/R but could prove counterproductive in cardioprotection and in oxygen sensing. Second, the existence of redundant signaling pathways that trigger HF also poses challenges for therapeutic intervention. Clearly, our improved understanding of the precise order of intracellular events, their downstream consequences, the overall interrelatedness, and regulation of these pathways will be necessary for the discovery of new therapies enabling the development and characterization of reagents with high specificity to heart signaling pathways and the arrival of new technologies for the inhibition of stress signaling (e.g., specific kinase inhibitors) in cardiac and vascular cells.

SUMMARY

- Multiple signaling pathways regulate cardiomyocyte growth and proliferation. These redundant mechanisms typically involve serine/threonine kinases that stimulate a series of sequential molecular and cellular events (cascades).
- These cascades usually initiated by signals received at the plasma membrane receptors are transmitted by means of coupled G-proteins or other second messengers to a wide spectrum of protein kinases and phosphatases. These include protein kinase A, B, C, and G, the latter which operates in conjunction with cyclic GMP.
- Several G-protein–coupled receptors (GPCRs), including the α- and β-adrenergic receptors, angiotensin II, and endothelin-1, are able to activate these signaling cascades. Other receptors pivotal in cardiac and cardiovascular signaling include protease-activated, Toll-like receptors, receptor tyrosine kinases (RTKS), and muscarinic receptors.
- These activated protein modifiers can lead to downstream activation or deactivation of specific transcription factors (e.g., NF-κB), which modulate specific gene expression affecting a broad spectrum of cellular events. The modulation of gene expression can be fostered as well by effectors that modulate chromatin structure and histone proteins (e.g., acetylases and deacetylases).
- These effectors also can directly target proteins involved in metabolic pathways, ion transport, Ca^{++} regulation, and handling, which affect contractility and excitability, as well as cardiomyocyte apoptosis and cell survival.
- Adrenoceptors (ARs) are members of the GPCR superfamily, interfacing between the sympathetic nervous system and the cardiovascular system, with integral roles in the rapid regulation of myocardial function.
- Stimulation of β1-adrenergic receptors (the predominant subtype found in both neonatal and adult ventricular myocytes) results in the activation of stimulatory G-protein and adenylyl cyclase (AC) with increased intracellular cAMP and induction of protein kinase A (PKA). This leads to the downstream phosphorylation of key target proteins, including L-type calcium channels, phospholamban, troponin I, and ryanodine receptors, resulting in the modulation of cardiac contractility in both neonatal and adult ventricular myocytes.
- Regulatory control of the receptors is exerted by phosphorylation (by either PKA or other kinases), which can lead to receptor internalization and desensitization to the

agonist. Chronic β-AR stimulation causes down-regulation of the receptors.
- The α1-adrenergic receptor family (α1-AR) each contains a seven transmembrane-spanning domain and is linked to G-proteins. On stimulation with agonists such as noradrenaline and adrenaline, α1-ARs activate G_q-proteins and, subsequently, activate phospholipase Cβ, resulting in increased levels of second messengers IP3 and DAG, which promote an increase in intracellular Ca^{++} levels and protein PKC activation. They are involved in both modulating cardiac contractility and cardiomyocyte hypertrophy.
- A variety of G-proteins play an essential role in mediating signal transduction from receptor to either kinase or effector and can either be stimulatory or inhibitory. These are trimeric structures, with the α-subunit binding guanine nucleotides, affecting the overall association of the G-protein subunits. GTP binding leads to inactivation/dissociation, whereas GDP binding leads to activation.
- Small GTP binding proteins, as well as effectors of GTP hydrolysis, can modulate the activity of G-proteins.
- Caveolae and specific proteins can organize the signaling complexes within microdomains at the plasma membrane. Other proteins serve to enhance and orchestrate the signaling complexes, including anchoring, adaptor and scaffolding proteins interacting with diverse stimuli.
- ER signaling involves calcium cycling by means of activation of the IP3 and ryanodine receptors, modulation of the calcium ATPase (SERCA), as well as control of protein synthesis function.
- Mitochondrial signaling occurs by means of ROS generation (a by-product of ETC).
- The global nuclear receptors/transcription factors (e.g., PPAR-α, PGC-1), calcium flux (derived mainly from ER), fatty acids, and hormones (e.g., thyroid hormone) regulate the mitochondrial bioenergetic process. This organelle also houses an antioxidant armamentarium, including MnSOD, glutathione, and glutathione peroxidase. Mitochondria are pivotal elements in the early events and progression of apoptosis; they are targets of proapoptotic proteins (e.g., Bad, Bax, Bid) that can lead to the permeabilization of both outer and inner mitochondrial membranes releasing mitochondrial apoptogenic proteins (e.g., cytochrome c, Smac, AIF, endoG) to the cytosol to stimulate caspase activation, DNA degradation, and apoptotic cell death. Mitochondrial membranes also contain the permeability transition pore, a nonselective megachannel operative during early apoptosis, responsive to ROS and calcium levels and the mitoK_{ATP} channel, which is thought to play a significant role in cardioprotective pathways during pharmacological and ischemic preconditioning.
- ROS can also be generated by nonmitochondrial enzymatic pathways present in both heart and endothelial cells, including cytochrome p450, NAD(P)H oxidase, and xanthine oxidase. Cytosolic neutralization of ROS and oxidative stress is achieved by enzymes such as catalase and CuSOD.
- ROS can be a potent insult to the cell, promoting a wide range of oxidative damage, including lipid peroxidation, DNA, and protein damage. Mitochondria are a primary target (e.g., ETC, mtDNA and cardiolipin). ROS can also be a stimulus for myocardial apoptotic progression and plays a fundamental role in the myocardial damage associated with ischemia and reperfusion (I/R). It also serves as an important regulatory signal (e.g., activating protein kinases, specific transcription factors, and survival pathways) and is an intrinsic element of cardioprotective pathways.
- Stress, metabolic, and mitotic stimuli can act through signaling pathways and share a variety of signaling components. Survival pathways involving Akt and PI3K, PKA, and PKC can be used by these different stimuli.
- Signaling pathways play an important role in morphogenesis and the regulation of gene expression in the developing heart. Signaling stimuli such as BMP receptors (e.g., BMPR) and specific transcription factors (e.g., NKX2-5 and GATA) are critical factors in early cardiac development.

References

1. Wallukat, G. (2002). The beta-adrenergic receptors. *Herz* **27**, 683–690.
2. Rybin, V. O., Pak, E., Alcott, S., and Steinberg, S. F. (2003). Developmental changes in beta2-adrenergic receptor signaling in ventricular myocytes: the role of Gi proteins and caveolae microdomains. *Mol. Pharmacol.* **63**, 1338–1348.
3. Rockman, H. A., Koch, W. J., Milano, C. A., and Lefkowitz, R. J. (1996). Myocardial beta-adrenergic receptor signaling in vivo: insights from transgenic mice. *J. Mol. Med.* **74**, 489–495.
4. Brodde, O. E., and Michel, M. C. (1999). Adrenergic and muscarinic receptors in the human heart. *Pharmacol. Rev.* **51**, 651–690.
5. O'Rourke, M. F., Iversen, L. J., Lomasney, J. W., and Bylund, D. B. (1994). Species orthologs of the alpha-2A adrenergic receptor: the pharmacological properties of the bovine and rat receptors differ from the human and porcine receptors. *J. Pharmacol. Exp. Ther.* **27**, 735–740.
6. Flordellis, C., Manolis, A., Scheinin, M., and Paris, H. (2004). Clinical and pharmacological significance of alpha2-adrenoceptor polymorphisms in cardiovascular diseases. *Int. J. Cardiol.* **97**, 367–372.
7. Ness, J. (1996). Molecular biology of muscarinic acetylcholine receptors. *Crit. Rev. Neurobiol.* **10**, 69–99.
8. Wang, Z., Shi, H., and Wang, H. (2004). Functional M3 muscarinic acetylcholine receptors in mammalian hearts. *Br. J. Pharmacol.* **142**, 395–408.
9. Shi, H., Wang, H., Yang, B., Xu, D., and Wang, Z. (2004). The M3 receptor-mediated K(+) current (IKM3), a G(q) protein-coupled K(+) channel. *J. Biol. Chem.* 279, 21774–21778.
10. Shi, H., Wang, H., Li, D., Nattel, S., and Wang, Z. (2004). Differential alterations of receptor densities of three

muscarinic acetylcholine receptor subtypes and current densities of the corresponding K+ channels in canine atria with atrial fibrillation induced by experimental congestive heart failure. *Cell Physiol. Biochem.* **14,** 31–40.

11. Giannessi, D., Del Ry, S., and Vitale, R. L. (2001). The role of endothelins and their receptors in heart failure. *Pharmacol. Res.* **43,** 111–126.
12. Sugden, P. H. (2003). An overview of endothelin signaling in the cardiac myocyte. *J. Mol. Cell. Cardiol.* **35,** 871–886.
13. Dinh, D. T., Frauman, A. G., Johnston, C. I., and Fabiani, M. E. (2001). Angiotensin receptors: distribution, signalling and function. *Clin. Sci. (Lond).* **100,** 481–492.
14. Saito, Y., and Berk, B. C. (2002(. Angiotensin II-mediated signal transduction pathways. *Curr. Hypertens. Rep.* **4,** 167–171.
15. Brasier, A. R., Jamaluddin, M., Han, Y., Patterson, C., and Runge, M. S. (2000). Angiotensin II induces gene transcription through cell-type-dependent effects on the nuclear factor-kappaB (NF-kappaB) transcription factor. *Mol. Cell. Biochem.* **212,** 155–169.
16. Chen, Y., Arrigo, A. P., and Currie, R. W. (2004). Heat shock treatment suppresses angiotensin II-induced activation of NF-kappaB pathway and heart inflammation: a role for IKK depletion by heat shock? *Am. J. Physiol. Heart Circ. Physiol.* **287,** H1104–H11014.
17. Clerk, A., Aggeli, I. K., Stathopoulou, K., and Sugden, P. H. (2006). Peptide growth factors signal differentially through protein kinase C to extracellular signal-regulated kinases in neonatal cardiomyocytes. *Cell Signal* **18,** 225–235.
18. Frantz, S., Kobzik, L., Kim, Y. D., Fukazawa, R., Medzhitov, R., Lee, R. T., and Kelly, R. A. (1999). Toll4 (TLR4) expression in cardiac myocytes in normal and failing myocardium. *J. Clin. Invest.* **104,** 271–280.
19. Satoh, M., Nakamura, M., Akatsu, T., Shimoda, Y., Segawa, I., and Hiramori, K. (2004). Toll-like receptor 4 is expressed with enteroviral replication in myocardium from patients with dilated cardiomyopathy. *Lab. Invest.* **84,** 173–181.
20. Oyama, J., Blais, C., Jr., Liu, X., Pu, M., Kobzik, L., Kelly, R. A., and Bourcier, T. (2004). Reduced myocardial ischemia-reperfusion injury in toll-like receptor 4-deficient mice. *Circulation* **109,** 784–789.
21. Wang, Y. P., Sato, C., Mizoguchi, K., Yamashita, Y., Oe, M., and Maeta, H. (2002). Lipopoly-saccharide triggers late preconditioning against myocardial infarction via inducible nitric oxide synthase. *Cardiovasc. Res.* **56,** 33–42.
22. Tavener, S. A., Long, E. M., Robbins, S. M., McRae, K. M., Van Remmen, H., and Kubes, P. (2004). Immune cell Toll-like receptor 4 is required for cardiac myocyte impairment during endotoxemia. *Circ. Res.* **95,** 700–707.
23. Frantz, S., Kelly, R. A., and Bourcier, T. (2001). Role of TLR-2 in the activation of nuclear factor kappaB by oxidative stress in cardiac myocytes. *J. Biol. Chem.* **276,** 5197–5203.
24. Coughlin, S. R., and Camerer, E. (2003). PARticipation in inflammation. *J. Clin. Invest.* **111,** 25–27.
25. Sabri, A., Muske, G., Zhang, H., Pak, E., Darrow, A., Andrade-Gordon, P., and Steinberg, S. F. (2000). Signaling properties and functions of two distinct cardiomyocyte protease-activated receptors. *Circ. Res.* **86,** 1054–1061.
26. Barnes, J. A., Singh, S., and Gomes, A. V. (2004). Protease activated receptors in cardiovascular function and disease. *Mol. Cell. Biochem.* **263,** 227–239.
27. Grassot, J., Mouchiroud, G., and Perriere, G. (2003). RTKdb: database of receptor tyrosine kinase. *Nucleic Acids Res.* **31,** 353–358.
28. Zhao, Y. Y., Feron, O., Dessy, C., Han, X., Marchionni, M. A., and Kelly, R. A. (1999). Neuregulin signaling in the heart. Dynamic targeting of erbB4 to caveolar microdomains in cardiac myocytes. *Circ. Res.* **84,** 1380–1387.
29. Zhao, Y. Y., Sawyer, D. R., Baliga, R. R., Opel, D. J., Han, X., Marchionni, M. A., and Kelly, R. A. (1998). Neuregulins promote survival and growth of cardiac myocytes. Persistence of ErbB2 and ErbB4 expression in neonatal and adult ventricular myocytes. *J. Biol. Chem.* **273,** 10261–10269.
30. Lee, H. J., and Koh ,G. Y. (2003). Shear stress activates Tie2 receptor tyrosine kinase in human endothelial cells. *Biochem. Biophys. Res. Commun.* **304,** 399–404.
31. Becker, E., Huynh-Do, U., Holland, S., Pawson, T., Daniel, T. O., and Skolnik, E. Y. (2000). Nck-interacting Ste20 kinase couples Eph receptors to c-Jun N-terminal kinase and integrin activation. *Mol. Cell. Biol.* **20,** 1537–1545.
32. Hamm, H. E. (1998). The many faces of G protein signaling. *J. Biol. Chem.* **273,** 669–672.
33. Adams, J. W., and Brown, J. H. (2001). G-proteins in growth and apoptosis: lessons from the heart. *Oncogene* **20,** 1626–1634.
34. Dorn, G. W., 2nd., and Brown, J. H. (1999). Gq signaling in cardiac adaptation and maladaptation. *Trends Cardiovasc. Med.* **9,** 26–34.
35. Clerk, A., and Sugden, P. H. (2000). Small guanine nucleotide-binding proteins and myocardial hypertrophy. *Circ. Res.* **86,** 1019–1023.
36. Meszaros, J. G., Gonzalez, A. M., Endo-Mochizuki, Y., Villegas, S., Villarreal, F., and Brunton, L. L. (2000). Identification of G protein-coupled signaling pathways in cardiac fibroblasts: cross talk between G(q) and G(s). *Am. J. Physiol. Cell. Physiol.* **278,** C154–C162.
37. Gao, M. H., Lai, N. C., Roth, D. M., Zhou, J., Zhu, J., Anzai, T., Dalton, N., and Hammond, H. K. (1999). Adenylyl cyclase increases responsiveness to catecholamine stimulation in transgenic mice. *Circulation* **99,** 1618–1622.
38. Lai, N. C., Roth, D. M., Gao, M. H., Tang, T., Dalton, N., Lai, Y. Y., Spellman, M., Clopton, P., and Hammond, H. K. (2004). Intracoronary adenovirus encoding adenylyl cyclase VI increases left ventricular function in heart failure. *Circulation* **110,** 330–336.
39. Roth, D. M., Gao, M. H., Lai, N. C., Drumm, J., Dalton, N., Zhou, J. Y., Zhu, J., Entrikin, D., an Hammond, H. K. (1999). Cardiac-directed adenylyl cyclase expression improves heart function in murine cardiomyopathy. *Circulation* **99,** 3099–3102.
40. Iwami, G., Kawabe, J., Ebina, T., Cannon, P. J., Homcy, C. J., and Ishikawa, Y. (1995). Regulation of adenylyl cyclase by protein kinase A. *J. Biol. Chem.* **270,** 12481–12484.
41. Rhee, S. G. (2001). Regulation of phosphoinositide-specific phospholipase C. *Annu. Rev. Biochem.* **70,** 281–312.
42. Wing, M. R., Bourdon, D. M., and Harden, T. K. (2003). PLC-epsilon: a shared effector protein in Ras-, Rho-, and

G alpha beta gamma-mediated signaling. *Mol. Interv.* **3,** 273–280.

43. Rybin, V. O., Xu, X., and Steinberg, S. F. (1999). Activated protein kinase C isoforms target to cardiomyocyte caveolae: Stimulation of local protein phosphorylation. *Circ. Res.* **84,** 980–988.

44. Head, B. P., Patel, H. H., Roth, D. M., Lai, N. C., Niesman, I. R., Farquhar, M. G., and Insel, P. A. (2005). G-protein coupled receptor signaling components localize in both sarcolemmal and intracellular caveolin-3-associated microdomains in adult cardiac myocytes. *J. Biol. Chem.* **280,** 31036–31044.

45. Williams, T. M., and Lisanti, M. P. (2004). The caveolin genes: From cell biology to medicine. *Ann. Med.* **36,** 584–595.

46. Ruehr, M. L., Russell, M. A., and Bond, M. (2004). A-kinase anchoring protein targeting of protein kinase A in the heart. *J. Mol. Cell Cardiol.* **37,** 653–656

47. Dodge, K. L., Khouangsathiene, S., Kapiloff, M. S., Mouton, R., Hill, E. V., Houslay, M. D., Langeberg, L. K., and Scott, J. D. (2001). mAKAP assembles a protein kinase A/PDE4 phosphodiesterase cAMP signaling module. *EMBO J.* **20,** 1921–1930.

48. Hulme, J. T., Scheuer, T., and Catterall, W. A. (2004). Regulation of cardiac ion channels by signaling complexes: role of modified leucine zipper motifs. *J. Mol. Cell. Cardiol.* **37,** 625–631.

49. Dorn, G. W., II, and Force, T. (2005). Protein kinase cascades in the regulation of cardiac hypertrophy. *J. Clin. Invest.* **115,** 527–537.

50. Sugden, P. H. (2003). Ras, Akt, and mechanotransduction in the cardiac myocyte. *Circ. Res.* **93,** 1179–1192.

51. Aikawa, R., Nawano, M., Gu, Y., Katagiri, H., Asano, T., Zhu, W., Nagai, R., and Komuro, I. (2000). Insulin prevents cardiomyocytes from oxidative stress-induced apoptosis through activation of PI3 kinase/Akt. *Circulation* **102,** 2873–2879.

52. Gao, F., Gao, E., Yue, T. L., Ohlstein, E. H., Lopez, B. L., Christopher, T. A., and Ma, X. L. (2002). Nitric oxide mediates the antiapoptotic effect of insulin in myocardial ischemia-reperfusion: the roles of PI3-kinase, Akt, and endothelial nitric oxide synthase phosphorylation. *Circulation* **105,** 1497–1502.

53. Matsui, T., Li, L., del Monte, F., Fukui, Y., Franke, T. F., Hajjar, R. J., and Rosenzweig, A. (1999). Adenoviral gene transfer of activated phosphatidylinositol 3-kinase and Akt inhibits apoptosis of hypoxic cardiomyocytes in vitro. *Circulation* **100,** 2373–2379.

54. Yamashita, K., Kajstura, J., Discher, D. J., Wasserlauf, B. J., Bishopric, N. H., Anversa, P., and Webster, K. A. (2001). Reperfusion-activated Akt kinase prevents apoptosis in transgenic mouse hearts overexpressing insulin-like growth factor-1. *Circ. Res.* **88,** 609–614.

55. Craig, R., Wagner, M., McCardle, T., Craig, A. G., and Glembotski, C. C. (2001). The cytoprotective effects of the glycoprotein 130 receptor-coupled cytokine, cardiotrophin-1, require activation of NF-kappa B. *J. Biol. Chem.* **276,** 37621–37629.

56. Negoro, S., Oh, H., Tone, E., Kunisada, K., Fujio, Y., Walsh, K., Kishimoto, T., and Yamauchi-Takihara, K. (2001). Glycoprotein 130 regulates cardiac myocyte survival in doxorubicin-induced apoptosis through phosphatidylinositol 3-kinase/Akt phosphorylation and Bcl-xL/caspase-3 interaction. *Circulation* **103,** 555–561.

57. Chesley, A., Lundberg, M. S., Asai, T., Xiao, R. P., Ohtani, S., Lakatta, E. G., and Crow, M. T. (2000). The beta(2)-adrenergic receptor delivers an antiapoptotic signal to cardiac myocytes through G(i)-dependent coupling to phosphatidylinositol 3′-kinase. *Circ. Res.* **87,** 1172–1179.

58. Tian, B., Liu, J., Bitterman, P., and Bache, R. J. (2003). Angiotensin II modulates nitric oxide-induced cardiac fibroblast apoptosis by activation of AKT/PKB. *Am. J. Physiol. Heart Circ. Physiol.* **285,** H1105–H1112.

59. Clerk, A., and Sugden, P. H. (1999). Activation of protein kinase cascades in the heart by hypertrophic G protein-coupled receptor agonists. *Am. J. Cardiol.* **83,** 64H–69H.

60. Krieg, T., Landsberger, M., Alexeyev, M. F., Felix, S. B., Cohen, M. V., and Downey, J. M. (2003). Activation of Akt is essential for acetylcholine to trigger generation of oxygen free radicals. *Cardiovasc. Res.* **58,** 196–202.

61. Yin, H., Chao, L., and Chao, J. (2004). Adrenomedullin protects against myocardial apoptosis after ischemia/reperfusion through activation of Akt-GSK signaling. *Hypertension* **43,** 109–116.

62. Naga Prasad, S. V., Esposito, G., Mao, L., Koch, W. J., and Rockman, H. A. (2000). Gbetagamma-dependent phosphoinositide 3-kinase activation in hearts with in vivo pressure overload hypertrophy. *J. Biol. Chem.* **275,** 4693–4698.

63. Mockridge, J. W., Marber, M. S., and Heads, R. J. (2000). Activation of Akt during simulated ischemia/reperfusion in cardiac myocytes. *Biochem. Biophys. Res. Commun.* **270,** 947–952.

64. Matsui, T., Tao, J., del Monte, F., Lee, K. H., Li, L., Picard, M., Force, T. L., Franke, T. F., Hajjar, R. J., and Rosenzweig, A. (2001). Akt activation preserves cardiac function and prevents injury after transient cardiac ischemia in vivo. *Circulation* **104,** 330–335.

65. Brunet, A., Bonni, A., Zigmond, M. J., Lin, M. Z., Juo, P., Hu, L. S., Anderson, M. J., Arden, K. C., Blenis, J., and Greenberg, M. E. (1999). Akt promotes cell survival by phosphorylating and inhibiting a Forkhead transcription factor. *Cell* **96,** 857–868.

66. Shioi, T., McMullen, J. R., Kang, P. M., Douglas, P. S., Obata, T., Franke, T. F., Cantley, L. C., and Izumo, S. (2002). Akt/protein kinase B promotes organ growth in transgenic mice. *Mol. Cell. Biol.* **22,** 2799–2809.

67. Matsui, T., Li, L., Wu, J. C., Cook, S. A., Nagoshi, T., Picard, M. H., Liao, R., and Rosenzweig, A. (2002). Phenotypic spectrum caused by transgenic overexpression of activated Akt in the heart. *J. Biol. Chem.* **277,** 22896–22901.

68. Wang, L., Wang, X., and Proud, C. G. (2000). Activation of mRNA translation in rat cardiac myocytes by insulin involves multiple rapamycin-sensitive steps. *Am. J. Physiol. Heart Circ. Physiol.* **278,** H1056–H1068.

69. Gao, X., Zhang, Y., Arrazola, P., Hino, O., Kobayashi, T., Yeung, R. S., Ru, B., and Pan, D. (2002). Tsc tumour suppressor proteins antagonize amino-acid-TOR signalling. *Nat. Cell Biol.* **4,** 699–704.

70. Marygold, S. J., and Leevers, S. J. (2002). Growth signaling: TSC takes its place. *Curr. Biol.* **12,** R785–R787.

71. Zhang, Y., Gao, X., Saucedo, L. J., Ru, B., Edgar, B. A., and Pan, D. (2003). Rheb is a direct target of the tuberous sclerosis tumour suppressor proteins. *Nat. Cell Biol.* **5,** 578–581.

72. Tee, A. R., Manning, B. D., Roux, P. P., Cantley, L. C., and Blenis, J. (2003). Tuberous sclerosis complex gene products, Tuberin and Hamartin, control mTOR signaling by acting as a GTPase-activating protein complex toward Rheb. *Curr. Biol.* **13,** 1259–1268.
73. Badorff, C., Ruetten, H., Mueller, S., Stahmer, M., Gehring, D., Jung, F., Ihling, C., Zeiher, A. M., and Dimmeler, S. (2002). Fas receptor signaling inhibits glycogen synthase kinase 3 beta and induces cardiac hypertrophy following pressure overload. *J. Clin. Invest.* **109,** 373–381.
74. Haq, S., Choukroun, G., Kang, Z. B., Ranu, H., Matsui, T., Rosenzweig, A., Molkentin, J. D., Alessandrini, A., Woodgett, J., Hajjar, R., Michael, A., and Force, T. (2000). Glycogen synthase kinase-3beta is a negative regulator of cardiomyocyte hypertrophy. *J. Cell Biol.* **151,** 117–130.
75. Morisco, C., Zebrowski, D., Condorelli, G., Tsichlis, P., Vatner, S. F., and Sadoshima, J. (2000). The Akt-glycogen synthase kinase 3beta pathway regulates transcription of atrial natriuretic factor induced by beta-adrenergic receptor stimulation in cardiac myocytes. *J. Biol. Chem.* **275,** 14466–14475.
76. Antos, C. L., McKinsey, T. A., Frey, N., Kutschke, W., McAnally, J., Shelton, J. M., Richardson, J. A., Hill, J. A., and Olson, E. N. (2002). Activated glycogen synthase-3 beta suppresses cardiac hypertrophy in vivo. *Proc. Natl. Acad. Sci. USA* **99,** 907–912.
77. Proud, C. G. (2004). Ras, PI3-kinase and mTOR signaling in cardiac hypertrophy. *Cardiovasc. Res.* **63,** 403–413.
78. Haq, S., Michael, A., Andreucci, M., Bhattacharya, K., Dotto, P., Walters, B., Woodgett, J., Kilter, H., and Force, T. (2003). Stabilization of beta-catenin by a Wnt-independent mechanism regulates cardiomyocyte growth. *Proc. Natl. Acad. Sci. USA* **100,** 4610–4615.
79. Morisco, C., Seta, K., Hardt, S. E., Lee, Y., Vatner, S. F., and Sadoshima, J. (2001). Glycogen synthase kinase 3beta regulates GATA4 in cardiac myocytes. *J. Biol. Chem.* **276,** 28586–28597.
80. Xiao, G., Mao, S., Baumgarten, G., Serrano, J., Jordan, M. C., Roos, K. P., Fishbein, M. C., and MacLellan, W. R (2001). Inducible activation of c-Myc in adult myocardium in vivo provokes cardiac myocyte hypertrophy and reactivation of DNA synthesis. *Circ. Res.* **89,** 1122–1129.
81. Michael, A., Haq, S., Chen, X., Hsich, E., Cui, L., Walters, B., Shao, Z., Bhattacharya, K., Kilter, H., Huggins, G., Andreucci, M., Periasamy, M., Solomon, R. N., Liao, R., Patten, R., Molkentin, J. D., and Force, T. (2004). Glycogen synthase kinase-3beta regulates growth, calcium homeostasis, and diastolic function in the heart. *J. Biol. Chem.* **279,** 21383–21393.
82. Liao, W., Wang, S., Han, C., and Zhang, Y. (2005). 14-3-3 proteins regulate glycogen synthase 3beta phosphorylation and inhibit cardiomyocyte hypertrophy. *FEBS J.* **272,** 1845–1854.
83. Chan, A. Y., Soltys, C. L., Young, M. E., Proud, C. G., and Dyck, J. R. (2004). Activation of AMP-activated protein kinase inhibits protein synthesis associated with hypertrophy in the cardiac myocyte. *J. Biol. Chem.* **279,** 32771–32779.
84. Arsham, A. M., Howell, J. J., and Simon, M. C. (2003). A novel hypoxia-inducible factor-independent hypoxic response regulating mammalian target of rapamycin and its targets. *J. Biol. Chem.* **278,** 29655–29660.
85. van Noort, M., Meeldijk, J., van der Zee, R., Destree, O., and Clevers, H. (2002). Wnt signaling controls the phosphorylation status of beta-catenin. *J. Biol. Chem.* **277,** 17901–17905.
86. Sabri, A., and Steinberg, S. F. (2003). Protein kinase C isoform-selective signals that lead to cardiac hypertrophy and the progression of heart failure. *Mol. Cell Biochem.* **251,** 97–101.
87. Mochly-Rosen, D., Wu, G., Hahn, H., Osinska, H., Liron, T., Lorenz, J. N., Yatani, A., Robbins, J., and Dorn, G. W., 2nd. (2000). Cardiotrophic effects of protein kinase C epsilon: analysis by in vivo modulation of PKCepsilon translocation. *Circ. Res.* **86,** 1173–1179.
88. Zhuang, D., Ceacareanu, A. C., Ceacareanu, B., and Hassid, A. (2005). Essential role of protein kinase G and decreased cytoplasmic Ca2+ levels in NO-induced inhibition of rat aortic smooth muscle cell motility. *Am. J. Physiol. Heart Circ. Physiol.* **288,** H1859–H1866.
89. Airhart, N., Yang, Y. F., Roberts, C. T., Jr., and Silberbach, M. (2003). Atrial natriuretic peptide induces natriuretic peptide receptor-cGMP-dependent protein kinase interaction. *J. Biol. Chem.* **278,** 38693–38698.
90. Fiedler, B., Lohmann, S. M., Smolenski, A., Linnemuller, S., Pieske, B., Schroder, F., Molkentin, J. D., Drexler, H., and Wollert, K. C. (2002). Inhibition of calcineurin-NFAT hypertrophy signaling by cGMP-dependent protein kinase type I in cardiac myocytes. *Proc. Natl. Acad. Sci. USA* **99,** 11363–11368.
91. Begum, N., Sandu, O. A., and Duddy, N. (2002). Negative regulation of rho signaling by insulin and its impact on actin cytoskeleton organization in vascular smooth muscle cells: role of nitric oxide and cyclic guanosine monophosphate signaling pathways. *Diabetes* **51,** 2256–2263.
92. Suzuki, Y. J., Nagase, H., Day, R. M., and Das, D. K. (2004). GATA-4 regulation of myocardial survival in the preconditioned heart. *J. Mol. Cell. Cardiol.* **37,** 1195–1203.
93. Gudi, T., Chen, J. C., Casteel, D. E., Seasholtz, T. M., Boss, G. R., and Pilz, R. B. (2002). cGMP-dependent protein kinase inhibits serum-response element-dependent transcription by inhibiting rho activation and functions. *J. Biol. Chem.* **277,** 37382–37393.
94. Gudi, T., Huvar, I., Meinecke, M., Lohmann, S. M., Boss, G. R., and Pilz, R. B. (1996). Regulation of gene expression by cGMP-dependent protein kinase. Transactivation of the c-fos promoter. *J. Biol. Chem.* **271,** 4597–4600.
95. Immenschuh, S., Hinke, V., Ohlmann, A., Gifhorn-Katz, S., Katz, N., Jungermann, K., and Kietzmann, T. (1998). Transcriptional activation of the haem oxygenase-1 gene by cGMP via a cAMP response element/activator protein-1 element in primary cultures of rat hepatocytes. *Biochem. J.* **334,** 141–146.
96. Mery, P. F., Lohmann, S. M., Walter, U., and Fischmeister, R. (1991). Ca2+ current is regulated by cyclic GMP-dependent protein kinase in mammalian cardiac myocytes. *Proc. Natl. Acad. Sci. USA* **88,** 1197–1201.
97. Kaye, D. M., Wiviott, S. D., and Kelly, R. A. (1999). Activation of nitric oxide synthase (NOS3) by mechanical activity alters contractile activity in a Ca2+-independent manner in cardiac myocytes: role of troponin I phosphorylation. *Biochem. Biophys. Res. Commun.* **256,** 398–403.
98. Layland, J., Li, J. M., and Shah, A. M. (2002). Role of cyclic GMP-dependent protein kinase in the contractile response to

exogenous nitric oxide in rat cardiac myocytes. *J. Physiol.* **540,** 457–467.

99. Becker, E. M., Schmidt, P., Schramm, M., Schroder, H., Walter, U., Hoenicka, M., Gerzer, R., and Stasch, J. P. (2000). The vasodilator-stimulated phosphoprotein (VASP): target of YC-1 and nitric oxide effects in human and rat platelets. *J. Cardiovasc. Pharmacol.* **35,** 390–397.

100. Sporbert, A., Mertsch, K., Smolenski, A., Haseloff, R. F., Schonfelder, G., Paul, M., Ruth, P., Walter, U., and Blasig, I. E. (1999). Phosphorylation of vasodilator-stimulated phosphoprotein: a consequence of nitric oxide- and cGMP-mediated signal transduction in brain capillary endothelial cells and astrocytes. *Brain Res. Mol. Brain Res.* **67,** 258–266.

101. Nicol, R. L., Frey, N., and Olson, E. N. (2000). From the sarcomere to the nucleus: role of genetics and signaling in structural heart disease. *Annu. Rev. Genomics Hum. Genet.* **1,** 179–223.

102. Chin, D., and Means, A. R. (2000). Calmodulin: a prototypical calcium sensor. *Trends Cell Biol.* **10,** 322–328.

103. Frey, N., McKinsey, T. A., and Olson, E. N. (2000). Decoding calcium signals involved in cardiac growth and function. *Nat. Med.* **6,** 1221–1227.

104. Zhang, T., Johnson, E. N., Gu, Y., Morissette, M. R., Sah, V. P., Gigena, M. S., Belke, D. D., Dillmann, W. H., Rogers, T. B., Schulman, H., Ross, J., Jr., and Brown, J. H. (2002). The cardiac-specific nuclear delta(B) isoform of Ca2+/calmodulin-dependent protein kinase II induces hypertrophy and dilated cardiomyopathy associated with increased protein phosphatase 2A activity. *J. Biol. Chem.* **277,** 1261–1267.

105. Zhu, W., Zou, Y., Shiojima, I., Kudoh, S., Aikawa, R., Hayashi, D., Mizukami, M., Toko, H., Shibasaki, F., Yazaki, Y., Nagai, R., and Komuro, I. (2000). Ca2+/calmodulin-dependent kinase II and calcineurin play critical roles in endothelin-1-induced cardiomyocyte hypertrophy. *J. Biol. Chem.* **275,** 15239–15245.

106. Liang, F., Wu, J., Garami, M., and Gardner, D. G. (1997). Mechanical strain increases expression of brain natriuretic peptide gene in rat cardiac myocytes. *J. Biol. Chem.* **272,** 28050–28056.

107. Zhang, T., Maier, L. S., Dalton, N. D., Miyamoto, S., Ross, J., Jr., Bers, D. M., and Brown, J. H. (2003). The deltaC isoform of CaMKII is activated in cardiac hypertrophy and induces dilated cardiomyopathy and heart failure. *Circ. Res.* **92,** 912–919.

108. Passier, R., Zeng, H., Frey, N., Naya, F. J., Nicol, R. L., McKinsey, T. A., Overbeek, P., Richardson, J. A., Grant, S. R., and Olson, E. N. (2000). CaM kinase signaling induces cardiac hypertrophy and activates the MEF2 transcription factor in vivo. *J. Clin. Invest.* **105,** 1395–1406.

109. Molkentin, J. D. (2004). Calcineurin-NFAT signaling regulates the cardiac hypertrophic response in coordination with the MAPKs. *Cardiovasc. Res.* **63,** 467–475.

110. Olson, E. N., and Molkentin, J. D. (1999). Prevention of cardiac hypertrophy by calcineurin inhibition: hope or hype? *Circ. Res.* **84,** 623–632.

111. Bueno, O. F., Brandt, E. B., Rothenberg, M. E., and Molkentin, J. D. (2002). Defective T cell development and function in calcineurin A beta-deficient mice. *Proc. Natl. Acad. Sci. USA* **99,** 9398–9403.

112. McKinsey, T. A., and Olson, E. N. (1999). Cardiac hypertrophy: sorting out the circuitry. *Curr. Opin. Genet. Dev.* **9,** 267–274.

113. Haq, S., Choukroun, G., Lim, H., Tymitz, K. M., del Monte, F., Gwathmey, J., Grazette, L., Michael, A., Hajjar, R., Force, T., and Molkentin, J. D. (2001). Differential activation of signal transduction pathways in human hearts with hypertrophy versus advanced heart failure. *Circulation* **103,** 670–677.

114. McKinsey, T. A., and Olson, E. N. (2005). Toward transcriptional therapies for the failing heart: chemical screens to modulate genes. *J. Clin. Invest.* **115,** 538–546.

115. Chen, M., Li, X., Dong, Q., Li, Y., and Liang, W. (2005). Neuropeptide Y induces cardiomyocyte hypertrophy via calcineurin signaling in rats. *Regul. Pept.* **125,** 9–15.

116. Pi, M., Oakley, R. H., Gesty-Palmer, D., Cruickshank, R. D., Spurney, R. F., Luttrell, L. M., and Quarles, L. D. (2005). Beta-arrestin- and G protein receptor kinase-mediated calcium-sensing receptor desensitization. *Mol. Endocrinol.* **19,** 1078–1087.

117. Hata, J. A., Williams, M. L., and Koch, W. J. (2004). Genetic manipulation of myocardial beta-adrenergic receptor activation and desensitization. *J. Mol. Cell. Cardiol.* **37,** 11–21.

118. Koch, W. J., Rockman, H. A., Samama, P., Hamilton, R. A., Bond, R. A., Milano, C. A., and Lefkowitz, R. J. (1995). Cardiac function in mice overexpressing the beta-adrenergic receptor kinase or a beta ARK inhibitor. *Science* **268,** 1350–1353.

119. Metaye, T., Gibelin, H., Perdrisot, R., and Kraimps, J. L. (2005). Pathophysiological roles of G-protein-coupled receptor kinases. *Cell Signal* **17,** 917–928.

120. Vinge, L. E., Oie, E., Andersson, Y., Grogaard, H. K., Andersen, G., and Attramadal, H. (2001). Myocardial distribution and regulation of GRK and beta-arrestin isoforms in congestive heart failure in rats. *Am. J. Physiol. Heart Circ. Physiol.* **281,** H2490–H2499.

121. Jones, W. K., Brown, M., Ren, X., He, S., and McGuinness, M. (2003). NF-kappaB as an integrator of diverse signaling pathways: the heart of myocardial signaling? *Cardiovasc. Toxicol.* **3,** 229–254.

122. Huss, J. M., and Kelly, D. P. (2004). Nuclear receptor signaling and cardiac energetics. *Circ. Res.* **95,** 568–578.

123. Chawla, A., Repa, J. J., Evans, R. M., and Mangelsdorf, D. J. (2001). Nuclear receptors and lipid physiology: Opening the X-files. *Science* **294,** 1866–1870.

124. van Bilsen, M., van der Vusse, G. J., Gilde, A. J., Lindhout, M., and van der Lee, K. A. (2002). Peroxisome proliferator-activated receptors: lipid binding proteins controlling gene expression. *Mol. Cell Biochem.* **239,** 131–138.

125. Brandt, J. M., Djouadi, F., and Kelly, D. P. (1998). Fatty acids activate transcription of the muscle carnitine palmitoyltransferase I gene in cardiac myocytes via the peroxisome proliferator-activated receptor alpha. *J. Biol. Chem.* **273,** 23786–23792.

126. Liu, H. R., Tao, L., Gao, E., Lopez, B. L., Christopher, T. A., Willette, R. N., Ohlstein, E. H., Yue, T. L., and Ma, X. L. (2004). Anti-apoptotic effects of rosiglitazone in hypercholesterolemic rabbits subjected to myocardial ischemia and reperfusion. *Cardiovasc. Res.* **62,** 135–144.

127. Muoio, D. M., MacLean, P. S., Lang, D. B., Li, S., Houmard, J. A., Way, J. M., Winegar, D. A., Corton, J. C., Dohm, G. L.,

and Kraus, W. E. (2002). Fatty acid homeostasis and induction of lipid regulatory genes in skeletal muscles of peroxisome proliferator-activated receptor (PPAR) alpha knock-out mice. Evidence for compensatory regulation by PPAR delta. *J. Biol. Chem.* **277,** 26089–26097.

128. Lehman, J. J., Barger, P. M., Kovacs, A., Saffitz, J. E., Medeiros, D. M., and Kelly, D. P. (2000). Peroxisome proliferator-activated receptor gamma coactivator-1 promotes cardiac mitochondrial biogenesis. *J. Clin. Invest.* **106,** 847–856.

129. Puigserver, P., and Spiegelman, B. M. (2003). Peroxisome proliferator-activated receptor-gamma coactivator 1 alpha (PGC-1 alpha): Transcriptional coactivator and metabolic regulator. *Endocr. Rev.* **24,** 78–90.

130. Clabby, M. L., Robison, T. A., Quigley, H. F., Wilson, D. B., and Kelly, D. P. (2003). Retinoid X receptor alpha represses GATA-4-mediated transcription via a retinoid-dependent interaction with the cardiac-enriched repressor FOG-2. *J. Biol. Chem.* **278,** 5760–5767.

131. Kahaly, G. J., and Dillmann, W. H. (2005). Thyroid hormone action in the heart. *Endocr. Rev.* **26,** 704–728.

132. Bahouth, S. W., Cui, X., Beauchamp, M. J., and Park, E. A. (1997). Thyroid hormone induces beta1-adrenergic receptor gene transcription through a direct repeat separated by five nucleotides. *J. Mol. Cell. Cardiol.* **29,** 3223–3237.

133. Chen, S., Nakamura, K., and Gardner, D. G. (2005). 1,25-dihydroxyvitamin D inhibits human ANP gene promoter activity. *Regul. Pept.* **128,** 197–202.

134. Sladek, R., Bader, J. A., and Giguere, V. (1997). The orphan nuclear receptor estrogen-related receptor alpha is a transcriptional regulator of the human medium-chain acyl coenzyme A dehydrogenase gene. *Mol. Cell. Biol.* **17,** 5400–5409.

135. Booth, E. A., Obeid, N. R., and Lucchesi, B. R. (2005). Activation of estrogen receptor alpha protects the in vivo rabbit heart from ischemia-reperfusion injury. *Am. J. Physiol. Heart Circ. Physiol.* **289,** H2039–H2047.

136. Jankowski, M., Rachelska, G., Donghao, W., McCann, S. M., and Gutkowska, J. (2001). Estrogen receptors activate atrial natriuretic peptide in the rat heart. *Proc. Natl. Acad. Sci. USA* **98,** 11765–11770.

137. Nuedling, S., Karas, R. H., Mendelsohn, M. E., Katzenellenbogen, J. A., Katzenellenbogen, B. S., Meyer, R., Vetter, H., and Grohe, C. (2001). Activation of estrogen receptor beta is a prerequisite for estrogen-dependent upregulation of nitric oxide synthases in neonatal rat cardiac myocytes. *FEBS Lett.* **502,** 103–108.

138. Yang, S. H., Liu, R., Perez, E. J., Wen, Y., Stevens, S. M., Jr., Valencia, T., Brun-Zinkernagel, A. M., Prokai, L., Will, Y., Dykens, J., Koulen, P., and Simpkins, J. W. (2004). Mitochondrial localization of estrogen receptor beta. *Proc. Natl. Acad. Sci. USA* 101, 4130–4135.

139. Romeih, M., Cui, J., Michaille, J. J., Jiang, W., and Zile, M. H. (2003). Function of RARgamma and RARalpha2 at the initiation of retinoid signaling is essential for avian embryo survival and for distinct events in cardiac morphogenesis. *Dev. Dyn.* **228,** 697–708.

140. Sato, A., Sheppard, K. E., Fullerton, M. J., and Funder, J. W. (1996). cAMP modulates glucocorticoid-induced protein accumulation and glucocorticoid receptor in cardiomyocytes. *Am. J. Physiol.* **271,** E827–E833.

141. Marsh, J. D., Lehmann, M. H., Ritchie, R. H., Gwathmey, J. K., Green, G. E., and Schiebinger, R. J. (1998). Androgen receptors mediate hypertrophy in cardiac myocytes. *Circulation* **98,** 256–261.

142. Perrier, E., Kerfant, B. G., Lalevee, N., Bideaux, P., Rossier, M. F., Richard, S., Gomez, A. M., and Benitah, J. P. (2004). Mineralocorticoid receptor antagonism prevents the electrical remodeling that precedes cellular hypertrophy after myocardial infarction. *Circulation* **110,** 776–783.

143. Le Menuet, D., Viengchareun, S., Muffat-Joly, M., Zennaro, M. C., and Lombes, M. (2004). Expression and function of the human mineralocorticoid receptor: lessons from transgenic mouse models. *Mol. Cell. Endocrinol.* **217,** 127–136.

144. Colombo, F., Gosselin, H., El-Helou, V., and Calderone, A. (2003). Beta-adrenergic receptor-mediated DNA synthesis in neonatal rat cardiac fibroblasts proceeds via a phosphatidyl-inositol 3-kinase dependent pathway refractory to the antiproliferative action of cyclic AMP. *J. Cell. Physiol.* **195,** 322–330.

145. Tuxworth, W. J., Jr., Saghir, A. N., Spruill, L. S., Menick, D. R., and McDermott, P. J. (2004). Regulation of protein synthesis by eIF4E phosphorylation in adult cardiocytes: the consequence of secondary structure in the 5-untranslated region of mRNA. *Biochem. J.* **378,** 73–82.

146. Pham, F. H., Sugden, P. H., and Clerk, A. (2000). Regulation of protein kinase B and 4E-BP1 by oxidative stress in cardiac myocytes. *Circ. Res.* **86,** 1252–1258.

147. Hannan, R. D., Luyken, J., and Rothblum, L. I. (1995). Regulation of rDNA transcription factors during cardiomyocyte hypertrophy induced by adrenergic agents. *J. Biol. Chem.* **270,** 8290–8297.

148. Hannan, K. M., Brandenburger, Y., Jenkins, A., Sharkey, K., Cavanaugh, A., Rothblum, L., Moss, T., Poortinga, G., McArthur, G. A., Pearson, R. B., and Hannan, R. D. (2003). mTOR-dependent regulation of ribosomal gene transcription requires S6K1 and is mediated by phosphorylation of the carboxy-terminal activation domain of the nucleolar transcription factor UBF. *Mol. Cell. Biol.* **23,** 8862–8877.

149. Lindemann, S. W., Weyrich, A. S., and Zimmerman, G. A. (2005). Signaling to translational control pathways: diversity in gene regulation in inflammatory and vascular cells. *Trends Cardiovasc. Med.* **15,** 9–17.

150. Flagg, T. P., and Nichols, C. G. (2005). Sarcolemmal K(ATP) channels: what do we really know? *J. Mol. Cell. Cardiol.* **39,** 61–70.

151. McLennan, H. R., and Degli Esposti, M. (2000). The contribution of mitochondrial respiratory complexes to the production of reactive oxygen species. *J. Bioenerg. Biomembr.* **32,** 153–162.

152. Pryor, W. A. (1986). Oxy-radicals and related species: Their formation, lifetimes, and reactions. *Annu. Rev. Physiol.* **48,** 657–667.

153. Han, D., Antunes, F., Canali, R., Rettori, D., and Cadenas, E. (2003). Voltage-dependent anion channels control the release of the superoxide anion from mitochondria to cytosol. *J. Biol. Chem.* **278,** 5557–5563.

154. Stadtman, E. R., and Berlett, B. S. (1998). Reactive oxygen-mediated protein oxidation in aging and disease. *Drug Metab. Rev.* **30,** 225–243.

155. Choksi, K. B., Boylston, W. H., Rabek, J. P., Widger, W. R., and Papaconstantinou, J. (2004). Oxidatively damaged pro-

teins of heart mitochondrial electron transport complexes. *Biochim. Biophys. Acta* **1688,** 95–101.
156. Vasquez-Vivar, J., Kalyanaraman, B., and Kennedy, M. C. (2000). Mitochondrial aconitase is a source of hydroxyl radical. An electron spin resonance investigation. *J. Biol. Chem.* **275,** 14064–14069.
157. Paradies, G., Petrosillo, G., Pistolese, M., and Ruggiero, F. M. (2002). Reactive oxygen species affect mitochondrial electron transport complex I activity through oxidative cardiolipin damage. *Gene* **286,** 135–141.
158. Petrosillo, G., Ruggiero, F. M., Pistolese, M., and Paradies, G. (2001). Reactive oxygen species generated from the mitochondrial electron transport chain induce cytochrome c dissociation from beef-heart submitochondrial particles via cardiolipin peroxidation: Possible role in the apoptosis. *FEBS Lett.* **509,** 435–438.
159. Wolin, M. S., Ahmad, M., and Gupte, S. A. (2005). Oxidant and redox signaling in vascular oxygen sensing mechanisms: basic concepts, current controversies, and potential importance of cytosolic NADPH. *Am. J. Physiol. Lung Cell. Mol. Physiol.* **289,** L159–L173.
160. Brown, G. C. (1999). Nitric oxide and mitochondrial respiration. *Biochim. Biophys. Acta* **1411,** 351–369.
161. Murray, J., Taylor, S. W., Zhang, B., Ghosh, S. S., and Capaldi, R. A. (2003). Oxidative damage to mitochondrial complex I due to peroxynitrite: Identification of reactive tyrosines by mass spectrometry. *J. Biol. Chem.* **278,** 37223–37230.
162. Riobo, N. A., Clementi, E., Melani, M., Boveris, A., Cadenas, E., Moncada, S, and Poderoso, J. J. (2001). Nitric oxide inhibits mitochondrial NADH:ubiquinone reductase activity through peroxynitrite formation. *Biochem. J.* **359,** 139–145.
163. Cassina, A. M., Hodara, R., Souza, J. M., Thomson, L., Castro, L., Ischiropoulos, H., Freeman, B. A., and Radi, R. (2000). Cytochrome c nitration by peroxynitrite. *J. Biol. Chem.* **275,** 21409–21415.
164. Castro, L., Rodriguez, M., and Radi, R. (1994). Aconitase is readily inactivated by peroxynitrite, but not by its precursor, nitric oxide. *J. Biol. Chem.* **269,** 29409–29415.
165. Packer, M. A., Scarlett, J. L., Martin, S. W., and Murphy, M. P. (1997). Induction of the mitochondrial permeability transition by peroxynitrite. *Biochem. Soc. Trans.* **25,** 909–914.
166. Brookes, P. S., and Darley-Usmar, V. M. (2004). Role of calcium and superoxide dismutase in sensitizing mitochondria to peroxynitrite-induced permeability transition. *Am. J. Physiol. Heart Circ. Physiol.* **286,** H39–H46.
167. Griendling, K. K., Minieri, C. A., Ollerenshaw, J. D., and Alexander, R. W. (1994). Angiotensin stimulates NADH and NADPH oxidase activity in cultured vascular smooth muscle cells. *Circ. Res.* **74,** 1141–1148.
168. De Keulenaer, G. W., Alexander, R. W., Ushio-Fukai, M., Ishizaka, N., and Griendling, K. K. (1998). Tumor necrosis factor activates a p22phox-based NADH oxidase in vascular smooth muscle. *Biochem. J.* **329,** 653–657.
169. Patterson, C., Ruef, J., Madamanchi, N. R., Barry-Lane, P., Hu, Z., Horaist, C., Ballinger, C. A., Brasier, A. R., Bode, C., and Runge, M. S. (1999). Stimulation of a vascular smooth muscle cell NAD(P)H oxidase by thrombin: evidence that p47phox may participate in forming this oxidase in vitro and in vivo. *J. Biol. Chem.* **274,** 19814–19822.
170. Hellsten-Westing, Y. (1993). Immunohistochemical localization of xanthine oxidase in human cardiac and skeletal muscle. *Histochemistry* **100,** 215–222.
171. Cappola, T. P., Kass, D. A., Nelson, G. S., Berger, R. D., Rosas, G. O., Kobeissi, Z. A., Marban, E., and Hare, J. M. (2001). Allopurinol improves myocardial efficiency in patients with idiopathic dilated cardiomyopathy. *Circulation* **104,** 2407–2411.
172. Nemoto, S., Takeda, K., Yu, Z. X., Ferrans, V. J., and Finkel, T. (2000). Role for mitochondrial oxidants as regulators of cellular metabolism. *Mol. Cell. Biol.* **20,** 7311–7318.
173. Cadenas, E. (2004). Mitochondrial free radical production and cell signaling. *Mol. Aspects Med.* **25,** 17–26.
174. Bogoyevitch, M. A., Ng, D. C., Court, N. W., Draper, K. A., Dhillon, A., and Abas, L. (2000). Intact mitochondrial electron transport function is essential for signalling by hydrogen peroxide in cardiac myocytes. *J. Mol. Cell. Cardiol.* **32,** 1469–1480.
175. Archer, S. L., Wu, X. C., Thebaud, B., Moudgil, R., Hashimoto, K., and Michelakis, E. D. (2004). O_2 sensing in the human ductus arteriosus: redox-sensitive K+ channels are regulated by mitochondria-derived hydrogen peroxide. *Biol. Chem.* **385,** 205–216.
176. Yamamura, T., Otani, H., Nakao, Y., Hattori, R., Osako, M., Imamura, H., and Das, D. K. (2001). Dual involvement of coenzyme Q10 in redox signaling and inhibition of death signaling in the rat heart mitochondria. *Antioxid. Redox. Signal* **3,** 103–112.
177. Brookes, P. S., Levonen, A. L., Shiva, S., Sarti, P., and Darley-Usmar, V. M . (2002). Mitochondria: Regulators of signal transduction by reactive oxygen and nitrogen species. *Free Radic. Biol. Med.* **33,** 755–764.
178. Boveris, A., D'Amico, G., Lores-Arnaiz, S., and Costa, L. E. (2003). Enalapril increases mitochondrial nitric oxide synthase activity in heart and liver. *Antioxid. Redox. Signal* **5,** 691–697.
179. O'Rourke, B. (2000). Myocardial KATP channels in preconditioning. *Circ. Res.* **87,** 845–855.
180. Lebuffe, G., Schumacker, P. T., Shao, Z. H., Anderson, T., Iwase, H., and Vanden Hoek, T. L. (2003). ROS and NO trigger early preconditioning: relationship to mitochondrial KATP channel. *Am. J. Physiol. Heart Circ. Physiol.* **284,** H299–H308.
181. Oldenburg, O., Cohen, M. V., Yellon, D. M., and Downey, J. M. (2002). RMitochondrial K(ATP) channels: role in cardioprotection. *Cardiovasc. Res.* **55,** 429–437.
182. Ardehali, H., and O'Rourke, B. (2005). Mitochondrial K(ATP) channels in cell survival and death. *J. Mol. Cell. Cardiol.* **39,** 7–16.
183. Garlid, K. D., Dos Santos, P., Xie, Z. J., Costa, A. D, and Paucek, P. (2003). Mitochondrial potassium transport: the role of the mitochondrial ATP-sensitive K(+) channel in cardiac function and cardioprotection. *Biochim. Biophys. Acta* **1606,** 1–21.
184. Das, M., Parker, J. E., and Halestrap, A. P. (2003). Matrix volume measurements challenge the existence of diazoxide/glibencamide-sensitive KATP channels in rat mitochondria. *J. Physiol.* **547,** 893–902.
185. Akao, M., Teshima, Y., and Marban, E. (2002). Antiapoptotic effect of nicorandil mediated by mitochondrial atp-sensitive

potassium channels in cultured cardiac myocytes. *J. Am. Coll. Cardiol.* **40,** 803–810.
186. Nagata, K., Obata, K., Odashima, M., Yamada, A., Somura, F., Nishizawa, T., Ichihara, S., Izawa, H., Iwase, M., Hayakawa, A., Murohara, T., and Yokota, M. (2003). Nicorandil inhibits oxidative stress-induced apoptosis in cardiac myocytes through activation of mitochondrial ATP-sensitive potassium channels and a nitrate-like effect. *J. Mol. Cell. Cardiol.* **35,** 1505–1512.
187. Xu, W., Liu, Y., Wang, S., McDonald, T., Van Eyk, J. E., Sidor, A., and O'Rourke, B. (2002). Cytoprotective role of Ca2+- activated K+ channels in the cardiac inner mitochondrial membrane. *Science* 298,1029–1033.
188. Hanley, P. J., and Daut, J. (2005). K(ATP) channels and preconditioning: a re-examination of the role of mitochondrial K(ATP) channels and an overview of alternative mechanisms. *J. Mol. Cell. Cardiol.* **39,** 17–50.
189. Halestrap, A. P., Clarke, S. J., and Javadov, S. A. (2004). Mitochondrial permeability transition pore opening during myocardial reperfusion–a target for cardioprotection. *Cardiovasc. Res.* **61,** 372–385.
190. Shanmuganathan, S., Hausenloy, D. J., Duchen, M. R., and Yellon, D. M. (2005). Mitochondrial permeability transition pore as a target for cardioprotection in the human heart. *Am. J. Physiol. Heart Circ. Physiol.* **289,** H237–H242.
191. Thomson, M. (2002). Evidence of undiscovered cell regulatory mechanisms: phospho-proteins and protein kinases in mitochondria. *Cell Mol. Life Sci.* **59,** 213–219.
192. Sugden, M. C., Orfali, K. A., Fryer, L. G., Holness, M. J., and Priestman, D. A. (1997). Molecular mechanisms underlying the long-term impact of dietary fat to increase cardiac pyruvate dehydrogenase kinase: regulation by insulin, cyclic AMP and pyruvate. *J. Mol. Cell. Cardiol.* **29,** 1867–1875.
193. Technikova-Dobrova, Z., Sardanelli, A. M., Stanca, M. R., and Papa, S. (1994). cAMP-dependent protein phosphorylation in mitochondria of bovine heart. *FEBS Lett.* **350,** 187–191.
194. Wang, Y., Hirai, K., and Ashraf, M. (1999). Activation of mitochondrial ATP-sensitive K(+) channel for cardiac protection against ischemic injury is dependent on protein kinase C activity. *Circ. Res.* **85,** 731–741.
195. Baines, C. P., Zhang, J., Wang, G. W., Zheng, Y. T., Xiu, J. X., Cardwell, E. M., Bolli, R., and Ping, P. (2002). Mitochondrial PKCepsilon and MAPK form signaling modules in the murine heart: enhanced mitochondrial PKCepsilon-MAPK interactions and differential MAPK activation in PKCepsilon-induced cardioprotection. *Circ. Res.* **90,** 390–397.
196. Garlid, K. D., Costa, A. D., Cohen, M. V., Downey, J. M., and Critz, S. D. (2004). Cyclic GMP and PKG activate mito K(ATP) channels in isolated mitochondria. *Cardiovasc. J. S. Afr.* **15,** S5.
197. He, H., Li, H. L., Lin, A., and Gottlieb, R. A. (1999). Activation of the JNK pathway is important for cardiomyocyte death in response to simulated ischemia. *Cell Death Differ.* **6,** 987–991.
198. Aoki, H., Kang, P. M., Hampe, J., Yoshimura, K., Noma, T., Matsuzaki, M., and Izumo, S. (2002). Direct activation of mitochondrial apoptosis machinery by c-Jun N-terminal kinase in adult cardiac myocytes. *J. Biol. Chem.* **277,** 10244–10250.
199. Court, N. W., Kuo, I., Quigley, O., and Bogoyevitch, M. A. (2004). Phosphorylation of the mitochondrial protein Sab by stress-activated protein kinase 3. *Biochem. Biophys. Res. Commun.* **319,** 130–137.
200. Baines, C. P., Song, C. X., Zheng, Y. T., Wang, G. W., Zhang, J., Wang, O. L., Guo, Y., Bolli, R., Cardwell, E. M., and Ping, P. (2003). Protein kinase Cepsilon interacts with and inhibits the permeability transition pore in cardiac mitochondria. *Circ. Res.* **92,** 873–880.
201. Papa, S. (2002). The NDUFS4 nuclear gene of complex I of mitochondria and the cAMP cascade. *Biochim. Biophys. Acta* **1555,** 147–153.
202. Lee, I., Bender, E., and Kadenbach, B. (2002). Control of mitochondrial membrane potential and ROS formation by reversible phosphorylation of cytochrome c oxidase. *Mol. Cell. Biochem.* **234–235,** 63–70.
203. Schulenberg, B., Aggeler, R., Beechem, J. M., Capaldi, R. A., and Patton, W. F. (2003). Analysis of steady-state protein phosphorylation in mitochondria using a novel fluorescent phosphosensor dye. *J. Biol. Chem.* **278,** 27251–27255.
204. Hood, D. A., and Joseph, A. M. (2004). Mitochondrial assembly: protein import. *Proc. Nutr. Soc.* 63, 293–300.
205. Colavecchia, M., Christie, L. N., Kanwar, Y. S., and Hood, D. A. (2003). Functional consequences of thyroid hormone-induced changes in the mitochondrial protein import pathway. *Am. J. Physiol. Endocrinol. Metab.* **284,** E29–E35.
206. Biswas, G., Guha, M., and Avadhani, N. G. (2005). Mitochondria-to-nucleus stress signaling in mammalian cells: Nature of nuclear gene targets, transcription regulation, and induced resistance to apoptosis. *Gene* 354, 132–139.
207. Berridge, M. J. (2002). The endoplasmic reticulum: a multifunctional signaling organelle. *Cell Calcium* **32,** 235–249.
208. Crow, M. T., Mani, K., Nam, Y. J., and Kitsis, R. N. (2004). The mitochondrial death pathway and cardiac myocyte apoptosis. *Circ. Res.* **95,** 957–970.
209. Danial, N. N., and Korsmeyer, S. J. (2004). Cell death: Critical control points. *Cell* **116,** 205–219.
210. Epand, R. F., Martinou, J. C., Montessuit, S., Epand, R. M., and Yip, C. M. (2002). Direct evidence for membrane pore formation by the apoptotic protein Bax. *Biochem. Biophys. Res. Commun.* **298,** 744–749.
211. Chipuk, J. E., Kuwana, T., Bouchier-Hayes, L., Droin, N. M., Newmeyer, D. D., Schuler, M., and Green, D. R. (2004). Direct activation of Bax by p53 mediates mitochondrial membrane permeabilization and apoptosis. *Science* **303,** 1010–1014.
212. Regula, K. M., Ens, K., and Kirshenbaum, L. A. (2003). Mitochondria-assisted cell suicide: A license to kill. *J. Mol. Cell. Cardiol.* **35,** 559–567.
213. Kroemer, G. (2003). Mitochondrial control of apoptosis: An introduction. *Biochem. Biophys. Res. Commun.* **304,** 433–435.
214. Scorrano, L., Ashiya, M., Buttle, K., Weiler, S., Oakes, S., Mannella, C. A., and Korsmeyer, S. J. (2002). A distinct pathway remodels mitochondrial cristae and mobilizes cytochrome c during apoptosis. *Dev. Cell* **2,** 55–67.
215. Belzacq, A. S., Vieira, H. L., Verrier, F., Vandecasteele, G., Cohen, I., Prevost, M. C., Larquet, E., Pariselli, F., Petit, P. X., Kahn, A., Rizzuto, R., Brenner, C., and Kroemer, G. (2003). Bcl-2 and Bax modulate adenine nucleotide translocase activity. *Cancer Res.* **63,** 541–546.
216. Engel, F. B., Schebesta, M., Duong, M. T., Lu, G., Ren, S., Madwed, J. B., Jiang, H., Wang, Y., and Keating, M. T.

(2005). p38 MAP kinase inhibition enables proliferation of adult mammalian cardiomyocytes. *Genes Dev.* **19**, 1175–1187.

217. Tamamori-Adachi, M., Ito, H., Sumrejkanchanakij, P., Adachi, S., Hiroe, M., Shimizu, M., Kawauchi, J., Sunamori, M., Marumo, F., Kitajima, S., and Ikeda, M. A. (2003). Critical role of cyclin D1 nuclear import in cardiomyocyte proliferation. *Circ. Res.* **92**, e12–e19.

218. Pasumarthi, K. B., Kardami, E., and Cattini, P. A. (1996). High and low molecular weight fibroblast growth factor-2 increase proliferation of neonatal rat cardiac myocytes but have differential effects on binucleation and nuclear morphology. Evidence for both paracrine and intracrine actions of fibroblast growth factor-2. *Circ. Res.* **78**, 126–136.

219. Sheikh, F., Jin, Y., Pasumarthi, K. B., Kardami, E., and Cattini, P. A. (1997). Expression of fibroblast growth factor receptor-1 in rat heart H9c2 myoblasts increases cell proliferation. *Mol. Cell Biochem.* **176**, 89–97.

220. Sheng, Z., Pennica, D., Wood, W. I., and Chien, K. R. (1996). Cardiotrophin-1 displays early expression in the murine heart tube and promotes cardiac myocyte survival. *Development* **122**, 419–428.

221. Kuwahara, K., Saito, Y., Kishimoto, I., Miyamoto, Y., Harada, M., Ogawa, E., Hamanaka, I., Kajiyama, N., Takahashi, N., Izumi, T., Kawakami, R., and Nakao, K. (2000). Cardiotrophin-1 phosphorylates akt and BAD, and prolongs cell survival via a PI3K-dependent pathway in cardiac myocytes. *J. Mol. Cell. Cardiol.* **32**, 1385–1394.

222. Condorelli, G., Drusco, A., Stassi, G., Bellacosa, A., Roncarati, R, Iaccarino, G., Russo, M. A., Gu, Y., Dalton, N., Chung, C, Latronico, M. V., Napoli, C., Sadoshima, J., Croce, C. M., and Ross, J., Jr. (2002). Akt induces enhanced myocardial contractility and cell size in vivo in transgenic mice. *Proc. Natl. Acad. Sci. USA* **99**, 12333–12338.

223. Liu, T. J., Lai, H. C., Wu, W., Chinn, S., and Wang, P. H. (2001). Developing a strategy to define the effects of insulin-like growth factor-1 on gene expression profile in cardiomyocytes. *Circ. Res.* **88**, 1231–1238.

224. Cook, S. A., Matsui, T., Li, L., and Rosenzweig, A. (2002). Transcriptional effects of chronic Akt activation in the heart. *J. Biol. Chem.* **277**, 22528–22533.

225. Bialik, S., Cryns, V. L., Drincic, A., Miyata, S., Wollowick, A. .L, Srinivasan, A., and Kitsis, R. N. (1999). The mitochondrial apoptotic pathway is activated by serum and glucose deprivation in cardiac myocytes. *Circ. Res.* **85**, 403–414.

226. Edinger, A. L., and Thompson, C. B. (2002). Akt maintains cell size and survival by increasing mTOR-dependent nutrient uptake. *Mol. Biol. Cell.* **13**, 2276–2288.

227. Juhaszova, M., Zorov, D. B., Kim, S. H., Pepe, S., Fu, Q., Fishbein, K. W., Ziman, B. D., Wang, S., Ytrehus, K., Antos, C. L., Olson, E. N., and Sollott, S. J. (2004). Glycogen synthase kinase-3beta mediates convergence of protection signaling to inhibit the mitochondrial permeability transition pore. *J. Clin. Invest.* **113**, 1535–1549.

228. Chandel, N. S., and Schumacker, P. T. (2000). Cellular oxygen sensing by mitochondria: old questions, new insight. *J. Appl. Physiol.* **88**, 1880–1889.

229. Waypa, G. B., Marks, J. D., Mack, M. M., Boriboun, C., Mungai, P. T., and Schumacker, P. T. (2002). Mitochondrial reactive oxygen species trigger calcium increases during hypoxia in pulmonary arterial myocytes. *Circ. Res.* **91**, 719–726.

230. Duranteau, J., Chandel, N. S., Kulisz, A., Shao, Z., and Schumacker, P. T. (1998). Intracellular signaling by reactive oxygen species during hypoxia in cardiomyocytes. *J. Biol. Chem.* **273**, 11619–11624.

231 Kacimi, R., Long, C. S., and Karliner, J. S. (1997). Chronic hypoxia modulates the interleukin-1β stimulated inducible nitric oxide synthase pathway in cardiac myocytes. *Circulation* **96**, 1937–1943.

232. French, S., Giulivi, C., and Balaban, R. S. (2001). Nitric oxide synthase in porcine heart mitochondria: evidence for low physiological activity. *Am. J. Physiol. Heart Circ. Physiol.* **280**, H2863–H2867.

233. Kulisz, A., Chen, N., Chandel, N. S., Shao, .Z, and Schumacker, P. T. (2002). Mitochondrial ROS initiate phosphorylation of p38 MAP kinase during hypoxia in cardiomyocytes. *Am. J. Physiol. Lung Cell. Mol. Physiol.* **282**, L1324–L1329.

234. Enomoto, N., Koshikawa, N., Gassmann, M., Hayashi, J., and Takenaga, K. (2002). Hypoxic induction of hypoxia-inducible factor-1alpha and oxygen-regulated gene expression in mitochondrial DNA-depleted HeLa cells. *Biochem. Biophys. Res. Commun.* **297**, 346–352.

235. Lopaschuk, G. D., Collins-Nakai, R. L., and Itoi, T. (1992). Developmental changes in energy substrate use by the heart. *Cardiovasc. Res.* **26**, 1172–1180.

236. Tripp, M. E. (1989). Developmental cardiac metabolism in health and disease. *Pediatr. Cardiol.* **10**, 150–158.

237. Bonnet, D., Martin, D., De Lonlay, P., Villian, E., Jouvet, P., Rabier, D., Brivet, M., and Saudubray, J. M. (1999). Arrhythmias and conduction defects as presenting symptoms of fatty acid oxidation disorders in children. *Circulation* **100**, 2248–2253.

238. Sparagna, G. C., Hickson-Bick, D. L., Buja, L. M., and McMillin, J. B. (2001). Fatty acid-induced apoptosis in neonatal cardiomyocytes: redox signaling. *Antioxid. Redox. Signal* **3**, 71–79.

239. Lanni, A., De Felice, M., Lombardi, A., Moreno, M., Fleury, C., Ricquier, D., and Goglia, F. (1997). Induction of UCP2 mRNA by thyroid hormones in rat heart. *FEBS Lett.* **418**, 171–174.

240. Boehm, E. A., Jones, B. E., Radda, G. K., Veech, R. L., and Clarke, K. (2001). Increased uncoupling proteins and decreased efficiency in palmitate-perfused hyperthyroid rat heart. *Am. J. Physiol Heart Circ. Physiol.* **280**, H977–H983.

241. Young, M. E., Patil, S., Ying, J., Depre, C., Ahuja, H. S., Shipley, G. L., Stepkowski, S. M., Davies, P. J., and Taegtmeyer, H. (2001). Uncoupling protein 3 transcription is regulated by peroxisome proliferator-activated receptor (alpha) in the adult rodent heart. *FASEB J.* **15**, 833–845.

242. Spinale, F. G. (2002). Bioactive peptide signaling within the myocardial interstitium and the matrix metalloproteinases. *Circ. Res.* **91**, 1082–1084.

243. Ross, R. S., and Borg, T. K. (2001). Integrins and the myocardium. *Circ. Res.* **88**, 1112–1119.

244. Iwami, K., Ashizawa, N., Do, Y. S., Graf, K, and Hsueh, W. A. (1996). Comparison of ANG II with other growth factors on Egr-1 and matrix gene expression in cardiac fibroblasts. *Am. J. Physiol.* **270**, H2100–H2107.

245. Temsah, R., and Nemer, M. (2005). GATA factors and transcriptional regulation of cardiac natriuretic peptide genes. *Regul. Pept.* **128,** 177–185.
246. Nemer, G., and Nemer, M. (2001). Regulation of heart development and function through combinatorial interactions of transcription factors. *Ann. Med.* **33,** 604–610.
247. Kim, T. G., Chen, J., Sadoshima, J., and Lee, Y. (2004). Jumonji represses atrial natriuretic factor gene expression by inhibiting transcriptional activities of cardiac transcription factors. *Mol. Cell Biol.* **24,** 10151–10160.
248. Small, E. M., and Krieg, P. A. (2003). Transgenic analysis of the atrial natriuretic factor (ANF) promoter: Nkx2-5 and GATA-4 binding sites are required for atrial specific expression of ANF. *Dev. Biol.* **261,** 116–131.
249. Pu, W. T., Ishiwata, T., Juraszek, A. L., Ma, Q., and Izumo, S. (2004). GATA4 is a dosage-sensitive regulator of cardiac morphogenesis. *Dev. Biol.* **275,** 235–244.
250. Heicklen-Klein, A., McReynolds, L. J., and Evans, T. (2005). Using the zebrafish model to study GATA transcription factors. *Semin. Cell Dev. Biol.* **16,** 95–106.
251. Kostetskii, I., Jiang, Y., Kostetskaia, E., Yuan, S., Evans, T., and Zile, M. (1999). Retinoid signaling required for normal heart development regulates GATA-4 in a pathway distinct from cardiomyocyte differentiation. *Dev. Biol.* **206,** 206–218.
252. Pehlivan, T., Pober, B. R., Brueckner, M., Garrett, S., Slaugh, R., Van Rheeden, R., Wilson, D. B., Watson, M. S., and Hing, A. V. (1999). GATA4 haploinsufficiency in patients with interstitial deletion of chromosome region 8p23.1 and congenital heart disease. *Am. J. Med. Genet.* **83,** 201–206.
253. Hirayama-Yamada, K., Kamisago, M., Akimoto, K., Aotsuka, H., Nakamura, Y., Tomita, H., Furutani, M., Imamura, S., Takao, A., Nakazawa, M., and Matsuoka, R. (2005). Phenotypes with GATA4 or NKX2.5 mutations in familial atrial septal defect. *Am. J. Med. Genet. A* **135,** 47–52.
254. Kelley, C., Blumberg, H., Zon, L. I., and Evans, T. (1993). GATA-4 is a novel transcription factor expressed in endocardium of the developing heart. *Development* **118,** 817–827.
255. Zeisberg, E. M., Ma, Q., Juraszek, A. L., Moses, K., Schwartz, R. J., Izumo, S., and Pu, W. T. (2005). Morphogenesis of the right ventricle requires myocardial expression of Gata4. *J. Clin. Invest.* **115,** 1522–1531.
256. Kawamura, T., Ono, K., Morimoto, T., Wada, H., Hirai, M., Hidaka, K., Morisaki, T., Heike, T., Nakahata, T., Kita, T., and Hasegawa, K. (2005). Acetylation of GATA-4 is involved in the differentiation of embryonic stem cells into cardiac myocytes. *J. Biol. Chem.* **280,** 19682–19688.
257. Dai, Y. S., and Markham, B. E. (2001). p300 functions as a coactivator of transcription factor GATA-4. *J. Biol. Chem.* **276,** 37178–37185.
258. Crispino, J. D., Lodish, M. B., Thurberg, B. L., Litovsky, S. H., Collins, T., Molkentin, J. D., and Orkin, S. H. (2001). Proper coronary vascular development and heart morphogenesis depend on interaction of GATA-4 with FOG cofactors. *Genes Dev.* **15,** 839–844.
259. Morin, S., Paradis, P., Aries, A., and Nemer, M. (2001). Serum response factor-GATA ternary complex required for nuclear signaling by a G-protein-coupled receptor. *Mol. Cell Biol.* **21,** 1036–1044.
260. Monzen, K., Shiojima, I., Hiroi, Y., Kudoh, S., Oka, T., Takimoto, E., Hayashi, D., Hosoda, T., Habara-Ohkubo, A., Nakaoka, T., Fujita, T., Yazaki, Y., and Komuro, I. (1999). Bone morphogenetic proteins induce cardiomyocyte differentiation through the mitogen-activated protein kinase kinase kinase TAK1 and cardiac transcription factors Csx/Nkx-2.5 and GATA-4. *Mol. Cell Biol.* **19,** 7096–7105.
261. Ventura, C., Zinellu, E., Maninchedda, E., and Maioli, M. (2003). Dynorphin B is an agonist of nuclear opioid receptors coupling nuclear protein kinase C activation to the transcription of cardiogenic genes in GTR1 embryonic stem cells. *Circ. Res.* **92,** 623–629.

SECTION II

Pediatric Cardiology in the Post-Genomic Era

CHAPTER **6**

Cardiac Development: Molecular and Genetic Analysis

OVERVIEW

Congenital heart disease (CHD) occurs in 1 of every 120 infants born each year, making it the most frequent form of birth defects in humans. Their causes are still an enigma with an imprecise number of factors; morphogenetic and hemodynamic events are apparently involved. Although our understanding of the pathology has grown rapidly in recent years, the basic underlying mechanisms of many specific pediatric cardiovascular diseases remain largely obscure. Given the manifold technical breakthroughs associated with the sequencing of the human genome, databases, and a number of model organisms, research has begun to provide increased understanding of specific molecular defects and to identify the specific "players" that contribute to the cardiac disorders, many of which are a consequence of defects in the development of the heart.

INTRODUCTION

A functional heart is of necessity the first organ to form for continuing embryonic organogenesis. The formation of the heart can be viewed as a process involving the synthesis and integration of several precisely patterned modular elements (e.g., atria, ventricles, septa, and valves). Explaining the pathways involved in both the genesis and integration of the individual modular elements in the developing heart is essential in furthering our understanding of CHD, which largely results from defects in specific structural components of the developing heart. Recently, it has become established that formation and integration of myocardial modular elements requires the complex interplay of a multitude of genes whose cell type-specific expression is highly organized and precisely regulated at the spatial and temporal levels by numerous transcription factors (Fig. 1). These critical regulatory factors can operate differently in various cell types and respond to a variety of intracellular and extracellular signals to modulate the precise integration of gene expression and morphological development. Their importance is underscored by the fact that most of the mutations thus far identified, which lead to specific congenital heart defects, reside in genes encoding transcription factors.

OVERVIEW OF CARDIAC MORPHOGENESIS

Gastrulation with its series of cell movements transforms the bilaminar germ disc (epiblast and hypoblast) into a three-layered embryo (ectoderm, mesoderm, and endoderm). These three germ layers and cells also become committed to endodermal or mesodermal lineages during this process. The different destiny of the mesodermal cell populations are determined by (1) the point of entrance of the epiblast cells into the primitive streak and (2) the direction of their subsequent migration. On the basis of these two events, mesodermal cells can form tissues as varied as muscle, heart, kidney, or bone.

The heart originates from the splanchnic mesoderm. It is formed from two crescent-like plates that through a process of differential growth are transformed into a tubular structure. As shown in Fig. 1, the overall developmental stages of the heart encompass:

1. A tubular stage, Carnegie stage 9—human: embryonic day (E) 20; mouse: E8; rat: E10; chicken: stage 10 (Hamburger and Hamilton (H/H)
2. A looped, segmented stage, Carnegie stage 12—human: E28; mouse: E10.5; rat: E12; chicken: stage 18 (H/H)
3. The septated fetal heart stage, Carnegie stage 23—human: E60; mouse: E16; rat: E17.5; chicken: stage 38 (H/H)

Before folding of the tubular embryo and approximately on day 18 or 19, angiogenic cell clusters on either side of the neural crest fuse to initially form capillaries and later blood vessels on either side of the neural tube joining at their cranial end. Progressively, the tubular heart becomes constricted in several regions delineating the future heart structures (i.e., the *bulbus cordis* that extends into the *truncus arteriosus*, the primitive ventricle, and atrium). Looping of the primitive heart occurs approximately on day 23 of

Post-Genomic Cardiology

Cardiac crescent →	Linear heart tube →	Looping heart →	Chamber formation →	Chamber maturation/ septation/valve formation
Nkx2.5	GATA4	Nkx2.5	Nkx2.5	Nkx2.5
Myocardin	Mesp1	MEF2c	Irx4	RXR-a
SRF	Mesp2	Tbx5	Tbx5	Tbx5
Mesp1		dHAND	dHAND	NFATc3
Mesp2		eHAND	eHAND	GATA4
		Irx4	Pitx2	Fog-2
			Smad6	ErbB
				Sox4
				HF-1B
				PitX2
				CITED2
				PAX3
Mouse: E7.5	E8	E8.5-9.5	E10-12	E12-birth

FIGURE 1 Formation of the heart. The heart is formed from two crescent-like plates, which through a process of differential growth are transformed into a tubular structure, followed by looping, chamber formation and maturation, septation, and valve formation. Factors responsible for each stage are shown.

embryo development.[1] Subsequently, the heart loops toward the right side, the bulboventricular groove becomes visible, and the atrial and ventricular segments become recognizable. The atrioventricular (AV) valves form during the fifth to eighth week of development.[2] The AV junction has two masses of endocardial cushions (ECs) that will meet in the middle, dividing the common AV canal into a right and a left AV orifice. Significantly, the AV cushions play a crucial role in the closure of the interatrial communication at the edge of the atrial *septum primum*. This septum grows toward the AV endocardial cushion and fuses with it.[3] With further development, septation of the ventricles starts approximately on day 30 and the atria approximately on day 35.[4] Right and left components can be recognized in the atrial and ventricular segments. The latest event concerning septation of the heart occurs in the outflow tract region, where two endocardial ridges fuse and each ventricle, right and left, acquires an independent outlet connection. At the end of the septation process, part of the outflow tract myocardium disappears, and part becomes incorporated into the right ventricular infundibulum[5]; subsequently, the AV canal and the inflow tract finally become incorporated into the atrial cavities.

The primary myocardium, found in the early heart tube, gives rise to the contracting myocardium (of the atria and ventricles) and the conducting myocardium (nodal and ventricular conducting tissue). The factors that direct the development of the myocardium phenotype, either as a pumping myocardium or a conduction system, are not known. Conducting myocardial tissue is frequently referred to as being a highly specialized tissue, implying that it has a uniform function. Actually, some portions, such as the nodal tissue, are slow conducting and represent a less developed primary myocardium, whereas other portions, such as the ventricular conduction tissue, are fast conducting.[6] The formation of the ventricular conduction system is better known than the sinus and AV nodal tissue, whose embryological origin and formation have not been clearly established.[7]

MOLECULAR PLAYERS IN HEART DEVELOPMENT

Role of Gene Expression

During embryonic development, the genesis of somatic, visceral and heart muscle are all derived from mesoderm progenitor cells, and they require the coordinated participation of multiple genes and signaling pathways. Subsequently, several distinctive transcriptional programs, as seen in the patterning of expression of a number of genes, will direct the compartmentalization of the forming heart.

Cell Differentiation and Mesoderm Development

Although the initiation of cardiac differentiation has been a topic of extensive investigation, no single transcription factor has yet been identified that is solely responsible for the differentiation of lateral plate mesoderm into cardiac cells. Rather, several factors seem to play a central role.

The *tinman* gene, originally discovered in the fruit fly *Drosophila melanogaster*, encodes an NK-class homeodomain-containing transcription factor that is required for the generation of myocardial cells in *Drosophila*.[8] In vertebrates, its homolog, the cardiac gene *Nkx2.5*, is crucial in cardiac differentiation processes, including the establishment or maintenance of a ventricular gene expression program. Its very early pattern of expression in the developing myocardium suggested that Nkx2.5 was a primary regulator of cardiogenesis. However, *Nkx2.5* is not an essential gene for the specification of heart cell lineage, nor is it required for heart tube formation (which is normal in Nkx2.5-deficient mice); however, it is required for completion of the looping of the heart.[9]

Factors that regulate the expression of Nkx2.5 (as well as other transcriptional factors) are important in early cardiac differentiation. Serum-response factor (SRF) has been shown to be an obligatory transcription factor required for the formation of vertebrate mesoderm leading to the origin of the cardiovascular system. SRF is necessary for serum induction of immediate early genes such as c-fos and for the downstream expression of many muscle and cardiac-specific genes.

In transgenic mice, the difficulty of studying factors such as SRF, which are required for mesoderm formation, is formidable, because null alleles for these factors typically result in embryonic lethality.

This limitation has been overcome by generating a conditional mutant of SRF using a Cre-LoxP strategy.[10] Mice containing a heart-specific deletion of SRF displayed lethal cardiac defects between embryonic day 10.5 (E10.5) and E13.5, as indicated by abnormally thin myocardium, dilated cardiac chambers, poor trabeculation, and a disorganized interventricular septum; this phenotype is preceded at E9.5 (before overt maldevelopment) by a marked reduction in the expression of transcription regulators including Nkx2.5, GATA4, and myocardin. In addition, SRF null embryonic stem cells used as a model system to investigate the specification of multiple embryonic lineages, including cardiac myocytes, exhibited absence of myogenic α-actin and myocardin expression and failure to form beating cardiac myocytes.[11]

Myocardin belongs to the SAP (SAF-A/B, Acinus, PIAS) domain family of nuclear proteins and activates cardiac muscle promoters by association with SRF. Experiments in Xenopus embryos using a dominant-negative myocardin molecule indicate that it may be necessary for early stages of cardiac differentiation, including high-level expression of Nkx2-5.[12] The myocardin-related family of transcription factors play a critical role in the transcriptional program regulating smooth muscle cell (SMC) differentiation. The function of myocardin-related transcription factor (MRTF)-B has been examined in mice harboring a conditional insertional mutation.[13] These studies found that MRTF-B plays a critical role in regulating differentiation of cardiac neural crest cells into smooth muscle and showed that neural crest–derived smooth muscle differentiation is specifically required for normal cardiovascular morphogenesis. A role for myocardin in cardiac differentiation is also supported by recent observations in which inhibition of myocardin function in the teratocarcinoma cell line P19CL6 prevented differentiation into cardiac myocytes.[14] On the other hand, forced expression of myocardin was not found to be sufficient for induction of SMC differentiation in multipotential embryonic cells, because overexpression of myocardin induced only a subset of SMC marker genes.[15]

Recently, a novel SRF cofactor, called p49/STRAP, for SRF-dependent transcription regulation–associated protein, has been identified.[16] This protein interacts mainly with the transcriptional activation domain of the SRF protein and binds to SRF or to the complex of SRF and another cofactor, such as myocardin or Nkx2.5. The expression of p49/STRAP differentially affects the promoter activity of SRF target genes, activating MLC2v and cardiac actin promoters, on cotransfection with SRF, while repressing atrial natriuretic factor (ANF) gene expression, which was strongly induced by myocardin. Although strongly expressed in the fetal heart, the role of this cofactor in early cardiac differentiation has not yet been determined.

The GATA family of zinc finger–containing transcription factors makes up another important group of transcriptional factors that seem to contribute to the activation of the cardiac-specific gene program involved in cardiac cell differentiation. Expression of three GATA family genes has been identified in the developing heart: GATA4, GATA5, and GATA6. A characteristic feature shared by the GATA factors (Fig. 2) is a central domain composed of two adjacent zinc fingers; the N–terminal zinc finger is involved in binding of some protein cofactors (e.g., FOG), whereas the C-terminal zinc finger contains the DNA sequence recognition domain (binding to specific promoters) and the binding site for most cofactors (e.g., dHAND, MEF2, Nkx2.5, p300, and SRF).[17]

GATA family members have been implicated as key regulators of cardiogenesis in several model systems.[18] Transgenic mice with inactivation of the GATA4 gene die during embryonic development because of failure of ventral morphogenesis and heart tube formation.[19] Embryos of mice homozygous for GATA4 null alleles developed splanchnic mesoderm differentiated into primitive cardiac myocytes that expressed contractile protein but failed to form a linear heart tube, indicating that GATA4 is not essential for the specification of the cardiac cell lineages, while suggesting a critical role for GATA4 in early cardiac morphogenesis.[20] These studies also found that up-regulation of endogenous GATA6 mRNA could potentially compensate for the lack of GATA4. By use of the cardiac differentiation model of pluripotent P19 embryonal carcinoma cells, which can differentiate into beating cardiac muscle cells, inhibition of GATA4 expression with antisense oligonucleotides blocked differentiation to beating cardiac muscle cells and interfered with the downstream expression of cardiac muscle markers.[21] In the absence of GATA4, differentiation is blocked at the precardiac (cardioblast) stage, and cells are lost through extensive apoptosis; however, these studies also demonstrated that mesoderm commitment was not affected in GATA4-deficient cells as gauged by specific marker levels (e.g., brachyury and goosecoid). Moreover, GATA4 overexpression increased differentiation of P19 cells to beating cardiomyocytes; however, this effect required cell aggregation, suggesting that GATA4 may relay cell contact–gener-

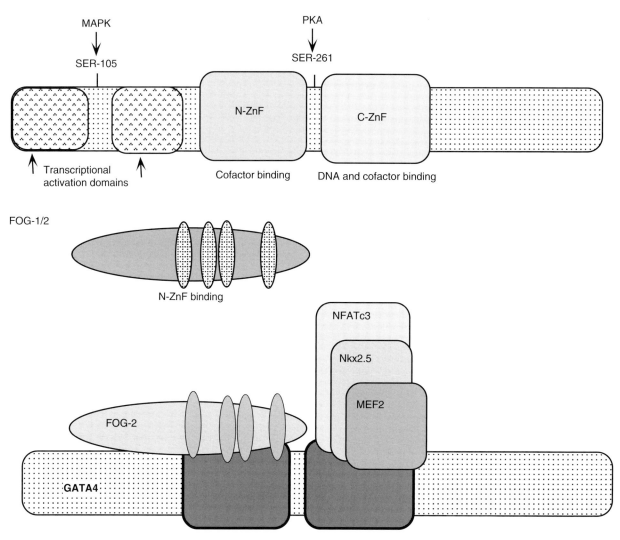

FIGURE 2 GATA family of transcription factors.

ated signals required for cardiomyocyte differentiation and survival.[22] There is evidence that GATA function is shared by other members of the redundant GATA family, suggesting that GATA5 and GATA6 can at least partially compensate for a lack of GATA4.[23,24] The mechanisms determining GATA factor specificity are not fully understood and may involve interaction among GATA factors or interactions between GATA factors and other cofactors differentially controlled at various stages of cardiogenesis. A number of highly conserved cofactors (present in nonvertebrates, as well as human) that interact with GATA have been identified, many by means of the yeast 2–hybrid technology. These include the *GATA4* homolog *pannier* required for normal proliferation of cardiogenic precursors and the multi-type zinc–finger proteins *U-shaped (USH)* and *friend of GATA (fog1* and *fog2)*; the latter are also required for cardiac development, more likely in the outflow tract (OFT) and AV valves, as we will shortly discuss.[25]

Post-translational modification of the transcriptional factors can modulate cardiac differentiation programming, a potential form of epigenetic control of differentiation. The histone acetyltransferase p300 is required for the acetylation and DNA binding of GATA4, modulating its transcriptional activity, as well as for promotion of a transcriptionally active chromatin configuration. In an embryonic stem (ES) cell model of developing embryoid bodies, the acetylated form of GATA4 and its DNA binding were increased, along with the expression of p300 during the differentiation of ES cells into cardiac myocytes.[26] Treatment of ES cells with trichostatin A, a specific histone deacetylase inhibitor, further elevated levels of acetylated forms of GATA4 and its DNA binding, as well as increased acetylation of histones 3 and 4 bound near the GATA binding site on the ANF gene promoter during ES cell differentiation; trichostatin A treatment also increased the expression of green fluorescence protein (GFP) under the control of the cardiac-specific Nkx2.5 promoter and of endogenous cardiac β-MHC. These data suggested that GATA4 acetylation is involved in cardiac cell differentiation.

Differentiation of Cardiac Precursors

Formation of the embryonic heart tube requires the determination or specification of heart precursor cells within defined areas of the mesoderm, their intracellular organization (often in an epithelial monolayer), medial migration, and the merger of bilateral precursor populations. Interestingly, the mechanisms and molecular players involved in these highly complex processes are extraordinarily well conserved.[27,28] With the availability of new molecular and cellular technologies to study the events of early cardiac development, many of the apparently well-established facts have had to be either discarded or substantially revised (i.e., the concept of the symmetrical nature of the bilateral cardiac precursors). Furthermore, anterior and posterior polarity seem to be established at the early stages of cardiac differentiation, with the anterior heart progenitors differing from the posterior heart progenitors in their myosin isoform gene expression.[29]

The heart is derived from the anterior splanchnic mesoderm. It forms from two crescent-like cardiogenic plates that already express cardiac-specific genes like *Nkx2.5* and *GATA4*. It is well established that heart induction in both vertebrates and nonvertebrates such as *Drosophila* is primarily dependent on the interaction of the heart primordia with the Spemann organizer. Signals from the Spemann organizer during gastrulation are a primary determinant of dorsal mesoderm specification, with several genes encoding transcription factors expressed specifically in the Spemann organizer region of the gastrula. Expression of one of these genes, the homeobox gene *goosecoid*, has been shown to be sufficient to elicit the formation of a dorsal axis in the embryo.[30] Other highly expressed genes in the organizer encode the bone morphogenetic protein (BMP) antagonists, Noggin,[31,32] a small polypeptide involved in dorsal development in *Xenopus* embryos, and Chordin, which are primary dorsalizing signals from the organizer.[32] In *Xenopus*, as well as other organisms, TGF-β signals of the Nodal family including Xnr-1, a homolog of the mouse Nodal protein, play a key role in organizer formation and function.[33–35] Evidence from studies in several organisms suggests that the gastrula organizer is populated by a succession of cell populations with different fates as opposed to a population with a single lineage.[36] Activation of the BMP pathway before the onset of gastrulation can promote the suppression of the Spemann organizer formation in *Xenopus* embryos.[37]

In addition to signals from the organizer, inductive signals derived from surrounding tissues (e.g., endoderm and ectoderm) during early gastrulation are contributory to mesoderm differentiation. For instance, the presence of deep endoderm can dramatically enhance heart formation in explants of heart primordia, both in the presence and absence of organizer,[38] whereas ablation of the entire endoderm can decrease the frequency of heart formation in embryos that retain organizer activity.[39]

An early and important event in the regional subdivision of the mesoderm is the restriction of *Nkx2.5* expression to dorsal mesodermal cells. An inductive signal originating from dorsal ectodermal cells is required for the activation of *Nkx2.5* in the underlying dorsal mesoderm in *Drosophila*; Decapentaplegic (Dpp), a member of the TGF-β family, serves as a pivotal signaling molecule in this process.[40] A second secreted signaling molecule, Wingless (Wg), has also been shown to be involved in cardiac mesoderm specification.[41] Various combinations of inductive signals and mesoderm-intrinsic transcription factors can cooperate to induce the progenitors of heart precursors at precisely defined positions within the mesoderm layer. Positive Dpp signals and antagonistic Wg inputs are integrated by virtue of combinatorial binding sites on a mesoderm-specific NK homeobox gene that functions as an early regulator of mesoderm development. Binding sites for Dpp-activated Smad proteins have been identified, as have adjacent binding sites for FoxG forkhead transcription factors, which are direct targets of the Wg signaling cascade. Binding to the second site blocks the activity of activated Smads.[42] Recently, it has been shown that the GATA transcription factor, pannier, mediates, as well as maintains, the cardiogenic Dpp signal in both germ layers.[43]

In chick embryos, endoderm-secreted activin-A and FGF2 have been shown to regulate early cardiac differentiation.[44] Combined treatment with BMP2 and fibroblast FGF4 can induce cardiogenic events, culminating in full cardiac differentiation of nonprecardiac mesoderm explanted from stage 6 avian embryos.[45] BMP2, like its *Drosophila* homolog Dpp, is an important signaling molecule for specification of cardiogenic mesoderm in vertebrates. The induction of cardiac lineage markers in central mesendoderm (i.e., Nkx2.5, GATA4, eHAND, MEF2A, and vMHC expression) exhibited a distinct time course with respect to BMP2 induction; Nkx2.5, GATA4, and MEF2A were induced within 6 h of BMP2 treatment, eHAND and dHAND required 12 h, whereas structural markers vMHC and titin were induced at significant levels only after 48 h of BMP2 addition.[46] BMPs apparently function, in part, by affecting the levels of these transcription factors and work in parallel to FGF signaling from the cardiac mesoderm.

FGF8 signaling also contributes to the heart-inducing properties of the endoderm.[47] FGF8 is highly expressed in endoderm adjacent to the precardiac mesoderm. Supplying exogenous FGF8 can reverse the rapid down-regulation of cardiac markers, including Nkx2.5 and MEF2c, which result from the removal of endoderm. Expression of cardiac markers is increased only in regions where BMP signaling is also present, suggesting that cardiogenesis occurs in regions exposed to both FGF and BMP signaling. These studies also showed that FGF8 expression is regulated by levels of BMP signaling with low levels of BMP2, resulting in ectopic expression of FGF8, whereas higher levels of BMP2 result in repression of FGF8 expression. The Heartless (Htl) FGF receptor is required for the differentiation of a variety of mesodermal tissues in the *Drosophila* embryo.[48] Null *Htl* mutant embryos display irregular migration and spreading of the mesoderm over the ectoderm. A common role for Htl in both directional mesoderm cell migration and pattern formation underlies the pleiotropic defects of the *Htl* mutation. Studies have indicated that the GATA-related U-shaped (USH) cofactor interacts directly with the Htl receptor in determining mesoderm migration.[49]

Two novel *FGF* genes, *Thisbe* (*Ths*) and *Pyramus* (*Pyr*), have been recently identified that may encode the elusive ligands for the Htl receptor.[50] The two genes exhibit dynamic patterns of expression in epithelial tissues adjacent to Htl-expressing mesoderm. Embryos that lack *Ths* and *Pyr* display embryonic defects related to those seen in *Htl* mutants, including delayed mesodermal migration during gastrulation and a loss of cardiac tissues.

The involvement of the canonical Wnt pathway has also been shown to play a significant role in mesoderm specification. The presence of Wnt antagonists Dickkopf-1 (Dkk-1) and Crescent (a Frizzled-related protein that inhibits Wnt-8) can induce heart formation in explants of ventral marginal zone mesoderm. Wnt3A and Wnt8, but not Wnt5A or Wnt11, inhibited endogenous heart induction.[51] Others have shown that the inhibition of Wnt signaling promotes heart formation in the anterior lateral mesoderm, whereas active Wnt signaling in the posterior lateral mesoderm promotes blood vessel development.[52] Little is presently known about the downstream effectors of these secreted Wnt antagonists or the mechanism by which they activate heart formation. A screen for downstream mediators has revealed that Dkk-1 and other inhibitors of the Wnt pathway induce the homeodomain transcription factor Hex, normally expressed in endoderm underlying the presumptive cardiac mesoderm in amphibian, bird, and mammalian embryos. Loss of Hex function blocks both endogenous heart development and heart induction by ectopic Dkk-1. As with the Wnt pathway antagonists, ectopic Hex induces expression of cardiac markers. Thus, to initiate cardiogenesis, Wnt antagonists act on endoderm to up-regulate *Hex*, which, in turn, mediates production of diffusible heart-inducing factors that convey critical patterning information to anterior mesoderm. This novel function for Hex suggests a potential etiology for cardiac malformations in *Hex* mutant mice and should enable the isolation of factors that specify heart induction directly in the mesoderm.[53]

These findings suggest that heart precursor specification is a result of multiple tissue and cell–cell interactions involving both temporal and spatial integrated programs of inductive signaling events.

Migration of the Cardiac Precursors

The basic helix-loop-helix (bHLH) transcription factors, Mesp1 and Mesp2, are required in the migration of cardiac precursors. Mesp1 is expressed in the early mesoderm that is destined to become the cranial-cardiac mesoderm and is one of the earliest molecular markers expressed in cardiac precursor cells, as revealed by lineage study.[54] Unlike other aforementioned transcription factors involved in heart morphogenesis such as GATA4 and Nkx2.5, Mesp1 and Mesp2 are expressed in the nascent heart precursor cell and not during later heart morphogenesis; therefore, they have a limited role in heart tube formation but are involved in mesodermal specification. Disruption of the *Mesp1* gene resulted in a morphogenetic abnormality of the heart, cardia bifida.[54] In the absence of *Mesp1*, mesodermal cells fated to become cardiac myocytes fail to migrate normally out of the primitive streak during gastrulation and, consequently, fall behind the morphogenetic movements of the rest of the embryo, resulting in complete or partial cardia bifida. Somatogenesis is not disrupted in these embryos because of normal expression of the related *Mesp2* gene. Mice lacking both *Mesp1* and *Mesp2* exhibit complete block in migration of the mesoderm from the primitive streak, resulting in a complete lack of cardiac and other mesodermal derivatives, and die at approximately 9.5 days. A major defect in this *Mesp1/Mesp2* double knock-out embryo is the absence of any mesodermal layer between the endoderm and ectoderm.[55]

The transcriptional pathway involving GATA factors may also participate in the movements of the paired progenitor pools that coalesce to form the linear heart tube. During early mouse development, whereas *GATA4* is expressed in cardiogenic splanchnic mesoderm and associated endoderm, endodermal *GATA4* expression is required for ventral morphogenesis.[56] Both GATA4 and GATA5 have been implicated in the regulation of normal formation of the endoderm underlying the myocardial precursors.[57] With antisense-mediated reduction of GATA4/5/6 function in chick embryos,[58] endodermal cells do not normally differentiate, their ventral migration is inhibited, preventing the concomitant movement of myocardial cells. This severe block in GATA expression results in a high percentage of the embryos developing abnormal hearts, including the development of cardia bifida. A Mix-like homeodomain transcription factor (Bon) has also been reported in conjunction with the impaired endodermal differentiation observed in zebrafish cardia bifida mutants.[59]

Cardiac precursor migration also involves formation of a coherent epithelial layer. Specifically targeting *GATA4* directly in cardiac mesodermal cells with siRNA resulted in the development of cardia bifida in chick embryo and selective suppression of the cell adhesion protein N-cadherin expression.[60] Recent studies with zebrafish cardiogenesis implicate cell adhesion molecules as playing an important role in maintaining cardiac epithelial integrity.[61] As precardiac mesoderm cells epithelialize, they become stably committed by the activation of several cell–matrix and intracellular signaling transduction pathways. Two different families of cell adhesion molecules are primarily involved, the calcium-dependent cadherins, specifically N-cadherin, and the extracellular matrix (ECM) glycoprotein, fibronectin.[62] N-cadherin acts by binding to the intracellular molecule β-catenin, and fibronectin acts by binding to integrins at focal adhesion sites. Both are involved in the regulation of gene expression by their association with the cytoskeleton and through signal transduction pathways. Cross-talk between the adhesion signaling pathways initiates a number of characteristic phenotypic changes associated with cardiomyocyte differentiation, electrical activity and myofibrillar organization.

Cardiac Progenitor Cells and the Anterior Field

Recent observations have demonstrated the contribution to the developing heart by a population of progenitor cells in pharyngeal mesoderm that gives rise to myocardium at the arterial pole. Lineage-tracing experiments have outlined the extent to which pharyngeal progenitor cells colonize the heart, revealing their contribution to venous, as well as arterial, pole myocardium.[63] In pharyngeal myocardial progenitor cells, transcriptional regulators including Forkhead, GATA, LIM, MEF2, Smad, and T-box transcription factors have been implicated.[1] Isl1, a LIM homeodomain transcription factor, has become an important marker for a distinct population of undifferentiated cardiac progenitors that proliferate before differentiation and gives rise to the cardiac segments.[64]

Tube Looping and Segmentation

Cardiovascular development involves the looping of the tubular heart with tightening of the inner curvature and the completion of an arterial and a venous pole, together with a complicated set of septa separating veins, atria, ventricles, and great arteries. The differentiation in chamber myocardium and myocardium of the transitional or intersegmental zones takes place during rightward looping, a process regulated by a cascade of genes that is essential for left-right programming. Disturbances in this transcriptional programming will lead to abnormal heart looping that can vary from random anterior to leftward looping. The consequences for human development are found in the abnormal atrial *situs* (*situs inversus* or isomerism), dextrocardia, and heterotaxy, which will be discussed in more detail in Chapter 7.

In the myocardium of the looped heart tube (Fig. 3), several transitional zones (TZ) and intervening primitive

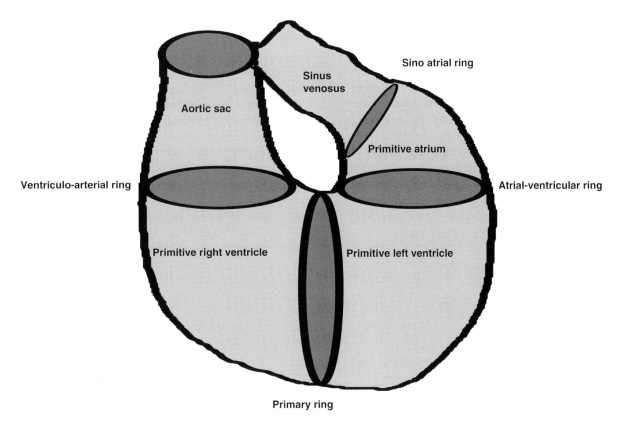

FIGURE 3 The looped heart tube with the primitive cardiac chambers and the transitional zones. After the blood flow from venous to arterial, one can distinguish the sinus venosus, the sinoatrial ring, the primitive atrium, the AV ring encircling the AV canal, the primitive left ventricle, the primary fold or ring, the primitive right ventricle, the outflow tract ending at the ventriculoarterial ring, and the aortic sac.

cardiac chambers (CC) can be discerned. The transitional zones will become part of the septa, valves, conduction system, and fibrous heart skeleton. They will be partly incorporated in the cardiac chambers during formation of the definitive right and left atrium and their ventricular counterparts. These include the sinus venosus and the sinoatrial ring (TZ), the primitive atrium (CC), the AV canal or ring (TZ), the primitive left ventricle (CC), the primary fold or ring (TZ), the primitive right ventricle (CC), and the ventricular outflow tract (OFT) with a proximal and a distal part, also referred to as the ventriculoarterial ring (TZ).[65] Several of the transitional zones, such as the AVC and OFT, develop endocardial cushions (EC), whereas others do not (e.g., the *sinus venosus* and primary fold). The process of looping aligns the primordial chambers such that they face the outer curvature, whereas the transitional zones are brought together in the inner curvature of the heart tube. Myocardium of the inner curvature, as well as that of the inflow tract, AVC, and OFT, retains the molecular signature originally found in linear heart tube myocardium. Specific gene defects affecting the morphological characteristics can result in deficient remodeling of the inner curvature, leading to a leftward movement of the OFT over the AVC, resulting in a spectrum of OFT abnormalities: for example, the formation of a double-outlet right ventricle with an obligatory ventricular septal defect (VSD) or, in less extreme cases, just a VSD, malformations more commonly found in transgenic mice.

The formation of the cardiac chambers from the primary heart tube has recently received a great deal of attention. The initial linear heart, composed of primary myocardium, shows a polarity both in phenotype and gene expression along its anteroposterior and dorsoventral axes. Specialized ventricular chamber myocardium is specified at the ventral surface of the linear heart tube, whereas distinct left and right atrial myocardium form more caudally on laterodorsal surfaces. Each chamber has specific biochemical and physiological properties important for heart function, which are established and maintained by chamber-specific gene expression. The spatial pattern of gene expression bears a strong relationship to morphogenesis. This involves various promoter elements of genes for ANF, sarcoplasmic reticulum calcium ATPase (*SERCA2a*), *MLC2v*, and β-myosin heavy chain (β-MHC), and the restriction of specific gene expression to the atrial or ventricular compartments, mediated primarily by transcription factors responsible for compartment-restricted gene expression.

Several factors contribute to patterning and chamber formation in the developing heart by regulating gene programs at specific sites within the heart tube and integrating positional information with respect to the anteroposterior (AP) patterning. These include retinoic acid, Irx4, T-box genes, including Tbx5, Tbx2, and Tbx20, and GATA factors.

Retinoic acid (RA), the active derivative of vitamin A, is a powerful effector of AP patterning, chamber specification, and morphogenesis. Transient exposure of zebrafish to RA either during or shortly after gastrulation results in heart tube truncation, initiating in the OFT and ventricle at low doses, and further along the AP axis with higher doses.[66] In several species, including chicken and mouse, excess RA promotes increased expression in anterior regions of genes normally expressed only in the posterior region. Conversely, RA deficiency, brought about by the use of an anti-RA monoclonal antibody, causes underdevelopment of the posterior structures of the heart, most notably the *sinus venosus* and atria.[67]

To further assess the role of RA in early cardiac development, the expression of a key enzyme involved in endogenous RA synthesis, retinaldehyde dehydrogenase type 2 (Raldh2), was investigated.[68] In avian embryos, Raldh2 expression was observed exclusively in the posterior mesoderm and in the posterior heart precursors. In mice, Raldh2 expression is initiated in the posterior mesoderm shortly after gastrulation, and when the synthesis of endogenous RA is blocked as a result of disrupted *Raldh2* alleles, the animal exhibited severe heart abnormalities and died *in utero*. Although the embryonic heart tube formed normally, the embryos showed failure in upward right looping and in the development of the posterior chambers (atria, *sinus venosus*). In addition, these embryos displayed an anteriorly based defect and premature differentiation of ventricular cardiomyocytes.

Numerous molecular and functional studies have revealed that RA regulates the expression of a variety of cardiogenic transcription factors, including GATA4, as well as several heart asymmetry genes.[69,70] During the crucial RA-requiring developmental window, RA transduces its signals to genes for heart morphogenesis by means of the retinoic acid receptors RARα, RXRα2, and RARγ. RARs and RXRs, ligand-activated transcription factors, have been implicated in many aspects of heart development, including ventricular maturation and cardiac septation.[71–73] In avian embryos, blocking the expression of *RARα, RXRα2,* and *RARγ* using antisense oligonucleotides recapitulates the complete RA-deficient phenotype.[74] Similar to mice whose synthesis of RA is blocked, mice lacking either the RAR coreceptor RXR or RAR have defective ventricular maturation, apparently related to accelerated cardiomyocyte differentiation.[75]

The *Iroquois homeobox gene4* (*Irx4*) is a member of the *Iroquois* family of homeodomain-containing transcription factor genes, which have been implicated in chamber-specific gene expression. When the tubular heart is formed, Irx4 expression is confined to a restricted segment of the linear heart tube and the AVC canal and ventricular myocardium, including the inner curvature after looping, resembling the pattern of MLC2V.[76] In later stages, in all species examined, Irx4 expression is largely confined to ventricular myocardium.[77,78]

Irx4 expression is reduced in mice lacking Nkx2-5 or dHAND, in which ventricular differentiation is compromised, suggesting that it operates as a downstream mediator of ventricular development relative to Nkx2.5 and dHAND.[79] Irx4 modulates specific ventricular gene expression as shown by its activation of the expression of the ventricle myosin heavy chain-1 (*VMHC1*) and suppression of the expression of the atrial myosin heavy chain-1 (*AMHC1*) in the ventricles of chick embryos.[77] How Irx4 functions as a positive activator of *VMHC1* is not known, because most observations have indicated that transcription factors belonging to the Iroquois family primarily act as repressors.[80] That Irx4 plays a role in chamber-specific gene expression in the ventricles has become evident from the analysis of the promoter of the *AMHC1* homolog, atrium-specific slow myosin heavy chain 3 (*SMyHC3*), which is repressed by Irx4 in chicken ventricular myocardium and up-regulated in *Irx4* −/− embryonic ventricles.[81] The regulatory elements of the *SMyHC3* gene are functional in quail and mouse,[82,83] and in both species the transcriptional elements controlling the chamber specificity of this promoter are under the control of Irx4. Interestingly, this promoter also contains both a functional GATA factor binding element and a vitamin D–response element (VDRE). These findings coupled with the fact that Irx4 does not bind directly to the *SMyHC3* promoter elements, required for ventricular repression, strongly suggest the involvement of additional factors in this compartment-specific transcriptional control. In fact, interaction of Irx4 with a RAR/vitamin D receptor complex of proteins has been reported.[84] Available information does not support the notion that Irx4 is a global regulator of ventricle-specific gene expression, because mice with a targeted disruption of *Irx4* have only a partial defect of ventricle-specific gene expression.[81] Moreover, the identification of five additional *Irx* genes present in the developing heart indicated the probability of some genetic redundancy.[85] However, their spatial and temporal patterns of expression during development are markedly different from *Irx4*, suggesting that they can only partially compensate for *Irx4*. *Irx4*-deficient mice have impaired cardiac function and develop cardiomyopathy, underscoring the important role of chamber-specific gene expression for proper cardiac function.[81]

Present in all metazoans, the T-box family of transcription factors is involved in early embryonic cell fate decisions, regulatory development of extraembryonic structures, embryonic patterning, and many aspects of organogenesis. T-box genes include *Tbx1*, *Tbx2*, *Tbx3*, *Tbx5*, *Tbx6*, *Tbx18*, and *Tbx20*, all of which exhibit complex patterns of temporal and spatial regulation in the developing cardiac structures. T-box transcription factors function in many different signaling pathways, notably BMP and FGF pathways, and the downstream target genes that have been identified thus far indicate a wide range of downstream effectors. Furthermore, mutations in the *T-box* genes are responsible for developmental dysmorphic syndromes with severe cardiac abnormalities in humans (as we will further discuss in Chapter 7) and have also been implicated in the regulation of cell proliferation in cancer.

Tbx5 is expressed initially throughout the cardiac mesoderm in its earliest stages and is colocalized with other cardiac transcription factors such as NKX-2.5 and GATA4. In the linear heart tube, its expression pattern displays a PA gradient, stronger near the posterior end and weaker at the anterior end.[86,87] At mid-gestation, it is restricted to the atria and LV. *Tbx5* mRNA levels decrease in the LV during subsequent stages of development in such a way that by late gestation and adulthood, low levels of *Tbx5* transcripts can be detected similarly in both LV and RV in mice and humans.[88] In zebrafish, *Tbx5* deficiency results in the *heartstrings* mutation, which is characterized by failure in heart tube looping and significant abnormalities of both the atrium and ventricle.[89] Lack of Tbx5 in mice results in severe hypoplasia of posterior structures such as the atria, whereas the RV and OFT growth remain intact.[90] This suggests that Tbx5 is required for formation of the posterior heart. *Tbx5* deficiency also results in a marked down-regulation of *Nkx2.5* and *GATA4*, as well as anteriorly expressed genes, including *Irx4*, *Mlcv*, and *Hey2*. *Tbx5* haploinsufficiency also markedly decreased ANF and connexin (*Cx40*) transcription.

Overall ventricular differentiation is impaired in *Tbx5*-deficient embryos, including decreased expression of the ventricle-specific genes *Mlc2v*, *Irx4*, and *Hey2*. This is perhaps because of the early pleiotropic effects of the absence of *Tbx5* on cardiac differentiation, including decreased GATA4 and Nkx2-5 expression. Consistent with these observations, Tbx5 has been shown to accelerate cardiac differentiation of P19CL6 cell lines, including increased *Nkx2-5* expression,[91] and the inhibition of *Tbx5* in *Xenopus* embryos also led to hypoplasia of the cardiac tissues and decreased *Nkx2-5* mRNA levels.[92]

In embryos in which *Tbx5* is misexpressed, the ventricular septum was not formed, resulting in a single ventricle.[93] In such hearts, the LV-specific *Nppa* gene for ANF was induced. Transgenic overexpression of *Tbx5* in tubular hearts of chicken or mice results in thinned and hypoproliferative ventricular myocardium, with retardation of ventricular chamber morphogenesis and loss of anterior gene expression.[87,94] These findings are consistent with the role of Tbx5 in AP patterning, imposing a posterior identity on the heart tube. Tbx5-mediated inhibition of myocardial growth is suggestive that Tbx5 is involved in the down-regulation of cell proliferation. Also to be noted is that *Tbx5* haploinsufficiency in mice causes distinct morphological and functional defects in the AV and bundle branch conduction systems,[95] and its critical role in the development and maturation of the cardiac conduction system will be further addressed later in this chapter.

Other T-Box Factors

Tbx2 functions as a primary determinant in the local repression of chamber-specific gene expression and chamber differentiation.[96] The pattern of *Tbx2* mRNA and protein expression in both human and mouse hearts displayed a temporal and spatial profile parallel to that found with chamber myocardium-specific genes *Nppa*, *Cx40*, *Cx43*, and *Chisel*.

In vitro, Tbx2 repressed the activity of regulatory fragments of *Cx40*, *Cx43*, and *Nppa*. Hearts of transgenic embryos that expressed Tbx2 in the prechamber myocardium completely failed to form chambers and to express the chamber myocardium-specific genes *Nppa*, *Cx40*, and *Chisel*, whereas other cardiac genes were normally expressed. Moreover, it has been suggested that Tbx2 inhibits chamber-specific gene expression by competition with the positive factor Tbx5.

The murine T-box transcription factor Tbx20 also plays a central role in the genetic hierarchy-guiding lineage and chamber specification. Disruption of *Tbx20* in mice results in a severely abnormal cardiac transcriptional program, reduced expansion of cardiac progenitors, and block chamber differentiation, leading to grossly abnormal heart morphogenesis and embryonic death at mid-gestation.[97] Furthermore, *Tbx20*-null embryos display activation of the repressive Tbx2 across the entire heart myogenic field, which is largely responsible for the cardiac phenotype in *Tbx20*-null mice, placing Tbx20 upstream of Tbx2. In addition, in the developing heart, Tbx2 directly binds the promoter and represses expression of the cell proliferation gene *N-myc1*. This gene is required for growth and development of multiple organs, including the heart, and likely contributes to the observed cardiac hypoplasia in *Tbx20* mutants.[98] This suggests a model by which the T-box proteins regulate regional differences in *N-myc1* expression and proliferation affecting organ morphogenesis. In normal chamber myocardium, Tbx20 represses Tbx2, preventing repression of *N-myc1* and resulting in relatively high proliferation.

Two mutant alleles of *Tbx6* have suggested that this factor is involved in both the specification and patterning of the somites along the entire length of the embryo. The null allele, *Tbx6 (tm1Pa)*, causes abnormal patterning of the cervical somites and improper specification of the more posterior paraxial mesoderm.[99]

MEF2c and HAND Proteins

eHAND (also termed Hand 1) and dHAND (Hand 2) are basic helix-loop-helix transcription factors that play critical roles in cardiac development (Fig. 4). *HAND* gene expression

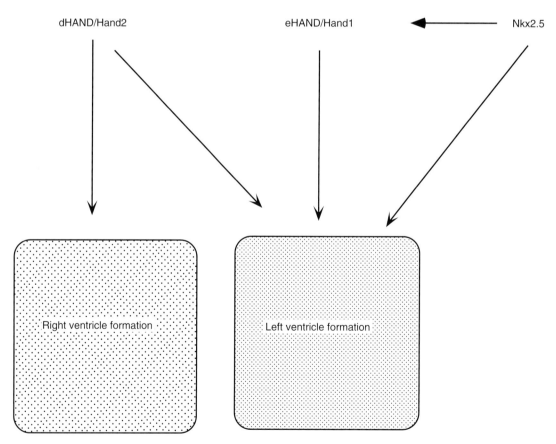

FIGURE 4 Asymmetrical pattern of HAND expression and its role in the developing myocardium. Knock-out mice lacking dHAND/Hand2 have demonstrated the essential role of this gene in formation of the RV. Knock-out mice lacking Nkx2.5 show a loss of the LV, as well as down-regulation of eHAND/Hand1. Knock-out mice lacking eHAND/Hand1 show relatively minor abnormalities in the LV, suggesting that the loss of eHAND/Hand1 alone is insufficient to account for the more severe LV defects in Nkx2.5 mutant embryos. Knock-out mice lacking both eHAND/Hand1 and dHAND/Hand2 show severe ventricular hypoplasia, suggesting redundant functions of the HAND genes in the developing LV.

is down-regulated in adult mouse. In the developing myocardium, these genes exhibit a complementary left-right (LR) pattern of asymmetrical expression, with *Hand2/dHAND* predominantly on the right side and *Hand1/eHAND* on the left side of the looped heart tube.[100]

Hand2/dHAND is expressed in cardiac precursors throughout the cardiac crescent and the linear heart tube before becoming restricted to the right ventricular chamber at the onset of looping morphogenesis. It has been reported that *Hand2/dHAND* is a direct transcriptional target of GATA and NKX2.5 factors during development of RV.[101,102] Because GATA factors are not chamber restricted, these findings suggest the existence of positive and/or negative coregulators that cooperate with GATA factors to control RV-specific gene expression in the developing heart. Mice that lack *Hand2/dHAND* die at embryonic day 10.5 from right ventricular hypoplasia and vascular defects. The RV and OFT are presently considered to arise from the secondary (or anterior) heart field in contrast to the atria and LV that are thought to arise from the primary heart field. Absence of *Hand2/dHAND* resulting in the deletion of the RV suggests that it is an essential component of the pathway for development of the secondary heart field.[103]

Although *Hand1/eHAND* is asymmetrically expressed along the AP and dorsoventral embryonic axes, it is symmetrically expressed along the left-right axis at early stages of embryonic and cardiac development. After cardiac looping, expression of *Hand1/eHAND* is restricted largely to the LV and the atria, a pattern identical to that of the *Nppa* gene encoding ANF, the major secretory product of the heart. Mice that lack *Hand1/eHAND* die at embryonic day 8.5 from placental and extraembryonic abnormalities that preclude analysis of its potential role in later stages of heart development.[100] To examine the role of *Hand1/eHAND* on chamber specification and morphogenesis, *Hand1/eHAND* knock-in mice were generated in which the *Hand1/eHAND* cDNA was placed under the control of the *MLC2V* promoter, which is fully expressed in ventricular myocardium throughout development.[104] Embryos with knock-in *eHAND* exhibited a morphologically single ventricle, but they showed distinctive LV and RV expression at the molecular level. Forced expression of *Hand1/eHAND* resulted in marked expansion of the outer curvature of the RV and LV with limited interventricular groove or septum formation between the two ventricles. Furthermore, these mice displayed abnormal expression patterns of molecular markers of the working myocardium (e.g., Chisel and ANF), as well as *Hand2/dHAND* in the RV, but did not affect *Tbx5* expression. These findings indicate that Hand1/eHAND is involved in ventricular wall expansion and is unlikely to act as a master regulatory gene required to specify LV myocyte lineage. Furthermore, *Hand1/eHAND* expression may be critical in the normal formation of the interventricular groove and septum. Recently, mice were generated that contained a conditional *Hand1/eHAND*-null allele flanked by Cre recombinase loxP recognition sites permitting the specific deletion of this allele in the developing heart by cardiac-specific expression of Cre recombinase.[105] Embryos homozygous for the cardiac-specific *Hand1/eHAND* gene deletion displayed defects in the LV and EC and exhibited deregulated ventricular gene expression but survived until the perinatal period when they succumbed to a spectrum of CHD. Combination of the conditional *Hand1/eHAND* mutation with a *Hand2/dHAND* loss-of-function mutation revealed pronounced effects of HAND gene dosage on the control of cardiac morphogenesis and ventricular gene expression. The phenotype resulting from cardiac-specific deletion of *Hand1/eHAND* was much less severe than that of *Hand2/dHAND* mutant embryos in which the entire RV was absent. *Hand1/eHAND* is expressed specifically in the outer curvatures of the embryonic LV, RV, and OFT in contrast to *Hand2/dHAND* expression that occurs in both RV and LV, although highest is in the RV.[103,106,107] In the absence of *Hand1/eHAND*, residual *Hand2/dHAND* expression in the LV and OFT probably compensates in part for the loss of *Hand1/eHAND*. In contrast, absence of *Hand2/dHAND* results in a complete lack of Hand factors in the presumptive RV.[103]

Also the less severe phenotypes of either the *Hand1/eHAND* or *Hand1/eHAND/Hand2/dHAND* double knock-out mice markedly contrasts with the phenotype of mice lacking *Nkx2.5*, in which the LV fails to grow after cardiac looping, and expression of several markers of cardiac differentiation is decreased throughout the remaining myocardium.[108] *Hand1/eHAND1* expression is abolished in the hearts of *Nkx2.5* mutant embryos,[106] and the loss of *Hand1/eHAND* contributes to the abnormal cardiac morphogenesis of *Nkx2.5* mutant hearts. These findings demonstrate that Hand factors play both pivotal, albeit partially, redundant roles in cardiac morphogenesis, cardiomyocyte differentiation, and cardiac-specific transcription.

It is worth noting that a single cardiac *Hand* gene is found closely related to *Hand2/dHAND* in lower vertebrates, such as frogs and fish, and it is responsible for the apparent single ventricular chamber found in these species.[109,110] Similarly, *Tbx5* is expressed in the single ventricle and is involved in the morphogenesis of the entire heart in these species.[109] This suggests that with the acquisition of pulmonary circulation and a RV, gene duplication and specialization of *Hand* and *Tbx5* function evolved along with their restricted chamber-specific expression.

In the human adult heart, *Hand2/dHAND* expression has been observed in all four chambers, but it was diminished in the right atrium. In contrast *Hand1/eHAND*, which is expressed in the RV and LV, is down-regulated in both atria. Expression of *Hand1/eHAND* and not *Hand2/dHAND* has been reported to be significantly down-regulated in ischemic hearts and DCM, suggesting a correlation between *Hand1/eHAND* dysregulation and the evolution of a subset of cardiomyopathies.[111]

As previously noted, the RV and OFT are derived primarily from a population of progenitors elements/genes known as the anterior heart field. These regions of the heart are severely hypoplastic in mutant mice lacking either the myocyte enhancer factor 2C (MEF2C) and BOP transcription factors, suggesting that these cardiogenic regulatory factors may act in a common pathway for development of the anterior heart field and its derivatives.[112,113] *Bop* expression in the developing heart depends on the direct binding of MEF2C to a MEF2-response element in the *Bop* promoter that is necessary and sufficient to recapitulate endogenous Bop expression in the anterior heart field and its cardiac derivatives during mouse development.[113] *Bop* has been identified as an essential downstream effector gene of *MEF2C* in the developing heart. Moreover, MEF2C also interacts with the HAND and GATA proteins.[114]

Members of the myocyte-specific enhancer factor 2 (MEF2) family of transcription factors are known to be important for cardiac muscle formation. In mice, there are four *MEF2* genes, *MEF2-A, B, C,* and *D*, each of which is expressed at some stage of cardiac development.[115] Several downstream myogenic genes, including *MLC2*, α*-MHC*, and *Nppa*, have promoter elements or enhancers that bind these proteins. In mice lacking *MEF2C*, these proteins are not expressed, and although the linear heart tube forms, cardiac looping is defective, and anterior heart structures, but not posterior structures, are malformed and hypoplastic. This latter feature is similar to the cardiac phenotypes obtained by targeted deletion of genes for *Nkx2-5*, *Hand2/dHAND*, and *Bop* in which embryos are affected more at the anterior than at the posterior side. The resulting defect in ventricular chamber morphogenesis and expansion is likely because of a failure in downstream cardiac gene expression, including the failure to activate the *HAND* genes with which MEF2 proteins interact. In mice with null mutations of *MEF2C*, the expression of *Hand2/dHAND* is absent, whereas *Hand1/eHAND* is present in both the LV and RV.[107,116] Because of its ventricular specificity requirement, MEF2C is a necessary cofactor for ventricle-specific factors, and a cooperative interaction of *Hand2/dHAND* with MEF2C has been suggested to be a pivotal event in formation of the anterior region of the heart.[117] Recently, the relationship between MEF2 and HAND factors has been further established with the demonstration that *Hand1/eHAND* is recruited to the cardiac *Nppa* promoter by means of physical interaction with MEF2 proteins.[118] This interaction results in a synergistic activation of MEF2-dependent promoters, and MEF2 binding sites are sufficient to mediate this synergy. Together with a variety of cofactors and coactivators that seem to play a role in MEF2C transactivation, post-transcriptional modification also seems to play a significant role in the MEF2's regulatory pathway. In skeletal and cardiac muscle, negative regulation of MEF2 function by histone deacetylases (HDACs) has been revealed to be an important mechanism modulating MEF2 activity.[119] The E1A-binding protein p300 that functions as an acetylase is involved in the regulation of MEF2C cardiac transcription during development.[120] Moreover, MEF2C activity has been shown to be regulated by phosphorylation,[121] and activity of a MEF2-dependent transgene in the heart is stimulated by calmodulin kinase activation.[122]

Dorsoventral Specification and Chamber Formation

The dorsal side of the forming linear heart tube is connected to the body wall by the dorsal mesocardium. During looping, the ventral side of the anterior ventricular region and the posterior dorsal side of the atrial region will define the outer curvatures from where the chambers will expand. The chambers form as an integrated response to dorsoventral, as well as anteroposterior, patterning. A limited number of genes have been found to discriminate between the ventral and dorsal sides of the heart tube, with *Hand1/eHAND* specifically expressed at the ventral side of the linear heart tube at E8-8.5.[76,106] Genes for Chisel, ANF, and Cited1 (formerly Msg1) are expressed selectively at the ventral side of the future ventricular region of the linear heart tube of E8.25 embryos.[76,123,124] After looping, expression is confined to the chambers. Also, the onset of *Irx3* expression in the myocardium and *Irx5* expression in the endocardium at the ventral side of the E9 mouse heart indicates dorsoventral differences in transcription regulation in the early tubular heart.[85] These genes are specifically expressed at the outer curvature and, except for *Hand1*, are restricted to the forming chambers underlining the relation between early dorsoventral patterning and chamber formation.

Left-Right Identity

The acquisition of left versus right (L-R) identity in the developing heart is presently an area of intensive investigation. The L-R identity of the heart-forming regions and the direction of looping of the atrial and ventricular regions of the tubular heart are highly conserved in evolution, and both are dependent on L-R signaling in the embryo.[125] Defects in the determination of laterality can result in severe cardiac malformations, a topic that will be covered in depth in Chapter 7.

The importance of L-R signaling is particularly evident in the atria, which are initially positioned in a left-right arrangement, unlike the ventricles for which the left and right derive from an initial anteroposterior orientation. The analysis of transcription profiles using reporter transgenes further supports the notion that molecular specification of left and right atrial chambers, but not ventricular chambers, is dependent on L-R signaling cues.[126] Atrial identity is essential for the proper alignment of septal and valve structures and for the normal connections of pulmonary veins. Abnormalities in these processes lead to severe defects, such as common AV canal or total anomalous pulmonary venous return.

A cascade of signaling molecules that regulate the establishment of L-R identity of the embryo reaches a decisive point in the expression of the paired-domain homeodomain transcription factor *Pitx2* on the left side of the visceral organs, including the heart. The mouse *Nodal* gene, and its homologs in chick and *Xenopus*, are among the first genes known to be asymmetrically expressed along the L-R axis. A key event in this pathway is the restriction of *Nodal* expression to the left side of the lateral plate mesoderm and its repression in the right lateral plate mesoderm.[127] In addition to *Nodal* involvement, several TGF-β family signaling proteins, including secreted extracellular factors (i.e., *lefty-1, lefty-2, BMP4*), membrane receptors (e.g., activin receptor type IIB), members of the membrane-associated proteins encoded by *EFC-CFC* genes (e.g., *cryptic, Notch* receptor, BMP type I receptor *ACVRI*), intracellular mediators (e.g., sonic hedgehog [*SHH*], Smad), and an array of transcription factors (e.g., Zic3, SnR, Hand1, NKX2.5, Cited, β-catenin) have been implicated in L-R axis determination.[128–142]

Asymmetrical *Pitx2* expression seems to be sufficient for establishing L-R identity of the heart and represents a major determinant in establishing atrial identity. Mice lacking *Pitx2* have a single large atrium with right atrial morphology and abnormal drainage of the venae cavae and pulmonary veins.[143,144] The resulting defects resemble complete AVC with single AV valve, ventricular, and atrial septal defects (ASDs), and double outlet right ventricle (DORV). Decreased dosage of *Pitx2* leads to relatively normal chamber formation, but septal and valve defects occur, probably because of misalignment of structures during development.[143] In addition, RA that was previously discussed in reference to its involvement in early mesoderm differentiation and AP patterning is required for the L-R asymmetry pathway that is responsive to either RA excess or deficiency.[145,146] RA controls both the level and location of expression of components of the L-R signaling pathway (e.g., *lefty, nodal, Pitx2*).

Chamber Growth and Maturation

Maturation of the heart into fully functional trabeculated chambers and septation of the atria and ventricles from one another and between their left and right sides are important processes that require precise integration of growth and differentiation signals.

There has been considerable effort to identify the components of the pathways involved in cardiomyocyte growth during embryogenesis, especially because these pathways are at least partly recapitulated during both physiological and pathological cardiac hypertrophy in the adult heart, as well as involved in the proliferation of potentially regenerative cells. Growth of the heart from a thin-walled structure with the atrial and ventricular chambers molecularly specified involves proliferation of myocytes along the walls of the heart tube and within the developing interventricular septum. The most highly proliferative cardiomyocytes are situated along the outer surface of the heart in a highly mitotic area designated the compact zone. As the tube wall thickens, cardiomyocytes along the inner wall become organized into finger-like projections, or trabeculae, which enhance oxygen and nutrient exchange and force generation. The thin epicardial layer of cells surrounding the heart provides a source of mitogenic signals (e.g., RA) that is necessary and sufficient to stimulate cardiomyocyte proliferation. Mice lacking the RA receptor (*RXR*) die during embryogenesis from failure in the proliferative expansion of ventricular cardiomyocytes, resulting in a thin-walled ventricle.[147,148] Interestingly, this effect is observed only with *RXR* deletions that are epicardium specific but not cardiomyocyte specific, indicating that the effects of RA on cardiac growth are primarily nonmyocyte autonomous.[149,150] The epicardium also expresses high levels of the RA synthetic enzyme RALDH2.[151] Furthermore, another epicardial-derived trophic signal affecting cardiomyocyte proliferation is erythropoietin (epo); blockade of either RA or epo signaling from the epicardium inhibits cardiac myocyte proliferation and survival.[152] Several observations have suggested that these epicardial-derived signals do not act directly on the myocardium but instead may regulate the production of an unidentified soluble epicardial-derived mitogen.[149,150] Downstream signaling pathways that are elicited in response to this epicardial-derived factor include the activation of phosphoinositol kinase (PI3K) and Erk pathways that are required for a proliferative response.[153]

The mitogenic signals emanating from the epicardium are essential for the maintenance of the correct amount of myocyte proliferation in the compact myocardium. By use of microsurgical-mediated inhibition of epicardium formation in the embryonic chick, it was noted that levels of myocyte expression of FGF2 and its receptor FGFR-1 is dependent on the presence of epicardial-derived signals.[154] Recently, FGFs expressed in the epicardium have been identified, including Fgf9, Fgf16, and Fgf20, that are RA-inducible and that contribute to the regulation of cardiomyocyte proliferation during mid-gestation.[155] These findings have led to the suggestion that FGFs contribute to the epicardial signal regulating myocardial growth and differentiation.

Growth signals originating from the endocardium, the specialized endothelial lining of the heart, are critical as well. The neuregulin family of peptide growth factors and their tyrosine kinase receptors (ErbBs), have been shown to promote growth of embryonic cardiomyocytes *in vivo*.[156] Knock-out mice lacking *ErbB2, ErbB4*, or *Neuregulin-1* die from cardiac growth defects characterized by the absence of trabeculae.[157,158] This abnormality can be ascribed to the lack of signaling between the endocardium and myocardium. In adult animals, cardiac-specific deletion of *ErbB2* results in DCM with thinning of the ventricular wall.[159] Interestingly, downstream effectors, including activated PI3K and Erk pathways, are similar in the endocardial and epicardial signaling of cardiomyocyte proliferation, although they are different signaling entities.[153]

Nuclear Regulators of Chamber Growth and Maturation

During embryogenesis, a number of transcription factors and nuclear regulatory factors participate in the control of cardiac growth and chamber maturation. For example, the forkhead transcription factor *Foxp1* plays a role in myocyte proliferation and maturation, and *Foxp1*-deleted embryos display a thin ventricular myocardium caused by defects in myocyte maturation and proliferation.[160] The role of GATA4 regulation in ventricular maturation has been demonstrated using Cre/loxP technology to conditionally delete *GATA4* in the myocardium of mice at an early embryonic stage. The *GATA4* deletion resulted in hearts with striking myocardial thinning and with reduced cardiomyocyte proliferation, more so in the RV than the LV, leading to specific hypoplasia of the RV.[161] Another previously discussed transcription factor, Tbx5, has a contributory role in cardiomyocyte proliferation during embryonic development. *Tbx5* overexpression in embryonic chick hearts *in vivo* inhibits myocardial growth and trabeculation largely as a result of suppressing embryonic cardiomyocyte proliferation.[162] Mice with targeted disruption of both *NFATc3* and *NFATc4* genes demonstrated early embryonic lethality (after embryonic day 10.5) and exhibited thin ventricles and a reduction in ventricular myocyte proliferation.[163] Furthermore, a role for mitochondrial energy metabolism in early cardiomyocyte proliferation was suggested by pronounced defects in mitochondrial structure (e.g., abnormal cristae) and function (e.g., respiratory complex II and IV activities) in mice containing these deleted genes. The cardiac-specific expression of constitutively active *NFATc4* in *NFATc3−/− NFATc4−/−* embryos prolonged viability to embryonic day 12 and preserved ventricular myocyte proliferation, compact zone density, and trabecular formation, with enhanced cardiac mitochondrial ultrastructure and complex II enzyme activity in the rescued embryos.

HOP (homeodomain only protein) is a small divergent protein that lacks certain conserved residues required for DNA binding, whose expression early in cardiogenesis is dependent on cardiac-restricted homeodomain protein Nkx2.5 and is involved in the control of cardiac growth during embryogenesis and early prenatal development.[164,165] During mid-embryogenesis, HOP is expressed predominantly in the trabecular region of the myocardium (where cardiomyocyte proliferation is diminished). HOP modulates cardiac growth and proliferation by inhibiting the transcriptional activity of serum response factor (SRF) in cardiomyocytes by recruiting HDAC activity, forming a complex that includes HDAC2 and affecting chromatin remodeling.[166]

Mice deficient in the protooncogene transcription factor N-myc also have defective trabeculation and thinned ventricular myocardium.[167] The negative regulation of *N-myc* (and cardiomyocyte proliferation) by *Tbx2* discussed earlier is relevant in this context.[98] Repression of *N-myc1* by aberrantly regulated *Tbx2* accounts in part for the observed cardiac hypoplasia in *Tbx20* mutants.

Chamber Septation

Normally, septation occurs at three levels: the atrium, the ventricle, and the arterial pole, and requires correct looping. Cell populations extrinsic to the developing heart, including the neural crest, influence the process of ventricular septation and OFT formation through inductive interactions with neighboring tissues. With proper septation at the various levels, the previously described transitional zones are incorporated into the chambers, leading to the formation of the definitive cardiac atria and ventricles. The molecular mechanisms responsible for various stages of ventricular septation remain largely unknown, particularly the processes that mediate morphogenetic movements and fusion between opposing structures.

The primary heart tube consists of a myocardial outer mantle with an endocardial inner lining. Between these two concentric epithelial cell layers, an acellular matrix is found that is generally referred to as the cardiac jelly. During cardiac looping, the cardiac jelly basically disappears from the chamber-forming regions of the cardiac tube but accumulates in the junction between the atria and ventricles, the AV junction, as well as in the developing OFT. This results in the formation of the endocardial cushion (EC) tissues in the AV junction and OFT. Subsequent maturation of these ECs is achieved when a subpopulation of endothelial cells overlying the cushions, triggered by growth factor signaling, undergoes a transformation from epithelial to mesenchymal cells followed by migration into the extracellular matrix of the cushions. Neural crest cells also contribute to the mesenchyme of the OFT, whereas epicardial-derived cells contribute to the AV cushions. These EC tissues constitute the major building blocks of the septal structures in the heart. The EC tissues in the AV junction contribute to the formation of AV septal structures and AV valves, and the EC tissues in the OFT participate in its septation and in the formation of the semilunar valves (aortic and pulmonary). Defects in the formation of these important developmental entities play a critical role in the etiology of a variety of CHD. Mouse models of EC defects have been generated by mutations in *neurofibromin-1*,[168] *hyaluronan synthase-2 (Has2)*,[169] and *RXR* knock-out mice[170] and are generally associated with fetal lethality. Although defects effecting EC formation in mice generally affect both AV and OFT cushions, some defects (e.g., *NFATc* and *SOX-4*) affect only the OFT cushions.[171]

AV Junction and the Formation of the AV Cushions

The endocardial jelly in the AV junction forms the basic material for the AV cushions. Initially, two prominent ECs develop at opposing sides of the common AV canal. These make up the inferior (or dorsal) AV cushion and the superior (or ventral) AV cushion. As development proceeds, the leading edges of the cushions fuse, thereby separating the common AV canal into a left and right AV orifice. Subsequently, smaller ECs develop in the lateral AV junction. These lateral cushions contribute to valvuloseptal morphogenesis and are involved in the formation of the anterosuperior leaflet of the tricuspid valve and the mural leaflet of the mitral valve.

After fusion, the major AV cushion-derived tissue basically forms a large mesenchymal "bridge" contiguous with the mesenchymal cap (also EC material) that covers the leading edge of the forming atrial *septum primum*. As development progresses, the mesenchymal cap on the *septum primum* and the fused cushions merge completely, thereby closing this primary foramen. The mesenchymal remodeling at this point basically consists of the fused major AV cushions, and the mesenchymal cap eventually leads to the formation of the membranous AV septum, the septal leaflet of the tricuspid valve, and the aortic leaflet of the mitral valve. Although the EC tissues are important in the formation of the valves, AV valve morphogenesis also involves a number of myocardial remodeling steps.[172]

Clinical Studies

Defects in the genesis, interaction, and fate of the EC account for most congenital heart malformations in humans, including ASDs and VSDs, respectively, tetralogy of Fallot (TOF), AV canal, and DORV. Although much information has been provided from animal models, the genetic analysis of cardiac septation defects has proved to be a major stimulus to demonstrate that specific transcription factors (e.g., Nkx2-5 and Tbx5) are required at both specific times and dosage as major factors in defining septal morphogenesis.

Dominant mutations in *Nkx2-5* have been found in patients with ASDs, VSDs, TOF, and Ebstein's anomaly of the tricuspid valve with associated conduction defects.[173–175] *Nkx2-5* haploinsufficiency seems to underlie these defects,[176] although mice lacking one copy of *Nkx2-5* do not exhibit as severe cardiac defects as found with human mutations.[177]

The identification of *Tbx5* mutations in Holt–Oram syndrome (HOS), a rare inherited disease characterized mainly by upper limb defects and CHD, has also provided insight into cardiac septation.[178,179] The cardiac defects in HOS are similar to those caused by *Nkx2-5* mutations: ASDs and VSDs with occasional cases of TOF are often associated with defects in the conduction system. As noted with *Nkx2-5* mutations, *Tbx5* haploinsufficiency is a primary aspect of HOS,[178,179] a notion supported by the observation that a deletion of one copy of *Tbx5* in the mouse recapitulates a HOS phenotype.[90] Patients with missense *Tbx5* mutations have variable phenotypes, some mutations associated with severe cardiac defects, whereas others are associated with milder defects.[180,181] Functional studies of *Tbx5* missense mutations indicate that some result in a nonfunctional protein, whereas others have altered Tbx5 function (e.g., binding to its DNA-binding sites).[182]

Formation of the AV Valves

The formation of the EC is a complex event characterized by endothelial-mesenchymal transdifferentiation (EMT) of a subset of endothelial cells specified in the cushion-forming regions to invade the cardiac jelly, where they subsequently proliferate and complete their differentiation into mesenchymal cells. The cushions protrude from the underlying myocardium, and by a complex and still unknown mechanism form thin, tapered leaflets with a single endothelial cell layer and a central matrix comprised of collagen, elastin, and glycosaminoglycans.[183] These delamination and remodeling events depend on further cell differentiation, apoptosis, and ECM remodeling. The final AV valves (mitral and tricuspid) are derived entirely from EC tissue.[184] Recent lineage analysis has documented that the leaflets and tendinous cords of the mitral and tricuspid valves, as well as the AV fibrous continuity, and the leaflets of valves are exclusively generated from mesenchyme derived from the endocardium, with no substantial contribution from cells of the myocardial and neural crest lineages.

A number of signaling molecules originating from both the AV myocardium and the endothelium participate in the formation of the AV valves and in EMT, a critical step of EC formation. A variety of techniques have implicated numerous genes and transcription factors in AV valve formation, including gene disruption in mice (over a dozen genes cause AV valve defective phenotype), spatiotemporal gene expression profiles, and the use of specific inhibitors. The signaling factors identified (Table I) include numerous ligands, membrane receptors, and transcriptional regulators involved in signaling pathways such as BMP/TGF-β, Notch, VEGF, NFATc1, Wnt/β-catenin, ErbB, SOX, Fog1/Gata, and NF1/Ras. We highlight several of the more well-established factors involved in AV valve formation in this section.

As we have noted elsewhere, the BMPs are members of the TGF-β cytokine superfamily of which BMP2, BMP4, TGF-β2, and TGF-β3 have been implicated in heart development. All TGF-β family members are homodimeric proteins that interact with transmembrane TGF-β receptors. Ligand binding activates type II receptors to transphosphorylate type I receptors within the ligand-receptor complexes. The phosphorylated type I receptor then acts as a serine/threonine kinase to phosphorylate and activate cytosolic Smad proteins, which are the major intracellular mediators of TGF-β signaling. TGF-β and BMP are the most extensively studied signaling partners in EC formation, and their pathways are depicted in Fig. 5.

Expression levels of *BMP2* in myocardial cells parallel the segmental pattern of EC formation.[185] BMP2 protein is localized in the AV myocardium in mice before the onset of AV mesenchymal cell formation but absent from

TABLE I Factors Implicated in Cardiac AV Valve Formation

Factor	Phenotype
NFATc1	Disruption of NFATc1 gene leads to selective absence of the aortic, pulmonary, and semilunar valves.
VEGF	Myocardial overexpression of VEGF results in failure of AVC and OFT cardiac cushion formation.
Connexin (Cx45)	*Cx45* knockout leads to decreased and delayed EC formation.
Notch 1	Disruptions of *notch1* leads to hypoplastic EC
Neurofibromin (NF1)	Disruption of *nf1* leads to markedly enlarged EC.
ErbB3	*ErbB3-/-* embryos exhibit EC abnormalities and defective valve formation.
BMP ligand	*Bmp6Bmp7* double mutants result in hypoplasia of the EC and delayed OFT formation.
BMPR (Alk3)	*Alk3* disruption leads to hypoplastic EC.
Smad6	*Smad6* disruption results in thickened and gelatinous AV and semilunar valves
Wnt/β-catenin/APC	*APC* disruption leads to thickened valves.
HB-EGF	*HB-EGF-/-* mice have enlarged AVC and OFT valves.
Hyaluronan synthase-2 (HAS-2)	*Has2-/-* mice are unable to form EC.
Hesr2	Disruption of the notch-signaling target *Hesr2* results in dysplastic AV valves.
TGF-β	*TGF-β1* inhibits valve myofibroblast / proliferation; *TGF-β2* knock-out results in AV and semilunar valve thickening.

EC, Endocardial cushion; OFT, outflow tract; AV, atrioventricular; APC, adenomatous polyposis coli (gene); AVC, atrioventricular canal; HB-EGF, heparin-binding epidermal growth factor; TGF-β, transforming growth factor-β.

ventricular myocardium throughout these stages. After the subsequent cellularization of the AV cushion, BMP2 protein expression is reduced in AV myocardium, while initiated and maintained in cushion mesenchymal cells even during later stages of development. During valvulogenesis, there is intense *BMP2* expression in the valve tissue that is maintained in adult mice. *In vitro* studies using cultured AV endocardial endothelium showed that addition of BMP2 protein promoted the formation of cushion mesenchymal cells in the absence of AV myocardium, with enhanced expression of the mesenchymal marker, smooth muscle (SM) α-actin, loss of the endothelial marker, PECAM-1, and elevated levels of TGF-β2. On treatment with *noggin,* a specific antagonist to BMPs, applied together with BMP2 to the culture medium, AV endothelial cells remained as an epithelial monolayer with reduced expression of SM α-actin and TGF-β2, and normal expression of PECAM-1, thereby blocking EMT. These data indicated that BMP signaling is necessary and sufficient for myocardial segmental regulation of AV cushion mesenchymal cell formation in mice.

Further evidence that BMP signaling plays an important role in the AV myocardium during the maturation of AV valves from the cushions has been obtained with the Cre/lox technique to target a null allele of *Alk3*, the type IA receptor for BMPs, to cardiac myocytes of the AV canal (AVC).[186] Cardiac myocytes of the AVC were shown by lineage analysis to contribute to the formation of the tricuspid mural and posterior leaflets, the mitral septal leaflet, and the atrial border of the annulus fibrosus. With *Alk3* deletion in these cells, defects were seen in the tricuspid mural leaflet and mitral septal leaflet, the tricuspid posterior leaflet was displaced, and the annulus fibrous was disrupted, resulting in ventricular preexcitation. In addition to providing further support for the role of BMP/Alk3 signaling in AV myocardium during the development of AV valves, these findings provided support for the potential role of *Alk3* in human CHD, such as Ebstein's anomaly.

The foregoing observations showed that the BMP ligand and/or receptor disruption results in specific phenotypes with abnormal valve formation. The Smad proteins are intracellular mediators of signaling initiated by TGF-β superfamily ligands, which either can positively trigger further downstream transcriptional responses (Smads 1, 2, 3, 5 and 8) or function as inhibitory transcriptional regulators (Smad 5

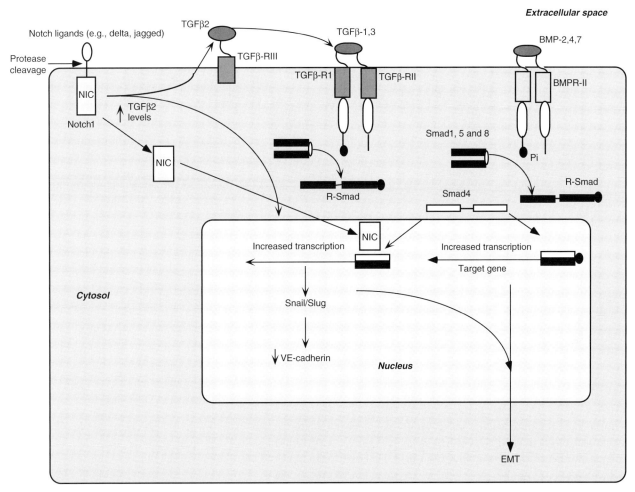

FIGURE 5 Pathway of TGFβ/BMP/Notch signaling in the initiation of EMT. Type III TGFβ receptor (TGFβ-RIII) presents TGFβ2 to TGFβ-RII in the developing heart valve. TGFβ signaling is mediated through snail/slug transcription factors and can result in reduced expression of VE-cadherin. BMPs (e.g., BMP2, 4 and 7) in the developing valve signal by means of the BMP-RII receptors, which are crucial for valve development, and by intracellular Smad mediators. Smad6 antagonizes the interaction of Smad1 with Smad4, thereby decreasing BMP signaling. Synergy between TGFβ and BMP signaling in cardiac cushion explants has been shown to facilitate EMT. Also shown is the Notch signaling pathway in which the transmembrane receptor Notch, on binding specific ligands, is cleaved to form NIC, which subsequently translocates to the nucleus activating specific gene transcription leading to EMT.

and 6). Smad6, an inhibitor of Smad signaling downstream of Alk2, inhibited EMT in AV cushion endocardial cells.[187] Mice with disrupted *Madh6*, which encodes Smad6, display marked hyperplasia of the cardiac valves and also have systemic hypertension and decreased endothelial cell–mediated vasodilation.[188] Inhibitory Smad6 signaling may also play a physiological feedback role in development by limiting the number of endocardial cells that undergo EMT.

In addition to the Smad transducers, the downstream effectors of TGF-β and BMP signaling in developing ECs are still not entirely defined but likely include factors of the Snail/Slug family, a group of zinc-finger transcription factors that primarily act as transcriptional repressors.[189] Inactivation of *Slug* with antisense oligonucleotides impairs epithelial-to-mesenchymal transdifferentiation.[190] The expression patterns of *Slug/Snail* in the developing heart suggest that these transcription factors could be involved in the regulation of EMT in EC formation. In chick hearts, Slug protein levels are highly expressed within the mesenchyme of developing cushions and in a subset of endocardial cells overlying the cushions,[191] where it is thought to be a target of TGF-β2 during EMT.[192] Similarly, in the embryonic mouse, *Snail* is expressed in the mesenchyme of many regions that undergo epithelial-to-mesenchymal transdifferentiation and in the endocardium and mesenchyme of developing heart valves.[193] VE-cadherin expression seems to be reciprocal to Snail, as revealed in studies of *Notch* signaling mutants.

Data derived from zebrafish studies suggest that the evolutionarily conserved Notch signaling pathway may have a significant role in endocardial EMT, because expression of the receptor *Notch1b* is localized within the endocardium of the presumptive AV valve region, both before and during the stages when EMT occurs.[194] Notch is a transmembrane protein that on binding specific ligands (e.g., Jagged or Delta) is proteolytically cleaved, and its intracellular domain (NIC) is translocated to the nucleus. The cleavage of Notch converts this transmembrane receptor to a transcriptional coactivator. Embryos that lack Notch signaling elements manifest severely attenuated cardiac *Snail* expression, abnormal maintenance of intercellular endocardial adhesion complexes, and abortive EMT. Transient ectopic expression of activated *Notch1* in zebrafish embryos led to hypercellular cardiac valves, whereas *Notch* inhibition prevented valve development.[195]

In mice, disruption of *Hesr2*, a downstream target of Notch signaling was found to result in dysplastic AV valves with tricuspid and mitral valve regurgitation, a perimembranous VSD, and a secundum ASD, with most mice dying from resultant congestive heart failure (CHF).[196] These observations support the view that the Notch signaling pathway, including the targeted *Hesr2*, plays an important role in the formation and function of the AV valves. *Notch1* mutants retain strong VE-cadherin expression, fail to express Snail, and fail to undergo EMT.[195] Notch activity promotes EMT during normal cardiac development, partially by means of the transcriptional induction of the Snail repressor, a potent and evolutionarily conserved mediator of EMT in many tissues types. In the embryonic heart, Notch functions by means of lateral induction to promote a selective TGF-β–mediated EMT that leads to cellularization of the developing cardiac valvular primordia. Embryos that entirely lack Notch signaling elements exhibit severely attenuated cardiac *snail* expression, abnormal maintenance of intercellular endocardial adhesion complexes, and abortive endocardial EMT *in vivo* and *in vitro*. A possible model to explain these findings would be that TGF-β induced by Notch signaling in the developing cushion activates *snail*, which decreases the expression of cell adhesion molecules and thereby down-regulates endocardial cell–cell adhesion, promoting endocardial cells to initiate invasion into the cardiac jelly.

The biological effects of VEGF are mediated by two receptor tyrosine kinases (RTKs), VEGF-R1 and VEGF-R2, which differ considerably in signaling properties. Cell proliferation, vascular permeability, chemotaxis, and survival in endothelial cells in the developing embryo are regulated by VEGF.[197] The downstream mediators of VEGF-R signaling are not completely known but may include inositol 1,4,5 phosphate/diacylglycerol, ERK, and MAPK pathways, and in valve endothelial cells, NFATc1.[198,199]

Although broadly expressed in early endocardial cells, VEGF expression becomes restricted to a subset of endocardial cells lining the AVC, an indication that VEGF has a role in EC formation.[200] It remains unclear whether the VEGF-expressing endothelial cells in the cushion-forming region represent a unique subpopulation of endothelial cells predetermined to undergo EMT or whether the VEGF-producing cells induce proliferation or increased permeability of adjacent endothelial cells in the developing EC to undergo EMT.[183]

Selective myocardial overexpression of *VEGF* in the early embryo (between E 3.5 and E 9.5) resulted in failure of EC formation at the AVC and OFT.[201] These embryos also had multilayered endocardium, suggesting a dysregulation of the differentiation process and overexpression of an endothelial phenotype. These findings were confirmed *ex vivo* using the collagen explant system, in which myocardium and endocardium of the developing cushion are explanted onto a type I collagen gel to recapitulate EMT events.[202] The addition of exogenous VEGF inhibited EMT in the forming AVC cushions. Taken together, these findings confirm that VEGF levels are tightly regulated during normal heart development and that even moderate increases in VEGF expression can have profound developmental consequences, possibly by inhibiting endothelial cell differentiation and thereby negatively regulating EMT.

Both hypoxia and hyperglycemia can regulate VEGF expression in the developing cardiac valves. Studies using tissue explants have shown that hypoxia decreases cushion EMT, which can be reversed by the addition of soluble VEGF-R1.[203] Moreover, under hypoxic conditions, VEGF expression was increased nearly 10-fold in EC. These results suggest that fetal hypoxia may increase VEGF expression in the cushion-forming areas inhibiting EMT and may contribute to CHD, including valves and interatrial septum.

In the developing mouse, hyperglycemia reduces VEGF expression.[204] By use of the tissue explant system, it has been shown that elevated glucose inhibits AVC cushion ability to undergo EMT.[205] Adding back exogenous recombinant VEGF-A165 abrogated the effect of hyperglycemia by allowing normal cushion EMT. These results suggest that decreased VEGF expression during development inhibits cushion formation, potentially by inhibiting endothelial migration into the cardiac jelly, and underscores the importance of highly controlled VEGF levels during EC formation, because either overexpression or underexpression of VEGF causes hypoplastic EC. Interestingly, neonates born to diabetic mothers have an approximately threefold increase in the incidence of CHD, with roughly a 10- to 20-fold increased risk of rare and complex abnormalities such as DORV and truncus arteriosus.[206] Dramatic reduction in the occurrence of CHD in children born to diabetic mothers with strict glycemic control during pregnancy suggests that hyperglycemia has a direct teratogenic effect.[207]

Members of the nuclear factors of activated T-cell (NFAT) family, which mediate transcriptional responses of the Ca^{++}/calmodulin-dependent protein phosphatase calcineurin, have been implicated in cardiovascular development almost exclusively in vertebrates, and a model of its pathway in endothelial cells is shown in Fig. 6.[208]

Genetic inactivation of *NFATc1* in the mouse demonstrated that *NFATc1* is required for cardiac valve formation,[209,210] with some studies finding greater effect on the pulmonary

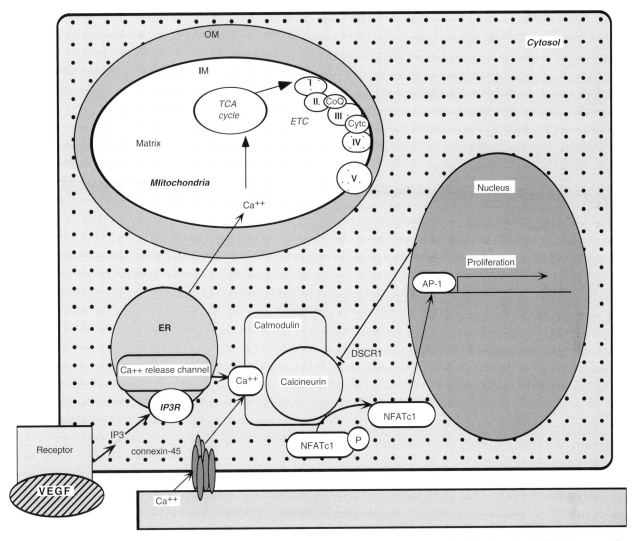

FIGURE 6 Model for NFATc1 as a transcriptional regulator of endothelial cell fate. VEGF signaling through NFATc1 increases the proliferation of pulmonary valve endothelial cells. In the developing cardiac cushion, Ca++ may enter the endothelial cell through connexin-45 gap junctions and activate calcineurin. Calcineurin, in turn, dephosphorylates NFAT family isoforms, including NFATc1. NFATc1 is then transported into the nucleus, where it interacts with transcriptional regulators, including AP-1, to affect gene transcription. The endogenous calcineurin inhibitor DSCR1, a target of NFATc1, may establish a negative feedback loop by inhibiting calcineurin.

and aortic valves.[209] Consistent with a function in the formation of these endocardial-derived structures, *NFATc1* has been found to be exclusively expressed in the endocardium from the initiation of endocardial differentiation in the primary heart-forming field. Within the endocardium, specific inductive events seem to activate NFATc. It is localized in the nucleus only in endocardial cells that are adjacent to the interface with the cardiac jelly and myocardium, which are thought to give the inductive stimulus to the valve primordia. Treatment of wild-type embryos with FK506, a specific calcineurin inhibitor, prevents nuclear localization of NFATc.[210] Studies have also shown that NFATc1 mediates VEGF signaling of proliferation of human pulmonary valve endothelial cells.[211] The calcineurin-specific peptide inhibitor reduced both VEGF-induced human pulmonary valve endothelial cell proliferation and abrogated VEGF-induced *NFATc1* nuclear translocation, suggesting a functional role for *NFATc1* in endothelial growth. Others have shown that the initiation of heart valve morphogenesis in mice requires calcineurin/NFAT interaction to repress VEGF expression in the myocardium underlying the site of prospective valve formation. This repression of VEGF at E9 is essential for endocardial cells to transform into mesenchymal cells.[212] Interestingly, an enhancer element located within the first intron of the mouse *NFATc1* gene has been recently identified that is responsible for the restricted and high-level *NFATc1* gene expression found in provalve endocardial cells of both the AVC and OFT during valvulogenesis.[213]

It has also been reported that the DSCR1 protein (Down syndrome critical region) alias MCIP1 (modulatory calcineurin interacting protein 1), whose gene maps to region 21q22.1–q22.2, may play a contributory role in the cardiac defects (mainly involving EC tissue) present in Down syndrome. Overexpression of *DSCR1* in cardiomyocytes inhibits Ca^{++}-dependent nuclear translocation of NFAT.[214] In addition, DSCR1 functions as an endogenous calcineurin inhibitor.[215] DSCR1 may play an important role in the development of EC defects as suggested by its increased expression in regions that correlate with areas of defective EC development.[216]

Gap junction channels are necessary during early cardiogenesis. Connexins are a group of transmembrane proteins that form gap junctions between cells. Products of several connexin genes have been identified in the mammalian heart (e.g., *Cx45, Cx43, Cx40, and Cx37*), and their expression is regulated during the development of the myocardium. *Cx45* is the first connexin expressed in the developing mammalian heart. In the mouse E9.5 developmental stage, *Cx45* is markedly up-regulated in cells in the AVC and OFT, suggesting a potential role in the development of the EC.[217] Whereas disruption of cardiac connexin *Cx40* and *Cx43* genes causes conduction defects and dysrhythmias but not embryonic death, disruption of *Cx45* gene does.[218,219] *Cx45*-deficient mice died of HF at around embryonic day 10 and have defective EC caused by impairment of the EMT of the cardiac endothelium.[220] Activation of this endothelium depends on the presence of the Cx45 gap junctions, because signaling through Ca^{++}/calcineurin pathway and *NFATc1* was disrupted in the mutant mice hearts. These findings indicate the need for gap junction channels during early cardiogenesis and implicate Cx45 in the development of a number of CHD. On the other hand, caution is necessary in this mice model, because it often features extensive AV conduction defects in addition to EC defects. It has been noted that *Cx 45*-deficient mice die at E10 3 days earlier than mice with CHF secondary to lack or failure of valve development, suggesting that the formation of the EC in this model may, in fact, be delayed rather than disrupted and that the cause of death may be secondary to a hypocontractile heart.

As previously discussed, the involvement of Wnt signaling in early cardiac development has been demonstrated by studies showing the requirement of *Wnt* inhibition in early cardiac specification. These observations revealed the critical role of Wnt signaling in the posterior mesoderm to repress cardiogenesis and patterning in the chick embryo.[221] Specific inhibition of canonical Wnt/β-catenin signaling by Dickkopf-1 (Dkk-1) or Crescent was used, and it resulted in the initiation of cardiogenesis in vertebrate embryos. Furthermore, Wnt signaling may also closely regulate valve development.

Induction of EMT in the endocardial cells requires β-catenin transcriptional activity. In mice, in which the *cat* gene encoding β-catenin was selectively inactivated in endothelial/endocardial cells, the EC fail to develop because endocardial cells do not undergo EMT.[222] Recently, EC-specific expression of Wnt/β-catenin signaling has been identified in the developing mouse heart. At E11.5, Wnt signaling is restricted to a subset of cells in the AVC and OFT, with only a subset of mesenchymal cells staining for Wnt/β-catenin signaling.[223] This pattern of expression, which is conserved in mammals, further supports the role of the Wnt/β-catenin pathway in EC development. The involvement of this pathway in valve formation has been also demonstrated in zebrafish, in which an early missense mutation in the *adenomatous polyposis coli* (APC) gene product resulted in embryonic lethality 96 h after fertilization, timing that corresponds with the cardiac valve formation. Lack of function mutations in the *APC* gene lead to increased β-catenin signaling and result in excessive and abnormal EC formation, suggesting that endocardial cells lining most of the heart had undergone EMT.[224] Moreover, in contrast to the intranuclear location of β-catenin found in wild-type embryos that is restricted to endocardial valve–forming cells overlying the cushion-forming areas, homozygous *APC* truncation mutants (which constitutively activate the Wnt signaling pathway) exhibited nuclear-localized β-catenin throughout the heart concomitant with a marked up-regulation of valve markers. Concomitantly, proliferation and EMT, normally restricted to EC, occurred throughout the endocardium. Overexpression of *APC* or inhibition of *Wnt* by Dickkopf 1 reversed this phenotype. These findings support the concept that the Wnt/ß-catenin pathway plays an important role in determining endocardial cell fate and EMT progression.

Other components of the Wnt signaling pathway have been shown to mediate mesenchymal cell proliferation leading to proper AVC cushion outgrowth and remodeling in the developing heart. Within the chick heart, *Wnt-9a* expression is restricted to AV cushions, primarily in the AVC endocardial cells, whereas the secreted *Wnt* antagonist, the Frizzled-related protein (*Frzb*) gene expression is detected in both endocardial and transformed mesenchymal cells of the developing AV cardiac cushions.[225] *Wnt-9a* stimulates β-catenin–responsive transcription and promotes cell proliferation in the AVC cells; overexpression of *Wnt-9a* results in enlarged EC and AV inlet obstruction. Functional studies also revealed that *Frzb* inhibits *Wnt-9a*–mediated cell proliferation in EC, indicating that a dynamic balance between *Wnt-9a* and *Frzb* is involved in the regulation of mesenchymal cell proliferation in EC formation. Studies directed at developing a set of molecular markers, encompassing all stages of cardiac valve development, and providing an analysis of their gene expression profile have identified 13 EC markers, including the demonstration of active Wnt/β-catenin signaling components, as well as the expression of the transcription factor Fog1 in developing EC.[223]

In addition to its signaling functions, β-catenin acts as a structural link between actin and VE-cadherin to form the adherens junction, a molecular scaffold that mediates cell–cell adhesion and cell polarity in endothclial cells.[226] Phosphorylated β-catenin also associates with platelet-endothelial cell adhesion molecule (PECAM-1/CD31), a transmembrane protein of the immunoglobulin (Ig) family involved in cell–cell contact. The association of β-catenin with cadherins and PECAM-1/CD31 in dynamic junctional complexes may provide a cellular reservoir for phosphorylated β-catenin to regulate the level of free, cytosolic β-catenin and modulate its availability for nuclear translocation.[227]

The interactions between ß-catenin, PECAM-1/CD31, and VEGF may be critical in the control of EMT during EC formation. VEGF signaling increases the phosphorylation of β-catenin in a time-dependent and dose-dependent manner, leading to increased association of β-catenin with PECAM-1/CD31.[227] Sequestration of β-catenin by VEGF signaling may represent one means by which increased VEGF levels can decrease EC formation. During normal EMT, PECAM-1/CD31 is down-regulated and smooth muscle α-actin is upregulated as endocardial cells differentiate into a mesenchymal phenotype.[228] If EMT is disrupted, PECAM-1/CD31 levels persist.[205] It has been proposed that as PECAM-1/CD31 is down-regulated, the cytosolic levels of β-catenin increase and activate proliferation of cells undergoing EMT; β-catenin could thus serve as the link between activation of the mesenchymal program and population of the cardiac jelly with mesenchymal cells. Furthermore, reduced PECAM-1/CD31 levels promote cell motility, enabling their ability to migrate and penetrate the ECM. Consistent with this hypothesis, in PECAM-1/ cd31$^{-/-}$ mice, the cushion-forming areas remain competent to undergo EMT, even in the presence of hyperglycemia, which reduces the level of VEGF.[205]

ErbB: Integration of Extracellular Matrix Signals

Several studies have demonstrated a role for the ErbB family of receptors in regulating cushion remodeling and valve formation (Fig. 7).[229] ErbB proteins mediate cell proliferation, migration, differentiation, adhesion, and apoptosis in numerous cell types.[230] The ErbB family proteins are RTKs and include ErbB1/EGFR/HER1, ErbB2/Neu/HER2, ErbB3/HER3, and ErbB4/HER4. The four ErbB proteins bind a wide range of ligands with varying affinities that include epidermal growth factor (EGF), members of the heregulin/neuregulin (HRG) family, heparin-binding epidermal growth factor (HB-EGF), TGF, amphiregulins, betacellulin, and epiregulin.

The requirement for ErbB/EGFR signaling in cushion development has been confirmed by observations on the cardiac-specific effects of ligand knockouts. HB-EGF is a widely expressed growth factor of the EGF family that can bind ErbB1 and ErbB4.[231] In the developing mouse heart, HB-EGF is strongly expressed in the endocardium overlying the cushion-forming area.[232] *HB-EGF* $^{-/-}$ mice display markedly enlarged and malformed semilunar and AV valves and OFT valves and die shortly after birth. Consistent with this finding, mice deficient for tumor necrosis factor–*c*onverting *e*nzyme (TACE/ADAM-17, a processing enzyme that cleaves HB-EGF from a precursor pro-HB-EGF polypeptide into its active form) also possess enlarged AV and semilunar valves. *HB-EGF* $^{-/-}$ and *TACE* $^{-/-}$ mice display normal cushion development through E13.5 but have thickened valves by E14.5. These findings suggested that *HB-EGF* $^{-/-}$ and *TACE* $^{-/-}$ mice exhibit normal EMT initiation but uncontrolled mesenchymal proliferation during remodeling that resulted in defective cardiac valvulogenesis. *HB-EGF* $^{-/-}$ mice also displayed dramatic increases in activated Smad1, 5, and 8, suggesting that ErbB/EGFR signaling is required to regulate the BMP pathway. Moreover, developing valves in *HB-EGF* $^{-/-}$ and *TACE* $^{-/-}$ mice show increased staining for bromodeoxyuridine (BrdU, a marker of cell proliferation) but display little evidence of cellular apoptosis. It seems likely that HB-EGF, after activation by TACE, signals through EGFR to limit mesenchymal cell proliferation, because *erbB4* knock-out mice display only moderately decreased EC size but have no defects in fully developed valves.[233]

The crucial importance of the cardiac jelly in providing a signal that initiates endocardial differentiation is well recognized. Although Krug and associates focused on soluble factors present in the jelly,[234] recent observations have centered on the involvement of the ECM in regulating growth factor activity. A specific area of investigation has been on the role of hyaluronic acid (HA) in mediating ErbB signaling.[235]

HA is a glycosaminoglycan composed of alternating glucuronic acid and *N*-acetylglucosamine (NAG) residues and is present as a hydrated gel in the ECM to expand the extracellular space and regulate ligand availability. Furthermore, HA interacts directly with numerous ECM proteins, including the proteoglycan versican, a major constituent of the cardiac jelly.[236] A variety of sulfated proteoglycans in ECM have been considered to be involved in EC remodeling into valvuloseptal tissue.[237] These molecules accumulate at the cell surface or pericellular matrix of the migrating cushion cells that secrete them and seem to play a role in the final positioning and proper spacing of cells within the differentiated tissue.

Three HAS genes, *has1*, *has2*, and *has3*, are present in mammals.[238] *Has2* encodes the major enzyme responsible for HA synthesis during development. *Has2* $^{-/-}$ mice exhibit severe cardiac and vascular abnormalities, including pericardial effusion, abnormal vessel growth, and complete absence of cardiac jelly and die by mid-gestation (between E9.5 and 10).[239] In the absence of cardiac jelly, no EMT occurs, and the EC are unable to form. In addition to its structural role, HA can modulate cell-signaling events.[240] Endocardial cells overlying the cushion-forming region in *Has2* $^{-/-}$ mice display reduced EMT and migration, an effect that can be blocked by gene rescue, adding back exogenous HA or by transfection with constitutively active Ras.[239] Transfection with a dominant-negative Ras shuts off the ability of HA to promote EMT. Interestingly, heregulin (a ligand for ErbB3) rescued the *Has2* $^{-/-}$ phenotype in *ex vivo* cushion explant

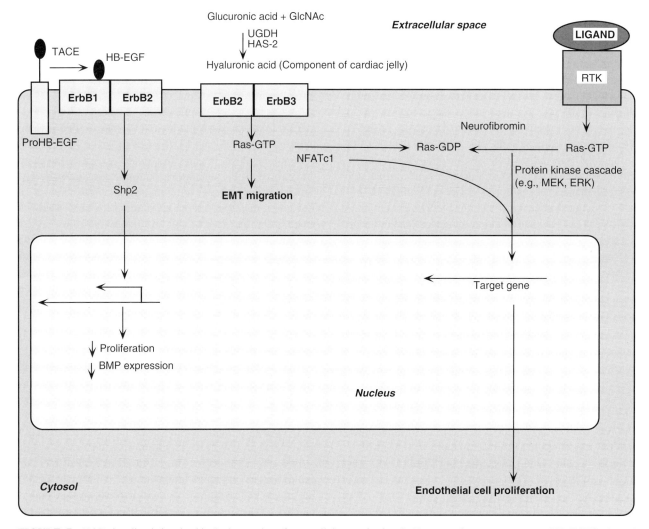

FIGURE 7 ErbB signaling is involved in the integration of extracellular matrix signals. Transmembrane precursor pro-HB-EGF is cleaved by TACE to HB-EGF, a ligand for membrane-localized ErbB1 and ErbB4. Binding of HB-EGF to ErbB1, and possibly formation of a heterodimer with ErbB2, seems to reduce downstream EMT, proliferation, and BMP expression. Activation of ErbB2/3 heterodimers by the extracellular matrix polysaccharide hyaluronic acid (HA) increases EMT and migration, an effect that is likely mediated by Ras signaling. Synthesis of HA from glucuronic acid and N-acetylglucosamine (GlcNAc) depends on the enzymes UDP-glucose dehydrogenase (UGDH) and hyaluronic acid synthase-2 (HAS-2). Also depicted is the involvement of neurofibromin in EMT regulation. Neurofibromin is a Ras-specific GTPase-activating protein (GAP) that cycles Ras from an active GTP-bound state to an inactive GDP-bound, state. Ras signaling is activated by receptor tyrosine kinases (RTKs), which bind a wide range of ligands and tranduce the activation of downstream targets to increase mesenchymal proliferation. Downstream signaling targets of Ras interact with NFATc1 to alter gene transcription. Neurofibromin may, therefore, decrease endothelial and/or mesenchymal cell proliferation by modulating Ras signaling.

models.[241] Furthermore, the *Has2* −/− mice possess decreased ErbB2/ErbB3 phosphorylation in EC compared with wild-type embryos. Addition of HA to *Has2* −/− tissue explants restored ErbB3 phosphorylation. *ErbB* −/− mice die by E13.5 and have cardiac cushions defects, with decreased mesenchymal content.[242] Notably, ErbB3 is expressed by EC cells and mesenchymal cells undergoing EMT. In contrast, EGF, ErbB2, and ErbB4 expression is limited largely to cardiomyocytes during the critical cushion-forming window (E9.5 to E10.5). How HA signaling interacts with other ErbB ligands, such as neuregulin and HB-EGF, remains to be determined.

Stainier and associates have identified 58 mutations that affect morphogenesis and function of the cardiovascular system in zebrafish. The *jekyll* mutant results in failure of EC formation and deficient heart valve formation.[243,244] The *jekyll* mutant contains a point mutation in the *ugdh* gene (encoding UDP-glucose dehydrogenase, UGDH) residing in the enzyme's active site. UGDH is required for heparan sulfate, chondroitin sulfate, and HA production.[245] The close phenotypic correlation between the *jekyll* mutant and *Has2* −/− mice suggests that these two mutations interrupt the same pathway affecting cushion formation and subsequent valvulogenesis.

The expression of the extracellular matrix proteoglycan versican is associated with valvulogenesis in the developing mouse heart.[246] Versican is a chondroitin sulfate proteoglycan expressed in the pathways of neural crest cell migration and in pre-chondrogenic areas of the developing chick and mouse.[247,248] This protein is nonpermissive for cell migration and appears in association with slow cell proliferation and cytodifferentiation. Versican may be a key participant in cardiogenesis, responding to the many diffusible signals that mediate interactions between the developing endocardium and myocardium. Localization of versican protein parallels versican mRNA during heart development in the mouse embryo. In mice at E8.5 to E9.5, the protein is found surrounding both the endocardial and myocardial cells in the atrial and ventricular regions.[246] However, by E10.5 as the heart loops, versican expression becomes chamber-specific with marked down-regulation in the atrium but is increased in the trabeculations of the ventricles (before formation of the ventricular septum is apparent) in the crest of the developing atrial and ventricular septa and throughout the development of the EC and valves. This protein is strongly expressed in the EC of the AV valves as cells delaminate from the endocardium and migrate into the cardiac jelly and continues to be expressed in the developing valves into the neonatal period. Versican is well known for its antiadhesive properties,[249,250] and its expression at sites of EMT may reflect a role in the detachment and migration of mesenchymal cells rather than play a role in the transformation process itself. In addition, versican binds to HA,[251] and their interaction may mediate the downstream effects of HA.

Immediately before EC tissue formation, small particulates accumulate within the cardiac jelly of only the AV canal and ventricular outflow tract regions of the heart.[252] These molecular complexes referred to as *adherons* can be visualized by use of antibodies to fibronectin, which is one of the constituent proteins of the particles. These proteins are expressed in regions where epithelial–mesenchymal interactions occur[253] and that coexpress BMP4, *msx-1*, and *msx-2* and are important for the formation of cushion mesenchyme involved in EMT.

The final stages of valvulogenesis involve the transition from EC to embryonic valves and septa and make up the least well-understood stage of valvuloseptal development.[183] These stages involve the further differentiation of regions of cushion mesenchyme into the fibrous connective tissue of both inlet and outlet valvular leaflets.[254] Subsequently, the inlet valve, particularly the tricuspid valve, as well as the septa formed by the fusion of the cushions, are partially "myocardialized" to give a muscular component to the terminally differentiated tissues. The differentiation of the cushion mesenchymal cell into a valvuloseptal fibroblast correlates with the expression of the two ECM microfibrillar proteins, fibrillin and fibulin, which may function as a scaffolding for cell adhesion as these cells differentiate, and in proper tissue remodeling.[255,256]

A number of regulatory pathways have been described in this section that contribute to early mesenchyme differentiation, cardiac jelly, migration and formation of EC, EMT, as well as the final remodeling of the mesenchymal cushions into atrioventricular valves. Although presented separately in this chapter, recent observations have suggested that there is considerable interaction and integration between these signaling pathways as shown in Fig. 8. Other regulatory factors will be identified in the future that will allow the integration of these pathways and will shed further light on their function. For instance, gene expression of the regulator

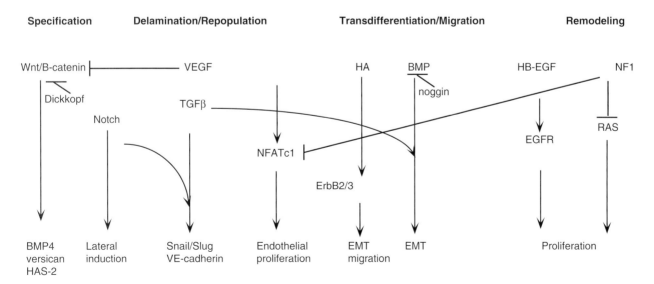

FIGURE 8 Interaction of signaling factors in valve development and remodeling. Numerous signaling pathways and transcriptional regulators act to coordinately regulate the process of heart valve formation. Black arrows denote positive/synergistic interactions between pathways. Blunt arrows denote inhibitory effects between pathways.

Sox9 is activated when endocardial endothelial cells undergo mesenchymal transformation and migrate into the ECM to form EC. This expression pattern suggests that *Sox9* might play a contributory role in the pathway that controls the formation of cardiac valves and septa. In *Sox9*-null mutants, EC are markedly hypoplastic.[257] In these mutants, *NFATc1* is ectopically expressed and no longer restricted to endothelial cells, and *Sox9*-deficient endocardial mesenchymal cells fail to express ErbB3, which is required for EC cell differentiation and proliferation. Another class of regulators recently identified in conjunction with AV valve defects are the FOG-1 and FOG-2 zinc-finger proteins, GATA- interacting cofactors. These cofactors can act either as enhancers or repressors of GATA transcriptional activity, depending on the cell and promoter context.[258] Mice containing disrupted FOG-2 alleles display tricuspid atresia, suggesting a genetic basis for this CHD.[259] Other cardiovascular defects occurring in these mice include AVC, TOF, and general failure of coronary vessel formation.[260] Transgenic reexpression of *FOG-2* in cardiomyocytes rescues the *FOG-2* −/− vascular phenotype, demonstrating that FOG-2 function in myocardium is required and is sufficient for coronary vessel development. *FOG-1* −/− mice die at E 14.5 with defects that include DORV and common AV valve.[261] Conditional inactivation of *FOG-1* established that endocardial-derived, rather than neural crest–derived tissues, are the site of expression for this regulatory factor. These findings revealed a succession of molecular steps in the pathway of EC development and suggest novel epistatic relationships within these interfacing pathways.

Formation of Semilunar Valves (Aortic and Pulmonic)

Although the number of animal models and the information available regarding the embryonic development of the AV valves is rather impressive, information on the development of the semilunar valves has been more slowly forthcoming. Previously, the mechanisms that regulate the development of the semilunar and AV valves were thought to be similar; however, specific defects found in the semilunar valves suggest that their development may be differently regulated.

The OFT in the looped heart tube develops as a transitional zone lined on the inside by two endocardial OFT cushions containing proximal (*bulbar*) and distal (*truncal*) regions that are often difficult to distinguish in humans and mammals. The distal cushions take part in semilunar valve formation, whereas the proximal cushions eventually become the muscular OFT septum.[262] The subsequent fusion of the cushions from distal to proximal has been shown to involve the participation of extracardiac neural crest cells that contribute to the central (or condensed) mesenchyme incorporated within the OFT cushion mass. At the proximal OFT cushions, the neural crest cells are in close contact with the myocardial cells and, in part, invade the myocardium, and most of these cells undergo apoptosis as a feature of the remodeling needed for valvulogenesis and are replaced during "myocardialization."[263]

The development of the semilunar valves is more controversial than the AV valves. OFT development is mediated by the involvement of a subpopulation of ectodermally derived neural crest cells derived from the branchial arches migrating to the distal OFT that is required for aortopulmonary septation.[264] However, recent cell lineage studies have demonstrated the endothelial origin of resident cells throughout the leaflets of aortic and pulmonic valves, as well as the AV valves.[265,266] Although the precise role of the neural crest cells in the development of the semilunar valves and OFT septation remains to be determined, critical elements of neural crest signaling are necessary for normal OFT and valve development, indicating that these cells play an important inductive role. Either ablation of the neural crest cells or neural crest–targeted mutations in specific resident signaling genes can lead to abnormal OFT and valve development, as well as to CHD. Defects in the secreted factor FGF15 and neural crest localized BMPR cause abnormal cardiac neural crest cell migration and OFT malformation.[267-269] Moreover, neural crest–produced neurotrophin-3 (NT-3) is present within the OFT throughout cardiac development, even when neural crest cells are not detectable, suggesting that NT-3 interacts with cells in the OFT that are not of neural crest origin.[270] It has been proposed that neural crest signaling (e.g., NT-3) may function by its interaction with cells in the OFT, such as those arising from the secondary heart field, which have been shown to provide both essential myocardium and smooth muscle to the developing OFT.[271-273] This hypothesis has been strongly supported by demonstration that in cardiac neural crest–ablated embryos, the secondary heart field (derived from splanchnic mesoderm) fails to add myocardial cells to the OFT, and elongation of the tube is deficient.[274] Defects in other signaling factors (not produced in neural crest cells), including SHH and Tbx1, lead to defects in neural crest migration and positioning, which at least partly mediates their overall effects on OFT and valve development.[275,276]

In addition to its role in OFT development, cardiac neural crest cells also contribute extensively to vascular smooth muscle differentiation. Recent observations have shown that the myocardin-related transcription factor (MRTF-B) targeted in neural crest cells plays a critical role in regulating the differentiation of cardiac neural crest cells into smooth muscle and is required for normal cardiovascular morphogenesis.[13]

Other significant signaling influences on OFT and semilunar valve development have been identified by use of techniques such as modification of gene expression *in vivo*. *Sox* is a family of transcription factors expressed at the neural plate border in response to neural crest–inducing signals, and Sox proteins (e.g., Sox8, Sox9, Sox10, LSox5, Sox4 and Sox11) seem to function in the regulation of multiple aspects of neural crest development in many tissues,[277] albeit in cardiac valve development little information clarifying this interaction is available. Expression of *Sox8*, *Sox9*, and *Sox10* in the developing heart correlates with heart septation and with the differentiation of the connective tissue of the valve leaflets.[278] As noted previously, Sox9 function is involved in mesenchymal transformation from endocardial endothelial cells required for cushion and valve formation. In the *Sox9*-null mutants, the EC are markedly hypoplastic. Moreover, in embryos in which *Sox9* inactivation was specifically targeted to cardiac neural crest cells, the EC defects appear only in the distal part of the OFT, whereas the proximal portion of the OFT and the AV cushions are normal. Thus, Sox9 is required for the development of EC derived from both neural crest cells and endothelial cells of the heart. Sox4 is abundantly present in the developing heart of chicks, mice, and humans; its expression was also abundantly detected in tissues of neural crest origin, including the pharyngeal arch.[279] Using targeted gene disruption, *Sox4* -/- embryos were found to succumb to circulatory failure at day E14.[280] This resulted from impaired development of the endocardial ridges (a specific site of *Sox4* expression) into the semilunar valves and the outlet portion of the muscular ventricular septum. These findings have suggested the existence of a *Sox4-deficiency syndrome* defined as a dysfunction of endocardial tissue of the OFT, leading to a lack of development and/or fusion of the endocardial ridges and the semilunar valves, with an arrangement of the ventriculoarterial connection corresponding with transposition of the great arteries (TGA).[281] The *Sox4*-deficiency syndrome is due to a defective function of the endocardially derived tissue and structures whose fate could be followed thanks to the presence of the highly characteristic rods of condensed mesenchyme. Restriction of these defects to the arterial pole, even though *Sox4* is equally expressed in endocardially derived tissue of the AVC, suggests that interaction between the endocardially derived tissue of the OFT and the neural crest–derived myofibroblasts may determine the proper development of the arterial pole. The abundant expression of Sox4 in neural crest tissues supports its potential role in the cardiovascular abnormalities detected in *Sox4* mutant mice.

Four genes are known to encode proteins belonging to the nuclear factor of activated T-cells (NFAT) complex (*NFATc*, *NFATp*, *NFAT3*, and *NFAT4*). These genes are expressed in several tissues, including the immune system, in which an enhanced response in mice lacking the transcription factor NFAT1 has been reported without definitive abnormalities in cardiac development.[282] In two independent studies, it has been shown that *NFATc3*-mutant mice embryos played a particular role in the morphogenesis of the semilunar valves and septa.[283,284] Interestingly, the specificity of *NFATc* expression in the endothelial lining is rather unique in heart development, because *NFATc* is the only endocardial-specific transcription factor thus far described. By E13.5 the embryos displayed abnormalities of valve structure, leading *in utero* to CHF and death. Although septation into right and left ventricle OFT was normal, both semilunar valves were underdeveloped. Vascular endothelium, myocardium, and ECM appeared unaffected in *NFATc3* mutant embryos, as indicated by the normal expression of PECAM-1, cardiac α-actin, and fibronectin. Because *NFATp* does not seem to be expressed in the heart, the cardiac-specific expression of *NFATc* may be under the control of different transcriptional molecules. Although the neural crest participates in the formation of the semilunar valves, *NFATc* is exclusively expressed in the endocardium; the relationship between neural crest differentiation and abnormal valve development of *NFATc3*-mutant embryos remains unclear. In addition, *Sox4* expression was normal in these embryos, indicating that *Sox4* is not likely to be a target for *NFATc*, although both genes may have converging roles in heart development. Interestingly, blocking calcineurin activity results in inhibition of *NFATc* nuclear translocation in these cardiac structures, suggesting that *NFATc* may be responsive to calcium signaling. Furthermore, the involvement of NFAT transcription factors as downstream effectors of calcineurin signaling has also been reported in cardiac hypertrophy.[285]

Similar to the effects on the OFT and valve phenotype observed in other mouse models, including *Sox4* and *NFATc1* null embryos, *Foxp1*-mutant embryos have severe defects in cardiac morphogenesis, including cardiac myocyte proliferation and maturation of OFT septation and cushion defects, which lead to embryonic death at E14.5.[160] *Foxp1*, is a member of a new *Foxp1/2/4* subfamily of *Fox* gene family of winged-helix/forkhead DNA-binding proteins that act as transcriptional repressors and are thought to regulate important aspects of cardiovascular development.[286,287] The phenotype in *Sox4*-null embryos bears striking similarities to the cardiac defects found in *Foxp1*-null embryos, including a similar embryonic stage of lethality, VSD, defective development of the EC in the OFT, marked increased cellularity within the cushion mesenchyme, and DORV in a subset of embryos.[281] Interestingly, *Sox4* expression in the OFT and cushions of *Foxp1*-null embryos is significantly reduced, suggesting that *Fox1p* may be upstream of *Sox4* in the pathways of OFT and valve development. Further studies involving the tissue-specific inactivation of *Foxp1* will be required to determine which cell type confers these cushion defects, as well as where and when *Sox4* expression is effected (e.g., by early defective EMT or as a result of defective tissue remodeling processes in the EC). Disruption in cushion remodeling is also a prominent feature of these embryos as demonstrated by reduced apoptosis and persistent *NFATc1* expression in the cushion mesenchyme. Semilunar valve defects were 100% penetrant, whereas defects in the AV valves were less frequently observed (45%) in *Foxp1* -/- embryos. Decreased apoptosis may contribute to the increased cellularity observed in the cushion mesenchyme of the OFT of *Foxp1* -/- hearts at E14.5. On the other hand, the atrioventricular cushions showed little difference in cell proliferation or apoptosis between *Foxp1* −/− and their wild-type litter mates.

As previously noted in connection with AV valves, ErbB signaling is clearly involved in semilunar valve development. Interestingly, in the analysis of genetic interaction between *EGFR*, encoding the epidermal growth factor receptor, and *Ptpn11*, encoding the protein-tyrosine-phosphatase Shp2, it was found that *EGFR* is required for semilunar, but not for AV valve development.[288] Mice homozygous for the hypomorphic *EGFR* allele waved-2 (*EGRF wa2/wa2*) that possess only 10–20% of the intrinsic tyrosine kinase activity of *EGFR*, exhibit semilunar valve enlargement resulting from overabundant mesenchymal cells in the leaflets. A similar phenotype of hyperplastic semilunar valves was found in *EGFR* −/− mice with a specific background (CD1). The penetrance and severity of the defects in *EGFR wa2/wa2* mice are enhanced by heterozygosity for a targeted mutation of exon 2 of *Ptpn11*, underscoring the interaction between these two proteins.

In zebrafish, the addition of EGFR kinase inhibitors or the targeted knock-down of *EGFR* expression with morpholinos resulted in a narrowed OFT with massive ventricular and atrial enlargement, likely secondary to increased afterload.[289] It is not yet clear whether the narrowed OFT results from thickened OFT cushions or an alteration of neural crest cell migration. However, it seems that the cardiac-specific phenotype of *EGFR* disruption is conserved across multiple experimental models.

A similar phenotype of hyperplastic semilunar valves present in the *EGF* receptor mutants (and, as previously noted, in animals deficient in HB-EGF) has been recently reported in mice containing disrupted alleles of phospholipase Cε (PLCε), an isoform of phosphoinositide-specific PLC, a downstream effector of RAS signaling.[290] Mice homozygous for the targeted *PLC* allele exhibit congenital malformations of both semilunar valves, with a varying degree of regurgitation, mild stenosis, and ventricular dilation. These malformations involve marked thickening of the valve leaflets, likely caused by a defect in valve remodeling at the late stages of semilunar valvulogenesis, and are accompanied by inhibition of Smad activation and BMP signaling.

Null mutation in the *Nf1* gene in mice causes markedly enlarged EC, DORV, and other noncardiac defects, including sympathetic ganglia, as well as liver and kidney abnormalities, and death by E14.5.[291] Mutations in *Nf1* cause neurofibromatosis, an autosomal-dominant disorder characterized by café au-lait spots, neurofibromas, an increased risk of neural crest cell–derived malignancy, and a sixfold increased risk of pulmonic stenosis. *Nf1* encodes neurofibromin, a Ras-specific GTPase activating protein (GAP), which inactivates Ras by cycling Ras from its active GTP-bound conformation to an inactive GDP-bound conformation (Fig. 7). By use of tissue-specific gene inactivation, it was demonstrated that endothelial-specific *Nf1* inactivation recapitulates key aspects of the complete null phenotype with multiple cardiovascular abnormalities involving the ECs and myocardium.[292] This phenotype is associated with an elevated level of Ras signaling in *Nf1* −/− endothelial cells and greater nuclear localization of the transcription factor NFATc1 (placing NF1 upstream of NFAT). In contrast, *Nf1* inactivation targeted to neural crest cells did not promote cardiac defects but resulted in tumors of neural crest origin resembling those seen in humans with neurofibromatosis. Therefore, neurofibromin plays an essential role in endothelial cells and is also required in the neural crest. Although the physiological role of neurofibromin in cardiac cushion development is not yet fully established, it may be to limit the extent of NFAT signaling and, thereby, to attenuate expression of *NFAT* gene targets, and availability of transcriptional coactivators.[183]

Cardiac Conduction System

The cardiac conduction system (CCS) is a heterogenous complex of cells responsible for establishing and maintaining the rhythmic excitation of the mature heart. Although substantial progress has been made in the understanding of the molecular pathways regulating many aspects of cardiogenesis, the molecular mechanisms that regulate the formation of the CCS are still poorly understood. However, as the number of mouse mutants with defects in CCS has increased, the potential to decipher the molecular pathways controlling the formation and function of this cellular network has increased.

The primary heart tube is a peristaltic pump that moves blood ahead as a result of a unidirectional wave of contractions along the tube. Within this slow-conducting heart tube, fast-conducting and synchronously contracting atrial and ventricular chambers develop, flanked by the slow-conducting primary myocardium of the inflow tract, AVC, and OFT.[6] The slow-conducting sinoatrial node, which contains the leading pacemaker and generates the impulse, and the AV node, which conducts that impulse to the bundle branches and Purkinje fibers, both originate from the slow-conducting myocardium of OFT and AVC, sometimes referred to as "cardiac specialized tissues." The fast-conducting AV bundle (or His bundles), bundle branches, and their ramifications develop from the ventricular segment.

Elements of the CCS are formed from the differentiation of cardiac cells into specialized conduction cells.[293] A role for Nkx2-5 and Tbx5 in the formation of a functional conduction system has been suggested from the AV node defects observed in humans and mice with dominant mutations in these transcription factors.[90,173–178] Increased expression of *Nkx2-5* in specialized conduction fibers is found (compared with myocardium) and may be the basis of the high sensitivity of these cells to decreased *Nkx2-5* dosage.[294] A direct downstream target of Tbx5 and Nkx2-5 that has been identified is connexin 40 (Cx40), with decreased expression levels in *Tbx5*-deficient mice and in mice expressing a mutant Nkx2-5 protein.

Electrical coupling of myocytes is mediated by gap junctions, which establish through membrane channels, a connection between the cytoplasm of adjacent cells permitting intercellular current flow and the transfer of depolarizing action potentials, a prerequisite for CCS. These gap junction channels are formed by alignment of two hexameric hemichannels, each composed of connexin (Cx) protein subunits, of which there are at least five different types expressed in the mammalian heart (e.g., Cx37, Cx40, Cx43, Cx45, and Cx46). Gap junctions and their connexin components are scarce in the developing sinoatrial and AV nodes, corresponding with the slow conduction in these regions. Cx37 is primarily found in endocardium,[295] and Cx40 is mainly found in the atrium and conduction system.[296–300] Cx43 is located in the atrial and ventricular myocardium and in the distal parts of the CCS,[298–300] and Cx45 is present throughout the heart in small amounts, with some evidence of increased expression in the conduction system.[301,302] The specific association of ventricular connexin45 with connexin40-expressing myocytes has been reported in the mouse and rat and may indicate that connexin45 contributes to the modulation of electrophysiological properties in the ventricular conduction system and needs to be further investigated in other species.[303] Connexin45 seems to be the isoform most continuously expressed by conduction tissues and constitutes a defining feature of the heterogenous nature of the tissues that make up the CCS of the rodent heart.[304]

In mice, a continuity between the common bundle and the septum is present, and *Cx40* deficiency results in right bundle-branch block and impaired left bundle branch.[305] Moreover, mice with null *Cx40* alleles exhibit reduced atrial, but not ventricular, conduction velocity.[306] *Cx40* knock-out mice have prolonged P waves on ECG, suggesting reduced atrial conduction velocity. Also, the PR interval is prolonged, suggesting AV nodal dysfunction, and the QRS complexes are prolonged.[307–310] Impaired function of the specific conduction system could explain the prolonged QRS duration, because *Cx40* is normally not expressed in the working ventricular myocardium.

In addition to the aforementioned regulatory control of the connexin channels and the conduction system exerted by Tbx5 and Nkx2.5, several signaling pathways including endothelin, neuregulin, Wnts, and BMPs,[311–318] as well as several other transcriptional regulators including additional T-box family members such as Tbx-3 and Tbx-2, the vertebrate muscle segment related homeobox factor Msx-2, the SP1-related factor HF-1b, and Hop, are operative in the development and functioning of the CCS.

The zinc-finger SP1-related transcription factor HF-1b has been shown to be critical in establishing conduction system identity. Mice lacking HF-1b exhibit a wide variety of abnormal conduction system function defects, including spontaneous ventricular tachycardia and a high incidence of AV block.[319] In the absence of HF-1b, molecular analysis of the myocardium surrounding specialized conduction fibers has revealed a heterogenous expression profile of genes resulting in decreased levels and mislocalization of connexins, as well as a marked increase in action potential heterogeneity. HF-1b, therefore, seems to be required to establish a molecular or physical identity between conduction and nonconduction cells in the developing myocardium, although the direct targets and precise mechanism of HF-1b remain to be determined. Interestingly, HF-1b, Cx40, and Cx43 have been found misexpressed and/or mislocalized in mice homozygous for the *LMNA* allele *N195K/N195K*, resulting in death at an early age because of cardiac dysrhythmia.[320] In addition, clinical laminopathies are associated with missense mutations in the *LMNA* gene encoding lamina, a nuclear intermediate filament protein serving as a major component of the nuclear envelope, and that result in DCM with significant conduction defects as discussed in the following chapter.[321]

The T-box transcription factor Tbx3 has been found to be an accurate marker for the murine CCS, and the regulatory function of Tbx3 in *Nppa* and *Cx40* promoter activity has been studied *in vitro*.[322] In the formed heart, *Tbx3* is expressed in the sinoatrial and AV node, bundle, and proximal bundle branches, as well as the internodal regions and the AV region. Throughout cardiac development, *Tbx3* is expressed in an uninterrupted myocardial region that extends from the sinoatrial node to the AV region, and this expression domain is present in the looping heart tube from E8.5 onward. Tbx3 is able to repress *Nppa* and *Cx40* promoter activity and abolish the synergistic activation of the *Nppa* promoter by Tbx5 and Nkx2.5. Therefore, Tbx3 has a significant role repressing a chamber-specific program of gene expression in regions from which the diverse components of the CCS are subsequently formed.

In addition to its critical role in the morphogenesis of the heart, the transcription factor Hop is a molecular component of the CCS regulation. Hop, the unusual homeodomain-only regulatory protein that does not bind DNA, apparently functions downstream of Nkx2.5 and has been implicated in the regulation of myocyte growth and proliferation by means of its antagonism of SRF activity in a process involving the recruitment of histone deacetylase activity. In the mouse heart, the *Hop* transcript is strongly expressed throughout the developing myocardium before 11.5 days post-coitum and, subsequently, becomes restricted to the trabecular zone.[323] Recently, using a knock-in strategy to place a *lacZ* reporter gene under the transcriptional control of the Hop locus, Ismat and associates reported that the adult heart expression of *Hop* was restricted to the AV node, His bundle, and bundle branches, as well as more broadly within the atria.[324] *Hop* inactivation in adult mice leads to conduction defects below the AV node as determined by invasive electrophysiological testing, including prolonged atrial refractoriness, widening of the QRS complex, and prolongation of the HV interval, and is associated with decreased *Cx40* expression. Unlike the case with *Nkx2.5*-deficiency, the AV node and

His bundle do not seem to be atrophied in surviving homozygous Hop mutant mice, suggesting Hop plays a primary role in the maintenance of CCS function rather than in CCS specification.

In addition to the Hop factor, which, with its involvement in regulating histone deacetylase, suggests an epigenetic developmental element, the impact of the environment on cardiac embryonic gene expression signaling and CCS has also been recently suggested. Biophysical factors that play a significant role in the differentiation of specialized CCS include the effects of physical conditioning[325,326] and of hemodynamic-induced molecular signaling cascades.[327,328] For instance, the timing of apex-first activation in chick embryo shows a striking dependence on hemodynamic load; conversion from immature to a mature pattern of ventricular activation was accelerated by increased loading and delayed by decreased load, reinforcing the importance of biophysical forces in the differentiation of the His–Purkinje system (HPS) in vivo.[329] Endothelin-1 (ET-1), a shear stress–sensitive cytokine prominently expressed by endothelial tissues, induces embryonic chick myocytes to express specific markers of Purkinje fiber differentiation.[330,331]

The ET-1 secretion by endothelial cells is particularly increased by a combination of pulsatile shear stress and increased blood pressure[332,333] and hemodynamic changes particularly pronounced in conotruncal banding hearts.[329] Additional aspects of ET-1 signaling, including alterations in ET-1 secretion and in other ET-1 signaling pathway components such as endothelin-converting enzyme-1 (ECE-1) in response to altered hemodynamic load, remain to be determined. Evidence has also been presented that a second factor secreted by endothelial cells, neuregulin-1 (NG-1), also plays key roles in both trabeculae formation and the differentiation of CCS cells.[334–336] However, at present, there is no definitive evidence that NG-1 expression or secretion is directly modulated by mechanical factors such as strain or fluid shear stress and pressure. Interestingly, it has been reported that ET-1 treatment increases *NG-1* expression in cultured endothelial cells,[337] indicating cross-talk between these two pathways and suggesting an indirect mechanism for *NG-1* upregulation by physical force. However, the transcriptional and molecular signaling processes controlling the functional maturation of the HPS remain to be explained.

What are the best markers for studying the early embryonic development of CCS? Most of the conduction system, including both the sinoatrial and AV nodes, in early embryonic hearts is composed of small myocytes with poorly developed actin and myosin filaments and sarcoplasmic reticulum, difficult to distinguish from the surrounding myocardium.[338]

Current techniques that facilitate delineation of CCS development include the use of specific antibody markers and transgenic mouse lines specifically expressing reporter genes. Several studies have used the monoclonal antibody HNK1 that reacts with a carbohydrate epitope in cell surface glycoproteins and glycolipids. Expression of HNK1 epitopes in cardiomyocytes during development has been reported in various species and has generally been thought to be a marker for the origin of CCS; however, the presence of HNK1 epitopes in migrating neural crest cells has led to questions concerning its involvement. Analysis of the spatiotemporal expression pattern of HNK1 in early chick cardiogenesis revealed HNK1 expression in premyocardium and that precardiac mesoderm generates HNK1-positive cardiomyocytes with morphological features similar to those of conduction cardiomyocytes.[339] Antibodies to GlN2 have also been used to delineate the developing conduction system in a number of species; in humans and rats, HNK-1 and GlN2 share almost the same spatiotemporal distribution in the heart.[340,341] The GlN2 epitope reacts with an antibody raised against an extract from the chicken nodose ganglion and was originally used to identify migrating neural crest cells.[342] Antibodies directed against the cell adhesion protein, NCAM, have also been found to be useful markers of CCS components. Furthermore, in the developing human heart, NCAM is highly expressed in the nodal areas.[343]

It is important to note that although HNK1 and GlN2 have proved to be informative as identifying markers of CCS during development, at present there is little information concerning their functional significance in mediating cell–cell interaction, adhesion, or differentiation, in contrast to the well-defined involvement of NCAM.[338,344]

LacZ reporter constructs are informative developmental markers available to monitor the expression of the *lacZ* transgene under the direction of various cardiac promoter constructs (e.g., *GATA6, mink, engrailed, cardiac troponin 1*).[345,346] Analysis of *lacZ* expression during sequential stages of cardiogenesis has provided a detailed view of the maturation of the conductive network and demonstrated that CCS patterning occurred very early in embryogenesis, beginning in transgenic mice 8.25 days after conception.[347] In addition, optical mapping of cardiac electrical activity, using a voltage-sensitive dye, confirmed that cells identified by the *lacZ* reporter gene were CCS components and that a murine His–Purkinje system is functional well before septation has completed. By use of the same CCS-*lacZ* strain, the developing conduction system has been identified in regions previously not formerly thought of, including Bachmann's bundle, the pulmonary veins, and sinus venosus–derived internodal structures, notably regions associated with the occurrence of cardiac dysrhythmias in adult patients.[348] Similar constructs have been used to examine the relationship of neural crest cells with the developing conduction system. *Wnt1-Cre/R26R* conditional reporter mice were used that express β-galactosidase from ROSA26 on Cre-mediated recombination.[349] In this study, two subpopulation of neural crest cells were found in the myocardium of the early embryonic heart, one adjacent to the bundle branches at the arterial pole, the other positioned contiguous to the nodes at the venous pole. The expression of a *minK-lacZ* construct has also been applied to the detection of conduction components in the early embryonic heart, as well as in the post-natal and adult heart.[350]

What can be inferred about the origin of the conducting system? At present, there is little evidence to support the idea of primary extracardiac contributions to the CCS; the primary myocardium can generate and conduct adult-like ECGs well before neural crest cells enter the embryonic heart. There is a consensus that both nodes arise from existing myocardium, and the ventricular conduction system is generated from a distinct region/transcriptional domain, the trabecular component of the ventricle, which is distinguished from the compact myocardium.[338] This latter conclusion has been drawn from molecular, functional, and morphological findings[338] and has been underscored by the demonstration that loss-of-function mutations in genes encoding the peptide growth factor neuregulin and its receptors ErbB2 and ErbB4 effectively shut down the initiation of trabecular development and result in early embryonic death, while not affecting the development of the atrial myocardium and the compact myocardium.[157,158,335] Furthermore, the trabecular ventricular component has a substantially different gene expression profile than the compact ventricular component with higher levels of atrial isoforms (e.g., ANF, α-MHC), as well as displaying different electrophysiological function. As embryonic development proceeds, the ventricular area involving the trabecular component markedly decreases, presumably by a gradual remodeling into compact myocardium. Finally, using replication-defective retroviruses encoding recombinant β-galactosidase, cell lineage analysis demonstrated that cells of the peripheral Purkinje conduction system have a myogenic origin.[351]

Development of the Pharyngeal Arch Arteries

As soon as the cardiogenic plates fuse, they connect to the primitive vascular system, which develops by both vasculogenesis (i.e., the formation of isolated proendothelial cells that merge into vessels) and angiogenesis (i.e., generation of the existing endothelial vessel network). The heart tube contacts the bilateral dorsal aorta through the first set of pharyngeal arch arteries (PAAs). These are subsequently followed by a second, third, fourth, and sixth set (the fifth does not develop).

Mesodermal cells and migratory neural crest cells populate the pharyngeal arches. As they form, each arch includes a blood vessel that connects the heart to the dorsal aorta. During development, the PAAs undergo extensive remodeling, leading to selective arterial growth and regression, a process that leads to the formation of the mature aortic arch and to connections with the great arteries.

Physical ablation of a discrete subset of neural crest cells before emergence from the neural tube results in predictable cardiac malformations, including persistent truncus arteriosus, DORV, interrupted aortic arch, and mispatterning of the great vessels.[352] These malformations have in common either outflow and/or inflow tract malalignment, suggesting that hemodynamic parameters during early cardiac morphogenesis may be disrupted causing cardiac dysmorphogenesis. Further studies have shown that neural crest cells are not required for PAA formation but are critically involved in their growth and maintenance.[353]

In mice, mutations in the paired-box–containing gene *Pax3* result in a similar array of cardiac malformations, strikingly reminiscent of those in human patients with DiGeorge syndrome.[354] Although mutations in *Pax3* have not been shown to cause DiGeorge syndrome in humans, it is likely that similar molecular and developmental pathways are affected, making *Splotch* mice a potentially useful model for the study of neural crest–related cardiac defects. In addition, targeted disruption of the secreted semaphorin signaling molecule also leads to interruption of the aortic arch and incorrect septation of the cardiac OFT in mice.[355] The precise mechanism of *Pax3* and semaphorin involvement on pharyngeal arch development remains to be determined. However, semaphorins act as ligands that bind plexin receptors; these plexin receptors have been reported to be expressed in neural crest cells[356] and, more recently, in endothelial cells.[357] Although semaphorin may direct the migration of neural crest cells, *Pax3* has been shown not to be a primary determinant in neural crest migration but is more likely involved in its fine-tuning[358]; in combination with other factors it may be a modulator of critical cell-surface interactions.[359]

Environmental Factors Affecting Cardiac Development

The mechanisms by which epigenetic factors, to which the developing heart is exposed, affect the morphogenesis and function of the cardiovascular system are not yet known. It is clear that a large number of environmental factors can affect cardiac development. In this section, we briefly review a number of the better-characterized factors impacting cardiac development, including environmental toxins and teratogens, drugs, alcohol and cocaine, micronutrients, diseases, and physiological stressors (e.g., hypoxia).

The significance of environmental effects on cardiovascular development extends well beyond the potential impact on structural and functional cardiac malformations in the embryo. It has become increasingly recognized that exposure of an embryo or fetus to a suboptimal environment increases its risk of acquiring coronary artery disease and HF in adult life through a process known as programming. As an example, stress experienced *in utero* and during early post-natal life imparts an increased vulnerability for adult-onset CVD. Programming involves a change in gene expression pattern that occurs in response to a stressor and leads to altered growth of specific organs during their most critical times of development. Known stressors include improper nourishment, hypoxia, and excess glucocorticoids.[360]

Among the cardiovascular problems that may be caused by environmental exposures are abnormal anatomical development, dysrhythmias, conduction defects, myocardial dysfunction, and derangements in blood pressure and cholesterol metabolism. Depending on when the exposure occurs and the magnitude of the exposure, it may cause transient or permanent effects, and the effects may change over time. Our focus herein will be on fetal exposure.

Drugs

Drugs that are known to be teratogenic to the fetal cardiovascular system when ingested by the mother include lithium, retinoids, thalidomide, and trimethadione.[361]

The teratogenic effect of erythromycin therapy in women who had taken erythromycin in early pregnancy significantly enhanced the risk of their infants developing cardiovascular malformations and an increased risk for pyloric stenosis.[362] Erythromycin inhibits a specific cardiac potassium channel (I_{Kr}) that play a major role in cardiac rhythm regulation in the early embryo.

Class III antidysrhythmics and phenytoin can result in a broad spectrum of cardiac abnormalities, including conduction defects as well as VSD.[363] Class III antidysrhythmics decrease the excitability of cardiac cells by selectively blocking the rapid component of the delayed rectified potassium channel (I_{Kr}), resulting in prolongation of the repolarization phase of the action potential. Phenytoin, which decreases the excitability of neurones, has recently also been shown to block I_{Kr}, in addition to its known blockade of sodium channels.

In chick embryos, exposure to phenobarbital can promote cardiovascular malformations, including isolated VSD and VSD associated with dextroposition of the aorta, DORV, and several types of aortic arch anomalies.[364] In addition, significant bradycardia and cardiac dysrhythmias were observed in embryos treated with teratogenic doses of phenobarbital.

Toxins and the Environment

Exposure to a variety of solvents during pregnancy can result in increased cardiac defects. The risk of VSD was increased with exposure to organic solvents during the first trimester.[365] Prenatal exposure to dyes, lacquers, and paints is associated with conal malformations[366] and mineral oil products with coarctation of the aorta.[367]

Exposure to halogenated hydrocarbons, specifically trichloroethylene (TCE), during pregnancy has been associated with CHD in animal models and also in retrospective epidemiology studies.[368,369] An association between TCE-contaminated drinking water and an increase in the incidence of cardiac malformations was reported in children of mothers who resided in the areas of contamination.[370]

Multiple animal studies support trichloracetic acid (TCA), a TRI metabolite, as the more potent teratogen.[371] Genetic and phenotypic differences in the key enzymes responsible for the formation of TCA may be associated with differences in the offspring susceptibility.

Another potential environmental hazard is exposure to the herbicide nitrofen (2,4-dichloro-4'-nitrodiphenyl ether). When administered to pregnant rats, it results in CHD, including, VSD, DORV, TOF, and TGA, as well as diaphragm and kidney abnormalities.[372,373] The incidence of cardiovascular abnormalities was dose related and was greatest with nitrofen treatment on day 11 of gestation.[374] Nitrofen-exposed fetal rats display heart hypoplasia,[375] associated with reduced endothelin-1,[376] reduced total DNA and cell proliferation,[377] and increased apoptosis.[378] This nitrofen-associated cardiotoxic effect during cardiogenesis can be minimized or reversed by therapy with corticosteroids administered prenatally and antioxidants or vitamin A provided during late gestation, which stimulates myogenesis.[377,378]

A proposed mechanism of action of nitrofen involves abnormal neural crest signaling. As previously noted, *Pax3* functions as a modulator of cardiac neural crest proliferation and/or migration to the developing heart. In addition, *Pax3* mRNA is decreased in the hearts of rats with experimental diaphragmatic hernia, and myocardial *Pax3* expression was significantly decreased in nitrofen-treated embryos.[379] Moreover, rats with nitrofen-induced hypoplasia of the heart exhibited decreased cardiac gene expression of growth factors including IGF-I and EGF, suggesting that nitrofen-induced hypoplasia may be due to reduced synthesis of IGF-I and EGF by cardiomyocytes in the developing heart.[380]

Deficiencies and Excesses in Vitamins and Micronutrients

RA, an analog of vitamin A, is known to be teratogenic in laboratory animals, as well as in humans. Human fetal exposure to isotretinoin (a retinoid prescribed for severe cystic acne) is associated with an increased risk for a selected set of malformations, including craniofacial, thymic, central nervous system, and cardiac defects, including conotruncal heart defects and aortic-arch abnormalities.[381] The pattern of malformation closely resembled that observed in animal studies of retinoid teratogenesis. Significantly, a potential target of isotretinoin teratogenesis is neural crest cell activity, whose deficiency would result in the observed craniofacial, cardiac, and thymic malformations.

Retinoid compounds exert pleiotropic effects on cellular differentiation, morphogenesis, and metabolism. The crucial role of retinoids in controlling differentiation processes has become evident from studies conducted in a variety of *in vivo* and *in vitro* system.[382] Deletions of individual retinoid receptors in the embryo induce several developmental abnormalities in the cardiovascular system.[383,384] On the other hand, overexpression of a constitutively active RAR in developing atria and/or in post-natal ventricles is relatively benign, whereas ventricular expression during gestation can lead to DCM and significant cardiac dysfunction. Retinoid-dependent pathways are involved in promoting the ventricular phenotype of the embryonic heart and indicate that retinoids may have an important role in maintaining the normal ventricular phenotype in the post-natal state. Nevertheless, the signaling mechanisms of RA action are not well understood.[385] RXR-α represses GATA4-mediated transcription by means of a retinoid-dependent interaction with the cardiac-enriched repressor FOG-2.[386] Several observations have also revealed a novel mechanism by which retinoids regulate cardiogenic gene expression through direct interaction with GATA4 and its corepressor FOG-2.

Vitamin A replacement at different stages of embryonic development in rodents and birds can modulate the severity and type of heart defects seen at birth, indicating that retinoid signaling pathways play key roles in a variety of developmental programs.[387] In addition, retinoids can exert teratogenic effects. Human offspring born of mothers who ingested isotretinoin, a vitamin A analog, during pregnancy exhibited a high incidence of VSD and complex cardiac malformations, including transposition of the great vessels, TOF, hypoplastic aortic arch, and VSD.[381] Moreover, in the developing chick, local application of high concentrations of RA disrupts the migration of the precardiac mesoderm, an effect that is dose-dependent and developmental stage-specific.[388]

Inadequate maternal vitamin intake during pregnancy has been suggested to be a risk factor for cleft lip with or without cleft palate and CHD. How folate reduces the risks of congenital anomalies is not known.

The potential role of genetic polymorphism in the *RFC1* gene encoding a folate carrier on the risks of developing orofacial clefts or conotruncal heart defects was recently assessed in a recent case-controlled study.[389] This study found modest evidence for a gene-nutrient interaction between infant *RFC1* genotype and periconceptional intake of vitamins on the risk of conotruncal defects.

ALCOHOL AND SUBSTANCE ABUSE

Alcohol (Fetal Alcohol Syndrome)

Cardiac abnormalities are present in an increased number of neonates with fetal alcohol syndrome.[390,391] The type of cardiac abnormalities varied with the extent of intrauterine exposure and the resulting severity of the fetal alcohol syndrome.[391] First trimester maternal use of alcohol has been reported to double the risk of ASD.[392]

Animal studies examining the effect of dose on the occurrence of specific abnormalities have shown that incubation of the chick embryo with a low ethanol concentration caused a significant incidence (45%) of VSD, and at a higher dose, an increased incidence (74%) of aortic and ventricular septal abnormalities was found.[393]

Cocaine

Women who use cocaine while pregnant are more likely to give birth to children with CHD,[394] including abnormalities of ventricular structure and function, dysrhythmias, and intracardiac conduction abnormalities, resulting in some cases in congestive HF, cardiorespiratory arrest, and death.[395] Electrocardiographic abnormalities, including sustained dysrhythmias, have also been documented in infants and children with prenatal exposure to cocaine.[396] These dysrhythmias manifest widely varied rhythms and are markedly resistant to conventional therapy.

Experimental studies have suggested that cocaine may exert direct toxic effects on fetal myocytes. Exposure of cocaine to primary cardiomyocyte cultures derived from near-term fetal rats resulted in apoptotic cell death.[397] Administration of cocaine to pregnant rats *in vivo* also promoted a dose-dependent myocyte cell death in the fetal heart. This was accompanied by both an up-regulation of the Bax/Bcl-2 ratio and an increase in myocyte caspase activities.[398]

Maternal Diseases

Maternal lupus provides an important illustration of how an altered fetal environment can promote the development of dysrhythmias. Immune-mediated damage to fetal AV nodal tissue is associated with transplacental passage of specific maternal autoantibodies,[399,400] with damage becoming evident at 16–30 weeks' gestation. By that time, an intrauterine myocarditis has damaged the AV node, resulting in both its destruction and replacement with fibrotic tissue. The resulting AV block causes fetal dysrhythmia that may be accompanied by CHF and pericarditis.

Dysrhythmias, including ventricular tachycardia, and conduction disturbances are also frequent among HIV-infected children.[401] Clinical effects of other viral diseases on early cardiac development are further discussed in the following chapter on pediatric cardiovascular disease.

STRESSORS

Hypoxia

Tissue hypoxia plays a critical role in cardiogenesis, and VEGF is an intrinsic link between hypoxia and cardiac septation defects by its function as a negative regulator of endocardial-to-mesenchymal transformation, which underlies EC formation.[402] Studies with the hypoxia indicator EF5 have suggested a potential regulatory role for hypoxia in cardiac morphogenesis. Recently, areas of the embryonic chicken heart that were intensely positive for EF5 and for nuclear-localized hypoxia-inducible factor 1alpha (HIF-1α) were identified in discrete regions of the atrial wall and the interventricular septum (IVS), including the developing CCS.[403] During development, the CCS and the patterning of coronary vessels may be subject to regulation by differential oxygen concentration within the cardiac tissues and subsequent HIF-1 regulation of gene expression. Hypoxia-responsive signaling also regulates the apoptosis-dependent remodeling of the embryonic avian cardiac OFT.[404] Furthermore, the hypoxia-dependent expression of VEGF-R2 in the distal OFT myocardium may be protective, because cardiomyocyte apoptosis in the early stages of OFT remodeling was absent from this region. Adenovirus-mediated forced expression of *VEGF*-165 induced conotruncal malformation such as DORV and VSD, similar to the defects observed when apoptosis-dependent remodeling of the OFT was specifically targeted. Normal developmental remodeling of the embryonic cardiac OFT involves hypoxia/HIF-1–dependent signaling, cardiomyocyte apoptosis, and regulated VEGF signaling. Interestingly, signaling through VEGF/VEGF-R2 provides survival signals for the hypoxic OFT cardiomyocytes.

Future Prospects

Cardiogenesis is an active and complex process with multiple and precisely coordinated phases of pattern formation and morphogenesis. Taking advantage of new technology and the availability of a number of animal models, key regulators of cardiac patterning and morphogenesis have been progressively discovered. As the specific genetic basis of the cardiac defect becomes known, the hope of early gene therapy may become a reality. However, to achieve this a number of pitfalls/shortcomings need to be addressed (e.g., delivery of a corrected gene to a target tissue and understanding the regulation of gene expression, because a defective gene in one animal may result in different cardiac phenotype in another, even within the same strain). To fully understand the process of normal gene regulation, the role of modifier genes and their interaction with the environment (including changes and effects brought about by environmental toxins, micronutrients, and medications), further research is necessary. Because most cardiac defects occur early in embryogenesis, the sooner a defect is diagnosed, the potential for treatment before the forming of the heart will increase.

SUMMARY

- The formation and integration of the myocardial modular elements requires the complex interplay of a multitude of genes whose cell type–specific expression is highly organized and precisely regulated at the spatial and temporal levels by numerous transcription factors. These critical regulatory factors can operate differently in various cell types and respond to a variety of intracellular and extracellular signals to modulate the precise integration of gene expression and morphological development.
- Most of the mutations identified that lead to specific CHD reside in genes encoding transcription factors.
- Factors that play a central role in the specification and differentiation of lateral plate mesoderm into cardiac cells include tinman (i.e., *Nkx2.5*), SRF, myocardin, and GATA. These factors interact with each other and with a variety of cofactors to affect the programming of cardiac differentiation by means of transcriptional regulatory changes.
- Signaling pathways are contributed from a number of surrounding cell types, including the endoderm and epicardium and from the anterior field constituting a complex milieu of inductive signaling. Elements of the TGFβ/BMP and Wnt families play crucial roles as does FGF signaling.
- A number of transcription factors (e.g., Mesp1, Mesp2, and GATA) are involved in the migration of the cardiac precursors; both intracellular signaling and cell adhesion pathways are activated in this process with evidence of substantial cross-talk between the components involved in their regulation.
- Cardiac development involves looping of the heart, definition of a left-right identity, and formation of the cardiac chambers.
- An early defined anteroposterior polarity in the expression of specific genes contributes to the development of chamber-specific expression. Key factors in defining this early polarity of expression include retinoic acid, *Irx4*, T-box genes, including Tbx5, Tbx2, and Tbx20, and GATA factors.
- Downstream markers of chamber specificity include ANF and MLC. Further definition of chamber-specific maturation, as well as establishing a dorsoventral polarity, involves regulation by dHAND and eHAND bHLH transcription factors, with considerable interaction by Nkx2.5 and MEF2.
- Left-right (L-R) identity is achieved by a cascade of signaling molecules culminating in the expression of the paired-domain homeodomain transcription factor *Pitx2* on the left side of the visceral organs, including the heart. In addition to *Nodal* involvement, this pathways involves several TGF-β family signaling proteins, including secreted extracellular factors (i.e., *lefty-1*, *lefty-2*, *BMP4*), membrane receptors (e.g., activin receptor type IIB), members of the membrane-associated proteins encoded by *EFC-CFC* genes (e.g., *cryptic*, *Notch* receptor, BMP type I receptor, *ACVRI*), intracellular mediators (e.g., sonic hedgehog [*SHH*] *Smad*s), and an array of transcription factors (e.g., *Zic3*, *SnR*, *Hand1*, *NKX2.5*, *cited*, β-*catenin*).
- In addition to its profound effect on anteroposterior polarity, retinoic acid controls both levels and location of expression of L-R signaling pathway components. Both retinoic acid (RA) and FGF are critical signalers of the proliferative events, with both epicardium and endocardium serving as important growth inducers.
- The neuregulin family of peptide growth factors and their tyrosine kinase receptors (ErbBs), the forkhead

transcription factor *Foxp1*, GATA4, Tbx5, the Hop regulator, and n-myc play critical roles in myocyte proliferation and maturation.
- The formation of cardiac septa and the valves is linked, involving many of the same molecular players.
- Preceding the formation of valves and septa is the pivotal formation of the endocardial cushions (EC), a complex event characterized by endothelial-mesenchymal transdifferentiation (EMT) of a subset of endothelial cells specified in the cushion-forming regions to invade the cardiac jelly, where they subsequently proliferate and complete their differentiation into mesenchymal cells.
- This is followed by a series of delamination and remodeling events requiring further cell differentiation, apoptosis, and remodeling of the extracellular matrix. The signaling factors involved include BMP/TGF-β, Notch, VEGF, NFATc1, Wnt/β-catenin, ErbB, Sox, Fog1/GATA, and NF1/Ras. The formation of the semilunar valves and outflow tract (OFT) involves some of the aforementioned signaling pathways, including Sox and NFAT, as well as neural crest cells.
- Development of the conduction system involves Nkx2.5, Hop, Tbx5, the connexin genes, and ErbB.
- Defining the molecular markers of the early developing conduction system in a work is progress of great significance.
- Development of the pharyngeal arches into the aorta and major vessels involves both the intervention/induction of neural crest and molecules such as PAX3 and semaphorin, whose precise role remains unknown.
- Environmental factors, including toxins and teratogens, drugs, alcohol, cocaine, micronutrients (e.g., retinoic acid and folate), diseases, and physiological stressors (e.g., hypoxia) might have severe effects on the development of the heart.

References

1. Kathiriya, I. S., and Srivastava, D. (2000). Left-right asymmetry and cardiac looping: implications for cardiac development and congenital heart disease. *Am. J. Med. Genet.* **97,** 271–279.
2. Larsen, W. J. (1997b). "Human Embryology," 2nd ed. pp. 151–188. Churchill Livingstone, New York.
3. Van Mierop, L. H. S. (1979). Morphological development of the heart. In "Handbook of Physiology, the Cardiovascular System." (R. M. Berne, Ed), pp. 1–28. American Physiology Society, Bethesda, MD.
4. Steding, G., and Seidl, W. (1984). Cardiac septation in normal development. In "Congenital Heart Disease: Causes and Processes." (J. J. Nora, and A. Talao, Eds.), pp. 481–500. Futura, New York.
5. De la Cruz, M. V., Sánchez-Gómez, C., and Palomino, M. (1989). The primitive cardiac regions in the straight tube heart (stage 9) and their anatomical expression in the mature heart: an experimental study in the chick embryo. *J. Anat.* **165,** 121–131.
6. Moorman, A. F., de Jong, F., Denyn, M. M., and Lamers, W. H. (1998). Development of the cardiac conduction system. *Circ. Res.* **82,** 629–644.
7. Abdulla, R., Blew, G. A., and Holterman, M. J. (2004). Cardiovascular embryology. *Pediatr. Cardiol.* **25,** 191–200.
8. Harvey, R. P. (1996). NK-2 homeobox genes and heart development. *Dev. Biol.* **178,** 203–216.
9. Lyons, I., Parsons, L. M., Hartley, L., Li, R., Andrews, J. E., Robb, L., and Harvey, R. P. (1995). Myogenic and morphogenetic defects in the heart tubes of murine embryos lacking the homeo box gene Nkx2-5. *Genes Dev.* **9,** 1654–1666.
10. Parlakian, A., Tuil, D., Hamard, G., Tavernier, G., Hentzen, D., Concordet, J. P., Paulin, D., Li, Z., and Daegelen, D. (2004). Targeted inactivation of serum response factor in the developing heart results in myocardial defects and embryonic lethality. *Mol. Cell Biol.* **24,** 5281–5289.
11. Niu, Z., Yu, W., Zhang, S. X., Barron, M., Belaguli, N. S., Schneider, M. D., Parmacek, M., Nordheim, A., and Schwartz, R. J. (2005). Conditional mutagenesis of the murine serum response factor gene blocks cardiogenesis and the transcription of downstream gene targets. *J. Biol. Chem.* **280,** 32531–32538.
12. Wang, D., Chang, P. S., Wang, Z., Sutherland, L., Richardson, J. A., Small, E., Krieg, P. A., and Olson, E. N. (2001). Activation of cardiac gene expression by myocardin, a transcriptional cofactor for serum response factor. *Cell* **105,** 851–862.
13. Li, J., Zhu, X., Chen, M., Cheng, L., Zhou, D., Lu, M. M., Du, K., Epstein, J. A., and Parmacek, M. S. (2005). Myocardin-related transcription factor B is required in cardiac neural crest for smooth muscle differentiation and cardiovascular development. *Proc. Natl. Acad. Sci. USA* **102,** 8916–8921.
14. Ueyama, T., Kasahara, H., Ishiwata, T., Nie, Q., and Izumo, S. (2003). Myocardin expression is regulated by Nkx2.5, and its function is required for cardiomyogenesis. *Mol. Cell Biol.* **23,** 9222–9232.
15. Yoshida, T., Kawai-Kowase, K., and Owens, G. K. (2004). Forced expression of myocardin is not sufficient for induction of smooth muscle differentiation in multipotential embryonic cells. *Arterioscler. Thromb. Vasc. Biol.* **24,** 1535–1537.
16. Zhang, X., Azhar, G., Zhong, Y., and Wei, J. Y. (2004). Identification of a novel serum response factor cofactor in cardiac gene regulation. *J. Biol. Chem.* **279,** 55626–55632.
17. Pikkarainen, S., Tokola, H., Kerkelä, R., and Ruskoaho, H. (2004). GATA transcription factors in the developing and adult heart. *Cardiovasc. Res.* **63,** 196–207.
18. Charron, F., and Nemer, M. (1999). GATA transcription factors and cardiac development. *Semin. Cell Dev. Biol.* **10,** 85–91.
19. Molkentin, J. D., Lin, Q., Duncan, S. A., and Olson, E. N. (1997). Requirement of the transcription factor GATA4 for heart tube formation and ventral morphogenesis. *Genes Dev.* **11,** 1061–1072.
20. Kuo, C. T., Morrisey, E. E., Anandappa, R., Sigrist, K., Lu, M. M., Parmacek, M. S., Soudais, C., and Leiden, J. M. (1997). GATA4 transcription factor is required for ventral morphogenesis and heart tube formation. *Genes Dev.* **11,** 1048–1060.

21. Grepin, C., Robitaille, L., Antakly, T., and Nemer, M. (1995). Inhibition of transcription factor GATA-4 expression blocks in vitro cardiac muscle differentiation. *Mol. Cell Biol.* **15**, 4095–4102.
22. Grepin, C., Nemer, G., and Nemer, M. (1997). Enhanced cardiogenesis in embryonic stem cells overexpressing the GATA-4 transcription factor. *Development* **124**, 2387–2395.
23. Narita, N., Bielinska, M., and Wilson, D. B. (1997). Cardiomyocyte differentiation by GATA-4-deficient embryonic stem cells. *Development* **124**, 3755–3764.
24. Charron, F., Paradis, P., Bronchain, O., Nemer, G., and Nemer, M. (1999). Cooperative interaction between GATA-4 and GATA-6 regulates myocardial gene expression. *Mol. Cell Biol.* **19**, 4355–4365.
25. Katz, S. G., Williams, A., Yang, J., Fujiwara, Y., Tsang, A. P., Epstein, J. A., and Orkin, S. H. (2003). Endothelial lineage-mediated loss of the GATA cofactor Friend of GATA 1 impairs cardiac development. *Proc. Natl. Acad. Sci. USA* **100**, 14030–14035.
26. Kawamura, T., Ono, K., Morimoto, T., Wada, H., Hirai, M., Hidaka, K., Morisaki, T., Heike, T., Nakahata, T., Kita, T., and Hasegawa, K. (2005). Acetylation of GATA-4 is involved in the differentiation of embryonic stem cells into cardiac myocytes. *J. Biol. Chem.* **280**, 19682–19688.
27. Zaffran, S., Astier, M., Gratecos, D., Guillen, A., and Semeriva, M. (1995). Cellular interactions during heart morphogenesis in the *Drosophila* embryo. *Biol. Cell* **84**, 13–24.
28. Olson, E. N., and Srivastava, D. (1996). Molecular pathways controlling heart development. *Science* **272**, 671–676.
29. Yutzey, C., Rhee, J. T., and Bader, D. (1994). Expression of the atrial-specific myosin heavy chain AMHC1 and the establishment of anteroposterior polarity in the developing chicken heart. *Development* **120**, 871–883.
30. Dawid, I. B. (1992). Mesoderm induction and axis determination in *Xenopus laevis*. *Bioessays* **14**, 687–691.
31. Smith, W. C., and Harland, R. M. (1992). Expression cloning of noggin, a new dorsalizing factor localized to the Spemann organizer in *Xenopus* embryos. *Cell* **70**, 829–840.
32. Smith, W. C., Knecht, A. K., Wu, M., and Harland, R. M. (1993). Secreted noggin protein mimics the Spemann organizer in dorsalizing *Xenopus* mesoderm. *Nature* **361**, 547–549.
33. Schier, A. F., and Talbot, W. S. (2001). Nodal signaling and the zebrafish organizer. *Int. J. Dev. Biol.* **45**, 289–297.
34. Smith, W. C., McKendry, R., Ribisi, S., Jr., and Harland, R. M. (1995). A nodal-related gene defines a physical and functional domain within the Spemann organizer. *Cell* **82**, 37–46.
35. Vincent, S. D., Dunn, N. R., Hayashi, S., Norris, D. P., and Robertson, E. J. (2003). Cell fate decisions within the mouse organizer are governed by graded Nodal signals. *Genes Dev.* **17**, 1646–1662.
36. Kinder, S. J., Tsang, T. E., Wakamiya, M., Sasaki, H., Behringer, R. R., Nagy, A., and Tam, P. P. (2001). The organizer of the mouse gastrula is composed of a dynamic population of progenitor cells for the axial mesoderm. *Development* **128**, 3623–3634.
37. Marom, K., Levy, V., Pillemer, G., and Fainsod, A. (2005). Temporal analysis of the early BMP functions identifies distinct anti-organizer and mesoderm patterning phases. *Dev. Biol.* **282**, 442–454.
38. Schultheiss, T. M., Xydas, S., and Lassar, A. B. (1995). Induction of avian cardiac myogenesis by anterior endoderm. *Development* **121**, 4203–4214.
39. Nascone, N., and Mercola, M. (1995). An inductive role for the endoderm in *Xenopus* cardiogenesis. *Development* **121**, 515–523.
40. Frasch, M. (1995). Induction of visceral and cardiac mesoderm by ectodermal Dpp in the early *Drosophila* embryo. *Nature* **374**, 464–467.
41. Lockwood, W. K., and Bodmer, R. (2002). The patterns of wingless, decapentaplegic, and tinman position the *Drosophila* heart. *Mech. Dev.* **114**, 13–26.
42. Lee, H. H., and Frasch, M. (2005). Nuclear integration of positive Dpp signals, antagonistic Wg inputs and mesodermal competence factors during *Drosophila* visceral mesoderm induction. *Development* **132**, 1429–1442.
43. Klinedinst, S. L., and Bodmer, R. (2003). Gata factor Pannier is required to establish competence for heart progenitor formation. *Development* **130**, 3027–3038.
44. Ladd, A. N., Yatskievych, T. A., and Antin, P. B. (1998). Regulation of avian cardiac myogenesis by activin/TGFbeta and bone morphogenetic proteins. *Dev. Biol.* **204**, 407–419.
45. Kruithof, B. P., van Wijk, B., Somi, S., Kruithof-de Julio, M., Perez Pomares, J. M., Weesie, F., Wessels, A. Moorman, A. F., vanden Hoff, M. J. (2006). BMP and FGF regulate the differentiation of multipotential pericardial mesoderm into the myocardial or epicardial lineage. *Dev Biol.* **295**, 507–522.
46. Schlange, T., Andree, B., Arnold, H. H., and Brand, T. (2000). BMP2 is required for early heart development during a distinct time period. *Mech. Dev.* **91**, 259–270.
47. Alsan, B. H., and Schultheiss, T. M. (2002). Regulation of avian cardiogenesis by Fgf8 signaling. *Development* **129**, 1935–1943.
48. Beiman, M., Shilo, B. Z., and Volk, T. (1996). Heartless, a *Drosophila* FGF receptor homolog, is essential for cell migration and establishment of several mesodermal lineages. *Genes Dev.* **10**, 2993–3002.
49. Fossett, N., Zhang, Q., Gajewski, K., Choi, C. Y., Kim, Y., and Schulz, R. A. (2000). The multi-type zinc-finger protein U-shaped functions in heart cell specification in the *Drosophila* embryo. *Proc. Natl. Acad. Sci. USA* **97**, 7348–7353.
50. Stathopoulos, A., Tam, B., Ronshaugen, M., Frasch, M., and Levine, M. (2004). Pyramus and thisbe: FGF genes that pattern the mesoderm of *Drosophila* embryos. *Genes Dev.* **18**, 687–699.
51. Schneider, V. A., and Mercola, M. (2001). Wnt antagonism initiates cardiogenesis in *Xenopus laevis*. *Genes Dev.* **15**, 304–315.
52. Marvin, M. J., Di Rocco, G., Gardiner, A., Bush, S. M., and Lassar, A. B. (2001). Inhibition of Wnt activity induces heart formation from posterior mesoderm. *Genes Dev.* **15**, 316–327.
53. Foley, A. C., and Mercola, M. (2005). Heart induction by Wnt antagonists depends on the homeodomain transcription factor Hex. *Genes Dev.* **19**, 387–396.
54. Saga, Y., Miyagawa-Tomita, S., Takagi, A., Kitajima, S., Miyazaki, J., and Inoue, T. (1999). MesP1 is expressed in the heart precursor cells and required for the formation of a single heart tube. *Development* **126**, 3437–3447.
55. Kitajima, S., Takagi, A., Inoue, T., and Saga, Y. (2000). MesP1 and MesP2 are essential for the development of cardiac mesoderm. *Development* **127**, 3215–3226.
56. Narita, N., Bielinska, M., and Wilson, D. B. (1997). Wild-type endoderm abrogates the ventral developmental defects

associated with GATA-4 deficiency in the mouse. *Dev. Biol.* **189,** 270–274.
57. Reiter, J. F., Alexander, J., Rodaway, A., Yelon, D., Patient, R., Holder, N., and Stainier, D. Y. (1999). Gata5 is required for the development of the heart and endoderm in zebrafish. *Genes Dev*. **13,** 2983–2995.
58. Jiang, Y., Tarzami, S., Burch, J. B., and Evans, T. (1998). Common role for each of the cGATA-4/5/6 genes in the regulation of cardiac morphogenesis. *Dev. Genet*. **22,** 263–277.
59. Kikuchi, Y., Trinh, L. A., Reiter, J. F., Alexander, J., Yelon, D., and Stainier, D. Y. (2000). The zebrafish bonnie and clyde gene encodes a Mix family homeodomain protein that regulates the generation of endodermal precursors. *Genes Dev*. **14,** 1279–1289.
60. Zhang, H., Toyofuku, T., Kamei, J., and Hori, M. (2003). GATA-4 regulates cardiac morphogenesis through transactivation of the N-cadherin gene. *Biochem. Biophys. Res. Commun*. **312,** 1033–1038.
61. Glickman, N. S., and Yelon, D. (2004). Coordinating morphogenesis: epithelial integrity during heart tube assembly. *Dev. Cell* **6,** 311–312.
62. Linask, K. K., Manisastry, S., and Han, M. (2005). Cross talk between cell–cell and cell–matrix adhesion signaling pathways during heart organogenesis: implications for cardiac birth defects. *Microsc. Microanal*. **11,** 200–208.
63. Kelly, R. G. (2005). Molecular inroads into the anterior heart field. *Trends Cardiovasc. Med*. **15,** 51–56.
64. Cai, C. L., Liang, X., Shi, Y., Chu, P. H., Pfaff, S. .L, Chen, J., and Evans, S. (2003). Isl1 identifies a cardiac progenitor population that proliferates prior to differentiation and contributes a majority of cells to the heart. *Dev. Cell* **5,** 877–889.
65. Gittenberger-de Groot, A. C., Bartelings, M. M., Deruiter, M. C., and Poelmann, R. E. (2005). Basics of cardiac development for the understanding of congenital heart malformations. *Pediatr. Res*. **57,** 169–176.
66. Stainier, D. Y., and Fishman, M. C. (1992). Patterning the zebrafish heart tube: acquisition of anteroposterior polarity. *Dev. Biol*. **153,** 91–101.
67. Twal, W., Roze, L., and Zile, M. H. (1995). Anti-retinoic acid monoclonal antibody localizes all-trans-retinoic acid in target cells and blocks normal development in early quail embryo. *Dev. Biol*. **168,** 225–234.
68. Niederreither, K., Vermot, J., Messaddeq, N., Schuhbaur, B., Chambon, P., and Dolle, P. (2001). Embryonic retinoic acid synthesis is essential for heart morphogenesis in the mouse. *Development* **128,** 1019–1031.
69. Kostetskii, I., Jiang, Y., Kostetskaia, E., Yuan, S., Evans, T., and Zile, M. (1999). Retinoid signaling required for normal heart development regulates GATA-4 in a pathway distinct from cardiomyocyte differentiation. *Dev. Biol*. **206,** 206–218.
70. Zile, M. H., Kostetskii, I., Yuan, S., Kostetskaia, E., St. Amand, T. R., Chen, Y., and Jiang, W. (2000). Retinoid signaling is required to complete the vertebrate cardiac left/right asymmetry pathway. *Dev. Biol*. **223,** 323–338.
71. Sucov, H. M., Dyson, E., Gumeringer, C. L., Price, J., Chien, K. R., and Evans, R. M. (1994). RXR mutant mice establish a genetic basis for vitamin A signaling in heart morphogenesis. *Genes Dev*. **8,** 1007–1018.
72. Kastner, P., Grondona, J. M., Mark, M., Gansmuller, A., LeMeur, M., Decimo, D., Vonesch, J. L., Dolle, P., and Chambon, P. (1994). Genetic analysis of RXR developmental function: convergence of RXR and RAR signaling pathways in heart and eye morphogenesis. *Cell* **78,** 987–1003.
73. Gruber, P. J., Kubalak, S. W., Pexieder, T., Sucov, H. M., Evans, R. M., and Chien, K. R. (1996). RXR deficiency confers genetic susceptibility for aortic sac, conotruncal, atrioventricular cushion, and ventricular muscle defects in mice. *J. Clin. Invest*. **98,** 1332–1343.
74. Romeih, M., Cui, J., Michaille, J. J., Jiang, W., and Zile, M. H. (2003). Function of RARγ and RARα2 at the initiation of retinoid signaling is essential for avian embryo survival and for distinct events in cardiac morphogenesis. *Dev. Dyn*. **228,** 697–708.
75. Kastner, P., Messaddeq, N., Mark, M., Wendling, O., Grondona, J. M., Ward, S., Ghyselinck, N., and Chambon. P. (1997). Vitamin A deficiency and mutations of RXR, RXRß and RAR lead to early differentiation of embryonic ventricular cardiomyocytes. *Development* **124,** 4749–4758.
76. Christoffels, V. M., Habets, P. E., Franco, D., Campione, M., de Jong, F., Lamers, W. H., Bao, Z. Z., Palmer, S., Biben, C., Harvey, R. P., and Moorman, A. F. (2000). Chamber formation and morphogenesis in the developing mammalian heart. *Dev. Biol*. **223,** 266–278.
77. Bao, Z.-Z., Bruneau, B. G., Seidman, J. G., Seidman, C. E., and Cepko, C. L. (1999). Irx4 regulates chamber-specific gene expression in the developing heart. *Science* **283,** 1161–1164.
78. Garriock, R. J., Vokes, S. A., Small, E. M., Larson, R., and Krieg, P. A. (2001). Developmental expression of the *Xenopus* Iroquois-family homeobox genes, Irx4 and Irx5. *Dev. Genes Evol*. **211,** 257–260.
79. Bruneau, B. G., Bao, Z. Z., Tanaka, M., Schott, J. J., Izumo, S., Cepko, C. L., Seidman, J. G., and Seidman, C. E. (2000). Cardiac expression of the ventricle-specific homeobox gene Irx4 is modulated by Nkx2-5 and dHand. *Dev. Biol*. **217,** 266–277.
80. Cavodeassi, F., Modolell, J., and Gomez-Skarmeta, J. L. (2001). The Iroquois family of genes: from body building to neural patterning. *Development* **128,** 2847–1255.
81. Bruneau, B. G., Bao, Z. Z., Fatkin, D., Xavier-Neto, J., Georgakopoulos, D., Maguire, C. T., Berul, C. I., Kass, D. A., Kuroski-de Bold, M. L., de Bold, A. J., Conner, D. A., Rosenthal, N., Cepko, C. L., Seidman, C. E., and Seidman, J. G. (2001). Cardiomyopathy in Irx4-deficient mice is preceded by abnormal ventricular gene expression. *Mol. Cell Biol*. **21,** 1730–1736.
82. Xavier-Neto, J., Neville, C. M., Shapiro, M. D., Houghton, L., Wang, G. F., Nikovits, W., Stockdale, F. E., and Rosenthal, N. (1999). A retinoic acid-inducible transgenic marker of sino-atrial development in the mouse heart. *Development* **126,** 2677–2687.
83. Wang, G. F., Nikovits, W. Jr., Schleinitz, M., and Stockdale, F. E. (1996). Atrial chamber-specific expression of the slow myosin heavy chain 3 gene in the embryonic heart. *J. Biol. Chem*. **271,** 19836–19845.
84. Wang, G. F., Nikovits, W. Jr., Bao, Z. Z., and Stockdale, F. E. (2001). Irx4 Forms an inhibitory complex with the vitamin D and retinoic X receptors to regulate cardiac chamber-specific slow MyHC3 expression. *J. Biol. Chem*. **276,** 28835–28841.
85. Christoffels, V. M., Keijser, A. G., Houweling, A. C., Clout, D. E., and Moorman, A. F. (2000). Patterning the embryonic

heart: identification of five mouse Iroquois homeobox genes in the developing heart. *Dev. Biol.* **224**, 263–274.
86. Bruneau, B. G., Logan, M., Davis, N., Levi, T., Tabin, C. J., Seidman, J. G., and Seidman, C. E. (1999). Chamber-specific cardiac expression of Tbx5 and heart defects in Holt-Oram syndrome. *Dev. Biol.* **211**, 100–108.
87. Liberatore, C. M., Searcy-Schrick, R. D., and Yutzey, K. E. (2000). Ventricular expression of tbx5 inhibits normal heart chamber development. *Dev. Biol.* **223**, 169–180.
88. Hatcher, C. J., Goldstein, M. M., Mah, C. S., Delia, C. S., and Basson, C. T. (2000). Identification and localization of TBX5 transcription factor during human cardiac morphogenesis. *Dev. Dyn.* **219**, 90–95.
89. Garrity, D. M., Childs, S., and Fishman, M. C. (2002). The heartstrings mutation in zebrafish causes heart/fin Tbx5 deficiency syndrome. *Development* **129**, 4635–4645.
90. Bruneau, B. G., Nemer, G., Schmitt, J. P., Charron, F., Robitaille, L., Caron, S., Conner, D., Gessler, M., Nemer, M., Seidman, C. E., and Seidman, J. G. (2001). A murine model of Holt-Oram syndrome defines roles of the T-box transcription factor Tbx5 in cardiogenesis and disease. *Cell* **106**, 709–721.
91. Hiroi, Y., Kudoh, S., Monzen, K., Ikeda, Y., Yazaki, Y., Nagai, R., and Komuro, I. (2001). Tbx5 associates with Nkx2-5 and synergistically promotes cardiomyocyte differentiation. *Nat. Genet.* **28**, 276–280.
92. Horb, M. E., and Thomsen, G. H. (1999). Tbx5 is essential for heart development. *Development* **126**, 1739–1751.
93. Takeuchi, J. K., Ohgi, M., Koshiba-Takeuchi, K., Shiratori, H., Sakaki, I., Ogura, K., Saijoh, Y., and Ogura, T. (2003). Tbx5 specifies the left/right ventricles and ventricular septum position during cardiogenesis. *Development* **130**, 5953–5964.
94. Hatcher, C. J., Kim, M. S, Mah, C. S., Goldstein, M. M., Wong, B., Mikawa, T., and Basson, C. T. (2001). TBX5 transcription factor regulates cell proliferation during cardiogenesis. *Dev. Biol.* **230**, 177–188.
95. Moskowitz, I. P., Pizard, A., Patel, V. V., Bruneau, B. G., Kim, J. B., Kupershmidt, S., Roden, D., Berul, C. I., Seidman, C. E., and Seidman, J. G. (2004). The T-Box transcription factor Tbx5 is required for the patterning and maturation of the murine cardiac conduction system. *Development* **131**, 4107–4116.
96. Christoffels, V. M., Hoogaars, W. M., Tessari, A., Clout, D. E, Moorman, A. F., and Campione, M. (2004). T-box transcription factor Tbx2 represses differentiation and formation of the cardiac chambers. *Dev. Dyn.* **229**, 763–770.
97. Stennard, F. A., Costa, M. W., Lai, D., Biben, C., Furtado, M. B., Solloway, M. J., McCulley, D. J., Leimena, C., Preis, J. I., Dunwoodie, S. L., Elliott, D. E., Prall, O. W., Black, B. L., Fatkin, D., and Harvey, R. P. (2005). Murine T-box transcription factor Tbx20 acts as a repressor during heart development, and is essential for adult heart integrity, function and adaptation. *Development* **132**, 2451–2462.
98. Cai, C. L., Zhou, W., Yang, L., Bu, L., Qyang, Y., Zhang, X., Li, X., Rosenfeld, M. G., Chen, J., and Evans, S. (2005). T-box genes coordinate regional rates of proliferation and regional specification during cardiogenesis. *Development* **132**, 2475–2487.
99. Chapman, D. L., Cooper-Morgan, A, Harrelson, Z., and Papaioannou, V. F. (2003). Critical role for Tbx6 in mesoderm specification in the mouse embryo. *Mech. Dev.* **120**, 837–847.
100. Srivastava, D. (1999). HAND proteins: molecular mediators of cardiac development and congenital heart disease. *Trends Cardiovasc. Med.* **9**, 11–18.
101. McFadden, D. G., Charite, J., Richardson, J. A., Srivastava, D., Firulli, A. B., and Olson, E. N. (2000). A GATA-dependent right ventricular enhancer controls dHAND transcription in the developing heart. *Development* **127**, 5331–5341.
102. Yamagishi, H., Yamagishi, C., Nakagawa, O., Harvey, R. P., Olson, E. N., and Srivastava, D. (2001). The combinatorial activities of Nkx2.5 and dHAND are essential for cardiac ventricle formation. *Dev. Biol.* **239**, 190–203.
103. Srivastava, D., Thomas, T., Lin, Q., Kirby, M. L., Brown, D., and Olson, E. N. (1997). Regulation of cardiac mesodermal and neural crest development by the bHLH transcription factor, dHAND. *Nat. Genet.* **16**, 154–160.
104. Togi, K., Kawamoto, T., Yamauchi, R., Yoshida, Y., Kita, T., and Tanaka, M. (2004). Role of Hand1/eHAND in the dorsoventral patterning and interventricular septum formation in the embryonic heart. *Mol. Cell Biol.* **24**, 4627–4635.
105. McFadden, D. G., Barbosa, A. C., Richardson, J. A., Schneider, M. D., Srivastava, D., and Olson, E. N. (2005). The Hand1 and Hand2 transcription factors regulate expansion of the embryonic cardiac ventricles in a gene dosage-dependent manner. *Development* **132**, 189–201.
106. Biben, C., and Harvey, R. P. (1997). Homeodomain factor Nkx2-5 controls left/right asymmetric expression of bHLH gene eHand during murine heart development. *Genes Dev.* **11**, 1357–1369.
107. Thomas, T., Yamagishi, H., Overbeek, P. A., Olson, E. N., and Srivastava, D. (1998). The bHLH factors, dHAND and eHAND, specify pulmonary and systemic cardiac ventricles independent of left-right sidedness. *Dev. Biol.* **196**, 228–236.
108. Lyons, I., Parsons, L. M., Hartley, L., Li, R., Andrews, J. E., Robb, L., and Harvey, R. P. (1995). Myogenic and morphogenetic defects in the heart tubes of murine embryos lacking the homeo box gene Nkx2-5. *Genes Dev.* **9**, 1654–1666.
109. Yelon, D., Ticho, B., Halpern, M. E., Ruvinsky, I., Ho, R. K., Silver, L. M., and Stainier, D. Y. (2000). The bHLH transcription factor hand2 plays parallel roles in zebrafish heart and pectoral fin development. *Development* **127**, 2573–2582.
110. Angelo, S., Lohr, J., Lee, K. H., Ticho, B. S., Breitbart, R. E., Hill, S., Yost, H. J., and Srivastava, D. (2000). Conservation of sequence and expression of *Xenopus* and zebrafish dHAND during cardiac, branchial arch and lateral mesoderm development. *Mech. Dev.* **95**, 231–237.
111. Natarajan, A., Yamagishi, H., Ahmad, F., Li, D., Roberts, R., Matsuoka, R., Hill, S., and Srivastava, D. (2001). Human eHAND, but not dHAND, is down-regulated in cardiomyopathies. *J. Mol. Cell. Cardiol.* **33**, 1607–1614.
112. Lin, Q., Schwarz, J., Bucana, C., and Olson, E. N. (1997). Control of mouse cardiac morphogenesis and myogenesis by transcription factor MEF2C. *Science* **276**, 1404–1407.
113. Phan, D., Rasmussen, T. L., Nakagawa, O., McAnally, J., Gottlieb, P. D., Tucker, P. W., Richardson, J. A., Bassel-Duby, R., and Olson, E. N. (2005). BOP, a regulator of right ventricular heart development, is a direct transcriptional target of MEF2C in the developing heart. *Development* **132**, 2669–2678.

114. Morin, S., Charron, F., Robitaille, L., and Nemer, M. (2000). GATA-dependent recruitment of MEF2 proteins to target promoters. *EMBO J.* **19,** 2046–2055.
115. Naya, F. J., Wu, C., Richardson, J. A., Overbeek, P., and Olson, E. N. (1999). Transcriptional activity of MEF2 during mouse embryogenesis monitored with a MEF2-dependent transgene. *Development* **126,** 2045–2052.
116. Overbeek, P. A. (1997). Right and left go dHAND and eHAND. *Nat. Genet.* **16,** 119–121.
117. Olson, E. N., and Black, B. L. (1999). Control of cardiac development by the MEF2 family of transcription factors. *In* "Heart Development." (R. P. Harvey and N. Rosenthal, Eds.), chap. 8, pp. 131–154. Academic Press, San Diego, CA.
118. Morin, S., Pozzulo, G., Robitaille, L., Cross, J., and Nemer, M. (2005). MEF2-dependent recruitment of the HAND1 transcription factor results in synergistic activation of target promoters. *J. Biol. Chem.* **280,** 32272–32278.
119. McKinsey, T. A., Zhang, C. L., and Olson, E. N. (2001). Control of muscle development by dueling HATs and HDACs. *Curr. Opin. Genet. Dev.* **11,** 497–504.
120. Slepak, T. I., Webster, K. A., Zang, J., Prentice, H., O'Dowd, A., Hicks, M. N., and Bishopric, N. H. (2001). Control of cardiac-specific transcription by p300 through myocyte enhancer factor-2D. *J. Biol. Chem.* **276,** 7575–7585.
121. Molkentin, J. D., Li, L., and Olson, E. N. (1996). Phosphorylation of the MADS-Box transcription factor MEF2C enhances its DNA binding activity. *J. Biol. Chem.* **271,** 17199–17204.
122. Passier, R., Zeng, H., Frey, N., Naya, F. J., Nicol, R. L., McKinsey, T. A., Overbeek, P., Richardson, J. A., Grant, S. R., and Olson, E. N. (2000). CaM kinase signaling induces cardiac hypertrophy and activates the MEF2 transcription factor in vivo. *J. Clin. Invest.* **105,** 1395–1406.
123. Dunwoodie, S. L., Rodriguez, T. A., and Beddington, R. S. (1998). Msg1 and Mrg1, founding members of a gene family, show distinct patterns of gene expression during mouse embryogenesis. *Mech. Dev.* **72,** 27–40.
124. Palmer, S., Groves, N., Schindeler, A., Yeoh, T., Biben, C., Wang, C. C., Sparrow, D., Barnett, L., Jenkins, N. A., Copeland, N. G., Koentgen, F., Mohun, T., and Harvey, R. P. (2001). The small muscle-specific protein Csl modifies cell shape and promotes myocyte fusion in an insulin-like growth factor 1-dependent manner. *J. Cell Biol.* **153,** 985–997.
125. Capdevila, J., Vogan, K. J., Tabin, C. J., and Izpisua Belmonte, J. C. (2000). Mechanisms of left-right determination in vertebrates. *Cell* **101,** 9–21.
126. Franco, D., Kelly, R., Moorman, A. F., Lamers, W. H., Buckingham, M., and Brown, N. A. (2001). MLC3F transgene expression in iv mutant mice reveals the importance of left-right signalling pathways for the acquisition of left and right atrial but not ventricular compartment identity. *Dev. Dyn.* **221,** 206–215.
127. Lohr, J. L., Danos, M. C., and Yost, H. J. (1997). Left-right asymmetry of a nodal-related gene is regulated by dorsoanterior midline structures during *Xenopus* development. *Development* **124,** 1465–1472.
128. Bamforth, S. D., Braganca, J., Farthing, C. R., Schneider, J. E., Broadbent, C., Michell, A. C., Clarke, K., Neubauer, S., Norris, D., Brown, N. A., Anderson, R. H., and Bhattacharya, S. (2004). Cited2 controls left-right patterning and heart development through a Nodal-Pitx2c pathway. *Nat. Genet.* **36,** 1189–1196.
129. Lowe, L. A., Yamada, S., and Kuehn, M. R. (2001). Genetic dissection of nodal function in patterning the mouse embryo. *Development* **128,** 1831–1843.
130. Tsukui, T., Capdevila, J., Tamura, K., Ruiz-Lozano, P., Rodriguez-Esteban, C., Yonei-Tamura, S., Magallon, J., Chandraratna, R. A., Chien, K., Blumberg, B., Evans, R. M., and Belmonte, J. C. (1999). Multiple left-right asymmetry defects in Shh(−/−) mutant mice unveil a convergence of the shh and retinoic acid pathways in the control of Lefty-1. *Proc. Natl. Acad. Sci. USA* **96,** 11376–11381.
131. Kishigami, S., Yoshikawa, S., Castranio, T., Okazaki, K., Furuta, Y., and Mishina, Y. (2004). BMP signaling through ACVRI is required for left-right patterning in the early mouse embryo. *Dev. Biol.* **276,** 185–193.
132. Raya, A., Kawakami, Y., Rodriguez-Esteban, C., Buscher, D., Koth, C. M., Itoh, T., Morita, M., Raya, R. M., Dubova, I., Bessa, J. G., de la Pompa, J. L., and Belmonte, J. C. (2003). Notch activity induces Nodal expression and mediates the establishment of left-right asymmetry in vertebrate embryos. *Genes Dev.* **17,** 1213–1218.
133. Gaio, U., Schweickert, A., Fischer, A., Garratt, A. N., Muller, T., Ozcelik, C., Lankes, W., Strehle, M., Britsch, S., Blum, M., and Birchmeier, C. (1999). A role of the cryptic gene in the correct establishment of the left-right axis. *Curr. Biol.* **9,** 1339–1342.
134. Chang, H., Zwijsen, A., Vogel, H., Huylebroeck, D., and Matzuk, M. M. (2000). Smad5 is essential for left-right asymmetry in mice. *Dev. Biol.* **219,** 71–78.
135. Grinberg, I., and Millen, K. (2005). The ZIC gene family in development and disease. *Clin. Genet.* **67,** 290–296.
136. Schilling, T. F., Concordet, J. P., and Ingham, P. W. (1999). Regulation of left-right asymmetries in the zebrafish by Shh and BMP4. *Dev. Biol.* **210,** 277–287.
137. Thomas, T., Yamagishi, H., Overbeek, P. A., Olson, E. N., and Srivastava, D. (1998). The bHLH factors, dHAND and eHAND, specify pulmonary and systemic cardiac ventricles independent of left-right sidedness. *Dev. Biol.* **196,** 228–236.
138. Chen, Y., Mironova, E., Whitaker, L. L., Edwards, L., Yost, H. J., and Ramsdell AF. (2004). ALK4 functions as a receptor for multiple TGF beta-related ligands to regulate left-right axis determination and mesoderm induction in *Xenopus*. *Dev. Biol.* **268,** 280–294.
139. Rodriguez-Esteban, C., Capdevila, J., Kawakami, Y., and Izpisua Belmonte, J. C. (2001). Wnt signaling and PKA control Nodal expression and left-right determination in the chick embryo. *Development* **128,** 3189–3195.
140. Isaac, A., Sargent, M. G., and Cooke, J. (1997). Control of vertebrate left-right asymmetry by a snail-related zinc finger gene. *Science* **275,** 1301–1304.
141. Biben, C., and Harvey, R. P. (1997). Homeodomain factor Nkx2-5 controls left/right asymmetric expression of bHLH gene eHand during murine heart development. *Genes Dev.* **11,** 1357–1369.
142. Meno, C., Shimono, A., Saijoh, Y., Yashiro, K., Mochida, K., Ohishi, S., Noji, S., Kondoh, H., and Hamada, H. (1998). Lefty-1 is required for left-right determination as a regulator of lefty-2 and nodal. *Cell* **94,** 287–297.

143. Gage, P. J., Suh, H., and Camper, S. A. (1999). Dosage requirement of Pitx2 for development of multiple organs. *Development* **126,** 4643–4651.
144. Kitamura, K., Miura, H., Miyagawa-Tomita, S., Yanazawa, M., Katoh-Fukui, Y., Suzuki, R., Ohuchi, H., Suehiro, A., Motegi, Y., Nakahara, Y., Kondo, S., and Yokoyama, M. (1999). Mouse Pitx2 deficiency leads to anomalies of the ventral body wall, heart, extra- and periocular mesoderm and right pulmonary isomerism. *Development* **126,** 5749–5758.
145. Zile, M. H., Kostetskii, I., Yuan, S., Kostetskaia, E., St. Amand, T. R., Chen, Y., and Jiang, W. (2000). Retinoid signaling is required to complete the vertebrate cardiac left/right asymmetry pathway. *Dev. Biol.* **223,** 323–338.
146. Chazaud, C., Chambon, P., and Dolle, P. (1999). Retinoic acid is required in the mouse embryo for left-right asymmetry determination and heart morphogenesis. *Development* **126,** 2589–2596.
147. Sucov, H. M., Dyson, E., Gumeringer, C. L., Price, J., Chien, K. R., and Evans, R. M. (1994). RXR α mutant mice establish a genetic basis for vitamin A signaling in heart morphogenesis. *Genes Dev.* **8,** 1007–1018.
148. Kastner, P., Messaddeq, N., Mark, M., Wendling, O., Grondona, J. M., Ward, S., Ghyselinck, N., and Chambon, P. (1997). Vitamin A deficiency and mutations of RXRα, RXRβ and RARα lead to early differentiation of embryonic ventricular cardiomyocytes. *Development* **124,** 4749–4758.
149. Chen, J., Kubalak, S. W., and Chien, K. R. (1998). Ventricular muscle-restricted targeting of the RXRα gene reveals a non-cell-autonomous requirement in cardiac chamber morphogenesis. *Development* **125,** 1943–1949.
150. Chen, T. H., Chang, T. C., Kang, J. O., Choudhary, B., Makita, T., Tran, C. M., Burch, J. B., Eid, H., and Sucov, H. M. (2002). Epicardial induction of fetal cardiomyocyte proliferation via a retinoic acid-inducible trophic factor. *Dev. Biol.* **250,** 198–207.
151. Moss, J. B., Xavier-Neto, J., Shapiro, M. D., Nayeem, S. M., McCaffery, P., Drager, U. C., and Rosenthal, N. (1998). Dynamic patterns of retinoic acid synthesis and response in the developing mammalian heart. *Dev. Biol.* **199,** 55–71.
152. Stuckmann, I., Evans, S., and Lassar, A. B. (2003). Erythropoietin and retinoic acid, secreted from the epicardium, are required for cardiac myocyte proliferation. *Dev. Biol.* **255,** 334–349.
153. Kang, J. O., and Sucov, H. M. (2005). Convergent proliferative response and divergent morphogenic pathways induced by epicardial and endocardial signaling in fetal heart development. *Mech. Dev.* **122,** 57–65.
154. Pennisi, D. J., Ballard, V. L., and Mikawa, T. (2003). Epicardium is required for the full rate of myocyte proliferation and levels of expression of myocyte mitogenic factors FGF2 and its receptor, FGFR-1, but not for transmural myocardial patterning in the embryonic chick heart. *Dev. Dyn.* **228,** 161–172.
155. Lavine, K. J., Yu, K., White, A. C., Zhang, X., Smith, C., Partanen, J., and Ornitz, D. M. (2005). Endocardial and epicardial derived FGF signals regulate myocardial proliferation and differentiation in vivo. *Dev. Cell* **8,** 85–95.
156. Zhao, Y. Y., Sawyer, D. R., Baliga, R. R., Opel, D. J., Han, X., Marchionni, M. A., and Kelly, R. A. (1998). Neuregulins promote survival and growth of cardiac myocytes. Persistence of ErbB2 and ErbB4 expression in neonatal and adult ventricular myocytes. *J. Biol. Chem.* **273,** 10261–10269.
157. Lee, K. F., Simon, H., Chen, H., Bates, B., Hung, M. C., and Hauser, C. (1995). Requirement for neuregulin receptor erbB2 in neural and cardiac development. *Nature* **378,** 394–398.
158. Gassmann, M., Casagranda, F., Orioli, D., Simon, H., Lai, C., Klein, R., and Lemke, G. (1995). Aberrant neural and cardiac development in mice lacking the ErbB4 neuregulin receptor. *Nature* **378,** 390–394.
159. Crone, S. A., Zhao, Y. Y., Fan, L., Gu, Y., Minamisawa, S., Liu, Y., Peterson, K. L., Chen, J., Kahn, R., Condorelli, G., Ross, J. Jr., Chien, K. R., and Lee, K. F. (2002). ErbB2 is essential in the prevention of dilated cardiomyopathy. *Nat. Med.* **8,** 459–465.
160. Wang, B., Weidenfeld, J., Lu, M. M., Maika, S., Kuziel, W. A., Morrisey, E. E., and Tucker, P. W. (2004). Foxp1 regulates cardiac outflow tract, endocardial cushion morphogenesis and myocyte proliferation and maturation. *Development* **131,** 4477–4487.
161. Zeisberg, E. M., Ma, Q., Juraszek, A. L., Moses, K., Schwartz, R. J., Izumo, S., and Pu, W. T. (2005). Morphogenesis of the right ventricle requires myocardial expression of Gata4. *J. Clin. Invest.* **115,** 1522–1531.
162. Hatcher, C. J., Kim, M. S., Mah, C. S., Goldstein, M. M., Wong, B., Mikawa, T., and Basson, C. T. (2001). TBX5 transcription factor regulates cell proliferation during cardiogenesis. *Dev. Biol.* **230,** 177–188.
163. Bushdid, P. B., Osinska, H., Waclaw, R. R., Molkentin, J. D., and Yutzey, K. E. (2003). NFATc3 and NFATc4 are required for cardiac development and mitochondrial function. *Circ. Res.* **92,** 1305–1313.
164. Shin, C. H., Liu, Z. P., Passier, R., Zhang, C. L., Wang, D. Z., Harris, T. M., Yamagishi, H., Richardson, J. A., Childs, G., and Olson, E. N. (2002). Modulation of cardiac growth and development by HOP, an unusual homeodomain protein. *Cell* **110,** 725–735.
165. Chen, F., Kook, H., Milewski, R., Gitler, A. D., Lu, M. M., Li, J., Nazarian, R., Schnepp, R., Jen, K., Biben, C., Runke, G., Mackay, J. P., Novotny, J., Schwartz, R. J., Harvey, R. P., Mullins, M. C., and Epstein, J. A. (2002). Hop is an unusual homeobox gene that modulates cardiac development. *Cell* **110,** 713–723.
166. Kook, H., Lepore, J. J., Gitler, A. D., Lu, M. M., Wing-Man Yung, W., Mackay, J., Zhou, R., Ferrari, V., Gruber, P., and Epstein, J. A. (2003). Cardiac hypertrophy and histone deacetylase-dependent transcriptional repression mediated by the atypical homeodomain protein Hop. *J. Clin. Invest.* **112,** 863–871.
167. Tran, C. M., and Sucov, H. M. (1998). The RXRα gene functions in a non-cell-autonomous manner during mouse cardiac morphogenesis. *Development* **125,** 1951–1956.
168. Lakkis, M. L., and Epstein, J. A. (1998). Neurofibromin modulation of ras activity is required for normal endocardial-mesenchymal transformation in the developing heart. *Dev. Biol.* **125,** 4359–4367.

169. Camenisch, T. D., Spicer, A. P., Brehm-Gibson, T., Biesterfeldt, J., Augustine, M. L., Calabro, A. Jr., Kubalak, S., Klewer, S. E., and McDonald, J. A. (2000). Disruption of hyaluronan synthase-2 abrogates normal cardiac morphogenesis and hyaluronan-mediated transformation of epithelium to mesenchyme. *J. Clin. Invest.* **106,** 349–360.
170. Gruber, P., Kubalak, S., Pexieder, T., Sucov, H., Evans, R., and Chien, K. (1996). RXRα deficiency confers genetic susceptibility for aortic sac, conotruncal, atrioventricular cushion, and ventricular muscle defects in mice. *J. Clin. Invest.* **98,** 1332–1343.
171. Ya, J., Schilham, M. W., de Boer, P. A., Moorman, A. F., Clevers, H., and Lamers, W. H. (1998). Sox-4 deficiency syndrome in mice is an animal model for common trunk. *Circ. Res.* **83,** 986–994.
172. Wessels, A., and Sedmera, D. (2003). Developmental anatomy of the heart: a tale of mice and man. *Physiol. Genomics* **15,** 165–176.
173. Schott, J.-J., Benson, D. W., Basson, C. T., Pease, W., Silberbach, G. M., Moak, J. P., Maron, B., Seidman, C. E., and Seidman, J. G. (1998). Congenital heart disease caused by mutations in the transcription factor *NKX2-5*. *Science* **281,** 108–111.
174. Benson, D. W., Silberbach, G. M., Kavanaugh-McHugh, A., Cottrill, C., Zhang, Y., Riggs, S., Smalls, O., Johnson, M. C., Watson, M. S., Seidman, J. G., Seidman, C. E., Plowden, J., and Kugler, J. D. (1999). Mutations in the cardiac transcription factor NKX2.5 affect diverse cardiac developmental pathways. *J. Clin. Invest.* **104,** 1567–1573.
175. Goldmuntz, E., Geiger, E., and Benson, D. W. (2001). NKX2.5 mutations in patients with tetralogy of Fallot. *Circulation* **104,** 2565–2568.
176. Kasahara, H., Lee, B., Schott, J. J., Benson, D. W., Seidman, J. G., Seidman, C. E., and Izumo, S. (2000). Loss of function and inhibitory effects of human CSX/NKX2.5 homeoprotein mutations associated with congenital heart disease. *J. Clin. Invest.* **106,** 299–308.
177. Biben, C., Weber, R., Kesteven, S., Stanley, E., McDonald, L., Elliott, D. A., Barnett, L., Koentgen, F., Robb, L., Feneley, M., and Harvey, R. P. (2000). Cardiac septal and valvular dysmorphogenesis in mice heterozygous for mutations in the homeobox gene Nkx2-5. *Circ. Res.* **87,** 888–895.
178. Basson, C. T., Bachinsky, D. R., Lin, R. C., Levi, T., Elkins, J. A., Soults, J., Grayzel, D, Kroumpouzou, E., Traill, T. A., Leblanc-Straceski, J., Renault, B., Kucherlapati, R., Seidman, J. G., and Seidman, C. E. (1997). Mutations in human TBX5 cause limb and cardiac malformation in Holt-Oram syndrome. *Nat. Genet.* **15,** 30–35.
179. Li, Q. Y., Newbury-Ecob, R. A., Terrett, J. A., Wilson, D. I., Curtis, A. R., Yi, C. H., Gebuhr, T., Bullen, P. J., Robson, S. C., Strachan, T., Bonnet, D., Lyonnet, S., Young, I. D., Raeburn, J. A., Buckler, A. J., Law, D. J., and Brook, J. D. (1997). Holt-Oram syndrome is caused by mutations in TBX5, a member of the Brachyury (T) gene family. *Nat. Genet.* **15,** 21–29.
180. Basson, C. T., Huang, T., Lin, R. C., Bachinsky, D. R., Weremowicz, S., Vaglio, A., Bruzzone, R., Quadrelli, R., Lerone, M., Romeo, G., Silengo, M., Pereira, A., Krieger, J., Mesquita, S. F., Kamisago, M., Morton, C. C., Pierpont, M. E., Muller, C. W., Seidman, J. G., and Seidman, C. E. (1999). Different TBX5 interactions in heart and limb defined by Holt-Oram syndrome mutations. *Proc. Natl. Acad. Sci. USA* **96,** 2919–2924.
181. Cross, S. J., Ching, Y.-H., Li, Q. Y., Armstrong-Buisseret, L., Spranger, S., Lyonnet, S., Bonnet, D., Penttinen, M., Jonveaux, P., Leheup, B., Mortier, G., Van Ravenswaaij, C., Gardiner, C.-A., Brook, J. D., and Newbury-Ecob, R. A. (2000). The mutation spectrum in Holt-Oram syndrome. *J. Med. Genet.* **37,** 785–787.
182. Ghosh, T. K., Packham, E. A., Bonser, A. J., Robinson, T. E., Cross, S. J., and Brook, J. D. (2001). Characterization of the TBX5 binding site and analysis of mutations that cause Holt-Oram syndrome. *Hum. Mol. Genet.* **10,** 1983–1994.
183. Armstrong, E. J., and Bischoff, J. (2004). Heart valve development: endothelial cell signaling and differentiation. *Circ. Res.* **95,** 459–470.
184. de Lange, F. J., Moorman, A. F., Anderson, R. H., Manner, J., Soufan, A. T, de Gier-de Vries, C., Schneider, M. D., Webb, S., van den Hoff, M. J., and Christoffels, V. M. (2004). Lineage and morphogenetic analysis of the cardiac valves. *Circ. Res.* **95,** 645–654.
185. Sugi, Y., Yamamura, H., Okagawa, H., and Markwald, R. R. (2004). Bone morphogenetic protein-2 can mediate myocardial regulation of atrioventricular cushion mesenchymal cell formation in mice. *Dev. Biol.* **269,** 505–501.
186. Gaussin, V., Morley, G. E., Cox, L., Zwijsen, A., Vance, K. M., Emile, L., Tian, Y., Liu, J., Hong, C., Myers, D., Conway, S. J., Depre, C., Mishina, Y., Behringer, R. R., Hanks, M. C., Schneider, M. D., Huylebroeck, D., Fishman, G. I., Burch, J. B., and Vatner, S. F. (2005). Alk3/Bmpr1a receptor is required for development of the atrioventricular canal into valves and annulus fibrosus. *Circ. Res.* **97,** 219–226.
187. Desgrosellier, J. S., Mundell, N. A., McDonnell, M. A., Moses, H. L., and Barnett, J V. (2005). Activin receptor-like kinase 2 and Smad6 regulate epithelial-mesenchymal transformation during cardiac valve formation. *Dev. Biol.* **280,** 201–210.
188. Galvin, K. M., Donovan, M. J., Lynch, C. A., Meyer, R. I., Paul, R. J., Lorenz, J. N., Fairchild-Huntress, V., Dixon, K. L., Dunmore, J. H., Gimbrone, M. A., Jr., Falb, D., and Huszar, D. (2000). A role for smad6 in development and homeostasis of the cardiovascular system. *Nat. Genet.* **24,** 171–174.
189. Nieto, M. A. (2002). The snail superfamily of zinc-finger transcription factors. *Nat. Rev. Mol. Cell Biol.* **3,** 155–166.
190. Nieto, M. A., Sargent, M. G., Wilkinson, D. G., and Cooke, J. (1994). Control of cell behavior during vertebrate development by Slug, a zinc finger gene. *Science* **264,** 835–839.
191. Carmona, R., Gonzalez-Iriarte, M., Macias, D., Perez-Pomares, J. M., Garcia-Garrido, L., and Munoz-Chapuli, R. (2000). Immunolocalization of the transcription factor Slug in the developing avian heart. *Anat. Embryol. (Berl)* **201,** 103–109.
192. Romano, L. A., and Runyan, R. B. (2000). Slug is an essential target of TGFß2 signaling in the developing chicken heart. *Dev. Biol.* **223,** 91–102.
193. Carver, E. A., Jiang, R., Lan, Y., Oram, K. F., and Gridley, T. (2001). The mouse snail gene encodes a key regulator of the epithelial-mesenchymal transition. *Mol. Cell Biol.* **21,** 8184–8188.
194. Hurlstone, A. F., Haramis, A. P., Wienholds, E., Begthel, H., Korving, J., Van Eeden, F., Cuppen, E., Zivkovic, D., Plasterk,

R. H., and Clevers, H. (2003). The Wnt/beta-catenin pathway regulates cardiac valve formation. *Nature* **425,** 633–637.

195. Timmerman, L. A., Grego-Bessa, J., Raya, A., Bertran, E., Perez-Pomares, J. M., Diez, J., Aranda, S., Palomo, S., McCormick, F., Izpisua-Belmonte, J. C., and de la Pompa, J. L. (2004). Notch promotes epithelial-mesenchymal transition during cardiac development and oncogenic transformation. *Genes Dev.* **18,** 99–115.

196. Kokubo, H., Miyagawa-Tomita, S., Tomimatsu, H., Nakashima, Y., Nakazawa M., Saga, Y., and Johnson, R. L. (2004). Targeted disruption of hesr2 results in atrioventricular valve anomalies that lead to heart dysfunction. *Circ. Res.* **95,** 540–547.

197. Ferrara, N., Gerber, H. P., and LeCouter, J. (2003). The biology of VEGF and its receptors. *Nat. Med.* **9,** 669–676.

198. Zachary, I., and Gliki, G. (2001). Signaling transduction mechanisms mediating biological actions of the vascular endothelial growth factor family. *Cardiovasc. Res.* **49,** 568–581.

199. Johnson, E. N., Lee, Y. M., Sander, T. L., Rabkin, E., Schoen, F. J., Kaushal, S., and Bischoff, J. (2003). NFATc1 mediates vascular endothelial growth factor-induced proliferation of human pulmonary valve endothelial cells. *J. Biol. Chem.* **278,** 1686–1692.

200. Miquerol, L., Gertsenstein, M., Harpal, K., Rossant, J., and Nagy, A. (1999). Multiple developmental roles of VEGF suggested by a LacZ-tagged allele. *Dev. Biol.* **212,** 307–322.

201. Dor, Y., Camenisch, T. D., Itin, A., Fishman, G. I., McDonald, J. A., Carmeliet, P., and Keshet, E. (2001). A novel role for VEGF in endocardial cushion formation and its potential contribution to congenital heart defects. *Development* **128,** 1531–1538.

202. Bernanke, D. H., and Markwald, R. R. (1982). Migratory behavior of cardiac cushion tissue cells in a collagen-lattice culture system. *Dev. Biol.* **91,** 235–245.

203. Dor, Y., Klewer, S. E., McDonald, J. A., Keshet, E., and Camenisch, T. D. (2003). VEGF modulates early heart valve formation. *Anat. Rec.* **271A,** 202–208.

204. Pinter, E., Haigh, J., Nagy, A., and Madri, J. A. (2001). Hyperglycemia-induced vasculopathy in the murine conceptus is mediated via reductions of VEGF-A expression and VEGF receptor activation. *Am. J. Pathol.* **158,** 1199–1206.

205. Enciso, J. M., Gratzinger, D., Camenisch, T. D., Canosa, S., Pinter, E., and Madri, J. A. (2003). Elevated glucose inhibits VEGF-A-mediated endocardial cushion formation: modulation by PECAM-1 and MMP-2. *J. Cell Biol.* **160,** 605–615.

206. Ferencz, C., Rubin, J. D., McCarter, R. J., and Clark, E. B. (1990). Maternal diabetes and cardiovascular malformations: predominance of double outlet right ventricle and truncus arteriosus. *Teratology* **41,** 319–326.

207. Kitzmiller, J. L., Gavin, L. A., Gin, G. D., Jovanovic-Peterson, L., Main, E K., and Zigrang, W. D. (1991). Preconception care of diabetes. Glycemic control prevents congenital anomalies. *JAMA* **265,** 731–736.

208. Crabtree, G. R., and Olson, E. N. (2002). NFAT signaling: choreographing the social lives of cells. *Cell* **109,** S67–S79.

209. Ranger, A. M., Grusby, M. J., Hodge, M. R., Gravallese, E M., de la Brousse, F. C., Hoey, T., Mickanin, C., Baldwin, H. S., and Glimcher, L. H. (1998). The transcription factor NF-ATc is essential for cardiac valve formation. *Nature* **392,** 186–190.

210. de la Pompa, J. L., Timmerman, L. A., Takimoto, H., Yoshida, H., Elia, A. J., Samper, E., Potter, J., Wakeham, A., Marengere, L., Langille, B. L., Crabtree, G. R., and Mak, T. W. (1998). Role of the NF-ATc transcription factor in morphogenesis of cardiac valves and septum. *Nature* **392,** 182–186.

211. Johnson, E. N., Lee, Y. M., Sander, T. L., Rabkin, E., Schoen, F. J., Kaushal, S., and Bischoff, J. (2003). NFATc1 mediates vascular endothelial growth factor-induced proliferation of human pulmonary valve endothelial cells. *J. Biol. Chem.* **278,** 1686–1692.

212. Chang, C. P., Neilson, J. R., Bayle, J. H., Gestwicki, J. E., Kuo, A., Stankunas, K., Graef, I. A., and Crabtree, G. R. (2004). A field of myocardial-endocardial NFAT signaling underlies heart valve morphogenesis. *Cell* **118,** 649–663.

213. Zhou, B., Wu, B., Tompkins, K. L., Boyer, K. L., Grindley, J. C., and Baldwin, H. S. (2005). Characterization of Nfatc1 regulation identifies an enhancer required for gene expression that is specific to pro-valve endocardial cells in the developing heart. *Development* **132,** 1137–1146.

214. Rothermel, B., Vega, R. B., Yang, J., Wu, H., Bassel-Duby, R., and Williams, R. S. (2000). A protein encoded within the Down syndrome critical region is enriched in striated muscles and inhibits calcineurin signaling. *J. Biol. Chem.* **275,** 8719–8725.

215. Yang, J., Rothermel, B., Vega, R. B, Frey, N., McKinsey, T. A., Olson, E. N., Bassel-Duby, R., and Williams, R. S. (2000). Independent signals control expression of the calcineurin inhibitory proteins MCIP1 and MCIP2 in striated muscles. *Circ. Res.* **87,** E61–E68.

216. Lange, A. W., Molkentin, J. D., and Yutzey K. E. (2004). DSCR1 gene expression is dependent on NFATc1 during cardiac valve formation and colocalizes with anomalous organ development in trisomy 16 mice. *Dev. Biol.* **266,** 346–360.

217. Delorme, B., Dahl, E., Jarry-Guichard, T., Briand, J. P., Willecke, K., Gros, D., and Theveniau-Ruissy, M. (1997). Expression pattern of connexin gene products at the early developmental stages of the mouse cardiovascular system. *Circ. Res.* **81,** 423–437.

218. Reaume, A. G., de Sousa, P. A., Kulkarni, S., Langille, B. L., Zhu, D., Davies, T. C., Juneja, S. C., Kidder, G. M., and Rossant, J. (1995). Cardiac malformation in neonatal mice lacking connexin43. *Science* **267,** 1831–1834.

219. Kirchhoff, S., Nelles, E., Hagendorff, A., Kruger, O., Traub, O., and Willecke, K. (1998). Reduced cardiac conduction velocity and predisposition to arrhythmias in connexin40-deficient mice. *Curr. Biol.* **8,** 299–302.

220. Kumai, M., Nishii, K., Nakamura, K., Takeda, N., Suzuki, M., and Shibata, Y. (2000). Loss of connexin45 causes a cushion defect in early cardiogenesis. *Development* **127,** 3501–3512.

221. Marvin, M. J., Di Rocco, G., Gardiner, A., Bush, S. M., and Lassar, A. B. (2001). Inhibition of Wnt activity induces heart formation from posterior mesoderm. *Genes Dev.* **15,** 316–327.

222. Liebner, S., Cattelino, A., Gallini, R., Rudini, N., Iurlaro, M., Piccolo, S., and Dejana, E. (2004). Beta-catenin is required for endothelial-mesenchymal transformation during heart cushion development in the mouse. *J. Cell Biol.* **166,** 359–367.

223. Gitler, A. D., Lu, M. M., Jiang, Y. Q., Epstein, J. A., and Gruber, P. J. (2003). Molecular markers of cardiac endocardial cushion development. *Dev. Dyn.* **228,** 643–650.

224. Hurlstone, A. F., Haramis, A. P., Wienholds, E., Begthel, H., Korving, J., Van Eeden, F., Cuppen, E., Zivkovic, D., Plasterk, R. H., and Clevers, H. (2003). The Wnt/beta-catenin pathway regulates cardiac valve formation. *Nature* **425**, 633–637.

225. Person, A. D., Garriock, R. J., Krieg, P. A., Runyan, R. B., and Klewer, S. E. (2005). Frzb modulates Wnt-9a-mediated beta-catenin signaling during avian atrioventricular cardiac cushion development. *Dev. Biol.* **278**, 35–48.

226. Perez-Moreno, M., Jamora, C., and Fuchs, E. (2003). Sticky business: orchestrating cellular signals at adherens junctions. *Cell* **112**, 535–548.

227. Ilan, N., Mahooti, S., Rimm, D. L., and Madri, J. A. (1999). PECAM-1 (CD31) functions as a reservoir for and a modulator of tyrosine-phosphorylated beta-catenin. *J. Cell Sci.* **112**, 3005–3014.

228. Nakajima, Y., Mironov, V., Yamagishi, T., Nakamura, H., and Markwald, R. R. (1997). Expression of smooth muscle -actin in mesenchymal cells during formation of avian endocardial cushion tissue: a role for transforming growth factor ß3. *Dev. Dyn.* **209**, 296–309.

229. Schroeder, J. A., Jackson, L. F., Lee, D. C., and Camenisch, T. D. (2003). Form and function of developing heart valves: coordination by extracellular matrix and growth factor signaling. *J. Mol. Med.* **81**, 392–403.

230. Yarden, Y., and Sliwkowski, M. X. (2001). Untangling the ErbB signalling network. *Nat. Rev. Mol. Cell Biol.* **2**, 127–137.

231. Raab, G., and Klagsbrun, M. (1997). Heparin-binding EGF-like growth factor. *Biochim. Biophys. Acta* **1333**, F179–F199.

232. Jackson, L. F., Qiu, T. H., Sunnarborg, S. W., Chang, A., Zhang, C., Patterson, C., and Lee, D. C. (2003). Defective valvulogenesis in HB-EGF and TACE-null mice is associated with aberrant BMP signaling. *EMBO J.* **22**, 2704–2716.

233. Gassmann, M., Casagranda, F., Orioli, D., Simon, H., Lai, C., Klein, R., and Lemke, G. (1995). Aberrant neural and cardiac development in mice lacking the ErbB4 neuregulin receptor. *Nature* **378**, 390–394.

234. Krug, E. L., Mjaatvedt, C. H., and Markwald, R. R. (1987). Extracellular matrix from embryonic myocardium elicits an early morphogenetic event in cardiac endothelial differentiation. *Dev. Biol.* **120**, 348–355.

235. McDonald, J. A., and Camenisch, T. D. (2002). Hyaluronan: genetic insights into the complex biology of a simple polysaccharide. *Glycoconj. J.* **19**, 331–339.

236. Day, A. J., and Prestwich, G. D. (2002), Hyaluronan-binding proteins: tying up the giant. *J. Biol. Chem.* **277**, 4585–4588.

237. Eisenberg, L. M., and Markwald, R. R. (1995). Molecular regulation of atrioventricular valvuloseptal morphogenesis. *Circ. Res.* **77**, 1–6.

238. Weigel, P. H., Hascall, V. C., and Tammi, M. (1997). Hyaluronan synthases. *J. Biol. Chem.* **272**, 13997–4000.

239. Camenisch, T. D., Spicer, A. P., Brehm-Gibson, T., Biesterfeldt, J., Augustine, M. L., Calabro, A., Jr., Kubalak, S., Klewer, S. E., and McDonald, J. A. (2000). Disruption of hyaluronan synthase-2 abrogates normal cardiac morphogenesis and hyaluronan-mediated transformation of epithelium to mesenchyme. *J. Clin. Invest.* **106**, 349–360.

240. Turley, E. A., Noble, P. W., and Bourguignon, L. Y. (2002). Signaling properties of hyaluronan receptors. *J. Biol. Chem.* **277**, 4589–4592.

241. Camenisch, T. D., Schroeder, J. A., Bradley, J., Klewer, S. E., and McDonald, J. A. (2002). Heart-valve mesenchyme formation is dependent on hyaluronan-augmented activation of ErbB2-ErbB3 receptors. *Nat. Med.* **8**, 850–855.

242. Erickson, S. L., O'Shea, K. S., Ghaboosi, N., Loverro, L., Frantz, G., Bauer, M., Lu, L. H., and Moore, M. W. (1997). ErbB3 is required for normal cerebellar and cardiac development: a comparison with ErbB2-and heregulin-deficient mice. *Development* **124**, 4999–5011.

243. Stainier, D. Y., Fouquet, B., Chen, J. N., Warren, K. S., Weinstein, B. M., Meiler, S. E., Mohideen, M. A., Neuhauss, S. C., Solnica-Krezel, L., Schier, A. F., Zwartkruis, F., Stemple, D. L., Malicki, J., Driever, W., and Fishman, M. C. (1996). Mutations affecting the formation and function of the cardiovascular system in the zebrafish embryo. *Development* **123**, 285–292.

244. Walsh, E. C., and Stainier, D. Y. (2001). UDP-glucose dehydrogenase required for cardiac valve formation in zebrafish. *Science* **293**, 1670–1673.

245. Lander, A. D., and Selleck, S. B. (2000). The elusive functions of proteoglycans: in vivo veritas. *J. Cell. Biol.* **148**, 227–232.

246. Henderson, D. J., and Copp, A. J. (1998). Versican expression is associated with chamber specification, septation, and valvulogenesis in the developing mouse heart. *Circ. Res.* **83**, 523–532.

247. Landolt, R. M., Vaughan, L., Winterhalter, K. H., and Zimmermann, D. R. (1995). Versican is selectively expressed in embryonic tissues that act as barriers to neural crest cell migration and axon outgrowth. *Development* **121**, 2303–2312.

248. Henderson, D. J., Ybot-Gonzalez, P., and Copp, A. J. (1997). Over-expression of the chondroitin sulphate proteoglycan versican is associated with defective neural crest migration in the Pax3 mutant mouse (splotch). *Mech. Dev.* **69**, 39–51.

249. Yamagata, M., Saga, S., Kato, M., Bernfield, M., and Kimata, K. (1993). Selective distributions of proteoglycans and their ligands in pericellular matrix of cultured fibroblasts: implications for their roles in cell-substratum adhesion. *J. Cell Sci.* **106**, 55–65.

250. Yamagata, M., and Kimata, K. (1994). Repression of a malignant cell-substratum adhesion phenotype by inhibiting the production of the anti-adhesive proteoglycan, PG-M/versican. *J. Cell Sci.* **107**, 2581–2590.

251. LeBaron, R. G., Zimmermann, D. R., and Ruoslahti, E. (1992). Hyaluronate binding properties of versican. *J Biol. Chem.* **267**, 10003–10010.

252. Mjaatvedt, C. H., and Markwald, R. R. (1989). Induction of an epithelial-mesenchymal transition by an in vivo adheron-like complex. *Dev. Biol.* **136**, 118–128.

253. Rezaee, M., Isokawa, K., Halligan, N., Markwald, R. R., and Krug, E. L. (1993). Identification of an extracellular 130 kDa protein involved in early cardiac morphogenesis. *J. Biol. Chem.* **268**, 14404–14411.

254. Lamers, W. H., Virágh, S., Wessels, A., Moorman, A. F. M., and Anderson, R. H. (1995). Formation of the tricuspid valve in the human heart. *Circulation* **91**, 111–121.

255. Spence, S. G., Argraves, W. S., Walters, L., Hungerford, J. E., and Little, C. D. (1992). Fibulin is localized at sites of epithelial-mesenchymal transitions in the early avian embryo. *Dev. Biol.* **151,** 473–484.
256. Wunsch, A., Little, C. D., and Markwald, R. R. (1994). Cardiac endothelial heterogeneity is demonstrated by the diverse expression of JB3 antigen, a fibrillin-like protein of the endocardial cushion tissue. *Dev. Biol.* **165,** 585–601.
257. Akiyama, H., Chaboissier, M. C., Behringer, R. R., Rowitch, D. H., Schedl, A., Epstein, J. A., and de Crombrugghe, B. (2004). Essential role of Sox9 in the pathway that controls formation of cardiac valves and septa. *Proc. Natl. Acad. Sci. USA* **101,** 6502–6507.
258. Robert, N. M., Tremblay, J. J., and Viger, R. S. (2002). Friend of GATA (FOG)-1 and FOG-2 differentially repress the GATA-dependent activity of multiple gonadal promoters. *Endocrinology* **143,** 3963–3973.
259. Svensson, E. C., Huggins, G. S., Lin, H., Clendenin, C., Jiang, F., Tufts, R., Dardik, F. B., and Leiden, J. M. (2000). A syndrome of tricuspid atresia in mice with a targeted mutation of the gene encoding Fog-2. *Nat. Genet.* **25,** 353–356.
260. Tevosian, S. G., Deconinck, A. E., Tanaka, M., Schinke, M., Litovsky, S. H., Izumo, S., Fujiwara, Y., and Orkin, S. H. (2000). FOG-2, a cofactor for GATA transcription factors, is essential for heart morphogenesis and development of coronary vessels from epicardium. *Cell* **101,** 729–739.
261. Katz, S. G., Williams, A., Yang, J., Fujiwara, Y., Tsang, A. P., Epstein, J. A., and Orkin, S. H. (2003). Endothelial lineage-mediated loss of the GATA cofactor Friend of GATA 1 impairs cardiac development. *Proc. Natl. Acad. Sci. USA* **100,** 14030–14035.
262. Gittenberger-de Groot, A. C., Bartelings, M. M., Deruiter, M. C., and Poelmann, R. E. (2005). Basics of cardiac development for the understanding of congenital heart malformations. *Pediatr. Res.* **57,** 169–176.
263. Poelmann, R. E., and Gittenberger-de Groot, A. C. (1999). A subpopulation of apoptosis-prone cardiac neural crest cells targets to the venous pole: multiple functions in heart development? *Dev. Biol.* **207,** 271–286.
264. Kirby, M. L., Gale, T. F., and Stewart, D. E. (1983). Neural crest cells contribute to normal aorticopulmonary septation. *Science* **220,** 1059–1061.
265. Lincoln, J., Alfieri, C. M., and Yutzey, K. E. (2004). Development of heart valve leaflets and supporting apparatus in chicken and mouse embryos. *Dev. Dyn.* **230,** 239–250.
266. de Lange, F. J., Moorman, A. F., Anderson, R. H., Manner, J., Soufan, A. T., de Gier-de Vries, C., Schneider, M. D., Webb, S., van den Hoff, M. J., and Christoffels, V. M. (2004). Lineage and morphogenetic analysis of the cardiac valves. *Circ. Res.* **95,** 645–654.
267. Vincentz, J. W., McWhirter, J. R., Murre, C. , Baldini, A., and Furuta, Y. (2005). Fgf15 is required for proper morphogenesis of the mouse cardiac outflow tract. *Genesis* **41,** 192–201.
268. Kaartinen, V., Dudas, M., Nagy, A., Sridurongrit, S., Lu, M. M., and Epstein, J. A. (2004). Cardiac outflow tract defects in mice lacking ALK2 in neural crest cells. *Development* **131,** 3481–3490.
269. Stottmann, R. W., Choi, M., Mishina, Y., Meyers, E. N., and Klingensmith, J. (2004). BMP receptor IA is required in mammalian neural crest cells for development of the cardiac outflow tract and ventricular myocardium. *Development* **131,** 2205–2218.
270. Bernd, P., Miles, K., Rozenberg, I., Borghjid, S., and Kirby, M. L. (2004). Neurotrophin-3 and TrkC are expressed in the outflow tract of the developing chicken heart. *Dev. Dyn.* **230,** 767–772.
271. Waldo, K., Kumiski, D. H., Wallis, K. T., Stadt, H. A., Hutson, M. R., Platt, D. H., and Kirby, M. L. (2001). Conotruncal myocardium arises from a secondary heart field. *Development* **128,** 3179–3188.
272. Kelly, R. G., and Buckingham, M. E. (2002). The anterior heart-forming field: voyage to the arterial pole of the heart. *Trends Genet.* **18,** 210–216.
273. Waldo, K. L., Hutson, M. R., Ward, C. C., Zdanowicz, M., Stadt, H. A., Kumiski, D., Abu-Issa, R., and Kirby, M. L. (2005). Secondary heart field contributes myocardium and smooth muscle to the arterial pole of the developing heart. *Dev. Biol.* **281,** 78–90.
274. Waldo, K. L., Hutson, M. R., Stadt, H. A., Zdanowicz, M., Zdanowicz, J., and Kirby, M. L. (2005). Cardiac neural crest is necessary for normal addition of the myocardium to the arterial pole from the secondary heart field. *Dev. Biol.* **281,** 66–77.
275. Washington Smoak, I., Byrd, N. A., Abu-Issa, R., Goddeeris, M. M., Anderson, R., Morris, J., Yamamura, K., Klingensmith, J., and Meyers, E. N. (2005). Sonic hedgehog is required for cardiac outflow tract and neural crest cell development. *Dev. Biol.* **283,** 357–372.
276. Moraes, F., Novoa, A., Jerome-Majewska, L. A., Papaioannou, V. E., and Mallo, M. (2005). Tbx1 is required for proper neural crest migration and to stabilize spatial patterns during middle and inner ear development. *Mech. Dev.* **122,** 199–212.
277. Hong, C. S., and Saint-Jeannet, J. P. (2005). Sox proteins and neural crest development. *Semin. Cell Dev. Biol.* **16,** 694–703.
278. Montero, J. A., Giron, B., Arrechedera, H., Cheng, Y. C., Scotting, P., Chimal-Monroy, J., Garcia-Porrero, J. A., and Hurle, J. M. (2002). Expression of Sox8, Sox9 and Sox10 in the developing valves and autonomic nerves of the embryonic heart. *Mech. Dev.* **118,** 199–202.
279. Maschhoff, K. L., Anziano, P. Q., Ward, P., and Baldwin, H. S. (2003). Conservation of Sox4 gene structure and expression during chicken embryogenesis. *Gene* **320,** 23–30.
280. Schilham, M. W., Oosterwegel, M. A., Moerer, P., Jing, Y., de Boer, P. A. J., Verbeek, S., Lamers, W. H., Kruisbeek, A. M., Cumano, A., and Clevers, H. (1996). Sox-4 gene is required for cardiac outflow tract formation and pro-B lymphocyte expansion. *Nature* **380,** 711–714.
281. Ya, J., Schilham, M. W., de Boer, P. A., Moorman, A. F., Clevers, H., and Lamers, W. H. (1998). Sox4-deficiency syndrome in mice is an animal model for common trunk. *Circ. Res.* **83,** 986–994.
282. Xanthoudakis, S., Viola, J. P., Shaw, K. T., Luo, C., Wallace, J. D., Bozza, P. T., Luk, D. C., Curran, T., and Rao, A. (1996). An enhanced immune response in mice lacking the transcription factor NFAT1. *Science* **272,** 892–895.
283. de la Pompa, J. L., Timmerman, L. A., Takimoto, H., Yoshida, H., Elia, A. J., Samper, E., Potter, J., Wakeham A., Marengere, L., Langille, B. L., Crabtree, G. R., and Mak, T. W. (1998). Role of the NF-ATc transcription factor in morphogenesis of cardiac valves and septum. *Nature* **392,** 182–186.

284. Ranger, A. M., Grusby, M. J., Hodge, M. R., Gravallese, E. M., de la Brousse, F. C., Hoey, T., Mickanin, C., Baldwin, H. S., and Glimcher, L. H. (1998). The transcription factor NF-ATc is essential for cardiac valve formation. *Nature* **392**, 129–130.
285. Wilkins, B. J., De Windt, L. J., Bueno, O. F., Braz, J. C., Glascock, B. J., Kimball, T. F., and Molkentin, J. D. (2002). Targeted disruption of NFATc3, but not NFATc4, reveals an intrinsic defect in calcineurin-mediated cardiac hypertrophic growth. *Mol. Cell Biol.* **22**, 7603–7613.
286. Shu, W., Yang, H., Zhang, L., Lu, M. M., and Morrisey, E. E. (2001). Characterization of a new subfamily of winged-helix/forkhead (Fox) genes that are expressed in the lung and act as transcriptional repressors. *J. Biol. Chem.* **276**, 27488–27497.
287. Li, S., Weidenfeld, J., and Morrisey, E. E. (2004). Transcriptional and DNA binding activity of the Foxp1/2/4 family is modulated by heterotypic and homotypic protein interactions. *Mol. Cell Biol.* **24**, 809–822.
288. Chen, B., Bronson, R. T., Klaman, L. D., Hampton, T. G., Wang, J. F., Green, P. J., Magnuson, T., Douglas, P. S., Morgan, J. P., and Neel, B. G. (2000). Mice mutant for Egfr and Shp2 have defective cardiac semilunar valvulogenesis. *Nat. Genet.* **24**, 296–299.
289. Goishi, K., Lee, P., Davidson, A. J., Nishi, E., Zon, L. I., and Klagsbrun, M. (2003). Inhibition of zebrafish epidermal growth factor receptor activity results in cardiovascular defects. *Mech. Dev.* **120**, 811–822.
290. Tadano, M., Edamatsu, H., Minamisawa, S., Yokoyama, U., Ishikawa, Y., Suzuki, N., Saito, H., Wu, D., Masago-Toda, M., Yamawaki-Kataoka, Y., Setsu, T., Terashima, T., Maeda, S., Satoh, T., and Kataoka, T. (2005). Congenital semilunar valvulogenesis defect in mice deficient in phospholipase C ε. *Mol. Cell Biol.* **25**, 2191–2199.
291. Brannan, C. I., Perkins, A. S., Vogel, K. S., Ratner, N., Nordlund, M. L., Reid, S. W., Buchberg, A. M., Jenkins, N. A., Parada, L. F., and Copeland, N. G. (1994). Targeted disruption of the neurofibromatosis type-1 gene leads to developmental abnormalities in heart and various neural crest-derived tissues. *Genes Dev.* **8**, 1019–1029.
292. Gitler, A. D., Zhu, Y., Ismat, F A., Lu, M. M., Yamauchi, Y., Parada, L. F., and Epstein, J. A. (2003). Nf1 has an essential role in endothelial cells. *Nat. Genet.* **33**, 75–79.
293. Gourdie, R. G., Kubalak, S., and Mikawa. T. (1999). Conducting the embryonic heart: orchestrating development of specialized cardiac tissues. *Trends Cardiovasc. Med.* **9**, 18–26.
294. Thomas, P. S., Kasahara, H., Edmonson, A. M., Izumo, S., Yacoub, M. H., Barton, P. J., and Gourdie, R. G. (2001). Elevated expression of Nkx-2.5 in developing myocardial conduction cells. *Anat. Rec.* **263**, 307–313.
295. Verheule, S., van Kempen, M. J., te Welscher, P. H., Kwak, B. R., and Jongsma, H. J. (1997). Characterization of gap junction channels in adult rabbit atrial and ventricular myocardium. *Circ. Res.* **80**, 673–681.
296. Bastide, B., Neyses, L., Ganten, D., Paul, M., Willecke, K., and Traub, O. (1993). Gap junction protein connexin40 is preferentially expressed in vascular endothelium and conductive bundles of rat myocardium and is increased under hypertensive conditions. *Circ. Res.* **73**, 1138–1149.
297. Gros, D., Jarry-Guichard, T., Ten Velde, I., de Maziere, A., van Kempen, M. J., Davoust, J., Briand, J. P., Moorman, A. F., and Jongsma, H. J. (1994). Restricted distribution of Connexin40, a gap junctional protein, in mammalian heart. *Circ. Res.* **74**, 839–851.
298. Gourdie, R. G., Severs, N. J., Green, C. R., Rothery, S., Germroth, P., and Thompson, R. P. (1993). The spatial distribution and relative abundance of gap-junctional connexin40 and connexin43 correlate to functional properties of components of the cardiac atrioventricular conduction system. *J. Cell Sci.* **105**, 985–991.
299. Davis, L. M., Rodefeld, M. E., Green, K., Beyer, E. C., and Saffitz, J. E. (1995). Gap junction protein phenotypes of the human heart and conduction system. *J. Cardiovasc. Electrophysiol.* **6**, 813–822.
300. Van Kempen, M. J., Vermeulen, J. L., Moorman, A. F., Gros, D., Paul, D. L., and Lamers, W. H. (1996). Developmental changes of connexin40 and connexin43 mRNA distribution patterns in the rat heart. *Cardiovasc. Res.* **32**, 886–900.
301. Kanter, H. L., Saffitz, J. E., and Beyer, E. C. (1992). Cardiac myocytes express multiple gap junction proteins. *Circ. Res.* **70**, 438–444.
302. Darrow, B. J., Laing, J. G., Lampe, P. D., Saffitz, J. E., and Beyer, E. C. (1995). Expression of multiple connexins in cultured neonatal rat ventricular myocytes. *Circ. Res.* **76**, 381–387.
303. Coppen, S. R., Dupont, E., Rothery, S., and Severs, N. J. (1998). Connexin45 expression is preferentially associated with the ventricular conduction system in mouse and rat heart. *Circ. Res.* **82**, 232–243.
304. Coppen, S. R, Severs, N. J., and Gourdie, R. G. (1999). Connexin45 (α6) expression delineates an extended conduction system in the embryonic and mature rodent heart. *Dev. Genet.* **24**, 82–90.
305. van Rijen, H. V., van Veen, T. A., van Kempen, M. J., Wilms-Schopman, F. J., Potse, M., Krueger, O., Willecke, K., Opthof, T., Jongsma, H. J., and de Bakker, J. M. (2001). Impaired conduction in the bundle branches of mouse hearts lacking the gap junction protein connexin40. *Circulation* **103**, 1591–1598.
306. Verheule, S., van Batenburg, C. A., Coenjaerts, F. E., Kirchoff, S., Willecke, K., and Jongsma, H. J. (1999) Cardiac conduction abnormalities in mice lacking the gap junction protein connexin40. *J. Cardiovasc. Electrophysiol.* **10**, 1380–1389.
307. Hagendorff, A., Schumacher, B., Kirchhoff, S., Luderitz, B., and Willecke, K. (1999). Conduction disturbances and increased atrial vulnerability in Connexin40-deficient mice analyzed by transesophageal stimulation. *Circulation* **99**, 1508–1515.
308. Kirchhoff, S., Nelles, E., Hagendorff, A., Kruger, O., Traub, O. and Willecke, K. (1998). Reduced cardiac conduction velocity and predisposition to arrhythmias in connexin40-deficient mice. *Curr. Biol.* **8**, 299–302.
309. Simon, A. M., Goodenough, D. A., and Paul, D. L. (1998). Mice lacking connexin40 have cardiac conduction abnormalities characteristic of atrioventricular block and bundle branch block. *Curr. Biol.* **8**, 295–298.
310. van Rijen, H. V., van Veen, T. A., van Kempen, M. J., Wilms-Schopman, F. J., Potse, M. , Krueger, O., Willecke, K., Opthof, T., Jongsma, H. J., and de Bakker, J. M. (2001). Impaired conduction in the bundle branches of mouse hearts lacking the gap junction protein connexin40. *Circulation* **103**, 1591–1598.
311. Gourdie, R. G., Wei, Y., Kim, D., Klatt, S. C., and Mikawa, T. (1998). Endothelin-induced conversion of embryonic heart

muscle cells into impulse-conducting Purkinje fibers. *Proc. Natl. Acad. Sci. USA* **95,** 6815–6818.

312. Takebayashi-Suzuki, K., Yanagisawa, M., Gourdie, R. G., Kanzawa, N., and Mikawa, T. I. (2000). In vivo induction of cardiac Purkinje fiber differentiation by coexpression of preproendo-thelin-1 and endothelin converting enzyme-1. *Development* **127,** 3523–3532.

313. Hyer, J., Johansen, M., Prasad, A., Wessels, A., Kirby, M. L., Gourdie, R. G., and Mikawa, T. (1999). Induction of Purkinje fiber differentiation by coronary arterialization. *Proc. Natl. Acad. Sci. USA* **96,** 13214–13218.

314. Kanzawa, N., Poma, C. P., Takebayashi-Suzuki, K., Diaz, K. G., Layliev, J., and Mikawa, T. (2002). Competency of embryonic cardiomyocytes to undergo Purkinje fiber differentiation is regulated by endothelin receptor expression. *Development* **129,** 3185–3194.

315. Hall, C. E., Hurtado, R., Hewett, K. W., Shulimovich, M., Poma, C. P., Reckova, M., Justus, C., Pennisi, D. J., Tobita, K., Sedmera, D., Gourdie, R. G., and Mikawa, T. (2004). Hemodynamic-dependent patterning of endothelin converting enzyme 1 expression and differentiation of impulse-conducting Purkinje fibers in the embryonic heart. *Development* **131,** 581–592.

316. Rentschler, S., Zander, J., Meyers, K., France, D., Levine, R., Porter, G., Rivkees, S. A., Morley, G. E., and Fishman, G. I. (2002). Neuregulin-1 promotes formation of the murine cardiac conduction system. *Proc. Natl. Acad. Sci. USA* **99,** 10464–10469.

317. Patel, R., and Kos, L. (2005). Endothelin-1 and neuregulin-1 convert embryonic cardiomyocytes into cells of the conduction system in the mouse. *Dev. Dyn.* **233,** 20–28.

318. Bond, J., Sedmera, D., Jourdan, J., Zhang, Y., Eisenberg, C. A., Eisenberg, L. M., and Gourdie, R. G. (2003). Wnt11 and Wnt7a are up-regulated in association with differentiation of cardiac conduction cells in vitro and in vivo. *Dev. Dyn.* **227,** 536–543.

319. Nguyen-Tran, V. T., Kubalak, S. W., Minamisawa, S., Fiset, C., Wollert, K. C., Brown, A. B., Ruiz-Lozano, P., Barrere-Lemaire, S., Kondo, R., Norman, L. W., Gourdie, R. G., Rahme, M. M., Feld, G. K., Clark, R. B., Giles, W. R., and Chien, K. R. (2000). A novel genetic pathway for sudden cardiac death via defects in the transition between ventricular and conduction system cell lineages. *Cell* **102,** 671–682.

320. Mounkes, L. C., Kozlov, S. V., Rottman, J. N., and Stewart, C. L. (2005). Expression of an LMNA-N195K variant of A-type lamins results in cardiac conduction defects and death in mice. *Hum. Mol. Genet.* **14,** 2167–2180.

321. Hershberger, R. E., Hanson, E. L., Jakobs, P. M., Keegan, H., Coates, K., Bousman, S., and Litt, M. (2002). A novel lamin A/C mutation in a family with dilated cardiomyopathy, prominent conduction system disease, and need for permanent pacemaker implantation. *Am. Heart. J.* **144,** 1081–1086.

322. Hoogaars, W. M., Tessari, A., Moorman, A. F., de Boer, P. A., Hagoort, J., Soufan, A. T., Campione, M., and Christoffels, V. M. (2004). The transcriptional repressor Tbx3 delineates the developing central conduction system of the heart. *Cardiovasc. Res.* **62,** 489–499.

323. Fishman, G. I. (2005). Understanding conduction system development: a hop, skip and jump away? *Circ. Res.* **96,** 809–811.

324. Ismat, F. A., Zhang, M., Kook, H., Huang, B., Zhou, R., Ferrari, V. A., Epstein, J. A., and Patel, V. V. (2005). Homeobox protein Hop functions in the adult cardiac conduction system. *Circ. Res.* **96,** 898–903.

325. Thompson, R. P., Lindroth, J. R., and Wong, Y. M. (1990). Regional differences in DNA-synthetic activity in the preseptation myocardium of the chick. *In* "Developmental Cardiology: Morphogenesis and Function." (E. B. Clark, and A. Takao, Eds.), pp. 219–234. Futura Publishing, Mount Kisco, NY.

326. Thompson, R. P., Reckova, M., DeAlmeida, A., Bigelow, M., Stanley, C. P., Spruill, J. B., Trusk, T., and Sedmera, D. (2003). The oldest, toughest cells in the heart. *In* "Development of the Cardiac Conduction System." (D. Chadwick and J. Goode, Eds.), Novartis Foundation Symposium, pp. 157–176. Wiley, Chichester, UK

327. Gourdie R. G., Kubalak S., Mikawa T. (1999). Conducting the embryonic heart: orchestrating development of specialized cardiac tissues. *Trends. Cardiovasc. Med.* **9,** 18–26.

328. Gourdie, R. G., Harris, B. S., Bond, J., Justus, C., Hewett, K. W., O'Brien, T. X., Thompson, R. P., and Sedmera, D. (2003). Development of the cardiac pacemaking and conduction system. *Birth Defects Res.* **69C,** 46–57.

329. Reckova, M., Rosengarten, C., deAlmeida, A., Stanley, C. P., Wessels, A., Gourdie, R. G., Thompson, R. P., and Sedmera, D. (2003). Hemodynamics is a key epigenetic factor in development of the cardiac conduction system. *Circ. Res.* **93,** 77–85.

330. Gourdie, R. G., Wei, Y., Kim, D., Klatt, S. C., and Mikawa, T. (1998). Endothelin-induced conversion of embryonic heart muscle cells into impulse-conducting Purkinje fibers. *Proc. Natl. Acad. Sci. USA* **95,** 6815–6818.

331. Takebayashi-Suzuki, K., Yanagisawa, M., Gourdie, R. G., Kanzawa, N., and Mikawa, T. (2000). In vivo induction of cardiac Purkinje fiber differentiation by coexpression of preproendothelin-1 and endothelin converting enzyme-1. *Development* **127,** 3523–3532.

332. Ziegler, T., Bouzourene, K., Harrison, V. J., Brunner, H. R., and Hayoz, D. (1998). Influence of oscillatory and unidirectional flow environments on the expression of endothelin and nitric oxide synthase in cultured endothelial cells. *Arterioscler. Thromb. Vasc. Biol.* **18,** 686–692.

333. Markos, F., Hennessy, B. A., Fitzpatrick, M., O'Sullivan, J., and Snow, H. M. (2002). The effect of tezosentan, a non-selective endothelin receptor antagonist, on shear stress-induced changes in arterial diameter of the anaesthetized dog. *J. Physiol.* **544,** 913–918.

334. Rentschler, S., Zander, J., Meyers, K., France, D., Levine, R., Porter, G., Rivkees, S. A., Morley, G. E., and Fishman, G. I. (2002). Neuregulin-1 promotes formation of the murine cardiac conduction system. *Proc. Natl. Acad. Sci. USA* **99,** 10464–10469.

335. Meyer, D., and Birchmeier, C. (1995). Multiple essential functions of neuregulin in development. *Nature* **378,** 386–390.

336. Hertig, C. M., Kubalak, S. W., Wang, Y., and Chien, K. R. (1999). Synergistic roles of neuregulin-1 and insulin-like growth factor-I in activation of the phosphatidylinositol 3-kinase pathway and cardiac chamber morphogenesis. *J. Biol. Chem.* **274,** 37362–37369.

337. Zhao, Y. Y., Sawyer, D. R., Baliga, R. R., Opel, D. J., Han, X., Marchionni, M. A., and Kelly, R. A. (1998). Neuregulins

promote survival and growth of cardiac myocytes: persistence of ErbB2 and ErbB4 expression in neonatal and adult ventricular myocytes. *J. Biol. Chem.* **273,** 10261–10269.

338. Moorman, A. F., de Jong, F., Denyn, M. M., and Lamers, W. H. (1998). Development of the cardiac conduction system. *Circ. Res.* **82,** 629–644.

339. Nakajima, Y., Yoshimura, K., Nomura, M., and Nakamura, H. (2001). Expression of HNK1 epitope by the cardiomyocytes of the early embryonic chick: in situ and in vitro studies. *Anat. Rec.* **263,** 326–333.

340. Wessels, A., Vermeulen, J. L. M., Verbeek, F. J., Virágh, S., Kálmán, F., Lamers, W. H., and Moorman, A. F. (2001). Spatial distribution of 'tissue-specific' antigens in the developing human heart and skeletal muscle, III: an immunohistochemical analysis of the distribution of the neural tissue antigen G1N2 in the embryonic heart: implications for the development of the atrioventricular conduction system. *Anat. Rec.* **232,** 97–111.

341. Ito, H., Iwasaki, K., Ikeda, T., Sakai, H., Shimokawa, I., and Matsuo, T. (1992). HNK-1 expression pattern in normal and bis-diamine induced malformed developing rat heart: three dimensional reconstruction analysis using computer graphics. *Anat. Embryol.* **186,** 327–334.

342. Barbu, M., Ziller, C., Rong, P. M., and Le Douarin, N. M. (1986). Heterogeneity in migrating neural crest cells revealed by a monoclonal antibody. *J. Neurol. Sci.* **6,** 2215–2225.

343. Gordon, L., Wharton, J., Moore, S. E., Walsh, F. S., Moscoso, J. G., Penketh, R., Wallwork, J., Taylor, K. M., Yacoub, M. H., and Polak, J. M. (1990). Myocardial localization and isoforms of neural cell adhesion molecule (N-CAM) in the developing and transplanted human heart. *J. Clin. Invest.* **86,** 1293–1300.

344. Hoffman, S., Crossin, K. L., Prediger, E. A., Cunningham, B. A., and Edelman G. M. (1990). Expression and function of cell adhesion molecules during the early development of the heart: embryonic origin of defective heart development. *Ann. N. Y. Acad. Sci.* **588,** 73–86.

345. Davis, D. L., Edwards, A. V., Juraszek, A. L., Phelps, A., Wessels, A., and Burch, J. B. (2001). A GATA-6 gene heart-region-specific enhancer provides a novel means to mark and probe a discrete component of the mouse cardiac conduction system. *Mech. Dev.* **108,** 105–119.

346. Wessels, A., Phelps, A., Trusk, T. C., Davis, D. L., Edwards, A. V., Burch, J. B., and Juraszek, A. L. (2003). Mouse models for cardiac conduction system development. *Novartis Found. Symp.* **250,** 44–59.

347. Rentschler, S., Vaidya, D. M., Tamaddon, H., Degenhardt, K., Sassoon, D., Morley, G. E., Jalife, J., and Fishman, G. I. (2001). Visualization and functional characterization of the developing murine cardiac conduction system. *Development* **128,** 1785–1792.

348. Jongbloed, M. R., Schalij, M. J., Poelmann, R. E., Blom, N. A., Fekkes, M. L., Wang, Z., Fishman, G I., and Gittenberger-De Groot, A. C. (2004). Embryonic conduction tissue: a spatial correlation with adult arrhythmogenic areas. *J. Cardiovasc. Electrophysiol.* **15,** 349–355.

349. Poelmann, R. E., Jongbloed, M. R., Molin, D. G., Fekkes, M. L., Wang, Z., Fishman, G. I., Doetschman, T., Azhar, M., and Gittenberger-de Groot, A. C. (2004). The neural crest is contiguous with the cardiac conduction system in the mouse embryo: a role in induction? *Anat. Embryol.* **20,** 389–393.

350. Kondo, R. P., Anderson, R. H., Kupershmidt, S., Roden, D. M., and Evans, S. M. (2003). Development of the cardiac conduction system as delineated by minK-lacZ. *J. Cardiovasc. Electrophysiol.* **14,** 383–391.

351. Gourdie, R. G., Mima, T., Thompson, R. P., and Mikawa, T. (1995). Terminal diversification of the myocyte lineage generates Purkinje fibers of the cardiac conduction system. *Development* **121,** 1423–1431.

352. Kirby, M. L. (1987). Cardiac morphogenesis—recent research advances. *Pediatr. Res.* **21,** 219–224.

353. Bockman, D. E., Redmond, M. E., Waldo, K., Davis, H., and Kirby, M. L. (1987). Effect of neural crest ablation on development of the heart and arch arteries in the chick. *Am. J. Anat.* **180,** 332–341.

354. Epstein, J. A., and Buck, C. A. (2000). Transcriptional regulation of cardiac development: implications for congenital heart disease and DiGeorge syndrome. *Pediatr. Res.* **48,** 717–724.

355. Feiner, L., Webber, A. L., Brown, C. B., Lu, M. M., Jia, L., Feinstein, P., Mombaerts, P., Epstein, J. A., and Raper, J. A. (2001). Targeted disruption of semaphorin 3C leads to persistent truncus arteriosus and aortic arch interruption. *Development* **128,** 3061–30670.

356. Brown, C. B., Feiner, L., Lu, M. M., Li, J., Ma, X., Webber, A. L., Jia, L., Raper, J. A., and Epstein, J. A. (2001). PlexinA2 and semaphorin signaling during cardiac neural crest development. *Development* **128,** 3071–3080.

357. Gitler, A. D., Lu, M. M., and Epstein, J. A. (2004). PlexinD1 and semaphorin signaling are required in endothelial cells for cardiovascular development. *Dev. Cell* **7,** 107–116.

358. Epstein, J. A., Li, J., Lang, D., Chen, F., Brown, C. B., Jin, F., Lu, M. M., Thomas, M., Liu, E., Wessels, A., and Lo, C. W. (2000). Migration of cardiac neural crest cells in Splotch embryos. *Development* **127,** 1869–1878.

359. Mansouri, A., Pla, P., Larue, L., and Gruss, P. (2001). Pax3 acts cell autonomously in the neural tube and somites by controlling cell surface properties. *Development* **128,** 1995–2005.

360. Louey, S., and Thornburg, K. L. (2005). The prenatal environment and later cardiovascular disease. *Early Hum. Dev.* **81,** 745–751.

361. Mone, S. M., Gillman, M. W., Miller, T. L., Herman, E. H., and Lipshultz, S. E. (2004). Effects of environmental exposures on the cardiovascular system: prenatal period through adolescence. *Pediatrics* **113,** 1058–1069.

362. Kallen, B. A., Otterblad Olausson, P., and Danielsson, B. R. (2005). Is erythromycin therapy teratogenic in humans? *Reprod. Toxicol.* **20,** 209–214.

363. Danielsson, B. R., Skold, A. C., and Azarbayjani, F. (2001). Class III antiarrhythmics and phenytoin: teratogenicity because of embryonic cardiac dysrhythmia and reoxygenation damage. *Curr. Pharm. Des.* **7,** 787–802.

364. Nishikawa, T., Bruyere, H. J. Jr., Takagi, Y., Gilbert, E. F., and Matsuoka, R. (1986). The teratogenic effect of phenobarbital on the embryonic chick heart. *J. Appl. Toxicol.* **6,** 91–94.

365. Tikkanen, J., and Heinonen, O. P. (1991). Risk factors for ventricular septal defect in Finland. *Public Health* **105,** 99–112.

366. Tikkanen, J., and Heinonen, O. P. (1992). Risk factors for conal malformations of the heart. *Eur. J. Epidemiol.* **8,** 48–57.

367. Tikkanen, J., and Heinonen, O. P. (1993). Risk factors for coarctation of the aorta. *Teratology* **47,** 565–572.
368. Goldberg, S. J., Lebowitz, M. D., and Graver, E. J. (1990). An association of human congenital cardiac malformations and drinking water contaminants. *J. Am. Coll. Cardiol.* **16,** 155–164.
369. Loeber, C. P., Hendrix, M. J. C., Diez de Pinos, S., and Goldberg, S. J. (1988). Trichloroethylene: cardiac teratogen in developing chick embryos. *Pediatr. Res.* **74,** 740–744.
370. Yauck, J., Malloy, M., Blair, K., Simpson, P., and McCarver, D. (2003). Closer residential proximity to trichloroethylene-emitting sites increases risk of offspring congenital heart defects among older women. *Toxicologist* **72,** 327.
371. Johnson, P. D., Dawson, B. V., and Goldberg, S. J. (1998). Cardiac teratogenicity of trichloroethylene metabolites. *J. Am. Coll. Cardiol.* **32,** 540–545.
372. Costlow, R. D., and Manson, J. M. (1981). The heart and diaphragm: target organs in the neonatal death induced by nitrofen (2,4-dichloro phenyl-p-nitrophenyl ether). *Toxicology* **20,** 209–227.
373. Lau, C., Cameron, A. M., Irsula, O., and Robinson, K. S. (1986). Effects of prenatal nitrofen exposure on cardiac structure and function in the rat. *Toxicol. Appl. Pharmacol.* **86,** 22–32.
374. Kim, W. G., Suh, J. W., and Chi, J. G. (1999). Nitrofen-induced congenital malformations of the heart and great vessels in rats. *J. Pediatr. Surg.* **34,** 1782–1786.
375. Migliazza, L., Xia, H., Alvarex, J. I., Arnaiz, A., Diez-Pardo, J. A., Alfonso, L. F., and Tovar, J. A. (1999). Heart hypoplasia in experimental congenital diaphragmatic hernia. *J. Pediatr. Surg.* **34,** 706–710.
376. Guarino, N., and Puri, P. (2002). Antenatal dexamethasone enhances endothelin-1 synthesis and gene expression in the heart in congenital diaphragmatic hernia in rats. *J. Pediatr. Surg.* **37,** 1563–1567.
377. Yu, J., Gonzalez, S., Diez-Pardo, J. A., and Tovar, J. A. (2002). Effects of vitamin A on malformations of neural crest-controlled organs induced by nitrofen in rats. *Pediatr. Surg. Int.* **18,** 600–605.
378. Guarino, N., Shima, H., and Puri, P. (2001). Cardiac gene expression and synthesis of atrial natriuretic peptide in the nitrofen model of congenital diaphragmatic hernia in rats: effect of prenatal dexamethasone treatment. *J. Pediatr. Surg.* **36,** 1497–1501.
379. Gonzalez-Reyes, S., Fernandez-Dumont, V., Martinez-Calonge, W., Martinez, L., Hernandez, F., and Tovar, J. (2005). Pax3 mRNA is decreased in the hearts of rats with experimental diaphragmatic hernia. *Pediatr. Surg. Int.* **21,** 203–207.
380. Teramoto, H., and Puri, P. (2001). Gene expression of insulin-like growth factor-1 and epidermal growth factor is downregulated in the heart of rats with nitrofen-induced diaphragmatic hernia. *Pediatr. Surg. Int.* **17,** 284–287.
381. Lammer, E. J., Chen, D. T., Hoar, R M., Agnish, N. D., Benke, P. J., Braun, J. T., Curry, C. J., Fernhoff, P. M., Grix, A. W. Jr., Lott, I. T., Richard, J. M., and Sun, S. C. (1985). Retinoic acid embryopathy. *N. Engl. J. Med.* **313,** 837–841.
382. Aboseif, S. R., Dahiya, R., Narayan, P., and Cunha, G. R. (1997). Effect of retinoic acid on prostatic development. *Prostate* **31,** 161–167.
383. Sucov, H. M., Dyson, E., Gumeringer, C. L., Price, J., Chien, K. R., and Evans, R. M. (1994). RXR α mutant mice establish a genetic basis for vitamin A signaling in heart morphogenesis. *Genes Dev.* **8,** 1007–1018.
384. Dyson, E., Sucov, H. M., Kubalak, S. W., Schmid-Schonbein, G. W., DeLano, F. A., Evans, R. M., Ross, J. Jr., and Chien, K. R. (1995). Atrial-like phenotype is associated with embryonic ventricular failure in retinoid X receptor α −/− mice. *Proc. Natl. Acad. Sci. USA* **92,** 7386–7390.
385. Palm-Leis, A., Singh, U. S., Herbelin, B. S., Olsovsky, G. D., Baker, K. M., and Pan, J. (2004). Mitogen-activated protein kinases and mitogen-activated protein kinase phosphatases mediate the inhibitory effects of all-trans retinoic acid on the hypertrophic growth of cardiomyocytes. *J. Biol. Chem.* **279,** 54905–54917.
386. Clabby, M. L., Robison, T. A., Quigley, H. F., Wilson, D. B., and Kelly, D. P. (2003). Retinoid X receptor represses GATA-4-mediated transcription via a retinoid-dependent interaction with the cardiac-enriched repressor FOG-2. *J. Biol. Chem.* **278,** 5760–5767.
387. Niederreither, K., Subbarayan, V., Dolle, P., and Chambon, P. (1999). Embryonic retinoic acid synthesis is essential for early mouse post-implantation development. *Nat. Genet.* **21,** 444–448.
388. Osmond, M. K., Butler, A. J., Voon, F. C., and Bellairs, R. (1991). The effects of retinoic acid on heart formation in the early chick embryo. *Development* **113,** 1405–1417.
389. Shaw, G. M., Zhu, H., Lammer, E. J., Yang, W., and Finnell, R. H. (2003). Genetic variation of infant reduced folate carrier (A80G) and risk of orofacial and conotruncal heart defects. *Am. J. Epidemiol.* **158,** 747–752.
390. Loser, H., and Majewski, F. (1977). Type and frequency of cardiac defects in embryo fetal alcohol syndrome. Report of 16 cases. *Br. Heart J.* **39,** 1374–1379.
391. Shillingford, A. J., and Weiner, S. (2001). Maternal issues affecting the fetus. *Clin. Perinatol.* **28,** 31–64.
392. Tikkanen, J., and Heinonen, O. P. (1992). Risk factors for atrial septal defect. *Eur. J. Epidemiol.* 8, 509–512.
393. Fang, T., Bruyere, H. T. Jr., Kargas, T., Nisikawa, T., Takagi, Y., and Gilbert, E. F. (1987). Ethyl alcohol induced cardiovascular malformations in the chick embryo. *Teratology* **35,** 95–103.
394. Lipshultz, S. E., Frassica, J. J., and Orav, E. J. (1991). Cardiovascular abnormalities in infants prenatally exposed to cocaine. *J. Pediatr.* **118,** 44–52.
395. Gintautiene, K., Longmore, A., Abadir, A. R., Goy, G. R., Berman, S., Levendoglu, H., and Gintautas, J. (1990). Cocaine-induced deaths in pediatric population. *Proc. West Pharmacol. Soc.* **33,** 247–248.
396. Frassica, J. J., Orav, E. J., Walsh, E. P., and Lipshultz, S. E. (1994). Arrhythmias in children prenatally exposed to cocaine. *Arch. Pediatr. Adolesc. Med.* **148,** 1162–1169.
397. Xiao, Y.-H., He, J., Gilbert, R. D., and Zhang, L. (2000). Cocaine induces apoptosis in fetal myocardial cells through a mitochondria-dependent pathway. *J. Pharmacol. Exp. Ther.* **292,** 8–14.
398. Xiao, Y., Xiao, D., He, J., and Zhang, L. (2001). Maternal cocaine administration during pregnancy induces apoptosis in fetal rat heart. *J. Cardiovasc. Pharmacol.* **37,** 639–648.

399. Chameides, L., Truex, R. C., Vetter, V., Rashkind, W. J., Galioto, F. M. Jr., and Noonan, J. A. (1977). Association of maternal systemic lupus erythematosus with congenital complete heart block. *N. Engl. J. Med.* **297,** 1204–1207.
400. Kalush, F., Rimon, E., Keller, A., and Mozes, E. (1994). Neonatal lupus erythematosus with cardiac involvement in offspring of mothers with experimental systemic lupus erythematosus. *J. Clin. Immunol.* **14,** 314–322.
401. Lipshultz, S. E., Chanock, S., Sanders, S. P., Colan, S. D., and McIntosh, K. (1989). Cardiovascular manifestations of human immunodeficiency infection in infants and children. *Am. J. Cardiol.* **63,** 1489–1497.
402. Dor, Y., Camenisch, T. D., Itin, A., Fishman, G. I., McDonald, J. A., Carmeliet, P., and Keshet, E. (2001). A novel role for VEGF in endocardial cushion formation and its potential contribution to congenital heart defects. *Development* **128,** 1531–1538.
403. Wikenheiser, J., Doughman, Y. Q., Fisher, S. A., and Watanabe, M. (2005). Differential levels of tissue hypoxia in the developing chicken heart. *Dev. Dyn.* **235,** 115–123.
404. Sugishita, Y., Leifer, D. W., Agani, F., Watanabe, M., and Fisher, S. A. (2004). Hypoxia-responsive signaling regulates the apoptosis-dependent remodeling of the embryonic avian cardiac outflow tract. *Dev. Biol.* **273,** 285–296.

CHAPTER 7

Congenital and Acquired Heart Disease

OVERVIEW

Although structural, congenital, and acquired cardiac diseases, cardiomyopathy, and dysrhythmias are common causes of mortality and morbidity in infants and children, the basic underlying genetic and molecular mechanisms remain undetermined. Breakthroughs in molecular genetic technology have just begun to be applied in pediatric cardiology stemming from the use of chromosomal mapping and the identification of genes involved in both the primary etiology and as significant risk factors in the development of cardiac and vascular abnormalities. This chapter will focus on information obtained thus far by molecular genetic analysis in the diagnosis, treatment, and overall understanding of pediatric cardiovascular disease (CVD) pathogenesis by examining the more prevalent congenital heart defects (CHD), dysrhythmias, and cardiomyopathies, as well as sporadic and acquired disorders. In addition, a survey of the pediatric cardiologist's armamentarium with regard to molecular and genetic analysis is presented, highlighting the current use of molecular diagnostic methods, including microarray, gene mapping, proteomic, transgenic and stem cell technologies, as well as future directions in both clinical application and research.

INTRODUCTION

CHD, cardiomyopathy, and dysrhythmias are common causes of mortality and morbidity in infants and children, particularly during the perinatal period. Cardiovascular abnormalities represent the most common class of birth defect, affecting approximately 1 in every 120 infants each year. The high incidence of cardiovascular defects in infants and children represents an enormous burden and cost borne by the families, health-care providers, and society at large.

Although the clinical applicability of genetic and molecular techniques shows great promise in the diagnosis, management, and treatment of pediatric heart disease, their present use in the clinical setting has been generally limited, in part because of the high costs and resources involved and in part because of the complexities posed by genetic heterogeneity.[1]

CONGENITAL HEART DISEASE

Recent advances in molecular genetics have shown that specific genetic and molecular factors are linked to congenital heart disease (CHD), allowing their identification on the human chromosome map as depicted in Fig. 1 and providing a great opportunity for improving genetic diagnostics and future gene therapy.

Single gene mutations have been implicated in the pathogenesis of a variety of CHD (Table I), and new evidence suggests that these mutations, (more common than previously thought) are present in a broad spectrum of genes involved in cardiac structure and function. The level of cardiac specificity for these mutations is highly variable. Many single-gene mutation–associated syndromes have neuromuscular and systemic presentation associated with cardiac involvement (e.g., Friedreich ataxia, Duchenne muscular dystrophy). A wide range of cardiovascular defects results from these genetic mutations, including abnormalities in electrophysiological function (e.g., conduction defects and dysrhythmias); extracellular matrix proteins, enzymes, and membrane transporters involved in fatty acid and mitochondrial biosynthesis; cardiac oxidative phosphorylation (OXPHOS) metabolism; sarcomeric structural and contractile proteins; and nuclear transcription factors that govern myocardial gene expression and developmental programming, as well as the architecture of the outflow tracts. Pleiotropic cardiac malformations can result from discrete mutations in specific nuclear transcription factors, proteins recognized as playing key regulatory roles during cardiovascular development and morphogenesis as discussed in Chapter 6.[2–4] Factors such as GATA4, Nkx2.5, dHAND, TFAP2, and Tbx5 are among the earliest transcription factors expressed in the developing heart and are crucial in the activation of cardiac-specific genes. Mutations in each of these genes results in cardiac abnormalities, including cardiac septal defects (*GATA4*), conduction defects (*NKX2-5*), right ventricular hypoplasia (*HAND2*), patent ductus arteriosus (PDA) in Char syndrome (*TFAP2B*), and Holt–Oram syndrome (*TBX5*), underscoring the critical role played by the disruption of early heart development and morphogenesis in the genesis of CHD.[5–9]

Genetic defects in proteins involved in the multiple signaling pathways that modulate cell proliferation, migration, and differentiation in early cardiovascular development have

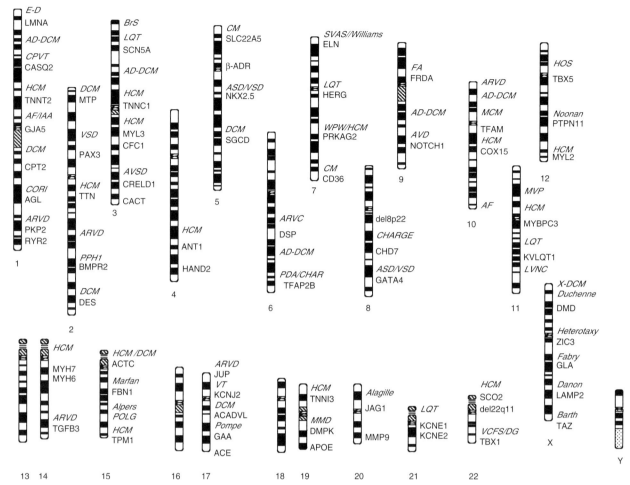

FIGURE 1 Chromosomal map of inherited pediatric cardiovascular disorders. Disorders (in italics) and affected genes are localized on ideograms of each of the human chromosomes. These include ACE, angiotensin-converting enzyme; ACTC, cardiac α-actin; AD-DCM, autosomal-dominant DCM; AF, *familial atrial fibrillation*; AGL, *glycogen-debranching enzyme*; Alagille, *Alagille syndrome*; Alpers, *Alpers syndrome*; ANT1, adenine nucleotide translocator 1; APOE, apolipoprotein E; ASD/VSD, *atrial and ventricular septal defects*; ARVD/ARVC, *arrhythmogenic right ventricular dysplasia or cardiomyopathy*; AVD, *aortic valve defect*; AVSD, *atrioventricular septal defect*; GAA, α-glucosidase; Barth, *Barth syndrome*; BMPR2, bone morphogenetic protein receptor; BrS, *Brugada syndrome*; MYH6, α-myosin heavy chain; TPM1, α-tropomyosin; β-ADR, β-adrenergic receptor; MYH7, β-myosin heavy chain; CACT, carnitine-acylcarnitine translocase (also SLC25A20); CASQ2, calsequestrin; CD36, fatty acid translocase (FAT); CFC1, cryptic; CHARGE, *CHARGE syndrome*; CHD7, chromo-domain gene 7; CORI, *Cori's syndrome*; COX15, cytochrome *c* oxidase assembly protein; CM, *cardiomyopathy*; CPT2, carnitine palmitoyltransferase II; CPVT, *catecholaminergic polymorphic ventricular tachycardia*; CRELD1, cysteine rich with EGF domains-1; SGCD, α-sarcoglycan; Danon, *Danon syndrome*; DCM, *dilated cardiomyopathy*; del8p22, deletion in chromosome 8p22; del22q11, deletion in chromosome 22q11; DES, desmin; DMD, dystrophin; DMPK, myotonin protein kinase; DSP, desmoplakin; Duchenne, *Duchenne syndrome*; ELN, elastin; E-D, *Emery–Dreifuss syndrome*; FA, *Friedreich's ataxia*; Fabry, *Fabry syndrome*; FBN1, fibrillin I; FRDA, frataxin; GATA4, GATA binding protein 4; GJA5, connexin 43; GLA, α-galactosidase; HCM, *hypertrophic cardiomyopathy*; HAND2, dhand/Hand 2; HERG, *human ether-a-go-go related syndrome*; Heterotaxy, *Heterotaxy syndrome*; HOS, *Holt–Oram syndrome*; IAA, *interrupted aortic arch*; JAG1, jagged-1; JUP, plakoglobin; KCNE1, potassium channel voltage-gated, Isk-related member 1; KCNE2, potassium channel, voltage-gated Isk-related member 2; KCNJ2, inwardly rectifying potassium Kir2.1 channel; KVLQT1, *potassium voltage-gated LQT syndrome 1 channel*; LAMP2, lysosome-associated membrane glycoprotein; LMNA, lamin A/C; LQT, *long QT syndrome*; LVNC, *left ventricular non-compaction*; Marfan, *Marfan syndrome*; MCM, *mitochondrial cardiomyopathy*; MMD, *myotonic muscular dystrophy*; MMP9, matrix metalloproteinase 9; MTP, mitochondrial trifunctional protein; NKX2.5, NK 2 transcription factor locus 5 (also NKX2-5); TFAM, mitochondrial transcription factor A (mtTFA); MVP, *mitral valve prolapse*; MYBPC3, myosin-binding protein C; MYL2, regulatory myosin light chain; MYL3, essential myosin light chain; Noonan, *Noonan syndrome*; NOTCH1, Notch 1 homolog; PAX3, paired domain transcription factor; PDA, *patent ductus arteriosus/Char syndrome*; PKP2, plakophilin; POLG, DNA polymerase γ; Pompe, *Pompe's syndrome*; PPH1, *primary pulmonary hypertension*; PRKAG2, AMP-activated protein kinase γ2; PTPN11, protein tyrosine phosphatase; RYR2, ryanodine receptor 2; SCO2, synthesis of cytochrome *c* oxidase (COX assembly protein); SCN5A, sodium channel voltage-gated α-polypeptide; SLC22A5, organic cation carnitine transporter 2 (OCTN2); SVAS, *supravalvular aortic stenosis/Williams syndrome*; TAZ (also G4.5), tafazzin; TBX1, T-box 1 transcription factor; TBX5, T-box 5 transcription factor; TFAP2B, transcription factor of the AP-2 family; TGFβ3, transforming growth factors-beta3; TNNT2, cardiac troponin T; TNNI3, cardiac troponin I; TNNC1, cardiac troponin c; TTN, titin; VCFS/DG, *velocardiofacial/DiGeorge syndrome*; ACADVL, very long-chain acyl CoA dehydrogenase; VSD, *ventricular septal defect*; VT, *ventricular tachycardia*; WPW, *Wolff–Parkinson–White syndrome*; X-DCM, *X-linked dilated cardiomyopathy*; ZIC3, zinc finger protein.

TABLE I Inborn Errors Causing Congenital Heart Defects

Genes affected (loci)	Cardiac phenotype/ (syndrome)
Cardiac voltage-gated sodium channel α-subunit (*SCN5A*)	Dysrhythmia, ventricular tachycardia, and fibrillation; LQT, and Brugada syndrome (BrS)
HERG (*KCNH2*)	Cardiac dysrhythmias, SD (LQT)
MinK (*KCNE1*)	Cardiac dysrhythmias, SD (LQT)
MiRP-1 (*KCNE2*)	Cardiac dysrhythmias, SD (LQT)
KVLQT1 (*KCNQ1*)	Cardiac dysrhythmias, SD (LQT)
Nuclear envelope protein lamin A/C (*LMNA*)	Conduction defects, muscular dystrophy (Emery-Dreifuss)
Cardiac ryanodine receptor (*RYR2*)	Ventricular tachycardia
Fibrillin-1 (*FBN1*)	Mitral or aortic valve regurgitation, SD (Marfan)
Elastin (*ELN*)	Aortic and systemic arterial stenoses (Williams)
GATA4	Cardiac septal defects
TBX1	DiGeorge/velocardiofacial
TBX5	Holt-Oram
CSX/NKX2-5	ASD/VSD; AV block
HAND 2 (dHAND)	Right ventricle hypoplasia
TFAP2	PDA (Char)
SHP-2 (*PTPN11*), protein tyrosine phosphatase	Conduction defects, pulmonary stenosis, (NS and LS)
Jagged 1 (*JAG1*)	Pulmonary artery stenosis; TOF (Alagille)
Myotonin protein kinase (*DMPK*)	Dysrhythmias and conduction defects (myotonic muscular dystrophy)

SD, sudden death; ASD, atrial septal defect; BrS, Brugada syndrome; VSD, ventricular septal defect; LQT, long QT; LS, Leopard syndrome; NS, Noonan syndrome; TOF, tetralogy of Fallot.

also been identified. Mutations in *JAG1* have been identified in kindred studies in association with Alagille syndrome, a complex autosomal-dominant disorder presenting with CHD, including pulmonary artery stenosis and tetralogy of Fallot (TOF).[10] *JAG1* encodes a ligand that binds the Notch receptor, an evolutionarily conserved signaling pathway involved in cell fate specification. Mutations in the signaling regulator *Notch1* have recently been implicated in aortic valve disease.[11] Mutations in *PTPN11* encoding a protein tyrosine–phosphatase (SHP-2) have been proposed to play a role in the pathogenesis of Noonan syndrome characterized by conduction defects, pulmonary stenosis, and hypertrophic cardiomyopathy (HCM)[12] and have been also recently implicated in the pathogenesis of LEOPARD syndrome, which likely represents an allelic disorder.[13]

SPECIFIC CARDIAC MALFORMATIONS

Although in the maternal environment multiple teratogens, infectious agents, and factors have been identified, which can contribute to the incidence of human congenital heart malformations, an increasing number of malformations have been shown to have a genetic basis as predicted by findings of similar isolated cardiac malformations in other species, many with heritable components.[14] Table II illustrates the most common cardiac malformations present in human subjects and provides information concerning their incidence and genetic etiology when known. Nearly one third of the congenital heart abnormalities are VSD, but ASD, atrioventricular canal (AVC), and TOF are not uncommon. It is of interest that clinically distinct malformations can arise from single genetic defects, suggesting that unrelated cardiac structures likely share similar developmental pathways.

Ventricular Septal Defect

Ventricular septal defect (VSD) is considered to be the most common type of CHD, occurring in more than 1 in 300 live births, as well as comprising a frequent component of more complex lesions.[15] VSD are of variable size and may occur anywhere in the ventricular septum and are primarily classified by their location. VSDs are either isolated small defects or larger defects associated with pulmonary

TABLE II Incidence of CHD in Human

CHD	Frequency	Gene defects
VSD	1:280	*TBX5, TBX1, NKX2-5*
ASD	1:1062	*TBX5, NKX2-5*
AVC	1:1372	*TBX5, NKX2-5*
Tetralogy of Fallot	1:2375	*JAG1, TBX1, NKX2-5*
Pulmonary stenosis	1:2645	*JAG1*
Transposition of great arteries	1:3175	
Hypoplastic left heart syndrome	1:3759	*NKX2-5*
Double-outlet right ventricle	1:6369	
Laterality/looping	1:6944	*ZIC3, LEFTYA*
Coarctation of aorta	1:7142	
Pulmonary atresia	1:7576	
Ebstein's anomaly	1:8772	*NKX2-5*
Truncus arteriosus	1:9346	
PDA	1:11,111	*TFAP2B*
Aortic valve stenosis	1:12,395	
Tricuspid atresia	1:12 658	
Bicuspid aortic valve	1:13,513	
Interrupted aortic arch	1:17,291	

stenosis, pulmonary hypertension, or aortic regurgitation. Only a small fraction of patients with VSDs ever become symptomatic. Left-to-right shunts through moderate to large VSDs become hemodynamically significant in the first 2–6 weeks of life. The direction and volume of the shunt most often dictates the clinical presentation of VSDs.

The heterogeneous structural composition of the ventricular septum suggests a variety of possible developmental mechanisms leading to VSD ontogeny. Although not present in the final anatomy, transitory structures are important in cardiac septation (e.g., proximal portions of the conal cushions) and are critical to its development. Conal structure position can also be a critical factor in the development of TOF and interrupted aortic arch.[16] Moreover, cell populations extrinsic to the developing heart, including the neural crest, seem to influence the process of ventricular septation through inductive interactions with neighboring tissues.

Some insights have been provided by animal models in which cell surface receptors including α4 integrin or vascular cell adhesion molecule (VCAM-1) are disrupted by homologous recombination. These molecules mediate adhesion and are expressed in complementary patterns in critical regions of the epicardium and the closing ventricular septum. Transgenic mice harboring null mutations of these genes exhibit multiple cardiac and noncardiac defects attributable to failure of tissue fusion, indicative of their critical role in the developing heart.[17,18]

A variety of hereditary syndromes are associated with VSDs. Individuals with chromosomal abnormalities such as trisomy 21 (Down syndrome), the most common genetic cause of CHD, display VSD. Similarly, VSD is commonly present in hearts of mice with trisomy 16, the murine model of human trisomy 21.

In animal models, mutations in a large number of genes have been associated with VSDs, usually in association with other complex heart defects. Human syndromic and sporadic cases of VSD have been associated with *NKX2-5*, *TBX5*, and *GATA4* mutations[5,19,20] and generally display an autosomal-dominant pattern of inheritance.

In addition, the signaling function of the Notch receptors, members of a gene family encoding transmembrane receptors and ligands involved in cell fate decisions (discussed in Chapter 6), may be critical for ventricular septation. Both missense or null defects in *JAG1*, encoding a Notch ligand, jagged 1, can result in VSD in individuals with Alagille syndrome, which produces a wide spectrum of developmental anomalies targeting other tissues in addition to heart or in children with TOF, who lack other features of Alagille syndrome.[21,22]

In addition, transgenic inactivation in mice of the basic helix-loop-helix transcription factor gene *Chf1/Hey2,* which acts as a nuclear effector of Notch signaling, results in VSDs.[23] Targeted disruption of many other genes participating in signaling pathways have been implicated in animal models that produce VSDs. A partial list includes mutations in the retinoic acid X receptor gene (*RXR*),[24] the type 1 neurofibromatosis gene (*Nf1*),[25] *Pax3*,[26] and *TGFβ-2*[27] all result in VSDs, although the etiology is unlikely to be related in each case. *RXR* defects may primarily relate to an epicardial abnormality in trophic signaling required for cardiomyocyte proliferation and ventricular morphogenesis.[24] *Nf1* cardiac defects are thought to be primarily due to the role of neurofibromin in endocardial cells as shown by the presence of cardiac defects in endothelial-specific inactivation of *Nf1*.[25] *Pax3* is expressed and functions in neural crest migration. Hence, diverse mechanisms in multiple cell types can converge to result in a phenotype that includes VSD.

Atrial Septal Defect

Atrial septal defect (ASD) is a common form of CHD, affecting more than 1 in 1000 live births, and often occurs in association with other types of more complex CHD.[1] ASDs are anatomically classified into four categories: *ostium secundum* (a defect constituting 85% of all ASDs and 10% of all CHD), *ostium primum* (a defect comprising 10% of ASDs), *sinus venosus* (a defect in the right horn of *sinus venosus* representing 5% of ASDs), and *coronary sinus* (a very rare defect in left horn of *sinus venosus*). These defects result in a communication between the right and left atria and clinically result in a shunt. Although there is little shunting evident early in post-natal life, over time, persistent left-to-right shunting of blood leads to enlargement of the right atrium and ventricle, atrial dysrhythmias, ventricular dysfunction, and pulmonary overload, which eventually can result in irreversible pulmonary vascular obstructive disease. Patients with significant shunts have an average life expectancy of 45 years. Therefore, it is recommended that children with ASDs should have them closed either surgically or through catheter-assisted devices, usually by age 3–4 y.

Gene loci leading to the genesis of ASDs have recently been identified. Some of these genetic factors have been grossly mapped to chromosomes (e.g., 5p) with the precise gene defect undetermined, whereas other genetic defects have been more fully characterized. For instance, patients with Holt–Oram syndrome frequently have ASDs in association with limb deformities. This disorder is due to mutations in the T-box transcription factor gene *TBX5*.[9] Homozygous null mutations of *Tbx5* in mice results in embryonic death, whereas heterozygous mice have a variety of ASDs and/or VSD and limb abnormalities, with specific missense mutations in *TBX5* leading to distinct phenotypes.[28]

Mutations in the transcription factor, GATA4 have been reported in some patients with isolated ASDs. GATA4 and Tbx5 can physically interact, suggesting that the cooperative activity of these two factors within a transcriptional complex may contribute to atrial septal formation.[5] Defects in NKX2-5, another transcription factor that can physically interact with GATA4, have also been associated with secundum ASDs in humans, as well as with a range of other cardiac abnormalities, including severe and progressive AV conduction block, VSD, TOF, double outlet right ventricle (DORV), subvalvular aortic stenosis, tricuspid valve abnormality, and Ebstein's anomaly.[6] *NKX2-5* (also *NKX2.5*) encodes a homeodomain-containing DNA binding protein that is homologous to the *Drosophila* tinman protein, so named because in its absence the fly has no heart.[29] Complete loss of *Nkx2-5* in mouse models results in embryonic lethality and a failure of cardiac development at the looping stage, defects that are more severe than the cardiac abnormalities found in human *NKX2-5* mutations.[30] Heterozygous loss of only one copy of *NKX2-5* results in ASD in some species.[31] In mouse models, haploinsufficiency for *Nkx2-5* and *Tbx5* resulted in an increased incidence of structural heart disease, confirming that normal heart development is sensitive to small changes in expression levels of *Nkx2-5*, *GATA4*, and *Tbx5*.[32] The variable expression of the ASD phenotype in different genetic backgrounds suggests the involvement of interacting modulators, whether of genetic or environmental origin. Studies in a wide variety of organisms suggest that *Tbx5*, *GATA4*, and *Nkx2-5* loci function at many stages of cardiac development affecting a variety of cardiac loci.[33]

In addition to *Nkx2-5*, *Tbx5*, and *GATA4*, other genes have been implicated in ASD etiology using animal models, although mutations in these other genes have not yet been identified in patients. ASD is a primary finding along with other cardiovascular abnormalities, including tricuspid atresia in mouse models in which the transcription factors *Cited2* and *Fog2/Zfpm2*, growth factor neurotrophin 3 (*NT-3*), and the *TrKc* receptor genes have been inactivated.[34–38]

It is noteworthy that development of the AV node and cardiac conduction system also depends on similar transcriptional programs, and conduction defects are often found in association with ASD; this association is attributed to common underlying genetic and developmental processes.[39] For instance, *Nkx2-5* mutations in both humans and mice cause ASD, and conduction defects, including heart block consistent with *Nkx2-5*, playing a significant role in both the regulation of septation during cardiac morphogenesis and in the maturation and maintenance of AV node function.[40]

Atrioventricular Canal

Atrioventricular canal (AVC) has been estimated to affect more than 1 in 2800 live births.[15] Although associated with extensive anatomical variations, the most prevalent form of AVC consists of a combination of defects in the atrial septum primum and the inlet portion of the ventricular septum. During early neonatal stages, patients may have mild cyanosis and relatively high pulmonary vascular resistance. If these defects are not detected in the neonatal period, these infants typically are seen with CHF resulting from the increased left-to-right shunt, which increases as the pulmonary vascular resistance falls.

Developmental formation of the AVC results from complex interactions of components of the extracellular matrix (ECM), and a series of endothelial/mesenchymal interactions in response to a number of signaling molecules that are crucial to its normal development. Atrioventricular septal defects (AVSDs) that include AVC result from the arrest or interruption of normal endocardial cushion (EC) development and have been associated with chromosome abnormalities and a variety of genetic syndromes. Early clinical studies also suggested the presence of a susceptibility gene for AVSD identified in a large kindred with many affected members.[41]

In mice, mutation of the *Friend of GATA2* (*Fog2*) transcription cofactor gene results in a phenotype with features of common AVC (CAVC), including a common AV valve, but coronary artery defects are also present, suggesting that the physical interactions of FOG2 and GATA4 factors are critical determinants in early heart morphogenesis.[42,43] In addition, the *RXR* transgenic mouse demonstrates rodent-human phenotype convergence, with *RXR* heterozygotes exhibiting a broad spectrum of defects ranging from mild defects to severe CAVC.[44]

Recently, the cardiac-specific inactivation of the gene for bone morphogenetic protein 4 (*Bmp4*) in mice resulted in defects of AV septation that closely resemble the range of AVSD phenotypes seen in human CAVC.[45] Also mutant embryonic mice containing diminished levels (approximately 30% of wild-type levels) GATA4 displayed CAVC, DORV, hypoplastic ventricular myocardium, and normal coronary vasculature.[32]

Nearly 20% of individuals with the common human chromosomal disorder trisomy 21/Down syndrome exhibit AVC defects.[46] Similarly, mice with an extra copy of the syntenic mouse region, resulting in trisomy 16 (Ts16) display an AVC defect.[47] However, Ts16 mice are markedly different phenotypically from patients and typically show an unbalanced AV connection with separate superior leaflets compared with a more balanced common AV junction as found in humans with Down syndrome, limiting their value as an informative model. Specific ECM molecules have been proposed as playing an intrinsic role in normal EC differentiation into valves and septa. Type VI collagen (COL6) is of particular interest, because COL6 genes encoding the α1 and α2 chains are located on chromosome 21, are expressed in the developing AVC ECM and have been associated with trisomy 21 AVC defects in human genetic studies.[48] Although the molecular mechanisms linking COL6 and trisomy 21 AVC defects are presently undetermined, studies have found that fibroblasts from trisomic patients have significantly increased cell adhesiveness that may be contributory to the development of the AVC defect in Down syndrome.[49]

AVSD not associated with trisomy 21 usually occurs as a sporadic trait with no indication of the genetic basis. The discovery of the cell adhesion molecule, CRELD1 (cysteine rich with EGF domains), as the first recognized genetic risk factor for AVSD has provided new insight into the genetic basis of sporadically occurring and syndromic AVSD.[50] Missense mutations of the CRELD1 gene increase susceptibility to AVSD but are not sufficient to cause the defect, indicating that AVSD is multigenic. CRELD1 was identified as a candidate gene from the AVSD2 locus (3p25-pter), and analysis of 50 patients indicated a significant association of the CRELD1 mutations with partial AVSD.

Valve Defects

The developmental formation of the thinly tapered cardiac valves, produced by the remodeling of regional swellings of both endothelial and mesenchymal cells, is a highly complex process, which is relatively poorly understood. This process involves the complex interaction of residential myocardial cells, ECM, and endothelium, featuring an array of reciprocal signaling events,[51] the transformation and migration of endocardial cells into mesenchymal cells and differentiation into the valvular fibrous tissue. Septa and valves of the AVC (i.e., mitral and tricuspid) are derived entirely from EC tissue, whereas the end stages in development of the outflow tract (OFT) valves (i.e., aortic and pulmonic) that are likely derived from endothelial cells are more controversial.

Defects in cardiac valves and associated structures are a common class of CHD, accounting by some estimates for approximately 25% of all cardiovascular malformations. Congenital valve defects have been identified as features of well-defined clinical syndromes, including Down syndrome (with EC defects, including incomplete septation of the AV valves), LEOPARD syndrome (pulmonic stenosis), chromosome 22 microdeletion syndromes, Holt–Oram, and Noonan syndrome (pulmonic stenosis). However, in a significant proportion of cases, defects in cardiac cushion development occur separate from any defined syndrome or genetic cause.[52] Considerable excitement was generated by the discovery that a specific gene, DSCR1, residing within the Down syndrome chromosome 21, had the appropriate temporal pattern of expression for valve formation, was regulated by VEGF, and interacted with the Wnt pathway.[53] However, further observations demonstrated that the cardiac defects were not corrected by returning the DSCR1 gene from trisomy to disomy, indicating that the cardiac developmental phenotype is likely mediated by a more complex mechanism.[54] Furthermore, the complexity of the multiple signaling pathways involved in cardiac valve formation is underscored by findings from mouse genetics that disruptions of a large number of genes alter the valve phenotypes (Table III).

It is also noteworthy that mitral valve prolapse (MVP), a common cardiovascular abnormality that occurs in approximately 2.4% of the general population, has a genetic component, that although heterogenous, tends to display an autosomal-dominant pattern of inheritance with variable penetrance. Although in many cases, MVP is clinically benign, it can be associated with significant sequelae, including mitral regurgitation, bacterial endocarditis, congestive heart failure (CHF), atrial fibrillation (AF), and even sudden death. Recently, a locus for MVP was mapped to chromosome 11p15.4.[55]

Transposition of the Great Arteries

Transposition of the great arteries (TGA) is the most common cyanotic lesion presenting in the first week of life, affecting approximately 1 in 3100 live births. In TGA, the aorta arises from the right ventricle and the pulmonary artery from the left ventricle, resulting in separate parallel systemic and pulmonary circulations. Although abnormalities such as VSD, ASDs, aortic arch abnormalities, and tricuspid valve abnormalities may accompany TGA, coronary artery malpositioning is the most frequently associated anomaly, occurring in 30% of patients. Children with TGA presenting in the first few days of life with cyanosis and tachypnea can be treated with surgical correction by an arterial switch procedure in which the great arteries are divided and proper ventriculoarterial alignment restored. Untreated, half of the children die in 1 mo, and 90% in 1 y.

Transgenic inactivation of the murine gene (HSPG2) encoding perlecan, the basement membrane heparin-sulfate proteoglycan, results in cardiac malformations, including a high incidence of complete TGA, and represents the closest genetic animal model to the common type of TGA.[72] Perlecan binds several growth factors (including FGF) and interacts with various ECM proteins and cell adhesion mol-

TABLE III Mouse Genes Implicated in Cardiac Valve Formation

Genes	Effect on myocardial valves	Ref
NFATc1	Disruption of NFATc1 gene leads to selective absence of the aortic and pulmonary valves; NFATc1 is also necessary for semilunar valve formation.	56
VEGF	Myocardial overexpression of VEGF results in failure of AVC and OFT cushion formation.	57
Connexin (Cx45)	Cx45 knockout leads to decreased and delayed EC formation.	58
Notch 1	Disruption of notch1 leads to hypoplastic EC.	59
Neurofibromin (NF1)	Disruption of nf1 leads to markedly enlarged EC.	60
ErbB3	ErbB3 -/- embryos exhibit EC abnormalities and defective valve formation.	61
BMP ligand	Bmp6Bmp7 double mutants result in hypoplastic EC and delay in OFT formation.	62
BMPR (Alk3)	Alk3 disruption leads to hypoplastic EC.	63
Smad6	Smad6 disruption results in thickened and gelatinous AV and semilunar valves.	64
Wnt/β-catenin/ APC	APC disruption leads to very large valves.	65, 66
HB-EGF	HB-EGF -/- mice have large AVC and semilunar valves.	67
Hyaluronan synthase-2 (HAS-2)	Has2 -/- mice are unable to form EC.	68
HESR2	Disruption of the Notch signaling target HESR2 results in dysplastic AV valves.	69
TGF-β	TGF-β1 inhibits valve myofibroblast proliferation; TGF-β2 knockout results in AV and semilunar valve thickening.	70, 71

ecules. However, individuals harboring mutations of *HSPG2* exhibit a range of skeletal disorders but do not manifest CHD.[73]

Inactivation of the gene for the type II activin receptor in mice also results in TGA,[74] although this phenotype is only one of several OFT defects (e.g., double outlet right ventricle [DORV] and persistent truncus arteriosus [PTA]).

Mice containing null mutations in *Dvl2*,[75] the EGF-related *cryptic* (*CFC*),[76] and *Neuropilin-1*[77] also exhibit a series of vascular defects in mice, including TGA, suggesting a role for neural crest–derived tissues in the pathogenesis of TGA. Mutations in the human homolog of *cryptic*, *CFC1*, a gene also extensively associated with laterality defects and heterotaxy syndrome[78,79] (as discussed later), have been reported in subjects with TGA and DORV in the absence of laterality defects. Furthermore, treatment with the teratogenic agent retinoic acid (RA) promotes a dose-dependent differential induction of complete TGA at high doses and dextroposition of the aorta at low dose.[80] The modulating effect of RA signaling, the interplay of growth factor signaling pathway elements in conotruncal maturation, and the effected movement and growth of tissues resulting in the various forms of TGA, PTA, and DORV remain to be determined.

Hypoplastic Left Heart Syndrome

Hypoplastic left heart syndrome (HLHS) is a relatively common (1 in 4000) heterogeneous group of abnormalities in which there is a small or absent left ventricle, with hypoplastic mitral and aortic valves rendering the left ventricle nonfunctional.[15] This syndrome is characterized by a functional single right ventricle and systemic outflow obstruction. Palliative surgical therapy of this lesion is necessary for survival. Although its etiology is largely unknown, there is considerable evidence of a genetic component.

Left ventricle outflow tract obstruction (LVOTO) encompasses a spectrum of defects that is developmentally related and can be caused by a single gene defect. The more severe form, HLHS, may occur in families showing left-sided cardiac disease of variable severity, including mild aortic stenosis and bicuspid aortic valve.[81] This association suggests a potential genetic etiology. Moreover, HLHS rarely occurs with trisomy 21 but has been reported in other trisomies such as 13 and 18, as well as in Turner's syndrome. Other HLHS associations with defects in chromosomal loci have been reported. A female newborn was reported with deletion of the short arm of the chromosome 18 (del 18p) and HLHS with intact atrial septum.[82]

Single Ventricle

Single ventricle (SV) is a relatively rare CHD, occurring in 1.25% of infants with CHD in the Baltimore–Washington Infant Study.[83] This complex malformation manifests a genetic component. In a relatively large study of patients seen with SV, approximately 3% of the patients' siblings harbored CHD.[84] The incidence of cardiac defects was significantly greater in siblings of patients having a SV with a single or common inlet, or a common/SV of right ventricular morphology (5%), compared with siblings of patients with a double-inlet left ventricle (0.5%). SV with aortic outlet from a rudimentary cavity was reported in a patient with trisomy 18.[85] In a recent study of fetuses having structural heart disease as detected by echocardiography, univentricular heart or SV was frequently associated with intrauterine HF and neonatal mortality.[86]

Defects in the Cardiac Outflow Tract and Aortic Arch

Defects in the cardiac outflow tract (OFT), including TOF, PTA, DORV, and in the aortic arch (e.g., coarctation of the aorta, interrupted aortic arch [IAA], PDA) make up 20–30% of all CHD. Large chromosomal deletions have been implicated in developmental/structural cardiac malformations, including conotruncal abnormalities, AVC defects, VSDs, and ASDs.[15] The most common human gene deletion syndrome, referred to as the *22q11* deletion syndrome, is the second most common genetic cause of CHD after trisomy 21. OFT (or conotruncal) defects and defects in the aortic arch (both of which are derived from the cardiac neural crest) are present in more than 75% of individuals harboring this deletion. In addition to pharyngeal arch defects, including, cleft palate, dysmorphic facial features, and thymic hypoplasia, these cardiovascular abnormalities are manifestations of the complex genetic disorder termed velocardiofacial syndrome/DiGeorge syndrome, also known as CATCH-22. Most patients are hemizygous for a 1.5- to 3.0-Mb deleted region of chromosome 22 (22q11), which is suspected to be critical for normal pharyngeal arch development and contains more than 30 genes. The del22q11 deletion is a relatively common event occurring in approximately 1 in 4000 live births.

The gene for the T-box transcription factor Tbx1 derived from the central area of the deleted region is the critical factor in the development of this congenital defect.[87] Tbx1, a member of a phylogenetically conserved family of genes that share a common DNA-binding domain (i.e., the T-box), is highly expressed in the pharyngeal arches and involved in the regulation of cardiac development. Reduction in murine *Tbx1* expression (which occurs in the deleted hemizygous state) greatly impacts on the early gene expression involved in cardiac morphogenesis.[88] Mice heterozygous for a single null allele of *Tbx1* exhibit a high incidence of cardiac OFT anomalies, as well as other developmental abnormalities characteristic of DiGeorge syndrome. Although mutations in humans have been difficult to identify, *TBX1* mutations have been detected in three unrelated patients without the 22q11 deletion.[89] It remains unclear whether *TBX1* haploinsufficiency is sufficient to bring about human cardiac abnormalities. Additional neighboring genes within the 22q11 region may modify *TBX1* function or may independently contribute to the del22q11 phenotype.[90] Furthermore, genes that encode transcription factors of the forkhead class (e.g., *Fox*) directly activate *Tbx1* transcription in the pharyngeal endoderm,[91] and *Foxc1* and *Foxc2* mutations in mice have similar phenotypes with respect to cardiac abnormalities as *Tbx1* mutants.

Other large-scale chromosomal deletions have been reported, leading to a wide spectrum of congenital OFT obstruction, septal, and valve disorders.[92] A distal 8p deletion encompassing a 5-centimorgan region at chromosome 8p23[93] has been described, as well as a deletion located at chromosome 10p14.[94] The specific genes involved in these regions whose haploinsufficiency is likely responsible for these cardiac abnormalities have not yet been determined. A number of cardiac defects, with obstruction of the aortic arch in particular, seem to be a particular feature of chromosome 1q21.1 contiguous gene deletion (which encompasses a region of 1.5–3 Mb inclusive of at least seven genes).[95] It is also possible that a number of microdeletions may have been previously overlooked because of smaller size and chromosomal location. At present, newer molecular cytogenetic techniques with high resolution, such as fluorescence *in situ* hybridization or FISH, are routinely used to confirm the clinical diagnoses of chromosomal damage, such as chromosomal microdeletions and small translocations.

It is important to note that such large genetic deletions are commonly associated with a wide spectrum of extracardiac malformations associated to CHD and have been estimated at more than 30% of cases. Significantly, these chromosomal defects are more prevalent in patients with cardiac anomalies than in the general population. Although neonatal cardiac malformations resulting from trisomies 13, 18, and 21, as well as monosomy XO (Turner syndrome), are well recognized, the precise molecular basis by which the gene dosage imbalance in these patients causes the cardiac phenotype is not known.

Another significant cause for CHD is uniparental disomy (UPD) and mosaic trisomy in which certain segments of both copies of specific chromosomes are derived from the same parent, resulting in aberrant development or prenatal lethality. A wide range of CHDs have been found in patients with uniparental disomy of chromosome 7,[96] 10,[97] 12,[98,99] 14,[100] 15,[101] 16,[102,103] 20,[104] and 22.[105] Furthermore, in a number of cases, the parental genomes have undergone epigenetic modifications during gametogenesis. These epigenetic modifications result in parent-of-origin specific expression for some genes, a phenomenon known as genomic imprinting. The best-known and most thoroughly studied epigenetic mechanisms include DNA methylation and chromatin/histone protein modifications (e.g., acetylation, methylation), both of which provide a basis for the switching of gene activities, allowing the alteration and maintenance of stable phenotypes (without affecting the DNA sequence) and leading to the basis of the imprinted character of several genes. Further discussion on DNA methylation and histone epigenetic modifications and their role in the development of cardiovascular defects was presented in Chapter 1.

Patent Ductus Arteriosus

Patent ductus arteriosus (PDA) is a relatively common CHD that results when the ductus arteriosus, a muscular artery connecting the pulmonary artery to the descending aorta, fails to remodel and close after birth, resulting in a left-to-right shunt.

A syndromic form of this disorder, Char syndrome, mapped to a single locus on chromosome 6p12–p21.[106] Subsequent positional cloning and mutation analysis demonstrated that this phenotype is caused by mutations in *TFAP2B*, the gene encoding a transcription factor AP2β, which is highly expressed in neural crest cells.[8] In addition to PDA, established features of this autosomal-dominant syndrome include facial dysmorphology and fifth-finger malformation. Several disease-causing *TFAP2B* missense mutations have been identified, which result in heterodimeric transcription factors unable to bind target DNA sequence, adversely affecting the transactivation of gene expression and consistently behave as dominant negative mutations.[8,107] Because only a small number of families have been reported, there is, at present, limited information on the spectrum of mutations and resulting phenotypes.

Two kindreds with Char syndrome containing 22 and 5 affected members, respectively, have been reported.[108] Sequencing of *TFAP2B* revealed mutations changing highly conserved bases in introns required for normal splicing in both kindreds, and these mutations were not found among control chromosomes. Transcripts generated from abnormal splicing events contained frameshift mutations and are degraded by nonsense-mediated mRNA decay, resulting in haploinsufficiency. Other phenotypes segregating with the *TFAP2B* mutations included dental and occipital bone abnormalities, as well as a sleep disorder (parasomnia), implicating *TFAP2B*-dependent functions in the normal regulation of sleep.

Interestingly, ethnicity may be an important factor in the incidence of PDA. For example, PDA accounts for a higher fraction of all CHD in Iran (15%) than in the United States (2–7%),[109] and there is documented information showing a marked increase of parental consanguinity (63%) associated with PDA cases compared with controls (25%). Linkage analysis in Iran has revealed the presence of a recessive component to PDA and implicated a single locus, *PDA1*, mapping to chromosome 12q24 in one third or more of all PDA cases.

Interrupted Aortic Arch

Interrupted aortic arch (IAA) is an extremely rare CHD defined as the loss of luminal continuity between the ascending and descending aorta. Its clinical presentation, including cardiovascular shock, acidemia, or severe CHF in the first 2 weeks of life, is similar to severe coarctation of the aorta, although the developmental mechanisms involved in their generation seem to be different. IAA is characterized by a narrowed left ventricular OFT, with resulting severe obstruction associated with decreased growth, hypoplasia, and interruption of the aortic arch. In addition to VSD and narrowed subaortic area, there is complete atresia of a segment of the aortic arch. This anomaly entails a very poor prognosis without prompt surgical treatment.

Although IAA is associated with chromosome 22 deletions, specific gene mutations in humans have not yet been described in patients with IAA. Research in mice has shown that the critical gene in the chromosome 22 region implicated in cardiac abnormalities, *Tbx1*, is likely involved in triggering diffusible signals from the pharyngeal arches to support the growth contributing to the mature aortic arch.[110] In addition, other large-scale genetic defects have been reported in association with the interrupted or obstructed aortic arch phenotype. Several unrelated patients containing the previously noted deletion on chromosome 1q21.1 spanning a 1.5- to 3-Mb region had IAA.[95] Interestingly, this contiguous gene deletion included the connexin 40 gene (*GJA5*), which previously was found to exhibit markedly reduced expression levels in patients with chronic atrial fibrillation.[111] Mouse studies have revealed that a *connexin 40* haploinsufficiency results in a number of cardiac malformations, including aortic arch abnormality.[112] Also, interrupted aortic arch has been reported in isolated cases of children with trisomy 5q31.1q35.1, resulting from a maternal balanced insertion.[113]

A number of candidate genes have been identified in animal models in which there is considerable phenotypic overlap between IAA and PTA. Defects in several different components of the endothelin signaling pathway also result in isolated IAA.[114–117] Mutations in two related forkhead transcription factors, *Foxc1* and *Foxc2*, can also result in IAA.[118,119] Null mutations in *CBP*/p300-interacting transactivator with ED-rich tail 2 (*CITED2*) result in mouse embryos with several cardiac abnormalities, including IAA, which derive in large part from its critical role as a TFAP2 coactivator.[34]

Abnormalities in the aortic arch also can result from RA deficiency such as that occurring with vitamin A deficiency in pregnancy or with excess RA exposure often termed RA or isotretinoin (a vitamin A analog) embryopathy.[120] Fetal alcohol exposure during pregnancy causes a similar cardiac phenotype to RA deficiency and has been proposed to reduce the amount of available RA at the level of formation and function of Spemann's organizer during early embryonic development.[121]

Persistent Truncus Arteriosus

Persistent truncus arteriosus (PTA) is an uncommon form of CHD, affecting 1 in 10,000 live births. PTA consists of a single great artery arising from the heart. The truncal valve is frequently anatomically abnormal, and associated VSD is present in most cases. Most infants with PTA are first seen with symptoms of CHF in the first weeks of life. Although infants may be cyanotic, CHF symptoms and signs tend to predominate. A prognosis of irreversible pulmonary vascular disease and hemodynamic instability makes prompt surgical repair a critical choice.

Microdeletions of chromosome 22, commonly associated with DiGeorge syndrome, have been found in as many as one third of patients with PTA. However, PTA can also occur as a result of this deleted chromosomal abnormality even in the absence of other signs of DiGeorge syndrome.

Of particular significance is that neural crest cells play a critical role in truncal septation. This multipotent cell population is specified in the dorsal neural tube, and a subpopulation migrates through the pharyngeal arches to the developing cardiac OFT. Physical ablation of neural crest cells in chick embryos demonstrated the critical importance of these cells in the patterning of the pharyngeal arches and aortic arch and in the development of the cardiac OFT and the outflow septum.[122] In particular, it is the failure of the outflow septum to form that results in PTA, which is a single outflow vessel with a single valve.

Several genes that are expressed in neural crest cells have been implicated in the development of PTA. Mutation of the murine *Pax3* gene, which is highly expressed in pre-migratory neural crest cells, results in PTA. In addition, members of the semaphorin guidance molecule family contribute to the migratory behavior of neural crest cells. Mice embryos containing targeted mutations in *Sema3C* exhibit a cardiac phenotype that includes PTA and IAA similar to that found with ablated neural crest cells.[123] Other signaling pathways required for the induction, migration, and differentiation of cardiac neural crest have been identified involving Wnt/β-catenin signaling pathway, bone morphogenetic protein (BMP), and vascular endothelial growth factor (VEGF), which have been implicated in OFT septation.[124–126]

Understanding the role of the BMP receptor signaling on cardiac development has been hampered, because *Bmpr* knock-outs often lead to very early embryonic lethality, with few organized structures formed.[125] By engineering the expression of a BMP type II receptor lacking half of the ligand-binding domain, an altered receptor with reduced signaling capability can be expressed in mice at levels comparable with the wild-type allele.[126] Unlike *Bmpr2*-null mice, mice homozygous for this defective receptor undergo normal gastrulation, providing evidence of the dose-dependent effects of BMPs during mammalian development. Mutant embryos, however, die at midgestation with specific cardiovascular and skeletal defects, demonstrating the importance of wild-type levels of BMP signaling in the development of these tissues. The most striking defects occurred in the OFT, with PTA and IAA. In addition, the semilunar valves do not form in mutants, whereas AV valves are unaffected.

Mouse embryos with neural crest–specific deletion of the type I BMP receptor, *ALK2*, displayed cardiovascular defects, including PTA and abnormal maturation of the aortic arch.[127] These studies found in the *ALK2* mutant mice impaired migration of mutant neural crest cells to the OFT, deficient differentiation to smooth muscle around aortic arch arteries, and reduced expression of *Msx1*, one of the major effectors of BMP signaling in distal OFT, leading to the abnormal development of the OFT and aortic arch derivatives.

Further observations have provided evidence that BMPs, particularly BMP2 and BMP4, are regulators of neural crest cell induction, maintenance, migration, differentiation, and survival. Moreover, BMP2/4 signaling in mammalian neural crest derivatives is essential for OFT development and plays a contributory role in proliferation signaling pathway for the ventricular myocardium.[128] The TGFβ signaling pathway is critical to conotruncal development. *TGFβ2*−/− mice have PTA, VSD, and DORV.[129] In addition, BMPs that are part of the TGFβ superfamily also play a role in conotruncal development, because mice with targeted disruption of both *Bmp6* and *Bmp7* exhibit PTA.[130]

Double Outlet Right Ventricle

Double outlet right ventricle (DORV) is a rare group of cardiac anomalies characterized by both great arteries (pulmonary and aorta) arising primarily from the right ventricle. It includes a broad spectrum of anatomical variants and associated malformations. A variety of surgical techniques have allowed the complete correction of even the more complex forms of biventricular DORV.

A number of mutations in specific murine genes may result in a DORV phenotype. Mutations in the *ActRIIB* gene encoding the type IIB activin A receptor result in a complicated array of cardiac defects, which include DORV.[74] Mutations in *Cited2*,[34] multiple members of the endothelin signaling pathway,[116–117] *PDGFR*,[131] *Pitx2*,[132] *TGFβ2*,[70] connexin (*Cx40*),[112] the high-mobility–group transcription factor *SOX-11*,[133] and mutants of the retinoic acid receptor family (*RAR, RARβ*) also result in DORV phenotype.[134] In patients with DORV, mutations have been reported in the *GJA1* gene encoding the connexin (Cx43) protein,[135] and DORV has been described in patients with deletions in chromosome 22q11.

Laterality/Heterotaxy

The position of the heart and viscera is strictly regulated and highly conserved. The normal left-right anatomical position is termed *situs solitus*; the failure to correctly establish left-right patterning during embryogenesis results in the clinical phenotype of heterotaxy or laterality defect. These include both *situs inversus*, involving a mirror-image reversal of all asymmetrical structures, and *situs ambiguus*, in which the entire anatomical left-right (L-R) axis is neither normal nor mirror-image reverse. Although both laterality defects occur with similar frequencies (1.4 in 10,000 births), *situs ambiguus* is much more deleterious and is primarily associated with severe cardiac malformation and dysfunction not present with *situs inversus*.

Laterality defects can occur as a function of environmental or teratogenic influences, as well as by strong genetic determinants.[136] High doses of RA can induce laterality defects in vertebrates, including humans.[137] Also, an increased risk of laterality defects has been documented in infants of mothers with maternal type-1 diabetes mellitus.[138]

A number of animal models of laterality, including mouse, chicken, *Xenopus,* and zebrafish, have provided additional insights into the genetic regulation of left-right positioning and determination. A plurality of genes involved in signaling (Table IV) has been identified, including upstream regulators such as the *TGFβ* family members *Nodal* and *lefty-1, cited-2, sonic hedgehog* (*Shh*), and the cofactor for Nodal, the *EGF-CFC cryptic* gene, and downstream factors including the bicoid-type homeobox gene *Pitx2, Nkx2.5,* and *SnR*.[139–158] The cascade of signaling molecules regulating the establishment of L-R identity of the embryo, positioning and restricting the expression of *Nodal* on the left side of the lateral plate mesoderm, culminates into the downstream expression of the homeodomain factor *Pitx2* in the left side of the visceral organs, including the heart. Asymmetrical *Pitx2* expression seems to be sufficient for establishing L-R identity of the heart.

Different approaches including the analysis of altered patterns of asymmetric gene expression have been necessary in defining these genes' roles in the development of laterality and L-R asymmetry in the development of the heart. In a number of cases, gene disruption of mouse alleles did not prove useful.[159]

The establishment of the left-right asymmetry starts at the Hensen node. Here, the initial embryonic symmetry is broken by cascades of gene activation that confer specific properties on the left and right sides of the embryo. Although there are variations between species, some basic patterns of

TABLE IV Genes Implicated in Left-Right Positioning in Animal Studies

Gene	Function	Ref
Cited2	Transcriptional coactivator	139, 140
Nodal	TGF-β extracellular protein signal	141, 142
Pit2x	Bicoid-type homeobox transcription factor	143
SSH	Intercellular signaling protein	144
BMP4	TGF-β family growth factor	145
Hand1	Basic helix-loop-helix transcription factors	146
Activin receptor ALK4	TGF-β cell surface receptor	147
β-catenin	Transcriptional activator	148
Snail (cSnR)	Zinc finger transcription factor	149
Nkx2.5	Homeobox gene	150
Caronte (Car)	Multifunctional extracellular protein	151
Lefty-1	TGF-β extracellular protein signal	152
Lefty-2	TGF-β extracellular protein signal	153
HFH-4	Forkhead transcription factor	154
BMP type I receptor ACVRI	TGF-β cell-surface receptor	155
Notch	Notch receptor	156
CFC (cryptic)	EGF-CFC membrane protein	157
Smad5	Downstream intracellular signal mediator of BMP/TGFβ-family	158

gene expression (*Nodal*, *Pitx2*) seem to be maintained along the phylogenetic scale. Anomalous expression of these genes induces the heterotaxy syndrome, which usually is associated with CHD.

The most extensively studied mouse model of L-R development is the spontaneous mutant *iv* (*situs inversus viscerum*). A candidate gene for *iv*, *left-right dynein* (*lrd*), has been identified in which a missense amino acid substitution appears only in *iv* alleles.[160] The identification of the *iv* gene as a dynein implicates the complicity of microtubule arrays in the generation of left-right asymmetry. Molecular data linking cilia dysfunction and left-right asymmetry was established by the identification of a mutation in the gene of heavy chain dynein (*LRD*), a protein that acts as a molecular motor in cilia motility in a well-characterized mouse with randomized *situs*. Subsequently, mutations in the human dynein axonemal heavy chain *DNAH5* were reported in individuals with Kartagener syndrome.[161] To date, mutations in *DNAH5* and in the axonemal dynein intermediate chain gene 1 (*DNAI1*) have been found to result in Kartagener syndrome.[162] A mutation in the human LRD homolog axonemal heavy chain type 11 (*DNAH11*) is also associated with *situs inversus* and ciliary dyskinesia.[163]

Several cases of *situs inversus* have been associated with primary ciliary dyskinesia (PCD). Sensory cilia responsive to a variety of stimuli generate a direct left-right asymmetry by signaling (by a calcium-dependent response) within the embryonic node soon after anteroposterior and proximal-distal axes are established. This response triggers a program that results in a left-sided asymmetrical heart and asymmetrical patterning of visceral organs. Thus, L-R axis (LRA) formation depends on proper node cell differentiation and cilia function.

Situs abnormalities in human have shown autosomal-dominant, autosomal-recessive, and X-linked patterns of inheritance.[164] Several genes, and a number of single gene mutations, have been implicated in heterotaxy and related isolated CHD, with evidence of extensive *locus* heterogeneity. Furthermore, a number of chromosomal abnormalities associated with *situs ambiguus* have been reported.[165–167] With the exception of *Nodal*, none of the candidate LRA genes identified in other organisms has been shown to map to one of the human cytogenetic breakpoints. Further evidence that mutations in the human homolog of *Nodal* may contribute to the pathogenesis in cases of heterotaxy has been presented. A *de novo* interstitial deletion encompassing the *NODAL* gene on human chromosome 10q21–q23 has been reported in an individual with *situs ambiguus* and midline malformations.[168]

Mutations in genes that control early L-R patterning and the earliest steps in cardiogenesis have been shown to cause isolated CHD in humans. Mutations in three genes that function in the TGFβ signaling pathway–activin receptor type IIB (*ACVR2B*), the EGF-CFC family member *CRYPTIC,* and *LEFTYA* have been found in a small number of patients with classic heterotaxy.[169–173] Recently, mutations in *CRELD1*, a cell adhesion molecule, and *NKX2-5* have been associated with laterality defects.[174,50]

Mutations in the zinc finger transcription factor *ZIC3* gene result in *situs ambiguus* with severe heart malformations as a consequence of the inability of the embryo to establish the appropriate left-right asymmetry in early development.[175] Point mutations, including frameshift, missense, and nonsense alleles, have been identified in four X-linked familial cases, in one sporadic case, and in one case with isolated CHD.[176–179] On the other hand, the association of *ZIC3* deletion with *situs ambiguus* suggests that *ZIC3* loss of function may underlie the pathogenesis seen in patients with point mutations.[177]

CHARGE Syndrome

The acronym CHARGE stands for the major features of this syndrome: *c*oloboma of the eye, *h*eart defects, *a*tresia of the choanae, *r*etarded growth and development, *g*enital hypoplasia, and *e*ar anomalies and/or deafness. CHD associated with this syndrome includes a high incidence of cases with TOF and AVSD, with an estimated prevalence of 1/10,000.[180] The combination of malformations in CHARGE syndrome supports the view that this multiple congenital anomalies/mental retardation syndrome is a developmental field defect involving the neural tube and neural crest cells. Although many cases seemed to be sporadic, the existence of rare familial cases have suggested the involvement of genetic factors such as *de novo* mutations of a dominant gene or subtle submicroscopic chromosome rearrangement.[181]

By use of a microarray approach, a 2.3-Mb *de novo* overlapping microdeletion on chromosome 8q12 was identified, by comparative genomic hybridization, in two individuals with CHARGE syndrome.[182] Positional cloning and sequence analysis revealed *CHD7* as a likely candidate gene. This was further supported by the detection of *CHD7* mutations in 10 of 17 individuals with CHARGE syndrome without microdeletions, accounting for the disease in most affected individuals. Recent observations have revealed that most of the *CHD7* mutations are unique and scattered throughout the gene.[183] A putative role for *CHD7* as a general regulator of developmental gene expression involved in chromatin structure and gene expression has been suggested.[184] Moreover, disruption of this developmental regulation leading to the altered disruption of mesenchymal–epithelial interaction, which is necessary for proper formation of the heart and other tissues, has been suggested in the pathogenesis of CHARGE syndrome.[185]

EXPRESSION PROFILING IN CHD

Expression profiling has been sporadically used in children with CHD. Recently, myocardial gene profiling was carried out in cardiomyocytes derived from individuals with HLHS. Transcript analysis was conducted by quantitative PCR, and protein analysis by two-dimensional gel electrophoresis and mass spectroscopy.[186] Transcriptome analysis revealed that both the left and right ventricles expressed the fetal or "heart failure" gene expression pattern, and proteomic analysis indicated that although differentiation was evident in the ventricle, the fetal isoforms of some cardiac-specific proteins were prevalently expressed. Furthermore, cardiomyocytes from each HLHS sample ($n = 21$) showed a unique (and inappropriate) expression of the platelet-endothelial cell adhesion molecule-1 (PECAM-1, CD31), a member of the cell adhesion molecule (CAM) family that has a primary role in the regulation of tissue morphogenesis.

A microarray study documented myocardial gene expression in samples collected from six normal subjects and 55 young patients (at surgery) with a variety of CHD.[187] After adjusting for age variation between the older children and infants, distinct gene expression profiles were identified for TOF, VSDs, and right ventricular hypertrophy (RVH). The cases with TOF featured differential expression of genes involved in cell cycle and cardiac development (e.g., up-regulation of *SNIP*, *A2BP1*, and *KIAA1437*, and down-regulation of *STK33*, *BRDG1*, and *TEKT2*) and exhibited up-regulation of ribosomal proteins encoded by *S6*, *L37a*, *S3A*, *S14*, and *L13A*. The RVH group displayed the expression profile of genes primarily involved in stress response, cell proliferation, and metabolism and included the up-regulation of *ADD2*. VSD cases exhibited a specific signature consisting of marked down-regulation (primarily in right atrium) of genes encoding ribosomal proteins *S11*, *L18A*, *L36*, *LP0*, *L31*, and *MRPS7* and genes involved in cell proliferation, differentiation, and apoptosis (e.g., *AMD1*, *RIPK3*, *EGLN1*, *SIAHBP1*, and *ARVCF*). Several ion channel genes, including *SLC26A8*, *SLC16A5*, *SLC4A7*, *KCNS2*, and *KCNN3*, were also found to be differentially expressed in patients with VSD.

CONGENITAL VASCULAR DEFECTS

Besides CHD, defective genes to account for congenital vascular disease have been also found. Molecular genetic defects have been found underlying the autosomal-dominant vasculopathies of Marfan syndrome, supravalvular aortic stenosis, and Williams' syndrome, indicating the critical role that microfibrils and extracellular matrix defects play in the pathogenesis and pathophysiology of these disorders.[188,189]

Marfan Syndrome

Marfan syndrome is an autosomal-dominant disorder of connective tissue and presents with abnormalities in the skeletal, ocular, and cardiovascular systems. Premature death may occur in Marfan syndrome, primarily because of progressive dilation of the aortic root, fatal aortic dissection, and aneurysm or aortic insufficiency. This syndrome is associated with high neonatal mortality because of polyvalvular involvement and subsequent severe CHF. Most cases of Marfan syndrome with cardiovascular involvement have mutations in the fibrillin (*FBN1*) gene, and more than 500 mutations have been identified. Affected individuals usually exhibit a distinct type of mutation,[190] and those with the same mutation may display marked clinical heterogeneity, making genotype-phenotype correlation a difficult task.

Fibrillin is a glycoprotein constituent of a multiprotein complex (including elastin) present in the microfibril component of the large-vessel vascular wall. It is encoded by a large complex gene containing 65 exons. Marfan syndrome with a variety of defects, such as aortic root dilatation, aortic dissection, and floppy mitral or aortic valve, shows characteristic histological findings (in valves and aorta) of disorganization and fragmentation of elastic fibers; these changes are similar to those found in non-Marfan dissected aorta and idiopathic floppy valves. Critical information about the role of fibrillin in microfibrillar assembly and vascular integrity has been largely derived from studies of animal models. Homologous gene-targeting experiments in the mouse demonstrated that *FBN1* disruption recapitulated the vascular effects of Marfan syndrome, primarily by effecting global tissue homeostasis rather than by disrupting elastic matrix assembly and cross-linking that have been suggested to occur in early embryogenesis.[190] Furthermore, studies performed on *mgR/mgR* mice, which contain a mutation resulting in the underexpression of *FBN1*, demonstrated that aortic dilation was due to failure by the microfibrillar array to sustain wall integrity in the face of hemodynamic stress. The resulting increase in wall stress is associated with localized calcium deposition, macrophage infiltration, and metalloproteinases release in the media, leading to fragmentation of the medial elastic network and aortic dilation.[191] These findings are similar to data collected in both animal models and clinical studies of patients with *FBN1* defects, showing that elastic fiber fragmentation occurs in association with up-regulation of the synthesis of metalloproteinases, and with increased susceptibility of the fiber to metalloproteinase activity.[192]

Neonatal Marfan syndrome, a more severe form of this disorder, is generally characterized by mitral and tricuspid valve involvement and aortic root dilatation, and patients often die of CHF during the first year of life. Although mutations in classical Marfan syndrome have been observed along the entire length of the *FBN1* gene, mutations in the more severe forms of Marfan syndrome, including the neonatal type, tend to cluster in a relatively small region of the *FBN1* gene, mostly between exons 24 and 34.[193,194] Furthermore, mutations leading to neonatal Marfan syndrome have been found in both the *FBN1* coding region (e.g., amino acid substitutions I1071S and E1073D) and in several specific splicing sites. An 11-year-old with an infantile onset of Marfan syndrome had a mutation in exon 28, resulting in a cysteine substitution.[194]

In patients with severe Marfan syndrome, substitutions at conserved cysteine residues (that disrupt disulfide bonding) particularly situated in the calcium-binding epidermal growth factor (cbEGF)-like domains (e.g., present in exon 28) are prevalent.[195] Other hot spots have been identified with missense mutations in exon 24–27 and splicing mutations causing skipping of exon 31 or 32.[196] In a recent case study of an infant with neonatal Marfan syndrome with pulmonic stenosis, AV valve insufficiency, and dilated aortic root, an intron mutation was found in a splicing region (IVS31-2A > G) in the *FBN1* gene.[197] In addition to the many missense mutations reported in *FBN1*, a number of frameshift mutations in *FBN1* have been identified, resulting in striking variations in the severity of the disease.[198] Nevertheless, it remains undetermined whether a dominant-negative effect of mutant *FBN1* proteins or a haploinsufficiency in *FBN1* is the cause of this disorder. Support for a dominant-negative model of *FBN1*-encoded proteins comes from several findings, including, a number of heterozygous mutations in patients with Marfan syndrome, the autosomal dominance of inheritance, the aggregation of multimeric fibrillin-1 molecules to form complex extracellular structures termed microfibrils, and the dramatic scarcity of extracellular microfibrils in patient-derived tissues (less than the 50% level that would be expected from a simple loss of contribution from the mutant allele). Some patients with nonsense *FBN1* alleles display a mild clinical picture (consistent with a dominant negative mechanism), whereas others have a severe disease.[198] Interestingly, mice engineered to overexpress a human transgene containing a mutant allele of *FBN1* (found in a number of patients with severe Marfan syndrome) did not display the diseased vascular phenotype in mice that contained normal endogenous wild-type *FBN1* genes, whereas an endogenous mutant *FBN1* allele mimicked the Marfan phenotype and could be rescued by introduction of a wild-type allele. These data suggest that in some cases, haploinsufficiency, rather than a specific mutation, can contribute to the Marfan phenotype.[199] The wide range of clinical presentations seen in Marfan syndrome are unlikely caused by either *FBN1* haploinsufficiency or by mutant alleles alone and suggest the existence of additional environmental and genotype-modified events in the progression of disease. Potential modifiers include the susceptibility of established microfibrillar matrices (and different mutant fibrillin-1peptides) to proteolytic degradation and the protective effect of calcium binding to EGF-like domains of fibrillin-1 shielding the wild-type protein from the activity of proteases.

Williams Syndrome

Mutations in the *ELN* gene encoding a component of the extracellular matrix (i.e., elastin) are responsible for familial supravalvular aortic stenosis (discrete narrowing of the ascending aorta) and stenoses of systemic and/or pulmonary arteries. Patients with Williams syndrome (WS) frequently have a deletion in chromosome 7,[200] spanning more than 20 genes, including the *ELN* gene. The constellation of behavioral, morphological, and developmental manifestations of WS seems to be a contiguous gene syndrome. In addition to the *ELN* gene, the *LIM-kinase 1* gene encoding a cytoskeletal regulatory protein is also located at 7q11.23.[201] Deletions in that gene may also contribute to the developmental and behavioral abnormalities seen in WS.

Friedreich Ataxia and Myotonic Dystrophy

In addition to point mutations in coding regions of specific genes, a number of inherited neuromuscular disorders that effect cardiac structure and function, referred to as Triplet Repeat syndromes, including, Friedreich ataxia (FA) and myotonic muscular dystrophy (MMD), are caused by expanded repeats of trinucleotide sequences within specific genes (e.g., the frataxin protein [*FRDA*] and myotonin protein kinase [*DMPK*]), respectively.[202,203] In both disorders, affected individuals exhibit severe cardiac abnormalities, including HCM, cardiac dysrhythmias, and conduction defects. In both disorders, the severity of the clinical phenotype correlates with the number of nucleotide repeats (i.e., > 200 repeats of GAA are found in affected individuals with FA, whereas > 50 copies of CTG are detected in cases with MMD).

FA is inherited as a recessive disorder characterized by progressive neurological disability and heart abnormalities. The *FRDA* gene located on chromosome 9q13 encodes a soluble mitochondrial protein of 18-kDa, frataxin, which is produced in insufficient amounts in the disease as a consequence of GAA triplet repeat expansion in the first intron of the gene.[202] The GAA repeat causes frataxin deficiency by interfering with the transcription elongation of the *FRDA* gene. Most patients (more than 98%) are homozygous for this repeat expansion. The size of the GAA expansion in the shorter of the two expanded alleles correlates with the clinical severity and is inversely related to age at onset.[204] The earlier FA develops, the more frequently left ventricular hypertrophy occurs.[205] However, the correlation between GAA repeat size and clinical phenotype is not invariable, suggesting that other genetic or environmental factors may significantly modify disease severity in FA.

Frataxin deficiency leads to excessive free radical production and dysfunction of Fe-S center containing mitochondrial enzymes (in particular respiratory complexes I, II, and III, and aconitase) as shown in Fig. 2, with effects on cardiac bioenergetic supply correlating with myocardial hypertrophy.[206] In humans, mutations in the *FRDA* gene can lead to progressive iron accumulation in heart and liver mitochondria.[204] In yeast, frataxin serves as a mitochondrial iron-binding protein that prevents this metal from participating in the Fenton reaction to generate toxic hydroxyl radicals. Further evidence that the generation of oxidative stress is coupled to *FRDA* mutations has emerged with the targeted use of antioxidants such

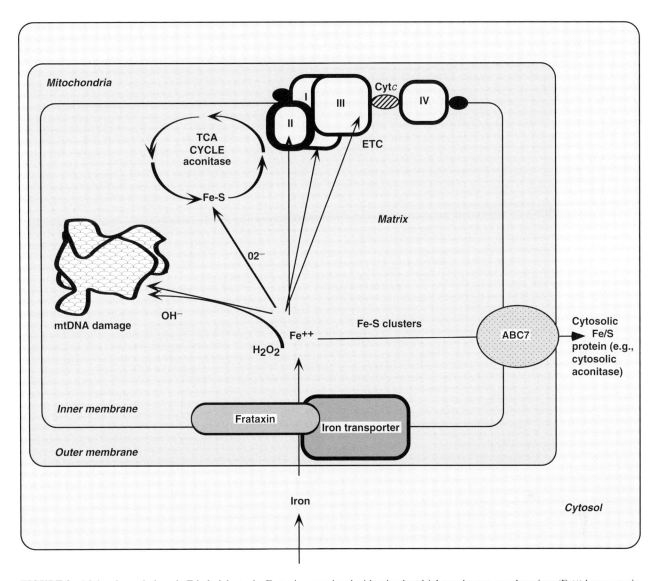

FIGURE 2 Molecular pathology in Friedreich ataxia. Frataxin, associated with mitochondrial membranes, regulates iron (Fe^{++}) homeostasis in this organelle. FRDA disruption results in the inactivation of mitochondrial iron-sulfur (Fe-S) proteins contained in complexes I, II, and III of the electron transport chain (ETC) involved in oxidative phosphorylation and in mitochondrial aconitase of the TCA cycle, as well as mitochondrial iron accumulation that leads to production of potent free radicals ($OH°$ and $O_2°-$) that, in turn, can lead to mtDNA damage. FRDA disruption also results in cytosolic Fe–S inactivation, as evidenced by cytosolic aconitase dysfunction. The protein ABC7 is thought to be involved in export of Fe–S clusters.

as idebenone to eliminate the cardiac hypertrophy accompanying FA.[207] Animal models with a range of phenotypes similar to FA are available, including mice containing null alleles for the *FRDA* gene encoding mitochondrial frataxin. *FRDA* knock-out mice die during embryonic life, further suggesting that frataxin is necessary for normal development.[208] Two conditional knock-out models for FA, a cardiac and a neuronal-cardiac model, recapitulate important pathophysiological features of the human disease. These models that develop severe cardiac dysfunction and HCM (without skeletal muscle involvement) also displayed intramitochondrial iron accumulation, decreased activity of respiratory complexes I and III, and also of the aconitases.[209] In addition, to further study the phenotypic effect of *FRDA* mutations, a transgenic mouse that contains the entire human *FRDA* gene within a YAC clone of 370 kb was generated.[210] When bred into a null *FRDA* background, human frataxin was expressed in the appropriate tissues at levels comparable to the endogenous mouse frataxin, correctly processed and imported into mitochondria, where it rescued the abnormal enzyme activities. These data showed that human frataxin could effectively substitute for endogenous murine frataxin in the *FRDA* null mutant.

CARDIAC DYSRHYTHMIAS

Although post-genomic analysis of cardiac dysrhythmias will be comprehensively discussed in Chapter 17, here we will briefly outline the genetic and metabolic factors that contribute to their etiology in infants and children, because cardiac dysrhythmias are a frequent complication of pediatric heart disease and, by themselves, can be a primary cause of sudden cardiac death (SCD). Mutations in specific genes encoding a variety of cardiac proteins involved in critical functions, such as ion transport, storage, and metabolism, have been identified as critical risk factors in the pathogenesis of lethal and nonlethal dysrhythmias in children (Table V). For instance, mutations in *SCN5A*, a gene that encodes the α-subunit of cardiac sodium channels responsible for initiating action potentials, are associated with prolongation of the QT interval or long QT syndrome (LQTS), which predisposes to syncope and SCD.[211,212] The phenotypic hallmark of LQTS is abnormal ventricular repolarization and can result in ventricular fibrillation, ventricular tachycardia (VT), cardiac conduction defects, and Brugada syndrome (BrS).[213,214] The BrS is a distinct form of idiopathic ventricular fibrillation that, in children, can be inherited as an autosomal-dominant trait with variable penetrance and can lead to SCD in healthy young individuals. Approximately 20% of cases have *SCN5A* mutations; however, the cause of the defect in most patients with BrS remains unknown.[215]

Mutations in four other genes (i.e., *HERG*, *KCNE1*, *KCNE2*, *KVLQT1*) involved in the formation of cardiac K+ channels have also been associated with the onset of LQTS and in atrial fibrillation.[216] These congenital defects resulting in LQTS are primarily autosomal dominant and are generally characterized by significant genetic heterogeneity and unpredictable expression of phenotypes with variable penetrance, even within single families. More than 500 mutations have been identified in the five genes encoding key ion channel subunits.[217] Swimming is a relatively genotype-specific dysrhythmogenic trigger for type 1 long-QT syndrome (LQT1).[218] In addition to LQTS, individuals with mutations in these genes can also be initially seen with other cardiac phenotypes. Families with hereditary short-QT syndrome and a high incidence of ventricular dysrhythmias and SCD were found to harbor mutations in the HERG (*KCNH2*) gene.[219]

A case of familial LQT interval and congenital sensorineural hearing loss was reported emphasizing the complex diagnostic and management implications. Jervell and Lange-Nielsen syndrome, an autosomal-recessive disorder that features LQT with recurrent attacks of syncope, is associated with SCD in children, who also display congenital sensorineural deafness. It is known to be associated with mutations of the genes *KCNQ1* (*KVLQT1*) and *KCNE1*.[220]

Polymorphic ventricular tachycardia (PVT) also called *torsade de pointes,* can result from mutations in genes encoding the K+ channel subunits described previously, as well as mutation in other genes, including heterozygous mutations in *KCNJ2* recently identified in two adolescents with PVT.[221] Anderson syndrome, a rare, inherited disorder characterized by LQT, ventricular dysrhythmias, periodic paralysis, and skeletal developmental abnormalities, is

TABLE V Mutations in Cardiac Dysrhythmias

Gene	Function	Cardiac phenotype
SCN5A also LQT3	Cardiac voltage-gated α-subunit sodium channel (I_{Na})	Dysrhythmia, VT and fibrillation, LQTS, BrS
HERG (KCNH2) also LQT2	Cardiac α-subunit of K+ channel (I_{Kr})	Dysrhythmias, LQTS
minK (KCNE1) also LQT5	Cardiac β-subunit of K+ channel (I_{Kr})	Dysrhythmias, LQTS
MiRP-1 (KCNE2) also LQT6	Cardiac β-subunit of K+ channel (I_{Kr})	Dysrhythmias, LQTS
KVLQT1 (KCNQ1) also LQT1	Cardiac α-subunit of K+ channel (I_{Kr})	Dysrhythmias, LQTS
LMNA (lamin A/C)	Nuclear envelope protein	Conduction defects, muscular dystrophy (Emery-Dreifuss)
RYR2 (RyR2)	Cardiac ryanodine receptor	VT, CPVT
KCNJ2 also LQT7	Inward rectifier K+ channel Kir2.1	PVT, Anderson syndrome
CASQ2	Calsequestrin	CPVT
PKP2 (Plakophilin-2)	Desmosomal protein	ARVC
JUP (Plakoglobin)	Desmosomal protein	ARVC
DSP (Desmoplakin)	Desmosomal protein	ARVC
DSG2 (Desmoglein-2)	Desmosomal protein	ARVC
TGFB3 (TGF-β3)	Transforming growth factor	ARVC

ARVC, Arrhythmogenic right ventricular dysplasia; BrS, Brugada syndrome; LQTS, long QT syndrome; CPVT, catecholaminergic polymorphic ventricular tachycardia; VT, ventricular tachycardia.

caused by mutations in *KCNJ2*, which encodes the inward rectifier K+ channel Kir2.1.[222]

Mutations in an assortment of membrane transporters operating at cellular loci other than the myocardial plasma membrane have been implicated in AV conduction defects, broadening substantially the concept of cardiac channelopathies. Missense mutations in the ryanodine-receptor calcium release channel (RyR2), involved in excitation–contraction coupling of the sarcomere, have been identified in stress-induced calcium overload cardiomyocytes leading to VT.[223]

Catecholaminergic polymorphic ventricular tachycardia (CPVT) is characterized by episodes of syncope, seizures, or SCD in response to physical activity or emotional stress and affects mainly young children with morphologically normal hearts. Mutations in RyR2 have been identified in autosomal-dominant pedigrees manifesting CPVT or exercise-induced PVT.[224,225] In addition, defects in the *CASQ2* gene encoding calsequestrin 2, another regulator of Ca++ storage and release in the cardiac myocyte, have been found that lead to an autosomal-recessive form of CPVT.[226]

Discrete mutations in the *LMNA* gene encoding the nuclear envelope proteins lamin A and lamin C are present in individuals affected with the autosomal-dominant form of Emery Dreifuss muscular dystrophy who display familial partial lipodystrophy, DCM, AV conduction defects, and atrial fibrillation.[227]

Severe cardiac dysrhythmias and conduction defects in the neonate can also arise from abnormal metabolism. Conduction disorders and atrial tachycardias have been reported in children with inborn errors of fatty acid oxidation (FAO). These include defects of long-chain fatty acid transport across the inner mitochondrial membrane (carnitine palmitoyltransferase type II deficiency and carnitine acylcarnitine translocase deficiency) and in patients with trifunctional protein deficiency. VT, although fortunately rare, may have significant deleterious effect on cardiac function and may be followed by ventricular fibrillation. Dysrhythmias have been reported in children with any type of FAO deficiency.[228] The accumulation of intermediary metabolites of fatty acids, such as long-chain acylcarnitines, arising from the dysfunction of fatty acid metabolism incurred by mutations in carnitine transport and FAO pathways can lead to dysrhythmias. In particular, FAO disorders often are first seen in infancy with myocardial dysfunction and dysrhythmias after exposure to stresses such as fasting, exercise, or recurrent viral illness. Infants and young children are particularly vulnerable to fasting and defective FAO, with rapid fatal consequences. In children, FAO defects, which generally are autosomal recessively inherited, also encompass a spectrum of clinical disorders, including recurrent hypoglycemic, hypoketotic encephalopathy, or Reye-like syndrome with secondary seizures and potential developmental delay, progressive lipid storage myopathy, recurrent myoglobinuria, neuropathy, and progressive cardiomyopathy. Metabolic defects attributed to mutations in the genes involved in the FAO pathway, including several mutations in the *ACADM* gene encoding the medium-chain acyl-CoA dehydrogenase (MCAD), can lead to sudden death in infants.[229,230] Both defective FAO genes and polymorphic variants have been closely examined in cases of sudden infant death syndrome (SIDS). The prevailing view is that any single mutation or polymorphism is unlikely to be the predisposing factor in all SIDS cases; rather, it has been proposed that there are "SIDS genes" operating as polygenic factors, predisposing infants to SIDS in combination with environmental risk factors.[231]

A familial type of Wolff-Parkinson-White (WPW) syndrome, characterized by ventricular preexcitation and tachy-dysrhythmias, progressive conduction system disease, and cardiac hypertrophy, was found to be due to mutations in the *PRKAG2* gene encoding the γ-2 regulatory subunit of AMP-activated protein kinase (AMPK), a key regulator of the glucose metabolic pathway in skeletal muscle and heart and cellular sensor of bioenergetic need.[232]

Another disorder, arrhythmogenic right ventricular cardiomyopathy (ARVC), a familial heart muscle disease characterized by structural, electrical, and pathological abnormalities of the right ventricle, primarily is seen with dysrhythmia, HF, and SCD and has a genetic basis.[233] Although in most families, ARVC shows autosomal-dominant inheritance with incomplete penetrance, autosomal-recessive ARVC has also been described.[233,234] In association with ARVC, defective genes have been identified that encode proteins of desmosomes, subcellular protein complexes that anchor intermediate filaments to the cytoplasmic membrane in adjoining cells. These include mutations in the plakoglobin,[235] desmoplakin,[236] and plakophilin[237] genes. Molecular defects in regulatory regions of the *TGFB3* gene encoding TGFβ3 have also been recently linked to ARVC in several families.[238] In addition, *RYR2* mutations have been identified in four independent families with ARVC.[239] These reported mutations in *RYR2* cluster in two highly conserved regions. By use of a quantitative yeast two-hybrid system, it was demonstrated that these ARVC-associated point mutations in *RYR2* markedly influence the binding of RyR2-calcium channel to its gating protein FKBP12.6, resulting in increasing the RyR2-mediated calcium release to cytoplasm.[240]

CARDIOMYOPATHIES

A large proportion of the clinical cases with HCM, conservatively estimated at more than 60%, are familial, primarily autosomal dominant. Familial occurrence of DCM, mostly as an autosomal dominant trait, is more common than generally believed and is responsible for 20–30% of all cases of DCM.[241] In addition, a variety of cardiomyopathies, including but not limited to DCM and HCM disorders, have been shown to occur in the young, and frequently have displayed genetic/familial components. Mutations causing human

TABLE VI Genetic Defects in Pediatric Cardiomyopathy

Gene product affected (gene locus)	Cardiac phenotype	Reference
Structural/contractile proteins		
β-Myosin heavy chain (β-MHC/MYH7)	HCM	248
α-Myosin heavy chain (α-MHC/MYH6)	HCM	247
Regulatory myosin light chain (MYL2)	HCM	381
Actin (ACTC)	DCM, HCM	249, 262
α-Tropomyosin (TPM1)	HCM	245
Cardiac troponin T (TNNT2)	HCM	245
Cardiac troponin I (TNNI3)	HCM	246
Desmin (DES)	DCM	265
δ-Sarcoglycan (SGCD)	DCM	264
Myosin-binding protein c (MYBPC)	HCM	244
Titin (TTN)	HCM	250
Dystrophin (DMD)	DCM (Duchenne/Becker muscular dystrophy)	260, 266
Metabolism and bioenergetics		
Mitochondrial trifunctional protein (HADHA)	Dysrhythmias, DCM	277–279
Carnitine palmitoyl transferase II (CPT2)	Dysrhythmias, SCD, CM	293–295
Carnitine-acylcarnitine translocase (CACT/SLC25A20)	Dysrhythmias, SCD, CM	289–291
Carnitine transport (OCTN2/SLC22A5)	HCM, DCM	286–288
Tafazzin (TAZ/G4.5)	DCM (Barth)	261
Mitochondrial Fe^{++} metabolism-frataxin (FRDA)	HCM (Friedreich ataxia)	202, 204
Very-long-chain acyl-CoA dehydrogenase (ACADVL)	HCM, SCD	276
Lysosomal α-glucosidase, acid maltase/glycogen storage (GAA)	Ventricular pre-excitation; HCM (Pompe's disease)	382–383
Glycogen-debranching enzyme (AGL)	HCM (Cori's disease)	384–385
α-Galactosidase (GLA)	HCM (Fabry's disease)	386–387
Mitochondrial heme metabolism (COX15)	Early-onset fatal HCM	255
γ-2 Subunit of AMP-activated protein kinase (PRKAG2)	HCM, conduction defects (WPW)	232, 258
Mitochondrial DNA		
$tRNA^{leu}$, $tRNA^{lys}$, $tRNA^{ile}$, $tRNA^{gly}$	HCM (MELAS, MERRF)	306–309
ATPase6	HCM (Leigh)	388
Sporadic mtDNA deletions	Conduction defects (KSS)	317

HCM, hypertrophy cardiomyopathy; DCM, dilated cardiomyopathy; SCD, sudden cardiac death; WPW, Wolff-Parkinson-White; KSS, Kearns–Sayre syndrome; MELAS, mitochondrial myopathy, encephalopathy, lactic acidosis, and strokelike episodes; MERRF, myoclonic epilepsy and ragged red fibers.

cardiomyopathies have been identified in a broad spectrum of nuclear genes encoding myocardial contractile proteins and structural proteins involved in sarcomeric structure, enzymes involved in glycogen storage (Pompe's and Cori's diseases), and mucopolysaccharide degradation (Fabry disease), lipid metabolism (FAO and carnitine deficiency), and in both nuclear and mitochondrial DNA (mtDNA) encoded genes essential for cardiac energy production (Table VI).

Studies directed at explaining the genetic basis of HCM have particular urgency, because HCM represents the most frequent cause of SCD in children and adolescents.[1] Most cases of familial HCM exhibit a pattern of autosomal-dominant transmission (the exception being those cases of pathogenic mtDNA mutations that are maternally inherited). Mutations causing HCM have been found in 11 genes encoding different sarcomeric proteins, including β-myosin heavy chain (*β-MHC* or *MYH7*), α-myosin heavy chain (*α-MHC* or *MYH6*), myosin-binding protein C (*MYBPC3*), cardiac troponin T (*TNNT2*), cardiac troponin C (*TNNC1*), cardiac troponin I (*TNNI3*), α-tropomyosin (*TPM1*), regulatory myosin light chains (*MYL2*), titin (*TTN*), and cardiac α-actin (*ACTC*).[242–250] Comprehensive screening studies of mutant alleles of sarcomere protein genes in diverse patient populations indicate that *MYBPC3*, *TNNT2*, and *MYH7* mutations are the most frequent mutant genes in HCM, containing more than 75% of the mutations identified.[251,252]

Other genetic loci affecting cardiac function lying outside the sarcomere have been implicated in HCM. Recently, a familial X-linked form of HCM (which frequently is accompanied by skeletal myopathy and WPW syndrome) occurred as a result of an X-linked lysosomal storage disorder (termed Danon disorder) caused by deficiency of lysosome-associated membrane protein-2 (*LAMP2*).[253,254] In addition, specific defects in genes involved in mitochondrial heme and Fe^{++} metabolism (e.g., nuclear encoded *FRDA* and *COX15*)[255,256] and in mitochondrial bioenergetics (e.g., mutations in mtDNA-encoded tRNAs and ATPase6)[257] have been detected in patients with HCM (albeit more rarely than the sarcomeric protein mutations). As previously noted, mutations in *PRKAG2* encoding the regulatory subunit of AMP-activated protein kinase (AMPK), a pivotal mediator of cellular energy metabolism resulting in glycogen storage disease, are present in a subset of cases of HCM.[258] Taken together, these findings indicate that cardiac mitochondrial energy depletion can be an underlying cause of HCM in some patients, rather than depressed sarcomeric contraction, and could be helpful in understanding a number of clinical observations in HCM such as its heterogeneity, variable onset, severity, and hypertrophic asymmetry.

In a large-scale population-based study of pediatric cardiomyopathy, DCM made up 51% of the cases, whereas HCM represented 42% of cases.[259] Several defective genes have been identified in association with DCM. Genes for X-linked familial DCM (*DMD, TAZ/G4.5*) have been identified,[260,261] and several genes for the autosomal-dominant form of DCM (*ACTC, DES, SGCD*) have been reported.[262-265] In cases of X-linked DCM attributed to defect in the *DMD* gene encoding dystrophin (a large cytoskeletal protein associated with the sarcolemma), the defect is manifested only in cardiac myocytes; the site of the mutation is primarily located in the *DMD* promoter regulatory region, which is consistent with its tissue-specific expression.[266] Mutations in the *DMD* gene can also lead to both Duchenne (DMD) and Becker (BMD) muscular dystrophies, affecting both skeletal and cardiac function.[260] Typically, patients with the more severe DMD lack detectable dystrophin protein in skeletal muscles caused by the presence of either deletion mutations in *DMD* alleles that disrupt the translational reading frame or specific point mutations that create stop codons. Male patients with X-linked DCM (because of dystrophin defect) tend to be asymptomatic in early childhood and have syncope and rapidly progressive congestive HF in late adolescence; affected females generally display a later onset. In DMD, skeletal muscle weakness is present at an early age (3–6 y). Subsequently, more than 30% of the patients have signs of cardiac dysfunction by age 14, and virtually all DMD patients have DCM by age 18.

Barth syndrome, an X-linked cardioskeletal myopathy with neutropenia and DCM, often is initially seen in infancy. The protein tafazzin responsible for Barth syndrome is encoded by the *TAZ* (also known as *G4.5*) gene and belongs to a family of acyltransferases involved in phospholipid synthesis.[261,267] In patients harboring *TAZ* mutations, saturated fatty acid levels increase, whereas unsaturated fatty acid and cardiolipin levels are markedly reduced, affecting cardiac membrane fluidity and function.

Arrhythmogenic right ventricular dysplasia (ARVD or ARVC) is a form of cardiomyopathy characterized by progressive degeneration and replacement of the right ventricular myocardium by adipose and fibrous tissue, dysrhythmias, and increased risk of SCD. Initial genetic linkage analysis demonstrated an association to *loci* on several chromosomes including, 2, 10, 14, and 17.[268] Molecular analysis of the genetic *locus* mapped to chromosome 14q24.3 resulted in the identification of a nucleotide substitution (-36G > A) in the 5′ untranslated region of the *TGFB3* gene, which was invariably associated with the typical ARVC clinical phenotype in the affected family members.[238] Another mutation residing in the 3′ untranslated region of *TGFB3* was identified in an unrelated patient with ARVC. Other genetic defects have been found in genes encoding desmosome proteins as previously discussed[235-237] and in the *RYR2* gene encoding the cardiac ryanodine receptor (RyR2).[239]

Another type of congenital pediatric cardiomyopathy, isolated non-compaction of left ventricular myocardium (INLVM) has been found in association with mitochondrial myopathy.[269] This disorder seems to be genetically heterogenous, linked to defects at a variety of genetic *loci*, including X-linked *TAZ*. Autosomal-dominant INLVM has recently been mapped to an unidentified gene at chromosome 11p15.[270]

FATTY ACID OXIDATION AND CARDIOMYOPATHY

Specific inborn deficiencies in fatty acid metabolism are associated with cardiomyopathy and HF. In Table VII, we present a list of disorders with defective fatty acid metabolism that may result in cardiomyopathy or HF and their

TABLE VII Fatty Acid Metabolism Defects in Cardiomyopathy and HF

Fatty acid metabolism disorders	Affected loci
CPT-II deficiency	*CPT2*
Barth syndrome	*TAZ/G4.5*
SCAD deficiency	SCAD (*ACADS*)
MTP deficiency (includes LCHAD defect)	MTPα (*HADHA*) + MTPβ (*HADHB*) subunits
VLCAD deficiency	VLCAD (*ACADVL*)
CPT-I deficiency	L-CPT-I (*CPT1A*)
Carnitine transport	OCTN2 (*SLC22A5*)
Carnitine translocase deficiency	CACT (*SLC25A20*)
MCAD deficiency	MCAD (*ACADM*)
ETF deficiency	*ETF*
ETF dehydrogenase deficiency	*ETFDH*
FAT deficiency	*CD36*

characterized genetic *loci*. Heritable mitochondrial defects in acyl-CoA dehydrogenase have been reported in cardiomyopathy and HF.[271] Generally, defects in the oxidation of long-chain fatty acids (LCFAs) are more likely to cause cardiomyopathy than defects in medium-chain or short-chain fatty acids. Specific defects in enzymes involved in the oxidation of short-chain, long-chain, and very long-chain fatty acids have been identified.[272–275]

Severe infantile cardiomyopathy is the most common clinical phenotype of very long chain acyl dehydrogenase (VLCAD) deficiency and is found in more than 67% of cases, often resulting in sudden death.[274] Mutation analysis of the *ACADVL* gene revealed a large number of different mutant *loci* (21 in 19 patients) with few repeated mutations.[276]

Mitochondrial trifunctional protein (MTP) is a complex protein that catalyzes the last three steps of LCFA oxidation. MTP defects have emerged recently as important inborn errors of metabolism that can cause cardiomyopathy. MTP deficiency has been classified into two different biochemical phenotypes.[277] In one, both α- and β-subunits are present, and only the 3-hydroxyacyl-CoA dehydrogenase (LCHAD) activity is affected. The most common mutation associated with MTP deficiency (G1528C) is associated with this phenotype. The second phenotype that is frequently present in patients with neonatal cardiomyopathy is characterized by an absence of both subunits and the complete lack of all three MTP enzymatic activities.[278] This phenotype can be caused by mutations localized to the 5′ donor splicing site of the β-subunit gene, which can result in the entire loss of an exon in the mRNA (exon 3). A mutation (976G->C) within the β-subunit of MTP (*HADHB*) has been found, which destabilizes the protein leading to complete deficiency and HF.[279] This mutation caused the loss of all three MTP enzymatic activities and near complete loss of the protein, as assayed by Western immunoblot analysis. Both DNA and enzymatic testing can be performed in fetal screening of this, often devastating, disease.[280]

Defects in malonyl CoA decarboxylase caused by mutations in the *MLYCD* gene can also lead to cardiomyopathy and neonatal death.[281] Also, as previously noted, HCM results from specific mutations in *PRKAG2* gene encoding the regulatory subunit of AMPK, a critical bioenergetic sensor that activates FAO by adjusting malonyl CoA levels and increasing glucose uptake, particularly during chronic low-energy states, as those that occur in cardiac hypertrophy and failure.[282]

Carnitine deficiency has been frequently associated with severe cardiomyopathy. Mutations in proteins that participate in carnitine transport and metabolism may cause either DCM or HCM as a recessive trait.[283–285] One of the genetic *loci* affected in carnitine-associated cardiac pathology encodes the plasma membrane–localized carrier that transports carnitine into the cell, and its deficiency has been described as primary carnitine deficiency.[286] This transport deficiency is due to specific defects in the *SLC22A5* gene encoding the plasma membrane–localized organic cation/carnitine transporter OCTN2.[287] Recent studies have described the involvement of homozygous mutations in *SLC22A5* in children with severe cardiomyopathy and in a number of sudden infant deaths.[288] The carnitine deficiency caused by defects at this *locus*, and its resultant pathology, which can prove lethal in childhood, can be dramatically reversed by intake of a high dosage of oral carnitine supplementation.[289] Defects in a second *locus*, *SLC25A20* encoding the mitochondrial membrane localized carnitine-acylcarnitine translocase (CATC), also lead to autosomal-recessive carnitine deficiency, cardiomyopathy, and HF,[290] and carnitine supplementation has been effective in some cases.[291]

Although generally not found in association with cardiomyopathy or HF, recent observations suggest that CPT-I deficiency can result in cardiac pathology.[292] In contrast, there is a consensus that deficiencies in CPT-II (specifically infantile CPT-II deficiency), an autosomal-recessive disorder, is associated with cardiac dysfunction and sudden death.[293] Infantile CPT-II deficiency has been associated with several *CPT2* mutations, including a homozygous mutation at A2399C causing a Tyr→Ser substitution at residue 628. This mutation produces a marked decrease in CPT-II activity in fibroblasts.[294] Another mutation has been found at C1992T predicting an Arg→Cys substitution at residue 631, which is associated with drastic reduction of CPT-II catalytic activity.[295]

Other mutations in fatty acid metabolism have been reported in association with cardiomyopathy in children. Reduced myocardial uptake of LCFA, a preferential energy substrate used by the heart, has been demonstrated in patients with DCM and HCM.[296] Myocardial uptake of LCFAs occurs by means of a specific fatty acid translocase, which is homologous with human CD36, an integral membrane protein with multiple ligands, and is highly expressed in the blood vessel walls and the heart. In addition to its role as a crucial transporter for LCFA, CD36 seems to play an important role as a major scavenger receptor for oxidized low-density lipoproteins.[297,298] CD36 deficiency has been linked to the phenotypic expression of the metabolic syndrome, often associated with atherosclerotic and diabetic cardiomyopathy.[299,300] Given the central importance of LCFAs in states such as fasting, infection, stress, and vigorous exercise, it is possible that defects in the transport of LCFAs, resulting from CD36 deficiency, may act as a contributory factor in disrupting the normal cardiac adaptation to these stressors and increase susceptibility to heart dysfunction in children.[301]

Electrons are transferred from acyl CoA dehydrogenases by means of the electron-transfer flavoprotein (ETF), ETF dehydrogenase, and ubiquinone (or coenzyme Q) to the mitochondrial respiratory enzymes. Individuals with deficiencies in the ETF pathway display impaired FAO and abnormal intramitochondrial accumulation of fatty acids and glutaric acid and may develop a fatal cardiomyopathy.[302,303]

Numerous studies have substantiated critical roles for the global transcription regulators PPAR and PGC-1 in

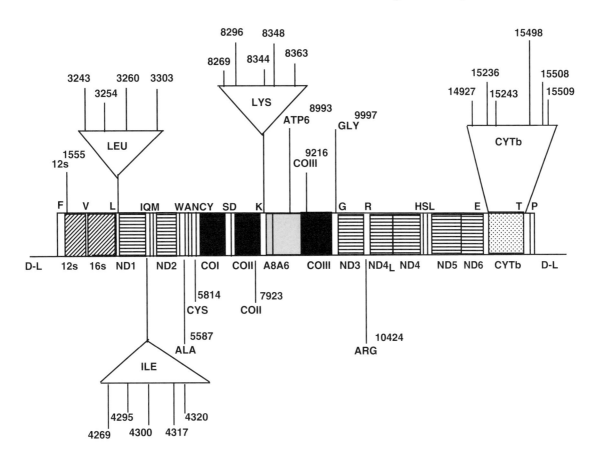

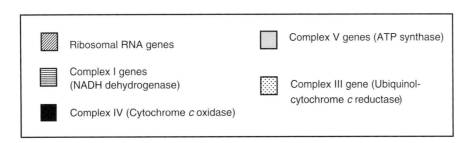

FIGURE 3   Pathogenic mitochondrial DNA mutations associated with pediatric cardiac disease. A linear representation of the circular 16,569 base pair human mtDNA molecule showing the relative location of all 13 protein-encoding genes (ND1-ND6, COI-COIII, cytb, and ATP6 and ATP8), 22 tRNAs identified by their cognate amino acid using single letter code (F, V, L, Y, Q, M, W, A, N, C, Y, S, D, K, G, R, H, S, L, E, T, P), the two rRNA genes (12s and 16s), and the noncoding D-loop region (D-L). The nucleotide location and genes containing pathogenic mtDNA mutations associated with pediatric cardiac disease are indicated as shown.

controlling fatty acid metabolism and for their modulation in ischemia and HF, a topic that will be comprehensively discussed in later chapters. However, to the best of our knowledge, defects in these factors, and specific mutations, have not yet been reported in association with HF or cardiomyopathy in children.

MITOCHONDRIAL CARDIOMYOPATHY

Cardiomyopathy in neonates and children can result from underlying deficiencies of energy production because of both genetic and sporadic defects at a wide spectrum of loci.[257] Genetic disorders in energy metabolism leading to either specific FAO/carnitine deficiencies or OXPHOS/respiration abnormalities may also result in either a DCM or a HCM phenotype, leading to the so called mitochondrial cardiomyopathy (MCM), which is characterized by abnormal cardiac mitochondria either in number, structure, or function. Furthermore, a number of MCMs have been described in association with neurological disorders and multisystemic defects, including MELAS (mitochondrial myopathy, encephalopathy, lactic acidosis and strokelike episodes), MERRF (myoclonic epilepsy and ragged red fibers), and Leigh syndromes, with specific pathogenic mutations identified in several mtDNA genes needed for mitochondrial function (Fig. 3)[304] and, more recently in nuclear genes involved in the assembly of mitochondrial respiratory complexes.[305] These disorders may be seen early in childhood, whereas others manifest themselves later. Mitochondrial enzyme and DNA defects have also been noted in cases of fatal infantile cardiomyopathy.[306–309] Molecular studies of patients with either HCM or DCM have resulted in the further identification of novel pathogenic mtDNA mutations prevalent in cardiac tissues.[310,311]

MCM can also arise in a sporadic fashion. Agents that cause damage to cardiac mitochondria and mtDNA, such as adriamycin (also doxorubicin) and alcohol, can result in cardiomyopathy.[312,313] Somatically generated (sporadic) deletion/mutations in cardiac mtDNA have been shown to increase during myocardial ischemia,[314] and their increased presence (although in low overall abundance) has been reported in the cardiomyopathic heart, probably arising from increased oxidative stress.[315,316] In addition, Kearns-Sayre syndrome (KSS), a neuromuscular disorder with AV conduction defects and cardiomyopathy, is commonly associated with abundant large-scale mtDNA deletions whose generation is thought to arise spontaneously, because they are rarely detected in mothers or siblings.[317]

In contrast, DCM associated with multiple, abundant mtDNA deletions have been reported as a distinct phenotype caused by genetic nuclear DNA defects either dominantly or recessively inherited.[318] Linkage analysis in families with DCM revealed abundant mtDNA deletions in affected individuals in association with specific mutations in proteins that participate in mtDNA replication (e.g., mtDNA polymerase γ gene [Pol] and the *Twinkle* gene, a putative mitochondrial helicase) and in mitochondrial nucleotide metabolism (e.g., adenine nucleotide translocator [*ANT*]).[319] Interestingly, mutations in *ANT* were found in patients with HCM (and skeletal myopathy) harboring markedly increased levels of mtDNA deletions in skeletal muscle biopsy samples.[320]

Depletion in cardiac mtDNA levels has also been reported in young children with isolated cardiomyopathy, either DCM or HCM.[321] Recently, several nuclear *loci* have been identified as likely responsible for mtDNA depletion, a phenotype that is rarely assessed. Autosomal-recessive mutations in factors that play a role in mitochondrial nucleotide metabolism (e.g., thymidine kinase 2, thymidine phosphorylase, and deoxyguanosine kinase) have been found in a subset of patients (and their families) with mtDNA depletion.[319] In addition, depletion of cardiac mtDNA levels can be specifically induced by zidovudine (AZT), which inhibits both the viral DNA polymerase and mitochondrial DNA polymerase (polγ), thereby stopping mtDNA replication.[322] On the other hand, recent observations do not support the association of AZT treatment with the development of cardiomyopathy in infants or children.[323]

Other disorders presenting with cardiomyopathy have been linked to mitochondrial defects. Congenital HCM and defective OXPHOS, in association with infantile cataracts and mitochondrial myopathy, have been described in children with Sengers syndrome.[324,325] Although patients with Sengers syndrome exhibit reduced levels of ANT in skeletal muscle and abnormal mitochondrial structure,[324–326] the underlying genetic defect is not known. Also, patients with Alpers syndrome display HCM in association with reduced mitochondrial respiration, including cytochrome *c* oxidase (COX) deficiency and extensive mtDNA depletion.[327] Moreover, pronounced mtDNA polymerase γ deficiency[328] and specific lesions in the Polγ gene (*POLG*) have been found in patients with Alpers syndrome.[329]

Wilson's disease, an autosomal-recessive disorder of copper homeostasis, is characterized by abnormal accumulation of copper in several tissues and has been reported to display a range of cardiac abnormalities, including HCM, supraventricular tachycardia, and autonomic dysfunction.[330,331] The disease-associated gene *ATP7B* encodes a copper-transporting P-type ATPase, the WND protein. Excessive accumulation of copper is particularly toxic to mitochondria, resulting in increased ROS formation, inhibition of mitochondrial dehydrogenases (e.g., pyruvate dehydrogenase), and reduced mitochondrial respiratory activitiy.[332–334]

CELLULAR TARGETS IN GENETIC-BASED CARDIAC DEFECTS

A broad range of structural and functional cardiac phenotypes, including cardiomyopathy, conduction defects, and dysrhythmias, may arise from genetic defects in a diverse set of molecular targets within the cardiomyocyte. Specific targets have been localized in a plurality of subcellular compartments, including the nucleus, mitochondria, lysosome, cytoplasm, endoplasmic reticulum, and plasma membrane (Fig. 4). In addition, these molecules (whether receptors, enzymes, channels, or kinases) often play multiple roles in interacting signaling pathways involved in the cell cycle, metabolic, developmental, and physiological transitions. The close intersection and communication signaling between these diverse pathways has made the unraveling of cardiac events highly informative, although arguably more complex, and has important ramifications for treatments focused on any specific target.

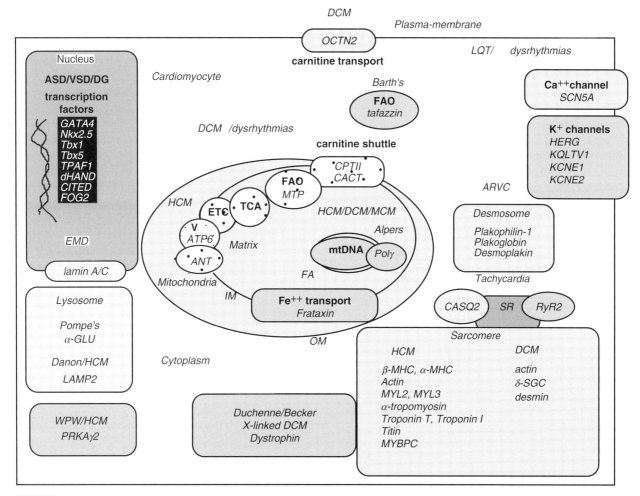

FIGURE 4 Subcellular location within the cardiomyocyte of defective gene products found in pediatric cardiac disease. Organelles and subcellular complexes such as nucleus, lysosome, mitochondria, sarcomere/sarcomeric reticulum (SR), desmosome, and plasma membrane sites are shown. The subcompartments of the mitochondria are indicated, including matrix, outer membrane (OM), and inner membrane (IM). Specific cardiac defects associated with specific cardiomyocyte-localized defects are indicated, including atrial and ventricular septal defect (ASD/VSD), ARVC, arrhythmogenic right ventricular cardiomyopathy or dysplasia, DiGeorge syndrome (DG), Barth's syndrome, Pompe's disease, Danon syndrome with HCM, Emery–Dreifuss muscular dystrophy (EMD), Duchenne/Becker muscular dystrophy, Wolff–Parkinson–White syndrome with HCM, Friedreich ataxia (FA), tachycardia, LQT syndrome, dysrhythmias, HCM, DCM and MCM (hypertrophic, dilated and mitochondrial cardiomyopathy). Functional pathways are indicated in smaller plain font, including transcription factors, membrane transporters, shuttles and channels, sarcomeric proteins, fatty acid oxidation (FAO), citric acid cycle (TCA), electron transport chain (ETC) with associated complex V (V), and adenine nucleotide translocator (ANT). Specific proteins/gene products involved are displayed in smallest plain italic font. Abbreviations used are as described in Figure 1.

HEART FAILURE IN CHILDREN

Although the topic of HF will be mainly discussed in Chapter 15, several comments can be added here in reference to specifically pediatric HF. In pediatrics, a large number of heterogeneous conditions cause HF. More common in neonates than older children, or adults, is HF caused by CHD. On the other hand, DCM and HCM can cause HF in neonates, infants, and children with a pathophysiology similar to adults. Neonates and young infants may be more susceptible to a number of factors triggering HF and sudden death, including metabolic stimuli, stresses, and acquired pathogens. However, in defining treatment options, more information is needed concerning the differential extent and susceptibility of infants and children to left ventricular dysfunction and cardiac dysrhythmias associated with HF, stemming from either genetic-based or metabolic defects and also from acquired pathogens. Specific indications for heart transplant in children may include metabolic and genetic forms of cardiomyopathy, as well structural CHD. Before transplant, screening should include a metabolic workup (to rule out mitochondrial disorders) and, thorough genetic studies, if indicated, by phenotypic appearance or family pedigree.

ACQUIRED CARDIAC DISEASES

Acquired diseases in neonates and children include Kawasaki disease, acute and chronic rheumatic cardiac disease, infective endocarditis, and myocarditis. Molecular genetic technology has been applied in a limited fashion in the analysis of these diseases and could improve the clinical diagnosis.

Kawasaki disease, an acute self-limited vasculitis of infancy and early childhood, is the leading cause of acquired heart disease in children in the United States and Japan.[335] Its etiology remains unknown, and although extensive molecular analysis has thus far been unable to detect viral or bacterial involvement, numerous clinical and epidemiological features strongly suggest an infectious cause. A self-limited, generally nonrecurring illness could be the trigger to produce clinically evident disease only in certain genetically predisposed individuals.[336] The rareness of disease occurrence in the first few months of life and in adults suggests an infectious agent for which older individuals have immunity and from which young infants are protected by passive maternal antibodies. Differing genetic susceptibility might explain the low levels of person-to-person transmission.

Although such genetic susceptibility factors to Kawasaki disease have not yet been fully identified, there are several reasons to believe that they exist. Although occurring in infants and children of all races, Kawasaki disease is markedly more prevalent in Japan and in children of Japanese ancestry, with an annual incidence of 112 cases per 100,000 children younger than 5 years of age.[337] A low but significant proportion of patients have a positive family history, including parents who themselves had the illness in childhood.[338,339] In certain populations (e.g., Japanese), rates of occurrence in siblings of infected subjects is 10-fold higher than in control populations; moreover, the risk of occurrence in twins is 13%.[340] These findings suggest a genetic predisposition that interacts with exposure to the etiologic agent or environmental factors.

Furthermore, a number of striking immune perturbations occur in acute Kawasaki disease and in the pathways leading to coronary arteritis, including marked cytokine cascade and endothelial cell activation, monocyte/macrophages recruitment, and a cytotoxic lymphocyte response.[341,342] Matrix metalloproteinases capable of damaging arterial wall integrity have been implicated in the development of coronary artery aneurysmal dilatation.[343] In addition, VEGF, monocyte chemotactic and activating factor (MCAF or MCP-1), tumor necrosis factor-α, and various interleukins also seem to play a contributory role in the vasculitis.[344–347] Although Kawasaki disease is a generalized systemic vasculitis involving blood vessels throughout the body, if untreated, 25% of children have aneurysms of the main coronary arteries, with potentially fatal consequences. The coronary arteries are virtually always involved in autopsy cases. Because treatment is generally only effective if administered within the first 10 days of illness (to prevent coronary artery involvement), it poses a diagnostic challenge for the pediatric cardiologist who must distinguish Kawasaki disease from other diseases within a relatively limited time frame. Although intravascular ultrasound has the promise of improving assessment of coronary arteries, molecular markers of the disease potentially identifiable by microarray analysis could prove a valuable asset in the diagnosis.

Viral and bacterial infections have been implicated in a wide range of CVDs that affect infants and children, including diseases such as myocarditis, DCM, endocardial fibroelastosis, and atherosclerosis. It is important to note that in a number of these diseases the infective agent has been shown to be causal in the development of the disease, whereas in other cases, only the presence of the infectious agent in the affected tissues has been established and, although potentially contributing to the pathogenesis, may represent an effect of the disease. Nevertheless, the persistence of cardiotropic viral infection can be associated with disease deterioration, progressive cardiac dysfunction, and poor prognosis.[348,349]

Myocarditis, a common inflammatory disease of the myocardium in children and young adults, can be associated with cardiac dysfunction and the development of DCM. Viral infection is a common etiology or trigger of this disorder. Immunological and molecular analysis have evolved to enable the sensitive quantitative detection of viral genomes in the myocardium in diagnosing viral myocarditis and have implicated a large variety of viruses in the pathogenesis of pediatric myocarditis.[348] The most frequent cardiotropic viruses associated with myocarditis, as detected on analysis of tissue obtained by endomyocardial biopsy, are enteroviruses, particularly the coxsackieviruses group B (CVB), parvovirus B19, adenoviruses, cytomegalovirus, and less frequently Epstein–Barr virus, and influenza virus.[350] Molecular screening by polymerase chain reaction (PCR) has proved to be a useful and highly sensitive tool for detecting and quantitating viral levels in the heart. The technique is highly adaptable, can be used in the screening of paraffin-embedded tissues,[351] and is able to determine viral load and copy number by using real-time PCR.[352] Such studies have shown a significantly increased presence of adenovirus and parvovirus B19 (PVB19) in patients with myocarditis.[353,354] Moreover, increased levels of PVB19 have been detected in the heart of children with CHD,[355] sudden infant death,[351] end-stage HF,[356] and with cardiac allograft rejection.[357] Despite concerns over the specificity of this technique, most of which can be addressed with appropriate controls, the use of PCR can play a powerful role in virus detection and has provided a new dimension to screen the viral etiology of myocarditis and DCM. Furthermore, PCR enhances the ability to detect the state of viral replicative activity by demonstrating the presence of enteroviral minus-strand RNA.[358]

A major determinant of the ability of viruses to infect the heart is the viral receptor (which has myocyte selective expression). Both coxsackie and adenovirus, two of the most common viral agents associated with acquired myocarditis and DCM, share a host receptor, the coxsackie-adenovirus receptor (CAR). This receptor mediates cell attachment and entry of cardiotropic viruses[359] working in collaboration with a second receptor, the decay-accelerating factor (DAF). CAR is predominantly expressed in intercalated discs, whereas DAF is abundantly expressed in epithelial and endothelial cells.[360] CAR has been shown to be up-regulated in patients with DCM, potentially making the expression of this protein a marker of susceptibility to virus infection.[361] Recombinant expression of human *CAR* in cardiomyocytes strongly enhanced their virus uptake rate, suggesting that CAR induction enhances cardiac vulnerability to viral disease, whereas healthy myocardium is relatively resistant to CAR-dependent viruses.[362] The notion that viral-host receptor induction may significantly aggravate the clinical course of viral heart disease opens new therapeutic possibilities by using the blockade of receptor expression or receptor-virus interactions. Similarly, identification of the host receptors for other potentially cardiotropic viruses (e.g., cytomegalovirus, Epstein–Barr virus, HIV, PVB19) and analysis of their expression patterns in cardiac cells might lead to a better understanding of individual risk factors for a variety of viral heart diseases, shedding light on their highly variable clinical courses and may as well provide new therapeutic options. It also should be pointed out that these viruses likely operate by distinct pathogenic pathways, which at present remain to be fully explained. Although there is considerable evidence that enteroviruses and adenoviruses can induce an acute inflammation of the myocardium without cardiac dysfunction (i.e., myocarditis) or with cardiac dysfunction (i.e., inflammatory cardiomyopathy), the subsequent development of DCM (with an attendant autoimmune response) often found in patients infected with these viruses is less likely to be found with cardiotropic infections by parvovirus and cytomegalovirus.[363] Moreover, enteroviruses damage the heart primarily by promoting the direct lysis of infected cardiomyocytes, whereas PVB19 may more limitedly attack cardiomyocytes directly. PVB19 infects endothelial cells of small intracardiac arterioles and venules mainly during acute infection, resulting in impairment of myocardial microcirculation with secondary myocyte necrosis.[364] A more complete understanding of the process will also likely allow the identification of host susceptibility factors needed for effective viral infection.

Host immune responses may be operative both in limiting or promoting viral entry and maintenance. CVB infection of the human myocardium is associated with a selective, yet polyclonal activation, of different T-cell subsets in genetically susceptible individuals.[365] This immune response may play a critical role in modulating the disease progression after chronic viral infections. Moreover, the host immune response can be subverted during later stages of viral-based myocarditis and cardiomyopathy into an autoimmune response, which can further exacerbate cardiac damage. A pivotal role for autoimmunity is supported by the presence of cardiac-specific autoantibodies, inflammatory cell infiltration (e.g., cytolytic T-cell lymphocytes), and pro-inflammatory cytotoxic cytokines found in both myocardium and sera of patients and animal models of myocarditis with cardiomyopathy.[366] Several observations have indicated a commonality of host susceptibility factors (e.g., MHC class II HLA-DQ8) that play a role in both viral-mediated myocarditis and cardiomyopathy and other autoimmune diseases.[367] Moreover, cross-reactivity between viral and cardiac protein epitopes (e.g., myosin, adenine nucleotide translocator) or antigenic mimicry contributes to both cellular and humoral autoimmunity in myocarditis.

Although the precise pathogenic mechanism of streptococcal-induced rheumatic fever and rheumatic heart disease has not yet been fully explained, molecular analysis has provided significant insight into critical autoimmune aspects of the disease. Structural and immunological mimicry between the streptococcal M protein and cardiac myosin have been directly demonstrated.[368] In rheumatic heart disease, streptococcal M protein–specific T-cells migrate to the cardiac valves where cardiac myosin is an important autoantigen. Further gene linkage/association analysis may provide key information about the genetic factors involved in host susceptibility.

GENETIC DIAGNOSTIC APPROACHES

Many of the nuclear gene defects implicated in cardiomyopathies were originally detected and mapped by linkage analyses in affected families, allowing the subsequent identification of candidate genes (and mutant alleles) by positional cloning and subsequent nucleotide sequence analysis. A variety of molecular techniques, including PCR, restriction fragment length polymorphism (RFLP), and single-strand conformation polymorphism (SSCP), have been used in screening defective alleles from the proband and family members to establish inheritance patterns. In most cases, detection of novel mutation by itself is an immense undertaking involving the comprehensive analysis of large and multiple coding regions (exons) of one, if not more, candidate genes. Furthermore, in the relatively well-characterized cases of familial HCM gene screening, the consensus has been that each specific HCM-causing mutation is rare, challenging the view of common mutations, because most families have "private" or novel mutations. Nevertheless, correlation of the clinical course and prognosis with specific mutations has in some cases proved informative; for instance, specific mutations in the *MYH7* gene encoding β-MHC *in* HCM are associated with a high incidence of SCD, whereas other *MYH7* mutations are associated with a better prognosis. Recent advances in the speed and sensitivity of mutation detection by applying high-throughput analytical techniques, such as denaturing high-performance liquid chromatography (DHPLC) or high-throughput capillary array electrophoresis, should improve the use of molecular genetic analysis in clinical and preclinical diagnosis and provide specific targeted treatments in pediatric cardiac disorders. Moreover, the availability of gene chip technology may allow in the near future automated and rapid screening of mtDNA and nuclear gene mutations.

Although modern imaging techniques are helpful in defining cardiac phenotypes in children, both extensive genetic heterogeneity and intrafamilial variability have made the precise molecular explanation of many cardiac defects, as well as the correlation of genotype with cardiac phenotype, particularly difficult. These difficulties may arise from the involvement of undefined multifactorial or polygenic factors, which can contribute to the expression of specific cardiac gene defect(s), as well as to a variety of epigenetic or acquired influences. Progress is gradually being made in defining these polygenic and epigenetic factors, some of which are also amenable to molecular analysis.

The genetic background in which deleterious mutations occur can significantly modulate their phenotypic expression. The presence of modifier genes in the genetic background that influence the phenotypic expression and severity of pathogenic HCM genes has been well established.[369] The identification of modifier genes that will markedly improve the identification of genetic risk factors has been assisted by large-scale genome-wide approaches to identify polymorphism variants correlated with disease severity. Single nucleotide polymorphism association studies have identified several candidate modifier genes for various cardiac disorders. A number of specific genetic polymorphisms have been found in association with myocardial infarction (MI), coronary artery disease (CAD), and HCM. With increased cataloging of single nucleotide polymorphisms, either alone or within a larger chromosomal region (haplotypes) in available shared databases, these modifier *loci* can be evaluated for their effects in predisposing to specific cardiac defects and may have an impact on the choice of diagnostic and treatment options.

The association of defective genes with specific cardiac disorder revealed by genomic analysis needs to be followed by proteomic analysis to establish the precise function and pathophysiological role played by the mutant protein and to reveal interacting modulators. Once the implicated genes and their gene products have been fully identified, sequence and subsequent bioinformatics analysis can be used to identify common structural and functional motifs and homologies with known proteins. The potentially significant functional interaction of proteins (which can be an important determinant of the cardiac phenotype) can be further determined by yeast two-hybrid analysis. This approach has been productive in establishing that mutant titin proteins (derived from patients with HCM) had reduced binding affinities for other specific sarcomeric proteins (e.g., α-actinin), as well as in characterizing the synergistic interactions of transcription factors NKX2-5 and TBX5 in early cardiac development.[370]

PRENATAL DIAGNOSIS

Three-dimensional reconstruction of heart defects by use of ultrasound, X-ray, or MRI has dramatically improved the diagnosis therapeutic strategies of cardiac diseases. Most forms of CHD can be detected *in utero*. After the diagnosis of CHD, further evaluation for extracardiac anomalies and chromosomal abnormalities is recommended, because these are found in up to 62% and 38% of cases, respectively. Counseling based on the prenatal evaluation can provide realistic information about the incidence, diagnosis, and prognosis of the fetal heart defects. Prenatal diagnosis of congenital heart malformations and their molecular correlates (e.g., microdeletions of 22q11 in DiGeorge syndrome and 7q in Williams syndrome),[371,372] detectable by cytogenetic and molecular techniques subsequent to amniocentesis, has proved to be a critical adjunct in the management of life-threatening malformations of the neonate such as TGA and HLHS.

Interestingly, the genetic analysis of severe LQT syndrome presenting with ventricular tachycardia in the neonatal period has revealed the presence of a novel homozygous *HERG* missense mutation (R752Q).[373] Recurrent late-term fetal loss or sudden infant death can result from LQT-associated mutations; their identification at the earliest time point possible can be critical. A heterozygous mutation in *SCN5A* (R1623Q) caused neonatal ventricular arrhythmia and LQTS, and the mutant allele was detected in the fetus cord blood.[374]

FUTURE FRONTIERS

Despite significant advances in the diagnosis and treatment of cardiac diseases in children, many fundamental questions remain unanswered regarding their pathophysiology and pathogenesis. Breakthroughs in molecular genetic technology have just begun to be applied in studies of CVD by allowing chromosomal mapping and the identification of many genes involved in both the primary etiology and also as significant risk factors in the development of these anomalies. Identification of novel genes involved in cardiac organogenesis and vascular development will serve as an important foundation for our understanding of how specific congenital gene defects generate their cardiac phenotypes. Furthermore, bioinformatics methods can be used to search existing databases with the use of reverse genetics techniques, with subsequent cloning of novel genes/cDNAs of interest, followed by the characterization of spatiotemporal patterns of specific gene expression in the developing embryo (using *in situ* hybridization).

Although not yet precisely known, the mechanisms governing the early specification of cardiac chambers in the developing heart tube seem to involve novel cell-to-cell signaling among migrating cells, as well as the triggering of chamber-specific gene expression programs mediated by specific transcription factors and growth factors. Future research will focus on explaining the role of network modules of signaling molecules using conditional gene knock-outs (in a variety of genetic backgrounds) and accessing their interaction with critical transcription factors. These approaches may become important tools in the early diagnosis of cardiac defects during embryogenesis, increasing the possibilities of treatment (e.g., gene delivery) before the forming of the heart. Evaluation of pediatric CVD at the genomic level may allow a more effective stratification of patient subclasses, as well as targeting and optimization of patient-specific therapy. The related fields of pharmacogenomics and pharmacogenetics hold the promise of improved drug development and the tailoring of therapies on the basis of the individual's ability to metabolize drugs, which are determined only in part by age and influenced by the disease, environmental factors (e.g., diet), concurrent medications, and variant genetic factors specifying the transport, metabolism, and targets of the drug. For example, a subset of SNPs identified in human genes (e.g., β-adrenergic receptor [β-AR] and angiotensin-converting enzyme [ACE]) have been associated with significant changes in the metabolism and/or effects of the medications used in the treatment of CVD and may be valuable in predicting the clinical response.[375] Individualization of therapy may be particularly critical in the calculation of drug dosages and efficacies in children with CVD, a population for which pharmacokinetics has proven to be poorly defined and often unpredictable. Importantly, both immunological and genetic phenotyping of pediatric patients can provide a more effective therapeutic strategy, either by inhibiting or stimulating specific responses.

Cardioprotection can be elicited by either ischemic preconditioning or by pharmacological means (e.g., nicorandil and diazoxide) and can potentially be harnessed as a strategy for organ and tissue protection in ischemic heart disease and hypoxic insult, albeit currently there is limited information regarding the cardioprotection responses in infants and children. Extensive work in several animal models has established that the molecular basis of the cardioprotection mechanism(s) involves a network of signal transduction pathways mediated by cell-surface receptors, the activation and subcellular translocation of specific protein kinases (e.g., PKCε, p38 MAP kinase, and JUN kinase), and the opening of both sarcolemmal and mitochondrial K_{ATP} channels.[376] Infants with cyanotic heart defects and hypoxia were found to have activated myocardial protein kinase levels of PKCε, p38 MAP kinase, and JUN kinase that were not present in either infants with acyanotic defects or normal individuals, indicating that the cardioprotective signal transduction pathway is, at least partially, operative in hypoxic infants.[377] Cardioprotection associated with stress proteins and mitochondrial signaling has also been demonstrated during brief periods of hypothermia before a prolonged ischemic insult.[378] Further research in this area might reveal potential target molecules (e.g., receptor, signaling kinase or channel) for highly specific pharmacological intervention. However, as a cautionary note, this may take time and a better understanding of the network of interacting pathways. Despite recent achievements in the identification of precise genetic and signaling defects causing cardiac dysrhythmias, the development of effective drugs (e.g., specific ion channel blockers), which can substantially reduce the mortality associated with these disorders, has shown so far strikingly little success, underscoring the complex cardiac circuitry, multiple causal, genotypic, and risk factors involved in these evolving disease phenotypes.[379]

A critical area of research is the identification of molecular regulators that control cardiomyocyte proliferation. Cardiomyocytes are mitotically active during embryogenesis and generally cease proliferation shortly after birth. Understanding the molecular basis of cardiomyocyte proliferation could have a significant impact on our attempts to repair the damaged heart. Furthermore, the mechanisms of cell growth regulation may be investigated by careful comparison of comprehensive gene expression profiles of embryonic and post-natal myocytes, as well as by the generation of myocyte cell culture lines with the capacity to respond to proliferative inducers. Alternately, cellular transplantation is a mechanism with which to augment myocyte number in diseased or ischemia-damaged hearts. Interestingly, recent observations showed that a subpopulation of adult cardiac stem cells injected into an ischemic heart were able to fully reconstitute well-differentiated myocardium, differentiating into both cardiomyocytes and new blood vessels.[380] New research efforts will be necessary to further define the optimal conditions necessary for cardiomyocyte differentiation, proliferation, the fully functional integration of stem cells in the myocardium of children, as well as to investigate the ability of transplanted stem cells to repair the damaged heart. Moreover, it will be critical to learn whether severe cardiac abnormalities such as cardiomyopathies (e.g., DCM), Kawasaki disease with myocardial damage, and ARVD can be rescued by stem cell transplantation.

Insight into the cardiovascular consequences of abnormal gene function and expression should ultimately impact on the development of targeted therapeutic strategies and disease management for children with congenital and acquired heart disorders and may replace less effective treatment modalities directed solely at rectifying structural cardiac defects and temporal improvement of function. As our knowledge of the genes that are responsible for both early and late cardiac abnormalities increases, we may be able to modify the effects of mutated genes by modifying the environment in which genes act. It is hoped that new drugs may be found to alter or prevent the adverse effect of some genes, and cardiac transplantation will hopefully be replaced by the direct use of myocardial cells for patients with cardiomyopathy and/or HF. Bioengineered valves will be developed to replace mechanical valves and homografts for the surgical treatment of some of the cardiac malformations in children. Valves made from the patient's own cells will grow as the patient grows and will not need to be replaced. Both pulmonary hypertension and cardiomyopathy will be better understood and treated. Finally, fetal intervention to treat and/or prevent a cardiac defect will become more available and successful.

SUMMARY

- CHD, cardiomyopathy, and dysrhythmias are common causes of mortality and morbidity in infants and children, particularly during the perinatal period. Cardiovascular abnormalities represent the most common class of birth defect affecting approximately 1 in every 120 infants each year.
- Recent advances in molecular genetics have shown that specific genetic and molecular factors are linked to CHD, allowing their identification on the human chromosome map.
- Single gene mutations have been implicated in the pathogenesis of a variety of CHD, and new evidence suggests that these mutations are present in a broad spectrum of genes involved in cardiac structure and function.
- Pleiotropic cardiac malformations can result from discrete mutations in specific nuclear transcription factors, proteins recognized as playing key regulatory roles during cardiovascular development and morphogenesis.
- Genetic defects in proteins involved in the multiple signaling pathways that modulate cell proliferation, migration, and differentiation in early cardiovascular development have been identified.
- In animal models, mutations in a large number of genes have been associated with VSDs, usually in association with other complex heart defects.
- Human syndromic and sporadic cases of VSD have been associated with *NKX2-5*, *TBX5*, and *GATA4* mutations and generally display an autosomal-dominant pattern of inheritance.
- Gene loci leading to the genesis of ASDs have recently been identified. Some of these genetic factors have been grossly mapped to chromosomes (e.g., 5p).
- The variable expression of the ASD phenotype in different genetic backgrounds suggests the involvement of interacting modulators, whether of genetic or environmental origin.
- Conduction defects are often found in association with ASD, and this association is attributed to common underlying genetic and developmental processes.
- Developmental formation of the AVC results from complex interactions of components of the extracellular matrix, and endothelial/mesenchymal interactions in response to a number of signaling molecules are crucial to its normal development.
- The discovery of the cell adhesion factor (*CRELD1*) as the first recognized genetic risk factor for atrioventricular septal defects (AVSD) has provided new insight into the genetic basis of sporadically occurring and syndromic AVSD.
- The developmental formation of cardiac valves, produced by the remodeling of regional swellings of both endothelial and mesenchymal cells is a highly complex process, which is relatively poorly understood.
- Transgenic inactivation of the *HSPG2* gene encoding the basement membrane heparin-sulfate proteoglycan perlecan results in cardiac malformations, including a high incidence of complete TGA.
- Treatment with the teratogenic agent retinoic acid (RA) promotes a dose-dependent differential induction of complete TGA at high doses and dextroposition of the aorta at low dose.
- HLHS may occur in families showing left-sided cardiac disease of variable severity, including mild aortic stenosis and bicuspid aortic valve.
- In fetuses having structural heart disease, as detected by echocardiography, univentricular heart was frequently detected in association with intrauterine HF and neonatal mortality.
- Large chromosomal deletions (e.g., 22q11) have been implicated in an array of developmental/structural malformations of the heart, including conotruncal abnormalities, AVC defects, VSDs, and ASDs. Detection of chromosomal microdeletions may prove informative and will undoubtedly improve with new advances.
- Parental genomes may undergo epigenetic modifications during gametogenesis, and these epigenetic modifications result in parent-of-origin specific expression for some genes, a phenomenon known as genomic imprinting.
- A syndromic form of PDA, Char syndrome, was mapped to a single locus on chromosome 6p12–p21.[106] Subsequent positional cloning and mutation analysis demonstrated that this phenotype is caused by mutations residing in *TFAP2B*, the gene encoding a transcription factor AP2β, highly expressed in neural crest cells.
- Specific gene mutations in humans have not yet been described in patients with IAA, although it is associated with chromosome 22q11 deletion.
- A number of candidate genes have been identified in animal models in which there is considerable phenotypic overlap between IAA and PTA. Microdeletion of chromosome 22 has been found in as many as one third of patients with PTA.
- A large number of mouse mutations may present with a DORV phenotype.
- Laterality defects can occur as a function of environmental or teratogenic influences, as well as by strong genetic determinants. High doses of retinoic acid (RA) can induce laterality defects in vertebrates, including humans.
- Also, an increased risk of laterality defects has been documented in infants of mothers with maternal type-1 diabetes mellitus.

- S*itus* abnormalities in humans have shown autosomal-dominant, autosomal-recessive, and X-linked patterns of inheritance.
- Mutations in the zinc finger transcription factor *ZIC3* gene result in *situs ambiguus* with severe heart malformations as a consequence of the inability of the embryo to establish in early development the appropriate left-to-right asymmetry.
- The combination of malformations in CHARGE syndrome supports the view that this multiple congenital anomalies/mental retardation syndrome is a developmental field defect involving the neural tube and the neural crests cells.
- CHD associated with CHARGE syndrome includes a high incidence of cases with TOF and AVSD, with an estimated prevalence of 1/10,000.
- Expression profiling of myocardial genes in cases with HLHS revealed that both left and right ventricles expressed the fetal or "heart failure" gene pattern. Proteomic analysis indicated that whereas differentiation was evident in the ventricle, the fetal isoform of some cardiac specific proteins was prevalently expressed.
- Most cases of Marfan syndrome with cardiovascular involvement have mutations in the fibrillin (*FBN1*) gene. Affected individuals usually exhibit a distinct type of mutation, and those with the same mutation may display marked clinical heterogeneity, making genotype-phenotype correlation a difficult task.
- Mutations in the ELN gene encoding a component of the extracellular matrix (i.e., elastin) are responsible for familial supravalvular aortic stenosis and stenoses of systemic and/or pulmonary arteries that may be present in Williams syndrome.
- Triplet repeat syndromes, including Friedreich ataxia (FA) and myotonic muscular dystrophy, are caused by expanded repeats of trinucleotide sequences within specific genes (e.g., *frataxin* [*FRDA*] and myotonin protein kinase [*DMPK*], respectively).
- Frataxin deficiency leads to excessive free radical production and dysfunction of Fe-S center containing enzymes, with effects on cardiac bioenergetic supply correlating with myocardial hypertrophy.
- Mutations in specific genes encoding a variety of cardiac proteins involved in critical functions, such as ion transport, storage, and metabolism, have been identified as risk factors in the pathogenesis of lethal and nonlethal dysrhythmias in children.
- In a population-based study of pediatric cardiomyopathy, DCM made up 51% of the cases, whereas HCM represented 42% of cases.
- Specific defects in genes involved in mitochondrial heme and Fe^{++} metabolism (e.g., nuclear encoded frataxin and COX15) and in mitochondrial bioenergetics (e.g., mutations in mtDNA-encoded tRNAs and ATPase6) have been detected in patients with HCM (although more rarely than the sarcomeric mutations).
- Severe infantile cardiomyopathy is the most common clinical phenotype of very long chain acyl dehydrogenase (VLCAD) deficiency and is found in more than 67% of cases, often resulting in sudden death. Defects in malonyl CoA decarboxylase caused by mutations in the *MCD* gene can also lead to cardiomyopathy and neonatal death.
- Carnitine deficiency has been frequently associated with severe cardiomyopathy. Mutations in proteins that participate in carnitine transport and metabolism may cause either DCM or HCM as a recessive trait.
- Genetic disorders in energy metabolism, leading to either specific FAO/carnitine deficiencies or OXPHOS/respiration abnormalities, may also result in either a DCM or a HCM phenotype leading to mitochondrial cardiomyopathy, which is characterized by abnormal heart mitochondria either in number, structure, or function.
- More information is needed concerning the differential extent and susceptibility of infants and children to left ventricular dysfunction and cardiac dysrhythmias associated with HF, stemming from either genetic-based or metabolic defects and also from acquired pathogens.
- Specific indications for heart transplant in children may include metabolic and genetic forms of cardiomyopathy, as well as structural CHD. Before transplant, screening should include a metabolic workup (to rule out mitochondrial disorders) and thorough genetic studies, if indicated by phenotypic appearance or family pedigree.
- Acquired diseases in neonates and children include Kawasaki disease, acute and chronic rheumatic cardiac disease, infective endocarditis, and myocarditis. Molecular genetic technology is progressively being applied in the analysis of these diseases and could be valuable in improving the clinical diagnosis.
- A number of striking immune perturbations occur in acute Kawasaki disease and in the pathways leading to coronary arteritis. These include cytokine cascade and endothelial cell activation, monocyte/macrophages recruitment, and a cytotoxic lymphocyte response.
- Myocarditis is a common inflammatory disease of the myocardium in children and young adults and can be associated with cardiac dysfunction and DCM development.
- Viral infection is a common etiology or trigger of myocarditis. Immunological and molecular analyses have evolved to enable the sensitive quantitative detection of viral genomes in the myocardium and have implicated a large variety of viruses in the pathogenesis of pediatric myocarditis.
- Advances in the speed and sensitivity of mutation detection by applying high-throughput analytical techniques, such as denaturing high-performance liquid chromatography or high-throughput capillary array electrophoresis, should improve the use of molecular genetic analysis in clinical and preclinical diagnosis.

- The availability of gene chip technology may allow in the near future automated and rapid screening of mtDNA and nuclear gene mutations.
- The association of defective genes with specific cardiac disorder, uncovered by genomic analysis, needs to be followed by proteomic analysis to establish the precise function and pathophysiological role played by the mutant protein and to reveal interacting modulators.
- Many forms of CHD can be detected *in utero*. After CHD diagnosis, further evaluation for extracardiac anomalies and chromosomal abnormalities is recommended, because these are found in up to 62% and 38% of cases, respectively.
- Cardioprotection can be elicited by either ischemic preconditioning or by pharmacological means (e.g., nicorandil and diazoxide) and can potentially be used as a strategy for organ and tissue protection in ischemic heart disease and hypoxic insult; however, limited information regarding the cardioprotection responses in infants and children is currently available.
- Critical areas of research include the identification of molecular regulators that control cardiomyocyte proliferation and further definition of the optimal conditions necessary for cardiomyocyte differentiation, proliferation, the fully functional integration of stem cells in the myocardium of children, as well as to investigate the ability of the transplanted stem cells to repair the damaged heart.
- New drugs may be found to alter or prevent the adverse effect of some genes, and cardiac transplantation will hopefully be replaced by the direct use of myocardial cells in patients with cardiomyopathy and/or HF.

References

1. Maron, B. J., Moller, J. H., Seidman, C. E., Vincent, G. M., Dietz, H. C., Moss, A. J., Towbin, J. A., Sondheimer, H. M., Pyeritz, R. E., McGee, G., and Epstein, A. E. (1998). Impact of laboratory molecular diagnosis on contemporary diagnostic criteria for genetically transmitted cardiovascular diseases: hypertrophic cardiomyopathy, long-QT syndrome, and Marfan syndrome. *Circulation* **98,** 1460–1471.
2. Benson, D. W. (2000). Advances in cardiovascular genetics and embryology: role of transcription factors in congenital heart disease. *Curr. Opin. Pediatr.* **12,** 497–500.
3. Srivastava, D. (1999). HAND proteins: molecular mediators of cardiac development and congenital heart disease. *Trends Cardiovasc. Med.* **9,** 11–18.
4. Benson, D. W., Silberbach, G. M., Kavanaugh-McHugh, A., Cottrill, C., Zhang, Y., Riggs, S., Smalls, O., Johnson, M. C., Watson, M. S., Seidman, J. G., Seidman, C. E., Plowden, J., and Kugler, J. D. (1999). Mutations in the cardiac transcription factor NKX2.5 affect diverse cardiac developmental pathways. *J. Clin. Invest.* **104,** 1567–1573.
5. Garg, V., Kathiriya, I. S., Barnes, R., Schluterman, M. K., King, I. N., Butler, C. A., Rothrock, C. R., Eapen, R. S., Hirayama-Yamada, K., Joo, K., Matsuoka, R., Cohen, J. C., and Srivastava, D. (2003). GATA4 mutations cause human congenital heart defects and reveal an interaction with TBX5. *Nature* **424,** 443–447.
6. Schott, J. J., Benson, D. W., Basson, C. T., Pease, W., Silberbach, G. M., Moak, J. P., Maron, B. J., Seidman, C. E., and Seidman, J. G. (1998). Congenital heart disease caused by mutations in the transcription factor NKX2-5. *Science* **281,** 108–111.
7. Jay, P. Y., Berul, C. I., Tanaka, M., Ishii, M., Kurachi, Y., and Izumo, S. (2003). Cardiac conduction and arrhythmia: insights from Nkx2.5 mutations in mouse and humans. *Novartis Found. Symp.* **250,** 227–238.
8. Satoda, M., Zhao, F., Diaz, G. A., Burn, J., Goodship, J., Davidson, H. R., Pierpont, M. E., and Gelb, B. D. (2000). Mutations in TFAP2B cause Char syndrome, a familial form of patent ductus arteriosus. *Nat. Genet.* **25,** 42–46.
9. Bruneau, B. G., Nemer, G., Schmitt, J. P., Charron, F., Robitaille, L., Caron, S., Conner, D. A., Gessler, M., Nemer, M., Seidman, C. E., and Seidman, J. G. (2001). A murine model of Holt-Oram syndrome defines roles of the T-box transcription factor Tbx5 in cardiogenesis and disease. *Cell* **106,** 709–721.
10. Krantz, I. D., Piccoli, D. A., and Spinner, N. B. (1999). Clinical and molecular genetics of Alagille syndrome. *Curr. Opin. Pediatr.* **11,** 558–564.
11. Garg, V., Muth, A. N., Ransom, J. F., Schluterman, M. K., Barnes, R., King, I. N., Grossfeld, P. D., and Srivastava, D. (2005). Mutations in NOTCH1 cause aortic valve disease. *Nature* **437,** 270–274.
12. Tartaglia, M., Mehler, E. L., Goldberg, R., Zampino, G., Brunner, H. G., Kremer, H., van der Burgt, I., Crosby, A. H., Ion, A., Jeffery, S., Kalidas, K., Patton, M. A,. Kucherlapati, R. S., and Gelb, B. D. (2001). Mutations in PTPN11, encoding the protein tyrosine phosphatase SHP-2, cause Noonan syndrome. *Nat. Genet.* **29,** 465–468.
13. Legius, E., Schrander-Stumpel, C., Schollen, E., Pulles-Heintzberger, C., Gewillig, M., and Fryns, J. P. (2002). PTPN11 mutations in LEOPARD syndrome. *J. Med. Genet.* **39,** 571–574.
14. Taussig, H. B. (1988). Evolutionary origin of cardiac malformations. *J. Am. Coll. Cardiol.* **12,** 1079–1086.
15. Hoffman, J. I., and Kaplan, S. (2002). The incidence of congenital heart disease. *J. Am. Coll. Cardiol.* **39,** 1890–1900.
16. Aranega, A., Egea, J., Alvarez, L., and Arteaga, M. (1985). Tetralogy of Fallot produced in chick embryos by mechanical interference with cardiogenesis. *Anat. Rec.* **213,** 560–565.
17. Kwee, L., Baldwin, H. S., Shen, H. M., Stewart, C. L., Buck, C., Buck, C. A., and Labow, M. A. (1995). Defective development of the embryonic and extraembryonic circulatory systems in vascular cell adhesion molecule (VCAM-1) deficient mice. *Development* **121,** 489–503.
18. Yang, J. T., Rayburn, H., and Hynes, R. O. (1995). Cell adhesion events mediated by 4 integrins are essential in placental and cardiac development. *Development* **121,** 549–560.
19. Kasahara, H., Lee, B., Schott, J. J., Benson, D. W., Seidman, J. G., Seidman, C. E., and Izumo, S. (2000). Loss of function and inhibitory effects of human CSX/NKX2.5 homeoprotein mutations associated with congenital heart disease. *J. Clin. Invest.* **106,** 299–308.

20. Basson, C. T., Bachinsky, D. R., Lin, R. C., Levi, T., Elkins, J. A., Soults, J., Grayzel, D., Kroumpouzou, E., Traill, T. A., Leblanc-Straceski, J., Renault, B., Kucherlapati, R., Seidman, J. G., and Seidman, C. E. (1997). Mutations in human TBX5 cause limb and cardiac malformation in Holt-Oram syndrome. *Nat. Genet.* **15,** 30–35.
21. Li, L., Krantz, I. D., Deng, Y., Genin, A., Banta, A. B., Collins, C. C., Qi, M., Trask, B. J., Kuo, W. L., Cochran, J., Costa, T., Pierpont, M. E., Rand, E. B., Piccoli, D. A., Hood, L., and Spinner, N. B. (1997). Alagille syndrome is caused by mutations in human Jagged1, which encodes a ligand for Notch1. *Nat. Genet.* **16,** 243–251.
22. Eldadah, Z. A., Hamosh, A., Biery, N. J., Montgomery, R. A., Duke, M., Elkins, R., and Dietz, H. C. (2001). Familial tetralogy of Fallot caused by mutation in the jagged1 gene. *Hum. Mol. Genet.* **10,** 163–169.
23. Sakata, Y., Kamei, C. N., Nakagami, H., Bronson, R., Liao, J. K., and Chin, M. T. (2002). Ventricular septal defect and cardiomyopathy in mice lacking the transcription factor CHF1/Hey2. *Proc. Natl. Acad. Sci. USA* **99,** 16197–16202.
24. Sucov, H. M., Dyson, E., Gumeringer, C. L., Price, J., Chien, K. R., and Evans, R. M. (1994). RXR mutant mice establish a genetic basis for vitamin A signaling in heart morphogenesis. *Genes Dev.* **8,** 1007–1018.
25. Gitler, A. D., Zhu, Y., Ismat, F. A., Lu, M. M., Yamauchi, Y., Parada, L. F., and Epstein, J. A. (2003). Nf1 has an essential role in endothelial cells. *Nat. Genet.* **33,** 75–79.
26. Li, J., Liu, K. C., Jin, F., Lu, M. M., and Epstein, J. A. (1999). Transgenic rescue of congenital heart disease and spina bifida in Splotch mice. *Development* **126,** 2495–2503.
27. Bartram, U., Molin, D. G., Wisse, L. J., Mohamad, A., Sanford, L. P., Doetschman, T., Speer, C. P., Poelmann, R. E., and Gittenberger-de Groot, A. C. (2001). Double-outlet right ventricle and overriding tricuspid valve reflect disturbances of looping, myocardialization, endocardial cushion differentiation, and apoptosis in TGF-beta(2)-knockout mice. *Circulation* **103,** 2745–2752.
28. Basson, C. T., Huang, T., Lin, R. C., Bachinsky, D. R., Weremowicz, S., Vaglio, A., Bruzzone, R., Quadrelli, R., Lerone, M., Romeo, G., Silengo, M., Pereira, A., Krieger, J., Mesquita, S. F., Kamisago, M., Morton, C. C., Pierpont, M. E., Muller, C. W., Seidman, J. G., and Seidman, C. E. (1999). Different TBX5 interactions in heart and limb defined by Holt-Oram syndrome mutations. *Proc. Natl. Acad. Sci. USA* **96,** 2919–2924.
29. Bodmer, R. (1993). The gene tinman is required for specification of the heart and visceral muscles in *Drosophila*. *Development* **118,** 719–729.
30. Lyons, I., Parsons, L. M., Hartley, L., Li, R., Andrews, J. E., Robb, L., and Harvey, R. P. (1995). Myogenic and morphogenetic defects in the heart tubes of murine embryos lacking the homeo box gene Nkx2-5. *Genes Dev.* **9,** 1654–1666.
31. Biben, C., Weber, R., Kesteven, S., Stanley, E., McDonald, L., Elliott, D. A., Barnett, L., Koentgen, F., Robb, L., Feneley, M., and Harvey, R. P. (2000). Cardiac septal and valvular dysmorphogenesis in mice heterozygous for mutations in the homeobox gene Nkx2-5. *Circ. Res.* **87,** 888–895.
32. Pu, W. T., Ishiwata, T., Juraszek, A. L., Ma, Q., and Izumo, S. (2004). GATA4 is a dosage-sensitive regulator of cardiac morphogenesis. *Dev. Biol.* **275,** 235–244.
33. Komuro, I., and Izumo, S. (1993). Csx: a murine homeobox-containing gene specifically expressed in the developing heart. *Proc. Natl. Acad. Sci. USA* **90,** 8145–8149.
34. Bamforth, S. D., Braganca, J., Eloranta, J. J., Murdoch, J. N., Marques, F. I., Kranc, K. R., Farza, H., Henderson, D. J., Hurst, H. C., and Bhattacharya, S. (2001). Cardiac malformations, adrenal agenesis, neural crest defects and exencephaly in mice lacking Cited2, a new Tfap2 co-activator. *Nat. Genet.* **29,** 469–474.
35. Svensson, E. C., Huggins, G. S., Lin, H., Clendenin, C., Jiang, F., Tufts, R., Dardik, F. B., and Leiden, J. M. (2000). A syndrome of tricuspid atresia in mice with a targeted mutation of the gene encoding Fog-2. *Nat. Genet.* **25,** 353–356.
36. Tevosian, S. G., Deconinck, A. E., Tanaka, M., Schinke, M., Litovsky, S. H., Izumo, S., Fujiwara, Y., and Orkin, S. H. (2000). FOG-2, a cofactor for GATA transcription factors, is essential for heart morphogenesis and development of coronary vessels from epicardium. *Cell* **101,** 729–739.
37. Donovan, M. J., Hahn, R., Tessarollo, L., and Hempstead, B. L. (1996). Identification of an essential nonneuronal function of neurotrophin 3 in mammalian cardiac development. *Nat. Genet.* **14,** 210–213.
38. Tessarollo, L., Tsoulfas, P., Donovan, M. J., Palko, M. E., Blair-Flynn, J., Hempstead, B. L., and Parada, L. F. (1997). Targeted deletion of all isoforms of the trkC gene suggests the use of alternate receptors by its ligand neurotrophin-3 in neuronal development and implicates trkC in normal cardiogenesis. *Proc. Natl. Acad. Sci. USA* **94,** 14776–14781.
39. Kasahara, H., Wakimoto, H., Liu, M., Maguire, C. T., Converso, K. L., Shioi, T., Huang, W. Y., Manning, W. J., Paul, D., Lawitts, J., Berul, C. I., and Izumo, S. (2001). Progressive atrioventricular conduction defects and heart failure in mice expressing a mutant Csx/Nkx2.5 homeoprotein. *J. Clin. Invest.* **108,** 189–201.
40. Pashmforoush, M., Lu, J. T., Chen, H., Amand, T. S, Kondo, R., Pradervand, S., Evans, S. M., Clark, B., Feramisco, J. R., Giles, W., Ho, S. Y., Benson, D. W., Silberbach, M., Shou, W., and Chien, K. R. (2004). Nkx2-5 pathways and congenital heart disease; loss of ventricular myocyte lineage specification leads to progressive cardiomyopathy and complete heart block. *Cell* **117,** 373–386.
41. Pierpont, M. E., Markwald, R. R., and Lin, A. E. (2000). Genetic aspects of atrioventricular septal defects. *Am. J. Med. Genet.* **97,** 289–296.
42. Crispino, J. D., Lodish, M. B., Thurberg, B. L., Litovsky, S. H., Collins, T., Molkentin, J. D., and Orkin, S. H. (2001). Proper coronary vascular development and heart morphogenesis depend on interaction of GATA-4 with FOG cofactors. *Genes Dev.* **15,** 839–844.
43. Tevosian, S. G., Deconinck, A. E., Tanaka, M., Schinke, M., Litovsky, S. H., Izumo, S., Fujiwara, Y., and Orkin, S. H. (2000). FOG-2, a cofactor for GATA transcription factors, is essential for heart morphogenesis and development of coronary vessels from epicardium. *Cell* **101,** 729–739.
44. Gruber, P. J., Kubalak, S. W., Pexieder, T., Sucov, H. M., Evans, R. M., and Chien, K. R. (1996). RXR deficiency confers genetic susceptibility for aortic sac, conotruncal, atrioventricular cushion, and ventricular muscle defects in mice. *J. Clin. Invest.* **98,** 1332–1343.
45. Jiao, K., Kulessa, H., Tompkins, K., Zhou, Y., Batts, L., Baldwin, H. S., and Hogan, B. L. (2003). An essential role

of Bmp4 in the atrioventricular septation of the mouse heart. *Genes Dev.* **17,** 2362–2367.
46. Maslen, C. L. (2004). Molecular genetics of atrioventricular septal defects. *Curr. Opin. Cardiol.* **19,** 205–210.
47. Webb, S., Brown, N. A., and Anderson, R. H. (1997). Cardiac morphology at late fetal stages in the mouse with trisomy 16: consequences for different formation of the atrioventricular junction when compared to humans with trisomy 21. *Cardiovasc. Res.* **34,** 515–524.
48. Gittenberger-de Groot, A. C., Bartram, U., Oosthoek, P. W., Bartelings, M. M., Hogers, B., Poelmann, R. E., Jongewaard, I. N., and Klewer, S. E. (2003). Collagen type VI expression during cardiac development and in human fetuses with trisomy 21. *Anat. Rec. A Discov. Mol. Cell Evol. Biol.* **275,** 1109–1116.
49. Jongewaard, I. N., Lauer, R. M., Behrendt, D. A., Patil,. S, and Klewer, S. E. (2002). Beta 1 integrin activation mediates adhesive differences between trisomy 21 and non-trisomic fibroblasts on type VI collagen. *Am. J. Med. Genet.* **109,** 298–305.
50. Robinson, S. W., Morris, C. D., Goldmuntz, E., Reller, M. D., Jones, M. A., Steiner, R. D., and Maslen, C. L. (2003). Missense mutations in CRELD1 are associated with cardiac atrioventricular septal defects. *Am. J. Hum. Genet.* **72,** 1047–1052.
51. Armstrong, E. J., and Bischoff, J. (2004). Heart valve development: endothelial cell signaling and differentiation. *Circ. Res.* **95,** 459–470.
52. Gelb, B. D. (2004). Genetic basis of congenital heart disease. *Curr. Opin. Cardiol.* **19,** 110–115.
53. Lange, A. W., Molkentin, J. D., and Yutzey, K. E. (2004). DSCR1 gene expression is dependent on NFATc1 during cardiac valve formation and colocalizes with anomalous organ development in trisomy 16 mice. *Dev. Biol.* **266,** 346–360.
54. Lange, A. W., Rothermel, B. A., and Yutzey, K. E. (2005). Restoration of DSCR1 to disomy in the trisomy 16 mouse model of Down syndrome does not correct cardiac or craniofacial development anomalies. *Dev. Dyn.* **233,** 954–963.
55. Freed, L. A., Acierno, J. S. Jr., Dai, D., Leyne, M., Marshall, J. E., Nesta, F., Levine, R. A., and Slaugenhaupt, S. A. (2003). A locus for autosomal dominant mitral valve prolapse on chromosome 11p15.4. *Am. J. Hum. Genet.* **72,** 1551–1559.
56. Ranger, A. M., Grusby, M. J., Hodge, M. R, Gravallese, E. M., de la Brousse, F. C., Hoey, T., Mickanin, C., Baldwin, H. S., and Glimcher, L. H. (1998). The transcription factor NF-ATc is essential for cardiac valve formation. *Nature* **392,** 186–190.
57. Dor, Y., Camenisch, T. D., Itin, A., Fishman, G. I., McDonald, J. A., Carmeliet, P., and Keshet, E. (2001). A novel role for VEGF in endocardial cushion formation and its potential contribution to congenital heart defects. *Development* **128,** 1531–1538.
58. Kumai, M., Nishii, K., Nakamura, K., Takeda, N., Suzuki, M., and Shibata, Y. (2000). Loss of connexin45 causes a cushion defect in early cardiogenesis. *Development* **127,** 3501–3512.
59. Timmerman, L. A., Grego-Bessa, J., Raya, A., Bertran, E., Perez-Pomares, J. M., Diez, J., Aranda, S., Palomo, S., McCormick, F., Izpisua-Belmonte, J. C., and de la Pompa, J. L. (2004). Notch promotes epithelial-mesenchymal transition during cardiac development and oncogenic transformation. *Genes Dev.* **18,** 99–115.
60. Brannan, C. I., Perkins, A. S., Vogel, K. S., Ratner, N., Nordlund, M. L., Reid, S. W., Buchberg, A. M., Jenkins, N. A., Parada, L. F., and Copeland, N. G. (1994). Targeted disruption of the neurofibromatosis type-1 gene leads to developmental abnormalities in heart and various neural crest-derived tissues. *Genes Dev.* **8,** 1019–1029.
61. Erickson, S. L., O'Shea, K. S., Ghaboosi, N., Loverro, L., Frantz, G., Bauer, M., Lu, L. H., and Moore, M. W. (1997). ErbB3 is required for normal cerebellar and cardiac development: a comparison with ErbB2-and heregulin-deficient mice. *Development* **124,** 4999–5011.
62. Kim, R. Y., Robertson, E. J., and Solloway, M. J. (2001). Bmp6 and Bmp7 are required for cushion formation and septation in the developing mouse heart. *Dev. Biol.* **235,** 449–466.
63. Gaussin, V., Van de Putte, T., Mishina, Y., Hanks, M. C., Zwijsen, A., Huylebroeck, D., Behringer, R. R., and Schneider, M. D. (2002). Endocardial cushion and myocardial defects after cardiac myocyte-specific conditional deletion of the bone morphogenetic protein receptor ALK3. *Proc. Natl. Acad. Sci. USA* **99,** 2878–2883.
64. Galvin, K. M., Donovan, M. J., Lynch, C. A., Meyer, R. I., Paul, R. J., Lorenz, J. N., Fairchild-Huntress, V., Dixon, K. L., Dunmore, J. H., Gimbrone, M. A., Jr., Falb, D., and Huszar, D. (2000). A role for smad6 in development and homeostasis of the cardiovascular system. *Nat. Genet.* **24,** 171–174.
65. Hurlstone, A. F., Haramis, A. P., Wienholds, E., Begthel, H., Korving, J., Van Eeden, F., Cuppen, E., Zivkovic, D., Plasterk, R. H., and Clevers, H. (2003). The Wnt/beta-catenin pathway regulates cardiac valve formation. *Nature* **425,** 633–637.
66. Hasegawa, S., Sato, T., Akazawa, H., Okada, H., Maeno, A., Ito, M., Sugitani, Y., Shibata, H., Miyazaki Ji. J., Katsuki, M., Yamauchi, Y., Yamamura, Ki. K., Katamine, S., and Noda, T. (2002). Apoptosis in neural crest cells by functional loss of APC tumor suppressor gene. *Proc. Natl. Acad. Sci. USA* **99,** 297–302.
67. Jackson, L. F., Qiu, T. H., Sunnarborg, S. W., Chang A., Zhang, C., Patterson, C., and Lee, D. C. (2003). Defective valvulogenesis in HB-EGF and TACE-null mice is associated with aberrant BMP signaling. *EMBO J.* **22,** 2704–2716.
68. Camenisch, T. D., Spicer, A. P., Brehm-Gibson, T., Biesterfeldt, J., Augustine, M. L., Calabro, A., Jr., Kubalak, S., Klewer, S. E., and McDonald, J. A. (2000). Disruption of hyaluronan synthase-2 abrogates normal cardiac morphogenesis and hyaluronan-mediated transformation of epithelium to mesenchyme. *J. Clin. Invest.* **106,** 349–360.
69. Kokubo, H., Miyagawa-Tomita, S., Tomimatsu, H., Nakashima, Y., Nakazawa, M., Saga, Y., and Johnson, R. L. (2004). Targeted disruption of hesr2 results in atrioventricular valve anomalies that lead to heart dysfunction. *Circ. Res.* **95,** 540–547.
70. Sanford, L. P., Ormsby, I., Gittenberger-de Groot, A., Sariola, H., Friedman, R., Boivin, G. P., Cardell, E. L., and Doetschman, T. (1997). TGF2 knockout mice have multiple developmental defects that are non-overlapping with other TGF knockout phenotypes. *Development* **124,** 2659–2670.
71. Walker, G. A., Masters, K. S., Shah, D. N., Anseth, K. S., and Leinwand, L. A. (2004). Valvular myofibroblast activation by transforming growth factor-beta: implications for pathological extracellular matrix remodeling in heart valve disease. *Circ. Res.* **95,** 253–260.
72. Costell, M., Carmona, R., Gustafsson, E., Gonzalez-Iriarte, M., Fassler, R., and Munoz-Chapuli, R. (2002). Hyperplastic

conotruncal endocardial cushions and transposition of great arteries in perlecan-null mice. *Circ. Res.* **91,** 158–164.
73. Arikawa-Hirasawa, E., Le, A. H., Nishino, I., Nonaka, I., Ho, N. C., Francomano, C. A., Govindraj, P., Hassell, J. R., Devaney, J. M., Spranger, J., Stevenson, R. E., Iannaccone, S., Dalakas, M. C., and Yamada, Y. (2002). Structural and functional mutations of the perlecan gene cause Schwartz-Jampel syndrome, with myotonic myopathy and chondrodysplasia. *Am. J. Hum. Genet.* **70,** 1368–1375.
74. Oh, S. P., and Li, E. (1997). The signaling pathway mediated by the type IIB activin receptor controls axial patterning and lateral asymmetry in the mouse. *Genes Dev.* **11,** 1812–1826.
75. Hamblet, N. S., Lijam, N., Ruiz-Lozano, P., Wang, J., Yang, Y., Luo, Z., Mei, L., Chien, K. R., Sussman, D. J., and Wynshaw-Boris, A. (2002). Dishevelled 2 is essential for cardiac outflow tract development, somite segmentation and neural tube closure. *Development* **129,** 5827–5838.
76. Gaio, U., Schweickert, A., Fischer, A., Garratt, A. N., Muller, T., Ozcelik, C., Lankes, W., Strehle, M., Britsch, S., Blum, M., and Birchmeier, C. (1999). A role of the cryptic gene in the correct establishment of the left-right axis. *Curr. Biol.* **9,** 1339–1342.
77. Kawasaki, T., Kitsukawa, T., Bekku, Y., Matsuda, Y., Sanbo, M., Yagi, T., and Fujisawa, H. (1999). A requirement for neuropilin-1 in embryonic vessel formation. *Development* **126,** 4895–4902.
78. Bamford, R. N., Roessler, E., Burdine, R. D., Saplakoglu, U., dela Cruz, J., Splitt, M., Goodship, J. A., Towbin, J., Bowers, P., Ferrero, G. B., Marino, B., Schier, A. F., Shen, M. M., Muenke, M., and Casey, B. (2000). Loss-of-function mutations in the EGF-CFC gene CFC1 are associated with human left-right laterality defects. *Nat. Genet.* **26,** 365–369.
79. Goldmuntz, E., Bamford, R., Karkera, J. D., dela Cruz, J., Roessler, E., and Muenke, M. (2002). CFC1 mutations in patients with transposition of the great arteries and double-outlet right ventricle. *Am. J. Hum. Genet.* **70,** 776–780.
80. Yasui, H., Morishima, M., Nakazawa, M., Ando, M, and Aikawa, E. (1999). Developmental spectrum of cardiac outflow tract anomalies encompassing transposition of the great arteries and dextroposition of the aorta: pathogenic effect of extrinsic retinoic acid in the mouse embryo. *Anat. Rec.* **254,** 253–260.
81. Wessels, M. W., Berger, R.M., Frohn-Mulder, I. M., Roos-Hesselink, J. W., Hoogeboom, J. J., Mancini, G. S., Bartelings, M. M., Krijger, R., Wladimiroff, J. W., Niermeijer, M. F., Grossfeld, P., and Willems, P. J. (2005). Autosomal dominant inheritance of left ventricular outflow tract obstruction. *Am. J. Med. Genet. A.* **134,** 171–179.
82. Vasquez, J. C., Rabah, R., Delius, R. E., and Walters, H. L. (2003). Hypoplastic left heart syndrome with intact atrial septum associated with deletion of the short arm of chromosome 18. *Cardiovasc. Pathol.* **12,** 102–104.
83. Steinberger, E. K., Ferencz, C., and Loffredo, C. A. (2002). Infants with single ventricle: a population-based epidemiological study. *Teratology* **65,** 106–115.
84. Weigel, T. J., Driscoll, D. J., and Michels, V. V. (1989). Occurrence of congenital heart defects in siblings of patients with univentricular heart and tricuspid atresia. *Am. J. Cardiol.* **64,** 768–771.
85. Lizarraga, M. A., Mintegui, S., Sanchez Echaniz, J., Galdeano, J. M., Pastor, E., and Cabrera, A. (1991). Heart malformations in trisomy 13 and trisomy 18. *Rev. Esp. Cardiol.* **44,** 605–610.
86. Boldt, T., Andersson, S., and Eronen, M. (2002). Outcome of structural heart disease diagnosed in utero. *Scand. Cardiovasc. J.* **36,** 73–79.
87. Merscher, S., Funke, B., Epstein, J. A, Heyer, J., Puech, A., Lu, M. M., Xavier, R. J., Demay, M. B., Russell, R. G., Factor, S., Tokooya, K., Jore, B. S., Lopez, M., Pandita, R. K., Lia, M., Carrion, D., Xu, H., Schorle, H., Kobler, J. B., Scambler P., Wynshaw-Boris, A., Skoultchi, A. I., Morrow, B. E., and Kucherlapati, R. (2001). TBX1 is responsible for cardiovascular defects in velo-cardio-facial/DiGeorge syndrome. *Cell* **104,** 619–629.
88. Lindsay, E. A., Vitelli, F., Su, H., Morishima, M., Huynh, T., Pramparo, T., Jurecic, V., Ogunrinu, G., Sutherland, H. F., Scambler, P. J., Bradley, A., and Baldini, A. (2001). Tbx1 haploinsufficiency in the DiGeorge syndrome region causes aortic arch defects in mice. *Nature* **410,** 97–101.
89. Yagi, H., Furutani, Y., Hamada, H., Sasaki, T., Asakawa, S., Minoshima, S., Ichida, F., Joo, K., Kimura, M, Imamura, S., Kamatani, N., Momma, K., Takao, A., Nakazawa, M., Shimizu, N., and Matsuoka, R. (2003). Role of TBX1 in human del22q11.2 syndrome. *Lancet* **362,** 1366–1373.
90. Baldini, A. (2002). DiGeorge syndrome: the use of model organisms to dissect complex genetics. *Hum. Mol. Genet.* **11,** 2363–2369.
91. Yamagishi, H., Maeda, J., Hu, T., McAnally, J., Conway, S. J., Kume, T., Meyers, E. N., Yamagishi, C., and Srivastava, D. (2003). Tbx1 is regulated by tissue-specific forkhead proteins through a common Sonic hedgehog-responsive enhancer. *Genes Dev.* **17,** 269–281.
92. Strauss, A. W. (1998). The molecular basis of congenital cardiac disease. *Semin. Thorac. Cardiovasc. Surg. Pediatr. Card. Surg. Annu.* **1,** 179–188.
93. Giglio, S., Graw, S. L., Gimelli, G., Pirola, B., Varone, P., Voullaire, L., Lerzo, F., Rossi, E., Dellavecchia, C., Bonaglia, M. C., Digilio, M. C., Giannotti, A., Marino, B., Carrozzo, R., Korenberg, J. R., Danesino, C., Sujansky, E., Dallapiccola, B., and Zuffardi, O. (2000). Deletion of a 5-cM region at chromosome 8p23 is associated with a spectrum of congenital heart defects. *Circulation* **102,** 432–437.
94. Skrypnyk, C., Goecke, T. O., Majewski, F., and Bartsch, O. (2002). Molecular cytogenetic characterization of a 10p14 deletion that includes the DGS2 region in a patient with multiple anomalies. *Am. J. Med. Genet.* **113,** 207–212.
95. Christiansen, J., Dyck, J. D., Elyas, B. G., Lilley, M., Bamforth, J. S., Hicks, M., Sprysak, K. A., Tomaszewski, R., Haase, S. M., Vicen-Wyhony, L. M., and Somerville, M. J. (2004). Chromosome 1q21.1 contiguous gene deletion is associated with congenital heart disease. *Circ. Res.* **94,** 1429–1435.
96. von Beust, G., Sauter, S. M., Liehr, T., Burfeind, P., Bartels, I., Starke, H., von Eggeling, F., and Zoll, B. (2005). Molecular cytogenetic characterization of a de novo supernumerary ring chromosome 7 resulting in partial trisomy, tetrasomy, and hexasomy in a child with dysmorphic signs, congenital heart defect, and developmental delay. *Am. J. Med. Genet. A* **137,** 59–64.
97. Hahnemann, J. M., Nir, M., Friberg, M., Engel, U., and Bugge, M. (2005). Trisomy 10 mosaicism and maternal uniparental

disomy 10 in a liveborn infant with severe congenital malformations. *Am. J. Med. Genet. A* **138A,** 150–154.
98. Villar, A. J., Carlson, E. J., Gillespie, A. M., Ursell, P. C., and Epstein, C. J. (2001). Cardiomyopathy in mice with paternal uniparental disomy for chromosome 12. *Genesis* **30,** 274–279.
99. DeLozier-Blanchet, C. D., Roeder, E., Denis-Arrue, R., Blouin, J. L., Low, J., Fisher, J., Scharnhorst, D., and Curry C. J. (2000). Trisomy 12 mosaicism confirmed in multiple organs from a liveborn child. *Am. J. Med. Genet.* **95,** 444–449.
100. Iglesias, A., McCurdy, L. D., Glass, I. A., Cotter, P. D., Illueca, M., Perenyi, A., and Sansaricq, C. (1997). Mosaic trisomy 14 with hepatic involvement. *Ann. Genet.* **40,** 104–108.
101. Olander, E., Stamberg, J., Steinberg, L., and Wulfsberg, E. A. (2000). Third Prader-Willi syndrome phenotype due to maternal uniparental disomy 15 with mosaic trisomy 15. *Am. J. Med. Genet.* **93,** 215–218.
102. Hsu, W. T., Shchepin, D. A., Mao, R., Berry-Kravis, E., Garber, A. P., Fischel-Ghodsian, N., Falk, R. E., Carlson, D. E., Roeder, E. R., Leeth, E. A., Hajianpour, M. J., Wang, J. C., Rosenblum-Vos, L. S., Bhatt, S. D., Karson, E. M., Hux, C. H., Trunca, C., Bialer, M. G., Linn, S. K., and Schreck, R. R. (1998). Mosaic trisomy 16 ascertained through amniocentesis: evaluation of 11 new cases. *Am. J. Med. Genet.* **80,** 473–480.
103. O'Riordan, S., Greenough, A., Moore, G. E., Bennett, P., and Nicolaides, K. H. (1996). Case report: uniparental disomy 16 in association with congenital heart disease. *Prenat. Diagn.* **16,** 963–965.
104. Venditti, C. P., Hunt, P., Donnenfeld, A., Zackai, E., and Spinner, N. B. (2004). Mosaic paternal uniparental (iso)disomy for chromosome 20 associated with multiple anomalies. *Am. J. Med. Genet. A* **124,** 274–279.
105. de Pater, J. M., Schuring-Blom, G. H., van den Bogaard, R., van der Sijs-Bos, C. J., Christiaens, G. C., Stoutenbeek, P., and Leschot, N. J. (1997). Maternal uniparental disomy for chromosome 22 in a child with generalized mosaicism for trisomy 22. *Prenat. Diagn.* **17,** 81–86.
106. Satoda, M., Pierpont, M. E., Diaz, G. A., Bornemeier, R. A., and Gelb, B. D. (1999). Char syndrome, an inherited disorder with patent ductus arteriosus, maps to chromosome 6p12-p21. *Circulation* **99,** 3036–3042.
107. Zhao, F., Weismann, C. G., Satoda, M., Pierpont, M. E., Sweeney, E., Thompson, E. M., and Gelb, B. D. (2001). Novel TFAP2B mutations that cause Char syndrome provide a genotype-phenotype correlation. *Am. J. Hum. Genet.* **69,** 695–703.
108. Mani, A., Radhakrishnan, J., Farhi, A., Carew, K. S., Warnes, C. A., Nelson-Williams, C., Day, R. W., Pober, B., State, M. W., and Lifton, R. P. (2005). Syndromic patent ductus arteriosus: evidence for haploinsufficient TFAP2B mutations and identification of a linked sleep disorder. *Proc. Natl. Acad. Sci. USA* **102,** 2975–2979.
109. Mani, A., Meraji, S. M., Houshyar, R., Radhakrishnan, J., Mani, A., Ahangar, M., Rezaie, T. M., Taghavinejad, M. A., Broumand, B., Zhao, H., Nelson-Williams, C., and Lifton, R. P. (2002). Finding genetic contributions to sporadic disease: a recessive locus at 12q24 commonly contributes to patent ductus arteriosus. *Proc. Natl. Acad. Sci. USA* **99,** 15054–15059.
110. Morishima, M., Yanagisawa, H., Yanagisawa, M., and Baldini, A. (2003). Ece1 and Tbx1 define distinct pathways to aortic arch morphogenesis. *Dev. Dyn.* **228,** 95–104.
111. Nao, T., Ohkusa, T., Hisamatsu, Y., Inoue, N., Matsumoto, T., Yamada, J., Shimizu, A., Yoshiga, Y., Yamagata, T., Kobayashi, S., Yano, M., Hamano, K., and Matsuzaki, M. (2003). Comparison of expression of connexin in right atrial myocardium in patients with chronic atrial fibrillation versus those in sinus rhythm. *Am. J. Cardiol.* **91,** 678–683.
112. Gu, H., Smith, F. C., Taffet, S. M., and Delmar, M. (2003). High incidence of cardiac malformations in connexin40-deficient mice. *Circ. Res.* **93,** 201–206.
113. Martin, D. M., Mindell, M. H., Kwierant, C. A., Glover, T. W., and Gorski, J. L. (2003). Interrupted aortic arch in a child with trisomy 5q31.1q35.1 due to a maternal (20;5) balanced insertion. *Am. J. Med. Genet. A* **116,** 268–271.
114. Kurihara, Y., Kurihara, H., Oda, H., Maemura, K., Nagai, R., Ishikawa, T., and Yazaki, Y. (1995). Aortic arch malformations and ventricular septal defect in mice deficient in endothelin-1. *J. Clin. Invest.* **96,** 293–300.
115. Yanagisawa, H., Hammer, R. E., Richardson, J. A., Williams, S. C., Clouthier, D. E., and Yanagisawa, M. (1998). Role of endothelin-1/endothelin-A receptor-mediated signaling pathway in the aortic arch patterning in mice. *J. Clin. Invest.* **102,** 22–33.
116. Yanagisawa, H., Yanagisawa, M., Kapur, R. P., Richardson, J. A., Williams, S. C., Clouthier, D. E., de Wit, D., Emoto, N., and Hammer, R. E. (1998). Dual genetic pathways of endothelin-mediated intercellular signaling revealed by targeted disruption of endothelin converting enzyme-1 gene. *Development* **125,** 825–836.
117. Clouthier, D. E., Hosoda, K., Richardson, J. A., Williams, S. C., Yanagisawa, H., Kuwaki, T., Kumada, M., Hammer, R. E., and Yanagisawa, M. (1998). Cranial and cardiac neural crest defects in endothelin-A receptor-deficient mice. *Development* **125,** 813–824.
118. Kume, T., Jiang, H., Topczewska, J. M., and Hogan, B. L. (2001). The murine winged helix transcription factors, Foxc1 and Foxc2, are both required for cardiovascular development and somatogenesis. *Genes Dev.* **15,** 2470–2482.
119. Kanzaki-Kato, N., Tamakoshi, T., Fu, Y., Chandra, A., Itakura, T., Uezato, T., Tanaka, T., Clouthier, D. E., Sugiyama, T., Yanagisawa, M., and Miura, N. (2005). Roles of forkhead transcription factor Foxc2 (MFH-1) and endothelin receptor A in cardiovascular morphogenesis. *Cardiovasc. Res.* **65,** 711–718.
120. Lammer, E. J., Chen, D. T., Hoar, R. M., Agnish, N. D., Benke, P. J., Braun, J. T., Curry, C. J., Fernhoff, P. M., Grix, A. W. Jr., and Lott, I. T. (1985). Retinoic acid embryopathy. *N. Engl. J. Med.* **313,** 837–841.
121. Yelin, R., Schyr, R. B., Kot, H., Zins, S., Frumkin, A., Pillemer, G., and Fainsod, A. (2005). Ethanol exposure affects gene expression in the embryonic organizer and reduces retinoic acid levels. *Dev. Biol.* **279,** 193–204.
122. Waldo, K., Zdanowicz, M., Burch, J., Kumiski, D. H., Stadt, H. A., Godt, R. E., Creazzo, T. L., and Kirby, M. L. (1999). A novel role for cardiac neural crest in heart development. *J. Clin. Invest.* **103,** 1499–1507.
123. Feiner, L., Webber, A. L., Brown, C. B., Lu, M. M., Jia, L., Feinstein, P., Mombaerts, P., Epstein, J. A., and Raper,

J. A. (2001). Targeted disruption of semaphorin 3C leads to persistent truncus arteriosus and aortic arch interruption. *Development* **128**, 3061–3070.

124. Kioussi, C., Briata, P., Baek, S. H., Rose, D. W., Hamblet, N. S., Herman, T., Ohgi, K. A., Lin, C., Gleiberman, A., Wang, J., Brault, V., Ruiz-Lozano, P., Nguyen, H. D., Kemler, R., Glass, C. K., Wynshaw-Boris, A., and Rosenfeld, M. G. (2002). Identification of a Wnt/Dvl/ß-cateninPitx2 pathway mediating cell-type-specific proliferation during development. *Cell* **111**, 673–685.

125. Beppu, H., Kawabata, M., Hamamoto, T., Chytil, A., Minowa, O., Noda, T., and Miyazono, K. (2000). BMP type II receptor is required for gastrulation and early development of mouse embryos. *Dev. Biol.* **221**, 249–258.

126. Delot, E. C., Bahamonde, M. E., Zhao, M., and Lyons, K. M. (2003). BMP signaling is required for septation of the outflow tract of the mammalian heart. *Development* **130**, 209–220.

127. Kaartinen, V., Dudas, M., Nagy, A., Sridurongrit, S., Lu, M. M., and Epstein, J. A. (2004). Cardiac outflow tract defects in mice lacking ALK2 in neural crest cells. *Development* **131**, 3481–3490.

128. Stottmann, R. W., Choi, M., Mishina, Y. Meyers, E. N., and Klingensmith, J. (2004). BMP receptor IA is required in mammalian neural crest cells for development of the cardiac outflow tract and ventricular myocardium. *Development* **131**, 2205–2218.

129. Bartram, U., Molin, D. G., Wisse, L. J., Mohamad, A., Sanford, L. P., Doetschman, T., Speer, C. P., Poelmann, R. E., and Gittenberger-de Groot, A. C. (2001). Double-outlet right ventricle and overriding tricuspid valve reflect disturbances of looping, myocardialization, endocardial cushion differentiation, and apoptosis in TGF-β(2)-knockout mice. *Circulation* **103**, 2745–2752.

130. Kim, R. Y., Robertson, E. J., and Solloway, M. J. (2001). Bmp6 and Bmp7 are required for cushion formation and septation in the developing mouse heart. *Dev. Biol.* **235**, 449–466.

131. Schatteman, G. C., Motley, S. T., Effmann, E. L., and Bowen-Pope, D. F. (1995). Platelet-derived growth factor receptor subunit deleted Patch mouse exhibits severe cardiovascular dysmorphogenesis. *Teratology* **51**, 351–366.

132. Franco, D., and Campione, M. (2003). The role of Pitx2 during cardiac development. Linking left-right signaling and congenital heart diseases. *Trends Cardiovasc. Med.* **13**, 157–163.

133. Sock, E., Rettig, S. D., Enderich, J., Bosl, M. R., Tamm, E. R., and Wegner, M. (2004). Gene targeting reveals a widespread role for the high-mobility-group transcription factor Sox11 in tissue remodeling. *Mol. Cell Biol.* **24**, 6635–6644.

134. Mendelsohn, C., Lohnes, D., Decimo, D., Lufkin, T., LeMeur, M., Chambon, P., and Mark, M. (1994). Function of the retinoic acid receptors (RARs) during development (II): multiple abnormalities at various stages of organogenesis in RAR double mutants. *Development* **120**, 2749–2771.

135. Chen, P., Xie, L. J., Huang, G. Y., Zhao, X. Q., and Chang, C. (2005). Mutations of connexin43 in fetuses with congenital heart malformations. *Chin. Med. J.* **118**, 971–976.

136. Belmont, J. W., Mohapatra, B., Towbin, J. A., and Ware, S. M. (2004). Molecular genetics of heterotaxy syndromes. *Curr. Opin. Cardiol.* **19**, 216–220.

137. Smith, S. M., Dickman, E. D., Thompson, R. P., Sinning, A. R., Wunsch, A. M., and Markwald, R. R. (1997). Retinoic acid directs cardiac laterality and the expression of early markers of precardiac asymmetry. *Dev. Biol.* **182**, 162–171.

138. Loffredo, C. A., Wilson, P. D., and Ferencz, C. (2001). Maternal diabetes: an independent risk factor for major cardiovascular malformations with increased mortality of affected infants. *Teratology* **64**, 98–106.

139. Weninger, W. J., Floro, K. L., Bennett, M. B., Withington, S. L., Preis, J. I., Barbera, J. P., Mohun, T. J., and Dunwoodie, S. L. (2005). Cited2 is required both for heart morphogenesis and establishment of the left-right axis in mouse development. *Development* **132**, 1337–1348.

140. Bamforth, S. D., Braganca, J., Farthing, C. R., Schneider, J. E., Broadbent, C., Michell, A. C., Clarke, K., Neubauer, S., Norris, D., Brown, N. A., Anderson, R. H., and Bhattacharya, S. (2004). Cited2 controls left-right patterning and heart development through a Nodal-Pitx2c pathway. *Nat. Genet.* **36**, 1189–1196.

141. Lowe, L. A., Yamada, S., and Kuehn, M. R. (2001). Genetic dissection of nodal function in patterning the mouse embryo. *Development* **128**, 1831–1843.

142. Saijoh, Y., Oki, S., Ohishi, S., and Hamada, H. (2003). Left-right patterning of the mouse lateral plate requires Nodal produced in the node. *Dev. Biol.* **256**, 161–173.

143. Gage, P. J., Suh, H., and Camper, S. A. (1999). Dosage requirement of Pitx2 for development of multiple organs. *Development* **126**, 4643–4651.

144. Tsukui, T., Capdevila, J., Tamura, K., Ruiz-Lozano, P., Rodriguez-Esteban, C., Yonei-Tamura, S., Magallon, J., Chandraratna, R. A., Chien, K., Blumberg, B., Evans, R. M., and Belmonte, J. C. (1999). Multiple left-right asymmetry defects in Shh(-/-) mutant mice unveil a convergence of the shh and retinoic acid pathways in the control of Lefty-1. *Proc. Natl. Acad. Sci. USA* **96**, 11376–11381.

145. Schilling, T. F., Concordet, J. P, and Ingham, P. W. (1999). Regulation of left-right asymmetries in the zebrafish by Shh and BMP4. *Dev. Biol.* **210**, 277–287.

146. Thomas, T., Yamagishi, H., Overbeek, P. A., Olson, E. N., and Srivastava, D. (1998). The bHLH factors, dHAND and eHAND, specify pulmonary and systemic cardiac ventricles independent of left-right sidedness. *Dev. Biol.* **196**, 228–236.

147. Chen, Y., Mironova, E., Whitaker, L. L., Edwards, L. Yost, H. J., and Ramsdell, A. F. (2004). ALK4 functions as a receptor for multiple TGF beta-related ligands to regulate left-right axis determination and mesoderm induction in *Xenopus*. *Dev. Biol.* **268**, 280–294.

148. Rodriguez-Esteban, C., Capdevila, J., Kawakami, Y., and Izpisua Belmonte, J. C. (2001). Wnt signaling and PKA control Nodal expression and left-right determination in the chick embryo. *Development* **128**, 3189–3195.

149. Isaac, A., Sargent, M. G., and Cooke, J. (1997). Control of vertebrate left-right asymmetry by a snail-related zinc finger gene. *Science* **275**, 1301–1304

150. Biben, C., and Harvey, R. P. (1997). Homeodomain factor Nkx2–5 controls left/right asymmetric expression of bHLH

gene eHand during murine heart development. *Genes Dev.* **11,** 1357–1369.

151. Rodriguez-Esteban, C., Capdevila, J., Economides, A. N., Pascual, J., Ortiz, A., and Izpisua Belmonte, J. C. (1999). The novel Cer-like protein Caronte mediates the establishment of embryonic left-right asymmetry. *Nature* **401,** 243–251.

152. Meno, C., Shimono, A., Saijoh, Y., Yashiro, K., Mochida, K., Ohishi, S., Noji, S., Kondoh, H., and Hamada, H. (1998). Lefty-1 is required for left-right determination as a regulator of lefty-2 and nodal. *Cell* **94,** 287–297.

153. Yost, H. J. (2001). Establishment of left-right asymmetry. *Int. Rev. Cytol.* **203,** 357–381.

154. Brody, S. L., Yan, X. H., Wuerffel, M. K., Song, S. K., and Shapiro, S. D. (2000). Ciliogenesis and left-right axis defects in forkhead factor HFH-4-null mice. *Am. J. Respir. Cell Mol. Biol.* **23,** 45–51.

155. Kishigami, S., Yoshikawa, S., Castranio, T., Okazaki, K., Furuta, Y., and Mishina, Y. (2004). BMP signaling through ACVRI is required for left-right patterning in the early mouse embryo. *Dev. Biol.* **276,** 185–193.

156. Raya, A., Kawakami, Y., Rodriguez-Esteban, C., Buscher, D., Koth, C. M., Itoh, T., Morita, M., Raya, R. M., Dubova, I., Bessa, J. G., de la Pompa, J. L., and Belmonte, J. C. (2003). Notch activity induces Nodal expression and mediates the establishment of left-right asymmetry in vertebrate embryos. *Genes Dev.* **17,** 1213–1218.

157. Gaio, U., Schweickert, A., Fischer, A., Garratt, A. N., Muller, T., Ozcelik, C., Lankes, W., Strehle M., Britsch, S., Blum, M., and Birchmeier, C. (1999). A role of the cryptic gene in the correct establishment of the left-right axis. *Curr. Biol.* **9,** 1339–1342.

158. Chang, H., Zwijsen, A., Vogel, H., Huylebroeck, D., and Matzuk, M. M. (2000). Smad5 is essential for left-right asymmetry in mice. *Dev. Biol.* **219,** 71–78.

159. Casey, B. (1998). Two rights make a wrong: human left-right malformations. *Hum. Mol. Genet.* **7,** 1565–1571.

160. Supp, D. M., Witte, D. P., Potter, S. S., and Brueckner, M. (1997). Mutation of an axonemal dynein affects left-right asymmetry in inversus viscerum mice. *Nature* **389,** 963–966.

161. Olbrich, H., Haffner, K., Kispert, A., Volkel, A., Volz, A., Sasmaz, G., Reinhardt, R., Hennig, S., Lehrach, H., Konietzko, N., Zariwala, M., Noone, P. G., Knowles, M., Mitchison, H. M., Meeks, M., Chung, E. M., Hildebrandt, F., Sudbrak, R., and Omran, H. (2002). Mutations in DNAH5 cause primary ciliary dyskinesia and randomization of left-right asymmetry. *Nat. Genet.* **30,** 143–144.

162. Guichard, C., Harricane, M. C., Lafitte, J. J., Godard, P., Zaegel, M., Tack, V., Lalau, G., and Bouvagnet, P. (2001). Axonemal dynein intermediate-chain gene (DNAI1) mutations result in situs inversus and primary ciliary dyskinesia (Kartagener syndrome). *Am. J. Hum. Genet.* **68,** 1030–1035.

163. Bartoloni, L., Blouin, J. L., Pan, Y., Gehrig, C., Maiti, A. K., Scamuffa, N., Rossier, C., Jorissen, M., Armengot, M., Meeks, M., Mitchison, H. M., Chung, E. M., Delozier-Blanchet, C. D., Craigen, W. J., and Antonarakis, S. E. (2002). Mutations in the DNAH11 (axonemal heavy chain dynein type 11) gene cause one form of situs inversus totalis and most likely primary ciliary dyskinesia. *Proc. Natl. Acad. Sci. USA* **99,** 10282–10286.

164. Carmi, R., Boughman, J. A., and Rosenbaum, K. R. (1992). Human situs determination is probably controlled by several different genes. *Am. J. Med. Genet.* **44,** 246–249.

165. Genuardi, M., Gurrieri, F., and Neri, G. (1994). Genes for split hand/split foot and laterality defects on 7q22.1 and Xq24-q27.1. *Am. J. Med. Genet.* **50,** 101.

166. Peeters, H., Debeer, P., Groenen, P., Van Esch, H., Vanderlinden, G., Eyskens, B., Mertens, L., Gewillig, M., Van de Ven, W., Fryns, J. P., an Devriendt, K. (2001). Recurrent involvement of chromosomal region 6q21 in heterotaxy. *Am. J. Med. Genet.* **103,** 44–47.

167. Wilson, G. N., Stout, J. P., Schneider, N. R., Zneimer, S. M., and Gilstrap, L. C. (1991). Balanced translocation 12/13 and situs abnormalities: homology of early pattern formation in man and lower organisms? *Am. J. Med. Genet.* **38,** 601–607.

168. Digilio, M. C., Marino, B., Giannotti, A., Di Donato, R., ad Dallapiccola, B. (2000). Heterotaxy with left atrial isomerism in a patient with deletion 18p. *Am. J. Med. Genet.* **94,** 198–200.

169. Kosaki, R., Gebbia, M., Kosaki, K., Lewin, M., Bowers, P., Towbin, J. A., ad Casey, B. (1999). Left-right axis malformations associated with mutations in ACVR2B, the gene for human activin receptor type IIB. *Am. J. Med. Genet.* **82,** 70–76.

170. Kosaki, K., Bassi, M. T., Kosaki, R., Lewin, M., Belmont, J., Schauer, G., and Casey, B. (1999). Characterization and mutation analysis of human LEFTY A and LEFTY B, homologues of murine genes implicated in left-right axis development. *Am. J. Hum. Genet.* **64,** 712–721.

171. Bamford, R. N., Roessler, E., Burdine, R. D., Saplakoglu, U., de la Cruz, J., Splitt, M., Goodship, J. A., Towbin, J., Bowers, P., Ferrero, G. B., Marino, B., Schier, A. F., Shen, M. M., Muenke, M., and Casey, B. (2000) Loss-of-function mutations in the EGF-CFC gene CFC1 are associated with human left-right laterality defects. *Nat. Genet.* **26,** 365–369.

172. Chen, C., and Shen, M. M. (2004). Two modes by which Lefty proteins inhibit nodal signaling. *Curr. Biol.* **14,** 618–624

173. Goldmuntz, E., Bamford, R., Karkera, J. D., dela Cruz,. J, Roessler, E., and Muenke, M. (2002). CFC1 mutations in patients with transposition of the great arteries and double-outlet right ventricle. *Am. J. Hum. Genet.* **70,** 776–780.

174. Watanabe, Y., Benson, D. W., Yano, S., Akagi, T., Yoshino, M., and Murray, J. C. (2002).Two novel frameshift mutations in NKX2.5 result in novel features including visceral inversus and sinus venosus type ASD. *J. Med. Genet.* **39,** 807–811.

175. Grinberg, I., and Millen, K. (2005). The ZIC gene family in development and disease. *Clin. Genet.* **67,** 290–296.

176. Casey, B., Devoto, M., Jones, K., and Ballabio, A. (1993). Mapping a gene for familial situs abnormalities to human chromosome Xq24–q27.1. *Nat. Genet.* **5,** 403–407.

177. Ferrero, G. B., Gebbia, M., Pilia, G., Witte, D., Peier, A., Hopkin, R. J., Craigen, W. J., Shaffer, L. G., Schlessinger, D., Ballabio, A., and Casey, B. (1997). A submicroscopic deletion in Xq26 associated with familial situs ambiguus. *Am. J. Hum. Genet.* **61,** 395–401.

178. Gebbia, M., Ferrero, G. B., Pilia, G., Bassi, M. T., Aylsworth, A., Penman-Splitt, M., Bird, L. M., Bamforth, J. S, Burn, J., Schlessinger, D., Nelson, D. L., and Casey, B. (1997). X-linked situs abnormalities result from mutations in ZIC3. *Nat. Genet.* **17,** 305–308.

179. Megarbane, A., Salem, N., Stephan, E., Ashoush, R., Lenoir, D., Delague, V., Kassab, R., Loiselet, J., and Bouvagnet, P. (2000). X-linked transposition of the great arteries and incomplete penetrance among males with a nonsense mutation in ZIC3. *Eur. J. Hum. Genet.* **8,** 704–708.
180. Wyse, R. K., al-Mahdawi, S., Burn, J., and Blake, K. (1993). Congenital heart disease in CHARGE association. *Pediatr. Cardiol.* **14,** 75–81.
181. Tellier, A. L., Cormier-Daire, V., Abadie, V., Amiel, J., Sigaudy, S., Bonnet, D., de Lonlay-Debeney, P., Morrisseau-Durand, M. P., Hubert, P., Michel, J. L, Jan, D., Dollfus, H., Baumann, C., Labrune, P., Lacombe, D., Philip, N., LeMerrer, M., Briard, M. L., Munnich, A., and Lyonnet, S. (1998). CHARGE syndrome: report of 47 cases and review. *Am. J. Med. Genet.* **76,** 402–409.
182. Vissers, L. E., van Ravenswaaij, C. M, Admiraal, R., Hurst, J. A., de Vries, B. B., Janssen, I. M., van der Vliet, W. A., Huys, E. H., de Jong, P. J., Hamel, B. C., Schoenmakers, E. F, Brunner, H. G., Veltman, J. A., and van Kessel, AG. (2004). Mutations in a new member of the chromodomain gene family cause CHARGE syndrome. *Nat. Genet.* **36,** 955–957.
183. Jongmans, M., Admiraal, R., van der Donk, K., Vissers, L., Baas, A., Kapusta, L, van Hagen, A., Donnai, D., de Ravel, T., Veltman, J., Geurts van Kessel, A., de Vries, B., Brunner, H., Hoefsloot, L., and van Ravenswaaij, C. (2006). CHARGE syndrome: the phenotypic spectrum of mutations in the CHD7 gene. *J. Med. Genet.* **43,** 306–314.
184. Brunner, H. G., and van Bokhoven, H. (2005). Genetic players in esophageal atresia and tracheoesophageal fistula. *Curr. Opin. Genet. Dev.* **15,** 341–34.7
185. Williams, M. S. (2005). Speculations on the pathogenesis of CHARGE syndrome. *Am. J. Med. Genet. A* **133,** 318–325
186. Bohlmeyer, T. J., Helmke, S., Ge, S., Lynch, J., Brodsky, G, Sederberg, J. H., Robertson, A. D., Minobe, W., Bristow, M. R., and Perryman, M. B. (2003). Hypoplastic left heart syndrome myocytes are differentiated but possess a unique phenotype. *Cardiovasc. Pathol.* **12,** 23–31.
187. Kaynak, B., von Heydebreck, A., Mebus, S., Seelow, D., Hennig, S., Vogel, J., Sperling, H. P., Pregla, R., Alexi-Meskishvili, V., Hetzer, R., Lange, P. E., Vingron, M., Lehrach, H., and Sperling, S. (2003). Genome-wide array analysis of normal and malformed human hearts. *Circulation* **107,** 2467–2474.
188. Dietz, H. C., Cutting, G. R., Pyeritz, R. E., Maslen, C. L., Sakai, L. Y., Corson, G. M., Puffenberger, E. G., Hamosh, A., Nanthakumar, E. J., Curristin, S. M., Stetten, G., Meyers, D. A., and Francomano, C. A. (1991). Marfan syndrome caused by a recurrent de novo missense mutation in the fibrillin gene. *Nature* **352,** 337–339.
189. Ewart, A. K., Morris, C. A., Atkinson, D., Jin, W., Sternes, K., Spallone, P., Stock, A. D., Leppert, M., and Keating, M. T. (1993). Hemizygosity at the elastin locus in a developmental disorder, Williams syndrome. *Nat. Genet.* **5,** 11–16.
190. Pereira, L., Andrikopoulos, K., Tian, J., Lee, S. Y., Keene, D. R., Ono, R., Reinhardt, D. P., Sakai, L. Y., Biery, N. J., Bunton, T., Dietz, H. C., and Ramirez, F. (1997). Targeting of the gene encoding fibrillin-1 recapitulates the vascular aspect of Marfan syndrome. *Nat. Genet.* **17,** 218–222
191. Pereira, L., Lee, S. Y., Gayraud, B., Andrikopoulos, K., Shapiro, S. D., Bunton, T., Biery, N. J., Dietz, H. C., Sakai L. Y., and Ramirez, F. (1999). Pathogenetic sequence for aneurysm revealed in mice underexpressing fibrillin-1. *Proc. Natl. Acad. Sci. USA* **96,** 3819–3823.
192. Segura, A. M., Luna, R. E., Horiba, K., Stetler-Stevenson, W. G., McAllister, H. A., Willerson, J. T., and Ferrans, V. J. (1998). Immunohistochemistry of matrix metalloproteinases and their inhibitors in thoracic aortic aneurysms and aortic valves of patients with Marfan's syndrome. *Circulation* **98,** 331–337.
193. Wang, M., Price, C., Han, J., Cisler, J., Imaizumi, K., Van Thienen, M. N., DePaepe, A., and Godfrey, M. (1995). Recurrent mis-splicing of fibrillin exon 32 in two patients with neonatal Marfan syndrome. *Hum. Mol. Genet.* **4,** 607–613.
194. Pepe, G., Giusti, B., Attanasio, M., Comeglio, P.,, Porciani M. C., Giurlani, L., Montesi, G. F., Calamai, G. C., Vaccari, M., Favilli, S., Abbate, R., and Gensini, G. F. (1997). A major involvement of the cardiovascular system in patients affected by Marfan syndrome: novel mutations in fibrillin 1 gene. *J. Mol. Cell Cardiol.* **29,** 1877–1884.
195. Schrijver, I., Liu, W., Brenn, T., Furthmayr, H., and Francke, U. (1999). Cysteine substitutions in epidermal growth factor-like domains of fibrillin-1: distinct effects on biochemical and clinical phenotypes. *Am. J. Hum. Genet.* **65,** 1007–1020.
196. Booms, P., Cisler, J., Mathews, K. R., Godfrey, M., Tiecke, F., Kaufmann, U. C., Vetter, U., Hagemeier, C., and Robinson, P. N. (1999). Novel exon skipping mutation in the fibrillin-1 gene: two 'hot spots' for the neonatal Marfan syndrome. *Clin. Genet.* **55,** 110–117.
197. Shinawi, M., Boileau, C., Brik, R., Mandel, H., and Bentur, L. (2005). Splicing mutation in the fibrillin-1 gene associated with neonatal Marfan syndrome and severe pulmonary emphysema with tracheobronchomalacia. *Pediatr. Pulmonol.* **39,** 374–378.
198. Pepe, G., Giusti, B., Evangelisti, L., Porciani, M.C., Brunelli, T., Giurlani, L., Attanasio, M., Fattori, R., Bagni, C., Comeglio, P., Abbate, R., and Gensini, G. F. (2001). Fibrillin-1 (FBN1) gene frameshift mutations in Marfan patients: genotype-phenotype correlation. *Clin. Genet.* **59,** 444–450.
199. Judge, D. P., Biery, N. J., Keene, D. R., Geubtner, J., Myers, L., Huso, D. L., Sakai, L. Y., and Dietz, H. C. (2004). Evidence for a critical contribution of haploinsufficiency in the complex pathogenesis of Marfan syndrome. *J. Clin. Invest.* **114,** 172–181.
200. Wouters, C. H., Meijers-Heijboer, H. J., Eussen, B. J., van der Heide, A. A., van Luijk, R. B., van Drunen, E., Beverloo, B. B., Visscher, F., and Van Hemel, J. O. (2001). Deletions at chromosome regions 7q11.23 and 7q36 in a patient with Williams syndrome. *Am. J. Med. Genet.* **102,** 261–265.
201. Hoogenraad, C.C., Akhmanova, A., Galjart, N., and De Zeeuw, C.I. (2004). LIMK1 and CLIP-115: linking cytoskeletal defects to Williams syndrome. *Bioessays* **26,** 141–150.
202. Palau, F. (2001). Friedreich's ataxia and frataxin: molecular genetics, evolution and pathogenesis. *Int. J. Mol. Med.* **7,** 581–589.
203. Korade-Mirnics, Z., Tarleton, J., Servidei, S., Casey, R. R., Gennarelli, M., Pegoraro, E., Angelini, C., and Hoffman, E. P. (1999). Myotonic dystrophy: tissue-specific effect of somatic CTG expansions on allele-specific DMAHP/SIX5 expression. *Hum. Mol. Genet.* **8,** 1017–1023.

204. Puccio, H., and Koenig, M. (2000). Recent advances in the molecular pathogenesis of Friedreich ataxia. *Hum. Mol. Genet.* **9,** 887–892.
205. Bit-Avragim, N., Perrot, A., Schols, L., Hardt, C., Kreuz, F. R., Zuhlke, C., Bubel, S., Laccone, F., Vogel, H. P., Dietz, R., and Osterziel, K. J. (2001). The GAA repeat expansion in intron 1 of the frataxin gene is related to the severity of cardiac manifestation in patients with Friedreich's ataxia. *J. Mol. Med.* **78,** 626–632.
206. Bunse, M., Bit-Avragim, N., Rieffin, A., Perrot, A., Schmidt, O., Kreuz, F. R., Dietz, R., Jung, W. I., and Osterziel, K. J. (2003). Cardiac energetics correlates to myocardial hypertrophy in Friedreich's ataxia. *Ann. Neurol.* **53,** 121–123.
207. Hausse, A. O., Aggoun, Y., Bonnet, D., Sidi, D., Munnich, A., Rotig, A., and Rustin, P. (2002). Idebenone and reduced cardiac hypertrophy in Friedreich's ataxia. *Heart* **87,** 346–349.
208. Cossee, M., Puccio, H., Gansmuller, A., Koutnikova, H., Dierich, A., LeMeur, M., Fischbeck, K., Dolle, P., and Koenig, M. (2000). Inactivation of the Friedreich ataxia mouse gene leads to early embryonic lethality without iron accumulation. *Hum. Mol. Genet.* **9,** 1219–1226.
209. Puccio, H., Simon, D., Cossee, M., Criqui-Filipe, P., Tiziano, F., Melki, J., Hindelang, C., Matyas, R., Rustin, P., and Koenig, M. (2001). Mouse models for Friedreich ataxia exhibit cardiomyopathy, sensory nerve defect and Fe-S enzyme deficiency followed by intramitochondrial iron deposits. *Nat. Genet.* **27,** 181–186.
210. Pook, M. A., Al-Mahdawi, S., Carroll, C. J., Cossee, M., Puccio, H., Lawrence, L., Clark, P., Lowrie, M. B., Bradley, J. L., Cooper, J. M., Koenig, M., and Chamberlain, S. (2001). Rescue of the Friedreich's ataxia knockout mouse by human YAC transgenesis. *Neurogenetics* **3,** 185–193.
211. Wang, D. W., Yazawa, K., George, A. L. Jr., and Bennett, P. B. (1996). Characterization of human cardiac Na+ channel mutations in the congenital long QT syndrome. *Proc. Natl. Acad. Sci. USA* **93,** 13200–13205.
212. Towbin, J. A., Wang, Z., and Li, H. (2001). Genotype and severity of long QT syndrome. *Drug Metab. Dispos.* **29,** 574–579.
213. Chen, Q., Kirsch, G. E., Zhang, D., Brugada, R., Brugada, J., Brugada, P., Potenza, D., Moya, A., Borggrefe, M., Breithardt, G., Ortiz-Lopez, R., Wang, Z., Antzelevitch, C., O'Brien, R. E., Schulze-Bahr, E., Keating, M. T., Towbin, J. A., and Wang Q. (1998). Genetic basis and molecular mechanism for idiopathic ventricular fibrillation. *Nature* **392,** 293–296.
214. Bezzina, C., Veldkamp, M. W., van Den Berg, M. P., Postma, A. V., Rook, M. B., Viersma, J. W., van Langen, I. M., Tan-Sindhunata, G., Bink-Boelkens, M. T., van Der Hout, A. H., Mannens, M. M., and Wilde, A. A. (1999). A single Na(+) channel mutation causing both long-QT and Brugada syndromes. *Circ. Res.* **85,** 1206–1213.
215. Grant, A. O. (2005). Electrophysiological basis and genetics of Brugada syndrome. *J. Cardiovasc. Electrophysiol.* **16,** S3-S7.
216. Splawski, I., Shen, J., Timothy, K. W., Lehmann, M. H., Priori, S., Robinson, J. L., Moss, A. J., Schwartz, P. J. Towbin, J. A., Vincent, G. M., and Keating, M. T. (2000). Spectrum of mutations in long-QT syndrome genes. KVLQT1, HERG, SCN5A, KCNE1, and KCNE2. *Circulation* **102,** 1178–1185.
217. Tester, D. J., Will, M. L., Haglund, C. M., and Ackerman, M. J. (2005). Compendium of cardiac channel mutations in 541 consecutive unrelated patients referred for long QT syndrome genetic testing. *Heart Rhythm* **2,** 507–517.
218. Choi, G., Kopplin, L. J., Tester, D. J., Will, M. L., Haglund, C. M., and Ackerman, M. J. (2004). Spectrum and frequency of cardiac channel defects in swimming-triggered arrhythmia syndromes. *Circulation* **110,** 2119–2124.
219. Brugada, R., Hong, K., Dumaine, R., Cordeiro, J., Gaita, F., Borggrefe, M., Menendez, T. M., Brugada, J., Pollevick, G. D., Wolpert, C., Burashnikov, E., Matsuo, K., Wu, Y. S., Guerchicoff, A., Bianchi, F., Giustetto, C., Schimpf, R., Brugada, P., and Antzelevitch, C. (2004). Sudden death associated with short-QT syndrome linked to mutations in HERG. *Circulation* **109,** 30–35.
220. Huang, L., Bitner-Glindzicz, M., Tranebjaerg, L., and Tinker, A. (2001). A spectrum of functional effects for disease causing mutations in the Jervell and Lange-Nielsen syndrome. *Cardiovasc. Res.* **51,** 670–680.
221. Chun, T. U., Epstein, M. R., Dick, M. 2nd., Andelfinger, G., Ballester, L., Vanoye, C. G., George, A. L. Jr., and Benson, D. W. (2004). Polymorphic ventricular tachycardia and KCNJ2 mutations. *Heart Rhythm* **1,** 235–241.
222. Tristani-Firouzi, M., Jensen, J. L., Donaldson, M. R., Sansone, V., Meola, G., Hahn, A., Bendahhou, S., Kwiecinski, H., Fidzianska, A., Plaster, N., Fu, Y. H., Ptacek, L. J., and Tawil, R. (2002). Functional and clinical characterization of KCNJ2 mutations associated with LQT7 (Andersen syndrome). *J. Clin. Invest.* **110,** 381–388.
223. Laitinen, P. J., Brown, K. M., Piippo, K., Swan, H., Devaney, J. M., Brahmbhatt, B., Donarum, E. A., Marino, M., Tiso, N., Viitasalo, M., Toivonen, L., Stephan, D. A., and Kontula, K. (2001). Mutations of the cardiac ryanodine receptor (RyR2) gene in familial polymorphic ventricular tachycardia. *Circulation* **103,** 485–490.
224. Laitinen, P. J., Swan, H., and Kontula, K. (2003). Molecular genetics of exercise-induced polymorphic ventricular tachycardia: identification of three novel cardiac ryanodine receptor mutations and two common calsequestrin 2 amino-acid polymorphisms. *Eur. J. Hum. Genet.* **11,** 888–891.
225. Laitinen, P. J., Swan, H., Piippo, K., Viitasalo, M., Toivonen, L., and Kontula, K. (2004). Genes, exercise and sudden death: molecular basis of familial catecholaminergic polymorphic ventricular tachycardia. *Ann. Med.* **36,** 81–86.
226. Lahat, H., Pras, E., and Eldar, M. (2004). A missense mutation in CASQ2 is associated with autosomal recessive catecholamine-induced polymorphic ventricular tachycardia in Bedouin families from Israel. *Ann. Med.* **36,** 87–91.
227. Bonne, G., Di Barletta, M. R., Varnous, S., Becane, H. M., Hammouda, E. H., Merlini, L., Muntoni, F., Greenberg, C. R., Gary, F., Urtizberea, J. A., Duboc, D., Fardeau, M., Toniolo, D., and Schwartz, K. (1999). Mutations in the gene encoding lamin A/C cause autosomal dominant Emery-Dreifuss muscular dystrophy. *Nat. Genet.* **21,** 285–288.
228. Bonnet, D., Martin, D., de Lonlay, P., Villain, E., Jouvet, P., Rabier, D., Brivet, M., and Saudubray, J. M. (1999). Arrhythmias and conduction defects as presenting symptoms

of fatty acid oxidation disorders in children. *Circulation* **100,** 2248–2253.
229. Kelly, D. P., Hale, D. E., Rutledge, S. L., Ogden, M. L., Whelan, A. J., Zhang, Z., and Strauss, A. W. (1992). Molecular basis of inherited medium-chain acyl-CoA dehydrogenase deficiency causing sudden child death. *J. Inherit. Metab. Dis.* **15,** 171–180.
230. Korman, S. H., Gutman, A., Brooks, R., Sinnathamby, T., Gregersen, N., and Andresen, B. S. (2004). Homozygosity for a severe novel medium-chain acyl-CoA dehydrogenase (MCAD) mutation IVS3-1G > C that leads to introduction of a premature termination codon by complete missplicing of the MCAD mRNA and is associated with phenotypic diversity ranging from sudden neonatal death to asymptomatic status. *Mol. Genet. Metab.* **82,** 121–129.
231. Opdal, S. H., and Rognum, T. O. (2004). The sudden infant death syndrome gene: does it exist? *Pediatrics* **114,** e506–e512.
232. Gollob, M. H., Green, M. S., Tang, A. S., and Roberts, R. (2002). PRKAG2 cardiac syndrome: familial ventricular preexcitation, conduction system disease, and cardiac hypertrophy. *Curr. Opin. Cardiol.* **17,** 229–234.
233. Sen-Chowdhry, S., Syrris, P., and McKenna, W. J. (2005). Genetics of right ventricular cardiomyopathy. *J. Cardiovasc. Electrophysiol.* **16,** 927–935.
234. Dokuparti, M. V., Pamuru, P. R., Thakkar, B., Tanjore, R. R., and Nallari, P. (2005). Etiopathogenesis of arrhythmogenic right ventricular cardiomyopathy. *J. Hum. Genet.* **50,** 375–381.
235. Gerull, B., Heuser, A., Wichter, T., Paul, M., Basson, C. T., McDermott, D. A., Lerman, B. B., Markowitz, S. M., Ellinor, P. T., MacRae, C. A., Peters, S., Grossmann. K. S., Drenckhahn, J., Michely, B., Sasse-Klaassen, S., Birchmeier, W., Dietz, R., Breithardt, G., Schulze-Barr, E., and Thierfelder, L. (2004). Mutations in the desmosomal protein plakophilin-2 are common in arrhythmogenic right ventricular cardiomyopathy. *Nat. Genet.* **36,** 1162–1164.
236. Norman, M., Simpson, M., Mogensen, J., Shaw, A., Hughes, S., Syrris, P., Sen-Chowdhry, S., Rowland, E., Crosby, A., and McKenna, W. J. (2005). Novel mutation in desmoplakin causes arrhythmogenic left ventricular cardiomyopathy. *Circulation* **112,** 636–642.
237. McKoy, G., Protonotarios, N., Crosby, A., Tsatsopoulou, A., Anastasakis, A., Coonar, A., Norman, M., Baboonian, C., Jeffery, S., and McKenna, W. J. (2000). Identification of a deletion in plakoglobin in arrhythmogenic right ventricular cardiomyopathy with palmoplantar keratoderma and woolly hair (Naxos disease). *Lancet* **355,** 2119–2124.
238. Beffagna, G., Occhi, G., Nava, A., Vitiello, L., Ditadi, A., Basso, C., Bauce, B., Carraro, G., Thiene, G., Towbin, J. A., Danieli, G. A., and Rampazzo, A. (2005). Regulatory mutations in transforming growth factor-beta3 gene cause arrhythmogenic right ventricular cardiomyopathy type 1. *Cardiovasc. Res.* **65,** 366–373.
239. Tiso, N., Stephan, D. A., Nava, A., Bagattin, A., Devaney, J. M., Stanchi, F., Larderet, G., Brahmbhatt, B., Brown, K., Bauce, B., Muriago, M., Basso, C., Thiene, G., Danieli, G. A., and Rampazzo, A. (2001). Identification of mutations in the cardiac ryanodine receptor gene in families affected with arrhythmogenic right ventricular cardiomyopathy type 2 (ARVD2). *Hum. Mol. Genet.* **10,** 189–194.
240. Tiso, N., Salamon, M., Bagattin, A., Danieli, G. A., Argenton, F., and Bortolussi, M. (2002). The binding of the RyR2 calcium channel to its gating protein FKBP12.6 is oppositely affected by ARVD2 and VTSIP mutations. *Biochem. Biophys. Res. Commun.* **299,** 594–598.
241. Maisch, B., Richter, A., Sandmoller, A., Portig, I., and Pankuweit, S. (2005). Inflammatory dilated cardiomyopathy (DCMI). *Herz* **30,** 535–544.
242. Roberts, R., and Sidhu, J. (2003). Genetic basis for hypertrophic cardiomyopathy: implications for diagnosis and treatment. *Am. Heart. Hosp. J.* **1,** 128–134.
243. Ramirez, C. D., and Padron, R. (2004). Familial hypertrophic cardiomyopathy: genes, mutations and animal models. A review. *Invest. Clin.* **45,** 69–99.
244. Bonne, G., Carrier, L., Bercovici, J., Cruaud, C., Richard, P., Hainque, B., Gautel, M., Labeit, S., James, M., Beckman, J., Weissenbach, J., Vosberg, H. P., Fiszman, M., Komajda, M., and Schwartz, K. (1995). Cardiac myosin binding protein-C gene splice acceptor site mutation is associated with familial hypertrophic cardiomyopathy. *Nat. Genet.* **11,** 438–440.
245. Thierfelder, L., Watkins, H., MacRae, C., Lamas, R., McKenna, W., Vosberg, H. P., Seidman, J. G., and Seidman, C. E. (1994). α-tropomyosin and cardiac troponin T mutations cause familial hypertrophic cardiomyopathy: a disease of the sarcomere. *Cell* **77,** 701–712.
246. Kimura, A., Harada, H., Park, J. E., Nishi, H., Satoh, M., Takahashi, M., Hiroi, S., Sasaoka, T., Ohbuchi, N., Nakamura, T., Koyanagi, T., Hwang, T. H., Choo, J. A., Chung, K. S., Hasegawa, A., Nagai, R., Okazaki, O., Nakamura, H., Matsuzaki, M., Sakamoto, T., Toshima, H., Koga, Y., Imaizumi, T., and Sasazuki, T. (1997). Mutations in the cardiac troponin I gene associated with hypertrophic cardiomyopathy. *Nat. Genet.* **16,** 379–382.
247. Berul, C. I., Christe, M. E., Aronovitz, M. J., Seidman, C. E., Seidman, J. G., and Mendelsohn, M. E. (1997). Electrophysiological abnormalities and arrhythmias in alpha MHC mutant familial hypertrophic cardiomyopathy mice. *J. Clin. Invest.* **99,** 570–576.
248. Anan, R., Greve, G., Thierfelder, L., Watkins, H., McKenna, W. J., Solomon, S., Vecchio, C., Shono, H., Nakao, S., Tanaka, H., Mares, A., Towbin, J. A., Spirito, P., Roberts, R., Seidman, J. G., and Seidman, C. E. (1994). Prognostic implications of novel β cardiac myosin heavy chain gene mutations that cause familial hypertrophic cardiomyopathy. *J. Clin. Invest.* **93,** 280–285.
249. Olson, T. M., Doan, T. P., Kishimoto, N. Y., Whitby, F. G., Ackerman, M. J., and Fananapazir, L. (2000). Inherited and de novo mutations in the cardiac actin gene cause hypertrophic cardiomyopathy. *J. Mol. Cell Cardiol.* **32,** 1687–1694.
250. Satoh, M., Takahashi, M., Sakamoto, T., Hiroe, M., Marumo, F., and Kimura, A. (1999). Structural analysis of the titin gene in hypertrophic cardiomyopathy: identification of a novel disease gene. *Biochem. Biophys. Res. Commun.* **262,** 411–417.
251. Richard, P., Charron, P., Carrier, L., Ledeuil, C., Cheav, T., Pichereau, C., Benaiche, A., Isnard, R., Dubourg, O., Burban, M., Gueffet, J. P., Millaire, A., Desnos, M., Schwartz, K., Hainque, B., and Komajda, M., and the EUROGENE Heart Failure Project. (2003). Hypertrophic cardiomyopathy:

distribution of disease genes, spectrum of mutations, and implications for a molecular diagnosis strategy. *Circulation* **107**, 2227–2232.

252. Morita, H., DePalma, S. R., Arad, M., McDonough, B., Barr, S., Duffy, C., Maron, B. J., Seidman, C. E., and Seidman, J. G. (2002). Molecular epidemiology of hypertrophic cardiomyopathy. *Cold Spring Harb. Symp. Quant. Biol.* **67**, 383–388.

253. Yang, Z., McMahon, C. J., Smith, L. R., Bersola, J., Adesina, A. M., Breinholt, J. P., Kearney, D. L., Dreyer, W. J., Denfield, S. W., Price, J. F., Grenier, M., Kertesz, N. J., Clunie, S. K., Fernbach, S. D., Southern, J. F., Berger, S., Towbin, J. A., Bowles, K. R., and Bowles, N. E. (2005). Danon disease as an underrecognized cause of hypertrophic cardiomyopathy in children. *Circulation* **112**, 1612–1617.

254. Balmer, C., Ballhausen, D., Bosshard, N. U., Steinmann, B., Boltshauser, E., Bauersfeld, U., and Superti-Furga, A. (2005). Familial X-linked cardiomyopathy (Danon disease): diagnostic confirmation by mutation analysis of the LAMP2 gene. *Eur. J. Pediatr.* **164**, 509–514.

255. Antonicka, H., Mattman, A., Carlson, C. G., Glerum, D. M., Hoffbuhr, K. C., Leary, S. C., Kennaway, N. G., and Shoubridge, E. A. (2003). Mutations in COX15 produce a defect in the mitochondrial heme biosynthetic pathway, causing early-onset fatal hypertrophic cardiomyopathy. *Am. J. Hum. Genet.* **72**, 101–114.

256. Van Driest, S. L., Gakh, O., Ommen, S. R., Isaya, G., and Ackerman, M. J. (2005). Molecular and functional characterization of a human frataxin mutation found in hypertrophic cardiomyopathy. *Mol. Genet. Metab.* **85**, 280–285.

257. Marin-Garcia, J., and Goldenthal, M. J. (2002). Understanding the impact of mitochondrial defects in cardiovascular disease: a review. *J. Card. Fail.* **8**, 347–361.

258. Arad, M., Benson, D. W., Perez-Atayde, A. R., McKenna, W. J., Sparks, E. A., Kanter, R. J., McGarry, K, Seidman, J. G., and Seidman, C. E. (2002). Constitutively active AMP kinase mutations cause glycogen storage disease mimicking hypertrophic cardiomyopathy. *J. Clin. Invest.* **109**, 357–362.

259. Lipshultz, S. E., Sleeper, L. A., Towbin, J. A., Lowe, A. M., Orav, E. J., Cox, G. F., Lurie, P. R., McCoy, K L., McDonald, M. A., Messere, J. E., and Colan, S. D. (2003). The incidence of pediatric cardiomyopathy in two regions of the United States. *N. Engl. J. Med.* **348**, 1647–1655.

260. Beggs, A. H. (1997). Dystrophinopathy, the expanding phenotype. Dystrophin abnormalities in X-linked dilated cardiomyopathy. *Circulation* **95**, 2344–2347.

261. D'Adamo, P., Fassone, L., Gedeon, A., Janssen, E. A., Bione, S., Bolhuis, P. A., Barth, P. G., Wilson, M., Haan, E., Orstavik, K. H., Patton, M. A., Green, A. J., Zammarchi, E., Donati, M. A., and Toniolo, D. (1997). The X-linked gene G4.5 is responsible for different infantile dilated cardiomyopathies. *Am. J. Hum. Genet.* **61**, 862–867.

262. Olson, T. M., Michels, V. V., Thibodeau, S. N., Tai, Y. S., and Keating, M. T. (1998). Actin mutations in dilated cardiomyopathy, a heritable form of heart failure. *Science* **280**, 750–752.

263. Dalakas, M. C., Park, K. Y., Semino-Mora, C., Lee, H. S., Sivakumar, K., and Goldfarb, L. G. (2000). Desmin myopathy, a skeletal myopathy with cardiomyopathy caused by mutations in the desmin gene. *N. Engl. J. Med.* **342**, 770–780.

264. Tsubata, S., Bowles, K. R., Vatta, M., Zintz, C., Titus, J., Muhonen, L., Bowles, N. E., and Towbin, J. A. (2000). Mutations in the human Δ-sarcoglycan gene in familial and sporadic dilated cardiomyopathy. *J. Clin. Invest.* **106**, 655–662.

265. Zhang, J., Kumar, A., Stalker, H. J., Virdi, G., Ferrans, V. J., Horiba, K., Fricker, F. J., and Wallace, M. R. (2001). Clinical and molecular studies of a large family with desmin-associated restrictive cardiomyopathy. *Clin. Genet.* **59**, 248–256.

266. Towbin, J. A., Hejtmancik, J. F., Brink, P., Gelb, B., Zhu, X. M., Chamberlain, J. S., McCabe, E. R., and Swift, M. (1993). X-linked dilated cardiomyopathy. Molecular genetic evidence of linkage to the Duchenne muscular dystrophy (dystrophin) gene at the Xp21 locus. *Circulation* **87**, 1854–1865.

267. Neuwald, A. F. (1997). Barth syndrome may be due to an acyltransferase deficiency. *Curr. Biol.* **7**, 465–466.

268. Rampazzo, A., Beffagna, G., Nava, A., Occhi, G., Bauce, B., Noiato, M., Basso, C., Frigo, G., Thiene, G., Towbin, J., and Danieli, G. A. (2003). Arrhythmogenic right ventricular cardiomyopathy type 1 (ARVD1): confirmation of locus assignment and mutation screening of four candidate genes. *Eur. J. Hum. Genet.* **11**, 69–76.

269. Pignatelli, R. H., McMahon, C. J., Dreyer, W. J., Denfield, S. W., Price, J., Belmont, J. W., Craigen, W. J., Wu, J., El Said, H., Bezold, L. I., Clunie, S., Fernbach, S., Bowles, N. E., and Towbin, J. A. (2003). Clinical characterization of left ventricular noncompaction in children: A relatively common form of cardiomyopathy. *Circulation* **108**, 2672–2678.

270. Sasse-Klaassen, S., Probst, S., Gerull, B., Oechslin, E., Nürnberg, P., Heuser, A., Jenni, R., Hennies, H. C., and Thierfelder, L. (2004). Novel gene locus for autosomal dominant left ventricular noncompaction maps to chromosome 11p15. *Circulation* **109**, 2720–2723.

271. Kelly, D. P., and Strauss, A. W. (1994). Inherited cardiomyopathies. *N. Engl. J. Med.* **330**, 913–919.

272. Hale, D. E., Batshaw, M. L., Coates, P. M., Frerman, F. E., Goodman, S. I., Singh, I., and Stanley, C. A. (1985). Long-chain acyl coenzyme A dehydrogenase deficiency: An inherited cause of nonketotic hypoglycemia. *Pediatr. Res.* **19**, 666–671.

273. Rocchiccioli, F., Wanders, R. J., Aubourg, P., Vianey-Liaud, C., Ijlst, L., Fabre, M., Cartier, N., ad Bougneres, P. F. (1990). Deficiency of long-chain 3-hydroxyacyl-CoA dehydrogenase: A cause of lethal myopathy and cardiomyopathy in early childhood. *Pediatr. Res.* **28**, 657–662.

274. Strauss, A. W., Powell, C. K., Hale, D. E., Anderson, M. M., Ahuja, A., Brackett, J. C., and Sims, H. F. (1995). Molecular basis of human mitochondrial very-long-chain acyl-CoA dehydrogenase deficiency causing cardiomyopathy and sudden death in childhood. *Proc. Natl. Acad. Sci. USA* **92**, 10496–10500.

275. Tein, I., Haslam, R. H., Rhead, W. J., Bennett, M. J., Becker, L. E., and Vockley, J. (1999). Short-chain acyl-CoA dehydrogenase deficiency: A cause of ophthalmoplegia and multicore myopathy. *Neurology* **52**, 366–372.

276. Mathur, A., Sims, H. F., Gopalakrishnan, D., Gibson, B., Rinaldo, P., Vockley, J., Hug, G., and Strauss, A. W. (1999). Molecular heterogeneity in very-long-chain acyl-CoA dehydrogenase deficiency causing pediatric cardiomyopathy and sudden death. *Circulation* **99**, 1337–1343.

277. Angdisen, J., Moore, V. D., Cline, J. M., Payne, R. M., and Ibdah, J. A. (2005). Mitochondrial trifunctional protein defects: molecular basis and novel therapeutic approaches. *Curr. Drug Targets Immune Endocr. Metabol. Disord.* **5**, 27–40.

278. Spiekerkoetter, U., Khuchua, Z., Yue, Z., Bennett, M. J., and Strauss, A. W. (2004). General mitochondrial trifunctional protein (TFP) deficiency as a result of either α- or β-subunit mutations exhibits similar phenotypes because mutations in either subunit alter TFP complex expression and subunit turnover. *Pediatr. Res.* **55,** 190–196.

279. Schwab, K. O., Ensenauer, R., Matern, D., Uyanik, G., Schnieders, B., Wanders, R. A., and Lehnert, W. (2003). Complete deficiency of mitochondrial trifunctional protein due to a novel mutation within the β-subunit of the mitochondrial trifunctional protein gene leads to failure of long-chain fatty acid β-oxidation with fatal outcome. *Eur. J. Pediatr.* **162,** 90–95.

280. Ibdah, J. A., Zhao, Y., Viola, J., Gibson, B., Bennett, M. J., and Strauss, A. W. (2001). Molecular prenatal diagnosis in families with fetal mitochondrial trifunctional protein mutations. *J. Pediatr.* **138,** 396–399.

281. Surendran, S., Sacksteder, K. A., Gould, S. J., Coldwell, J. G., Rady, P. L., Tyring, S. K., and Matalon, R. (2001). Malonyl CoA decarboxylase deficiency: C to T transition in intron 2 of the MCD gene. *J. Neurosci. Res.* **65,** 591–594.

282. Oliveira, S. M., Ehtisham, J., Redwood, C. S., Ostman-Smith, I., Blair, E. M., and Watkins, H. (2003). Mutation analysis of AMP-activated protein kinase subunits in inherited cardiomyopathies: Implications for kinase function and disease pathogenesis. *J. Mol. Cell Cardiol.* **35,** 1251–1255.

283. Roe, C. R., and Ding, J. H. (2001). *In* "Metabolic and Molecular Basis of Inherited Disease." Vol. 2. (C. Scriver, C. R. Scriver, and W. L. Sly, Eds.), pp. 2297–2326. McGraw-Hill.

284. Engel, A. G., and Angelini, C. (1973). Carnitine deficiency of human skeletal muscle with associated lipid storage myopathy: A new syndrome. *Science* **179,** 899–902.

285. Stanley, C. A. (2004). Carnitine deficiency disorders in children. *Ann. N. Y. Acad. Sci.* **1033,** 42–51.

286. Stanley, C. A., Treem, W. R., Hale, D. E., and Coates, P. M. (1990). A genetic defect in carnitine transport causing primary carnitine deficiency. *Prog. Clin. Biol. Res.* **321,** 457–464.

287. Nezu, J., Tamai, I., Oku, A., Ohashi, R., Yabuuchi, H., Hashimoto, N., Nikaido, H., Sai, Y., Koizumi, A., Shoji, Y., Takada, G., Matsuishi, T., Yoshino, M., Kato, H., Ohura, T., Tsujimoto, G., Hayakawa, J., Shimane, M., and Tsuji, A. (1999). Primary systemic carnitine deficiency is caused by mutations in a gene encoding sodium ion-dependent carnitine transporter. *Nat. Genet.* **21,** 91–94.

288. Melegh, B., Bene, J., Mogyorosy, G., Havasi, V., Komlosi, K., Pajor, L., Olah, E., Kispal, G., Sumegi, B., and Mehes, K. (2004). Phenotypic manifestations of the OCTN2 V295X mutation: sudden infant death and carnitine-responsive cardiomyopathy in Roma families. *Am. J. Med. Genet. A.* **131,** 121–126.

289. Tein, I. (2003). Carnitine transport: pathophysiology and metabolism of known molecular defects. *J. Inherit. Metab. Dis.* **26,** 147–169.

290. Roschinger, W., Muntau, A. C., Duran, M., Dorland, L., Ijlst, L., Wanders, R. J., and Roscher, A. A. (2000). Carnitine-acylcarnitine translocase deficiency: Metabolic consequences of an impaired mitochondrial carnitine cycle. *Clin. Chim. Acta* **298,** 55–68.

291. Iacobazzi, V., Pasquali, M., Singh, R., Matern, D., Rinaldo, P., Amat di San Filippo, C., Palmieri, F., and Longo, N. (2004). Response to therapy in carnitine/acylcarnitine translocase (CACT) deficiency due to a novel missense mutation. *Am. J. Med. Genet.* **126A,** 150–155.

292. Olpin, S. E., Allen, J., Bonham, J. R., Clark, S., Clayton, P. T., Calvin, J., Downing, M., Ives, K., Jones, S., Manning, N. J., Pollitt, R. J., Standing, S. J., and Tanner, M. S. (2001). Features of carnitine palmitoyltransferase type I deficiency. *J. Inherit. Metab. Dis.* **24,** 35–42.

293. Demaugre, F., Bonnefont, J. P., Colonna, M., Cepanec, C., Leroux, J. P., and Saudubray, J. M. (1991). Infantile form of carnitine palmitoyltransferase II deficiency with hepatomuscular symptoms and sudden death: Physiopathological approach to carnitine palmitoyltransferase II deficiencies. *J. Clin. Invest.* **87,** 859–864.

294. Bonnefont, J. P., Taroni, F., Cavadini, P., Cepanec, C., Brivet, M., Saudubray, J. M., Leroux, J. P., and Demaugre, F. (1996). Molecular analysis of carnitine palmitoyltransferase II deficiency with hepatocardiomuscular expression. *Am. J. Hum. Genet.* **58,** 971–978.

295. Taroni, F., Verderio, E., Fiorucci, S., Cavadini, P., Finocchiaro, G., Uziel, G., Lamantea, E., Gellera, C., and DiDonato, S. (1992). Molecular characterization of inherited carnitine palmitoyltransferase II deficiency. *Proc. Natl. Acad. Sci. USA* **89,** 8429–8433.

296. Okamoto, F., Tanaka, T., Sohmiya, K., and Kawamura, K. (1998). CD36 abnormality and impaired myocardial long-chain fatty acid uptake in patients with hypertrophic cardiomyopathy. *Jpn. Circ. J.* **7,** 499–504.

297. Hirano, K., Kuwasako, T., Nakagawa-Toyama, Y., Janabi, M., Yamashita, S., and Matsuzawa, Y. (2003). Pathophysiology of human genetic CD36 deficiency. *Trends Cardiovasc. Med.* **13,** 136–141.

298. Brinkmann, J. F., Abumrad, N. A., Ibrahimi, A., van der Vusse, G. J., and Glatz, J. F. (2002). New insights into long-chain fatty acid uptake by heart muscle: a crucial role for fatty acid translocase/CD36. *Biochem. J.* **367,** 561–570.

299. Febbraio, M., Guy, E., Coburn, C., Knapp, F. F. Jr., Beets, A. L., Abumrad, N. A., and Silverstein, R. L. (2002). The impact of overexpression and deficiency of fatty acid translocase (FAT)/CD36. *Mol. Cell Biochem.* **239,** 193–197.

300. Treem, W. R. (2000). New developments in the pathophysiology, clinical spectrum, and diagnosis of disorders of fatty acid oxidation. *Curr. Opin. Pediatr.* **12,** 463–468.

301. Teraguchi, M., Ikemoto, Y., Unishi, G., Ohkohchi, H., and Kobayashi, Y. (2004). Influence of CD36 deficiency on heart disease in children. *Circ. J.* **68,** 435–438.

302. Salazar, D., Zhang, L., deGala, G. D., and Frerman, F. E. (1997). Expression and characterization of two pathogenic mutations in human electron transfer flavoprotein. *J. Biol. Chem.* **272,** 26425–26433.

303. Gregersen, N. (1985). Riboflavin-responsive defects of beta-oxidation. *J. Inherit. Metab. Dis.* **8,** 65–69.

304. Wallace, D. C. (1992). Diseases of mitochondrial DNA. *Ann. Rev. Biochem.* **61,** 1175–1212.

305. Papadopoulou, L. C., Sue, C. M., Davidson, M. M., Tanji, K., Nishino, I., Sadlock, J. E., Krishna, S., Walker, W., Selby, J., Glerum, D. M., Coster, R. V., Lyon, G., Scalais, E., Lebel, R., Kaplan, P., Shanske, S., De Vivo, D. C., Bonilla, E., Hirano, M., DiMauro, S., and Schon, E. A. (1999). Fatal infantile cardioencephalo-myopathy with COX deficiency and mutations in SCO2, a COX assembly gene. *Nat. Genet.* **23,** 333–337.

306. Tanaka, M., Ino, H., and Ohno, K. (1990). Mitochondrial mutation in fatal infantile cardiomyopathy. *Lancet* **336,** 1452.
307. Taniike, M., Fukushima, H., Yanagihara, I., Tsukamoto, H., Tanaka, J., Fujimura, H., Nagai, T., Sano, T., Yamoka, K., and Innui, K. (1992). Mitochondrial tRNAIle mutation in fatal cardiomyopathy. *Biochem. Biophys. Res. Commun.* **186,** 47–53.
308. Silvestri, G., Santorelli, F. M., Shanske, S., Whitley, C. B., Schimmenti, L. A., Smith, S. A., and DiMauro, S. (1994). A new mtDNA mutation in the tRNALEU(UUR) gene associated with maternally inherited cardiomyopathy. *Hum. Mutat.* **3,** 37–43.
309. van den Bosch, B. J., de Coo, I. F., Hendrickx, A. T., Busch, H. F., de Jong, G., Scholte, H. R., and Smeets, H. J. (2004). Increased risk for cardiorespiratory failure associated with the A3302G mutation in the mitochondrial DNA encoded tRNALeu(UUR) gene. *Neuromuscul. Disord.* **14,** 683–688.
310. Marín-García, J., Goldenthal, M. J., Ananthakrishnan, R., and Pierpont, M. E. (2000). The complete sequence of mtDNA genes in idiopathic dilated cardiomyopathy shows novel missense and tRNA mutations. *J. Card. Fail.* **6,** 321–329.
311. Marín-García, J., Ananthakrishnan, R., Goldenthal, M. J., and Pierpont, M. E. (2000). Biochemical and molecular basis for mitochondrial cardiomyopathy in neonates and children. *J. Inherit. Metab. Dis.* **23,** 625–633.
312. Serrano, J., Palmeira, C. M., Kuehl, D. W., and Wallace, K. B. (1999). Cardioselective and cumulative oxidation of mitochondrial DNA following subchronic doxorubicin administration. *Biochim. Biophys. Acta.* **1411,** 201–205.
313. Schoppet, M., and Maisch, B. (2001). Alcohol and the heart. *Herz* **26,** 345–352.
314. Corral-Debrinski, M., Shoffner, J. M., Lott, M. T., and Wallace, D. C. (1992). Association of mitochondrial DNA damage with aging and coronary atherosclerotic heart disease. *Mutat. Res.* **275,** 169–180.
315. Marín-García, J., Goldenthal, M. J., Ananthakrishnan, R., Pierpont, M. E., Fricker, F. J., Lipshultz, S. E., and Perez-Atayde, A. (1996). Specific mitochondrial DNA deletions in idiopathic dilated cardiomyopathy. *Cardiovasc. Res.* **31,** 306–313.
316. Li, Y. Y., Hengstenberg, C., and Maisch, B. (1995). Whole mitochondrial genome amplification reveals basal level multiple deletions in mtDNA of patients with dilated cardiomyopathy. *Biochem. Biophys. Res. Commun.* **210,** 211–218.
317. Holt, I. J., Harding, A. E., and Morgan-Hughes, J. A. (1988). Deletions of mtDNA in patients with mitochondrial myopathies. *Nature* **331,** 717–719.
318. Carrozzo, R., Hirano, M., Fromenty, B., Casali, C., Santorelli, F. M., Bonilla, E., DiMauro, S., Schon, E. A., and Miranda, A. F. (1998). Multiple mtDNA deletions features in autosomal dominant and recessive diseases suggest distinct pathogeneses. *Neurology* **50,** 99–106.
319. Zeviani, M., Spinazzola, A., and Carelli, V. (2003). Nuclear genes in mitochondrial disorders. *Curr. Opin. Genet. Dev.* **13,** 262–270.
320. Palmieri, L., Alberio, S., Pisano, I., Lodi, T., Meznaric-Petrusa, M., Zidar, J., Santoro, A., Scarcia, P., Fontanesi, F., Lamantea, E., Ferrero, I., and Zeviani, M. (2005). Complete loss of function of the heart/muscle specific adenine nucleotide translocator is associated with mitochondrial myopathy and cardiomyopathy. *Hum. Mol. Genet.* **14,** 3079–3088.
321. Marín-García, J., Ananthakrishnan, R., Goldenthal, M. J., Filiano, J. J., and Pérez-Atayde, A. (1997). Cardiac mitochondrial dysfunction and DNA depletion in children with hypertrophic cardiomyopathy. *J. Inherit. Metab. Dis.* **20,** 674–680.
322. Lewis, W., and Dalakas, M. C. (1995). Mitochondrial toxicity of antiviral drugs. *Nature Med.* **1,** 417–422.
323. Lipshultz, S. E., Easley, K. A., Orav, E. J., Kaplan, S., Starc, T. J., Bricker, J. T., Lai, W. W., Moodie, D. S., Sopko, G., McIntosh, K., and Colan, S. D. (2000). Absence of cardiac toxicity of zidovudine in infants. Pediatric Pulmonary and Cardiac Complications of Vertically Transmitted HIV Infection Study Group. *N. Engl. J. Med.* **343,** 759–766.
324. Sengers, R. C., Stadhouders, A. M., Jaspar, H. H., Trijbels, J. M., and Daniels, O. (1976). Cardiomyopathy and short stature associated with mitochondrial and/or lipid storage myopathy of skeletal muscle. *Neuropadiatrie* **7,** 196–208.
325. Morava, E., Sengers, R., Ter Laak, H., Van Den Heuvel, L., Janssen, A., Trijbels, F., Cruysberg, H., Boelen, C., and Smeitink, J. (2004). Congenital hypertrophic cardiomyopathy, cataract, mitochondrial myopathy and defective oxidative phosphorylation in two siblings with Sengers-like syndrome. *Eur. J. Pediatr.* **163,** 467–471.
326. Jordens, E. Z., Palmieri, L., Huizing, M., van den Heuvel, L. P., Sengers, R. C., Dorner, A., Ruitenbeek, W., Trijbels, F. J., Valsson, J., Sigfusson, G., Palmieri, F., and Smeitink, J. A. (2002). Adenine nucleotide translocator 1 deficiency associated with Sengers syndrome. *Ann. Neurol.* **52,** 95–99.
327. Rasmussen, M., Sanengen, T., Skullerud, K., Kvittingen, E. A., and Skjeldal, O. H. (2000). Evidence that Alpers-Huttenlocher syndrome could be a mitochondrial disease. *J. Child. Neurol.* **15,** 473–477.
328. Naviaux, R. K., Nyhan, W. L., Barshop, B. A., Poulton, J., Markusic, D., Karpinski, N. C., and Haas, R. H. (1999). Mitochondrial DNA polymerase Γ deficiency and mtDNA depletion in a child with Alpers' syndrome. *Ann. Neurol.* **45,** 54–58.
329. Naviaux, R. K., and Nguyen, K. V. (2004). POLG mutations associated with Alpers' syndrome and mitochondrial DNA depletion. *Ann. Neurol.* **55,** 706–712.
330. Hlubocka, Z., Marecek, Z., Linhart, A., Kejkova, E., Pospisilova, L., Martasek, P., and Aschermann, M. (2002). Cardiac involvement in Wilson disease. *J. Inherit. Metab. Dis.* **25,** 269–277.
331. Kuan, P. (1987). Cardiac Wilson's disease. *Chest* **91,** 579–583.
332. Gu, M., Cooper, J. M., Butler, P., Walker, A. P., Mistry, P. K., Dooley, J. S., and Schapira, A. H. (2000). Oxidative-phosphorylation defects in liver of patients with Wilson's disease. *Lancet* **356,** 469–474.
333. Davie, C. A., and Schapira, A. H. (2002). Wilson disease. *Int. Rev. Neurobiol.* 53, 175–190.
334. Sheline, C. T., and Choi, D. W. (2004). Cu^{2+} toxicity inhibition of mitochondrial dehydrogenases *in vitro* and *in vivo*. *Ann. Neurol.* **55,** 645–653.
335. Singh, G. K. (1998). Kawasaki disease: an update. *Indian J. Pediatr.* **65,** 231–241.
336. Newburger, J. W., Takahashi, M., Gerber, M. A., Gewitz, M. H., Tani, L. Y., Burns, J. C., Shulman, S. T., Bolger, A. F., Ferrieri, P., Baltimore, R. S., Wilson, W. R., Baddour, L. M., Levison, M. E., Pallasch, T. J., Falace, D. A., Taubert, K. A., and the Committee on Rheumatic Fever, Endocarditis, and

Kawasaki Disease, Council on Cardiovascular Disease in the Young, American Heart Association. (2004). Diagnosis, treatment, and long-term management of Kawasaki disease: a statement for health professionals from the Committee on Rheumatic Fever, Endocarditis, and Kawasaki Disease, Council on Cardiovascular Disease in the Young, American Heart Association. *Pediatrics* **114**, 1708–1733.

337. Burns, J. C., Kushner, H. I., Bastian, J. F., Shike, H., Shimizu, C., Matsubara, T., and Turner, C. L. (2000). Kawasaki disease: A brief history. *Pediatrics* **106**, E27.

338. Fujita, Y., Nakamura, Y., Sakata, K., Hara, N., Kobayashi, M., Nagai, M., Yanagawa, H., and Kawasaki, T. (1989). Kawasaki disease in families. *Pediatrics* **84**, 666–669.

339. Uehara, R., Yashiro, M., Nakamura, Y., and Yanagawa, H. (2003). Kawasaki disease in parents and children. *Acta Paediatr.* **92**, 694–697.

340. Harada, F., Sada, M., Kamiya, T., Yanase, Y., Kawasaki, T., and Sasazuki, T. (1986). Genetic analysis of Kawasaki syndrome. *Am. J. Hum. Genet.* **39**, 537–539.

341. Rowley, A. H., Shulman, S. T., Spike, B. T., Mask, C. A., and Baker, S. C. (2001). Oligoclonal IgA response in the vascular wall in acute Kawasaki disease. *J. Immunol.* **166**, 1334–1343.

342. Brown, T. J., Crawford, S. E., Cornwall, M. L., Garcia, F., Shulman, S. T., and Rowley, A. H. (2001). CD8 T lymphocytes and macrophages infiltrate coronary artery aneurysms in acute Kawasaki disease. *J. Infect. Dis.* **184**, 940–943.

343. Takeshita, S., Tokutomi, T., Kawase, H., Nakatani, K., Tsujimoto, H., Kawamura, Y., and Sekine, I. (2001). Elevated serum levels of matrix metalloproteinase-9 (MMP-9) in Kawasaki disease. *Clin. Exp. Immunol.* **125**, 340–344.

344. Yasukawa, K., Terai, M., Shulman, S. T., Toyozaki, T., Yajima, S., Kohno, Y., and Rowley, A. H. (2002). Systemic production of vascular endothelial growth factor and fms-like tyrosine kinase-1 receptor in acute Kawasaki disease. *Circulation* **105**, 766–769.

345. Asano, T., and Ogawa, S. (2000). Expression of monocyte chemoattractant protein-1 in Kawasaki disease: the anti-inflammatory effect of gamma globulin therapy. *Scand. J. Immunol.* **51**, 98–103.

346. Furukawa, S., Matsubara, T., Umezawa, Y., Okumura, K., and Yabuta, K. (1994). Serum levels of p60 soluble tumor necrosis factor receptor during acute Kawasaki disease. *J. Pediatr.* **124**, 721–725.

347. Lin, C. Y., Lin, C. C., Hwang, B., and Chiang, B. (1992). Serial changes of serum interleukin-6, interleukin-8, and tumor necrosis factor α among patients with Kawasaki disease. *J. Pediatr.* **121**, 924–926.

348. Pauschinger, M., Chandrasekharan, K., Noutsias, M., Kuhl, U., Schwimmbeck, L. P., and Schultheiss, H. P. (2004). Viral heart disease: molecular diagnosis, clinical prognosis, and treatment strategies. *Med. Microbiol. Immunol.* **193**, 65–69.

349. Kuhl, U., Pauschinger, M., Seeberg, B., Lassner, D., Noutsias, M., Poller, W., and Schultheiss, H. P. (2005). Viral persistence in the myocardium is associated with progressive cardiac dysfunction. *Circulation* **112**, 1965–1970.

350. Calabrese, F., Rigo, E., Milanesi, O., Boffa, G. M., Angelini, A., Valente, M., and Thiene, G. (2002). Molecular diagnosis of myocarditis and dilated cardiomyopathy in children: clinico-pathologic features and prognostic implications. *Diagn. Mol. Pathol.* **11**, 212–221.

351. Baasner, A., Dettmeyer, R., Graebe, M., Rissland, J., and Madea, B. (2003). PCR-based diagnosis of enterovirus and parvovirus B19 in paraffin-embedded heart tissue of children with suspected sudden infant death syndrome. *Lab. Invest.* **83**, 1451–1455.

352. Vliegen, I., Herngreen, S., Grauls, G., Bruggeman, C., and Stassen, F. (2003). Improved detection and quantification of mouse cytomegalovirus by real-time PCR. *Virus Res.* **98**, 17–25.

353. Klein, R. M., Jiang, H., Niederacher, D., Adams, O., Du, M., Horlitz, M., Schley, P., Marx, R., Lankisch, M. R., Brehm, M. U., Strauer, B. E., Gabbert, H. E., Scheffold, T., and Gulker, H. (2004). Frequency and quantity of the parvovirus B19 genome in endomyocardial biopsies from patients with suspected myocarditis or idiopathic left ventricular dysfunction. *Z. Kardiol.* **93**, 300–309.

354. Bowles, N. E., Ni, J., Kearney, D. L., Pauschinger, M., Schultheiss, H. P., McCarthy, R., Hare, J., Bricker, J. T., Bowles, K. R., and Towbin, J. A. (2003). Detection of viruses in myocardial tissues by polymerase chain reaction: evidence of adenovirus as a common cause of myocarditis in children and adults. *J. Am. Coll. Cardiol.* **42**, 466–472.

355. Wang, X., Zhang, G., Liu, F., Han, M., Xu, D., and Zang, Y. (2004). Prevalence of human parvovirus B19 DNA in cardiac tissues of patients with congenital heart diseases indicated by nested PCR and in situ hybridization. *J. Clin. Virol.* **31**, 20–24.

356. Francalanci, P., Chance, J. L., Vatta, M., Jimenez, S., Li, H., Towbin, J. A., and Bowles, N. E. (2004). Cardiotropic viruses in the myocardium of children with end-stage heart disease. *J. Heart Lung Transplant.* **23**, 1046–1052.

357. Schowengerdt, K. O., Ni, J., Denfield, S. W., Gajarski, R. J., Bowles, N. E., Rosenthal, G., Kearney, D. L., Price, J. K., Rogers, B. B., Schauer, G. M., Chinnock, R. E., and Towbin, J. A. (1997). Association of parvovirus B19 genome in children with myocarditis and cardiac allograft rejection: diagnosis using the polymerase chain reaction. *Circulation* **96**, 3549–3554.

358. Deguchi, H., Fujioka, S., Terasaki, F., Ukimura, A., Hirasawa, M., Kintaka, T., Kitaura, Y., Kondo, K., Sasaki, S., Isomura, T., and Suma, H. (2001). Enterovirus RNA replication in cases of dilated cardiomyopathy: light microscopic in situ hybridization and virological analyses of myocardial specimens obtained at partial left ventriculectomy. *J. Card. Surg.* **16**, 64–71.

359. Bergelson, J. M., Cunningham, J. A., Droguett, G., Kurt-Jones, E. A., Krithivas, A., Hong, J. S., Horwitz, M. S., Crowell, R. L., and Finberg, R. W. (1997). Isolation of a common receptor for Coxsackie B viruses and adenoviruses 2 and 5. *Science* **275**, 1320–1323.

360. Selinka, H. C., Wolde, A., Sauter, M., Kandolf, R., and Klingel, K. (2004). Virus-receptor interactions of coxsackie B viruses and their putative influence on cardiotropism. *Med. Microbiol. Immunol. (Berl)* **193**, 127–131.

361. Noutsias, M., Fechner, H., de Jonge, H., Wang, X., Dekkers, D., Houtsmuller, A. B., Pauschinger, M., Bergelson, J., Warraich, R., Yacoub, M., Hetzer, R., Lamers, J., Schultheiss, H. P., and Poller, W. (2001). Human coxsackie-adenovirus receptor is colocalized with integrins α(v)β(3) and α(v)β(5) on the cardiomyocyte sarcolemma and upregulated

in dilated cardiomyopathy: implications for cardiotropic viral infections. *Circulation* **104**, 275–280.

362. Poller, W., Fechner, H., Noutsias, M., Tschoepe, C., and Schultheiss, H. P. (2002). Highly variable expression of virus receptors in the human cardiovascular system. Implications for cardiotropic viral infections and gene therapy. *Z. Kardiol.* **91**, 978–991.

363. Figulla, H. R. (2004). Transformation of myocarditis and inflammatory cardiomyopathy to idiopathic dilated cardiomyopathy: facts and fiction. *Med. Microbiol. Immunol. (Berl)* **193**, 61–64,

364. Klingel, K., Sauter, M., Bock, C. T., Szalay, G., Schnorr, J. J., and Kandolf, R. (2004). Molecular pathology of inflammatory cardiomyopathy. *Med. Microbiol. Immunol.* **193**, 101–107.

365. Luppi, P., Rudert, W., Licata, A., Riboni, S., Betters, D., Cotrufo, M., Frati, G., Condorelli, G., and Trucco, M. (2003). Expansion of specific αβ+ T-cell subsets in the myocardium of patients with myocarditis and idiopathic dilated cardiomyopathy associated with Coxsackievirus B infection. *Hum. Immunol.* **64**, 194–210.

366. Pankuweit, S., Portig, I., and Maisch, B. (2002). Pathophysiology of cardiac inflammation: molecular mechanisms. *Herz* **27**, 669–676.

367. Taylor, J. A., Havari, E., McInerney, M. F., Bronson, R., Wucherpfennig, K. W., and Lipes, M. A. (2004). A spontaneous model for autoimmune myocarditis using the human MHC molecule HLA-DQ8. *J. Immunol.* **172**, 2651–2658.

368. Cunningham, M. W. (2004). T cell mimicry in inflammatory heart disease. *Mol. Immunol.* **40**, 1121–1127.

369. Marian, A. J. (2002). Modifier genes for hypertrophic cardiomyopathy. *Curr. Opin. Cardiol.* **17**, 242–252.

370 Hiroi, Y., Kudoh, S., Monzen, K., Ikeda, Y., Yazaki, Y., Nagai, R., and Komuro, I. (2001). Tbx5 associates with Nkx2-5 and synergistically promotes cardiomyocyte differentiation. *Nat. Genet.* **28**, 276–280.

371. Machlitt, A., Tennstedt, C., Korner, H., Bommer, C., and Chaoui, R. (2002). Prenatal diagnosis of 22q11 microdeletion in an early second-trimester fetus with conotruncal anomaly presenting with increased nuchal translucency and bilateral intracardiac echogenic foci. *Ultrasound Obstet. Gynecol.* **19**, 510–513.

372. Driscoll, D. A. (2001). Prenatal diagnosis of the 22q11.2 deletion syndrome. *Genet. Med.* **3**, 14–18.

373. Johnson, W. H. Jr., Yang, P., Yang, T., Lau, Y. R., Mostella, B. A., Wolff, D. J., Roden, D. M., and Benson, D. W. (2003). Clinical, genetic, and biophysical characterization of a homozygous HERG mutation causing severe neonatal long QT syndrome. *Pediatr. Res.* **53**, 744–748.

374. Miller, T. E., Estrella, E., Myerburg, R. J., Garcia de Viera, J., Moreno, N., Rusconi, P., Ahearn, M. E., Baumbach, L., Kurlansky, P., Wolff, G., and Bishopric, N. H. (2004). Recurrent third-trimester fetal loss and maternal mosaicism for long-QT syndrome. *Circulation* **109**, 3029–3034.

375. Daley, G. Q., and Cargill, M. (2001). The heart SNPs a beat: polymorphisms in candidate genes for cardiovascular disease. *Trends Cardiovasc. Med.* **11**, 60–66.

376. O'Rourke, B. (2000). Myocardial KATP channels in preconditioning. *Circ. Res.* **87**, 845–855.

377. Rafiee, P., Shi, Y., Kong, X., Pritchard, K. A. Jr., Tweddell, J. S., Litwin, S. B., Mussatto, K., Jaquiss, R. D., Su, J., and Baker, J. E. (2002). Activation of protein kinases in chronically hypoxic infant human and rabbit hearts: role in cardioprotection. *Circulation* **106**, 239–245.

378. Ning, X. H., Xu, C. S., Song, Y. C., Xiao, Y., Hu, Y. J., Lupinetti, F. M., and Portman, M. A. (1998). Hypothermia preserves function and signaling for mitochondrial biogenesis during subsequent ischemia. *Am. J. Physiol.* **274**, H786–H793.

379. Sanguinetti, M. C., and Bennett, P. B. (2003). Antiarrhythmic drug target choices and screening. *Circ. Res.* **93**, 491–499.

380. Beltrami, A. P., Barlucchi, L., Torella, D., Baker, M., Limana, F., Chimenti, S., Kasahara, H., Rota, M., Musso, E., Urbanek, K., Leri, A., Kajstura, J., Nadal-Ginard, B., and Anversa, P. (2003). Adult cardiac stem cells are multipotent and support myocardial regeneration. *Cell* **114**, 763–776.

381. Flavigny, J., Richard, P., Isnard, R., Carrier, L., Charron, P., Bonne, G., Forissier, J. F., Desnos, M., Dubourg, O., Komajda, M., Schwartz, K., and Hainque, B. (1998). Identification of two novel mutations in the ventricular regulatory myosin light chain gene (MYL2) associated with familial and classical forms of hypertrophic cardiomyopathy. *J. Mol. Med.* **76**, 208–214.

382. Zhong, N., Martiniuk, F., Tzall, S., and Hirschhorn, R. (1991). Identification of a missense mutation in one allele of a patient with Pompe disease, and use of endonuclease digestion of PCR-amplified RNA to demonstrate lack of mRNA expression from the second allele. *Am. J. Hum. Genet.* **49**, 635–645.

383. Teng, Y. T., Su, W. J., Hou, J. W., and Huang, S. F. (2004). Infantile-onset glycogen storage disease type II (Pompe disease): report of a case with genetic diagnosis and pathological findings. *Chang Gung Med. J.* **27**, 379–384.

384. Shen, J. J., and Chen, Y. T. (2002). Molecular characterization of glycogen storage disease type III. *Curr. Mol. Med.* **2**, 167–175.

385. Lucchiari, S., Fogh, I., Prelle, A., Parini, R., Bresolin, N., Melis, D., Fiori, L., Scarlato, G., and Comi, G. P. (2002). Clinical and genetic variability of glycogen storage disease type IIIa: seven novel AGL gene mutations in the Mediterranean area. *Am. J. Med. Genet.* **109**, 183–190.

386. Yoshitama, T., Nakao, S., Takenaka, T., Teraguchi, H., Sasaki, T., Kodama, C., Tanaka, A., Kisanuki, A., and Tei, C. (2001). Molecular genetic, biochemical, and clinical studies in three families with cardiac Fabry's disease. *Am. J. Cardiol.* **87**, 71–75.

387. Germain, D. P. (2002). Fabry's disease (α-galactosidase-A deficiency): physiopathology, clinical signs, and genetic aspects. *J. Soc. Biol.* **196**, 161–173.

388. Pastores, G. M., Santorelli, F. M., Shanske, S., Gelb, B. D., Fyfe, B., Wolfe, D., and Willner, J. P. (1994). Leigh syndrome and hypertrophic cardiomyopathy in an infant with a mitochondrial DNA point mutation (T8993G). *Am. J. Med. Genet.* **50**, 265–271.

SECTION III

Post-Genomic Analysis of Coronary Artery Disease, Angiogenesis, and Hypertension

CHAPTER 8

Molecular Basis of Lipoprotein Disorders, Atherogenesis, and Thrombosis

OVERVIEW

Atherosclerosis is a complex disease caused by multiple genetic and environmental factors and complex gene–environment interactions. Coronary artery disease (CAD) is the most common cause of death in the Western hemisphere and by the year 2020 is expected to become the leading cause of morbidity and mortality in the world. The molecular mechanisms of atherosclerosis is a complex web of cellular events that is only gradually becoming explained. These mechanisms involve lipid metabolism, inflammatory signaling, and interaction with the complex vascular system involved in thrombosis. Risk factors involved in these areas (e.g., dyslipidemia and diabetes, procoagulant, and anticoagulant factors) have provided information about important genes that seem to play a role in establishing the risk of atherosclerosis developing. Although common, atherosclerosis is clearly polygenic; critical molecular information concerning the process has been gleaned from rare monogenic forms of atherosclerosis and thrombosis. In this chapter, we will discuss candidate genes, their pathogenic role in atherosclerotic and thrombotic pathways, and potential therapies.

INTRODUCTION

Coronary artery disease (CAD), the most common cause of death in the Western hemisphere, is expected to become the leading cause of morbidity and mortality in the world by the year 2020. Atherosclerosis, the primary cause of CAD, involves a wide spectrum of cell types, organs, and diverse physiological processes that underlie its complex genetic basis. More than 100 genes have been identified thus far that influence the development of atherosclerotic lesions. Moreover, known risk factors in the atherosclerotic disease process such as dyslipidemia, diabetes, and hypertension have both significant genetic and environmental components, leading to the further complexity that commonly arises from CAD resulting from the interaction of environmental and genetic risk factors. Known risk factors are presented in Table I.

The molecular mechanisms and genetic factors that contribute to the susceptibility to atherosclerosis will be discussed in this chapter. Also, we will analyze the progress made by applying this information in the clinical diagnosis and therapeutics in the current post-genomic era and what is expected to be achieved in the foreseeable future.

Clinical genetic studies have established a major role for genetic factors in the susceptibility to coronary atherosclerosis. Studies in both twins and families have documented the heritability of CAD, which exceeds 50% in most studies of twins and families.[1] Although the evidence from most cases supports the view that CAD is a polygenic disorder, relatively few single-gene defects leading to atherosclerosis have been identified, and most are related to dyslipidemia.[2] These atherosclerotic disorders comprise single gene traits inherited in a Mendelian fashion as an autosomal-dominant or recessive or X-linked disorder. A representative list of monogenic forms of dyslipidemia leading to coronary atherosclerosis is shown in Table II. Understanding the molecular basis of the Mendelian forms of atherosclerosis not only has provided insights into the pathogenesis of specific monogenic disorders but has also allowed critical information and technical resources in delineating the pathogenesis of common non-Mendelian forms of atherosclerosis.

LIPOPROTEINS, CHOLESTEROL, AND DYSLIPIDEMIAS

Familial Hypercholesterolemia

Familial hypercholesterolemia (FH) is characterized by a triad of elevated cholesterol, tendon xanthomas, and premature CAD and provided strong evidence for an association between blood lipids and atherosclerosis. The studies of Goldstein and Brown[3] implicated receptor-mediated endocytosis and revealed that FH is the result of mutations that affect the low-density lipoprotein (LDL) receptor responsible for the binding of LDL, a cholesteryl-rich particle containing the apoB100 lipoprotein, and control of its levels. These mutations perturb LDLR function by a variety of mechanisms,

TABLE I Genetic and Environmental Risk Factors for CAD

Risk factors with a significant genetic component
Elevated LDL and VLDL cholesterol
Low HDL cholesterol
Elevated triglycerides
Hypertension
Obesity
Elevated lipoprotein(a) levels
Elevated homocysteine levels
Type 2 diabetes mellitus
Elevated C-reactive protein
Gender
Age
Family history
Environmental risk factors
Smoking
Diet
Exercise
Alcohol consumption

including the synthesis, transportation, and affinity to bind LDL cholesterol, internalization, recycling, and degradation of the receptors. FH is a relatively rare disease that affects 1 in 500 cases of the population in the heterozygous form and 1 1,000,000 in homozygotes. Affected individuals show severe elevation of plasma levels of total and LDL cholesterol that can range between 350 and 500 mg/dl in the heterozygous individuals and much higher levels in the homozygous subjects. The disorder is generally heterogeneous, with more than 800 mutations in *LDLR* having been identified and displays a pattern of dominant inheritance.[4–6]

Familial Defective Apolipoprotein B

Familial defective apolipoprotein B or apoB (also called FLB), another relatively common hypercholesterolemia (the heterozygous form is present in approximately 1 in 800 cases), is the result of the mutations of apoB, the major protein of LDL, which reduces its affinity and binding to LDLR. In contrast to FH, this disorder is homogeneous with most cases resulting from a single nucleotide substitution at codon 3500 in *APOB*.[7] Although the phenotype is somewhat milder than FH, cholesterol levels are still markedly elevated. Like FH, the disorder exhibits a dominant inheritance.

Other rare forms of autosomal dominant hypercholesterolemia have been described.[2] A novel molecular defect has been identified with a clinical phenotype indistinguishable from patients with heterozygous FH and FLB, and the disease does not segregate with either *LDLR* or *APOB*. Mutations responsible for this disorder have been identified in the *PCSK9* gene encoding neural apoptosis-regulated convertase 1 (NARC-1), a member of the proteinase K family of subtilases.[8]

Autosomal-recessive hypercholesterolemia (ARH) is a rare disorder with a clinical phenotype similar to that of the homozygous FH caused by *LDLR* defects but more variable, less severe, and more responsive to lipid-lowering therapy.[9] Mutations in the *ARH* gene on chromosome 1p35-36 have

TABLE II Single Gene Defects in CAD

Gene	Disease	Lipid profile	Frequency
LDLR	Familial hypercholesterolemia dominant	Excess LDL-C	1/500-heterozygote
			1/10⁶-homozygote
APOB	Familial defective apoB dominant	Excess LDL-C	1/1000-heterozygote
			1/10⁶-homozygote
ARH	Autosomal recessive familial hypercholesterolemia	Excess LDL-C	Rare
PCSK9	Familial hypercholesterolemia dominant	Excess LDL-C	1/2500 heterozygous
APOC2	Recessive fasting hypertriglyceridemia	Elevated triglycerides, chylomicrons + VLDL	Rare
LPL	LPL deficiency	Elevated triglycerides Reduced HDL-C	Rare
APOA1	Premature CAD	Elevated triglycerides Reduced HDL-C	Rare
ABCA1	Tangier disease	Reduced HDL-C	Rare
APOE	Hyperlipoproteinemia (type III)	Excess VLDL	1/10,000
LCAT	Fish-eye disease	Reduced HDL-C	Rare
ABCG5	Sitosterolemia	Excess LDL-C	Rare
ABCG8	Sitosterolemia	Excess LDL-C	Rare
CYP7A	Recessive hypercholesterolemia	Excess LDL-C	Rare
ND	Familial combined hyperlipidemia	Excess VLDL, LDL-C and triglycerides	1/100

been identified that prevent the normal internalization of the LDLR in patients' cultured lymphocytes and monocyte-derived macrophages but not in skin fibroblasts. ARH encodes an adaptor protein containing a phosphotyrosine binding domain, which binds a specific motif (NPXY) in LDLR.[10] Interestingly, the LDLR protein accumulates at the cell surface in the affected cells from either patients or in models of transgenic mice with null mutations in ARH.[11,12]

Other rare recessive forms of hypercholesterolemia have been described in several patients exhibiting a genetic deficiency of cholesterol 7-hydroxylase (CYP7A1), the first enzyme in the classical pathway for bile acid biosynthesis.[13] Cholesterol 7-hydroxylase deficiency presumably causes hypercholesterolemia by reducing hepatic LDLR activity, and the disposal of cholesterol by means of the bile acid biosynthetic pathway is likely to be essential for the maintenance of normal plasma cholesterol levels in humans.

Other significant causes of dyslipidemia have been found to be closely associated with atherosclerosis. Familial combined hyperlipidemia (FCHL) is the most common discrete hyperlipidemia and is a common cause of premature atherosclerosis.[14,15] This condition is characterized by elevated levels of plasma triglycerides and LDL cholesterol, VLDL cholesterol, or both. Although the identity of most genes involved in the etiology of FCHL has proved elusive, a number of promising genetic leads have been identified. Several independent studies have shown linkage of FCHL with a locus on chromosome 1q21-q23.[16,17] In addition, recent linkage studies have identified a major locus for FCHL at chromosome 1q21-23 containing the upstream transcription factor-1 (USF-1) gene.[18] In addition, the mouse strain, HcB-19/Dem (HcB-19), shares features with FCHL, including hypertriglyceridemia, hypercholesterolemia, elevated plasma apolipoprotein B, and increased secretion of triglyceride-rich lipoproteins.[19] Murine hyperlipidemia with a similar phenotype to FCHL results from spontaneous mutation at the Hyplip1 locus mapped to mouse chromosome 3.[20]

Hyperlipoproteinemia can present in several forms. It has long been recognized that the synthesis of an abnormal isoform of apolipoprotein E (i.e., apoE2) is the most common genetic defect underlying type III hyperlipoproteinemia that is phenotypically manifested as elevated levels of VLDL, xanthoma, premature CAD, and peripheral artery disease.[21] ApoE, a 229 amino acid polypeptide, encoded by the APOE gene located on the long arm of chromosome 19, is classified into three major isoform (E2, E3, E4) according to the differences of amino acids at positions 112 and 158.[22,23] In the normal population, the apoE3 isoform is most prevalent, and individuals who are homozygous for the APOE2 allele have a strong association with type III hyperlipoproteinemia. Although many of these individual have a cysteine replacing an arginine at position 158, other APOE2 allele variants have been reported (e.g., cysteine replacing arginine at position 136).[24] Allelic variants in the APOE2 allele underlying both type III hyperlipoproteinemia, as well as dysbetalipoproteinemia, may be either dominantly or recessively inherited and are defective in binding to lipoprotein receptors. The receptor-defective APOE allele results in an impaired clearance of remnant atherogenic lipoproteins (e.g., β-VLDL), leading to the their accumulation.

In normal individuals, cholesterol constitutes more than 99% of circulating sterols; noncholesterol sterols, such as sitosterol, are generally present in only trace amounts. In normal metabolism, both sitosterol and cholesterol are taken up into the enterocytes in the small intestine, and between 20 and 80% of dietary cholesterol is incorporated into chylomicrons. In contrast, less than 5% of dietary sitosterol is absorbed, and the small amount of sitosterol transported to the liver is preferentially secreted into the bile. In sitosterolemia, plasma levels of sitosterol are elevated more than 50-fold, and approximately 15% of circulating and tissue sterols are derived from plants and shellfish.[25] Sitosterolemia (also a hypercholesteremic disorder) is a rare autosomal-recessive disease in which patients can have plasma LDL-C levels as high as those seen in FH homozygotes (>500 mg/dl). Like patients with FH, these patients can develop aortic stenosis and premature CAD. In addition, elevated levels of plasma sitosterol and of the sitosterol/cholesterol ratio have been significantly associated with increased occurrence of major coronary events (MI or sudden coronary death) in males with high risk of CAD in the prospective PROCAM case-control study.[26]

Patients with sitosterolemia also display increased absorption of dietary sterols and have a defect in the ability to secrete sterols into the bile, resulting in the accumulation of dietary animal and plant sterols in the blood and body tissues. Whereas the primary metabolic defect in FH, FLB, and ARH is in the receptor-mediated uptake of circulating LDL, sitosterolemia results from a defect in sterol efflux from cells.

Sitosterolemia is caused by mutations in either of two adjacent genes that encode ABC half-transporters, ABCG5 and ABCG8 (also termed sterolin-1 and sterolin-2 respectively).[27,28] More than 25 different mutations in these genes have been identified that cause sitosterolemia. Sitosterolemic patients invariably have two mutant alleles of ABCG5 or two mutant alleles of ABCG8 alleles; none of the patients identified has a single mutation in ABCG5 and in ABCG8.[29]

In addition to atherosclerosis stemming from the accumulation of the "bad" cholesterol lipoproteins (i.e., aberrant LDL levels), reduced HDL levels have also been associated with atherosclerosis, particularly in several rare disorders. Tangier disease (TD) is an autosomal co-dominant disease characterized by the absence of HDL and very low plasma levels of apoA-I. The deposition of cholesterol esters results in hypertrophic orange-color tonsils, hepatosplenomegaly, and premature CAD. TD is caused by mutations in the ATP binding cassette transporter (ABCA1) gene, located on chromosome 9q31.[30-33] The ABCA1 gene encodes an integral membrane protein with 12 transmembrane domains involved in cholesterol and phospholipid efflux at the membrane. In the absence of ABCA1, free cholesterol is not transported extracellularly.

Another rare autosomal-dominant disorder affecting HDL levels is termed Fish eye disease.[34,25] This disease is due to a deficiency in lecithin/cholesterol acyltransferase (LCAT).

LCAT is located on chromosome 16q22.1 and encodes a protein involved in the synthesis of HDL3 from pre-lipoprotein A-I and its conversion to HDL2 cholesterol. The deficiency of LCAT leads to premature CAD, proteinuria, anemia, and renal failure.

Although the mechanism(s) by which reduced HDL levels are associated with an increased CAD risk in most patients are not completely known, several beneficial or protective properties of HDL-cholesterol are potentially involved: (1) HDL particles have the ability to carry excess cholesterol from peripheral tissues back to the liver, a process termed reverse cholesterol transport (2) apolipoprotein A-I (apo A-I), the major protein component of HDL, is associated with two antioxidant enzymes on HDL, paraoxonase, and platelet-activating factor acetylhydrolase (also known as plasma lipoprotein-associated phospholipase A2, Lp-PLA2).[36,37] These enzymes help diminish the formation of the highly atherogenic oxidized LDL, reducing its proatherogenic potential. Moreover, HDL lipoproteins function as a protective factor in endothelial cells by providing anti-inflammatory and anti-apoptotic function. For instance, HDL has been implicated in down-regulating the expression of cellular adhesion molecules (e.g., VCAM, ICAM, and E-selectin) and cytokines (e.g., IL-8) involved in events in leukocyte-mediated CAD.[38] In addition, HDL mediates inhibition of caspase activation and prevents subsequent endothelial apoptosis, as well as activates protein kinase Akt, a mediator of antiapoptotic signaling.[39] Also, HDL induces endothelial nitric oxide synthase (eNOS) activation, NO release, and vasorelaxatory effects.[40-42]

Increased triglyceride levels have long been associated with atherosclerosis, although usually in association with increased LDL levels. Increased triglyceride levels in CAD have been mapped to a gene cluster on chromosome 11 termed the apoAI-CIII-AIV-AV locus.[43] The apolipoprotein CIII (apoCIII) impedes triglyceride hydrolysis and triglyceride remnant clearance and, as such, may exert pro-atherogenic activities. Several apo*AV* gene variants have been implicated in increased plasma triglyceride accumulation.[44] Variants in the newly identified apolipoprotein *APOA5* gene were found to be strongly associated with elevated triglyceride levels in different racial groups.[45] Modifications in lipoprotein lipase (LPL) levels lead to elevated triglycerides and reduced HDL, both risk factors for CAD. The dynamics of lipoprotein formation and modifications are diagrammed in Fig. 1.

Recent studies have focused on the association of CAD and profound lipid and metabolic dysregulation in the metabolic syndrome (MetSyn) further described in Chapter 16.[46] This disorder is characterized by visceral adiposity, insulin resistance, low HDL level, hypertriglyceridemia, a systemic pro-inflammatory state, and small dense LDL and is a strong predictor of both CAD and type 2 diabetes. Genetic variations have been reported in the nuclear receptor peroxisome proliferator-activator gamma (PPAR-γ) receptor in both MetSyn and type 2 diabetes mellitus.[47,48] Common variants of the lipoprotein lipase (LPL) gene have also been implicated in MetSyn and insulin resistance.[49,50]

Lipoprotein Oxidation and Modification

Lipid oxidation plays a central role in atherogenesis.[51] Specific proinflammatory oxidized phospholipids result from the oxidation of LDL phospholipids containing arachidonic acid, and these are recognized by the innate immune system of animals and humans. These oxidized phospholipids are largely generated by oxidants produced by the lipoxygenase and myeloperoxidase pathways. The levels of specific oxidized lipids in plasma and lipoproteins may serve as useful markers of the susceptibility to atherogenesis. The reduction of oxidized lipids promoted by the administration of apolipoprotein A-I, and mimetic peptides may have therapeutic potential.[52]

Oxidized LDL is present in the lesions of atherosclerosis in humans. In animals with hypercholesterolemia, antioxidants can reduce the size of the atherosclerotic lesions and also can reduce the presence of fatty streaks. The latter observation suggests that antioxidants have an anti-inflammatory effect, perhaps by preventing the up-regulation of adhesion molecules by monocytes. Antioxidants can increase the resistance of human LDL to oxidation generated *ex vivo* in proportion to the vitamin E content of the plasma.

Although the genetic differences contributing to CAD and atherosclerosis are greatest in lipid metabolism, a large array of candidate genes has been examined in population-association studies. A number of these genes have common variations with significant (and convincing) association to CAD. A partial list of candidate genes is presented in Table III.

Inflammation and Atherosclerosis

Over the last decade, there has been a large accumulation of data supporting the view that atherosclerosis is an inflammatory disease. Although high levels of plasma cholesterol, in particular LDL-cholesterol, represent a primary risk factor in patients with atherosclerosis (approximately 50% of which exhibit hypercholesteremia), the process of atherogenesis clearly constitutes much more than the accumulation of lipids within the artery wall. In fact, one of the earliest type of atherosclerotic lesions, the fatty streak, particularly common in infants and young children, is primarily an inflammatory lesion, consisting only of monocyte-derived macrophages and T lymphocytes. An imbalance between proinflammatory and anti-inflammatory cellular and molecular elements constitutes a primary element in the evolution of the atherosclerotic plaque. Not unlike current models of chronic wound healing or ischemia-reperfusion, the atherosclerotic lesion at any given time is transient, reflecting a dynamic balance of numerous local and circulating inflammatory and lipid factors.[53] Therefore, it can be stated that atherosclerosis represents a series of highly specific cellular and molecular responses characterizing an inflammatory disease, and we will discuss some of the key molecular players acting in those pathways.

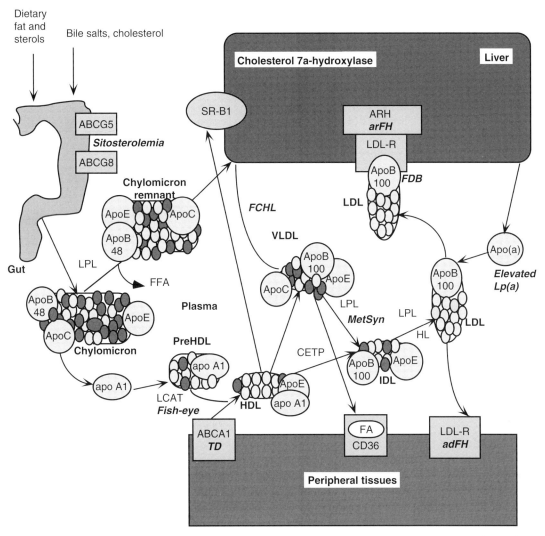

FIGURE 1 Overview of LDL metabolism in humans. Dietary cholesterol and triglycerides are packaged with apolipoproteins in enterocytes of the small intestine, secreted into the lymphatic system as chylomicrons. As chylomicrons circulate, the core triglycerides are hydrolyzed by lipoprotein lipase (LPL), resulting in the formation of chylomicron remnants that are rapidly removed by the liver. In liver, dietary cholesterol has several possible fates: it can be esterified and stored, packaged into VLDL particles and secreted into plasma or into bile; or converted into bile acids and secreted into bile. VLDL particles secreted into the plasma undergo lipolysis to form VLDL remnants, including IDL. VLDL remnants are also removed by the liver by the LDLR, and the remainder mature into LDL, the major cholesterol transport particle in the blood. ABCG5 and ABCG8 are located predominantly in the enterocytes of the duodenum and jejunum, the sites of uptake of dietary sterols. Mutations in either transporter cause an increase in delivery of dietary sterols to the liver and a decrease in secretion of sterols into the bile (see color insert).

The inflammatory events associated with atherosclerosis have been frequently described as a response to injury or endothelial dysfunction. Possible causes of endothelial dysfunction leading to atherosclerosis include elevated and modified LDL, hypertension, smoking, diabetes mellitus, elevated plasma homocysteine, infectious microorganisms such as *Chlamydia pneumoniae*, and various combinations of these or other factors.[53]

The endothelial dysfunction that results from the injury leads to a series of compensatory responses that alter the normal homeostatic properties of the endothelium, such as modifying its adhesiveness with respect to leukocytes or platelets, as well as its permeability. The injury also induces the endothelium to form vasoactive molecules, cytokines, and growth factors. If the offending agents are not effectively neutralized or removed, the inflammatory response may continue indefinitely. This inflammatory response stimulates the migration and proliferation of smooth muscle cells (SMCs) that become embedded within the area of inflammation to form an intermediate lesion. If these responses continue unabated, they can thicken the artery wall, which compensates by gradual dilation, so that up to a point, the lumen remains unaltered, a phenomenon referred to as remodeling. This response is largely mediated by monocyte-derived macrophages and specific subtypes of T lymphocytes.[54]

TABLE III Candidate Genes Associated with CAD

Gene

Apo A-I *(APOA1)*
Apo A-V *(APOA5)*
ApoB *(APOB)*
ApoE *(APOE)*
Liver x receptor *(LXR)*
Myocyte enhancer factor 2A *(MEF2A)*
Lymphotoxin-α *(LTA)*
Endothelial nitric oxide synthase *(eNOS)*
Angiotensin-converting enzyme *(ACE)*
Cholesteryl ester transfer protein *(CETP)*
Paraoxonase-1 *(PON-1)*
Lipoprotein lipase *(LPL)*
Upstream transcription factor-1 *(USF-1)*
Peroxisome proliferator-activator receptor-γ *(PPAR-γ)*
Plasminogen activator inhibitor *(PAI)*
Methylenetetrahydrofolate reductase *(MTHFR)*
5-lipoxygenase activating protein *(ALOX5AP)*
5-lipoxygenase *(5-LO)*
Hepatic lipase
LDL receptor *(LDLR)*
Phosphodiesterase 4D *(PDE4D)*

Continued inflammation results in increased numbers of macrophages and lymphocytes, which both emigrate to and multiply at the lesion. Activation of these cells leads to the release of numerous hydrolytic enzymes, cytokines, chemokines, and growth factors, which further induce damage and can eventually lead to necrosis. The accumulation of mononuclear cells, migration, and proliferation of SMCs and the formation of fibrous tissue promote a further enlargement and restructuring of the lesion, enabling the generation of a fibrous cap that overlies a core of lipid and necrotic tissue. When, at some point, the artery can no longer compensate by dilation, the atherosclerotic lesion may become occlusive, compromising blood flow.

LDL that is subject to modification by oxidation, glycation, aggregation, association with proteoglycans, or incorporation into immune complexes is a major stimulus for injury to the endothelium and underlying smooth muscle. LDL particles entrapped in an artery often undergo progressive oxidation and are targeted by scavenger receptors on the surfaces of macrophages and become internalized. The internalization of LDL leads to the formation of lipid peroxides with the subsequent accumulation of cholesteryl esters, resulting in the formation and activation of foam cells. The removal and sequestration of modified LDL represents an intrinsic, protective role of the macrophage in the inflammatory response to minimize the effects of modified LDL on endothelial cells and SMCs. In addition to its ability to injure these cells, modified LDL is chemotactic for other monocytes, can up-regulate the expression of genes for macrophage colony-stimulating factor (MCSF),[55] and monocyte chemotactic protein (MCP)[56] and may help further promote the inflammatory response by stimulating monocyte-derived macrophage replication and the entry of new monocytes into lesions.[53]

There are other interactions between LDL and the inflammatory pathway. Mediators of inflammation such as tumor necrosis factor (TNF), interleukin-1, and MCSF increase binding of LDL to endothelium and smooth muscle and increase the transcription of the *LDLR* gene.[57] After binding to scavenger receptors *in vitro*, modified LDL induces the expression of inflammatory cytokines such as interleukin-1, thereby promoting a vicious circle of inflammation, modification of lipoproteins, and further inflammation in the artery.[58]

Both the initial and later stages in the development of atherosclerosis involve the recruitment of inflammatory cells from the circulation and their transendothelial migration. This process is largely mediated by a variety of cellular adhesion molecules, which are expressed in the endothelial cell, as well as on circulating leukocytes in response to several inflammatory stimuli.[59] Endothelial cell adhesion molecules, including several selectins, intercellular adhesion molecules, and vascular-cell adhesion molecules, serve as receptors for glycoconjugates and integrins present on monocytes and T cells. Molecules associated with the migration of leukocytes across the endothelium act in conjunction with chemoattractant molecules (e.g., such as MCP1 generated by the endothelium, smooth muscle, and monocytes), as well as oxidized LDL, interleukin-8, platelet-derived growth factor (PDGF), MCSF, and osteopontin to attract monocytes and T cells into the artery. Selectins (P, E and L) and their ligands are involved in the tethering of leukocytes on the vascular wall. Intercellular adhesion molecules (ICAMs) and vascular cell adhesion molecules (VCAM-1), as well as some of the integrins, induce firm adhesion of inflammatory cells at the vascular surface. Interestingly, the expression of specific adhesion molecules (e.g., VCAM-1, ICAM-1 and L-selectin) in endothelial cells has been consistently observed in atherosclerotic plaques and, in some cases, has been demonstrated to be regulated by properties of blood flow (e.g., decreased shear stress and increased turbulence).[60] A large number of common polymorphisms have been identified in the genes encoding the different adhesion molecules, but their relationship with CAD awaits further investigation.[61]

Several chemokine factors are responsible for the chemotaxis and accumulation of macrophages, lipid-laden monocytes, T lymphocytes, and SMCs in fatty streaks. Activation of monocytes and T cells leads to the up-regulation of receptors on their surfaces, including molecules that bind selectins, integrins that bind adhesion molecules, and receptors that bind chemoattractant molecules, interactions that profoundly contribute to localizing the lesion and defining the extent of the inflammatory response.

Scavenging monocyte-related macrophages and antigen-presenting cells accumulate that secrete cytokines (e.g.,

TNF-α, interleukin-1, and transforming growth factor ß), chemokines, growth factors (e.g., PDGF and IGF-1), and proteolytic enzymes (e.g., metalloproteinases). The continuing incorporation, survival, and replication of monocytes in lesions depend on factors such as MCSF and granulocyte-macrophage colony-stimulating factor (GMCSF) for monocytes and interleukin-2 for lymphocytes.[62] Continued exposure to MCSF permits macrophages to survive *in vitro* and possibly to multiply within the lesions. In contrast, inflammatory cytokines can activate macrophages and, in some cases, can induce them to undergo apoptosis.

Cells and molecules that mediate both adaptive immunity, which depends on antigen-specific immunological memory, and innate immunity, which is characteristically antigen and memory independent, contribute extensively to the atherosclerotic lesions. The principal effector cell with innate immunity is the macrophage that can express receptors that recognize a broad range of molecular patterns commonly found on pathogens and can act as a first line of defense. These pattern-recognition receptors include various scavenger (SR) and Toll-like receptors (TLR). A number of ligands bound to these receptors are shown in Table IV.

Pathogenic microorganisms such as *Chlamydia pneumoniae*, cytomegalovirus, and *Helicobacter pylori* have been detected with high frequency in atherosclerotic lesions and have been demonstrated to aggravate atherosclerosis in experimental models.[63] However, because neither infection nor TLR expression is sufficient to induce atherosclerosis in animal models, it is unlikely that microbes and/or TLR signaling play a causative role.[64]

A family of mammalian TLRs has recently been defined as a key component of pathogen-associated molecular pattern recognition machinery.[65] A variety of bacterial and fungal components are known TLR ligands, including peptidoglycan for TLR2, lipopolysaccharide (LPS) for TLR4, flagellin for TLR5, and unmethylated CpG motifs in bacterial DNA for TLR9. It seems that the TLRs may collectively be responsible for recognizing a large repertoire of microbial pathogens. The binding of the TLRs to oxidized LDL is also accompanied by up-regulated TLR expression in macrophages by oxidized LDL *in vitro*.[66]

Whereas binding of the recognized particles to scavenger receptors leads to endocytosis and lysosomal degradation, engagement of TLR transmits transmembrane signals that activate NF-κB,[67] and MAPK pathways.[68] TLR binding induces the expression of a wide variety of genes such as those encoding proteins involved in leukocyte recruitment, production of reactive oxygen species (ROS), and phagocytosis. Activation of TLRs can also elicit the production of cytokines that augment local inflammation. Finally, TLR ligation may also directly induce apoptosis.[69]

It is noteworthy that endothelial cells also express TLRs and SRs that, on binding the appropriate ligand, induce the expression of leukocyte adhesion molecules, iNOS2, endothelin, interleukin-1, and other inflammatory molecules. Their activation causes leukocyte recruitment, increased permeability, edema, and other characteristic features of inflammation.

The more specific response of adaptive immunity involves the recognition of specific molecular structures and depends on the generation of large numbers of antigen receptors (i.e., T-cell receptors and immunoglobulins) generated by somatic rearrangement processes in blast cells. These responses include direct attack of antigen-bearing cells by cytotoxic T-cell lymphocytes, stimulation of B cells to produce antibodies against the antigens, and induction of inflammation, with enhanced innate responses in the area where the antigen is present. The adaptive response tends to be a more delayed response.

The activated macrophages express MHC class II antigens, which allow the incorporation of T cells and ensure the involvement of both an innate and an adaptive cell-mediated immune response at the atherosclerotic lesion. In the artery wall, adaptive immune recognition mainly leads to Th1 effector responses, characterized by the secretion of proinflammatory cytokines and by the activation of macrophages and vascular cells. The activation of T cells by the antigens presented by the macrophages results in the secretion of cytokines (e.g., Interferon-γ and TNF-α) that amplify the inflammatory response. Further immune activation is furnished by the up-regulation of CD40 ligand, a potent immunoregulatory molecule, and the CD40 receptor, in macrophages, T cells, endothelial, and SMCs in atherosclerotic lesions *in vivo*. Another signal that is recognized in chronic atherosclerosis (perhaps as a danger signal) by both T cells and macrophages is HSP60 eliciting increased cytokine expression.[70,71] In animal models, activation of innate immunity leads to changes in lipoproteins, enzymes, transfer proteins, and receptors with an increase in atherogenic lipoprotein particles.[72] Adaptive immunity is largely initiated by the recognition of disease-related antigens, which include oxidatively modified lipoproteins, heat-shock proteins (HSPs),

TABLE IV Ligands for Pattern Recognition Receptors

Ligand	Scavenger Receptor	Toll-Like Receptor
LPS	SR-A	TLR2, TLR4
Lipoteichoic acid	SR-A	TLR2, TLR4
Acetyl-LDL	SR-A, MARCO, SR-EC	?
Oxidized LDL	SR-A, CD36, SR-PSOX, LOX-1	?
HSP60	?	TLR2, TLR4
CpG DNA	?	TLR9

SR, scavenger receptor; TLR, toll-like receptor; LPS, lipopolysaccharide; MARCO, macrophage scavenger receptor; EC, endothelial cell; LOX, lectin-like oxidized LDL receptor; SR-PSOX, scavenging receptor that binds phosphatidylserine and oxidized lipoprotein.

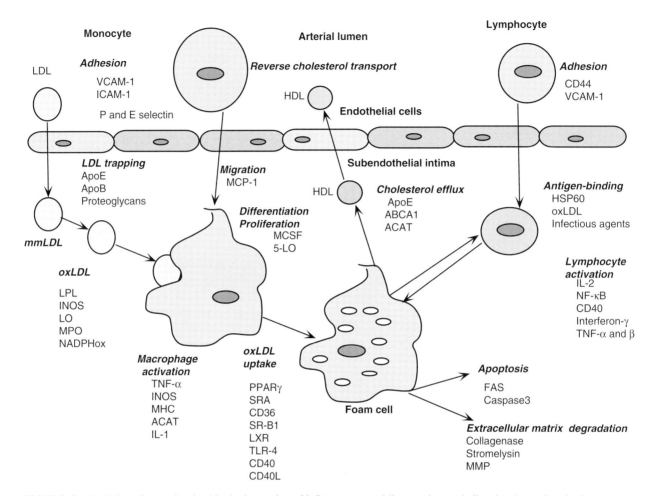

FIGURE 2 Multiple cell types involved in the interaction of inflammatory and lipoprotein metabolism in atherosclerosis. Formation of lesions of atherosclerosis initiated at the endothelium. These changes include increased endothelial permeability to lipoproteins and other plasma constituents; up-regulation of leukocyte adhesion molecules, including L-selectin, integrin, and PECAM 1; up-regulation of endothelial adhesion molecules, which include E-selectin, P-selectin, ICAM-1, and vascular-cell adhesion molecule 1; migration of leukocytes into the artery wall, which is mediated by oxidized low-density lipoprotein (ox-LDL), monocyte chemotactic protein 1, interleukin-8, platelet-derived growth factor, and macrophage colony-stimulating factor. Monocytes and macrophages (foam cells) together with T lymphocytes are recruited to form the fatty streak. Factors associated with macrophage accumulation include macrophage colony-stimulating factor, monocyte chemotactic protein 1 (MCP-1), and ox-LDL. The fatty streaks progress to intermediate and advanced lesions, with the formation of a fibrous cap that covers a mixture of leukocytes, lipid, and debris and may form a necrotic core. The fibrous cap forms as a result of increased activity of platelet-derived growth factor, transforming growth factor ß, interleukin-1, tumor necrosis factor, and osteopontin and of decreased connective-tissue degradation. The necrotic core represents the results of apoptosis and necrosis, increased proteolytic activity, and lipid accumulation. Specific markers are shown in plain text.

and microbial macromolecules. Furthermore, endothelial cells can activate adaptive immunity by presenting foreign antigens to specific T cells. A diagram illustrating the cross-talk between these different cell-based inflammatory and immune responses in atherosclerosis is shown in Fig. 2.

In addition to their role in the innate immune system, macrophages are essential modulators of lipid metabolism. The induction of both lipid metabolic and inflammatory pathways in activated macrophages are central to the pathogenesis of atherosclerosis.[73] Up-regulation of expression of genes involved in oxidized lipid uptake (e.g., CD36) and cholesterol efflux in macrophages is mediated by PPAR-γ and liver X receptors (LXRs).[74–76] The same receptors have also been implicated in negatively modulating macrophage inflammatory gene expression.[77,78] These findings suggest that specific nuclear receptors may serve to integrate pathways of lipid metabolism and inflammation with control of macrophage gene expression and may provide viable targets for therapies with high-affinity ligands such as the thiazolidinedione class of drugs to treat atheromatous lesions. It is also noteworthy that PPAR-γ agonism has been found to be an important target in the therapeutic treatment of metabolic syndrome-related conditions.[79]

The Cap of the Atherosclerotic Lesion and Its Rupture

Myocardial infarction (MI) primarily occurs as a result of erosion or uneven thinning and rupture of the atherosclerotic lesion's fibrous cap, often at the shoulders of the lesion site where macrophages enter, accumulate, and are activated and where apoptosis may occur.[80] Degradation of the fibrous cap may result from the increased expression and activity of metalloproteinases such as collagenases, elastases, and stromelysins. Activated T cells may stimulate metalloproteinase production by the macrophages in the lesions, promoting plaque instability. Matrix metalloproteinases, cathepsins, and mast cell proteases can impair the integrity of the fibrous cap by degrading its collagen cap. Proinflammatory cytokines can regulate the release of these matrix-degrading proteinases. Moreover, LPS, TNF-α, IL-1, and interferon-γ all induce tissue-factor procoagulant gene expression in human endothelial cells. The production of tissue-factor procoagulant and other hemostatic factors is considered to be a principal factor in the thrombosis of the atherosclerotic lesion.[81]

Stable advanced lesions usually have uniformly dense fibrous caps. The potentially more dangerous lesions are often nonocclusive and thus may be difficult to detect by angiography, yet at autopsy active inflammation is generally evident in the accumulation of macrophages at sites of plaque rupture. Macrophage accumulation may be associated with increased plasma concentrations of both fibrinogen and C-reactive protein, two markers of inflammation thought to be early signs of atherosclerosis. Plaque rupture and thrombosis may be responsible for as many as 50% of cases of acute coronary syndromes (ACS) and MI.

NULL MUTATIONS IN TRANSGENIC MICE AS ATHEROSCLEROSIS MODELS

Among the many animal models that have been used in studying the genetics of atherosclerosis, the generation and use of transgenic mice with null mutations in genes such as *LDLR* and apolipoprotein E (*APOE*) have proved to be highly informative.[82–84]

In the absence of apolipoprotein E, lipoproteins are not carried to the liver, their normal site of metabolism, and the mice become an excellent atherogenic model with hypercholesterolemic and atherosclerotic lesions developing in a diet-responsive manner similar to those found in humans. Null mutations in specific genes have been introduced into *apoE*-deficient strains and examined for the effect on the phenotype and progression of atherosclerosis mediated by the absence of their specific function. The effect on atherosclerotic progression in strains harboring specific gene knock-outs is presented in Table V. Although some gene knock-outs increase the genesis of atherosclerotic lesions, others reduce the atherosclerotic phenotype. Also, the effects on atherosclerotic progression (in the *apoE* and *LDLR* deficient mice) from overexpressed genes can also be assessed. For instance, overexpression of protective genes such as *PON-1* in *apoE*-deficient mice markedly reduced atherosclerosis.[85] In contrast, overexpression of apoC1 markedly increased hyperlipidemia and atherosclerosis in the *apoE*-deficient mouse.[86] A similar approach has been recently reported with the recombinant apoA1 overexpression,[87] as well as by introducing A-1 mimetic peptides (e.g., D-4F).[88]

In addition to enhancing our understanding of the underlying processes, the transgenic mice can be crossed with strains containing a variety of overexpressed genes and examined for their responses to drugs directed against the atherosclerotic process.

Studies in transgenic mice have shown that Lp (a) lipoprotein, CETP, and apolipoprotein A (the principal apoprotein subunit of the high-density lipoproteins), have little effect on atherogenesis, whereas null mutations in *MCSF* and *5-LO* seem to be important in the regulation of the numbers of monocytes and macrophages and in lesion formation.[89,90] Transplantation of *PPAR-γ* null bone marrow into *LDLR* −/− mice results in a significant increase in atherosclerosis, consistent with the hypothesis that the regulation by this nuclear receptor is protective *in vivo*.[75] The knock-out of serum paraoxonase (*PON1*), an esterase associated with plasma HDLs that seems to confer protection against CAD by reducing pro-inflammatory oxidized LDLs, produced mice that were more susceptible to atherosclerosis than their wild-type litter mates when fed a high-fat, high-cholesterol diet.[91]

Studies with genetically altered mice, particularly the *apoE* and *LDLR* knock-out strains, have also provided critical information concerning the components of innate and adaptive immunity in atherosclerosis. Hyperlipidemic mice

TABLE V Atherosclerosis in Knock-out Mouse Models

Gene	Disease Model	Atherosclerotic Lesion
MCSF	apoE	↓
P-selectin	apoE	↓
E-selectin	apoE	↓
VCAM-1	apoE	↓
ICAM-1	apoE	↓
SR-A	apoE	↓
CD 36	apoE	↓
IFN-γR	apoE	↓
CD40L	apoE	↓
TLR-4	apoE	—
LTA	apoE	↓
5-LO	LDLR	↑
PON1	apoE	↑
IL-10	apoE	↑
SRB-1	apoE	↑

lacking scavenger receptor *SR-A* or *CD36* alleles exhibit a reduced incidence of atherosclerosis, implicating these pattern-recognition receptors in CAD pathogenesis.[92,93] In contrast, deficiency in TLR-4 does not significantly affect the incidence/severity of atherosclerosis in *apoE −/−* mice.[94] In *apoE −/−* mice lacking interferon-γ signaling, there is also a reduction in the incidence and severity of atherosclerosis, with enhanced atherosclerotic lesions occurring on administration of exogenous interferon-γ.[95] The immune mediator CD40 and its ligand CD40L, which are expressed on endothelial and SMCs, macrophages, and T lymphocytes within human atherosclerotic lesions, have been implicated in atherogenesis, including regulation of matrix metalloproteinases and cytokines.[96] In vivo interruption of CD40 signaling stemmed the initiation and progression of atherosclerotic lesion formation in hypercholesterolemic mice. These findings suggest a contributory proatherogenic role for these signaling elements.

MOLECULAR MARKERS OF CAD

As previously noted, analysis of lipid/lipoprotein levels (HDL, VLDL, and LDL) can be predictive of atherosclerosis in a large subset of patients with CAD. Interestingly, the measurement of CRP has been shown to have a significant prognostic value, indicating that inflammation likely has a contributory role in atherosclerosis, even in cases not caused by dyslipidemias.

Gene Variation and Disease

Many genetic variations have been associated with increased risk of atherosclerosis. Although many of these variations influence traditional risk factors, they also can provide information useful for assessing the risk to relatives or guiding therapy (e.g., LDLR, apoAV, apoB, apoE, CETP, LPL, upstream transcription factor-1).

As we have previously noted, a number of genes that are associated with an increased risk for atherosclerosis have been recently identified. Some of these genes such as lymphotoxin-α (*LTA*), *PDE4D*, and *ALOX5AP* exhibit common variations that influence the risk of CAD independently of the traditional risk factors.[97–99] Other gene variants are rare, but they are highly penetrant (e.g., myocyte enhancer factor 2A [*MEF2A*]), and still others are important markers of familial disease (e.g., upstream transcription factor-1 [*USF-1*]).

Prospective large-scale trials to evaluate the effectiveness of genetic testing to individual risk stratification and to guide the effectiveness of therapy are needed before these tests can be used as a routine clinical tool. Rapid progress in this area will follow further studies of gene–gene, gene–phenotype, and gene–treatment interactions in large patient populations and should provide useful information on the clinical application of individual's genotyping for the assessment, prevention, and treatment of atherosclerosis.

Rationale for DNA Testing

DNA is stable and simple to isolate, and with multiplex analysis a large number of tests can be performed on the DNA derived from a few milliliters of blood. Until recently, the cost of typing DNA genotyping was >$1/genotype, but emerging technologies are reducing this amount by more than a factor of 10. Nevertheless, the number of tests that will be required per gene is a critical issue. If risk is determined largely by genetic variations that are common in human populations, then genotyping can be relatively inexpensive. An example of a significant common polymorphic variant affecting the atherosclerotic phenotype occurs with specific *APOE* alleles.[100] On the other hand, if susceptibility to CAD results largely from many different rare mutations, such as those found in the highly variable mutations of the *LDLR* locus in FH, then genotyping becomes a more difficult, less reliable, and expensive undertaking. However, newer techniques are emerging that use rapid, high-throughput genetic screens that can detect a significant fraction of FH mutations.[101,102]

Notwithstanding, considerable obstacles remain in the routine clinical application of genetic analysis for atherosclerosis. For instance, information is lacking to assess differences in DNA sequence because of ethnic groups. In addition, many of the previously reported gene associations with CAD require further confirmation. Gene association studies of complex traits, such as coronary atherosclerosis, caused by the interaction of a multiplicity of genetic and nongenetic factors and complex gene–environment interactions, can be largely affected by the genetic heterogeneity within the population studied and incomplete penetrance of risk alleles and are notoriously difficult to replicate. It has also been argued that genetic testing may not have much value, because the presently known genetic differences may not predict the risk over and above that of measured traits, such as plasma lipids. Such values reflect both genotype and exposure, and they may be better predictors of the clinical outcome. The value of genetic testing in the analysis of atherosclerosis will likely become more evident only when implemented on a large scale.

Several potential benefits to genetic testing exist: (1) a number of recently identified genes are likely to be independent of measured traits; (2) they may help to guide therapy as discussed later; and (3) genetic analysis may reveal highly penetrant disorders that could significantly influence the risk for individuals and family members. In view of these considerations, it is suggested that multilocus testing be evaluated in large trials to determine its overall value and to identify the most important risk factors. Finally, the identification of DNA sequence variations associated with coronary atherosclerosis may provide windows into new therapeutic targets. However, it remains to be seen whether assaying single or even hundreds of DNA sequence polymorphisms will provide useful predictive information regarding CAD risk, which reflects the complexity of the atherosclerotic lesion and highlights the difficulties associated with predicting the clinical outcome of individual patients.

Pharmacogenomics

Pharmacogenetics provides the basis to understand the variability in response to drugs as a function of an individual's genetic makeup. Genetic polymorphisms may influence drug response through a number of mechanisms, including pharmacokinetic interactions, pharmacodynamic gene–drug interactions that involve gene products expressed as receptors, and genes that are in the causal pathway of disease. The modified gene products can affect the transport or metabolism of specific drugs.

A number of different genetic polymorphisms with a potential role in the different responses to lipid-lowering therapy have been recently identified and may have predictive value in assessing individual successes in hypolipemic drug treatment. Significant interactions between an individual's *APOE* genotype and plasma lipoprotein levels in patients on statin therapy have been reported.[103] Individuals with the *E2* allele are more likely to respond to statin therapy, with a favorable reduction in total cholesterol and LDL-C, than subjects harboring at least 1 *E4* allele. Variation in response to statins on disease progression also has been associated with genetic polymorphisms. A specific polymorphism in the *CETP* gene has been correlated with the effectiveness of pravastatin on disease progression. In a recent clinical trial evaluating statin therapy in atherosclerosis (REGRESS), approximately 16% of the trial population had the *CETP B2B2* genotype and exhibited no benefit from the use of statins on the atherosclerosis progression, whereas individuals with either the *B1B2* or *B1B1* genotypes significantly benefited by statin treatment.[104] Similarly, in pravastatin-treated patients in the REGRESS trial, the risk of clinical events was associated with a common polymorphism in the promoter of the matrix metalloproteinase stromelysin-1 gene.[105] Interestingly, these effects were independent of the effects of pravastatin on the lipid levels.

Genetic testing may also reveal significant dietary interactions. In the case of 5-LO, dietary arachidonic acid intake significantly enhanced the proatherogenic effect of 5-LO variants, whereas intake of n-3 polyunsaturated fatty acids (PUFA) attenuated this effect.[106] Because n-3 PUFA decreases the formation of leukotrienes by competing with arachidonic acid as substrates for 5-LO, the antiatherogenic effects of n-3 PUFA may be limited to or may be more prominent in individuals with genotypes that favor increased 5-LO activity.

Several observations have suggested that specific polymorphic variants in the angiotensin-1 converting enzyme (*ACE*) are more prevalent in individuals with CAD, as well as in association with hypertension and MI. These findings are further discussed in Chapters 9 and 12.

Further understanding of the genetic basis of atherosclerosis may provide mechanisms for creating personalized designer therapies for an individual or groups of patients with similar pathology on the basis of individual genotype. Pharmacogenetics may allow for a more targeted approach to identify effective therapies to which individual patients will respond best.

Gene Profiling in Atherosclerosis

To understand fundamental pathobiological mechanisms in atherogenesis and to develop and target new therapies, information on gene expression patterns (atherogenomics) and protein expression patterns (atheroproteomics) are crucially needed. Despite the relatively recent advent of this technology, a large number of studies have been performed in animal models, and several global gene profiling studies have been performed in tissues and cells derived from patients.[107–110] Gene arrays provide a valuable approach for analysis of atherosclerotic plaque composition and for the identification of candidate markers of plaque progression. Further studies will be able to examine how gene profiles may vary in response to diet and specific pharmacological treatments and may be a useful adjunct in tailoring individual therapies.

THROMBOSIS

Blood coagulation and thrombosis

The human blood coagulation system is composed of a series of linked plasma glycoproteins that on activation induce the generation of downstream enzymes ultimately forming fibrin primarily involved in arresting bleeding (hemostasis). Also involved is the mediation of activated platelets and endothelial factors.

Hemostasis involves the macromolecular assembly of substrates, enzymes, protein cofactors, and calcium ions on a phospholipid surface that markedly accelerates the rate of coagulation. Excess and pathological coagulation activity occurs in thrombosis, the formation of an intravascular clot, which in the most dramatic form precipitates in the microvasculature as disseminated intravascular coagulation. Thrombosis has been likened to a biochemical machine[111] operative in the case of atherothrombosis on a ruptured atherosclerotic plaque but that may develop at a slower rate in venous thrombosis, illustrating that the coagulation machinery can function at different velocities.

Coagulation Pathways

A highly simplified general scheme of the coagulation clotting cascade is shown in Fig. 3. After vascular injury, clotting is initiated by the binding of plasma FVII/FVIIa to tissue factor (TF), also known as coagulation factor III. In the extrinsic pathway, the TF–FVIIa complex initiates blood coagulation by activating both FX (forming FXa) and FIX. Alternately, by means of the intrinsic

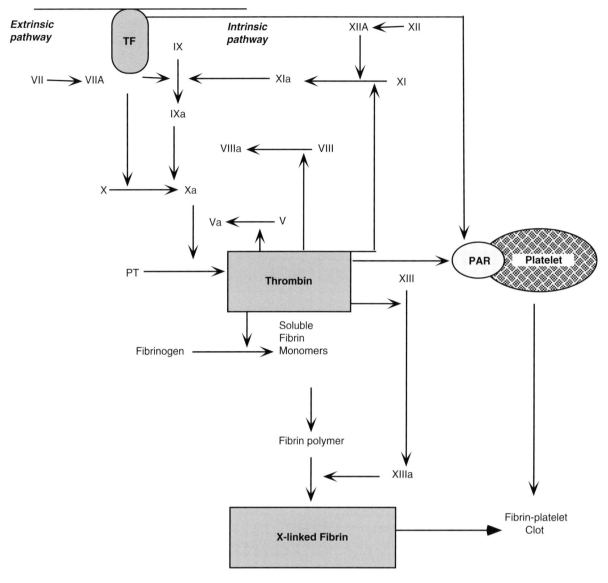

FIGURE 3 Coagulation pathway. The intrinsic, extrinsic, and common pathways of the coagulation (clotting) cascade are shown as described in text.

pathway, FXI can activate FIXa, which, on forming a complex FVIIIa–FIXa, generates FXa, The prothrombinase complex (FVa–FXa) activates prothrombin (PT) to form thrombin, a multifunctional serine-protease cast as a central player in the coagulation protease cascade by virtue of its activation of various proteases and cofactors. Thrombin mediates the cleavage of fibrinogen to soluble monomers (SFM), which are cross-linked by FXIIIa to form fibrin, activates FXI, which is an alternate way to generate FIXa, and activates several G-protein–coupled protease-activated receptors (PARs) on platelets, which, in association with fibrin production, leads to the formation of a clot. These receptors (which are also functional in endothelial cells) mediate thrombin signaling, triggering a variety of phenotypic changes including alterations in cell shape, permeability, vasomotor tone, leukocyte trafficking, migration, and hemostasis. Of the four PAR isoforms, thrombin activates three, PAR-1, PAR-3, and PAR-4.

Platelets also accelerate the activation of the coagulation cascade by binding FXI by means of the receptor glycoprotein Ib-IX-V and by providing a thrombogenic surface for the assembly of the prothrombinase complex (FVa–FXa).

Tissue Factor

Tissue factor (TF) is a potent initiator of the coagulation cascade. Recent evidence has indicated that the transmembrane cellular receptor TF not only is critical to triggering the coagulation by its interaction with FVIIa but also induces cell signaling by means of activation of specific PARs, as do several of the other coagulation proteins, although different PAR isoforms are targeted. The TF–FVIIa complex activates

PAR-2, and FXa activates both PAR-1 and PAR-2. Crosstalk between coagulation and inflammation in endotoxemia is mediated by PAR-1 and PAR-2.[112]

ROLE OF TF IN THROMBOSIS ASSOCIATED WITH ATHEROTHROMBOSIS

Aberrant TF expression triggers intravascular thrombosis associated with various diseases, such as atherosclerosis and in sepsis. In atherosclerosis, TF is expressed by macrophage-derived foam cells within atherosclerotic plaques.[113] Moreover, there is considerable evidence that several mediators such as inflammatory cytokines and oxidized lipids involved in the process of atherosclerotic plaque formation are capable of inducing TF expression in cells such as monocytes, macrophages, and endothelial cells, which under normal physiological conditions do not express TF. TF levels were higher in atheroma from patients with unstable angina compared with stable angina.[114] Moreover, in coronary and carotid plaques, TF content was found to be higher in plaques from symptomatic than asymptomatic patients. Results have strongly suggested that high levels of TF exposed on plaque rupture trigger thrombosis and MI. Studies have also demonstrated an important contribution of TF in the pathogenesis of thrombosis and restenosis after balloon angioplasty.[113]

Although the classical view of TF held that it is primarily expressed locally within an atherosclerotic lesion, recent studies have suggested that an additional source of TF, known as blood-borne or plasma TF, may also contribute to thrombosis. This form of TF is mainly associated with microparticles originating from endothelial cells, SMCs, platelets, or leukocytes.[115–117] Importantly, several studies have shown that levels of blood-borne TF are increased in atherosclerosis.[116–118] Tissue factor is also found in adventitia of blood vessels, as well as the lipid core of atherosclerotic plaques. Thrombogenic TF on cell-derived microparticles of both monocyte and platelet origin is also present in the circulating blood of patients with ACS. Plasma levels of TF have also been found to be significantly elevated in patients with cardiovascular risk factors, including hypertension, dyslipidemia, hyperglycemia, and diabetes.[115] Some studies show a correlation between the levels of blood-borne TF and acute myocardial infarction (AMI).[119] Interestingly, TF is constitutively expressed in cardiac myocytes but not in skeletal myocytes.[120] A likely function of TF in the heart is to provide additional hemostatic protection.

Studies in transgenic mice have also suggested another role for TF and its interaction with factor VII in mediating the migration and proliferation of vascular SMCs, a critical event in the vascular remodeling integral to the formation of atherosclerotic lesions. Vascular remodeling in response to injury is markedly reduced in transgenic mice lacking the cytoplasmic domain of TF.[121]

There has been substantial interest in the development of specific and nonspecific inhibitors of TF or the TF–factor VIIa complex to reduce thrombosis. This antithrombotic strategy can target the initial steps of the coagulation pathway leaving the downstream effectors and mediators intact.

An endogenous tissue factor pathway inhibitor (TFPI) present in plasma modulates the initiation of coagulation induced by TF. This serine protease inhibitor binds directly and inhibits the TF–factor VIIa complex, as well as with factor Xa formation. This inhibition is mediated by three tandemly located domains in TFPI, the first and second of which inhibit the TF–factor VIIa complex and factor Xa, respectively. Several studies have suggested that the C-terminus of TFPI has an additional function (i.e., the modulation of cell proliferation).[122,123] TFPI was recently found to inhibit the proliferation of vascular wall cells by inducing apoptosis.[122] Although found in plasma both as a full-length molecule and as C-termini truncated forms, TFPI also circulates bound with plasma lipoproteins. *In vivo* administration of recombinant TFPI (rTFPI) or of the *TFPI* gene in experimental animal models has shown marked anticoagulant effects, including the elimination of shear stress–induced thrombosis,[124] reducing fibrin formation and neointimal development after repeated balloon injury of rabbit aorta,[125] and preventing fibrin deposition on subendothelial human matrix in an *ex vivo* model.[126] More recent studies have shown that TFPI can provide protection against the widespread systemic activation of coagulation promoted by disseminated intravascular coagulation (DIC), a severe disorder resulting in diffuse fibrin deposition in small and midsize vessels that occurs frequently in septic patients, associated with increased mortality. Moreover, treatment with TFPI has been shown to attenuate thrombin generation and to reduce mortality in experimental sepsis models in mice, rabbits, and baboons.[127,128] Concerns have been raised that the therapeutic effects and potential clinical application of TFPI could be limited because of its short half-life *in vivo*. By use of computer homology modeling and molecule docking, two recombinant long half-life human *TFPI* mutants coined *TFPI-Mut1* and *TFPI-Mut4* were designed and expressed in *E. coli*.[129] Both purified re-engineered TFPI proteins displayed similar bioactivities of wild-type TFPI with significantly prolonged half-life *in vivo* and are expected to provide better clinical value and therapeutic effect.

A phase II clinical trial of the efficacy of TFPI in patients with severe sepsis demonstrated a mortality reduction in TFPI-treated compared with placebo-treated patients and an improvement of organ dysfunctions.[130] However, findings with larger, multinational phase III clinical trials in patients with severe sepsis have not demonstrated significant reductions in overall 28-day mortality rates,[131,132] although some did suggest the presence of specific definable patient subgroups that may have benefited from TFP.[131]

TF and TFPI Genes and Variants

The human *TF* gene spans 12.4 kb and is organized into six exons separated by five introns; its promoter sequences identified and transcriptional activation in human monocytes and endothelial cells have been well characterized.[133,134] In endothelial cells, mononuclear phagocytes, and SMCs, the induction of *TF* transcription is mediated by growth factors,[135] mechanical injury,[136] myocardial ischemia and reperfusion,[137] and shear stress,[138] which suggest the *TF* gene acts as a primary response gene. Genotype analysis of a large population of patients with MI has revealed that the *TF* gene promoter exists in two major haplotype variants with similar prevalence designated as 1208 D and 1208 I.[139] These haplotypes differ at four sites, including an 18 bp deletion/insertion at -1208, and three single nucleotide polymorphisms (-603A/G, -1322C/T, -1812C/T). Although neither type was found to be associated with coronary thrombosis, the 1208 D haplotype is associated with reduced plasma TF levels and a lower risk of venous thrombosis. Moreover, a recent study demonstrated that the 1208 D-allele is associated with elevated TF expression in human umbilical vein endothelial cells (HUVEC), which may be due to variant transcription factor binding in the -1208 region.[140] Another single nucleotide polymorphism in *TF* (at 5466 A>G) was recently analyzed for a potential role in patients with ACS at both the genotype level and compared with healthy patients for its effect on *TF* mRNA expression and activity.[141] These studies concluded that the 5466 AG genotype was a significant novel predictor of cardiovascular death in ACS, and that whereas *TF* mRNA and basal TF activity were significantly lower among 5466 AG carriers, there was a significant elevation in monocyte TF activity on LPS stimulation.

The human *TFPI* gene localized to chromosome 2 consists of nine exons with a promoter region.[142,143] Although the nucleotide sequence of the promotor region containing several binding sites for transcription factors has been determined (including the identification of three GATA sites),[144] it remains to be established precisely how endothelial cells synthesize TFPI and how the expression is regulated. TFPI expression is abundant in endothelial cells (supporting the view of the endothelium as the primary site of TFPI synthesis) and less so in monocytes; expression is markedly elevated in endothelial cells treated with TNF-α or phorbol esters and in monocytes treated with endotoxin.[145] Some cell types such as vascular SMCs that constitutively express TF also synthesize TFPI, providing dual components necessary for the regulation of clotting in their microenvironment; these cells show a marked up-regulation in serum and more specifically by its constituent growth factors, epidermal growth factor (EGF) and PDGF-B.[146] Similarly, human pulmonary artery SMCs that constitutively synthesize and secrete TFPI show marked up-regulation of TFPI activity and content when grown in the presence of serum but not with phorbol esters, LPS, or TNF-α.[147]

A second gene *TFPI2* has been identified containing a similar overall domain organization and considerable primary amino acid sequence homology to *TFPI*.[148] It is primarily expressed transcribed in umbilical vein endothelial cells, liver, and placenta. However, the genomic structure of the 7-kb human *TFPI2* gene is different than that found with *TFPI1* consisting of five exons and four introns.[149]

Mutations and polymorphism of *TFPI* have been reported. In the 342 patients with venous thrombosis, 4 individuals were identified as being heterozygous for a P151L mutation in *TFPI*.[150,151] However, a Spanish group has recently reported that this mutation is not associated with an increased risk for venous and arterial thrombosis.[152]

Another mutation of *TFPI*, V264M, was examined in patients with ACS, but no association between the ACS phenotype and the mutation was found.[153] To assess the effect of this mutation on plasma TFPI levels, patients with different genotypes (i.e., 5 V/M and 13 V/V) were examined. Levels of total and free TFPI antigen were markedly lower in the heterozygous variants than in the wild type; patients' TFPI activity levels were also lower, but the difference was not significant. Further studies by the same group found no link between this mutation and venous thromboembolic disease or restenosis after coronary angioplasty.[154] Polymorphism in the promoter region of *TFPI* was not associated with the plasma level of TFPI or with venous thrombosis.[155]

Factor VIIA

The human factor VII gene (*F7*) encoded on chromosome 13 consists of nine exons and spans approximately 12 kb.[156,157] Factor VII is an inactive vitamin K–dependent zymogen synthesized in the liver and secreted as a single-chain glycoprotein of 50 kDa. All vitamin K–dependent coagulation zymogens share a similar protein domain structure consisting of an N-terminal γ-carboxyglutamic acid (Gla) domain with 9–12 residues, C-terminal serine protease domain (catalytic domain), and 2 EGF-like domains. On contact with TF, it is converted to the two-chain active form by selective proteolysis at a single peptide bond at Arg152-Ile153. Factor VIIa, the activated form of factor VII, consists of a light chain (20 kDa) and a heavy chain (30 kDa) covalently linked by a disulfide bond. Rapid activation occurs when factor VII is combined with TF cofactor in the presence of calcium. Conversion of factor VII to factor VIIa is catalyzed by a number of proteases, including thrombin, factor IXa, factor Xa, factor XIa, and factor XIIa. Comparison of these proteins has shown that factor Xa, in association with phospholipids, exhibited the highest potential to activate factor VII.[158]

Elevated factor VII plasma levels was found to be associated with myocardial ischemia in an early study with a large patient population.[159] However, subsequent studies have been unable to show a clear-cut relationship between coronary events and factor VII levels and did not confirm that factor VII is an independent risk factor or predictor of coronary disease.[160,161]

Inherited factor VII (FVII) deficiency is a rare autosomal-recessive monogenic disorder. Over the last decade, more than 30 mutations and polymorphisms of the *F7* gene have been characterized in patients with FVII deficiency.[162] Most mutations thus far identified are missense mutations and affect the protease domain, indicating that loss of protease function is the most common cause of the clinical phenotype.[163] Severe forms of deficiency can be associated with intracranial hemorrhages occurring close to birth with a high mortality rate. Two cases of neonatal intracerebral bleeding were associated with FVII activity levels less than 1% of normal. *F7* genotyping investigations revealed genotypes. including the deleterious Cys135Arg mutation and a novel Ser52 nonsense mutation, at the homozygous state.[164]

Although complete absence of FVII activity in plasma has been shown to be generally lethal, most individuals with *F7* mutations are either asymptomatic or the clinical phenotype is unknown. Generally, only homozygote or compound heterozygote patients with factor VII deficiency are symptomatic.

An early study demonstrated a single base substitution (A/G) at codon 353 (Arg/Gln) in exon 8 associated with a 20–30% variance in factor VII levels in males and females and in different ethnic groups.[165] Subsequent studies have confirmed that carriers of the *F7* allele coding for Gln353 have reduced factor VII levels and clotting activity.[166] One study of Italian patients with familial arterial disease has also shown that individuals homozygous for the Gln353 allele had significant protection from MI, whereas individuals with the more common Arg353 allele had a higher risk, suggesting that in some populations the *F7* gene variants may attenuate atherosclerotic risk. These results remain somewhat controversial, because a larger case-controlled study from The Netherlands while finding the expected association of reduced factor VII levels with Gln353 also reported that Arg353 was associated with a reduced rather than an increased risk for MI. This study concluded that a genetic propensity to high factor VII levels is not associated with a risk for MI and that an elevated factor VII level itself is not a causal determinant. It remains unclear whether these difference in outcomes are population-specific or related to the different study methods used; the selection of patients with familial arterial disease in the Italian study may also select for associated genetic factors that might strengthen an associated risk.

Moreover, a recent multicenter European study of patients who are congenitally factor VII deficient showed that clinical symptoms did not vary with the frequency of functional polymorphisms and that homozygotes with the same mutation presented with striking differences in severity of bleeding,[167] suggesting that genotype–phenotype relationships indicate the presence of major environmental and/or extragenic components modulating the expression of FVII deficiency.

It is noteworthy that specific mutations and polymorphisms in *F7* are known to occur in some ethnic populations. Among Iranian and Moroccan Jews, a missense Ala244Val mutation is responsible for frequent occurrences of disease. The highest frequencies of the polymorphism, an Arg353Gln substitution, are observed in Gujaratis (25%) and Dravidian Indians (29%) compared with northern Europeans (9%) and Japanese (3%), resulting in decreases in factor VII levels.[168]

Recent attention has been directed to factor VIIa as a promising anticoagulation target, because of its role in complex with TF in initiating the coagulation cascade after blood vessel damage. Studies have revealed novel inhibitors of the TF/factor VIIa complex that can prevent thrombosis with a lower bleeding risk than other types of inhibitors. Ideally, these studies will result in the generation of potent and selective small-molecule factor VIIa inhibitors that can be safely administered once or twice daily in an oral formulation with no need for routine coagulation monitoring.

INTRINSIC PATHWAY

Factor XI

Factor XI (FXI) is the zymogen of a serine protease enzyme in the intrinsic pathway of blood coagulation and is an important factor in the creation of a stable fibrin clot. The FXI protein is composed of five domains: four tandem repeat domains of approximately 80 residues known as Apple domains, and the catalytic serine protease (Sp) domain. The gene for human factor XI mapped to the distal end of the long arm of chromosome 4 is 23 kb in length and consists of 15 exons and 14 introns.[169,170]

Deficiency of FXI leads to an injury-related bleeding disorder and is remarkable for the lack of correlation between bleeding symptoms and FXI coagulant activity (FXI:C). Factor XI (FXI) deficiency is a rare inherited disorder in the general population, although it is commonly found in individuals of Ashkenazi Jewish ancestry. It can cause bleeding complications, especially in the case of hemostatic challenge and/or in tissues with high fibrinolytic activity. The bleeding tendency associated with plasma FXI deficiency in patients is variable, with approximately 50% of patients exhibiting excessive post-traumatic or post-surgical bleeding.

A large number of mutations (more than 65) throughout the FXI gene (*F11*) have been reported in FXI-deficient patients.[171] The disorder is primarily inherited as an autosomal-recessive trait manifesting in homozygotes or compound heterozygotes and infrequently in heterozygotes.[172] A number of causative mutations that have been detected by various screening methods have been described in FXI-deficient individuals. In particular, two mutations—a stop mutation (type II) in Glu117stop and a missense mutation (type III) at Phe283Leu—that are responsible for FXI deficiency are prevalent in the Jewish population. Other mutations in different populations have been identified. Mutation in exon 4 resulting in Q88X can result in a severely truncated polypeptide.[173] Homozygous Q88X was found in a severely affected patient and in three other unrelated families as homozygous, heterozygous, or compound heterozygous states. Other identified mutations include two nonsense mutations in the *F11* gene, in exon 7 and 15, resulting in R210X and C581X, respectively, identified in three families and a novel insertion in exon 3 (nucleotide 137 + G) creating a stop codon.[174-176] A novel nonsense mutation (CAA Gly263→TAA stop) was identified in exon 8 of the *F11* gene in a 42-year-old female patient without bleeding tendency and with undetectable levels of coagulant activity and FXI antigen in her plasma. Missense mutations resulting in G336R and G350A (exon 10) and T575M (exon 15) accompanied by decreased antigen and functional values are consistent with most of the described *F11* mutations associated with the absence of secreted protein.

A role of *F11* mutations or deficiency in arterial thrombosis has not been found. Mutations in *F11* leading to FXI deficiency were found to be similarly distributed in patients with and without AMI.[177] In contrast, high levels of factor XI are a risk factor for deep venous thrombosis, with a doubling of the risk at levels that are present in 10% of the population.[178] The genetic basis for increased levels of factor XI remains undetermined.

Factor XII

The human factor XII gene (*F12*) is approximately 12 kilobase pairs in size and is composed of 13 introns and 14 exons and has been mapped to 5q33-qter.[179,180] The coding sequence for factor XII consists of multiple domains homologous to domains found in fibronectin and tissue-type plasminogen activator.

Coagulation factor XII deficiency is rarely found to be associated with bleeding, but several case studies have reported thromboembolic events in FXII-deficient patients. A common C→T polymorphism in the *F12* promoter region at nucleotide 46 is associated with lower plasma FXII activity levels in Orientals. Evaluation of plasma FXII activity and *F12* genotype in 80 randomly selected and unrelated European subjects also revealed a highly statistically significant association of the *F12* 46T allele with reduced FXII plasma activity.[181] Kanaji and associates[182] reported that this mutation decreased the translation efficiency and led to low plasma levels of FXII activity and antigen, probably as a result of the creation of an alternate ATG codon and/or the impairment of the consensus sequence for the translation-initiation scanning model. Several *F12* missense mutations resulting in lower factor XII levels have been described that are thought to affect intracellular processing of the FXII protein. These include a mutation at W486C and an Ala392 to Thr in the catalytic domain and a Tyr34 to Cys substitution in the N-terminal domain of factor XII.[183-185] All patients homozygous for these alleles displayed extremely low levels of factor XII.

The relationship of these *F12* mutations, FXII deficiency, and thrombosis has been examined. An early study found that FXII deficiency was significantly higher in patients with either recurrent arterial thromboembolism and/or MI and that nearly 70% of the patients with FXII deficiency had a positive family history of thrombosis.[186] Another study found no difference between the prevalence of the C46T polymorphism in patients with MI compared with those without MI and controls, nor was any association found between the extent of CAD and *F12* genotype.[187] However, in a study in which C46T allele, FXII levels, and high cholesterol were examined as risk factors in CAD, patients homozygous for the T allele (TT) harbored lower FXII levels and had an increased risk of CAD in the presence of pravastatin (whereas individuals with the other allele combinations were responsive to pravastatin), suggesting an interaction between these effectors.[188] Another recent genetic case-controlled study analyzing 250 unrelated consecutive Spanish patients with venous thrombotic disease and 250 Spanish subjects matched for gender and age as a control group found that the C46T allele in the homozygous state (genotype T/T) was significantly associated with an increased risk of thrombosis, suggesting that the polymorphism itself is an independent risk factor for venous thromboembolism in a Spanish population.[189] The same group recently reported a relationship of the same TT genotype with an increased risk of ischemic stroke.[190]

The GAIT (Genetic Analysis of Idiopathic Thrombophilia) project sought to quantify the genetic component(s) of susceptibility to thrombosis and related phenotypes.[191,192] Among the clotting factors studied, factor XII levels exhibited one of the highest heritabilities (67%) and a significant positive genetic correlation with thrombotic disease,[191] suggesting that some of the genes that influence variation in this physiological risk factor also influence liability to thrombosis. Subsequent studies using a genomewide linkage screen to localize genetic loci that influence variation in FXII levels detected two loci: one on chromosome 5 in the 5q33-5ter region, near the *F12* gene, and another on chromosome 10.[193] Further examination using bivariate linkage analysis of FXII activity and thrombosis improved the linkage signal. The addition of a 46C/T mutation in the *F12* gene significantly increased the multipoint LOD score but was found to contribute to only a part of the variance in FXII activity, suggesting that there are additional polymorphic loci responsible for determining FXII levels.

Factor IX

On activation, and in the presence of calcium ions and phospholipid surfaces, the serine-protease factor IX forms an active complex with its protein cofactor factor VIII to form the tenase complex, which activates factor X. Deficiency or dysfunction of either factor VIII or factor IX will abrogate the activation of factor X, so that the ensuing steps of the coagulation cascade are compromised, leading to either inefficient or non-existent fibrin deposition. The insufficiency of the activity of the tenase complex is the fundamental biochemical lesion underlying the hemophilias brought about either by a deficiency of coagulation factor VIII cofactor activity (hemophilia A) or coagulation factor IX enzyme activity (hemophilia B).

Hemophilia B is an X-linked recessive, bleeding disorder caused by mutations in the factor IX (*F9*) gene and the resultant deficiency of its serine-protease activity. A wide range of mutations, showing large molecular heterogeneity, has been described in hemophilia B patients. The *F9* gene located on the long arm of the X chromosome at Xq27 is approximately 34 kb in length and contains eight exons.[194] Factor IX consists of a pre- and pro-sequence (27 and 19 amino acids, respectively) and a mature peptide of 415 amino acids containing GLA domain, in which 12 glutamic acid residues undergo post-translational γ-carboxylation by a vitamin K–dependent carboxylase, two EGF-like domains, an activation peptide released after proteolytic cleavages at arginine 145 and arginine 180, and a catalytic serine protease domain. After proteolytic activation and cleavage, activated factor IX is composed of a N-terminal light chain containing the GLA and both EGF domains and a C-terminal heavy chain containing the catalytic domain held together by a disulfide bridge between cysteine residues 132 and 279. The pre-sequence directs factor IX for secretion, and the pro-sequence provides a binding domain for vitamin K–dependent carboxylase to carboxylate specific glutamic acid residues in the adjacent GLA domain. The proteolytic cleavages resulting in factor IX activation are mediated by activated factor XI generated through the intrinsic pathway[195] or through the TF/activated factor VII complex of the extrinsic pathway.[196,197] It is noteworthy that severe factor VII deficiency is characterized by a severe bleeding disorder, whereas severe factor IX deficiency is typically associated with a mild bleeding disorder, which may be indicative of their relative importance.

Although a large number of *F9* mutations have been identified and a limited number of polymorphic loci detected, no relationship of these mutations has been found with either arterial or venous thrombosis.

Factor VIII

The factor VIII (*F8*) gene is also X-linked, located toward the end of the long arm at Xq28, a considerable genetic distance (35 cM) away from factor IX. It is extremely large (approximately 180 kb) and structurally complex (26 exons). Factor VIII is synthesized from a 9-kb transcript as a single-chain protein of more than 2300 amino acid residues that constitutes the circulating inactive pro-cofactor (normally carried and protected by von Willebrand factor)[198,199] and a 19 aa signal peptide (which directs factor VIII passage through the cell). The mature factor VIII peptide is composed of the A1-A3, B, C1, and C2 domains. The activation of factor VIII is mediated by proteolytic cleavage at several arginine residues and is primarily catalyzed by thrombin and by factor X. The activation cleavage sites flank the A and B domains, the latter of which is released from factor VIII on activation and results in two N-terminal heavy chains (one with A1, the other with A2) present as a dimer and a C-terminal light chain (containing A3, C1 and C2). This activated heterotrimer is held together by a metal Ca^{++} bridge. Proteolytic cleavage of factor VIII at the time of activation simultaneously releases it from its complex with von Willebrand factor.[200]

As noted, factor VIII deficiency results in hemophilia A, the most common inherited bleeding disorder. An enormous number of *F8* mutations have been identified that lead to hemophilia A. These include nonsense mutations at CGA (arginine) codons, missense mutations at activation cleavage sites, mutations affecting factor VIII binding to von Willebrand factor, mutations affecting factor VIII secretion and factor IX binding to factor VIII.

There is epidemiological evidence suggesting that patients with hemophilia have modified atherothrombosis.[201] In an early large-scale study of Dutch patients with hemophilia following patients over a 20-year period, the number of deaths caused by ischemic heart disease was significantly lower (80% reduction) than expected, suggesting that hemophilia may provide protection against ischemic heart disease. A follow-up study from the same group confirmed this finding. Moreover, subjects with either factor VIII defective hemophilia and von Willebrand disease had fewer atherosclerotic plaques in the aorta and in the leg arteries, directly correlated to the disease severity.[202] Although some protection with hemophilia-related coagulation defects has been suggested relative to venous thrombosis, thrombosis that is rare in such patients can result from the co-existence of prothrombotic risk factors and from central venous catheter (CVC)–associated thrombosis.[203] Conversely, there are numerous studies showing that factor VIII levels are a significant risk factor of arterial thrombosis, carotid atherosclerosis, and venous thromboembolism.[204–206] Moreover, a substantial proportion of patients (approximately 30%) with deep vein thrombosis were found to have high levels of plasma factor VIII, a number with increasing plasma levels unbound to von Willebrand factor.[207] Modified thrombosis mediated by elevated factor VIII levels[208] or by reduced factor VIII activity (effected by an inhibitory monoclonal antibody)[209] has also been described in experimental mouse models.

COMMON PATHWAY FACTORS

Factor V

Factor V (FV) is a large single-chain glycoprotein of 330 kDa present both in free form in plasma, as well as in platelets. It principal site of biosynthesis is the liver, where human FV is synthesized as a single-chain molecule, undergoing extensive post-translational modifications before being secreted into the blood. The *F5* gene mapped to chromosome 1q23 spans more than 80 kb and contains 25 exons. The isolated cDNA has a length of 6672 bp and encodes a preprotein of 2224 amino acids, including the 28 amino-acid residue long signal peptide.[210,211] FV has a domain organization (A1-A2-B-A3-C1-C2) similar to that of factor VIII. Moreover, its progression from circulating single-chain inactive pro-cofactor to an activated essential cofactor for enzyme-activated factor X (FXa) is triggered by the site-directed cleavage of several peptide bonds in FV mediated by procoagulant enzymes such as thrombin and FXa, similar to factor VIII activation.[212,213] This results in the formation of the activated prothrombinase complex comprising FXa and FVa, which in the presence of calcium ions assemble on negatively charged phospholipid membranes. FVa is considered an essential FXa cofactor, inasmuch as its presence in the prothrombinase complex enhances the rate of prothrombin activation by several orders of magnitude, and FXa is virtually ineffective in its absence.[214]

FV deficiency is inherited as an autosomal-recessive disorder with an estimated frequency of 1 in 1 million.[215,216] Heterozygous cases are usually asymptomatic, whereas homozygous individuals show variable bleeding symptoms, suggesting that FV deficiency in humans is not a lethal disease.

In addition to its procoagulant potential, FV possesses an anticoagulant cofactor capacity functioning in synergy with protein S and activated protein C (APC) to mediate the APC-catalyzed inactivation of the activated factor VIII. Moreover, the expression of anticoagulant cofactor function of FV is dependent on APC-mediated proteolysis of FV at positions Arg306, Arg506, and Arg679. The regulation of the procoagulant activity of FVa by APC together with its cofactor protein S is a central pathway for down-regulating FVa cofactor activity leading to less coagulation. Therefore, failure to control FVa activity can result in thrombotic complications.

An increased interest in FV is linked to its relationship to activated protein C (APC) resistance characterized by a reduced anticoagulant response to APC, which has been found to be a common and significant risk factor for thrombosis. APC resistance is caused by a single point mutation in the *F5* gene (designated factor V Leiden) that not only renders FVa less susceptible to the proteolytic inactivation by APC but also impairs the anticoagulant properties of FV. This G1691A point mutation in *F5* results in an arginine to glutamine substitution at amino acid 506, the site where APC cleaves FVa.[217] The factor V Leiden mutation associated with APC resistance results in the continued activation of thrombin in the prothrombinase complex. Although a significant association has been established between the incidence of venous thrombosis and the factor V Leiden mutation, the influence of this mutation on the arterial circulatory system remains controversial. A recent meta-analysis found that the factor V Leiden mutation modestly increased the risk for MI and ischemic stroke, particularly among younger patients and women.[218] In addition, an increasing prevalence of the factor V Leiden mutation has been reported in children with arterial ischemic stroke.[219] However, other studies have concluded that factor V Leiden does not represent a risk factor for arterial thrombosis.[220,221]

Other allelic variants of the *F5* gene have been described in regard to APC resistance and thrombosis. Two variants, F5 Arg306Thr (FV Cambridge) and F5 Arg306Gly (FV Hong Kong), have been identified in patients with thrombosis, resulting in the loss of the APC cleavage site at Arg306, an important cleavage site in FVa.[222,223] Their relationship to venous thrombosis, however, is not as clear-cut as that provided by factor V Leiden. In addition, an *R2* haplotype characterized by several linked *F5* mutations (both missense and silent) residing in exons 13, 16, and 25 encoding the B, A3, and C2 domains, respectively,[224] has been reported in association with a slightly increased risk of venous thrombosis. No consensus has been reached regarding whether the *R2* allele is really a risk factor for thrombosis.[225] The *R2* allele has been associated with decreased levels of circulating FV in some populations. Moreover, the *R2* allele of *F5* promotes reduced cofactor activity in APC-catalyzed FVIII(a) inactivation. By reducing its cofactor activity of FV in APC-catalyzed FVIII(a) inactivation, the *R2* allele, if combined with FV Leiden, may enhance the APC-resistance phenotype and increase the risk of thrombosis.[226]

Of related interest is the recent identification of modifier genes that can affect phenotype associated with factor V abnormalities. Specific mutations in the ER/Golgi intermediate compartment proteins LMAN1 and MCFD2 promote a combined factor V and VIII deficiency, manifested by an autosomal-recessive bleeding disorder characterized by coordinate reduction of both clotting proteins.[227] These proteins form a cargo receptor that facilitates the transport of factors V and VIII, and presumably other proteins, from the ER to the Golgi.[228]

Factor X

Factor X (FX) circulates as a vitamin K–dependent serine protease that is converted to the active form at the point of convergence of the intrinsic and extrinsic coagulation pathways. The factor X zymogen serves as a substrate for both the extrinsic (TF/FVIIa) and the intrinsic (FVIIIa/FIXa) tenase enzyme complexes, which cleave the Arg15-Ile16 peptide bond in FX, releasing a 52-amino acid activation peptide generating factor Xa. Subsequently, factor Xa interacts with its cofactor, factor Va, phospholipids, and calcium to form a macromolecular prothrombinase complex that converts prothrombin into thrombin.

The human *F10* gene (27 kb) encoding factor X maps to the long arm of chromosome 13, approximately 2.8 kb downstream of the factor VII *F7* gene. Structural analysis shows the *F10* gene consists of eight exons, each of which encodes a specific functional modular domain within the protein.[229] These structural domains in factor X are responsible for specific functional properties, including γ-carboxylase recognition, calcium binding, phospholipid surface interaction, as well as cofactor and substrate binding. The *F10* gene shares a number of structural and organizational features in common with the other vitamin K–dependent coagulation proteins, suggesting that they have evolved from a common ancestral gene. In plasma, FX circulates as a two-chain glycoprotein (~59 kDa) composed of a 306-residue heavy chain (42 kDa) that is covalently linked by a disulfide bond to a 139-residue light chain (16 kDa).[230]

Factor X deficiency, which is inherited as an autosomal recessive trait, is one of the rarest of the inherited coagulation disorders.[231] The clinical phenotype is of a variable bleeding tendency, with heterozygotes often clinically asymptomatic. Homozygous factor X deficiency has an incidence of 1:1,000,000 in the general population. Acquired factor X deficiency is rare, but when it occurs, it is usually in association with amyloidosis.

A modest association of increased factor X levels with venous thrombosis was reported, although no relationship between FX levels and specific *F10* gene polymorphisms was found.[232]

Fibrinogen

The genes for the three polypeptide chains making up fibrinogen are clustered located together in a 50-kb region on the distal arm of chromosome 4, bands q23-q32.[233] The α gene, which consists of five exons, is located at the middle of the fibrinogen cluster. It produces a polypeptide chain that is 625 residues long. The gene for the β chain is located 13 kb downstream of that of the α, contains eight exons, is transcribed in the opposite direction to those of the α gene, and codes for a 461-residue polypeptide.[234] The third and last gene of the fibrinogen cluster is located 10 kb upstream of the α gene and codes for a 411-residue chain.[233]

Among the components of the coagulation system, elevated fibrinogen has been most consistently associated with arterial thrombotic disorders. Numerous prospective studies of both healthy subjects and patients with pre-existing vascular disease, such as the Northwick Park Heart study,[235] the Prospective Cardiovascular Munster (PROCAM) study,[236] and the Prospective Epidemiological Study of Myocardial Infarction (PRIME),[237] demonstrated an association between fibrinogen levels and MI, ischemic stroke, and peripheral vascular disease. Several mechanisms explain the association of increased fibrinogen with arterial thrombotic disease, including increased fibrin formation, blood viscosity, platelet aggregation, and vascular endothelial and SMC proliferation.[238] In addition, high fibrinogen concentrations lead to the formation of a fibrin clot with thin and tightly packed fibers that has high thrombogenicity, possibly because the small pore size restricts access of fibrinolytic enzymes.[239] Fibrinogen levels are strongly correlated with traditional vascular risk factors, including age, physical inactivity, hypertension, smoking, and features of the insulin resistance syndrome. Furthermore, fibrinogen is an acute-phase reactant, in part owing to its up-regulation by means of activation of IL-6–responsive elements in the promoter of all three fibrinogen chains; the acute-phase response arising from viral infection, inflammatory stimuli, and smoking in particular is strongly implicated in the development of arterial disease. Another perspective is that elevated fibrinogen might be a result of the inflammation associated with atherosclerosis rather than being a causal risk factor.[240]

Genetic factors contribute to the total variability in fibrinogen levels. The most common fibrinogen polymorphisms are summarized in Table VI. Although polymorphisms have been identified in genes encoding the three pairs of fibrinogen polypeptide chains, largely because the synthesis of the β-chain is rate-limiting *in vitro*, most studies have focused on this gene (*FIBB*). The main β-chain variants include the Arg448Lys, *Bcl*I, -148C/T, -455G/A (*Hae*III), and -854G/A polymorphisms. The -455G/A and -854G/A substitutions are the most physiologically relevant, because the respective alleles have distinct nuclear protein–binding properties, and reporter gene studies in isolated cells showed an increased rate of basal transcription with the less common *-455A* and *-854A* alleles.[241,242]

TABLE VI Fibrinogen Polymorphisms

Polymorphism	Intermediate phenotype	Disease association
Fibrinogen β -455 G/A	Elevated plasma fibrinogen	AT established, possibly not causal
Fibrinogen β -854 G/A	Elevated plasma fibrinogen	AT established, possibly not causal
Fibrinogen β -*Bcl*1	Elevated plasma fibrinogen	AT established, possibly not causal
Fibrinogen α-chain Thr312Ala	Influences α-chain cross-linking and clot stability	AT suggestive; AF/ pulmonary embolism
Fibrinogen β-448G/A	Alters nuclear protein binding	More common in CAD
Fibrinogen β-148C/T	None detected	None detected
Fibrinogen β-*Bcl*1	Increases fibrinogen level	More common in CAD
Fibrinogen β Arg448Lys	Changes fibrin structure/function	Macrovascular disease

AT, arterial thrombosis; AF, atrial fibrillation.

Of the β-chain *FIBB* polymorphisms, the -455G/A has been the most extensively studied clinically. The *-455AA* genotype is present in 10–20% of the population and is correlated with fibrinogen levels that are 10% higher than in individuals with the *GG* genotype. Nevertheless, the relation between the -455G/A variant and the risk of arterial thrombotic disease is controversial, with some case-controlled studies indicating an association,[243] whereas other large studies reported none.[244] In a pooled analysis of inherited hemostatic risk factors and the risk of AMI, homozygosity for the fibrinogen *-455A* allele was significantly, although marginally, associated with a decreased risk of MI.[245] In addition, the *-455A* allele has been associated with the progression of atheroma.[246] In a recent cohort of elderly patients with stroke, the presence of the *-455A* allele was associated with a 2.5-fold increase in risk of multiple infarcts but not with large-artery strokes; the authors suggested that elevated fibrinogen levels might predispose to the development of thrombosis primarily in small arteries.[247] The association of other β-chain polymorphisms and arterial thrombosis remains unclear.

In addition to the β-chain polymorphisms, a variant in the α-chain contains a Thr312Ala substitution within its C-terminal end,[248] a region important for factor XIII–dependent processes, including α chain cross-linking. Clots generated *in vitro* in the presence of the Ala312 fibrinogen isoform have more extensive α-chain cross-linking and, as a consequence, thicker fibers.[249] Although in one study the Ala312 variant had a gene dose-related influence on post-stroke mortality rates in subjects with atrial fibrillation,[250] these findings were not confirmed among patients with MI in the ECTIM study.[251] More recent studies indicate that this α-chain polymorphism has a more relevant role in the pathogenesis of venous thromboembolism.[252] Taken together, these results suggest that Ala312-induced changes in clot structure predispose to embolization in both the arterial and venous vascular systems.

Prothrombin

The gene for human prothrombin, or factor II (*F2*), has been assigned to 11p11-q12.[253] The 21-kb gene for human prothrombin contains 14 exons separated by 13 introns that comprise 90% of the gene.[254] Human prothrombin is a single-chain glycoprotein composed of 579 amino acid residues containing five functional domains that include a pre-pro leader sequence required for secretion and γ-carboxylation, the γ-carboxyglutamic acid containing Gla domain, two kringle domains for interaction with cofactors, and the catalytically active serine protease domain. On the initiation of coagulation, Factor Xa in the presence of Factor Va, calcium ions, and membrane phospholipids cleaves prothrombin at Arg271-Thr272 and Arg320-Ile321 to release the catalytic domain from the C-terminal of the protein. Human thrombin contains an A-chain of 36 amino acids and a B-chain of 259 amino acids connected by a disulfide bond.

Prothrombin deficiency is a rare (1:200,000) autosomal-recessive disorder associated with a moderately severe bleeding tendency caused by diverse mutations in the *F2* gene. Prothrombin deficiency and lower thrombin activity are typified by two types of congenital disorders, hypoprothrombinemia and dysprothrombinemia, which result in excessive bleeding. Prothrombin antigen levels in plasma decrease significantly in hypoprothrombinemia, whereas in dysprothrombinemia normal prothrombin antigen levels are detectable. More than 20 mutations (mostly missense) in the *F2* gene have been identified that cause prothrombin deficiency.[255]

Elevated levels of prothrombin have been associated with a risk of arterial and venous thrombosis. A mutation in the 3′-untranslated region of the prothrombin *F2* gene, a G to A transition at position 20210, has been associated with increased prothrombin levels.[256,257] The molecular mechanism by which this base change confers an increased risk of thrombosis compared with noncarriers is undefined. However, there is evidence suggesting that the *F2* 20210A variant has a more effective poly (A) site, leading to increased mRNA and protein expression, irrespective of the promoter and gene.[258] A recent proteomic analysis of patients' plasma proteins suggests that the G20210A mutation is associated with increased glycosylation of prothrombin, which confers greater stability to the protein.[259] Interestingly, two recently identified *F2* gene mutations (20209C>T and 20221C>T) located near the common variant G20210A in the 3-untranslated region have been detected in non-Caucasian patients with thrombosis.[260]

It is known that elevated prothrombin results in increased thrombin generation, which affects fibrin clot structure. Clots produced in conditions of low thrombin concentration are composed of thicker fibers and are more porous, whereas those formed at higher thrombin concentrations have thinner and more tightly cross-linked fibers.[261] It has not been determined whether the altered clot structure produced at different thrombin concentrations *in vivo* has clinical relevance, but many previous studies have shown that thin-fibered, tightly cross-linked clots have increased resistance to fibrinolysis.

Moreover, thrombin seems to participate in atherosclerotic heart disease in ways that do not directly involve thrombus formation. Several nonthrombotic pathways have been identified involving thrombin as a signaling molecule primarily acting through thrombin receptors such as the protease-activated receptors (PAR).[262,263] These signaling events concern virtually all aspects of vascular biology, including vessel tone maintenance, cell migration and proliferation, inflammation, and angiogenesis. Thrombin also has role in anticoagulation as the thrombomodulin-associated activator of protein C. Thrombin is also responsible for the activation of the thrombosis regulator, thrombin-activated fibrinolysis inhibitor (TAFI) that plays a major role in down-regulating fibrinolysis as discussed further in the following section. A large variety of direct thrombin inhibitors that block its catalytic activity, including hirudin, bivalirudin, and newer synthetic small-peptides (e.g., ximelagatran, melagatran), have proved valuable as antithrombotic agents as an alternative to heparin and vitamin K antagonists in recent clinical studies. These thrombin inhibitors have shown benefit in the treatment of ACS and venous and arterial thrombosis and the prevention of cardioembolic events in high-risk patients.

Thrombin and the Protein C Pathway

The protein C pathway converts the coagulation signal generated by thrombin into an anticoagulant response by means of the activation of protein C. Protein C (PC) is a vitamin K-dependent plasma glycoprotein that is synthesized by the liver and circulates as a two-chain biologically inactive species. It is transformed to its active form, APC, by thrombin-mediated cleavage of protein C at the N-terminus sequence Arg169-Leu170. Effective activation of PC by thrombin requires the transmembrane glycoprotein, thrombomodulin (TM), as a cofactor for thrombin, amplifying this event >1000-fold. When complexed with TM, thrombin has reduced procoagulant activity as exhibited by its reduced ability to cleave fibrinogen, activate factor V, and trigger platelet activation. Thus, thrombin's substrate specificity is entirely switched by TM.

As part of its overall anticoagulant program, APC also suppresses further thrombin generation by proteolytically destroying coagulation factors Va and VIIIa, facilitated by the cofactor for APC, protein S (PS), and increasing fibrinolytic activity by neutralizing plasminogen activator inhibitor 1 (PAI-1). It also interferes with thrombin-induced proinflammatory activities that include platelet activation, cytokine-induced chemotaxis for monocytes and neutrophils, and up-regulation of leukocyte adhesion molecules, largely by APC-mediated proteolytic cleavage of protease activated receptor 1 (PAR1).

The protein C gene (PROC) (12 kb) localized on chromosome 2 has nine exons. The amino acid sequence indicates that protein C is synthesized as a 62-kDa single-chain polypeptide containing the light chain and the heavy chain connected by a dipeptide of Lys-Arg. Cleavage of two or more internal peptide bonds results in the formation of the light and heavy chains that in plasma are linked together by a disulfide bond.[264–266] The N-terminus contains metal-binding GLA residues involved in lipid membrane binding. Binding to specific membrane phospholipids (e.g., cardiolipin) primarily by the GLA domain can modulate the anticoagulant protein C pathway and enhance the APC-mediated enzymatic degradation of factor VA. Antiphospholipid or anticardiolipin antibodies stem PC stimulation, resulting in a clinical thrombophilia.

Although the clinical manifestations of hereditary protein C deficiency are highly variable, both homozygosity and compound heterozygosity have been linked to severe thrombotic complications.[267] Although rare, lethal thromboembolic complications can occur, especially in the homozygous type. Children born with severe homozygous protein C deficiency often will not survive beyond the neonatal period unless they receive protein C replacement.[268] Numerous studies have demonstrated that genetic variants in PROC are independent risk factors for venous thrombosis.[269–271] Individuals heterozygous for a variety of PROC alleles display a moderate form of the disease with deep venous thrombosis during adulthood. Some PROC mutations are associated with a nonfunctional circulating protein; most of them are located in the GLA domain and in the serine protease domain. Although most studies have reported that inherited protein C defects do not play a significant role in arterial thrombosis,[272,273] others have found a significant association.[274–276] One report found that congenital protein C deficiency hastened the onset of arterial occlusive diseases, especially AMI, in Japanese subjects.[274]

A high incidence (estimated at nearly 80%) of children and adults with severe sepsis develop acquired deficiency of protein C because of factor consumption.[277] This deficiency is associated with poor outcomes in patients with septic shock, including multiple organ failure and mortality. Recently, recombinant activated protein C was shown to significantly reduce the mortality of adults with severe sepsis and is now approved for such use in the United States and Europe.[277]

Thrombomodulin

Thrombomodulin is a cell surface–expressed glycoprotein, predominantly synthesized by vascular endothelial cells serving as a high-affinity receptor for thrombin that accelerates thrombin-induced activation of protein C acting as a critical cofactor in the anticoagulant APC formation. Encoded by an intronless gene on chromosome 20, the mature single-chain TM glycoprotein in the human is 557 amino acids long, structurally organized into five distinct regions, including the cytoplasmic and transmembrane domains, a serine/threonine-rich region with an attached chondroitin sulfate moiety, six consecutive EGF-like repeats, and the N-terminal lectin-like domain.[278–280] One group of the EGF-like repeats (domains 4–6) provides essential cofactor function for thrombin-mediated activation of PC, which can be further amplified by its binding to the endothelial cell protein C receptor (EPCR).[281] A different EGF repeat (domain 3) is required for TM cofactor function, enabling the thrombin-mediated activation of TAFI, a plasma procarboxypeptidase B.[282] Activated TAFI (TAFIa) cleaves basic C-terminal amino acid residues of its substrates, including fibrin, and thereby interferes with the efficient transformation of plasminogen to plasmin and, therefore, down-regulates fibrinolysis.

In the presence of cytokines, activated neutrophils, and macrophages, endothelial TM is cleaved enzymatically, releasing soluble fragments that circulate in the blood and are eliminated in urine. Plasma TM level is regarded as a primary molecular marker reflecting injury of endothelial cells. It is often increased in response to diffuse endothelial damage such as occurs in disseminated intravascular coagulation and diabetic microangiopathy.[283] As with protein C, decreased TM occurs in sepsis. Gene expression analysis of human endothelial cells demonstrated that TNF-α significantly decreases expression of TM and EPCR at both the mRNA and protein levels.[284] EPCR is likely involved in the enhancement of protein C binding to TM on endothelial cell surfaces. A large number of recent findings support the view that several components of the protein C pathway, including TM, APC, and EPCR, have activities that impact not only coagulation but also inflammation, fibrinolysis, and cell proliferation.

Gene variants in *TM* with associated changes in plasma thrombomodulin levels have been linked with thrombosis. In a large case-controlled study, decreased TM levels associated with two polymorphic variants in the *TM* gene coding sequence (i.e., Ala455Val and Ala25Thr) were identified with an increased risk of MI.[285] Another report suggested an association between the Val455 and MI,[286] but this was not subsequently confirmed.[287] In the Study of Myocardial Infarctions Leiden (SMILE), the Ala25Thr substitution was found to increase the risk of MI, particularly in smokers[288]; however, no evidence that the 25Thr isoform alters protein function has yet been presented. A recent study found no evidence that *TM* genotypes or haplotypes were associated with venous thrombosis.[289]

Studies have found an increased level of specific polymorphisms residing within the *TM* promoter in patients with thrombosis. One study found five different variants more prevalent in a cohort of patients with MI than in controls.[290] The frequency of one such promoter polymorphism (G-33A) has been subsequently reported to be more significantly prevalent in patients with CAD than controls; this genotype was accompanied by differences in the level of soluble TM.[291] Another report from the same group found that the G-33A genotype was significantly associated with AMI in Chinese patients.[292]

Thrombin-Activatable Fibrinolysis Inhibitor

Thrombin-activatable fibrinolysis inhibitor (TAFI) is the precursor of an enzyme (TAFIa) with basic carboxypeptidase activity that attenuates the lysis of fibrin clots by removal of the C-terminal lysine residues from partially degraded fibrin that mediates positive feedback in the fibrinolytic cascade. The plasma concentration of TAFI varies substantially (up to approximately 10-fold) in the human population and may constitute a novel risk factor for thrombotic disorders.

The complete *TAFI* gene contains 11 exons, spans approximately 48 kb of genomic DNA, and maps to chromosome 13.[293–295] Sixteen SNPs have been identified in the 5′-flanking, protein coding, and 3′-untranslated regions of the *TAFI* gene.[296,297] The polymorphisms all have been shown to be associated with variations in plasma TAFI concentrations. One amino acid substitution has been found to directly alter the properties of the TAFIa enzyme. Two naturally occurring variants of *TAFI* (Thr-325 and Ile-325) have been shown to differ substantially with respect to thermal stability and antifibrinolytic activity of the enzyme.[298] In clot lysis assays with thrombomodulin and the TAFI variant proteins, the Ile-325 variants exhibited an antifibrinolytic effect that was 60% greater than the Thr-325 variants. Similarly, without thrombomodulin, the Ile-325 variants exhibited an antifibrinolytic effect that was 30–50% greater than the Thr-325 variants. These findings suggest that individuals homozygous for the Ile-325 variant of TAFI would likely have a longer-lived and more potent TAFIa enzyme than those homozygous for the Thr-325 variant. A recent study assessed TAFI activity levels and TAFI Thr325Ile polymorphism in a relatively young adult patient group (<51 y) with MI. Irrespective of the genotype, patients had higher TAFI activity levels; increased stability of TAFI seems to be related to a fibrinolytic hypofunction.[299]

Plasma TAFI levels in Behçet's disease, a systemic vasculitis frequently complicated by arterial and venous thrombosis, were significantly higher than in healthy controls.[300] However, no difference was found in patients with and without thrombosis. In a large study of six *TAFI* polymorphisms, a single *TAFI* polymorphism (Ala147Thr) the Thr 147 allele was significantly associated with CAD in a French population, whereas the Ala147 allele was more prevalent in Irish patients assessed with CAD.[301]

Single-locus and haplotype analyses revealed that two additional polymorphisms, C-2599G and T + 1583A, had additive effects on TAFI levels and explained a significant proportion of *TAFI* variability.[301] In studies to evaluate the risk of deep venous thrombosis (DVT) caused by the polymorphisms in the *TAFI* gene, no significant association was found between investigated "TAFI-increasing" alleles *TAFI* 505A (Thr147) and *TAFI* +1542C and the risk of venous thrombosis, rather a moderately increased thrombotic risk of with *TAFI* +1542GG carriers (low TAFI level) was found.[302]

Factor XIIIa

Although a critical mass of fibrin is polymerizing, thrombin simultaneously activates factor XIII by a calcium-dependent mechanism. Both fibrinogen and factor XIII are unusual among clotting factors in that neither is a serine protease. Once activated, factor XIIIa is involved in cross-linking of the fibrin clot by transglutaminase reactions between glutamine and lysine residues on fibrin. The result of fibrin polymerization and FXIII-induced cross-linking is the formation of thick fibrin bundles and a complex branched network conferring strength, rigidity, and resistance to lysis to the fibrin clot. In addition, activated factor XIII cross-links several other proteins to fibrin molecule, including 2-antiplasmin, the major physiological inhibitor of plasmin, TAFI, plasminogen activator inhibitor-2, and fibronectin, which increases fiber thickness and makes the clot less susceptible to lysis.[303,304] Collagen is also cross-linked to fibrin, which may stabilize the extracellular matrix (ECM) forming at tissue injury sites.[305]

Factor XIII (plasma transglutaminase, fibrin stabilizing factor) is a glycoprotein that circulates in blood as a tetramer consisting of two α- and two β-subunits (α2β2); the α-subunit contains the catalytic active site. The α-subunit gene (F13a) maps to chromosome 6p24-25,[306] spans more than 160 kb, and contains 15 exons separated by 14 introns.[307,308] It is composed of 731 amino acids, including an activation peptide (37 amino acids), an active site (-Tyr-Gly-Gln-Cys-Glu-), two putative calcium binding sites, and a thrombin-inactivation site; these functional regions are located in separate exons. The 28-kb β-subunit gene (F13b) mapped to human chromosome 1 q31-q32.1 encodes a peptide of 641 amino acids, including 10 tandem repeats of 60 amino acids known as GP-I structures that are homologous with those in at least 13 other proteins.[309,310] The β-subunit (sometimes termed the carrier subunit) is also composed of 10 modular domains, called Sushi domains, that are involved in protein–protein interactions.

A large number of missense and nonsense mutations and deletions/insertions with or without out-of-frame shift/premature termination and splicing abnormalities have been identified in the genes for α- and β-subunits in factor XIII deficiency. In some cases, the mRNA levels of the α- or β-subunit were severely reduced. In most cases, impaired folding and conformational changes in mutant α- or β-subunits led to altered factor XIII protein stability (either intracellularly or extracellularly), which is responsible for factor XIII deficiency.[311]

Genetic variations in factor XIII can lead to modified thrombotic risk stemming from changes in the process of fibrin clot formation.[312,313] For instance, a F13a polymorphism in the factor XIII α-subunit (Val34Leu), which codes for an amino acid change near the thrombin cleavage site at position 37, has been associated with moderately reduced venous thrombotic risk and with higher FXIII activity.[314,315] In individuals containing the F13a Leu34 allele, cleavage of the FXIII activation peptide by thrombin is enhanced two-fold to threefold, with significant impact on fibrin structure and function dependent on the concentration of fibrinogen.[316] Fibrin clots formed in the presence of F13a Leu34 form quicker and have thinner fibers, smaller pores, and reduced permeability compared with the Val34 variant. At high fibrinogen concentrations, plasma samples homozygous for the Leu34 allele form clots with increased permeability and looser structures than do clots formed from plasma samples homozygous for the Val allele.[317] The smaller and tighter clot structure associated with the Val34 allele is more resistant to fibrinolysis, possibly because the small pores within the clot restrict the access of fibrinolytic enzymes. Therefore, a protective effect of the Leu34 allele should emerge. The F13a Leu34 allele has also been found to be less prevalent in patients with MI and cerebral infarction and to provide protection against MI in younger subjects.[318]

Because fibrinogen concentrations are often increased in cardiovascular disease, it is possible that environmental factors altering fibrinogen concentrations will influence the structure of the clot formed in the presence of the Leu34 allele to provide a protective effect. In a study to assess the contributions of both environmental and genetic factors on fibrin structure and clot formation in a large group of twins (66 monozygotic, 71 dizygotic), environmental factors were found to have a considerable influence on fibrin clot structure, mediated in large part by an modulated fibrinogen levels. Moreover, plasma levels of fibrinogen and FXIII β-subunit levels correlated with fibrin clot pore size and fiber size and an overlap of both genetic and environmental influences were shown to contribute to these correlations.[319]

FIBRINOLYTIC SYSTEM

Activation of the fibrinolytic system is dependent on the conversion of the plasma zymogen, plasminogen, to the serine protease plasmin by the physiological activators urokinase-type plasminogen activator (uPA) or tissue-type plasminogen activator (tPA). The primary *in vivo* function of plasmin is to regulate vascular patency by degrading fibrin-containing thrombi. Plasminogen activator inhibitors 1 and 2 (PAI-1 and PAI-2) are the primary physiological inhibitors of plasminogen activation *in vivo* and are integral regulators of the fibrinolytic system. Significant physiological inhibition of the fibrinolytic system is also provided by α2-antiplasmin (α2-AP). In addition to playing a role in clot dissolution as shown in Fig. 4, the PA/plasmin system has been implicated in fibrin removal, as well as in the tissue remodeling and cell migration occurring during both physiological and pathological processes. In this section, we review information concerning the genes involved and genetic variants identified, with particular focus on their role in thrombosis.

Plasminogen

Plasminogen is a single-chain glycoprotein of 92 kDa consisting of 792 amino acids.[320] The gene for human plasminogen (*PLG*) spans about 52.5 kb of DNA and consists of 19 exons separated by 18 introns.[321] It contains five "kringles," triple-looped structures that contain lysine-binding and aminohexyl-binding sites that mediate its specific binding to fibrin, as well as its interaction with α2-antiplasmin and, therefore, play an integral role in fibrinolysis. Genetic studies and *in situ* hybridization have strongly demonstrated that the *PLG* gene coding for plasminogen is closely linked to apolipoprotein(a) and maps to chromosome 6 at band 6q26-27.[322–324]

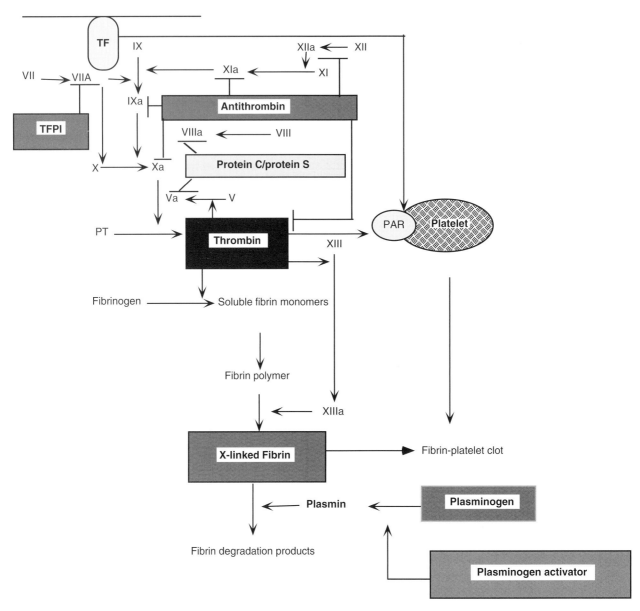

FIGURE 4 Antithrombotic pathways. The sites of action of the four major antithrombotic pathways are illustrated superimposed on the coagulation pathway. These include antithrombin, protein C/protein S, the TF pathway inhibitor (TFPI), and the fibrinolytic system consisting of plasminogen, plasminogen activator (PA), and plasmin.

Plasminogen deficiency in man is extremely rare and is associated with thrombolic complications.[325] Several polymorphic variants of the plasminogen gene have been reported, albeit not implicated in disease.[326] Plasminogen-deficient mice survive embryonic development but develop spontaneous fibrin deposition because of an impaired thrombolytic potential and suffer retarded growth and reduced fertility and survival.[327]

Tissue-Type Plasminogen Activator

Tissue-type plasminogen activator (tPA), the main endothelial cell–derived blood activator of the fibrinolytic system, is a 70-kDa serine protease activator of plasminogen whose catalytic efficiency is greatly increased by the presence of fibrin. Because of its fibrin specificity, tPA is primarily involved in clot dissolution, although it has also been implicated in ovulation, bone remodeling, and brain function. Cellular receptors for tPA and plasminogen have been identified that might localize plasmin proteolysis to the cell surface.

In striking contrast to most other members of the chymotrypsin family of serine proteases, including the closely related uPA, tPA is not synthesized and secreted as a true zymogen and is enzymatically active as a single-chain peptide.[328] The 527-amino acid tPA protein is composed of four different putative structural domains with homologies to fibronectin finger-like structures, EGF, two kringle structures, located at the N-terminal heavy chain and the active site of serine proteases (located at C-terminal light chain). Only the finger and EGF domains are each entirely encoded by unique single exons. The stimulatory effect of fibrin on the plasminogen activator activity of tPA was shown to be mediated by the kringle domain and, to a lesser extent, by the finger domain.[329]

The tPA gene *PLAT* localized on chromosome 8 encompasses more than 20 kb with 14 exons and 13 introns.[330] A 2.4-kb DNA fragment, located 7.1 kb upstream from the human *PLAT* gene acts as an enhancer that is activated by glucocorticoids, progesterone, androgens, and mineralocorticoids.[331]

For many years it was believed that impaired fibrinolysis caused by the decreased function of tPA might be a risk factor for thrombosis. Recent interest has been on elevated, rather than reduced, levels of tPA as a risk marker. The prospective ECAT study found an increased risk for future MI in individuals with elevated levels of tPA.[332]

Of the several nucleotide sequence changes that have been identified in the tPA gene *PLAT*, the most studied is a 311-bp Alu insertion/deletion in intron 8.[333] In a population-based cohort study of almost 8000 subjects, the presence of one insertion allele was associated with an 50% increase in risk of MI, whereas homozygous carriers had a greater than twofold adjusted increase in risk, suggesting an association between the number of Alu repeats and arterial thrombosis.[334] Other studies, however, did not confirm this association.[335]

Urokinase-Type Plasminogen Activator

The gene for urokinase-plasminogen activator (uPA) *PLAU* mapped to chromosome 10q24-qter encodes a peptide of Mr of 50,000 with 411 residues that undergoes processing by plasmin resulting in an A chain (157 amino acids) and B chain (253 amino acids).[336,337] Single-chain pro-urokinase is an inactive proenzyme form of human urokinase. It contains an N-terminal growth factor domain and a single kringle domain. On its secretion, pro-uPA binds to its cell receptor (uPAR) anchored to the plasma membrane by a C-terminal glycosylphosphatidylinositol residue. Subsequent proteolytic cleavage of pro-uPA into its two-chain activated form is mediated not only by plasmin but also by kallikrein, trypsin, cathepsin B, and thermolysin. The uPAR plays a fundamental role in directing the enzymatic activity to focal and cell–cell contacts, as well as regulating it by internalizing and degrading only the inhibited form of urokinase, and contributes to regulating cell processes, including cell adhesion and migration by providing a platform for an inducible, transient, and localized cell surface proteolytic activity.[338]

Recombinant tissue-type plasminogen activator (r-tPA) and single-chain urokinase-type plasminogen activator (scu-PA) are thrombolytic agents, characterized by a high, but not absolute, degree of fibrin specificity that is mediated through different molecular mechanisms.[339] Both activators are still under clinical investigation, but it has become apparent that their therapeutic dose in humans is high and associated with a variable degree of systemic activation of the fibrinolytic system and fibrinogen breakdown. For instance, prourokinase (proUK) at pharmacological doses is prone to nonspecific activation to urokinase, which has handicapped therapeutic exploitation of its fibrin-specific physiological properties. Reengineering this therapeutic protein by introducing a Lys300→His mutation attenuates the susceptibility to nonspecific activation without compromising specific activation of proUK on a fibrin clot.[340] The site-directed mutation designed to improve the stability of proUK in blood at therapeutic concentrations effectively induced superior clot lysis both *in vitro* and *in vivo* without causing significant interference with hemostasis.

Plasminogen Activator Inhibitors

The family of serine protease inhibitors termed serpins are key regulators of numerous biological pathways that initiate inflammation, coagulation, angiogenesis, apoptosis, ECM composition, and complement activation responses. The mammalian serpin, plasminogen activator inhibitor-1 (PAI-1), targets thrombotic and thrombolytic proteases and can induce a prothrombotic response.

Plasminogen Activator Inhibitor-1

Plasminogen activator inhibitor-1 (PAI-I) rapidly inhibits either tPA or uPA in normal blood plasma. Although highly variable plasma levels of PAI-I are found in healthy adults, PAI-I levels are markedly elevated during arterial injury and thromboembolic disease, including ischemic heart disease, DIC, and type 2 diabetes.[341,342] There is also evidence that elevated PAI-1 levels seem to increase the risk of atherothrombotic events and may also promote the progression of vascular disease.[343] PAI-1 exists in the circulation in great excess over tPA primarily to prevent systemic bleeding while permitting local clot lysis; thus, most circulating tPA is inactive and complexed with PAI-1.

The mature human PAI-1 is 379 amino acids long (not including its 23 residue signal peptide) encoding a 52-kDa glycoprotein.[344] Extensive homology with alpha 1-antitrypsin and antithrombin III suggested that PAI-1 is a member of the serpin superfamily. The amino acid sequence of PAI-1 includes three potential asparagine-linked glycosylation sites and lacks cysteine residues. The *PAI-I* gene spans approximately 12 kb, contains eight introns, and has been localized to chromosome 7.[345–347]

The most frequently studied of the *PAI-1* genetic variants is the 4G/5G insertion/deletion polymorphism located at position -675 of the promoter.[348] The *4G* allele has been correlated with higher levels of gene transcription and elevated

PAI-1 plasma levels compared with its more common *5G* counterpart. Interestingly, this promoter site has genotype-specific responses to triglycerides, which lead to the highest levels of PAI-1 in carriers of the *4G/4G* genotype, who are also hypertriglyceridemic.[349] This interaction has been explained by a triglyceride-responsive region adjacent to the 4G/5G site.[350] Although several case-controlled studies have demonstrated an increased risk of MI, CAD, and ischemic stroke in carriers of the *4G* allele, these findings have not been confirmed in several larger studies.[351] A meta-analysis of nine studies that included 1500 cases and >2000 controls found an overall slight increase in the risk of MI associated with the *4G* allele that was confined to subgroups of high-risk populations.[352] In addition to the promoter 4G/5G polymorphism, a CA(n) dinucleotide repeat in intron 3 and a 3′ *Hin*dIII site has been identified, variants for which no role in arterial thrombosis has yet been established.

In addition to inhibiting tPA and uPA, PAI-1 interacts with different components of the extracellular matrix, including fibrin, heparin, and vitronectin (Vn).[353,354] PAI-1 binding to Vn facilitates migration and invasion of tumor cells. The most important determinants of the Vn-binding site of PAI-1 seem to reside between amino acids 110 and 147, which includes alpha helix E at amino acids 109–118. Interestingly, in the presence of vitronectin, PAI-I displays a 200-fold elevation of thrombin inhibition, possibly as a result of conformation changes arising from vitronectin binding to the reactive site of PAI-I.[355] A similar increase in PAI-I thrombin-inhibitory properties was promoted by the glycosaminoglycan heparin.

Plasminogen Activator Inhibitor-2

Another member of the serpin family, plasminogen activator inhibitor type-2 (PAI-2), is a nonconventional serine protease inhibitor considered to be an authentic and physiological inhibitor of urokinase.[356] The fact that only a small percentage of PAI-2 is secreted suggests alternative roles for this serpin. Indeed, PAI-2 exhibits a variety of intracellular roles: it can alter gene expression, modulate the rate of cell proliferation and differentiation, and inhibit apoptosis in a manner independent of urokinase inhibition. Its normally low level in the blood is markedly up-regulated in pregnancy. In selective cell types, including activated monocytes and macrophages, high levels of PAI-2 gene expression result from stimulation by phorbol esters, LPS, TNF-α, retinoic acid, lipoprotein (a), and interferon-γ.

Nucleotide sequence analysis indicated that the *PAI-2* cDNA encodes a protein containing 415 amino acids with a predicted unglycosylated Mr of 46,543 and shows extensive homology with members of the serpin superfamily.[357] Interestingly, PAI-2 was found to be more homologous to ovalbumin than the endothelial plasminogen activator inhibitor, PAI-1. In addition, both ovalbumin and PAI-2 have no typical N-terminal signal sequence. PAI-2 exists in both an intracellular, nonglycosylated form of 47 kDa and a secreted, glycosylated form of approximately 60 kDa. The absence of a signal peptide might explain why PAI-2 remains primarily intracellular. The 3′ untranslated region of the *PAI-2* cDNA contains a putative regulatory sequence that has been associated with inflammatory mediators including TNF-α, interferons, and interleukins. The *PAI-2* promoter tightly regulates *PAI-2* gene expression in a cell-specific manner, and this control is mediated, in part, by the upstream silencer element, PAUSE-1.[358] The chromosomal assignment of the human *PAI-2* gene to 18q21.3→18q22.1, a location corresponding to a "hotspot" for common somatic deletions found in colorectal carcinomas is suggestive of a potential involvement of PAI-2 in determining metastatic phenotype.[359]

Two polymorphic haplotypes variants of *PAI-2*, designated A and B, are associated with the formation of large molecular PAI-2 complexes. Variant A consists of Asn120, Asn404, and Ser413, and variant B consists of Asp120, Lys404, and Cys413.[360] One study reported that the *AA* genotype of the *PAI-2* gene was more frequent in Turkish subjects with MI, suggesting a potential role as a risk factor for MI.[361]

α2-Antiplasmin

The plasma protein α2-antiplasmin (α2AP or α2PI) or plasmin inhibitor is the main physiological inhibitor of the serine protease plasmin, which is responsible for the dissolution of fibrin clots. Mature human α2-antiplasmin is a single-chain glycoprotein of 70 kDa containing 464 amino acids, homologous with five other proteins belonging to the serpin superfamily[362] and is encoded by the α2AP (*PI*) gene containing 10 exons and 9 introns distributed over approximately 16 kb.[363] The α2AP protein contains an N-terminal hydrophobic prepeptide (signal sequence) of 27 amino acids, followed by a hydrophilic propeptide of 12 amino acids. Two forms of α2AP circulate in human plasma: a 464-residue protein with methionine as N-terminus (Met-α2AP) and an N-terminally shortened 452-residue form with asparagine as the N-terminus (Asn-α2AP).[364] Its reactive site (i.e., the peptide bond cleaved by reaction with its primary target enzyme, plasmin) consists of Arg364-Met365, corresponding to the reactive site Met358-Ser359 of the archetypal serpin, α1-antitrypsin. The *PI* gene has been localized to chromosome 17p13.[365]

Although α2-antiplasmin exhibits most characteristics of other inhibitory serpins, it also possesses unique N- and C-terminal extensions that significantly modify its activities. The N-terminus of the major form (Asn-α2AP) is rapidly cross-linked to fibrin during blood clotting by activated coagulation factor XIII, and, as a consequence, fibrin becomes more resistant to fibrinolysis with incorporation of antiplasmin into a clot as it is formed. The highly conserved C-terminal portion of antiplasmin contains several charged amino acids, including four lysines with one of these at the C-terminus, which mediates the initial interaction with plasmin and is a key component of antiplasmin's rapid and efficient inhibitory mechanism.

A large number of isolated clinical cases of congenital α-antiplasmin deficiency have been reported in the literature, usually associated with bleeding tendency, although cases have been reported with severe hemorrhagic diathesis.[366] Two families with congenital α2PI deficiency were found to have changes of nucleotide sequence in an exon coding for plasma α2AP, resulting in productions of variant proteins. These proteins are largely retained within the cells, resulting in a deficiency of α2AP in a circulating blood plasma.[362]

Platelets

As previously noted, a significant role of platelets in regulating thrombosis and hemostasis has long been recognized. Platelet plugs are an integral part of the response to vascular injury, including plaque rupture. Moreover, it has become evident that platelets also have a relevant role in the inflammatory process.[367] The expression and release of various mediators from platelets such as thrombin, CD40L, and ADP have significant effects on events in the vessel wall. These trigger a number of autocrine and paracrine activation processes that lead to leukocyte recruitment into the vascular wall.[368–374]

Formation of the platelet plug involves a cascade of events, including platelet activation, aggregation, and subsequent thrombus formation generally initiated by platelet adhesion to the damaged vessel wall. The subendothelial components responsible for triggering platelet reactivity include different types of collagen, von Willebrand factor (vWF), fibronectin, and other adhesive proteins such as vitronectin and thrombospondin. The extent of this response is largely dependent on the multiple factors, including level of damage, the specific matrix proteins exposed, and the determinants of blood flow. Several collagen-binding proteins that are expressed on the platelet surface (see Table VII) also regulate collagen-induced platelet adhesion, specifically under flow conditions.[367] These receptors include GP IV, GP VI, and integrin 2β1.

The platelet-expressed GP Ib/IX/V complex adhesive receptor is central to both platelet adhesion and activation and normal hemostatic function. Extensive damage to the blood vessel wall such as occurs in complex plaque rupture exposes a number of subendothelial proteins, including vWF, fibronectins, and collagen, to the circulating blood. The GP Ib/IX/V complex expressed on the platelet surface binds to the exposed vWF, causing platelets to adhere to the subendothelium at the site of injury (Fig. 5).

In addition to anchoring the platelet to the injured vessel wall, engagement of the GP Ib/IX/V complex leads to the transduction of signals contributing to platelet activation, including triggering the synthesis and release of thromboxane (TXA2), serotonin, and ADP, and causes activation of various platelet receptors. The use of aspirin as an inhibitor of platelet aggregation is largely based on its inhibition of TXA2 synthesis. As a member of the class of drugs known as the thienopyridines, clopidogrel, a relatively new antiplatelet agent, irreversibly prevents platelet activation by blocking one of the three ADP receptors on its surface. Clopidogrel reduces the incidence of recurrent ischemic events in patients with ACS and after coronary stenting. The development of patient resistance to clopidogrel was recently shown to be associated with a higher risk of recurrent cardiovascular events including AMI.[368] Although the molecular basis of clopidogrel resistance has not yet been determined, possible explanations include modulation of clopidogrel metabolite activation by cytochrome p450 activity, genetic polymorphisms in the ADP receptor, and alterations in the downstream postreceptor signaling pathway.

Thrombin is also a pivotal agonist of platelet activation and provides a vital link between the previously described humoral coagulation pathway and cellular-based hemostasis mediated by platelets. Thrombin mediates platelet adhesion by its interaction with multiple PARs localized on the platelet surface. This interaction triggers a complex signaling cascade, including tyrosine kinase activation, serine/threonine kinase activation, and lipid kinase activities, which generate TXA2 and arachidonic acid.

Von Willebrand factor–bound GP Ib/IX/V induces a conformational change in the platelet GP IIb/IIIa receptor, transforming it from an inactive low-affinity state to an active receptor that binds additional vWF or fibrinogen with high affinity that can result in thrombus formation. The binding of vWF to the platelet GP IIb/IIIa receptor occurs at its RGD sequence, a ligand-binding domain on the N-terminal of both chains of the glycoprotein receptor. Other ligands, including fibronectin, vitronectin, and fibrinogen, also bind at this site. The IIb/IIIa receptor also plays a critical role in platelet aggregation that occurs when fibrinogen and vWF are bound, resulting in platelet cross-linking. Given its central role in platelet adhesion and aggregation, the GP Ib/IX/V complex has become an attractive target for the development of antiplatelet drugs. A number of therapeutics for preventing arterial thrombosis, including anti-vWF monoclonal antibodies and the GP Ib/IX/V antagonists isolated from snake venoms, have been developed that target the interaction between the GP Ib/IX/V complex and vWF.

Platelets are highly organized and complex cell fragments that share many of the same biological mechanisms as other cells, including cytoskeletal structure, housekeeping enzymes, and signal transduction components. Unlike most cells, platelets lack a nucleus and, therefore, need an alternative mechanism to adapt to changing biological settings other than by regulating nuclear transcription. Platelets maintain some protein synthetic capacity from megakaryocyte-derived mRNA, as well as the proteins and molecular machinery necessary for translation.[375] Moreover, platelets can respond to physiological stimuli by use of biosynthetic processes that are regulated at the level of protein translation, demonstrat-

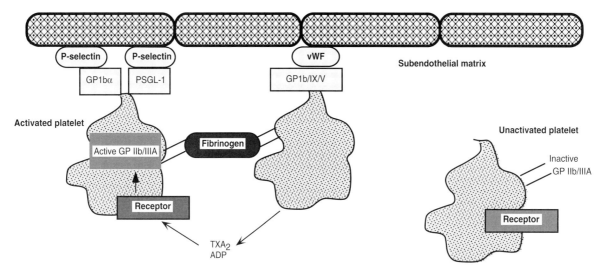

FIGURE 5 Platelet adhesion and aggregation. Damage or vascular injury exposes subendothelial von Willebrand factor (vWF) or collagen to the circulating blood. Platelets adhere to the site of injury when GP Ib/IX/V complex expressed on the platelet surface binds to the subendothelial vWF. This event triggers the synthesis and release of TXA2 and ADP and causes activation of various platelet receptors. These events cause a conformational change in GP IIb/IIIa, enabling the high-affinity binding of fibrinogen and resulting in thrombus formation. Other interactions between the platelet membrane proteins (PSGL-1 and GP 1bα) with the P-selectin of endothelial cells are shown.

ing a functional role for platelet mRNA.[376,377] Interestingly, microarray and SAGE analyses of platelet RNA (the platelet transcriptome) have revealed that the platelet is not entirely transcriptionally silent; mitochondrial transcription is rather abundant.[378]

In addition, a large number of molecules needed to respond to various physiological and pathological stimuli are present in secretory storage granules and membranes. Because a primary function of platelets is regulation of hemostasis, the membrane receptors (see Table VII) play a direct role in the processes regulating adhesion or thrombus formation. In addition, platelet secretion of adhesive proteins, cytokines, enzymes, growth factors, inflammatory modulators, and activating agents has been implicated in these processes, as well as in signaling inflammatory pathways. Moreover, the recent use of proteomic analysis has been effectively applied to analyze specific platelet signaling and regulatory subproteomes, including the phosphotyrosine proteome, the platelet releasate (also termed the platelet secretome), and, more recently, the proteome of membrane lipid rafts so as to assess levels and qualitative changes in biologically relevant proteins of low abundance.[379–381] The analysis of secretome from thrombin-activated platelets has identified more than 300 proteins that are secreted on activation.[382, 383] In addition to tyrosine phosphorylation, the regulation of other post-transcriptional modifications of platelet proteins such as glycosylation will be targeted in the near future.

A large number of disorders of platelet structure and function have been shown to have clinical relevance.[384] Although some of these are congenital with increasingly identified genetic loci, others have an acquired origin. The latter include triggering by various drugs with anti-platelet action (e.g., aspirin, NSAIDS, the thienopyridines clopidogrel and ticlopidine), clinical conditions, including uremia, hepatic cirrhosis, myeloma, and related disorders, polycythemia vera, essential thrombocythemia, and

Table VII Major Platelet Membrane Receptors in Thrombosis

Protein	Ligand	Platelet effect
GP Ib-IX-V (CD42)	vWF, thrombin, P-selectin	Adhesion
GP Ia-IIa	Collagen	Adhesion
GP IIb-IIIa	vWF, fibrinogen, fibronectin, CD40L	Aggregation
GP IV (CD36)	Collagen, thrombospondin	Adhesion
GP VI	Collagen	Adhesion
GP Ic-IIa	Fibronectin	Adhesion
PAR-1	Thrombin	Adhesion
PAR-4	Thrombin	Adhesion
P-selectin (CD62)	Neutrophils/monocytes	Platelet attachment to neutrophils/monocytes
Thromboxane A2 receptor	TXA2	Activation
P2Y1	ADP	Activation
P2Y12	ADP	Activation
PSGL-1	P-selectin	Adhesion

cardiopulmonary bypass.[385] Thrombotic complications involving virtually any site of the venous, arterial, and/or microcirculation are frequently observed in patients with myeloproliferative disorders such as polycythemia vera and essential thrombocythemia.[386] Thrombotic events such as left ventricular failure after an MI are often described as the presenting manifestation and are a major cause of death. Ischemic stroke and ACS may occur in the presence of normal arteriograms and in the absence of traditional cardiovascular risk factors, whereas microvascular disturbances may lead to ocular migraine, Raynaud's phenomenon, or erythromelalgia. Heparin-induced thrombocytopenia (HIT), a prothrombotic disorder can develop with any dose or form of heparin (including heparin-coated indwelling vascular catheters), resulting in aberrant platelet activation that may lead to thrombotic complications involving the venous, arterial, and microvascular circulation, although venous thromboembolism is more common than arterial thrombosis.[387] The HIT disorder stems from an immunoglobulin G antibody–mediated reaction to an antigenic complex consisting of heparin and platelet factor 4; the demonstration of antibodies directed against the heparin-platelet factor 4 (H-PF4) complex is an important component of the diagnosis of HIT.[388]

In von Willebrand disease, the most common human congenital bleeding disorder, both quantitative and qualitative abnormalities in the vWF glycoprotein have been implicated.[389,390] Types 1 and 3 are characterized by a quantitative defect of vWF. The molecular basis of type I von Willebrand disorders has only been recently determined. Recessive type I disease can either be severe (caused by homozygosity or double heterozygosity for a missense vWF mutation) or moderate (caused by missense mutations), whereas dominant type 1 mutations are caused by a heterozygous missense mutation in the VWF gene that produces a mutant vWF protein that exerts its dominant effect on the wild-type allele by defective processing, storage secretion, and/or proteolysis of vWF in endothelial cells. Type 3 von Willebrand disease is a recessive severe hemophilia-like bleeding disorder caused by homozygosity or double heterozygosity for two nonsense mutations (null alleles) characterized by a strongly prolonged bleeding time, absence of platelet aggregation, absence of vWF protein, and prolonged activated partial thromboplastin time (APTT) because of factor VIII deficiency.

Type 2 vWF disorders, comprising subtypes 2A, 2B, 2M and 2N, involve molecular variants with a qualitative defect of vWF.[390] Type 2A disorder is characterized by a decreased platelet-dependent function of vWF associated with the absence of vWF multimers. Type 2A mutations have been identified in the A2 domain of vWF that contains a proteolytic site and in the propeptide and the C-terminal part of vWF that are involved in its multimerization and dimerization, respectively.[391] In type 2B disorder typified by increased vWF for platelet glycoprotein Ib (GP Ib), and in type 2M, defined by a decreased platelet-dependent function not caused by the absence of vWF multimers, several mutations have been localized within the A1 domain containing the GP Ib binding site. The type 2B defect is unusual in that it comprises a gain-of-function phenotype characterized by an enhanced vWF protein–binding affinity to platelet GP Ib, causing the platelets to disappear from the circulation, resulting in thrombocytopenia, which can be further triggered by stress or desmopressin.[392] In type 2N, characterized by a defective binding of vWF to factor VIII, several mutations have been identified within the factor VIII–binding domain in the N-terminal part of vWF.

Proteolytic cleavage of vWF multimers normally limits platelet thrombus growth, and failure to cleave vWF seems to encourage microvascular thrombosis. The vWF cleaving protease has recently been shown to be a member of the ADAMTS family of metalloproteases, designated ADAMTS13. Mutations in the *ADAMTS13* gene cause an autosomal-recessive form of chronic relapsing thrombotic thrombocytopenic purpura (TTP), a severe thrombotic microangiopathy characterized by marked thrombocytopenia, systemic platelet aggregation, erythrocyte fragmentation, and organ ischemia.[393,394] Autoantibodies that inhibit ADAMTS13 can also cause sporadic TTP.

Glanzmann's thrombasthenia, literally translated as weak platelets, is a rare disorder in which platelets can carry out most biochemical reactions and are present in normal number, size, and morphology but fail to form aggregates.[395] Their failure to aggregate is due to the loss or dysfunction of the platelet integrin receptor complex GP IIb/IIIa or CD41. This rare disorder is inherited as an autosomal-recessive trait and is genetically heterogenous arising from multiple mutations that perturb the biosynthesis and assembly of the multisubunit GP IIb/IIIa complex.[396–398] The carrier state, in which there is a 50% reduction in the number of GP IIb/IIIa molecules on the platelets, is asymptomatic, and most patients with thrombasthenia are compound heterozygotes who carry two independent mutations. Patients with Glanzmann's disease have lifelong mucosal bleeding and may require platelet transfusions for severe bleeding episodes.

The Bernard-Soulier syndrome (BSS) is a rare autosomal-recessive disorder caused by mutations in various polypeptides in the GP Ib/IX/V complex, the principal platelet receptor for vWF.[395,399] In the absence of a functioning vWF receptor, platelets cannot adhere to vascular subendothelium under high shear stress. Patients with BSS also frequently have giant platelets, sometimes approaching the size of lymphocytes, mild to moderate thrombocytopenia, a prolonged bleeding time, and the absence of ristocetin-induced platelet aggregations. Thus far, more than 30 mutations of the GP Ibα, GP Ibβ, or GP IX proteins have been described in BSS, although no mutations have yet been found for the GP V protein.[400] Recent studies also have shown that the phenotypes caused by mutations in the subunits of the GP Ib/IX/V complex span a wide spectrum, from the normal phenotype to isolated giant platelet disorders/macrothrombocytopenia, to full-blown BSS and platelet-type von Willebrand disease.[401] The recurrent occurrence of the Asn45→Ser mutation in the GP IX protein has suggested that it represents an ancient mutation shared by northern and central European populations.

Numerous studies have examined the association of polymorphic variants of various platelet integrin genes with thrombosis and ischemic events with often interesting, as well as controversial, findings. In contrast, no studies have demonstrated any relationship between the presence of these variants and venous thrombosis, reinforcing the primary role of platelets in arterial and not venous thrombosis.

A common T/C polymorphism at position 1565 in exon 2 of the *GPIIIa* gene leads to a substitution of proline for leucine at amino acid 33 (Leu33Pro), resulting in a conformational change in the N-terminal disulfide loop important for fibrinogen binding.[402] This polymorphism is found in 15% of whites and 5–8% of blacks but is virtually absent in Asians; the more common *GPIIIa* isoform is known as *PIA1* allele (HPA-1a) and the 33Pro allele, as the *PIA2* allele (HPA-1b).[403] The homozygous *A2/A2* form is known to be associated with post-transfusional purpura and neonatal thrombocytopenia, conditions in which antibodies are formed against the *PIA1* allele. A significant association of *PIA2* was noted with acute coronary thrombosis, most strongly in patients younger than 60 y.[404] Several subsequent studies confirmed a significant association of the *PIA2* allele with CAD and ischemic stroke,[405,406] but others did not.[407,408] In contrast, a case-controlled study of 200 young survivors of MI found a modest, but significant, 1.8-fold increase in risk among carriers of the *PIA2* allele, which was increased further to 13.7 in carriers who smoked.[409] Almost 50% of premature MIs in that study were attributable to the interaction between these two risk factors.

In the gene encoding the GP Ib subunit, a length polymorphism with a variable number of tandem repeats of 39 bp and a linked C/T polymorphism at nucleotide 3550 that leads to a Thr145Met substitution have been found.[410] Several reports have correlated these genetic variants with CAD and ischemic stroke, particularly in younger patients.[409,411] A silent exonic C/T nucleotide substitution at position 807 in the gene coding for the GP Ia/IIa 2-peptide has been found to increase receptor density and might be associated with MI, as well as in younger patients.[412]

Other Factors

Once thrombosis has been initiated, variations in the activity of coagulation proteins and endogenous anticoagulants, as well as the kinetics of platelet aggregation, may alter the effectiveness of thrombus formation. Epidemiological studies have identified several acquired states (which also may have genetic components) that may result in endothelial damage or altered hemostatic equilibrium, thereby predisposing patients to arterial thrombosis. These include hyperhomocysteinemia, elevated C-reactive protein (CRP), antiphospholipid antibodies, and elevated fibrinogen levels.

Homocysteine

Homocysteine is a reactive amino acid intermediate in methionine metabolism. Once formed, homocysteine can either be remethylated by methionine synthetase to form methionine, or, alternately, it is enzymatically modified by cystathionine-ß-synthase, a B6-dependent enzyme, to form cysteine. The adverse effects of homocysteine are manifold but ultimately lead to endothelial dysfunction with associated platelet activation and thrombus formation. *In vitro*, clots formed in the presence of homocysteine have thicker, shorter fibers with a more compact structure.[413] Clots formed with fibrinogen from homocystinemic plasma are more resistant to lysis.[414]

The potential role of plasma total homocysteine in the development of atherosclerotic vascular disease was first reported in fibrous vascular plaques in two children, each with different enzymatic defects that resulted in severe hyperhomocysteinemia.[415] Impaired homocysteine metabolism was reported in a cohort of patients with premature CAD.[416] Many studies have supported a modest, but consistent, association between homocysteine and arterial vascular outcomes.[417]

A meta-analysis was performed of 72 prospective cohort studies that examined the association between the methylenetetrahydrofolate reductase (*MTHFR*) C677T gene variants, the most common form of genetic hyperhomocysteinemia, serum homocysteine levels, CAD, stroke, and venous thromboembolism.[418] In the 46 studies that examined the end point of CAD, individuals with the *MTHFR* 677 TT genotype had a significant association with CAD compared with the CC wild type. Moreover, significant associations were found between increased homocysteine levels and either stroke or deep vein thrombosis with or without pulmonary embolism. Other studies reported a similar, but less robust, association of the TT genotype with CAD compared with individuals harboring the CC genotype, particularly in the setting of low folate status.[419]

There is substantial evidence that significant variability in serum homocysteine levels is found in the general population. Moreover, folate and vitamin B_{12} status, vitamin cofactors required for homocysteine metabolism, are negatively correlated with fasting serum homocysteine concentration, and serum homocysteine has been positively correlated with increased age, serum creatinine, systolic blood pressure, male gender, cigarette smoking, and caffeine and alcohol intake.[420,421] The presently undetermined mechanisms by which elevated levels of homocysteine lead to atherosclerotic vascular disease may include impaired endothelium-dependent vasomotor regulation, increased endoplasmic reticulum stress, the formation of oxidized LDL, enhanced production of inflammatory mediators, the stimulation of SMC proliferation, and induction of a prothrombotic state through enhanced platelet activation, increased levels of TF expression, and inhibition of protein C.[422–425]

Despite evidence of an association between serum homocysteine and vascular end points, there are currently no convincing data demonstrating that therapies aimed at reducing serum homocysteine effectively result in favorable outcomes. A recent trial found that the moderate reduction of total homocysteine resulting from folic acid, vitamin B_6 and B_{12} therapy in subjects after nondisabling cerebral infarction had no effect on vascular outcomes during the 2-year follow-up.[426]

C-Reactive Protein

C-reactive protein (CRP) is an acute-phase reactant and binds with high affinity to numerous endogenous and exogenous ligands, including modified lipids, apoptotic cells, and microbial polysaccharides.[427,428] Once bound, CRP is a potent stimulator of the classical complement pathway; it promotes macrophage phagocytosis, as well as TGF-ß production.[427] CRP is increasingly recognized as a significant cardiovascular risk factor, with numerous studies demonstrating that serum CRP levels are positively correlated with adverse cardiovascular outcomes, independent of traditional cardiac risk factors.[429,430]

There are several mechanisms by which CRP may contribute to the development of atherothrombotic vascular events. Elevated serum CRP levels tend to cluster with traditional cardiovascular risk factors, including obesity, smoking, hypertension, type 2 diabetes, and exercise.[431–434] CRP also likely promotes atherogenesis through direct effects on monocytes and endothelial cells, including the up-regulation of vascular adhesion molecules, stimulation of proinflammatory mediators, and the impairment of vascular NO-dependent vasodilation.[435]

In addition, CRP induces the production of the procoagulant TF in monocytes, increasing thrombogenicity.[436] In human aortic endothelial cells, CRP also induces the expression and activity of PAI-1, a marker of atherothrombosis, also elevated in the metabolic syndrome and in diabetes. The CRP-mediated increase in PAI-1 expression is even more pronounced under hyperglycemic conditions.[437]

Considerable evidence has shown that serum CRP levels are predictive of future cardiovascular events. Several studies have demonstrated the prognostic value of serum CRP levels in the development of peripheral vascular disease[438] and recurrent atherothrombotic events, including ischemic stroke,[439] unstable angina, and MI.[440,441] Further demonstration that CRP concentrations can be predictive of future arterial vascular events was provided in a large case-controlled study of 1086 males in which baseline plasma CRP levels were significantly elevated in subjects who went on to have MI or ischemic stroke but not venous thrombosis. The increased risks of stroke and MI were stable over a long term and were independent of other cardiovascular risk factors, including smoking, hyperlipidemia, diabetes, and hypertension.[442,443]

Baseline CRP levels were found to be a stronger predictor than serum LDL cholesterol of CAD, stroke, and death from other cardiovascular causes in a large population of women studied over an 8-year period, a finding that persisted after adjustment for traditional cardiac risk factors.[444,445]

Epidemiological data unequivocally support CRP as a strong independent risk factor for the development of atherothrombotic events. However, it remains unclear whether CRP is simply a marker of the systemic inflammation that is part of atherogenesis or is actually an instigator of vascular disease. Recent studies of the use of statins in patients with vascular disease have provided further insight. The use of 3-hydroxy-3-methylglutaryl coenzyme A reductase inhibitors or statins can reduce plasma levels of CRP by 13–50%.[446] A case-controlled study examined statin therapy's effect on CRP levels and the combined end points of recurrent MIs or death from coronary causes in a large cohort of patients with ACS.[447] Patients with low CRP levels after statin therapy had significantly lower adverse cardiovascular events than patients with higher CRP. Reduction in CRP levels was associated with a significant improvement in event-free survival that was additive and independent of the achieved LDL cholesterol level. A decreased rate of coronary artery plaque progression by statin-mediated reduction in CRP levels (in addition to its reduction of atherogenic lipoproteins) has also been demonstrated.[448]

Preliminary data show that lowering CRP levels improves cardiovascular outcomes, but it is still not clear that the reduction in CRP levels completely explains the improvement in cardiovascular outcomes. Statins are known to have a variety of pleiotropic effects, and it is possible that the lower CRP levels are a reflection of other immunomodulatory effects on the vascular endothelium.

Antiphospholipid Antibodies

Antiphospholipid antibodies (APLA) are a heterogeneous group of immunoglobulins directed against epitopes that result from the interaction of phospholipids and proteins such as annexin V, prothrombin, cardiolipin, and β2-glycoprotein I (β2-GPI).[449–451] The antiphospholipid syndrome is another acquired thrombophilic state of hypercoagulability in which several of these autoantibodies are associated with an increased risk for the development of venous and arterial thromboembolism. Several categories of APLA have been defined with considerable overlap between subtypes, including those directed against cardiolipin (anticardiolipin antibodies), lupus anticoagulants (LA), and those that bind β2-GPI. Interestingly, anticardiolipin antibodies from patients with systemic lupus erythematosus (SLE) or the antiphospholipid syndrome, but not from patients with syphilis or other infectious diseases, require the presence of the plasma phospholipid–binding protein 2-glycoprotein I to bind to cardiolipin.[449,450,452–454] The demonstration that autoimmune anticardiolipin antibodies are directed against a phospholipid-binding protein rather than against a phospholipid led to the discovery that some autoantibodies bind directly to 2-glycoprotein I in the absence of phospholipids.[455] This has resulted in a change of focus from phospholipids to phospholipid-binding proteins.

The precise role and significance of anticardiolipin antibodies in the development of arterial thrombosis remains unclear, primarily because of the lack of uniformity in assays and interpretation, lack of distinction in earlier studies between venous and arterial thrombosis, and the preponderance of data derived from patients with connective tissue disease. A systematic review of prospective, case-controlled and cross-sectional studies on anticardiolipin antibodies and the risk of thrombosis revealed that only 50% of studies finding an association between anticardiolipin antibodies and thrombosis reached statistical significance.[456] In a separate analysis, it was found that elevated anticardiolipin antibodies' IgG titers were associated with an increased risk of arterial thrombosis and that the higher the titer, the higher the statistical significance. Although it seems that mild to moderate elevations in anticardiolipin antibodies are associated with arterial thrombosis, more rigorous research on this topic is indicated.

Interestingly, APLAs can interfere with both anticoagulant and procoagulant pathways. Lupus anticoagulant antibodies (LA) are somewhat of a misnomer, because they are associated with thromboembolic events rather than clinical bleeding. LA makes up a subgroup of APLA that disrupt the *in vitro* assembly of the prothrombinase complex (factors Xa, Va, and prothrombin in complex on phospholipid membranes), leading to prolongation of the APTT. It has been proposed that β2-GPI antibodies are the primary LA subtype associated with thrombotic outcomes.[457] A recent study reported that the frequency and titers of anti-β2-GPI antibodies were significantly higher in patients with ACS than in controls,[458] consistent with previous studies on patients with stable and unstable angina.[459] A review of 28 studies found an overall association between β2-GPI antibodies and the risk of thrombosis, although an association between anti-β2-GPI antibodies and arterial events was considerably weaker than with venous thrombosis.[460]

Considerable interest has centered on the extent to which interactions between APLA and epitopes associated with oxidized LDL may contribute to atherogenesis.[461,462] The presence of β2-GPI in LDL and oxidized LDL has led to the hypothesis that β2-GPI antibodies may facilitate the phagocytosis of oxidized lipid by foam cells and macrophages in developing atheromas. Some patients with SLE possess elevated levels of autoantibodies against oxidized LDL in association with anticardiolipin antibodies, and some anticardiolipin antibodies cross-react with oxidized LDL.[463] Moreover, anticardiolipin antibodies have been found to bind to oxidized, but not reduced, cardiolipin, suggesting that anticardiolipin antibodies recognize oxidized phospholipids, phospholipid-binding proteins, or both.[464] Preliminary studies in patients with accelerated atherosclerosis suggest that the presence of elevated titers of antibodies against oxidized LDL predict future cardiovascular events.[465] Although it is likely that these antibodies play a role in atherogenesis, the mechanisms remains undefined.

Although the literature supports a role for APLA in the development of arterial thrombotic events, this connection remains to be firmly established, because most studies have been observational or retrospective and subject to multiple confounders, the diffuse nature of the SLE disorder with which APLA is often associated, and the wide variability in the designation of APLA subtypes, assays, and titers.

FUTURE THERAPIES OF ATHEROSCLEROSIS AND THROMBOSIS

The list of potential therapeutic targets for treating atherosclerosis is very large and growing. Treatment of the hyperlipidemias often uses drugs such as statins, fibric acid derivatives, niacin, and ezetimibe, which can be applied to specific defects in lipid metabolism. This type of therapy will likely become more effective with the improved delineation of genetic biomarkers and gene profiling (e.g., using microarrays) in the clinical setting.

Preclinical studies have shown that gene therapy with specific genes can be effective in reducing atherosclerosis in various animal models. Some of these therapies are now undergoing clinical evaluation in patients with cardiovascular disease. However, as previously noted, the application of these therapies to clinical medicine awaits improvements to vector and delivery tools and further documentation of safety and efficacy through large-scale randomized trials. Studies with peripheral artery disease have also shown, both in numerous preclinical studies and in limited preliminary clinical trials, successes in the transfer of genes for growth factors (e.g., VEGF and FGF) or regulatory genes inducing angiogenesis and neovascularization or inhibiting restenosis. The use and incorporation of stem cell transplantation may also prove useful both as a vehicle and as a platform in treating atherosclerosis. Recent preclinical studies have implemented novel retroviral and lentiviral vectors for retroviral gene transfer into hematopoietic stem cells to express therapeutic genes into targeted cell types within atherosclerotic lesions.[466] The vector-driven expression of *ApoE* in macrophages was sufficient to reverse both hypercholesterolemia and prevent atherosclerotic lesion development on introduction in the ApoE-deficient mice.[467] Preliminary studies have demonstrated that bone marrow–derived precursors can give rise to vascular cells contributing to repair, remodeling lesion formation in the arterial wall.[468] Moreover, the use of vascular precursor cells might enable targeting the migration and proliferation of SMCs to prevent occlusive vascular remodeling.

SUMMARY

- Clinical genetic studies have established a major role for genetic factors in the susceptibility to coronary atherosclerosis.
- Although CAD is a polygenic disorder, a few monogenic gene defects leading to cases of atherosclerosis have been identified, and most are related to dyslipidemia. These disorders include familial hypercholesterolemia (FH) characterized by elevated cholesterol and premature CAD, resulting from mutations that affect the LDL receptor (LDLR), furnishing strong evidence for an association between blood lipids and atherosclerosis.
- Familial defective apolipoprotein B (apoB), another relatively common hypercholesterolemia, is the result of the mutations of apoB, the major protein of LDL.
- Autosomal-recessive hypercholesterolemia (ARH) is a rare disorder with a clinical phenotype similar to that of the homozygous FH caused by LDLR defects because of mutations in the ARH gene that prevent the normal internalization of the LDL receptor.
- Other lipoprotein defects resulting in atherosclerosis from both accumulation of bad cholesterol, defective lipoprotein transport, and HDL deficiency include Fisheye disease, Tangier disease, and sitosterolemia.
- Modifications in lipoprotein lipase (LPL) levels lead to elevated triglycerides and reduced HDL, both risk factors for CAD.
- Aberrant triglyceride metabolism and accumulation are defining markers of metabolic syndrome and diabetes, and pathways leading to these disorders intersect with those of the atherosclerotic pathway(s).
- Lipid oxidation plays a central role in atherogenesis. Oxidation of LDL phospholipids, largely generated by oxidants produced by the lipoxygenase and myeloperoxidase pathways, are recognized by the innate immune system of animals and humans. These are present in atherosclerotic lesions in humans and are useful markers of the susceptibility to atherogenesis.
- Activation of the inflammatory pathways is a critical element of atherosclerosis, whose lesions contain abundant immune cells, particularly macrophages and T cells. These cells mediate both adaptive immunity, which depends on antigen-specific immunological memory through toll-receptor activation, and innate immunity, which is characteristically antigen and memory independent.
- Variations in the activity of coagulation proteins and endogenous anticoagulants, as well as the kinetics of platelet aggregation, may alter the effectiveness of thrombus formation.
- In venous thrombosis, primary hypercoagulable states reflecting defects of the proteins of coagulation and fibrinolysis can lead to thrombosis; arterial thrombosis results primarily from events at the vessel wall and platelet involvement.
- Arterial disease develops with high-pressure, lipid deposition and smooth muscle hyperplasia to form an atherosclerotic plaque. Subsequently, the plaque becomes unstable and ruptures, exposing a prothrombotic lipid core activating the clotting cascade and leading to the development of a platelet-rich fibrin blood clot and arterial thrombotic occlusion.
- Epidemiological studies have identified several acquired states that may result in endothelial damage or altered hemostatic equilibrium, thereby predisposing patients to thrombosis. These include hyperhomocysteinemia, elevated C-reactive protein, antiphospholipid antibodies, and elevated fibrinogen levels.
- Hyperhomocysteinemia, elevated C-reactive protein, and lupus anticoagulants are risk factors for arterial thrombosis.

References

1. Lusis, A. J. (2000). Atherosclerosis. *Nature* **407**, 233–241.
2. Rader, D. J., Cohen, J., and Hobbs, H. H. (2003). Monogenic hypercholesterolemia: new insights in pathogenesis and treatment. *J. Clin. Invest.* **111**, 1795–1803.
3. Goldstein, J. L., and Brown, M. S. (19287). Regulation of low-density lipoprotein receptors: implications for pathogenesis and therapy of hypercholesterolemia and atherosclerosis. *Circulation* **76**, 504–507.
4. Villeger, L., Abifadel M., Allard, D., Rabes, J. P., Thiart, R., Kotze, M. J., Beroud, C., Junien, C., Boileau, C., and Varret, M. (2002). The UMD-LDLR database: Additions to the software and 490 new entries to the databbase. *Hum. Mutat.* **20**, 81–87.
5. Austin, M. A., Hutter, C. M., Zimmern, R. L., and Humphries, S. E. (2004). Genetic causes of monogenic heterozygous familial hypercholesterolemia: a HuGE prevalence review. *Am J Epidemiol.* **160**, 407–420.
6. van Aalst-Cohen, E. S., Jansen, A. C., Tanck, M. W., Defesche, J. C., Trip, M. D., Lansberg P. J., Stalenhoef, A. F., and Kastelein, J. J. (2006). Diagnosing familial hypercholesterolaemia: The relevance of genetic testing. *Eur Heart J.* **27**, 2240–2246.
7. Soria, L. F., Ludwig, E. H., Clarke, H. R, Vega, G. L., Grundy, S. M., and McCarthy, B. J. (1989). Association between a specific apolipoprotein B mutation and familial defective apolipoprotein B-100. *Proc. Natl. Acad. Sci. USA* **86**, 587–591.
8. Abifadel, M., Varret, M., Rabes, J. P., Allard, D., Ouguerram, K., Devillers, M., Cruaud, C., Benjannet, S., Wickham, L., Erlich, D., Derre, A., Villeger, L., Farnier, M., Beucler, I., Bruckert, E., Chambaz, J., Chanu, B., Lecerf, J. M., Luc, G., Moulin, P., Weissenbach, J., Prat, A., Krempf, M., Junien, C., Seidah, N. G., and Boileau, C. (2003). Mutations in PCSK9 cause autosomal dominant hypercholesterolemia. *Nat. Genet.* **34**, 154–156.
9. Soutar, A. K., Naoumova, R. P., and Traub. L. M. (2003). Genetics, clinical phenotype, and molecular cell biology of autosomal recessive hypercholesterolemia. *Arterioscler. Thromb. Vasc. Biol.* **23**, 1963–1970.
10. Garcia, C. K., Wilund, K., Arca, M., Zuliani, G., Fellin, R., Maioli, M., Calandra, S., Bertolini, S., Cossu, F., Grishin, N., Barnes, R., Cohen, J. C., and Hobbs, H. H. (2001). Autosomal recessive hypercholesterolemia caused by mutations in a putative LDL receptor adaptor protein. *Science* **292**, 1394–1398.

11. Wilund, K. R., Yi, M., Campagna, F., Arca, M., Zuliani, G., Fellin, R., Ho, Y. K., Garcia, J. V., Hobbs, H. H., and Cohen, J. C. (2002). Molecular mechanisms of autosomal recessive hypercholesterolemia. *Hum. Mol. Genet.* **11,** 3019–3030.

12. Jones, C., Hammer, R. E., Li, W. P., Cohen, J. C., Hobbs, H. H., and Herz, J. (2003). Normal sorting but defective endocytosis of the low density lipoprotein receptor in mice with autosomal recessive hypercholesterolemia. *J. Biol. Chem.* **278,** 29024–29030.

13. Pullinger, C. R., Eng, C., Salen, G., Shefer, S., Batta, A. K., Erickson, S. K., Verhagen, A., Rivera, C. R., Mulvihill, S. J., Malloy, M. J., and Kane, J. P. (2002). Human cholesterol 7α-hydroxylase (CYP7A1) deficiency has a hypercholesterolemic phenotype. *J. Clin. Invest.* **110,** 109–117.

14. Shoulders, C. C., Jones, E. L., and Naoumova, R. P. (2004). Genetics of familial combined hyperlipidemia and risk of coronary heart disease. *Hum Mol Genet.* **13,** R149–R160.

15. Ayyobi, A. F., and Brunzell, J. D. (2003). Lipoprotein distribution in the metabolic syndrome, type 2 diabetes mellitus, and familial combined hyperlipidemia. *Am J Cardiol.* **92,** 27J–33J.

16. Coon, H., Myers, R. H., Borecki, I. B., Arnett, D. K., Hunt, S. C., Province, M. A., Djousse, L., and Leppert, M. F. (2000). Replication of linkage of familial combined hyperlipidemia to chromosome 1q with additional heterogeneous effect of apolipoprotein A-I/C-III/A-IV locus. The NHLBI Family Heart Study. *Arterioscler. Thromb. Vasc. Biol.* **20,** 2275–2280.

17. Pajukanta, P., Nuotio, I., Terwilliger, J. D., Porkka, K. V., Ylitalo, K., Pihlajamaki, J., Suomalainen, A. J., Syvanen, A. C., Lehtimaki, T., Viikari, J. S., Laakso, M., Taskinen, M. R., Ehnholm, C., and Peltonen, L. (1998). Linkage of familial combined hyperlipidaemia to chromosome 1q21-q23. *Nat. Genet.* **18,** 369–373.

18. Pajukanta, P., Lilja, H. E., Sinsheimer, J. S., Cantor, R. M., Lusis, A. J., Gentile, M., Duan, X. J., Soro-Paavonen, A., Naukkarinen, J., Saarela, J., Laakso, M., Ehnholm, C., Taskinen, M. R., and Peltonen, L. (2004). Familial combined hyperlipidemia is associated with upstream transcription factor 1 (USF1). *Nat. Genet.* **36,** 371–376.

19. Castellani, L. W., Weinreb, A., Bodnar, J., Goto, A. M., Doolittle, M., Mehrabian, M., Demant, P., and Lusis, A.J. (1998). Mapping a gene for combined hyperlipidemia in a mutant mouse strain. *Nat Genet.* **18,** 374–377.

20. Bodnar, J. S., Chatterjee, A., Castellani, L. W., Ross, D. A., Ohmen, J., Cavalcoli, J., Wu, C., Dains, K. M., Catanese, J., Chu, M., Sheth, S. S., Charugundla, K., Demant, P., West, D. B., de Jong, P., and Lusis, A. J. (2002). Positional cloning of the combined hyperlipidemia gene Hyplip1. *Nat. Genet.* **30,** 110–116.

21. Mahley, R. W., Innerarity, T. L., Rall, S. C. Jr., and Weisgraber, K. H. (1985). Lipoproteins of special significance in atherosclerosis. Insights provided by studies of type III hyperlipoproteinemia. *Ann. N. Y. Acad. Sci.* **454,** 209–221.

22. Myklebost, O., Rogne, S., Olaisen, B., Gedde-Dahl, T., Jr, and prydz, H. (1984). The locus for apolipoprotein CII is closely linked to the apolipoprotein E locus on chromosome 19 in man. *Hum Genet.* **67,** 309–312.

23. Weisgraber, K. H., Rall, S. C., Jr, and Mahley, R. W. (1981). Human E apoprotein heterogeneity. Cysteine-arginine intechanges in the amino acid sequence of the apo-E isoforms. *J Biol Chem.* **256,** 9077–9083.

24. Vrablik, M., Horinek, A., Ceska, R., Stulc, T., and Kvasnicka, T. (2003). Familial dysbetalipoproteinemia in three patients with apoE 2*(Arg136–>Cys) gene variant. *Physiol. Res.* **52,** 647–650.

25. Lee M. H., Lu, K., and Patel, S. B. (2001). Genetic basis of sitosterolemia. *Curr Opin Lipidol.* 12141–12149.

26. Assmann, G., Cullen, P., Erbey, J., Ramey, D. R., Kannenberg, F., and Schulte, H. (2006). Plasma sitosterol elevations are associated with an increased incidence of coronary events in men: results of a nested case-control analysis of the Prospective Cardiovascular Munster (PROCAM) study. *Nutr Metab Cardiovasc Dis.* **16,** 13–21.

27. Berge, K. E., Tian, H., Graf, G. A., Yu, L., Grishin, N. V., Schultz, J., Kwiterovich, P., Shan, B., Barnes, R., and Hobbs, H. H. (2000). Accumulation of dietary cholesterol in sitosterolemia caused by mutations in adjacent ABC transporters. *Science* **290,** 1771–1775.

28. Lu, K., Lee, M. H., Hazard, S., Brooks-Wilson, A., Hidaka, H., Kojima, H., Ose, L., Stalenhoef, A. F., Mietinnen, T., Bjorkhem, I., Bruckert, E., Pandya, A., Brewer, H. B. Jr., Salen, G., Dean, M., Srivastava, A., and Patel, S. B. (2001). Two genes that map to the STSL locus cause sitosterolemia: genomic structure and spectrum of mutations involving sterolin-1 and sterolin-2, encoded by ABCG5 and ABCG8, respectively. *Am. J. Hum. Genet.* **69,** 278–290.

29. Hubacek, J. A., Berge, K. E., Cohen, J. C., and Hobbs, H. H. (2001). Mutations in ATP-cassette binding proteins G5 (ABCG5) and G8 (ABCG8) causing sitosterolemia. *Hum. Mutat.* **18,** 359–360.

30. Brooks-Wilson, A., Marcil, M., Clee, S. M., Zhang, L. H., Roomp, K., van Dam, M., Yu, L., Brewer, C., Collins, J. A., Molhuizen, H. O., Loubser, O., Ouelette, B. F., Fichter, K., Ashbourne-Excoffon, K. J., Sensen, C. W., Scherer, S., Mott, S., Denis, M., Martindale, D., Frohlich, J., Morgan, K., Koop, B., Pimstone, S., Kastelein, J. J., Genest, J. Jr., and Hayden, M. R. (1999). Mutations in ABC1 in Tangier disease and familial high-density lipoprotein deficiency. *Nat. Genet.* **22,** 336–345.

31. Bodzioch, M., Orso, E., Klucken, J., Langmann, T., Bottcher, A., Diederich, W., Drobnik, W., Barlage, S., Buchler, C., Porsch-Ozcurumez, M., Kaminski, W. E., Hahmann, H. W., Oette, K., Rothe, G., Aslanidis, C., Lackner, K. J., and Schmitz, G. (1999). The gene encoding ATP-binding cassette transporter 1 is mutated in Tangier disease. *Nat. Genet.* **22,** 347–351.

32. Brousseau, M. E., Schaefer, E. J., Dupuis, J., Eustace, B., Van Eerdewegh, P., Goldkamp, A. L., Thurston, L. M., FitzGerald, M. G., Yasek-McKenna, D., O'Neill, G., Eberhart, G. P., Weiffenbach, B., Ordovas, J. M., Freeman, M. W., Brown, R. H. Jr., and Gu, J. Z. (2000). Novel mutations in the gene encoding ATP-binding cassette 1 in four tangier disease kindreds. *J. Lipid. Res.* **41,** 433–441.

33. Remaley, A. T., Rust, S., Rosier, M., Knapper, C., Naudin, L., Broccardo, C., Peterson, K. M., Koch, C., Arnould, I., Prades, C., Duverger, N., Funke, H., Assman, G., Dinger, M., Dean, M., Chimini, G., Santamarina-Fojo, S., Fredrickson, D. S., Denefle, P., and Brewer, H. B. Jr. (1999). Human ATP-binding cassette transporter 1 (ABC1): genomic organization and identification of the genetic defect in the original Tangier disease kindred. *Proc. Natl. Acad. Sci. USA* **96,** 12685–12690.

34. Klein, H. G., Santamarina-Fojo, S., Duverger, N., Clerc, M., Dumon, M. F., Albers, J. J., Marcovina, S., and Brewer, H. B.

Jr. (1993). Fish eye syndrome: a molecular defect in the lecithin-cholesterol acyltransferase (LCAT) gene associated with normal alpha-LCAT-specific activity. Implications for classification and prognosis. *J. Clin. Invest.* **92,** 479–485.
35. Kuivenhoven, J. A., Stalenhoef, A. F., Hill, J. S., Demacker, P. N., Errami, A., Kastelein, J. J., and Pritchard, P. H. (1996). Two novel molecular defects in the LCAT gene are associated with fish eye disease. *Arterioscler. Thromb. Vasc. Biol.* **16,** 294–303.
36. Assmann, G., and Gotto, A. M. Jr. (2004). HDL cholesterol and protective factors in atherosclerosis. *Circulation* **109,** III8–III14.
37. Caslake, M. J., and Packard, C. J. (2003). Lipoprotein-associated phospholipase A2 (platelet-activating factor acetylhydrolase) and cardiovascular disease. *Curr. Opin. Lipidol.* **14,** 347–352.
38. Cockerill, G. W., Rye, K. A., Gamble, J. R., Vadas, M. A., and Barter, P. J. (1995). High-density lipoproteins inhibit cytokine-induced expression of endothelial cell adhesion molecules. *Arterioscler. Thromb. Vasc. Biol.* **15,** 1987–1994.
39. Nofer, J. R., Levkau, B., Wolinska, I., Junker, R., Fobker, M., von Eckardstein, A., Seedorf, U., and Assmann, G. (2001). Suppression of endothelial cell apoptosis by high density lipoproteins (HDL) and HDL-associated lysosphingolipids. *J. Biol. Chem.* **276,** 34480–34485.
40. Nofer, J. R., van der Giet, M., Tolle, M., Wolinska, I., von Wnuck Lipinski, K., Baba, H. A., Tietge, U. J., Godecke, A., Ishii, I., Kleuser, B., Schafers, M., Fobker, M., Zidek, W., Assmann, G., Chun, J., and Levkau, B. (2004). HDL induces NO-dependent vasorelaxation via the lysophospholipid receptor S1P3. *J. Clin. Invest.* **113,** 569–581.
41. Li, X. A., Titlow, W. B., Jackson, B. A., Giltiay, N., Nikolova-Karakashian, M., Uittenbogaard, A., and Smart, E. J. (2002). High density lipoprotein binding to scavenger receptor, Class B, type I activates endothelial nitric-oxide synthase in a ceramide-dependent manner. *J. Biol. Chem.* **277,** 11058–11063.
42. Yuhanna, I. S., Zhu, Y., Cox, B. E., Hahner, L. D., Osborne-Lawrence, S., Lu, P., Marcel, Y. L., Anderson, R. G., Mendelsohn, M. E., Hobbs, H. H., and Shaul, P. W. (2001). High-density lipoprotein binding to scavenger receptor-BI activates endothelial nitric oxide synthase. *Nat. Med.* **7,** 853–857.
43. Rees, A., Stocks, J., Sharpe, C. R., Vella, M. A., Shoulders, C. C., Katz, J., Jowett, N. I., Baralle, F. E., and Galton, D. J. (1985). Deoxyribonucleic acid polymorphism in the apolipoprotein A-1-C-III gene cluster. Association with hypertriglyceridemia. *J. Clin. Invest.* **76,** 1090–1095.
44. Hubacek, J. A., Adamkova, V., Ceska, R., Poledne, R., Horinek, A., and Vrablik, M. (2004). New variants in the apolipoprotein AV gene in individuals with extreme triglyceride levels. *Physiol. Res.* **53,** 225–228.
45. Bi, N., Yan, S. K., Li, G. P., Yin, Z. N., and Chen, B. S. (2004). A single nucleotide polymorphism -1131T>C in the apolipoprotein A5 gene is associated with an increased risk of coronary artery disease and alters triglyceride metabolism in Chinese. *Mol. Genet. Metab.* **83,** 280–286.
46. Corella, D., and Ordovas, J. M. (2004). The metabolic syndrome: a crossroad for genotype-phenotype associations in atherosclerosis. *Curr. Atheroscler. Rep.* **6,** 186–196.
47. Savage, D. B., Tan, G. D., Acerini, C. L., Jebb, S. A., Agostini, M., Gurnell, M., Williams, R. L., Umpleby, A. M., Thomas, E. L., Bell, J. D., Dixon, A. K., Dunne, F., Boiani, R., Cinti, S., Vidal-Puig, A., Karpe, F., Chatterjee, V. K., and O'Rahilly, S. (2003). Human metabolic syndrome resulting from dominant-negative mutations in the nuclear receptor peroxisome proliferator-activated receptor-gamma. *Diabetes* **52,** 910–917.
48. van Raalte, D. H., Li, M., Pritchard, P. H., and Wasan, K. M. (2004). Peroxisome proliferator-activated receptor (PPAR)-α: a pharmacological target with a promising future. *Pharm. Res.* **21,** 1531–1538.
49. Goodarzi, M. O., Wong, H., Quinones, M. J., Taylor, K. D., Guo, X., Castellani, L. W., Antoine, H. J., Yang, H., Hsueh, W. A., and Rotter, J. I. (2005). The 3′ untranslated region of the lipoprotein lipase gene: haplotype structure and association with post-heparin plasma lipase activity. *J. Clin. Endocrinol. Metab.* **90,** 4816–4823.
50. Holzl, B., Iglseder, B., Sandhofer, A., Malaimare, L., Lang, J., Paulweber, B., and Sandhofer, F. (2002). Insulin sensitivity is impaired in heterozygous carriers of lipoprotein lipase deficiency. *Diabetologia* **45,** 378–384.
51. Navab, M., Ananthramaiah, G. M., Reddy, S. T., Van Lenten, B. J., Ansell, B. J., Fonarow, G. C., Vahabzadeh, K., Hama, S., Hough, G., Kamranpour, N., Berliner, J. A., Lusis, A. J., and Fogelman, A. M. (2004). The oxidation hypothesis of atherogenesis: the role of oxidized phospholipids and HDL. *J. Lipid. Res.* **45,** 993–1007.
52. Navab, M., Anantharamaiah, G. M., Hama, S., Garber, D. W., Chaddha, M., Hough, G., Lallone, R., and Fogelman, A. M. (2002). Oral administration of an Apo A-I mimetic Peptide synthesized from D-amino acids dramatically reduces atherosclerosis in mice independent of plasma cholesterol. *Circulation* **105,** 290–292.
53. Ross, R. (1999). Atherosclerosis: an inflammatory disease. *N. Engl. J. Med.* **340,** 115–126.
54. Jonasson, L., Holm, J., Skalli, O., Bondjers, G., and Hansson, G. K. (1986). Regional accumulations of T cells, macrophages, and smooth muscle cells in the human atherosclerotic plaque. *Arteriosclerosis* **6,** 131–138.
55. Rajavashisth, T. B., Andalibi, A., Territo, M. C., Berliner, J. A., Navab, M., Fogelman, A. M., and Lusis, A. J. (1990). Induction of endothelial cell expression of granulocyte and macrophage colony-stimulating factors by modified low-density lipoproteins. *Nature* **344,** 254–257.
56. Cushing, S. D., Berliner, J. A., Valente, A. J., Territo, M. C., Navab, M., Parhami, F., Gerrity, R., Schwartz, C. J., and Fogelman, A. M. (1990). Minimally modified low density lipoprotein induces monocyte chemotactic protein 1 in human endothelial cells and smooth muscle cells. *Proc. Natl. Acad. Sci. USA* **87,** 5134–5138.
57. Ruan, X. Z., Varghese, Z., Powis, S. H., and Moorhead, J. F. (2001). Dysregulation of LDL receptor under the influence of inflammatory cytokines: a new pathway for foam cell formation. *Kidney Int.* **60,** 1716–1725.
58. Palkama, T. (1991). Induction of interleukin-1 production by ligands binding to the scavenger receptor in human monocytes and the THP-1 cell line. **74,** 432–438.
59. Hwang, S. J., Ballantyne C. M., Sharrett. A. R., Smith L. C., Davis, C. E., Gotto, A. M. Jr., and Boerwinkle, E. (1997). Circulating adhesion molecules VCAM-1, ICAM-1, and E-selectin in carotid Atherosclerosis Risk In Communities (ARIC) study. *Circulation* **96,** 4219–4225.
60. Cunningham, K. S., and Gotlieb, A. I. (2005). The role of shear stress in the pathogenesis of atherosclerosis. *Lab. Invest.* **85,** 9–23.

61. Auer, J., Weber, T., Berent, R., Lassnig, E., Lamm, G., and Eber, B. (2003). Genetic polymorphisms in cytokine and adhesion molecule genes in coronary artery disease. *Am. J. Pharmacogenomics* **3**, 317–328.
62. Clinton, S. K., Underwood, R., Hayes, L., Sherman, M. L., Kufe, D. W., and Libby, P. (1992). Macrophage colony-stimulating factor gene expression in vascular cells and in experimental and human atherosclerosis. *Am. J. Pathol.* **140**, 301–316.
63. Blessing, E., Campbell, L. A., Rosenfeld, M. E., Chough, N., and Kuo, C. C. (2001). *Chlamydia pneumoniae* infection accelerates hyperlipidemia induced atherosclerotic lesion development in C57BL/6J mice. *Atherosclerosis* **158**, 13–17.
64. Caligiuri, G., Rottenberg, M., Nicoletti, A., Wigzell, H., and Hansson, G. K. (2001). *Chlamydia pneumoniae* infection does not induce or modify atherosclerosis in mice. *Circulation* **103**, 2834–2838.
65. Johnson, G. B., Brunn, G. J., and Platt, J. L. (2003). Activation of mammalian Toll-like receptors by endogenous agonists. *Crit. Rev. Immunol.* **23**, 15–44.
66. Xu, X. H., Shah, P. K., Faure, E., Equils, O., Thomas, L., Fishbein, M. C., Luthringer, D., Xu, X. P., Rajavashisth, T. B., Yano, J., Kaul, S., and Arditi, M. (2001). Toll-like receptor-4 is expressed by macrophages in murine and human lipid-rich atherosclerotic plaques and upregulated by oxidized LDL. *Circulation* **104**, 3103–3108.
67. Edfeldt, K., Swedenborg, J., Hansson, G. K., and Yan, Z. Q. (2002). Expression of toll-like receptors in human atherosclerotic lesions: a possible pathway for plaque activation. *Circulation* **105**, 1158–1161.
68. Uematsu, S., Sato, S., Yamamoto, M., Hirotani, T., Kato, H., Takeshita, F., Matsuda, M., Coban, C., Ishii, K. J., Kawai, T., Takeuchi, O., and Akira, S. (2005). Interleukin-1 receptor-associated kinase-1 plays an essential role for Toll-like receptor (TLR)7- and TLR9-mediated interferon-α induction. *J. Exp. Med.* **201**, 915–923.
69. Ruckdeschel, K., Pfaffinger, G., Haase, R., Sing, A., Weighardt, H., Hacker, G., Holzmann, B., and Heesemann, J. (2004). Signaling of apoptosis through TLRs critically involves toll/IL-1 receptor domain-containing adapter inducing IFN-β, but not MyD88, in bacteria-infected murine macrophages. *J. Immunol.* **173**, 3320–3328.
70. Ford, P., Gemmell, E., Walker, P., West, M., Cullinan, M., and Seymour, G. (2005). Characterization of heat shock protein-specific T cells in atherosclerosis. *Clin. Diagn. Lab. Immunol.* **12**, 259–267.
71. Kol, A., Lichtman, A. H., Finberg, R. W., Libby, P., and Kurt-Jones, E. A. (2000). Cutting edge: heat shock protein (HSP) 60 activates the innate immune response: CD14 is an essential receptor for HSP60 activation of mononuclear cells. *J. Immunol.* **164**, 13–17.
72. Hansson, G. K., Libby, P., Schönbeck, U., and Yan, Z. (2002). Innate and adaptive immunity in the pathogenesis of atherosclerosis. *Circ. Res.* **91**, 281–291.
73. Castrillo, A., and Tontonoz, P. (2004). Nuclear receptors in macrophage biology, at the crossroads of lipid metabolism and inflammation. *Annu. Rev. Cell Dev. Biol.* **20**, 455–480.
74. Laffitte, B. A., Repa, J. J., Joseph, S. B., Wilpitz, D. C., Kast, H. R., Mangelsdorf, D. J., and Tontonoz, P. (2001). LXRs control lipid-inducible expression of the apolipoprotein E gene in macrophages and adipocytes. *Proc. Natl. Acad. Sci. USA* **98**, 507–512.
75. Chawla, A., Boisvert, W. A., Lee, C. H., Laffitte, B. A., Barak, Y., Joseph, S. B., Liao, D., Nagy, L., Edwards, P. A., Curtiss, L. K., Evans, R. M., and Tontonoz, P. (2001). A PPAR gamma-LXR-ABCA1 pathway in macrophages is involved in cholesterol efflux and atherogenesis. *Mol. Cell* **7**, 161–171.
76. Chinetti, G., Lestavel, S., Bocher, V., Remaley, A. T., Neve, B., Torra, I. P., Teissier, E., Minnich, A., Jaye, M., Duverger, N., Brewer, H. B., Fruchart, J. C., Clavey, V., and Staels, B. (2001). PPAR-alpha and PPAR-gamma activators induce cholesterol removal from human macrophage foam cells through stimulation of the ABCA1 pathway. *Nat. Med.* **7**, 53–58.
77. Joseph, S. B., Castrillo, A., Laffitte, B. A., Mangelsdorf, D. J., and Tontonoz, P. (2003). Reciprocal regulation of inflammation and lipid metabolism by liver X receptors. *Nat. Med.* **9**, 213–219.
78. Ricote, M., Li, A. C., Willson, T. M., Kelly, C. J., and Glass, C. K. (1998). The peroxisome proliferator-activated receptor-gamma is a negative regulator of macrophage activation. *Nature* **391**, 79–82.
79. Campbell, I. W. (2005). The clinical significance of PPAR gamma agonism. *Curr. Mol. Med.* **5**, 349–363.
80. Gutstein, D. E., and Fuster, V. (1999). Pathophysiology and clinical significance of atherosclerotic plaque rupture. *Cardiovasc. Res.* **41**, 323–333.
81. Corti, R., Hutter, R., Badimon, J. J., and Fuster, V. (2004). Evolving concepts in the triad of atherosclerosis, inflammation and thrombosis. *J. Thromb. Thrombolysis* **17**, 35–44.
82. Plump, A. S., Smith, J. D., Hayek, T., Aalto-Setala, K., Walsh, A., Verstuyft, J. G., Rubin, E. M., and Breslow, J. L. (1992). Severe hypercholesterolemia and atherosclerosis in apolipoprotein E-deficient mice created by homologous recombination in ES cells. *Cell* **71**, 343–353.
83. Ishibashi, S., Brown, M. S., Goldstein, J. L., Gerard, R. D., Hammer, R. E., and Herz, J. (1993). Hypercholesterolemia in low density lipoprotein receptor knockout mice and its reversal by adenovirus-mediated gene delivery. *J. Clin. Invest.* **92**, 883–893.
84. Chien, K. R. (1996). Genes and physiology: molecular physiology in genetically engineered animals. *J. Clin. Invest.* **97**, 901–909.
85. Tward, A., Xia, Y. R., Wang, X. P., Shi, Y. S., Park, C., Castellani, L. W., Lusis, A. J., and Shih, D. M. (2002). Decreased atherosclerotic lesion formation in human serum paraoxonase transgenic mice. *Circulation* **106**, 484–490.
86. Jong, M. C., Gijbels, M. J., Dahlmans, V. E., Gorp, P. J., Koopman, S. J., Ponec, M., Hofker, M. H., and Havekes, L. M. (1998). Hyperlipidemia and cutaneous abnormalities in transgenic mice overexpressing human apolipoprotein C1. *J. Clin. Invest.* **101**, 145–152.
87. Kaul, S., Coin, B., Hedayiti, A., Yano, J., Cercek, B., Chyu, K. Y., and Shah, P. K. (2004). Rapid reversal of endothelial dysfunction in hypercholesterolemic apolipoprotein E-null mice by recombinant apolipoprotein A-I(Milano)-phospholipid complex. *J. Am. Coll. Cardiol.* **44**, 1311–1319.
88. Navab, M., Anantharamaiah, G. M., Reddy, S. T., Hama, S., Hough, G., Grijalva, V. R., Yu, N., Ansell, B. J., Datta, G., Garber, D. W., and Fogelman, A. M. (2005). Apolipoprotein A-I mimetic peptides. *Arterioscler. Thromb. Vasc. Biol.* **25**, 1325–1331.

89. Qiao, J. H., Tripathi, J., Mishra, N. K., Cai, Y., Tripathi, S., Wang, X. P., Imes, S., Fishbein, M. C., Clinton, S. K., Libby, P., Lusis, A. J., and Rajavashisth, T. B. (1997). Role of macrophage colony-stimulating factor in atherosclerosis: studies of osteopetrotic mice. *Am. J. Pathol.* **150,** 1687–1699.
90. Mehrabian, M., Allayee, H., Wong, J., Shi, W., Wang, X. P., Shaposhnik, Z., Funk, C. D., and Lusis, A. J. (2002). Identification of 5-lipoxygenase as a major gene contributing to atherosclerosis susceptibility in mice. *Circ. Res.* **91,** 120–126.
91. Shih, D. M., Gu, L., Xia, Y. R., Navab, M., Li, W. F., Hama, S., Castellani, L. W., Furlong, C. E., Costa, L. G., Fogelman, A. M., and Lusis, A. J. (1998). Mice lacking serum paraoxonase are susceptible to organophosphate toxicity and atherosclerosis. *Nature* **394,** 284–287.
92. Suzuki, H., Kurihara, Y., Takeya, M., Kamada, N., Kataoka, M., Jishage, K., Ueda, O., Sakaguchi, H., Higashi, T., Suzuki, T., Takashima, Y., Kawabe, Y., Cynshi, O., Wada, Y., Honda, M., Kurihara, H., Aburatani, H., Doi, T., Matsumoto, A., Azuma, S., Noda, T., Toyoda, Y., Itakura, H., Yazaki, Y., Horiuchi, S., Takahashi, K., Kruijt, J. K., van Berkel, T. J. C., Steinbrecher, U. P., Ishibashi, S., Maeda, N., Gordon, S., and Kodama, T. (1997). A role for macrophage scavenger receptors in atherosclerosis and susceptibility to infection. *Nature* **386,** 292–296.
93. Febbraio, M., Podrez, E. A., Smith, J. D., Hajjar, D. P., Hazen, S. L., Hoff, H. F., Sharma, K., and Silverstein, R. L. (2000). Targeted disruption of the class B scavenger receptor CD36 protects against atherosclerotic lesion development in mice. *J. Clin. Invest.* **105,** 1049–1056.
94. Wright, S. D., Burton, C., Hernandez, M., Hassing, H., Montenegro, J., Mundt, S., Patel, S., Card, D. J., Hermanowski-Vosatka, A., Bergstrom, J. D., Sparrow, C. P., Detmers, P. A., and Chao, Y. S. (2000). Infectious agents are not necessary for murine atherogenesis. *J. Exp. Med.* **191,** 1437–1442.
95. Whitman, S. C., Ravisankar, P., Elam, H., and Daugherty, A. (2000). Exogenous interferon-gamma enhances atherosclerosis in apolipoprotein E-/– mice. *Am. J. Pathol.* **157,** 1819–1824.
96. Mach, F., Schonbeck, U., and Libby, P. (1998). CD40 signaling in vascular cells: a key role in atherosclerosis? *Atherosclerosis* **137,** S89–S95.
97. Suzuki, G., Izumi, S., Hakoda, M., and Takahashi, N. (2004). LTA 252G allele containing haplotype block is associated with high serum C-reactive protein levels. *Atherosclerosis* **176,** 91–94.
98. Helgadottir, A., Gretarsdottir, S., St. Clair, D., Manolescu, A., Cheung, J., Thorleifsson, G., Pasdar, A., Grant, S. F., Whalley, L. J., Hakonarson, H., Thorsteinsdottir, U., Kong, A., Gulcher, J., Stefansson, K., and MacLeod, M. J. (2005). Association between the gene encoding 5-lipoxygenase-activating protein and stroke replicated in a Scottish population. *Am. J. Hum. Genet.* **76,** 505–509.
99. Gretarsdottir, S., Thorleifsson, G., Reynisdottir, S. T., Manolescu, A., Jonsdottir, S., Jonsdottir, T., Gudmundsdottir, T., Bjarnadottir, S. M., Einarsson, O. B., Gudjonsdottir, H. M., Hawkins, M., Gudmundsson, G., Gudmundsdottir, H., Andrason, H., Gudmundsdottir, A. S., Sigurdardottir, M., Chou, T. T., Nahmias, J., Goss, S., Sveinbjornsdottir, S., Valdimarsson, E. M., Jakobsson, F., Agnarsson, U., Gudnason, V., Thorgeirsson, G., Fingerle, J., Gurney, M., Gudbjartsson, D., Frigge, M. L., Kong, A., Stefansson, K., and Gulcher, J. R. (2003). The gene encoding phosphodiesterase 4D confers risk of ischemic stroke. *Nat. Genet.* **35,** 131–138.
100. Ordovas, J. M., and Mooser, V. (2002). The APOE locus and the pharmacogenetics of lipid response. *Curr. Opin. Lipidol.* **13,** 113–117.
101. Wang, J., Ban, M. R., and Hegele, R. A. (2005). Multiplex ligation-dependent probe amplification of LDLR enhances molecular diagnosis of familial hypercholesterolemia. *J. Lipid Res.* **46,** 366–372.
102. Kotze, M. J., Kriegshauser, G., Thiart, R., de Villiers, N. J., Scholtz, C. L., Kury, F., Moritz, A., and Oberkanins, C. (2003). Simultaneous detection of multiple familial hypercholesterolemia mutations facilitates an improved diagnostic service in South African patients at high risk of cardiovascular disease. *Mol. Diagn.* **7,** 169–174.
103. Hagberg, J. M., Wilund, K. R., and Ferrell, R. E. (2000). APO E gene and gene-environment effects on plasma lipoprotein-lipid levels. *Physiol. Genomics* **4,** 101–108.
104. Kuivenhoven, J. A., Jukema, J. W., Zwinderman, A. H., de Knijff, P., McPherson, R., Bruschke, A. V., Lie, K. I., and Kastelein, J. J. (1998). The role of a common variant of the cholesteryl ester transfer protein gene in the progression of coronary atherosclerosis. The Regression Growth Evaluation Statin Study Group. *N. Engl. J. Med.* **338,** 86–93.
105. de Maat, M. P., Jukema, J. W., Ye, S., Zwinderman, A. H., Moghaddam, P. H., Beekman, M., Kastelein, J. J., van Boven, A. J., Bruschke, A. V., Humphries, S. E., Kluft, C., and Henney, A. M. (1999). Effect of the stromelysin -1 promoter on efficacy of pravastatin in coronary atherosclerosis and restenosis. *Am. J. Cardiol.* **83,** 852–856.
106. Dwyer, J. H., Allayee, H., Dwyer, K. M., Fan, J., Wu, H., Mar, R., Lusis, A. J., and Mehrabian, M. (2004). Arachidonate 5-lipoxygenase promoter genotype, dietary arachidonic acid, and atherosclerosis. *N. Engl. J. Med.* **350,** 29–37.
107. Satterthwaite, G., Francis, S. E., Suvarna, K., Blakemore, S., Ward, C., Wallace, D., Braddock, M., and Crossman, D. (2005). Differential gene expression in coronary arteries from patients presenting with ischemic heart disease: further evidence for the inflammatory basis of atherosclerosis. *Am. Heart J.* **150,** 488–499.
108. Gagarin, D., Yang, Z., Butler, J., Wimmer, M., Du, B., Cahan, P., and McCaffrey, T. A. (2005). Genomic profiling of acquired resistance to apoptosis in cells derived from human atherosclerotic lesions: potential role of STATs, cyclinD1, BAD, and Bcl-XL. *J. Mol. Cell Cardiol.* **39,** 453–465.
109. Randi, A. M., Biguzzi, E., Falciani, F., Merlini, P., Blakemore, S., Bramucci, E., Lucreziotti, S., Lennon, M., Faioni, E. M., Ardissino, D., and Mannucci, P. M. (2003). Identification of differentially expressed genes in coronary atherosclerotic plaques from patients with stable or unstable angina by cDNA array analysis. *J. Thromb. Haemost.* **1,** 829–835.
110. Csoka, A. B., English, S. B., Simkevich, C. P., Ginzinger, D. G., Butte, A. J., Schatten, G. P., Rothman, F. G., and Sedivy, J. M. (2004). Genome-scale expression profiling of Hutchinson-Gilford progeria syndrome reveals widespread transcriptional misregulation leading to mesodermal/mesenchymal defects and accelerated atherosclerosis. *Aging Cell* **3,** 235–243.
111. Spronk, H. M., Govers-Riemslag, J. W., and ten Cate, H. (2003). The blood coagulation system as a molecular machine. *Bioessays* **25,** 1220–1228.

112. Mackman, N. (2004). Role of tissue factor in hemostasis, thrombosis, and vascular development. *Arterioscler. Thromb. Vasc. Biol.* **24,** 1015–1022.
113. Wilcox, J. N., Smith, K. M., Schwartz, S. M., and Gordon, D. (1989). Localization of tissue factor in the normal vessel wall and in the atherosclerotic plaque. *Proc. Natl. Acad. Sci. USA* **86,** 2839–2843.
114. Annex, B. H., Denning, S. M., Channon, K. M., Sketch, M. H., Stack, R. S., Morrissey, J. H., and Peters, K. G. (1995). Differential expression of tissue factor protein in directional atherectomy specimens from patients with stable and unstable coronary syndromes. *Circulation* **91,** 619–622.
115. Steffel, J., Luscher, T. F., and Tanner, F. C. (2006). Tissue factor in cardiovascular diseases: molecular mechanisms and clinical implications. *Circulation* **113,** 722–731.
116. Diamant, M., Nieuwland, R., Pablo, R. F., Sturk, A., Smit, J. W., and Radder, J. K. (2002). Elevated numbers of tissue-factor exposing microparticles correlate with components of the metabolic syndrome in uncomplicated type 2 diabetes mellitus. *Circulation* **106,** 2442–2447.
117. Nieuland, R., Berckmans, R.,J., Rotteveel-Eijkman, R. C., Maquelin, K. N., Roozendaal, K. J., Jansen, P. G. M., ten Have, K., Eijsman, L., Hack, C. E., and Sturk, A. (1997). Cell-derived microparticles generated in patients during cardiopulmonary bypass are highly procoagulant. *Circulation* **96,** 3534–3541.
118. Soejima, H., Ogawa, H., Yasue, H., Kaikita, K., Nishiyama, K., Misumi, K., Takazoe, K., Miyao, Y., Yoshimura, M., Kugiyama, K., Nakamura, S., Tsuji, I., and Kumeda, K. (1999). Heightened tissue factor associated with tissue factor pathway inhibitor and prognosis in patients with unstable angina. *Circulation* **99,** 2908–2913.
119. Seljeflot, I., Hurlen, M., Hole, T., and Arnesen, H. (2003). Soluble tissue factor as predictor of future events in patients with acute myocardial infarction. *Thromb. Res.* **111,** 369–372.
120. Bajaj, M. S., Steer, S., Kuppuswamy, M. N., Kisiel, W., and Bajaj, S. P. (1999). Synthesis and expression of tissue factor pathway inhibitor by serum-stimulated fibroblasts, vascular smooth muscle cells and cardiac myocytes. *Thromb. Haemost.* **82,** 1663–1672.
121. Ott, I., Michaelis, C., Schuermann, M., Steppich, B., Seitz, I., Dewerchin, M., Zohlnhofer, D., Wessely, R., Rudelius, M., Schomig, A., and Carmeliet, P. (2005). Vascular remodeling in mice lacking the cytoplasmic domain of tissue factor. *Circ. Res.* **97,** 293–298.
122. Hamuro, T., Kamikubo, Y., Nakahara, Y., Miyamoto, S., and Funatsu, A. (1998). Human recombinant tissue factor pathway inhibitor induces apoptosis in cultured human endothelial cells. *FEBS Lett.* **421,** 197–202.
123. Kamikubo, Y., Nakahara, Y., Takemoto, S., Hamuro, T., Miyamoto, S., and Funatsu, A. (1997). Human recombinant tissue factor pathway inhibitor prevents the proliferation of cultured human neonatal aortic smooth muscle cells. *FEBS Lett.* **407,** 116–120.
124. Nishida, T., Ueno, H., Atsuchi, N., Kawano, R., Asada, Y., Nakahara, Y., Kamikubo, Y., Takeshita, A., and Yasui, H. (1999). Adenovirus-mediated local expression of human tissue factor pathway inhibitor eliminates shear stress-induced recurrent thrombosis in the injured carotid artery of the rabbit. *Circ. Res.* **84,** 1446–1452.
125. Asada, Y., Hara, S., Tsuneyoshi, A., Hatakeyama, K., Kisanuki, A., Marutsuka, K., Sato, Y., Kamikubo, Y., and Sumiyoshi, A. (1998). Fibrin-rich and platelet-rich thrombus formation on neointima: recombinant tissue factor pathway inhibitor prevents fibrin formation and neointimal development following repeated balloon injury of rabbit aorta. *Thromb. Haemost.* **80,** 506–511.
126. Lindahl, A. K., Sandset, P. M., Thune-Wiiger, M., Nordfang, O., and Sakariassen, K. S. (1994). Tissue factor pathway inhibitor prevents thrombus formation on procoagulant subendothelial matrix. *Blood Coagul. Fibrinolysis* **5,** 755–760.
127. Creasey, A. A., Chang, A. C., Feigen, L., Wun, T. C., Taylor, F. B. Jr., and Hinshaw, L. B. (1993). Tissue factor pathway inhibitor reduces mortality from *Escherichia coli* septic shock. *J. Clin. Invest.* **91,** 2850–2860.
128. Opal, S. M., Palardy, J. E., Parejo, N. A., and Creasey, A. A. (2001). The activity of tissue factor pathway inhibitor in experimental models of superantigen-induced shock and polymicrobial intra-abdominal sepsis. *Crit. Care Med.* **29,** 13–17.
129. Bai, H., Ma, D., Zhang, Y. G., Zhang, N., Kong, D. S., Guo, H. S., Mo, W., Tang, Q. Q., and Song, H. Y. (2005). Molecular design and characterization of recombinant long half-life mutants of human tissue factor pathway inhibitor. *Thromb. Haemost.* **93,** 1055–1060.
130. Kaiser, B., Hoppensteadt, D. A., and Fareed, J. (2001). Tissue factor pathway inhibitor, an update of potential implications in the treatment of cardiovascular disorders. *Expert Opin. Invest. Drugs* **10,** 1925–1935.
131. LaRosa, S. P., and Opal, S. M. (2005). Tissue factor pathway inhibitor and antithrombin trial results. *Crit. Care Clin.* **21,** 433–448.
132. Abraham, E., Reinhart, K., Opal, S., Demeyer, I., Doig, C., Rodriguez, A. L., Beale, R., Svoboda, P., Laterre, P. F., Simon, S., Light, B., Spapen, H., Stone, J., Seibert, A., Peckelsen, C., De Deyne, C., Postier, R., Pettila, V., Artigas, A., Percell, S. R., Shu, V., Zwingelstein, C., Tobias, J., Poole, L., Stolzenbach, J. C., and Creasey, A. A. (2003). OPTIMIST Trial Study Group. Efficacy and safety of tifacogin (recombinant tissue factor pathway inhibitor) in severe sepsis: a randomized controlled trial. *JAMA* **290,** 238–247.
133. Mackman, N., Morrissey, J. H., Fowler, B., and Edgington, T. S. (1989). Complete sequence of the human tissue factor gene, a highly regulated cellular receptor that initiates the coagulation protease cascade. *Biochemistry* **28,** 1755–1762.
134. Mackman, N. (1995). Regulation of the tissue factor gene. *FASEB J.* **9,** 883–889.
135. Samad, F., Pandey, M., and Loskutoff, D. J. (1998). Tissue factor gene expression in the adipose tissues of obese mice. *Proc. Natl. Acad. Sci. USA* **95,** 7591–7596.
136. Marmur, J. D., Rossikhina, M., Guha, A., Fyfe, B., Friedrich, V., Mendlowitz, M., Nemerson, Y., and Taubman, M. B. (1993). Tissue factor is rapidly induced in arterial smooth muscle after balloon injury. *J. Clin. Invest.* **91,** 2253–2259.
137. Golino, P., Ragni, M., Cirillo, P., Avvedimento, V. E., Feliciello, A., Esposito, N., Scognamiglio, A., Trimarco, B., Iaccarino, G., Condorelli, M., Chiariello, M., and Ambrosio, G. (1996). Effects of tissue factor induced by oxygen free

138. Lin, M. C., Almus-Jacobs, F., Chen, H. H., Parry, G. C., Mackman, N., Shyy, J. Y., and Chien, S. (1997). Shear stress induction of the tissue factor gene. *J. Clin. Invest.* **99,** 737–744.
139. Arnaud, E., Barbalat, V., Nicaud, V., Cambien, F., Evans, A., Morrison, C., Arveiler, D., Luc, G., Ruidavets, J. B., Emmerich, J., Fiessinger, J. N., and Aiach, M. (2000). Polymorphisms in the 5′ regulatory region of the tissue factor gene and the risk of myocardial infarction and venous thromboembolism: the ECTIM and PATHROS studies. Etude Cas-Temoins de l'Infarctus du Myocarde. Paris Thrombosis case-control Study. *Arterioscler. Thromb. Vasc. Biol.* **20,** 892–898.
140. Terry, C. M., Kling, S. J., Cheang, K. I., Hoidal, J. R., and Rodgers, G. M. (2004). Polymorphisms in the 5′-UTR of the tissue factor gene are associated with altered expression in human endothelial cells. *J. Thromb. Haemost.* **2,** 1351–1358.
141. Malarstig, A., Tenno, T., Johnston, N., Lagerqvist, B., Axelsson, T., Syvanen, A. C., Wallentin, L., and Siegbahn, A. (2005). Genetic variations in the tissue factor gene are associated with clinical outcome in acute coronary syndrome and expression levels in human monocytes. *Arterioscler. Thromb. Vasc. Biol.* **25,** 2667–2672.
142. Enjyoji, K., Emi, M., Mukai, T., Imada, M., Leppert, M. L., Lalouel, J. M., and Kato, H. (1993). Human tissue factor pathway inhibitor (TFPI) gene: complete genomic structure and localization on the genetic map of chromosome 2q. *Genomics* **17,** 423–428.
143. Girard, T. J., Eddy, R., Wesselschmidt, R. L., MacPhail, L. A., Likert, K. M., Byers, M. G., Shows, T. B., and Broze, G. J. Jr. (1991). Structure of the human lipoprotein-associated coagulation inhibitor gene. Intron/exon gene organization and localization of the gene to chromosome 2. *J. Biol. Chem.* **266,** 5036–5041.
144. Bajaj, M. S., Tyson, D. R., Steer, S. A., and Kuppuswamy, M. N. (2001). Role of GATA motifs in tissue factor pathway inhibitor gene expression in malignant cells. *Thromb. Res.* **101,** 203–211.
145. Iochmann, S., Reverdiau-Moalic, P., Beaujean, S., Rideau, E., Lebranchu, Y., Bardos, P., and Gruel, Y. (1999). Fast detection of tissue factor and tissue factor pathway inhibitor messenger RNA in endothelial cells and monocytes by sensitive reverse transcription-polymerase chain reaction. *Thromb. Res.* **94,** 165–173.
146. Caplice, N. M., Mueske, C. S., Kleppe, L. S., Peterson, T. E., Broze, G. J. Jr., and Simari, R. D. (1998). Expression of tissue factor pathway inhibitor in vascular smooth muscle cells and its regulation by growth factors. *Circ. Res.* **83,** 1264–1270.
147. Pendurthi, U. R., Rao, L. V., Williams, J. T., and Idell, S. (1999). Regulation of tissue factor pathway inhibitor expression in smooth muscle cells. *Blood* **94,** 579–586.
148. Sprecher, C. A., Kisiel, W., Mathewes, S., and Foster, D. C. (1994). Molecular cloning, expression, and partial characterization of a second human tissue-factor-pathway inhibitor. *Proc. Natl. Acad. Sci. USA* **91,** 3353–3357.
149. Kamei, S., Kazama, Y., Kuijper, J. L., Foster, D. C., and Kisiel, W. (2001). Genomic structure and promoter activity of the human tissue factor pathway inhibitor-2 gene. *Biochim. Biophys. Acta* **1517,** 430–435.
150. Kleesick, K., Schmidt, M., Goting, C., Brinkmann, T., and Prohaska, W. (1998). A first mutation in the human tissue factor pathway inhibitor gene encoding [P151L]TFPI. *Blood* **92,** 3976–3977.
151. Kleesick, K., Schmidt, M., Goting, C., Schwenz, B., Lange, S., Muller-Berghaus, G., Brinkmann, T., and Prohaska, W. (1999). The 536C-T transition in the human tissue factor pathway inhibitor (TFPI) gene is statistically associated with a higher risk for venous thrombosis. *Thromb. Haemost.* **82,** 1–5.
152. Gonzalez-Conejero, R., Lozano, M. L., Corral, J., Martinez, C., and Vicente, V. (2000). The TFPI C536T mutation is not associated with increased risk for venous or arterial thrombosis. *Thromb. Haemost.* **83,** 787–788.
153. Armaud, E., Moatti, D., Emmerich, J., Aiach, M., and de Prost, D. (1999). No link between the TFPI V264M mutation and venous thromboembolic disease. *Thromb. Haemost.* **82,** 159–160.
154. Moatti, D., Meirhaeghe Ollivier, V., Bauters, C., Amouyel, P., and de Prost, D. (2001). Polymorphisms of the tissue factor pathway inhibitor gene and the risk of restenosis after coronary angioplasty. *Blood Coagul. Fibrinolysis* **12,** 317–323.
155. Miyata, T., Sakata, T., Kumeda, K., Uchida, K., Tsushima, M., Fujimura, H., Kawasaki, T., and Kato, H. (1998). C-399 polymorphism in the promoter region of human tissue factor pathway inhibitor (TFPI) gene does not change the plasma TFPI antigen level and does not cause venous thrombosis. *Thromb. Haemost.* **80,** 345–346.
156. O'Hara, P. J., Grant, F. J., Haldeman, B. A., Gray, C. L., Insley, M. Y., Hagen, F. S., and Murray, M. J. (1987). Nucleotide sequence of the gene coding for human factor VII, a vitamin K-dependent protein participating in blood coagulation. *Proc. Natl. Acad. Sci. USA* **84,** 5158–5162.
157. Hagen, F. S., Gray, C. L., O'Hara, P., Grant, F. J., Saari, G. C., Woodbury, R. G., Hart, C. E., Insley, M., Kisiel, W., Kurachi, K., and Davie, E. W. (1986). Characterization of a cDNA coding for human factor VII. *Proc. Natl. Acad. Sci. USA* **83,** 2412–2416.
158. Laurian, Y. (2002). Treatment of bleeding in patients with platelet disorders: is there a place for recombinant factor VIIa? *Pathophysiol. Haemost. Thromb.* **32,** 37–40.
159. Meade, T. W., Mellows, S., Brozovic, M., Miller, G. J., Chakrabarti, R. R., North, W. R., Haines, A. P., Stirling, Y., Imeson, J. D., and Thompson, S. G. (1986). Haemostatic function and ischaemic heart disease: principal results of the Northwick Park Heart Study. *Lancet* **2,** 533–537.
160. Moor, E., Silveira, A., van't Hooft, F., Suontaka, A. M., Eriksson, P., Blomback, M., and Hamsten, A. (1995). Coagulation factor VII mass and activity in young men with myocardial infarction at a young age. Role of plasma lipoproteins and factor VII genotype. *Arterioscler. Thromb. Vasc. Biol.* **15,** 655–664.
161. Lowe, G. D., Rumley, A., McMahon, A. D., Ford, I., O'Reilly, D. S., Packard, C. J., and the West of Scotland Coronary Prevention Study Group. (2004). Interleukin-6, fibrin D-dimer, and coagulation factors VII and XIIa in prediction of

coronary heart disease. *Arterioscler. Thromb. Vasc. Biol.* **24,** 1529–1534.

162. McVey, J. H., Boswell, E., Mumford, A. D., Kemball-Cook, G., and Tuddenham, E. G. (2001). Factor VII deficiency and the FVII mutation database. *Hum. Mutat.* **17,** 3–17.

163. Herrmann, F. H., Wulff, K., Auberger, K., Aumann, V., Bergmann, F., Bergmann, K., Bratanoff, E., Franke, D., Grundeis, M., Kreuz, W., Lenk, H., Losonczy, H., Maak, B., Marx, G., Mauz-Korholz, C., Pollmann, H., Serban, M., Sutor, A., Syrbe, G., Vogel, G., Weinstock, N., Wenzel, E., and Wolf, K. (2000). Molecular biology and clinical manifestation of hereditary factor VII deficiency. *Semin. Thromb. Hemost.* **26,** 393–400.

164. Giansily-Blaizot, M., Aguilar-Martinez, P., Briquel, M. E., d'Oiron, R., De Maistre, E., Epelbaum, S., and Schved, J. F. (2003). Two novel cases of cerebral haemorrhages at the neonatal period associated with inherited factor VII deficiency, one of them revealing a new nonsense mutation (Ser52Stop). *Blood. Coagul. Fibrinolysis* **14,** 217–220.

165. Bernardi, F., Arcieri, P., Bertina, R. M., Chiarotti, F., Corral, J., Pinotti, M., Prydz, H., Samama, M., Sandset, P. M., Strom, R., Garcia, V. V., and Mariani, G. (1997). Contribution of factor VII genotype to activated FVII levels. Differences in genotype frequencies between northern and southern European populations. *Arterioscler. Thromb. Vasc. Biol.* **17,** 2548–2553.

166. Hunault, M., Arbini, A. A., Lopaciuk, S., Carew, J. A., and Bauer, K. A. (1997). The Arg353Gln polymorphism reduces the level of coagulation factor VII. In vivo and in vitro studies. *Arterioscler. Thromb. Vasc. Biol.* **17,** 2825–2829.

167. Mariani, G., Herrmann, F. H., Dolce, A., Batorova, A., Etro, D., Peyvandi, F., Wulff, K., Schved, J. F., Auerswald, G., Ingerslev, J., Bernardi, F., and the International Factor VII Deficiency Study Group. (2005). Clinical phenotypes and factor VII genotype in congenital factor VII deficiency. *Thromb. Haemost.* 93, 481–487.

168. Fromovich-Amit, Y., Zivelin, A., Rosenberg, N., Tamary, H., Landau, M., and Seligsohn, U. (2004). Characterization of mutations causing factor VII deficiency in 61 unrelated Israeli patients. *J. Thromb. Haemost.* **2,** 1774–1781.

169. Kato, A., Asakai, R., Davie, E. W., and Aoki, N. (1989). Factor XI gene (F11) is located on the distal end of the long arm of human chromosome 4. *Cytogenet. Cell Genet.* **52,** 77–78.

170. Asakai, R., Davie, E. W., and Chung, D. W. (1987). Organization of the gene for human factor XI. *Biochemistry* **26,** 7221–7228.

171. Saunders, R. E., O'Connell, N. M., Lee, C. A., Perry, D. J., and Perkins, S. J. (2005). Factor XI deficiency database: an interactive web database of mutations, phenotypes, and structural analysis tools. *Hum. Mutat.* **26,** 192–198.

172. Salomon, O., Steinberg, D. M., Dardik, R., Rosenberg, N., Zivelin, A., Tamarin, I., Ravid, B., Berliner, S., and Seligsohn, U. (2003). Inherited factor XI deficiency confers no protection against acute myocardial infarction. *J. Thromb. Haemost.* **1,** 658–661.

173. Quelin, F., Trossaert, M., Sigaud, M., Mazancourt, P. D., and Fressinaud, E. (2004). Molecular basis of severe factor XI deficiency in seven families from the west of France. Seven novel mutations, including an ancient Q88X mutation. *J. Thromb. Haemost.* **2,** 71–76.

174. Bezak, A., Kaczanowski, R., Dossenbach-Glaninger, A., Kucharczyk, K., Lubitz, W., and Hopmeier, P. (2005). Detection of single nucleotide polymorphisms in coagulation factor XI deficient patients by multitemperature single-strand conformation polymorphism analysis. *J. Clin. Lab. Anal.* **19,** 233–240.

175. Sato, E., Kawamata, N., Kato, A., and Oshimi, K. (2000). A novel mutation that leads to a congenital factor XI deficiency in a Japanese family. *Am. J. Hematol.* **63,** 165–169.

176. Asakai, R., Chung, D. W., Davie, E. W., and Seligsohn, U. (1991). Factor XI deficiency in Ashkenazi Jews in Israel. *N. Engl. J. Med.* **325,** 153–158.

177. Salomon, O., and Seligsohn, U. (2004). New observations on factor XI deficiency. *Haemophilia* **10,** 184–187.

178. Meijers, J. C., Tekelenburg, W. L., Bouma, B. N., Bertina, R. M., and Rosendaal, F. R. (2000). High levels of coagulation factor XI as a risk factor for venous thrombosis. *N. Engl. J. Med.* **342,** 696–701.

179. Royle, N. J., Nigli, M., Cool, D., MacGillivray, R. T., and Hamerton, J. L. (1988). Structural gene encoding human factor XII is located at 5q33-qter. *Somat. Cell Mol. Genet.* **14,** 217–221.

180. Cool, D. E., and MacGillivray, R. T. (1987). Characterization of the human blood coagulation factor XII gene. Intron/ exon gene organization and analysis of the 5′-flanking region. *J. Biol. Chem.* 262, 13662–13673.

181. Endler, G., Exner, M., Mannhalter, C., Meier, S., Ruzicka, K., Handler, S., Panzer, S., Wagner, O., and Quehenberger, P. (2001). A common C→T polymorphism at nt 46 in the promoter region of coagulation factor XII is associated with decreased factor XII activity. *Thromb. Res.* **101,** 255–260.

182. Kanaji, T., Okamura, T., Osaki, K., Kuroiwa, M., Shimoda, K., Hamasaki, N., and Niho, Y. (1998). A common genetic polymorphism (46 C to T substitution) in the 5′-untranslated region of the coagulation factor XII gene is associated with low translation efficiency and decrease in plasma factor XII level. *Blood* **91,** 2010–2014.

183. Kondo, S., Tokunaga, F., Kawano, S., Oono, Y., Kumagai, S., and Koide, T. (1999). Factor XII Tenri, a novel cross-reacting material negative factor XII deficiency, occurs through a proteasome-mediated degradation. *Blood* **993,** 4300–4308.

184. Oguchi, S., Ishii, K., Moriki, T., Takeshita, E., Murata, M., Ikeda, Y., and Watanabe, K. (2005). Factor XII Shizuoka, a novel mutation (Ala392Thr) identified and characterized in a patient with congenital coagulation factor XII deficiency. *Thromb. Res.* **115,** 191–197.

185. Ishii, K., Oguchi, S., Moriki, T., Yatabe, Y., Takeshita, E., Murata, M., Ikeda, Y., and Watanabe, K. (2004). Genetic analyses and expression studies identified a novel mutation (W486C) as a molecular basis of congenital coagulation factor XII deficiency. *Blood Coagul. Fibrinolysis* **15,** 367–373.

186. Halbmayer, W.-M., Mannhalter, C., Feichtinger, C., Rubi, K., and Fischer, M. (1992). The prevalence of factor XII deficiency in 103 orally anticoagulated outpatients suffering from recurrent venous and/or arterial thromboembolism. *Thromb. Haemost.* **68,** 285–290.

187. Kohler, H. P., Futers, T. S., and Grant, P. J. (1999). FXII (46C→T) polymorphism and in vivo generation of FXII

188. Zito, F., Lowe, G. D., Rumley, A., McMahon, A. D., Humphries, S. E., and the WOSCOPS Study Group West of Scotland Coronary Prevention Study. (2002). Association of the factor XII 46C>T polymorphism with risk of coronary heart disease (CHD) in the WOSCOPS study. *Atherosclerosis* **165,** 153–158.
189. Tirado, I., Soria, J. M., Mateo, J., Oliver, A., Souto, J. C., Santamaria, A., Felices, R., Borrell, M., and Fontcuberta, J. (2004). Association after linkage analysis indicates that homozygosity for the 46C→T polymorphism in the F12 gene is a genetic risk factor for venous thrombosis. *Thromb. Haemost.* **91,** 899–904.
190. Santamaria, A., Mateo, .J, Tirado, I., Oliver, A., Belvis, R., Marti-Fabregas, J., Felices, R., Soria, J. M., Souto, J. C., and Fontcuberta, J. (2004). Homozygosity of the T allele of the 46 C→T polymorphism in the F12 gene is a risk factor for ischemic stroke in the Spanish population. *Stroke* **35,** 1795–1799.
191. Souto, J. C., Almasy, L., Borrell, M., Blanco-Vaca, F., Mateo, J., Soria, J. M., Coll, I., Felices, R., Stone, W., Fontcuberta, J., and Blangero, J. (2000). Genetic susceptibility to thrombosis and its relationship to physiological risk factors: the GAIT study. *Am. J. Hum. Genet.* **67,** 1452–1459.
192. Souto, J. C., Almasy, L., Borrell, M., Garí, M., Martínez, E., Mateo, J., Stone, W., Blangero, J., and Fontcuberta, J. (2000). Genetic determinants of hemostasis phenotypes in Spanish families. *Circulation* **101,** 1546–1551.
193. Soria, J. M., Almasy, L., Souto, J. C., Bacq, D., Buil, A., Faure, A., Martinez-Marchan, E., Mateo, J., Borrell, M., Stone, W., Lathrop, M., Fontcuberta, J., and Blangero, J. (2002). A quantitative-trait locus in the human factor XII gene influences both plasma factor XII levels and susceptibility to thrombotic disease. *Am. J. Hum. Genet.* **70,** 567–574.
194. Yoshitake, S., Schach, B. G., Foster, D. C., Davie, E. W., and Kurachi, K. (1985). Nucleotide sequence of the gene for human factor IX (antihemophilic factor B). *Biochemistry* **24,** 3736–3750.
195. Walsh, P. N., Bradford, H., Sinha, D., Piperno J. R., and Tusznski, G. P. (1984). Kinetics of the factor XIa catalyzed activation of human blood coagulation factor IX. *J. Clin. Invest.* **73,** 1392–1399.
196. Jesty, J., and Morrison, S. A. (1983). The activation of factor IX by tissue factor-factor VII in a bovine plasma system lacking factor X. *Thromb. Res.* **32,** 171–181.
197. Osterud, B., and Rapaport, S. I. (1977). Activation of factor IX by the reaction product of tissue factor and factor VII: additional pathway for initiating blood coagulation. *Proc. Natl. Acad. Sci. USA* **74,** 5260–5364.
198. Weiss, H. J., Sussman, I. I., and Hoyer, L. W. (1977). Stabilization of factor VIII in plasma by the von Willebrand factor. Studies on posttransfusion and dissociated factor VIII and in patients with von Willebrand's disease. *J. Clin. Invest.* **60,** 390–404.
199. Tuddenham, E. G., Lane, R. S., Rotblat, F., Johnson, A. J., Snape, T. J., Middleton, S., and Kernoff, P. B. (1982). Response to infusions of polyelectrolyte fractionated human factor VIII concentrate in human haemophilia A and von Willebrand's disease. *Br. J. Haematol.* **52,** 259–267.
200. Saenko, E. L., Shima, M., and Sarafanov, A. G. (1999). Role of activation of the coagulation factor VIII in interaction with vWf, phospholipid, and functioning within the factor Xase complex. *Trends Cardiovasc. Med.* **9,** 185–192.
201. Triemstra, M., Rosendaal, F. R., Smit, C., Van der Ploeg, H. M., and Briet, E. (1995). Mortality in patients with hemophilia. Changes in a Dutch population from 1986 to 1992 and 1973 to 1986. *Ann. Intern. Med.* **123,** 823–827.
202. Bilora, F., Boccioletti, V., Zanon, E., Petrobelli, F., and Girolami, A. (2001). Hemophilia A, von Willebrand disease, and atherosclerosis of abdominal aorta and leg arteries: factor VIII and von Willebrand factor defects seem to protect abdominal aorta and leg arteries from atherosclerosis. *Clin. Appl. Thromb. Hemost.* **7,** 311–313.
203. Franchini, M. (2004). Thrombotic complications in patients with hereditary bleeding disorders. *Thromb. Haemost.* **92,** 298–304.
204. Wells, P. S., Langlois, N. J., Webster, M. A., Jaffey, J., and Anderson, J. A. (2005). Elevated factor VIII is a risk factor for idiopathic venous thromboembolism in Canada—Is it necessary to define a new upper reference range for factor VIII? *Thromb. Haemost.* **93,** 842–846.
205. Folsom, A. R., Wu, K. K., Shahar, E., and Davis, C. E. (1993). Association of hemostatic variables with prevalent cardiovascular disease and asymptomatic carotid artery atherosclerosis. The Atherosclerosis Risk in Communities (ARIC) Study Investigators. *Arterioscler. Thromb.* **13,** 1829–1836.
206. Bank, I., Libourel, E. J., Middeldorp, S., Hamulyak, K., van Pampus, E. C., Koopman, M. M., Prins, M. H., van der Meer, J., and Buller, H. R. (2005). Elevated levels of FVIII:C within families are associated with an increased risk for venous and arterial thrombosis. *J. Thromb. Haemost.* **3,** 79–84.
207. Schambeck, C. M., Grossmann, R., Zonnur, S., Berger, M., Teuchert, K., Spahn, A., and Walter, U. (2004). High factor VIII (FVIII) levels in venous thromboembolism: role of unbound FVIII. *Thromb. Haemost.* **92,** 42–61.
208. Kawasaki, T., Kaida, T., Arnout, J., Vermylen, J., and Hoylaerts, M. F. (1999). A new animal model of thrombophilia confirms that high plasma factor VIII levels are thrombogenic. *Thromb. Haemost.* **81,** 306–311.
209. Singh, I., Smith, A., Vanzieleghem, B., Collen, D., Burnand, K., Saint-Remy, J. M., and Jacquemin, M. (2002). Antithrombotic effects of controlled inhibition of factor VIII with a partially inhibitory human monoclonal antibody in a murine vena cava thrombosis model. *Blood* **99,** 3235–3240.
210. Kane, W. H., Ichinose, A., Hagen, F. S., and Davie, E. W. (1987). Cloning of cDNAs coding for the heavy chain region and connecting region of human factor V, a blood coagulation factor with four types of internal repeats. *Biochemistry* **26,** 6508–6514.
211. Jenny, R. J., Pittman, D. D., Toole, J. T., Kriz, R. W., Aldape, R. A., Hewick, M. H., Kaufman, R. J., and Mann, K. G. (1987). Complete cDNA and derived amino acid sequence of human factor V. *Proc. Natl. Acad. Sci. USA* **84,** 4846–4850.
212. Monkovic, D., and Tracy, P. (1990). Activation of human factor V by factor Xa and thrombin. *Biochemistry* **29,** 1118–1128.

213. Suzuki, K., Dahlbäck, B., and Stenflo, J. (1982). Thrombin-catalyzed activation of human coagulation factor V. *J. Biol. Chem.* **257,** 6556–6564.
214. Nesheim, M. E., Taswell, J. B., and Mann, K. G. (1979). The contribution of bovine factor V and factor Va to the activity of the prothrombinase. *J. Biol. Chem.* **254,** 10952–10962.
215. van Wijk, R., Nieuwenhuis, K., van den Berg, M., Huizinga, E. G., van der Meijden, B. B., Kraaijenhagen, R. J., and van Solinge, W. W. (2001). Five novel mutations in the gene for human blood coagulation factor V associated with type I factor V deficiency. *Blood* **98,** 358–367.
216. Nicolaes, G. A., and Dahlback, B. (2002). Factor V and thrombotic disease: description of a janus-faced protein. *Arterioscler. Thromb. Vasc. Biol.* **22,** 530–538.
217. Dahlbäck, B. (1997). Resistance to activated protein C caused by the R(506)Q mutation in the gene for factor V is a common risk factor for venous thrombosis. *J. Intern. Med.* **242,** 1–8.
218. Kim, R. J., and Becker, R. C. (2003). Association between factor V Leiden, prothrombin G20210A, and methylenetetrahydrofolate reductase C677T mutations and events of the arterial circulatory system: a meta-analysis of published studies. *Am. Heart. J.* **146,** 948–957.
219. Barnes, C., and Deveber, G. (2006). Prothrombotic abnormalities in childhood ischaemic stroke. *Thromb. Res.* **118,** 67–74.
220. Juul, K., Tybjaerg-Hansen, A., Steffensen, R., Kofoed, S., Jensen, G., and Nordestgaard, B. G. (2002). Factor V Leiden: The Copenhagen City Heart Study and 2 meta-analyses. *Blood* **100,** 3–10.
221. Ridker, P. M., Hennekens, C. H., Lindpaintner, K., Stampfer, M. J., Eisenberg, P. R., and Miletich, J. P. (1995). Mutation in the gene coding for coagulation factor V and the risk of myocardial infarction, stroke, and venous thrombosis in apparently healthy men. *N. Engl. J. Med.* **332,** 912–917.
222. Williamson, D., Brown, K., Luddington, R., Baglin, C., Baglin, T. (1998). Factor V Cambridge: a new mutation (Arg306Thr) associated with resistance to activated protein C. *Blood* **91,** 1140–1144.
223. Chan, W. P., Lee, C. K., Kwong, Y. L., Lam, C. K., and Liang, R. (1998). A novel mutation of Arg306 of factor V gene in Hong Kong Chinese. *Blood* **91,** 1135–1139.
224. Faioni, E. M., Franchi, F., Bucciarelli, P., Margaglione, M., De Stefano, V., Castaman, G., Finazzi, G., and Mannucci, P. M. (1999). Coinheritance of the HR2 haplotype in the factor V gene confers an increased risk of venous thromboembolism to carriers of factor V R506Q (factor V Leiden). *Blood* **94,** 3062–3066.
225. Luddington, R., Jackson, A., Pannerselvam, S., Brown, K., and Baglin, T. (2000). The factor V R2 allele: risk of venous thromboembolism, factor V levels and resistance to activated protein C. *Thromb. Haemost.* **83,** 204–208.
226. Hoekema, L., Castoldi, E., Tans, G., Girelli, D., Gemmati, D., Bernardi, F., and Rosing, J. (2001). Functional properties of factor V and factor Va encoded by the R2-gene. *Thromb. Haemost.* **85,** 75–81.
227. Zhang, B., McGee, B., Yamaoka, J. S., Guglielmone, H., Downes, K. A., Minoldo, S., Jarchum, G., Peyvandi, F., de Bosch, N. B., Ruiz-Saez, A., Chatelain, B., Olpinski, M., Buckenstedt, P., Sperl, W., Kauman, R. J., Nichols, W. C., Egd, T., and Ginsburg, D. (2006). Combined deficiency of factor V and factor VIII is due to mutations in either LMAN1 or MCFD2. *Blood* **107,** 1903–1907.
228. Zhang, B., Cunningham, M. A., Nichols, W. C., Bernat, J. A., Seligsohn, U., Pipe, S. W., McVey, J. H., Schulte-Overberg, U., de Bosch, N. B., Ruiz-Saez, A., White, G. C., Tuddenham, E. G., Kaufman, R. J., and Ginsburg, D. (2003). Bleeding due to disruption of a cargo-specific ER-to-Golgi transport complex. *Nat. Genet.* **34,** 220–225.
229. Uprichard, J., and Perry, D. J. (2002). Factor X deficiency. *Blood Rev.* 16, 97–110. Leytus, S. P., Foster, D. C., Kurachi, K., and Davie, E. W. (1986). Gene for human factor X: a blood coagulation factor whose gene organization is essentially identical with that of factor IX and protein C. *Biochemistry* **25,** 5098–5102.
230. Hertzberg, M. (1994). Biochemistry of factor X. *Blood Rev.* **8,** 56–62.
231. Cooper, D. N., Millar, D. S., Wacey, A., Pemberton, S., and Tuddenham, E. G. (1997). Inherited factor X deficiency: molecular genetics and pathophysiology. *Thromb. Haemost.* **78,** 161–172.
232. de Visser, M. C., Poort, S. R., Vos, H. L., Rosendaal, F. R., and Bertina, R. M. (2001). Factor X levels, polymorphisms in the promoter region of factor X, and the risk of venous thrombosis. *Thromb. Haemost.* **85,** 1011–1017.
233. Kant, J. A., Fornace, A. J. Jr., Saxe, D., Simon, M. I., McBride, O. W., and Crabtree, G. R. (1985). Evolution and organization of the fibrinogen locus on chromosome 4: gene duplication accompanied by transposition and inversion. *Proc. Natl. Acad. Sci. USA* **82,** 2344–2348.
234. Tuddenham, E. G. D., and Cooper, D. N. (1994). "The Molecular Genetics of Haemostasis And Its Inherited Disorders." Oxford University Press, Oxford.
235. Meade, T. W., Mellows, S., Brozovic, M., Miller, G. J., Chakrabarti, R. R., North, W. R., Haines, A. P., Stirling, Y., Imeson, J. D., and Thompson, S. G. (1986). Haemostatic function and ischaemic heart disease: principal results of the Northwick Park Heart Study. *Lancet* **2,** 533–537.
236. Heinrich, J., Balleisen, L., Schulte, H., Assmann, G., and van de Loo, J. (1994). Fibrinogen and factor VII in the prediction of coronary risk: results from the PROCAM study in healthy men. *Arterioscler. Thromb.* **14,** 54–59.
237. Scarabin, P. Y., Arveiler, D., Amouyel, P., Dos Santos, C., Evans, A., Luc, G., Ferrieres, J., and Juhan-Vague, I. (2003). Plasma fibrinogen explains much of the difference in risk of coronary heart disease between France and Northern Ireland. The PRIME study. *Atherosclerosis* **166,** 103–109.
238. Folsom, A. R. (2001). Hemostatic risk factors for atherothrombotic disease: an epidemiologic view. *Thromb. Haemost.* **86,** 366–373.
239. Collet, J. P., Soria, J., Mirshahi, M., Hirsch, M., Dagonnet, F. B., Caen, J., and Soria, C. (1993). Dusart syndrome: a new concept of the relationship between fibrin clot architecture and fibrin clot degradability, hypofibrinolysis related to an abnormal clot structure. *Blood* **82,** 2462–2469.
240. Sakkinen, P. A., Wahl, P., Cushman, M., Lewis, M. R., and Tracy, R. P. (2000). Clustering of procoagulation, inflammation, and fibrinolysis variables with metabolic factors in insulin resistance syndrome. *Am. J. Epidemiol.* **152,** 897–907.

241. Brown, E. T., and Fuller, G. M. (1998). Detection of a complex that associates with the Bß fibrinogen G-455-A polymorphism. *Blood* **92,** 3286–3293.
242. van 't Hooft, F. M., von Bahr, S. J., Silveira, A., Iliadou, A., Eriksson, P., and Hamsten, A. (1999). Two common, functional polymorphisms in the promoter region of the ß-fibrinogen gene contribute to regulation of plasma fibrinogen concentration. *Arterioscler. Thromb. Vasc. Biol.* **19,** 3063–3070.
243. Behague, I., Poirier, O., Nicaud, V., Evans, A., Arveiler, D., Luc, G., Cambou, J. P., Scarabin, P. Y., Bara, L., Green, F., and Cambien, F. (1996). ß-Fibrinogen gene polymorphisms are associated with plasma fibrinogen and coronary artery disease in patients with myocardial infarction: The ECTIM Study. Etude Cas-Temoins sur l'Infarctus du Myocarde. *Circulation* **93,** 440–449.
244. Endler, G., and Mannhalter, C. (2003). Polymorphisms in coagulation factor genes and their impact on arterial and venous thrombosis. *Clin. Chim. Acta* **330,** 31–55.
245. Boekholdt, S. M., Bijsterveld, N. R., Moons, A. H., Levi, M., Buller, H. R., and Peters, R. J. (2001). Genetic variation in coagulation and fibrinolytic proteins and their relation with acute myocardial infarction: a systematic review. *Circulation* **104,** 3063–3068.
246. de Maat, M. P., Kastelein, J. J., Jukema, J. W., Zwinderman, A. H., Jansen, H., Groenemeier, B., Bruschke, A. V., and Kluft, C. (1998). -455G/A polymorphism of the ß-fibrinogen gene is associated with the progression of coronary atherosclerosis in symptomatic men: proposed role for an acute-phase reaction pattern of fibrinogen. REGRESS group. *Arterioscler. Thromb. Vasc. Biol.* **18,** 265–271.
247. Martiskainen, M., Pohjasvaara, T., Mikkelsson, J., Mantyla, R., Kunnas, T., Laippala, P, Ilveskoski, E., Kaste, M., Karhunen, P. J., and Erkinjuntti, T. (2003). Fibrinogen gene promoter -455 A allele as a risk factor for lacunar stroke. *Stroke* **34,** 886–891.
248. Baumann, R. E., and Henschen, A. H. (1993). Human fibrinogen polymorphic site analysis by restriction endonuclease digestion and allele-specific polymerase chain reaction amplification: identification of polymorphisms at positions A 312 and Bß 448. *Blood* **82,** 2117–2124.
249. Standeven, K. F., Grant, P. J., Carter, A. M., Scheiner, T., Weisel, J. W., and Ariens, R. A. (2003). Functional analysis of the fibrinogen α Thr312Ala polymorphism: effects on fibrin structure and function. *Circulation* **107,** 2326–2330.
250. Carter, A. M., Catto, A. J., and Grant, P. J. (1999). Association of the α-fibrinogen Thr312Ala polymorphism with poststroke mortality in subjects with atrial fibrillation. *Circulation* **99,** 2423–246.
251. Curran, J. M., Evans, A., Arveiler, D., Luc, G., Ruidavets, J. B., Humphries, S. E., and Green, F. R. (1998). The α-fibrinogen T/A312 polymorphism in the ECTIM study. *Thromb. Haemost.* **79,** 1057–1058.
252. Carter, A. M., Catto, A. J., Kohler, H. P., Ariens, R. A., Stickland, M. H., and Grant, P. J. (2000). α-fibrinogen Thr312Ala polymorphism and venous thromboembolism. *Blood* **96,** 1177–1179.
253. Royle, N. J., Irwin, D. M., Koschinsky, M. L., MacGillivray, R. T., and Hamerton, J. L. (1987). Human genes encoding prothrombin and ceruloplasmin map to 11p11-q12 and 3q21-24, respectively. *Somat. Cell Mol. Genet.* **13,** 285–292.
254. Degen, S. J., and Davie, E. W. (1987). Nucleotide sequence of the gene for human prothrombin. *Biochemistry* **26,** 6165–6177.
255. Sun, W. Y., Burkart, M. C., Holahan, J. R., and Degen, S. J. (2000). Prothrombin San Antonio: A single amino acid substitution at a factor Xa activation site (Arg320 to His) results in dysprothrombinemia. *Blood* **95,** 711–714.
256. Rosendaal, F. R., Siscovick, D. S., Schwartz, S. M., Psaty, B. M., Raghunathan, T. E., and Vos, H. L. (1997). A common prothrombin variant (20210 G to A) increases the risk of myocardial infarction in young women. *Blood* **90,** 1747–1750.
257. Poort, S. R., Rosendaal, F. R., Reitsma, P. H., and Bertina, R. M. (1996). A common genetic variation in the 3′-untranslated region of the prothrombin gene is associated with elevated plasma prothrombin levels and an increase in venous thrombosis. *Blood* **88,** 3698–3703.
258. Ceelie, H., Spaargaren-van Riel, C. C., Bertina, R. M., and Vos, H. L. (2004). G20210A is a functional mutation in the prothrombin gene effect on protein levels and 3′-end formation. *J. Thromb. Haemost.* **2,** 119–127.
259. Gelfi, C., Vigano, A., Ripamonti, M., Wait, R., Begum, S., Biguzzi, E., Castaman, G., and Faioni, E. M. (2004). A proteomic analysis of changes in prothrombin and plasma proteins associated with the G20210A mutation. *Proteomics* **4,** 2151–2159.
260. Schrijver, I., Lenzi, T. J., Jones, C. D., Lay, M. J., Druzin, M. L., and Zehnder, J. L. (2003). Prothrombin gene variants in non-Caucasians with fetal loss and intrauterine growth retardation. *J. Mol. Diagn.* **5,** 250–253.
261. Wolberg, A. S., Monroe, D. M., Roberts, H. R., and Hoffman, M. (2003). Elevated prothrombin results in clots with an altered fiber structure: a possible mechanism of the increased thrombotic risk. *Blood* **101,** 3008–3013.
262. Minami, T., Sugiyama, A., Wu, S. Q., Abid, R., Kodama, T., and Aird, W. C. (2004). Thrombin and phenotypic modulation of the endothelium. *Arterioscler. Thromb. Vasc. Biol.* **24,** 41–53.
263. Tracy, R. P. (2003). Thrombin, inflammation, and cardiovascular disease: an epidemiologic perspective. *Chest* **124,** 49S–57S.
264. Foster, D., and Davie, E. W. (1984). Characterization of a cDNA coding for human protein C. *Proc. Natl. Acad. Sci. USA* **81,** 4766–4770.
265. Foster, D. C., Yoshitake, S., and Davie, E. W. (1985). The nucleotide sequence of the gene for human protein C. *Proc. Natl. Acad. Sci. USA* **82,** 4673–4677.
266. Rocchi, M., Roncuzzi, L., Santamaria, R., Archidiacono, N., Dente, L., and Romeo, G. (1986). Mapping through somatic cell hybrids and cDNA probes of protein C to chromosome 2, factor X to chromosome 13, and α 1-acid glycoprotein to chromosome 9. *Hum. Genet.* **74,** 30–33.
267. Romeo, G., Hassan, H. J., Staempfli, S., Roncuzzi, L., Cianetti, L., Leonardi, A., Vicente, V., Mannucci, P. M., Bertina, R., Peschle, C., and Cortese, R. (1987). Hereditary thrombophilia: identification of nonsense and missense mutations in the protein C gene. *Proc. Natl. Acad. Sci. USA* **84,** 2829–2832.
268. Mathias, M., Khair, K., Burgess, C., and Liesner, R. (2004). Subcutaneous administration of protein C concentrate. *Pediatr. Hematol. Oncol.* **21,** 551–556.

269. Wautrecht, J. C. (2005). Venous thromboembolic disease: which coagulation screening, for whom, when? *Rev. Med. Brux.* **26,** S315–S319.
270. Grundy, C. B., Schulman, S., Tengborn, L., Kakkar, V. V., and Cooper, D. N. (1992). Two different missense mutations at Arg 178 of the protein C (PROC) gene causing recurrent venous thrombosis. *Hum. Genet.* **89,** 685–686.
271. Brenner, B., Zivelin, A., Lanir, N., Greengard, J. S., Griffin, J. H., and Seligsohn, U. (1996). Venous thromboembolism associated with double heterozygosity for R506Q mutation of factor V and for T298M mutation of protein C in a large family of a previously described homozygous protein C-deficient newborn with massive thrombosis. *Blood* **88,** 877–880.
272. Wu, K. K. (1997). Genetic markers: genes involved in thrombosis. *J. Cardiovasc. Risk* **4,** 347–352.
273. Martinelli, I., Mannucci, P. M., De Stefano, V., Taioli, E., Rossi, V., Crosti, F., Paciaroni, K., Leone, G., and Faioni, E. M. (1998). Different risks of thrombosis in four coagulation defects associated with inherited thrombophilia: a study of 150 families. *Blood* **92,** 2353–2358.
274. Sakata, T., Kario, K., Katayama, Y., Matsuyama, T., Kato, H., and Miyata, T. (2000). Studies on congenital protein C deficiency in Japanese: prevalence, genetic analysis, and relevance to the onset of arterial occlusive diseases. *Semin. Thromb. Hemost.* **26,** 11–16.
275. Bovill, E. G., Tomczak, J. A., Grant, B., Bhushan, F., Pillemer, E., Rainville, I. R., and Long, G. L. (1992). Protein C Vermont: symptomatic type II protein C deficiency associated with two GLA domain mutations. *Blood* **79,** 1456–1465.
276. Takahashi, T., Shinohara, K., Nawata, R., Wakiyama, M., and Hamasaki, N. (1999). A novel mutation of the protein C gene with a frameshift deletion of 3 base pair F (3380)AGG in exon 6 in type 1 deficiency associated with arterial and venous thrombosis. *Am. J. Hematol.* **62,** 260–261.
277. Giroir, B. P. (2003). Recombinant human activated protein C for the treatment of severe sepsis: is there a role in pediatrics? *Curr. Opin. Pediatr.* **15,** 92–96.
278. Espinosa, R., III, Sadler, J. E., and Le Beau, M. M. (1989). Regional localization of the human thrombomodulin gene to 20p12-cen. *Genomics* **5,** 649–650.
279. Wen, D. Z., Dittman, W. A., Ye, R. D., Deaven, L. L., Majerus, P. W., and Sadler, J. E. (1987). Human thrombomodulin: complete cDNA sequence and chromosome localization of the gene. *Biochemistry* **26,** 4350–4357.
280. Jackman, R. W., Beeler, D. L., Fritze, L., Soff, G., and Rosenberg, R. D. (1987). Human thrombomodulin gene is intron depleted: nucleic acid sequences of the cDNA and gene predict protein structure and suggest sites of regulatory control. *Proc. Natl. Acad. Sci. USA* **84,** 6425–6429.
281. Suzuki, K., Hayashi, T., Nishioka, J., Kosaka, Y., Zushi, M., Honda, G., and Yamamoto, S. (1989). A domain composed of epidermal growth factor-like structures of human thrombomodulin is essential for thrombin binding and for protein C activation. *J. Biol. Chem.* **264,** 4872–4876.
282. Kokame, K., Zheng, X., and Sadler, J. (1998). Activation of thrombin-activatable fibrinolysis inhibitor requires epidermal growth factor-like domain 3 of thrombomodulin and is inhibited competitively by protein C. *J. Biol. Chem.* **273,** 12135–12139.
283. Boffa, M. C., and Karmochkine, M. (1998). Thrombomodulin: an overview and potential implications in vascular disorders. *Lupus* **7,** S120–S125.
284. Nan, B., Lin, P., Lumsden, A. B., Yao, Q., and Chen, C. (2005). Effects of TNF-alpha and curcumin on the expression of thrombomodulin and endothelial protein C receptor in human endothelial cells. *Thromb. Res.* **115,** 417–426.
285. Salomaa, V., Matei, C., Aleksic, N., Sansores-Garcia, L., Folsom, A. R., Juneja, H., Chambless, L. E., and Wu, K. K. (1999). Soluble thrombomodulin as a predictor of incident coronary heart disease and symptomless carotid artery atherosclerosis in the Atherosclerosis Risk in Communities (ARIC) Study: A case-cohort study. *Lancet* **353,** 1729–1734.
286. Norlund, L., Holm, J., Zoller, B., and Ohlin, A. K. (1997). A common thrombomodulin amino acid dimorphism is associated with myocardial infarction. *Thromb. Haemost.* **77,** 248–251.
287. Ohlin, A. K., Holm, J., and Hillarp, A. (2004). Genetic variation in the human thrombomodulin promoter locus and prognosis after acute coronary syndrome. *Thromb. Res.* **113,** 319–326.
288. Doggen, C. J., Kunz, G., Rosendaal, F. R., Lane, D. A., Vos, H. L., Stubbs, P. J., Manger Cats, V., and Ireland, H. (1998). A mutation in the thrombomodulin gene, 127G to A coding for Ala25Thr, and the risk of myocardial infarction in men. *Thromb. Haemost.* **80,** 743–748.
289. Heit, J. A., Petterson, T. M., Owen, W. G., Burke, J. P., De Andrade, M., and Melton, L. J., 3rd. (2005). Thrombomodulin gene polymorphisms or haplotypes as potential risk factors for venous thromboembolism: a population-based case-control study. *J. Thromb. Haemost.* **3,** 710–717.
290. Ireland, H., Kunz, G., Kyriakoulis, K., Stubbs, P. J., and Lane, D. A. (1997). Thrombomodulin gene mutations associated with myocardial infarction. *Circulation* **96,** 15–18.
291. Li, Y. H., Chen, J. H., Wu, H. L., Shi, G. Y., Huang, H. C., Chao, T. H., Tsai, W. C., Tsai, L. M., Guo, H. R., Wu, W. S., and Chen, Z. C. (2000). G-33A mutation in the promoter region of thrombomodulin gene and its association with coronary artery disease and plasma soluble thrombomodulin levels. *Am. J. Cardiol.* **85,** 8–12.
292. Chao, T. H., Li, Y. H., Chen, J. H., Wu, H. L., Shi, G. Y., Tsai, W. C., Chen, P. S. and, Liu, P. Y. (2004). Relation of thrombomodulin gene polymorphisms to acute myocardial infarction in patients ≤50 years of age. *Am. J. Cardiol.* **93,** 204–207.
293. Vanhoof, G., Wauters, J., Schatteman, K., Hendriks, D., Goossens, F., Bossuyt, P., and Scharpe, S. (1996). The gene for human carboxypeptidase U (CPU)—a proposed novel regulator of plasminogen activation—maps to 13q14.11. *Genomics* 38, 454–455.
294. Boffa, M. B., Reid, T. S., Joo, E., Nesheim, M. E., and Koschinsky, M. L. (1999). Characterization of the gene encoding human TAFI (thrombin-activatable fibrinolysis inhibitor plasma procarboxypeptidase B). *Biochemistry* **38,** 6547–6558.
295. Bouma, B. N., and Meijers, J. C. (2003). Thrombin-activatable fibrinolysis inhibitor (TAFI, plasma procarboxypeptidase B, procarboxypeptidase R, procarboxypeptidase U). *J. Thromb. Haemost.* **1,** 1566–1574.

296. Boffa, M. B., Nesheim, M. E., and Koschinsky, M. L. (2001). Thrombin activatable fibrinolysis inhibitor (TAFI): molecular genetics of an emerging potential risk factor for thrombotic disorders. Curr. *Drug Targets Cardiovasc. Haematol. Disord.* **1,** 59–74.
297. Henry, M., Aubert, H., Morange, P. E., Nanni, I., Alessi, M. C., Tiret, L., and Juhan-Vague, I. (2001). Identification of polymorphisms in the promoter and the 3′ region of the TAFI gene: evidence that plasma TAFI antigen levels are strongly genetically controlled. *Blood* **97,** 2053–2058.
298. Schneider, M., Boffa, M., Stewart, R., Rahman, M., Koschinsky, M., and Nesheim, M. (2002). Two naturally occurring variants of TAFI (Thr-325 and Ile-325) differ substantially with respect to thermal stability and antifibrinolytic activity of the enzyme. *J. Biol. Chem.* **277,** 1021–1030.
299. Zorio, E., Castello, R., Falco, C., Espana, F., Osa, A., Almenar, L., Aznar, J., and Estelles, A. (2003). Thrombin-activatable fibrinolysis inhibitor in young patients with myocardial infarction and its relationship with the fibrinolytic function and the protein C system. *Br. J. Haematol.* **122,** 958–965.
300. Donmez, A., Aksu, K., Celik, H. A., Keser, G., Cagirgan, S., Omay, S. B., Inal, V., Aydin, H. H., Tombuloglu, M., and Doganavsargil, E. (2005). Thrombin activatable fibrinolysis inhibitor in Behçet's disease. *Thromb. Res.* **115,** 287–292.
301. Morange, P. E., Tregouet, D. A., Frere, C., Luc, G., Arveiler, D., Ferrieres, J., Amouyel, P., Evans, A., Ducimetiere, P., Cambien, F., Tiret, L., Juhan-Vague, I., and The Prime Study Group. (2005). TAFI gene haplotypes, TAFI plasma levels and future risk of coronary heart disease: the PRIME Study. *J. Thromb. Haemost.* **3,** 1503–1510.
302. Kostka, H., Kuhlisch, E., Schellong, S., and Siegert, G. (2003). Polymorphisms in the TAFI gene and the risk of venous thrombosis. *Clin. Lab.* **49,** 645–647.
303. Bereczky, Z., Katona, E., and Muszbek, L. (2003). Fibrin stabilization (factor XIII), fibrin structure and thrombosis. *Pathophysiol. Haemost. Thromb.* **33,** 430–437.
304. Greenberg, C. S., Birckbichler, P. J., and Rice, R. H. (1991). Transglutaminases: multifunctional cross-linking enzymes that stabilize tissues. *FASEB J.* **5,** 3071–3077.
305. Panetti, T. S., Kudryk, B. J., and Mosher, D. F. (1999). Interaction of recombinant procollagen and properdin modules of thrombospondin-1 with heparin and fibrinogen/fibrin. *J. Biol. Chem.* **274,** 430–437.
306. Board, P. G., Webb, G. C., McKee, J., and Ichinose, A. (1988). Localization of the coagulation factor XIII A subunit gene (F13A) to chromosome bands 6p24–p25. *Cytogenet. Cell Genet.* **48,** 25–27.
307. Ichinose, A., and Davie, E. W. (1988). Characterization of the gene for the a subunit of human factor XIII (plasma transglutaminase), a blood coagulation factor. *Proc. Natl. Acad. Sci. USA* **85,** 5829–5833.
308. Ichinose, A., and Davie, E. W. (1988). Primary structure of human coagulation factor XIII. *Adv. Exp. Med. Biol.* **231,** 15–27.
309. Bottenus, R. E., Ichinose, A., and Davie, E. W. (1990). Nucleotide sequence of the gene for the b subunit of human factor XIII. *Biochemistry* **29,** 11195–11209.
310. Webb, G. C., Coggan, M., Ichinose, A., and Board, P. G. (1989). Localization of the coagulation factor XIII B subunit gene (F13B) to chromosome bands 1q31-32.1 and restriction fragment length polymorphism at the locus. *Hum. Genet.* **81,** 157–160.
311. Ichinose, A. (2001). Physiopathology and regulation of factor XIII. *Thromb. Haemost.* **86,** 57–65.
312. Scott, E. M., Ariens, R. A., and Grant, P. (2004). Genetic and environmental determinants of fibrin structure and function: relevance to clinical disease. *Arterioscler. Thromb. Vasc. Biol.* **24,** 1558–1566.
313. Ariens, R. A., Lai, T. S., Weisel, J. W., Greenberg, C. S., and Grant, P. J. (2002). Role of factor XIII in fibrin clot formation and effects of genetic polymorphisms. *Blood* **100,** 743–754.
314. Van Hylckama Vlieg, A., Komanasin, N., Ariens, R. A., Poort, S. R., Grant, P. J., Bertina, R. M., and Rosendaal, F. R. (2002). Factor XIII Val34Leu polymorphism, factor XIII antigen levels and activity and the risk of deep venous thrombosis. *Br. J. Haematol.* **119,** 169–175.
315. Franco, R. F., Reitsma, P. H., Lourenco, D., Maffei, F. H., Morelli, V., Tavella, M. H., Araujo, A. G., Piccinato, C. E., and Zago, M. A. (1999). Factor XIII Val34Leu is a genetic factor involved in the etiology of venous thrombosis. *Thromb. Haemost.* **81,** 676–679.
316. Vossen, C. Y., and Rosendaal, F. R. (2005). The protective effect of the factor XIII Val34Leu mutation on the risk of deep venous thrombosis is dependent on the fibrinogen level. *J. Thromb. Haemost.* **3,** 1102–1103.
317. Lim, B. C., Ariens, R. A., Carter, A. M., Weisel, J. W., and Grant, P. J. (2003). Genetic regulation of fibrin structure and function: complex gene-environment interactions may modulate vascular risk. *Lancet* **361,** 1424–1431.
318. Franco, R. F., Pazin-Filho, A., Tavella, M. H., Simoes, M. V., Marin-Neto, J. A., and Zago, M. A. (2000). Factor XIII val34leu and the risk of myocardial infarction. *Haematologica* **85,** 67–71.
319. Dunn, E. J., Ariens, R. A., de Lange, M., Snieder, H., Turney, J. H., Spector, T. D., and Grant, P. J. (2004). Genetics of fibrin clot structure: a twin study. *Blood* **103,** 1735–1740.
320. Forsgren, M., Raden, B., Israelsson, M., Larsson, K., and Heden, L. O. (1987). Molecular cloning and characterization of a full-length cDNA clone for human plasminogen. *FEBS Lett.* **213,** 254–260.
321. Petersen, T. E., Martzen, M. R., Ichinose, A., and Davie, E. W. (1990). Characterization of the gene for human plasminogen, a key proenzyme in the fibrinolytic system. *J. Biol. Chem.* **265,** 6104–6111.
322. Murray, J. C., Buetow, K. H., Donovan, M., Hornung, S., Motulsky, A. G., Disteche, C., Dyer, K., Swisshelm, K., Anderson, J., Giblett, E., Sadler, E., Eddy, R., and Shows, T. B. (1987). Linkage disequilibrium of plasminogen polymorphisms and assignment of the gene to human chromosome 6q26-6q27. *Am. J. Hum. Genet.* **40,** 338–350.
323. Frank, S. L., Klisak, I., Sparkes, R. S., Mohandas, T., Tomlinson, J. E., McLean, J. W., Lawn, R. M., and Lusis, A. J. (1988). The apolipoprotein(a) gene resides on human chromosome 6q26-27, in close proximity to the homologous gene for plasminogen. *Hum. Genet.* **79,** 352–356.
324. Ichinose, A. (1992). Multiple members of the plasminogen-apolipoprotein(a) gene family associated with thrombosis. *Biochemistry* **31,** 3113–3118.

325. Lijnen, H. R. (1996). Pathophysiology of the plasminogen/plasmin system. *Int. J. Clin. Lab. Res.* **26,** 1–6.
326. Kida, M., H-Kawabata, M., Yamazaki, T., and Ichinose, A. (1998). Presence of two plasminogen alleles in normal populations. *Thromb. Haemost.* **79,** 150–154.
327. Ploplis, V. A., Carmeliet, P., Vazirzadeh, S., Van Vlaenderen, I., Moons, L., Plow, E. F., and Collen, D. (1995). Effects of disruption of the plasminogen gene on thrombosis, growth, and health in mice. *Circulation* **92,** 2585–2593.
328. Tachias, K., and Madison, E. L. (1996). Converting tissue-type plasminogen activator into a zymogen. *J. Biol. Chem.* **271,** 28749–28752.
329. van Zonneveld, A. J., Veerman, H., MacDonald, M. E., van Mourik, J. A., and Pannekoek, H. (1986). Structure and function of human tissue-type plasminogen activator (t-PA). *J. Cell Biochem.* **32,** 169–178.
330. Pennica, D., Holmes, W. E., Kohr, W. J., Harkins, R. N., Vehar, G. A., Ward, C. A., Bennett, W. F., Yelverton, E., Seeburg, P. H., Heyneker, H. L., Goeddel, D. V., and Collen, D. (1983). Cloning and expression of human tissue-type plasminogen activator cDNA in *E. coli. Nature* **301,** 214–221.
331. Bulens, F., Merchiers, P., Ibanez-Tallon, I., De Vriese, A., Nelles, L., Claessens, F., Belayew, A., and Collen, D. (1997). Identification of a multihormone responsive enhancer far upstream from the human tissue-type plasminogen activator gene. *J. Biol. Chem.* **272,** 663–671.
332. Juhan-Vague, I., Pyke, S. D., Alessi, M. C., Jespersen, J., Haverkate, F., and Thompson, S. G. (1996). Fibrinolytic factors and the risk of myocardial infarction or sudden death in patients with angina pectoris. ECAT Study Group. European Concerted Action on Thrombosis and Disabilities. *Circulation* **94,** 2057–2063.
333. Ludwig, M., Wohn, K. D., Schleuning, W. D., and Olek, K. (1992). Allelic dimorphism in the human tissue-type plasminogen activator (t-PA) gene as a result of an Alu insertion/deletion event. *Hum. Genet.* **88,** 388–392.
334. van der Bom, J. G., de Knijff, P., Haverkate, F., Bots, M. L., Meijer, P., de Jong, P. T., Hofman, A., Kluft, C., and Grobbee, D. E. (1997). Tissue plasminogen activator and risk of myocardial infarction: the Rotterdam Study. *Circulation* **95,** 2623–2627.
335. Ridker, P. M., Baker, M. T., Hennekens, C. H., Stampfer, M. J., and Vaughan, D. E. (1997). Alu-repeat polymorphism in the gene coding for tissue-type plasminogen activator (t-PA) and risks of myocardial infarction among middle-aged men. *Arterioscler. Thromb. Vasc. Biol.* **17,** 1687–1690.
336. Stein, P. M., Stass, S. A., and Kagan, J. (1993). The human urokinase-plasminogen activator gene (PLAU) is located on chromosome 10q24 centromeric to the HOX11 gene. *Genomics* **16,** 301–302.
337. Kasai, S., Arimura, H., Nishida, M., and Suyama, T. (1985). Primary structure of single-chain pro-urokinase. *J. Biol. Chem.* **260,** 12382–12389.
338. Blasi, F. (1993). Urokinase and urokinase receptor: a paracrine/autocrine system regulating cell migration and invasiveness. *Bioessays* **15,** 105–111.
339. Collen, D. (1987). Molecular mechanism of action of newer thrombolytic agents. *J. Am. Coll. Cardiol.* **10,** 11B–15B.
340. Liu, J. N., Liu, J. X., Liu, B. F., Sun, Z., Zuo, J. L., Zhang, P. X., Zhang, J., Chen, Y. H., and Gurewich, V. (2002). Prourokinase mutant that induces highly effective clot lysis without interfering with hemostasis. *Circ. Res.* **90,** 757–763.
341. Watanabe, R., Wada, H., Miura, Y., Murata, Y., Watanabe, Y., Sakakura, M., Okugawa, Y., Nakasaki, T., Mori, Y., Nishikawa, M., Gabazza, E. C., Shiku, H., and Nobori, T. (2001). Plasma levels of total plasminogen activator inhibitor-I (PAI-I) and tPA/PAI-1 complex in patients with disseminated intravascular coagulation and thrombotic thrombocytopenic purpura. *Clin. Appl. Thromb. Hemost.* 7, 229–233.
342. Dunn, E. J., and Grant, P. J. (2005). Type 2 diabetes: an atherothrombotic syndrome. *Curr. Mol. Med.* **5,** 323–332.
343. Vaughan, D. E. (2005). PAI-1 and atherothrombosis. *J. Thromb. Haemost.* **3,** 1879–1883.
344. Pannekoek, H., Veerman, H., Lambers, H., Diergaarde, P., Verweij, C. L., van Zonneveld, A. J., and van Mourik, J. A. (1986). Endothelial plasminogen activator inhibitor (PAI): a new member of the Serpin gene family. *EMBO J.* **5,** 2539–2544.
345. Ny, T., Sawdey, M., Lawrence, D., Millan, J. L., and Loskutoff, D. J. (1986). Cloning and sequence of a cDNA coding for the human β-migrating endothelial-cell-type plasminogen activator inhibitor. *Proc. Natl. Acad. Sci. USA* **83,** 6776–6780.
346. Ginsburg, D., Zeheb, R., Yang, A. Y., Rafferty, U. M., Andreasen, P. A., Nielsen, L., Dano, K., Lebo, R. V., and Gelehrter, T. D. (1986). cDNA cloning of human plasminogen activator-inhibitor from endothelial cells. *J. Clin. Invest.* **78,** 1673–1680.
347. Follo, M., and Ginsburg, D. (1989). Structure and expression of the human gene encoding plasminogen activator inhibitor, PAI-1. *Gene* **84,** 447–453.
348. Dawson, S. J., Wiman, B., Hamsten, A., Green, F., Humphries, S., and Henney, A. M. (1993). The two allele sequences of a common polymorphism in the promoter of the plasminogen activator inhibitor-1 (PAI-1) gene respond differently to interleukin-1 in HepG2 cells. *J. Biol. Chem.* **268,** 10739–10745.
349. Panahloo, A., Mohamed-Ali, V., Lane, A., Green, F., Humphries, S. E., and Yudkin, J. S. (1995). Determinants of plasminogen activator inhibitor 1 activity in treated NIDDM and its relation to a polymorphism in the plasminogen activator inhibitor 1 gene. *Diabetes* **44,** 37–42.
350. Eriksson, P., Nilsson, L., Karpe, F., and Hamsten, A. (1998). Very-low-density lipoprotein response element in the promoter region of the human plasminogen activator inhibitor-1 gene implicated in the impaired fibrinolysis of hypertriglyceridemia. *Arterioscler. Thromb. Vasc. Biol.* **18,** 20–26.
351. Simmonds, R. E., Hermida, J., Rezende, S. M., and Lane, D. A. (2001). Haemostatic genetic risk factors in arterial thrombosis. *Thromb. Haemost.* **86,** 374–385.
352. Iacoviello, L., Burzotta, F., Di Castelnuovo, A., Zito, F., Marchioli, R., and Donati, M. B. (1998). The 4G/5G polymorphism of PAI-1 promoter gene and the risk of myocardial infarction: a meta-analysis. *Thromb. Haemost.* **80,** 1029–1030.
353. Ehrlich, H. J., Gebbink, R. K., Preissner, K. T., Keijer, J., Esmon, N. L., Mertens, K., and Pannekoek, H. (1991). Thrombin neutralizes plasminogen activator inhibitor 1 (PAI-1) that is complexed with vitronectin in the endothelial cell matrix. *J. Cell Biol.* **115,** 1773–1781.

354. Gebbink, R. K., Reynolds, C. H., Tollefsen, D. M., Mertens, K., and Pannekoek, H. (1993). Specific glycosaminoglycans support the inhibition of thrombin by plasminogen activator inhibitor 1. *Biochemistry* **32,** 1675–1680.
355. Arroyo De Prada, N., Schroeck, F., Sinner, E. K., Muehlenweg, B., Twellmeyer, J., Sperl, S., Wilhelm, O. G., Schmitt, M., and Magdolen, V. (2002). Interaction of plasminogen activator inhibitor type-1 (PAI-1) with vitronectin. *Eur. J. Biochem.* **269,** 184–192.
356. Medcalf, R. L., and Stasinopoulos, S. J. (2005). The undecided serpin. The ins and outs of plasminogen activator inhibitor type 2. *FEBS J.* **272,** 4858–4867.
357. Antalis, T. M., Clark, M. A., Barnes, T., Lehrbach, P. R., Devine, P. L., Schevzov, G., Goss, N. H., Stephens, R. W., and Tolstoshev, P. (1988). Cloning and expression of a cDNA coding for a human monocyte-derived plasminogen activator inhibitor. *Proc. Natl. Acad. Sci. USA* **85,** 985–989.
358. Ogbourne, S. M., and Antalis, T. M. (2001). Characterisation of PAUSE-1, a powerful silencer in the human plasminogen activator inhibitor type 2 gene promoter. *Nucleic Acids Res.* **29,** 3919–3927.
359. Webb, G., Baker, M. S., Nicholl, J., Wang, Y., Woodrow, G., Kruithof, E., and Doe, W. F. (1994). Chromosomal localization of the human urokinase plasminogen activator receptor and plasminogen activator inhibitor type-2 genes: implications in colorectal cancer. *J. Gastroenterol. Hepatol.* **9,** 340–343.
360. Foy, C. A., and Grant, P. J. (1997). PCR-RFLP detection of PAI-2 gene variants: prevalence in ethnic groups and disease relationship in patients undergoing coronary angiography. *Thromb. Haemost.* **77,** 955–958.
361. Buyru, N., Altinisik, J., Gurel, C. B., and Ulutin, T. (2003). PCR-RFLP detection of PAI-2 variants in myocardial infarction. *Clin. Appl. Thromb. Hemost.* **9,** 333–336.
362. Aoki, N. (1990). Molecular genetics of α 2 plasmin inhibitor. *Adv. Exp. Med. Biol.* **281,** 195–200.
363. Holmes, W. E., Nelles, L., Lijnen, H. R., and Collen, D. (1987). Primary structure of human α 2-antiplasmin, a serine protease inhibitor (serpin). *J. Biol. Chem.* **262,** 1659–1664.
364. Lee, K. N., Jackson, K. W., Christiansen, V. J., Chung, K. H., and McKee, P. A. (2004). α2-antiplasmin: potential therapeutic roles in fibrin survival and removal. *Curr. Med. Chem. Cardiovasc. Hematol. Agents* **2,** 303–310.
365. Kato, A., Hirosawa, S., Toyota, S., Nakamura, Y., Nishi, H., Kimura, A., Sasazuki, T., and Aoki, N. (1993). Localization of the human α 2-plasmin inhibitor gene (PLI) to 17p13. *Cytogenet. Cell Genet.* **62,** 190–191.
366. Yoshinaga, H., Nakahara, M., Koyama, T., Shibamiya, A., Nakazawa, F., Miles, L. A., Hirosawa, S., and Aoki, N. (2002). A single thymine nucleotide deletion responsible for congenital deficiency of plasmin inhibitor. *Thromb. Haemost.* **88,** 144–148.
367. Gawaz, M., Langer, H., and May, A. E. (2005). Platelets in inflammation and atherogenesis. *J. Clin. Invest.* **115,** 3378–3384.
368. Matetzky, S., Shenkman, B., Guetta, V., Shechter, M., Bienart, R., Goldenberg, I., Novikov, I., Pres, H., Savion, N., Varon, D., and Hod, H. (2004). Clopidogrel resistance is associated with increased risk of recurrent atherothrombotic events in patients with acute myocardial infarction. *Circulation* **109,** 3171–3175.
369. Becker, R. C. (2001). Markers of platelet activation and thrombin generation. *Cardiovasc. Toxicol.* **1,** 141–145.
370. Walsh, P. N. (2004). Platelet coagulation-protein interactions. *Semin. Thromb. Hemost.* **30,** 461–471.
371. Packham, M. A., and Mustard, J. F. (2005). Platelet aggregation and adenosine diphosphate/adenosine triphosphate receptors: a historical perspective. *Semin. Thromb. Hemost.* **31,** 129–138.
372. Lisman, T., Weeterings, C., and de Groot, P. G. (2005). Platelet aggregation: involvement of thrombin and fibrin(ogen). *Front. Biosci.* **10,** 2504–2517.
373. Danese, S., and Fiocchi, C. (2005). Platelet activation and the CD40/CD40 ligand pathway: mechanisms and implications for human disease. *Crit. Rev. Immunol.* **25,** 103–121.
374. Yip, J., Shen, Y., Berndt, M. C., and Andrews, R. K. (2005). Primary platelet adhesion receptors. *IUBMB Life* **57,** 103–108.
375. Kieffer, N., Guichard, J., Farcet, J. P., Vainchenker, W., and Breton-Gorius, J. (1987). Biosynthesis of major platelet proteins in human blood platelets. *Eur. J. Biochem.* **164,** 189–195.
376. Booyse, F. M., and Rafelson, M. E. Jr. (1968). Studies on human platelets. I. Synthesis of platelet protein in a cell-free system. *Biochim. Biophys. Acta* **166,** 689–697.
377. Weyrich, A. S., Dixon, D. A., Pabla, R., Elstad, M. R., McIntyre, T. M., Prescott, S. M., and Zimmerman, G. A. (1998). Signal-dependent translation of a regulatory protein, Bcl-3, in activated human platelets. *Proc. Natl. Acad. Sci. USA* **95,** 5556–5561.
378. Gnatenko, D. V., Dunn, J. J., McCorkle, S. R., Weissmann, D., Perrotta, P. L., and Bahou, W. F. (2003). Transcript profiling of human platelets using microarray and serial analysis of gene expression. *Blood* **101,** 2285–2293.
379. Maguire, P. B., Wynne, K. J., Harney, D. F., O'Donoghue, N. M., Stephens, G., and Fitzgerald, D. J. (2002). Identification of the phosphotyrosine proteome from thrombin activated platelets. *Proteomics* **2,** 642–648.
380. Maguire, P. B. (2004). Platelet proteomics: identification of potential therapeutic targets. *Pathophysiol. Haemost. Thromb.* **33,** 481–486.
381. Macaulay, I. C., Carr, P., Gusnanto, A., Ouwehand, W. H., Fitzgerald, D., and Watkins, N. A. (2005). Platelet genomics and proteomics in human health and disease. *J. Clin. Invest.* **115,** 3370–3377.
382. McRedmond, J. P., Park, S. D., Reilly, D. F., Coppinger, J. A., Maguire, P. B., Shields, D. C., and Fitzgerald, D. J. (2004). Integration of proteomics and genomics in platelets: a profile of platelet proteins and platelet-specific genes. *Mol. Cell Proteomics* **3,** 133–144.
383. Coppinger, J. A., Cagney, G., Toomey, S, Kislinger, T., Belton, O., McRedmond, J. P., Cahill, D. J., Emili, A., Fitzgerald, D. J., and Maguire, P. B. (2004). Characterization of the proteins released from activated platelets leads to localization of novel platelet proteins in human atherosclerotic lesions. *Blood* **103,** 2096–2104.
384. Andreotti, F., and Becker, R. C. (2005). Atherothrombotic disorders: new insights from hematology. *Circulation* **111,** 1855–1863.

385. Hassan, A. A., and Kroll, M. H. (2005). Acquired disorders of platelet function. *Hematology (Am Soc Hematol Educ Program)* 403–408.
386. Landolfi, R., Rocca, B., and Patrono, C. (1995). Bleeding and thrombosis in myeloproliferative disorders: mechanisms and treatment. *Crit. Rev. Oncol. Haematol.* **20**, 203–222.
387. Poncz, M. (2005). Mechanistic basis of heparin-induced thrombocytopenia. *Semin. Thorac. Cardiovasc. Surg.* **17**, 73–79.
388. Francis, J. L. (2005). Detection and significance of heparin-platelet factor 4 antibodies. *Semin. Hematol.* **42**, S9–S14.
389. Sadler, J. E. (2005). von Willebrand factor: two sides of a coin. *J. Thromb. Haemost.* **3**, 1702–1709.
390. Mazurier, C., Ribba, A. S., Gaucher, C., Meyer, D. (1998). Molecular genetics of von Willebrand disease. *Ann. Genet.* **41**, 34–43.
391. Fressinaud, E., Mazurier, C., and Meyer, D. (2002). Molecular genetics of type 2 von Willebrand disease. *Int. J. Hematol.* **75**, 9–18.
392. Gomez Garcia, E. B., Brouwers, G. J., and Leebeek, F. W. (2002). From gene to disease from mutations in the Von Willebrand factor gene to hemorrhagic diathesis and thrombocytopenia. *Ned. Tijdschr. Geneeskd.* **146**, 1180–1182.
393. Levy, G. G., Nichols, W. C., Lian, E. C., Foroud, T., McClintick, J. N., McGee, B. M., Yang, A. Y., Siemieniak, D. R., Stark, K. R., Gruppo, R., Sarode, R., Shurin, S. B., Chandrasekaran, V., Stabler, S. P., Sabio, H., Bouhassira, E. E., Upshaw, J. D., Jr., Ginsburg, D, and Tsai, H. M. (2001). Mutations in a member of the ADAMTS gene family cause thrombotic thrombocytopenic purpura. *Nature* **413**, 488–494.
394. Tsai, H. M. (2003). Platelet activation and the formation of the platelet plug: deficiency of ADAMTS13 causes thrombotic thrombocytopenic purpura. *Arterioscler. Thromb. Vasc. Biol.* **23**, 388–396.
395. Handin, R. I. (2005). Inherited platelet disorders. *Hematology (Am Soc Hematol Educ Program)* **396**, 96–402.
396. D'Andrea, G., Colaizzo, D., Vecchione, G., Grandone, E., Di Minno, G., and Margaglione, M. (2002). Glanzmann's thrombasthenia: identification of 19 new mutations in 30 patients. *Thromb. Haemost.* **87**, 1034–1042.
397. Bellucci, S., and Caen, J. (2002). Molecular basis of Glanzmann's thrombasthenia and current strategies in treatment. *Blood Rev.* **16**, 193–202.
398. Lanza, F., Stierle, A., Fournier, D., Morales, M., Andre, G., Nurden, A. T., and Cazenave, J. P. (1992). A new variant of Glanzmann's thrombasthenia (Strasbourg I). Platelets with functionally defective glycoprotein IIb-IIIa complexes and a glycoprotein IIIa 214Arg-> 214Trp mutation. *J. Clin. Invest.* **89**, 1995–2004.
399. Lopez, J. A., Andrews, R. K., Afshar-Kharghan, V., and Berndt, M. C. (1998). Bernard-Soulier syndrome. *Blood* **91**, 4397–4418.
400. Kunishima, S., Kamiya, T., and Saito, H. (2002). Genetic abnormalities of Bernard-Soulier syndrome. *Int. J. Hematol.* **76**, 319–327.
401. Liang, H. P., Morel-Kopp, M. C., Clemetson, J. M., Clemetson, K. J., Kekomaki, R., Kroll, H., Michaelides, K., Tuddenham, E. G., Vanhoorelbeke, K., and Ward, C. M. (2005). A common ancestral glycoprotein (GP) 9 1828AG (Asn45Ser) gene mutation occurring in European families from Australia and Northern Europe with Bernard-Soulier Syndrome (BSS). *Thromb. Haemost.* **94**, 599–605.
402. Honda, S., Honda, Y., Bauer, B., Ruan, C., and Kunicki, T. J. (1995). The impact of three-dimensional structure on the expression of PlA alloantigens on human integrin beta3. *Blood* **86**, 234–242.
403. Newman, P. J. (1997). Platelet alloantigens: cardiovascular as well as immunological risk factors? *Lancet* **349**, 370–371.
404. Weiss, E. J., Bray, P. F., Tayback, M., Schulman, S. P., Kickler, T. S., Becker, L. C., Weiss, J. L., Gerstenblith, G., and Goldschmidt-Clermont, P. J. (1996). A polymorphism of a platelet glycoprotein receptor as an inherited risk factor for coronary thrombosis. *N. Engl. J. Med.* **334**, 1090–1094.
405. Carter, A. M., Ossei-Gerning, N., Wilson, I. J., and Grant, P. J. (1997). Association of the platelet Pl(A) polymorphism of glycoprotein IIb/IIIa and the fibrinogen β 448 polymorphism with myocardial infarction and extent of coronary artery disease. *Circulation* **96**, 1424–1431.
406. Wagner, K. R., Giles, W. H., Johnson, C. J., Ou, C. Y., Bray, P. F., Goldschmidt-Clermont, P. J., Croft, J. B., Brown, V. K., Stern, B. J., Feeser, B. R., Buchholz, D. W., Earley, C. J., Macko, R. F., McCarter, R. J., Sloan, M. A., Stolley, P. D., Wityk, R. J., Wozniak, M. A., Price, T. R., and Kittner, S. J. (1998). Platelet glycoprotein receptor IIIa polymorphism P1A2 and ischemic stroke risk: the Stroke Prevention in Young Women Study. *Stroke* **29**, 581–585.
407. Ridker, P. M., Hennekens, C. H., Schmitz, C., Stampfer, M. J., and Lindpaintner, K. (1997). PIA1/A2 polymorphism of platelet glycoprotein IIIa and risks of myocardial infarction, stroke, and venous thrombosis. *Lancet* **349**, 385–388.
408. Herrmann, S. M., Poirier, O., Marques-Vidal, P., Evans, A., Arveiler, D., Luc, G., Emmerich, J., and Cambien, F. (1997). The Leu33/Pro polymorphism (PlA1/PlA2) of the glycoprotein IIIa (GPIIIa) receptor is not related to myocardial infarction in the ECTIM Study. Etude Cas-Temoins de l'Infarctus du Myocarde. *Thromb. Haemost.* **77**, 1179–1181.
409. Ardissino, D., Mannucci, P. M., Merlini, P. A., Duca, F., Fetiveau, R., Tagliabue, L., Tubaro, M., Galvani, M., Ottani, F., Ferrario, M., Corral, J., and Margaglione, M. (1999). Prothrombotic genetic risk factors in young survivors of myocardial infarction. *Blood* **94**, 46–51.
410. Bray, P. F. (1999). Integrin polymorphisms as risk factors for thrombosis. *Thromb. Haemost.* **82**, 337–344.
411. Gonzalez-Conejero, R., Lozano, M. L., Rivera, J., Corral, J., Iniesta, J. A., Moraleda, J. M., and Vicente, V. (1998). Polymorphisms of platelet membrane glycoprotein Ib associated with arterial thrombotic disease. *Blood* **92**, 2771–2776.
412. Santoso, S., Kunicki, T. J., Kroll, H., Haberbosch, W., and Gardemann, A. (1999). Association of the platelet glycoprotein Ia C807T gene polymorphism with nonfatal myocardial infarction in younger patients. *Blood* **93**, 2449–2453.
413. Lauricella, A. M., Quintana, I. L., and Kordich, L. C. (2002). Effects of homocysteine thiol group on fibrin networks: another possible mechanism of harm. *Thromb. Res.* **107**, 75–79.
414. Sauls, D. L., Wolberg, A. S., and Hoffman, M. (2003). Elevated plasma homocysteine leads to alterations in fibrin clot structure and stability: implications for the mechanism of thrombosis in hyperhomocysteinemia. *J. Thromb. Haemost.* **1**, 300–306.

415. McCully, K. S. (1969). Vascular pathology of homocysteinemia: implications for the pathogenesis of arteriosclerosis. *Am. J. Pathol.* 56, 111–128.
416. Wilcken, D. E., and Wilcken, B. (1976). The pathogenesis of coronary artery disease. A possible role for methionine metabolism. *J. Clin. Invest.* 57, 1079–1082.
417. Welch, G. N., and Loscalzo, J. (1998). Homocysteine and atherothrombosis. *N. Engl. J. Med.* 338, 1042–1050.
418. Wald, D. S., Law, M., and Morris, J. K. (2002). Homocysteine and cardiovascular disease, evidence on causality from a meta-analysis. *BMJ* 325, 1202.
419. Klerk, M., Verhoef, P., Clarke, R., Blom, H. J., Kok, F. J., and Schouten, E. G. (2002). MTHFR 677C→T polymorphism and risk of coronary heart disease: a meta-analysis. *JAMA* 288, 2023–2031.
420. Ganji, V., and Kafai, M. R. (2003). Demographic, health, lifestyle, and blood vitamin determinants of serum total homocysteine concentrations in the third National Health and Nutrition Examination Survey, 1988–1994. *Am. J. Clin. Nutr.* 77, 826–833.
421. Jacques, P. F., Bostom, A. G., Wilson, P. W., Rich, S., Rosenberg, I. H., and Selhub, J. (2001). Determinants of plasma total homocysteine concentration in the Framingham Offspring cohort. *Am. J. Clin. Nutr.* 73, 613–621.
422. De Bree, A., Verschuren, W. M. M., Kromhout, D., Kluijtmans, L. A. J., and Blom, H. J. (2002). Homocysteine determinants and the evidence to what extent homocysteine determines the risk of coronary heart disease. *Pharmacol. Rev.* 54, 599–618.
423. Thambyrajah, J., and Townend, J. N. (2000). Homocysteine and atherothrombosis-mechanisms for injury. *Eur. Heart. J.* 21, 967–974.
424. Woo, K. S., Chook, P., Lolin, Y. I., Cheung, A. S., Chan, L. T., Sun, Y. Y., Sanderson, J. E., Metreweli, C., Celermajer, D. S., Tawakol, A., Omland, T., Gerhard, M., Wu, J. T., and Creager, M. A. (1997). Hyperhomocysteinemia is a risk factor for arterial endothelial dysfunction in humans. *Circulation* 96, 2542–2544.
425. Welch, G. N., Upchurch, G. R., Jr., and Loscalzo, J. (1997). Homocysteine, oxidative stress, and vascular disease. *Hosp. Pract. (Off Ed)* 32, 81–92.
426. Toole, J. F., Malinow, M. R., Chambless, L. E., Spence, J. D., Pettigrew, L. C., Howard, V. J., Sides, E. G., Wang, C. H., and Stampfer, M. (2004). Lowering homocysteine in patients with ischemic stroke to prevent recurrent stroke, MI, and death: the Vitamin Intervention for Stroke Prevention (VISP) randomized controlled trial. *JAMA* 291, 565–575.
427. Gershov, D., Kim, S., Brot, N., and Elkon, K. B. (2000). C-reactive protein binds to apoptotic cells, protects the cells from assembly of the terminal complement components, and sustains an antiinflammatory innate immune response: implications for systemic autoimmunity. *J. Exp. Med.* 192, 1353–1364.
428. Pepys, M. B., and Hirschfield, G. M. (2003). C-reactive protein: a critical update. *J. Clin. Invest.* 111, 1805–1812.
429. Pearson, T. A., Mensah, G. A., Alexander, R. W., Anderson, J. L., Cannon, R. O., III, Criqui, M., Fadl, Y. Y., Fortmann, S. P., Hong, Y., Myers, G. L., Rifai, N., Smith, S. C., Jr., Taubert, K., Tracy, R. P., and Vinicor, F. (2003). Markers of inflammation and cardiovascular disease: application to clinical and public health practice: a statement for healthcare professionals from the Centers for Disease Control and Prevention and the American Heart Association. *Circulation* 107, 499–511.
430. Torres, J. L., and Ridker, P. M. (2003). Clinical use of high sensitivity C-reactive protein for the prediction of adverse cardiovascular events. *Curr. Opin. Cardiol.* 18, 471–478.
431. Ridker, P. M. (2003). Clinical application of C-reactive protein for cardiovascular disease detection and prevention. *Circulation* 107, 363–369.
432. Mendall, M. A., Patel, P., Ballam, L., Strachan, D., and Northfield, T. C. (1996). C reactive protein and its relation to cardiovascular risk factors: a population based cross sectional study. *BMJ* 312, 1061–1065.
433. Visser, M., Bouter, L. M., McQuillan, G. M., Wener, M. H., and Harris, T. B. (1999). Elevated C-reactive protein levels in overweight and obese adults. *JAMA* 282, 2131–2135.
434. Freeman, D. J., Norrie, J., Caslake, M. J., Gaw, A., Ford, I., Lowe, G. D., O'Reilly, D. S., Packard, C. J., and Sattar, N. (2002) C-reactive protein is an independent predictor of risk for the development of diabetes in the West of Scotland Coronary Prevention Study. *Diabetes* 51, 1596–1600.
435. Jialal, I., Devaraj, S., and Venugopal, S. K. (2004). C-reactive protein: risk marker or mediator in atherothrombosis? *Hypertension* 44, 6–11.
436. Cermak, J., Key, N. S., Bach, R. R., Balla, J., Jacob, H. S., and Vercellotti, G. M. (1993). C-reactive protein induces human peripheral blood monocytes to synthesize tissue factor. *Blood* 82, 513–520.
437. Devaraj, S., Xu, D. Y., and Jialal, I. (2003). C-reactive protein increases plasminogen activator inhibitor-1 expression and activity in human aortic endothelial cells: implications for the metabolic syndrome and atherothrombosis. *Circulation* 107, 398–404.
438. Ridker, P. M., Stampfer, M. J., and Rifai, N. (2001). Novel risk factors for systemic atherosclerosis: a comparison of C-reactive protein, fibrinogen, homocysteine, lipoprotein(a), and standard cholesterol screening as predictors of peripheral arterial disease. *JAMA* 285, 2481–2485.
439. Di Napoli, M., Papa, F., and Bocola, V. (2001). Prognostic influence of increased C-reactive protein and fibrinogen levels in ischemic stroke. *Stroke* 32, 133–138.
440. Biasucci, L. M., Liuzzo, G., Grillo, R. L., Caligiuri, G., Rebuzzi, A.G., Buffon, A., Summaria, F., Ginnetti, F., Fadda, G., and Maseri, A. (1999). Elevated levels of C-reactive protein at discharge in patients with unstable angina predict recurrent instability. *Circulation* 99, 855–860.
441. Tommasi, S., Carluccio, E., Bentivoglio, M., Buccolieri, M., Mariotti, M., Politano, M., and Corea, L. (1999). C-reactive protein as a marker for cardiac ischemic events in the year after a first, uncomplicated myocardial infarction. *Am. J. Cardiol.* 83, 1595–1599.
442. Ridker, P. M., Cushman, M., Stampfer, M. J., Tracy, R. P., and Hennekens, C. H. (1997). Inflammation, aspirin, and the risk of cardiovascular disease in apparently healthy men. *N. Engl. J. Med.* 336, 973–979.
443. Steering Committee of the Physicians' Health Study Research Group. (1989). Final report on the aspirin component of the ongoing Physicians' Health Study. *N. Engl. J. Med.* 321, 129–135.

444. Ridker, P. M., Rifai, N., Rose, L., Buring, J. E., and Cook, N. R. (2002). Comparison of C-reactive protein and low-density lipoprotein cholesterol levels in the prediction of first cardiovascular events. *N. Engl. J. Med.* **347,** 1557–1565.
445. Rexrode, K. M., Lee, I. M., Cook, N. R., Hennekens, C. H., and Buring, J. E. (2000). Baseline characteristics of participants in the Women's Health Study. *J. Womens Health Gend. Based Med.* **9,** 19–27.
446. Balk, E. M., Lau, J., Goudas, L. C., Jordan, H. S., Kupelnick, B., Kim, L. U., and Karas, R. H. (2003). Effects of statins on nonlipid serum markers associated with cardiovascular disease: a systematic review. *Ann. Intern. Med.* **139,** 670–682.
447. Ridker, P. M., Cannon, C. P., Morrow, D., Rifai, N., Rose, L. M., McCabe, C. H., Pfeffer, M. A., and Braunwald, E. (2005). C-reactive protein levels and outcomes after statin therapy. *N. Engl. J. Med.* **352,** 20–28.
448. Nissen, S. E., Tuzcu, E. M., Schoenhagen, P., Crowe, T., Sasiela, W. J., Tsai, J, Orazem, J., Magorien, R. D., O'Shaughnessy, C., and Ganz, P. (2005). Reversal of Atherosclerosis with Aggressive Lipid Lowering (REVERSAL) Investigators. Statin therapy, LDL cholesterol, C-reactive protein, and coronary artery disease. *N. Engl. J. Med.* **352,** 29–38.
449. de Laat, H. B., Derksen, R. H., Urbanus, R. T., Roest, M., and De Groot, P. G. (2004). β 2-glycoprotein I dependent lupus anticoagulant highly correlates with thrombosis in the antiphospholipid syndrome. *Blood* **104,** 3598–3602.
450. Rauch, J. (1998). Lupus anticoagulant antibodies: recognition of phospholipid-binding protein complexes. *Lupus* **7,** S29–S31.
451. Horkko, S., Miller, E., Dudl, E., Reaven, P., Curtiss, L. K., Zvaifler, N. J., Terkeltaub, R., Pierangeli, S. .S, Branch, D. W., Palinski, W., and Witztum, J. L. (1996). Antiphospholipid antibodies are directed against epitopes of oxidized phospholipids. Recognition of cardiolipin by monoclonal antibodies to epitopes of oxidized low density lipoprotein. *J. Clin. Invest.* **98,** 815–825.
452. McNeil, H. P., Simpson, R. J., Chesterman, C. N., and Krilis, S. A. (1990). Anti-phospholipid antibodies are directed against a complex antigen that includes a lipid-binding inhibitor of coagulation: 2-glycoprotein I (apolipoprotein H). *Proc. Natl. Acad. Sci. USA* **87,** 4120–4124.
453. Hunt, J. E., McNeil, H. P., Morgan, G. J., Crameri, R. M., and Krilis, S. A. (1992). A phospholipid-β 2-glycoprotein I complex is an antigen for anticardiolipin antibodies occurring in autoimmune disease but not with infection. *Lupus* **1,** 75–81.
454. Arvieux, J., Roussel, B., Jacob, M. C., and Colomb, M. G. (1991). Measurement of anti-phospholipid antibodies by ELISA using β 2-glycoprotein I as an antigen. *J. Immunol. Methods* **143,** 223–229.
455. Galli, M., Comfurius, P., Maassen, C., Hemker, H. C., de Baets, M. H., van Breda-Vriesman, P. J., Barbui, T., Zwaal, R. F., and Bevers, E. M. (1990). Anticardiolipin antibodies (ACA) directed not to cardiolipin but to a plasma protein cofactor. *Lancet* **335,** 1544–1547.
456. Galli, M., Luciani, D., Bertolini, G., and Barbui, T. (2003). Lupus anticoagulants are stronger risk factors for thrombosis than anticardiolipin antibodies in the antiphospholipid syndrome: a systematic review of the literature. *Blood* **101,** 1827–1832.
457. de Laat, H. B., Derksen, R. H. W. M., Urbanus, R. T., Roest, M., and de Groot, P. G. (2004). β2-glycoprotein I-dependent lupus anticoagulant highly correlates with thrombosis in the antiphospholipid syndrome. *Blood* **104,** 3598–3602.
458. Veres, K., Lakos, G., Kerenyi, A., Szekanecz, Z., Szegedi, G., Shoenfeld, Y., and Soltesz, P. (2004). Antiphospholipid antibodies in acute coronary syndrome. *Lupus* **13,** 423–427.
459. Farsi, A., Domeneghetti, M. P., Fedi, S., Capanni, M., Giusti, B., Marcucci, R., Giurlani, L., Prisco, D., Passaleva, A., Gensini, G. F., and Abbate, R. (1999). High prevalence of anti-β2 glycoprotein I antibodies in patients with ischemic heart disease. *Autoimmunity* **30,** 93–98.
460. Galli, M., Luciani, D., Bertolini, G., and Barbui, T. (2003). Anti-β 2-glycoprotein I, antiprothrombin antibodies, and the risk of thrombosis in the antiphospholipid syndrome. *Blood* **102,** 2717–2723.
461. Lehtimaki, T., Lehtinen, S., Solakivi, T., Nikkila, M., Jaakkola, O., Jokela, H., Yla-Herttuala, S., Luoma, J. S., Koivula, T., and Nikkari, T. (1999). Autoantibodies against oxidized low density lipoprotein in patients with angiographically verified coronary artery disease. *Arterioscler. Thromb. Vasc. Biol.* **19,** 23–27.
462. Meraviglia, M. V., Maggi, E., Bellomo, G., Cursi, M., Fanelli, G., and Minicucci, F. (2002). Autoantibodies against oxidatively modified lipoproteins and progression of carotid restenosis after carotid endarterectomy. *Stroke* **33,** 1139–1141.
463. Vaarala, O., Alfthan, G., Jauhiainen, M., Leirisalo-Repo, M., Aho, K., and Palosuo, T. (1993). Crossreaction between antibodies to oxidised low-density lipoprotein and to cardiolipin in systemic lupus erythematosus. *Lancet* **341,** 923–925.
464. Hörkkö, S., Miller, E., Dudl, E., Reaven, P., Curtiss, L. K., Zvaifler, N. J., Terkeltaub, R., Pierangeli, S. S., Branch, D. W., Palinski, W., and Witztum, J. L. (1996). Antiphospholipid antibodies are directed against epitopes of oxidized phospholipids, recognition of cardiolipin by monoclonal antibodies to epitopes of oxidized low density lipoprotein. *J. Clin. Invest.* **98,** 815–825.
465. Salonen, J. T., Yla-Herttuala, S., Yamamoto, R., Butler, S., Korpela, H., Salonen, R., Nyyssonen, K., Palinski, W., and Witztum, J. L. (1992). Autoantibody against oxidised LDL and progression of carotid atherosclerosis. *Lancet* **339,** 883–887.
466. Gough, P. J., and Raines, E. W. (2003). Advances in retroviral transduction of hematopoietic stem cells for the gene therapy of atherosclerosis. *Curr. Opin. Lipidol.* **14,** 491–497.
467. Gough, P. J., and Raines, E. W. (2003). Gene therapy of apolipoprotein E-deficient mice using a novel macrophage-specific retroviral vector. *Blood* **101,** 485–491.
468. Sata, M., Saiura, A., Kunisato, A., Tojo, A., Okada, S., Tokuhisa, T., Hirai, H., Makuuchi, M., Hirata, Y., and Nagai, R. (2002). Hematopoietic stem cells differentiate into vascular cells that participate in the pathogenesis of atherosclerosis. *Nat. Med.* **8,** 403–409.

CHAPTER **9**

Ischemia and Myocardial Infarction: A Post-Genomic Analysis

OVERVIEW

Coronary artery disease and myocardial infarction are severe endemic health problems all over the world. Despite considerable clinical and research effort during the past two decades and the availability of new drugs and surgical modalities of therapy, the mortality and morbidity remain very high. In myocardial infarct (MI), the coronary blood flow is acutely disrupted because of rupture of an endothelial atherosclerotic plaque and vascular thrombotic occlusion. In assessing the severity of this syndrome, several factors need to be considered, including the extension of collateral circulation, vasoconstriction, and spontaneous thrombosis. Risk factors such as diabetes mellitus, arterial hypertension, and hypercholesterolemia contribute to the development of the disease, and although each risk factor by itself is partly under genetic control, a positive family history is an independent predictor, which suggests that there are additional susceptibility genes.

A number of MI-associated genes have been identified in both human and animal studies, suggesting that the gene profiling analysis of animal MI and ischemia models may have relevance for human MI. Moreover, the availability of the human genome is already changing our understanding of global and specific patterns of gene expression, which will greatly impact the diagnosis and treatment of these conditions. Identification of those individuals at risk for the disease may allow its prevention by a systematic genetic screening of family members of the proband. In addition, modification of certain environmental factors should be a high priority.

INTRODUCTION

Coronary artery disease (CAD), including its most important manifestation, myocardial infarction (MI), is a multifactorial polygenic disorder caused by multiple genetic and environmental factors and complex gene–environment interactions modulated by modifiers such as diabetes, dyslipidemias, and hypertension as depicted in Fig. 1. Post-genomic studies of myocardial ischemia and infarction are currently underway to identify those genes that contribute to the polygenic nature of CAD and MI, their expression and involvement in the pathophysiological mechanism and phenotype of both MI and events that occur after MI, and their interaction(s) with a number of genetic and environmental risk factors.[1,2]

To identify genes involved in increased risk of CAD and susceptibility to MI, genetic case-control studies have proved increasingly powerful.[3] Genes associated with MI include genes involved in thrombosis described in the previous chapter such as thrombospondin (*TSP*) genes, *TSP-1*, *TSP-2*, and *TSP-4*, plasminogen activator inhibitor (*PAI-1*), platelet glycoprotein IIIa, fibrinogen, factor V and VII), encoding inflammatory mediators (including various cytokines, the *LTA* gene encoding lymphotoxin-α), other atherosclerotic factors like stromelysin-1, apolipoprotein E (*ApoE*), cholesterol ester transfer protein (*CETP*), and other factors, including angiotensin-converting enzyme (*ACE*), angiotensinogen, and endothelial nitric oxide synthase (*eNOS*). Although this genetic information is primarily derived from human/clinical studies, the availability of animal models of ischemia/MI has allowed the identification of novel genes, the temporal studies of gene expression, the testing of the effects of null mutations in the genetic loci, and finally pre-clinical therapeutic trials. Expression of these and other genes in myocardial ischemia and MI has been assessed in both limited clinical studies and in animal models using both global gene profiling and specific gene expression approaches. The gene profiling has resulted in the identification of genes that predispose to MI and ischemia susceptibility and in defining useful biomarkers for diagnosis. The impact of gene variants on pharmaceutical treatment and the use of pharmacogenomic information may allow for a more personalized and potentially more effective type of medicine that can be tailored to the individual's genomic profile.[1]

LINKAGE ASSOCIATION: GENOME-WIDE APPROACH

A genome-wide approach to identify susceptibility genes for complex traits such as MI is based on analyzing cosegregation of polymorphic DNA markers with inheritance of a phenotype. The principle behind the techniques of genome-wide

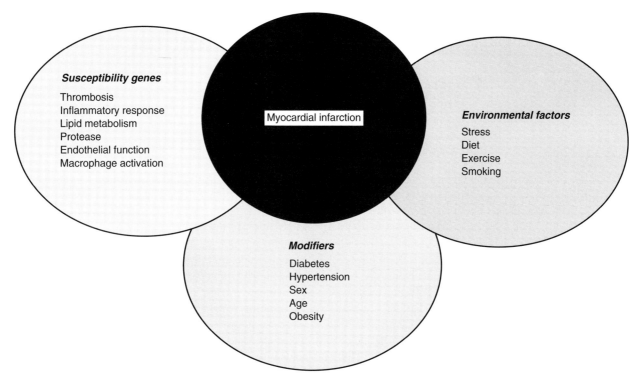

FIGURE 1 Intersection of susceptibility genes, environmental factors, and modifiers in the genesis of myocardial infarction.

search is based on the likelihood of sharing a susceptibility allele between the relatives who also have inherited the trait. The studies of family pedigrees (typically using hundreds of families) for polymorphic chromosomal loci that cosegregate with disease phenotype has proved successful in implicating several chromosomal loci with increased susceptibility to MI and stroke (Table I).[1,4,5]

There is increasing recognition that conventional linkage techniques have limited power to map susceptibility genes with small or moderate effects. However, with the identification of hundreds of thousands of genetic differences, such as single-nucleotide repeats or single-nucleotide polymorphisms (SNPs) that are present throughout the human genome, the power of linkage association studies has been greatly accelerated by the abundance of markers for mapping. Genetic analysis also has increasingly made use of the identification of haplotypes (specific combinations of genetic differences in blocks on chromosomes) present throughout the genome. Unfortunately, given the large genotypic heterogeneity and variability in human populations, the complex multifactorial nature of a polygenic disorder such as CAD/MI with both numerous interacting genetic and environmental components, studies using SNPs to detect an association with the MI phenotype can also generate spurious results, and a number of association studies using SNPs have not been concordant.

TABLE I Chromosomal Loci Linked to MI and Related Disorders

Phenotype	Loci identified	References
Premature acute MI	1p34-36	6
Stroke	5q12	7
MI and stroke	13q12-13	8
MI	Chromosome 14	9
MI	22q12-q13	10
MI	6p21.3	11
Ischemic heart disease	16p13-pter	12
Acute coronary syndrome (ACS)	2 q36-q37.3	13
Premature CAD/MI	Chromosome 2	170

CANDIDATE GENE APPROACH

The Human Genome Project has ushered in new opportunities for studying the complex disorders with genetic components such as MI through construction of polymorphism maps and genome-wide association studies. Although there is a considerable debate regarding the best approach for genome-wide association studies, a candidate gene approach has emerged as a highly informative approach.[1,14,15] Because polymorphisms often do not exist in isolation, a comprehensive analysis of the selected candidate genes is necessary. Because a positive association does not necessarily establish causality and often indicates linkage disequilibrium with the actual mutation, the relationship of the variant gene with the phenotype must be confirmed through independent tests.

Unlike genome-wide studies, the candidate gene approach generally uses an *a priori* understanding of the potential involvement of the specific gene under investigation with regard to the trait of interest. Candidate genes are often selected on the basis of evidence derived from biochemical and molecular biology studies implicating a specific molecule in disease pathogenesis.[2] In the post-genomic era, the identification of novel candidate genes has been greatly assisted by our enhanced ability to rapidly and efficiently survey expression of genes implicated in pathogenetic pathways.

The association of biologically functional polymorphisms in the candidate gene with the specific phenotype is commonly analyzed by either case-control studies or by prospective study, which tends to be more robust. The subjects of such studies may or may not be related, and the analysis is often targeted to ethnically homogenous populations. The objective of an association study is to examine whether a variant of the candidate gene is associated with the presence or severity of the phenotype, clinical outcome, or response to treatment. We will discuss a variety of candidate genes examined with regard to their association with ischemic heart disease, MI, and stroke in the following sections, which are presented and organized broadly by phenotypic function (e.g., thrombosis, inflammatory response).

Genes involved in coronary thrombosis and MI

Platelet aggregation and coronary thrombosis have a central role in the development of acute coronary syndromes (ACS) and MI. A number of "prothrombotic" genetic factors that can influence the individual thrombotic risk have been identified, as discussed in depth in the preceding chapter, and have been reported as candidate genes contributory to the risk of MI (Table II). For instance, genetic variation in platelet surface receptors mediating thrombus formation has been suggested to be associated with platelet hyperreactivity and with increased risk of MI.

Roldan and associates recently reported that polymorphisms in five hemostatic genes, when analyzed independently, had no significant effect on the incidence of premature MI in a group of 281 patients compared with 530 control subjects. However, the simultaneous combinations of polymorphisms in the genes for factor XIII (*F8*) and prothrombin (*F2*) significantly increased the risk of MI, supporting the occurrence of gene–gene interactions in MI.[40]

Platelet glycoprotein IIb/IIIa is a membrane receptor for fibrinogen and von Willebrand factor (vWF) and plays an important role in platelet aggregation. It is known to be involved in the pathogenesis of ACS. Several studies have found that the Pl A2 polymorphism of the *ITGB3* gene encoding platelet glycoprotein IIIa (PGIIIA) was associated with a high incidence of premature MI.[17,18] Interestingly, this association was strongest in patients younger than 60 years of age who had experienced previous coronary vascular events.[41]

In addition to the Pl A1/A2 variants of *ITGB3*, a modest, but significant, association between polymorphisms in the prothrombin *F2* gene (G20210A) and acute MI (AMI) in young subjects[19] has been found in several studies, including a meta-analysis of 19 studies.[37] Although some studies have reported a modest association of the factor V (G1691A) variant with AMI limited to women and smokers,[19] other studies have found no evidence of an association with MI in general with variants of either the factor V Leiden[34–36] or glycoprotein 1B (C3550T).[19] Interestingly, some polymorphic alleles in Factor VII (*F7*) alleles have been reported to be protective, whereas others increase the susceptibility to MI.[23,24] The presence of the Pl A2 polymorphic allele in PGIIIA was the only prothrombic genetic factor associated with the risk of MI in an analysis of 200 survivors of MI who had experienced the event before the age of 45 years.[18]

Evidence of an association of specific polymorphic variants of the plasminogen activator inhibitor type-I (*PAI-1*) gene and MI has been reported.[20–22] A large-scale study of 2819 unrelated Japanese patients with MI and 2242 unrelated Japanese controls showed that the risk of MI was significantly associated with the 4G-668 5G polymorphism in the *PAI-1* promoter region.[20] In a population of subjects with end-stage renal disease, the risk of occurrence of fatal and nonfatal MI was associated with *PAI-1* 4G/5G gene polymorphism[22]; however, the association of this allele with MI risk was not found in other studies in various populations without end-stage renal disease (Japanese and Caucasian, respectively).[42,43] Interestingly, in a recent European case-control study comparing 598 men with MI and 653 age-matched controls, Juhan-Vague and associates found that the 4G *PAI-1* allele was associated with higher plasma PAI-1 levels in patients than in controls, but this association did not change the risk of MI.[21] On the other hand, that study demonstrated a significant interaction with both insulin or proinsulin and the risk of MI, which was observed only in the carriers of the 4G/4G genotype. The effect of PAI-1 in the incidence of MI may be related to its interaction with underlying insulin resistance.

On univariate analysis, no significant association was found between the Ala 387 Pro polymorphism in the thrombospondin-4 (*TSP-4* or *THBS4*) gene and MI.[32] However, such an association became apparent only after taking into account other variables that seem to modify the effect of *TSP-4* on MI, including waist-to-hip ratio, diabetes mellitus, and hypertension. This association was significant regardless of the gender or age of onset.

These findings suggested that the Ala 387 Pro variant of the *TSP-4* gene might be an important determinant in the

TABLE II Thrombotic Gene Variants and MI

Gene loci	Variant/mutation	Associated phenotype	References
GPVI	T 13254 C	Increased MI risk	16
ITGB3 (PGIIIa)	Pl A2	Increased MI risk (Scandinavia)	17
ITGB3 (PGIIIa)	Pl A2	Increased MI risk (young adults)	18
PAI	Allele 4G in promoter	AMI	19
PAI-1	4G -668/ 5G	Increased MI risk	20
PAI-1	-675 4G/5G promoter 4G/4G genotype	Increased MI risk (insulin-involved)	21,22
PFA-AH	Ala 379 Val	Premature MI risk	171
F7	Arg 353 Gln A2 promoter allele	Protective	23
F7	-402 GA promoter	Increased MI risk in men	24
F7	-323 A2 allele promoter	Decreased MI risk in men	24
FGA/FGG	-58G>A/ 1299 +79T>C (haplotype 1)	Increased MI	25
FGA/FGG	FGA 2224A/ FGG 9340C (haplotype 4)	Protective	25
TF	A -603 G promotor	Increased MI risk	26
TM	Ala 455 Val	Premature MI risk (smoking enhanced)	27, 28
TM	G–33 A promoter	Premature MI	29
TSP-1	Asn 700 Ser	Increased MI	30,33
TSP-2	T>G substitution in 3'-untranslated region	Reduced premature MI	31
TSP-4	Ala 387 Pro	Familial premature MI	30,32
Contradictory findings			
Factor V Leiden:	G1691A	No MI	19, 34–36
F2 (prothrombin)	G20210A	Increased MI risk	19, 37
		No increased MI risk	35, 38
PG IIIa	Pl A1/A2	No premature MI risk	39

FGA/FGG, fibrinogen alpha/fibrinogen gamma; *F7*, factor VII; GP VI, glycoprotein VI; PAI, plasminogen activator inhibitor; *ITGB3*, PGIIIA or platelet glycoprotein IIIA; TSP, thrombospondin; TM, thrombomodulin; TF, tissue factor; Factor V, factor V Leiden; *PFA-AH*, platelet-activating factor-acetylhydrolase; AMI, Acute myocardial infarction.

development of MI at any age. A role for thrombospondin gene variants as significant risk factors in CAD and MI is highly plausible, given the pivotal role that the thrombospondin family of extracellular matrix glycoproteins play in platelet adhesion, in the modulation of vascular injury, coagulation, and angiogenesis, as a key ligand for CD36, an oxidized LDL receptor, and for integrins, as well as their involvement in atherosclerotic plaque formation.[30]

In addition to TSP-4, two other members of the thrombospondin gene family encoding the thrombospondin protein (*TSP-1 and TSP -2*) have been associated with the modulation of familial, premature MI,[30] an aggressive form of MI occurring before age 40 in men and age 45 in women. In addition to the Ala 387 Pro variant of the *TSP-4* gene discussed previously, the *TSP-1* SNP causing a Asn 700 Ser missense variant is a rare recessive polymorphism (1% frequency) but displays the highest association with the disease. In a recent large case-control study of 1425 individuals who survived MI before age 45, the Asn 700 Ser polymorphism was identified as a significant risk factor for MI in both homozygous and heterozygous carriers of the S700 allele.[33] Biochemical studies have shown that the Ser-700 variant of *TSP-1* has enhanced interactions with fibrinogen, which increases the susceptibility to proteolysis and denaturants and contributes to its prothrombic phenotype.[44] In contrast, homozygosity for the *TSP-2* variant allele (a T>G substitution in the 3'-untranslated region of TSP-2) was significantly associated with a reduced risk of premature MI and seems to be protective against MI.[30,31]

Tissue factor (TF), the main initiator of the extrinsic coagulation cascade as discussed in Chapter 8 is expressed in atherosclerotic lesions and contributes to the formation of thrombus in the pathogenesis of MI. A number of polymorphic variants have been found in the upstream promoter region of the *TF* gene.[26] Although previous findings concerning the association of these 5' regulatory variant promoter polymorphisms with the incidence of MI were rather contradictory,[45] more recently, Ott and associates have confirmed that one of these *TF* promoter variants (the -603 G allele) is associated with a significant increase in the incidence of

MI.[26] Because higher plasma TF concentrations are present in the -603 G carriers, enhanced TF expression could be the primary mechanism underlying this association.

A high incidence in -455 G/A β-fibrinogen gene promoter polymorphism, plasma fibrinogen levels, and the development of CAD and MI have been previously reported.[46,47] However, the present consensus is that although there is a strong association between polymorphisms of the fibrinogen β-chain gene and fibrinogen plasma concentration, a significant association of these variants (acting as independent risk factors) with the incidence of MI has not been demonstrated in most populations studied.

Genes Involved in the Inflammatory Response in MI

Inflammation pathways play a prominent role in plaque initiation that may result in stroke and MI.[3] Elevated plasma levels of inflammatory marker proteins, such as C-reactive protein (CRP) and CD40 ligand/CD40, in concert with increased expression of adhesion molecules, chemokines, cytokines, matrix metalloproteinases (MMP), and plaque inflammatory cells, characterize the atherothrombotic state as discussed in the previous chapter. It is, therefore, not surprising that several genes that contribute to the inflammatory response have variants that can modulate the susceptibility to MI as shown in Table III.

Matrix metalloproteinases (MMPs) play an important role in the pathogenesis of MI, contributing to the weakening and rupture of the fibrous cap of an atherosclerotic plaque, a key event that predisposes to AMI. Stromelysin (*MMP-3*) has been identified in human atherosclerotic lesions and can degrade constituents of the extracellular matrix (ECM), contributing to vascular remodeling and activating other MMPs, including collagenase and gelatinase. A common variant (5A/6A at -1171) in the promoter of the *MMP-3* gene has been identified in association with MI independent of other risk factors.[48] This genetic polymorphism has been shown to have allele-specific effects on the transcriptional

TABLE III Gene Variants in Inflammatory Response and MI

Gene loci	Variant	Associated phenotype	References
MMP-3	5A/6A -1171 promoter	AMI and smoking	20, 48, 53
MMP-3	5A/6A–1612 promoter	MI	54
PECAM-1/CD31	Gly 670 Arg	AMI	55
		MI (Japanese)	56
PECAM-1/CD31	Asn 563 Ser	MI (Japanese)	56
TLR-4	Asp 299 Gly	Statin therapy MI	57
TLR-4	Thr 399 Ile	MI in men	58
IL-1β	TT genotype of -11C/T	Decreased MI risk	59
PON1	Gln 192 Arg	MI in diabetes/obese MI (Chinese)	60, 61
PON1/PON2	Haplotype: met55 + gln192 (ON1); cys311 (PON2)	MI protective	15
PON2	Ser 311 Cys	MI (in smokers)	62
PPAR-γ	C161T allele in exon 6	AMI	63
PPAR-γ2	Pro 12 Ala	Reduced risk of MI	64
LGALS2	—	MI-Japanese	10
LTA LTA/TNFβ	Thr 26 Asn intron 1 A252G	MI	11
CD14	C-260T promoter	AMI	19, 65, 66
CCL11	Ala 23 Thr	Increased MI	67
IL-1β, IL-6, and APOE4 (in combination)	IL-1β (allele C); IL-6 (allele C)	MI	68
Contradictory findings			
LTA	A 252G and C804A	No MI (Japanese)	20
IL-6	G -174 C promoter	No MI	69
TNFα TNFβ	-308 TNFα promoter A252G	No MI	70
IL-10	-1082G/A, -819C/T and -592C/A,	No MI	71

PON, paraoxonase; LGALS2, galectin-2; LTA, lymphotoxin A; IL, interleukin; MMP-3, stromelysin-1 (also matrix metalloproteinase); PECAM, platelet-endothelial cell adhesion molecule; TLR, toll-like receptor 4; CCL11, eotaxin; CD14, cell-surface bacterial lipopolysaccharide receptor; PPARγ, peroxisome proliferator-activated receptor gamma; AMI, acute myocardial infarction.

activity of *MMP-3* gene promoter.[49,50] This *MMP-3* variant has also been associated with the development of coronary artery aneurysms[51] and stable angina.[52] The involvement of variant *MMP-3* as a significant risk factor for MI in a variety of populations suggests that its association with AMI is enhanced in women[20] and is markedly strengthened in young individuals who smoke.[53]

Cytokines, such as the proinflammatory interleukins IL-1β and IL-6, which are significantly increased early after AMI in both patients and animal models, are promising candidates as potential MI susceptibility genes.[72,73] The presence of 511 C/T *IL-1β* gene polymorphism seems to increase the incidence of MI and ischemic stroke in young patients and also serves as a contributory factor in the response of mononuclear cells to inflammatory stimuli.[59] Data have shown an independent association of a specific *IL-6* promoter polymorphism (e.g., -174 G/C) with a mild or moderate increased risk of MI in elderly men,[68,74] whereas several studies reported no evidence of an association between *IL-6* variants and MI in other populations.[69,72,75] Interestingly, a significantly increased risk of MI was found in older patients harboring combinations of several cytokine (i.e. *IL-1β* and *IL-6*) and *APOE* variants.[68] In addition, an A23T polymorphic variant of the *CCL11* gene encoding the chemokine, eotaxin, involved in transendothelial immune cell migration during inflammatory disorders is significantly associated with an increased risk for MI.[67]

Another inflammation-mediating cytokine lymphotoxin-alpha (*LTA*) has been implicated in plaque formation inside the coronary artery wall and may be associated with its acute rupture.[76] Studies have shown a significant association of each of two SNPs in the *LTA* gene, also termed tumor necrosis factor β (*TNF-β*), with MI when present in homozygous state in a large Japanese population.[11] One variant, the Thr26Asn is located within the *LTA* coding region and mediated a twofold increase in induction of several cell-adhesion molecules, including VCAM1, in vascular smooth muscle cells (SMCs) of human coronary artery as determined by functional studies. A second polymorphism (252G/A) resides within intron 1 and is associated with increased *LTA* transcription as shown by *in vitro* studies. However, the association of *LTA* polymorphisms with MI could not be demonstrated in a German population.[71] Other interacting environmental factors, including smoking, dyslipidemia, and obesity, significantly increase *LTA* as a risk factor.[77] Recent studies have confirmed a strong association between the *LTA* Thr26Asn polymorphism and the extent of coronary atherosclerosis in a large cohort (n=1082) of well-documented patients with CAD. Multiple-vessel disease was significantly elevated in male patients of the Asn/Asn genotype compared with patients of the Thr/Thr or Thr/Asn genotype.[78] Recently, a single nucleotide polymorphism in the *LGALs2* gene encoding an LTA-interacting galactose-binding protein (galectin-2) has also been found in a Japanese case-control study associated with significant susceptibility to MI.[10]

As previously noted, macrophage activation is a central component of the inflammatory process in atherosclerosis, and genes involved in this pathway are potential candidates for MI susceptibility loci. A polymorphic variant (161T allele in exon 6) in the transcriptional regulatory factor PPAR-γ, a pivotal regulator of macrophage activation, has been associated with premature AMI and increased lipid peroxidation.[63] Other studies have shown that a Pro12Ala polymorphism in codon 12 of PPAR-γ is significantly associated with both a modulation of the risk of developing type 2 diabetes and a reduced risk of MI in a type 2 diabetic population.[79,80] This PPAR-γ variant was also associated with significantly reduced incidence of MI in a large cohort of American males.[64] A second PPAR-γ polymorphism (C1431T) that influenced Ala12-associated diabetes risk and significantly increased MI risk in individuals with the T allele has also been reported.[79,80]

Oxidative damage is a major cause of atherosclerosis, and oxLDL is a primary substrate for macrophage activation. Because paraoxonase (encoded by either *PON1* or *PON2*) has been postulated to be an antioxidant factor that plays a role in protection from LDL oxidation, several recent studies have addressed whether *PON* gene polymorphisms are contributory risk factors for MI. Although previous data on the role of *PON1* polymorphisms in MI have been conflicting, it has recently been reported that the *PON1* Q192R polymorphic variant was not independently associated with MI but rather was a significant risk factor in association with either of two modifying factors (i.e., diabetes or obesity); individuals with the *PON1* RR genotype had a high risk for MI.[60,81] There is also an effect of the *PON* genotypes with aging; individuals homozygous for the *PON1* Q genotype have significantly lower PON1 activity as a function of age. In subjects carrying the low-activity QQ *PON1* genotype, the risk of MI has been shown to be elevated along with advancing age.[82] Another recent study reported that a *PON2* polymorphism (Ser311Cys) had no independent association with MI but did significantly affect MI risk in smokers.[62] Moreover, a haplotype, which includes 3 *PON* gene polymorphisms, comprising met55 and gln192 in *PON1* and cys311 in *PON2*, was recently shown to be associated with a protective effect on MI.[15] These data illustrate the delineation of several diverse modifying influences on the genotype determining MI risk.

The genetic predisposition for MI can be modulated by the interactions of several components of the inflammatory pathway. Variants in several genes involved in cell adhesion function in atherosclerotic process have been associated with increased MI risk. Specific haplotypes of the P-selectin gene (defined by specific polymorphisms located in the coding region of the P-selectin gene [*SELP*]) have been associated with increased risk for MI; the association with MI risk was significantly modulated by different polymorphic variants within the P-selectin coding region.[83] A Ser128Arg polymorphism has been reported in the E-selectin gene (*SELE*) in association with MI in Japanese patients; this variant seems to modulate kinase activities in endothelial signaling pathways and may functionally alter leukocyte–endothelial interactions.[84] Platelet endothelial cell adhesion molecule-1 (*PECAM1/CD31*) plays a pivotal role in the migration of circulating leukocytes during vascular inflammation. Polymorphisms in the *PECAM1* gene, Asn563Ser and Gly670Arg, were associated with AMI, with a stronger association in young male Japanese patients.[56] Others have reported a significantly higher frequency of Gly670Arg *PECAM1* variant in young adult male Italian patients with AMI.[55]

The CD14 receptor of monocytes is an important mediator for the activation of monocytes/macrophages by endotoxins and can lead to increased cytokine expression. Individuals harboring both a common polymorphism (C-260T) in the promoter of the *CD14* gene and high plasma IL-6 levels have enhanced risk for MI.[65] Several studies have found that this CD14 promoter variant is significantly prevalent in survivors of MI compared with controls,[66] associated with MI in a Japanese population,[85] and prevalent in elderly men with ACS.[86] Other large-scale studies have provided data indicating that screening for *CD14* genotypes as independent risk factors in MI is unproductive.[87,88] These contradictory findings likely reflect the variable interactions of genetic risk factors with the ethnic and environmental backgrounds, other gene variants, as well as gender and age differences.

Genes encoding the toll-like receptors (*TLR*) involved in macrophage innate immunity and in cardiac remodeling after MI have been examined in regard to their association with MI risk.[89] In a large-scale study of a Swedish population with MI, men with the Asp299Gly and the Thr399Ile *TLR4* genotypes exhibited an increased risk of MI, whereas no association was observed for women.[58] This study also found a synergistic interaction in men between the *TLR4* polymorphism and smoking. However, in a recent large-scale study of American men, there was no association of the Asp299Gly *TLR4* polymorphism with either MI or stroke.[90] Interestingly, individuals carrying the *TLR4* Asp299Gly variant have a lower risk of cardiovascular events, including MI, when treated with pravastatin compared with noncarriers.[91] The *TLR* variant modified the effect of pravastatin in preventing cardiovascular events in such a way that carriers of the variant Gly299 allele have significantly more benefit from pravastatin treatment. A synergistic effect of *TLR4* variation and statin efficacy in reducing MI has been suggested by a recent meta-analysis combining the data of three large independent studies.[57]

A polymorphic haplotype (HapA) in the *ALOX5AP* gene located on chromosome 13q12-1, encoding the 5-lipoxygenase activating protein, which is involved in the synthesis of potent inflammatory mediators (leukotrienes) by macrophages at the atherosclerotic lesion, was significantly associated with the risk of MI and stroke in a population from Iceland[8] and with ischemic stroke in a Scottish population.[92]

A different *ALOX5AP* haplotype (HapB) variant was associated with MI in subjects from England.[8] The finding that *ALOX5AP* variants contribute to susceptibility for MI was further validated in a Japanese population in a recent large-scale study of 1875 subjects (871 men, 1004 women).[93] In addition, analysis indicated that the homozygous genotype of the (CA) haplotype comprising two SNPs (A162C and T8733A) in the *ALOX5AP* gene was significantly associated with a reduced risk for MI.

Interestingly, a polymorphic allele at a completely unrelated loci, in the *PDE4D* gene encoding phosphodiesterase 4D located on chromosome 5q12, was also associated with an increased risk of ischemic stroke in the same Icelandic population described previously.[94] In addition, the -765G→C polymorphism of the cyclooxygenase (*COX-2*) gene promoter has been recently associated with a significantly lower risk of MI and stroke.[95] This relationship likely stems from cyclooxygenase-2 involvement in the induction of macrophage metalloproteinases (e.g., MMP-2 and MMP-9) leading to rupture of atherosclerotic plaques; the expression levels of both *COX-2* and *MMPs* were significantly lower in plaques from subjects carrying the -765C allele. The identification of genes (Table IV) with strong association to common forms of stroke, including carotid and cardiogenic stroke (forms of stroke related to atherosclerosis), may clarify its pathogenesis and may eventually allow novel therapies.

Other Gene Loci Associated With MI

A number of other gene loci previously associated with cardiovascular diseases other than MI have been examined for their relationship to MI (see Table V). There is increasing evidence to suggest a synergistic effect between atherogenic and thrombogenic risk factors in the pathogenesis of AMI.[102] In subjects with hypertension, hypercholesterolemia, or diabetes, any one of several polymorphic gene variants, including factor V Leiden (G1691GA), F2 prothrombin (G20210A), and the *APOE4* allele substantially increased the risk of MI, nearly 10-fold. Moreover, the combination of prothrombotic and *APOE4* polymorphisms in individuals who smoke increased their risk of MI 25-fold. The *APOE4* variant has been reported to be associated with MI in young adults as an independent risk factor.[102] Moreover, evidence has been presented that *APOE4* is significantly associated with MI in men but not in women.[104] Carriers of *APOE2* alleles may have a modestly reduced risk for MI, although the data presently available are rather limited.[105,106] Variants in other genes involved in the lipid pathways in atherosclerosis have been reported in association with MI. These include the cholesterol ester transfer (*CETP*) protein -629A variant associated with a protective effect in MI,[15] several apolipoprotein C-III (*APOC3*) haplotypes associated with an increased risk of MI,[15] and a polymorphism (R219K) in the

TABLE IV Gene Variants Involved in MI and Stroke

Gene loci	Variant	Associated phenotype	References
ALOX5AP	HapA encompassing 4SNPs:SG13S25, SG13S114, SG13S89, and SG13S32	↑ Risk of MI and stroke in Icelandic + Scottish subjects	8, 92
PDE4D	Combination of SNP45 + SNP41 and microsatellite marker (AC008818-1)	↑ Ischemic stroke in Icelandic subjects	94
COX-2	G765C	↓ risk of MI and stroke	95
vWF	Sma I polymorphism in intron 2	↑ Ischemic stroke risk but not AMI (Chinese)	96
IL-1β	-511C/T TT genotype	↓ Risk of MI and stroke in young patients	59
MTHFR	C677T -TT genotype	↑ MI and ischemic stroke	97
	TT genotype	↑ Stroke risk with increased T allele dosage	98
ESR1	T –397 C CC genotype	↑ MI and stroke risk	99, 101

ALOX5AP, arachidonate 5-lipoxygenase activating protein (FLAP); PDE, phosphodiesterase; COX-2, cyclooxygenase; vWF, von Willebrand factor; MTHFR, methylenetetrahydrofolate reductase; ESR1, estrogen receptor alpha; IL-1β, interleukin-1β.

TABLE V Other Cardiovascular Gene Variants and MI

Gene loci	Variant	Associated phenotype	References
Positive correlation			
APOE 4	Presence of epsilon 4 allele	AMI in young adults	103
APOC3	Haplotypes CCTTCG and ATCCCG at -641, 482, -455, 1100, 3175, and 3206	MI	15
ESR1	Haplotype 1	MI in women—no men	99–101, 108
CX37 (GJA4)	C1019T	MI in men	20
ADD1	Gly 460 Trp	MI protective	15
CETP	C-629A promoter	MI protective	15
ABCA1	Arg 219 Lys	Reduced risk of MI	107
ANV	-1C>T	Lower risk of MI in young	109
β1-AR	Arg 389 Gly.	AMI	110
Contradictory			
eNOS	4a/4a variant homozygosity - rare	AMI and ACS, No AMI	115, 116
eNOS	G894T Glu 298 Asp	MI ↑ Risk premature MI + IHD No MI (Australian)	116–119, 120, 121, 122
eNOS	T-786C	↑ MI risk (Korean) ↑ ACS No IHD	115, 123 / 120
MTHFR	C677T	↑ MI (Turkish) ↑ CAD (Japanese)	124 / 125
MTHFR	C677T	No MI	35, 126
ACE	DD genotype (but not CC genotype of AT1R)	Lower risk of MI	127
ACE	DD genotype	Increased risk MI in men	128, 129
CETP	TaqI B polymorphic site	No MI in middle-aged men	130
AT1R	A1166C	↑ AMI	131

ESR1, estrogen receptor alpha; *AT1R*, angiotensin II type-1 receptor; *CETP*, cholesteryl ester transfer protein; *ACE*, angiotensin-converting enzyme; *AGT*, angiotensinogen; *MTHFR*, methylenetetrahydrofolate reductase; *APOE4*, apolipoprotein E4; *eNOS*, endothelial nitric oxide synthase; *ADD*, α-adducin; *APOC3*, apolipoprotein C-III; *CX37*, connexin 37; *ANV*, annexin V; *ABCA1*, ATP-binding cassette transporter 1; IHD, ischemic heart disease; AMI, acute myocardial infarction; AR, adrenergic receptor.

ABCA1 gene encoding the ATP-binding cassette transporter involved in reverse cholesterol efflux and HDL uptake with the K219 allele associated with a decreased risk of MI.[107]

Gene variants involved in cardiovascular physiology outside of the atherosclerotic pathways have been implicated in MI. A common Arg389Gly polymorphism in the *β1-AR* gene has been associated with AMI; the prevalence of the Arg389 homozygote genotype was significantly higher in patients with AMI than in control subjects.[110] The site of the polymorphism lies within a region important for receptor–G_s-protein coupling and subsequent agonist-stimulated adenylyl cyclase activation. A suggested mechanism underlying the *β1-AR*-mediated AMI is that augmented sympathetic activity might trigger AMI by means of enhanced hemodynamic or mechanical forces through β-AR activation. The Arg 389 polymorphism, when present with an $α_{2c}$-AR variant, is a risk factor for human HF.[111] Individuals carrying a combination of the Arg389 allele and the $α_{2c}$-AR variant (α2CDel322-325) exhibit a 10-fold risk for development of HF. Moreover, variants at this position have also been associated with increased diastolic blood pressure and heart rate (in individuals homozygous for the Arg389 allele)[112] and differential therapeutic responses to β1-AR antagonists (greater in patients with Arg389 compared with those carrying Gly389 alleles).[113] Functional studies of transgenic mice containing either the Arg389 or Gly389 variants demonstrated that the Arg389 variant provided enhanced ventricular function and blunted β-AR signaling and elicited improved myocardial recovery after ischemic injury, which in the mouse model, was age-dependent.[114]

A common variant (a single Alu insertion/deletion polymorphism) of the angiotensin-converting enzyme (*ACE*) has an effect on MI susceptibility. Petrovic and associates reported that the *ACE D* allele is frequently found in association with MI.[128,129] However, this was not the case in several larger studies[132] and meta-analyses.[133] The overall risk for MI as a result of ACE D is not clear, because small, underpowered studies can erroneously implicate polymorphisms as genetic risk factors.[134] Another possibility is the involvement of interacting environmental factors and modulating genetic influences of which there are many candidates.

The association of specific variants of the eNOS (endothelial nitric oxide synthase) gene with MI has been documented. This marker–disease association may be due to the impaired effects of nitric oxide on the cardiovascular system (i.e., dysregulation of vascular tone, platelet aggregation and leukocyte adhesion, and SMC proliferation), all of which promote coronary atherosclerosis and thrombosis. A tandem 27-bp repeat in intron 4 of the *eNOS* gene is polymorphic (i.e., *eNOS4a* allele has 4 and *eNOS4b* has 5 tandem repeats), and an association between *eNOS4a* and MI has been reported.[135] In addition, a missense Glu298Asp mutation in exon 7 of the *eNOS* gene has been found to be a risk factor for MI.[116,117] Moreover, the GG genotype of the *eNOS* Glu298Asp polymorphism exerts a beneficial reduction in the risk of ACS.[136] A third polymorphism in the promoter of the eNOS gene (T-786C) has been found in association with ACS; C/C *eNOS* variants were found significantly more often in ACS patients than in controls.[115] Interestingly, this study also documented the contemporary presence of hyperhomocysteinemia, and allelic NOS variants influenced the predisposition to ACS. A recent large-scale meta-analysis confirmed that individuals homozygous for the Asp298 and intron-4a alleles of eNOS are at moderately increased risk of MI but was unable to confirm the effect of T-786C on MI risk.[120]

The interaction of gene variations governing hyperhomocysteinemia in relation to the susceptibility and development of MI has received considerable attention. The accumulation of homocysteine (hyperhomocysteinemia) in blood has a multifactorial etiology, of both hereditary and acquired origins. A C->T677 polymorphism in the *MTHFR* gene encoding methylenetetrahydrofolate reductase has been identified as a cause of mild hyperhomocysteinemia, a risk factor for arterial thrombosis. Several studies have reported an association of the homozygous *MTHFR* TT genotype as a risk factor with premature CAD onset,[124,137] MI,[102,125,138] and ischemic stroke,[96,98] although others have found no association of MI with this *MTHFR* allele.[35,126,139] There are indications that other genes involved in homocysteine metabolism may need to be screened. Interestingly, a recent genome-wide search for genes affecting variation in plasma homocysteine levels using linkage scanning has succeeded in identifying a locus on chromosome 11q23 containing a candidate gene involved in the metabolism of homocysteine (the nicotinamide *N*-methyltransferase gene, *NNMT*).[140] This gene may prove a viable candidate to screen for MI risk.

Connexin represents an essential protein forming myocardial channels and gap junctions in the myocardium and has been implicated in dysrhythmia susceptibility and cardiac conduction defects. Connexin-formed channels are also involved in ischemia/reperfusion injury, in ischemic hypercontracture, infarct development, post-MI remodeling, and in ischemic preconditioning. The risk of MI in a large case-control study of the Japanese population was associated with the C1019T polymorphism in the *GJA-4* gene encoding connexin 37, most significantly in men.[20] Interestingly, the MI risk-associated TT genotype was found to be more prevalent in the African-American population than European Americans.[141]

A T-413A variant of the promoter region of *HO-1*, the gene encoding the antioxidant heme oxygenase, which promotes greater *HO-1* promoter activity and transcription, has been associated with a decreased risk of MI and the AA genotype of *HO-1* significantly reduced the incidence of ischemic heart disease.[142]

GENES, GENDER, AND MI

It is well established that estrogens have a variety of vasodilatory, anti-inflammatory, and anti-proliferative effects on the cardiovascular system, as well as favorable effects on the lipid profile. Despite these beneficial effects, recent clinical trials have failed to demonstrate a lower rate of CAD in postmenopausal women receiving hormone replacement therapy. Estrogen effects on the vascular system are primarily mediated by two distinct estrogen receptors (ERs), ER-α and ER-β, encoded by two separate genes (*ESR1* and *ESR2*), expressed in endothelial cells and vascular SMCs in both genders.

The strikingly lower incidence of MI in premenopausal women compared with men of the same age suggests a potential role for sex hormones in the etiology of MI. Conversely, the increased incidence of cardiovascular events in postmenopausal women suggests that estrogen deficiency may predispose women to a higher cardiovascular risk. Selective estrogen receptor modulators (e.g., tamoxifen) have been reported to affect the onset of myocardial ischemia and atherosclerosis in both animal studies and in postmenopausal women, primarily by reduced arterial intima–media thickness and improved endothelial function.[143] These findings have been suggestive that *ESR* genes might be examined as candidate genes in association with MI risk.

In one large-scale study, postmenopausal women who carry the *ESR1* haplotype 1 (c.454-397 T allele and c.454-351 A allele) either in the heterozygous or homozygous state had an increased risk of MI and ischemic heart disease, independent of known cardiovascular risk factors.[108] No association of this haplotype with MI was observed in men in this study. However, a previous study examining only the 454-397 T allele found that homozygous individuals carrying the CC genotype exhibited a significantly increased risk of MI regardless of gender.[100] In several other studies, a polymorphic sequence repeat of a (TA) dinucleotide of differing lengths located within the ER-1α encoding *ESR1* gene promoter region (upstream of exon 1) was studied; individuals homozygous for longer TA alleles were associated with increased angiographic severity of CAD in both Finnish men[144] and in a mixed-gender population of patients <55 y of age, with little significant gender difference.[145] More recent studies have demonstrated that the longer TA repeat genotype of the *ESR1* gene was associated with lower adenosine-stimulated coronary flow in men.[146] These studies suggest that the *ESR* variants can significantly modulate MI risk in both genders.

ETHNICITY, RACE, AND MI

There is little question that various ethnic groups have a differential distribution of risk factors for the development of CAD and MI. Although some of these differences are thought to be attributable to a combination of environmental (e.g., diet and smoking) and noncardiovascular risk factors (e.g., diabetes/obesity), numerous genetic components seem to be involved.

The identification of these genetic determinants or susceptibility factors will not only contribute to improved diagnosis but will also likely impact treatment of this disorder. Although a cataloging of different ethnic group susceptibilities for CAD and MI is beyond the scope of this chapter, we briefly discuss two examples.

It is generally well recognized that South Asians (e.g., Indian, Pakistani, Sri Lankan) seem to have a high incidence of CAD, whereas East Asians (e.g., Chinese, Japanese) have a very low incidence. Although the genetic factors to explain this differential distribution have not yet been determined, several potentially relevant observations have been made. In a genotype analysis of two ethnic groups in Singapore examined for distribution of a polymorphic site in exon 2 of the FXIIIa subunit *F13a* gene resulting in a valine to leucine (V34L) change, a significantly higher frequency of the L34 allele was found among the Asian Indians than the Chinese. Mean FXIII plasma levels were also significantly higher among the Asian Indians.[147] A study of risk factors for AMI in a Southeast Asian population (in Singapore) comprising different ethnic groups showed that the odds of having AMI was higher for subjects with hypertension, smoking habit, lower plasma folate and vitamin B_{12} levels, and non-Chinese ethnic group, whereas plasma homocysteine levels were not significantly associated with AMI.[148] Differences in *APOE* alleles have been reported in different Chinese ethnic groups (Uygar vs, Han); CAD patients, particularly in Uygar, have higher incidence of *APOE4* allele (and less of the protective *APOE2* allele).[149] In another study of groups of Chinese and Indian men in Singapore, there was no significant association of ACE gene variants with CAD or MI in the Chinese or Indians, either in the entire sample or in different risk groups.[150]

Increasing attention has been focused on CAD/MI in African-Americans, a population with a high incidence of CAD, with higher mortality rates particularly at a young age. Progress has been made in defining potential genetic determinants for CAD/MI in this group. In a recent study of healthy populations of African-Americans (AA) and European Americans (EA), the frequencies of multiple "risk-associated" genotypes, including variants in genes encoding connexin-37 (*GJA-4*), plasminogen activator inhibitor-1 (*PAI-1*), and stromelysin-1 (*MMP-3*), were significantly higher in the AA population than the EA population.[141] Nearly 9.1% of AA had all three high-risk genotypes, compared with 0% among the EA group. Other studies have revealed that the *ACE* DD genotype was also associated with MI in African-American men but not in women. In addition, allele frequencies of the A1166C variant of the angiotensin II type I receptor were different in African-Americans compared with Caucasians.[151] A haplotype (HapK) spanning the *LTA4H* gene encoding leukotriene A4 hydrolase, a protein involved in the same biochemical pathway as aforementioned *ALOX5AP*, was recently identified that confers increased risk to MI, an effect that was ethnic-specific with an approximately three-fold higher risk in African-Americans than in Caucasians.[152] As genetic differences between ethnic groups become better understood, targeted interventions to prevent and treat CAD/MI in specific populations can be developed.

GENE PROFILING IN MYOCARDIAL ISCHEMIA/INFARCT

In addition to genetic analyses that have provided significant information about the involvement of specific genes as both risk factors, as well as their potential application as diagnostic markers of myocardial ischemia and infarction, the profiling of gene expression can provide complementary information concerning the activity of genes preceding, during, and after MI. Gene profiling of myocardial ischemia and MI using both global approaches (e.g., microarray and SAGE analysis) and specifically targeted gene analyses has been applied in both clinical studies and in a variety of animal models of ischemia and MI. There have been limited studies of global gene profiling specific to MI in humans, a situation that will undoubtedly be remedied in the near future. Difficulties with the access or availability of tissues (e.g., human coronary arteries), as well as the highly variable nature of genes in human populations, have made this undertaking a difficult proposition. Given these constraints, most studies have been performed to assess gene expression at a single time point (usually at a more advanced stage in disease progression), which provides only a limited snapshot of the continuous and complex processes involved. For obvious reasons, there has also been little study addressing the contemporaneous gene expression events at the artery site (e.g., the primary lesions involved), including macrophage leukocytes, endothelial cells, and SMCs, as well as in myocardial cell response in MI.

One approach has been to comprehensively assess and compare gene expression profiles in coronary arteries from normal and diseased patients using microarray analysis. This profiling analysis has revealed primarily a pattern of transcriptional up-regulation for 56 genes, whereas one gene was down-regulated, glutathione-S-transferase (*GST*).[153] Among the genes that were up-regulated in response to CAD were several previously linked to CAD, such as osteopontin expressed in SMCs, the vascular cell adhesion molecule (*VCAM-1*) expressed in endothelial cells, and matrix metalloprotease (*MMP-9*) expressed in macrophages.

Another approach has been to profile gene expression specifically at the site of the ruptured human atherosclerotic lesions in human coronary arteries. This type of analysis was done using subtraction hybridization analysis that can detect low abundant mRNAs more sensitively than microarray analysis.[154] Among the genes that exhibited a significant difference in expression in ruptured compared with stable plaques was the up-regulated perilipin gene, whereas genes-encoding β-actin, fibronectin, and immunoglobulin chain were down-regulated in ruptured plaques. Another profiling study that used microarray analysis evaluated gene expression in atherosclerotic plaques located in a variety of vessels (e.g., aorta, the common, internal and external iliac arteries) but not the coronary arteries. This study identified more than 75 genes, including several novel factors such as JAK-1 and VEGF receptor-2, but interestingly most plaque-related genes were not identified in other studies.[155]

A larger number of studies have examined specific gene expression in clinical MI focusing on the individual trees rather than the forest. Both the numbers of studies and genes involved are beyond the scope of this chapter. A partial list of such studies have included the assessment of *eNOS* gene expression in coronary plaques from patients,[156] hypoxia-inducible factor-1 (*HIF-1*) in myocardium,[157] myocardial *ACE2* expression,[158] cardiac matrix metalloproteinase-9 (*MMP-9*) expression,[159] apoptotic-inducing factors in serum of patients in early-stage AMI,[160] TNF-α in circulating leukocytes of AMI patients,[161] and the induction of cytokines in leukocytes from patients with AMI.[162] One recent study examined a transcript profile of 35 inflammatory genes in leukocytes from patients with a history of MI compared with controls and found that MI patients had significantly enhanced levels for most inflammatory genes. Two of the most prominent up-regulated factors included transcripts for the *MIF* gene encoding secreted protein macrophage migration inhibitory factor and the *PI9* gene encoding the intracellular regulator proteinase inhibitor 9.[163]

In addition to transcriptional analysis, a number of studies have examined global and specific changes in protein levels occurring during MI. Given that the cellular composition of the plaques is highly heterogenous, a proteomic approach of entire lesions seems unlikely to be informative. One proteomic study analyzed proteins secreted from human carotid atherosclerotic plaques obtained by endarterectomy and demonstrated a marked increase in the number and kinds of secreted proteins with increased complexity of the lesion.[164]

A proteomic analysis of plasma from patients with AMI compared with controls found that specific alpha1-antitrypsin (AAT) isoforms 1, 5, 6, and 7 were markedly reduced as were five apolipoprotein A-I isoforms in plasma from AMI patients compared with controls, whereas fibrinogen γ chain isoforms 1 and 2 and γ-immunoglobulin heavy chains were increased in AMI patients.[165] Studies are currently underway to incorporate the use of mass spectroscopy in the proteomic analysis of cardiac myocytes/tissue or serum/plasma from patients with MI for further biomarker discovery.[166]

ANIMAL STUDIES

A number of studies have used gene profiling in the examination of animal models of MI and ischemia. The study of animal models of MI and ischemia is highly amenable to the gene profiling approach circumventing difficulties with profiling expression in MI patients, including extensive genetic heterogeneity, the presence in most patients of a large-number of risk factors, difficult to control variables (e.g., diet), and the confounding effects of medicines and treatments, both previously and concurrently administered. However, it remains unclear whether the events in animal and human infarction are the same; for instance, until recently, most murine models of atherosclerosis could not develop fibrous and unstable atherosclerotic plaques. Nevertheless, a number of genes affected in animal models also seem to be affected in the human studies as noted in the following.

One informative animal model involved expression profiling of reversible ischemia caused by one or several brief transient episodes of complete coronary occlusion or with a more prolonged, but partial, coronary ligation in the rat. Many up-regulated genes related to cell survival in myocardium were analyzed using an Affymetrix Gene Chip microarray. This analysis showed the up-regulation of myocardial genes encoding mitogen-activated protein kinase-activated protein kinase 3 *(MAPKAPK 3)*, heat shock proteins 70, 70, 27, 105, 86 and 40 kDa, β-crystalline, vascular endothelial growth factor (VEGF), inducible NOS, and plasminogen activator inhibitors 1 and 2 and suggested the presence of an early protective myocardial gene program against cellular injury.[167,168] With a longer term coronary occlusion lasting from 24h to several weeks and resulting in a true MI, the up-regulated genes tended to include those related to remodeling (e.g., collagens I and III, fibronectin, laminin) and apoptosis (Bax), whereas down-regulated genes were related to major myocardial energy-generating pathways including fatty acid metabolism.

Using an *in vivo* rat MI model generated by permanent coronary occlusion over 16 weeks, a microarray study found a total of 731 genes differentially expressed in two regions of the heart, the left ventricular free wall and the interventricular septum.[169] Expression was up-regulated for genes encoding atrial natriuretic peptide *(ANP)*, sarcoplasmic/endoplasmic reticulum Ca^{++}-ATPase *(SERCA)*, collagen, fibronectin, tissue inhibitor of metalloproteinase-3 (TIMP3), fibrillin, laminin, and osteoblast-specific factor-2. These genes were classified as belonging to a cardiac remodeling pathway in the post-MI period. Most of the up-regulated genes were cytoskeletal and ECM proteins, whereas down-regulated genes tended to be contractile proteins or fatty acid metabolism–related genes, suggesting a significant change in the energy-generating processes in the ischemic and injured myocardial tissue.

It is noteworthy that a number of MI-associated genes have been identified in both the human and animal model studies (e.g., *HSP 70*, *fibronectin*, *VCAM-1* and *GST*), further suggesting that the gene profiling analysis of animal MI and ischemia models has considerable relevance for human MI.

CONCLUSION

Data from both genetic and profiling studies of myocardial ischemia and MI have proved useful in the identification of genes that contribute to the susceptibility to this highly complex pathogenic process of MI. These data once confirmed should be useful in the further explanation of the elaborate pathway(s) and in defining the temporal sequence of events involved in MI. Clearly, the reconstruction of such events occurring at a variety of time points and spatial *loci* require more than the reductionist science of the molecular biologist or biochemist but likely will need extensive contributions from an integrative system biology approach using a combined approach of proteomics, genomics, physiological, and statistical modeling. The data as confirmed may allow the development of an array of biomarkers that can be applied to screening of disease susceptibility (e.g., which plaques might be most vulnerable for rupture), as well as to assess progressive ischemic damage, the viability of the cellular response to myocardial ischemic injury, and the ability to repair the damage. Finally, there is increasing evidence that our ability to develop more effective drugs, as well as more efficiently treat MI either pharmaceutically or with cell and gene-based therapies, will be greatly impacted by the information generated by this post-genomic analysis.

SUMMARY

- Myocardial infarction (MI) is a multifactorial polygenic disorder caused by multiple genetic and environmental factors and complex gene–environment interactions modulated by modifiers such as diabetes, dyslipidemias, and hypertension. A diverse variety of environmental factors such as diet and smoking can modulate the effects of specific genetic loci on MI risk.
- Factors such as gender, age, and ethnic background seem to play a substantial role in the expression of the MI phenotype.
- Linkage studies have provided evidence of genetic susceptibility loci involved in MI, as well as the identification of specific genes that seem to be involved in MI. The use of single-nucleotide polymorphisms and haplotype analysis has allowed for the identification of gene variants that contribute to MI.

- To identify genes involved in increased risk of coronary artery disease and susceptibility to MI, genetic case-control studies have proved increasingly powerful. Candidate genes significantly associated with MI include genes involved in thrombosis such as thrombospondin genes *TSP-1*, *TSP-2*, and *TSP-4*, plasminogen activator inhibitor (*PAI-1*), platelet glycoprotein IIIa, fibrinogen, factor V and VII, encoding inflammatory mediators (including various cytokines, the *LTA* gene encoding lymphotoxin-α), other atherosclerotic factors like stromelysin-1 (*MMP-3*), apolipoprotein E (*ApoE*), cholesterol ester transfer protein (*CETP*), and other factors, including angiotensin converting enzyme (*ACE*), angiotensinogen (*AGT*), and endothelial nitric oxide synthase (*eNOS*). A number of specific genotypes have also been shown to result in protective phenotypes with reduced MI risk.
- There is some controversy concerning the role of many of the polymorphic variants of these genes in modulating MI, which may be a result of different genetic backgrounds and environmental factors.
- The availability of animal models of ischemia/MI has allowed both the identification of novel genes by use of gene profiling techniques, the temporal studies of gene expression, the testing of the effects of null mutations in the genetic loci, and finally pre-clinical therapeutic trials. These models circumvent many of the problems associated with human studies, including genetic heterogeneity and difficult to control variables such as diet, medicines, and other diseases. Some events in the phenotype and progression of MI may be different in animals and humans. Nevertheless, some specific gene expression patterns have been found to be similar in both animal models of MI and patient profiles.
- Gene profiling studies with clinical MI are just beginning. These may increase the identification of susceptibility and effector genes that are significantly up-regulated (or down-regulated) in MI either as a promoter or consequence of the disorder, as well as potential biomarkers useful in diagnosis and assessing treatment.

References

1. Winkelmann, B. R., and Hager, J. (2000). Genetic variation in coronary heart disease and myocardial infarction: methodological overview and clinical evidence. *Pharmacogenomics* **1**, 73–94.
2. Nordlie, M. A., Wold, L. E., and Kloner, R. A. (2005). Genetic contributors toward increased risk for ischemic heart disease. *J. Mol. Cell Cardiol.* **39**, 667–679.
3. Jefferson, B. K., and Topol, E. J. (2005). Molecular mechanisms of myocardial infarction. *Curr. Probl. Cardiol.* **30**, 333–374.
4. Wang, Q. (2005). Molecular genetics of coronary artery disease. *Curr. Opin. Cardiol.* **20**, 182–188.
5. Rosand, J., and Altshuler, D. (2003). Human genome sequence variation and the search for genes influencing stroke. *Stroke* **34**, 2512–2516.
6. Wang, Q., Rao, S., Shen, G. Q., Li, L., Moliterno, D. J., Newby, L. K., Rogers, W. J., Cannata, R., Zirzow. E., Elston, R. C., and Topol, E. J. (2004). Premature myocardial infarction novel susceptibility locus on chromosome 1P34-36 identified by genomewide linkage analysis. *Am. J. Hum. Genet.* **74**, 262–271.
7. Gretarsdottir, S., Sveinbjornsdottir, S., Jonsson, H. H., Jakobsson, F., Einarsdottir, E., Agnarsson, U., Shkolny, D., Einarsson, G., Gudjonsdottir, H. M., Valdimarsson, E. M., Einarsson, O. B., Thorgeirsson, G., Hadzic, R., Jonsdottir, S., Reynisdottir, S. T., Bjarnadottir, S. M., Gudmundsdottir, T., Gudlaugsdottir, G. J., Gill, R., Lindpaintner, K., Sainz, J., Hannesson, H. H., Sigurdsson, G. T., Frigge, M. L., Kong, A., Gudnason, V., Stefansson, K., and Gulcher, J. R. (2002). Localization of a susceptibility gene for common forms of stroke to 5q12. *Am. J. Hum. Genet.* **70**, 593–603.
8. Helgadottir, A., Manolescu, A., Thorleifsson, G., Gretarsdottir, S., Jonsdottir, H., Thorsteinsdottir, U., Samani, N. J., Gudmundsson, G., Grant, S. F., Thorgeirsson, G., Sveinbjornsdottir, S., Valdimarsson, E. M., Matthiasson, S. E., Johannsson, H., Gudmundsdottir, O., Gurney, M. E., Sainz, J., Thorhallsdottir, M., Andresdottir, M., Frigge, M. L., Topol, E. J., Kong, A., Gudnason, V., Hakonarson, H., Gulcher, J. R., and Stefansson, K. (2004). The gene encoding 5-lipoxygenase activating protein confers risk of myocardial infarction and stroke. *Nat. Genet.* **36**, 233–239.
9. Broeckel, U., Hengstenberg, C., Mayer, B., Holmer, S., Martin, L. J., Comuzzie, A. G., Blangero, J., Nurnberg, P., Reis, A., Riegger, G. A., Jacob, H. J., and Schunkert, H. (2002). A comprehensive linkage analysis for myocardial infarction and its related risk factors. *Nat. Genet.* **30**, 210–214.
10. Ozaki, K., Inoue, K., Sato, H., Iida, A., Ohnishi, Y., Sekine, A., Sato, H., Odashiro, K., Nobuyoshi, M., Hori, M., Nakamura, Y., and Tanaka, T. (2004). Functional variation in LGALS2 confers risk of myocardial infarction and regulates lymphotoxin-alpha secretion in vitro. *Nature* **429**, 72–75.
11. Ozaki, K., Ohnishi, Y., Iida, A., Sekine, A., Yamada, R., Tsunoda, T., Sato, H., Sato, H., Hori, M., Nakamura, Y., and Tanaka, T. (2002). Functional SNPs in the lymphotoxin-alpha gene that are associated with susceptibility to myocardial infarction. *Nat. Genet.* **32**, 650–65.4
12. Francke, S., Manraj, M., Lacquemant, C., Lecoeur, C., Lepretre, F., Passa, P., Hebe, A., Corset, L., Yan, S. L., Lahmidi, S., Jankee, S., Gunness, T. K., Ramjuttun, U. S., Balgobin, V., Dina, C., and Froguel, P. (2001). A genome-wide scan for coronary heart disease suggests in Indo-Mauritians a susceptibility locus on chromosome 16p13 and replicates linkage with the metabolic syndrome on 3q27. *Hum. Mol. Genet.* **10**, 2751–2765.
13. Harrap, S. B., Zammit, K. S., Wong, Z. Y., Williams, F. M., Bahlo, M., Tonkin, A. M., and Anderson, S. T. (2002). Genome-wide linkage analysis of the acute coronary syndrome suggests a locus on chromosome 2. *Arterioscler. Thromb. Vasc. Biol.* **22**, 874–878.
14. Chiodini, B. D., Barlera, S., Franzosi, M. G., and Tognoni, G. (2001). Susceptibility gene to infarct: a review of the literature. *Ital. Heart J. Suppl.* **2**, 935–944.

15. Tobin, M. D., Braund, P. S., Burton, P. R., Thompson, J. R., Steeds, R., Channer, K., Cheng, S., Lindpaintner, K., and Samani, N. J. (2004). Genotypes and haplotypes predisposing to myocardial infarction: a multilocus case-control study. *Eur. Heart J.* **25**, 459–467.
16. Ollikainen, E., Mikkelsson, J., Perola, M., Penttila, A., and Karhunen, P. J. (2004). Platelet membrane collagen receptor glycoprotein VI polymorphism is associated with coronary thrombosis and fatal myocardial infarction in middle-aged men. *Atherosclerosis* **176**, 95–99.
17. Grove, E. L., Orntoft, T. F., Lassen, J. F., Jensen, H. K., and Kristensen, S. D. (2004). The platelet polymorphism PlA2 is a genetic risk factor for myocardial infarction. *J. Intern. Med.* **255**, 637–644.
18. Ardissino, D., Mannucci, P. M., Merlini, P. A., Duca, F., Fetiveau, R., Tagliabue, L., Tubaro, M., Galvani, M., Ottani, F., Ferrario, M., Corral, J., and Margaglione, M. (1999). Prothrombotic genetic risk factors in young survivors of myocardial infarction. *Blood* **94**, 46–51.
19. Incalcaterra, E., Hoffmann, E., Averna, M. R., and Caimi, G. (2004). Genetic risk factors in myocardial infarction at young age. *Minerva Cardioangiol.* **52**, 287–312.
20. Yamada, Y., Izawa, H., Ichihara, S., Takatsu, F., Ishihara, H., Hirayama, H., Sone, T., Tanaka, M., and Yokota, M. (2002). Prediction of the risk of myocardial infarction from polymorphisms in candidate genes. *N. Engl. J. Med.* **347**, 1916–1923.
21. Juhan-Vague, I., Morange, P. E., Frere, C., Aillaud, M. F., Alessi, M. C., Hawe, E., Boquist, S., Tornvall, P., Yudkin, J. S., Tremoli, E., Margaglione, M., Di Minno, G., Hamsten, A., Humphries, S. E., and the HIFMECH Study Group. (2003). The plasminogen activator inhibitor-1 -675 4G/5G genotype influences the risk of myocardial infarction associated with elevated plasma proinsulin and insulin concentrations in men from Europe: the HIFMECH study. *J. Thromb. Haemost.* **1**, 2322–2329.
22. Aucella, F., Margaglione, M., Vigilante, M., Gatta, G., Grandone, E., Forcella, M., Ktena, M., De Min, A., Salatino, G., Procaccini, D. A., and Stallone, C. (2003). PAI-1 4G/5G and ACE I/D gene polymorphisms and the occurrence of myocardial infarction in patients on intermittent dialysis. *Nephrol. Dial. Transplant.* **18**, 1142–1146.
23. Girelli, D., Russo, C., Ferraresi, P., Olivieri, O., Pinotti, M., Friso, S., Manzato, F., Mazzucco, A., Bernardi, F., and Corrocher, R. (2000). Polymorphisms in the factor VII gene and the risk of myocardial infarction in patients with coronary artery disease. *N. Engl. J. Med.* **343**, 774–780.
24. Bozzini, C., Girelli, D., Bernardi, F., Ferraresi, P., Olivieri, O., Pinotti, M., Martinelli, N., Manzato, F., Friso, S., Villa, G., Pizzolo, F., Beltrame, F., and Corrocher, R. (2004). Influence of polymorphisms in the factor VII gene promoter on activated factor VII levels and on the risk of myocardial infarction in advanced coronary atherosclerosis. *Thromb. Haemost.* **92**, 541–549.
25. Mannila, M. N., Eriksson, P., Lundman, P., Samnegard, A., Boquist, S., Ericsson, C. G., Tornvall, P., Hamsten, A., and Silveira, A. (2005). Contribution of haplotypes across the fibrinogen gene cluster to variation in risk of myocardial infarction. *Thromb. Haemost.* **93**, 570–577.
26. Ott, I., Koch, W., von Beckerath, N., de Waha, R., Malawaniec, A., Mehilli, J., Schomig, A., and Kastrati, A. (2004). Tissue factor promotor polymorphism -603 A/G is associated with myocardial infarction. *Atherosclerosis* **177**, 189–191.
27. Norlund, L., Holm, J., Zoller, B., and Ohlin, A. K. (1997). A common thrombomodulin amino acid dimorphism is associated with myocardial infarction. *Thromb. Haemost.* **77**, 248–251.
28. Ranjith, N., Pegoraro, R. J., and Rom, L. (2003). Haemostatic gene polymorphisms in young Indian Asian subjects with acute myocardial infarction. *Med. Sci. Monit.* **9**, CR417–CR421.
29. Chao, T. H., Li, Y. H., Chen, J. H., Wu, H. L., Shi, G. Y., Tsai, W. C., Chen, P. S., and Liu, P. Y. (2004). Relation of thrombomodulin gene polymorphisms to acute myocardial infarction in patients ≤50 years of age. *Am. J. Cardiol.* **93**, 204–207.
30. Topol, E. J., McCarthy, J., Gabriel, S., Moliterno, D. J., Rogers, W. J., Newby, L. K., Freedman, M., Metivier, J., Cannata, R., O'Donnell, C. J., Kottke-Marchant, C., Murugesan, G., Plow, E. F., Stenina, O., and Daley, G. Q. (2001). Single nucleotide polymorphisms in multiple novel thrombospondin genes may be associated with familial premature myocardial infarction. *Circulation* **104**, 2641–2644.
31. Boekholdt, S. M., Trip, M. D., Peters, R. J., Engelen, M., Boer, J. M., Feskens, E. J., Zwinderman, A. H., Kastelein, J. J., and Reitsma, P. H. (2002). Thrombospondin-2 polymorphism is associated with a reduced risk of premature myocardial infarction. *Arterioscler. Thromb. Vasc. Biol.* **22**, e24–e27.
32. Wessel, J., Topol, E. J., Ji, M., Meyer, J., and McCarthy, J. J. (2004). Replication of the association between the thrombospondin-4 A387P polymorphism and myocardial infarction. *Am. Heart J.* **147**, 905–909.
33. Zwicker, J. I., Pevandi, F., Palla, R., Lombardi, R., Canciani, M. T., Cairo, A., Ardissino, D., Bernardinelli, L., Bauer, K. A., Lawler, J., and Mannucci, P. M. (2006). The thrombospondin-1 N700S polymorphism is associated with early myocardial infarction without altering von Willebrand Factor multimer size. *Blood* **108**, 1280–1283.
34. Ridker, P. M., Hennekens, C. H., Lindpaintner, K., Stampfer, M. J., Eisenberg, P. R., and Miletich, J. P. (1995). Mutation in the gene coding for coagulation factor V and the risk of myocardial infarction, stroke, and venous thrombosis in apparently healthy men. *N. Engl. J. Med.* **332**, 912–917.
35. Ucar, F., Celik, S., Ovali, E., Karti, S. S., Pakdemir, A., Yilmaz, M., and Onder, E. (2004). Coexistence of prothrombic risk factors and its relation to left ventricular thrombus in acute myocardial infarction. *Acta Cardiol.* **59**, 33–39.
36. Doix, S., Mahroussseh, M., Jolak, M., Laurent, Y., Lorenzini, J. L., Binquet, C., Zeller, M., Cottin, Y., and Wolf, J. E. (2003). Factor V Leiden and myocardial infarction, a case, review of the literature with a meta-analysis. *Ann. Cardiol. Angeiol. (Paris)* **52**, 143–149.
37. Burzotta, F., Paciaroni, K., De Stefano, V., Crea, F., Maseri, A., Leone, G., and Andreotti, F. (2004). G20210A prothrombin gene polymorphism and coronary ischaemic syndromes: a phenotype-specific meta-analysis of 12,034 subjects. *Heart* **90**, 82–86.
38. Russo, C., Girelli, D., Olivieri, O., Guarini, P., Manzato, F., Pizzolo, F., Zaia, B., Mazzucco, A., and Corrocher, R. (2001). G20210A prothrombin gene polymorphism and prothrombin activity in subjects with or without angiographically documented coronary artery disease. *Circulation* **103**, 2436–2440.

39. Lagercrantz, J., Bergman, M., Lundman, P., Tornvall, P., Hjemdahl, P., Hamsten, A., and Eriksson, P. (2003). No evidence that the PLA1/PLA2 polymorphism of platelet glycoprotein IIIa is implicated in angiographically characterized coronary atherosclerosis and premature myocardial infarction. *Blood Coagul. Fibrinolysis* **14**, 749–753.

40. Roldan, V., Gonzalez-Conejero, R., Marin, F., Pineda, J., Vicente, V., and Corral, J. (2005). Five prothrombotic polymorphisms and the prevalence of premature myocardial infarction. *Haematologica* **90**, 421–423.

41. Weiss, E. J., Bray, P. F., Tayback, M., Schulman, S. P., Kickler, T. S., Becker, L. C., Weiss, J. L., Gerstenblith, G., and Goldschmidt-Clermont, P. J. (1996). A polymorphism of a platelet glycoprotein receptor as an inherited risk factor for coronary thrombosis. *N. Engl. J. Med.* **334**, 1090–1094.

42. Sugano, T., Tsuji, H., Masuda, H., Nakagawa, K., Nishimura, H., Kasahara, T., Yoshizumi, M., Nakahara, Y., Kitamura, H., Yamada, K., Yoneda, M., Maki, K., Tatsumi, T., Azuma, A., and Nakagawa, M. (1998). Plasminogen activator inhibitor-1 promoter 4G/5G genotype is not a risk factor for myocardial infarction in a Japanese population. *Blood Coagul. Fibrinolysis* **9**, 201–204.

43. Anderson, J. L., Muhlestein, J. B., Habashi, J., Carlquist, J. F., Bair, T. L., Elmer, S. P., and Davis, B. P. (1999). Lack of association of a common polymorphism of the plasminogen activator inhibitor-1 gene with coronary artery disease and myocardial infarction. *J. Am. Coll. Cardiol.* **34**, 1778–1783.

44. Narizhneva, N. V., Byers-Ward, V. J., Quinn, M. J., Zidar, F. J., Plow, E. F., Topol, E. J, and Byzova, T. V. (2004). Molecular and functional differences induced in thrombospondin-1 by the single nucleotide polymorphism associated with the risk of premature, familial myocardial infarction. *J. Biol. Chem.* **279**, 21651–21657.

45. Arnaud, E., Barbalat, V., Nicaud, V., Cambien, F., Evans, A., Morrison, C., Arveiler, D., Luc, G., Ruidavets, J. B., Emmerich, J., Fiessinger, J. N., and Aiach, M. (2000). Polymorphisms in the 5′ regulatory region of the tissue factor gene and the risk of myocardial infarction and venous thromboembolism: the ECTIM and PATHROS studies. Etude Cas-Temoins de l'Infarctus du Myocarde. Paris Thrombosis case-control Study. *Arterioscler. Thromb. Vasc. Biol.* **20**, 892–898.

46. Zito, F., Di Castelnuovo, A., Amore, C., D'Orazio, A., Donati, M. B., and Iacoviello, L. (1997). Bcl I polymorphism in the fibrinogen beta-chain gene is associated with the risk of familial myocardial infarction by increasing plasma fibrinogen levels. A case-control study in a sample of GISSI-2 patients. *Arterioscler. Thromb. Vasc. Biol.* **17**, 3489–3494.

47. Behague, I., Poirier, O., Nicaud, V., Evans, A., Arveiler, D., Luc, G., Cambou, J. P., Scarabin, P. Y., Bara, L., Green, F., and Cambien, F. (1996). Beta fibrinogen gene polymorphisms are associated with plasma fibrinogen and coronary artery disease in patients with myocardial infarction. The ECTIM Study. Etude Cas-Temoins sur l'Infarctus du Myocarde. *Circulation* **93**, 44044–9.

48. Terashima, M., Akita, H., Kanazawa, K., Inoue, N., Yamada, S., Ito, K., Matsuda, Y., Takai, E., Iwai, C., Kurogane, H., Yoshida, Y., and Yokoyama, M. (1999). Stromelysin promoter 5A/6A polymorphism is associated with acute myocardial infarction. *Circulation* **99**, 2717–2719.

49. Ye, S. (2000). Polymorphism in matrix metalloproteinase gene promoters: implication in regulation of gene expression and susceptibility of various diseases. *Matrix Biol.* **19**, 623–629.

50. Ye, S., Eriksson, P., Hamsten, A., Kurkinen, M., Humphries, S. E., and Henney, A. M. (1996). Progression of coronary atherosclerosis is associated with a common genetic variant of the human stromelysin-1 promoter which results in reduced gene expression. *J. Biol. Chem.* **271**, 13055–13060.

51. Lamblin, N., Bauters, C., Hermant, X., Lablanche, J. M., Helbecque, N., and Amouyel, P. (2002). Polymorphisms in the promoter regions of MMP-2, MMP-3, MMP-9 and MMP-12 genes as determinants of aneurysmal coronary artery disease. *J. Am. Coll. Cardiol.* **40**, 43–48.

52. Kim, J. S., Park, H. Y., Kwon, J. H., Im, E. K., Choi, D. H., Jang, Y. S., and Cho, S. Y. (2002). The roles of stromelysin-1 and the gelatinase B gene polymorphism in stable angina. *Yonsei Med. J.* **43**, 473–481.

53. Liu, P. Y., Chen, J. H., Li, Y. H., Wu, H. L., and Shi, G. Y. (2003). Synergistic effect of stromelysin-1 (matrix metalloproteinase-3) promoter 5A/6A polymorphism with smoking on the onset of young acute myocardial infarction. *Thromb. Haemost.* **90**, 132–139.

54. Samnegard, A., Silveira, A., Lundman, P., Boquist, S., Odeberg, J., Hulthe, J., McPheat, W., Tornvall, P., Bergstrand, L., Ericsson, C. G., Hamsten, A., and Eriksson, P. (2005). Serum matrix metalloproteinase-3 concentration is influenced by MMP-3 -1612 5A/6A promoter genotype and associated with myocardial infarction. *J. Intern. Med.* **258**, 411–419.

55. Listi, F., Candore, G., Lio, D., Cavallone, L., Colonna-Romano, G., Caruso, M., Hoffmann, E., and Caruso, C. (2004). Association between platelet endothelial cellular adhesion molecule 1 (PECAM-1/CD31) polymorphisms and acute myocardial infarction: a study in patients from Sicily. *Eur. J. Immunogenet.* **31**, 175–178.

56. Sasaoka, T., Kimura, A., Hohta, S. A., Fukuda, N., Kurosawa, T., and Izumi, T. (2001). Polymorphisms in the platelet-endothelial cell adhesion molecule-1 (PECAM-1) gene, Asn563Ser and Gly670Arg, associated with myocardial infarction in the Japanese. *Ann. N. Y. Acad. Sci.* **947**, 259–269.

57. Holloway, J. W., Yang, I. A., and Ye, S. (2005). Variation in the toll-like receptor 4 gene and susceptibility to myocardial infarction. *Pharmacogenet. Genomics* **15**, 15–21.

58. Edfeldt, K., Bennet, A. M., Eriksson, P., Frostegard, J., Wiman, B., Hamsten, A., Hansson, G. K., de Faire, U., and Yan, Z. Q. (2004). Association of hypo-responsive toll-like receptor 4 variants with risk of myocardial infarction. *Eur. Heart J.* **25**, 1447–1453.

59. Iacoviello, L., Di Castelnuovo, A., Gattone, M., Pezzini, A., Assanelli, D., Lorenzet, R., Del Zotto, E., Colombo, M., Napoleone, E., Amore, C., D'Orazio, A., Padovani, A., de Gaetano, G., Giannuzzi, P., and Donati, M. B., and the IGIGI Investigators. (2005). Polymorphisms of the interleukin-1beta gene affect the risk of myocardial infarction and ischemic stroke at young age and the response of mononuclear cells to stimulation in vitro. *Arterioscler. Thromb. Vasc. Biol.* **25**, 222–227.

60. Li, J., Wang, X., Huo, Y., Niu, T., Chen, C., Zhu, G., Huang, Y., Chen, D., and Xu, X. (2005). PON1 polymorphism, diabetes mellitus, obesity, and risk of myocardial infarction: Modifying effect of diabetes mellitus and obesity on the association

between PON1 polymorphism and myocardial infarction. *Genet. Med.* **7,** 58–63.

61. Baum, L., Ng, H. K., Woo, K. S., Tomlinson, B., Rainer, T. H., Chen, X., Cheung, W. S., Yin Chan, D. K., Thomas, G. N., Wai Tong, C. S., and Wong, K. S. (2006). Paraoxonase 1 gene Q192R polymorphism affects stroke and myocardial infarction risk. *Clin. Biochem.* **39,** 191–195.

62. Martinelli, N., Girelli, D., Olivieri, O., Stranieri, C., Trabetti, E., Pizzolo, F., Friso, S., Tenuti, I., Cheng, S., Grow, M. A., Pignatti, P. F., and Corrocher, R. (2004). Interaction between smoking and PON2 Ser311Cys polymorphism as a determinant of the risk of myocardial infarction. *Eur. J. Clin. Invest.* **34,** 14–20.

63. Chao, T. H., Li, Y. H., Chen, J. H., Wu, H. L., Shi, G. Y., Liu, P. Y., Tsai, W. C., and Guo, H. R. (2004). The 161TT genotype in the exon 6 of the peroxisome-proliferator-activated receptor γ gene is associated with premature acute myocardial infarction and increased lipid peroxidation in habitual heavy smokers. *Clin. Sci. (Lond)* **107,** 461–466.

64. Ridker, P. M., Cook, N. R., Cheng, S., Erlich, H. A., Lindpaintner, .K, Plutzky, J., and Zee, R. Y. (2003). Alanine for proline substitution in the peroxisome proliferator-activated receptor γ-2 (PPARG2) gene and the risk of incident myocardial infarction. *Arterioscler. Thromb. Vasc. Biol.* **23,** 859–863.

65. Morange, P. E., Saut, N., Alessi, M. C., Frere, C., Hawe, E., Yudkin, J. S., Tremoli, E., Margaglione, M., Di Minno, G., Hamsten, A., Humphries, S. E., Juhan-Vague, I., and the HIFMECH Study Group. (2005). Interaction between the C-260T polymorphism of the CD14 gene and the plasma IL-6 concentration on the risk of myocardial infarction, the HIFMECH study. *Atherosclerosis* **179,** 317–323.

66. Hubacek, J. A., Rothe, G., Pit'ha, J., Skodova, Z., Stanek, V., Poledne, R., and Schmitz, G. C. (1999). (-260)→T polymorphism in the promoter of the CD14 monocyte receptor gene as a risk factor for myocardial infarction. *Circulation* **99,** 3218–3220.

67. Zee, R. Y., Cook, N. R., Cheng, S., Erlich, H. A., Lindpaintner, K., Lee, R. T., and Ridker, P. M. (2004). Threonine for alanine substitution in the eotaxin (CCL11) gene and the risk of incident myocardial infarction. *Atherosclerosis* **175,** 91–94.

68. Licastro, F., Chiappelli, M., Caldarera, C. M., Tampieri, C., Nanni, S., Gallina, M., and Branzi, A. (2004). The concomitant presence of polymorphic alleles of interleukin-1β, interleukin-6 and apolipoprotein E is associated with an increased risk of myocardial infarction in elderly men. Results from a pilot study. *Mech. Ageing Dev.* **125,** 575–579.

69. Lieb, W., Pavlik, R., Erdmann, J., Mayer, B., Holmer, S. R., Fischer, M., Baessler, A., Hengstenberg, C., Loewel, H., Doering, A., Riegger, G. A., and Schunkert, H. (2004). No association of interleukin-6 gene polymorphism (-174 G/C) with myocardial infarction or traditional cardiovascular risk factors. *Int. J. Cardiol.* **97,** 205–212.

70. Keso, T., Perola, M., Laippala, P., Ilveskoski, E., Kunnas, T. A., Mikkelsson, J., Penttila, A., Hurme, M., and Karhunen, P. J. (2001). Polymorphisms within the tumor necrosis factor locus and prevalence of coronary artery disease in middle-aged men. *Atherosclerosis* **154,** 691–697.

71. Koch, W., Tiroch, K., von Beckerath, N., Schomig, A., and Kastrati, A. (2003). Tumor necrosis factor-alpha, lymphotoxin-alpha, and interleukin-10 gene polymorphisms and restenosis after coronary artery stenting. *Cytokine* **24,** 161–171.

72. Bennermo, M., Held, C., Green, F., Strandberg, L. E., Ericsson, C. G., Hansson, L. O., Watkins, H., Hamsten, A., and Tornvall, P. (2004). Prognostic value of plasma interleukin-6 concentrations and the -174 G > C and -572 G > C promoter polymorphisms of the interleukin-6 gene in patients with acute myocardial infarction treated with thrombolysis. *Atherosclerosis* **174,** 157–163.

73. Deten, A., Volz, H. C., Briest, W., and Zimmer, H. G. (2003). Differential cytokine expression in myocytes and non-myocytes after myocardial infarction in rats. *Mol. Cell Biochem.* **242,** 47–55.

74. Georges, J. L., Loukaci, V., Poirier, O., Evans, A., Luc, G., Arveiler, D., Ruidavets, J. B., Cambien, F., and Tiret, L. (2001). Interleukin-6 gene polymorphisms and susceptibility to myocardial infarction: the ECTIM study. Etude Cas-Temoin de l'Infarctus du Myocarde. *J. Mol. Med.* **79,** 300–305.

75. Nauck, M., Winkelmann, B. R., Hoffmann, M. M., Bohm, B. O., Wieland, H., and Marz, W. (2002). The interleukin-6 G(-174)C promoter polymorphism in the LURIC cohort: no association with plasma interleukin-6, coronary artery disease, and myocardial infarction. *J. Mol. Med.* **80,** 507–513.

76. Schreyer, S. A., Vick, C. M., and LeBoeuf, R. C. (2002). Loss of lymphotoxin-alpha but not tumor necrosis factor-α reduces atherosclerosis in mice. *J. Biol. Chem.* **277,** 12364–12368.

77. Padovani, J. C., Pazin-Filho, A., Simoes, M. V., Marin-Neto, J. A., Zago, M. A., and Franco, R. F. (2000). Gene polymorphisms in the TNF locus and the risk of myocardial infarction. *Thromb. Res.* **100,** 263–269.

78. Laxton, R., Pearce, E., Kyriakou, T., and Ye, S. (2005). Association of the lymphotoxin-alpha gene Thr26Asn polymorphism with severity of coronary atherosclerosis. *Genes Immun.* **6,** 539–541.

79. Doney, A. S., Fischer, B., Leese, G., Morris, A. D., and Palmer, C. N. (2004). Cardiovascular risk in type 2 diabetes is associated with variation at the PPARG locus: a Go-DARTS study. *Arterioscler. Thromb. Vasc. Biol.* **24,** 2403–2407.

80. Doney, A. S., Fischer, B., Cecil, J. E., Boylan, K., McGuigan, F. E., Ralston, S. H., Morris, A. D., and Palmer, C. N. (2004). Association of the Pro12Ala and C1431T variants of PPARG and their haplotypes with susceptibility to Type 2 diabetes. *Diabetologia* **47,** 555–558.

81. Aubo, C., Senti, M., Marrugat, J., Tomas, M., Vila, J., Sala, J., and Masia, R. (2000). Risk of myocardial infarction associated with Gln/Arg 192 polymorphism in the human paraoxonase gene and diabetes mellitus. The REGICOR Investigators. *Eur. Heart J.* **21,** 33–38.

82. Senti, M., Tomas, M., Vila, J., Marrugat, J., Elosua, R., Sala, J., and Masia, R. (2001). Relationship of age-related myocardial infarction risk and Gln/Arg 192 variants of the human paraoxonase1 gene: the REGICOR study. *Atherosclerosis* **156,** 443–449.

83. Tregouet, D. A., Barbaux, S., Escolano, S., Tahri, N., Golmard, J. L., Tiret, L., and Cambien, F. (2002). Specific haplotypes of the P-selectin gene are associated with myocardial infarction. *Hum. Mol. Genet.* **11,** 2015–2023.

84. Yoshida, M., Takano, Y., Sasaoka, T., Izumi, T., and Kimura, A. (2003). E-selectin polymorphism associated with myocardial infarction causes enhanced leukocyte-endothelial interac-

tions under flow conditions. *Arterioscler. Thromb. Vasc. Biol.* **23,** 783–788.

85. Hohda, S., Kimura, A., Sasaoka, T., Hayashi, T., Ueda, K., Yasunami, M., Okabe, M., Fukuta, N., Kurosawa, T., and Izumi, T. (2003). Association study of CD14 polymorphism with myocardial infarction in a Japanese population. *Jpn. Heart J.* **44,** 613–622.

86. Arroyo-Espliguero, R., El-Sharnouby, K., Vazquez-Rey, E., Kalidas, K., Jeffery, S., and Kaski, J. C. (2005). CD14 C(-260)T promoter polymorphism and prevalence of acute coronary syndromes. *Int. J. Cardiol.* **98,** 307–312.

87. Zee, R. Y., Lindpaintner, K., Struk, B., Hennekens, C. H., and Ridker, P. M. (2001). A prospective evaluation of the CD14 C(-260)T gene polymorphism and the risk of myocardial infarction. *Atherosclerosis* **154,** 699–702.

88. Nauck, M., Winkelmann, B. R., Hoffmann, M. M., Bohm, B. O., Wieland, H., and Marz, W. C. (2002). (-260)T polymorphism in the promoter of the CD14 gene is not associated with coronary artery disease and myocardial infarction in the Ludwigshafen Risk and Cardiovascular Health (LURIC) study. *Am. J. Cardiol.* **90,** 1249–1252.

89. Shishido, T., Nozaki, N., Yamaguchi, S., Shibata, Y., Nitobe, J., Miyamoto, T., Takahashi, H., Arimoto, T., Maeda, K., Yamakawa, M., Takeuchi, O., Akira, S., Takeishi, Y., and Kubota, I. (2003). Toll-like receptor-2 modulates ventricular remodeling after myocardial infarction. *Circulation* **108,** 2905–2910.

90. Zee, R. Y., Hegener, H. H., Gould, J., and Ridker, P. M. (2005). Toll-like receptor 4 Asp299Gly gene polymorphism and risk of atherothrombosis. *Stroke* **36,** 154–157.

91. Boekholdt, S. M., Agema, W. R., Peters, R. J., Zwinderman, A. H., van der Wall, E. E., Reitsma, P. H., Kastelein, J .J, Jukema, J. W., and the Regression Growth Evaluation Statin Study Group. (2003). Variants of toll-like receptor 4 modify the efficacy of statin therapy and the risk of cardiovascular events. *Circulation* **107,** 2416–2421.

92. Helgadottir, A., Gretarsdottir, S., St. Clair, D., Manolescu, A., Cheung, J., Thorleifsson, G., Pasdar, A., Grant, S. F., Whalley, L. J., Hakonarson, H., Thorsteinsdottir, U., Kong, A., Gulcher, J., Stefansson, K., and MacLeod, M. J. (2005). Association between the gene encoding 5-lipoxygenase-activating protein and stroke replicated in a Scottish population. *Am. J. Hum. Genet.* **76,** 505–509.

93. Kajimoto, K., Shioji, K., Ishida, C., Iwanaga, Y., Kokubo, Y., Tomoike, H., Miyazaki, S., Nonogi, H., Goto, Y., and Iwai, N. (2005). Validation of the association between the gene encoding 5-lipoxygenase-activating protein and myocardial infarction in a Japanese population. *Circ. J.* **69,** 1029–1034.

94. Gretarsdottir, S., Thorleifsson, G., Reynisdottir, S. T., Manolescu, A., Jonsdottir, S., Jonsdottir, T., Gudmundsdottir, T., Bjarnadottir, S. M., Einarsson, O. B., Gudjonsdottir, H. M., Hawkins, M., Gudmundsson, G., Gudmundsdottir, H., Andrason, H., Gudmundsdottir, A. S., Sigurdardottir, M., Chou, T. T., Nahmias, J., Goss, S., Sveinbjornsdottir, S., Valdimarsson, E. M., Jakobsson, F., Agnarsson, U., Gudnason, V., Thorgeirsson, G., Fingerle, J., Gurney, M., Gudbjartsson, D., Frigge, M. L., Kong, A., Stefansson, K., and Gulcher, J. R. (2003). The gene encoding phosphodiesterase 4D confers risk of ischemic stroke. *Nat. Genet.* **35,** 131–138.

95. Cipollone, F., Toniato, E., Martinotti, S., Fazia, M., Iezzi, A., Cuccurullo, C., Pini, B., Ursi, S., Vitullo, G., Averna, M., Arca, M., Montali, A., Campagna, F., Ucchino, S., Spigonardo, F., Taddei, S., Virdis, A., Ciabattoni, G., Notarbartolo, A., Cuccurullo, F., and Mezzetti, A., and the Identification of New Elements of Plaque Stability (INES) Study Group. (2004). A polymorphism in the cyclooxygenase 2 gene as an inherited protective factor against myocardial infarction and stroke. *JAMA* **291,** 2221–2228.

96. Dai, K., Gao, W., and Ruan, C. (2001). The Sma I polymorphism in the von Willebrand factor gene associated with acute ischemic stroke. *Thromb. Res.* **104,** 389–395.

97. Kim, R. J., and Becker, R. C. (2003). Association between factor V Leiden, prothrombin G20210A, and methylenetetrahydrofolate reductase C677T mutations and events of the arterial circulatory system: a meta-analysis of published studies. *Am. Heart J.* **146,** 948–957.

98. Cronin, S., Furie, K. L., and Kelly, P. J. (2005). Dose-related association of MTHFR 677T allele with risk of ischemic stroke. Evidence from a cumulative meta-analysis. *Stroke* **36,** 1581–1587.

99. Shearman, A. M., Cupples, L. A., Demissie, S., Peter, I., Schmid, C. H., Karas, R. H., Mendelsohn, M. E., Housman, D. E., and Levy, D. (2003). Association between estrogen receptor α gene variation and cardiovascular disease. *JAMA* **290,** 2263–2270.

100. Shearman, A. M., Cooper, J. A., Kotwinski, P. J., Miller, G. J., Humphries, S. E., Ardlie, K. G., Jordan, B., Irenze, K., Lunetta, K. L., Schuit, S. C., Uitterlinden, A. G., Pols, H. A., Demissie, S., Cupples, L. A., Mendelsohn, M. E., Levy, D., and Housman, D. E. (2006). Estrogen receptor α gene variation is associated with risk of myocardial infarction in more than seven thousand men from five cohorts. *Circ. Res.* **98,** 590–592.

101. Shearman, A. M., Cooper, J. A., Kotwinski, P. J., Humphries, S. E., Mendelsohn, M. E., Housman, D. E., and Miller, G. J. (2005). Estrogen receptor α gene variation and the risk of stroke. *Stroke* **36,** 2281–2282.

102. Inbal, A., Freimark, D., Modan, B., Chetrit, A., Matetzky, S., Rosenberg, N., Dardik, R., Baron, Z., and Seligsohn, U. (1999). Synergistic effects of prothrombotic polymorphisms and atherogenic factors on the risk of myocardial infarction in young males. *Blood* **93,** 2186–2190.

103. Brscic, E., Bergerone, S., Gagnor, A., Colajanni, E., Matullo, G., Scaglione, L., Cassader, M., Gaschino, G., Di Leo, M., Brusca, A., Pagano, G. F., Piazza, A., and Trevi, G. P. (2000). Acute myocardial infarction in young adults: prognostic role of angiotensin-converting enzyme, angiotensin II type I receptor, apolipoprotein E, endothelial constitutive nitric oxide synthase, and glycoprotein IIIa genetic polymorphisms at medium-term follow-up. *Am. Heart J.* **139,** 979–984.

104. Scuteri, A., Bos, A. J., Zonderman, A. B., Brant, L. J., Lakatta, E. G., and Fleg, J. L. (2001). Is the apoE4 allele an independent predictor of coronary events? *Am. J. Med.* **110,** 28–32.

105. Frikke-Schmidt, R., Tybjaerg-Hansen, A., Steffensen, R., Jensen, G., and Nordestgaard, B. (2000). Apolipoprotein E genotype: epsilon32 women are protected while epsilon43 and epsilon44 men are susceptible to ischemic heart disease:

the Copenhagen City Heart Study. *J. Am. Coll. Cardiol.* **35**, 1192–1199.
106. Yamamura, T., Dong, L. M., and Yamamoto, A. (1992). Apolipoprotein E polymorphism and coronary heart disease. *Chin. Med. J. (Engl)* **105**, 738–741.
107. Tregouet, D. A., Ricard, S., Nicaud, V., Arnould, I., Soubigou, S., Rosier, M., Duverger, N., Poirier, O., Mace, S., Kee, F., Morrison, C., Denefle, P., Tiret, L., Evans, A., Deleuze, J. F., and Cambien, F. (2004). In-depth haplotype analysis of ABCA1 gene polymorphisms in relation to plasma ApoA1 levels and myocardial infarction. *Arterioscler. Thromb. Vasc. Biol.* **24**, 775–781.
108. Schuit, S. C., Oei, H. H., Witteman, J. C., Geurts van Kessel, C. H., van Meurs, J. B., Nijhuis, R. L., van Leeuwen, J. P., de Jong, F. H., Zillikens, M. C., Hofman, A., Pols, H. A., and Uitterlinden, A. G. (2004). Estrogen receptor alpha gene polymorphisms and risk of myocardial infarction. *JAMA* **291**, 2969–2977.
109. Gonzalez-Conejero, R., Corral, J., Roldan, V., Martinez, C., Marin, F., Rivera, J., Iniesta, J. A., Lozano, M. L., Marco, P., and Vicente, V. (2002). A common polymorphism in the annexin V Kozak sequence (-1C>T) increases translation efficiency and plasma levels of annexin V, and decreases the risk of myocardial infarction in young patients. *Blood* **100**, 2081–2086.
110. Iwai, C., Akita, H., Kanazawa, K., Shiga, N., Terashima, M., Matsuda, Y., Takai, E., Miyamoto, Y., Shimizu, M., Kajiya, T., Hayashi, T., and Yokoyama, M. (2003). Arg389Gly polymorphism of the human β1-adrenergic receptor in patients with nonfatal acute myocardial infarction. *Am. Heart J.* **146**, 106–109.
111. Small, K. M., Wagoner, L. E., Levin, A. M., Kardia, S. L., and Liggett, S. B. (2002). Synergistic polymorphisms of β1- and α2C-adrenergic receptors and the risk of congestive heart failure. *N. Engl. J. Med.* **347**, 1135–1142.
112. Bengtsson, K., Melander, O., Orho-Melander, M., Lindblad, U., Ranstam, J., Rastam, L., and Groop, L. (2001). Polymorphism in the β(1)-adrenergic receptor gene and hypertension. *Circulation* **104**, 187–190.
113. Liu, J., Liu, Z. Q., Tan, Z. R., Chen, X. P., Wang, L. S., Zhou, G., and Zhou, H. H. (2003). Gly389Arg polymorphism of β1-adrenergic receptor is associated with the cardiovascular response to metoprolol. *Clin. Pharmacol. Ther.* **74**, 372–379.
114. Akhter, S. A., D'Souza, K. M., Petrashevskaya, N. N., Mialet-Perez, J., and Liggett, S. B. (2006). Myocardial β1-adrenergic receptor polymorphisms affect functional recovery after ischemic injury. *Am. J. Physiol. Heart Circ. Physiol.* **290**, H1427–H1432.
115. Fatini, C., Sofi, F., Sticchi, E., Gensini, F., Gori, A. M., Fedi, S., Lapini, I., Rostagno, C., Comeglio, M., Brogi, D., Gensini, G., and Abbate, R. (2004). Influence of endothelial nitric oxide synthase gene polymorphisms (G894T, 4a4b, T-786C) and hyperhomocysteinemia on the predisposition to acute coronary syndromes. *Am. Heart J.* **147**, 516–521.
116. Hibi, K., Ishigami, T., Tamura, K., Mizushima, S., Nyui, N., Fujita, T., Ochiai, H., Kosuge, M., Watanabe, Y., Yoshii, Y., Kihara, M., Kimura, K., Ishii, M., and Umemura, S. (1998). Endothelial nitric oxide synthase gene polymorphism and acute myocardial infarction. *Hypertension* **32**, 521–526.
117. Shimasaki, Y., Yasue, H., Yoshimura, M., Nakayama, M., Kugiyama, K., Ogawa, H., Harada, E., Masuda, T., Koyama, W., Saito, Y., Miyamoto, Y., Ogawa, Y., and Nakao, K. (1998). Association of the missense Glu298Asp variant of the endothelial nitric oxide synthase gene with myocardial infarction. *J. Am. Coll. Cardiol.* **31**, 1506–1510.
118. Antoniades, C., Tousoulis, D., Vasiliadou, C., Pitsavos, C., Toutouza, M., Tentolouris, C., Marinou, K., and Stefanadis, C. (2006). Genetic polymorphisms G894T on the eNOS gene is associated with endothelial function and vWF levels in premature myocardial infarction survivors. *Int. J. Cardiol.* **107**, 95–100.
119. Antoniades, C., Tousoulis, D., Vasiliadou, C., Pitsavos, C., Chrysochoou, C., Panagiotakos, D., Tentolouris, C., Marinou, K., Koumallos, N., and Stefanadis, C. (2005). Genetic polymorphism on endothelial nitric oxide synthase affects endothelial activation and inflammatory response during the acute phase of myocardial infarction. *J. Am. Coll. Cardiol.* **46**, 1101–1109.
120. Casas, J. P., Bautista, L. E., Humphries, S. E., and Hingorani, A. D. (2004). Endothelial nitric oxide synthase genotype and ischemic heart disease: meta-analysis of 26 studies involving 23028 subjects. *Circulation* **109**, 1359–1365.
121. Hingorani, A. D., Liang, C. F., Fatibene, J., Lyon, A., Monteith, S., Parsons, A., Haydock, S., Hopper, R. V., Stephens, N. G., O'Shaughnessy, K. M., and Brown, M. J. (1999). A common variant of the endothelial nitric oxide synthase (Glu298→Asp) is a major risk factor for coronary artery disease in the UK. *Circulation* **100**, 1515–1520.
122. Granath, B., Taylor, R. R., van Bockxmeer, F. M., and Mamotte, C. D. (2001). Lack of evidence for association between endothelial nitric oxide synthase gene polymorphisms and coronary artery disease in the Australian Caucasian population. *J. Cardiovasc. Risk* **8**, 235–241.
123. Jo, I., Moon, J., Yoon, S., Kim, H. T., Kim, E., Park, H. Y., Shin, C., Min, J., Jin, Y. M., Cha, S. H., and Jo, S. A. (2006). Interaction between -786TC polymorphism in the endothelial nitric oxide synthase gene and smoking for myocardial infarction in Korean population. *Clin. Chim. Acta* **365**, 86–92.
124. Gulec, S., Aras, O., Akar, E., Tutar, E., Omurlu, K., Avci, F., Dincer, I., Akar, N., and Oral, D. (2001). Methylenetetrahydrofolate reductase gene polymorphism and risk of premature myocardial infarction. *Clin. Cardiol.* **24**, 281–284.
125. Nakai, K., Itoh, C., Nakai, K., Habano, W., and Gurwitz, D. (2001). Correlation between C677T MTHFR gene polymorphism, plasma homocysteine levels and the incidence of CAD. *Am. J. Cardiovasc. Drugs* **1**, 353–361.
126. Brugada, R., and Marian, A. J. (1997). A common mutation in methylenetetrahydrofolate reductase gene is not a major risk of coronary artery disease or myocardial infarction. *Atherosclerosis* **128**, 107–112.
127. Andrikopoulos, G. K., Richter, D. J., Needham, E. W., Tzeis, S. E., Zairis, M. N., Gialafos, E. J., Vogiatzi, P. G., Papasteriadis, E. G., Kardaras, F. G., Foussas, S. G., Gialafos, J. E., Stefanadis, C. I., Toutouzas, P. K., Mattu, R. K., and the GEMIG study investigators. (2004). The paradoxical association of common polymorphisms of the renin-angiotensin system genes with risk of myocardial infarction. *Eur. J. Cardiovasc. Prev. Rehabil.* **11**, 477–483.
128. Petrovic, D., and Peterlin, B. (2004). Pharmacogenomic considerations of the insertion/deletion gene polymorphism

of the Angiotensin I-converting enzyme and coronary artery disease. *Curr. Vasc. Pharmacol.* **2**, 271–279.

129. Petrovic, D., Bregar, D., Guzic-Salobir, B., Skof, E., Span, M., Terzic, R., Petrovic, M. G., Keber, I., Letonja, M., Zorc, M., Podbregar, M., and Peterlin, B. (2004). Sex difference in the effect of ACE-DD genotype on the risk of premature myocardial infarction. *Angiology* **55**, 155–158.

130. Liu, S., Schmitz, C., Stampfer, M. J., Sacks, F., Hennekens, C. H., Lindpaintner, K., and Ridker, P. M. (2002). A prospective study of TaqIB polymorphism in the gene coding for cholesteryl ester transfer protein and risk of myocardial infarction in middle-aged men. *Atherosclerosis* **161**, 469–474.

131. Araujo, M. A., Goulart, L. R., Cordeiro, E. R., Gatti, R. R., Menezes, B. S., Lourenco, C., and Silva, H. D. (2005). Genotypic interactions of renin-angiotensin system genes in myocardial infarction. *Int. J. Cardiol.* **103**, 27–32.

132. Holmer, S. R., Bickeboller, H., Hengstenberg, C., Rohlmann, F., Engel, S., Lowel, H., Mayer, B., Erdmann, J., Baier, C., Klein, G., Riegger, G. A., and Schunkert, H. (2003). Angiotensin converting enzyme gene polymorphism and myocardial infarction a large association and linkage study. *Int. J. Biochem. Cell Biol.* **35**, 955–962.

133. Agerholm-Larsen, B., Nordestgaard, B. G., and Tybjaerg-Hansen, A. (2000). ACE gene polymorphism in cardiovascular disease: meta-analyses of small and large studies in whites. *Arterioscler. Thromb. Vasc. Biol.* **20**, 484–492.

134. Morgan, T. M., Coffey, C. S., and Krumholz, H. M. (2003). Overestimation of genetic risks owing to small sample sizes in cardiovascular studies. *Clin. Genet.* **64**, 7–17.

135. Ichihara, S., Yamada, Y., Fujimura, T., Nakashima, N., and Yokota, M. (1998). Association of a polymorphism of the endothelial constitutive nitric oxide synthase gene with myocardial infarction in the Japanese population. *Am. J. Cardiol.* **81**, 83–86.

136. Park, K. W., You, K. H., Oh, S., Chae, I. H., Kim, H. S., Oh, B. H., Lee, M. M., and Park, Y. B. (2004). Association of endothelial constitutive nitric oxide synthase gene polymorphism with acute coronary syndrome in Koreans. *Heart* **90**, 282–285.

137. Mager, A., Lalezari, S., Shohat, T., Birnbaum, Y., Adler, Y., Magal, N., and Shohat, M. (1999). Methylenetetrahydrofolate reductase genotypes and early-onset coronary artery disease. *Circulation* **100**, 2406–2410.

138. Margaglione, M., Colaizzo, D., Cappucci, G., del Popolo, A., Vecchione, G., Grandone, E., and Di Minno, G. (1999). Genetic polymorphism of 5,10-MTHFR reductase gene in offspring of patients with myocardial infarction. *Thromb. Haemost.* **82**, 19–23.

139. Schmitz, C., Lindpaintner, K., Verhoef, P., Gaziano, J. M., and Buring, J. (1996). Genetic polymorphism of methylenetetrahydrofolate reductase and myocardial infarction. A case-control study. *Circulation* **94**, 1812–1814.

140. Souto, J. C., Blanco-Vaca, F., Soria, J. M., Buil, A., Almasy, L., Ordonez-Llanos, J., Martin-Campos, J. M., Lathrop, M., Stone, W., Blangero, J., and Fontcuberta, J. (2005). A genomewide exploration suggests a new candidate gene at chromosome 11q23 as the major determinant of plasma homocysteine levels: results from the GAIT project. *Am. J. Hum. Genet.* **76**, 925–933.

141. Lanfear, D. E., Marsh, S., Cresci, S., Shannon, W. D., Spertus, J. A., and McLeod, H. L. (2004). Genotypes associated with myocardial infarction risk are more common in African Americans than in European Americans. *J. Am. Coll. Cardiol.* **44**, 165–167.

142. Ono, K., Goto, Y., Takagi, S., Baba, S., Tago, N., Nonogi, H., and Iwai, N. (2004). A promoter variant of the heme oxygenase-1 gene may reduce the incidence of ischemic heart disease in Japanese. *Atherosclerosis* **173**, 315–319.

143. Stamatelopoulos, K. S., Lekakis, J. P., Poulakaki, N. A., Papamichael, C. M., Venetsanou, K., Aznaouridis, K., Protogerou, A. D., Papaioannou, T. G., Kumar, S., and Stamatelopoulos, S. F. (2004). Tamoxifen improves endothelial function and reduces carotid intima-media thickness in postmenopausal women. *Am. Heart J.* **147**, 1093–1099.

144. Kunnas, T. A., Laippala, P., Penttila, A., Lehtimaki, T., and Karhunen, P. J. (2000). Association of polymorphism of human alpha oestrogen receptor gene with coronary artery disease in men: a necropsy study. *BMJ* **321**, 273–274.

145. Pollak, A., Rokach, A., Blumenfeld, A., Rosen, L. J., Resnik, L., and Dresner Pollak, R. (2004). Association of oestrogen receptor alpha gene polymorphism with the angiographic extent of coronary artery disease. *Eur. Heart J.* **25**, 240–245.

146. Kunnas, T. A., Lehtimaki, T., Karhunen, P. J., Laaksonen, R., Janatuinen, T., Vesalainen, R., Nuutila, P., Knuuti, J., and Nikkari, S. T. (2004). Estrogen receptor genotype modulates myocardial perfusion in young men. *J. Mol. Med.* **82**, 821–825.

147. Heng, C. K., Lal, S., Saha, N., Low, P. S., and Kamboh, M. I. (2004). The impact of factor XIIIa V34L polymorphism on plasma factor XIII activity in the Chinese and Asian Indians from Singapore. *Hum. Genet.* **114**, 186–191.

148. Ng, K. C., Yong, Q. W., Chan, S. P., and Cheng, A. (2002). Homocysteine, folate and vitamin B12 as risk factors for acute myocardial infarction in a Southeast Asian population. *Ann. Acad. Med. Singapore* **31**, 636–640.

149. Yang, S. L., He, B. X., Liu, H. L., He, Z. Y., Zhang, H., Luo, J. P., Hong, X. F., and Zou, Y. C. (2004). Apolipoprotein E gene polymorphisms and risk for coronary artery disease in Chinese Xinjiang Uygur and Han population. *Chin. Med. Sci. J.* **19**, 150–154.

150. Saha, N., Talmud, P. J., Tay, J. S., Humphries, S. E., and Basair, J. (1996). Lack of association of angiotensin-converting enzyme (ACE). Gene insertion/deletion polymorphism with CAD in two Asian populations. *Clin. Genet.* **50**, 121–125.

151. Hooper, W. C., Dowling, N. F., Wenger, N. K., Dilley, A., Ellingsen, D., and Evatt, B. L. (2002). Relationship of venous thromboembolism and myocardial infarction with the renin-angiotensin system in African-Americans. *Am. J. Hematol.* **70**, 1–8.

152. Helgadottir, A., Manolescu, A., Helgason, A., Thorleifsson, G., Thorsteinsdottir, U., Gudbjartsson, D. F., Gretarsdottir, S., Magnusson, K. P., Gudmundsson, G., Hicks, A., Jonsson, T., Grant, S. F., Sainz, J., O'Brien, S. J., Sveinbjornsdottir, S., Valdimarsson, E. M., Matthiasson, S. E., Levey, A. I., Abramson, J. L., Reilly, M. P., Vaccarino, V., Wolfe, M. L., Gudnason, V., Quyyumi, A. A., Topol, E. J., Rader, D. J., Thorgeirsson, G., Gulcher, J. R., Hakonarson, H., Kong, A., and Stefansson, K. (2006). A variant of the gene encoding

leukotriene A4 hydrolase confers ethnicity-specific risk of myocardial infarction. *Nat. Genet.* **38,** 68–74.
153. Archacki, S., and Wang, Q. (2004). Expression profiling of cardiovascular disease. *Hum. Genomics* **1,** 355–370.
154. Faber, B. C., Cleutjens, K. B., Niessen, R. L., Aarts, P. L., Boon, W., Greenberg, A. S., Kitslaar, P. J., Tordoir, J. H., and Daemen, M. J. (2001). Identification of genes potentially involved in rupture of human atherosclerotic plaques. *Circ. Res.* **89,** 547–554.
155. Hiltunen, M. O., Tuomisto, T. T., Niemi, M., Brasen, J. H., Rissanen, T. T., Toronen, P., Vajanto, I., and Yla-Herttuala, S. (2002). Changes in gene expression in atherosclerotic plaques analyzed using DNA array. *Atherosclerosis* **165,** 23–32.
156. Rossi, M. L., Marziliano, N., Merlini, P. A., Bramucci, E., Canosi, U., Presbitero, P., Arbustini, E., Mannucci, P. M., and Ardissino, D. (2005). Phenotype commitment in vascular smooth muscle cells derived from coronary atherosclerotic plaques: differential gene expression of endothelial nitric oxide synthase. *Eur. J. Histochem.* **49,** 39–46.
157. Parisi, Q., Biondi-Zoccai, G. G., Abbate, A., Santini, D., Vasaturo, F., Scarpa, S., Bussani, R., Leone, A. M., Petrolini, A., Silvestri, F., Biasucci, L. M., and Baldi, A. (2005). Hypoxia inducible factor-1 expression mediates myocardial response to ischemia late after acute myocardial infarction. *Int. J. Cardiol.* **99,** 337–339.
158. Burrell, L. M., Risvanis, J., Kubota, E., Dean, R. G., MacDonald, P. S., Lu, S., Tikellis, C., Grant, S. L., Lew, R. A., Smith, A. I., Cooper, M. E., and Johnston, C. I. (2005). Myocardial infarction increases ACE2 expression in rat and humans. *Eur. Heart J.* **26,** 369–375.
159. Kawakami, R., Saito, Y., Kishimoto, I., Harada, M., Kuwahara, K., Takahashi, N., Nakagawa, Y., Nakanishi, M., Tanimoto, K., Usami, S., Yasuno, S., Kinoshita, H., Chusho, H., Tamura, N., Ogawa, Y., and Nakao, K. (2004). Overexpression of brain natriuretic peptide facilitates neutrophil infiltration and cardiac matrix metalloproteinase-9 expression after acute myocardial infarction. *Circulation* **110,** 3306–3312.
160. Hasegawa, R., Kita, K., Hasegawa, R., Fusejima, K., Fukuzawa, S., Wano, C., Watanabe, S., Saisho, H., Masuda, Y., Nomura, F., and Suzuki, N. (2003). Induction of apoptosis and ubiquitin hydrolase gene expression by human serum factors in the early phase of acute myocardial infarction. *J. Lab. Clin. Med.* **141,** 168–178.
161. Akatsu, T., Nakamura, M., Satoh, M., and Hiramori, K. (2003). Increased mRNA expression of tumour necrosis factor-alpha and its converting enzyme in circulating leucocytes of patients with acute myocardial infarction. *Clin. Sci. (Lond)* **105,** 39–44.

162. Marx, N., Neumann, F. J., Ott, I., Gawaz, M., Koch, W., Pinkau, T., and Schomig, A. (1997). Induction of cytokine expression in leukocytes in acute myocardial infarction. *J. Am. Coll. Cardiol.* **30,** 165–170.
163. Wettinger, S. B., Doggen, C. J., Spek, C. A., Rosendaal, F. R., and Reitsma, P. H. (2005). High throughput mRNA profiling highlights associations between myocardial infarction and aberrant expression of inflammatory molecules in blood cells. *Blood* **105,** 2000–2006.
164. Duran, M. C., Mas, S., Martin-Ventura, J. L., Meilhac, O., Michel, J. B., Gallego-Delgado, J., Lazaro, A., Tunon, J., Egido, J., and Vivanco, F. (2003). Proteomic analysis of human vessels: application to atherosclerotic plaques. *Proteomics* **3,** 973–978.
165. Mateos-Caceres, P. J., Garcia-Mendez, A., Lopez Farre, A., Macaya, C., Nunez, A., Gomez, J., Alonso-Orgaz, S., Carrasco, C., Burgos, M. E., de Andres, R., Granizo, J. J., Farre, J., and Rico, L. A. (2004). Proteomic analysis of plasma from patients during an acute coronary syndrome. *J. Am. Coll. Cardiol.* **44,** 1578–1583.
166. Stanley, B. A., Gundry, R. L., Cotter, R. J., and Van Eyk, J. E. (2004). Heart disease, clinical proteomics and mass spectrometry. *Dis. Markers* **20,** 167–178.
167. Simkhovich, B. Z., Marjoram, P., Poizat, C., Kedes, L., and Kloner, R. A. (2003). Brief episode of ischemia activates protective genetic program in rat heart: a gene chip study. *Cardiovasc. Res.* **59,** 450–459.
168. Simkhovich, B. Z., Kloner, R. A., Poizat, C., Marjoram, P., and Kedes, L. H. (2003). Gene expression profiling-a new approach in the study of myocardial ischemia. *Cardiovasc. Pathol.* **12,** 180–185.
169. Stanton, L. W., Garrard, L. J., Damm, D., Garrick, B. L., Lam, A., Kapoun, A. M., Zheng, Q., Protter, A. A., Schreiner, G. F., and White, R. T. (2000). Altered patterns of gene expression in response to myocardial infarction. *Circ. Res.* **86,** 939–945.
170. Samani, N. J., Burton, P., Mangino, M., Ball, S. G., Balmforth, A. J., Barrett, J., Bishop, T., Hall, A., and the BHF Family Heart Study Research Group. (2005). A genomewide linkage study of 1,933 families affected by premature coronary artery disease: The British Heart Foundation (BHF) Family Heart Study. *Am. J. Hum. Genet.* **77,** 1011–1020.
171. Liu, P. Y., Li, Y. H., Wu, H. L., Chao, T. H., Tsai, L. M., Lin, L. J., Shi, G. Y., and Chen, J. H. (2006). Platelet-activating factor-acetylhydrolase A379V (exon 11) gene polymorphism is an independent and functional risk factor for premature myocardial infarction. *J. Thromb. Haemost.* **4,** 1023–1028.

CHAPTER **10**

Cellular Pathways and Molecular Events in Cardioprotection

OVERVIEW

Acute ischemia and myocardial infarct as a consequence of coronary artery disease are major public health problems, with significantly increasing numbers of heart failure secondary to myocardial ischemia/myocardial infarct diagnosed each year in adults and aging individuals. Moreover, acute myocardial infarction poses a high risk for recurrent cardiovascular events, heart failure, and increased mortality. Present treatment options, including early reperfusion by thrombolysis or percutaneous coronary intervention, are the most effective strategies for limiting the size of an evolving infarct after acute myocardial infarct, although mortality remains significant caused in part by the lethal injury that occurs on reperfusing the ischemic myocardium.

Novel cardioprotective strategies are required to target injury stemming from both ischemic and reperfusion damage. Ischemic pre-conditioning (IPC) by single or multiple brief nonlethal periods of ischemia and reperfusion protects the heart against a more prolonged ischemic insult, reducing infarct size, and can restore cardiac function and diminish myocardial arrhythmias. This robust form of cardioprotection has been widely documented in animal models, as well as in cardiomyocytes *in vitro*, and seems to occur in humans in response to balloon angioplasty and angina. IPC has been shown to have an acute and a delayed cardioprotective phase, both involving the activation within the myocyte of specific triggering pathways and mediating signals and end effectors. The complex molecular mechanism and cellular targets of both phases of IPC are presented.

In addition, a similar regimen of brief ischemia/reperfusion can be applied at the onset of reperfusion after ischemia, resulting in cardioprotection. This *ischemic post-conditioning* shares commonality in a number of the signaling pathways with IPC, thereby offering a common target for cardioprotection. In addition, treatment with a variety of chemicals can substitute for either IPC or post-conditioning; this *pharmacological pre-conditioning* includes the targeted use of volatile anesthetics, potassium channel openers, nitric oxide donors, and modulators of some of the downstream components, including erythropoietin, statins, insulin, and pyruvate.

The cardioprotective potential of IPC and post-conditioning has been difficult to realize in clinical practice, because they necessitate highly invasive interventions applied either before the onset of ischemic insult, which is difficult to predict, or during reperfusion. Therefore, pharmacological pre-conditioning at the time of myocardial reperfusion may have greater applicability. We discuss current clinical trials and future cardioprotective approaches.

INTRODUCTION

Acute ischemia and myocardial infarct (MI) occurring as a consequence of coronary artery disease (CAD) pose major public health problems. The incidence of heart failure (HF) secondary to myocardial ischemia/MI has been reported to be increasing, and, in the United States alone, more than a half million new cases are diagnosed each year in adults and aging individuals. Moreover, after an acute myocardial infarction (AMI), patients are at high risk for recurrent cardiovascular events, new onset HF, and increased mortality. Treatment that uses early reperfusion by thrombolysis or primary percutaneous coronary intervention after AMI has been the most effective strategy for limiting the size of an evolving infarct. Nevertheless, the mortality resulting from AMI remains significant, caused in part by both the lethal ischemia and reperfusion injury that often occur on reperfusing the ischemic myocardium.

Myocardial ischemia has a large number of effects on cardiac physiology.[1] Its lethality (although not completely delineated) most likely stems from its marked perturbation of metabolism ultimately depriving the cardiomyocyte of the needed bioenergy to provide pumping energy and electrical signaling. Early ischemic damage shares with other physiological stresses (e.g., heat shock) a marked perturbation of the mitochondrial phenotype, including increased mitochondrial swelling and the uncoupling of respiration and oxidative phosphorylation (OXPHOS). Myocardial ischemia also results in the selective depletion of the mitochondrial inner membrane phospholipid, cardiolipin (CL), involved in cytochrome *c* insertion, retention, and electron transport function.[2] In addition, the onset of ATP depletion and

Post-Genomic Cardiology

subsequent de-energization of the cell found in sustained ischemia results in both necrotic cell death and in signaling apoptosis (programmed cell death).[3,4]

Reperfusion of the ischemic myocardium, until recently, was the only intervention available to restore the various cellular functions affected by myocardial ischemia, including preventing cell death by necrosis or apoptosis. Unfortunately, reperfusion can result in extensive myocardial damage, including myocardial stunning, dysrhythmias, and myocyte cell death, and the functional recovery of the heart may appear only after a period of cardiac contractile dysfunction that may last for several hours or days. Paradoxically, functional mitochondria can exacerbate ischemic damage, especially at the onset of reperfusion. A pronounced increase in fatty acid influx and unbalanced fatty acid β-oxidation (FAO) predominates during reperfusion, producing an excess of acetyl-CoA saturating the TCA cycle at the expense of glucose and pyruvate oxidation, which is eventually inhibited. Increased mitochondrial OXPHOS causes elevated ROS accumulation with associated increased lipid peroxidation; this results in lower cardiolipin levels in the inner membrane with a consequent effect on respiratory complex activities. At present, there is also evidence that reperfusion injury involves apoptotic cell death in contrast to ischemic injury, which primarily involves necrotic cell death.[5]

When exposed to an ischemic insult, cardiomyocytes are able to mount a cardioprotective response. This has been best documented with respect to IPC, in which short periods of ischemia protect the heart against a more prolonged ischemic insult. Chemical, metabolic, and even physical stressors have also been shown to generate cardioprotective responses that share aspects of the IPC model and in these responses recruit a variety of cell signaling pathways that seem to converge on several cardiomyocyte end effectors. Among those so far identified are the targeting of several aspects of mitochondria function, including the generation of ROS (as a potent signaling component), calcium flux, bioenergetic function, and the regulation of pivotal early events in the apoptotic pathway. Although our understanding of cardioprotection has increased substantially within the past few years, the precise cellular, biochemical, and molecular events involved remain to be delineated. This chapter presents a discussion of our present cardioprotective armamentarium from a primarily cellular/molecular point of view.

ISCHEMIC PRE-CONDITIONING

Ischemic pre-conditioning (IPC), by single or multiple brief periods of ischemia, protects the heart against a more prolonged ischemic insult. Two distinct pathways of IPC cardioprotection have been demonstrated. Acute protection, resulting from brief periods of ischemia applied 1–2 h before a longer ischemic insult (termed the index ischemia), occurs within a few minutes after the initial stimulus, and lasting for 2–3 h, is considered the model of classic pre-conditioning. This early phase of IPC is protective against MI but not myocardial stunning. Delayed pre-conditioning often referred to as a second window of protection occurs approximately 12–24 h after the pre-conditioning event and lasts several days.[6] The delayed pre-conditioning pathway is generally recognized as having greater clinical relevance, with a longer protective phase and greater effectiveness against both MI and stunning.

First demonstrated by Murry and colleagues in a canine model,[7] IPC has subsequently been successfully carried out in other animals studied. A variety of interventions and cardioprotective drugs have been tested with these animal models (Table I) in an effort to identify the mechanisms involved in cardioprotection. IPC-like effects have also been observed in cultured cells exposed to hypoxia and metabolic inhibition and can reverse the defects in both the functioning and the structure of mitochondria in a manner similar to the animal studies. It is also important to note that although IPC has been demonstrated in every species in which it has been tested, a number of conflicting findings have suggested that fundamental interspecific differences may underlie the cellular and molecular mechanisms of cardioprotection.

The duration of the IPC, the index ischemia, and the subsequent reperfusion are critical determinants of both the signaling pathways involved, as well as the cardioprotective outcome. The transient pre-conditioning state has been described for periods ranging from one cycle of ischemia/reperfusion (I/R) of 1.25 min to 5 cycles of 5 min ischemia/5 min reperfusion.[8] If the duration of the index ischemia is extended beyond 3 h (with no intermediary reperfusion), cardioprotection is abolished, suggesting that reperfusion after a damaging (but not prolonged) index ischemia is required for cardioprotection.[9] These early findings also strongly supported the view that IPC delays the lethal effects (i.e., cell death) of ischemia rather than entirely preventing it.

CELLULAR AND MOLECULAR EVENTS IN IPC

The mechanism of pre-conditioning has been characterized as being composed of triggers, mediators, memory, and end effectors.[10] Although some events can be clearly distinguished as to their temporal role, others are less well defined and, in fact, may act at several places in the pathway as triggers, mediators, and end effectors. In rendering the heart resistant to infarction, the pre-conditioning ischemia triggers a change in myocardial physiology of the heart, involving a series of signal transduction pathways that carry the signal for cardioprotection and subsequently converge on one or more end effectors. Triggers exert their effect before the

TABLE I Interventions/Drugs in Cardioprotection

Class	Specific drug chemical
K channel openers	Nicorandil
	Diazoxide
	Pinacidil
	Cromakalim
K channel blockers	Glibenclamide
	5-HD
Inhibitors of mitochondrial PT pore	Cyclosporin A
Receptor-mediated signaling pathways/ligands in IPC-PC	Adenosine
	Opioids
	Bradykinin
	Acetylcholine
	Endothelin
Sphingolipid/ceramide signaling	Ceramide
	Sphingosine
	Chelerythrine
	(pan-PKC inhibitors)
	PKCε-TIP
Mitochondria-generated ROS and lipid peroxidation	Idebenone
	FCCP
	DNP
	Coenzyme Q
	Carvedilol
	Quercetin
Other pharmacological preconditioning agents	Erythropoietin (EPO)
	Statin (Avastatin)
	Glucose-insulin-potassium (GIK)
	Insulin
	Nitroglycerin (NTG)
	Pyruvate
	Glucagon-like peptide
	Sildenafil
	Monophosphoryl lipid A (MLA)

index ischemia; after the index ischemia has been initiated, mediators convey the signal to the end effectors that actually promotes the cardioprotection during the lethal ischemic insult (index ischemia) and/or the subsequent reperfusion period. There is evidence that somewhere within the signal transduction pathways between the trigger signal and the end effector resides a memory element that is set by the pre-conditioning protocol and that keeps the myocardium in a pre-conditioned state.

Triggering Early IPC

It is well established that adenosine and its G-protein–coupled receptors represent an important triggering stimulus, as well as a locus for feedback control of IPC.[11] Adenosine is generated at high levels during myocardial ischemia from ATP metabolism. IPC-induced cardioprotection was found to be abrogated by treatment with adenosine receptor antagonists, suggesting that adenosine produced during IPC acting on cell-surface receptors is a critical event in triggering IPC.[12] Further studies extended these findings with the demonstration that intravenous administration of adenosine A1 receptor agonists could be substituted for the IPC regimen, resulting in the same cardioprotection, an early example of pharmacological pre-conditioning.[13]

Other triggering ligands have been identified, including opioids, bradykinin, norepinephrine, and endothelin, all of which the heart can release during a brief ischemic period, which bind specific G-protein–coupled receptors and elicit IPC-mediated cardioprotection. At least three surface receptors acting in parallel can trigger pre-conditioning; these receptors are thought to represent parallel and redundant pathways.[14] Whether these multiple receptor pathways act in concert or synergistically may have important clinical significance impacting on the development of a more effective therapeutic approach by use of smaller drug dosages, thereby avoiding potential toxicity and adverse side effects. Different signaling pathways are triggered by ligand occupation of these specific receptors.

The binding of specific ligands to their membrane-bound receptors can lead to the subsequent activation of phospholipase C (PLC) and production of diacylglycerol (DAG), which in turn stimulates the pivotal signal transducer protein kinase C (PKC). There is also recent evidence implicating tyrosine kinases as signaling cardioprotective responses in conjunction with nonadenosine receptors, as well as the involvement of sphingolipid signaling pathway in IPC-mediated cardioprotection.[15,16] These signaling events will be further discussed later in the mediator section. As a consequence of the receptor redundancy, blockade of a single receptor type often does not completely block cardioprotection but rather raises the ischemic threshold required for protection.[17]

A number of studies have identified the generation of ROS and NO as triggering elements in the early IPC pathway.[18] Treatment with a free radical scavenger (e.g., N-2-mercaptopropionylglycine or MPG) can raise the threshold of pre-conditioning, abolishing cardioprotection from a single cycle of IPC but not from a more extensive four-cycle regimen.[19] Moreover, a free radical generator (i.e., xanthine oxidase) exposed to isolated rabbit hearts for 5 min before index ischemia and reperfusion could trigger a pre-conditioned state on the basis of its subsequent cardioprotective effect.[20] There is evidence that the triggering ROS acts to directly activate protective kinases[21,22] and to trigger the activation of the mitoK$_{ATP}$ channel as a cardioprotective mediator,[18] resulting in a further generation of mitochondrial-produced ROS. However, the precise source of the triggering ROS has not yet been determined.

Some studies have shown the involvement of NO as an early IPC triggering event, although this remains largely controversial. Data have shown that exogenous NO (derived from perfusion with an NO donor S-nitroso-N-acetylpenicillamine (SNAP) provided to isolated rabbit hearts before I/R exhibited cardioprotection similar to that with IPC.[23] This cardioprotection was found to be PKC dependent; however, no evidence for endogenous NO generation or triggering was found in this study. Other studies have found that endogenous NO synthesis does not play a contributory role in early IPC in either rat or swine models.[24,25] Studies with eNOS knock-out mice revealed that the loss of endogenous NO had no obvious effect on the development of IPC cardioprotection with higher levels of IPC. Lower levels of IPC (which were effective in the development of cardioprotection in wild-type mice) resulted in significant attenuation of cardioprotection in the eNOS null mice.[26] These results suggest that although eNOS or endogenous NO is not likely required for robust early IPC, NO may contribute to early IPC by lowering the ischemic threshold for protection. As we shall shortly see, NO's involvement in delayed IPC is more clearcut.

Mediators of Early IPC

A broad array of signal transducing elements has been identified as mediators within the cardioprotective pathway. Several intracellular kinases have been implicated as have transducers located at both the cell membrane (e.g., sarcolemmal K$_{ATP}$ channels) and within the mitochondrial membranes (e.g., mitoK$_{ATP}$ channels).[10] In acute cardioprotection, a primary mechanism by which activation of the trigger/mediator signaling pathways leads to activation of the end effectors seems to involve post-translational modification such as the phosphorylation mediated by kinases. Despite an abundance of information concerning the involvement of these cardioprotective transducing elements, the sequence of events in these intracellular cascades remains not well defined. A diagram of the major signaling components in early or acute IPC pathway is shown in Fig. 1.

The Phosphatidylinositol 3-Kinase (PI3K) Pathway

One of the earliest mediators of ischemic cardioprotection (triggering a variety of other mediating events) is PI3K.[27] Activation of the PI3K pathway has been demonstrated in a large number of experimental pre-conditioning models to be cardioprotective. In addition, IPC leads to the activation of protein kinase B (also called Akt) and p70S6K, kinases downstream of PI3K, consistent with a significant role for PI3K in IPC.[28,29] Inhibitors of PI3K, such as wortmannin and LY294002, attenuate the cardioprotection afforded by IPC.[28,29]

These studies also demonstrated that using isolated perfused rat heart, a standard IPC stimulus induced Akt activation and Akt-mediated phosphorylation of its downstream GSK3β target immediately before the index ischemic period.[30] Because there are presently no inhibitors of Akt, it is unclear whether the activation of Akt, which has been shown to accompany IPC, is in fact required for IPC.

PI3K activation (which has been previously discussed in Chapter 5) occurs after the triggering ligands (e.g., adenosine, bradykinin or opioids) bind the G-protein–coupled receptors and initiate a signaling cascade. Krieg and associates reported that activation of the G-protein–coupled receptor in response to acetylcholine-induced IPC leads to the transactivation of the epidermal growth factor receptor (EGRF) and associated tyrosine kinases, which then activate the PI3K–Akt pathway.[31]

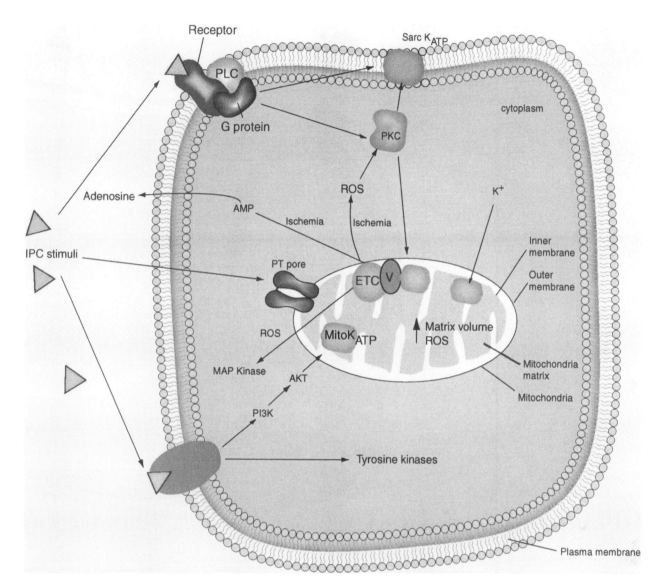

FIGURE 1 Schematic representation depicting cellular events occurring in early ischemic preconditioning. Agonists of G-protein–coupled receptors, including adenosine and bradykinin, can stimulate parallel signaling pathways by stimulating either phosphoinositide 3-kinase (PI3K) or tyrosine kinase-dependent cascades, or a PKC-dependent cascade involving activation of phospholipase C (PLC) can confer cardioprotection. Multiple mitochondrial events integral to cardioprotection include the increased modulation of mitoK$_{ATP}$ channels (and to a lesser extent, sarcolemmal K$_{ATP}$ channels), mitochondrial reactive oxygen species (ROS) generation, modulation of mitochondrial calcium intake, permeability transition (PT) pore opening, and electron transport chain (ETC) activity. MitoK$_{ATP}$ channel opening also provides positive feedback by altering components such as ROS, mitogen-activated protein kinase (MAPK), or protein kinase C (PKC) activation. PI3K activation results in further signaling by downstream targets such as Akt, eNOS, GSK-3ß, PKC, mTOR, and p70S6K. PI3K activation also plays a role in the activation of PKC and perhaps other signaling events (such as the activation of mitoK$_{ATP}$) (see color insert).

Protein Kinase C

A considerable number of studies conducted in a variety of pre-conditioning models have shown that protein kinase C (PKC) activation is essential for IPC. Blockade of PKC with PKC inhibitors eliminated the cardioprotection resulting from IPC.[32,33] Moreover, the activation of PKC by phorbol esters can mimic IPC in the isolated rabbit heart.[33]

PKC is a serine/threonine kinase that is activated by lipid cofactors derived from breakdown of membrane lipids by PLC and has multiple isoforms in the heart, each with a similar substrate specificity. The PKC isoforms achieve specificity by their physical translocation to specific docking sites located on specific intracellular organelles. The activation of phosphorylation activity of the PKC isoforms at those sites results in the further mediation of the cardioprotective signal, although the precise intracellular targets have not yet been completely identified.

Studies have revealed a potential contributory role of PKC-δ and PKC-α isoforms and a primary role for myocardial PKC-ε in IPC. Although there are data showing a relationship between

PKC-δ and delayed IPC,[34,35] there are considerable data challenging a positive role for PKC-δ in early IPC.[36-38] Limited data have reported PKC-α translocation to sarcolemmal membrane in IPC.[39] In contrast, several studies have demonstrated a positive role for PKC-ε in early IPC.[37,38,40,41] Targeted disruption of the PKC-ε gene abolishes the cardioprotective infarct size reduction resulting from IPC in isolated perfused mouse hearts.[42] The specific activation and intracellular translocation of PKC-ε during IPC has been shown; interestingly, PKC-ε translocates to the mitochondria, where it interacts in modular functional signaling complexes with MAPK. Moreover, PKC-ε/MAPK complex formation can lead to the activation of mitochondrial signaling ERKs and subsequently to the phosphorylation of the proapoptotic Bad protein that could reduce or delay cell death.[43]

Other potential mitochondrial targets for PKC-ε that may be impacted in pre-conditioning as both mediators and end effectors of the pathway include the mitochondrial permeability transition pore (PT pore) and the mitoK$_{ATP}$ channel. A direct protein–protein interaction between PKC-ε and several components of the PT pore (e.g., voltage-dependent anion channel [VDAC], adenine nucleotide translocase [ANT], and hexokinase II) has recently been shown by Baines and associates.[44] This interaction may inhibit the pathological opening of the pore (including Ca^{++}-induced opening and subsequent mitochondrial swelling) contributing to PKC-ε–induced cardioprotection. Although such a direct physical relationship between PKC-ε and the mitoK$_{ATP}$ channel has not yet been demonstrated, two recent studies suggested that myocardial PKC-ε is involved in IPC upstream of mitoK$_{ATP}$ channels.[45,46]

It has been reported that PKC is activated by PI3K.[27,29] Wortmannin, the PI3K inhibitor, was shown to block the PC-induced translocation of PKC. The precise role of PI3K in activation of PKC has not been determined; in this respect, it is noteworthy that phosphoinositide products have been reported to activate several isoforms of PKC.[47] Recent studies conducted by Okada and coworkers have suggested that the interplay between PI3K and PKC activation may be involved in the establishment of cardioprotective memory.[48]

Tyrosine and MAP Kinases

Activation of a tyrosine kinase pathway seems to be recruited in acute IPC as an early mediating event. This pathway has been reported by some to be downstream of the PKC pathway,[49] whereas others have assigned it as a parallel pathway.[50,51] There is also considerable evidence that subfamilies of the MAPKs including p42/p44 extracellular signal–regulated kinases (Erk1/2), the 52–54kDa c-JUN kinase, and the 38-kDa p38 MAPK may be involved in IPC. The role of Erk1/2 as a potential signaling component of IPC is controversial, with some demonstrating that classic IPC results in Erk1/2 activation occurring before the index ischemic period,[52] whereas other studies reported no change in Erk1/2 phosphorylation in response to an IPC stimulus, suggesting that Erk1/2 may have no significant role in early IPC.[28] Most evidence supports a downstream role for p38 MAPK, which we will discuss shortly.

K$_{ATP}$ Channels: IPC, KCOs, and the Mitochondrial K$_{ATP}$ Channel

Another essential component of the early cardioprotection pathway in IPC is the ATP-sensitive potassium channel (K$_{ATP}$ channel). This channel opens when the intracellular levels of ATP decline as occurs in ischemia. Initially, sarcolemmal K$_{ATP}$ (sarcK$_{ATP}$) channels were implicated in IPC, but more recently the focus has shifted to the mitochondrial K$_{ATP}$ (mitoK$_{ATP}$) channels. Information about IPC and the K$_{ATP}$ channels has also been derived from the use of potassium channel openers (KCOs) such as nicorandil, diazoxide, pinacidil, and cromakalim and also by the use of K$_{ATP}$ channel blockers (e.g., glibenclamide and 5-hydroxydecanoic acid [5-HD]), which diminish the beneficial effects of short ischemic events on cardiac tissue.[53,54]

Garlid and associates have shown that the KCO diazoxide was 1000–2000 times more potent in opening the mitoK$_{ATP}$ channel than the sarcK$_{ATP}$ channel.[55] The negative effect of 5-HD (abolishing diazoxide's protective effect entirely) further suggested that mitoK$_{ATP}$ mediated IPC, because 5-HD is highly selective in interacting with the mitoK$_{ATP}$ compared with the sarcolemmal channel. However, 5-HD treatment has been recently shown to also modulate mitochondrial respiration and fatty acid β-oxidation independently of its interaction with the mitoK$_{ATP}$ channel.[56] Nevertheless, recent data support the view that both mitochondrial and sarcK$_{ATP}$ channels have contributory roles in cardioprotection.[57,58] Under specific conditions, diazoxide and 5HD can modulate sarcK$_{ATP}$ channel opening,[57] whereas mitoK$_{ATP}$ can bind molecules previously thought to interact solely with sarcolemmal K$_{ATP}$ channels, and in some species (e.g., dogs) both the mitochondrial and sarcK$_{ATP}$ channels must be blocked to entirely abolish IPC cardioprotection.[58] Another ion channel previously thought to be exclusive to the myocardial sarcolemmal membrane (i.e., the calcium-activated K$^+$ channel) has recently been located on the mitochondrial inner membrane and demonstrated to be functionally cardioprotective against MI.[59]

Biochemical structural analysis of the proteins comprising the K$_{ATP}$ channels has been performed with functional channels isolated from both myocardial sarcolemmal and mitochondrial membranes.[60,61] The channel activities of these preparations derived from either heart or brain have been reconstituted in proteoliposomes and shown to be regulated by the same ligands as *in vivo*. Although the sarcK$_{ATP}$ channels are heteromultimeric complexes of sulfonylurea receptors (SUR) and potassium inward rectifier (Kir) gene products,[62] the precise molecular composition of the mitoK$_{ATP}$ channel has yet to be identified.[63] By use of immunoblot analysis, no evidence for the presence of either known sulfonylurea receptor (SUR1 or SUR2) in mitochondria has been found.[64] Although there is evidence for the presence of Kir6.1 and 6.2 in cardiac mitochondria,[65] gene knock-out studies demonstrated no discernible functional role for these proteins in the mitoK$_{ATP}$ channel.[63] A recent study has indicated that a multiprotein complex can be purified from the mitochondrial inner membrane that functions as a mitoK$_{ATP}$ channel, fully sensitive to channel activators and blockers on reconstitution into proteoliposomes and lipid bilayers.[66] This complex contains five mitochondrial proteins, including ANT, the mitochondrial ATP-binding cassette protein 1 (mABC1), the phosphate carrier (PIC), ATP synthase, and SDH. However, at this time, the pore-forming component of the mitoK$_{ATP}$ channel remains undetermined, as do the potential identity of other proteins within this complex. The identity of major bioenergetic proteins (ATP synthase and SDH) within the mitoK$_{ATP}$ channel and their overlap with structural components of the PT pore (e.g., ANT) have important ramifications for the mechanism of cardioprotection and its relatedness to both bioenergetic function and apoptosis regulation. However, the lack of an agreed on, defined structure for the mitoK$_{ATP}$ channel has increasingly led to questions regarding both its significance and even its existence.

A number of the pharmacological treatments used to promote cardioprotection may have compromised "specificity" (particularly when used at high dosage) impacting on more than a single cellular event, thereby complicating the interpretation of their action. Recent findings have shown that diazoxide and 5-HD have distinctive effects on mitochondria, independent of the mitoK$_{ATP}$ channel targets,[67-70] making the evaluation of upstream and downstream events rather difficult. For example, respiratory complex II (SDH) activity seems to be directly inhibited by some KCOs (e.g., diazoxide),[71] which may complicate the interpretation of experiments that use diazoxide to elicit mitoK$_{ATP}$ channel opening (a standard procedure in this field). This finding, however, takes on added significance with the recent discovery that SDH may be a primary component of the mitoK$_{ATP}$ channel.[66] This is also consistent with the finding that specific SDH inhibitors such as 3-nitropropionic acid (3-NPA) can result in increased K$^+$ transport by the mitoK$_{ATP}$ channel and can mimic the cardioprotection provided by IPC.[66,72] Moreover, K$_{ATP}$ channel blockers, which entirely abolish the cardioprotective effect of diazoxide, also reverse the effect of complex II inhibitors.

Mitochondrial ETC Activity and ATP Levels in IPC

The relationship of mitochondrial respiratory activity and ATP levels to IPC and to mitochondrial K$_{ATP}$ channel opening has been proposed as important in the overall regulation of myocardial bioenergetics, as well as in the survival of the ischemic cardiomyocyte. A modest, but significant, level of uncoupling of respiration and OXPHOS occurs with IPC.[73] The uncoupling of OXPHOS, under certain circumstances, can lead to increased bioenergetic efficiency and has been also proposed as a primary *raison d'etre* for mitochondrial uncoupling proteins. The importance of uncoupling has been further supported by studies in which treatment with uncouplers (e.g., DNP) mimics IPC.[74] Moreover, in cardiomyocytes it has been demonstrated that a crucial determinant in opening of both sarcK$_{ATP}$ and mitoK$_{ATP}$ channels is the cytoplasmic ATP pool, in the main produced by mitochondrial OXPHOS. Uncoupling of mitochondrial OXPHOS results in the opening of cardiac K$_{ATP}$ channels, even in the presence of glucose.[75] In addition, myocardial ATP levels are elevated in cardiomyocytes treated with IPC or with diazoxide compared with untreated ischemic cells.[76-78] IPC also promotes inhibition of cardiac mitochondrial ATPase activity and reduced ATP depletion during early stages of ischemia that persist during the critical early phase of reperfusion.[79,80] Inhibition of ATP hydrolysis conserves high-energy phosphates and improves the energy state during ischemia, contributing to postischemic myocardial recovery. On the other hand, other studies have concluded that mitochondrial ATP depletion and declining mitochondrial ATPase activity are unlikely to be responsible for IPC cardioprotection, because ATP depletion can remain unchanged and even be enhanced during IPC.[81,82]

As we have noted, it is generally accepted that the mitoK$_{ATP}$ channel closes when the mitochondrial ATP level is high and opens under conditions of low ATP levels, responding to the ATP/ADP ratio as well as to localized changes in adenosine content derived from adenine nucleotide metabolism.[83] These nucleotide levels are largely regulated by the adenine nucleotide translocator (ANT) and creatine kinase, important factors in regulating ATP and ADP levels in mitochondria and both PT pore components. IPC has also been reported to induce a significant decrease in mitochondrial NADH-supported respiration that can be reproduced by pharmacologically inhibiting the activity of complex I, a primary site of mitochondrial ROS generation.[84]

In addition, IPC and treatment with KCOs causes significant increases in cardiomyocyte mitochondrial volume, presumably because of K$^+$ influx.[85] *In vivo* and *in vitro* studies have demonstrated that mitochondrial matrix swelling may result in improved mitochondrial energy production with activated FAO, mitochondrial respiration, and ATP production.[86,87] However, controversy exists concerning the extent, cause, and significance of mitochondrial matrix changes during IPC.[88] Another reported effect of mitoK$_{ATP}$ channel opening is the preservation of low outer membrane permeability to nucleotides and cytochrome *c*, beneficial effects that can be abrogated by the mitoK$_{ATP}$ channel inhibitor 5-HD.[89] Opening of mitoK$_{ATP}$ channels during ischemia can also contribute to the tight structure of the intermembrane space needed for preserving low outer membrane permeability to adenine nucleotides. Consequently, mitoK$_{ATP}$

channel opening is considered both a signaling trigger and a mediator/effector of the cardioprotection pathway.[90]

IPC and ROS Generation

Increasing evidence has supported the fundamental involvement of ROS generation and signaling in the cardioprotective pathway(s). Diazoxide-mediated pre-conditioning results in ROS production, and blocking of ROS production with antioxidants blunts cardioprotection.[91] Moreover, it has been established that mitochondrial ROS production occurs downstream and as a consequence of mitoK$_{ATP}$ channel opening. ROS generation may function as a downstream intracellular signal as a second messenger leading to further activation of protein kinases, G-proteins, and the nuclear poly (ADP-ribose) polymerase (PARP) or directly modulate mitoK$_{ATP}$ channel opening.[92-94]

A burst of ROS occurs during the first moments of reperfusion and is associated with changes in mitochondria (e.g., PT pore opening) and myocardial injury.[95] The source of ROS generation during early reperfusion has not been determined and may be of either mitochondrial or cytoplasmic origin. In contrast, the source of ROS generated during ischemia (and likely in the early/acute pathway of IPC) more clearly involves the mitochondrial ETC and may be different than the source of ROS generated in early reperfusion.[96,97] The large increase in ROS accumulation that occurs in ischemic cardiomyocytes in response to reperfusion can be reduced either by diazoxide exposure or by hypoxic pre-conditioning.[98,99] Moreover, ROS levels are reduced in pre-conditioned tissues after index ischemia at the time of reperfusion, likely as a consequence of inhibiting PT pore opening.[88]

These findings have prompted the characterization of ROS as dual-sided with respect to cardioprotection, both involved in signaling or triggering of pre-conditioning and as a consequence of I/R damage mediating ischemic cell death (which cardioprotection can stem).[88] Although it is generally accepted that the initial signaling mechanism of ROS generation lies downstream of mitoK$_{ATP}$ channel opening, it has not yet been satisfactorily explained how ROS generation occurs as a consequence of K$_{ATP}$ channel opening.[100] Its role in the generation of ROS positions the mitoK$_{ATP}$ channel as a signal mediator in IPC, furnishing free radicals that can activate the downstream kinases that ultimately modulate the end effector, although this channel also can be considered an end effector as well.

End Effectors of IPC Cardioprotection

At this time, there has not been a definitive identification of the end effector(s) in IPC-mediated cardioprotection. Multiple end targets for the cardioprotection pathways seem to be likely. These include metabolic effects (i.e., an improved cardiomyocyte energy balance), modulation of mitoK$_{ATP}$ channel, mitochondrial PT pore opening, Na$^+$/H$^+$ exchanger, osmotic swelling, decreased cytoskeletal fragility, changes in gap junction function, reduced levels of TNF-α and ROS, and probably, most importantly, changes in apoptotic remodeling.[10] Clearly, the end effector(s) of IPC cardioprotection must be involved in the reduction of cell death (either apoptotic or necrotic and likely both), which is paramount to its success. Because necrotic cell death (and not apoptosis) is characterized by a loss of plasma membrane integrity, the end effector(s) of IPC must address and correct this defect. It is interesting to note that overexpression of antiapoptotic factors such as Bcl-2 has been shown to inhibit necrosis arising from myocardial I/R injury.[101]

A number of the potential candidates for end effectors listed previously are associated with the mitochondria. This is not surprising, because mitochondria, besides being the primary organelles involved in ATP production, have been implicated in maintaining cell integrity, and mitochondrial events are pivotally linked to the initiation of apoptosis. A key participant in the trigger and mediation of cardioprotective responses, mitochondria house and regulate pivotal early events in the apoptotic pathway. Mitochondrial plasticity enables its functioning both as a target and as a player in myocardial signal transduction events, in the generation of ROS in response to a variety of cellular insults, and providing an appropriate antioxidant response (Table II). Data are available suggesting a role for several mitochondrial proteins as end effectors; the mitoK$_{ATP}$ channel, apoptotic proteins such as Bad and the mitochondrial PT pore.[102-104] Because of their common mitochondrial location, it is possible that the mechanisms involved in the regulation of these different mitochondrial proteins are related.

As a potential unifying mechanism, modulating the activation of mitoK$_{ATP}$ channels could reduce cell death in several ways. These include the aforementioned modulation of ROS, inhibition of mitochondrial Ca^{++} uptake, and the regulation of mitochondrial volume that could alter mitochondrial permeability control by VDAC or the PT pore. There is considerable evidence from studies of simulated ischemia and reperfusion in rabbit ventricular myocytes and in isolated

TABLE II Mitochondrial Functional Plasticity

Regulatable mitochondrial functions

Respiration and oxidative phosphorylation

Ion (K$^+$, Ca^{++}) transport and storage

ATP production

Cytochrome c release

Redox state

ROS generation

Permeability transition (PT) pore opening

Antioxidant response (e.g., MnSOD, glutathione)

Role in apoptotic progression

Responsive to K$_{ATP}$ channel openers (e.g., diazoxide, nicorandil)

mitochondria that K_{ATP} channel openers such as diazoxide make the mitochondria resistant to Ca^{++} overload by reducing Ca^{++} influx and activating Ca^{++} release.[105-107] Therefore, the mitoK_{ATP} channel may provide cardioprotective effects on end effectors by acting against myocardial necrosis and stunning through alleviation of oxidative stress and Ca^{++} overload.

A relationship between the mitochondrial PT pore and the mitoK_{ATP} channel in IPC has been suggested partially by their close physical proximity, as well as by their response to potassium channel openers, although a direct interaction has not yet been reported.[104] Changes in mitochondrial Ca^{++} could inhibit the mitochondrial PT pore opening. Studies that used isolated mitochondria have reported that diazoxide prevents opening of the PT pore induced by Ca^{++} addition.[108] They additionally reported that PKC activation mimicked the effect of diazoxide on the PT pore and was blocked by 5-HD. The PT pore opens in the first minutes of reperfusion after ischemia,[109] in association with elevated levels of mitochondrial Ca^{++}, increased oxidative stress, and increased matrix pH.[110] PT pore opening allows water, foreign substances, and solutes to enter the mitochondria, increasing matrix volume, disrupting mitochondrial function, and rupturing the outer membrane and may result in cell death by either apoptosis or necrosis. Inhibiting PT pore opening at reperfusion using cyclosporin A, which has a high affinity for binding the PT pore component cyclophilin D, has been found to be cardioprotective.[74] Although the role of the PT pore in IPC has not been directly demonstrated, cardioprotection afforded from IPC, diazoxide, or an adenosine agonist treatment could be blocked by pre-treatment with atractyloside, an opener of the PT pore.[88,111]

Moreover, the VDAC component of the PT pore is affected by events at the mitoK_{ATP} channel and interacts with some of the proapoptotic proteins, including the Bcl-2 family proteins.[112,113] Opening the mitoK_{ATP} channel can result in mitochondrial swelling and promotes a low permeability state for the outer mitochondrial membrane protein VDAC/porin.[89] Reduced levels of adenine nucleotides would effectively reduce ATP entry into the mitochondria during ischemia and diminish the consumption of glycolytic ATP by the mitochondrial ATPase. The hypothesis that activation of the mitoK_{ATP} channel induces protection by means of modulation of VDAC is interesting in light of a recent study indicating that overexpression of Bcl-2 modulates cardioprotection by inhibition of VDAC.[114] The cardiac-specific overexpression of Bcl-2 has been shown to reduce myocyte death after I/R, reducing the rate of decline in ATP during ischemia and attenuated ischemia-mediated acidification.[115] Similar to the effects of cardiac Bcl-2 overexpression, reduction of VDAC activity might explain many of the intracellular changes associated with pre-conditioning, such as the attenuated rate of decline in ATP, reduced cardiomyocyte acidification, and reduced cell death. Diazoxide addition has also been reported to reduce acidification during ischemia,[116] and perhaps this effect is mediated by means of K_{ATP}-dependent closure of VDAC.[89]

Moreover, the interaction of VDAC with proapoptotic proteins is potentially contributory to IPC cardioprotection. Recent data have suggested that early apoptotic progression involves proapoptotic family members such as Bax targeting the outer mitochondrial membrane and forming a large conductance channel to promote cytochrome c release from the mitochondria.[117-119] Although the precise mechanism by which Bax mediates release of cytochrome c is not well understood, the association of channel-forming Bax with other mitochondrial membrane proteins, such as VDAC or the larger PT pore complex, remains to be examined. Interestingly, pre-conditioning has been reported to lower the levels of myocardial proapoptotic Bax protein.[120] Antiapoptotic family members such as Bcl-2 oppose release of cytochrome c either by binding and sequestering proapoptotic members[117] or by binding to proteins such as VDAC and blocking the formation or opening of a cytochrome c release pathway.[118,119] Bad is another pro-apoptotic protein that binds and sequesters Bcl-2, such that Bcl-2 can no longer inhibit apoptosis. Phosphorylation of Bad releases it, thereby freeing Bcl-2 to inhibit apoptosis. As previously noted, a mitochondrial-localized signaling complex containing PKC-ε and MAPK has been isolated and characterized; this signaling complex was associated with increased mitochondrial phosphorylation of Bad.[43] Similarly, PKC-ε forms a functional signaling complex with components of the PT pore, including VDAC, ANT, and hexokinase.[44]

IPC not only protects against necrotic cell death occurring with I/R but also blunts the progression of apoptosis, a process that is increased with aging. The prevailing view is that the cardiomyocyte apoptosis occurring during I/R is primarily reperfusion triggered.[121] Reperfusion rapidly restores intracellular ATP levels required for apoptosis. Altering the expression of apoptogenic proteins (e.g., Bax) and PKC activity by IPC can also reduce apoptosis. Early hallmark events of apoptosis occur at the mitochondrial inner membrane, including PT pore opening and cytochrome c release from mitochondria (Fig. 2); on cardioprotection, cytochrome c release is inhibited along with PT pore opening and mitochondrial Ca^{++} flux stemmed. Modulation of myocardial apoptotic events has been documented in early, but not delayed, IPC.[122]

Nonmitochondrial Targets

In addition to the substantial mitochondrial involvement in the IPC pathways, there are numerous nonmitochondrial targets in the cardiomyocyte. These include cytoskeletal proteins and membrane proteins (e.g., Na^+/H^+ exchangers), which may represent plausible targets for cardioprotection because they might assist in maintaining cell integrity.

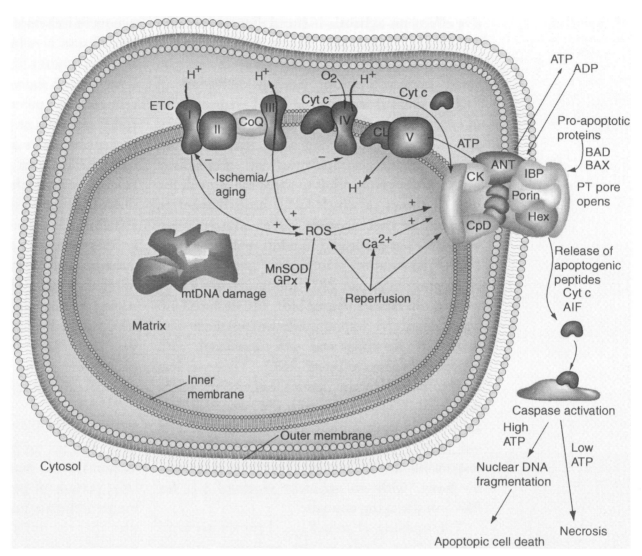

FIGURE 2 The mitochondrial pathway of apoptosis. An array of cellular signals, including ROS, trigger the apoptotic pathway regulated by proapoptotic proteins (e.g., Bax, Bid, and Bad) binding to the outer mitochondrial membrane and leading to mitochondrial outer-membrane permeabilization (MOMP) and PT pore opening. Elevated levels of mitochondrial Ca^{++}, as well as ETC-generated ROS, also promote PT pore opening. This is followed by the release of cytochrome c (Cyt c), Smac, endonuclease G (Endo G), and apoptosis-inducing factor (AIF) from the mitochondrial intermembrane space to the cytosol and apoptosome formation leading to caspase and endonuclease activation, DNA fragmentation, and cell death. Association of Bax with mitochondria is prevented by antiapoptotic proteins (e.g., Bcl-2). Also shown are proteins making up the PT pore, including hexokinase (Hex), adenine nucleotide translocator (ANT), creatine kinase (CK), cyclophilin D (CpD), and porin (VDAC), as well as the inner membrane phospholipid cardiolipin (CL) (see color insert).

It is also possible that the same signaling kinases that target mitochondrial proteins can also modify nonmitochondrial proteins and modulate other aspects of cardiomyocyte metabolism and function. Notably, IPC has been reported to alter connexins (a major component of gap junctions)[123,124] and Ca^{++} handling by sarcoplasmic reticulum.[125,126] Opening of $mitoK_{ATP}$ channels has also been shown to activate p38 MAPK,[127,128] and MAPK has been proposed to act as a downstream mediator in IPC signaling by its involvement in the phosphorylation and redistribution of heat shock proteins (e.g., HSP27), thereby impacting on the integrity and stability of myocardial cytoskeleton.[129] Mitochondrial ROS generation by means of electron flux at complex I and III is a necessary prerequisite leading to the downstream activation of p38 MAPK during hypoxia[130] and plays a mediating role in cardioprotective signaling,[131] although the mechanism for achieving this activation has not been fully delineated.

Transcriptional/Nuclear Events

Although it is well established that delayed cardioprotection involves up-regulation of gene expression with *de novo* protein synthesis, the involvement of gene expression in early or acute pre-conditioning has generally been excluded. The time for pre-conditioning seemed to be too short and the pre-conditioned state after a pre-conditioning protocol too transient (1–2 h) for new mRNA and protein synthesis to fully kick in to provide early cardioprotection. Early studies found that protein synthesis inhibition with either actinomycin D or cycloheximide did not block pre-conditioning's protection precluding gene expression as a possible mechanism of IPC memory.[132] Other studies showed that higher concentrations of cycloheximide (but not actinomycin D) could abrogate pre-conditioning protection, suggesting that a *de novo* protein synthesized from a pre-existing transcript might be involved.[133] However, more recent studies show that actinomycin D transcriptional inhibition abolished early IPC-induced cardioprotection with dramatic reductions in the activation of several MAP kinases and subsequent transcription and phosphorylation of specific transcription factors.[134]

Microarray analysis to profile myocardial gene expression in pre-conditioned tissues has been used to extend these findings. Large-scale changes in the program of gene expression with classic IPC have been reported in several studies.[135,136] Among the identified genes with altered myocardial expression caused by ischemia and/or pre-conditioning include a number of encoding proteins functioning in controlling protein degradation, stress responses, apoptosis, metabolic enzymes, regulatory proteins, as well as several with unknown cellular functions.[135] Moreover, a subset of the genes identified as having altered expression with pre-conditioning (e.g., oligoadenylate synthase, chaperonin subunit epsilon, a cGMP phosphodiesterase [PDE9A1]), a secretory carrier membrane protein, an amino acid transporter, and protease 28 subunit had not been previously implicated in cardioprotection and may reveal new end effectors of IPC. Interestingly, distinct patterns of myocardial gene expression (with some shared elements) were found with pharmacological pre-conditioning with anesthetics compared with IPC.[136] Moreover, further support for the view that transcription is likely required for both classical ischemic and pharmacological pre-conditioning was provided by the demonstration that both types of pre-conditioning were found to be entirely abolished using isolated hearts of mice deficient in the transcription factor, signal transducer, and activator of transcription-3 (STAT-3).[137]

Delayed Pre-Conditioning

Delayed pre-conditioning or the "second window of protection" (SWOP) lasts for up to 72 h, depending on the species and end point used for cardioprotection evaluation. Recent evidence indicates that in addition to enhanced tolerance to lethal ischemic injury, delayed pre-conditioning confers cardioprotection against other end points of I/R injury, including I/R-induced ventricular dysrhythmias[138] and post-ischemic myocardial dysfunction (stunning).[139,140] By use of isolated cardiomyocytes subjected to simulated ischemia or hypoxia, the cytoprotective effects of delayed pre-conditioning have also been demonstrated *in vitro*.[6]

A similar spectrum of stimuli, receptors, and transducers (e.g., the mitoK$_{ATP}$ channels, protein kinases) and pathophysiological mediators (e.g., ROS) are involved in delayed IPC cardioprotection as used in early IPC.[6] The triggering elements tend to be freely diffusible molecules (e.g., adenosine, opioids, NO and bradykinin) or free radicals generated during the pre-conditioning period. These promote the subsequent activation of a protein kinase signal cascade (e.g., protein kinase C, tyrosine kinases and various mitogen- and stress-activated protein kinases). The activated kinases phosphorylate specific substrate proteins.

In delayed PC, it is thought that the phosphorylation of transcription factors, initiating the synthesis of late-appearing effector proteins that promote cell survival during subsequent ischemia, may be a crucial event. Gene expression is tailored in a regulated fashion to induce new proteins that promote cell repair and to protect against subsequent I/R insult. The nature of the pre-conditioning stimulus determines the activation of a variety of transcription factors, regulating a large number of target genes. Among several transcription factors, NF-κB and AP-1 have consistently been shown to be activated in delayed IPC.[141] Both IPC and NO donors are able to induce NF-κB activation.[142] Conversely, direct inhibition of NF-κB abrogates delayed IPC cardioprotection.[142]

Gene profiling has identified several genes as up-regulated in both delayed ischemic and heat-shock pre-conditioning, including the TGFβ receptor interacting protein 1, the alpha isoform of the A subunit of PP2, and the cap binding protein NCBP,1 as well as a number of novel genes.[143]

Unlike the acute protection pathway, the delayed IPC pathway seems to be more dependent on the involvement of NO and its synthase (NOS), heat-shock proteins (e.g., HSP27, HSP72), manganese superoxide dismutase (MnSOD), cytokine induction, and cyclooxygenase-2 (COX2) (Fig. 3).[6]

During delayed IPC, NO exerts a more profound influence on the development of the protective pre-conditioned phenotype.[6] In the trigger phase of delayed IPC, eNOS-derived NO serves as the initiator of a molecular cascade, culminating in the subsequent activation of inducible nitric oxide synthase (iNOS), which then confers protection.[144] Therefore, eNOS is activated within an early time frame, and iNOS is activated later, and both events seem to be absolute requirements for the generation of delayed IPC-mediated cardioprotection. The ability of NO to act in concert with ROS signaling (generated after cell exposure to KCOs and other pre-conditioning stimuli) may be a critical factor in targeting the mitochondrial respiratory chain and modulating mitoK$_{ATP}$ channels to provide protection against ischemia.[145] This is supported by data showing that (1)

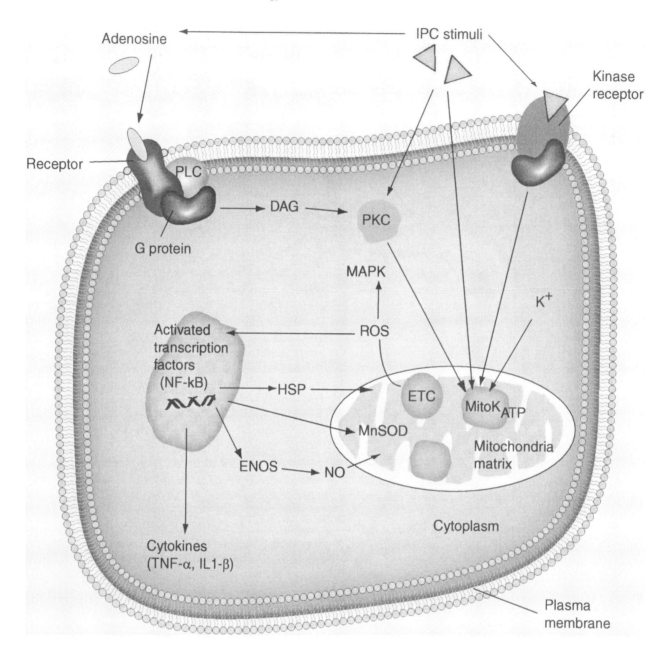

FIGURE 3 Schematic representation depicting cellular events occurring in delayed IPC. In delayed IPC, preconditioning stimuli impact sarcolemmal receptors by use of a variety of agonists and triggering a series of parallel signal transduction pathways involving G-coupled proteins and protein kinases, including PLC, PKC, and production of diacylglycerol (DAG). In addition, modulation of mitoK$_{ATP}$ channel opening and extensive ROS involvement occur as noted with the acute preconditioning model. Also involved is a protein kinase–mediated activation of specific nuclear transcription factors (e.g., NF-κB), leading to increased gene expression and synthesis of protective proteins to enable sustained cardioprotection. These include mitochondrial Mn-superoxide dismutase (MnSOD), eNOS, and several stress-activated, heat shock proteins (HSPs), ETC proteins, and the activation of cytokines (including TNF-α and IL-1β). Both eNOS and iNOS produce increased NO that provides cardioprotective stimulation of mitochondrial metabolism. These proteins provide a variety of levels of cytoprotection by acting at a variety of cellular sites including mitochondria (see color insert).

inhibitors of NOS abrogate delayed IPC-mediated protection;[146] (2) NO donors can mimic IPC in conferring CP,[145] and (3) free radical scavengers can reduce NOS activation, NO generation, and the induction of cardioprotective signaling.[6]

As reported with early pre-conditioning, delayed pre-conditioning also promotes cardioprotection by augmenting cardiomyocyte bioenergetic capacity in response to ischemic and anoxic insult. The modulation of mitochondrial respiration in response to delayed pre-conditioning has recently been shown to occur primarily as a consequence of the induction of ETC and ANT transcription, promoted by the ROS-dependent up-regulation of nuclear transcription factors NRF-1 and PGC-1α.[147]

In addition to the contributory roles that the mitoK$_{ATP}$ channels, mitochondrial-generated ROS, and ETC induction play in the events of both types of pre-conditioning, a mitochondrial-specific antioxidant response has been demonstrated, primarily with respect to the events of delayed pre-conditioning.[148] The time course of MnSOD induction, a mitochondrial-specific protective response to ROS mediated damage, correlates well with the appearance of ischemic tolerance. The activities of other cellular antioxidants (i.e., catalase and CuSOD) were not similarly affected. That MnSOD induction plays a role in delayed IPC is further supported by findings that treatment with antisense oligonucleotides to MnSOD applied immediately after IPC completely abolished both the IPC cardioprotective effects and MnSOD induction.[148] MnSOD induction is likely downstream of both ROS generation and ROS-mediated activation of specific transcription factors (e.g., NF-κB) and the subsequent induction of cytokines (e.g., TNF-α and IL-1β).[16,149] It is noteworthy that the myocyte intracellular pathways affected by TNF-α include the rapid activation of the sphingolipid-ceramide signaling pathway generating endogenous second-messengers ceramide and sphingosine.[16]

Thus, in the delayed IPC pathway, the mitochondrial response to relatively short-term metabolic stresses contributes to a dynamic intracellular cross-talk and signaling between multiple cellular compartments, enabling the cardiomyocyte to adopt a stress-tolerant state. Moreover, in addition to IPC, several physical stresses can serve to promote MnSOD-associated delayed cardioprotective induction, including, whole-body hyperthermia and exercise.[148,149]

PHARMACEUTICAL PRE-CONDITIONING: BEYOND KCOS AND CARDIOPROTECTION

A rather large assortment of drugs has been shown to mimic IPC when either substituted for the pre-conditioning period or applied at reperfusion. These are listed in Table I. In this section, we discuss a number of the better characterized cardioprotective agents.

Anesthetics

Transient administration of volatile anesthetics, such as halothane, isoflurane, and sevoflurane, before a prolonged ischemic episode reduces myocardial infarct size to a degree comparable to that observed during IPC. Many components of the signal transduction pathways responsible for cardioprotection are shared by both anesthetic pre-conditioning (APC) and IPC, including the activation of protein kinase C and tyrosine kinases and the involvements of both sarcolemmal and mitochondrial K$_{ATP}$ channels (inward rectifiers).[150] Exposure to volatile anesthetics generates small "triggering" quantities of ROS that likely arise from the direct inhibition of enzymes of the mitochondrial electron transport chain (e.g., complex III) or indirectly through a signaling cascade in which G-protein–coupled receptors, protein kinases, and mitoK$_{ATP}$ channels play important roles.[151,152] Experiments showing that pre-treatment with the mitoK$_{ATP}$ blocker 5-HD before isoflurane abolished ROS production, whereas administration of 5-HD after isoflurane only partially attenuated ROS generation produced by the volatile anesthetic agent, suggesting that mitoK$_{ATP}$ channel opening

TABLE III Anesthetic Preconditioning-Signaling Events

Signaling event	References
Attenuated mitochondrial electron transport	168, 169
Increased ROS generation with overall decrease in ROS accumulation after ischemia and reperfusion	170–173
PKC activation	154, 155, 157, 162
MitoK$_{ATP}$ channel opening	174–176
Differential gene regulation	136, 165
MAPK activation	158
Mitochondrial PT pore	177
Inhibition of glycogen synthase kinase 3β.	163, 164
Sarcolemmal K$_{ATP}$ channel	178
Up-regulation of nitric oxide synthase (as both trigger and mediator)	156, 161
ERK1/2 activation	179
Attenuation of NF-κB activation and down-regulation of NF-κB-dependent inflammatory gene expression	180
Cyclooxygenase-2 mediation	181
Phosphatidylinositol-3-kinase/Akt signaling	182
Reduction of intracellular calcium	183
Attenuation of mitochondrial calcium overload	184, 185

likely acts as a trigger for isoflurane-induced pre-conditioning by generating ROS *in vivo*.[153] Moreover, APC-mediated cardioprotection can be abolished by NOS inhibitors or by deployment of ROS scavengers. Notably, all halogenated anesthetics can produce an anesthetic pre-conditioning effect. Evidence for the involvement of ROS, mitoK$_{ATP}$ channels, and protein kinase activation has been obtained in both separate experimental models and in a single model with sevoflurane-induced cardioprotection in the rat.[154] APC leads to improved mitochondrial electron transport chain function and cardiac function during reperfusion.[155] In addition, APC has both an acute and a delayed cardioprotective phase, the latter involving NOS activation.[156]

Despite substantial similarities in structural and functional cardioprotection elicited by APC compared with IPC, some differences with respect to key signaling steps have been reported.[157,158] These include the differential activation and translocation of protein kinase C isoforms to subcellular targets,[157] as well as the role of other intracellular kinases (e.g., MAPK) in triggering and mediating pre-conditioning–induced cardioprotection.[15] Salient features of the signaling pathway in APC cardioprotection are listed in Table III.

Transcriptional profiling using genome chips of APC and IPC in rat hearts while identifying many commonalities in the gene expression profiles induced by these very different forms of cardioprotection also revealed several significant distinct features.[136,154] A high number of commonly regulated genes were found in IPC and APC (39 up-regulated, 17 down-regulated), including genes associated with cell defense (heat shock protein 10, aldose reductase, Bcl-xS). Conversely, a pool of protective and antiprotective genes was differentially regulated in APC versus IPC (heat shock protein 27/70, programmed cell death 8), indicating trigger-dependent transcriptome variability. Moreover, the gene profile of the IPC shared more commonality with the program found with unprotected myocardium, indicating less "protection" (and conversely more ischemic damage) with IPC compared with the distinct pattern of a cardioprotective genomic response found with APC.

Recent studies have also demonstrated that volatile anesthetics applied during early reperfusion after myocardial ischemia can produce cardioprotection and termed this phenomenon "anesthetic post-conditioning."[160,161] As with ischemic pre-conditioning, anesthetic post-conditioning is mediated by the involvement of PI3K activation, mitoK$_{ATP}$ channels, and mitochondrial PT pore closing.[161–163] The closing of the PT pore in both anesthetic post-conditioning and classical pre-conditioning is largely mediated by the inhibition of glycogen synthase kinase 3B by means of the P13K and Akt pathways.[163–164] However, there are presently no other reports concerning glycogen synthase kinase mediation in the relatively nascent field of post-conditioning. Interestingly, transcription profiling of the anesthetic post-conditioned heart compared with the anesthetic pre-conditioned myocardium reveal strikingly different genomic responses in these otherwise highly similar cardioprotective approaches.[165]

Several clinical studies have suggested that pre-conditioning by volatile anesthetics exerts beneficial effects in patients undergoing cardiac surgery.[166] However, a variety of factors, including old age, coexisting conditions such as diabetes mellitus, and the use of oral hypoglycemic drugs or cyclooxygenase inhibitors, as well as the timing and duration of myocardial ischemia, may limit the benefits of APC when applied under clinical conditions.[167]

Erythropoietin

Erythropoietin (EPO), the principal hematopoietic cytokine that regulates mammalian erythropoiesis and erythrocyte survival, exhibits diverse cellular effects in nonhematopoietic tissues, including protection of neuronal tissues from ischemic-induced apoptosis. The physiological functions of EPO are directly mediated by its specific cell-surface receptor EPOR, which has been recently found to be expressed at both the RNA and protein levels in both fetal and adult cardiomyocytes.[186,187]

Several studies have demonstrated that EPO can serve a cardioprotective function in both H9C2 cardiac myoblasts and cardiac fibroblasts *in vitro*[188] and in the myocardium *in vivo* in response to the injury stemming from I/R in rat and rabbits.[186,188,189] *In vivo* studies indicate that treatment with recombinant EPO either before or during ischemia significantly enhances cardiac function and recovery, including left ventricular contractility, after myocardial I/R.[186,188,190] These studies have all demonstrated that pre-conditioning with erythropoietin activates the same cell survival pathways, including the PI3K/Akt pathway, as found to be operative in IPC and that EPO cardioprotection was fundamentally associated with reduced levels of myocyte apoptosis and less ventricular dysfunction. Rafiee and associates found that multiple cardioprotective signaling pathways are activated by pre-ischemic EPO treatment in the rabbit heart, including the phosphorylation and activation of JAK1/2, STAT3 and STAT5A, the phosphorylation and activation of PI3K and its downstream kinases Akt and Rac, and the activation of PKC-ε, Raf, MEK1/2, p42/44 MAPK, and p38 MAPK.[191] Moreover, Hanlon and associates demonstrated that EPO treatment of perfused hearts induced the translocation of PKC-ε isoform to the membrane fraction and that the protective effect of EPO was significantly inhibited by the PKC catalytic inhibitor chelerythrine added before and concomitant with EPO, suggesting that EPO-mediated activation of the PKC signaling pathway before or during ischemia is required for the cardioprotective effect of EPO during I/R injury.[190] This study also found that post-ischemic EPO treatment (applied at the time of reperfusion) also resulted in significantly improved recovery of LVDP and reduced infarct size and that this cardioprotection required the PI3K pathway but was not affected by inhibition of PKC at the time of EPO treatment. These findings indicate that EPO has direct action on cardiac fibroblast and myocyte signaling to alter survival and ventricular remodeling in response to ischemic injury and may be a novel treatment for myocardial ischemic disease.

Glucose, Insulin, and Metabolic Conditioning

It has been known for some time that a metabolic "cocktail" of glucose, insulin, and potassium (GIK) can provide beneficial pre-conditioning effects to injured myocardium. Administration of GIK significantly reduced mortality in patients with AMI, supporting the concept that metabolic interventions may be an important adjunctive therapy for protecting ischemic myocardium.[192] GIK provided either during the entire I/R period or at the onset of the post-ischemic reperfusion provided similar levels of cardioprotection.[193] The basis for this cardioprotection has generally been considered to be primarily metabolic in origin, acting by modulating cardiac and circulating metabolites to provide the heart with an optimal metabolic milieu to resist I/R injury.[194] The use of carbohydrates rather than fatty acids for bioenergetic fuel is metabolically more efficient and may improve the coupling between glycolysis and pyruvate oxidation. Therefore, promoting a shift in metabolic fuel substrate use toward glucose oxidation during times of reduced oxygen availability may represent a cardioprotective strategy. A variety of metabolic interventions that promote glucose use during ischemia have been shown to protect ischemic myocardium and improve functional recovery on reperfusion. Enhanced glucose uptake, glucose metabolism, and glycolysis have been proposed as contributing to an anti-ischemic cardioprotection. These metabolic processes are also enhanced by the use of fatty acid inhibitors such as trimetazidine and ranolazine that stimulate carbohydrate oxidation either by enhancing oxidation at the pyruvate dehydrogenase complex or by limiting FAO.

Recent data have suggested that insulin acts as the major component mediating cardioprotection primarily by promoting tolerance against ischemic cell death and enhancing cardiomyocyte survival by activating cell-survival signaling pathways.[195] Insulin's cytoprotective capacity was initially demonstrated in rat neonatal cardiomyocytes in a simulated model of ischemia and reoxygenation.[196] Administration of insulin at the moment of reoxygenation enhanced myocardial cell viability. This cardioprotection was found to be due, in part, to a reduction in myocyte apoptosis and was completely abolished by addition of the tyrosine kinase inhibitor lavendustin A and by the PI3K- inhibitor wortmannin.

Insulin's cardioprotective effect was further confirmed in isolated perfused rat hearts in which insulin administered at the onset of reperfusion attenuated infarct size.[197,198] Protection was abrogated if insulin administration was delayed until 15 min into reperfusion. Co-administration of insulin with lavendustin A, wortmannin, and the mTOR/p70S6 kinase inhibitor rapamycin completely abolished cardioprotection. Interestingly, insulin-mediated cardioprotection was found to be independent of the presence of glucose at reperfusion. This study also demonstrated that insulin treatment affected downstream prosurvival targets of Akt, including p70S6 kinase and the proapoptotic protein Bad, which is maintained in its inert phosphorylated state. Further studies with an *in vitro* cell system have provided confirmation of p70S6 kinase as a critical target of insulin-activated Akt cardiomyocyte-survival signaling.[199] In human cardiac–derived Girardi cells subjected to 6 h of simulated ischemia and 2 h of reoxygenation, insulin provided at reoxygenation enhanced cell viability with attenuated LDH release and significantly increased p70S6 kinase levels and activity compared with controls. The cytoprotection afforded by insulin and P70S6 kinase activity was abolished by pre-treatment with antisense oligonucleotides targeting p70S6 kinase but not by the sense or scrambled oligonucleotide constructs.

Given the critical role that insulin seems to play in stemming cardiomyocyte apoptosis, current evidence that insulin may play a contributory role in preserving the integrity of mitochondria, a key player in the early events of myocardial apoptosis, is not entirely surprising. Recent studies have demonstrated that several cardioprotective interventions, including insulin treatment, opioid treatment, and classic IPC can promote the myocardial redistribution of the glycolytic enzyme hexokinase from the cytosol to the mitochondria in the heart.[200] The significance of this putative end effector mechanism common to several disparate cardioprotective mechanisms is not yet known; the association of hexokinase with mitochondria may be involved in promoting inhibition of the mitochondrial PT pore and thereby reducing cell death and apoptosis.

Adult rat ventricular cardiomyocytes with induced oxidative stress (promoted by laser illumination of the fluorophore, tetramethyl rhodamine methyl ester, TMRM) exhibit global mitochondrial depolarization stemming from PT pore opening followed by ATP depletion and rigor contracture.[201] Insulin treatment of the stressed cardiomyocytes significantly delayed the mitochondrial PT pore opening.[201] Moreover, the effect of insulin on the PT pore was prevented by wortmannin and by LY-294002, inhibitors of the PI3K pathway, and could be mediated by different Akt constructs (e.g., a dominant negative Akt construct abolished the delayed PT pore opening), suggesting that the insulin activation of the PI3K/Akt pathway occurs upstream of the mitochondrial PT pore opening in insulin's cardioprotective pathway.

Although insulin can provide potent cardioprotection, particularly against myocardial apoptosis, its widespread use as an applicable therapy may be limited because of its association with a variety of metabolic perturbations, the development of pharmacotherapeutic agents that target downstream cell-survival insulin-activated signaling molecules has been proposed as an alternate approach to promote clinical cardioprotection.

Glucagon-like peptide 1 (GLP-1), a gut hormone that stimulates insulin secretion and also activates antiapoptotic signaling pathways such as PI3K and MAP kinase initially described in pancreatic cells, has recently been shown to be cardioprotective by eliciting the activation of these pro-survival signaling pathways.[202] Experiments in pre-clinical models using either isolated perfused rat heart or whole animal models of I/R showed that GLP-1 added before ischemia promoted a significant reduction in infarction. This cardioprotection could be abolished by the addition of the GLP-1 receptor antagonist exendin, the PI3K inhibitor LY294002, and the p42/44 MAP kinase inhibitor UO126, implicating the involvement of multiple survival kinases and expanding the possibility for pharmacological therapeutic agents. Recent clinical studies have revealed that GLP-1 infusion in patients with AMI having left ventricular dysfunction undergoing primary angioplasty significantly improved LVEF global wall motion score indexes and regional wall motion score indexes.[203] Infusion of GLP-1 also enhanced recovery from ischemic myocardial stunning after brief coronary occlusion and reperfusion in dogs.[204]

Pyruvate, a natural metabolic fuel and antioxidant in myocardium and other tissues, exerts a variety of cardioprotective actions when provided at very high concentrations.[205,206] Provision of pyruvate alone or as a cosubstrate (along with glucose) markedly reduced ischemic contracture and enhanced post-ischemic recovery in mouse hearts subjected to I/R. Cardioprotection did not require the presence of pyruvate during ischemia or reperfusion, and the effects of pyruvate pre-treatment could be mimicked by pre-treatment with dichloroacetate (DCA), an activator of pyruvate dehydrogenase. Myocardial adenosine efflux and Ca^{++} content were elevated after pre-treatment with pyruvate, potentially triggering a pre-conditioned state. The response seems unrelated to glycolytic inhibition but may be mediated by transient changes in adenosine levels and/or cellular Ca^{++} mitigated by pyruvate enhancement of sarcoplasmic reticular Ca^{++} transport. Moreover, there is evidence that pyruvate also inhibits the mitochondrial PT pore opening at reperfusion and enhances subsequent pore closure to provide cardioprotection, although the mechanism by which it accomplishes this has not been determined.[207]

Diabetes and Cardioprotection

The importance of insulin in promoting cardioprotection is also of particular of interest, because it has been frequently reported both in clinical and animal studies that the diabetic heart is refractory to cardioprotection, although some have challenged this view. While a more extensive discussion of diabetes and myocardial disease will be presented elsewhere in this volume, a few points relevant to the present discussion of cardioprotection can be made here.

Most animal models of diabetes, including streptozotocin-induced or alloxan-induced diabetic rats, rabbits, or dogs, have demonstrated a pronounced reduction in cardioprotection from either IPC or pharmacological pre-conditioning. Several recent examples (among numerous studies) include the demonstration that alloxan-induced diabetes (as well as acute hyperglycemia) blocks the cardioprotection induced by ischemic delayed pre-conditioning in rabbits and that the cardioprotection is not restored by short-term insulin treatment.[208] Similarly, the cardioprotective action of GIK administered in a dog model of coronary artery occlusion was abolished in animals with alloxan-induced diabetes (or with hyperglycemia preceding the ischemic insult).[209] IPC in hearts from obese Zucker diabetic fatty (ZDF) and lean Goto-Kakizaki (GK) type 2 diabetic rats failed to protect hearts from reperfusion injury during reperfusion in both diabetic animal models.[210] This latter study is particularly of interest, given the increasing worldwide prevalence of type 2 diabetes mellitus and its well-documented association with increased cardiovascular risk.

Although the cellular signaling pathways that specifically mediate the effects of cardioprotection in the diabetic myocardium have not been fully explained, diabetes-associated changes in a variety of specific pathways have been described.[211] Defects in the insulin receptor β, insulin receptor substrate-1, GLUT-4 protein,[212] basal and insulin-stimulated Akt,[213] and extracellular signal–related kinase and PI3K[214] have been reported in diabetic animal models. Moreover, sarcolemmal K_{ATP} channels have been shown to be altered in diabetic animals.[215] A recent study found that cardioprotection signaling defect(s) in the diabetic rat (type 2) can be bypassed by using an increased IPC stimulus to achieve the threshold for cardioprotection and a critical level of Akt phosphorylation to mediate myocardial protection.[211]

Susceptibility of the type 2 diabetic myocardium to ischemic damage is lower than in the nondiabetic heart. Previous studies have reported that in the acute phase of diabetes in streptozotocin-induced diabetes (which is most analogous to type I diabetes), diabetic hearts are more resistant to irreversible cell damage.[216] Potential reasons that ischemic sensitivity might be lower in diabetic animals may be substrate related, including a decreased accumulation of glycolytic products during ischemia (decreased lactate and protons), as well as alterations in the regulation of intracellular pH in the diabetic heart.[217] Others have reported the reverse (i.e., increased ischemic susceptibility with diabetes [218] or no differences in ischemic susceptibility [in a model of type 2 diabetes]).[219] These conflicting findings might be attributed to a variety of other factors, including species- and model-specific differences, different genetic backgrounds, and metabolic factors/levels including the presence of fatty acids. Interestingly, gender-specific differences in ischemic susceptibility have recently been described in which female diabetic rats are significantly more susceptible to ischemic damage than males.[220]

Moreover, the role of high blood glucose that often accompanies diabetes also may have an effect on cardioprotection and on susceptibility to ischemia. High blood glucose levels have been shown to antagonize IPC and anesthetic pre-conditioning in the presence and absence of diabetes in a number of animal models.[221–223] The onset of hyperglycemia may also provide some protection of the heart to ischemic injury. Interestingly, high glucose incubation of isolated myocytes causes up-regulation of the antiapoptotic factor Bcl-2 and inactivation of the proapoptotic factor Bad, which leads to increased tolerance to hypoxia.[224] Moreover, hyperglycemia also activates PKC,[225] which has been shown to have a protective effect against ischemia.[226]

Clinical studies have also generally supported the diminished capability of diabetic patients to undergo cardioprotection. Interestingly, early studies proposed that the failure to pre-condition the diabetic heart in patients is primarily the result of dysfunction of the mitochondrial K_{ATP} channels.[227] Analysis of mitochondrial preparations from nondiabetic and diabetic myocardium of patients revealed that diazoxide-mediated pre-conditioning resulted in impaired mitochondrial depolarization in diabetic myocardium compared with controls, as well in decreased superoxide production in diabetic mitochondria presumably mediated by a dysfunctional mitoK_{ATP} channel.[228] Also of relevance is the inhibitory effect of a number of oral sulfonylureas (e.g., glibenclamide), a widely used antidiabetic hypoglycemic therapy, on myocardial ATP-sensitive potassium K_{ATP} channels, an integral player in IPC as we have noted. Inhibition of these channels prevents IPC and likely explains the finding that patients taking sulfonylurea drugs have increased cardiovascular mortality.[229,230]

Statins

There is accumulating evidence that statins (HMG-CoA reductase inhibitors) exert lipid-independent cardioprotective effects. Direct cytoprotective effects of statins on cardiomyocytes (independent of endothelial cells) in response to hypoxic injury have also been demonstrated. Human ventricular cardiomyocytes treated with pravastatin were markedly protected against cell death after simulated hypoxia and reoxygenation.[231] This cytoprotection was shown to be mediated by an increase in NO release, a decrease in myocyte ET-1 production/action, and an increase in Akt activation. In a study featuring neonatal rat cardiomyocytes subjected to H_2O_2-induced oxidative stress, Jones and associates showed that simvastatin pre-treatment significantly attenuated mitochondrial membrane depolarization after exposure to oxidant stress.[232] This study also demonstrated in experiments using the NOS inhibitor (L-NAME) and mitoK_{ATP} channel blocker (5-HD) that the NO pathway and mitoK_{ATP} channel are necessary components of the statin-induced cytoprotection.

Long-term pre-treatment with statins reduces myocardial injury after acute I/R by increasing the expression of eNOS.[233] To evaluate whether statins act rapidly enough to protect the myocardium from I/R injury when administered at the beginning of the reperfusion period, activated simvastatin was given intravenously 3 min before starting the reperfusion after temporary coronary artery occlusion in anesthetized rats. Simvastatin significantly increased myocardial PI3K activity, phosphorylation of both AktSer473 and eNOS Ser1177, and reduced infarct size by 42%. Infarct size reduction and the activation of PI3K/Akt/eNOS pathways were not observed in rats cotreated with the PI3K inhibitor wortmannin. The contribution of eNOS was further confirmed in experiments that used the NOS inhibitor L-NAME, which could completely block cardioprotection by the statin treatment.[234]

Recent evidence indicates that cyclooxygenase-2 (COX2), which mediates delayed IPC, also mediates statin-induced cardioprotection.[235] In rats, administration of valdecoxib, a selective COX2 inhibitor, attenuated the cardioprotective effect of atorvastatin when administered together before 30 min of myocardial ischemia followed 4 h of reperfusion. Although atorvastatin treatment resulted in smaller infarct size and significantly increased levels of myocardial PGF1a and PDE2, as well as myocardial content of cytosolic phospholipase A2 (cPLA2), COX2, PGI2 synthase, and PGE2 synthase, coadministration of atorvastatin and valdecoxib did not. These results suggest that the prostaglandins are essential for mediating the cardioprotective effects of atorvastatin and also further indicate that their production is downstream to eNOS phosphorylation and iNOS, because atorvastatin-mediated changes in these NO signaling molecules were unchanged in the presence of valdecoxib.

Studies have also indicated that atorvastatin administered to isolated perfused mouse hearts at the onset of reperfusion after a 35-min period of global ischemia resulted in a marked dose-dependent reduction of infarct size, with a concomitant and rapid (occurring within 5 min) activation of both Akt and eNOS phosphorylation.[236] The involvement of both the pI3K/Akt and eNOS pathways in this statin-mediated cardioprotection was further confirmed by the abrogation of this protection by the administration of wortmannin and in eNOS knock-out mice. Subsequent studies that used the same model system have shown that atorvastatin applied (as an acute treatment) at reperfusion attenuated lethal reperfusion-induced injury that occurs with the activation of other prosurvival signaling pathways besides PI3K/Akt, including the significant phosphorylation of p44/42 MAPK, p38 MAPK, and HSP27.[237]

Because many patients are receiving chronic statin therapy for their well-documented provision of beneficial lipid-lowering effects, recent studies have also examined the important question of whether chronic oral therapy with atorvastatin also provides protection against I/R injury. Although rats treated for 1–3 days with atorvastatin before global I/R displayed significant reduction of infarct size, longer treatment of 1–2 weeks resulted in no protection coincident with increased levels of PTEN, an endogenous PI3K inhibitor. However, the waning protection can be re-elicited by the administration of a supplementary dose of atorvastatin provided just before the ischemic insult.[238]

The clinical benefits of statin treatment for preventing coronary injury have been well documented in several large-scale prospective clinical trials.[239] A large-scale clinical study documented the significant presence of a cardioprotective effect of statin treatment in response to AMI both in patients under chronic treatment and in patients in which statin treatment was applied within the first 24 h of admission with AMI.[240] Mortality was significantly decreased in patients with early or new statin treatment compared with either untreated patients or patients with discontinued statin treatment. Early statin use was also associated with a lower incidence of cardiogenic shock, dysrhythmias, cardiac arrest, rupture, but not recurrent MI. Other clinical studies have also suggested that statin treatment can lower the risk of procedural myocardial injury. In the clinical trial designated ARMYDA (*A*torvastatin for *R*eduction of *MY*ocardial *D*amage during *A*ngioplasty), pre-treatment with atorvastatin (40 mg/d for 7 days) significantly reduced procedural myocardial injury in patients with chronic stable angina undergoing elective coronary intervention.[241] Detection of markers of myocardial injury (e.g., creatine kinase-MB, troponin I, and myoglobin levels) above the upper normal limit was significantly lower in the statin group than the placebo group. Moreover, MI was detected after coronary intervention in 5% of patients in the statin group and in 18% of those in the placebo group.

Other Pharmacological Agents

Several other pharmacological agents have shown promise in promoting cardioprotection. These include agents such as sildenafil, nitroglycerin, and monophosphoryl lipid A.[242] Significantly, both nitroglycerin and sildenafil are associated with increased NO bioavailability.

As we have previously noted, NO is a critical component of the delayed cardioprotection pathway; nitroglycerin (NTG), which serves as a potent NO donor, has been shown to provide cardioprotection in a mouse model of occlusion/reperfusion.[243] This cardioprotection was entirely abolished in iNOS knock-out (iNOS$^-$/iNOS$^-$) mice, indicating that iNOS function is required for NTG cardioprotection. Rat studies have also linked NTG with the endogenous myocardial protective factor calcitonin gene-related peptide (CGRP)–mediated delayed cardioprotection[244] and iNOS activation.[245] Pre-treatment with NTG causes a significant increase in α-CGRP mRNA expression, concomitant with an increase in plasma concentrations of cGMP and CGRP. These effects of NTG were completely abolished by pre-treatment with aminoguanidine, a selective inhibitor of iNOS activity suggestive that delayed cardioprotection afforded by NTG is mediated by the α-CGRP isoform by generation of NO derived from iNOS.

Clinically, NTG has been effectively used as a cardioprotective supplement in coronary angioplasty[246] and can alleviate exercise-induced ischemia 24 h after administration, substantially enhancing exercise capacity in patients with CAD.[247] Recent studies have also reported that transdermal NTG-induced myocardial pre-conditioning promotes enhanced global cardiac performance and hemodynamic improvements in patients with stable angina, mimicking exercise-induced pre-conditioning.[248]

Considerable evidence has indicated that the phosphodiesterase type-5 (PDE5) inhibitor, sildenafil, initially developed for the treatment of erectile dysfunction, has a pre-conditioning-like cardioprotective effect against I/R in the intact heart.[249] Pre-clinical studies have demonstrated that sildenafil exerts cardioprotection through NO generated from eNOS/iNOS, activation of PKC/ERK signaling, and opening of mitoK$_{ATP}$ channels.[250,251] Additional studies show that the drug attenuates cell death resulting from necrosis and apoptosis and increases the Bcl-2/Bax ratio through NO signaling in adult cardiomyocytes.[252] Clinical studies have also suggested that sildenafil may be applied in the treatment of pulmonary arterial hypertension and endothelial dysfunction.

Monophosphoryl lipid A (MLA), a nontoxic derivative of the endotoxin pharmacophore lipid A, when given as a single dose pre-treatment in various canine and rabbit models 12–24 h before ischemia, limits infarct size and reduces regional and global contractile dysfunction.[253] Priming of the K$_{ATP}$ channel for enhanced opening during ischemia may be a prerequisite for the cardioprotective activity of MLA, establishing a similarity between MLA and IPC, although how MLA may regulate the K$_{ATP}$ channel is undetermined. Pre-treatment with MLA displays a time course for development similar to that of the second window of IPC, and the delayed cardioprotection afforded by MLA is largely mediated by endogenous NO and is related to stimulation of CGRP release.[254] Pre-treatment of isolated rat hearts with MLA for 24 h before hypothermic ischemia improved cardiac function, reduced creatine kinase release concomitant with decreased cardiac TNF-α levels, and increased plasma concentrations of CGRP and NO. Pre-treatment with either L-NAME or capsaicin abolished both the cardioprotection and the increased release of CGRP induced by MLA, confirming the involvement of CGRP in the cardioprotective pathway.

POST-CONDITIONING AND CARDIOPROTECTION

Post-conditioning is a series of brief mechanical interruptions of reperfusion after a specific prescribed algorithm

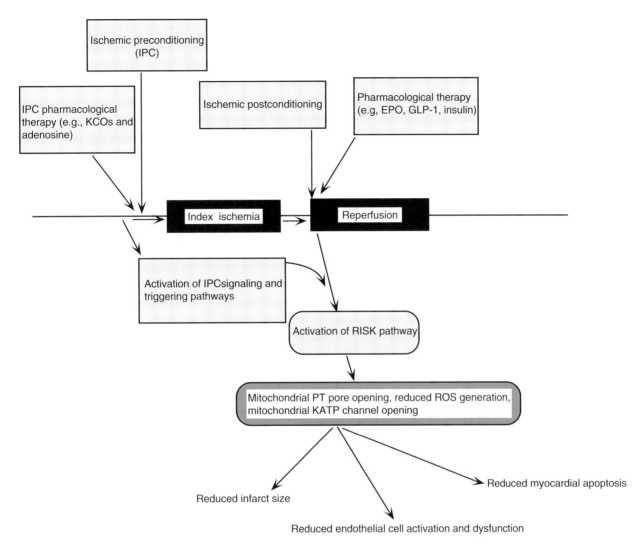

FIGURE 4 Flowchart comparing post-conditioning and pre-conditioning protocols.

applied at the very onset of reperfusion. This algorithm lasts only from 1–3 min, depending on species. Post-conditioning has been observed to reduce infarct size and apoptosis as the end points in myocardial therapeutics; salvage of infarct size was similar to that achieved by IPC. Post-conditioning cardioprotection has been also associated with a reduction in endothelial cell activation and dysfunction, tissue superoxide anion generation, neutrophil activation, and accumulation in reperfused myocardium, microvascular injury, tissue edema, intracellular and mitochondrial calcium accumulation.[255]

In contrast to pre-conditioning, which exerts its effects both during the index ischemia and in the reperfusion stage, post-conditioning seems to exert its effects during reperfusion alone. Post-conditioning modifies the early phase of reperfusion by reducing the oxidant burden and consequent oxidant-induced injury by attenuating the local inflammatory response to reperfusion and by engaging end effectors and signaling pathways implicated in ischemic and pharmacological pre-conditioning. As with IPC, the post-conditioning phenomenon has been observed in both large and small animal in vivo models, as well as in ex vivo and cell culture models. A comparison between these two models or protocols of cardioprotection is shown in Fig. 4.

Post-conditioning sets in motion triggers and signals that are functionally related to reduced cell death, in part by promoting the up-regulation of survival kinases known to attenuate the pathogenesis of apoptosis and possibly necrosis. Adenosine has been implicated in the cardioprotection of post-conditioning, as has eNOS, NO, and guanylyl cyclase, opening of K_{ATP} channels, and closing of the mitochondrial PT pore.[256–258] Cardioprotection by post-conditioning has also been associated with the activation of intracellular survival pathways such as ERK1/2 and PI3K–Akt pathways.[259] Other pathways have yet to be identified. Although many of the pathways involved in post-conditioning have also been identified in IPC, some may not be involved in pre-conditioning (ERK1/2). The timing of action of these pathways and other

TABLE IV Comparison of IPC and Post-Conditioning

	IPC	Post-conditioning
Time of intervention	Before onset of ischemia	After onset of ischemia at start of reperfusion
Physiological phenotype		
Reduced infarct size	+	+
Reduced apoptosis	+	+
Reduced tissue edema	+	+
Reduced endothelial dysfunction	+	+
Signaling components		
Adenosine-mediated	+	+
NO, eNOS, and guanylyl cyclase	+	+
Opening of mitoK$_{ATP}$ channels	+	+
Closing of the mitochondrial PT pore	+	+
Activation of PI3K-Akt survival pathway +	+	+
Activation of ERK1/2	+	+
Opioid mediated	+	?
Reduced expression of leukocyte adhesion molecules and cytokines	+	+
Reduced lipid peroxidation	+	+
Reduced ROS generation coincident with reduced cell death	+	+
Genomic response (mostly with delayed IPC)	+	−
Pharmacological agents for clinical application	Adenosine, nicorandil	Insulin, GLP1, erythropoietin, atorvastatin

mediators of protection in post-conditioning differs from that of pre-conditioning.[260,261] In contrast to pre-conditioning, which requires a foreknowledge of the ischemic event, post-conditioning can be applied at the onset of reperfusion at the point of clinical service (i.e., angioplasty, stenting, cardiac surgery, and transplantation).[262]

A comparison of key signaling components used in pre-conditioning and post-conditioning is shown in Table IV.

REMOTE CONDITIONING

Pre-conditioning is not confined to one organ but can also limit infarct size in remote, non-pre-conditioned organs ("*remote* pre-conditioning").

The first evidence of remote, pre-conditioning–induced protection was obtained with a canine model. Brief occlusions of the circumflex (Cx) artery protected not only the Cx vascular bed but pre-conditioned as well the left anterior descending (LAD) vascular bed.[263] Subsequent studies have demonstrated that brief episodes of ischemia in peripheral sites (i.e., kidney, hind limb, and, most recently, the brain) can protect the heart against infarction.[264–266] Moreover, it has been reported that isolated buffer-perfused hearts release cardioprotective factor(s) during brief I/R, which, when administered to virgin acceptor hearts by means of transfer of coronary effluent, elicit a reduction in infarct size comparable to that achieved with IPC.[267] Remote pre-conditioning has been reported in many other species, including rats, mice, rabbits, and pigs, in most conferring cardioprotection (usually assessed as limiting myocardial infarct size) to a similar extent as classical IPC. At present, there is limited evidence documenting remote pre-conditioning in humans.[268]

As with classical IPC, both early and delayed (or second window of protection) components have been described in remote pre-conditioning.[8] Several lines of evidence have pointed to the contribution of opioids,[269] adenosine,[270] and stimulation of myocardial adenosine receptors,[266] PKC-ε activation,[271] ROS involvement,[272] as well as the activation of the mitoK$_{ATP}$ channel[273] in the signaling pathway of remote pre-conditioning. Gene profiling studies that used microarray analysis found that remote pre-conditioning modified myocardial gene expression by up-regulating cardioprotective genes, including genes protecting against oxidative stress (e.g., Hadhsc, Prdx4, and Fabp4) and cytoprotection (Hsp73) and suppressing genes potentially involved in the pathogenesis of I/R injury such as proinflammatory genes (e.g., Egr-1, Dusp 1 and 6).[274] These findings support earlier studies showing that remote IPC (a brief forearm ischemia) provided a stimulus for mediating inflammatory gene transcription in circulating human leukocytes.[275] In that study, genes encoding key proteins involved in cytokine synthesis, leukocyte chemotaxis, adhesion and migration, exocytosis, innate immunity signaling pathways, and apoptosis were all suppressed within 15 min (early phase IPC) and more so after 24 h (second window IPC). Other studies have reported that remote hind limb ischemic pre-conditioning promoted myocardial nuclear translocation of the transcription factor NF-κB within 2 h of the pre-conditioning stimulus, whereas cardiac mRNA for iNOS gradually increased in a 24-h time course after remote hind limb pre-conditioning, an example of delayed cardioprotection.[276]

However, the mechanism of remote conditioning is presently less well defined than classical IPC; this includes both the temporal sequence of events mediating the cardioprotection and delineating the elusive identity of the diffusible protective factor(s), apparently released from the heart and other organs, that initiates protection at remote sites. It remains unclear whether this factor is a humoral protein, a neurogenic factor, or both.[8] Even recent attempts that used sophisticated proteomic analysis were unable to identify a specific protein factor.[277] Recent studies have suggested that the calcitonin gene–related peptide (CGRP) may be a viable candidate, because CGRP levels in the plasma are significantly increased in remote pre-conditioning, and it is an important mediator of sensory neurons. Moreover, recent studies have shown that CGRP infusion markedly reduces infarct size in rat heart and activates PKC-ε. Both infarct size reduction and PKC-ε activation were significantly attenuated by CGRP (8–37), a specific CGRP receptor antagonist.[278]

CLINICAL APPLICATION

Human Studies

Numerous studies support the concept that cardioprotection mediated by ischemic and pharmacological conditioning occurs in humans. When isolated human myocardial cells in culture are exposed to severe hypoxia for 90 min followed by reoxygenation, the resulting cell death can be significantly reduced by exposing the cells to short periods of hypoxia and reoxygenation before the longer 90-min index hypoxia.[279] This suggested that human myocardial cells can be pre-conditioned and that other cell types are not required for pre-conditioning. Benefits of IPC, therefore, depend on primary biochemical changes within the myocyte itself and not on changes in the vasculature. These findings have been extended by the demonstration that human ventricular cardiomyocytes also exhibited delayed cardioprotection 24 h after a short period of simulated ischemia in addition to classic pre-conditioning.[280]

There is provocative clinical evidence suggesting that human myocardial tissue can be pre-conditioned. Studies of patients undergoing procedures that involve brief periods of ischemia such as coronary angioplasty (PTCA) are suggestive of IPC. When repetitive balloon inflations (usually on the order of 60–90 sec) and deflations are performed with intervening periods of perfusion, the following clinical observations have been made, including a lessening of the severity of chest pain, reduction in degree of ST-segment elevation, reduced lactate production, and attenuated release of myocardial markers such as creatine kinase-MB with subsequent balloon inflations constituting a myocardial cardioprotection.[8]

This sequential angioplasty balloon inflation model has proved to be a useful screening tool, whereby pharmacological agents that mimic pre-conditioning but do not cause ischemia can be tested. Blockade of K_{ATP} channels with oral glibenclamide administered before angioplasty abolishes the reduction in the ischemia-mediated events observed during subsequent balloon inflations, implying a role for these channels in mediating this form of cardioprotection.[281] Conversely, administration of the KCO nicorandil reduces the subsequent amount of ST changes. Intracoronary infusion of adenosine before PTCA attenuates ischemic indices during the first balloon inflation, whereas inhibition of adenosine receptors by aminophylline abolishes myocardial adaptation during the second balloon inflation.

Given that many patients experience brief episodes of ischemia before an acute MI, it seems to be possible that this pre-infarct angina may offer the potential to pre-condition the myocardium, thereby reducing infarct size and improving survival. A retrospective analysis of the TIMI-4 trial[282] revealed that the presence of pre-infarct angina was associated with smaller infarct size on the basis of creatine kinase release, improved left ventricular function with reduced incidence of congestive heart failure and shock, and reduced mortality, a finding confirmed in other studies.[283] Interestingly, evidence suggesting that the interval between the last episode of angina and the index MI is a critical factor was shown by the finding that prodromal angina is only protective if it occurs within 24–72 h of MI, a time course remarkably similar to that of the delayed IPC in animal models.[284,285]

The clinical use of IPC has also been applied in the setting of coronary artery bypass grafting (CABG) in protecting the myocardium from global ischemia. Evidence that IPC conferred protection in CABG was obtained in studies that used pre-treatment with two 3-min episodes of aortic cross-clamping and reperfusion before a 10-min period of global ischemia and ventricular fibrillation, resulting in an improved preservation of left ventricular content and reduced post-operative troponin T release (an index of myocardial injury).[286,287] Techniques other than intermittent cross-clamping such as cardioplegia arrest have, in some cases, shown beneficial results and not in others. The use of anesthetics may have a confounding effect in these surgical studies. The use of pharmacological pre-conditioning with adenosine A1 receptor agonists was shown to confer some myocardial protection in CABG but not as much as obtained with cross-clamping IPC.[288]

Clinical Trials

Clinical testing of IPC has been initiated in studies involving cardiac surgery, coronary angioplasty or treadmill exercise. However, in most circumstances, the clinical application of IPC is not desirable or feasible. Repeated cross-clamping of the aorta poses potential risks such as atheroembolism from calcifications, and IPC can prolong the length of surgery. Protecting the heart of patients with CAD likely requires a pharmacological approach using drugs that either enhance or mimic IPC. An early clinical trial (e.g., the IONA trial) provided evidence of the clinical usefulness of pharmacological pre-conditioning, with its demonstration that the chronic administration of the KCO nicorandil significantly improved the cardiovascular prognosis in patients with CAD.[289] A limiting factor in the efficacious use of either ischemic or pharmacological pre-conditioning pertains to the temporal boundaries of its protection; it is required that they be applied before the ischemic insult, which in most clinical situations (other than CABG or PTCA) is difficult to predict. In this respect, the use of pharmacological post-conditioning (administered at reperfusion) may circumvent the unpredictable nature of the ischemic insult and may be more clinically applicable. Pre-clinical studies have shown robust cardioprotection with the application of agents such as insulin, erythropoietin, and glucagon-like peptide 1 during the first few minutes of reperfusion, as acute pharmacological activators of myocardial survival kinases. However, there is presently limited clinical data concerning pharmacological pre-conditioning. In the AMISTAD clinical trial, adenosine as an adjunct to thrombolytic therapy resulted in decreased infarct size, most probably an example of post-conditioning cardioprotection.[290]

A critical caveat has been raised in regard to the clinical application of findings from pre-clinical studies that have largely been conducted with young and healthy animals to be used primarily with an older and often compromised patient population harboring CAD, diabetes, and other cardiovascular complications. In an earlier section of this chapter, we addressed data that suggest that diabetes may decrease the effectiveness of pre-conditioning, although recent studies suggest that increased strength of the pre-conditioning stimulus might overcome this limitation. Similarly, although there is limited evidence that aging may also decrease cardioprotective efficiency,[291–293] large-scale clinical studies are not available, and several recent studies have challenged that conclusion.[294–296] Moreover, the effects of hypercholesterolemia and atherosclerosis have not been clinically evaluated in human cardioprotective studies, although pre-clinical studies suggest that these will not be a major factor in limiting cardioprotection.[294]

FUTURE PERSPECTIVE

Data accumulated thus far from animal models show that ischemic and pharmacological pre-conditioning and post-conditioning may offer a desirable alternative and/or adjunct to the currently available drug armamentarium and surgical interventions for CAD. The use of pharmacological post-conditioning may provide the most attractive approach. Clearly, large-scale trials are needed to test their clinical efficacy. However, before these interventions can be routinely clinically used as therapy, the following questions need to be answered:

1. Because most IPC events have been studied primarily in the healthy heart of experimental young and middle-aged animals, are these cardioprotective interventions applicable successfully in human myocardial ischemia, including the aging heart?
2. Besides myocardial ischemia, could we achieve IPC cardioprotection in patients with primary cardiomyopathies and cardiomyopathy of aging?
3. Can pre-conditioning or cardioprotection be successfully applied in treating AMI, after heart transplant, and after bypass surgery for CAD?[297]
4. Can the period of cardioprotection (whether IPC, pharmacological, or post-conditioning) be extended even beyond the 72-h period of delayed pre-conditioning without losing effect?
5. Can combinations of various pharmacological triggers improve cardioprotection?

CONCLUSIONS

Understanding the cellular and molecular mechanisms of cardioprotection is essential in developing new therapies to improve the welfare of so many patients with myocardial diseases. Evidently, unequivocal data and more refined techniques to evaluate the molecular events of cardioprotection are required before clinical trials with drugs and regimens whose specificity of action including their overall metabolic and physiological role and potential benefits are not absolutely clear. Fully understanding the signals, transducing mediators, and effectors in the temporal pathway(s) of cardioprotection will be instrumental in the development of novel pharmaceutical agents with increased efficacy and enhanced specificity of action. Further search for alternative methods to improve mitochondrial cardioprotective responses are also warranted. For instance, the findings that PT pore opening during reperfusion can be modulated by either mitochondrial Ca^{++} levels or oxidative stress suggests that controlled management of mitochondrial Ca^{++} and/or ROS levels before reperfusion might be effective in providing cardioprotection.

Finally, advancing our knowledge of the cellular and molecular events involved in IPC and cardioprotection should stimulate further collaboration among cardiologists and other researchers in the field of drug discovery to successfully treat human cardiac diseases.

SUMMARY

- Cardioprotection shields the heart from damage secondary to insults such as ischemia.
- Ischemic pre-conditioning, by use of single or multiple brief periods of ischemia, protects the heart against prolonged ischemic insult (index ischemia).
- The myocardial signaling pathway of ischemic pre-conditioning, including numerous triggering and mediating components, is increasingly being identified, with a number of the components showing redundancy (i.e., parallel pathways). The end effectors, however, are not well defined. Early and delayed stages of cardioprotection have differences in their signaling components, as well as in their effects on cardiac phenotype.
- Pharmacological pre-conditioning can be effectively used to mimic ischemic pre-conditioning and seems to be more applicable to clinical settings. The use of erythropoietin, statins, sildenafil, nitroglycerin, potassium channel openers, adenosine receptor agonists, and insulin has provided cardioprotection in protecting against ischemia/reperfusion injury in pre-clinical models.
- The aging heart may have a diminished capacity to tolerate and respond to various forms of stresses, and the likelihood of myocardial ischemia and cardiac dysfunction increases.
- Prospective interventions and new drugs should be evaluated not only in young and middle-aged animals but also in the aging or senescent heart, a population in which cardioprotection is most relevant.
- Cardioprotective interventions await successfully application in humans.
- Post-conditioning with either brief ischemia/reperfusion episodes or pharmacological agents applied within the first few minutes of reperfusion provides similar cardioprotective benefits to ischemic pre-conditioning and uses many of the same signaling pathway components. This cardioprotective approach is highly attractive, because it can be applied with thrombolytic perfusion after AMI and does not have to precede the infarct/ischemic insult, which is, in most cases, difficult to predict.

References

1. Lesnefsky, E. J., Moghaddas, S., Tandler, B., Kerner, J., and Hoppel, C. L. (2001). Mitochondrial dysfunction in cardiac disease: Ischemia-reperfusion, aging, and heart failure. *J. Mol. Cell Cardiol.* **33,** 1065–1089.
2. Paradies, G., Petrosillo, G., Pistolese, M., Di Venosa, N., Serena, D., and Ruggiero, F. M. (1999). Lipid peroxidation and alterations to oxidative metabolism in mitochondria isolated from rat heart subjected to ischemia and reperfusion. *Free Radic. Biol. Med.* **27,** 42–50.
3. Eefting, F., Rensing, B., Wigman, J., Pannekoek, W. J., Liu, W. M., Cramer, M. J., Lips, D. J., and Doevendans, P. A. (2004). Role of apoptosis in reperfusion injury. *Cardiovasc. Res.* **61,** 414–426.
4. Honda, H. M., Korge, P., and Weiss, J. N. (2005). Mitochondria and ischemia/ reperfusion injury. *Ann. N. Y. Acad. Sci.* **1047,** 248–258.
5. Gottlieb, R. A., Burleson, K. O., Kloner, R. A., Babior, B. M., and Engler, R. L. (1994). Reperfusion injury induces apoptosis in rabbit cardiomyocytes. *J. Clin. Invest.* **94,** 1621–1628.
6. Bolli, R. (2000). The late phase of preconditioning. *Circ. Res.* **87,** 972–983.
7. Murry, C. E., Jennings, R. B., and Reimer, K. A. (1986). Preconditioning with ischemia: a delay of lethal cell injury in ischemic myocardium. *Circulation* **74,** 1124–1136.
8. Riksen, N. P., Smits, P., and Rongen, G. A. (2004). Ischaemic preconditioning: from molecular characterisation to clinical application—part I. *Neth. J. Med.* **62,** 353–363.
9. Murry, C. E., Jennings, R. B., and Reimer, K. A. (1986). Preconditioning with ischemia: a delay of lethal cell injury in ischemic myocardium. *Circulation* **74,** 1124–1136.
10. Yellon, D. M., and Downey, J. M. (2003). Preconditioning the myocardium: from cellular physiology to clinical cardiology. *Physiol. Rev.* **83,** 1113–1151.
11. Mubagwa, K., and Flameng, W. (2001). Adenosine, adenosine receptors and myocardial protection: an updated overview. *Cardiovasc. Res.* **52,** 25–39.
12. Liu, G. S., Thornton, J., Van Winkle, D. M., Stanley, A. W., Olsson, R. A., and Downey, J. M. (1991). Protection against infarction afforded by preconditioning is mediated by A1 adenosine receptors in rabbit heart. *Circulation* **84,** 350–356.
13. Thornton, J. D., Liu, G. S., Olsson, R. A., and Downey, J. M. (1992). Intravenous pretreatment with A1-selective adenosine analogues protects the heart against infarction. *Circulation* **85,** 659–665.
14. Cohen, M. V., Baines, C. P., and Downey, J. M. (2000). Ischemic preconditioning: from adenosine receptor of K_{ATP} channel. *Annu. Rev. Physiol.* **62,** 79–109.
15. Qin, Q., Downey, J. M., and Cohen, M. V. (2003). Acetylcholine but not adenosine triggers preconditioning through PI3-kinase and a tyrosine kinase. *Am. J. Physiol. Heart Circ. Physiol.* **284,** H727–H734.
16. Lecour, S., Smith, R. M., Woodward, B., Opie, L. H., Rochette, L., and Sack, M. N. (2002). Identification of a novel role for sphingolipid signaling in TNF alpha and ischemic preconditioning mediated cardioprotection. *J. Mol. Cell Cardiol.* **34,** 509–518.
17. Goto, M., Liu, Y., Yang, X. M., Ardell, J. L., Cohen, M. V., and Downey, J. M. (1995). Role of bradykinin in protection of ischemic preconditioning in rabbit hearts. *Circ. Res.* **77,** 611–621.
18. Lebuffe, G., Schumacker, P. T., Shao, Z. H., Anderson, T., Iwase, H., and Vanden Hoek, T. L. (2003). ROS and NO trigger early preconditioning: relationship to mitochondrial K_{ATP} channel. *Am. J. Physiol. Heart Circ. Physiol.* **284,** H299–H308.

19. Baines, C. P., Goto, M., and Downey, J. M. (1997). Oxygen radicals released during ischemic preconditioning contribute to cardioprotection in the rabbit myocardium. *J. Mol. Cell Cardiol.* **29,** 207–216.
20. Tritto, I., D'Andrea, D., Eramo, N., Scognamiglio, A., De Simone, C., Violante, A., Esposito, A., Chiariello, M., and Ambrosio, G. (1997). Oxygen radicals can induce preconditioning in rabbit hearts. *Circ. Res.* **80,** 743–748.
21. Gopalakrishna, R., and Anderson, W. B. (1989). Ca^{2+} and phospholipid-independent activation of protein kinase C by selective oxidative modification of the regulatory domain. *Proc. Natl. Acad. Sci. USA* **86,** 6758–6762.
22. Von Ruecker, A. A., Han-Jeon, B.-G., Wild, M., and Bidlingmaier, F. (1989). Protein kinase C involvement in lipid peroxidation and cell membrane damage induced by oxygen-based radicals in hepatocytes. *Biochem. Biophys. Res. Commun.* **163,** 836–842.
23. Nakano, A., Liu, G. S., Heusch, G., Downey, J. M., and Cohen, M. V. (2000). Exogenous nitric oxide can trigger a preconditioned state through a free radical mechanism, but endogenous nitric oxide is not a trigger of classical ischemic preconditioning. *J. Mol. Cell Cardiol.* **32,** 1159–1167.
24. Weselcouch, E. O., Baird, A. J., Sleph, P., and Grover, G. J. (1995). Inhibition of nitric oxide synthesis does not affect ischemic preconditioning in isolated perfused rat hearts. *Am. J. Physiol. Heart Circ. Physiol.* **268,** H242–H249.
25. Post, H., Schulz, R., Behrends, M., Gres, P., Umschlag, C., and Heusch, G. (2000). No involvement of endogenous nitric oxide in classical ischemic preconditioning in swine. *J. Mol. Cell Cardiol.* **32,** 725–733.
26. Bell, R. M., and Yellon, D. M. (2001). The contribution of endothelial nitric oxide synthase to early ischaemic preconditioning: the lowering of the preconditioning threshold. An investigation in eNOS knockout mice. *Cardiovasc. Res.* **52,** 274–280.
27. Murphy, E. (2004). Primary and secondary signaling pathways in early preconditioning that converge on the mitochondria to produce cardioprotection. *Circ. Res.* **94,** 7–16.
28. Mocanu, M. M., Bell, R. M., and Yellon, D. M. (2002). PI3 kinase and not p42/p44 seems to be implicated in the protection conferred by ischemic preconditioning. *J. Mol. Cell Cardiol.* **34,** 661–668.
29. Tong, H., Chen, W., Steenbergen, C., and Murphy, E. (2000). Ischemic preconditioning activates phosphatidylinositol-3-kinase upstream of protein kinase C. *Circ. Res.* **87,** 309–315.
30. Tong, H., Imahashi, K., Steenbergen, C., and Murphy, E. (2002). Phosphorylation of glycogen synthase kinase-3β during preconditioning through a phosphatidylinositol-3-kinase–dependent pathway is cardioprotective. *Circ. Res.* **90,** 377–379.
31. Krieg, T., Qin, Q., McIntosh, E. C., Cohen, M. V., and Downey, J. M. (2002). ACh and adenosine activate PI3-kinase in rabbit hearts through transactivation of receptor tyrosine kinases. *Am. J. Physiol. Heart Circ. Physiol.* **283,** H2322–H2330.
32. Mitchell, M. B., Meng, X., Ao, L., Brown, J. M., Harken, A. H., and Banerjee, A. (1995). Preconditioning of isolated rat heart is mediated by protein kinase C. *Circ. Res.* **76,** 73–78.
33. Ytrehus, K., Liu, Y., and Downey, J. M. (1994). Preconditioning protects ischemic rabbit heart by protein kinase C activation. *Am. J. Physiol.* **266,** H1145–H1152.
34. Kudo, M., Wang, Y., Xu, M., Ayub, A., and Ashraf, M. (2002). Adenosine A(1) receptor mediates late preconditioning via activation of PKC-delta signaling pathway. *Am. J. Physiol. Heart Circ. Physiol.* **283,** H296–H301.
35. Zhao, T. C., and Kukreja, R. C. (2003). Protein kinase C-delta mediates adenosine A3 receptor-induced delayed cardioprotection in mouse. *Am. J. Physiol. Heart Circ. Physiol.* **285,** H434–H441.
36. Fryer, R. M., Hsu, A. K., Wang, Y., Henry, M., Eells, J., and Gross, G. J. (2002). PKC-delta inhibition does not block preconditioning-induced preservation in mitochondrial ATP synthesis and infarct size reduction in rats. *Basic Res. Cardiol.* **97,** 47–54.
37. Liu, H., McPherson, B. C., and Yao, Z. (2001). Preconditioning attenuates apoptosis and necrosis: role of protein kinase C epsilon and -delta isoforms. *Am. J. Physiol. Heart Circ. Physiol.* **281,** H404–H410.
38. Gray, M. O., Zhou, H. Z., Schafhalter-Zoppoth, I., Zhu, P., Mochly-Rosen, D., and Messing, R. O. (2004). Preservation of base-line hemodynamic function and loss of inducible cardioprotection in adult mice lacking protein kinase C epsilon. *J. Biol. Chem.* **279,** 3596–3604.
39. Wang, Y., and Ashraf, M. (1998). Activation of 1-adrenergic receptor during Ca^{2+} preconditioning elicits strong protection against Ca^{2+} overload injury via protein kinase C signaling pathway. *J. Mol. Cell Cardiol.* **30,** 2423–2435.
40. Liu, G. S., Cohen, M. V., Mochly-Rosen, D., and Downey, J. M. (1999). Protein kinase C-ε is responsible for the protection of preconditioning in rabbit cardiomyocytes. *J. Mol. Cell Cardiol.* **31,** 1937–1948.
41. Ping, P., Zhang, J., Qiu, Y., Tang, X.-L., Manchikalapudi, S., Cao, X., and Bolli, R. (1997). Ischemic preconditioning induces selective translocation of protein kinase C isoforms and in the heart of conscious rabbits without subcellular redistribution of total protein kinase C activity. *Circ. Res.* **81,** 404–414.
42. Saurin, A. T., Pennington, D. J., Raat, N. J., Latchman, D. S., Owen, M. J., and Marber, M. S. (2002). Targeted disruption of the protein kinase C epsilon gene abolishes the infarct size reduction that follows ischaemic preconditioning of isolated buffer-perfused mouse hearts. *Cardiovasc. Res.* **55,** 672–680.
43. Baines, C. P., Zhang, J., Wang, G. W., Zheng, Y. T., Xiu, J. X., Cardwell, E. M., Bolli, R., and Ping, P. (2002). Mitochondrial PKCepsilon and MAPK form signaling modules in the murine heart: enhanced mitochondrial PKCepsilon-MAPK interactions and differential MAPK activation in PKCepsilon-induced cardioprotection. *Circ. Res.* **90,** 390–397.
44. Baines, C. P., Song, C. X., Zheng, Y. T., Wang, G. W., Zhang, J., Wang, O. L., Guo, Y., Bolli, R., Cardwell, E. M., and Ping, P. (2003). Protein kinase Cepsilon interacts with and inhibits the permeability transition pore in cardiac mitochondria. *Circ. Res.* **92,** 873–880.
45. Hassouna, A., Matata, B. M., and Galinanes, M. (2004). PKC-epsilon is upstream and PKC-alpha is downstream of mitoK_{ATP} channels in the signal transduction pathway of ischemic preconditioning of human myocardium. *Am. J. Physiol. Cell Physiol.* **287,** C1418–C1425.
46. Ohnuma, Y., Miura, T., Miki, T., Tanno, M., Kuno, A., Tsuchida, A., and Shimamoto, K. (2002). Opening of mitochondrial $K_{(ATP)}$ channel occurs downstream of PKC-ε activation in the

mechanism of preconditioning. *Am. J. Physiol. Heart Circ. Physiol.* **283**, H440–H447.
47. Toker, A., Meyer, M., Reddy, K. K., Falck, J. R., Aneja, R., Aneja, S., Sparra, A., Burns, D. J., Ballas, L. M., and Cantley, L. C. (1994). Activation of protein kinase C family members by the novel polyphosphoinositides PtdIns-3,4-P2 and PtdIns-3,4,5-P3. *J. Biol. Chem.* **269**, 32358–32367.
48. Okada, T., Otani, H., Wu, Y., Uchiyama, T., Kyoi, S., Hattori, R., Sumida, T., Fujiwara, H., and Imamura, H. (2005). Integrated pharmacological preconditioning and memory of cardioprotection: role of protein kinase C and phosphatidylinositol 3-kinase. *Am. J. Physiol. Heart Circ. Physiol.* **289**, H761–H767.
49. Baines, C. P., Wang, L., Cohen, M. V., and Downey, J. M. (1998). Protein tyrosine kinase is downstream of protein kinase C for ischemic preconditioning's anti-infarct effect in the rabbit heart. *J. Mol. Cell Cardiol.* **30**, 383–392.
50. Fryer, R. M., Schultz, J. E. J., Hsu, A. K., and Gross, G. J. (1999). Importance of PKC and tyrosine kinase in single or multiple cycles of preconditioning in rat hearts. *Am. J. Physiol. Heart Circ. Physiol.* **276**, H1229–H1235.
51. Vahlhaus, C., Schulz, R., Post, H., Rose, J., and Heusch, G. (1998). Prevention of ischemic preconditioning only by combined inhibition of protein kinase C and protein tyrosine kinase in pigs. *J. Mol. Cell Cardiol.* **30**, 197–209.
52. Fryer, R. M., Pratt, P. F., Hsu, A. K., and Gross, G. J. (2001). Differential activation of extracellular signal regulated kinase isoforms in preconditioning and opioid-induced cardioprotection. *J. Pharmacol. Exp. Ther.* **296**, 642–649.
53. O'Rourke, B. (2000). Myocardial K_{ATP} channels in preconditioning. *Circ. Res.* **87**, 845–855.
54. Szewczyk, A., and Wojtczak, L. (2002). Mitochondria as a pharmacological target. *Pharmacol. Rev.* **54**, 101–127.
55. Garlid, K. D., Paucek, P., Yarov-Yarovoy, V., Murray, H. N., Darbenzio, R., D'Alonzo, A. J., Lodge, N. J., Smith, M. A., and Grover, G. J. (1997). Cardioprotective effect of diazoxide and its interaction with mitochondrial ATP-sensitive K^+ channels. Possible mechanism of cardioprotection. *Circ. Res.* **81**, 1072–1082.
56. Hanley, P. J., Mickel, M., Loffler, M., Brandt, U., and Daut, J. (2002). $K(_{ATP})$ channel–independent targets of diazoxide and 5-hydroxydecanoate in the heart. *J. Physiol.* **542**, 735–741.
57. D'hahan, N., Moreau, C., Prost, A. L., Jacquet, H., Alekseev, A. E., Terzic, A., and Vivaudou, M. (1999). Pharmacological plasticity of cardiac ATP-sensitive potassium channels toward diazoxide revealed by ADP. *Proc. Natl. Acad. Sci. USA* **96**, 12162–12167.
58. Sanada, S., Kitakaze, M., Asanuma, H., Harada, K., Ogita, H., Node, K., Takashima, S., Sakata, Y., Asakura, M., Shinozaki, Y., Mori, H., Kuzuya, T., and Hori, M. (2001). Role of mitochondrial and sarcolemmal $K(_{ATP})$ channels in ischemic preconditioning of the canine heart. *Am. J. Physiol. Heart Circ. Physiol.* **280**, H256–H263.
59. Xu, W., Liu, Y., Wang, S., McDonald, T., Van Eyk, J. E., Sidor, A., and O'Rourke, B. (2003). Cytoprotective role of Ca^{2+}-activated K^+ channels in the cardiac inner mitochondrial membrane. *Science* **298**, 1029–1033.
60. Paucek, P., Mironova, G., Mahdi, F., Beavis, A. D., Woldegiorgis, G., and Garlid, K. D. (1992). Reconstitution and partial purification of the glibenclamide-sensitive, ATP-dependent K^+ channel from rat liver and beef heart mitochondria. *J. Biol. Chem.* **267**, 26062–26069.
61. Bajgar, R., Seetharaman, S., Kowaltowski, A. J., Garlid, K. D., and Paucek, P. (2001). Identification and properties of a novel intracellular (mitochondrial) ATP-sensitive potassium channel in brain. *J. Biol. Chem.* **276**, 33369–33374.
62. Babenko, A. P., Gonzalez, G., Aguilar-Bryan, L., and Bryan, J. (1998). Reconstituted human cardiac K_{ATP} channels: Functional identity with the native channels from the sarcolemma of human ventricular cells. *Circ. Res.* **83**, 1132–1143.
63. Seharaseyon, J., Ohler, A., Sasaki, N., Fraser, H., Sato, T., Johns, D. C., O'Rourke, B., and Marban, E. (2000). Molecular composition of mitochondrial ATP-sensitive potassium channels probed by viral Kir gene transfer. *J. Mol. Cell Cardiol.* **32**, 1923–1930.
64. Lacza, Z., Snipes, J. A., Kis, B., Szabo, C., Grover, G., and Busija, D. W. (2003). Investigation of the subunit composition and the pharmacology of the mitochondrial ATP-dependent K^+ channel in the brain. *Brain Res.* **994**, 27–36.
65. Lacza, Z., Snipes, J. A., Miller, A. W., Szabo, C., Grover, G., and Busija, D. W. (2003). Heart mitochondria contain functional ATP-dependent K^+ channels. *J. Mol. Cell Cardiol.* **35**, 1339–1347.
66. Ardehali, H., Chen, Z., Ko, Y., Mejia-Alvarez, R., and Marban, E. (2004). Multiprotein complex containing succinate dehydrogenase confers mitochondrial ATP-sensitive K^+ channel activity. *Proc. Natl. Acad. Sci. USA* **101**, 11880–11885.
67. Hanley, P. J., Gopalan, K. V., Lareau, R. A., Srivastava, D. K., von Meltzer, M., and Daut, J. (2003). Beta-oxidation of 5-hydroxydecanoate, a putative blocker of mitochondrial ATP-sensitive potassium channels. *J. Physiol.* **547**, 387–393.
68. Lim, K. H., Javadov, S. A., Das, M., Clarke, S. J., Suleiman, M. S., and Halestrap, A. P. (2002). The effects of ischaemic preconditioning, diazoxide and 5-hydroxydecanoate on rat heart mitochondrial volume and respiration. *J. Physiol.* **545**, 961–974.
69. Das, M., Parker, J. E., and Halestrap, A. P. (2003). Matrix volume measurements challenge the existence of diazoxide/glibencamide-sensitive K_{ATP} channels in rat mitochondria. *J. Physiol.* **547**, 893–902.
70. Javadov, S. A., Clarke, S., Das, M., Griffiths, E. J., Lim, K. H., and Halestrap, A. P. (2003). Ischaemic preconditioning inhibits opening of mitochondrial permeability transition pores in the reperfused rat heart. *J. Physiol.* **549**, 513–524.
71. Grimmsmann, T., and Rustenbeck, I. (1998). Direct effects of diazoxide on mitochondria in pancreatic B-cells and isolated liver mitochondria. *Br. J. Pharmacol.* **123**, 781–788.
72. Akao, M., O'Rourke, B., Kusuoka, H., Teshima, Y., Jones, S. P., and Marban, E. (2003). Differential actions of cardioprotective agents on the mitochondrial death pathway. *Circ. Res.* **92**, 195–202.
73. Opie, L. H., and Sack, M. N. (2002). Metabolic plasticity and the promotion of cardiac protection in ischemia and ischemic preconditioning. *J. Mol. Cell Cardiol.* **34**, 1077–1089.
74. Minners, J., van den Bos, E. J., Yellon, D. M., Schwalb, H., Opie, L. H., and Sack, M. N. (2000). Dinitrophenol, cyclosporin A, and trimetazidine modulate preconditioning in the isolated rat heart: support for a mitochondrial role in cardioprotection. *Cardiovasc. Res.* **47**, 68–73.

75. Knopp, A., Thierfelder, S., Doepner, B., and Benndorf, K. (2001). Mitochondria are the main ATP source for a cytosolic pool controlling the activity of ATP-sensitive K+ channels in mouse cardiac myocytes. *Cardiovasc. Res.* **52,** 236–245.
76. Miura, T., Liu, Y., Kita, H., Ogawa, T., and Shimamoto, K. (2000). Roles of mitochondrial ATP-sensitive K channels and PKC in anti-infarct tolerance afforded by adenosine A1 receptor activation. *J. Am. Coll. Cardiol.* **35,** 238–245.
77. Kobara, M., Tasumi, T., Matoba, S., Yamahara, Y., Nakagawa, C., Ohta, B., Matsumoto, T., Inoue, D., Asayama, J., and Nakagawa, M. (1997). Effect of ischemic preconditioning on mitochondrial oxidative phosphorylation and high energy phosphates in rat hearts. *J. Mol. Cell Cardiol.* **28,** 417–428.
78. Wang, Y., Hirai, K., and Ashraf, M. (1999). Activation of mitochondrial ATP-sensitive K(+) channel for cardiac protection against ischemic injury is dependent on protein kinase C activity. *Circ. Res.* **85,** 731–741.
79. Bosetti, F., Yu, G., Zucchi, R., Ronca-Testoni, S., and Solaini, G. (2000). Myocardial ischemic preconditioning and mitochondrial F1F0-ATPase activity. *Mol. Cell Biochem.* **215,** 31–37.
80. Vuorinen, K., Ylitalo, K., Peuhkurinen, K., Raatikainen, P., Ala-Rami, A., and Hassinen, I. E. (1995). Mechanisms of ischemic preconditioning in rat myocardium. Roles of adenosine, cellular energy state, and mitochondrial F1F0-ATPase. *Circulation* **91,** 2810–2818.
81. Green, D. W., Murray, H. N., Sleph, P. G., Wang, F. L., Baird, A. J., Rogers, W. L., and Grover, G. J. (1998). Preconditioning in rat hearts is independent of mitochondrial F1F0 ATPase inhibition. *Am. J. Physiol.* **274,** H90–H97.
82. Asimakis, G. K., Inners-McBride, K., Medellin, G., and Conti, V. R. (1992). Ischemic preconditioning attenuates acidosis and postischemic dysfunction in isolated rat heart. *Am. J. Physiol.* **263,** H887–H894.
83. Laclau, M. N., Boudina, S., Thambo, J. B., Tariosse, L., Gouverneur, G., Bonoron-Adele, S., Saks, V. A., Garlid, K. D., and Dos Santos, P. (2001). Cardioprotection by ischemic preconditioning preserves mitochondrial function and functional coupling between adenine nucleotide translocase and creatine kinase. *J. Mol. Cell Cardiol.* **33,** 947–956.
84. Da Silva, M. M., Sartori, A., Belisle, E., and Kowaltowski, A. J. (2003). Ischemic preconditioning inhibits mitochondrial respiration, increases H_2O_2 release and enhances K+ transport. *Am. J. Physiol. Heart Circ. Physiol.* **285,** H154–H162.
85. Garlid, K. D. (2000). Opening mitochondrial K($_{ATP}$) in the heart—what happens, and what does not happen. *Basic Res. Cardiol.* **95,** 275–279.
86. Halestrap, A. P. (1989). The regulation of the matrix volume of mammalian mitochondria in vivo and in vitro and its role in the control of mitochondrial metabolism. *Biochim. Biophys. Acta* **973,** 355–382.
87. Halestrap, A. P. (1994). Regulation of mitochondrial metabolism through changes in matrix volume. *Biochem. Soc. Trans.* **22,** 522–529.
88. Hausenloy, D. J., Maddock, H. L., Baxter, G. F., and Yellon, D. M. (2002). Inhibiting mitochondrial permeability transition pore opening: a new paradigm for myocardial preconditioning? *Cardiovasc. Res.* **55,** 534–543.
89. Dos Santos, P., Kowaltowski, A. J., Laclau, M. N., Seetharaman, S., Paucek, P., Boudina, S., Thambo, J. B., Tariosse, L., and Garlid, K. D. (2002). Mechanisms by which opening the mitochondrial ATP-sensitive K+ channel protects the ischemic heart. *Am. J. Physiol. Heart Circ. Physiol.* **283,** H284–H295.
90. Oldenburg, O., Cohen, M., Yellon, D., and Downey, J. (2002). Mitochondrial K($_{ATP}$) channels: role in cardioprotection. *Cardiovasc. Res.* **55,** 429–437.
91. Pain, T., Yang, X. M., Critz, S. D., Yue, Y., Nakano, A., Liu, G. S., Heusch, G., Cohen, M. V., and Downey, J. M. (2000). Opening of mitochondrial K($_{ATP}$) channels triggers the preconditioned state by generating free radicals. *Circ. Res.* **87,** 460–466.
92. Krenz, M., Oldenburg, O., Wimpee, H., Cohen, M. V., Garlid, K. D., Critz, S. D., Downey, J. M., and Benoit, J. N. (2002). Opening of ATP-sensitive potassium channels causes generation of free radicals in vascular smooth muscle cells. *Basic Res. Cardiol.* **97,** 365–373.
93. Cohen, M. V., Yang, X. M., Liu, G. S., Heusch, G., and Downey, J. M. (2001). Acetylcholine, bradykinin, opioids, and phenylephrine, but not adenosine, trigger preconditioning by generating free radicals and opening mitochondrial K($_{ATP}$) channels. *Circ. Res.* **89,** 273–278.
94. Halmosi, R., Berente, Z., Osz, E., Toth, K., Literati-Nagy, P., and Sumegi, A. (2001). Effect of poly (ADP-ribose) polymerase inhibitors on the ischemia-reperfusion-induced oxidative cell damage and mitochondrial metabolism in Langendorff heart perfusion system. *Mol. Pharmacol.* **59,** 1497–1505.
95. Hess, M. L., and Manson, N. H. (1984). Molecular oxygen, Friend and foe. The role of the oxygen free radical system in the calcium paradox, the oxygen paradox and ischemia/reperfusion injury. *J. Mol. Cell Cardiol.* **16,** 969–985.
96. Becker, L. B. (2004). New concepts in reactive oxygen species and cardiovascular reperfusion physiology. *Cardiovasc. Res.* **61,** 461–470.
97. Becker, L. B., vanden Hoek, T. L., Shao, Z. H., Li, C. Q., and Schumacker, P. T. (1999). Generation of superoxide in cardiomyocytes during ischemia before reperfusion. *Am. J. Physiol.* **277,** H2240–H2246.
98. Van den Hoek, T. L., Becker, L. B., Shao, Z., Li, C., and Schumacker, P. T. (1998). Reactive oxygen species released from mitochondria during brief hypoxia induce preconditioning in cardiomyocytes. *J. Biol. Chem.* **273,** 18092–18098.
99. Ozcan, C., Bienengraeber, M., Dzeja, P. P., and Terzic, A. (2002). Potassium channel openers protect cardiac mitochondria by attenuating oxidant stress at reoxygenation. *Am. J. Physiol. Heart Circ. Physiol.* **282,** H531–H539.
100. Yue, Y., Qin, Q., Cohen, M. V., Downey, J. M., and Critz, S. D. (2002). The relative order of mK ($_{ATP}$) channels, free radicals and p38 MAPK in preconditioning's protective pathway in rat heart. *Cardiovasc. Res.* **55,** 681–689.
101. Chen, Z., Chua, C. C., Ho, Y. S., Hamdy, R. C., and Chua, B. H. (2001). Overexpression of bcl-2 attenuates apoptosis and protects against myocardial I/R injury in transgenic mice. *Am. J. Physiol. Heart Circ. Physiol.* **280,** H2313–H23120.
102. Liu, Y., Sato, T., O'Rourke, B., and Marbán, E. (1998). Mitochondrial ATP-dependent potassium channels: novel effectors of cardioprotection? *Circulation* **97,** 2463–2469.

103. Schultz, Je.-J., Hsu, A. K., Nagase, H., and Gross, G. J. (1998). TAN-67, a 1-opioid receptor agonist, reduces infarct size via activation of Gi/o proteins and K_{ATP} channels. *Am. J. Physiol.* **274,** H909–H914.
104. Weiss, J. N., Korge, P., Honda, H. M., and Ping, P. (2003). Role of the mitochondrial permeability transition in myocardial disease. *Circ. Res.* **93,** 292–301.
105. Holmuhamedov, E. L., Wang, L., and Terzic, A. (1999). ATP-sensitive K^+ channel openers prevent Ca^{2+} overload in rat cardiac mitochondria. *J. Physiol.* **519,** 347–360.
106. Murata, M., Akao, M., O'Rourke, B., and Marbán, E. (2001). Mitochondrial ATP-sensitive potassium channels attenuate matrix Ca^{2+} overload during simulated ischemia and reperfusion: possible mechanism of cardioprotection. *Circ. Res.* **89,** 891–898.
107. Ishida, H., Hirota, Y., Genka, C., Nakazawa, H., Nakaya, H., and Sato, T. (2001). Opening of mitochondrial K_{ATP} channels attenuates the ouabain-induced calcium overload in mitochondria. *Circ. Res.* **89,** 856–858.
108. Korge, P., Honda, H. M., and Weiss, J. N. (2002). Protection of cardiac mitochondria by diazoxide and protein kinase C: implications for ischemic preconditioning. *Proc. Natl. Acad. Sci. USA* **99,** 3312–3317.
109. Griffiths, E. J., and Halestrap, A. P. (1995). Mitochondrial non-specific pores remain closed during cardiac ischaemia, but open on reperfusion. *Biochem. J.* **307,** 93–98.
110. Crompton, M., and Costi, A. (1988). Kinetic evidence for a heart mitochondrial pore activated by Ca^{2+}, inorganic phosphate and oxidative stress. A potential mechanism for mitochondrial dysfunction during cellular Ca^{2+} overload. *Eur. J. Biochem.* **178,** 489–501.
111. Hausenloy, D., Wynne, A., Duchen, M., and Yellon, D. (2004). Transient mitochondrial permeability transition pore opening mediates preconditioning-induced protection. *Circulation* **109,** 1714–1717.
112. Shoshan-Barmatz, V., Israelson, A., Brdiczka, D., and Sheu, S. S. (2006). The voltage-dependent anion channel (VDAC): Function in intracellular signalling, cell life, and cell death. *Curr. Pharm. Des.* **12,** 2249–2270.
113. Shimizu, S., Narita, M., and Tsujimoto, Y. (1999). Bcl-2 family proteins regulate the release of apoptogenic cytochrome c by the mitochondrial channel VDAC. *Nature* **399,** 483–487.
114. Imahashi, K., Schneider, M. D., Steenbergen, C., and Murphy, E. (2004). Transgenic expression of Bcl-2 modulates energy metabolism and prevents cytosolic acidification during ischemia and reduces ischemia-reperfusion injury. *Circ. Res.* **95,** 734–741.
115. Chatterjee, S., Stewart, A. S., Bish, L. T., Jayasankar, V., Kim, E. M., Pirolli, T., Burdick, J., Woo, Y. J., Gardner, T. J., and Sweeney, H. (2002). Viral gene transfer of the antiapoptotic factor Bcl-2 protects against chronic postischemic heart failure. *Circulation* **106,** 212–217.
116. Forbes, R. A., Steenbergen, C., and Murphy, E. (2001). Diazoxide-induced cardioprotection requires signaling through a redox-sensitive mechanism. *Circ. Res.* **88,** 802–809.
117. Cheng, E. H., Wei, M. C., Weiler, S., Flavell, R. A., Mak, T. W., Lindsten, T., and Korsmeyer, S. J. (2001). BCL-2, BCL-XL sequester BH3 domain-only molecules preventing BAX- and BAK-mediated mitochondrial apoptosis. *Mol. Cell Biochem.* **8,** 705–711.
118. Shimizu, S., Narita, M., and Tsujimoto, Y. (1999). Bcl-2 family proteins regulate the release of apoptogenic cytochrome c by the mitochondrial channel VDAC. *Nature* **399,** 483–487.
119. Shimizu, S., Ide, T., Yanagida, T., and Tsujimoto, Y. (2000). Electrophysiological study of a novel large pore formed by Bax and the voltage-dependent anion channel that is permeable to cytochrome c. *J. Biol. Chem.* **275,** 12321–12325.
120. Nakamura, M., Wang, N. P., Zhao, Z. Q., Wilcox, J. N., Thourani, V., Guyton, R. A., and Vinten-Johansen, J. (2000). Preconditioning decreases Bax expression, PMN accumulation and apoptosis in reperfused rat heart. *Cardiovasc. Res.* **45,** 661–670.
121. Zhao, Z. Q., Nakamura, M., Wang, N. P., Wilcox, J. N., Shearer, S., Ronson, R. S., Guyton, R. A., and Vinten-Johansen, J. (2000). Reperfusion induces myocardial apoptotic cell death. *Cardiovasc. Res.* **45,** 651–660.
122. Zhao, Z. Q., and Vinten-Johansen, J. (2002). Myocardial apoptosis and ischemic preconditioning. *Cardiovasc. Res.* **55,** 438–455.
123. Schwanke, U., Konietzka, I., Duschin, A., Schulz, R., and Heusch, G. (2002). No ischemic preconditioning in heterozygous connexin43-deficient mice. *Am. J. Physiol.* **283,** H1740–H1742.
124. Jain, S. K., Schuessler, R. B., and Saffitz, J. E. (2003). Mechanisms of delayed electrical uncoupling induced by ischemic preconditioning. *Circ. Res.* **92,** 1138–1144.
125. Zucchi, R., Ronca, F., and Ronca-Testoni, S. (2001). Modulation of sarcoplasmic reticulum function: a new strategy in cardioprotection? *Pharmacol. Ther.* **89,** 47–65.
126. Chen, W., London, R., Murphy, E., and Steenbergen, C. (1998). Regulation of the Ca^{2+} gradient across the sarcoplasmic reticulum in perfused rabbit heart: a 19F nuclear magnetic resonance study. *Circ. Res.* **83,** 898–907.
127. Zhao, T. C., Hines, D. S., and Kukreja, R. C. (2001). Adenosine-induced late preconditioning in mouse hearts: role of p38 MAP kinase and mitochondrial K(ATP) channels. *Am. J. Physiol. Heart Circ. Physiol.* **280,** H1278–H1285.
128. Fryer, R. M., Schultz, J. E., Hsu, A. K., and Gross, G. J. (1999). Importance of PKC and tyrosine kinase in single or multiple cycles of preconditioning in rat hearts. *Am. J. Physiol. Heart Circ. Physiol.* **276,** H1229–H1235.
129. Schulz, R., Cohen, M. V., Behrends, M., Downey, J. M., and Heusch, G. (2001). Signal transduction of ischemic preconditioning. *Cardiovasc. Res.* **52,** 181–198.
130. Kulisz, A., Chen, N., Chandel, N. S., Shao, Z., and Schumacker, P. T. (2002). Mitochondrial ROS initiate phosphorylation of p38 MAP kinase during hypoxia in cardiomyocytes. *Am. J. Physiol. Lung Cell Mol. Physiol.* **282,** L1324–L1329.
131. Das, D. K., Maulik, N., Sato, M., and Ray, P. S. (1999). Reactive oxygen species function as second messenger during ischemic preconditioning of heart. *Mol. Cell Biochem.* **196,** 59–67.
132. Thornton, J., Striplin, S., Liu, G. S., Swafford, A., Stanley, A. W. H., Van Winkle, D. M., and Downey, J. M. (1990). Inhibition of protein synthesis does not block myocardial protection afforded by preconditioning. *Am. J. Physiol. Heart Circ. Physiol.* **259,** H1822–H1825.
133. Matsuyama, N., Leavens, J. E., McKinnon, D., Gaudette, G. R., Aksehirli, T. O., and Krukenkamp, I. B. (2000).

Ischemic but not pharmacological preconditioning requires protein synthesis. *Circulation* **102,** III312–III318.

134. Strohm, C., Barancik, M., von Bruehl, M., Strniskova, M., Ullmann, C., Zimmermann, R., and Schaper, W. (2002). Transcription inhibitor actinomycin-D abolishes the cardioprotective effect of ischemic reconditioning. *Cardiovasc. Res.* **55,** 602–618.

135. Onody, A., Zvara, A., Hackler, L., Jr., Vigh, L., Ferdinandy, P., and Puskas, L. G. (2003). Effect of classic preconditioning on the gene expression pattern of rat hearts: a DNA microarray study. *FEBS Lett.* **536,** 35–40.

136. Sergeev, P., da Silva, R., Lucchinetti, E., Zaugg, K., Pasch, T., Schaub, M. C., and Zaugg, M. (2004). Trigger-dependent gene expression profiles in cardiac preconditioning: evidence for distinct genetic programs in ischemic and anesthetic preconditioning. *Anesthesiology* **100,** 474–488.

137. Smith, R. M., Suleman, N., Lacerda, L., Opie, L. H., Akira, S., Chien, K. R., and Sack, M. N. (2004). Genetic depletion of cardiac myocyte STAT-3 abolishes classical preconditioning. *Cardiovasc. Res.* **63,** 611–616.

138. Vegh, A., Komori, S., Szekeres, L., and Parratt, J. R. (1992). Antiarrhythmic effects of preconditioning in anaesthetised dogs and rats. *Cardiovasc. Res.* **26,** 487–495.

139. Sun, J.-Z., Tang, X.-L., Knowlton, A. A., Park, S.-W., Qiu, Y., and Bolli, R. (1995). Late preconditioning against myocardial stunning: an endogenous protective mechanism that confers resistance to post-ischemic dysfunction 24 h after brief ischemia in conscious pigs. *J. Clin. Invest.* **95,** 388–403.

140. Bolli, R., Bhatti, Z. A., Tang, X.-L., Qiu, Y., Zhang, Q., Guo, Y., and Jadoon, A. K. (1997). Evidence that late preconditioning against myocardial stunning in conscious rabbits is triggered by the generation of nitric oxide. *Circ. Res.* **81,** 42–52.

141. Jancso, G., Lantos, J., Borsiczky, B., Szanto, Z., and Roth, E. (2004). Dynamism of NF-κB and AP-1 activation in the signal transduction of ischaemic myocardial preconditioning. *Eur. Surg. Res.* **36,** 129–135.

142. Xuan, Y.-T., Tang, X.-L., Banerjee, S., Takano, H., Li, R. C., Han, H., Qiu, Y., Li, J. J., and Bolli, R. (1999). Nuclear factor κB plays an essential role in the late phase of preconditioning in conscious rabbits. *Circ. Res.* **84,** 1095–1109.

143. Fauchon, M. A., Pell, T. J., Baxter, G. F., Yellon, D. M., Latchman, D. S., Hubank, M. F., and Mayne, L. V. (2005). Representational difference analysis of cDNA identifies novel genes expressed following preconditioning of the heart. *Exp. Mol. Med.* **37,** 311–322.

144. Bolli, R., Dawn, B., Tang, X. L., Qiu, Y., Ping, P., Xuan, Y. T., Jones, W. K., Takano, H., Guo, Y., and Zhang, J. (1998). The nitric oxide hypothesis of late preconditioning. *Basic Res. Cardiol.* **93,** 325–338.

145. Rakhit, R. D., Mojet, M. H., Marber, M. S., and Duchen, M. R. (2001). Mitochondria as targets for nitric oxide-induced protection during simulated ischemia and reoxygenation in isolated neonatal cardiomyocytes. *Circulation* **103,** 2617–2623.

146. Ockaili, R., Emani, V. R., Okubo, S., Brown, M., Krottapalli, K., and Kukreja, R. C. (2000). Opening of mitochondrial K_{ATP} channel induces early and delayed cardioprotective effect: Role of nitric oxide. *Am. J. Physiol.* **277,** H2425–H2434.

147. McLeod, C. J., Jeyabalan, A. P., Minners, J. O., Clevenger, R., Hoyt, R. F., Jr., and Sack, M. N. (2004). Delayed ischemic preconditioning activates nuclear-encoded electron-transfer-chain gene expression in parallel with enhanced postanoxic mitochondrial respiratory recovery. *Circulation* **110,** 534–539.

148. Hoshida, S., Yamashita, N., Otsu, K., and Hori, M. (2002). The importance of manganese superoxide dismutase in delayed preconditioning. Involvement of reactive oxygen species and cytokines. *Cardiovasc. Res.* **55,** 495–505.

149. Yamashita, N., Hoshida, S., Otsu, K., Taniguchi, N., Kuzuya, T., and Hori, M. (2000). The involvement of cytokines in the second window of ischaemic preconditioning. *Br. J. Pharmacol.* **131,** 415–422.

150. Zaugg, M., and Schaub, M. C. (2003). Signaling and cellular mechanisms in cardiac protection by ischemic and pharmacological preconditioning. *J. Muscle Res. Cell Motil.* **24,** 219–249.

151. Ludwig, L. M., Tanaka, K., Eells, J. T., Weihrauch, D., Pagel, P. S., Kersten, J. R., and Warltier, D. C. (2004). Preconditioning by isoflurane is mediated by reactive oxygen species generated from mitochondrial electron transport chain complex III. *Anesth. Analg.* **99,** 1308–1315.

152. Stowe, D. F., and Kevin, L. G. (2004). Cardiac preconditioning by volatile anesthetic agents: a defining role for altered mitochondrial bioenergetics. *Antioxid. Redox. Signal* **6,** 439–448.

153. Tanaka, K., Weihrauch, D., Kehl, F., Ludwig, L. M., LaDisa, J. F., Jr., Kersten, J. R., Pagel, P. S., and Warltier, D. C. (2002). Mechanism of preconditioning by isoflurane in rabbits: a direct role for reactive oxygen species. *Anesthesiology* **97,** 1485–1490.

154. de Ruijter, W., Musters, R. J., Boer, C., Stienen, G. J., Simonides, W. S., and de Lange, J. J. (2003). The cardioprotective effect of sevoflurane depends on protein kinase C activation, opening of mitochondrial K(+) (ATP) channels, and the production of reactive oxygen species. *Anesth. Analg.* **397,** 1370–1376.

155. Novalija, E., Kevin, L. G., Camara, A. K., Bosnjak, Z. J., Kampine, J. P., and Stowe, D. F. (2003). Reactive oxygen species precede the epsilon isoform of protein kinase C in the anesthetic preconditioning signaling cascade. *Anesthesiology* **99,** 421–428.

156. Wakeno-Takahashi, M., Otani, H., Nakao, S., Imamura, H., and Shingu, K. (2005). Isoflurane induces second window of preconditioning through upregulation of inducible nitric oxide synthase in rat heart. *Am. J. Physiol. Heart Circ. Physiol.* **289,** H2585–H2591.

157. Uecker, M., da Silva, R., Grampp, T., Pasch, T., Schaub, M. C., and Zaugg, M. (2003). Translocation of protein kinase C isoforms to subcellular targets in ischemic and anesthetic preconditioning. *Anesthesiology* **99,** 138–147.

158. da Silva, R., Grampp, T., Pasch, T., Schaub, M. C., and Zaugg, M. (2004). Differential activation of mitogen-activated protein kinases in ischemic and anesthetic preconditioning. *Anesthesiology* **100,** 59–69.

159. da Silva, R., Lucchinetti, E., Pasch, T., Schaub, M. C., and Zaugg, M. (2004). Ischemic but not pharmacological preconditioning elicits a gene expression profile similar to unprotected myocardium. *Physiol. Genomics* **20,** 117–130.

160. Chiari, P. C., Bienengraeber, M. W., Pagel, P. S., Krolikowski, J. G., Kersten, J. R., and Warltier, D. C. (2005). Isoflurane

protects against myocardial infarction during early reperfusion by activation of phosphatidylinositol-3-kinase signal transduction: Evidence for anesthetic-induced postconditioning in rabbits. *Anesthesiology* **102**, 102–109.

161. Chiari, P. C., Bienengraeber, M. W., Weihrauch, D., Krolikowski, J. G., Kersten, J. R., Warltier, D. C., and Pagel, P. S. (2005). Role of endothelial nitric oxide synthase as a trigger and mediator of isoflurane-induced delayed preconditioning in rabbit myocardium. *Anesthesiology* **103**, 74–83.

162. Obal, D., Dettwiler, S., Favoccia, C., Scharbatke, H., Preckel, B., and Schlack, W. (2005). The influence of mitochondrial K_{ATP}-channels in the cardioprotection of preconditioning and postconditioning by sevoflurane in the rat in vivo. *Anesth. Analg.* **101**, 1252–1260.

163. Feng, J., Lucchinetti, E., Ahuja, P., Pasch, T., Perriard, J. C., and Zaugg, M. (2005). Isoflurane postconditioning prevents opening of the mitochondrial permeability transition pore through inhibition of glycogen synthase kinase 3beta. *Anesthesiology* **103**, 987–995.

164. Juhaszova, M., Zorov, D. B., Kim, S. H., Pepe, S., Fu, Q., Fishbein, K. W., Ziman, B. D., Wang, S., Ytrehus, K., Antos, C. L., Olson, E. N., and Sollott, S. J. (2004). Glycogen synthase kinase-3β mediates convergence of protection signaling to inhibit the mitochondrial permeability transition pore. *J. Clin. Invest.* **113**, 1535–1549.

165. Lucchinetti, E., da Silva, R., Pasch, T., Schaub, M. C., and Zaugg, M. C. (2005). Anaesthetic preconditioning but not postconditioning prevents early activation of the deleterious cardiac remodelling programme: evidence of opposing genomic responses in cardioprotection by pre- and postconditioning. *Br. J. Anaesth.* **95**, 140–152.

166. Bienengraeber, M. W., Weihrauch, D., Kersten, J. R., Pagel, P. S., and Warltier, D. C. (2005). Cardioprotection by volatile anesthetics. *Vascul. Pharmacol.* **42**, 243–252.

167. Riess, M. L., Stowe, D. F., and Warltier, D. C. (2004). Cardiac pharmacological preconditioning with volatile anesthetics: from bench to bedside? *Am. J. Physiol. Heart Circ. Physiol.* **286**, H1603–H1607.

168. Riess, M. L., Eells, J. T., Kevin, L. G., Camara, A. K. S., Henry, M. M., and Stowe, D. F. (2003). Attenuation of mitochondrial respiration by sevoflurane in isolated cardiac mitochondria is mediated in part by reactive oxygen species. *Anesthesiology* **100**, 498–505.

169. Kissin, I., Aultman, D. F., and Smith, L. R. (1983). Effects of volatile anesthetics on myocardial oxidation-reduction status assessed by NADH fluorometry. *Anesthesiology* **59**, 447–452.

170. Kevin, L. G., Novalija, E., Riess, M. L., Camara, A. K., Rhodes, S. S., and Stowe, D. F. (2003). Sevoflurane exposure generates superoxide but leads to decreased superoxide during ischemia and reperfusion in isolated hearts. *Anesth. Analg.* **96**, 949–955.

171. Novalija, E., Varadarajan, S. G., Camara, A. K., An, J., Chen, Q., Riess, M. L., Hogg, N., and Stowe, D. F. (2002). Anesthetic preconditioning: triggering role of reactive oxygen and nitrogen species in isolated hearts. *Am. J. Physiol. Heart Circ. Physiol.* **283**, H44–H52.

172. Müllenheim, J., Ebel, D., Frassdorf, J., Preckel, B., Thamer, V., and Schlack, W. I. (2002). Isoflurane preconditions myocardium against infarction via release of free radicals. *Anesthesiology* **96**, 934–940.

173. Novalija, E., Kevin, L. G., Eells, J. T., Henry, M. M., and Stowe, D. F. (2003). Anesthetic preconditioning improves adenosine triphosphate synthesis and reduces reactive oxygen species formation in mitochondria after ischemia by a redox dependent mechanism. *Anesthesiology* **98**, 1155–1163.

174. Zaugg, M., Lucchinetti, E., Spahn, D. R., Pasch, T., and Schaub, M. C. (2002). Volatile anesthetics mimic cardiac preconditioning by priming the activation of mitochondrial K_{ATP} channels via multiple signaling pathways. *Anesthesiology* **97**, 4–14.

175. Kwok, W. M., Martinelli, A. T., Fujimoto, K., Suzuki, A., Stadnicka, A., and Bosnjak, Z. J. (2002). Differential modulation of the cardiac adenosine triphosphate-sensitive potassium channel by isoflurane and halothane. *Anesthesiology* **97**, 50–56.

176. Tanaka, K., Weihrauch, D., Ludwig, L. M., Kersten, J. R., Pagel, P. S., and Warltier, D. C. (2003). Mitochondrial adenosine triphosphate-regulated potassium channel opening acts as a trigger for isoflurane-induced preconditioning by generating reactive oxygen species. *Anesthesiology* **98**, 935–943.

177. Piriou, V., Chiari, P., Gateau-Roesch, O., Argaud, L., Muntean, D., Salles, D., Loufouat, J., Gueugniaud, P. Y., Lehot, J. J., and Ovize, M. (2004). Desflurane-induced preconditioning alters calcium-induced mitochondrial permeability transition. *Anesthesiology* **100**, 581–588.

178. Marinovic, J., Bosnjak, Z. J., and Stadnicka, A. (2005). Preconditioning by isoflurane induces lasting sensitization of the cardiac sarcolemmal adenosine triphosphate-sensitive potassium channel by a protein kinase C-Δ-mediated mechanism. *Anesthesiology* **103**, 540–547.

179. Toma, O., Weber, N. C., Wolter, J. I., Obal, D., Preckel, B., and Schlack, W. (2004). Desflurane preconditioning induces time-dependent activation of protein kinase C ε and extracellular signal-regulated kinase 1 and 2 in the rat heart in vivo. *Anesthesiology* **101**, 1372–1380.

180. Zhong, C., Zhou, Y., and Liu, H. (2004). Nuclear factor κB and anesthetic preconditioning during myocardial ischemia-reperfusion. *Anesthesiology* **100**, 540–546.

181. Alcindor, D., Krolikowski, J. G., Pagel, P. S., Warltier, D. C., and Kersten, J. R. (2004). Cyclooxygenase-2 mediates ischemic, anesthetic, and pharmacologic preconditioning in vivo. *Anesthesiology* **100**, 547–554.

182. Raphael, J., Rivo, J., and Gozal, Y. (2005). Isoflurane-induced myocardial preconditioning is dependent on phosphatidylinositol-3-kinase/Akt signalling. *Br. J. Anaesth.* **95**, 756–763.

183. An, J., Varadarajan, S. G., Novalija, E., and Stowe, D. F. (2001). Ischemic and anesthetic preconditioning reduces cytosolic [Ca^{2+}] and improves Ca(2+) responses in intact hearts. *Am. J. Physiol. Heart Circ. Physiol.* **281**, H1508–H1523.

184. Liu, H., Wang, L., Eaton, M., and Schaefer, S. (2005). Sevoflurane preconditioning limits intracellular/ mitochondrial Ca^{2+} in ischemic newborn myocardium. *Anesth. Analg.* **101**, 349–355.

185. Riess, M. L., Camara, A. K., Novalija, E., Chen, Q., Rhodes, S. S., and Stowe, D. F. (2002). Anesthetic preconditioning attenuates mitochondrial Ca^{2+} overload during ischemia in Guinea pig intact hearts: reversal by 5-hydroxydecanoic acid. *Anesth. Analg.* **95**, 1540–1546.

186. Wright, G. L., Hanlon, P., Amin, K., Steenbergen, C., Murphy, E., and Arcasoy, M. O. (2004). Erythropoietin receptor expression in adult rat cardiomyocytes is associated with an acute cardioprotective effect for recombinant erythropoietin during ischemia-reperfusion injury. *FASEB J.* **18,** 1031–1033.
187. Juul, S. E., Yachnis, A. T., and Christensen, R. D. (1998). Tissue distribution of erythropoietin and erythropoietin receptor in the developing human fetus. *Early Hum. Dev.* **52,** 235–249.
188. Parsa, C. J., Matsumoto, A., Kim, J., Riel, R. U., Pascal, L. S., Walton, G. B., Thompson, R. B., Petrofski, J. A., Annex, B. H., Stamler, J. S., and Koch, W. J. (2003). A novel protective effect of erythropoietin in the infarcted heart. *J. Clin. Invest.* **112,** 999–1007.
189. Calvillo, L., Latini, R., Kajstura, J., Leri, A., Anversa, P., Ghezzi, P., Salio, M., Cerami, A., and Brines, M. (2003). Recombinant human erythropoietin protects the myocardium from ischemia-reperfusion injury and promotes beneficial remodeling. *Proc. Natl. Acad. Sci. USA* **100,** 4802–4806.
190. Hanlon, P. R., Fu, P., Wright, G. L., Steenbergen, C., Arcasoy, M. O., and Murphy, E. (2005). Mechanisms of erythropoietin-mediated cardioprotection during ischemia-reperfusion injury: role of protein kinase C and phosphatidylinositol 3-kinase signaling. *FASEB J.* **19,** 1323–1325.
191. Rafiee, P., Shi, Y., Su, J., Pritchard, K. A., Jr., Tweddell, J. S., and Baker, J. E. (2005). Erythropoietin protects the infant heart against ischemia-reperfusion injury by triggering multiple signaling pathways. *Basic Res. Cardiol.* **100,** 187–197.
192. Fath-Ordoubadi, F., and Beatt, K. J. (1997). Glucose-insulin-potassium therapy for treatment of acute myocardial infarction: an overview of randomized placebo-controlled trials. *Circulation* **96,** 1152–1156.
193. Jonassen, A. K., Aasum, E., Riemersma, R. A., Mjos, O. D., and Larsen, T. S. (2000). Glucose-insulin-potassium reduces infarct size when administered during reperfusion. *Cardiovasc. Drugs Ther.* **14,** 615–623.
194. Opie, L. H., and Sack, M. N. (2002). Metabolic plasticity and the promotion of cardiac protection in ischemia and ischemic preconditioning. *J. Mol. Cell Cardiol.* **34,** 1077–1089.
195. Sack, M. N., and Yellon, D. M. (2003). Insulin therapy as an adjunct to reperfusion after acute coronary ischemia: a proposed direct myocardial cell survival effect independent of metabolic modulation. *J. Am. Coll. Cardiol.* **41,** 1404–1407.
196. Jonassen, A. K., Brar, B. K., Mjos, O. D., Sack, M. N., Latchman, D. S., and Yellon, D. M. (2000). Insulin administered at reoxygenation exerts a cardioprotective effect in myocytes by a possible anti-apoptotic mechanism. *J. Mol. Cell Cardiol.* **32,** 757–764.
197. Jonassen, A. K., Sack, M. N., Mjos, O. D., and Yellon, D. M. (2001). Myocardial protection by insulin at reperfusion requires early administration and is mediated via Akt and p70s6 kinase cell-survival signaling. *Circ. Res.* **89,** 1191–1198.
198. Fischer-Rasokat, U., Beyersdorf, F., and Doenst, T. (2003). Insulin addition after ischemia improves recovery of function equal to ischemic preconditioning in rat heart. *Basic Res. Cardiol.* **98,** 329–336.
199. Jonassen, A. K., Mjos, O. D., and Sack, M. N. (2004). p70s6 kinase is a functional target of insulin activated Akt cell-survival signaling. *Biochem. Biophys. Res. Commun.* **315,** 160–165.
200. Zuurbier, C. J., Eerbeek, O., and Meijer, A. J. (2005). Ischemic preconditioning, insulin, and morphine all cause hexokinase redistribution. *Am. J. Physiol. Heart Circ. Physiol.* **289,** H496–H499.
201. Davidson, S. M., Hausenloy, D., Duchen, M. R., and Yellon, D. M. (2006). Signalling via the reperfusion injury signalling kinase (RISK) pathway links closure of the mitochondrial permeability transition pore to cardioprotection. *Int. J. Biochem. Cell Biol.* **38,** 414–419.
202. Bose, A. K., Mocanu, M. M., Carr, R. D., Brand, C. L., and Yellon, D. M. (2005). Glucagon-like peptide 1 can directly protect the heart against ischemia/reperfusion injury. *Diabetes* **54,** 146–151.
203. Nikolaidis, L. A., Mankad, S., Sokos, G. G., Miske, G., Shah, A., Elahi, D., and Shannon, R. P. (2004). Effects of glucagon-like peptide-1 in patients with acute myocardial infarction and left ventricular dysfunction after successful reperfusion. *Circulation* **109,** 962–965.
204. Nikolaidis, L. A., Doverspike, A., Hentosz, T., Zourelias, L., Shen, Y. T., Elahi, D., and Shannon, R. P. (2005). Glucagon-like peptide-1 limits myocardial stunning following brief coronary occlusion and reperfusion in conscious canines. *J. Pharmacol. Exp. Ther.* **312,** 303–308.
205. Flood, A., Hack, B. D., and Headrick, J. P. (2003). Pyruvate-dependent preconditioning and cardioprotection in murine myocardium. *Clin. Exp. Pharmacol. Physiol.* **30,** 145–152.
206. Mallet, R. T., Sun, J., Knott, E. M., Sharma, A. B., and Olivencia-Yurvati, A. H. (2005). Metabolic cardioprotection by pyruvate: recent progress. *Exp. Biol. Med. (Maywood)* **230,** 435–443.
207. Halestrap, A. P., Clarke, S. J., and Javadov, S. A. (2004). Mitochondrial permeability transition pore opening during myocardial reperfusion—a target for cardioprotection. *Cardiovasc. Res.* **61,** 372–385.
208. Ebel, D., Mullenheim, J., Frassdorf, J., Heinen, A., Huhn, R., Bohlen, T., Ferrari, J., Sudkamp, H., Preckel, B., Schlack, W., and Thamer, V. (2003). Effect of acute hyperglycaemia and diabetes mellitus with and without short-term insulin treatment on myocardial ischaemic late preconditioning in the rabbit heart in vivo. *Pflugers Arch.* **446,** 175–182.
209. LaDisa, J. F., Jr., Krolikowski, J. G., Pagel, P. S., Warltier, D. C., and Kersten, J. R. (2004). Cardioprotection by glucose-insulin-potassium: dependence on K_{ATP} channel opening and blood glucose concentration before ischemia. *Am. J. Physiol. Heart Circ. Physiol.* **287,** H601–H607.
210. Kristiansen, S. B., Lofgren, B., Stottrup, N. B., Khatir, D., Nielsen-Kudsk, J. E., Nielsen, T. T., Botker, H. E., and Flyvbjerg, A. (2004). Ischaemic preconditioning does not protect the heart in obese and lean animal models of type 2 diabetes. *Diabetologia* **47,** 1716–1721.
211. Tsang, A., Hausenloy, D. J., Mocanu, M. M., Carr, R. D., and Yellon, D. M. (2005). Preconditioning the diabetic heart: the importance of Akt phosphorylation. *Diabetes* **54,** 2360–2364.
212. Desrois, M., Sidell, R. J., Gauguier, D., King, L. M., Radda, G. K., and Clarke, K. (2004). Initial steps of insulin signaling and glucose transport are defective in the type 2 diabetic rat heart. *Cardiovasc. Res.* **61,** 288–296.

213. Huisamen, B. (2003). Protein kinase B in the diabetic heart. *Mol. Cell Biochem.* **249,** 31–38.
214. Steiler, T. L., Galuska, D., Leng, Y., Chibalin, A. V., Gilbert, M., and Zierath, J. R. (2003). Effect of hyperglycemia on signal transduction in skeletal muscle from diabetic Goto-Kakizaki rats. *Endocrinology* **144,** 5259–5267.
215. del Valle, H. F., Lascano, E. C., Negroni, J. A., and Crottogini, A. J. (2003). Absence of ischemic preconditioning protection in diabetic sheep hearts: role of sarcolemmal K_{ATP} channel dysfunction. *Mol. Cell Biochem.* **249,** 21–30.
216. Ravingerova, T., Neckar, J., and Kolar, F. (2003). Ischemic tolerance of rat hearts in acute and chronic phases of experimental diabetes. *Mol. Cell Biochem.* **249,** 167–174.
217. Feuvray, D., and Lopaschuk, G. D. (1997). Controversies on the sensitivity of the diabetic heart to ischemic injury: the sensitivity of the diabetic heart to ischemic injury is decreased. *Cardiovasc. Res.* **34,** 113–120.
218. Paulson, D. J. (1997). The diabetic heart is more sensitive to ischemic injury. *Cardiovasc. Res.* **34,** 104–112.
219. Wang, P., and Chatham, J. C. (2004). Onset of diabetes in Zucker diabetic fatty (ZDF) rats leads to improved recovery of function after ischemia in the isolated perfused heart. *Am. J. Physiol. Endocrinol. Metab.* **286,** E725–E736.
220. Desrois, M., Sidell, R. J., Gauguier, D., Davey, C. L., Radda, G. K., and Clarke, K. (2004). Gender differences in hypertrophy, insulin resistance and ischemic injury in the aging type 2 diabetic rat heart. *J. Mol. Cell Cardiol.* **37,** 547–555.
221. Ebel, D., Mullenheim, J., Frassdorf, J., Heinen, A., Huhn, R., Bohlen, T., Ferrari, J., Sudkamp, H., Preckel, B., Schlack, W., and Thamer, V. (2003). Effect of acute hyperglycaemia and diabetes mellitus with and without short-term insulin treatment on myocardial ischaemic late preconditioning in the rabbit heart in vivo. *Pflugers Arch.* **446,** 175–182.
222. Kehl, F., Krolikowski, J. G., Mraovic, B., Pagel, P. S., Warltier, D. C., and Kersten, J. R. (2002). Hyperglycemia prevents isoflurane-induced preconditioning against myocardial infarction. *Anesthesiology* **96,** 183–188.
223. Kersten, J. R., Montgomery, M. W., Ghassemi, T., Gross, E. R., Toller, W. G., Pagel, P. S., and Warltier, D. C. (2001). Diabetes and hyperglycemia impair activation of mitochondrial $K_{(ATP)}$ channels. *Am. J. Physiol. Heart Circ. Physiol.* **280,** H1744–H1750.
224. Schaffer, S. W., Croft, C. B., and Solodushko, V. (2000). Cardioprotective effect of chronic hyperglycemia: effect on hypoxia-induced apoptosis and necrosis. *Am. J. Physiol. Heart Circ. Physiol.* **278,** H1948–H1954.
225. Brownlee, M. (2001). Biochemistry and molecular cell biology of diabetic complications. *Nature* **414,** 813–820.
226. Schulz, R., Cohen, M. V., Behrends, M., Downey, J. M., and Heusch, G. (2001). Signal transduction of ischemic preconditioning. *Cardiovasc. Res.* **52,** 181–198.
227. Ghosh, S., Standen, N. B., and Galinianes, M. (2001). Failure to precondition pathological human myocardium. *J. Am. Coll. Cardiol.* **37,** 711–718.
228. Hassouna, A., Loubani, M., Matata, B. M., Fowler, A., Standen, N. B., and Galinanes, M. (2006). Mitochondrial dysfunction as the cause of the failure to precondition the diabetic human myocardium. *Cardiovasc. Res.* **69,** 450–458.
229. Brady, P. A., and Terzic, A. (1998). The sulfonylurea controversy: more questions from the heart. *J. Am. Coll. Cardiol.* **31,** 950–956.
230. Cleveland, J. C., Jr., Meldrum, D. R., Cain, B. S., Banerjee, A., and Harken, A. H. (1997). Oral sulfonylurea hypoglycemic agents prevent ischemic preconditioning in human myocardium. Two paradoxes revisited. *Circulation* **96,** 29–32.
231. Verma, S., Rao, V., Weisel, R. D., Li, S. H., Fedak, P. W., Miriuka, S., and Li, R. K. (2004). Novel cardioprotective effects of pravastatin in human ventricular cardiomyocytes subjected to hypoxia and reoxygenation: beneficial effects of statins independent of endothelial cells. *J. Surg. Res.* **119,** 66–71.
232. Jones, S. P., Teshima, Y., Akao, M., and Marban, E. (2003). Simvastatin attenuates oxidant-induced mitochondrial dysfunction in cardiac myocytes. *Circ. Res.* **93,** 697–699.
233. Di Napoli, P., Antonio Taccardi, A., Grilli, A., Spina, R., Felaco, M., Barsotti, A., De Caterina, R. (2001). Simvastatin reduces reperfusion injury by modulating nitric oxide synthase expression: an ex vivo study in isolated working rat hearts. *Cardiovasc. Res.* **51,** 283–293.
234. Wolfrum, S., Dendorfer, A., Schutt, M., Weidtmann, B., Heep, A., Tempel, K., Klein, H. H., Dominiak, P., and Richardt, G. (2004). Simvastatin acutely reduces myocardial reperfusion injury in vivo by activating the phosphatidylinositide 3-kinase/Akt pathway. *J. Cardiovasc. Pharmacol.* **44,** 348–355.
235. Birnbaum, Y., Ye, Y., Rosanio, S., Tavackoli, S., Hu, Z. Y., Schwarz, E. R., and Uretsky, B. F. (2005). Prostaglandins mediate the cardioprotective effects of atorvastatin against ischemia-reperfusion injury. *Cardiovasc. Res.* **65,** 345–355.
236. Bell, R. M., and Yellon, D. M. (2003). Atorvastatin, administered at the onset of reperfusion, and independent of lipid lowering, protects the myocardium by up-regulating a pro-survival pathway. *J. Am. Coll. Cardiol.* **41,** 508–515.
237. Efthymiou, C. A., Mocanu, M. M., and Yellon, D. M. (2005). Atorvastatin and myocardial reperfusion injury: new pleiotropic effect implicating multiple prosurvival signaling. *J. Cardiovasc. Pharmacol.* **45,** 247–252.
238. Mensah, K., Mocanu, M. M., and Yellon, D. M. (2005). Failure to protect the myocardium against ischemia/reperfusion injury after chronic atorvastatin treatment is recaptured by acute atorvastatin treatment: a potential role for phosphatase and tensin homolog deleted on chromosome ten? *J. Am. Coll. Cardiol.* **45,** 1287–1291.
239. Scalia, R. (2005). Statins and the response to myocardial injury. *Am. J. Cardiovasc. Drugs* **5,** 163–170.
240. Fonarow, G. C., Wright, R. S., Spencer, F. A., Fredrick, P. D., Dong, W., Every, N., French, W. J., and the National Registry of Myocardial Infarction 4 Investigators. (2005). Effect of statin use within the first 24 hours of admission for acute myocardial infarction on early morbidity and mortality. *Am. J. Cardiol.* **96,** 611–616.
241. Pasceri, V., Patti, G., Nusca, A., Pristipino, C., Richichi, G., and Di Sciascio, G., and the ARMYDA Investigators. (2004). Randomized trial of atorvastatin for reduction of myocardial damage during coronary intervention: results from the ARMYDA (Atorvastatin for Reduction of MYocardial Damage during Angioplasty) study. *Circulation* **110,** 674–678.
242. Riksen, N. P., Smits, P., and Rongen, G. A. (2004). Ischaemic preconditioning: from molecular characterisation to clinical application—Part II. *Neth. J. Med.* **62,** 409–423.

243. Guo, Y., Stein, A. B., Wu, W. J., Zhu, X., Tan, W., Li, Q., and Bolli, R. (2005). Late preconditioning induced by NO donors, adenosine A1 receptor agonists, and delta-opioid receptor agonists is mediated by iNOS. *Am. J. Physiol. Heart Circ. Physiol.* **289,** H2251–H2257.
244. Li, Y. J., Song, Q. J., and Xiao, J. (2000). Calcitonin gene-related peptide: an endogenous mediator of preconditioning. *Acta Pharmacol. Sin.* **21,** 865–869.
245. Du, Y. H., Peng, J., Huang, Z. Z., Jiang, D. J., Deng, H. W., and Li, Y. J. (2004). Delayed cardioprotection afforded by nitroglycerin is mediated by α-CGRP via activation of inducible nitric oxide synthase. *Int. J. Cardiol.* **93,** 49–54.
246. Leesar, M. A., Stoddard, M. F., Dawn, B., Jasti, V. G., Masden, R., and Bolli, R. (2001). Delayed preconditioning-mimetic action of nitroglycerin in patients undergoing coronary angioplasty. *Circulation* **103,** 2935–2941.
247. Jneid, H., Chandra, M., Alshaher, M., Hornung, C. A., Tang, X. L., Leesar, M., and Bolli, R. (2005). Delayed preconditioning-mimetic actions of nitroglycerin in patients undergoing exercise tolerance tests. *Circulation* **111,** 2565–2571.
248. Crisafulli, A., Melis, F., Tocco, F., Santoboni, U. M., Lai, C., Angioy, G., Lorrai, L., Pittau, G., Concu, A., and Pagliaro, P. (2004). Exercise-induced and nitroglycerin-induced myocardial preconditioning improves hemodynamics in patients with angina. *Am. J. Physiol. Heart Circ. Physiol.* **287,** H235–H242.
249. Kukreja, R. C., Salloum, F., Das, A., Ockaili, R., Yin, C., Bremer, Y. A., Fisher, P. W., Wittkamp, M., Hawkins, J., Chou, E., Kukreja, A. K., Wang, X., Marwaha, V. R., and Xi, L. (2005). Pharmacological preconditioning with sildenafil: Basic mechanisms and clinical implications. *Vascul. Pharmacol.* **42,** 219–232.
250. Das, A., Xi, L., and Kukreja, R. C. (2005). Phosphodiesterase-5 inhibitor sildenafil preconditions adult cardiac myocytes against necrosis and apoptosis. Essential role of nitric oxide signaling. *J. Biol. Chem.* **280,** 12944–12955.
251. Salloum, F., Yin, C., Xi, L., and Kukreja, R. C. (2003). Sildenafil induces delayed preconditioning through inducible nitric oxide synthase-dependent pathway in mouse heart. *Circ. Res.* **92,** 595–597.
252. Das, A., Ockaili, R., Salloum, F., and Kukreja, R. C. (2004). Protein kinase C plays an essential role in sildenafil-induced cardioprotection in rabbits. *Am. J. Physiol. Heart Circ. Physiol.* **286,** H1455–H1460.
253. Elliott, G. T. (1996). Monophosphoryl lipid A: A novel agent for inducing pharmacologic myocardial preconditioning. *J. Thromb. Thrombolysis* **3,** 225–237.
254. He, S. Y., Deng, H. W., and Li, Y. J. (2001). Monophosphoryl lipid A-induced delayed preconditioning is mediated by calcitonin gene-related peptide. *Eur. J. Pharmacol.* **420,** 143–149.
255. Vinten-Johansen, J., Zhao, Z. Q., Zatta, A. J., Kin, H., Halkos, M. E., and Kerendi F. (2005). Postconditioning—A new link in nature's armor against myocardial ischemia–reperfusion injury. *Basic Res. Cardiol.* **100,** 295–310.
256. Zhao, Z. Q., Corvera, J. S., Halkos, M. E., Kerendi, F., Wang, N. P., Guyton, R. A., and Vinten-Johansen, J. (2003). Inhibition of myocardial injury by ischemic postconditioning during reperfusion: comparison with ischemic preconditioning. *Am. J. Physiol. Heart Circ. Physiol.* **285,** H579–H588.
257. Argaud, L., Gateau-Roesch, O., Raisky, O., Loufouat, J., Robert, D., and Ovize, M. (2005). Postconditioning inhibits mitochondrial permeability transition. *Circulation* **111,** 194–197.
258. Kin, H., Zatta, A. J., Lofye, M. T., Amerson, B. S., Halkos, M. E., Kerendi, F., Zhao, Z. Q., Guyton, R. A., Headrick, J. P., and Vinten-Johansen, J. (2005). Postconditioning reduces infarct size via adenosine receptor activation by endogenous adenosine. *Cardiovasc. Res.* **67,** 124–133.
259. Hausenloy, D. J., Tsang, A., and Yellon, D. M. (2005). The reperfusion injury salvage kinase pathway: a common target for both ischemic preconditioning and postconditioning. *Trends Cardiovasc. Med.* **15,** 69–75.
260. Vinten-Johansen, J., Zhao, Z. Q., Jiang, R., and Zatta, A. J. (2005). Myocardial protection in reperfusion with postconditioning. *Expert Rev. Cardiovasc. Ther.* **3,** 1035–1045.
261. Kin, H., Zhao, Z. Q., Sun, H. Y., Wang, N. P., Corvera, J. S., Halkos, M. E., Kerendi, F., Guyton, R. A., and Vinten-Johansen, J. (2004). Postconditioning attenuates myocardial ischemia-reperfusion injury by inhibiting events in the early minutes of reperfusion. *Cardiovasc. Res.* **62,** 74–85.
262. Yellon, D. M., and Hausenloy, D. J. (2005). Realizing the clinical potential of ischemic preconditioning and postconditioning. *Nat. Clin. Pract. Cardiovasc. Med.* **2,** 568–575.
263. Przyklenk, K., Bauer, B., Ovize, M., Kloner, R. A., and Whittaker, P. (1993). Regional ischemic 'preconditioning' protects remote virgin myocardium from subsequent sustained coronary occlusion. *Circulation* **87,** 893–899.
264. Pell, T. G., Baxter, G. F., Yellon, D. M., and Drew, G. M. (1998). Renal ischemia preconditions myocardium: role of adenosine receptors and ATP-sensitive potassium channels. *Am. J. Physiol.* **275,** H1542–H1547.
265. Tokuno, S., Hinokiyama, K., Tokuno, K., Lowbeer, C., Hansson, L. O., and Valen, G. (2002). Spontaneous ischemic events in the brain and heart adapt the hearts of severely atherosclerotic mice to ischemia. *Arterioscler. Thromb. Vasc. Biol.* **22,** 995–1001.
266. Schulte, G., Sommerschild, H., Yang, J., Tokuno, S., Goiny, M., Lovdahl, C., Johansson, B., Fredholm, B. B., and Valen, G. (2004). Adenosine A receptors are necessary for protection of the murine heart by remote, delayed adaptation to ischaemia. *Acta Physiol. Scand.* **182,** 133–143.
267. Dickson, E. W., Lorbar, M., Porcaro, W. A., Fenton, R. A., Reinhardt, C. P., Gysembergh, A., and Przyklenk, K. (1999). Rabbit heart can be 'preconditioned' via transfer of coronary effluent. *Am. J. Physiol.* **277,** H2451–H2457.
268. Gunaydin, B., Cakici, I., Soncul, H., Kalaycioglu, S., Cevik, C., Sancak, B., Kanzik, I., and Karadenizli, Y. (2000). Does remote organ ischaemia trigger cardiac preconditioning during coronary artery surgery? *Pharmacol. Res.* **41,** 493–496.
269. Peart, J. N., Gross, E. R., and Gross, G. J. (2005). Opioid-induced preconditioning: recent advances and future perspectives. *Vascul. Pharmacol.* **42,** 211–218.
270. Takaoka, A., Nakae, I., Mitsunami, K., Yabe, T., Morikawa, S., Inubushi, T., and Kinoshita, M. (1999). Renal ischemia/reperfusion remotely improves myocardial energy metabolism during myocardial ischemia via adenosine receptors in rabbits, effects of "remote preconditioning." *J. Am. Coll. Cardiol.* **33,** 556–564.

271. Wolfrum, S., Schneider, K., Heidbreder, M., Nienstedt, J., Dominiak, P., and Dendorfer, A. (2002). Remote preconditioning protects the heart by activating myocardial PKCepsilon-isoform. *Cardiovasc. Res.* **55,** 583–589.

272. Weinbrenner, C., Schulze, F., Sarvary, L., and Strasser, R. H. (2004). Remote preconditioning by infrarenal aortic occlusion is operative via delta1-opioid receptors and free radicals in vivo in the rat heart. *Cardiovasc. Res.* **61,** 591–599.

273. Kristiansen, S. B., Henning, O., Kharbanda, R. K., Nielsen-Kudsk, J. E., Schmidt, M. R., Redington, A. N., Nielsen, T. T., and Botker, H. E. (2005). Remote preconditioning reduces ischemic injury in the explanted heart by a K_{ATP} channel-dependent mechanism. *Am. J. Physiol. Heart Circ. Physiol.* **288,** H1252–H1256.

274. Konstantinov, I. E., Arab, S., Li, J., Coles, J. G., Boscarino, C., Mori, A., Cukerman, E., Dawood, F., Cheung, M. M., Shimizu, M., Liu, P. P., and Redington, A. N. (2005). The remote ischemic preconditioning stimulus modifies gene expression in mouse myocardium. *J. Thorac. Cardiovasc. Surg.* **130,** 1326–1332.

275. Konstantinov, I. E., Arab, S., Kharbanda, R. K., Li, J., Cheung, M. M., Cherepanov, V., Downey, G. P., Liu, P. P., Cukerman, E., Coles, J. G., and Redington, A. N. (2004). The remote ischemic preconditioning stimulus modifies inflammatory gene expression in humans. *Physiol. Genomics* **19,** 143–150.

276. Li, G., Labruto, F., Sirsjo, A., Chen, F., Vaage, J., and Valen, G. (2004). Myocardial protection by remote preconditioning: the role of nuclear factor kappa-B p105 and inducible nitric oxide synthase. *Eur. J. Cardiothorac. Surg.* **26,** 968–973.

277. Lang, S. C., Elsasser, A., Scheler, C., Vetter, S., Tiefenbacher, C. P., Kubler, W., Katus, H. A., and Vogt, A. M. (2006). Myocardial preconditioning and remote renal preconditioning Identifying a protective factor using proteomic methods? *Basic Res. Cardiol.* **101,** 149–158.

278. Wolfrum, S., Nienstedt, J., Heidbreder, M., Schneider, K., Dominiak, P., and Dendorfer, A. (2005). Calcitonin gene related peptide mediates cardioprotection by remote preconditioning. *Regul. Pept.* **127,** 217–224.

279. Ikonomidis, J. S., Tuniati, L. C., Weisel, R. D., Mickle, D. A., and Li, R. K. (1994). Preconditioning human ventricular cardiomyocytes with brief periods of simulated ischaemia. *Cardiovasc. Res.* **28,** 1285–1291.

280. Arstall, M. A., Zhao, Y. Z., Hornberger, L., Kennedy, S. P., Buchholtz, R. A., Osathanondh, R., and Kelly, R. A. (1998). Human ventricular myocytes in vitro exhibit early and delayed preconditioning responses to simulated ischemia. *J. Mol. Cell Cardiol.* **30,** 210–214.

281. Forlani, S., Tomai, F., De Paulis, R., Turani, F., Colella, D. F., Nardi, P., De Notaris, S., Moscarelli, M., Magliano, G., Crea, F., and Chiariello, L. (2004). Preoperative shift from glibenclamide to insulin is cardioprotective in diabetic patients undergoing coronary artery bypass surgery. *J. Cardiovasc. Surg. (Torino)* **45,** 117–122.

282. Kloner, R. A., Shook, T., Przyklenk, K., Davis, V. G., Junio, L., Matthews, R. V., Burstein, S., Gibson, C. M., Poole, W. K., Cannon, C. P., McCabe, C. H., and Braunwald, E. (1995). Previous angina alters in-hospital outcome in TIMI 4: a clinical correlate to preconditioning? *Circulation* **591,** 37–45.

283. Iwasaka, T., Nakamura, S., Karakawa, M., Sugiura, T., and Inada, M. (1994). Cardioprotective effect of unstable angina prior to acute anterior myocardial infarction. *Chest* **105,** 57–61.

284. Kloner, R. A., Shook, T., Antman, E. M., Cannon, C. P., Przyklenk, K., Yoo, K., Mccabe, C. H., Braunwald, E., and The Timi-9b Investigators. (1998). Prospective temporal analysis of the onset of preinfarction angina versus outcome: an ancillary study in TIMI-9B. *Circulation* **97,** 1042–1045.

285. Yamagishi, H., Akioka, K., Hirata, K., Sakanoue, Y., Toda, I., Yoshiyama, M., Teragaki, M., Takeuchi, .K, Yoshikawa, J., and Ocji, H. (2000). Effects of preinfarction angina on myocardial injury in patients with acute myocardial infarction: a study with resting 123I-BMIPP and 201TI myocardial SPECT. *J. Nucl. Med.* **41,** 830–836.

286. Yellon, D. M., Alkhulaifi, A. M., and Pugsley, W. B. (1993). Preconditioning the human myocardium. *Lancet* **342,** 276–277.

287. Jenkins, D. P., Pugsley, W. B., Alkhulaifi, A. M., Kemp, M., Hooper, J., and Yellon, D. M. (1997). Ischaemic preconditioning reduces troponin T release in patients undergoing coronary artery bypass surgery. *Heart* **77,** 314–318.

288. Teoh, L. K. K., Grant, R., Hulf, J. A., Pugslry, W. B., and Yellon, D. M. (2002). The effect of preconditioning (ischaemic and pharmacological) on myocardial necrosis following coronary artery bypass surgery. *Cardiovasc. Res.* **53,** 175–180.

289. Argaud, L., and Ovize, M. (2004). How to use the paradigm of ischemic preconditioning to protect the heart? *Med. Sci. (Paris)* **20,** 521–525.

290. Mahaffey, K. W., Puma, J. A., Barbagelata, N. A., DiCarli, M. F., Leesar, M. A., Browne, K. F., Eisenberg, P. R., Bolli, R., Casas, A. C., Molina-Viamonte, V., Orlandi, C., Blevins, R., Gibbons, R. J., Califf, R. M., and Granger, C. B. (1999). Adenosine as an adjunct to thrombolytic therapy for acute myocardial infarction: results of a multicenter, randomized, placebo-controlled trial: the Acute Myocardial Infarction STudy of ADenosine (AMISTAD) trial. *J. Am. Coll. Cardiol.* **34,** 1711–1720.

291. Sniecinski, R., and Liu, H. (2004). Reduced efficacy of volatile anesthetic preconditioning with advanced age in isolated rat myocardium. *Anesthesiology* **100,** 589–597.

292. Bartling, B., Friedrich, I., Silber, R. E., and Simm, A. (2003). Ischemic preconditioning is not cardioprotective in senescent human myocardium. *Ann. Thorac. Surg.* **76,** 105–111.

293. Lee, T. M., Su, S. F., Chou, T. F., Lee, Y. T., and Tsai, C. H. (2002). Loss of preconditioning by attenuated activation of myocardial ATP-sensitive potassium channels in elderly patients undergoing coronary angioplasty. *Circulation* **105,** 334–340.

294. Shinmura, K., Nagai, M., Tamaki, K., and Bolli, R. (2004). Gender and aging do not impair opioid-induced late preconditioning in rats. *Basic Res. Cardiol.* **99,** 46–55.

295. Loubani, M., Ghosh, S., and Galinanes, M. (2003). The aging human myocardium: tolerance to ischemia and responsiveness to ischemic preconditioning. *J. Thorac. Cardiovasc. Surg.* **126,** 143–147.

296. Peart, J. N., and Gross, G. J. (2004). Chronic exposure to morphine produces a marked cardioprotective phenotype in aged mouse hearts. *Exp. Gerontol.* **39,** 1021–1026.

297. Valen, G., and Vaage, J. (2005). Pre- and postconditioning during cardiac surgery. *Basic Res. Cardiol.* **100,** 179–186.

CHAPTER **11**

Cardiac Neovascularization: Angiogenesis, Arteriogenesis, and Vasculogenesis

OVERVIEW

From vasculogenic, angiogenic, and arteriogenic mechanisms, the heart vasculature develops in the embryo and post-natally. These three different pathways in the formation of new blood vessels or neovascularization can be defined as follows: (1) vasculogenesis, the *de novo* formation of the first primitive vascular plexus and post-natal vascularization; (2) angiogenesis, the formation of new vessels from pre-existing ones and, in particular, the sprouting of new capillaries from post-capillary venules; and (3) arteriogenesis, the process of maturation and/or *de novo* growth of specifically collateral arteries, which mainly occurs after ischemic vascular diseases. During embryonic vasculogenesis, endothelial cells precursors (hemangioblasts) start to proliferate, migrate into avascular areas, and aggregate to form a primitive network of vessels, because this vascular system is required to ensure adequate blood flow to provide the cells with sufficient supply of nutrients and oxygen, as well as the removal of toxic metabolic byproducts. Angiogenesis, a complex biological process, requires the precise coordination of multiple steps, being initiated by vasodilatation and an increased vascular permeability. In arteriogenesis, flow-related remodeling of existing collateral vessels occurs. In this chapter, we will analyze the potential molecular mechanisms involved in neovascularization, with particular emphasis on transcriptional factors that may function as activators, regulators, and inhibitors of the processes, restenosis, available animal models of angiogenesis, and current therapeutic modalities.

INTRODUCTION

Angiogenesis is a complex process involving the participation and communication of multiple cell types and is an important element in both physiological and pathological processes. The molecular basis of blood and lymph vessel formation remains incompletely understood. Nonetheless, understanding how these vessels grow offers novel therapeutic opportunities to treat ischemic cardiac disease by stimulating revascularization or to treat cancer by blocking lymph/angiogenesis.

The embryo responding to a partially hypoxic environment develops a circulatory system that allows the delivery of nutrients and oxygen and the removal of metabolic waste products. Blood vessels develop by vasculogenesis, a process that entails the differentiation of precursor cells (angioblasts) to endothelial cells, forming a complex and tortuous structure. Hypoxia, as the primary stimulator of vascular development, will trigger the expression of angiogenic factors, such as vascular endothelial growth factor (VEGF). Hypoxia-inducible factor 1 (HIF-1), a heterodimeric basic helix-loop-helix protein that activates transcription of the human erythropoietin gene in hypoxic cells, is involved in the activation of transcription of VEGF-A. After the developing embryo has formed a primary vascular plexus, further blood vessels are generated by both sprouting and nonsprouting angiogenesis, which are progressively growing and remodeling into a complete, functional circulatory system.

The present availability of animal models lacking some of the transcription factors/signaling pathways involved in vasculogenesis and angiogenesis has advanced our understanding of these complex processes. In this chapter, the cellular and molecular mechanisms that participate in the formation of endothelium-lined channels (angiogenesis) and their maturation by recruitment of smooth muscle cells (SMCs) and vessel remodeling (arteriogenesis) during physiological and pathological conditions will be presented, and potential therapeutic applications will be also discussed.

MECHANISMS OF NEOVASCULARIZATION

Although the processes that are involved in neovascularization are also not yet fully understood, it seems that the most important determinants in the formation of new vessels and their growth are in angiogenesis, local tissue ischemia or hypoxia; in arteriogenesis, nonspecific factors such as shear stress and blood flow; and in vasculogenesis, the release of endothelial progenitor cells from the bone marrow. We agree with Simons[1] that the best understood of these processes in molecular terms is hypoxia-induced angiogenesis, which is examined in the following section.

The Role of Specific Genes/Gene Products and Modulatory Factors

Angiogenesis is a complex process in which numerous genes/transcription factors are involved either as activators or regulators. During vasculogenesis, mesoderm-inducing factors of the fibroblast growth factor (FGF) family are crucial in inducing paraxial and lateral plate mesoderm to form angioblasts and hematopoietic cells. Also, because a variety of transcription factors are expressed in endothelial cells during angiogenesis, the cross-regulation of transcription factors is important. The downstream-regulated processes encompass the evolution from progenitor to mature endothelial cells in vasculogenesis and the involvement of extensive endothelial cell proliferation and tube formation as central events in the angiogenesis pathway (Fig. 1) in which SMCs play a key role.

In the formation of angioblasts, VEGF and the VEGF receptors (VEGF-R) play essential roles (thru VEGF-R2 expression) and their assembly into functional blood vessels (by VEGF-R1). Both VEGF-R2 and basic FGF (bFGF) also play key roles in the differentiation of angioblasts.[2] Embryos lacking VEGF-R2 develop defects in both hematopoietic and angioblastic lineages, and after differentiation, down-regulation of VEGF-R2 occurs in hematopoietic but not in endothelial cells.[3] Later in development VEGF-R1 (Flt-1) also provides a regulatory role, because mice lacking VEGF-R1 produce angioblasts, but their assembly into functional blood vessels is impaired.[4]

In contrast, transgenic overexpression of VEGF alone in mice results in increased numbers of primarily leaky vascular vessels with tissue edema and inflammation,[5,6] suggesting that VEGF needs to work in conjunction with other angiogenic factors to produce a healthy vasculature.[7] There is considerable evidence that in a variety of conditions, such as malignant tumors, wound healing, and myocardial ischemia, hypoxia is a fundamental stimulus for angiogenesis.[8] The upstream regulator HIF-1, a PAS domain transcription factor, mediates activation of hypoxia-responsive genes including VEGF.[9,10] HIF-1 mediates the angiogenic response to hypoxia by up-regulating the expression of multiple angiogenic factors involved in the assembly of newly formed vasculature and the maintenance of vascular integrity. HIF-1 is composed of two subunits, HIF-1α and HIF-1β (aryl hydrocarbon nuclear translocator). Whereas the β-subunit protein is constitutively present, the stability of the α-subunit and its transcriptional activity are regulated by various post-translational modifications, hydroxylation, acetylation, and phosphorylation precisely controlled by the intracellular oxygen concentration.[11–13] Under nonhypoxic conditions, HIF-1α is rapidly ubiquitinated and degraded by proteosomes triggered by the hydroxylation of prolines and the acetylation of lysine within a polypeptide segment known as the oxygen-dependent degradation (ODD) domain. On the other hand, in hypoxia, HIF-1α subunit becomes stable, and in cells subjected to decreasing O_2 concentrations, levels of HIF-1 protein increase exponentially.[14]

Gene disruption of both *HIF-1α* alleles in mice results in developmental arrest and embryonic lethality at E11 with vascular regression.[15] Transfection of naked plasmid DNA containing a hybrid protein composed of the DNA binding and dimerization domains from the HIF-1α subunit and the transactivation domain from herpes simplex virus VP16 protein (to provide robust constitutive expression) increases angiogenesis and blood supply in both animal models of hind limb ischemia and myocardial infarct (MI).[16,17] Similarly, transgenic expression of a constitutively stable HIF-1α mutant in mice epidermis resulted in a marked hypervascularity characterized by increased numbers of dermal capillaries and a greater than 10-fold increase in VEGF levels without the excessive permeability and inflammatory phenotype observed in transgenic mice overexpressing VEGF.[18] However, the underlying molecular mechanism for the differences in the quality of the vasculature resulting from overexpression of VEGF or HIF-1α remains to be explained.

Angiopoietins comprise a family of ligands that bind the endothelium-specific receptor tyrosine kinase Tie-2 and effect blood vessel development, primarily interactions between endothelial cells. Angiopoietin-1 (Ang-1), the major physiological ligand for Tie-2, does not directly promote the mitogenic growth of cultured endothelial cells, nor by itself does it produce neovascularization *in vivo*. However, its expression in close proximity with developing blood vessels implicates this factor in endothelial developmental processes.[19,20] Transgenic mice overexpressing a combination of VEGF and Ang-1 exhibit marked induction of angiogenesis and collateral formation without excessive permeability and the inflammatory events associated with increased VEGF expression. This highlights the importance of Ang-1 in the formation of normal vessels and in the maintenance of an anti-inflammatory phenotype.[6] On the other hand, angiopoietin-2 (Ang-2) competes for binding to a common receptor and, therefore, serves as a natural antagonist of Ang-1 often acting as an angiogenic antagonist as well.[21,22] The role of Ang-2 in angiogenesis is less defined, because it is highly dependent on the presence of other angiogenic factors, particularly VEGF.[23,24] Ang-2 antagonizes the activation of Tie-2, disrupting blood vessel formation in part by promoting endothelial cell apoptosis and vascular regression.[23,25] In the presence of VEGF, however, Ang-2 destabilizes the pre-existing vasculature and consequently makes it more responsive to angiogenic stimuli and is, in fact, associated with neovascularization characterized by increased vessel length.[24,25] When coadministered with VEGF, both Ang-1 and Ang-2 are, therefore, capable of augmenting angiogenesis.[25,26] Although its precise function has not been documented, Ang-4, similar to Ang-1, binds to Tie-2 as an agonist.

In addition, hypoxia has been shown to regulate the expression of angiopoietins and Tie-2, suggesting their participation in the angiogenic response to hypoxia.[27–30]

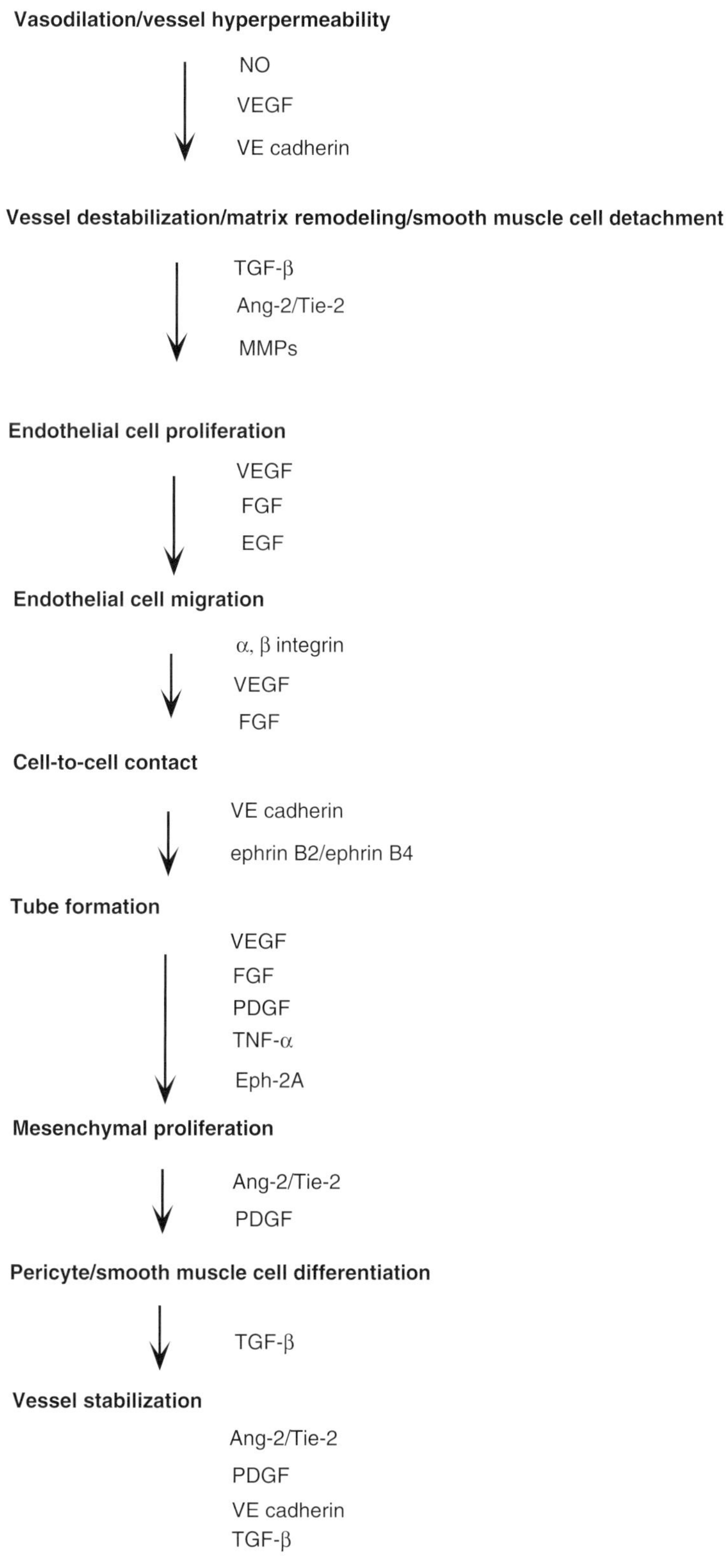

Vasodilation/vessel hyperpermeability

↓   NO
      VEGF
      VE cadherin

Vessel destabilization/matrix remodeling/smooth muscle cell detachment

↓   TGF-β
      Ang-2/Tie-2
      MMPs

Endothelial cell proliferation

↓   VEGF
      FGF
      EGF

Endothelial cell migration

↓   α, β integrin
      VEGF
      FGF

Cell-to-cell contact

↓   VE cadherin
      ephrin B2/ephrin B4

Tube formation

↓   VEGF
      FGF
      PDGF
      TNF-α
      Eph-2A

Mesenchymal proliferation

↓   Ang-2/Tie-2
      PDGF

Pericyte/smooth muscle cell differentiation

↓   TGF-β

Vessel stabilization

      Ang-2/Tie-2
      PDGF
      VE cadherin
      TGF-β

FIGURE 1 The angiogenesis pathway illustrating key molecular factors.

Recent data showing that HIF-1–deficient hepatoma cells (Hepa1 c4), as well as its wild-type counterpart (Hepa1 c1c4), up-regulate Ang-2 during hypoxia. In addition, pharmacological induction of HIF-1 with cobalt chloride or desferrioxamine (DFO) in cultured endothelial cells had no effect on Ang-2 expression, suggesting that hypoxia-driven Ang-2 expression may be independent of the HIF pathway.[31] However, to address whether HIF-1 mediates hypoxic induction of the other angiopoietins, and whether Ang-4 participates in the angiogenic response to hypoxia, the role of HIF-1 in the angiogenic response to hypoxia was studied in primary culture human artery endothelial cells.[32] Treatment with DFO or infection with *Ad2/HIF-1alpha/VP16*, a constitutively stable hybrid form of HIF-1α present on an adenoviral vector, mimicked the hypoxic induction of multiple angiogenic factors, including VEGF, Ang-2, and Ang-4 (Ang-1 was not effected in these cells). In addition, expression of HIF-1α stimulated the proliferation and tube formation of cultured endothelial cells, which was partially inhibited by a VEGF-R2 or Tie-2 inhibitor, further suggesting that HIF-1 mediates the angiogenic response to hypoxia by activating both the VEGF and angiopoietin/Tie-2 system. This study also demonstrated that hypoxia stimulated the mRNA levels of Ang-4, but not Ang-1 and Ang-2, in human fetal cardiac cells and that cells treated with recombinant Ang-4 protein were protected against starvation-induced apoptosis and exhibited increased migration and tube formation compared with untreated cells or cells treated with recombinant Ang-2 protein. Activation of the angiopoietin/Tie-2 system by HIF-1, in particular the up-regulation of Ang-4, may explain the difference in the quality of the vasculature resulting from overexpression of VEGF compared with HIF-1α. Hence, these data implicate HIF-1 as a global mediator of the angiogenic response to hypoxia in cultured endothelial cells by inducing multiple angiogenic factors, including VEGF and angiopoietins.

There is increasing evidence that the effect of hypoxia-associated gene regulation in angiogenesis is dependent on several transcription factors for the activation of targeting genes. HIF-1 is one of the transcriptional factors that has been shown to regulate VEGF expression (either *in vivo* or *in vitro*) with the responding element located between -975 and -968 on *VEGF* promoter (related to the 5' transcriptional start site).[33] Recent studies have reported that various stimuli such as oxidative stress[34] and oxidants[35] may also regulate the expression of VEGF by binding to a GC-rich region (-114 to -50) on the *VEGF* promoter sequence, suggesting that other transcription factors independent of HIF-1 are likely operative.

The transcriptional enhancer factor-1 (TEF-1) is a transcriptional factor family that can regulate many target genes expressed in cardiac and skeletal muscle cells by binding to myocyte-specific chloramphenicol acetyltransferase (M-CAT) heptamer sequence CATN(T/C)(T/C) located in the promoters of these target genes. Shie and associates[36] reported that related TEF-1 (RTEF-1), a member of the TEF-1 family, is up-regulated in cultured hypoxic bovine aortic endothelial cells and that overexpression of transfected *RTEF-1* increases *VEGF* promoter activity and *VEGF* expression. Sequential deletion and site-directed mutation analyses of the *VEGF* promoter demonstrated that a GC-rich region containing four Sp1 response elements, located between -114 and -50, was essential for RTEF-1 function. This region is beyond the HIF-1α binding site and does not consist of M-CAT related elements. Using an electrophoretic mobility shift assay, RTEF-1 was found to interact with the first Sp1 residue (-97 to -87) of the four consecutive Sp1 elements. Binding activity of RTEF-1 to *VEGF* promoter was also confirmed by chromatin immunoprecipitation. In addition, induction of *VEGF* promoter activity by RTEF-1 results in an increase of angiogenic processes, including endothelial cell proliferation and vascular structure formation. These data indicate that RTEF-1 acts as a transcriptional stimulator of *VEGF* by regulating *VEGF* promoter activity through binding to Sp1 site in a GC-rich region of the promoter. In addition, RTEF-1–induced *VEGF* promoter activity was enhanced in hypoxia, indicating that RTEF-1 may play an important role in the regulation of VEGF under hypoxia and, therefore, in angiogenesis.

Furthermore, given the significant roles that HIF plays in mouse development and physiology (e.g., homeostasis regulation and lung development), outside vascular development, an important question to be addressed is whether its role in vascular development was a direct one or an indirect derivative of its functions in other tissues.

To understand whether specific gene(s) such as *HIF-2α* regulate vascular development, either directly from within the vascular system (e.g., endothelial cell) or indirectly from nonvascular cells, the targeted expression of *HIF-2α* cDNA (under the control of an endothelial cell-specific promoter) in the vascular endothelium of *HIF-2 α-/-* embryos by an embryonic stem (ES) cell–mediated transgenic approach was used and assessment of whether endothelial-specific re-expression of *HIF-2α* could rescue vascular development.[37] Although ES cell–derived *HIF-2α-/-* embryos developed severe vascular defects by E11.5 and died *in utero* before E12.5, endothelial cell–specific *HIF-2α* expression restored normal vascular development at all stages examined (up to E14.5) and allowed *HIF-2α-/-* embryos to survive at a frequency comparable to that of *HIF-2α+/-* embryos. Furthermore, Tie-2 expression was significantly reduced in *HIF-2α-/-* mutants but was restored by *HIF-2α* expression, demonstrating an intrinsic requirement for HIF-2α by endothelial cells and implying that hypoxia may control endothelial functions directly by means of *HIF-2α*–regulated Tie-2 expression.

A number of factors have been implicated in vasculogenesis, the initial events in vascular growth in which endothelial cell precursors (angioblasts) migrate to discrete locations, differentiate *in situ*, and assemble into solid endothelial cords, later forming a plexus with endocardial tubes, which are finally stabilized by pericytes and SMCs and angiogenesis, including the subsequent growth, expansion, and remodeling of these primitive vessels into a mature vascular network.[38] Numerous soluble growth factors and inhibitors, cytokines, proteases, and extracellular matrix (ECM) proteins and adhesion molecules strictly control these multi-step processes. These include angiogenic molecules such as VEGFs, FGFs, angiopoietins, PDGF, angiogenin, angiotropin, HGF, CXC chemokines (with ELR motif), PECAM-1, integrins, and VE-cadherin as shown in Table I. Key angiostatic players such as angiostatin, endostatin, thrombospondin, and CXC chemokines (without ELR motif) also are involved in the regulation of vasculogenesis and angiogenesis—a number of which will be discussed further in a later section of this chapter.

The highly regulated process of blood vessel growth is characterized by a combination of sprouting of new vessels from the sides and ends of pre-existing ones or by longitudinal division of existing vessels with periendothelial cells (intussusception), either of which may then split and branch into precapillary arterioles and capillaries.[39] Depending on the ultimate fate with respect to the type of vessel (artery, vein, capillary) and vascular bed, activated endothelial cells that are migrating and proliferating to form new vessels, forming anastomotic connections with each other, become variably surrounded by layers of periendothelial cells—pericytes for small vessels and SMCs for large vessels ("vascular myogenesis").[40] During this dynamic period, extracellular matrix produced by mural cells serves to stabilize the network. Finally, further functional modifications of larger arteries occur during "arteriogenesis" as a thick muscular coat is added, concomitant with acquisition of viscoelastic and vasomotor properties.

An important element in the mobilization and migration of endothelial precursor cells is PDGF-CC, a recently discovered member of the VEGF/PDGF superfamily, whose contribution to the angiogenic mechanism remains incompletely characterized.

PDGF-CC is expressed in actively angiogenic tissues such as placenta, some embryonic tissues, and tumors and can stimulate neovascularization in the mouse cornea and induce branch sprouts from established blood vessels in the developing chick embryo; these PDGF-CC–mediated angiogenic responses are most likely transduced by PDGF-$\alpha\alpha$ and PDGF-$\alpha\beta$ receptors.[41] In recent studies conducted to examine the direct effects of PDGF-CC on endothelial cells, recombinant PDGF-CC delivered to mouse heart subjected to ischemic conditions (achieved by coronary ligation) stimulated vessel growth and maturation.[42] In achieving this revascularization (also demonstrated with ischemic limb preparation), PDGF-CC enhanced endothelial progenitor cell mobilization, promoted differentiation of bone marrow progenitor cells into endothelial cells, and stimulated the migration of endothelial cells and microvessel sprouting. In addition, PDGF-CC

TABLE I Activators of Angiogenesis

Molecule	Function
VEGF, VEGF-C, PlGF	Stimulate angiogenesis by ↑ EC proliferation, migration, and apoptosis, ↑ vascular permeability and matrix degradation; VEGF-C stimulates lymphangiogenesis; PlGF role is limited to pathologic angiogenesis.
Angiopoietin-1 (Ang-1) and Tie-2-receptor	Ang1 stabilizes vessels by tightening EC and smooth muscle cell interaction; ↑ EC sprouting.
NO and NOS	↑ Vasodilation and permeability; ↑ EC proliferation.
Cyclooxygenase-2	Proteolytic fragments of plasminogen inhibit EC proliferation, migration, and survival.
Plasminogen activators, matrix metalloproteinases	Proteinases involved in cellular migration and matrix remodeling; liberate FGF and VEGF from the matrix; activate TGF-β; generate angiostatin.
Ephrins	Regulate arterial/venous specification.
VE-cadherin	Endothelial junctional molecules; essential for endothelial survival effect; decreased EC apoptosis.
PECAM	↑ EC aggregation and migration, tube formation, and vessel stabilization.
Integrins $\alpha_v\beta_3$, $\alpha_v\beta_5$	Receptors for matrix macromolecules and proteinases (MMP-2), ↑ EC migration and attachment.
VEGF receptor	VEGF-R2, angiogenic signaling receptor; VEGF-R3, (lymph)angiogenic signaling receptor.
FGF, HGF, MCP-1	Stimulate angiogenesis (FGF, HGF) by ↑ EC proliferation and migration, ↑ PAs, integrins, and adhesion molecules and arteriogenesis (FGF, MCP-1).
TGF-β, endoglin, TGF-β receptors	Stabilize vessels by stimulating extracellular matrix production.
Erythropoietin	↑ EC proliferation.
PDGF-BB and receptors	Recruit SMCs; ↑vessel stabilization.

EC, endothelial cell.

induced the differentiation of bone marrow cells into SMCs and stimulated their growth during vessel sprouting. These findings have suggested that modulating the activity of PDGF-CC may prove clinically useful as an angiogenic factor and provide novel opportunities for treating ischemic diseases.

The evolutionary conserved *forkhead box O (Foxo)* subclass of transcription factors seems to play an important role in the regulation of vessel formation in adults.[43] Genetic analysis of DAF-16, a Foxo ortholog, in *Caenorhabditis elegans* revealed essential roles in longevity, energy metabolism, and reproduction. Mammalian forkhead transcription factors Foxo1 (Fkhr), Foxo3a (Fkhrl1), and Foxo 4 (Afx), all of which are regulated by growth factor–dependent activation of the PI3K pathway, have been implicated in regulating cell processes (in a large variety of cell types), including cell cycle arrest, apoptosis, and stress responses *in vitro*. However, the physiological roles of the specific Foxo isoforms *in vivo* are largely unknown. To explain their role in normal development and physiology, the Foxo genes were disrupted in mice.[43] *Foxo1*-null embryos died on E10.5 as a consequence of incomplete vascular development. *Foxo1*-null embryonic and yolk sac vessels were not well developed at E9.5, and *Foxo1* expression was found in a variety of embryonic vessels, suggesting a crucial role of this transcription factor in vascular formation. On the other hand, both *Foxo3a*- and *Foxo4*-null mice were viable and grossly indistinguishable from their litter mate controls, indicating dispensability of these two members of the Foxo transcription factor family for normal vascular development. *Foxo3a*-null females showed age-dependent infertility and had abnormal ovarian follicular development. In contrast, histological analyses of *Foxo4*-null mice did not identify any consistent abnormalities. These results demonstrate that the physiological roles of *Foxo* genes are functionally diverse in mammals

Foxo1 and *Foxo3a* encode the most abundant Foxo isoforms in mature endothelial cells, and overexpression of constitutively active Foxo1 or Foxo3a, but not Foxo4, significantly inhibits endothelial cell migration and tube formation *in vitro*. Foxo3a overexpression also enhances apoptosis and stems the growth factor–induced endothelial cell proliferation. Silencing of either *Foxo1* or *Foxo3a* gene expression with siRNA led to a marked increase in the migratory and sprout-forming capacity of endothelial cells.[44] Gene expression profiling showed that *Foxo1* and *Foxo3a* specifically shared a common subset of target genes with angiogenesis and vascular remodeling function, including *eNOS*, which is essential for post-natal neovascularization, and the proangiogenic transcription factor ETS domain protein Elk-3 (ELK-3, also known as Net). Other genes such as Ang-2 (exclusively regulated by Foxo1) were only under the regulation of a specific Foxo isoform. Consistent with these findings, constitutively active Foxo1 and Foxo3a repressed eNOS protein expression and bound to the *eNOS* promoter. *In vivo*, Foxo3a deficiency increased eNOS expression and enhanced post-natal vessel formation and maturation.

Focal adhesion kinase (FAK), a cytoplasmic (nonreceptor) tyrosine kinase that plays a major role in integrin-mediated signal transduction in several different cell types including cardiomyocytes and endothelial cells, has recently been shown to have a significant role in angiogenesis and vascular development in late embryogenesis. Integrins cluster in focal adhesion complexes, where the extracellular matrix is connected to the cytoskeleton and FAK is located. Moreover, recent studies have suggested that FAK is a critical mediator and a converging point in the signaling pathways triggered by integrin, G-protein–coupled, and growth factor receptors.

Consistent with its critical roles as revealed by *in vitro* studies, *FAK* gene inactivation in mice resulted in a relatively early embryonic lethality (E8.5) with major defects in the mesoderm development, reduced cell motility and a defective cardiovascular system, including a lack of fully developed blood vessels and heart.[45]

In contrast, in *FAK* knock-out mice specifically targeted to endothelial cells using a Cre-loxP approach, deletion of *FAK* did not affect early embryonic development, including normal vasculogenesis, but rather showed multiple defects in angiogenesis in late embryogenesis (by E13.5), including a reduced blood vessel network in the superficial vasculature, hemorrhage, edema, developmental delay in the embryos, abnormalities of blood vessels in both yolk sac and placental labyrinth, and embryonic lethality.[46] Moreover, endothelial cells derived from the endothelial cell–targeted FAK exhibited increased apoptosis, reduced proliferation and migration, and reduced ability to form capillaries on Matrigel when cultured *in vitro*. These findings strongly suggest a role of FAK in angiogenesis and vascular development because of its essential function in the regulation of multiple endothelial cell activities, which are required in later, but not early, embryogenesis.

Also, the effect of FAK overexpression on angiogenesis has been demonstrated in models of skin wound healing and ischemia in skeletal muscle by the same group of investigators.[47] Transgenic mice with chicken FAK overexpressed in their vascular endothelial cells under the control of the Tie-2 promoter and enhancer displayed high levels of FAK transgene expression detected by RT-PCR, immunoprecipitation and Western blots in all vessel-rich tissues. Vessel number in the granulation tissue of healing wounds in the wound-induced angiogenesis model and capillary density of the ischemic skeletal tissues were significantly increased in the transgenic mouse compared with that of wild-type control mice, further suggesting that FAK plays an important role in the promotion of angiogenesis *in vivo*. Eventually, targeted overexpression of FAK may be used in human cardiovascular angiogenesis.

It has also been recently demonstrated that FAK plays a pivotal role in NO-mediated flow-induced coronary arteriole dilation. Earlier studies have found that flow-induced regulation of *eNOS* and arteriole dilation were dependent on integrin signaling function and tyrosine kinases as mechanosensors, VEGF-R2 as mechanotransducers and downstream activation of Akt/PI3K.[48–55]

Given FAK's central role in integrin signaling, the role of FAK was further explored using an anti-FAK phospho-specific (Tyr397) antibody (FAKab) to inactivate FAK-mediated signaling. These studies found that FAK signaling inhibition with FAKab significantly impaired flow-induced dilation and reduced phosphorylation of FAK, Akt, and eNOS. These data suggest that the upstream activation of FAK is central to flow-induced dilation by means of regulation of activation of Akt and eNOS.[56]

Another transcription factor that participates in angiogenesis is Net. This ETS ternary complex transcription factor, normally a transcriptional repressor, is activated by phosphorylation by Ras and Src signaling and converted into a positive regulator. It is expressed at sites of vasculogenesis and angiogenesis during early mouse development, suggestive of a function in blood vessel formation.[57] Recent *in vivo* and *in vitro* evidence has shown that in the adult, Net is a primary regulator of the angiogenic switch, in which growth factors, generated by surrounding normal or tumor cells, induce quiescent endothelial cells to proliferate and differentiate to form new blood vessels. These studies show that endogenous Net is phosphorylated on critical amino acids in response to FGF-2 and Ras-V12, growth factors and oncogenes that are known to activate angiogenesis and MAP kinase–signaling cascades. Phosphorylation results in converting Net from repressor to P-Net, a transcriptional activator stimulating specifically VEGF expression. Conversely, down-regulation of Net inhibits VEGF expression and decreased angiogenesis *in vivo* and assayed *in vitro*. Both P-Net and VEGF are coexpressed in angiogenic processes in wild-type mouse tissues and human tumors.[58]

Available data support an important paracrine role for endogenously produced NO in endothelial cell migration and differentiation *in vitro*, and suggest that cell-based eNOS gene transfer may be a useful approach to increase new blood vessel formation *in vivo*. In addition to increasing angiogenic growth factor expression,[59] NO can also have direct effects stimulating endothelial cell migration and capillary-like tube formation. New blood vessel formation is a complex process and requires coordinate regulation of multiple mechanisms, including modifications of cell–cell interactions and their extracellular matrix, endothelial cell migration/invasion and proliferation, and finally maturation into neovessels.[60]

Shear stress regulates endothelial iNOS expression through c-Src signaling pathway,[61] as well as by the PI3K/Janus kinase 2/MEK-1–dependent pathways.[62] NO production can be stimulated by shear stress, is known to inhibit NF-κB activation, and plays a critical role in the endothelial cell inflammatory state. Grumbach and associates[63] reported that exposure of bovine aortic endothelial cells to laminar shear stimulated *eNOS* mRNA expression and *eNOS* promoter activity as measured by an *eNOS* promoter/CAT construct. The shear effects were enhanced by the NOS inhibitor L-NAME and decreased by the NO-donor DPTA-NO by 30–50%. In addition, the NF-κB inhibitor panepoxydone prevented the increase in *eNOS* mRNA caused by shear, confirming a role of NF-κB in this response and underscoring the critical role of NO in the modulation of the endothelial cell inflammatory state. In pathological conditions such as hypercholesterolemia, diabetes, and hypertension, all associated with eNOS dysfunction, decreased NO may result in sustained activation of NF-κB secondary to shear and endothelial cell inflammation.

Regulation of VEGF and angiogenesis by the tumor suppressor *PTEN* gene has been shown in several studies.[64] Mice heterozygous for an endothelial-specific mutation of *PTEN* (generated using Cre-loxP system) displayed an increase in angiogenesis driven by vascular growth factors. Endothelial cells derived from these mice showed enhanced proliferation/migration. Mice homozygous for this endothelial-targeted *PTEN* mutation died before E11.5 because of bleeding and heart failure (HF) caused by impaired recruitment of pericytes and vascular SMCs to blood vessels and of cardiomyocytes to the endocardium. These phenotypes were dependent on the PI3K p110γ subunit and were associated with decreased expression of Ang-1, VCAM-1, connexin 40, and ephrinB2 but increased expression of Ang-2, VEGF-A, VEGF-R1, and VEGF-R2.

It is well established that the *PTEN*-encoded phosphatase can mediate the down-regulation of PI3K signaling, which largely explains *PTEN* effects on VEGF-mediated signaling and angiogenesis. The inactivation of the tumor suppressor gene *PTEN* and overexpression of VEGF are two common events observed in high-grade malignant gliomas. Promoter activity assays in human glioma cells indicated that the *VEGF* promoter region containing the HIF-1α binding site is necessary and sufficient for PTEN-mediated down-regulation of VEGF.[65] Moreover, glioblastoma cell lines containing mutant *PTEN* infected with adenovirus expressing wild-type *PTEN* exhibited decreased levels of both V*EGF* mRNA and phosphorylated Akt.[66] Experiments with PI3K inhibitors and kinase assays suggested that PI3K is mediating the effect of PTEN on VEGF and not the p42/p48 or p38 MAP kinases.

These results indicate that the restoration of PTEN function in gliomas may induce therapeutic effects by down-regulating VEGF. A *PTEN* adenoviral vector has been reported to effectively treat bladder cancers in nude mice that have genomic alterations in *PTEN* in part by promoting the marked down-regulation of VEGF and decreased microvessel density associated with tumorigenic growth.[67] Because many glioblastomas, the most common brain cancer of adults, contain mutations and deletions of tumor suppressor gene PTEN, targeted molecular therapies using pathway inhibitors, monoclonal antibodies, and viral-mediated gene therapy to increase PTEN or decrease PI3K activation are currently in early stages of application.[68]

Angiogenesis regulation mediated by mitochondrial-generated hydrogen peroxide (H_2O_2) and PTEN oxidation has been recently reported by Connor and associates.[69]

It is well established that PI3K activity, the level of PIP3, and the activity of Akt are increased in cells exposed to H_2O_2. This H_2O_2-mediated activation of PI3K/Akt pathway has been found to be largely the result of the reversible oxidation and inactivation of protein tyrosine phosphatase family proteins such as PTEN that have a critical cysteine residue in the active site. Recent studies show that cell lines engineered by MnSOD overexpression to have elevated levels of mitochondrial-derived H_2O_2 manifest PTEN oxidation that can be reversed by coexpression of the H_2O_2-detoxifying enzyme catalase. The accumulation of an oxidized PTEN favored the formation of plasma membrane PIP3, increased Akt activation, and modulation of its downstream targets. Moreover, the mitochondrial H_2O_2-induced PTEN oxidation leading to enhanced PI3K signaling resulted in increased VEGF expression. Wild-type PTEN overexpression abrogated the H_2O_2-dependent stimulation of *VEGF* promoter activity, whereas overexpression of a mutant PTEN (G129R), lacking phosphatase activity, did not. In addition, MnSOD-mediated mitochondrial generation of H_2O_2 promoted endothelial cell sprouting in a three-dimensional *in vitro* angiogenesis assay that could be attenuated by either catalase coexpression or the PI3K inhibitor LY2949002. The involvement of mitochondrial H_2O_2 in regulating PTEN function and the angiogenic switch was further supported by the finding that MnSOD overexpression resulted in increased *in vivo* blood vessel formation that was H_2O_2 dependent as assessed by the chicken chorioallantoic membrane assay.

ARTERIOGENESIS

Although there is some controversy regarding whether arteriogenesis involves the development of neovascularization (collateral) or is limited to enlargement of existing vessels with remodeling, the consensus is that collateral artery growth to compensate for the loss of an occluded artery is a central feature of arteriogenesis. Mechanisms of collateral artery growth differ from angiogenic capillary growth, which is largely driven by ischemia or hypoxia. A comparison of some of the salient differences among angiogenesis, arteriogenesis, and vascularization is shown in Fig. 2. Although these processes share common endothelial cell regulators, the arteriogenic formation of larger more complex blood vessels consisting of endothelium and incorporating SMCs, ECM to form media, elastin fibers and adventitia requires a more elaborate mechanism with complex spatial cues than

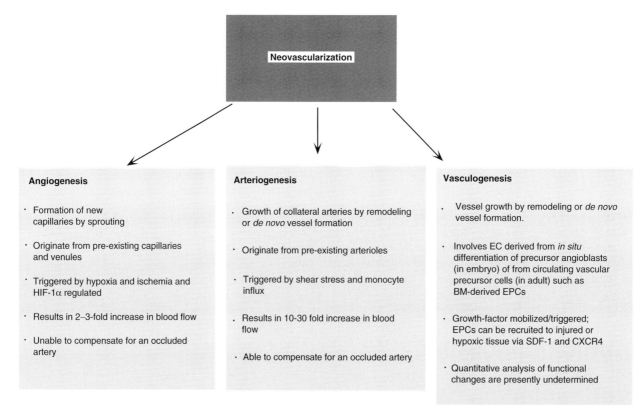

FIGURE 2 Comparison of key features of angiogenesis, arteriogenesis, and vasculogenesis.

the formation of a capillary composed primarily of an endothelium, basement membrane, and mural cells.

The distinct processes contributing to arteriogenesis involve a complex interplay between molecular stimuli and factors and cellular events, most often initiated on a pre-existing substrate of collateral anastomoses. Early stages of arteriogenesis involve stimulation of endothelial cell activation by hemodynamic/physical forces, including shear and wall stress,[70] release from activated endothelial cells of chemokines and monocyte-attracting factors, and attraction of blood leukocytes to the vessel wall followed by their adhesion and subsequent transmigration. Activation of vascular endothelial surfaces induces, in an NF-κB–dependent fashion, the up-regulation of adhesion molecules including selectins, vascular cell, and intercellular cell adhesion molecules. A subsequent stage of proliferation and vessel remodeling is initiated by the recruitment of blood-derived mononuclear cells capable of secreting a number of cytokines and growth factors, including FGFs, PDGF, PlGF, TGF-β, VEGF, and MCP-1, as well as numerous matrix-degrading protease enzymes mediating digestion of the external elastic lamina and ECM and mediating endothelial cell and SMC proliferation of vascular wall.[71] Cytokines that attract monocytes or prolong the life span of monocytes (MCP-1, GM-CSF) are strong arteriogenic factors. A final maturation phase of the vessel wall involves rearrangement of wall cells and a recovery of extracellular structures occurring with positive remodeling and diameter increases. Interestingly, arteriogenesis and atherosclerosis share many elements, in particular the stimulation of inflammatory pathways, up-regulation of adhesion molecules and matrix proteases, and SMC recruitment, although collateral vessels enlarge in diameter, whereas atherosclerotic vessels shrink. An overview of the major stages of arteriogenesis highlighting the key molecular players is depicted in Fig. 3.

The critical role of specific growth factors in arteriogenesis remains not as clearly defined as in angiogenesis. This is particularly relevant to potential therapeutic application. A multitude of growth factors, such as PDGF and hepatocyte growth factor, have been found to contribute to arteriogenic responses. Some growth factors such as placenta growth factor (PlGF), a member of the VEGF family, and MCP are more clearly identified with arteriogenesis than with angiogenesis, likely because of their action attracting inflammatory cells, including monocytes.[72] In monocyte-depleted animals, the ability of PlGF to enhance collateral growth in the rabbit model and to rescue impaired arteriogenesis in *PlGF* gene-deficient mice was abrogated. Numerous preclinical studies have not only indicated the involvement of factors including FGF, PlGF, VEGF, and TGF-β in arteriogenesis but that FGF and VEGF could stimulate collateral flow in animal models, albeit duration of exposure could be limiting. As discussed in more detail later, recent clinical trials with isolated growth factors such as FGF-2 or VEGF showed little beneficial effect on patients with severe myocardial ischemia[73,74] nor

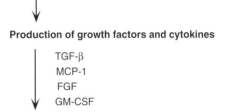

FIGURE 3 Flowchart depicting salient events of arteriogenesis with major molecular factors.

did GM-CSF with patients with peripheral artery disease.[75] These findings may reflect the inability of individual growth factors to orchestrate the entire spectrum of events required for growth of new blood vessels, and there is some indication that synergistic effects with more successful outcomes might occur with administration of a mixture of factors.[76]

Although some studies of the interactions of these multiple factors and synergy between particular growth factor combinations have been conducted, further analysis seems warranted.[77] Another explanation for the present inability of these factors to display clinical utility may arise from difficulties with local delivery, peptide half-life, and/or clinical trial design and execution. Moreover, the application of sensitive methods for imaging and noninvasively monitoring the quantitative effects of angiogenic and arteriogenic therapy in clinical trials, including CT, MRI, and PET and "molecular imaging" targeting specific endothelial cell–specific antigenes has been limited thus far.[1]

As noted elsewhere in this volume, there is increasing evidence that bone marrow stem cells can provide vasculogenesis after birth. It is known that bone marrow cells (BMCs) give rise to most of the SMCs that contribute to arterial remodeling in models of post-angioplasty restenosis, graft vasculopathy, and hyperlipidemia-induced atherosclerosis. Somatic stem cells contribute to pathological remodeling of remote organs and may provide the basis for the development of new therapeutic strategies for vascular diseases through targeting mobilization, homing, differentiation, and proliferation of bone marrow–derived vascular progenitor cells.[78] The role of BMCs in arteriogenesis remains ill defined. Recent observations have shown no evidence of bone marrow–derived cells incorporated into vessels of mice with hind limb ischemia treated with transplanted BMC (labeled with GFP protein).[79] No colocalization of endothelial cell or SMC markers with GFP was found in association with growing collateral arteries, although evidence of GFP-containing fibroblasts and leukocytes (BMC-derived) surrounding the collateral arteries was found, indicating a supportive function. Other studies have suggested that paracrine signaling by marrow stromal cells augmenting collateral remodeling through release of several cytokines such as VEGF and bFGF is an important mediator of collateral remodeling and does not require cell incorporation.[80] Local proliferation of tissue-resident precursors, including the accumulation of macrophages and vascular cells, play a prominent role in the initial phase of vascular growth in arteriogenesis compared with cells derived from migrating or circulating blood-borne cells.[81] In this study, the depletion of circulating cells affected neither macrophage accumulation nor collateral growth. Other studies have shown that transplanted monocytes can localize to areas of collateral growth and create a highly arteriogenic environment through secretion of multiple growth factors. Transplanted rabbit monocytes that were either *ex vivo* stimulated or engineered by adenoviral transduction to express a transgene encoding an arteriogenic growth factor (e.g., GM-CSF) were examined for their therapeutic potential in a rabbit model of arteriogenesis.[82] In particular, when autologous monocytes were used as vehicles to deliver GM-CSF as therapeutic transgene, a strong promotion of arteriogenesis resulted. Interestingly, transplantation of autologous cells (without re-engineered GM-CSF) provided no augmentation of collateral growth. Transplantation of autologous monocytes modified *ex vivo* may represent a potential therapeutic approach in treating ischemic disease, where arteriogenesis might be required. Recently, it has been shown that targeted delivery of bone marrow mononuclear cells by ultrasound-mediated destruction of contrast microbubbles can enhance blood flow restoration by stimulating both angiogenesis and arteriogenesis.[83]

CARDIAC NEOVASCULARIZATION

Although VEGF and FGF-2 have proved to be successful in the angiogenesis of animal models of ischemia, their potential benefits in humans are still questionable.[73] VIVA (Vascular endothelial growth factor in Ischemia for Vascular Angiogenesis), a double-blind, placebo-controlled trial was designed to evaluate the safety and efficacy of intracoronary and intravenous infusion of recombinant human vascular endothelial growth factor protein (rhVEGF). In this trial, 178 patients with stable exertional angina, unsuitable for standard revascularization, were randomly assigned to receive placebo, low-dose rhVEGF, or high-dose rhVEGF by intracoronary infusion on day 0, followed by intravenous infusions on days 3, 6, and 9. Angina class and quality of life were significantly improved within each group, with no difference between groups. By day 120, placebo-treated patients demonstrated a reduced benefit in all three measures, with no significant difference compared with low-dose rhVEGF. In contrast, high-dose rhVEGF resulted in significant improvement in angina class and nonsignificant trends in exercise treadmill test (ETT) time and angina frequency compared with placebo. Therefore, although rhVEGF seems to be safe and well tolerated, it offered no improvement beyond placebo in all measurements by day 60. Only by day 120, high-dose rhVEGF resulted in significant improvement in angina and favorable trends in ETT time and angina frequency.

It has been suggested that therapeutic angiogenesis (and arteriogenesis) in clinical settings may be promoted more successfully by the restoration of effective endothelial signaling rather than by supraphysiological administration of a growth factor.[1] VEGF and FGF-2 operate through the release of NO by means of the activation of tyrosine kinase receptors,[84] and patients with advanced CAD have been found to have significant endothelial dysfunction and diminished coronary NO release, which are determinant pathological features of their disease.[85] Moreover, *in vitro* data and murine studies in peripheral organs suggested that the effects of VEGF and FGF-2 may depend on local NO availability.[86,87] Furthermore, in hypercholesterolemic swine models with endothelial dysfunction, it has been shown that the cardiac angiogenic response to FGF-2 compared with normocholesterolemic animals is markedly inhibited.[88] Because endothelial dysfunction and diminished coronary NO release constitute key features in the pathophysiology of CAD, therapies such as chronic oral L-arginine supplementation can clinically improve coronary endothelial function,[89–91] and this may have significant implications in clinical trials of angiogenesis, where the evaluation of potential endothelial modulation may be an important adjunct to other therapeutic angiogenesis modalities.

Another endothelial cell mediator function that can modulate angiogenesis is Big MAP kinase 1 (BMK1 or ERK5), because abnormal embryonic angiogenesis and vascular collapse occurred in *BMK1* knockout mice.[92] The findings that BMK1 activity regulates angiogenesis (by HIF-1 and VEGF expression), endothelial cell migration, and apoptosis suggests that BMK1 is a potential target for treatment that can modulate wound repair, tumor angiogenesis, and atherosclerosis. In addition, stimulating BMK1 activity would have beneficial effects to limit edema formation, to promote endothelial cell survival, and to decrease cancer angiogenesis. Furthermore, BMK1-mediated proliferation of cultured rat aortic SMCs is stimulated by aldosterone, an effect that is inhibited by splerone (a selective mineralocorticoid receptor antagonist).[93] This aldosterone-mediated event was ascribed to a nongenomic effect, because cycloheximide failed to inhibit aldosterone-induced BMK1 activation. Transfection of dominant-negative MAP kinase/ERK kinase 5 (MEK5), which is an upstream regulator of BMK1, partially inhibited aldosterone-induced rat aortic SMC proliferation, which was almost completely inhibited by MEK inhibitor PD98059. Consequently, in addition to its known steroid activity, the nongenomic effects induced by aldosterone by BMK1 may represent an alternative etiology for vascular diseases, including systemic hypertension.

Several sources of endothelial precursor cells (EPCs) have been identified in the neonate and the adult heart that contribute to physiological and pathological vascularization consistent with the concept of *de novo* vasculogenesis after birth. New vascularization after cardiac ischemia is induced by infusion of EPCs; however, the mechanisms of homing of EPCs to sites of ischemia are not known. *Ex vivo*–expanded EPCs, as well as murine hematopoietic Sca-$1^+/Lin^-$ progenitor cells, express β2-integrins, which mediate the adhesion of EPCs to endothelial cell monolayers and their chemokine-induced transendothelial migration *in vitro*. In a murine model of hind limb ischemia, Sca-$1^+/Lin^-$ hematopoietic progenitor cells from β2-integrin–deficient mice are less capable of homing to sites of ischemia and of improving neovascularization; however, pre-activation of the β2-integrins expressed on EPCs by activating antibodies augments the EPC-induced neovascularization *in vivo* and provide evidence for a novel function of β2-integrins in postnatal vasculogenesis.[94] Furthermore, the protease cathepsin L (CathL) has been found to be highly expressed in EPCs as opposed to mature endothelial cells and also that CathL-deficient mice have impaired recovery after hind limb ischemia. Interestingly, forced expression of CathL in mature endothelial cells significantly enhanced their capacity to improve neovascularization *in vivo*, suggesting that CathL plays a critical role in the incorporation of EPCs into the ischemic tissue and is involved in the formation of new vessels.[95]

Given the increasingly evident role of endothelial cell loss, largely through apoptosis, in the pathogenesis of vascular diseases, strategies achieving rapid endothelial recovery seem to be warranted. Circulating EPCs originating in the bone marrow play a significant role in endogenous neovascularization of ischemic tissue[96–98] and in the re-endothelialization of injured vessels.[99,100] Moreover, EPC mobilization and proliferation also seems to contribute to the beneficial effects of 3-hydroxy-3-methylglutaryl (HMG)-CoA reductase inhibitors, or statins[101,102] and estrogen.[103] Transplantation of EPCs induced new vessel formation in ischemic myocardium and hind limb[104,105] and accelerated the re-endothelialization of injured vessels and prosthetic vascular grafts in humans and in various animal models.[106] Therefore, EPCs may be used as a cell-based strategy to rescue and repair ischemic tissues and injured blood vessels, and because EPCs are amenable to genetic manipulation they may be used as vectors for local delivery of therapeutic genes.[107,108] However, with aging, there is an obsolescence and functional impairment of EPCs that normally repair and rejuvenate the arteries and that correlate with risk factors for atherosclerosis and CAD.[109,110]

VASCULAR REMODELING DURING ANGIOGENESIS

Angiogenic vascular remodeling is the process whereby the primitive vascular network develops into the mature vasculature. This process is an adaptive one that occurs in response to chronic changes in hemodynamic conditions.[111] Changes in cell growth, cell migration, cell death, and extracellular matrix composition operative in remodeling lead to a compensatory adjustment in vessel diameter and lumen area. Pathological remodeling occurs during the development of atherosclerosis and restenosis after balloon angioplasty, where the migration of proliferating phenotypically modified vascular SMCs from the medial to the luminal side of the vessel is an important mechanism in the intimal thickening that narrows the vessel lumen and leads to ischemia, or even anoxia, in the downstream tissues.[112]

The master regulatory genes of the Homeobox (Hox) family, besides their direct participation in organogenesis and differentiated tissue function, promotes angiogenesis thru *HoxD3* and *HoxB3* genes; however, overexpression of *HoxD10* gene in endothelial cells revealed a nonangiogenic phenotype,[113] because sustained expression of *HoxD10*-impaired endothelial cell migration and blocked angiogenesis induced by basic FGF and VEGF in the chick chorioallantoic membrane *in vivo*. Furthermore, HoxD10-overexpressing human endothelial cells also failed to form new vessels when implanted into immunocompromised mice.

The homeobox genes have multiple functions in many cell types, including their role in cell proliferation or growth arrest, cell differentiation, and de-differentiation.[114] Because the endothelial cells and vascular SMCs are highly moldable, the specific characteristics of the homeobox proteins make these important transcription factors potential candidates as regulators of cellular differentiation and remodeling of the vascular system not only during normal development but also in a number of pathological states.

Recently, it has been demonstrated that angiogenic vascular remodeling is also under the control of the Notch signaling pathway.[115] Of the two mammalian Notch receptors expressed in vascular endothelium, Notch1 is broadly expressed in diverse cell types, whereas Notch4 is preferentially expressed in endothelial cells. Analysis of *NOTCH4* primary transcripts in human umbilical vein endothelial cells by RNA fluorescence *in situ hybridization* showed that 36% of the cells transcribed one or both *NOTCH4* alleles. The *NOTCH4* promoter was sufficient to confer endothelial cell–specific transcription in transfection assays, but intron 1 or upstream sequences were required for expression in the vasculature of transgenic mouse embryos. The cell type–specific activator protein 1 (AP-1) complexes occupied *NOTCH4* chromatin and conferred endothelial cell–specific transcription. In HeLa cells, vascular angiogenic factors activated AP-1 and reprogrammed the endogenous *NOTCH4* gene from a repressed to a transcriptionally active state, implying the presence of an AP-1/Notch4 pathway, which may be crucial for transducing angiogenic signals and may be deregulated on aberrant signal transduction in cancer.

INHIBITORS OF ANGIOGENESIS

Numerous molecules with anti-angiogenic effects after administration in various animal models have been described (Table II). These include natural or endogenous inhibitors of endothelial cell proliferation, survival, or function. Some of the endogenous angiogenesis inhibitors that are being evaluated for clinical use include angiostatin, endostatin, antithrombin III, interferons, and platelet factor-4. The precise physiological role of most of these molecules during embryonic, post-natal, and adult vessel growth remains largely unknown. Several factors such as TGF-β, TNF-α, and IL-4 also are bifunctional modulators of angiogenesis, acting either as stimulators or inhibitors, depending on a variety of factors including their amount, the lesion site, the microenvironment, and the presence of other cytokines.

Synthetic exogenous inhibitors are also currently being used in clinical trials, including inhibitors of endothelial proliferation and survival (e.g., TNP470 ([synthetic derivative of fumagillin]), captopril (inhibitor of angiotensin-I–converting enzyme), inhibitors of endothelial cell adhesion (e.g., Vitaxin [humanized monoclonal antibody against the integrin $\alpha v \beta 3$ present on the surface of "activated" endothelial cells]), inhibitors of ECM proteolysis, and endothelial cell invasion (e.g., synthetic MMP inhibitors: marimastat, AG-3340, COL-3, and Neovastat).

The specific factors that mediate and maintain collateral formation in animal models and humans with coronary vessel occlusion are not known. In dogs undergoing repetitive coronary occlusions under control conditions or during antagonism of NOS using L-NAME, it was found that inhibition of NOS increased expression of angiostatin, an inhibitor of tumor angiogenesis produced by the actions of matrix

TABLE II Inhibitors of Angiogenesis

Molecule	Function
Angiopoietin-2	Antagonist of Ang1: induces vessel regression.
Endostatin	Fragment of type XVIII collagen; inhibits EC proliferation, survival and migration; promotes EC apoptosis.
Vasostatin, calreticulin	Calreticulin and N-terminal fragment (vasostatin) inhibit endothelial growth.
Angiostatin and related plasminogen kringles	Proteolytic fragments of plasminogen inhibit EC proliferation, migration, and survival.
Thrombospondin-1 (TSP-1)	Extracellular matrix protein; type I repeats inhibit EC migration, proliferation, adhesion, and survival; related TSP-2 also inhibits angiogenesis.
Platelet factor-4	Heparin-binding CXC chemokine inhibits binding of bFGF and VEGF and EC proliferation.
Anti-thrombin III fragment	Fragments of the hemostatic factor suppress EC proliferation.
Tissue-inhibitors of MMP (TIMPs)	Suppress pathologic angiogenesis.
VEGF receptor	Sink for VEGF, VEGF-B, and PlGF.
Interferon (IFN) α, β, γ	Cytokines and chemokines, inhibiting EC proliferation and migration; IFNα down-regulates FGF.
Prolactin (16-kDa fragment)	Attenuates NO production by inhibiting EC iNOS activity and expression and induces EC cell cycle arrest.
Plasminogen activator inhibitor 1 (PAI-1)	Primarily by its antiproteolytic activity (inhibiting plasmin production).
Wild-type p53	Down-regulates angiogenesis inducers, neutralizes FGF, and maintains expression of inhibitors such as thrombospondin.

metalloproteinases (MMPs) on plasminogen. Activities of tissue MMP-2 and MMP-9, which generate angiostatin, were increased in the L-NAME group, suggesting that angiostatin inhibits coronary angiogenesis during compromised NO production and may underscore the impairment of coronary angiogenesis during endothelial dysfunction.[116] Also, angiostatin levels were increased in a model of inadequate coronary collateral growth and angiogenesis in response to ischemia (despite high levels of VEGF), suggesting that therapies attempting to provoke coronary collateral growth should incorporate approaches that limit or neutralize the effects of growth inhibitors.[117] Furthermore, patients with CAD and grade 3 collateralization have endostatin levels in pericardial fluid nearly 40% lower than those lacking angiographic evidence of collaterals. Interestingly, VEGF levels were not significantly different between the groups, suggesting that pericardial fluid levels of endostatin, but not VEGF, are associated with the presence or absence of collaterals in patients with CAD and that endostatin levels may locally modulate coronary collateral formation.[118]

Negative vascular modulators may have a function in embryogenesis, as well as in the revascularization process that follows acute myocardial infarction (AMI). The endothelial-monocyte activating polypeptide II (EMAP II) is an anti-angiogenic protein whose expression has been detected in a number of systems of the mouse embryo, including the respiratory and cardiovascular systems, and it may play an important role in vertebrate morphogenesis.[119] In a rat model of MI, the temporal and spatial distribution of EMAP II has been examined by *in situ* hybridization and protein expression by Western analysis over a 6-week period.[120] EMAP II protein was increased above control levels and changed its location of transcription from the inflammatory cell population to that of the fibroblasts located in the relative avascular scar tissue and resumed perivascular stromal distribution in the viable peri-infarct tissue. This suggests that EMAP II, besides its role in morphogenesis, has a negative modulatory effect in the angiogenesis occurring after AMI.

Another negative factor of angiogenesis is the hematopoietically expressed homeobox gene (*HEX*) that is transiently expressed in endothelial cells during vascular formation in the mouse embryo. The HEX homeobox protein significantly reduces the expression of genes, including VEGF-R1, TIE-1, TIE-2, and neuropilin-1, and abrogates the angiogenic properties of endothelial cells. *In vitro* experiments suggest that HEX acts as a negative regulator of angiogenesis.[121] However, whether HEX has the same effects *in vivo* is not known. Furthermore, HEX can act as a transcription activator as well[122]; EX overexpression in human umbilical vein endothelial cells results in increased expression of endoglin.[126] In association with a number of genes with increased expression, endoglin is a binding target of TGF-β,[123] and is highly expressed in endothelial cells during angiogenesis.[124] Endoglin is involved in vascular maturation, because targeted disruption of the *endoglin* gene in mice resulted in defective vascular maturation, poor SMC accumulation, and arrested endothelial remodeling.[125] Consequently, HEX may not act only as negative regulator of angiogenesis but also promotes vascular maturation by inducing endoglin in endothelial cells.[126]

DYSREGULATORS OF ANGIOGENESIS

Angiogenesis is dysregulated in coronary artery (CA) aneurysms in the chronic phase of Kawasaki disease (KD), and neovascularization may occur in inflammatory-related vascular diseases, because many angiogenesis mediators are secreted by inflammatory cells. Recently, it has been shown that in acute KD, marked inflammation and angiogenesis occurred in both CA aneurysms and myocardium, with the highest microvessel density seen in patients who died 2–3 weeks after onset of the disease.[127] Expression of proangiogenic proteins was higher than expression of inhibitors in KD CA aneurysms and myocardium. Angiogenesis mediators were localized to inflammatory cells in the myointima, adventitia, and myocardium, suggesting that significant neovascularization occurs in acute KD CA aneurysms and myocardium soon after onset of the disease with the involvement of multiple angiogenesis factors, and dysregulation of angiogenesis may contribute to the vasculopathy of KD.

A causal relationship between elevated plasminogen activator inhibitor 1 (PAI-1) levels and poor outcome in patients with MI through mechanisms that directly inhibit bone marrow–dependent neovascularization has been reported,[128] and reduction of myocardial PAI-1 expression seems to be capable of enhancing cardiac neovascularization, regeneration, and functional recovery after ischemic insult.

In acute myocardial ischemia, down-regulation of PAI-1 besides increasing the microvasculature also improves cardiomyocyte (CM) survival. Recently, Xiang and associates[129] developed a sequence-specific catalytic DNA enzyme used to reduce PAI-1 levels in cultured endothelial cells and in ischemic myocardium. At 2 weeks, hearts from experimental rats had more than fivefold greater capillary density, 70% reduction in apoptotic CMs, and fourfold greater functional recovery than controls, implying a causal relationship between elevated PAI-1 levels in the ischemic heart and adverse outcomes, suggesting that strategies to reduce cardiac PAI-1 activity may augment neovascularization and improve functional recovery.

Serine protease inhibitors (SPI), termed serpins, are important regulators of numerous biological pathways, including angiogenesis. In vascular injury models, Serp-1 altered tissue plasminogen activator, the inhibitor PAI-1, and receptor (urokinase-type plasminogen activator) expression. Serp-1, but not a reactive center loop mutant, up-regulated PAI-1 serpin expression in human endothelial cells.[130] Treatment of endothelial cells with antibody to urokinase-type plasminogen activator and vitronectin blocked Serp-1-induced changes. Serp-1 blocked intimal hyperplasia after aortic allograft transplant in PAI-1–deficient mice and also blocked plaque growth after aortic isograft transplant and after wire-induced injury in PAI-1–deficient mice, indicating that increase in PAI-1 expression is not required for Serp-1 to block vasculopathy development.

RESTENOSIS

Arterial injury such as occurs in patients with CAD undergoing percutaneous coronary intervention or placement of intravascular stents frequently results in restenosis of the lumen because of constrictive remodeling of the vessel wall and intimal accumulation of SMCs and matrix.[131]

Research in injured animal artery models has helped in our understanding of angioplasty and stenting mechanisms, as well as the testing of drug-eluting stent (DES) technologies.[132] Prediction of adverse clinical outcomes in patients with DESs can be forecast from suboptimal results in the animal model studies, and similarly stent thrombosis in animal models suggests stent thrombogenicity in human patients. Nevertheless, equivocal animal model results occasionally have contrasted with excellent clinical outcomes in patients. Although the reasons for these differences have not been explained, they may result from variations in method, including less arterial injury than originally described in the models. Further research into animal models will likely resolve these apparent dissimilarities with human clinical trials and will advance our understanding of how to use animal models of clinical stenting in the era of DESs.

Voisine and associates[133] reported significant endothelial dysfunction in the coronary microvessel of a swine model of chronic ischemia by the administration of a cholesterol-rich diet, whereas there was a markedly decreased response to the angiogenic (vasodilatation) effects of administered VEGF. In addition, key molecules in the VEGF-signaling pathways were affected. These changes did not occur in pigs treated with the same VEGF therapy but kept on a normal diet. Moreover, in agreement with the association between VEGF and eNOS in the growth and stabilization of vascular cells in vessel development, the high-cholesterol-diet animals exhibited decreased eNOS and VEGF expression than the normal diet group, which displayed an increase in both factors. In an earlier study, it was shown that patients with stable CAD and patients with acute myocardial ischemia experienced a diminished release of NO in response to VEGF.[134] Later Bussolati and associates[135] also reported that VEGF-stimulated NO release is inhibited by blockade of VEGF-R1 and that VEGF-R1 mediated NO release negatively regulates VEGF-R2–mediated proliferation and promotes formation of capillary networks in human umbilical vein endothelial cells. This indicates that VEGF-R1 is a signaling receptor that promotes endothelial cell differentiation into vascular tubes, in part by limiting VEGF-R2–mediated endothelial cell proliferation by NO, and likely contributes to the molecular switch for endothelial cell differentiation.

Early growth response factor-1 (Egr-1) controls the expression of a number of genes involved in the pathogenesis of atherosclerosis and post-angioplasty restenosis. Egr-1 is activated by diverse proatherogenic stimuli. As such, this transcription factor represents a key molecular target in efforts to control vascular lesion formation in humans. Lowe and associates[136] generated catalytic DNAzymes targeting specific sequences in human *EGR-1* mRNA cleaving *in vitro* transcribed *EGR-1* mRNA efficiently at preselected sites, inhibiting EGR-1 protein expression in human aortic SMCs, blocking serum-inducible cell proliferation, and abrogating cellular regrowth after mechanical injury *in vitro*. These DNAzymes also selectively inhibited EGR-1 expression and proliferation of porcine arterial SMCs and reduced intimal thickening after stenting pig coronary arteries *in vivo*. These findings demonstrated that endoluminally delivered DNAzymes targeting *EGR-1* might serve as inhibitors of in-stent restenosis. In addition, this group of investigators, using a rat model of neointima (NI) formation involving the complete ligation of the common carotid artery, demonstrated the importance of the immediate-early gene and zinc finger transcription factor EGR-1 in this process. Acute cessation of common carotid blood flow by vessel ligation was followed by the expression of *EGR-1* in the arterial media within 3 h and NI formation proximal to the point of ligation at 18 days. Local delivery of catalytic oligodeoxynucleotides (ODN) targeting rat *EGR-1* mRNA at the time of ligation reduced both EGR-1 expression and NI formation in this model. In contrast, a scrambled version of this ODN had no inhibitory effect. These findings demonstrate for the first time that arterial intimal thickening after artery ligation is critically dependent on the activation of EGR-1.[137] On the other hand, EGR-1 can bind and activate the promoters of many genes, whose products influence vascular repair. Antisense RNA overexpression has been used to inhibit EGR-1 protein synthesis, without affecting levels of the immediate early gene product, *c-fos*. Furthermore, antisense *EGR-1* RNA overexpression inhibited SMC regrowth after mechanical injury *in vitro*, whereas sense *EGR-1* RNA had no effect on SMC repair, *EGR-1* mRNA expression, or protein synthesis. An antisense RNA approach may be useful to control SMC growth in the injured artery wall.

TRANSGENIC MOUSE MODELS

Several transgenic mouse models of perturbed vascular development have been developed over the past several years.[138] Abnormal embryonic vascular development, resulting from defects in the formation of a primitive capillary plexus, has been observed in mice lacking VEGF, FLT-1, FLK-1, TGF-β, fibronectin, or VE-cadherin. Defects in the expansion and remodeling of the embryonic vasculature occur in mice deficient in TIE-1, TIE-2, angiopoietin-1, Braf, ARNT, VHL, VCAM-1, integrin, or in mice overexpressing neuropilin, or angiopoietin-2. Impaired recruitment and investment of mural cells has been observed in mice with disruption of the genes encoding PDGF-B, PDGF-B receptor, tissue factor, and the transcription factor lung Kruppel-like factor (LKLF), whereas reduced ECM formation occurs in mice lacking collagen type I. Abnormal blood vessel development or function after birth occurs in mice lacking collagen type III (vessel fragility with bleeding), P-selectin (reduced leukocyte rolling), or fibrillin-1 (aneurysmal dilatation and rupture). Mice overexpressing VEGF in the retina or mice with sickle cell disease develop retinal and choroidal neovascularization, whereas overexpression of the Fps/Fes tyrosine kinase or the polyoma middle-T antigen results in the formation of hemangiomas. Overexpression of VEGF-C in the skin induces lymphangiomas, whereas overexpression of the bovine polyoma virus-1 genome, the simian virus large T-antigen, or conditional overexpression of VEGF results in the formation of vascularized tumors.

MOLECULAR PROFILING OF CARDIAC ENDOTHELIAL CELLS

Endothelial cells lining the blood vessels differ among different tissues, and molecular profiling of specific markers for the cardiac vasculature could be very useful for targeting specific drugs. Zhang and associates[139] have profiled mice heart vasculature by phage screening and by matching heart-homing peptides against a heart cDNA library in a bacterial two-hybrid system. They found a set of four heart-homing peptides and their receptors, expressed in high levels in endothelial cells of the coronary vasculature and endocardium (in some cases in cardiomyocytes), and mostly absent in other major organs. Although these protein receptors were also expressed in lung and skeletal muscles, they were at significantly lower levels than in the heart. Three of these peptides are novel heart endothelial markers, and the fourth one (CRIP2) had also been previously expressed in the heart endothelium during development and in the adult heart. Homing peptides can be used to deliver concentrated therapeutic drugs in the chosen target organ/tissue[140–142] and eventually may be used for imaging of the vasculature and as pro-angiogenic and antithrombotic tools. Nevertheless, further research is necessary to fully understand the functional role of these peptides and receptors, their cell localization (intracellular versus cell surface), mechanisms for the effects of cell delivery, and, most importantly, their usefulness in human cardiovascular pathology.

THERAPY

A number of physical and biological modalities are being used in an attempt to increase blood flow in ischemic tissues. A better understanding of how vessels sprout and reach maturation has made the use of gene therapy, to promote angiogenesis in ischemic cardiac and skeletal muscle, a reality; nevertheless, and as previously discussed in earlier chapters, new and better ways of delivery and vectors are necessary before widespread clinical use. Furthermore, approaches that include cell transplant, with their pool of growth factors and cytokines and transfer of transcription factors, are in progress.

Gene Therapy

On the basis of studies of various animal models, gene therapy has emerged as a potential modality to treat complex diseases such as cardiomyopathy, vascular diseases such as hypertension, atherosclerosis, and myocardial ischemia. Some of these experimental therapies have moved into clinical trials to assess their feasibility, safety, and their potential in humans with CVD. However, the biology of coronary artery restenosis is incompletely understood, catheter-based gene delivery is still poorly adapted to the coronary circulation, and the safety and effectiveness of current gene transfer vectors to the coronary artery wall is still an important, if not unsolvable, problem. On the other hand, the availability of adult progenitor cells with the capacity to differentiate into endothelial and cardiomyocyte phenotypes offers the promise of further progress in achieving the vascularization and repair of ischemic tissues and injured blood vessels; furthermore, genetic engineering of EPCs to develop these cells into successful therapeutic tools seems to have great potential. However, successful translation of these experimental therapies into clinical practice will require safer and more effective vectors and delivery tools, increasing understanding of the biology of progenitor cell, as well as the absolute documentation of the efficacy and safety of the method through multicenter randomized trials. For a more comprehensive discussion on this subject the reader is referred to Chapter 4.

siRNA/Antisense RNA

Recently, siRNA-mediated silencing of lactosylceramide (LacCer) synthase expression (GalT-V) has been used in human umbilical vein endothelial cells and showed the involvement of LacCer in VEGF-induced angiogenesis.[143] This gene silencing markedly inhibited VEGF-induced platelet endothelial cell adhesion molecule-1 (PECAM-1) expression, and angiogenesis. The deployment of D-PDMP, an inhibitor of LacCer synthase and glucosylceramide synthase, was found to significantly reduce both VEGF-induced PECAM-1 expression and angiogenesis. Interestingly, these phenotypic changes could be reversed by LacCer but not by structurally related compounds such as glucosylceramide and ceramide. Moreover, VEGF/LacCer failed to stimulate PECAM-1 expression and tube formation/angiogenesis in a human endothelial cell line (REN) that lacks the endogenous expression of PECAM-1.[143] In contrast, REN cells expressing human PECAM-1 gene/protein displayed both VEGF and LacCer-induced PECAM-1 protein expression and tube formation/angiogenesis. Notably, VEGF-induced, but not LacCer-induced, angiogenesis was abrogated by exposure to SU-1498, a VEGF receptor tyrosine kinase inhibitor. In addition, VEGF/LacCer-induced PECAM-1 expression and angiogenesis were diminished by treatment with protein kinase C and phospholipase A2 inhibitors. These results indicate that LacCer generated in VEGF-treated endothelial cells may serve as an important signaling molecule for the regulation of PECAM-1 expression in angiogenesis. This finding and the reagents developed in these studies may prove useful as anti-angiogenic drugs for further studies to be conducted *in vitro* and *in vivo*.

Antisense RNA methods may be useful in the gene therapy approach to control SMC growth in the injured arterial wall. Furthermore, inhibition of SMC migration and apoptosis induction may provide novel targets for prevention and treatment of vascular diseases.

PDGF-CC

The platelet-derived growth factor (PDGF) family of ligands and receptors plays a pivotal role in angiogenesis. This family consists of four members (PDGF-AA, PDGF-BB, PDGF-CC, and PDGF-DD) and must form dimers to be physiologically active.[144] Their receptors have been detected on the microvascular endothelium, suggesting a direct role for PDGFs on endothelial cells and vascular SMCs. PDGF-CC has been shown to stimulate microvessel growth and induces branching of pre-existing vessels in the developing embryo[145] but did not affect endothelial cell outgrowth from the aorta.[146] PDGF-CC also stimulates vascular SMC growth[147,148] and induces the release of VEGF,[149] probably because angiogenesis is induced indirectly by affecting mural cells. However, whether PDGF-CC is capable of affecting endothelial cells directly or indirectly was not clearly established until the recent report from Li and associates.[150] They showed that PDGF-CC stimulated vessel growth and improved blood flow in two mouse models of myocardial and hind limb ischemia. Apparently, besides inducing the release of VEGF, PDGF-CC also mobilized vascular progenitors and promoted their differentiation into endothelial cells and SMCs, stimulating endothelial cell migration and affecting SMCs. These effects of PDGF-CC on vascular and muscle regeneration, together with its safety profile and activity in ischemic conditions, may have significant implications for developing novel strategies for the treatment of heart and limb ischemia, either alone or in combination therapy with other growth factors, such as VEGF or FGF, both of which failed in phase 2 clinical trials to improve neovascularization in patients with ischemic heart disease.[151]

Gene Transfer/Overexpression

Proteases of the plasminogen activator (PA) and matrix metalloproteinase (MMP) system play an important role in SMCs migration and neointima formation after vascular injury, such as occurs in arterial restenosis after balloon angioplasty. Inhibition of either PAs or MMPs has been shown to result in decreased neointima formation *in vivo*. The availability of gene transfer technology has opened the door to genetically manipulate the endothelium. An increase in expression of biologically active and immunoreactive TIMP-1 has been reported *in vitro* after infection of rat SMCs with *TIMP-1*, an adenoviral vector containing the human *TIMP-1* cDNA; expression of tissue inhibitor of MMP-1 inhibited SMC migration and reduced neointimal hyperplasia in the rat model of vascular balloon injury.[152] On the other hand, Lamfers and associates[153] using gene transfer technology have demonstrated that the receptor-binding amino terminal fragment (ATF) of urokinase suppresses neointima formation more efficiently than tissue inhibitor of MMP-1. They constructed a novel hybrid protein, TIMP-1.ATF, consisting of TIMP-1 domain, as MMP inhibitor, linked to ATF of urokinase. By binding to u-PA receptor, this protein not only anchors the TIMP-1 moiety directly to the cell surface, it also prevents the local activation of plasminogen by blocking the binding of u-PA to its receptor. Adenoviral expression of TIMP-1.ATF was used to inhibit SMC migration and neointima formation in human saphenous vein segments *in vitro*. Binding of TIMP-1.ATF hybrid protein to the u-PA receptor at the cell surface strongly enhances the inhibitory effect of TIMP-1 on neointima formation in human saphenous vein cultures. Also, it has been demonstrated that the tissue inhibitor of MMP-4 (TIMP-4) can inhibit SMC migration and induce apoptosis *in vitro* and *in vivo*, which may generate new targets for prevention and treatment of vascular diseases.[154] Human *TIMP-4* cDNA transduced into rat aortic SMCs using adenoviral vector constructs resulted in *TIMP-4* overexpression blocking the conversion of pro-MMP-2 to the active form and markedly inhibiting basic FGF-induced migration. Overexpression of *TIMP-4* significantly increased apoptotic cell death without changing their proliferation. Importantly, overexpression of human *TIMP-4* in the wall of balloon-injured rat carotid artery also increased SMC apoptosis.

Cell Transplant

Early EPCs and late EPCs, also called outgrowth endothelial cells (OECs), have been shown to have angiogenic properties. OECs are like mature endothelial cells in phenotype but have proliferative, migrating, and tube-forming capabilities.[155] On the other hand, OECs are different from mature endothelial cells in terms of caveolae,[156] integrin expression,[157] oxidative stress resistance,[158] and angiogenic potency *in vivo*.[159,160] In contrast to OECs, early EPCs secrete large amounts of

VEGF and IL-8. Both of these cytokines are proangiogenic molecules that increase endothelial cell proliferation, tube formation, and migration.[161,162] Rehman and associates[163] reported that EPCs isolated from human blood are primarily of monocyte/macrophage origin and are capable of secreting angiogenic peptides, including VEGF and IL-8. That EPCs have a limited proliferative potential was previously shown by Murasawa and associates,[164] a finding not consistent with *in vitro* generation of highly proliferative OECs.[165] IL-8 and VEGF are also known to increase MMP secretion in endothelial cells.[166,167] Yoon and associates[168] using athymic nude mice with hind limb ischemia, demonstrated that transplantation of a mixture of these cells (EPCs and OECs) resulted in synergistic neovascularization through cytokines and MMPs. Early EPCs seem to contribute to neovascularization mainly by secretion of cytokines and MMP-9, whereas OECs differentiated into endothelial cells and contributed to angiogenesis by providing building blocks and secreting MMP-2. These findings are of significance in relation to future work on stem cells for tissue regeneration, because synergistic interaction may also occur in other types of stem or progenitor cells.

Recently, the effects of intramyocardial injection of hepatocyte growth factor (HGF), a novel angiogenic factor, have been compared with those of transmyocardial laser revascularization (TMLR) by evaluating the improvement in regional blood flow and regional function in a canine heart model of chronic ischemia.[169] The intramural injection of recombinant human HGF resulted in therapeutic angiogenesis and is likely a better alternative to TMLR in the treatment of chronic ischemic heart disease. Furthermore, overexpression of HGF resulted in enhanced angiogenesis 3 weeks after MI in a rat model of experimental MI, as well as promoting significantly reduced myocyte apoptosis, greater preservation of ventricular geometry, and preservation of cardiac contractile function, suggesting that HGF could also prevent postinfarction HF.[170]

Although enhanced angiogenesis, mediated by several factors mentioned previously, seems a viable and an attractive approach at least in animal models of CAD, the opposite (i.e., inhibition of angiogenesis in the treatment of atherosclerotic disease) is rather questionable. It has been reported that antiangiogenic therapies may prevent the growth of atherosclerotic and neointimal lesions, particularly in vein graft stenosis and restenosis after angioplasty, although any benefit is likely to be nullified by the harmful effects of inhibiting endothelial function and regeneration.[171] Although the main paradigm for VEGF cardiovascular therapy is the stimulation of "therapeutic angiogenesis" in ischemic myocardial and peripheral vascular limb disease, VEGF also acts as a vascular protective factor by means of increased NO and PGI2 production, maintenance of the antiapoptotic signaling pathways to enhance endothelial integrity, inhibit SMC proliferation, and enhance the antithrombogenic, and anti-inflammatory properties of the endothelium. This vascular protection paradigm may be mainly relevant in cases in which stenosis occurs in previously normal vessels characterized by relatively undamaged or undiseased endothelia.[172] In addition, in hypercholesterolemic rabbits, periadventitial *VEGF* gene delivery inhibits intimal thickening, macrophage accumulation, and endothelial VCAM-1 expression induced by collar placement around the carotid artery.[173] VEGF may exert a local arterioprotective effect in the presence of high blood cholesterol, which may be therapeutically valuable for the treatment of vasculoproliferative disease. We agree with Khurana and associates[171] that the many biological roles that VEGF plays and the importance of endothelial integrity for vascular function are both strong arguments against an antiangiogenic approach for the treatment of CVD. Proangiogenic therapy for ischemic heart disease seems to be a more desirable approach that is waiting to be validated by solid clinical studies.

CONCLUSION

Although neovascularization or the generation of new blood vessels plays an essential role in embryonic growth and development, as well as in a number of human pathological conditions, the role of therapeutic angiogenesis in CVDs, such as ischemia and atherosclerosis, is not yet completely established. Modulation of neovascularization in CVDs is an extremely attractive goal, but the role that angiogenesis (either pro- or anti-angiogenesis) plays in cardiovascular disease is still an unresolved issue requiring more unbiased research. Therapies that stimulate or inhibit neovascularization have provided important insights into the diverse role that growth factors play in angiogenesis, as well as on their undesirable side effects, such as tumor growth with proangiogenic drugs. *Therapeutic angiogenesis* is the terminology that has been adopted when using cytokines and growth factors for the development of collateral vessels in cardiac and limb ischemia;[174] however, despite promising results obtained with animals models, available data from human clinical trials are rather disappointing.

Although cytokine therapy has been regarded as an attractive approach, both to treat ischemic heart disease and to enhance the arterioprotective functions of the endothelium, we should keep in mind that neovascularization may contribute to the growth of atherosclerotic lesions and may also be a key factor in plaque destabilization and, finally, plaque rupture. Although there is evidence supporting a role for angiogenesis and angiogenic factors in atherosclerosis and neointima formation, significant questions raised by some landmark clinical studies, together with the problematic suitability of some of the animal models of atherosclerosis and neointimal thickening used, indicate that further research must be carried out before pro-angiogenic therapy can be successfully used in human. Furthermore, because a number of studies have been focused on the specific role of VEGF in angiogenesis, it is important to consider VEGF in the wider context of its biology, as well as the collected experience from clinical trials of VEGF together with other angiogenic cytokines in ischemic heart disease. Even though angiogenesis may contribute to neointimal growth, it is not a necessary requirement for the initiation of intimal thickening. It remains unclear, however, whether angiogenesis plays a central role in the development of atherosclerosis or is responsible for plaque instability. Although data from clinical trials of both pro-angiogenic and anti-angiogenic therapies do not suggest that inhibition of angiogenesis is likely to be a viable therapeutic strategy in the treatment of ischemic heart disease in humans, pro-angiogenic therapies look like a more promising venue. Notwithstanding, comprehensive and unbiased clinical trials must be the prerequisite for its final application in human.

SUMMARY

- Angiogenesis is a complex process involving the participation and communication of multiple cell types and is elementary to both physiological and pathological processes.
- Blood vessels will develop by *vasculogenesis*, a process that entails the differentiation of precursors cells (angioblasts) to endothelial cells forming a complex and tortuous structure.
- Because a variety of transcription factors are expressed in endothelial cells during angiogenesis, the cross-regulation of transcription factors is important. The processes encompass the evolution from progenitor endothelial cells to mature endothelial cells and tube formation in which SMCs play a key role.
- VEGF receptor 2 (VEGF-R2) and basic FGF (bFGF) also play key roles in the differentiation of angioblasts. Transgenic overexpression of VEGF alone in mice results in increased numbers of primarily leaky vascular vessels with tissue edema and inflammation, suggesting that VEGF needs to work in conjunction with other angiogenic factors to produce a healthy vasculature.
- Hypoxia-inducible factor-1 mediates the angiogenic response to hypoxia by up-regulating the expression of multiple angiogenic factors.
- Overexpression of a DNA plasmid encoding HIF-1α/VP16, a constitutively stable hybrid of HIF-1α, increases angiogenesis and blood supply in animal models of ischemic hind limb and myocardium.
- Transgenic mice overexpressing a combination of VEGF and angiopoietin-1 exhibit marked induction of hypervascularity without excessive permeability highlighting the importance of Ang-1 in the formation of normal vessels.
- Although the angiogenic mechanism and therapeutic potential of PDGF-CC, a recently discovered member of the VEGF/PDGF superfamily, remain incompletely characterized, PDGF-CC mobilizes endothelial progenitor cells in ischemic conditions, induces differentiation of bone marrow cells into endothelial cells, and stimulates their migration.
- FOXO transcription factors may play an important role in the regulation of vessel formation in the adult.
- Data support an important paracrine role for endogenously produced NO in endothelial cell migration and differentiation *in vitro* and suggest that the cell-based eNOS gene transfer may be a useful approach to increase new blood vessel formation *in vivo*.
- The focal adhesion kinase (FAK) plays a significant role in angiogenesis and in the vascular development of late embryogenesis.
- Mitochondrial generation of H_2O_2 by SOD2 promoted endothelial cell sprouting in a three-dimensional *in vitro* angiogenesis assay that was attenuated by catalase coexpression or the PI3K inhibitor LY2949002.
- Arteriogenesis is stimulated by hemodynamics forces and is mainly dependent on shear stress and local activation of endothelium; however, other undefined factors may play roles as well.
- New vascularization after cardiac ischemia is induced by infusion of progenitor endothelial cells (EPCs); however, the mechanisms of homing of EPCs to sites of ischemia are not known.
- With aging, there is an obsolescence and functional impairment of EPCs that normally repair and rejuvenate the arteries and that correlate with risk factors for atherosclerosis and CAD.
- Changes in cell growth, cell migration, cell death, and in extracellular matrix composition, all occurring in remodeling, lead to a compensatory adjustment in vessel diameter and lumen area.
- Negative vascular modulators may have a function in embryogenesis and in the revascularization process that follows AMI.
- Serine protease inhibitors, termed serpins, are important regulators of numerous biological pathways, including angiogenesis.
- Arterial injury such as occurs in patients with CAD undergoing percutaneous coronary intervention or placement of intravascular stents frequently results in restenosis of the lumen because of constrictive remodeling of the vessel wall and intimal accumulation of smooth-muscle cells and matrix.
- Early growth response factor-1 controls the expression of a number of genes involved in the pathogenesis of atherosclerosis and postangioplasty restenosis.
- Several transgenic mouse models of perturbed vascular development have been developed over the past several years.
- Endothelial cells lining the blood vessels differ among different tissues, and molecular profiling of specific markers for the cardiac vasculature could be very use-

ful for targeting specific drugs. Mice heart vasculature has been profiled by phage screening and by matching heart-homing peptides against a heart cDNA library in a bacterial two-hybrid system.
- The availability of adult progenitor cells with the capacity to differentiate into endothelial and cardiomyocyte phenotypes promises further progress in achieving the vascularization and repair of ischaemic tissues and injured blood vessels
- PDGF-CC stimulates vessel growth and improves blood flow in mouse models of myocardial and hind limb ischemia.
- Tissue inhibitor of metalloproteinases-4 (TIMP-4) can inhibit SMC migration and induce apoptosis *in vitro* and *in vivo*, which may generate new targets for prevention and treatment of vascular diseases.
- Transplantation of a mixture of these early EPCs and late EPCs has resulted in synergistic neovascularization through cytokines and matrix metalloproteinases.
- The intramural injection of recombinant human hepatocyte growth factor resulted in therapeutic angiogenesis and appears to be a better alternative to transmyocardial laser revascularization in the treatment of chronic ischemic heart disease.
- Proangiogenic therapy for ischemic heart disease seems to be a desirable approach that is waiting to be validated by solid clinical studies.

References

1. Simons, M. (2005). Angiogenesis: where do we stand now? *Circulation* **111,** 1556–1566.
2. Carmeliet, P. (2000). Mechanisms of angiogenesis and arteriogenesis. *Nat. Med.* **6,** 389–395.
3. Risau, W. (1997). Mechanisms of angiogenesis. *Nature* **386,** 671–674.
4. Fong, G. H., Rossant, J., Gertsenstein, M., and Breitman, M. L. (1995). Role of the Flt-1 receptor tyrosine kinase in regulating the assembly of vascular endothelium. *Nature* **376,** 66–70.
5. Larcher, F., Murillas, R., Bolontrade, M., Conti, C. J., and Jorcano, J. L. (1998). VEGF/VPF overexpression in skin of transgenic mice induces angiogenesis, vascular hyperpermeability and accelerated tumor development. *Oncogene* **17,** 303–311.
6. Thurston, G., Suri, C., Smith, K., McClain, J., Sato, T. N., Yancopoulos, G. D., and McDonald, D. M. (1999). Leakage-resistant blood vessels in mice transgenically overexpressing angiopoietin-1. *Science* **286,** 2511–2514.
7. Blau, H. M., and Banfi, A. (2001). The well-tempered vessel. *Nat. Med.* **7,** 532–534.
8. Folkman, J., and D'Amore, P. A. (1996). Blood vessel formation: what is its molecular basis? *Cell* **87,** 1153–1155.
9. Wang, G. L., and Semenza, G. L. (1995). Purification and characterization of hypoxia-inducible factor 1. *J. Biol. Chem.* **270,** 1230–1237.
10. Semenza, G. L. (1999). Regulation of mammalian O_2 homeostasis by hypoxia-inducible factor 1. *Annu. Rev. Cell Dev. Biol.* **15,** 551–578.
11. Van, M., Kondo, K., Yang, H., Kim, W., Valiando, J., Ohh, M., Salic, A., Asara, J. M., Lane, W. S., and Kaelin, W. G., Jr. (2001). HIF-1α targeted for VHL-mediated destruction by proline hydroxylation: implications for O_2 sensing. *Science* **292,** 464–468.
12. Jaakkola, P., Mole, D. R., Tian, Y. M., Wilson, M. I., Gielbert, J., Gaskell, S. J., Kriegsheim, A. V., Hebestreit, H. F., Mukherji, M., Schofield, C. J., Maxwell, P. H., Pugh, C. W., and Ratcliffe, P. J. (2001). Targeting of HIF-α to the von Hippel-Lindau ubiquitylation complex by O_2-regulated prolyl hydroxylation. *Science* **292,** 468–472.
13. Epstein, A. C., Gleadle, J. M., McNeill, L. A., Hewitson, K. S., O'Rourke, J., Mole, D. R., Mukherji, M., Metzen, E., Wilson, M. I., Dhanda, A., Tian, Y. M., Masson, N., Hamilton, D. L., Jaakkola, P., Barstead, R., Hodgkin, J., Maxwell, P. H., Pugh, C. W., Schofield, C. J., and Ratcliffe, P. J. (2001). *C. elegans* EGL-9 and mammalian homologs define a family of dioxygenases that regulate HIF by prolyl hydroxylation. *Cell* **107,** 43–54.
14. Jiang, B. H., Semenza, G. L., Bauer, C., and Marti, H. H. (1996). Hypoxia-inducible factor 1 levels vary exponentially over a physiologically relevant range of O_2 tension. *Am. J. Physiol.* **271,** C1172–C1180.
15. Iyer, N. V., Kotch, L. E., Agani, F., Leung, S. W., Laughner, E., Wenger, R. H., Gassmann, M., Gearhart, J. D., Lawler, A. M., Yu, A. Y., and Semenza, G. L. (1998). Cellular and developmental control of O_2 homeostasis by hypoxia-inducible factor 1α. *Genes Dev.* **12,** 149–162.
16. Vincent, K. A., Shyu, K., Luo, Y., Magner, M., Tio, R. A., Jiang, C., Goldberg, M. A., Akita, G. Y., Gregory, R. J., and Isner, J. M. (2000). Angiogenesis is induced in a rabbit model of hind limb ischemia by naked DNA encoding a HIF-1α/VP16 hybrid transcriptional factor. *Circulation* **102,** 2255–2261.
17. Shyu, K. G., Wang, M. T., Wang, B. W., Chang, C. C., Leu, J. G., Kuan, P., and Chang, H. (2002). Intramyocardial injection of naked DNA encoding HIF-1α/VP16 hybrid to enhance angiogenesis in an acute myocardial infarction model in the rat. *Cardiovasc. Res.* **54,** 576–583.
18. Elson, D. A., Thurston, G., Huang, L. E., Ginzinger, D. G., McDonald, D. M., Johnson, R. S., and Arbeit, J. M. (2001). Induction of hypervascularity without leakage or inflammation in transgenic mice overexpressing hypoxia-inducible factor–1α. *Genes Dev.* **15,** 2520–2532.
19. Davis, S., Aldrich, T. H., Jones, P. F., Acheson, A., Compton, D. L., Jain, V., Ryan, T. E., Bruno, J., Radziejewski, C., Maisonpierre, P. C., and Yancopoulos, G. D. (1996). Isolation of angiopoietin-1, a ligand for the TIE2 receptor, by secretion-trap expression cloning. *Cell* **87,** 1161–1169.
20. Suri, C., Jones, P. F., Patan, S., Bartunkova, S., Maisonpierre, P. C., Davis, S., Sato, T. N., and Yancopoulos, G. D. (1996). Requisite role of angiopoietin-1, a ligand for the TIE2 receptor, during embryonic angiogenesis. *Cell* **87,** 1171–1180.
21. Maisonpierre, P. C., Suri, C., Jones, P. F., Bartunkova, S., Wiegand, S. J., Radziejewski, C., Compton, D., McClain, J., Aldrich, T. H., Papadopoulos, N., Daly, T. J., Davis, S., Sato, T. N., and Yancopoulos, G. D. (1997). Angiopoietin-2, a natural antagonist for Tie2 that disrupts in vivo angiogenesis. *Science* **277,** 55–60.
22. Holash, J., Maisonpierre, P. C., Compton, D., Boland, P., Alexander, C. R., Zagzag, D., Yancopoulos, G. D., and Wiegand, S. J. (1999). Vessel cooption, regression, and growth

in tumors mediated by angiopoietins and VEGF. *Science* **284,** 1994–1998.

23. Holash, J., Wiegand, S. J., and Yancopoulos, G. D. (1999). New model of tumor angiogenesis: dynamic balance between vessel regression and growth mediated by angiopoietins and VEGF. *Oncogene* **18,** 5356–5362.

24. Davis, S., and Yancopoulos, G. D. (1999). The angiopoietins: yin and yang in angiogenesis. *Curr. Top. Microbiol. Immunol.* **237,** 173–185.

25. Asahara, T., Chen, D., Takahashi, T., Fujikawa, K., Kearney, M., Magner, M., Yancopoulos, G. D., and Isner, J. M. (1998). Tie2 receptor ligands, angiopoietin-1 and angiopoietin-2, modulate VEGF-induced postnatal neovascularization. *Circ. Res.* **83,** 233–240.

26. Visconti, R. P., Richardson, C. D., and Sato, T. N. (2002). Orchestration of angiogenesis and arteriovenous contribution by angiopoietins and vascular endothelial growth factor (VEGF). *Proc. Natl. Acad. Sci. USA* **99,** 8219–8224.

27. Mandriota, S. J., and Pepper, M. S. (1998). Regulation of angiopoietin-2 mRNA levels in bovine microvascular endothelial cells by cytokines and hypoxia. *Circ. Res.* **83,** 852–859.

28. Krikun, G., Schatz, F., Finlay, T., Kadner, S., Mesia, A., Gerrets, R., and Lockwood, C. J. (2000). Expression of angiopoietin-2 by human endometrial endothelial cells: regulation by hypoxia and inflammation. *Biochem. Biophys. Res. Commun.* **275,** 159–163.

29. Willam, C., Koehne, P., Jurgensen, J. S., Grafe, M., Wagner, K. D., Bachmann, S., Frei, U., and Eckardt, K. U. (2000). Tie2 receptor expression is stimulated by hypoxia and proinflammatory cytokines in human endothelial cells. *Circ. Res.* **87,** 370–377.

30. Ray, P. S., Estrada-Hernandez, T., Sasaki, H., Zhu, L., and Maulik, N. (2000). Early effects of hypoxia/reoxygenation on VEGF, Ang-1, Ang-2 and their receptors in the rat myocardium: implications for myocardial angiogenesis. *Mol. Cell Biochem.* **213,** 145–153.

31. Pichiule, P., Chavez, J. C., and LaManna, J. C. (2004). Hypoxic regulation of angiopoietin-2 expression in endothelial cells. *J. Biol. Chem.* **279,** 12171–12180.

32. Yamakawa, M., Liu, L. X., Date, T., Belanger, A. J., Vincent, K. A., Akita, G. Y., Kuriyama, T., Cheng, S. H., Gregory, R. J., and Jiang, C. (2003). Hypoxia-inducible factor-1 mediates activation of cultured vascular endothelial cells by inducing multiple angiogenic factors. *Circ. Res.* **93,** 664–673.

33. Forsythe, J. A., Jiang, B. H., Iyer, N. V., Agani, F., Leung, S. W., Koos, R. D., and Semenza, G. L. (1996). Activation of vascular endothelial growth factor gene transcription by hypoxia-inducible factor 1. *Mol. Cell Biol.* **16,** 4604–4613.

34. Schafer, G., Cramer, T., Suske, G., Kemmner, W., Wiedenmann, B., and Hocker, M. (2003). Oxidative stress regulates vascular endothelial growth factor-A gene transcription through Sp1- and Sp3-dependent activation of two proximal GC-rich promoter elements. *J. Biol. Chem.* **278,** 8190–8198.

35. Sen, C. K., Khanna, S., Babior, B. M., Hunt, T. K., Ellison, E. C., and Roy, S. (2002). Oxidant-induced vascular endothelial growth factor expression in human keratinocytes and cutaneous wound healing. *J. Biol. Chem.* **277,** 33284–33290.

36. Shie, J. L., Wu, G., Wu, J., Liu, F. F., Laham, R. J., Oettgen, P., and Li, J. (2004). RTEF-1, a novel transcriptional stimulator of vascular endothelial growth factor in hypoxic endothelial cells. *J. Biol. Chem.* **279,** 25010–25016.

37. Duan, L. J., Zhang-Benoit, Y., and Fong, G. H. (2005). Endothelium-intrinsic requirement for Hif-2α during vascular development. *Circulation* **111,** 2227–2232.

38. Distler, J. H., Hirth, A., Kurowska-Stolarska, M., Gay, R. E., Gay, S., and Distler, O. (2003). Angiogenic and angiostatic factors in the molecular control of angiogenesis. *Q. J. Nucl. Med.* **47,** 149–161.

39. Conway, E. M., Collen, D., and Carmeliet, P. (2001). Molecular mechanisms of blood vessel growth. *Cardiovasc. Res.* **49,** 507–521.

40. Hirschi, K. K., and D'Amore, P. A. (1996). Pericytes in the microvasculature. *Cardiovasc. Res.* **32,** 687–698.

41. Cao, R., Brakenhielm, E., Li, X., Pietras, K., Widenfalk, J., Ostman, A., Eriksson, U., and Cao, Y. (2002). Angiogenesis stimulated by PDGF-CC, a novel member in the PDGF family, involves activation of PDGFR-$\alpha\alpha$ and -$\alpha\beta$ receptors. *FASEB J.* **16,** 1575–1583.

42. Li, X., Tjwa, M., Moons, L., Fons, P., Noel, A., Ny, A., Zhou, J. M., Lennartsson, J., Li, H., Luttun, A., Ponten, A., Devy, L., Bouche, A., Oh, H., Manderveld, A., Blacher, S., Communi, D., Savi, P., Bono, F., Dewerchin, M., Foidart, J. M., Autiero, M., Herbert, J. M., Collen, D., Heldin, C. H., Eriksson, U., and Carmeliet, P. (2005). Revascularization of ischemic tissues by PDGF-CC via effects on endothelial cells and their progenitors. *J. Clin. Invest.* **115,** 118–127.

43. Potente, M., Urbich, C., Sasaki, K. I., Hofmann, W. K., Heeschen, C., Aicher, A., Kollipara, R., Depinho, R. A., Zeiher, A. M., and Dimmeler, S. (2005). Involvement of Foxo transcription factors in angiogenesis and postnatal neovascularization. *J. Clin. Invest.* **115,** 2382–2392.

44. Hosaka, T., Biggs, W. H., 3rd., Tieu, D., Boyer, A. D., Varki, N. M., Cavenee, W. K., and Arden, K. C. (2004). Disruption of forkhead transcription factor (FOXO) family members in mice reveals their functional diversification. *Proc. Natl. Acad. Sci. USA* **101,** 2975–2980.

45. Ilic, D., Furuta, Y., Kanazawa, S., Takeda, N., Sobue, K., Nakatsuji, N., Nomura, S., Fujimoto, J., Okada, M., and Yamamoto, T. (1995). Reduced cell motility and enhanced focal adhesion contact formation in cells from FAK-deficient mice. *Nature* **377,** 539–544.

46. Shen, T. L., Park, A. Y., Alcaraz, A., Peng, X., Jang, I., Koni, P., Flavell, R. A., Gu, H., and Guan, J. L. (2005). Conditional knockout of focal adhesion kinase in endothelial cells reveals its role in angiogenesis and vascular development in late embryogenesis. *Cell Biol.* **169,** 941–952.

47. Peng, X., Ueda, H., Zhou, H., Stokol, T., and Shen, T. L. A. (2004). Overexpression of focal adhesion kinase in vascular endothelial cells promotes angiogenesis in transgenic mice. *Cardiovasc. Res.* **64,** 421–430.

48. Chen, K. D., Li, Y. S., Kim, M., Li, S., Yuan, S., Chien, S., and Shyy, J. Y. (1999). Mechanotransduction in response to shear stress: roles of receptor tyrosine kinases, integrins, and Shc. *J. Biol. Chem.* **274,** 18393–18400.

49. Jin, Z. G., Ueba, H., Tanimoto, T., Lungu, A. O., Frame, M. D., and Berk, B. C. (2003). Ligand-independent activation of vascular endothelial growth factor receptor 2 by fluid shear stress regulates activation of endothelial nitric oxide synthase. *Circ. Res.* *93,* 354–363.

50. Shyy, J. Y., and Chien, S. (2002). Role of integrins in endothelial mechanosensing of shear stress. *Circ. Res.* **91**, 769–775.
51. Muller, J. M., Davis, M. J., and Chilian, W. M. (1996). Coronary arteriolar flow-induced vasodilation signals through tyrosine kinase. *Am. J. Physiol.* **270**, H1878–H1884.
52. Dimmeler, S., Fleming, I., Fisslthaler, B., Hermann, C., Busse, R., and Zeiher, A. M. (1999). Activation of nitric oxide synthase in endothelial cells by Akt-dependent phosphorylation. *Nature* **399**, 601–605.
53. Corson, M. A., James, N. L., Latta, S. E., Nerem, R. M., Berk, B. C., and Harrison, D. G. (1996). Phosphorylation of endothelial nitric oxide synthase in response to fluid shear stress. *Circ. Res.* **79**, 984–991.
54. Martinez-Lemus, L. A., Sun, Z., Trache, A., Trzciakowski, J. P., and Meininger, G. A. (2005). Integrins and regulation of the microcirculation: from arterioles to molecular studies using atomic force microscopy. *Microcirculation* **12**, 99–112.
55. Ali, M. H., and Schumacker, P. T. (2002). Endothelial responses to mechanical stress: where is the mechanosensor? *Crit. Care Med.* **30**, S198–S206.
56. Koshida, R., Rocic, P., Saito, S., Kiyooka, T., Zhang, C., and Chilian, W. M. (2005). Role of focal adhesion kinase in flow-induced dilation of coronary arterioles. *Arterioscler. Thromb. Vasc. Biol.* **25**, 2548–2553.
57. Ayadi, A., Suelves, M., Dolle, P., and Wasylyk, B. (2001). Net, an Ets ternary complex transcription factor, is expressed in sites of vasculogenesis, angiogenesis, and chondrogenesis during mouse development. *Mech. Dev.* **102**, 205–208.
58. Zheng, H., Wasylyk, C., Ayadi, A., Abecassis, J., Schalken, J. A., Rogatsch, H., Wernert, N., Maira, S. M., Multon, M. C., and Wasylyk, B. (2003). The transcription factor Net regulates the angiogenic switch. *Genes Dev.* **17**, 2283–2297.
59. Dulak, J., Jozkowicz, A., Dembinska-Kiec, A., Guevara, I., Zdzienicka, A., Zmudzinska-Grochot, D., Florek, I., Wojtowicz, A., Szuba, A., and Cooke, J. P. (2000). Nitric oxide induces the synthesis of vascular endothelial growth factor by rat vascular smooth muscle cells. *Arterioscler. Thromb. Vasc. Biol.* **20**, 659–666.
60. Babaei, S., and Stewart, D. J. (2002). Overexpression of endothelial NO synthase induces angiogenesis in a co-culture model. *Cardiovasc. Res.* **55**, 190–200.
61. Davis, M. E., Cai, H., Drummond, G. R., and Harrison, D. G. (2001). Shear stress regulates endothelial nitric oxide synthase expression through c-Src by divergent signaling pathways. *Circ. Res.* **89**, 1073–1080.
62. Cieslik, K., Abrams, C. S., and Wu, K. K. (2001). Up-regulation of endothelial nitric-oxide synthase promoter by the phosphatidylinositol 3-kinase γ/Janus kinase 2/MEK-1-dependent pathway. *J. Biol. Chem.* **276**, 1211–1219.
63. Grumbach, I. M., Chen, W., Mertens, S. A., and Harrison, D. G. (2005). A negative feedback mechanism involving nitric oxide and nuclear factor χ-B modulates endothelial nitric oxide synthase transcription. *J. Mol. Cell Cardiol.* **39**, 595–603.
64. Hamada, K., Sasaki, T., Koni, P. A., Natsui, M., Kishimoto, H., Sasaki, J., Yajima, N., Horie, Y., Hasegawa, G., Naito, M., Miyazaki, J., Suda, T., Itoh, H., Nakao, K., Mak, T. W., Nakano, T., and Suzuki, A. (2005). The PTEN/PI3K pathway governs normal vascular development and tumor angiogenesis. *Genes Dev.* **19**, 2054–2065.
65. Pore, N., Liu, S., Haas-Kogan, D. A., O'Rourke, D. M., and Maity, A. (2003). PTEN mutation and epidermal growth factor receptor activation regulate vascular endothelial growth factor (VEGF) mRNA expression in human glioblastoma cells by transactivating the proximal VEGF promoter. *Cancer Res.* **63**, 236–2.41
66. Gomez-Manzano, C., Fueyo, J., Jiang, H., Glass, T. L., Lee, H. Y., Hu, M., Liu, J. L., Jasti, S. L., Liu, T. J., Conrad, C. A., and Yung, W. K. (2003). Mechanisms underlying PTEN regulation of vascular endothelial growth factor and angiogenesis. *Ann. Neurol.* **53**, 109–117.
67. Tanaka, M., and Grossman, H. B. (2003). In vivo gene therapy of human bladder cancer with PTEN suppresses tumor growth, downregulates phosphorylated Akt, and increases sensitivity to doxorubicin. *Gene Ther.* **10**, 1636–1642.
68. Mischel, P. S., and Cloughesy, T. F. (2003). Targeted molecular therapy of GBM. *Brain Pathol.* **13**, 52–61.
69. Connor, K. M., Subbaram, S., Regan, K. J., Nelson, K. K., Mazurkiewicz, J. E., Bartholomew, P. J., Aplin, A. E., Tai, Y. T., Aguirre-Ghiso, J., Flores, S. C., and Melendez, J. A. (2005). Mitochondrial H_2O_2 regulates the angiogenic phenotype via PTEN oxidation. *J. Biol. Chem.* **280**, 16916–16924.
70. Resnick, N., Einav, S., Chen-Konak, L., Zilberman, M., Yahav, H., and Shay-Salit, A. (2003). Hemodynamic forces as a stimulus for arteriogenesis. *Endothelium* **10**, 197–206.
71. Helisch, A., and Schaper, W. (2003). Arteriogenesis: the development and growth of collateral arteries. *Microcirculation* **10**, 83–97; Schaper, W., and Scholz, D. (2003). Factors regulating arteriogenesis. *Arterioscler. Thromb. Vasc. Biol.* **23**, 1143–1151.
72. Pipp, F., Heil, M., Issbrucker, K., Ziegelhoeffer, T., Martin, S., van den Heuvel, J., Weich, H., Fernandez, B., Golomb, G., Carmeliet, P., Schaper, W., and Clauss, M. (2003). VEGFR-1-selective VEGF homologue PlGF is arteriogenic: Evidence for a monocyte-mediated mechanism. *Circ. Res.* **92**, 378–385.
73. Henry, T. D., Annex, B. H., McKendall, G. R., Azrin, M. A., Lopez, J. J., Giordano, F. J., Shah, P. K., Willerson, J. T., Benza, R. L., Berman, D. S., Gibson, C. M., Bajamonde, A., Rundle, A. C., Fine, J., and McCluskey, E. R. (2003). The VIVA trial: Vascular endothelial growth factor in ischemia for vascular angiogenesis. *Circulation* **107**, 1359–1365.
74. Simons, M. (2001). Therapeutic coronary angiogenesis: a fronte praecipitium a tergo lupi? *Am. J. Physiol. Heart Circ. Physiol.* **280**, H1923–H1927.
75. van Royen, N., Schirmer, S. H., Atasever, B., Behrens, C. Y., Ubbink, D., Buschmann, E. E., Voskuil, M., Bot, P., Hoefer, I., Schlingemann, R. O., Biemond, B. J., Tijssen, J. G., Bode, C., Schaper, W., Oskam, J., Legemate, D. A., Piek, J. J., and Buschmann, I. (2005). START Trial: a pilot study on STimulation of ARTeriogenesis using subcutaneous application of granulocyte-macrophage colony-stimulating factor as a new treatment for peripheral vascular disease. *Circulation* **112**, 1040–1046.
76. Cao, R., Brakenhielm, E., Pawliuk, R., Wariaro, D., Post, M. J., Wahlberg, E., Leboulch, P., and Cao, Y. (2003). Angiogenic synergism, vascular stability and improvement of hind-limb ischemia by a combination of PDGF-BB and FGF-2. *Nat. Med.* **9**, 604–613.
77. Carmeliet, P., Moons, L., Luttun, A., Vincenti, V., Compernolle, V., De Mol, M., Wu, Y., Bono, F., Devy, L., Beck, H.,

Scholz, D., Acker, T., DiPalma, T., Dewerchin, M., Noel, A., Stalmans, I., Barra, A., Blacher, S., Vandendriessche, T., Ponten, A., Eriksson, U., Plate, K. H., Foidart, J. M., Schaper, W., Charnock-Jones, D. S., Hicklin, D. J., Herbert, J. M., Collen, D., and Persico, M. G. (2001). Synergism between vascular endothelial growth factor and placental growth factor contributes to angiogenesis and plasma extravasation in pathological conditions. *Nat. Med.* **7,** 575–583.

78. Hristov, M., Erl, W., and Weber, P.C. (2003). Endothelial progenitor cells: mobilization, differentiation, and homing. *Arterioscler. Thromb. Vasc. Biol.* **23,** 1185–1189.

79. Ziegelhoeffer, T., Fernandez, B., Kostin, S., Heil, M., Voswinckel, R., Helisch, A., and Schaper, W. (2004). Bone marrow-derived cells do not incorporate into the adult growing vasculature. *Circ. Res.* **94,** 230–238.

80. Kinnaird, T., Stabile, E., Burnett, M. .S, Lee, C. W., Barr, S., Fuchs, S., and Epstein, S. E. (2004). Marrow-derived stromal cells express genes encoding a broad spectrum of arteriogenic cytokines and promote in vitro and in vivo arteriogenesis through paracrine mechanisms. *Circ. Res.* **94,** 678–685.

81. Khmelewski, E., Becker, A., Meinertz, T., and Ito, W. D. (2004). Tissue resident cells play a dominant role in arteriogenesis and concomitant macrophage accumulation. *Circ. Res.* **95,** E56–E64.

82. Herold, J., Pipp, F., Fernandez, B., Xing, Z., Heil, M., Tillmanns, H., and Braun-Dullaeus, R. C. (2004). Transplantation of monocytes: a novel strategy for in vivo augmentation of collateral vessel growth. *Hum. Gene Ther.* **15,** 1–12.

83. Imada, T., Tatsumi, T., Mori, Y., Nishiue, T., Yoshida, M., Masaki, H., Okigaki, M., Kojima, H., Nozawa, Y., Nishiwaki, Y., Nitta, N., Iwasaka, T., and Matsubara, H. (2005). Targeted delivery of bone marrow mononuclear cells by ultrasound destruction of microbubbles induces both angiogenesis and arteriogenesis response. *Arterioscler. Thromb. Vasc. Biol.* **25,** 2128–2134.

84. Sellke, F. W., Wang, S. Y., Stamler, A., Lopez, J. J., Li, J., and Simons, M. (1996). Enhanced microvascular relaxations to VEGF and bFGF in chronically ischemic porcine myocardium. *Am. J. Physiol.* **271,** H713–H720.

85. Ludmer, P. L., Selwyn, A. P., Shook, T. L., Wayne, R. R., Mudge, G. H., Alexander, R. W., and Ganz, P. (1986). Paradoxical vasoconstriction induced by acetylcholine in atherosclerotic coronary arteries. *N. Engl. J. Med.* **315,** 1046–1051.

86. Babaei, S., Teichert-Kuliszewska, K., Monge, J. C., Mohamed, F., Bendeck, M. P., and Stewart, D. J. (1998). Role of nitric oxide in the angiogenic response in vitro to basic fibroblast growth factor. *Circ. Res.* **82,** 1007–1015.

87. Jang, J. J., Ho, H. K., Kwan, H. H., Fajardo, L. F., and Cooke, J. P. (2000). Angiogenesis is impaired by hypercholesterolemia: role of asymmetric dimethylarginine. *Circulation* **102,** 1414–1419.

88. Ruel, M., Wu, G. F., Khan, T. A., Voisine, P., Bianchi, C., Li, J., Laham, R. J., and Sellke, F. W. (2003). Inhibition of the cardiac angiogenic response to surgical FGF-2 therapy in a Swine endothelial dysfunction model. *Circulation* **108,** II335–II340.

89. Voisine, P., Li, J., Bianchi, C., Khan, T. A., Ruel, M., Xu, S. H., Feng, J., Rosinberg, A., Malik, T., Nakai, Y., and Selke, F. W. (2005). Effects of L-arginine on fibroblast growth factor-2 induced angrogenesis in a model of endothelial dysfunction. *Circulation* **30,** 1202–1207.

90. Quyyumi, A. A., Dakak, N., Diodati, J. G., Gilligan, D. M., Panza, J. A., and Cannon, R. O., III. (1997). Effect of L-arginine on human coronary endothelium-dependent and physiologic vasodilation. *J. Am. Coll. Cardiol.* **30,** 1220–1227.

91. Lerman, A., Burnett, J. C., Jr., Higano, S. T., McKinley, L. J., and Holmes, D. R., Jr. (1999). Long-term L-arginine supplementation improves small-vessel coronary endothelial function in humans. *Circulation* **99,** 1648–1649.

92. Pi, X., Garin, G., Xie, L., Zheng, Q., Wei, H., Abe, J., Yan, C., and Berk, B. C. (2005). BMK1/ERK5 is a novel regulator of angiogenesis by destabilizing hypoxia inducible factor 1alpha. BMK1/ERK5 is a novel regulator of angiogenesis by destabilizing hypoxia inducible factor 1alpha. *Circ. Res.* **96,** 1145–1151.

93. Ishizawa, K., Izawa, Y., Ito, H., Miki, C., Miyata, K., Fujita, Y., Kanematsu, Y., Tsuchiya, K., Tamaki, T., Nishiyama, A., and Yoshizumi, M. (2005). Aldosterone stimulates vascular smooth muscle cell proliferation via big mitogen-activated protein kinase 1 activation. *Hypertension* **46,** 1046–10.52

94. Chavakis, E., Aicher, A., Heeschen, C., Sasaki, K., Kaiser, R., El Makhfi, N., Urbich, C., Peters, T., Scharffetter-Kochanek, K., Zeiher, A. M., Chavakis, T., and Dimmeler, S. (2005). Role of β2-integrins for homing and neovascularization capacity of endothelial progenitor cells. *J. Exp. Med.* **201,** 63–72.

95. Urbich, C., Heeschen, C., Aicher, A., Sasaki, K., Bruhl, T., Farhadi, M. R., Vajkoczy, P., Hofmann, W. K., Peters, C., Pennacchio, L. A., Abolmaali, N. D., Chavakis, E., Reinheckel, T., Zeiher, A. M., and Dimmeler, S. (2005). Cathepsin L is required for endothelial progenitor cell-induced neovascularization. *Nat. Med.* **11,** 206–213.

96. Kalka, C., Tehrani, H., Laudernberg, B., Vale, P., Isner, J. M., Asahara, T., and Symes, J. F. (2000). Mobilization of endothelial progenitor cells following gene therapy with VEGF165 in patients with inoperable coronary disease. *Ann. Thorac. Surg.* **70,** 829–834.

97. Shintani, S., Murohara, T., Ikeda, H., Ueni, T., Honma, T., Katoh, A., Sasaki, K., Shimada, T., Oike, Y., and Imaizumi, T. (2001). Mobilization of endothelial progenitor cells in patients with acute myocardial infarction. *Circulation* **103,** 2776–2779.

98. Raffi, S., and Lyden, D. (2003). Therapeutic stem and progenitor cell transplantation for organ vascularization and regeneration. *Nat. Med.* **9,** 702–712.

99. Kong, D., Melo, L. G., Gnecchi, M., Zhang, L., Mostoslavski, G., Liew, C. C., Pratt, R. E., and Dzau, V. J. (2004). Cytokine-induced mobilization of circulating endothelial progenitor cells enhances repair of injured arteries. *Circulation* **110,** 2039–2046.

100. Werner, N., Junk, S., Laufs, L., Link, A., Walenta, K., Bohm, M., and Nickenig, G. (2003). Intravenous transfusion of endothelial progenitor cells reduces neointima formation after vascular injury. *Circ. Res.* **93,** e17–e24.

101. Vasa, M., Fichtlscherer, S., Adler, K., Aicher, A., Martin, H., Zeiher, A. M., and Dimmeler, S. (2001). Increase in circulating endothelial progenitor cells by statin therapy in patients with stable coronary artery disease. *Circulation* **103,** 2885–2890.

102. Walter, D. H., Rittig, K., Bahlmann, F. H., Kirchmair, R., Silver, M., Murayama, T., Nishimura, H., Losordo, D. W., Asahara, T., and Isner, J. M. (2002). Statin therapy accelerates reendothelialization: a novel effect involving mobilization and

incorporation of bone marrow–derived endothelial progenitor cells. *Circulation* **105,** 3017–3024.
103. Strehlow, K., Werner, N., Berweiler, J., Link, A., Dirnagl, U., Priller, J., Laufs, K., Ghaeni, L., Milosevic, M., Bohm, M., and Nickenig, G. (2003). Estrogen increases bone-marrow derived endothelial progenitor cell production and diminishes neointima formation. *Circulation* 107, 3059–3065.
104. Kalka, C., Masuda, H., Takahashi, T., Kalka-Moll, W. M., Silver, M., Kearney, M., Li, T., Isner, J. M., and Asahara, T. (2000). Transplantation of ex vivo expanded endothelial progenitor cells for therapeutic neovascularization. *Proc. Natl. Acad. Sci. USA* **97,** 3422–3427.
105. Kocher, A. A., Schuster, M. D., Szabolcs, M. J., Burkhoff, D., Wang, J., Homma, S., Edwards, N. M., and Itescu, S. (2001). Neovascularization of ischemic myocardium by human bone-marrow–derived angioblasts prevents cardiomyocyte apoptosis, reduces remodeling and improves cardiac function. *Nat. Med.* **7,** 430–436.
106. Griese, D. P., Ehsan, A., Melo, L. G., Kong, D., Zhang, L., Mann, M. J., Pratt, R. E., Mulligan, R. C., and Dzau, V. J. (2003). Isolation and transplantation of autologous circulating endothelial cells into denuded vessels and prosthetic grafts: implications for cell–based vascular therapy. *Circulation* **108,** 2710–2715.
107. Kong, D., Melo, L. G., Mangi, A. A., Zhang, L., Lopez-Ilasaca, M., Perrella, M. A., Liew, C. C., Pratt, R. E., and Dzau, V. J. (2004). Enhanced inhibition of neointimal hyperplasia by genetically engineered endothelial progenitor cells. *Circulation* **109,** 1769–1775.
108. Iwaguro, H., Yamaguchi, J., Kalka, C., Murasawa, S., Masuda, H., Hayashi, S., Silver, M., Li, T., Isner, J. M., and Asahara, T. (2002). Endothelial progenitor cell vascular endothelial growth factor gene transfer for vascular regeneration. *Circulation* **105,** 732–738.
109. Rauscher, F. M., Goldschmidt-Clermont, P. J., Davis, B. H., Wang, T., Gregg, D., Ramaswami, P., Pippen, A. M., Annex, B. H., Dong, C., and Taylor, D. A. (2003). Aging, progenitor cell exhaustion, and atherosclerosis. *Circulation* **108,** 457–463.
110. Vasa, M., Fichtlscherer, S., Aicher, A., Adler, K., Urbich, C., Martin, H., Zeiher, A. M., and Dimmeler, S. (2001). Number and migratory activity of circulating endothelial progenitor cells inversely correlates with risk factors for coronary artery disease. *Circ. Res.* **89,** E1–E7.
111. Cowan, D. B., and Langille, B. L. (1996). Cellular and molecular biology of vascular remodeling. *Curr. Opin. Lipidol.* **7,** 94–100.
112. Ross, R. (1993). The pathogenesis of atherosclerosis: a perspective for the 1990s. *Nature* **362,** 801–809.
113. Myers, C., Charboneau, A., Cheung, I., Hanks, D., and Boudreau, N. (2002). Sustained expression of homeobox D10 inhibits angiogenesis. *Am. J. Pathol.* **161,** 2099–2109.
114. Gorski, D. H., and Walsh, K. (2000). The role of homeobox genes in vascular remodeling and angiogenesis. *Circ. Res.* **87,** 865–872.
115. Wu, J., Iwata, F., Grass, J. A., Osborne, C. S., Elnitski, L., Fraser, P., Ohneda, O., Yamamoto, M., and Bresnick, E. H. (2005). Molecular determinants of NOTCH4 transcription in vascular endothelium. *Mol. Cell Biol.* **25,** 1458–1474.
116. Matsunaga, T., Weihrauch, D. W., Moniz, M. C., Tessmer, J., Warltier, D. C., and Chilian, W. M. (2002). Angiostatin inhibits coronary angiogenesis during impaired production of nitric oxide. *Circulation* **105,** 2185–2191.
117. Matsunaga, T., Chilian, W. M., and March, K. (2005). Angiostatin is negatively associated with coronary collateral growth in patients with coronary artery disease. *Am. J. Physiol. Heart Circ. Physiol.* **288,** H2042–H2046.
118. Panchal, V. R., Rehman, J., Nguyen, A. T., Brown, J. W., Turrentine, M. W., Mahomed, Y., and March, K. L. (2004). Reduced pericardial levels of endostatin correlate with collateral development in patients with ischemic heart disease. *J. Am. Coll. Cardiol.* **43,** 1383–1387.
119. Zhang, F., and Schwarz, M. A. (2000). Temporo-spatial distribution of endothelial-monocyte activating polypeptide II, an anti-angiogenic protein, in the mouse embryo. *Dev. Dyn.* **218,** 490–498.
120. Thompson, J. L., Ryan, J. A., Barr, M. L., Franc, B., Starnes, V. A., and Schwarz, M. A. (2004). Potential role for antiangiogenic proteins in the myocardial infarction repair process. *J. Surg. Res.* **116,** 156–164.
121. Nakagawa, T., Abe, M., Yamazaki, T., Miyashita, H., Niwa, H., Kokubun, S., and Sato, Y. (2003). HEX acts as a negative regulator of angiogenesis by modulating the expression of angiogenesis-related gene in endothelial cells in vitro. *Arterioscler. Thromb. Vasc. Biol.* **23,** 231–237.
122. Sekiguchi, K., Kurabayashi, M., Oyama, Y., Aihara, Y., Tanaka, T., Sakamoto, H., Hoshino, Y., Kanda, T., Yokoyama, T., Shimomura, Y., Iijima, H., Ohyama, Y., and Nagai, R. (2001). Homeobox protein Hex induces SMemb/nonmuscle myosin heavy chain-B gene expression through the cAMP-responsive element. *Circ. Res.* **88,** 52–58.
123. Cheifetz, S., Bellon, T., Cales, C., Vera, S., Bernabeu, C., Massague, J., and Letarte, M. (1992). Endoglin is a component of the transforming growth factor-β receptor system in human endothelial cells. *J. Biol. Chem.* **267,** 19027–19030.
124. Raab, U., Lastres, P., Arevalo, M. A., Lopez-Novoa, J. M., Cabanas, C., de la Rosa, E. J., and Bernabeu, C. (1999). Endoglin is expressed in the chicken vasculature and is involved in angiogenesis. *FEBS Lett.* **459,** 249–254.
125. Li, D. Y., Sorensen, L. K., Brooke, B. S., Urness, L. D., Davis, E. C., Taylor, D. G., Boak, B. B., and Wendel, D. P. (1999). Defective angiogenesis in mice lacking endoglin. *Science* **284,** 1534–1537.
126. Nakagawa, T., Abe, M., Yamazaki, T., Miyashita, H., Niwa, H., Kokubun, S., and Sato, Y. (2003). HEX acts as a negative regulator of angiogenesis by modulating the expression of angiogenesis-related gene in endothelial cells in vitro. *Arterioscler. Thromb. Vasc. Biol.* **23,** 231–237.
127. Freeman, A. F., Crawford, S. E., Cornwall, M. L., Garcia, F. L., Shulman, S. T., and Rowley, A. H. (2005). Angiogenesis in fatal acute Kawasaki disease coronary artery and myocardium. *Pediatr. Cardiol.* **26,** 578–584.
128. Xiang, G., Schuster, M. D., Seki, T., Kocher, A. A., Eshghi, S., Boyle, A., and Itescu, S. (2004). Down-regulation of plasminogen activator inhibitor 1 expression promotes myocardial neovascularization by bone marrow progenitors. *J. Exp. Med.* **200,** 1657–1666.
129. Xiang, G., Schuster, M. D., Seki, T., Witkowski, P., Eshghi, S., and Itescu, S. (2005). Downregulated expression of plasminogen activator inhibitor-1 augments myocardial neovascularization and reduces cardiomyocyte apoptosis

after acute myocardial infarction. *J. Am. Coll. Cardiol.* **46,** 536–541.

130. Dai, E., Guan, H., Liu, L., Little, S., McFadden, G., Vaziri, S., Cao, H., Ivanova, I. A., Bocksch, L., and Lucas, A. (2003). Serp-1, a viral anti-inflammatory serpin, regulates cellular serine proteinase and serpin responses to vascular injury. *J. Biol. Chem.* **278,** 18563–18572.

131. Schwartz, S. M., Reidy, M. A., and O'Brien, E. R. (1995). Assessment of factors important in atherosclerotic occlusion and restenosis. *Thromb. Haemost.* **74,** 541–551.

132. Schwartz, R. S., Chronos, N. A., and Virmani, R. (2004). Preclinical restenosis models and drug-eluting stents: still important, still much to learn. *J. Am. Coll. Cardiol.* **44,** 1373–1385.

133. Voisine, P., Bianchi, C., Ruel, M., Malik, T., Rosinberg, A., Feng, J., Khan, T. A., Xu, S. H., Sandmeyer, J., Laham, R. J., and Sellke, F. W. (2004). Inhibition of the cardiac angiogenic response to exogenous vascular endothelial growth factor. *Surgery* **136,** 407–415.

134. Metais, C., Li, J., Simons, M., and Sellke, F. W. (1998). Effects of coronary artery disease on expression and microvascular response to VEGF. *Am. J. Physiol.* **275,** H1411–H1418.

135. Bussolati, B., Dunk, C., Grohman, M., Kontos, C. D., Mason, J., and Ahmed, A. (2001). Vascular endothelial growth factor receptor-1 modulates vascular endothelial growth factor-mediated angiogenesis via nitric oxide. *Am. J. Pathol.* **159,** 993–1000.

136. Lowe, H. C., Fahmy, R. G., Kavurma, M. M., Baker, A., Chesterman, C. N., and Khachigian, L. M. (2001). Catalytic oligodeoxynucleotides define a key regulatory role for early growth response factor-1 in the porcine model of coronary in-stent restenosis. *Circ. Res.* **89,** 670–677.

137. Lowe, H. C., Chesterman, C. N., and Khachigian, L. M. (2002). Catalytic antisense DNA molecules targeting Egr-1 inhibit neointima formation following permanent ligation of rat common carotid arteries. *Thromb. Haemost.* **87,** 134–140.

138. Carmeliet, P., Moons, L., and Collen, D. (1998). Mouse models of angiogenesis, arterial stenosis, atherosclerosis and hemostasis. *Cardiovasc. Res.* **39,** 8–33.

139. Zhang, L., Hoffman, J. A., and Ruoslahti, E. (2005). Molecular profiling of heart endothelial cells. *Circulation* **112,** 1601–1611.

140. Ruoslahti, E. (2002). Specialization of tumour vasculature. *Nat. Rev. Cancer* **2,** 83–90.

141. Arap, W., Pasqualini, R., and Ruoslahti, E. (1998). Cancer treatment by targeted drug delivery to tumor vasculature in a mouse model. *Science* **279,** 377–380.

142. Laakkonen, P., Akerman, M. E., Biliran, H., Yang, M., Ferrer, F., Karpanen, T., Hoffman, R. M., and Ruoslahti, E. (2004). Antitumor activity of a homing peptide that targets tumor lymphatics and tumor cells. *Proc. Natl. Acad. Sci. USA* **101,** 9381–9386.

143. Rajesh, M., Kolmakova, A., and Chatterjee, S. (2005). Novel role of lactosylceramide in vascular endothelial growth factor-mediated angiogenesis in human endothelial cells. *Circ. Res.* **97,** 796–804.

144. Betsholtz, C. (2004). Insight into the physiological functions of PDGF through genetic studies in mice. *Cytokine Growth Factor Rev.* **15,** 215–228.

145. Cao, R., Brakenhielm, E., Li, X., Pietras, K., Widenfalk, J., Ostman, A., Eriksson, U., and Cao, Y. (2002). Angiogenesis stimulated by PDGF-CC, a novel member in the PDGF family, involves activation of PDGFR-alphaalpha and -alphabeta receptors. *FASEB J.* **16,** 1575–1583.

146. Gilbertson, D. G., Duff, M. E., West, J. W., Kelly, J. D., Sheppard, P. O., Hofstrand, P. D., Gao, Z., Shoemaker, K., Bukowski, T. R., Moore, M., Feldhaus, A. L., Humes, J. M., Palmer, T. E., and Hart, C. E. (2001). Platelet-derived growth factor C (PDGF-C), a novel growth factor that binds to PDGF alpha and beta receptor. *J. Biol. Chem.* **276,** 27406–27414.

147. Li, X., Ponten, A., Aase, K., Karlsson, L., Abramsson, A., Uutela, M., Backstrom, G., Hellstrom, M., Bostrom, H., Li, H., Soriano, P., Betsholtz, C., Heldin, C. H., Alitalo, K., Ostman, A., and Eriksson, U. (2000). PDGF-C is a new protease-activated ligand for the PDGF alpha-receptor. *Nat. Cell Biol.* **2,** E78–79.

148. Uutela, M., Lauren, J., Bergsten, E., Li, X., Horelli-Kuitunen, N., Eriksson, U., and Alitalo, K. (2001). Chromosomal location, exon structure, and vascular expression patterns of the human PDGFC and PDGFD genes. *Circulation* **103,** 2242–2247.

149. Li, H., Fredriksson, L., Li, X., and Eriksson, U. (2003). PDGF-D is a potent transforming and angiogenic growth factor. *Oncogene* **22,** 1501–1510.

150. Li, X., Tjwa, M., Moons, L., Fons, P., Noel, A., Ny, A., Zhou, J. M., Lennartsson, J., Li, H., Luttun, A., Ponten, A., Devy, L., Bouche, A., Oh, H., Manderveld, A., Blacher, S., Communi, D., Savi, P., Bono, F., Dewerchin, M., Foidart, J. M., Autiero, M., Herbert, J. M., Collen, D., Heldin, C. H., Eriksson, U., and Carmeliet, P. (2005). Revascularization of ischemic tissues by PDGF-CC via effects on endothelial cells and their progenitors. *J. Clin. Invest.* **115,** 118–127.

151. Dimmeler, S. (2005). Platelet-derived growth factor CC–a clinically useful angiogenic factor at last? *N. Engl. J. Med.* **352,** 1815–1816.

152. Dollery, C. M., Humphries, S. E., McClelland, A., Latchman, D. S., and McEwan, J. R. (1999). Expression of tissue inhibitor of matrix metalloproteinases 1 by use of an adenoviral vector inhibits smooth muscle cell migration and reduces neointimal hyperplasia in the rat model of vascular balloon injury. *Circulation* **99,** 3199–3205.

153. Lamfers, M. L., Grimbergen, J. M., Aalders, M. C., Havenga, M. J., de Vries, M. R., Huisman, L. G., van Hinsbergh, V. W., and Quax, P. H. (2002). Gene transfer of the urokinase-type plasminogen activator receptor-targeted matrix metalloproteinase inhibitor TIMP-1.ATF suppresses neointima formation more efficiently than tissue inhibitor of metalloproteinase-1. *Circ. Res.* **91,** 945–952.

154. Guo, Y. H., Gao, W., Li, Q., Li, P. F., Yao, P. Y., and Chen, K. (2004). Tissue inhibitor of metalloproteinases-4 suppresses vascular smooth muscle cell migration and induces cell apoptosis. *Life Sci.* **75,** 2483–2493.

155. Yoon, C.-H., Hur, J., Park, K.-W., Kim, J.-H., Lee, C.-S., Oh, I.-Y., Kim, T.-Y., Cho, H.-J., Kang, H.-J., Chae, I.-H., Yang, H.-K., Oh, B.-H., Park, Y.-B., and Kim, H.-S. (2005). Synergistic neovascularization by mixed transplantation of early endothelial progenitor cells and late outgrowth endothelial cells: the role of angiogenic cytokines and matrix metalloproteinases. *Circulation* **112,** 1618–1627.

156. Gulati, R., Jevremovic, D., Peterson, T. E., Chatterjee, S., Shah, V., Vile, R. G., and Simari, R. D. (2003). Diverse origin and function of cells with endothelial phenotype obtained from adult human blood. *Circ. Res.* **93,** 1023–1025.

157. Deb, A., Skelding, K. A., Wang, S., Reeder, M., Simper, D., and Caplice, N. M. (2004). Integrin profile and in vivo homing of human smooth muscle progenitor cells. *Circulation* **110,** 1–5.
158. He, T., Peterson, T. E., Holmuhamedov, E. L., Terzic, A., Caplice, N. M., Oberley, L. W., and Katusic, Z. S. (2004). Human endothelial progenitor cells tolerate oxidative stress due to intrinsically high expression of manganese superoxide dismutase. *Arterioscler. Thromb. Vasc. Biol.* **24,** 2021–2027.
159. Reyes, M., Dudek, A., Jahagirdar, B., Koodie, L., Marker, P. H., and Verfaillie, C. M. (2002). Origin of endothelial progenitors in human postnatal bone marrow. *J. Clin. Invest.* **109,** 337–346.
160. Hur, J., Yoon, C. H., Kim, H. S., Choi, J. H., Kang, H. J., Hwang, K. K., Oh, B. H., Lee, M. M., and Park, Y. B. (2004). Characterization of two types of endothelial progenitor cells and their different contributions to neovasculogenesis. *Arterioscler. Thromb. Vasc. Biol.* **24,** 288–293.
161. Carmeliet, P. (2003). Angiogenesis in health and disease. *Nat. Med.* **9,** 653–660.
162. Li, A., Dubey, S., Varney, M. L., Dave, B. J., and Singh, R. K. (2003). IL-8 directly enhanced endothelial cell survival, proliferation, and matrix metalloproteinases production and regulated angiogenesis. *J. Immunol.* **170,** 3369–3376.
163. Rehman, J., Li, J., Orschell, C. M., and March, K. L. (2003). Peripheral blood "endothelial progenitor cells" are derived from monocyte/macrophages and secrete angiogenic growth factors. *Circulation* **107,** 1164–1169.
164. Murasawa, S., Llevadot, J., Silver, M., Isner, J. M., and Losordo, D. W. (2002). Constitutive human telomerase reverse transcriptase expression enhances regenerative properties of endothelial progenitor cells. *Circulation* **106,** 1133–1139.
165. Lin, Y., Weisdorf, D. J., Solovey, A., and Hebbel, R. P. (2000). Origins of circulating endothelial cells and endothelial outgrowth from blood. *J. Clin. Invest.* **105,** 71–77.
166. Li, A., Dubey, S., Varney, M. L., Dave, B. J., and Singh, R. K. (2003). IL-8 directly enhanced endothelial cell survival, proliferation, and matrix metalloproteinases production and regulated angiogenesis. *J. Immunol.* **170,** 3369–3376.
167. Zucker, S., Mirza, H., Conner, C. E., Lorenz, A. F., Drews, M. H., Bahou, W. F., and Jesty, J. (1998). Vascular endothelial growth factor induces tissue factor and matrix metalloproteinase production in endothelial cells: conversion of prothrombin to thrombin results in progelatinase A activation and cell proliferation. *Int J Cancer* **75,** 780–786.
168. Yoon, C.-H., Hur, J., Park, K.-W., Kim, J.-H., Lee, C.-S., Oh, I.-Y., Kim, T.-Y., Cho, H.-J., Kang, H.-J., Chae, I.-H., Yang, H.-K., Oh, B.-H., Park, Y.-B., and Kim, H.-S. (2005). Synergistic neovascularization by mixed transplantation of early endothelial progenitor cells and late outgrowth endothelial cells: the role of angiogenic cytokines and matrix metalloproteinases. *Circulation* **112,** 1618–1627.
169. Yamaguchi, T., Sawa, Y., Miyamoto, Y., Takahashi, T., Jau, C. C., Ahmet, I., Nakamura, T., and Matsuda, H. (2005). Therapeutic angiogenesis induced by injecting hepatocyte growth factor in ischemic canine hearts. *Surg. Today* **35,** 855–860.
170. Jayasankar, V., Woo, Y. J., Pirolli, T. J., Bish, L. T., Berry, M. F., Burdick, J., Gardner, T. J., and Sweeney, H. L. (2005). Induction of angiogenesis and inhibition of apoptosis by hepatocyte growth factor effectively treats postischemic heart failure. *J. Card. Surg.* **20,** 93–101.
171. Khurana, R., Simons, M., Martin, J. F., and Zachary, I. C. (2005). Role of angiogenesis in cardiovascular disease: a critical appraisal. *Circulation* **112,** 1813–1824.
172. Zachary, I., Mathur, A., Yla-Herttuala, S., and Martin, J. (2000). Vascular protection: A novel nonangiogenic cardiovascular role for vascular endothelial growth factor. *Arterioscler. Thromb. Vasc. Biol.* **20,** 1512–1520.
173. Khurana, R., Shafi, S., Martin, J., and Zachary, I. (2004). Vascular endothelial growth factor gene transfer inhibits neointimal macrophage accumulation in hypercholesterolemic rabbits. *Arterioscler. Thromb. Vasc. Biol.* **24,** 1074–1080.
174. Simons, M., and Ware, J. A. (2003). Therapeutic angiogenesis in cardiovascular disease. *Nat. Rev. Drug Discov.* **2,** 863–871.

CHAPTER **12**

Systemic and Pulmonary Hypertension

OVERVIEW

Hypertension affects up to 30% of the adult population in Western societies and is a major risk factor for kidney disease, stroke, and coronary heart disease. It is a complex multifactorial trait with demonstrable genetic and environmental components, whose precise etiology remains undetermined. Several approaches and different technologies, including genome-wide scans, case-control association studies, experiments on inbred animal models, and expression profiling using both transcriptomic and proteomic analysis, have been used to identify chromosomal regions harboring blood pressure loci and to evaluate candidate genes involved in hypertension susceptibility. These have proved to be most successful with several monogenic hypertension conditions and with primary pulmonary arterial hypertension, while unraveling the genetic etiology of the more complex, polygenic essential hypertension has shown promising albeit inconclusive results thus far.

INTRODUCTION

Hypertension afflicts more than 65 million Americans and poses an increased risk for cardiovascular morbidity such as stroke, myocardial infarction (MI), and end-stage renal disease, resulting in significant mortality. A number of family and epidemiological studies have supported the view that hypertension arises from a complex interplay between genetic and environmental factors, including dietary sodium intake, excess alcohol consumption, and body weight.[1]

In this chapter, we review the findings of specific gene defects that underlie specific conditions associated with hypertension, including both systemic and pulmonary arterial hypertension and several more rare hypertensive phenotypes that are inherited as monogenic Mendelian traits such as Gordon and Liddle syndromes. We discuss the findings of several recent genome-wide scans implicating specific chromosomal loci in the pathogenesis of essential systemic hypertension, the use of case-association studies to evaluate the complicity of specific candidate genes in hypertension susceptibility in clinical studies, and the use of microarray, SNP genotyping, and proteomic analysis to complement the clinical genetic analysis. We also discuss the use of animal models to identify both the genetic loci and to investigate the functional mechanisms of the suspected candidate genes involved in hypertensive phenotypes and comment on the pharmacogenomic strategies available for enhanced diagnosis and therapies of hypertension.

PRIMARY RISKS

Long-term caloric intake in excess of energy expenditures, chronic intake of dietary sodium, excessive alcohol consumption, and psychosocial stressors all are contributory factors in the development of hypertension. In addition, body weight and obesity are positively associated with increased blood pressure (BP). A substantial body of evidence strongly supports the concept that lifestyle modification, including increased physical activity, reduced salt intake, weight loss, moderation of alcohol intake, increased potassium intake, and an overall healthy dietary pattern can effectively lower BP. The interaction of these primarily environmental factors with genetic factors has not yet been determined but may prove important in the overall expression and penetrance of specific gene action on the development of hypertensive phenotypes. Exposure to a variety of environmental stresses (both acute and chronic) may impact genes underlying the physiological systems mediating the stress response of heart, vasculature, and kidney (i.e., the sympathetic nervous system, renin-angiotensin-aldosterone system and sodium reabsorption, and the endothelial system), as well neural and hormonal pathways (e.g., the parasympathetic nervous system, the serotonergic system, and the hypothalamus-pituitary-adrenal axis) conferring enhanced susceptibility to the development of essential hypertension.[1-3]

Another significant factor with effects on the development of hypertension is age; in Western countries, hypertension is prevalent in the elderly. This may relate to any of several of the aforementioned environmental factors (e.g., diet, stress), because the elderly from primitive societies do not exhibit increased BP,[4] and immigrants from rural areas who formally showed no age-related increase in BP

Post-Genomic Cardiology

develop hypertension when placed in urban Westernized settings.[5]

Unlike essential hypertension for which there is no recognized primary cause, secondary hypertension is elevated BP resulting from an underlying, identifiable, often correctable cause, and is thought to constitute about 5–10% of hypertension cases. These include obstructive sleep apneas, primary aldosteronism caused by an adrenal adenoma or bilateral hyperplasia of the adrenal glands, renal artery stenosis and parenchymal disease, excess catecholamines (e.g., pheochromocytoma, acute stress), coarctation of the aorta, Cushing's syndrome, and endocrine disorders, including diabetes, acromegaly, hypothyroidism, and hyperthyroidism. Other secondary causes of hypertension include a wide variety of drug therapies, including immunosuppressive agents, nonsteroidal anti-inflammatory drugs and cyclooxygenase-2 (COX-2) inhibitors, stimulants, and steroids. Dramatic progress has been made in our understanding of several forms of secondary hypertension as discussed later in which a number of single genes and defined mutations have been implicated.

Elevated BP, particularly systolic hypertension, is well recognized as a key determinant of risk for morbidity and mortality for multiple clinical disorders, including stroke, heart failure (HF), MI, renal insufficiency/failure, peripheral vascular disease, retinopathy, dementia, and premature mortality. However, there is increasing evidence to suggest that susceptibility to these outcomes is influenced by genetic factors that are likely separate and distinct from the genetic factors involved in hypertension risk.

EVIDENCE OF GENETIC CONTRIBUTION(S) TO HYPERTENSION

Family segregation analysis and twin studies show that between 30 and 60% of BP variation is determined by genetic factors. Previous twin studies have found heritabilities around 50% for both systolic (SBP) and diastolic (DBP) blood pressure in either white or black individuals.[1,2,6,7]

In addition to a large array of genetic components involved in primary essential hypertension indicated by both clinical and animal models as discussed later, hypertension can be also regarded as a trait associated with a variety of inherited disorders with their own genetic components (which may have considerable overlap with the hypertensive genes), including type II diabetes, atherosclerosis, dyslipidemias, obesity, and metabolic syndrome. Moreover, several disorders with large-scale chromosomal abnormalities, including Turners syndrome (XO) and Williams–Beuren syndrome (e.g., deletion on chromosome 7q) can present with increased risk of hypertension.[8] In addition, point mutations in the elastin gene, a primary defect in Williams syndrome, mapping to chromosome 7q11.2, can lead to supravalvular aortic stenosis and hypertension.[9,10]

MENDELIAN FORMS OF HYPERTENSION

A number of relatively rare monogenic Mendelian disorders associated with hypertensive phenotypes (Table I) have been reported.[11]

Substantial progress has been made in the explanation of genes and specific mutations that underlie several Mendelian hypertensive traits, a number of which can present early in life with distinct phenotypes. Most of the causative genes contributing to these uncommon forms of hypertension are involved in the renin-angiotensin pathway or are related components acting further downstream, underscoring their well-established importance in BP regulation. It remains a challenging, but undetermined, hypothesis whether "milder" variants of these same genes contribute and are relevant in the more common type of essential hypertension and whether the same physiological pathways are operative in both forms of hypertension.

In glucocorticoid-remediable aldosteronism (GRA), two flanking genes (*CYP11B1*) and (*CYP11B2*) are recombined to form a unique chimeric gene composed of the promoter-containing regulatory region of the *CYP11B1* gene encoding the ACTH-responsive 11β-hydroxylase (the terminal enzymatic step in glucocorticoid synthesis) and the structural portion of the *CYP11B2* gene encoding aldosterone synthase (the terminal enzymatic step of mineralocorticoid biosynthesis). Affected individuals, therefore, produce a functional aldosterone in excess in response to an ACTH-responsive promoter, generating an autosomal-dominant hypertension and lower plasma renin activity (PRA).[12] The syndrome of apparent mineralocorticoid excess (SAME) can occur as a rare, autosomal-recessive form of hypertension, usually presenting in infancy or childhood or as an acquired syndrome resulting from the inhibition of 11β-hydroxysteroid dehydrogenase type 2, an enzyme responsible for converting cortisol to cortisone, as shown in Fig. 1. This enzyme normally serves a protective function, preventing cortisol from occupying the mineralocorticoid receptor (MR) and triggering enhanced Na$^+$ reabsorption (ENaC, Na-K-ATPase) and K$^+$ excretion. Mutations in the renal-specific isoform gene encoding 11β-hydroxysteroid dehydrogenase (*11β-HSD2*) have been found that cause an increase in intracellular cortisol levels, activation of MR resulting in volume-dependent salt-sensitive hypertension, hypokalemia, and reduced aldosterone and PRA levels.[13] Mutations in the mineralocorticoid receptor (*MR*) gene have also provided insights into mineralocorticoid-induced hypertension.[14] By screening for *MR* single-strand conformation polymorphisms (SSCP) in seven unrelated patients with suspected monogenic hypertension, a heterozygous mutation at codon 810 was detected, resulting in a leucine for serine substitution within the *MR* hormone-binding domain.[15] This mutation causes early-onset hypertension that is markedly exacerbated in pregnancy. It also results in constitutive MR activity and alters receptor specificity, with groups such as progesterone steroids

TABLE I Mendelian Forms of Hypertension

Condition	Inh	Markers	Genetic defect
GRA (FH1)	AD	PRA↓ ALD↑	*CYP11B1/CYP11B2* hybrid
Liddle's syndrome	AD	PRA↓ ALD↓	Activating βγENaC mutations
PHA1	AD	ALD↑	Inactivating βγENaC mutations
SAME	AR	PRA↓ ALD↓	*11β-HSD2* mutation
MR mutations	AD	PRA↓ ALD↓	*MR* S810L
Gordon's syndrome (PHAII)	AD	PRA↓ ALD↓	*WNK1, WNK4*
Brachydactyly	AD	Normal	Maps to chromosome 12p

AR, autosomal dominant; AD, autosomal recessive; PRA, plasma renin activity; ALD, aldosterone; MR, mineralocorticoid receptor; GRA, glucocorticoid-remediable aldosteronism; SAME, syndrome of apparent mineralocorticoid excess; PHA1, pseudohypoaldosteronism type I; Inh, inheritance; FH, familial hyperaldosteronism.

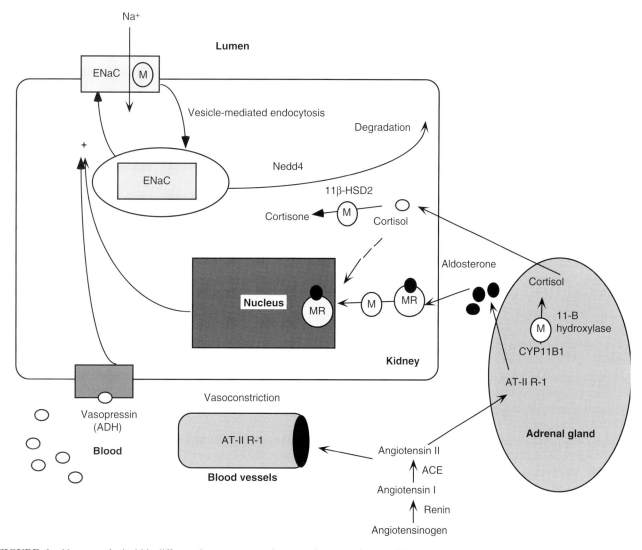

FIGURE 1 Numerous loci within different sites can promote hypertension. Mutations in different proteins are indicated by M. MR, mineralocorticoid receptor.

lacking 21-hydroxyl, which normally behave as MR antagonists becoming potent agonists.

In Liddle's syndrome, an autosomal-dominant form of hypertension is present with both hypokalemia and increased sodium reabsorption. The genetic defect was mapped to chromosome 16 and localized further to specific defects in either the β- or γ-subunit of the epithelial sodium channel (*ENaC*).[16,17] Gain-of-function mutations in the β- and γ-subunits of *ENaC* were shown to cause Liddle's syndrome by promoting an unregulated activation of the sodium channel at the cell surface, resulting in the channel remaining inappropriately open even with high salt intake, explaining the salt-sensitive hypertension. Loss-of-function mutations in all three subunits of ENaC can cause hypotension (pseudohypoaldosteronism [PHA] type I).

Gordon's syndrome or pseudohypoaldosteronism type II (PHA type II) is a rare Mendelian trait characterized by familial hypertension with increased renal salt reabsorption, impaired K+ and H+ excretion, and low renin activity. Genetic loci at chromosomes 1q31-42 and 17p11-q21 have been linked to PHA type II.[18] Two genes of the *WNK* (with no lysine K) serine-threonine kinase family, *WNK1* on chromosome 12 and *WNK4* localized to chromosome 17, respectively, have been identified, which can cause PHA type II.[19] Both genes encode proteins that localize to the distal nephron, a kidney segment involved in salt, K+, and pH homeostasis; the WNK1 protein is cytoplasmic, whereas WNK4 localizes to tight junctions. WNK4 is involved in electrolyte homeostasis control by regulating the surface activity of several proteins involved in ion transport, including the thiazide-sensitive sodium chloride cotransporter (NCCT) and the renal outer medullary potassium channel (ROMK); WNK 4 directly interacts with the NCCT *in vitro* and inhibits its activity by reducing the numbers of the receptors on the cell surface and directly interacts and suppresses ROMK potassium channel activity and endocytotic regulation.[20,21] WNK1 seems to interact directly with WNK4, mediating the WNK4-suppression of NCCT.[22,23] Recent studies have also implicated WNK1 in the activation of the serum- and glucocorticoid-inducible protein kinase SGK1, leading to activation of the epithelial sodium channel.[24]

Disease-causing mutations in *WNK1* are large intronic deletions that increase *WNK1* expression. In contrast, missense mutations in *WNK4* have been found that cluster in a short, highly conserved segment of the encoded protein and that relieve the inhibition of NCCT, increase the inhibition of the renal outer medullary potassium ion channel, and further increase paracellular chloride ion flux.[25] Some of the WNKs' effects seem to be independent of their kinase function, suggesting that they may occur as a result of specific protein–protein interactions, functioning as part of much larger protein scaffolds with effects in cells transcending ion transport.[26]

WNKs, in particular *WNK4*, have been suggested as candidate genes for more common essential hypertension itself, but this has not yet been demonstrated. The connection between variations in WNK1 function and essential hypertension was suggested by recent findings showing a significant association of a common *WNK1* variant with hypertension.[27] Common *WNK1* haplotypes, including a SNP at rs1468326, located 3 kb from the *WNK1* promoter, were found to be associated with the severity of hypertension, with both SBP and DBP.

Unlike the aforementioned Mendelian hypertension phenotypes, the syndrome of autosomal-dominant hypertension with brachydactyly features normal sodium and renin-angiotensin-aldosterone responses and in several ways is the form of monogenic hypertension most closely resembling primary hypertension. The condition is interesting because it may represent a novel neural form of hypertension suggested by a significant pattern of neurovascular compression of the ventrolateral medulla at the root entry zone of cranial nerves IX and X in affected patients, with deregulated sympathetic responses.[28] The gene has been mapped to chromosome 12p,[29] and the evaluation of potential candidate genes is underway and should prove of great interest. Interestingly, evidence for a more generalized importance of the chromosome 12p locus with regard to BP regulation and essential hypertension has recently emerged from studies of Chinese kindreds from the relatively isolated rural province of Shijingshan. A total genome scan conducted in these families indicated that the essential hypertensive trait was linked to chromosome 12p with a highly significant LOD score exceeding 3.0.[30] This genomic region overlaps with the assigned locus, causing severe autosomal-dominant hypertension and brachydactyly. Further studies of this genomic region, spanning 18 annotated genes, may prove of broad significance in explaining new mechanisms for primary hypertension.

GENOMIC ANALYSIS OF PRIMARY HYPERTENSION

As previously noted, the primary method for identifying genetic effectors independent of the knowledge of *a priori* candidate genetic loci is genome-wide linkage analysis. This technique allows the broad identification of novel genetic loci and their linkage to a trait essentially featuring a broad mapping of the loci within the chromosomes; the finer resolution of the specific gene loci involved often uses candidate association studies, which are discussed in a later section. As of 2005, there have been more than 20 studies conducted describing the results of genome-wide screens for genes controlling BP. Most of these studies have reported numerous chromosomal regions with suggestive evidence of linkage. Each genome scan featured different numbers, family types, ethnic populations, and design and phenotyping strategy, making comparisons difficult. In summarizing most of studies conducted thus far, most human chromosomes have at least one locus with suggestive linkage to hypertension (Table II). Some chromosomes such as 2 and 6 have several loci, whereas others such as chromosomes 13 and 20 have none. Most of the identified loci show a suggestive level of linkage, with only a few loci showing a highly significant LOD score. In many cases, linked regions identified in one study were not found in others. However, genetic loci associated with a hypertensive phenotype in one ethnic population may simply not be present in a different population. Increasing the size of the sampled group has not necessarily increased the resolution of loci. In a recent series of large-scale genome-wide studies conducted by the National Heart Lung and Blood Institute–funded Family Blood Pressure Program (FBPP) found no chromosomal locus with genome-wide significant level of linkage to hypertension.[31–35]

Several of these studies did identify chromosomal loci with a suggestive relationship to hypertension, including loci at chromosome 2p and 1p, as well as providing evidence for a quantitative trait locus on chromosome 18q21, significantly influencing postural change in SBP.[31,33,36] The large-scale MRC-funded BRIGHT study involving more than 3599 members of sibling pairs identified a principle locus on chromosome 6q, with a LOD score over 3 and three other loci at 2q, 5q, and 9q with LOD scores higher than 1.5 that all displayed genome-wide significance, although showing no overlap with the loci from the FBPP study.[37]

Smaller genome screens with a variety of more homogenous populations have shown other chromosomal loci with association with hypertension. A genome scan of 91 Scandinavian families with early-onset hypertension[38] found one region on chromosome 14 with genome-wide significance and a locus on chromosome 2 (found in previous studies of Mexican-Americans)[39] with suggestive linkage. The α2β-adrenergic receptor gene, a key component in the sympathetic nervous system, is located in this region of chromosome 2 and represents a viable candidate gene. Functional variants of the α-AR gene (ADRA2B) have been significantly associated with hypertension.[40] A novel susceptibility gene for essential hypertension on chromosome 18q was identified with a genome-wide scan using 120 extended Icelandic families with 490 hypertensive patients.[41]

A recent genome-wide scan to identify loci linked to essential hypertension performed with Anglo-Celtic Australian sib pairs identified potential loci on chromosomes 1 and 4.[42] The chromosome 4 locus coincided with a previously known QTL for SBP, and the locus on chromosome 1 contains the chloride channel gene CLCNKB and tumor necrosis factor receptor 2 gene TNFRSF1B, which have each shown association with hypertension. A genome-wide linkage analysis for SBP and DBP with 1109 white female dizygotic twin pairs from the Twins UK registry replicated the locations of previously reported linkage loci on chromosome 16, 17, and 22. Results from multipoint analysis showed one novel suggestive linkage for SBP on chromosome 11.[43]

A recent study examining genetic loci influencing the long-term levels and trends of BP over time conducted with 775 white siblings, ages 13–43 y, found significant linkage (LOD>3.0) of DBP with a region on chromosome 2 spanning from 44–103 cM. In addition, suggestive linkage for SBP and DBP on chromosome 4, and DBP on chromosome 18 was noted. Several hypertension candidate genes encoding proteins such as α-adducin, β-adducin, the sodium bicarbonate co-transporter, and G-protein–coupled receptor kinase 4 are located in these regions.[44]

These genome-scan data also further reinforce the notion that essential hypertension is polygenic. However, limited data are presently available for delineating the actual number of genes involved to explain specific gene–gene interactions or gene–environment interactions involved in hypertension phenotype. Several studies have recently suggested that BP is governed by multiple genetic loci, each with a relatively modest effect.[33,45] The number of human hypertension loci

TABLE II Genomic Loci Associated with Essential HT

Human chromosomal loci	QTL (LOD>3)
1q23.1–24.2	9
2p11.2–q12.1	1
2q31.3–32.3	7
3p21.31	10
3p26.3–25.3	
3q24–25.1	2
4p16.1–15.3	3
5p15.2–12	11
6p22.1	
6q14.1–16.1	8
6q27	
7q11.21–23	
7q21.3–22.1	
7q31.32–32.2	
8q24.21–24.3	12
9p21.2	
10q24.33	
11q12.1	
12q21.31–21.33	
12p12.2	
14q11.2	
15q26.1–26.3	4
16p12.3	
16q12.1	
17q11.2–23.2	5
18q21.2–22.1	6
19p13.3–12	
21q22.13	
22q12.3	
Xp11.4	

has recently been suggested to be in the order of tens of genes each with a relatively modest effect on hypertension in the population at large (e.g., relative risk of 1.2–1.5).[46]

IDENTIFICATION OF PRIMARY RISK GENES AND CANDIDATE GENES

Genome-wide screens have revealed a number of chromosomal regions that likely harbor candidate genes involved in hypertension. Several positional candidate genes have been suggested by these studies, including the angiotensin II type I receptor (*AGTR1*) gene on chromosome 3, the β2 adrenergic receptor (see later), and the lipoprotein lipase (*LPL*) gene on chromosome 8. Unfortunately, there has been little commonality or overlap indicating the involvement of specific candidate genes in the different genome studies performed thus far.[31]

The analysis of candidate genes by case association study simply assesses whether a suspected genetic variant is significantly present more commonly in individuals with raised BP compared with normotensive subjects. A diverse range of genes that could represent viable candidates for affecting BP if mutated has emerged from physiological studies in both rodents and man (Table III).

TABLE III Candidate Genes in Primary Hypertension

Renin-angiotensin
Angiotensinogen (*AGT*)
Renin (*REN*)
ACE
AGTR1
Sympathetic nervous system
Adrenergic receptors (*ADRB2, ADRB3, ADRA2B*)
Endothelin system
ET-A receptor
ET-B receptor
ET-1
NO pathway
eNOS
iNOS
Adducin
Natriuretic peptides and their receptors
ANP
BNP
NPRA
NPRB
NPRC
Aldosterone action
Aldosterone synthase
ENaC α, β, and *γ*
Guanine nucleotide binding protein b3 (GNb3)

Several studies (including meta-analyses) have reported that variation in the angiotensinogen gene (*AGT*), particularly in the case of a coding polymorphism (met235thr), is significantly associated with essential hypertension.[47,48] A mutation (G-6A) in the *AGT* promoter region affecting *AGT* transcription *in vitro* also has been associated (independently of the met235thr variation) with essential hypertension.[49] The level of association has been found to vary with gender (increased in females) and with specific ethnic groups.[50,51] Conflicting data have been found concerning the association of hypertension with specific variants of the angiotensin-1 converting enzyme (*ACE*) gene, another important player in the renin-angiotensin pathway.[52,53] A moderate level of linkage of hypertension with *ACE* was found to be dependent in one study on gender (e.g., present in males)[52] or several contexts that included ethnicity, gender, and body weight.[53]

Polymorphisms of the β2-adrenergic receptor subtype (*ADRB2*) have been reported,[54] and studies have shown a suggestive association with essential hypertension, but these findings vary with the populations studied.[55] A chromosomal locus of 5q containing the *ADRB2* gene has shown suggestive linkage and further supports its status as a viable candidate gene for hypertension susceptibility.[56]

In summary, although a number of studies have shown some preliminary positive relationship between specific candidate gene variants with essential hypertension, the findings have often not been replicated in different populations; only genetic variants in angiotensinogen (*AGT*) and the β2 adrenergic receptor (*ADRB2*) have garnered consistent support in association with a mild hypertensive effect. No candidate gene has yet been implicated with a major effect on the risk of human essential hypertension. Where effects have been found, most studies have suggested the interaction of the candidate gene with other factors including genetic background and other variables (e.g., body weight and diabetes).

ANIMAL MODELS OF HYPERTENSION

Genetic models of essential hypertension are available such as the spontaneously hypertensive stroke-prone rat (SHRSP) and the Dahl-salt sensitive rat, which have assisted the analysis of this complex phenotype by providing the element of genetic homogeneity not possible with generally, highly heterogeneous human populations, as well as allowing controlled environmental factors (e.g., diet) particularly useful in the search for causative genes. A principal strategy in the rat has been the identification of quantitative trait loci (QTL) responsible for BP regulation by genome-wide scanning. Linkage studies have identified many QTLs, each tens of centimorgans in length, on most rat chromosomes.[57]

Traditional genetic crosses between hypertensive strains (e.g., the Dahl salt-sensitive rat and the spontaneously hypertensive rat) and inbred nonhypertensive strains (e.g., the Wistar Kyoto rat, Lewis, and Milan normotensive strain) revealed linkage to BP with loci on all but three rat chromosomes; the large number of genomic loci revealed by these genetic studies of rodent hypertension was similar to the findings in human clinical studies. Crosses of different hypertensive strains have been used to identify whether different hypertensive strains share loci in common.[58] A large proportion of QTLs loci was shared (13 of 16) by two different hypertensive strains, whereas three QTLs were dissimilar between them.

The use of designer rats such as congenic strains, which are produced by repeated backcrosses to an inbred strain, and consomic or chromosome substitution strains, which are produced by repeated backcrossing of an entire chromosome onto an inbred strain, for the confirmation of QTLs has significantly enabled the genetic dissection of the implicated regions.[59,60] Several studies have generated large panels of consomic rat strains, in which single autosomes are replaced one at a time to evaluate specific chromosome effects on hypertensive phenotype.[61,62] One of the first studies using consomic strains indicated that the substitution of chromosome 13, carrying the renin gene (*REN*), derived from a Brown Norway rat into a Dahl salt-sensitive genetic background resulted in significantly reduced arterial pressure and associated proteinuria with a high-salt diet.[62]

To further identify susceptible variants within these large QTLs, the congenic mapping approach has been used involving the substitution of small, defined segments of chromosome from one strain into another. The construction of panels of congenic strains, each of which carries a different hypertensive QTL allele with a genetic composition that is otherwise similar to that of the hypertensive Dahl salt-sensitive (S) rat strain, has been useful in helping to dissect the effects of each QTL.[57] Congenic strains have also been used to detect strong epistatic interactions between different QTLs contributing to SBP (e.g., QTLs residing on chromosomes 1 and 10).[63,64]

Although congenic mapping has proved to be a formidable approach in further delineating QTLs, its level of resolution is limited, because it is unlikely to further resolve QTLs to a region smaller than 200 kb in size. QTLs of moderate size can contain upward of tens to hundreds of potential candidate genes, making fine mapping a difficult process. In one of the better-case scenarios, the molecular dissection of the 177-kb segment of chromosome 7,[65] although a relatively small region, revealed the presence of eight positional candidate genes, one of which was aldosterone synthase *Cyp11b1*, the gene involved in GRA associated hypertension.[12]

Given the increasing potential of genetic analysis with animal models, interest has been heightened in the comparative mapping of loci in rats, mice, and man (syntenic analysis). A number of studies have reported that specific QTLs in rats or mice also play a role in hypertension in humans. For instance, *in silico* maps have been generated by translating large QTLs between rats and humans, predicting 26 chromosomal locations in the human genome that may harbor genes influencing BP.[66] Human chromosome loci at 1q21-32, 2q11-q14, 5p14, 7q21, 8p23, 10q11, and 11q14 have QTL homologs in both rat and mouse, whereas 3p21, 4q25, 16p11, 17q23, and 18q21 have syntenic homologs in just the rat. Although there is uncertainty about the precise relationship in the mechanisms of rodent and human hypertension, these findings of synteny corroborate many of the candidate regions thus far identified.[67] This comparative approach has been extended to findings with specific candidate genes at some of these loci. For example, variations in the α-adducin gene, which encodes a membrane protein associated with renal sodium transport have been found in both rat hypertensive strains and show a suggestive level of association with human essential hypertension (but only in some populations).[68,69]

The deployment of post-genomic methods such as RNA interference, knock-out, and knock-in transgenic models, is just beginning to be gainfully applied to the analysis of highly suspected individual genes contained within identified QTL regions to further understand their contribution to hypertension. For instance, an siRNA approach has been used to block the expression of Nedd4 and Nedd4-2, both regulators of the renal epithelial sodium channel (*ENaC*), the mutant gene implicated in Liddle's syndrome.[70] Endogenous Nedd4-2 negatively regulates *ENaC* in epithelia and seems to function as a critical component of the regulatory signaling pathway by which steroid hormones regulate EnaC. These studies also reported that the effect of *Nedd4-2* siRNA was ameliorated in epithelial cell lines expressing a variant allele of *ENaC* from an individual with Liddle's syndrome, suggesting that a defect in *ENaC* regulation by Nedd4-2 contributes to the pathogenesis of this inherited form of hypertension.

Numerous mouse knock-out strains have been produced in which the targeted gene deletion modulates BP or hypertension susceptibility in the presence of interacting environmental factors (e.g., elevated salt intake). The diversity of such genes (Table IV) further reinforces the hypothesis that a large number of genes possess alleles that contribute to BP and underscores the extensive degree of genetic heterogeneity in susceptibility to clinical hypertension.

Transgenic mice with overexpression or with multiple copies of specific alleles of candidate genes have also proved informative (e.g., angiotensinogen resulting in increased BP).[71]

GENE EXPRESSION AND PROFILING IN HYPERTENSION

Global and Specific Analysis

High-throughput differential gene expression profiling, a transcriptomic approach, is another strategy that has been

TABLE IV Targeted Gene Deletion and Blood Pressure in Mice

Gene (symbol)	Interacting	BP	References
Adrenomedullin (*AM*)		↑	72
Bradykinin receptor type 2 (*Bdkrb2*)	High sodium	↑	73
ANP receptor guanylyl cyclase-A (*GC-A*)		↑	74
Pro-atrial natriuretic peptide (*ANP*)	Salt	↑	75
Angiotensinogen (*AGT*)		↓	76
Angiotensin II type 1 receptor (*AT1R*)		↓	76
Angiotensin II type 2 receptor (*AT2R*)	Angiotensin II	↑	77
Apolipoprotein E (*Apoe*)		↑	78
β-adducin (*ADD2*)		↑	79
α-Calcitonin gene-related peptide (*α-CGRP*)		↑	80
D5 dopamine receptor (*D1* and D5)		↑	81
Heme oxygenase 1 (*HO-1*)	Hypoxia	↑	82
β-2 Adrenergic receptor (*β2AR*)	Exercise	↑	83
11-β-Hydroxysteroid dehydrogenase type II (*11 HSDβ2*)		↑	84
Endothelial nitric oxide synthase (*eNOS3*)	High-fat diet	↑	85,86
Insulin-like growth factor (*IGF-1*)		↑	87
Endothelin-1 (*ET-1*)		↑	88

increasingly applied in the search for the molecular genetic basis of complex diseases such as hypertension. The choice of tissue for gene expression profiling is critical; several studies that used renal transplantation have suggested that the kidney is an ideal choice in studies of essential hypertension.[89] A recent study examining rat kidney gene expression in three spontaneously hypertensive rat (SHR) substrains at defined time points, including periods of pre-hypertension, rapid BP elevation, and sustained hypertension, identified a limited group of genes whose expression profile was shared among all three substrains (and which significantly varied from the normotensive strain profile).[90] At present, few microarray profiling studies have been undertaken and replicated in tissues other than kidney in either animal hypertensive models or clinical hypertension. In those studies available, significant differences in gene expression profiles have been recently reported in cardiac tissue,[91] leukocytes,[92] vascular smooth muscle cells (SMCs),[93] adipose tissues[94] and in the adrenal gland.[95] In an experimental animal model in which hypertension is induced by the administration of angiotensin II (ANG II), cardiac gene expression was decreased for mitochondrial bioenergetic enzymes, including electron transport chain, Krebs cycle, and fatty acid oxidative activities with acute treatment (i.e., a 24-h ANG II administration), whereas growth-related genes encoding ribosome protein and protein translation components were up-regulated. The gene expression of proteins involved in cardiac remodeling, including cytoskeletal and extracellular matrix (ECM) proteins, increased with chronic exposure to ANG II (i.e., a 14-day treatment) as did increased cardiac hypertrophy.[91] The reader is referred to Chapter 15, which is dedicated to the topic of cardiac remodeling. In blood leukocytes from patients with hypertension, there is significant modulation of expression of a broad array of genes involved in inflammatory pathways; these changes are absent in patients after hypertensive treatment.[92]

As in the case of genetic mapping, however, investigators applying this technology (particularly in human studies) as a stand-alone approach have not identified single primary genes that modulate hypertension. This is in part due to the very large number of genes that are differentially expressed in tissues or organs of contrasting populations, many of which do not bear relevance to the disease under study, as well as being consistent with the notion that a large number of genes contribute to the hypertensive phenotype.

An emerging alternative strategy to the search for the genetic basis of complex diseases consists of integrating the genomic approach with the analysis of the transcriptome. This strategy has the considerable advantage of combining the power of genetic mapping with that of differential gene expression profiling. Although hundreds of differentially expressed genes would be expected in comparing two different unrelated parental strains, this number is significantly reduced in the profiling analysis of congenic and parental strains. A combination of congenic mapping and transcription profiling has been used to construct a transcriptional profile in kidneys from the SHR stroke-prone rat and a congenic strain containing a 22-cM region of WKY chromosome 2,[96] and identified the marked down-regulation of the transcript level of glutathione *S*-transferase μ-type 2 (*Gstm2*), an antioxidant, in the hypertensive strains. It remains to be shown whether this effect contributes to the hypertensive phenotype or whether it is a secondary phenotype. Recently, this integrative approach successfully led to the reduction of the number of genes incorporated within the two QTLs on chromosome 1 from more than 1000 to just seven genes, novel candidate genes involved in salt-sensitive hypertension in the rat.[97] Studies that use proteomic analysis and targeted functional analysis by RNA interference in these animal models of essential hypertension are presently underway.

PULMONARY HYPERTENSION

Primary pulmonary hypertension (PPH) is a rare autosomal-dominant disease with incomplete penetrance characterized by distinctive changes in pulmonary arterioles that lead to increased pulmonary artery pressures, right ventricular failure, and death. PPH is a potentially lethal disorder, because the elevation of the pulmonary arterial pressure can result in right-heart failure, and it is also commonly referred to as pulmonary arterial hypertension (PAH). Histologically, PPH is characterized by pulmonary artery smooth muscle and endothelial cell proliferation. In particular, plexiform lesions characterized by arterial lumen occlusion, aneurysmal dilatation, proliferation of interconnected vascular channels, and endothelial and SMC proliferation are present in a large proportion of cases (ranging from one third to two thirds). PPH occurs with greater incidence in adult women, although it can be present in both males and neonates, and the age of onset is variable.[98] Although PPH is rare, cases secondary to known etiologies are more common and include those associated with appetite-suppressant drugs, including phentermine-fenfluramine, responses to hemodynamic changes associated with congenital or acquired cardiac defects, pulmonary emboli, or hypoxia. A variety of congenital cardiac defects (e.g., atrial and ventricular septal defects) with left-to-right shunt leading to increased pulmonary blood flow can cause pulmonary hypertension, including some that have been repaired by surgery. In addition, secondary pulmonary hypertension (SPH), which is essentially indistinguishable from PPH with regard to clinical course and histopathology, can also arise in association with connective tissue disorders (e.g., collagen vascular disease including scleroderma or systemic sclerosis) and may be triggered by human immunodeficiency virus infection or portal hypertension. Considerable progress has been made in the clinical treatment of pulmonary hypertension whether of primary or secondary etiology with the use of vasodilators such as prostacyclin analogs (e.g., epoprostenol, beraprost, or iloprost) and the endothelin receptor antagonists (e.g., oral bosentan), the phosphodiesterase-5 inhibitor sildenafil, and calcium-channel blockers. It has also been suggested that there may be a genetic predisposition underlying several forms of secondary pulmonary hypertension and in persistent neonatal pulmonary hypertension. Understanding the molecular genetic and cellular basis for pulmonary hypertension has just begun to reveal the commonalities shared (as well as the differences) between the primary and secondary forms of pulmonary hypertension.

The aforementioned plexiform lesion, a hallmark of irreversible vessel damage in pulmonary hypertension, as well as a significant clinical indicator of poor prognosis, can also be found in patients with PPH. However, one of the few histopathological features that can be used to distinguish between the primary and secondary forms of pulmonary hypertension relates to the clonal nature of the plexiform lesion. The proliferation of endothelial cells in plexiform lesions has been shown to be primarily monoclonal in PPH but polyclonal in secondary pulmonary hypertension.[99] These findings underscore the important role that vascular cell proliferation plays in the development of familial PPH, as well as the differences in the pathogenic mechanisms used by primary and secondary disorders.

Although originally believed to be primarily acquired disorders, genetic analysis of PPH has documented significant linkage to a 3-cM region on chromosome 2q33 (locus PPH1).[100] By use of a positional cloning approach, several candidate genes residing within this region were investigated with denaturing high-performance liquid chromatography. In a number of unrelated affected patients, the *BMPR2* gene encoding the bone morphogenetic protein receptor type II (BMPR-II) was found to contain mutations, including five that predicted premature termination of the protein product and two missense mutations.[101,102] These data and those from an *in vitro* expression analysis demonstrating loss of BMPR-II function for a number of the identified mutations are consistent with the premise that BMPR-II haploinsufficiency constitutes a critical molecular mechanism in PPH.[103] Mutations in *BMPR2* exons are found in approximately 50% of patients with familial PPH. Because familial PPH is highly linked to chromosome 2q33, it seems probable that the many of the remaining 50% of familial cases may harbor either intronic *BMPR2* abnormalities or alterations within the promoter or regulatory genes. Moreover, varying estimates ranging between 10 and 25% of patients with "sporadic" idiopathic PPH have identifiable *BMPR2* mutations, suggesting that other loci might be operative. The findings that PPH does not develop in all individuals with *BMPR2* mutations and that *BMPR2* mutations confer a 15–20% chance of developing PPH in a carrier's lifetime[104] suggest that gene–gene or gene–environment interactions may contribute to either enhance or prevent the development of the vascular disease. It is also noteworthy that the clinical and pathological manifestations tend to be identical whether the disease is familial or acquired.

The *BMPR2* gene is a member of the transforming growth factor TGF-ß superfamily. Members of the TGF-β superfamily transduce signals by binding to heteromeric complexes of type I and II receptors, which can act as kinases phosphorylating a group of cytoplasmic signaling proteins (Smads) that

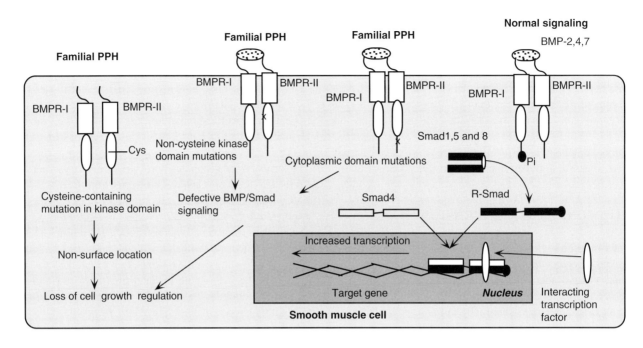

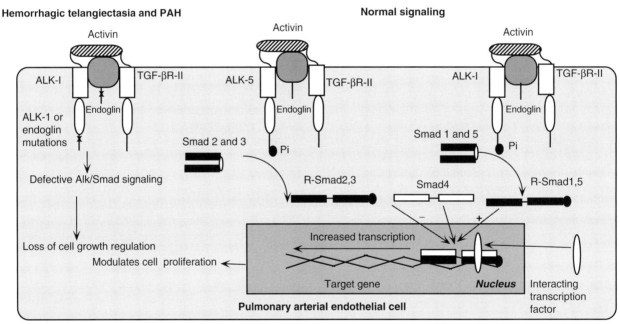

FIGURE 2 Model depicting the role of mutations in BMPR2 and ALK-1 in the development of primary pulmonary hypertension. As depicted, in vascular smooth-muscle cells, the binding of the BMP ligand to the complex of BMPR-I and BMPR-II results in receptor-dependent phosphorylation of Smad 1, 2, and 8. These associate with Smad 4 and an interacting transcription factor, which translocates to the nucleus to modulate transcription that results (depending on cell-type, ligand etc.) in growth arrest. In primary pulmonary hypertension, mutant BMPR-II has defective Smad signaling (as shown in the top diagram) with mutations at different domains of the protein, resulting in increased cell proliferation. The bottom portion of the diagram depicts mutations in both ALK-1 and endoglin that operate by means of a similar pathway in pulmonary endothelial cells to promote hypertension in combination with hemorrhagic telangiectasia. This pathway uses different ligands (e.g., activin), Smads (2 and 3), receptor molecules, and acts in dynamic opposition to a third pathway, the ALK-5–dependent pathway, as indicated. Interacting epigenetic and genetic factors are thought to be necessary to modulate the effects of these mutations on hypertension phenotype.

translocate to the nucleus and directly regulate gene transcription (Fig. 2). In addition to Smad-mediated signaling, BMPs and TGF-ßs also activate mitogen-activating protein (MAP) kinases, including ERK, JNK, and p38MAPK. The net result of TGF-ß signaling on vascular growth, proliferation, and structure is rather complex, depending on the ligand, the formation of heteromeric receptor complexes, cell type, downstream signals, and the ongoing transcriptional program.[105]

Levels of the BMPR-II protein are markedly reduced in patients with PPH and more moderately in patients with sporadic disease. It is thought that the down-regulation of the BMP pathway reduces a growth arrest regulation contributing to the extensive vascular proliferation and remodeling that characterizes the condition. Mutations in *BMPR2* have been found that result in increased mRNA decay and lower protein levels yielding haploinsufficiency. By use of multiplex ligation-dependent probe amplification, enabling a full characterization of the *BMPR2* gene at the exon level, more than nine novel *BMPR2* rearrangements were recently identified in 12% of familial PPH cases and roughly 5% of idiopathic PPH cases.[106] Importantly, this analysis also described two *BMPR2* deletions that encompassed all functional protein domains and were predicted to result in null mutations, providing further support that the predominant molecular mechanism for PPH disease predisposition is *BMPR2* haploinsufficiency. Other mutations (as depicted in Fig. 2) have been detected that affect the ligand binding, effecting the trafficking of the receptor to the cell surface that disrupts Smad signaling and that enhances p38 activation.[107]

Mutations within other receptors of the transforming growth factor-β family have recently been reported in patients with PPH, suggesting that the genetic etiology of PPH is heterogeneous in nature.[108,109] Mutations in the *ALK1* gene encoding the activin-like kinase type-1 (ALK-1) have been found in several patients with hereditary hemorrhagic and co-existent PPH. Mutations have been identified within the *ENG* gene encoding endoglin, a type III receptor responsive to TGF-β ligand signaling like *ALK*-1 in vascular endothelial cells in patients with hereditary hemorrhagic telangiectasia with associated PPH.[110] The observation that mutations in different, but mechanistically related, genes, *ALK1*, *ENG*, and *BMPR2*, can produce the same clinical phenotype points to TGF-β receptors as an important molecular pathway at the origin of pulmonary vascular remodeling.

Screening of other families with PPH has suggested the involvement of other genetic loci in PPH, including one at chromosome 2q.31 designated PPH2.[111,112] In addition, Grunig and associates recently found no evidence of *BMPR2* mutations or linkage of PPH to either chromosome 2 or 12 in the analysis of 13 unrelated children with PPH and their families and have proposed a substantially different pattern of inheritance (i.e., recessive) in the PPH phenotype in children compared with adults.[113] To explain the genetic basis of PPH further, genome scanning for major and minor genes (Table V), and the analysis of genetic profiles of patients for candidate genes that potentially modify the risk for PPH disease (e.g., serotonin transporter *5-HTT* alleles, nitric oxide synthases) are underway as are studies of PPH that use proteomics, transgenic mice, and studies of altered signal transduction. For instance, studies investigating the molecular processes that underlie the complex vascular changes associated with PPH have identified major pathways involved in constriction and proliferation of pulmonary vascular SMCs,

dysfunction of endothelial cells, and ECM remodeling that may be involved either in initiating or in perpetuating the disease.

Pathophysiological studies of patients with PPH have indicated a critical role for the L-variant in the promoter of the *5-HTT* gene encoding the serotonin transporter, which is associated with 5-HTT overexpression and increased proliferative growth of pulmonary artery SMCs.[114] The L-allelic variant of *5-HTT* was present in homozygous form in 65% of patients but in only 27% of controls, suggesting that this *5-HTT* polymorphism confers susceptibility to PPH. The role of *5-HTT* overexpression in the genesis of PPH has recently been confirmed in studies of transgenic mice overexpressing 5-*HTT* under the control of the SM22 promoter.[115] Compared with wild-type mice, mice containing the *SM22-5-HTT* construct exhibited a threefold to fourfold increase in lung *5-HTT* mRNA and protein, together with increased lung 5-HT uptake activity. By 8 weeks of age, the transgenic mice exhibited PPH, with marked increases in right ventricular systolic pressure and muscularization of distal pulmonary vessels, but no changes in systemic arterial pressure, a phenotype that worsened with age.

The integral role of matrix metalloproteinases (MMPs) in the arterial wall remodeling occurring in the pulmonary artery SMCs hyperplasia accompanying PPH suggested that MMPs might be potential factors contributing to PPH susceptibility or modifiers of the PPH phenotype.[116] Both *in situ* and *in vitro* studies performed on pulmonary arteries from patients undergoing lung transplantation for PAH found increased levels of the MMP-tissue inhibitor of metalloproteinase (TIMP-1) and altered levels of MMP-2 and MMP-3. Increasing focus has been directed to therapies that target factors that contribute to the vascular remodeling and the resultant thrombosis occurring in small and medium-sized pulmonary arteries and arterioles in patients with PPH, including growth and inflammatory mediators such as

TABLE V Candidate Genes for Pulmonary Hypertension

TGF-βRII
Bax
Endothelial nitric oxide synthase (eNOS)
5-Lipoxygenase (5LO)
Prostacyclin synthase (PGIS)
Serotonin transporter (5-HTT)
Bone morphogenetic protein receptor type-2 (BMPR2)
Activin-like kinase type-1 (ALK1)
IQ-GTPase-activating protein-1 (IQGAP1)
Decorin
Insulin-like growth factor binding protein-3 (IGFBP3)
Lactotransferrin
Endoglin (ENG)

TABLE VI Gene Profiling in Familial and Sporadic Human Pulmonary Hypertension[a]

Oncogene and cancer-related genes	Transcript level	Etiology
v-myc	↑	F/S
jun D proto-oncogene	↑	F/S
vav 1 oncogene	↑	F/S
Laminin receptor 1 (67 kDa)	↑	S
Pre–B-cell leukemia transcription factor 2	↑	S
Replication protein A2 (32 kDa)	↑	S
Apoptosis resistance		
Bcl-2	↑	F/S
Inositol 1,4,5-triphosphate receptor, type 3	↓	F/S
Caspase 9	↑	S
Mitochondrial proteins		
Mitochondrial outer membrane protein 19	↓	F/S
Cytochrome oxidase subunit X	↑	F/S
Mitochondrial citrate transport protein	↓	F/S
Mitochondrial H+-transporting ATP synthase F1	↓	F/S
Ubiquinol—cytochrome c reductase core protein 1	↓	F/S
Genes coding for ion channels		
Inward rectifier K+ channel	↑	F/S
Voltage-gated Na+-channel type Iβ polypeptide	↑	F/S
Voltage-gated, shaker-related K+ channel	↓	S
Chloride channel 1	↑	F/S
Genes encoding proteins of the TGF-β signaling superfamily		
TGF-β3	↓	S
TGF-βRIII (B-glycan)	↓	S
BMP4	↑	S
BMP2	↓	S
Smad1	↓	S
Genes encoding kinases		
TYK tyrosine kinase	↑	S
p21/Cdc 42/Rac 1–activated kinase	↑	S
G protein–coupled receptor kinase 6	↑	S
MAP/microtubule affinity regulating kinase 3	↑	S
CaM kinase II	↓	S
Cyclin-dependent kinase 9	↓	S
MAPK-activated protein kinase 3	↓	S
Casein kinase 2 1 polypeptide	↑	S
MAPKK5	↑	S

[a]Data from Geraci et al.[118]; S, sporadic; F/S, familial and sporadic; ↑ and ↓, increased and decreased specific transcript levels relative to normal patients.

prostaglandin-1 (prostacyclin), nitric oxide (NO), endothelin-1, serotonin, cytokines, chemokines, and TGF-β.[117]

Although evidence has been presented that the sporadic and familial form of PPH have different gene expression profiles[118] (Table VI), as well as different levels of microsatellite instability,[119] it remains unclear whether they are distinct and separable disorders. Other studies have shown that a high incidence of patients (81%) with PPH had relatives with active or latent disease, indicating a primary genetic role in determining disease.[120] Moreover, a considerable percentage of patients previously considered to have the "sporadic" form of PPH were found to contain specific PPH-related mutations (e.g., *BMPR2*) and patients harboring specific genetic defects have been identified who may not manifest the disease because of their low penetrance, whose basis also remains to be determined. In this context, it is noteworthy that transgenic mice heterozygous for *BMPR2* display pulmonary hemodynamics and vascular morphometry similar to wild-type litter mate controls under normoxic or chronic hypoxic conditions.[121] However, chronic infusion of serotonin promotes increased pulmonary artery systolic pressure, right ventricular hypertrophy, and pulmonary artery remodeling in the *BMPR2* heterozygotes compared with wild-type mice, suggesting a significant linkage between the serotonin and the *BMPR2* pathways that may modulate the penetrance of *BMPR2* mutations. Finally, familial PPH is characterized by genetic anticipation (i.e., the disease becomes increasingly severe or occurs at an earlier age with successive generations) suggestive that disease-modifying genes or environmental factors influence the expression of the PPH phenotype.[122] The molecular basis for genetic anticipation in pulmonary hypertension remains undetermined. In other diseases in which genetic anticipation is operative (e.g., fragile X syndrome, Huntington disease, myotonic dystrophy), the mechanism most often documented is an expansion of nucleotide repeats (e.g., trinucleotide or pentanucleotide repeats associated with spinocerebellar ataxias).

Gene Profiling in Pulmonary Hypertension

Gene expression profiling of pulmonary hypertension in both animal models (e.g., mice) and in patients has been performed in both lung tissue[118,123,124] and in peripheral blood mononuclear cells.[125] Interestingly, although patients with sporadic and familial forms of pulmonary hypertension shared some gene profiling features, a considerable number of genes displayed distinguishable patterns of expression in these hypertensive phenotypes with many more being affected in the sporadic cases of PPH. These studies have also identified a relatively large group of genes (approximately 300) whose differential expression defines a molecular signature for pulmonary hypertension, including several genes coding for proteins of the TGF-β signaling superfamily, ion channel genes, protein kinases and phosphatases, ribosomal proteins, mitochondrial metabolic proteins, transcription factors, and oncoproteins, a number of which are shown in Table VI. The findings of differential expression of growth-related genes are consistent with the important role that abnormal cell growth and phenotypic alterations of pulmonary arterial endothelial cells and vascular SMCs play in the pathobiology of PPH. Similarly, changes in the ECM, cytoskeletal, and apoptosis-related proteins found in patients with pulmonary hypertension are consistent with the essential role of tissue remodeling that occurs with pulmonary hypertension.[126] Interestingly, there has been limited global proteomic analysis of PPH in either animal models or clinical studies.

In addition, subsequent studies have documented decreased levels of the global transcription regulator, PPAR-γ, in both lung tissues and endothelial cells of patients with PPH.[127] Ligand activators of PPAR-γ have been reported to exert antiproliferative, anti-inflammatory, antifibrotic, and antioxidant effects on vascular cells, including smooth SMCs, both *in vivo* and *in vitro*[128]; therefore, lower PPAR-γ levels could be a potential factor contributing to pulmonary hypertension.

FUTURE DIRECTIONS

The increasing use of pharmacogenetic studies with larger study groups to evaluate specific genotypes and phenotypes in response to specific treatment regimens may provide extremely important information about hypertension. These also offer the long-term goal of tailoring medication on the basis of genotype to maximize drug efficiency and minimize individual adverse reactions.

The delineation of the chromosomal loci of both identified QTLs in animal models and candidate genetic loci by use of molecular technologies may eventually contribute to the understanding of the genes (and interacting factors) involved in hypertension. To unravel the web of interacting genetic factors impacting the hypertensive phenotype, the concerted use of several of the recent advances in SNP and haplotype genotyping techniques may be adapted to perform genome-wide multiplex SNP screening to simultaneously evaluate multiple genetic loci involved in hypertension. Also, there is a recognized need for further prospective clinical studies to incorporate stringent evaluation of patient histories and thoroughly catalog potential interacting (and confounding) environmental factors in their subjects.

Information from these findings may also result not only in enhanced effectiveness of known drugs but also in the design of novel therapeutic regimens and in the use of gene therapies, as we have previously described.

SUMMARY

- Hypertension affects up to 30% of the adult population in Western societies and is a major risk factor for kidney disease, stroke, and coronary heart disease. The most common form, essential hypertension, is a complex multifactorial trait with demonstrable genetic (polygenic) and environmental components, whose precise etiology remains undetermined.
- Rare hypertensive phenotypes inherited as monogenic Mendelian traits include Gordon and Liddle syndromes. Most of the genes contributing to these rare forms of hypertension are involved in the renin-angiotensin pathway or are related components. Some of the signaling components (e.g., WNK) implicated in these rare phenotypes may also play a role in the more common hypertensive phenotype.
- Long-term caloric intakes, dietary sodium, excessive alcohol consumption, and stresses all are contributory

factors and primary risk factors in the development of hypertension. In addition, body weight and obesity are positively associated with increased blood pressure (BP) and hypertension.
- Lifestyle modification, including increased physical activity, a reduced salt intake, weight loss, moderation of alcohol intake, increased potassium intake, and healthy dietary pattern can effectively lower BP.
- Genome-wide screening has indicated a number of chromosomal loci with a moderate association with essential hypertension, although findings of identified loci are often not replicable in studies with different populations. Linkage analysis supports the notion that essential hypertension is a polygenic trait whose phenotypic expression is influenced by ethnic and gender background.
- Analysis of candidate genes by case-control studies has identified specific gene loci with moderate effect on hypertension expression such as angiotensin and β-adrenergic receptor.
- Animal models of hypertension (particularly rats and transgenic mice) have proved informative in identifying genes with contributory effects on hypertension. Linkage studies have revealed specific chromosomal loci (QTLs), which have a strong association with hypertension phenotypes, and, although rather large, offer the potential for identifying specific candidates genes involved in those phenotypes. Some of these QTLs are syntenic with human chromosomal loci.
- The use of designer rats such as congenic and consomic strains for the confirmation of QTLs has significantly enabled the genetic dissection of the implicated regions. These allow the studies of isolated chromosomes or segments in the same genetic background.
- Analysis of gene profiling in both animal studies and patient tissues has further identified genes whose expression contributes to hypertension. Expression profiling in congenic hypertensive rat strains has been particularly informative.
- Primary pulmonary hypertension (PPH) is a rare autosomal-dominant disease with incomplete penetrance. Mutations in the BMP and TGF-β signaling pathway, including the BMPR2, endoglin, and activin receptors, have been implicated in a large number of patients with both familial and sporadic disease. Changes in the metalloproteinases affecting tissue remodeling and in the serotonin signaling pathway have also been implicated in the pathogenesis of PPH.

References

1. Snieder, H., Harshfield, G. A., and Treiber, F. A. (2003). Heritability of blood pressure and hemodynamics in African- and European-American youth. *Hypertension* **41,** 1196–1201.
2. Snieder, H., and Treiber, F. A. (2002). The Georgia Cardiovascular Twin Study. *Twin Res.* **5,** 497–498.
3. Imumorin, I. G., Dong, Y., Zhu, H., Poole, J. C., Harshfield, G. A., Treiber, F. A., and Snieder, H. (2005). A gene-environment interaction model of stress-induced hypertension. *Cardiovasc. Toxicol.* **5,** 109–132.
4. Truswell, A. S., Kennelly, B. M., Hansen, J. D., and Lee, R. B. (1972). Blood pressures of Kung bushmen in Northern Botswana. *Am. Heart. J.* **84,** 5–12.
5. Poulter, N. R., Khaw, K. T., Mugambi, M., Peart, W. S., Rose, G., and Sever, P. (1985). Blood pressure patterns in relation to age, weight and urinary electrolytes in three Kenyan communities. *Trans. R. Soc. Trop. Med. Hyg.* **79,** 389–392.
6. Bielen, E. C., Fagard, R., and Amery, A. K. (1991). Inheritance of blood pressure and haemodynamic phenotypes measured at rest and during supine dynamic exercise. *J. Hypertens.* **9,** 655–663.
7. Kotchen, T. A., Kotchen, J. M., Grim, C. E., Varghese, G., Kaldunski, M. L., Cowley, A. W., Hamet, P., and Chelius, T. H. (2000). Genetic determinants of hypertension: identification of candidate phenotypes. *Hypertension* **36,** 7–13.
8. Sylos, C., Pereira, A. C., Azeka, E., Miura, N., Mesquita, S. M., and Ebaid, M. (2002). Arterial hypertension in a child with Williams-Beuren syndrome (7q11.23 chromosomal deletion). *Arq. Bras. Cardiol.* **79,** 173–180.
9. Urban, Z., Riazi, S., Seidl, T. L., Katahira, J., Smoot, L. B., Chitayat, D., Boyd, C. D., and Hinek, A. (2002). Connection between elastin haploinsufficiency and increased cell proliferation in patients with supravalvular aortic stenosis and Williams-Beuren syndrome. *Am. J. Hum. Genet.* **71,** 30–44.
10. D'Armiento, J. (2003). Decreased elastin in vessel walls puts the pressure on. *J. Clin. Invest.* **112,** 1308–1310.
11. Toka, H. R., and Luft, F. C. (2002). Monogenic forms of human hypertension. *Semin. Nephrol.* **22,** 81–88.
12. Lifton, R. P., Gharavi, A. G., and Geller, D. S. (2001). Molecular mechanisms of human hypertension. *Cell* **104,** 545–556.
13. Mune, T., Rogerson, F. M., Nikkilä, H., Agarwal, A. K., and White, P. C. (1995). Human hypertension caused by mutations in the kidney isozyme of 11 ß-hydroxysteroid dehydrogenase. *Nat. Genet.* **10,** 394–399.
14. Geller, D. S., Rodriguez-Soriano, J., Vallo Boado, A., Schifter, S., Bayer, M., Chang, S. S., and Lifton, R. P. (1998). Mutations in the mineralocorticoid receptor gene cause autosomal dominant pseudohypoaldosteronism type I. *Nat. Genet.* **19,** 279–281.
15. Geller, D. S., Farhi, A., Pinkerton, N., Fradley, M., Moritz, M., Spitzer, A., Meinke, G., Tsai, F. T., Sigler, P. B., and Lifton, R. P. (2000). Activating mineralocorticoid receptor mutation in hypertension exacerbated by pregnancy. *Science* **289,** 119–123.
16. Chang, S. S., Grunder, S., Hanukoglu, A., Rosler, A., Mathew, P. M., Hanukoglu, I., Schild, L., Lu, Y., Shimkets, R. A., Nelson-Williams, C., Rossier, B. C., and Lifton, R. P. (1996). Mutations in subunits of the epithelial sodium channel cause salt wasting with hyperkalaemic acidosis, pseudohypoaldosteronism type 1. *Nat. Genet.* **12,** 248–2.53
17. Hansson, J. H., Nelson-Williams, C., Suzuki, H., Schild, L., Shimkets, R., Lu, Y., Canessa, C., Iwasaki, T., Rossier, B., and Lifton, R. P. (1995). Hypertension caused by a truncated epithelial sodium channel gamma subunit: genetic heterogeneity of Liddle syndrome. *Nat. Genet.* **11,** 76–82.

18. Mansfield, T. A., Simon, D. B., Farfel, Z., Bia, M., Tucci, J. R., Lebel, M., Gutkin, M., Vialettes, B., Christofilis, M. A., Kauppinen-Makelin, R., Mayan, H., Risch, N., and Lifton, R. P. (1997). Multilocus linkage of familial hyperkalaemia and hypertension, pseudohypo-aldosteronism type II, to chromosomes 1q31-42 and 17p11-q21. *Nat. Genet.* **16,** 202–205.
19. Wilson, F. H., Disse-Nicodeme, S., Choate, K. A., Ishikawa, K., Nelson-Williams, C., Desitter, I., Gunel, M., Milford, D. V., Lipkin, G. W., Achard, J. M., Feely, M. P., Dussol, B., Berland, Y., Unwin, R. J., Mayan, H., Simon, D. B., Farfel, Z., Jeunemaitre, X., and Lifton, R. P. (2001). Human hypertension caused by mutations in WNK kinases. *Science* **293,** 1107–1112.
20. Wilson, F. H., Kahle, K. T., Sabath, E., Lalioti, M. D., Rapson, A. K., Hoover, R. S., Hebert, S. C., Gamba, G., and Lifton, R. P. (2003). Molecular pathogenesis of inherited hypertension with hyperkalemia: the Na-Cl cotransporter is inhibited by wild-type but not mutant WNK4. *Proc. Natl. Acad. Sci. USA* **100,** 680–684.
21. Kahle, K. T., Wilson, F. H., Leng, Q., Lalioti, M. D., O'Connell, A. D., Dong, K., Rapson, A. K., MacGregor, G. G., Giebisch, G., Hebert, S. C., and Lifton, R. P. (2003). WNK4 regulates the balance between renal NaCl reabsorption and K+ secretion. *Nat. Genet.* **35,** 372–376.
22. Wang, Z., Yang, C. L., and Ellison, D. H. (2004). Comparison of WNK4 and WNK1 kinase and inhibiting activities. Biochem. Biophys. Res. Commun. **317,** 939–944.
23. Yang, C. L., Angell, J., Mitchell, R., and Ellison, D. H. (2003). WNK kinases regulate thiazide-sensitive Na-Cl cotransport. *J. Clin. Invest.* **111,** 1039–1045.
24. Xu, B. E., Stippec, S., Chu, P. Y., Lazrak, A., Li, X. J., Lee, B. H., English, J. M., Ortega, B., Huang, C. L., and Cobb, M. H. (2005). WNK1 activates SGK1 to regulate the epithelial sodium channel. *Proc. Natl. Acad. Sci. USA* **102,** 10315–10320.
25. Kahle, K. T., Wilson, F. H., Lalioti, M., Toka, H., Qin, H., and Lifton, R. P. (2004). WNK kinases: molecular regulators of integrated epithelial ion transport. *Curr. Opin. Nephrol. Hypertens.* **13,** 557–562.
26. Cope, G., Golbang, A., and O'Shaughnessy, K. M. (2005). WNK kinases and the control of blood pressure. *Pharmacol. Ther.* **106,** 221–231.
27. Newhouse, S. J., Wallace, C., Dobson, R., Mein, C., Pembroke, J., Farrall, M., Clayton, D., Brown, M., Samani, N., Dominiczak, A., Connell, J. M., Webster, J., Lathrop, G. M., Caulfield, M., and Munroe, P. B. (2005). Haplotypes of the WNK1 gene associate with blood pressure variation in a severely hypertensive population from the British Genetics of Hypertension study. *Hum. Mol. Genet.* **14,** 1805–1814.
28. Naraghi, R., Schuster, H., Toka, H. R., Bahring, S., Toka, O., Oztekin, O., Bilginturan, N., Knoblauch, H., Wienker, T. F., Busjahn, A., Haller, H., Fahlbusch, R., and Luft, F. C. (1997). Neurovascular compression at the ventrolateral medulla in autosomal dominant hypertension and brachydactyly. *Stroke* **28,** 1749–1754.
29. Schuster, H., Wienker, T. E., Bähring, S., Bilginturan, N., Toka, H. R., Neitzel, H., Jeschke, E., Toka, O., Gilbert, D., Lowe, A., Ott, J., Haller, H., and Luft, F. C. (1996). Severe autosomal dominant hypertension and brachydactyly in a unique Turkish kindred maps to human chromosome 12. *Nat. Genet.* **13,** 98–100.
30. Gong, M., Zhang, H., Schulz, H., Lee, Y. A., Sun, K., Bähring, S., Luft, F. C., Nürnberg, P., Reis, A., Rohde, K., Ganten, D., Hui, R., and Hübner, N. (2003). Genome-wide linkage reveals a locus for human essential (primary) hypertension on chromosome 12p. *Hum. Mol. Genet.* **12,** 1273–1277.
31. Rao, D. C., Province, M. A., Leppert, M. F., Oberman, A., Heiss, G., Ellison, R. C., Arnett, D. K., Eckfeldt, J. H., Schwander, K., Mockrin, S. C., and Hunt, S. C., and the HyperGEN Network. (2003). A genome-wide affected sibpair linkage analysis of hypertension: the HyperGEN network. *Am. J. Hypertens.* **16,** 148–150.
32. Kardia, S. L., Rozek, L. S., Krushkal, J., Ferrell, R. E., Turner, S. T., Hutchinson, R., Brown, A., Sing, C. F., and Boerwinkle, E. (2003). Genome-wide linkage analyses for hypertension genes in two ethnically and geographically diverse populations. *Am. J. Hypertens.* **16,** 154–157.
33. Thiel, B. A., Chakravarti, A., Cooper, R. S., Luke, A., Lewis, S., Lynn, A., Tiwari, H., Schork, N. J., and Weder, A. B. (2003). A genome-wide linkage analysis investigating the determinants of blood pressure in whites and African Americans. *Am. J. Hypertens.* **16,** 151–153.
34. Province, M. A., Kardia, S. L., Ranade, K., Rao, D. C., Thiel, B. A., Cooper, R. S., Risch, N., Turner, S. T., Cox, D. R., Hunt, S. C., Weder, A. B., Boerwinkle, E., and the National Heart, Lung and Blood Institute Family Blood Pressure Program. (2003). A meta-analysis of genome-wide linkage scans for hypertension: the National Heart, Lung and Blood Institute Family Blood Pressure Program. *Am. J. Hypertens.* **16,** 144–147.
35. Ranade, K., Hinds, D., Hsiung, C. A., Chuang, L. M., Chang, M. S., Chen, Y. T., Pesich, R., Hebert, J., Chen, Y. D., Dzau, V., Olshen, R., Curb, D., Botstein, D., Cox, D. R., and Risch, N. (2003). A genome scan for hypertension susceptibility loci in populations of Chinese and Japanese origins. *Am. J. Hypertens.* **16,** 158–162.
36. Pankow, J. S., Dunn, D. M., Hunt, S. C., Leppert, M. F., Miller, M. B., Rao, D. C., Heiss, G., Oberman, A., Lalouel, J. M., and Weiss, R. B. (2005). Further evidence of a quantitative trait locus on chromosome 18 influencing postural change in systolic blood pressure: the Hypertension Genetic Epidemiology Network (HyperGEN) Study. *Am. J. Hypertens.* **18,** 672–678.
37. Caulfield, M., Munroe, P., Pembroke, J., Samani, N., Dominiczak, A., Brown, M., Benjamin, N., Webster, J., Ratcliffe, P., O'Shea, S., Papp, J., Taylor, E., Dobson, R., Knight, J., Newhouse, S., Hooper, J., Lee, W., Brain, N., Clayton, D., Lathrop, G. M., Farrall, M., Connell, J. and the MRC British Genetics of Hypertension Study. (2003). Genome-wide mapping of human loci for essential hypertension. *Lancet* **361,** 2118–2123.
38. von Wowern, F., Bengtsson, K., Lindgren, C. M., Orho-Melander, M., Fyhrquist, F., Lindblad, U., Rastam, L., Forsblom, C., Kanninen, T., Almgren, P., Burri, P., Katzman, P., Groop, L., Hulthen, U. L., and Melander, O. (2003). A genome wide scan for early onset primary hypertension in Scandinavians. *Hum. Mol. Genet.* **12,** 2077–2081.
39. Atwood, L. D., Samollow, P. B., Hixson, J. E., Stern, M. P., and MacCluer, J. W. (2001). Genome-wide linkage analysis of

blood pressure in Mexican Americans. *Genet. Epidemiol.* **20**, 373–382.
40. Von Wowern, F., Bengtsson, K., Lindblad, U., Rastam, L., and Melander, O. (2004). Functional variant in the (alpha)2B adrenoceptor gene, a positional candidate on chromosome 2, associates with hypertension. *Hypertension* **43**, 592–597.
41. Kristjansson, K., Manolescu, A., Kristinsson, A., Hardarson, T., Knudsen, H., Ingason, S., Thorleifsson, G., Frigge, M., Kong, A., Glucher, J., and Stefansson, K. (2002). Linkage of essential hypertension to chromosome 18q. *Hypertension* **39**, 1044–1049.
42. Benjafield, A. V., Wang, W. Y., Speirs, H. J., and Morris, B. J. (2005). Genome-wide scan for hypertension in Sydney Sibships: the GENIHUSS study. *Am. J. Hypertens.* **18**, 828–832.
43. de Lange, M., Spector, T. D., and Andrew, T. (2004). Genome-wide scan for blood pressure suggests linkage to chromosome 11 and replication of loci on 16, 17, and 22. *Hypertension* **44**, 872–877.
44. Chen, W., Li, S., Srinivasan, S. R., Boerwinkle, E., and Berenson, G. S. (2005). Autosomal genome scan for loci linked to blood pressure levels and trends since childhood: the Bogalusa Heart Study. *Hypertension* **45**, 954–959.
45. Koivukoski, L., Fisher, S. A., Kanninen, T., Lewis, C. M., von Wowern, F., Hunt, S., Kardia, S. L., Levy, D., Perola, M., Rankinen, T., Rao, D. C., Rice, T., Thiel, B. A., and Melander, O. (2004). Meta-analysis of genome-wide scans for hypertension and blood pressure in Caucasians shows evidence of susceptibility regions on chromosomes 2 and 3. *Hum. Mol. Genet.* **13**, 2325–2332.
46. Mein, C. A., Caulfield, M. J., Dobson, R. J., and Munroe, P. B. (2004). Genetics of essential hypertension. *Hum. Mol. Genet.* **13**, R169–R175.
47. Jeunemaitre, X., Inoue, I., Williams, C., Charru, A., Tichet, J., Powers, M., Sharma, A. M., Gimenez-Roqueplo, A. P., Hata, A., Corvol, P., and Lalouel, J. M. (1997). Haplotypes of angiotensinogen in essential hypertension. *Am. J. Hum. Genet.* **60**, 1448–1460.
48. Kato, N., Sugiyama, T., Morita, H., Kurihara, H., Yamori, Y., and Yazaki, Y. (1999). Angiotensinogen gene and essential hypertension in the Japanese: extensive association study and meta-analysis on six reported studies. *J. Hypertens.* **17**, 757–763.
49. Inoue, I., Nakajima, T., Williams, C. S., Quackenbush, J., Puryear, R., Powers, M., Cheng, T., Ludwig, E. H., Sharma, A. M., Hata, A., Jeunemaitre, X., and Lalouel, J. M. (1997). A nucleotide substitution in the promoter of human angiotensinogen is associated with essential hypertension and affects basal transcription in vitro. *J. Clin. Invest.* **99**, 1786–1797.
50. Sethi, A. A., Nordestgaard, B. G., Gronholdt, M. L., Steffensen, R., Jensen, G., and Tybjaerg-Hansen, A. (2003). Angiotensinogen single nucleotide polymorphisms, elevated blood pressure, and risk of cardiovascular disease. *Hypertension* **41**, 1202–1211.
51. Zhu, X., Chang, Y. P., Yan, D., Weder, A., Cooper, R., Luke, A., Kan, D., and Chakravarti, A. (2003). Associations between hypertension and genes in the renin-angiotensin system. *Hypertension* **41**, 1027–1034.
52. Higaki, J., Baba, S., Katsuya, T., Sato, N., Ishikawa, K., Mannami, T., Ogata, J., and Ogihara, T. (2000). Deletion allele of angiotensin-converting enzyme gene increases risk of essential hypertension in Japanese men: the Suita Study. *Circulation* **101**, 2060–2065.
53. Turner, S. T., Boerwinkle, E., and Sing, C. F. (1999). Context-dependent associations of the ACE I/D polymorphism with blood pressure. *Hypertension* **34**, 773–778.
54. Bray, M. S., Krushkal, J., Li, L., Ferrell, R., Kardia, S., Sing, C. F., Turner, S. T., and Boerwinkle, E. (2000). Positional genomic analysis identifies the $\beta(2)$-adrenergic receptor gene as a susceptibility locus for human hypertension. *Circulation* **101**, 2877–2882.
55. Herrmann, S. M., Nicaud, V., Tiret, L., Evans, A., Kee, F., Ruidavets, J. B., Arveiler, D., Luc, G., Morrison, C., Hoehe, M. R., Paul, M., and Cambien, F. (2002;). Polymorphisms of the $\beta 2$-adrenoceptor (ADRB2) gene and essential hypertension: the ECTIM and PEGASE studies. *J. Hypertens.* **20**, 229–235.
56. Krushkal, J., Xiong, M., Ferrell, R., Sing, C. F., Turner, S. T., and Boerwinkle, E. (1998). Linkage and association of adrenergic and dopamine receptor genes in the distal portion of the long arm of chromosome 5 with systolic blood pressure variation. *Hum. Mol. Genet.* **7**, 1379–1383.
57. Rapp, J. P. (2000). Genetic analysis of inherited hypertension in the rat. *Physiol. Rev.* **80**, 135–172.
58. Garrett, M. R., Joe, B., Dene, H., and Rapp, J. P. (2002). Identification of blood pressure quantitative trait loci that differentiate two hypertensive strains. *J. Hypertens.* **20**, 2399–2406.
59. Kwitek-Black, A. E., and Jacob, H. J. (2001). The use of designer rats in the genetic dissection of hypertension. *Curr. Hypertens. Rep.* **3**, 12–18.
60. Graham, D., McBride, M. W., Brain, N. J., and Dominiczak, A. F. (2004). Congenic/consomic models of hypertension. *Methods Mol. Med.* **108**, 3–16.
61. Negrin, C. D., McBride, M. W., Carswell, H. V. O., Graham, D., Carr, F. J., Clark, J. S., Jeffs, B., Anderson, N. H., Macrae, I. M., and Dominiczak, A. F. (2001). Reciprocal consomic strains to evaluate Y chromosome effects. *Hypertension* **37**, 391–397.
62. Cowley, A. W., Roman, R. J., Kaldunski, M. L., Dumas, P., Dickout, J. D., Greene, A. S., and Jacob, H. J. (2001). Brown Norway chromosome 13 confers protection from high salt to consomic Dahl S rats. *Hypertension* **37**, 456–461.
63. Rapp, J. P., Garrett, M. R., and Deng, A. Y. (1998). Construction of a double congenic strain to prove epistatic interaction on blood pressure between rat chromosomes 2 and 10. *J. Clin. Invest.* **101**, 1591–1595.
64. Monti, J., Plehm, R., Schultz, H., Ganten, D., Kreutz, R., and Hubner, N. (2003). Interaction between blood pressure QTL in rats in which trait variation at chromosome 1 is conditional upon a specific allele at chromosome 10. *Hum. Mol. Genet.* **12**, 435–439.
65. Garrett, M. R., and Rapp, J. P. (2003). Defining the blood pressure QTL on chromosome 7 in Dahl rats by a 177-kb congenic segment containing Cyp11b1. *Mamm. Genome* **14**, 268–273.
66. Stoll, M., Kwitek-Black, A. E., Cowley, A. W., Harris, E. L., Harrap, S. B., Krieger, J. E., Printz, M. P., Provoost, A. P., Sassard, J., and Jacob, H. J. (2000). New target regions for human hypertension via comparative genomics. *Genome Res.* **10**, 473–482.

67. Kato, N. (2002). Genetic analysis in human hypertension. *Hypertens. Res.* **25,** 319–327.
68. Manunta, P., Barlassina, C., and Bianchi, G. (1998). Adducin in essential hypertension. *FEBS Lett.* **430,** 41–44.
69. Lanzani, C., Citterio, L., Jankaricova, M., Sciarrone, M. T., Barlassina, C., Fattori, S., Messaggio, E., Serio, C. D., Zagato, L., Cusi, D., Hamlyn, J. M., Stella, A., Bianchi, G., and Manunta, P. (2005). Role of the adducin family genes in human essential hypertension. *J. Hypertens.* **23,** 543–549.
70. Snyder, P. M., Steines, J. C., and Olson, D. R. (2004). Relative contribution of Nedd4 and Nedd4-2 to ENaC regulation in epithelia determined by RNA interference. *J. Biol. Chem.* **279,** 5042–5046.
71. Kim, H. S., Krege, J. H., Kluckman, K. D., Hagaman, J. R., Hodgin, J. B., Best, C. F., Jennette, J. C., Coffman, T. M., Maeda, N., and Smithies, O. (1995). Genetic control of blood pressure and the angiotensinogen locus. *Proc. Natl. Acad. Sci. USA* **92,** 2735–2739.
72. Shindo, T., Kurihara, Y., Nishimatsu, H., Moriyama, N., Kakoki, M., Wang, Y., Imai, Y., Ebihara, A., Kuwaki, T., Ju, K. H., Minamino, N., Kangawa, K., Ishikawa, T., Fukuda, M., Akimoto, Y., Kawakami, H., Imai, T., Morita, H., Yazaki, Y., Nagai, R., Hirata, Y., and Kurihara, H. (2001). Vascular abnormalities and elevated blood pressure in mice lacking adrenomedullin gene. *Circulation* **104,** 1964–1971.
73. Alfie, M. E., Sigmon, D. H., Pomposiello, S.I., and Carretero, O. A. (1997). Effect of high salt in-take in mutant mice lacking bradykinin-B2 receptors. *Hypertension* **29,** 483–487.
74. Lopez, M. J., Wong, S. K., Kishimoto, I., Dubois, S., Mach, V., Friesen, J., Garbers, D. L., and Beuve, A. (1995). Salt-resistant hypertension in mice lacking the guanylyl cyclase-A receptor for atrial natriuretic peptide. *Nature* **378,** 65–68.
75. Melo, L. G., Veress, A. T., Chong, C. K., Ackermann, U., and Sonnenberg, H. (1999). Salt-sensitive hypertension in ANP knockout mice is prevented by AT1 receptor antagonist losartan. *Am. J. Physiol.* **277,** R624–R630.
76. Matsusaka, T., Fogo, A., and Ichikawa, I. (1997). Targeting the genes of angiotensin receptors. *Semin. Nephrol.* **17,** 396–403.
77. Siragy, H. M., Inagami, T., Ichiki, T., and Carey, R. M. (1999). Sustained hypersensitivity to angiotensin II and its mechanism in mice lacking the subtype-2 (AT2) angiotensin receptor. *Proc. Natl. Acad. Sci. USA* **96,** 6506–6510.
78. Yang, R., Powell-Braxton, L., Ogaoawara, A. K., Dybdal, N., Bunting, S., Ohneda, O., and Jin, H. (1999). Hypertension and endothelial dysfunction in apolipoprotein E knockout mice. *Arterioscler. Thromb. Vasc. Biol.* **19,** 2762–2768.
79. Marro, M. L., Scremin, O. U., Jordan, M. C., Huynh, L., Porro, F., Roos, K. P., Gajovic, S., Baralle, F. E., and Muro, A. F. (2000). Hypertension in β-adducin-deficient mice. *Hypertension* **36,** 449–453.
80. Gangula, P. R., Zhao, H., Supowit, S. C., Wimalawansa, S. J., Dipette, D. J., Westlund, K. N., Gagel, R. F., and Yallampalli, C. (2000). Increased blood pressure in α-calcitonin gene-related peptide/calcitonin gene knockout mice. *Hypertension* **35,** 470–475.
81. Yang, Z., Sibley, D. R., and Jose, P. A. (2004). D5 dopamine receptor knockout mice and hypertension. *J. Recept. Signal Transduct. Res.* **24,** 149–164.
82. Yet, S. F., Perrella, M. A., Layne, M. D., Hsieh, C. M., Maemura, K., Kobzik, L., Wiesel, P., Christou, H., Kourembanas, S., and Lee, M. E. (1999). Hypoxia induces severe right ventricular dilatation and infarction in heme oxygenase-1 null mice. *J. Clin. Invest.* **103,** R23–R29.
83. Chruscinski, A. J., Rohrer, D. K., Schauble, E., Desai, K. H., Bernstein, D., and Kobilka, B. K. (1999). Targeted disruption of the β2 adrenergic receptor gene. *J. Biol. Chem.* **274,** 16694–16700.
84. Holmes, M. C., Kotelevtsev, Y., Mullins, J. J., and Seckl, J. R. (2001). Phenotypic analysis of mice bearing targeted deletions of 11β-hydroxysteroid dehydrogenases 1 and 2 genes. *Mol. Cell Endocrinol.* **171,** 15–20.
85. Shesely, E. G., Maeda, N., Kim, H. S., Desai, K. M., Krege, J. H., Laubach, V. E., Sherman, P. A., Sessa, W. C., and Smithies, O. (1996). Elevated blood pressures in mice lacking endothelial nitric oxide synthase. *Proc. Natl. Acad. Sci. USA* **93,** 13176–13181.
86. Cook, S., Hugli, O., Egli, M., Menard, B., Thalmann, S., Sartori, C., Perrin, C., Nicod, P., Thorens, B., Vollenweider, P., Scherrer, U., and Burcelin, R. (2004). Partial gene deletion of endothelial nitric oxide synthase predisposes to exaggerated high-fat diet-induced insulin resistance and arterial hypertension. *Diabetes* **53,** 2067–2072.
87. Lembo, G., Rockman, H. A., Hunter, J. J., Steinmetz, H., Koch, W. J., Ma, L., Prinz, M. P., Ross, J., Chien, K. R., and Powell-Braxton, L. (1996). Elevated blood pressure and enhanced myocardial contractility in mice with severe IGF-1 deficiency. *J. Clin. Invest.* **98,** 2648–2655.
88. Maemura, K., Kurihara, H., Kurihara, Y., Kuwaki, T., Kumoda, M., and Yazaki, Y. (1995). Gene expression of endothelin isoforms and receptors in endothelin-1 knockout mice. *J. Cardiovasc. Pharmacol.* **26,** S17–S21.
89. Rettig, R. (1993). Does the kidney play a role in the aetiology of primary hypertension? Evidence from renal transplantation studies in rats and humans. *J. Hum. Hypertens.* **7,** 177–180.
90. Hinojos, C. A., Boerwinkle, E., Fornage, M., and Doris, P. A. (2005). Combined genealogical, mapping, and expression approaches to identify spontaneously hypertensive rat hypertension candidate genes. Hypertension **45,** 698–704.
91. Larkin, J. E., Frank, B. C., Gaspard, R. M., Duka, I., Gavras, H., and Quackenbush, J. (2004). Cardiac transcriptional response to acute and chronic angiotensin II treatments. *Physiol. Genomics* **18,** 152–166.
92. Chon, H., Gaillard, C. A., van der Meijden, B. B., Dijstelbloem, H. M., Kraaijenhagen, R. J., van Leenen, D., Holstege, F. C., Joles, J. A., Bluyssen, H. A., Koomans, H. A., and Braam, B. (2004). Broadly altered gene expression in blood leukocytes in essential hypertension is absent during treatment. *Hypertension* **43,** 947–951.
93. Hu, W. Y., Fukuda, N., and Kanmatsuse, K. (2002). Growth characteristics, angiotensin II generation, and microarray-determined gene expression in vascular smooth muscle cells from young spontaneously hypertensive rats. *J. Hypertens.* **20,** 1323–1333.
94. Aitman, T. J., Glazier, A. M., Wallace, C. A., Cooper, L. D., Norsworthy, P. J., Wahid, F. N., Al-Majali, K.M., Trembling, P. M., Mann, C. J., Shoulders, C. C., Graf, D., St. Lezin, E., Kurtz, T. W., Kren, V., Pravenec, M., Ibrahimi, A., Abumrad, N. A., Stanton, L. W., and Scott, J. (1999). Identification of Cd36 (Fat) as an insulin-resistance gene causing defective fatty

acid and glucose metabolism in hypertensive rats. *Nat. Genet.* **21,** 76–83.
95. Fries, R. S., Mahboubi, P., Mahapatra, N. R., Mahata, S. K., Schork, N. J., Schmid-Schoenbein, G. W., and O'Connor, D. T. (2004). Neuroendocrine transcriptome in genetic hypertension: multiple changes in diverse adrenal physiological systems *Hypertension* **43,** 1301–1311.
96. McBride, M. W., Carr, F. J., Graham, D., Anderson, N. H., Clark, J. S., Lee, W. K., Charchar, F. J., Brosnan, M. J., and Dominiczak, A. F. (2003). Microarray analysis of rat chromosome 2 congenic strains. *Hypertension* **41,** 847–853.
97. Yagil, C., Hubner, N., Monti, J., Schulz, H., Sapojnikov, M., Luft, F. C., Ganten, D., and Yagil, Y. (2005). Identification of hypertension-related genes through an integrated genomic-transcriptomic approach. *Circ. Res.* **96,** 617–625.
98. Loyd, J. E. (2002). Genetics and pulmonary hypertension. *Chest* **122,** 284S–286S.
99. Lee, S. D., Shroyer, K. R., Markham, N. E., Cool, C. D., Voelkel, N. F., and Tuder, R. M. (1998). Monoclonal endothelial cell proliferation is present in primary but not secondary pulmonary hypertension. *J. Clin. Invest.* **101,** 927–934.
100. Deng, Z., Haghighi, F., Helleby, L., Vanterpool, K., Horn, E. M., Barst, R. J., Hodge, S. E., Morse, J. H., and Knowles, J. A. (2000). Fine mapping of PPH1, a gene for familial primary pulmonary hypertension, to a 3-cM region on chromosome 2q33. *Am. J. Respir. Crit. Care Med.* **161,** 1055–1059.
101. Deng, Z., Morse, J. H., Slager, S. L., Cuervo, N., Moore, K. J., Venetos, G., Kalachikov, S., Cayanis, E., Fischer, S. G., Barst, R. J., Hodge, S. E., and Knowles, J. A. (2000). Familial primary pulmonary hypertension (gene PPH1) is caused by mutations in the bone morphogenetic protein receptor-II gene. *Am. J. Hum. Genet.* **67,** 737–744.
102. Newman, J. H., Wheeler, L., Lane, K. B., Loyd, E., Gaddipati, R., Phillips III, J. A., and Loyd, J. E. (2001). Mutations in the gene for bone morphogenetic protein receptor II as a cause of primary pulmonary hypertension in a large kindred. *N. Engl. J. Med.* **5,** 319–324.
103. Machado, R. D., Pauciulo, M. W., Thomson, J. R., Lane, K. B., Morgan, N. V., Wheeler, L., Phillips III, J. A., Newman, J., Williams, D., Galie, N., Manes, A., McNeil, K., Yacoub, M., Mikhail, G., Rogers, P., Corris, P., Humbert, M., Donnai, D., Martensson, G., Tranebjaerg, L., Loyd, J. E., Trembath, R. C., and Nichols, W. C. (2001). BMPR2 haploinsufficiency as the inherited molecular mechanism for primary pulmonary hypertension. *Am. J. Hum. Genet.* **68,** 92–102.
104. Newman, J. H., Trembath, R. .C, Morse, J. A., Grunig, E., Loyd, J. E., Adnot, S., Coccolo, F., Ventura, C., Phillips, J. A., III., Knowles, J. A., Janssen, B., Eickelberg, O., Eddahibi, S., Herve, P., Nichols, W. C., and Elliott, G. (2004). Genetic basis of pulmonary arterial hypertension: current understanding and future directions. *J. Am. Coll. Cardiol.* **43,** 33S–39S.
105. Eddahibi1, S., Morrell, N., d'Ortho, M.-P., R., Naeije, R., and Adnot, S. (2002). Pathobiology of pulmonary arterial hypertension. *Eur. Respir. J.* **20,** 1559–1572.
106. Aldred, M. A., Vijayakrishnan, J., James, V., Soubrier, F., Gomez-Sanchez, M. A., Martensson, G., Galie, N., Manes, A., Corris, P., Simonneau, G., Humbert, M., Morrell, N. W., and Trembath, R. C. (2006). BMPR2 gene rearrangements account for a significant proportion of mutations in familial and idiopathic pulmonary arterial hypertension. *Hum. Mutat.* **27,** 212–213.
107. Rudarakanchana, N., Flanagan, J. A., Chen, H., Upton, P. D., Machado, R., Patel, D., Trembath, R. C., and Morrell, N. (2002). Functional analysis of bone morphogenetic protein type II receptor mutations underlying primary pulmonary hypertension. *Hum. Mol. Genet.* **11,** 1517–1525.
108. Trembath, R. C., Thomson, J. R., Machado, R. D., Morgan, N. V., Atkinson, C., Winship, I., Simonneau, G., Galie, N., Loyd, J. E., Humbert, M., Nichols, W. C., Morrell, N. W., Berg, J., Manes, A., McGaughran, J., Pauciulo, M., and Wheeler, L. (2001). Clinical and molecular genetic features of pulmonary hypertension in patients with hereditary hemorrhagic telangiectasia. *N. Engl. J. Med.* **345,** 325–334.
109. Harrison, R. E., Berger, R., Haworth, S. G., Tulloh, R., Mache, C. J., Morrell, N. W., Aldred, M. A., and Trembath, R. C. (2005). Transforming growth factor-β receptor mutations and pulmonary arterial hypertension in childhood. *Circulation* **111,** 435–441.
110. Chaouat, A., Coulet, F., Favre, C., Simonneau, G., Weitzenblum, E., Soubrier, F., and Humbert, M. (2004). Endoglin germline mutation in a patient with hereditary haemorrhagic telangiectasia and dexfenfluramine associated pulmonary arterial hypertension. *Thorax* **59,** 446–448.
111. Janssen, B., Rindermann, M., Barth, U., Miltenberger-Miltenyi, G., Mereles, D., Abushi, A., Seeger, W., Kübler, W., Bartram, C. R., and Grünig, E. (2002). Linkage analysis in a large family with primary pulmonary hypertension: genetic heterogeneity and a second PPH locus on 2q31-32. *Chest* **121,** 54S–56S.
112. Rindermann, M., Grünig, E., von Hippel, A., Koehler, R., Miltenberger-Miltenyi, G., Mereles, D., Arnold, K., Pauciulo, M., Nichols, W., Olschewski, H., Hoeper, M. M., Winkler, J., Katus, H. A., Kübler, W., Bartram, C. R., and Janssen, B. (2003). Primary pulmonary hypertension may be a heterogeneous disease with a second locus on chromosome 2 q31. *J. Am. Coll. Cardiol.* **12,** 2237–2244.
113. Grunig, E., Koehler, R., Miltenberger-Miltenyi, G., Zimmermann, R., Gorenflo, M., Mereles, D., Arnold, K., Naust, B., Wilkens, H., Benz, A., von Hippel, A., Ulmer, H. E., Kubler, W., Katus, H. A., Bartram, C. R., Schranz, D., and Janssen, B. (2004). Primary pulmonary hypertension in children may have a different genetic background than in adults. *Pediatr. Res.* **56,** 571–578.
114. Eddahibi, S., Humbert, M., Fadel, E., Raffestin, B., Darmon, M., Capron, F., Simonneau, G., Dartevelle, P., Hamon, M., and Adnot, S. (2001). Serotonin transporter overexpression is responsible for pulmonary artery smooth muscle hyperplasia in primary pulmonary hypertension. *J. Clin. Invest.* **108,** 1141–1150.
115. Guignabert, C., Izikki, M., Tu, L. I., Li, Z., Zadigue, P., Barlier-Mur, A. M., Hanoun, N., Rodman, D., Hamon, M., Adnot, S., and Eddahibi, S. (2001). Transgenic mice overexpressing the 5-hydroxytryptamine transporter gene in smooth muscle develop pulmonary hypertension. *Circ. Res.* **98,** 1323–1330.
116. Lepetit, H., Eddahibi, S., Fadel, E., Frisdal, E., Munaut, C., Noel, A., Humbert, M., Adnot, S., D'Ortho, M. P., and Lafuma, C. (2005). Smooth muscle cell matrix metalloproteinases in idiopathic pulmonary arterial hypertension. *Eur. Respir. J.* **25,** 834–842.

117. Perros, F., Dorfmuller, P., and Humbert, M. (2005). Current insights on the pathogenesis of pulmonary arterial hypertension. *Semin. Respir. Crit. Care Med.* **26**, 355–364.
118. Geraci, M. W., Moore, M., Gesell, T., Yeager, M. E., Alger, L., Golpon, H., Gao, B., Loyd, J. E., Tuder, R. M., and Voelkel, N. F. (2001). Gene expression patterns in the lungs of patients with primary pulmonary hypertension: a gene microarray analysis. *Circ. Res.* **88**, 555–562.
119. Yeager, M. E., Halley, G. R., Golpon, H. A., Voelkel, N. F., and Tuder, R. M. (2001). Microsatellite instability of endothelial cell growth and apoptosis genes within plexiform lesions in primary pulmonary hypertension. *Circ. Res.* **88**, E2–E11.
120. Grunig, E., Mereles, D., Arnold, K., Benz, A., Olschewski, H., Miltenberger-Miltenyi, G., Borst, M. M., Abushi, A., Seeger, W., Winkler, J., Hoper, M. M., Bartram, C. R., Kubler, W., and Janssen, B. (2002). Primary pulmonary hypertension is predominantly a hereditary disease. *Chest* **121**, 81S–82S.
121. Long, L., MacLean, M. R., Jeffery, T. K., Morecroft, I., Yang, X., Rudarakanchana, N., Southwood, M., James, V., Trembath, R. C., and Morrell, N. W. (2006). Serotonin increases susceptibility to pulmonary hypertension in BMPR2-deficient mice. *Circ. Res.* **98**, 818–827.
122. Loyd, J. E., Butler, M. G., Foroud, T. M., Conneally, P. M., Phillips, J. A. III, and Newman, J. H. (1995). Genetic anticipation and abnormal gender ratio at birth in familial primary pulmonary hypertension. *Am. J. Respir. Crit. Care Med.* **152**, 93–97.
123. Merklinger, S. L., Wagner, R. A., Spiekerkoetter, E., Hinek, A., Knutsen, R. H., Kabir, M. G., Desai, K., Hacker, S., Wang, L., Cann, G. M., Ambartsumian, N. S., Lukanidin, E., Bernstein, D., Husain, M., Mecham, R. P., Starcher, B., Yanagisawa, H., and Rabinovitch, M. (2005). Increased fibulin-5 and elastin in S100A4/Mts1 mice with pulmonary hypertension. *Circ. Res.* **97**, 596–604.
124. Golpon, H. A., Geraci, M. W., Moore, M. D., Miller, H. L., Miller, G. J., Tuder, R. M., and Voelkel, N. F. (2001). HOX genes in human lung: altered expression in primary pulmonary hypertension and emphysema. *Am. J. Pathol.* **158**, 955–966.
125. Bull, T. M., Coldren, C. D., Moore, M., Sotto-Santiago, S. M., Pham, D. V., Nana-Sinkam, S. P., Voelkel, N. F., and Geraci, M. W. (2004). Gene microarray analysis of peripheral blood cells in pulmonary arterial hypertension. *Am. J. Respir. Crit. Care Med.* **170**, 911–919.
126. Huang, W., Sher, Y. P., Delgado-West, D., Wu, J. T., Peck, K., and Fung, Y. C. (2001). Tissue remodeling of rat pulmonary artery in hypoxic breathing. I. Changes of morphology, zero-stress state, and gene expression. *Ann. Biomed. Eng.* **29**, 535–551.
127. Ameshima, S., Golpon, H., Cool, C. D., Chan, D., Vandivier, R. W., Gardai, S. J., Wick, M., Nemenoff, R. A., Geraci, M. W., and Voelkel, N. F. (2003). Peroxisome proliferator-activated receptor α (PPARα) expression is decreased in pulmonary hypertension and affects endothelial cell growth. *Circ. Res.* **92**, 1162–1169.
128. Schiffrin, E. L., Amiri, F., Benkirane, K., Iglarz, M., and Diep, Q. N. (2003). Peroxisome proliferator-activated receptors: vascular and cardiac effects in hypertension. *Hypertension* **42**, 664–668.

SECTION IV

Post-Genomic Analysis of the Myocardium

CHAPTER 13

Cardiomyopathies

OVERVIEW

Cardiomyopathies, diseases of heart muscle, may result from an array of factors that damage the heart and other organs and impair myocardial function, including infections, toxins, and cardiac ischemia. Over the past decade, the significance of inherited gene defects in the pathogenesis of primary cardiomyopathies has been recognized, with numerous mutations identified as etiological factors in the more prevalent types of cardiomyopathy (i.e., hypertrophic cardiomyopathy [HCM] and dilated cardiomyopathy [DCM]) and more recently in more uncommon phenotypes such as restrictive cardiomyopathy (RCM) and arrhythmogenic right ventricular dysplasia/cardiomyopathy (ARVD/ARVC). Moreover, genetic defects are increasingly implicated in the pathogenesis of metabolic cardiomyopathies (often associated with extracardiac presentations), including the mitochondrial cardiomyopathies and the cardiomyopathy associated with diabetes. These genetic defects involve numerous intracellular pathways sharing a number of critical features, as well as displaying distinct elements, in fostering the various cardiomyopathic phenotypes, including those arising in sporadic and acquired cardiomyopathies resulting from infection (e.g., viral-induced), toxins (e.g., alcohol), and ischemic insult. The primary pathophysiological mechanisms implicated in cardiomyopathy include defective force generation caused by mutations in sarcomeric protein genes; defective force transmission caused by mutations in cytoskeletal protein genes; myocardial energy deficits caused by mutations in both nuclear and mitochondrial DNA encoded genes; and abnormal Ca^{++} homeostasis caused by altered availability of Ca^{++} and altered myofibrillar Ca^{++} sensitivity.

A leading premise of ongoing research on cardiomyopathy is defining the role(s) of a plurality of genes in cardiac function and explaining the mechanisms by which mutations in these genes lead to hypertrophy, dilation, and contractile failure and, hopefully, to discover successful therapeutic strategies.

INTRODUCTION

Cardiomyopathies are traditionally classified according to morphological and functional criteria into four categories: HCM, DCM, RCM, and ARVD/ARVC. As depicted in Table I, these cardiomyopathies exhibit a number of fundamentally different phenotypes and some shared features. These cardiomyopathies can be primary myocardial disorders or develop as a secondary complication of a variety of factors, including infection and inflammation, toxic agents, and myocardial ischemia. Over the past decade, the importance of gene defects in the etiology of primary cardiomyopathies has been recognized. Autosomal-dominant, autosomal-recessive, X-linked, and maternal patterns of inheritance have been observed. Families with inherited cardiomyopathies have provided a unique opportunity to study the genetic basis of these disorders. Molecular genetic screening performed to date has focused largely on monogenic inherited cardiomyopathies (i.e., caused by mutations in a single gene). Although relatively uncommon, these monogenic disorders enable the evaluation of pathophysiological processes applicable to a wide range of more commonly occurring cardiac diseases.

Substantial progress has been made in explaining further gene defects in HCM, DCM, RCM, and ARVD/ARVC. Although these genetic studies have enabled molecular triggers of cardiomyopathies to be identified, the functional consequences of gene mutations and precise details of the signaling pathways that lead to hypertrophy, dilation, and contractile failure have largely remained to be determined.

HYPERTROPHIC CARDIOMYOPATHY

Clinical Manifestations

Hypertrophic cardiomyopathy (HCM) is a primary myocardial disorder with an autosomal-dominant pattern of inheritance clinically characterized by hypertrophy of the ventricles, histological features of myocyte hypertrophy, myofibrillar disarray, and interstitial fibrosis. HCM is one of the most common inherited cardiac disorders, with prevalence in young adults of 1 in 500.[1] A wide spectrum of clinical manifestations of varying severity has been reported, with affected individuals exhibiting significant variability in their clinical presentation. Genotype-phenotype studies have demonstrated that genotype-positive individuals may be asymptomatic or present with symptoms ranging from

Post-Genomic Cardiology

TABLE I Classification of Cardiomyopathies

Class	Pathophysiology	Phenotype	Etiology
Dilated	Dilated ventricles with impaired contraction	Mild wall hypertrophy; initially normal diastolic compliance; ↓ systolic function; ↓ ejection fraction	Idiopathic, familial, ischemia, valvular, dystrophy, immune, hypertension, toxic myocarditis (e.g., cocaine, alcohol, and Adriamycin)
Hypertrophic	Left and/or right ventricular hypertrophy	Severe wall hypertrophy; ↓ diastolic compliance; normal systolic function; high/ normal ejection fraction; dysrhythmia; SD	Familial/genetic, hypertension, aortic stenosis
Restrictive	Restrictive filling with ↓ diastolic relaxation of one/both ventricles	Moderate wall hypertrophy; ↓ diastolic compliance; normal systolic function; ↓ ejection fraction	Idiopathic, sarcoidosis, amyloidosis, endomyocardial fibrosis
Arrhythmogenic right ventricular cardiomyopathy	Fibro-fatty replacement of right ventricular myocardium	No effect on LV function; SD; recurrent ventricular tachycardia with left bundle-branch block	Unknown; familial with incomplete penetrance.

SD, sudden death.

palpitations and dizziness to syncope and sudden death (SD), and the age of onset of symptoms varies between different HCM disease genes. Clinical diagnosis of HCM is primarily by transthoracic echocardiography, with findings of asymmetrical ventricular septum hypertrophy, with or without left ventricular outflow tract obstruction (occurring in < 25% of affected individuals). Differentiation of HCM from physiological left ventricular hypertrophy (LVH) can be difficult, particularly in competitive athletes. Individuals with different gene mutations/alleles may display differences in the extent of LVH. For example, individuals with mutations in the MYH7 gene encoding β-MHC usually develop moderate or severe hypertrophy with a high penetrance, whereas those with TNNT2 mutations in cardiac troponin T generally have only mild or clinically undetectable hypertrophy.[2,3] The extent of LVH may also vary between members of a single family with the same gene mutation, which suggests that the HCM phenotype is determined by additional modifying genetic and environmental factors (such as blood pressure, exercise, diet, and body mass), in addition to the effects of a disease-causing gene.[4]

The natural history of HCM is variable; some individuals remain asymptomatic throughout life, and others may develop progressive symptoms with or without heart failure (HF) or experience SD. Several observations have documented left ventricular remodeling and progressive increases in left ventricular wall thickness with age, as well and in other cases, an age-related reduction in left ventricular wall thickness associated with myocyte loss and fibrosis.

HCM is a frequent cause of SD, particularly in young individuals and competitive athletes.[5] Numerous mechanisms for SD have been proposed, including bradydysrhythmias caused by sinus node and AV conduction abnormalities and tachydysrhythmias, abnormal Ca^{++} homeostasis, myocardial ischemia, left ventricular diastolic dysfunction, and left ventricular outflow tract obstruction. No single risk factor has been identified that can precipitate SD. However, some mutant alleles have been associated with either high or low risk; for example, the Arg403Gln and Arg453Cys-β-MHC mutations, cardiac troponin T mutations, and some tropomyosin mutations (Ala63Val, Lys70Thr) have been characterized as "high-risk" mutations, with reduced life expectancy and high rates of SD, whereas the β-MHC Val606Met mutation and cMyBP-C mutations have a relatively benign course.[2,3,6,7] At this time, the mechanisms whereby HCM gene mutations influence the prognosis remain undetermined.

GENETICS OF HCM

Using standard genetic mapping techniques, 12 chromosomal *loci* have been detected in familial hypertrophic cardiomyopathy (FHCM).[8] A list of the genes identified thus far and their respective chromosomal loci is presented in Table II and depicted in Fig. 1.

Of the 15 HCM-causing genes identified, 10 encode protein components of the cardiac sarcomere. Mutations have been found in four genes that encode components of the thick filament (i.e., β-MHC, essential MLC, regulatory MLC, and cMyBP-C), five genes that encode thin filament proteins (i.e., cardiac actin, cardiac troponin T, cardiac troponin I, cardiac troponin C, and α-tropomyosin), and in the sarcomeric cytoskeletal protein titin. Mutations have also been reported in four genes encoding nonsarcomeric proteins, including the γ2-regulatory subunit of an AMP-activated protein kinase (AMPK), the cytoskeletal muscle

TABLE II Genes Implicated in Familial HCM

Gene	Protein	Loci	References
TNNT2	Cardiac troponin T	1q32	9
TTN	Titin	2q31	10
MYL3	Essential myosin light chain	3p21	11
TNNC1	Cardiac troponin C	3p21-p14	12
PRKAG2	AMP-activated protein kinase (regulatory subunit)	7q36	13
MYBPC3	Cardiac myosin binding protein C	11p11	14
CSRP3	Cardiac muscle LIM protein	11p15	15
MYL2	Regulatory myosin light chain	12q23-q24	16
MYH7	β-Myosin heavy chain	14q12	17
ACTC	Cardiac actin	15q14	18
TPM1	α-Tropomyosin	15q22	9
TNNI3	Cardiac troponin I	19q13	19
CAV3	Caveolin-3	3p25	20
LAMP2	Lysosome associated membrane protein 2	Xq24-25	21, 22
TCAP	T-Cap (telethonin)	17q12	54

LIM protein, LAMP-2, a lysosome-associated membrane glycoprotein, and caveolin-3, a contributor to the caveolin microdomains involved in cellular signaling. Interestingly, several of these nonsarcomeric pathogenic *loci* are also associated with other disease phenotypes; the same AMPK mutations involved in HCM cause an accumulation of cardiac glycogen, hence termed glycogen storage disease, and lead to severe electrophysiological abnormalities, particularly ventricular pre-excitation (Wolff-Parkinson-White syndrome). Defects of the LAMP-2 lysosomal structural protein cause an increase in intracytoplasmic vacuoles containing autophagic material and glycogen in skeletal and cardiac muscle cells (also leading to a glycogen storage disease) and promote myopathy and mental retardation in addition to HCM, constituting a syndrome known as Danon disease.[21,22] Furthermore, mutations in caveolin-3 cause limb-girdle muscular dystrophy (LGMD).[20]

In HCM there is significant genetic heterogeneity with more than 200 different mutations in these 15 genes so far identified[23]; however, approximately 75% of cases with defined mutations have been estimated to be due to mutations in just three genes, the β-myosin heavy chain *MYH7* gene, cardiac troponin T *TNNT2* gene, and the myosin binding protein-C *MYBPC3* gene.[24] Certain phenotypes are more common with certain gene variants, such as the *MYBPC3* mutations, which induce the disease predominantly in the fifth or sixth decade of life. Interestingly, the distribution of mutations in elderly onset disease is strikingly different from that found in familial early-onset HCM. Defects in β-myosin heavy chain, cardiac troponin T, and α-tropomyosin account for more than 45% of FHCM, whereas mutations in cardiac myosin binding protein-C, troponin I, and α-myosin heavy chain are more prevalent in elderly onset HCM.[25]

The mutations reported encompass both missense and nucleotide deletion/insertions. In general, most individuals with HCM-causing mutations are heterozygous at the disease *locus*, although two cases of homozygous HCM mutations have been reported, one an Arg869Gly point mutation in the β-MHC gene,[26] and the other, a Ser179Phe mutation in the cardiac troponin T gene.[27] Both homozygous mutations caused a particularly severe phenotype with onset in childhood and premature death. Moreover, because FHCM is inherited in an autosomal-dominant fashion, most mutations might be expected to exert a dominant-negative effect to produce the hypertrophic phenotype. In the following section, we will review some of the salient features of mutations thus far identified.

β-Myosin Heavy Chain

The *MYH7* gene encoding β-MHC is located in tandem with the *MYH6* gene (encoding α–MHC) on chromosome 14. In humans, β-MHC is present in the embryonic heart and in the adult atria and is the predominant isoform expressed in the adult ventricle. *MYH7* encompasses 23 kb of genomic DNA, including 41 exons, 38 of which encode a protein of 1935 amino acids. More than 60 mutations in the *MYH7* gene have been reported, accounting for approximately 30% of cases of HCM.[28] Although most are missense mutations, deletions and premature termination (nonsense) codons have also been

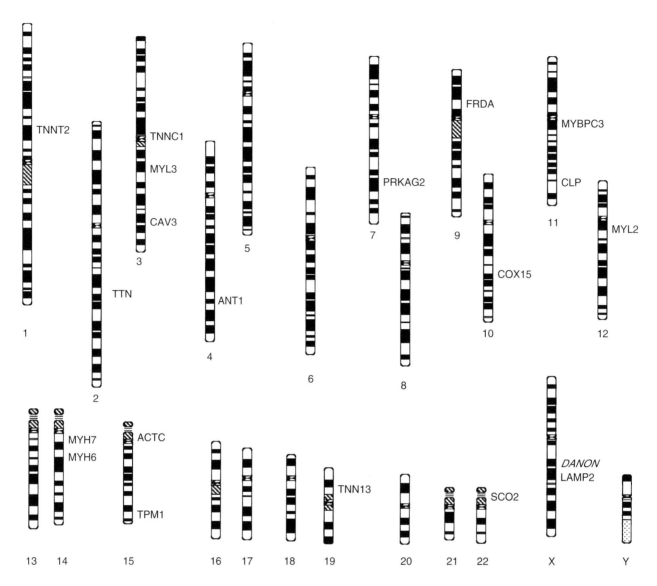

FIGURE 1 Human chromosome map of genetic loci involved in HCM.

identified. *MYH7* mutations seem to be distributed throughout the gene-coding sequence, and most of the *MYH7* mutations result in amino acid substitutions located in the globular head of the protein, involving the binding sites for ATP, actin, and essential or regulatory light chains. Because the myosin heads seem to be the target of most pathogenic mutations and are primarily responsible for the force transduction properties of myosin, it has been proposed that diminished force transmission from the head to the body of the thick filament is a primary mechanism by which these β-MHC mutations mediate HCM. However, a number of mutations have been recently localized to the β-MHC rod, suggesting that there may be alternative pathogenetic mechanisms.[29]

In families with a benign form of HCM, the prevalence of *MYBPC3* and *MYH7* gene mutations was reported to be similar (45% and 43%, respectively). In contrast, in families whose disorder exhibits a more malignant course and prognosis, the *MYH7* gene mutations were the most prevalent (45%), and in families with an intermediate prognosis, *MYBPC3* gene mutations were the most prevalent (70%).[30] In a number of cases, the correlation of genotype-phenotype has implicated a number of HCM mutations as "benign defects" associated with near-normal survival, whereas others seem to be more pathogenic (Table III). However, genotype-phenotype correlations are difficult to assess on the basis of findings from a limited number of families. Furthermore, there is also evidence that challenges the notion of mutation-specific clinical outcomes, with several benign mutations detected in patients with severe HCM.[31,32] Interestingly, other recent observations have suggested that the prognosis of specific β-MHC genotypes is determined by the affected structural-functional domain.[33]

Missense mutations of β-MHC that result in change of a charged amino acid residue likely confer conformational

TABLE III Benign and Malignant HCM Mutations in β-MHC

Benign mutations
N232S
G256E
F513C
V606M
R719Q
L908V

Malignant mutations
R403Q
R453C
G716R
R719W

---

structural changes and functional abnormalities of the sarcomeric protein and, for a limited number of *loci*, have been associated with a worse prognosis.[34] The nonconservative missense mutations, Arg403Gln (occurring in the actin binding region) and Arg719Cys (present within the light chain binding domain), have been characterized by high disease penetrance, a high incidence of sudden cardiac death (SCD), and a decreased life expectancy.[33] In contrast, patients with the Leu908Val defect (a conservative mutation occurring in the head-rod junction) have a low incidence of SCD and a more benign course.[6] The association of poor prognosis with the reported missense mutations affecting the actin binding and rod regions suggests that alteration in the actin-myosin interaction and in force transmission to the thick filament, respectively, can lead to a particularly malignant HCM phenotype.[35]

Myosin Binding Protein C

The *MYBPC3* gene comprises 24 kb of genomic DNA with 37 exons encoding a protein of 1274 amino acids. The gene contains multiple immunoglobulin C2–like and fibronectin type 3 domains, as well as a cardiac-specific region, a phosphorylation region, and overlapping myosin and titin binding sites.[36] The cardiac MyBP-C isoform is expressed exclusively in cardiac tissue. More than 30 mutations have been reported in the *MYBPC3* gene and represent approximately 15–20% of all the individuals with HCM. Most mutations are frameshift mutations (e.g., insertions, deletions, or splice site mutations) that result in the premature truncation of the cardiac MyBP-C protein with loss of the myosin and titin binding sites.[37] Alternately, missense mutations in *MYBPC3* that preserve the myosin and titin binding sites have also been found. It is also worth noting that mutations of the cardiac myosin binding protein C gene have been associated with a milder form of hypertrophy and delayed phenotypic expression until middle or older age.[7]

Cardiac Troponins

Cardiac troponin T encoded by the *TNNT2* gene is composed of 17 kb of genomic DNA, contains 15 exons, and is expressed in the embryonic heart, the adult heart, and in the developing skeletal muscle. Alternate splicing produces a number of different cardiac troponin T isoforms. The principal isoform in the adult heart consists of 288 amino acids and possesses two major domains: (1) an N-terminal domain that interacts with tropomyosin, and (2) a C-terminal domain that binds to tropomyosin, troponin C, and troponin I. Binding of troponin T to tropomyosin is responsible for the positioning of the troponin complex on the thin filaments.[38-41]

Although mutations in all three subunits of cardiac troponin (C, T and I), as well as tropomyosin, have been associated with FHCM, with approximately 60 different mutations thus far detected, more than 30 mutations have been reported in *TNNT2*, accounting for 5–10% of cases of HCM, the largest class among these regulatory proteins. Most of the cardiac troponin T mutations are missense. Interestingly, a deletion mutation has been reported within a 5′ splice donor site GA transition in residue 1 of intron 15, resulting in a truncated cardiac troponin T protein with loss of the terminal 28 amino acid residues.[9] The 5′ splice donor mutation is likely to function as a null allele and result in synthesis of a truncated troponin T that degrades rapidly. Many of the *TNNT2* mutations cluster between residue 79 and 179, a region essential for anchoring the troponin–tropomyosin complex onto the thin filament. Cardiac troponin T mutations, in particular, predispose affected patients to SCD (Table IV), and one of the most malignant mutations in FHCM is a missense mutation resulting in an exchange of a glutamine for an arginine at residue 92 (R92Q) in the tail portion of *TNNT2*.[2] Patients with this mutant allele also display a lower degree of LVH than patients with other pathogenic HCM alleles.

TABLE IV Troponin T Mutations and Phenotype

Mutation	Phenotype
Arg92Leu	SCD
Arg92Trp	SCD
Arg94Leu	SCD; low level of hypertrophy
Arg102Leu	Wide range of hypertrophy; no SCD; high penetrance
Arg102Gln	Mild hypertrophy; SCD; low penetrance
Ile79Asn	SCD
Intron15G1A	SCD
Pro77Leu	SCD
Ser69Phe	SCD
Phe110Ile	SCD; highly variable penetrance; more severe hypertrophy with increased gene dosage (homozygote vs. heterozygote)
Arg278Cys	Low SCD; hypertrophy occurs later in life; low penetrance

Several troponin T mutations have decreased disease penetrance and a milder degree of hypertrophy and fibrosis, but a significantly higher risk of SCD than patients with certain β-MHC missense mutations.[42,43] Although considerable data have shown that marked myocardial hypertrophy is a risk for SCD in HCM, patients with specific troponin T mutations (e.g., R94L) display a high rate of premature SCD even in the absence of hypertrophy.[44] The presence of extensive myocyte disarray in these patients suggests that it may be a determinant of troponin T–related SCD.[43]

Cardiac troponin I encoded by *TNNI3* is composed of 6.2 kb of genomic DNA with eight exons encoding a protein of 210 amino acids and is expressed solely in cardiac tissue. Its primary function is to serve as an inhibitory subunit of the troponin complex that prevents contraction in the absence of Ca^{++} binding to troponin C. More than 20 *TNNI3* mutations have been reported, representing between 3 and 5% of all cases of HCM. Most mutations seem to cluster in exons 7 and 8; their clinical expression is very broad and heterogeneous, ranging from asymptomatic to SCD and varying both within and between families. Many of the mutations disrupt functional interactions with troponin C and T and often affect the calcium sensitivity of force development, which may account for the increased severity of disease in these families. In addition, a mutation linked to FHCM has recently been detected in the N-terminal region (R21C) of *TNNI3* that results in a troponin I, which is phosphorylated by PKA at a lower rate than wild-type troponin I and manifests a markedly altered calcium sensitivity of force development.[45]

Troponin C has two isoforms that are expressed in cardiac and skeletal muscle. The *TNNC1* gene encodes the isoform present in the heart and slow skeletal muscle, and encompasses 3 kb of genomic DNA with six exons, encoding a protein of 161 amino acids. Calcium binding to cardiac troponin C induces conformational changes in the troponin–tropomyosin complex that initiates muscle contraction. Only one missense mutation in the *TNNC1* gene has been reported in association with FHCM.[12]

The α-tropomyosin protein functions to bridge the binding of the troponin protein complex to thin actin filament. The *TPM1* gene encoding α-tropomyosin is located on chromosome 15q2 and is composed of 15 exons with a corresponding mature mRNA of 1 kb. Fewer than 5% of cases of HCM are caused by *TPM1* mutations, six of which have been reported, and several are located near the calcium-dependent troponin T binding domain. Two missense mutations have been described in exon 5 of the *TPM1* gene in affected individuals of families with HCM. The first mutation alters the amino acid sequence from aspartic acid to asparagine at position 175 (Asp175Asn).[9] The second missense mutation changes the glutamic acid residue at amino acid 180 to a glycine residue.[3] A third *TPM1* missense mutation resulting in Glu62Gln was identified in a family, which presented a variable clinical phenotype, including both a malignant phenotype at an early age (e.g., five cases of SCD) and a milder form with variable penetrance in older individuals.[46]

Regulatory and Essential Myosin Light Chain Genes

The myosin light chains (MLC) belong to a superfamily of Ca^{++}-binding proteins, which contain a highly conserved helix-loop-helix region (termed EF hands), which are Ca^{++}-binding domains; other members of this family include troponin C and calmodulin. Cardiac muscle contains two regulatory MLC isoforms. *MYL2* encodes the MLC2 slow isoform, expressed in ventricle and in slow skeletal muscle. It has seven exons that encode a protein of 166 amino acids. Eight *MYL2* missense mutations have been reported in association with HCM.[47] In addition to the regulatory MLC isoforms, two essential MLC isoforms are present in cardiac muscle. The MLC-1 slow/ventricular isoform expressed in the ventricle and slow skeletal muscle is encoded by *MYL3*, comprising seven exons that encode a protein of 195 amino acids. Two *MYL3* missense mutations have been reported in patients with HCM.[48,49]

Interestingly, some mutations in both the regulatory *MYL2* (Glu22Lys, Asn47Lys) and essential (Met149Val) *MYL3* myosin light chain genes can result in an unusual pattern of hypertrophy termed midventricular hypertrophy (MVH) in which papillary muscle hypertrophy leads to obstruction of the midventricular cavity.[47] Interestingly, a novel missense mutation, the L277M in the *VCL* gene encoding the cytoskeletal protein vinculin, was recently reported in a patient with severely obstructive MVH.[50] Other mutations in both *MYL2* (e.g., Ala13Thr, Phe18Leu, Arg58Gln) and *MYL3* (Ala57Gly) have been reported to cause the more classical variety of HCM.[51] It is noteworthy that the Ala57Gly mutation in *MYL 3* is located in the EF-hand domain.

Cardiac Actin

Actin is a major constituent of the thin filaments, which in association with myosin, is involved in force generation within the sarcomere, as well as in force transmission from the sarcomere to the surrounding syncytium by the thin filament.

Cardiac actin is encoded by *ACTC* gene, which has six exons that encode 375 amino acids; its N-terminal domain contains the site of myosin cross-bridge attachment; whereas its C-terminal domain has binding sites for α-actinin and dystrophin. Five *ACTC* mutations have been identified in FHCM, albeit rarely, comprising less than 1% of all HCM.[18,52] Several of these are missense mutations that affect amino acid residues lying in close proximity to the myosin binding region (with side chains exposed to the actin surface) and that interact with myosin; these mutant actin residues likely promote HCM primarily by their effect on sarcomeric force generation. The clinical manifestations of these mutations are heterogeneous. In one study, patients heterozygous for *de novo* missense mutations resulting in Pro164Ala and Ala331 substitutions exhibited sporadic HCM with syncope in early childhood.[53] This study also reported a family in which several members developed ventricular dysrhythmias and apical ventricular hypertrophy harboring a missense *ACTC* mutation, resulting in Glu99Lys. In contrast, patients harboring a missense mutation in *ACTC* that causes Ala295Ser displayed high penetrance, diverse phenotypes, and variable age at onset with no evidence of SCD.[18,52]

Z Disc Proteins: Titin Gene (TTN, MLB, and T-cap)

Cardiac Z-discs are positioned at the junction between the cytoskeleton and the myofilaments, providing a physical connection between the sarcomere, nucleus, membrane, and sarcoplasmic reticulum (SR). Furthermore, numerous molecular messengers congregate at the Z-disc. This combination of physical and chemical signals moving through the Z-disc makes this element a vital switching station of the heart and plays a significant regulatory role in stretch sensing and in regulating cardiac hypertrophy. Mutations in genes encoding several protein components of the Z disc and interacting proteins result in FHCM further broadening the cast of players implicated in HCM pathogenesis.

Titin, a giant (more than 3 MDa) protein, the largest known polypeptide, anchors in the Z-disc and spans a considerable portion of the sarcomere, contributing to the maintenance of sarcomere organization and myofibrillar elasticity, as well as participating in myofibrillar cell signaling. Ninety percent of titin's mass is composed of up to 298 repeating immunoglobulin and fibronectin 3 domains. One HCM-causing *TTN* missense mutation resulted in an R740 L change located in the Z-disc binding region of titin.[10] This mutation increased the binding affinity of titin to the Z disc protein α-actinin in the yeast two-hybrid assay.

The Z disc protein, muscle LIM protein (MLP), plays a selective role in mechanical stretch sensing and its coupling to sarcomeric contraction. MLP interacts and colocalizes with telethonin (T-cap), a titin interacting protein. T-cap interacts with calsarcin, which tethers calcineurin to the Z-disc. Interestingly, mutations in the genes encoding both MLP (*CSRP3*) and T-cap (*TCAP*) have recently been reported in FHCM.[54] Two HCM-associated *TCAP* mutations in T137I and R153H resulted in altered binding among Z disc components, as gauged by an *in vitro* binding assay. The HCM-associated mutant T-cap proteins exhibited enhanced binding to titin and to calsarcin-1, in marked contrast with T-cap proteins associated with a DCM phenotype, which exhibit reduced binding to MLP, titin, and calsarcin-1.

MLP also acts as an essential nuclear regulator of myogenic differentiation, in addition to its role as an integrator of protein assembly of the actin-based cytoskeleton. Several different heterozygous missense mutations in the *CSRP3* gene have been found in three unrelated patients with FHCM, including Cys58Gly, Leu44Pro, and Ser54Arg/ Glu55Gly, which likely alter the electrostatic interactions, structural, and function properties of the MLP protein.[15] All mutations involved substitution at highly conserved amino acid residues in the functionally important LIM1 domain (present throughout the entire family of cysteine-rich proteins or CSRPs, including, CSRP1, CSRP2, and CSRP3), which is responsible for interaction with α-actinin and with muscle-specific transcription factors. Subsequent protein-binding studies demonstrated that the variant MLP proteins, resulting from Cys58Gly mutation in the *CSRP3* gene, have decreased binding to α-actinin.

Non-Sarcomeric Genes

As we have previously pointed out, a number of non-sarcomeric genes have been implicated in FHCM; however, it is too early to assess the overall influence in HCM of genes, such as *PRKAG2*, *LAMP2,* and *CAV3*.

AMPK is a heterotrimeric protein composed of a catalytic subunit α and two regulatory subunits β and γ, the latter with three isoforms that vary in length and tissue expression. The *PRKAG2* gene encodes the γ2-subunit, the predominant γ-isoform present in the heart. In myocytes, AMPK acts as a "metabolic sensor," responding to ATP depletion by regulating diverse intracellular pathways that use and generate ATP. In the absence of metabolic stress, AMPK activity is suppressed, and during periods of hypoxic or metabolic stress, ATP consumption results in an increase in the AMP/ATP ratio, and rising levels of AMP activate AMPK. In addition to its protein kinase activity, AMPK may have a transcriptional regulatory role, given its homology to the SNF1 kinase/transcription factor that regulates glucose metabolism in yeast. Six mutations in the *PRKAG2* gene have been reported in families with HCM associated with Wolff-Parkinson-White (WPW) syndrome (Table V), five were missense mutations, and one an in-frame single codon insertion.[13,55,56] The effect of the mutations on AMP kinase activity is rather controversial, because one group has shown that AMP kinase activity regulation becomes constitutively active with specific mutations, and another group found that the *PRKAG2* mutations can lead to decreased AMP kinase activity, suggesting that its loss-of-function may be an intrinsic part of the pathogenic mechanism. New observations

TABLE V *PRKAG2* Mutations Leading to HCM and WPW

Mutation	Reference
Arg531Gly	60
Arg302Glu	56
His142Arg	13
Exon5:InsLeu (at codon 110)	13
Thr400Asn	55
Asn488Ile	55

support the view that *PRKAG2* mutations (which often map to the nucleotide-binding region) render AMPK insensitive to the inhibitory and stimulatory effects of the regulatory nucleotides ATP and AMP, respectively, suggesting that the resulting cardiac pathogenesis may not be attributable to a simple loss or gain-of-function.[57]

At present, there are two diverse points of view in regard to AMPK, *PRKAG2,* and HCM. One suggests that AMPK mutations (as well as *LAMP-2* mutations) enlarge the perspective of HCM beyond its characterization as a disease of the sarcomere engendered by defective sarcomeric proteins, which result in defective force generation.[13] This view suggests that the underlying mechanism of disease may involve an inefficient ATP regulation or defective myocardial energetic homeostasis and signaling, consistent with the AMPK mutation. This argument also marshals evidence that consistent changes in sarcomeric contractility are not shared by many of the sarcomeric protein mutations, that most of the sarcomeric mutations result in deficient ATP use, and also that HCM can result from defective mitochondrial ATP production as a consequence of specific maternally inherited mutations in mitochondrial DNA. The second point of view is that the AMPK/*PRKAG2* mutation*s* (along with the *LAMP2*/Danon disease mutations) are essentially glycogen-storage diseases, which mimic HCM. This perspective notes that the HCM phenotype found with AMPK/*PRKAG2* mutations always is accompanied by WPW syndrome and conduction defects, and can be further distinguished from other HCM phenotypes by the absence of myocyte and myofibrillar disarray, a pathognomonic feature of HCM caused by sarcomere protein mutations and by the presence of pronounced vacuole formation in myocytes.[58] Recent evidence at the clinical level also supports the notion that the AMPK/*PRKAG2* and *LAMP2* Danon diseases may be distinct entities.[58,59] Furthermore, a novel *PRKAG2* mutation (Arg531Gly) has been reported in a family with WPW syndrome and conduction defects with onset in childhood and no evident HCM.[60] These observations suggest a primary role for AMPK/*PRKAG2* in cardiac ion channel regulation, and its role in "mimicking" cardiac hypertrophy (and that of *LAMP2*) remains to be explained.

ANIMAL MODELS OF GENETIC HCM

The effects of putative mutations on sarcomere structure and function, as well as on overall cardiac structure and function *in vivo,* have been examined in genetically engineered mouse models. A large number of transgenic studies have provided a wealth of information, including the identification of new genetic targets, the confirmation of pathogenic mutations described in clinical studies, and in some cases explanation of the role of these specific pathogenic mutations and resultant proteins in the progression of cardiac hypertrophy and onset of cardiomyopathy. Several specific alterations in genes that generate HCM in mice are presented in Table VI.

Sarcomeric Mutations

Major questions that need to be addressed include whether specific sarcomere protein mutations act by a dominant-negative mechanism or alter function by causing haploinsufficiency and what can the sarcomeric mutations tell us about the stimulus and pathway for hypertrophy in HCM. With the dominant-negative model, both wild-type and mutant proteins are presumably present at similar levels. Namely, the mutant peptide is stably incorporated into the sarcomere and acts as a poison polypeptide to perturb wild-type protein function.

Mutations can also result in null alleles or cause a reduction in the amount of wild-type protein, leading to an imbalance of sarcomere protein stoichiometry. In addition, mutations that truncate the encoded protein are thought to act by haploinsufficiency.

Heterozygous Arg 403Gln α-MHC (designated α-MHC403/+) mutant mice exhibited progressive LVH on transthoracic echocardiogram, together with histological evidence of myocyte hypertrophy, myofibrillar disarray, and fibrosis.[61] *In vivo* hemodynamic studies found that 6-week α-MHC403/+ mice had normal myocardial histology but altered contraction kinetics with accelerated systolic pressure rise, whereas by 20 weeks of age, α-MHC403/+ mice had developed myocardial hypertrophy, hyperdynamic left ventricular contraction, increased end-systolic chamber stiffness, left ventricular outflow tract pressure gradient, and decreased cardiac index.[62] Interestingly, the preserved and enhanced systolic function found in both patients and in animal models of HCM contrast with findings from a large number of *in vitro* studies showing that Arg403Gln α-MHC perturbs actin-myosin interaction and depresses motor function. However, the *in vitro* studies may miss the full interplay of the multiple regulatory factors present in the myocardial milieu. On the other hand, left ventricular diastolic function studies in α-MHC403/+ mouse hearts uniformly display prolongation of relaxation analogous to that observed in patients with HCM.[61,62] Several observations have suggested that enhanced diastolic relaxation may be a consequence of defective cross-bridge cycling resulting from a mutant myosin allele, changes in Ca^{++} transients, Ca^{++} sensitivity, and abnormal myocardial energetics.

As previously noted, most mutations in the cardiac myosin binding protein C (MyBP-C) linked to FHCM are predicted to encode truncated forms of the protein that lack portions of the C-terminus containing the titin and myosin binding domains. However, because truncated peptides have not been detected in patients with FHCM, it is unclear whether mutant proteins are incorporated into the sarcomeres, whether they are degraded, and whether disease results from N-terminal cMyBP-C fragments that exert dominant-negative effects or from reduced expression of cMyBP-C due to a null allele (i.e., haploinsufficiency).

Transgenic mice were generated in which varying amounts of a truncated MyBP-C, lacking the myosin and

TABLE VI Gene Alterations Causing HCM in Transgenic Models

Gene (protein)	Genetic alteration	References
Myosin heavy chain	Arg 403 Gln	91
Cardiac myosin binding protein C	Cardiac-specific knock-out (homozygous and heterozygous)	65, 66
Cardiac myosin binding protein C	Truncation of titin and myosin binding sites at the C-terminus	63, 64
Cardiac troponin T	Arg 92 Gln, Ile 79 Asn	68–69, 73, 91
Caveolin-1	Homozygous null	368
Caveolin-3	Homozygous null	369
Caveolin-3	Pro 104 Leu	370
DMPK	Overexpression	371
α-Tropomyosin	Glu 180 Gly	76
p38 Kinase	Cardiac-specific dominant-negative alleles	372

titin binding domains, were expressed in the heart.[63] The transgenic-encoded, truncated protein is stable, its overexpression compensated by a reduction in wild-type protein, and therefore not changing the overall stoichiometry of the protein pool. The mutant MyBP-C, although present in the sarcomere, is not incorporated efficiently into the contractile apparatus, showing a diffuse pattern in the cytosol. Histological analyses of myocardial sections from 25-week-old mice showed *foci* of degenerate myocytes and sarcomeric disorganization, whereas both *in vivo* and *in vitro* hemodynamic analyses showed no differences between mutant and wild-type mice in contraction or relaxation parameters. These findings were more consistent with a model of haploinsufficiency, in which the disease progresses very slowly because of decreased levels of endogenous protein (the "null allele" hypothesis) playing a primary role in the pathogenic process.

A second mouse model with a similar cMyBP-C truncation was generated by homologous recombination.[64] In contrast to α-MHC403/+ mice generated in the same study, heterozygous mice (MyBP-Ct/+) did not develop left ventricular hypertrophy until after 125 weeks of age. This late onset of hypertrophy is analogous to that observed in patients with HCM caused by cMyBP-C mutations. Dose-related effects were suggested by the findings of normal myocardial histology in heterozygous MyBP-Ct/+ mice compared with prominent left ventricular hypertrophy, disarray, and fibrosis in homozygous MyBP-Ct/t mice.

Gene targeting has been used to produce a knock-out mouse that lacks MyBP-C in the heart, including both mice heterozygous (+/−) and homozygous (−/−) for the knock-out allele.[65] Whereas *cMyBP-C* (+/−) mice were indistinguishable from wild-type litter mates, *cMyBP-C* (−/−) mice exhibited significant cardiac hypertrophy. Cardiac function, assessed by echocardiography, showed significantly depressed diastolic and systolic function only in *cMyBP-C* (−/−) mice. The Ca^{++} sensitivity of tension, measured in single-skinned myocytes, was reduced in *cMyBP-C* (−/−) but not *cMyBP-C* (+/−) mice. However, other studies of null mice have demonstrated that inactivation of one or two mouse *cMyBP-C* alleles leads to different cardiac changes within specific post-natal time windows.[66] Homozygous *cMyBP-C* null mice develop eccentric left ventricular hypertrophy with decreased fractional shortening at 3–4 months of age, markedly impaired relaxation after 9 months associated with myocardial disarray, and an increase of interstitial fibrosis. In contrast, heterozygous *cMyBP-C* null mice develop asymmetrical septal hypertrophy associated with fibrosis only after 10–11 months of age.

Cardiac Troponin T Gene Mutations

Transgenic mice expressing a truncated mouse *cTnT* allele similar to one found in FHCM expressed truncated cardiac troponin T at low (< 5%) levels, developed cardiomyopathy and have significantly smaller hearts (18–27%) compared with wild type.[67] These animals also exhibited significantly impaired diastolic relaxation with milder but significant systolic dysfunction. Animals that express higher levels of transgene troponin T protein die within 24 h of birth. Transgenic mouse hearts show myocyte disarray and have a reduced number of cardiac myocytes that are smaller in size. These observations suggested that multiple cellular mechanisms are involved in the troponin-T–mediated disease, which is generally characterized by mild hypertrophy, but also frequent SD.

In contrast, an Arg92Gln cardiac troponin T transgenic mouse model with transgene expression levels ranging from 1–10% of the cardiac troponin T protein exhibited normal left ventricular dimensions and systolic function but marked diastolic dysfunction accompanied by variable myocyte degeneration, necrosis, myofibril disarray, fibrosis with increased myocardial collagen; the effects seem to be the result of a dominant-negative troponin T allele.[68]

Transgenic mice with different levels of the Arg92Gln cardiac troponin T protein (e.g., 30, 67, and 92% of their total cardiac troponin levels) have also been generated with a dose-dependent myocardial disarray and fibrosis.[69] Similar to the effects of the cardiac troponin T truncation, hearts of all the Arg92Gln strains were smaller than control hearts, with a reduction in both the number and size of myocytes. Enhanced systolic function in the presence of diastolic dysfunction was found, consistent with the view that some troponin T mutant alleles result in a hypercontractile heart, and the presence of distinctly different phenotypes associated with the various troponin T alleles was also demonstrated. Hypercontractility was also noted in isolated myocytes derived from the mutant mouse hearts that manifested increased basal sarcomeric activation, impaired relaxation, and shorter sarcomere lengths. Mild degeneration of the myofibrils and gross alterations in mitochondrial morphology with increased mitochondrial number along lipid deposition were detected by electron microscopy. These findings stand in contrast with the primary findings of myofibrillar disarray with the truncated troponin T allele.

Cardiac energetics using ^{31}P NMR spectroscopy and contractile performance of the intact beating heart were evaluated in R92Q mice, and they showed both a decrease in the free energy of ATP hydrolysis available to support contractile work and a marked inability to increase contractile performance on acute inotropic challenge.[70] These findings suggest that defects in troponin T within thin filament protein structure and function can lead to significant changes in myocardial energetics and contractile reserve.

Cardiac troponin T mutations may also influence the inhibitory regulatory effect of the tropomyosin–troponin complex. The altered contractility observed has been attributed to increased Ca^{++} sensitivity and to abnormal basal levels of sarcomeric activation, resulting in the initiation of contraction at shorter, less optimal sarcomere lengths against a stronger passive restoring force.[70] Modifications in troponin–tropomyosin interactions, occurring as a consequence of mutant troponin T alleles, may be responsible for the increase in myofilament Ca^{++} sensitivity observed in mutant myofilaments.[71] Other contributory factors that influence the relationship of specific troponin T mutations to the overall extent of myofilament dysfunction include the concentration and distribution of the mutant protein within the sarcomere.[72]

It is worth noting that despite myocardial histopathology and functional changes, none of the cardiac troponin T mouse models exhibited left ventricular hypertrophy. This again is consistent with the clinical findings of FHCM caused by cardiac troponin T mutations, which typically have minimal left ventricular hypertrophy.

Cardiac troponin T I79N mutation has been linked to FHCM with a high incidence of SCD, despite causing little or no cardiac hypertrophy in human. Transgenic mice expressing mutant human troponin T (I79N) had increased cardiac contractility but exhibited no ventricular hypertrophy or fibrosis. Because enhanced cardiac function has been associated with myofilament Ca^{++} sensitization, altered cellular Ca^{++} handling, as well as several electrophysiological parameters, have been analyzed in isolated myocytes collected from perfused hearts, whole mutant mice, and controls. Although no significant differences were found either in L-type Ca^{++} or transient outward K^+ currents, inward rectifier K^+ current (I_{K1}) was significantly decreased. In addition, Ca^{++} transients of ventricular I79N myocytes were reduced, exhibiting slow decay kinetics, consistent with increased Ca^{++} sensitivity of I79N mutant fibers. Moreover, at higher pacing rates or in the presence of isoproterenol, diastolic Ca^{++} became significantly elevated in myocytes derived from I79N mice compared with controls. Stress-induced ventricular tachycardia was more prevalent in I79N mice, suggesting that specific troponin mutations (e.g., I79N) can mediate stress-induced ventricular tachycardia, even in absence of hypertrophy and/or fibrosis, in part by altered Ca^{++} transients and suppression of I_{K1}.[73]

Other mutant alleles of cardiac troponin T that cause clinical FHCM also display disparate phenotypes in the transgenic mice. Three to four-month-old transgenic mice expressing F110I and R278C troponin T did not develop significant hypertrophy or ventricular fibrosis, even after chronic exercise challenge.[74] However, the F110I allele had impaired acute exercise tolerance, whereas the R278C did not. Analysis of skinned papillary muscle fibers from transgenic mice expressing F110I demonstrated an increased Ca^{++} sensitivity of force and ATPase activity not found in muscle fibers from transgenic mice expressing R287C, which may underlie the more severe F110I phenotype.

Cardiac Troponin I Gene

The Arg-145 residue is located in the inhibitory region of cardiac troponin I, a highly negatively charged region that alternately binds to either actin or troponin C, depending on the intracellular concentration of calcium. This region is critical for the inhibition of actin-tropomyosin–activated myosin ATPase (when Ca^{++} levels are low) and for Ca^{++}-mediated binding to troponin C. A transgenic mouse with Arg145Gly cardiac troponin I exhibited myofibrillar disarray and interstitial fibrosis. In this model some of the molecular markers of cardiac hypertrophy are activated, although no overt hypertrophy, as measured by increased ventricle mass/body mass, was observed.[75] At the whole organ level (using isolated working heart preparations), mutant mice showed significantly enhanced contractility and prolonged relaxation. Mice with 3.5-fold transgene overexpression died by 17 days. In 10-day-old mice, the ventricles showed myocyte disarray, nuclear degeneration, and interstitial fibrosis, together with a reduced ability to inhibit the actin-tropomyosin–activated ATPase, as well as an increase in Ca^{++} sensitivity. The findings suggested that the mutant troponin I protein either does not interact appropriately with actin or perhaps binds strongly to troponin C. The net effect would be lack of ATPase inhibition, essentially leaving the molecular switch in the "on" position, analogous to the changes observed with cardiac troponin T gene mutations. The differences in contractile performance between mice with 3.5-fold transgene expression may reflect dose-related differences in basal sarcomere length or more severe myocardial damage.

α-Tropomyosin

Five distinct point mutations within α-tropomyosin are associated with the development of FHCM. Two of these mutations are found within a troponin T binding site, located at amino acid residues 175 and 180. Transgenic mice have been generated that encode a mutation in α-tropomyosin at codon 180 (Glu180Gly).[76] Expression of exogenous mutant tropomyosin leads to a concomitant decrease in endogenous α-tropomyosin without altering the expression of other contractile proteins. Histological analysis shows that initial pathological changes, which include ventricular concentric hypertrophy, fibrosis, and atrial enlargement, are detected within 1 month. The disease-associated changes progressively increase and result in death between 4 and 5 months. Physiological analyses of the Glu180Gly mice using echocardiography, work-performing heart analyses, and force measurements of cardiac myofibers demonstrated dramatic functional differences in diastolic performance and increased sensitivity to calcium. Mutations in α-tropomyosin can be severely disruptive of sarcomeric function, which consequently triggers a dramatic hypertrophic response culminating in death.

The Glu180Gly mice develop severe cardiac hypertrophy, substantial interstitial fibrosis, and increased heart weight/body weight ratio.[77] Furthermore, calcium-handling proteins associated with the SR exhibit decreased expression. These alterations in gene expression, coupled with the structurally altered tropomyosin, may contribute to the decreased physiological performance exhibited by these transgenic mice. A DNA microarray analysis of the transgenic versus control ventricular RNAs shows that 50 transcripts are differentially expressed during the onset of hypertrophy, many of which are associated with the extracellular matrix (ECM). It is apparent that mutations within tropomyosin can be severely disruptive of sarcomeric function, triggering a hypertrophic response coupled with a cascade of alterations in gene expression.

Myosin Light Chain

Mutations in both the regulatory (at Glu22Lys) and essential myosin light chains (at Met149Val) result in an unusual pattern of hypertrophy, which leads to obstruction of the midventricular cavity. When a human genomic fragment containing the Met149Val essential myosin light chain was used to generate transgenic mice, the HCM phenotype occurred.[78]

To further establish a causal relationship for the regulatory and essential light chain mutations in HCM, mice were generated that expressed either the wild-type or mutated forms, using cDNA clones encompassing only the coding regions of the mouse gene *loci*. However, expression of the mutant mouse MLC proteins did not lead to a hypertrophic response, even in senescent animals. In fact, transgenic mice had lower left ventricular mass and smaller myocytes than that observed in control mice.[79] Functional analyses in isolated working heart preparations showed that mutant mice displayed hypercontractility and impaired relaxation. In addition, at the myofilament and cellular levels, changes were noted, including dose-related myofibril disarray and fibrosis, myofibrils increased Ca^{++} sensitivity, and reduced power output. Although these observations revealed several significant changes in myofilament structure and function linked to MLC mutations, they also illustrate that transgenic mice do not always reproduce pivotal aspects of human hypertrophy.

HCM-linked mutations in the ventricular myosin regulatory light-chain have also been reported. The Glu22Lys mutation in *Myl2* has been associated with a form of cardiac hypertrophy defined by mid-left ventricular obstruction caused by papillary muscle hypertrophy. A transgenic mouse line has been generated with the Glu22Lys mutant of human ventricular MLC, and the functional consequences of this mutation were examined in skinned cardiac-muscle preparations. The Glu22Lys mutant hearts of 13-month-old animals showed ventricular *septum* hypertrophy and enlarged papillary muscles with no filament disarray using staining techniques, whereas no cardiac hypertrophy was detected by echocardiography. Functional studies that used skinned cardiac muscle preparations showed an increase in the Ca^{++} sensitivity of myofibrillar ATPase activity and force development in mutant mice compared with wild-type mice, suggesting that Glu22Lys-linked HCM is likely mediated through Ca^{++}-dependent events.[80]

Other Non-Sarcomeric Models of HCM

Several transgenic mice models containing the AMPK/*PRKAG2* mutations have proven informative. Transgenic mice were generated by the cardiac-restricted expression of the wild-type and mutant *PRKAG2* genes containing the cardiac-specific promoter of α-myosin heavy chain and showed that expression of the mutant allele *R302Q* in mice reproduced the clinical phenotypes of ventricular preexcitation, conduction abnormalities, and cardiac hypertrophy.[81] Not only were ventricular pre-excitation and prolonged QRS detected by ECG recordings and intracardiac electrophysiology studies, but a distinct AV accessory pathway was confirmed by electrical and pharmacological stimulation and by induction of orthodromic AV reentrant tachycardia. Both significant cardiac hypertrophy and excessive cardiac glycogen deposits were documented, as well as significant reduction of AMPK activity in the mutant heart.

Studies in transgenic mice in which the *PRKAG2* mutant *N488I* allele was overexpressed also demonstrated severe left ventricular hypertrophy, ventricular pre-excitation, sinus node dysfunction, and large accumulation of cardiac glycogen (30-fold above normal) in numerous vacuoles.[82] In contrast to the findings with the *R302Q* transgenic mouse, the *N488I* mutant mice exhibited elevated AMPK activity. Electrophysiological data confirmed the presence of alternative atrioventricular conduction pathways consistent with WPW, and histological findings revealed that the annulus fibrosis, which normally insulates the ventricles from inappropriate excitation by the atria, was disrupted by glycogen-filled myocytes. These anomalous atrioventricular connections have been proposed as providing the anatomical substrate for ventricular pre-excitation

Several transgenic models of concentric or symmetrical left ventricular hypertrophy have now been reported, including overexpression of the protooncogenes *ras* and *myc*, α-adrenergic receptors, the heterodimeric G-protein α subunit, and protein kinase C (PKC).[83–86] The mechanisms for the induction of increased ventricular wall thickness are diverse, with overexpression of some signaling factors such as *ras*, G_α, and *PKC* exhibiting true cellular hypertrophy with increase in cell size, whereas the *myc*-overexpression animal exhibits cardiac myocyte hyperplasia. Interestingly, the HCM phenotypes discussed earlier illustrate the principle that apparently diverse signals can culminate in the same phenotype, presumably by converging on final common pathways.

TRIGGERS AND EFFECTORS OF LVH

Triggers of LVH

The hypertrophic process may be initiated by factors both extrinsic and intrinsic to the cardiac myocyte. Extrinsic stimuli include vasoactive peptides (e.g., angiotensin II, endothelin-1), adrenergic agonists (e.g., norepinephrine, epinephrine, phenylephrine), activators of PKC (e.g., tumor-producing phorbol esters), peptide growth factors (e.g., insulin-like growth factor, fibroblast growth factor), cytokines (e.g., cardiotrophin-1), arachidonate metabolites (e.g., prostaglandin F2), mechanical stretch, and cell contact. Intrinsic stimuli include elevated $[Ca^{++}]_i$, the heterotrimeric G-protein G_q, as well as activated small G-proteins, kinases, phosphatases, and numerous transcriptional factors.[87,88] These extrinsic and intrinsic factors trigger a complex cascade of intracellular pathways, termed the "hypertrophic response," which results in increased myocardial mass, altered spatial relationships between myocytes and other cellular and extracellular components of the myocardium, reprogramming of myocardial gene expression, and apoptosis. A rapidly growing list of genes has been found to elicit cardiac hypertrophy when overexpressed in transgenic mice, and determination of which of these genes are clinically relevant in human hypertrophy will be important.

The stimulus for hypertrophy in HCM has not been definitively identified. Although numerous studies have demonstrated that sarcomere protein gene mutations perturb sarcomere structure and/or function, the precise consequences of sarcomere protein gene mutations differ according to the type of model studied with both reduced and augmented motor function as described for individual HCM disease genes. However, a shared feature among these models is mechanical dysfunction of the sarcomere, which also is a potential stimulus for hypertrophy.

Calcium

Transgenic mice expressing constitutively activated calcineurin develop LVH that can be prevented by administration of the calcineurin inhibitors cyclosporin and FK506. These findings promoted interest in the role of elevated Ca^{++} and the calcineurin-signaling pathway in the pathogenesis of LVH. Calcineurin is a cytoplasmic protein phosphatase that is activated by sustained $[Ca^{++}]_i$. Activated calcineurin dephosphorylates NFAT3 transcription factors, inducing their nuclear translocation and interaction with the cardiac-restricted transcription factor GATA4, resulting in the synergistic activation of embryonic cardiac genes and a hypertrophic response. Although transgenic mouse studies suggested that the calcineurin pathway might be critical in the development of myocardial hypertrophy, a number of models suggested that hypertrophy could be achieved by means of multiple signaling pathways.

Myocardial Energetics

The recent discovery of mutations in the *PRKAG2* gene has been somewhat unexpected, given the traditional dogma that sarcomere dysfunction was the fundamental defect in HCM. These *PRKAG2* mutations raise the intriguing possibility that defective myocardial energetics may be a common feature of HCM mutations. Alterations of myocardial metabolism had been observed previously in patients with HCM[89] and in α-*MHC403/+* mice.[90,91] Although these energetic changes might result from factors such as myocardial ischemia caused by increased oxygen demands of hypertrophied myocardium, it is also possible that alterations of myocardial metabolism may occur as primary effects of mutant proteins. Moreover, mutations that cause deficiency in mitochondrial ATP production (e.g., mtDNA mutations) can result in HCM.

Effectors of LVH

Considerable research has been focused on the characterization of the consequences of specific gene mutations on sarcomere structure and function and in the identification of factors that might trigger the hypertrophic response. Only recently has the cascade of signaling pathways that constitute the effectors of hypertrophy in HCM begun to be examined. An important, but unanswered, question is whether the various HCM gene mutations stimulate the same or different hypertrophic pathways and to what extent the hypertrophic response pathways in HCM might differ from those in other cardiac diseases.

Microarray gene profiling has been used to address differential gene expression occurring as a result of different HCM stimuli. Several observations have demonstrated that a diverse array of genes is up-regulated in the heart with HCM, which could account for the diversity of clinical and pathological phenotypes.[92–97] Markers of hypertrophy caused by secondary causes, such as pressure overload, have been found to be up-regulated in patients with HCM, suggesting some commonality of pathways are involved in HCM and the acquired forms of cardiac hypertrophy.[95] More than 190 genes have been found to be up-regulated in both DCM and HCM, and 51 down-regulated in both conditions. Also, genes encoding a diverse subset of proteins were differentially expressed between DCM and HCM (e.g., α B-crystallin, antagonizer of *myc* transcriptional activity, β-dystrobrevin, calsequestrin, lipocortin, and lumican).[96]

As we have noted, the severity of LVH often varies considerably between individuals with different sarcomere protein gene mutations, as well as among individuals in a family with the same gene mutations. Are these differences caused by varying severity of the stimulus for hypertrophy or do other factors influence the expression of the hypertrophic phenotype? Observations in mouse models do not support a direct correlation between the extent of sarcomere dysfunction and LVH. For example, although cardiac function is impaired in α-MHC403/+ mice and normal in mice with truncated cMyBP-C, both mouse models develop LVH of similar severity.[64] In both these mouse models, heterozygous mice have LVH, but homozygous mice exhibit DCM rather than severe hypertrophy. These findings suggest that the extent to which LVH can compensate for a cardiac motor deficit may be related to gene dosage and that severe dysfunction is less likely to be "rescued." More studies are also needed to determine the effects of additional genetic factors and environmental factors as modifiers of the hypertrophic phenotype.

In evaluating the effects of genes on HCM in mouse models, it is important to consider that transgenic models can show entirely different effects in different developmental stages. In the pathogenesis of myocardial hypertrophy, hundreds of signaling molecules have been assigned critical roles largely on the basis of their cardiac phenotypes in transgenic mouse models. The phenotypes are the combined results of transgene effects and normal trophic influences, which vary developmentally, physiologically, as well as with different genetic backgrounds. For instance, expression of the proapoptotic Nix protein caused ventricular dilation and severely decreased contractility with extensive apoptosis in neonates, with minimal effects in adults.[98]

DILATED CARDIOMYOPATHY

Clinical Manifestations

DCM is characterized by cardiac dilatation and contractile dysfunction of the left and/or right ventricles. DCM can arise from diverse etiologies (Table I) that cause cardiomyocyte dysfunction, whereas in a significant proportion of cases, a specific precipitating factor cannot be identified, termed idiopathic DCM (IDCM). Although the overall incidence of DCM is 36 per 100,000, the incidence of idiopathic DCM has been estimated to be 5–8 cases per 100,000. DCM has traditionally been regarded as a sporadic nongenetic disorder, but recent studies of large families with DCM have suggested that inherited gene defects, discussed in depth later, comprise a significant subset of idiopathic DCM cases.

The prevalence of familial DCM (FDCM) has varied according to the methods of detection and diagnostic criteria used. The diagnosis of DCM has mostly been based on the World Health Organization criteria (i.e., the presence of both ventricular dilation and contractile dysfunction),[99] whereas studies solely on the basis of a positive clinical history of DCM in relatives of probands have yielded a relatively low prevalence (<10%). Nevertheless, there is a consensus that the true prevalence of FDCM is likely to be underestimated. Abnormalities of either ventricular dimensions or contractile function have been observed in asymptomatic relatives of probands with DCM, but it is not clear whether these findings represent early disease or a more benign form of disease. Revised estimates gauged by careful screening of relatives of patients has documented a prevalence of FDCM, reflecting the contribution of inherited genes, ranging from 25–50% of the cases of IDCM. It is also interesting to note that the incidence of IDCM increases with age, and males are afflicted at a higher rate than females.

Several factors, including the diversity of clinical presentation, the variability in penetrance and expression, and significant heterogeneity in the genetic loci involved, have contributed to the under-recognition of this disease entity as an inherited disorder. Age-related penetrance (i.e., absence of disease manifestations in genotype-positive individuals until after a particular age) and nonpenetrance (i.e., genotype-positive but phenotype-negative) can lead to the underestimation of the prevalence of familial disease. Moreover, the diagnosis of FDCM can also be complicated by the comorbid conditions that can contribute to cardiac dysfunction, such as myocardial ischemia, viral infection, or alcohol excess. The extent to which genetic factors may increase the susceptibility to recognized DCM precipitating factors has only just begun to be assessed.

The diagnosis of FDCM and nonfamilial DCM is primarily made using transthoracic echocardiography. The echocardiographic images can be used to quantitatively assess ventricular structure, systolic function, valvular morphology, and atrial size. Color flow mapping and Doppler echocardiography can assess ventricular diastolic function, as well as the severity of mitral and tricuspid valve regurgitation. Radionuclide scans may be helpful in quantitating ventricular volumes and contractile performance, particularly in individuals in whom echocardiography is technically difficult. Invasive techniques, such as cardiac catheterization may also be used to assess ventricular function and visualize the coronary anatomy, particularly in individuals concurrently at risk for coronary artery disease (CAD).

Endomyocardial biopsy generally shows nonspecific histological findings such as myocyte hypertrophy, necrosis, nuclear abnormalities, and interstitial fibrosis. Cardiac mass is usually increased, and findings of moderate eccentric ventricular hypertrophy frequently are accompanied by severe atrial and ventricular dilatation.

Unlike HCM, myocyte disarray is not a feature of DCM. ECM components, such as fibronectin, laminin, and collagen, are often increased, and in some cases, the presence of interstitial T cells and endothelial cells indicates that an inflammatory process is contributing to DCM.

The electrocardiogram in DCM may reveal sinus tachycardia, intraventricular conduction delays, or nonspecific S-T segment and T-wave changes. Individuals with FDCM accompanied by defects in the conduction system may exhibit sinus bradycardia, AV block, and atrial fibrillation or flutter. Electrocardiographic monitoring of individuals with DCM often reveals premature ventricular beats and/or nonsustained ventricular tachycardia.

The natural history of FDCM is variable both between and within families. Affected individuals may have a relatively benign clinical course, may develop progressive HF requiring cardiac transplantation, or may experience SCD often resulting from ventricular dysrhythmias. In FDCM, SCD may occur at any age, irrespective of ventricular function. In subgroups of families with DCM and conduction system defects, affected individuals generally experience progressive AV conduction abnormalities in the second to fourth decades, with subsequent development of DCM. Implantation of permanent pacemakers may be required for advanced AV conduction block.

Apart from the family history, no single clinical parameters have been found to reliably distinguish FDCM from nonfamilial DCM. In some cases, genetically affected individuals may be asymptomatic, with an incidental finding of cardiomegaly. Most individuals have symptoms attributable to ventricular failure or cardiac dysrhythmias, including weakness, fatigue, dyspnea, palpitations, and exercise intolerance. It has also been suggested that individuals with FDCM can be diagnosed at an earlier age than individuals with nonfamilial DCM.[100] Although this may reflect the presence of a relatively malignant DCM phenotype in some families, it also may be attributable to more intensive screening of individuals deemed to be at high risk because of their family history.

With increasing understanding of DCM and the routine use of ACE inhibitors and β-adrenergic blockade for its treatment, the 5-y survival rate after initial diagnosis has improved beyond the 50% previously reported.[101] Although multiple clinical parameters have been proposed in DCM as predictors of mortality, it is likely that family genotype may be the strongest determinant of outcome.

Genetics of FDCM

Studies of families with DCM have revealed autosomal-dominant, autosomal-recessive, X-linked, and maternal modes of inheritance, with the autosomal-dominant pattern observed most frequently.[100] FDCM exhibits marked clinical variability, with several distinct phenotypic groups: DCM alone, DCM with conduction system defects/dysrhythmias (sinus bradycardia, AV block, atrial tachydysrhythmias), DCM with skeletal myopathy with or without AV conduction defects, or DCM with sensorineural deafness.[102] In addition, in some families, DCM may occur as part of a multisystem inherited disorder. The relative prevalence of each of these phenotypic groups has not yet been determined.

Chromosomal Loci

Adult-onset autosomal-dominant DCM without accompanying AV conduction defects has been associated with 12 loci, seven of which were identified by genome-wide linkage analyses in large families (chromosomes 1q32, 2q31, 6q12–q16, 9q13–q22, 9q22–q31, 10q22–q23, and 14q12).[102] The remaining five loci were identified by use of a candidate gene approach in small families. The identification of DCM-causing mutations (Table VII) at a wide variety of genetic loci associated with clinically diverse features confirms the assessment of DCM as highly heterogeneous both genetically and clinically. Their locations on the human chromosome map are shown in Fig. 2.

The first gene identified by candidate gene analysis in DCM was cardiac actin (chromosome 15q14). Five other genes encoding sarcomere proteins cause DCM (all of which can also cause HCM) including β-MHC (chromosome 14q12), cMyBP-C (chromosome 11p11), cardiac troponin T (chromosome 1q32), α-tropomyosin (chromosome 15q22), and titin (chromosome 2q31). Although a subset of HCM patients (estimated as ranging between 10 and 20%) develop a dilated phenotype, DCM arising from sarcomeric gene mutations often occurs without previous HCM and can occur in very early childhood.

Autosomal-dominant DCM-only mutations have also been found in genes encoding the cytoskeletal proteins desmin (chromosome 2q35), δ-sarcoglycan (chromosome 5q33), and metavinculin (chromosome 10q22–q23). These proteins stabilize the myofibrillar apparatus and link the cytoskeleton to the contractile apparatus.

Linkage studies performed in families with autosomal-dominant DCM and conduction defects have identified three *loci*: on chromosomes 1p1–q21,[125] 2q14–q22,[126]

TABLE VII Genes Implicated in FDCM

Gene	Protein	Chromosomal locus	Inheritance	References
ABCC9	SUR2A, regulatory subunit of cardiac K_{ATP} channel	12p12.1	AD	103
ACTC	Cardiac α-actin	15q14	AD	104
ACTN2	α-Actinin-2	1q43	AD	105
DES	Desmin	2q35	AD	106
DMD	Dystrophin	Xp21	X-R	107
TAZ (G4.5)	Tafazzin	Xq28	X-R	108
LMNA	Lamins A and C	1p1-q21	AD	109
CSRP3	Muscle LIM protein	11p15	AD	105
MYBPC3	Cardiac myosin binding protein C	11p11	AD	110
MYH6	α-Myosin heavy chain	14q12	AD	111
MYH7	β-Myosin heavy chain	14q12	AD	112
PLN	Phospholamban	6q22.1	AD	113
SCN5A	Cardiac sodium channel	3p22-25	AD	114, 115
SGCD	δ-Sarcoglycan	5q33	AD	116
STA	Emerin	Xq28	X-R	117
TCAP	Tcap/telethonin	17q12	AD	54
TNNC1	Cardiac troponin C	3p21.3-3p14.3	AD	118
TNNI3	Cardiac troponin I	19q13.4	AR	119
TNNT2	Cardiac troponin T	1q32	AD	120
TPM1	α-Tropomyosin	15q22	AD	121
TTN	Titin	2q31	AD	122
VCL	Metavinculin	10q22-q23	AD	123
ZASP	Cypher/Zasp	10q22.3-q23.2	AD	124

and 3p22–p25.[127] A disease-causing gene has been found in one of these *loci* (the lamin A and C encoding *LMNA* gene at chromosome 1p1–q21). Autosomal-dominant DCM with conduction defects and skeletal myopathy mapped to two *loci* (chromosomes 1p1–q21[128] and 6q23).[129] Mutations in *LMNA* have been associated with two skeletal myopathies with cardiac involvement, Emery–Dreifuss muscular dystrophy and autosomal-dominant limb girdle muscular dystrophy 1B.[109] Autosomal-dominant DCM with sensorineural hearing loss has been mapped to chromosome 6q23–q24.[130] Autosomal-recessive inheritance has been associated with infantile forms of DCM. A mutation in the cardiac troponin I (*TNNI3*) has been reported in association with autosomal recessive DCM.[118] Three disease genes have been found to cause X-linked DCM. Mutations in the genes encoding dystrophin and emerin have been found in Duchenne and Becker muscular dystrophy[107,117] but have also been shown to cause an adult-onset X-linked DCM without skeletal myopathy.[131] Mutations in the *TAZ* gene (also known as *G4.5*) encoding tafazzin cause Barth's syndrome as well as a childhood-onset X-linked DCM.[108]

Mutations in mitochondrial DNA have been found in maternally inherited mitochondrial cardiomyopathies,[132] and these will be addressed in further depth later in this chapter.

Cardiac Actin Gene

The cardiac actin gene (ACTC) gene, the first gene associated with adult-onset autosomal-dominant DCM, encodes the sarcomere thin filament protein cardiac actin. Because cardiac actin has an integral role in cardiac muscle contraction, it was considered to be a promising candidate for DCM. Missense mutations in the *ACTC* gene including G867A and A1014G were reported in two small unrelated families with DCM and, subsequently in several families with HCM.[104] Although the pathophysiological basis for the differences in phenotypic expression of *ACTC* mutations has not yet been established, these differences may be attributable to the location of mutations in distinct functional domains of the *ACTC* molecule. HCM-causing mutations identified to date have been located in regions that are involved in actin-myosin interaction and force generation, whereas DCM-causing mutations have been located in the immobilized end of the actin molecule at the Z line, which is likely involved in enabling the transmission of force from the sarcomere to the extra-sarcomeric cytoskeleton. Nevertheless, pathogenic *ACTC* mutations in association with DCM are exceedingly rare and have not been found in several large population screenings.[133–136]

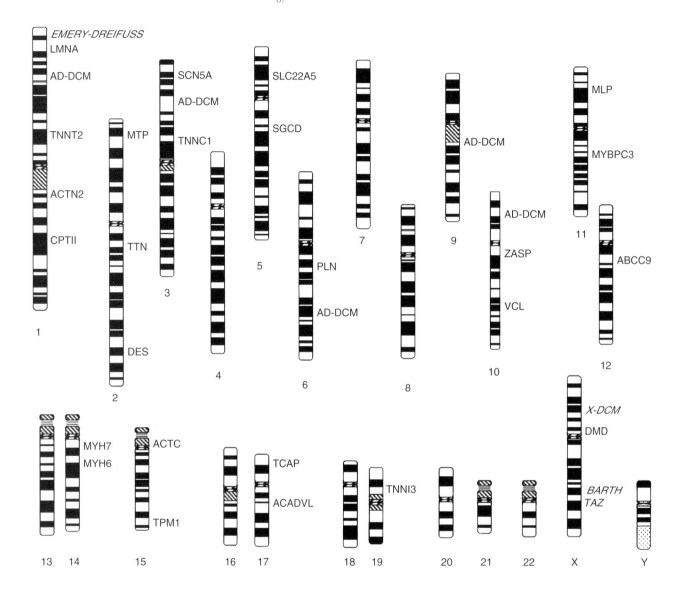

FIGURE 2 Human chromosome map of genetic *loci* involved in DCM.

MYOSIN HEAVY CHAIN GENES

In a linkage analysis performed in a multigeneration family the chromosome 14q12 DCM *locus* was associated with early-onset (<25 y) DCM in ~50% of affected members.[137] Subsequent screening of the myosin heavy chain *MYH7* gene, encoding the sarcomeric β-myosin heavy chain previously discussed in reference to HCM, in probands from the index family and from 20 additional families with DCM revealed two missense mutations, one mutation in a large family and a second mutation in a small unlinked family. The phenotype of the second family was also characterized by early onset of ventricular dilation and dysfunction, including SCD of a 2-month-old infant.

HCM-causing *MYH7* mutations have been reported to cluster predominantly in four regions of the molecule and are thought to perturb actin-myosin interaction and force generation. One of the two DCM-causing *MYH7* mutations (Ser532Pro) was located in a conserved helical structure of the myosin domain and predicted to disrupt actin-myosin binding. The other mutation (Phe764Leu), located in a "hinge" region between the myosin head and neck, is predicted to alter either the magnitude or polarity of cross-bridge movement, and thereby modulate contraction efficiency.

A recent study suggested that *MYH7* may represent a "hot spot" for DCM mutations.[138] Seven novel heterozygous missense mutations in *MYH7* were identified in 7 of 96 unrelated patients with DCM, 5 with familial disease, and 2 sporadic. The mutant amino acid residues were located either in the head (I201T, T412N, A550V) or tail domains (T1019N, R1193S, E1426K, R1634S) of the protein. The DCM presented by these patients showed delayed onset, incomplete penetrance, and was not associated with skeletal myopathy or conduction defects. In addition, in a study of 46 young patients with DCM, 2 patients harbored novel missense *MYH7* mutations (Ala223Thr and Ser642Leu).[110] Ser642Leu is part of the actin-myosin interface, whereas Ala223Thr affects a residue near the ATP binding site. It is noteworthy that although mutations in *MYH7* are present in high incidence in both DCM and HCM, the site of the mutations detected seems to be distinct in each form, supporting the view that the underlying mechanism of the defective protein to elicit each respective mutant phenotype is likely different. Recently, a large-scale study of the myosin heavy chain *MYH6* gene encoding α-MHC in patients with DCM identified three heterozygous missense mutations (P830L, A1004S, and E1457K) in six affected patients associated with delayed-onset DCM.[139]

cMyBP-C Gene

The *cMyBP-C* gene *(MYBPC3)* described previously in association with HCM was selected as a potential candidate gene for DCM subsequent to findings that a number of HCM-causing genes could also cause DCM. Several heterozygous missense *MYBPC3* mutations, for example, Asn948Thr[110] and R820Q,[140] have been reported in individuals with DCM, although this is less prevalent than with HCM. The missense mutation (Arg820Gln) previously found in several patients with HCM has been recently reported in elderly patients with DCM.[141] Furthermore, a novel mutation in *MYBPC3* causing the loss of titin and myosin binding sites has been reported in a patient with HCM who progressed to DCM.[142]

Cardiac Troponin T and Tropomyosin Genes

The cardiac troponin T (*TNNT2*) gene has also been considered a potential candidate gene for DCM because of its previous consideration as a HCM gene. On screening 19 probands from families with DCM, a deletion in cardiac troponin T (Lys210) was identified in two unrelated families in this cohort, with early-onset DCM (average age 24 y) and no evidence of LVH or HCM.[137] Sudden deaths of two infants with infantile cardiomyopathy and three young adults were also identified in these families. The DCM-causing *TNNT2* deletion was located in a C-terminal region (residues 207–234), which has been implicated in Ca^{++}-sensitive troponin C binding. Moreover, the functional consequences of this DCM mutation (such as decreased calcium sensitivity) are qualitatively different from those of the *R92Q* HCM troponin T mutation, further suggesting that DCM and HCM may be triggered by distinct primary stimuli.[143] Another novel troponin T mutation located at a highly conserved arginine residue at codon 141 (*R141W*) was reported in the tropomyosin-binding site of troponin T in a large family consisting of 72 members; 14 presented with DCM, mainly in the second decade of life. Before genotyping studies, 17 members of this family had died of HF.[144]

The *TPM1* gene encoding the thin filament protein α-tropomyosin has been screened in individuals with both familial and sporadic DCM and HCM. Two missense mutations in highly conserved residues (E40K and E54K) have been identified in probands with familial DCM.[121] As previously noted, HCM-causing *TPM1* mutations have been found in regions of the α-tropomyosin molecule.[9] These mutations may impair troponin T binding, functionally may increase the Ca^{++} sensitivity of myofibrillar ATPase activity, and exhibit deregulated activation of ATPase activity, which subsequently will affect contractility and force generation.[145] In contrast, the two DCM-causing *TPM1* mutations change the surface charge of α-tropomyosin, functionally reducing the Ca^{++} sensitivity of ATPase activity, and likely depressing myofibrillar contractility.[145,146]

Titin Gene

The 2q31 *locus* had been identified by linkage analysis in a multigeneration family with autosomal-dominant DCM.[147] The titin gene (*TTN*) encoding the giant sarcomeric cytoskeletal protein titin, an important element in the extensible scaffold for the contractile machinery and a crucial determinant of myofibrillar elasticity and integrity, maps to this *locus* and was screened as a candidate gene in two large families with FDCM. Screening of more than 340 exons of the *TTN* resulted in finding two novel mutations, a 2-bp insertion that caused a frameshift and premature stop codon predicted to truncate the titin protein and a missense mutation (Trp930Arg) in a highly conserved hydrophobic core sequence at the Z-disc/I-band transition zone.[122] Four mutations in *TTN* have been reported in another study on DCM.[148] Two of these mutations, Val54Met and Ala743Val in the Z-line region of titin, resulted in decreased binding affinities of titin to the Z-line proteins T-cap/telethonin and α-actinin, respectively, as assessed by yeast two-hybrid assays. The other two mutations were found in the cardiac-specific N2-B region of titin, and one of them was a nonsense mutation, Glu4053ter, presumably encoding a truncated nonfunctional molecule.

Mutations have also been identified in genes encoding several of the proteins known to interact with titin at the Z-disc. Mutations in T-cap (*TCAP*),[54] α-actinin (*ACTN2*),[105] αB-crystallin (*CRYAB*),[149] muscle LIM protein (*CSRP3*),[105,150] and *Zasp*[124] genes, which result in altered Z disc protein interactions, have all been reported in association with FDCM.

Desmin Gene

Considerable evidence suggests that desmin, a muscle-specific cytoskeletal protein that belongs to the intermediate filament family, plays a significant role in cardiac growth and development and contributes to force transmission from the sarcomere, as well as resistance to mechanical stress in mature muscle. Several families with myopathy and cardiomyopathy and an unusual distribution of desmin in skeletal and cardiac muscle have been reported.[151,152] These findings suggested that the desmin gene (*DES*) encoding desmin might be a candidate gene in DCM. The *DES* gene, a single-copy gene located on chromosome 2, is composed of 8.4 kb of genomic DNA containing nine exons that encode a protein of 470 amino acids. A missense mutation (Ile451Met) located in the desmin tail domain of the *DES* gene has been reported to segregate with FDCM in a four-generation family, without clinical evidence of skeletal myopathy.[153] *DES* mutations are exceedingly rare in DCM and have not been detected in screened European patient populations.[137] However, three unrelated young male patients with the Ile451Met mutation have been identified in a group of 265 Japanese patients; these patients were younger and manifested a more severe phenotype (lower left ventricle shortening and ejection fraction) than other sporadic cases.[154]

In addition, familial cases of RCM caused by desmin accumulation and characterized by severe cardiac conduction defects have been reported.[155] Desmin accumulation has been detected in a small subset of RCMs and was further characterized by the presence of two isoforms of desmin, one of normal and another of lower molecular weight in cardiac and skeletal muscle of familial cases.[156] Furthermore, two new families with desmin-related cardioskeletal myopathy (a familial disorder characterized by skeletal myopathy, conduction defects, RCM, and intracytoplasmic accumulation of desmin-reactive deposits) were associated with mutations in the highly conserved C-terminal end of the desmin rod domain. A heterozygous A337P mutation was identified in a family with adult-onset skeletal myopathy and mild cardiac involvement. Two other mutations, A360P and N393I, were detected in a second family characterized by childhood-onset aggressive course of cardiac and skeletal myopathy.[106] The mechanism by which some *DES* mutations cause RCM and others cause DCM is unknown.

δ-Sarcoglycan Gene

The δ-sarcoglycan gene (*SGCD*) encodes the sarcolemmal transmembrane glycoprotein, δ-sarcoglycan (δ-SG), one of four sarcoglycan proteins (α-, β-, γ-, and δ-SG) highly expressed in cardiac and skeletal muscle. The sarcoglycan proteins associate in stoichiometrically equal proportions to form the sarcoglycan complex, a component of the dystrophin-associated glycoprotein transmembrane complex. The *SGCD* gene located at chromosome 5q33 is composed of nearly 100 kb of genomic DNA with eight exons encoding a protein of 290 amino acids (35-kDa protein) and has a small intracellular domain, a single transmembrane hydrophobic domain, and a large extracellular C-terminus. The expression of δ-SG is abundant in striated and smooth muscles, with the highest levels of expression found in skeletal and cardiac muscle.

A mutation in the *SGCD* gene was first reported in families with autosomal-recessive limb girdle muscular dystrophy (LGMD2F), a skeletal muscle disorder with a highly variable phenotype that often includes AV conduction defects. The *SGCD* mutation was homozygous and comprised a single nucleotide deletion that alters its reading frame.[157] Other LGMD disorders arise from mutations in the α, β, and γ sarcoglycan genes, may present with cardiac involvement and skeletal myopathy, and primarily display a pattern of autosomal-recessive inheritance. Interestingly, an autosomal-dominant form of LGMD results from mutations in two other genes, *CAV3* and *LMNA*.[158,159]

A missense mutation (S151A) in the *SGCD* gene was found in a family with autosomal-dominant DCM. All affected members displayed both congestive HF and a high prevalence of SCD at an early age.[116] The change from a polar to nonpolar amino acid is significant and likely alters the secondary structure of δ-sarcoglycan. Moreover, in 2 of 50 individuals with sporadic DCM, a 3 base pair deletion of the Lys238 codon has also been identified. These patients presented with HF at less than 20 y of age, and none of them showed clinical evidence of skeletal myopathy. In addition, in a group of 52 patients with DCM, recently screened for δ-SG/*SGCD* mutations, a novel mutation, Arg71Thr, was found in two patients who presented with a relatively mild DCM phenotype and late onset.[160]

Metavinculin Gene

The metavinculin gene (*VCL*) encodes the cytoskeletal proteins vinculin and its cardiac isoform, the splice variant, metavinculin. Vinculin is ubiquitously expressed, whereas metavinculin is coexpressed with vinculin exclusively in cardiac, skeletal, and smooth muscle.[161] In the heart, vinculin and metavinculin are localized to the sarcolemmal-sarcomere attachment sites (structures termed costameres), where they interact with α-actinin and actin to form a microfilamentous network. Vinculin and metavinculin are also present in adherens junctions in intercalated discs, structures involved in anchoring thin filaments, cell–cell adhesion, and force transmission between myocytes. Metavinculin seems to be specifically involved in the connection of actin microfilaments to the intercalated disk and membrane costameres of the heart. Because skeletal muscle lacks intercalated disks, it is possible that the localization of metavinculin to this highly specialized cardiac structure is related to a specific function conferred by metavinculin, a function that vinculin alone cannot support.

The *VCL* gene is composed of more than 75 kb of genomic DNA and has 22 exons that encode the two isoforms vinculin (1066 amino acids) and metavinculin (1134 amino acids). The larger isoform, metavinculin, is generated by alternate splicing of exon 19 and contains an additional 68 amino acids in the C-terminal region. Vinculin and metavinculin are composed of identical globular head regions, which interact with membrane-associated ligand proteins and a unique rod-shaped tail fragment, which harbors actin filament–binding sites.

A metavinculin deficiency has been reported in one individual (from 28 examined) with idiopathic DCM.[123] Transcript analysis indicated the presence of normal vinculin mRNA, but not metavinculin mRNA, reflecting abnormal splicing. This cytoskeletal defect was associated with abnormalities in ventricular function, cardiac dilatation, and immunohistological defects at the sites of metavinculin absence (i.e., the cardiomyocyte membrane and intercalated disk).

Recently, SSCP screening was used to identify 2 *VCL* mutations in a cohort of 350 unrelated individuals with familial and sporadic DCM, including one missense mutation (Arg975Trp) and one 3-bp deletion (Leu954del) in highly conserved residues.[162] Heart tissue from the individual with an Arg975Trp *VCL* mutation showed grossly abnormal intercalated discs, and *in vitro* assays demonstrated that both the Arg975Trp substitution and the Leu-95 deletion significantly altered metavinculin-mediated cross-linking of actin filaments.

Interestingly, the same Arg975Trp mutation in *VCL* has recently been reported in a patient with HCM exhibiting severe obstruction, midventricular, and apical hypertrophy.[50] Immunohistochemical analysis revealed a scarcity of vinculin/metavinculin in the intercalated discs. It is worth noting that mutations in functionally distinct regions of certain cardiomyopathy-associated genes often have a dominant effect in determining a remodeling pathway of either maladaptive hypertrophy or dilation. In contrast, the discovery that the same *VCL* mutation can yield either cardiomyopathic phenotype, underscores a critical role for modifier genes, or environmental stressors, in determining these pathways.

Lamin A/C Gene and Emerin

The lamin A/C gene (*LMNA*) located on chromosome 1q11-q23 encodes the intermediate filament proteins lamins A and C. The lamins are located in the nuclear lamina at the nucleoplasmic side of the inner nuclear membrane and have a structural role in maintaining membrane integrity. In dividing cells, the binding of lamins to M-phase chromatin is important for the reassembly of nuclear membranes in daughter cells during mitosis. Lamin binding to interphase chromatin may maintain chromatin architecture and contribute to the regulation of the cell cycle and gene transcription.

Lamins are classified broadly as A type and B type. The A-type lamins include four transcripts produced by alternate splicing of the *LMNA* gene: lamins A, C, A10, and C2 and are expressed in differentiated cells in a wide range of tissues. The B-type lamins include lamins B1 and B2 and are expressed throughout development and are the only lamins present in the embryo. Lamins A and C are expressed in heart and skeletal muscle, and specific functions of lamins A and C have not been precisely determined.

The coding sequence of the *LMNA* gene spans 24 kb of genomic DNA with 12 exons. The principal transcripts, lamins A and C, have identical sequence for the first 566 amino acids (exons 1–10) but differ in their C-terminal regions, which contain 98 and 6 unique amino acids, respectively. Lamins A and C consist of a central rod domain flanked by globular head and tail domains (Fig. 3).

Nineteen *LMNA* mutations have been found in families with autosomal-dominant DCM with cardiac conduction defects.[109,163] The defects include missense mutations, deletions, and frameshift mutations that result in premature termination codons. These *LMNA* mutations have been located predominantly in the central rod domain common to lamins A and C. Interestingly, although the mutations in the lamin A/C rod found in families with autosomal-dominant DCM with conduction defects typically are not associated with clinical evidence of skeletal myopathy and abnormal creatine kinase levels, each *LMNA* mutation was associated with progressive conduction defects, including sinus bradycardia and AV block, atrial dysrhythmias, and DCM. There was also frequent occurrence of HF and SCD within these families, and the age at the onset of the cardiac involvement averaged 30 y. In one family with a mutation in the lamin C–specific region of the tail domain, no clinical features of skeletal myopathy were present; however, in several family members, creatine kinase levels were mildly elevated.[109]

It is important to note that the occurrence of conduction defects and DCM in individuals with *LMNA* mutations seems to be strongly linked. Recent observations have shown no *LMNA* mutation in cases with isolated DCM while revealing several mutations in association with conduction defects and atrial fibrillation.[164]

LMNA mutations have also been found to cause two forms of autosomal-dominant skeletal myopathy, Emery–Dreifuss muscular dystrophy and limb girdle muscular dystrophy type 1B, as well as familial partial lipodystrophy, Charcot-Marie-Tooth disorder type 2B, and Hutchinson–Gilford progeria.[158,163,165]

Emery–Dreifuss muscular dystrophy is characterized by a childhood onset of contractures of the elbows, Achilles tendons, and post-cervical muscles, together with slowly progressive wasting and weakness of humeroperoneal muscles. In later life, affected individuals may develop cardiac conduction defects, most commonly AV block. A number of patients also develop DCM and present left ventricular dysfunction. Creatine kinase levels are typically elevated. The clinical manifestations of the autosomal-dominant form of Emery–Dreifuss muscular dystrophy (EDMD) are similar to those of the X-linked recessive form, resulting from mutations in the *STA* gene located at Xq28 encoding emerin, a serine-rich integral protein in the inner nuclear membrane.[166] These *STA* mutations, most of which involve early translation termination, result in the loss of all or part of the emerin protein.[167] A number of *STA* mutations have also been identified in EDMD patients in which emerin is targeted to the

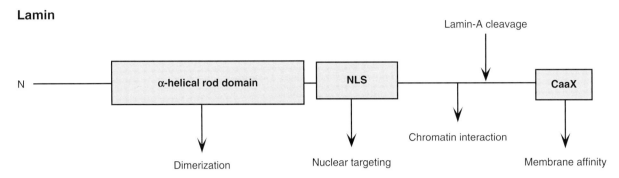

FIGURE 3 The structure of the lamin protein. Domains are shown in rectangles with arrows indicating function.

nuclear membrane, but to a lesser extent than wild-type emerin. Moreover, mutant emerin proteins are unstable if they are unable to integrate into a membrane.[168] Emerin defects have also been reported in patients presenting with a predominantly cardiac-defective phenotype (e.g., conduction defects and cardiomyopathy) with no apparent skeletal muscle involvement.[169] Emerin seems to affect multiple functions in the nucleus, including gene and cell-cycle regulation and maintenance of nuclear integrity. Furthermore, the emerin protein has been shown to have specific interactions with a variety of nuclear proteins, including barrier-to-autointegration factor (BAF; a regulator of higher order chromatin structure), lamin A, transcription factors (e.g., GCL), an mRNA splicing regulator, a nuclear membrane protein named nesprin, nuclear myosin, and F-actin isoforms.[170]

Mutations causing autosomal-dominant Emery–Dreifuss muscular dystrophy were first identified in the globular head and tail domains of lamins A and C.[171] Recent studies have found that Emery–Dreifuss muscular dystrophy-causing mutations may also be located in the rod domain of lamins A and C. In an analysis of large French kindred, among which 17 members presented with an autosomal-dominant cardiomyopathy, and 5 with manifestations of Emery–Dreifuss muscular dystrophy, a nonsense mutation in the *LMNA* gene was identified.[172]

Moreover, this family had a high incidence of SD in individuals with history of dysrhythmias. This is strikingly similar to findings with a family harboring EDMD caused by a nonsense *STA* mutation exhibiting high incidence of cardiac dysrhythmias in all male carriers and SD.[173]

Another important finding with the *LMNA* gene is the high incidence of *de novo* mutations that have been detected. One study found that *de novo LMNA* mutations were found in 76% of the patients screened.[171] The high proportion of *de novo* mutations together with the large phenotypic spectrum of both *LMNA* mutations indicates that screening for *LMNA* may prove informative in both familial and sporadic cases of EDMD, as well as in DCM associated with conduction defects.

Autosomal-dominant limb girdle muscular dystrophy type 1B is a slowly progressive skeletal myopathy without contractures and with age-related conduction disturbances. Mutations causing this disorder have been located in the rod and tail domains of lamins A and C.[158] Cardiac muscle involvement has not been reported in any of the families with the other disorders (e.g., Charcot-Marie-Tooth disorder type 2B and Hutchinson–Gilford progeria) associated with *LMNA* mutations. However, *LMNA* variants have been reported in two families with familial partial lipodystrophy (Dunnigan variety) who also had cardiac conduction defects and cardiomyopathy.[174] The multisystemic dystrophy in these families resulted from two novel missense mutations in exon 1 of *LMNA*. One mutation, R28W (CGG→TGG), affected the N-terminal head domain, and the other, R62G (CGC→GGC), affected the α-helical rod domain.

The mechanisms by which mutations in the A-type lamin proteins cause tissue-specific diseases are not well understood. It remains unknown why defects in these ubiquitous proteins can result in abnormalities of specific and highly differentiated tissues, such as skeletal and cardiac muscle, composed of differentiated cells. Immunofluorescence studies have demonstrated abnormal localization of mutant lamin A proteins,[175,176] a partial disruption of the endogenous lamina and altered localization of the nuclear envelope protein, emerin. In addition to structural changes, these mutations can also affect processes such as gene regulation. As we will discuss shortly, functional studies of the effects of these mutations in transgenic animal models have provided further information.

Dystrophin Gene

The dystrophin gene (*DMD*) encodes dystrophin, a large cytoskeletal protein expressed predominantly in skeletal, heart, and smooth muscle. The dystrophin protein interacts with both cytoskeletal actin and the dystrophin-associated glycoprotein complex, which provide a strong mechanical link from the intracellular cytoskeleton to the ECM. Dystrophin contributes to intracellular organization, force transduction, and membrane stability. The *DMD* gene, the largest known human gene (spanning more than 3 Mb), contains 79 exons (accounting for about 0.6% of the gene) that encode a protein of 3685 amino acids, and it is transcribed to form a 14-kb mRNA. Transcription regulation of the *DMD* gene is complex, with at least seven independent, tissue-specific promoters. Alternate splicing gives rise to a range of different transcripts and protein isoforms. Four domains in the dystrophin protein have been identified (Fig. 4): (1) a 240-amino acid N-terminal domain that is homologous to the actin-binding domain of α–actinin, responsible for anchoring dystrophin to cytoskeletal actin; (2) a large central rod domain composed of 24 triple-helical segments, similar to the repeat domains of spectrin; (3) a cysteine-rich segment that anchors dystrophin to the sarcolemma; and (4) a 420-amino acid α-helical C-terminal domain.

Mutations in the *DMD* gene were initially reported in both familial and sporadic cases of Duchenne and Becker muscular dystrophies.[177,178] Duchenne muscular dystrophy (DMD) is an X-linked recessive disorder, affecting 1 in 3500 male births. During childhood, affected males frequently develop gait disturbances, progressive skeletal muscle weakness, mainly of the large proximal muscle groups, and generally by early adolescence loss of ambulation. DCM and conduction defects tend to present late in the disease, and the severity of the cardiac involvement is unrelated to the severity of the skeletal muscle disease. Most patients die in their twenties, most commonly as a result of respiratory failure caused by diaphragm muscle weakness.

Becker muscular dystrophy (BMD) is a milder, allelic form of DMD with an incidence of 1 in 14,000 live male births, but it is almost as common as DMD because of longer patient survival. Affected males present later in life and have a milder course. Patients with BMD also have a high incidence of cardiac involvement despite their milder skeletal muscle disease; the most common cause of death in BMD is HF. Interestingly, female carriers of DMD and BMD experience a high incidence of cardiac involvement that progresses with age and manifests primarily as cardiomyopathy.

In general, individuals with Duchenne muscular dystrophy have frameshift or nonsense *DMD* mutations that result in premature termination of translation and a reduction or absence of dystrophin protein. Typically, patients with DMD lack any detectable dystrophin expression in their skeletal muscles. In contrast, individuals with BMD usually have *DMD* deletions that result in truncation or reduced levels of expression of dystrophin.[177] Muscle from patients with BMD contains dystrophin proteins of altered size and/or reduced abundance.

DMD mutations have also been reported to cause X-linked DCM, a disorder in which affected males present with HF generally in late adolescence, often with rapid progression to death within 1–2 years; female carriers display a later onset of DCM with milder symptoms.[179,180] In contrast to either DMD or BMD, no evidence of significant skeletal myopathy is manifested. The presentation, age at onset, severity, and clinical course of the cardiomyopathy are variable, ranging from an early onset and fatal cardiomyopathy to milder forms with better prognosis.

DMD gene defects associated with X-linked DCM were initially noted to cluster at the 5' end, including the promoter, first exon, and a "hinge" region between N-terminal actin–binding domain and the central rod domain. Mutations in other regions further downstream have been also identified.[181,183] Point mutations, deletions, insertions, and inversions have been reported[181] and recently grouped in two classes (Table VIII): Class A includes mutations affecting transcription or splicing of the dystrophin gene, resulting in a more severe cardiac involvement; Class B includes mutations in which specific protein domains of dystrophin are affected.

In contrast to group A patients, in whom dystrophin is totally absent within the heart, in group B patients the dystrophin concentration is reduced, and no major difference in promoter use or splicing between skeletal and cardiac muscle has been reported. Thus, in group B, the mutations tend to cause a qualitative rather than a quantitative defect in perturbing the expression of dystrophin.[182,183] Most patients belonging to this group carry mutations in the spectrin-like domain of dystrophin.

Tafazzin Gene

The tafazzins are a group of proteins produced by alternate splicing of the tafazzin gene (*TAZ*) or *G4.5*, which is expressed at high levels in cardiac and skeletal muscle. Although the protein product of the *TAZ* gene shows sequence homology

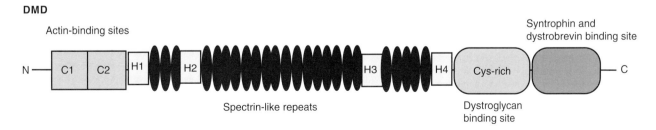

FIGURE 4 Domains of the dystrophin protein. The N-terminus contains the primary actin binding sites, whereas the C-terminus contains the β-dystroglycan, dystrobrevin, and syntrophin binding sites. The N- and C-terminal domains are connected by 24 spectrin-like repeats, some of which have been shown to bind actin. The four "hinge" regions are also shown (H1–H4).

TABLE VIII Mutations in DMD Resulting in X-Linked DCM

Location	Type
Group A	
Muscle exon/intron 1 junction	Deletion
Muscle exon/intron 1 junction	Deletion
Muscle exon 1 3' splice site	Point mutation
Muscle exon 1 3' splice site	Point mutation
Splicing in intron 1	Point mutation
Muscle exon 1	Insertion of L element
Exons 2–7	Duplication
Exon 9	Point mutation
Intron 11	Deletion
Group B	
Exon 29	Point mutation
Exon 35	Missense mutation
Exons 45–48	Deletion
Exons 45–55	Deletion
Exon 48	Deletion
Exon 48	Deletion
Exons 48–49	Deletion
Exons 48–51	Deletion
Exons 48–53	Deletion
Exons 49–51	Deletion

to the glycerolipid acyltransferase family of enzymes, its precise biochemical function remains to be explained.

TAZ is a single-copy gene composed of ~11 kb of genomic DNA with 11 exons, which maps to Xq28, and two ATG initiation sites. Up to 10 different mRNAs can be produced by alternate splicing at exons 5–7, resulting in tafazzin proteins ranging from 129–292 amino acids in length that differ at the N-terminal and central regions. Two putative functional domains have been identified: (1) a highly hydrophobic segment of 30 residues at the N-terminus, which likely acts as a membrane anchor; and (2) a hydrophilic segment in the central region that forms an exposed loop interacting with other proteins.

Mutations in *TAZ* have been associated with Barth's syndrome, an X-linked disorder characterized by infantile-onset DCM, skeletal myopathy, short stature, neutropenia, and abnormal mitochondria.[108] Histopathological findings include ventricular hypertrophy, dilation, and/or endocardial fibroelastosis. On ultrastructural examination, mitochondria exhibit concentric, tightly packed cristae with occasional inclusion bodies. In addition to mitochondrial dysfunction, biochemical analysis has shown the occurrence of 3-methylglutaconic aciduria and a deficiency of cardiolipin, especially its tetralinoleoyl form.[184,185]

A variety of *TAZ* mutations, including splice site mutations, insertions, deletions, nonsense, and missense mutations, located throughout the *TAZ* gene have been reported to cause Barth's syndrome.[186] No correlation has been found between genotype and either severity or age at onset of the cardiac phenotype. In addition, the genetic defect usually differs among families. Because the same *TAZ* defect within families can cause varied clinical phenotypes, ranging from mild to SD, it is likely that modifier genes are involved in determining the phenotype and the clinical severity.[187] Interestingly, despite the X-linked inheritance of Barth's syndrome and the identification of many females carrying *TAZ* mutations, there have been no reports of females with classic Barth's syndrome. Mutations in *TAZ* have subsequently been associated with three other myocardial disorders: X-linked endocardial fibroelastosis,[188] an X-linked form of noncompaction of the left ventricle,[187] and X-linked infantile DCM.

DCM, neutropenia, mitochondrial abnormalities, and infantile death characterize X-linked endocardial fibroelastosis. The same missense mutation in a conserved region of exon 10 of the *TAZ* gene has been found in two unrelated families.[188]

X-linked isolated noncompaction of the left ventricular myocardium (INVM) may be a recessive disorder with infantile DCM, cardiac dysrhythmias, and SD. Post-mortem examination shows numerous prominent trabeculations and deep intertrabecular recesses in the ventricular myocardium, as well as ventricular hypertrophy, dilation, and endocardial fibroelastosis. A missense mutation in a conserved region of exon 8 of the *TAZ* gene (G197R) has been reported in one family and prompted the hypothesis that INVM is a severe allelic variant of Barth's syndrome, with a specific effect on the heart.[189] However, recent observations indicated that most INVM cases are autosomally dominant and are rarely caused by *TAZ* mutations.[190]

X-linked DCM is a recessive disorder in which male infants develop DCM and SD. A deletion in exon 8 of the *TAZ* gene was identified in one family, resulting in the complete absence of all tafazzin proteins.[188] Recently, other *TAZ* mutations in association with X-linked DCM, including a stop codon (E188X) and a nonsense mutation, have been detected in one family and an amino acid substitution (G240R), a missense mutation, in a second kindred.[191]

Cardiac biopsy specimens of both kindreds with X-linked DCM showed markedly enlarged mitochondria containing large bundles of stacked and disarrayed cristae and decreased unsaturated and increased saturated fatty acid concentrations. Impaired acyltransferase function could result in increased fatty acid saturation that would decrease membrane fluidity.

Additional Genes

Additions to the growing list of DCM genes can be expected. Of the 15 chromosomal *loci* mapped for autosomal-dominant DCM, with or without conduction defects, 10 disease-causing genes have been identified. Moreover, for most of these pathogenic genes, only a very small number of mutations in each gene have been reported. At present, available data suggest that these known disease genes account for a relatively small proportion of all cases of DCM and suggest that a large number of different genes may cause DCM, as well as intimating that important disease-causing genes remain to be discovered. Linkage studies in large families with DCM will continue to be an invaluable method for further identification of DCM genes.

A major limitation of the candidate gene approach alone, without linkage or functional analyses, is that apparent mutations may in fact be rare polymorphisms.

TRANSGENIC MODELS OF DILATED CARDIOMYOPATHY

In the past few years, there has been a proliferation of mouse models (as well as other animal models) generated by overexpression or deletion of a variety of cardiac proteins, including sarcomeric, cytoskeletal, and cell signaling proteins, generating a DCM phenotype. These studies have demonstrated that perturbation of the normal function of diverse proteins, which contribute to cardiac structure and function, may cause DCM. A number of these models have preceded the findings with subsequent human/clinical studies and presaged their involvement in DCM. For instance, studies with mice containing missense mutations in desmin demonstrated the critical importance of this intermediate filament in maintaining myocardial integrity.[192] In addition, both naturally occurring models of inherited DCM in the hamster and transgenic models of δ-sarcoglycan mutants have suggested, and confirmed, the role of this component of the dystrophin/dystroglycan complex in cardiomyopathy and demonstrated that destabilization of the cytoskeleton leads to membrane fragility and loss of membrane integrity, with resulting degeneration of skeletal muscle and cardiomyocytes.[193–195] Transgenic mice containing null mutations in MLP were instrumental in defining the role of these genes in DCM.[196] Transgenic studies have also shown that overexpression or constitutive activation of an array of signaling components, including β-adrenergic receptor,[197,198] the catalytic subunit of PKA,[199] retinoic acid receptor,[200] and the stimulatory G_q α subunit[201] (as well as G_i subunit),[202] all can lead to DCM in the mouse. A list of transgenic mice with alterations in specific genes leading to DCM is shown in Table IX.

A number of concerns have been raised as to the relevance of these models. Murine models of human cardiac disease, although exhibiting important facets of the phenotype, often fail to demonstrate a complete spectrum of the characteristic human pathological conditions.[203] Significant differences exist between the human and murine electrophysiological and contractile protein isoforms, which underlie the functional differences in normal cardiac function, as well as the responses to specific genetic mutations. A limited number of genes such as phospholamban have shown one effect in transgenic mice (a salutary effect of overexpression relative to DCM development) and an entirely different effect in humans.[204]

Although the use of genetically manipulated models with larger animals, such as rabbits and pigs, has been proposed as a solution, the continued and cautious use, and extrapolation of findings, with mouse models to human findings has continued to provide a valuable treasure trove of insights into DCM.

PATHOPHYSIOLOGICAL MECHANISMS OF DCM

"Defective Force Transmission" Hypothesis

The "defective force transmission" hypothesis is based on the premise that the cytoskeleton provides intracellular scaffolding that is essential for the efficient transmission of force from the sarcomere to the ECM and for the protection of the myocyte from external mechanical stress. Defects in cytoskeletal proteins are presumed to predispose to DCM by reducing force transmission and/or resistance to mechanical stress. Support for the defective force transmission hypothesis for DCM has come from the identification of mutations in the genes encoding cardiac actin, desmin, metavinculin, the δ-sarcoglycan subunit of the dystrophin-associated glycoprotein complex, and dystrophin. Cardiac dysfunction may be related to effects of the protein itself or from abnormal interactions with other proteins. In addition, there is evidence implicating two other cytoskeletal proteins in familial DCM, the muscle LIM protein MLP and α-actinin.

"Defective Force Generation" Hypothesis

The identification of mutations in the genes encoding sarcomere proteins (e.g., β-MHC and cardiac troponin T), previously found to be involved in HCM and in FDCM raised the alternative hypothesis that defects in force generation might, in fact, be responsible for FDCM. Although DCM can occur as a late complication of chronic HCM, individuals with these gene mutations present with a primary DCM with no clinical or histological evidence of antecedent HCM.

The factors that determine the phenotypic expression of sarcomere protein gene mutations have not as yet been identified. One possibility is that the HCM-causing and DCM-causing mutations may occur in distinct functional domains of the encoded proteins. For instance, HCM-causing mutations in cardiac actin occur in regions involved in force generation, whereas DCM-causing mutations have been identified in regions putatively involved in force transmission.[104] A DCM-causing mutation, but not HCM-causing mutation, was found in the troponin C–binding domain of cardiac troponin T, suggesting that these two disease phenotypes might arise by different pathophysiological mechanisms.[120] In contrast, both

TABLE IX Gene Alterations Causing DCM in Transgenic Animals

Protein (gene)	Gene Alteration	Function	References
Angiotensinogen (*AGT*)	Null	Signaling	205
MLP	Null	Muscle LIM cytoskeletal protein	196
Bradykinin receptor B2 (*Bdkrb2*)	Null	Bradykinin receptor signaling	206
Angiotensin II type 2 receptor (*AT2R*)	Ventricle-targeted overexpression	Signaling	207
CHF1 (*Hey2*)	Null	Helix-loop-helix transcription factor	208
Cardiac Ena-VASP (*VASP-EVH1*)	Null	Vasodilator-stimulated phosphoprotein	209
Lamin A/C (*Lmna*)	Null	Nuclear membrane protein	210
TIMP-3	Null	Matrix metalloproteinases (MMP) inhibitor	211
CELF	Cardiac-specific dominant-negative allele	RNA binding proteins implicated in alternative splicing regulation	212
ABCA5	Null	Lysosomal ABC transporter	213
N-cadherin (*Cdh2*)	Cardiac-specific null	Cell adhesion molecule involved in intercalated disc	214
SC35	Cardiac-specific null	Trans-acting splicing factor	215
ErbB2 (*Her2*)	Ventricle-specific conditional allele	Receptor tyrosine kinase	216
Caveolin-1 (*Cav-1*)	Null	Signaling protein	217
MnSOD	Null	Antioxidant	219
CREB	Dominant negative	Transcription factor	218

HCM- and DCM-causing mutations in the α-MHC gene were identified in regions involved in actin-myosin interaction and force generation.[120] This might be explained by a second possibility, that is whether a mutation resulting in HCM or DCM is determined by the extent of contractile deficit incurred. If myocardial hypertrophy is a compensatory response to a certain degree of contractile dysfunction, it is feasible that this compensatory response might be inadequate to rescue more severe dysfunction, and DCM may ensue. Dosage effects of mutant protein may lead to specific phenotypes and are to an extent consistent with this hypothesis. For example, heterozygous mice expressing Arg403Gln-MHC develop HCM, whereas homozygous mice develop rapidly progressive DCM and neonatal death.[6] Similarly, heterozygous mice expressing a truncated cMyBP-C exhibit HCM; homozygous mice develop DCM.[64] These observations in mice suggest that there might be a continuum between the extent of sarcomere dysfunction and the presence of hypertrophy or dilation, in contrast to the hypothesis that independent pathways are triggered that lead either to HCM or DCM. Resolving the relationship of these different phenotypes will undoubtedly have broad implications for the therapeutic management of individuals with sarcomere protein gene mutations.

Role of Nuclear Lamina Proteins

The finding that DCM could be mediated by mutations in the nuclear envelope proteins (e.g., lamin A/C and emerin) has greatly expanded our ideas concerning genes involved in the pathogenesis of cardiac and skeletal myopathies. Although some data collected from mouse studies suggested that lamins A and C may be involved in force transmission in cardiac myocytes by virtue of their interaction with the desmin intermediate filament network, it has also been suggested that *LMNA* mutations might promote DCM by disruption of nuclear functions, including transcription, DNA replication, RNA splicing, and nuclear protein import. Lamins have been found to interact with a wide variety of proteins, including chromatin components, cell-cycle markers, transcription, and splicing factors. Moreover, it has been noted that DCM, skeletal myopathies, and partial lipodystrophy caused by *LMNA* mutations are not present at birth but develop later in life. Their progressive onset is consistent with the hypothesis that mutant lamins may confer an increased susceptibility to processes such as apoptosis, biomechanical stress, or other degenerative processes. Although deregulated apoptosis has been implicated in several cardiovascular disorders, including HF, cardiac conduction defects, myocardial infarction, and ARVD, its role in these disorders and in DCM, whether primary or secondary, has not been determined.

Role of Altered Gene Expression in the Progression of DCM

Microarray analyses of myocardial transcripts from DCM patients have shown that expression of a large number of genes may be up-regulated or down-regulated as a secondary response to HF.[96,220–222] This molecular response is composed of factors that may exacerbate progression of HF, as well as factors that form part of the compensatory mechanisms.

In one study, genes with altered expression profiles were clustered into several groups: (1) cytoskeletal and myofibrillar genes (e.g., striated muscle LIM protein-1, regulatory MLC-2, α-actin); (2) genes responsible for degradation and disassembly of myocardial proteins (e.g., 1-antichymotrypsin, ubiquitin, gelsolin); (3) genes responsible for protein synthesis (elongation factor-2, eukaryotic initiation factor-4AII); (4) genes encoding stress proteins; and (5) genes involved in metabolism and energetics (e.g., ATP synthase α-subunit, succinate dehydrogenase flavoprotein subunit, aldose reductase). It is relevant that progressive myocardial energy deficit resulting from cumulative mutations in mitochondrial genes may also contribute to the DCM phenotype.[132]

Finally, available data suggest that mutations in a number of individual genes can cause DCM by multiple distinct pathophysiological mechanisms. Given the clinical variability of autosomal-dominant DCM found both between and within families, it is likely that the phenotypic expression of DCM disease genes can be modified by other genetic factors and/or environmental factors. In determining the role of genetic factors in the pathogenesis of DCM, it will be important to establish whether genetic variants (polymorphisms) in various genes increase the susceptibility to other genetic and/or environmental factors.

RESTRICTIVE CARDIOMYOPATHY

Clinical Manifestations

Restrictive cardiomyopathy (RCM), the rarest form of cardiomyopathy, involves impaired ventricular filling and reduced diastolic volume in the presence of normal systolic function and normal or near normal myocardial thickness. It is most frequently caused by pathological conditions that stiffen the myocardium by promoting infiltration or fibrosis, including eosinophilic endomyocardial disease, amyloidosis, sarcoidosis, scleroderma, carcinoid, storage diseases (e.g., hemochromatosis, Gaucher's disease, Fabry disease, glycogen storage disease), metastatic malignancy, anthracycline toxicity, or radiation damage. Several of the infiltrative diseases resulting in RCM may be inherited, including familial amyloidosis, hemochromatosis, Gaucher's disease, and glycogen storage disease. Although most idiopathic RCM has not been found to be familial, there have been several reports of small families with variable phenotypes, including isolated RCM, RCM with AV conduction block and skeletal myopathy, and RCM with skeletal myopathy manifesting either autosomal-dominant or autosomal-recessive patterns of inheritance.[223,224]

Individuals with RCM may present with dyspnea, weakness, or exercise intolerance because of their inability to increase cardiac output by tachycardia without further reducing ventricular filling. In some cases, there may be signs of right HF, including hepatomegaly, ascites, and peripheral edema. Transthoracic echocardiography may show thickening of the ventricular walls, with variable effects on chamber diameter and contractility. A classic feature of cardiac amyloidosis is the presence of sparkling echodense spots in the myocardial walls giving a speckled "ground-glass" appearance. The mitral and aortic valve leaflets may also be thickened.

Both RCM and constrictive pericarditis, which has similar clinical and hemodynamic features, manifest impairment of diastolic filling with elevation of left and right ventricular pressures and reduced stroke volume and cardiac output.[225] Distinguishing between them is critical, because patients with constrictive pericarditis may benefit from surgical intervention. In RCM, the diastolic filling pressure in the left ventricle usually exceeds that in the right ventricle by at least 5 mmHg; in constrictive pericarditis, diastolic filling pressures in the left and right ventricles are equivalent. In RCM, endomyocardial biopsy can be useful to evaluate myocardial histopathology; transthoracic or transesophageal echocardiography, computed tomography scanning, and magnetic resonance imaging may demonstrate pericardial thickening or calcification in constrictive pericarditis.

The prognosis in RCM varies according to the underlying cause, but the overall prognosis tends to be poor, especially when the onset is in childhood, and patients often require cardiac transplantation. Most individuals experience progressive deterioration because of congestive HF with a high incidence of early mortality. Symptomatic improvement in RCM may be achieved with diuretic or vasodilator therapy, and specific therapies have proved beneficial in some cases, such as iron chelation and immunosuppressive drugs in individuals with hemochromatosis and primary amyloidosis, respectively.

Chromosomal Loci and Disease Genes

There is little evidence of genetic linkage in studies of large families with idiopathic RCM. A recent study has described a large kindred with an autosomal-dominant RCM linked to a genetic *locus* on chromosome 10;[226] none of the genes at this *locus* have been previously associated with DCM. Also, as previously noted, RCM has been described in association with abnormal accumulations of desmin in both cardiac and skeletal muscle.[227,228] Although desmin accumulation in skeletal and cardiac muscle can be a relatively nonspecific finding and has been observed in the absence of mutations in the *DES* gene encoding desmin, several missense mutations in the *DES* on chromosome 2q35 have been found in families with RCM. Heterozygous mutations have been identified in the highly conserved C-terminal end of the desmin rod domain at A337P, A360P, and N393I.[106]

Interestingly, mutations in the sarcomeric protein cardiac troponin I gene (*TNNI3*) can result in idiopathic restrictive cardiomyopathy. In a large family in which individuals were affected by either idiopathic RCM or HCM, linkage and subsequent mutation analyses revealed a novel missense mutation (Asp190His) in the *TNNI3* gene encoding cardiac troponin I (cTnI), co-segregating with the presence of either cardiomyopathic phenotype in affected individuals in the family.[229] In addition, mutations in conserved and functionally important domains of *TNNI3* were identified in six of nine unrelated RCM patients. Two of the RCM mutations identified, Leu144Gln and Arg145Trp, were localized to a domain required for inhibition of human cardiac troponin I actomyosin ATPase activity. Ala171Thr and Lys178Glu mutations, which map to the region involved in actin binding and may influence the inhibitory function through actin binding, were identified. Furthermore, Asp190His and Arg192His mutations that are localized within the conserved C-terminal region of the protein are also required for normal inhibitory function of troponin I. These findings suggested that *TNNI3* mutation might be a prevalent contributory factor in this disease. Further studies to assess actomyosin ATPase activity in skinned fibers from patients with five of the RCM mutations (i.e., L144Q, R145W, A171T, K178E, and R192H) showed that all five *TNNI3* mutations exhibited an increase in the Ca^{++} sensitivity of force development compared with wild-type cTnI, as well as that observed for most HCM mutations.[230] Moreover, the two mutations associated with the worst clinical phenotype (K178E and R192H) showed large increases in Ca^{++} sensitivity. In addition, all of the mutations investigated caused a decrease in the ability of cTnI to inhibit actomyosin ATPase activity and force development. Hence, these mutations in RCM can result in severe diastolic dysfunction.

Amyloidosis is the most prevalent underlying cause of RCM.[231] Restrictive cardiomyopathy in cardiac amyloidosis results from the replacement of normal myocardial contractile elements by infiltration and interstitial deposits of amyloid, leading to alterations in cellular metabolism, calcium transport, receptor regulation, and cellular edema. Injury can also occur from circulating light chains in the absence of amyloid fibril formation. Amyloid myocardium becomes firm, rubbery, and noncompliant and can also involve the cardiac conduction system presenting with different types of conduction defects and dysrhythmias.

Familial amyloidosis, or hereditary amyloidosis, although overall less common than immunoglobin amyloidosis (AL), is more frequently associated with RCM and is most often caused by an autosomal-dominant mutation in the serum protein transthyretin encoded by the *TTR* gene. This gene encodes 127-amino acid residues of four identical, non-covalently linked subunits that form a pair of dimers in the plasma protein complex. More than 60 distinct amino acid substitutions distributed throughout the *TTR* sequence have been found to be correlated with an increased amyloidogenicity of *TTR*.[231] With familial amyloidosis, the pattern of myocardial involvement varies according to the specific mutation; for example, patients with the Met 30 transthyretin variant, the most prevalent *TTR* mutation, primarily display conduction defects and often require pacemaker implantation.[232]

The Tyr77 mutation, the second most prevalent mutation, was found in a large family with 12 affected individuals over four generations; the phenotype is associated with an initial and sometimes prolonged carpal tunnel syndrome, beginning between the sixth and seventh decade, with subsequent RCM.[233] A large number of these mutations have been associated with late-onset amyloid cardiomyopathy, as well as with polyneuropathy. In addition, different kindred groups have been shown to have varying degrees of susceptibility to cardiac amyloid deposition, whereas other mutational groups do not have cardiac involvement. Substitution of isoleucine for valine at position 122 of the transthyretin gene has been reported in high prevalence in African-Americans (estimated to be present in approximately 4% of the black population)[234] and is associated with the occurrence of late-onset RCM.[235] Molecular analysis shows that this residue substitution shifts the equilibrium toward monomer (indicating lower tetramer stability) and favors tetramer dissociation required for amyloid fibril formation.[236]

Iron-overload cardiomyopathy is a RCM that manifests systolic or diastolic dysfunction secondary to increased deposition of iron in the heart and occurs with common genetic disorders such as primary hemochromatosis.[237] Although the exact mechanism of iron-induced HF has not yet been determined, the toxicity of iron in biological systems is believed to be attributed to its ability to catalyze the generation of oxygen-free radicals. Hereditary hemochromatosis is one of the most common autosomal-recessive disorders among Caucasians, because the genotype at risk for hemochromatosis accounts for 1:200–400 individuals of Northern European ancestry.[238] Clinical complications appear late in life and may include cardiomyopathy (although it is not entirely clear whether it is limited to the restrictive-type) developing only in homozygotes. The phenotypic expression of the disease is variable even within the same family because of the effect of modifier genes or environmental factors. Hereditary iron overload has been linked to pathogenic mutations of the gene coding for *HFE*, an atypical HLA class I molecule on chromosome 6 (6p21.3), hemojuvelin (*HJV* or *HFE2*) on chromosome 1 (most often associated with the juvenile form of hemochromatosis), and, more rarely, the gene coding for hepcidin (*HAMP*) on chromosome 19 and the gene encoding serum transferrin receptor 2.[238–240] Two mutations in *HFE* have been well-characterized, C282Y and H63D; the C282Y mutation has a higher penetrance than the H63D mutation and seems to result in a greater loss of HFE protein function.[237]

ARRHYTHMOGENIC RIGHT VENTRICULAR DYSPLASIA

Clinical Manifestations

Arrhythmogenic right ventricular dysplasia or cardiomyopathy (ARVD or ARVC) is a primary heart muscle disease characterized by myocyte loss caused by necrosis and/or apoptosis, with fatty or fibrotic replacement. The pathological process predominantly affects the right ventricle and results in wall thinning and chamber dilation. Extensive wall thinning may give rise to a parchment-thin appearance (Uhl's anomaly). Left ventricular involvement has also been found, including the Carvajal variant, although of a more variable and lesser degree. Electrocardiographic abnormalities include inverted T waves in the right precordial leads, late potentials, and right ventricular dysrhythmias with left bundle-branch block (LBBB). Sick sinus syndrome and complete AV block are late complications observed in some older patients with severe disease phenotype. As a significant cause of SCD, it is particularly common in areas such as Italy.

The prevalence of ARVD has been estimated to be 1:5,000, although higher rates (4.4:1,000) have been reported in some areas of northern Italy.[241] ARVD is thought to be familial in at least 30% of cases, primarily with autosomal-dominant inheritance with variable expression and penetrance.[242] In addition, an autosomal-recessive form of ARVD has been reported in association with palmoplantar keratoderma and woolly hair (Naxos disease).

By use of clinical criteria, a definitive diagnosis of ARVD may be difficult to establish, because the disease primarily affects only the right ventricle. A right ventricular biopsy may be definitive, but often can produce false-negative results, because the disease initiates in the epicardium and spreads to the endocardium of the right ventricular free wall, making it inaccessible to biopsy. The age of onset, degree of penetrance, and clinical features are highly variable both between and within families. Patients typically present with recurrent ventricular tachycardia of LBBB morphology and, less commonly, HF.[243] Affected individuals present most commonly with ventricular and supraventricular dysrhythmias, which can be well tolerated but can result in syncope or SCD. Ventricular tachycardia, particularly in the young, is frequently precipitated by exercise as is SCD, which occurs at a rate of 2% year. In Italy, ARVD accounts for 20% of SDs in individuals aged less than 35 y and over 20% of deaths in competitive athletes.

Genetics of ARVD

Chromosomal Loci

Eleven chromosomal *loci* have been identified in families with ARVD (Table X). Seven of these *loci* were initially found in autosomal-dominant ARVD, on chromosomes 14q23-q24 (ARVD1),[244,245] 1q42-q43 (ARVD2),[246] 14q12-q22 (ARVD3),[247] 2q32 (ARVD4),[248] 3p23 (ARVD5),[249] and 10p12-p14 (ARVD6).[250] The seventh *locus* (ARVD7) (chromosome 10q22) was mapped in a family with autosomal-dominant disease, characterized mainly by skeletal muscle disease with ARVD. The muscle biopsy specimens revealed accumulation of desmin, dystrophin, and sarcoglycan.[251]

Disease-causing genes have been identified at several of these *loci*, including ARV1 (*TGFB3*),[244] ARVD2 (cardiac *RYR2* gene),[246] two newly discovered *loci*, ARVD8[252–255] and ARVD9,[256] which encode the desmosome proteins, desmoplakin and plakophilin-2, respectively. In addition, a tenth *locus* specifying autosomal-recessive ARVD (Naxos disease) has been mapped to chromosome 17q21, and mutations in the *plakoglobin* gene have been identified.[257] Furthermore, several mutations in desmoglein-2 (*DSG2*) have been recently reported in ARVC probands.[258]

Cardiac Ryanodine Receptor Gene

The cardiac ryanodine receptor gene (*RYR2*) encodes the cardiac isoform of the ryanodine receptor, which is the major Ca^{++} release channel in cardiac muscle. Primarily expressed in the heart and to a lesser degree in the brain, the cardiac ryanodine receptor is one of three ryanodine receptor isoforms (RYR1-3), each encoded by a separate gene, with *RYR1* and *RYR3* primarily expressed in skeletal muscle and brain, respectively. The *RYR2*-encoded proteins associate as homotetramers with four FK506-binding proteins (FKBP12.6) to form a membrane-spanning complex in the SR.

Stimulation of voltage-sensitive L-type Ca^{++} channels on the outer myocardial cell membrane permits small amounts of Ca^{++} to enter the cell, which activates the ryanodine receptors in the SR, resulting in the release of large levels of Ca^{++} responsible for initiating myocardial contraction. In addition to this depolarization-induced Ca^{++} release, spontaneous SR Ca^{++} release by RYR can occur under conditions of SR Ca^{++} overload, termed store-overload induced–Ca^{++} release, which can alter the membrane potential by generating delayed after depolarizations, which in turn can trigger dysrhythmias.

Phosphorylation of *RYR2* by protein kinase A dissociates FKBP12.6 and regulates the SR channel opening. *RYR2* is composed of 105 exons that encode a large protein of 4967 amino acids (565 kDa). Several distinct transcripts of *RYR2* arise by alternate splicing. The cardiac ryanodine receptor contains a large N-terminal cytoplasmic domain that contains interaction sites for FKBP12.6, ATP, and calmodulin, and a C-terminal transmembrane domain that is anchored to the SR by 4–10 hydrophobic motifs. Four missense mutations in *RYR2* have been identified in four unrelated families that map to the *ARVD2 locus*.[246] These four mutations were located in two clusters in highly conserved regions in the cytosolic portion of the molecule, including one cluster at the N-terminal end and a second cluster in the central FKBP12.6 interacting domain. All *RYR2* mutations occurred in domains critical for the regulation of the calcium channel. One of the *RYR2* mutations in the N-terminus corresponds to a mutation in *RYR1*, which has been previously linked to malignant hyperthermia and core disease.

These mutations are believed to unblock the SR calcium channel, resulting in hyperactivation/hypersensitization. This is postulated to result in "leaky" calcium channels, particularly evident during sympathetic stimulation. Under

TABLE X Genes Associated with Familial ARVD

Gene	Protein	Loci	Inheritance	References
ARVD1	TGF-β3	14q24.3	AD	244, 245
ARVD2 (RYR2)	Cardiac ryanodine receptor	1q42-q43	AD	246
ARVD3	ND	14q11-q12	AD	247
ARVD4	ND	2q32	AD	248
ARVD5	ND	3p23	AD	249
ARVD6	ND	10p12-p14	AD	250
ARVD7	ND	10q22	AD	251
ARVD8 (DSP)	Desmoplakin	6p24	AD/AR	252–255
ARVD9 (PKP2)	Plakophilin-2	12p11	AD	256
JUP (Naxos disease)	Plakoglobin	17q21	AR	257
DSG2	Desmoglein-2	18	AD	258

ND, not determined

conditions in which SR calcium content is rapidly increased, such as exercise, emotional stressor, and catecholamine infusion, there will be an increased Ca^{++} spillover, leading to delayed after-polarizations and dysrhythmia.

As previously discussed in the Chapter 7, *RYR2* mutations have also recently been reported in families with sporadic cases of catecholamine-sensitive polymorphic ventricular tachycardia (CPVT), indicating that this disorder is allelic to ARVD.[259,260] *RYR2* mutations were found in two clusters, in the cytosolic central FKBP12.6-interacting domain and in the C-terminal transmembrane region, respectively.[259,261] Furthermore, recent observations have shown that pathogenic *RYR2* mutations leading either to ARVD2 or ventricular tachycardia and SCD (regardless of their location in the gene) acted similarly in response to store-overload–induced Ca^{++} release (SOICR) and channel sensitivity to luminal Ca^{++} activity.[262] However, the relationship between the different phenotypes engendered by *RYR2* mutations is mostly unknown. Interestingly, screening of a family harboring a novel missense mutation of *RYR2* at A77V, in which there were affected members with ARVD and others with CPVT, suggested that the two entities correspond to varying degrees with phenotypic expression of the same disease.[263]

Desmosomal Genes

Desmosomes are highly organized intercellular junctions that provide mechanical integrity to tissues by anchoring intermediate filaments to sites of strong adhesion. Several mutations in structural protein components of desmosomes have been reported that have been implicated in the pathogenesis of ARVD. Mutations in plakophilin 2, desmoplakin, desmoglein2, and plakoglobin have been reported in ARVD, often in conjunction with several unique phenotypes.

Mutations in desmoplakin have recently been isolated in both autosomal-dominant and autosomal-recessive forms of ARVC, and primary left ventricular variants of the disease are being increasingly recognized. In a large family with autosomally dominant disease with high incidence of ventricular dysrhythmia, SD, and fibrofatty replacement of the left ventricular myocardium, linkage analysis revealed cosegregation of disease with desmoplakin.[255] A heterozygous, single adenine insertion mutation (2034insA) in the desmoplakin (*DSP*) gene was identified in affected individuals only, resulting in a truncated terminus of the desmoplakin protein; the truncated protein resulting from this frameshift mutation was detectable by Western blot analysis and likely disrupts intermediate filament binding.

Another study of 38 subjects belonging to four families with ARVD/ARVC identified four different dominant *DSP* mutations, three of which were missense mutation, and one in the intron-exon splicing region. Affected individuals had a high incidence of effort-induced polymorphic ventricular dysrhythmias and SD.[252] In addition, in an Italian family with ARVD/ARVC, a dominant mutation (S299R) in exon 7 of *DSP*, which modifies a putative phosphorylation site in the N-terminal domain binding plakoglobin, has been reported.[254] Recessive ARVD phenotypes have also been identified with mutations of desmoplakin; one patient had a homozygous missense *DSP* mutation (Gly2375Arg) in the C-terminus, where the intermediate filament binding site is located.[253] This patient also presented with woolly hair and a pemphigus-like skin disorder, common features of desmosomal mutations. Four related heterozygous carriers of the mutation showed no hair, skin, or cardiac abnormalities.

The *JUP* gene encodes the cytoplasmic protein plakoglobin that belongs to the catenin family of proteins. In many tissues, this protein is a key component of desmosomes and adherens junctions, including the heart, skin, and hair. Plakoglobin is associated with desmoglein (one of the transmembrane desmosomal proteins) and is a component of the cadherin–catenin complex involved in tight adhesion and signaling between cells. Mutations in plakoglobin cause a recessive form of ARVD accompanied by palmoplantar keratoderma and woolly hair (Naxos disease). Nineteen affected individuals were found in a Greek cohort with a homozygous 2 base pair deletion in the *plakoglobin* gene.[257] Heterozygous individuals were not affected.

This deletion causes a frameshift and premature termination of the protein, as shown by Western blot analysis. Subsequently, in a larger study of 12 families (all adults homozygous for this recessive mutation fulfilled the diagnostic criteria for ARVD),[264] 92% developed ventricular dysrhythmia and 100% right ventricular structural defects. Furthermore, autosomal-recessive ARVC caused by a mutation in plakoglobin was 100% penetrant by adolescence.

Gerull and associates have identified heterozygous mutations in *PKP2* (which encodes plakophilin-2, an essential armadillo-repeat protein of the cardiac desmosome) in 32 of 120 unrelated individuals with ARVC and in two kindreds with ARVC; the disease was incompletely penetrant in most of the carriers of *PKP2* mutations.[256]

Transforming Growth Factor-β3

Although the first chromosomal *locus* found to be linked to ARVD was ARVD1 on chromosome 14q24.3, until very recently the gene(s) involved has proved to be elusive. The identification of the candidate gene transforming growth factor-β3 (*TGFβ3*) was confirmed by the demonstration of a nucleotide substitution (C-36G>A) in the 5′ untranslated region of the *TGFβ3* gene, which in affected family members was invariably associated with the ARVC clinical phenotype.[244] The mechanism by which TGF-β3 mutations modulate these severe cardiac phenotypes, which likely involves its functions in signaling and apoptotic regulation, remains to be determined.

Pathophysiological Mechanisms in ARVD

Linkage studies and genetic studies have demonstrated that ARVD is a genetically heterogeneous disorder. Although six disease-causing genes have been identified to date, it is clear that diverse pathophysiological pathways can culminate in the ARVD phenotype. Mutations in the cardiac ryanodine receptor affect Ca^{++} channel properties, with either hyperactivation or hypersensitivity to activating levels of Ca^{++}. The subsequent imbalance of intracellular Ca^{++} homeostasis disrupts normal excitation-contraction coupling and promotes cardiac dysrhythmias.

In addition, changing the intracellular calcium levels can modulate cell death in nearly all cell types. In contrast, mutant plakoglobin might render cells more susceptible to the effects of mechanical stress, resulting in detachment of intercalated discs and subsequent remodeling. In ARVD, remodeling of gap junctions seems to be a significant consequence of plakoglobin mutations.[265] Impaired functioning of cell adhesion junctions during exposure to shear stress may lead to myocyte detachment and death, accompanied by inflammation and fibrofatty repair. Findings of the *TGF-β* involvement are also suggestive of modulation of apoptotic signaling effects. It is worth noting that in patients with ARVD, increased apoptotic cell death in the right ventricular myocardium has been recently documented,[266] and it may prove interesting and revealing to find which of the known genes are involved.

METABOLIC CARDIOMYOPATHY

That metabolic dysfunction may lead to cardiomyopathy has been recognized. Some defects in fatty acid metabolism, mitochondrial bioenergetics and signaling, diabetes/glucose metabolism may arise as a consequence of inherited mutations, whereas other defects occurring more sporadically have been identified in a number of cardiomyopathic disorders. Furthermore, cardiomyopathy can arise as a manifestation of systemic metabolic diseases caused by deficiencies of specific enzymes in a variety of metabolic pathways. Examples of these classes of metabolic cardiomyopathies include glycogen storage diseases, such as glycogenosis type II and III, and lysosomal storage diseases, including Niemann-Pick disease, Gaucher disease, I-cell disease, various types of mucopolysaccharidoses, GM1 gangliosidosis, galactosialidosis, carbohydrate-deficient glycoprotein syndromes, and Sandhoff's disease. Given the striking heterogeneity of the implicated genetic *loci*, their variable expression, and penetrance, as well as their interaction and modulation by a variety of environmental and genetic influences, the prevalence of these defects in cardiomyopathy has been extremely difficult to assess.

As discussed in Chapter 7, defects in fatty acid metabolism can lead to severe cardiomyopathy in infants and children, although they are not entirely limited to these groups. These include defects in carnitine palmitoyltransferase, carnitine acetyltransferase, the cardiomyocyte carnitine transporter, the mitochondrial trifunctional protein, and fatty acid acyl transporters. Moreover, defects in fatty acid metabolism can lead to a series of abnormalities labeled as lipotoxic cardiomyopathy. Changes in fatty acid metabolism are likely important mediators of the DCM associated with ischemia. Similarly, abnormalities in mitochondria caused by point mutations, deletions, and depletion in mtDNA can result in a variety of cardiomyopathies. Although a subset of these defects occurs in children, defects in mitochondrial oxidative phosphorylation may also be present in adults with cardiomyopathy. A large number of mutations in mtDNA have been identified in association with either HCM or DCM (Table XII).

The phenotypes can be quite variable; some are associated with multisystemic and neurological diseases, including a wide spectrum of associated disorders, such as hypotonia, myopathies, muscle weakness, lactic acidosis, deafness, ophthalmic disease, and diabetes). These multisystemic disorders (e.g., Leigh, MELAS, and MERRF syndromes) are generally

TABLE XII MtDNA Mutations in Cardiomyopathy

Gene	Mutation	Phenotype-cardiac/other
Val	1644 (G→A)	HCM, MELAS
Leu	3243 (A→G)	DCM, MELAS
Leu	3260 (A→G)	CM, tachycardia
Leu	3302 (A→G)	CM, fatal dysrhythmia
Leu	3303 (C→T)	Fatal CM
Leu	3310 (C→T)	HCM, diabetes
Ile	4300 (A→G)	HCM, adult onset
Ile	4320 (C→T)	Fatal CM
Ile	4269 (A→G)	CM, HF
Ile	4295 (A→G)	HCM
Ile	4284 (G→A)	Fatal CM
Lys	8296 (A→G)	HCM, MELAS
Lys	8334 (A→G)	HCM, MERRF
Lys	8363 (G→A)	HCM, deafness, LS
Lys	8348 (A→G)	HCM, DCM
Lys	8296 (A→G)	Fatal HCM
Gly	9997 (T→C)	Dysrhythmia, HCM
Ala	5587 (T→C)	DCM
Arg	10415 (T→C)	DCM
Arg	10424 (T→C)	Fatal DCM
16S rRNA	3093 (C→G)	MELAS, CM
12S rRNA	1555 (A→G)	CM
Cys	5814 (A→G)	HCM
Val	1644 (G→A)	HCM, MELAS
Val	1624 (C→T)	HCM
Leu	12297 (T→C)	DCM
Cytb	14927 (A→G) Thr→Ala	DCM, HCM
Cytb	15236 (A→G) Ile→Val	DCM
Cytb	15508 (C→G) Asp→Glu	DCM
Cytb	15509 (A→C) Asn→His	PPCM
Cytb	15498 (G→A) Gly→Asp	HiCM
Cytb	15243 (G→A) Gly→Glu	HCM
COII	7923 (A→G) Tyr→Cys	DCM
COIII	9216 (A→G) Gln→Glu	DCM
COI	6860 (A→C) Lys→Asn	DCM
ND1	3310 (C→T) Pro→Ser	HCM
ND5	14609 (C→T) Ser→Leu	DCM
ATPase6	8993 (T→G) Leu→Arg	LS, HCM

CM, cardiomyopathy; PPCM, post-partum cardiomyopathy; LS, Leigh syndrome; HiCM, histiocytoid cardiomyopathy.

maternally inherited (caused by mtDNA mutation). Other cardiac phenotypes, including ventricular dysrhythmias, hypertrophy, cardiomegaly, conduction defects, as well as isolated cases of cardiomyopathy, may occur. Furthermore, a rare entity called histiocytoid cardiomyopathy has also been associated with specific mtDNA mutations.

As noted with the cardiomyopathy of children, point mutations in mitochondrial tRNA genes are frequently found in adult cardiomyopathies, resulting in abnormal mitochondrial protein synthesis and deficiencies in specific respiratory enzyme activities (mainly in HCM).[132] Most of the pathogenic cardiomyopathy-related mitochondrial tRNA mutations have been detected in highly conserved nucleotides, frequently present in a heteroplasmic fashion (a mixed population of both mutant and wild-type mtDNA genomes) and with a particularly high incidence in tRNAs for Leu, Lys, and Ile.[267–277] Mutations in mitochondrial rRNAs have also been found in cardiomyopathy.[278,279] Although most pathogenic mtDNA mutations are heteroplasmic and reside in highly conserved sequences,[280–283] recent observations have also implicated specific homoplasmic mutations in mtDNA as a possible, if not frequent, cause of cardiomyopathy.[284,285]

One of the more puzzling aspects in the area of mtDNA pathogenesis relates to the frequent finding that specific mtDNA mutations, found in association with primary cardiomyopathy, can also be found in patients with a plurality of neurological disorders. For instance, the 8363 mutation has been found not only in patients with HCM but also in patients and relatives with severe encephalomyopathies, including Leigh syndrome,[286] ataxia, or sensorineural deafness with or without cardiomyopathy.[270] Similarly, a mutation at nt 3423 in tRNA^{Leu} (probably the best characterized, and most prevalent mtDNA mutation reported in mitochondrial disorders) has been specifically associated with the MELAS phenotype and has also been recently detected in patients with isolated cardiomyopathy without neuromuscular involvement.[287] The involvement of other genetic or environmental cofactors, which can modulate the effect of mtDNA mutations, may contribute to the difficulties in correlating genotypic mutation to phenotypic manifestation.[285]

In addition to maternally inherited disorders, sporadic mitochondrial defects with associated cardiac manifestations have been reported, some of which involve somatic large-scale mtDNA deletions, such as KSS. Patients with KSS may display cardiomyopathy, mitral valve prolapse, and/or cardiac conduction defects, and most of the mtDNA deletions identified in this syndrome are of a single type, not inherited, and primarily found in skeletal muscle.[288]

On the other hand, defects in autosomal nuclear *loci* may be the cause of multiple mtDNA deletion phenotypes; however, the precise genetic defect has not yet been identified[289] and can be either dominantly or recessively inherited.[290,291] MtDNA deletions, found in KSS and in autosomal disorders, tend to be very abundant (up to 95% of the total mtDNA), and because of their abundance, they are often detected by Southern blot analysis. A second type of large-scale, less abundant (<0.1%) mtDNA deletion has been found in cardiac tissue of many primary, nonfamilial cardiomyopathies and may be detectable only by PCR analysis. Although these deletions may be evidence of specific mtDNA damage and tend to occur in an age-dependent manner, their significance in cardiac pathogenesis is not clear.[292–294] Whether mtDNA deletions precede the cardiomyopathy or whether they represent secondary somatic mutations arising from cardiac dysfunction and resulting metabolic changes (e.g., increased ROS-mediated damage to mtDNA) remains unresolved.[292] Evidence that increased levels of oxidative stress can result in increased levels of myocardial mtDNA deletions, in association with elevated ROS production, has been provided by observations in transgenic mice containing defective *ANT1* alleles[295] and in patients with specific *ANT1* mutations.[296]

Cardiac mtDNA depletion has also been found in patients with cardiomyopathy. Depletion of cardiac mtDNA, with concomitant reduction in mitochondrial respiratory enzyme activities, accompanies DCM in both animal models and patients treated with AZT (or zidovudine), an inhibitor of both DNA polymerase of the HIV-virus and mitochondrial DNA polymerase γ activities.[297,298.] Reduced cardiac mtDNA levels have also been reported in patients with severe cardiomyopathy and decreased cardiac respiratory enzyme activities.[299] Animals treated with Adriamycin and with ethanol have also been found to have decreased levels of cardiac mtDNA (decreased copy number).

CLINICAL EVIDENCE OF NUCLEAR MUTATIONS IN MITOCHONDRIAL COMPONENTS

Mutations in a wide spectrum of nuclear genes encoding mitochondrial proteins can also cause cardiomyopathy (Table XIII). For example, mutations in mitochondrial transport proteins (e.g., carnitine-acylcarnitine translocase), which facilitate the passage of critical metabolites across the inner mitochondrial membrane, are implicated in cardiomyopathy.[300] Friedreich ataxia (FA), which often presents with HCM, is caused by mutations in a nuclear-encoded mitochondrial transport protein frataxin, encoded by the *FRDA* gene, that is involved in mitochondrial iron accumulation and compromised respiratory enzyme activities.[301–303] Recently, mutations in a novel protein DNAJC19, localized to cardiac myocyte mitochondria, have been reported in the Canadian Dariusleut Hutterite population. Individuals with these mutations presented a syndrome characterized by autosomal-recessive dilated cardiomyopathy with ataxia (DCMA).[304] This disorder, characterized by early-onset DCM with cardiac conduction defects, nonprogressive cerebellar ataxia, growth failure, and 3-methylglutaconic aciduria, was mapped by use of a homozygosity mapping approach to a 2.2-Mb region of chromosome 3q26.3. Mutation analysis performed on positional candidate genes in this chromosomal region identified a disease-associated mutation in the *DNAJC19* gene, which shares extensive sequence and organizational similarity to yeast mitochondrial inner membrane proteins Tim14 and translocase involved in mitochondrial protein import.

Mutations in several of the 36 nuclear-encoded subunits of respiratory complex I have been identified as contributory to the onset of Leigh disease with associated cardiomyopathy and to HCM (e.g., *NDUFS2*).[305,306] In addition, mutations in nuclear genes encoding factors required for the assembly and function of the multiple-subunit enzyme respiratory complexes have been implicated in mitochondrial-based

TABLE XIII Nuclear Mutations Causing Clinical Mitochondrial Cardiomyopathy

Gene	Cardiac phenotype	Protein function
SCO2	HCM	COX assembly
NDUFV2	Early onset HCM	Complex I subunit
NDUFS2	HCM	Complex I subunit
ANT	HCM	Adenine nucleotide transporter/mtDNA maintenance
ACADVL	DCM/HCM	VLCAD activity (FAO)
DNAJC19	DCM/ataxia	Mitochondrial protein import
FRDA/frataxin	HCM/FA	Iron import
MTP	Neonatal CM	Mitochondrial trifunctional protein (FAO)
COX 10	HCM	COX assembly
SLC22A4	HCM, DCM	Carnitine transporter (OCTN2)
COX15	Early onset HCM	COX assembly
CPTII	CM, dysrhythmias,	Carnitine palmitoyl transferase II/FAO
SLC25A20	CM, dysrhythmias,	Carnitine-acylcarnitine translocase (CACT)/FAO
TAZ (G4.5)	DCM (Barth)	Tafazzin/FAO

FA, Friedreich ataxia.

diseases, such as Leigh syndrome. These include mutations in the *SCO2* gene encoding a copper chaperone involved in cytochrome *c* oxidase (COX) assembly, which can result in fatal infantile HCM with complex IV deficiency.[307] It is noteworthy that the clinical phenotype in patients with *SCO2* mutations is distinct from that found with mutations in other COX assembly factors (e.g., SURF1), which typically present without cardiac involvement. Recently, deleterious mutations in *COX15*, a heme A farnesyl-transferase involved in the synthesis of the prosthetic heme A group in COX, were identified in HCM.[308]

As previously stated, a number of nuclear gene defects in the mitochondrial fatty-acid β-oxidation pathway can lead to cardiomyopathy.[309] Cardiomyopathy results from defects in the genes encoding very long-chain acyl-Co A dehydrogenase (VLCAD)[310] and long-chain 3-hydroxylacyl CoA dehydrogenase (LCAD),[311] and from mutations affecting carnitine metabolism.[312] Moreover, mutations in the tafazzin *TAZ* gene result in reduced mitochondrial phospholipid levels, including cardiolipin, leading to Barth syndrome, an X-linked disorder characterized by DCM, cardiac dysrhythmias, and HF.[313–315]

CONTRIBUTION OF TRANSGENIC MODELS TO THE STUDY OF MITOCHONDRIA IN HEART DYSFUNCTION

In mouse transgenic models, disruption of mitochondrial bioenergy at a variety of specific *loci* or pathways can cause cardiomyopathy and HF. Targeting a relatively wide spectrum of nuclear genes encoding specific mitochondrial proteins using gene ablation results in severe cardiomyopathy in mice. Targeted genes include *ANT*,[316] *MnSOD*,[219] *CK*[317], *GPx*,[318] proteins involved in fatty acid metabolism (e.g., MTP subunits),[319] mtTFA or TFAM,[320] and frataxin, the protein responsible for Friedreich ataxia (FA).[321] *ANT*-deficient mice develop progressive HCM with marked proliferation of mitochondria containing COX and SDH activation, whereas MnSOD-deficient mice develop DCM, yet both types of null mutation cause severe cardiac ATP deficiency, which is believed to underlie the resulting cardiac phenotype(s). Another major contribution of the transgenic mouse model has been to advance our understanding of the family of transcriptional coactivators and factors, including PPAR-α, PGC-1α, and MEF-2, which coordinately regulate myocardial energy metabolism and of their essential role in the developing embryonic heart. Furthermore, these models have helped to delineate the order of biochemical and molecular events in the metabolic and transcriptional cascade governing energy regulation in both the normal and abnormal heart.[322]

Ablation of the intermediate filament protein desmin, which is involved in cytoskeletal function, causes murine DCM (characterized by extensive cardiomyocyte death, fibrosis, calcification, and eventual HF) and promotes aberrant myocardial mitochondrial proliferation, defects in mitochondrial structure, and mitochondrial dysfunction. Proteomic analysis in desmin-null mouse strains has documented a relatively large number of changes in myocardial proteins and in metabolic pathways involved in calcium homeostasis, fibrosis, and apoptosis, and most markedly in proteins of the mitochondrial proteome, including NADH shuttle proteins, amino-acid metabolism proteins, and respiratory enzymes.[323] Disruption of another cytoskeletal component, the muscle LIM protein

(MLP) can also result in DCM with abnormal mitochondrial number and organization, leading to energy depletion.[324]

In addition to examining the effects on cardiac phenotype by eliminating specific nuclear genes regulating mitochondrial function, tissue-specific knock-out mice with mitochondrial cardiomyopathy (MCM) have been used to identify modifying genes of potential therapeutic value.[325] Although there is limited information about the impact on myocardium of knocking-out nuclear genes involved directly in mitochondrial OXPHOS, cardiac dysfunction in mice lacking cytochrome c oxidase subunit VIa-H, the heart isoform, has been reported.[326] The knock-out approach has not yet been accomplished with mtDNA genes because of the formidable technical difficulty involved in direct gene replacement or ablation of a multicopy gene in the setting of the multicopy mitochondrial genome.

The use of cardiac-specific overexpression of specific genes has also advanced our understanding of the role of mitochondria in cardiac dysfunction. Overexpression of global regulatory factors that mediate the expression and control of cardiac energy metabolism (e.g., PGC-1α and PPAR-α) have been shown to lead to severe cardiac dysfunction and marked changes in mitochondrial structure and function.[327–329] Similarly, transgenic mice containing cardiac-specific overexpression of calcineurin exhibited severe cardiac hypertrophy (that progresses to HF), marked mitochondrial respiratory dysfunction, and superoxide generation.[330] Overexpression of the G_q-coupled 5-HT2B receptor in murine heart leads to compensated HCM associated with proliferation of the mitochondria organelle and altered bioenergetic function.[331]

The development of animal models of mitochondrial-based cardiac dysfunction may offer the possibility of direct testing for potential treatments. For example, the demonstration that MnSOD-deficient animals developed ROS toxicity and DCM prompted speculation that effective treatment with antioxidants could ameliorate the cardiac phenotype. Indeed, peritoneal injection of MnSOD-deficient mice with the MnSOD mimetic, MnTBAP, eliminated the cardiac dysfunction and reversed ROS accumulation.[332] Interestingly, in the frataxin-null mice (animal model of FA) treated with the antioxidant MnTBAP, no effect on the cardiomyopathy was observed, suggesting that contrary to popular belief, increased oxidative stress and damage may not be a direct result of mitochondrial iron accumulation or associated with the early stages of the neuronal and cardiac FA pathology as previously suggested.[333]

By use of a different approach, overexpression of the mitochondrial-localized antiapoptotic protein Bcl-2 in the murine heart was shown to ameliorate the mitochondrial defects found in desmin-null mice, resulting in marked changes in cardiac phenotype, including reduced occurrence of myocardial fibrotic lesions, prevention of cardiac hypertrophy, restoration of cardiomyocyte ultrastructure, and significant improvement of cardiac function.[334] These findings suggest a potential therapeutic role for Bcl-2 to target inherited cardiomyopathies, including the desmin-induced CM.

Diabetic Cardiomyopathy

Numerous clinical studies have confirmed the association of diabetes with left ventricular dysfunction, independent of hypertension, CAD, and other heart diseases and have provided evidence of extensive myocardial structural and functional changes. The most important mechanisms of diabetic cardiomyopathy are metabolic disturbances (i.e., depletion of glucose transporters [e.g., GLUT4], increased free fatty acids, carnitine deficiency, changes in Ca^{++} homeostasis, myocardial fibrosis [with increases in angiotensin II, IGF-I, and inflammatory cytokines], small vessel disease [microangiopathy, impaired coronary flow reserve, and endothelial dysfunction], cardiac autonomic neuropathy [denervation and alterations in myocardial catecholamine levels], and insulin resistance [hyperinsulinemia and reduced insulin sensitivity]).[335]

Defects in the ability of insulin to regulate GLUT4 translocation can lead to insulin resistance and noninsulin-dependent type 2 diabetes. Patients with diabetic cardiomyopathy exhibit marked down-regulation of myocardial GLUT transporters, limiting both glucose uptake and oxidation and contributing to the heart's inability to generate much needed ATP.[336] In diabetic cardiomyopathy, decreased glucose use results in an almost exclusive use of fatty acids as the myocardial energy source.[337,338]

Diabetes and hyperglycemia exert profound effect on cardiac structure and function. Insulin deficiency or resistance is associated with LV hypertrophy, as well as cardiomyopathy. In addition to effects in myocardial glucose uptake and use, deficits in insulin signaling lead to an increased reliance of the heart on FAO for energy generation. This contributes to an increased accumulation of potentially toxic lipid intermediates, elevated cellular acidosis, decreased cardiac efficiency, and contractile dysfunction.[339] Moreover, hyperglycemia and diabetes can affect cardiac mitochondria function directly. In animals treated with streptozocin to induce diabetes, cardiac mitochondria show pronounced swelling, marked structural changes, and targeting by lysosomes.[340] Mitochondria from a variety of diabetic animal models display diminished respiratory control, as well as increased oxidative stress.[341,342] These changes in cardiac structure and function are reversed with insulin administration. Conversely, the activity of the PPAR-α gene regulatory pathway is increased in the diabetic heart, which relies primarily on FAO for energy production, providing further stimulus for excessive fatty acid import and oxidation, underlying the cardiac remodeling of the diabetic heart.

The relationship between abnormal cardiac mitochondria and diabetes is also relevant to cardiomyopathy. Increasingly, specific nuclear and mtDNA mutations associated with both diabetes and cardiomyopathy are being identified. Defective genes and cardiac phenotypes associated with diabetes are shown in Table XIV.

Diabetes has been found in patients with large rearrangements (e.g., deletions) in mtDNA or in association with specific point mutations,[343,344] such as those identified in mitochondrial tRNAs, including sites previously associated with MELAS (e.g., nt 3243), MERRF (e.g., nt 8344), protein genes (e.g., ND1 at nt 3310), and 16s rRNA.[345–347] In addition, mitochondria are primary targets of oxidative damage in the diabetic heart,[348] which could lead to increased generation of mtDNA mutations at specific hotspots. Nevertheless, the causal relationship between diabetes and mtDNA mutation remains unclear.

ACQUIRED CARDIOMYOPATHY

Alcohol

Alcoholic cardiomyopathy is characterized by cardiomegaly, disruptions of myofibrillary architecture, reduced myocardial contractility, decreased ejection fraction, and enhanced risk of stroke and hypertension. Although several mechanisms have been postulated for alcoholic cardiomyopathy, including oxidative damage, accumulation of triglycerides, altered fatty acid extraction, decreased myofilament Ca^{++} sensitivity, and impaired protein synthesis, neither the mechanism nor the ultimate toxin has been revealed.[349,350] Primary candidates acting as specific toxins of myocardial tissue are ethanol, its first and major metabolic product, acetaldehyde, and fatty acid ethyl esters. Acetaldehyde has been demonstrated to impair directly cardiac contractile function, disrupt cardiac excitation-contractile coupling, and contribute to oxidative damage and lipid peroxidation. Acetaldehyde-elicited cardiac dysfunction may be mediated through cytochrome P450 oxidase, xanthine oxidase, and the stress-signaling cascade.[351,352]

Myocyte atrophy and death are the main pathological findings. A clear dose-related effect has been established with ethanol consumption, with gender, and some specific gene polymorphisms being factors of increased susceptibility to alcohol-induced muscle damage.[353,354] Pathogenic mechanisms are pleiotropic, the most relevant being disturbances in carbohydrate, protein, and energy cell turnover, signal transduction, and induction of apoptosis and gene dysregulation. A number of these molecular/cellular mechanisms of alcohol-mediated myocardial damage have been substantiated with animal models of alcoholic cardiomyopathy.[355–357] Overexpression of alcohol dehydrogenase in a murine transgenic model has further explained effects on alcohol-mediated myocardial protein damage, lipid peroxidation, and cardiac contractile defects.[358,359]

Chagas' Disease

Chagas' disease is caused by *Trypanosoma cruzi* (*T. cruzi*), a flagellated protozoan parasite related to the African trypanosome that causes sleeping sickness. It is spread by reduvid bugs and is one of the major health problems in South and Central America, where more than 10 million people are infected. Because of immigration, a significant population residing in the United States is believed to be infected. Moreover, risk factors for Chagas' disease include receiving a blood transfusion from a person who carries the parasite but does not have active Chagas' disease.

The acute phase of Chagas' disease may have no symptoms or have very mild symptoms, including swelling and reddening at the site of infection, swelling on one eye, and fever and are particularly prevalent in children living in endemic areas. Some cases display a distinct myocarditis, primarily promoted by mononuclear cells, which tends to be mild and reversible. The disease goes into remission after the acute phase and becomes chronic in approximately 30% of infected and/or untreated subjects, with no presentation of further symptoms for many years. When symptoms finally develop, they appear most commonly as a DCM, often with a fulminant myocarditis, and can be accompanied by occasional swallowing difficulties, digestive abnormalities, neurological sequelae including dementia, and severe dysrhythmias that may cause SCD.

Gene expression profiles of human Chagas' cardiomyopathy were examined to identify selective disease pathways and potential therapeutic targets.[360] By use of cDNA microarray, it was shown that the immune response, lipid metabolism, and mitochondrial OXPHOS genes were selectively up-regulated in myocardial tissue, with interferon (IFN)-γ–inducible genes representing 15% of the genes specifically up-regulated. Fetal mouse cardiomyocytes increase ANF expression 15-fold in response to IFN-γ, 400-fold in response to combined IFN-γ and monocyte chemoattractant

TABLE XIV Defective Genes in Diabetic Cardiomyopathy

Gene defect	Phenotype
Nuclear loci	
LMNA	DCM
ALMS1	DCM
PKCb2	HCM
Mitochondrial loci	
Ile G4284A	CM
Leu 3243	HCM
ND1 3310	HCM
MtDNA deletions	KSS
Lys A8296G	HCM/MELAS
D-loop T16189C	DCM
16s rRNA C3093G	CM

KSS, Kearns–Sayre syndrome; CM, cardiomyopathy.

protein-1, suggesting that IFN-γ and chemokine signaling can directly up-regulate cardiomyocyte expression of genes involved in myocardial hypertrophy, which ultimately may lead to HF in patients with Chagas' disease.

Molecular analysis suggests the existence of circulating antibodies in Chagas' disease, which bind to β-adrenergic and muscarinic cholinergic receptors of myocardium.[361,362] The interaction of agonist-like antibodies with myocardial neurotransmitter receptors, triggers in the cells intracellular signal transduction that changes the physiological behavior of the target organs, promoting the conversion of normal cells into pathologically active cells leading to extensive tissue damage. Analysis of the prevalence and distribution of these antibodies reveals a strong association with cardiac and esophageal autonomic dysfunction in seropositive patients and may partially explain the cardiomyoneuropathy and achalasia of Chagas' disease, in which the sympathetic and parasympathetic systems are affected. The behavior of auto-antibodies acting as agonists on neurotransmitter receptors could induce receptor desensitization and/or down-regulation, triggering a progressive blockade of neurotransmitter receptors, with sympathetic and parasympathetic denervation, as it has described during the course of Chagas' cardioneuropathy and achalasia.[363] These findings suggested the potential use of peptides corresponding to the amino acid sequence of the second extracellular loop of human M2 muscarinic acetylcholine receptor to target the circulating antibodies.[364] These peptides have proven useful in stemming myocardial contractile dysfunction in a mouse model of Chagas' disease. Furthermore, molecular analysis has also permitted the identification of the specific *T. cruzi* antigen(s) and shown that specific surface antigens (e.g., TC13) on the parasite have affinity for β1-adrenergic receptors on target organs, resulting in synthesis of cyclic adenosine monophosphate (cAMP) and an increase in cardiac contractility.

The heterogeneity in the clinical expression of Chagas' disease has suggested the involvement of the host genetic factors that modulate its pathogenesis. Several observations have provided evidence for different markers of host genetic susceptibility to Chagas' cardiomyopathy.

Specific HLA variants seem to confer either increased susceptibility or decreased susceptibility to *T. cruzi* infection in endemic areas.[365] Specific MHC alleles have been proposed to be associated with the development of chronic infection and with cardiac involvement in Chagas' disease. Cruz-Robles and associates found increased frequencies of HLA-A68 and HLA-B39 in asymptomatic individuals compared with patients with cardiomyopathy. Also, patients with cardiomyopathy exhibited increased frequency of HLA-B35 compared with healthy controls, and increased frequency of HLA-DR16 compared with asymptomatic individuals. These findings suggest that MHC alleles might be associated with the development of chronic infection and with heart damage in Chagas' disease. Although HLA-DR4 and HLA-B39 could be associated directly with the infection by *T. cruzi*, HLA-DR16 could be a marker of susceptibility to heart damage, and HLA-A68 might confer protection against developing cardiomyopathy.[366] More recently, evidence of an association was shown between complement C3 and BF allotypes and the susceptibility to Chagas' disease and the development of cardiomyopathy.[367] A significantly increased frequency of C3F was observed in patients with the Chagas' cardiomyopathy, whereas a negative association was found with the BFS allotype, suggesting that it may play a protective role against the severe cardiomyopathy.

CONCLUSION

The use of genomic and post-genomic technologies has expanded our view of the pathology of cardiomyopathy, including the identification of many molecular elements shared between the many diverse phenotypes involved. This technology has also enabled the identification and characterization of many genes involved in the pathogenesis of both familial and, increasingly, the idiopathic forms of cardiomyopathy, in particular the dilated type. The considerable number and variation of genes and pathways involved in the pathogenesis of cardiomyopathy makes diagnostic screening a difficult, time-consuming task, although new methods will surely solve this problem. Also, increased understanding of the gene profiles of the subtypes of cardiomyopathy may contribute as well. Furthermore, combined use of pharmacogenomics and gene profiling will be useful in ascertaining not only the expression of specific genes correlated to the presence of specific genotypes but also the dynamic of their responses to various stimuli (in animal models) and to specific therapeutic regimens. Understanding the contribution of environmental factors, as well as the interactions between genes, will be necessary to unravel the high heterogeneity and multi-layered profiles of these pathologies. Moreover, there is a need to refine techniques that presently exist for a more precise monitoring of cellular and molecular events in the myocardium, hopefully in a less invasive manner than endomyocardial biopsy.

An essential adjunct to the clinical studies of cardiomyopathy is the transgenic models of cardiomyopathy that allow not only the discovery of new genes involved in pathogenesis but also can advance our understanding of the precise roles that specific genes have in determining the phenotype. These models also provide a highly informative substrate for testing new therapies, as well as novel hypotheses, which cannot be tested with clinical models. Although a dramatic increase in information concerning cardiomyopathy has emerged from these transgenic models, considerably more should be expected. New treatments for cardiomyopathy are needed, and currently a number of new drugs are under development, including modulators of myocardial remodeling. Future strategies will also include antiapoptotic drugs, and gene therapy.

SUMMARY

- There is increasing recognition of the involvement of inherited gene defects in the pathogenesis of primary cardiomyopathies.
- Numerous genes mutations have been identified as a common etiological factor in the more prevalent varieties of cardiomyopathy, HCM, and DCM and also in the more rarely found phenotypes such as RCM and ARVD/ARVC.
- Specific defects in 14 genes have been linked to familial HCM, most encoding sarcomeric proteins, including β-MHC, α-MHC, the regulatory and essential MLC genes, the cardiac troponins (T, I and C), actin, and α-tropomyosin. These implicate defects in sarcomeric function and force generation as a major cause of pathology in HCM.
- Non-sarcomeric genes found in familial HCM include defects in metabolic regulatory genes (*PRAKG2*), signaling proteins (caveolin), and nuclear membrane proteins (Lamin A/C).
- Data from transgenic models of HCM (primarily mouse) have confirmed the involvement of most of the preceding genes, further explaining their role in cardiomyopathy and allowing the identification of other genetic *loci* involved in the pathways leading to HCM.
- The combination of the highly diverse contributors to HCM, and the extremely variable penetrance and expression of these genes in diseased individuals, has made genetic diagnosis of familial HCM a difficult task at present; the involvement of a veritable phalanx of environmental and genetic modifiers is suggested.
- DCM is a highly heterogeneous disorder with more than 20 different defective genes thus far characterized, many in cases previously labeled as idiopathic.
- In DCM, most of the identified genes are involved with the cytoskeleton and the dystrophin–glycoprotein complex and are considered to implicate force transmission as the major pathogenic mechanism.
- The molecular mechanism in DCM also seems to involve myocardial energy deficits (caused by mutations in both nuclear and mitochondrial DNA encoded genes which can cause both DCM and HCM), as well as abnormal Ca^{++} homeostasis caused by altered availability of Ca^{++} and altered myofibrillar Ca^{++} sensitivity.
- A number of defective sarcomeric genes implicated in HCM also can lead to DCM, although it remains uncertain whether different mutations, gene dosage, or interacting factors are responsible for deciding the cardiac phenotype.
- The availability of animal models has enabled both the characterization of novel pathogenic genetic *loci* and the explanation of signaling pathways involved in the development of this complex disorder.
- The genes implicated in RCM are highly diverse, including troponin I and desmin, as well as genetic factors that influence the pathogenesis (and/or susceptibility) to secondary causes of RCM (e.g., amyloidosis).
- The genes implicated in ARVD encode desmosomal proteins (plakophilin 2, desmoplakin, desmoglein and plakoglobin), TGF-β3, and the cardiac ryanodine receptor. Their involvement in myocardial remodeling and apoptosis regulation has been suggested as central to the pathogenic mechanism of ARVD.
- Genetic defects are increasingly implicated in the pathogenesis of metabolic cardiomyopathies, including mitochondrial cardiomyopathy and the cardiomyopathy associated with diabetes.
- Genetic and post-genomic methods of analysis have also been applied to acquired entities such as Chagas' disease and alcoholic cardiomyopathies. Both types of cardiomyopathy involve numerous intracellular pathways, sharing a number of critical features as well as displaying distinct elements.
- New techniques are needed for a more precise monitoring of the cellular and molecular events occurring in the myocardium and, hopefully, in a less invasive manner than endomyocardial biopsy.

References

1. Maron, B. J., Gardin, J. M., Flack, J. M., Gidding, S. S., Kurosaki, T. T., and Bild, D. E. (1995). Prevalence of hypertrophic cardiomyopathy in a general population of young adults. Echocardiographic analysis of 4111 subjects in the CARDIA study. *Circulation* **92**, 785–789.
2. Moolman, J. C., Corfield, V. A., Posen, B., Ngumbela, K., Seidman, C. E., Brink, P. A., and Watkins, H. (1997). Sudden death due to troponin T mutations. *J. Am. Coll. Cardiol.* **29**, 549–555.
3. Watkins, H., McKenna, W. J., Thierfelder, L., Suk, H. J., Anan, R., O'Donoghue, A., Spirito, P., Matsumori, A., Moravec, C. S., Seidman, J. G., and Seidman, C. E. (1995). Mutations in the genes for cardiac troponin T and α-tropomyosin in hypertrophic cardiomyopathy. *N. Engl. J. Med.* **332**, 1058–1064.
4. Marian, A. J. (2002). Modifier genes for hypertrophic cardiomyopathy. *Curr. Opin. Cardiol.* **17**, 242–252.
5. Ly, H. Q., Greiss, I., Talakic, M., Guerra, P. G., Macle, L., Thibault, B., Dubuc, M., and Roy, D. (2005). Sudden death and hypertrophic cardiomyopathy: A review. *Can. J. Cardiol.* **21**, 441–448.
6. Watkins, H., Rosenzweig, A., Hwang, D. S., Levi, .T, McKenna, W., Seidman, C. E., and Seidman, J. G. (1992). Characteristics and prognostic implications of myosin missense mutations in familial hypertrophic cardiomyopathy. *N. Engl. J. Med.* **326**, 1108–1114.
7. Niimura, H., Bachinski, L. L., Sangwatanaroj, S., Watkins, H., Chudley, A. E., McKenna, W., Kristinsson, A., Roberts, R., Sole, M., Maron, B. J., Seidman, J. G., and Seidman, C. E. (1998). Mutations in the gene for cardiac myosin-binding protein C and late-onset familial hypertrophic cardiomyopathy. *N. Engl. J. Med.* **338**, 1248–1257.
8. Fatkin, D., and Graham. R. M. (2002). Molecular mechanisms of inherited cardiomyopathies. *Physiol.* Rev. **82**, 945–980.

9. Thierfelder, L., Watkins, H., MacRae, C., Lamas, R., McKenna, W., Vosberg, H. P., Seidman, J. G., and Seidman, C. E. (1994). Alpha-tropomyosin and cardiac troponin T mutations cause familial hypertrophic cardiomyopathy: a disease of the sarcomere. *Cell* **77,** 710–712.
10. Satoh, M., Takahashi, M., Sakamoto, T., Hiroe, M., Marumo, F., and Kimura, A. (1999). Structural analysis of the titin gene in hypertrophic cardiomyopathy: identification of a novel disease gene. *Biochem. Biophys. Res. Commun.* **262,** 411–417.
11. Epstein, N. D. (1998). The molecular biology and pathophysiology of hypertrophic cardiomyopathy due to mutations in the beta myosin heavy chains and the essential and regulatory light chains. *Adv. Exp. Med. Biol.* **453,** 105–114.
12. Hoffmann, B., Schmidt-Traub, H., Perrot, A., Osterziel, K. J., and Gessner, R. (2001). First mutation in cardiac troponin C, L29Q, in a patient with hypertrophic cardiomyopathy. *Hum. Mutat.* **17,** 524.
13. Blair, E., Redwood, C., Ashrafian, H., Oliveira, M., Broxholme, J., Kerr, B., Salmon, A., Ostman-Smith, I., and Watkins, H. (2001). Mutations in the gamma(2) subunit of AMP-activated protein kinase cause familial hypertrophic cardiomyopathy: evidence for the central role of energy compromise in disease pathogenesis. *Hum. Mol. Genet.* **10,** 1215–1220.
14. Watkins, H., Conner, D., Thierfelder, L., Jarcho, J. A., MacRae, C., McKenna, W. J., Maron, B. J., Seidman, J. G., and Seidman, C. E. (1995). Mutations in the cardiac myosin binding protein-C gene on chromosome 11 cause familial hypertrophic cardiomyopathy. *Nat. Genet.* **11,** 434–437.
15. Geier, C., Perrot, A., Ozcelik, C., Binner, P., Counsell, D., Hoffmann, K., Pilz, B., Martiniak, Y., Gehmlich, K., van der Ven, P. F., Furst, D. O., Vornwald, A., von Hodenberg, E., Nurnberg, P., Scheffold, T., Dietz, R., and Osterziel, K. J. (2003). Mutations in the human muscle LIM protein gene in families with hypertrophic cardiomyopathy. *Circulation* **107,** 1390–1395.
16. Flavigny, J., Richard, P., Isnard, R., Carrier, L., Charron, P., Bonne, G., Forissier, J. F., Desnos, M., Dubourg, O., Komajda, M., Schwartz, K., and Hainque, B. (1998). Identification of two novel mutations in the ventricular regulatory myosin light chain gene (MYL2) associated with familial and classical forms of hypertrophic cardiomyopathy. *J. Mol. Med.* **76,** 208–214.
17. Geisterfer-Lowrance, A. A. T., Kass, S., Tanigawa, G., Vosberg, H. P., McKenna, W., Seidman, C. .E, and Seidman, J. G. (1990). A molecular basis for familial hypertrophic cardiomyopathy: a cardiac myosin heavy chain missense mutation. *Cell* **62,** 999–1006.
18. Mogensen, J., Klausen, I. C., Pedersen, A. K., Egeblad, H., Bross, P., Kruse, T. A., Gregersen, N., Hansen, P. S., Baandrup, U., and Borglum, A. D. (1999). Alpha-cardiac actin is a novel disease gene in familial hypertrophic cardiomyopathy. *J. Clin. Invest.* **103,** R39–R43.
19. Elliott, K., Watkins, H., and Redwood, C. S. (2000). Altered regulatory properties of human cardiac troponin I mutants that cause hypertrophic cardiomyopathy. *J. Biol. Chem.* **275,** 22069–22074.
20. Hayashi, T., Arimura, T., Ueda, K., Shibata, H., Hohda, S., Takahashi, M., Hori, H., Koga, Y., Oka, N., Imaizumi, T., Yasunami, M., and Kimura, A. (2004). Identification and functional analysis of a caveolin-3 mutation associated with familial hypertrophic cardiomyopathy. *Biochem. Biophys. Res. Commun.* **313,** 178–184.
21. Horvath, J., Ketelsen, U. P., Geibel-Zehender, A., Boehm, N., Olbrich, H., Korinthenberg, R., and Omran, H. (2003). Identification of a novel LAMP2 mutation responsible for X-chromosomal dominant Danon disease. *Neuropediatrics* **34,** 270–273.
22. Yang, Z., McMahon, C. J., Smith, L. R., Bersola, J., Adesina, A. M., Breinholt, J. P., Kearney, D. L., Dreyer, W. J., Denfield, S. W., Price, J. F., Grenier, M., Kertesz, N. J., Clunie, S. K., Fernbach, S. D., Southern, J. F., Berger, S., Towbin, J. A., Bowles, K. R., and Bowles, N. E. (2005). Danon disease as an underrecognized cause of hypertrophic cardiomyopathy in children. *Circulation* **112,** 1612–1617.
23. Taylor, M. R., Carniel, E., and Mestroni, L. (2004). Familial hypertrophic cardiomyopathy: clinical features, molecular genetics and molecular genetic testing. *Exp. Rev. Mol. Diagn.* **4,** 99–113.
24. Roberts, R., and Sidhu, J. (2003). Genetic basis for hypertrophic cardiomyopathy: implications for diagnosis and treatment. *Am. Heart Hosp. J.* **1,** 128–134.
25. Niimura, H., Patton, K. K., McKenna, W. J., Soults, J., Maron, B. J., Seidman, J. G., and Seidman, C. E. (2002). Sarcomere protein gene mutations in hypertrophic cardiomyopathy of the elderly. *Circulation* **105,** 446–451.
26. Richard, P., Charron, P., Leclercq, C., Ledeuil, C., Carrier, L., Dubourg, O., Desnos, M., Bouhour, J.B., Schwartz, K., Daubert, J. C., Komajda, M., and Hainque, B. (2000). Homozygotes for a R869G mutation in the beta-myosin heavy chain gene have a severe form of familial hypertrophic cardiomyopathy. *J. Mol. Cell Cardiol.* **32,** 1575–1583.
27. Ho, C. Y., Lever, H. M., DeSanctis, R., Farver, C. F., Seidman, J. G., and Seidman, C. E. (2000). Homozygous mutation in cardiac troponin T. Implications for hypertrophic cardiomyopathy. *Circulation* **102,** 1950–1955.
28. Perrot, A., Schmidt-Traub, H., Hoffmann, B., Prager, M., Bit-Avragim, N., Rudenko, R. I., Usupbaeva, D. A., Kabaeva, Z., Imanov, B., Mirrakhimov, M. M., Dietz, R., Wycisk, A., Tendera, M., Gessner, R., and Osterziel, K. J. (2005). Prevalence of cardiac beta-myosin heavy chain gene mutations in patients with hypertrophic cardiomyopathy. *J. Mol. Med.* **83,** 468–477.
29. Blair, E., Redwood, C., de Jesus Oliveira, M., Moolman-Smook, J. C., Brink, P., Corfield, V. A., Ostman-Smith, I., and Watkins, H. (2002). Mutations of the light meromyosin domain of the β-myosin heavy chain rod in hypertrophic cardiomyopathy. *Circ. Res.* **90,** 263–269.
30. Richard, P., Charron, P., Carrier, L., Ledeuil, C., Cheav, T., Pichereau, C., Benaiche, A., Isnard, R., Dubourg, O., Burban, M., Gueffet, J. P., Millaire, A., Desnos, M., Schwartz, K., Hainque, B., Komajda, M., and the EUROGENE Heart Failure Project. (2003). Hypertrophic cardiomyopathy: distribution of disease genes, spectrum of mutations, and implications for a molecular diagnosis strategy. *Circulation* **107,** 2227–2232.
31. Van Driest, S. L., Ackerman, M. J., Ommen, S. R., Shakur, R., Will, M. L., Nishimura, R. A., Tajik, A. J., and Gersh, B. J. (2002). Prevalence and severity of "benign" mutations in the beta-myosin heavy chain, cardiac troponin T, and alpha-tropomyosin genes in hypertrophic cardiomyopathy. *Circulation* **106,** 3085–3090.

32. Van Driest, S. L., Maron, B. J., and Ackerman, M. J. (2004). From malignant mutations to malignant domains: the continuing search for prognostic significance in the mutant genes causing hypertrophic cardiomyopathy. *Heart* **90,** 7–8.
33. Woo, A., Rakowski, H., Liew, J., Zhao, M. S., Liew, C. C., Parker, T. G., Zeller, M., Wigle, E. D., and Sole, M. J. (2003). Mutations of the β-myosin heavy chain gene in hypertrophic cardiomyopathy: critical functional sites determine prognosis. *Heart* **89,** 1179–1185.
34. Fananapazir, L., and Epstein, N. D. (1994). Genotype-phenotype correlations in hypertrophic cardiomyopathy: insights provided by comparisons of kindreds with distinct and identical β-myosin heavy chain gene mutations. *Circulation* **89,** 22–32.
35. Rayment, I., Holden, H. M., Sellers J, Fananapazir, L., and Epstein, N. D. (1995). Structural interpretation of the mutations in the β-cardiac myosin that have been implicated in familial hypertrophic cardiomyopathy. *Proc. Natl. Acad. Sci. USA* **92,** 3864–3868.
36. Flashman, E., Redwood, C., Moolman-Smook, J., and Watkins, H. (2004). Cardiac myosin binding protein C: its role in physiology and disease. *Circ. Res.* **94,** 1279–1289.
37. Bonne, G., Carrier, L., Bercovici, J., Cruaud, C., Richard, P., Hainque, B., Gautel, M., Labeit, S., James, M., Beckmann, J., Weissenbach, J., Vosberg, H. P., Fiszman, M., Komajda, M., and Schwartz, K. (1995). Cardiac myosin binding protein-C gene splice acceptor site mutation is associated with familial hypertrophic cardiomyopathy. *Nat. Genet.* **11,** 438–440.
38. Lin, T., Ichihara, S., Yamada, Y., Nagasaka, T., Ishihara, H., Nakashima, N., and Yokota, M. (2000). Phenotypic variation of familial hypertrophic cardiomyopathy caused by the Phe(110)→Ile mutation in cardiac troponin T. *Cardiology* **93,** 155–162.
39. Anan, R., Shono, H., Kisanuki, A., Arima, S., Nakao, S., and Tanaka, H. (1998). Patients with familial hypertrophic cardiomyopathy caused by a Phe110Ile missense mutation in the cardiac troponin T gene have variable cardiac morphologies and a favorable prognosis. *Circulation* **98,** 391–397.
40. Elliott, P. M., D'Cruz, L., and McKenna, W. J. (1999). Late-onset hypertrophic cardiomyopathy caused by a mutation in the cardiac troponin T gene. *N. Engl. J. Med.* **341,** 1855–1856.
41. Torricelli, F., Girolami, F., Olivotto, I., Passerini, I., Frusconi, S., Vargiu, D., Richard, P., and Cecchi, F. (2003). Prevalence and clinical profile of troponin T mutations among patients with hypertrophic cardiomyopathy in Tuscany. *Am. J. Cardiol.* **92,** 1358–1362.
42. Maass, A. H., and Leinwand, L. A. (2003). Mechanisms of the pathogenesis of troponin T-based familial hypertrophic cardiomyopathy. *Trends Cardiovasc. Med.* **13,** 232–237.
43. Varnava, A. M., Elliott, P. M., Baboonian, C., Davison, F., Davies, M. J., and McKenna, W. J. (2001). Hypertrophic cardiomyopathy: histopathological features of sudden death in cardiac troponin T disease. *Circulation* **104,** 1380–1384.
44. Varnava, A., Baboonian, C., Davison, F., de Cruz, L., Elliott, P. M., Davies, M. J., and McKenna, W. J. (1999). A new mutation of the cardiac troponin T gene causing familial hypertrophic cardiomyopathy without left ventricular hypertrophy. *Heart* **82,** 621–624.
45. Gomes, A. V., Harada, K., and Potter, J. D. (2005). A mutation in the N-terminus of Troponin I that is associated with hypertrophic cardiomyopathy affects the Ca(2+)-sensitivity, phosphorylation kinetics and proteolytic susceptibility of troponin. *J. Mol. Cell Cardiol.* **39,** 754–765.
46. Jongbloed, R. J., Marcelis, C. L., Doevendans, P. A., Schmeitz-Mulkens, J. M., Van Dockum, W. G., Geraedts, J. P., and Smeets, H. J. (2003). Variable clinical manifestation of a novel missense mutation in the alpha-tropomyosin (TPM1) gene in familial hypertrophic cardiomyopathy. *J. Am. Coll. Cardiol.* **41,** 981–986.
47. Poetter, K., Jiang, H., Hassanzadeh, S., Master, S. R., Chang, A., Dalakas, M. C., Rayment, I., Sellers, J. R., Fananapazir, L., and Epstein, N. D. (1996). Mutations in either the essential or regulatory light chains of myosin are associated with a rare myopathy in human heart and skeletal muscle. *Nat. Genet.* **13,** 63–69.
48. Flavigny, J., Richard, P., Isnard, R., Carrier, L., Charron, P., Bonne, G., Forissier, J. F., Desnos, M., Dubourg, O., Komajda, M., Schwartz, K., and Hainque, B. (1998). Identification of two novel mutations in the ventricular regulatory myosin light chain gene (MYL2) associated with familial and classical forms of hypertrophic cardiomyopathy. *J. Mol. Med.* **76,** 208–214.
49. Andersen, P. S., Havndrup, O., Bundgaard, H., Moolman-Smook, J. C., Larsen, L. A., Mogensen, J., Brink, P. A., Borglum, A. D., Corfield, V. A., Kjeldsen, K., Vuust, J., and Christiansen, M. (2001). Myosin light chain mutations in familial hypertrophic cardiomyopathy: phenotypic presentation and frequency in Danish and South African populations. *J. Med. Genet.* **38,** E43.
50. Vasile, V. C., Ommen, S. R., Edwards, W. D., and Ackerman, M. J. (2006). A missense mutation in a ubiquitously expressed protein, vinculin, confers susceptibility to hypertrophic cardiomyopathy. *Biochem. Biophys. Res. Commun.* **345,** 998–1003.
51. Lee, W., Hwang, T. H., Kimura, A., Park, S. W., Satoh, M., Nishi, H., Harada, H., Toyama, J., and Park, J. E. (2001). Different expressivity of a ventricular essential myosin light chain gene Ala57Gly mutation in familial hypertrophic cardiomyopathy. *Am. Heart J.* **141,** 184–189.
52. Mogensen, J., Perrot, A., Andersen, P. S., Havndrup, O., Klausen, I. C., Christiansen, M., Bross, P., Egeblad, H., Bundgaard, H., Osterziel, K. J., Haltern, G., Lapp, H., Reinecke, P., Gregersen, N., and Borglum, A. D. (2004). Clinical and genetic characteristics of alpha cardiac actin gene mutations in hypertrophic cardiomyopathy. *J. Med. Genet.* **41,** e10.
53. Olson, T. M., Doan, T. P., Kishimoto, N. Y., Whitby, F. G., Ackerman, M. J., and Fananapazir, L. (2000). Inherited and de novo mutations in the cardiac actin gene cause hypertrophic cardiomyopathy. *J. Mol. Cell Cardiol.* **32,** 1687–1694.
54. Hayashi, T., Arimura, T., Itoh-Satoh, M., Ueda, K., Hohda, S., Inagaki, N., Takahashi, M., Hori, H., Yasunami, M., Nishi, H., Koga, Y., Nakamura, H., Matsuzaki, M., Choi, B. Y., Bae, S. W., You, C. W., Han, K. H., Park, J. E., Knoll, R., Hoshijima, M., Chien, K. R., and Kimura, A. (2004). Tcap gene mutations in hypertrophic cardiomyopathy and dilated cardiomyopathy. *J. Am. Coll. Cardiol.* **44,** 2192–2201.

55. Arad, M., Benson, D. W., Perez-Atayde, A. R., McKenna, W. J., Sparks, E. A., Kantor, R. J., McGarry, K., Seidman, J. G., and Seidman, C. E. (2002). Constitutively active AMP kinase mutations cause glycogen storage disease mimicking hypertrophic cardiomyopathy. *J. Clin. Invest.* **109**, 357–362.

56. Gollob, M. H., Green, M. S., Tang, A. S. L., Gollob, T., Karibe, A., Hassan, A. S., Ahmad, F., Lozado, R., Shah, G., Fananapazir, L., Bachinski, L., and Roberts, R. (2001). Identification of a gene responsible for familial Wolff-Parkinson-White syndrome. *N. Engl. J. Med.* **344**, 1823–1831.

57. Zou, L., Shen, M., Arad, M., He, H., Lofgren, B., Ingwall, J. S., Seidman, C. E., Seidman, J. G., and Tian, R. (2005). N488I mutation of the gamma2-subunit results in bidirectional changes in AMP-activated protein kinase activity. *Circ. Res.* **97**, 323–328.

58. Murphy, R. T., Mogensen, J., McGarry, K., Bahl, A., Evans, A., Osman, E., Syrris, P., Gorman, G., Farrell, M., Holton, J. L., Hanna, M. G., Hughes, S., Elliott, P. M., Macrae, C. A., and McKenna, W. J. (2005). Adenosine monophosphate-activated protein kinase disease mimicks hypertrophic cardiomyopathy and Wolff-Parkinson-White syndrome: natural history. *J. Am. Coll. Cardiol.* **45**, 922–930.

59. Arad, M., Maron, B. J., Gorham, J. M., Johnson, W. H., Jr., Saul, J. P., Perez-Atayde, A. R., Spirito, P., Wright, G. B., Kanter, R. J., Seidman, C. E., and Seidman, J. G. (2005). Glycogen storage diseases presenting as hypertrophic cardiomyopathy. *N. Engl. J. Med.* **352**, 362–372.

60. Gollob, M. H., Seger, J. J., Gollob, T. N., Tapscott, T., Gonzales, O., Bachinski, L., and Roberts, R. (2001). Novel PRKAG2 mutation responsible for the genetic syndrome of ventricular preexcitation and conduction system disease with childhood onset and absence of cardiac hypertrophy. *Circulation* **104**, 3030–3033.

61. Geisterfer-Lowrance, A. A. T., Christe, M., Conner, D. A., Ingwall, J. S., Schoen, F. J., Seidman, C. E., and Seidman, J. G. (1996). A mouse model of familial hypertrophic cardiomyopathy. *Science* **272**, 731–734.

62. Georgakopoulos, D., Christe, M. E., Giewat, M., Seidman, C. E., Seidman, J. G., and Kass, D. A. (1999). The pathogenesis of familial hypertrophic cardiomyopathy: early and evolving effects from an alpha-cardiac myosin heavy chain missense mutation. *Nat. Med.* **5**, 327–330.

63. Yang, Q., Sanbe, A., Osinska, H., Hewett, T. E., Klevitsky, R., and Robbins, J. (1998). A mouse model of myosin binding protein C human familial hypertrophic cardiomyopathy. *J. Clin. Invest.* **102**, 1292–1300.

64. McConnell, B. K., Fatkin, D., Semsarian, C., Jones, K. A., Georgakopoulos, D., Maguire, C. T., Healey, M. J., Mudd, J. O., Moskowitz, I. P. G., Conner, D. A., Giewat, M., Wakimoto, H., Berul, C. I., Schoen, F. J., Kass, D. A., Seidman, C. E., and Seidman, J. G. (2001). Comparison of two murine models of familial hypertrophic cardiomyopathy. *Circ. Res.* **88**, 383–389.

65. Harris, S. P., Bartley, C. R., Hacker, T. A., McDonald, K. S., Douglas, P. S., Greaser, M. L., Powers, P. A., and Moss, R. L. (2002). Hypertrophic cardiomyopathy in cardiac myosin binding protein-C knockout mice. *Circ. Res.* **90**, 594–601.

66. Carrier, L., Knoll, R., Vignier, N., Keller, D. I., Bausero, P., Prudhon, B., Isnard, R., Ambroisine, M. L., Fiszman, M., Ross, J., Jr., Schwartz, K., and Chien, K. R. (2004). Asymmetric septal hypertrophy in heterozygous cMyBP-C null mice. *Cardiovasc. Res.* **63**, 293–304.

67. Tardiff, J. C., Factor, S. M., Tompkins, B. D., Hewett, T. E., Palmer, B. M., Moore, R. L., Schwartz, S., Robbins, J., and Leinwand, L. A. (1998). A truncated cardiac troponin T molecule in transgenic mice suggests multiple cellular mechanisms for familial hypertrophic cardiomyopathy. *J. Clin. Invest.* **101**, 2800–2811.

68. Oberst, L., Zhao, G., Park, J. T., Brugada, R., Michael, L. H., Entman, M. L., Roberts, R., and Marian, A. J. (1998). Dominant-negative effect of a mutant cardiac troponin T on cardiac structure and function in transgenic mice. *J. Clin. Invest.* **102**, 1498–1505.

69. Tardiff, J. C., Hewett, T. E., Palmer, B. M., Olsson, C., Factor, S. M., Moore, R. L., Robbins, J., and Leinwand, L. A. (1999). Cardiac troponin T mutations result in allele-specific phenotypes in a mouse model for hypertrophic cardiomyopathy. *J. Clin. Invest.* **104**, 469–481.

70. Javadpour, M. M., Tardiff, J. C., Pinz, I., and Ingwall, J. S. (2003). Decreased energetics in murine hearts bearing the R92Q mutation in cardiac troponin T. *J. Clin. Invest.* **112**, 768–775.

71. Chandra, M., Rundell, V. L., Tardiff, J. C., Leinwand, L. A., De Tombe, P. P., and Solaro, R. J. (2001). Ca(2+) activation of myofilaments from transgenic mouse hearts expressing R92Q mutant cardiac troponin T. *Am. J. Physiol. Heart Circ. Physiol.* **280**, H705–H713.

72. Montgomery, D. E., Tardiff, J. C., and Chandra, M. (2001). Cardiac troponin T mutations: correlation between the type of mutation and the nature of myofilament dysfunction in transgenic mice. *J. Physiol.* **536**, 583–592.

73. Knollmann, B. C., Kirchhof, P., Sirenko, S. G., Degen, H., Greene, A. E., Schober, T., Mackow, J. C., Fabritz, L., Potter, J. D., and Morad, M. (2003). Familial hypertrophic cardiomyopathy-linked mutant troponin T causes stress-induced ventricular tachycardia and Ca2+-dependent action potential remodeling. *Circ. Res.* **92**, 428–436.

74. Hernandez, O., Szczesna-Cordary, D., Knollmann, B. C., Miller, T., Bell, M., Zhao, J., Sirenko, S. G., Diaz, Z., Guzman, G., Xu, Y., Wang, Y., Kerrick, W. G., and Potter, J. D. (2005). F110I and R278C troponin T mutations that cause familial hypertrophic cardiomyopathy affect muscle contraction in transgenic mice and reconstituted human cardiac fibers. *J. Biol. Chem.* **280**, 37183–37194

75. James, J., Zhang, Y., Osinska, H., Sanbe, A., Klevitsky, R., Hewett, T. E., and Robbins, J. (2000). Transgenic modeling of a cardiac troponin I mutation linked to familial hypertrophic cardiomyopathy. *Circ. Res.* **87**, 805–811.

76. Prabhakar, R., Boivin, G. P., Grupp, I. L., Hoit, B., Arteaga, G., Solaro, J. R., and Wieczorek, D. F. (2001). A familial hypertrophic cardiomyopathy alpha-tropomyosin mutation causes severe cardiac hypertrophy and death in mice. *J. Mol. Cell Cardiol.* **33**, 1815–1828.

77. Prabhakar, R., Petrashevskaya, N., Schwartz, A., Aronow, B., Boivin, G. P., Molkentin, J. D., and Wieczorek. D. F. (2003). A mouse model of familial hypertrophic cardiomyopathy caused by an alpha-tropomyosin mutation. *Mol. Cell Biochem.* **251**, 33–42.

78. Vemuri, R., Lankford, E. B., Poetter, K., Hassanzadeh, S., Takeda, K., Yu, Z. X., Ferrans, V. J., and Epstein, N. D.

(1999). The stretch-activation response may be critical to the proper functioning of the mammalian heart. *Proc. Natl. Acad. Sci. USA* **96**, 1048–1053.

79. Sanbe, A., Nelson, D., Gulick, J., Setser, E., Osinska, H., Wang, X., Hewett, T. E., Klevitsky, R., Hayes, E., Warshaw, D. M., and Robbins, J. (2000). In vivo analysis of an essential myosin light chain mutation linked to familial hypertrophic cardiomyopathy. *Circ. Res.* **87**, 296–302.

80. Szczesna-Cordary, D., Guzman, G., Zhao, J., Hernandez, O., Wei, J., and Diaz-Perez, Z. (2005). The E22K mutation of myosin RLC that causes familial hypertrophic cardiomyopathy increases calcium sensitivity of force and ATPase in transgenic mice. *J. Cell Sci.* **118**, 3675–3683.

81. Sidhu, J. S., Rajawat, Y. S., Rami, T. G., Gollob, M. H., Wang, Z., Yuan, R., Marian, A. J., DeMayo, F. J., Weilbacher, D., Taffet, G. E., Davies, J. K., Carling, D., Khoury, D. S., and Roberts, R. (2005). Transgenic mouse model of ventricular preexcitation and atrioventricular reentrant tachycardia induced by an AMP-activated protein kinase loss-of-function mutation responsible for Wolff-Parkinson-White syndrome. *Circulation* **111**, 21–29.

82. Arad, M., Moskowitz, I. P., Patel, V. V., Ahmad, F., Perez-Atayde, A. R., Sawyer, D. B., Walter, M., Li, G. H., Burgon, P. G., Maguire, C. T., Stapleton, D., Schmitt, J. P., Guo, X. X., Pizard, A., Kupershmidt, S., Roden, D. M., Berul, C. I., Seidman, C. E., and Seidman, J.G. (2003). Transgenic mice overexpressing mutant PRKAG2 define the cause of Wolff-Parkinson-White syndrome in glycogen storage cardiomyopathy. *Circulation* **107**, 2850–2856.

83. Tanaka, N., Dalton, N., Mao, L., Rockman, H. A., Peterson, K. L., Gottshall, K. R., Hunter, J. J., Chien, K. R., and Ross, J. Jr. (1996). Transthoracic echocardiography in models of cardiac disease in the mouse. *Circulation* **94**, 1109–1117.

84. Robbins, R. J., and Swain, J. L. (1992). C-myc protooncogene modulates cardiac hypertrophic growth in transgenic mice. *Am. J. Physiol.* **262**, H590–H597.

85. Takeishi, Y., Ping, P., Bolli, R., Kirkpatrick, D. L., Hoit, B. D., and Walsh, R. A. (2000). Transgenic overexpression of constitutively active protein kinase C epsilon causes concentric cardiac hypertrophy. *Circ. Res.* **86**, 1218–1223.

86. Mende, U., Kagen, A., Cohen, A., Aramburu, J., Schoen, F. J., and Neer, E. J. (1998). Transient cardiac expression of constitutively active Galphaq leads to hypertrophy and dilated cardiomyopathy by calcineurin-dependent and independent pathways. *Proc. Natl. Acad. Sci. USA* **95**, 13893–13898.

87. Dorn, G. W., II, and Hahn, H. S. (2004). Genetic factors in cardiac hypertrophy. *Ann. N.Y. Acad. Sci.* **1015**, 225–237.

88. Frey, N., Katus, H. A., Olson, E. N., and Hill, J. A. (2004). Hypertrophy of the heart: A new therapeutic target? *Circulation* **109**, 1580–1589.

89. Jung, W. I., Sieverding, L., Breuer, J., Hoess, T., Widmaier, S., Schmidt, O., Bunse, M., van Erckelens, F., Apitz, J., Lutz, O., and Dietze, G. J. (1998). 31P NMR spectroscopy detects metabolic abnormalities in asymptomatic patients with hypertrophic cardiomyopathy. *Circulation* **97**, 2536–2542.

90. Spindler, M., Saupe, K. W., Christe, M. E., Sweeney, H. L., Seidman, C. E., Seidman, J. G., and Ingwall, J. S. (1998). Diastolic dysfunction and altered energetics in the MHC403/+ mouse model of familial hypertrophic cardiomyopathy. *J. Clin. Invest.* 101, 1775–1783.

91. Lucas, D. T., Aryal, P., Szweda, L. I., Koch, W. J., and Leinwand, L. A. (2003). Alterations in mitochondrial function in a mouse model of hypertrophic cardiomyopathy. *Am. J. Physiol. Heart Circ. Physiol.* **284**, H575–H583.

92. Aronow, B. J., Toyokawa, T., Canning, A., Haghighi, K., Delling, U., Kranias, E., Molkentin, J. D., and Dorn, G. W., 2nd. (2001). Divergent transcriptional responses to independent genetic causes of cardiac hypertrophy. *Physiol. Genomics* **6**, 19–28.

93. Kong, S. W., Bodyak, N., Yue, P., Liu, Z., Brown, J., Izumo, S., and Kang, P. M. (2005). Genetic expression profiles during physiological and pathological cardiac hypertrophy and heart failure in rats. *Physiol. Genomics* **21**, 34–42.

94. Strom, C. C., Aplin, M., Ploug, T., Christoffersen, T. E., Langfort, J., Viese, M., Galbo, H., Haunso, S., and Sheikh, S. P. (2005). Expression profiling reveals differences in metabolic gene expression between exercise-induced cardiac effects and maladaptive cardiac hypertrophy. *FEBS J.* **272**, 2684–2695.

95. Lim, D. S., Roberts, R., and Marian, A. J. (2001). Expression profiling of cardiac genes in human hypertrophic cardiomyopathy: insight into the pathogenesis of phenotypes. *J. Am. Coll. Cardiol.* **38**, 1175–1180.

96. Hwang, J. J., Allen, P. D., Tseng, G. C., Lam, C. W., Fananapazir, L., Dzau, V. J., and Liew, C. C. (2002). Microarray gene expression profiles in dilated and hypertrophic cardiomyopathic end-stage heart failure. *Physiol. Genomics* **10**, 31–44.

97. Ohki, R., Yamamoto, K., Ueno, S., Mano, H., Misawa, Y., Fuse, K., Ikeda, U., and Shimada, K. (2004). Transcriptional profile of genes induced in human atrial myocardium with pressure overload. *Int. J. Cardiol.* **96**, 381–387.

98. Syed, F., Odley, A., Hahn, H. S., Brunskill, E. W., Lynch, R. A., Marreez, Y., Sanbe, A., Robbins, J., and Dorn, G. W. 2nd. (2004). Physiological growth synergizes with pathological genes in experimental cardiomyopathy. *Circ. Res.* **95**, 1200–1206.

99. Report of the 1995 World Health Organization/International Society and Federation of Cardiology Task Force. (1996). The definition and classification of cardiomyopathies. *Circulation* **93**, 841–842.

100. Mestroni, L., Rocco, C., Gregori, D., Sinagra, G., Di Lenarda, A., Miocic, S., Vatta, M., Pinamonti, B., Muntoni, F., Caforio, A. L. P., McKenna, W. J., Falaschi, A., Giacca, M., and Camerini, F. (2000). Familial dilated cardiomyopathy: Evidence for genetic and phenotypic heterogeneity. *J. Am. Coll. Cardiol.* **34**, 181–190.

101. Dec, G. W., and Fuster, V. (1994). Idiopathic dilated cardiomyopathy. *N. Engl. J. Med.* **331**, 1564–1575.

102. Burkett, E. L., and Hershberger, R. E. (2005). Clinical and genetic issues in familial dilated cardiomyopathy. *J. Am. Coll. Cardiol.* **45**, 969–981.

103. Bienengraeber, M., Olson, T. M., Selivanov, V. A., Kathmann, E. C., O'Cochlain, F., Gao, F., Karger, A. B., Ballew, J. D., Hodgson, D. M., Zingman, L. V., Pang, Y. P., Alekseev, A. E., and Terzic, A. (2004). ABCC9 mutations identified in human dilated cardiomyopathy disrupt catalytic KATP channel gating. *Nat. Genet.* **36**, 382–387.

104. Olson, T. M., Michels, V. V., Thibodeau, S. N., Tai, Y. S., and Keating, M. T. (1998). Actin mutations in dilated cardiomyopathy, a heritable form of heart failure. *Science* **280,** 750–752.
105. Mohapatra, B., Jimenez, S., Lin, J. H., Bowles, K. R., Coveler, K. J., Marx, J. G., Chrisco, M. A., Murphy, R. T., Lurie, P. R., Schwartz, R. J., Elliott, P. M., Vatta, M., McKenna, W., Towbin, J. A., and Bowles, N. E. (2003). Mutations in the muscle LIM protein and alpha-actinin-2 genes in dilated cardiomyopathy and endocardial fibroelastosis. *Mol. Genet. Metab.* **80,** 207–215.
106. Goldfarb, L. G., Park, K. Y., Cervenakova, L., Gorokhova, S., Lee, H. S., Vasconcelas, O., Nagle, J. W., Semino-Mora, C., Sivakumar, K., and Dalakas, M. C. (1998). Missense mutations in desmin associated with familial cardiac and skeletal myopathy. *Nat. Genet.* **19,** 402–403.
107. Muntoni, F., Di Lenarda, A., Porcu, M., Sinagra, G., Mateddu, A., Marrosu, G., Ferlini, A., Cau, M., Milasin, J., Melis, M. A., Marrosu, M. G., Cianchetti, C., Sanna, A., Falaschi, A., Camerini, F., Giacca, M., and Mestroni, L. (1997). Dystrophin gene abnormalities in two patients with idiopathic dilated cardiomyopathy. *Heart* **78,** 608–612.
108. Bione, S., D'Adamo, P., Maestrini, E., Gedeon, A. K., Bolhuis, P. A., and Toniolo, D. (1996). A novel X-linked gene, G4.5. is responsible for Barth syndrome. *Nat. Genet.* **12,** 385–389.
109. Fatkin, D., MacRae, C., Sasaki, T., Wolff, M. R., Porcu, M., Frenneaux, M., Atherton, J., Vidaillet, H. J., Spudich, S., De Girolami, U., Seidman, J. G., and Seidman, C. E. (1999). Missense mutations in the rod domain of the lamin A/C gene as causes of dilated cardiomyopathy and conduction-system disease. *N. Engl. J. Med.* **341,** 1715–1724.
110. Daehmlow, S., Erdmann, J., Knueppel, T., Gille, C., Froemmel, C., Hummel, M., Hetzer, R., and Regitz-Zagrosek, V. (2002). Novel mutations in sarcomeric protein genes in dilated cardiomyopathy. *Biochem. Biophys. Res. Commun* **298,** 116–120.
111. Carniel, E., Taylor, M. R., Sinagra, G., Di Lenarda, A., Ku, L., Fain, P. R., Boucek, M. M., Cavanaugh, J., Miocic, S., Slavov, D., Graw, S. L., Feiger, J., Zhu, X. Z., Dao, D., Ferguson, D. A., Bristow, M. R., and Mestroni, L. (2005). Alpha-myosin heavy chain: a sarcomeric gene associated with dilated and hypertrophic phenotypes of cardiomyopathy. *Circulation* **112,** 54–59.
112. Villard, E., Duboscq-Bidot, L., Charron, P., Benaiche, A., Conraads, V., Sylvius, N., and Komajda, M. (2005). Mutation screening in dilated cardiomyopathy: prominent role of the beta myosin heavy chain gene. *Eur. Heart J.* **26,** 794–803.
113. Schmitt, J. P., Kamisago, M., Asahi, M., Li, G. H., Ahmad, F., Mende, U., Kranias, E. G., MacLennan, D. H., Seidman, J. G., and Seidman, C. E. (2003). Dilated cardiomyopathy and heart failure caused by a mutation in phospholamban. *Science* **299,** 1410–1413.
114. Olson, T. M., Michels, V. V., Ballew, J. D., Reyna, S. P., Karst, M. L., Herron, K. J., Horton, S. C., Rodeheffer, R. J., and Anderson, J. L. (2005). Sodium channel mutations and susceptibility to heart failure and atrial fibrillation. *JAMA* **293,** 447–454.
115. McNair, W. P., Ku, L., Taylor, M. R., Fain, P. R., Dao, D., Wolfel, E., and Mestroni, L. (2004). Familial Cardiomyopathy Registry Research Group. SCN5A mutation associated with dilated cardiomyopathy, conduction disorder, and arrhythmia. *Circulation* **110,** 2163–2167.
116. Tsubata, S., Bowles, K. R., Vatta, M., Zintz, C., Titus, J., Muhonen, L., Bowles, N. E., and Towbin, J. A. (2000). Mutations in the human delta-sarcoglycan gene in familial and sporadic dilated cardiomyopathy. *J. Clin. Invest.* **106,** 655–662.
117. Bione, S., Maestrini, E., Rivella, S., Mancini, M., Regis, S., Romeo, G., and Toniolo, D. (1994). Identification of a novel X-linked gene responsible for Emery-Dreifuss muscular dystrophy. *Nat. Genet.* **8,** 323–327.
118. Mogensen, J., Murphy, R. T., Shaw, T., Bahl, A., Redwood, C., Watkins, H., Burke, M., Elliott, P. M., and McKenna, W. J. (2004). Severe disease expression of cardiac troponin C and T mutations in patients with idiopathic dilated cardiomyopathy. *J. Am. Coll. Cardiol.* **44,** 2033–4200.
119. Murphy, R. T., Mogensen, J., Shaw, A., Kubo, T., Hughes, S., and McKenna, W. J. (2004). Novel mutation in cardiac troponin I in recessive idiopathic dilated cardiomyopathy. *Lancet* **363,** 371–372.
120. Kamisago, M., Sharma, S. D., DePalma, S. R., Solomon, S., Sharma, P., McDonough, B., Smoot, L., Mullen, M. P., Woolf, P. K., Wigle, E. D., Seidman, J. G., and Seidman, C. E. (2000). Mutations in sarcomere protein genes as a cause of dilated cardiomyopathy. *N. Engl. J. Med.* **343,** 1688–1696.
121. Olson, T. M., Kishimoto, N. Y., Whitby, F. G., and Michels, V. V. (2001). Mutations that alter the surface charge of alpha-tropomyosin are associated with dilated cardiomyopathy. *J. Mol. Cell Cardiol.* **33,** 723–732.
122. Gerull, B., Gramlich, M., Atherton, J., McNabb, M., Trombitas, K., Sasse-Klaassen, S., Seidman, J. G., Seidman, C. E., Granzier, H., Labeit, S., Frenneaux, M., and Thierfelder, L. (2002). Mutations of *TTN*, encoding the giant muscle filament titin, cause familial dilated cardiomyopathy. *Nat. Genet.* **30,** 201–204.
123. Maeda, M., Holder, E., Lowes, B., Valent, S., and Bies, R. D. (1997). Dilated cardiomyopathy associated with deficiency of the cytoskeletal protein metavinculin. *Circulation* **95,** 17–20.
124. Vatta, M., Mohapatra, B., Jimenez, S., Sanchez, X., Faulkner, G., Perles, Z., Sinagra, G., Lin, J. H., Vu, T. M., Zhou, Q., Bowles, K. R., Di Lenarda, A., Schimmenti, L., Fox, M., Chrisco, M. A., Murphy, R. T., McKenna, W., Elliott, P., Bowles, N. E., Chen, J., Valle, G., and Towbin, J. A. (2003). Mutations in Cypher/ZASP in patients with dilated cardiomyopathy and left ventricular non-compaction. *J. Am. Coll. Cardiol.* **42,** 2014–2027.
125. Kass, S., MacRae, C., Graber, H. L., Sparks, E. A., McNamara, D., Boudoulas, H., Basson, C. T., Baker, P. B., Cody, R. J., Fishman, M. C., Cox, N., Kong, A., Wooley, C. F., Seidman, J. G., and Seidman, C. E. (1994). A gene defect that causes conduction system disease and dilated cardiomyopathy maps to chromosome 1p1-1q1. *Nat. Genet.* **7,** 546–551.
126. Jung, M., Poepping, I., Perrot, A., Ellmer, A. E., Wienker, T. F., Dietz, R., Reis, A., and Osterziel, K. J. (1999). Investigation of a family with autosomal dominant dilated cardiomyopathy defines a novel locus on chromosome 2q14-q22. *Am. J. Hum. Genet.* **65,** 1068–1077.
127. Olson, T. M., and Keating, M. T. (1996). Mapping a cardiomyopathy locus to chromosome 3p22-p25. *J. Clin. Invest.* **97,** 528–532.

128. Muchir, A., Bonne, G., van der Kooi, A. J., van Meegen, M., Baas, F., Bolhuis, P. A., de Visser, M., and Schwartz, K. (2000). Identification of mutations in the gene encoding lamins A/C in autosomal dominant limb girdle muscular dystrophy with atrioventricular conduction disturbances (LGMD1B). *Hum. Mol. Genet.* **9,** 1453–1459.

129. Messina, D. N., Speer, M. C., Pericak-Vance, M. A., and McNally, E. M. (1997). Linkage of familial dilated cardiomyopathy with conduction defect and muscular dystrophy to chromosome 6q23. *Am. J. Hum. Genet.* **61,** 909–917.

130. Schonberger, J., Levy, H., Grunig, E., Sangwatanaroj, S., Fatkin, D., MacRae, C., Stacker, H., Halpin, C., Eavey, R., Philbin, E. F., Katus, H., Seidman, J. G., and Seidman, C. E. (2000). Dilated cardiomyopathy and sensorineural hearing loss. A heritable syndrome that maps to chromosome 6q23-q24. *Circulation* **101,** 1812–1818.

131. Towbin, J. A., Hejtmancik, J. F., Brink, P., Gelb, B., Zhu, X. M., Chamberlain, J. S., McCabe, E. R. B., and Swift, M. (1993). X-linked dilated cardiomyopathy. Molecular genetic evidence of linkage to the Duchenne muscular dystrophy (dystrophin) gene at the Xp21 locus. *Circulation* **87,** 1854–1865.

132. Marin-Garcia, J., and Goldenthal, M. J. (2002). Understanding the impact of mitochondrial defects in cardiovascular disease: a review. *J. Card. Fail.* **8,** 347–361.

133. Karkkainen, S., Peuhkurinen, K., Jaaskelainen, P., Miettinen, R., Karkkainen, P., Kuusisto, J., and Laakso, M. (2002). No variants in the cardiac actin gene in Finnish patients with dilated or hypertrophic cardiomyopathy. *Am. Heart J.* **143,** E6.

134. Shimizu, M., Ino, H., Yasuda, T., Fujino, N., Uchiyama, K., Mabuchi, T., Konno, T., Kaneda, T., Fujita, T., Masuta, E., Katoh, M., Funada, A., and Mabuchi, H. (2005). Gene mutations in adult Japanese patients with dilated cardiomyopathy. *Circ. J.* **69,** 150–153.

135. Takai, E., Akita, H., Shiga, N., Kanazawa, K., Yamada, S., Terashima, M., Matsuda, Y., Iwai, C., Kawai, K., Yokota, Y., and Yokoyama, M. (1999). Mutational analysis of the cardiac actin gene in familial and sporadic dilated cardiomyopathy. *Am. J. Med. Genet.* **86,** 325–327.

136. Tesson, F., Sylvius, N., Pilotto, A., Dubosq-Bidot, L., Peuchmaurd, M., Bouchier, C., Benaiche, A., Mangin, L., Charron, P., Gavazzi, A., Tavazzi, L., Arbustini, E., and Komajda, M. (2000). Epidemiology of desmin and cardiac actin gene mutations in a European population of dilated cardiomyopathy. *Eur. Heart J.* **21,** 1872–1876.

137. Kamisago, M., Sharma, S. D., DePalma, S. R., Solomon, S., Sharma, P., McDonough, B., Smoot, L., Mullen, M. P., Woolf, P. K., Wigle, E. D., Seidman, J. G., and Seidman, C. E. (2000). Mutations in sarcomere protein genes as a cause of dilated cardiomyopathy. *N. Engl. J. Med.* **343,** 1688–1696.

138. Villard, E., Duboscq-Bidot, L., Charron, P., Benaiche, A., Conraads, V., Sylvius, N., and Komajda, M. (2005). Mutation screening in dilated cardiomyopathy: prominent role of the beta myosin heavy chain gene. *Eur. Heart J.* **26,** 794–803.

139. Carniel, E., Taylor, M. R., Sinagra, G., Di Lenarda, A., Ku, L., Fain, P. R., Boucek, M. M., Cavanaugh, J., Miocic, S., Slavov, D., Graw, S. L., Feiger, J., Zhu, X. Z., Dao, D., Ferguson, D. A., Bristow, M. R., and Mestroni, L. (2005). Alpha-myosin heavy chain: a sarcomeric gene associated with dilated and hypertrophic phenotypes of cardiomyopathy. *Circulation* **112,** 54–59.

140. Shimizu, M., Ino, H., Yasuda, T., Fujino, N., Uchiyama, K., Mabuchi, T., Konno, T., Kaneda, T., Fujita, T., Masuta, E., Katoh, M., Funada, A., and Mabuchi, H. (2005). Gene mutations in adult Japanese patients with dilated cardiomyopathy. *Circ. J.* **69,** 150–153.

141. Konno, T., Shimizu, M., Ino, H., Matsuyama, T., Yamaguchi, M., Terai, H., Hayashi, K., Mabuchi, T., Kiyama, M., Sakata, K., Hayashi, T., Inoue, M., Kaneda, T., and Mabuchi, H. (2003). A novel missense mutation in the myosin binding protein-C gene is responsible for hypertrophic cardiomyopathy with left ventricular dysfunction and dilation in elderly patients. *J. Am. Coll. Cardiol.* **41,** 781–786.

142. Nanni, L., Pieroni, M., Chimenti, C., Simionati, B., Zimbello, R., Maseri, A., Frustaci, A., and Lanfranchi, G. (2003). Hypertrophic cardiomyopathy: two homozygous cases with "typical" hypertrophic cardiomyopathy and three new mutations in cases with progression to dilated cardiomyopathy. *Biochem. Biophys. Res. Commun.* **309,** 391–398.

143. Robinson, P., Mirza, M., Knott, A., Abdulrazzak, H., Willott, R., Marston, S., Watkins, H., and Redwood, C. (2002). Alterations in thin filament regulation induced by a human cardiac troponin T mutant that causes dilated cardiomyopathy are distinct from those induced by troponin T mutants that cause hypertrophic cardiomyopathy. *J. Biol. Chem.* **277,** 40710–40716.

144. Li, D., Czernuszewicz, G. Z., Gonzalez, O., Tapscott, T., Karibe, A., Durand, J. B., Brugada, R., Hill, R., Gregoritch, J. M., Anderson, J. L., Quinones, M., Bachinski, L. .L, and Roberts, R. (2001). Novel cardiac troponin T mutation as a cause of familial dilated cardiomyopathy. *Circulation* **104,** 2188–2193.

145. Chang, A. N., Harada, K., Ackerman, M. J., and Potter, J. D. (2005). Functional consequences of hypertrophic and dilated cardiomyopathy-causing mutations in alpha-tropomyosin. *J. Biol. Chem.* **280,** 34343–34349.

146. Mirza, M., Marston, S., Willott, R., Ashley, C., Mogensen, J., McKenna, W., Robinson, P., Redwood, C., and Watkins, H. (2005). Dilated cardiomyopathy mutations in three thin filament regulatory proteins result in a common functional phenotype. *J. Biol. Chem.* **280,** 28498–28506.

147. Siu, B. L., Niimura, H., Osborne, J. A., Fatkin, D., MacRae, C., Solomon, S., Benson, D. W., Seidman, J. G., and Seidman, C. E. (1999). Familial dilated cardiomyopathy locus maps to chromosome 2q31. *Circulation* **99,** 1022–1026.

148. Itoh-Satoh, M., Hayashi, T., Nishi, H., Koga, Y., Arimura, T., Koyanagi, T., Takahashi, M., Hohda, S., Ueda, K., Nouchi, T., Hiroe, M., Marumo, F., Imaizumi, T., Yasunami, M., and Kimura, A. (2002). Titin mutations as the molecular basis for dilated cardiomyopathy. *Biochem. Biophys. Res. Commun.* **291,** 385–393.

149. Inagaki, N., Hayashi, T., Arimura, T., Koga, Y., Takahashi, M., Shibata, H., Teraoka, K., Chikamori, T., Yamashina, A., and Kimura, A. (2006). Alpha B-crystallin mutation in dilated cardiomyopathy. *Biochem. Res. Commun.* **342,** 379–386.

150. Knoll, R., Hoshijima, M., Hoffman, H. M., Person, V., Lorenzen-Schmidt, I., Bang, M. L., Hayashi, T., Shiga, N., Yasukawa, H., Schaper, W., McKenna, W., Yokoyama, M., Schork, N. J., Omens, J. H., McCulloch, A. D., Kimura,

A., Gregorio, C. C., Poller, W., Schaper, J., Schultheiss, H. P., and Chien, K.R. (2002). The cardiac mechanical stretch sensor machinery involves a Z disc complex that is defective in a subset of human dilated cardiomyopathy. *Cell* **111**, 943–955.

151. Vajsar, J., Becker, L. E., Freedom, R. M., and Murphy, E. G. (1993). Familial desminopathy: myopathy with accumulation of desmin-type intermediate filaments. *J. Neurol. Neurosurg. Psychiatry* **56**, 644–648.

152. Muntoni, F., Catani, G., Mateddu, A., Rimoldi, M., Congiu, T., Faa, G., Marrosu, M. G., Cianchetti, C., and Porcu, M. (1994). Familial cardiomyopathy, mental retardation and myopathy associated with desmin-type intermediate filaments. *Neuromuscul. Disord.* **4**, 233–241.

153. Li, D., Tapscoft, T., Gonzalez, O., Burch, P. E., Quinones, M. A., Zoghbi, W. A., Hill, R., Bachinski, L. L., Mann, D. L., and Roberts, R. (1999). Desmin mutation responsible for idiopathic dilated cardiomyopathy. *Circulation* **100**, 461–464.

154. Miyamoto, Y., Akita, H., Shiga, N., Takai, E., Iwai, C., Mizutani, K., Kawai, H., Takarada, A., and Yokoyama, M. (2001). Frequency and clinical characteristics of dilated cardiomyopathy caused by desmin gene mutation in a Japanese population. *Eur. Heart J.* **22**, 2284–2289.

155. Zachara, E., Bertini, E., Lioy, E., Boldrini, R., Prati, P. L., and Bosman, C. (1997). Restrictive cardiomyopathy due to desmin accumulation in a family with evidence of autosomal dominant inheritance. *G. Ital. Cardiol.* **27**, 436–442.

156. Arbustini, E., Morbini, P., Grasso, M., Fasani, R., Verga, L., Bellini, O., Dal Bello, B., Campana, C., Piccolo, G., Febo, O., Opasich, C., Gavazzi, A., and Ferrans, V. J. (1998). Restrictive cardiomyopathy, atrioventricular block and mild to subclinical myopathy in patients with desmin-immunoreactive material deposits. *J. Am. Coll. Cardiol.* **31**, 645–653.

157. Nigro, V., de Sa Moreira, E., Piluso, G., Vainzof, M., Belsito, A., Politano, L., Puca, A. A., Passos-Bueno, M. R., and Zatz, M. (1996). Autosomal recessive limb-girdle muscular dystrophy, LGMD2F, is caused by a mutation in the delta-sarcoglycan gene. *Nat. Genet.* **14**, 195–198.

158. Muchir, A., Bonne, G., van der Kooi, A. J., van Meegen, M., Baas, F., Bolhuis, P. A., de Visser, M., and Schwartz, K. (2000). Identification of mutations in the gene encoding lamins A/C in autosomal dominant limb girdle muscular dystrophy with atrioventricular conduction disturbances (LGMD1B). *Hum. Mol. Genet.* **9**, 1453–1459.

159. Minetti, C., Sotgia, F., Bruno, C., Scartezzini, P., Broda, P., Bado, M., Masetti, E., Mazzocco, M., Egeo, A., Donati, M. A., Volonte, D., Galbiati, F., Cordone, G., Bricarelli, F. D., Lisanti, M. P., and Zara, F. (1998). Mutations in the caveolin-3 gene cause autosomal dominant limb-girdle muscular dystrophy. *Nat. Genet.* **18**, 365–368.

160. Karkkainen, S., Miettinen, R., Tuomainen, P., Karkkainen, P., Helio, T., Reissell, E., Kaartinen, M., Toivonen, L., Nieminen, M. S., Kuusisto, J., Laakso, M., and Peuhkurinen, K. (2003). A novel mutation, Arg71Thr, in the delta-sarcoglycan gene is associated with dilated cardiomyopathy. *J. Mol. Med.* **81**, 795–800.

161. Rudiger, M., Korneeva Schwienbacher, C., Weiss, E. E., and Jockusch, B. M. (1998). Differential actin organization by vinculin isoforms; implications for cell type-specific microfilament anchorage. *FEBs Lett.* **431**, 49–54.

162. Olson, T. M., Illenberger, S., Kishimoto, N. Y., Huttelmaier, S., Keating, M. T., and Jockusch, B. M. (2002). Metavinculin mutations alter actin interaction in dilated cardiomyopathy. *Circulation* **105**, 431–437.

163. Genschel, J., and Schmidt, H. H. J. (2000). Mutations in the LMNA gene encoding lamin A/C. *Hum. Mutat.* **16**, 451–459.

164. Sebillon, P., Bouchier, C., Bidot, L. D., Bonne, G., Ahamed, K., Charron, P., Drouin-Garraud, V., Millaire, A., Desrumeaux, G., Benaiche, A., Charniot, J. C., Schwartz, K., Villard, E., and Komajda, M. (2003). Expanding the phenotype of LMNA mutations in dilated cardiomyopathy and functional consequences of these mutations. *J. Med. Genet.* **40**, 560–567.

165. Bonne, G., Mercuri, E., Muchir, A., Urtizberea, A., Becane, H. M., Recan, D., Merlini, L., Wehnert, M., Boor, R., Reuner, U., Vorgerd, M., Wicklein, E. M., Eymard, B., Duboc, D., Penisson-Besnier, I., Cuisset, J. M., Ferrer, X., Desguerre, I., Lacombe, D., Bushby, K., Pollitt, C., Toniolo, D., Fardeau, M., Schwartz, K., and Muntoni, F. (2000). Clinical and molecular genetic spectrum of autosomal dominant Emery-Dreifuss muscular dystrophy due to mutations of the lamin A/C gene. *Ann. Neurol.* **48**, 170–180.

166. Wehnert, M. S., and Bonne, G. (2002). The nuclear muscular dystrophies. *Semin. Pediatr. Neurol.* **9**, 100–107.

167. Manilal, S., Recan, D., Sewry, C. A., Hoeltzenbein, M., Llense, S., Leturcq, F., Deburgrave, N., Barbot, J., Man, N., Muntoni, F., Wehnert, M., Kaplan, J., and Morris, G. E. (1998). Mutations in Emery-Dreifuss muscular dystrophy and their effects on emerin protein expression. *Hum. Mol. Genet.* **7**, 855–864.

168. Fairley, E. A., Riddell, A., Ellis, J. A., and Kendrick-Jones, J. (2002). The cell cycle dependent mislocalisation of emerin may contribute to the Emery-Dreifuss muscular dystrophy phenotype. *J. Cell Sci.* **115**, 341–354.

169. Vohanka, S., Vytopil, M., Bednarik, J., Lukas, Z., Kadanka, Z., Schildberger, J., Ricotti, R., Bione, S., and Toniolo, D. (2001). A mutation in the X-linked Emery-Dreifuss muscular dystrophy gene in a patient affected with conduction cardiomyopathy. *Neuromuscul. Disord.* **11**, 411–413.

170. Wilson, K. L., Holaska, J. M., de Oca, R. M., Tifft, K., Zastrow, M., Segura-Totten, M., Mansharamani, M., and Bengtsson, L. (2005). Nuclear membrane protein emerin: roles in gene regulation, actin dynamics and human disease. *Novartis Found. Symp.* **264**, 51–58.

171. Bonne, G., Di Barletta, M. R., Varnous, S., Becane, H. M., Hammouda, E. H., Merlini, L., Muntoni, F., Greenberg, C. R., Gary, F., Urtizberea, J. A., Duboc, D., Fardeau, M., Toniolo, D., and Schwartz, K. (1999). Mutations in the gene encoding lamin A/C cause autosomal dominant Emery-Dreifuss muscular dystrophy. *Nat. Genet.* **21**, 285–288.

172. Becane, H. M., Bonne, G., Varnous, S., Muchir, A., Ortega, V., Hammouda, E. H., Urtizberea, J. A., Lavergne, T., Fardeau, M., Eymard, B., Weber, S., Schwartz, K., and Duboc, D. (2000). High incidence of sudden death with conduction system and myocardial disease due to lamins A and C gene mutation. *Pacing Clin. Electrophysiol.* **23**, 1661–1666.

173. Sakata, K., Shimizu, M., Ino, H., Yamaguchi, M., Terai, H., Fujino, N., Hayashi, K., Kaneda, T., Inoue, M., Oda, Y., Fujita, T., Kaku, B., Kanaya, H., and Mabuchi, H. (2005). High incidence of sudden cardiac death with conduction

disturbances and atrial cardiomyopathy caused by a nonsense mutation in the STA gene. *Circulation* **111**, 3352–3358.
174. Garg, A., Speckman, R. A., and Bowcock, A. M. (2002). Multisystem dystrophy syndrome due to novel missense mutations in the amino-terminal head and alpha-helical rod domains of the lamin A/C gene. *Am. J. Med.* **112**, 549–555.
175. Raharjo, W. H., Enarson, P., Sullivan, T., Stewart, C. L., and Burke, B. (2001). Nuclear envelope defects associated with LMNA mutations cause dilated cardiomyopathy and Emery-Dreifuss muscular dystrophy. *J. Cell Sci.* **114**, 4447–4457.
176. Ostlund, C., Bonne, G., Schwartz, K., and Worman, H. J. (2001). Properties of lamin A mutants found in Emery-Dreifuss muscular dystrophy, cardiomyopathy and Dunnigan-type partial lipodystrophy. *J. Cell Sci.* **114**, 4435–4445.
177. Hart, K. A., Hodgson, S., Walker, A., Cole, C. G., Johnson, L., Dubowitz, V., and Bobrow, M. (1987). DNA deletions in mild and severe Becker muscular dystrophy. *Hum. Genet.* **75**, 281–285.
178. Koenig, M., Hoffman, E. P., Bertelson, C. J., Monaco, A. P., Feener, C., and Kunkel, L. M. (1987). Complete cloning of the Duchenne muscular dystrophy (DMD) cDNA and preliminary genomic organization of the DMD gene in normal and affected individuals. *Cell* **50**, 509–517.
179. Towbin, J. A., Hejtmancik, J. .F, Brink, P., Gelb,. B, Zhu, X. M., Chamberlain, J. S., McCabe, E. R. B., and Swift, M. (1993). X-linked dilated cardiomyopathy. Molecular genetic evidence of linkage to the Duchenne muscular dystrophy (dystrophin) gene at the Xp21 locus. *Circulation* **87**, 1854–1865.
180. Arbustini, E., Diegoli, M., Morbini, P., Dal Bello, B., Banchieri, N., Pilotto, A., Magani, F., Grasso, M., Narula, J., Gavazzi, A., Vigano, M., and Tavazzi, L. (2000). Prevalence and characteristics of dystrophin defects in adult male patients with dilated cardiomyopathy. *J. Am. Coll. Cardiol.* **35**, 1760–1768.
181. Muntoni, F., Torelli, S., and Ferlini, A. (2003). Dystrophin and mutations: one gene, several proteins, multiple phenotypes. *Lancet Neurol.* **2**, 731–740.
182. Franz, W. M., Muller, M., Muller, O. J., Herrmann, R., Rothmann, T., Cremer, M., Cohn, R. D., Voit, T., and Katus, H. A. (2000). Association of nonsense mutation of dystrophin gene with disruption of sarcoglycan complex in X-linked dilated cardiomyopathy. *Lancet* **355**, 1781–1785.
183. Muntoni, F., Wilson, L., Marrosu, G., Marrosu, M. G., Cianchetti, C., Mestroni, L., Ganau, A., Dubowitz, V., and Sewry, C. (1995). A mutation in the dystrophin gene selectively affecting dystrophin expression in the heart. *J. Clin. Invest.* **96**, 693–699.
184. Barth, P. G., Valianpour, F., Bowen, V. M., Lam, J., Duran, M., Vaz, F. M., and Wanders, R. J. (2004). X-linked cardioskeletal myopathy and neutropenia (Barth syndrome): an update. *Am. J. Med. Genet. A* **126**, 349–354.
185. Vreken, P., Valianpour, F., Nijtmans, L. G., Grivell, L. A., Plecko, B., Wanders, R. J., and Barth, P. G. (2000). Defective remodeling of cardiolipin and phosphatidylglycerol in Barth syndrome. *Biochem. Biophys. Res. Commun.* **279**, 378–382.
186. Johnston, J., Kelley, R. I., Feigenbaum, A., Cox, G. F., Iyer, G. S., Funanage, V. L., and Proujansky, R. (1997). Mutation characterization and genotype-phenotype correlation in Barth syndrome. *Am. J. Hum. Genet.* **61**, 1053–105.8

187. Ichida, F., Tsubata, S., Bowles, K. R., Haneda, N., Uese, K., Miyawaki, T., Dreyer, W. J., Messina, J., Li, H., Bowles, N. E., and Towbin, J. (2001). A. Novel gene mutations in patients with left ventricular noncompaction or Barth syndrome. *Circulation* **103**, 1256–1263.
188. D'Adamo, P., Fassone, L., Gedeon, A., Janssen, E. A., Bione, S., Bolhuis, P. A., Barth, P. G., Wilson, M., Haan, E., Orstavik, K. H., Patton, M. A., Green, A. J., Zammarchi, E., Donati, M. A., and Toniolo, D. (1997). The X-linked gene G4.5 is responsible for different infantile dilated cardiomyopathies. *Am. J. Hum. Genet.* **61**, 862–867.
189. Bleyl, S. B., Mumford, B. R., Thompson, V., Carey, J. C., Pysher, T. J., Chin, T. K., and Ward, K. (1997). Neonatal lethal noncompaction of the left ventricular myocardium is allelic with Barth syndrome. *Am. J. Hum. Genet.* **61**, 868–872.
190. Sasse-Klaassen, S., Gerull, B., Oechslin, E., Jenni, R., and Thierfelder, L. (2003). Isolated noncompaction of the left ventricular myocardium in the adult is an autosomal dominant disorder in the majority of patients. *Am. J. Med. Genet. A* **119**, 162–167.
191. Bissler, J. J., Tsoras, M., Goring, H. H., Hug, P., Chuck, G., Tombragel, E., McGraw, C., Schlotman, J., Ralston, M. A., and Hug, G. (2002). Infantile dilated X-linked cardiomyopathy, G4.5 mutations, altered lipids, and ultrastructural malformations of mitochondria in heart, liver, and skeletal muscle. *Lab. Invest.* **82**, 335–344.
192. Milner, D. J., Weitzer, G., Tran, D., Bradley, A., ad Capetanaki, Y. (1996). Disruption of muscle architecture and myocardial degeneration in mice lacking desmin. *J. Cell Biol.* **134**, 1255–1270.
193. Nigro, V., Okazaki, Y., Belsito, A., Piluso, G., Matsuda, Y., Politano, L., Nigro, G., Ventura, C., Abbondanza, C., Molinari, A. M., Acampora, D., Nishimura, M., Hayashizaki, Y., and Puca, G. A. (1997). Identification of the Syrian hamster cardiomyopathy gene. *Hum. Mol. Genet.* **6**, 601–607.
194. Sakamoto, A., Ono, K., Abe, M., Jasmin, G., Eki, T., Murakami, Y., Masaki, T., Toyo-oka, T., and Hanaoka, F. (1997). Both hypertrophic and dilated cardiomyopathies are caused by mutation of the same gene, delta-sarcoglycan, in hamster: an animal model of disrupted dystrophin-associated glycoprotein complex. *Proc. Natl. Acad. Sci. USA* **94**, 13873–13878.
195. Coral-Vazquez, R., Cohn, R. D., Moore, S. A., Hill, J. A., Weiss, R. M., Davisson, R. L., Straub, V., Barresi, R., Bansal, D., Hrstka, R. F., Williamson, R., and Campbell, K. P. (1999). Disruption of the sarcoglycan-sarcospan complex in vascular smooth muscle: a novel mechanism for cardiomyopathy and muscular dystrophy. *Cell* **98**, 465–474.
196. Arber, S., Hunter, J. J., Ross, J. Jr., Hongo, M., Sansig, G., Borg, J., Perriard, J. C., Chien, K. R., and Caroni, P. (1997). MLP-deficient mice exhibit a disruption of cardiac cytoarchitectural organization, dilated cardiomyopathy, and heart failure. *Cell* **88**, 393–403.
197. Liggett, S. B., Tepe, N. M., Lorenz, J. N., Canning, A. M., Jantz, T. D., Mitarai, S., Yatani, A., and Dorn, G. W. 2nd. (2000). Early and delayed consequences of beta(2)-adrenergic receptor overexpression in mouse hearts: critical role for expression level. *Circulation* **101**, 1707–1714.
198. Lemire, I., Ducharme, A., Tardif, J. C., Poulin, F., Jones, L. R., Allen, B. G., Hebert, T. E., and Rindt, H. (2001).

Cardiac-directed overexpression of wild-type alpha1B-adrenergic receptor induces dilated cardiomyopathy. *Am. J. Physiol. Heart Circ. Physiol.* **281,** H931–H938.

199. Antos, C. L., Frey, N., Marx, S. O., Reiken, S., Gaburjakova, M., Richardson, J. A., Marks, A. R., and Olson, E. N. (2001). Dilated cardiomyopathy and sudden death resulting from constitutive activation of protein kinase a. *Circ. Res.* **89,** 997–1004.

200. Colbert, M. C., Hall, D. G., Kimball, T. R., Witt, S. A., Lorenz, J. N., Kirby, M. L., Hewett, T. E., Klevitsky, R., and Robbins, J. (1997). Cardiac compartment-specific overexpression of a modified retinoic acid receptor produces dilated cardiomyopathy and congestive heart failure in transgenic mice. *J. Clin. Invest.* **100,** 1958–1968.

201. D'Angelo, D. D., Sakata, Y., Lorenz, J. N., Boivin, G. P., Walsh, R. A., Liggett, S. B., and Dorn, G. W., 2nd. (1997). Transgenic Galphaq overexpression induces cardiac contractile failure in mice. *Proc. Natl. Acad. Sci. USA* **94,** 8121–8126.

202. Baker, A. J., Redfern, C. H., Harwood, M. D., Simpson, P. .C, and Conklin, B. R. (2001). Abnormal contraction caused by expression of G(i)-coupled receptor in transgenic model of dilated cardiomyopathy. *Am. J. Physiol. Heart Circ. Physiol.* **280,** H1653–H1659.

203. James, J. F., Hewett, T. E., and Robbins, J. (1998). Cardiac physiology in transgenic mice. *Circ. Res.* **82,** 407–415.

204. Haghighi, K., Kolokathis, F., Pater, L., Lynch, R. A., Asahi, M., Gramolini, A. O., Fan, G. C., Tsiapras, D., Hahn, H. S., Adamopoulos, S., Liggett, S. B., Dorn, G. W., II., MacLennan, D. H., Kremastinos, D. T., and Kranias, E. G. (2003). Human phospholamban null results in lethal dilated cardiomyopathy revealing a critical difference between mouse and human. *J. Clin. Invest.* **111,** 869–876.

205. Walther, T., Steendijk, P., Westermann, D., Hohmann, C., Schulze, K., Heringer-Walther, S., Schultheiss, H.P., and Tschope, C. (2004). Angiotensin deficiency in mice leads to dilated cardiomyopathy. *Eur. J. Pharmacol.* **493,** 161–165.

206. Emanueli, C., Maestri, R., Corradi, D., Marchione, R., Minasi, A., Tozzi, M. G., Salis, M. B., Straino, S., Capogrossi, M. C., Olivetti, G., and Madeddu, P. (1999). Dilated and failing cardiomyopathy in bradykinin B(2) receptor knockout mice. *Circulation* **100,** 2359–2365.

207. Yan, X., Price, R. L., Nakayama, M., Ito, K., Schuldt, A. J., Manning, W. J., Sanbe, A., Borg, T. K., Robbins, J., and Lorell, B. H. (2003). Ventricular-specific expression of angiotensin II type 2 receptors causes dilated cardiomyopathy and heart failure in transgenic mice. *Am. J. Physiol. Heart Circ. Physiol.* **285,** H2179–H2187.

208. Sakata, Y., Kamei, C. N., Nakagami, H., Bronson, R., Liao, J. K., and Chin, M. T. (2002). Ventricular septal defect and cardiomyopathy in mice lacking the transcription factor CHF1/Hey2. *Proc. Natl. Acad. Sci. USA* **99,** 16197–16202.

209. Eigenthaler, M., Engelhardt, S., Schinke, B., Kobsar, A., Schmitteckert, E., Gambaryan, S., Engelhardt, C. M., Krenn, V., Eliava, M., Jarchau, T., Lohse, M. J., Walter, U., and Hein, L. (2003). Disruption of cardiac Ena-VASP protein localization in intercalated disks causes dilated cardiomyopathy. *Am. J. Physiol. Heart Circ. Physiol.* **285,** H2471–H2481.

210. Nikolova, V., Leimena, C., McMahon, A. C., Tan, J. C., Chandar, S., Jogia, D., Kesteven, S. H., Michalicek, J., Otway, R., Verheyen, F., Rainer, S., Stewart, C. L., Martin, D., Feneley, M. P., and Fatkin D. (2004). Defects in nuclear structure and function promote dilated cardiomyopathy in lamin A/C-deficient mice. *J. Clin. Invest.* **113,** 357–369.

211. Fedak, P. W., Smookler, D. S., Kassiri, Z., Ohno, N., Leco, K. J., Verma, S., Mickle, D. A., Watson, K., Hojilla, C. V., Cruz, W., Weisel, R. D., Li, R. K., and Khokha, R. (2004). TIMP-3 deficiency leads to dilated cardiomyopathy. *Circulation* **110,** 2401–2409.

212. Ladd, A. N., Taffet, G., Hartley, C., Kearney, D. L., and Cooper, T. A. (2005). Cardiac tissue-specific repression of CELF activity disrupts alternative splicing and causes cardiomyopathy. *Mol. Cell Biol.* **25,** 6267–6278.

213. Kubo, Y., Sekiya, S., Ohigashi, M., Takenaka, C., Tamura, K., Nada, S., Nishi, T., Yamamoto, A., and Yamaguchi, A. (2005). ABCA5 resides in lysosomes, and ABCA5 knockout mice develop lysosomal disease-like symptoms. *Mol. Cell Biol.* **25,** 4138–4149.

214. Kostetskii, I., Li, J., Xiong, Y., Zhou, R., Ferrari, V. A., Patel, V. V., Molkentin, J. D., and Radice, G. L. (2005). Induced deletion of the N-cadherin gene in the heart leads to dissolution of the intercalated disc structure. *Circ. Res.* **96,** 346–354.

215. Ding, J. H., Xu, X., Yang, D., Chu, P. H., Dalton, N. D., Ye, Z., Yeakley, J. M., Cheng, H., Xiao, R. P., Ross, J., Chen, J., and Fu, X. D. (2004). Dilated cardiomyopathy caused by tissue-specific ablation of SC35 in the heart. *EMBO J.* **23,** 885–896.

216. Ozcelik, C., Erdmann, B., Pilz, B., Wettschureck, N., Britsch, S., Hubner, N., Chien, K. R., Birchmeier, C., and Garratt, A. N. (2002). Conditional mutation of the ErbB2 (HER2) receptor in cardiomyocytes leads to dilated cardiomyopathy. *Proc. Natl. Acad. Sci. USA* **99,** 8880–8885.

217. Zhao, Y. Y., Liu, Y., Stan, R. V., Fan, L., Gu, Y., Dalton, N., Chu, P. H., Peterson, K., Ross, J. Jr., and Chien, K. R. (2002). Defects in caveolin-1 cause dilated cardiomyopathy and pulmonary hypertension in knockout mice. *Proc. Natl. Acad. Sci. USA* **99,** 11375–11380.

218. Fentzke, R. C., Korcarz, C. E., Lang, R. M., Lin, H., and Leiden, J. M. (1998). Dilated cardiomyopathy in transgenic mice expressing a dominant-negative CREB transcription factor in the heart. *J. Clin. Invest.* **101,** 2415–2426.

219. Li, Y., Huang, T. T., Carlson, E. J., Melov, S., Ursell, P. C., Olson, J. L., Noble, L. J., Yoshimura, M. P., Berger, C., Chan, P. H., Wallace, D. C., and Epstein, C. J. (1995). Dilated cardiomyopathy and neonatal lethality in mutant mice lacking manganese superoxide dismutase. *Nat. Genet.* **11,** 376–381.

220. Kaab, S., Barth, A. S., Margerie, D., Dugas, M., Gebauer, M., Zwermann, L., Merk, S., Pfeufer, A., Steinmeyer, K., Bleich, M., Kreuzer, E., Steinbeck, G., and Nabauer, M. (2004). Global gene expression in human myocardium-oligonucleotide microarray analysis of regional diversity and transcriptional regulation in heart failure. *J. Mol. Med.* **82,** 308–316.

221. Barrans, J. D., Allen, P. D., Stamatiou, D., Dzau, V. J., and Liew, C. C. (2002). Global gene expression profiling of end-stage dilated cardiomyopathy using a human cardiovascular-based cDNA microarray. *Am. J. Pathol.* **160,** 2035–2043.

222. Yang, J., Moravec, C. S., Sussman, M. A., DiPaola, N. R., Fu, D., Hawthorn, L., Mitchell, C. A., Young, J. B., Francis, G. S., McCarthy, P. M., and Bond, M. (2000). Decreased

SLIM1 expression and increased gelsolin expression in failing human hearts measured by high-density oligonucleotide arrays. *Circulation* **102**, 3046–3052.
223. Thiene, G., Basso, C., Calabrese, F., Angelini, A., and Valente, M. (2005). Twenty years of progress and beckoning frontiers in cardiovascular pathology: cardiomyopathies. *Cardiovasc. Pathol.* **14**, 165–169.
224. Hughes, S. E., and McKenna, W. J. (2005). New insights into the pathology of inherited cardiomyopathy. *Heart* **91**, 257–264.
225. Yazdani, K., Maraj, S., and Amanullah, A. M. (2005). Differentiating constrictive pericarditis from restrictive cardiomyopathy. *Rev. Cardiovasc. Med.* **6**, 61–71.
226. Zhang, J., Kumar, A., Kaplan, L., Fricker, F. J., and Wallace, M. R. (2005). Genetic linkage of a novel autosomal dominant restrictive cardiomyopathy locus. *J. Med. Genet.* **42**, 663–665.
227. Arbustini, E., Morbini, P., Grasso, M., Fasani, R., Verga, L., Bellini, O., Dal Bello, B., Campana, C., Piccolo, G., Febo, O., Opasich, C., Gavazzi, A., and Ferrans, V. J. (1998). Restrictive cardiomyopathy, atrioventricular block and mild to subclinical myopathy in patients with desmin-immunoreactive material deposits. *J. Am. Coll. Cardiol.* **31**, 645–653.
228. Zachara, E., Bertini, E., Lioy, E., Boldrini, R., Prati, P. L., and Bosman, C. (1997). Restrictive cardiomyopathy due to desmin accumulation in a family with evidence of autosomal dominant inheritance. *G. Ital. Cardiol.* **27**, 436–442.
229. Mogensen, J., Kubo, T., Duque, M., Uribe, W., Shaw, A., Murphy, R., Gimeno, J. R., Elliott, P., and McKenna, W. J. (2003). Idiopathic restrictive cardiomyopathy is part of the clinical expression of cardiac troponin I mutations. *J. Clin. Invest.* **111**, 209–216.
230. Gomes, A. V., Liang, J., and Potter, J. D. (2005). Mutations in human cardiac troponin I that are associated with restrictive cardiomyopathy affect basal ATPase activity and the calcium sensitivity of force development. *J. Biol. Chem.* **280**, 30909–30915.
231. Hassan, W., Al-Sergani, H., Mourad, W., and Tabbaa, R. (2005). Amyloid heart disease. New frontiers and insights in pathophysiology, diagnosis, and management. *Tex. Heart Inst. J.* **32**, 178–184.
232. Koike, H., Misu, K., Sugiura, M., Iijima, M., Mori, K., Yamamoto, M., Hattori, N., Mukai, E., Ando, Y., Ikeda, S., and Sobue, G. (2004). Pathology of early- vs late-onset TTR Met30 familial amyloid polyneuropathy. *Neurology* **63**, 129–138.
233. Blanco-Jerez, C. R., Jimenez-Escrig, A., Gobernado, J. M., Lopez-Calvo, S., de Blas, G., Redondo, C., Garcia Villanueva, M., and Orensanz, L. (1998). Transthyretin Tyr77 familial amyloid polyneuropathy: A clinicopathological study of a large kindred. *Muscle Nerve* **21**, 1478–1485.
234. Hamidi Asl, K., Nakamura, M., Yamashita, T., and Benson, M. D. (2001). Cardiac amyloidosis associated with the transthyretin Ile122 mutation in a Caucasian family. *Amyloid* **8**, 263–269.
235. Yamashita, T., Asl, K. H., Yazaki, M., and Benson, M. D. (2005). A prospective evaluation of the transthyretin Ile122 allele frequency in an African-American population. *Amyloid* **12**, 127–130.
236. Jiang, X., Buxbaum, J. N., and Kelly, J. W. (2001). The V122I cardiomyopathy variant of transthyretin increases the velocity of rate-limiting tetramer dissociation, resulting in accelerated amyloidosis. *Proc. Natl. Acad. Sci. USA* **98**, 14943–14948.
237. Burke, W., Press, N., and McDonnell, S. M. (1998). Hemochromatosis: genetics helps to define a multifactorial disease. *Clin. Genet.* **54**, 1–9.
238. Hanson, E. H., Imperatore, G., and Burke, W. (2001). HFE gene and hereditary hemochromatosis: a HuGE review. Human Genome Epidemiology. *Am. J. Epidemiol.* **154**, 193–206.
239. Papanikolaou, G., Samuels, M. E., Ludwig, E. H., MacDonald, M. L., Franchini, P. L., Dube, M. P., Andres, L., MacFarlane, J., Sakellaropoulos, N., Politou, M., Nemeth, E., Thompson, J., Risler, J. K., Zaborowska, C., Babakaiff, R., Radomski, C. C., Pape, T. D., Davidas, O., Christakis, J., Brissot, P., Lockitch, G., Ganz, T., Hayden, M. R., and Goldberg, Y. P. (2004). Mutations in HFE2 cause iron overload in chromosome 1q-linked juvenile hemochromatosis. *Nat. Genet.* **36**, 77–82.
240. Limdi, J. K., and Crampton, J. R. (2004). Hereditary haemochromatosis. *QJM* **97**, 315–324.
241. Thiene, G., Basso, C., Danieli, G., Rampazzo, A., Corrado, D., and Nava, A. (1997). Arrhythmogenic right ventricular cardiomyopathy. *Trends Cardiovasc. Med.* **7**, 84–90.
242. Hulot, J. S., Jouven, X., Empana, J. P., Frank, R., and Fontaine, G. (2004). Natural history and risk stratification of arrhythmogenic right ventricular dysplasia/cardiomyopathy. *Circulation* **110**, 1879–1884.
243. Corrado, D., Basso, C., Thiene, G., McKenna, W. J., Davies, M. J., Fontaliran, F., Nava, A., Silvestri, F., Blomstrom-Lundqvist, C., Wlodarska, E. K., Fontaine, G., and Camerini, F. (1997). Spectrum of clinicopathologic manifestations of arrhythmogenic right ventricular cardiomyopathy/dysplasia: a multicenter study. *J. Am. Coll. Cardiol.* **30**, 1512–1520.
244. Beffagna, G., Occhi, G., Nava, A., Vitiello, L., Ditadi, A., Basso, C., Bauce, B., Carraro, G., Thiene, G., Towbin, J. A., Danieli, G. A., and Rampazzo, A. (2005). Regulatory mutations in transforming growth factor-beta3 gene cause arrhythmogenic right ventricular cardiomyopathy type 1. *Cardiovasc. Res.* **65**, 366–373.
245. Rampazzo, A., Beffagna, G., Nava, A., Occhi, G., Bauce, B., Noiato, M., Basso, C., Frigo, G., Thiene, G., Towbin, J., and Danieli, G. A. (2003). Arrhythmogenic right ventricular cardiomyopathy type 1 (ARVD1): confirmation of locus assignment and mutation screening of four candidate genes. *Eur. J. Hum. Genet.* **11**, 69–76.
246. Tiso, N., Stephan, D. A., Nava, A., Bagattin, A., Devaney, J. M., Stanchi, F., Larderet, G., Brahmbhatt, B., Brown, K., Bauce, B., Muriago, M., Basso, C., Thiene, G., Danieli, G. A., and Rampazzo, A. (2001). Identification of mutations in the cardiac ryanodine receptor gene in families affected with arrhythmogenic right ventricular cardiomyopathy type 2 (ARVD2). *Hum. Mol. Genet.* **10**, 189–194.
247. Severini, G. M., Krajinovic, M., Pinamonti, B., Sinagra, G., Fioretti, P., Brunazzi, M. C., Falaschi, A., Camerini, F., Giacca, M., and Mestroni, L. (1996). A new locus for arrhythmogenic right ventricular dysplasia on the long arm of chromosome 14. *Genomics* **31**, 193–200.
248. Rampazzo, A., Nava, A., Miorin, M., Fonderico, P., Pope, B., Tiso, N., Livolsi, B., Zimbello, R., Thiene, G., and Danieli, G. A. (1997). ARVD4, a new locus for arrhythmogenic right

ventricular cardiomyopathy, maps to chromosome 2 long arm. *Genomics* **45**, 259–263.
249. Ahmad, F., Li, D., Karibe, A., Gonzalez, O., Tapscott, T., Hill, R., Weilbaecher, D., Blackie, P., Furey, M., Gardner, M., Bachinski, L. L., and Roberts, R. (1998). Localization of a gene responsible for arrhythmogenic right ventricular dysplasia to chromosome 3p23. *Circulation* **98**, 2791–2795.
250. Li, D., Ahmad, F., Gardner, M. J., Weilbaecher, D., Hill, R., Karibe, A., Gonzalez, O., Tapscott, T., Sharratt, G. P., Bachinski, L. L., and Roberts, R. (2000). The locus of a novel gene responsible for arrhythmogenic right-ventricular dysplasia characterized by early onset and high penetrance maps to chromosome 10p12-p14. *Am. J. Hum. Genet.* **66**, 148–156.
251. Melberg, A., Oldfors, A., Blomstrom-Lundqvist, C., Stalberg, E., Carlsson, B., Larsson, E., Lidell, C., Eeg-Olofsson, K. E., Wikstrom, G., Henriksson, K. G., and Dahl, N. (1999). Autosomal dominant myofibrillar myopathy with arrhythmogenic right ventricular cardiomyopathy linked to chromosome 10q. *Ann. Neurol.* **4**, 684–692.
252. Bauce, B., Nava, A., Rampazzo, A., Daliento, L., Muriago, M., Basso, C., Thiene, G., and Danieli, G. A. (2000). Familial effort polymorphic ventricular arrhythmias in arrhythmogenic right ventricular cardiomyopathy map to chromosome 1q42-43. *Am. J. Cardiol.* **85**, 573–579.
253. Alcalai, R., Metzger, S., Rosenheck, S., Meiner, V., and Chajek-Shaul, T. (2003). A recessive mutation in desmoplakin causes arrhythmogenic right ventricular dysplasia, skin disorder, and woolly hair. *J. Am. Coll. Cardiol.* **42**, 319–327.
254. Rampazzo, A., Nava, A., Malacrida, S., Beffagna, G., Bauce, B., Rossi, V., Zimbello, R., Simionati, B., Basso, C., Thiene, G., Towbin, J. A., and Danieli, G. A. (2002). Mutation in human desmoplakin domain binding to plakoglobin causes a dominant form of arrhythmogenic right ventricular cardiomyopathy. *Am. J. Hum. Genet.* **71**, 1200–1206.
255. Norman, M., Simpson, M., Mogensen, J., Shaw, A., Hughes, S., Syrris, P., Sen-Chowdhry, S., Rowland, E., Crosby, A., and McKenna, W. J. (2005). Novel mutation in desmoplakin causes arrhythmogenic left ventricular cardiomyopathy. *Circulation* **112**, 636–642.
256. Gerull, B., Heuser, A., Wichter, T., Paul, M., Basson, C. T., McDermott, D. A., Lerman, B. B., Markowitz, S. M., Ellinor, P. T., MacRae, C. A., Peters, S., Grossmann, K. S., Drenckhahn, J., Michely, B., Sasse-Klaassen, S., Birchmeier, W., Dietz, R., Breithardt, G., Schulze-Bahr, E., and Thierfelder, L. (2004). Mutations in the desmosomal protein plakophilin-2 are common in arrhythmogenic right ventricular cardiomyopathy. *Nat. Genet.* **36**, 1162–1164.
257. McKoy, G., Protonotarios, N., Crosby, A., Tsatsopoulou, A., Anastasakis, A., Coonar, A., Norman, M., Baboonian, C., Jeffery, S., and McKenna, W. J. (2000). Identification of a deletion in plakoglobin in arrhythmogenic right ventricular cardiomyopathy with palmoplantar keratoderma and woolly hair (Naxos disease). *Lancet* **355**, 2119–2124.
258. Pilichou, K., Nava, A., Basso, C., Beffagna, G., Bauce, B., Lorenzon, A., Frigo, G., Vettori, A., Valente, M., Towbin, J., Thiene, G., Danieli, G. A., and Rampazzo, A. (2006). Mutations in desmoglein-2 gene are associated with arrhythmogenic right ventricular cardiomyopathy. *Circulation* **113**, 1171–1179.
259. Marks, A. R., Priori, S., Memmi, M., Kontula, K., and Laitinen, P. J. (2002). Involvement of the cardiac ryanodine receptor/calcium release channel in catecholaminergic polymorphic ventricular tachycardia. *J. Cell Physiol.* **190**, 1–6.
260. Laitinen, P. J., Swan, H., Piippo, K., Viitasalo, M., Toivonen, L., and Kontula, K. (2004). Genes, exercise and sudden death: molecular basis of familial catecholaminergic polymorphic ventricular tachycardia. *Ann. Med.* **36**, 81–86.
261. Lehnart, S. E., Wehrens, X. H., Laitinen, P. J., Reiken, S. R., Deng, S. X., Cheng, Z., Landry, D. W., Kontula, K., Swan, H., and Marks, A. R. (2004). Sudden death in familial polymorphic ventricular tachycardia associated with calcium release channel (ryanodine receptor) leak. *Circulation* **109**, 3208–3214.
262. Jiang, D., Wang, R., Xiao, B., Kong, H., Hunt, D. J., Choi, P., Zhang, L., and Chen, S. R. (2005). Enhanced store overload-induced Ca2+ release and channel sensitivity to luminal Ca2+ activation are common defects of RyR2 mutations linked to ventricular tachycardia and sudden death. *Circ. Res.* **97**, 1173–1181.
263. d'Amati, G., Bagattin, A., Bauce, B., Rampazzo, A., Autore, C., Basso, C., King, K., Romeo, M. D., Gallo, P., Thiene, G., Danieli, G. A., and Nava, A. (2005). Juvenile sudden death in a family with polymorphic ventricular arrhythmias caused by a novel RyR2 gene mutation: evidence of specific morphological substrates. *Hum. Pathol.* **36**, 761–767.
264. Protonotarios, N., Tsatsopoulou, A., Anastasakis, A., Sevdalis, E., McKoy, G., Stratos, K., Gatzoulis, K., Tentolouris, K., Spiliopoulou, C., Panagiotakos, D., McKenna, W., and Toutouzas, P. (2001). Genotype-phenotype assessment in autosomal recessive arrhythmogenic right ventricular cardiomyopathy (Naxos disease) caused by a deletion in plakoglobin. *J. Am. Coll. Cardiol.* **38**, 1477–1484.
265. Kaplan, S. R., Gard, J. J., Protonotarios, N., Tsatsopoulou, A., Spiliopoulou, C., Anastasakis, A., Squarcioni, C. P., McKenna, W. J., Thiene, G., Basso, C., Brousse, N., Fontaine, G., and Saffitz, J. E. (2004). Remodeling of myocyte gap junctions in arrhythmogenic right ventricular cardiomyopathy due to a deletion in plakoglobin (Naxos disease). *Heart Rhythm* **1**, 3–11.
266. Yamaji, K., Fujimoto, S., Ikeda, Y., Masuda, K., Nakamura, S., Saito, Y., and Yutani, C. (2005). Apoptotic myocardial cell death in the setting of arrhythmogenic right ventricular cardiomyopathy. *Acta Cardiol.* **60**, 465–470.
267. Taniike, M., Fukushima, H., Yanagihara, I., Tsukamoto, H., Tanaka, J., Fujimura, H., Nagai, T., Sano, T., Yamaoka, K., and Inui, K. (1992). Mitochondrial tRNAIle mutation in fatal cardiomyopathy. *Biochem. Biophys. Res. Commun.* **186**, 47–53.
268. Silvestri, G., Santorelli, F. M., Shanske, S., Whitley, C. B., Schimmenti, L. A., Smith, S. A., and DiMauro, S. (1994). A new mtDNA mutation in the tRNALEU(UUR) gene associated with maternally inherited cardiomyopathy. *Hum. Mutat.* **3**, 37–43.
269. Santorelli, F. M., Mak, S. C., Vazquez-Acevedo, M., Gonzalez-Astiazaran, A., Ridaura-Sanz, C., Gonzalez-Halphen, D., and DiMauro, S. (1995). A novel mtDNA point mutation associated with mitochondrial encephalocardiomyopathy. *Biochem. Biophys. Res. Commun.* **216**, 835–840.
270. Santorelli, F. M., Mak, S. C., El-Schahawi, M., Casali, C., Shanske, S., Baram, T. Z., Madrid, R. E., and DiMauro, S. (1996). Maternally inherited cardiomyopathy and hearing loss

associated with a novel mutation in mitochondrial tRNALys gene (G8363). *Am. J. Hum. Genet.* **58,** 933–939.

271. Merante, F., Tein, I., Benson, L., and Robinson, B. H. (1994). Maternally inherited cardiomyopathy due to a novel T-to-C transition at nt 9997 in the mitochondrial tRNAGly gene. *Am. J. Hum. Genet.* **55,** 437–446.

272. Zeviani, M., Gellera, C., Antozzi, C., Rimoldi, M., Morandi, L., Villani, F., Tiranti, V., and DiDonato, S. (1991). Maternally inherited myopathy and cardiomyopathy: Association with mutation in mitochondrial DNA tRNALeu. *Lancet* **338,** 143–147.

273. Casali, C., Santorelli, F. M., D'Amati, G., Bernucci, P., DeBiase, L., and DiMauro, S. (1995). A novel mtDNA point mutation in maternally inherited cardiomyopathy. *Biochem. Biophys. Res. Commun.* **213,** 588–593.

274. Terasaki, F., Tanaka, M., Kawamura, K., Kanzaki, Y., Okabe, M., Hayashi, T., Shimomura, H., Ito, T., Suwa, M., Gong, J. S., Zhang, J., and Kitaura, Y. (2001). A case of cardiomyopathy showing progression from the hypertrophic to the dilated form: Association of Mt8348A→G mutation in the mitochondrial tRNA(Lys) gene with severe ultra-structural alterations of mitochondria in cardiomyocytes. *Jpn. Circ. J.* **65,** 691–694.

275. Akita, Y., Koga, Y., Iwanaga, R., Wada, N., Tsubone, J., Fukuda, S., Nakamura, Y., and Kato, H. (2000). Fatal hypertrophic cardiomyopathy associated with an A8296G mutation in the mitochondrial tRNA(Lys) gene. *Hum. Mutat.* **15,** 382.

276. Merante, F., Myint, T., Tein, I., Benson, L., and Robinson, B. H. (1996). An additional mitochondrial tRNA(Ile) point mutation (A-to-G at nucleotide 4295) causing hypertrophic cardiomyopathy. *Hum. Mutat.* **8,** 216–222.

277. Schon, E. A., Bonilla, E., and DiMauro, S. (1997). Mitochondrial DNA mutations and pathogenesis. *J. Bioenerg. Biomembr.* **29,** 131–149.

278. Santorelli, F. M., Tanji, K., Manta, P., Casali, C., Krishna, S., Hays, A. P., Mancini, D. M., DiMauro, S., and Hirano, M. (1999). Maternally inherited cardiomyopathy: An atypical presentation of the mtDNA 12S rRNA gene A1555G mutation. *Am. J. Hum. Genet.* **64,** 295–300.

279. Hsieh, R. H., Li, J. Y., Pang, C. Y., and Wei, Y. H. (2001). A novel mutation in the mitochondrial 16S rRNA gene in a patient with MELAS syndrome, diabetes mellitus, hyperthyroidism and cardiomyopathy. *J. Biomed. Sci.* **8,** 328–335.

280. Li, Y. Y., Maisch, B., Rose, M. L., and Hengstenberg, C. (1997). Point mutations in mitochondrial DNA of patients with dilated cardiomyopathy. *J. Mol. Cell Cardiol.* **29,** 2699–2709.

281. Ozawa, T., Katsumata, K., Hayakawa, M., Tanaka, M., Sugiyama, S., Tanaka, T., Itoyama, S., Nunoda, S., and Sekiguchi, M. (1995). Genotype and phenotype of severe mitochondrial cardiomyopathy: A recipient of heart transplantation and the genetic control. *Biochem. Biophys. Res. Commun.* **207,** 613–619.

282. Arbustini, E., Diegoli, M., Fasani, R., Grasso, M., Morbini, P., Banchieri, N., Bellini, O., Dal Bello, B., Pilotto, A., Magrini, G., Campana, C., Fortina, P., Gavazzi, A., Narula, J., and Vigano, M. (1998). Mitochondrial DNA mutations and mitochondrial abnormalities in dilated cardiomyopathy. *Am. J. Pathol.* **153,** 1501–1510.

283. Marín-García, J., Goldenthal, M. J., Ananthakrishnan, R., and Pierpont, M. E. (2000). The complete sequence of mtDNA genes in idiopathic dilated cardiomyopathy shows novel missense and tRNA mutations. *J. Card. Fail.* **6,** 321–329.

284. Taylor, R. W., Giordano, C., Davidson, M. M., d'Amati, G., Bain, H., Hayes, C. M., Leonard, H., Barron, M. J., Casali, C., Santorelli, F. M., Hirano, M., Lightowlers, R. N., DiMauro, S., and Turnbull, D. M. (2003). A homoplasmic mitochondrial transfer ribonucleic acid mutation as a cause of maternally inherited hypertrophic cardiomyopathy. *J. Am. Coll. Cardiol.* **41,** 1786–1796.

285. Carelli, V., Giordano, C., and d'Amati, G. (2003). Pathogenic expression of homoplasmic mtDNA mutations needs a complex nuclear-mitochondrial interaction. *Trends Genet.* **19,** 257–262.

286. Marín-García, J., and Goldenthal, M. J. (2000). Mitochondrial biogenesis defects and neuromuscular disorders. *Pediatr. Neurol.* **22,** 122–129.

287. Silvestri, G., Bertini, E., Servidei, S., Rana, M., Zachara, E., Ricci, E., and Tonali, P. (1997). Maternally inherited cardiomyopathy: A new phenotype associated with the A to G at nt 3243 in mitochondrial DNA. *Muscle Nerve* **20,** 221–225.

288. Zeviani, M., Moraes, C. T., DiMauro, S., Nakase, H., Bonilla, E., Schon, E. A., and Rowland, L. P. (1988), Deletions of mitochondrial DNA in Kearns-Sayre syndrome. *Neurology* **38,** 1339–1346.

289. Suomalainen, A., Kaukonen, J., Amati, P., Timonen, R., Haltia, M., Weissenbach, J., Zeviani, M., Somer, H., and Peltonen, L. (1995). An autosomal locus predisposing to deletions of mitochondrial DNA. *Nat. Genet.* **9,** 146–151.

290. Bohlega, S., Tanji, K., Santorelli, F. M., Hirano, M., al-Jishi, A., and DiMauro, S. (1996). Multiple mitochondrial DNA deletions associated with autosomal recessive ophthalmoplegia and severe cardiomyopathy. *Neurology* **46,** 1329–1334.

291. Suomalainen, A., Paetau, A., Leinonen, H., Majander, A., Peltonen, L., and Somer, H. (1992). Inherited idiopathic dilated cardiomyopathy with multiple deletions of mitochondrial DNA. *Lancet* **340,** 1319–1320.

292. Marín-García, J., Goldenthal, M. J., Ananthakrishnan, R., Pierpont, M., Fricker, F. J., Lipshultz, S. E., and Perez-Atayde, A. (1996). Specific mitochondrial DNA deletions in idiopathic dilated cardiomyopathy. *Cardiovasc. Res.* **31,** 306–314.

293. Li, Y. Y., Hengstenberg, C., and Maisch, B. (1995). Whole mitochondrial genome amplification reveals basal level multiple deletions in mt-DNA of patients with dilated cardiomyopathy. *Biochem. Biophys. Res. Commun.* **210,** 211–218.

294. Corral-Debrinski, M., Stepien, G., Shoffner, J. M., Lott, M. T., Kanter, K., and Wallace, D. C. (1991). Hypoxemia is associated with mitochondrial DNA damage and gene induction: Implications for cardiac disease. *JAMA* **266,** 1812–1816.

295. Esposito, L. A., Melov, S., Panov, A., Cottrell, B. A., and Wallace, D. C. (1999). Mitochondrial disease in mouse results in increased oxidative stress. *Proc. Natl. Acad. Sci. USA* **96,** 4820–4825.

296. Kaukonen, J., Juselius, J. K., Tiranti, V., Kyttala, A., Zeviani, M., Comi, G. P., Keranen, S., Peltonen, L., and Suomalainen, A. (2000). Role of adenine nucleotide translocator 1 in mtDNA maintenance. *Science* **289,** 782–785.

297. Lewis, W., and Dalakas, M. C. (1995). Mitochondrial toxicity of antiviral drugs. *Nat. Med.* **1,** 417–422.
298. Herskowitz, A., Willoughby, S. B., Baughman, K. L., Schulman, S. P., and Bartlett, J. D. (1992). Cardiomyopathy associated with antiretroviral therapy in patients with HIV infection: A report of 6 cases. *Ann. Intern. Med.* **116,** 311–313.
299. Marín-García, J., Ananthakrishnan, R., and Goldenthal, M. J. (1998). Hypertrophic cardiomyopathy with mitochondrial DNA depletion and respiratory enzyme defects. *Pediatr. Cardiol.* **19,** 266–268.
300. Huizing, M., Iacobazzi, V., Ijlst, L., Savelkoul, P., Ruitenbeek, W., van den Heuvel, L., Indiveri, C., Smeitink, J., Trijbels, F., Wanders, R., and Palmieri, F. (1997). Cloning of the human carnitine-acylcarnitine carrier cDNA and identification of the molecular defect in a patient. *Am. J. Hum. Genet.* **61,** 1239–1245.
301. Babcock, M., de Silva, D., Oaks, R., Davis-Kaplan, S., Jiralerspong, S., Montermini, L., Pandolfo, M., and Kaplan, J. (1997). Regulation of mitochondrial iron accumulation by Yfh1p, a putative homolog of frataxin. *Science* **276,** 1709–1712.
302. Campuzano, V., Montermini, L., Molto, M. D., Pianese, L., Cossee, M., Cavalcanti, F., Monros, E., Rodius, F., Duclos, F., and Monticelli, A. (1996). Friedreich's ataxia: Autosomal recessive disease caused by an intronic GAA triplet repeat expansion. *Science* **271,** 1423–1427.
303. Lodi, R., Rajagopalan, B., Blamire, A. M, Cooper, J. M., Davies, C. H., Bradley, J. L., Styles, P., and Schapira, A. H. (2001). Cardiac energetics are abnormal in Friedreich ataxia patients in the absence of cardiac dysfunction and hypertrophy: An *in vivo* 31P magnetic resonance spectroscopy study. *Cardiovasc. Res.* **52,** 111–119.
304. Davey, K. M., Parboosingh, J. S., McLeod, D. R., Chan, A., Casey, R., Ferreira, P., Snyder, F. F., Bridge, P. J., and Bernier, F. P. (2006). Mutation of DNAJC19, a human homolog of yeast inner mitochondrial membrane co-chaperones, causes DCMA syndrome, a novel autosomal recessive Barth syndrome-like condition. *J. Med. Genet.* **43,** 385–393.
305. Benit, P., Beugnot, R., Chretien, D., Giurgea, I., De Lonlay-Debeney, P., Issartel, J. P., Corral-Debrinski, M., Kerscher, S., Rustin, P., Rotig, A., and Munnich, A. (2003). Mutant NDUFV2 subunit of mitochondrial complex I causes early onset hypertrophic cardiomyopathy and encephalopathy. *Hum. Mutat.* **21,** 582–586.
306. Loeffen, J., Elpeleg, O., Smeitink, J., Smeets, R., Stockler-Ipsiroglu, S., Mandel, H., Sengers, R., Trijbels, F., and van den Heuvel, L. (2001). Mutations in the complex I NDUFS2 gene of patients with cardiomyopathy and encephalomyopathy. *Ann. Neurol.* **49,** 195–201.
307. Papadopoulou, L. C., Sue, C. M., Davidson, M. .M, Tanji, K., Nishino, I., Sadlock, J. E., Krishna, S., Walker, W., Selby, J., Glerum, D. M., Coster, R. V., Lyon, G., Scalais, E., Lebel, R., Kaplan, P., Shanske, S., De Vivo, D. C., Bonilla, E., Hirano, M., DiMauro, S., and Schon, E. A. (1999). Fatal infantile cardioencephalomyopathy with COX deficiency and mutations in SCO2, a COX assembly gene. *Nat. Genet.* **23,** 333–337.
308. Antonicka, H., Leary, S. C., Guercin, G. H., Agar, J. N., Horvath, R., Kennaway, N. G., Harding, C. O., Jaksch, M., and Shoubridge, E. A. (2003). Mutations in COX10 result in a defect in mitochondrial heme A biosynthesis and account for multiple, early-onset clinical phenotypes associated with isolated COX deficiency. *Hum. Mol. Genet.* **12,** 2693–2702.
309. Kelly, D. P., and Strauss, A. W. (1994). Inherited cardiomyopathies. *N. Engl. J. Med.* **330,** 913–919.
310. Strauss, A. W., Powell, C. K., Hale, D. E., Anderson, M. M., Ahuja, A., Brackett, J. C., and Sims, H. F. (1995). Molecular basis of human mitochondrial very-long-chain acyl CoA dehydrogenase deficiency causing cardiomyopathy and sudden death in childhood. *Proc. Natl. Acad. Sci. USA* **92,** 10496–10500.
311. Rocchiccioli, F., Wanders, R. J., Aubourg, P., Vianey-Liaud, C., Ijlst, L., Fabre, M., Cartier, N., and Bougneres, P. F. (1990). Deficiency of long-chain 3-hydroxyacyl CoA dehydrogenase: A cause of lethal myopathy and cardiomyopathy in early childhood. *Pediatr. Res.* **28,** 657–662.
312. Taroni, F., Verderio, E., Fiorucci, S., Cavadini, P., Finocchiaro, G., Uziel, G., Lamantea, E., Gellera, C., and DiDonato, S. (1992). Molecular characterization of inherited carnitine palmitoyltransferase II deficiency. *Proc. Natl. Acad. Sci. USA* **89,** 8429–8433.
313. D'Adamo, P., Fassone, L., Gedeon, A., Janssen, E. A., Bione, S., Bolhuis, P. A., Barth, P. G., Wilson, M., Haan, E., Orstavik, K. H., Patton, M. A., Green, A. J., Zammarchi, E., Donati, M. A., and Toniolo, D. (1997). The X-linked gene G4.5 is responsible for different infantile dilated cardiomyopathies. *Am. J. Hum. Genet.* **61,** 862–867.
314. Schlame, M., Kelley, R. I., Feigenbaum, A., Towbin, J. A., Heerdt, P. M., Schieble, T., Wanders, R. J., DiMauro, S., and Blanck, T. J. (2003). Phospholipid abnormalities in children with Barth syndrome. *J. Am. Coll. Cardiol.* **42,** 1994–1999.
315. Vreken, P., Valianpour, F., Nijtmans, L. G., Grivell, L. A., Plecko, B., Wanders, R. J., and Barth, P. G. (2000). Defective remodeling of cardiolipin and phosphatidylglycerol in Barth syndrome. *Biochem. Biophys. Res. Commun.* **279,** 378–382.
316. Graham, B. H., Waymire, K. G., Cottrell, B., Trounce, I. A., MacGregor, G. R., and Wallace, D. C. (1997). A mouse model for mitochondrial myopathy and cardiomyopathy resulting from a deficiency in heart/muscle isoform of the adenine nucleotide translocator. *Nat. Genet.* **16,** 226–234.
317. Nahrendorf, M., Spindler, M., Hu, K., Bauer, L., Ritter, O., Nordbeck, P., Quaschning, T., Hiller, K. H., Wallis, J., Ertl, G., Bauer, W. R., and Neubauer, S. (2005). Creatine kinase knockout mice show left ventricular hypertrophy and dilatation, but unaltered remodeling post-myocardial infarction. *Cardiovasc. Res.* **65,** 419–427.
318. Wallace, D. C. (2002). Animal models for mitochondrial disease. *Methods Mol. Biol.* **197,** 3–54.
319. Ibdah, J. A., Paul, H., Zhao, Y., Banford, S., Salleng, K., Cline, M., Matern, D., Bennett, M. J., Rinaldo, P., and Strauss, A. W. (2001). Lack of mitochondrial trifunctional protein in mice causes neonatal hypoglycemia and sudden death. *J. Clin. Invest.* **107,** 1403–1409.
320. Wang, J., Wilhelmsson, H., Graff, C., Li, H., Oldfors, A., Rustin, P., Bruning, J. C., Kahn, C. R., Clayton, D. A., Barsh, G. S., Thoren, P., and Larsson, N. G. (1999). Dilated cardiomyopathy and atrioventricular conduction blocks induced by heart-specific inactivation of mitochondrial gene expression. *Nat. Genet.* **21,** 133–137.

321. Puccio, H., Simon, D., Cossee, M., Criqui-Filipe, P., Tiziano, F., Melki, J., Hindelang, C., Matyas, R., Rustin, P., and Koenig, M. (2001). Mouse models for Friedreich ataxia exhibit cardiomyopathy, sensory nerve defect and Fe-S enzyme deficiency followed by intramitochondrial iron deposits. *Nat. Genet.* **27**, 181–186.
322. Ingwall, J. S. (2004). Transgenesis and cardiac energetics: new insights into cardiac metabolism. *J. Mol. Cell Cardiol.* **37**, 613–623.
323. Fountoulakis, M., Soumaka, E., Rapti, K., Mavroidis, M., Tsangaris, G., Maris, A., Weisleder, N., and Capetanaki, Y. (2005). Alterations in the heart mitochondrial proteome in a desmin null heart failure model. *J. Mol. Cell Cardiol.* **38**, 461–474.
324. van den Bosch, B. J., van den Burg, C. M., Schoonderwoerd, K., Lindsey, P. J., Scholte, H. R., de Coo, R. F., van Rooij, E., Rockman, H. A., Doevendans, P. A., and Smeets, H. J. (2005). Regional absence of mitochondria causing energy depletion in the myocardium of muscle LIM protein knockout mice. *Cardiovasc. Res.* **65**, 411–418.
325. Li, H., Wilhelmsson, H., Hanson, A., Thoren, P., Duffy, J., Rustin, P., and Larsson, N. (2000). Genetic modification of survival in tissue-specific knockout mice with mitochondrial cardiomyopathy. *Proc. Natl. Acad. Sci. USA* **97**, 3467–3472.
326. Radford, N. B., Wan, B., Richman, A., Szczepaniak, L. S., Li, J. L., Li, K., Pfeiffer, K., Schagger, H., Garry, D. J., and Moreadith, R. W. (2002). Cardiac dysfunction in mice lacking cytochrome-c oxidase subunit VIaH. *Am. J. Physiol. Heart Circ. Physiol.* **282**, H726–H733.
327. Finck, B. N., Lehman, J. J., Leone, T. C., Welch, M. J., Bennett, M. J., Kovacs, A., Han, X., Gross, R. W., Kozak, R., Lopaschuk, G. D., and Kelly, D. P. (2002). The cardiac phenotype induced by PPARalpha overexpression mimics that caused by diabetes mellitus. *J. Clin. Invest.* **109**, 121–130.
328. Russell, L. K., Mansfield, C. M., Lehman, J. J., Kovacs, A., Courtois, M., Saffitz, J. E., Medeiros, D. M., Valencik, M. L., McDonald, J. A., and Kelly, D. P. (2004). Cardiac-specific induction of the transcriptional coactivator peroxisome proliferator-activated receptor gamma coactivator-1alpha promotes mitochondrial biogenesis and reversible cardiomyopathy in a developmental stage-dependent manner. *Circ. Res.* **94**, 525–533.
329. Lehman, J. J., Barger, P. M., Kovacs, A., Saffitz, J. E., Medeiros, D. M., and Kelly, D. P. (2000). Peroxisome proliferator-activated receptor gamma coactivator-1 promotes cardiac mitochondrial biogenesis. *J. Clin. Invest.* **106**, 847–856.
330. Sayen, M. R., Gustafsson, A. B., Sussman, M. A., Molkentin, J. D., and Gottlieb, R. A. (2003). Calcineurin transgenic mice have mitochondrial dysfunction and elevated superoxide production. *Am. J. Physiol. Cell Physiol.* **284**, C562–C570.
331. Nebigil, C. G., Jaffre, F., Messaddeq, N., Hickel, P., Monassier, L., Launay, J. M., and Maroteaux, L. (2003). Overexpression of the serotonin 5-HT2B receptor in heart leads to abnormal mitochondrial function and cardiac hypertrophy. *Circulation* **107**, 3223–3229.
332. Melov, S., Schneider, J. A., Day, B. J., Hinerfeld, D., Coskun, P., Mirra, S. S., Crapo, J. D., and Wallace, D. C. (1998). A novel neurological phenotype in mice lacking mitochondrial manganese superoxide dismutase. *Nat. Genet.* **18**, 159–163.
333. Seznec, H., Simon, D., Bouton, C., Reutenauer, L., Hertzog, A., Golik, P., Procaccio, V., Patel, M., Drapier, J. C., Koenig, M., and Puccio, H. (2005). Friedreich ataxia: the oxidative stress paradox. *Hum. Mol. Genet.* **14**, 463–474.
334. Weisleder, N., Taffet, G. E., and Capetanaki, Y. (2004). Bcl-2 overexpression corrects mitochondrial defects and ameliorates inherited desmin null cardiomyopathy. *Proc. Natl. Acad. Sci. USA* **101**, 769–774.
335. Fang, Z. Y., Prins, J. B., and Marwick, T. H. (2004). Diabetic cardiomyopathy: evidence, mechanisms, and therapeutic implications. *Endocr. Rev.* **25**, 543–567.
336. Razeghi, P., Young, M. E., Ying, J., Depre, C., Uray, I. P., Kolesar, J., Shipley, G. L., Moravec, C. S., Davies, P. J., Frazier, O. H., and Taegtmeyer, H. (2002). Downregulation of metabolic gene expression in failing human heart before and after mechanical unloading. *Cardiology* **97**, 203–209.
337. Razeghi, P., Young, M. E., Cockrill, T. C., Frazier, O. H., and Taegtmeyer, H. (2002). Downregulation of myocardial myocyte enhancer factor 2C and myocyte enhancer factor 2C-regulated gene expression in diabetic patients with nonischemic heart failure. *Circulation* **106**, 407–411.
338. Stanley, W. C., Lopaschuk, G. D., and McCormack, J. G. (1997). Regulation of energy substrate metabolism in the diabetic heart. *Cardiovasc. Res.* **34**, 25–33.
339. Avogaro, A., Vigili de Kreutzenberg, S., Negut, C., Tiengo, A., and Scognamiglio, R. (2004). Diabetic cardiomyopathy: A metabolic perspective. *Am. J. Cardiol.* **93**, 13A–16A.
340. Seager, M. J., Singal, P. K., Orchard, R., Pierce, G. N., and Dhalla, N. S. (1984). Cardiac cell damage: A primary myocardial disease in streptozotocin-induced chronic diabetes. *Br. J. Exp. Pathol.* **65**, 613–623.
341. Mokhtar, N., Lavoie, J. P., Rousseau-Migneron, S., and Nadeau, A. (1993). Physical training reverses defect in mitochondrial energy production in heart of chronically diabetic rats. *Diabetes* **42**, 682–687.
342. Tomita, M., Mukae, S., Geshi, E., Umetsu, K., Nakatani, M., and Katagiri, T. (1996). Mitochondrial respiratory impairment in streptozotocin induced diabetic rat heart. *Jpn. Circ. J.* **60**, 673–682.
343. Rotig, A., Bonnefont, J. P., and Munnich, A. (1996). Mitochondrial diabetes mellitus. *Diabetes Metab.* **22**, 291–298.
344. Maassen, J. A., 'T Hart, L. M., Van Essen, E., Heine, R. J., Nijpels, G., Jahangir Tafrechi, R. S., Raap, A. K., Janssen, G. M., and Lemkes, H. H. (2004). Mitochondrial diabetes: Molecular mechanisms and clinical presentation. *Diabetes* **53**, S103–S109.
345. Hattori, Y., Nakajima, K., Eizawa, T., Ehara, T., Koyama, M., Hirai, T., Fukuda, Y., and Kinoshita, M. (2003). Heteroplasmic mitochondrial DNA 3310 mutation in NADH dehydrogenase subunit 1 associated with type 2 diabetes, hypertrophic cardiomyopathy, and mental retardation in a single patient. *Diabetes Care* **26**, 952–953.
346. Suzuki, S., Oka, Y., Kadowaki, T., Kanatsuka, A., Kuzuya, T., Kobayashi, M., Sanke, T., Seino, Y., and Nanjo, K. (2003). Clinical features of diabetes mellitus with the mitochondrial DNA 3243 (A-G) mutation in Japanese: Maternal inheritance and mitochondria-related complications. *Diabetes Res. Clin. Pract.* **59**, 207–217.
347. Hsieh, R. H., Li, J. Y., Pang, C. Y., and Wei, Y. H. (2001). A novel mutation in the mitochondrial 16S rRNA gene in a patient with MELAS syndrome, diabetes mellitus, hyperthyroidism and cardiomyopathy. *J. Biomed. Sci.* **8**, 328–335.

348. Shen, X., Zheng, S., Thongboonkerd, V., Xu, M., Pierce, Jr., W. M., Klein, J. B., and Epstein, P. N. (2004). Cardiac mitochondrial damage and biogenesis in a chronic model of type I diabetes. *Am. J. Physiol. Endocrinol. Metab.* **287,** E896–E905.
349. Spies, C. D., Sander, M., Stangl, K., Fernandez-Sola, J., Preedy, V. R., Rubin, E., Andreasson, S., Hanna, E. Z., and Kox, W. J. (2001). Effects of alcohol on the heart. *Curr. Opin. Crit. Care* **7,** 337–343.
350. Piano, M. R. (2002). Alcoholic cardiomyopathy: incidence, clinical characteristics, and pathophysiology. *Chest* **121,** 1638–16350.
351. Eriksson, C. J. (2001). The role of acetaldehyde in the actions of alcohol (update 2000). *Alcohol Clin. Exp. Res.* **25,** 15S–32S.
352. Zhang, X., Li, S. Y., Brown, R. A., and Ren, J. (2004). Ethanol and acetaldehyde in alcoholic cardiomyopathy: from bad to ugly en route to oxidative stress. *Alcohol* **32,** 175–186.
353. Fernandez-Sola, J., Nicolas, J. M., Oriola, J., Sacanella, E., Estruch, R., Rubin, E., and Urbano-Marque, A. (2002). Angiotensin-converting enzyme gene polymorphism is associated with vulnerability to alcoholic cardiomyopathy. *Ann. Intern. Med.* **137,** 321–326.
354. Kajander, O. A., Kupari, M., Perola, M., Pajarinen, J., Savolainen, V., Penttila, A., and Karhunen, P. J. (2001). Testing genetic susceptibility loci for alcoholic heart muscle disease. *Alcohol Clin. Exp. Res.* **25,** 1409–1413.
355. Jones, W. K. (2005). A murine model of alcoholic cardiomyopathy: a role for zinc and metallothionein in fibrosis. *Am. J. Pathol.* **167,** 301–304.
356. Wang, L., Zhou, Z., Saari, J. T., and Kang, Y. J. (2005). Alcohol-induced myocardial fibrosis in metallothionein-null mice: prevention by zinc supplementation. *Am. J. Pathol.* **167,** 337–344.
357. Jankala, H., Eriksson, P. C., Eklund, K., Sarviharju, M., Harkonen, M., and Maki, T. (2005). Effect of chronic ethanol ingestion and gender on heart left ventricular p53 gene expression. *Alcohol Clin. Exp. Res.* **29,** 1368–1373.
358. Hintz, K. K., Relling, D. P., Saari, J. T., Borgerding, A. J., Duan, J., Ren, B. H., Kato, K., Epstein, P. N., and Ren, J. (2003). Cardiac overexpression of alcohol dehydrogenase exacerbates cardiac contractile dysfunction, lipid peroxidation, and protein damage after chronic ethanol ingestion. *Alcohol Clin. Exp. Res.* **27,** 1090–1098.
359. Duan, J., McFadden, G. E., Borgerding, A. J., Norby, F. L., Ren, B. H., Ye, G., Epstein, P. N., and Ren, J. (2002). Overexpression of alcohol dehydrogenase exacerbates ethanol-induced contractile defect in cardiac myocytes. *Am. J. Physiol. Heart Circ. Physiol.* **282,** H1216–H1222.
360. Cunha-Neto, E., Dzau, V. J., Allen, P. D., Stamatiou, D., Benvenutti, L., Higuchi, M. L., Koyama, N. S., Silva, J. S., Kalil, J., and Liew, C. C. (2005). Cardiac gene expression profiling provides evidence for cytokinopathy as a molecular mechanism in Chagas' disease cardiomyopathy. *Am. J. Pathol.* **167,** 305–313.
361. Sterin-Borda, L., Gorelik, G., Postan, M., Gonzalez Cappa, S., and Borda, E. (1999). Alterations in cardiac beta-adrenergic receptors in chagasic mice and their association with circulating beta-adrenoceptor-related autoantibodies. *Cardiovasc. Res.* **41,** 116–225.
362. Joensen, L., Borda, E., Kohout, T., Perry, S., Garcia, G., and Sterin-Borda, L. (2003). *Trypanosoma cruzi* antigen that interacts with the beta1-adrenergic receptor and modifies myocardial contractile activity. *Mol. Biochem. Parasitol.* **127,** 169–177.
363. Garcia, G. A., Joensen, L. G., Bua, J., Ainciart, N., Perry, S. J., and Ruiz, A. M. (2003). *Trypanosoma cruzi:* Molecular identification and characterization of new members of the Tc13 family. Description of the interaction between the Tc13 antigen from Tulahuen strain and the second extracellular loop of the beta(1)-adrenergic receptor. *Exp. Parasitol.* **103,** 112–119.
364. Sterin-Borda, L., Joensen, L., Bayo-Hanza, C., Esteva, M., and Borda, E. (2002). Therapeutic use of muscarinic acetylcholine receptor peptide to prevent mice chagasic cardiac dysfunction. *J. Mol. Cell Cardiol.* **34,** 1645–1654.
365. Nieto, A., Beraun, Y., Collado, M. D. Caballero, A., Alonso, A., Gonzalez, A., and Martin, J. (2000). HLA haplotypes are associated with differential susceptibility to Trypanosoma cruzi infection. *Tissue Antigens* **55,** 195–198.
366. Cruz-Robles, D., Reyes, P. A., Monteon-Padilla, V. M., Ortiz-Muniz, A. R., and Vargas-Alarcon, G. (2004). MHC class I and class II genes in Mexican patients with Chagas disease. *Hum. Immunol.* **65,** 60–65.
367. Messias-Reason, I. J., Urbanetz, L., and Pereira da Cunha, C. (2003). Complement C3 F and BF S allotypes are risk factors for Chagas disease cardiomyopathy. *Tissue Antigens* **62,** 308–312.
368. Cohen, A. W., Park, D. S., Woodman, S. E., Williams, T. M., Chandra, M., Shirani, J., Pereira de Souza, A., Kitsis, R. N., Russell, R. G., Weiss, L. M., Tang, B., Jelicks, L. A., Factor, S. M., Shtutin, V., Tanowitz, H. B., and Lisanti, M. P. (2003). Caveolin-1 null mice develop cardiac hypertrophy with hyperactivation of p42/44 MAP kinase in cardiac fibroblasts. *Am. J. Physiol. Cell Physiol.* **284,** C457–C474.
369. Woodman, S. E., Park, D. S., Cohen, A. W., Cheung, M. W., Chandra, M., Shirani, J., Tang, B., Jelicks, L. A., Kitsis, R. N., Christ, G. J., Factor, S. M., Tanowitz, H. B., and Lisanti, M. P. (2002). Caveolin-3 knock-out mice develop a progressive cardiomyopathy and show hyperactivation of the p42/44 MAPK cascade. *J. Biol. Chem.* **277,** 38988–38997.
370. Ohsawa, Y., Toko, H., Katsura, M., Morimoto, K., Yamada, H., Ichikawa, Y., Murakami, T., Ohkuma, S., Komuro, I., and Sunada, Y. (2004). Overexpression of P104L mutant caveolin-3 in mice develops hypertrophic cardiomyopathy with enhanced contractility in association with increased endothelial nitric oxide synthase activity. *Hum. Mol. Genet.* **13,** 151–157.
371. O'Cochlain, D. F., Perez-Terzic, C., Reyes, S., Kane, G. C., Behfar, A., Hodgson, D. M., Strommen, J. A., Liu, X. K., van den Broek, W., Wansink, D. G., Wieringa, B., and Terzic, A. (2004). Transgenic overexpression of human DMPK accumulates into hypertrophic cardiomyopathy, myotonic myopathy and hypotension traits of myotonic dystrophy. *Hum. Mol. Genet.* **13,** 2505–2518.
372. Braz, J. C., Bueno, O. F., Liang, Q., Wilkins, B. J., Dai, Y. S., and Parsons, S. (2003). Targeted inhibition of p38 MAPK promotes hypertrophic cardiomyopathy through upregulation of calcineurin-NFAT signaling. *J. Clin. Invest.* **111,** 1475–1486.

CHAPTER 14

Heart Response to Inflammation and Infection

OVERVIEW

Myocarditis is an inflammatory disease of the myocardium associated with cardiac dysfunction and can be a precursor to dilated cardiomyopathy (DCM). Because of its variable clinical presentation, ranging from latent subacute presentation to congestive heart failure (HF) and sudden death (SD), its prevalence is still unknown. A wide range of pathogenic agents, as well as toxic chemicals and drugs, can cause myocarditis, often acting in concert with autoimmune and inflammatory mechanisms within the host. Its diagnosis depends largely on the immunohistochemical detection of inflammatory infiltrates in endomyocardial biopsy specimens, according to the Dallas criteria, in combination with clinical data. The use of molecular techniques, such as PCR, gene sequencing, *in situ* hybridization, and real-time PCR, often applied to the same endomyocardial biopsied specimen, represents important breakthroughs in diagnosis, providing rapid, specific, and sensitive identification of infective agents and important information about pathogenesis that may allow a finely targeted treatment.

Infective endocarditis is an infection of the endocardium usually involving valves and adjacent structures, and it may be caused by a wide variety of bacteria and fungi. Molecular information available concerning the pathogenesis of endocarditis, disease progression, its diagnosis and treatment, as well as available information concerning the pathogenesis, epidemiology, molecular diagnosis, and treatment of myocarditis, will be presented in this chapter.

INTRODUCTION

In the preceding chapter, we examined primary and metabolic cardiomyopathies, focusing primarily on the familial/genetic causes and the genomic/post-genomic approach to their analysis. In this chapter, we will discuss the molecular underpinnings of cardiac sequelae resulting from specific myocardial infection and inflammation (i.e., endocarditis and myocarditis).

Although myocarditis is considered here as a separate entity, we are fully aware that there can be a considerable overlap between the myocarditis occurring from inflammatory damage and the DCM that can result. In this chapter, we will seek to identify the molecular correlates and pathways in the onset of both the inflammatory and immune response in myocarditis, endocarditis, and other disorders involving viral/bacterial infections of the heart, as well as the identification of the potential role of susceptibility genes leading to these processes. In addition, we will critically examine the information coming from numerous clinical and animal studies, as well as which diagnostic and therapeutic strategies need to be applied for in the different stages of the diseases.

MYOCARDITIS

Myocarditis is an inflammatory disorder of the myocardium that is associated with cardiac dysfunction. It can be associated with SD, particularly in younger individuals, and is a frequent precursor of DCM. The clinical presentation of myocarditis is highly variable, with some patients exhibiting fulminant disease, including acute heart failure (HF) and severe dysrhythmias, whereas many present with minimal symptoms or are entirely asymptomatic. The heterogeneity of symptoms is a primary factor, leading to the uncertainty regarding the true prevalence of myocarditis, which is most likely underestimated. The wide spectrum of clinical presentation reported with myocarditis may be explained by the interaction of several determining factors, including infectious agents, genetic susceptibility, age, gender, and the immunocompetence of the host.[1] Given the variability in clinical presentation, histopathology is, therefore, a cornerstone of diagnosis, albeit more recently molecular analysis has proved to be an important tool.

There are a large variety of agents, both infectious and noninfectious, that can cause myocarditis.[2] Among the infectious agents that are listed in Table I, the most prevalent in infectious myocarditis are viral pathogens, including the human enteroviruses (i.e., coxsackieviruses) and the adenoviruses. In noninfectious myocarditis, inflammation occurs in the absence of myocardial infection and may be a result of autoimmune disease, drug-induced hypersensitivity, neoplasia, and other systemic disorders. Furthermore, cases involving infectious

TABLE I	Etiological Agents of Infectious Myocarditis
Virus	Picornavirus (enteroviruses), adenoviruses, influenza, paramyxovirus (rubeola, mumps), herpes (cytomegalovirus, Epstein–Barr virus, varicella-zoster), retrovirus (HIV), hepatitis C virus, parvovirus
Bacteria	Chlamydiae (*C. pneumoniae, C. trachomatis*), *Mycoplasma (pneumoniae), Borrelia, Salmonella, Shigella, Streptococcus, Meningococcus, Staphylococcus, Campylobacter, Corynebacterium diphtheriae*
Mycobacteria	Tuberculosis
Fungi	Candida, aspergillus, cryptococcus
Protozoa	*Trypanosoma cruzi, Toxoplasma gondii*
Parasites	*Trichinella spinalis*
Rickettsiae	*Rickettsia rickettsii, R. conorii, Orientia tsutsugamushi, Coxiella burnetii*

myocarditis often have an autoimmune component or stage in their progression.

Giant cell myocarditis is a rare idiopathic inflammatory disease characterized by a history of severe, rapidly progressive HF, and refractory ventricular dysrhythmias. It is associated with a poor prognosis, a distinct histological presentation, and is most commonly featured in patients with latent or symptomatic autoimmune disease.

By reviewing both the commonalities and unique features of the various forms of infectious myocarditis, insights concerning the pathogenetic mechanism(s) can be confirmed, and this can lead to better diagnosis and treatment. Myocarditis can be considered a progressive disease with three distinct successive stages.[3] In the first phase, an initial myocardial insult occurs, often because of a viral infection; HF rarely occurs during this phase, and the initial insult may go unnoticed. The second phase develops as a result of autoimmunity triggered by the initial injury and often is accompanied by the development of congestive HF, resulting from additional injury. In the third phase, DCM may develop and proceed despite the absence/cessation of the first two processes. On the other hand, spontaneous recovery of cardiac function may also occur.

Viral Myocarditis

ENTEROVIRUSES

The key viruses that have been linked to myocarditis and DCM are the human enteroviruses. The viruses most commonly found associated with this disease are the Coxsackie B family of single-stranded RNA viruses and, in particular, the plus-strand RNA virus Coxsackievirus B3 and Coxsackievirus B5.[2] Early events in viral infection include the attachment of the virus onto cell surface receptors. Coxsackieviruses of group B (CBV) specifically interact with at least two receptor proteins, the coxsackievirus-adenovirus receptor (CAR) and its coreceptor, the decay-accelerating factor (DAF) encoded by *CD55*, and cause a broad spectrum of clinical presentations, including acute and chronic myocarditis; as the name implies, this receptor is shared by both the coxsackievirus B and adenovirus pathogens.[4,5] In the human heart, CAR is predominantly expressed in the intercalated discs, regions of utmost importance for the functional integrity of the organ. Because DAF is abundantly expressed in epithelial and endothelial cells, interaction of cardiotropic CVB with the DAF coreceptor protein, in addition to CAR, enhances viral internalization.[6,7]

Although, DAF and Coxsackievirus adenovirus receptor proteins (CAR) have been identified as receptors for Coxsackievirus B3, the exact mechanisms that Coxsackievirus B3 and B5 use to infect the cardiac muscle are not yet known, neither is the natural function of the receptor. Recent attempts to inhibit Coxsackievirus B3 and Coxsackievirus B5 infectivity of cardiac cells by use of CAR and DAF specific antibodies have demonstrated that these antibodies could not completely inhibit Coxsackievirus B3 and Coxsackievirus B5 binding or infectivity.[8] Moreover, five unrelated proteins have been identified that are used by Coxsackieviruses for binding to cardiac tissue, which are distinct from CAR or DAF, indicating that these viruses may use a number of receptors for infection of cardiac muscle.

Although a broad spectrum of cardiotropic viruses clearly can replicate in the heart, they also share another common feature; they are notoriously difficult to isolate in infectious form during chronic infection (a failure in fulfilling the Koch postulates for rigorously identifying a disease-causing entity). However, recent methodological advances for the analysis of nucleic acids using nested, real-time PCR and RT-PCR have enabled the sensitive and specific detection of viral genomes in endomyocardial-biopsied tissues. In a recent study, viral persistence in patient tissues of enteroviral RNA, adenovirus DNA, parvovirus B19 DNA, and human herpesvirus type 6 DNA genomes was significantly associated with a progressive impairment of left ventricular contractile function, whereas spontaneous viral elimination (which occurred in approximately 50% of patients) was associated with a significant improvement in LV function.[9]

Although these viruses are known to be cytopathic viruses that can cause death of the host cell, their viral RNA has been shown to persist in cardiac muscle, contributing to a chronic inflammatory cardiomyopathy. Both the severity of viral myocarditis and the different age-related clinical phenotypes have been associated with viral-triggered inflammatory responses.[5]

The precise mechanism by which coxsackie B viruses (CBVs) trigger this inflammatory response is poorly understood. Studies of the involvement of Toll-like receptors (TLRs) in the recognition of CBV virions, as well as CBV single-stranded RNA, have shown that the CBV-induced inflammatory response is mediated through TLR8, and to a lesser extent through TLR7.[10]

Investigations at the molecular level have revealed that a concerted interference of coxsackievirus replication with the host's cellular metabolism is achieved by cleavage of host cell proteins by virus-encoded proteinases. A pivotal aspect of the pathogenetic mechanism by which enteroviruses operate involves the targeting of the cytoskeletal protein dystrophin, whose genetic defects we have previously described as leading to familial DCM. The enteroviral protease 2A directly cleaves dystrophin in the hinge 3 region, leading to functional dystrophin impairment.[11] On infection of mice with coxsackievirus B3, the myocardial dystrophin–glycoprotein complex is disrupted, and sarcolemmal integrity is lost in virus-infected cardiomyocytes. Moreover, dystrophin deficiency markedly increases enterovirus-induced cardiomyopathy *in vivo*, suggesting a pathogenetic role of the dystrophin cleavage in this type of cardiomyopathy.[12] These findings have been extended to isolated cases of DCM resulting from coxsackievirus B2 myocarditis. Analysis of endomyocardial biopsy specimens showed an inflammatory infiltrate and myocytolysis. In addition, immunostaining for the enteroviral capsid antigen VP1 revealed virus-infected cardiomyocytes. Focal areas of cardiomyocytes displayed a loss of the sarcolemmal staining pattern for dystrophin and β-sarcoglycan identical to previous findings in virus-infected mouse hearts. *In vitro* studies have also demonstrated that coxsackievirus B3 protease 2A cleaved human dystrophin.[13] Interestingly, an interaction between the genetic and acquired forms of cardiomyopathy has recently been described in which mice with dystrophin mutation are predisposed to enteroviral-induced cardiomyopathy. On viral infection, these mice exhibited enhanced viral replication and more severe cardiac dysfunction, presumably as a function of greater viral release.[14] Furthermore, expression of a wild-type dystrophin or of a cleavage-resistant dystrophin significantly decreased both the cytopathic effect and release of the virus. These findings suggest that viral infection can impact both the severity and penetrance of inherited dystrophin-deficient cardiomyopathy.

OTHER VIRUSES

With the increased use of molecular techniques, other viruses associated with myocarditis have been detected in high prevalence. For example, endomyocardial biopsy samples from patients with peri-partum cardiomyopathy have revealed that a significant proportion (more than 30%) of patients harbored viral genomes, including parvovirus B19, human herpes virus 6, Epstein–Barr virus, and human cytomegalovirus, that immunohistologically were associated with interstitial inflammation.[15]

Gathered observations had suggested a link of mumps infection and endocardial fibroelastosis (EFE), previously a significant cause of infant mortality often resulting in HF and death, together with myocarditis in its early stages. This suggestion was based on the dramatic decline of the incidence of EFE, which corresponded chronologically to the availability of the mumps vaccine, and the decline in mumps-related disease. In addition, it was supported by positive serological studies of mumps in EFE, although direct evidence of a viral etiology for EFE was lacking, because an infectious virus could not be cultured in these patients. This observation has been further supported by a molecular study of biopsied myocardial tissue in a group of children with EFE, in which more than 70% harbored mumps viral RNA.[16]

Furthermore, myocarditis associated with respiratory tract viruses such as adenoviruses, Epstein–Barr viruses, and influenza has also been reported with variable frequency. Interestingly, whereas in several studies adenovirus infection (which shares the same CAR receptor with coxsackie virus) seems to be significantly prevalent in myocarditis and DCM in both the young and adults,[9,17] other have not confirmed this finding.[18] It is also worthy to note that the detection of a virus/viral genome is not in itself proof of a causal relationship with myocarditis, because such agents may be present as "innocent bystanders"[2] or as a result of secondary infections. It is evident that a more rigorous proof of their pathogenicity is required before assigning them a role in the etiology of myocarditis.

Increasingly, parvovirus B19 (PVB19) has been identified as a possible cause of myocarditis and HF in both children and adults. Several lines of evidence support the assignment of cardiotropic properties to PVB19, because PVB19 DNA has been found in fetal myocardial cells. Moreover, parvovirus infection during pregnancy has been shown to be an important cause of hydrops fetalis, as demonstrated by histological and serological evidence.[19–21] Furthermore, transmission electron microscopy of cardiac tissue revealed the presence of intranuclear virions in both erythroid precursor cells and in cardiac myocytes. Parvovirus has been found not only in patients with suspected myocarditis but also in patients with cardiac allograft rejection after transplantation[22,23] and in patients with idiopathic DCM.[24] It has been suggested that this virus plays a pivotal role in the induction of endothelial dysfunction and coronary vasospasms in patients with acute myocarditis, mimicking acute myocardial infarction (AMI).[25] Most observations have suggested that endothelial cells may be the primary targets for PVB19 infection in small cardiac vessels, although presently there is little evidence that cardiomyocytes are the direct target for this virus type.[26] Endothelial parvoviral infection may result in the impairment of myocardial microcirculation with secondary myocyte necrosis occurring during acute infection.[27] Parvovirus B19 has recently been found to be the most frequent pathogen present in 31 of 37 patients with unexplained isolated LV diastolic dysfunction.[28] Also, patients infected with parvovirus B19 exhibited a high incidence of endothelial dysfunction, suggesting that endothelial parvovirus-induced dysfunction may underlie cardiac diastolic dysfunction.

HIV

Human immunodeficiency virus (HIV) disease is increasingly recognized as an important cause of myocarditis and DCM. In the pre-antiretroviral therapy era, the predominant cardiac pathology was associated with the localization of opportunistic infections, including *Toxoplasma*, nontuberculous mycobacteria, cytomegalovirus, and *Cryptococcus*. In patients with HIV infection, cardiac involvement is frequently associated with tuberculosis, malignant neoplasms, non-antiretroviral drug-related cardiotoxicity, endocarditis in IV drug users, and direct HIV infection of the blood vessels.[29] The use of highly active antiretroviral therapy (HAART) has substantially modified the course of HIV by lengthening patient survival and significantly reducing the incidence of DCM-related myocarditis although posing additional cardiovascular complications, particularly when incorporating protease inhibitors, such as activation of the metabolic syndrome, including hyperlipidemia, insulin resistance, diabetes mellitus, and atherosclerotic heart disease.[30,31]

The direct role and mechanism of HIV in promoting myocarditis and DCM remains unclear. An early study demonstrated that both DCM and myocarditis were present in more than 83% of a large population of asymptomatic HIV-positive patients, with 60% having detectable HIV in the myocardium.[32] On the other hand, several studies have failed to detect HIV proviral DNA in myocardial samples obtained post-mortem from HIV-infected children, despite the presence of distinctive signs of myocarditis.[33] Using molecular techniques (sequence amplification and hybridization), HIV-1 DNA sequences have been found in the patients' myocardial cells in a patchy distribution; however, there was not direct association between the presence of the virus and myocyte dysfunction.[34] Furthermore, *in vitro* studies have reported that HIV did not infect fetal cardiomyocytes.[35]

Despite observations indicating that HIV infection and viral replication in cardiomyocytes may be limited, there is evidence that HIV infection has significant impact on the heart. *In vitro* studies with neonatal rat cardiomyocytes suggest that HIV-infected myocardium may contribute to the pathogenesis by activating multifunctional cytokines (i.e., tumor necrosis factor-α) and the inducible form of nitric oxide synthase (NOS) that plays a role in the progressive and late myocardial tissue damage.[36] Rabbit ventricular cells perfused with a solution containing the HIV surface envelope protein, glycoprotein 120 (gp120) sustained marked reduction in electric field–stimulated contractions and L-type Ca^{++} current.[37] Furthermore, in heart tissues of HIV-infected patients, the presence of HIV-1 DNA and RNA has been detected in inflammatory cells but not endothelial cells or cardiomyocytes, although the viral envelope protein gp120 was found in macrophages, lymphocytes, cardiomyocytes, and endothelial cells.[38] Of interest, macrophages invading the myocardium expressed TNF ligands, whereas adjacent cardiomyocytes underwent apoptosis. These data support a model of HIV pathogenesis by which HIV infects and replicates in inflammatory cells, leading to the induction of cardiomyocyte apoptosis through apoptotic ligands and by gp120-proapoptotic signaling. Other studies have suggested that NOS and mitochondria may also play a significant role in the triggering of apoptosis by HIV/gp120. Ethyl isothiourea, an inhibitor of NOS, inhibited apoptosis induction by gp120 in rat neonatal cardiomyocytes.[39] Observations in isolated cardiomyocytes and in transgenic mice have shown that HIV, the myocardial-specific targeted expression of several of its component peptides (gp120, and the transactivator TAT), and therapeutic AZT can cause significant damage to myocardial mitochondria with resulting cardiac dysfunction and cardiomyocyte apoptosis.[38,40,41]

NONVIRAL INFECTIVE MYOCARDITIS

Myocarditis caused by nonviral infectious agents is rather uncommon, although more prevalent in immunosuppressed or compromised individuals, and is a cause of morbidity and mortality mainly in the developing world. An example in case is diphtheria-mediated myocarditis, which is rare in the United States because of the availability of vaccination. The production of an exotoxin by *Corynebacterium diphtheriae* induces myocarditis with dysrhythmias and can be stemmed by antibiotic and antitoxin therapy if initiated without delay. Both nonrheumatic and rheumatic myocarditis can arise from streptococcal infection and are primarily due to bacterial exotoxins and autoimmunity.[2] Immune responses to streptococcal antigens resulting in antibodies and immunocompetent cells, which cross-react with myocardial antigens, are thought to be central to the pathogenesis of rheumatic carditis, which often targets the cardiac valves and peri-myocardium. A bacterial endotoxin has been attributed a role in the etiology of meningococcal myocarditis, and more than 70% of cases of meningococcal infection have marked endotoxinemia.[42]

Although not common, it is well documented that infection with the spirochete *Borrelia burgdorferi*, the causative agent of Lyme disease, can result in myocarditis in addition to a wide spectrum of other manifestations. This is often accompanied by varying degrees of intermittent atrioventricular block, myopericarditis, intraventricular conduction defects, bundle branch block, and in some cases DCM.[43] The myocarditis can resolve either spontaneously or with antimicrobial therapy. Animal models of *Borrelia* infection have confirmed the role of this spirochete in causing myocarditis.[44] Interestingly, infected mice have histopathological and ultrastructural findings similar to those observed in human Lyme carditis and shown that *Borrelia* spirochetes had a predilection for connective tissue in the base of the heart, especially around the aorta, epicardium of the upper ventricles and atria, myocardial interstitium, and endocardium.[45]

Observations in mouse models have also indicated that *Borrelia*-induced myocarditis may result from autoimmunity mediated by molecular mimicry between the bacterium *Borrelia burgdorferi* and self-components (e.g., myosin).[46]

Studies that used *Borrelia*-infected mouse models have directly demonstrated a contributory role for CD4[+] T cells in mediating and exacerbating myocarditis.[47] Specific mouse strains often with marked immunodeficiency (e.g., SCID, NIH-3) can be more extensively affected, indicating that host susceptibility plays an important role.[48] Recently, *Borrelia*-mediated myocarditis in specific immunocompromised animals has been found in a nonhuman primate model (*Macaca mulatta*) and suggests that *Borrelia*-mediated myocarditis in nonhuman primates is frequent and can persist for years but is mild unless the host is immunosupressed.[49]

By use of gene-profiling techniques, progress has been gained in defining the genes of the *Borrelia* spirochete that are preferentially expressed during mammalian infection and are responsible for the adaptation and gene expression required for *B. burgdorferi* to effectively colonize the host, evade humoral responses, and cause disease.[50]

Generally, untreated patients with Lyme disease have high levels of serum immunoglobulin G antibodies and sometimes low levels of immunoglobulin M antibodies to *Borrelia burgdorferi* antigens by ELISA and Western blot. These responses can persist for many years after antibiotic treatment, and serology does not accurately distinguish between active or past infection. Detection of *Borrelia*, which initially was difficult because specific and dependable serological tests were often lacking, has evolved with the increasing acquisition of knowledge of the infectious antigens. The use of recombinant protein antigens and PCR (most informative with synovial fluid) have improved the detection and confirmation of ongoing infections.[51] However, it might be expected that active *Borrelia* infection, as gauged by the presence of *Borrelia* DNA, would be negative in later stages of *Borrelia*-mediated myocarditis once an autoimmune response has been triggered.

Another tickborne pathogenic organism *Rickettsiae* has been found linked to myocarditis in a significant number of patients with Rocky Mountain spotted fever.[52,53] Cardiac involvement includes multifocal myocarditis, cardiac dysfunction, chamber enlargement, pericardial effusion, vasculitis, secondary thrombosis, and evidence of tissue necrosis. The pathogenetic mechanism was further suggested by specific immunofluorescent analysis demonstrating the presence of *Rickettsia rickettsii* in myocardial capillaries, venules, and arterioles correlating well with the patchy distribution of interstitial mononuclear myocarditis.[54] The related *Rickettsia Helvetica* transmitted by *Ixodes ricinus* ticks has been causally linked to myocarditis in several young Swedish men who died suddenly during exercise and who had signs of myocarditis.[55] PCR-positive tissue showed chronic interstitial inflammation and the presence of *Rickettsia*-like organisms predominantly located in the endothelium that reacted with antisera. Electron microscopy confirmed that the size and form of the organisms was similar to that observed with spotted fever *Rickettsia*.

Isolated cases of myocarditis have been causally related to infections by *Salmonella*, *Campylobacter*, *Yersinia*, *Mycoplasma*, and *Chlamydia*,[1,2,56–58] and in some of these cases, enteroviral RNA coinfection has been noted, suggesting either a viral etiology or function as a cofactor. It is possible that increased molecular surveillance for viral infection may establish the etiology.

PARASITES

In Chapter 13, we described in some detail the cardiomyopathy that may occur after infection with *Trypanosome cruzi* (*T. cruzi*), a parasite that causes Chagas' disease and is endemic in both rural and urban areas of South America. In this chapter, we will comment on the involvement of *T. cruzi* in the development of myocarditis and progression into the chronic stages of the disease, both of which involve the presence of mononuclear cells. Frequently, patients with acute Chagas' disease display myocarditis (evaluated by myocardial fiber injury), whose extent is related to the presence of detectable *T. cruzi* antigens and increased numbers of CD4[+] (helper) and CD8[+] (cytotoxic-suppressor) T cells in endomyocardial biopsies.[59] Diagnosis of acute disease with high mortality is particular prevalent in children. Fatal cases of myocarditis with disseminated and diffuse foci are primarily located in contracting myocytes and in the conduction system of the heart. These lesions feature enhanced parasite multiplication, with myocyte-contained *T. cruzi* causing disruption and liberation of several inflammatory mediators.[60] High levels of CC chemokines, especially CCL5/RANTES (regulated on activation, normal T cell expressed, and secreted) and CCL3/macrophage inflammatory protein-1 (MIP-1) have been found in experimental *T. cruzi*–elicited chronic myocarditis.[61,62] In mice infected with *T. cruzi* during early stages of infection, most of the inflammatory cells invading the heart tissue were CD8[+] cells, as well as peripheral CD8[+] leukocytes, which exhibited enhanced expression of CCR5, a CCL5/RANTES, and CCL3/MIP1- receptor.[63] By use of selective CCR1 and CCR5 antagonists, termed Met-RANTES, to modulate the acute *T. cruzi*–elicited myocarditis, significant reduction in the numbers of CD4[+] and CD8[+] T cells, CCR5[+], and interleukin-4[+] cells invading the heart occurred, resulting in increased survival of the infected animals. Moreover, enhanced expression of the chemokine receptor CCR5 on leukocytes from patients with mild Chagas' disease has also been detected,[64] and these findings indicate that CC-chemokine receptors may serve as an attractive therapeutic target in Chagas' disease and deserve further evaluation.

A dramatic up-regulation of NCAM expression in acute and chronic Chagas' myocarditis has been reported in the intercalated discs of cardiomyocytes and within the intracellular nests of the amastigote forms of the parasites, suggesting a potential contributory role of NCAM as a receptor for tissue targeting and cellular invasion by *T. cruzi*.[65]

African trypanosomiasis (e.g., *Trypanosoma gambiense*) may also cause myocarditis, primarily in Europeans, albeit rarely. Another parasite that may cause myocarditis is the *Toxoplasma gondii*, which is mainly common in immunocompromised hosts and in cardiac transplant recipients. More than 50% of transplanted patients (without *T. cruzi* antibodies) were reported to develop toxoplasmosis-associated myocarditis. Also, it has been associated with HIV infection.

Severe infection with the nematode *Trichinella spiralis* may cause myocarditis mostly associated with electrocardiographic changes.[66] DNA analysis in a rat model of *Trichinella* infection showed that from 21 days after infection onwards, the extensive morphological and functional myocardial changes detected could not be ascribed to the presence of the parasite, whereas its antigens, and the attendant host immunopathological response, likely played a role in the induction of myocardial damage and dysfunction.[67]

Toxic Myocarditis

A number of toxic agents, including commonly used drugs (Table II), can promote myocarditis, in some by exerting a direct toxic effect on the myocardium, although others act by eliciting immune-mediated mechanisms. Toxic myocarditis (such as found with cyclophosphamide) can manifest with dose-related morphological abnormalities, including myocyte necrosis and vasculitis. Hypersensitivity myocarditis, such as found with thiazide diuretics and clozapine, often are not dose-related in their effects and are associated with eosinophilic infiltrate in the myocardium and eosinophil degranulation.[68] Myocardial damage can occur as a direct result of tissue injury produced by released toxic eosinophil granule proteins, which can act as selective activators of cardiac mast cells.[69,70] These conditions are often reversible on withdrawal of the toxic stimuli and often can be resolved after treatment with steroids. In rare cases, giant cell myocarditis can result from drug hypersensitivity.[71] Interestingly, eosinophilia and increased deposition of toxic eosinophil granule protein on cardiac myofibers has been detected in patients with both acute and chronic Chagas' disease.[72]

There are also drugs (e.g., anthracyclines, particularly the antineoplastic drug doxorubicin) that promote an early myocarditis and an eventual (chronic) cardiomyopathy with ventricular dilatation and failure. This group of drugs causes a gradual myofibrillar loss within the cardiac myocytes (by stimulating myocyte apoptosis)[73] and a pronounced dilatation of the sarcoplasmic reticulum manifested by cytoplasmic vacuolization.[74] Effects on cardiac mitochondria (e.g., mtDNA damage, respiratory enzyme dysfunction and lipid peroxidation) potentiated, at least partly, by the damaging effects of free radicals are contributory to the pathogenesis-mediated doxorubicin cardiotoxicity.[75-77] The damaged cells are replaced by fibrosis, and ventricular failure ensues.

Some toxic agents such as cocaine seem to engender all three types of myocarditis, including an acute insult to the heart (by blocking sodium-channels and potentiation of cardiotoxic catecholamine effects, by inhibition of the presynaptic uptake carrier), increased hypersensitivity myocarditis, and degenerative/inflammatory myocarditis, leading to cardiomyopathy.[78,79]

Diagnosis

The Dallas criteria for the histological diagnosis of active myocarditis introduced in 1986 included the use of routine light microscopy to demonstrate both infiltrating lymphocytes and myocytolysis in endomyocardial biopsy specimens. However, the complex and highly variable clinical phenotype of myocarditis in combination with mostly negative or inconsistent results of the histological evaluation according to the Dallas criteria has contributed to the frequent failure to confirm the clinical suspicion of myocarditis. The persistence of cardiotropic viruses (enterovirus, adenovirus) and anticardiac autoimmunity constitute the predominant etiopathogenic pathways of DCM. The diagnosis of inflammatory cardiomyopathy requires a level of sensitivity and specificity not fulfilled by the histological Dallas Criteria. The immunohistological quantification and characterization of immunocompetent infiltrates and cell adhesion molecule (CAM) expression has been crucial in the establishment of the diagnostic entity of inflammatory cardiomyopathy as acknowledged by the World Health Organization (WHO), reported in approximately 50% of

TABLE II Toxic Myocarditis

Amphetamines
Anthracyclines
Cocaine
Ethanol
Fluorouracil
Interleukin-2
Lithium
Trastuzumab
Heavy metals—copper, iron, lead
Arsenic
Azide
Inhalants
Carbon monoxide
Bites (snake, spider) and stings (wasp/bee)
Catecholamines (norepinephrine, epinephrine)
Tricyclic antidepressants
Clozapine
Dopamine
Phenothiazines
Organophosphate (carbamate, cyclophosphamide)

patients with DCM. The use of cellular immunohistochemical analysis can include the overall analysis of inflammatory cells (e.g., T lymphocytes, leukocytes and macrophages), markers of necrosis and apoptotic remodeling, complement-associated protein complex, and the presence of viral antigens (e.g., viral capsid proteins [VP1]).[80]

The use of molecular techniques has emerged as an important tool in myocarditis diagnosis. Viral detection by RT-PCR and *in situ* hybridization, as well as serological studies detecting viral antibodies, revealed an association between enteroviral Coxsackie group B viruses infection and myocarditis.[80]

The finding of the enteroviral genome in a significant proportion of patients with DCM has been increasingly reported, and other viral genomes (e.g., adenovirus, parvovirus, herpes virus) have been detected. Data showing that some viruses associated with pneumonia (and respiratory disease) can also be involved in myocarditis have prompted the use of targeted PCR amplification for screening suspected viral genomes in less invasively obtained tracheal aspirate samples in patients; one study showed an excellent correlation of detected virus in both tracheal isolates and endomyocardial biopsy specimens.[81]

Enteroviral persistence tends to be associated with an adverse prognosis.[82] The induction of the CAR exclusively in 63% of DCM patients, but not in other cardiomyopathies, suggests that CAR may constitute a key molecular determinant for cardiotropic viral infections in DCM. Moreover, the ability to detect the state of viral replicative activity by demonstrating the presence of enteroviral minus-strand RNA has contributed an additional dimension to studies on viral etiology of myocarditis and DCM.

In tissues that have enteroviral plus-strand RNA, the identification and quantitative assessment of the minus-strand RNA intermediate serves as an indicator of active viral replication that has ramifications for both the prognosis and appropriate treatment. Furthermore, high levels of active enteroviral replication have been reported both in patients with myocarditis, and in patients with end-stage DCM, in which active viral replication was associated with a poor prognosis.[82] Similar findings showing the presence of both positive- and negative-strand RNA in myocardium of patients with DCM have been reported for the hepatitis C viral–mediated myocarditis and suggest that hepatitis C virus can replicate in human myocardium.[83]

It is worth noting that the evaluation of viral genome persistence with adequate sensitivity and specificity has been made possible by sensitive molecular biological techniques, such as *in situ* hybridization and nested polymerase chain reaction; unfortunately, these tests are not routinely available to the clinician, although delineation of the pathogenic agent may impact on the initiation of specific therapy. Currently, antiviral therapies (e.g., β-interferon) are undergoing randomized testing in the BICC Study (Betaferon In Patients with Chronic Viral Cardiomyopathy).[84]

ANIMAL MODELS

Our knowledge regarding the pathophysiology of infective myocarditis has been derived in part from animal studies (e.g., myocarditis can be triggered by inoculation with CVB3 in genetically predisposed mice).[85] This model has revealed a large array of mechanisms by which a cardiotropic virus like coxsackie can cause myocarditis and HF (Table III).

Animal models have also documented direct enterovirus-induced myocyte damage in the absence of a specific immune response.[86] Immunohistological analysis has revealed evidence of multiple foci of cardiomyocyte necrosis in the infected myocardium, as well as the presence of clusters of macrophages that could be recruited as phagocytic or cytolytic effectors. *In vivo* studies of viral-infected mice have shown that immune cells (e.g., B lymphocytes) are significant early targets of enteroviral infection, providing a noncardiac reservoir for viral RNA during acute and persistent myocardial enterovirus infection.[87,88]

In some strains of mice, an initial noninflammatory stage is followed by severe myocarditis occurring 4–10 days after infection (Fig. 1). This second inflammatory stage involves the infiltration of the myocardium by inflammatory cells, including natural killer (NK) cells and macrophages, with activated proinflammatory cytokine expression. The NK cells and activated macrophages play a protective function by eliminating virally infected cells and stemming viral replication. Cytokines such as interferon and TNF-α serve both a protective function (assisting with viral clearance)[89] and also provoke chronic inflammatory pathways, including the activation of endothelial cells and recruitment of inflammatory cells with attendant negative inotropic effects.[90] Cytokines also activate NOS, which provides both cardioprotection[91] contributing to recovery and worsens injury by inflammatory producing effects.[92]

Animal models have also been useful in defining the cell-mediated immune response that occurs with infective myocarditis. Woodruff and Woodruff,[93] have shown a marked decline in myocarditis in mice depleted of T cells and then infected with coxsackie virus B3. In untreated infected mice, antigen-specific T cells infiltrate the myocardium by 7 days of infection with activated cytotoxic T cells, recognizing

TABLE III Viral-Mediated Mechanisms of Myocarditis Pathogenesis

Mechanism
Direct cell lysis by replicating virus
Destruction of cellular and humoral immunity
Activation of cellular and humoral autoimmunity
Latent viral infection-replication
Viral protease cleavage of host proteins
Latent viral infection-immune response
Latent viral infection
Viral triggered remodeling

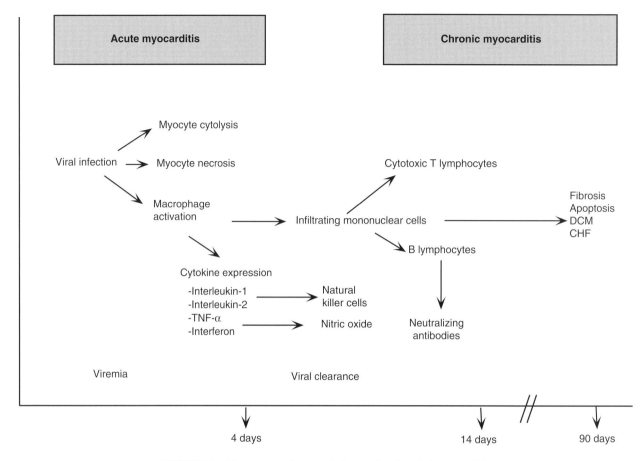

FIGURE 1 Time course of myocardial events in mice viral myocarditis.

degraded viral proteins and are able to lyse viral-infected cells. This activation involves a highly regulated interaction of numerous antigens, cofactors, and cell-to-cell contact mediated by adhesion molecules.[94]

Enteroviral-infected mice have also suggested an autoimmune component in myocarditis. Chronic myocyte damage is associated with the viral-triggered production of circulating heart-specific autoantibodies and autoreactive lymphocytes. This is comparable to the finding of circulating organ-specific autoantibodies reported in patients with myocarditis and idiopathic DCM.[95] Moreover, persistent activation of the infiltrating T cells during viral infection can cause long-term inflammatory tissue damage, leading to DCM.[96,97] In addition, the inflammatory process is largely mediated by an autoimmune response to myocardial antigens that are exposed to the immune system after viral-induced myocyte necrosis.

Animal models have been established to explain how infections initiate autoimmunity and how autoimmune mediators cause death or transient dysfunction of myocytes. This also suggested the development of a virus-free model of autoimmune myocarditis in which cardiac inflammation can be induced by immunization with cardiac myosin.[98] This was further supported by the development of autoimmune myocarditis in mice immunized with cardiac C protein[99] or with streptococcal M protein peptide (which strongly cross-reacts with cardiac myosin).[100]

Mouse models of cardiac myosin–induced experimental autoimmune myocarditis (EAM) and cardiac α-myosin heavy chain peptide–induced myocarditis have been widely used in which cardiac myosin, or the relevant peptide in Freund's complete adjuvant, is injected subcutaneously. The immune response, the histological changes, and the genetic susceptibility seen in EAM are similar to those of CVB3-induced myocarditis.[101] Furthermore, investigations that used gene-targeted mice have provided insights into viral pathogenicity and host factors involved in the control of viral replication. Studies with transgenic animals and cultured cells have revealed that remarkably low persisting levels of replication-restricted coxsackie viral genomes are associated with an induction of a cytopathic effect in cardiac myocytes, leading to a DCM phenotype. In addition, these studies have revealed the involvement of both innate and acquired immunity, as well as the role of viral receptors in disease phenotype.[104] For instance, the impact of the innate immune system on coxsackie viral replication has been demonstrated by studies in gene-targeted mice deficient of either type I or type II interferon signaling, with findings implicating type I, but not type II, interferons as essential for the control of early viral replication and survival of coxsackie viral infection.[102]

The pathogenesis of viral myocarditis involves contributions from the virus, the host immune system, and myocytes. In defining the specific molecular pathways underlying myocarditis, modulations of the components of the immune system through models of transgenic knock-out, knock-in, and overexpression have provided significant insights.[103,104] An example is the mouse null for the IRF-1 (interferon regulatory factor) transcription factor that exhibits myocarditis even though there is a complete absence of the inducible form of NOS in the tissues. These IRF-1–deficient animals are exquisitely sensitive to coxsackie viral infection, with extremely high mortality, and have increased cardiac lesions (compared with wild-type mice) in response to reovirus infection, suggesting a role for IRF-1 in protection against viral myocarditis. On the other hand, *CD4* knock-out strains seem to have myocarditis in an autoimmune myocarditis model, although p56lck knock-outs (targeting the T-cell tyrosine kinase signaling molecule) seem to be free of viral myocarditis.

Treatment

In the not so distant past, myocarditis therapy primarily used supportive treatment against HF and dysrhythmias with ACE inhibitors, angiotensin receptor blockers, β-blockers, diuretics, and amiodarone, and, in severe cases, heart transplantation (particularly in cases of pediatric giant cell myocarditis). In treatment of the more aggressive giant cell myocarditis that may present with rapid hemodynamic deterioration, heart transplantation can be accompanied by ECMO (extracorporeal membrane oxygenation). At present, the establishment of a definite diagnosis of myocarditis and identification of the infective/pathogenic agent has begun to allow the initiation of more specific therapeutic strategies.

The available armamentarium against viral infection is presently limited compared with the wealth of antibiotics that can be directed against bacterial or intracellular pathogens such as *Chlamydia* and rickettsiae, although this trend may be rapidly changing. Despite several complications, HAART therapy with HIV infection is a powerful example of a relatively successful therapy against a constantly evolving pathogen. Moreover, nucleoside analogs and interferon have been used to treat many of the signs and symptoms of common viral infections, ranging from herpes, hepatitis C, and mumps to parvoviral infections, which can be associated with myocarditis.[105,106]

Because no specific treatment for CVB3 infections is available to date, RNA interference (RNAi) approaches have been initiated to prevent virus propagation. Initially, fusion constructs of a reporter (green fluorescent protein) and viral subgenomic fragments were used to select active siRNAs against the virus. Moreover, in an attempt to achieve sustained virus silencing and reduce the risk of generating escape mutants, only highly efficient siRNAs directed against regions of the viral genome, which are unlikely to tolerate mutations, were considered for virus inhibition. Two siRNAs directed against the RNA-dependent RNA polymerase were found to inhibit virus propagation by 80–90%.[107] The protective effect of the efficient siRNAs lasted for several days. Furthermore, inhibition of the cellular CAR by RNAi also reduces the virus titer.

Recently, particular attention has centered on the use of immunomodulation and immunosuppression as optional therapies for defined subgroups of patients with myocarditis or DCM.[108] On the basis of previous data from animal studies showing an anti-coxsackie viral effect of interferon treatment,[109] new observations have documented that treatment of viral-mediated DCM with subcutaneously administered interferon B (by weekly injections) over a 6-month period resulted in a dramatic clearance of viral genomes (both enteroviral and adenoviral) paralleled by a significant improvement in left ventricular function.[110] Despite the significant clinical benefits obtained in this relatively small phase II study with the well-tolerated interferon B treatment, similar approaches that used TNF-α antagonists, such as etanercept and infliximab, will likely not be undertaken in clinical myocarditis, because there is no evidence that these agents provide a beneficial effect in patients with chronic HF resulting in the premature termination of two large-scale randomized trials, the RENAISSANCE and ATTACH studies and confirmed by the RECOVER clinical trial. Moreover, evidence of increased morbidity and mortality with infliximab was found in the ATTACH study.[111]

Although recent large randomized clinical trials did not indicate that immunosuppression (using prednisone with either cyclosporine or azathioprine) improves patient survival or ameliorated left ventricular function in myocarditis,[112] the use of immunosuppression has been shown to provide benefits in smaller well-delineated subgroups of patients with myocarditis. For instance, in eight patients with fulminant (primarily lymphocytic) myocarditis, immunosuppression in combination with mechanical circulatory support allowed improved ventricular function and enhanced survival.[113] The successful treatment of a small group of patients with active myocarditis because of hepatitis C virus infection (by extensive molecular and immunochemical analysis) with prednisone and azathioprine for 6 months resulted in complete recovery of cardiac volume and function.[114] In addition, screening of patients with active lymphocytic myocarditis showed that those with circulating cardiac autoantibodies and no viral genome in the myocardium (excluding hepatitis C) are the most likely to benefit from immunosuppression treatment (i.e., prednisone and azathioprine for 6 months).[115] Moreover, IgG antibodies, which are cardiac and disease-specific (myocarditis/ DCM), can be used as autoimmune markers for identifying patients in whom immunosuppression may be beneficial and to identify relatives at risk.

The use of gene therapy that has been applied with some success in a number of defined animal models of myocarditis is discussed in Chapter 3.

ENDOCARDITIS

Infective endocarditis, a microbial infection of the endocardial surface of the heart, has been classified as "acute," "subacute," and "chronic" on the basis of the timing and severity of the clinical presentation and the progression of the disease when left untreated. The characteristic lesion, termed vegetation, is composed of a collection of platelets, fibrin, microorganisms, and inflammatory cells. It most commonly involves the heart valves, but it may also occur at other sites, including that of a septal defect, or on the mural endocardium.[116]

Despite improvements in health care, the incidence of infective endocarditis has not significantly decreased over the past decade, and it remains a disease that is associated with considerable morbidity and mortality. This paradox has been largely explained by a progressive evolution in risk factors. Although classic predisposing conditions such as rheumatic heart disease have been significantly reduced, mainly in Western countries, albeit still a significant morbidity factor in developing countries, new risk factors for infective endocarditis have emerged (Table IV). These include intravenous drug use, hemodialysis, sclerotic valve disease in elderly patients, use of prosthetic valves, and nosocomial disease.[117] Furthermore, a large number of infectious agents have now been implicated in its pathogenesis, many of which are difficult to cultivate by standard blood culture techniques, requiring molecular techniques for identification, such as PCR and sequence analysis. Several newly identified, difficult-to-cultivate, and likely under-reported pathogens including *Bartonella* spp. and *Tropheryma whipplei* (Whipple's disease), have been increasingly detected in specific cases of endocarditis. The ongoing evolution of antibiotic-resistant organisms presents an enormous challenge to conventional antimicrobial therapy. Keeping up with these myriad changes depends on a comprehensive approach, requiring a better understanding of disease pathogenesis, as well as the development of new drugs. This understanding may be greatly enhanced by dissection at the molecular level of both the infectious agents and the host response.[118]

MAJOR MICROBIAL PATHOGENS

Currently, *Staphylococcus aureus*, *Streptococcus* species, and *Enterococci* account for more than 80% of the agents causing infective endocarditis.

Streptococci, particularly viridans streptococci, are the most common cause of native valve endocarditis. Affected patients usually have underlying cardiac disease. Viridans streptococci are part of a normal oral flora and generally gain access to the bloodstream through breaches in the oral mucosa (in dental surgical procedures, for example). The most common streptococci isolated from patients with endocarditis are *Streptococcus sanguis*, *S. bovis*, *S. mutans*, and *S. mitis*.

In recent years, staphylococci, and particularly *S. aureus*, have challenged viridans streptococci as the most common pathogen implicated in infective endocarditis. The highly virulent *S. aureus*, the only coagulase-positive staphylococcus, is associated with high morbidity and mortality rates, especially in mitral and aortic valve disease in both nosocomial and community-acquired settings. Other staphylococci species that colonize humans are coagulase-negative, among which *S. epidermidis* has emerged as an important pathogen in the setting of implanted devices and in hospitalized patients and is almost always resistant to methicillin or oxacillin. Complications are more common in *S. aureus* infective endocarditis than in endocarditis caused by other bacteria and include conduction defects, myocardial abscess, valve ring abscess, purulent pericarditis, and peripheral emboli. In addition to causing endocarditis, *S. aureus* infection may result in severe sepsis syndrome with fulminating coagulopathy. In up to 40% of patients, a variety of complications may result from metastatic foci of infection, including lung abscess (from tricuspid valve involvement), CNS abscess, and splenic abscess, and contribute to the high mortality. Endocarditis can also result from infection with *S. lugdunensis*, a coagulase-negative organism that can be confused with *S. aureus*. However, despite being susceptible to penicillin, this pathogen is extremely virulent, yielding high mortality unless the infected valves are removed.

Infective endocarditis caused by *Enterococci* infection has been increasingly reported and is estimated to account for approximately 10% of the cases. Enterococci are frequently implicated in nosocomial bacteremias and infective endocarditis that is resistant to medical therapy, because some strains may be not only be resistant to penicillin but also to vancomycin. This type of infective endocarditis has been most frequently seen in elderly men, it has relatively low short-term mortality, frequently involves the aortic valve, and tends to produce HF rather than embolic events.

TABLE IV Primary Risk Factors in Endocarditis

Use of prosthetic valves
Mitral valve prolapse
Intravenous drug use
Atherosclerotic valve disease in elderly patients
Nosocomial disease.
Rheumatic heart disease
Long-term hemodialysis
Poor dental hygiene
Diabetes mellitus
Congenital heart disease

In 5–15% of patients with presumed infective endocarditis, no etiologic microorganism can be isolated from blood cultures. The most common factors behind this inability to identify the etiologic microorganism are prior/ongoing antimicrobial therapy, right-sided endocarditis, and a variety of organisms (Table V). These include fastidious/slow-growing organisms (such as the gram-negative HACEK group of bacteria) that resist culturing in routine microbiological media because of the lack of appropriate growth factors, intracellular organisms (such as *Rickettsiae, Coxiella, Bartonella*), and fungi.

The HACEK group of fastidious gram-negative aerobic bacteria also can cause endocarditis. HACEK is an all-inclusive term for infective endocarditis caused by *Haemophilus, Actinobacillus, Cardiobacterium, Eikenella,* and *Kingella* species of bacteria. Clinically, a subacute or chronic course, frequent embolic lesions, large vegetations, and frequent need for valve replacement characterize these cases.

Infective endocarditis, although more prevalent in adults, can also occur in children, with increasing incidence reported over the last few years. This may, in part, be related to improved survival of infants with CHD, the primary predisposing factor.[119,120] Different infectious agents seem to be associated with infective endocarditis in patients of different ages (Table VI). In neonates, infection with staphylococcal species, including coagulation-negative staphylococci, is significantly increased, whereas in individuals older than 60 years of age, infection with streptococci and with enterococci are more prevalent.[116] Furthermore, the incidence of fungal infection seems to be significantly higher in neonates, mainly in recent years, perhaps as a consequence of the overuse of broad-spectrum antibiotics.[121]

It is also notable that significant differences have been found in the incidence, clinical presentation, pathophysiology, and etiology in infections causing prosthetic valve endocarditis (engendering the term PVE) compared with native valve endocarditis (NVE). PVE accounts for 7–25% of cases of infective endocarditis in most developed countries. Patients with enterococcal PVE have a higher rate of myocardial abscess formation and lower rates of new regurgitation

TABLE V Infectious Agents (Blood Culture-Negative) of Endocarditis

Bacterium	Clinical presentation	Method of diagnosis	Underlying disease/condition
Coxiella burnetii	Fever; HF; hepatomegaly; splenomegaly; exposure to risk factors	IFA; PCR; serological tests; cultivation of blood and resected valve in tissue culture;	Prosthetic valve; previous valve injury; rheumatic HD; immunocompromise (cancer, HIV, renal dialysis, corticosteroid therapy, organ transplant)
Mycoplasma spp.	Fever; CHF; prosthetic valve dysfunction	Cultivation of annulus of resected valve; serology	Lupus; prosthetic valve; prednisone; rheumatic HD
Bartonella henselae	Fever; acute HF; cardiac murmur; vegetation on echo	Serology: IFA	Previous valve injury cat owner
Bartonella quintana	Fever; acute HF; cardiac murmur; body lice; vegetation on echo	Culture and PCR; DNA sequence analysis	Alcoholism; homelessness
Mycobacteria other than *M. tuberculosis*	Fever; acute HF	Blood culture; valve culture; acid-fast bacilli on valves	Late prosthetic valve infection; contaminated prosthesis, or water
Mycobacterium tuberculosis	Fever; acute HF	Necropsy; acid-fast bacilli on valves; valve culture	Miliary tuberculosis; congenital heart defects
Chlamydia spp.	Fever; cardiac murmur	Serology; blood culture; immunohistochemistry and monoclonal antibody	Pneumonia
Legionella spp.	Fever; mild CHF; vegetation on echo	Serological testing; valve cultures; blood culture; fluorescent antibody	Valve prosthesis; surgery; rheumatic fever
Whipple's disease bacillus	Fever; diarrhea; acute HF; cardiac murmur; vegetation on echo	PCR and sequence analysis from valve; PAS stain of valve	Not known
Brucella	Fever; cardiac murmur; large vegetation on echo	Prolonged incubation; culture of valves; serology	Underlying valvular disease, including prosthetic valves
HACEK	Fever; vegetation on echo; emboli; CHF	Prolonged culture required	Upper respiratory tract infection; dental infection
Abiotrophia (nutritionally variant streptococci)	Fever; vegetations; cardiac murmur; emboli; CHF	Growth in thioglycolate medium or as satellite colonies around *S. aureus* on blood agar	Underlying valve injury

HACEK, *Haemophilus* species, *Actinobacillus actinomycetemcomitans*, *Cardiobacterium hominis*, *Eikenella corrodens*, and *Kingella* species.

TABLE VI Infectious Agents Associated with Infective Endocarditis

	Native-valve endocarditis			Prosthetic-valve endocarditis		
	Neonate	2–60 y	>60 y	Early (<60 d)	Intermediate (60 d–12 mo)	Late (>12 mo)
Streptococcus spp.	15–20%	40–65%	30–45%	1%	7–10%	30–32%
Staphylococcus spp.	40–50%	22–40%	25–30%	20–24%	10–14%	15–20%
Coagulation-negative staphylococci	8–12%	4–8%	3–5%	30–35%	30–35%	10–12%
Enterococci spp.	<1%	3–8%	14–17%	5–10%	10–15%	8–12%
Gram-negative bacilli	8–12%	4–10%	5%	10–15%	2–4%	4–7%
Fungi	8–12%	1–3%	1–2%	5–10%	10–15%	1%
HACEK	2–6%	0–15%	5%	3–7%	3–7%	3–8%
Diphtheroids	<1%	<1%	<1%	5–7%	2–5%	2–3%

than patients with enterococcal NVE.[122] A recent comparative survey of a large patient population found that mitral valve involvement seems to be more common in PVE, although multivalvular involvement is more prevalent in patients with NVE.[123] In addition, patients with PVE tended to have worse early and long-term outcomes than patients with NVE, including a higher incidence of recurrence and worse prognosis especially if the mitral valve is involved. Previously, early-onset PVE had an extremely high mortality (as high as 90%); however, with advanced use of surgical and standard perioperative antibiotic prophylaxis and targeted antimicrobial therapy, early-onset mortality from PVE has been reduced to approximately 25%.[124]

As shown in Table VI, coagulase-negative staphylococci are the most common pathogens in early PVE, which usually occurs within 2 months of surgery and is most often acquired although the patient is in the hospital. Cases of late PVE (occurring 12 months or more after surgery) tend to be community-acquired and are most prevalently secondary to *Streptomyces* infection, exhibiting a pattern similar to NVE. Gathered observations reveal that the microbiology of both early and late PVE endocarditis has changed markedly over the past few decades, with a marked reduction of gram-negative bacilli infection in early PVE and increases in enterococcal and *Staphylococcus* spp. infection in later PVE.[125]

Pathogenesis

Cardiac valves and other endocardial surfaces become infected after exposure to microemboli from bacteria or fungi circulating in the bloodstream (Fig. 2). Dextran-producing bacteria, like *Streptococcus mutans*, have a virulence factor that promotes their adherence to endovascular surfaces.

Coagulase-negative staphylococci can produce a biofilm that also promotes adherence on prosthetic surfaces. β-Hemolytic streptococci and enteric gram-negative bacteria lack recognized adherence factors and seem less likely to cause endocarditis. Previously damaged valves, defective endocardial surfaces, previous endocarditis, surgery, and pacemaker wires provide a favorable environment for thrombus formation. Over time, microorganisms proliferate in the thrombus, resulting in classic vegetation.

The characteristic vegetative lesion found with infective endocarditis, primarily composed of bacteria surrounded by a platelet/fibrin layer attached to the underlying endothelium, has long been believed to hinder host defenses in effective clearing of bacteria. Although this has not been directly demonstrated, the ability of the vegetations to exclude *in vivo* host antibodies specific for the bacterial surface protein aggregation factors has been shown in a model of experimental endocarditis caused by *Enterococcus faecalis*.[126] These findings demonstrated that once the vegetation encloses bacteria, they are no longer accessible to high-titer bacterial–specific host antibodies, consistent with a protective role of the vegetation on bacteria from the humoral immune response.

Damaged tissue in susceptible valves provides a coagulum platform consisting of fibrin, fibronectin, plasma proteins, and platelet proteins to which bacteria can adhere. This complex induces the production of cytokines and procoagulant factors, which in turn results in enlargement of the vegetation. Molecular techniques have allowed the identification of several surface structures of staphylococci (also streptococci and enterococci) as both markers and factors of virulence.[127] For instance, in the presence of fibronectin, binding proteins on the coagulum enhance the attachment of some bacteria, such as *S. aureus* to the substrate. The fibronectin-binding proteins encoded by the *S. aureus* gene (*FnbA*) have been shown to be involved in adherence to damaged heart valves but also seem to play a fundamental role in the invasion and persistence within such cells and triggering of host cell apoptosis. FnBPs expressed on the *S. aureus* surface can be degraded by extracellular proteases, suggesting that such enzymes participate in the transition of *S. aureus* cells from an adhesive to an invasive phenotype.[128,129] A role for FnBP in staphylococcal virulence and invasiveness is supported

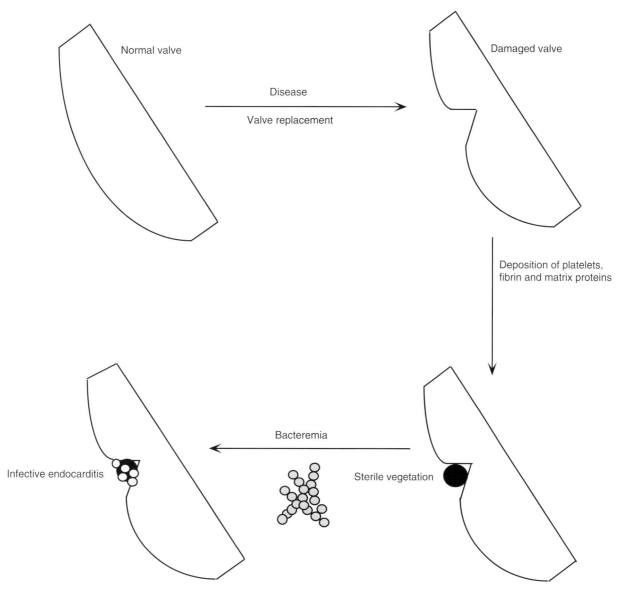

FIGURE 2 The pathogenesis of infective endocarditis.

by findings showing that FnBP deletion mutants of invasive laboratory strains lost invasiveness for epithelial cells, endothelial cells, and fibroblasts and that expression of FnBPs in noninvasive strains conferred invasiveness.[130] Moreover, in *S. aureus*, staphylococcal global regulatory elements such as the staphylococcal accessory regulator, *sarA*, and the accessory gene regulator *agr* have been shown to co-regulate genes that encode a variety of exoenzymes and exotoxins, including those responsible for cytolysis of host tissues (e.g., the α-hemolysin gene, *hla*) and enterotoxins needed for further invasion.[117] In addition, they play a critical role in regulating the expression of the antiphagocytic polysaccharide capsules that represent an important pathogenetic step in establishing *S. aureus* infections.[131] In particular, *agr* seems to positively regulate the synthesis of the capsular polysaccharide 5 gene (*cap5*), and *sarA* positively up-regulates the expression of a number of adhesin (e.g., FnBPs) and exotoxin (e.g., *hla*) genes, the latter with which *agr* is also involved as a co-regulator. Co-activation by these regulators is rare, because *sarA* and *agr* more often have opposing effects on select target genes. Expression of the *agr* effector molecule, RNAIII, correlates with down-regulation of select genes encoding cell surface proteins, such as protein A and fibronectin-binding proteins (FnBPA), and up-regulation of genes encoding secreted proteins, such as α- and β-toxins, lipases, and proteases.[132] In contrast, *sarA* up-regulates cell surface proteins (e.g., FnBPA) and down-regulates secreted proteins such as lipase, hemolysins, and proteases. A number of proteins with homology to the SarA regulatory factor have been identified in *S. aureus* as well. One of these SarA homologs, Rot (repressor of toxins), was shown by transcriptional profiling to also act as a global regulator of virulence genes and, like SarA, has opposing effects on select target genes.[132] Because *S. aureus* strains contain multiple virulence factors, studying their pathogenic role by single-gene inactivation generates equivocal results.

Interestingly, aspirin (salicylic acid) has been shown to reduce the *in vivo* virulence of *S. aureus* in experimental endocarditis, primarily by reducing the expression of the *hla* gene promoter and the *fnbA* gene promoter.[133] This leads to the attenuation of two important virulence phenotypes in both clinical and laboratory *S. aureus* strains: α-hemolysin secretion and fibronectin binding *in vitro*. Moreover, salicyclic acid likely attenuates the virulent phenotype by mediating the activation of the stress-response gene *sigB*, which causes down-regulation of the expression of the sigB-repressible *sarA* and *agr*, corresponding to the reduced expression of the *hla* and *fnbA* genes *in vitro*. Therefore, aspirin may eventually be used as an adjuvant therapy against *S. aureus* mediated–endocardial infections by down-modulating key staphylococcal global regulatory and structural genes *in vivo*.

In some cases, these factors have been purified and used as immunogens in animal models of experimental endocarditis and have been shown to induce protective antibody responses.[134,135] Gathered observations support the notion that the interactions of Gram-positive cocci with platelets and the organism's capacity to evade the antimicrobial host defense properties of platelets are pivotal in the production and persistence of endocardial infections.[136] Platelet microbicidal proteins (PMPs) are small cationic antimicrobial peptides secreted by mammalian platelets; a subgroup of these proteins are thrombin-induced and are termed thrombin-induced platelet-microbicidal proteins (tPMPs). These staphylocidal peptides play a key role in host defense against infective endocarditis. Although many *S. aureus* strain are susceptible to these proteins, strains of *S. aureus* with resistance to PMPs have been found, and these produce in experimental models a more severe form of infective endocarditis.[137] Interestingly, loss of *agr* function, vancomycin exposure, and abnormal cell lysis have been linked with the phenotypic development of resistance *in vitro* to tPMPs.[138]

Two cell surface proteins, the DltABCD and MprF proteins, have been recently identified that modify the surface charge of the *S. aureus* cell envelope (contributing a positive charge), markedly affecting wall teichoic acid molecules, and modulate *in vitro* resistance profiles of *S. aureus* to a number of endogenous cationic antimicrobial peptides, including tPMPs.[139] Bacterial strains containing mutations in genes encoding these proteins displayed substantial increases in the *in vitro* susceptibility to tPMP-1 and a significantly attenuated virulence. Mutation in the DltABCD, but not MprF, protein resulted in a significantly reduced capacity to bind to endothelial cells *in vitro*. These findings may identify a potential therapeutic target to enhance the efficacy of platelet protection against infective endocarditis.

Interestingly, recent studies have demonstrated that *S. aureus* virulence gene regulation profiles, defined *in vitro*, are often not precisely mirrored *in vivo*.[131,140] This suggests that host environmental cues likely play a major role in the activation of key *S. aureus* virulence genes.

Although *Streptococcus* and *S. aureus* seem to share the same primary site of infection in infective endocarditis, there are several major differences in their pathogenic mechanism. Although *Streptococcus* primarily adheres to cardiac valves with pre-existing endothelial lesions, *S. aureus* can colonize either damaged endothelium or invade physically intact endothelial cells. *Streptococcus* virulence in endocarditis involves factors that promote infectivity and pathogenicity. Multiple surface adhesins and exopolysaccharide (glycocalyx) contribute to infectivity. *Streptococcus* produces surface glucans (gtf and ftf), ECM adhesins (e.g., fibronectin-binding proteins, FimA), and platelet aggregating factors (phase I and phase II antigens, pblA, pblB, and pblT).[141]

Although many factors contribute to the pathogenicity, the platelet aggregation–associated protein (PAAP) of *Streptococcus sanguis* directly contributes to the development of experimental endocarditis. PAAP is synthesized as a cell wall glycoprotein of 115 kDa containing a collagen-like platelet-interactive domain and interacts with a signal-transducing receptor complex on platelets. On injured heart valves, PAAP first enhances platelet accumulation into a fibrin-enmeshed thrombus (vegetation), within which *Streptococcus sanguis* colonizes, only if successfully resisting platelet microbicidal protein (PMPs). The expression of PAAP can be modified in response to heat shock or collagen *in vitro* and potentially during infection, because collagen is exposed on damaged heart valves, and fever (and presumably heat shock) occurs during endocarditis. After colonization, proteases and other enzymes from *Streptococcus* and host sources may directly destroy the heart valves. When PAAP is neutralized with specific antibodies, experimental endocarditis exhibits a milder course, and vegetations are smaller. Finally, PAAP also contributes to the formation of the characteristic septic mural thrombus (vegetation) of infective endocarditis.

Streptococcus agalactiae is another important microbial cause of human endocarditis; it produces FbsA, a fibrinogen-binding protein that is a crucial virulent factor in *S. agalactiae*–induced platelet aggregation.[142,143] *S. agalactiae* clinical isolates bearing FbsA attach to fibrinogen eliciting a fibrinogen-dependent aggregation of platelets.

Infection by the gram-positive enterococci is among the leading causes of nosocomial bacteremia and of community-acquired subacute endocarditis. A number of enterococcal factors that contribute to the pathogenesis of disease have been identified. *E. faecalis* produces the plasmid-encoded extracellular toxin cytolysin (also called hemolysin) that contributes to enterococcal virulence in all models studied and increases mortality in combination with aggregation substance in an endocarditis model.[144]

Aggregation substance is an enterococcal surface protein, encoded by several pheromone-responsive plasmids that contribute to cardiac vegetation size *in vivo*.[145] The endocarditis antigen (EfaA) of *E. faecalis* bears similarity to adhesins encoded by genes in other *Streptococcus* spp.[146] The biological role of EfaA is undetermined, although more is known concerning the regulation of its expression (see later).

Gelatinase, an extracellular zinc metalloproteinase secreted by *E. faecalis*, may contribute to its virulence in some animal models.[147] The production of gelatinase seems to be regulated in a cell-density–dependent manner. Genes for gelatinase, hemolysin, and aggregation substance were found in 50–55% of isolates of *E. faecalis* obtained from patients with endocarditis. Their absence in nearly 45% of endocarditis-associated *E. faecalis* isolates suggests that although these traits may play a role in virulence, other properties contribute to pathogenicity.[148]

Biological cues in serum may play an important role in modulating enterococcal virulence at sites of infection. Expression of the *E. faecalis* endocarditis-associated virulence factor EfaA is induced by growth in serum. In addition, growth of endocarditis isolates of *E. faecalis* in serum induces the expression of carbohydrate ligands responsible for adhesion to Girardi heart cells.[149] Serum has also been reported to induce the expression of aggregation substance,[150] but its effect on the expression of other enterococcal virulence traits is not known. Furthermore, the extent to which biological cues affect enterococcal virulence-associated gene expression has been evaluated by quantitative real-time PCR to compare mRNA levels in *E. faecalis* cultures grown in serum. Both environmental and growth phase–specific variations were identified.

In addition, expression of the *efaCBA* operon encoding a putative ABC-type transporter has been shown to be regulated by Mn^{++} levels.[151] An Mn^{++}-responsive transcriptional regulatory protein, EfaR, sharing identity with the *Corynebacterium diphtheriae* toxin repressor (DtxR), has been identified. Analysis of the *E. faecalis* V583 genome revealed 10 additional putative EfaR-binding sites, suggesting that manganese availability may have a significant regulatory role in infection and also suggesting that a Mn^{++}-sensing element in *Enterococcus* is involved in regulating the expression of virulence factors in enterococcal endocarditis.

Besides the virulence factors described previously, the capacity of the *Enterococcus* spp. to readily acquire, accumulate, and share extrachromosomal elements (i.e., plasmids), encoding virulence traits or antibiotic resistance genes, provides numerous advantages to their survival under unusual environmental stresses and has undoubtedly contributed to their increasing prevalence as nosocomial pathogens. Other factors associated with enterococcal plasmids, including pheromones and plasmid-encoded pheromone inhibitors, which are chemotactic for polymorphonuclear leukocytes, may also contribute to the host inflammatory response often associated with *Enterococcus* infection.[152]

ANIMAL MODELS OF ENDOCARDITIS

From previous discussion on pathogenetic mechanisms, it should be evident that animal models have greatly contributed to our understanding of the pathogenesis of infective endocarditis.

Key findings regarding microbial adherence and persistence were obtained in animal models that used a plastic catheter that induced left-sided enterococcal endocarditis in rabbits, guinea pigs, or rats. Analysis of the host response to endocardial infection has revealed the complex issues operative in the pathogenesis of endocarditis. Furthermore, experimental models that closely simulate the characteristics of endocarditis in humans have helped to improve potential therapeutic and prophylactic regimens, bypassing the difficulties encountered in clinical trials. A primary advantage of the experimental model is the provision of clear end points, which allow statistical comparisons among different therapeutic regimens and parameters that are difficult to obtain in humans, such as number of infective organisms per gram of tissue, frequency of resistance, positivity of blood cultures, death versus survival rates, multiple sampling points, and percentage of relapses after treatment has been terminated.[153] In addition, data from animal models have established that bactericidal therapy is warranted and that *in vitro* susceptibility tests, especially those evaluating killing rate, are helpful in determining the efficacy and dosages required of specific treatment. These data have a good predictive value regarding therapeutic outcome and clinical relevance.[154]

Diagnosis

In 1994, the Duke criteria were established for assessing patients with suspected infective endocarditis and have proved to be useful as a guide for the clinician. These criteria integrated a variety of factors predisposing patients to infective endocarditis, including the use of blood-culture isolates and persistence of bacteremia, echocardiographic findings, and other clinical and laboratory information. Subsequently, they have been modified, and currently the consensus is that transesophageal echocardiography is preferred to the transthoracic approach, because it provides superior imaging for detecting vegetations and abscesses.

A major criterion and method for the diagnosis of endocarditis is the isolation of an endocarditis-associated microorganism from two to three separate blood cultures within 24 h. If bacteremia is present, the first two blood cultures will yield the etiologic agent in more than 90% of the cases. Additional blood cultures may be needed if the patient received antibiotics in the preceding 2 weeks. In most cases, the pathogen is identified by routine microbiological tests. In addition, a positive valve culture can be used.

If culture and biochemical tests fail to identify the organism, molecular methods, such as broad-range prokaryotic PCR in conjunction with DNA sequencing, can be used. By amplifying and sequencing a target gene from the causal agent, pathogens can be accurately identified, and this may allow sufficient discrimination among different bacterial species. The 16S rRNA gene is most commonly used in the identification of fastidious and uncultivatable bacteria, whereas the 18S, 28S, and 5.8S rRNA genes are used for fungal identification. This molecular approach can be applied with a variety of clinical samples, such as blood or infected valve tissue or the unidentified bacterial colony on solid medium. The DNA sequence of the target gene can then be compared with the available sequences in GenBank, which results in a list of bacterial names with a sequence identity index score for the query sequence. A typical search yielding a sequence identity of >97% of the nucleotides with the 16S rRNA gene from a known bacterium is considered a good match. If the query sequence yields more than one species with >97% sequence identity obtained through the BLAST search feature, further discriminating tests (e.g., staining, serology) need to be taken into account before a final identification is made. High-quality and nearly full-length consensus sequence data from both sense and antisense strands are required for reliable interpretation, because partial sequence data can lead to improper identification. The demonstration that this molecular diagnostic technique could unambiguously identify etiological microbial agent(s) in a large series of blood cultures[155]; its highly specific, sensitive, and definitive identification of the causative infective agents in valve tissues from a group of patients with infective endocarditis[156]; and its rapidity and increased availability throughout the biomedical community have all strongly supported its inclusion in the Duke criteria.[118]

A further extension of this molecular diagnostic approach has been its application in the rapid screening and identification in clinical specimens of common antibiotic-resistance genes specific to streptococci, staphylococci, and enterococci organisms causing infective endocarditis.[157] This technique can be used to supplement more classically derived data, assist in the most efficacious use of antibiotics, and allow for an earlier identification of resistant organisms, particularly critical in cases of culture-negative endocarditis.

The use of serology, with or without cell culture, has shown a diagnostic value in cases of infective endocarditis caused by *Coxiella burnetii*, species of *Bartonella*, *Brucella*, *Chlamydia*, *Legionella*, and *Mycoplasma*. Bacteria grown in culture could be assayed with immunofluorescent antibodies. Although rarely used in clinical practice, histological staining of biopsy material provides a direct observation of the inflammatory response and can be useful in the detection of the etiological agent. A number of stains, such as the Gram stain and periodic acid–Schiff (PAS), each with its respective advantages and disadvantages with specific organisms, are routinely used.

Treatment

ANTIMICROBIAL THERAPY

The penicillins, often in combination with gentamicin, remain the cornerstones in the therapy of endocarditis, when caused by susceptible streptococci (Table VII).

For penicillin-allergic patients, vancomycin is an alternative. Intravenous ceftriaxone (Rocephin), given once daily for 4 weeks, also works, and even a 2-week course in combination with gentamicin has proven successful. However,

TABLE VII Antimicrobial Therapy for Endocarditis

Pathogen	Antimicrobial therapy (regimen)
Viridans streptococci, penicillin-susceptible	Penicillin G (4 wk) or ceftriaxone (4 wk)
Streptococci, moderately penicillin-resistant	Penicillin G (4 wk) and gentamycin (2 wk)
Streptococci, highly penicillin-resistant	Penicillin G (6 wk) or ceftriaxone (4–6 wk)
Staphylococci, methicillin-susceptible	Nafcillin or oxacillin (4–6 wk) ± gentamycin (3–5 d)
Staphylococci methicillin-resistant	Vancomycin ± gentamycin (3–5 d)
HACEK	Ceftriaxone (4–6 wk) or ciprofloxan (4 wk)
Enterococci, penicillin-sensitive	Ampicillin (4–6 wk) or penicillin G (4–6 wk) plus gentamycin (4–6 wk)
Enterococci, penicillin-sensitive, gentamycin resistant	Ampicillin (4–6 wk) or penicillin G (4–6 wk) plus streptomycin (6 wk) or vancomycin (6 wk) plus streptomycin (6 wk)
Enterococci, penicillin-resistant, vancomycin resistant	Linezolid (8 wk) or imipenem/cilastatin (8 wk) plus ampicillin (8 wk)
Culture-negative endocarditis (including *Bartonella*)	Ampicillin (4–6 wk) plus gentamycin (4–6 wk) or vancomycin (4–6wk) plus gentamycin (4–6 wk) plus ciprofloxan (4–6 wk)
Fungal (*Candida* or *Aspergillus*)	Amphotericin B/surgical valve replacement

short-course therapy is not indicated for patients who have PVE or symptoms for longer than 2 months.

For relatively penicillin-insensitive streptococci (minimal inhibitory concentration 0.1–0.5 μg/ml), penicillin dosage is increased and continued for 4 weeks. Gentamicin is given for the first 2 weeks. Treatment of *Enterococcus* endocarditis is longer; both penicillin and gentamicin are given for 6 weeks.

The preferred treatment for NVE caused by methicillin-susceptible *Staphylococcus* is oxacillin or cefazolin for 4–6 weeks.[12] If the organism is methicillin-resistant, vancomycin is used. As a supplement, gentamicin may be given for the first 3–5 days to reduce the duration of bacteremia.

Antibiotic therapy for *Staphylococcus* PVE must be more aggressive because of the greater likelihood of treatment failure or relapse. When the isolate is methicillin-susceptible, oxacillin plus rifampin is given for 6 weeks; and gentamicin for the first 2 weeks. When the isolate is methicillin-resistant, vancomycin is substituted for oxacillin. The preferred treatment for the HACEK group of bacterial is ceftriaxone alone or ampicillin plus gentamicin for 4 weeks.

In approximately 10% of patients with clinically suspected endocarditis, who have negative blood cultures, the most effective regimen is ampicillin plus gentamicin for NVE or vancomycin plus ampicillin and gentamicin for PVE. Although the usual organisms are not recovered from blood, they may be seen on smear or cultured from vegetations taken at surgery. Approximately 50% of patients with negative blood culture will respond to medical treatment.

SURGICAL INTERVENTION

Death from infective endocarditis is primarily due to HF, often accompanied by valve dysfunction. The use of surgery has been a critical advance in therapy. Valve damage/dysfunction causing moderate to severe HF (NYHA Class III or IV) is a strong indication for urgent surgery. An endocardial abscess involving the aortic root, valve ring, or ventricular septum is another indication for surgical intervention as are vegetations larger than 1 cm in diameter, a major embolic event, and failure or relapse after medical therapy.

CONCLUSION

In this chapter we attempted to provide a view of both myocarditis and endocarditis from a largely molecular and cellular perspective, including information concerning clinical diagnostics and therapeutics, as well as information from studies that used animal models, and *in vitro* systems, which have furthered defined the pathways of the underlying pathogenesis. Concerning the latter, it should be remembered that the development of multifactorial, and often heterogenous, myocarditis and its progression to DCM, as demarcated into discrete stages in the experimental animal models, misses the human host side of the equation and clearly cannot be entirely extrapolated to humans. However, these models have proved highly informative in the establishment of a conceptual framework to understand the events of pathogenesis, as well as to test out therapeutic options.

Future studies will further address the components of both the infectious elements (e.g., ligands, receptors and virulence) at the molecular level, as well as further unravel the sequence of molecular and cellular events involved in the autoimmune and inflammatory aspects of both myocarditis and endocarditis. The increased use of gene profiling, proteomic, and pharmacogenomic analysis will further identify novel components that play a role in these processes and regulators that might interface between infective and host responses, including the identification of genetic variants that might act as susceptibility loci. Understanding these events will undoubtedly contribute to the development of novel therapeutic approaches and may provide an early and more definitive diagnosis.

SUMMARY

- Myocarditis is an inflammatory disease of the myocardium associated with cardiac dysfunction, and frequently is a precursor of DCM.
- The clinical presentation of myocarditis is highly variable, ranging from asymptomatic to acute HF, severe dysrhythmias, and sudden death, and this is partly responsible for the undetermined disease prevalence.
- A broad spectrum of infectious agents, including viral, bacterial, fungal, and parasitic pathogens and noninfectious toxic agents (e.g., chemicals and drugs), can cause myocarditis. These agents are modulated by host-specific factors, including genetic susceptibility, age, gender, and immunocompetence. Viral agents (especially the enteroviruses) are the best characterized pathogens and have been increasingly identified in patients with myocarditis, mostly using molecular approaches.
- The diagnosis of myocarditis is made by a careful clinical and histochemistry analysis. Molecular screening is a highly useful adjunct.
- Pathogenesis of myocarditis has been approached by use of a variety of highly informative animal models. These have established that there are clearly autoimmune events and inflammatory pathways in the host that can attack both the infectious agent and also can cause further myocardial tissue damage and disease progression in the host, particularly in the development of chronic disease including DCM.
- Pathogenesis also proceeds by the molecular mimicry and immune response triggered by viral and self-antigens.
- Enhanced therapies for myocarditis will make use of this information and will not only potentiate the targeting of the infective agent but by the use of immunomodulation

- and suppression can modulate the host response and stem disease progression.
- Infective endocarditis is a microbial infection of the endocardial surface of the heart and most commonly is involved in either native or prosthetic heart valves but may also occur at other sites.
- The pathogen in infective endocarditis is most often bacterial, although a variety of other agents, including fungal pathogens, can produce this disorder. These can be both community and nosocomially acquired.
- Identification of the infective pathogen in infective endocarditis generally derives from blood and valve culture. Some pathogens require immunohistochemical and serological tests. The advent of molecular diagnostics (PCR and sequencing) can provide rapid, early, and definitive pathogen identification, particularly important for pathogens resistant to culturing.
- In infective endocarditis, pathogenesis requires the complicity of host susceptibility factors. An effective immune response in the host is necessary to promote adequate defense; therefore, immunocompromised hosts are more susceptible to infection.
- Molecular and genetic analyses have allowed the identification and characterization of virulence and adhesive factors of the pathogens; however, less is known about host susceptibility factors.
- Treatment of infective endocarditis with antibiotics is targeted to the infection and also can be used prophylactically (e.g., before oral/dental surgery); however, a considerable problem has been found with increasing multidrug resistance, particularly endemic with specific infections and with specific host populations (e.g., IV drug users).

References

1. Burian, J., Buser, P., and Eriksson, U. (2005). Myocarditis: the immunologist's view on pathogenesis and treatment. *Swiss Med. Wkly.* **135,** 359–364.
2. Calabrese, F., and Thiene, G. (2003). Myocarditis and inflammatory cardiomyopathy: Microbiological and molecular biological aspects. *Cardiovasc. Res.* **60,** 11–25.
3. Mason, J. W. (2003). Myocarditis and dilated cardiomyopathy: an inflammatory link. *Cardiovasc. Res.* **60,** 5–10.
4. Bergelson, J. M., Cunningham, J. A., Droguett, G., Kurt-Jones, E. A., Krithivas, A., Hong, J. S., Horwitz, M. S., Crowell, R. L., and Finberg, R. W. (1997). Isolation of a common receptor for Coxsackie B viruses and adenoviruses 2 and 5. *Science* **275,** 1320–1323.
5. Liu, P. P., and Opavsky, M. A. (2000). Viral myocarditis: Receptors that bridge the cardiovascular with the immune system? *Circulation. Res.* **86,** 253–254.
6. Selinka, H. C., Wolde, A., Sauter, M., Kandolf, R., and Klingel, K. (2004). Virus-receptor interactions of coxsackie B viruses and their putative influence on cardiotropism. *Med. Microbiol. Immunol. (Berl)* **193,** 127–131.
7. Martino, T. A., Petric, M., Brown, M., Aitken, K., Gauntt, C. J., Richardson, C. D., Chow, L. H., and Liu, P. P. (1998). Cardiovirulent coxsackieviruses and the decay-accelerating factor (CD55) receptor. *Virology* **244,** 302–314.
8. Orthopoulos, G., Triantafilou, K., and Triantafilou, M. (2004). Coxsackie B viruses use multiple receptors to infect human cardiac cells. *J. Med. Virol.* **74,** 291–299.
9. Kuhl, U., Pauschinger, M., Seeberg, B., Lassner, D., Noutsias, M., Poller, W., and Schultheiss, H. P. (2005). Viral persistence in the myocardium is associated with progressive cardiac dysfunction. *Circulation* **112,** 1965–1970.
10. Triantafilou, K., Orthopoulos, G., Vakakis, E., Ahmed, M. A., Golenbock, D. T., Lepper, P. M., and Triantafilou, M. (2005). Human cardiac inflammatory responses triggered by Coxsackie B viruses are mainly Toll-like receptor (TLR) 8-dependent. *Cell Microbiol.* **7,** 1117–1126.
11. Badorff, C., and Knowlton, K. U. (2004). Dystrophin disruption in enterovirus-induced myocarditis and dilated cardiomyopathy: from bench to bedside. *Med. Microbiol. Immunol. (Berl)* **193,** 121–126.
12. Badorff, C., Lee, G. H., Lamphear, B. J., Martone, M. E., Campbell, K. P., Rhoads, R. E., and Knowlton, K. U. (1999). Enteroviral protease 2A cleaves dystrophin: evidence of cytoskeletal disruption in an acquired cardiomyopathy. *Nat. Med.* **5,** 320–326.
13. Badorff, C., Berkely, N., Mehrotra, S., Talhouk, J. W., Rhoads, R. E., and Knowlton, K. U. (2000). Enteroviral protease 2A directly cleaves dystrophin and is inhibited by a dystrophin-based substrate analogue. *J. Biol. Chem.* **275,** 11191–11197.
14. Xiong, D., Lee, G. H., Badorff, C., Dorner, A., Lee, S., Wolf, P., and Knowlton, K. U. (2002). Dystrophin deficiency markedly increases enterovirus-induced cardiomyopathy: a genetic predisposition to viral heart disease. *Nat. Med.* **8,** 872–877.
15. Bultmann, B. D., Klingel, K., Nabauer, M., Wallwiener, D., and Kandolf, R. (2005). High prevalence of viral genomes and inflammation in peripartum cardiomyopathy. *Am. J. Obstet. Gynecol.* **193,** 363–365.
16. Ni, J., Bowles, N. E., Kim, Y. H., Demmler, G., Kearney, D., Bricker, J. T., and Towbin, J. A. (1997). Viral infection of the myocardium in endocardial fibroelastosis. Molecular evidence for the role of mumps virus as an etiologic agent. *Circulation* **95,** 133–139.
17. Bowles, N. E., Ni, J., Kearney, D. L., Pauschinger, M., Schultheiss, H. P., McCarthy, R., Hare, J., Bricker, J. T., Bowles, K. R., and Towbin, J. A. (2003). Detection of viruses in myocardial tissues by polymerase chain reaction. Evidence of adenovirus as a common cause of myocarditis in children and adults. *J. Am. Coll. Cardiol.* **42,** 466–472.
18. Fujioka, S., Kitaura, Y., Deguchi, H., Shimizu, A., Isomura, T., Suma, H., and Sabbah, H. N. (2004). Evidence of viral infection in the myocardium of American and Japanese patients with idiopathic dilated cardiomyopathy. *Am. J. Cardiol.* **94,** 602–605.
19. O'Malley, A., Barry-Kinsella, C., Hughes, C., Kelehan, P., Devaney, D., Mooney, E., and Gillan, J. (2003). Parvovirus infects cardiac myocytes in hydrops fetalis. *Pediatr. Dev. Pathol.* **6,** 414–420.
20. Morey, A. L., Keeling, J. W., Porter, H. J., and Fleming, K. A. (1992). Clinical and histopathological features of parvovirus B19 infection in the human fetus. *Br. J. Obstet. Gynaecol.* **99,** 566–574.

21. Katz, V. L., Chescheir, N. C., and Bethea, M. (1990). Hydrops fetalis from B19 parvovirus infection. *J. Perinatol.* **10,** 366–368.
22. Schowengerdt, K. O., Ni, J., Denfield, S. W., Gajarski, R. J., Bowles, N. E., Rosenthal, G., Kearney, D. L., Price, J. K., Rogers, B. B., Schauer, G. M., Chinnock, R. E., and Towbin, J. A. (1997). Association of parvovirus B19 genome in children with myocarditis and cardiac allograft rejection, diagnosis using the polymerase chain reaction. *Circulation* **96,** 3549–3554.
23. Heegaard, E. D., Eiskjaer, H., Baandrup, U., and Hornsleth, A. (1998). Parvovirus B19 infection associated with myocarditis following adult cardiac transplantation. *Scand. J. Infect. Dis.* **30,** 607–610.
24. Pankuweit, S., Moll, R., Baandrup, U., Portig, I., Hufnagel, G., and Maisch, B. (2003). Prevalence of the parvovirus B19 genome in endomyocardial biopsy specimens. *Hum. Pathol.* **34,** 497–503.
25. Kuhl, U., Pauschinger, M., Bock, T., Klingel, K., Schwimmbeck, C. P., Seeberg, B., Krautwurm, L., Poller, W., Schultheiss, H. P., and Kandolf, R. (2003). Parvovirus B19 infection mimicking acute myocardial infarction. *Circulation* **108,** 945–950.
26. Bultmann, B. D., Klingel, K., Sotlar, K., Bock, C. T., Baba, H. A., Sauter, M., and Kandolf, R. (2003). Fatal parvovirus B19-associated myocarditis clinically mimicking ischemic heart disease: an endothelial cell-mediated disease. *Hum. Pathol.* **34,** 92–95.
27. Klingel, K., Sauter, M., Bock, C. T., Szalay, G., Schnorr, J. J., and Kandolf, R. (2004). Molecular pathology of inflammatory cardiomyopathy. *Med. Microbiol. Immunol. (Berl)* **193,** 101–107.
28. Tschope, C., Bock, C. T., Kasner, M., Noutsias, M., Westermann, D., Schwimmbeck, P. L., Pauschinger, M., Poller, W. C., Kuhl, U., Kandolf, R., and Schultheiss, H. P. (2005). High prevalence of cardiac parvovirus B19 infection in patients with isolated left ventricular diastolic dysfunction. *Circulation* **111,** 879–886.
29. Barbaro, G. (2005). HIV-associated cardiomyopathy etiopathogenesis and clinical aspects. *Herz* **30,** 486–492.
30. Zareba, K. M., Miller, T. L., and Lipshultz, S. E. (2005). Cardiovascular disease and toxicities related to HIV infection and its therapies. *Exp. Opin. Drug Saf.* **4,** 1017–1025.
31. Barbaro, G. (2005). Reviewing the cardiovascular complications of HIV infection after the introduction of highly active antiretroviral therapy. *Curr. Drug Targets Cardiovasc. Haematol. Disord.* **5,** 337–343.
32. Barbaro, G., Di Lorenzo, G., Grisorio, B., and Barbarini, G. (1998). Incidence of dilated cardiomyopathy and detection of HIV in myocardial cells of HIV-positive patients. Gruppo Italiano per lo Studio Cardiologico dei Pazienti Affetti da AIDS. *N. Engl. J. Med.* **339,** 1093–1099.
33. Bowles, N. E., Kearney, D. L., Ni, J., Perez-Atayde, A. R., Kline, M. W., Bricker, J. T., Ayres, N. A., Lipshultz, S. E., Shearer, W. T., and Towbin, J. A. (1999). The detection of viral genomes by polymerase chain reaction in the myocardium of pediatric patients with advanced HIV disease. *J. Am. Coll. Cardiol.* **34,** 857–865.
34. Rodriguez, E. R., Nasim, S., Hsia, J., Sandin, R. L., Ferreira, A., Hilliard, B. A., Ross, A. M., and Garrett, C. T. (1991). Cardiac myocytes and dendritic cells harbor human immunodeficiency virus in infected patients with and without cardiac dysfunction: detection by multiplex, nested, polymerase chain reaction in individually microdissected cells from right ventricular endomyocardial biopsy tissue. *Am. J. Cardiol.* **68,** 1511–1520.
35. Rebolledo, M. A., Krogstad, P., Chen, F., Shannon, K. M., and Klitzner, T. S. (1998). Infection of human fetal cardiac myocytes by a human immunodeficiency virus-1-derived vector. *Circ. Res.* **83,** 738–742.
36. Kan, H., Xie, Z., and Finkel, M. S. (2000). HIV gp120 enhances NO production by cardiac myocytes through p38 MAP kinase-mediated NF-kappaB activation. *Am. J. Physiol. Heart Circ. Physiol.* **279,** H3138–H3143.
37. Chen, F., Shannon, K., Ding, S., Silva, M. E., Wetzel, G. T., Klitzner, T. S., and Krogstad, P. (2002). HIV type 1 glycoprotein 120 inhibits cardiac myocyte contraction. *AIDS Res. Hum. Retroviruses* **18,** 777–784.
38. Fiala, M., Popik, W., Qiao, J. H., Lossinsky, A. S., Alce, T., Tran, K., Yang, W., Roos, K. P., and Arthos, J. (2004). HIV-1 induces cardiomyopathy by cardiomyocyte invasion and gp120, Tat, and cytokine apoptotic signaling. *Cardiovasc. Toxicol.* **4,** 97–107.
39. Fiala, M., Murphy, T., MacDougall, J., Yang, W., Luque, A., Iruela-Arispe, L., Cashman, J., Buga, G., Byrns, R. E., Barbaro, G., and Arthos, J. (2004). HAART drugs induce mitochondrial damage and intercellular gaps and gp120 causes apoptosis. *Cardiovasc. Toxicol.* **4,** 327–337.
40. Raidel, S. M., Haase, C., Jansen, N. R., Russ, R. B., Sutliff, R. L., Velsor, L. W., Day, B. J., Hoit, B. D., Samarel, A. M., and Lewis, W. (2002). Targeted myocardial transgenic expression of HIV Tat causes cardiomyopathy and mitochondrial damage. *Am. J. Physiol. Heart Circ. Physiol.* **282,** H1672–H1678.
41. Lewis, W. (2003). Use of the transgenic mouse in models of AIDS cardiomyopathy. *AIDS* **17,** S36–S45.
42. Hardman, J. M., and Earle, K. M. (1969). Myocarditis in 200 fatal meningococcal infections. *Arch. Pathol.* **87,** 318–325.
43. Nagi, K. S., Joshi, R., and Thakur, R. K. (1996). Cardiac manifestations of Lyme disease: A review. *Can. J. Cardiol.* **12,** 503–506.
44. Defosse, D. L., Duray, P. H., and Johnson, R. C. (1992). The NIH-3 immunodeficient mouse is a model for Lyme borreliosis myositis and carditis. *Am. J. Pathol.* **141,** 3–10.
45. Armstrong, A. L., Barthold, S. W., Persing, D. H., and Beck, D. S. (1992). Carditis in Lyme disease susceptible and resistant strains of laboratory mice infected with *Borrelia burgdorferi*. *Am. J. Trop. Med. Hyg.* **47,** 249–258.
46. Raveche, E. S., Schutzer, S. E., Fernandes, H., Bateman, H., McCarthy, B. A., Nickell, S. P., and Cunningham, M. W. (2005). Evidence of *Borrelia* autoimmunity-induced component of Lyme carditis and arthritis. *J. Clin. Microbiol.* **43,** 850–856.
47. McKisic, M. D., Redmond, W. L., and Barthold, S. W. (2000). Cutting edge: T cell-mediated pathology in murine Lyme borreliosis. *J. Immunol.* **164,** 6096–6099.
48. Schaible, U. E., Kramer, M. D., Museteanu, C., Zimmer, G., Mossmann, H., and Simon, M. M. (1989). The severe combined immunodeficiency (scid) mouse. A laboratory model for the analysis of Lyme arthritis and carditis. *J. Exp. Med.* **170,** 1427–1432.

49. Cadavid, D., Bai, Y., Hodzic, E., Narayan, K., Barthold, S. W., and Pachner, A. R. (2004). Cardiac involvement in non-human primates infected with the Lyme disease spirochete *Borrelia burgdorferi*. *Lab. Invest.* **84,** 1439–1450.
50. Anguita, J., Samanta, S., Revilla, B., Suk, K., Das, S., Barthold, S. W., and Fikrig, E. (2000). *Borrelia burgdorferi* gene expression in vivo and spirochete pathogenicity. *Infect. Immun.* **68,** 1222–1230.
51. Aguero-Rosenfeld, M. E., Wang, G., Schwartz, I., and Wormser, G. P. (2005). Diagnosis of lyme borreliosis. *Clin. Microbiol. Rev.* **18,** 484–509.
52. Marin-Garcia, J., and Mirvis, D. M. (1984). Myocardial disease in Rocky Mountain spotted fever: Clinical, functional, and pathologic findings. *Pediatr. Cardiol.* **5,** 149–154.
53. Marin-Garcia, J., Gooch, W. M. 3rd., and Coury, D. L. (1981). Cardiac manifestations of Rocky Mountain spotted fever. *Pediatrics* **67,** 358–361.
54. Walker, D. H., Paletta, C. E., and Cain, B. G. (1980). Pathogenesis of myocarditis in Rocky Mountain spotted fever. *Arch. Pathol. Lab. Med.* **104,** 171–174.
55. Nilsson, K., Lindquist, O., and Pahlson, C. (1999). Association of *Rickettsia helvetica* with chronic perimyocarditis in sudden cardiac death. *Lancet* **354,** 1169–1177.
56. Uzoigwe, C. (2005). *Campylobacter* infections of the pericardium and myocardium. *Clin. Microbiol. Infect.* 11, 253–255.
57. Saikku, P. (1996). *Chlamydia pneumoniae* and cardiovascular diseases. *Clin. Microbiol. Infect.* **1 Suppl 1,** S19–S22.
58. Paz, A., and Potasman, I. (2002). Mycoplasma-associated carditis. Case reports and review. *Cardiology* **97,** 83–88.
59. Fuenmayor, C., Higuchi, M. L., Carrasco, H., Parada, H., Gutierrez, P., Aiello, V., and Palomino, S. (2005). Acute Chagas' disease: immunohistochemical characteristics of T cell infiltrate and its relationship with *T. cruzi* parasitic antigens. *Acta Cardiol.* **60,** 33–37.
60. Pinto, A. Y., Valente, S. A., ad Valente, Vda C. (2004). Emerging acute Chagas disease in Amazonian Brazil: Case reports with serious cardiac involvement. *Braz. J. Infect. Dis.* **8,** 454–460.
61. Talvani, A., Ribeiro, C. S., Aliberti, J. C. S., Michailowsky, V., Santos, P. V., Murta, S. M., Romanha, A. J., Almeida, I. C., Farber, J., Lannes-Vieira, J., Silva, J. S., and Gazzinelli, R. T. (2000). Kinetics of cytokine gene expression in experimental chagasic cardiomyopathy: tissue parasitism and endogenous IFN-gamma as important determinants of chemokine mRNA expression during infection with *Trypanosoma cruzi*. *Microbes Infect.* **2,** 851–866.
62. dos Santos, P. V.A., Roffê, E., Santiago, H. C., Torres, R. A., Marino, A. P., Paiva, C. N., Silva, A. A., Gazzinelli, R. T., and Lannes-Vieira, J. (2001). Prevalence of CD8+ T cells in *Trypanosoma cruzi*-elicited myocarditis is associated with acquisition of CD62L (Low) HFA-1 (High) VLA-4 (High) activation phenotype and expression of IFN-gamma inducible adhesion and chemoattractant molecules. *Microbes Infect.* **3,** 971–984.
63. Marino, A. P., da Silva, A., dos Santos, P., Pinto, L. M., Gazzinelli, R. T., Teixeira, M. M., and Lannes-Vieira, J. (2004). Regulated on activation, normal T cell expressed and secreted (RANTES) antagonist (Met-RANTES) controls the early phase of *Trypanosoma cruzi*-elicited myocarditis. *Circulation* **110,** 1443–1449.
64. Talvani, A., Rocha, M. O. C., Ribeiro, A .L., Correa-Oliveira, R., and Teixeira, M. M. (2004). Chemokine receptor expression on the surface of peripheral blood mononuclear cells in Chagas disease. *J. Infect. Dis.* **189,** 214–220.
65. Soler, A. P., Gilliard, G., Xiong, Y., Knudsen, K. A., Martin, J. L., De Suarez, C. B., Mota Gamboa, J. D., Mosca, W., and Zoppi, L. B. (2001). Overexpression of neural cell adhesion molecule in Chagas' myocarditis. *Hum. Pathol.* **32,** 149–155.
66. Puljiz, I., Beus, A., Kuzman, I., and Seiwerth, S. (2005). Electrocardiographic changes and myocarditis in trichinellosis: a retrospective study of 154 patients. *Ann. Trop. Med. Parasitol.* **99,** 403–411.
67. Paolocci, N., Sironi, M., Bettini, M., Bartoli, G., Michalak, S., Bandi, C., Magni, F., and Bruschi, F. (1998). Immunopathological mechanisms underlying the time-course of *Trichinella spiralis* cardiomyopathy in rats. *Virchows Arch.* **432,** 261–266.
68. Pieroni, M., Cavallaro, R., Chimenti, C., Smeraldi, E., and Frustaci, A. (2004). Clozapine-induced hypersensitivity myocarditis. *Chest* **126,** 1703–1705.
69. Spry, C. J., Take, M., and Tai, P. C. (1985). Eosinophilic disorders affecting the myocardium and endocardium: a review. *Heart Vessels Suppl.* **1,** 240–242.
70. Patella, V., de Crescenzo, G., Marino, I., Genovese, A., Adt, M., Gleich, G. J., and Marone, G. (1997). Eosinophil granule proteins are selective activators of human heart mast cells. *Int. Arch. Allergy Immunol.* **113,** 200–202.
71. Daniels, P. R., Berry, G. J., Tazelaar, H. D., and Cooper, L. T. (2000). Giant cell myocarditis as a manifestation of drug hypersensitivity. *Cardiovasc. Pathol.* **9,** 287–291.
72. Molina, H. A., and Kierszenbaum, F. (1989). Eosinophil activation in acute and chronic chagasic myocardial lesions and deposition of toxic eosinophil granule proteins on heart myofibers. *J. Parasitol.* **75,** 129–133.
73. Kumar, D., Kirshenbaum, L., Li, T., Danelisen, I., and Singal, P. (1999). Apoptosis in isolated adult cardiomyocytes exposed to Adriamycin. *Ann. N. Y. Acad. Sci.* **874,** 156–168.
74. Ferrans, V. J., Clark, J. R., Zhang, J., Yu, Z.X., and Herman, E. H. (1997). Pathogenesis and prevention of doxorubicin cardiomyopathy. *Tsitologiia* **39,** 928–937.
75. Berthiaume, J. M., Oliveira, P. J., Fariss, M. W., and Wallace, K. B. (2005). Dietary vitamin E decreases doxorubicin–induced oxidative stress without preventing mitochondrial dysfunction. *Cardiovasc. Toxicol.* **5,** 257–268.
76. Lebrecht, D., Kokkori, A., Ketelsen, U. P., Setzer, B., and Walker, U. A. (2005). Tissue-specific mtDNA lesions and radical-associated mitochondrial dysfunction in human hearts exposed to doxorubicin. *J. Pathol.* **207,** 436–444.
77. Chaiswing, L., Cole, M. P., St. Clair, D. K., Ittarat, W., Szweda, L. I., and Oberley, T. D. (2004). Oxidative damage precedes nitrative damage in adriamycin-induced cardiac mitochondrial injury. *Toxicol. Pathol.* **32,** 536–547.
78. Rump, A. F., Theisohn, M., and Klaus, W. (1995). The pathophysiology of cocaine cardiotoxicity. *Forensic Sci. Int.* **71,** 103–115.
79. Kloner, R. A., Hale, S., Alker, K., and Rezkalla, S. (1992). The effects of acute and chronic cocaine use on the heart. *Circulation* **85,** 407–419.
80. Zhang, H., Li, Y., Peng, T., Aasa, M., Zhang, L., Yang, Y., and Archard, L. C. (2000). Localization of enteroviral antigen in

myocardium and other tissues from patients with heart muscle disease by an improved immunohistochemical technique. *J. Histochem. Cytochem.* **48,** 579–484.

81. Akhtar, N., Ni, J., Stromberg, D., Rosenthal, G. L., Bowles, N. E., and Towbin, J. A. (1999). Tracheal aspirate as a substrate for polymerase chain reaction detection of viral genome in childhood pneumonia and myocarditis. *Circulation* **99,** 2011–2018.

82. Calabrese, F., Rigo, E., Milanesi, O., Boffa, G. M., Angelini, A., Valente, M., and Thiene, G. (2002). Molecular diagnosis of myocarditis and dilated cardiomyopathy in children: clinicopathologic features and prognostic implications. *Diagn. Mol. Pathol.* **11,** 212–221.

83. Okabe, M., Fukuda, K., Arakawa, K., and Kikuchi, M. (1997). Chronic variant of myocarditis associated with hepatitis C virus infection. *Circulation* **96,** 22–24.

84. Pauschinger, M., Noutsias, M., Kuhl, U., and Schultheiss, H. P. (2004). Antiviral therapy in viral heart disease. *Herz* **29,** 618–623.

85. Wolfgram, L. J., Beisel, K. W., Hershkowitz, A., and Rose, N. R. (1986). Variations in the susceptibility to coxsackie B3-induced myocarditis among different strains of mice. *J. Immunol.* **136,** 1848–1852.

86. Chow, L. H., Beisel, K. W., and McManus, B. M. (1992). Enteroviral infection of mice with severe combined immunodeficiency. Evidence for direct viral pathogenesis of myocardial injury. *Lab. Invest.* **66,** 24–31.

87. Klingel, K., Stephan, S., Sauter, M., Zell, R., McManus, B. M., Bultmann, B., and Kandolf, R. (1996). Pathogenesis of murine enterovirus myocarditis: virus dissemination and immune cell targets. *J. Virol.* **70,** 8888–8895.

88. Mena, I., Perry, C. M., Harkins, S., Rodriguez, F., Gebhard, J., and Whitton, J. L. (1999). The role of B lymphocytes in coxsackievirus B3 infection. *Am. J. Pathol.* **155,** 1205–1215.

89. Matsumori, A., Tomioka, N., and Kawai, C. (1988). Protective effect of recombinant alpha interferon on coxsackievirus B3 myocarditis in mice. *Am. Heart. J.* **115,** 1229–1232.

90. Lane, J. R., Neumann, D. A., Lafond-Walker, A., Herskowitz, A., and Rose, N. R. (1992). Interleukin 1 or tumor necrosis factor can promote Coxsackie B3-induced myocarditis in resistant B10.A mice. *J. Exp. Med.* **175,** 1123–1129.

91. Zaragoza, C., Ocampo, C., Saura, M., Leppo, M., Wei, X. Q., Quick, R., Moncada, S., Liew, F. Y., and Lowenstein, C. J. (1998). The role of inducible nitric oxide synthase in the host response to Coxsackievirus myocarditis. *Proc. Natl. Acad. Sci. USA* **95,** 2469–2474.

92. Gluck, B., Schmidtke, M., Merkle, I., Stelzner, A., and Gemsa, D. (2001). Persistent expression of cytokines in the chronic stage of CVB3-induced myocarditis in NMRI mice. *J. Mol. Cell Cardiol.* **33,** 1615–1626.

93. Woodruff, J. F., and Woodruff, J. J. (1974). Involvement of T lymphocytes in the pathogenesis of coxsackie virus B3 heart disease. *J. Immunol.* **113,** 1726–1734.

94. Feldman, A. M., and McNamara, D. (2000). Myocarditis. *N. Engl. J. Med.* **343,** 1388–1398.

95. Caforio, A. L., Bonifacio, E., Keeling, P. J., Grazzini, M., Schiaffino, S., Bottazzo, G F., and McKenna, W. J. (1992). Idiopathic dilated cardiomyopathy: a persistent viral infection or an organ-specific autoimmune disease? The trial of 2 major pathogenetic hypotheses. *G. Ital. Cardiol.* **22,** 63–72.

96. Kawai, C. (1999). From myocarditis to cardiomyopathy: mechanisms of inflammation and cell death: learning from the past for the future. *Circulation* **99,** 1091–1100.

97. Opavsky, M. A., Penninger, J., Aitken, K., Wen, W. H., Dawood, F., Mak, T., and Liu, P. (1999). Susceptibility to myocarditis is dependent on the response of αβ T lymphocytes to coxsackieviral infection. *Circ. Res.* **85,** 551–558.

98. Neu, N., Rose, N. R., Beisel, K. W., Herskowitz, A., Gurri-Glass, G., and Craig, S. W. (1987). Cardiac myosin induces myocarditis in genetically predisposed mice. *J. Immunol.* **139,** 3630–3636.

99. Kasahara, H., Itoh, M., Sugiyama, T., Kido, N., Hayashi, H., Saito, H., Tsukita, S., and Kato, N. (1994). Autoimmune myocarditis induced in mice by cardiac C-protein. Cloning of complementary DNA encoding murine cardiac C-protein and partial characterization of the antigenic peptides. *J. Clin. Invest.* **94,** 1026–1036.

100. Huber, S. A., and Cunningham, M. W. (1996). Streptococcal M protein peptide with similarity to myosin induces CD4+ T cell-dependent myocarditis in MRL/++ mice and induces partial tolerance against coxsakieviral myocarditis. *J. Immunol.* **156,** 3528–3534.

101. Cihakova, D., Sharma, R. B., Fairweather, D., Afanasyeva, M., and Rose, N. R. (2004). Animal models for autoimmune myocarditis and autoimmune thyroiditis. *Methods Mol. Med.* **102,** 175–193.

102. Wessely, R. (2004). Coxsackieviral replication and pathogenicity: lessons from gene modified animal models. *Med. Microbiol. Immunol. (Berl)* **193,** 71–74.

103. Liu, P., Penninger, J., Aitken, K., Sole, M., and Mak, T. (1995). The role of transgenic knockout models in defining the pathogenesis of viral heart disease. *Eur. Heart J.* **16,** 25–27.

104. Ayach, B., Fuse, K., Martino, T., and Liu, P. (2003). Dissecting mechanisms of innate and acquired immunity in myocarditis. *Curr. Opin. Cardiol.* **18,** 175–181.

105. Chakrabarty, A., and Beutner, K. (2004). Therapy of other viral infections: herpes to hepatitis. *Dermatol. Ther.* **17,** 465–490.

106. Matsumori, A. (2000). Hepatitis C virus and cardiomyopathy. *Herz* **25,** 249–254.

107. Werk, D., Schubert, S., Lindig, V., Grunert, H. P., Zeichhardt, H., Erdmann, V. A., and Kurreck, J. (2005). Developing an effective RNA interference strategy against a plus-strand RNA virus: silencing of coxsackievirus B3 and its cognate coxsackievirus-adenovirus receptor. *Biol. Chem.* **386,** 857–863.

108. Burian, J., Buser, P., and Eriksson, U. (2005). Myocarditis: the immunologist's view on pathogenesis and treatment. *Swiss Med. Wkly.* **135,** 359–364.

109. Matsumori, A., Tomioka, N., and Kawai, C. (1988). Protective effect of recombinant alpha interferon on coxsackievirus B3 myocarditis in mice. *Am. Heart* J. **115,** 1229–1232.

110. Kuhl, U., Pauschinger, M., Schwimmbeck, P. L., Seeberg, B., Lober, C., Noutsias, M., Poller, W., and Schultheiss, H. P. (2003). Interferon-beta treatment eliminates cardiotropic viruses and improves left ventricular function in patients with myocardial persistence of viral genomes and left ventricular dysfunction. *Circulation* **107,** 2793–2798.

111. Mann, D. L., McMurray, J. J., Packer, M., Swedberg, K., Borer, J. S., Colucci, W. S., Djian, J., Drexler, H., Feldman, A.,

Kober, L., Krum, H., Liu, P., Nieminen, M., Tavazzi, L., van Veldhuisen, D. J., Waldenstrom, A., Warren, M., Westheim, A., Zannad, F., and Fleming, T. (2004). Targeted anticytokine therapy in patients with chronic heart failure: results of the Randomized Etanercept Worldwide Evaluation (RENEWAL). *Circulation* **109**, 1594–1602.

112. Mason, J. W., O'Connell, J. B., Herskowitz, A., Rose, N. R., McManus, B. M., Billingham, M. E., and Moon, T. E. (1995). A clinical trial of immunosuppressive therapy for myocarditis. The Myocarditis Treatment Trial Investigators. *N. Engl. J. Med.* **333**, 269–275.

113. Chau, E. M., Chow, W. H., Chiu, C. S., and Wang, E. (2006). Treatment and outcome of biopsy-proven fulminant myocarditis in adults. *Int. J. Cardiol.* **110**, 405–406.

114. Frustaci, A., Calabrese, F., Chimenti, C., Pieroni, M., Thiene, G., and Maseri, A. (2002). Lone hepatitis C virus myocarditis responsive to immunosuppressive therapy. *Chest* **122**, 1348–1356.

115. Frustaci, A., Chimenti, C., Calabrese, F., Pieroni, M., Thiene, G., and Maseri, A. (2003). Immunosuppressive therapy for active lymphocytic myocarditis: virological and immunologic profile of responders versus nonresponders. *Circulation* **107**, 857–863.

116. Mylonakis, E., and Calderwood, S. B. (2001). Infective endocarditis in adults. *N. Engl. J. Med.* **345**, 1318–1330.

117. Moreillon, P., and Que, Y. A. (2004). Infective endocarditis. *Lancet* **363**, 139–149.

118. Tak, T., and Shukla, S. K. (2004). Molecular diagnosis of infective endocarditis: A helpful addition to the Duke criteria. *Clin. Med. Res.* **2**, 206–208.

119. Saiman, L., Prince, A., and Gersony, W. M. (1993). Pediatric infective endocarditis in the modern era. *J. Pediatr.* **122**, 847–853.

120. Martin, J. M., Neches, W. H., and Wald, E. R. (1997). Infective endocarditis: 35 years of experience at a children's hospital. *Clin. Infect. Dis.* **24**, 669–675.

121. Tissieres, P., Jaeggi, E. T., Beghetti, M., and Gervaix, A. (2005). Increase of fungal endocarditis in children. *Infection* **33**, 267–272.

122. Anderson, D. J., Olaison, L., McDonald, J. R., Miro, J. M., Hoen, B., Selton–Suty, C., Doco-Lecompte, T., Abrutyn, E., Habib, G., Eykyn, S., Pappas, P. A., Fowler, V. G., Sexton, D. J., Almela, M., Corey, G. R., and Cabell, C. H. (2005). Enterococcal prosthetic valve infective endocarditis: report of 45 episodes from the International Collaboration on Endocarditis-merged database. *Eur. J. Clin. Microbiol. Infect. Dis.* **24**, 665–670.

123. Romano, G., Carozza, A., Della Corte, A., De Santo, L. S., Amarelli, C., Torella, M., De Feo, M., Cerasuolo, F., and Cotrufo, M. (2004). Native versus primary prosthetic valve endocarditis: comparison of clinical features and long-term outcome in 353 patients. *J. Heart Valve Dis.* **13**, 200–208.

124. Keys, T. F. (2000). Infective endocarditis: prevention, diagnosis, treatment, referral. *Cleve. Clin. J. Med.* **67**, 353–360.

125. Rivas, P., Alonso, J., Moya, J., de Gorgolas, M., Martinell, J., and Fernandez Guerrero, M. L. (2005). The impact of hospital-acquired infections on the microbial etiology and prognosis of late-onset prosthetic valve endocarditis. *Chest* **128**, 764–771.

126. McCormick, J. K., Tripp, T. J., Dunny, G. M., and Schlievert, P. M. (2002). Formation of vegetations during infective endocarditis excludes binding of bacterial-specific host antibodies to *Enterococcus faecalis*. *J. Infect. Dis.* **185**, 994–997.

127. Minhas, T., Ludlam, H. A., Wilks, M., and Tabaqchali, S. (1995). Detection by PCR and analysis of the distribution of a fibronectin-binding protein gene (fbn) among staphylococcal isolates. *J. Med. Microbiol.* **42**, 96–101.

128. Dubin, G. (2002). Extracellular proteases of Staphylococcus spp. *Biol. Chem.* **383**, 1075–1086.

129. McGavin, M. J., Zahradka, C., Rice, K., and Scott, J. E. (1997). Modification of the *Staphylococcus aureus* fibronectin binding phenotype by V8 protease. *Infect. Immun.* **65**, 2621–2628.

130. Sinha, B., Francois, P. P., Nusse, O., Foti, M., Hartford, O. M., Vaudaux, P., Foster, T. J., Lew, D. P., Herrmann, M., and Krause, K. H. (1999). Fibronectin-binding protein acts as *Staphylococcus aureus* invasin via fibronectin bridging to integrin α5β1. *Cell Microbiol.* **1**, 101–117.

131. van Wamel, W., Xiong, Y. Q., Bayer, A. S., Yeaman, M. R., Nast, C. C., and Cheung, A. L. (2002). Regulation of *Staphylococcus aureus* type 5 capsular polysaccharides by agr and sarA in vitro and in an experimental endocarditis model. *Microb. Pathog.* **33**, 73–79.

132. Said-Salim, B., Dunman, P. M., McAleese, F. M., Macapagal, D., Murphy, E., McNamara, P. J., Arvidson, S., Foster, T. J., Projan, S. J., and Kreiswirth, B. N. (2003). Global regulation of Staphylococcus aureus genes by Rot. *J. Bacteriol.* **185**, 610–619.

133. Kupferwasser, L. I., Yeaman, M. R., Nast, C. C., Kupferwasser, D., Xiong, Y. Q., Palma, M., Cheung, A. L., and Bayer, A. S. (2003). Salicylic acid attenuates virulence in endovascular infections by targeting global regulatory pathways in Staphylococcus aureus. *J. Clin. Invest.* **112**, 222–233.

134. Baddour, L. M., Sullam, P. M., and Bayer, A. S. (2001). The pathogenesis of infective endocarditis. *In* "Molecular Medical Microbiology." (M. Sussman, Ed.), pp. 999–1020. Academic Press, San Diego, CA.

135. Baddour, L. M. (1999). Immunization for prevention of infective endocarditis. *Curr. Infect. Dis. Rep.* **1**, 126–128.

136. Yeaman, M. R., and Bayer, A. S. (1999). Antimicrobial peptides from platelets. *Drug Resist. Updat.* **2**, 116–126.

137. Kupferwasser, L. I., Yeaman, M. R., Shapiro, S. M., Nast, C. C., and Bayer, A. S. (2002). In vitro susceptibility to thrombin-induced platelet microbicidal protein is associated with reduced disease progression and complication rates in experimental *Staphylococcus aureus* endocarditis: microbiological, histopathologic, and echocardiographic analyses. *Circulation* **105**, 746–752.

138. Sakoulas, G., Eliopoulos, G. M., Fowler, V. G., Jr., Moellering, R. C., Jr., Novick, R. P., Lucindo, N., Yeaman, M. R., and Bayer, A. S. (2005). Reduced susceptibility of *Staphylococcus aureus* to vancomycin and platelet microbicidal protein correlates with defective autolysis and loss of accessory gene regulator (agr) function. *Antimicrob. Agents Chemother.* **49**, 2687–2692.

139. Weidenmaier, C., Peschel, A., Kempf, V. A., Lucindo, N., Yeaman, M. R., and Bayer, A. S. (2005). DltABCD- and MprF-mediated cell envelope modifications of *Staphylococcus aureus* confer resistance to platelet microbicidal proteins and contribute to virulence in a rabbit endocarditis model. *Infect. Immun.* **73**, 8033–8038.

140. Xiong, Y. Q., Bayer, A. S., Yeaman, M. R., Van Wamel, W., Manna, A. C., and Cheung, A. L. (2004). Impacts of sarA and agr in *Staphylococcus aureus* strain Newman on fibronectin-binding protein A gene expression and fibronectin adherence capacity in vitro and in experimental infective endocarditis. *Infect. Immun.* **72,** 1832–1836.

141. Moreillon, P., Que, Y. A., and Bayer, A. S. (2002). Pathogenesis of streptococcal and staphylococcal endocarditis. *Infect. Dis. Clin. North Am.* **16,** 297–318.

142. Herzberg, M. C. (1996). Platelet-streptococcal interactions in endocarditis. *Crit. Rev. Oral Biol. Med.* **7,** 222–236.

143. Pietrocola, G., Schubert, A., Visai, L., Torti, M., Fitzgerald, J. R., Foster, T. J., Reinscheid, D. J., and Speziale, P. (2005). FbsA, a fibrinogen-binding protein from *Streptococcus agalactiae*, mediates platelet aggregation. *Blood* **105,** 1052–1059.

144. Chow, J. W., Thal, L. A., Perri, M. B., Vazquez, J. A., Donabedian, S. M., Clewell, D. B., and Zervos, M. J. (1993). Plasmid-associated hemolysin and aggregation substance production contribute to virulence in experimental enterococcal endocarditis. *Antimicrob. Agents Chemother.* **37,** 2474–2477.

145. Schlievert, P. M., Gahr, P. J., Assimacopoulos, A. P., Dinges, M. M., Stoehr, J. A., Harmala, J. W., Hirt, H., and Dunny, G. M. (1998). Aggregation and binding substances enhance pathogenicity in rabbit models of *Enterococcus faecalis* endocarditis. *Infect. Immun.* **66,** 218–223.

146. Lowe, A. M., Lambert, P. A., and Smith, A. W. (1995). Cloning of an *Enterococcus faecalis* endocarditis antigen: homology with adhesins from some oral streptococci. *Infect. Immun.* **63,** 703–706.

147. Shepard, B. D., and Gilmore, M. S. (2002). Differential expression of virulence-related genes in *Enterococcus faecalis* in response to biological cues in serum and urine. *Infect. Immun.* **70,** 4344–4352.

148. Coque, T. M., Patterson, J. E., Steckelberg, J. M., and Murray, B. E. (1995). Incidence of hemolysin, gelatinase, and aggregation substance among enterococci isolated from patients with endocarditis and other infections and from feces of hospitalized and community-based persons. *J. Infect. Dis.* **171,** 1223–1229.

149. Guzmàn, C. A., Pruzzo, C., Plate, M., Guardati, M. C., and Calegari, L. (1991). Serum dependent expression of *Enterococcus faecalis* adhesins involved in the colonization of heart cells. *Microb. Pathog.* **11,** 399–409.

150. Kreft, B., Marre, R., Schramm, U., and Wirth, R. (1992). Aggregation substance of *Enterococcus faecalis* mediates adhesion to cultured renal tubular cells. *Infect. Immun.* **60,** 25–30.

151. Low, Y. L., Jakubovics, N. S., Flatman, J. C., Jenkinson, H. F., and Smith, A. W. (2003). Manganese-dependent regulation of the endocarditis-associated virulence factor EfaA of *Enterococcus faecalis*. *J. Med. Microbiol.* **52,** 113–119.

152. Johnson, A. P. (1994). The pathogenicity of enterococci. *J. Antimicrob. Chemother.* **33,** 1083–1089.

153. Carbon, C. (1994). Animal models of endocarditis. *Int. J. Biomed. Comput.* **36,** 59–67.

154. Gutschik, E. (1999). New developments in the treatment of infective endocarditis infective cardiovasculitis. *Int. J. Antimicrob. Agents* **13,** 79–92.

155. Millar, B., Moore, J., Mallon, P., Xu, J., Crowe, M., Mcclurg, R., Raoult, D., Earle, J., Hone, R., and Murphy, P. (2001). Molecular diagnosis of infective endocarditis—a new Duke's criterion. *Scand. J. Infect. Dis.* **33,** 673–680.

156. Kotilainen, .P, Heiro, M., Jalava, J., Rantakokko, V., Nikoskelainen, J., Nikkari, S., and Rantakokko-Jalava, K. (2006). Aetiological diagnosis of infective endocarditis by direct amplification of rRNA genes from surgically removed valve tissue. An 11-year experience in a Finnish teaching hospital. *Ann. Med.* **38,** 263–273.

157. Moore, J. E., Millar, B. C., Yongmin, X., Woodford, N., Vincent, S., Goldsmith, C. E., McClurg, R. B., Crowe, M., Hone, R., and Murphy, P. G. (2001). A rapid molecular assay for the detection of antibiotic resistance determinants in causal agents of infective endocarditis. *J. Appl. Microbiol.* **90,** 719–726.

SECTION V

The Failing Heart

CHAPTER **15**

Molecular Analysis of Heart Failure and Remodeling

OVERVIEW

Heart failure (HF), an endemic health problem of great magnitude in the Western world, is essentially the final and common pathway of cardiovascular diseases (CVD) that results in impairment of myocardial systolic and/or diastolic function. Triggers of HF include impaired pump function after myocyte death from MI, stimuli resulting from myocardial ischemia (e.g., oxidative and nitrosative stress), and abnormal hemodynamic loading conditions such as found in hypertension. Interestingly, HF can be elicited by mutations in nuclear and mitochondrial DNA encoded proteins, leading to defects in a variety of myocardial functions, including signal transduction, transcriptional activation and bioenergetic metabolic remodeling, oxidative stress, calcium handling, and sarcomeric/contractile protein function. These stimuli activate a series of biomechanical stress-dependent signaling cascades involving cardiomyocyte sensors, leading to altered signal transduction pathways and resulting in the activation of transcription factors, coactivators, and corepressors orchestrating myocardial gene expression. However, despite all this knowledge and considerable clinical and research advances, the morbidity and mortality of HF remain high. Accordingly, there is an urgent need to develop new paradigms and to identify novel therapeutic targets for HF.

In this chapter, we will discuss the molecular basis of HF, as well as its application in current and prospective diagnostic and therapeutic strategies.

INTRODUCTION

Heart failure (HF) is a condition associated with significant mortality and morbidity.[1] In the United States, HF affects approximately 5 million people, accounting for nearly 50,000 deaths each year and an estimated direct and indirect cost of near $28 billion.[2,3] There is evidence to indicate that both the prevalence and incidence of HF is increasing, with projected further increases over the next decade.[4,5] The hallmark of HF is that of relentless clinical *progression*, which is most often manifested as repeated hospitalizations.[4] Although over the past decade neurohormonal inhibition, including the use of angiotensin-converting enzyme (ACE) inhibitors, with or without angiotensin receptor blockers (ARBs), β-blockers, and aldosterone receptor blockers, has resulted in improved clinical outcomes in these patients, the disappointing results of several recent large clinical trials have led to the suggestion that neurohormonal inhibition may have reached its limit for efficacy.[6]

Common underlying disorders in HF include cardiomyopathy, primary/idiopathic or acquired from previous myocardial infarctions (MI), chronic myocardial ischemia, hypertension, diabetes, valvular disorders, dysrhythmias, and congenital defects. In response to these initial insults and also critical to the progression of most forms of HF that are encountered in clinical practice is the process of cardiac remodeling.[7,8] Cardiac remodeling involves changes in both the myocytes and the extracellular matrix (ECM), the latter includes the activation of proteolytic enzymes, leading to the degradation and reorganization of collagen.[9–12]

In Fig. 1, a working dynamic model is depicted encompassing key biological circuits involved in the onset and progression of HF. Primary triggers include impaired pump function after myocyte death from MI, stimuli resulting from myocardial ischemia (e.g., oxidative and nitrosative stress), and abnormal hemodynamic loading conditions such as those found in hypertension.[13] These stimuli activate a biomechanical stress-dependent signaling cascade involving a series of sensors, leading to altered signal transduction pathways and resulting in the activation of transcription factors, coactivators, and corepressors for myocardial gene expression, as well as cardiac cell effector mechanisms like calcium cycling, mitochondrial metabolism, and myocyte growth and loss (primarily by apoptosis) that underlie cardiac remodeling. These cellular events may cooperatively interact with each other and can culminate in extensive ventricular structural changes and in contractile dysfunction, as well as leading to secondary neurohumoral responses (such as adrenergic drive) that can further promote or amplify the signal transduction changes.[13] Mutations that lead to cardiomyopathy can affect proteins at various stages of this pathway (from the initial stimuli to the progression of signal transduction pathways, transcription factors, and mitochondrial metabolism) and can contribute to the HF phenotype.[14] This model also suggests discrete steps by which HF might be targeted by use of gene-based and cell-based therapies to stem HF progression (e.g., replacement of dead myocytes) and to augment cardiac self-repair.

The morphological patterns that accompany cardiac remodeling include hypertrophy and chamber dilatation with associated progressive impairment in function.[8] The molecular basis and cellular triggers for the underlying structural changes associated with cardiac remodeling and forms of HF commonly encountered in clinical practice will be the subject of review in this chapter. Discussion will focus on HF in general rather than detailed discussions of the specific form of cardiomyopathies, which were covered in Chapter 13.

MOLECULAR AND CELLULAR BASIS OF HF

Relevant Pre-Clinical Models of HF

As should be evident from several chapters in this volume, many advances in cardiovascular research have been made possible by the use of animal models. Given the complicated pathophysiology of HF, the design of animal models that would provide HF phenotypes that closely mimic human HF has captured the imaginations of generations of investigators.[15,16] A number of animal models have been developed to study both the pathophysiology of HF and new therapeutic approaches to the complex syndrome of HF. For the general purpose of explanation of disease mechanisms, widely used preparations include models of rapid ventricular pacing in the dog and pig[17] and MI in the rat. Other common approaches include surgically induced pressure or volume overload and toxic myocardial depression. There is no ideal animal model that both perfectly mimics human HF and is also technically feasible for use in the laboratory. Each model has advantages and limitations in terms of extrapolations from experimental to clinical HF. As will be apparent in the subsequent discussions, there are frequent discrepancies between the abundant experimental data supporting a key mechanistic role for a biological system in evolving HF compared with disappointing results of therapies that inhibit or augment the system in patients. Nevertheless, animal models have provided new

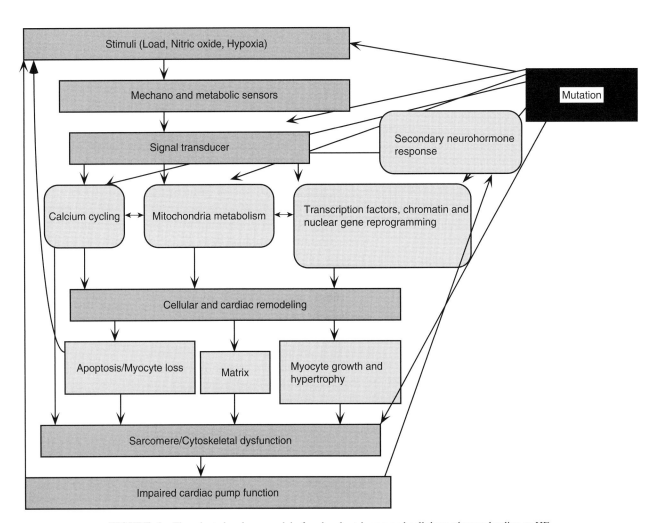

FIGURE 1 Flowchart showing a model of molecular triggers and cellular pathways leading to HF.

insights into many aspects of the complex pathophysiology of the syndrome of HF and have helped to investigate the efficacy of new therapeutic interventions.

For genetic and molecular analyses in HF, the mouse transgenic models that effect gain or loss of function have become the method of choice to study genotype-phenotype relationships.[18] As will be reviewed in later sections of this chapter, these models have been particularly valuable to the study of the relationship between mitochondrial and myocardial function in HF.[19]

Alterations in Myocardial Energetics

The heart relies predominantly on fatty acids as its key fuel supply, whereas glucose and lactate provide a relatively small percentage. However, in pathological conditions, this balance becomes altered such that the heart relies more on glucose, as seen in cardiac hypertrophy, or may rely almost solely on fatty acids, as observed in cardiac tissue of animal models of diabetes.[20] Initially, this switch in metabolic substrate provides adequate energy to maintain normal cardiac function; however, over time, diastolic dysfunction and HF occur, which is associated with depletion in high-energy phosphates.

Mitochondrial oxidative catabolism of carbohydrates and fatty acids is the major source of energy for the adult mammalian heart and crucial to proper functioning of the myocardium. As described more extensively in Chapter 16, fatty acids and pyruvate from glycolysis are oxidized in the mitochondrial fatty acid β-oxidation (FAO) and pyruvate dehydrogenase (PDH) pathways, respectively, to generate acetyl-CoA, which enters the tricarboxylic acid (TCA) cycle. NADH and $FADH_2$, the reducing equivalents that transfer electrons to the electron transport chain (ETC), are produced by the TCA cycle and during fatty acid and glucose oxidation (Fig. 2). The ETC consists of five multi-subunit complexes that receive electrons from NADH and $FADH_2$, ultimately producing ATP through oxidative phosphorylation (OXPHOS). Subsequently, ATP is transported from the mitochondrial matrix to the cytoplasm in exchange for ADP through the adenine nucleotide transporter (ANT), making energy available for cellular work. Cardiac mitochondrial capacity is dynamically regulated at both the transcriptional and post-transcriptional levels. Changes in substrate use are mediated by signaling pathways, metabolite concentrations, substrates, and precursors, and by allosteric regulation of key enzymes. In addition to the signaling and metabolic regulatory networks that mediate acute response to changes in substrate availability, there is also substantial regulation of mitochondrial number and substrate selection at the gene transcriptional level.[21]

Although the mitochondrial genome encodes the mitochondrial ribosomal and transfer RNA genes and 13 protein subunits of the ETC,[22] most mitochondrial proteins are encoded by nuclear genes, and many of these are essential for mitochondrial function. Mutations in either the nuclear DNA or mtDNA can, therefore, cause mitochondrial disease.[23,24] Several mtDNA disorders with cardiac involvement, including Kearns–Sayre syndrome, Leber hereditary optic neuropathy, and Leigh syndrome, result in a global impairment of mitochondrial respiratory function. Mitochondrial disease may also be associated with mutations in specific metabolic pathways. For example, mutations in the mitochondrial genome frequently result in skeletal myopathy in association with cardiac defects, including DCM, HCM, and conduction abnormalities.[25] In addition, several inherited FAO disorders caused by mutations in nuclear genes encoding FAO enzymes involve impaired mitochondrial catabolism of fatty acids.[26] Inborn errors in the mitochondrial FAO pathways can manifest as HCM.[26] Cardiomyopathy in these patients usually appears during childhood and often presents as sudden onset of HF or ventricular dysrhythmia induced by stress. Mitochondrial dysfunction has also been reported in acquired cardiomyopathies. In pressure overload hypertrophy and end-stage HF, several studies have demonstrated that the FAO and OXPHOS pathway are impaired.[27–29] It is possible that in early stages of HF progression, higher rates of oxidative flux lead to mitochondrial oxidative damage and eventual impairment of respiration. Moreover, discordance between rates of FAO and OXPHOS could lead to toxic intermediate accumulation and perpetuate a continued cycle of mitochondrial damage.

Abnormal regulation of the nuclear receptor transcription factor peroxisome proliferator–activated receptor (PPAR-α), and the co-activator (PGC-1α) and their control of cardiac fatty acid use have been reported in several cardiomyopathic disease states in both humans and mouse models (Fig. 3). In animal models of HF induced by pressure overload or ischemia, the expression of PPAR-α, PGC-1α, and downstream target genes encoding FAO enzymes is diminished.[30] These observations have been confirmed in patients with HF, and collectively they suggest that deactivation of the PPAR-α/PGC-1α axis is involved in the metabolic switch from fatty acid catabolism in the failing heart.[31,32] It is quite likely that abnormal regulation of the PPAR-α/PGC-1α axis in these disease states is involved in the metabolic remodeling that occurs in HF. However, the exact role of these metabolic "switches" in the development of HF has not been fully explained.

Oxidative Stress and HF

Oxygen serves as the critical terminal electron acceptor in the ETC; in its absence, the ETC shuts down, and cardiac demands for ATP are not met. Molecular oxygen is also central in both the formation of nitric oxide (NO), a primary determinant of both vascular tone and cardiac contractility, and in the generation of ROS (and subsequent oxidative stress) as a significant byproduct of energy metabolism during its sequential acceptance of electrons in the mitochondrial ETC. Because

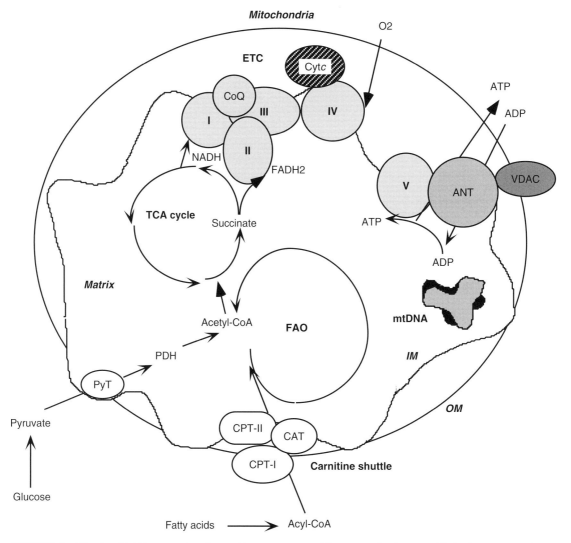

FIGURE 2 Mitochondrial bioenergetic pathways. Pyruvate generated from cytosolic glucose oxidation is transported into mitochondria by the pyruvate transporter (PyT), oxidized to acetyl CoA by matrix-localized pyruvate dehydrogenase (PDH), and fed into the TCA cycle. Fatty acids are transported into mitochondria by the carnitine shuttle composed of several transporters (CAT, CPT-I, and CPT-II) further oxidized by fatty acid β-oxidation (FAO) to acetyl CoA and sent to the TCA cycle. Reducing equivalents (NADH and $FADH_2$) generated by the TCA cycle enter the electron transport chain (ETC) composed of five mitochondrial inner-membrane (IM) localized-respiratory complexes (I–V) with associated electron-transfer components CoQ and Cyt c, with oxygen (O_2) serving as the terminal electron acceptor to ultimately produce ATP by oxidative phosphorylation. ATP is delivered to the cytosol in exchange for ADP by the adenine nucleotide transporter (ANT), closely associated with the outer membrane (OM) VDAC/porin protein. Several subunits of complex I, III, IV, and V are encoded by the matrix-localized mtDNA.

these short-lived intermediates can act either as an important signaling molecules or induce irreversible oxidative damage to proteins, lipids, and nucleic acids, both ROS and oxygen exert both beneficial and deleterious effects. It is important to note that ROS can be generated in the heart and endothelial tissues by non-mitochondrial reactions, as well including the involvement of xanthine oxidase (XO), NAD(P)H oxidases, and cytochrome P450.[33]

Ischemia causes alterations in the endogenous defense mechanisms against oxygen free radicals, mainly through a reduction in the activity of mitochondrial superoxide dismutase (SOD) and a decline in tissue content of reduced glutathione.[34] Moreover, increased ROS production in both the mitochondria and leukocytes and toxic oxygen metabolite production are exacerbated by readmission of oxygen during post-ischemic reperfusion as extensively discussed in Chapter 10. It is well documented that oxidative stress, resulting from both increased ROS generation and diminished antioxidant protection promotes the oxidation of thiol groups in proteins and lipid peroxidation, leading first to reversible damage and eventually to necrosis.

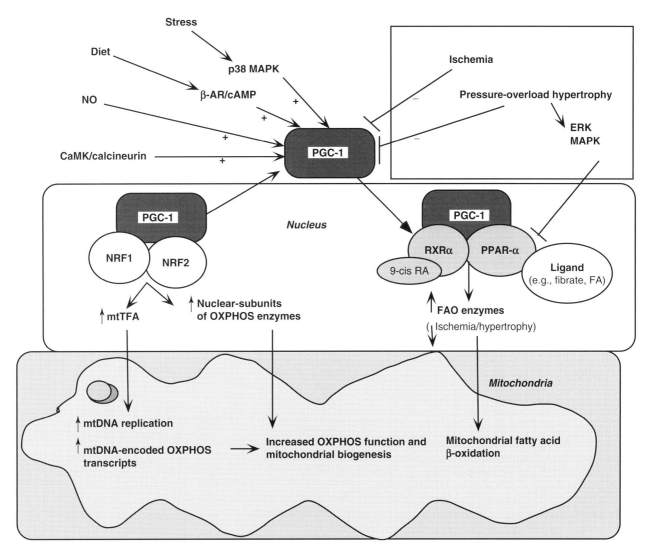

FIGURE 3 PGC-1 and its metabolic pathway are affected in HF. PGC-1 transduces cell signals associated with physiological stimuli to regulate cardiac metabolic genes. Numerous signaling pathways downstream of physiological stimuli, including fasting and exercise, activate the PGC-1 cascade either by increasing PGC-1 expression or activity, whereas pressure-overload hypertrophy and ischemia down-regulate. PGC-1, in turn, coactivates transcriptional partners, including nuclear respiratory factor-1 and 2 (NRF-1/2) and PPAR-α, resulting in downstream activation of mitochondrial biogenesis and fatty acid oxidation (FAO) pathways, respectively.

Because coronary artery disease (CAD), with consequent myocardial ischemia and necrosis, is a well-recognized leading cause of HF worldwide, it is relevant to note here that ROS plays an integral role in the genesis and progression of CAD. In the vessel wall, ROS contributes to the formation of oxidized LDL, a major player in the pathogenesis of atherosclerosis, and ROS-mediated activation of matrix metalloproteinases (MMPs) may play a contributory role in vessel plaque rupture, initiating coronary thrombosis and occlusion.[35–37]

A significant number of *in vitro* and animal studies have demonstrated that in the failing heart, ROS influences several components of the cardiac phenotype and its remodeling, including contractile function, interstitial fibrosis, endothelial dysfunction, and myocyte hypertrophy. As we shall further discuss later in this chapter, ROS contributes to the remodeling processes in a number of ways, including activating MMPs that participate in reconfiguration of the ECM; acting as signaling molecules in the development of compensatory hypertrophy; and contributing to myocyte loss by apoptosis signaling.

Excessive production of NO has also been implicated in the pathogenesis of chronic HF.[38,39] Uncoupling of constitutive nitric oxide synthase (NOS) leads to overproduction of superoxide (O_2^-) and peroxynitrite ($ONOO^-$), two extremely potent oxidants. Peroxynitrite produced from the reaction of highly reactive NO with the superoxide anion impairs cardiovascular function through multiple mechanisms, including activation of MMPs and nuclear enzyme poly (ADP-ribose) polymerase (PARP). Increased oxidative stress resulting from the overproduction of superoxide also mediates the dysregulation of S-nitrosylation of proteins at specific cysteine residues by reactive nitrogen species, a more selective modification of proteins than found with protein oxidation. This redox mechanism has been demonstrated to lead to altered myocardial excitation-contractility and vascular reactivity.[40,41]

INSIGHTS FROM TRANSGENIC MODELS

As referred to earlier, the creation of transgenic mice with altered expression of genes involved in carbohydrate and lipid metabolism has provided unique insights into the fine balance within the mouse heart to maintain energy status and cardiac function, as well as to explore the cause–effect relationships between mitochondrial function and myocardial disease.[42] A list of transgenic models of metabolic modification in the heart that are associated with cardiac dysfunction and/or HF phenotype is shown in Table I.

Loss-of-function model systems, which disrupt mitochondrial metabolism, exhibit cardiac phenotypes. One genetically engineered mouse model demonstrating a causal relationship between a mitochondrial energetic defect and cardiomyopathy is the *Ant1* null mouse. The adenine nucleotide translocators (ANT) are a family of proteins that exchange mitochondrial ATP for cytosolic ADP, providing new ADP substrate to the mitochondria while delivering ATP to the cytoplasm for cellular work. Mice express two isoforms of this enzyme; with tissue-specific expression patterns.[43] Ant1 is expressed in skeletal muscle, heart, and the brain, whereas Ant2 is expressed in all tissues except skeletal muscle. The *Ant1* gene–deficient mice exhibit mitochondrial abnormalities, including a partial deficit in ADP-stimulated respiration consistent with impairment in the translocation of ADP into mitochondria in both skeletal muscle and the heart. The skeletal muscle respiratory defect was profound, presumably because Ant2 can partially compensate for the loss of Ant1 in heart, but not in skeletal muscle. *Ant1$^{-/-}$* mice exhibit a progressive cardiac hypertrophic phenotype coincident with the proliferation of mitochondria.[44] The mitochondrial biogenic response has, therefore, been hypothesized to be a compensatory mechanism to correct the energy deficit but could also contribute to cardiac remodeling. Transgenic models of specific defects in the mitochondrial FAO pathways have also been established (listed in Table II).

TABLE I Transgenic Models of Metabolopathies in HF

Gene	Manipulation	Cardiac phenotype
GLUT4	Global KO	Hypertrophy, ↑ expression of MCAD and LCAD, ↑ glycogen synthesis, interstitial fibrosis
	Heterozygous KO	Diabetic cardiomyopathy
	Cardiac-specific KO	Hypertrophy, ↑BNP, glucose uptake
IR	Cardiac-specific KO	↓ Myocyte size, ↓ basal glucose uptake, impaired cardiac function
IGF-R	Cardiac overexpression	Cardiac hypertrophy
IGF-1	Overexpression	Cardiac hypertrophy, attenuation of cardiac dysfunction after myocardial infarct
PI3Kα	Constitutively active	Cardiac hypertrophy with no alterations in cardiac function or fibrosis
	Dominant negative	Smaller hearts with normal cardiac function
PTEN	Cardiac-specific KO	Cardiac hypertrophy and impaired contractility
PTEN	Constitutively active	Concentric LV hypertrophy, ↓ infarct size after I/R, altered contractility
PDK1	Cardiac-specific KO	Sudden death from HF
AMPKα2	Inactive kinase	↓ Heart weight and *in vivo* LV dP/dt, impaired glucose uptake, glycolysis and FAO, ↑ apoptosis and impaired LV recovery in response to ischemia
PFK2	Cardiac-specific kinase deficient	Multiple cardiac pathological conditions, fibrosis, ↓ contractility, impaired glycolysis
G6PDH	Global KO	Myocardial dysfunction

KO, knock-out; IR, insulin receptor; IGF, insulin-like growth factor; MCAD, medium chain acyl-CoA dehydrogenase; PDK, pyruvate dehydrogenase kinase; I/R, ischemia/reperfusion; LCAD, long chain acyl CoA dehydrogenase; G6PDH, glucose-6-phosphate dehydrogenase; PFK2, phosphofructokinase 2; AMPK, adenosine monophosphate kinase; PTEN, phosphatase and tensin homolog deleted on chromosome 10; GLUT, glucose transporter; BNP, B-type or brain natriuretic peptide; PI3K, phosphatidylinositol-3-kinase; FAO, fatty acid oxidation.

Two distinct mouse models with genetic deletion of the second step in the mitochondrial FAO pathway, a fatty acid chain-length–specific dehydrogenase enzyme (VLCAD and LCAD), display a cardiomyopathic phenotype.[45,46] Furthermore, mice null for the PPAR-α gene also exhibit diminished myocardial fat catabolic capacity and mild cardiomyopathic phenotype that accompanies aging.[47]

Besides "loss of function" models, "gain of function" overexpression strategies have also shed light on the relation between mitochondrial dysfunction and cardiac dysfunction, particularly in a poorly characterized entity of cardiomyopathy associated with diabetes. The healthy heart generates ATP by use of both carbohydrate utilization and mitochondrial FAO pathways as shown in transgenic mice (Table II), the diabetic heart derives a preponderance of ATP from FAO.[48] The cardiac PPAR-α/PGC-1α system is activated in diabetes mellitus.[49] This metabolic shift associated with high level of fatty acid import and oxidation may eventually lead to pathological mitochondrial and cardiac remodeling. Transgenic mice were generated with cardiac-restricted overexpression of PPAR-α enabled by fusion of the promoter of the cardiac α myosin heavy chain (MHC) with the PPAR-α cDNA. The MHC-PPAR-α mice exhibit increased expression of genes encoding enzymes involved in multiple steps of mitochondrial FAO with strong reciprocal down-regulation of glucose transporter (GLUT4) and glycolytic enzyme gene expression. Myocardial fatty acid uptake and mitochondrial fatty acid β-oxidation are markedly increased in MHC-PPAR-α hearts, whereas glucose uptake and oxidation are profoundly diminished in MHC-PPAR-α mice.[50] Echocardiographic assessment identified LV hypertrophy and dysfunction in the MHC-PPAR-α mice in a transgene expression–dependent manner. Sequential studies showed that both HF diet and insulinopenia induced further remodeling accompanied by signs of HF. The HF phenotype caused by the HF chow was completely reversed by resumption of standard chow.

CELLULAR AND MOLECULAR BASIS OF CARDIAC REMODELING

A critical step in the evolution of LV dysfunction and HF is cardiac remodeling.[7,8] Cardiac remodeling involves changes in both the myocytes and the ECM, the latter includes the activation of proteolytic enzymes leading to the degradation and reorganization of collagen.[9–12] The morphological patterns that accompany cardiac remodeling include hypertrophy and subsequent chamber dilatation with associated progressive impairment in function.[8] Progressive and worsening cardiac remodeling results in the onset of LV dilatation with relatively proportional increases in LV mass, a process sometimes termed eccentric hypertrophy.[51] These events are more characteristic of patients with compensated LV dysfunction. The development of overt HF is largely triggered by the marked ventricular dilatation that occurs once the resource from myocardial hypertrophic response is exhausted.[52] With end-stage disease and clinically overt HF, profound LV dilatation occurs without comparable levels of hypertrophy, resulting in relative wall thinning from the disproportional increases in LV chamber size. In addition, excessive LV dilatation alters the overall shape of the LV from the normal ellipsoid configuration to a more spherical shape.

Hypertrophy is an adaptive response to increased workload and is accompanied by alterations in metabolism.[53] At least some of the cytoplasmic signaling pathways thought to be responsible for pathological hypertrophy are mediated through increased levels of growth factors signaling through G-protein– coupled cell-surface receptors (GPCR).[54] Atrial natri-

TABLE II Transgenic Models of Fatty Acid Defects in HF

PPAR-α	Global KO	↑ Myocardial fibrosis; ↓ LV fractional shortening; ↓ cardiac contractile performance under basal and under stimulation of β1 adrenergic receptors; ↓ cardioprotection to I/R
	Cardiac-specific overexpression	Diabetic cardiomyopathy; ↑ FAO, ↓ glucose uptake and oxidation
FAT/CD36	Global KO	DCM; ↓ cardiac fatty acid uptake and TG
FAT/CD36	Overexpression	↑ FAO; ↓ plasma TG + fatty acid; no cardiac pathology; ↑ glucose
H-FABP	Global KO	Hypertrophy, ↑ANF expression, ↓ LCFA utilization
FATP-1	Cardiac-specific overexpression	Lipotoxic CM
ACS1	Cardiac-specific overexpression	Cardiomyopathy, hypertrophy, LV dysfunction, HF, intramyocellular TG accumulation
LpL	Cardiac-specific KO	DCM; ↑ TG; ↓ FAO; ↑ Glut expression and glucose uptake
LCAD	Global ablation	Cardiomyopathy, lipid accumulation, myocardial fibrosis
Leptin (ob)	Null	↑ FAO, fatty acid uptake, TG + lipid accumulation; ↑ diastolic dysfunction

FATP, fatty acid transport protein; ACS, long-chain acyl-CoA synthetase; TG, triglyceride; Lp, lipoprotein lipase; LCAD, long-chain acyl-coenzyme A dehydrogenase; FAT/CD36, fatty acid translocase; H-FABP, heart-type fatty acid binding protein; LCFA, long chain fatty acids; FAO, fatty acid oxidation; Glut, glucose transporter; I/R, ischemia/reperfusion; ANF, atrial natriuretic factor; PPAR, peroxisome proliferator-activated receptor.

uretic peptide (ANP), through its guanylyl cyclase-A (GC-A) receptor, locally moderates cardiomyocyte growth. To characterize the anti-hypertrophic effects of ANP, the possible contribution of Na$^+$/H$^+$ exchanger (NHE-1) to cardiac remodeling was recently examined in a model of GC-A–deficient (GC-A [-/-]) mice.[55] Fluorometric measurements in the cardiomyocytes demonstrated that cardiac hypertrophy in GC-A (-/-) mice was associated with enhanced NHE-1 activity, alkalinization of intracellular pH, and increased Ca^{++} levels. Chronic treatment of GC-A (-/-) mice with the NHE-1 inhibitor cariporide normalized cardiomyocyte pH and Ca^{++} levels and regressed cardiac hypertrophy and fibrosis. Activity of four pro-hypertrophic signaling pathways—the mitogen-activated protein kinases (MAPK), the serine-threonine kinase Akt, calcineurin, and Ca^{++}/calmodulin–dependent kinase II (CaMKII)—were activated in GC-A (-/-) mice, but only CaMKII and Akt activity regressed on reversal of the hypertrophic phenotype after cariporide. By contrast, the MAPK and calcineurin/NFAT signaling pathways remained activated during regression of hypertrophy. These observations suggest that the ANP/GC-A system modulates the cardiac growth response to pressure overload during remodeling by preventing excessive activation of NHE-1 and subsequent increases in cardiomyocyte intracellular pH, Ca^{++}, and CaMKII, as well as Akt activity.

Progression of Cardiac Remodeling and Transition to Overt HF

The mechanisms of cardiac remodeling, particularly those underlying the transition from stable hypertrophy to cardiac dilatation and ultimately to overt HF, remain unclear. Many factors, including neurohormonal and cytokine activation and impaired Ca^{2+} handling, are thought to play a contributory role, particularly after MI (Fig. 4).[56,57] Pressure and volume overload have been known to be accompanied by increased glucose uptake and glycolysis and decreased FAO.[58] Alterations in energy status and metabolism are proposed to play a role in the transition from stable cardiac hypertrophy to overt HF at least

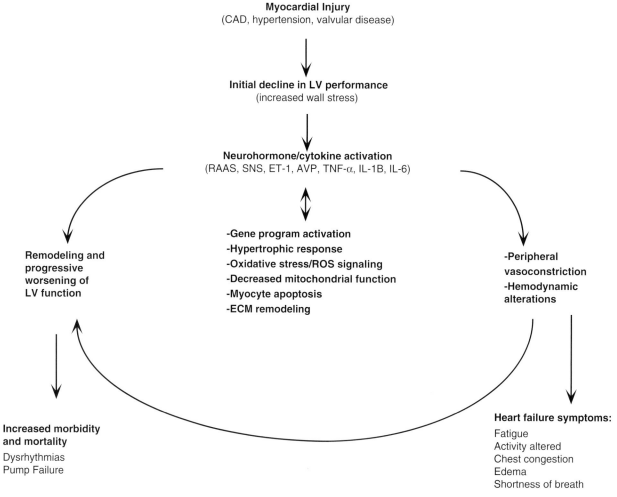

FIGURE 4 Flowchart model showing the complicity of neurohormonal and cytokine activation in HF. CAD, Coronary artery disease; LV, left ventricle; RAAS, renin-angiotensin-aldosterone system; SNS, sympathetic nervous system; ET-1, endothelin-1; TNF-α, tumor necrosis factor-α; IL-1β, interleukin-1β; AVP, arginine vasopressin.

on the basis of observations from animal studies.[28,53,59] Some of the recent research conducted on the relationship between myocardial metabolism and cardiac remodeling using transgenic models have been discussed in the previous section.

Diverse neurohormonal signals acting through interwoven signal transduction pathways can lead to pathological cardiac hypertrophy and HF. Many such agonists act through cell surface receptors coupled with G-proteins to mobilize intracellular calcium, with consequent activation of downstream kinases and the calcium- and calmodulin-dependent phosphatase calcineurin. MAPK signaling pathways are also interconnected at multiple levels with calcium-dependent kinases and calcineurin.[60] The details of these signaling pathways have been reviewed in other chapters.

β-Adrenergic agonists influence cardiac growth and function through the generation of cAMP, which activates protein kinase A (PKA) and other downstream effectors.[61] These signaling pathways target a variety of substrates in the cardiomyocyte, including components of the contractile apparatus, calcium channels, and their regulatory proteins.

CONTRACTILE ELEMENTS

The intrinsic contractile ability of individual cardiomyocytes is impaired during the evolving HF process and reflects important cellular changes that include altered expression of membrane and contractile proteins, altered energy metabolism, and impaired excitation-contraction coupling.[62] Cell isolation and function studies have demonstrated that the reduction in contractile performance in these failing cells is not simply a consequence of associated wall stresses, ischemia, or myocardial perfusion defects.[63,64] However, there is no substantial direct evidence for a causative role for depressed contractility in the initiation and progression of human HF, and cardiac remodeling with HF can occur in the absence of depressed myocyte contractility.[62,65] Although cardiomyocyte contractile dysfunction contributes to cardiac dysfunction, it cannot adequately account for the profound structural changes in both the cells and the surrounding ECM that characterize cardiac remodeling.

CELLULAR HYPERTROPHY

The hypertrophy of cardiomyocytes is a well-recognized remodeling response to increased hemodynamic load.[66] Myocyte hypertrophy is likely an adaptive mechanism designed to improve pump function by expanding the number of contractile units while simultaneously reducing wall stress by increasing wall thickness. When excessive or prolonged, myocyte hypertrophy is maladaptive. Hypertrophy can directly result in HF, as evidenced by patients with HCM.[67] In the failing heart, excessive LV hypertrophy is associated with reduced myocardial compliance, myocardial fibrosis, and lethal dysrhythmias.[68,69] However, the time frame of transition from adaptive to maladaptive hypertrophy is not known. Experimental evidence indicates that targeted increases in wall thickness in the absence of concomitant increases in LV volume can have beneficial effects on cardiac performance and on retarding the progression of HF.[70]

The first stimulus to induce myocyte hypertrophy is usually mechanical, such as hemodynamic overload, although neurohormones and cytokines also play an important role in its maintenance.[71,72] Transduction of mechanical stress and other environmental signals is believed to occur through integrin proteins, transmembrane receptors that couple extracellular matrix components directly to the intracellular cytoskeleton and nucleus.[73,74] In general, the signal for hypertrophy is mediated by a complex cascade of signaling systems within the cardiomyocyte, resulting in gene reprogramming,[75] and these signaling mechanisms have been discussed in other chapters. In brief, diverse neurohumoral signals acting through many interrelated signal transduction pathways lead to pathological cardiac hypertrophy and HF.[75] Moreover, activated hypertrophy-related genes induce the synthesis of new contractile proteins that are organized into new sarcomeres. Thus, extensive remodeling of the complete intracellular contractile apparatus is characteristic of the failing heart.

CELL DEATH AND RENEWAL

Environmental stress and injury produce cell death by apoptosis, necrosis, and perhaps autophagy. Little is known about role of necrosis and autophagy in HF. On the other hand, a conceptual framework has been developed for apoptosis, a highly regulated cell suicide process that is hard-wired into all metazoan cells. Chronic cardiac remodeling response and transition to overt HF have been associated with modestly increased apoptosis.[76–78] The actual burden of chronic cell loss attributable to apoptosis is not clear. Measures of actual rates are highly variable and depend on the species, type of injury, timing, location, and method of assessment. When viewed in absolute terms, the rate of apoptosis is quite low[79]; however, when the relatively low rates are viewed in the context of months or years, it is entirely plausible that the apoptotic burden could be substantial. Unfortunately, the timing of the apoptotic process is not well defined, and the assessment of the true rates and their consequences is still quite limited.

The regulation of apoptosis is complex and not completely defined. An important signaling pathway for apoptosis in the heart during the transition to HF is mediated by the balance of the proapoptotic protein Bax relative to the antiapoptotic protein Bcl-2.[80] These proteins can stimulate or suppress the action of the caspases, which carry out the characteristic biochemical and morphological changes of apoptosis. Other key pathways include the expression of the death receptor,

Fas, which is up-regulated in failing cardiomyocytes and activates downstream caspases, resulting in apoptosis.[81,82]

In brief, apoptosis is mediated by two evolutionarily conserved central death pathways: the extrinsic pathway, which uses cell surface death receptors; and the intrinsic pathway, involving mitochondria and the ER (Fig. 5).[83] In the extrinsic pathway, death ligands (e.g., FasL) initiate apoptosis by binding their cognate receptors.[84] This stimulates the recruitment of the adaptor protein Fas-associated by death domain (FADD), which then recruits procaspase-8 into the death-inducing signaling complex (DISC).[85] Procaspase-8 is activated by dimerization within this complex and subsequently cleaves and activates procaspase-3 and other downstream procaspases.[86]

The intrinsic pathway transduces a wide variety of extracellular and intracellular stimuli, including loss of survival/trophic factors, toxins, radiation, hypoxia, oxidative stress, myocardial ischemia/reperfusion (I/R) injury, and DNA damage. Although a number of peripheral pathways connect these signals with the central death machinery, each ultimately feeds into a variety of proapoptotic Bcl-2–related proteins that possess only Bcl-2 homology domain 3 (BH3-only proteins) and the proapoptotic multidomain Bcl-2 proteins Bax and Bak.[87] These proteins undergo activation through diverse mechanisms

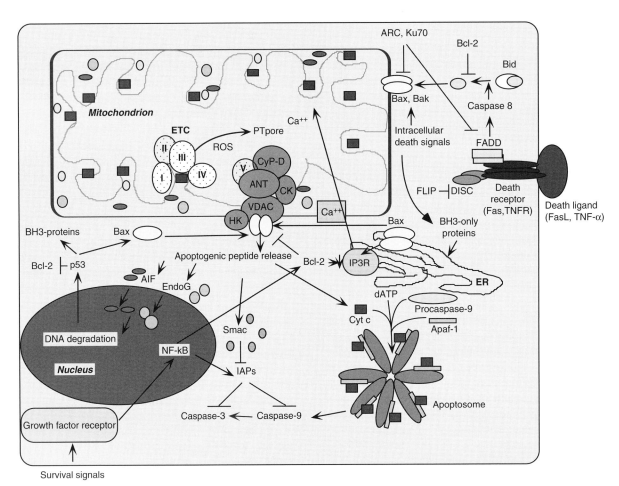

FIGURE 5 The intrinsic and extrinsic pathways of apoptosis. An array of extracellular and intracellular signals trigger the intrinsic apoptotic pathway regulated by proapoptotic proteins (e.g., Bax, Bid, and Bak) binding to outer mitochondrial membrane leading to mitochondrial outer-membrane permeabilization and PT pore opening. Elevated levels of mitochondrial Ca^{++}, as well as ETC-generated ROS, also promote PT pore opening. This is followed by the release of cytochrome c (Cytc), Smac, endonuclease G (Endo G), and apoptosis-inducing factor (AIF) from the mitochondrial intermembrane space to the cytosol. This leads to apoptosome formation (with Cytc) leading to caspase 9 activation, DNA fragmentation stimulated by the nuclear translocation of AIF and EndoG, whereas inhibition of IAP (by Smac) further promotes the activation of caspases 9 and 3. Bax and Bid mediate mitochondrial membrane permeabilization, and antiapoptogenic proteins (e.g., Bcl-2) prevent apoptogen release. Also shown are major proteins making up the PT pore, including hexokinase (Hx), adenine nucleotide translocator (ANT), creatine kinase (CK), cyclophilin D (CyP-D), and porin (VDAC). The extrinsic pathway is initiated by ligand binding to death receptors, leading to recruitment of FADD and DISC, which stimulate the activation of caspase 8, resulting in caspase 3 activation and Bid cleavage (a C-terminal fragment of Bid targets mitochondria). FLIP, ARC, and Ku70 can stem this pathway's progression at specific points. Intracellular stimuli trigger ER release of Ca^{++} through both Bax and BH3-protein interactions. Also shown is the survival pathway triggered by survival stimuli, mediated by growth factor receptors, transcription factor activation (e.g., NF-κB), and enhanced expression of IAPs and Bcl-2 (see color insert).

to trigger the release of mitochondrial apoptogens, such as cytochrome *c*, Smac, EndoG, and AIF into the cytoplasm.[88–91] Once in the cytoplasm, cytochrome *c* binds Apaf-1 along with dATP. This stimulates Apaf-1 to homo-oligomerize and recruit procaspase-9 into the multiprotein complex called the apoptosome.[92–95] Within the apoptosome, procaspase-9 is activated by dimerization, after which it cleaves and activates downstream procaspases. Bid, a BH3-only protein, unites the extrinsic and intrinsic pathways; after cleavage by caspase-8, Bid's C-terminal portion translocates to the mitochondria and triggers further apoptogen release.[96,97]

The extrinsic and intrinsic pathways are regulated by a variety of endogenous inhibitors of apoptosis (Fig. 5). FLICE-like (Fas-associated death domain protein-like–interleukin-1–converting enzyme-like) inhibitory protein (FLIP), whose expression is highly enriched in striated muscle, binds to and inhibits procaspase-8 in the DISC.[98] Antiapoptotic Bcl-2 proteins, such as Bcl-2 and Bcl-xL, inhibit mitochondrial apoptogen release through biochemical mechanisms that are still incompletely understood. Ku-70 and humanin bind Bax and block its conformational activation and translocation to the mitochondria.[99] X-linked inhibitor of apoptosis (XIAP) and related proteins that contain baculovirus inhibitor of apoptosis repeats bind to and inhibit already activated caspases-9, -3, and -7, as well as interfering with procaspase-9 dimerization and activation.[100,101] Each of these inhibitors acts on circumscribed portions of either the extrinsic or intrinsic pathway. By contrast, the apoptosis repressor with a CARD (ARC), which is expressed preferentially in striated muscle and some neurons, antagonizes both the intrinsic and extrinsic apoptosis pathways.[102] The extrinsic pathway is inhibited by ARC's direct interactions with Fas, FADD, and procaspase-8, which prevent DISC assembly, whereas the intrinsic pathway is inhibited by ARC's direct binding and inhibition of Bax's interaction with the mitochondrial membrane.[102,103] More recently, the endoplasmic reticulum (ER) has been recognized as an important organelle in the intrinsic pathway. In addition to its role in mediating cellular responses to traditional ER stresses, such as misfolded proteins, this organelle seems to be critical in mediating cell death elicited by a subset of stimuli originating outside of the ER, such as oxidative stress.[104] Similar to their roles in transducing upstream signals to the mitochondria, BH3-only proteins seem to relay upstream death signals to the ER.[105]

In the failing human heart, the expression of protooncogenes that regulate programmed cell death is increased and is associated with increased cardiomyocyte apoptosis.[106] To assess the role of apoptosis in human HF, patients with acute MI (AMI) were assessed for the rate of cardiomyocyte apoptosis relative to indices of structural LV remodeling. Within the infarct area and in areas remote from the site of injury, the rate of myocardial apoptosis was increased and was strongly associated with maladaptive LV remodeling, in addition to adverse clinical outcomes.[107] The most common form of HF that has been associated with significant levels of apoptosis is DCM.[108] Apoptosis is also correlated with the clinical severity of deterioration in DCM and the subsequent need for cardiac transplantation.[109] Despite the demonstration of apoptosis in cardiac remodeling and HF, it is not clear whether apoptosis is a cause or consequence of transition to HF. Experimental models of progressive LV dysfunction and failure provide an opportunity to regulate apoptosis and examine the resultant influence on LV structure and function. In rabbits with coronary ligation, the administration of a viral antiapoptotic Bcl-2 gene to the heart limited apoptosis and simultaneously attenuated LV dilatation and dysfunction.[110] In addition, bioactive peptides such as insulin-like growth factor-1 (IGF-1) can protect cardiomyocytes from apoptosis.[111] The overexpression of IGF-1 in a mouse coronary ligation model limited cell death and correspondingly attenuated maladaptive remodeling and the transition to HF.[112] In a canine model of HF, exogenously administered IGF-1 attenuated wall thinning and improved function in association with reduced apoptosis.[113] These observations in animal models lend support to the concept that apoptosis may play a contributory role in cardiac remodeling and the transition to HF.

Investigators recently questioned the widely held belief that human cardiomyocytes do not divide after birth and propose the intriguing new concept of cardiomyocyte regeneration. These concepts are based on the premise that myocyte death and regeneration are homeostatic mechanisms intrinsic to both the normal and diseased heart.[114,115] Estimates of the rates of cell death would suggest that an innate system of cardiomyocyte replacement might be present to account for the maintenance of cardiac muscle mass over a lifetime.[116] Other investigators believe that significant regeneration is not observed in conditions in which myocytes are extensively lost; myocardial tumors are rare, and the current evidence supporting cell renewal in the heart is preliminary and controversial.[117] Findings of cyclin and cyclin-dependent kinase up-regulation in both normal and pathological myocardium, as well as increased telomerase activity in cardiomyocytes, have been proposed to be evidence of myocyte renewal. Opposing viewpoints dismiss these observations as nonspecific biochemical events of hypertrophy and not as evidence of cardiomyocyte hyperplasia.[117] Direct evidence for the concept of myocyte regeneration was obtained by Anversa and associates by examining normal and post-MI left ventricular myocardium for the presence of cytological and biochemical markers of cell replication in cardiomyocytes. Measurements of Ki67 and the mitotic index indicated that cardiomyocyte replication occurred in normal myocardium and was increased in diseased myocardium, particularly in the border zone between infarcted and viable myocardium.[118–120] These investigators hypothesize that most adult cardiomyocytes are indeed terminally differentiated, and only a small proportion of cardiomyocytes retains the ability to reenter the cell cycle, becoming particularly prolific when the myocardium is injured and cell losses were incurred.

On the basis of the magnitude and rate at which these events are hypothesized to occur, the proponents of the cell renewal theory estimate that the entire LV could be regenerated in a period of 6 months, and as such, the generation of new myocytes likely contributes to remodeling the failing heart.[114] Unfortunately, these mechanisms of cell renewal do not seem capable of regenerating the burden of cells lost within an area of infarction and are limited to the restoration of cells lost in the remaining viable myocardium.

Although there exists evidence in support of a limited population of cardiomyocyte stem cells in the myocardium, the possibility exists that a population of extracardiac stem cells, perhaps from the bone marrow, is capable of regenerating cardiomyocytes lost to injury.[121-124] In transplanted human hearts, chimerism has recently been observed, supporting the possibility that extracardiac progenitor cells exist and serve to regenerate cardiomyocytes in the heart.[125,126] Chimerism has also been demonstrated in the heart after bone-marrow transplantation.[127]

THE EXTRACELLULAR MATRIX

The myocardial extracellular matrix (ECM) is an integral part of the growth and development of the heart, as well as in response to pathophysiological stimuli.[128] The myocardial ECM surrounds and interconnects cardiac myocytes, myofibrils, muscle fibers, and the coronary circulation.[129,130] Because of the high tensile strength of fibrillar collagen and its close association with the functioning components of the myocardium, alterations in the ECM can significantly influence the size, shape, and function of the cardiac chamber. The classical perspective that the ECM is an inert biomaterial and passive structural support mechanism has been replaced by the new concept that the ECM is a dynamic entity that is in equilibrium with the cellular components of the heart and that it alters its composition and organization in response to environmental cues and tissue injury.[128,131] These changes in both form and function have been termed ECM remodeling. Remodeling of the ECM, therefore, involves the dynamic interaction of both cellular and acellular components, soluble factors, and mechanical signals, providing both positive and negative signals generated by various physiological conditions.

The myocardial ECM has considerable plasticity such that the components of the ECM are actively degraded and replaced.[130,132,133] Physiological degradation of matrix elements is primarily the result of the coordinated activity of the proteolytic enzymes. A family of zinc-dependent enzymes capable of degrading the ECM of biological tissues and that play a fundamental role in ECM remodeling in normal and disease states are the MMPs.[134,135] Among the 20 species currently reported, MMPs can be divided into two principal types: those that are secreted into the extracellular space and those that are membrane bound.

The secreted MMPs make up most known MMP species and are released into the extracellular space in a latent or proenzyme state (proMMP). Activation of these latent MMPs is required for proteolytic activity, achieved through enzymatic cleavage of the propeptide domain. Serine proteases such as plasmin, as well as other MMP species, can convert proMMPs to active enzyme. Rapid amplification of MMP activity, therefore, occurs after an initial enzymatic step. The secreted MMPs bind to specific ECM proteins on the basis of the sequence of the C-terminus of the enzyme, and, therefore, they are in very close juxtaposition to the future proteolytic substrate, providing a means for rapid induction of proteolytic activity. The activated MMPs subsequently undergo autocatalysis, resulting in lower molecular weight forms and ultimately inactive protein fragments. The classification of MMPs was originally determined by substrate specificity, but as the characterization of this enzyme system proceeded, a great deal of substrate crossover between MMP classes and species has been identified. Nevertheless, a general classification has been developed for the MMPs and summarized in several recent reviews.[136,137] In brief, the interstitial collagenases such as MMP-1, MMP-13, the stromelysins, MMP-3, and the gelatinases, MMP-2 and MMP-9 have been observed within myocardium.[138-140]

A second and novel class of MMPs is the membrane-type MMPs (MT-MMPs), which are membrane bound and, therefore, provide a focalized area for ECM proteolytic degradation.[141] During trafficking to cell membrane, MT-MMPs undergo intracellular activation through a proprotein convertase pathway. Thus, unlike other classes of MMPs, MT-MMPs are proteolytically active once inserted into the cell membrane. Finally, MT-MMPs contain substrate recognition site for other MMP species and, therefore, constitute an important pathway for activation of other MMPs.[142,143] Type 1 MT-MMP (MT1-MMP) proteolytically processes the proforms of the gelatinase MMP-2 and the interstitial collagenase MMP-13. MT-MMPs do not seem to be under the influence of local inhibitory control, because the tissue inhibitors of the MMPs (TIMPs), including TIMP-1, fail to effectively bind to MT-MMPs.[141,144] Six different MT-MMPs have been cloned and seem to be expressed in both normal and diseased tissue. A number of cell types within the myocardium express MT-MMPs, including fibroblasts, vascular smooth muscle, and cardiac myocytes. The most well-characterized MT-MMP is MT1-MMP, and it has been the focus of several studies.[141] It has been shown that MT1-MMP degrades fibrillar collagens and a wide range of ECM glycoproteins and proteoglycans. MT-MMPs, therefore, likely play a key role in ECM degradation localized to basement membrane and cell–cell contact points.

An important control point of MMP activity is through the presence of an endogenous class of low molecular weight TIMPs.[144] To date, four different TIMP species have been identified and are known to bind to activated

MMPs in a 1:1 stoichiometric ratio. The four TIMP species all bind and inactivate the various MMPs, including MMP-2 and MMP-9, but with different affinities. TIMP-3 is a more potent inhibitor of MMP-9 than the other TIMPs.[145] Certain TIMPs bind to proMMPs and thereby form MMP–TIMP complexes. The functional significance of these proMMP–TIMP complexes is not completely understood but may actually facilitate MMP activation. One of the better-characterized TIMPs is TIMP-1, which binds with great affinity to activated MMPs, as discussed previously, and does not effectively bind and inhibit MT-MMPs. TIMP-4 seems to have a predominant distribution within the myocardium.[146] However, the significance of the myocardial expression of TIMP-4 remains to be determined. In addition to binding to MMPs, TIMPs seem also to influence cell growth and metabolism *in vitro*.[144] Thus, TIMPs may have multiple biological effects with respect to MMP activity within the myocardium that would be relevant to the cardiac remodeling process. An important emerging concept in the biology of MMPs is that MMPs can alter the bioactive properties and signaling capacity of the ECM through the activation and release of growth factors and other biologically active matrix elements, including the collagen degradation products.[142]

The ADAMs (a disintegrin and metalloproteinase) are a unique and novel class of enzymes from the metalloproteinase family that degrade ECM components and simultaneously activate key biopeptides in the interstitial microenvironment.[147] ADAMs are believed to be able to modify cell–cell and cell–matrix interactions by integrin receptors and may, therefore, influence tissue architecture and remodeling by reorganizing cells within their matrix. ADAMs may also activate biologically active peptides, cytokines, and growth factor, resulting in biological effects.[147] Although ADAMs have been implicated in cardiac disease and remodeling,[148–151] their specific profile and their role in HF remain to be evaluated.

There is currently ample evidence to indicate a pathophysiological role for MMPs in LV remodeling and HF; the topic has been the subject of several recent reviews.[152,153] In brief, increased MMP expression has been reported in patients with end-stage HF and in animal models of HF caused by LV dysfunction.[139,146,154–157] A cause and effect relationship between MMPs and LV remodeling seems to have been established through the use of transgenic models and through the use of pharmacological MMP inhibitors.[11,156,158–162] For example, a loss of MMP inhibitory control through TIMP-1 gene deletion has been shown to produce LV chamber dilation in mice.[161] Cardiac-restricted overexpression of the MMP-1 (interstitial collagenase) resulted in a loss of collagen abundance with transition to LV dysfunction.[159] Deletion of the MMP-9 gene in mice reduces the rate of cardiac rupture and alters the course of LV remodeling after MI.[156]

INSIGHTS FROM PATIENTS WITH LEFT VENTRICULAR ASSIST DEVICE

Mechanical left-ventricular assist devices (LVADs) are used as a "bridge to transplant" for some patients with HF. These devices can induce profound cardiac unloading and have been shown to be associated with reverse cardiac remodeling on a structural and functional level.[163] They offer an opportunity to study the molecular mechanisms underlying cardiac remodeling. To analyze this reverse remodeling on a transcriptional level, a series of studies were performed on cardiac tissue from patients before and after placement of an LVAD.[164,165] One study analyzed six patients with HF with different etiologies using oligonucleotide microarrays. Paired *t* test analysis revealed numerous genes that were regulated in a statistically significant fashion, including a down-regulation of several previously studied genes. Further analysis revealed that the overall gene expression profiles could significantly distinguish pre- and post-LVAD status. Importantly, the data also identified two distinct groups among the pre-LVAD failing hearts, in which there was blind segregation of patients on the basis of HF etiology. In addition to the substantial divergence in genomic profiles for these two HF groups, there were significant differences in their corresponding LVAD-mediated regulation of gene expression, with an association between the process of reverse remodeling and changes in cellular metabolic pathways.[164] Another study analyzed seven patients with DCM. On average, 1374 (± 155) genes were reported as "increased" and 1629 (± 45) as "decreased" after LVAD support. Up-regulated genes included a large proportion of transcription factors, genes related to cell growth/apoptosis/DNA repair, cell structure proteins, metabolism, and cell signaling/communication. LVAD support resulted in down-regulation of genes for a group of cytokines.[166]

In another study, analysis of a gene expression library of 19 paired human failing heart samples from different etiologies harvested at the time of LVAD implant and again at explant revealed a high percentage of genes involved in the regulation of vascular networks, including neuropilin-1 (a VEGF receptor), FGF9, Sprouty1, stromal-derived factor 1, and endomucin, suggesting that mechanical unloading alters the regulation of vascular organization and migration in the heart. In addition to vascular signaling networks, GATA4 binding protein, a critical mediator of myocyte hypertrophy, was significantly down-regulated after mechanical unloading. In summary, these findings may have important implications for defining the role of mechanical stretch and load on autocrine/paracrine signals directing vascular organization in the failing human heart and the role of GATA4 in orchestrating reverse myocardial remodeling.[165]

GLOBAL ANALYSIS OF GENE FUNCTION IN HF

Gene Expression Profiling: Transcriptome and Proteome

The emergence of microarray technology in early 1990s paved the way to simultaneously assess the expression of tens of thousands of gene transcripts in a single experiment, providing a resolution and precision of phenotypic characterization not previously possible.[167,168] The diverse etiologies and multiple consequences of HF make it attractive to analyze by use of gene array technology, especially when HF is idiopathic in nature. At present, microarray studies include two major applications: gene discovery and molecular signature analysis. The first application identifies differentially expressed genes characteristic of different disease states through which novel pathways and therapeutic targets can be identified. Using this application, many microarray evaluations in HF have made comparisons between failing and nonfailing hearts,[169,170] between DCM and HCM,[171] as well as before and after the placement of LVAD, as discussed in the previous section.[165] This approach of gene discovery has also provided new insights into more rare diseases such as giant cell myocarditis.[172] All these studies have the potential of providing insights into novel genetic pathways and therapeutic targets for common and uncommon conditions associated with HF.

The second application arises from the fact that the state of the transcriptome in a given disease tissue may contain a highly accurate representation of key biological phenomena. Patterns of gene expression or molecular signatures, therefore, have the potential to also identify biomarkers useful for diagnostic, prognostic, and therapeutic purposes. In this approach, the goal is to use molecular signature analysis to identify a pattern of gene expression that is associated with a clinical parameter such as etiology, prognosis, or response to treatment, thus providing diagnostic or prognostic precision otherwise not possible with standard clinical information. The basic principle of molecular signature analysis is as follow. First, samples are divided into groups on the basis of a clinically relevant parameter such as disease etiology, prognosis, or response to therapy. Then a molecular signature is created by choosing genes whose expression is associated with the parameter in question, by weighting the genes on the basis of their individual predictive strengths.[173–175] For example, using prediction analysis of microarrays, Kittleson and associates[176] have recently identified a gene expression profile that differentiates the two major forms of cardiomyopathy, ischemic and nonischemic, a classification commonly used in clinical practice and clinical trials.

It is generally accepted that both post-translational and translational processes play key roles in determining the cellular proteome.[177,178] Techniques commonly used in proteomic-based analysis include a combination of two-dimensional gel electrophoresis (2-DE), mass spectrometry, and autoradiography to map changes in myocardial protein expression. An example of the application of these techniques is provided by a recent report of an analysis of 27 proteins in the right ventricle (RV) and 21 proteins in the LV in load-induced and catecholamine-induced RV and LV hypertrophy, respectively.[179] The study describes a potentially novel pathway (BRAP2/BRCA1) that is involved in myocardial hypertrophy. Increased afterload-induced hypertrophy leads to striking changes in the energy metabolism with down-regulation of pyruvate dehydrogenase (subunit βE1), isocitrate dehydrogenase, succinyl coenzyme A ligase, NADH dehydrogenase, ubiquinol-cytochrome c reductase, and propionyl coenzyme A carboxylase. These changes go in parallel with alterations of the thin filament proteome (troponin T, tropomyosin), probably associated with Ca^{++} sensitization of the myofilaments. By contrast, neurohumoral stimulation of the LV increases the abundance of proteins relevant for energy metabolism. These techniques, therefore, allow for in-depth analysis of global proteome alterations in a controlled animal model of pressure overload–induced myocardial hypertrophy.

Global and Specific Analysis of Expression

Within the heart, numerous examples of genetic and protein changes correlated with functional alterations have been noted both during normal development and during the development of numerous pathologies.[180] One of the earliest correlations that was rigorously investigated was the association with pathology or cardiac hypertrophy of β-myosin heavy chain (MHC) expression in the ventricles of small rodents whose normal MHC complement consists largely of α-MHC.[181] Similarly, different congenital heart diseases are associated with certain shifts in the motor proteins.[182] Expression levels of 6000–15,000 genes have been compared between failing and non-failing hearts. In general, up-regulation or down-regulation of different effectors/modulators has been documented in a number of experimental models and sometimes with conflicting results.[183–185] The progressive cytoskeletal stiffness, the contractile dysfunction, and fibrosis typical for failing hearts may be partly explained by the up-regulation of many genes encoding cytoskeletal proteins, sarcomeric proteins, and ECM proteins. Remodeling on the transcriptional and translational level is represented by the up-regulation of elongation factors and ribosomal proteins. Most studies also showed increased expression of genes encoding stress proteins. For example, in the spontaneous hypertensive rats (SHR), during the transition from LV hypertrophy to frank HF, cardiac α-MHC mRNA was reduced, whereas β-MHC, α-cardiac actin, and myosin light chain-2 mRNAs were not significantly altered. ANP and B-type (brain) natriuretic peptide (BNP) mRNA was also significantly increased.[185] Possible limitations of the preceding, nevertheless important, observations include occasional high and variable level of ANP and BNP gene expression in the "nonfailing hearts" used as controls,[186] as well as a lack of proof that the new or mutated protein(s) expressed were responsible for the phenotype.

ENERGY METABOLISM PROFILING

As discussed in a previous section, observations from animal models have generally suggested a "metabolic switch" from fatty acid to glucose oxidation during evolving HF. There has not, however, been a consensus that the same pathological process is operative in human HF.[170,187] In general, the studies demonstrated up-regulation of genes involved in OXPHOS, which might reflect a decrease in activity of mitochondrial respiratory pathways during developing HF.[171,186,188] The different studies did not show consistent up-regulation or down-regulation of genes involved in FAO, highlighting the difficulty of reproducing findings from animal models to humans. When only considering end-stage failing hearts, there seems to be a trend toward the down-regulation of glycolysis, contradicting the hypothesis that failing hearts switch from fatty acid to glucose oxidation, but rather favoring a decreased use of glucose as an energy substrate.[189,190] None of the transcriptional regulators of cardiac mitochondrial biogenesis and respiratory function were found to be deregulated. This may be related to the fact that most studies lacked an experimental design that would allow the detection of relatively small expression changes.[190]

INTRACELLULAR CALCIUM CYCLING PROFILING

Intracellular Ca^{++} cycling is a key regulator of contraction in the human heart.

During membrane depolarization, Ca^{++} enters the cardiomyocyte through the L-type Ca^{++} channel. The influx of Ca^{++} into the cell triggers Ca^{++} release from the sarcoplasmic reticulum (SR) through the ryanodine receptor. This then triggers muscle contraction through the actin–myosin complex. Relaxation is initiated by Ca^{++} reuptake in the SR by Ca^{++} ATPase-2 (SERCA2). Although not all studies have yielded consistent results, altered functional properties of the L-type Ca^{++} channel, the ryanodine receptor, ATPase-2, and related regulatory proteins have been linked to human HF.[191] These changes in calcium cycling factors are thought to contribute to the reduced velocity of contraction in HF.[192] Gene profiling studies of human HF have consistently revealed a pattern of down-regulation of SERCA2 expression in failing hearts.[169,186,193] This is concordant with the results at both the mRNA and protein levels.[191] This down-regulation could be related to the decreased SR Ca^{++} content found in cardiomyocytes from failing hearts. In most gene profiling studies in human failing hearts, the negative regulator of ATPase-2—phospholamban—was not differentially expressed between failing and nonfailing hearts. The only exception was the study by Grzeskowiak and associates who found that phospholamban was up-regulated in failing hearts.[194] These findings corroborate the proposed mechanism that HF is associated with a decreased ATPase-2/phospholamban ratio and, therefore, decreased ATPase-2. Phospholamban will only inhibit ATPase-2 activity when it is hypophosphorylated. The dephosphorylation of phospholamban is accomplished by type 1 phosphatase, which is inhibited by protein phosphatase inhibitor PPP1R1A. This inhibitor was down-regulated in failing hearts in one transcriptomal study.[195] This may point toward an activation of phospholamban. In addition, down-regulation of PPP1R1A may also increase the dephosphorylation of other Ca^{++} cycling proteins, such as the L-type Ca^{++} channel and the ryanodine receptor.[196] In general, protein dephosphorylation is associated with impaired cardiac function, and increased type 1 phosphatase levels and activity have been found in human failing hearts.[197] Furthermore, ablation of PPP1R1A in murine hearts is associated with impaired β-adrenergic contractile responses.[198] Other protein phosphatase inhibitors were up-regulated in failing hearts: PPP1R14C,[194] PPP1R12A,[186] and PPP1R15A.[169] However, it remains unclear whether the encoded proteins play a role in the regulation of intracellular Ca^{++} cycling proteins. The fact that relatively few proteins involved in intracellular Ca^{++} cycling were differentially expressed in failing hearts in the transcriptomal studies indicates that either disturbed Ca^{++} cycling is mostly regulated on the protein level or that it is regulated by small expression changes that were below detection in these studies.

THE GENETICS OF HUMAN HF

Although the genesis of most forms of adult HF does not seem to be primarily genetic, HF, like other forms of adult heart disease, can be precipitated by an underlying genetic condition that results in the expression of a causative protein from birth with inherited DCM and adult HCMs particularly salient examples.[199] There exists substantial cross-talk between genetic and acquired processes in both adult and congenital heart disease. Environmental factors can profoundly disturb normal cardiac development, and it is increasingly apparent that common genetic variations can act as important modifiers of acquired adult heart disease, influencing susceptibility or progression and determining the response to therapy. As discussed in other sections, the development of cardiac hypertrophy and subsequent HF are accompanied by the reprogramming of cardiac gene expression and the activation of "fetal" cardiac genes that encode proteins involved in contraction, calcium handling, and metabolism. This transcriptional reprogramming is related to loss of cardiac function and as described earlier, improvement in cardiac function after LVAD is frequently accompanied by normalization of cardiac gene expression.[164,200]

From the perspective of etiology in HF, the greatest insight has been gained from the discovery that HCM and DCM can result from mutations in genes encoding an astonishingly broad range of cardiac proteins. The vast spectrum of causative mutations and new insights into their mechanisms of action has been extensively reviewed[199] and discussed in other chapters (in particular in Chapter 13). To date, hundreds of mutations that cause HCM have been pinpointed, with varying phenotypes resulting from missense mutations in a given protein. An intriguing handful of mutations can promote aggressive clinical courses. Determining the way in which human myosin heavy chain mutations alter force-generation in single-molecule motility studies is a triumph of the reductionist approach. In contrast, it remains unclear how hyperdynamic properties at this level of organization "trickle up" to myofiber disarray, sporadic cell death, and reactive fibrosis at the tissue level; this illustrates a gap in the present understanding of pathogenesis, as opposed to etiology. From one point of view, many hereditary cardiomyopathies may differ little from acquired cardiac disorders; the instigating signal is known in both cases, whether it is a mutation, long-standing hypertension, or past ischemic injury, yet the mechanisms of disease progression are still cryptic. Because genetic defects alter cells in highly precise and defined ways, it is reasonable to hope that dissecting their effector pathways might prove simpler than dissecting those of other forms of heart disease. Human mutations are an experiment of nature through which extraordinary details of cardiac proteins structure and function have come to be unmasked—not only for the sarcomeric and cytoskeletal proteins that are most familiar in hereditary heart disease but also for genetic defects in energy-generating mechanisms, calcium-cycling mechanisms, and transcriptional control, some of which were discussed earlier.

Recent progress in genomic applications has led to a better understanding of the relationship between genetic background and HF. A considerable component of the variability in HF outcome is due to modifier genes (i.e., genes that are not involved in the genesis of a disease but that modify the severity of the phenotypic expression once the disease has developed).[201] The strategy most commonly used to identify modifier genes, the candidate strategy, is based on association studies correlating the severity of the phenotype of the disease (morbidity and/or mortality) and the sequence variation(s) of selected candidate gene(s). Single nucleotide polymorphisms (SNPs) are the most common among these genetic variations and are widely used in association studies. This strategy has showed that, for example, several polymorphisms of the β-1 and β-2 adrenergic receptors genes and the angiotensin-converting enzyme gene are correlated to the prognosis of patients with HF. For example, a polymorphism at the nucleotide position 145 leads to a missense mutation of amino-acid residue 49 of the β-1 adrenergic receptor (β1-AR) and replacement of a serine (Ser-49) by a glycine (Gly-49). The consequence of this Ser49Gly polymorphism on the risk to develop HF and the time course of the disease has been examined in a study of 184 patients and 77 healthy controls.[202] The allele frequency of the Gly49 variant was 0.13 in controls and 0.18 in patients. At the time of the 5-year follow-up, 62% of the patients with the wild-type gene and 39% of the patients with the Ser49Gly variant had died or had experienced hospitalization. Patients without the mutation had significantly poorer survival than those with the mutation.

More recently, an experimental strategy (i.e., genome mapping) has been used for the identification of HF modifier genes.[201] Genome mapping has previously been used with success to identify the genes involved in the development of both monogenic and multifactorial diseases. It has been shown that the prognosis of mice with HF, induced through overexpressing calsequestrin, is linked to two Quantitative Trait Loci (QTL) localized on chromosome 2 and 3.[203] Using the two strategies in combination, candidate gene and genome mapping, investigators may in the near future identify an increasing number of modifier genes that may provide a more rational approach to identify patients at risk for disease and their response to therapy.

APPLICATION TO CLINICAL PRACTICE

Markers for Diagnosis, Prognostication, Disease Monitoring, and Clinical Decision-Making

The diagnosis of the HF syndrome, in general, is usually not difficult by use of conventional tools such as history, physical examination, radiographic, and echocardiographic evaluations. In more difficult cases, the use of biomarkers, most notably BNP, proves to be extremely helpful.[204] Indeed, the cardiac natriuretic peptides, particularly BNP, have also evolved to be useful biomarkers of cardiac function and prognosis in HF and other CVDs. Multiple observational studies have established the close association between plasma BNP, as well as the N-terminal fragment of the BNP prohormone (NT-proBNP) with the diagnosis of HF and an independent prediction of short-term mortality and HF events.[204–208] When they are used in appropriate clinical settings, BNP or NT-proBNP testing can be extremely useful. Furthermore, preliminary data from randomized controlled trials suggest that knowledge of BNP and/or NT-proBNP level may optimize the management of patients with HF. Large-scale randomized controlled trials that evaluate BNP/NT-proBNP–guided therapy are ongoing.

It has been appreciated for many years that HF is accompanied by the activation of neurohormone and pro-inflammatory cytokines, which likely mediate the progression of HF (i.e., the neurohormone and cytokine hypotheses) (Fig. 4).[209,210] Although circulating markers of activation are increased and predict prognosis in patients with HF,[209,211] none of these biomarkers, except for BNP and NT-proBNP, have been used extensively in clinical practice, particularly as they relate to clinical decision-making.

More specific genetic markers are required, however, for the diagnosis of specific cardiomyopathy (as opposed to HF in general) or specific etiologies of HF. For example, DNA analysis remains the only definitive test for HCM,[67] a cardiac disease frequently caused by a variety of mutations in genes encoding sarcomeric proteins, accompanied by a broad and expanding clinical spectrum, and now considered the most common genetic CVD. Likewise, the finding that mutations in a gene that encodes for contractile, cytoskeletal, nuclear membrane and other proteins could contribute to the development of DCM reminds clinicians that a considerable number of so-called "idiopathic" DCM is likely to be familial.[212,213] The first study that demonstrated a difference in cardiac gene-expression profiles from HF patients with different etiologies was reported by Hwang and associates.[171] They analyzed the expression profiles from patients with DCM and HCM and reported that functional categories "cell and organism defense" (up-regulated) and "metabolism" (down-regulated) were more strongly associated with the DCM than with the hypertrophic phenotype. Down-regulation of genes involved in "cell signaling and communication" and "cell structure and motility" and up-regulation of ribosomal protein encoding genes were found more often in the hypertrophic patients. Because it is known that both forms of cardiomyopathy result in end-stage HF through different remodeling and molecular pathways, these results are not unexpected. Tan and associates were able to distinguish six DCM heart profiles from the profiles of two failing hearts with alcoholic cardiomyopathy and familial cardiomyopathy.[195] The gene clusters responsible for the different phenotypes contained genes encoding proteins involved in signal transduction or transcription and certain cardiac-specific genes (e.g., myosin heavy chain). These genes were up-regulated in the six DCM hearts, but only slightly in the two hearts of different etiology. As noted in a previous section, Blaxall and associates were the first to classify DCM patients separately from ischemic cardiomyopathy patients.[164] After analysis of six HF patients before and after LVAD placement, they found that the pre-LVAD ischemic cardiomyopathy patients separated from the pre-LVAD nonischemic patients in their classification. In addition, the pre-LVAD ischemic patients were closer to all post-LVAD patients than the pre-LVAD nonischemic patients. These and other data suggest that non-ischemic patients undergo more extensive remodeling in the development of HF and, therefore, more extensive reverse remodeling after LVAD. Recently, Kittleson and associates described the identification of a 90-gene expression profile that specifically differentiates between ischemic and nonischemic.[176] This expression profile included many genes involved in signal transduction, metabolism, and cell growth/maintenance, with higher expression evident in ischemic than in nonischemic cardiomyopathic hearts.

There are interesting preliminary data to suggest that establishing a molecular gene-profiling portrait may help in clinical decision as it relates to the urgency of intervention. In a recent study, expression profiling was conducted on 15 failing and two non-failing hearts using a cDNA microarray containing cardiac-relevant genes.[170] Patients were classified according to expression profile on the basis of differentially expressed genes. Three patient subgroups were identified ("1", "2", and "3" in the patient classification tree), each with a specific molecular portrait. These patient subgroups did not coincide with a clinical classification on the basis of etiology. However, when the patients were annotated according to their United Network for Organ Sharing Status,[214] there seemed to be an association with the molecular patient classification. All patients with status 1A (i.e., the highest medical urgency status) clustered together in subgroup 2. This subgroup was characterized by a relatively low expression level of sarcomeric genes (e.g., titin) and metabolic genes (e.g., NADH dehydrogenase) and a relatively high expression level of natriuretic peptides. Athough the clinical relevance of this study remains to be established, it opens up the intriguing possibility that human failing hearts can be distinguished by their molecular portrait.

THERAPEUTIC OPTIONS

Cardiac transplantation is the cornerstone of surgical treatment of patients with end-stage HF, but its applicability is limited by the availability of donor organs and complications such as graft rejection and allograft coronary vasculopathy.[215] Other surgical interventions for HF that have evolved include cardiac restraint devices, ventricular remodeling procedures, and mechanical assist devices.[215,216] Ventricular remodeling procedures involve the direct surgical resection and physical restructuring of the heart to reverse the remodeling process. Most of these procedures have not yet been fully validated, and the benefits are often unpredictable and transient. On the other hand, the mechanical unloading of end-stage failing hearts with ventricular assist devices (e.g., LVAD as described earlier) has resulted in a reversal of remodeling in some patients.[217,218] Although reverse remodeling is uncommon, these striking observations indicate that cardiac remodeling can be reversed under the appropriate conditions, particularly with a reduction in mechanical load. However, the specific conditions to induce reverse remodeling are not clear, and the reversal process may be only transient once normal loading conditions are reestablished in the damaged heart. To provide significant ventricular restoration, adjunctive interventions may be required to sustain the benefits of these new surgical procedures.

Gene Therapy

Given the fact that patients with HF have an altered pattern of myocardial gene expression with a characteristic genetic fingerprint,[195] the use of genes themselves as the vehicle for replacing or altering the expression of defective genes to treat patients at the molecular level is an attractive approach to therapy. After a decade of preclinical and early phase clinical investigations, gene therapy has emerged as a genuine therapeutic option with the potential to alter the way clinicians manage patients with coronary heart disease and with HF. In the case of HF, some investigators believe that gene therapy may still be a long time away.[219] Indeed, gene therapy had multiple "initial" successes followed by the realization that much of the enthusiasm for each success has been perhaps premature. Furthermore, much of the early excitement with regard to this approach has been dampened after the death of a young male in one clinical trial. However, as with all discoveries and new fields, problems will arise, and they need to be and likely will be identified and overcome with time.

Muscular dystrophies constitute a good model to assess the novel approach of gene therapy in the setting of cardiomyopathy. These conditions are common and debilitating genetic diseases that commonly arise from single gene mutations. In humans and hamsters, mutations in the *SGCD* gene encoding δ-sarcoglycan (δ–SG) can cause limb girdle muscular dystrophy (LGMD) with accompanying DCM. The success of gene therapy requires widespread and stable gene delivery with minimal trauma. A recent study reported the therapeutic effect of systemic delivery of adeno-associated virus (AAV) vectors carrying human *SGCD* gene in TO-2 hamsters, a HF and muscular dystrophy model with a δ-SG mutation.[220] A single injection of a double-stranded AAV vector carrying the *SGCD* gene, without the need of physical or pharmaceutical interventions achieved near complete gene transfer and expression in the heart and skeletal muscles. Sustained restoration of the missing δ-SG gene in the TO-2 hamsters corrected muscle cell membrane leakiness throughout the body and normalized serum creatine kinase levels. Echocardiography revealed significantly increased percent fractional shortening and decreased LV end-diastolic and end-systolic dimensions.

CELL TRANSPLANTATION

In contrast to gene therapy, cell-based approaches to remodel the failing heart are in the preliminary stage of clinical application as discussed in Chapter 4, and the results are thus far quite encouraging.[221,222]

Stem Cell Therapy

Stem cell therapy offers potential help to those who develop severe HF after MI. Two potentially major advances in stem cell therapy for HF were published recently. The first report demonstrates how bone marrow stem cells can regenerate myocardium in the infarct area of a mouse heart,[121] whereas the second report describes the use of a subgroup of bone marrow stem cells to stimulate neovascularization and prevent remodeling in the infarct area of a rat heart.[223] In both studies, improved cardiac function was reported.

Cardiomyogenic differentiation of marrow stem cells (MSCs) can occur *in vivo*.[224] In this study, isogeneic cultured MSCs were labeled with 4′, 6-diamidino-2-phenylindole (DAPI) and implanted into the LV wall of recipient rats. After 4 weeks, DAPI-labeled donor MSCs demonstrated myogenic differentiation with the expression of sarcomeric myosin heavy chain in the cytoplasm. Orlic and associates reported that they have identified bone marrow stem cells that, when injected into MI mouse, migrate specifically into the infarct area, replenish it with cardiomyocytes, endothelial cells, and smooth muscle cells, and partially restore cardiac function.[121] In this study, the stem cells were isolated from donor mice and injected into recipients in viable myocardium bordering a 3- to 5-hour-old infarction. Nine days after injection, the researchers observed new myocardium complete with cardiomyocytes and vascular structures filling almost 70% of the infarcted region in 12 of the 30 mice. The treated mice showed significant improvement in LV function. At 9 days, the new myocardial cells were still proliferating and maturing.

An alternative approach to transplanting new myocardium to the infarcted heart is to help the heart in its own recovery by preventing remodeling. This approach was pursued by Kocher and associates,[223] who identified a subpopulation of bone marrow stem cells with hemangioblast-like properties that, when injected into MI rats, migrate to the infarct zone, generate new blood vessels, and keep the hypertrophied cardiomyocytes viable as LV function restores. Left ventricular end-diastolic pressure declined by 36% and LV-dP/dt increased by 41%. At 2-week follow-up, these investigators observed significant increases in microvascularity and in the number of capillaries and feeding vessels both within the infarct zone and at its perimeter. The revascularization of the infarct tissue resulted in a 6-fold reduction in myocyte apoptosis and scar and a significant restoration of cardiac function. The benefit was sustained up to 15 weeks. LV ejection fraction increased by 34%.[223,225] In the first human study, Strauer and associates transplanted the patient's bone marrow cells into his myocardium after an MI and found a significant improvement of the patient's heart function.[226] Ten weeks after stem cell implantation, there was a reduction of the infarct area and an increase of ejection fraction, cardiac index, and stroke volume. The investigators assume that the adult stem cells, which were transplanted into the necrotic areas of the myocardium, had differentiated into cardiomyocytes, which regenerated the heart wall. Although at this point these investigators could not prove the hypothesis, because tissue samples from the myocardium could not be extracted and analyzed, there is no other explanation for the improvement of the patient's heart function. After the first successful implantation, the team has treated six more patients with autologous bone marrow cells.

Larger clinical studies on stem cell therapy have so far been less compelling. A new randomized, double-blind, placebo-controlled study of stem cells in MI patients has failed to show an improvement in LV function, although there was a suggestion that such treatment could favorably affect infarct remodeling.[227] The randomized, open-label BOOST study has previously shown improvement of LV systolic function after transfer of stem cells, but this trial was not blinded. However, as bone-marrow aspiration and intracoronary injection can both induce cytokine release and, therefore, affect subsequent infarct healing and functional recovery, the control group must undergo exactly the same procedures as the active treatment group if the true benefit of cell transfer is to be fully appreciated. The investigators, therefore, followed with a study in 67 patients, who all underwent bone-marrow aspiration 1 day after successful percutaneous coronary intervention for ST-elevation MI (STEMI). Patients were then assigned to intracoronary infusion of stem cells or placebo. The primary end point was the increase in LV ejection fraction, and secondary end points were change in infarct size and regional LV function at 4-month follow-up, all assessed by magnetic resonance imaging. Results showed no difference in LV function improvements between the two groups, although stem-cell infusion was associated with a significant reduction in myocardial infarct size and a better recovery of regional systolic function. No complications were associated with the treatment.

The first double-blind, randomized, placebo-controlled trial of granulocyte-colony stimulating factor (G-CSF) in patients with acute MI has just been published and failed to show any benefit.[228] The Stem Cells in Myocardial Infarction (STEMMI) clinical trial was assessed with G-CSF to mobilize stem cells in patients with STEMI after successful primary percutaneous coronary stent intervention. A total of 78 STEMI patients were evenly randomly assigned to receive either a once-daily under-the-skin injection of G-CSF (10 μg/kg of body weight) or placebo for 6 days. Patients underwent angioplasty and stenting less than 12 h after the onset of symptoms, and 85% of patients were treated within 48 h. Although G-CSF proved safe, at 6-month follow-up, there was no evidence of any significant benefit in the primary end point change in systolic wall thickening in the infarct area from baseline, determined by cardiac magnetic resonance imaging (MRI), which improved 17% in the G-CSF group and 17% in the placebo group. Comparable results were found in the infarct border and noninfarcted myocardium. LVEF improved similarly in the two groups, as measured by both MRI and echocardiography. These findings differ from those observed in previous open-label trials of G-CSF and underscore the importance of blinding and placebo controls in evaluating new, potentially antiangiogenic, stem cell therapies.

Autologous Skeletal Myoblast Transfer

Skeletal myoblasts are endogenous skeletal muscle progenitor cells that serve to regenerate functional skeletal muscle after injury. These cells are dormant in healthy skeletal tissues and will proliferate and differentiate into mature skeletal muscle fibers when new tissue is required. Adult autologous skeletal myoblasts can be harvested from small peripheral muscle biopsy, and, if successfully isolated from the donor tissue, they will grow rapidly in culture. In a rat coronary artery ligation, Scorcin and Ménasché provided evidence that skeletal muscle myoblasts improved post-MI LV function to a similar degree as cardiomyocytes.[229] Ménasché and associates have reported the first clinical use of autologous skeletal myoblast transplantation in a 72-year-old patient with HF after multiple previous MIs. These investigators injected 800 million skeletal myoblasts cultured from biopsy specimens taken from the patient's leg into infarcted tissue on the posterior wall of the heart during double-bypass surgery. Echocardiography and positron-emission tomography scanning before the transplant had shown the area metabolically nonviable. After 5 months of follow-up, these studies reveal contraction in the area of the transplant, the magnitude of which increases when challenged with dobutamine.[230] French investigators, who have now undertaken autologous skeletal myoblast transplantation in more patients with severe ischemic HF, report that the technique seems feasible, is generally safe, and has shown some early signs of efficacy. Most patients have had symptomatic improvement, and echocardiographic studies showed new systolic thickening in implanted areas of the previous infarct, indicating contractility in implanted segments. A phase II randomized trial of this strategy, the Myoblast Autologous Grafting in Ischemic Cardiomyopathy (MAGIC) trial, is currently underway. Only at the end of the randomized trial will one know whether this technique really holds the promise that is currently anticipated.

PHARMACOGENETICS AND PHARMACOGENOMICS IN HF

Pharmacogenetics is the study of the effect of a drug as it relates to single or defined sets of genes. With increasing understanding of the human genome and the development of wide assortment of pharmacological and mechanical therapies in HF, a potentially realizable goal will be to integrate pharmacogenetics, pharmacogenomics, and therapy so that more tailored treatment can be delivered.[231] More detailed discussions on method and the general development, including limitations in this field, are covered in other chapters. Only some examples from drugs used widely and specifically in the treatment of HF will be discussed in this chapter. A list of examples is shown in Table III.

DRUG METABOLISM

A number of drug-metabolizing enzymes have inactivating mutations, leading to absent or nonfunctional protein.

TABLE III Examples of Pharmacological Agents Used in the Treatment of HF with Evidence of Association Between Genetic and Efficacy or Adverse Effects

Drug class	Genes associated with efficacy or toxicity
β-Agonists/blockers	β1-Adrenergic receptors
	ACE
	G_s protein α subunit
	CYP2D6
ACE inhibitors	ACE
	Angiotensinogen
	AT_1 receptor subtype
	Bradykinin B_2
AT_1 receptor blockers	AT_1 receptor subtype
Digoxin	P-glycoprotein drug transporter (MDR1)
Antihypertensives:	
Diuretics	G protein β3-subunit (GNβ3), Adducin (ADD 1–3)
Hydralazine	N-acetyltransferase (NAT2)

For drugs with a high dependence on a particular enzyme for their elimination, these polymorphisms may have clinical consequence. In the case of drugs used in the treatment of HF, there are a few substrates for the enzymes that exhibit genetic variability, but most do not have important clinical significance. For example, approximately 70% of the metabolism of the β-blocker metoprolol, a drug used widely in the treatment of HF associated with systolic function, is controlled by the polymorphic cytochrome P450 2D6 (CYP2D6) enzyme, and patients with inactivating mutations on both alleles have no functional protein. The "poor metabolizers" have up to five times the usual drug concentrations, and yet neither drug plasma concentration nor CYP2D6 genotype are determinants of tolerability or adverse effects in patients with HF.[232,233]

GENETIC INFLUENCES ON DRUG TARGETS

Genetic variability in the proteins involved in drug pharmacodynamics has proved to be informative in understanding the variability in responses to cardiovascular drugs. At present, there are data supporting genetic associations between drug targets and response for β-blockers, ACE inhibitors, and angiotensin receptor blockers.[231,234] Of most relevance to the treatment of HF is the emerging data on the genetic determinants of response to β-blockers. Several studies have now reported that Arg389Arg genotype of the β1-adrenergic receptor is associated with the greatest response, both in terms of LV systolic function as well as improvement in survival.[233,235,236] On the other hand, there are now genetic markers that can potentially predict cardiac toxicity of drugs. A recent study of genotyped patients with non-Hodgkin lymphoma treated with doxorubicin was undertaken and followed for the development of HF.[237] Single-nucleotide polymorphisms were selected from 82 genes with conceivable relevance to anthracycline-induced cardiotoxicity. Of 1697 patients, 55 developed acute and 54 developed chronic toxicity. Five significant associations with polymorphisms of the NAD(P)H oxidase and doxorubicin efflux transporters were detected. Chronic toxicity was associated with a variant of the NAD(P)H oxidase subunit NCF4 (-212AG). Acute toxicity was associated with the His72Tyr polymorphism in the p22phox subunit and with the variant 7508TA of the RAC2 subunit of the same enzyme. Mice deficient in NAD(P)H oxidase activity, unlike wild-type mice, were resistant to chronic doxorubicin treatment. In addition, acute toxicity was associated with the Gly671Val variant of the doxorubicin efflux transporter multidrug resistance protein 1 (MRP1) and with the Val1188Glu and Cys1515Tyr haplotype of the functionally similar MRP2, a drug transporter gene. In this study, polymorphisms in adrenergic receptors previously demonstrated to be predictive of HF were not associated with anthracycline cardiac toxicity. Genetic variants in doxorubicin transport and free radical metabolism may, therefore, modulate the individual risk to develop doxorubicin-induced cardiotoxicity.

CONCLUSION

HF is essentially the final common pathway of cardiovascular disorders that results in impairment of myocardial systolic and/or diastolic function. Since the publication of the human genome, there have been many changes in the field of genetic research, and it has provided investigators with a vast new array of potential therapeutic genes to test in animal models *in vivo*. Many cardiovascular pathological conditions that may culminate in

HF are associated with disturbances in energy metabolism. The genetic modification of key steps in the uptake and metabolism of glucose and fatty acids, by the use of transgenic technology, has enhanced the understanding of how these pathways influence the pathogenesis of cardiac disorders. In general, these results would suggest that the metabolic shifts, which accompany HF, contribute to the development of cardiac dysfunction.

A critical step in the evolution of LV dysfunction and HF is cardiac remodeling. The structural changes in cardiac remodeling include hypertrophy and subsequent chamber dilatation with associated progressive impairment in function. The process involves changes in both the myocytes and the ECM, the latter including the activation of proteolytic enzymes leading to the degradation and reorganization of collagen. Neurohumoral signals acting through many interrelated signal transduction pathways lead to pathological cardiac hypertrophy and subsequent progression of HF. Many such agonists act through cell surface receptors coupled with G-proteins to mobilize intracellular calcium, with consequent activation of downstream kinases and the calcium- and calmodulin-dependent phosphatase calcineurin. MAPK signaling pathways are also interconnected at multiple levels with calcium-dependent kinases and calcineurin.

The knowledge of the significance of myocyte loss during HF has increased substantially since cardiac pathologists initially observed this phenomenon. Molecular and genetic studies demonstrate clearly that cardiomyocyte apoptosis is a critical process in the pathogenesis of HF in rodent models. If this paradigm is proved to extend to humans, apoptosis will be a logical target for novel therapies. However, despite success in establishing a mechanistic link between apoptosis and HF, the knowledge of the precise molecular mechanisms that regulate cell death specifically in this syndrome remains preliminary.

The emergence of microarray technology has paved the way to simultaneously assess the expression of tens of thousands of gene transcripts in a single experiment, providing a resolution and precision of phenotypic characterization not previously possible. Within the heart, many examples of genetic and protein changes correlated with functional alterations have been noted both during normal development and during the development of HF from diverse etiologies. Detailed profiling has been performed on structural changes, as well as functional alterations, including energy metabolism and intracellular calcium handling.

Knowledge about the genetic and molecular changes in HF has begun to find clinical applications, including their use as diagnostic and prognostic markers, as well as therapy in conditions associated with the development of HF.

SUMMARY

- HF is the final common pathway of cardiovascular disorders resulting from impairment of myocardial systolic and/or diastolic function.
- Primary triggers of HF include impaired pump function after myocyte death from MI, stimuli resulting from myocardial ischemia (e.g., oxidative and nitrosative stress), and abnormal hemodynamic loading conditions such as found in hypertension.
- These stimuli activate a series of biomechanical stress-dependent signaling cascades involving cardiomyocyte sensors, leading to altered signal transduction pathways and resulting in the activation of transcription factors, coactivators, and corepressors orchestrating myocardial gene expression. Both neurohormonal and cytokine activation further promote and amplify these cascades.
- These pathways directly impact cardiomyocyte effector mechanisms such as calcium cycling, mitochondrial metabolism, myocyte growth and loss (primarily by apoptosis) that underlie cardiac remodeling, a critical feature of HF progression.
- Central to cardiac remodeling is the activation of proteolytic enzymes involved in extracellular matrix remodeling.
- HF can be elicited by mutations in nuclear and mitochondrial DNA–encoded proteins, leading to defects in a variety of myocardial functions, including signal transduction, transcriptional activation and bioenergetic metabolic remodeling, oxidative stress, calcium handling, and sarcomeric/contractile protein function.
- The use of transgenic models (primarily murine) has allowed the identification of critical genetic *loci* involved in HF and provided useful data concerning both gene targeting and therapeutic modulation of HF.
- Global and specific gene profiling in preclinical studies has proved informative about the involvement of specific genes and the development of useful biomarkers in the progression of HF.
- Gene profiling studies have shown distinctive profiles of gene expression in association with different types of cardiomyopathy leading to HF (e.g., ischemic, DCM or HCM).
- Early clinical studies suggest that the integrated use of genetic technologies (gene profiling, pharmacogenomics) with clinical medicine may enable a highly effective individually tailored approach to treating HF.
- Modifier *loci* and SNPs have been identified to impact HF progression.
- Gene therapies and cell transplantation studies have just begun to be used in clinical studies of HF with modest but encouraging preliminary results.

References

1. Braunwald E. (2001). Congestive heart failure: A half century perspective. *Eur. Heart* **22,** 825–836.
2. The American Heart Association. (2004). "Heart Disease and Stroke Statistics—2004 Update." American Heart Association, Dallas, Texas.

3. The American Heart Association. (2005). Heart disease and stroke statistics—2005 Update. American Heart Association, Dallas, Texas.
4. Johansen, H., Strauss, B., Arnold, M., and Moe, G. (2003). On the rise: the current and projected future burden of congestive heart failure hospitalization in Canada. *Can. J. Cardiol.* **19,** 430–435.
5. O'Connell, J. B. (2000). The economic burden of heart failure. *Clin. Cardiol.* **23,** III6–III10.
6. Massie, B. M. (2002). Treating heart failure: It's time for new paradigms and novel approaches. *J. Card. Fail.* **8,** 117–119.
7. Cohn, J. N. (2004). New therapeutic strategies for heart failure: Left ventricular remodeling as a target. *J. Card. Fail.* **10,** S200–S201.
8. Cohn, J. N., Ferrari, R., and Sharpe, N. (2000). Cardiac remodeling—concepts and clinical implications: A consensus paper from an international forum on cardiac remodeling. Behalf of an International Forum on Cardiac Remodeling. *J. Am. Coll. Cardiol.* **35,** 569–582.
9. Gunja-Smith, Z., Morales, A. R., Romanelli, R., and Woessner, J. F., Jr. (1996). Remodeling of human myocardial collagen in idiopathic dilated cardiomyopathy. Role of metalloproteinases and pyridinoline cross-links. *Am. J. Pathol.* **148,** 1639–1648.
10. Spinale, F. G., Tomita, M., Zellner, J. L., Cook, J. C., Crawford, F. A., and Zile, M. R. (1991). Collagen remodeling and changes in LV function during development and recovery from supraventricular tachycardia. *Am. J. Physiol.* **261,** H308–H318.
11. Spinale, F. G., Coker, M. L., Thomas, C. V., Walker, J. D., Mukherjee, R., and Hebbar, L. (1998). Time-dependent changes in matrix metalloproteinase activity and expression during the progression of congestive heart failure: relation to ventricular and myocyte function. *Circ. Res.* **82,** 482–495.
12. Weber, K. T., Pick, R., Janicki, J. S., Gadodia, G., and Lakier, J. B. (1988). Inadequate collagen tethers in dilated cardiopathy. *Am. Heart J.* **116,** 1641–1646.
13. Benjamin, I. J., and Schneider, M. D. (2005). Learning from failure: congestive heart failure in the postgenomic age. *J. Clin. Invest.* **115,** 495–499.
14. Morita, H., Seidman, J., and Seidman, C. E. (2005). Genetic causes of human heart failure. *J. Clin. Invest.* **115,** 518–526.
15. Armstrong, P. W., Howard, R. J., and Moe, G. W. (1989). Clinical lessons learned from experimental heart failure. *Int. J. Cardiol.* **24,** 133–136.
16. Elsner, D., and Riegger, G. A. (1995). Characteristics and clinical relevance of animal models of heart failure. *Curr. Opin. Cardiol.* **10,** 253–259.
17. Moe, G. W., and Armstrong, P. (1999). Pacing-induced heart failure: a model to study the mechanism of disease progression and novel therapy in heart failure. *Cardiovasc. Res.* **42,** 591–599.
18. Robbins, J. (2004). Genetic modification of the heart: exploring necessity and sufficiency in the past 10 years. *J. Mol. Cell Cardiol.* **36,** 643–652.
19. Russell, L. K., Finck, B. N., and Kelly, D. P. (2005). Mouse models of mitochondrial dysfunction and heart failure. *J. Mol. Cell Cardiol.* **38,** 81–91.
20. Carvajal, K., and Moreno-Sanchez, R. (2003). Heart metabolic disturbances in cardiovascular diseases. *Arch. Med. Res.* **34,** 89–99.
21. Kelly, D. P., and Scarpulla, R. C. (2004). Transcriptional regulatory circuits controlling mitochondrial biogenesis and function. *Genes Dev.* **18,** 357–368.
22. Anderson, S., Bankier, A. T., Barrell, B. G., de Bruijn, M. H., Coulson, A. R., Drouin, J., Eperon, I. C., Nierlich, D. P., Roe, B. A., Sanger, F., Schreier, P. H., Smith, A. J., Staden, R., and Young, I. G. (1981). Sequence and organization of the human mitochondrial genome. *Nature* **290,** 457–465.
23. Smeitink, J., van den, H. L., and DiMauro, S. (2001). The genetics and pathology of oxidative phosphorylation. *Nat. Rev. Genet.* **2,** 342–352.
24. Wallace, D. C. (1999). Mitochondrial diseases in man and mouse. *Science* **283,** 1482–1488.
25. Larsson, N. G., and Oldfors, A. (2001). Mitochondrial myopathies. *Acta Physiol. Scand.* **171,** 385–393.
26. Kelly, D. P., and Strauss, A. W. (1994). Inherited cardiomyopathies. *N. Engl. J. Med.* **330,** 913–919.
27. Garnier, A., Fortin, D., Delomenie, C., Momken, I., Veksler, V., and Ventura-Clapier, R. (2003). Depressed mitochondrial transcription factors and oxidative capacity in rat failing cardiac and skeletal muscles. *J. Physiol.* **551,** 491–501.
28. van Bilsen, M. (2004). "Energenetics" of heart failure. *Ann. N. Y. Acad. Sci.* **1015,** 238–249.
29. Ventura-Clapier, R., Garnier, A., and Veksler, V. (2004). Energy metabolism in heart failure. *J. Physiol.* **555,** 1–13.
30. Barger, P. M., Brandt, J. M., Leone, T. C., Weinheimer, C. J., and Kelly, D. P. (2000). Deactivation of peroxisome proliferator-activated receptor-alpha during cardiac hypertrophic growth. *J. Clin. Invest.* **105,** 1723–1730.
31. Razeghi, P., Young, M. E., Alcorn, J. L., Moravec, C. S., Frazier, O. H., and Taegtmeyer, H. (2001). Metabolic gene expression in fetal and failing human heart. *Circulation* **104,** 2923–2931.
32. Razeghi, P., Young, M. E., Cockrill, T. C., Frazier, O. H., and Taegtmeyer, H. (2002). Downregulation of myocardial myocyte enhancer factor 2C and myocyte enhancer factor 2C-regulated gene expression in diabetic patients with nonischemic heart failure. *Circulation* **106,** 407–411.
33. Giordano, F. J. (2005). Oxygen, oxidative stress, hypoxia, and heart failure. *J. Clin. Invest.* **115,** 500–508.
34. Ferrari, R., Guardigli, G., Mele, D., Percoco, G. F., Ceconi, C., and Curello, S. (2004). Oxidative stress during myocardial ischaemia and heart failure. *Curr. Pharm. Des.* **10,** 1699–1711.
35. Witztum, J. L., and Steinberg, D. (1991). Role of oxidized low density lipoprotein in atherogenesis. *J. Clin. Invest.* **88,** 1785–1792.
36. Rajagopalan, S., Meng, X. P., Ramasamy, S., Harrison, D. G., and Galis, Z. S. (1996). Reactive oxygen species produced by macrophage–derived foam cells regulate the activity of vascular matrix metalloproteinases in vitro. Implications for atherosclerotic plaque stability. *J. Clin. Invest.* **98,** 2572–2579.
37. Khatri, J. J., Johnson, C., Magid, R., Lessner, S. M., Laude, K. M., Dikalov, S. I., Harrison, D. G., Sung, H. J., Rong, Y., and Galis, Z. S. (2004). Vascular oxidant stress enhances progression and angiogenesis of experimental atheroma. *Circulation* **109,** 520–525.

38. Malinski, T. (2005). Understanding nitric oxide physiology in the heart: a nanomedical approach. *Am. J. Cardiol.* **96**, 13i–24i.
39. Ungvari, Z., Gupte, S. A., Recchia, F. A., Batkai, S., and Pacher, P. (2005). Role of oxidative-nitrosative stress and downstream pathways in various forms of cardiomyopathy and heart failure. *Curr. Vasc. Pharmacol.* **3**, 221–229.
40. Hare, J. M., and Stamler, J. S. (2005). NO/redox disequilibrium in the failing heart and cardiovascular system. *J. Clin. Invest.* **115**, 509–517.
41. Martinez-Ruiz, A., and Lamas, S. (2004). S-nitrosylation: a potential new paradigm in signal transduction. *Cardiovasc. Res.* **62**, 43–52.
42. Hartil, K., and Charron, M. J. (2005). Genetic modification of the heart; transgenic modification of cardiac lipid and carbohydrate utilization. *J. Mol. Cell Cardiol.* **39**, 581–593.
43. Stepien, G., Torroni, A., Chung, A. B., Hodge, J. A., and Wallace, D. C. (199). Differential expression of adenine nucleotide translocator isoforms in mammalian tissues and during muscle cell differentiation. *J. Biol. Chem.* **2267**, 14592–14597.
44. Graham, B. H., Waymire, K. G., Cottrell, B., Trounce, I. A., MacGregor, G. R., and Wallace, D. C. (1997). A mouse model for mitochondrial myopathy and cardiomyopathy resulting from a deficiency in the heart/muscle isoform of the adenine nucleotide translocator. *Nat. Genet.* **16**, 226–234.
45. Exil, V. J., Roberts, R. L., Sims, H., McLaughlin, J. E., Malkin, R. A., Gardner, C. D., Ni, G., Rottman, J. N., and Strauss, A. W. (2003). Very-long-chain acyl-coenzyme A dehydrogenase deficiency in mice. *Circ. Res.* **93**, 448–455.
46. Kurtz, D. M., Rinaldo, P., Rhead, W. J., Tian, L., Millington, D. S., Vockley, J., Hamm, D. A., Brix, A. E., Lindsey, J. R., Pinkert, C. A., O'Brien, W. E., and Wood, P. A. (1998). Targeted disruption of mouse long-chain acyl-CoA dehydrogenase gene reveals crucial roles for fatty acid oxidation. *Proc. Natl. Acad. Sci. USA* **95**, 15592–15597.
47. Watanabe, K., Fujii, H., Takahashi, T., Kodama, M., Aizawa, Y., Ohta, Y., Ono, T., Hasegawa, G., Naito, M., Nakajima, T., Kamijo, Y., Gonzalez, F. J., and Aoyama, T. (2000). Constitutive regulation of cardiac fatty acid metabolism through peroxisome proliferator-activated receptor alpha associated with age-dependent cardiac toxicity. *J. Biol. Chem.* **275**, 22293–22299.
48. Lopaschuk, G. D. (2002). Metabolic abnormalities in the diabetic heart. *Heart Fail. Rev.* **7**, 149–159.
49. Finck, B. N., Lehman, J. J., Leone, T. C., Welch, M. J., Bennett, M. J., Kovacs, A., Han, X., Gross, R. W., Kozak, R., Lopaschuk, G. D., and Kelly, D. P. (2002). The cardiac phenotype induced by PPARα overexpression mimics that caused by diabetes mellitus. *J. Clin. Invest.* **109**, 121–130.
50. Finck, B. N., Han, X., Courtois, M., Aimond, F., Nerbonne, J. M., Kovacs, A., Gross, R. W., and Kelly, D. P. (2003). A critical role for PPARα-mediated lipotoxicity in the pathogenesis of diabetic cardiomyopathy: modulation by dietary fat content. *Proc. Natl. Acad. Sci. USA* **100**, 1226–1231.
51. Carabello, B. A. (2002). Concentric versus eccentric remodeling. *J. Card. Fail.* **8**, S258–S263.
52. Brower, G. L., and Janicki, J. S. (2001). Contribution of ventricular remodeling to pathogenesis of heart failure in rats. *Am. J. Physiol. Heart Circ. Physiol.* **280**, H674–H683.
53. Sambandam, N., Lopaschuk, G. D., Brownsey, R. W., and Allard, M. F. (2002). Energy metabolism in the hypertrophied heart. *Heart Fail. Rev.* **7**, 161–173.
54. Molkentin, J. D., and Dorn, II G. W. (2001). Cytoplasmic signaling pathways that regulate cardiac hypertrophy. *Annu. Rev. Physiol.* **63**, 391–426.
55. Kilic, A., Velic, A., De Windt, L. J., Fabritz, L., Voss, M., Mitko, D., Zwiener, M., Baba, H. A., van, E. M., Schlatter, E., and Kuhn, M. (2005). Enhanced activity of the myocardial Na+/H+ exchanger NHE-1 contributes to cardiac remodeling in atrial natriuretic peptide receptor-deficient mice. *Circulation* **112**, 2307–2317.
56. del, M. F., and Hajjar, R. J. (2003). Targeting calcium cycling proteins in heart failure through gene transfer. *J. Physiol.* **546**, 49–61.
57. Fedak, P. W., Verma, S., Weisel, R. D., and Li, R. K. (2005). Cardiac remodeling and failure: From molecules to man (Part II). **14**, 49–60.
58. Leong, H. S., Brownsey, R. W., Kulpa, J. E., and Allard, M. F. (2003). Glycolysis and pyruvate oxidation in cardiac hypertrophy—why so unbalanced? *Comp. Biochem. Physiol. A Mol. Integr. Physiol.* **135**, 499–513.
59. Ananthakrishnan, R., Moe, G. W., Goldenthal, M. J., and Marin-Garcia, J. (2005). Akt signaling pathway in pacing-induced heart failure. *Mol. Cell Biochem.* **268**, 103–110.
60. Sugden, P. H., and Clerk, A. (1998). "Stress-responsive" mitogen-activated protein kinases (c-Jun N-terminal kinases and p38 mitogen-activated protein kinases) in the myocardium. *Circ. Res.* **83**, 345–352.
61. Rockman, H. A., Koch, W. J., and Lefkowitz, R. J. (2002). Seven-transmembrane-spanning receptors and heart function. *Nature* **415**, 206–212.
62. Houser, S. R., and Margulies, K. B. (2003). Is depressed myocyte contractility centrally involved in heart failure? *Circ. Res.* **92**, 350–358.
63. Davies, C. H., Davia, K., Bennett, J. G., Pepper, J. R., Poole-Wilson, P. A., and Harding, S. E. (1995). Reduced contraction and altered frequency response of isolated ventricular myocytes from patients with heart failure. *Circulation* **92**, 2540–2549.
64. Harding, S. E., MacLeod, K. T., Davies, C. H., Wynne, D. G., and Poole-Wilson, P. A. (1995). Abnormalities of the myocytes in ischaemic cardiomyopathy. *Eur. Heart J.* **16**, 74–81.
65. Anand, I. S., Liu, D., Chugh, S. S., Prahash, A. J., Gupta, S., John, R., Popescu, F., and Chandrashekhar, Y. (1997). Isolated myocyte contractile function is normal in postinfarct remodeled rat heart with systolic dysfunction. *Circulation* **96**, 3974–3984.
66. Sugden, P. H. (1999). Signaling in myocardial hypertrophy: life after calcineurin? *Circ. Res.* **84**, 633–646.
67. Maron, B. J. (2002). Hypertrophic cardiomyopathy: a systematic review. *JAMA* **287**, 1308–1320.
68. Swynghedauw, B. (1999). Molecular mechanisms of myocardial remodeling. *Physiol. Rev.* **79**, 215–262.
69. Weber, K. T., Jalil, J. E., Janicki, J. S., and Pick, R. (1989). Myocardial collagen remodeling in pressure overload hypertrophy. A case for interstitial heart disease. *Am. J. Hypertens.* **2**, 931–940.
70. Litwin, S. E., Raya, T. E., Anderson, P. G., Litwin, C. M., Bressler, R., and Goldman, S. (1991). Induction of myocardial

hypertrophy after coronary ligation in rats decreases ventricular dilatation and improves systolic function. *Circulation* **84,** 1819–1827.

71. Asakura, M., Kitakaze, M., Takashima, S., Liao, Y., Ishikura, F., Yoshinaka, T., Ohmoto, H., Node, K., Yoshino, K., Ishiguro, H., Asanuma, H., Sanada, S., Matsumura, Y., Takeda, H., Beppu, S., Tada, M., Hori, M., and Higashiyama, S. (2002). Cardiac hypertrophy is inhibited by antagonism of ADAM12 processing of HB–EGF: metalloproteinase inhibitors as a new therapy. *Nat. Med.* **8,** 35–40.
72. Liao, J. K. (2002). Shedding growth factors in cardiac hypertrophy. *Nat. Med.* **8,** 20–21.
73. Ross, R. S., and Borg, T. K. (2001). Integrins and the myocardium. *Circ. Res.* **88,** 1112–1119.
74. Terracio, L., Rubin, K., Gullberg, D., Balog, E., Carver, W., Jyring, R., and Borg, T. K. (1991). Expression of collagen binding integrins during cardiac development and hypertrophy. *Circ. Res.* **68,** 734–744.
75. Frey, N., and Olson, E. N. (2003). Cardiac hypertrophy: the good, the bad, and the ugly. *Annu. Rev. Physiol.* **65,** 45–79.
76. Li, Z., Bing, O. H., Long, X., Robinson, K. G., and Lakatta, E. G. (1997). Increased cardiomyocyte apoptosis during the transition to heart failure in the spontaneously hypertensive rat. *Am. J. Physiol.* **272,** H2313–H2319.
77. Mani, K., and Kitsis, R. N. (2003). Myocyte apoptosis: programming ventricular remodeling. *J. Am. Coll. Cardiol.* **41,** 761–764.
78. Moe, G. W., Naik, G., Konig, A., Lu, X., and Feng, Q. (2002). Early and persistent activation of myocardial apoptosis, bax and caspases: insights into mechanisms of progression of heart failure. *Pathophysiology* **8,** 183–192.
79. Kang, P. M., and Izumo, S. (2000). Apoptosis and heart failure: A critical review of the literature. *Circ. Res.* **86,** 1107–1113.
80. Condorelli, G., Morisco, C., Stassi, G., Notte, A., Farina, F., Sgaramella, G., de, R. A., Roncarati, R., Trimarco, B., and Lembo, G. (1999). Increased cardiomyocyte apoptosis and changes in proapoptotic and antiapoptotic genes bax and bcl-2 during left ventricular adaptations to chronic pressure overload in the rat. *Circulation* **99,** 3071–3078.
81. Schulze-Osthoff, K., Ferrari, D., Los, M., Wesselborg, S., and Peter, M. E. (1998). Apoptosis signaling by death receptors. *Eur. J. Biochem.* **254,** 439–459.
82. Yue, T. L., Ma, X. L., Wang, X., Romanic, A. M., Liu, G. L., Louden, C., Gu, J. L., Kumar, S., Poste, G., Ruffolo, R. R., Jr., and Feuerstein, G. Z. (1998). Possible involvement of stress-activated protein kinase signaling pathway and Fas receptor expression in prevention of ischemia/reperfusion-induced cardiomyocyte apoptosis by carvedilol. *Circ. Res.* **82,** 166–174.
83. Danial, N. N., and Korsmeyer, S. J. (2004). Cell death: critical control points. *Cell* **116,** 205–219.
84. Ashkenazi, A., and Dixit, V. M. (1998). Death receptors: signaling and modulation. *Science* **281,** 1305–1308.
85. Muzio, M., Chinnaiyan, A. M., Kischkel, F. C., O'Rourke, K., Shevchenko, A., Ni, J., Scaffidi, C., Bretz, J. D., Zhang, M., Gentz, R., Mann, M., Krammer, P. H., Peter, M. E., and Dixit, V. M. (1996). FLICE, a novel FADD-homologous ICE/CED-3-like protease, is recruited to the CD95 (Fas/APO-1) death—inducing signaling complex. *Cell* **85,** 817–827.
86. Boatright, K. M., Renatus, M., Scott, F. L., Sperandio, S., Shin, H., Pedersen, I. M., Ricci, J. E., Edris, W. A., Sutherlin, D. P., Green, D. R., and Salvesen, G. S. (2003). A unified model for apical caspase activation. *Mol. Cell* **11,** 529–541.
87. Crow, M. T., Mani, K., Nam, Y. J., and Kitsis, R. N. (2004). The mitochondrial death pathway and cardiac myocyte apoptosis. *Circ. Res.* **95,** 957–970.
88. Du, C., Fang, M., Li, Y., Li, L., and Wang, X. (2000). Smac, a mitochondrial protein that promotes cytochrome c-dependent caspase activation by eliminating IAP inhibition. *Cell* **102,** 33–42.
89. Li, L. Y., Luo, X., and Wang, X. (2001). Endonuclease G is an apoptotic DNase when released from mitochondria. *Nature* **412,** 95–99.
90. Liu, X., Kim, C. N., Yang, J., Jemmerson, R., and Wang, X. (1996). Induction of apoptotic program in cell-free extracts: requirement for dATP and cytochrome c. *Cell* **86,** 147–157.
91. Susin, S. A., Lorenzo, H. K., Zamzami, N., Marzo, I., Snow, B. E., Brothers, G. M., Mangion, J., Jacotot, E., Costantini, P., Loeffler, M., Larochette, N., Goodlett, D. R., Aebersold, R., Siderovski, D. P., Penninger, J. M., and Kroemer, G. (1999). Molecular characterization of mitochondrial apoptosis-inducing factor. *Nature* **397,** 441–446.
92. Acehan, D., Jiang, X., Morgan, D. G., Heuser, J. E,. Wang, X., and Akey, C. W. (2002). Three-dimensional structure of the apoptosome: implications for assembly, procaspase-9 binding, and activation. *Mol. Cell* **9,** 423–432.
93. Hu, Y., Ding, L., Spencer, D. M., and Nunez, G. (1998). WD-40 repeat region regulates Apaf-1 self-association and procaspase-9 activation. *J. Biol. Chem.* **273,** 33489–33494.
94. Qin, H., Srinivasula, S. M., Wu, G., Fernandes-Alnemri, T., Alnemri, E. S., and Shi, Y. (1999). Structural basis of procaspase-9 recruitment by the apoptotic protease-activating factor 1. *Nature* **399,** 549–557.
95. Zou, H., Henzel, W. J., Liu, X., Lutschg, A., and Wang, X. (1997). Apaf-1, a human protein homologous to C. elegans CED-4, participates in cytochrome c-dependent activation of caspase-3. *Cell* **90,** 405–413.
96. Gross, A., Yin, X. M., Wang, K., Wei, M. C., Jockel, J., Milliman, C., Erdjument-Bromage, H., Tempst, P., and Korsmeyer, S. J. (1999). Caspase cleaved BID targets mitochondria and is required for cytochrome c release, while BCL-XL prevents this release but not tumor necrosis factor-R1/Fas death. *J. Biol. Chem.* **274,** 1156–1163.
97. Luo, X., Budihardjo, I., Zou, H., Slaughter, C., and Wang, X. (1998). Bid, a Bcl2 interacting protein, mediates cytochrome c release from mitochondria in response to activation of cell surface death receptors. *Cell* **94,** 481–490.
98. Peter, M. E. (2004). The flip side of FLIP. *Biochem. J.* **382,** e1–e3.
99. Guo, B., Zhai, D., Cabezas, E., Welsh, K., Nouraini, S., Satterthwait, A. C., and Reed, J. C. (2003). Humanin peptide suppresses apoptosis by interfering with Bax activation. *Nature* **423,** 456–461.
100. Shiozaki, E. N., Chai, J., Rigotti, D .J, Riedl, S. J., Li, P., Srinivasula, S. M., Alnemri, E. S., Fairman, R., and Shi, Y. (2003). Mechanism of XIAP-mediated inhibition of caspase-9. *Mol. Cell* **11,** 519–527.
101. Sun, C., Cai, M., Meadows, R. P., Xu, N., Gunasekera, A. H., Herrmann, J., Wu, J. C., and Fesik, S. W. (2000).

NMR structure and mutagenesis of the third Bir domain of the inhibitor of apoptosis protein XIAP. *J. Biol. Chem.* **275**, 33777–33781.
102. Nam, Y. J., Mani, K., Ashton, A. W., Peng, C. F., Krishnamurthy, B., Hayakawa, Y., Lee, P., Korsmeyer, S. J., and Kitsis, R. N. (2004). Inhibition of both the extrinsic and intrinsic death pathways through nonhomotypic death-fold interactions. *Mol. Cell* **15**, 901–912.
103. Gustafsson, A. B., Tsai, J. G., Logue, S. E., Crow, M. T., and Gottlieb, R. A. (2004). Apoptosis repressor with caspase recruitment domain protects against cell death by interfering with Bax activation. *J. Biol. Chem.* **279**, 21233–21238.
104. Scorrano, L., Oakes, S. A., Opferman, J. T., Cheng, E. H., Sorcinelli, M. D., Pozzan, T., and Korsmeyer, S. J. (2003). BAX and BAK regulation of endoplasmic reticulum Ca2+: a control point for apoptosis. *Science* **300**, 135–139.
105. Morishima, N., Nakanishi, K., Tsuchiya, K., Shibata, T., and Seiwa, E. (2004). Translocation of Bim to the endoplasmic reticulum (ER) mediates ER stress signaling for activation of caspase-12 during ER stress-induced apoptosis. *J. Biol. Chem.* **279**, 50375–50381.
106. Olivetti, G., Abbi, R., Quaini, F., Kajstura, J., Cheng, W., Nitahara, J. A., Quaini, E., Di, L. C., Beltrami, C. A., Krajewski, S., Reed, J. C., and Anversa, P. (1997). Apoptosis in the failing human heart. *N. Engl. J. Med.* **336**, 1131–1141.
107. Abbate, A., Biondi-Zoccai, G. G., Bussani, R., Dobrina, A., Camilot, D., Feroce, F., Rossiello, R., Baldi, F., Silvestri, F., Biasucci, L. M., and Baldi, A. (2003).Increased myocardial apoptosis in patients with unfavorable left ventricular remodeling and early symptomatic post-infarction heart failure. *J. Am. Coll. Cardiol.* **41**, 753–760.
108. Schaper, J., Lorenz-Meyer, S., and Suzuki, K. (1999). The role of apoptosis in dilated cardiomyopathy. *Herz* **24**, 219–224.
109. Saraste, A., Pulkki, K., Kallajoki, M., Heikkila, P., Laine, P., Mattila, S., Nieminen, M. S., Parvinen, M., and Voipio-Pulkki, L. M. (1999). Cardiomyocyte apoptosis and progression of heart failure to transplantation. *Eur. J. Clin. Invest.* **29**, 380–386.
110. Chatterjee, S., Stewart, A. S., Bish, L. T., Jayasankar, V., Kim, E. M., Pirolli, T., Burdick, J., Woo, Y. J., Gardner, T. J., and Sweeney, H. L. (2002). Viral gene transfer of the antiapoptotic factor Bcl-2 protects against chronic postischemic heart failure. *Circulation* **106**, I212–1217.
111. Ren, J., Samson, W. K., and Sowers, J. R. (1999). Insulin-like growth factor I as a cardiac hormone: physiological and pathophysiological implications in heart disease. *J. Mol. Cell Cardiol.* **31**, 2049–2061.
112. Li, Q., Li, B., Wang, X., Leri, A., Jana, K. P., Liu, Y., Kajstura, J., Baserga, R., and Anversa, P. (1997). Overexpression of insulin-like growth factor-1 in mice protects from myocyte death after infarction, attenuating ventricular dilation, wall stress, and cardiac hypertrophy. *J. Clin. Invest.* **100**, 1991–1999.
113. Lee, W. L., Chen, J. W., Ting, C. T., Ishiwata, T., Lin, S. J., Korc, M., and Wang, P. H. (1999). Insulin-like growth factor I improves cardiovascular function and suppresses apoptosis of cardiomyocytes in dilated cardiomyopathy. *Endocrinology* **140**, 4831–4840.
114. Nadal-Ginard, B., Kajstura, J., Leri, A., and Anversa, P. (2003). Myocyte death, growth, and regeneration in cardiac hypertrophy and failure. *Circ. Res.* **92**, 139–150.
115. Nadal-Ginard, B., Kajstura, J., Anversa, P., and Leri, A. (2003). A matter of life and death: cardiac myocyte apoptosis and regeneration. *J. Clin. Invest.* **111**, 1457–1459.
116. Anversa, P., and Nadal-Ginard, B. (2002). Myocyte renewal and ventricular remodelling. *Nature* **415**, 240–243.
117. Soonpaa, M. H., and Field, L. J. (1998). Survey of studies examining mammalian cardiomyocyte DNA synthesis. *Circ. Res.* **83**, 15–26.
118. Anversa, P., and Kajstura, J. (1998). Ventricular myocytes are not terminally differentiated in the adult mammalian heart. *Circ. Res.* **83**, 1–14.
119. Beltrami, A. P., Urbanek, K., Kajstura, J., Yan, S. M., Finato, N., Bussani, R., Nadal-Ginard, B., Silvestri, F., Leri, A., Beltrami, C. A., and Anversa, P. (2001). Evidence that human cardiac myocytes divide after myocardial infarction. *N. Engl. J. Med.* **344**, 1750–1757.
120. Kajstura, J., Leri, A., Finato, N., Di, L. C., Beltrami, C. A., and Anversa, P. (1998). Myocyte proliferation in end-stage cardiac failure in humans. *Proc. Natl. Acad. Sci. USA* **95**, 8801–8805.
121. Orlic, D., Kajstura, J., Chimenti, S., Jakoniuk, I., Anderson, S. M., Li, B., Pickel, J., McKay, R., Nadal-Ginard, B., Bodine, D. M., Leri, A., and Anversa, P. (2001). Bone marrow cells regenerate infarcted myocardium. *Nature* **410**, 701–705.
122. Orlic, D., Kajstura, J., Chimenti, S., Limana, F., Jakoniuk, I., Quaini, F., Nadal-Ginard, B., Bodine, D. M., Leri, A., and Anversa, P. (2001). Mobilized bone marrow cells repair the infarcted heart, improving function and survival. *Proc. Natl. Acad. Sci. USA* **98**, 10344–10349.
123. Orlic, D., Kajstura, J., Chimenti, S., Bodine, D. M., Leri, A., and Anversa, P. (2001). Transplanted adult bone marrow cells repair myocardial infarcts in mice. *Ann. N. Y. Acad. Sci.* **938**, 221–229.
124. Orlic, D., Hill, J. M., and Arai, A. E. (2002). Stem cells for myocardial regeneration. *Circ. Res.* **91**, 1092–1102.
125. Laflamme, M. A., Myerson, D., Saffitz, J. E., and Murry, C. E. (2002). Evidence for cardiomyocyte repopulation by extracardiac progenitors in transplanted human hearts. *Circ. Res.* 90, 634–640.
126. Quaini, F., Urbanek, K., Beltrami, A. P., Finato, N., Beltrami, C. A., Nadal-Ginard, B., Kajstura, J., Leri, A., and Anversa, P. (2002). Chimerism of the transplanted heart. *N. Engl. J. Med.* **346**, 5–15.
127. Deb, A., Wang, S., Skelding, K. A., Miller, D., Simper, D., and Caplice, N. M. (2003). Bone marrow-derived cardiomyocytes are present in adult human heart: A study of gender-mismatched bone marrow transplantation patients. *Circulation* **107**, 1247–1249.
128. Goldsmith, E. C., and Borg, T. K. (2002). The dynamic interaction of the extracellular matrix in cardiac remodeling. *J. Card. Fail.* **8**, S314–S318.
129. Janicki, J. S., and Brower, G. L. (2002). The role of myocardial fibrillar collagen in ventricular remodeling and function. *J. Card. Fail.* **8**, S319–S325.
130. Weber, K. T., Anversa, P., Armstrong, P. W., Brilla, C. G., Burnett, J. C., Jr., Cruickshank, J. M., Devereux, R. B., Giles, T. D., Korsgaard, N., and Leier, C. V. (1992). Remodeling and reparation of the cardiovascular system. *J. Am. Coll. Cardiol.* **20**, 3–16.
131. Libby, P., and Lee, R. T. (2000). Matrix matters. *Circulation* **102**, 1874–1876.

132. Bonnin, C. M., Sparrow, M. P., and Taylor, R. R. (1981). Collagen synthesis and content in right ventricular hypertrophy in the dog. *Am. J. Physiol.* **241,** H708–H713.
133. Weber, K. T. (1989). Cardiac interstitium in health and disease: The fibrillar collagen network. *J. Am. Coll. Cardiol.* **13,** 1637–1652.
134. Benjamin, I. J. (2001). Matrix metalloproteinases: from biology to therapeutic strategies in cardiovascular disease. *J. Invest. Med.* **49,** 381–397.
135. Vu, T. H., and Werb, Z. (2000). Matrix metalloproteinases: effectors of development and normal physiology. *Genes Dev.* **14,** 2123–2133.
136. Nagase, H. (1997). Activation mechanisms of matrix metalloproteinases. *Biol. Chem.* **378,** 151–160.
137. Nelson, A. R., Fingleton, B., Rothenberg, M. L., and Matrisian, L. M. (2000). Matrix metalloproteinases: biologic activity and clinical implications. *J. Clin. Oncol.* **18,** 1135–1149.
138. Gunja-Smith, Z., Morales, A. R., Romanelli, R., and Woessner, J. F., Jr. (1996). Remodeling of human myocardial collagen in idiopathic dilated cardiomyopathy. Role of metalloproteinases and pyridinoline cross-links. *Am. J. Pathol.* **148,** 1639–1648.
139. Spinale, F. G., Coker, M. L., Heung, L. J., Bond, B. R., Gunasinghe, H. R., Etoh, T., Goldberg, A. T., Zellner, J. L., and Crumbley, A. J. (2000). A matrix metalloproteinase induction/activation system exists in the human left ventricular myocardium and is upregulated in heart failure. *Circulation* **102,** 1944–1949.
140. Thomas, C. V., Coker, M. L., Zellner, J. L., Handy, J. R., Crumbley, A. J., III, and Spinale, F. G. (1998). Increased matrix metalloproteinase activity and selective upregulation in LV myocardium from patients with end-stage dilated cardiomyopathy. *Circulation* **97,** 1708–1715.
141. Knauper, V., and Murphy, G. (1998). Membrane-type matrix metalloproteinases and cell surface-associated activation cascades for matrix metalloproteinases. *In* "Matrix Metalloproteinases." (W. C. Parks, and R. P. Mecham, Eds.), pp. 199–218. Academic Press, San Diego, CA.
142. Nagase, H. (1997). Activation mechanisms of matrix metalloproteinases. *Biol. Chem.* **378,** 151–160.
143. Woessner, J. F., and Nagase, H. (2003). Activation of the zymogen forms of MMPs. *In* "Matrix Metalloproteinases and TIMPs." (J. F. Woessner, and H. Nagase, Eds.), pp. 72–86. Oxford University Press, Oxford.
144. Woessner, J. F., and Nagase, H. (2000). "Function of the TIMPs. Matrix Metalloproteinases and TIMPs." pp. 130–135. Oxford University Press, Oxford.
145. Sternlicht, M. D., and Werb, Z. (2001). How matrix metalloproteinases regulate cell behavior. *Annu. Rev. Cell Dev. Biol.* **17,** 463–516.
146. Li, Y. Y., Feldman, A. M., Sun, Y., and McTiernan, C. F. (1998). Differential expression of tissue inhibitors of metalloproteinases in the failing human heart. *Circulation* **98,** 1728–1734.
147. Schlondorff, J., and Blobel, C. P. (1999). Metalloprotease-disintegrins: modular proteins capable of promoting cell-cell interactions and triggering signals by protein-ectodomain shedding. *J. Cell Sci.* **112,** 3603–3617.
148. Amour, A., Slocombe, P. M., Webster, A., Butler, M., Knight, C. G., Smith, B. J., Stephens, P. E., Shelley, C., Hutton, M., Knauper, V., Docherty, A. J., and Murphy, G. (1998). TNF-alpha converting enzyme (TACE) is inhibited by TIMP-3. *FEBS Lett.* **435,** 39–44.
149. Asakura, M., Kitakaze, M., Takashima, S., Liao, Y., Ishikura, F., Yoshinaka, T., Ohmoto, H., Node, K., Yoshino, K., Ishiguro, H., Asanuma, H., Sanada, S., Matsumura, Y., Takeda, H., Beppu, S., Tada, M., Hori, M., and Higashiyama, S. (2002). Cardiac hypertrophy is inhibited by antagonism of ADAM12 processing of HB-EGF: Metalloproteinase inhibitors as a new therapy. *Nat. Med.* **8,** 35–40.
150. Leco, K. J., Khokha, R., Pavloff, N., Hawkes, S. P., and Edwards, D. R. (1994). Tissue inhibitor of metalloproteinases-3 (TIMP-3) is an extracellular matrix-associated protein with a distinctive pattern of expression in mouse cells and tissues. *J. Biol. Chem.* **269,** 9352–9360.
151. Woessner, J. F., Jr. (2001). That impish TIMP: The tissue inhibitor of metalloproteinases-3. *J. Clin. Invest.* **108,** 799–800.
152. Shastry, S., Hayden, M. R., Lucchesi, P. A., ad Tyagi, S. C. (2003). Matrix metalloproteinase in left ventricular remodeling and heart failure. *Curr. Cardiol. Rep.* **5,** 200–204.
153. Sierevogel, M. J., Pasterkamp, G., De Kleijn, D. P., and Strauss, B. H. (2003). Matrix metalloproteinases: a therapeutic target in cardiovascular disease. *Curr. Pharm. Des.* **9,** 1033–1040.
154. Bradham, W. S., Moe, G., Wendt, K. A., Scott, A. A., Konig, A., Romanova, M., Naik, G., and Spinale, F. G. (2002). TNF-alpha and myocardial matrix metalloproteinases in heart failure: relationship to LV remodeling. *Am. J. Physiol. Heart Circ. Physiol.* **282,** H1288–H1295.
155. Coker, M. L., Thomas, C. V., Clair, M. J., Hendrick, J. W., Krombach, R. S., Galis, Z. S., and Spinale, F. G. (1998). Myocardial matrix metalloproteinase activity and abundance with congestive heart failure. *Am. J. Physiol.* **274,** H1516–H1523.
156. Rohde, L. E., Ducharme, A., Arroyo, L. H., Aikawa, M., Sukhova, G. H., Lopez-Anaya, A., McClure, K. F., Mitchell, P. G., Libby, P., and Lee, R. T. (1999). Matrix metalloproteinase inhibition attenuates early left ventricular enlargement after experimental myocardial infarction in mice. *Circulation* **99,** 3063–3070.
157. Spinale, F. G., Coker, M. L., Krombach, S. R., Mukherjee, R., Hallak, H., Houck, W. V., Clair, M. J., Kribbs, S. B., Johnson, L. L., Peterson, J. T., and Zile, M. R. (1999). Matrix metalloproteinase inhibition during the development of congestive heart failure: effects on left ventricular dimensions and function. *Circ. Res.* **85,** 364–376.
158. Bauvois, B. (2001). Transmembrane proteases in focus: diversity and redundancy? *J. Leukoc. Biol.* **70,** 11–17.
159. Kim, H. E., Dalal, S. S., Young, E., Legato, M. J., Weisfeldt, M. L., and D'Armiento, J. (2000). Disruption of the myocardial extracellular matrix leads to cardiac dysfunction. *J. Clin. Invest.* **106,** 857–866.
160. Peterson, J. T., Hallak, H., Johnson, L., Li, H., O'Brien, P. M., Sliskovic, D. R., Bocan, T. M., Coker, M. L., Etoh, T., and Spinale, F. G. (2001). Matrix metalloproteinase inhibition attenuates left ventricular remodeling and dysfunction in a rat model of progressive heart failure. *Circulation* **103,** 2303–2309.
161. Roten, L., Nemoto, S., Simsic, J., Coker, M. L., Rao, V., Baicu, S., Defreyte, G., Soloway, P. J., Zile, M. R., and Spinale, F. G.

(2000). Effects of gene deletion of the tissue inhibitor of the matrix metalloproteinase-type 1 (TIMP-1) on left ventricular geometry and function in mice. *J. Mol. Cell Cardiol.* **32,** 109–120.

162. Heymans, S., Luttun, A., Nuyens, D., Theilmeier, G., Creemers, E., Moons, L., Dyspersin, G. D., Cleutjens, J. P., Shipley, M., Angellilo, A., Levi, M., Nube, O., Baker, A., Keshet, E., Lupu, F., Herbert, J. M., Smits, J. F., Shapiro, S. D., Baes, M., Borgers, M., Collen, D., Daemen, M. J., and Carmeliet, P. (1999). Inhibition of plasminogen activators or matrix metalloproteinases prevents cardiac rupture but impairs therapeutic angiogenesis and causes cardiac failure. *Nat. Med.* **5,** 1135–1142.

163. Margulies, K. B. (2002). Reversal mechanisms of left ventricular remodeling: lessons from left ventricular assist device experiments. *J. Card. Fail.* **8,** S500–S505.

164. Blaxall, B. C., Tschannen-Moran, B. M., Milano, C. A., and Koch, W. J. (2003). Differential gene expression and genomic patient stratification following left ventricular assist device support. *J. Am. Coll. Cardiol.* **41,** 1096–1106.

165. Hall, J. L., Grindle, S., Han, X., Fermin, D., Park, S., Chen, Y., Bache, R. J., Mariash, A., Guan, Z., Ormaza, S., Thompson, J., Graziano, J., Sam Lazaro, S. E., Pan, S., Simari, R. D., and Miller, L. W. (2004). Genomic profiling of the human heart before and after mechanical support with a ventricular assist device reveals alterations in vascular signaling networks. *Physiol. Genomics* **17,** 283–291.

166. Chen, Y., Park, S., Li, Y., Missov, E., Hou, M., Han, X., Hall, J. L., Miller, L. W., and Bache, R. J. (2003). Alterations of gene expression in failing myocardium following left ventricular assist device support. *Physiol. Genomics* **14,** 251–260.

167. Kittleson, M. M., and Hare, J. M. (2005). Molecular signature analysis: using the myocardial transcriptome as a biomarker in cardiovascular disease. *Trends Cardiovasc. Med.* **15,** 130–138.

168. Sanoudou, D., Vafiadaki, E., Arvanitis, D. A., Kranias, E., and Kontrogianni-Konstantopoulos, A. (2005). Array lessons from the heart: focus on the genome and transcriptome of cardiomyopathies. *Physiol. Genomics* **21,** 131–143.

169. Kaab, S., Barth, A. S., Margerie, D., Dugas, M., Gebauer, M., Zwermann, L., Merk, S., Pfeufer, A., Steinmeyer, K., Bleich, M., Kreuzer, E., Steinbeck, G., and Nabauer, M. (2004). Global gene expression in human myocardium-oligonucleotide microarray analysis of regional diversity and transcriptional regulation in heart failure. *J. Mol. Med.* **82,** 308–316.

170. Steenman, M., Lamirault, G., Le Meur, N., Le Cunff, M., Escande, D., and Leger, J. J. (2005). Distinct molecular portraits of human failing hearts identified by dedicated cDNA microarrays. *Eur. J. Heart Fail.* **7,** 157–165.

171. Hwang, J. J., Allen, P. D., Tseng, G. C., Lam, C. W., Fananapazir, L., Dzau, V. J., and Liew, C. C. (2002). Microarray gene expression profiles in dilated and hypertrophic cardiomyopathic end-stage heart failure. *Physiol. Genomics* **10,** 31–44.

172. Kittleson, M. M., Minhas, K. M., Irizarry, R. A., Ye, S. Q., Edness, G., Breton, E., Conte, J. V., Tomaselli, G., Garcia, J. G., and Hare, J. M. (2005). Gene expression in giant cell myocarditis: Altered expression of immune response genes. *Int. J. Cardiol.* **102,** 333–340.

173. Carey, V. J., Gentry, J., Whalen, E., and Gentleman, R. (2005). Network structures and algorithms in Bioconductor. *Bioinformatics* **21,** 135–136.

174. Simon, R., Radmacher, M. D., Dobbin, K., and McShane, L. M. (2003). Pitfalls in the use of DNA microarray data for diagnostic and prognostic classification. *J. Natl. Cancer Inst.* **95,** 14–18.

175. Tibshirani, R., Hastie, T., Narasimhan, B., and Chu, G. (2002). Diagnosis of multiple cancer types by shrunken centroids of gene expression. *Proc. Natl. Acad. Sci. USA* **99,** 6567–6572.

176. Kittleson, M. M., Ye, S. Q., Irizarry, R. A., Minhas, K. M., Edness, G., Conte, J. V., Parmigiani, G., Miller, L. W., Chen, Y., Hall, J. L., Garcia, J. G., and Hare, J. M. (2004). Identification of a gene expression profile that differentiates between ischemic and nonischemic cardiomyopathy. *Circulation* **110,** 3444–3451.

177. Song, Q., Schmidt, A. G., Hahn, H. S., Carr, A. N., Frank, B., Pater, L., Gerst, M., Young, K., Hoit, B. D., McConnell, B. K., Haghighi, K., Seidman, C. E., Seidman, J. G., Dorn, G. W., and Kranias, E. G. (2003). Rescue of cardiomyocyte dysfunction by phospholamban ablation does not prevent ventricular failure in genetic hypertrophy. *J. Clin. Invest.* **111,** 859–867.

178. Wilkie, G. S., Dickson, K. S., and Gray, N. K. (2003). Regulation of mRNA translation by 5′- and 3′-UTR-binding factors. *Trends Biochem. Sci.* **28,** 182–188.

179. Schott, P., Singer, S. S., Kogler, H., Neddermeier, D., Leineweber, K., Brodde, O. E., Regitz-Zagrosek, V., Schmidt, B., Dihazi, H., and Hasenfuss, G. (2005). Pressure overload and neurohumoral activation differentially affect the myocardial proteome. *Proteomics* **5,** 1372–1381.

180. Figueredo, V. M., and Camacho, S. A. (1995). Basic mechanisms of myocardial dysfunction: cellular pathophysiology of heart failure. *Curr. Opin. Cardiol.* **10,** 246–252.

181. Barany, M. (1967). ATPase activity of myosin correlated with speed of muscle shortening. *J. Gen. Physiol.* **50,** 197–218.

182. Morano, M., Zacharzowski, U., Maier, M., Lange, P. E., Alexi-Meskishvili, V., Haase, H., and Morano, I. (1996). Regulation of human heart contractility by essential myosin light chain isoforms. *J. Clin. Invest.* **98,** 467–473.

183. Braz, J. C., Bueno, O. F., Liang, Q., Wilkins, B. J., Dai, Y. S., Parsons, S., Braunwart, J., Glascock, B. J., Klevitsky, R., Kimball, T. F., Hewett, T. E., and Molkentin, J. D. (2003). Targeted inhibition of p38 MAPK promotes hypertrophic cardiomyopathy through upregulation of calcineurin-NFAT signaling. *J. Clin. Invest.* **111,** 1475–1486.

184. Dorn, G. W., and Molkentin, J. D. (2004). Manipulating cardiac contractility in heart failure: data from mice and men. *Circulation* **109,** 150–158.

185. Boluyt, M. O., O'Neill, L., Meredith, A. L., Bing, O. H., Brooks, W. W., Conrad, C. H., Crow, M. T., and Lakatta, E. G. (1994). Alterations in cardiac gene expression during the transition from stable hypertrophy to heart failure. Marked upregulation of genes encoding extracellular matrix components. *Circ. Res.* **75,** 23–32.

186. Yung, C. K., Halperin, V. L., Tomaselli, G. F., and Winslow, R. L. (2004). Gene expression profiles in end-stage human idiopathic dilated cardiomyopathy: altered expression of apoptotic and cytoskeletal genes. *Genomics* **83,** 281–297.

187. Huss, J. M., and Kelly, D. P. (2005). Mitochondrial energy metabolism in heart failure: a question of balance. *J. Clin. Invest.* **115,** 547–555.

188. Yang, J., Moravec, C. S., Sussman, M. A., DiPaola, N. R., Fu, D., Hawthorn, L., Mitchell, C. A., Young, J. B., Francis,

G. S., McCarthy, P. M., and Bond, M. (2000). Decreased SLIM1 expression and increased gelsolin expression in failing human hearts measured by high-density oligonucleotide arrays. *Circulation* **102**, 3046–3052.

189. Paolisso, G., Gambardella, A., Galzerano, D., D'Amore, A., Rubino, P., Verza, M., Teasuro, P., Varricchio, M., and D'Onofrio, F. (1994). Total-body and myocardial substrate oxidation in congestive heart failure. *Metabolism* **43**, 174–179.

190. Steenman, M., Lamirault, G., Le, M. N., and Leger, J. J. (2005). Gene expression profiling in human cardiovascular disease. *Clin. Chem. Lab. Med.* **43**, 696–701.

191. Hasenfuss, G. (1998). Alterations of calcium-regulatory proteins in heart failure. *Cardiovasc. Res.* **37**, 279–289.

192. Yano, M., Ikeda, Y., and Matsuzaki, M. (2005). Altered intracellular Ca2+ handling in heart failure. *J. Clin. Invest.* **115**, 556–564.

193. Barrans, J. D., Allen, P. D., Stamatiou, D., Dzau, V. J., and Liew, C. C. (2002). Global gene expression profiling of end-stage dilated cardiomyopathy using a human cardiovascular-based cDNA microarray. *Am. J. Pathol.* **160**, 2035–2043.

194. Grzeskowiak, R., Witt, H., Drungowski, M., Thermann, R., Hennig, S., Perrot, A., Osterziel, K. J., Klingbiel, D., Scheid, S., Spang, R., Lehrach, H., and Ruiz, P. (2003). Expression profiling of human idiopathic dilated cardiomyopathy. *Cardiovasc. Res.* **59**, 400–411.

195. Tan, F. L., Moravec, C. S., Li, J., Apperson-Hansen, C., McCarthy, P. M., Young, J. B., and Bond, M. (2002). The gene expression fingerprint of human heart failure. *Proc. Natl. Acad. Sci. USA* **99**, 11387–11392.

196. Rapundalo, S. T. (1998). Cardiac protein phosphorylation: functional and pathophysiological correlates. *Cardiovasc. Res.* **38**, 559–588.

197. Neumann, J., Eschenhagen, T., Jones, L. R., Linck, B., Schmitz, W., Scholz, H., and Zimmermann, N. (1997). Increased expression of cardiac phosphatases in patients with end-stage heart failure. *J. Mol. Cell Cardiol.* **29**, 265–272.

198. Carr, A. N., Schmidt, A. G., Suzuki, Y., del, M. F., Sato, Y., Lanner, C., Breeden, K., Jing, S. L., Allen, P. B., Greengard, P., Yatani, A., Hoit, B. D., Grupp, I. L., Hajjar, R. J., Paoli-Roach, A. A., and Kranias, E. G. (2002). Type 1 phosphatase, a negative regulator of cardiac function. *Mol. Cell Biol.* **22**, 4124–4135.

199. Morita, H., Seidman, J., and Seidman, C. E. (2005). Genetic causes of human heart failure. *J. Clin. Invest.* **115**, 518–526.

200. Abraham, W. T., Gilbert, E. M., Lowes, B. D., Minobe, W. A., Larrabee, P., Roden, R. L., Dutcher, D., Sederberg, J., Lindenfeld, J. A., Wolfel, E. E., Shakar, S. F., Ferguson, D., Volkman, K., Linseman, J. V., Quaife, R. A., Robertson, A. D., and Bristow, M. R. (2002). Coordinate changes in Myosin heavy chain isoform gene expression are selectively associated with alterations in dilated cardiomyopathy phenotype. *Mol. Med.* **8**, 750–760.

201. Le, C. P., Park, H. Y., and Rockman, H. A. (2003). Modifier genes and heart failure. *Minerva Cardioangiol.* **51**, 107–120.

202. Borjesson, M., Magnusson, Y., Hjalmarson, A., and Andersson, B. (2000). A novel polymorphism in the gene coding for the beta(1)-adrenergic receptor associated with survival in patients with heart failure. *Eur. Heart J.* **21**, 1853–1858.

203. Wheeler, F. C., Fernandez, L., Carlson, K. M., Wolf, M. J., Rockman, H. A., and Marchuk, D. A. (2005). QTL mapping in a mouse model of cardiomyopathy reveals an ancestral modifier allele affecting heart function and survival. *Mamm. Genome* **16**, 414–423.

204. Moe, G. W. (2005). BNP in the Diagnosis and Risk Stratification of Heart Failure. *Heart Fail. Monit.* **4**, 116–122.

205. Januzzi, J. L., van, K. R., Lainchbury, J., Bayes-Genis, A., Ordonez-Llanos, J., Santalo-Bel, M., Pinto, Y. M., and Richards, M. (2006). NT-proBNP testing for diagnosis and short-term prognosis in acute destabilized heart failure: an international pooled analysis of 1256 patients The International Collaborative of NT-proBNP Study. *Eur. Heart J.* **27**, 330–33.7

206. Januzzi, J. L., Jr., Camargo, C. A., Anwaruddin, S., Baggish, A. L., Chen, A. A., Krauser, D. G., Tung, R., Cameron, R., Nagurney, J. T., Chae, C. U., Lloyd-Jones, D. M., Brown, D. F., Foran-Melanson, S., Sluss, P. M., Lee-Lewandrowski, E., and Lewandrowski, K. B. (2005). The N-terminal Pro-BNP investigation of dyspnea in the emergency department (PRIDE) study. *Am. J. Cardiol.* **95**, 948–954.

207. Maisel, A. S., Krishnaswamy, P., Nowak, R. M., McCord, J., Hollander, J. E., Duc, P., Omland, T., Storrow, A. B., Abraham, W. T., Wu, A. H., Clopton, P., Steg, P. G., Westheim, A., Knudsen, C. W., Perez, A., Kazanegra, R., Herrmann, H. C., and McCullough, P. A. (2002). Rapid measurement of B-type natriuretic peptide in the emergency diagnosis of heart failure. *N. Engl. J. Med.* **347**, 161–167.

208. McDonagh, T. A., Robb, S. D., Murdoch, D. R., Morton, J. J., Ford, I., Morrison, C. E., Tunstall-Pedoe, H., McMurray, J. J., and Dargie, H. J. (1998). Biochemical detection of left-ventricular systolic dysfunction. *Lancet* **351**, 9–13.

209. Armstrong, P. W., and Moe, G. W. (1993). Medical advances in the treatment of congestive heart failure. *Circulation* **88**, 2941–2952.

210. Seta, Y., Shan, K., Bozkurt, B., Oral, H., and Mann, D. L. (1996). Basic mechanisms in heart failure: the cytokine hypothesis. *J. Card. Fail.* **2**, 243–249.

211. Vidal, B., Roig, E., Perez-Villa, F., Orus, J., Perez, J., Jimenez, V., Leivas, A., Cuppoletti A., Roque, M., and Sanz, G. (2002). Prognostic value of cytokines and neurohormones in severe heart failure. *Rev. Esp. Cardiol.* **55**, 481–486.

212. Michels, V. V., Moll, P. P., Miller, F. A., Tajik, A. J., Chu, J. S., Driscoll, D. J., Burnett, J. C., Rodeheffer, R. J., Chesebro, J. H., and Tazelaar, H. D. (1992). The frequency of familial dilated cardiomyopathy in a series of patients with idiopathic dilated cardiomyopathy. *N. Engl. J. Med.* **326**, 77–82.

213. Olson, T. M., Illenberger, S., Kishimoto, N. Y., Huttelmaier, S., Keating, M. T., and Jockusch, B. M. (2002). Metavinculin mutations alter actin interaction in dilated cardiomyopathy. *Circulation* **105**, 431–437.

214. Renlund, D. G., Taylor, D. .O, Kfoury, A. G., and Shaddy, R. S. (1999). New UNOS rules: historical background and implications for transplantation management. United Network for Organ Sharing. *J. Heart Lung Transplant.* **18**, 1065–1070.

215. Zeltsman, D., and Acker, M. A. (2002). Surgical management of heart failure: An overview. *Annu. Rev. Med.* **53**, 383–391.

216. Dor, V. (2001). The endoventricular circular patch plasty ("Dor procedure") in ischemic akinetic dilated ventricles. *Heart Fail. Rev.* **6**, 187–193.

217. Barbone, A., Oz, M. C., Burkhoff, D., and Holmes, J. W. (2001). Normalized diastolic properties after left ventricular

assist result from reverse remodeling of chamber geometry. *Circulation* **104,** I229–I232.
218. Barbone, A., Holmes, J. W., Heerdt, P. M., The', A. H., Naka, Y., Joshi, N., Daines, M., Marks, A. R., Oz, M. C., and Burkhoff, D. (2001). Comparison of right and left ventricular responses to left ventricular assist device support in patients with severe heart failure: A primary role of mechanical unloading underlying reverse remodeling. *Circulation* **104,** 670–675.
219. Towbin, J. A., and Bowles, N. E. (2002). The failing heart. *Nature* **415,** 227–233.
220. Zhu, T., Zhou, L., Mori, S., Wang, Z., McTiernan, C. F., Qiao, C., Chen, C., Wang, D. W., Li, J., and Xiao, X. (2005). Sustained whole-body functional rescue in congestive heart failure and muscular dystrophy hamsters by systemic gene transfer. *Circulation* **112,** 2650–2659.
221. Menasche, P. (2002). Myoblast transplantation: feasibility, safety and efficacy. *Ann. Med.* **34,** 314–315.
222. Fedak, P. W., Verma, S., Weisel, R. D., Skrtic, M., and Li, R. K. (2005). Cardiac remodeling and failure: from molecules to man (Part III). *Cardiovasc. Pathol.* **14,** 109–119.
223. Kocher, A. A., Schuster, M. D., Szabolcs, M. J., Takuma, S., Burkhoff, D., Wang, J., Homma, S., Edwards, N. M., and Itescu, S. (2001). Neovascularization of ischemic myocardium by human bone-marrow–derived angioblasts prevents cardiomyocyte apoptosis, reduces remodeling and improves cardiac function. *Nat. Med.* **7,** 430–436.
224. Wang, J. S., Shum-Tim, D., Galipeau, J., Chedrawy, E., Eliopoulos, N., and Chiu, R. C. (2000). Marrow stromal cells for cellular cardiomyoplasty: Feasibility and potential clinical advantages. *J. Thorac. Cardiovasc. Surg.* **120,** 999–1005.
225. Rosenthal, N., and Tsao, L. (2001). Helping the heart to heal with stem cells. *Nat. Med.* **7,** 412–413.
226. Strauer, B. E., Brehm, M., Zeus, T., Gattermann, N., Hernandez, A., Sorg, R. V., Kogler, G., and Wernet, P. (2001). Intracoronary, human autologous stem cell transplantation for myocardial regeneration following myocardial infarction. *Dtsch. Med. Wochenschr.* **126,** 932–938.
227. Janssens, S., Dubois, C., Bogaert, J., Theunissen, K., Deroose, C., Desmet, W., Kalantzi, M., Herbots, L., Sinnaeve, P., Dens, J., Maertens, J., Rademakers, F., Dymarkowski, S., Gheysens, O., Van, C. J., Bormans, G., Nuyts, J., Belmans, A., Mortelmans, L., Boogaerts, M., and Van de Werf, F. (2006). Autologous bone marrow-derived stem-cell transfer in patients with ST-segment elevation myocardial infarction: Double-blind, randomised controlled trial. *Lancet* **367,** 113–121.
228. Ripa, R. S., Jorgensen, E., Wang, Y., Thune, J. J., Nilsson, J. C., Sondergaard, L., Johnsen, H. E., Kober, L., Grande, P., and Kastrup, J. (2006). Stem cell mobilization induced by subcutaneous granulocyte-colony stimulating factor to improve cardiac regeneration after acute ST-elevation myocardial infarction. Result of the double-blind, randomized, placebo-controlled Stem Cells in Myocardial Infarction (STEMMI) trial. *Circulation* **113,** 1983–1992.
229. Scorsin, M., Hagege, A. A., Marotte, F., Mirochnik, N., Copin, H., Barnoux, M., Sabri, A., Samuel, J. .L, Rappaport, L., and Menasche, P. (1997). Does transplantation of cardiomyocytes improve function of infarcted myocardium? *Circulation* **96,** II–93.
230. Menasche, P. (2002). Cell transplantation for the treatment of heart failure. *Semin. Thorac. Cardiovasc. Surg.* **14,** 157–166.
231. Johnson, J. A., and Cavallari, L. H. (2005). Cardiovascular pharmacogenomics. *Exp. Physiol.* **90,** 283–289.
232. McGourty, J. C., Silas, J. H., Lennard, M. S., Tucker, G. T., and Woods, H. F. (1985). Metoprolol metabolism and debrisoquine oxidation polymorphism—population and family studies. *Br. J. Clin. Pharmacol.* **20,** 555–566.
233. Terra, S. G., Pauly, D. F., Lee, C. R., Patterson, J. H., Adams, K. F., Schofield, R. S., Belgado, B. S., Hamilton, K. K., Aranda, J. M., Hill, J. A., Yarandi, H. N., Walker, J. R., Phillips, M. S., Gelfand, C. A., and Johnson, J. A. (2005). Beta-adrenergic receptor polymorphisms and responses during titration of metoprolol controlled release/extended release in heart failure. *Clin. Pharmacol. Ther.* **77,** 127–137.
234. Roden, D. M. (2003). Cardiovascular pharmacogenomics. *Circulation* **108,** 3071–3074.
235. Mialet, P. J., Rathz, D. A., Petrashevskaya, N. N., Hahn, H. S., Wagoner, L. E., Schwartz, A., Dorn, G. W., and Liggett, S. B. (2003). Beta 1-adrenergic receptor polymorphisms confer differential function and predisposition to heart failure. *Nat. Med.* **9,** 1300–1305.
236. Terra, S. G., Hamilton, K. K., Pauly, D. F, Lee, C. R., Patterson, J. H., Adams, K. F., Schofield, R. S., Belgado, B. S., Hill, J. A., Aranda, J. M., Yarandi, H. N., and Johnson, J. A. (2005). Beta1-adrenergic receptor polymorphisms and left ventricular remodeling changes in response to beta-blocker therapy. *Pharmacogenet. Genomics* **15,** 227–234.
237. Wojnowski, L., Kulle, B., Schirmer, M., Schluter, G., Schmidt, A., Rosenberger, A., Vonhof, S., Bickeboller, H., Toliat, M. R., Suk, E. K., Tzvetkov, M., Kruger, A., Seifert, S., Kloess, M., Hahn, H., Loeffler, M., Nurnberg, P., Pfreundschuh, M., Trumper, L., Brockmoller, J., and Hasenfuss, G. (2005). NAD(P)H oxidase and multidrug resistance protein genetic polymorphisms are associated with doxorubicin-induced cardiotoxicity. *Circulation* **112,** 3754–3762.

SECTION VI

Molecular and Genetic Analysis of Metabolic Disorders

CHAPTER 16

Fatty Acid and Glucose Metabolism in Cardiac Disease

OVERVIEW

Fatty acid and glucose metabolism have been extensively studied in the heart; however, information regarding the role(s) that specific genes and their interactions with the environment play in this metabolism and in both normal myocardium and in the pathogenesis and progression of cardiac disease is rather limited. In this chapter, our understanding of the mechanisms, diagnosis, and treatment of metabolic defects occurring in human cardiac disorders and the experimental findings observed in animal models are discussed.

INTRODUCTION

Fatty acids and associated lipids play an integral role in cardiomyocyte structure and function. In the post-natal and adult mammalian heart, fatty acid β-oxidation is the preferred pathway for the energy required for normal cardiac function. Glucose metabolism, which provides the bulk of ATP during prenatal growth, contributes significantly to the ATP production in the adult heart (up to 30% of myocardial ATP can be generated by glucose oxidation). In myocardial ischemia and hypertrophy, profound changes in both glucose and fatty acid metabolism occur, with glucose metabolism taking on greater importance. In addition, specific abnormalities in myocardial fatty acid oxidation (FAO) metabolism, caused by either inherited or acquired physiological stresses, and in cardiac glucose use, caused by chronic insulin deficiency or resistance, may result in dysrhythmias, cardiomyopathy, and HF. Moreover, abnormalities in glucose and fatty acid metabolism associated with diabetes and metabolic syndrome (MetSyn), chronic diseases that are becoming increasingly prevalent in the urbanized world, also have associated cardiovascular abnormalities, ranging from hypertension to increased cardiomyopathy and HF. Research to identify contributory genes and susceptibility loci and to delineate the molecular pathways involved in these disorders is just at its inception, as is the important dissection of the modulation of the genetic background by interacting environmental features such as diet and stress.

In this chapter, the molecular, genetic, and cellular basis of fatty acid, lipid, and glucose metabolic defects that can lead to cardiac disease are presented. In this context, the focus is on the molecular, cellular, and biochemical events that can lead to myocardial hypertrophy, cardiomyopathy, dysrhythmias, and HF. For an extensive discussion concerning acquired and inherited lipid disorders (e.g., cholesterol, the apolipoproteins, and HDL/LDL) related to coronary artery disease and stroke, the reader is referred to Chapter 8.

CELLULAR PERSPECTIVE

Fatty Acids and Their Metabolism Play an Important Role in Normal Cardiomyocytes

Fatty acids play an integral role in the structure and function of the cardiac cell plasma and mitochondrial membranes. Their influence on the fluidity and stability of membrane structure markedly impacts on membrane functions such as the transport of ions and substrates and electrophysiology intrinsic to cardiac function and excitability. In addition to their multiple structural and functional roles within the cardiac cell membrane, fatty acids and associated lipids are also potent regulatory molecules acting as second messengers in cell signaling, as effectors in apoptotic cell death, and in response to oxidative and ischemic damage.

Fatty Acid Transporters

In the heart, the oxidation of long-chain fatty acids (LCFAs) provides much of the energy needed for proper function. A continuous supply of exogenous plasma nonesterified LCFAs primarily derived from adipose tissue is required; under conditions of metabolic stress such as ischemia, diabetes, or starvation, the levels of plasma LCFAs are elevated. Fatty acids derived from adipocytes are released from triglyceride hydrolysis mediated by hormone-sensitive lipases activated by catecholamines and inhibited by insulin. Therefore, fasting states in which insulin levels are reduced and catecholamines elevated result in increased plasma free fatty acid, a slight misnomer, because hydrophobic fatty acids are not truly free *in vivo* but are associated with proteins (e.g.,

Post-Genomic Cardiology

albumin), covalently bound to coenzyme A, carnitine or triglycerides, or contained within lipoprotein particles such as chylomicrons or very low-density lipoproteins.[1]

Entry of fatty acids into the cardiomyocyte is primarily mediated by several proteins, including the 43-kDa plasma membrane–associated fatty acid binding proteins (FABPpm), and a myocardial-specific 88-kDa integral membrane transporter (fatty acid translocase or FAT/CD36), a homolog of human CD36. The nonenzymatic FABPpm also serves as a facilitator of intracellular transport of relatively insoluble LCFAs to sites of metabolic use (e.g., mitochondria). In mammals, the FABP content in skeletal and cardiac muscle is related to the FAO capacity of the tissue.[2] Although FABPpm is present constitutively in the sarcolemma, FAT/CD36 is present in an intracellular compartment from which it can be translocated to the plasma membrane within minutes of stimulation by either muscle contraction or insulin treatment, leading to elevated myocardial LCFA uptake. Interestingly, two distinct signaling pathways are used by these different translocation-inducing stimuli: insulin acting through the phosphatidylinositol 3-kinase (PI3K) pathway, and contraction-induced FAT/CD36 translocation mediated through AMP-activated protein kinase (AMPK) signaling.[3,4] Insulin also rapidly stimulates FAT/CD36 protein expression in either cultured cardiomyocytes or in isolated perfused hearts but has no effect on FABPpm expression.[5] The involvement in LCFA transport of other integral membrane proteins, including the fatty acid transport proteins (FATP1 and FATP6) in cardiac muscle has also been reported.

Of related interest, the intracellular recycling of the myocardial FAT/CD36 transporter and its induction by insulin and contraction are features shared with two myocardial glucose transporters GLUT1 and GLUT4. In rodent models of obesity and type 1 diabetes, LCFA uptake into heart and muscle is increased, either by permanently relocating FAT/CD36 to the plasma membrane without altering its expression (in obesity) or by increasing the expression of both FAT/CD36 and FABPpm (in type 1 diabetes).[6,7]

Once the LCFAs are transported through the plasma membrane, the 15-kDa cytoplasmic fatty acid–binding proteins (FABPc or H-FABP) are involved in the cytosolic diffusion of fatty acids to the intracellular sites of conversion, such as the mitochondria for oxidation and subsequent energy production, or to sites where they will be esterified into phospholipids and intracellular triacylglycerol (TAG). Insulin predominantly directs intracellular LCFAs toward esterification,[8] whereas during cellular contractions, LCFAs are efficiently used for energy production by mitochondrial β-oxidation.[9] A strong relationship has been reported in both skeletal and cardiac muscle between the accumulation of intracellular TAGs and the development of insulin resistance, characterized by a diminished ability of insulin-sensitive tissues to take up and metabolize glucose in response to insulin.[10–13]

Individuals with obesity and type 2 diabetes exhibit increased plasma LCFA levels and intramuscular accumulation of TAG (as well as related lipid intermediates such as diacylglycerol, fatty acyl-CoA, and ceramides); whereas the latter is not in itself harmful, it interferes with the insulin-signaling pathway (by activating a serine/threonine kinase cascade with PKC involvement), leading to phosphorylation of serine/threonine sites on insulin receptor substrates (IRS-1 and IRS-2). This, in turn, reduces the IRS-activated PI3K activity, resulting in the downstream reduction of glucose transport.[14] As discussed later, other intracellular stimuli can trigger these cascades resulting in insulin resistance.

Before transport into the mitochondria or TAG synthesis, fatty acids must be activated by their conversion into acyl-CoA in the cytoplasm. This activation process results in the consumption of ATP, requires CoA-SH, and is catalyzed by fatty acyl-CoA synthetases, associated with either the endoplasmic reticulum or the outer membrane of the mitochondria. At least three different acyl-CoA synthetase enzymes have been described whose specificities depend on fatty acid chain length. The fatty acid transport protein (FATP1) has been shown to be a bifunctional protein with both a fatty acid transport function and an intrinsic fatty acyl-CoA synthetase activity with broad substrate specificity, including both long-chain and very long-chain fatty acids.[15] Some evidence supports the view that increased fatty acid uptake mediated by FATP1 is likely driven by increased conversion of fatty acids to fatty acyl-CoA; mutation of a key residue that is predicted to abolish catalytic synthetase activity results in a loss of FATP1-mediated fatty acid uptake.[16]

Moreover, a subset of fatty acids taken up through FATP1 are preferentially channeled into triglyceride synthesis, suggesting a functional link between FATP1-mediated fatty acid uptake and lipid storage.[17] Recent studies have demonstrated a significant role for FATP1 and intramuscular fatty acid metabolite accumulation in causing insulin resistance. Transgenic mice deleted for FATp1 were protected from fat-induced insulin resistance and intramuscular accumulation of fatty acyl-CoA without alteration in whole-body adiposity compared with wild-type mice.[18]

For the β-oxidation pathway to proceed, the fatty acyl-CoA has to be transported across the inner mitochondrial membrane. Long-chain fatty acyl-CoA molecules cannot pass directly across the inner mitochondrial membrane and need to be transported as carnitine esters, whereas short-chain and medium-chain fatty acids can be easily transported without the assistance of carnitine. The transport of long-chain fatty acyl-CoA into the mitochondria depicted in Fig. 1 is accomplished by an acyl-carnitine intermediate, which itself is generated by the action of carnitine palmitoyltransferase I (CPT-I), an enzyme residing in the inner face of the outer mitochondrial membrane. The resulting acyl-carnitine molecule is subsequently transported into the mitochondria by the carnitine translocase, a transmembrane protein residing in the inner membrane that delivers acyl-carnitine in exchange for free carnitine from the mitochondrial matrix. CPT-II

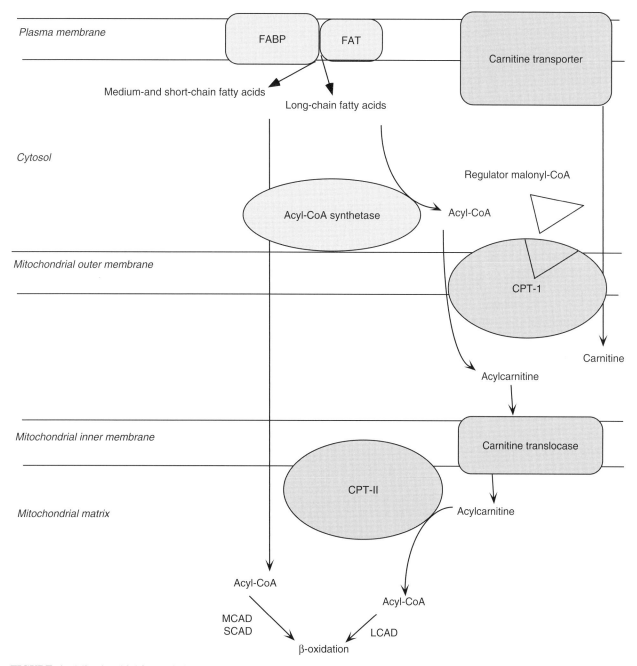

FIGURE 1 Mitochondrial fatty acid import and oxidation. Fatty acid and carnitine, after entry into the cardiomyocyte, are transported into the mitochondria for oxidation. FABP, fatty acid binding protein; FAT, fatty acid translocase; CPT-I, carnitine palmitoyltransferase I; CPT-II, carnitine palmitoyltransferase II; MCAD, medium-chain acyl CoA dehydrogenase; SCAD, short-chain acyl CoA dehydrogenase; LCAD, long-chain acyl CoA dehydrogenase.

located within the inner mitochondrial membrane catalyzes the regeneration of the fatty acyl-CoA molecule, with the acyl group transferred back to CoA from carnitine. Once inside the mitochondrion, the fatty acid-CoA is a substrate for the FAO machinery. Recent evidence supports an interaction of CPTI with FAT/CD36 at the mitochondrial outer membrane, suggesting a potential role for this protein as either an accessory or an additional fatty acid transporter at the mitochondrial membrane.[19]

The uptake of fatty acids by heart mitochondria is also regulated by the levels of the metabolite malonyl-CoA, which functions as a potent allosteric inhibitor of CPT-I and which binds to CPT-I on the cytosolic side of the enzyme; the heart isoform of CPT-I (i.e., CPT-IB) is 30 times more sensitive to malonyl-CoA inhibition than the liver isoform CPT-IA.[20]

Malonyl-CoA is synthesized by the enzyme acetyl-CoA carboxylase (ACC) from cytoplasmic and peroxisomal acetyl-CoA. The enzyme malonyl-CoA decarboxylase (MCD) is involved in the regulation of malonyl-CoA turnover. Levels of malonyl-CoA can be affected by changes in extramitochondrial acetyl-CoA levels or by modulation of the ACC and MCD activities. ACC activity is allosterically regulated by citrate and by kinase-mediated phosphorylation (e.g., AMPK). In the ischemic heart, AMP-activated protein kinase (AMPK) seems to be pivotally involved in promoting the phosphorylation and inhibition of ACC and the activation of MCD, resulting in lower myocardial malonyl-CoA levels and increasing FAO.[21] Decreasing the ischemic-induced activation of AMPK or preventing the downstream decrease in malonyl-CoA levels to reduce FAO may be an effective therapeutic approach in treating ischemic heart disease and HF. Because AMPK also activates glucose metabolism, increasing AMPK activity, which occurs with increased myocardial metabolic rate such as occurs with exercise, should ensure ample acetyl-CoA production, substrate oxidation, and energy. Therefore, malonyl-CoA production linked to altered metabolic demand or use of carbohydrate resources can, in turn, impact on either the down-regulation or up-regulation of fatty acid import into mitochondria and myocardial FAO.

Glucose Carriers

As in all other cells, the entry of glucose into cardiac myocytes depends on the transmembrane glucose gradient and is facilitated by members of the GLUT family of facilitative glucose transporters.[22] The GLUT1 transporter, which is localized on the plasma membrane under basal conditions, is thought to be the primary mediator of basal glucose uptake in the heart.[23] Its myocardial expression is stably increased within hours of ischemia or induction of hypertrophy. The most abundant glucose transporter in the heart is the insulin-responsive GLUT4 transporter. Insulin mediates the translocation of GLUT4 to the plasma membrane from a pool of intracellular vesicles and represents a critical control point by which the net flux of glucose is regulated. A variety of physiological stimuli, including hypoxia, ischemia, and cardiac work overload, can induce this translocation, thereby increasing glucose uptake and glycolytic metabolism.[24] It has also been reported that other stimuli, including catecholamines, calcium, and exercise, and contraction-induced stimulation of GLUT4 translocation and enhanced glucose uptake can occur in cardiac tissues and myocytes that are insulin independent.[25-27] Moreover, studies with mice containing muscle-specific deletions in the insulin receptor gene revealed that normal expression of muscle insulin receptors is not needed for the exercise-mediated increase in glucose uptake and glycogen synthase activity *in vivo*.[28] Moreover, GLUT4 translocation to the plasma membrane can also be stimulated by activation of AMPK activity that occurs during exercise stress.[24] Studies with transgenic mice containing an inactive form of AMPK have shown normal GLUT4 levels and glucose uptake but no increase in glucose uptake, glycolysis, or FAO during ischemia.[24] These findings have led to the conclusion that AMPK is involved in mediating glucose uptake and glycolysis during ischemia, presumably as a protective adaptation. Gathered observations have also demonstrated that changes in myocardial AMPK activity in exercise-trained rats (which increased in proportion to exercise intensity) were associated with physiological AMPK effects (e.g., GLUT4 translocation to the myocardial sarcolemma) and are consistent with AMPK as a key mediator in the cardiac response to exercise, as previously demonstrated with skeletal muscle.[29]

Defects in the ability of insulin to regulate GLUT4 translocation may be contributory (as one of multiple effects of insulin) to the development of insulin resistance and non-insulin-dependent type 2 diabetes. Mice heterozygous for a null *GLUT4* allele display reduced muscle glucose uptake, insulin-resistance, and diabetes.[30] In contrast, mice homozygous for a null *GLUT4* were growth retarded and exhibited decreased longevity associated with cardiac hypertrophy but displayed little effect on muscle glucose uptake in either fasted or fed state; however, *GLUT4*-deleted animals had postprandial hyperinsulinemia, indicating possible insulin resistance. A more recent study targeting muscle-specific *GLUT4* demonstrated a profound reduction in basal glucose transport and near-absence of stimulation by insulin or contraction, demonstrating severe insulin resistance and glucose intolerance from an early age.[31]

In a model of type 2 diabetes that used the Goto-Kakizaki (GK) rat, insulin-stimulated glucose uptake was 50% ($p<0.03$) lower in GK rat hearts than their Wistar controls with marked GLUT4 protein depletion.[32] Moreover, these animals exhibited significant decreases in levels of the myocardial insulin receptor substrate-1 (IRS-1), as well as reduced IRS-1 association with PI3K, all key upstream events in the insulin signaling pathway. Although this study also revealed decreased levels of the myocardial insulin receptor in GK rats, other studies have found conflicting information about its role in muscle insulin resistance.[33] Interestingly, whereas most transgenic studies addressing the complicity of the insulin signaling pathway on the development of insulin resistance have used targeting of muscle-specific components, such as GLUT4 and the insulin receptor, there has been limited examination of the myocardial-specific inactivation of these critical genes. In transgenic mice containing a Cre/LoxP generated construct of a null *GLUT4* allele directed to the heart, reduction of the GLUT4 to a level as low as 15% of wild-type levels was sufficient to allow normal levels of insulin-stimulated glucose uptake, which was markedly reduced with further reduction of GLUT4 levels. If GLUT4 levels were reduced to 5% of wild type, cardiac hypertrophy resulted.[34]

Clinical studies of patients with HF revealed a marked down-regulation of myocardial GLUT transporters, limiting both glucose uptake and oxidation and contributing to the heart's inability to generate much needed ATP.[35] An earlier study by this group demonstrated that GLUT4 transcripts were markedly down-regulated in the human failing heart.[36] In diabetic cardiomyopathy, decreased glucose use results in an almost exclusive use of fatty acids as the myocardial energy source.[37,38] The relatively recent finding of increased myocardial insulin resistance accompanying advanced dilated cardiomyopathy (DCM) limiting both glucose uptake and oxidation may provide critical targets for therapeutic intervention.[39,40]

Insulin resistance characterized by lower myocardial insulin-sensitive glucose uptake and a reduced GLUT4 protein level was reported in patients with severe cardiac hypertrophy in the absence of hypertension, diabetes, and CAD.[41,42] Interestingly, myocardial insulin-independent glucose uptake (and basal glucose uptake) and glycolytic metabolism are enhanced in patients with hypertension or in experimental animals with cardiac hypertrophy, the latter in striking contrast to insulin-stimulated glucose uptake that is depressed in these animals and suggests that hearts subjected to pressure-overload seem to be resistant to the metabolic effects of insulin.[43,44] This may have important therapeutic consequences in the setting of individuals with hypertension/hypertrophy, in whom treatment with glucose-potassium-insulin aimed at altering myocardial glucose use may be less effective.[45]

The expression of myocardial GLUT4 is present throughout embryonic development, albeit at low levels, whereas GLUT1 is highly expressed in the prenatal heart.[46] At birth, the expression of genes that control myocardial glucose transport and oxidation is down-regulated. In the adult heart, GLUT4 becomes the main glucose transporter, although GLUT1 is expressed at a considerable level. Regulation of myocardial glucose transporter levels is primarily exerted transcriptionally.[47] Data collected from numerous studies suggest that in the failing heart a fetal metabolic gene profile is established largely by down-regulating adult gene transcripts of metabolic proteins (e.g., GLUT4) rather than by up-regulating fetal genes (e.g., GLUT1).[48] Similar findings of re-activation of a fetal metabolic program involving the down-regulation of adult but not fetal isoforms and a re-expression of growth factors and protooncogenes has been described in both the hypertrophied heart and with mechanical unloading of the heart.[49-51] Reactivation of these fetal genes includes a pivotal metabolic switch from fat to glucose oxidation, which, although initially adaptive, ultimately results in a loss of insulin sensitivity and, hence, a loss of metabolic flexibility. Recent evidence suggests that this loss of flexibility then becomes an early feature of metabolic dysregulation in the failing heart, which also exhibits all the features of insulin resistance.[48]

BIOENERGETICS OF FATTY ACID AND GLUCOSE OXIDATION

The mitochondrial fatty acid β-oxidation pathway contains four reaction steps, including acyl-CoA dehydrogenases (short-chain, SCAD, medium-chain, MCAD, long-chain, LCAD, and very long-chain, VLCAD), short-chain enoyl-CoA hydratase, 3-hydroxyacyl CoA dehydrogenase, and 3-ketoacyl CoA thiolase (Fig. 2).

In the initial acyl-CoA dehydrogenase reaction, VLCAD and LCAD are responsible for the enzymatic dehydrogenation of long-chain C8 to C22 fatty acids (e.g., palmitate and linoleic acids), with VLCAD having greater activity with the longer chain substrates (e.g., C22 and C24 acyl-CoA esters). MCAD is active with the C4–C12 straight-chain fatty acids (e.g., decanoic acid), and SCAD primarily is active with C2–C4 fatty acids (e.g., butyryl CoA). The remaining enzymatic reactions of FAO are performed by a highly organized single enzymatic complex known as the *mitochondrial trifunctional protein* (MTP or TFP) bound to the inner membrane of mitochondria. Recently, a different set of enzymes, localized in the mitochondrial matrix, has been reported to be responsible for the β-oxidation of medium-chain and short-chain fatty acids.[52,53]

The process of FAO is termed β-oxidation, because it occurs through the sequential removal of two-carbon units by oxidation at the β-carbon position of the fatty acyl-CoA molecule. Each round of β-oxidation produces NADH, FADH$_2$, and acetyl-CoA. Acetyl-CoA, the end product of each round of β-oxidation, enters the TCA cycle, where it is further oxidized to CO_2 with the concomitant generation of NADH, FADH$_2$, and ATP. In the well-perfused heart, between 60 and 90% of the acetyl-CoA comes from β-oxidation of fatty acids, whereas 10–40% is derived from the oxidation of pyruvate and lactate. The NADH and FADH$_2$ generated during FAO and acetyl-CoA oxidation in the TCA cycle will subsequently enter the respiratory pathway for the production of ATP. Consequently, the oxidation of fatty acids yields more energy per carbon atom than does the oxidation of carbohydrates. However, although fatty acids produce more ATP during complete aerobic oxidation than glucose, this occurs at the expense of a higher rate of oxygen consumption. The supply of oxygen can be an important determinant of myocardial fuel use. As we will shortly see, there is a critical balance between the myocardial use of these substrates; disruption of this balance can comprise a primary defect resulting in cardiac disease.

The carbohydrate substrates for myocardial glycolytic oxidation are provided by exogenous glucose (obtained by transport) and internal glycogen stores. On entry into the cytosol, glucose is rapidly phosphorylated by hexokinase to glucose-6-phosphate and either catabolized by subsequent cytosolic glycolysis or converted into glycogen. The glycolytic pathway produces ATP (2 molecules/every glucose molecule oxidized), NADH, and pyruvate, which can either be converted into NAD and cytosolic lactate (often secreted) or transported into the mitochondria for further oxidation by the TCA cycle and oxidative phosphorylation (OXPHOS). The myocardium only generates significant lactate levels as a net producer under conditions of accelerated glycolysis

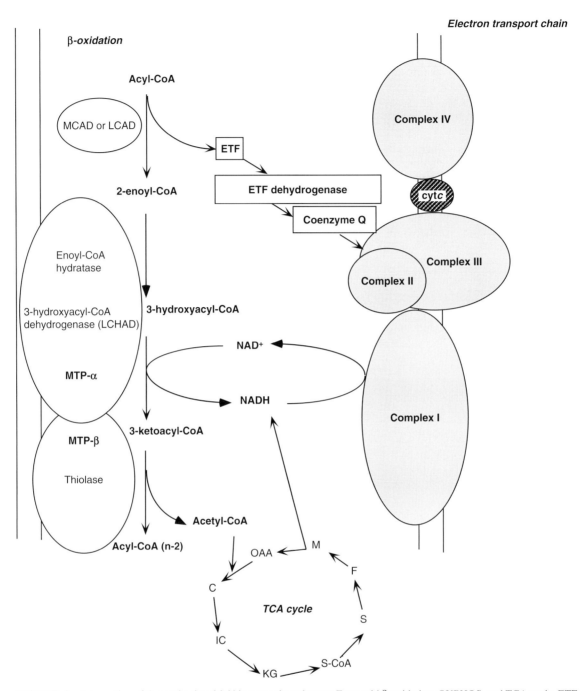

FIGURE 2 Intersection of three mitochondrial bioenergetic pathways: Fatty acid β-oxidation, OXPHOS, and TCA cycle. ETF, electron transfer flavoprotein; cyt*c*, cytochrome *c*; MTP, mitochondrial trifunctional protein; MCAD, medium-chain acyl CoA dehydrogenase; LCAD, long-chain acyl CoA dehydrogenase; LCHAD, long-chain 3-hydroxylacyl-CoA dehydrogenase; OAA, oxaloacetate; M, malate; C, citrate; F, fumarate; IC, isocitrate; KG, ketoglutarate; S, succinate; S-CoA, succinyl CoA.

and impaired pyruvate oxidation, which occurs in ischemia and often in diabetes.[1] The oxidative metabolism of pyruvate once transported to the mitochondria (by a monocarboxylate carrier)[54] generates acetyl-CoA for fueling the TCA cycle and is often termed glucose oxidation.

The regulation of cardiac glycolysis occurs at several steps. Glucose phosphorylation by hexokinase comprises the first regulatory step that commits glucose to further metabolism. Two different isozymes of hexokinase are present in the heart, hexokinases I and II (HKI and HKII), with HKI predominant in the fetal and newborn heart and the insulin-regulated HKII prevalent in the adult heart.[55]

The first regulatory site that commits glucose to the glycolytic pathway occurs at the level of phosphofructokinase

(PFK1), catalyzing the phosphorylation of fructose 6-phosphate to fructose 1,6-bisphosphate. The PFK1-mediated conversion of fructose 6-phosphate into fructose 1,6-bisphosphate is a rate-limiting step of glycolysis. Negative allosteric effectors include ATP, citrate, and protons,[56] whereas positive effectors consist of AMP and fructose 2,6-bisphosphate (the main activator of PFK-1 in normoxic heart) and stimulate glycolysis.[57,58] Levels of this later effector increase when the glycolytic flux is stimulated and decrease when the heart oxidizes competing substrates.[58] Fructose 1,6-bisphosphate generated by PFK-1 also stimulates pyruvate kinase, which catalyzes the transformation of phosphoenolpyruvate into pyruvate, indicating a further role of PFK-1 in synchronizing several glycolytic reactions, allowing an acceleration of the glycolytic pathway without glycolytic intermediate accumulation.[59]

The pyruvate dehydrogenase complex (PDC), an allosteric enzyme that transforms pyruvate into acetyl-CoA by a process called oxidative decarboxylation, is a key determinant in the rate of glucose oxidation. This large multiprotein complex located in the mitochondrial matrix includes 132 subunits (30 E1 dimers, 60 E2 monomers, and 6 E3 dimers) with a variety of coenzymes, including thiamine pyrophosphate, lipoamide, CoA, FAD, and NAD and also contains three catalytic subunits responsible for different enzymatic reactions. The PDC complex is highly regulated by its substrates and products and by the activation/inactivation of its catalytic components (i.e., E1, E2, and E3) by their dephosphorylation/phosphorylation. Phosphorylation of PDC by the associated enzyme pyruvate dehydrogenase kinase (PDK) inactivates the PDC complex, whereas pyruvate dehydrogenase phosphatase (stimulated primarily by Ca^{++}) dephosphorylates and reactivates the enzyme. Both the PDC-activating phosphatase and PDC-deactivating kinase have several cardiac-specific isoforms subject to regulation by diverse developmental, dietary, and hormonal stimuli.[60,61] The PDK-1 isoform is up-regulated in the adult compared with the neonatal heart and is primarily involved in regulating glucose oxidation through the inhibitory phosphorylation of PDC. Expression of the cardiac PDH kinase 4 isoform is responsive to changes in myocardial lipid supply; its up-regulation in the post-natal heart contributes to the perinatal developmental switch to fatty acids as the primary myocardial energy source. This myocardial kinase isoform is also up-regulated in response to thyroid hormone and high-fat diet. Starvation and diabetes decrease myocardial PDC activity levels by both activating PDK gene expression and by the inactivation of myocardial PDH phosphatase activity, affected largely by promoting a reduced expression of the *PDHP2* gene.[62]

In addition to its regulation by reversible phosphorylation, PDC is regulated through negative feedback by acetyl-CoA and NADH, end products of both the PDC reaction and mitochondrial FAO. Acetyl-CoA and NADH accumulation (produced primarily by FAO) also activate PDK, which phosphorylates and inhibits PDC, thereby decreasing glucose oxidation. In addition, increased pyruvate supply can inhibit PDK, thereby stimulating PDC, a process that may occur in isolated hearts perfused with insulin.

When circulating glucose and insulin levels are high, as occurs in the postprandial state, glucose is a primary contributor to cardiac energy metabolism,[63] and during the fasting state, free fatty acids become the dominant fuel. With increased FAO and the mitochondrial production of acetyl-CoA, NADH, and citrate, cytosolic glycolysis and glucose oxidation are inhibited through inactivation of the PDC and phosphofructokinase activities. During oxygen deprivation and anoxia, the inhibition of glucose use is removed, and glycolysis is accelerated. A significant increase in carbohydrate oxidation also occurs in the adult heart in response to an acute increase in cardiac work. Because an increase in glucose uptake is delayed in this cardiac response, the increase in glucose oxidation is initiated by rapid glycogen breakdown.[64]

Glycogen turnover has been proposed as a control site for myocardial glucose metabolism. The myocardial glycogen pool in the adult heart is relatively small, occupying approximately 2% of the cell volume and is more prominent in the fetal and newborn cardiomyocyte comprising 30% of the cell volume; it has a relatively rapid turnover.[65] Unlike liver and skeletal muscle, heart muscle increases its glycogen content with fasting, consistent with the conclusion that fatty acids, the predominant fuel for the heart during fasting, inhibit glycolysis more than glucose uptake, thereby rerouting glucose toward glycogen synthesis.[66] Myocardial glycogen levels are also increased by insulin from the concerted stimulation of both glucose transport and glycogen synthase activity.[67]

At the other end of the spectrum, glycogen is rapidly broken down when glycogen phosphorylase, the main regulator of glycogenolysis, is stimulated by adrenergic stimulation (e.g., epinephrine) or glucagon.[68] It is activated by phosphorylation, either by cAMP-dependent protein kinase or by Ca^{++}-activated phosphorylase kinase.[69] Glycogen breakdown is also rapidly stimulated during sudden increases of heart work, decreased tissue content of ATP, and increased levels of inorganic phosphate accompanying ischemia or intense exercise.[70,71]

A role of AMPK in glycogen turnover has recently been identified after the discovery that mutations in the γ-2 regulatory subunit (PRKAγ2) of the AMPK enzyme. These mutations elicited a constitutively active AMPK resulting in extensive myocardial glycogen accumulation and HCM and Wolff-Parkinson-White (WPW) syndrome, as discussed in Chapters 13 and 17.[72] The relationship between the glycogen storage diseases and the development of cardiomyopathy and pre-excitation syndrome remains unclear. In contrast, acute AMPK activation leads to increased glycogenolysis. Isolated working hearts perfused with 5-aminoimidazole-4-carboxamide 1-β-D-ribofuranoside (AICAR), an adenosine analog and cell-permeable activator of AMPK, caused an allosteric activation of glycogen phosphorylase responsible for glycogenolysis.[73] In addition, AMPK contains a putative glycogen-binding site in its β-subunit and has been reported to associate with specific subcellular structures containing both glycogen and glycogen phosphorylase.[74]

CELLULAR LOCATION OF FAO AND GLUCOSE OXIDATION

Both peroxisomes and mitochondria have multiple enzymes involved in fatty acid β-oxidation. The peroxisomal enzymes include palmitoyl-CoA oxidase, L-functional protein, and 3-ketoacyl oxidase, which are all inducible enzymes acting on straight-chain substrates. In addition, peroxisomes contain branched-chain acyl-CoA oxidase, D-functional protein, and sterol-carrier protein X, which are noninducible and primarily use branched-chain substrates. The inducible enzymes increase in response to the peroxisomal proliferating activating receptor (PPAR), resulting in increased peroxisomal biogenesis. In addition, recent evidence has shown that myocardial peroxisomal fatty acid is a primary supplier of acetyl-CoA used for the synthesis of malonyl-CoA.[75] Peroxisomal β-oxidation, therefore, contributes to the regulation of mitochondrial FAO.

It is important to note that although specific deficiencies in the mitochondrial-located enzymes involved in FAO may result in cardiomyopathy (as discussed later), defects in the peroxisomal FAO enzymes primarily result in neurological pathology, including seizures, hypotonia, and psychomotor retardation. Cardiac abnormalities have been rarely described in peroxisomal deficiencies. This is also true of diseases involving general peroxisomal biogenesis abnormalities such as Zellweger syndrome and neonatal adrenoleukodystrophy, where there is little or no cardiac involvement. PPAR plays a pivotal role in both mitochondrial FAO and mitochondrial biogenesis, not only in normal cardiac growth and development but also in HF. This suggests an important interrelationship between the two cellular compartments and further underscores the mitochondrial compartment as a critical effector of cardiac homeostasis. The commonality of biogenesis and potential feedback between these two cellular organelles needs further investigation in both normal growth and development and in cardiac disease (both FAO and mitochondrial OXPHOS disorders).

After entering the plasma membrane, glucose is oxidized by the glycolytic enzymes located primarily in the cytosol but often in association with specific organelles. Glyceraldehyde-3-phosphate dehydrogenase and pyruvate kinase bind to sarcolemmal and sarcoplasmic reticulum membranes. *In silico* (i.e., performed on computer or by computer simulation) studies to simulate the glycolytic burst associated with the onset of ischemia *in vivo* have suggested the presence of compartmentation of glycolytic function to a defined cytosolic subdomain.[76] Similarly, several studies have supported the organization of the TCA enzymes within the mitochondrial matrix as constituting a supercomplex or metabolon, providing a kinetic advantage in concentrating intermediates and channeling substrates within the supercomplex.[77] In addition, an association of specific glycolytic enzymes with mitochondria has been noted.[78] Both the first enzyme in the glycolytic pathway (i.e., hexokinase) and the primary enzyme in glucose oxidation (i.e., PDC) are associated with the mitochondria. Hexokinase binds at the outer membrane to peripheral protein complexes such as the PT pore, whereas PDC (which determines the fate of the glycolytic product pyruvate) is entirely located within the mitochondrial matrix.

Regulatory Elements in FAO

Although cardiac energy conversion capacity and metabolic flux are modulated at many levels, an important mechanism of regulation occurs at the level of gene expression, which is subject to modulation by developmental, physiological, and pathophysiological stimuli. Much attention has focused on nuclear transcription factors that oversee and orchestrate the regulation of specific and global gene expression. These ligand-activated factors can function as metabolite sensors to rapidly program gene expression in response to fluctuating substrate levels. In this section, we describe several such factors, members of the nuclear receptor superfamily (described in depth in our signaling chapter) and include the fatty acid–activated peroxisome proliferator–activated receptors (PPARs) and the nuclear receptor coactivator, PPAR gamma coactivator-1 (PGC-1), which modulate the transcriptional regulation of genes involved in key energy metabolic pathways.

PPAR

Over the past decade, PPARs have been identified as playing a central role in the transcriptional regulation of genes involved in intracellular lipid and energy metabolism, including FAO enzymes.[79–81] Three isoforms of the PPAR subfamily (α, β, and γ) are enriched in tissues that depend on lipid use for energy metabolism (e.g., heart, liver, brown adipose tissue, and all critical vascular cells) and have been implicated in the rapid mobilization of bioenergetic stores in response to physiological stresses.

All three PPAR isoforms are activated by fatty acids. Each factor acts in concert with the nuclear retinoid X receptor (RXR) as a heterodimer binding to a consensus DNA response element with the sequence AGGTCANAGGTCA (direct repeat with a single nucleotide spacing) contained within the regulatory regions of target genes. This is followed by the transcriptional activation and increased gene expression of these target genes, including a constellation of genes encoding enzymes involved in both peroxisome and mitochondrial FAO (e.g., mitochondrial MCAD, CPT-I, and peroxisomal acyl-CoA oxidase). The functional specificity of the PPARs is determined by isoform-specific tissue distribution, specific interaction with activating ligands (e.g., prostaglandins, eicosanoids, and long-chain unsaturated fatty acids), and cofactor interactions (i.e., coactivators and corepressors).[82,83] Although several lines of evidence suggest that all three isoforms modulate cardiac energy metabolism, PPAR-α has been generally characterized as the central regulator of myocardial mitochondrial fatty acid catabolism, whereas PPAR-γ is thought to be involved in myocardial lipid storage regulation. Studies have recently shown the myocardial function of PPAR-β to have considerable overlap with PPAR-α, although indicating that these isoforms are not redundant.[84]

PPAR-α has been shown largely by transgenic overexpression, as well as by knock-out gene analysis, to be intrinsically involved in nearly every step of myocardial fatty acid metabolism, including fatty acid uptake, esterification, transport into mitochondria, and both mitochondrial and peroxisomal β-oxidation. A list of the activated gene products involved in these functions is presented in Table I and includes MCAD, frequently used as a marker of PPAR–α activation. The activity of PPAR-α, which is highly expressed in the heart, is a critical determinant of myocardial energy production and can serve to match cardiac lipid delivery to oxidative capacity.[100]

Although the primary endogenous ligand for PPAR has not yet been precisely determined, PPAR-α is activated by a number of lipid-derived molecules, including naturally occurring long-chain fatty acids, eicosanoids, and leukotriene B4.[101,102] In addition, the fibrate class of anti-hyperlipidemic drugs, that is widely used clinically, are synthetic PPAR ligands and include ciprofibrate, bezafibrate, fenofibrate and gemfibrozil.[101] Interestingly, although some of the synthetic agonists are highly specific for PPAR-α activation, others show dual specificity, activating both PPAR-α and PPAR-γ (e.g., glitazars) and some activate equally all PPARs (e.g., bezafibrate).[103] To precisely define the specific endogenous ligands used by these different PPARs, their specific coactivators, and the full set of targeted genes for each receptor isoforms, further research is needed. Eventually, this information may allow a finely tuned pharmacological modulation of cardiac metabolism.

Cardiac metabolic gene expression is activated by PPAR-α regulation during post-natal development, during short-term starvation, and in response to exercise training.[104] Conversely, pressure-overload hypertrophy results in the deactivation of PPAR-α with lower FAO expression, abnormal cardiac lipid homoeostasis, and reduced energy production.[105] This suggests that PPAR-α plays a contributory role in the energy substrate switch away from fatty acid use in the hypertrophied heart. Lower nuclear levels of the PPAR-α protein in these tissues have been largely explained to occur as a function of the negative regulation of PPAR-α mediated at the transcriptional level during ventricular overload in mice. In addition, PPAR activity is altered at the post-transcriptional level by phosphorylation by several kinases, including PKA, PKC, MAPKs, and AMPK.[106–111] Ventricular overload, therefore, results in hypertrophied myocytes with intracellular fat accumulation (in response to oleate loading).

In addition to the effects on FAO and glucose metabolism, there have been some reports that PPAR-α activation may provide cardioprotection against ischemia/reperfusion (I/R) injury.[112,113] However, several recent studies have challenged that conclusion. One recent study found that in wild-type mice, the down-regulation of GLUT4 levels and glucose uptake resulting from PPAR-α activation was particularly detrimental in the ischemic heart, resulting in poor post-ischemic functional recovery.[114] In contrast, *PPAR-α* null mice, although having elevated plasma free fatty acids, display no reduction in cardiac GLUT4 levels or glucose uptake during ischemia and, consequently, did not have poor recovery during reperfusion, suggesting that PPAR-α could enhance susceptibility to ischemic injury.[114] Other recent studies have confirmed that chronic activation of PPAR-α seems to be detrimental to the cardiac recovery during reperfusion after ischemia.[115]

Chemically distinct ligands of PPAR-γ include the thiazolidines (TZDs), including rosiglitazone, ciglitazone, and pioglitazone, as well as the cyclopentanone prostaglandins 15D-PGJ2 and PGA1. PPAR-γ agonists, such as the glitazones, increase insulin sensitivity and decrease plasma glucose levels in patients with diabetes and are used therapeutically.[116] Similarly, thiazolidinediones, acting by PPAR-γ, influence free fatty acid flux and thus reduce insulin resistance and blood glucose levels. The levels of PPAR-γ are barely detectable in the heart (much more abundant in adipose tissue), supportive against a direct role for PPAR-γ in cardiac lipid metabolism.[117] However, numerous studies have shown that PPAR-γ–specific agonists provide cardioprotection in response to ischemia and HF,[113,116] even in isolated myocytes exposed to PPAR-γ agonists or fatty acid ligands.

TABLE I PPAR-α Regulated Genes in Cardiac Metabolism

Function	Gene product	References
Fatty acid uptake/esterification	LPL	85
	CD36/FAT, FABP	86
	FATP1, FACS-1	87
Glucose oxidation	PDK4	88, 89
	GLUT4	84
Mitochondrial and peroxisomal β-oxidation	MCAD,	96
	LCAD, VLCAD	86
	M-CPT I (CPT-Iβ)	90, 91
	Bifunctional enzyme	92, 93
	ACO	94, 95
	MCD	97
Uncoupling proteins	UCP3	98, 99
	UCP2	84, 99

Finally, the less characterized myocardial PPAR-β isoform is similar in abundance to PPAR-α, is fatty acid inducible, and promotes myocardial PPAR target gene expression.

Peroxisome Proliferator-Activated Receptor γ Coactivator (PGC-1α)

PGC-1α was initially identified as a PPAR-γ coactivator involved in the regulation of energy metabolic pathways in tissues specialized for thermogenesis (e.g., brown adipose tissue and skeletal muscle).[118] PGC-1 is selectively expressed in highly oxidative tissues such as heart, skeletal muscle, brown adipose, and liver. Its tissue expression and inducibility in response to a variety of physiological stimuli (e.g., exercise, cold exposure, and fasting) are intricately linked to its role as a regulator of cellular and mitochondrial energy metabolism, including myocardial mitochondrial biogenesis and oxidation, hepatic gluconeogenesis, and skeletal muscle glucose uptake.[119–121] A number of key signaling pathways, which activate PGC-1 have been identified in association with these stimuli in these tissues, including p38 MAP kinase, β-adrenergic/cAMP, NO, AMPK, and Ca^{++}-calmodulin kinase, which seem to act either by increasing PGC-1 expression or its transcriptional activating function.[122–127] A structurally related homolog PGC-1β with similar tissue distribution has also been identified and cloned and shown to play a role in regulating metabolic pathways, albeit by activating a distinct array of target genes.[128,129] Also, a homologous PGC-1–released coactivator protein (PRC) has been reported that has functional similarities to PGC-1, although it is serum-induced and seems to be distinct in its metabolic activation in response to specific cell proliferative signals.[130] The similarity in domain composition between these three related genes is shown in Fig. 3.

The mechanism of PGC activation is somewhat complex.[131] As a coactivator, PGC works as a docking station for a variety of transcription factors and nuclear receptors that are loaded at the promoter region through transcription factor docking and recruitment of histone acetyltransferase (HAT) complexes. In some instances, PGC confers promoter specificity to the transcription factor; others have demonstrated a role for PGC-1 in RNA elongation and processing and described interactions of PGC-1 with splicing and elongation factors.

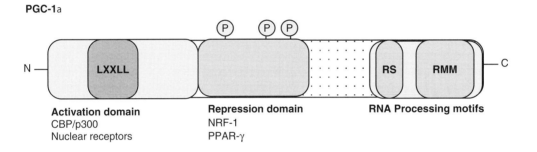

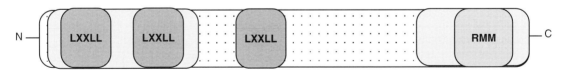

FIGURE 3 PGC-1 gene family's functional domains. PGC-1α contains a potent activation domain at its N-terminus that interacts with other transcriptional coactivators. The LXXLL motif is responsible for ligand-dependent interaction with certain hormone nuclear receptors. A central suppression domain contains several p38 MAPK phosphorylation sites. The C-terminus contains RNA processing motifs such as RS and RMM domains. Also shown is its alignment with other members of the PGC-1 gene family (PGC-1β and PRC) with the more conserved domains at the N- and C-termini.

Of interest is the finding that the N-terminus of the PGC-1α and β proteins contains a transcriptional activation domain that binds to DNA-binding proteins and can recruit proteins with HAT activity, and contains the major nuclear hormone receptor-interacting sequence (LXXLL), a motif known to be responsible for ligand-dependent interaction of other coactivators with nuclear hormone receptors.[132] The LXXLL motif on PGC-1 is absolutely required for the ligand-dependent interaction with ER, PPAR, RXR, glucocorticoid receptor, and probably other nuclear hormone receptors.[131,133–135] Interestingly, HNF4 also interacts with this motif of PGC-1 without addition of any ligand, suggesting that this nuclear hormone receptor is in an active conformation even without the addition of exogenous ligand.[136] PGC-1 also uses different non-LXXLL domains to interact with certain other transcription factors. A domain between amino acid residues 200–400 interacts with PPAR and NRF-1, and a region between amino acid residues 400–500 interacts with MEF2-C.[137,138] PGC-1 can interact with several nuclear hormone receptors bound to a target DNA sequence (e.g., *UCP1* enhancer). In that particular case, the interactions could be ligand dependent, as in the case of RAR or TR, or ligand independent as with PPAR. Although many of the previously described coactivators potentiate transcriptional activity by having specific enzymatic functions involved in chromatin remodeling and transcriptional initiation,[139,140] neither such an enzymatic activity or sequence homology with HAT domains present in other coactivators has been seen with PGC-1 thus far; however, the N-terminal activation domain of PGC-1 has been documented to recruit proteins that contain HAT activity, such as SRC-1 and CREB-binding protein (CBP/p300).[141] Moreover, the binding of SRC-1 and CBP/p300 to the N-terminus of PGC-1 depends on docking of transcription factors such PPAR and NRF-1 to PGC-1.

Residing within the C-terminus of the PGC-1 protein are two conserved protein domains (see Fig. 3), including an RNA-binding motif with homology to proteins involved in RNA processing and likely contribute to its participation in post-transcriptional processes.[131]

PGC-1α gene expression is induced in the mouse heart after birth coincident with the perinatal shift from glucose metabolism to FAO and can be subsequently induced in response to short-term fasting, conditions known to increase cardiac mitochondrial energy production.[119] Expression of PGC-1α in cardiac myocytes has been reported to induce nuclear and mitochondrial gene expression involved in multiple mitochondrial bioenergetic pathways, increased mitochondrial biogenesis, and respiration. Similarly, gain-of-function studies that used cardiac-specific overexpression of PGC-1α in transgenic mice resulted in uncontrolled mitochondrial proliferation in cardiac myocytes, leading to loss of sarcomeric structure and DCM and to enhanced mitochondrial FAO and overall mitochondrial oxidative capacity, including ATP-generating, coupled respiration.

Moreover, gene profiling of cardiac myocytes overexpressing PGC-1 has shown that PGC-1 activates the expression of genes encoding enzymes at every level of oxidative energy metabolism, including fatty acid uptake, mitochondrial β-oxidation, TCA cycle, electron transport, and OXPHOS.[119] Recent studies have confirmed PGC-1's importance in cardiac metabolism by use of loss-of-function analysis.[142] *PGC-1* knock-out mice exhibit markedly diminished expression of myocardial OXPHOS genes, resulting in reduced mitochondrial enzymatic activities and decreased levels of ATP and exhibit a diminished ability to increase myocardial work output in response to chemical or electrical stimulation. These strains develop cardiac dysfunction with progressive age. Thus, PGC-1 has a critical role in mediating the myocardium's response to meeting increased demands for ATP and energy in response to physiological stimuli. A flowchart depicting myocardial PGC-1's responses to a variety of stimuli is shown in Fig. 4.

Studies of cardiac hypertrophy caused by pressure overload in mice have shown that a shift from FAO to glycolytic metabolism is accompanied by a dramatic decline in PGC-1 expression as FAO drops, suggesting that this decrease in oxidative metabolism is secondary to the loss of PGC-1 expression.[143]

PGC-1 mediates its broad metabolic regulatory effects through coactivation of numerous transcription factor partners, including many nuclear receptors. Initial studies showed that PGC-1 coactivates PPAR-α and enhances fatty acid–dependent regulation of PPAR-α responsive genes involved in the FAO pathway,[144] and more recent studies have found that PPAR-β is also a transcriptional partner of PGC-1.[145] Other studies have implicated additional orphan nuclear receptors in mediating PGC-1 regulation of cardiac energy metabolism. Recently, two members of the estrogen-related receptor (ERR) family, ERRα and ERRγ, were identified as partners for the cardiac PGC-1 family of coactivators.[146] The ERRs generally are considered to be constitutively active receptors that interact with coactivator proteins in the absence of exogenous ligands.[147] These factors are abundantly expressed in adult tissues that rely primarily on mitochondrial oxidative metabolism for ATP production, such as heart and slow-twitch skeletal muscle;[148] myocardial ERRα is markedly increased at birth coincident with PGC-1α and cellular FAO levels.[146]

There is also accumulating evidence that the role of PGC-1 in promoting mitochondrial biogenesis and oxidative metabolism is linked to the array of its interacting transcription-regulatory factor partners.[149] In addition to the PPARs and the ERRs, PGC-1 coactivates transcription factors involved in regulating mitochondrial biogenesis and oxidative metabolism, including nuclear respiratory factor-1 and -2 (NRF-1 and NRF-2). The NRFs directly regulate downstream nuclear genes involved in mitochondrial respiratory function and biogenesis, including mitochondrial transcription factor A (mtTFA or TFAM), which is involved in mitochondrial DNA maintenance, replication, and transcription,[150,151] and nuclear-encoded components

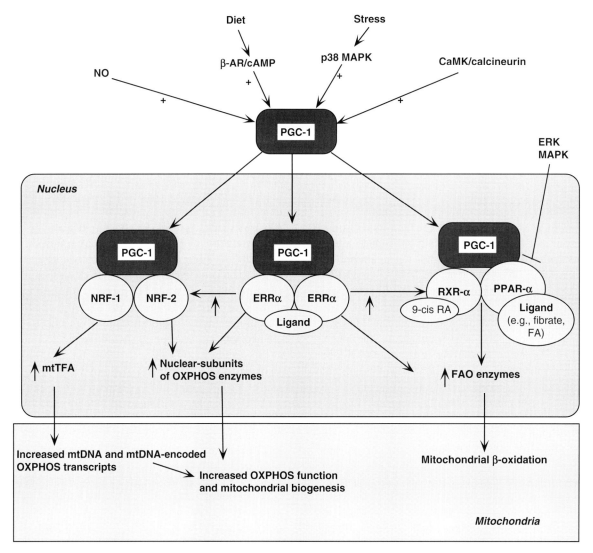

FIGURE 4 PGC-1 and its metabolic pathway. PGC-1 transduces cell signals associated with physiological stimuli to regulate cardiac metabolic genes. Numerous signaling pathways downstream of physiological stimuli, like fasting and exercise, activate the PGC-1 cascade either by increasing PGC-1 expression or activity. PGC-1, in turn, coactivates transcriptional partners, including nuclear respiratory factor-1 and 2 (NRF-1/2), estrogen-related receptor (ERR), and PPAR-α, resulting in downstream activation of mitochondrial biogenesis and FAO pathways, respectively.

of mitochondrial electron transport and OXPHOS such as subunits of cytochrome *c* oxidase and ATP synthase.[152] Recent studies have also suggested that the effects of thyroid hormone on cardiac energy mitochondria are at least partly mediated through transcriptional changes exerted by PGC1 and PPAR-α.[153–155]

Interestingly, ERRα has been recently found to mediate PGC-1 activation of the NRF pathway through regulation of the GA repeat binding protein alpha (Gabpa), a subunit of the NRF-2 complex,[156] and, in concert with its coactivator PGC-1, is involved in the direct transcriptional activation of mitochondrial oxidative metabolism, including activating promoters of genes encoding MCAD, cytochrome *c*, and ATP synthase β genes.[157,158]

The involvement of PGC in regulating adaptive glucose and fat oxidation in myocardium, muscle, and fat tissue; gluconeogenesis in liver; and even glucose-regulated insulin secretion in pancreatic beta cells has led to its investigation in association with diabetes and hyperglycemia.[159] Gene profiling in patients with diabetes, and in insulin-resistant ("pre-diabetic") individuals with normal glucose tolerance has revealed the reduced expression of genes encoding key enzymes in oxidative metabolism and mitochondrial function, most of which are regulated by the NRF-1 and PGC-1. These findings are suggestive that genes and environmental stresses that modify the risk of diabetes may do so by modifying the expression or activity of these transcriptional regulators. In addition, these studies revealed that muscle from individuals with pre-diabetes and type 2 diabetes have reduced *PGC-1α* and *PGC-1β* mRNA levels[160] and the mitochondrial genes they regulate.[161] In contrast, the activity of the *PPAR-α* gene regulatory pathway is increased in the diabetic heart, which relies primarily on FAO for energy production.[162] Moreover, as we will shortly see, a variety of *PGC-1α* polymorphisms have been reported in association with diabetes and insulin resistance.[163]

Effects of Dysfunctional or Perturbations in Fatty Acid and Glucose Metabolism Cause Changes in Cardiac Structure/Function

USE OF CELLULAR AND ANIMAL MODELS

Studies that use the isolated cardiomyocyte have been useful particularly in the study of rapid and transient signal transduction events occurring in the metabolic signaling pathways in response to specific defined stimuli. An important limitation with such studies is that most metabolic stressors and insults endured by the heart are less well defined and often feature a broad range of effects, including neuroendocrine, intercellular signaling involving several cell types, and circulatory/hemodynamic effects extending well beyond oxygen availability. Nevertheless, hypoxic simulations of ischemia or electrical stimulation studies with isolated cardiomyocytes can be useful in delineating a subset of the signaling pathways and molecules involved in the *in vivo* response.

The use of transgenic mouse models has been extremely helpful in our understanding of the initiation, severity, and progression of the cardiac phenotypes associated with specific abnormalities in fatty acid and glucose metabolism.[164] Although some of these studies have used global knock-out of specific genes of interest, the use of cardiac-specific gene disruption generally achieved by the Cre/loxP technology has proved to be useful in several instances in preventing embryonic or fetal lethality that may result from global gene ablation, as well as more readily allowing the distinction of primarily cardiac-specific events from systemic events occurring secondary to altered gene expression in other tissues. Moreover, the addition of inducible elements to the gene of interest allows the manipulation of gene expression within specific developmental or environmental conditions. A number of the genes examined affecting metabolic pathways and cardiac phenotypes are presented in Table II.

Both global and cardiac-specific disruption of *GLUT4* in transgenic mice result in marked cardiac hypertrophy.[30,165] Insulin-mediated glucose uptake is abolished in mice with the cardiac-specific *GLUT4* disruption that display a likely compensatory increase in GLUT1 and basal glucose uptake; moreover, decreased FAO suggests a switch from fatty acid to glucose metabolism. Both models show marked eventual cardiac dysfunction.

Interestingly, male heterozygous *GLUT4* +/- knock-out mice become diabetic and hypertensive with features of diabetic cardiomyopathy.[166] These mice exhibit decreased GLUT4 expression in adipose tissue and skeletal muscle, leading to increased serum glucose and insulin, reduced muscle glucose uptake, hypertension, and diabetic histopathology in the heart and liver, similar to humans with noninsulin-dependent diabetes.

In mice knock-outs of *PPAR-α*, there is reduced myocardial expression of genes involved in fatty acid uptake (CD36/FAT, FATP, FACS-1), mitochondrial transport (CPT I, CPT II), ß-oxidation (MCAD, VLCAD, SCAD, SCHAD, MTP), and myocardial LCFA uptake and oxidation rates.[167,168] Although unstressed hearts are relatively normal in these knock-out strains, a fasting stress, which in wild-type mice induces cardiac FAO enzyme gene expression, promotes hypoglycemia and hepatic and cardiac triglyceride accumulation. In contrast, transgenic mice containing cardiac-specific *PPAR-α* overexpression display activated myocardial gene expression for fatty acid use, including fatty acid uptake and FAO, whereas glucose use is reciprocally decreased.[84, 169] Reduced glucose use is due to the aforementioned inhibitory effects of acetyl-CoA and NADH (increased during high rates of FAO) on PDC activity and on direct gene down-regulation of the glucose transporter, GLUT4 and the glycolytic enzyme phosphofructokinase, and up-regulation of PDK4 (which decreases glucose oxidation through PDH inhibition) in hearts with increased PPAR-α levels. These results suggested that PPAR-α provides a critical link in circuits involved in myocardial fatty acid and glucose use.

Transgenic mice null for the MTP α allele lack both MTP α and β subunits and display necrosis and acute degradation of the cardiac myocytes. They also accumulate LCFA metabolites, have low birth weight, and develop neonatal hypoglycemia, with sudden death occurring between 6 and 36 h after birth.[170]

PGC-1α overexpression in the myocardium can display strikingly developmental-specific differences. Although constitutive cardiac-specific overexpression of PGC-1α proved to be lethal, resulting in the early development of DCM, marked fibrosis, and the accumulation of high numbers of large mitochondria,[119] inducible cardiac-specific PGC-1α overexpression (driven by a tetracycline-responsive promoter) was useful in examining developmental stage-specific responses.[171] For instance, in the neonatal myocardium, PGC-1α overexpression leads to increased cardiac mitochondrial size and number, concurrent with the increased gene expression associated with mitochondrial biogenesis and with no overt effect on neonatal cardiac function. However, in the adult mouse, PGC-1α, overexpression resulted in a moderate increase in mitochondrial number, with many organelles containing striking abnormalities in structure (e.g., vacuoles, inclusions) and within 2 weeks development of cardiomyopathy, featuring increased ventricular mass, wall thinning, and chamber dilation. Cessation of PGC-1α expression reverses the cardiac phenotype over time.

To further test the hypothesis that a disturbance in myocardial fatty acid uptake and use leads to the accumulation of cardiotoxic lipid species and to establish a mouse model of metabolic cardiomyopathy, transgenic mouse lines that overexpress long-chain acyl-CoA synthetase in the heart were generated. These mice showed cardiac-restricted expression of the transgene and marked cardiac myocyte

TABLE II Cardiac Phenotype Resulting from Altered Metabolic Genes in Transgenic Mice

Protein (gene)	Alteration	Myocardial phenotype	References
Fatty acid transport protein 1 (FATP1)	Cardiac-specific overexpression	Lipotoxic CM	184
Insulin-sensitive glucose transporter (Glut4)	Null	Cardiac hypertrophy; ↓ FAO	166
Long-chain acyl-CoA synthetase (ACS)	Cardiac-specific overexpression	Lipotoxic CM, hypertrophy, TG accumulation	172
Peroxisome proliferator-activated receptor-α (PPAR-α)	Cardiac-specific overexpression	↑ FAO; ↓ glucose uptake and oxidation	84
Peroxisome proliferator-activated receptor-δ (PPAR-δ)	Cardiac-specific deletion	Lipotoxic CM with ↓ FAO	185
Leptin (Lep)	Null	↑ FAO, FA uptake, TG + lipid accumulation; ↑ diastolic dysfunction	186
Very-long-chain acyl-coenzyme A dehydrogenase (VLCAD)	Null	Lipid accumulation, CM, mitochondrial proliferation, facilitated PVT induction	187
Mitochondrial transcription factor A (Tfam)	Cardiac-specific deletion	↓ MtDNA, ETC and ATP; ↓ FAO + ↑ glycolytic expression	188
Lipoprotein lipase (LpL)	Cardiac-specific deletion	↑ TG; ↓ FAO and ↑ Glut expression + glucose uptake	189
Fatty acid translocase (FAT/CD36)	Null	↓ Cardiac FA uptake +TG; DCM	190
Fatty acid translocase (FAT/CD36)	Overexpression	↑ FAO; ↓ plasma TG + FA; no cardiac pathology; ↑ glucose	191
Insulin receptor (Insr)	Cardiac-specific deletion	↓ FAO; ↓ Glut1 + basal glucose uptake; ↑ Glut4, insulin-glucose uptake + oxidation; smaller heart	192
Heart-type fatty acid binding protein (H-FABP)	Null	↓ Free LCFA uptake; ↑ Glucose use; hypertrophy	193
Mitochondrial trifunctional protein (Mtpa)	Null	Accumulate LCFA metabolites; growth retardation, neonatal hypoglycemia + SD	170
Peroxisome proliferator-activated receptor γ coactivator-1 (PGC-1α)	Null	↓ ETC + ATP; lower work output in response to chemical or electrical stimulation	142
Peroxisome proliferator-activated receptor γ coactivator-1 (PGC-1α)	Cardiac-specific overexpression	Mitochondrial proliferation, loss of sarcomeric structure and DCM in adults	119

Lipotoxic CM, cardiac hypertrophy; cardiac dysfunction, progressive myocardial lipid accumulation, and congestive HF with decreased survival; PVT, polymorphic ventricular tachycardia; FA, fatty acid; SD, sudden death; DCM, dilated cardiomyopathy; ETC, electron transport chain; TG, triglyceride.

triglyceride accumulation, similar to that found in mice with disrupted PPAR-α. Lipid accumulation was associated with initial cardiac hypertrophy, followed by the development of left-ventricular dysfunction and premature death.[172] Lipotoxic cardiomyopathy is a significant entity associated with alterations in cellular lipid metabolism. It can occur as a result of lipid accumulation stemming from defective fat oxidation with leptin deficiency,[10] increased oxidation, increased expression of fatty acid transporters caused by PPAR-α overexpression,[84] and accumulation of more fatty acids because of expression of long-chain fatty acyl-CoA synthase.[172] In addition, mouse hearts with targeted overexpression of anchored lipoprotein lipase (LpL), the primary enzyme responsible for conversion of lipoprotein triglyceride (TG) into free fatty acids, also develop lipotoxic cardiomyopathy because of increased myocardial lipid uptake and accumulation.[173,174] Increased lipid uptake in LpL-overexpressed strains is also associated with the increased expression of genes mediating FAO (e.g., ACO, CPT-I), "heart failure" markers such as atrial natriuretic factor (ANF), brain natriuretic factor (BNP), myosin, apoptosis markers (e.g., caspase-3, cytochrome c release), a decrease in glucose transporters (GLUT1, GLUT4), and AMPK expression.[174]

Lipotoxicity and lipid accumulation may be also at least partly responsible for the development of the myocardial dysfunction in strains with type 2 diabetes and diabetic cardiomyopathy, as well as with FAO defects. The myocardial lipid accumulation that occurs with high-fat diets in the

PPAR-α null strains has been temporally correlated with cardiac contractile dysfunction, and both can be reversed by diet change. Notably, among patients with nonischemic HF, a significant subset (approximately 30%) exhibited intramyocardial lipid accumulation associated with an up-regulation of PPAR-α -regulated genes, as well as β-*MHC*, and *TNF-α*. Similarly, intramyocardial lipid and triglyceride overload in the hearts of Zucker diabetic fatty (ZDF) rats was found to be associated with contractile dysfunction and similar changes in gene expression as found in failing human hearts with lipid overload.[175] An extract derived from *Salacia oblonga* root (SOE), an anti-diabetic and anti-obesity medicine, improves hyperlipidemia in ZDF rats and possesses PPAR-α–activating properties. Chronic oral SOE administration in ZDF rats reduced myocardial triglyceride and fatty acid content and suppressed cardiac overexpression of fatty acid transporter protein-1 mRNA and protein, suggesting inhibition of increased cardiac fatty acid uptake as the basis for decreased cardiac fatty acid levels.[176] SOE treatment also inhibited the overexpression in ZDF rat heart of *PPAR-α* mRNA and protein and CPT-1, acyl-CoA oxidase, and AMPK mRNAs, thereby down-regulating myocardial FAO. This suggests that SOE ameliorates excess cardiac lipid accumulation and increased cardiac FAO in diabetes and obesity primarily by reducing cardiac fatty acid uptake and, thereby, modulating cardiac PPAR-α–mediated FAO gene transcription.

In most cases, both the precise cellular mechanism of cardiac remodeling and the molecular identity of the lipotoxic stimulus remain to be determined. Lipotoxicity has frequently been postulated to result from excess lipid oxidation,[177] or from the accumulation of a toxic lipid intermediate.[178] Cellular studies, as discussed later, have shown that palmitate, but not oleate, leads to apoptosis.[179] It remains to be determined whether this saturated fatty acid, its metabolic product, or cellular processes affected by palmitate lead to cellular dysfunction or death.

The role of RXR-α in HF has been examined in the transgenic mouse as well. *RXR-α* null mutant mice display ocular and cardiac malformations and liver developmental delay and die from HF in early embryo life. A large percentage (more than 50%) of the downstream target genes, identified by subtractive hybridization, encode proteins involved in fatty acid metabolism and electron transport, suggesting energy deficiency in the null *RXR-α* embryos. ATP content and *MCAD* mRNA were significantly lower in *RXR-α* mutant hearts than in wild-type mice. These findings suggest that defects in intermediary metabolism may be a causative factor in the *RXR-α* −/− phenotype, an embryonic form of DCM.[180]

Our understanding of the role that mitochondrial gene expression and function play in defining cardiac metabolism has also been greatly enhanced by transgenic studies. The previously discussed studies with PGC-1 overexpression[171] suggest that increased mitochondrial function and biogenesis may be fine in the neonatal heart but may result in cardiomyopathy in the adult. Disturbing mitochondrial bioenergetic function by creating null mutations in *ANT1* can also lead to mitochondrial proliferation with dysfunctional mitochondria and cardiomyopathy.[181] Similarly, null mutations in *mtTFA* in the mouse cause progressive cardiac dysfunction with depleted mtDNA and an associated decline in OXPHOS ATP production.[182] An early feature in the progression of this cardiac mitochondrial dysfunction is the activation of a fetal metabolic gene-expression program characterized by the decreased expression of FAO genes and increased expression of glycolytic genes.[183] The switch in the programming of cardiac metabolism was followed by increased myocardial mitochondrial biogenesis, which could not compensate in ATP production but rather contributes to the progression of HF, probably in a way similar to the PCG-1α overexpression in the adult heart.

Myocardial Apoptosis and Metabolism

MOLECULAR MECHANISM: FATTY ACIDS, GLUCOSE, CARDIAC APOPTOSIS, AND REMODELING

There is increasing evidence of a relationship between perturbation of fatty acid and glucose metabolism and the development of the cardiac remodeling processes. In this section, we survey some factors that likely play a contributory role in fostering this relationship.

The phospholipid cardiolipin is a component of the inner membrane associated with the mitochondrial permeability transition (PT) pore and its constituent protein ANT. Cardiolipin mediates the targeting of the proapoptotic protein tBid to mitochondria, implicating cardiolipin in the pathway for cytochrome *c* release.[194] It may also play a role in membrane permeability and proton conductance, as well as in the functioning of cytochrome *c* oxidase (complex IV).

During ischemia, oxidation of the saturated fatty acid palmitate is associated with diminished myocyte function.[195] Saturated long-chain fatty acid substrates such as palmitate (but not monounsaturated fatty acids such as oleate) readily induce apoptosis in rat neonatal cardiomyocytes.[196,197] As an early feature of palmitate-induced cardiomyocyte apoptosis, palmitate diminishes the content of the mitochondrial cardiolipin by causing a marked reduction in cardiolipin synthesis. Decreased levels of cardiolipin synthesis and cytochrome *c* release have been reported to be temporally correlated, suggesting that cardiolipin modulates the association of cytochrome *c* with the mitochondrial inner membrane.[198]

Palmitate also decreases the oxidative metabolism of fatty acids and respiratory complex III activity and has been associated with an increase in the intracellular second messenger ceramide.[195] However, recent data have shown that the ceramide apoptotic pathway, which involves ROS production, is distinct from the palmitate-induced pathway, which involves neither ROS production nor oxidative stress.[179,199,200] The modulations in FAO metabolism (e.g., CPT-I activity decline) and complex III activity in fatty acid–induced apoptosis have been shown to be downstream occurring well after cytochrome *c* release.[195,201,202]

Glucose and glucose uptake can play an important role in modulating myocardial apoptosis. Glucose uptake in cardiomyocytes reduces hypoxia-induced apoptosis.[203] Overexpression of GLUT1, to promote increased glucose uptake, also blocked the progression of apoptosis in hypoxia-treated cardiomyocytes.[204] A protective antiapoptotic role of glucose is further supported by studies in which glucose deprivation promoted myocardial apoptosis. Insulin administration attenuates cardiac I/R-induced apoptosis by activation of Akt-mediated cell-survival signaling.[205,206] In contrast to normal myocardial glucose uptake and signaling, which promote cell survival pathways, there is evidence that defective glucose uptake and hyperglycemia, as found in diabetic cardiomyopathy, can lead to increased myocardial apoptosis. Hyperglycemia induces myocyte apoptosis, cytochrome *c* release, and high levels of ROS in cardiomyocytes in culture, as well as in a mouse model of diabetes produced by streptozocin treatment.[207]

Metabolic Dysfunction and Specific Cardiovascular Disease Phenotypes

Cardiac Failure

In HF after cardiac hypertrophy, there is a major switch in myocardial bioenergetic substrate used from fatty acid to glucose. A key component and marker of the switch is the coordinate down-regulation of FAO enzymes and mRNA levels (>40%) in the human left ventricle with a concomitant increase in glucose uptake and oxidation.[208] This switch is thought to represent a reversal to a fetal energy substrate preference pattern of glucose oxidation in the heart. During the development of cardiac hypertrophy, a fetal metabolic gene program is initiated through the complicity of transcription factors that bind to regulatory elements, reducing gene expression of FAO enzymes (e.g., MCAD and CPT-I β), and is often accompanied by increased expression of ANF, BNP, and β-MHC. Although the molecular mechanisms mediating this down-regulation are not entirely understood, the participation of several nuclear receptors, intermediate metabolites, and transcription factors (e.g., SP1 and PPAR) has been implicated in the programmatic change in myocardial gene expression.[209]

The hypertrophied and failing heart becomes increasingly dependent on glucose as energy substrate. However, in the failing heart, it is unlikely that increased anaerobic glycolysis can compensate in ATP production for the decline in FAO, together with the diminished levels of high-energy phosphates, characterized by a decreased phosphocreatine (PCr):ATP ratio, resulting from declining PCr content, diminished creatine kinase activity, and mitochondrial OXPHOS dysfunction. Moreover, despite the rise in glycolysis, the rate of mitochondrial-localized pyruvate oxidation does not keep up with the increased pyruvate levels.[210] This has led to the conclusion that the failing heart is energetically, severely compromised.[211]

Cardiomyopathy

In earlier chapters, we have described defects in FAO enzymes that result in cardiomyopathy. With the exception of defects in the MTP that affect long-chain L-3 hydroxyl-acyl-CoA activity and in the protein tafazzin encoded by the *G4.5/TAZ* gene in the X-linked Barth syndrome, which are associated with DCM,[212] most of the disturbances in fatty acid metabolism are found in patients with HCM, and many of the reported mutations in fatty acid β-oxidation pathway, result in HCM rather than DCM.[213]

Specific heritable (inborn) deficiencies in fatty acid and glucose metabolism are associated with cardiomyopathy and HF. Table III details identified defects affecting fatty acid

TABLE III Inherited Fatty Acid Metabolism Defects in Clinical Cardiac Abnormalities

Specific fatty acid defect	Affected loci	Human chromosome location	References
CPT-II	*CPT2*	1p32	226
Barth syndrome	*G4.5/TAZ*	Xq28	256
MTP (also called TFP)	MTP-α (*HADHA*)	2p24.1-23.3	259
(includes LCHAD)	MTP-β (*HADHB*)	2p23	260
VLCAD	*ACADVL*	17p13	258,218
CPT-I	*CPT1*	22qter	257
Carnitine transport	OCTN2 (*SLC22A4*)	5q31	230
Carnitine translocase	CACT (*SLC25A20*)	3p21.31	236,237
MCAD	MCAD (*ACADM*)	1p31	250
FAT/CD36	*CD36*	7q11.2.	242
PGC-1α variant	PGC-1 (*PGC1*)	4p15.1	248
PPAR-α variant	PPAR-α (*PPARA*)	22q12.2–13.1	247

metabolism, which can result in clinical cardiomyopathy or HF, including both their characterized genetic loci and the specific mutation. Within the group of defects reported in the mitochondrial acyl-CoA dehydrogenase in association with cardiomyopathy and HF,[214] those affecting the oxidation of LCFAs tend to be more likely to cause cardiomyopathy than defects in medium-chain or short-chain fatty acids, although the reason underlying this is unclear.

Although severe cardiomyopathy is unusual in patients with MCAD deficiency, sudden death in children is a common outcome, and its pathogenetic mechanism is presently undetermined. MCAD deficiency is autosomally recessive and associated with sudden death and severe cardiac dysrhythmias. More than 90% of cases of MCAD deficiency are associated with a homozygous mutation at nt 985 (A985G), a mutation that directs a glutamate replacement of lysine at residue 304 in the mature MCAD subunit, resulting in a defect in MCAD tetramer assembly leading to increased protein instability.[215] MCAD deficiency represents the most frequent inborn metabolic disorder in populations of northwestern European origin.[216]

Defects in long-chain and very long-chain acyl dehydrogenases (LCHAD and VLCAD respectively) leading to cardiomyopathy are more common, albeit still rare overall.[217,218] Pediatric cardiomyopathy is the most common clinical phenotype of VLCAD deficiency, with sudden death a frequent outcome. VLCAD deficiency generally is accompanied by a marked reduction of VLCAD mRNA and decreased levels of VLCAD enzyme activity. A large number of different mutations in the *ACADVL* gene encoding VLCAD, (21 in 19 patients) with few repeated mutations have been reported thus far. At this time, the critical distinction between truly pathogenic mutations and polymorphic variations in VLCAD remains difficult to assess.

The last three steps of long-chain FAO are catalyzed by MTP. This is a multienzyme complex composed of four molecules of the α-subunit (encoded by *HADHA*) that contains both the enoyl-CoA hydratase and 3-hydroxyacyl-CoA dehydrogenase domains and four molecules of the β-subunit (encoded by *HADHB*) containing the 3-ketoacyl-CoA thiolase domain.[219]

Recessively inherited defects in MTP have recently emerged as important inborn errors of metabolism, which can cause cardiomyopathy and sudden death, particularly among infants, and severe hepatic dysfunction and neuropathy.[220] In one biochemical phenotype associated with the most common MTP mutation (G1528C), both α- and β-subunits are present, and only the long-chain 3-hydroxyacyl-CoA dehydrogenase (LCHAD) activity is affected. A second biochemical phenotype is frequently displayed by patients with neonatal cardiomyopathy and is characterized by an absence of both subunits (as detected by Western blot) and the complete lack of all three enzymatic activities of MTP.[221] This phenotype can be caused by mutations localized to the 5′ donor splicing site of the β-subunit gene, which can result in the entire loss of an exon in the mRNA (exon 3). There is considerable overlap between the clinical features found in each molecular/biochemical phenotype. In addition, a homozygous missense mutation (976G→C) within the β-subunit of MTP was reported in patients, leading to complete deficiency, acidosis, and HF.[222] This mutation destabilized the MTP protein and caused the loss of all three MTP activities and loss of the protein, as assessed by Western immunoblot. Neonatal screening for MTP defects by use of tandem mass spectrometry has been shown to be extremely effective for detecting these deficiencies that can be treated with appropriate low-fat diet (with restriction of LCFA intake and substitution with medium-chain fatty acids) and fasting avoidance.[223–225]

As we have previously noted, cardiac abnormalities, including cardiomyopathy and neonatal death, can result, albeit rarely, from mutations in the genes encoding carnitine palmitoyltransferases CPT-I and the infantile-type CPT-II as well as in *MLYCD* encoding the malonyl-CoA decarboxylase.[226] In addition, HCM can result from specific *PRKAG2* mutations in the regulatory subunit of AMPK, a critical bioenergetic sensor that activates FAO by adjusting malonyl-CoA levels and increasing glucose uptake, particularly during chronic low-energy states, such as those that occur in cardiac hypertrophy and failure as has been previously discussed.

Carnitine deficiency has been frequently associated with severe cardiomyopathy. Mutations in proteins that participate in carnitine transport and metabolism may cause either DCM or HCM as an autosomal-recessive trait.[227,228] A primary genetic *locus* affected in carnitine-associated cardiac involvement encodes the plasma membrane–localized carnitine transporter. Mutations in the *SLC22A5* gene encoding the high affinity plasma membrane–localized organic cation/carnitine transporter OCTN2 lead to primary carnitine deficiency.[229,230] Several studies have described the involvement of homozygous mutations in *SLC22A5* in children with severe cardiomyopathy associated with myopathy and in a number of sudden infant death subjects.[231,232] A homozygous deletion of 17081C of the *SLC22A5* gene resulting in a frameshift at R282D and leading ultimately to a premature stop codon (V295X) in the OCTN2 carnitine transporter was described in two non-consanguineous Hungarian gypsy children who presented with cardiomyopathy and decreased plasma carnitine levels. Interestingly, analysis of the family history revealed four sudden deaths, two of them corresponded to the classic SIDS phenotype; post-mortem tissue specimens available in three cases contained the homozygous mutation. Carnitine treatment resulted in dramatic improvement of the cardiac symptoms, echocardiographic, and ECG findings in both children. Another study documented 11 different OCTN2 mutations in 11 affected individuals and could show no correlation between residual uptake and severity of clinical presentation, suggesting that the wide phenotypic variability is likely related to exogenous stressors exacerbating carnitine deficiency.[233]

Findings have also shown that cardiac manifestations can develop even in heterozygotes most likely determined by gene-dosage levels. A heterozygous C→T mutation within exon V of OCTN2 was found in a serine 280 phenylalanine exchange in a patient with cardiomyopathy and ischemic heart disease.[231]

Defects in a second locus, the mitochondrial membrane localized carnitine-acylcarnitine translocase (CACT) encoded by *SLC25A20*, also can lead to an autosomal-recessive carnitine deficiency, cardiomyopathy, and HF.[234] Thus far, nine mutations have been identified in the CACT gene. Combined analysis of clinical, biochemical, and molecular data failed to indicate a correlation between the phenotype and the genotype.[235,236] The carnitine deficiency caused by both these gene defects, which can prove lethal in childhood, has recently been found to be reversed by intake of high-dose oral carnitine supplementation, albeit more dramatically with the *SCL22A5* mutations.[237,238] Because of its key role in FAO, carnitine has long been thought to be of benefit in other genetic or acquired disorders of energy production by enhancing FAO, removing accumulated toxic fatty acyl-CoA metabolites, or restoring the balance between free and acyl-CoA.

Other genetic defects in fatty acid metabolism have been reported in association with cardiomyopathy. A number of reports have described reduced myocardial LCFA uptake in patients with DCM and HCM.[239] Myocardial scintigraphy with iodine-123-15-(p-iodophenyl)-3-(R,S)-methylpentadecanoic acid (BMIPP) has been used to detect impairment in LCFA metabolism. Because fatty acid translocase (FAT/CD36) has been associated with a variety of normal and pathological processes, including scavenger receptor functions (uptake of apoptotic cells and of oxidized low-density lipoproteins), lipid metabolism and fatty acid transport, adhesion, angiogenesis, modulation of inflammation and atherosclerosis,[240] its potential role in cardiomyopathy has been of considerable interest. In one study of 47 patients with HCM, 11 individuals with HCM (and asymmetric septal hypertrophy) manifested defects in myocardial FAT/CD36 activity and low LCFA uptake. Another study found that CD36 was not expressed in the myocardial capillary endothelial cells in 11 patients with type I CD36 deficiency, all of whom had no myocardial BMIPP accumulation, suggesting that type I CD36 deficiency may be closely related to lack of myocardial LCFA accumulation and metabolism in the myocardium.[241] Others have found a major subset of HCM patients (40%) with either negligible (<5%) or reduced (<50%) levels of CD36 expression in patients' platelets. A number of these patients were either homozygous or heterozygous for a C-478→T substitution in the *FAT/CD36* gene.[242] In addition, a recent case study of a rare form of cardiomyopathy involving a transient left ventricular apical ballooning (termed "Takotsubo-shaped" cardiomyopathy) reported a type I CD36 deficiency, with no myocardial BMPIPP accumulation, no CD36 expression in platelets, and the heterozygous C-478T allele of *FAT/CD36*.[243]

On the other hand, other recent studies have found no linkage of reduced FAT/36 deficiency (either type I or type II) in either individuals with HCM or DCM or in individuals with a family history of cardiomyopathy, indicating that CD36 deficiency is not a uniformly characteristic factor of most inherited forms of HCM or DCM.[244,245]

Studies have also suggested that CD36 deficiency is linked with the phenotypic expression of MetSyn, frequently associated with atherosclerotic cardiovascular diseases and with diabetic cardiomyopathy.[246] These findings are consistent with the view of CD36 deficiency resulting in defective myocardial LCFA transport as a contributory risk factor that might enhance susceptibility to myocardial damage in stress (e.g., fasting), exercise, or infection.

As noted previously, numerous studies have substantiated critical roles for the global transcription regulators, PPAR-α and PGC-1, in controlling fatty acid metabolism, and for their modulation in the ischemic heart and in HF. Although defects or genetic variants in these factors and specific mutations have not yet been reported in association with cardiomyopathy, a common L162V variant in PPAR-α has recently been described in association with myocardial infarction,[247] and variants in both PPAR-α and PGC-1 genes have been described in association with MetSyn and diabetes,[248] which we will shortly examine.

Dysrhythmias

As we have previously noted, conduction defects and cardiac dysrhythmias frequently occur in patients with certain FAO defects.[249] Specifically, these electrocardiographic abnormalities have been found in patients with deficiencies in CPT-II, carnitine translocase, and MTP enzyme activities, all of which are associated with aberrant myocardial LCFA accumulation. On the other hand, patients with deficiencies in CPT-I, the primary carnitine carrier, and MCAD generally do not exhibit cardiac dysrhythmias or have LCFA accumulation (although a recent report found that MCAD defects can lead to severe dysrhythmia in adults).[250] These findings further support the notion that the accumulation of dysrhythmogenic intermediary metabolites of fatty acids (e.g., long-chain acylcarnitines) play a role in cardiac dysrhythmias and are potentially contributory to HF and sudden death.

Amphiphilic long-chain acylcarnitines possess detergent-like properties, can extensively modify membrane proteins and lipids, and have a variety of toxic effects on the electrophysiological function of the cardiac membranes, including ion transport (Na^+, Ca^{++}) and impaired gap junction activity. This is further supported by the demonstration that patients with cardiomyopathy, caused by inborn defects in carnitine translocase, have an increased incidence of cardiac dysrhythmias.[251] Moreover, accumulation of LCFA intermediates (e.g., acylcarnitine) has been implicated in the genesis of ventricular dysrhythmias during myocardial ischemia.[252] Selective blocking of CPT-I activity prevents the accumulation of potentially toxic long-chain esters during hypoxia/

ischemia, thereby reducing the risk of electrophysiological disturbance and membrane disruption.[253] Studies to define the precise site of action of these toxic long-chain intermediates within the cardiomyocytes (e.g., at the level of the plasma membrane, mitochondrial matrix, or inner membrane) may prove to be of great significance in the development of new therapies.

It is noteworthy that other metabolic storage diseases can present with cardiac dysrhythmias. The accumulation of glycosphingolipids in myocardial lysosomes resulting from an inherited deficiency in the enzyme α-galactosidase A in the X-linked Fabry disease causes a multisystemic disorder with common cardiac involvement, including dysrhythmias, conduction abnormalities, HCM, and valvular abnormalities.[254] In addition, myocardial glycogen accumulation caused by mutations in the *PRKAG2* gene resulting in a constitutively activated AMPK also produces in a familial arrhythmogenic syndrome characterized by ventricular pre-excitation (WPW syndrome), tachydysrhythmias, and progressive cardiac conduction defects.[255] The molecular mechanism by which glycogen accumulation causes this cardiac phenotype has not been determined.

Diabetes and Hyperglycemia

As should be clear from the previous discussion, the effects of diabetes and hyperglycemia on cardiac structure and function are also profound. Insulin deficiency or resistance is frequently associated with LV hypertrophy and cardiomyopathy. Deficits in insulin signaling lead to diminished myocardial glucose uptake and use, resulting in an increased reliance of the heart on FAO for energy generation. This contributes to an increased accumulation of lipid intermediates, elevated cellular acidosis, decreased cardiac efficiency, and contractile dysfunction.[261]

Moreover, hyperglycemia and diabetes affect cardiac mitochondria function directly. In animals treated with streptozocin to induce diabetes, cardiac mitochondria show pronounced swelling, increased damage, and targeting by lysosomes.[262] Mitochondria from a variety of diabetic animal models show diminished respiratory control and increased oxidative stress.[263,264] These changes in cardiac structure and function are reversed with insulin administration. Conversely, the activity of the *PPAR-α* gene regulatory pathway is increased in the diabetic heart, which relies primarily on FAO for energy production, providing further stimulus for the excessive fatty acid import and oxidation underlying the cardiac remodeling of the diabetic heart.

The genetic basis for diabetes has been a matter of tremendous interest. In several rare forms of diabetes, including neonatal diabetes and maturity-onset diabetes, a monogenic etiology has been demonstrated. As with commons forms of hypertension, the common types of diabetes (type 1 and 2) are polygenic; a large number of genes seem to be involved with extensive modulation by environmental and epigenetic factors.

Recent reports have identified discrete activating mutations in the *KCNJ11* gene, which encodes the Kir6.2 subunit of the sarcolemmal K_{ATP} channels that prevent its closure in the pancreatic β cells (and affect insulin secretion) as a primary cause of neonatal diabetes mellitus. The mutated K_{ATP} channels do not close in the presence of metabolically generated ATP, so the β-cell membrane is hyperpolarized and insulin secretion does not occur. The degree of K_{ATP} channel overactivity has been shown to correlate with the severity of the diabetic phenotype. Mutations in channel properties of Kir6.2 that underlie transient neonatal diabetes (I182V) or more severe forms of permanent neonatal diabetes (V59M, Q52R, and I296L) have been identified, which in all cases result in a significant decrease in sensitivity to inhibitory ATP correlating with channel overactivity in intact cells and increasing the K_{ATP} current, which inhibits β-cell electrical activity and insulin secretion.[265-268]

The targeted ATP-sensitive potassium channel couples membrane excitability to cellular metabolism and is a critical mediator in the process of glucose-stimulated insulin secretion. These findings have also proven applicable to studies of type 2 diabetes. A number of K_{ATP} channel polymorphisms have been described and linked to altered insulin secretion, indicating that genes encoding this ion channel could be susceptibility markers for type 2 diabetes and that genetic variants of K_{ATP} channels may underlie altered β-cell electrical activity and glucose homeostasis, in addition to increased susceptibility to type-2 diabetes. In particular, the Kir6.2 E23K polymorphism has been linked to increased susceptibility to type 2 diabetes in Caucasian populations. As with many of the genetic polymorphisms associated with diabetes (to be discussed), this polymorphism has also been associated with weight gain and obesity, both of which constitute major diabetes risk factors. Mechanistically, it has been proposed that the LCFA accumulation in the plasma and in pancreatic β cells in obese and type 2 diabetic patients elicit an enhanced stimulation of K_{ATP} channels containing subunits encoded by the polymorphic Kir6.2 E23K allele.[269-271]

In addition, loss-of-function mutations in the genes encoding the two subunits of K_{ATP} channels have been identified that reduce K_{ATP} channel activity and lead to the most common form of congenital hyperinsulinism, resulting in persistent and severe hypoglycemia in the neonatal and infancy period. Moreover, sulfonylureas, which inhibit K_{ATP} channels, can enhance insulin secretion in type 2 diabetics. This has led to their widespread use in treating patients who were insulin dependent and provides an important alternative treatment to the use of insulin injections with improved glycemic control.[272]

A second type of monogenic subtype of diabetes mellitus, maturity-onset diabetes of the young (MODY), is characterized by an early onset of type 2 diabetes (usually in children or young adults), including some abnormalities of the β-cell function and an autosomal-dominant inheritance with high penetrance.[273] MODY types represent fewer than 5%

of all cases of type 2 diabetes. Mutations in six genes have been described thus far; these different gene mutations are associated with different clinical forms of the disease. For instance, mutation in the *GCK* gene encoding glucokinase, a key regulatory glycolytic enzyme of the β cell, is found in MODY 2 patients. The other mutant loci present in MODY 1,3,4,5,6, respectively, include defects in specific transcription factors, including the hepatocyte nuclear factors-1α, -4 α, and -1β (HNF-1α, HNF-4α and HNF-1β), the insulin promoter factor-1 (IPF-1), and NeuroD.[274–276] Individuals harboring either of the two most frequently found forms of MODY 2 display a more benign clinical prognosis with an elevated threshold for glucose sensing, resulting in mild, regulated hyperglycemia or impaired glucose tolerance with relatively few cardiovascular complications. In contrast, subjects with MODY 3 (with defective HNF-1α) exhibit a much more severe disorder, more typical of type 2 diabetes and frequently require treatment with sulfonylureas or insulin.

Other monogenic forms of type 2 diabetes characterized by severe insulin resistance are the result of mutations in genes encoding PPAR-γ (*PPARG*), Akt (*AKT2*), and the insulin receptor (*INSR*).[277–280] These patients will sometimes develop discrete extrapancreatic phenotypes; for example, lipid abnormalities or a variety of cystic renal diseases.

Efforts to identify genes responsible for more common, polygenic forms of type 1 and type 2 diabetes have been less fruitful. Despite intensive research, there is still no definitive genetic test to diagnose type 1 or type 2 diabetes. Data from both animal and clinical studies have revealed multiple overlaps in the genes implicated in both types of diabetes, as well as in the pathogenic pathways including apoptotic remodeling. Moreover, both types of diabetes have been shown to be associated with significant damage to mitochondria in pancreatic β cells, liver, and heart.[281,282]

In type 2 diabetes, which accounts for approximately 85% of all diabetic patients, the body either produces too little insulin or does not respond well to it. Forms of type 2 diabetes tend to have a middle/late age of onset and occur with both impaired insulin secretion and insulin resistance. Although this type of diabetes might be predicted to be induced by multiple gene defects specifically involved in insulin action and/or insulin secretion, a variety of other genes have been at least partially implicated in its genesis as shown in Table IV. Moreover, other genetically influenced traits like obesity and hyperlipidemia are strongly associated with type 2 diabetes. Interestingly, many of the risk factors implicated in the development of type 2 diabetes, including weight gain, lack of physical exercise, and increasing age, are associated with an impaired mitochondrial function. Furthermore, recent studies have suggested that mitochondrial bioenergetic dysfunction largely underlies the defects in insulin responsiveness found in skeletal muscle and liver responsible for insulin resistance[283] as well as defects in glucose-stimulated insulin secretion by pancreatic β cells responsible for the progression to hyperglycemia.[284–287]

Genes and genetic variants thus far identified in mediating type 2 diabetes include calpain 10, PPAR-γ, *KCJN11*, and insulin. In addition, some evidence exists that genes, such as adiponectin, IRS-1, and some others, may also influence the susceptibility to type 2 diabetes. The clinical manifestations and course of this disease is fostered by the interaction of environmental and genetic factors, including frequent polymorphisms of many genes, not just one. These polymorphisms may be localized in the coding or regulatory parts of the genes and are present, although with different frequencies, in both patients and healthy individuals. A large number of pathogenic mechanisms for type 2 diabetes have been proposed, including increased nonesterified fatty acids, inflammatory cytokines, adipokines, mitochondrial dysfunction for insulin resistance, glucotoxicity, lipotoxicity, and amyloid formation for β-cell dysfunction.[288]

Multifactorial and polygenic type 1 diabetes is strongly influenced by multiple genes controlling the immune system, within the major histocompatibility complex primarily HLA-DQ and DR.[289] Another well-characterized susceptibility locus is the insulin gene, including the variable nucleotide tandem repeat locus within the regulatory region of the gene. This genetic variation affects the expression of insulin in the thymus and may play a role in the modulation of tolerance to this molecule.[290] Moreover, a significant autoimmune component has been identified in a large subset of type 1 diabetes cases, with measurable autoantibodies a useful diagnostic and prognostic marker.[291] In addition, several predisposition loci (Table IV), interacting with each other, have significant influence on the susceptibility to type 1 diabetes. These include polymorphic variants of genes involved in signal transduction, including the cytotoxic T lymphocyte–associated molecule 4 (*CTLA-4*)[292] and the *PTPN22* gene encoding the lymphoid protein tyrosine phosphatase (LYP).[293] In some populations, genomic variations of vitamin D metabolism and in target cell action predispose to type 1 diabetes. There is increasing epidemiological evidence suggesting that vitamin D deficiency in early life increases the incidence of later onset autoimmune diseases, such as type 1 diabetes, in genetically predisposed individuals, and that high-dose vitamin D supplementation can be protective against its development in both animals and humans[294,295]

METABOLIC SYNDROME AND INSULIN RESISTANCE

MetSyn includes a clustering of metabolic derangements that cause affected subjects to have an increased risk for developing diabetes, cardiovascular disease (CVD), and, according to recent epidemiological studies, chronic kidney disease.

MetSyn is characterized by visceral obesity, insulin resistance, hypertension, chronic inflammation, and thrombotic disorders contributing to endothelial dysfunction and, subsequently, to accelerated atherosclerosis. The atherogenic dyslipidemia includes the combination of hypertriglyceridemia, low levels of high-density lipoprotein cholesterol, and a preponderance of small, dense low-density lipoprotein (LDL)

TABLE IV Polymorphisms Associated with Diabetes, Insulin Resistance, and Metabolic Syndrome

Phenotype	Variants (gene)	Polymorphism	References
Type 2 diabetes	Methylenetetrahydrofolate reductase (*MTHFR*)	677C>T	296
Type 2 diabetes	Transcription-factor-activating protein 2β (*TFAP2B*)	Variable # of tandem repeats	297
Type 2 diabetes	*PGC-1α*	Thr394Thr	298
		Gly482Ser	299
Type 2 diabetes	*TNF-α*	G-308 A allele	300, 301
Type 2 diabetes	SUR1/Kir6.2/ (KCNJ11)	E23K	302
Type 2 diabetes	Calpain 10 (*CAPN10*)	SNP44	303, 304
Type 2 diabetes	*LMNA*	1908C/T	305
Type 2 diabetes	*IRS-1*	Gly972Arg	306, 307
Type 2 diabetes	Adiponectin (*APM1*)	SNP exon 2 (45T/G)	308
Type 2 diabetes	Adiponectin (*APM1*)	SNP intron 2 (276G/T)	309
Type 2 diabetes	Adiponectin (*APM1*)	Tyr111His	310
Type 2 diabetes	PPAR-γ (*PPARG*)	Pro12Ala	311–313
Type 1 diabetes	Lymphoid protein tyrosine phosphatase, (*PTPN22*)	1858T	293
Type 1 diabetes	Catatonic T-lymphocyte–associated antigen-4 (*CTLA-4*)	A49G	292, 314, 315
Type 1 diabetes	HLA class II DQ and DR alleles	Several haplotypes in DQb, DQa, and DRB1	316, 317
Type 1 diabetes	Insulin (*INS*)	Polymorphism 5′ tandem repeats	318, 319
Insulin resistance, FPLD	*LMNA*	1908C/T	320–322
		R482Q	
		R133L	
Body fat/insulin sensitivity	Protein tyrosine phosphatase 1B (*PTPN1*)	Pro387Leu	323
Insulin resistance	PPAR-α (*PPARA*)	L162V	324
Insulin resistance/premature CHD/obesity	β3-adrenoceptor (*ADRB3*)	Trp64Arg	325, 326
Insulin resistance, FPLD MetSyn	PPAR-γ (*PPARG*)	Pro12Ala	248, 327–330
Insulin resistance/MetSyn	β2-adrenoceptor (ADRB2)	Arg16Gly	331
MetSyn	*PGC-1α*	Gly482 Ser	248
MetSyn	Adiponectin (*APM1*)	I164T	332

MetSyn, metabolic syndrome; FPLD, familial partial lipodystrophy.

particles.[334] Inflammation in the vasculature has been posited as an important pathogenic link between CVDs and MetSyn.[335] Obesity is also thought to be a central component in development of MetSyn, and it is becoming increasingly clear that a primary factor is the production by adipose cells of bioactive substances that directly influence insulin sensitivity and vascular injury.[335–337]

Inflammation can be reduced by a variety of approaches, including diet, exercise, cardiovascular drugs, and insulin sensitizers. These different measures can improve vascular function and reduce inflammation by distinct mechanisms. Combination therapy along with lifestyle modifications and multiple drugs might produce additive beneficial outcomes, including the reduction of inflammation, improvement of endothelial dysfunction, and reduction of insulin resistance and hypertension in insulin-resistant states, including diabetes, obesity, and MetSyn.[338]

The confluence of several cardiovascular conditions comprising MetSyn may provide an important opportunity to fully investigate the dynamic interactions between the various components comprising the phenotype(s) and to identify the pathways leading to this syndrome. However, increasing concerns have been raised that many of those components may or may not be invariably involved in this phenotype (in particular, insulin resistance) and have suggested that the MetSyn entity may simply complicate the genetic analysis (which already is highly complex).[339] Nevertheless, many groups are seeking to identify key genes involved as risk factors in MetSyn, with many of the common polymorphic genetic variants showing some complicity, and to establish presymptomatic disease biomarkers.[340] Table IV also contains information concerning several of the candidate genetic loci that have shown some significant association with MetSyn.[248,327–332]

A significant recent development in understanding MetSyn has been the emergence of familial partial lipodystrophy (FPLD), a rare monogenic form of insulin resistance proposed as a potential model of MetSyn, with its gradual evolution and marked recapitulation of key clinical and biochemical features of MetSyn. FPLD can be caused by mutations in either *LMNA*, encoding nuclear lamin A/C (subtype FPLD2),[320–322] or in *PPAR-γ* (subtype FPLD3),[327–330] a transcription factor with a key role in adipocyte differentiation.[341,342]

Alterations in the abundance and activity of transcription factors can lead to complex dysregulation of gene expression, which is pivotal in the generation of insulin resistance–associated and its clustering of coronary risk factors at the cellular or gene regulatory level. Members of the nuclear hormone receptor superfamily, for example, peroxisome proliferator–activated receptors (PPARs), PGC-1, RXR-α, and sterol regulatory element-binding proteins (SREBPs), have all been implicated in both insulin resistance and MetSyn.[343–345] In addition to their regulation by a host of metabolites and nutrients, these transcription factors are also targets of hormones (like insulin and leptin), growth factors, inflammatory signals, and drugs. Extracellular stimuli are coupled to transcription factors by a variety of signaling pathways, including the MAP kinase cascades. For instance, SREBPs seem to be substrates of MAP kinases and have been proposed to play a contributory role in the development of cellular features belonging to lipid toxicity and MetSyn.[343] Therefore, MetSyn seems to be not only a disease or state of altered glucose tolerance, plasma lipid levels, blood pressure, and body fat distribution, but rather a complex clinical phenomenon of dysregulated gene expression.

These findings have suggested an important approach for management and treatment of MetSyn.[346–348] Both PPAR-α activators such as fibric acid class of hypolipidemic drugs and PPAR-γ agonists, including antidiabetic thiazolidinediones (TZDs), have proved to be effective for improving MetSyn. PPAR-α agonists, such as the fibrates, correct dyslipidemia, thus decreasing CVD risk, whereas PPAR-γ agonists, such as the glitazones, increase insulin sensitivity and decrease plasma glucose levels in patients with diabetes. Moreover, both PPAR-α and PPAR-γ agonists exert anti-inflammatory activities in liver, adipose, and vascular tissues.

Diagnosis and Therapy

DIAGNOSTIC ADVANCES

At the biochemical level, the diagnostic evaluation of fatty acid and glucose metabolic defects and determination of carnitine levels are easily performed. Rapid and correct diagnosis (including newborn screening that uses a noninvasive, highly sensitive method profiling specific metabolic intermediates such as acylcarnitines and carnitines by tandem mass spectrometry, on blood spot collected on a Guthrie card) is particularly critical, because dramatic recovery from or prevention of dysrhythmias and HF have been demonstrated in a number of these disorders (e.g., VLCAD deficiency).[349,350] The use of noninvasive techniques for evaluating intermediates has greatly progressed in the past few years, including the use of nuclear magnetic resonance (NMR) spectroscopy to assess glucose transport, glycogen levels, and cellular levels of triglyceride content (in skeletal muscle).[351–353] These methods allow the noninvasive evaluation of specific lipid levels, of accumulated "toxic" metabolites, and of insulin resistance. The use of genetic analysis is also available, although the presence of nonrepeating mutations makes this analysis rather problematic.

CURRENT AND FUTURE GUIDELINES: METABOLIC THERAPIES

Treatment of disorders of mitochondrial long-chain FAO is based on the avoidance of fasting, low-fat diet with the restriction of LCFA intake, frequent carbohydrate-rich feeding, and the replacement of normal dietary fat by medium-chain triglycerides. The activation of PPAR-α using lipid-lowering fibrates has also proved an effective supplement in some cases of FAO deficiency.[354]

Information regarding the precise site of the biochemical or molecular defect can be of critical importance regarding the choice of the therapeutic modality to be applied. For instance, deficiencies in CPT-II, carnitine acylcarnitine translocase, or MTP can be treated with drugs targeted to enhance glucose use and pyruvate oxidation energy, at the expense of FAO, to prevent the accumulation of long-chain acylcarnitines that can result in cardiac conduction defects and dysrhythmias.[355] In contrast, acute cardiomyopathy associated with VLCAD deficiency, which can be diagnosed by acylcarnitine analysis even in the neonatal period, can be effectively treated with the aforementioned dietary therapy, including medium-chain triglycerides.[356] The accumulation of LCFA and their side effects can also be effectively reversed by inhibition of CPT-I activity with perhexiline and amiodarone.[357] Perhexiline has also been used to reduce ventricular ectopic beats associated with chronic ischemic injury and in chronic HF.[358,359] As previously discussed, the use of carnitine supplementation in patients with carnitine deficiency may be of significant benefit in the prevention and treatment of potentially lethal disorders. In addition, gene and cell-based therapies for some of the fatty acid disturbances of cardiac function and structure to decrease LCFA intermediates and redirect metabolic programs hold great promise, albeit studies are largely limited at this time to animals.[223] In a rat model of MetSyn caused by leptin insufficiency and abnormal fat accumulation, recent studies have shown that a single central administration of recombinant adeno-associated virus vector containing the gene encoding leptin severely depletes fat and ameliorates the major symptoms of MetSyn for extended periods in rodents.[360] Recent work suggests that the deployment of endogenous hematopoietic stem cells to regenerate pancreatic β cells and produce insulin may be helpful in targeting autoimmune diabetes.[361] Other cell-based approaches include the engineering of extrapancreatic cells (using a patient's somatic cells) to secrete insulin.[362]

Metabolic therapies that use modulation of myocardial glucose and fatty acid metabolism are recognized as a potential approach in treating the failing and ischemic heart. Treatment of patients in congestive HF with carvedilol, a β-adrenoreceptor blocker, results in marked improvement in myocardial energy efficiency and a reduction in myocardial oxygen consumption by shifting myocardial oxidative substrates from fatty acid to glucose.[363] Earlier studies with β-blocker treatment indicated that along with some cardiovascular improvement in regard to both HF and hypertension, some components of the MetSyn (including glycemic control) seem to be worsened, largely because of negative effects on carbohydrate metabolism and, therefore, their use in diabetics was restricted.[364] However, nonselective β-blocker vasodilators, such as carvedilol, do not exhibit the same negative metabolic consequences with glucose use seen with the use of earlier generation β-blockers and may be used in patients with CVD and diabetes.

Free fatty acids are a primary source of energy during myocardial ischemia and can also serve to uncouple OXPHOS and increase myocardial O_2 consumption. On the other hand, inhibitors of FAO can increase glucose oxidation and may serve to improve cardiac efficiency. It is noteworthy that inhibitors of β-FAO can help to prevent the hyperglycemia that occurs in noninsulin dependent diabetes. Because the inhibition of FAO is effective in controlling abnormalities in diabetes, FAO inhibitors targeting enzymes such as CPT-I may also prove useful in the treatment of diabetic cardiomyopathy. FAO inhibition can be achieved with a number of enzymatic inhibitors, such as etomoxir, oxfenicine, perhexiline, aminocarnitine, trimetazidine, ranolazine, and dichloroacetic acid (DCA).[365–367] In animal models, etomoxir, an inhibitor of CPT-I, reversed changes in myocardial fetal gene expression, preserved cardiac function, and prevented ventricular dilatation.[367] Other studies using pacing-induced HF have shown that CPT-I inhibition mediated by oxfenicine treatment not only attenuates ventricular dilation but also significantly slows left ventricular remodeling and delays the time to end-stage failure.[368] Clinical studies of patients in HF treated with etomoxir revealed improved systolic ventricular function, increased ejection fraction, and decreased pulmonary capillary pressure.[369]

Partial inhibition of FAO has also proved to ameliorate many of the hemodynamic abnormalities associated with myocardial ischemia and HF.[370] Treatment with ranolazine mediates a partial inhibition of fatty acid oxidation (it is therefore termed a *pFOX inhibitor*), because it reduces cellular acetyl-CoA content and activates PDH activity. In dogs in which chronic HF is induced by intracoronary microembolizations, intravenous administration of ranolazine significantly increased LV ejection fraction and systolic function.[371] Clinically, ranolazine has been used to treat both ischemia and angina.[372] This metabolic switch from FAO to glucose oxidation increases ATP production, reduces the rise in lactic acidosis, and improves myocardial function under conditions of reduced myocardial oxygen delivery leading to a reduced gluconeogenesis and improved economy of cardiac work. In addition, trimetazidine treatment has been demonstrated to provide cardioprotective effects against myocardial ischemia, diabetic cardiomyopathy, and exercise-induced angina in numerous clinical and experimental investigations.[365,366,373] Particularly notable is its amelioration of cardiac function and exercise performance in diabetic patients with ischemic heart disease.[374] Although initially trimetazidine was thought to be an inhibitor of the activity of the long-chain isoform of the last enzyme involved in mitochondrial fatty acid β-oxidation, 3-ketoacyl coenzyme A thiolase,[375] recent studies have cast doubt on FAO inhibition as being the primary mechanism by which trimetazidine mediates cardiac recovery.[376] Other related cardioprotective effects of trimetazidine, which may contribute to its anti-ischemic action, are its acceleration of phospholipid synthesis and turnover with significant consequences for α-adrenergic signaling[377] and its inhibition of mitochondrial PT pore opening, resulting in the prevention of lethal ischemia-reperfusion injury.[378]

Clinical studies have suggested that polyunsaturated fatty acids (e.g., N-3 PUFA) or fish oil supplementation seems to reduce mortality and sudden death, as well as dysrhythmias, associated with HF.[379] Its effects on mortality and morbidity have recently been gauged in the GISSI HF project, a large-scale, randomized, double-blind study and shown to significantly reduce sudden death incidence, particularly evident in patients with left ventricular systolic dysfunction[380,381] A smaller study confirmed that N-3 PUFA treatment markedly reduces the incidence of both atrial and ventricular dysrhythmias.[382] Among a large assortment of PUFA-mediated effects on cardiomyocyte membrane lipid organization and function, the incorporation of n-3 PUFA (normally associated with reduced arachidonic acid) induces a reduction of mitochondrial β-FAO and oxygen consumption in the heart. These effects on mitochondrial metabolism are manifested primarily during post-ischemic reperfusion as improved metabolic and ventricular function. Both aging and ischemia markedly decrease levels of N-3 PUFA and cardiolipin in myocardial membranes, effects that have been correlated to increased mitochondrial Ca^{++} levels and the effects of Ca^{++} on mitochondrial enzymatic activities.[383]

SUMMARY

- Fatty acid use is a critical source of bioenergy for the adult heart. Fatty acid transport into the cardiomyocytes and within the cell is a highly regulated process involving specific transport proteins located in the plasma membrane, fatty acid–binding proteins, and transporters within the mitochondrial membranes.

- Defects in these protein functions can result in a variety of cardiac abnormalities, resulting in deficiencies of FAO or an accumulation of fatty acid in metabolic intermediates, which can promote a lipotoxic cardiomyopathy.
- Animal models, particularly transgenic mice, have been highly informative in the molecular analysis of metabolic disorders.
- Fatty acid oxidation, to provide energy for the downstream TCA and oxidative phosphorylation cycles, is largely localized to the mitochondrial organelle and is also present in the peroxisome. This process is governed by substrate levels, a variety of physiological stimuli, and regulators, including AMP kinase, several members of the nuclear receptor transcription factor family including the PPARs, and associated coactivators such as PGC-1.
- Regulatory factors are modulated by many physiological stimuli, as well as by pathophysiological conditions such as myocardial ischemia, HF, and diabetes.
- Mutations in genes encoding fatty acid oxidation proteins can lead to cardiomyopathy and cardiac dysrhythmias.
- Glucose transport in the heart is also facilitated by the function of specific membrane transporters (GLUT1 and GLUT4) regulated by a complex signaling pathway, including insulin and its receptor, a variety of kinase transducers, by the fatty acid oxidation process, and by other physiological stimuli. Subsequent glycolytic metabolism and glucose oxidation (most effective in the mitochondria) is also highly regulated by many of the same signals and transducers.
- The dysregulation of both glucose transport and oxidation, as well as changes in fatty acid levels and oxidation, are primary events in both in diabetes and insulin resistance.
- Rare forms of diabetes (e.g., neonatal and MODY) are a result of monogenic defects in proteins, including PPAR-γ, LMNA, Akt2, glucokinase, and a number of transcription factors. Both type 1 and 2 diabetes are polygenic disorders that involve the interaction of several polymorphic gene variants that are gradually being identified with environmental risk factors, including stressors, neurohormones, and dietary factors.
- Further pharmacogenomic studies will evaluate the interaction of these variants and their response to specific stressors, genetic backgrounds, and drugs. Insulin resistance has also been shown to be a polygenic disorder for which polymorphic variants are being identified.
- Mitochondrial pathways are increasingly being identified as significant mediators of cardiovascular damage in diabetes and insulin resistance.
- The increasingly prevalent MetSyn involves the clustering of several phenotypes, including hypertension, insulin resistance, and obesity. The identification of contributory genes and biomarkers is underway and should allow the identification of the pathways involved as well as therapeutic targets.
- Present therapeutic approaches for these metabolic defects include dietary modification, elimination of stresses (e.g., fasting), and supplementation with carnitine. Shifts in metabolic programming (from fatty acid to glucose use) using specific metabolic inhibitors and cardioprotective agents have proven beneficial in treating both specific metabolic defects and more global defects caused by myocardial ischemia, HF, and diabetes.
- Metabolic therapy that uses fibrates and agonists of PPAR has shown promise in redirecting bioenergetic pathways with beneficial cardiovascular effects. Gene and cell-mediated therapy hold promise in treating fatty acid and metabolic storage disorders, as well as diabetes and insulin resistance.

References

1. Stanley, W. C., Recchia, F. A., and Lopaschuk, G. D. (2005). Myocardial substrate metabolism in the normal and failing heart. *Physiol. Rev.* **85**, 1093–1129.
2. Glatz, J. F., and Storch, J. (2001). Unravelling the significance of cellular fatty acid binding-protein. *Curr. Opin. Lipidol.* **12**, 267–274.
3. Luiken, J. J., Coort, S. L., Willems, J., Coumans, W. A., Bonen, A., van der Vusse, G. J., and Glatz, J. F. (2003). Contraction-induced fatty acid translocase/CD36 translocation in rat cardiac myocytes is mediated through AMP-activated protein kinase signaling. *Diabetes* **52**, 1627–1634.
4. Luiken, J. J., Koonen, D. P., Willems, J., Zorzano, A., Becker, C., Fischer, Y., Tandon, N. N., Van Der Vusse, G. J., Bonen, A., and Glatz, J. F. (2002). Insulin stimulates long-chain fatty acid utilization by rat cardiac myocytes through cellular redistribution of FAT/CD36. *Diabetes* **51**, 3113–3119.
5. Chabowski, A., Coort, S. L., Calles-Escandon, J., Tandon, N. N., Glatz, J. F., Luiken, J. J., and Bonen, A. (2004). Insulin stimulates fatty acid transport by regulating expression of FAT/CD36 but not FABPpm. *Am. J. Physiol. Endocrinol. Metab.* **287**, E781–E789.
6. Bonen, A., Benton, C. R., Campbell, S. E., Chabowski, A., Clarke, D. C., Han, X. X., Glatz, J. F., and Luiken, J. J. (2003). Plasmalemmal fatty acid transport is regulated in heart and skeletal muscle by contraction, insulin and leptin, and in obesity and diabetes. *Acta Physiol. Scand.* **178**, 347–356.
7. Luiken, J. J., Arumugam, Y., Bell, R C., Calles-Escandon, J., Tandon, N. N., Glatz, J. F., and Bonen, A. (2002). Changes in fatty acid transport and transporters are related to the severity of insulin deficiency. *Am. J. Physiol. Endocrinol. Metab.* **283**, E612–E621.
8. Muoio, D. M., Dohm, G. L., Tapscott, E. B., and Coleman, R. A. (1999). Leptin opposes insulin's effects on fatty acid partitioning in muscles isolated from obese ob/ob mice. *Am. J. Physiol.* **1276**, E913–E921.
9. Luiken, J. J., Willems, J., Coort, S. L., Coumans, W. A., Bonen, A., Van Der Vusse, G. J., and Glatz, J. F. (2002). Effects of cAMP modulators on long-chain fatty-acid uptake and utilization by electrically stimulated rat cardiac myocytes. *Biochem. J.* **367**, 881–887.
10. Zhou, Y. T., Grayburn, P., Karim, A., Shimabukuro, M., Higa, M., Baetens, D., Orci, L., and Unger, R. H. (2000). Lipotoxic

heart disease in obese rats: Implications for human obesity. *Proc. Natl. Acad. Sci. USA* **97,** 1784–1789.
11. Krssak, M., Falk Petersen, K., Dresner, A., DiPietro, L., Vogel, S. M., Rothman, D. L., Roden, M., and Shulman, G. I. (1999). Intramyocellular lipid concentrations are correlated with insulin sensitivity in humans: a 1H NMR spectroscopy study. *Diabetologia* **42,** 113–116.
12. Turcotte, L. P., Swenberger, J. R., Zavitz Tucker, M., and Yee, A. J. (2001). Increased fatty acid uptake and altered fatty acid metabolism in insulin-resistant muscle of obese Zucker rats. *Diabetes* **50,** 1389–1396.
13. Young, M. E., Guthrie, P. H., Razeghi, P., Leighton, B., Abbasi, S., Patil, S., Youker, K. A., and Taegtmeyer, H. (2002). Impaired long-chain fatty acid oxidation and contractile dysfunction in the obese Zucker rat heart. *Diabetes* **51,** 2587–2595.
14. Shulman, G. I. (2000). Cellular mechanisms of insulin resistance. *J. Clin. Invest.* **106,** 171–176.
15. Hall, A. M., Smith, A. J., and Bernlohr, D. A. (2003). Characterization of the acyl CoA synthetase activity of purified murine fatty acid transport protein 1. *J. Biol. Chem.* **278,** 43008–43013.
16. Stuhlsatz-Krouper, S. M., Bennett, N. E., and Schaffer, J. E. (1998). Substitution of alanine for serine 250 in the murine fatty acid transport protein inhibits long chain fatty acid transport. *J. Biol. Chem.* **273,** 28642–28650.
17. Hatch, G. M., Smith, A. J., Xu, F. Y., Hall, A. M., and Bernlohr, D. A. (2002). FATP1 channels exogenous FA into 1,2,3-triacyl-sn-glycerol and downregulates sphingomyelin and cholesterol metabolism in growing 293 cells. *J. Lipid Res.* **43,** 1380–1389.
18. Kim, J. K., Gimeno, R. E., Higashimori, T., Kim, H. J., Choi, H., Punreddy, S., Mozell, R. L., Tan, G., Stricker-Krongrad, A., Hirsch, D. J., Fillmore, J. J., Liu, Z. X., Dong, J., Cline, G., Stahl, A., Lodish, H. F, and Shulman, G. I. (2004). Inactivation of fatty acid transport protein 1 prevents fat-induced insulin resistance in skeletal muscle. *J. Clin. Invest.* **113,** 756–763.
19. Campbell, S. E., Tandon, N. N., Woldegiorgis, G., Luiken, J. J., Glatz, J. F., and Bonen, A. (2004). A novel function for fatty acid translocase (FAT)/CD36: involvement in long chain fatty acid transfer into the mitochondria. *J. Biol. Chem.* **279,** 36235–36241.
20. McGarry, J. D., and Brown, N. F. (1997). The mitochondrial carnitine palmitoyltransferase system. From concept to molecular analysis. *Eur. J. Biochem.* **244,** 1–14.
21. Hopkins, T. A., Dyck, J. R., and Lopaschuk, G. D. (2003). AMP-activated protein kinase regulation of fatty acid oxidation in the ischaemic heart. *Biochem. Soc. Trans.* **31,** 207–212.
22. Abel, E. D. (2004). Glucose transport in the heart. *Front. Biosci.* **9,** 201–215.
23. Flier, J. S., Mueckler, M. M., Usher, P., and Lodish, H. F. (1987). Elevated levels of glucose transport and transporter messenger RNA are induced by ras or src oncogenes. *Science* **235,** 1492–1495.
24. Russell, R. R., 3rd., Li, J., Coven, D. L., Pypaert, M., Zechner, C., Palmeri, M., Giordano, F. J., Mu, J., Birnbaum, M. J., and Young, L. H. (2004). AMP-activated protein kinase mediates ischemic glucose uptake and prevents postischemic cardiac dysfunction, apoptosis, and injury. *J. Clin. Invest.* **114,** 495–503.
25. Chou, S. W., Chiu, L. L., Cho, Y. M., Ho, H. Y., Ivy, J. L., Ho, C. F., and Kuo, C. H. (2004). Effect of systemic hypoxia on GLUT4 protein expression in exercised rat heart. *Jpn. J. Physiol.* **54,** 357–363.
26. Till, M., Kolter, T., and Eckel, J. (1997). Molecular mechanisms of contraction-induced translocation of GLUT4 in isolated cardiomyocytes. *Am. J. Cardiol.* **80,** 85A–89A.
27. Rattigan, S., Appleby, G. J., and Clark, M. G. (1991). Insulin-like action of catecholamines and Ca2+ to stimulate glucose transport and GLUT4 translocation in perfused rat heart. *Biochim. Biophys. Acta* **1094,** 217–223.
28. Wojtaszewski, J. F., Higaki, Y., Hirshman, M. F., Michael, M. D., Dufresne, S. D., Kahn, C. R., and Goodyear, L. J. (1999). Exercise modulates postreceptor insulin signaling and glucose transport in muscle-specific insulin receptor knockout mice. *J. Clin. Invest.* **104,** 1257–1264.
29. Coven, D. L., Hu, X., Cong, L., Bergeron, R., Shulman, G. I., Hardie, D. G., and Young, L. H. (2003). Physiological role of AMP-activated protein kinase in the heart: graded activation during exercise. *Am. J. Physiol. Endocrinol. Metab.* **285,** E629–E636.
30. Katz, E. B., Stenbit, A. E., Hatton, K., DePinho, R., and Charron, M. J. (1995). Cardiac and adipose tissue abnormalities but not diabetes in mice deficient in GLUT4. *Nature* **377,** 151–155.
31. Zisman, A., Peroni, O. D., Abel, E. D., Michael, M. D., Mauvais-Jarvis, F., Lowell, B. B., Wojtaszewski, J. F., Hirshman, M. F., Virkamaki, A., Goodyear, L. J., Kahn, C. R., and Kahn, B. B. (2000). Targeted disruption of the glucose transporter 4 selectively in muscle causes insulin resistance and glucose intolerance. *Nat. Med.* **6,** 924–928.
32. Desrois, M., Sidell, R. J., Gauguier, D., King, L. M., Radda, G. K., and Clarke, K. (2004). Initial steps of insulin signaling and glucose transport are defective in the type 2 diabetic rat heart. *Cardiovasc. Res.* **61,** 288–296.
33. Bruning, J. C., Michael, M. D., Winnay, J. N., Hayashi, T., Horsch, D., Accili, D., Goodyear, L. J., and Kahn, C.R. (1998). A muscle-specific insulin receptor knockout exhibits features of the metabolic syndrome of NIDDM without altering glucose tolerance. *Mol. Cell* **2,** 559–569.
34. Kaczmarczyk, S. J., Andrikopoulos, S., Favaloro, J., Domenighetti, A. A., Dunn, A., Ernst, M., Grail, D., Fodero-Tavoletti, M., Huggins, C. E., Delbridge, L. M., Zajac, J. D., and Proietto, J. (2003). Threshold effects of glucose transporter-4 (GLUT4) deficiency on cardiac glucose uptake and development of hypertrophy. *J. Mol. Endocrinol.* **31,** 449–459.
35. Razeghi, P., Young, M. E., Ying, J., Depre, C., Uray, I. P., Kolesar, J., Shipley, G. L., Moravec, C. S., Davies, P. J., Frazier, O. H., and Taegtmeyer, H. (2002). Downregulation of metabolic gene expression in failing human heart before and after mechanical unloading. *Cardiology* **97,** 203–209.
36. Razeghi, P., Young, M. E., Alcorn, J. L., Moravec, C. S., Frazier, O. H., and Taegtmeyer, H. (2001). Metabolic gene expression in fetal and failing human heart. *Circulation* **104,** 2923–2931.
37. Razeghi, P., Young, M. E., Cockrill, T. C., Frazier, O. H., and Taegtmeyer, H. (2002). Downregulation of myocardial myocyte enhancer factor 2C and myocyte enhancer factor 2C-regulated gene expression in diabetic patients with nonischemic heart failure. *Circulation* **106,** 407–411.

38. Stanley, W. C., Lopaschuk, G. D., and McCormack, J. G. (1997). Regulation of energy substrate metabolism in the diabetic heart. *Cardiovasc. Res.* **34**, 25–33.
39. Nikolaidis, L. A., Sturzu, A., Stolarski, C., Elahi, D., Shen, Y. T., and Shannon, R. P. (2004). The development of myocardial insulin resistance in conscious dogs with advanced dilated cardiomyopathy. *Cardiovasc. Res.* **61**, 297–306.
40. Shah, A., and Shannon, R. P. (2003). Insulin resistance in dilated cardiomyopathy. *Rev. Cardiovasc. Med.* **4**, S50–S57.
41. Paternostro, G., Pagano, D., Gnecchi-Ruscone, T., Bonser, R. S., and Camici, P. G. (1999). Insulin resistance in patients with cardiac hypertrophy. *Cardiovasc. Res.* **42**, 246–253.
42. Nuutila, P., Maki, M., Laine, H., Knuuti, M. J., Ruotsalainen, U., Luotolahti, M., Haaparanta, M., Solin, O., Jula, A., Koivisto, V. A., Voipio-Pulkki, L. M., and Yki-Jarvinen, H. (1995). Insulin action on heart and skeletal muscle glucose uptake in essential hypertension. *J. Clin. Invest.* **96**, 1003–1009.
43. Paternostro, G., Clarke, K., Heath, J., Seymour, A. M., and Radda, G. K (1995). Decreased GLUT-4 mRNA content and insulin-sensitive deoxyglucose uptake show insulin resistance in the hypertensive rat heart. *Cardiovasc. Res.* **30**, 205–211.
44. Nascimben, L., Ingwall, J. S., Lorell, B. H., Pinz, I., Schultz, V., Tornheim, K., and Tian, R. (2004). Mechanisms for increased glycolysis in the hypertrophied rat heart. *Hypertension* **44**, 662–667.
45. Brownsey, R. W., Boone, A. N., and Allard, M. F. (1997). Actions of insulin on the mammalian heart: metabolism, pathology and biochemical mechanisms. *Cardiovasc. Res.* **34**, 3–24.
46. Vannucci, S. J., Rutherford, T., Wilkie, M. B., Simpson, I. A., and Lauder, J. M (2000). Prenatal expression of the GLUT4 glucose transporter in the mouse. *Dev. Neurosci.* **22**, 274–282.
47. Santalucia, T., Boheler, K. R., Brand, N. J., Sahye, U., Fandos, C., Vinals, F., Ferre, J., Testar, X., Palacin, M., and Zorzano, A. (1999). Factors involved in GLUT-1 glucose transporter gene transcription in cardiac muscle. *J. Biol. Chem.* **274**, 17626–17634.
48. Taegtmeyer, H., Sharma, S., Golfman, L., Razeghi, P., and van Arsdall, M. (2004). Linking gene expression to function: metabolic flexibility in normal and diseased heart. *Ann. N. Y. Acad. Sci.* **1015**, 1–12.
49. Young, L. H., Coven, D. L., and Russell, R. R., 3rd. (2000). Cellular and molecular regulation of cardiac glucose transport. *J. Nucl. Cardiol.* **7**, 267–276.
50. Depre, C., Shipley, G. L., Chen, W., Han, Q., Doenst, T., Moore, M. L., Stepkowski, S., Davies, P. J., and Taegtmeyer, H. (1998). Unloaded heart in vivo replicates fetal gene expression of cardiac hypertrophy. *Nat. Med.* **4**, 1269–1275.
51. Doenst, T., Goodwin, G. W., Cedars, A. M., Wang, M., Stepkowski, S., and Taegtmeyer. H. (2001). Load-induced changes in vivo alter substrate fluxes and insulin responsiveness of rat heart in vitro. *Metabolism* **50**, 1083–1090.
52. Liang, X., Le, W., Zhang, D., and Schulz, H. (2001). Impact of the intramitochondrial enzyme organization on fatty acid organization. *Biochem. Soc. Trans.* **29**, 279–282.
53. Jackson, S., Schaefer, J., Middleton, B., and Turnbull, D. M. (1995). Characterization of a novel enzyme of human fatty acid beta-oxidation: A matrix-associated, mitochondrial 2-enoyl CoA hydratase. *Biochem. Biophys. Res. Commun.* **214**, 247–253.
54. Poole, R. C., and Halestrap, A. P. (1993). Transport of lactate and other monocarboxylates across mammalian plasma membranes. *Am. J. Physiol.* **264**, C761–C782.
55. Printz, R. L., Koch, S., Potter, L. R., O'Doherty, R. M., Tiesinga, J. J., Moritz, S., and Granner, D. K. (1993). Hexokinase II mRNA and gene structure, regulation by insulin, and evolution. *J. Biol. Chem.* **268**, 5209–5219.
56. Uyeda, K. (1979). Phosphofructokinase. *Adv. Enzymol. Relat. Areas Mol. Biol.* **48**, 193–244.
57. Hue, L., and Rider, M. H. (1987). Role of fructose 2,6-bisphosphate in the control of glycolysis in mammalian tissues. *Biochem. J.* **245**, 313–324.
58. Depre, C., Rider, M. H., Veitch, K., and Hue, L. (1993). Role of fructose 2,6-bisphosphate in the control of heart glycolysis. *J. Biol. Chem.* **268**, 13274–13279.
59. Kiffmeyer, W. R., and Farrar, W. W. (1991). Purification and properties of pig heart pyruvate kinase. *J. Protein Chem.* **10**, 585–591.
60. Sugden, M. C., Langdown, M. L., Harris, R. A., and Holness, M. J. (2000). Expression and regulation of pyruvate dehydrogenase kinase isoforms in the developing rat heart and in adulthood: Role of thyroid hormone status and lipid supply. *Biochem. J.* **352**, 731–738.
61. Denton, R. M., McCormack, J. G., Rutter, G. A., Burnett, P., Edgell, N. J., Moule, S. K., and Diggle, T. A. (1996). The hormonal regulation of pyruvate dehydrogenase complex. *Adv. Enzyme Regul.* **36**, 183–198.
62. Huang, B., Wu, P., Popov, K. M., and Harris, R. A. (2003). Starvation and diabetes reduce the amount of pyruvate dehydrogenase phosphatase in rat heart and kidney. *Diabetes* **52**, 1371–1376.
63. Opie, L. H., and Sack, M. N. (2002). Metabolic plasticity and the promotion of cardiac protection in ischemia and ischemic preconditioning. *J. Mol. Cell Cardiol.* **34**, 1077–1089.
64. Goodwin, G. W., and Taegtmeyer, H. (2000). Improved energy homeostasis of the heart in the metabolic state of exercise. *Am. J. Physiol. Heart Circ. Physiol.* **279**, H1490–H1501.
65. Shelley, H. J. (1961). Cardiac glycogen in different species before and after birth. *Br. Med. Bull.* **17**, 137–156.
66. Schneider, C. A., Nguyên, V. T. B., and Taegtmeyer, H. (1991). Feeding and fasting determine postischemic glucose utilization in isolated working rat hearts. *Am. J. Physiol.* **260**, H542–H548.
67. Moule, S. K., and Denton, R. M. (1997). Multiple pathways involved in the metabolic effects of insulin. *Am. J. Cardiol.* **80**, 41A–49A.
68. Goodwin, G. W., Arteaga, J. R., and Taegtmeyer, H. (1995). Glycogen turnover in the isolated working rat heart. *J. Biol. Chem.* **270**, 9234–9240.
69. Morgan, H. E., and Parmeggiani, A. (1964). Regulation of glycogenolysis in muscle, II: control of glycogen phosphorylase reaction in isolated perfused heart. *J. Biol. Chem.* **239**, 2435–2439.
70. Goodwin, G., Ahmad, F., and Taegtmeyer, H. (1996). Preferential oxidation of glycogen in isolated working rat heart. *J. Clin. Invest.* **97**, 1409–1416.
71. Stanley, W. C., Lopaschuk, G. D., Hall, J. L., and McCormack, J. G. (1997). Regulation of myocardial carbohydrate metabolism under normal and ischaemic conditions. Potential for pharmacological interventions. *Cardiovasc. Res.* **33**, 243–257.

72. Gollob, M. H. (2003). Glycogen storage disease as a unifying mechanism of disease in the PRKAG2 cardiac syndrome. *Biochem. Soc. Trans.* **31,** 228–231.

73. Longnus, S. L., Wambolt, R. B., Parsons, H. L., Brownsey, R. W., and Allard, M. F. (2003). 5-Aminoimidazole-4-carboxamide 1-beta-D-ribofuranoside (AICAR) stimulates myocardial glycogenolysis by allosteric mechanisms. *Am. J. Physiol. Regul. Integr. Comp. Physiol.* **284,** R936–R944.

74. Polekhina, G., Gupta, A., Michell, B. J., van Denderen, B., Murthy, S., Feil, S. C., Jennings, I. G., Campbell, D. J., Witters, L. A., Parker, M. W., Kemp, B. E., and Stapleton, D. (2003). AMPK beta subunit targets metabolic stress sensing to glycogen. *Curr. Biol.* **13,** 867–871.

75. Reszko, A. E., Kasumov, T., David, F., Jobbins, K. A., Thomas, K. R., Hoppel, C. L., Brunengraber, H., and Des Rosiers, C. (2004). Peroxisomal fatty acid oxidation is a substantial source of the acetyl moiety of malonyl-CoA in rat heart. *J. Biol. Chem.* **279,** 19574–19579.

76. Zhou, L., Salem, J. E., Saidel, G. M., Stanley, W. C., and Cabrera, M. E. (2005). Mechanistic model of cardiac energy metabolism predicts localization of glycolysis to cytosolic subdomain during ischemia. *Am. J. Physiol. Heart Circ. Physiol.* **288,** H2400–H2411.

77. Srere, P. A., Sumegi, B., and Sherry, A. D. (1987). Organizational aspects of the citric acid cycle. *Biochem. Soc. Symp.* **54,** 173–178.

78. Ishibashi, S. (1999). Cooperation of membrane proteins and cytosolic proteins in metabolic regulation—involvement of binding of hexokinase to mitochondria in regulation of glucose metabolism and association and complex formation between membrane proteins and cytosolic proteins in regulation of active oxygen production. *Yakugaku. Zasshi.* **119,** 16–34.

79. Barger, P. M., and Kelly, D. P. (2000). PPAR signaling in the control of cardiac energy metabolism. *Trends Cardiovasc. Med.* **10,** 238–245.

80. Finck, B. N., and Kelly, D. P. (2002). Peroxisome proliferator-activated receptor alpha (PPARalpha) signaling in the gene regulatory control of energy metabolism in the normal and diseased heart. *J. Mol. Cell Cardiol.* **34,** 1249–1257.

81. Huss, J. M., and Kelly, D. P. (2004). Nuclear receptor signaling and cardiac energetics. *Circ. Res.* **95,** 568–578.

82. Braissant, O., Foufelle, F., Scotto, C., Dauca, M., and Wahli, W. (1996). Differential expression of peroxisome proliferator-activated receptors (PPARs): tissue distribution of PPAR-alpha, -beta, and -gamma in the adult rat. *Endocrinology* **137,** 354–366.

83. Escher, P., Braissant, O., Basu-Modak, S., Michalik, L., Wahli, W., and Desvergne, B. (2001). Rat PPARs: Quantitative analysis in adult rat tissues and regulation in fasting and refeeding. *Endocrinology* **142,** 4195–4202.

84. Finck, B., Lehman, J. J., Leone, T. C., Welch, M. J., Bennett, M. J., Kovacs, A., Han, X., Gross, R. W., Kozak, R., Lopaschuk, G. D., and Kelly, D. P. (2002). The cardiac phenotype induced by PPARalpha overexpression mimics that caused by diabetes mellitus. *J. Clin. Invest.* **109,** 121–130.

85. Schoonjans, K., Peinado-Onsurbe, A. M., Heyman, R. A., Briggs, M., Deeb, S., Staels, B., and Auwerx, J. (1996). PPARalpha and PPARgamma activators direct a distinct tissue-specific transcriptional response via a PPRE in the lipoprotein lipase gene. *EMBO J.* **15,** 5336–5348.

86. van der Lee, K. A. J. M., Vork, M. M., de Vries, J E., Willemsen, P. H. M., Glatz, J. F. C., Reneman, R. S., Van der Vusse, G. J., and Van Bilsen, M. (2000). Long-chain fatty acid-induced changes in gene expression in neonatal cardiac myocytes. *J. Lipid Res.* **41,** 41–47.

87. Martin, G., Schoonjans, K., Lefebvre, A. M., Staels, B., and Auwerx, J. (1997). Coordinate regulation of the expression of the fatty acid transport protein and acyl-CoA synthetase genes by PPARalpha and PPARgamma activators. *J. Biol. Chem.* **272,** 28210–28217.

88. Sugden, M. C., Bulmer, K., Gibbons, G. F., and Holness, M. J. (2001). Role of peroxisome proliferator-activated receptor-alpha in the mechanism underlying changes in renal pyruvate dehydrogenase kinase isoform 4 protein expression in starvation and after refeeding. *Arch. Biochem. Biophys.* **395,** 246–252.

89. Wu, P., Peters, J. M., and Harris, R. A. (2001). Adaptive increase in pyruvate dehydrogenase kinase 4 during starvation is mediated by peroxisome proliferator-activated receptor alpha. *Biochem. Biophys. Res. Commun.* **287,** 391–396.

90. Brandt, J. M., Djouadi, F., and Kelly, D. P. (1998). Fatty acids activate transcription of the muscle carnitine palmitoyltransferase I gene in cardiac myocytes via the peroxisome proliferator-activated receptor alpha. *J. Biol. Chem.* **273,** 23786–23792.

91. Mascaro, C., Acosta, E., Ortiz, J. A., Marrero, P. F., Hegardt, F. G., and Haro, D. (1998). Control of human muscle-type carnitine palmitoyltransferase Iß gene promoters by fatty acid enzyme substrate. *J. Biol. Chem.* **273,** 32901–32909.

92. Zhang, B. W., Marcus, S. L., Sajjadi, F. G., Alvares, K., Reddy, J. K., Subramani, S., Rachubinski, R. A., and Capone, J. P. (1992). Identification of a peroxisome proliferator-responsive element upstream of the gene encoding rat peroxisomal enoyl-CoA hydratase/3-hydroxyacyl-CoA dehydrogenase. *Proc. Natl. Acad. Sci. USA* **85,** 7541–7545.

93. Bardot, O., Aldridge, T. C., Latruffe, N., and Green, S. (1993). PPAR-RXR heterodimer activates a peroxisome proliferator response element upstream of the bifunctional enzyme gene. *Biochem. Biophys. Res. Commun.* **19,** 237–245.

94. Osumi, T., Wen, J. K., and Hashimoto, T. (1991). Two cis-acting regulatory sequences in the peroxisome proliferator-responsive enhancer region of the rat acyl-CoA oxidase gene. *Biochem. Biophys. Res. Commun.* **175,** 866–871.

95. Dreyer, C., Krey, G., Keller, H., Givel, F., Helftenbein, G., and Wahli, W. (1992). Control of the peroxisomal ß-oxidation pathway by a novel family of nuclear hormone receptors. *Cell* **68,** 879–887.

96. Gulick, T., Cresci, S., Caira, T., Moore, D. D., and Kelly, D. P. (1994). The peroxisome proliferator activated receptor regulates mitochondrial fatty acid oxidative enzyme gene expression. *Proc. Natl. Acad. Sci. USA* **91,** 11012–11016.

97. Campbell, F. M., Kozak, R., Wagner, A., Altarejos, J. Y., Dyck, J. R. B., Belke, D. D., Severson, D. L., Kelly, D. P., and Lopaschuk, G. D. (2002). A role for PPARalpha in the control of cardiac malonyl-CoA levels: reduced fatty acid oxidation rates and increased glucose oxidation rates in the hearts of mice lacking PPARalpha are associated with higher concentrations of malonyl-CoA and reduced expression of malonyl-CoA decarboxylase. *J. Biol. Chem.* **277,** 4098–4103.

98. Young, M. E., Patil, S., Ying, J., Depre, C., Singh Ahuja, H., Shipley, G. L., Stepkowski, S. M., Davies, P. J. A., and

Taegtmeyer, H. (2001). Uncoupling protein 3 transcription is regulated by peroxisome proliferator-activated receptor alpha in the adult rodent heart. *FASEB J.* **15**, 833–845.

99. Murray, A. J., Panagia, M., Hauton, D., Gibbons, G. F., and Clarke, K. (2005). Plasma free fatty acids and peroxisome proliferator-activated receptor {alpha} in the control of myocardial uncoupling protein levels. *Diabetes* **54**, 3496–3502.

100. Kelly, D. P. (2003). PPARs of the heart: Three is a crowd. *Circ. Res.* **92**, 482–484.

101. Forman, B. M., Chen, J., and Evans, R. M. (1997). Hypolipidemic drugs, polyunsaturated fatty acids, and eicosanoids are ligands for peroxisome proliferator-activated receptors alpha and delta. *Proc. Natl. Acad. Sci. USA* **94**, 4312–4317.

102. Devchand, P. R., Keller, H., Peters, J. M., Vazquez, M., Gonzalez, F. J., and Wahli, W. (1996). The PPARalpha-leukotriene B4 pathway to inflammation control. *Nature* **384**, 39–43.

103. Tenenbaum, A., Motro, M., and Fisman, E. Z. (2005). Dual and pan-peroxisome proliferator-activated receptors (PPAR) co-agonism: the bezafibrate lessons. *Cardiovasc. Diabetol.* **4**, 14.

104. Lehman, J. J., and Kelly, D. P. (2002). Gene regulatory mechanisms governing energy metabolism during cardiac hypertrophic growth. *Heart Fail. Rev.* **7**, 175–185.

105. Barger, P. M., Brandt, J. M., Leone, T. C., Weinheimer, C. J., and Kelly, D. P. (2000). Deactivation of peroxisome proliferator-activated receptor-alpha during cardiac hypertrophic growth. *J. Clin. Invest.* **105**, 1723–1730.

106. Diradourian, C., Girard, J., and Pegorier, J. P. (2005). Phosphorylation of PPARs: from molecular characterization to physiological relevance. *Biochimie.* **87**, 33–38.

107. Bronner, M., Hertz, R., and Bar-Tana, J. (2004). Kinase-independent transcriptional co-activation of peroxisome proliferator-activated receptor alpha by AMP-activated protein kinase. *Biochem. J.* **384**, 295–305.

108. Blanquart, C., Mansouri, R., Paumelle, R., Fruchart, J. C., Staels, B., and Glineur, C. (2004). The protein kinase C signaling pathway regulates a molecular switch between transactivation and transrepression activity of the peroxisome proliferator-activated receptor alpha. *Mol. Endocrinol.* **18**, 1906–1918.

109. Oberkofler, H., Esterbauer, H., Linnemayr, V., Strosberg, A. D., Krempler, F., and Patsch, W. (2002). Peroxisome proliferator-activated receptor (PPAR) gamma coactivator-1 recruitment regulates PPAR subtype specificity. *J. Biol. Chem.* **277**, 16750–16757.

110. Lazennec, G., Canaple, L., Saugy, D., and Wahli, W. (2000). Activation of peroxisome proliferator-activated receptors (PPARs) by their ligands and protein kinase A activators. *Mol. Endocrinol.* **14**, 1962–1975.

111. Passilly, P., Schohn, H., Jannin, B., Cherkaoui Malki, M., Boscoboinik, D., Dauca, M., and Latruffe, N. (1999). Phosphorylation of peroxisome proliferator-activated receptor alpha in rat Fao cells and stimulation by ciprofibrate. *Biochem. Pharmacol.* **58**, 1001–1008.

112. Yue, T. L., Bao, W., Jucker, B. M., Gu, J. L., Romanic, A. M., Brown, P. J., Cui, J., Thudium, D. T., Boyce, R., Burns-Kurtis, C. L., Mirabile, R. C., Aravindhan, K., and Ohlstein, E. H. (2003). Activation of peroxisome proliferator-activated receptor-alpha protects the heart from ischemia/reperfusion injury. *Circulation* **108**, 2393–2399.

113. Wayman, N. S., Hattori, Y., McDonald, M. C., Mota-Filipe, H., Cuzzocrea, S., Pisano, B., Chatterjee, P. K., and Thiemermann, C. (2002). Ligands of the peroxisome proliferator-activated receptors (PPAR-gamma and PPAR-alpha) reduce myocardial infarct size. *FASEB J.* **16**, 1027–1040.

114. Panagia, M., Gibbons, G. F., Radda, G. K., and Clarke, K. (2005). PPAR-alpha activation required for decreased glucose uptake and increased susceptibility to injury during ischemia. *Am. J. Physiol. Heart Circ. Physiol.* **288**, H2677–H2683.

115. Sambandam, N., Morabito, D., Wagg, C., Finck, B. N., Kelly, D. P., and Lopaschuk, G. D. (2005). Chronic activation of PPAR{alpha} is detrimental to cardiac recovery following ischemia. *Am. J. Physiol. Heart Circ. Physiol.* **290**, H87–H95.

116. Thiemermann, C. (2004). Ligands of the peroxisome proliferator-activated receptor-gamma and heart failure. *Br. J. Pharmacol.* **142**, 1049–1051.

117. Meng, H., Li, H., Zhao, J. G., and Gu, Z. L. (2005). Differential expression of peroxisome proliferator-activated receptors alpha and gamma gene in various chicken tissues. *Domest. Anim. Endocrinol.* **28**, 105–110.

118. Puigserver, P., Wu, Z., Park, C. W., Graves, R., Wright, M., and Spiegelman, B. M. (1998). A cold-inducible coactivator of nuclear receptors linked to adaptive thermogenesis. *Cell* **92**, 829–839.

119. Lehman, J. J., Barger, P. M., Kovacs, A., Saffitz, J. E., Medeiros, D., and Kelly, D. P. (2000). PPARgamma coactivator-1 (PGC-1) promotes cardiac mitochondrial biogenesis. *J. Clin. Invest.* **106**, 847–856.

120. Herzig, S., Long, F., Jhala, U. S., Hedrick, S., Quinn, R., Bauer, A., Rudolph, D., Schutz, G., Yoon, C., Puigserver, P., Spiegelman, B., and Montminy, M. (2001). CREB regulates hepatic gluconeogenesis through the coactivator PGC-1. *Nature* **413**, 179–183.

121. Michael, L. F., Wu, Z., Cheatham, R. B., Puigserver, P., Adelmant, G., Lehman, J. J., Kelly, D. P., and Spiegelman, B. M. (2001). Restoration of insulin-sensitive glucose transporter (GLUT4) gene expression in muscle cells by the transcriptional coactivator PGC-1. *Proc. Natl. Acad. Sci. USA* **98**, 3820–3825.

122. Puigserver, P., Rhee, J., Lin, J., Wu, Z., Yoon, J. C., Zhang, C.-Y., Krauss, S., Mootha, V. K., Lowell, B. B., and Spiegelman, B. M. (2001). Cytokine stimulation of energy expenditure through p38 MAP kinase activation of PPARgamma coactivator-1. *Mol. Cell* **8**, 971–982.

123. Wu, H., Kanatous, S. B., Thurmond, F. A., Gallardo, T., Isotani, E., Bassel-Duby, R., and Williams, R. S. (2002). Regulation of mitochondrial biogenesis in skeletal muscle by CaMK. *Science* **296**, 349–352.

124. Handschin, C., Rhee, J., Lin, J., Tam, P. T., and Spiegelman, B. M. (2003). An autoregulatory loop controls peroxisome proliferator-activated receptor gamma coactivator 1 expression in muscle. *Proc. Natl. Acad. Sci. USA* **100**, 7111–7116.

125. Boss, O., Bachman, E., Vidal-Puig, A., Zhang, C.-Y., Peroni, O., and Lowell, B. B. (1999). Role of the beta(3)-adrenergic receptor and/or a putative beta(4)-adrenergic receptor on the expression of uncoupling proteins and peroxisome proliferator-activated receptor-gamma coactivator-1. *Biochem. Biophys. Res. Commun.* **261**, 870–876.

126. Nisoli, E., Clementi, E., Paolucci, C., Cozzi, V., Tonello, C., Sciorati, C., Bracale, R., Valerio, A., Francolini, M., Moncada, S., and Carruba, M. O. (2003). Mitochondrial biogenesis in mammals: The role of endogenous nitric oxide. *Science* **299,** 896–899.

127. Ojuka, E. O. (2004). Role of calcium and AMP kinase in the regulation of mitochondrial biogenesis and GLUT4 levels in muscle. *Proc. Nutr. Soc.* **63,** 275–278.

128. Lin, J., Puigserver, P., Donovan, J., Tarr, P., and Spiegelman, B. M. (2002). Peroxisome proliferator-activated receptor gamma coactivator 1beta (PGC-1beta), a novel PGC-1-related transcription coactivator associated with host cell factor. *J. Biol. Chem.* **277,** 1645–1648.

129. Lin, J., Tarr, P. T., Yang, R., Rhee, J., Puigserver, P., Newgard, C. B., and Spiegelman, B. M. (2003). PGC-1beta in the regulation of hepatic glucose and energy metabolism. *J. Biol. Chem.* **278,** 30843–30848.

130. Andersson, U., and Scarpulla, R. C. (2001). Pgc-1-related coactivator, a novel, serum-inducible coactivator of nuclear respiratory factor 1-dependent transcription in mammalian cells. *Mol. Cell Biol.* **21,** 3738–3749.

131. Puigserver, P., and Spiegelman, B. M. (2003). Peroxisome proliferator-activated receptor-gamma coactivator 1 alpha (PGC-1 alpha): transcriptional coactivator and metabolic regulator. *Endocr. Rev.* **24,** 78–90.

132. Heery, D. M., Kalkhoven, E., Hoare, S., and Parker, M. G. (1997). A signature motif in transcriptional co-activators mediates binding to nuclear receptors. *Nature* **387,** 733–736.

133. Tcherepanova, I., Puigserver, P., Norris, J. D., Spiegelman, B. M., and McDonnell, D. P. (2000). Modulation of estrogen receptor-alpha transcriptional activity by the coactivator PGC-1. *J. Biol. Chem.* **275,** 16302–16308.

134. Delerive, P., Wu, Y., Burris, T. P., Chin, W. W., and Suen, C. S. (2001). PGC-1 functions as a transcriptional coactivator for the retinoid X receptors. *J. Biol. Chem.* **277,** 3913–3917.

135. Knutti, D., Kaul, A., and Kralli, A. (2000). A tissue-specific coactivator of steroid receptors, identified in a functional genetic screen. *Mol. Cell Biol.* **20,** 2411–2422.

136. Yoon, J. C., Puigserver, P., Chen, G., Donovan, J., Wu, Z., Rhee, J., Adelmant, G., Stafford, J., Kahn, C. R., Granner, D. K., Newgard, C. B., and Spiegelman, B. M. (2001). Control of hepatic gluconeogenesis through the transcriptional coactivator PGC-1. *Nature* **413,** 131–138.

137. Wu, Z., Puigserver, P., Andersson, U., Zhang, C., Adelmant, G., Mootha, V., Troy, A., Cinti, S., Lowell, B. B., Scarpulla, R. C., and Spiegelman, B. M. (1999). Mechanisms controlling mitochondrial biogenesis and respiration through the thermogenic coactivator PGC-1. *Cell* **98,** 115–124.

138. Michael, L. F., Wu, Z., Cheatham, R. B., Puigserver, P., Adelmant, G., Lehman, J. J., Kelly, D. P., and Spiegelman, B. M. (2001). Restoration of insulin-sensitive glucose transporter (GLUT4) gene expression in muscle cells by the transcriptional coactivator PGC-1. *Proc. Natl. Acad. Sci. USA* **98,** 3820–3825.

139. Struhln K. (1998). Histone acetylation and transcriptional regulatory mechanisms. *Genes Dev.* **12,** 599–606.

140. Naar, A. M., Lemon, B. D., and Tjian, R. (2001). Transcriptional coactivator complexes. *Annu. Rev. Biochem.* **70,** 475–501.

141. Puigserver, P., Adelmant, G., Wu, Z., Fan, M., Xu, J., O'Malley, B., and Spiegelman, B. M. (1999). Activation of PPARgamma coactivator-1 through transcription factor docking. *Science* **286,** 1368–1371.

142. Arany, Z., He, H., Lin, J., Hoyer, K., Handschin, C., Toka, O., Ahmad, F., Matsui, T., Chin, S., Wu, P. H., Rybkin, I. I., Shelton, J. M., Manieri, M., Cinti, S., Schoen, F. J., Bassel-Duby, R., Rosenzweig, A., Ingwall, J. S., and Spiegelman, B. M. (2005). Transcriptional coactivator PGC-1 alpha controls the energy state and contractile function of cardiac muscle. *Cell Metab.* **1,** 259–271.

143. Sack, M. N., Disch, D. L., Rockman, H. A., and Kelly, D. P. (1997). A role for Sp and nuclear receptor transcription factors in a cardiac hypertrophic growth program. *Proc. Natl. Acad. Sci. USA* **94,** 6438–6443.

144. Vega, R. B., Huss, J. M., and Kelly, D. P. (2000). The coactivator PGC-1 cooperates with peroxisome proliferator-activated receptor alpha in transcriptional control of nuclear genes encoding mitochondrial fatty acid oxidation enzymes. *Mol. Cell Biol.* **20,** 1868–1876.

145. Planavila, A., Sanchez, R. M., Merlos, M., Laguna, J. C., and Vazquez-Carrera, M. (2005). Atorvastatin prevents peroxisome proliferator-activated receptor gamma coactivator-1 (PGC-1) downregulation in lipopolysaccharide-stimulated H9c2 cells. *Biochim. Biophys. Acta* **1736,** 120–127.

146. Huss, J. M., Kopp, R. P., and Kelly, D. P. (2002). PGC-1 coactivates the cardiac-enriched nuclear receptors estrogen-related receptor-alpha and -gamma. *J. Biol. Chem.* **277,** 40265–40274.

147. Willy, P. J., Murray, I. R., Qian, J., Busch, B. B., Stevens, W. C., Jr., Martin, R., Mohan, R., Zhou, S., Ordentlich, P., Wei, P., Sapp, D. W., Horlick, R. A., Heyman, R. A., and Schulman, I. G. (2004). Regulation of PPARgamma coactivator 1alpha (PGC-1alpha) signaling by an estrogen-related receptor alpha (ERRalpha) ligand. *Proc. Natl. Acad. Sci. USA* **101,** 8912–8917.

148. Heard, D. J., Norbu, P. L., Holloway, J., and Vissing, H. (2000). Human ERR, a third member of the estrogen receptor-related receptor (ERR) subfamily of orphan nuclear receptors: Tissue-specific isoforms are expressed during development and in the adult. *Mol. Endocrinol.* 14, 383–392.

149. Kelly, D. P., and Scarpulla, R. C. (2004). Transcriptional regulatory circuits controlling mitochondrial biogenesis and function. *Genes Dev.* **18,** 357–368.

150. Larsson, N.-G., Wang, J. M., Wilhelmsson, H., Oldfors, A., Rustin, P., Lewandoski, M., Barsh, G. S., and Clayton, D. A. (1998). Mitochondrial transcription factor A is necessary for mtDNA maintenance and embryogenesis in mice. *Nat. Genet.* **18,** 231–236.

151. Garesse, R., and Vallejo, C. G. (2001). Animal mitochondrial biogenesis and function: a regulatory cross-talk between two genomes. *Gene* **263,** 1–16.

152. Scarpulla, R. C. (2002). Nuclear activators and coactivators in mammalian mitochondrial biogenesis. *Biochim. Biophys. Acta* **1576,** 1–14.

153. McClure, T. D., Young, M. E., Taegtmeyer, H., Ning, X. H., Buroker, N. E., Lopez-Guisa, J., and Portman, M. A. (2005). Thyroid hormone interacts with PPARalpha and PGC-1 during mitochondrial maturation in sheep heart. *Am. J. Physiol. Heart Circ. Physiol.* **289,** H2258–H2264.

154. Goldenthal, M. J., Ananthakrishnan, R., and Marin-Garcia, J. (2005). Nuclear-mitochondrial cross-talk in cardiomyocyte

T3 signaling: a time-course analysis. *J. Mol. Cell Cardiol.* **39**, 319–326.

155. Irrcher, I., Adhihetty, P. J., Sheehan, T., Joseph, A. M., and Hood, D. A. (2003). PPARgamma coactivator-1alpha expression during thyroid hormone- and contractile activity-induced mitochondrial adaptations. *Am. J. Physiol. Cell Physiol.* **284**, C1669–C1677.

156. Mootha, V. K., Handschin, C., Arlow, D., Xie, X., St. Pierre, J., Sihag, S., Yang, W., Altshuler, D., Puigserver, P., Patterson, N., Willy, P. J., Schulman, I. G., Heyman, R. A., Lander, E. S., and Spiegelman, B. M. (2004). Errα and Gabpa/b specify PGC-1α-dependent oxidative phosphorylation gene expression that is altered in diabetic muscle. *Proc. Natl. Acad. Sci. USA* **101**, 6570–6575.

157. Schreiber, S. N., Emter, R., Hock, M. B., Knutti, D., Cardenas, J., Podvinec, M., Oakeley, E. J., and Kralli, A. (2004). The estrogen-related receptor alpha (ERRalpha) functions in PPARgamma coactivator 1alpha (PGC-1alpha)-induced mitochondrial biogenesis. *Proc. Natl. Acad. Sci. USA* **101**, 6472–6477.

158. Dressel, U., Allen, T. L., Pippal, J. B., Rohde, P. R., Lau, P., and Muscat, G. E. (2003). The peroxisome proliferator-activated receptor beta/delta agonist, GW501516, regulates the expression of genes involved in lipid catabolism and energy uncoupling in skeletal muscle cells. *Mol. Endocrinol.* **17**, 2477–2493.

159. Shuldiner, A. R., and McLenithan, J. C. (2004). Genes and pathophysiology of type 2 diabetes: more than just the Randle cycle all over again. *J. Clin. Invest.* **114**, 1414–1417.

160. Patti, M. E., Butte, A. J., Crunkhorn, S., Cusi, K., Berria, R., Kashyap, S., Miyazaki, Y., Kohane, I., Costello, M., Saccone, R., Landaker, E. J., Goldfine, A. B., Mun, E., DeFronzo, R., Finlayson, J., Kahn, C. R., and Mandarino, L. J. (2003). Coordinated reduction of genes of oxidative metabolism in humans with insulin resistance and diabetes: Potential role of PGC1 and NRF1. *Proc. Natl. Acad. Sci. USA* **100**, 8466–8471.

161. Mootha, V. K., Lindgren, C. M., Eriksson, K. F., Subramanian, A., Sihag, S., Lehar, J., Puigserver, P., Carlsson, E., Ridderstrale, M., Laurila, E., Houstis, N., Daly, M. J., Patterson, N., Mesirov, J. P., Golub, T. R., Tamayo, P., Spiegelman, B., Lander, E. S., Hirschhorn, J. N., Altshuler, D., and Groop, L. C. (2003). PGC-1alpha-responsive genes involved in oxidative phosphorylation are coordinately downregulated in human diabetes. *Nat. Genet.* **34**, 267–273.

162. Finck, B. N., and Kelly, D. P. (2002). Peroxisome proliferator-activated receptor alpha (PPARalpha) signaling in the gene regulatory control of energy metabolism in the normal and diseased heart. *J. Mol. Cell Cardiol.* **34**, 1249–1257.

163. Oberkofler, H., Linnemayr, V., Weitgasser, R., Klein, K., Xie, M., Iglseder, B., Krempler, F., Paulweber, B., and Patsch, W. (2004). Complex haplotypes of the PGC-1alpha gene are associated with carbohydrate metabolism and type 2 diabetes. *Diabetes* **53**, 1385–1393.

164. Hartil, K., and Charron, M. J. (2005). Genetic modification of the heart: transgenic modification of cardiac lipid and carbohydrate utilization. *J. Mol. Cell Cardiol.* **39**, 581–593.

165. Abel, E. D., Kaulbach, H. C., Tian, R., Hopkins, J. C., Duffy, J., Doetschman, T., Minnemann, T., Boers, M. E., Hadro, E., Oberste-Berghaus, C., Quist, W., Lowell, B. B., Ingwall, J. S., and Kahn, B. B. (1999). Cardiac hypertrophy with preserved contractile function after selective deletion of GLUT4 from the heart. *J. Clin. Invest.* **104**, 1703–1714.

166. Stenbit, A. E., Tsao, T. S., Li, J., Burcelin, R., Geenen, D. L., Factor, S. M., Houseknecht, K., Katz, E. B., and Charron, M. J. (1997). GLUT4 heterozygous knockout mice develop muscle insulin resistance and diabetes. *Nat. Med.* **3**, 1096–1101.

167. Watanabe, K., Fujii, H., Takahashi, T., Kodama, M., Aizawa, Y., Ohta, Y., Ono, T., Hasegawa, G., Naito, M., Nakajima, T., Kamijo, Y., Gonzalez, F. J., and Aoyama, T. (2000). Constitutive regulation of cardiac fatty acid metabolism through peroxisome proliferator-activated receptor alpha associated with age-dependent cardiac toxicity. *J. Biol. Chem.* **275**, 22293–22299.

168. Djouadi, F., Brandt, J., Weinheimer, C. J., Leone, T. C., Gonzalez, F. J., and Kelly, D. P. (1999). The role of the peroxisome proliferator-activated receptor alpha (PPAR alpha) in the control of cardiac lipid metabolism. *Prostaglandins Leukot. Essent. Fatty Acids* **60**, 339–343.

169. Hopkins, T. A., Sugden, M. C., Holness, M. J., Kozak, R., Dyck, J. R. B., and Lopaschuk, G. D. (2003). Control of cardiac pyruvate dehydrogenase activity in peroxisome proliferator-activated receptor-alpha transgenic mice. *Am. J. Physiol. Heart Circ. Physiol.* **285**, H270–H276.

170. Ibdah, J. A., Paul, H., Zhao, Y., Binford, S., Salleng, K., Cline, M., Matern, D., Bennett, M. J., Rinaldo, P., and Strauss, A. W. (2001). Lack of mitochondrial trifunctional protein in mice causes neonatal hypoglycemia and sudden death. *J. Clin. Invest.* **107**, 1403–1409.

171. Russell, L. K., Mansfield, C. M., Lehman, J. J., Kovacs, A., Courtois, M., Saffitz, J. E., Medeiros, D. M., Valencik, M. L., McDonald, J. A., and Kelly, D. P. (2004). Cardiac-specific induction of the transcriptional coactivator peroxisome proliferator-activated receptor gamma coactivator-1alpha promotes mitochondrial biogenesis and reversible cardiomyopathy in a developmental stage-dependent manner. *Circ. Res.* **94**, 525–533.

172. Chiu, H. C., Kovacs, A., Ford, D., Hsu, F. F., Garcia, R., Herrero, P., Saffitz, J. E., and Schaffer, J. E. (2001). A novel mouse model of lipotoxic cardiomyopathy. *J. Clin. Invest.* **107**, 813–822.

173. Yagyu, H., Chen, G., Yokoyama, M., Hirata, K., Augustus, A., Kako, Y., Seo, T., Hu, Y., Lutz, E. P., Merkel, M., Bensadoun, A., Homma, S., and Goldberg, I. J. (2003). Lipoprotein lipase (LpL) on the surface of cardiomyocytes increases lipid uptake and produces a cardiomyopathy. *J. Clin. Invest.* **111**, 419–426.

174. Yokoyama, M., Yagyu, H., Hu, Y., Seo, T., Hirata, K., Homma, S., and Goldberg, I. J. (2004). Apolipoprotein B production reduces lipotoxic cardiomyopathy: studies in heart-specific lipoprotein lipase transgenic mouse. *J. Biol. Chem.* **279**, 4204–4211.

175. Sharma, S., Adrogue, J. V., Golfman, L., Uray, I., Lemm, J., Youker, K., Noon, G. P., Frazier, O. H., and Taegtmeyer, H. (2004). Intramyocardial lipid accumulation in the failing human heart resembles the lipotoxic rat heart. *FASEB J.* **18**, 1692–1700.

176. Huang, T. H., Yang, Q., Harada, M., Uberai, J., Radford, J., Li, G. Q., Yamahara, J., Roufogalis, B. D., and Li, Y. (2006). Salacia oblonga root improves cardiac lipid metabolism

in Zucker diabetic fatty rats: Modulation of cardiac PPAR-alpha-mediated transcription of fatty acid metabolic genes. *Toxicol. Appl. Pharmacol.* **210,** 78–85.

177. Finck, B. N., Han, X., Courtois, M., Aimond, F., Nerbonne, J. M., Kovacs, A., Gross, R. W., and Kelly, D. P. (2003). A critical role for PPARalpha-mediated lipotoxicity in the pathogenesis of diabetic cardiomyopathy: modulation by dietary fat content. *Proc. Natl. Acad. Sci. USA* **100,** 1226–1231.

178. Pillutla, P., Hwang, Y. C., Augustus, A., Yokoyama, M., Yagyu, H., Johnston, T. P., Kaneko, M., Ramasamy, R., and Goldberg, I. J. (2005). Perfusion of hearts with triglyceride-rich particles reproduces the metabolic abnormalities in lipotoxic cardiomyopathy. *Am. J. Physiol. Endocrinol. Metab.* **288,** E1229–E1235.

179. Listenberger, L. L., Ory, D. S., and Schaffer, J. E. (2001). Palmitate-induced apoptosis can occur through a ceramide-independent pathway. *J. Biol. Chem.* **276,** 14890–14895.

180. Ruiz-Lozano, P., Smith, S. M., Perkins, G., Kubalak, S. W., Boss, G. R., Sucov, H. M., Evans, R. M., and Chien, K. R. (1998). Energy deprivation and a deficiency in downstream metabolic target genes during the onset of embryonic heart failure in RXR alpha −/− embryos. *Development* **125,** 533–544.

181. Graham, B. H., Waymire, K. G., Cottrell, B., Trounce, I. A., MacGregor, G. R., and Wallace, D. C. (1997). A mouse model for mitochondrial myopathy and cardiomyopathy resulting from a deficiency in the heart/muscle isoform of the adenine nucleotide translocator. *Nat. Genet.* **16,** 226–234.

182. Wang, J., Wilhelmsson, H., Graff, C., Li, H., Oldfors, A., Rustin, P., Bruning, J. C., Kahn, C. R., Clayton, D. A., Barsh, G. S., Thoren, P., and Larsson, N. G. (1999). Dilated cardiomyopathy and atrioventricular conduction blocks induced by heart-specific inactivation of mitochondrial DNA gene expression. *Nat. Genet.* **21,** 133–137.

183. Hansson, A., Hance, N., Dufour, E., Rantanen, A., Hultenby, K., Clayton, D. A., Wibom, R., and Larsson, N. G. (2004). A switch in metabolism precedes increased mitochondrial biogenesis in respiratory chain-deficient mouse hearts. *Proc. Natl. Acad. Sci. USA* **101,** 3136–3141.

184. Chiu, H. C., Kovacs, A., Blanton, R. M., Han, X., Courtois, M., Weinheimer, C. J., Yamada, K. A., Brunet, S., Xu, H., Nerbonne, J. M., Welch, M. J., Fettig, N. M., Sharp, T. L., Sambandam, N., Olson, K. M., Ory, D. S., and Schaffer, J. E. (2005). Transgenic expression of fatty acid transport protein 1 in the heart causes lipotoxic cardiomyopathy. *Circ. Res.* **96,** 225–233.

185. Cheng, L., Ding, G., Qin, Q., Huang, Y., Lewis, W., He, N., Evans, R. M., Schneider, M. D., Brako, F. A., Xiao, Y., Chen, Y. E., and Yang, Q. (2004). Cardiomyocyte-restricted peroxisome proliferator-activated receptor-delta deletion perturbs myocardial fatty acid oxidation and leads to cardiomyopathy. *Nat. Med.* **10,** 1245–1250.

186. Christoffersen, C., Bollano, E., Lindegaard, M. L., Bartels, E. D., Goetze, J. P., Andersen, C. B., and Nielsen, L. B. (2003). Cardiac lipid accumulation associated with diastolic dysfunction in obese mice. *Endocrinology* **144,** 3483–3490.

187. Exil, V. J., Roberts, R. L., Sims, H., McLaughlin, J. E., Malkin, R. A., Gardner, C. D., Ni, G., Rottman, J. N., and Strauss, A. W. (2003). Very-long-chain acyl-coenzyme a dehydrogenase deficiency in mice. *Circ. Res.* **93,** 448–455.

188. Hansson, A., Hance, N., Dufour, E., Rantanen, A., Hultenby, K., Clayton, D. A., Wibom, R., and Larsson, N. G. (2004). A switch in metabolism precedes increased mitochondrial biogenesis in respiratory chain-deficient mouse hearts. *Proc. Natl. Acad. Sci. USA* **101,** 3136–3141.

189. Augustus, A., Yagyu, H., Haemmerle, G., Bensadoun, A., Vikramadithyan, R. K., Park, S. Y., Kim, J. K., Zechner, R., and Goldberg, I. J. (2004). Cardiac-specific knock-out of lipoprotein lipase alters plasma lipoprotein triglyceride metabolism and cardiac gene expression. *J. Biol. Chem* **279,** 25050–25057.

190. Coburn, C. T., Hajri, T., Ibrahimi, A., and Abumrad, N. A. (2001). Role of CD36 in membrane transport and utilization of long-chain fatty acids by different tissues. *J. Mol. Neurosci.* **16,** 117–121.

191. Ibrahimi, A., Bonen, A., Blinn, W. D., Hajri, T., Li, X., Zhong, K., Cameron, R., and Abumrad, N. A. (1999). Muscle-specific overexpression of FAT/CD36 enhances fatty acid oxidation by contracting muscle, reduces plasma triglycerides and fatty acids, and increases plasma glucose and insulin. *J. Biol. Chem.* **274,** 26761–26766.

192. Belke, D. D., Betuing, S., Tuttle, M. J., Graveleau, C., Young, M. E., Pham, M., Zhang, D., Cooksey, R. C., McClain, D. A., Litwin, S. E., Taegtmeyer, H., Severson, D., Kahn, C. R., and Abel, E. D. (2002). Insulin signaling coordinately regulates cardiac size, metabolism, and contractile protein isoform expression. *J. Clin. Invest.* **109,** 629–639.

193. Binas, B., Danneberg, H., McWhir, J., Mullins, L., and Clark, A. J. (1999). Requirement for the heart-type fatty acid binding protein in cardiac fatty acid utilization. *FASEB J.* **13,** 805–812.

194. Lutter, M., Fang, M., Luo, X., Nishijima, M., Xie, X., and Wang, X. (2000). Cardiolipin provides specificity for targeting of tBid to mitochondria. *Nat. Cell. Biol.* **2,** 754–761.

195. Hickson-Bick, D. L., Buja, M. L., and McMillin, J. B. (2000). Palmitate-mediated alterations in the fatty acid metabolism of rat neonatal cardiac myocytes. *J. Mol. Cell. Cardiol.* **32,** 511–519.

196. De Vries, J. E., Vork, M. M., Roemen, T. H., de Jong, Y. F., Cleutjens, J. P., van der Vusse, G. J., and van Bilsen, M. (1997). Saturated but not monounsaturated fatty acids induce apoptotic cell death in neonatal rat ventricular myocytes. *J. Lipid Res.* **38,** 1384–1394.

197. Sparagna, G. C., Hickson-Bick, D. L., Buja, L. M., and McMillin, J. B. (2001). Fatty acid-induced apoptosis in neonatal cardiomyocytes: Redox signaling. *Antioxid. Redox. Signal.* **3,** 71–79.

198. Ostrander, D. B., Sparagna, G. C., Amoscato, A. A., McMillin, J. B., and Dowhan, W. (2001). Decreased cardiolipin synthesis corresponds with cytochrome c release in palmitate-induced cardiomyocyte apoptosis. *J. Biol. Chem.* **276,** 38061–38067.

199. Kong, J. Y., and Rabkin, S. W. (2003). Mitochondrial effects with ceramide-induced cardiac apoptosis are different from those of palmitate. *Arch. Biochem. Biophys.* **412,** 196–206.

200. Hickson-Bick, D. L., Sparagna, G. C., Buja, L. M., and McMillin, J. B. (2002). Palmitate-induced apoptosis in neonatal cardiomyocytes is not dependent on the generation of ROS. *Am. J. Physiol. Heart Circ. Physiol.* **282,** H656–H664.

201. Sparagna, G. C., Hickson-Bick, D. L., Buja, L. M., and McMillin, J. B. (2000). A metabolic role for mitochondria in

palmitate-induced cardiac myocyte apoptosis. *Am. J. Physiol. Heart Circ. Physiol.* **279,** H2124–H2132.
202. Sparagna, G. C., and Hickson-Bick, D. L. (1999). Cardiac fatty acid metabolism and the induction of apoptosis. *Am. J. Med. Sci.* **318,** 15–21.
203. Malhotra, R., and Brosius, F. C. (1999). Glucose uptake and glycolysis reduce hypoxia-induced apoptosis in cultured neonatal rat cardiac myocytes. *J. Biol. Chem.* **274,** 12567–12575.
204. Lin, Z., Weinberg, J. M., Malhotra, R., Merritt, S. E., Holzman, L. B., and Brosius, F. C. (2000). GLUT-1 reduces hypoxia-induced apoptosis and JNK pathway activation. *Am. J. Physiol. Endocrinol. Metab.* **278,** E958–E966.
205. Fujio, Y., Nguyen, T., Wencker, D., Kitsis, R. N., and Walsh, K. (2000). Akt promotes survival of cardiomyocytes in vitro and protects against ischemia-reperfusion injury in mouse heart. *Circulation* **101,** 660–667.
206. Aikawa, R., Nawano, M., Gu, Y., Katagiri, H., Asano, T., Zhu, W., Nagai, R., and Komuro, I. (2000). Insulin prevents cardiomyocytes from oxidative stress-induced apoptosis through activation of PI3 kinase/Akt. *Circulation* **102,** 2873–2879.
207. Cai, L., Li, W., Wang, G., Guo, L., Jiang, Y., and Kang, Y. J. (2002). Hyperglycemia-induced apoptosis in mouse myocardium: Mitochondrial cytochrome c-mediated caspase-3 activation pathway. *Diabetes* **51,** 1938–1948.
208. Sack, M. N., Rader, T. A., Park, S., Bastin, J., McCune, S. A., and Kelly, D. P. (1996). Fatty acid oxidation enzyme gene expression is downregulated in the failing heart. *Circulation* **94,** 2837–2842.
209. Kanda, H., Nohara, R., Hasegawa, K., Kishimoto, C., and Sasayama, S. (2000). A nuclear complex containing PPARa/RXR is markedly downregulated in the hypertrophied rat left ventricular myocardium with normal systolic function. *Heart Vessels* **15,** 191–196.
210. Leong, H. S., Brownsey, R. W., Kulpa, J. E., and Allard, M. F. (2003). Glycolysis and pyruvate oxidation in cardiac hypertrophy: Why so unbalanced? *Comp. Biochem. Physiol. A Mol. Integr. Physiol.* **135,** 499–513.
211. van Bilsen, M., Smeets, P. J., Gilde, A. J., and van der Vusse, G. J. (2004). Metabolic remodelling of the failing heart: The cardiac burn-out syndrome? *Cardiovasc. Res.* **61,** 218–226.
212. Brackett, J. C., Sims, H. F., Rinaldo, P., Shapiro, S., Powell, C. K., Bennett, M. J., and Strauss, A. W. (1995). Two alpha-subunit donor splice site mutations cause human trifunctional protein deficiency. *J. Clin. Invest.* **95,** 2076–2082.
213. Schonberger, J., and Seidman, C. E. (2001). Many roads lead to a broken heart: The genetics of dilated cardiomyopathy. *Am. J. Hum. Genet.* **69,** 249–260.
214. Kelly, D. P., and Strauss, A. W. (1994). Inherited cardiomyopathies. *N. Engl. J. Med.* **330,** 913–919.
215. Kelly, D. P., Hale, D. E., Rutledge, S. L., Ogden, M. L., Whelan, A. J., Zhang, Z., and Strauss, A. W. (1992). Molecular basis of inherited medium chain acyl-CoA dehydrogenase deficiency causing sudden child death. *J. Inherit. Metab. Dis.* **15,** 171–180.
216. Tanaka, K., Yokota, I., and Coates, P. M. (1992). Mutations in the medium chain acyl-CoA dehydrogenase (MCAD) gene. *Hum. Mutat.* **1,** 271–279.

217. Andresen, B. S., Bross, P., Vianey-Saban, C., Divry, P., Zabot, M. T., Roe, C. R., Nada, M. A., Byskov, A., Kruse, T. A., Neve, S., Kristiansen, K., Knudsen, I., Corydon, M. J., and Gregersen, N. (1996). Cloning and characterization of human very-long-chain acyl-CoA dehydrogenase cDNA, chromosomal assignment of the gene and identification in four patients of nine different mutations within the VLCAD gene. *Hum. Mol. Genet.* **5,** 461–472.
218. Mathur, A., Sims, H. F., Gopalakrishnan, D., Gibson, B., Rinaldo, P., Vockley, J., Hug, G., and Strauss, A. W. (1999). Molecular heterogeneity in very-long-chain acyl-CoA dehydrogenase deficiency causing pediatric cardiomyopathy and sudden death. *Circulation* **99,** 1337–1343.
219. Orii, K. E., Aoyama, T., Wakui, K., Fukushima, Y., Miyajima, H., Yamaguchi, S., Orii, T., Kondo, N., and Hashimoto, T. (1997). Genomic and mutational analysis of the mitochondrial trifunctional protein beta-subunit (HADHB) gene in patients with trifunctional protein deficiency. *Hum. Mol. Genet.* **6,** 1215–1224.
220. den Boer, M. E., Dionisi-Vici, C., Chakrapani, A., van Thuijl, A. O., Wanders, R. J., and Wijburg, F. A. (2003). Mitochondrial trifunctional protein deficiency: A severe fatty acid oxidation disorder with cardiac and neurologic involvement. *J. Pediatr.* **142,** 684–689.
221. Spiekerkoetter, U., Khuchua, Z., Yue, Z., Bennett, M. J., and Strauss, A. W. (2004). General mitochondrial trifunctional protein (TFP) deficiency as a result of either alpha- or beta-subunit mutations exhibits similar phenotypes because mutations in either subunit alter TFP complex expression and subunit turnover. *Pediatr. Res.* **55,** 190–196.
222. Schwab, K. O., Ensenauer, R., Matern, D., Uyanik, G., Schnieders, B., Wanders, R. A., and Lehnert, W. (2003). Complete deficiency of mitochondrial trifunctional protein due to a novel mutation within the beta-subunit of the mitochondrial trifunctional protein gene leads to failure of long-chain fatty acid beta-oxidation with fatal outcome. *Eur. J. Pediatr.* **162,** 90–95.
223. Angdisen, J., Moore, V. D., Cline, J. M., Payne, R. M., and Ibdah, J. (2005). A. Mitochondrial trifunctional protein defects: molecular basis and novel therapeutic approaches. *Curr. Drug Targets Immune. Endocr. Metabol. Disord.* **5,** 27–40.
224. Sander, J., Sander, S., Steuerwald, U., Janzen, N., Peter, M., Wanders, R. J., Marquardt, I., Korenke, G. C., and Das, A. M. (2005). Neonatal screening for defects of the mitochondrial trifunctional protein. *Mol. Genet. Metab.* **85,** 108–114.
225. Olpin, S. E., Clark, S., Andresen, B. S., Bischoff, C., Olsen, R. K., Gregersen, N., Chakrapani, A., Downing, M., Manning, N. J., Sharrard, M., Bonham, J. R., Muntoni, F., Turnbull, D. N., and Pourfarzam, M. (2005). Biochemical, clinical and molecular findings in LCHAD and general mitochondrial trifunctional protein deficiency. *J. Inherit. Metab. Dis.* **28,** 533–544.
226. Bonnefont, J. P., Djouadi, F., Prip-Buus, C., Gobin, S., Munnich, A., and Bastin, J. (2004). Carnitine palmitoyltransferases 1 and 2: biochemical, molecular and medical aspects. *Mol. Aspects Med.* **25,** 495–520.
227. Engel, A. G., and Angelini, C. (1973). Carnitine deficiency of human skeletal muscle with associated lipid storage myopathy: A new syndrome. *Science* **179,** 899–902.

228. Stanley, C. A. (2004). Carnitine deficiency disorders in children. *Ann. N. Y. Acad. Sci.* **1033,** 42–51.
229. Stanley, C. A., Treem, W. R., Hale, D. E., and Coates, P. M. (1990). A genetic defect in carnitine transport causing primary carnitine deficiency. *Prog. Clin. Biol. Res.* **321,** 457–464.
230. Nezu, J., Tamai, I., Oku, A., Ohashi, R., Yabuuchi, H., Hashimoto, N., Nikaido, H., Sai, Y., Koizumi, A., Shoji, Y., Takada, G., Matsuishi, T., Yoshino, M., Kato, H., Ohura, T., Tsujimoto, G., Hayakawa, J., Shimane, M., and Tsuji, A. (1999). Primary systemic carnitine deficiency is caused by mutations in a gene encoding sodium ion-dependent carnitine transporter. *Nat. Genet.* **21,** 91–94.
231. Melegh, B. (2004). The human OCTN2 carnitine transporter and its mutations. *Orv. Hetil.* **145,** 679–686.
232. Melegh, B., Bene, J., Mogyorosy, G., Havasi, V., Komlosi, K., Pajor, L., Olah, E., Kispal, G., Sumegi, B., and Mehes, K. (2004). Phenotypic manifestations of the OCTN2 V295X mutation: sudden infant death and carnitine-responsive cardiomyopathy in Roma families. *Am. J. Med. Genet. A* **131,** 121–126.
233. Lamhonwah, A. M., Olpin, S. E., Pollitt, R. J., Vianey-Saban, C., Divry, P., Guffon, N., Besley, G. T., Onizuka, R., De Meirleir, L. J., Cvitanovic-Sojat, L., Baric, I., Dionisi-Vici, C., Fumic, K., Maradin, M., and Tein, I. (2002). Novel OCTN2 mutations: no genotype-phenotype correlations: early carnitine therapy prevents cardiomyopathy. *Am. J. Med. Genet.* **111,** 271–284.
234. Roschinger, W., Muntau, A. C., Duran, M., Dorland, L., Ijlst, L., Wanders, R. J., and Roscher, A. A. (2000). Carnitine-acylcarnitine translocase deficiency: Metabolic consequences of an impaired mitochondrial carnitine cycle. *Clin. Chim. Acta* **298,** 55–68.
235. Rubio-Gozalbo, M. E., Bakker, J. A., Waterham, H. R., and Wanders, R. J. (2004). Carnitine-acylcarnitine translocase deficiency, clinical, biochemical and genetic aspects. *Mol. Aspects Med.* **25,** 521–532.
236. Iacobazzi, V., Invernizzi, F., Baratta, S., Pons, R., Chung, W., Garavaglia, B., Dionisi-Vici, C., Ribes, A., Parini, R., Huertas, M. D., Roldan, S., Lauria, G., Palmieri, F., and Taroni, F. (2004). Molecular and functional analysis of SLC25A20 mutations causing carnitine-acylcarnitine translocase deficiency. *Hum. Mutat.* **24,** 312–320.
237. Iacobazzi, V., Pasquali, M., Singh, R., Matern, D., Rinaldo, P., Amat di San Filippo, C., Palmieri, F., and Longo, N. (2004). Response to therapy in carnitine/acylcarnitine translocase (CACT) deficiency due to a novel missense mutation. *Am. J. Med. Genet.* **126,** 150–155.
238. Tein, I. (2003). Carnitine transport: pathophysiology and metabolism of known molecular defects. *J. Inherit. Metab. Dis.* **26,** 147–169.
239. Tanaka, T., Sohmiya, K., and Kawamura, K. (1997). Is CD36 deficiency an etiology of hereditary hypertrophic cardiomyopathy? *J. Mol. Cell Cardiol.* **29,** 121–127.
240. Febbraio, M., Guy, E., Coburn, C., Knapp, F. F., Jr., Beets, A. L., Abumrad, N. A., and Silverstein, R. L. (2002). The impact of overexpression and deficiency of fatty acid translocase (FAT)/CD36. *Mol. Cell Biochem.* **239,** 193–197.
241. Watanabe, K., Ohta, Y., Toba, K., Ogawa, Y., Hanawa, H., Hirokawa, Y., Kodama, M., Tanabe, N., Hirono, S., Ohkura, Y., Nakamura, Y., Kato, K., Aizawa, Y., Fuse, I., Miyajima, S., Kusano, Y., Nagamoto, T., Hasegawa, G., and Naito, M. (1998). Myocardial CD36 expression and fatty acid accumulation in patients with type I and II CD36 deficiency. *Ann. Nucl. Med.* **12,** 261–266.
242. Okamoto, F., Tanaka, T., Sohmiya, K., and Kawamura, K. (1998). CD36 abnormality and impaired myocardial long-chain fatty acid uptake in patients with hypertrophic cardiomyopathy. *Jpn. Circ. J.* **62,** 499–504.
243. Kushiro, T., Saito, F., Kusama, J., Takahashi, H., Imazeki, T., Tani, S., Kikuchi, S., Imai, S., Matsudaira, K., Watanabe, I., Hino, T., Sato, Y., Nakayama, T., Nagao, K., and Kanmatsuse, K. (2005). Takotsubo-shaped cardiomyopathy with type I CD36 deficiency. *Heart Vessels* **20,** 123–125.
244. Nakamura, T., Sugihara, H., Inaba, T., Kinoshita, N., Adachi, Y., Hirasaki, S., Matsuo, A., Azuma, A., and Nakagawa, M. (1999). CD36 deficiency has little influence on the pathophysiology of hypertrophic cardiomyopathy. *J. Mol. Cell Cardiol.* **31,** 1253–1259.
245. Pohl, J., Fitscher, B. A., Ring, A., Ihl-Vahl, R., Strasser, R. H., and Stremmel, W. (2000). Fatty acid transporters in plasma membranes of cardiomyocytes in patients with dilated cardiomyopathy. *Eur. J. Med. Res.* **5,** 438–442.
246. Hirano, K., Kuwasako, T., Nakagawa-Toyama, Y., Janabi, M., Yamashita, S., and Matsuzawa, Y. (2003). Pathophysiology of human genetic CD36 deficiency. *Trends Cardiovasc. Med.* **13,** 136–141.
247. Doney, A. S., Fischer, B., Lee, S. P., Morris, A. D., Leese, G., and Palmer, C. N. (2005). Association of common variation in the PPARA gene with incident myocardial infarction in individuals with type 2 diabetes: A Go-DARTS study. *Nucl. Recept.* **3,** 4.
248. Sookoian, S., Garcia, S. I., Porto, P. I., Dieuzeide, G., Gonzalez, C. D., and Pirola, C. J. (2005). Peroxisome proliferator-activated receptor gamma and its coactivator-1 alpha may be associated with features of the metabolic syndrome in adolescents. *J. Mol. Endocrinol.* **35,** 373–380.
249. Bonnet, D., Martin, D., De Lonlay, P., Villain, E., Jouvet, P., Rabier, D., Brivet, M., and Saudubray, J. M. (1999). Arrhythmias and conduction defects as presenting symptoms of fatty acid oxidation disorders in children. *Circulation* **100,** 2248–2253.
250. Feillet, F., Steinmann, G., Vianey-Saban, C., de Chillou, C., Sadoul, N., Lefebvre, E., Vidailhet, M., and Bollaert, P. E. (2003). Adult presentation of MCAD deficiency revealed by coma and severe arrhythmias. *Intensive Care Med.* **29,** 1594–1597.
251. Stanley, C. A., Hale, D. E., Berry, D. T., Deleeuw, S., Boxer, J., Bonnefont, J. P. (1992). A deficiency of carnitine-acylcarnitine translocase in the inner mitochondrial membrane. *N. Engl. J. Med.* **327,** 19–23.
252. Corr, P. B., Creer, M. H., Yamada, K. A., Saffitz, J. E., and Sobel, B. E. (1989). Prophylaxis of early ventricular fibrillation by inhibition of acylcarnitine accumulation. *J. Clin. Invest.* **83,** 927–936.
253. Tripp, M. E. (1989). Developmental cardiac metabolism in health and disease. *Pediatr. Cardiol.* **10,** 150–158.
254. Perrot, A., Osterziel, K. J., Beck, M., Dietz, R., and Kampmann, C. (2002). Fabry disease: focus on cardiac manifestations and molecular mechanisms. *Herz* **27,** 699–702.
255. Gollob, M. H., Green, M. S., Tang, A. S., and Roberts, R. (2002). PRKAG2 cardiac syndrome: familial ventricular

preexcitation, conduction system disease, and cardiac hypertrophy. *Curr. Opin. Cardiol.* **17,** 229–234.

256. D'Adamo, P., Fassone, L., Gedeon, A., Janssen, E. A., Bione, S., Bolhuis, P. A., Barth, P. G., Wilson, M., Haan, E., Orstavik, K. H., Patton, M. A., Green, A. J., Zammarchi, E., Donati, M. A., and Toniolo, D. (1997). The X-linked gene G4.5 is responsible for different infantile dilated cardiomyopathies. *Am. J. Hum. Genet.* **61,** 862–867.

257. Brivet, M., Boutron, A., Slama, A., Costa, C., Thuillier, L., Demaugre, F., Rabier, D., Saudubray, J. M., and Bonnefont, J. P. (1999). Defects in activation and transport of fatty acids. *J. Inherit. Metab. Dis.* **22,** 428–441.

258. Strauss, A. W., Powell, C. K., Hale, D. E., Anderson, M. M., Ahuja, A., Brackett, J. C., and Sims, H. F. (1995). Molecular basis of human mitochondrial very-long-chain acyl-CoA dehydrogenase deficiency causing cardiomyopathy and sudden death in childhood. *Proc. Natl. Acad. Sci. USA* **92,** 10496–10500.

259. Sims, H. F., Brackett, J. C., Powell, C. K., Treem, W. R., Hale, D. E., Bennett, M. J., Gibson, B., Shapiro, S., and Strauss, A. W. (1995). The molecular basis of pediatric long chain 3-hydroxyacyl-CoA dehydrogenase deficiency associated with maternal acute fatty liver of pregnancy. *Proc. Natl. Acad. Sci. USA* **92,** 841–845.

260. Matern, D., Strauss, A. W., Hillman, S. L., Mayatepek, E., Millington, D. S., and Trefz, F. K. (1999). Diagnosis of mitochondrial trifunctional protein deficiency in a blood spot from the newborn screening card by tandem mass spectrometry and DNA analysis. *Pediatr. Res.* **46,** 45–49.

261. Avogaro, A., Vigili de Kreutzenberg, S., Negut, C., Tiengo, A., and Scognamiglio, R. (2004). Diabetic cardiomyopathy: A metabolic perspective. *Am. J. Cardiol.* **93,** 13A–16A.

262. Seager, M. J., Singal, P. K., Orchard, R., Pierce, G. N., and Dhalla, N. S. (1984). Cardiac cell damage: A primary myocardial disease in streptozotocin-induced chronic diabetes. *Br. J. Exp. Pathol.* **65,** 613–623.

263. Mokhtar, N., Lavoie, J. P., Rousseau-Migneron, S., and Nadeau, A. (1993). Physical training reverses defect in mitochondrial energy production in heart of chronically diabetic rats. *Diabetes* **42,** 682–687.

264. Tomita, M., Mukae, S., Geshi, E., Umetsu, K., Nakatani, M., and Katagiri, T. (1996). Mitochondrial respiratory impairment in streptozotocin induced diabetic rat heart. *Jpn. Circ. J.* **60,** 673–682.

265. Koster, J. C., Permutt, M. A., and Nichols, C. G. (2005). Diabetes and Insulin Secretion: The ATP-Sensitive K+ Channel (KATP) Connection. *Diabetes* **54,** 3065–3072.

266. Hattersley, A. T., and Ashcroft, F. M. (2005). Activating mutations in Kir6.2 and neonatal diabetes: new clinical syndromes, new scientific insights, and new therapy. *Diabetes* **54,** 2503–2513.

267. Gloyn, A. L., Pearson, E. R., Antcliff, J. F., Proks, P., Bruining, G. J., Slingerland, A. S., Howard, N., Srinivasan, S., Silva, J. M., Molnes, J., Edghill, E. L., Frayling, T. M., Temple, I. K., Mackay, D., Shield, J. P., Sumnik, Z., van Rhijn, A., Wales, J. K., Clark, P., Gorman, S., Aisenberg, J., Ellard, S., Njolstad, P. R., Ashcroft, F. M., and Hattersley, A. T. (2004). Activating mutations in the gene encoding the ATP-sensitive potassium-channel subunit Kir6.2 and permanent neonatal diabetes. *N. Engl. J. Med.* **350,** 1838–1849.

268. Sperling, M. A. (2005). Neonatal diabetes mellitus: from understudy to center stage. *Curr. Opin. Pediatr.* **17,** 512–518.

269. Gloyn, A. L., Weedon, M. N., Owen, K. R., Turner, M. J., Knight, B. A., Hitman, G., Walker, M., Levy, J. C., Sampson, M., Halford, S., McCarthy, M. I., Hattersley, A. T., and Frayling, T. M. (2003). Large-scale association studies of variants in genes encoding the pancreatic beta-cell KATP channel subunits Kir6.2 (KCNJ11) and SUR1 (ABCC8) confirm that the KCNJ11 E23K variant is associated with type 2 diabetes. *Diabetes* **52,** 568–572.

270. Riedel, M. J., Steckley, D. C., and Light, P. E. (2005). Current status of the E23K Kir6.2 polymorphism: implications for type-2 diabetes. *Hum. Genet.* **116,** 133–145.

271. Riedel, M. J., Boora, P., Steckley, D., de Vries, G., and Light, P. E. (2003). Kir6.2 polymorphisms sensitize beta-cell ATP-sensitive potassium channels to activation by acyl CoAs: a possible cellular mechanism for increased susceptibility to type 2 diabetes? *Diabetes* **52,** 2630–2635.

272. Slingerland, A. S., and Hattersley, A. T. (2005). Mutations in the Kir6.2 subunit of the KATP channel and permanent neonatal diabetes: new insights and new treatment. *Ann. Med.* **37,** 186–195.

273. Malecki, M. T. (2005). Genetics of type 2 diabetes mellitus. *Diabetes Res. Clin. Pract.* **68,** S10–S21.

274. Gupta, R. K., and Kaestner, K. H. (2004). HNF-4alpha: from MODY to late-onset type 2 diabetes. *Trends Mol. Med.* **10,** 521–524.

275. Gloyn, A. L. (2003). Glucokinase (GCK) mutations in hyper- and hypoglycemia: maturity-onset diabetes of the young, permanent neonatal diabetes, and hyperinsulinemia of infancy. *Hum. Mutat.* **22,** 353–362.

276. Mitchell, S. M., and Frayling, T. M. (2002). The role of transcription factors in maturity-onset diabetes of the young. *Mol. Genet. Metab.* **77,** 35–43.

277. George, S., Rochford, J. J., Wolfrum, C., Gray, S. L., Schinner, S., Wilson, J. C., Soos, M. A., Murgatroyd, P. R., Williams, R M., Acerini, C. L., Dunger, D. B., Barford, D., Umpleby, A. M., Wareham, N. J., Davies, H. A., Schafer, A. J., Stoffel, M., O'Rahilly, S., and Barroso, I. (2004). A family with severe insulin resistance and diabetes due to a mutation in AKT2. *Science* **304,** 1325–1328.

278. Hone, J., Accili, D., al-Gazali, L. I., Lestringant, G., Orban, T., and Taylor, S. I. (1994). Homozygosity for a new mutation (Ile119→Met) in the insulin receptor gene in five sibs with familial insulin resistance. *J. Med. Genet.* **31,** 715–716.

279. Kusari, J., Takata, Y., Hatada, E., Freidenberg, G., Kolterman, O., and Olefsky, J. M. (1991). Insulin resistance and diabetes due to different mutations in the tyrosine kinase domain of both insulin receptor gene alleles. *J. Biol. Chem.* **266,** 5260–5267.

280. Musso, C., Cochran, E., Moran, S. A., Skarulis, M. C., Oral, E. A., Taylor, S., and Gorden, P. (2004). Clinical course of genetic diseases of the insulin receptor (type A and Rabson-Mendenhall syndromes): a 30-year prospective. *Medicine* **83,** 209–222.

281. Shen, X., Zheng, S., Thongboonkerd, V., Xu, M., Pierce, W. M., Jr., Klein, J. B., and Epstein, P. N. (2004). Cardiac mitochondrial damage and biogenesis in a chronic model of type 1 diabetes. *Am. J. Physiol. Endocrinol. Metab.* **287,** E896–E905.

282. Ferreira, F. M., Seica, R., Oliveira, P. J., Coxito, P. M., Moreno, A. J., Palmeira, C. M., and Santos, M. S. (2003). Diabetes induces metabolic adaptations in rat liver mitochondria: role of coenzyme Q and cardiolipin contents. *Biochim. Biophys. Acta* **1639**, 113–118.
283. Ritov, V. B., Menshikova, E. V., He, J., Ferrell, R. E., Goodpaster, B. H., and Kelley, D. E. (2005). Deficiency of subsarcolemmal mitochondria in obesity and type 2 diabetes. *Diabetes* **54**, 8–14.
284. Petersen, K. F., Dufour, S., Befroy, D., Garcia, R., and Shulman, G. I. (2004). Impaired mitochondrial activity in the insulin-resistant offspring of patients with type 2 diabetes. *N. Engl. J. Med.* **350**, 664–671.
285. Silva, J. P., Kohler, M., Graff, C., Oldfors, A., Magnuson, M. A., Berggren, P. O., and Larsson, N. G. (2000). Impaired insulin secretion and beta-cell loss in tissue-specific knockout mice with mitochondrial diabetes. *Nat. Genet.* **26**, 336–340.
286. Brownlee, M. (2003). A radical explanation for glucose-induced beta cell dysfunction. *J. Clin. Invest.* **112**, 1788–1790.
287. Lowell, B. B., and Shulman, G. I. (2005). Mitochondrial dysfunction and type 2 diabetes. *Science* **307**, 384–387.
288. Stumvoll, M., Goldstein, B. J., and van Haeften, T. W. (2005). Type 2 diabetes: principles of pathogenesis and therapy. *Lancet* **365**, 1333–1346.
289. Malecki, M. T. (2005). Genetics of type 2 diabetes mellitus. *Diabetes Res. Clin. Pract.* **68**, S10–S21.
290. Kelly, M. A., Mijovic, C. H., and Barnett, A. H. (2001). Genetics of type 1 diabetes. *Best Pract. Res. Clin. Endocrinol. Metab.* **15**, 279–291.
291. Achenbach, P., Bonifacio, E., and Ziegler, A. G. (2005). Predicting type 1 diabetes. *Curr. Diab. Rep.* **5**, 98–103.
292. Kavvoura, F. K., and Ioannidis, J. P. (2005). CTLA-4 gene polymorphisms and susceptibility to type 1 diabetes mellitus: a HuGE Review and meta-analysis. *Am. J. Epidemiol.* **162**, 3–16.
293. Bottini, N., Musumeci, L., Alonso, A., Rahmouni, S., Nika, K., Rostamkhani, M., MacMurray, J., Meloni, G. F., Lucarelli, P., Pellecchia, M., Eisenbarth, G. S., Comings, D., and Mustelin, T. (2004). A functional variant of lymphoid tyrosine phosphatase is associated with type I diabetes. *Nat. Genet.* **36**, 337–338.
294. Mathieu, C., and Badenhoop, K. (2005). Vitamin D and type 1 diabetes mellitus: state of the art. *Trends Endocrinol. Metab.* **16**, 261–266.
295. Luong, K., Nguyen, L. T., and Nguyen, D. N. (2005). The role of vitamin D in protecting type 1 diabetes mellitus. *Diabetes Metab. Res. Rev.* **21**, 338–346.
296. Pollex, R. L., Mamakeesick, M., Zinman, B., Harris, S. B., Hanley, A. J., and Hegele, R. A. (2005). Methylenetetrahydrofolate reductase polymorphism 677C>T is associated with peripheral arterial disease in type 2 diabetes. *Cardiovasc. Diabetol.* **4**, 17.
297. Maeda, S., Tsukada, S., Kanazawa, A., Sekine, A., Tsunoda, T., Koya, D., Maegawa, H., Kashiwagi, A., Babazono, T., Matsuda, M., Tanaka, Y., Fujioka, T., Hirose, H., Eguchi, T., Ohno, Y., Groves, C. J., Hattersley, A. T., Hitman, G. A., Walker, M., Kaku, K., Iwamoto, Y., Kawamori, R., Kikkawa, R., Kamatani, N., McCarthy, M. I., and Nakamura, Y. (2005). Genetic variations in the gene encoding TFAP2B are associated with type 2 diabetes mellitus. *J. Hum. Genet.* **50**, 283–292.
298. Vimaleswaran, K. S., Radha, V., Ghosh, S., Majumder, P. P., Deepa, R., Babu, H. N., Rao, M. R., and Mohan, V. (2005). Peroxisome proliferator-activated receptor-gamma co-activator-1alpha (PGC-1alpha) gene polymorphisms and their relationship to Type 2 diabetes in Asian Indians. *Diabet. Med.* **22**, 1516–1521.
299. Ek, J., Andersen, G., Urhammer, S. A., Gaede, P. H., Drivsholm, T., Borch-Johnsen, K., Hansen, T., and Pedersen, O. (2001). Mutation analysis of peroxisome proliferator-activated receptor-gamma coactivator-1 (PGC-1) and relationships of identified amino acid polymorphisms to Type II diabetes mellitus. *Diabetologia* **44**, 2220–2226.
300. Nicaud, V., Raoux, S., Poirier, O., Cambien, F., O'Reilly, D. S., and Tiret, L. (2002). The TNF alpha/G-308A polymorphism influences insulin sensitivity in offspring of patients with coronary heart disease: the European Atherosclerosis Research Study II. *Atherosclerosis* **161**, 317–325.
301. Vendrell, J., Fernandez-Real, J. M., Gutierrez, C., Zamora, A., Simon, I., Bardaji, A., Ricart, W., and Richart, C. (2003). A polymorphism in the promoter of the tumor necrosis factor-alpha gene (-308) is associated with coronary heart disease in type 2 diabetic patients. *Atherosclerosis* **167**, 257–264.
302. Florez, J. C., Burtt, N., de Bakker, P. I., Almgren, P., Tuomi, T., Holmkvist, J., Gaudet, D., Hudson, T. J., Schaffner, S. F., Daly, M. J., Hirschhorn, J. N., Groop, L., and Altshuler, D. (2004). Haplotype structure and genotype-phenotype correlations of the sulfonylurea receptor and the islet ATP-sensitive potassium channel gene region. *Diabetes* **53**, 1360–1368.
303. Hayes, M. G., Del Bosque-Plata, L., Tsuchiya, T., Hanis, C. L., Bell, G. I., and Cox, N. J. (2005). Patterns of linkage disequilibrium in the type 2 diabetes gene calpain-10. *Diabetes* **54**, 3573–3576.
304. Evans, J. C., Frayling, T. M., Cassell, P. G., Saker, P. J., Hitman, G. A., Walker, M., Levy, J. C., O'Rahilly, S., Rao, P. V., Bennett, A. J., Jones, E. C., Menzel, S., Prestwich, P., Simecek, N., Wishart, M., Dhillon, R., Fletcher, C., Millward, A., Demaine, A., Wilkin, T., Horikawa, Y., Cox, N. J., Bell, G. I., Ellard, S., McCarthy, M. I., and Hattersley, A. T. (2001). Studies of association between the gene for calpain-10 and type 2 diabetes mellitus in the United Kingdom. *Am. J. Hum. Genet.* **69**, 544–552.
305. Liang, H., Murase, Y., Katuta, Y., Asano, A., Kobayashi, J., and Mabuchi, H. (2005). Association of LMNA 1908C/T polymorphism with cerebral vascular disease and diabetic nephropathy in Japanese men with type 2 diabetes. *Clin. Endocrinol.* **63**, 317–322.
306. Armstrong, M., Haldane, F., Taylor, R. W., Humphriss, D., Berrish, T., Stewart, M. W., Turnbull, D. M., Alberti, K. G., and Walker, M. (1996). Human insulin receptor substrate-1: variant sequences in familial non-insulin-dependent diabetes mellitus. *Diabet. Med.* **13**, 133–138.
307. Jellema, A., Zeegers, M. P., Feskens, E. J., Dagnelie, P. C., and Mensink, R. P. (2003). Gly972Arg variant in the insulin receptor substrate-1 gene and association with Type 2 diabetes: a meta-analysis of 27 studies. *Diabetologia* **46**, 990–995.
308. Zacharova, J., Chiasson, J. L., Laakso, M., and the STOP-NIDDM Study Group. (2005). The common polymorphisms

(single nucleotide polymorphism [SNP] +45 and SNP +276) of the adiponectin gene predict the conversion from impaired glucose tolerance to type 2 diabetes: the STOP-NIDDM trial. *Diabetes* **54,** 893–899.

309. Bacci, S., Menzaghi, C., Ercolino, T., Ma, X., Rauseo, A., Salvemini, L., Vigna, C., Fanelli, R., Di Mario, U., Doria, A., and Trischitta, V. (2004). The +276 G/T single nucleotide polymorphism of the adiponectin gene is associated with coronary artery disease in type 2 diabetic patients. *Diabetes Care* **27,** 2015–2020.

310. Ukkola, O., Santaniemi, M., Rankinen, T., Leon, A. S., Skinner, J. S., Wilmore, J. H., Rao, D. C., Bergman, R., Kesaniemi, Y. A., and Bouchard, C. (2005). Adiponectin polymorphisms, adiposity and insulin metabolism: HERITAGE family study and Oulu diabetic study. *Ann. Med.* **37,** 141–150.

311. Mori, H., Ikegami, H., Kawaguchi, Y., Seino, S., Yokoi, N., Takeda, J., Inoue, I., Seino, Y., Yasuda, K., Hanafusa, T., Yamagata, K., Awata, T., Kadowaki, T., Hara, K., Yamada, N., Gotoda, T., Iwasaki, N., Iwamoto, Y., Sanke, T., Nanjo, K., Oka, Y., Matsutani, A., Maeda, E., and Kasuga, M. (2001). The Pro12→Ala substitution in PPAR-gamma is associated with resistance to development of diabetes in the general population: possible involvement in impairment of insulin secretion in individuals with type 2 diabetes. *Diabetes* **50,** 891–894.

312. Doney, A. S., Fischer, B., Leese, G., Morris, A. D., and Palmer, C. N. (2004). Cardiovascular risk in type 2 diabetes is associated with variation at the PPARG locus, a Go-DARTS study. *Arterioscler. Thromb. Vasc. Biol.* **24,** 2403–2407.

313. Altshuler, D., Hirschhorn, J. N., Klannemark, M., Lindgren, C. M., Vohl, M. C., Nemesh, J., Lane, C. R., Schaffner, S. F., Bolk, S., Brewer, C., Tuomi, T., Gaudet, D., Hudson, T. J., Daly, M., Groop, L., and Lander, E. S. (2000). The common PPARgamma Pro12Ala polymorphism is associated with decreased risk of type 2 diabetes. *Nat. Genet.* **26,** 76–80.

314. Nistico, L., Buzzetti, R., Pritchard, L. E., Van der Auwera, B., Giovannini, C., Bosi, E., Larrad, M. T., Rios, M. S., Chow, C. C., Cockram, C. S., Jacobs, K., Mijovic, C., Bain, S. C., Barnett, A. H., Vandewalle, C. L., Schuit, F., Gorus, F. K., Tosi, R., Pozzilli, P., and Todd, J. A. (1996). The CTLA-4 gene region of chromosome 2q33 is linked to, and associated with, type 1 diabetes. Belgian Diabetes Registry. *Hum. Mol. Genet.* **5,** 1075–1080.

315. Van der Auwera, B. J., Vandewalle, C. L., Schuit, F. C., Winnock, F., De Leeuw, I. H., Van Imschoot, S., Lamberigts, G., and Gorus, F. K. (1997). CTLA-4 gene polymorphism confers susceptibility to insulin-dependent diabetes mellitus (IDDM) independently from age and from other genetic or immune disease markers. The Belgian Diabetes Registry. *Clin. Exp. Immunol.* **110,** 98–103.

316. Redondo, M. J., Fain, P. R., and Eisenbarth, G. S. (2001). Genetics of type 1A diabetes. *Recent Prog. Horm. Res.* **56,** 69–89.

317. Erlich, H. A. (1991). HLA class II sequences and genetic susceptibility to insulin dependent diabetes mellitus. *Baillieres Clin. Endocrinol. Metab.* **5,** 395–411.

318. Kennedy, G. C., German, M. S., and Rutter, W. J. (1995). The minisatellite in the diabetes susceptibility locus IDDM2 regulates insulin transcription. Nat. Genet. **9,** 293–298.

319. Bell, G. I., Horita, S., and Karam, J. H. (1984). A polymorphic locus near the human insulin gene is associated with insulin-dependent diabetes mellitus. *Diabetes* **33,** 176–183.

320. Caux, F., Dubosclard, E., Lascols, O., Buendia, B., Chazouilleres, O., Cohen, A., Courvalin, J. C., Laroche, L., Capeau, J., Vigouroux, C., and Christin-Maitre, S. (2003). A new clinical condition linked to a novel mutation in lamins A and C with generalized lipoatrophy, insulin-resistant diabetes, disseminated leukomelanodermic papules, liver steatosis, and cardiomyopathy. *J. Clin. Endocrinol. Metab.* **88,** 1006–1013.

321. Haque, W. A., Oral, E. A., Dietz, K., Bowcock, A. M., Agarwal, A. K., and Garg, A. (2003). Risk factors for diabetes in familial partial lipodystrophy, Dunnigan variety. *Diabetes Care* **26,** 1350–1355.

322. Cao, H., and Hegele, R. A. (2000). Nuclear lamin A/C R482Q mutation in Canadian kindreds with Dunnigan-type familial partial lipodystrophy. *Hum. Mol. Genet.* **9,** 109–112.

323. Ukkola, O., Rankinen, T., Lakka, T., Leon, A. S., Skinner, J. S., Wilmore, J. H., Rao, D. C., Kesaniemi, Y. A., and Bouchard, C. (2005). Protein tyrosine phosphatase 1B variant associated with fat distribution and insulin metabolism. *Obes. Res.* **13,** 829–834.

324. Tai, E. S., Collins, D., Robins, S. J., O'connor, J. J., Jr., Bloomfield, H. E., Ordovas, J. M., Schaefer, E. J., and Brousseau, M. E. (2005). The L162V polymorphism at the peroxisome proliferator activated receptor alpha locus modulates the risk of cardiovascular events associated with insulin resistance and diabetes mellitus: The Veterans Affairs HDL Intervention Trial (VA-HIT). *Atherosclerosis* **187,** 153–160.

325. Manraj, M., Francke, S., Hebe, A., Ramjuttun, U. S., and Froguel, P. (2001). Genetic and environmental nature of the insulin resistance syndrome in Indo-Mauritian subjects with premature coronary heart disease: contribution of beta3-adrenoreceptor gene polymorphism and beta blockers on triglyceride and HDL concentrations. *Diabetologia* **44,** 115–122.

326. Strazzullo, P., Iacone, R., Siani, A., Cappuccio, F. P., Russo, O., Barba, G., Barbato, A., D'Elia, L., Trevisan, M., and Farinaro, E. (2001). Relationship of the Trp64Arg polymorphism of the beta3-adrenoceptor gene to central adiposity and high blood pressure: interaction with age. Cross-sectional and longitudinal findings of the Olivetti Prospective Heart Study. *J. Hypertens.* **19,** 399–406.

327. Pischon, T., Pai, J. K., Manson, J. E., Hu, F. B., Rexrode, K. M., Hunter, D., and Rimm, E. B. (2005). Peroxisome proliferator-activated receptor-gamma2 P12A polymorphism and risk of coronary heart disease in US men and women. *Arterioscler. Thromb. Vasc. Biol.* **25,** 1654–1658.

328. Li, S., Chen, W., Srinivasan, S. R., Boerwinkle, E., and Berenson, G. S. (2003). The Bogalusa Heart Study The peroxisome proliferator-activated receptor-gamma2 gene polymorphism (Pro12Ala) beneficially influences insulin resistance and its tracking from childhood to adulthood: the Bogalusa Heart Study. *Diabetes* **52,** 1265–1269.

329. Kahara, T., Takamura, T., Hayakawa, T., Nagai, Y., Yamaguchi, H., Katsuki, T., Katsuki, K., Katsuki, M., and Kobayashi, K. (2003). PPARgamma gene polymorphism is associated with exercise-mediated changes of insulin resistance in healthy men. *Metabolism* **52,** 209–212.

330. Meirhaeghe, A., Cottel, D., Amouyel, P., and Dallongeville, J. (2005). Association between peroxisome proliferator-activated receptor gamma haplotypes and the metabolic syndrome in French men and women. *Diabetes* **54,** 3043–3048.

331. Masuo, K., Katsuya, T., Fu, Y., Rakugi, H., Ogihara, T., and Tuck, M. L. (2005). Beta2-adrenoceptor polymorphisms relate to insulin resistance and sympathetic overactivity as early markers of metabolic disease in nonobese, normotensive individuals. *Am. J. Hypertens.* **18,** 1009–1014.
332. Ohashi, K., Ouchi, N., Kihara, S., Funahashi, T., Nakamura, T., Sumitsuji, S., Kawamoto, T., Matsumoto, S., Nagaretani, H., Kumada, M., Okamoto, Y., Nishizawa, H., Kishida, K., Maeda, N., Hiraoka, H., Iwashima, Y., Ishikawa, K., Ohishi, M., Katsuya, T., Rakugi, H., Ogihara, T., and Matsuzawa, Y. (2004). Adiponectin I164T mutation is associated with the metabolic syndrome and coronary artery disease. *J. Am. Coll. Cardiol.* **43,** 1195–1200.
334. Menuet, R., Lavie, C. J., and Milani, R. V. (2005). Importance and management of dyslipidemia in the metabolic syndrome. *Am. J. Med. Sci.* **330,** 295.
335. Zambon, A., Pauletto, P., and Crepaldi, G. (2005). Review article: The metabolic syndrome—A chronic cardiovascular inflammatory condition. *Aliment Pharmacol. Ther.* **22,** 20–23.
336. Chan, J. C., Tong, P. C., and Critchley, J. A. (2002). The insulin resistance syndrome: mechanisms of clustering of cardiovascular risk. *Semin. Vasc. Med.* **2,** 45–57.
337. Hutley, L., and Prins, J. B. (2005). Fat as an endocrine organ: Relationship to the metabolic syndrome. *Am. J. Med. Sci.* **330,** 280–289.
338. Scheen, A. J. (2004). Management of the metabolic syndrome. *Minerva Endocrinol.* **29,** 31–45.
339. Kim, S. H., and Reaven, G. M. (2004). The metabolic syndrome: one step forward, two steps back. *Diab. Vasc. Dis. Res.* **1,** 68–75.
340. Koh, K. K., Han, S. H., and Quon, M. J. (2005). Inflammatory markers and the metabolic syndrome: insights from therapeutic interventions. *J. Am. Coll. Cardiol.* **46,** 1978–1985.
341. Hegele, R. A., and Pollex, R. L. (2005). Genetic and physiological insights into the metabolic syndrome. *Am. J. Physiol. Regul. Integr. Comp. Physiol.* **289,** R663–R669.
342. Savage, D. B., Tan, G. D., Acerini, C. L., Jebb, S. A., Agostini, M., Gurnell, M., Williams, R. L., Umpleby, A. M., Thomas, E. L., Bell, J. D., Dixon, A. K., Dunne, F., Boiani, R., Cinti, S., Vidal-Puig, A., Karpe, F., Chatterjee, V. K., and O'Rahilly S. (2003). Human metabolic syndrome resulting from dominant-negative mutations in the nuclear receptor peroxisome proliferator-activated receptor-gamma. *Diabetes* **52,** 910–917.
343. Kotzka, J., and Muller-Wieland, D. (2004). Sterol regulatory element-binding protein (SREBP)-1: Gene regulatory target for insulin resistance? *Expert Opin. Ther. Targets* **8,** 141–149.
344. Koo, S. H., Satoh, H., Herzig, S., Lee, C. H., Hedrick, S., Kulkarni, R., Evans, R. M., Olefsky, J., and Montminy, M. (2004). PGC-1 promotes insulin resistance in liver through PPAR-alpha-dependent induction of TRB-3. *Nat. Med.* **10,** 530–534.
345. Shulman, A. I., and Mangelsdorf, D. J. (2005). Retinoid x receptor heterodimers in the metabolic syndrome. *N. Engl. J. Med.* **353,** 604–615.
346. Berger, J. P., Akiyama, T. E., and Meinke, P. T. (2005). PPARs: therapeutic targets for metabolic disease. *Trends Pharmacol. Sci.* **26,** 244–251.
347. Han, S. H., Quon, M. J., and Koh, K. K. (2005). Beneficial vascular and metabolic effects of peroxisome proliferator-activated receptor-alpha activators. *Hypertension* **46,** 1086–1092.
348. Chinetti-Gbaguidi, G., Fruchart, J. C., and Staels, B. (2005). Role of the PPAR family of nuclear receptors in the regulation of metabolic and cardiovascular homeostasis: new approaches to therapy. *Curr. Opin. Pharmacol.* **5,** 177–183.
349. Cavedon, C. T., Bourdoux, P., Mertens, K., Van Thi, H. V., Herremans, N., de Laet, C., and Goyens, P. (2005). Age-related variations in acylcarnitine and free carnitine concentrations measured by tandem mass spectrometry. *Clin. Chem.* **51,** 745–752.
350. Delolme, F., Vianey-Saban, C., Guffon, N., Favre-Bonvin, J., Guibaud, P., Becchi, M., Mathieu, M., and Divry, P. (1997). Study of plasma acylcarnitines using tandem mass spectrometry. Application to the diagnosis of metabolism hereditary diseases. *Arch. Pediatr.* **4,** 819–826.
351. Sinha, R., Dufour, S., Petersen, K. F., LeBon, V., Enoksson, S., Ma, Y. Z., Savoye, M., Rothman, D. L., Shulman, G. I., and Caprio, S. (2002). Assessment of skeletal muscle triglyceride content by (1)H nuclear magnetic resonance spectroscopy in lean and obese adolescents: relationships to insulin sensitivity, total body fat, and central adiposity. *Diabetes* **51,** 1022–1027.
352. Laurent, D., Hundal, R. S., Dresner, A., Price, T. B., Vogel, S. M., Petersen, K. F., and Shulman, G. I. (2000). Mechanism of muscle glycogen autoregulation in humans. *Am. J. Physiol. Endocrinol. Metab.* **278,** E663–E668.
353. Petersen, K. F., Dufour, S., Befroy, D., Lehrke, M., Hendler, R. E., and Shulman, G. I. (2005). Reversal of nonalcoholic hepatic steatosis, hepatic insulin resistance, and hyperglycemia by moderate weight reduction in patients with type 2 diabetes. *Diabetes* **54,** 603–608.
354. Olpin, S. E. (2005). Fatty acid oxidation defects as a cause of neuromyopathic disease in infants and adults. *Clin. Lab.* **51,** 289–306.
355. Saudubray, J. M., Martin, D., de Lonlay, P., Touati, G., Poggi-Travert, F., Bonnet, D., Jouvet, P., Boutron, M., Slama, A., Vianey-Saban, C., Bonnefont, J. P., Rabier, D., Kamoun, P., and Brivet, M. (1999). Recognition and management of fatty acid oxidation defects: A series of 107 patients. *J. Inherit. Metab. Dis.* **22,** 488–502.
356. Brown-Harrison, M. C., Nada, M. A., Sprecher, H., Vianey-Saban, C., Farquhar, J., Jr., Gilladoga, A. C., and Roe, C. R. (1996). Very long-chain acyl-CoA dehydrogenase deficiency: Successful treatment of acute cardiomyopathy. *Biochem. Mol. Med.* **58,** 59–65.
357. Kennedy, J. A., Unger, S. A., and Horowitz, J. D. (1996). Inhibition of carnitine palmitoyltransferase-1 in rat heart and liver by perhexiline and amiodarone. *Biochem. Pharmacol.* **52,** 273–280.
358. Myburgh, D. P., and Goldman, A. P. (1978). The anti-arrhythmic efficacy of perhexiline maleate, disopyramide and mexiletine in ventricular ectopic activity. *S. Afr. Med. J.* **54,** 1053–1055.
359. Lee, L., Campbell, R., Scheuermann-Freestone, M., Taylor, R., Gunaruwan, P., Williams, L., Ashrafian, H., Horowitz, J., Fraser, A. G., Clarke, K., and Frenneaux, M. (2005). Metabolic modulation with perhexiline in chronic heart

failure: a randomized, controlled trial of short-term use of a novel treatment. *Circulation* **112,** 3280–3288.

360. Kalra, S. P., and Kalra, P. S. (2005). Gene-transfer technology: a preventive neurotherapy to curb obesity, ameliorate metabolic syndrome and extend life expectancy. *Trends Pharmacol. Sci.* **26,** 488–495.

361. Shah, R., and Jindal, R. M. (2003). Reversal of diabetes in the rat by injection of hematopoietic stem cells infected with recombinant adeno-associated virus containing the preproinsulin II gene. *Pancreatology* **3,** 422–428.

362. Sasaki, T., Fujimoto, K., Sakai, K., Nemoto, M., Nakai, N., and Tajima, N. (2003). Gene and cell-based therapy for diabetes mellitus: endocrine gene therapeutics. *Endocr. Pathol.* **14,** 141–144.

363. Wallhaus, T. R., Taylor, M., DeGrado, T. R., Russell, D. C., Stanko, P., Nickles, R. J., and Stone, C. K. (2001). Myocardial free fatty acid and glucose use after carvedilol treatment in patients with congestive heart failure. *Circulation* **103,** 2441–2446.

364. Bell, D. S. (2005). Optimizing treatment of diabetes and cardiovascular disease with combined alpha,beta-blockade. *Curr. Med. Res. Opin.* **21,** 1191–1200.

365. Rupp, H., Zarain-Herzberg, A., and Maisch, B. (2002). The use of partial fatty acid oxidation inhibitors for metabolic therapy of angina pectoris and heart failure. *Herz* **27,** 621–636.

366. Stanley, W. C. (2002). Partial fatty acid oxidation inhibitors for stable angina. *Expert Opin. Investig. Drugs* **11,** 615–629.

367. Zarain-Herzberg, A., and Rupp, H. (1999). Transcriptional modulators targeted at fuel metabolism of hypertrophied heart. *Am. J. Cardiol.* **83,** 31H–37H.

368. Lionetti, V., Linke, A., Chandler, M. P., Young, M. E., Penn, M. S., Gupte, S., d'Agostino, C., Hintze, T. H., Stanley, W. C., and Recchia, F. A. (2005). Carnitine palmitoyl transferase-I inhibition prevents ventricular remodeling and delays decompensation in pacing-induced heart failure. *Cardiovasc. Res.* **66,** 454–461.

369. Schmidt-Schweda, S., and Holubarsch, C. (2000). First clinical trial with etomoxir in patients with chronic congestive heart failure. *Clin. Sci.* **99,** 27–35.

370. Stanley, W. C. (2004). Myocardial energy metabolism during ischemia and the mechanisms of metabolic therapies. *J. Cardiovasc. Pharmacol. Ther.* **9,** S31–S45.

371. Sabbah, H. N., Chandler, M. P., Mishima, T., Suzuki, G., Chaudhry, P., Nass, O., Biesiadecki, B. J., Blackburn, B., Wolff, A., and Stanley, W. C. (2002). Ranolazine, a partial fatty acid oxidation (pFOX) inhibitor, improves left ventricular function in dogs with chronic heart failure. *J. Card. Fail.* **8,** 416–422.

372. Pepine, C. J., and Wolff, A. A. (1999). A controlled trial with a novel anti-ischemic agent, ranolazine, in chronic stable angina pectoris that is responsive to conventional antianginal agents. *Am. J. Cardiol.* **84,** 46–50.

373. Fragasso, G., Piatti Md, P. M., Monti, L., Palloshi, A., Setola, E., Puccetti, P., Calori, G., Lopaschuk, G. D., and Margonato, A. (2003). Short- and long-term beneficial effects of trimetazidine in patients with diabetes and ischemic cardiomyopathy. Am. Heart J. **146,** E18.

374. Stanley, W. C. (2005). Rationale for a metabolic approach in diabetic coronary patients. *Coron. Artery Dis.* **16,** S11–S15.

375. Kantor, P. F., Lucien, A., Kozak, R., and Lopaschuk, G. D. (2000). The antianginal drug trimetazidine shifts cardiac energy metabolism from fatty acid oxidation to glucose oxidation by inhibiting mitochondrial long-chain 3-ketoacyl coenzyme A thiolase. *Circ. Res.* **86,** 580–588.

376. MacInnes, A., Fairman, D. A., Binding, P., Rhodes, J., Wyatt, M. J., Phelan, A., Haddock, P. S., and Karran, E. H. (2003). The antianginal agent trimetazidine does not exert its functional benefit via inhibition of mitochondrial long-chain 3-ketoacyl coenzyme A thiolase. *Circ. Res.* **93,** e26–e32.

377. Tabbi-Anneni, I., Helies-Toussaint, C., Morin, D., Bescond-Jacquet, A., Lucien, A., and Grynberg, A. (2003). Prevention of heart failure in rats by trimetazidine treatment: a consequence of accelerated phospholipid turnover? *J. Pharmacol. Exp. Ther.* **304,** 1003–1009.

378. Argaud, L., Gomez, L., Gateau-Roesch, O., Couture-Lepetit, E., Loufouat, J., Robert, D., and Ovize, M. (2005). Trimetazidine inhibits mitochondrial permeability transition pore opening and prevents lethal ischemia-reperfusion injury. *J. Mol. Cell Cardiol.* **39,** 893–899.

379. Chung, M. K. (2004). Vitamins, supplements, herbal medicines, and arrhythmias. *Cardiol. Rev.* **12,** 73–84.

380. Tavazzi, L., Tognoni, G., Franzosi, M. G., Latini, R., Maggioni, A. P., Marchioli, R., Nicolosi, G. L., and Porcu, M. (2004). Rationale and design of the GISSI heart failure trial: A large trial to assess the effects of n-3 polyunsaturated fatty acids and rosuvastatin in symptomatic congestive heart failure. *Eur. J. Heart Fail.* **6,** 635–641.

381. Macchia, A., Levantesi, G., Franzosi, M. G., Geraci, E., Maggioni, A. P., Marfisi, R., Nicolosi, G. L., Schweiger, C., Tavazzi, L., Tognoni, G., Valagussa, F., Marchioli, R., and the GISSI-Prevenzione Investigators. (2005). Left ventricular systolic dysfunction, total mortality, and sudden death in patients with myocardial infarction treated with n-3 polyunsaturated fatty acids. *Eur. J. Heart Fail.* **7,** 904–909.

382. Singer, P., and Wirth, M. (2004). Can n-3 PUFA reduce cardiac arrhythmias? Results of a clinical trial. *Prostaglandins Leukot. Essent. Fatty Acids* **71,** 153–159.

383. Pepe, S., Tsuchiya, N., Lakatta, E. G., and Hansford, R. G. (1999). PUFA and aging modulate cardiac mitochondrial membrane lipid composition and Ca2+ activation of PDH. *Am. J. Physiol.* **276,** H149–H158.

SECTION VII

Molecular Genetics of Dysrhythmias

CHAPTER **17**

Dysrhythmias and Sudden Death

OVERVIEW

During the past 15 years, increasing interaction between clinical cardiologists, molecular biologists, and geneticists allowed the detection of a number of gene mutations in patients with life-threatening dysrhythmias, particularly mutations in genes encoding most of the proteins forming the ion channels and transporters. In many cases, subsequent genetic screening has revealed significant genetic diversity in the channel genes, with allelic heterogeneity spread over the entire gene, and often with more than one mutation present. Although genetic heterogeneity represents a tremendous challenge for mutation identification, the availability of new tools, including high-throughput technologies, to map the human genome, together with the availability of large databases of single nucleotide polymorphisms (SNPs) and haplotype markers is facilitating the progressive discovery of new mutations and numerous SNPs. Furthermore, the availability of genotype–phenotype correlation in large families is being exploited to evaluate the intergene, interfamilial, and intrafamilial differences in the clinical phenotype, the gene specificity, and the effects of a given mutation. Heterogeneity, even within mutation carriers in the same family, has increased the focus on the significance of modulatory/modifying factors, as well as in the discovery of new gene mutations.

Currently available data on the molecular genetics and genotype–phenotype correlations in rhythm disorders, as well as their implications in diagnosis and treatment, will be discussed in this chapter.

INTRODUCTION

Cardiac dysrhythmias represent an important cause of mortality, and in the United States alone more than 300,000 cases of sudden death occur each year because of ventricular dysrhythmias. The availability of the Human Genome Project (HGP) has led to a collection of new data and enabled discernment of the polygenic nature of some of the cardiac dysrhythmias. At the same time, it has been possible by genetic screening of rare monogenic dysrhythmogenic disorders to obtain specific new and crucial information that has advanced our understanding of the molecular and genetic basis of many inherited cardiac dysrhythmias, in particular the channelopathies.

Looking back, it is clear that progress in the molecular genetic analysis of cardiac dysrhythmias has been rather delayed in part because of the difficulties of carrying out traditional genetic linkage studies (which are based on the genetic information obtained from large multigeneration families) and because of the high mortality associated with these disorders. Despite this difficulty, it is significant that increasing communication and interaction between clinicians and molecular biologists resulted in the screening of a significant number of individuals with genotyped dysrhythmias, and this led to the discovery of numerous mutated genes, in particular those encoding subunits of proteins that constitute the cardiac ion channels. In addition, silent mutations and functional DNA polymorphisms have also been found to play a significant role in increasing the susceptibility for these dysrhythmogenic disorders, together with nongenetic or environmental factors such as gender, aging, the presence of cardiac structural defects, and ethnicity. Although molecular genetic analysis has revealed an unforeseen genetic diversity within the inherited dysrhythmias, making its genetic analysis rather complex, the availability of the HGP and evolving technologies seem to offer the promise of continued progress in this important field of cardiac pathology.

CAUSALITY OF CARDIAC DYSRHYTHMIAS

Until recently, many of the cardiac pathologies, including hypertension, cardiomyopathies, and, of course, dysrhythmias, were considered to be primary or idiopathic. However, with the availability of the human genome sequence and accelerated progress in computer and molecular genetic technologies, our understanding of the role that genetics and molecular biology play in cardiovascular pathology in general, and specifically in dysrhythmias, has grown and greatly modified our perspective. Analysis of the causality of cardiac dysrhythmias (either of the inherited or acquired type), like in other cardiovascular pathologies, has revealed their multifactorial origin, encompassing individual molecular and genetic changes together with a number of epigenetic factors, including age, gender, and environment. The patient's clinical findings often are a representation of aggregated manifestations in cardiac pathology, including cardiac structural defects (e.g., coronary artery disease [CAD] and cardiomyopathies) and triggered/associated rhythm disorders. Furthermore, to identify the precise factors involved in the causality of cardiac dysrhythmias, it will be necessary to carry out further larger population studies that will integrate both genetic and molecular findings to specific variables such as ethnicity and gender.

VENTRICULAR ACTION POTENTIAL AND ION CURRENTS

Myocytes have a characteristically long action potential (AP) that is orchestrated by the highly organized activity and interaction of multiple voltage-dependent ion channels and membrane transporters. The properties of AP (e.g., duration, configuration) show striking variation, depending on the region of the heart (e.g., atrium compared with ventricle).[1] Cells of the sinoatrial node have very simple APs with no stable resting potential in contrast to ventricular myocytes that have more complex APs. In the SA node, after repolarization, the membrane potential slowly depolarizes spontaneously, reflecting its pacemaker role and automaticity. Purkinje and mid-myocardial cells display APs that are much longer than those in epicardium. These physiological heterogeneities are thought to reflect variations in the expression or function of the ion channels and other proteins that constitute cardiac ion currents. Further modulation of these APs by changes in rate, mutations in ion channel genes, or drug exposures can promote re-entrant excitation, a common basis for many cardiac dysrhythmias. The impact of exogenous stressors such as drugs, myocardial ischemia, or autonomic activation on the acute electrophysiological response of a myocyte likely reflects changes in the activity of specific ion channels, including channels activated by specific stimuli such as ATP depletion, muscarinic stimulation, or mechanical stretch.[2] Moreover, exogenous stressors may also include long-term changes in gene expression, leading to chronic responses in myocardium and to further electrophysiological remodeling.[1,3]

The model generally used to describe the cardiac action potential (AP) is the action potential of the ventricular myocyte (Fig. 1). Depolarizing currents, primarily sodium and calcium, are responsible for the AP upstroke and maintenance of the AP plateau, whereas repolarizing currents, primarily potassium, in combination with a reduction in depolarizing currents, determine the restoration of the resting membrane potential. The AP is propagated as an impulse serving as an electrical stimulation to the cells that lie adjacent to it and can be conducted from one cell to all the cells of the heart.

There are five phases in the ventricular myocyte action potential (numbered 0–4). The cell AP depicted in Fig. 1 encompasses both the cellular depolarization (phase 0) and the repolarization phase (phases 1, 2, and 3) that correspond with systole. At the end of systole, the AP returns to its initial level (phase 4) with polarized ventricular myocytes at the resting membrane potential, the membrane potential state in the absence of cell stimuli and are in diastole; they have a stable diastolic transmembrane potential compared with a non-stable potential present in automatic cells.[4]

The resting potential is characterized by a myocyte membrane highly permeable to K^+, with specific K^+ channels (i.e., inward rectifier channel) open at rest; in contrast, cardiac Na^+ channels, which provide the route for Na^+ to enter cells, are closed, and the resting membrane is Na^+ impermeant despite the large electrochemical gradient favoring Na^+ entry.

Once the cell is electrically stimulated (by an electric current from an adjacent cell), a sequence of events begins involving the influx and efflux of multiple ions that together generate the AP. A change in the potential across the cell caused by a propagating impulse is sensed by the Na^+ channel protein, which alters its conformation to open. Inasmuch as Na^+ ions (and to a lesser extent Ca^{++} ions) are at higher concentrations extracellularly than intracellularly, openings of Na^+ (and Ca^{++}) channels cause these cations to enter the cell in a large, rapid Na^+ flux and rapidly depolarize the membrane (phase O). The resultant fast upstroke in ventricular (and atrial) myocytes is accomplished by the I_{Na} channel and inward current that is short-lived, with rapid inactivation occurring as a function of time and voltage. A slower channel capable of carrying both Ca^{++} and Na^+ currents (I_{si}) also contributes to the upstroke of the AP.

In some cells, a rapid phase 1 repolarization then ensues, because of outward movement of K^+ mediated by specific channels by transient outward currents (I_{to}). Two kinetic variants of the cardiac I_{to} have been described, a fast component ($I_{to,f}$) and a slow component ($I_{to,s}$), which extends beyond phase 1 into phase 2. Although epicardial myocytes have greater expression and activity of the overall I_{to} current, $I_{to,f}$ expression predominates (also observed in RV); in contrast, in the LV endocardium with its longer AP and septal cells, lower overall I_{to} currents are found, primarily the $I_{to,s}$ variety.

In addition to phase 1 repolarization being correlated with rapid voltage-dependent inactivation of I_{Na} during phase 0 and phase 1, Ca^{++} channels open. In phase 2, cardiac cells uniquely exhibit a characteristically long (hundreds of milliseconds) plateau phase of the AP. The plateau phase reflects a balance between two types of inward Ca^{++} currents (although largely mediated by the L-type channels) and several types of outward K^+ current, largely through delayed rectifier K^+ channels. The two distinct Ca^{++} currents include the low-voltage–activated transient Ca^{++} current (I_{Ca-T} T-type) and the high-voltage–activated long-lasting Ca^{++} current (I_{Ca-L} L-type) that bring in Ca^{++} needed for contraction. The T type Ca^{++} channels are primarily found in atrial pacemaker cells, Purkinje fibers, and coronary artery smooth muscle and contribute little to the Ca^{++} influx needed

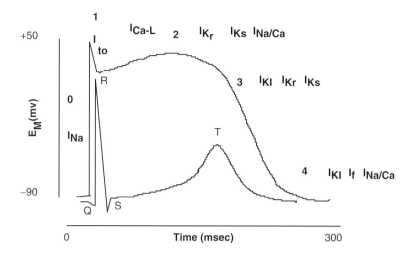

Figure 1 The action potential (AP) in the ventricular myocyte is mediated by ion currents. A representation of a ventricular AP is depicted with voltage on the Y-axis and time on the X-axis. Inward sodium current (I_{Na}) mediates the rapid phase 0 depolarization. Transient outward potassium currents (I_{to}) mediate rapid phase 1 repolarization (also termed the notch). The L-type inward calcium current contributes to the long plateau of phase 2. Several outward potassium currents are primarily responsible for the repolarization of the AP in phases 2 and 3, including I_{Kr}, I_{Ks}, and I_{K1}. Phase 4 represents the reestablishment of the resting potential. Also shown is a typical ECG profile denoting QRS depolarization and the ST-T repolarization.

for excitation contraction. In contrast, the L-type Ca^{++} current is found in virtually all cardiac cells and, as noted previously, is the primary Ca^{++} current operative during the plateau phase. Another current found in the terminal stages of the phase 2 plateau is an inward current through the electrogenic Na^+/Ca^{++} exchanger; it seems to play a pivotal role at this point in extruding Ca^{++} and serves a regulatory function in modulating myocyte Ca^{++} levels.

In the subsequent phase 3, with the rapid inactivation of the inward currents and Ca^{++} channels, the net outward currents accomplished by outward movement of K^+ occur by means of the delayed rectifier K^+ channels (I_K) and drive phase 3 repolarization. The delayed rectifying K^+ currents include three distinct channel populations: a rapid (I_{Kr}), a slow channel (I_{Ks}), and ultra-rapid channels (I_{Kur}).

As the I_{si} current becomes inactivated, the inflow of Na^+ and Ca^{++} stop. The existing ionic imbalance created in phase 3 will be corrected in phase 4 by an active ATPase energy-consuming ion pump, which allows Na^+ and Ca^{++} ions to leave and K^+ to enter the cell. Moreover, the involvement of electrogenic components such as ion exchangers (i.e., Na^+-Ca^{++} exchanger) and ion pumps (i.e., Na^+-K^+ ion exchange pump) expressed in cardiac tissue is contributory in maintaining intracellular ionic homeostasis and the electrical gradient in the face of large ion fluxes accompanying each AP.

Slow response cells in the sinus node and in the AV node exhibit slow depolarization during phase 4, a manifestation of pacemaker channel activity. Furthermore, a rapid phase 1 upstroke is absent, and initial depolarization is primarily accomplished by the opening of L-type Ca^{++} channels.

CARDIAC ION CHANNELS

Ion channels are multi-subunit transmembrane protein complexes that perform the task of mediating the selective flow of millions of ions per second across cell membranes and are the fundamental functional units of biological excitability. Although some ion channels open and close randomly at all membrane potentials (i.e., voltage-independent gating), other ion channels are normally closed, but their open probability can be greatly enhanced by a change in membrane potential (voltage-gated channels), by the binding of extracellular or intracellular ligands (ligand-gated channels), or by physical stimuli (mechanosensitive and heat-sensitive channels). The voltage-gated channels include Na^+ and Ca^{++} channels, transient outward (I_{to}) and delayed rectifier K^+ channels (I_k), and the pacemaker channels (I_f).

The opening and closing of specific ion channels resulting in the various currents as described previously can be individually modulated by a variety of specific activators and blockers as noted in Table I.

Most of the relevant cardiac ion channel genes are composed of pore-forming α-subunits (normally transmembrane proteins) and function-modifying auxiliary β-subunits that play modulatory, structural, or stabilizing roles and that can either be cytoplasmic proteins or transmembrane spanning proteins. These auxiliary subunits have been subdivided into two main classes.[5] One class consists of entirely cytoplasmic intracellular subunits with no transmembrane domains and includes the β-subunits of the voltage-gated K^+ and Ca^{++} channels (see Table II). The other class contains at least one transmembrane domain and includes the $α_2δ$-subunits of the Ca^{++} channel and the β-subunits of the Na^+ channel.[6] Although the expression of genes encoding the pore-forming α-subunit is sufficient to generate an ion current, recapitulation of all the physiological features of ion currents in myocytes requires the expression of both α- and β-subunits, as well as numerous gene products that regulate channel subunit trafficking, post-translational modifications including phosphorylation and dephosphorylation, assembly, and targeting, and anchoring of the channel subunits to specific subcellular domains.[7] This expands the spectrum of potential genetic loci (e.g., modifying genes) that can modulate ion channel

TABLE I Pharmacological and Physiological Modulators of Specific Ion Currents

Current	Blockers/inhibitors	Activators
Inward		
I_{Na}	Tetrodoxin, cadmium, zinc	
	Class IA: quinidine, procainamide	
	Class IB: lidocaine, tocainide, mexiletine	
	Class IC: encainide, flecainide, propafenone	
I_{Ca-L}	Nifedipine, diltiazem, verapamil	BayK 8644
I_{Ca-T}	Nickel, mibefradil	Endothelin-1
I_f, I_h (pacemaker)	Cesium	
$I_{Na/Ca}$		
Outward		
Delayed rectifier		
I_{Ks}	Amiodarone, clofilium, chromanol-293B	
I_{Kr}	Sotalol, propafenone, quinidine, clofilium, almokalant, dofetilide	
I_{Kur}	4-AP, quinidine, terfenadine	
Inward rectifier		
I_{K1}	TEA, cesium, barium, propafenone	
Other		
I_{K-ACh}	Barium	Acetylcholine, adenosine, somatostatin
I_{K-ATP}	ATP, phentolamine, glyburide, glipizide	Pinacidil, cromakalim, diazoxide, nicorandil, minoxidil
I_{to}	4-AP, quinidine, TEA, terfenadine	
I_{Kp}	Barium	

4-AP, 4-Aminopyridine; TEA, tetraethylammonium.

function. As we will shortly see, mutations that cause a variety of cardiac dysrhythmias have been identified primarily in the ion channel subunit genes but have also been found in other closely associated loci (e.g., *ankyrin*).

A typical α-subunit of the Na⁺ channel (shown in Fig. 2A) forms a functional channel by itself, and is composed of four homologous domains (I–IV), which each contain six helical transmembrane segments (S1–S6) and an additional membrane-associated pore loop. The S4 transmembrane segments in each domain contain an array of positively charged amino acid residues (arginine and lysine) serving as the major voltage sensors for channel activation[8,9]; the S5 and S6 segments and the pore loop between them form the transmembrane pore, and the highly conserved intracellular loop between domains III and IV is proposed to form an inactivation gate. The Na⁺ channel also contains two ancillary subunits, β1- and β2-subunits, which are single membrane–spanning glycoproteins that modulate channel gating.

As in voltage-gated Na⁺ channels, the α1-subunit of Ca⁺⁺ channels is composed of four homologous internal domains, each containing six transmembrane helices and a pore-forming loop with both N- and C-termini and hydrophilic peptide linking sequences (between domains) intracellularly located (Fig. 2B). The pore-forming α1-subunit forms an active L-type Ca⁺⁺ channel complex by interacting with an intracellular β-subunit and with a large polypeptide generated by expression of a single gene, comprising two protein products, α2- and δ-subunits linked by a disulfide bridge.[10] The α2-protein is large and located extracellularly, whereas the δ-subunit is small and includes a single membrane-spanning segment and an extracellular domain that links to the α2-subunit. In contrast, the α-subunit of the voltage-gated K⁺ channels is similar to the K⁺ channel initially cloned from the *Drosophila* mutant *shaker* and contains a single domain with six transmembrane segments (homologous to the Na⁺ channel). The segment between the fifth and sixth membrane-spanning α-helices (S5 and S6) re-enters the membrane to form the P-loop, which generates the outer pore. The pore region of nearly all the K⁺ channels includes a signature GXG (often GYG) motif comprising a critical element of the selectivity filter that confers ion selectivity (i.e., the capacity to discriminate between cations). That the GXG motif is necessary for K⁺ selectivity has been amply demonstrated both from studies showing that mutations or deletions of the motif can result in loss of K⁺ selectivity[11] and consistent with

TABLE II Major Cardiac Currents and Their Ion Channel Subunits

Currents	α-Subunit (gene)	Auxiliary subunit (gene)
Inward		
I_{Na}	SCN5A	β1 (*SCN1B*)
		β2 (*SCN2B*)
I_{Ca-L}	$α_1C$ (*CACNL1A1*)	β1 (*CACN1B*)
		β2 (*CACN2B*)
		$α_2γ$ (*CACNA2D1*)
I_{Ca-T}	$α_1H$ (*CACNA1H*)	
I_f, I_h (pacemaker current)	BCNG, HCN2, HCN4	
$I_{Na/Ca}$	SLC8A1 (*NCX1.1*)	
Outward		
Delayed rectifier		
I_{Ks}	KvLQT1 (*KCNQ1*)	minK/IsK (*KCNE1*)
I_{Kr}	HERG (*KCNH2*)	minK /IsK (*KCNE1*)
		MiRP1 (*KCNE2*)
I_{Kur}	Kv1.5 (*KCNA5*)	Kvβ1 (*KCNAB1*)
		Kvβ2 (*KCNAB2*)
Inward rectifier		
I_{K1}	Kir2.1 (*KCNJ2*)	
	Kir2.2 (*KCNJ2*)	
Other		
I_{K-ACh}	GIRK1, Kir3.1 (*KCNJ3*)	
	GIRK2, Kir3.4 (*KCNJ5*)	
I_{K-ATP}	KIR6.2, BIR (*KCNJ11*)	SUR2 (*ABCC9*)
I_{to}	Kv4.3 (*KCND3*)	
	Kv1.4 (*KCNA4*)	
I_{Kp}	TWIK1 (*KCNK1*)	
	CFTR (*ABCC7*)	MiRP$_1$ (*KCNE2*)
	KvLQT1 (*KCNQ1*)	

findings from high-resolution crystallographic studies of the K+ channel.[12,13] However, it is also clear that additional structural features are necessary for the GXG motif to confer K+ selectivity.[14,15] For example, cyclic nucleotide–gated (*HCN*) pacemaker channels contain a GYG motif in the P-loop but show relatively poor selectivity for K+ over Na+.[15]

As previously noted with the Na+ channel, the S4 transmembrane segments of the voltage-gated K+ channel serves as the voltage sensor (see Fig. 3A). Interestingly, the α-subunit of voltage-gated K+ channels (like the voltage-gated Na+ and Ca++ channels) also forms a structure with fourfold symmetry, albeit with a single internal domain; they do so by assembling as tetramers to generate a pore-forming structure (Fig. 3B–3C).

This basic channel architecture of the α-subunit is shared by a variety of cardiac K+ channels, including the primary repolarization channel(s) with I_{to} current (i.e., *Kv4.3* and *Kv1.4*), the channels associated with the delayed rectifier currents I_{kr} (i.e., HERG), I_{Ks} (i.e., KvLQT1 or *KCNQ1*), and I_{Kur} (Kv1.5), and the two kinetically different I_f pacemaker channels encoded by *HCN2* (fast) and *HCN4* (slow). The latter two are 90% homologous with each other in the S1–S6 transmembrane sequences, voltage sensor, and pore and contain a cyclic nucleotide–binding domain (cNBD) but are divergent in their N-and C-termini.[16,17]

The auxiliary β-subunits that are recruited both impact K+ channel surface expression and confer additional regulatory kinetic control to specific K+ channels. The assignment of specific regulatory function to specific subunits with K+ channels is often difficult given the tetrameric nature of the α-subunits that can be either heteromultimeric or homomultimers and the diversity of specific β-subunits. Much information can be gleaned from the functional coexpression of specific α-subunit genes with β-subunits and the biophysical and pharmacological characterization of the resulting current. These functional expression studies have primarily been conducted using specific transcripts generated *in vitro* from the

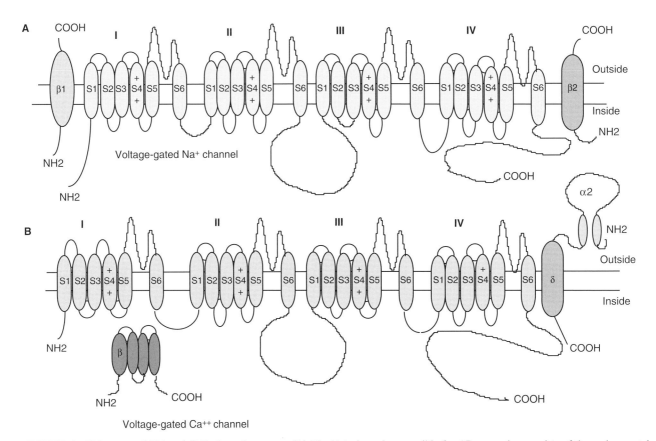

FIGURE 2 Voltage-gated Na+ and Ca++ channel structure. (A) The Na+ channel responsible for AP generation consists of three glycoprotein subunits: a pore-forming α-subunit associated with β1- and β2-subunits. The α-subunit is composed of four homologous domains (I–IV), which contain six transmembrane segments (S1–S6) and a pore loop. The β-subunits are single-membrane–spanning glycoproteins that modulate channel gating. S4 segment in each domain serves as a voltage sensor for channel activation, S5 and S6 segments and the intervening pore loop form the transmembrane pore, and the intracellular loop between domains III and IV forms the inactivation gate. (B) Ca++ channels are composed of a pore forming α-subunit, a transmembrane δ-subunit covalently linked to an extracellular α2-subunit, and an intracellular β-subunit. The α1-subunit of Ca++ channels is composed of four homologous membrane-spanning internal domains, each with six transmembrane helices and a pore-forming P loop. The α2δ complex is derived from a precursor protein that is cleaved to yield separate α2 and δ proteins linked by disulfide bonds.

gene(s) of interest after injection in *Xenopus laevis* oocytes; currents in the frog oocytes are subsequently measured by voltage-clamp technique and compared with the profiles obtained from *in vivo* cardiac studies. For instance, coexpression of the *HERG* α-subunit (encoded by *KCNH2*) with the MiRP1 β-subunit (encoded by *KCNE2*) recapitulates the biophysical and pharmacological properties of cardiac I_{Kr}.[18] Similarly, coexpression in COS cells of another β-subunit minK (encoded by *KCNE1*) with *HERG* increased I_{Kr} amplitude compared with *HERG* alone.[19] Coexpression and coassembly of the *KvLQT1* α-subunit (encoded by *KCNQ1*) and the single transmembrane-spanning minK β-subunit recapitulates the slow delayed rectifier current (I_{Ks}), whereas expression of *KvLQT1* alone gave rise to a current without correlate in cardiac myocytes.[20] Heterologous expression of the α-subunit Kv1.5 results in a delayed rectifier current with kinetics similar to I_{Kur}[21]; moreover, antisense oligonucleotide inhibition of Kv1.5 in cultured human atrial cardiomyocytes reduces I_{Kur}.[22] A number of cytosolic β-subunits have been identified whose coexpression with *Kv1.5* accelerates inactivation of the I_{Kur} channel and also seems to alter the channel activity regulation by phosphorylation by PKC or PKA.[23,24]

The cardiac inward rectifying K+ selective channels (IRKs) encoded by the *Kir* superfamily possess a simpler α structure consisting of two membrane-spanning domains (M1 and M2) that flank a highly conserved pore region containing the conserved H5 segment (Fig. 3D). The H5 and M2 segments, in conjunction with the C-terminal hydrophilic domain, are critical for K+ permeation. As in the voltage-gated K+ channels, four α-subunits assemble to form a functional IRK channel. However, these channels do not display intrinsic voltage sensitivity, presumably because they lack the S4 voltage sensor. As previously noted, these channels (I_{K1}) play a primary role in establishing the resting potential, enabling the formation of the plateau phase, and contribute to repolarization, playing a role in slowing heart rate. Cardiac I_{K1} is recapitulated by the expression of Kir2.1 (*IRK1*).[25] Heteromultimeric channel formation and interactions with a variety of other proteins increase the diverse range of regulatory functions with this channel class. This group of K+ channels conducts current

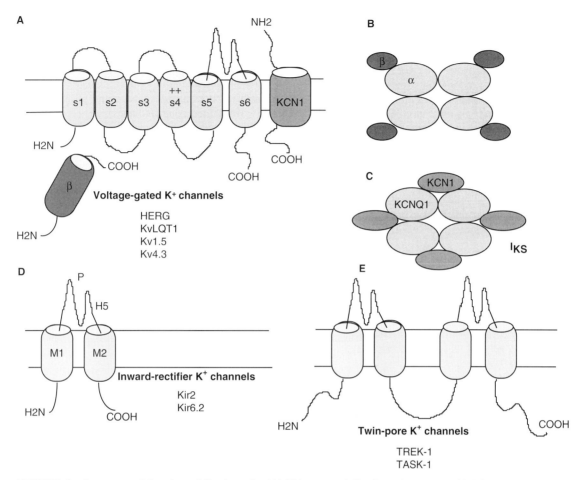

FIGURE 3 Structure and function of K⁺ channels. (A) Voltage-gated K⁺ channels are assembled from α-subunits (containing 6 transmembrane domains [s1–6]) and auxiliary β-subunits. (B) A functional voltage-gated K⁺ channel is formed by four subunits clustered around the central pore. Physical association of identical or different subunits can form a tetrameric channel complex. (C) The voltage-gated K⁺ channel KCNQ1 associates with the KCNE1 β-subunit to form the slow cardiac outward delayed rectifier K⁺ current (I_{Ks}). KCNQ1 can form functional homotetrameric channels with different biophysical properties than heteromeric KCNQ1/KCNE1 channels. (D) The structure of inwardly rectifying K⁺ selective channels (IRKs) consists of two membrane-spanning domains (M1 and M2) flanking a conserved pore (P) region with the H5 segment. Four channel subunits assemble to form a functional IRK channel (E) Two pore K⁺ channels contain four transmembrane domains and have been characterized as involved in "background" K⁺-selective conductance and are gated by pH, mechanical stretch, heat, and coupling to G-proteins but not voltage.

much more effectively into the cell than out of it and largely determines the transmembrane voltage of most cells at rest, because they are open in the steady state.

In addition to I_{K1}, several important modulating cardiac ion channels, including the K_{ATP} and the K_{ACh} channels, are inward rectifying K⁺ channels with a common structure. The K_{ATP} channel is an octameric complex composed of four IRK subunits of Kir6.2 (*KCNJ11*) and four subunits of the cardiac sulfonylurea receptor (*SUR2A*), a member of the ATP-binding cassette superfamily that generally features 12 membrane-spanning segment and 2 intracellular ATP-binding sites. A common trafficking motif (RKR) has been identified in both subunit types relating to the retention of these subunits by the ER and regulation of their cell surface expression as a functional K_{ATP} channel.[26] Coexpression of Kir6.2 with *SUR2* in COS cells recapitulates the cardiac K_{ATP} channel current sensitive to ATP and the sulfonylurea glibenclamide inhibition activated by cromakalim and pinacidil (but not by diazoxide); neither subunit expressed alone generates significant current.[27] At this time, the role of the K_{ATP} channel in the heart is largely undetermined. Its regulation by potassium channel openers such as nicorandil (which likely reduces its affinity for the negative-acting ATP, thereby promoting channel activation), its linkage with the metabolic activity of the cardiomyocyte, its location both in the plasma membrane and in the mitochondrial membrane, and its putative role in cardioprotection have been reported,[28,29] and are further discussed in Chapter 10.

The K_{ACh} is a heteromultimeric complex of two G-protein–activated inwardly rectifying K⁺ channels (GIRK), GIRK 1 (Kir 3.1) and GIRK4 (Kir 3.4).[30] Although oocyte expression of single subunits was infrequent and modest in intensity, coexpression of GIRK1 and four mixtures leads to robust expression of K$^+$ channels, suggesting that heterologous subunit assembly is required for activity.[31] Heart rate is slowed in part by the vagal-secreted acetylcholine (ACh)-dependent activation of the I$_{ACh}$ channel mediated by ACh-bound muscarinic receptors activating G-protein beta gamma-subunits (G$_{\beta\gamma}$). Released G$_{\beta\gamma}$ directly stimulate channel activity[31] by binding to multiple regions of the GIK1 and GIRK4 subunits.[32] This channel is a critical regulator of atrial, SA, and AV node excitability, primarily by promoting the hyperpolarization of pacemaker cells. Adenosine similarly produces pacemaker hyperpolarization by producing G$_{\beta\gamma}$ and activating I$_{ACh}$ channels. Interestingly, the I$_{ACh}$ channels are primary therapeutic targets for treating supraventricular tachycardia (SVT).

A relatively poorly characterized outward rectifying background or leak current has been widely reported in cardiac cells. Primary candidates responsible for this current include a number of two-pore domain K$^+$ channels (depicted in Fig. 3E), which are sensitive to a variety of stimuli, including mechanical stretch, pH, heat, and lipids. These include TREK-1,[33,34] TASK-1,[35] and cTBAK.[36] Inhibition of these channels can delay repolarization, cause prolongation of the AP duration, and, in some cases, promote early afterdepolarizations.

Many voltage-gated ion channels display the characteristic of inactivation, entry into a nonconducting state during depolarization. It is now clear that multiple types of inactivation, with varying time courses, play a role in gating of cardiac ion channels and that a single channel can exhibit entry into multiple inactivated states.[37] One primary mechanism of inactivation exemplified in some voltage-gated channels is mediated by a tethered N-terminal blocking cytoplasmically located particle (the ball and chain mechanism); this N-type inactivation is relatively fast and coupled both to the S4 voltage sensor and occlusion of the permeation pathway.[38] A second mechanism of inactivation termed C-type involves several residues in the C-terminus S6 transmembrane segment and involves rearrangement at the pore. Studies of channel proteins re-engineered by site-specific mutagenesis have demonstrated that a cluster of hydrophobic residues within the cytoplasmic linker connecting domains III and IV are critical for fast inactivation of the voltage-gated Na$^+$ channel.[39,40] Molecular chimeric proteins in which the Na$^+$ channel III–IV linker was attached to the N-terminus of a non-inactivating K$^+$ channel recapitulated the fast inactivation of the K$^+$ channel, and a mutation that abolishes the Na$^+$ channel inactivation also completely abolished inactivation in the chimera, suggesting a functional relationship between the fast inactivation processes of Na$^+$ and K$^+$ channels.[41] Other studies have demonstrated that a single amino acid substitution (F1304Q) in Na$^+$ channels profoundly altered both fast and slow inactivation simultaneously, indicating that these processes share physical determinants in Na$^+$ channels.[42]

The number of cardiac ion channels identified continues to increase, as well as the genes whose expression results in the molecular components underlying the individual ion currents of the heart. An updated compendium of the presently known ion channel subunits with myocardial expression indicating the preferred nomenclatures of gene and protein (and indicating other published alternative names) is shown in Table III. This information is largely derived from a recent issue of IUPHAR featuring several articles listing all known ion channels.[43]

MOLECULAR AND CELLULAR CHARACTERIZATION OF DYSRHYTHMIAS

The molecular characterization of primary cardiac dysrhythmias has provided important information toward our understanding of their pathogenesis. Dysrhythmias associated with mutations in ion channel genes were first described a decade ago.[44,45] Since then, more mutations in ion channels, as well as their anchoring proteins, have been reported. These discoveries are significantly improving our understanding of the molecular mechanisms of cardiac dysrhythmias and advancing the search for novel therapeutic modalities.

INHERITED CARDIAC DYSRHYTHMIAS

Inherited or *primary* dysrhythmias are primary electrical disorders not associated with structural cardiac pathology and include long-QT syndrome (LQTS), short-QT syndrome (SQTS), catecholamine-induced polymorphic ventricular tachycardia (CPVT), and Brugada syndrome (BrS). Most of the inherited dysrhythmias, with a known genetic basis (mostly monogenic defects), have been found to be associated with cardiac ion channel defects (i.e., in ion channels subunits), as well as in the proteins essential for proper channel functioning. These mutations tend to affect the cardiomyocyte membrane ion flux, resulting in increasing duration of the ventricular AP and prolongation of the Q-T interval.

LQTS can be inherited either as an autosomal-recessive or autosomal-dominant disorder. The autosomal-recessive type, first described by Jervell and Lange-Nielsen in 1957[46] tended to present a more severe phenotype, which includes congenital neural deafness, syncope, and sudden death. Romano and colleagues in 1963[47] and Ward in 1964[48] reported children with remarkable clinical similarities, except that there was not neural deafness, and their transmission had an autosomal-dominant pattern. The autosomal-dominant form of LQTS also termed the Romano–Ward syndrome (RWS) has a prevalence of 1/10,000, whereas the autosomal-recessive form termed Jervell Lange Nielsen syndrome (JLNS) has an incidence of 1/250,000 (Table IV).

Since the initial reports on LQTS, there have been an overwhelming number of publications that have brought considerable insight both into the clinical aspects of both

TABLE III Cardiac Ion Channel Subunits: Gene, Protein, and Nomenclature

Gene	Protein	Subunit type	Other names
SCN5A	Nav1.5	Cardiac voltage-gated α-subunit Na$^+$ channel	Also LQT3
SCN1A	Nav1.1	Cardiac voltage-gated α-subunit Na$^+$ channel	R-I
SCN3A	Nav1.3	Cardiac voltage-gated α-subunit Na$^+$ channel	R-III
CACNA1C	Cav1.2	Cardiac voltage-gated α-subunit Ca^{++} channel	L-type; DHP receptor
CACNA1D	Cav1.3	Cardiac voltage-gated α-subunit Ca^{++} channel	L-type
CACNA1G	Cav1.2	Cardiac voltage-gated α-subunit Ca^{++} channel	T-type
CACNA1H	Cav1.2	Cardiac voltage-gated α-subunit Ca^{++} channel	T-type
KCNT2	Kca4.2	Ca^{++} activated K$^+$ channel	Slick, Slo2.1
KCNN3	Kca2.3	Ca^{++} activated K$^+$ channel	SK3
KCNN2	Kca2.2	Ca^{++} activated K$^+$ channel	SK2
KCNJ1	Kir1.1	Cardiac inward rectifier K$^+$ channel	ROMK
KCNJ2	Kir2.1	Cardiac inward rectifier K$^+$ channel	IRK1
KCNJ4	Kir2.3	Cardiac inward rectifier K$^+$ channel	IRK3
KCNJ8	Kir6.1	ATP-sensitive inward rectifier K$^+$ channel	—
KCNJ11	Kir6.2	ATP-sensitive inward rectifier K$^+$ channel	BIR
KCNJ12	Kir2.2	Cardiac inward rectifier K$^+$ channel	IRK2
KCNJ14	Kir2.4	Cardiac inward rectifier K$^+$ channel	IRK4
KCNJ5	Kir3.4	Cardiac inward rectifier K$^+$ channel	GIRK4
KCNA1	KV1.1	Voltage-gated K$^+$ channel, delayed rectifier	RBK1, HBK1
KCNA2	KV1.2	Voltage-gated K$^+$ channel, delayed rectifier	RcK5, HBK5
KCNA4	KV1.4	Voltage-gated K$^+$ channel, A-type, fast-inactivating	RcK4, MK4
KCNA5	Kv1.5	Voltage-gated α subunit K$^+$ channel	
KCNA6	Kv1.6	Voltage-gated K$^+$ channel, delayed rectifier	RcK2, HBK2
KCNA7	Kv1.7	Voltage-gated K$^+$ channel, delayed rectifier	None
KCNA10	Kv1.8	Voltage-gated K$^+$ channel, delayed rectifier	KV1.10, Kcn1
KCND1	Kv4.1	Voltage-gated K$^+$ channel, A-type K$^+$ current	mShal
KCND2	Kv4.2	Voltage-gated K$^+$ channel, A-type K$^+$ current	Shall, RK5
KCND3	Kv4.3	Voltage-gated K$^+$ channel, A-type K$^+$ current	None
KCNF1	Kv5.1	Modifier of the KV2 family of channels	KH1, IK8
KCNG1	Kv6.1	Modifier/silencer of KV2 family channels	KH2, K13
KCNG2	Kv6,2	Modifier/silencer of KV2 family channels	None
KCNQ1	Kv7.1	Voltage-gated K$^+$ channel, delayed rectifier	KVLQT1
KCNQ2	Kv7.2	Voltage-gated K$^+$ channel, delayed rectifier	KQT2
KCNS3	Kv9.3	Modifier/silencer	None
KCNH2	Kv11.1	Voltage-gated K$^+$ channel, inwardly rectifying	HERG, erg1, LQT2
KCNE1	MinK	Voltage-gated K$^+$ channel β-subunit	Also LQT5
KCNE2	MiRP-1	Voltage-gated K$^+$ channel β-subunit	Also LQT6
KCNE3	MiRP-2	Voltage-gated K$^+$ channel β-subunit	
KCNE4	KCNE4	Voltage-gated K$^+$ channel β-subunit, inhibitory subunit	
KCNE5	MiRP4	Voltage-gated K$^+$ channel β-subunit, inhibitory subunit	
HCN1	HCN1	Hyperpolarization-activated, CN-gated cation channel	HAC2, BCNG1
HCN2	HCN2	Hyperpolarization-activated, CN-gated cation channel	HAC1, BCNG2
HCN3	HCN3	Hyperpolarization-activated, CN-gated cation channel	HAC3, BCNG4
HCN4	HCN4	Hyperpolarization-activated, CN-gated cation channel	

congenital and acquired types and into the molecular and genetic basis of LQTS. Mutations in the genes encoding the protein subunits of cardiac channels leading to a distinct phenotype turned out to be not only a family-specific event but also when expressed in heterologous expression systems, these mutations permitted the further characterization of defective ion channels in a topological way, leading to a more specific understanding of the ion channel function.

TABLE IV LQT Channelopathies

Disease	Locus	Gene (alternate)	Protein	Mech	Freq (%)	Inh	Ref.
LQTS1 (Romano-Ward syndrome)	11p15.5	KCNQ1 (KvLQT1)	I_{Ks} K⁺ channel α-subunit	Abn rep	40–55	AD	49, 50
LQTS2	7q35–36	KCNH2 (HERG)	I_{Kr} K⁺ channel α-subunit	Abn rep	35–45	AD	44
LQTS3	3p21–24	SCN5A	I_{Na} Na⁺ channel α-subunit	Abn rep	5–10	AD	45
LQTS4	4q25–27	ANK2	Ankyrin B	Abn Ca⁺⁺ homeostasis	Rare	AD	51, 52
LQTS5	21q22.1–22.2	KCNE1 (minK)	I_{Ks} K⁺ channel β-subunit	Abn rep	Rare	AD, AR	53
LQTS6	21q22.1–22.2	KCNE2 (MiRP1)	I_{Kr} K⁺ channel β-subunit	Abn rep	Rare	AD	54
LQTS7 (Andersen syndrome)	17q23	KCNJ2 (IK1)	Kir2.1	↓ PIP$_2$ channel act	Rare	AD	55, 56
LQTS8 (Timothy syndrome)		CACNA1c (Cav1.2)	L-type Ca⁺⁺ channel	↑ AP; ↑ DAD	Rare	SP	57, 58
Long QT: JLNS							
JLN1	11p15.5	KCNQ1 (KVLQT1)	I_{Ks} K⁺ channel α-subunit	Abn rep	90% of JLNS	AR	59
JLN2	21q22.1–22.2	KCNE1 (minK)	I_{Ks} K⁺ channel β-subunit	Abn rep	10% of JLNS	AR	60

Abn Rep, abnormal repolarization; Freq, frequency; Inh, inheritance; Ref, references; AD, autosomal dominant; AR, autosomal recessive; Mech, mechanism; SP, sporadic; Act, activity; JLNS, Jervell Lange Nielsen syndrome; AP, action potential; DAD, delayed afterdepolarization.

Cardiac Channelopathies

The genes of the most important ion channel players (e.g., the pore-forming α-subunits and a number of β-subunits) have been characterized, and a rather impressive number of inherited and acquired dysrhythmias have been demonstrated to be secondary to mutations in these genes.

Long QT Syndrome

The discovery of the genetic basis of the LQTS has ushered in a new methodological paradigm. Classical genetic linkage analysis, (i.e., positional candidate strategies) has been useful in both the identification of causative genes and in a number of functional components, with a previously unknown but fundamental role for a normal repolarization process.

The phenotypic hallmark of the LQTS is abnormally prolonged ventricular repolarization that can result in ventricular tachycardia (VT), ventricular fibrillation (VF), and sudden death. Most, if not all, of the primary electrical cardiac disorders have shown a high degree of genetic diversity, and this diversity is well manifested in LQTS in which eight disease loci and the responsible genes have thus far been discovered (see Table IV).

In 1995, genetic linkage between LQTS2 and polymorphisms within the KCNH2 gene (encoding the human ether a-go-go related gene or HERG), as well as between LQTS3 and polymorphisms within the SCN5A gene (encoding the voltage-gated sodium-channel-α subunit) were simultaneously reported.[44,45] A short time later, a SCN5A mutation was characterized using the heterologous expression of recombinant human heart sodium channel, revealing that mutant channels exhibited sustained inward current during membrane depolarization, and the persistence of the inward current could explain the prolongation of the AP.[61] These breakthroughs were followed by the identification of several genes and mutations responsible for LQT, including KCNQ1 gene encoding the LQTS1 voltage-dependent K⁺ channel KvLQT1,[49,50] the ankyrin B (ANK2) gene causing LQTS4,[52] KCNE1 (minK) encoding for the subunit IsK of the K⁺ channel and causing LQTS5,[53] and the KCNJ2 (or MiRP1) encoding the β-subunit of the K⁺ channel I_{Kr} and causing LQTS6.[54]

The genetic basis for two LQT-related syndromes has also been recently explained with the demonstration that the KCNJ2 (IK1) gene coding the K⁺ channel inward rectifier Kir2.1 causes Andersen syndrome (LQTS7)[56–62] and a missense mutation (G406R) found in the *CACNA1c* gene, encoding the L-type Ca^{++} α-subunit Cav1.2 causes the Timothy syndrome (TS) or LQTS8.[57,58]

In TS, functional expression analysis revealed that the G406R mutation acts as a gain-of-function mutation, producing a maintained inward Ca^{++} current in multiple cell types and inducing intracellular Ca^{2+} overload.[58] In the heart, this prolonged Ca^{++} current delays cardiomyocyte repolarization and extends the AP with increased delayed after-depolarization events, leading to an increased risk for lethal dysrhythmia. TS encompasses multiorgan dysfunction, including lethal cardiac dysrhythmias, webbing of fingers and toes, congenital heart disease, immune deficiency, intermittent hypoglycemia, cognitive abnormalities, and autism.

Andersen syndrome (also known as Andersen–Tawil syndrome) was first reported in 1971 in a patient with muscle weakness, extrasystoles, and developmental delay.[63] On subsequent evaluation of a larger number of patients with comparable symptoms, periodic paralysis and more severe cardiac dysrhythmias were added to the syndrome. Patients may exhibit a variety of ventricular and bidirectional ventricular tachycardias. Andersen syndrome or LQTS type 7 is inherited as an autosomal-dominant entity (see Table IV) with significant phenotypic and genotypic variability, because at least 21 different mutations have been found that are distributed throughout the *KCNJ2* gene. A common feature of these mutations is that they all resulted in loss of function and dominant-negative suppression of Kir2.1 channel function.[56] When coexpressed with Kir2.1 wild-type (WT) channels, two missense mutations (C154F and T309I) exerted a dominant-negative effect, leading to a loss of the inward rectifying K$^+$ current.[64] An unexplained aspect of Andersen syndrome is that mutations in *KCNJ2* do not account for all cases of this disorder; more than 30% of patients with clinically indistinguishable presentation do not harbor Kir2.1 mutations.[65] Because the Kir2.1 mutations that do lead to Andersen syndrome have a dominant-negative effect, it is unlikely that promoter mutations resulting in lower Kir2.1 levels or haploinsufficiency will cause these remaining unexplained cases, but, more likely, these cases involve a defective gene essential or related to Kir2.1 channel function. Candidate proteins include aberrant scaffolding or adaptor proteins, which anchor the Kir2.1 protein to regions of the plasma membrane or are involved in its trafficking through the cell and targeting to specific subcellular locations regulating its channel function. A proteomics approach recently applied to identify proteins associated with cardiac Kir2.1 channels found a number of high-affinity interactions with scaffolding proteins, including SAP97 and the Lin7/calcium/calmodulin-dependent serine protein kinase (CASK).[66] Further studies by this group have revealed that the targeting of Kir2.1 channels to the basolateral membrane in polarized epithelial cells and their colocalization with SAP97 and CASK could be abrogated by a dominant interfering form of CASK that caused the channels to mislocalize, implicating CASK as a central protein of a macromolecular complex mediating the trafficking and plasma membrane localization of Kir2 channels.[67] Interestingly, transgenic mice harboring nonfunctional SAP97 and CASK exhibit a similar phenotype to *KCNJ2* knock-out mice; unfortunately, all are nonviable after birth.[68–70] It remains to be seen whether mutations in these genes (e.g., Kir2.1-interacting domains) are involved in Andersen syndrome.

Mutations in the scaffolding protein ankyrin B were initially implicated in the etiology of long QT syndrome (LQTS4). Mutations in the *ANK2* gene encoding ankyrin B in humans resulted in a loss-of function of ankyrin B and development of fatal cardiac dysrhythmias.[52] Moreover, transgenic mice heterozygous for a null *ANK2* mutation in ankyrin B display dysrhythmias similar to humans. The importance of the ubiquitously expressed ankyrin adaptor protein is underscored by the pleiotropic effects of ankyrin B mutations. These include the disruption in the cellular organization of the Na$^+$-K^{+-}ATPase pump, the Na$^+$–Ca^{++} exchanger, and inositol-1,4,5-trisphosphate (IP3) receptors (all ankyrin B-binding proteins), as well as the reduced targeting of these proteins to the transverse tubules and diminished levels of overall protein. Moreover, ankyrin B mutations also lead to altered Ca^{++} signaling in adult cardiomyocytes that results in extrasystoles, providing a significant rationale for dysrhythmia.

Mammals express three ankyrins proteins (ankyrin R, ankyrin B, and ankyrin G), which are encoded by three different genes (*ANK1*, *ANK2*, and *ANK3*, respectively). These proteins interface between diverse membrane ion channels and transporters and the cytoskeleton. Although most vertebrate tissues harbor multiple alternatively spliced forms of each ankyrin, cardiomyocytes are exceptional in expressing a single form of ankyrin B (220 kDa) and one isoform of ankyrin G (190 kDa).

Mohler and Bennett have recently identified a number of loss-of-function mutations in ankyrin B in patients who display cardiac dysrhythmias in association with sinus node dysfunction, suggesting a novel class of channelopathy distinct from LQTS.[71] These patients exhibited severe sinus bradycardia, ventricular and atrial fibrillation, catecholaminergic-induced ventricular tachycardia, syncope, and had an increased risk of sudden cardiac death. Studies in mice lacking ankyrin B expression suggested that altered calcium homeostasis resulting from defective or reduced ankyrin B function likely underlie spontaneous calcium-based depolarization events in response to catecholaminergic stimulation leading to ventricular dysrhythmias.

The other myocardial ankyrin isoform, ankyrin G, is directly associated with the principal voltage-gated cardiac channel, Nav1.5, responsible for I$_{Na}$. As we will discuss, loss of function in the Nav1.5 channel resulting from *SCN5A* mutations can lead to BrS, characterized by right bundle branch block, dysrhythmia, and increased risk of sudden cardiac death.[72] A *SCN5A* mutation (E1053) in the critical nine amino acid motif required for interaction with ankyrin G has recently been reported in a proband with BrS.[73] These *SCN5A* mutations disrupt the interaction of human Nav1.5 and 190-kDa ankyrin G, leading to abnormal Nav1.5 membrane surface expression (i.e., not properly localized at intercalated discs or at T tubules) and Nav1.5 loss of function.

Recently, an ankyrin B–based macromolecular complex including the Na$^+$/K$^+$– ATPase, Na$^+$/Ca^{++} exchanger, and IP3 receptor localized in cardiomyocyte T tubules in discrete microdomains has been reported as distinct from classic dihydropyridine receptor/ryanodine receptor "dyads."[74] Besides its mediation of long QT, the E1425G *ANK2* mutation also blocks binding of ankyrin B to all three components of the complex. The ankyrin B–based complex has been proposed to be a specialized adaptation of cardiomyocytes with a role for cytosolic Ca^{++} modulation because a T-tubule–associated complex of ankyrin B, Na$^+$/K$^+$– ATPase and the Na$^+$/Ca^{++} exchanger is not present in skeletal muscle, in which ankyrin B is expressed at 10-fold lower levels than in heart, and, moreover, ankyrin B is not abundantly expressed in smooth muscle.

The variability of phenotypic expression present in cases with ankyrin mutation–based dysrhythmias, as well as their heterogeneity, have been recently examined in a study screening 541 patients and 200 control subjects with targeted mutational analysis of 10 *ANK2* exons on genomic DNA by PCR, denaturing HPLC, and direct DNA sequencing.[75] Fourteen distinct variants (including 10 novel) were observed in 9 (3.3%) of 269 genotype-negative LQTS patients, 5 (1.8%) of 272 genotype-positive LQTS cases, 4 (4%) of 100 white controls, and 9 (9%) of 100 black controls. Four variants found in controls (L1622I, T1626N, R1788W, and E1813K) have been implicated previously as LQTS4-associated mutations and displayed functional abnormalities *in vitro*. All genotype-negative LQTS cases hosting *ANK2* variants were diagnosed as "atypical" or "borderline" cases, most presenting with normal QTc, non-exertional syncope, U waves, and/or sinus bradycardia. Nonsynonymous ankyrin B variants were detected in nearly 3% of unrelated LQTS patients but also in 7% of healthy control subjects. Genotype-negative LQTS patients with a single *ANK2* variant displayed non-exertional syncope, U waves, sinus bradycardia, and extracardiac findings. We concur with the authors that further research is necessary to definitively establish whether the identification of previously reported cases with functionally significant *ANK2* variants (residing in 2% of apparently healthy subjects) suggests pro-dysrhythmic potential.

On the basis of recent findings and the ubiquitous expression of ankyrin gene products, dysfunction in ankyrins and their interacting proteins in other excitable tissues may be increasingly linked with additional new classes of human disease.

It is well established that the abnormal lengthening of the cardiac action potential present in each type of LQTS1-8 can result in a polymorphic ventricular tachycardia known also as *torsade de pointes* (TdP). This dysrhythmia may be self-limited, but it may also be followed by syncope and sudden death. Physical activities such as swimming and emotional excitement can precipitate the LQTS, including syncope and sudden death. In Table V, we show a listing of suggested criteria based on a scoring of clinical parameters, as developed by Schwartz and associates, to predict the individual risk for the development of LQTS.[76]

In mutational analysis, it is important to keep in mind that mutations within the preceding genes may be distributed over the entire gene (allelic heterogeneity), and more than one mutation may be present in the same gene. This complexity requires a thorough mutation analysis of all LQT genes before treatment recommendations can be given. Moreover, genotype-phenotype correlations in large families should be used to evaluate the intergene, interfamilial, and intrafamilial differences in the clinical phenotype; the specific gene locus involved; and the individual consequences of a given mutation. The presence of extensive phenotypic heterogeneity, even within mutation carriers in the same family, raises the importance of modifying factors and genes that remain to be identified. Furthermore, the reduced penetrance and variable expressivity associated with the LQT mutations still remain to be explained.

Brugada Syndrome

Brugada syndrome (BrS) is characterized by idiopathic ventricular fibrillation, ST segment elevation in the right precordial leads (V_1–V_3), often followed by negative T waves,[77] right bundle branch block, and sudden unexpected death at a young age in the absence of cardiac structural abnormalities. Three ECG repolarization patterns in the right precordial leads are recognized (Table VI). Type 1 is diagnostic of BrS and is characterized by a coved ST-segment elevation of 2 mm (0.2 mV) followed by a negative T wave. These ECG changes may be present all the time (congenital form) or can be elicited after the administration of specific Na$^+$ channel blockers (e.g., Ajmaline, flecainide, and procainamide).

A syndrome characterized by sudden death and ST-segment elevation in the right precordial leads (V_1–V_3) with no structural heart disease was described in 1953 by Osher and Wolff,[78] but it was first described as a distinct clinical entity associated with a high risk of sudden cardiac death (SCD) in 1992.[79] Geographically, the incidence of BrS varies, estimated to be between 5 and 50 per 10,000. In endemic areas such as East/Southeast Asia (particularly Japan and Thailand), BrS is a leading cause of death among young and otherwise healthy adult males.[80,81]

Although the intermittent (and often hidden) characteristics of the ECG pattern in BrS likely has led to an underestimation of the true incidence of the syndrome in

TABLE V LQTS Diagnostic Criteria

Electrocardiogram findings[a]	Points
QTc[b]	
≥ 480 msec$^{1/2}$	3
460–470 msec$^{1/2}$	2
450 msec$^{1/2}$ (in males)	1
Torsades de pointes[c]	2
T-wave alternant	1
Notched T waves in three leads	1
Low heart rate for age[d]	0.5
Clinical history	
Syncope[e]	
With stress	2
Without stress	1
Congenital deafness	0.5
Family history[f]	
Family member with definitive LQTS[g]	1
Unexplained SCD of immediate family member younger than age 30	0.5

[a]In the absence of medications or disorders known to affect these ECG findings.
[b]QTc calculated by Bazett's formula where QTc = QT/square root of RR.
[c]Mutually exclusive.
[d]Resting heart rate below the second percentile for age.
[e]The same family member can not be counted in both categories.
[f]LQTS defined by an LQTS score ≥4. Scoring: ≤1 point, low probability of LQTS; 2–3 points, intermediate probability of LQTS; ≥4 points, high probability of LQTS.

TABLE VI ST-Segment Abnormalities in Leads V1–V3

	Type 1	Type 2	Type 3
J wave amplitude	≥2 mm	≥2 mm	≥2 mm
T wave	Negative	Positive or positive biphasic	Positive
ST-T configuration	Coved type	Saddleback	Saddleback
ST segment (terminal portion)	Gradually descending	Elevated ≥1 mm	Elevated <1 mm

1 mm = 0.1 mV. The terminal portion of the ST segment refers to the latter half of the ST segment.

asymptomatic patients, it is, nevertheless, estimated that 12% of all SCD are secondary to BrS. The determination of SCD in a large population of patients with the electrocardiographic characteristics of BrS and no previous cardiac arrests was reported by Brugada and associates in 2003.[82] In this study, a diagnosis of BrS on ECG was made spontaneously in 391 patients (of a total of 547), and in the remaining 156 individuals, the ECG was abnormal only after the administration of an antidysrhythmic drug. One hundred twenty-four patients had experienced at least one episode of syncope. During programmed ventricular stimulation, a sustained ventricular dysrhythmia was induced in 163 of 408 patients, and at follow-up (24 ± 32 months), 45 patients (8%) suffered sudden death or documented VF. Inducibility of a sustained ventricular dysrhythmia and a history of syncope were found to be strong predictors of events. Thus, patients with a spontaneously abnormal ECG, a previous history of syncope, and inducible sustained ventricular dysrhythmias had a probability of 27.2% of suffering an event at follow-up.

BrS is inherited as an autosomal-dominant trait (Table VII). The only gene so far implicated in BrS is *SCN5A*, the gene encoding the α-subunit of the cardiac Na+ channel gene located on the short arm of the third chromosome (3p21).[83] More than 80 mutations in *SCN5A* have been linked to the syndrome so far.[86–89] Unlike those *SCN5A* mutations causing long QT syndrome that are associated with gain-of-function,

TABLE VII Channelopathies/ BrS

Disease	Chromosomal Locus	Gene (alternate)	Protein	Mech	Inherit	Ref.
BrS1	3p21–24	*SCN5A (Nav1.5)*	I_{Na} Na$^+$ channel α-subunit	Abn. Rep	AD	83
BrS2	3p22–25	?	?	?	AD	84
Idiopathic VF	3p21–24	*SCN5A (Nav1.5)*	I_{Na} Na$^+$ channel α-subunit	?	AD	85

AD, autosomal dominant; ?, not known.

SCN5A mutations leading to BrS are invariably associated with loss of function. More than two dozen of these mutations have been studied in functional expression systems and shown to result in loss of function related to multiple mechanisms, including failure of Na$^+$ channel protein expression; changes in the voltage and time dependence of Na$^+$ channel (I_{Na}) current activation, inactivation, or reactivation, or accelerated inactivation of the sodium channel.[81] Interestingly, a higher incidence of *SCN5A* mutations has been reported in familial than in sporadic cases.[90]

Less than a third of the patients with this disorder screened for these mutations have mutations in *SCN5A*, suggesting that there are other unaccounted for mutations in the *SCN5A* gene or that other genes are involved. Moreover, studies for promoter mutations, cryptic splicing mutations, or the presence of gross rearrangements in *SCN5A* have rarely been investigated. A loss of current in the ion channels secondary to *SCN5A* mutations could also occur by creating a truncated SCN5A protein. A second locus on chromosome 3, close to, but distinct from, the *SCN5A* locus, was linked recently to BrS in a large pedigree in which the syndrome is autosomal-dominant inherited and associated with progressive conduction disease, a low sensitivity to procainamide, and a relatively benign prognosis.[84]

The primary cause of ST elevation in BrS and its strong linkage to VF remains largely unresolved.[91] The most widely accepted mechanism supported by numerous experimental and clinical studies ascribes the ST elevation in this disorder as essentially a repolarization disorder involving an unequal repolarizing current response to reduced I_{Na} in the epicardium relative to endocardium; the nonuniform loss of monophasic AP (MAP) in the right ventricular epicardium compared with the endocardium results in a transmural voltage gradient.[80] This electrophysiological mechanism was corroborated from the MAP analysis carried out by Kurita and associates.[92] A spike-and-dome configuration was documented by the authors on the epicardial sites of the right ventricle outflow tract in patients with BrS but not in controls. MAP recordings from the endocardium did not demonstrate morphological defects in the patients with BrS. A deep-notched AP in the RV epicardium, but not in endocardium, may induce a transmural current that would contribute to the elevation of the ST-segment in the right precordial leads. The spike-and-dome configuration may also prolong the epicardial AP, contributing to a rapid reversal of the transmural gradients and inscription of an inverted T wave. Others have proposed that conduction delays in the right ventricular outflow tract[93] underlie both the unique ECG signature and VF characteristic of BrS and have, therefore, termed it a depolarization disorder.[91] As in many diseases, BrS may not be fully explained by a single mechanism.

Previously, we have stated that inherited dysrhythmias are primarily electrical disorders not associated with structural cardiac pathology, and within this group, BrS was included together with LQTS. But there might be some exceptions. For example, in an interesting editorial, Saffitz noted that it is increasingly recognized that some types of myocardial pathology may occur in patients with the clinical phenotype of BrS, even when conventional clinical assessment does not identify obvious signs of structural heart disease.[94] That myocardial pathology can be present in BrS has been confirmed in a group of 18 patients by Frustaci and associates.[95] Each patient in this cohort (15 men and 3 women; mean age, 42 ± 12 years) exhibited typical ECG features characteristic of BrS, and on the basis of conventional noninvasive analysis with 2D echocardiography, they were considered not to have structural cardiac defects. Moreover, VF was documented in seven patients, sustained PVT in seven, and syncope in four at the time of initial clinical presentation. However, most of the patients had evidence of myocarditis at biopsy, which was diffuse or localized in the right ventricle in 14 patients (78%). Viral genomes were detected by polymerase chain reaction in four of them (Coxsackie B3 in two patients, and Epstein–Barr virus and Parvovirus B-19 in single patients). Furthermore, the typical ST-segment abnormalities of BrS disappeared a few weeks after hospital discharge and were not observed at follow-up in eight of the patients with myocarditis at biopsy. In this group of patients, *SCN5A* gene mutations were not identified. Similar observations have been previously reported in a clinicopathological study of 273 young individuals who suffered sudden death in the Veneto region of Italy.[96] ST-elevation and right bundle branch block were present on ECG in 13 patients; at autopsy, 12 showed pathological features of arrhythmogenic right ventricular cardiomyopathy or dysplasia (ARVD); and 1 was found to have a normal heart. These findings provide further evidence that myocarditis and ARVD may mimic BrS, and both may account for the transient development of the ECG abnormalities that characterize this disorder in some patients. Therefore, myocarditis should be considered, mainly, if at follow-up the ECG landmarks of BrS disappear. On the other hand, it is possible that patients with the BrS phenotype and an underlying *SCN5A* mutation may also have structural heart disease. As noted by Saffitz, it remains to be determined whether the genetic defect itself contributes to the development of structural heart disease or whether the two defects arise independently and together contribute to the clinical phenotype.[94]

The prognosis of the asymptomatic form of BrS is not clear; although some investigators report an increased risk of sudden, unexpected death,[97] others found that the risk for fatal dysrhythmias is low.[98] This discrepancy may be related to a number of factors, including ethnicity, because the prevalence of BrS varies between and within geographical areas.

From all of the preceding and the fact that many patients with BrS are asymptomatic, it is becoming evident that genetic screening of this disease is of paramount significance. Because the inherited form of BrS has an autosomal-dominant pattern, we concur with Roberts that it is extremely important after a proband with the disease is identified to investigate which offspring have the mutation, because they are at risk for SCD.[99] Finally, none of the drugs currently available have been successful in the treatment of this disease, and only an implantable cardioverter-defibrillator seems to be an effective alternative. Further research is needed to be able to diagnose all cases of BrS, and to develop new drugs.

Catecholaminergic Polymorphic Ventricular Tachycardia

Mutations in two important components of the sarcoplasmic reticulum (SR), which are essential in excitation-contraction coupling, the ryanodine receptor type 2 (*RYR2*) and calsequestrin (*CASQ2*), have been found to be associated to catecholaminergic polymorphic ventricular tachycardia (CPVT) 1 and 2, respectively.[100] A third type of CPVT has also been described, but its genetic etiology and its inheritance pattern are not known (Table VIII).

CPVT1 is inherited in an autosomal-dominant pattern, whereas CPVT2 can be inherited as either an autosomal-dominant or autosomal-recessive trait. Both types 1 and 2 have been associated with enhanced SR Ca^{++} release that will cause a delayed after-depolarization activity. However, the precise molecular and pathophysiological mechanisms involved are not known. The disease is characterized by ventricular dysrhythmias, syncope, and seizures in response to physiological (exercise) or emotional stress, and it is an important cause of SCD in children and young adults (as noted previously in Chapter 7). The mortality rate tends to increase with age, reaching 30–35% by the time the patient is 30 years old.[102,107] The dysrhythmia may initially present as isolated premature ventricular beats or bigeminy with later evolution to polymorphic ventricular tachycardia. In CPVT1, increased adrenergic activity has been considered a primary culprit in triggering the attacks, which explains why β-adrenergic blockers are often used as the treatment of choice,[108,109] although not all patients with CPVT are protected by this therapy.[107] Interestingly, infusion of catecholamines may initiate the dysrhythmia, because during emotional and physical stresses, catecholamines are released activating Ca^{++}-induced Ca^{++} release by protein kinase A–mediated phosphorylation of RyR2.[110] *In vitro* functional characterization of mutant RyR2 channels have shown abnormal behavior when subjected to adrenergic stimulation and caffeine administration, resulting in enhanced calcium release from the SR. In a effort to demonstrate whether *RYR2* mutations can reproduce the dysrhythmias observed in CPVT patients in a laboratory animal setting, Cerrone and associates developed a conditional knock-in mouse model carrier of the R4496C mutation in the *RYR2* gene (the murine equivalent of the R4497C mutations identified in CPVT families).[111] This model was used to demonstrate whether the animals could develop a CPVT phenotype and whether β-blockers might prevent dysrhythmias. Twenty-six mice (12 wild-type [WT] and 14 *RYR* [R4496C] mutant) underwent exercise stress testing followed by epinephrine administration; none of the WT mice developed VT compared with 5 of the 14 *RYR* (R4496C) mice. Of 21 mice (8 WT, 8 *RYR* (R4496C) and 5 *RYR* (R4496C) mutants pretreated with β-blockers) treated with epinephrine and caffeine, none of the WT developed VT, whereas 4 of the 8 *RYR* (50%) (R4496C) and 4 of the 5 *RYR* (R4496C) mice pretreated with β-blocker did. These data provide the first experimental demonstration that the R4496C *RYR2* mutation predisposes the murine heart to VT and VF in response to caffeine and/or adrenergic stimulation. Similarly, to what has been observed in some patients[107] with *RYR2* -CPVT, β-blocker treatment seems to be ineffective in preventing life-threatening dysrhythmias in mice.

TABLE VIII Cardiac Channelopathies (CPVT)

Disease	Chromo	Gene (alternate)	Protein	Mech	Inher	Ref.
CPVT1	1q42–43	*RYR2*	RyR/Ca^{++} release channel	Ca^{++} overload	AD	101–104
CPVT2	1p13–21	*CASQ2*	Calsequestrin	Impaired Ca^{++} storage and release	AD, AR	105
CPVT3	?	?	?	?	AD	106

AD, autosomal dominant; AR, autosomal recessive; ?, not known.

Recently, 13 missense mutations in the cardiac *RYR2* gene (12 of which were novel) were identified in 12 probands with CPVT.[109] Another 11 patients were silent gene carriers, suggesting that some mutations are associated with low penetrance. Interestingly, a marked resting sinus bradycardia was found in all carriers in the absence of drug treatment. On β-blocker treatment, most (more than 95%) of the *RYR2* mutation carriers remained asymptomatic at follow-up (median, 2 years). CVPT patients with the *RYR2* mutation had bradycardia regardless of the site of the mutation, which may be important in establishing a molecular diagnosis in young patients presenting with syncopal episodes and bradycardia but normal QTc on resting ECG. Because the risk of SCD is high and β-blockers can be an effective therapy, early genetic screening to confirm the diagnosis and subsequent preventive strategies are warranted.

Another study demonstrated the effectiveness of treatment with β-blockers in 13 patients with type 2 CPVT (D307H mutation in the *CASQ2* gene) because 11 patients became completely asymptomatic, and the remaining 2 showed a significant reduction in the frequency of syncopal episodes.[105] Nevertheless, alternative therapies might be necessary in those cases that do not respond to β-adrenergic blockers.

Short QT Syndrome

Short QT syndrome is a rare, recently described, heritable disorder characterized by an abnormally short Q-T interval (<300 ms) and a propensity to atrial fibrillation (AF), SCD, or both. At this time, the precise cause of short QT syndrome is not known. A current hypothesis is that short QT syndrome is due to increased activity of outward potassium currents in phase 2 and 3 of the cardiac AP. This would cause a shortening of the plateau phase (phase 2), resulting in a shortening of the overall AP, and leading to a shortening of the refractory periods and the Q-T interval. In fact, during programmed electrical stimulation, atrial and ventricular effective refractory periods are shortened, and in a high percentage of cases, ventricular tachydysrhythmias are inducible.

Individuals with short QT syndrome frequently complain of palpitations and may have syncope. Individuals with this disorder exhibit an autosomal-dominant inheritance pattern, and most have a family history of unexplained or sudden death at a young age (even in newborns), palpitations, and AF. SCD in short QT is most likely due to VF. In families with short QT syndrome, missense mutations in three different genes encoding K+ channels involved in cardiac repolarization have been found (Table IX), including *KCNH2* (HERG, *KCNQ1* (KvLQT1), and most recently, *KCNJ2* (Kir2.1).

Two different mutations in the *HERG/KCNH2* gene result in the same amino acid change (N588K) in the cardiac delayed rectifier current I_{Kr} ion channel.[112,113] The mutated I_{Kr} exhibits dramatically increased activity (gain-of-function) compared with the normal ion channel, leading to a heterogeneous abbreviation of AP duration and refractoriness and reduced affinity of the channels to I_{Kr} blockers.[112] Similarly, missense mutation in the *KCNQ1* gene has recently been linked to short QT syndrome and exhibits a similar gain of function in the I_{Ks} channel in functional studies with predicted shortening of repolarization.[114] A defect in the gene coding for the inwardly rectifying Kir2.1 (I_{K1}) channel encoded by the *KCNJ2* gene has been recently linked to a novel variant of short QT syndrome (SQT3), displaying a unique ECG phenotype characterized by asymmetrical T waves.[116] The cause for the development of VT and VF with these mutations is thought to be a marked transmural dispersion or acceleration of the repolarization and a heterogenous reduction of the AP duration.

At present, the only effective treatment for individuals with short QT syndrome is the implantation of an implantable cardioverter-defibrillator (ICD). Nevertheless, antidysrhythmic agents, particularly quinidine (a class IA antidysrhythmic), may be of benefit in individuals with short QT syndrome because of their effects on prolonging the AP and by their action on the I_K channels. Although the use of these drugs alone is not indicated at present, there may be a benefit of adding these agents to individuals who have already had ICD implantation by decreasing the number of dysrhythmic events. Gaita and associates reported a group of six patients with short QT syndrome, five of whom had received a ICD, and one child in whom they have tried different antidysrhythmic drugs, including flecainide, sotalol, ibutilide, and hydroquinidine, to determine whether they could prolong the Q-T interval into the normal range and thus prevent symptoms and dysrhythmia recurrences.[117] Class IC and III antidysrhythmic drugs did not produce any significant QT prolongation. Only hydroquinidine produced a sig-

TABLE IX Molecular Genetic Data in Short QT Syndrome

Disease	Chrom Locus	Gene (alternate name)	Mutation (nucleotide change)	Amino acid change	Channel	Inher	Ref.
SQT1	7q35–36	*KCNH2 (HERG)*	C1764A	N588K	I_{Kr}	AD	112, 113
			C1764G	N588K			
SQT2	11p15.5	*KCNQ1 (KvLQT1)*	G919C	V307L	I_{Ks}	AD	114
				V141M	I_{Ks}	?	115
SQT3	17q23	*KCNJ2 (Kir 2.1)*	G514A	D172N	I_{Ki}	?	116

nificant QT prolongation (from 263 ± 12 ms to 362 ± 25 ms) followed by ventricular programmed stimulation increased the ventricular effective refractory period to ≥200 ms, and VF was no longer induced. These data suggest that quinidine has the potential to be an effective therapy for short QT patients, mainly in young children at risk for sudden death because ICD implantation is not feasible.

Also, in a recent review, Schimpf and associates[118] observed that quinidine has been effective in normalizing the Q-T interval, with normalization of the QT/heart rate ratio and ventricular effective refractory period. Moreover, this drug makes VT/VF noninducible and may serve as either an adjunct to ICD or as an alternative in the very young to treat AF and recurrent tachydysrhythmias. However, the authors advised caution, because the quinidine effects have only been shown in patients with mutations in *HERG*, and the response could be different in cases with a short Q-T interval secondary to other mutations.

Familial Atrial Fibrillation

Atrial fibrillation (AF) is the most common dysrhythmia seen in clinical cardiology, and although in some cases it may be inherited in a monogenic form, familial AF (FAF) seems to be genetically heterogenous. In other cases called lone AF, it is inherited, but the patient does not have a positive family history. Linkage to a locus in chromosome 10q22–q24 was reported in three families with FAF.[119] It was inherited with an autosomal-dominant pattern in 10 of 26 living relatives of a three-generation family, but a causative gene was not found. Another study reported that 5% of 914 patients with AF had a positive family history for this dysrhythmia, and four multi-generation families with autosomal-dominant AF (FAF 1–4) were tested for linkage to the chromosome 10 AF locus.[120] Fifty probands with lone AF (no apparent cause) and a positive family history were identified, but genotyping analysis of FAF 1–4 excluded a linkage to locus 10q22–q24 region. In addition, linkage was also excluded to the chromosome 3p22–p25 and *lamin A/C* loci associated with FAF, conduction system disease, and dilated cardiomyopathy.

Further validating the genetic heterogeneity of this rhythm disorder, Ellinor and associates, using linkage and haplotype analysis in a large family with AF, have mapped a novel locus to chromosome 6q14–16.[121] (Table X).

The inward rectifier K⁺ channel Kir2.1 mediates the potassium I_{K1} current in the heart. It is encoded by *KCNJ2* gene linked to Andersen's syndrome. Because Kir2.1 channels were associated with mouse atrial AF, it was hypothesized that *KCNJ2* was linked with clinical familial AF. In an analysis of 30 Chinese AF kindreds evaluated for mutations in *KCNJ2* gene, a highly conserved valine-to-isoleucine mutation at position 93 (V93I) of Kir2.1 was found in all affected members in one kindred.[125] Functional analysis of the V93I mutant demonstrated a gain-of-function in the Kir2.1 current in marked contrast to the loss-of-function effect of Kir2.1 mutations previously reported in Andersen's syndrome. Hence, Kir2.1 V93I mutation plays a role in initiating and/or maintaining AF by increasing the activity of the inward rectifier K⁺ channel.

Chen and associates screened a large Chinese family with autosomal-dominant inherited lone AF and demonstrated the presence of a mutation (S140G) located in the *KCNQ1* gene mapped to chromosome 11p15.5.[122] This gene encodes the pore forming α-subunit of the cardiac I_{Ks} potassium channel, and the mutation results in a gain of function effect on *KCNQ1/KCNE1* and *KCNQ1/KCNE2* currents. However, the significance of this mutation is unclear, because it has neither been confirmed in six other Chinese families nor in unselected patients with AF from the United Stattes.[127]

In addition, an arginine-to-cysteine mutation at residue 27 (R27C) of the *KCNE2* gene encoding the β-subunit of the KCNQ1-KCNE2 channel responsible for a background potassium current was identified in 2 of 28 Chinese probands.[124] This mutation was found in each affected member of the two kindreds, whereas it was absent in a large group of unrelated individuals. These findings also indicated that the *KCNE2* R27C is a gain-of-function mutation associated with the initiation or maintenance of AF.

Inheritance of AF with an autosomal-recessive pattern has also been reported.[126] In a large family with AF, the onset of AF presented at the fetal stage and was associated with

TABLE X Genes Involved in Familial Atrial Fibrillation

Disease	Chromo	Gene	Protein	Mutation	Inher	Ref.
Lone FAF	11p15.5	*KCNQ1*	KvLQT1	S140G	AD	122
FAF	10q22–24	?	?	?	AD	
		KCNE2	MiRP1	R27C	?	123–125
		KCNJ2	Kir2.1	V93I	?	
	6q14–16	?	?	?	AD	121
Neonatal AF	5p13	?	?	?	AR	126

AD, autosomal dominant; AR, autosomal recessive; Inher, inheritance; ?, not known.

neonatal sudden death and, in some cases, with ventricular tachydysrhythmias and cardiomyopathy of variable severity. A novel genetic locus for AF on chromosome 5p13 (arAF1) and a link between AF and prolonged P-wave duration were identified. Eventually, the cloning of the *arAF1* gene should further increase our understanding of the molecular mechanisms of AF.

Cardiac Conduction Defects

Lenegre–Lev disease was initially described as an acquired complete atrial-ventricular (AV) block with right (RBBB) or left bundle branch block (LBBB) and widening QRS complexes.[128] The disease is secondary to idiopathic fibrosis of the heart electrical conduction system and may cause syncope and sudden death. In 1999, Schott and associates reported the first mutation in the *SCN5A* gene that segregated with progressive conduction defect (PCCD) in an autosomal-dominant manner in a large French family and a second *SCN5A* mutation that co-segregated in a smaller Dutch family with familial nonprogressive conduction defect.[129] Fifteen patients from the French family were clinically and electrocardiographically affected (the mean QRS duration was 135 ± 7 ms). RBBB was present in five patients, LBBB in two, left anterior or posterior hemiblock in three, and long P-R interval (>210 ms) in eight. None of the patients had structural heart disease. Of significance, four patients received a pacemaker implantation because of syncope or complete AV block, and in a number of affected patients, the conduction defect increased in severity with age. On the other hand, in the Dutch family, the proband presented after birth with an asymptomatic first-degree AV block associated with RBBB. Three brothers were asymptomatic, one of whom had RBBB, and the asymptomatic mother had a nonspecific conduction defect with a QRS duration of 120 ms. By use of markers flanking *SCN5A* in the French family, these investigators demonstrated segregation of the disease with marker D3S1260 in every affected individual, and analyses with flanking markers of the region confirmed a linkage to the 3p21 locus. Sequencing the entire *SCN5A* coding region in this family identified a T→C substitution in the highly conserved +2 donor-splicing site of intron 22. This abnormal transcript predicts an in-frame skipping of exon 22 and an impaired gene product lacking the voltage-sensitive domain III S4 segment. Importantly, this mutation was found in all affected members, but not in 100 control chromosomes. In the Dutch family, sequence analysis of the *SNC5A* gene revealed a deletion of a single nucleotide (G) at position 5280, resulting in a frameshift and a premature stop codon. This mutation co-segregated with the phenotype in all affected family members.

These findings also indicated that with aging there is a progressive increase in cardiac fibrosis, which, in association with the *SNC5A* gene mutation, can slow the impulse along the electrical conduction system. In the Dutch family, the mutation conferring a premature stop codon and the presentation of PCCD at birth suggests that as a consequence of the sodium channel mutation a congenital phenotype can arise that may be either progressive or immediate.

It is worth noting that none of the affected individuals had LQTS or BrS, although heterozygous mutations in the cardiac *SCN5A* gene have been associated with LQTS, BrS, and conduction system disease. The same mutation in *SCN5A* can lead either to BrS or to an isolated cardiac conduction defect.[130] In a large family with both BrS and isolated cardiac conduction defects, a G-to-T mutation at position 4372 was found in 13 affected members and was predicted to change a glycine for an arginine (G1406R) between the S5 and S6 segments of domain III of the Na+ channel protein. Four individuals showed typical BrS phenotypes, including ST-segment elevation in the right precordial leads and RBBB, and seven individuals had isolated cardiac conduction defects but no BrS phenotype; one patient with an isolated cardiac conduction defect (CDD) had an episode of syncope and required pacemaker implantation. These findings suggest that modifier gene(s) may influence the phenotypic consequences of a *SCN5A* mutation.

Often a mutant cardiac sodium channel may be associated with multiple biophysical defects and concomitant clinical features of BrS and CCD. For example, LQTS3, which is caused by mutations in the human cardiac *SCN5A* gene, may present, in addition to LQT, with bradycardia and sinus pauses. In an interesting study, Veldkamp and associates reported the effect of the 1795insD Na+ channel mutation (previously characterized by the presence of a persistent inward current (I_{pst}) at −20 mV and a negative shift in voltage dependence of inactivation) on sinoatrial (SA) pacemaking.[131] By use of functional studies, I_{pst} was characterized over the complete voltage range of the SA node AP by measuring whole-cell Na+ currents (I_{Na}) in HEK-293 cells expressing either wild-type or 1795insD channels. I_{pst} for 1795insD channels varied between 0.8 ± 0.2% and 1.9 ± 0.8% of peak I_{Na}, and the activity of 1795insD channels during SA node pacemaking was confirmed by AP clamp experiments. When implemented into SA node AP models, the negative shift decreased sinus rate by decreasing diastolic depolarization rate, whereas I_{pst} decreased sinus rate by AP prolongation, despite a concomitant increase in diastolic depolarization rate. Furthermore, moderate I_{pst} together with the shift reduced sinus rate by approximately 10%. Further increase in I_{pst} could result in plateau oscillations and failure to repolarize completely. The authors concluded that Na+ channel mutations displaying an I_{pst} or a negative shift in inactivation may account for the bradycardia seen in LQTS3, whereas SA node pauses or arrest may result from failure of SA node cells to repolarize under conditions of extra net inward current.

On the other hand, a CCD such as complete atrial-ventricular block (AV block) or sick sinus syndrome (SSS) can be the only electrical rhythm disorder associated with *SCN5A* mutations. Wang and associates have reported the clinical, genetic, and biophysical features of two new *SCN5A* mutations that resulted in AV conduction block.[132] Molecular analysis demonstrated two G to A transition mutations that resulted in the substitution of serine for glycine (G298S) in the domain I S5-S6 loop and asparagine for aspartic acid (D1595N) within the S3 segment of domain IV. Both mutations impair fast inactivation but do not exhibit sustained non-inactivating currents. The mutations also reduce Na+ current density and enhance slower inactivation components. AP simulations predicted that this combination of biophysical abnormalities could significantly slow myocardial conduction velocity. In addition, Benson and associates have screened the α-subunit of *SCN5A* as a candidate gene in 10 pediatric patients from 7 families with congenital SSS.[133] Compound heterozygosity for six distinct *SCN5A* alleles, including two mutations previously associated with dominant disorders of cardiac excitability, was identified in probands from three kindreds. Among 27 heterozygotes, no individual exhibited ECG evidence of BrS. With heterologously expressed recombinant human heart sodium channels, biophysical characterization of mutant channels demonstrated either loss of function or significant impairment(s) in channel gating (inactivation) consistent with reduced myocardial excitability. These findings contribute to establishing a molecular basis for some forms of congenital SSS and in explaining a recessive disorder of a cardiac voltage-gated Na+ channel.

A novel Na+ channel mutation in *SCN5A*, E161K, has recently been identified in individuals of two nonrelated families with symptoms of bradycardia, sinus node dysfunction, generalized conduction disease, and BrS, or combinations thereof.[134] Functional studies of mutant Na+ channels were performed with wild-type or E161K Na+ channel α-subunit and β-subunit co-transfected into tsA201 cells. Whole-cell sodium current (I_{Na}) gauged using whole cell patch-clamp technique from cells containing the E161K mutation exhibited an almost threefold reduction in current density and an 11.9-mV positive shift of the voltage-dependence of activation, whereas the inactivation properties of wild-type and mutant Na+ channels were similar. These data suggested an overall reduction of I_{Na} in E161K mutants. Computational models demonstrated a marked atrial and ventricular conduction slowing, as well as a reduction in sinus rate stemming from slowing of the diastolic depolarization rate and upstroke velocity of the sinus node AP. This reduction in sinus rate was further aggravated by application of acetylcholine, simulating the dominant vagal tone during the night.

Mutations and their genetic loci implicated in CCD and sinus node disease are shown in Table XI.

SUDDEN DEATH INFANT SYNDROME

It is unlikely that one mutation or polymorphism is the predisposing factor in each sudden death infant syndrome (SIDS) case, but rather, there is a strong probability that there are "SIDS genes" operating as a polygenic inheritance and predisposing the infant to sudden death, in combination with a number of environmental risk factors.[136]

Inherited LQTS have been considered to be the cause of death in some cases of SIDS, and in some cases, mutations in *KvLQT1* and *SCN5A* genes have been identified. In addition, several polymorphisms in these two genes and in the *HERG* gene have been reported (see Table XII).

Inherited and Sporadic Wolff–Parkinson–White Syndrome

In general, the Wolff–Parkinson–White (WPW) syndrome ECG is characterized by an abnormal short P-R interval with prolonged QRS duration. Paroxysmal atrial tachycardia is frequently present in WPW secondary to reentry by means of an anomalous pathway, which in the atria becomes a retrograde conduction pathway. In some patients, episodic atrial fibrillation and flutter may occur,

TABLE XI Gene defects in Cardiac Conduction Defects (CCD)

Disease	Chromos. locus	Gene	Protein	Mutation	References
CCD	3p21–24	SCN5A	hNav1.5	E161K	134
CCD/Brugada	3p21–24	SCN5A	hNav1.5	G1406R	130
CCD/AV block	3p21–24	SCN5A	hNav1.5	G298S	132
CCD/AV block	3p21–24	SCN5A	hNav1.5	D1595N	132
Sick sinus syndrome	15q24–23	HCN4	HCN4–573X	1631delC	135
	p21–24	SCN5A	hNav1.5	E161K	134
	3p21–24	SCN5A	hNav1.5	Multiple mutant alleles	133

TABLE XII Cardiac Channelopathies and SIDS

Disease	Chromosomal locus	Gene (alternate name)	Protein	Frequency	References
LQTS/SIDS in neonates					
	11p15.5	KCNQ1 (KvLQT1)	KvLQT1 (Kv7.1)	Rare	137
	3p21–24	SCN5A	hNaV1.5	Rare	138
	7q35–36	KCNH2 (HERG)	HERG (Kv11.1)	Rare	139

as well as VT and VF. Clinically, the patient presents with palpitations and syncope. Sudden death, although infrequent, has been reported. In symptomatic patients, radiofrequency ablation of the anomalous pathway is indicated, and its success rate is close to 95%. WPW can present as an inherited (familial) or as an acquired cardiac dysrhythmia. The familial form is rare and exhibits an autosomal-dominant pattern, often associated with hypertrophic cardiomyopathy (HCM), and has been linked to mutations in the PRKAG2 gene encoding the γ2 regulatory subunit of AMP-activated protein kinase (AMPK). Mutations in PRKAG2 have been reported in two families with severe HCM, and some individuals had WPW.[140] These mutations occurred in highly conserved regions, one a missense mutation, and one, an in-frame single codon insertion. In addition, Gollob and associates mapped the gene responsible to 7q34–q36 (a locus previously identified to be responsible for an inherited form of WPW) in two large families with WPW, which segregated as an autosomal-dominant disorder.[141] Candidate genes were identified, sequenced, and analyzed in normal and affected family members to identify the disease-causing gene. A total of 31 members had WPW, and the linkage of the gene was confirmed in both families to 7q34–q36. Haplotype analysis suggested that the two families do not have a common founder. A missense mutation was identified, resulting in the substitution of glutamine for arginine at residue 302 in the PRKAG2 gene. These findings may have important implications in understanding the pathogenesis of ventricular pre-excitation and suggest a common molecular basis in some patients for several phenotypes, including ventricular preexcitation, conduction abnormalities, and cardiac hypertrophy.

Studies in transgenic mice were generated by the cardiac-restricted expression of the wild-type (TG-WT) and mutant (TG-R302Q) PRKAG2 gene by use of the cardiac-specific promoter α-myosin heavy chain.[142] Both ECG and electrophysiological data demonstrated the TG (R302Q) mice to have ventricular pre-excitation and a prolonged QRS. Enzymatic activity of AMPK in the mutant heart was significantly reduced, probably a result of secondary disruption of the AMP binding site by the mutation. This genetic animal model of WPW expresses a mutation responsible for a familial form of WPW syndrome with a phenotype identical to that of the human. The defect seems to be due to loss of function of AMPK.

Other causes for WPW have been described. WPW has also been reported in patients with mutations in a 1,4-α-glucosidase.[143] A study of 24 Finnish families with Leber's hereditary optic neuropathy (LHON), showed that 5 affected individuals with both LHON and WPW pre-excitation syndrome had a mutation at nt 11778 in mitochondrial DNA.[144] Vaughan and associates have screened 26 patients with sporadic (non-familial) WPW for the PRKAG2 mutations using denaturing HPLC and automated sequencing.[145] Cardiac hypertrophy was ruled out by echocardiography in each case, and only one patient had an associated congenital heart disease. Two polymorphisms in PRKAG2 [inv6+36insA] were found in intron 6 in four patients and [inv10+10delT] in intron 10 in one patient, and no mutations in PRKAG2 coding sequence were found. Unlike familial WPW syndrome, this study showed that PRKAG2 mutation is not commonly associated with sporadic WPW syndrome. Moreover, there is no evidence that the identified polymorphisms predispose to accessory pathway formation, because their incidence was similar in both WPW and normal individuals.

ACQUIRED DYSRHYTHMIAS

Acquired or secondary dysrhythmias are mainly those caused by diseases of the myocardium (e.g., CAD, HCM, and DCM), or drug related.

Atrial Fibrillation and Cardiac Structural Defects

AF is a dysrhythmia frequently associated with HF of diverse etiologies, CAD, hyperthyroidism, and systemic hypertension, although in these clinical settings, some genetic factors may influence its occurrence as previously noted. It has been recently reported that AF in parents significantly increases the future risk for offspring AF, not only in lone AF but also when structural defects are present, further underscoring the presence of genetic factors predisposing to AF.[146] Moreover, in the Framingham Heart study, 681 individuals of a total of 2242 had at least one parent with AF. At follow-up 10% of the 681 individuals developed AF.[147] As shown in Table XIII, a number of polymorphism have been reported in patients with AF and underlying cardiac defects.

An association between the KCNE1/minK 38G allele and AF has been reported in a Chinese population.

TABLE XIII Genotype and Phenotype of AF with Structural Cardiac Defects

Gene	Variant	Type AF	Race	Sympt	UHD	Substrate	Ref.
KCNE1/minK	38G	CAF	Chinese	NA	VHD, HT	I_{Ks}	147
ACE	M235T, G217A, G-6A	CAF	Chinese	Yes	VHD	Possible	148
Cx40	−44AA	PAF	Caucasian	Yes	WPW	Possible	149

AF, atrial fibrillation; CAF, chronic AF; NA, not available; PAF, paroxysmal AF; UHD, underlying heart disease; VHD, valvular heart disease; HT, hypertension; WPW, Wolff–Parkinson–White syndrome.

Compared with patients without *KCNE1* 38G allele, the odds ratios for AF in patients with 1 and 2 *KCNE1* 38G alleles were 2.16 and 3.58, suggesting a possible genetic control in the pathogenesis of AF. More recently, the same group of investigators found, in a group of 250 patients with structural heart disease and AF, that the expression of polymorphisms M235T, G-6A, G-217A of the *ACE* gene encoding angiotensin-converting enzyme were significantly increased in patients with AF compared with matched controls (n = 250).[148] The odds ratio for AF was 2.5 with M235/M235 + M235/T235 genotype, 3.3 for the G6/G6 genotype, and 2.0 for the G217/G217 genotype. Significant associations have also been found in cases of AF with polymorphisms in the *Cx40* gene (encoding the cardiac gap junction protein connexin40), which mediates cardiomyocyte electro-coupling.[149] After identification of two linked polymorphisms within the regulatory regions of *Cx40* at nucleotides −44 G→A and +71 A→G (which were found to be associated with familial atrial standstill), an evaluation of whether the *Cx40* polymorphisms were associated with increased atrial vulnerability *in vivo* and dysrhythmia susceptibility was conducted. Carriers of the −44AA genotype exhibited a significantly higher coefficient of spatial dispersion of refractoriness (CD) compared with those with −44GG genotype, whereas heterozygotes exhibited intermediate values. All individuals with increased CD had a history of idiopathic AF compared with only one subject with normal CD. Moreover, the −44A allele and −44AA genotype were more often present in subjects with prior AF than in those without. This study provided evidence linking *Cx40* polymorphisms to increased atrial vulnerability and risk of AF. However, as the authors pointed out, the precise mechanism by which these *Cx40* gene variations may modulate atrial electrophysiological properties conferring dysrhythmia risk is currently undetermined. In addition, these are case-control association studies, and a larger sample size is necessary to validate any definitive conclusion.

Transcriptional profiling of genes modulated in the atrium of AF patients using DNA microarray technology was recently carried out in right atrial appendage tissues collected from 10 patients with normal sinus rhythm and 7 patients with chronic AF who underwent cardiac surgery.[150] The diameter of the left atrium in the AF group was greater than that in the sinus rhythm group. Affymetrix GeneChip analysis of 12,000 human genes used for each specimen, revealed 33 AF-specific genes that were significantly activated compared with the sinus rhythm group. The genes included an ion channel, an antioxidant, inflammatory factors, three cell growth/cell cycle factors, three transcription factors, several cell signaling proteins, and seven expressed sequence tags (ESTs). On the other hand, 63 sinus rhythm–specific genes were identified, including several with cell signaling/communication function such as SR Ca^{++}-ATPase 2 (SERCA2), cellular respiratory and energy production function, and two antiproliferative or negative regulator of cell growth genes, as well as 22 ESTs. This study showed that more than 100 genes were modulated in the atria of AF patients and suggests that these genes may play a critical role in the initiation or perpetuation of AF and in the pathophysiology of atrial remodeling.

Furthermore, in a experimental study in a model of persistent AF over several weeks, electrical remodeling was accompanied by structural remodeling.[151] Ultrastructural analysis by electron microscopy of samples from the two atria showed extensive mitochondrial changes, accumulation of glycogen, deficit in myofibrils, redistribution of the nuclear chromatin, and a reduction of SR with marked changes in protein structure. This atrial structural remodeling seems to be an adaptation reaction similar to that observed in hibernating myocardium during ischemia, prolonging cellular viability by decreasing atrial contractility. Remodeling also involves the activation of fibroblasts and promotion of fibrosis, resulting in extensive heterogeneity of the conduction tissue. An increase in ACE levels and angiotensin II concentrations was also demonstrated, and irbesartan (an angiotensin II antagonist) significantly reduced fibroblast growth. These findings have practical applications, because a reduction in the recurrence of AF after cardioversion occurred when administered in association with amiodarone. Nevertheless, to determine the full effect of atrial remodeling, which might also act on the foci responsible for inducing the AF, further research is indicated.

In another study of gene expression in noninherited paroxysmal AF (PAF), Barth and associates compared atrial mRNA expression in patients with PAF to individuals in sinus rhythm and to ventricular gene expression by use of Affymetrix U133 arrays.[152] In PAF cases, 1434 genes were found deregulated in the atrial myocardium (most of them down-regulated). Significant up-regulation of transcripts involved in metabolic activities was found, suggesting an adaptive response to increased metabolic demand in AF. Ventricular-predominant genes were five times more likely to be up-regulated, whereas atrial-specific transcripts were predominantly down-regulated. Dedifferentiation, with the adoption of a ventricular-like signature, seems to be a general feature of AF.

Ventricular Dysrhythmias and Sudden Dysrhythmia Death Syndrome

Premature ventricular contractions (PVCs) occur in healthy individuals, as well as those with structural heart defects, mainly in cardiomyopathies and ischemic heart disease. External (environmental)/epigenetic factors, such as electrolyte imbalance, sympathetic overdrive, medications, anxiety, and toxins (e.g., alcohol, tobacco, and coffee), may facilitate their appearance. The mechanism involved in PVCs is mainly reentry, but it also may be an ectopic focus with increased automaticity. The same factors can produce VT defined by the presence of one or more runs of three or more PVCs, and these can be either monomorphic or polymorphic. In the absence of structural heart defects, sustained VT (defined as a long run of >30 PVCs) is rare. In an analysis of 230 patients who died suddenly while undergoing Holter monitoring, Bayés de Luna found that in more than 80% of the cases, sudden death was due to ventricular tachydysrhythmias, including 10% primary VF, 50% VF preceded by classic VT, and 20% *torsades de pointes* tachycardia.[153] The latter phenotype appeared mainly in patients without structural heart disease and was mostly related to antidysrhythmic therapy and/or electrolyte and metabolic imbalances. As pointed out by Bayés de Luna, in the context of ventricular tachydysrhythmias, the risk for sudden death is mainly related to the degree of myocardial vulnerability, which in post-MI patients is primarily manifested by the presence of markers such as structural defects, decreased myocardial contractility, residual ischemia, and imbalanced autonomic nervous system.

In general, HF and myocardial hypertrophy account for most cases of lethal dysrhythmias, and these may be related to down-regulation of repolarizing K$^+$ currents, as well as changes in cardiomyocyte Ca^{++} handling, two mechanisms that can explain the prolongation of the AP duration (APD) and alterations of the AP dynamics (which occur in these conditions). Functional down-regulation of the major K$^+$ currents, I_{to}, I_{Kr}, I_{Ks} and I_{K1}, has been reported in HF, and their impact on the AP varies with species and the cause of CHF.[154] Among these currents, I_{to} is the most frequently altered, not only in HF but also in cardiac hypertrophy, especially in humans, and abnormal current channel density is caused by a diminished expression of the genes encoding either the ion channel subunits or regulatory proteins, such as KChIP2.[155]

Furthermore, dependence of APD and the Ca^{++} transient on pacing rate is a fundamental property of cardiac myocytes that, when altered, may promote life-threatening cardiac dysrhythmias. The ionic mechanisms underlying the calcium transient and APD rate dependence have been investigated by Hund and Rudy in a theoretical model of canine ventricular epicardial AP and Ca^{++} cycling.[156] They found that I_{to} creates a phase 1 notch that increases the driving force for I_{Ca-L} and facilitates the activation of a sustained component. In addition, the phase 1 notch decreases the repolarizing currents I_{Kr} and reverse-mode $I_{Na/Ca}$. Together, these processes prolong APD. Alterations in depolarizing currents during the AP plateau phase can affect the balance in favor of secondary depolarizations, which then will trigger dysrhythmias. Enhanced susceptibility to after-depolarization–mediated ventricular tachydysrhythmias had been confirmed in animal studies of HF in which APD prolongation was provoked with pharmacological agents.[157] Moreover, both intracellular Ca^{++} handling, which is defective in HF, and increased Na$^+$-Ca^{++} exchanger activity may promote ventricular dysrhythmias, the first by increasing cardiomyocyte electrical instability and abnormal mechanics and the latter by altering the AP configuration. Compared with controls, the pig hibernating myocardium exhibits myocyte hypertrophy, marked AP prolongation, and L-type Ca^{++} currents with decreased Ca^{++} release.[158] Furthermore, a model of regional ischemia in isolated perfused pig hearts showed that the incidence of VF was 38% after a 10-min period of ischemia.[159] After 80–90 sec of reperfusion, the extracellular K$^+$ was 0.8 mM less than in normal tissue in more than half of the reperfused tissue, especially in the border zone, and this was associated with TQ elevation and large peaked T waves. The T-wave changes were caused by an abrupt decrease of the APD in reperfused tissue. It is worth noting that reperfusion VF started with a closely coupled PVC. The closely coupled PVC that initiates reentry was not caused by delayed after-depolarizations but more likely by intramural re-entry.

In general, the mechanism of sudden dysrhythmia death syndrome (SDDS) is uncertain, although most often it seems to be secondary to a lethal cardiac dysrhythmia (mostly ventricular tachydysrhythmia) with or without structural cardiac abnormalities. Less often, SDDS is caused by bradydysrhythmias such as complete AV block, frequently in the presence of cardiac structural defects. Some recent data suggest that bradydysrhythmias and pulseless electrical activity may account for an increasing percentage of SDDSs, because the frequency of VT and VF may be decreasing. The WHO definition of sudden death has opened the possibility that death may result from a precipitous decline in the mechanical function of the heart, such as pulseless electrical activity. As pointed out by Tomaselli and Zipes, even witnessed sudden death may be produced by a sudden mechanical or vascular catastrophe (pulmonary embolus, cardiac, or vascular rupture) rather than a malignant dysrhythmia.[160]

Drug-Related Dysrhythmias (Pharmacogenetics)

During the past few years, lethal ventricular dysrhythmias have been reported associated with the use of a number of drugs, most of which subsequently were withdrawn from the market. The implicated drugs include not only antidysrhythmics but also a number of noncardiac drugs such as antihistaminics and antipsychotics. *Torsades de pointes* (TdP) ventricular tachycardia may be drug-induced (in addition to the type associated with the congenital form of LQTS), and this type is called acquired TdP. Often, an aberrantly prolonged AP may lead to TdP, although the term "decreased repolarization reserve" has also been proposed as a better description for the cause of this rhythm disorder, which electrophysiologically is probably a result of triggered automatic activity.[161] On ECG, TdP presents with short, self-limited bouts of rapid ventricular rhythm together with a changing morphology of the QRS complexes. In the absence of tachycardia, the QT is prolonged (500 ms or more) with the first complex of each run appearing relatively late. Most drugs with the potential to induce TdP are HERG blockers. These drugs inhibit the *KCNH2*-encoded HERG potassium channel, making it capable of developing delayed repolarization as indicated by LQT. Rare mutations in *KCNH2* have provided the pathogenic substrate for LQT2, placing this cardiac potassium channel exactly in the intersection between congenital LQTS (cLQTS) and acquired LQTS (aLQTS) (i.e., drug-induced TdP).[162]

Furthermore, polymorphisms residing in the LQTS-causing channel genes may also increase the dysrhythmogenic susceptibility of a vulnerable host. It is worth noting that besides mutations in LQTS genes, polymorphisms may also increase the risk for aLQTS. In a study of 98 patients with drug-induced aLQTS, Sesti and associates evaluated the *KCNE2* gene encoding the *minK*-related peptide 1 (MiRP1), a β-subunit of the cardiac K$^+$ channel I$_{Kr}$.[163] Three individuals with sporadic mutations were identified, as well as a patient with sulfamethoxazole-associated LQTS, who carried a SNP (detected in approximately 1.6% of the general population). Although mutant channels showed diminished K$^+$ flux at baseline and wild-type drug sensitivity, channels with the SNP-containing allele were normal at baseline but were inhibited by sulfamethoxazole at therapeutic levels that did not affect wild-type channels. Apparently, allelic variants of MiRP1 contribute to a significant fraction of cases of aLQTS through multiple mechanisms. Importantly, common sequence variations that increase the risk of life-threatening drug reactions could be clinically silent before drug exposure.

Furthermore, in a mutation analysis of 32 unrelated patients with drug-related LQTS (with confirmed TdP) and 32 healthy individuals, Paulussen and associates, after careful screening of five cLQTS genes (*KCNH2, KCNQ1, SCN5A, KCNE1, KCNE2*), identified highly significant missense mutations in four aLQTS patients, D85N in *KCNE1* (two cases), T8A in *KCNE2*, and P347S in *KCNH2*.[164] Three other missense variations were identified both in patients and controls, which are unlikely to significantly influence the aLQTS susceptibility. In addition, 13 silent and 6 intronic variations were detected, 4 of which were found in a single aLQTS patient but not in controls. Thus, it seems that missense mutations in cLQTS genes only explain a minority of aLQTS cases.

Besides genetic risk factors for aLQTS, it is significant that there are a number of environmental or nongenetic factors, which may contribute to increase the susceptibility to aLQTS/TdP, such as female gender (see Chapter 18), hypokalemia, and other heart diseases. Also, interindividual differences in drug metabolism, caused by functional SNPs in drug-metabolizing enzyme genes may be a significant risk factor for aLQTS, especially if multiple drugs are involved. Furthermore, because most drugs involved in LQTS are metabolized by the hepatic enzymes CYP2D6 and CYP3A, variants in the *CYP2D6* gene may abolish the enzyme function, leading to a poor drug metabolism and increasing the risk for TdP (secondary to higher drug plasma levels) compared with normal metabolizers.[165]

Dysrhythmias Associated with Defects in FAO and Mitochondrial Function

As noted elsewhere in this book, fatty acids play an integral role in determining the structural and functional nature of the plasma and mitochondrial membranes. Their influence on the fluidity and stability of membrane structure markedly impacts on cardiac membrane functions such as transport of ions and substrates and, therefore, in the cardiac electrophysiology, which is intrinsic to function and excitability. Conduction defects and cardiac dysrhythmias have been shown to be frequently present in patients with certain defects in FAO.[166] Specifically, these cardiac defects were present in patients with deficiencies in carnitine palmitoyltransferase (CPT-II), carnitine translocase, and mitochondrial trifunctional protein (MTP). However, cardiac dysrhythmias were notably absent in patients with other FAO defects, including deficiencies in CPT-I, the primary carnitine carrier and medium chain acyl-CoA dehydrogenase (MCAD). This strongly suggests that the accumulation of dysrhythmogenic intermediary metabolites of fatty acids (e.g., long-chain acylcarnitines) may be responsible for dysrhythmias and potentially contributory to HF and sudden death. This finding is consistent with the observation that long-chain acylcarnitines accumulate with the defects in CPT-II, carnitine translocase, and MTP, whereas MCAD, CPT-I, and carnitine carrier do not result in accumulation of these intermediates. It is well known that amphiphilic long-chain acylcarnitines possess detergent-like properties, can extensively modify membrane proteins and lipids, and have a variety of toxic effects on the electrophysiological function of cardiac membranes, including ion transport (Na$^+$, Ca^{++}) and impaired gap junction activity. Moreover, patients with cardiomyopathy caused by inborn defects in carnitine translocase often have an increased onset of cardiac dysrhythmias.[167] Furthermore, long-chain fatty acid intermediate accumulation (e.g., acyl-carnitine) has been implicated in the genesis of ventricular rhythm disorders during myocardial ischemia.[168] The selective blocking of CPT-I activity prevents the accumulation of potentially toxic long-chain esters during hypoxia/ischemia, thereby reducing the risk of electrophysiological disturbance and membrane disruption.[169]

Research to further define the site of action of these toxic long-chain intermediates within the cardiomyocytes (e.g., at the level of the plasma membrane, mitochondrial matrix, or membrane[s]) may be of use in development of new therapies. The rescue of mitochondrial inner membrane potential ($\Delta\psi_m$) is a key determinant of post-ischemic functional recovery of the heart. Mitochondrial ROS-induced ROS release causes the collapse of ($\Delta\psi_m$) and the instability of the AP, largely mediated by a mechanism involving a mitochondrial inner membrane anion channel (i.e., mitochondrial channelopathy) modulated by the mitochondrial benzodiazepine receptor. Recently, Akar and colleagues have shown how failure of the mitochondrial network at the muscle–cell level leads to differential electrical excitability in hearts subjected to ischemia and reperfusion.[170] This may constitute a new mechanism to explain post-ischemic dysrhythmias, in which mitochondrial dysfunction in the reperfused heart leads to a local fall of electrical currents preventing propagation. This study also showed that by stabilizing the mitochondrial membrane potential, the cellular AP was maintained, blunting the AP shortening during ischemia, and prevented ventricular dysrhythmias during reperfusion. These findings could explain the pathogenesis of ischemia-related dysrhythmias and suggest potential future treatment with specific drugs targeting the mitochondrial organelle.

ANIMAL MODELS FOR CARDIAC DYSRHYTHMIAS

Studies in animal models have provided extremely important insights about cardiac dysrhythmias. The mouse has become the most prevalent animal used in the genetic analysis, largely because of the available advances in transgene and gene targeting technology. As shown in Table XIV, numerous mouse strains have been generated exhibiting ion channelopathies and dysrhythmogenic disorders recapitulating human clinical phenotypes. These models have allowed the further examination of the genes and dissection of the molecular pathways involved in mechanism underlying these cardiac conduction disorders.

As shown in Table XV, other animal models such as the rat, dog, guinea pig, and pig are also increasingly being used, some of which might be more physiologically similar to humans.

POST-GENOMIC APPROACHES TO RHYTHM DISORDERS

Identification of each dysrhythmia gene relevant to humans is a great challenge ahead. Rare dysrhythmias, either associated or not to structural heart disease, pose a significant challenge to physiologically define genes that are necessary for the maintenance and propagation of normal heart rhythm, and in this vein, it may be possible to identify naturally occurring DNA polymorphisms. Most of the SNPs so far identified have a low (<10%) allele frequency of the minor allele, meaning that they are potentially not suitable for case-control studies, because small differences in the allele frequencies between patients and controls would require samples of a very large size to achieve statistically significant results.[200] The deciphering of haplotype block structure of human gene variation seems to provide a powerful tool for gene mapping in cases of inherited dysrhythmias. Of significance will be the further identification of frequently occurring, nonsilent SNPs with proven functional relevance *in vitro*, although it has been suggested that intrinsically gene expression is only partially under genetic control.[201–203]

Aydin and associates have screened a white population of 141 subjects for SNPs in five long QT syndrome genes, namely, *KCNQ1* (LQTS1), *HERG* (LQTS2), *SCN5A* (LQTS3), *KCNE1* (LQTS5), and *KCNE2* (LQTS6).[204] They found 35 SNPs, 10 of which have not been previously described. Ten SNPs were in *KCNE1*, six in *HERG*, eight in *KCNQ1*, four in *KCNE2*, and seven in *SCN5A*. Four SNPs were associated with QTc interval, 1 in *KCNE1*, 1 in *KCNE2*, and 2 in *SCN5A*. Therefore, these five long QT syndrome genes contain common variants, some of which are associated with the QTc interval in phenotypically normal individuals. The authors suggested that analysis of these SNPs in a much larger cohort would enable establishment of common haplotypes that are associated with QTc, and these haplotypes could facilitate prediction of dysrhythmia risk in the general population. In a similar study that used linkage disequilibrium–based SNP association, the KORA study reported common variants in four myocardial ion channel genes that modified the Q-T interval in the general population.[205] By use of a two-step analysis, 174 SNPs from the *KCNQ1*, *KCNH2*, *KCNE1*, and *KCNE2* genes were analyzed in 689 individuals and 14 SNPs with suggestive association to QTc were further analyzed in a larger population (n = 3,277) from the KORA study. A significant association with a gene variant in intron 1 of the *KCNQ1* gene and a weaker association with a variant upstream of the *KCNE1* gene were detected. In addition, association with two SNPs in

TABLE XIV Mouse Models of Dysrhythmias and Conduction Defects

Disease	Gene	Mutation	Phenotype	Ref.
LQTS1/JLN	KCNQ1	Loss-of-function	QT prolongation	171
		Dominant-negative	*Torsade de pointes*; long QT	172
LQTS2	HERG	Loss-of-function	Normal QT	173
LQTS3	SCN5A	Gain-of-function	QT prolongation; spontaneous VT and VF; sudden death	174
LQTS4	ANK2	Loss-of-function	QT prolongation; exercise-induced VT	175
LQTS5	KCNE1	Loss-of-function	Altered Qt/RR	176
Andersen syndrome	KCNJ2	Loss-of-function	QT prolongation	177
Brugada syndrome	SCN5A	Loss-of-function	Impaired conduction; Triggered VT	178
Inherited Lenegre/ Lev syndrome	SCN5A	Loss-of-function	Impaired conduction; Progressive impairment of conduction with age	179
AF	KCNE1 junctin	Loss-of-function	AF	180
		Overexpression	Bradycardia and AF	181
CPVT/ARVC	RYR2	Gain-of-function	VT in response to caffeine or adrenergic stimulation	182
WPW syndrome	PRKAG2	Loss-of-function	Ventricular preexcitation; prolonged QRS; HCM	183
AV block	HF-1b	Loss-of-function	AV block, sinus bradycardia, spontaneous VT	184
	Nkx2.5	Loss-of-function	Underdeveloped AV node	185
	DMPK	Loss-of-function	Onset of complete heart block	186
	Cx40	Loss-of-function	AV block	187
	ACE2	Overexpression	AV and bundle branch block	188
	INOS	Overexpression	Heart block; premature death	189
	Rho GDIα	Overexpression	AV block	190
	SCN5A	Loss-of-function	AV block; atrial dysrhythmia;	178

AV, atrioventricular; AF, atrial fibrillation; HCM, hypertrophic cardiomyopathy; LQTS, long QT syndrome; VT, ventricular tachycardia; VF, ventricular fibrillation; WPW, Wolff-Parkinson-White syndrome.

the *KCNH2* gene, a previously described K897T variant, and a gene variant that tags a different haplotype in the same block were found. These findings confirmed previous studies of heritability, indicating that repolarization is a complex trait with a significant genetic component and further demonstrated that high-resolution SNP mapping in large population samples could be useful in detecting and fine mapping quantitative trait loci even when locus-specific heritabilities were small.

The interrelationship between genomic variance in dysrhythmia genes, promoter variability, and alterations in gene transcription is mostly unknown to date.[206] Studies involving a large sample size of genotyped patients have demonstrated that epigenetic factors (e.g., gender, age, and ethnicity) are important modulators of disease phenotype and in the development of dysrhythmias (Fig. 4). Whether these modulators or genetic cofactors play a leading causative role is not known. Integration of a comprehensive and standardized clinical screening, together with a thorough genetic analysis (e.g., genetic subtype, polymorphic changes, haplotype blocks) and transcriptome/proteome mapping, may be successful in the risk stratification of patients with dysrhythmias.

Genetic Screening and Risk Stratification of Dysrhythmias

Genetic screening of patients with potentially lethal dysrhythmias is a recent strategy in the identification of markers or clinical predictors of SDDS. Although still controversial largely because of the phenotypic heterogeneity of cases with single gene mutation, and the fact that the technique is rarely available outside specialized medical centers, genetic assessment of patients and families can be of great value at least to define specific antidysrhythmic therapies.

Recently, Shah and associates have reported that genetic screening (e.g., FAMILION test) has become available to screen for selected mutations in Na^+ and K^+ channel genes

TABLE XV Genetic Models of Dysrhythmia and Conduction Defects in Other Animal Models

Gene	Animal	Mutation	Phenotype	Ref.
Hepatocyte growth factor	Rat	Overexpression	Decreased ischemia-induced ventricular dysrhythmias	191
SERCA2a	Rat	Overexpression	Decreased ischemia-induced VT and VF	192
HCN2	Dog	Overexpression targeted to left bundle-branch system	Induction of pacemaker activity with normal automatic rate and vagal stimulation	193
HERG	Guinea pig	Overexpression HERG-G628S	Slows cardiac repolarization, QT prolongation and frequent EADs	194
$g\alpha_i$	Pig	Overexpression in AV node of constitutively active gαi (cGi)	Decreased heart rates in animals with induced AF	195, 196
KCNE1	Guinea pig	KCNE1-D76N overexpression	Slows cardiac repolarization, QT prolongation and frequent EADs	194
ras-Related small G-protein Gem	Guinea pig	Overexpression in heart; Overexpression in AV node	Shortening of QTc interval; slowed AV nodal conduction (prolongation of PR and AH intervals) and reduced heart rate with induced AF	197
NOS-1	Guinea pig	Overexpression in right atrium	Enhanced heart rate in response to vagal stimulation	198
MiRP1	Pig	hMiRP1-Q9E overexpression in right atrium	Prolongation of repolarization of right atrial epicardial MAPs	199

MAP, monophasic action potential; AF, atrial fibrillation; VT, ventricular tachycardia; VF, ventricular fibrillation; EAD, early afterdepolarizations; AV, atrioventricular.

(e.g., SCN5A, KCNQ1, KCNH2, KCNE1, KCNE2). Mutations in these genes are considered to be responsible for 50–75% of cases of congenital LQTS and 15–30% of BrS cases.[154] Moreover, the yield of genetic testing, as well as the type and the prevalence of mutations in patients with LQTS, have been assessed by Napolitano and associates.[207] They examined whether the detection of a set of frequently mutated codons in the KCNQ1, KCNH2, and SCN5A genes could be translated into a novel strategy for rapid and efficient genetic testing of 430 patients referred to a medical center during an 8-year period. The entire coding regions of KCNQ1, KCNH2, SCN5A, KCNE1, and KCNE2 were screened with denaturing HPLC and DNA sequencing, and the frequency and the type of mutations were assessed to identify a set of recurring mutations. A separate cohort of 75 consecutive probands was used as a validation group to quantify prospectively the prevalence of the recurring mutations identified in the primary LQTS population. In this study, 235 different mutations were identified, 138 of which were novel in 72% of 430 probands (49% KCNQ1, 39% KCNH2, 10% SCN5A, 1.7% KCNE1, and 0.7% KCNE2). Non-private mutations were found in 58% of probands and were carried in 64 codons of KCNQ1, KCNH2, and SCN5A genes. A similar occurrence of mutations at these codons (52%) was confirmed in the prospective cohort of 75 probands, as well as in previously published LQTS cohorts. This study offers a novel approach to improve the efficiency of LQTS genetic screening and may enhance wider access to genotyping, resulting in better risk stratification in the prediction of SCD and treatment. Previously, this group of investigators led by Priori attempted to establish risk stratification of patients with LQTS. They prospectively followed up a large group of patients ($n = 647$) with mutations at the LQTS1, LQTS2, and LQTS3 genetic loci to identify their first cardiac event (e.g., syncope, cardiac arrest, or SCD) before their 40th birthday.[208] The incidence of a first cardiac event before age 40 was 30% for patients with mutations in the LQTS1 locus, 46% in the LQTS2, and 42% in the LQTS3. By use of a multivariate analysis, the investigators found that the genetic locus and the QTc, but not gender, were independent predictors of risk. The QTc was an independent

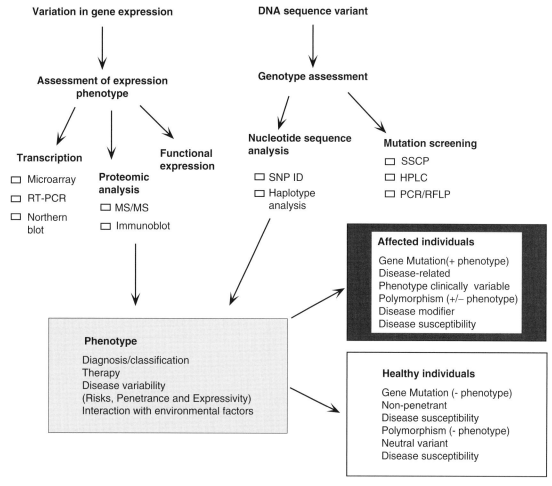

FIGURE 4 Molecular, genetic data, and risk stratification of patients. Genetic screening and risk stratification of dysrhythmias/cardiac events.

predictor of risk among patients with a mutation at either the LQTS1 or LQTS2 locus but not among those with a LQTS3 mutation; in contrast, gender was an independent predictor of events only among those with a mutation at the LQTS3 locus. On the basis of these data, it is apparent that the locus of the causative mutation affects the clinical course of the LQTS and modulates the effects of the QTc and gender on the clinical manifestations, therefore allowing for risk stratification. As the authors pointed out, their findings make possible a revision of previous recommendations for risk stratification and the management of asymptomatic LQTS whenever information on genotype is available.[209] Thus, they proposed that prophylactic treatment with β-blockers is warranted in male and female patients with a mutation at the LQTS1 locus who have a QTc of 500 msec or more, male patients with a mutation at the LQTS2 locus who have a QTc of 500 msec or more, all female patients with a mutation at the LQTS2 locus irrespective of the QTc, and all patients with a mutation at the LQTS3 locus.[210] By contrast, the decision to institute therapy in patients at lower risk of becoming symptomatic before the age of 40 y should be individualized. As the

authors noted, their study was based on the assumption that patients with LQTS with mutations at the same locus have a similar risk of cardiac events. If it could be confirmed that specific mutations are more malignant than others, it might possible to develop locus-specific risk-stratification schemes based either on specific mutations or on their functional effects.[211,212] Similarly, Chen and associates have described the genotype–phenotype characteristics in 10 families with mutations of *KCNQ1*, including 5 novel mutations.[213] By use of single-strand conformation polymorphism (SSCP) and DNA sequence analyses, they screened 102 families with a history of lethal cardiac events, including 55 LQTS, 9 BrS, 18 idiopathic VF, and 20 acquired LQTS. Genotypes of *KCNQ1* were correlated with phenotypes assessed by ECG and cardiac event history. No *KCNQ1* mutations were found in BrS, idiopathic VF, or acquired LQTS families. Ten of 55 LQTS families had *KCNQ1* mutations, with 62 carriers identified. Mutations included G269S in domain S5; W305X, G314C, Y315C, and D317N in the pore region; A341E and Q357R in domain S6; and 1338insC, G568A and T587M mutations in the C-terminus. W305X, G314C, Q357R, 1338insC, and

G568A, seemed to be novel mutations. Gene carriers were 26 ± 19 y (32 females). Baseline QTc was 0.47 ± 0.03 sec (range 0.40–0.57 sec), and 40% had normal to borderline QTc ≤0.46 sec). Typical LQTS1 T wave patterns were present in at least one affected member of each family and in 73% of all affected members. A history of cardiac events was present in 19 of 62 (31%), 18 with syncope, 2 with aborted cardiac arrest, and 6 SD. Two of six SDs (33%) occurred as the first symptom. No difference in phenotype was evident in pore versus nonpore mutations. *KCNQ1* mutations were limited to LQTS families, and all five novel mutations produced a typical LQTS1 phenotype. This study identified a reduced penetrance for QTc and symptoms; however, catastrophic SD (as the first symptom) represented 33% of those who died. Finally, genetic testing can be informative in the identification of gene carriers with reduced penetrance to provide adequate treatment and prevent lethal cardiac dysrhythmias.

The effects of age and gender on the clinical course of LQTS by genotype have been evaluated by Zareba and associates.[214] The probability of cardiac events (e.g., syncope, aborted cardiac arrest, or sudden death) was determined as a function of genotype, gender, and age (children up to 15 y, and adults from 16–40 y). In addition, the risk of SD and the lethality of cardiac events were evaluated in 1,075 LQTS1, 976 LQTS2, and 324 LQTS3 family members from families with known genotype. In children, the risk of cardiac events was significantly higher in males with LQTS1 than females, whereas there was no significant gender-related difference in the risk of cardiac events between LQTS2 and LQTS3 carriers. In adults, women with LQTS1 and LQTS2 had a significantly higher risk of cardiac events than men of similar age. The lethality of cardiac events was highest in men and women with LQTS3, and higher in men with LQTS1 and LQTS2 than in women with LQTS1 and LQTS2. These findings demonstrated that age and gender have different genotype-specific modulating effects on the probability of cardiac events in patients with LQTS.

Notwithstanding the preceding findings, we agree with Shah and associates[154] that only in rare cases are mutations in genes that affect cardiac excitability enough to cause a life-threatening dysrhythmia. For the most part, the generation of cardiac dysrhythmias will require a susceptible myocardial substrate (unlikely in structurally normal heart) and an appropriate trigger(s).

Pharmacogenomics/Pharmacogenetics and Other Therapeutic Options in Cardiac Dysrhythmias

In the past, many marketed drugs, including antidysrhythmics, have had to be withdrawn by the pharmaceutical companies because of the occurrence of pro-dysrhythmic effects, sometimes with lethal consequences. The related fields of pharmacogenomics and pharmacogenetics seem to have the potential to improve drug development and the tailoring of a drug therapy based mainly on the individual's ability to metabolize drugs that are determined only in part by age and influenced by disease, environmental factors (e.g., diet), concurrent medications, and variant genetic factors, specifying the transport, metabolism, and targets of the drug. New pharmacogenetic tests may allow pharmaceutical companies and clinical cardiologists to identify which individual are at risk for severe dysrhythmias.

Both inherited and, to a lesser degree, acquired cardiac dysrhythmias seem to have a genetic basis. Changes in channel protein expression, which play a significant role in cardiomyocyte AP and cellular electrical stability, may have a significant role in modifying the individual's pharmacological profile by altering the relative expression and contribution of different drug targets. For example, abnormal expression of the *KCNE2* (MiRP1)protein was noted in two different models of cardiac pathology.[215]

In the first model, canine myocardial ischemia secondary to coronary microembolizations, the rapid delayed rectifier current (I_{Kr}) density was increased, the protein level of ERG (I_{Kr}) pore-forming α-subunit was affected, whereas *KCNE2*-encoded protein level was found to be markedly reduced. These findings highlighted the effect of heterologously expressed *KCNE2* (MiRP1) on ERG and suggested that MiRP1 may associate with ERG and suppress its current amplitude. In the second model, aging rat ventricle, both the pacemaker current (I_f) density and the MiRP1 protein level were markedly increased, and no significant changes were found in the α-subunit (HCN2). Taken together, these findings show that the MiRP1 protein is expressed in the ventricles, and under different pathophysiological conditions, such as myocardial ischemia and aging, it can play diverse roles in modulating ventricular electrical activity. Careful use of certain drugs with variable interactions with the genetic background may be possible in the future if rapid and affordable genetic screening of defective genes becomes available; nonetheless, this may only apply to the more common polymorphisms predisposing to acquired dysrhythmias, such as T8A and Q9E in *KCNE2* or Y1102 in *SCN5A*.[216] On the other hand, specific DNA polymorphisms may make an individual insensitive to some medications, affecting altogether the response to therapy. For example, two variant polymorphisms (P532L and R578K) in the *KCNA5* gene that encodes the α-subunit of the I_{Kur} current, both residing in the C-terminus, were found to be resistant to block by quinidine.[217] Furthermore, a particular drug may affect more than just one gene or protein. In a review by Brendel and Peukert on the antidysrhythmic potential of the Kv1.5 channel blocker in atrial dysrhythmias, they found that these blockers were not selective because they blocked other ion channels.[218] On the other hand, it seems that the additional inhibition of Kv4.3 and K_{ACh} by these compounds may be beneficial in their antidysrhythmic effects, or at least does not affect the atrial selectivity of a Kv1.5 blocker. However, it is worth noting that marked block of I_{K1}, HERG, or Na$^+$ channels may lead to the loss of atrial selectivity, thus increasing the risk of lethal ventricular pro-dysrhythmia.

The forenamed investigators have also brought increased attention to the potential use of virtual screening approaches to improve the discovery of Kv1.5 blockers, because they can be targets for the treatment of AF.[219] Interestingly, by use of a protein-based pharmacophore derived from a homology model, they screened 244 molecules *in vitro*, and 19 of them (7.8%) were found to be active. The performance of this structure-based virtual screening protocol was compared with those of similarity and ligand-based pharmacophore searches, and this supports the use of as many virtual screening techniques as possible to increase the chances of identifying as many chemotypes as possible.

Gene/Cell-Based Therapies

The use of gene therapy for specific cardiac dysrhythmias, although promising, is still in its developmental stages, and only a few laboratories have reported a handful of successful cases in the experimental animals. For example, pacemaker implantation is the treatment of choice in patients with complete heart block and sinus node dysfunction. In an effort to develop biological pacemaker function, several approaches have been tested. By injection of plasmids to overexpress human β2-adrenergic receptors in the porcine right atrium, Edelberg and associates were able to produce heart rates that were 50% faster than those of controls, demonstrating the value of gene transfer to modify pacemaker function.[220] This study, however, focused only on modulating native pacemaker cells rather than the pacemaker itself. Subsequently, Marban's laboratory reported the *in vivo* viral gene transfer into the guinea pig ventricle of a dominant negative mutant of Kir 2.1 (the major subunit of the inward rectifier current I_{k1}) that caused cellular automaticity.[221] After successful gene transfer, they converted quiescent cardiomyocytes into pacemaker cells with the generation of spontaneous rhythmic electrical activity. These findings suggest that genetically engineered pacemakers could be developed as a possible alternative to implantable electronic devices. More recently, Qu and associates have used adenoviral constructs of mouse *HCN2* and green fluorescent protein (*GFP*) or *GFP* alone injected into the dog left atrium (LA).[222] After terminal studies 3 to 4 days later, the hearts were removed, and cardiomyocytes were examined for native and expressed pacemaker current (using the molecular correlate of I_f). Spontaneous LA rhythms occurred after vagal stimulation-induced sinus arrest in four of four *HCN2+GFP* dogs and none of three *GFP* dogs. Native I_f in nonexpressed atrial myocytes was 7 ± 4 pA at −130 mV (n = 5), whereas *HCN2+GFP* LA expressed pacemaker current (I_f) of 3823 ± 713 pA at −125 mV (n = 10) and 768 ± 365 pA at −85 mV. These data suggest that *HCN2* overexpression provides an I_f-based pacemaker sufficient to drive the heart when injected into a localized region of atrium, offering a promising gene therapy for pacemaker disease. Although considerable data suggest that the application of such a pacemaker can be further explored in animal models, significant issues need to be resolved before such constructs can be delivered into the human heart.[222] An issue of paramount significance is improvement in the delivery methods if we are to achieve the necessary density in gene transfer. With further development in gene transfer techniques, it may be possible to modify both conduction and repolarization to treat a number of specific dysrhythmias.

Although embryonic stem cells (ESCs) are considered to have great potential for cell-based therapies, their electrophysiological properties have not been yet fully characterized. Recent studies that used a combination of electrophysiological and imaging techniques have demonstrated that electrically active, donor cardiomyocytes (CMs) derived from human ESCs (hESCs) that had been stably genetically engineered by a recombinant lentivirus can functionally integrate with otherwise-quiescent, recipient, ventricular CMs to induce rhythmic electrical and contractile activities *in vitro*.[223] The integrated syncytium was responsive to the β-adrenergic agonist isoproterenol and to other pharmacological agents such as lidocaine and ZD7288. Similarly, a functional hESC-derived pacemaker could be implanted in the left ventricle *in vivo*. Mapping of the epicardial surface of guinea pig hearts transplanted with hESC-derived CMs showed successful spread of membrane depolarization from the site of injection to the surrounding myocardium. These findings prove that electrically active, hESC-derived CMs are capable of actively pacing quiescent, recipient, ventricular CMs *in vitro* and ventricular myocardium *in vivo*, and this may lead to an alternative or a supplemental method to correct defects in cardiac impulse generation.

CONCLUSION

Recently, breakthrough information on the molecular and molecular genetics basis of a large number of life-threatening dysrhythmias has become available. Mutated genes, in particular genes encoding the proteins forming the ion channels and transporters, as well as the discovery of numerous SNPs, have allowed the labeling of many of these dysrhythmias as inherited or familial. For example, genotyping single SNPs by use of haplotype block analysis by the human genome consortium has revealed that by the application of modern informatics tools to this vast collection of data, we could select from the approximately 10 million polymorphisms (or variants) present in the human genome, in particular those nonsilent SNPs that actually may have a functional role (by producing sequence changes in specific cardiac proteins) in many dysrhythmogenic syndromes and other cardiovascular pathological conditions.

As noted with the inherited cardiomyopathies, some dysrhythmogenic syndromes are genetically heterogenous (i.e., the same phenotype may result from a number of genetic defects). For example, the same LQTS phenotype is produced by several mutated genes, which account for 50% of cases clinically diagnosed. Heterogeneity is also present in patients with CPVT, in which two mutations account for approximately 50% of the cases with this phenotype. On the other hand, a single gene mutation has been found to result in BrS, yet only 20% of the cases with BrS phenotype seem to have mutations in that gene. It has been questioned whether it is appropriate to group all the genetic variants responsible for the same phenotype under the same name (e.g., LQTS) and whether it makes sense to group (under the same diagnosis) those individuals with the same electrocardiographic markers, knowing that each mutation has a distinctive clinical phenotype.[224] The answer is probably no; nevertheless, these and other questions may be soon appropriately addressed as progress continues in our understanding of the molecular, cellular, and genetic mechanisms of dysrhythmias.

Screening of several families with monogenic dysrhythmia has identified genes extremely valuable in our understanding of the variability of electrophysiological markers and in the response to exogenous stressors. It is apparent that individuals from the same kindred with the same disease-associated gene mutation or SNPs may display markedly different phenotypes. Moreover, SNPs are being continually identified in a number of genes that regulate cardiac excitability. According to Roden, these findings, combined with the increasing recognition that very common dysrhythmias/syndromes (such as AF or SDDS), include a genetic component increase the prospect that screening of the genomic variability among individuals and populations may in the future be used to manage patients.[225] The possibility of applying high-density genome-wide SNP analyses to examine genetic contributions to dysrhythmia susceptibility in community-based, case-control studies of common forms of SCD had been recently reviewed by Arking and associates, and they suggest that the development of novel strategies to identify contributors to susceptibility in common cardiac phenotypes is most likely to lead to new and relevant therapeutic targets for SCD.[226]

Notwithstanding the recent scientific advances in genetic analysis to assign a risk stratification for many of these dysrhythmogenic syndromes, we must also take into account numerous and important epigenetic modifiers/modulator factors, including family history, gender, repolarization abnormalities, and sympathetic tone, each of which may influence disease presentation and severity. Moreover, other environmental factors and modulators such as ethnicity and geographical distribution need to be closely evaluated in assessing causality and in establishing the specific diagnosis of the rhythm disorder. This can be facilitated not only by monitoring specific cardiac markers but also by close observation of the pharmacological responses and the electrophysiological phenotypes. Individualizing therapy may be particularly critical in establishing drug dosages and efficacy in children and aging patients with dysrhythmias and structural heart defects, a population for which pharmacokinetics has proven to be poorly defined and often unpredictable. Both immunological and genetic phenotyping can provide a more effective therapeutic strategy by either inhibiting or stimulating specific responses.

Parenthetically, in regard to the practicality and effectiveness of developing pharmacogenetic therapies, it is apparent that this modality has a promising future, and it may be the way to go in the treatment of cardiac dysrhythmias and probably other cardiac pathological conditions. Nevertheless, a recent report from United Kingdom's Royal Society (Ref: 18/05;21 Sep 2005) may have cooled down a little bit the currently growing, if rather impatient, general enthusiasm, because the report estimated that it may take another 15–20 y before so-called personalized medicine becomes widely available. Although the report estimates may be correct, with really focused and well-funded research on molecular genetics, it probably will not take more than 10 y before a patient's visit to the cardiologist's office results in the prescription of drug therapy tailored to the individual genetic makeup. After successful testing in sizable and well-designed clinical trials, those expected novel drugs should become a reality.

SUMMARY

- Cardiac dysrhythmias are an important cause of mortality in humans, and in the United States alone, more than 300,000 cases of sudden death occur each year because of ventricular dysrhythmias.
- Although molecular genetic analysis has shown an unforeseen genetic diversity on the inherited dysrhythmias, making its genetic differentiation rather complex, the availability of the Human Genome Project and evolving technologies seem to offer the promise of fast progress in this important field of cardiac pathology.
- Improving interaction between clinical cardiologists, molecular biologists, and geneticists is important for the detection of gene mutations in patients with life-threatening dysrhythmias, particularly mutations in genes encoding most of the regulatory proteins forming the ion channels and transporters.
- Genetic heterogeneity represents a tremendous challenge for mutation identification. However, the availability of new tools, including high-throughput technologies to map the human genome together with the availability of large databases of single nucleotide polymorphisms (SNPs) and haplotype markers, is facilitating the progressive discovery of new mutations and numerous SNPs in patients with cardiac dysrhythmias.
- Heterogeneity, even within mutation carriers in the same family, has increased the focus on the significance of environmental modulatory/modifying factors, as well as in the discovery of new gene mutations.

- Analysis of the causality of cardiac dysrhythmias (either of the inherited or acquired type), as in other cardiovascular pathological conditions, supports their multifactorial origin (i.e., encompassing individual molecular and genetic changes together with a number of epigenetic factors including age, gender, and environment). Thus, to identify causality of cardiac dysrhythmias, it is necessary to carry out larger population studies that will integrate the genetics and molecular findings with specific variables such as ethnicity and gender.
- Action potential (AP) results from the highly organized interaction of multiple voltage-dependent ion channels and membrane transporters. The model generally used to describe the cardiac AP is the ventricular myocyte.
- Inherited or primary dysrhythmias are primary electrical disorders not associated with structural cardiac pathology. Most of the inherited dysrhythmias, with a known genetic basis (mostly monogenic defects), have been found to be associated with cardiac ion channels defects (i.e., channelopathies).
- In LQTS, there is an abnormally prolonged ventricular repolarization that can result in ventricular tachycardia (VT), ventricular fibrillation (VF), and sudden death.
- Abnormal lengthening of the cardiac AP present in each type of LQTS1-8 can result in a polymorphic ventricular tachycardia known also as *torsade de pointes*. This dysrhythmia may be self-limited, but it may also be followed by syncope and sudden death.
- Physical activities such as swimming and emotional excitement can precipitate the LQTS, including syncope and sudden death.
- The presence of extensive phenotypic heterogeneity, even within mutation carriers in the same family, raises the importance of modifying factors and genes that are yet not known.
- Mutations identified as causative of Brugada syndrome (BrS) are missense mutations in the SCN5A gene that encodes the α-subunit of the cardiomyocytes Na^+ ion channel.
- Mutations in two important components of the sarcoplasmic reticulum, which are essential in excitation-contraction coupling, the ryanodine receptor type 2 (*RYR2*), and calsequestrin (*CASQ2*), have been found to be associated with catecholaminergic polymorphic ventricular tachycardia types 1 and 2, respectively.
- Short QT syndrome has an autosomal-dominant inheritance pattern, and most individuals will have a family history of unexplained or sudden death at a young age (even in newborns), palpitations, and atrial fibrillation (AF).
- AF is the most common dysrhythmia seen in clinical cardiology, and although in some cases it may be inherited in a monogenic form, familial AF seems to be genetically heterogenous. Nevertheless, AF is mostly acquired and associated with cardiac structural defects leading to heart failure.
- Acquired or secondary dysrhythmias are mainly those caused by diseases of the myocardium (e.g., CAD, HCM, and DCM) or are drug related.
- Of 230 patients who died suddenly while undergoing Holter monitoring, more than 80% of the cases of sudden death were due to ventricular tachydysrhythmias, including 10% primary VF, 50% VF preceded by classic VT, and 20% *torsades de pointes* tachycardia. The latter appeared mainly in patients without structural heart disease and were mostly related to antidysrhythmic therapy and/or electrolyte and metabolic imbalances.
- Heart failure and myocardial hypertrophy account for most cases of lethal dysrhythmias, and these may be related to down-regulation of repolarizing K^+ currents, as well as changes in cardiomyocyte Ca^{2+} handling, two mechanisms that can explain the prolongation of the AP duration and alterations of the AP dynamics.
- Lethal ventricular dysrhythmias, mainly *torsades de pointes,* have been reported to be associated not only with the use of a number of antidysrhythmic drugs but also with noncardiac drugs such as antihistaminics and antipsychotics.
- Most drugs with the potential to induce *torsades de pointes* are HERG blockers. These drugs inhibit the KCNH2-encoded HERG potassium channel, making it capable of developing delayed repolarization as indicated by LQT.
- Besides mutations in LQTS genes, DNA polymorphisms may also increase the risk for acquired LQTS.
- Failure of the mitochondrial network at the muscle–cell level leads to differential electrical excitability in hearts subjected to ischemia and reperfusion. This may be a new mechanism that could explain post-ischemic dysrhythmias, in which mitochondrial dysfunction in the reperfused heart leads to a local decline of electrical currents preventing propagation.
- Studies in animal models have provided extremely important insights about cardiac dysrhythmias. The mouse has become the most prevalent animal used in genetic analysis largely because of the advances in transgene and gene targeting technology available.
- Analysis of polymorphisms (SNPs) in large cohorts of individuals would enable the establishment of common haplotypes associated with changes in the Q-T interval, and these haplotypes could facilitate prediction of dysrhythmia risk in the general population.
- The interrelationship between genomic variance in dysrhythmia genes, promoter variability, and alterations in gene transcription is mostly unknown to date.
- Genetic screening of patients with potentially lethal dysrhythmias is a recent strategy in the identification of markers or clinical predictors of sudden dysrhythmia death syndrome.
- The related fields of pharmacogenomics and pharmacogenetics seem to have the potential to improve drug

development and the tailoring of a drug therapy based mainly on the individual's ability to metabolize drugs.
- The use of gene and cell-based therapies for specific cardiac dysrhythmias, although promising, is still in its developmental stages.

References

1. Roden, D. M., Balser, J. R., George, A. L., Jr., and Anderson, M. E. (2002). Cardiac ion channels. *Annu. Rev. Physiol.* **64**, 431–475.
2. Zeng, T., Bett, G. C., and Sachs, F. (2000). Stretch-activated whole cell currents in adult rat cardiac myocytes. *Am. J. Physiol. Heart Circ. Physiol.* **278**, H548–H557.
3. Pinto, J. M., and Boyden, P. A. (1999). Electrical remodeling in ischemia and infarction. *Cardiovasc. Res.* **42**, 284–297.
4. Roepke, T. K., and Abbott, G. W. (2006). Pharmacogenetics and cardiac ion channels. *Vascul. Pharmacol.* **44**, 90–106.
5. Isom, L. L., De Jongh, K. S., and Catterall, W. A. (1994). Auxiliary subunits of voltage-gated ion channels. *Neuron* **12**, 1183–1189.
6. Deal, K. K., England, S. K., and Tamkun, M. M. (1996). Molecular physiology of cardiac potassium channels. *Physiol. Rev.* **76**, 49–67.
7. Catterall, W. A. (1995). Structure and function of voltage-gated ion channels. *Annu. Rev. Biochem.* **64**, 493–531.
8. Goldstein, S. A. N. (1996). A structural vignette common to voltage sensors and conduction pores: canaliculi. *Neuron* **16**, 717–722.
9. Yang, N., George, A. L., Jr., and Horn, R. (1996). Molecular basis of charge movement in voltage-gated sodium channels. *Neuron* **16**, 113–122.
10. De Jongh, K. S., Warner, C., and Catterall, W. A. (1990). Subunits of purified calcium channels. Alpha 2 and delta are encoded by the same gene. *J. Biol. Chem.* **265**, 14738–14741.
11. Heginbotham, L., Lu, Z., Abramson, T., and MacKinnon, R. (1994). Mutations in the K+ channel signature sequence. *Biophys. J.* **66**, 1061–1067.
12. Doyle, D. A., Morais Cabral, J., Pfuetzner, R. A., Kuo, A., Gulbis, J. M., Cohen, S. L., Chait, B. T., and MacKinnon, R. (1998). The structure of the potassium channel: molecular basis of K+ conduction and selectivity. *Science* **280**, 69–77.
13. Kuo, A., Gulbis, J. M., Antcliff, J. F., Rahman, T., Lowe, E. D., Zimmer, J., Cuthbertson, J., Ashcroft, F. M., Ezaki, T., and Doyle, D. A. (2003). Crystal structure of the potassium channel KirBac1.1 in the closed state. *Science* **300**, 1922–1926.
14. Korn, S. J., and Ikeda, S. R. (1995). Permeation selectivity by competition in a delayed rectifier potassium channel. *Science* **269**, 410–412.
15. Dibb, K. M., Rose, T., Makary, S. Y., Claydon, T. W., Enkvetchakul, D., Leach, R., Nichols, C. G., and Boyett, M. R. (2003). Molecular basis of ion selectivity, block, and rectification of the inward rectifier Kir3.1/Kir3.4 K(+) channel. *J. Biol. Chem.* **278**, 49537–49548.
16. Kaupp, U. B., and Seifert, R. (2002). Cyclic nucleotide-gated ion channels. *Physiol. Rev.* **82**, 769–824.
17. Santoro, B., and Tibbs, G. R. (1999). The HCN gene family: molecular basis of the hyperpolarization-activated pacemaker channels. *Ann. N. Y. Acad. Sci.* **868**, 741–764.
18. Abbott, G. W., Sesti, F., Splawski, I., Buck, M. E., Lehmann, M. H., Timothy, K. W., Keating, M. T., and Goldstein, S. A. (1999). MiRP1 forms IKr potassium channels with HERG and is associated with cardiac arrhythmia. *Cell* **97**, 175–187.
19. McDonald, T. V., Yu, Z., Ming, Z., Palma, E., Meyers, M. B., Wang, K. W., Goldstein, S. A., and Fishman, G. I. (1997). A minK-HERG complex regulates the cardiac potassium current I(Kr). *Nature* **388**, 289–292.
20. Sanguinetti, M. C., Curran, M. E., Zou, A., Shen, J., Spector, P. S., Atkinson, D. L., and Keating, M. T. (1996). Coassembly of K(V)LQT1 and minK (IsK) proteins to form cardiac I(Ks) potassium channel. *Nature* **384**, 80–83.
21. Snyders, D. J., Tamkun, M. M., and Bennett, P. B. (1993). A rapidly activating and slowly inactivating potassium channel cloned from human heart. Functional analysis after stable mammalian cell culture expression. *J. Gen. Physiol.* **101**, 513–543.
22. Feng, J., Wible, B., Li, G. R., Wang, Z., and Nattel, S. (1997). Antisense oligodeoxynucleotides directed against Kv1.5 mRNA specifically inhibit ultrarapid delayed rectifier K+ current in cultured adult human atrial myocytes. *Circ. Res.* **80**, 572–579.
23. Kwak, Y. G., Hu, N., Wei, J., George, A. L., Jr., Grobaski, T. D., Tamkun, M. M., and Murray, K. T. (1999). Protein kinase A phosphorylation alters Kvbeta1.3 subunit-mediated inactivation of the Kv1.5 potassium channel. *J. Biol. Chem.* **274**, 13928–13932.
24. Williams, C. P., Hu, N., Shen, W., Mashburn, A. B., and Murray, K. T. (2002). Modulation of the human Kv1.5 channel by protein kinase C activation: role of the Kvbeta1.2 subunit. *J. Pharmacol. Exp. Ther.* **302**, 545–550.
25. Wible, B. A., De Biasi, M., Majumder, K., Taglialatela, M., and Brown, A. M. (1995). Cloning and functional expression of an inwardly rectifying K+ channel from human atrium. *Circ. Res.* **76**, 343–350.
26. Zerangue, N., Schwappach, B., Jan, Y. N., and Jan, L. Y. (1999). A new ER trafficking signal regulates the subunit stoichiometry of plasma membrane K(ATP) channels. *Neuron* **22**, 537–548.
27. Inagaki, N., Gonoi, T., Clement, J. P., Wang, C. Z., Aguilar-Bryan, L., Bryan, J., and Seino, S. (1996). A family of sulfonylurea receptors determines the pharmacological properties of ATP-sensitive K+ channels. *Neuron* **16**, 1011–1017.
28. Grover, G. J., and Garlid, K. D. (2000). ATP-sensitive potassium channels: A review of their cardioprotective pharmacology. *J. Mol. Cell Cardiol.* **32**, 677–695.
29. Sato, T., Sasaki, N., Seharaseyon, J., O'Rourke, B., and Marban, E. (2000). Selective pharmacological agents implicate mitochondrial but not sarcolemmal K(ATP) channels in ischemic cardioprotection. *Circulation* **101**, 2418–2423.
30. Krapivinsky, G., Gordon, E. A., Wickman, K., Velimirovic, B., Krapivinsky, L., and Clapham, D. E. (1995). The G-protein–gated atrial K+ channel IKACh is a heteromultimer of two inwardly rectifying K+-channel proteins. *Nature* **374**, 135–141.
31. Duprat, F., Lesage, F., Guillemare, E., Fink, M., Hugnot, J. P., Bigay, J., Lazdunski, M., Romey, G., and Barhanin, J. (1995). Heterologous multimeric assembly is essential for K+ channel

activity of neuronal and cardiac G-protein–activated inward rectifiers. *Biochem. Biophys. Res. Commun.* **212,** 657–663.

32. Huang, C. L., Jan, Y. N., and Jan, L. Y. (1997). Binding of the G protein betagamma subunit to multiple regions of G protein-gated inward-rectifying K+ channels. *FEBS Lett.* **405,** 291–298.

33. Hughes, S., Magnay, J., Foreman, M., Publicover, S. J., Dobson, J. P., and El Haj, A. J. (2006). Expression of the mechanosensitive 2PK+ channel TREK-1 in human osteoblasts. *J. Cell Physiol.* **206,** 738–748.

34. Terrenoire, C., Lauritzen, I., Lesage, F., Romey, G., and Lazdunski, M. (2001). A TREK-1–like potassium channel in atrial cells inhibited by beta-adrenergic stimulation and activated by volatile anesthetics. *Circ. Res.* **89,** 336–342.

35. Besana, A., Barbuti, A., Tateyama, M. A., Symes, A. J., Robinson, R. B., and Feinmark, S. J. (2004). Activation of protein kinase C epsilon inhibits the two-pore domain K+ channel, TASK-1, inducing repolarization abnormalities in cardiac ventricular myocytes. *J. Biol. Chem.* **279,** 33154–33160.

36. Kim, D., Fujita, A., Horio, Y., and Kurachi, Y. (1998). Cloning and functional expression of a novel cardiac two-pore background K+ channel (cTBAK-1). *Circ. Res.* **82,** 513–518.

37. Hoshi, T., Zagotta, W. N. and Aldrich, R. W. (1990). Biophysical and molecular mechanisms of Shaker potassium channel inactivation. *Science* **250,** 533–538.

38. Armstrong, C. M., and Bezanilla, F. (1977). Inactivation of the sodium channel II. Gating current experiments. *J. Gen. Physiol.* **70,** 567–590.

39. West, J. W., Patton, D. E., Scheuer, T., Wang, Y., Goldin, A. L., and Catterall, W. (1992). A cluster of hydrophobic amino acid residues required for fast Na(+)-channel inactivation. *Proc. Natl. Acad. Sci. USA* **89,** 10910–10914.

40. Patton, D. E., West, J. W., Catterall, W. A., and Goldin, A. L. (1992). Amino acid residues required for fast Na(+)-channel inactivation: charge neutralizations and deletions in the III-IV linker. *Proc. Natl. Acad. Sci. USA* **89,** 10905–10909.

41. Patton, D. E., West, J. W., Catterall, W. A., and Goldin, A. L. (1993). A peptide segment critical for sodium channel inactivation functions as an inactivation gate in a potassium channel. *Neuron* **11,** 967–974.

42. Nuss, H. B., Balser, J. R., Orias, D. W., Lawrence, J. H., Tomaselli, G. F., and Marban, E. (1996). Coupling between fast and slow inactivation revealed by analysis of a point mutation (F1304Q) in mu 1 rat skeletal muscle sodium channels. *J. Physiol.* **494,** 411–429.

43. IUPHAR compendium of voltage-gated ion channels 2005. (2005). *Pharmacol. Rev.* **57,** 385–540.

44. Curran, M. E., Splawski, I., Timothy, K. W., Vincent, G. M., Green, E. D., and Keating, M. T. (1995). A molecular basis for cardiac arrhythmia: HERG mutations cause long QT syndrome. *Cell* **80,** 795–803.

45. Wang, Q., Shen, J., Splawski, I., Atkinson, D., Li, Z., Robinson, J. L., Moss, A. J., Towbin, J. A., and Keating, M. T. (1995). SCN5A mutations associated with an inherited cardiac arrhythmia, long QT syndrome. *Cell* **80,** 805–810.

46. Jervell, A., and Lange-Nielsen, F. (1957). Congenital deaf-mutism, functional heart disease with prolongation of the QT and sudden death. *Am. Heart J.* **54,** 59–68.

47. Romano, C., Gemme, G., and Pongiglione, R. (1963). Rare cardiac arrhythmias of the pediatric age. Syncopal attacks due to paroxysmal ventricular fibrillation. *Clin. Pediatr.* **45,** 656–683.

48. Ward, O. (1964). A new familial cardiac syndrome in children. *J. Ir. Med. Assoc.* **54,** 103–106.

49. Keating, M. T., and Sanguinetti, M. C. (2001). Molecular and cellular mechanisms of cardiac arrhythmias. *Cell* **104,** 569–580.

50. Wang, Q., Curran, M. E., Splawski, I., Burn, T. C., Millholland, J. M., VanRaay, T. J., Shen, J., Timothy, K. W., Vincent, G. M., de Jager, T., Schwartz, P. J., Toubin, J. A., Moss, A. J., Atkinson, D. L., Landes, G. M., Connors, T. D., and Keating, M. T. (1996). Positional cloning of a novel potassium channel gene: KVLQT1 mutations cause cardiac arrhythmias. *Nat. Genet.* **12,** 17–23.

51. Schott, J. J., Charpentier, F., Peltier, S., Foley, P., Drouin, E., Bouhour, J. B., Donnelly, P., Vergnaud, G., Bachner, L., and Moisan, J. P. (1995). Mapping of a gene for long QT syndrome to chromosome 4q25-27. *Am. J. Hum. Genet.* **57,** 1114–1122.

52. Mohler, P. J., Schott, J. J., Gramolini, A. O., Dilly, K. W., Guatimosim, S., duBell, W. H., Song, L. S., Haurogne, K., Kyndt, F., Ali, M. E., Rogers, T. B., Lederer, W. J., Escande, D., Le Marec, H., and Bennett, V. (2003). Ankyrin-B mutation causes type 4 long-QT cardiac arrhythmia and sudden cardiac death. *Nature* **421,** 634–639.

53. Splawski, I., Tristani-Firouzi, M., Lehmann, M. H., Sanguinetti, M. C., and Keating, M. T. (1997). Mutations in the hminK gene cause long QT syndrome and suppress IKs function. *Nat. Genet.* **17,** 338–340.

54. Abbott, G. W., Sesti, F., Splawski, I., Buck, M. E., Lehmann, M. H., Timothy, K. W., Keating, M. T., and Goldstein, S. A. (1999). MiRP1 forms IKr potassium channels with HERG and is associated with cardiac arrhythmia. *Cell* **97,** 175–187.

55. Donaldson, M. R., Yoon, G., Fu, Y. H., and Ptacek, L. J. (2004). Andersen-Tawil syndrome: a model of clinical variability, pleiotropy, and genetic heterogeneity. *Ann. Med.* **36,** 92–97.

56. Tristani-Firouzi, M., Jensen, J. L., Donaldson, M. R., Sansone, V., Meola, G., Hahn, A., Bendahhou, S., Kwiecinski, H., Fidzianska, A., Plaster, N., Fu, Y. H., Ptacek, L. J., and Tawil, R. (2002). Functional and clinical characterization of KCNJ2 mutations associated with LQT7 (Andersen syndrome). *J. Clin. Invest.* **110,** 381–388.

57. Splawski, I., Timothy, K. W., Sharpe, L. M., Decher, N., Kumar, P., Bloise, R., Napolitano, C., Schwartz, P. J., Joseph, R. M., Condouris, K., Tager-Flusberg, H., Priori, S. G., Sanguinetti, M. C., and Keating, M. T. (2004). Ca(V)1.2 calcium channel dysfunction causes a multisystem disorder including arrhythmia and autism. *Cell* **119,** 19–31.

58. Splawski, I., Timothy, K. W., Decher, N., Kumar, P., Sachse, F. B., Beggs, A. H., Sanguinetti, M. C., and Keating, M. T. (2005). Severe arrhythmia disorder caused by cardiac L-type calcium channel mutations. *Proc. Natl. Acad. Sci. USA* **102,** 8089–8096.

59. Neyroud, N., Tesson, F., Denjoy, I., Leibovici, M., Donger, C., Barhanin, J., Faure, S., Gary, F., Coumel, P., Petit, C., Schwartz, K., and Guicheney, P. (1997). A novel mutation in the potassium channel gene KVLQT1 causes the Jervell and Lange-Nielsen cardioauditory syndrome. *Nat. Genet.* **15,** 186–189.

60. Schulze-Bahr, E., Wang, Q., Wedekind, H., Haverkamp, W., Chen, Q., Sun, Y., Rubie, C., Hordt, M., Towbin, J. A.,

Borggrefe, M., Assmann, G., Qu, X., Somberg, J. C., Breithardt, G., Oberti, C., and Funke, H. (1997). KCNE1 mutations cause Jervell and Lange-Nielsen syndrome. *Nat. Genet.* **17,** 267–268.

61. Bennett, P. B., Yazawa, K., Makita, N., and George, A. L., Jr. (1995). Molecular mechanism for an inherited cardiac arrhythmia. *Nature* **376,** 683–685.
62. Plaster, N. M., Tawil, R., Tristani-Firouzi, M., Canun, S., Bendahhou, S., Tsunoda, A., Donaldson, M. R., Iannaccone, S. T., Brunt, E., Barohn, R., Clark, J., Deymeer, F., George, A. L., Jr., Fish, F. A., Hahn, A., Nitu, A., Ozdemir, C., Serdaroglu, P., Subramony, S. H., Wolfe, G., Fu, Y. H., and Ptacek, L. J. (2001). Mutations in Kir2.1 cause the developmental and episodic electrical phenotypes of Andersen's syndrome. *Cell* **105,** 511–519.
63. Andersen, E. D., Krasilnikoff, P. A., and Overvad, H. (1971). Intermittent muscular weakness, extrasystoles, and multiple developmental anomalies. A new syndrome? *Acta Paediatr. Scand.* **60,** 559–564.
64. Bendahhou, S., Fournier, E., Sternberg, D., Bassez, G., Furby, A., Sereni, C., Donaldson, M. R., Larroque, M. M., Fontaine, B., and Barhanin, J. (2005). In vivo and in vitro functional characterization of Andersen's syndrome mutations. *J. Physiol.* **565,** 731–741.
65. Donaldson, M. R., Yoon, G., Fu, Y. H., and Ptacek, L. J. (2004). Andersen-Tawil syndrome: A model of clinical variability, pleiotropy, and genetic heterogeneity. *Ann. Med.* **36,** 92–97.
66. Leonoudakis, D., Conti, L. R., Anderson, S., Radeke, C. M., McGuire, L. M., Adams, M. E., Froehner, S. C., Yates, J. R., 3rd., and Vandenberg, C. A. (2004). Protein trafficking and anchoring complexes revealed by proteomic analysis of inward rectifier potassium channel (Kir2.x)-associated proteins. *J. Biol. Chem.* **279,** 22331–22346.
67. Leonoudakis, D., Conti, L. R., Radeke, C. M., McGuire, L. M., and Vandenberg, C. A. (2004). A multiprotein trafficking complex composed of SAP97, CASK, Veli, and Mint1 is associated with inward rectifier Kir2 potassium channels. *J. Biol. Chem.* **279,** 19051–19063.
68. Caruana, G., and Bernstein, A. (2001). Craniofacial dysmorphogenesis including cleft palate in mice with an insertional mutation in the discs large gene. *Mol. Cell Biol.* **21,** 1475–1483.
69. Laverty, H. G., and Wilson, J. B. (1998). Murine CASK is disrupted in a sex-linked cleft palate mouse mutant. *Genomics* **53,** 29–41.
70. Zaritsky, J. J., Eckman, D. M., Wellman, G. C., Nelson, M. T., and Schwarz, T. L. (2000). Targeted disruption of Kir2.1 and Kir2.2 genes reveals the essential role of the inwardly rectifying K(+) current in K(+)-mediated vasodilation. *Circ. Res.* **87,** 160–166.
71. Mohler, P. J., and Bennett, V. (2005). Ankyrin-based cardiac arrhythmias: a new class of channelopathies due to loss of cellular targeting. *Curr. Opin. Cardiol.* **20,** 189–193.
72. Priori, S. G., and Napolitano, C. (2004). Genetics of cardiac arrhythmias and sudden cardiac death. *Ann. N. Y. Acad. Sci.* **1015,** 96–110.
73. Mohler, P. J., Rivolta, I., Napolitano, C., LeMaillet, G., Lambert, S., Priori, S. G., and Bennett, V. (2004). Nav1.5 E1053K mutation causing Brugada syndrome blocks binding to ankyrin-G and expression of Nav1.5 on the surface of cardiomyocytes. *Proc. Natl. Acad. Sci. USA* **101,** 17533–17538.
74. Mohler, P. J., Davis, J. Q., and Bennett, V. (2005). Ankyrin-B coordinates the Na/K ATPase, Na/Ca exchanger, and InsP3 receptor in a cardiac T-tubule/SR microdomain. *PLoS Biol.* **3,** e423.
75. Sherman, J., Tester, D. J., and Ackerman, M. J. (2005). Targeted mutational analysis of ankyrin-B in 541 consecutive, unrelated patients referred for long QT syndrome genetic testing and 200 healthy subjects. *Heart Rhythm* **2,** 1218–1223.
76. Schwartz, P. J., Moss, A. J., Vincent, G. M., and Crampton, R. S. (1993). Diagnostic criteria for the long QT syndrome. *Circulation* **88,** 782–784.
77. Wilde, A. A., Antzelevitch, C., Borggrefe, M., Brugada, J., Brugada, R., Brugada, P., Corrado, D., Hauer, R. N., Kass, R. S., Nademanee, K., Priori, S. G., Towbin, J. A., and the Study Group on the Molecular Basis of Arrhythmias of the European Society of Cardiology. (2002). Proposed diagnostic criteria for the Brugada syndrome, consensus report. *Circulation* **106,** 2514–2519.
78. Osher, H. L., and Wolff, L. (1953). Electrocardiographic pattern simulating acute myocardial injury. *Am. J. Med. Sci.* **226,** 541–545.
79. Brugada, P., and Brugada, J. (1992). Right bundle branch block, persistent ST segment elevation and sudden cardiac death: a distinct clinical and electrocardiographic syndrome: a multicenter report. *J. Am. Coll. Cardiol.* **20,** 1391–1396.
80. Antzelevitch, C., Brugada, P., Brugada, J., Brugada, R., Shimizu, W., Gussak, I., and Perez Riera, A. R. (2002). Brugada syndrome: A decade of progress. *Circ. Res.* **91,** 1114–1118.
81. Antzelevitch, C., Brugada, P., Borggrefe, M., Brugada, J., Brugada, R., Corrado, D., Gussak, I., LeMarec, H., Nademanee, K., Perez Riera, A. R., Shimizu, W., Schulze-Bahr, E., Tan, H., and Wilde, A. (2005). Brugada syndrome: Report of the second consensus conference: endorsed by the Heart Rhythm Society and the European Heart Rhythm Association. *Circulation* **111,** 659–670.
82. Brugada, J., Brugada, R., and Brugada, P. (2003). Determinants of sudden cardiac death in individuals with the electrocardiographic pattern of Brugada syndrome and no previous cardiac arrest. *Circulation* **108,** 3092–3096.
83. Chen, Q., Kirsch, G. E., Zhang, D., Brugada, R., Brugada, J., Brugada,. P, Potenza, D., Moya, A., Borggrefe, M., Breithardt, G., Ortiz-Lopez, R., Wang, Z., Antzelevitch, C., O'Brien, R. E., Schulze-Bahr, E., Keating, M. T., Towbin, J. A., and Wang, Q. (1998). Genetic basis and molecular mechanism for idiopathic ventricular fibrillation. *Nature* **392,** 293–296.
84. Weiss, R., Barmada, M. M., Nguyen, T., Seibel, J. S., Cavlovich, D., Kornblit, C. A., Angelilli, A., Villanueva, F., McNamara, D. M., and London, B. (2002). Clinical and molecular heterogeneity in the Brugada syndrome: A novel gene locus on chromosome 3. *Circulation* **105,** 707–713.
85. Akai, J., Makita, N., Sakurada, H., Shirai, N., Ueda, K., Kitabatake, A., Nakazawa, K., Kimura, A., and Hiraoka, M. (2000). A novel SCN5A mutation associated with idiopathic ventricular fibrillation without typical ECG findings of Brugada syndrome. *FEBS Lett.* **479,** 29–34.
86. Priori, S. G., Napolitano, C., Gasparini, M., Pappone, C., Della Bella, P., Giordano, U., Bloise, R., Giustetto, C., De Nardis, R., Grillo, M., Ronchetti, E., Faggiano, G., and Nastoli, J. (2002). Natural history of Brugada syndrome: insights for risk stratification and management. *Circulation* **105,** 1342–1347.

87. Balser, J. R. (2001). The cardiac sodium channel: gating function and molecular pharmacology. *J. Mol. Cell Cardiol.* **33**, 599–613.
88. Tan, H. L., Bezzina, C. R., Smits, J .P, Verkerk, A. O., and Wilde, A. A. (2003). Genetic control of sodium channel function. *Cardiovasc. Res.* **57**, 961–973.
89. Fondazione Salvatore Maugeri, Istituto di Ricovero e Cura a Carattere Scientifico. Brugada syndrome (BrS) mutations. Gene: SCN5A. Available at: http://pc4.fsm.it:81/cardmoc/SCN5A_brugada_mut.htm. Accessed November 30, 2004.
90. Schulze-Bahr, E., Eckardt, L., Breithardt, G., Seidl, K., Wichter, T., Wolpert, C., Borggrefe, M., and Haverkamp, W. (2003). Sodium channel gene (SCN5A) mutations in 44 index patients with Brugada syndrome: Different incidences in familial and sporadic disease. *Hum. Mutat.* **21**, 651–652.
91. Meregalli, P. G., Wilde, A. A., and Tan, H. L. (2005). Pathophysiological mechanisms of Brugada syndrome: depolarization disorder, repolarization disorder, or more? *Cardiovasc. Res.* **67**, 367–378.
92. Kurita, T., Shimizu, W., Inagaki, M., Suyama, K., Taguchi, A., Satomi, K., Aihara, N., Kamakura, S., Kobayashi, J., and Kosakai, Y. (2002). The electrophysiologic mechanism of ST-segment elevation in Brugada syndrome. *J. Am. Coll. Cardiol.* **40**, 330–334.
93. Tukkie, R., Sogaard, P., Vleugels, J., de Groot, I. K., Wilde, A. A., and Tan, H. L. (2004). Delay in right ventricular activation contributes to Brugada syndrome. *Circulation* **109**, 1272–1277.
94. Saffitz, J. E. (2005). Structural heart disease, SCN5A gene mutations, and Brugada syndrome. *Circulation* **112**, 3672–3674.
95. Frustaci, A., Priori, S. G., Pieroni, M., Chimenti, C., Napolitano, C., Rivolta, I., Sanna, T., Bellocci, F., and Russo, M. A. (2005). Cardiac histological substrate in patients with clinical phenotype of Brugada syndrome. *Circulation* **112**, 3680–3687.
96. Corrado, D., Basso, C., Buja, G., Nava, A., Rossi, L., and Thiene, G. (2001). Right bundle branch block, right precordial ST-segment elevation, and sudden death in young people. *Circulation* **103**, 710–717.
97. Matsuo, K., Akahoshi, M., Nakashima, E., Suyama, A., Seto, S., Hayano, M., and Yano, K. (2001). The prevalence, incidence and prognostic value of the Brugada-type electrocardiogram: A population-based study of four decades. *J. Am. Coll. Cardiol.* **38**, 765–770.
98. Junttila, M. J., Raatikainen, M. J., Karjalainen, J., Kauma, H., Kesaniemi, Y. A., and Huikuri, H. V. (2004). Prevalence and prognosis of subjects with Brugada-type ECG pattern in a young and middle-aged Finnish population. *Eur. Heart J.* **25**, 874–878.
99. Roberts, R. (2006). Genomics and cardiac arrhythmias. *J. Am. Coll. Cardiol.* **47**, 9–21.
100. Laitinen, P. J., Swan, H., Piippo, K., Viitasalo, M., Toivonen, L., and Kontula, K. (2004). Genes, exercise and sudden death: molecular basis of familial catecholaminergic polymorphic ventricular tachycardia. *Ann. Med.* **36**, 81–86.
101. Lehnart, S. E., Wehrens, X. H., Laitinen, P. J., Reiken, S. R., Deng, S. X., Cheng, Z., Landry, D. W., Kontula, K., Swan, H., and Marks, A. R. (2004). Sudden death in familial polymorphic ventricular tachycardia associated with calcium release channel (ryanodine receptor) leak. *Circulation* **109**, 3208–3214.
102. Swan, H., Piippo, K., Viitasalo, M., Heikkila, P., Paavonen, T., Kainulainen, K., Kere, J., Keto, P., Kontula, K., and Toivonen, L. (1999). Arrhythmic disorder mapped to chromosome 1q42-q43 causes malignant polymorphic ventricular tachycardia in structurally normal hearts. *J. Am. Coll. Cardiol.* **34**, 2035–2042.
103. Priori, S. G., Napolitano, C., Tiso, N., Memmi, M., Vignati, G., Bloise, R., Sorrentino, V., and Danieli, G. A. (2000). Mutations in the cardiac ryanodine receptor gene (hRyR2) underlie catecholaminergic polymorphic ventricular tachycardia. *Circulation* **102**, 149–153.
104. Laitinen, P. J., Brown, K. M., Piippo, K., Swan, H., Devaney, J. M., Brahmbhatt, B., Donarum, E. A., Marino, M., Tiso, N., Viitasalo, M., Toivonen, L., Stephan, D. A., and Kontula, K. (2001). Mutations of the cardiac ryanodine receptor (RyR2) gene in familial polymorphic ventricular tachycardia. *Circulation* **103**, 485–490.
105. Lahat, H., Eldar, M., Levy-Nissenbaum, E., Bahan, T., Friedman, E., Khoury, A., Lorber, A., Kastner, D. L., Goldman, B., and Pras, E. (2001). Autosomal recessive catecholamine- or exercise-induced polymorphic ventricular tachycardia: clinical features and assignment of the disease gene to chromosome 1p13-21. *Circulation* **103**, 2822–2827.
106. Postma, A. V., Denjoy, I., Hoorntje, T. M., Lupoglazoff, J. M., Da Costa, A., Sebillon, P., Mannens, M. M., Wilde, A. A., and Guicheney, P. (2002). Absence of calsequestrin 2 causes severe forms of catecholaminergic polymorphic ventricular tachycardia. *Circ. Res.* **91**, e21–e26.
107. Priori, S. G., Napolitano, C., Memmi, M., Colombi, B., Drago, F., Gasparini, M., DeSimone, L., Coltorti, F., Bloise, R., Keegan, R., Cruz Filho, F. E., Vignati, G., Benatar, A., and DeLogu, A. (2002). Clinical and molecular characterization of patients with catecholaminergic polymorphic ventricular tachycardia. *Circulation* **106**, 69–74.
108. Bauce, B., Rampazzo, A., Basso,. C, Bagattin, A., Daliento, L., Tiso, N., Turrini, P., Thiene, G., Danieli, G. A., and Nava, A. (2002). Screening for ryanodine receptor type 2 mutations in families with effort-induced polymorphic ventricular arrhythmias and sudden death: early diagnosis of asymptomatic carriers. *J. Am. Coll. Cardiol.* **40**, 341–349.
109. Postma, A. V., Denjoy, I., Kamblock, J., Alders, M., Lupoglazoff, J. M., Vaksmann, G., Dubosq-Bidot, L., Sebillon, P., Mannens, M. M., Guicheney, P., and Wilde, A. A. (2005). Catecholaminergic polymorphic ventricular tachycardia: RYR2 mutations, bradycardia, and follow up of the patients. *J. Med. Genet.* **42**, 863–870.
110. Kontula, K., Laitinen, P. J., Lehtonen, A., Toivonen, L., Viitasalo, M., and Swan, H. (2005). Catecholaminergic polymorphic ventricular tachycardia: recent mechanistic insights. *Cardiovasc. Res.* **67**, 379–387.
111. Cerrone, M., Colombi, B., Santoro, M., di Barletta, M. R., Scelsi, M., Villani, L., Napolitano, C., and Priori, S. G. (2005). Bidirectional ventricular tachycardia and fibrillation elicited in a knock-in mouse model carrier of a mutation in the cardiac ryanodine receptor. *Circ. Res.* **96**, e77–e82.
112. Gaita, F., Giustetto, C., Bianchi, F., Wolpert, C., Schimpf, R., Riccardi, R., Grossi, S., Richiardi, E., and Borggrefe,

M. (2003). Short QT syndrome: A familial cause of sudden death. *Circulation* **108**, 965–970.

113. Brugada, R., Hong, K., Dumaine, R., Cordeiro, J., Gaita, F., Borggrefe, M., Menendez, T. M., Brugada, J., Pollevick, G. D., Wolpert, C., Burashnikov, E., Matsuo, K., Sheng, W. Y., Guerchicoff, A., Bianchi, F., Giustetto, C., Schimpf, R., Brugada, P., and Antzelevitch, C. (2003). Sudden death associated with short-QT syndrome linked to mutations in HERG. *Circulation* **109**, 30–35.

114. Bellocq, C., van Ginneken, A., Bezzina, C. R., Escande, D., Mannens, M., Baró, I., and Wilde, A. A. (2004). Mutation in the KCNQ1 gene leading to the short interval syndrome. *Circulation* **109**, 2394–2397.

115. Hong, K., Piper, D. R., Diaz-Valdecantos, A., Brugada, J., Oliva, A., Burashnikov, E., Santos-de-Soto, J., Grueso-Montero, J., Diaz-Enfante, E., Brugada, P., Sachse, F., Sanguinetti, M. C., and Brugada, R. (2005). De novo KCNQ1 mutation responsible for atrial fibrillation and short QT syndrome in utero. *Cardiovasc. Res.* **68**, 433–440.

116. Priori, S. G., Pandit, S. V., Rivolta, I., Berenfeld, O., Ronchetti, E., Dhamoon, A., Napolitano, C., Anumonwo, J., di Barletta, M. R., Gudapakkam, S., Bosi, G., Stramba-Badiale, M., and Jalife, J. (2005). A novel form of short QT syndrome (SQT3) is caused by a mutation in the KCNJ2 gene. *Circ. Res.* **96**, 800–807.

117. Gaita, F., Giustetto, C., Bianchi, F., Schimpf, R., Haissaguerre, M., Calo, L., Brugada, R., Antzelevitch, C., Borggrefe, M., and Wolpert, C. (2004). Short QT syndrome: pharmacological treatment. *J. Am. Coll. Cardiol.* **43**, 1494–1499.

118. Schimpf, R., Wolpert, C., Gaita, F., Giustetto, C., and Borggrefe, M. (2005). Short QT syndrome. *Cardiovasc. Res.* **67**, 357–366.

119. Brugada, R., Tapscott, T., Czernuszewicz, G. Z., Marian, A. J., Iglesias, A., Mont, L., Brugada, J., Girona, J., Domingo, A., Bachinski, L. L., and Roberts, R. (1997). Identification of a genetic locus for familial atrial fibrillation. *N. Engl. J. Med.* **336**, 905–911.

120. Darbar, D., Herron, K. J., Ballew, J. D., Jahangir, A., Gersh, B. J., Shen, W. K., Hammill, S. C., Packer, D. L., and Olson, T. M. (2003). Familial atrial fibrillation is a genetically heterogeneous disorder. *J. Am. Coll. Cardiol.* **41**, 2185–2192.

121. Ellinor, P. T., Shin, J. T., Moore, R. K., Yoerger, D. M., and MacRae, C. A. (2003). Locus for atrial fibrillation maps to chromosome 6q14–16. *Circulation* **107**, 2880–2883.

122. Chen, Y. H., Xu, S. J., Bendahhou, S., Wang, X. L., Wang, Y., Xu, W. Y., Jin, H. W., Sun, H., Su, X. Y., Zhuang, Q. N., Yang, Y. Q., Li, Y. B., Liu, Y., Xu, H. J., Li, X. F., Ma, N., Mou, C. P., Chen, Z., Barhanin, J., and Huang, W. (2003). KCNQ1 gain-of-function mutation in familial atrial fibrillation. *Science* **299**, 251–254.

123. Brugada, R., Tapscott, T., Czernuszewicz, G. Z., Marian, A. J., Iglesias, A., Mont, L., Brugada, J., Girona, J., Domingo, A., Bachinski, L. L., and Roberts, R. (1997). Identification of a genetic locus for familial atrial fibrillation. *N. Engl. J. Med.* **336**, 905–911.

124. Yang, Y., Xia, M., Jin, Q., Bendahhou, S., Shi, J., Chen, Y., Liang, B., Lin, J., Liu, Y., Liu, B., Zhou, Q., Zhang, D., Wang, R., Ma, N., Su, X., Niu, K., Pei, Y., Xu, W., Chen, Z., Wan, H., Cui, J., Barhanin, J., and Chen, Y. (2004). Identification of a KCNE2 gain-of-function mutation in patients with familial atrial fibrillation. *Am. J. Hum. Genet.* **75**, 899–905.

125. Xia, M., Jin, Q., Bendahhou, S., He, Y., Larroque, M. M., Chen, Y., Zhou, Q., Yang, Y., Liu, Y., Liu, B., Zhu, Q., Zhou, Y., Lin, J., Liang, B., Li, L., Dong, X., Pan, Z., Wang, R., Wan, H., Qiu, W., Xu, W., Eurlings, P., Barhanin, J., and Chen, Y. (2005). A Kir2.1 gain-of-function mutation underlies familial atrial fibrillation. *Biochem. Biophys. Res. Commun.* **332**, 1012–1019.

126. Oberti, C., Wang, L., Li, L., Dong, J., Rao, S., Du, W., and Wang, Q. (2004). Genome-wide linkage scan identifies a novel genetic locus on chromosome 5p13 for neonatal atrial fibrillation associated with sudden death and variable cardiomyopathy. *Circulation* **110**, 3753–3759.

127. Wiesfeld, A. C., Hemels, M. E., Van Tintelen, J. P., Van den Berg, M. .P, Van Veldhuisen, D. J., and Van Gelder, I. C. (2005). Genetic aspects of atrial fibrillation. *Cardiovasc. Res.* **67**, 414–418.

128. Lev, M., Kinare, S. G., and Pick, A. (1970). The pathogenesis of atrioventricular block in coronary disease. *Circulation* **42**, 409–425.

129. Schott, J. J., Alshinawi, C., Kyndt, F., Probst, V., Hoorntje, T. M., Hulsbeek, M., Wilde, A. A., Escande, D., Mannens, M. M., and Le Marec, H. (1999). Cardiac conduction defects associate with mutations in SCN5A. *Nat. Genet.* **23**, 20–21.

130. Kyndt, F., Probst, V., Potet, F., Demolombe, S., Chevallier, J. C., Baro, I., Moisan, J. P., Boisseau, P., Schott, J. J., Escande, D., and Le Marec, H. (2001). Novel SCN5A mutation leading either to isolated cardiac conduction defect or Brugada syndrome in a large French family. *Circulation* **104**, 3081–3086.

131. Veldkamp, M. W., Wilders, R., Baartscheer, A., Zegers, J. G., Bezzina, C. R., and Wilde, A. A. (2003). Contribution of sodium channel mutations to bradycardia and sinus node dysfunction in LQT3 families). *Circ. Res.* **92**, 976–983.

132. Wang, D. W., Viswanathan, P. C., Balser, J. R., George, A. L., Jr., and Benson, D. W. (2002). Clinical, genetic, and biophysical characterization of SCN5A mutations associated with atrioventricular conduction block. *Circulation* **105**, 341–346.

133. Benson, D. W., Wang, D. W., Dyment, M., Knilans, T. K., Fish, F. A., Strieper, M. J., Rhodes, T. H., and George, A. L., Jr. (2003). Congenital sick sinus syndrome caused by recessive mutations in the cardiac sodium channel gene (SCN5A). *J. Clin. Invest.* **112**, 1019–1028.

134. Smits, J. P., Koopmann, T. T., Wilders, R., Veldkamp, M. W., Opthof, T., Bhuiyan, Z. A., Mannens, M. M., Balser, J. R., Tan, H. L., Bezzina, C. R., and Wilde, A. A. (2005). A mutation in the human cardiac sodium channel (E161K) contributes to sick sinus syndrome, conduction disease and Brugada syndrome in two families. *J. Mol. Cell Cardiol.* **38**, 969–981.

135. Schulze-Bahr, E., Neu, A., Friederich, P., Kaupp, U. B., Breithardt, G., Pongs, O., and Isbrandt, D. (2003). Pacemaker channel dysfunction in a patient with sinus node disease. *J. Clin. Invest.* **111**, 1537–1545.

136. Opdal, S. H., and Rognum, T. O. (2004). The sudden infant death syndrome gene: Does it exist? *Pediatrics* **114**, e506–e512.

137. Schwartz, P. J., Priori, S. G., Dumaine, R., Napolitano, C., Antzelevitch, C., Stramba-Badiale, M., Richard, T. A., Berti, M. R., and Bloise, R. (2000). A molecular link between the

sudden infant death syndrome and the long-QT syndrome. *N. Engl. J. Med.* **343,** 262–267.

138. Ackerman, M. J., Siu, B. L., Sturner, W. Q., Tester, D. J., Valdivia, C. R., Makielski, J. C., and Towbin, J. A. (2001). Postmortem molecular analysis of SCN5A defects in sudden infant death syndrome. *JAMA* **286,** 2264–2269.

139. Lupoglazoff, J. M., Denjoy, I., Villain, E., Fressart, V., Simon, F., Bozio, A., Berthet, M., Benammar, N., Hainque, B., and Guicheney, P. (2004). Long QT syndrome in neonates: conduction disorders associated with HERG mutations and sinus bradycardia with KCNQ1 mutations. *J. Am. Coll. Cardiol.* **43,** 826–830.

140. Blair, E., Redwood, C., Ashrafian, H., Oliveira, M., Broxholme, J., Kerr, B., Salmon, A., Ostman-Smith, I., and Watkins, H. (2001). Mutations in the gamma(2) subunit of AMP-activated protein kinase cause familial hypertrophic cardiomyopathy: Evidence for the central role of energy compromise in disease pathogenesis. *Hum. Mol. Genet.* **10,** 1215–1220.

141. Gollob, M. H., Green, M. S., Tang, A. S., Gollob, T., Karibe, A., Ali Hassan, A. S., Ahmad, F., Lozado, R., Shah, G., Fananapazir, L., Bachinski, L. L., and Roberts, R. (2001). Identification of a gene responsible for familial Wolff-Parkinson-White syndrome. *N. Engl. J. Med.* **344,** 1823–1831.

142. Sidhu, J. S., Rajawat, Y. S., Rami, T. G., Gollob, M. H., Wang, Z., Yuan, R., Marian, A. J., DeMayo, F. J., Weilbacher, D., Taffet, G. E., Davies, J. K., Carling, D., Khoury, D. S., and Roberts, R. (2005). Transgenic mouse model of ventricular preexcitation and atrioventricular reentrant tachycardia induced by an AMP-activated protein kinase loss-of-function mutation responsible for Wolff-Parkinson-White syndrome. *Circulation* **111,** 21–29.

143. Raben, N., Plotz, P., and Byrne, B. J. (2002). Acid alpha-glucosidase deficiency (glycogenosis type II, Pompe disease). *Curr. Mol. Med.* **2,** 145–166.

144. Nikoskelainen, E. K., Savontaus, M. L., Huoponen, K., Antila, K., and Hartiala, J. (1994). Pre-excitation syndrome in Leber's hereditary optic neuropathy. *Lancet* **344,** 857–858.

145. Vaughan, C. J., Hom, Y., Okin, D. A., McDermott, D. A., Lerman, B. B., and Basson, C. T. (2003). Molecular genetic analysis of PRKAG2 in sporadic Wolff-Parkinson-White syndrome. *J. Cardiovasc. Electrophysiol.* **14,** 263–268.

146. Fox, C. S., Parise, H., D'Agostino, R. B., Lloyd-Jones, D. M., Vasan, R. S., Wang, T. J., Levy, D., Wolf, P. A., and Benjamin, E. J. (2004). Parental atrial fibrillation as a risk factor for atrial fibrillation in offspring. *JAMA* **29,** 2851–2855.

147. Lai, L. P., Su, M. J., Yeh, H. M., Lin, J. L., Chiang, F. T., Hwang, J. J., Hsu, K. L., Tseng, C. D., Lien, W. P., Tseng, Y. Z., and Huang, S. K. (2002). Association of the human minK gene 38G allele with atrial fibrillation: Evidence of possible genetic control on the pathogenesis of atrial fibrillation. *Am. Heart J.* **144,** 485–490.

148. Tsai, C. T., Lai, L. P., Lin, J. L., Chiang, F. T., Hwang, J. J., Ritchie, M. D., Moore, J. H., Hsu, K. L., Tseng, C. D., Liau, C. S., and Tseng, Y. Z. (2004). Renin-angiotensin system gene polymorphisms and atrial fibrillation. *Circulation* **109,** 1640–1646.

149. Firouzi, M., Ramanna, H., Kok, B., Jongsma, H. J., Koeleman, B. P., Doevendans, P. A., Groenewegen, W. A., and Hauer, R. N. (2004). Association of human connexin40 gene polymorphisms with atrial vulnerability as a risk factor for idiopathic atrial fibrillation. *Circ. Res.* **95,** e29–e33.

150. Ohki, R., Yamamoto, K., Ueno, S., Mano, H., Misawa, Y., Fuse, K., Ikeda, U., and Shimada, K. (2005). Gene expression profiling of human atrial myocardium with atrial fibrillation by DNA microarray analysis. *Int. J. Cardiol.* **102,** 233–238.

151. Levy, S., and Sbragia, P. (2005). Remodelling in atrial fibrillation. *Arch. Mal. Coeur. Vaiss.* **98,** 308–312.

152. Barth, A. S., Merk, S., Arnoldi, E., Zwermann, L., Kloos, P., Gebauer, M., Steinmeyer, K., Bleich, M., Kaab, S., Hinterseer, M., Kartmann, H., Kreuzer, E., Dugas, M., Steinbeck, G., and Nabauer, M. (2005). Reprogramming of the human atrial transcriptome in permanent atrial fibrillation: expression of a ventricular-like genomic signature. *Circ. Res.* **96,** 1022–1029.

153. Bayés de Luna, A. (1993). "Clinical Electrocardiography. A Textbook." pp. 388–389. Futura Publishing Company, Mount Kisco, NY.

154. Shah, M., Akar, F. G., and Tomaselli, G. F. (2005). Molecular basis of arrhythmias. *Circulation* **112,** 2517–2529.

155. Swynghedauw, B., Baillard, C., and Milliez, P. (2003). The long QT interval is not only inherited but is also linked to cardiac hypertrophy. *J. Mol. Med.* **81,** 336–345.

156. Hund, T. J., and Rudy, Y. (2004). Rate dependence and regulation of action potential and calcium transient in a canine cardiac ventricular cell model. *Circulation* **110,** 3168–3174.

157. Pak, P. H., Nuss, H. B., Tunin, R. S., Kaab, S., Tomaselli, G. F., Marban, E., and Kass, D. A. (1997). Repolarization abnormalities, arrhythmia and sudden death in canine tachycardia-induced cardiomyopathy. *J. Am. Coll. Cardiol.* **30,** 576–584.

158. Bito, V., Heinzel, F. R., Weidemann, F., Dommke, C., van der Velden, J., Verbeken, E., Claus, P., Bijnens, B., De Scheerder, I., Stienen, G. J., Sutherland, G. R., and Sipido, K. R. (2004). Cellular mechanisms of contractile dysfunction in hibernating myocardium. *Circ. Res.* **94,** 794–801.

159. Coronel, R., Wilms-Schopman, F. J., Opthof, T., Cinca, J., Fiolet, J. W., and Janse, M. J. (1992). Reperfusion arrhythmias in isolated perfused pig hearts: inhomogeneities in extracellular potassium, ST and TQ potentials, and transmembrane action potentials. *Circ. Res.* **71,** 1131–1142.

160. Tomaselli, G. F., and Zipes, D. P. (2004). What causes sudden death in heart failure? *Circ. Res.* **95,** 754–763.

161. Roden, D. M. (1998). Taking the "idio" out of "idiosyncratic": Predicting torsades de pointes. *Pacing Clin. Electrophysiol.* **21,** 1029–1034.

162. Fitzgerald, P. T., and Ackerman, M. J. (2005). Drug-induced torsades de pointes: The evolving role of pharmacogenetics. *Heart Rhythm* **2,** S30–S37.

163. Sesti, F., Abbott, G. W., Wei, J., Murray, K. T., Saksena, S., Schwartz, P. J., Priori, S. G., Roden, D. M., George, A. L., Jr., and Goldstein, S. A. (2000). A common polymorphism associated with antibiotic-induced cardiac arrhythmia. *Proc. Natl. Acad. Sci. USA* **97,** 10613–10618.

164. Paulussen, A. D., Gilissen, R. A., Armstrong, M., Doevendans, P. A., Verhasselt, P., Smeets, H. J., Schulze-Bahr, E., Haverkamp, W., Breithardt, G., Cohen, N., and Aerssens, J. (2004). Genetic variations of KCNQ1, KCNH2, SCN5A, KCNE1, and KCNE2 in drug-induced long QT syndrome patients. *J. Mol. Med.* **82,** 182–188.

165. Aerssens, J., and Paulussen, A. D. (2005). Pharmacogenomics and acquired long QT syndrome. *Pharmacogenomics* **6,** 259–270.

166. Bonnet, D., Martin, D., De Lonlay, P., Villain, E., Jouvet, P., Rabier, D., Brivet, M., and Saulubray, J. M. (1999). Arrhythmias and conduction defects as presenting symptoms of fatty acid oxidation disorders in children. *Circulation* **100,** 2248–2253.

167. Stanley, C. A., Hale, D. E., Berry, D. T., Deleeuw, S., Boxer, J., and Bonnefont, J. P. (1992). A deficiency of carnitine-acylcarnitine translocase in the inner mitochondrial membrane. *N. Engl. J. Med.* **327,** 19–23.

168. Corr, P. B., Creer, M. H., Yamada, K. A., Saffitz, J. E., and Sobel, B. E. (1989). Prophylaxis of early ventricular fibrillation by inhibition of acylcarnitine accumulation. *J. Clin. Invest.* **83,** 927–936.

169. Tripp, M. E. (1989). Developmental cardiac metabolism in health and disease. *Pediatr. Cardiol.* **10,** 150–158.

170. Akar, F. G., Aon, M. A., Tomaselli, G. F., and O'Rourke, B. (2005). The mitochondrial origin of postischemic arrhythmias. *J. Clin. Invest.* **115,** 3527–3535.

171. Casimiro, M. C., Knollmann, B. C., Ebert, S. N., Vary, J. C., Jr., Greene, A. E., Franz, M. R., Grinberg, A., Huang, S. P., and Pfeifer, K. (2001). Targeted disruption of the Kcnq1 gene produces a mouse model of Jervell and Lange-Nielsen syndrome. *Proc. Natl. Acad. Sci. USA* **98,** 2526–2531.

172. Demolombe, S., Lande, G., Charpentier, F., van Roon, M. A., van den Hoff, M. J., Toumaniantz, G., Baro, I., Guihard, G., Le Berre, N., Corbier, A., de Bakker, J., Opthof, T., Wilde, A., Moorman, A. F., and Escande, D. (2001). Transgenic mice overexpressing human KvLQT1 dominant-negative isoform. Part I: Phenotypic characterisation. *Cardiovasc. Res.* **50,** 314–327.

173. Babij, P., Askew, G. R., Nieuwenhuijsen, B., Su, C. M., Bridal, T. R., Jow, B., Argentieri, T. M., Kulik, J., DeGennaro, L. J., Spinelli, W., and Colatsky, T. J. (1998). Inhibition of cardiac delayed rectifier K+ current by overexpression of the long-QT syndrome HERG G628S mutation in transgenic mice. *Circ. Res.* **83,** 668–678.

174. Nuyens, D., Stengl, M., Dugarmaa, S., Rossenbacker, T., Compernolle, V., Rudy, Y., Smits, J. F., Flameng, W., Clancy, C. E., Moons, L., Vos, M. A., Dewerchin, M., Benndorf, K., Collen, D., Carmeliet, E., and Carmeliet, P. (2001). Abrupt rate accelerations or premature beats cause life-threatening arrhythmias in mice with long-QT3 syndrome. *Nat. Med.* **7,** 1021–1027.

175. Mohler, P. J., Gramolini, A. O., and Bennett, V. (2002). Ankyrins. *J. Cell Sci.* **115,** 1565–1566.

176. Vetter, D. E., Mann, J. R., Wangemann, P., Liu, J., McLaughlin, K. J., Lesage, F., Marcus, D. C., Lazdunski, M., Heinemann, S. F., and Barhanin, J. (2002). Inner ear defects induced by null mutation of the isk gene. *Neuron* **17,** 1251–1264.

177. Antzelevitch, C., Brugada, P., Brugada, J., Brugada, R., Shimizu, W., Gussak, I., and Perez Riera, A. R. (2002). Brugada syndrome: A decade of progress. *Circ. Res.* **91,** 1114–1118.

178. Papadatos, G. A., Wallerstein, P. M., Head, C. E., Ratcliff, R., Brady, P. A., Benndorf, K., Saumarez, R. C., Trezise, A. E., Huang, C. L., Vandenberg, J. I., Colledge, W. H., and Grace, A. A. (2002). Slowed conduction and ventricular tachycardia after targeted disruption of the cardiac sodium channel gene Scn5a. *Proc. Natl. Acad. Sci. USA* **99,** 6210–6215.

179. Royer, A., van Veen, T. A., Le Bouter, S., Marionneau, C., Griol-Charhbili, V., Leoni, A. L., Steenman, M., van Rijen, H. V., Demolombe, S., Goddard, C. A., Richer, C., Escoubet, B., Jarry-Guichard, T., Colledge, W. H., Gros, D., de Bakker, J. M., Grace, A. A., Escande, D., and Charpentier, F. (2005). Mouse model of SCN5A-linked hereditary Lenegre's disease: age-related conduction slowing and myocardial fibrosis. *Circulation* **111,** 1738–1746.

180. Temple, J., Frias, P., Rottman, J., Yang, T., Wu, Y., Verheijck, E. E., Zhang, W., Siprachanh, C., Kanki, H., Atkinson, J. B., King, P., Anderson, M. E., Kupershmidt, S., and Roden, D. M. (2005). Atrial fibrillation in KCNE1-null mice. *Circ. Res.* **97,** 62–69.

181. Hong, C. S., Cho, M. C., Kwak, Y. G., Song, C. H., Lee, Y. H., Lim, J. S., Kwon, Y. K., Chae, S. W., and Kim, D. H. (2002). Cardiac remodeling and atrial fibrillation in transgenic mice overexpressing junctin. *FASEB J.* **16,** 1310–1312.

182. Cerrone, M., Colombi, B., Santoro, M., di Barletta, M. R., Scelsi, M., Villani, L., Napolitano, C., and Priori, S. G. (2005). Bidirectional ventricular tachycardia and fibrillation elicited in a knock-in mouse model carrier of a mutation in the cardiac ryanodine receptor. *Circ. Res.* **96,** e77–e82.

183. Sidhu, J. S., Rajawat, Y. .S, Rami, T. G., Gollob, M. H., Wang, Z., Yuan, R., Marian, A. J., DeMayo, F. J., Weilbacher, D., Taffet, G. E., Davies, J. K., Carling, D., Khoury, D. S., and Roberts, R. (2005). Transgenic mouse model of ventricular preexcitation and atrioventricular reentrant tachycardia induced by an AMP-activated protein kinase loss-of-function mutation responsible for Wolff-Parkinson-White syndrome. *Circulation* **111,** 21–29.

184. Nguyen-Tran, V. T., Kubalak, S. W., Minamisawa, S., Fiset, C., Wollert, K. C., Brown, A. B., Ruiz-Lozano, P., Barrere-Lemaire, S., Kondo, R., Norman, L. W., Gourdie, R. G., Rahme, M. M., Feld, G. K., Clark, R. B., Giles, W. R., and Chien, K. R. (2000). A novel genetic pathway for sudden cardiac death via defects in the transition between ventricular and conduction system cell lineages. *Cell* **102,** 671–682.

185. Kasahara, H., Wakimoto, H., Liu, M., Maguire, C. T., Converso, K. L., Shioi, T., Huang, W. Y., Manning, W. J., Paul, D., Lawitts, J., Berul, C. I., and Izumo, S. (2001). Progressive atrioventricular conduction defects and heart failure in mice expressing a mutant Csx/Nkx2.5 homeoprotein. *J. Clin. Invest.* **108,** 189–201.

186. Berul, C. I., Maguire, C. T., Aronovitz, M. J., Greenwood, J., Miller, C., Gehrmann, J., Housman, D., Mendelsohn, M. E., and Reddy, S. (1999). DMPK dosage alterations result in atrioventricular conduction abnormalities in a mouse myotonic dystrophy model. *J. Clin. Invest.* **103,** R1–R7.

187. Simon, A. M., Goodenough, D. A., and Paul, D. L. (1998). Mice lacking connexin40 have cardiac conduction abnormalities characteristic of atrioventricular block and bundle branch block. *Curr. Biol.* **8,** 295–298.

188. Donoghue, M., Wakimoto, H., Maguire, C. T., Acton, S., Hales, P., Stagliano, N., Fairchild-Huntress, V., Xu, J., Lorenz, J. N., Kadambi, V., Berul, C. I., and Breitbart, R. E. (2003). Heart block, ventricular tachycardia, and sudden death in ACE2 transgenic mice with downregulated connexins. *J. Mol. Cell Cardiol.* **35,** 1043–1053.

189. Mungrue, I. N., Gros, R., You, X., Pirani, A., Azad, A., Csont, T., Schulz, R., Butany, J., Stewart, D. J., and Husain, M.

(2002). Cardiomyocyte overexpression of iNOS in mice results in peroxynitrite generation, heart block, and sudden death. *J. Clin. Invest.* **109,** 735–743.

190. Wei, L., Taffet, G. E., Khoury, D. S., Bo, J., Li, Y., Yatani, A., Delaughter, M. C., Klevitsky, R., Hewett, T. E., Robbins, J., Michael, L. H., Schneider, M. D., Entman, M. L., and Schwartz, R. J. (2004). Disruption of Rho signaling results in progressive atrioventricular conduction defects while ventricular function remains preserved. *FASEB J.* **18,** 857–859.

191. Yumoto, A., Fukushima Kusano, K., Nakamura, K., Hashimoto, K., Aoki, M., Morishita, R., Kaneda, Y., and Ohe, T. (2005). Hepatocyte growth factor gene therapy reduces ventricular arrhythmia in animal models of myocardial ischemia. *Acta Med. Okayama* **59,** 73–78.

192. del Monte, F., Lebeche, D., Guerrero, J. L., Tsuji, T., Doye, A. A., Gwathmey, J. K., and Hajjar, R. J. (2004). Abrogation of ventricular arrhythmias in a model of ischemia and reperfusion by targeting myocardial calcium cycling. *Proc. Natl. Acad. Sci. USA* **101,** 5622–5627.

193. Plotnikov, A. N., Sosunov, E. A., Qu, J., Shlapakova, I. N., Anyukhovsky, E. P., Liu, L., Janse, M. J., Brink, P. R., Cohen, I. S., Robinson, R. B., Danilo, P., Jr., and Rosen, M. R. (2004). Biological pacemaker implanted in canine left bundle branch provides ventricular escape rhythms that have physiologically acceptable rates. *Circulation* **109,** 506–512.

194. Hoppe, U. C., Marban, E., and Johns, D. C. (2001). Distinct gene-specific mechanisms of arrhythmia revealed by cardiac gene transfer of two long QT disease genes, HERG and KCNE1. *Proc. Natl. Acad. Sci. USA* **98,** 5335–5340.

195. Bauer, A., McDonald, A. D., Nasir, K., Peller, L., Rade, J. J., Miller, J. M., Heldman, A. W., and Donahue, J. K. (2004). Inhibitory G protein overexpression provides physiologically relevant heart rate control in persistent atrial fibrillation. *Circulation* **110,** 3115–3120.

196. Donahue, J. K., Heldman, A. W., Fraser, H., McDonald, A. D., Miller, J. M., Rade, J. J., Eschenhagen, T., and Marban, E. (2000). Focal modification of electrical conduction in the heart by viral gene transfer. *Nat. Med.* **6,** 1395–1398.

197. Murata, M., Cingolani, E., McDonald, A. D., Donahue, J. K., and Marban, E. (2004). Creation of a genetic calcium channel blocker by targeted gem gene transfer in the heart. *Circ. Res.* **95,** 398–405.

198. Mohan, R. M., Heaton, D. A., Danson, E. J., Krishnan, S. P., Cai, S., Channon, K. M., and Paterson, D. J. (2002). Neuronal nitric oxide synthase gene transfer promotes cardiac vagal gain of function. *Circ. Res.* **91,** 1089–1091.

199. Perlstein, I., Burton, D. Y., Ryan, K., Defelice, S., Simmers, E., Campbell, B., Connolly, J. M., Hoffman, A., and Levy, R. J. (2005). Posttranslational control of a cardiac ion channel transgene in vivo: clarithromycin-hMiRP1-Q9E interactions. *Hum. Gene Ther.* **16,** 906–910.

200. Au, W. W., Oh, H. Y., Grady, J., Salama, S. A., and Heo, M. Y. (2001). Usefulness of genetic susceptibility and biomarkers for evaluation of environmental health risk. *Environ. Mol. Mutagen.* **37,** 215–225.

201. Cheung, V. G., Spielman, R. S., Ewens, K. G., Weber, T. M., Morley, M., and Burdick, J. T. (2005). Mapping determinants of human gene expression by regional and genome-wide association. *Nature* **437,** 1365–1369.

202. Morley, M., Molony, C. M., Weber, T. M., Devlin, J. L., Ewens, K. G., Spielman, R. S., and Cheung, V. G. (2004). Genetic analysis of genome-wide variation in human gene expression. *Nature* **430,** 743–747.

203. Cheung, V. G., Conlin, L. K., Weber, T. M., Arcaro, M., Jen, K. Y., Morley, M., and Spielman, R. S. (2003). Natural variation in human gene expression assessed in lymphoblastoid cells. *Nat. Genet.* **33,** 422–425.

204. Aydin, A., Bahring, S., Dahm, S., Guenther, U. P., Uhlmann, R., Busjahn, A., and Luft, F. C. (2005). Single nucleotide polymorphism map of five long-QT genes. *J. Mol. Med.* **83,** 159–165.

205. Pfeufer, A., Jalilzadeh, S., Perz, S., Mueller, J. C., Hinterseer, M., Illig, T., Akyol, M., Huth, C., Schopfer-Wendels, A., Kuch, B., Steinbeck, G., Holle, R., Nabauer, M., Wichmann, H. E., Meitinger, T., and Kaab, S. (2005). Common variants in myocardial ion channel genes modify the QT interval in the general population: results from the KORA study. *Circ. Res.* **96,** 693–701.

206. Kaab, S., and Schulze-Bahr, E. (2005). Susceptibility genes and modifiers for cardiac arrhythmias. *Cardiovasc. Res.* **67,** 397–413.

207. Napolitano, C., Priori, S. G., Schwartz, P. J., Bloise, R., Ronchetti, E., Nastoli, J., Bottelli, G., Cerrone, M., and Leonardi, S. (2005). Genetic testing in the long QT syndrome: Development and validation of an efficient approach to genotyping in clinical practice. *JAMA* **294,** 2975–2980.

208. Priori, S. G., Schwartz, P. J., Napolitano, C., Bloise, R., Ronchetti, E., Grillo, M., Vicentini, A., Spazzolini, C., Nastoli, J., Bottelli, G., Folli, R., and Cappelletti, D. (2003). Risk stratification in the long-QT syndrome. *N. Engl. J. Med.* **348,** 1866–1874.

209. Priori, S. G., Aliot, E., Blomstrom-Lundqvist, C., Bossaert, L., Breithardt, G., Brugada, P., Camm, A. J., Cappato, R., Cobbe, S. M., Di Mario, C., Maron, B. J., McKenna, W. J., Pedersen, A. K., Ravens, U., Schwartz, P. J., Trusz-Gluza, M., Vardas, P., Wellens, H. J., and Zipes, D. P. (2001). Task Force on Sudden Cardiac Death of the European Society of Cardiology. *Eur. Heart J.* **22,** 1374–1450.

210. Moss, A. J., Zareba, W., Hall, W. J., Schwartz, P. J., Crampton, R. S., Benhorin, J., Vincent, G. M., Locati, E. H., Priori, S. G., Napolitano, C., Medina, A., Zhang, L., Robinson, J. L., Timothy, K., Towbin, J. A., and Andrews, M. L. (2000). Effectiveness and limitations of beta-blocker therapy in congenital long-QT syndrome. *Circulation* **101,** 616–623.

211. Donger, C., Denjoy, I., Berthet, M., Neyroud, N., Cruaud, C., Bennaceur, M., Chivoret, G., Schwartz, K., Coumel, P., and Guicheney, P. (1997). KVLQT1 C-terminal missense mutation causes a forme fruste long-QT syndrome. *Circulation* **96,** 2778–2781.

212. Moss, A. J., Zareba, W., Kaufman, E. S., Gartman, E., Peterson, D. R., Benhorin, J., Towbin, J. A., Keating, M. T., Priori, S. G., Schwartz, P. J., Vincent, G. M., Robinson, J. L., Andrews, M. L., Feng, C., Hall, W. J., Medina, A., Zhang, L., and Wang, Z. (2002). Increased risk of arrhythmic events in long-QT syndrome with mutations in the pore region of the human ether-a-go-go-related gene potassium channel. *Circulation* **105,** 794–799.

213. Chen, S., Zhang, L., Bryant, R. M., Vincent, G. M., Flippin, M., Lee, J. C., Brown, E., Zimmerman, F., Rozich, R.,

Szafranski, P., Oberti, C., Sterba, R., Marangi, D., Tchou, P. J., Chung, M. K., and Wang, Q. (2003). KCNQ1 mutations in patients with a family history of lethal cardiac arrhythmias and sudden death. *Clin. Genet.* **63,** 273–282.

214. Zareba, W., Moss, A. J., Locati, E. H., Lehmann, M. H., Peterson, D. R., Hall, W. J., Schwartz, P. J., Vincent, G. M., Priori, S. G., Benhorin, J., Towbin, J. A., Robinson, J. L., Andrews, M. L., Napolitano, C., Timothy, K., Zhang, L., Medina, A., and the International Long QT Syndrome Registry. (2003). Modulating effects of age and gender on the clinical course of long QT syndrome by genotype. *J. Am. Coll. Cardiol.* **42,** 103–109.

215. Jiang, M., Zhang, M., Tang, D. G., Clemo, H. F., Liu. J., Holwitt, D., Kasirajan, V., Pond, A. L., Wettwer, E., and Tseng, G. N. (2004). KCNE2 protein is expressed in ventricles of different species, and changes in its expression contribute to electrical remodeling in diseased hearts. *Circulation* **109,** 1783–1788.

216. Roepke, T. K., and Abbott, G. W. (2006). Pharmacogenetics and cardiac ion channels. *Vascul. Pharmacol.* **44,** 90–106.

217. Simard, C., Drolet, B., Yang, P., Kim, R. B., and Roden, D. M. (2005). Polymorphism screening in the cardiac K+ channel gene KCNA5. *Clin. Pharmacol. Ther.* **77,** 138–144.

218. Brendel, J., and Peukert, S. (2003). Blockers of the Kv1.5 channel for the treatment of atrial arrhythmias. *Curr. Med. Chem. Cardiovasc. Hematol. Agents* **1,** 273–287.

219. Pirard, B., Brendel, J., and Peukert, S. (2005). The discovery of Kv1.5 blockers as a case study for the application of virtual screening approaches. *J. Chem. Inf. Model* **45,** 477–485.

220. Edelberg, J. M., Huang, D. T., Josephson, M. E., and Rosenberg, R. D. (2001). Molecular enhancement of porcine cardiac chronotropy. *Heart* **86,** 559–562.

221. Miake, J., Marban, E., and Nuss, H. B. (2002). Biological pacemaker created by gene transfer. *Nature* **419,** 132–133.

222. Qu, J., Plotnikov, A. N., Danilo, P., Jr., Shlapakova, I., Cohen, I. S., Robinson, R. B., and Rosen, M. R. (2003). Expression and function of a biological pacemaker in canine heart. *Circulation* **107,** 1106–1109.

223. Xue, T., Cho, H. C., Akar, F. G., Tsang, S. Y., Jones, S. P., Marban, E., Tomaselli, G. F., and Li, R. A. (2005). Functional integration of electrically active cardiac derivatives from genetically engineered human embryonic stem cells with quiescent recipient ventricular cardiomyocytes: Insights into the development of cell-based pacemakers. *Circulation* **111,** 11–20.

224. Priori, S. G. (2004). Inherited arrhythmogenic diseases: the complexity beyond monogenic disorders. *Circ. Res.* **94,** 140–145.

225. Roden, D. M. (2004). Human genomics and its impact on arrhythmias. *Trends Cardiovasc. Med.* **14,** 112–116.

226. Arking, D. E., Chugh, S. S., Chakravarti, A., and Spooner, P. M. (2004). Genomics in sudden cardiac death. *Circ. Res.* **94,** 712–723.

SECTION VIII

Genes, Gender, and Environment

CHAPTER 18

Gender and Cardiovascular Disease

OVERVIEW

It is well established that there are gender differences in the incidence and severity of cardiac defects in human and that in a number of clinical and experimental studies, it has been found that estrogen may play a cardioprotective role. However, in large-scale randomized, double-blind, placebo-controlled studies of hormone replacement therapy (HRT) in menopausal women, both asymptomatic and those with known coronary artery disease, the results have been disappointing. Alternatives to HRT are been sought, and selective estrogen receptor modulators (SERMs) with both estrogen antagonist effects in the breast and agonist effects in the heart that seem to be clinically safe have recently began to be used. These include raloxifene, tamoxifen, and a third-generation derivative of SERMs, lasofoxifene. Notwithstanding the progress achieved so far, further understanding of the molecular and cellular physiology of these compounds and their receptors in the heart is warranted. Furthermore, the decision to use these compounds in healthy menopausal women should be individualized on the basis of an assessment of relative risks and benefits.

INTRODUCTION

Gender differences in cardiovascular disease (CVD) are mostly caused by sex hormones. Although the specific molecular mechanism(s) underlying the gender factor in the incidence and phenotypes of CVD remains poorly understood, an increasing number of studies, derived mainly from animal models, continue to generate information on basic issues vital to the understanding of the effect of gender on the heart, such as molecular signaling pathways, genetics, energy metabolism and post-injury repair. Understandably, one must be very cautious when translating research data from animal models, mainly rodents, to humans regarding the pathogenesis of CVD and the response to different modulatory factors and therapies. From a clinical standpoint, gender differences in morbidity and mortality have been mostly attributed to the earlier perceived protective actions of female hormones as suggested by the increased cardiovascular risk in women after menopause.[1] However, clinically and experimentally reported benefits of HRT have been challenged and kept on hold by the negative outcome in a number of recent clinical trials, including the Heart and Estrogen/Progestin Replacement Study (HERS) and the Women's Health Initiative (WHI) studies.[2,3] A number of issues need to be resolved to clearly understand why there is sexual dimorphism in regard to CVD in human. Although not denying the importance of research with animal models, it is also imperative to generate more investigative studies in humans in which the molecular and cellular mechanisms and differences in cardiac phenotypes can be properly addressed.

SEX STEROID HORMONES

Steroid hormones, including the three gonadal hormones estrogen, progesterone, and testosterone, and their receptors are the principal regulators of cardiovascular gender differences. Although estrogen has received most of the attention by clinicians and researchers, there has been much less focus on the role of the other steroid hormones in CVD. In this chapter, although most of the discussion will be centered on estrogen and its receptors, we will also present relevant laboratory and clinical data available on other steroid hormones, in particular testosterone (Fig. 1).

Estrogen and Its Receptors

Estrogen receptors (ERs) that exist in two forms, ERα and ERβ, are activated by estrogen and related sex hormones and have been implicated in both non-transcriptional pathways (often at the plasma membrane) and in specific target gene transcriptional activation. To function as transcriptional activators, the ERs must translocate into the nucleus and bind specific DNA sites. ERs are expressed in a number of cell types, including, SMCs, cardiac myocytes, fibroblasts, and tissues such as endothelium, myocardium, and coronary arteries. Lin and associates[4] have examined sexual dimorphism with respect to sex steroid hormone receptors in baboon myocardium. They found that the myocardial content and distribution of both the ER and progesterone receptor did not differ between the genders. However, the distribution of myocardial androgen receptors between the cytosolic and nuclear compartments was sexually dimorphic. Female baboon myocardial androgen receptors were restricted

Post-Genomic Cardiology

Copyright © 2007 by Academic Press.
All rights of reproduction in any form reserved.

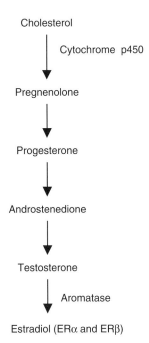

FIGURE 1 Flowchart depicting the synthesis of steroid hormones. Estrogen, progesterone, and testosterone are synthesized from pregnenolone. By the intermediate action of an aromatic ring, catalyzed by the enzyme aromatase, estrogen is produced from androgens.

to the cytosolic compartment, whereas male myocardial androgen receptors were distributed between the cytosolic and nuclear compartments. Therefore, gonadal steroid hormone receptor gene transcription occurs in cells of the baboon cardiovasculature, but a subset of these receptors may be restricted to specialized cells, and they also may be physiologically functional. Recently, Nordmeyer and associates[5] analyzed ER expression in human heart and its relation to hypertrophy-mediated gene expression. By use of real-time PCR in LV biopsies, they quantified ERα and ERβ, calcineurin A-β, and BNP mRNA in a group of patients with aortic valve stenosis ($n=14$) and controls ($n=17$). LV ERα and ERβ mRNA were increased 2.6-fold, and LV ERα protein was increased 1.7 fold in these patients. In addition, ERα and ERβ were found both in the cytoplasm and nuclei of human hearts, and both receptors seem to be up-regulated by myocardial pressure load. It is important to note that the sample size for this study was relatively small, and a larger population of patients with aortic stenosis, and probably with other causes of myocardial hypertrophy (e.g., hypertension), needs to be studied to definitely establish whether sexual dimorphism occurs in humans.

Non-Genomic Effects of Estrogen

The so-called non-genomic effects of estrogen are independent of gene transcription or protein synthesis and involve steroid-induced modulation of cytoplasmic or of cell membrane–bound regulatory proteins. Relevant biological actions of steroids have been associated with this signaling in different tissues.[6] Of significance is that signaling regulatory cascades such MAPK, PI3K, and tyrosine kinases are modulated through non-transcriptional mechanisms by steroid hormones. Furthermore, steroid hormone–receptor modulation of cell membrane–associated molecules such as ion channels and G-protein–coupled receptors has been detected in a number of tissues.

Non-genomic steroid hormone actions are particularly prominent in the vascular wall. For instance, estrogens and glucocorticoids trigger rapid vasodilatation because of rapid induction of endothelial nitric oxide synthase (eNOS) in the endothelial cells by means of the ER-dependent activation of MAPK and PI3K, with pathophysiological consequences, *in vitro* and *in vivo*. The rapid non-genomic actions of estrogen resulting in the activation of eNOS, several kinase cascades, and the modulation of ion channels are achieved from membrane signaling complexes that incorporate a variety of diverse proteins, a number of which have been identified, including ERα, G-coupled protein receptors and G-proteins, eNOS, and, in some cell types, caveolin-associated striatin, which serves as a molecular anchor.[7]

Furthermore, in ovariectomized female mice, 17β-estradiol (E2) reduces cardiomyocyte apoptosis both *in vivo* and *in vitro* by ER- and PI3K-Akt–dependent pathways, providing further support of the role and significance of these pathways in the observed estrogen-mediated reduction in myocardial injury.[8] The mechanism by which E2 activates PI3K-Akt signaling in cardiomyocytes is not fully understood. Patten and associates suggested that ERα specifically mediates this E2 effect,[8] consistent with findings in vascular endothelial cells in which E2 rapidly stimulates eNOS activity, in part by activation of PI3K-Akt pathways through a non-genomic mechanism.[9,10] Moreover, a direct interaction of ligand-activated ERα and the p85 regulatory subunit of PI3K coinciding with activation of PI3K activity in endothelial cells has been suggested.[11]

There is also increasing evidence to suggest that both transcriptional and non-transcriptional signaling mechanisms play a primary role in the generation of the effects of steroids on endothelial cells, which may turn out to be clinically relevant. After menopause, increased levels of proinflammatory cytokines, including IL-1, IL-6, and TNF-α, are found whose activity is blocked by estrogen. Although the mechanisms by which estrogen affects cytokine levels are not completely understood, these may include transcriptional regulation by interaction of ERs with other transcription factors (as found in the regulation of bone resorption, as shown in Fig. 2) and non-genomic actions, including the modulation of NO activity, antioxidative effects, plasma membrane actions, and changes in immune cell function.[12]

In addition to its multiple vascular effects, estrogen directly acts on the myocardium, which contains functional estrogen receptors. Myocardial hypertrophy resulting from a number of insults, including pressure overload (i.e., aortic valvular stenosis), is reduced by the action of estrogen.

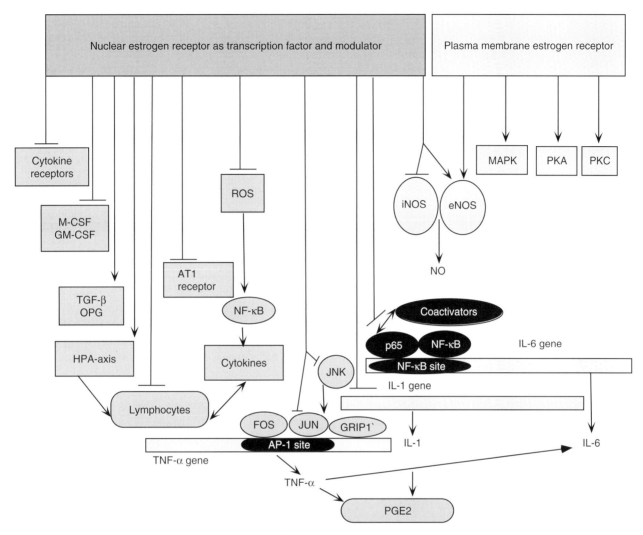

FIGURE 2 Overview of the multiple interactions by which estrogen bound to the estrogen receptor may regulate bone resorption. Both stimulatory (indicated by arrow) and inhibitory (by perpendicular line) effects of estrogen receptor and its cellular mediators (e.g., ROS, protein kinases, NO, cytokines) are shown. Macrophages, lymphocytes, and stromal cells/osteoblasts contribute to increased cytokine production in bone with estrogen deficiency. M-CSF, GM-CSF, and IL-6 facilitate osteoclast precursor proliferation. This suggests a model of the estrogen receptor in which both proinflammatory cytokines (e.g., IL-1, PGE2, TNF-α, and IL-6) and specific receptor production are attenuated, as are synthesis of M-CSF, GM-CSF, and IL-6 as shown, thought to occur primarily by estrogen receptor acting as an upstream transcription factor, as well as by its interaction with multiple transcription factors. Other effects of estrogen receptor are shown, including modulation of NO (by action on eNOS and iNOS), direct effects on immune cell function (e.g., lymphocytes), and rapid nongenomic effects mediated largely by plasma membrane ERs resulting in the activation of signaling kinases (e.g., MAPK, PKA, and PKC).

Female ovariectomized mice develop a more significant left ventricular hypertrophic response than ovariectomized mice with restoration of physiological levels of 17β-estradiol.[13] Furthermore, estrogen has been reported to modulate myocardial responses to ischemia/reperfusion (I/R) injury, with ER mediating the action of estrogen. In ovariectomized rabbits, Booth and associates[14] investigated the cardioprotection promoted by 17β-estradiol and 17α-estradiol by use of an *in vivo* occlusion-reperfusion model. The infarct size decreased in the 17β-estradiol–treated group compared with 17α-estradiol or vehicle-treated groups. In addition, pre-treatment with the ER antagonist fulvestrant markedly decreased the size of the infarct area, indicating that 17β-estradiol protects the heart against I/R injury and that this cardioprotection is ER mediated. In a more recent study, this group of investigators also reported that activation of ERα, but not ERβ, is required for the cardioprotective effects of 17β-estradiol.[15]

Genomic Effect of Estrogen

The direct genomic effect of estrogen is mediated by the interaction of its receptors with specific target sequences of DNA termed estrogen response elements (ERE), and this interaction of ER with ERE can be affected by differences in the ERE sequence, the ER subtype, and/or dimerization of ERs.[16] Both ERα and ERβ are able to form homodimers and heterodimers, and the variable composition of the dimeric ER complex differentially regulates gene expression.[17] Gabel and associates[18] have observed in mice that under hypercontractile conditions induced by increased intracellular calcium, male hearts sustain enhanced I/R injury compared with female hearts, and they attempted to identify the specific ER involved in this gender difference. After a brief period of treatment with isoproterenol, isolated mouse hearts subjected to I/R were assessed for post-ischemic contractile function and infarct size with wild-type male and female hearts and female hearts from transgenic knock-out (KO) strains lacking either functional ERα (αERKO), or ERβ (βERKO). Wild-type male hearts exhibited significantly less functional recovery and more necrosis than females. The αERKO female hearts exhibited I/R injury similar to that observed in wild-type females, whereas βERKO females exhibited significantly less functional recovery than wild-type females and were similar to wild-type males. These data suggested that estrogen, acting through ERβ, plays a role in the protection observed in the female heart. This study also identified genes that were expressed in βERKO female hearts and not in αERKO and WT female hearts, and found abnormal expression of a number of lipid metabolism genes, including ATP citrate lyase, stearoyl CoA desaturase, and fatty acid synthase, that may be important in ischemic injury. Furthermore, in female mice with HF secondary to experimental MI, systemic deletion of ERβ increased mortality, suggesting, although indirectly, that after MI, selective activation of ERβ may improve cardiac function.[19] Pelzer and associates[20] have also shown that ERα activation favorably affects cardiac hypertrophy, myocardial contractility, and gene expression in ovariectomized spontaneously hypertensive rats (SHR). The selective ERα agonist 16α-LE2 reduced cardiac hypertrophy and improved function in estrogen-deficient SHR, which was linked to differential expression patterns of cardiac myosin heavy chains. Although these findings are helpful in our understanding of the role(s) that ERs plays in cardiac structure and function, we and others[21,22] believe that further research is needed. A potential hypothesis for future testing might be that ERβ is protective by improving metabolic function, whereas ERα provides cardioprotection by acting in an antitrophic fashion.

Androgens

A number of publications in the medical literature have suggested that testosterone replacement therapy may produce a wide range of benefits for men with hypogonadism, including improvement in libido,[23,24] bone density,[25,26] muscle mass,[27] and erythropoiesis.[28] On the other hand, there is considerable controversy regarding the indications for testosterone (T) supplementation in aging men.[29]

It is generally accepted that human exposure to testosterone does not shorten the life span in either gender; however, androgen exposure in early life (perinatal androgen imprinting) may predispose men to premature atherosclerosis.[30] Although there is increasing interest in the potential effect of androgen replacement therapy (ART), randomized trials of ART are not yet available. Wu and von Eckardstein[31] have reviewed the effects of exogenous T on cardiovascular mortality or morbidity, and on the basis of preliminary data, there may be short-term improvements in the electrocardiogram of men with coronary artery disease (CAD) although prospective controlled studies are needed to further verify this. In addition, they found that in most animal experiments, exogenous T exerts either neutral or beneficial effects on the development of atherosclerosis. Exogenous androgens may induce both beneficial and deleterious effects on cardiovascular risk factors, including reducing serum levels of HDL-C, plasminogen activator type 1 (probably deleterious), lipoprotein (a), fibrinogen, insulin, leptin, and visceral fat mass (probably beneficial) in men and women. T exerts proatherogenic effects on macrophage function by facilitating the uptake of modified lipoproteins and an antiatherogenic effect by stimulating efflux of cellular cholesterol to HDL. The authors found the data collected were inconsistent (partly attributable to different dosage and types of androgens used), which precludes a meaningful assessment of the net effect of T on atherosclerosis. Nevertheless, on the basis of current evidence, therapeutic use of T in men may not need to be restricted by considerations of potential cardiovascular side effects, but at the same time, existing data do not justify the uncontrolled use of T or dehydroepiandrosterone for the prevention or treatment of CAD. It is important to keep in mind that, before new therapeutic use of androgens is initiated, it is necessary to have a clear understanding of their multiple effects, the function and location of their receptors and regulators, the explanation of the non-genomic androgen effect, and the specificity of tissue activation and sensitivity. Novel therapeutic targets may develop arising from improved insight into how androgens can enhance early plaque formation, how they may cause vasodilatation by means of non-genomic androgen effects on vascular smooth muscle, how tissue-specific androgen effects are modulated by androgen receptor (AR) coregulators, and insights on the metabolic activation of T to enhance the biological effects of T on the vasculature.[30]

Observational studies have shown that T concentrations in blood are lower in men with CVD,[32] implying that T therapy may have a possible preventive role, but again, this needs to be further confirmed by prospective studies. Furthermore, a definitive association between endogenous T levels and CAD in men and women has not been confirmed in several large, prospective studies, although cross-sectional data have suggested that CAD may be associated with low T in men. On the other hand, short-term interventional studies have shown that T, compared with placebo, produces a modest improvement in cardiac ischemia, similar to the effects associated with antianginal drugs.

Non-Genomic and Genomic Androgen Effects

Besides regulation of sexual function, male steroids play an important role in a variety of physiological homeostasis systems, including the cardiovascular system. T activates both

AR and ER (by aromatase conversion to E2) in cardiovascular tissues and mediates cardiomyocyte growth responses both in physiological situations and in hypertrophy-related cardiac diseases.[33] According to Liu and associates,[30] the classical pathway of androgen action involves its binding to the AR, a single copy member of the nuclear receptor superfamily that functions as a ligand-activated transcription factor acting on the genome. AR is almost ubiquitously expressed in tissues, and its genomic action is modulated by a diverse group of coregulators, which are proteins that fine-tune target gene expression by either enhancing (coactivator) or restraining (corepressor) transcription. Although the factors responsible for variation in tissue androgen sensitivity are not known, expression levels of AR in part define androgen sensitivity. However, it is not yet known how extensively ARs mediate vascular androgen effects, as well as the full spectrum of specific target genes that are regulated by these receptors.

The genomic effects of androgens can be initiated in either of two ways. The first pathway involves binding to and activation of AR by either T or other androgens such as dihydrotestosterone (DHT) shortly on entering the cell. In the second pathway, T or other aromatizable synthetic androgens cross the cell membrane to be converted to estradiol (or the aromatic synthetic analog) by aromatase (CYP19), which then binds to and activates ERα or ERβ. DHT or other nonaromatizable androgens are limited to the first pathway.[30] Aromatase gene expression, protein, and enzymatic activity have been widely reported in vascular tissues, including human coronary arteries, particularly in endothelium and smooth muscle. Interestingly, inhibition of aromatase disrupts normal vascular relaxation in healthy human males, and aromatase knock-out mice exhibit abnormal vascular relaxation, suggesting that conversion of T to estrogen in males by aromatase helps to maintain normal vascular tone.[34] Despite the widespread use of aromatase in breast cancer therapy, the effects of aromatase inhibition on vascular function in females have not yet been examined.[35]

The subsequent events are similar in both pathways; either the ligand-bound ER or AR dissociate their heat-shock protein (HSP), undergo conformational changes, dimerize, and translocate into the nucleus, where they bind to specific DNA sequences, including ERE or androgen response elements (ARE) primarily located within promoters of target nuclear genes to produce long-term genomic effects.[30]

The non-genomic effects of testosterone are triggered by binding to an unknown membrane receptor, followed by activation of second messengers and transducers, including Ca^{++}, protein kinases (A and C), and MAPK, which produce rapid responses leading to diverse cellular effects, including smooth muscle relaxation, neuromuscular and junctional signal transmission, and neuronal plasticity.[36] Most non-genomic effects of androgens are thought to be mediated by a membrane receptor (not fully characterized so far), to which putative binding has been described for all major classes of steroids including androgens.[36,37] In mammals, only a membrane receptor for progesterone (P) has been cloned, but its functional characterization is not yet known.[38,39]

Androgen's non-genomic effects have been particularly prevalent in vascular cells, including macrophages, endothelial cells, and SMCs. Testosterone has been implicated in the rapid blockade of calcium influx and potassium efflux by modulating specific channel opening and endothelial-derived NO levels, leading to coronary vasodilation and reduced contractility.[40–43] Moreover, non-genomic effects of T have been shown to modulate adenosine vasodilation in isolated heart.[44]

Recent studies have further examined whether T might improve the myocardium tolerance to ischemia secondary to activation of mitochondrial and/or sarcoplasmic K_{ATP} channels.[45] In a cellular model of ischemia, T significantly reduced cardiomyocyte death that could be prevented by 5-hydroxydecanoic acid but was not affected by the sarcoplasmic K_{ATP} blocker HMR1098 and the T antagonist flutamide. These findings suggest a relation between T-induced cytoprotection and activation of mitoK$_{ATP}$ and also that endogenous T might play an important role in the post-MI recovery.

ROLE OF GONADAL HORMONES IN CARDIAC PATHOLOGY

The possible protective action of estrogen in the cardiovascular system seems to be mediated indirectly by its effect on serum lipoprotein and triglyceride profiles, on the expression of coagulant and fibrinolytic proteins, and by a direct effect on the vessel wall itself. Although estrogen exerts both rapid effects on membrane ionic permeability, activation of membrane-bound enzymes, and increases in endothelial cell eNOS activity, it also has longer term effects on gene expression that are mediated, at least in part, by the ligand-activated transcription factors ERα and ERβ.[46] Preparations with pure antiestrogenic activity and selective estrogen receptor modulators (SERM), such as tamoxifen and raloxifene, that regulate ER function in a tissue-specific manner, have been developed in an effort to attain the cardioprotective effects of estrogens while minimizing the risks associated with hormone replacement therapy (e.g., endometrial and breast cancer). As we have pointed out, in addition to smooth muscle and endothelial cells,[47] the myocardium with its functional ERs seems to be a direct and important target for estrogens, because female gender and sex hormones modulate an array of myocardial signaling pathway elements, some of which can reduce cardiac hypertrophy in humans and a variety of animal models.[1,13,48–51]

Although our understanding of estrogen signaling with knockout animal models had been progressively improving, further insights into the function of ERα and ERβ have been revealed by the application of isotype-selective ER agonists. Steroid ligands were designed on the basis of the crystal structure of the ERα-ligand–binding domain, and a homology model of the ERβ-ligand–binding domain, taking advantage of the differences in size and flexibility of the two ligand-binding sites.[52] Compounds predicted to bind preferentially to either ERα or ERβ were synthesized and tested *in vitro* with radioligand competition and transactivation assays. These investigators also conducted *in vivo* experiments with rats to identify the physiological roles of the two receptors using the ERα- and ERβ-selective agonists compared with 17β-estradiol (E2). The ERα agonist induced uterine growth, caused bone-protective effects, reduced LH and FSH plasma levels, and increased angiotensin I, whereas the ERβ agonist did not at all or only at high doses lead to such effects, despite high plasma levels. On the basis of these concepts Pelzer and associates[20] have used the ERα agonist 16α-LE2 (which binds and transactivate ERα more efficiently than ERβ) to inhibit cardiac hypertrophy and to improve the cardiovascular hemodynamics of ovariectomized spontaneously hypertensive rats (SHR). They found that activation of ERα had a positive effect on cardiac hypertrophy and contractility, which has been linked to the differential expression of cardiac α-MHC.

Moreover, a novel role for ovarian hormones in the regulation of cardiac protein synthesis has been recently described.[53] In ovariectomized rats, P replacement resulted in significantly increased cardiac protein synthesis rates and increased plasma volume. Treatment of P-supplemented ovariectomized rats with the P receptor antagonist RU 486 or with E2 reduced protein synthesis rates to control levels, indicating that the progesterone stimulation of cardiac muscle protein synthesis was mediated through a receptor-dependent pathway and that estrogens may play an inhibitory role in regulating cardiac growth. Although the basis for this ovarian hormone action on the heart has not yet been established, the effect of P was probably related to changes in myocardial gene transcription or to circulatory volume homeostasis. Previous studies have found that the administration of medroxyprogesterone acetate, a synthetic progestin, to post-menopausal women increased left ventricular mass, end diastolic volume, plasma volume, and stroke volume.[54] It is, however, possible that these changes may not be comparable to a progestogenic effect but rather a function of the synthetic nature of progestin with potential androgenic activity.[55]

ANIMAL AND *IN VITRO* STUDIES

Differences in the Cardiovascular Phenotypic Response to Stimuli by Gender

Animal models exposed to different stress stimuli may respond with cardiovascular changes that are different between males and females as shown in Table I.

Modulation of the cardiovascular phenotype, either genetically and/or biochemically, in animal models has emerged as a powerful tool for cardiovascular research. Studies with these models have generated important information on the effect of gender in the development of diverse cardiovascular pathological conditions. For instance, by use of strains of mice with disruption of the ApoE low-density lipoprotein receptor, ER, aromatase, or follitropin receptor (disruption of the latter two lead to estrogen deficiency), several groups have shown that estrogen and ER-mediated signaling are protective against vascular dysfunction and atherosclerotic lesions.[56–60] Epidemiological and clinical data have suggested differences in the incidence and severity of atherosclerotic CAD in age-matched men versus premenopausal women. Herrera and associates[61] have studied a moderately expressed transgene for human cholesteryl ester transfer protein (CETP) in the Dahl salt-sensitive hypertensive rat strain, Tg25, which recapitulates the premenopausal female atheroresistance. After establishing identical genetic background, environmental factors, and equivalent CETP hepatic RNA levels, male Tg25 rats exhibited more hypercholesterolemia, hypertriglyceridemia, coronary plaques, and worse survival outcome thanTg25 female rats. These findings showed that the atherosclerosis phenotype is modulated by gender-specific factors.

Recent studies have examined the potential role of gender differences in the cardiomyopathic phenotype and of gonadal status in mice with cardiac-specific overexpression of β2-adrenergic receptors (β2-ARs).[62] Survival to 15 months was less in transgenic (TG) mice and lower for males than females. Echocardiogram analysis indicated progressive LV dilatation and reduced LV shortening fraction in male, and much less pronounced changes in female TG mice. TG males also had a higher incidence of atrial thrombosis, pleural effusion, LV fibrosis, and lung congestion. Deprivation of testicular hormones by castration improved survival and significantly ameliorated LV dysfunction, remodeling, and hypertrophy compared with intact TG males. In contrast, no significant effect, except for a trend of a better survival, was detected by ovariectomy in TG females. Thus, cardiac β2-AR overexpression in males leads to cardiomyopathy and HF with aging. Female mice exhibited less cardiac remodeling, dysfunction, and pathology and a marked survival advantage over male mice, and this was independent of prevailing levels of ovarian hormones. Also, TG males showed benefit from orchiectomy, suggesting a contribution by testicular hormones to the progression of the cardiomyopathic phenotype, an interesting contrast to the previously discussed findings of T's cardiovascular benefits.

Similarly, studies on the myocardial response to cardiac ischemia with the overexpression of β2-AR showed that females were protected from the detrimental effects of β2-AR overexpression compared with males, which exhibit significantly decreased myocardial ATP levels for energy use and increased I/R injury.[63] Protection in female β2-AR overexpressors was mediated by eNOS, suggesting that eNOS may be relevant to the protection from cardiovascular injury observed clinically in pre-menopausal females.

Recent studies using monkeys (*Macaca fascicularis*) as animal models found a marked gender difference by proteomic analysis of the aging heart.[64] These differences

TABLE I Differences on Cardiovascular Phenotypes by Animal Gender

Gene targeting	Model	Cardiac phenotype	Reference
Gene Disruption			
FKBP12.6	Mice	↑ Myocardial hypertrophy in males not females	69
GC-A	Mice	↑ Hypertrophy in males	79
α1A/βAR	Mice	Impaired developmental heart growth, ↓ exercise capacity and ↓ tolerance to pressure overload in males	144
Relaxin	Mice	Males with ↑ cardiac fibrosis, atrial weight, and diastolic dysfunction	70
PPAR-α	Mice	↑ Hypoglycemia, myocardial lipids accumulation, and death triggered by inhibition of fatty acid influx with etomoxir; more so in males	145
IGF-I	Mice	Retarded growth including the heart, more so in males	71
AT1aR	Mice	↑ Myocyte hypertrophy post-MI in females, partially dependent on activation of AT1aR	78
Transgene overexpression			
CETP	Rat	Females ↓ cholesterol level and better survival rate	61
R92Q allele of TnT	Mice	Males have delayed cardiac growth and development	72
Myostatin	Mice	Males have ↓ cardiac and skeletal muscle mass	146
β1AR	Mice	↑ Cardiomyopathy and premature death in males	147,148
β2AR	Mice	↑ Ischemia/reperfusion injury in males	62,63
NCX	Mice	↑ Ischemia/reperfusion injury in males	149
R403Q allele of MHC	Mice	Males ↑ severe cardiomyopathy	76
R403Q allele of MHC + deletion in actin-binding domain of α–MHC	Mice	Old males ↑ diastolic and systolic dysfunction	77

MHC, myosin heavy chain; TnT, troponin T; CETP, cholesteryl ester transfer protein; βAR, beta adrenergic receptor; NCX, Na^+-Ca^{++} exchanger; GC-A, guanylyl cyclase A; IGF, insulin-like growth factor; PPAR-α, peroxisome proliferator-activated receptor alpha; AT1R, angiotensin II type 1 receptor.

seem to be linked to previous findings of a distinctive desensitization to β-AR stimulation in old male (OM) monkeys, whereas the response to β-AR in old female (OF) monkeys was preserved. Furthermore, the OM monkeys showed evidence of early aging cardiomyopathy, which was not observed in OF monkeys. Although adrenergic stimulation enhances the contribution of glucose to overall metabolism in rat hearts,[65–67] OM monkeys showed a gender-specific reduction in glucose utilization (glycolysis and glucose oxidation), suggesting that the decreased expression of glycolytic enzymes observed in the aging males may be related to a gender-specific β-AR desensitization, and these two mechanisms acting together may be responsible for the impaired cardiac function detected in OM monkeys.[68]

Estrogen has also been reported to play a protective role in the hypertrophic response of the heart to Ca^{++} dysregulation.[69] Disruption of gene coding of FK506 binding protein 12.6 (*FKBP12.6*) associated with the cardiac ryanodine receptor results in cardiac hypertrophy in male but not in female mice. Male and female *FKBP12.6*-knockout mice displayed similar dysregulation of Ca^{++} release, but female hearts were normal. On the other hand, female *FKBP12.6*-null mice treated with tamoxifen, an ER antagonist, developed cardiac hypertrophy similar to that of male mice. Similarly, male-restricted phenotypes in mice were observed in the fibrotic cardiomyopathy that develops after disruption of relaxin-1 (an anti-fibrotic hormone),[70] in the delayed cardiac growth and development occurring after disruption of IGF-1[71] and in the expression of mutant troponin T.[72] In addition, after aortic banding, male rats exhibited an earlier transition to left ventricular chamber dilatation, eccentric hypertrophy, and diastolic dysfunction compared with females.[73,74] Also, Dash and associates[75] were able to demonstrate that catecholamine-induced hypertrophy was gender dependent and modulated by differential regulation of p38 mitogen-activated protein kinase (MAPK). Although male transgenic mice exhibited ventricular hypertrophy and mortality along with cardiac p38 MAPK activation at 15 months, female transgenics, despite similar contractile dysfunction, displayed a temporal delay in p38 MAPK activation, hypertrophy, and mortality (at 22 months), which was associated with sustained cardiac levels of MAPK phosphatase-1 (MKP-1), a potent inhibitor of MAPK p38. Subsequently, decreases in cardiac MKP-1 were accompanied by increased levels of p38 MAPK activation. Furthermore, *in vitro* studies have shown that pre-incubation with E2 induced high MKP-1 levels, which precluded norepinephrine-induced p38 MAPK activation. These findings indicated that norepinephrine-induced hypertrophy is linked closely with p38 MAPK activation, which can be endogenously modulated through estrogen-responsive regulation of MKP-1 expression, and that estrogen may protect the heart from structural decompensation through induction of MKP-1 with subsequent suppression of p38 MAPK activation.

In a mouse model of familial HCM generated by the introduction of an Arg403 Gln mutation into the cardiac α-myosin heavy chain (MHC) gene, it was found that young male mice showed more evidence of disease than did their female counterparts.[76] In addition, in a transgenic mouse model of HCM in which a R403Q missense allele and an actin-binding deletion in α-MHC were expressed in the heart, both male and female TG mice developed LV hypertrophy with diastolic dysfunction at 4 months of age, but LV systolic function was normal and supranormal in the young TG females and males, respectively.[77] At 10 months of age, females continued to present LV concentric hypertrophy, as well as impaired LV diastolic function without evidence of systolic dysfunction, whereas the males began to display LV dilation with worsened LV diastolic function, and systolic performance was significantly impaired.

Recently, the role that angiotensin II type 1a receptor (AT1aR) subtype activation plays in post-MI remodeling and potential gender differences in cardiac remodeling in AT1aR knock-out strains compared with wild-type mice was analyzed by Bridgman and associates.[78] No increases in LV mass/body weight ratio or in myocyte cross-sectional area after MI were observed in female AT1aR-KO mice compared with shams. After MI, both male and female wild-type mice showed higher ANP levels, with female levels higher than males. However, in AT1aR-KO mice with MI, both males and females showed no ANP gene overexpression, despite increases in LV mass and myocyte size in males. Taken together, these findings suggest that gender-specific patterns of LV and myocyte hypertrophy exist after MI in mice with a disrupted AT1aR gene and that myocyte hypertrophy after MI in females was partially dependent on activation of the AT1aR.

Androgens may also contribute to gender-related cardiac hypertrophy and fibrosis. In a guanylyl cyclase-A knock-out (GC-A KO) model, cardiac hypertrophy and fibrosis were significantly more pronounced in males than females, and these differences were not seen in wild-type mice.[79] Moreover, castration or the AR antagonist flutamide markedly decreased the hypertrophy and fibrosis in male GC-A KO mice, whereas ovariectomy had little effect. Furthermore, chronic T infusion increased cardiac mass and fibrosis in ovariectomized GC-A mice. In addition, castration markedly decreased the levels of ventricular angiotensin-converting enzyme (ACE) of male GC-A KO mice, and the gender differences were abolished by targeted deletion of *AT1aR*. As previously noted, T had been shown to act as a vasodilator *in vitro* and *in vivo*, and short-term administration of T acutely induced vasodilation in the systemic, coronary, and pulmonary vascular beds and inhibited the calcium-dependent elements of vascular contraction.[41] Furthermore, administration of T in orchiectomized rats improved recovery of myocardial function after I/R injury.[80] Moreover, the aforementioned findings of T's role in protecting the heart from ischemia should be mentioned in this context. These studies demonstrated that T acutely and directly depolarized and oxidized cardiac mitochondria in a K^+-dependent, ATP-sensitive, and AR-independent manner by use of a cellular model of ischemia (single ventricular cells isolated from Sprague-Dawley rats) and found that T induced mitoK_{ATP} channel activation, which could be inhibited by ATP, 5-HD, and glibenclamide.[45]

STUDIES IN HUMANS

Gender Differences on Cardiovascular Phenotype Development

Although the data collected so far in animals studies have been helpful in our understanding and the progressive dissection of the many variables involved in the mechanisms of cardiovascular gender differences, the precise effects of genetic variation in estrogen receptor α and β (*ER1* and *ER2*) in human and CVD risk are not yet known.

Using the Framingham Heart Study, in a prospective study of 1739 unrelated men and women, Shearman and associates[81] investigated whether a specific variant in the ERα gene promoter (*ER1* c.454-397T>C) is associated with CVD risk. Twenty percent of participants were homozygous for the *ER1* c.454-397C allele. After adjustment for covariates such as age, gender, body mass index, hypertension, diabetes mellitus, total cholesterol, high-density lipoprotein cholesterol, and cigarette smoking, the CC genotype was significantly associated with major atherosclerotic CVD (i.e., acute MI, coronary insufficiency, coronary heart disease, death, or thrombotic stroke [with an odds ratio of 2.0] compared with individuals with the CT or TT genotypes). Individuals with the CC genotype had 3.0-fold greater odds of MI compared with those with the CT or TT genotype. Moreover, individuals with the common *ER1* c.454-397CC genotype have a substantial increase in risk of MI. Although the *ER1* genotype association with MI risk was significant, analysis was restricted to males, since the lower number of women having events precluded evaluation. These findings support the concept that ERs play a significant role in CVD susceptibility in humans, especially in men, and that ER variation may explain the existing conflict regarding the effects of hormone therapy on CVD susceptibility in women.

The relationship of genetic variations in the estrogen receptor α (ERα) and the risk of ischemic heart disease (IHD) has also been evaluated by Schuitt and associates[82] in a group of 2617 men and 3791 post-menopausal women whose primary outcome was MI, IHD, revascularization procedures, and IHD-related death. In this prospective study, an increased risk of MI was found in post-menopausal women who carry ERα haplotype 1 (*c.454-397 T* allele and *c.454-351 A* allele). The risk estimates did not change after adjustment for clinically relevant cardiovascular risk factors, indicating that ERα1 haplotype 1 is an independent risk factor. Heterozygous carriers of haplotype 1 had a 2.2-fold increased risk of MI compared with noncarriers, whereas homozygous carriers had a 2.5-fold times increased risk. Analysis of the risk of IHD events by taking together MIs, revascularization procedures, and IHD mortality data showed similar increased risk in female carriers of haplotype 1. For men, no association between the ERα haplotypes and MI or IHD risk was observed. For women, the effect of haplotype 1 on fatal IHD was greater than on nonfatal IHD, and IHD resulted in death more often in carriers of haplotype 1, suggesting that post-menopausal women who carry ERα haplotype 1 have not only an increased risk of having an IHD event but also an increased risk of death from such an event. As pointed out by Newton-Cheh and O'Donnell,[83] some of the Schuit data seem to contradict the findings by Shearman, because Schuit found that having two copies of the *C* allele of *c.454-397* is protective in women and has no effect in men, whereas Shearman found that having two copies of the *C* allele is harmful in men. Further research in this area is needed to resolve this apparent conflict.

Regarding its significance in the incidence of CAD by gender, the genotypic distribution of ER polymorphisms is still a matter of debate. *Pvu*II and *Xba*I restriction fragment length polymorphisms (RFLPs) of the ERα (*ER1*) gene have been associated with receptor expression and function in nonatherosclerotic diseases like breast cancer and osteoporosis.[84–86] A number of studies have evaluated the role of these polymorphisms in CAD, but findings about the functional role of these polymorphisms in disease/phenotype expression seem rather at variance.[87–91] The association of CAD with three polymorphic changes in the ERα was examined by Matsubara and associates in a Japanese population.[88] Genotypes P1/P2, X1/X2, and B-wild type/B-variant type were assessed in 87 men and post-menopausal women with MI or angina pectoris (as confirmed by coronary angiography) and from 94 control individuals with no history of CAD. In contrast with the reported allele frequency for B-variant type (0.1) in Caucasians, all individuals examined had B-wild type. Genotype distribution and allele frequencies of *Pvu*II or *Xba*I polymorphisms were not significantly different between control subjects and patients. When the allele frequencies were analyzed separately by gender, there was still no statistically significant difference for both polymorphisms. No association was found between the polymorphisms and the angiographic severity of CAD. Similarly, total cholesterol, triglyceride, or HDL-cholesterol levels were not significantly different among ER genotypes. These findings suggest that these ER polymorphisms are not associated with the prevalence and severity of CAD and that the polymorphisms are unrelated to the serum lipid levels in both control subjects and patients.

In a recent study, the prevalence of mutations in ERs was similarly analyzed in patients with CAD and controls.[92] Although the *Pvu*II mutation in ERα was more prevalent in controls than in CAD patients, a mutation in the ERβ (*Alu*I polymorphism) was more prevalent in CAD patients. Homozygosity for this mutation was associated with increased body mass index, elevated serum triglycerides and apolipoprotein B, and reduced HDL cholesterol. On multivariate logistic regression analysis, dyslipidemia, low serum HDL levels, and the *Alu*I polymorphism in ERα were shown to be independent risk factors for CAD. In an analysis of the potential role that ERα polymorphisms (*Pvu*II and *Xba*I) may have in the angiographic outcome after coronary artery stenting,[93] a population of 858 patients (including 148 women), from which 955 lesions were treated with stent implantation, was evaluated. Findings revealed that women homozygous for the T allele of the *Pvu*II polymorphism of the ERα (*ER1*) gene treated with coronary stent implantation have a higher risk of angiographic in-stent restenosis than men.

Besides the effect of ER polymorphism in CAD that has been previously discussed, association studies have been conducted to identify other genes that may confer susceptibility to restenosis after plain old balloon angioplasty (POBA). Horibe and associates[94] evaluated by angiography a population of 730 individuals (424 men, 306 women) 6 months after they underwent successful POBA, in at least one major coronary artery. A total of 469 subjects (273 men, 196 women) exhibited no restenosis after POBA for any of the coronary lesions, whereas 261 subjects (151 men, 110 women) manifested restenosis for all lesions. The genotypes for 40 polymorphisms of 34 genes were determined. Multivariate logistic regression analysis with adjustment for age, body mass index, and the prevalence of smoking, hypertension, diabetes mellitus, hyperuricemia, and hypercholesterolemia revealed that two polymorphisms (242C→T in the NADH/NADPH oxidase p22 phox (*p22-PHOX*) gene and 2136C→T in the thrombomodulin (*TM*) gene) in men and two polymorphisms (584G→A) in the paraoxonase 1 (*PON1*) gene and 2445G→A in the fatty acid–binding protein 2 (*FABP2*) gene in women were significantly associated with restenosis after POBA. The effects of these polymorphisms on restenosis were statistically independent of conventional risk factors for CAD. Taken together, genotyping of these and other polymorphisms may prove informative in the assessment of genetic risk factors for CAD after or before therapeutic interventions.

GENDER DIFFERENCES IN THE OUTCOME OF CVDS IN HUMANS

At the outset it is important to point out that normally there are a number of differences between male and female hearts even in the absence of pathology. For example, everything being equal, contractility is higher in women than men, and women maintain better myocardial mass. That this is so probably relates to existing differences in the expression of glycolytic and mitochondrial metabolic enzymes as previously noted[64] and/or to the prosurvival effects of E2 ERs on cardiomyocytes, which, as discussed earlier, may be mediated by ERα and PI3K-Akt–dependent pathways.[8]

Heart Failure

The relationship of gender to phenotype and severity of CVD variably affects different groups of the population. The Beta-Blocker Evaluation of Survival Trial (BEST) study specifically targeted women in HF, because in previous clinical trials, women have been underrepresented, and limitation of data in women had impeded a clear understanding of gender-related differences in patients with HF.[95] The BEST study analyzed 2708 randomly assigned patients with NYHA class III/IV and left ventricle ejection fraction (LVEF)< or = 0.35 in response to bucindolol versus placebo. Women ($n=593$) were compared with men ($n=2115$), and significant differences in baseline clinical and laboratory characteristics were found. Women were younger, more likely to be black, had a higher prevalence of nonischemic etiology, had higher right and LVEF, higher heart rate, greater cardiothoracic ratio, higher prevalence of left bundle branch block, lower prevalence of atrial fibrillation, and lower plasma norepinephrine level. Ischemic etiology and measures of severity of HF were found to be predictors of prognosis in women and men. The predictive values of various variables showed some differences; mainly, CAD and LVEF seem to be stronger predictors of prognosis in women, and in the nonischemic patients, women had a significantly better survival rate than men.

In another study, female patients with similar degrees of advanced, nonischemic LV dysfunction exhibited improved survival rates over men.[96] To further understand why CVD is often delayed and less common in women than in men, studies have further probed the extent of cell death in HF as a function of gender. Analysis of a small group of patients of both genders in HF revealed that the level of myocyte necrosis was 7-fold greater than apoptosis.[97] However, cell death was 2-fold higher in men than in women. Apoptosis increased 35-fold in women and 85-fold in men. The lower degree of cell death in women was associated with a longer duration of the cardiomyopathy, a later onset of cardiac decompensation, and a longer interval between HF and transplantation. On the other hand, Crabbe and associates[98] did not observe gender-based differences in cardiac or cellular remodeling among a large cohort of patients with DCM. However, in patients with ischemic cardiomyopathy ($n=50$), the heart weight index was significantly greater in men, along with a strong trend toward increased LV mass index. These gender differences in cardiac and LV mass were paralleled by marked gender differences in myocyte volume (36% greater in men than in women) in association with a 14% increase in resting cell length, suggesting that gender differences in cardiac remodeling in ischemic cardiomyopathy are mainly related to differences in cellular remodeling. The different response between ischemic and DCM imply that gender may influence the myocardial responses to injury.

As discussed in Chapter 15, HF in both clinical and experimental studies is invariably accompanied by abnormalities in Ca^{++}-handling that will decrease cardiac contractility. However, it is not known how gender may affect these changes. Dash and associates[99] have studied the potential contributory role of gender to SERCA cycling alterations, the levels of SERCA, phospholamban, and calsequestrin, as well as the site-specific phospholamban phosphorylation status, in a mixed gender population of failing ($n=14$) and donor ($n=15$) hearts. Myocardial protein levels and Ca^{++}-uptake parameters were analyzed, demonstrating that decreases in phosphorylated serine-16 in phospholamban were specific to male failing hearts, reflecting increases in the apparent affinity values of SERCA uptake for Ca^{++} compared with donor males. Although decreased levels of SERCA protein and phospholamban phosphorylation contribute to depressed SERCA uptake and LV dysfunction in HF, the subcellular abnormalities that may underlie these effects may not be uniform with respect to gender.

Dysrhythmias

It is well known that there are gender-specific cardiac dysrhythmias, although the underlying reasons for this are not yet known. Moreover, because women, compared with men, have been highly underrepresented in past controlled studies for primary and secondary prevention, it is presently difficult to assess the distribution by gender of type, incidence, and severity of dysrhythmias. Notwithstanding, it is generally accepted that from puberty on women show a higher basic heart rate, as well as a longer Q-T interval.[100] Supraventricular dysrhythmias (in women sinus and AV-nodal–reentry tachycardias, less frequently Wolff–Parkinson–White tachycardias) may show cyclical differences. The incidence of atrial fibrillation (AF) by gender is still debatable; whereas some investigators have reported that it occurs more frequently in women,[101] others indicate that the incidence is higher in men.[102,103] However, if we consider that the incidence of AF increased with aging and there are a greater number of women older than 75 years of age than men, then the number of men and women with AF is likely equal.[104,105] Compared with men, women with AF tend to be more symptomatic, and because it is frequently associated with HF, the outcome in women is worse.[100,101] Rienstra and associates[106] using data from the RACE (rate control versus electrical cardioversion) study also found that women had more AF-related complaints and their quality of life was worse than men. However, the cardiovascular morbidity and mortality were equal after a follow-up of nearly 3 y. On the other hand, women randomly assigned to rhythm control developed more end points, mainly HF, thromboembolic complications, and adverse effects of antidysrhythmic drugs than rate control randomly assigned female patients. During follow-up, compared with men, the quality of life in women remained worse. Because treatment did not change the quality of life in women, it was suggested that a rate control strategy might be considered in these patients.

Ventricular dysrhythmias are more prevalent in men, mainly in association with CAD. *Torsade de pointes* tachycardias seem to occur more frequently in women, because they have a higher incidence of acquired and congenital long-QT syndrome. On the other hand, sudden cardiac death (SCD) occurs more often in men, and in women it occurs later.[101] In comparison with men, female survivors of SCD have a lower frequency of spontaneous or inducible ventricular tachycardia. Significantly, the incidence of dysrhythmias is increased during pregnancy, and the management of pregnant patients poses a significant challenge.[100]

Although the response to therapies (either pharmacological and nonpharmacological) seems not to be affected by gender, the risks of pharmacological therapy may, in fact, be different in men and women. Further controlled studies are necessary to specifically address gender-related solutions for adequate risk stratification and therapy.

Cardiac Hypertrophy and Hypertension

Cardiac hypertrophy, frequently accompanied by cardiac fibrosis, and myocardial dysfunction are associated with gender-based differences, with higher mortality in men. Female patients with aortic stenosis exhibit better preservation of systolic function and increased LV hypertrophy than men.[107-109] Nevertheless, it is still unclear whether sex hormones, and by how much, are responsible for these differences in LV function and structural remodeling/hypertrophy.[110] For women it may be particularly important to regulate the development of cardiac hypertrophy, because female patients with a similar degree of nonischemic hypertrophy as men exhibit higher mortality rates.[111]

As we have noted in several chapters, mutations in sarcomeric proteins, depending on their effects on the structural and functional properties of the contractile unit of the heart, can lead to either HCM or DCM. Stefanelli and associates[112] reported a novel cardiac troponin T mutation (A171S) leading to DCM and SCD. In contrast to prior described mutations, the A171S mutation results in a significant gender difference in the severity of the observed phenotype with adult men demonstrating more severe LV dilatation and dysfunction than adult women.

Gender has a significant effect on arterial blood pressure (ABP), with pre-menopausal women having a lower ABP than age-matched men. Also, compared with pre-menopausal women, post-menopausal women have higher ABP, implying a modulating role in ABP for ovarian hormones.[113] Despite that, it is not known whether sex hormones are responsible for the gender-associated differences in ABP and whether ovarian hormones account for differences in blood pressure in pre-menopausal compared with post-menopausal women. On the other hand, sex hormone receptors have been identified in vascular endothelium and smooth muscle, and sex hormone interaction with cytosolic/nuclear receptors initiates long-term genomic effects that stimulate endothelial cell growth but inhibit smooth muscle proliferation. As mentioned previously, and also as Khalil pointed out,[114] the activation of sex hormone receptors on the plasma membrane triggers non-genomic effects that stimulate endothelium-dependent vascular relaxation by means of NO-cGMP, prostacyclin-cAMP, and hyperpolarization pathways. In addition, sex hormones cause endothelium-independent inhibition of vascular smooth muscle contraction, $[Ca^{++}]_i$ and protein kinase C. These vasorelaxant/vasodilator effects suggested that sex hormone therapy might have a positive vascular effect in natural and surgically induced hypogonadism. Nevertheless, in some clinical trials, HRT showed none or minimal benefits in post-menopausal women with hypertension; whether this was due to the type/dose of sex hormone, subject's age, or other factors remains unclear.

The potential relationship between estrogen-related gene polymorphisms and blood pressure had been further evaluated recently in a Framingham Heart Study offspring cohort.[115] Untreated cross-sectional and longitudinal blood pressure was correlated with polymorphisms in genes encoding the ERα (*ER1*), ERβ (*ER2*), aromatase (*CYP19A1*), and nuclear receptor coactivator 1 (*NCOA1*). In men, systolic blood pressure and pulse pressure were associated with two polymorphisms in *ER1*, whereas pulse pressure was also associated with variations in *NCOA1* and *CYP19A1*. Polymorphisms in *ER1*, *CYP19A1*, and *NCOA1* were associated with diastolic blood pressure in women. Although the underlying relations between genes involved in estrogen action and hypertension are incompletely understood, these findings suggest that there is a gender-specific contribution of estrogen-related genes to blood pressure variation. Nevertheless, further studies are warranted to confirm these results.

Coronary Artery Disease

The diagnosis of CAD in women is more difficult than in men because of the lower specificity of symptoms and diagnostic accuracy of noninvasive testing. Wiviott and associates[116] have examined the relationship between gender and cardiac biomarkers in patients with unstable angina (UA) and non-ST–segment elevation myocardial infarction (NSTEMI). In patients with UA/NSTEMI, there was a gender-specific difference in presenting biomarkers. Men were more likely to have elevated creatine kinase-MB and troponins and markers of myocardial necrosis, whereas women were more likely to have elevated C-reactive protein (hs-CRP) and BNP. These findings imply that a multimarker approach may aid the initial risk assessment of UA/NSTEMI and that further research is necessary to explain whether gender-related pathophysiological differences exist in the presentation of acute coronary syndromes.

Increases in vascular, ventricular systolic, and ventricular diastolic elastance (stiffness) may contribute to the pathogenesis of HF with preserved ejection fraction (HFnlEF).[117] Advancing age and female gender are associated with increases in vascular ventricular systolic, and diastolic stiffness even in the absence of CVD. This combined ventricular-vascular stiffening probably contributes to the increased prevalence of HFnlEF in the elderly, particularly in women.

Recent randomized clinical trials in post-menopausal women have failed to demonstrate that hormone replacement therapy (HRT) is beneficial for CAD secondary or primary prevention. Endogenous T in men is correlated positively with HDL-C and negatively with LDL-C, triglycerides, fibrinogen, and plasminogen activator type 1 PAI-1. On the other hand, these relationships are reversed in women. In addition, hypoandrogenemia in men and hyperandrogenemia in women are often confounded by central obesity and insulin resistance, making these associations not helpful with respect to deciding whether androgens have a direct pro- or antiatherogenic effect. Although a relationship in men of T level with the progression of atherosclerosis, accumulation of visceral adipose tissue, and other risk factors for MI have been reported, neither the level of T nor of estrogen was found to be predictive of coronary events in any of the eight prospective studies that have been carried out.[118] Whereas gender difference in the incidence of MI tends to support the view that T promotes and/or estrogen prevents MI, it is surprising that cross-sectional hormone administration and prospective studies have suggested that in men T may prevent and estrogen may promote MI, a so-called estrogen–androgen paradox, meaning that endogenous sex hormones may relate both to atherosclerotic CVD and its risk factors oppositely in women and men.

In their evaluation of the medical literature (by MEDLINE searches), Wu and von Eckardstein[31] found that the gender difference in CAD could not be explained on the basis of endogenous sex hormone exposure (because none of the epidemiological studies in the literature have showed a positive association between T and CAD in men). In women also, there is not definitive evidence that endogenous T plays a causal or protective role for CAD, although patients with the polycystic ovarian syndrome (PCOS) have shown an adverse risk profile. Furthermore, observational studies on dehydroepiandrosterone (DHEA) do not support the hypothesis that DHEA deficiency is a risk factor for CAD in men or women.[119]

THERAPY

On the basis of numerous human and animal studies, it has been postulated that HRT has potent cardiovascular effects against ischemic or nonischemic injury of the heart and vessels and that the loss of estrogen/progesterone and/or unopposed androgen may promote CVD in post-menopausal women. However, recent randomized, double-blind controlled trials, using a number of clinical CAD end points, have shown that previous recommendations that most post-menopausal women should be treated with HRT may not hold true. Results of the previously mentioned WHI and HERS studies,[2,3] which were considered definitive in the search for answers about health benefits and risks of post-menopausal HRT, as well as other randomized trials of HRT, have failed to show any significant effect in lowering cardiovascular events. On the contrary, these studies have engendered confusion among clinicians, researchers, and the public in general because conjugated equine estrogens plus medroxyprogesterone acetate did not only fail to reduce the risk but rather increased the risk of MI in both primary (WHI) and secondary (HERS) prevention study. These results have led the American Heart Association to issue the statement that hormone therapy should not be used for the primary nor secondary prevention of CVD in menopausal women. On the other hand, and despite the results from the preceding clinical trials that HRT does not afford cardioprotection, there are numerous observational clinical and animal studies that found profound beneficial effects of HRT on CVD risk.

Treatment with the SERMs raloxifene and tamoxifen has been considered as possible alternatives to HRT, because they favorably affect several markers of cardiovascular risk in healthy post-menopausal women. Raloxifene has a potentially beneficial fibrinogen-lowering effect not seen with conventional HRT and does not have the potentially adverse effects of HRT on triglycerides and CRP; however, it seems to lack the potentially beneficial effects of HRT on HDL cholesterol levels and plasminogen activation inhibitor-1.[120] The significance of these differences is still unclear, and further evidence that raloxifene or HRT reduces the risk of CVD is critically needed.

In an excellent review on hormone therapy published in the aftermath of the WHI studies, Turgeon and associates[121] critically examined the roles of ERs and progesterone receptors (PRs) as potential therapeutic targets in menopausal women. Examination of the mechanism of action of these receptors reveals that the existing pharmaceuticals are relatively primitive and that by using mechanism-based approaches for drug discovery, such as ER subtype-specific compounds, improved drugs may emerge.

Selective receptor modulators (SRMs) are receptor ligands that exhibit agonistic or antagonistic properties in the cell. The prototype of SRM is tamoxifen, the first SERM that has been successfully tested for the prevention of breast cancer, capable of activating or inhibiting ER action dependent on context.[122]

SRMs, by mediating changes in the conformation of the ligand-binding domains of nuclear receptors, influence their abilities to interact with coregulatory proteins, such as coactivators and corepressors that are critical for gene transcription.[123] The relative balance of coactivator and corepressor molecules within a given target cell has been proposed to determine the relative agonist or antagonist action of the SRMs in a given cell environment. Recent evidence has also implicated the importance of cell signaling pathways in influencing the activity and subcellular localization of coactivators and corepressors and nuclear receptors, contributing to gene-, cell-, and tissue-specific responses to SRM ligands and SRM phenotypic effect.

Because SERMs act as either ER agonists or antagonists in a tissue-specific manner, it may be possible to selectively distinguish the cardiovascular effects of estrogen from unfavorable stimulatory effects on the breast and endometrium. Much of the data regarding tamoxifen's effects on the cardiovascular system in post-menopausal women were actually obtained from breast cancer trials. These studies showed fewer fatal myocardial events in women randomly assigned to tamoxifen compared with women assigned to placebo.[124] Raloxifene (a second-generation SERM) increases bone mineral density, has no effect on the endometrium, and holds high promise for the prevention of breast cancer and potential in CVD treatment, given its favorable effect on lipid profile as noted previously. Evidence from the MORE (Multiple Outcomes of Raloxifene) trial indicates that raloxifene offers some protection to women with CVD or to those who are at high risk.[124] Whether raloxifene significantly reduces the risk of CAD in post-menopausal women will require a clinical trial (currently underway) with hard clinical end points.

The addition of raloxifene, and probably soon of lasofoxifene (a third-generation SERM), a selective ER modulator replacing the progestin in HRT, to the armamentarium of new therapies may prove successful and clinically applicable because of their estrogenic activity in bone and action in the cardiovascular system while opposing estrogen action in the uterus and breast. Although current SERMs cannot treat vasomotor symptoms associated with menopause, there is evidence that the combination of a classical estrogen with a SERM may bring relief with fewer side effects than are seen with current treatments.[125]

According to Mendelsohn and Karas, both methodological and biological issues have contributed to the discordance between observational and randomized clinical trials, with both the age at which women initiate HRT and the combination of hormones used, particularly critical factors in explaining the seemingly contradictory results.[35] Studies in rabbits have provided supportive evidence that HRT is beneficial in the prevention of atherosclerosis, only if replacement is begun before the development of severe disease.[126,127] In addition, studies in mice also have demonstrated that estrogen inhibits the initiation of the fatty streak, although it does not alter the progression of established lesions through the stages of instability and healing.[128]

Interestingly, data from monkey studies have brought attention to the pre-menopausal estrogen deficiency that results in premature coronary artery atherosclerosis.[129] Findings in monkey studies were subsequently confirmed for pre-menopausal women in the NHLBI-sponsored Women's Ischemia Syndrome Evaluation (WISE) Study. Prevention of coronary atherosclerosis was achieved when estrogens were administered soon after the development of estrogen deficiency. Moreover, convincing data from monkey studies have shown the total loss of the beneficial effects of estrogen if treatment is delayed for a period equal to 6 post-menopausal years for women. The monkey model has been used to identify the most effective treatment regimen to prevent the progression of CAD; this included an estrogen-containing oral contraceptive initiated during the peri-menopausal transition, followed directly by post-menopausal treatment. The collected findings from the monkey studies suggest little or no effects of estrogen-only treatment; whereas a combined estrogen and progestin treatment clearly increased the risk of breast cancer.

Williams and Suparto[130] have also suggested that estrogen replacement therapy (ERT)/HRT is effective in inhibiting progression of early-stage (fatty streak) atherosclerosis in both human and monkey studies, whereas ERT/HRT is much less effective in advanced atherosclerosis. Moreover, the results from monkey studies are consistent with those studies in women wherein ERT/HRT was initiated in the post-menopausal period with different initial amounts of atherosclerosis. It may be possible that ERT/HRT is more cardioprotective in younger post-menopausal women with less CAD, and less effective in women with established CAD. Furthermore, it is important to identify the cardiovascular risk/benefit in large groups of post-menopausal women on the basis of a number of parameters such as age, severity of pre-existing atherosclerosis, and other risk factors.

In regard to T, short-term studies suggest that T treatment may improve exercise ECG in men with established CAD. On the other hand, in most animal studies, it had been found that exogenous T and DHEA exert neutral or beneficial effects on atherosclerosis in male, but have detrimental effects in female animals.

In regard to the use of androgens in the clinical prevention or treatment of CAD, no definitive course can be taken, because their effects on cardiovascular mortality or morbidity have not been evaluated in prospective, double-blind, controlled studies. According to Wu and von Eckardstein,[31] although preliminary data suggest that exogenous androgens may induce short-term improvements in the electrocardiographic changes of men with CAD, they induce both apparently beneficial and deleterious effects on cardiovascular risk factors by decreasing serum levels of HDL-C, plasminogen activator type 1 (apparently deleterious), lipoprotein (a), fibrinogen, insulin, leptin, and visceral fat mass (apparently beneficial) in men and in women. Suprapbysiological concentrations of T stimulate vasorelaxation, but at physiological concentrations, beneficial, neutral, and detrimental effects on vascular reactivity have been observed. In conclusion, the inconsistent data, which can only be partly explained by differences in dose and source of androgens, militate against a meaningful assessment of the net effect of T on atherosclerosis. Thus, on the basis of current evidence, there is not sufficient justification for uncontrolled use of T or DHEA for the prevention or treatment of CAD. Further pilot studies in humans are also needed to examine cardioprotective aspects of T treatment that have been demonstrated in both *in vivo* and *in vitro* animal and cell studies.

Compared with men, women with AMI have a higher risk for death.[131–137] Although it is presently not known which factors are responsible for this difference, there are numerous reports that significant gender differences exist in regard to specific therapies and interventions. Outcomes after intervention with angioplasty versus stenting, with and without abciximab (a chimeric mouse–human monoclonal antibody directed against the GP IIb/IIIa receptor), in a population of 2082 individuals (27% women) after AMI were assessed in the CADILLAC trial (controlled abciximab and device investigation to lower late angioplasty complications), in which women had higher mortality rates that men.[138] Although similar to other studies showing that women with AMI undergoing primary angioplasty have higher rates of morbidity and mortality than do men, this study may have limited value, because the number of subjects under study was too small to establish whether stenting versus glycoprotein IIb/IIIa inhibitors could be favorable or not in women. On the other hand, it suggested that the reduction of the time to reperfusion in women (with earlier administration of abciximab) might have a positive effect and improve the prognosis. Clearly, this is an important area for further research.

An important aspect of therapy that needs to be kept in mind regarding drug treatment is that not only are there gender differences in the clinical manifestations and progression of CVD but also that female gender has been shown to be a risk factor for significant adverse drug reactions. Anderson reported that of the 300 new drug applications received by the Food and Drug Administration (FDA) between 1995 and 2000, only 163 included an analysis according to gender, and 11 of these drugs showed a significant difference in pharmacokinetics between men and women.[139] The response of women and men to drugs seems to be related to lower body weight, smaller organ sizes, and a higher proportion of fat in women compared with men. Furthermore, hormone levels and differences in metabolism may affect the absorption and elimination of drugs in women.[140]

Administration of a number of important cardiovascular drugs can have entirely different effects on the basis of gender. For example, there is evidence that women are more prone than men to develop *torsades de pointes* during administration of cardiovascular drugs that prolong cardiac repolarization; the incidence of drug-induced *torsades de pointes* is higher in women than in men, with a female prevalence of 70% of the 332 cases reported in 24 published studies.[140]

Gender differences in salicylate metabolism have also been reported, although whether these differences account for dissimilar incidence of adverse or protective effects in men and women is not known.[141] Moreover, the WHI Study failed to show a beneficial effect for low-dosage aspirin in the prevention of MI in women. In addition, data from five prior trials involving 55,580 participants with no history of heart disease showed that although aspirin therapy was associated with a significant reduction in the risk of stroke, there was no reduction in the risk of MI in women.[141]

Furthermore, it is worth noting that genetic polymorphisms in the expression of several cytochrome P450 (CYP) enzymes involved in drug metabolism may influence the side effects and the efficacy of several agents. Gender differences in enzyme activities have been demonstrated for several CYP metabolized drugs, including diltiazem, nifedipine, and verapamil, with a significantly increased activity in women compared with men.[142]

CONCLUSION

In the Western world, CVD is the major cause of morbidity and mortality for both men and women, and although uncommon in the pre-menopausal women, the incidence significantly increases during menopause. Gender differences manifest long before cardiac pathology appears in men and women, and besides those related to health, life span, and cognitive abilities, both genders also exhibit different responses to diseases such as anemia, CAD, hypertension, and renal dysfunction. That there are gender differences in the molecular and cellular physiology of the heart and blood vessels in health and disease is unquestionable, and it is increasingly evident that these differences must be known and kept in mind when considering potential approaches to therapy

Our understanding of the action, biology, and pharmacology of sex hormones, SERMs, and their receptors in the cardiovascular system is rapidly progressing, but further basic research is required to pinpoint how selectivity is achieved in the cardiovascular system and how that can be exploited in the design of new estrogen and progestin analogs, as well as in treatment protocols. Moreover, when considering treatment, it is important to keep in mind that the effects of steroid hormones are subject to multiple variables such as preparation, dose, sequence of administration, duration of post-menopausal hormone deficit, and tissue-specific and situation-dependent context of treatment. Genetic background, age, and health status of women at the time of treatment may also influence their responses. Furthermore, future clinical trials that fully integrate the complex pleiotropic actions of estrogens and progesterone and their analogs may provide further insights regarding post-menopausal roles for hormone therapy.

Recent studies that used microarray analysis and other methods of gene profiling in animal models have made substantial progress toward explaining the molecular differences between mammalian sexes in a variety of tissues, including, brain, liver, and kidney, and toward identifying the transcription factors that regulate gender-biased gene expression.[143] Although only a few tissues in rodent models have been examined for gender-specific gene expression, it is likely that other tissues, including the human heart, may contain genes that are differentially expressed in each gender. Understanding the differences in protein levels and their alterations among men and women may shed more light on the cardiovascular dimorphism, because proteins and post-translational modifications may be important to gender-specific physiology.

Notwithstanding, the great and continuous contribution of research in animal models toward our understanding of the multifaceted mechanisms of sex hormones and the gender effect, we must point out, nonetheless, that it is difficult to faithfully extrapolate the results obtained from the bench to humans. We need more studies focused on human subjects per se and to improve communication and interactions between laboratory and clinical researchers to decipher further the cardiovascular biology of gender differences and to re-evaluate what was wrong in previous approaches to HRT. Because there are numerous gender-related differences in the outcome and susceptibility to a variety of CVDs, we need to identify those pathways that mediate these responses. Also, many of the differences seen in human males and females are not apparent until they reach adulthood, even old age. Therefore, this is a variable to keep in mind when comparing human to reported laboratory studies, where animals are comparatively much younger.[72] Also, it is important to underscore the notion that gender plays a significant role in the response to drug therapy, and this reinforces the need for further research and clinical trials. Gender-dependent pharmacodynamic effects have been identified. However, the role of pharmacokinetics versus pharmacodynamics is unclear, because of the impact of pharmacogenetics on both.[139] A thorough understanding of pharmacogenetics, as well as the molecular and cellular biology underlying these gender differences, is necessary if we are to similarly advance the diagnosis and treatment of CVDs in both genders.

SUMMARY

- Steroid hormones, including the three gonadal hormones estrogen, progesterone, and testosterone and their receptors, are the definitive regulators of cardiovascular gender differences.
- Modulation of cardiovascular phenotype, either genetically and/or biochemically, in animal models has emerged as a powerful tool for cardiovascular research.
- Caution is warranted when translating research data from animal models, mainly rodent, to humans regarding the pathogenesis of CVD and the response to different modulatory factors/therapies.
- Gender differences in cardiovascular morbidity and mortality have been attributed to protective actions of female hormones, as suggested by increased cardiovascular risk in menopausal women.
- The protective action of estrogen in the cardiovascular system seems to be mediated indirectly by its effect on serum lipoprotein and triglyceride profiles, on the expression of coagulant and fibrinolytic proteins, and by a direct effect on the vessel wall itself.
- Estrogen genomic effect seems to result by the interaction of its receptors with specific target sequences of DNA (estrogen response elements [ERE]).
- Estrogen non-genomic effects are independent of gene transcription or protein synthesis and involve steroid-induced modulation of cytoplasmic or of cell membrane–bound regulatory proteins.
- ERs have been found both in the cytoplasm and nuclei of human hearts, and both receptors seem to be up-regulated by myocardial pressure load.
- Human exposure to testosterone does not shorten life span in either gender, although androgen exposure in early life may predispose males to premature atherosclerosis.
- Testosterone activates both androgen and estrogen receptors in cardiovascular tissues and mediates cardiomyocyte responses both in physiological situations and in hypertrophy-related cardiac diseases.
- Recent randomized, double-blind, controlled clinical trials, using a number of clinical coronary heart disease end points, have shown that previous recommendations that most post-menopausal women should be treated with HRT may not be warranted.
- Selective estrogen receptor modulators (SERM) have been developed to attain the cardioprotective effects of estrogens while minimizing the risks associated with HRT.
- Gender-dependent pharmacodynamic effects have been identified. However, the role of pharmacokinetics versus pharmacodynamics is unclear, because of the impact of pharmacogenetics on both.

References

1. Babiker, F. A., De Windt, L. J., van Eickels, M., Grohe, C., Meyer, R., and Doevendans, P. A. (2002). Estrogenic hormone action in the heart: regulatory network and function. *Cardiovasc. Res.* **53,** 709–719.
2. Grady, D., Herrington, D., Bittner, V., Blumenthal, R., Davidson, M., Hlatky, M., Hsia, J., Hulley, S., Herd, A., Khan, S., Newby, L. K., Waters, D., Vittinghoff, E., and Wenger, N. (2002). Cardiovascular disease outcomes during 6.8 years of hormone therapy: heart and estrogen/progestin Replacement Study follow-up (HERS II). *JAMA* **288,** 49–57.

3. Manson, J. E., Hsia, J., Johnson, K. C., Rossouw, J. E., Assaf, A. R., Lasser, N. L., Trevisan, M., Black, H. R., Heckbert, S. R., Detrano, R., Strickland, O. L., Wong, N. D., Crouse, J. R., Stein, E., Cushman, M, and the Women's Health Initiative Investigators. (2003). Estrogen plus progestin and the risk of coronary heart disease. *N. Engl. J. Med.* **349,** 523–534.
4. Lin, A. L., Schultz, J. J., Brenner, R. M., and Shain, S. A. (1990). Sexual dimorphism characterizes baboon myocardial androgen receptors but not myocardial estrogen and progesterone receptors. *J. Steroid Biochem. Mol. Biol.* **37,** 85–95.
5. Nordmeyer, J., Eder, S., Mahmoodzadeh, S., Martus, P., Fielitz, J., Bass, J., Bethke, N., Zurbrugg, H. R., Pregla, R., Hetzer, R., and Regitz-Zagrosek, V. (2004). Upregulation of myocardial estrogen receptors in human aortic stenosis. *Circulation* **110,** 3270–3275.
6. Simoncini, T., Mannella, P., Fornari, L., Caruso, A., Varone, G., and Genazzani, A. R. (2004). Genomic and non-genomic effects of estrogens on endothelial cells. *Steroids* **69,** 537–542.
7. Lu, Q., Pallas, D. C., Surks, H. K., Baur, W. E., Mendelsohn, M. E., and Karas, R. H. (2004). Striatin assembles a membrane signaling complex necessary for rapid, nongenomic activation of endothelial NO synthase by estrogen receptor alpha. *Proc. Natl. Acad. Sci. USA* **101,** 17126–17131.
8. Patten, R. D., Pourati, I., Aronovitz, M. J., Baur, J., Celestin, F., Chen, X., Michael, A., Haq, S., Nuedling, S., Grohe, C., Force, T., Mendelsohn, M. E., and Karas, R. H. (2004). 17beta-estradiol reduces cardiomyocyte apoptosis in vivo and in vitro via activation of phospho-inositide-3 kinase/Akt signaling. *Circ. Res.* **95,** 692–699.
9. Haynes, M. P., Sinha, D., Russell, K. S., Collinge, M., Fulton, D., Morales-Ruiz, M., Sessa, W. C., and Bender, J. R. (2000). Membrane estrogen receptor engagement activates endothelial nitric oxide synthase via the PI3-kinase-Akt pathway in human endothelial cells. *Circ. Res.* **87,** 677–682.
10. Hisamoto, K., Ohmichi, M., Kurachi, H., Hayakawa, J., Kanda, Y., Nishio, Y., Adachi, K., Tasaka, K., Miyoshi, E., Fujiwara, N., Taniguchi, N., and Murata, Y. (2001). Estrogen induces the Akt-dependent activation of endothelial nitric-oxide synthase in vascular endothelial cells. *J. Biol. Chem.* **276,** 3459–3467.
11. Simoncini, T., Hafezi-Moghadam, A., Brazil, D. P., Ley, K., Chin, W. W., and Liao, J. K. (2000). Interaction of oestrogen receptor with the regulatory subunit of phosphatidylinositol-3-OH kinase. *Nature* **407,** 538–554.
12. Pfeilschifter, J., Koditz, R., Pfohl, M., and Schatz, H. (2002). Changes in proinflammatory cytokine activity after menopause. *Endocr. Rev.* **23,** 90–119.
13. van Eickels, M., Grohe, C., Cleutjens, J. P., Janssen, B. J., Wellens, H. J., and Doevendans, P. A. (2001). 17beta-estradiol attenuates the development of pressure-overload hypertrophy. *Circulation* **104,** 1419–1423.
14. Booth, E. A., Marchesi, M., Kilbourne, E. J., and Lucchesi, B. R. (2003). 17Beta-estradiol as a receptor-mediated cardioprotective agent. *J. Pharmacol. Exp. Ther.* **307,** 395–401.
15. Booth, E. A., Obeid, N. R., and Lucchesi, B. R. (2005). Activation of estrogen receptor-alpha protects the in vivo rabbit heart from ischemia-reperfusion injury. *Am. J. Physiol. Heart Circ. Physiol.* **289,** H2039–H2047.
16. Gruber, C. J., Gruber, D. M., Gruber, I. M., Wieser, F., and Huber, J. C. (2004). Anatomy of the estrogen response element. *Trends Endocrinol. Metab.* **15,** 73–78.
17. Mendelsohn, M. E. (2000). Mechanisms of estrogen action in the cardiovascular system. *J. Steroid Biochem. Mol. Biol.* **74,** 337–343.
18. Gabel, S. A., Walker, V. R., London, R. E., Steenbergen, C., Korach, K. S., and Murphy, E. (2005). Estrogen receptor beta mediates gender differences in ischemia/reperfusion injury. *J. Mol. Cell Cardiol.* **38,** 289–297.
19. Pelzer, T., Loza, P. A., Hu, K., Bayer, B., Dienesch, C., Calvillo, L., Couse, J. F., Korach, K. S., Neyses, L., and Ertl, G. (2005). Increased mortality and aggravation of heart failure in estrogen receptor-beta knockout mice after myocardial infarction. *Circulation* **111,** 1492–1498.
20. Pelzer, T., Jazbutyte, V., Hu, K., Segerer, S., Nahrendorf, M., Nordbeck, P., Bonz, A. W., Muck, J., Fritzemeier, K. H., Hegele-Hartung, C., Ertl, G., and Neyses, L. (2005). The estrogen receptor-alpha agonist 16alpha-LE2 inhibits cardiac hypertrophy and improves hemodynamic function in estrogen-deficient spontaneously hypertensive rats. *Cardiovasc. Res.* **67,** 604–612.
21. Schönfelder, G. (2005). The biological impact of estrogens on gender differences in congestive heart failure. *Cardiovasc. Res.* **67,** 573–574.
22. Meyer, M. R., Haas, E., and Barton, M. (2006). Gender differences of cardiovascular disease. New perspectives for estrogen receptor signaling. *Hypertension* **47,** 1019–1026.
23. Snyder, P. J., Peachey, H., Berlin, J. A., Hannoush, P., Haddad, G., Dlewati, A., Santanna, J., Loh, L., Lenrow, D. A., Holmes, J. H., Kapoor, S. C., Atkinson, L. E., and Strom, B. L. (2000). Effects of testosterone replacement in hypogonadal men. *J. Clin. Endocrinol. Metab.* **85,** 2670–2677.
24. Tenover, J. L. (1998). Male hormone replacement therapy including "andropause." *Endocrinol. Metab. Clin. North Am.* **27,** 969–987.
25. Kenny, A. M., Prestwood, K. M., Gruman, C. A., Marcello, K. M., and Raisz, L. G. (2001). Effects of transdermal testosterone on bone and muscle in older men with low bioavailable testosterone levels. *J. Gerontol. A Biol. Sci. Med. Sci.* **56,** M266–M272.
26. Snyder, P. J., Peachey, H., Hannoush, P., Berlin, J. A., Loh, L., Holmes, J. H., Dlewati, A., Staley, J., Santanna, J., Kapoor, S. C., Attie, M. F., Haddad, J. G., Jr., and Strom, B. L. (1999). Effect of testosterone treatment on bone mineral density in men over 65 years of age. *J. Clin. Endocrinol. Metab.* **84,** 1966–1972.
27. Sih, R., Morley, J. E., Kaiser, F. E., Perry, H. M., III, Patrick, P., and Ross, C. (1997). Testosterone replacement in older hypogonadal men: a 12-month randomized controlled trial. *J. Clin. Endocrinol. Metab.* **82,** 1661–1667.
28. Dobs, A. S., Meikle, A. W., Arver, S., Sanders, S. W., Caramelli, K. E., and Mazer, N. A. (1999). Pharmacokinetics, efficacy, and safety of a permeation-enhanced testosterone transdermal system in comparison with bi-weekly injections of testosterone enanthate for the treatment of hypogonadal men. *J. Clin. Endocrinol. Metab.* 84, 3469–3478.
29. Rhoden, E. L., and Morgentaler, A. (2004). Risks of testosterone-replacement therapy and recommendations for monitoring. *N. Engl. J. Med.* 350, 482–492.
30. Liu, P. Y., Death, A. K., and Handelsman, D. J. (2003). Androgens and cardiovascular disease. *Endocr. Rev.* **24,** 313–340.
31. Wu, F. C., and von Eckardstein, A. (2003). Androgens and coronary artery disease. *Endocr. Rev.* **24,** 183–217.

32. English, K. M., Mandour, O., Steeds, R. P., Diver, M. J., Jones, T. H., and Channer, K. S. (2000). Men with coronary artery disease have lower levels of androgens than men with normal coronary angiograms. *Eur. Heart J.* **21,** 890–894.
33. Smeets, L., and Legros, J. J. (2004). The heart and androgens. *Ann. Endocrinol. (Paris)* **65,** 163–170.
34. Mendelsohn, M. E., and Rosano, G. M. (2003). Hormonal regulation of normal vascular tone in males. *Circ. Res.* **93,** 1142–1145.
35. Mendelsohn, M. E., and Karas, R. H. (2005). Molecular and cellular basis of cardiovascular gender differences. *Science* **308,** 1583–1587.
36. Heinlein, C. A., and Chang, C. (2002). The roles of androgen receptors and androgen-binding proteins in nongenomic androgen actions. *Mol. Endocrinol.* **16,** 2181–2187.
37. Falkenstein, E., Tillmann, H. C., Christ, M., Feuring, M., and Wehling, M. (2000). Multiple actions of steroid hormones-a focus on rapid, nongenomic effects. *Pharmacol. Rev.* **52,** 513–556.
38. Gerdes, D., Wehling, M., Leube, B., and Falkenstein, E. (1998). Cloning and tissue expression of two putative steroid membrane receptors. *Biol. Chem.* **379,** 907–911.
39. Bernauer, S., Wehling, M., Gerdes, D., and Falkenstein, E. (2001). The human membrane progesterone receptor gene: genomic structure and promoter analysis. *DNA Seq.* **12,** 13–25.
40. Murphy, J. G., and Khalil, R. A. (1999). Decreased [Ca(2+)](i) during inhibition of coronary smooth muscle contraction by 17ß-estradiol, progesterone, and testosterone. *J. Pharmacol. Exp. Ther.* **291,** 44–52.
41. English, K. M., Jones, R. D., Jones, T. .H, Morice, A. H., and Channer, K. S. (2002). Testosterone acts as a coronary vasodilator by a calcium antagonistic action. *J. Endocrinol. Invest.* **25,** 455–458.
42. Ding, A. Q., and Stallone, J. N. (2001). Testosterone-induced relaxation of rat aorta is androgen structure specific and involves K+ channel activation. *J. Appl. Physiol.* **91,** 2742–2750.
43. Tep-areenan, P., Kendall, D. A., and Randall, M. D. (2002). Testosterone-induced vasorelaxation in the rat mesenteric arterial bed is mediated predominantly via potassium channels. *Br. J. Pharmacol.* **135,** 735–740.
44. Ceballos, G., Figueroa, L., Rubio, I., Gallo, G., Garcia, A., Martinez, A., Yanez, R., Perez, J., Morato, T., and Chamorro, G. (1999). Acute and nongenomic effects of testosterone on isolated and perfused rat heart. *J. Cardiovasc. Pharmacol.* **33,** 691–697.
45. Er, F., Michels, G., Gassanov, N., Rivero, F., and Hoppe, U. C. (2004). Testosterone induces cytoprotection by activating ATP-sensitive K+ channels in the cardiac mitochondrial inner membrane. *Circulation* **110,** 3100–3107.
46. Sanz-Gonzalez, S. M., Cano, A., Valverde, M. A., Hermenegildo, C., and Andres, V. (2004). Drug targeting of estrogen receptor signaling in the cardiovascular system: preclinical and clinical studies. *Curr. Med. Chem. Cardiovasc. Hematol. Agents* **2,** 107–122.
47. Venkov, C. D., Rankin, A. B., and Vaughan, D. E. (1996). Identification of authentic estrogen receptor in cultured endothelial cells. A potential mechanism for steroid hormone regulation of endothelial function. *Circulation* **94,** 727–733.
48. Pelzer, T., Shamim, A., Wolfges, S., Schumann, M., and Neyses, L. (1997). Modulation of cardiac hypertrophy by estrogens. *Adv. Exp. Med. Biol.* **432,** 83–89.
49. de Jager, T., Pelzer, T., Muller-Botz, S., Imam, A., Muck, J., and Neyses, L. (2001). Mechanisms of estrogen receptor action in the myocardium. Rapid gene activation via the ERK1/2 pathway and serum response elements. *J. Biol. Chem.* **276,** 27873–27880.
50. Nuedling, S., van Eickels, M., Allera, A., Doevendans, P., Meyer, R., Vetter, H., and Grohe, C. (2003). 17 Beta-estradiol regulates the expression of endothelin receptor type B in the heart. *Br. J. Pharmacol.* **140,** 195–201.
51. Pelzer, T., de Jager, T., Muck, J., Stimpel, M., and Neyses, L. (2002). Oestrogen action on the myocardium in vivo: specific and permissive for angiotensin-converting enzyme inhibition. *J. Hypertens.* **20,** 1001–1006.
52. Hall, J. M., Couse, J. F., and Korach, K. S. (2001). The multifaceted mechanisms of estradiol and estrogen receptor signaling. *J. Biol. Chem.* **276,** 36869–36872.
53. Goldstein, J., Sites, C. K., and Toth, M. J. (2004). Progesterone stimulates cardiac muscle protein synthesis via receptor-dependent pathway. *Fertil. Steril.* **82,** 430–436.
54. Sites, C. K., Tischler, M. D., Blackman, J. A., Niggel, J., Fairbank, J. T., O'Connell, M., and Ashikaga, T. (1999). Effect of short-term hormone replacement therapy on left ventricular mass and contractile function. *Fertil. Steril.* **71,** 137–143.
55. Darney, P. D. (1995). The androgenicity of progestins. *Am. J. Med.* **98,** 104S–110S.
56. Kimura, M., Sudhir, K., Jones, M., Simpson, E., Jefferis, A. M., and Chin-Dusting, J. P. (2003). Impaired acetylcholine-induced release of nitric oxide in the aorta of male aromatase-knockout mice, regulation of nitric oxide production by endogenous sex hormones in males. *Circ. Res.* **93,** 1267–1271.
57. Adams, M. R., Golden, D. L., Register, T. C., Anthony, M. S., Hodgin, J. B., Maeda, N., and Williams, J. K. (2002). The atheroprotective effect of dietary soy isoflavones in apolipoprotein E$^{-/-}$ mice requires the presence of estrogen receptor-alpha. *Arterioscler. Thromb. Vasc. Biol.* **22,** 1859–1864.
58. Zhu, Y., Bian, Z., Lu, P., Karas, R. H., Bao, L., Cox, D., Hodgin, J., Shaul, P. W., Thoren, P., Smithies, O., Gustafsson, J. A., and Mendelsohn, M. E. (2002). Abnormal vascular function and hypertension in mice deficient in estrogen receptor beta. *Science* **295,** 505–508.
59. Hodgin, J. B., Krege, J. H., Reddick, R. L., Korach, K. S., Smithies, O., and Maeda, N. (2001). Estrogen receptor alpha is a major mediator of 17beta-estradiol's atheroprotective effects on lesion size in Apoe$^{-/-}$ mice. *J. Clin. Invest.* **107,** 333–340.
60. Hodgin, J. B., and Maeda, N. (2002). Estrogen and mouse models of atherosclerosis. *Endocrinology* **143,** 4495–4501.
61. Herrera, V. L., Tsikoudakis, A., Didishvili, T., Ponce, L. R., Bagamasbad, P., Gantz, D., Herscovitz, H., Van Tol, A., and Ruiz-Opazo, N. (2004). Analysis of gender-specific atherosclerosis susceptibility in transgenic [hCETP] 25DS rat model. *Atherosclerosis* **177,** 9–18.
62. Gao, X. M., Agrotis, A., Autelitano, D. J., Percy, E., Woodcock, E. A., Jennings, G. L., Dart, A. M., and Du, X. J. (2003). Sex hormones and cardiomyopathic phenotype induced by cardiac beta 2-adrenergic receptor overexpression. *Endocrinology* **144,** 4097–4105.
63. Cross, H. R., Murphy, E., Koch, W. J., and Steenbergen, C. (2002). Male and female mice overexpressing the beta(2)-

adrenergic receptor exhibit differences in ischemia/reperfusion injury: role of nitric oxide. *Cardiovasc. Res.* **53**, 662–671.
64. Yan, L., Ge, H., Li, H., Lieber, S. C., Natividad, F., Resuello, R. R., Kim, S. J., Akeju, S., Sun, A., Loo, K., Peppas, A. P., Rossi, F., Lewandowski, E. D., Thomas, A. P., Vatner, S. F., and Vatner, D. E. (2004). Gender-specific proteomic alterations in glycolytic and mitochondrial pathways in aging monkey hearts. *J. Mol. Cell Cardiol.* **37**, 921–929.
65. Henning, S. L., Wambolt, R. B., Schonekess, B. O., Lopaschuk, G. D., and Allard, M. F. (1996). Contribution of glycogen to aerobic myocardial glucose utilization. *Circulation* **93**, 1549–1555.
66. Goodwin, G. W., Taylor, C. S., and Taegtmeyer, H. (1998). Regulation of energy metabolism of the heart during acute increase in heart work. *J. Biol. Chem.* **273**, 29530–29539.
67. Goodwin, G. W., Ahmad, F., Doenst, T., and Taegtmeyer, H. (1998). Energy provision from glycogen, glucose, and fatty acids on adrenergic stimulation of isolated working rat hearts. *Am. J. Physiol.* **274**, H1239–H1247.
68. Takagi, G., Asai, K., Vatner, S. F., Kudej, R. K., Rossi, F., Peppas, A., Takagi, I., Resuello, R. R., Natividad, F., Shen, Y. T., and Vatner, D. E. (2003). Gender differences on the effects of aging on cardiac and peripheral adrenergic stimulation in old conscious monkeys. *Am. J. Physiol. Heart Circ. Physiol.* **285**, H527–H534.
69. Xin, H. B., Senbonmatsu, T., Cheng, D. S., Wang, Y. X., Copello, J. A., Ji, G. J., Collier, M. L., Deng, K. Y., Jeyakumar, L. H., Magnuson, M. A., Inagami, T., Kotlikoff, M. I., and Fleischer, S. (2002). Oestrogen protects FKBP12.6 null mice from cardiac hypertrophy. *Nature* **416**, 334–338.
70. Du, X. J., Samuel, C. S., Gao, X. M., Zhao, L., Parry, L. J., and Tregear, G. W. (2003). Increased myocardial collagen and ventricular diastolic dysfunction in relaxin deficient mice: a gender-specific phenotype. *Cardiovasc. Res.* **57**, 395–404.
71. Holzenberger, M., Lenzner, C., Leneuve, P., Zaoui, R., Hamard, G., Vaulont, S., and Bouc, Y. L. (2001). Experimental IGF-I receptor deficiency generates a sexually dimorphic pattern of organ-specific growth deficits in mice, affecting fat tissue in particular. *Endocrinology* **142**, 4469–4478.
72. Leinwand, L. A. (2003). Sex is a potent modifier of the cardiovascular system. *J. Clin. Invest.* **112**, 302–307.
73. Douglas, P. S., Katz, S. E., Weinberg, E. O., Chen, M. H., Bishop, S. P., and Lorell, B. H. (1998). Hypertrophic remodeling: gender differences in the early response to left ventricular pressure overload. *J. Am. Coll. Cardiol.* **32**, 1118–1125.
74. Weinberg, E. O., Thienelt, C. D., Katz, S. E., Bartunek, J., Tajima, M., Rohrbach, S., Douglas, P. S., and Lorell, B. H. (1999). Gender differences in molecular remodeling in pressure overload hypertrophy. *J. Am. Coll. Cardiol.* **34**, 264–273.
75. Dash, R., Schmidt, A. G., Pathak, A., Gerst, M. J., Biniakiewicz, D., Kadambi, V. J., Hoit, B. D., Abraham, W. T., and Kranias, E. G. (2003). Differential regulation of p38 mitogen-activated protein kinase mediates gender-dependent catecholamine-induced hypertrophy. *Cardiovasc. Res.* **57**, 704–714.
76. Geisterfer-Lowrance, A. A., Christe, M., Conner, D. A., Ingwall, J. S., Schoen, F. J., Seidman, C. E., and Seidman, J. G. (1996). A mouse model of familial hypertrophic cardiomyopathy. *Science* **272**, 731–734.
77. Olsson, M. C., Palmer, B. M., Leinwand, L. A., and Moore, R. L. (2001). Gender and aging in a transgenic mouse model of hypertrophic cardiomyopathy. *Am. J. Physiol. Heart Circ. Physiol.* **280**, H1136–H1144.
78. Bridgman, P., Aronovitz, M. A., Kakkar, R., Oliverio, M. I., Coffman, T. M., Rand, W. M., Konstam, M. A., Mendelsohn, M. E., and Patten, R. D. (2005). Gender-specific patterns of left ventricular and myocyte remodeling after myocardial infarction in mice deficient in the angiotensin II type 1a receptor. *Am. J. Physiol. Heart Circ. Physiol.* **289**, H586–H592.
79. Li, Y., Kishimoto, I., Saito, Y., Harada, M., Kuwahara, K., Izumi, T., Hamanaka, I., Takahashi, N., Kawakami, R., Tanimoto, K., Nakagawa, Y., Nakanishi, M., Adachi, Y., Garbers, D. L., Fukamizu, A., and Nakao, K. (2004). Androgen contributes to gender-related cardiac hypertrophy and fibrosis in mice lacking the gene encoding guanylyl cyclase-A. *Endocrinology* **145**, 951–958.
80. Callies, F., Stromer, H., Schwinger, R. H., Bolck, B., Hu, K., Frantz, S., Leupold, A., Beer, S., Allolio, B., and Bonz, A. W. (2003). Administration of testosterone is associated with a reduced susceptibility to myocardial ischemia. *Endocrinology* **144**, 4478–4483.
81. Shearman, A. M., Cupples, L. A., Demissie, S., Peter, I., Schmid, C. H., Karas, R. H., Mendelsohn, M. E., Housman, D. E., and Levy, D. (2003). Association between estrogen receptor alpha gene variation and cardiovascular disease. *JAMA* **290**, 2317–2319.
82. Schuit, S. C., Oei, H. H., Witteman, J. C., Geurts van Kessel, C. H., van Meurs, J. B., Nijhuis, R. L., van Leeuwen, J. P., de Jong, F. H., Zillikens, M. C., Hofman, A., Pols, H. A., and Uitterlinden, A. G. (2004). Estrogen receptor alpha gene polymorphisms and risk of myocardial infarction. *JAMA* **291**, 2969–2977.
83. Newton-Cheh, C., and O'Donnell, C. J. (2004). Sex differences and genetic associations with myocardial infarction. *JAMA* **29**, 3008–3010.
84. Andersen, T. I., Heimdal, K. R., Skrede, M., Tveit, K., Berg, K., and Borresen, A. (1994). Oestrogen receptor (ESR) polymorphisms and breast cancer susceptibility. *Hum. Genet.* **94**, 665–670.
85. Kobayashi, S., Inoue, S., Hosoi, T., Ouchi, Y., Shiraki, Y., and Orimo, H. (1996). Association of bone mineral density with polymorphism of the estrogen receptor gene. *J. Bone Miner. Res.* **11**, 306–311.
86. Kobayashi, N., Fujino, T., Shirogane, T., Furuta, I., Kobamatsu, Y., Yaegashi, M., Sakuragi, N., and Fujimoto, S. (2002). Estrogen receptor alpha polymorphism as a genetic marker for bone loss, vertebral fractures and susceptibility to estrogen. *Maturitas* **41**, 193–201.
87. Sudhir, K., Chou, T. M., Chatterjee, K., Smith, E. P., Williams, T. C., Kane, J. P., Malloy, M. J., Korach, K. S., and Rubanyi, G. M. (1997). Premature coronary artery disease associated with a disruptive mutation in the estrogen receptor gene in a man. *Circulation* **96**, 3774–3777.
88. Matsubara, Y., Murata, M., Kawano, K., Zama, T., Aoki, N., Yoshino, H., Watanabe, G., Ishikawa, K., and Ikeda, Y. (1997). Genotype distribution of estrogen receptor polymorphisms in men and post-menopausal women from healthy and coronary populations and its relation to serum lipid levels. *Arterioscler. Thromb. Vasc. Biol.* **17**, 3006–3012.
89. Kunnas, T. A., Laippala, P., Penttila, A., Lehtimaki, T., and Karhunen, P. J. (2000). Association of polymorphism of human

oestrogen receptor gene with coronary artery disease in men: a necroscopy study. *BMJ* **321,** 273–274.

90. Herrington, D. M., Howard, T. D., Hawkins, G. A., Reboussin, D. M., Xu, J., Zheng, S. L., Brosnihan, B., Meyers, D. A., and Bleecker, E. R. (2002). Estrogen-receptor polymorphisms and effects of estrogen replacement on high-density lipoprotein cholesterol in women with coronary artery disease. *N. Engl. J. Med.* **346,** 1017–1018.

91. Lu, H., Higashikata, I., Inazu, A., Nohara, A., Yu, W., Shimuzu, M., and Mabuchi, H. (2002). Association of estrogen receptor-alpha gene polymorphisms with coronary artery disease in patients with familial hypercholesterolemia. *Arterioscler. Thromb. Vasc. Biol.* **22,** 817–823.

92. Mansur Ade, P., Nogueira, C. C., Strunz, C. M., Aldrighi, J. M., and Ramires, J. A. (2005). Genetic polymorphisms of estrogen receptors in patients with premature coronary artery disease. *Arch. Med. Res.* **36,** 511–517.

93. Ferrero, V., Ribichini, F., Matullo, G., Guarrera, S., Carturan, S., Vado, A., Vassanelli, C., Piazza, A., Uslenghi, E., and Wijns, W. (2003). Estrogen receptor-alpha polymorphisms and angiographic outcome after coronary artery stenting. *Arterioscler. Thromb. Vasc. Biol.* **23,** 2223–2228.

94. Horibe, H., Yamada, Y., Ichihara, S., Watarai, M., Yanase, M., Takemoto, K., Shimizu, S., Izawa, H., Takatsu, F., and Yokota, M. (2004). Genetic risk for restenosis after coronary balloon angioplasty. *Atherosclerosis* **174,** 181–187.

95. Ghali, J. K., Krause-Steinrauf, H. J., Adams, K. F., Khan, S. S., Rosenberg, Y. D., Yancy, C. W., Young, J. B., Goldman, S., Peberdy, M. A., and Lindenfeld, J. (2003). Gender differences in advanced heart failure: insights from the BEST study. *J. Am. Coll. Cardiol.* **42,** 2128–2134.

96. Adams, K. F., Jr., Sueta, C. A., Gheorghiade, M., O'Connor, C. M., Schwartz, T. A., Koch, G. G., Uretsky, B., Swedberg, K., McKenna, W., Soler-Soler, J., and Califf, R. M. (1999). Gender differences in survival in advanced heart failure. Insights from the FIRST study. *Circulation* **99,** 1816–1821.

97. Guerra, S., Leri, A., Wang, X., Finato, N., Di Loreto, C., Beltrami, C. A., Kajstura, J., and Anversa, P. (1999). Myocyte death in the failing human heart is gender dependent. *Circ. Res.* **85,** 867–869.

98. Crabbe, D. L., Dipla, K., Ambati, S., Zafeiridis, A., Gaughan, J. P., and Houser, S. R. (2003). Gender differences in post-infarction hypertrophy in end-stage failing hearts. *J. Am. Coll. Cardiol.* **41,** 300–306.

99. Dash, R., Frank, K. F., Carr, A. N., Moravec, C. S., and Kranias, E. G. (2001). Gender influences on sarcoplasmic reticulum Ca2+-handling in failing human myocardium. *J. Mol. Cell Cardiol.* **33,** 1345–1353.

100. Wolbrette, D., Naccarelli, G., Curtis, A., Lehmann, M., and Kadish, A. (2002). Gender differences in arrhythmias. *Clin. Cardiol.* **2,** 49–56.

101. Rogge, C., Geibel, A., Bode, C., and Zehender, M. (2004). Cardiac arrhythmias and sudden cardiac death in women. *Z. Kardiol.* **93,** 427–438.

102. Kannel, W. B., Wolf, P. A., Benjamin, E. J., and Levy, D. (1998). Prevalence, incidence, prognosis, and predisposing conditions for atrial fibrillation population-based estimates, *Am. J. Cardiol.* **82,** 2N–9N.

103. Kerr, C. R., and Humphries, K. (2005). Gender-related differences in atrial fibrillation. *J. Am. Coll. Cardiol.* **46,** 1307–1308.

104. Feinberg, W. M., Blackshear, J. L., Laupacis, A., Kronmal, R., and Hart, R. G. (1995). Prevalence, age distribution, and gender of patients with atrial fibrillation. Analysis and implications. *Arch. Intern. Med.* **155,** 469–473.

105. Levy, S., Maarek, M., Coumel, P., Guize, L., Lekieffre, J., Medvedowsky, J. L., and Sebaoun, A. (1999). Characterization of different subsets of atrial fibrillation in general practice in France: the ALFA study. The College of French Cardiologists. *Circulation* **99,** 3028–3035.

106. Rienstra, M., Van Veldhuisen, D. J., Hagens, V. E., Ranchor, A. V., Veeger, N. J., Crijns, H. J., Van Gelder, I. C., and the RACE Investigators. (2005). Gender-related differences in rhythm control treatment in persistent atrial fibrillation: data of the Rate Control Versus Electrical Cardioversion (RACE) study. *J. Am. Coll. Cardiol.* **46,** 1298–1306.

107. Aurigemma, G. P., Silver, K. H., McLaughlin, M., Mauser, J., and Gaasch, W. H. (1994). Impact of chamber geometry and gender on left ventricular systolic function in patients >60 years of age with aortic stenosis. *Am. J. Cardiol.* **74,** 794–798.

108. Douglas, P. S., Otto, C. M., Mickel, M. C., Labovitz, A., Reid, C. L., and Davis, K. B. (1995). Gender differences in left ventricle geometry and function in patients undergoing balloon dilatation of the aortic valve for isolated aortic stenosis. NHLBI Balloon Valvuloplasty Registry. *Br. Heart J.* **73,** 548–554.

109. Carroll, J. D., Carroll, E. P., Feldman, T., Ward, D. M., Lang, R. M., McGaughey, D., and Karp, R. B. (1992). Sex-associated differences in left ventricular function in aortic stenosis of the elderly. *Circulation* **86,** 1099–1107.

110. Bristow, M. R. (1998). Why does the myocardium fail? Insights from basic science. *Lancet* **352,** SI8–S14.

111. Liao, Y., Cooper, R. S., Mensah, G. A., and McGee, D. L. (1995). Left ventricular hypertrophy has a greater impact on survival in women than in men. *Circulation* **92,** 805–810.

112. Stefanelli, C. B., Rosenthal, A., Borisov, A. B., Ensing, G. J., and Russell, M. W. (2004). Novel troponin T mutation in familial dilated cardiomyopathy with gender-dependant severity. *Mol. Genet. Metab.* **83,** 188–196.

113. Dubey, R. K., Oparil, S., Imthurn, B., and Jackson, E. K. (2002). Sex hormones and hypertension. *Cardiovasc. Res.* **53,** 688–708.

114. Khalil, R. A. (2005). Sex hormones as potential modulators of vascular function in hypertension. *Hypertension* **46,** 249–254.

115. Peter, I., Shearman, A. M., Zucker, D. R., Schmid, C. H., Demissie, S., Cupples, L. A., Larson, M. G., Vasan, R. S., D'agostino, R. B., Karas, R. H., Mendelsohn, M. E., Housman, D. E., and Levy, D. (2005). Variation in estrogen-related genes and cross-sectional and longitudinal blood pressure in the Framingham Heart Study. *J. Hypertens.* **23,** 2193–2200.

116. Wiviott, S. D., Cannon, C. P., Morrow, D. A., Murphy, S. A., Gibson, C. M., McCabe, C. H., Sabatine, M. S., Rifai, N., Giugliano, R. P., DiBattiste, P. M., Demopoulos, L. A., Antman, E. M., and Braunwald, E. (2004). Differential expression of cardiac biomarkers by gender in patients with

unstable angina/non-ST-elevation myocardial infarction: a TACTICS-TIMI (Treat Angina with Aggrastat and determine Cost of Therapy with an Invasive or Conservative Strategy-Thrombolysis In Myocardial Infarction) substudy. *Circulation* **109,** 580–586.

117. Redfield, M. M., Jacobsen, S. J., Borlaug, B. A., Rodeheffer, R. J., and Kass, D. A. (2005). Age- and gender-related ventricular-vascular stiffening: a community-based study. *Circulation* **112,** 2254–2262.

118. Phillips, G. B. (2005). Is atherosclerotic cardiovascular disease an endocrinological disorder? The estrogen-androgen paradox. *J. Clin. Endocrinol. Metab.* **90,** 2708–2701.

119. Silvestri, A., Gambacciani, M., Vitale, C., Monteleone, P., Ciaponi, M., Fini, M., Genazzani, A. R., Mercuro, G., and Rosano, G. M. (2005). Different effect of hormone replacement therapy, DHEAS and tibolone on endothelial function in postmenopausal women with increased cardiovascular risk. *Maturitas* **50,** 305–311.

120. Walsh, B. W. (2001). The effects of estrogen and selective estrogen receptor modulators on cardiovascular risk factors. *Ann. N. Y. Acad. Sci.* **949,** 163–167.

121. Turgeon, J. L., McDonnell, D. P., Martin, K. A., and Wise, P. M. (2004). Hormone therapy: physiological complexity belies therapeutic simplicity. *Science* **304,** 1269–1273.

122. Smith, C. L., and O'Malley, B. W. (2004). Coregulator function: A key to understanding tissue specificity of selective receptor modulators. *Endocr. Rev.* **25,** 45–71.

123. Lewis, J. S., and Jordan, V. C. (2005). Selective estrogen receptor modulators (SERMs): Mechanisms of anticarcinogenesis and drug resistance. *Mutat. Res.* **591,** 247–263.

124. Vogelvang, T. E., van der Mooren, M. J., and Mijatovic, V. (2004). Hormone replacement therapy, selective estrogen receptor modulators, and tissue-specific compounds: cardiovascular effects and clinical implications. *Treat. Endocrinol.* **3,** 105–115.

125. Labrie, F., El-Alfy, M., Berger, L., Labrie, C., Martel, C., Belanger, A., Candas, B., and Pelletier, G. (2003). The combination of a novel selective estrogen receptor modulator with an estrogen protects the mammary gland and uterus in a rodent model: the future of postmenopausal women's health? *Endocrinology* **144,** 4700–4706.

126. Haarbo, J., and Christiansen, C. (1996). The impact of female sex hormones on secondary prevention of atherosclerosis in ovariectomized cholesterol-fed rabbits. *Atherosclerosis* **123,** 139–144.

127. Hanke, H., Kamenz, J., Hanke, S., Spiess, J., Lenz, C., Brehme, U., Bruck, B., Finking, G., and Hombach, V. (1999). Effect of 17-beta estradiol on pre-existing atherosclerotic lesions: role of the endothelium. *Atherosclerosis* **147,** 123–132.

128. Rosenfeld, M. E., Kauser, K., Martin-McNulty, B., Polinsky, P., Schwartz, S. M., and Rubanyi, G. M. (2002). Estrogen inhibits the initiation of fatty streaks throughout the vasculature but does not inhibit intra-plaque hemorrhage and the progression of established lesions in apolipoprotein E deficient mice. *Atherosclerosis* **164,** 251–259.

129. Clarkson, T. B., and Appt, S. E. (2005). Controversies about HRT—lessons from monkey models. *Maturitas* **51,** 64–74.

130. Williams, J. K., and Suparto, I. (2004). Hormone replacement therapy and cardiovascular disease: lessons from a monkey model of postmenopausal women. *ILAR J.* **45,** 139–146.

131. Weaver, W. D., White, H. D., Wilcox, R. G., Aylward, P. E., Morris, D., Guerci, A., Ohman, E. M., Barbash, G. I., Betriu, A., Sadowski, Z., Topol, E. J., and Califf, R. M. (1996). Comparisons of characteristics and outcomes among women and men with acute myocardial infarction treated with thrombolytic therapy: GUSTO-I investigators. *JAMA* **275,** 777–782.

132. Malacrida, R., Genoni, M., Maggioni, A. P., Spataro, V., Parish, S., Palmer, A., Collins, R., and Moccetti, T. (1998). A comparison of the early outcome of acute myocardial infarction in women and men. *N. Engl. J. Med.* **338,** 8–14.

133. Vaccarino, V., Berkman, L. F., and Krumholz, H. M. (2000). Long-term outcome of myocardial infarction in women and men: a population perspective. *Am. J. Epidemiol.* **152,** 965–973.

134. Watanabe, C. T., Maynard, C., and Ritchie, J. L. (2001). Comparison of short-term outcomes following coronary artery stenting in men versus women. *Am. J. Cardiol.* **88,** 848–852.

135. Stone, G. W., Grines, C. L., Browne, K. F., Marco, J., Rothbaum, D., O'Keefe, J., Hartzler, G. O., Overlie, P., Donohue, B., and Chelliah, N. (1995). Comparison of in-hospital outcomes in men and women treated by either thrombolytic therapy or primary coronary angioplasty for acute myocardial infarction. *Am. J. Cardiol.* **75,** 987–992.

136. Tamis-Holland, J. E., Palazzo, A., Stebbins, A. L., Slater, J. N., Boland, J., Ellis, S. G., and Hochman, J. S. (2004). (GUSTO II-B) Angioplasty Substudy. *Am. Heart J.* **147,** 133–139.

137. Zijlstra, F., Hoorntje, J. C., de Boer, M. J., Reiffers, S., Miedema, K., Ottervanger, J. P., van'T Hof, A. W., and Suryapranata, H. (1999). Long-term benefit of primary angioplasty as compared with thrombolytic therapy for acute myocardial infarction. *N. Engl. J. Med.* 341, 1413–1419.

138. Lansky, A. J., Pietras, C., Costa, R. A., Tsuchiya, Y., Brodie, B. R., Cox, D. A., Aymong, E. D., Stuckey, T. D., Garcia, E., Tcheng, J. E., Mehran, R., Negoita, M., Fahy, M., Cristea, E., Turco, M., Leon, M. B., Grines, C. L., and Stone, G. W. (2005). Gender differences in outcomes after primary angioplasty versus primary stenting with and without abciximab for acute myocardial infarction: results of the Controlled Abciximab and Device Investigation to Lower Late Angioplasty Complications (CADILLAC) trial. *Circulation* **111,** 1611–1618.

139. Anderson, G. D. (2005). Sex and racial differences in pharmacological response. Where is the evidence? Pharmacogenetics, pharmacokinetics, and pharmacodynamics. *J. Women's Health* **14,** 19–29.

140. Makkar, R. R., Fromm, B. S., Steinman, R. T., Meissner, M. D., and Lehmann, M. H. (1993). Female gender as a risk factor for torsade de pointe associated with cardiovascular drugs. *JAMA* **270,** 2590–2597.

141. Ridker, P. M., Cook, N. R., Lee, I. M., Gordon, D., Gaziano, J. M., Manson, J. A. E., Hennekens, C. H., and Buring, J. E. (2005). A randomized trial of low-dose aspirin in the primary prevention of cardiovascular disease in women. *N. Engl. J. Med.* **352,** 1293–1304.

142. Stramba-Badiale, M., and Priori, S. G. (2005). Gender-specific prescription for cardiovascular diseases? *Eur. Heart J.* **26,** 1571–1572.

143. Rinn, J. L., and Snyder, M. (2005). Sexual dimorphism in mammalian gene expression. *Trends Genet.* **21,** 298–305.

144. O'Connell, T. D., O'Connell, T. D., Ishizaka, S., Nakamura, A., Swigart, P. M., Rodrigo, M. C., Simpson, G. L., Cotecchia, S., Rokosh, D. G., Grossman, W., Foster, E., and Simpson, P. C. (2003). The alpha (1A/C)- and alpha (1B)-adrenergic receptors are required for physiological cardiac hypertrophy in the double-knockout mouse. *J. Clin. Invest.* **111,** 1783–1791.
145. Djouadi, F., Weinheimer, C. J., Saffitz, J. E., Pitchford, C., Bastin, J., Gonzalez, F. J., and Kelly, D. P. (1998). A gender-related defect in lipid metabolism and glucose homeostasis in peroxisome proliferator-activated receptor α-deficient mice. *J. Clin. Invest.* **102,** 1083–1091.
146. Reisz-Porszasz, S., Bhasin, S., Artaza, J. N., Shen, R., Sinha-Hikim, I., Hogue, A., Fielder, T. J., and Gonzalez-Cadavid, N. F. (2003). Lower skeletal muscle mass in male transgenic mice with muscle-specific overexpression of myostatin. *Am. J. Physiol. Endocrinol. Metab.* **285,** E876–E888.
147. Engelhardt, S., Hein, L., Wiesmann, F., and Lohse, M. J. (1999). Progressive hypertrophy and heart failure in beta1-adrenergic receptor transgenic mice. *Proc. Natl. Acad. Sci. USA* **96,** 7059–7064.
148. Engelhardt, S., Hein, L., Keller, U., Klambt, K., and Lohse, M. J. (2002). Inhibition of Na+-H+ exchange prevents hypertrophy, fibrosis, and heart failure in beta1-adrenergic receptor transgenic mice. *Circ. Res.* **90,** 814–819.
149. Cross, H. R., Lu, L., Steenbergen, C., Philipson, K. D., and Murphy, E. (1998). Overexpression of the cardiac Na+/Ca2+ exchanger increases susceptibility to ischemia/ reperfusion injury in male, but not female, transgenic mice. *Circ. Res.* **83,** 1215–1223.

SECTION IX
Aging and the Cardiovascular System

CHAPTER **19**

The Aging Heart: A Post-Genomic Appraisal

OVERVIEW

As the human population ages and the average life span increases, so does the burden of cardiovascular disease (CVD). Although risk factors such as lifestyle patterns, genetic traits, blood lipid levels, and diabetes can contribute to its development, advancing age unequivocally remains the most significant predictor of cardiac disease. Several parameters of left ventricular function may be affected with aging, including increased duration of systole, decreased sympathetic stimulation, and increased left ventricle ejection time, whereas compliance decreases. In addition, changes in cardiac phenotype with diastolic dysfunction, reduced contractility, left ventricular hypertrophy, and heart failure (HF) all increase in incidence with age. Given the limited capacity that the heart has for regeneration, reversing or slowing the progression of these abnormalities poses a major challenge.

During aging, a significant loss of cardiac myocytes occurs, probably related to programmed cell death (apoptosis), and the cumulative effect of this loss may result in a significant physiological decline, although by definition the decline occurring during aging is much subtler than that observed in the young and adults with cardiac disease. This loss in cardiomyocytes may be secondary to mitochondrial dysfunction, likely caused by chronic exposure to oxidative free radicals, damage to mtDNA (mutations and deletions), and mitochondrial membranes. In spite of the fact that mtDNA damage occurs with aging, mtDNA levels, even though decreased mainly in liver and skeletal muscle, are for the most part preserved in the aging heart. Genomic and proteomic analysis and recombinant DNA techniques are increasingly being applied to the study of aging in general, and to the aging heart in particular. In this chapter, the bioenergetics, biogenesis, and molecular changes occurring in the aging heart in human and animal models, together with current diagnostic and future therapeutic modalities, will be discussed.

INTRODUCTION

The association of age and decreasing heart function has been amply documented. For example, the Baltimore Longitudinal Study on Aging found that even in the absence of disease, there is still a significant loss of cardiac reserve manifested by a reduction in maximum achievable heart rate during stress. At rest in the sitting position, age-associated decline in heart rate (HR) and increased systolic blood pressure occurred in both genders. When hemodynamics were expressed as the change from rest to peak effort (upright cycle exercise) as an index of cardiovascular reserve function, both genders demonstrated age-associated increases in end-diastolic volume index and end-systolic volume index (ESVI), and reductions in ejection fraction, HR, and cardiac index (CI). However, the exercise-induced reduction in ESVI and the increases in ejection fraction, CI, and stroke work index at rest were greater in men than in women. Thus, age and gender each have a significant impact on the cardiac response to exhaustive upright cycle exercise.[1] In the Honolulu Heart Program, after each examination, 6 y of follow-up were available to assess the risk factor effects as the cohort aged from 45–93 y. It was found that in the absence of hypertension, the risk of coronary artery disease (CAD) developing is significantly higher in elderly men, with the incidence of CAD increasing from 1.8% in younger adults (45–54 y) to 8.1% in the elderly (75–93 y) adults. In the later group, alcohol intake was unrelated to CAD, whereas the effects of sedentary lifestyles on promoting CAD seemed to be stronger than in those who were younger.[2]

Notwithstanding, our current knowledge of age-associated cardiac pathological conditions has outpaced our understanding of the basic mechanisms underlying these processes. At present, with the availability of the Human Genome Project (HGP) and an ever increasing number of animals models, as well as new and exciting molecular technologies, the unraveling of the underlying basic mechanisms of cardiac aging have already begun.

Cardiovascular changes occurring with aging include decreased β-adrenergic sympathetic responsiveness,[3,4] slowed and delayed early diastolic filling,[5,6] increased vascular stiffness,[7,8] and endothelial dysfunction.[9,10] Compounding these problems, the cellular changes of aging are most pronounced in post-mitotic organs (e.g., brain and heart), and abnormalities in the structure and function of cardiac myocytes may be the definitive factors in the overall cardiac aging process. As the heart ages, myocytes undergo some degree of hypertrophy, and this may be accompanied by intracellular changes, including mitochondrial-derived oxidative injury, which will contribute to the overall cellular aging, as well as to ischemia-induced tissue damage. Although the aging heart suffers greater damage after an episode of ischemia and reperfusion than the adult heart, the occurrence and degree of aging-related defects remain uncertain. Interestingly, aging-related defects have been found to be limited to a subpopulation of cardiac mitochondria; COX activity and rate of oxidative phosphorylation (OXPHOS) were decreased in the interfibrillar mitochondria, whereas the subsarcolemmal mitochondria remained unaffected[11]; nevertheless, these findings need to be further validated.

Furthermore, during ischemia in the aging heart, ATP and tissue adenosine levels have been found to be decreased compared with adult normal controls.[12] However, reduction in cardiac ATP content in early ischemia is unlikely to provide the mechanism for the increased damage observed after more prolonged periods of ischemia in the aging heart. The interaction of mitochondria, endoplasmic reticulum (ER), peroxisomes, and lysosomes in cellular homeostasis may also be of significance, lysosomes and mitochondria being the organelles that suffer the most remarkable age-related alterations in post-mitotic cells.[13] Coupled mitochondrial and lysosomal defects contribute to irreversible myocyte functional impairment and cell death. Therefore, interactions among the different components of the cardiac cells must be regulated (e.g., mitochondria-nucleus for adequate expression of a number genes, the ER-mitochondria for Ca^{++} metabolism, and peroxisomes-mitochondria for the interchange of antioxidant enzymes essential in the production and decomposition of H_2O_2), because defects in communication between these organelles will accelerate aging. In this chapter, the molecular and genetic analysis of a number of pathways all contributing to the aging process, together with a discussion of new diagnostic modalities and therapeutic interventions will be presented.

UNDERSTANDING THE MECHANISMS OF CARDIAC AGING

Cardiac aging is a rather complex process unlikely to be dependent on a unique, singular pathway or determined gene(s) or gene products. On the contrary, a number of specific and nonspecific pathways and genes seem to play a role in the general regulation/modulation of microorganisms and mammals life span and the cardiac aging in particular. It is relevant to say that at the molecular level, multiple mechanisms interplay in aging, either in parallel or in series, including the involvement of somatic mutations, telomere loss, defects in protein turnover and protein functional decline with accumulation of defective proteins (i.e., impaired induction of heat shock proteins and decline in chaperone function), mitochondrial defects, and most of them produced significant damage to cardiac macromolecules.

From a cardiovascular standpoint, some of these mechanisms seem more at work than others, in particular, the molecular stresses that defective mitochondrial bioenergetics and biogenesis may bring to cardiomyocytes, hormonal and inflammatory signaling, and telomere shortening, as will be addressed later. Identification of the totality of the mechanisms/pathways contributing to cardiac aging is a rather difficult, probably unattainable, goal, although important work in this regard is in progress. A discussion is presented on some of the known mechanisms.

Role of Reactive Oxidative Species on Oxidative Stress and Signaling of the Aging Heart

A causative role for ROS in the aging process, known as the free radical theory of aging (FRTA),[14] presupposes that in biological systems, ROS attack molecules and cause a decline in the function of organ systems, eventually leading to failure and death. Although still controversial, at present, the FRTA is an important, if not *the* dominant, theory to explain the pathogenesis of aging in a vast number of animal species. As we have noted previously, the production of ROS and oxidative stress (OS) is a function of both the inefficiency of the transfer of electrons through the mitochondria respiratory chain and the overall level of antioxidant defenses in the cell.[15] Because ROS are the result of normal metabolic processes, the more active tissues, such as the heart, suffer the most damage. In addition, the bioenergetic dysfunction occurring with aging will further increase the accumulation of ROS. Among its many targets, ROS can reduce the inner-membrane fluidity by attacking polyunsaturated fatty acids and cardiolipin, which can substantially affect protein transport function and the electron transport chain (ETC).

There is evidence of increasing ROS-mediated oxidative damage to lipids and proteins in the aging heart, and both myocardial mtDNA and nuclear DNA damage will result in further accumulation of oxidative species. In particular, mtDNA damage will accumulate because of its inefficient repair machinery and its close proximity to the sources of ROS.

In the aging heart, neutralization of ROS by mitochondrial antioxidants such as superoxide dismutase (SOD), catalase, glutathione peroxidase (GPx), and glutathione becomes of critical significance (Fig. 1). Interestingly, in the aging heart a marked decline in mitochondrial ascorbate levels and reduced glutathione (GSH) has been found in interfibrillar (but not sarcolemmal) mitochondria,[16] and the levels of myocardial GPx and glutathione reductase,[17,18] as well as MnSOD, are increased,[19] possibly as an adaptive mechanism to cope with the increased ROS generation.

Mimetics of the antioxidant enzymes catalase and SOD are currently been considered as a way of reversing the accumulation of ROS and, therefore, aging. However, caution is required, because this modality of therapy may also produce undesirable effects on the "good role" of ROS as important signaling molecules.

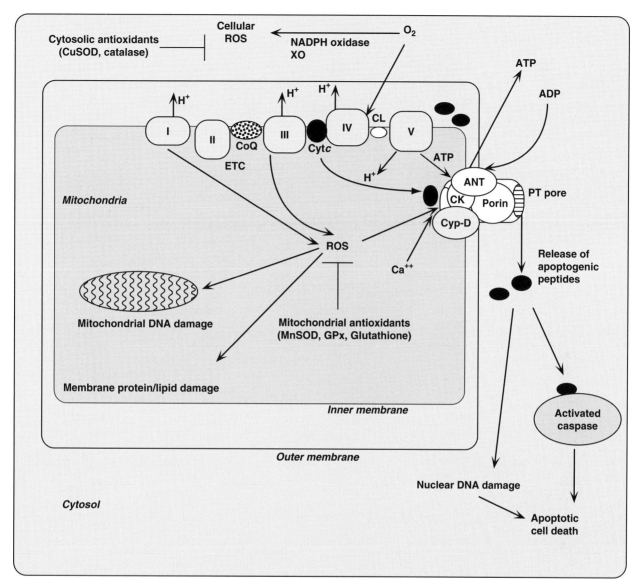

FIGURE 1 Location of antioxidant enzymes and ROS generation in the aging heart. Shown are the cytosolic pathways of ROS generation involving NADPH oxidase and xanthine oxidase (XO), the cytosolic antioxidant enzymes copper SOD (CuSOD), glutathione peroxidase (GPx), and catalase, as well as the mitochondrial pathway of ROS generation (primarily through complex I and III) and mitochondrial antioxidant response featuring MnSOD, GPx, and glutathione. Also depicted are the primary mitochondrial targets of ROS, including the PT pore and mitochondrial apoptotic pathway and mtDNA.

During aging, mitochondrial dysfunction and ROS generation may trigger increased apoptosis, with resultant cell loss. For instance, *in vitro* studies indicate that mitochondrial OS and declining mitochondrial energy production can lead to activation of apoptotic pathways, but whether this also occurs in the *in vivo* aging heart is not clear. Although the role and extent of apoptosis in normal myocardial aging is presently unknown, ample evidence of cardiomyocyte apoptosis is supported by studies demonstrating the release of cytochrome *c* in the aging rat heart mitochondria and decreased levels of Bcl-2 (an antiapoptotic protein), whereas Bax, a proapoptotic protein, remained unchanged.[19,20] Furthermore, myocytes derived from the hearts of old mice displayed increased levels of markers of cell death and senescence compared with myocytes from younger animals.[21]

Undoubtedly, mitochondrial ROS can be an important limiting factor in determining mammalian longevity. Further supporting the role of FRTA, Schriner and associates[22] provided an important advance in our understanding of the mechanisms controlling life span in mammalian species. A transgenic mouse strain was constructed (MCAT) with a 50-fold increase in its expression of the antioxidant catalase in cardiac and skeletal muscle mitochondria. The MCAT strain with increased catalase activity was found to have reduced severity of age-dependent arteriosclerosis, reduction in OS (e.g., H_2O_2 production and oxidative damage), and increased genomic stability, as indicated by reduced levels of mtDNA deletions in heart and muscle tissues. Interestingly, both median and maximum life span were increased approximately 20% compared with wild-type controls. Unfortunately, the Gompertz plot (to characterize the age-related changes in mortality of a genetically heterogeneous population, whether human or animal) of the MCAT strain ran parallel to the wild-type control, indicating a delay in the onset of aging rather than a decrease in aging rate. Nevertheless, these results reinforce the notion that mitochondrial OS has a determinant role in both health and life span. In addition, the relatively large increase in life span resulting from the up-regulation of a single gene involved in boosting antioxidant defenses suggests the possibility that similar up-regulation of several key longevity determinant genes may result in further increases in life span. Consequently, mitochondrial ROS seems to be an important limiting factor in determining mammalian longevity, and this may be proved in future research with new and combined antioxidant mouse models.

One of the major difficulties in arriving at a conclusive proof of the classical OS theory of aging has been the existence of confounding observations, such as noted in several models of OS mice deficient in various antioxidant pathways, leading to increased sensitivity to OS, which were not associated with reduced longevity.[23] On the other hand, new insights into the molecular interactions between ROS and reactive nitrogen species (RNS), such as NO, suggest that their dynamic interplay is involved in the critical modulation of *S*-nitrosylation of targeted proteins, a ubiquitous post-translational modification system involving the covalent attachment of NO to cysteine thiol moieties.[24] This NO-derived post-translational modification serves as a major effector of NO bioactivity and an important mode of cellular signal transduction, and ROS such as superoxide (O_2^-) and H_2O_2 can control the responsivity to S-nitrosylation. At low physiological concentrations, controlled superoxide/ROS and NO production facilitate protein *S*-nitrosylation, whereas high O_2^- levels disrupt *S*-nitrosylation, as well as inactivate NO. In this way, ROS may not only damage cells but also affect a number of signaling pathways.

Hare and associates have proposed a change in our way of thinking regarding the effect of free radicals on aging.[24,25] Although it is increasingly evident that O_2^- and NO both contribute, alone and in combination, to OS and aging, the relation between the degree of OS and the pathological consequences of aging is not linear, rather it is a disruption of the physiological balance between NO and O_2^- that will lead to pathology.[24-29] Achieving a physiological balance between NO and O_2^- (nitroso-redox balance) seems to be critical to the regulation of cardiac excitation–contraction coupling, mitochondrial respiration, and apoptotic cell death. Establishing the precise mechanisms that regulate the balance between OS and NO seems to be essential for understanding cardiovascular pathophysiology and pathogenesis and, ultimately, the aging process as a whole.

NO also plays a significant role in myocardial OS as confirmed by Li and associates. On examining the role of inducible nitric oxide synthase (iNOS) in aging-related myocardial ischemic injury, as well as its relation to β-AR stimulation, they found that iNOS is up-regulated in the aging rat heart.[30] Isolated perfused hearts from young (3–5 mo) and aging (24–25 mo) rats subjected to 30 min of myocardial ischemia resulted in cardiac dysfunction. Infusion of isoproterenol for 30 min caused a partial recovery of cardiac function in hearts from young rats receiving either vehicle or 1400W (a nonselective iNOS inhibitor). In striking contrast, isoproterenol infusion to hearts from aging animals receiving vehicle failed to improve ischemia-induced cardiac depression and worsened cardiac function, with a significant increase in myocardial NO production, peroxynitrite formation, caspase-3 activation, and CK release. Therefore, β-AR stimulation interacts with ischemia and triggers a marked increase in myocardial NO production, creates a nitrosative stress, generating toxic peroxynitrite, activates apoptosis, and eventually causes cardiac dysfunction and myocardial injury in the aging heart. Moreover, myocardial NO production, peroxynitrite formation, and caspase-3 activation were attenuated and LV function significantly improved in aging hearts treated with the iNOS inhibitor, 1400W. Significant increases in iNOS protein expression, activity, and immunoreactivity were found in aging compared with young rat hearts, confirming that aging induces a phenotypic up-regulation of myocardial iNOS. These findings revealed a critical link between iNOS-generated NO production and aging-associated myocardial ischemic injury.

Inflammatory Mechanisms/ Signaling

Besides ROS, NO, and iNOS, other molecules and signaling pathways such as *inflammatory signaling* are actively involved in aging. As noted in several earlier chapters, inflammatory markers increasingly have been identified as significant independent risk indicators for cardiovascular events. Although adults older than the age of 65 have experienced a high proportion of such events, the available epidemiological data comes mainly from middle-aged subjects. Kritchevsky and associates[31] have examined the role that inflammatory markers play in predicting the incidence of CVD, specifically in older adults. Interestingly, IL-6, TNF-α, and IL-10 levels were found to predict cardiovascular outcomes in adults <65 y. Data on C-reactive protein was rather inconsistent and seemed to be less reliable in old age than in middle age. In addition, fibrinogen levels have some value in predicting mortality but in a nonspecific manner. They concluded that in the elderly, inflammatory markers are nonspecific measures of health and may predict both disability and mortality, even in the absence of clinical CVD. Interventions designed to prevent CVD through the modulation of inflammation may be helpful in reducing disability and mortality. The role of increased inflammatory markers such as IL-6 and IL-1β as a risk factor in aging and in the development of MI has also been reported.[32] Analysis of polymorphisms in *IL-6* gene promoter (−174 G→C) has indicated that elderly patients with ACS carrying *IL-6* −174 GG genotypes exhibited a marked increase in 1 y follow-up mortality rate, suggesting that *IL-6* −174 G→C polymorphisms can be added to the other clinical markers such as CRP serum levels and a history of CAD, useful in identifying elderly male patients at higher risk of death after ACS.[33]

Adrenergic (and Muscarinic) Receptors in the Aging Heart

The decline in cardiac function that occurs with aging is due in part to decreased α- and β-AR–mediated contractility. Although impairment in β-AR signaling is known to occur in the aging heart, which components of the α1-AR signaling cascade are responsible for the aging-associated deficit in α1-AR contractile function have just began to be identified. These include PKC and associated anchoring proteins receptors for activated C kinase (RACKs).

Korzick and associates[34] have measured cardiac contractility (dP/dt) in Langendorff-perfused hearts isolated from 5-month-old adult and 24-month-old aging Wistar rats, after maximal α1-AR stimulation with phenylephrine. On assessing the subcellular distribution of PKCα and PKCε and their respective anchoring proteins RACK1 and RACK2 by Western blotting, they found that the subcellular translocation of PKCα and PKCε, in response to α1-AR stimulation, is disrupted in the aging myocardium. Moreover, age-related reductions in RACK1 and RACK2 levels were also observed, suggesting that alterations in PKC-anchoring proteins may contribute to impaired PKC translocation and defective α1-AR contraction in the aged rat heart. Interestingly, this group of investigators also sought to determine whether age-related defects in α1-AR contraction could be reversed by chronic exercise training (treadmill) in the adult and aged rat.[35] They found that the age-related decrease in α1-AR contractility in the rat heart can be partially reversed by exercise, suggesting that alterations in PKC levels underlie, at least in part, exercise training–induced improvements in α1-AR contraction.

Reperfusion of an isolated mammalian heart with a calcium-containing solution after a brief calcium-free perfusion results in irreversible cell damage (the calcium paradox). Activation of the α1-AR pathway has been shown to confer protection against the lethal injury of the Ca^{++} paradox by way of PKC-mediated signaling pathways, and this protection is shared by stimuli common with calcium pre-conditioning.[36] Age-related changes have been analyzed in the cardiac effects of α1-adrenergic stimulation, both in cardiomyocyte Ca^{++}-transient and cardiac PKC activity in 3-month-old and 24-month-old Wistar rats.[37] In a dose-response curve to phenylephrine, the response of Ca^{++} transient was maximal at 10^{-7} M. Although this concentration induced a significant increase in Ca^{++} transient in the young, a significant decrease was observed in the aging rat. Moreover, α-1 adrenergic stimulation led to a reduction of intracellular PKC translocation in the older but not in the younger rats. The negative effect of α1-adrenergic stimulation on cardiomyocyte Ca^{++} transient observed in old rats may be related to the absence of α1-adrenergic–induced PKC translocation.

Notwithstanding these findings, the effect of aging on the human sympathetic nervous system remains a controversial issue. At present, the interest in this subject has significantly increased, mainly because diverse cardiac pathological conditions, including essential hypertension, CAD, HF, and dysrhythmias, increase with age, and the sympathetic nervous system may be an important pathophysiological component.[38] Recently, in a study on the role of the sympathetic nervous system in aging and HF, Kaye and Esler[39] found no additive effect of aging in the activation of the sympathetic nervous system occurring in HF, suggesting that other factors such as CAD and MI may impact the increased incidence of HF with aging.

Cardiac G-Protein–Coupled Receptors

As previously noted, cardiac G-protein–coupled receptors (GPCRs) that function through stimulatory G-protein $G\alpha_s$, such as β1- and β2-ARs, play a key role in cardiac contractility. Several $G\alpha_s$-coupled receptors in the heart also activate $G\alpha_i$, including β2-ARs (but not β1-ARs); PKA- dependent phosphorylation of β2-AR can shift its coupling preference from $G\alpha_s$ to $G\alpha_i$.[40] Coupling of cardiac β2-ARs to $G\alpha_i$ inhibits adenylyl cyclase (AC) and opposes β1-AR–mediated apoptosis.[41] As found in studies of congestive HF, $G\alpha_{i_2}$ levels increases with age in both human atria[42] and in ventricles of old (24 mo) Fischer 344 rats, resulting in diminished AC activity.[41] These levels may subsequently increase the receptor-mediated activation of G_i through multiple GPCRs. Increased G_i activity is likely to have an adverse effect on heart function, because G_i-coupled signaling pathways in the heart reduce both the rate and force of contraction.[43]

Investigation of the effects of age on G-protein–coupled receptor signaling in human atrial tissue showed that the density of atrial muscarinic acetylcholine receptor (mAChR) increases with age but reaches statistical significance only in patients with diabetes.[44] Interestingly, in elderly subjects of similar ages, those with diabetes have 1.7-fold higher levels of $G\alpha_{i_2}$ and twofold higher levels of $G\beta_1$. Other studies have reported that right atrial mAChR density significantly decreased in advanced age.[45] The disparity between these studies could be explained by differences in age between patient groups; one study examined only adults with an age range from 41–85 y,[44] whereas the other study group's age ranged from 5 days to 76 y.[45] The finding that in humans, after 50 y of age, atrial mAChR density exhibits an upward trend with age[44] differs with findings in most animal studies, which have been less conclusive showing unchanged muscarinic receptor levels,[46–49] or indicate decreased mAChR density with age.[50]

SERCA and Thyroid Hormone in the Aging Heart

The cardiac sarcoplasmic reticulum Ca^{++}-ATPase (SERCA) pumps Ca^{++} from the cytosol back to the sarcoplasmic reticulum (SR) and is considered an important determinant of intracellular Ca^{++} signaling and cardiac contractility. The aging heart shows an increased susceptibility to HF, which may be, in part, mediated by abnormal Ca^{++} handling and decreased expression of the *SERCA* gene.[51] Although the molecular mechanisms that regulate cardiac *SERCA* expression in aging are still unclear, new findings have implicated decreased thyroid hormone (TH) responsiveness in the aging rat heart; this decrease in large part involves the binding of the TH receptor (TR) and retinoid X receptor (RXR) heterodimer to TH-responsive elements (TREs) located in the *SERCA* and cardiac myosin heavy chain (*MHC*) gene promoters.

Age-associated changes in the TR and RXRs could explain the age-associated changes in SERCA and MHC expression. Long and associates found that although no significant myocardial changes in RXRα or RXRβ mRNA levels occur with age in the rat heart, both α1 and α2 TR mRNA levels decreased significantly between 2 and 6 mo of age.[52] During this time, the mRNA levels for α-MHC declined by more than half, whereas β-MHC mRNA levels remained low and unchanged. On the other hand, between 6 and 24 mo, when mRNA levels for β-MHC increased and α-MHC continued to decrease, there was a significant decline in TRβ1 and RXRγ mRNA levels accompanied by a reduction in the TRβ1 and RXRγ protein levels. These findings suggest that the decline in α-MHC gene expression may be biphasic and, in part, due to a decline in α1 (and possibly α2) TR levels between 2 and 6 mo of age, and a decline in TRβ1 and RXRγ levels at later ages.

Recent observations have demonstrated that the aging-induced down-regulation of α-MHC and SERCA mediated by myocardial TH/TR signaling–triggered transcriptional control can be reversed with exercise.[53] Although the expression of myocardial TRα1 and TRβ1 proteins are significantly lower in sedentary aged rats than in sedentary young rats, their expression is significantly higher in exercise-trained aged rats than in sedentary aged rats. Furthermore, the activity of TR DNA binding to the TRE transcriptional regulatory region in the α-MHC and SERCA genes and the myocardial expression of α-MHC and SERCA (both mRNA and protein) were up-regulated with exercise training in the aging heart, in association with changes in the myocardial TR protein levels. In addition, plasma 3,3′-triiodothyronine (T3) and thyroid hormone levels, which decrease in aging,[54,55] are increased with exercise training. The reversal of aging-induced down-regulation of myocardial TR signaling–mediated transcription of α-MHC and SERCA genes by exercise training seems to be related to the functional improvement observed in trained aged hearts.

The identification of the specific mechanisms contributing to decreased TH signaling in the aging heart may provide significant insights into possible therapeutics, keeping in mind that in the aging heart decreased TH activity may be a physiological adaptation. It is possible that therapies that increase SERCA activity might improve cardiac function in the senescent heart. On the other hand, it has been shown that a decrease in SERCA activity contributes to the functional abnormalities observed in senescent hearts and that Ca^{++} cycling proteins can be targeted to improve cardiac function in senescence.[56] The well-established decline in myocardial SERCA content with age may also contribute to increased development of impaired function after I/R in aging subjects. Also, the ratio of SERCA to either phospholamban or calsequestrin decreased in the senescent human myocardium.[57] Decreased rates of Ca^{++} transport mediated by the SERCA isoform are responsible for the slower sequestration of cytosolic Ca^{++} and consequent prolonged muscle relaxation times in the aging heart. Knyushko and associates[58] have found that senescent Fischer 344 rat hearts showed a 60% decrease in SERCA activity relative to that of young adult hearts, and this functional reduction in activity could be attributed, in part, to both a lower abundance of SERCA protein and increased 3-nitrotyrosine modifications of multiple tyrosines within the cardiac SERCA protein. Nitration in the senescent heart was found to increase by more than two nitrotyrosines per Ca^{++}-ATPase, coinciding with the appearance of partial nitrated Tyr(294), Tyr(295), and Tyr(753) residues. In contrast, skeletal muscle SERCA exhibited a homogeneous pattern of nitration, with full-site nitration of Tyr(753) in the young, with additional nitration of Tyr(294) and Tyr(295) in the senescent muscle. The nitration of these latter sites correlates with diminished transport function in both types of muscle, suggesting that these sites have a potential role in the down-regulation of ATP use by the Ca^{++}-ATPase under conditions of nitrosative stress.

Growth Hormone (GH) and IGF-1

Reduced signaling of insulin and highly conserved insulin-like peptides can profoundly affect organismal life span. Mutation in genes involved in the insulin/insulin-like growth factor I (IGF-I) signal response pathway have been reported to significantly extend life span in diverse species, including yeast, nematodes, fruit flies, and rodents. Intriguingly, the long-lived mutants share important phenotypic characteristics, including reduced insulin signaling, enhanced sensitivity to insulin, and reduced IGF-I plasma levels. In the nematode and the fly, secondary hormones downstream of insulin-like signaling also seem to regulate aging. However, the relative order and significance in which the hormones act in mammals has been difficult to resolve, because there is a complex network of interacting and interdependent signaling molecules, including insulin, IGF-1, growth hormone (GH), and TH, affecting multiple interacting cellular pathways.[59] For instance, whereas endocrine manipulations in animal models can slow aging without concurrent costs in reproduction, these bring inevitable increases in stress resistance.[60]

Several mutant mouse strains have been instrumental in providing models of life span and aging modulation. These include GH-deficient/resistant animals that have a prolonged life span compared with their normal siblings. Studies have indicated that the Ames and Snell dwarf mouse strain and GH receptor/GH binding protein knock-out (GHR/BP-KO) do not experience aging at the same rate as their normal siblings but are subject to delayed aging. The Snell and Ames Dwarf mice are homozygous for recessive mutations at the pituitary-1 (pit1) *Pit-1*, or *Prop-1 locus*, respectively, which encode transcription factors controlling pituitary development.[61] Both Snell and Ames Dwarf mice demonstrate increased longevity (50% in males and 64% in females) compared with their wild-type controls, which has been generally attributed to GH/IGF-I deficiency.[62] Mice homozygous for such a mutation are deficient in serum GH, thyroid-stimulating hormone (TSH), and prolactin, as well as IGF-I.

Although the mechanism of increased life span has not been fully delineated, there is increased support for the centrality of insulin signaling in the control of mammalian aging and for the involvement of this pathway in extending the life span of IGF-I–deficient mice. In the Snell dwarf mouse, GH deficiency leads to reduced insulin release and alterations in insulin signaling, including a decreased IRS-2 pool level, a reduction in PI3K activity and its association with IRS-2, decreased docking of p85 to IRS-2, and preferential docking of IRS-2 to p85-p110, leading to reduced insulin levels, enhanced insulin sensitivity, alterations in carbohydrate and lipid metabolism, reduced generation of ROS, enhanced resistance to stress, reduced oxidative damage, and delayed onset of age-related disease.[61,63,64] These alterations would establish a physiological homeostasis that favors longevity. Mouse longevity is also increased by fat-specific disruption of the insulin receptor gene *FIRKO*.[65]

Although a lower level of circulating growth hormone and an enhanced life span was found in transgenic mice expressing bovine growth hormone (bGH),[66] mouse mutant models containing high plasma GH but a 90% lower IGF-I also live longer than wild-type mice. This suggests that reduction in plasma IGF-I levels may be primarily responsible for a major portion of the life span increase in dwarf, GH-deficient, and GHR/BP-null mice.[64,67] Further evidence for the direct role of IGF-I signaling in the control of mammalian aging has also been provided by mouse strains in which the loss of a single copy of the *igf1r* gene (encoding the IGF receptor) results in a 26% increase in mouse life span.[68]

Moreover, the IGF-I/GH/IGF-1 receptor pathway not only plays an important role in determining organism development and life span but is in itself affected by age. IGF-I decreased linearly with age in both genders, with significantly higher levels in men than women.[69] The decrease in GH-induced IGF-I secretion in the elderly suggests that resistance to the action of GH may be a secondary contributing factor in the low plasma IGF-1 concentrations.[70]

It has also been argued that decreased IGF-I levels with age may contribute to the increase in cardiac disease found in the elderly, including HF.[71] Findings from the Framingham Heart Study in a prospective, community-based investigation indicated that serum IGF-I level was inversely related to the risk for HF in elderly people without a previous MI, suggesting that the maintenance of an optimal IGF-I levels in the elderly may reduce the risk for HF.[72] In addition, this study revealed that greater levels or production of the catabolic cytokines TNF-α and IL-6 were associated with increased mortality in community-dwelling elderly adults, whereas IGF-I levels had the opposite effect.[73]

In aged animals and humans, the secretion of GH and the response of GH to the administration of GH-releasing hormone (GHRH) are lower than in young adults.[74] In rodents, a twofold increase in GH receptors has been observed with age, but this increase fails to compensate for the reduction in GH secretion[75,76] Further investigation revealed that the apparent size of the GH receptor was not altered with age, whereas the capacity of GH to induce *IGF-1* gene expression and secretion was 40–50% less in old than in young animal.[77]

There is considerable literature indicating that GH administration to old animals and humans raises plasma IGF-I levels and results in increases in skeletal muscle and lean body mass, a decrease in adiposity, increased immune function, improvements in learning and memory, and increases in cardiovascular function. Evidence has been presented suggesting that GH administration can induce improvement in hemodynamic and clinical status in some patients with chronic HF, largely resulting from the ability of GH to increase cardiac mass.[78] However, disappointing results were reported in patients with DCM undergoing infusion of GH.[79] These results could be related to the choice of the incorrect agent (GH instead of IGF-I) and/or the failure to selectively target patients with low IGF-I levels.[80] Other evidence has shown that pharmacological administration of GH to adults may also pose risks; mice transgenic for GH and acromegalic patients secreting high amounts of GH have premature death.[81] Similarly, caution must be used in the clinical administration of IGF-I in treating cardiac disease. Given its potent antiapoptotic role in proliferation, some association of high-dosage IGF-I with human cancers has been shown in early studies,[82,83] although a more recent review found that the overexpression of *IGF-1* in animals or the administration of rhIGF-I does not have a carcinogenic effect.[84]

Interestingly, the Klotho protein, which functions as a circulating hormone, binds to a cell-surface receptor and represses intracellular signals of insulin and IGF-I. Amelioration of the aging-like phenotypes in Klotho-deficient mice was observed by perturbing insulin and IGF-I signaling, suggesting that Klotho-mediated inhibition of insulin and IGF-I signaling contributes to its anti-aging properties.[85]

Cellular Damage/Cell Loss

Because cell damage occurs at random in any organ or tissue, including the heart, a population of damaged cells will always coexist with normal cells at any time in the process of aging, and an important unknown relates to the number of damaged cells required to impair organ/tissue function. Kirkwood[86] correctly pointed out that there is a significant difference in assessing cell damage in the *in vitro* state versus the *in vivo* system, with cells in culture reaching a limit in their potential for cell division/differentiation, which may not occur *in vivo*. Therefore, caution is called for regarding the interpretation of data that use different methods.

During aging, there is a significant loss of post-mitotic cells, such as cardiac myocytes, potentially triggered by the onset of mitochondrial dysfunction and ROS generation.[87] For instance, *in vitro* studies of H_2O_2-treated cardiomyocytes showed that increased mitochondrial OS and declining mitochondrial energy production lead to the activation of apoptotic pathways,[88,89] but whether this also occurs in the aging heart *in vivo* is not known. Although the role and extent of apoptosis in normal myocardial aging is presently under considerable debate, evidence of cardiomyocyte apoptosis has been corroborated by data demonstrating that the aging rat heart had significantly elevated levels of cytochrome *c* release from mitochondria, as well as decreased levels of the antiapoptotic protein Bcl-2, whereas levels of the proapoptotic protein Bax were unchanged.[20] Moreover, myocytes derived from the hearts of old mice displayed increased levels of markers of cell death and senescence compared with myocytes from younger animals.[21] It is possible that apoptosis, at least to a certain degree in cardiac aging, may be a protective mechanism to get rid of those damaged, potentially dangerous, cells in a mechanistic effort to incline the balance toward healthy cells, although we do not know where this balance is.

According to Uhrbom and associates,[90] glioma cells stained for senescence-associated β-galactosidase activity, apparently specific for senescent cells, showed that the enlarged cells gave a distinctive positive staining reaction. This senescence phenotype seems to be dependent on the continuous expression of p16INK4A. Thus, the induced expression of p16INK4A in these cells reverted their immortal phenotype and caused immediate cellular senescence. Furthermore, increased expression of p16INK4A also occurred in aging cardiomyocytes.[21] Proteins implicated in growth arrest and senescence, such as p27Kip1, p53, p16INK4a, and p19ARF, were also present in myocytes of young mice, and their expression increased with age. Moreover, DNA damage and myocyte death were found to exceed cell formation in older mice, leading to a decline in the number of myocytes and HF. Interestingly, this effect did not occur in transgenic mice in which cardiac stem cells mediated myocyte regeneration compensating for the extent of cell death and preventing ventricular dysfunction.

Telomeres and Telomere-Related Proteins

Telomeres are specialized DNA-protein structures located at the end of chromosomes, which shorten with each replication, unless they are preserved by the action of the enzyme telomerase reverse transcriptase (TERT). During aging, their length is progressively reduced in most somatic cells, including human, and after a number of cell cycles, telomere length reaches a critical size, cellular replication stops, and the cell becomes senescent (Fig. 2). Age-dependent telomere shortening in somatic cells, including vascular endothelial cells, SMCs, and cardiomyocytes, seems to impair cellular function and the viability of the aged organism. Although the association of telomere length and CVD seems likely, whether telomere shortening is a direct cause of the vascular pathology of aging or a consequence is not known.[91]

Notwithstanding, telomere dysfunction and reduction in telomere length have been observed with age in smooth muscle, endothelial, and white blood cells, and they may be the primary factor in predisposing vascular tissues to atherosclerosis and also to a decreasing capacity for neovascularization.[91] This attrition of telomere length is most prominent under conditions of high OS, but particularly prevalent in hypertensives, diabetics, and those with CAD.

In the preservation of telomere function in endothelial cells, a key role is played by glutathione-dependent redox homeostasis.[92] Under mild chronic OS, the loss of telomere integrity is a major trigger for the onset of premature senescence. Interestingly, antioxidants and statins can delay the replicative senescence of endothelial cells by inhibition of the nuclear export of telomerase reverse transcriptase into the cytosol.[93] Furthermore, recent studies have revealed telomere biology plays a significant role in the functional augmentation of endothelial progenitor cells (EPCs) by statins.[94] These studies found that the *ex vivo* culturing of EPCs leads to "uncapping" of telomeres, indicated by the loss of telomere repeat-binding factor (TRF2). Moreover, co-treatment with statins maintained the EPCs functional capacity by a TRF2-dependent, post-transcriptional mechanism.

Although the ability of EPCs to sustain ischemic tissue and repair may be limited in the aging/senescence heart, estrogens have been shown to accelerate recovery of the endothelium after vascular injury and significantly increased telomerase activity.[95] RT-PCR analysis showed that 17β-estradiol in a dose-dependent fashion increased levels of the telomerase catalytic subunit (TERT), an effect that was significantly inhibited by pharmacological PI3K blockers (either wortmannin or LY294002). In addition, EPCs treated with 17β-estradiol had significantly enhanced mitogenic potential and release of VEGF protein; moreover, EPCs treated with both 17β-estradiol and VEGF were more likely to integrate into the network formation than those treated with VEGF alone.

Telomerase inactivation during aging may also be related to the oxidized low-density lipoprotein-accelerated onset of EPC senescence, which leads to the impairment of proliferative

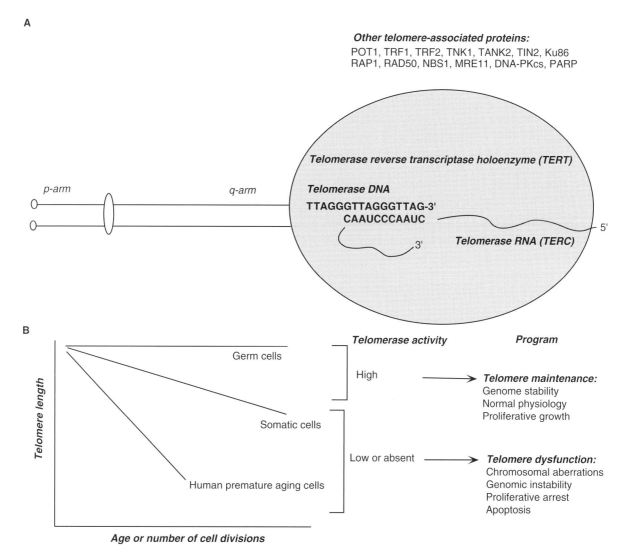

FIGURE 2 Telomeres, telomerase, and cell proliferation. (A) Telomeres present at the ends of the chromosomes (one end at the Q arm is shown here) are composed of a tandem repeated sequence in telomeric DNA, a telomerase ribonucleoprotein complex including a catalytic telomerase reverse transcriptase holoenzyme (TERT), a RNA component, and additional telomeric proteins (listed on top). (B) Telomere length decreases in aging in somatic cells but not in germ cells. Because telomerase activity is low or absent in most somatic cells, progressive telomere erosion occurs with each mitotic division during normal aging. In germ cells (and tumor cells) containing high telomerase activity, no change in telomere length is seen with aging or progressive divisions. Accelerated telomere attrition is associated with human premature aging. Some of the phenotypic changes associated with telomere length are shown on the bottom right.

capacity and network formation.[96] Interestingly, estrogen also stimulates NO production in vascular endothelial cells,[97] which, in turn, induces telomerase in these cells.[98] The cardioprotective effects of estrogens by indirect actions on lipoprotein metabolism and through direct effects on vascular endothelial cells and SMCs are likely to contribute to the lower incidence of CVD observed in premenopausal women compared with men, and, significantly, women have a decelerated rate of age-dependent telomere attrition over men.[99]

Most human somatic cells can undergo only limited replication *in vitro*, and senescence can be triggered when telomeres cannot carry out their normal protective functions. Furthermore, it has been observed that senescent human fibroblasts display molecular markers characteristic of cells bearing DNA double-strand breaks and that inactivation of DNA damage checkpoint kinases in senescent cells may restore the cell-cycle progression into S phase.[100] This telomere-initiated senescence may reflect a DNA damage checkpoint response that is activated with a direct contribution from dysfunctional telomeres.

A high rate of age-dependent telomere attrition has been noted in the human distal abdominal aorta, probably reflecting enhanced cellular turnover rate because of local factors, such as an increase in shear wall stress in this vascular segment.[101] These findings seem to contradict the findings in mice that short telomeres have a protective effect from atherosclerosis.[102] These apparent conflicting findings might be reconciled if cellular damage accumulation imposed by prolonged exposure to cardiovascular risk factors ultimately prevails over protective mechanisms, including telomere shortening.[99]

Several observations support the concept that the average telomere length is better maintained in conditions of low OS.[103–105] Selective targeting of antioxidants directly to the mitochondria can counteract telomere shortening and increase life span in fibroblasts under mild OS.[106] Mitochondrial dysfunction can lead to a loss of cellular proliferative capacity through telomere shortening (Fig. 3), and the generation of ROS may signal the nucleus to limit cell proliferation through telomere shortening with telomeres acting as sensors to damaged mitochondria.[107]

It is increasingly evident that telomere dysfunction is emerging as an important factor in the pathogenesis of human CVDs associated with aging, including hypertension, atherosclerosis, and HF (Fig. 4).

In the future, in addition to focussing on telomerase, research should also be oriented toward finding the role that additional telomere-associated proteins play in aging and to apply future discoveries to develop novel and more effective cardiovascular therapies.

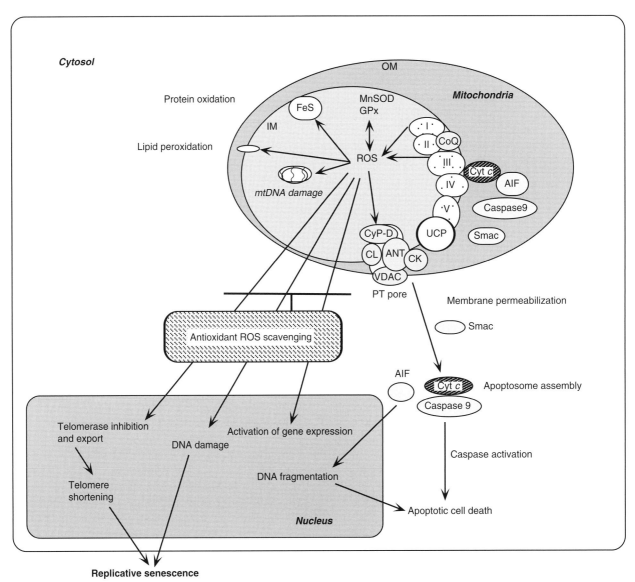

FIGURE 3 Mitochondrial ROS production contributes to telomere-dependent replicative senescence. Production of ROS by aberrant mitochondrial respiratory complexes thought to occur in aging leads to mtDNA and protein damage and lipid peroxidation, PT pore, and apoptotic progression (by cytochrome *c* release, membrane permeabilization, apoptosome formation, and nuclear DNA fragmentation). ROS also affects the nucleus by reducing telomerase activity and increasing its export from the nucleus, causing direct DNA damage and activating specific gene expression—all of which contribute to the signaling of replicative senescence. Also depicted is the involvement of antioxidant response both in the mitochondria (e.g. MnSOD and GpX) and in the cytosol, which can remove ROS by scavenging free radicals.

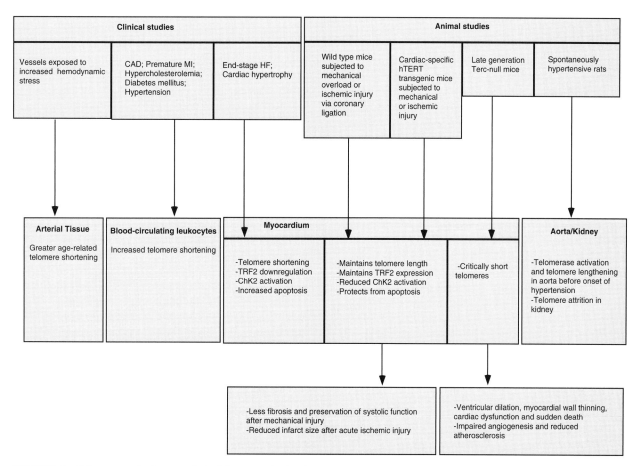

FIGURE 4 Telomeres and telomerase in both human and animal CVD. Telomere and telomerase phenotype in different tissues of patients with indicated CVD disorders and of selected animal models of CVD, including both wild-type and genetically modified mice (e.g., cardiac-specific hTERT and Terc-null strains) and in spontaneously hypertensive rats.

Autophagy and Cardiac Aging

Cells faced with a short supply of nutrients in their extracellular fluid begin to engulf specific, often defective, organelles (e.g., mitochondria) and to reuse their components. A number of steps are involved in autophagy including: (1) formation of a double membrane within the cell; (2) confinement of the material to be degraded into an autophagosome; (3) Fusion of the autophagosome with a lysosome, and (4) the enzymatic degradation of the materials.

Autophagy occurs in many types of cells during development, including cardiomyocytes. Moreover, in cardiac diseases associated with aging, such as ischemic heart disease and cardiomyopathy, intralysosomal degradation of cells plays an essential role in the renewal of cardiac myocytes. The interaction of mitochondria and lysosomes in cellular homeostasis is of great significance, because both organelles suffer significant age-related alterations in post-mitotic cells.[108] Many mitochondria undergo enlargement and structural disorganization, and because lysosomes responsible for mitochondrial turnover experience a loss of function, the rate of total mitochondrial protein turnover declines with age.[109] Coupled mitochondrial and lysosomal defects contribute to irreversible functional impairment and cell death, and similarly mitochondrial interaction with other functional compartments of the cardiac cell (e.g., the ER for Ca^{++} metabolism, peroxisomes for the interchange of antioxidant enzymes essential in the production and decomposition of H_2O_2) must be kept in check, because defects in communication between these organelles may accelerate the aging process.

Several mechanisms may potentially contribute to the age-related accumulation of damaged mitochondria after initial oxidative injury, including clonal expansion of defective mitochondria, reduction in the number of mitochondria targeted for autophagocytosis (secondary to mitochondrionmegaly or decreased membrane damage associated with decrease mitochondrial respiration), suppressed autophagy because of heavy lipofuscin loading of lysosomes, and decreased efficiency of Lon protease.[110]

Abnormal autophagic degradation of damaged macromolecules and organelles, known as biological "garbage," is also considered an important contributor to aging and the death of post-mitotic cells, including cardiomyocytes. Recently, Stroikin and associates compared the survival of density-dependent growth-arrested and proliferating human fibroblasts and astrocytes after inhibition of autophagic sequestration with 3-methyladenine (3MA).[111] Exposure of confluent fibroblast cultures to 3MA for 2 weeks resulted in an increased number of dying cells compared with both untreated confluent cultures and dividing cells with 3MA-inhibited autophagy. Similarly, autophagic degradation was suppressed by the protease inhibitor leupeptin. These findings suggest that lysosomal "garbage" accumulation plays an important role in the aging and death of post-mitotic cells and also supports the anti-aging role of cell division.

Gender Differences

In the functioning of the normal heart, there are differences among men and women, namely cardiac contractility is better preserved in women's hearts before hormone replacement therapy withdrawal, after which contractility decreases. Moreover, aging has a different effect on the integrity of the myocardium in the two genders, because aging is accompanied by appreciably less myocyte cell loss, myocyte cellular reactive hypertrophy, and myocardial remodeling in women than man.[112] The better preserved myocardial mass in women could be related to the effect of 17β-estradiol (E2) on the reduction in cardiomyocyte apoptosis. Recently, Patten and associates[113] have shown that physiological E2 replacement reduces cardiomyocyte apoptosis after MI in ovariectomized female mice, and E2 in vivo treatment increased activation of the prosurvival, serine-threonine kinase Akt, which preceded the reduction in cardiomyocyte apoptosis 24 and 72 h after MI.

Yan and associates have found a decreased expression of enzymes of glycolysis (e.g., pyruvate kinase, α-enolase, triosephosphate isomerase), glucose oxidation (e.g., pyruvate dehydrogenase E1 β-subunit), and the TCA cycle (e.g., 2-oxoglutarate dehydrogenase) in LV samples from old males (OM) monkeys; these changes were not observed either in young animals nor in old females (OF).[114] Moreover, gender differences in the reduced expression and function of proteins that are responsible for mitochondrial ETC and OXPHOS were only found in hearts from OM monkeys. The glycolytic and mitochondrial metabolic pathway changes in OM monkey hearts were similar to changes observed in hearts affected by diabetes or LV dysfunction, suggesting a potential involvement in the mechanism of the cardiomyopathy of aging. Because OF hearts did not undergo these abnormal metabolic changes, it was suggested that this may explain the delayed cardiovascular risk observed in OFs. For further discussion on gender differences see Chapter 18.

Susceptibility Genes: Genetic Makeup in Cardiac Aging

Gathered observations have suggested that a heritable genetic component(s) exists, which is an important determinant in the duration of life span in humans similar to the inherited factors operative in the incidence of certain cardiac diseases such as CHD, cardiomyopathics, and CAD. Genes that otherwise may remain in a quiescent state are stimulated with aging under adequate environmental conditions to express transcription factors/proteins that may facilitate the development of cardiac pathological conditions (e.g., on physiological stress families of stress-response genes are activated as natural defense mechanisms). Saito and associates[115] have recently reported that the induction of specific inflammatory genes is significantly deregulated and altered in the heart of aged versus young mice when challenged with the bacterial endotoxin lipopolysaccharide, suggesting that endotoxin-mediated induction of specific inflammatory genes in cardiovascular tissues is abnormal with aging, and this may be causally related to the increased susceptibility of aged animals to endotoxic stress.

The KLOTHO-deficient mice develop a syndrome resembling accelerated human aging, with significant and accelerated arteriosclerosis.[116] The KLOTHO gene encodes a single-pass transmembrane protein that functions in signaling pathways that suppress aging and that has β-glucuronidase activity. In humans, a functional variant of KLOTHO, termed KL-VS, has been found to be common in the general population (frequency 0.157), and individuals homozygous for KL-VS manifest reduced human longevity.[117] The KL-VS variant harbors three mutations in the coding region, one of which is silent, and two code for missense mutations F352V and C370S, which substantially alter KLOTHO metabolism. The KL-VS allele influences the trafficking and catalytic activity of KLOTHO, and the variations in KLOTHO function contributed to heterogeneity in the onset and severity of human age-related phenotypes[117] and early-onset of occult CAD.[118] Furthermore, recent cross-sectional and prospective studies have confirmed a genetic model in which the KL-VS allele confers a heterozygous advantage in conjunction with a marked homozygous disadvantage for HDL-C levels, systolic blood pressure, stroke, and longevity.[119]

Epigenetic/Environmental Factors Affecting Cardiac Aging

A number of epigenetic factors may contribute to the development of diverse cardiac pathological conditions (e.g., CAD, hypertension) in aging, including increased caloric intake, inadequate diet, alcohol intake, smoking, obesity, and lack of adequate aerobic exercise. Although a discussion on the effect of caloric intake/diet on the aging heart will be presented later in this chapter, at this time it may suffice to comment on some of the effects that aerobic exercise may have on cardiac aging. Intrinsically, in the normal human aging heart there is a significant decrease in the chronotropic and inotropic responses to catecholamine stimulation, compromising cardiac function. An age-associated reduction in cardiovascular β-adrenergic (β-AR) responsiveness has been noted in Fischer 344 rats, corresponding with alterations in post-receptor adrenergic signaling rather than with a decrease in LV β-AR receptor number.[120] Interestingly, chronic dynamic exercise partially attenuated these reductions through alterations in post-receptor elements of cardiac signal transduction. Moreover, exercise training improves the aging-induced down-regulation of myocardial

PPAR-α–mediated molecular system and contributes to an amelioration in fatty acid metabolic enzyme activity in rats.[121]

Moreover, endothelial function deteriorates with aging in human, and exercise training seems to improve the function of vascular endothelial cells. Regular aerobic-endurance exercise has been found to reduce plasma ET-1 concentration and to increase NO production in previously sedentary older women, with probable beneficial effects on the cardiovascular system (i.e., prevention of progression of hypertension and/or atherosclerosis by endogenous ET-1 and the potent vasodilatory effects of NO).[122,123] Also, regular aerobic exercise may prevent the age-associated loss in endothelium-dependent vasodilation and restore the levels in previously sedentary middle-aged and older healthy men. This may be an important mechanism by which aerobic exercise lowers the risk of cardiovascular disease in this population.[124] Furthermore, endothelial release of tissue-type plasminogen activator (t-PA), a primary regulator of fibrinolysis and part of the endogenous defense mechanism against thrombosis, decreases with age in sedentary men, and regular aerobic exercise may not only prevent it but could also reverse the age-related loss in endothelial fibrinolytic function.[125] In obesity, which is associated with an increased risk of atherothrombosis, significant endothelial fibrinolytic dysfunction may be present; but regular aerobic exercise can increase the capacity of the endothelium to release t-PA.[126] On the other hand, the endothelium release of NO was not compromised in overweight and obese adults under basal conditions.[127]

Endurance exercise provides cardioprotection (CP) against ischemia-reperfusion (I/R)–induced necrotic cell death, not only in young but also in aged Fischer 344 rats, by reducing I/R-induced myocardial apoptosis. The mechanisms for this exercise-induced CP against I/R-induced apoptosis may be mediated by improved myocardial antioxidant capacity and the prevention of calpain and caspase-3 activation.[128] Similarly, compared with sedentary animals, French and associates[129] found that exercise training prevented the I/R-induced rise in calpain activity and improved cardiac work in a working heart preparation from adult male rats. Pharmacological inhibition of calpain activity resulted in comparable CP against I/R injury. This exercise-induced protection against IR-induced calpain activation was not due to abnormal myocardial protein levels of calpain or calpastatin. Interestingly, exercise training was associated with increased levels of myocardial Mn-SOD, catalase, and a reduction in OS. In addition, exercise training also prevented the I/R-induced degradation of SERCA2a, apparently by increases in endogenous antioxidants.

Exercise intolerance has long been recognized as an important symptom of HF, but it also may develop in aged individuals without cardiac pathology. A number of nonspecific factors such as skeletal muscle dysfunction (likely secondary to mitochondrial bioenergetics defects), ventilatory abnormalities, and endothelial dysfunction, individually or in association, may contribute to limitation in exercise capacity. An important contributing factor for skeletal muscle catabolism (e.g., elevated cytokine expression) can be found in both normal, healthy aging, and in patients with HF.[130] This commonality of aging and HF-associated changes in the skeletal muscle may explain the more severe clinical presentation of the HF syndrome among elderly patients. A decline in maximal aerobic capacity and the ability to sustain submaximal exercise with advancing age was demonstrated in young (6–8 mo) and old (27–29 mo) Fischer 344 × Brown Norway rats.[131] Besides heart rate and mean arterial pressure, blood flow (BF) to different organs (kidneys, splanchnic organs, and 28 hind limb muscles) was measured at rest and during submaximal treadmill exercise with radiolabeled microspheres. BF to the total hind limb musculature increased during exercise but was similar for both young and old animals. However, in old compared with young rats, the BF was reduced in six (highly oxidative) and elevated in 8 (highly glycolytic) of the 28 individual hindquarter muscles or muscle parts examined, suggesting that although there were similar increases in total hind limb BF in young and old rats during submaximal exercise, there was a profound BF redistribution from highly oxidative to highly glycolytic muscles. With the same animal model, the effect of aging on muscle BF with similar degrees of MI-induced LV dysfunction was evaluated.[132] A significant age-related redistribution of BF from the highly oxidative to the highly glycolytic muscles of the hind limb was found during exercise in old rats compared with the young.

ARTERIAL AGING

Age is the main clinical determinant of stiffness in large arteries. Central arteries stiffen progressively with age, whereas peripheral muscular arteries change little with age. With aging, arterial stiffening is accompanied by elevated systolic blood pressure and pulse pressure, and these two signs are associated with higher cardiovascular morbidity and mortality.[133] Furthermore, increasing aortic stiffness with age occurs gradually and continuously and is similar in both men and women. On the other hand, cross-sectional studies have shown that aortic and carotid stiffness (evaluated by pulse wave velocity) increase with age by approximately 10–15% during a period of 10 y, and women always have 5–10% lower stiffness than men of the same age.

Abnormalities in a number of structural and functional properties of large arteries, including diameter, wall thickness, wall stiffness, and endothelial function, all occur with aging and contribute to the acceleration of arterial aging as a significant risk factor for the development of cardiovascular pathological conditions.[134] Genetic defects, as well as a multitude of biochemical, enzymatic, and cellular defects, participate in the arterial aging process. Locally derived and circulating factors, including NO, ET-1, and the natriuretic peptides, contribute in the short term to the functional regulation of large artery stiffness, changes in the balance between these factors, and, in particular, a reduction in NO production may well explain why conditions such as hypercholesterolemia and diabetes are themselves associated with arterial stiffening before the development of definitive atherosclerosis.[135] These vascular and biochemical alterations occurring with aging could be targeted by interventions aimed at preventing and/or attenuating the process (i.e., the modification of epigenetic factors such as diet and lifestyle, the development of new pharmacological agents, and potentially gene therapies). We concur with Najjar and associates[134] that new strategies should be developed to change or decrease the effects of molecules or signaling cascades that may accelerate intimal thickening (e.g., TGF-β), stiffening (e.g., NO bioavailability, deficits in elastin synthesis), protein degradation (e.g., MMP-2), arterial wall inflammation (e.g., MCP-1), fibrosis (e.g., Ang II), or injury (e.g., ROS). This should be the focus of future research.

THROMBOSIS IN AGING

Hypercoagulability and advanced vascular sclerotic changes may contribute to the increased incidence of thrombosis in the elderly. With aging, there are elevated plasma levels of clotting factors, including fibrinogen, factor VII, and factor VIII; increased levels of β-thromboglobulin and thromboxane A2 involved in platelet activation and aggregation; and reduced fibrinolysis primarily because of an increase in plasminogen activator inhibitor-1 (PAI-1), a principal regulator of fibrinolysis.[136,137]

Elevated PAI-1 plays a significant role in the increased incidence of thrombosis and vascular atherosclerosis observed in human cardiovascular aging, with cytokines and hormones, including TNF-α, TGF-α, angiotensin II, and insulin, positively regulating the gene expression of PAI-1. Although PAI-1 may also induce other pathological conditions associated with aging such as obesity, insulin resistance, immune responses, and vascular sclerosis/remodeling, the genetic mechanism of aging-associated PAI-1 induction is not yet known.[138]

CARDIOMYOPATHY OF AGING: REMODELING

The failing human myocardium (i.e., cardiomyopathy of aging) is a complication not uncommon in aging/senescent patients. HF will occur by a dynamic series of events set in motion after a threshold of physiological imbalance is reached. Multiple and complex interacting factors, including CAD, hypertension, and diabetes, may contribute to the process, resulting in progressive cardiac enlargement and dysfunction. Extracellular matrix and endothelial remodeling, with a potpourri of interacting molecular and metabolic changes, will occur.

Extracellular Remodeling

By use of echocardiogram analysis and the measurement of collagen, metalloproteinases (MMPs), and tissue inhibitor of the metalloproteinase (TIMP) levels, as well as fibroblast function, Lindsey and associates[139] have studied the effects of aging on LV geometry, collagen levels, MMPs, TIMP abundance, and myocardial fibroblast function in young (3 mo) and old (23 mo) CB6F1 mice. They found specific differences in cellular and extracellular processes likely contributing to the age-dependent extracellular matrix (ECM) remodeling.

Structural changes are mainly implicated in the development of LV diastolic dysfunction in the aging human heart, but changes in systolic function may also occur. De Santis and associates[140] measured the LV end-systolic elastance (Ees), a major determinant of systolic function and ventricular–arterial interaction, in a group of adult patients with either DCM or hypertensive cardiomyopathy compared with a control group of similar age and to an elderly group (mean age, 76.3 y). Ees was reduced in DCM and increased in patients with hypertensive cardiomyopathy compared with age-matched control subjects. Ees was significantly higher in the elderly group and was linearly correlated with age, suggesting that increased Ees may contribute to the cardiomyopathy and HF of the aging heart.

The increase in Ees is characterized by a significant reduction in stroke volume with little modification in afterload and also may worsen diastolic dysfunction in the aging heart through further limited filling and raised diastolic pressure.[141] These factors may contribute to decompensation of the aging heart in response to an episodic event such as acute increase in blood pressure or tachydysrhythmias. Furthermore, because atrial fibrillation, which is closely associated with increased age, involves extensive remodeling of the cardiomyocyte electrical properties, ECM, and fibrosis, it could lead to acute HF, without clinically apparent cardiac disease, in the aging heart.

Maladaptive cardiac tissue remodeling could be prevented by expression of TIMP-3, because its deficiency alone may be sufficient to cause progressive remodeling and dysfunction similar to human HF. Mice with a targeted TIMP-3 deficiency, analyzed with respect to aging and compared with age-matched wild-type litter mates, demonstrated that loss of TIMP-3 function triggered spontaneous LV dilatation, cardiomyocyte hypertrophy, and contractile dysfunction at 21 mo of age, consistent with human DCM.[142] TIMP-3 absence also resulted in interstitial matrix disruption with elevated MMP-9 activity and activation of TNF-α, a hallmark of human myocardial remodeling. TIMP-3 deficiency seems to disrupt matrix homeostasis and the balance of inflammatory mediators, eliciting the transition to cardiac dilation and dysfunction. It is possible that therapeutic restoration of myocardial TIMP-3 may limit the cardiac remodeling and progressive failure often developing in aging patients with DCM.

Endothelial Remodeling

Decline in endothelium-dependent regulation of vasodilation in aging is due, in part, to increased superoxide and peroxynitrite formation, diminished eNOS, estrogens, and increased inflammatory signaling that promotes iNOS.[143] In a recent review of the vascular consequences of menopause and the relevance of estrogen therapy, addressing the lack of protection against CVD provided by estrogen supplementation found in the disappointing large-scale clinical trials. Dubey and associates stressed the critical importance of timing, age, and the type of estrogen that would be required to effectively delay vascular remodeling.[144] Furthermore, aging Fischer 344 rats exhibited a decreased endothelium-dependant responsiveness of coronary arterioles to endothelin, potassium chloride, and pressure-induced myogenic responses.[145] In studies to determine the role of endothelin-1 (ET-1) and PKC signaling in age-related increases in coronary vascular resistance, Korzick and associates found enhanced PKC-mediated vasoconstriction in response to ET-1 in aged coronary arteries, the effects of which could not be explained by alterations in voltage-gated calcium channel (VGCC)–induced mechanisms.[146] Their data suggested that enhanced ETA receptor-mediated vasoconstriction in the coronary vasculature is, in part, mediated by PKC-dependent mechanisms localized in vascular smooth muscle, supported by immunoblot analyses revealing increased PKCβI, PKCβII, and PKCα levels in isolated coronary arteries of aged compared with adult rats. On the other hand, Donato and associates[147] reported an age-associated increase in gastrocnemius arteriole vasoconstrictor responsiveness and sensitivity to endothelin-1 but were unable to demonstrate that exercise training could reverse the age-associated effects of the vasoconstrictor responsiveness to endothelin-1.

Other studies have demonstrated other changes in vascular/endothelial remodeling with aging. Increases in large artery stiffness and resultant loss of capacitance and faster pulse wave velocity occur with aging.[148] Vascular remodeling that ultimately predisposes to hypertension includes loss of SMCs and elasticity, medial calcification, and accumulation of advanced glycation end products in collagen and elastin. Furthermore, remodeling of the ECM, under the regulation of MMPs, occurs also in cardiac aging.[139,149,150]

ENERGY METABOLISM IN THE AGED HEART

Experimental models of aging have shown that decline in myocardial function is at least partly associated with a pronounced shift in myocardial intermediary metabolism from primary fatty acid substrate to glucose. With advancing age, the changes noted in cardiac muscle are related to modifications in membrane fatty acid composition, decreasing levels of polyunsaturated fatty acids, and increasing levels of saturated fatty acids. Moreover, significant reduction in heart mitochondrial cardiolipin content occurs with aging. Cardiolipin, an anionic phospholipid, is the principal polyglycerophospholipid (carrying four acyl groups and two negative charges) found in the heart and the most unsaturated cellular phospholipid that is localized mainly in mitochondria. The age-related reduction in cardiolipin is considered to have a major impact on cardiac mitochondrial membrane transport function, fluidity, and stability. Studies with cultured adult cardiomyocytes isolated from rat hearts of a broad age range also exhibited changes in the fatty acid profile related to alterations in the mechanism of desaturation and elongation of essential fatty acids. The ability of heart cells to metabolize linoleic acid to higher and more unsaturated metabolites decreased with age, and the pattern of fatty acids of the cultured cardiomyocytes showed a gradual, but significant, shift, similar to that reported in the entire heart.[151]

Available data on cardiac carnitine levels in healthy aging people are rather limited. However, a marked reduction of carnitine and its derivatives in muscle, and of long-chain acyl-carnitine, has been reported in hearts of older mice and rats compared with younger animals.[152,153] In healthy humans, analysis of muscle samples showed drastic reduction of carnitine and acetyl carnitine in older subjects, with strong reverse correlation between age and carnitine levels.

Most of the information available concerning the switch from fatty acid to glucose metabolism in the failing aging heart (mild initially and significant in advanced HF) has been obtained from different animals models (e.g., rats and dogs). Data from humans, although currently limited, is slowly becoming available. Kates and associates,[154] using the technique of positron emission tomography (PET), measured myocardial blood flow, myocardial oxygen consumption (MVO_2), myocardial fatty acid use and oxidation, and myocardial glucose use in 17 healthy young normal men (mean age, 26 y) and 19 healthy older men (mean age, 67 y). BF was similar between the groups, but MVO_2 was higher in older men. Rates of fatty acid use and oxidation were significantly lower in older subjects, and the rates of glucose use did not differ between the groups. These findings show a decline in fatty acid use and oxidation in the aging heart, with a likely increase in the relative contribution of glucose use to substrate metabolism. On the other hand, Soto and associates[155] carried out a similar study evaluating the effect of dopamine infusion (β-adrenergic stimulation) on myocardial metabolism in young (mean age, 26 y) and older individuals (mean age, 69 y) by PET. Although the elevation in rate-pressure product, BF, and MVO_2 measurements was similar in both groups, glucose use increased only in the younger group. These findings suggest that lack of an increase in myocardial glucose use may lead to a state of energy deprivation, and this could partially explain the age-related decrease in contractile function during stress; they also suggest that if the aged heart under stress is more dependent on fatty acids as a source of energy, it may be more susceptible to manifestations of myocardial ischemia caused by inhibition of fatty acid β-oxidation. However, additional studies will be necessary to definitely clarify these issues.

It is a well-known phenomenon that in biological aging there is a decrease in stress tolerance; however, its biochemical basis is incompletely understood. AMP-activated protein kinase (AMPK), an important regulator of cellular metabolism during stress, may be decreased with aging, contributing to the poor stress tolerance of aged cardiac and skeletal muscle. Gonzalez and associates[156] found that old mice after 10 min of hypoxemia showed significantly higher activity of myocardial α2-AMPK but not α1-AMPK than young mice, probably related to differences in the phosphorylation of α2-AMPK. However, the hearts from the young mice did not show significant activation of AMPK until 30 min of hypoxemia, duration of stress that was poorly tolerated by old mice. In contrast, AMPK activity in the gastrocnemius muscle was not affected by age or hypoxemia. These findings suggest that an age-associated decline in cardiac and skeletal muscle hypoxic tolerance is not caused by changes in basal AMPK activity or a blunted AMPK response to hypoxia. On the other hand, Wang and associates[157] found evidence that increased AMPK activity directly contributed to the implementation of the senescent phenotype in cultured fibroblasts. Treatment of human fibroblasts with AMPK activators such as 5-amino-imidazole-4-carboxamide riboside, antimycin A, and sodium azide triggered a senescence phenotype in fibroblasts, such as the acquisition of senescence-associated β-galactosidase (β-gal) activity, as well as increased p16INK4a expression. Infection of cells with an adenoviral vector that expresses active AMPK increased senescence-associated β-galactosidase activity, whereas infection with an adenovirus that expresses dominant-negative AMPK decreased senescence-associated β-galactosidase activity. These findings suggest that AMPK activation can cause premature fibroblast senescence likely by mechanisms that involve a reduction in HuR (an AMP-regulated cytoplasmic RNA-binding protein) function. Nevertheless, the seemingly contradictory results may be accounted for by the use of different tissues/systems.

SKELETAL MUSCLE CHANGES

Progressive catabolism occurring with chronic HF and aging may result in skeletal muscle atrophy, a problem that may be compounded by the relatively poor regenerative properties of skeletal muscle. With aging, there is a steady decline in mass and functional performance with fibrotic invasion replacing the muscle contractile tissue and loss of muscle fibers.[158,159] Abnormalities in muscle structure and function are also found in neurodegenerative syndromes and disease-related cachexia.[160] Schulze and associates[161] have investigated the pathways of skeletal muscle proteolysis with an experimental model of chronic LV dysfunction. Twelve weeks after MI, skeletal muscle atrophy developed in wild-type mice accompanied by increased total protein ubiquitination and enhanced proteasome activity, activation of Foxo transcription factors, and marked induction of atrogin-1/MAFbx, a muscle-specific ubiquitin-ligase required for muscle atrophy. Further observations have identified skeletal muscle myosin as a specific target of ubiquitin-mediated degradation in muscle atrophy. In contrast, transgenic IGF-1 overexpression prevented muscle atrophy and proteasome activity stimulation, inhibited skeletal muscle Foxo4 activation, and blocked atrogin-1/MAFbx expression. These results suggest that skeletal muscle atrophy occurs through increased activity of the ubiquitin-proteasome pathway and that inhibition of muscle atrophy by local IGF-1 provides a promising therapeutic avenue for the prevention of skeletal muscle atrophy in chronic HF, and potentially other chronic diseases associated with skeletal muscle atrophy.

Musaro and associates[162] generated a model of persistent, functional myocyte hypertrophy with a tissue-restricted transgene encoding a locally acting isoform of IGF-1 that is expressed in skeletal muscle (mIGF-1). Transgenic embryos developed normally, and postnatal increases in muscle mass and strength were not accompanied by the additional pathological changes seen in other IGF-1 transgenic models and manifested expression of GATA2, a transcription factor normally undetected in skeletal muscle. This study demonstrated that the integrity and regenerative capacity of total body skeletal musculature is safely extended by local mIGF-1 supplementation into senescence. Even when expressed under the control of strong muscle regulatory elements, the mIGF-1 protein seems to remain in the muscle bed and does not enter the circulation, thereby avoiding hypertrophic changes in distal organs such as the heart, at the same time eliminating the risk of possible neoplasms induced by abnormally high levels of circulating IGF-1.[163] On the other hand, other observations[164] revealed that other IGF-1 constructs expressed specifically in skeletal muscle and heart, was also detected in the serum of transgenic animals at all ages and resulted initially in physiological cardiac hypertrophy that progressed to a pathological hypertrophic state.

The expression of two genes encoding ubiquitin-protein ligases, MAFbx/Atrogin-1 and MuRF1, is increased during muscle atrophy. Experiments with mouse knock-out models have demonstrated that MAFbx and MuRF1 are required for muscle atrophy and thus might be targets for clinical intervention. Other strategies for blocking muscle atrophy involve the stimulation of pathways leading to skeletal muscle hypertrophy such as with IGF-1 overexpression, which may induce skeletal muscle hypertrophy by activating the PI3K-Akt pathway[165] or by targeting caspase-3, which cleaves actin to facilitate its destruction by the ubiquitin-proteasome system for attenuating muscle wasting.[166]

On the other hand, according to Anversa,[167] the skeletal muscle–specific IGF-1 isoform may neutralize the decline in mass and functional performance occurring with old age and ventricular decompensation, protecting the pool of satellite cells, which may replace senescent dying cells, and further supporting the hypothesis that well-preserved functional cardiac stem cells and skeletal muscle satellite cells maintain the youth of the heart and skeletal muscle opposing aging effects and disease states. Nevertheless, caution is to be exercised in the translation of results from simpler organisms to large mammals, mainly to humans in whom the life and death of most somatic organs is regulated by a stem cell compartment.

At this point it is safe to say that these strategies aimed to reconstitute muscle mass with systemic administration of growth factors, and the potential implementation of cell therapies is a promising avenue.

REDUCED TOLERANCE TO MYOCARDIAL ISCHEMIA AND REPERFUSION INJURY

Most of the data thus far accumulated regarding the tolerance of the aging heart to I/R have been generated from different animals models. In rat models, compared with the young adult, senescent animals exhibit a reduced tolerance to myocardial I/R injury. I/R is associated with an increased number of circulating leukocytes and generation of superoxide in the peri-ischemic areas of the heart of young compared with aged animals, and it induces a significant decrease in cardiac index and stroke volume index of aged compared with young animals. Furthermore, aged rats exhibit an increase in the ratio of Bax mRNA to Bcl-2 mRNA and cardiomyocyte apoptosis after myocardial I/R, which may explain, at least in part, the increased myocardial dysfunction.[168] Although myocardial I/R can induce OS and an inflammatory response (i.e., increased infiltration of leukocytes and myeloperoxidase activity), with significantly higher plasma levels of TNF-α and IL-1β in the perinecrotic cardiac areas of young compared with aged rats, plasma 8-hydroxy-2′-deoxyguanosine levels and creatine kinase activity were increased in the aged compared with young rats exposed to I/R.[169] In addition, increased reperfusion damage was associated with a significant decrease in the plasma ratio of GSH to glutathione disulfide (GSSG) in the aged rats. These findings suggest that enhanced I/R injury in the aged rat heart may be due to reduced antioxidative capacity rather than to an increase in ROS production.

Although the consensus among investigators interested in aging is that the aged heart has a decreased functional and adaptive reserve capacity to tolerate and respond to various forms of stress with increased likelihood of myocardial ischemia and cardiac dysfunction, the bulk of studies with ischemia and CP have been conducted in young and middle-aged animals and cells. Research on the tolerance to ischemia and CP should be carried out mainly in the aging or senescent heart, because this will be most relevant to humans. Interestingly, in one of the few studies carried out in humans, age did not prove to be a factor in the tolerance to myocardial ischemia. Loubani and associates[170] studied a group of 128 patients undergoing elective heart surgery (age range, 30–90 y). In the collected right atrial specimens, tissue injury and viability were assessed after different settings of ischemia, reoxygenation, and preconditioning. They found that age did not influence myocardial tolerance to ischemia or the protective effect of ischemic preconditioning, indicating the need for a reevaluation of the importance of age in risk scoring in cardiac surgery. Interestingly, in a previous study it was found that the operative mortality in elderly patients has decreased significantly in recent years, despite an increase in the prevalence and severity of their risk factors.[171] The authors of this study pointed out that careful weighing of risk, rather than advanced age alone, should determine who is offered surgical revascularization, poor ventricular function and repeat coronary artery bypass graft surgery being the greatest risk factors impacting on the operative mortality in elderly patients.

Although the application of preconditioning/CP in human CVDs and the aged heart has been limited, it may be a useful adjunct in their treatment. Further research to ascertain the effect of age on the myocardial response to ischemia and CP is needed.

APPRAISAL OF MITOCHONDRIAL FUNCTION IN AGING

ROS, considered the pathogenic agent of many diseases and aging, is mainly the product of the mitochondrial respiratory chain. The mitochondrial theory of aging suggests that somatic mutations of mtDNA, induced by oxygen free radicals, are the primary cause of energy decline (Fig. 5). Although questions remain regarding the validity of this theory, it is supported by a large number of observations describing mtDNA instability and shortcomings of the mtDNA repair machine, as well as the absence of histones surrounding mtDNA, molecules that protect the nuclear genome from damage. This lack of stability is accompanied by an increasing number of mtDNA deletions with aging, probably secondary to increased oxygen free radicals.

Although the aging heart suffers greater damage after an episode of I/R than the adult heart, the occurrence and degree of aging-related defects in mitochondrial OXPHOS remain uncertain. Interestingly, aging-related OXPHOS defects have been found to be limited to interfibrillar mitochondria (IFM) in which complex III and IV activity and rate of OXPHOS were decreased, whereas the subsarcolemmal mitochondria (SSM) remained unaffected.[172–174] The selective alteration of IFM during aging raises the possibility that the consequences of aging-induced mitochondrial dysfunction will be enhanced in specific subcellular regions of the senescent cardiomyocyte. Recent studies suggest that mitochondrial ROS cause OS and impaired mitochondrial function in IFM to a greater degree than in SSM with age, and because of their proximity to myofibrils, IFM are probably the primary source of ATP for myosin ATPases, and, therefore, the OS in IFM may be the culprit for the myocardial dysfunction occurring with aging.[175] It is important to keep in mind that subfractionation of mitochondria may provide a mixture

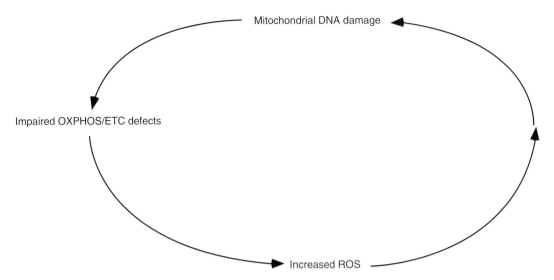

FIGURE 5 A vicious circle is initiated by somatic mtDNA damage, leading to eventual defects in the aging heart.

of organelles, of SSM and IFM that may complicate the assessment of the age-related changes in mitochondrial oxidant production and OS. Consequently, further research in this area is needed.

In postmitotic cells (e.g., cardiomyocytes), mitochondria have a limited life span of a few weeks. Their replacement during normal turnover requires an intergenomic coordination between both the mitochondrial and the nuclear genome. As mentioned earlier, age-induced mtDNA damage has been found to accumulate in cardiac tissue, and to some degree it may be a secondary phenomenon (e.g., ROS accumulation). Nevertheless, there are clonal accumulations of damaged/mutated mtDNA within individual cells, up to homoplasmy of mutated mtDNAs, which are either neutral with regard to phenotype or which cause substantial phenotype alterations. A deficient mitochondrial respiration phenotype (likely with less ROS accumulation and less ATP production) or a phenotype with a deficient ETC (i.e., defects within one or more respiratory enzymes with enhanced ROS formation proximal to the defect(s) and with enhanced susceptibility to OS-triggered apoptosis) may explain the progressive loss of cardiomyocytes occurring with advanced age. Thus, age-associated mtDNA alterations may explain important features of the aging heart such as myocyte loss and myocyte heterogeneity. However, a definitive proof of a genotype-phenotype correlation in aging heart is not yet available.[176]

In the past decade, the accumulation of a significant number of mtDNA point mutations has been reported in humans above a certain age. These mutations were found in fibroblasts and skeletal muscle mtDNA main control region (D-Loop), at critical sites for mtDNA replication.[177] More recently, Marin-Garcia and colleagues,[178] in their study on the incidence and location of D-loop mutations in myocardial mtDNA as a function of age in patients with cardiomyopathy and in normal controls, found no evidence of an age-related correlation in the accumulation of homoplasmic mutations in either group. Similar data were found for heteroplasmic mutations; because they did not accumulate in incidence with age and neither was there an age-related accumulation in cytochrome b structural gene mutations. In contrast, damage in mtDNA coding genes and a decline in bioenergetic generation have been found in human cardiac tissues from aging patients with and without cardiac disease.[179,180] In humans, considerable research interest has recently developed focused on the role that pathogenic mtDNA mutations may play in programmed cell death, even though mtDNA damage and defective mitochondrial respiration may not be essential factors for the process of apoptosis to occur. The integrity of mtDNA may influence the rate of apoptosis during aging, most probably by regulating ROS production. However, there is not agreement in the research community whether or not mitochondrial dysfunction really occurs with aging. A recent study in humans revealed that mitochondrial respiratory enzymes maintained normal activities in the aging heart, and thus defects in ETC cannot be considered the main cause of the increased oxidative damage associated with aging.[181] At odds with this study are the findings from rat experiments that the aging heart sustains greater injury during I/R than the adult heart, and, accordingly, aging decreases OXPHOS and the activity of complexes III and IV, although only in interfibrillar mitochondria[172–174,182]

Interestingly, as previously noted, cytochrome c was increasingly released from aging rat heart mitochondria, whereas the antiapoptotic Bcl-2 protein showed a strong tendency to decrease with age, and Bax, a proapoptotic protein, remained unchanged.[19,20] Moreover, indicative of the chronic OS that develops with age, heart mitochondria from old animals displayed increased MnSOD and GPx activity, as well as increases in lipid peroxidation.

Decreased mtDNA levels have been reported in aged human skeletal muscle, probably as a consequence of having fewer mtDNA copies per mitochondrion in older muscle, and mtDNA concentration was weakly related to physical fitness.[183] However, at present there is not clear evidence in humans that myocardial mtDNA copy number decreases with aging. Experiments with 6- and 27-month-old rats have shown an age-related reduction of mtDNA in skeletal muscle and liver, but not in the heart.[184] Also, a decline in mtDNA was not associated with reduced COX transcript levels in tissues with high oxidative capacities, such as red soleus muscle or liver, but transcript levels were reduced in the less oxidative mixed-fiber gastrocnemius muscle with aging. Consistent with transcript levels, COX activity also remained unchanged in aging liver and heart but declined with age in the lateral gastrocnemius. Thus, the effects of aging on mitochondrial gene expression are tissue specific, and mtDNA levels are preserved in the aging heart muscle, presumably because of its continual aerobic activity. Unfortunately, data on the aging human heart are rather limited. In our laboratory, we have found an increase in specific myocardial mtDNA deletions (the common 5- and 7-kb deletions) in aging patients with idiopathic DCM compared with the young patients.[179] In this group of patients, the age-related decline in ETC and the increased accumulation of mtDNA deletions may be the result of oxidative damage, which has been posited to increase with aging.[185] Moreover, in these studies, whereas age seems to play a role in the increased incidence of specific myocardial mtDNA deletions and in the development of multiple respiratory enzymes defects, it was found to be noncontributory to the severity or frequency of single enzyme activity defects.

Recently, the potential role of point mutations and deletions on aging has been addressed experimentally by creating homozygous knock-in mice that express a proof-reading-deficient version of Polγ, the nuclear-encoded catalytic subunit of mtDNA polymerase γ.[186] The knock-in mice developed a threefold to fivefold increase in the levels of point mutations, as well as an increased amount of deleted mtDNA. The increase in somatic mtDNA mutations was associated with reduced life span and premature onset of aging-related phenotypes, including heart enlargement, thus providing a causative link between mtDNA mutations and aging. Furthermore, the accumulation of age-dependent mtDNA point mutations and deletions in post-mitotic tissues has been found to be significantly elevated in individuals with either mutations in the mitochondrial helicase Twinkle or in Polγ, suggesting that the activity of proteins involved in mtDNA replication may be targeted in aging.[187]

In response to the increased level of ROS in the senescent heart, genes encoding mitochondrial stress proteins, including both heat shock and antioxidant response proteins, such as HSP60, HO-1, and GPx, are up-regulated.

Increased levels of HSP60 protein have been reported in association with increased levels of protein import into mitochondria,[188] indicating that defective mitochondrial protein import is not the primary cause of mitochondrial dysfunction. A significant increase in myocardial mitochondrial GPx activity has been found in the senescent heart,[189,190] and elevated GPx levels have been considered an important defensive mechanism in dealing with free radical damage. Increased expression of GPx and HO-1 has been found in the senescent Fischer 344 rat heart, and this suggests that a signaling mechanism for this transcriptional response can be identified (Marin-Garcia and associates, unpublished data). In addition to, and likely acting downstream of, the effects on mitochondria and ROS generation, myocardial aging can induce and activate nuclear transcription factors effecting cellular programming. In our laboratory, we found that the transcription factor NF-κB, which is involved in triggering a broad array of inflammatory cytokines and stress protein responses in myocardium, was increased in the senescent heart. Aging-related stimulation of NF-κB transcription factor activation is ROS-sensitive[191] and has been previously identified in the heart.[192] Interestingly, the activation of NF-κB by mitochondrial-produced ROS is a good example of nuclear-mitochondrial cross-talk.

Another consequence of increased OS and ROS accumulation is myocardial apoptosis. Cardiac remodeling and myocyte cell loss are well-established findings in the senescent Fischer rat model.[193,194] In the 29-month-old senescent rats, several studies have documented significant increases of myocardial apoptosis with DNA laddering, cytochrome c release, and TUNEL analysis.[194,195] Although recent findings of increased antiapoptotic Bcl-2 levels and procaspase3 levels in the senescent heart do not directly support an increase in myocardial apoptosis with aging (Marin-Garcia and associates, unpublished data), they may contribute to an increased age-related sensitivity to apoptotic stimulation along with proapoptotic changes in apoptosis programming.[196,197] Also, senescent hearts exhibited an up-regulation of the protein phosphatase (PP2A) regulatory subunit B56α, an abundantly expressed cardiac protein involved in multiple cell growth and signaling pathways, including dephosphorylation of the antiapoptotic protein Bcl-2.[198,199]

It is known that the opening of the mitochondrial permeability transition (PT) pore represents a landmark event of early myocardial apoptosis, as well as a result of OS in cardiomyocytes; surprisingly, there is little information on PT pore opening in the aging heart. In our laboratory, we detected an increased Ca^{++} sensitivity of the PT pore opening in the senescent compared with the young adult heart. Whether this inner membrane event contributes to the mitochondrial respiratory enzyme dysfunction observed in the aging heart, serves as a "gate" to the nuclear-mitochondrial cross-talk and signaling, represents a by-product for the production of mitochondrial ROS, or is an intrinsic component of the myocardial apoptosis triggered during aging remains to be determined. Di Lisa and Bernardi have noted that the causal relationships between dysfunctional mitochondria and age-related myocardial pathological conditions are not yet well defined, probably because of current methodological limitations that hamper the study of mitochondria and PT pore *in situ*.[200] Nevertheless, mitochondria seem to be involved in the increasing susceptibility to injury displayed by the aging heart, and under these conditions mitochondria become dysfunctional because of increased ROS formation and further ROS accumulation. This vicious circle is likely to favor PT pore opening. We concur with the aforementioned authors that other mitochondrial channels might contribute to the dysfunction of the organelle.

It has been suggested that the anion channel of the inner mitochondrial membrane (IMAC) causes the propagation of ROS from a subset of damaged mitochondria to the rest of the cell, generating oscillations of mitochondrial membrane potential.[201] However, at present, no information is available on the activity of IMAC or of the mitochondrial K_{ATP} channel (which is ROS sensitive) in the aging heart. Further investigation on the role of IMAC and mitoK$_{ATP}$ in ROS-induced aged-related defects are warranted.

Another interesting aspect of the mitochondrial changes during cardiac aging is their bioenergetics and biogenesis response to thyroid hormone (TH). Recently, evaluation of nuclear-mitochondrial cross-talk and the time course of events involved in cardiomyocyte T3 signaling have been carried out.[202] Neonatal cardiomyocytes responded to T3 in a similar manner to young adults, with increased mitochondrial enzyme activities; however, cardiomyocytes from senescent animals displayed a marked decline in their response to T3, indicating that the mechanism mediating stimulation of mitochondrial activities is inoperative with aging. The reduced T3-mediated mitochondrial response in senescent cardiomyocytes may be related to a decrease in nuclear TH receptor-α levels, which has been described in the aging cardiomyocyte[52] or, alternately, to a reduction in the mitochondrial membrane proteins (e.g., ANT, UCP) and lipids (e.g., cardiolipin), which are both altered and depleted in the senescent heart.[203–205]

That mitochondria play a significant role in the pathophysiology and pathogenesis of aging, and in particular of cardiac aging, seems unquestionable; however, a better understanding of the mechanisms through which mitochondria contribute to the aging process is necessary. This may facilitate the discovery of novel therapies to prevent the cardiac pathology and other diseases associated with aging and consequently to increase human life span.

GENETIC CONTRIBUTION TO CARDIAC AGING

The cracking of the genetic code has provided insights into the biological aspects of cardiac aging. Until recently, little information was available on the genetics of human cardiac aging, because most of the research had been carried out on several animal models, mainly mice. "Can mouse genetics teach us enough about the biology of aging?" is the question asked by Miller[206] and ourselves. If so, it will be possible "to guide the search for anti-aging medicines that can delay late-life illnesses," such as atherosclerosis and congestive HF. With the completion of the Human Genome Project (HGP), as well as the availability of complete DNA sequences from an increasing number of animal species, it is possible to say at this moment that genetics through the new science of comparative genomics is ready to open the way to positively answer the preceding question, and eventually new advances in scientific methodology and technology may allow the discovery of designed anti-aging drugs that can be safely used in humans. In addition, an understanding of the necessary changes to be made in our human lifestyle (see epigenetic factors) will be crucial in reaching the ultimate goal of longevity without disease, in another words, to die of age.

COMPARATIVE GENOMICS

The human genome is unintelligible by itself. To tie together the information from the human genome, we must use the tool of comparative genomics.[207] That is, we must compare the human genome to that of other organisms to understand which regions of the genome do what. In the wake of the HGP, high-throughput analyses in functional genomics are generating an immense amount of data. The functions of human genes and other DNA regions can be revealed by studying their parallels in non-humans, and, in principle, it could be very useful in the study of the human aging heart. The availability of complete genome sequences generated both inside and outside the HGP constitutes a major breakthrough in fundamental biology, because by comparing entire genomes research, advances in the evolutionary, biochemical, genetic, metabolic, and physiological pathways can be achieved. One important challenge is to distinguish genes that may influence cardiac aging, rather than being a consequence of aging. Although it is unlikely that a particular gene(s) may actively cause human aging/cardiac aging, it is probable that genes involved in the repair of DNA, including the "fragile" mitochondrial DNA, are down-regulated with age.

According to de Magalhaes and associates, mutations that either delay or accelerate aging in mice may provide some of the few clear hints on the genetics of aging.[208] On the other hand, both the clinical (as distinct from scientific) relevance and importance of basic animal research have been questioned, and a distinction has to be made whether a given gene affects the aging process or simply preserves health.[209] Moreover, to extrapolate whether a human homolog of a gene discovered in a model organism may be related to human aging in general, and cardiac aging in particular, is still rather difficult. To resolve this, the following criteria have been used[208]: (1) the influence of the gene in the model organism's aging process; (2) literature available suggesting that the human homolog has a similar function; (3) information on the phenotype of human variants or mutations of the gene; and (4) effects on aging of the genetic manipulation (e.g., overexpression or knock-out, of the gene's product[s] in mammals). Accordingly, evolutionary distant models such as invertebrates should have a much smaller impact than mammalian models, such as mice compared with human aging in general, and specifically cardiac aging.

Databases of genes related to human aging are available on the Internet. The Human Aging Genomic Resources (HAGR) is a collection of online resources for studying the biology of human aging. HAGR features two main databases: GenAge and AnAge. GenAge provides the most complete and comprehensive database of genes related to human aging on the Internet, as well as rendering an overview of the genetics of human aging. AnAge is an integrative database describing the aging process in several organisms and featuring, when available, maximum life span, taxonomy, developmental schedules, and metabolic rate, making AnAge a unique resource for the comparative biology of aging.

SINGLE NUCLEOTIDE POLYMORPHISMS/ CANDIDATE GENES/TRANSCRIPTION ANALYSIS

The discovery, description, and cataloging of single nucleotide polymorphisms (SNPs) and their involvement in age and disease may result in a better understanding of the pathophysiology and pathogenesis of CVDs in aging. In addition, determination of the sequence of the human genome and knowledge of the genetic code through which mRNA is translated can allow identification of mammalian proteins. However, less is known about the molecular mechanisms and environmental regulators that control expression of human genes and about the variations in gene expression that underlie many pathological states, such as diseases of the aging heart. This is caused in part by lack of information about the "second genetic code"–binding specificities of transcription factors. New tools to decipher this regulatory code may be critical for cardiac research to further explain the mechanisms by which known genetic defects induce transcriptional programs that control cell proliferation, survival, and angiogenesis. Furthermore, changes in the binding of transcription factors caused by regulatory SNPs may be a major factor in the detection of the familial predisposition to CVD.

Gene polymorphisms may influence drug response, and the genetic background may help to identify good responders and poor responders to the different treatments used in aging patients with cardiac disease. For example, β2 adrenergic receptor polymorphisms seem to influence the responsiveness to carvedilol in patients with HF. However, small sample size, differences in ethnic backgrounds, lack of result replication, and poorly defined functional significance of the genetic polymorphism are significant problems waiting to be overcome. Potentially, the availability of large population samples for collaborative studies may provide the answer to the preceding limitations.

GENE EXPRESSION IN CARDIAC AGING

The comparative analysis of developmental gene regulation between morphologically divergent animals, screening of intraspecific variation, and the response of organisms and genes to specific selection all support the assertion that regulatory DNA is the predominant source of the genetic diversity underlying morphological variation and evolution.[210] The differential expression of specific regulatory genes may determine the phenotypic variations among related organisms, as well the phenotypic changes in the same individual with aging. DNA microarrays are being used in cardiovascular aging research through generation and screening of panels of hundreds of transcriptional biomarkers, offering a new modality of measuring cardiac and skeletal muscles biological age, as well as the evaluation of procedures designed to postpone aging in these tissues. For instance, the transcriptional response to OS in the heart and how it changes with age has been studied by Edwards and associates.[211] Cardiac expression profiling showed genes associated with stress, inflammatory, immune, and growth factor responses were induced in young, middle-aged, and old C57BL/6 mice treated with a single intraperitoneal injection of paraquat. Only young mice displayed a significant increase in expression of all three isoforms of *GADD45*, a DNA damage-responsive gene; also, the number of immediate early genes found to be induced by paraquat was considerably higher in the younger animals. These findings demonstrated that, at the transcriptional level, specific inducible pathways in response to OS are impaired in the aged mouse heart.

High-density arrays have also been used to demonstrate that aging is associated with specific transcriptional alterations in other tissues such as skeletal muscle (e.g., gastrocnemius), cerebral cortex, and cerebellum of C57BL/6 mice, including an enhanced expression of inflammatory and stress genes (such as HSPs) and lower expression of metabolic and biosynthetic genes.[212,213] Similar gene profiling studies in mouse heart demonstrated that aging was associated with transcriptional alterations consistent with a metabolic shift from fatty acid to carbohydrate metabolism, increased expression of extracellular matrix genes, and reduced protein synthesis, albeit the enhanced stress transcriptional program seen in aging skeletal muscle and brain was not evident in the heart.[214] Furthermore, transcriptional expression patterns of genes required for energy metabolism in the aging heart are similar to those occurring in cardiac hypertrophy.

Caloric restriction (CR) but not antioxidant diets seems to stop the age-associated changes in gene expression suggesting that oxidative stress may only play a contributory role in cardiac aging. Studies in mice[212–214] have demonstrated that CR retarded or completely prevented the age-enhanced transcriptional changes, as well as several age-dependent physiological and biochemical changes, including increased steady-state levels of oxidative damage to lipids, DNA, and proteins. These gene-profiling studies in calorie-restricted animals suggested that CR retards the aging process by causing a metabolic shift toward increased protein turnover and decreased macromolecular damage. Interestingly, CR also resulted in alterations in myocardial gene expression consistent with preserved fatty acid metabolism, reduced endogenous DNA damage, decreased innate immune activity, apoptosis modulation, and a marked cytoskeletal reorganization.[214]

On the other hand, the profiling of gene expression in middle-aged monkeys (mean age, 20 y) subjected to CR in early adulthood (mean age, 11 y) resulted in an up-regulation of cytoskeletal protein-encoding genes and also a decrease in the expression of genes involved in mitochondrial bioenergetics.[215] Interestingly, no evidence was found for an inhibitory effect of adult-onset CR on age-related changes in gene expression such as reported in rodents. These findings indicated that, at least in rodents and primates, modulation of OS-induced transcriptional response might be a common feature of skeletal muscle aging and how much CR changes these responses may be species-specific.

Notwithstanding the attractiveness of microarrays as an important diagnostic technique, at present, there are still significant limitations associated with the reliability of the data generated by this method and with the interpretation. We are in agreement with Park and Prolla[216] that the diversity of cells present in the heart makes the gathering and interpretation of changes in gene expression rather difficult. Furthermore, there may not be a correlation between changes in RNA and proteins levels because, among a variety of reasons, changes in mRNA levels may be due to age-dependent changes in mRNA decay processes. Nevertheless, microarray gene profiling has revealed that in the mouse, aging results in a differential gene expression pattern specific to each tissue. It has also highlighted the important observation that CR can reverse most of the aging-induced alterations and retard the aging process by reducing endogenous damage and by inducing metabolic shifts associated with specific transcriptional profiles. Profiling of cardiac gene expression in conjunction with CR research might well be examined in humans (although noninvasive ways to approach this are sorely needed), and further research may open the way to find nutrients or drugs with effects similar to CR.

NEW APPROACHES IN THE ASSESSMENT OF CARDIAC INVOLVEMENT IN AGING

Proteomics

Proteomic screening still has a number of hurdles to be overcome, mainly protein family complexity and high individual variability, including age and gender (i.e., several proteins are differently expressed in male and female hearts). Furthermore, expressed proteins typically undergo a number of post-translational changes such as glycosylation, phosphorylation, and deamination that may complicate the understanding and significance of the pattern of their expression. At present, proteomic analysis seems to be a promising tool to advance our understanding of the pathogenesis and pathophysiology of the cardiac diseases occurring with aging, in particular HF. Using a proteomic approach in the aging monkey (*Macaca fascicularis*) heart, alterations in glycolytic and mitochondrial pathways were found together with down-regulation of specific proteins such a α-enolase, triosephosphate isomerase, pyruvate dehydrogenase E1 β-subunit, ATP-specific succinyl-CoA synthetase β-subunit, and 2-oxoglutarate dehydrogenase.[114] Interestingly, a significant gender difference in proteomic changes was noted in the aging monkeys.

Atomic Force Microscopy

Atomic force microscopy (AFM) is an imaging technique that has the potential to measure detailed micromechanical properties of soft biological samples, including changes in the mechanical properties of cells. This technique has been used to quantify the micromechanical properties of cultured rat atrial myocytes, in which alterations in cell contractile activity, with physiological perturbations and dynamic changes in cell stiffness during a single contraction, have been observed.[217] This technique offers the possibility of directly observing the relationship of the behavior and morphology of the cell with its biochemical and/or structural environment. A comparative study among different cells types has found that cardiac cells were the stiffest, the skeletal muscle cells were intermediate, and the endothelial cells were the softest with a range of elastic moduli depending on the location of the cell surface tested. Cardiac and skeletal muscle exhibited nonlinear elastic behavior, and these passive mechanical properties were generally consistent with the function of these different cell types.[218]

Recently, the mechanical properties of aging cardiomyocytes were determined by use of AFM by evaluating changes at nanoscale resolution in myocytes, from young (4 mon) and old (30 mo) male Fischer 344 × Brown Norway F1 hybrid rats. A significant increase in the apparent elastic modulus of single, aging cardiac myocytes was detected, supporting the concept that the mechanism mediating LV diastolic dysfunction in the aging heart may reside at the level of the myocyte.[219]

Recombinant DNA and Gene Expression Techniques

As in so many other areas of clinical and biological research, advances in recombinant DNA techniques are increasingly being used to study the aging heart. For some time it was assumed that mtDNA repair machinery was not present, poorly developed, or inefficient. Recently, the presence of mtDNA repair machinery in mammalian cells has been unequivocally demonstrated by the removal of a number of DNA lesions from mtDNA in cells exposed to various deleterious chemicals.[220] The activity of several proteins that process damaged DNA has been detected in mitochondria. Specific evidence has been found for mitochondrial base excision repair (BER), mismatch repair, and recombinational repair mechanisms, whereas mitochondrial nucleotide excision repair (NER) has not been detected.[221]

In aerobic cells, the most frequent type of injury encountered by nuclear DNA and even more so by mtDNA is damage arising from ROS. This damage includes single-strand breaks and oxidative base damage. Mitochondria are proficient organelles at removing oxidized and alkylated lesions generated by alloxan, acridine orange, and monofunctional alkylating agent, which are typically mended by base excision repair. These agents enter the cell, undergo redox cycling, and produce ROS similar to that normally generated in mitochondria. The addition of 5 mM alloxan to cultured rat cells increased the rate of oxidative base damage and by several fold the lesion frequency in mtDNA. After alloxan removal, the frequency of oxidized bases decreased rapidly, returning to levels found in mitochondria from untreated cells, indicating that mitochondrial repair of these lesions is extremely efficient.[222]

Animal models have been developed in which nuclear genes (coding for antioxidant enzymes active in mitochondria) have been inactivated, and deliberate, targeted mutations have been introduced into mitochondrial DNA for the purpose of observing the consequences.[223] Furthermore, nuclear DNA genes that govern energy production in mitochondria have been inactivated, and mice so treated developed heart defects that mimic human diseases. Analogous to the models developed for mitochondrial-associated neurodegenerative disorders, these animal models may provide the basis for investigating the etiology and potential therapies for a number of cardiac diseases of aging (i.e., cardiomyopathy).

Because down-regulation of proteins and factors that neutralize ROS can have negative consequences, it seems reasonable that increased expression of genes encoding these proteins might, in some situations, be used as a therapeutic modality. In the future, it may be possible, by raising antioxidant enzyme levels, to block the increased oxidative stress that accompanies aging. In addition, it is anticipated that in the near future mtDNA may be accessible to manipulation in an effort to stop or slow the organelle decay occurring with cardiac aging and other age-related diseases.

As noted earlier, telomere length gradually decreases as a function of age in most human tissues, including the heart. Evidence from epidemiology and genetic studies supports a contributory role for telomerase repression and short telomeres in a broad spectrum of human diseases, including CVDs. Cells in culture tend to lose telomeric DNA and undergo changes that mirror age-associated changes *in vivo*. However, telomerase-transduced cells exhibit extended replicative capacities, increased resistance to stress, and improved functional activities *in vitro* and *in vivo* without loss of differentiation capacity or growth control.[224] Thus, activation of telomerase may have significant potential for the treatment of a broad spectrum of chronic or degenerative diseases, including the cardiac pathologies of aging. Blasco and associates[225] demonstrated the role that telomerase plays in physiological and pathological processes in a mouse strain, in which the gene encoding the telomerase RNA component was deleted (Terc$^{-/-}$ mice). The complete loss of the Terc gene profoundly affected successive generations of mice. Importantly, defects in mice lacking the RNA component of telomerase involve apoptosis, not just proliferation defects,[226] suggesting that telomerase may protect cells against, at least some, of the causes of programmed cell death.[227] Furthermore, these findings bring up the possibility that telomerase replacement, or other telomerase-targeted therapies, may ameliorate or rescue the cardiac defects of the aged heart, including HF.

It has been pointed out that because telomerase-directed therapy would probably be administered relatively late in the course of a person's life, it is important to understand the effects of inducing telomerase expression in mostly nondividing cells, which represent a large fraction of the cardiomyocyte cell population at this stage of life.[228] Oh and associates[229] successfully demonstrated that the forced expression of exogenous TERT in mice cardiac muscle was sufficient to rescue telomerase activity and telomere length, even in adults. Increased ventricular myocyte density and DNA synthesis were initially present, indicating delay in cell-cycle exiting. However, by 12 weeks, cell cycling had largely subsided, and cell enlargement was present; this hypertrophic response was provoked at later ages, without mechanical dysfunction or fibrosis as initiating factors. Furthermore, *in vitro* studies have shown that viral delivery of TERT in cultured cardiac myocytes was sufficient for hypertrophy and that the introduction of TERT-containing virus to cardiac myocytes also conferred protection from apoptosis, both *in vitro* and *in vivo*. These studies have also demonstrated that TERT activity must be present for cell hypertrophy and survival, because it was not seen in constructs containing a catalytically inactive mutation in TERT. These findings confirmed that TERT can delay cell cycle exit in cardiac muscle, induce hypertrophy in postmitotic cells, and promote cardiac myocyte survival, but did not induce cell proliferation. However, the decrease in cell death may be enough to slow down the cardiac deterioration of the aged heart.

To further explain the role of telomeres in senescence, and its linkage with apoptotic signaling and cardiac dysfunction, additional studies by Oh and associates revealed that cardiac apoptosis in human HF is associated with the defective expression of the telomere repeat-binding factor TRF2, resulting in telomere shortening.[230] Furthermore, *in vivo* observations showed that mechanical stress was sufficient to down-regulate TRF2 levels, shorten telomeres, and activate the DNA damage checkpoint kinase, Chk2, in the mouse myocardium; therefore, transgenic TRF2 expression provided protection from all three responses. Interestingly, *in vivo* studies with cultured cardiomyocytes showed that interference with either TRF2 function or expression promoted extensive telomere erosion and myocyte apoptosis, indicating that cell death can occur by this pathway even in post-mitotic, non-cycling cells. Taken together, these findings support a role for telomere dysfunction in apoptosis in HF and suggest an essential role for TRF2 even in post-mitotic cells.

Similarly, Armstrong and associates,[231] by use of microarray analysis, have determined the effects of TERT overexpression in murine embryonic stem cells (ESCs). They found that TERT-overexpressing ESCs grow faster and are more resistant to apoptosis than are wild-type ESCs. Differentiated progeny with high levels of telomerase activity showed lower levels of spontaneous apoptosis, slower accumulation of peroxides, and significantly higher numbers of cells able to differentiate along hematopoietic lineages. In summary, these findings support the potential use of TERT overexpression to block the cell loss occurring in the aged human heart.

THERAPIES

Cardioprotection (CP) in Aging

The aging heart has a decreased capacity to tolerate and respond to various forms of stresses, and the likelihood of myocardial ischemia and cardiac dysfunction increases.

With aging, there is an age-associated loss of both TNF-α–induced platelet-derived growth factor (PDGF)-AB–mediated CP, as well as ischemic preconditioning (IPC)–mediated CP. Functional studies have confirmed that dysregulation of TNF-α receptor pathways, including the cardioprotective pathways, occurs in the senescent heart. TNF-α–induced PDGF-B is expressed in cardiac microvascular endothelial cells of young adults, but not in aging rats, suggesting that aging-associated defects in TNF-α receptor cardiac microvascular pathways likely contribute to the cardiovascular pathology of aging.[232] Strategies need to be developed to reconstitute the TNF-α receptor–mediated expression of PDGF-B to improve and prevent the cardiac microvascular dysfunction of the aging and senescent heart.

Data from a murine cardiac transplantation model showed that the synergistic interactions of combined PDGF-AB, VEGF and Angiopoietin-2 (PVA) can provide immediate restoration of senescent cardiac vascular function. PVA injection in young rat hearts, but not PDGF-AB alone or other cytokine combinations, at the time of coronary occlusion was able to suppress acute myocardial cell death by >50%.[233] PVA also reduced the extent of MI with an age-associated cardioprotective benefit. These findings reveal that synergistic cytokine pathways augmenting the actions of PDGF-AB are limited in older hearts. In contrast, Zheng and associates[234] reported that targeting of PDGF-AB–based pathways might restore CP by IPC in the aging heart. After IPC induction in 4- and 24-mo-old F344 rat, treatment of IPC-aging rats with PVA, but not PDGF-AB-alone, reversed IPC-induced mortality and also reduced myocardial injury, demonstrating that PDGF-AB–based strategies can reverse the senescent impairment in IPC-mediated CP.

Genomic Therapies to Improve Cardiac Function in Aging

Gene transfer and overexpression of parvalbumin in Fischer 344 rats improves myocardial calcium handling and diastolic function in aging, without augmenting energy consumption.[235] Furthermore, vascular calcification caused by reduced cellular expression of osteopontin can also be targeted by a gene-transfer approach. Recently, in a knock-out transgenic mouse model of osteopontin, Speer and associates[236] have used retroviral transduction of osteopontin cDNA to rescue osteopontin$^{-/-}$ aortic smooth muscle cells from inorganic phosphate-induced calcification.

Pharmacogenomics

Variability in the response to cardiovascular drugs fluctuates among patients. Although some achieve the desired therapeutic response, others do not. In addition, a subset of patients will experience variable adverse effects, ranging from mild to life threatening, and genetics may be an important contributor to this variable drug response.

Pharmacogenomics is a field focused on unraveling the genetic determinants of variable drug response. Although current research is largely focused on a limited candidate gene approach, which allows for the potential determination of significant genetic associations with variable response, it often does not explain the genetic basis of variable drug response enough to be useful clinically.[237] Given that most drug responses involve a large number of proteins, all of whose genes could have several polymorphisms, it seems unlikely that a single polymorphism in a single gene would explain a high degree of drug response variability in a consistent fashion, suggesting that a polygenic or genomic approach might be more appropriate.[238] Recently, Siest and associates,[239] in an excellent review, proposed a comprehensive pharmacogenomic approach in the field of cardiovascular therapy by considering five sources of variability: (1) the genetics of pharmacokinetics; (2) the genetics of pharmacodynamics (drug targets); (3) genetics linked to a defined pathology and its corresponding drug therapies; (4) the genetics of physiological regulation, and (5) environmental-genetic interactions.

In addition, they illustrated this five-tiered approach by use of examples of cardiovascular drugs in relation to genetic polymorphism.

At present, with a progressively aging population, the increasing pressure on public spending, and promises of individualized, safe, and effective treatment at lower cost, the public acceptance of pharmacogenomics is apparently high. Nevertheless, and despite the great benefits that this new approach may bring, there are a number of hurdles (e.g., ethical, social and legal concerns) to be overcome before the successful application of this modality of therapy.[240] Undoubtedly, cardiovascular pharmacogenomics has the potential to improve the use of cardiovascular drugs through the selection of the most appropriate drug therapy in an individual on the basis of their genetic information. However, it may take a decade or more before the available genetic information is widely used in making drug therapy decisions, although it is evident that new and important findings in this area will continue to appear, and the experimental approaches will continue to evolve.

Cell-Based Therapy

The dogma that "the birth of new cardiomyocytes is confined to the fetal and neonatal heart" has recently collapsed when researchers discovered that the heart of adult rat, mice, and human undergo significant cellular cardiac changes as a function of age. These include the demonstration of a capacity, albeit limited, of some post-mitotic cardiomyocytes to proliferate and form viable myocardium. These findings have set off a large number of parallel discoveries in rats, mice, and humans, with dramatic implications for how we think about cardiac plasticity and its potential role in rehabilitating individuals with acquired myocardial ischemia/infarct, HF, and different types of cardiomyopathies, including the cardiomyopathy of aging.

The human heart is composed of dividing and nondividing myocytes, with nearly 20% of the myocytes exhibiting telomeric shortening in the aging/senescent heart.[241] In the aging heart, there is not only a decrease in the functional reserve and capacity to adapt to sudden increases in pressure and volume loads but also myocyte loss, and when this loss reaches a threshold (variable with the presence or absence of other pathologies), HF will follow. Cell-based therapy in the aging failing heart is becoming a definitive alternative to other modalities (i.e., drugs and diet) and will likely become more prevalent, because currently heart transplant is mostly off limits beyond 60 y of age.

Therefore, replacement and regeneration of functional cardiac muscle is an important goal that could be achieved either by stimulation of autologous resident cardiomyocytes or by the transplantation of allogenic cells (e.g., embryonic stem cells, bone marrow mesenchymal cells, or skeletal myoblast). The use of these diverse stem cells, fetal or "newly differentiated cardiomyocytes" to proliferate, home, and integrate into damaged myocardium and restore myocardial function, as well as generate new tissue, has tremendous clinical possibility. However, important questions need to be answered, such as which type of cells to use and which route should be used to deliver them? As Ott and associates[242] have pointed out the choice seems to be based on availability and convenience, because a definitive scientific reason for the choice is not yet available.

A thorough discussion of the advantages and limitations of the cell types presently used in transplantation is presented more extensively in Chapter 4. Although no clear-cut choice has yet emerged regarding which cell type is best to transplant in myocardial repair, there are reasons to believe that the deployment of a multiplicity of approaches in the application of cell engineering will be required to develop novel therapies for different cardiac disorders. The approach to treat HF in aging may require the transplantation of cell types (e.g., skeletal myoblasts) that are different than those used in the targeted treatment of cardiac dysrhythmias, conduction disorders, and CHD. It is also possible that the long-term repair of a fully functioning myocardium may require more than a single cell type—for instance, cardiomyocytes, fibroblasts, and endothelial cells—in the generation and integration of a stable and responsive cardiac graft.

The refinement of nuclear transfer, cybrid, and cell fusion techniques may allow further engineering of stem cells to provide CP or stimulate antioxidant or antiapoptotic responses in the myocardium. Moreover, cell-engineering techniques might also allow the specific targeting of mitochondrial-based myopathies.[243] Furthermore, the combination of gene therapy and stem cell engineering seems to be an attractive alternative for treating cardiac disorders in the aging and senescent heart. Overexpression (and in some cases inhibition of expression) of specific proteins can result in striking changes in cardiomyocyte and in cardiac phenotype.

In summary, our increasing capacity to understand in detail the function of different cardiomyocytes/cardiac differentiation pathways will eventually allow the replacement of tissue, the transplant of others, and to shift imbalances on the molecular and the biochemistry of the aging heart. With the rapid progress being made in cell engineering, it is hoped that we may see the end of cardiac diseases that weaken human life in general, and in the aging population in particular, and that bankrupt the health care system.

Caloric Restriction

Dietary restriction is known to increase the life span of a number of organisms and mammals. Analysis of somatic mtDNA mutations and deletions in mouse heart and brain showed an age-related increase similar to humans. Old animals had a significant number of mtDNA deletions of variable size, and particularly large ones in the heart. Animals maintained on a CR diet had a significant reduction in mtDNA damage, mainly in the brain,[244] and gathered observations suggest that life span extension in animals kept on a CR diet is related to reduction in the generation of ROS and mtDNA damage, although this needs to be confirmed.[245] Furthermore, life-extended CR seems to reduce the age-related increases in lipid peroxidation, at least in the liver and kidney. F2-isoprostanes (F2-isoPs) have been found to be a sensitive and accurate biomarkers of lipid peroxidation for measuring the OS status of an organism. Age-related increases in esterified F2-isoPs levels correlated well with DNA oxidation, as measured by 8-oxodeoxyguanosine production, demonstrating that F2-isoPs are excellent biomarkers for age-related changes in oxidative damage to membranes.[246]

NO has been found to play an essential role in the mitochondrial changes induced by CR.[247] Male wild-type mice were fed either *ad libitum* (AL) or with a CR diet (food provided on alternate days) for 3 or 12 mo. Mice maintained on a CR feeding schedule consume 30–40% fewer calories over time compared with animals fed AL, have a lower body weight, and an extended life span.[248] The CR regimen, implemented for either 3 or 12 mo, induced the expression in several tissues of eNOS and 3′,5′-cyclic GMP, mtDNA, PGC-1α, nuclear respiratory factor-1 (NRF-1), mitochondrial transcription factor A, cytochrome *c* oxidase subunit IV (COX-IV), and cytochrome *c* (Cyt *c*). These findings were consistent with CR-mediated stimulation of mitochondrial biogenesis and mitochondrial gene expression. In addition, levels of ATP and the expression of sirtuin 1 were increased. In comparison, the *eNOS* null-mutant mice had much less significant changes with CR, suggesting that NO plays an important part in the response to CR.

CR is also known to influence the GH-IGF-1 axis. In a series of experiments by Sontag and associates,[249] age-related changes in GH secretory dynamics were compared in *ad libitum* fed and moderately caloric-restricted male Brown-Norway rats. They found that GH secretory dynamics decreased in young animals maintained on a moderate CR diet, but by 26 mo, GH pulse amplitude increased and was indistinguishable from young *ad libitum* fed animals. In addition, the moderately caloric-restricted animals failed to exhibit the decline in somatostatin mRNA characteristic of the *ad libitum* fed 25-month-old animals, suggesting that altered regulation of somatostatin mRNA may be a contributing factor in the decreased GH secretion observed in aging animals. These investigators also suggested that alterations in the GH-IGF-1 axis might underlie much of the life-prolonging and anti-aging actions of CR in the rat and mouse models.[250]

Regarding the anti-aging and life-prolonging effect of CR, Masoro[251] has proposed an interesting theory called "hormesis" defined as the beneficial action resulting from the response of an organism to a low-intensity stressor. An example of such low-intensity stressor is the moderate increase in the daily levels of plasma free corticosterone in rats under CR. The fact that it enhances the ability of rats and mice of any age to cope with insults such as surgery or exposure to toxic chemicals points to a "hormetic" action. Masoro suggests that if aging is primarily the result of injury caused by intrinsic living processes and environmental agents that are not repaired, then it is reasonable to propose that CR protects against this damage in a similar fashion. Moreover, because CR enhances the induction of stress proteins in response to damage,[252] and single-gene mutations have been shown to extend the life of invertebrate species and increase their ability to cope with damaging agents,[253] these are potential mechanisms supporting the theory. We agree with Masoro that with further progress in technology it may be possible to test within the *in situ* organ the validity of the proposed mechanism(s) of how CR may prolong life. At the same time it may shed further light on the existing controversy regarding the metabolic theory[254,255] and on the various evolutionary theories such as the "energy apportionment" hypothesis[256] and the "hibernation-like hypothesis" of the anti-aging action of CR.[257]

CR retards the aging processes, extends maximal life span, and consistently improves insulin resistance in lower species. However, the beneficial effect of CR in the prevention of atherosclerosis remains rather controversial. Data on the effects of CR on cardiovascular aging in cynomolgus monkeys showed that CR significantly improved insulin sensitivity and reduced intra-abdominal fat over the 4-y intervention, whereas no significant differences were seen in the lipid profile between groups.[258] In this study, there was improved insulin sensitivity with CR, but the extent of atherosclerosis did not differ between the *ad libitum*-fed or CR groups, and elevated plasma cholesterol concentrations were similar in both groups. On the other hand, data from *Rhesus* monkeys on the effect of CR on plasma lipoprotein Lp(a), an independent risk factor for the age-associated process of atherosclerosis, showed that plasma Lp(a) levels in control animals were almost twofold higher for males than females, and CR resulted in a reduction in circulating Lp(a) in males to levels similar to those measured in calorie-restricted females.[259] These findings suggest that in primates, CR may have a beneficial effect on the risk factors for development of atherosclerosis. Further research is needed to determine whether CR is beneficial in the prevention of age-related atherosclerosis and to unequivocally determine whether CR can be safely applied in humans for the prevention and/or amelioration of the diverse cardiac pathological conditions occurring with aging.

Drug Therapy and Metabolic Interventions

The technology of gene expression profiling and the use of DNA microarray may prove valuable in the discovery of nutrient or drug therapies that mimic the CR state.[260] This technique has been used by Lee and associates[261] to evaluate the effect of α-lipoic acid (LA), coenzyme Q_{10} (CoQ), and CR on the life span and patterns of gene expression in mice. They monitored the expression of 9977 genes in hearts from young (5 mo) and old (30 mo) animals. LA, CoQ, and CR inhibited age-related alterations in the expression of genes involved in the extracellular matrix, cellular structure, and protein turnover. However, unlike CR, LA and CoQ did not prevent age-related transcriptional alterations associated with energy metabolism. LA supplementation lowered the expression of genes encoding major histocompatibility complex components and of genes involved in protein turnover and folding. CoQ increased expression of genes involved in OXPHOS and reduced expression of genes involved in the complement pathway and several aspects of protein function. These observations suggest that supplementation with LA or CoQ results in transcriptional changes consistent with a state of reduced OS in the heart, but these therapeutic modalities are not as effective as CR in inhibiting the aging process in the heart.

Approaches to Rescue Mitochondrial Function

To reverse the mitochondrial dysfunction found in the aging heart, it has been suggested that supplements of acetyl-l-carnitine (ALCAR) and R-α-lipoic acid can improve myocardial bioenergetics and decrease the OS associated with aging.[262] Old rats fed with ALCAR exhibited a reverse of the age-related decline in carnitine levels and enhanced mitochondrial fatty acid β-oxidation in a number of tissues. However, ALCAR supplementation does not seem to reverse the age-related decline in the cardiac antioxidant status and may not improve the indices of OS. Lipoic acid, a potent thiol antioxidant and mitochondrial metabolite, seems to increase low molecular weight antioxidants, decreasing the age-associated oxidative insult. ALCAR, in combination with lipoic acid, may be effective supplemental regimens to maintain myocardial function in aging.[263] Moreover, the lipophilic antioxidant and mitochondrial redox coupler coenzyme Q_{10} may have the potential to improve energy production in the aging heart mitochondria by bypassing defective components in the respiratory chain, as well as by reducing the effects of OS. Recent studies in rats and humans suggest that coenzyme Q_{10} protects the aging heart against stress.

Judging by the popular media it would seem that CR and the use of antioxidants to slow/stop aging and aging-related cardiac defects are well-established facts. However, so far, there is a lack of meticulous and carefully done research to prove this contention. There is little doubt that mitochondria are desirable pharmacological targets, and treatments that modulate mitochondrial function (e.g., targeting OS) may be helpful in the aging heart; however, it is not clear if and how much of these antioxidant agents should be administered without jeopardizing the ROS "protector side" role in the heart signaling pathways. Recently, the antioxidant organoselenium compound ebselen, a mimic of GPx, has been shown to prevent the structural and functional changes occurring in mitochondria of aged red blood cells of the rainbow trout induced by OS. However, it did not prevent the swelling of the organelle.[264] If this compound can be of any significance in reversing the changes occurring in human cardiac aging is not known at this time.

The beneficial effects of moderate exercise on cardiac physiology and metabolism are well established. Moderate exercise on a treadmill increased the life span of male and female mice (age, 28–78 weeks).[265] Moreover, moderate exercise decreased the aging-associated ROS accumulation in a number of tissues, including the heart, probably by reducing the decrease in antioxidant enzymes, and in respiratory complex I and IV activities. However, these effects were not significant in the very old animals (78 weeks).

Contrary to simple organisms, such as yeast and worms, in which aging is easily altered through external stimuli, such as CR or through genetic manipulation (*SIR2* genes), the manipulation of aging in mammals is more complex, because it includes a closed relation with programmed cell death that does not tolerate well the effects of anti-aging therapy.[266] With advancing age, cells are much more susceptible to certain toxins or stresses and may not adequately respond to environmental and toxicological insults that would require increased ATP production to be successfully detoxified. Further research is needed to understand the role that mitochondrial dysfunction plays in aging, as well as to identify novel strategies to slow the progression of accumulative loss of cardiomyocytes (by apoptosis or programmed cell death) and the metabolic decline evident with age.

CONCLUSION

Cardiac aging, like aging in general, is a complex process, which involves numerous cellular and molecular changes, which along the way contribute to the expression of the multiple phenotypes of aging, "the different faces of cardiac aging." Of significance, aging in general and in particular cardiac aging are relative terms dependent on family history, prior cardiovascular history, and on a number of environmental/epigenetic factors. Individual cardiac aging may manifest earlier or later and is to some degree independent of the chronological time, with some individuals reaching aging at 60, whereas others do so at 75–80 y of age. Several plausible theories have been considered to explain aging (e.g., evolution, free radicals, somatic mutation); however, at this time, it is most likely that these different theories are intertwined with each other without a definitive "winner," reflecting a mixture of genetic and epigenetic elements found in most aged individuals with cardiovascular defects. Furthermore, and on the basis of the involved molecules and factors dominating at the time of the aging phenotypic expression, there may be a dominant "physiological," "biochemical," or molecular phenotype, with an integrated phenotype in the final stages of the process. In the final analysis, the aging heart encompasses an integrated pathology of cell damage, nuclear damage (including telomeres length), mitochondria, and the aforementioned signaling pathways, all-important focal points in the pathogenesis of cardiac aging. *A priori*, mitochondrial bioenergetic dysfunction is unlikely to be the cause of an aging-induced decline in mitochondrial transcription, mtDNA depletion, or defective mitochondrial biogenesis, but we are not certain. The mechanism of mitochondrial enzyme dysfunction needs to be further investigated, probably focusing on the post-translational modification of the component proteins, particularly mtDNA-encoded subunits. By a comprehensive examination of all the peptide subunits involved in each aging-affected respiratory enzyme (by blue native polyacrylamide gel electrophoresis, in which the catalytic activities can be maintained) we may able to distinguish among the possible mechanisms for enzymatic dysfunction.

Moreover, with increased ROS, we may see a strong mitochondrial stress response with compensatory gene expression of HSP60 and GPx, which parenthetically may be mediated by the increased transcription of NF-κB. The role(s) of both aging-dependent mtDNA damage and increased sensitivity of the PT pore are being actively investigated at present as indicators of oxidative damage and as potential stimuli or signaling events of downstream cellular transcriptional and apoptotic events in myocardial aging. Furthermore, the significance and extent of myocardial nuclear gene expression governing mitochondrial function needs to be thoroughly examined, given the limited data available on aging-gene regulation. A complete appraisal of the mitochondrial events will be very useful in determining both the individual pathway responses and the overall cardiac phenotype in aging.

CR, together with scheduled aerobic exercise plans, are appealing avenues of therapy in the prevention and treatment of the cardiac complications accompanying aging. They are attractive by both their simplicity and by their effect on mitochondrial function and are gaining increased support among clinicians and researchers alike. On the other hand, most of the available data have been collected from animals. For example, rodents maintained on a calorie-restricted diet had a significant increase in life span, reduction in the extension of mtDNA damage, and ROS. If this can be achieved in humans awaits to be demonstrated.

Independent of the many available theories trying to explain why and how we age (i.e., theological, evolutionary, disposable soma, free radical), aging is an inevitable, normal, and natural process for all living organisms and plants. How long humans can live (longevity) is a difficult question to answer at present. Improvements in health care and better control of a number of epigenetic factors (e.g., exercise and diet) seem to show that longevity is progressively increasing. On the basis of data from the World Health Organization (WHO) and the U. S. Census, in 2005 the average life span in the United States was estimated to be 77.7 y, and by the year 2050 is estimated to be in the mid-80s, eventually reaching the low 90s, excepting major scientific advances that can change the rate of human aging itself, as opposed to merely treating the effects of aging as is done today. The Census Bureau also predicted that the Unites States would have 5.3 million people aged older than 100 in 2100 (versus 50,000 in the 2000 census). It is interesting that approximately 150 years ago Arthur Schopenhauer in his book the *Wisdom of Life* comments that the *Vedic Upanishad* (Oupnekhat, II) characterized the natural length of human life at 100 y, although the Old Testament (Psalms xc. 10) put it at 70–80 y. This remarkable philosopher thought that "to die at seventy to eighty years will be due to disease, which is by no mean essential to old age, but rather something abnormal, and it is only when we are between 90 and 100 that people die of old age, meaning without suffering from any disease, simply put 'ceasing to live rather than die.'" To reach this important goal, the worldwide community of researchers, clinicians, pharmaceutical companies, and so forth are eagerly trying to prolong not only longevity but living disease-free, meaning preventing/lowering the suffering and misery currently associated with aging and senescence.

Because genetic factors are important variables in the current incidence of cardiac diseases, in the future these could be modified with the use of safe gene therapy. This together with improved, smarter drugs, and a better implementation/control of epigenetic factors, including a healthier lifestyle with appropriate diet, exercise, nonsmoking, prevention/treatment of obesity/diabetes, and so forth, together with the delivery of better health care for all, will significantly decrease the incidence and severity of cardiac diseases in aging. Finally, despite the potential effects that increased longevity may have on the world economy and ecology, all workable with our human ingenuity, increasing human life span with "healthy" aging is a desirable and attainable goal.

SUMMARY

- Cardiac aging is a rather complex process unlikely to be dependent on a unique, singular, pathway or determined gene(s) or gene products.
- As the human population ages and the average life span increases, so does the burden of CVD.
- Identification of the totality of the mechanisms/pathways contributing to cardiac aging is a rather difficult, probably unattainable, goal, although important work in this regard is in progress.
- Cardiovascular changes occurring with aging include decreased β-adrenergic sympathetic responsiveness, slowed and delayed early diastolic filling, increased vascular stiffness, and endothelial dysfunction.
- During aging, a significant loss of cardiac myocytes occurs, probably related to programmed cell death (apoptosis). The cumulative effect of this loss may result in significant physiological decline.
- Loss in cardiomyocytes may be secondary to mitochondrial dysfunction, likely caused by chronic

- exposure to oxidative free radicals, damage to mtDNA (mutations and deletions), and mitochondrial membranes.
- Besides cell loss, other mechanisms involved in cardiac aging are ROS and oxidative stress, inflammatory mechanisms/signaling, adrenergic (and muscarinic) receptors, cardiac G-protein–coupled receptors, SERCA, thyroid hormone, growth hormone and IGF-1, telomeres and telomere-related proteins, autophagy and cardiac aging, gender differences, genetic makeup, susceptibility genes, and epigenetic/environmental factors.
- Age is the main clinical determinant of stiffness in large arteries. Increasing aortic stiffness with age occurs gradually and continuously and is similar for men and women.
- Hypercoagulability and advanced vascular sclerotic changes may contribute to the increased incidence of thrombosis in the elderly.
- Multiple and complex interacting factors and pathological conditions, including CAD, hypertension, and diabetes may contribute to the cardiomyopathy of aging, resulting in progressive cardiac enlargement and dysfunction.
- Myocardial extracellular matrix and endothelial remodeling, with a plurality of interacting molecular and metabolic changes, occur with aging.
- A decline in myocardial function and a shift in myocardial intermediary metabolism from primary fatty acid substrate to glucose may also occur in the aging heart.
- Age-related reduction in cardiolipin is considered to have major impact on cardiac mitochondrial membrane transport function, fluidity, and stability.
- Progressive catabolism occurring with HF and aging may result in skeletal muscle atrophy, a problem that may be compounded by the relatively poor regenerative properties of skeletal muscle.
- Reduced tolerance to myocardial ischemia/reperfusion injury occurs in human cardiac aging; however, most of the published data in regard to the tolerance of the aging heart to ischemia and reperfusion have been generated from animal models.
- The mitochondrial theory of aging suggests that somatic mutations of mitochondrial DNA, induced by oxygen free radicals, are the primary cause of energy decline. In response to the increased level of ROS in the senescent heart, genes encoding mitochondrial stress proteins, including both heat shock and antioxidant response proteins such as HSP60, HO-1, and GPx, are upregulated.
- Another consequence of increased OS and ROS accumulation is myocardial apoptosis.
- To rescue mitochondrial function in the aging heart new approaches are being applied.
- The cracking of the genetic code has provided insights in the biological aspects of cardiac aging. However, we must compare the human genome to that of other organisms to understand which regions of the genome do what.
- The identification of single nucleotide polymorphisms (SNPs) as a function of age and diseases may result in a better understanding of the pathophysiology and pathogenesis of CVDs in aging.
- Differential expression of specific regulatory genes in the aging heart may determine the phenotypic variations among related organisms, as well as the phenotypic changes in the same individual with aging.
- New technological advances to assess cardiac pathology in aging include proteomics and atomic force microscopy.
- The aging heart has a decreased capacity to tolerate and respond to various forms of stresses, with increased likelihood for myocardial ischemia and dysfunction. Most studies on cardioprotection (CP) have been carried out in animal models. There is a need for CP to be further studied in the aging and senescent human heart.
- Gene therapies may soon become available to improve cardiac function in the aged heart.
- Pharmacogenomics is a new field mainly focused on unraveling the genetic determinants of variable drug responses. This approach may be very useful in aging.
- Replacement and regeneration of functional cardiac muscle is an important goal in aging. This could be achieved either by stimulation of autologous resident cardiomyocytes or by transplantation of allogenic cells.
- Caloric restriction (CR) increases the life span of a number of organisms and mammals. Animals maintained on a calorie-restricted diet had a significant reduction in the extension of mtDNA damage, mainly in the brain. More focused studies in humans will determine whether CR can be safely used for the prevention and/or amelioration of the pathologies occurring in the aging heart.

References

1. Fleg, J. L., O'Connor, F., Gerstenblith, G., Becker, L. C., Clulow, J., Schulman, S. P., and Lakatta, E. G. (1995). Impact of age on the cardiovascular response to dynamic upright exercise in healthy men and women. *J. Appl. Physiol.* **78,** 890–900.
2. Abbott, R. D., Curb, J. D., Rodriguez, B. L., Masaki, K. H., Yano, K., Schatz, I. J., Ross, G. W., and Petrovitch, H. (2002). Age-related changes in risk factor effects on the incidence of coronary heart disease. *Ann. Epidemiol.* **12,** 173–181.
3. Lakatta, E. G. (1993). Cardiovascular regulatory mechanisms in advanced age. *Physiol. Rev.* **73,** 413–467.
4. Lakatta, E. G., Gerstenblith, G., Angell, C. S., Shock, N. W., and Weisfeldt, M. L. (1975). Diminished inotropic response of aged myocardium to catecholamines. *Circ. Res.* **36,** 262–269.
5. Lakatta, E. G., Gerstenblith, G., Angell, C. S., Shock, N. W., and Weisfeldt, M. L. (1975). Prolonged contraction duration in aged myocardium. *J. Clin. Invest.* **55,** 61–68.
6. Schulman, S. P., Lakatta, E. G., Fleg, J. L., Lakatta, L., Becker, L. C., and Gerstenblith, G. (1992). Age-related decline in left

ventricular filling at rest and exercise. *Am. J. Physiol.* **263**, H1932–H1938.

7. Merillon, J. P., Motte, G., Masquet, C., Azancot, I., Aumont, M. C., Guiomard, A., and Gourgon, R. (1982). Changes in the physical properties of the arterial system and left ventricular performance with age and in permanent arterial hypertension: their interrelation. *Arch. Mal Coeur Vaiss.* **75**, 127–132.

8. Roman, M. J., Ganau, A., Saba, P. S., Pini, R., Pickering, T. G., and Devereux, R. B. (2000). Impact of arterial stiffening on left ventricular structure. *Hypertension* **36**, 489–494.

9. Taddei, S., Virdis, A., Mattei, P., Ghiadoni, L., Fasolo, C. B., Sudano, I., and Salvetti, A. (1997). Hypertension causes premature aging of endothelial function in humans. *Hypertension* **29**, 736–743.

10. Taddei, S., Virdis, A., Mattei, P., Ghiadoni, L., Gennari, A., Fasolo, C. B., Sudano, I., and Salvetti, A. (1995). Aging and endothelial function in normotensive subjects and patients with essential hypertension. *Circulation* **91**, 1981–1987.

11. Fannin, S. W., Lesnefsky, E. J., Slabe, T. J., Hassan, M. O., and Hoppel, C. L. (1999). Aging selectively decreases oxidative capacity in rat heart interfibrillar mitochondria. *Arch. Biochem. Biophys.* **372**, 399–407.

12. Ramani, K., Lust, W. D., Whittingham, T. S., and Lesnefsky, E. J. (1996). ATP catabolism and adenosine generation during ischemia in the aging heart. *Mech. Ageing Dev.* **89**, 113–124.

13. Brunt, U. T., and Terman, A. (2002). The mitochondrial-lysosomal axis theory of aging: accumulation of damaged mitochondria as a result of imperfect autophagocytosis. *Eur. J. Biochem.* **269**, 1996–2002.

14. Harman, D. (1956). Aging: A theory based on free radical and radiation chemistry. *J. Gerontol.* **11**, 298–300.

15. Melov, S. (2000). Mitochondrial oxidative stress. Physiologic consequences and potential for a role in aging. *Ann. N. Y. Acad. Sci.* **908**, 219–225.

16. Suh, J. H., Heath, S. H., and Hagen, T. (2003). Two subpopulations of mitochondria in the aging rat heart display heterogenous levels of oxidative stress. *Free Radic. Biol. Med.* **35**, 1064–1072.

17. Leichtweis, S., Leeuwenburgh, C., Bejma, J., and Ji, L. L. (2001). Aged rat hearts are not more susceptible to ischemia-reperfusion injury in vivo: Role of glutathione. *Mech. Aging Dev.* **122**, 503–518.

18. Vertechy, M., Cooper, M. B., Ghirardi, O., and Ramacci, M. T. (1989). Antioxidant enzyme activities in heart and skeletal muscle of rats of different ages. *Exp. Gerontol.* **24**, 211–218.

19. Phaneuf, S., and Leeuwenburgh, C. (2002). Cytochrome *c* release from mitochondria in the aging heart: A possible mechanism for apoptosis with age. *Am. J. Physiol. Integr. Comp. Physiol.* **282**, R423–R430.

20. Pollack, M., Phaneuf, S., Dirks, A., and Leeuwenburgh, C. (2002). The role of apoptosis in the normal aging brain, skeletal muscle, and heart. *Ann. N. Y. Acad. Sci.* **959**, 93–107.

21. Torella, D., Rota, M., Nurzynska, D., Musso, E., Monsen, A., Shiraishi, I., Zias, E., Walsh, K., Rosenzweig, A., Sussman, M. A., Urbanek, K., Nadal-Ginard, B., Kajstura, J., Anversa, P., and Leri, A. (2004). Cardiac stem cell and myocyte aging, heart failure, and insulin-like growth factor-1 overexpression. *Circ. Res.* **94**, 514–524.

22. Schriner, S. E., Linford, N. J., Martin, G. M., Treuting, P., Ogburn, C. E., Emond, M., Coskun, P. E., Ladiges, W., Wolf, N., Van Remmen, H., Wallace, D. C., and Rabinovitch, P. S. (2005). Extension of murine life span by overexpression of catalase targeted to mitochondria. *Science* **308**, 1909–1911.

23. Van Remmen, H., Qi, W., Sabia, M., Freeman, G., Estlack, L., Yang, H., Mao Guo, Z., Huang, T. T., Strong, R., Lee, S., Epstein, C. J., and Richardson, A. (2004). Multiple deficiencies in antioxidant enzymes in mice result in a compound increase in sensitivity to oxidative stress. *Free Radic. Biol. Med.* **36**, 1625–1634.

24. Hare, J. M., and Stamler, J. S. (2005). NO/redox disequilibrium in the failing heart and cardiovascular system. *J. Clin. Invest.* **115**, 509–517.

25. Raju, S. V., Barouch, L. A., and Hare, J. M. (2005). Nitric oxide and oxidative stress in cardiovascular aging. *Sci. Aging Knowledge Environ.* **2005**, re4.

26. Ide, T., Tsutsui, H., Kinugawa, S., Utsumi, H., Kang, D. C., Hattori, N., Uchida, K., Arimura, K., Egashira, K., and Takeshita, A. (1999). Mitochondrial electron transport complex I is a potential source of oxygen free radicals in the failing myocardium. *Circ. Res.* **85**, 357–363.

27. Clementi, E., Brown, G. C., Feelisch, M., and Moncada, S. (1998). Persistent inhibition of cell respiration by nitric oxide: Crucial role of S-nitrosylation of mitochondrial complex I and protective action of glutathione. *Proc. Natl. Acad. Sci. USA* **95**, 7631–7636.

28. Chandra, J., Samali, A., and Orrenius, S. (2000). Triggering and modulation of apoptosis by oxidative stress. *Free Radic. Biol. Med.* **9**, 323–333.

29. Ferdinandy, P., Panas, D., and Schulz, R. (1999). Peroxynitrite contributes to spontaneous loss of cardiac efficiency in isolated working rat hearts. *Am. J. Physiol.* **276**, H1861–H1867.

30. Li, D., Qu, Y., Tao, L., Liu, H., Hu, A., Gao, F., Sharifi-Azad, S., Grunwald, Z., Ma, X. L., and Sun, J. Z. (2005). Inhibition of iNOS protects the aging heart against beta-adrenergic receptor stimulation-induced cardiac dysfunction and myocardial ischemic injury. *J. Surg. Res.* **131**, 64–72.

31. Kritchevsky, S. B., Cesari, M., and Pahor, M. (2005). Inflammatory markers and cardiovascular health in older adults. *Cardiovasc. Res.* **66**, 265–275.

32. Deten, A., Marx, G., Briest, W., Volz, H. C., and Zimmer, H.-G. (2005). Heart function and molecular biological parameters are comparable in young adult and aged rats after chronic myocardial infarction. *Cardiovasc. Res.* **66**, 364–373.

33. Antonicelli, R., Olivieri, F., Bonafe, M., Cavallone, L., Spazzafumo, L., Marchegiani, F., Cardelli, M., Recanatini, A., Testarmata, P., Boemi, M., Parati, G., and Franceschi, C. (2005). The interleukin-6 -174 G>C promoter polymorphism is associated with a higher risk of death after an acute coronary syndrome in male elderly patients. *Int. J. Cardiol.* **103**, 266–271.

34. Korzick, D. H., Holiman, D. A., Boluyt, M. O., Laughlin, M. H., and Lakatta, E.G. (2001). Diminished alpha1-adrenergic-mediated contraction and translocation of PKC in senescent rat heart. *Am. J. Physiol. Heart Circ. Physiol.* **281**, H581–H589.

35. Korzick, D. H., Hunter, J. C., McDowell, M. K., Delp, M. .D, Tickerhoof, M. M., and Carson, L. D. (2004). Chronic exercise improves myocardial inotropic reserve capacity through alpha1-adrenergic and protein kinase C-dependent effects in senescent rats. *J. Gerontol. A Biol. Sci. Med. Sci.* **59**, 1089–1098.

36. Wang, Y., and Ashraf, M. (1998). Activation of alpha1-adrenergic receptor during Ca2+ pre-conditioning elicits strong protection against Ca2+ overload injury via protein kinase C signaling pathway. *J. Mol. Cell Cardiol.* **30,** 2423–2435.
37. Montagne, O., Le Corvoisier, P., Guenoun, T., Laplace, M., and Crozatier, B. (2005). Impaired alpha1-adrenergic responses in aged rat hearts. *Fundam. Clin. Pharmacol.* **19,** 331–339.
38. Esler, M., and Kaye, D. (2000). Sympathetic nervous system activation in essential hypertension, cardiac failure and psychosomatic heart disease. *J. Cardiovasc. Pharmacol.* **35,** S1–S7.
39. Kaye, D., and Esler, M. (2005) Sympathetic neuronal regulation of the heart in aging and heart failure. *Cardiovasc. Res.* **66,** 256–264.
40. Daaka, Y., Luttrell, L. M., and Lefkowitz, R. J. (1997). Switching of the coupling of the beta2-adrenergic receptor to different G proteins by protein kinase A. *Nature* **390,** 88–91.
41. Kilts, J. D., Akazawa, T., Richardson, M. D., and Kwatra, M. M. (2002). Age increases cardiac Galpha (i2) expression, resulting in enhanced coupling to G protein-coupled receptors. *J. Biol. Chem.* **277,** 31257–31262.
42. Kilts, J. D., Akazawa, T., El-Moalem, H. E., Mathew, J. P., Newman, M. F., and Kwatra, M. M. (2003). Age increases expression and receptor-mediated activation of Galpha i in human atria. *J. Cardiovasc. Pharmacol.* **42,** 662–670.
43. Brodde, O.-E., and Michel, M. C. (1999). Adrenergic and muscarinic receptors in the human heart. *Pharmacol. Rev.* **51,** 651–689.
44. Richardson, M. D., Kilts, J. D., and Kwatra, M. M. (2004). Increased expression of Gi-coupled muscarinic acetylcholine receptor and Gi in atrium of elderly diabetic subjects. *Diabetes* **53,** 2392–2396.
45. Brodde, O.-E., Konschack, U., Becker, K., Rüter, F., Poller, U., Jakubetz, J., Radke, J., and Zerkowski, H.-R. (1998). Cardiac muscarinic receptors decrease with age: in vitro and in vivo studies. *J. Clin. Invest.* **101,** 471–478.
46. Su, N., Duan, J., Moffat, M. P., and Narayanan, N. (1995). Age-related changes in electrophysiological responses to muscarinic receptor stimulation in rat myocardium. *Can. J. Physiol. Pharmacol.* **73,** 1430–1436.
47. Narayanan, N., and Derby, J. A. (1983). Effects of age on muscarinic cholinergic receptors in rat myocardium. *Can. J. Physiol. Pharmacol.* **61,** 822–829.
48. Hardouin, S., Mansier, P., Bertin, B., Dakhly, T., Swynghedauw, B. and, Moalic, J. M. (1997). ß-Adrenergic and muscarinic receptor expression are regulated in opposite ways during senescence in rat left ventricle. *J. Mol. Cell Cardiol.* **29,** 309–319.
49. Elfellah, M. S., Johns, A., and Shepherd, A. M. M. (1986). Effect of age on responsiveness of isolated rat atria to carbachol and on binding characteristics of atrial muscarinic receptors. *J. Cardiovasc. Pharmacol.* **8,** 873–877.
50. Lo, S.-H., Liu, I.-M., Huang, L. W., and Cheng, J.-T. (2001). Decrease of muscarinic m2 cholinoceptor gene expression in the heart of aged rat. *Neurosci. Lett.* **300,** 185–187.
51. Maciel, L. M., Polikar, R., Rohrer, D., Popovich, B. K., and Dillmann, W. H. (1990). Age-induced decreases in the messenger RNA coding for the sarcoplasmic reticulum Ca2+-ATPase of the rat heart. *Circ. Res.* **67,** 230–234.
52. Long, X., Boluyt, M. O., O'Neill, L., Zheng, J. S., Wu, G., Nitta, Y. K., Crow, M. T., and Lakatta, E. G. (1999). Myocardial retinoid X receptor, thyroid hormone receptor, and myosin heavy chain gene expression in the rat during adult aging. *J. Gerontol. A Biol. Sci. Med. Sci.* **54,** B23–B27.
53. Iemitsu, M., Miyauchi, T., Maeda, S., Tanabe, T., Takanashi, M., Matsuda, M., and Yamaguchi, I. (2004). Exercise training improves cardiac function-related gene levels through thyroid hormone receptor signaling in aged rats. *Am. J. Physiol. Heart Circ. Physiol.* **286,** H1696–H1705.
54. Tang, F. (1985). Effect of sex and age on serum aldosterone and thyroid hormones in the laboratory rat. *Horm. Metab. Res.* **17,** 507–509.
55. Buttrick, P., Malhotra, A., Factor, S., Greenen, D., Leinwand, L., and Scheuer, J. (1991). Effect of aging and hypertension on myosin biochemistry and gene expression in the rat heart. *Circ. Res.* **68,** 645–652.
56. Schmidt, U., del Monte, F., Miyamoto, M. I., Matsui, T., Gwathmey, J. K., Rosenzweig, A., and Hajjar, R. J. (2000). Restoration of diastolic function in senescent rat hearts through adenoviral gene transfer of sarcoplasmic reticulum Ca(2+)-ATPase. *Circulation* **101,** 790–796.
57. Cain, B. S., Meldrum, D. R., Joo, K. S., Wang, J. F., Meng, X., Cleveland, J. C., Jr., Banerjee, A., and Harken, A. H. (1998). Human SERCA2a levels correlate inversely with age in senescent human myocardium. *J. Am. Coll Cardiol.* **32,** 458–467.
58. Knyushko, T. V., Sharov, V. S., Williams, T. D., Schoneich, C., and Bigelow, D. J. (2005). 3-Nitro-tyrosine modification of SERCA2a in the aging heart: a distinct signature of the cellular redox environment. *Biochemistry* **44,** 13071–13081.
59. Tatar, M., Bartke, A., and Antebi, A. (2003). The endocrine regulation of aging by insulin-like signals. *Science* **299,** 1346–1351.
60. Muller, E. E., Cella, S. G., De Gennaro Colonna, V., Parenti, M., Cocchi, D., and Locatelli, V. (1993). Aspects of the neuroendocrine control of growth hormone secretion in ageing mammals *J. Reprod. Fertil. Suppl.* **46,** 99–114.
61. Bartke, A. (2005). Minireview: Role of the growth hormone/insulin-like growth factor system in mammalian aging. *Endocrinology* **146,** 3718–3723.
62. Brown-Borg, H. M., Borg, K. F., Meliska, C. J., and Bartke, A. (1996). Dwarf mice and the ageing process. *Nature* **384,** 33.
63. Hsieh, C. C., de Ford, J. .H, Flurkey, K., Harrison, D. E., and Papaconstantinou, J. (2002). Effects of the Pit1 mutation on the insulin signaling pathway: implication on the longevity of the long lived Snell dwarf mouse. *Mech. Ageing Dev.* **123,** 1254–1255.
64. Barbieri, M., Bonafe, M., Franceschi, C., and Paolisso, G. (2003). Insulin/IGF-I-signaling pathway: An evolutionarily conserved mechanism of longevity from yeast to humans. *Am. J. Physiol. Endocrinol. Metab.* **285,** E1064–E1071.
65. Bluher, M., Kahn, B. B., and Kahn, C. R. (2003). Extended longevity in mice lacking the insulin receptor in adipose tissue. *Science* **299,** 572–574.
66. Steger, R. W., Bartke, A., and Cecim, M. (1993). Premature ageing in transgenic mice expressing different growth hormone genes. *J. Reprod. Fertil. Suppl.* **46,** 61–75.
67. Muller, F. Growth hormone receptor knockout (Laron) mice. http://sageke.sciencemag.org/cgi/content/full/sageke 2002/8/tg1 (27 February 2002).
68. Holzenberger, M., Dupont, J., Ducos, B., Leneuve, P., Geloen, A., Even, P. C., Cervera, P., and Le Bouc, Y. (2003). IGF-1 receptor regulates life span and resistance to oxidative stress in mice. *Nature* **421,** 182–186.

69. Goodman-Gruen, D., and Barrett-Connor, E. (1997). Epidemiology of insulin-like growth factor-I in elderly men and women. The Rancho Bernardo Study. *Am. J. Epidemiol.* **145,** 970–976.
70. Lieberman, S. A., Mitchell, A. M., Marcus, R., Hintz, R. L., and Hoffman, A. R. (1994). The insulin-like growth factor I generation test: resistance to growth hormone with aging and estrogen replacement therapy. *Horm. Metab. Res.* **26,** 229–233.
71. Khan, A. S., Sane, D. C., Wannenburg, T., and Sonntag, W. E. (2002). Growth hormone, insulin-like growth factor-1 and the aging cardiovascular system. *Cardiovasc. Res.* **54,** 25–35.
72. Vasan, R. S., Sullivan, L. M., D'Agostino, R. B., Roubenoff, R., Harris, T., Sawyer, D. B., Levy, D., and Wilson, P. W. (2003). Serum insulin-like growth factor I and risk for heart failure in elderly individuals without a previous myocardial infarction: the Framingham Heart Study. *Ann. Intern. Med.* **139,** 642–648.
73. Roubenoff, R., Parise, H., Payette, H. A., Abad, L. W., D'Agostino, R., Jacques, P. F., Wilson, P. W., Dinarello, C. A. and Harris, T. B. (2003). Cytokines, insulin-like growth factor 1, sarcopenia, and mortality in very old community-dwelling men and women: The Framingham Heart Study. *Am. J. Med.* **115,** 429–435.
74. Ghigo, E., Arvat, E., Gianotti, L., Ramunni, J., DiVito, L., Maccagno, B., Grottoli, S., and Camanni, F. (1996). Human aging and the GH-IGF-I axis. *J. Pediatr. Endocrinol. Metab.* **9,** 271–278.
75. Takahashi, S., and Meites, J. (1987). GH binding to liver in young and old female rats: Relation to somatomedin-C secretion. *Proc. Soc. Exp. Biol. Med.* **186,** 229–233.
76. Xu, X., Bennett, S. A., Ingram, R. L., and Sonntag, W. E. (1995). Decreases in growth hormone receptor signal transduction contribute to the decline in insulin-like growth factor I gene expression with age. *Endocrinology* **136,** 4551–4557.
77. Khan, A. S., Sane, D. C., Wannenburg, T., and Sonntag, W. E. (2002). Growth hormone, insulin-like growth factor-1 and the aging cardiovascular system. *Cardiovasc. Res.* **54,** 25–35.
78. Colao, A., Marzullo, P., Di Somma, C., and Lombardi, G. (2001). Growth hormone and the heart. *Clin. Endocrinol.* **54,** 137–154.
79. Osterziel, K. J., Strohm, O., Schuler, J., Friedrich, M., Hänlein, D., Willenbrock. R., Anker, S. D., Poole-Wilson, P. A., Ranke, M. B., and Dietz, R. (1998). Randomised, double-blind, placebo-controlled trial of human recombinant growth hormone in patients with chronic heart failure due to dilated cardiomyopathy. *Lancet* **351,** 1233–1237.
80. Wang, P. H. (2001). Roads to survival: insulin-like growth factor-1 signaling pathways in cardiac muscle. *Circ. Res.* **88,** 552–554.
81. Laron, Z. (2005). Do deficiencies in growth hormone and insulin-like growth factor-1 (IGF-1) shorten or prolong longevity? *Mech. Ageing Dev.* **126,** 305–307.
82. Chan, J. M., Stampfer, M. J., Giovannucci, E., Gann, P. H., Ma, J., Wilkinson, P., Hennekens, C. H., and Pollak, M. (1998). Plasma insulin-like growth factor-I and prostate cancer risk, a prospective study. *Science* **279,** 563–566.
83. Hankinson, S. E., Willett, W. C., Colditz, G. A., Hunter, D. J., Michaud, D. S., Deroo, B., Rosner, B., Speizer, F. E., and Pollak, M. (1998). Circulating concentrations of insulin-like growth factor-I and risk of breast cancer. *Lancet* **351,** 1393–1396.
84. Clark, R. G. (2004). Recombinant human insulin-like growth factor I (IGF-I): risks and benefits of normalizing blood IGF-I concentrations. *Horm. Res.* **62,** 93–100.
85. Kurosu, H., Yamamoto, M., Clark, J. D., Pastor, J. V., Nandi, A., Gurnani, P., McGuinness, O. P., Chikuda, H., Yamaguchi, M., Kawaguchi, H., Shimomura, I., Takayama, Y., Herz, J., Kahn, C. R., Rosenblatt, K. P., and Kuro-o, M. (2005). Suppression of aging in mice by the hormone Klotho. *Science* **309,** 1829–1833.
86. Kirkwood, T. B. (2005). Understanding the odd science of aging. *Cell* **120,** 437–447.
87. Pollack, M., and Leeuwenburgh, C. (2001). Apoptosis and aging: Role of the mitochondria. *J. Gerontol. A. Biol. Sci. Med. Sci.* **56,** B475–B482.
88. Cook, S. A., Sugden, P. H., and Clerk, A. (1999). Regulation of Bcl-2 family proteins during development and in response to oxidative stress in cardiac myocytes: Association with changes in mitochondrial membrane potential. *Circ. Res.* **85,** 940–949.
89. Long, X., Goldenthal, M. J., Wu, G. M., and Marín-García, J. (2004). Mitochondrial Ca2+ flux and respiratory enzyme activity decline are early events in cardiomyocyte response to H_2O_2. *J. Mol. Cell Cardiol.* **37,** 63–70.
90. Uhrbom, L., Nister, M., and Westermark, B. (1997). Induction of senescence in human malignant glioma cells by p16INK4A. *Oncogene* **15,** 505–514.
91. Edo, M. D., and Andrés, V. (2005). Aging, telomeres, and atherosclerosis. *Cardiovasc. Res.* **66,** 213–221.
92. Kurz, D. J., Decary, S., Hong, Y., Trivier, E., Akhmedov, A., and Erusalimsky, J. D. (2004). Chronic oxidative stress compromises telomere integrity and accelerates the onset of senescence in human endothelial cells. *J. Cell Sci.* **117,** 2417–2426.
93. Haendeler, J., Hoffmann, J., Diehl, J. F., Vasa, M., Spyridopoulos, I., Zeiher, A. M., and Dimmeler, S. (2004). Antioxidants inhibit nuclear export of telomerase reverse transcriptase and delay replicative senescence of endothelial cells. *Circ. Res.* **94,** 768–775.
94. Spyridopoulos, I., Haendeler, J., Urbich, C., Brummendorf, T. H., Oh, H., Schneider, M. D., Zeiher, A. M., and Dimmeler, S. (2004). Statins enhance migratory capacity by upregulation of the telomere repeat-binding factor TRF2 in endothelial progenitor cells. *Circulation* **110,** 3136–3142.
95. Imanishi, T., Hano, T., and Nishio, I. (2005). Estrogen reduces endothelial progenitor cell senescence through augmentation of telomerase activity. *J. Hypertens.* **23,** 1699–1706.
96. Imanishi, T., Hano, T., Sawamura, T., and Nishio, I. (2004). Oxidized low-density lipoprotein induces endothelial progenitor cell senescence, leading to cellular dysfunction. *Clin. Exp. Pharmacol. Physiol.* **31,** 407–413.
97. Simoncini, T., Hafezi-Moghadam, A., Brazil, D. P., Ley, K., Chin, W. W., and Liao, J. K. (2000). Interaction of oestrogen receptor with the regulatory subunit of phosphatidylinositol-3-OH kinase. *Nature* **407,** 538–541.
98. Vasa, M., Breitschopf, K., Zeiher, A. M., and Dimmeler, S. (2000). Nitric oxide activates telomerase and delays endothelial cell senescence. *Circ. Res.* **87,** 540–542.
99. Serrano, A. L., and Andres, V. (2004). Telomeres and cardiovascular disease: does size matter? *Circ. Res.* **94,** 575–584.

100. d'Adda di Fagagna, F., Reaper, P. M., Clay-Farrace, L., Fiegler, H., Carr, P., Von Zglinicki, T., Saretzki, G., Carter, N. P., and Jackson, S. P. (2003). A DNA damage checkpoint response in telomere-initiated senescence. *Nature* **426,** 194–198.
101. Okuda, K., Khan, M. Y., Skurnick, J., Kimura, M., Aviv, H., and Aviv, A. (2000). Telomere attrition of the human abdominal aorta: relationships with age and atherosclerosis. *Atherosclerosis* **152,** 391–398.
102. Poch, E., Carbonell, P., Franco, S., Díez-Juan, A., Blasco, M. A., and Andrés, V. (2004). Short telomeres protect from diet-induced atherosclerosis in apolipoprotein E-null mice. *FASEB J.* **18,** 418–420.
103. von Zglinicki, T., Pilger, R., and Sitte, N. (2000). Accumulation of single-strand breaks is the major cause of telomere shortening in human fibroblasts. *Free Radic. Biol. Med.* **28,** 64–74.
104. Forsyth, N. R., Evans, A. P., Shay, J. W., and Wright, W. E. (2003). Developmental differences in the immortalization of lung fibroblasts by telomerase. *Aging Cell* **2,** 235–243.
105. Serra, V., von Zglinicki, T., Lorenz, M., and Saretzki, G. (2003). Extracellular superoxide dismutase is a major antioxidant in human fibroblasts and slows telomere shortening. *J. Biol. Chem.* **278,** 6824–6830.
106. Saretzki, G., Murphy, M. P., and von Zglinicki, T. (2003). MitoQ counteracts telomere shortening and elongates lifespan of fibroblasts under mild oxidative stress. *Aging Cell* **2,** 141–143.
107. Passos, J. F., and von Zglinicki, T. (2005). Mitochondria, telomeres and cell senescence. *Exp. Gerontol.* **40,** 466–472.
108. Brunt, U. T., and Terman, A. (2002). The mitochondrial-lysosomal axis theory of aging: Accumulation of damaged mitochondria as a result of imperfect autophagocytosis. *Eur. J. Biochem.* **269,** 1996–2002.
109. Rooyackers, O. E., Adey, D. B., Ades, P. A., and Nair, K. S. (1996). Effect of age on in vivo rates of mitochondrial protein synthesis in human skeletal muscle. *Proc. Natl. Acad. Sci. USA* **93,** 15364–15369.
110. Terman, A., and Brunk, U. T. (2004). Myocyte aging and mitochondrial turnover. *Exp. Gerontol.* **39,** 701–705.
111. Stroikin, Y., Dalen, H., Brunk, U. T., and Terman, A. (2005). Testing the "garbage" accumulation theory of aging: mitotic activity protects cells from death induced by inhibition of autophagy. *Biogerontology* **6,** 39–47.
112. Olivetti, G., Giordano, G., Corradi, D., Melissari, M., Lagrasta, C., Gambert, S. R., and Anversa, P. (1995). Gender differences and aging: effects on the human heart. *J. Am. Coll. Cardiol.* **26,** 1068–1079.
113. Patten, R. D., Pourati, I., Aronovitz, M. J., Baur, J., Celestin, F., Chen, X., Michael, A., Haq, S., Nuedling, S., Grohe, C., Force, T., Mendelsohn, M. E., and Karas, R. H. (2004). 17beta-estradiol reduces cardiomyocyte apoptosis in vivo and in vitro via activation of phospho-inositide-3 kinase/Akt signaling. *Circ. Res.* **95,** 692–699.
114. Yan, L., Ge, H., Li, H., Lieber, S. C., Natividad, F., Resuello, R. R., Kim, S. J., Akeju, S., Sun, A., Loo, K., Peppas, A. P., Rossi, F., Lewandowski, E. D., Thomas, A. P., Vatner, S. F., and Vatner, D. E. (2004). Gender-specific proteomic alterations in glycolytic and mitochondrial pathways in aging monkey hearts. *J. Mol. Cell Cardiol.* **37,** 921–929.
115. Saito, H., and Papaconstantinou, J. (2001). Age-associated differences in cardiovascular inflammatory gene induction during endotoxic stress. *J. Biol. Chem.* **276,** 29307–29312.
116. Kuro-o, M., Matsumura, Y., Aizawa, H., Kawaguchi, H., Suga, T., Utsugi, T., Ohyama, Y., Kurabayashi, M., Kaname, T., Kume, E., Iwasaki, H., Iida, A., Shiraki-Iida, T., Nishikawa, S., Nagai, R., and Nabeshima, Y. I. (1997). Mutation of the mouse klotho gene leads to a syndrome resembling aging. *Nature* **390,** 45–51.
117. Arking, D. E., Krebsova, A., Macek, M., Sr., Macek, M., Jr., Arking, A., Mian, I. S., Fried, L., Hamosh, A., Dey, S., McIntosh, I., and Dietz, H. C. (2002). Association of human aging with a functional variant of klotho. *Proc. Natl. Acad. Sci. USA* **99,** 856–861.
118. Arking, D. E., Becker, D. M., Yanek, L. R., Fallin, D., Judge, D. P., Moy, T. F., Becker, L. C., and Dietz, H. C. (2003). KLOTHO allele status and the risk of early-onset occult coronary artery disease. *Am. J. Hum. Genet.* **2,** 1154–1161.
119. Arking, D. E., Atzmon, G., Arking, A., Barzilai, N., and Dietz, H. C. (2005). Association between a functional variant of the KLOTHO gene and high-density lipoprotein cholesterol, blood pressure, stroke, and longevity. *Circ. Res.* **96,** 412–418.
120. Roth, D. A., White, C. D., Podolin, D. A., and Mazzeo, R. S. (1998). Alterations in myocardial signal transduction due to aging and chronic dynamic exercise. *J. Appl. Physiol.* **84,** 177–184.
121. Iemitsu, M., Miyauchi, T., Maeda, S., Tanabe, T., Takanashi, M., Irukayama-Tomobe, Y., Sakai, S., Ohmori, H., Matsuda, M., and Yamaguchi, I. (2002). Aging-induced decrease in the PPAR-alpha level in hearts is improved by exercise training. *Am. J. Physiol. Heart Circ. Physiol.* **283,** H1750–H1760.
122. Maeda, S., Tanabe, T., Miyauchi, T., Otsuki, T., Sugawara, J., Iemitsu, M., Kuno, S., Ajisaka, R., Yamaguchi, I., and Matsuda, M. (2003). Aerobic exercise training reduce plasma endothelin-1 concentration in older women. *J. Appl. Physiol.* **95,** 336–3.41
123. Maeda, S., Tanabe, T., Otsuki, T., Sugawara, J., Iemitsu, M., Miyauchi, T., Kuno, S., Ajisaka, R., and Matsuda, M. (2004). Moderate regular exercise increases basal production of nitric oxide in elderly women. *Hypertens. Res.* **27,** 947–953.
124. DeSouza, C. A., Shapiro, L. F., Clevenger, C. M., Dinenno, F. A., Monahan, K. D., Tanaka, H., and Seals, D. R. (2000). Regular aerobic exercise prevents and restores age-related declines in endothelium-dependent vasodilation in healthy men. *Circulation* **102,** 1351–1357.
125. Smith, D. T., Hoetzer, G. L., Greiner, J. J., Stauffer, B. L., and DeSouza, C. A. (2003). Effects of ageing and regular aerobic exercise on endothelial fibrinolytic capacity in humans. *J. Physiol.* **546,** 289–298.
126. DeSouza, C. A., Van Guilder, G. P., Greiner, J. J., Smith, D. T., Hoetzer, G. L., and Stauffer, B. L. (2005). Basal endothelial nitric oxide release is preserved in overweight and obese adults. *Obes. Res.* **13,** 1303–1306.
127. Van Guilder, G. P., Hoetzer, G. L., Smith, D. T., Irmiger, H. M., Greiner, J. J., Stauffer, B. L., and Desouza, C. A. (2005). Endothelial t-PA release is impaired in overweight and obese adults but can be improved with regular aerobic exercise. *Am. J. Physiol. Endocrinol. Metab.* **289,** E807–E813.

128. Quindry, J., French, J., Hamilton, K., Lee, Y., Mehta, J. L., and Powers, S. (2005). Exercise training provides cardioprotection against ischemia-reperfusion induced apoptosis in young and old animals. *Exp. Gerontol.* **40**, 416–425.

129. French, J. P., Quindry, J. C., Falk, D. J., Staib, J. L., Lee, Y., Wang, K. K., and Powers, S. K. (2005). Ischemia-reperfusion induced calpain activation and SERCA2a degradation are attenuated by exercise training and calpain inhibition. *Am. J. Physiol. Heart Circ. Physiol.* **290**, H128–H123.

130. Gielen, S., Adams, V., Niebauer, J., Schuler, G., and Hambrecht, R. (2005). Aging and heart failure—Similar syndromes of exercise intolerance? Implications for exercise-based interventions. *Heart Fail Monit.* **4**, 130–136.

131. Musch, T. I., Eklund, K. E., Hageman, K. S., and Poole, D. C. (2004). Altered regional blood flow responses to submaximal exercise in older rats. *J. Appl. Physiol.* **96**, 81–88.

132. Eklund, K. E., Hageman, K. S., Poole, D. C., and Musch, T. I. (2005). Impact of aging on muscle blood flow in chronic heart failure. *J. Appl. Physiol.* **99**, 505–514.

133. Benetos, A., Waeber, B., Izzo, J., Mitchell, G., Resnick, L., Asmar, R., and Safar, M. (2002). Influence of age, risk factors, and cardiovascular and renal disease on arterial stiffness: clinical applications. *Am. J. Hypertens.* **15**, 1101–1108.

134. Najjar, S. S., Scuteri, A., and Lakatta, E. G. (2005). Arterial aging: is it an immutable cardiovascular risk factor? *Hypertension* **46**, 454–462.

135. Wilkinson, I. B., and McEniery, C. M. (2004). Arterial stiffness, endothelial function and novel pharmacological approaches. Clin. *Exp. Pharmacol. Physiol.* **31**, 795–799.

136. Abbate, R., Prisco, D., Rostagno, C., Boddi, M., and Gensini, G. F. (1993). Age-related changes in the hemostatic system. *Int. J. Clin. Lab. Res.* **23**, 1–3.

137. Hashimoto, Y., Kobayashi, A., Yamazaki, N., Sugawara, Y., Takada, Y., and Takada, A. (1987). Relationship between age and plasma t-PA, PA-inhibitor, and PA activity. *Thromb. Res.* **46**, 625–633.

138. Yamamoto, K., Takeshita, K., Kojima, T., Takamatsu, J., and Saito, H. (2005). Aging and plasminogen activator inhibitor-1 (PAI-1) regulation: Implication in the patho-genesis of thrombotic disorders in the elderly. *Cardiovasc. Res.* **66**, 276–285.

139. Lindsey, M. L., Goshorn, D. K., Squires, C. E., Escobar, G. P., Hendrick, J. W., Mingoia, J. T., Sweterlitsch, S. E., and Spinale, F. G. (2005). Age-dependent changes in myocardial matrix metalloproteinase/tissue inhibitor of metalloproteinase profiles and fibroblast function. *Cardiovasc. Res.* **66**, 410–419.

140. de Santis, D., Abete, P., Testa, G., Cacciatore, F., Galizia, G., Leosco, D., Viati, L., Del Villano, V., Della Morte, D., Mazzella, F., Ferrara, N., and Rengo, F. (2005). Echocardiographic evaluation of left ventricular end-systolic elastance in the elderly. *Eur. J. Heart Fail.* **7**, 829–833.

141. Leite-Moreira, A. F., Correia-Pinto, J., and Gillebert, T. C. (1999). Afterload induced changes in myocardial relaxation: a mechanism for diastolic dysfunction. *Cardiovasc. Res.* **43**, 344–353.

142. Fedak, P. W., Smookler, D. S., Kassiri, Z., Ohno, N., Leco, K. J., Verma, S., Mickle, D. A., Watson, K. L., Hojilla, C. V., Cruz, W., Weisel, R. D., Li, R. K., and Khokha, R. (2004). TIMP-3 deficiency leads to dilated cardiomyopathy. *Circulation* **110**, 2401–2409.

143. Brandes, R. P., Fleming, I., and Busse, R. (2005). Endothelial aging. *Cardiovasc. Res.* **66**, 286–294.

144. Dubey, R. K., Imthurn, B., Barton, M., and Jackson, E. K. (2005). Vascular consequences of menopause and hormone therapy: importance of timing of treatment and type of estrogen. *Cardiovasc. Res.* **66**, 295–306.

145. Shipley, R. D., and Muller-Delp, J. M. (2005). Aging decreases vasoconstrictor responses of coronary resistance arterioles through endothelium-dependent mechanisms. *Cardiovasc. Res.* **66**, 374–383.

146. Korzick, D. H., Muller-Delp, J. M., Dougherty, P., Heaps, C. L., Bowles, D. K., and Krick, K. K. (2005). Exaggerated coronary vasoreactivity to endothelin-1 in aged rats: Role of protein kinase C. *Cardiovasc. Res.* **66**, 384–392.

147. Donato, A. J., Lesniewski, L. A., and Delp. M. D. (2005). The effects of aging and exercise training on endothelin-1 vasoconstrictor responses in rat skeletal muscle arterioles. *Cardiovasc. Res.* **66**, 393–401.

148. Dao, H. H., Essalihi, R., Bouvet, C., and Moreau, P. (2005). Evolution and modulation of age-related medial elastocalcinosis: impact on large artery stiffness and isolated systolic hypertension. *Cardiovasc. Res.* **66**, 307–317.

149. Defawe, O. D., Kenagy, R. D., Choi, C., Wan, S. Y. C., Deroanne, C., Nusgens, B., Sakalihasan, N., Colige, A., and Clowes, A. W. (2005). MMP-9 regulates both positively and negatively collagen gel contraction. A nonproteolytic function of MMP-9. *Cardiovasc. Res.* **66**, 402–409.

150. Robert, V., Besse, S., Sabri, A., Silvestre, J. S., Assayag, P., Nguyen, V. T., Swynghedauw, B., and Delcayre, C. (1997). Differential regulation of matrix metalloproteinases associated with aging and hypertension in the rat heart. *Lab. Invest.* **76**, 729–738.

151. Lopez Jimenez, J. A., Bordoni, A., Lorenzini, A., Rossi, C. A., Biagi, P. L., and Hrelia, S. (1997). Linoleic acid metabolism in primary cultures of adult rat cardiomyocytes is impaired by aging. *Biochem. Biophys. Res. Commun.* **237**, 142–145.

152. Costell, M., O'Connor, J. E., and Grisolia, S. (1989). Age-dependent decrease of carnitine content in muscle of mice and humans. *Biochem. Biophys. Res. Commun.* **161**, 1135–1143.

153. Hansford, R. G., and Castro, F. (1982). Age-linked changes in the activity of enzymes of the tricarboxylate cycle and lipid oxidation, and of carnitine content, in muscles of the rat. *Mech. Ageing Dev.* **19**, 191–200.

154. Kates, A. M., Herrero, P., Dence, C., Soto, P., Srinivasan, M., Delano, D. G., Ehsani, A., and Gropler, R. J. (2003). Impact of aging on substrate metabolism by the human heart. *J. Am. Coll. Cardiol.* **41**, 293–299.

155. Soto, P. F., Herrero, P., Kates, A. M., Dence, C. S., Ehsani, A. A., Davila-Roman, V., Schechtman, K. B., and Gropler, R. J. (2003). Impact of aging on myocardial metabolic response to dobutamine. *Am. J. Physiol. Heart Circ. Physiol.* **285**, H2158–H2164.

156. Gonzalez, A. A., Kumar, R., Mulligan, J. D., Davis, A. J., and Saupe, K. W. (2004). Effects of aging on cardiac and skeletal muscle AMPK activity: Basal activity, allosteric activation, and response to in vivo hypoxemia in mice. *Am. J. Physiol. Regul. Integr. Comp. Physiol.* **287**, R1270–R1275.

157. Wang, W., Yang, X., Lopez de Silanes, I., Carling, D., and Gorospe, M. (2003). Increased AMP:ATP ratio and AMP-activated protein kinase activity during cellular senescence

158. Brooks, S., and Faulkner, J. (1988). Contractile properties of skeletal muscles from young, adult and aged mice. *J. Physiol.* **404,** 71–82.
159. Musaro, A., Cusella De Angelis, M. G., Germani, A., Ciccarelli, C., Molinaro, M., and Zani, B. M. (1995). Enhanced expression of myogenic regulatory factors in aging skeletal muscle. *Exp. Cell Res.* **221,** 241–248.
160. Nelson, K. (2000). The cancer anorexia-cachexia syndrome. *Semin. Oncol.* **27,** 64–68.
161. Schulze, P. C., Fang, J., Kassik, K. A., Gannon, J., Cupesi, M., MacGillivray, C., Lee, R. T., and Rosenthal, N. (2005). Transgenic overexpression of locally acting insulin-like growth factor-1 inhibits ubiquitin-mediated muscle atrophy in chronic left-ventricular dysfunction. *Circ. Res.* **97,** 418–426.
162. Musaro, A., McCullagh, K., Paul, A., Houghton, L., Dobrowolny, G., Molinaro, M., Barton, E. R., Sweeney, H. L., and Rosenthal, N. (2001). Localized Igf-1 transgene expression sustains hypertrophy and regeneration in senescent skeletal muscle. *Nat. Genet.* **27,** 195–200.
163. Coleman, M. E., DeMayo, F., Yin, K. C., Lee, H. M., Geske, R., Montgomery, C., and Schwartz, R. J. (1995). Myogenic vector expression of insulin-like growth factor I stimulates muscle cell differentiation and myofiber hypertrophy in transgenic mice. *J. Biol. Chem.* **270,** 12109–12116.
164. Delaughter, M. C., Taffet, G. E., Fiorotto, M. L., Entman, M. L., and Schwartz, R. J. (1999). Local insulin-like growth factor I expression induces physiologic, then pathologic, cardiac hypertrophy in transgenic mice. *FASEB J.* **13,** 1923–1929.
165. Glass, D. J. (2003). Molecular mechanisms modulating muscle mass. *Trends Mol. Med.* **9,** 344–350.
166. Franch, H. A., and Price, S. R. (2005). Molecular signaling pathways regulating muscle proteolysis during atrophy. *Curr. Opin. Clin. Nutr. Metab. Care* **8,** 271–275.
167. Anversa, P. (2005). Aging and longevity, the IGF-1 enigma. *Circ. Res.* **97,** 411–414.
168. Liu, P., Xu, B., Cavalieri, T. A., and Hock, C. E. (2002). Age-related difference in myocardial function and inflammation in a rat model of myocardial ischemia-reperfusion. *Cardiovasc. Res.* **56,** 443–453.
169. Liu, P., Xu, B., Cavalieri, T. A., and Hock, C. E. (2004). Attenuation of antioxidative capacity enhances reperfusion injury in aged rat myocardium after MI/R. *Am. J. Physiol. Heart Circ. Physiol.* **287,** H2719–H2727.
170. Loubani, M., Ghosh, S., and Galinanes, M. (2003). The aging human myocardium: tolerance to ischemia and responsiveness to ischemic preconditioning. *J. Thorac. Cardiovasc. Surg.* **126,** 143–147.
171. Ivanov, J., Weisel, R. D., David, T. E., and Naylor, C. D. (1998). Fifteen-year trends in risk severity and operative mortality in elderly patients undergoing coronary artery bypass graft surgery. *Circulation* **97,** 673–680.
172. Fannin, S. W., Lesnefsky, E. J., Slabe, T. J., Hassan, M. O., and Hoppel, C. L. (1999). Aging selectively decreases oxidative capacity in rat heart interfibrillar mitochondria. *Arch. Biochem. Biophys.* **372,** 399–407.
173. Hoppel, C. L., Moghaddas, S., and Lesnefsky, E. J. (2002). Interfibrillar cardiac mitochondrial complex III defects in the aging rat heart. *Biogerontology* **3,** 41–44.
174. Suh, J. H., Heath, S. H., and Hagen, T. M. (2003). Two subpopulations of mitochondria in the aging rat heart display heterogenous levels of oxidative stress. *Free Radic. Biol. Med.* **35,** 1064–1072.
175. Judge, S., Jang, Y. M., Smith, A., Hagen, T., and Leeuwenburgh, C. (2005). Age-associated increases in oxidative stress and antioxidant enzyme activities in cardiac interfibrillar mitochondria: Implications for the mitochondrial theory of aging. *FASEB J.* **19,** 419–421.
176. Szibor, M., and Holtz, J. (2002). Mitochondrial ageing. *Basic Res. Cardiol.* **98,** 210–218.
177. Chomyn, A., and Attardi, G. (2003). MtDNA mutations in aging and apoptosis. *Biochem. Biophys. Res. Commun.* **304,** 519–529.
178. Marin-Garcia, J., Zoubenko, O., and Goldenthal, M. J. (2002). Mutations in the cardiac mtDNA control region associated with cardiomyopathy and aging. *J. Card. Fail.* **8,** 93–100.
179. Marin-Garcia, J., Goldenthal, M. J., Pierpont, M. E., Ananthakrishnan, R., and Perez-Atayde, A. (1999). Is age a contributory factor of mitochondrial bioenergetic decline and DNA defects in idiopathic dilated cardiomyopathy? *Cardiovasc. Pathol.* **8,** 217–222.
180. Marin-Garcia, J., Goldenthal, M. J., Ananthakrishnan, R., and Pierpont, M. E. (2000). The complete sequence of mtDNA genes in idiopathic dilated cardiomyopathy shows novel missense and tRNA mutations. *J. Card. Fail.* **6,** 321–329.
181. Miro, O., Casademont, J., Casals, E., Perea, M., Urbano-Marquez, A., Rustin, P., and Cardellach, F. (2000). Aging is associated with increased lipid peroxidation in human hearts, but not with mitochondrial respiratory chain enzyme defects. *Cardiovasc. Res.* **47,** 624–631.
182. Lesnefsky, E. J., and Hoppel, C. L. (2003). Ischemia-reperfusion injury in the aged heart: role of mitochondria. *Arch. Biochem. Biophys.* **420,** 287–297.
183. Welle, S., Bhatt, K., Shah, B., Needler, N., Delehanty, J. M., and Thornton, C. A. (2003). Reduced amount of mitochondrial DNA in aged human muscle. *J. Appl. Physiol.* **94,** 1479–1484.
184. Barazzoni, R., Short, K. R., and Nair, K. S. (2000). Effects of aging on mitochondrial DNA copy number and cytochrome c oxidase gene expression in rat skeletal muscle, liver, and heart. *J. Biol. Chem.* **275,** 3343–3347.
185. Wallace, D. C. (1992). Diseases of the mitochondrial DNA. *Annu. Rev. Biochem.* **61,** 1175–1212.
186. Trifunovic, A., Wredenberg, A., Falkenberg, M., Spelbrink, J. N., Rovio, A. T., Bruder, C. E., Bohlooly-Y, M., Gidlof, S., Oldfors, A., Wibom, R., Tornell, J., Jacobs, H. T., and Larsson, N. G. (2004). Premature ageing in mice expressing defective mitochondrial DNA polymerase. *Nature* **27,** 417–423.
187. Wanrooij, S., Luoma, P., van Goethem, G., van Broeckhoven, C., Suomalainen, A., and Spelbrink, J. N. (2004). Twinkle and POLγ defects enhance age-dependent accumulation of mutations in the control region of mtDNA. *Nucleic Acids Res.* **32,** 3053–3064.
188. Craig, E. E., and Hood, D. A. (1997). Influence of aging on protein import into cardiac mitochondria. *Am. J. Physiol.* **272,** H2983–H2988.
189. Ji, L. L., Dillon, D., and Wu, E. (1991). Myocardial aging: antioxidant enzyme systems and related biochemical properties. *Am. J. Physiol.* **261,** R386–R392.

190. Vertechy, M., Cooper, M. B., Ghirardi, O., and Ramacci, M. T. (1989), Antioxidant enzyme activities in heart and skeletal muscle of rats of different ages. *Exp. Gerontol.* **24**, 211–218.
191. Radak, Z., Chung, H. Y., Naito, H., Takahashi, R., Jung, K. J., Kim, H. J., and Goto, S. (2004). Age-associated increase in oxidative stress and nuclear factor kappaB activation are attenuated in rat liver by regular exercise. *FASEB J.* **18**, 749–750.
192. Helenius, M., Hanninen, M., Lehtinen, S. K., and Salminen, A. (1996). Changes associated with aging and replicative senescence in the regulation of transcription factor nuclear factor-kappa B. *Biochem. J.* **318**, 603–608.
193. Anversa, P., and Capasso, J. M. (1991). Cellular basis of aging in the mammalian heart. *Scanning Microsc.* **5**, 1065–1073.
194. Kajstura, J., Cheng, W., Sarangarajan, R., Li, P., Li, B., Nitahara, J. A., Chapnick, S., Reiss, K., Olivetti, G., and Anversa, P. (1996). Necrotic and apoptotic myocyte cell death in the aging heart of Fischer 344 rats. *Am. J. Physiol.* **271**, H1215–H1228.
195. Nitahara, J. A., Cheng, W., Liu, Y., Li, B., Leri, A., Li, P., Mogul, D., Gambert, S. R., Kajstura, J., and Anversa, P. (1998). Intracellular calcium, DNase activity and myocyte apoptosis in aging Fischer 344 rats. *J. Mol. Cell Cardiol.* **30**, 519–535.
196. Centurione, L., Antonucci, A., Miscia, S., Grilli, A., Rapino, M., Grifone, G., Di Giacomo, V., Di Giulio, C., Falconi, M., and Cataldi, A. (2002). Age-related death-survival balance in myocardium: An immunohistochemical and biochemical study. *Mech. Ageing Dev.* **123**, 341–350.
197. Kang, P. M., Yue, P., Liu, Z., Tarnavski, O., Bodyak, N., and Izumo, S. (2004). Alterations in apoptosis regulatory factors during hypertrophy and heart failure. *Am. J. Physiol. Heart Circ. Physiol.* **287**, H72–H80.
198. McCright, B., and Virshup, D. M. (1995). Identification of a new family of protein phosphatase 2A regulatory subunits. *J. Biol. Chem.* **270**, 26123–26128.
199. Ruvolo, P. P., Clark, W., Mumby, M., Gao, F., and May, W. S. (2002). A functional role for the B56 alpha-subunit of protein phosphatase 2A in ceramide-mediated regulation of Bcl2 phosphorylation status and function. *J. Biol. Chem.* **277**, 22847–22852.
200. Di Lisa, F., and Bernardi, P. (2005). Mitochondrial function and myocardial aging. A critical analysis of the role of permeability transition. *Cardiovasc. Res.* **66**, 222–232.
201. Aon, M. A., Cortassa, S., Marban, E., and O'Rourke, B. (2003). Synchronized whole cell oscillations in mitochondrial metabolism triggered by a local release of reactive oxygen species in cardiac myocytes. *J. Biol. Chem.* **278**, 44735–44744.
202. Goldenthal, M. J., Ananthakrishnan, R., and Marin-Garcia, J. (2005). Nuclear-mitochondrial cross-talk in cardiomyocyte T3 signaling: a time-course analysis. *J. Mol. Cell Cardiol.* **39**, 319–326.
203. Portman, M. A. (2002). The adenine nucleotide translocator: regulation and function during myocardial development and hypertrophy. *Clin. Exp. Pharmacol. Physiol.* **29**, 334–338.
204. Barazzoni, R., and Nair, K. S. (2001). Changes in uncoupling protein-2 and -3 expression in aging rat skeletal muscle, liver, and heart. *Am. J. Physiol. Endocrinol. Metab.* **280**, E413–E419.
205. Paradies, G., Ruggiero, F. M., Petrosillo, G., and Quagliariello, E. (1998). Peroxidative damage to cardiac mitochondria: Cytochrome oxidase and cardiolipin alterations. *FEBS Lett.* **424**, 155–158.
206. Miller, R. A. (2005). Genetic approaches to the study of aging. *J. Am. Geriatr. Soc.* **53**, S284–S286.
207. Ureta-Vidal, A., Ettwiller, L., and Birney, E. (2003). Comparative genomics: genome-wide analysis in metazoan eukaryotes. *Nat. Rev. Genet.* **4**, 251–262.
208. de Magalhaes, J. P., Costa, J., and Toussaint, O. (2005). HAGR: The Human Aging Genomic Resources. *Nucleic Acids Res.* **33**, D537–D543.
209. Pound, P., Ebrahim, S., Sandercock, P., Bracken, M. B., and Roberts, I. (2004). Where is the evidence that animal research benefits humans? *BMJ* **328**, 514–517.
210. Carroll, S. B. (2000). Endless forms: the evolution of gene regulation and morphological diversity. *Cell* **101**, 577–580.
211. Edwards, M. G., Sarkar, D., Klopp, R., Morrow, J. D., Weindruch, R., and Prolla, T. A. (2004). Impairment of the transcriptional responses to oxidative stress in the heart of aged C57BL/6 mice. *Ann. N. Y. Acad. Sci.* **1019**, 85–95.
212. Lee, C. K., Weindruch, R., and Prolla, T. A. (2000). Gene-expression profile of the aging brain in mice. *Nat. Genet.* **25**, 294–297.
213. Lee, C. K., Klopp, R. G., Weindruch, R., and Prolla, T. A. (1999). Gene expression profile of aging and its retardation by caloric restriction. *Science* **285**, 1390–1393.
214. Lee, C. K., Allison, D. B., Brand, J., Weindruch, R., and Prolla, T. A. (2002). Transcriptional profiles associated with aging and middle age-onset caloric restriction in mouse hearts. *Proc. Natl. Acad. Sci. USA* **99**, 14988–14993.
215. Kayo, T., Allison, D. B., Weindruch, R., and Prolla, T. A. (2001). Influences of aging and caloric restriction on the transcriptional profile of skeletal muscle from rhesus monkeys. *Proc. Natl. Acad. Sci. USA* **98**, 5093–5098.
216. Park, S. K., and Prolla, T. A. (2005). Gene expression profiling studies of aging in cardiac and skeletal muscles. *Cardiovasc. Res.* **66**, 205–212.
217. Shroff, S. G., Saner, D. R., and Lal, R. (1995). Dynamic micromechanical properties of cultured rat atrial myocytes measured by atomic force microscopy. *Am. J. Physiol.* **269**, C286–C292.
218. Mathur, A. B., Collinsworth, A. M., Reichert, W. M., Kraus, W. E., and Truskey, G. A. (2001). Endothelial, cardiac muscle and skeletal muscle exhibit different viscous and elastic properties as determined by atomic force microscopy. *J. Biomech.* **34**, 1545–1553.
219. Lieber, S. C., Aubry, N., Pain, J., Diaz, G., Kim, S. J., and Vatner, S. F. (2004). Aging increases stiffness of cardiac myocytes measured by atomic force microscopy nanoindentation. *Am. J. Physiol. Heart Circ. Physiol.* **287**, H645–H651.
220. Bogenhagen, D. F. (1999). Repair of mtDNA in vertebrates. *Am. J. Hum. Genet.* **64**, 1276–1281.
221. Bohr, V. A., and Dianov, G. L. (1999). Oxidative DNA damage processing in nuclear and mtDNA. *Biochimie* **81**, 155–160.
222. Driggers, W. J., Holmquist, G. P., LeDoux, S. P., and Wilson, G. L. (1997). Mapping frequencies of endogenous oxidative damage and the kinetic response to oxidative stress in a region of rat mtDNA. *Nucleic Acids Res.* **25**, 4362–4369.

223. Wallace, D. C. (2002). Animal models for mitochondrial disease. *Methods Mol. Biol.* **197,** 3–54.
224. Harley, C. B. (2005). Telomerase therapeutics for degenerative diseases. *Curr. Mol. Med.* **5,** 205–211.
225. Blasco, M. A., Lee, H. W., Handle, M. P., Samper, E., Lansdorp, P. M., DePinho, R. A., and Greider, C. W. (1997). Telomere shortening and tumor formation by mouse cells lacking telomerase RNA. *Cell* **91,** 25–34.
226. Goytisolo, F. A., Samper, E., Martin-Caballero, J., Finnon, P., Herrera, E., Flores, J. M., Bouffler, S. D., and Blasco, M. A. (2000). Short telomeres result in organismal hypersensitivity to ionizing radiation in mammals. *J. Exp. Med.* **192,** 1625–1636.
227. Herbert, B., Pitts, A. E., Baker, S. I., Hamilton, S. E., Wright, W. E., Shay, J. W., and Corey, D. R. (1999). Inhibition of human telomerase in immortal human cells leads to progressive telomere shortening and cell death. *Proc. Natl. Acad. Sci. USA* **96,** 14276–14281.
228. Heist, E. K., Huq, F., and Hajjar, R. (2003). Telomerase and the aging heart. *Sci. Aging Knowledge Environ.* **2003,** PE11.
229. Oh, H., Taffet, G. E., Youker, K. A., Entman, M. L., Overbeek, P. A., Michael, L. H., and Schneider, M. D. (2001). Telomerase reverse transcriptase promotes cardiac muscle cell proliferation, hypertrophy, and survival. *Proc. Natl. Acad. Sci. USA* **98,** 10308–10313.
230. Oh, H., Wang, S. C., Prahash, A., Sano, M., Moravec, C. S., Taffet, G. E., Michael, L. H., Youker, K. A., Entman, M. L., and Schneider, M. D. (2003). Telomere attrition and Chk2 activation in human heart failure. *Proc. Natl. Acad. Sci. USA* **100,** 5378–5383.
231. Armstrong, L., Saretzki, G., Peters, H., Wappler, I., Evans, J., Hole, N., von Zglinicki, T., and Lako, M. (2005). Overexpression of telomerase confers growth advantage, stress resistance, and enhanced differentiation of ESCs toward the hematopoietic lineage. *Stem Cells* **23,** 516–529.
232. Cai, D., Xaymardan, M., Holm, J. M., Zheng, J., Kizer, J. R., and Edelberg, J. M. (2003). Age-associated impairment in TNF-alpha cardioprotection from myocardial infarction. *Am. J. Physiol. Heart Circ. Physiol.* **285,** H463–H469.
233. Xaymardan, M., Zheng, J., Duignan, I., Chin, A., Holm, J. M., Ballard, V. L., and Edelberg, J. M. (2004). Senescent impairment in synergistic cytokine pathways that provide rapid cardioprotection in the rat heart. *J. Exp. Med.* **199,** 797–804.
234. Zheng, J., Chin, A., Duignan, I., Won, K. H., Hong, M. K., and Edelberg, J. M. (2005). Growth factor-mediated reversal of senescent dysfunction of ischemia-induced cardioprotection. *Am. J. Physiol Heart Circ. Physiol.* **290,** H525–H530.
235. Schmidt, U., Zhu, X., Lebeche, D., Huq, F., Guerrero, J. L., and Hajjar, R. J. (2005). In vivo gene transfer of parvalbumin improves diastolic function in aged rat hearts. *Cardiovasc. Res.* **66,** 318–323.
236. Speer, M. Y., Chien, Y.-C., Quan, M., Yang, H.-Y., Vali, H., McKee, M. D., and Giachelli, C. M. (2005). Smooth muscle cells deficient in osteopontin have enhanced susceptibility to calcification in vitro. *Cardiovasc. Res.* **66,** 324–333.
237. Johnson, J. A., and Cavallari, L. H. (2005). Cardiovascular pharmacogenomics. *Exp. Physiol.* **90,** 283–289.
238. Johnson, J. A. (2001). Drug target pharmacogenomics: An overview. *Am. J. Pharmacogenomics* **1,** 271–281.
239. Siest, G., Jeannesson, E., Berrahmoune, H., Maumus, S., Marteau, J. B., Mohr, S., and Visvikis, S. (2004). Pharmacogenomics and drug response in cardiovascular disorders. *Pharmacogenomics* **5,** 779–802.
240. Mahlknecht, U., and Voelter-Mahlknecht, S. (2005). Pharmacogenomics: questions and concerns. *Curr. Med. Res. Opin.* **21,** 1041–1047.
241. Anversa, P., Rota, M., Urbanek, K., Hosoda, T., Sonnenblick, E. H., Leri, A., Kajstura, J., and Bolli, R. (2005). Myocardial aging A stem cell problem. *Basic Res. Cardiol.* **100,** 482–493.
242. Ott, H. C., McCue, J., and Taylor, D. A. (2005). Cell-based cardiovascular repair. The hurdles and the opportunities. *Basic Res. Cardiol.* **100,** 504–517.
243. Zullo, S. J. (2001). Gene therapy of mitochondrial DNA mutations: A brief, biased history of allotopic expression in mammalian cells. *Semin. Neurol.* **21,** 327–335.
244. Melov, S., Hinerfeld, D., Esposito, L., and Wallace, D. C. (1997). Multi-organ characterization of mitochondrial genomic rearrangements in ad libitum and caloric restricted mice show striking somatic mitochondrial DNA rearrangements with age. *Nucl. Acids Res.* **25,** 974–982.
245. Masoro, E. J. (1993). Dietary restriction and aging. *J. Am. Geriatr. Soc.* **41,** 994–999.
246. Ward, W. F., Qi, W., Van Remmen, H., Zackert, W. E., Roberts, L. J., 2nd., and Richardson, A. (2005). Effects of age and caloric restriction on lipid peroxidation: measurement of oxidative stress by F2-isoprostane levels. *Gerontol. A Biol. Sci. Med. Sci.* **60,** 847–851.
247. Nisoli, E., Tonello, C., Cardile, A., Cozzi, V., Bracale, R., Tedesco, L., Falcone, S., Valerio, A., Cantoni, O., Clementi, E., Moncada, S., and Carruba, M. O. (2005). Calorie restriction promotes mitochondrial biogenesis by inducing the expression of eNOS. *Science* **310,** 314–317.
248. Goodrick, C. L., Ingram, D. K., Reynolds, M. A., Freeman, J. R., and Cider, N. (1990). Effects of intermittent feeding upon body weight and lifespan in inbred mice: Interaction of genotype and age. *Mech. Ageing Dev.* **55,** 69–87.
249. Sonntag, W. E., Xu, X., Ingram, R. L., and D'Costa, A. (1995). Moderate caloric restriction alters the subcellular distribution of somatostatin mRNA and increases growth hormone pulse amplitude in aged animals. *Neuroendocrinology* **61,** 601–608.
250. Sonntag, W. E., Lynch, C. D., Cefalu, W. T., Ingram, R. L., Bennett, S. A., Thornton, P. L., and Khan, A. S. (1999). Pleiotropic effects of growth hormone and insulin-like growth factor (IGF)-1 on biological aging: inferences from moderate caloric-restricted animals. *J. Gerontol. A Biol. Sci. Med. Sci.* **54,** B521–B538.
251. Masoro, E. J. (2003). Subfield history: Caloric restriction, slowing aging, and extending life. *Sci. Aging Knowledge Environ.* **2003,** RE2.
252. Heydari, A. R., Wu, B., Takahashi, R., Strong, R., and Richardson, A. (1993). Expression of heat shock protein 70 is altered by age and diet at the level of transcription. *Mol. Cell Biol.* **13,** 2909–2918.
253. Martin, G. M., Austad, S. N., and Johnson, T. E. (1996). Genetic analysis of ageing: role of oxidative damage and environmental stresses. *Nat. Genet.* **13,** 25–34.
254. Sacher, G. A. (1977). Life table modification and life prolongation. *In* "Handbook of the Biology of Aging." (C. E. Finch, and L. Hayflick, Eds.) pp. 582–638. van Nostrand-Reinhold, New York.

255. Masoro, E. J., Yu, B. P., and Bertrand, H. A. (1982). Action of food restriction in delaying the aging process. *Proc. Natl. Acad. Sci. USA* **79,** 4239–4241.
256. Holliday, R. (1989). Food, reproduction, and longevity: Is the extended lifespan of calorie-restricted animals an evolutionary adaptation? *BioEssays* **10,** 125–127.
257. Walford, R. L., and Spindler, S. R. (1997). The response to caloric restriction in mammals shows features also common to hibernation: A cross-adaptation hypothesis. *J. Gerontol. A Biol. Sci. Med. Sci.* **52,** B179–B183.
258. Cefalu, W. T., Wang, Z. Q., Bell-Farrow, A. D., Collins, J., Morgan, T., and Wagner, J. D. (2004). Caloric restriction and cardiovascular aging in cynomolgus monkeys (*Macaca fascicularis*): Metabolic, physiologic, and atherosclerotic measures from a 4-year intervention trial. *J. Gerontol. A Biol. Sci. Med. Sci.* **59,** 1007–1014.
259. Edwards, I. J., Rudel, L. L., Terry, J. G., Kemnitz, J. W., Weindruch, R., Zaccaro, D. J., and Cefalu, W. T. (2001). Caloric restriction lowers plasma lipoprotein (a) in male but not female rhesus monkeys. *Exp. Gerontol.* **36,** 1413–1418.
260. Weindruch, R., Keenan, K. P., Carney, J. M., Fernandes, G., Feuers, R. J., Floyd, R. A., Halter, J. B., Ramsey, J. J., Richardson, A., Roth, G. S., and Spindler, S. R. (2001). Caloric restriction mimetics: metabolic interventions. *J. Gerontol. A Biol. Sci. Med. Sci.* **56,** 20–33.
261. Lee, C. K., Pugh, T. D., Klopp, R. G., Edwards, J., Allison, D. B., Weindruch, R., and Prolla, T. A. (2004). The impact of alpha-lipoic acid, coenzyme Q10 and caloric restriction on life span and gene expression patterns in mice. *Free Radic. Biol. Med.* **36,** 1043–1057.
262. Hagen, T. M., Moreau, R., Suh, J. H., and Visioli, F. (2002). Mitochondrial decay in the aging rat heart: evidence for improvement by dietary supplementation with acetyl-L-carnitine and/or lipoic acid. *Ann. N. Y. Acad. Sci.* **959,** 491–507.
263. Rosenfeldt, F. L., Pepe, S., Linnane, A., Nagley, P., Rowland, M., Ou, R., Marasco, S., Lyon, W., and Esmore, D. (2002). Coenzyme Q10 protects the aging heart against stress: studies in rats, human tissues, and patients. *Ann. N. Y. Acad. Sci.* **959,** 355–359.
264. Tiano, L., Fedeli, D., Santoni, G., Davies, I., Wakabayashi, T., and Falcioni, G. (2003). Ebselen prevents mitochondrial ageing due to oxidative stress: In vitro study of fish erythrocytes. *Mitochondrion* **2,** 428–436.
265. Navarro, A., Gomez, C., Lopez-Cepero, J. M., and Boveris, M. (2004). Beneficial effects of moderate exercise on mice aging: Survival, behavior, oxidative stress and mitochondrial electron transfer. *Am. J. Physiol. Regul. Integr. Comp. Physiol.* **286,** R505–R511.
266. Guarente, L. P. (2004). Forestalling the great beyond with the help of SIR2. *The Scientist* **18,** 34–35.

SECTION X

Looking to the Future

CHAPTER **20**

Future of Post-Genomic Cardiology

OVERVIEW

New discoveries in molecular, genetic, and cellular cardiology have contributed greatly toward our understanding of the basic mechanisms involved in the pathogenesis of cardiovascular disease. This knowledge is currently being adapted into new tools for future clinical application in diagnostics and therapeutics, supplementing and possibly replacing current approaches to treatment. In this chapter, we will discuss future frontiers in post-genomic cardiology, including new technologies such as proteomics, biomarkers, systems biology, mitochondrial medicine, and new frontiers in cardiovascular diagnosis and therapy.

INTRODUCTION

Since the publication of the Human Genome Project (HGP), new discoveries in molecular biology and genetics are being applied in cardiology. Chromosomal mapping and the discovery of genes involved in both the primary etiology and as significant risks factors are playing an essential role in our understanding of the pathogenesis of cardiac and vascular defects. However, fundamental questions still remain unanswered regarding the underlying molecular and biochemical mechanisms involved and how this information can be used to improve cardiac diagnosis and treatment. To address these questions, emerging technologies are being recruited, some tested so far in animal models and others being investigated in clinical trials. These novel approaches include the use of molecular genetics, microarrays, proteomics, and integrated systems biology. This, together with assessment of the present status of an increasing number of novel cardiac biomarkers for cardiac diagnosis and their future clinical applications, and the use of gene and cell transplantation therapies represent some of the advances in cardiology to be further reviewed in this chapter.

CARDIAC DISEASE BIOMARKERS

A rather underdeveloped and yet extremely important area in cardiovascular medicine is the identification and use of markers in the diagnosis, prognosis, and risk stratification of various cardiovascular diseases (CVDs).

It is worth noting that cardiac biomarkers are mostly important during the early stages of myocardial infarction (MI), as well as in acute coronary syndrome (ACS), in which clinical signs may not be entirely informative and in which early diagnosis can be rapidly established and early treatment can be started to reduce the extent of myocardial injury, enhancing patient survival. Moreover, a number of these cardiac biomarkers have been shown to be excellent predictors of future cardiovascular events (including mortality) and, therefore, may be used not only in the diagnosis but also in developing strategies for prevention, therapy, and risk stratification. A list of biomarkers, their association with an array of pathological processes, their use in diagnosis, as well as in risk stratification of ACS is presented in Table I.

These markers range from markers of ischemia and resultant myocardial damage such as myocyte necrosis (e.g., cardiac troponins, myoglobin, creatine kinase-MB), hemodynamic markers of neurohumoral and vascular stress (e.g., brain natriuretic peptide [BNP] and its relative, NT-proBNP), markers of systemic inflammatory process (e.g., C-reactive protein, myeloperoxidase, matrix metalloproteinases and interleukins), markers of platelet function (e.g., the soluble CD40 ligand, P-selectin) and markers of hemostasis, including fibrinogen and soluble fibrin, vWF).[1] These markers are all present in the blood, allowing relatively low-cost and noninvasive screening. In addition, the assays of these biomarkers are all relatively rapid, show high sensitivity and specificity, can be performed easily in a highly reproducible and accurate fashion, and display a low coefficient of variation.

Among the most effective diagnostic protein markers to emerge in cardiovascular medicine, the most definitive and widely used in the diagnosis of MI are cardiac troponins, either cardiac TnI or, alternately, TnT, both cardiac-specific muscle contractile protein isoforms, because of their sensitivity and tissue-specificity.[2,3] The diagnosis of MI generally includes a finding of elevated cardiac marker (e.g., TnI > 1 ng/ml), indicative of destruction of cardiac muscle tissue surrounding the infarct area and typically leading to the initiation of reperfusion treatment. The troponin biomarkers are also useful in differentiating myocardial injury from skeletal muscle injury, which the next best alternative to troponin, the cardiac isozyme of creatine kinase (CK-MB), might not provide. Serum myoglobin, which has a high concentration in skeletal muscle, has a lower specificity for MI; however, it is useful as a very

TABLE I Biomarkers Function, and Their Use in Diagnosis and Risk Stratification in Myocardial Ischemia/ACS

Biomarker	Pathological Indication	Diagnostic Indication	Prognostic Indication
Troponin I (cTnI)	Ischemia/necrosis	Later MI/ACS	Death/MI
Troponin T (cTnT)	Ischemia/necrosis	Later MI/ACS	Death/MI/RI
Ischemia-modified albumin (IMA)	Early myocardial ischemia	ACS	—
Myoglobin	Myocardial ischemia/damage	Early MI/RI	—
Creatine kinase MB (CK-MB)	Myocardial ischemia/damage	Later MI	Death
Myosin light chain	Myocardial ischemia/damage	Later MI	—
Brain natriuretic peptide (BNP)	Hemodynamic stress	Acute HF	Death/CHF
N-terminal pro-brain natriuretic peptide (NT-proBNP)	Hemodynamic stress		Death/CHF
High sensitivity C-reactive protein (hsCRP)	Inflammation		Death/MI/RI
Myeloperoxidase (MPO)	Inflammation	ACS	Death/MI
Interleukin-10 (IL-10)	Inflammation		
Interleukin-6 (IL-6)	Inflammation		
Monocyte chemoattractant protein (MCP-1)	Inflammation		Potential risk predictor in ACS
Matrix metalloproteinases (MMPs)	Inflammation		Death/MI
Soluble CD40 ligand (sCD40L)	Platelet function		Death/MI
P-selectin	Platelet function		
von Willebrand factor (vWF)	Hemostasis		Death
Fibrinogen	Hemostasis		ACS
Plasminogen activator inhibitor 1 (PAI-1)	Hemostasis		Death
Unbound free fatty acids (FFAu)	Myocardial ischemia	Early indicator of ischemia	Death/ventricular dysrhythmias
Heart fatty acid binding protein (H-FABP)	Myocardial ischemia/necrosis	AMI/stroke	Death/RI/CHF
Fibrinopeptide A	Hemostasis		
Soluble fiber	Hemostasis	Early marker of ischemia in unstable angina	

ACS, acute coronary syndrome; CHF, congestive heart failure; AMI, acute myocardial infarct; RI, recurrent infarct.

early marker of MI because of its relatively small size and rapid release from cardiac cells in the setting of myocardial necrosis (usually within 1 h). In combination with cardiac troponin and/or CK-MB, multi-marker evaluation provides a more rapid and accurate assessment of MI or its exclusion.[4] Although troponins I and T (TnI and TnT) are cardiac-specific markers that yield diagnostic and prognostic value in patients with myocardial injury, troponins cannot be used in all clinical settings. It is notable that troponins have limited use for the diagnosis of early ischemia (because they become evident only 6 h or more after MI), as well as in preoperative MI.

Biomarkers that reliably detect myocardial ischemia in the absence of necrosis would be particularly important in the initial identification of patients with unstable angina and in the differentiation of chest pain from conditions other than coronary ischemia, as well as providing complementary clinical utility to that of cardiac troponins, the established markers of necrosis. Two such biomarkers that fit this description (although largely untested) include unbound free fatty acids (FFAu) and their intracellular binding protein, heart-type fatty acid-binding protein (H-FABP), which have shown potential clinical application as indicators of cardiac ischemia and necrosis, respectively.[5] Recent data suggest that serum FFAu concentrations increase well before other markers of cardiac necrosis and are sensitive indicators of ischemia in acute MI (AMI); this marker seems to be upstream of the biomarkers of necrosis (cardiac troponins I and T) and may provide an earlier assessment of the overall patient ischemic risk. H-FABP that is cardiac-specific behaves similarly to myoglobin, increasing within 3 h after MI and, therefore, is a useful gauge of cardiac injury in ACS and early MI.[6] It also allows detection of minor myocardial injury in patients with heart failure (HF) and unstable angina and can be used as a reliable marker of severe congestive heart failure (CHF).[7,8] Moreover, H-FABP levels in early ACS are prognostic of recurrent cardiac events.[6] Recently, Niizeki and associates have demonstrated that serum H-FABP levels could be effectively used for risk stratification and prediction of cardiac events in elderly patients with CHF.[9]

Ischemia-modified albumin is another biochemical marker of myocardial ischemia recently reported, and it is useful in the assessment of ACS in patients with acute chest pain but normal electrocardiograms.[10] Previous studies have indicated that albumin undergoes a significant reduction in its capacity to bind exogenous cobalt soon after transient coronary occlusion during human percutaneous transluminal coronary angioplasty (PTCA) and well before significant elevations of CK-MB, myoglobin, or cTnI, indicating its potential use in the detection of early myocardial ischemia.[11] Nevertheless, important questions remain as to why the modification of a ubiquitous circulating protein such as albumin would be specific for the local phenomenon of myocardial ischemia. The evaluation of other conditions that increase this modification (other than MI) merits additional exploration.

Other hemodynamic markers of myocardial ischemia include BNP and the N-terminal fragment of its prohormone (NT-proBNP), which are produced and released from cardiac myocytes in the left ventricle in response to elevated ventricular wall stress.[12] Measurement of BNP and N-proBNP has become a routine practice in the screening of patients with suspected HF, because their determination can facilitate the diagnosis and risk stratification of patients with CHF and also to guide the response to therapy.

Several reports have indicated that BNP is a marker of myocardial ischemia, and experimental studies have shown that BNP gene expression is markedly up-regulated in infarcted tissue and in surrounding ischemic tissue but not in viable myocardium.[13] Clinical studies of patients with coronary artery disease (CAD) have shown that after exercise, BNP levels increase proportionally to the size and severity of ischemia.[14,15] Increased BNP has also been found after coronary angioplasty and in patients with unstable angina.[16,17] Because of the relatively wide variety in stimuli that can elicit modest BNP elevation, it has been suggested that its diagnostic value may be limited in predicting early ischemic events;[18] others have argued that given the faster response of BNP (compared with troponin), it might be useful in assessing early MI (perhaps in conjunction with other markers of early MI, which remain to be identified).[1] Furthermore, BNP levels have significant prognostic value in patients with ACS.[1,18] Higher BNP levels are associated with worse clinical outcomes, and the relationship between natriuretic peptide levels and mortality is independent of risk factors that include age, renal insufficiency, and LV dysfunction and is complementary to cardiac troponin.

Compelling evidence has implicated inflammation in all phases of the atherosclerotic pathogenetic process, from plaque formation to the progression, and ultimately the thrombotic complications of atherosclerosis.[19] The composition of the atherosclerotic plaque is now recognized as a key feature in determining plaque vulnerability, and hence the risk of acute coronary ischemic events. Stemming from the increasing focus on the role of inflammation in atherogenesis, questions have been raised as to whether circulating levels of inflammatory biomarkers can help to identify those at risk for future cardiovascular events.

A growing body of evidence has confirmed that high-sensitivity C-reactive protein (hsCRP), is a reliable acute-phase inflammatory biomarker, and it is widely used as a marker of future cardiovascular events.[20] First described as a component of the inflammatory pathway in ACS, hsCRP has been consistently found to be associated with the risk of future events in both patients with ST elevation MI and with non-ST elevation ACS, independently of other risk factors.[21,22] Although a number of studies have demonstrated that hsCRP (and other inflammatory biomarkers) is elevated in patients with ACS, with or without evidence of myocardial damage and necrosis including troponin release,[23–25] recent observations in ACS have challenged that view, because both systemic and local markers of inflammatory activity may be associated with myocardial injury.[26] Moreover, in large populations of apparently healthy subjects, elevated hsCRP levels (presumably reflective of low-grade chronic inflammation) have been found to be a powerful marker of risk for future cardiovascular events, including AMI, stroke, and peripheral arterial disease, and are independent of traditional risk factors in the setting of low levels of low-density lipoprotein (LDL) cholesterol.[27] Similarly, it has been found that in older men and women, elevated CRP is associated with increased 10-y risk of CAD, regardless of the presence or absence of cardiac risk factors.[28] Whether a better inflammatory marker currently exists is not known; therefore, hsCRP continues to be the principle inflammatory marker in clinical use.

Another inflammatory marker, which seems to have predictive value in ACS, is soluble CD40 ligand (sCD40L).[29] This factor seems to be a marker of inflammation and platelet activation, which directly destabilize atherosclerotic plaques by stimulating proinflammatory T lymphocytes. Recently, it has been reported that elevated plasma levels of sCD40L can identify patients with ACS at heightened risk of death and recurrent MI independent of other predictive variables, including cTnI and CRP. Interestingly, the combined evaluation of sCD40L with cTnI complements the prognostic assessment for death and MI.[30]

Two other inflammatory markers (e.g., interleukin-6, myeloperoxidase) have also proven useful in ACS.[31] Circulating levels of interleukin-6 are elevated in unstable (but not stable) angina, AMI, and in ACS and tend to parallel CRP increases. Interestingly, coronary sinus IL-6 concentrations are also increased after percutaneous coronary intervention (PCI), and late restenosis correlates with an increase in IL-6 concentration after the procedure, suggesting that IL-6 expression may be not only related to instability of atheromatous plaques but also to the formation of restenotic lesions after PCI.[32]

USE OF PROTEOMICS TO DISCOVER NEW MARKERS

Despite the rapid strides made in the past few years in proteomics technology, there has been a conspicuous lull in the development of new plasma protein markers with widespread clinical application in the diagnosis of CVD.

Primarily, two major approaches have been used in proteomic research, the complete proteomic approach in which the entire proteome—all proteins—is characterized and a more limited proteomic analysis that is either targeted to specific candidate proteins or limited to specific classes of proteins or subproteomes.[33] Recently, Anderson[34] has presented a well-thought-out argument that a complete proteomic analysis of plasma proteins, although extremely desirable in the long-run (and a Holy Grail of a considerable number of laboratories), is likely unattainable with the presently available proteomic technology. This is largely due to the high sensitivity required to evaluate in depth the plasma proteome, reflecting both the dynamic range of its constituent specific protein levels, such as regulatory factors and cytokines simply undetectable by present proteomic technology, as well as the astonishing level of post-translational modifications that most proteins undergo. In strong advocacy of a targeted approach, particularly with regard to cardiovascular biomarkers, Anderson has presented a list of 177 candidate proteins, many of which can be simultaneously analyzed using available mass spectrometry (MS) technology (specifically SPE-LC-MS/MS). This is a widely used method for the precise quantitative determination of small molecules such as drugs, drug metabolites, and hormones. It also might be achieved by use of multiplex assessment with immunoblot microarrays; however, at this time, the use of antibody arrays, which could provide the high sensitivity and specificity required, are currently limited by the availability (and expense) of suitable antibodies. The analysis by MS/MS techniques focused on examining multiple small peptides generated (by proteolytic digestion) from the larger plasma proteins would have a high sensitivity, and also it would be extremely rapid and quantitative (compared with other proteomic techniques such as electrophoresis that tend to show great variability in quantitative assessment). Furthermore, the pre-selection or pre-fractionation of a group of candidate proteins may allow the use of assay technologies with higher sensitivity and a greater dynamic range than what is available with current proteomics. The simultaneous profiling analysis of multiple independent disease-related circulating marker proteins and their peptides, considered in the aggregate, would also be less prone to the influence of heterogenous genetic factors and disease processes, as well as environmental "noise" that might impact on the level of a single marker protein, giving a better fingerprint for a disease state. It has been noted that multiple biomarkers, considered as a composite, may provide better prediction of disease state than single markers (e.g., in the assessment of inflammation with a panel of weak acute phase reactants compared with a single marker such as CRP or serum amyloid A).[35] Similarly the relative risk of CAD is better predicted (by CRP and LDL-cholesterol) together than by either marker alone.[36]

Recently, a novel cardiac multi-marker approach has been used that is based on protein biochip array technology simultaneously assessing cTnI, CK-MB, myoglobin, carbonic anhydrase III (CAIII), and FABP in a single chip.[37] This method has been applied to the clinical diagnosis of ACS, and the data obtained have showed that FABP had a better diagnostic profile than myoglobin in the detection of AMI during the first hours after the onset of the chest pain. Furthermore, myoglobin/CAIII ratio, previously noted to be a sensitive and specific marker for perioperative MI,[38] significantly improved the myoglobin specificity.

The HUPO Plasma Proteome Project pilot phase has been initiated to accelerate the identification and development of novel cardiovascular biomarkers provided by proteomic profiling of accessible bodily fluids, such as plasma. Launched from multiple laboratories worldwide, this project incorporates data derived from the analyses of human plasma by use of a myriad of distinct proteomic approaches. A subset of the 3020 proteins thus far identified (by MS/MS spectra) are related to cardiovascular function and have been organized into eight groups: markers of inflammation and/or CVD, vascular and coagulation, signaling, growth and differentiation, cytoskeletal, transcription factors, channels/receptors and HF, and remodeling.[39] The functional annotation of these proteins and structural analyses are available as a shared database and constitute a significant resource for further development of molecular biosignatures for diseases such as myocardial ischemia and atherosclerosis.[40] This type of analysis has emerged in several areas of cardiovascular medicine/biology. For example, serum proteomic pattern diagnostics is a new type of proteomic platform in which patterns of proteomic signatures from MS data are used as a diagnostic classifier.[41] This approach has shown promise in early detection of cancers and recently has also been applied in the detection of anthracycline-induced cardiotoxicity by analysis of serum proteins from rats. This new technology emphasized the clinical usefulness of diagnostic proteomic patterns in which low molecular weight peptides and protein fragments may have higher accuracy than traditional biomarkers of cardiotoxicity such as troponin.

Endomyocardial biopsy is the most reliable method of detecting rejection after cardiac transplantation, but it is invasive. A reliable method to detect rejection noninvasively is not currently available; however, proteomics may be successful in the discovery of novel blood markers of rejection. Recently, sequential cardiac biopsies were analyzed by 2-D gel electrophoresis and classified according to whether they showed rejection ($n=16$) or no rejection ($n=17$).[42] In this analysis, more than 100 proteins were found to be up-regulated by between 2- and 50-fold during rejection; 13 of these were identified and found to be cardiac specific or HSPs. Levels of two of these proteins (i.e., α B-crystallin, tropomyosin) were measured by ELISA in the sera of 17 patients and followed for 3 mo after their transplants. These proteins were significantly higher in the sera of patients whose cardiac biopsies showed rejection compared with those with no rejection. These findings suggest that proteomic analysis can be used in the identification of novel serum markers of human cardiac allograft rejection.

MICROARRAY ANALYSIS AND BIOMARKERS

Gene expression profiling may also be used to identify a pattern of genes (a molecular signature) that serves as a biomarker of relevant clinical parameters of CVD (e.g., disease presence, progression, or response to therapy).[43] Although not as easy or convenient to perform as the screening of circulating biomarkers, the deployment of gene/transcriptome profiling targeted to specific tissues or cell populations can often provide important information unavailable with other approaches, as well as augmenting the growing list of significant (and novel) biomarkers.

Using transcriptional profiling by microarray technology of genes modulated in patients with atrial fibrillation (AF), a group of 33 AF-specific genes has been identified that were significantly activated compared with patients in sinus rhythm.[44] These included genes encoding proteins with multiple functions, including ion channel, antioxidant, inflammatory, cell growth/cell cycle, transcription factors (e.g., NF-κB), and cell signaling. In contrast, 63 sinus rhythm-specific genes were identified, including several genes with cell signaling function such as SERCA 2, cellular respiration, energy production, and antiproliferative or negative regulator of cell growth. This gene profiling data should allow further inroads in understanding the role that modulated expression of these genes plays in the initiation or perpetuation of AF and their specific contribution to the pathophysiological mechanism of atrial remodeling, as well as furnishing potential biomarkers that could be useful in the prediction of AF.

As we have noted, a large number of inflammatory mediators, including circulating levels of inflammatory cytokines (e.g., IL-6 and TNF-α) or the hepatic product, C-reactive protein, are useful biomarkers of long-term cardiovascular risk in both apparently healthy populations and in those with already established CAD. Although these circulating biomarkers are generally produced from blood-borne cells (e.g., macrophages and T cells) retained in the intima layer and activated resident cells of vascular origin (e.g., endothelial cells), it has been recently recognized that inflammatory mediators originating outside the coronary artery are also capable of inducing compositional changes in the intima layer. Such inflammatory mediators originating from remote extravascular sources can provide an explanation for the increased cardiovascular risk in certain patient populations, including not only patients with chronic infections or chronic inflammation (e.g., rheumatoid arthritis), but also insulin-resistant individuals who exhibit increased release of cytokines from adipose tissue. In fact, gene profiling has demonstrated increased expression of a panel of inflammatory genes (e.g., MCP-1, IL-6, IL-1β, and TNF-α) in epicardial adipose tissue in patients with established CAD. Importantly, these inflammatory signals present in epicardial adipose tissue were neither strongly correlated with plasma inflammatory biomarkers nor attenuated by chronic treatment with conventional cardiovascular therapies, including statins or ACE inhibitors/angiotensin II receptor blockers.[45] Although these findings are suggestive that the presence of inflammatory mediators and bioactive molecules in the localized tissues surrounding epicardial coronary arteries may profoundly alter arterial homeostasis (e.g., leading to amplification of vascular inflammation or plaque instability) in a manner not revealed by plasma biomarkers and refractory to some treatment options, they also suggest a fundamental limitation with biomarkers as presently configured. Recent studies with epicardial fat biopsies from patients undergoing coronary artery bypass grafting (CABG) have extended the genes profiled to include other inflammatory markers such as resistin (a recently identified adipocytokine), CRP, adiponectin and leptin, and the macrophage marker CD45 compared with both plasma levels and other fat depots.[45] These findings revealed increased macrophage infiltration (e.g., increased CD45) into epicardial fat in CABG patients, and they are indicative of significant local inflammation and evidence of a pathogenic gene profile in these tissues, including markedly decreased levels of adiponectin. Adiponectin exhibits both insulin sensitivity, anti-inflammatory, and anti-atherogenic properties, and its serum levels have been reported to be reduced in both type 2 diabetes mellitus and CAD.[46] Interestingly, the serum profile of CABG patients showed significantly higher levels of both CRP and resistin and significantly lower levels of adiponectin compared with matched controls.

Delineation of both plasma and genetic biomarkers should prove informative in the clinical setting of stent therapy in the treatment of atherosclerotic CAD, in particular to risk-stratify patients before therapy.[47] Prospective risk stratification may allow the rational selection of specialized treatments against the development of in-stent restenosis (ISR), such as might be promoted in some individuals by drug-eluting stents. A large-scale study is presently underway to both understand the molecular mechanisms of restenosis and to identify genetic biomarkers predictive of restenosis. In this study, a combined proteomic analysis of plasma and microarray profiling are being used to identify candidate genes that show differential expression. Furthermore, the screening of candidate genes to identify variants (e.g., in promoter regions, SNPs etc.) by genotype analysis will be carried out, as well as a genome-wide scan to identify genetic loci that are associated with ISR.

MITOCHONDRIAL MEDICINE

Abnormalities in mitochondrial function and structure have been increasingly recognized as a common feature in cardiac diseases, including cardiomyopathy and HF presenting with heterogeneous patterns of defects in mitochondrial respiratory enzymatic function, as well as frequent changes in mitochondrial structure and mtDNA.[48] Numerous animal studies have confirmed that changes in mitochondrial biogenesis because of targeted gene disruptions (e.g., mtTFA, ANT) can lead to DCM. Furthermore, mitochondria have also been shown to play a prominent role in the development of myocardial ischemia and in cardioprotection, partly through the generation of ROS and by its primary role in early apoptotic events, and through changes in mitochondrial metabolic function both in the use of fatty acids as substrate and in the effectiveness of oxidative phosphorylation.[49] It is worth noting that mutations of mtDNA and mitochondrial respiratory changes presumably resulting from oxidative damage may play an important role in the aging process (which was discussed in Chapter 19). Although there are several strong indications of mitochondrial involvement in cardiac dysfunction, the precise temporal location of mitochondria in the subcellular pathways involved has often not been determined. Questions related to whether mitochondrial defects represent a primary event or are secondary to other myocardial changes contributing to the pathophysiology of cardiac dysfunction have been largely unanswered.[50] Even in cases in which mitochondrial-based cardiac disease is a certainty, there are indications that mitochondrial genetic abnormalities are markedly influenced by other genetic factors (modifiers), as well as by environmental factors leading to heterogeneity of phenotype. Whether situated upstream or downstream, the broad spectrum of mitochondrial defects evident in both clinical and experimental studies have spurred increased interest in the potential application of mitochondrial medicine in both the diagnosis and the treatment of mitochondrial-associated cardiac defects.

Gene therapy to replace or repair defective mitochondrial genes could be an important adjunct in the treatment of mitochondrial-based CVD. However, mitochondrial genes with their own genetic code, as well as their multicopy status, have proved to be rather intractable to genetic manipulation compared with nuclear genes. Techniques of targeted mitochondrial gene transfection in animal studies have been rarely successful; electroporation and the use of the "gene gun" approach that uses biolistic transformation while successful in yeast and plants have not been extensively applied to incorporating genes into animal and human mitochondria.[51,52]

An alternate delivery system for nucleic acids into mitochondria involves the use of peptide nucleic acids (PNA).[53] Early experiments used PNA as a selective antisense inhibitor to target the replication of a pathogenic mtDNA allele *in vitro*, although this effect could not be demonstrated in cultured cells.[54] The difficulties associated with mitochondrial uptake of nucleic acids in living mammalian cells have been more recently surmounted by the addition of a mitochondrial-targeting leader peptide to the PNA-oligonucleotide molecule and the introduction of the PNA-oligonucleotide construct in cationic liposomes,[55,56] and, even more effectively, with cationic polyethylenimine.[57] The latter approach successfully allowed the import of PNA-oligonucleotides into the mitochondrial matrix of living cultured cells or isolated mitochondria, a critical step in potential mitochondrial gene-specific therapy. Similarly, a mitochondrial-specific delivery system has been recently developed that uses DQAsomes, liposome-like vesicles formed in aqueous medium from a dicationic amphiphile called *dequalinium*.[58] These DQAsomes can also bind and carry DNA (as well as drugs), are able to transfect cells with a high efficiency, and selectively accumulate in the mitochondrial organelle releasing their load.[59] Moreover, in addition to PNA-oligonucleotides, plasmid DNAs can be incorporated and condensed within the DQAsomes and exclusively delivered to the mitochondrial compartment.[59]

In addition to modifying mtDNA genes, the selective delivery of a variety of compounds (e.g., antiapoptotic drugs, antioxidants, and proton uncouplers) to the mitochondria could be used as a potential alternate strategy in the treatment of a number of mitochondrial-based disorders with cardiac involvement. The previously mentioned DQAsome can also deliver drugs that trigger apoptosis to mitochondria and inhibit carcinoma growth in mice.[60] A synthetic ubiquinone analog (termed *mitoQ*) has been selectively targeted to mitochondria by the addition of a lipophilic triphenylphosphate cation.[61] These positively charged lipophilic molecules can rapidly permeate the lipid bilayers and accumulate at high levels within negatively charged energized mitochondria.[62] Significant doses of these bioactive compounds can be administered safely by mouth to mice over long periods of time and accumulate within most organs, including the heart and brain. The incorporation of mitoQ within mitochondria can prevent apoptotic cell death and caspase activation induced by H_2O_2 (in isolated Jurkat cells) and can function as a potent antioxidant, preventing lipid peroxidation and protecting the mitochondria from oxidative damage. Feeding mitoQ to rats significantly decreased heart dysfunction, cell death, and mitochondrial damage after ischemia-reperfusion.[63]

This procedure of targeting bioactive molecules to mitochondria can be adapted to other neutral bioactive molecules, offering a potential vehicle for testing other mitochondrial-specific therapies. For instance, synthetic peptide antioxidants containing dimethyltyrosine, which are cell-permeable and concentrate 1000-fold in the mitochondria, can reduce intracellular ROS and cell death in a cell model. In ischemic hearts, these peptides potently improved contractile force in an *ex vivo* heart model.[64] In addition, the successful incorporation into the mitochondrial matrix of another modified antioxidant, a synthetic analog of vitamin E (MitoVitE), has been shown to significantly reduce mitochondrial lipid peroxidation and protein damage and can accumulate after oral administration at therapeutic concentrations within the cardiac tissue.[62]

Besides direct modulation of mitochondria by introduction of "therapeutic" nucleic acids, proteins, and antioxidants, a variety of therapeutic agents have been used to indirectly affect mitochondrial function. These act either through bypassing specific steps of mitochondrial dysfunction (e.g., vitamins and metabolic cofactors such as riboflavin, thiamine, succinate, ascorbate) augmenting the "general" antioxidant balance in the cell (e.g., vitamin E, coenzyme Q_{10}), supplementing metabolites that might be reduced because of either defective synthesis or transport in patients with mitochondrial-based disease (e.g., carnitine), or by redirecting the metabolic pathways (e.g., drugs targeted to enhance glucose use and pyruvate oxidation energy, at the expense of FAO). Although there have been multiple reports and case studies citing the beneficial use of many of these compounds in treating cardiomyopathies, there have been no large-scale rigorously controlled studies supporting their efficacy in patients with mitochondrial cardiomyopathy. In some cases, these pharmacological treatments are being used in various combinations as a therapeutic "cocktail." Knowledge of the precise site of the biochemical or molecular defect can be of critical importance regarding the choice of the therapeutic modality used.

In addition to the widening array of therapeutic strategies, advances are also being made in the area of diagnostic evaluation of mitochondrial defects (albeit more slowly than with other diagnostic problems). Furthermore, many clinicians seem not to be aware either of the significance of mitochondrial defects in CVD or how to screen for them. Although in some cases screening of different cell types (e.g., blood and biopsied skeletal muscle) can provide useful information about specific mtDNA mutations and bioenergetic defects (that often can exhibit a multisystemic phenotype), some defects (e.g., mtDNA deletions and some mtDNA point mutations) may not be present in active mitotic tissues such as blood. Unfortunately, direct examination of cardiac tissue, although important, can be limited by the invasive nature of a biopsy. Nevertheless, McDonell and associates[65] have recently found a strong correlation between the mutation load (of a specific mtDNA mutation) present in post-mitotic muscle and urinary epithelium, suggesting that urinary epithelial cells may be the tissue of choice in the noninvasive diagnosis.

The development of noninvasive methods to evaluate both mtDNA and *in vivo* bioenergetic dysfunction is of critical significance. In this regard, techniques involving nuclear magnetic resonance spectrometry have been used diagnostically (albeit in a limited fashion) as an *in vivo* diagnostic platform in animal models and in healthy and diseased patients to assess mitochondrial energy reserves, OXPHOS rates, metabolite levels, pH, and energy coupling[66–71] (e.g., insulin-resistant patients with type 2 diabetes[69]). Notably, most of these studies have only evaluated skeletal muscle bioenergetics, not the heart. A recent study using positron emission tomography (PET) was used as a noninvasive diagnostic method to evaluate myocardial mitochondrial Krebs cycle activity and dysfunction in children with cardiomyopathy.[72]

SYSTEMS BIOLOGY: INTEGRATING TRANSCRIPTOME, PROTEOME, AND PHYSIOME DATA AND MODELING FROM MUTANT GENE TO DYSFUNCTIONAL CELL, ORGAN, PHENOTYPE, AND DISEASE

A major challenge facing cardiovascular medicine is how to translate the wealth of reductionist detail about molecules, cells, and tissues into a real understanding of how these systems function and are perturbed in disease processes. There is an increased interest in integrating the large comprehensive and ever-growing databases containing genomic, proteomic, biochemical, anatomical, and physiological information that can be searched and retrieved through the Internet to further understand the functioning of the heart in health and disease. This could eventually lead to a model of cardiac function at the genetic, molecular, and physiological level, which can allow predictions about potential interventions to be made. It is anticipated that such simulated modeling would also highlight the gaps in our knowledge and elicit new perspectives on how to fill them.[73] On a limited scale, early work on this type of integrative approach (before the acquisition of much of the proteomic/genomic/molecular data) has used data on ion concentration and metabolite levels during cardiac ischemia to construct a simulated computer model that integrates cardiac energetics with electrophysiological changes (a novel approach to studying myocardial ischemia) that proved informative in predicting the effects of specific therapeutic interventions.[74] This model aided the identification of electrophysiological effects of therapeutic interventions such as Na^+-H^+ block and suggested an effective strategy to control cardiac dysrhythmias during calcium overload by regulating sodium–calcium exchange. In addition to myocardial ischemia, other types of cardiovascular disorders that have thus far been "modeled" by use of this type of approach include cardiac dysrhythmias and contractile disorders.[75–77]

The construction of these models, in addition to the array of data discussed previously, requires an extensive and sophisticated mathematical and computer algorithmic treatment.[78] Its undertaking is clearly a multidisciplinary approach involving mathematics, computer skills, molecular and cell biology, genetics, physiology, and anatomy. A central premise recapitulates the maxim of the noted molecular biologist, Sydney Brenner, that genes can only specify the properties of the proteins they code for, and any integrative properties of the system must be "computed" by their interactions.

Moreover, it is important to keep in mind that these models underscore that much critical post-genomic data remains to be obtained, including the identification of gene–environmental factor and protein–protein interactions, which seem to underlie many phenotypic changes and signaling events in the cardiomyocyte. Biomolecular interactions revealed by proteomic information will be important for unravelling metabolic and signaling pathways operating in the cell and, in particular, in response to disease and injury. The further identification of functional interactions between signaling pathways and genetic networks will also be of key interest, because they will provide a window into regulation of the coordinate expression of functional groups of genes, with a few key pathways switching between alternative cell fates. Also underlying these models is the important recognition that the heart and vascular system are dynamic and more than just electrical circuitry or mechanical pumps, and they have the ability to grow and remodel in response to changing environments partly determined by genes and partly by their physical environment.[79] Such a systems approach in the near future may need to be applied to metabolic events in the cardiac mitochondria, as well as to events occurring in both myocardial aging and early development.

TRANSLATION OF FINDINGS FROM PRE-CLINICAL/CELL STUDIES INTO CLINICAL MEDICINE

In this section, we will briefly discuss current and future applications involving the translation of the increasing myriad of physiological, genetic, molecular, biochemical, and cellular findings (mostly derived from pre-clinical studies) into effective cardiovascular medicine for the 21st century.

Post-genomic biology has not only substantially increased our understanding of the mechanisms underlying CVD (as well as the events of aging) but has also provided an armamentarium of new approaches, which are gradually being adapted to clinical medicine, and over the next decade should provide both improved diagnostic tools and gene- and cell-based therapies. Moreover, the biological information derived from increasing human-based studies may eventually be used in the context of enhanced pharmacogenetic/genomic medicine, allowing a more individually tailored gauging of genetic susceptibility, environmental stresses, toxic insults, and even dietary requirements that might be required or avoided and, in addition, delineating what could be the most effective treatment regimens for specific CVDs.

A primary step in the realization of this vision of post-genomic cardiovascular medicine is the further acquisition of genetic data from human studies. The continued development of methods for increasingly powerful haplotype and SNP mapping throughout the human genome needs to be more widely examined throughout the general population.

Data gathered from these techniques for genotyping may be helpful in enhancing the identification of candidate genes (both known and novel loci) and in testing their association with specific targeted CVD disorders. Pharmacogenetic analysis has recently allowed the identification of genetic factors that increase susceptibility to dysrhythmia and MI.[80,81] Moreover, the underlying responsiveness to specific drugs therapies targeting CVDs has been revealed by pharmacogenomic studies of statin responsiveness.[82] Critical factors that should also be stressed include the involvement and interest of the physician and a better educated lay public for understanding the significant ramifications of gathering and applying this information.

Unfortunately, despite the clear inroads made with respect to both identification of specific gene involvement in CVDs and cardioprotective treatments (both using pharmacological and nonpharmacological approaches) as reported in numerous animal studies, few of these studies have been translated into human clinical practice. Some of the difficulties arise from the nature of the pre-clinical studies themselves; these are often conducted with models that do not approximate the human model (e.g., isolated cell/heart studies, variable levels of ischemia) and generally with young healthy animals and are primarily focused on understanding the molecular and cellular mechanisms of injury and protection, rather than establishing the potential clinical efficacy of the interventions tested.[83] In addition, end points in human studies are more difficult to precisely measure and are often indirectly compared with animal studies (in which the duration of ischemia and size of infarct can be controlled) and seem to be subject to a myriad of confounding factors (e.g., concurrent use of medications, environmental stressors). Problems inherent in measuring patient infarct size may be alleviated by the use of novel imaging techniques, including delayed contrast-enhanced MRI, which provide higher resolution and highly quantitative and noninvasive measures of infarct size in the clinical setting.[84–86] Also as noted in our chapter on ischemia and cardioprotection, there is a high level of non-reproducibility found in the cardioprotection literature.

Despite a rather large number of cardioprotective interventions, several of which proved to be disappointing in clinical trials of patients with AMI, a number of recent studies have shown promising results. In patients with AMI, two large trials of adenosine administered at the time of reperfusion (AMISTAD I and AMISTAD II) have demonstrated a marked reduction in the size of anterior wall MI.[87,88] The beneficial effect in AMISTAD II seemed to be associated with a trend (although not statistically significant) toward improved clinical outcome. Furthermore, small studies have shown that the K_{ATP} channel opener and NO donor nicorandil was beneficial in patients with AMI when administered at the time of reperfusion.[89,90] Also, glucose-insulin-potassium (GIK) infusion has been found to reduce mortality in AMI, although some studies have suggested that the mortality was reduced only in subsets of patients.[91]

In the setting of high-risk CABG procedures, it has been reported that adenosine administered before, during, and after aortic clamping reduces perioperative MI incidence, and it improves a composite end point consisting of the need for mechanical or inotropic support, MI, or death.[92] Likewise, inhibitors of Na^+/H^+ exchange (e.g., cariporide) have been shown to protect ischemic myocardium in cardiac surgery, although the neurological complications observed with high dosages of cariporide precludes its clinical use.[93,94]

There is an increasing sense that large clinical trials need to be more highly targeted while inclusive of more perspectives in their design. A key suggestion that has been recently offered involves the increased use of patient subgroups. For instance, the rational selection of patient subgroups most likely to benefit from such cardioprotective therapies should enable the design of adequately powered studies with relatively small numbers of patients. However, there is uncertainty as to which subgroups of patients will have postoperative complications. This is an important problem that may hinder further studies of cardioprotection.

The selection of drugs for specific subgroups of patients has also been proposed in other areas of medicine. More than half the withdrawal of drugs (ranging from antihistamines, anti-cancer compounds, anti-emetics, antibiotics, to anti-migraine drugs) mandated by the FDA since 1998 has been attributable to cardiac side effects, most of them dysrhythmias.[95] This costly problem is largely attributable to the high degree of receptivity of one of the channel proteins, I_{Kr}, on which cardiac repolarization depends. A heart that is already prone to dysrhythmia, because of slowed conduction and/or failed repolarization, as a consequence of genetic or disease disturbance of sodium, or other channels and transporters, may therefore be tipped over into a fatal state by even a modest amount of I_{Kr} block. The advent of gene array technologies and proteomic analyses may be highly useful, not only in the identification of high-risk patients who are most likely to benefit by a specific treatment but also in the identification of patients who are genetically prone to dysrhythmia. If these goals can be achieved, the benefits should be very significant.

Another relevant subgroup type analysis that may arise from a concerted pharmacogenetic analysis also includes the potential role of ethnicity, race, and gender. Evidence of racial/ethnic differences in CVD has been widely recognized for some time. For instance, the cause of HF is predominantly ischemic disease in non-blacks but is related primarily to hypertension, which tends to be both more frequent and severe in blacks;[96] blacks seem to be stroke-prone, but relatively protected from CAD.[97,98]

In clinical trials, ethnic differences in the antihypertensive responses to β-blockers and ACE inhibitors as (therapeutic agents) have been reported; blacks exhibit somewhat reduced blood pressure responses to monotherapy with β-blockers and ACE inhibitors compared with diuretics.[99] Recently, retrospective analyses of HF trials have suggested differences in the response to ACE inhibitors (e.g., enalapril).[100,101] Initial trials with the β-blocker carvedilol, when retrospectively reanalyzed by ethnicity, did not reveal differences in the benefit between blacks and whites.[102]

In contrast, in the BEST trial conducted with a diverse group of patients with NYHA class III and IV HF, treatment with the β-blocker bucindolol provided a significant increase in survival benefit in non-black patients.[103] Moreover, clinical studies have demonstrated that treatment with the NO donor isosorbide dinitrate, in combination with the antioxidant hydralazine compared with placebo or prazosin, conferred a survival advantage for black but not for white subjects.[100] More recently, the African-American Heart Failure Trial (A-HeFT), a randomized, placebo-controlled, double-blind trial involving 1050 black patients with NYHA class III and IV HF, confirmed the finding that a fixed dose of isosorbide dinitrate plus hydralazine, in addition to standard therapy for HF including neurohormonal blockers, was efficacious and increased the survival of black patients with advanced HF.[104,105]

It remains to be seen whether a discrete DNA difference might underlie these different phenotypic responses to treatment. It also has been argued that the treatment of HF in blacks, which has a hypertensive component, might be more effective in targeting that pathology. Interestingly, the benefit of the combined isosorbide dinitrate and hydralazine in the treatment of HF has not been provided by other antihypertensive medications. Although the interpretation of these findings and their implications have been hotly debated, most clinicians agree that the inclusion of diverse groups in significant numbers (for subgroup analysis) should be an intrinsic part of future clinical trials.

Diagnostic Challenges And Genetic Counseling

Although the application of post-genomic technology in improving cardiovascular diagnosis has been discussed in several previous chapters, its use is of particular interest in pre-natal diagnosis and in pre-implantation genetic diagnosis (PGD). The use of pre-natal testing with DNA markers has been successful in identifying cases of X-linked cardiomyopathy (because of dystrophin mutations),[106] severe neonatal long QT syndrome (HERG mutations),[107] specific mutations associated with HCM,[108,109] and in Marfan syndrome (elastin mutations).[110] It seems to be of a greater benefit when a specific mutation is suspected (because of family history) and can be directly screened.

Specific mtDNA mutations leading to Leigh disease[111,112] and respiratory deficiencies[113,114] have been screened pre-natally, in some cases using chorionic villi as the source of the genetic material, and in other cases with amniotic fluid.[114,115] Analysis of mutant loads of the nt8993 mutations in fetal and adult tissues confirmed that there is no substantial tissue variation and suggested that the mutant load in a pre-natal sample will represent the mutant load in other fetal tissues; these 8993 mutations show a strong correlation between mutant load and symptom severity and between maternal blood mutant load and risk of a severe outcome.[116]

As an adjunct to *in vitro* fertilization (IVF) technology, the application of PGD will be increasingly available to prevent the transmission of devastating diseases from affected parents to their children. Although this technique has not yet received widespread attention, the detection of mutant alleles of the TBX5 transcription factor resulting in Holt–Oram syndrome (HOS) (which presents with multiple malformations, including congenital heart defect) has allowed the successful identification of fertilized eggs affected by HOS for potential embryo selection. With molecular genotyping techniques, blastocysts containing wild-type genotypes were distinguishable from those containing mutant alleles, and their transfer to the mother resulted in the delivery of a normal child.[117]

The use of trans-mitochondrial oocytes in human studies has a limited but controversial history. Ooplasmic transplantation has been reported in several studies in conjunction with IVF.[118,119] The addition of a small amount of injected ooplasm, derived from fertile donor oocytes, into developmentally compromised oocytes from patients with recurrent pre-implantation failure was reported to enhance embryo viability and led to the birth of 15 children. The mtDNA from the donor and the recipient cell mtDNA were found to be present in the blood of the child, resulting from the transplanted oocyte at 1 y of age. Excluding the numerous ethical considerations provoked by this human germ-line genetic modification (going somewhat beyond screening for the wild-type genotype), several concerns have been raised by these studies, including the potential for long-term harm in chromosomal segregation and aberrant division (predicted by similar studies conducted in lower organisms) and negative epigenetic influences of foreign cytoplasm, as has been demonstrated in numerous studies of cytoplasmic transfer in mice.[120] In fact, 2 of the 15 pregnancies resulted in chromosomal abnormalities, including Turner syndrome. Moreover, the long-term deleterious influence of heteroplasmic mtDNA has also been considered an additional problem of this technique.[121,122] If we are to fully appreciate the potential outcomes associated with embryo manipulation, extensive investigations with animal models that incorporate genetic, biochemical, and physiological analyses are mandated. Also, these investigations should be accompanied by careful clinical monitoring to demonstrate the suitability of these techniques for human use.

New and Future Therapeutic Options in Cardiovascular Medicine; Post-Genomic Contributions

Advances in post-genomic cardiology may make possible a bright future for new cardiovascular therapies, which can offer greater specificity and breadth, some of which are already coming to fruition. Molecular genetic analysis has substantially improved our understanding of the structure and functioning of the heart, both in early development and in aging, and has opened the door to further unraveling the order of molecular/cellular events and the principal molecules involved in both normal and malformed/dysfunctional hearts. Information derived from transgenic models has been instrumental in defining numerous therapeutic targets in signaling pathways in which the heart and cardiovascular system respond to stresses and insults, the exposition of both apoptotic and survival/proliferative pathways affecting cardiomyocyte growth, oxidative stress, hypertrophy, aging and cell death, and metabolic pathways essential for energy transduction necessary for contractile function and electrical excitability. Models of specific gene dysfunction have been highly informative in defining the roles that specific contractile and ion channel proteins play in both the normal and diseased heart. Some of this information has been useful in the design of pharmaceutical strategies with a variety of novel therapeutic targets in the treatment of CVDs, ranging from stroke, MI, acute, and chronic inflammatory diseases.[123] Potentially, the use of therapies aimed at metabolic remodeling may be developed to effectively supplement the treatment of myocardial ischemia, HF, the more obvious metabolic-based CVDs (i.e., metabolic syndrome), and underlying therapies such as calorie restriction aimed at modulating overall longevity and the aging processes.[124–126]

The potential of combining genetic and cell engineering has been demonstrated in a number of animal models, and in limited clinical studies, to be a useful adjunct in repairing damaged hearts and vessels and in enhancing the heart's regenerative potential. Later in this chapter, we will briefly review new developments in gene therapy, cell-based therapy, and tissue reengineering, and we will also discuss possible future applications.

Gene Engineering

It should be evident from much of the material discussed throughout this book that the introduction of specific genes could be successfully used to correct a number of specific myocardial defects resulting in cardiac dysfunction. As previously noted, the limiting factor in the clinical applications of gene therapy at this time is the timely development of safe and effective vectors/carriers for the transgene and its clinical testing. The alternatives have been reviewed in earlier chapters.

Cardiac Cell Engineering

The use of both embryonic and adult stem cells offers an exciting platform for repairing the damage to the heart. Direct transplantation of isolated myoblasts or bone marrow mononuclear cells and recruitment of stem cells from bone marrow by cytokine administration (e.g., G-CSF) have been already clinically performed as discussed in Chapter 4. Nevertheless, numerous and important unanswered questions have been raised concerning stem cell biology, including their differentiation process, the heterogeneity within stem cell populations, what effects are due to recruitment and activation of endogenous stem cell populations compared with circulating stem cells originating in bone marrow (such as recently described in cardiac cells), what effects are due to cytokines and angiogenesis, and what factors impact in their differentiation and homing properties.

There is considerable interest in combining the use and fabrication of biopolymers as a scaffolding matrix for rebuilding damaged cardiac tissue with proliferating cells.[127] The importance of re-engineering the cardiac milieu so stem cells can effectively adhere and be better functionally incorporated into the injured myocardium has been increasingly recognized. Cardiac tissue engineering research has also centered on fabricating 3-D cardiac grafts and biodegradable scaffolds as alternatives of extracellular matrix.[128] Other alternatives include the use of beating myocardial tissue by layering cell sheets that are harvested from cultured cells grafted with temperature-responsive polymer.[129] The transplantation of cardiomyocyte cell sheets in the treatment of HF, which eliminate the need for a donor, have reached the stage of pre-clinical testing. It is worth noting that skeletal myoblasts have been grown on polymer sheets in culture and implanted into coronary artery–ligated rat hearts. This resulted in the repair of the damaged myocardium with markedly reduced fibrosis (compared with skeletal myoblast cell injection) and prevention of remodeling in association with a recruitment of hematopoietic stem cells through the release of stromal-derived factor 1 and other growth factors, suggesting their potential use in the treatment of patients with severe HF.[130] In addition, it is possible that the transplant-associated dysrhythmias occurring with skeletal myoblasts in the heart (likely caused by the absence of gap junctions) might be eliminated by the *ex vivo* genetic modification of the myoblasts containing expression of the gap junction protein connexin43.[131] This antidysrhythmic engineering (achieved thus far only in cells in culture) may increase the safety (and perhaps the efficacy) of myoblast transplantation in patients.

Particular interest has been centered on the reengineering of heart valves.[128] The use of collagen scaffolds produced by a novel processs termed rapid prototyping, in which valve interstitial cells isolated from three human aortic valves seeded on the scaffolds and cultured for up to 4 weeks remained viable and proliferated, is an important step in the tissue engineering of an aortic valve. Repopulation of a scaffold of a decellularized valve matrix (usually porcine) *in vitro* with stem cells, in particular mesenchymal stem cells, has also been an area of intensive investigation.[132,133] Although this approach has not yet proved successful in either animal models or human clinical testing,[134] conditions for the optimization of cell seeding and repopulation with either valve interstitial cells[135] or mesenchymal cells have recently been optimized. Recently, the creation of autologous semilunar heart valves *in vitro* with mesenchymal stem cells and a biodegradable scaffold has been reported after their implantation, under cardiopulmonary bypass, into the pulmonary valve position in sheep. These valves underwent extensive remodeling *in vivo*, resembling the native heart valves, and functioned satisfactorily for periods of >4 mon.[136]

Targeting various regions of the heart associated with specific function has become another interesting approach. This has been focused on the possibility of grafting pacemaking cells, either derived from differentiating human embryonic stem cells or engineered from mesenchymal stem cells, into the myocardium.[137] Moreover, an initial proof that the use of gene therapy may create a biological pacemaker was provided by Miake and associates using an adenoviral gene transfer approach in guinea pig hearts.[138] By modifying the Kir2.1 gene encoding the inward rectifying potassium current K_1 (which pacemaker cells lack) and mutating it (i.e., obtaining a dominant-negative allele) to make it a dysfunctional channel, a subset of transfected ventricular myocytes were converted to cells with pacemaker activity. A limitation of these studies was that the induced automaticity was threefold slower than normal, and genetic suppression of I_{K1} does not provide a direct means to modulate the induced rhythm. In contrast, Xue and associates found that gene transfer of an engineered HCN1 construct to quiescent adult guinea pig ventricular cardiomyocytes can induce pacing with a normal firing rate.[139] Previous studies have reported that plasmids harboring the human β-2 adrenergic receptor injected into the right atria of Yorkshire pig hearts significantly enhanced porcine cardiac chronotropy up-regulating heart rate by 50%.[140]

Other studies have found that transient overexpression of the channel protein HCN2 in the left atrium or bundle-branch system (in a canine model) could generate an ectopic biological pacemaker.[141,142] This construct used an adenoviral vector, and vagal suppression was required to observe the effect. Moreover, it has been reported that transplanted stem cells (either MSCs or hESCs) can serve as platforms for the delivery of these pacemaker-conferring genes.[143,144] In addition, it has been noted that hESC-derived cardiomyocytes (generated *in vitro* using the embryoid body differentiating system), when introduced into swine hearts with complete AV block, stably integrated, restoring myocardial electromechanical properties acting as a rate-responsive biological pacemaker.[145]

Vessel Engineering

As previously discussed in Chapter 11, it is well recognized that therapeutic angiogenesis/vasculogenesis can be mediated by supplementation of a variety of growth factors or transplantation of vascular progenitor cells. This approach fosters the formation of arterial collaterals and promotes the regeneration of damaged tissues and, therefore, may be useful in treating ischemia.[146,147] Although angiogenic factors can be delivered in the form of recombinant proteins or by gene transfer with viral vectors, novel nonviral methods, including liposomes, naked plasmid vectors, or cell-mediated gene transfer, are promising alternatives with a safer profile.[148–150] Although growth factors offer distinct advantages in terms of efficacy, a number of approaches featuring the combination of several growth factors with cell transplantation are currently being explored to both initiate growth and stabilize vessels. Some angiogenic factors not only stimulate the growth of arterioles and capillaries but also inhibit vascular destabilization triggered by metabolic and oxidative stress. Endothelial progenitor cells (EPCs) for the treatment of peripheral or myocardial ischemia can be transplanted either without any preliminary conditioning or after *ex vivo* genetic manipulation.[151,152] Also, delivery of genetically modified autologous progenitor cells eliminates the drawback of immune response against viral vectors.

The formation of a microvascular network can be achieved by promoting vasculogenesis *in situ*, using seeding vascular endothelial cells within a biopolymeric scaffolding construct; the inclusion of human smooth muscle cells seeded with EPC-derived endothelial cells can form capillary-like microvessel structures throughout the scaffold.[153]

SUMMARY

- Cardiac biomarkers widely used for diagnosis of MI include troponin I and T, CK-MB, and myoglobin. The combined use of biomarkers (i.e., multi-markers) increases the reliability of disease diagnosis.
- Although other markers seem to better suited for detecting early myocardial ischemia and infarct, including unbound free fatty acids and ischemia-modified albumin, these have not been widely tested or used.
- Biomarkers of inflammation (e.g., CRP, cytokines, MCP) and hemodynamic function (BNP, NT-proBNP) are important prognostic indicators of future cardiovascular risk in both healthy individuals and in patients with acute coronary syndrome (ACS). BNP is also an excellent diagnostic marker of HF.
- Proteomics and gene profiling techniques are critical techniques that can be used in the identification of novel biomarkers. Also, proteomic techniques such as mass spectroscopy and protein microarrays offer improved methods for rapid and highly accurate screening of multiple markers.
- Mitochondrial-based defects have been demonstrated in a variety of CVDs because of the primary role of mitochondria in generating metabolic energy and its role in oxidative stress signaling and ROS generation and in the early events of apoptotic cell death. Mitochondrial also play a pivotal role in myocardial ischemia, the development of cardioprotective responses, and aging.
- The present therapies for mitochondrial-based diseases are limited, involving primarily metabolic bypasses of specific defects and antioxidant treatment (i.e., coenzyme Q_{10}). Direct genetic modulation of mitochondria involves novel gene therapeutic approaches, because those used for nuclear gene therapies have not been successful. A variety of lipophilic carriers for bringing molecules into the mitochondria (modified nucleic acids, proteins, drugs, inhibitors, and antioxidants) are being developed. Targeting their apoptotic function will likely involve new pharmacological development.
- Noninvasive diagnostic methods to detect mitochondrial defects in the cardiovascular system are presently being developed, including magnetic resonance spectroscopy.
- Genetic screening has been used to clinically detect prenatal defects in patients with congenital heart disease, cardiac dysrhythmias, HCM, and Marfan syndrome. Similar tests will likely be used in pre-implantation genetic diagnosis in association with *IVF* procedures.
- The use of gene therapy to target specific genetic defects or supplement deficiencies has an enormous potential with many possible useful applications; however, this is still awaiting the development of safe, effective, and easily testable vector systems.
- Cardiac targeted therapies may involve the direct modulation of myocardial cells with transfected genes or injected gene products or the transplantation of cells or groups of cells (cell sheets).
- Cell-based therapies will likely play a critical role in the future. A number of cell types have shown the ability to be recruited into damaged myocardium with potential benefits. These cells include several types of stem cells (either embryonic or adult), neonatal cardiomyocytes, and skeletal myocytes. These cell types can be modified *ex vivo* to express (or not express) specific genes; the cell types, once incorporated into the myocardium, can serve as platforms for introducing specific gene products. Early studies have shown that pacemaking activity can be affected by the introduction of specific genes either directly or by transplanted stem cells.
- Both stem cell biology and the cardiac milieu optimal for their homing, integration, and long-term survival need to be better characterized. In addition to cardiac transplantation, both stem cells and a combination of angiogenic factors can be used in therapeutic angiogenesis with its potential application in myocardial ischemia.
- A variety of biomaterials have been used to simulate the cardiac environment, and they have been useful in tissue reengineering of valves (e.g., aortic) and vessel remodeling.

References

1. Maisel, A. S., Bhalla, V., and Brunwald, E. (2005). Cardiac biomarkers; a contemporary status report. *Nature* **3**, 24–34.
2. Katus, H. A., Remppis, A., Neumann, F. J., Scheffold, T., Diederich, K. W., Vinar, G., Noe, A., Matern, G., and Kuebler, W. (1991). Diagnostic efficiency of troponin T measurements in acute myocardial infarction. *Circulation* **83**, 902–912.
3. Adams, J. E., 3rd., Sicard, G. A., Allen, B. T., Bridwell, K. H., Lenke, L. G., Davila-Roman, V. G., Bodor, G. S., Ladenson,

J. H., and Jaffe, A. S. (1994). Diagnosis of perioperative myocardial infarction with measurement of cardiac troponin I. *N. Engl. J. Med.* **330,** 670–674.

4. Newby, L. K., Storrow, A. B., Gibler, W. B., Garvey, J. L., Tucker, J. F., Kaplan, A. L., Schreiber, D. H., Tuttle, R. H., McNulty, S. E., and Ohman, E. M. (2001). Bedside multimarker testing for risk stratification in chest pain units: The chest pain evaluation by creatine kinase-MB, myoglobin, and troponin I (CHECKMATE) study. *Circulation* **103,** 1832–1837.

5. Azzazy, H. M., Pelsers, M. M., and Christenson, R. H. (2006). Unbound free fatty acids and heart-type fatty acid-binding protein: diagnostic assays and clinical applications. *Clin. Chem.* **52,** 19–29.

6. Ishii, J., Ozaki, Y., Lu, J., Kitagawa, F., Kuno, T., Nakano, T., Nakamura, Y., Naruse, H., Mori, Y., Matsui, S., Oshima, H., Nomura, M., Ezaki, K., and Hishida, H. (2005). Prognostic value of serum concentration of heart-type fatty acid-binding protein relative to cardiac troponin T on admission in the early hours of acute coronary syndrome. *Clin. Chem.* **51,** 1397–1404.

7. Pelsers, M. M., Hermens, W. T., and Glatz, J. F. (2005). Fatty acid-binding proteins as plasma markers of tissue injury. *Clin. Chim. Acta* **352,** 15–35.

8. Goto, T., Takase, H., Toriyama, T., Sugiura, T., Sato, K., Ueda, R., and Dohi, Y. (2003). Circulating concentrations of cardiac proteins indicate the severity of congestive heart failure. *Heart* **89,** 1303–1307.

9. Niizeki, T., Takeishi, Y., Arimoto, T., Okuyama, H., Takabatake, N., Tachibana, H., Nozaki, N., Hirono, O., Tsunoda, Y., Miyashita, T., Fukui, A., Takahashi, H., Koyama, Y., Shishido, T., and Kubota, I. (2005). Serum heart-type fatty acid binding protein predicts cardiac events in elderly patients with chronic heart failure. *J. Cardiol.* **46,** 9–15.

10. Roy, D., Quiles, J., Aldama, G., Sinha, M., Avanzas, P., Arroyo-Espliguero, R., Gaze, D., Collinson, P., and Carlos Kaski, J. (2004). Ischemia modified albumin for the assessment of patients presenting to the emergency department with acute chest pain but normal or non-diagnostic 12-lead electrocardiograms and negative cardiac troponin T. *Int. J. Cardiol.* **97,** 297–301.

11. Bar-Or, D., Winkler, J. V., Vanbenthuysen, K., Harris, L., Lau, E., and Hetzel, F. W. (2001). Reduced albumin-cobalt binding with transient myocardial ischemia after elective percutaneous transluminal coronary angioplasty: a preliminary comparison to creatine kinase-MB, myoglobin, and troponin I. *Am. Heart J.* **141,** 985–991.

12. de Lemos, J. A., McGuire, D. K., and Drazner, M. H. (2003). B-type natriuretic peptide in cardiovascular disease. *Lancet* **362,** 316–322.

13. Hama, N., Itoh, H., Shirakami, G., Nakagawa, O., Suga, S., Ogawa, Y., Masuda, I., Nakanishi, K., Yoshimasa, T., Hashimoto, Y., Yamaguchi, M., Hori, R., Yasue, H., and Nakao, K. (1995). Rapid ventricular induction of brain natriuretic peptide gene expression in experimental acute myocardial infarction. *Circulation* **92,** 1558–1564.

14. Sabatine, M. S., Morrow, D. A, de Lemos, J. A., Omland, T., Desai, M. Y., Tanasijevic, M., Hall, C., McCabe, C. H., and Braunwald, E. (2004). Acute changes in circulating natriuretic peptide levels in relation to myocardial ischemia. *J. Am. Coll. Cardiol.* **44,** 1988–1995.

15. Marumoto, K., Hamada, M., and Hiwada, K. (1995). Increased secretion of atrial and brain natriuretic peptides during acute myocardial ischaemia induced by dynamic exercise in patients with angina pectoris. *Clin. Sci.* **88,** 551–556.

16. Sadanandan, S., Cannon, C. P., Chekuri, K., Murphy, S. A., Dibattiste, P. M., Morrow, D. A., de Lemos, J. A., Braunwald, E., and Gibson, C. M. (2004). Association of elevated B-type natriuretic peptide levels with angiographic findings among patients with unstable angina and non-ST-segment elevation myocardial infarction. *J. Am. Coll. Cardiol.* **44,** 564–568.

17. Tateishi, J., Masutani, M., Ohyanagi, M., and Iwasaki, T. (2000). Transient increase in plasma brain (B-type) natriuretic peptide after percutaneous transluminal coronary angioplasty. *Clin. Cardiol.* **23,** 776–780.

18. de Lemos, J. A., and Morrow, D. A. (2002). Brain natriuretic peptide measurement in acute coronary syndromes: ready for clinical application? *Circulation* **106,** 2868–2870.

19. Libby, P., Ridker, P. M., and Maseri, A. (2002). Inflammation and atherosclerosis. *Circulation* **105,** 1135–1143.

20. Ridker, P. M., and Morrow, D. A. (2003). C-reactive protein, inflammation, and coronary risk. *Cardiol. Clin.* **21,** 315–325.

21. Anzai, T., Yoshikawa, T., Shiraki, H., Asakura, Y., Akaishi, M., Mitamura, H., and Ogawa, S. (1997). C-reactive protein as a predictor of infarct expansion and cardiac rupture after a first Q-wave acute myocardial infarction. *Circulation* **96,** 778–784.

22. James, S. K., Armstrong, P., Barnathan, E., Califf, R., Lindahl, B., Siegbahn, A., Simoons, M. L., Topol, E. J., Venge, P., Wallentin, L., and the GUSTO-IV-ACS Investigators. (2003). Troponin and C-reactive protein have different relations to subsequent mortality and myocardial infarction after acute coronary syndrome: A GUSTO-IV substudy. *J. Am. Coll. Cardiol.* **41,** 916–924.

23. Biasucci, L. M., Vitelli, A., Liuzzo, G., Altamura, S., Caligiuri, G., Monaco, C., Rebuzzi, A. G., Ciliberto, G., and Maseri, A. (1996). Elevated levels of interleukin-6 in unstable angina. *Circulation* **94,** 874–877.

24. Blake, G. J., and Ridker, P. M. (2003). C-reactive protein and other inflammatory risk markers in acute coronary syndromes. *J. Am. Coll. Cardiol.* **41,** 37S–42S.

25. Morrow, D. A., Rifai, N., Antman, E. M., Weiner, D. L., McCabe, C. H., Cannon, C. P., and Braunwald, E. (1998). C-reactive protein is a potent predictor of mortality independently of and in combination with troponin T in acute coronary syndromes: A TIMI 11A substudy. Thrombolysis in Myocardial Infarction. *J. Am. Coll. Cardiol.* **31,** 1460–1465.

26. Manginas, A., Bei, E., Chaidaroglou, A., Degiannis, D., Koniavitou, K., Voudris, V., Pavlides, G., Panagiotakos, D., and Cokkinos, D. V. (2005). Peripheral levels of matrix metalloproteinase-9, interleukin-6, and C-reactive protein are elevated in patients with acute coronary syndromes: correlations with serum troponin I. *Clin. Cardiol.* **28,** 182–186.

27. Koenig, W., Sund, M., Frohlich, M., Fischer, H. G., Lowel, H., Doring, A., Hutchinson, W. L., and Pepys, M. B. (1999). C-Reactive protein, a sensitive marker of inflammation, predicts future risk of coronary heart disease in initially healthy middle-aged men: results from the MONICA (Monitoring Trends and Determinants in Cardiovascular Disease) Augsburg Cohort Study, 1984 to 1992. *Circulation* **99,** 237–242.

28. Cushman, M., Arnold, A. M., Psaty, B. M., Manolio, T. A., Kuller, L. H., Burke, G. L., Polak, J. F., and Tracy, R. P. (2005). C-reactive protein and the 10-year incidence of coronary heart

disease in older men and women: the cardiovascular health study. *Circulation* **112,** 25–31.

29. Melanson, S. F., and Tanasijevic, M. J. (2005). Laboratory diagnosis of acute myocardial injury. *Cardiovasc. Pathol.* **14,** 156–161.

30. Varo, N., de Lemos, J. A., Libby, P., Morrow, D. A., Murphy, S. A., Nuzzo, R., Gibson, C. M., Cannon, C. P., Braunwald, E., and Schonbeck, U. (2003). Soluble CD40L: risk prediction after acute coronary syndromes. *Circulation* **108,** 1049–1052.

31. Manten, A., de Winter, R. J., Minnema, M. C., ten Cate, H., Lijmer, J. G., Adams, R., Peters, R. J., and van Deventer, S. J. (1998). Procoagulant and proinflammatory activity in acute coronary syndromes. *Cardiovasc. Res.* **40,** 389–395.

32. Ikeda, U., Ito, T., and Shimada, K. (2001). Interleukin-6 and acute coronary syndrome. *Clin. Cardiol.* **24,** 701–704.

33. Stanley, B. A., Gundry, R. L., Cotter, R J., and Van Eyk, J. E. (2004). Heart disease, clinical proteomics and mass spectrometry. *Dis. Markers* **20,** 167–178.

34. Anderson, L. (2005). Candidate-based proteomics in the search for biomarkers of cardiovascular disease. *J. Physiol.* **563,** 23–60.

35. Doherty, N. S., Littman, B. H., Reilly, K., Swindell, A. C., Buss, J. M., and Anderson, N. L. (1998). Analysis of changes in acute-phase plasma proteins in an acute inflammatory response and in rheumatoid arthritis using two-dimensional gel electrophoresis. *Electrophoresis* **19,** 355–363.

36. Ridker, P. M., Rifai, N., Rose, L., Buring, J. E., and Cook, N. R. (2002). Comparison of C-reactive protein and low-density lipoprotein cholesterol levels in the prediction of first cardiovascular events. *N. Engl. J. Med.* **347,** 1557–1565.

37. Di Serio, F., Amodio, G., Ruggieri, E., De Sario, R., Varraso, L., Antonelli, G., and Pansini, N. (2005). Proteomic approach to the diagnosis of acute coronary syndrome: Preliminary results. *Clin. Chim. Acta* **357,** 226–235.

38. Vuotikka, P., Ylitalo, K., Vuori, J., Vaananen, K., Kaukoranta, P., Lepojarvi, M., and Peuhkurinen, K. (2003). Serum myoglobin/ carbonic anhydrase III ratio in the diagnosis of perioperative myocardial infarction during coronary bypass surgery. *Scand. Cardiovasc. J.* **37,** 23–29.

39. Berhane, B. T., Zong, C., Liem, D. A., Huang, A., Le, S., Edmondson, R. D., Jones, R. C., Qiao, X., Whitelegge, J. P., Ping, P., and Vondriska, T. M. (2005). Cardiovascular-related proteins identified in human plasma by the HUPO Plasma Proteome Project pilot phase. *Proteomics* 5, 3520–3530.

40. Gallego-Delgado, J., Lazaro, A., Osende, J. I., Barderas, M. G., Blanco-Colio, L. M., Duran, M. C., Martin-Ventura, J. L., Vivanco, F., and Egido, J. (2005). Proteomic approach in the search of new cardiovascular biomarkers. *Kidney Int. Suppl.* **99,** S103–S107.

41. Petricoin, E. F., Rajapaske, V., Herman, E. H., Arekani, A. M., Ross, S., Johann, D., Knapton, A., Zhang, J., Hitt, B. A., Conrads, T. P., Veenstra, T. D., Liotta, L. A., and Sistare, F. D. (2004). Toxicoproteomics: serum proteomic pattern diagnostics for early detection of drug induced cardiac toxicities and cardioprotection. *Toxicol. Pathol.* **32,** 122–130.

42. Borozdenkova, S., Westbrook, J. A., Patel, V., Wait, R., Bolad, I., Burke, M. M., Bell, A. D., Banner, N. R., Dunn, M. J., and Rose, M. L. (2004). Use of proteomics to discover novel markers of cardiac allograft rejection. *J. Proteome Res.* **3,** 282–288.

43. Kittleson, M. M., and Hare, J. M. (2005). Molecular signature analysis: Using the myocardial transcriptome as a biomarker in cardiovascular disease. *Trends Cardiovasc. Med.* **15,** 130–138.

44. Ohki, R., Yamamoto, K., Ueno, S., Mano, H., Misawa, Y., Fuse, K., Ikeda, U., and Shimada, K. (2005). Gene expression profiling of human atrial myocardium with atrial fibrillation by DNA microarray analysis. *Int. J. Cardiol.* **102,** 233–238.

45. Mazurek, T., Zhang, L., Zalewski, A., Mannion, J. D., Diehl, J. T., Arafat, H., Sarov-Blat, L., O'Brien, S., Keiper, E. A., Johnson, A. G., Martin, J., Goldstein, B. J., and Shi, Y. (2003). Human epicardial adipose tissue is a source of inflammatory mediators. *Circulation* **108,** 2460–2466.

46. Baker, A. R., Silva, N. F., Quinn, D. W., Harte, A. L., Pagano, D., Bonser, R. S., Kumar, S., and McTernan, P. G. (2006). Human epicardial adipose tissue expresses a pathogenic profile of adipocytokines in patients with cardiovascular disease. *Cardiovasc. Diabetol.* **5,** 1–7.

47. Ganesh, S. K., Skelding, K. A., Mehta, L., O'Neill, K., Joo, J., Zheng, G., Goldstein, J., Simari, R., Billings, E., Geller, N. L., Holmes, D., O'Neill, W. W., and Nabel, E. G. (2004). Rationale and study design of the CardioGene Study: Genomics of in-stent restenosis. *Pharmacogenomics* **5,** 952–1004.

48. Marin-Garcia, J., and Goldenthal, M. J. (2002). Understanding the impact of mitochondrial defects in cardiovascular disease: a review. *J. Card. Fail.* **8,** 347–361.

49. Marin-Garcia, J., and Goldenthal, M. J. (2004). Mitochondria play a critical role in cardioprotection. *J. Card. Fail.* **10,** 55–66.

50. Marin-Garcia, J., Goldenthal, M. J., and Moe, G. W. (2001). Mitochondrial pathology in cardiac failure. *Cardiovasc. Res.* **49,** 17–26.

51. McGregor, A., Temperley, R., Chrzanowska-Lightowlers, Z. M., and Lightowlers, R. N. (2001). Absence of expression from RNA internalised into electroporated mammalian mitochondria. *Mol. Genet. Genomics* **265,** 721–729.

52. Johnston, S. A., Anziano, P. Q., Shark, K., Sanford, J. C., and Butow, R. A. (1988). Mitochondrial transformation in yeast by bombardment with microprojectiles. *Science* **240,** 1538–1541.

53. Chinnery, P. F., Taylor, R. W., Diekert, K., Lill, R., Turnbull, D. M., and Lightowlers, R. N. (1999). Peptide nucleic acid delivery to human mitochondria. *Gene Ther.* **6,** 1919–1928.

54. Taylor, R. W., Chinnery, P. F., Turnbull, D. M., and Lightowlers, R. N. (1997). Selective inhibition of mutant human mitochondrial DNA replication in vitro by peptide nucleic acids. *Nat. Genet.* **15,** 212–215.

55. Muratovska, A., Lightowlers, R. N., Taylor, R. W., Turnbull, D. M., Smith, R. A., Wilce, J. A., Martin, S. W., and Murphy, M. P. (2001). Targeting peptide nucleic acid (PNA) oligomers to mitochondria within cells by conjugation to lipophilic cations: Implications for mitochondrial DNA replication, expression and disease. *Nucleic Acids Res.* **29,** 1852–1863.

56. Geromel, V., Cao, A., Briane, D., Vassy, J., Rotig, A., Rustin, P., Coudert, R., Rigaut, J. P., Munnich, A., and Taillandier, E. (2001). Mitochondria transfection by oligonucleotides containing a signal peptide and vectorized by cationic liposomes. *Antisense Nucleic Acid Drug Dev.* **11,** 175–180.

57. Flierl, A., Jackson, C., Cottrell, B., Murdock, D., Seibel, P., and Wallace, D. C. (2003). Targeted delivery of DNA to the mitochondrial compartment via import sequence-conjugated peptide nucleic acid. *Mol. Ther.* **7,** 550–557.

58. Weissig, V., Lasch, J., Erdos, G., Meyer, H. W., Rowe, T. C., and Hughes, J. (1998). DQAsomes: A novel potential drug and gene delivery system made from Dequalinium. *Pharm. Res.* **15,** 334–337.
59. D'Souza, G. G., Rammohan, R., Cheng, S. M., Torchilin, V. P., and Weissig, V. (2003). DQAsome-mediated delivery of plasmid DNA toward mitochondria in living cells. *J. Control Release* **92,** 189–197.
60. Weissig, V., Cheng, S. M., and D'Souza, G. G. (2004). Mitochondrial pharmaceutics. *Mitochondrion* **3,** 229–244.
61. Kelso, G. F., Porteous, C. M., Coulter, C. V., Hughes, G., Porteous, W. K., Ledgerwood, E. C., Smith, R. A., and Murphy, M. P. (2001). Selective targeting of a redox-active ubiquinone to mitochondria within cells: Antioxidant and antiapoptotic properties. *J. Biol. Chem.* **276,** 4588–4596.
62. Smith, R. A., Porteous, C. M., Gane, A. M., and Murphy, M. P. (2003). Delivery of bioactive molecules to mitochondria in vivo. *Proc. Natl. Acad. Sci. USA* **100,** 5407–5412.
63. Adlam, V. J., Harrison, J. C., Porteous, C. M., James, A. M., Smith, R. A., Murphy, M. P., and Sammut, I. A. (2005). Targeting an antioxidant to mitochondria decreases cardiac ischemia-reperfusion injury. *FASEB J.* **19,** 1088–1095.
64. Zhao, K., Zhao, G. M., Wu, D., Soong, Y., Birk, A. V., Schiller, P. W., and Szeto, H. H. (2004). Cell-permeable peptide antioxidants targeted to inner mitochondrial membrane inhibit mitochondrial swelling, oxidative cell death and reperfusion injury. *J. Biol. Chem.* **279,** 34682–34690.
65. McDonnell, M. T., Schaefer, A. M., Blakely, E. L., McFarland, R., Chinnery, P. F., Turnbull, D. M., and Taylor, R. W. (2004). Noninvasive diagnosis of the 3243A > G mitochondrial DNA mutation using urinary epithelial cells. *Eur. J. Hum. Genet.* **12,** 778–781.
66. Mattei, J. P., Bendahan, D., and Cozzone, P. (2004). P-31 magnetic resonance spectroscopy. A tool for diagnostic purposes and pathophysiological insights in muscle diseases. *Reumatismo* **56,** 9–14.
67. Jucker, B. M., Dufour, S., Ren, J., Cao, X., Previs, S. F., Underhill, B., Cadman, K. S., and Shulman, G. I. (2000). Assessment of mitochondrial energy coupling in vivo by 13C/31P NMR. *Proc. Natl. Acad. Sci. USA* **97,** 6880–6884.
68. Lebon, V., Dufour, S., Petersen, K. F., Ren, J., Jucker, B. M., Slezak, L. A., Cline, G. W., Rothman, D. L., and Shulman, G. I. (2001). Effect of triiodothyronine on mitochondrial energy coupling in human skeletal muscle. *J. Clin. Invest.* **108,** 733–737.
69. Petersen, K. F., Dufour, S., Befroy, D., Garcia, R., and Shulman, G. I. (2004). Impaired mitochondrial activity in the insulin-resistant offspring of patients with type 2 diabetes. *N. Engl. J. Med.* **350,** 664–671.;
70. Petersen, K. F., Befroy, D., Dufour, S., Dziura, J., Ariyan, C., Rothman, D. L., DiPietro, L., Cline, G. W., and Shulman, G. I. (2003). Mitochondrial dysfunction in the elderly: Possible role in insulin resistance. *Science* **300,** 1140–1142.
71. Ingwall, J. S., Atkinson, D. E., Clarke, K., and Fetters, J. K. (1990). Energetic correlates of cardiac failure: changes in the creatine kinase system in the failing myocardium. *Eur. Heart J.* **11,** 108–115.
72. Leont'eva, I. V., Litvinova, I. S., Litvinov, M. M., Sebeleva, I. A., Sukhorukov, V. S., Tumanian, M. R., and Koledinskii, D. G. (2002). The use of positron emission tomography for noninvasive diagnosis of mitochondrial dysfunction and assessment of myocardial compensatory reserve in children with cardiomyopathies. *Kardiologiia* **42,** 80–85.
73. Bassingthwaighte, J. B., Qian, H., and Li, Z. (1999). The Cardiome Project. An integrated view of cardiac metabolism and regional mechanical function. *Adv. Exp. Med. Biol.* **471,** 541–553.
74. Ch'en, F., Clarke, K., Vaughan-Jones, R., and Noble, D. (1997). Modeling of internal pH, ion concentration, and bioenergetic changes during myocardial ischemia. *Adv. Exp. Med. Biol.* **430,** 281–290.
75. Noble, D. (2002). Modeling the heart—from genes to cells to the whole organ. *Science* **295,** 1678–1682.
76. Rudy, Y. (2000). From genome to physiome: integrative models of cardiac excitation. *Ann. Biomed. Eng.* **28,** 945–950.
77. Hunter, P., Smith, N., Fernandez, J., and Tawhai, M. (2005). Integration from proteins to organs: the IUPS Physiome Project. *Mech. Ageing Dev.* **126,** 187–192.
78. Crampin, E. J., Halstead, M., Hunter, P., Nielsen, P., Noble, D., Smith, N., and Tawhai, M. (2004). Computational physiology and the Physiome Project. *Exp. Physiol.* **89,** 1–26.
79. Noble, D. (2002). Modelling the heart: Insights, failures and progress. *Bioessays* **24,** 1155–1163.
80. Roden, D. M. (2005). Proarrhythmia as a pharmacogenomic entity: a critical review and formulation of a unifying hypothesis. *Cardiovasc. Res.* **67,** 419–425.
81. Holloway, J. W., Yang, I. A., and Ye, S. (2005). Variation in the toll-like receptor 4 gene and susceptibility to myocardial infarction. *Pharmacogenet. Genomics* **15,** 15–21.
82. Kajinami, K., Akao, H., Polisecki, E., and Schaefer, E. J. (2005). Pharmacogenomics of statin responsiveness. *Am. J. Cardiol.* **96,** 65K–70K.
83. Bolli, R., Becker, L., Gross, G., Mentzer, R., Jr., Balshaw, D., Lathrop, D. A., and the NHLBI Working Group on the Translation of Therapies for Protecting the Heart from Ischemia. (2004). Myocardial protection at a crossroads: the need for translation into clinical therapy. *Circ. Res.* **95,** 125–134.
84. Simonetti, O. P., Kim, R. J., Fieno, D. S., Hillenbrand, H. B., Wu, E., Bundy, J. M., Finn, J. P., and Judd, R. M. (2001). An improved MR imaging technique for the visualization of myocardial infarction. *Radiology* **218,** 215–223.
85. Klein, C., Nekolla, S. G., and Schwaiger, M. (2001). The role of magnetic resonance imaging in the diagnosis of coronary disease. *Z. Kardiol.* **90,** 208–217.
86. Mahrholdt, H., Wagner, A., Holly, T. A., Elliott, M. D., Bonow, R. O., Kim, R. J., and Judd, R. M. (2002). Reproducibility of chronic infarct size measurement by contrast-enhanced magnetic resonance imaging. *Circulation* **106,** 2322–2327.
87. Mahaffey, K. W., Puma, J. A., Barbagelata, N. A., DiCarli, M. F., Leesar, M. A., Browne, K. F., Eisenberg, P. R., Bolli, R., Casas, A. C., Molina-Viamonte, V., Orlandi, C., Blevins, R., Gibbons, R. J., Califf, R. M., and Granger, C. B. (1999). Adenosine as an adjunct to thrombolytic therapy for acute myocardial infarction: results of a multicenter, randomized, placebo-controlled trial: The Acute Myocardial Infarction Study of Adenosine (AMISTAD) trial. *J. Am. Coll. Cardiol.* **34,** 1711–1120.
88. Ross, A. M., Gibbons, R. J., Stone, G. W., Kloner, R. A., Alexander, R. W., and the AMISTAD-II Investigators. (2005). A randomized, double-blinded, placebo-controlled multicenter trial of adenosine as an adjunct to reperfusion in the treatment of acute myocardial infarction (AMISTAD-II). *J. Am. Coll. Cardiol.* **45,** 1775–1780.

89. Sakata, Y., Kodama, K., Komamura, K., Lim, Y.-J., Ishikura, F., Hirayama, A., Kitakaze, M., Masuyama, T., and Hori, M. (1997). Salutary effect of adjunctive intracoronary nicorandil administration on restoration of myocardial blood flow and functional improvement in patients with acute myocardial infarction. *Am. Heart J.* **133**, 616–621.

90. Ito, H., Taniyama, Y., Iwakura, K., Nishikawa, N., Masuyama, T., Kuzuya, T., Hori, M., Higashino, Y., Fujii, K., and Minamino, T. (1999). Intravenous nicorandil can preserve microvascular integrity and myocardial viability in patients with reperfused anterior wall myocardial infarction. *J. Am. Coll. Cardiol.* **33**, 654–660.

91. Fath-Ordoubadi, F., and Beatt, K. J. (1997). Glucose-insulin-potassium therapy for treatment of acute myocardial infarction: An overview of randomized placebo-controlled trials. *Circulation* **96**, 1152–1156.

92. Mentzer, R. M., Jr., Birjiniuk, V., Khuri, S., Lowe, J. E., Rahko, P. S., Weisel, R. D., Wellons, H. A., Barker, M. L., and Lasley, R. D. (1999). Adenosine myocardial protection: preliminary results of a phase II clinical trial. *Ann. Surg.* **229**, 643–649.

93. Theroux, P., Chaitman, B. R., Danchin, N., Erhardt, L., Meinertz, T., Schroeder, J. S., Tognoni, G., White, H. D., Willerson, J. T., and Jessel, A. (2000). Inhibition of the sodium-hydrogen exchanger with cariporide to prevent myocardial infarction in high-risk ischemic situations. Main results of the GUARDIAN trial. Guard during ischemia against necrosis (GUARDIAN) Investigators. *Circulation* **102**, 3032–3038.

94. Mentzer, R. M., Jr., and the EXPEDITION Study Investigators. (2003). Effects of Na+/H+ exchange inhibition by cariporide on death and nonfatal myocardial infarction in patients undergoing coronary artery bypass graft surgery: The EXPEDITION study. *Circulation* **108**, 3M.

95. Noble, D. (2002). Unraveling the genetics and mechanisms of cardiac arrhythmia. *Proc. Natl. Acad. Sci. USA* **99**, 5755–5756.

96. Gillum, R. F. (1979). Pathophysiology of hypertension in blacks and whites: a review of the basis of racial blood pressure differences. *Hypertension* **1**, 468–475.

97. Yancy, C. W. (2000). Heart failure in African-Americans: a cardiovascular enigma. *J. Card. Fail.* **6**, 183–186.

98. Watkins, L. O. (1984). Epidemiology of coronary heart disease in black populations. *Am. Heart J.* **108**, 635–640.

99. Chobanian, A. V., Bakris, G. L., Black, H. R., Cushman, W. C., Green, L. A., Izzo, J. L., Jr., Jones, D. W., Materson, B. J., Oparil, S., and Wright, J. T., Jr. (2003). The Seventh Report of the Joint National Committee on Prevention, Detection, Evaluation, and Treatment of High Blood Pressure: The JNC 7 report. *JAMA* **289**, 2560–2572.

100. Carson, P., Ziesche, S., Johnson, G., and Cohn, J. N. (1999). Racial differences in response to therapy for heart failure: analysis of the vasodilator-heart failure trials. *J. Card. Fail.* **5**, 178–187.

101. Exner, D. V., Dries, D. L., Domanski, M. J., and Cohn, J. N. (2001). Lesser response to angiotensin-converting-enzyme inhibitor therapy in black as compared with white patients with left ventricular dysfunction. *N. Engl. J. Med.* **344**, 1351–1357.

102. Yancy, C. W., Fowler, M. B., Colucci, W. S., Gilbert, E. M., Bristow, M. R., Cohn, J. N., Lukas, M. A., Young, S. T., Packer, M., and the US Carvedilol Heart Failure Study Group. (2001). Race and the response to adrenergic blockade with carvedilol in patients with chronic heart failure. *N. Engl. J. Med.* **344**, 1358–1365.

103. The Beta-Blocker Evaluation of Survival Trial Investigators. (2001). A trial of the beta-blocker bucindolol in patients with advanced chronic heart failure. *N. Engl. J. Med.* **344**, 1659–1667.

104. Taylor, A. L. (2005). The African American Heart Failure Trial: A clinical trial update. *Am. J. Cardiol.* **96**, 44–48.

105. Taylor, A. L., Ziesche, S., Yancy, C., Carson, P., D'Agostino, R. Jr., Ferdinand, K., Taylor, M., Adams, K., Sabolinski, M., Worcel, M., Cohn, J. N., and the African-American Heart Failure Trial Investigators. (2004). Combination of isosorbide dinitrate and hydralazine in blacks with heart failure. *N. Engl. J. Med.* **351**, 2049–2057.

106. Rimessi, P., Gualandi, F., Duprez, L., Spitali, P., Neri, M., Merlini, L., Calzolari, E., Muntoni, F., and Ferlini, A. (2005). Genomic and transcription studies as diagnostic tools for a prenatal detection of X-linked dilated cardiomyopathy due to a dystrophin gene mutation. *Am. J. Med. Genet. A* **132**, 391–394.

107. Johnson, W. H. Jr., Yang, P., Yang, T., Lau, Y. R., Mostella, B. A., Wolff, D. J., Roden, D. M., and Benson, D. W. (2003). Clinical, genetic, and biophysical characterization of a homozygous HERG mutation causing severe neonatal long QT syndrome. *Pediatr. Res.* **53**, 744–748.

108. Charron, P., Heron, D., Gargiulo, M., Feingold, J., Oury, J. F., Richard, P., and Komajda, M. (2004). Prenatal molecular diagnosis in hypertrophic cardiomyopathy: Report of the first case. *Prenat. Diagn.* **24**, 701–703.

109. Charron, P., Heron, D., Gargiulo, M., Richard, P., Dubourg, O., Desnos, M., Bouhour, J. B., Feingold, J., Carrier, L., Hainque, B., Schwartz, K., and Komajda, M. (2002). Genetic testing and genetic counselling in hypertrophic cardiomyopathy: The French experience. *J. Med. Genet.* **39**, 741–746.

110. Loeys, B., Nuytinck, L., Van Acker, P., Walraedt, S., Bonduelle, M., Sermon, K., Hamel, B., Sanchez, A., Messiaen, L., and De Paepe, A. (2002). Strategies for prenatal and preimplantation genetic diagnosis in Marfan syndrome (MFS). *Prenat. Diagn.* **22**, 22–28.

111. Jacobs, L. J., de Coo, I. F., Nijland, J. G., Galjaard, R. J., Los, F. J., Schoonderwoerd, K., Niermeijer, M. F., Geraedts, J. P., Scholte, H. R., and Smeets, H. J. (2005). Transmission and prenatal diagnosis of the T9176C mitochondrial DNA mutation. *Mol. Hum. Reprod.* **11**, 223–228.

112. Leshinsky-Silver, E., Perach, M., Basilevsky, E., Hershkovitz, E., Yanoov-Sharav, M., Lerman-Sagie, T., and Lev, D. (2003). Prenatal exclusion of Leigh syndrome due to T8993C mutation in the mitochondrial DNA. *Prenat. Diagn.* **23**, 31–33.

113. Amiel, J., Gigarel, N., Benacki, A., Benit, P., Valnot, I., Parfait, B., Von Kleist-Retzow, J. C., Raclin, V., Hadj-Rabia, S., Dumez, Y., Rustin, P., Bonnefont, J. P., Munnich, A., and Rotig, A. (2001). Prenatal diagnosis of respiratory chain deficiency by direct mutation screening. *Prenat. Diagn.* **21**, 602–604.

114. Niers, L., van den Heuvel, L., Trijbels, F., Sengers, R., Smeitink, J.; the Nijmegen Centre for Mitochondrial Disorders, The Netherlands. (2003). Prerequisites and strategies for prenatal diagnosis of respiratory chain deficiency in chorionic villi. *J. Inherit. Metab. Dis.* **26**, 647–658.

115. Thorburn, D. R., and Dahl, H. H. (2001). Mitochondrial disorders: genetics, counseling, prenatal diagnosis and reproductive options. *Am. J. Med. Genet.* **106**, 102–114.

116. Dahl, H. H., Thorburn, D. R., and White, S. L. (2000). Towards reliable prenatal diagnosis of mtDNA point mutations: studies of nt8993 mutations in oocytes, fetal tissues, children and adults. *Hum. Reprod.* **15**, 246–255.
117. He, J., McDermott, D. A., Song, Y., Gilbert, F., Kligman, I., and Basson, C. T. (2004). Preimplantation genetic diagnosis of human congenital heart malformation and Holt-Oram syndrome. *Am. J. Med. Genet. A* **126**, 93–98.
118. Barritt, J. A., Brenner, C. A., Malter, H. E., and Cohen, J. (2001). Mitochondria in human offspring derived from ooplasmic transplantation. *Hum. Reprod.* **16**, 513–516.
119. Malter, H. E., and Cohen, J. (2002). Ooplasmic transfer: Animal models assist human studies. *Reprod. Biomed. Online* **5**, 26–35.
120. Hawes, S. M., Sapienza, C., and Latham, K. E. (2002). Ooplasmic donation in humans: The potential for epigenic modifications. *Hum. Reprod.* **17**, 850–852.
121. St. John, J. C. (2002). Ooplasm donation in humans: The need to investigate the transmission of mitochondrial DNA following cytoplasmic transfer. *Hum. Reprod.* **17**, 1954–1958.
122. Poulton, J., and Marchington, D. R. (2002). Segregation of mitochondrial DNA (mtDNA) in human oocytes and in animal models of mtDNA disease: Clinical implications. *Reproduction* **123**, 751–755.
123. Kreuter, M., Langer, C., Kerkhoff, C., Reddanna, P., Kania, A. L., Maddika, S., Chlichlia, K., Bui, T. N., and Los, M. (2004). Stroke, myocardial infarction, acute and chronic inflammatory diseases: Caspases and other apoptotic molecules as targets for drug development. *Arch. Immunol. Ther. Exp.* **52**, 141–155.
124. Roth, G. S., Lane, M. A., and Ingram, D. K. (2005). Caloric restriction mimetics: The next phase. *Ann. N. Y. Acad. Sci.* **1057**, 365–371.
125. Ingram, D. K., Anson, R. M., de Cabo, R., Mamczarz, J., Zhu, M., Mattison, J., Lane, M. A., and Roth, G. S. (2004). Development of calorie restriction mimetics as a prolongevity strategy. *Ann. N. Y. Acad. Sci.* **1019**, 412–423.
126. Dirks, A. J., and Leeuwenburgh, C. (2006). Caloric restriction in humans: Potential pitfalls and health concerns. *Mech. Ageing Dev.* **127**, 1–7.
127. Davis, M. E., Hsieh, P. C., Grodzinsky, A. J., and Lee, R. T. (2005). Custom design of the cardiac microenvironment with biomaterials. *Circ. Res.* **97**, 8–15.
128. Taylor, P. M., Sachlos, E., Dreger, S. A., Chester, A. H., Czernuszka, J. T., and Yacoub, M. H. (2006). Interaction of human valve interstitial cells with collagen matrices manufactured using rapid prototyping. *Biomaterials* **27**, 2733–2737.
129. Fukuda, K. (2005). Progress in myocardial regeneration and cell transplantation. *Circ. J.* **69**, 1431–1461.
130. Memon, I. A., Sawa, Y., Fukushima, N., Matsumiya, G., Miyagawa, S., Taketani, S., Sakakida, S. K., Kondoh, H., Aleshin, A. N., Shimizu, T., Okano, T., and Matsuda, H. (2005). Repair of impaired myocardium by means of implantation of engineered autologous myoblast sheets. *J. Thorac. Cardiovasc. Surg.* **130**, 1333–1341.
131. Abraham, M. R., Henrikson, C. A., Tung, L., Chang, M. G., Aon, M., Xue, T., Li, R. A., O'Rourke, B., and Marban, E. (2005). Antiarrhythmic engineering of skeletal myoblasts for cardiac transplantation. *Circ. Res.* **97**, 159–167.
132. Knight, R. L., Booth, C., Wilcox, H. E., Fisher, J., and Ingham, E. (2005). Tissue engineering of cardiac valves: Re-seeding of acellular porcine aortic valve matrices with human mesenchymal progenitor cells. *J. Heart Valve Dis.* **14**, 806–813.
133. Nagy, R. D., Tsai, B. M., Wang, M., Markel, T. A., Brown, J. W., and Meldrum, D. R. (2005). Stem cell transplantation as a therapeutic approach to organ failure. *J. Surg. Res.* **129**, 152–160.
134. Vesely, I. (2005). Heart valve tissue engineering. *Circ. Res.* **97**, 743–755.
135. Cushing, M. C., Jaeggli, M. P., Masters, K. S., Leinwand, L. A., and Anseth, K. S. (2005). Serum deprivation improves seeding and repopulation of acellular matrices with valvular interstitial cells. *J. Biomed. Mater Res. A* **75**, 232–241.
136. Sutherland, F. W. H., Perry, T. E., Yu, .Y, Sherwood, M. C., Rabkin, E., Masuda, Y., Garcia, G. A., McLellan, D. L., Engelmayr, G. C., Jr., Sacks, M. S., Schoen, F. J., and Mayer, J. E., Jr. (2005). From stem cells to viable autologous semilunar heart valve. *Circulation* **111**, 2783–2791.
137. Gepstein, L. (2005). Stem cells as biological heart pacemakers. *Expert Opin. Biol. Ther.* **5**, 1531–1537.
138. Miake, J., Marban, E., and Nuss, H. B. (2002). Biological pacemaker created by gene transfer. *Nature* **419**, 132–133.
139. Xue, T., Cho, H. C., and Li, R. A. Induction of ventricular automaticity by overexpression of an engineered HCN1 construct. *Biophys. J.* In press.
140. Edelberg, J. M., Huang, D. T., Josephson, M. E., and Rosenberg, R. D. (2001). Molecular enhancement of porcine cardiac chronotropy. *Heart* **86**, 559–562.
141. Qu, J., Itskovitz-Eldor, J., Shapiro, S. S., Waknitz, M. A., Cohen, I. S., Robinson, R. B., and Rosen, M. R. (2003). Expression and function of a biological pacemaker in canine heart. *Circulation* **107**, 1106–1109.
142. Plotnikov, A. N., Sosunov, E. A., Qu, J., Shlapakova, I., Anyukhovsky, E. P., Liu, L., Janse, M. J., Brink, P. R., Cohen, I. S., Robinson, R. B., Danilo, P. Jr., and Rosen, M. R. (2004). A biological pacemaker implanted in the canine left bundle branch provides ventricular escape rhythms having physiologically acceptable rates. *Circulation* **109**, 506–512.
143. Potapova, I., Plotnikov, A., Lu, Z., Danilo, P., Jr., Valiunas, V., Qu, J., Doronin, S., Zuckerman, J., Shlapakova, I. N., Gao, J., Pan, Z., Herron, A. J., Robinson, R. B., Brink, P. R., Rosen, M. R., and Cohen, I. S. (2004). Human mesenchymal stem cells as a gene delivery system to create cardiac pacemakers. *Circ. Res.* **94**, 952–959.
144. Xue, T., Cho, H. C., Akar, F. G., Tsang, S. Y., Jones, S. P., Marban, E., Tomaselli, G. F., and Li, R. A. (2005). Functional integration of electrically active cardiac derivatives from genetically engineered human embryonic stem cells with quiescent recipient ventricular cardiomyocytes: Insights into the development of cell-based pacemakers. *Circulation* **111**, 11–20.
145. Kehat, I., Khimovich, L., Caspi, O., Gepstein, A., Shofti, R., Arbel, G., Huber, I., Satin, J., Itskovitz-Eldor, J., and Gepstein, L. (2004). Electromechanical integration of cardiomyocytes derived from human embryonic stem cells. *Nat. Biotechnol.* **22**, 1282–1289.
146. Madeddu. P. (2005). Therapeutic angiogenesis and vasculogenesis for tissue regeneration. *Exp. Physiol.* **90**, 315–326.
147. Hughes, G. C., Post, M. J., Simons, M., and Annex, B. H. (2003). Translational physiology: Porcine models of human

coronary artery disease: Implications for preclinical trials of therapeutic angiogenesis. *J. Appl. Physiol.* **94,** 1689–1701.
148. Shimamura, M., Sato, N., Yoshimura, S., Kaneda, Y., and Morishita, R. (2006). HVJ-based non-viral gene transfer method: Successful gene therapy using HGF and VEGF genes in experimental ischemia. *Front. Biosci.* **11,** 753–759.
149. Shah, P. B., and Losordo, D. W. (2005). Non-viral vectors for gene therapy: Clinical trials in cardiovascular disease. *Adv. Genet.* **54,** 339–361.
150. Lei, Y., Haider, H. Kh., Shujia, J., and Sim, E. S. (2004). Therapeutic angiogenesis. Devising new strategies based on past experiences. *Basic Res. Cardiol.* **99,** 121–132.
151. Riha, G. M., Lin, P. H., Lumsden, A. B., Yao, Q., and Chen, C. (2005). Review: Application of stem cells for vascular tissue engineering. *Tissue Eng.* **11,** 1535–1552.
152. Sales, K. M., Salacinski, H. J., Alobaid, N., Mikhail, M., Balakrishnan, V., and Seifalian, A. M. (2005). Advancing vascular tissue engineering: the role of stem cell technology. *Trends Biotechnol.* **23,** 461–467.
153. Wu, X., Rabkin-Aikawa, E., Guleserian, K. J., Perry, T. E., Masuda, Y., Sutherland, F. W., Schoen, F. J., Mayer, J. E., Jr., and Bischoff, J. (2004). Tissue-engineered microvessels on three-dimensional biodegradable scaffolds using human endothelial progenitor cells. *Am. J. Physiol. Heart Circ. Physiol.* **287,** H480–H487.

SECTION XI
Glossary

Glossary

AAV Adeno-associated virus vector. A defective human parvovirus with potential as a vector with long-term gene expression for human gene therapy of cardiovascular disorders.

Acetyl-CoA Small water-soluble molecule that carries acetyl groups linked to coenzyme A (CoA) by a thioester bond.

AC Adenylyl cyclase, a membrane-bound enzyme that catalyzes the synthesis of the second messenger cyclic AMP from ATP in conjunction with specific signaling ligands (e.g., adrenergic), receptors, and G-proteins.

ACC Acetyl-CoA carboxylase, an enzyme that synthesizes malonyl-CoA from cytoplasmic and peroxisomal acetyl-CoA.

ACE Angiotensin-converting enzyme, a central element of the renin-angiotensin system, converts the decapeptide angiotensin I to the potent pressor octapeptide angiotensin II (Ang II), mediating peripheral vascular tone, as well as glomerular filtration in the kidney.

ACH Acetylcholine.

ACM Alcoholic cardiomyopathy. Characterized by cardiomegaly, disruptions of myofibrillar architecture, reduced myocardial contractility, decreased ejection fraction, and enhanced risk of stroke and hypertension.

ACS Acute coronary syndrome.

ACS cells Adult cardiac stem cells.

Adenovirus Common vector for gene transfer with high efficiency of transfection *in vivo* but limited by transient transgene expression and host immunogenic response.

ADP Adenosine diphosphate.

Adrenoceptors Members of the G-protein–coupled receptor superfamily linking adrenergic signaling from the sympathetic nervous system and the cardiovascular system, with integral roles in the rapid regulation of myocardial function.

AF Atrial fibrillation, the most common dysrhythmia seen in clinical cardiology, can be familial, with both monogenic and more heterogenous genetic cases reported.

Affinity tag Biochemical indicator, which, appended to recombinant expressed proteins, can serve several functions (e.g., purification of proteins, solubilization of a fusion partner, indicator of fusion protein folding, and providing a common epitope to allow a single antibody to recognize each fusion protein).

AFM Atomic force microscopy can determine cellular mechanical property changes at nanoscale resolution.

AIF Apoptosis-inducing factor. Released from mitochondrial intermembrane space in early apoptosis and subsequently involved in nuclear DNA fragmentation.

AKAP Specific PKA anchoring proteins; regulators of PKA function and signaling by directing and concentrating PKA at specific subcellular sites.

Akt Protein kinase B (PKB). Myocardial Akt phosphorylates a number of downstream targets, including cardioprotective factors involved in glucose and mitochondrial metabolism, apoptosis, and regulators of protein synthesis.

Alagille syndrome An autosomal-dominant disorder with a wide spectrum of developmental anomalies and commonly presenting with TOF caused by mutations in the JAG1 gene encoding a notch ligand (jagged1).

ALCAR Acetyl-l-carnitine, supplementation with lipoic acid (LA) seems to improve myocardial bioenergetics and decrease oxidative stress associated with aging.

Allele One of several alternate forms of a single gene occupying a given locus on a chromosome or mtDNA.

Allotopic expression Alternate method of mitochondrial gene therapy in that a mitochondrial gene is reengineered for expression from the nucleus and targeting its translation product to the mitochondria.

AMI Acute myocardial infarction.

Amphipathic Molecule with distinct hydrophobic and hydrophilic domains (e.g., phospholipids and detergents).

AMPK AMP-activated protein kinase involved in myocardial metabolic energy sensing.

Amplification Generation of many copies of a specific region of DNA.

Aneurysm A sac-like protrusion from a blood vessel or the heart, resulting from a weakening of the vessel wall or heart muscle.

Angiogenesis Formation of new vessels from pre-existing ones, and in particular the sprouting of new capillaries from post-capillary venules.

Annulus The ring around a heart valve where the valve leaflet merges with the heart muscle.

Andersen syndrome Also known as Andersen–Tawil syndrome or LQT7, an autosomal dominant characterized by a heterogenous phenotype, including a variety of cardiac dysrhythmias, with many patients having mutations in the KCNJ2 gene coding the K^+ channel inward rectifier (IK1) Kir2 channel.

ANP Atrial natriuretic peptide (also **ANF**).

ANT Adenine nucleotide translocator. A mitochondrial inner membrane carrier protein of ADP and ATP and constituent of the PT pore.

Antimycin A Specific inhibitor of complex III activity.

Antisense RNA RNA complementary to a specific transcript of a gene that can hybridize to the specific RNA and block its function.

AP Action potential.

APC Anesthetic pre-conditioning.

APLA Antiphospholipid antibodies.

Apoptosis Programmed cell death.

Apoptosome Cytosolic complex involved in the activation of apoptotic caspases.

APTT Activated partial thromboplastin time.

AR Androgen receptor.

α2-AR α2-Adrenergic receptor, receptor for endogenous catecholamine agonists (e.g., norepinephrine and epinephrine) that mediate a number of physiological and pharmacological responses such as changes in blood pressure and heart rate.

β-AR Beta-adrenergic receptor, G-protein–coupled receptor containing a seven-transmembrane domain involved in signaling pathways of diverse cardiovascular functions, including blood pressure control and cardiac contractility.

ARC protein Apoptosis repressor with a caspase recruitment domain (CARD), inhibitor of both the intrinsic and extrinsic apoptosis pathways.

ARE Androgen response element, specific regulatory sites within the DNA of target nuclear genes, which, bound by AR, produce long-term genomic effects of testosterone.

ARH Autosomal-recessive familial hypercholesterolemia; a rare disorder with a clinical phenotype similar to homozygous FH, but less severe, more variable, and responsive to lipid-lowering therapy.

β-ARK Beta-adrenergic receptor kinase, a GRK that mediates the desensitization of the β-adrenergic receptor by phosphorylation of agonist-occupied receptors.

Aromatase Enzyme that converts testosterone or other aromatizable synthetic androgens to estradiol.

ART Androgen replacement therapy.

Arteriogenesis Process of maturation and/or *de novo* growth of specifically collateral arteries, which mainly occurs after ischemic vascular disease.

ARVC Arrhythmogenic right ventricular cardiomyopathy (same as ARVD).

ARVD Arrhythmogenic right ventricular dysplasia; the most common symptoms are ventricular dysrhythmias, heart palpitations, fainting or loss of consciousness (syncope), and sudden death.

ASO Antisense oligonucleotides. These short, synthetic DNA molecules can reduce specific gene expression by acting either directly or as decoys of transcription factors.

ATP Adenosine triphosphate.

Atractyloside Inhibitor of the adenine nucleotide translocator and PT pore opener

Autosomal inheritance pattern The gene of interest is present on any of the non-sex chromosomes.

AV Atrioventricular.

AVC Atrioventricular canal.

AV node Atrioventricular node; a group of specialized cells located between the atria and ventricles that regulate electrical current passing to the ventricles.

Azacytidine Demethylating agent; can induce ES cell cardiomyocyte-specific differentiation.

AZT Zidovudine. Used to treat AIDS. An inhibitor of DNA polymerase that can cause mtDNA depletion.

Bacteriophage A virus that infects bacteria; useful as a vector for gene transfer.

Barth syndrome An X-linked recessive cardioskeletal myopathy with neutropenia and DCM with childhood onset caused by mutations in TAZ/G4.5 gene encoding the tafazzin protein.

BER Base excision repair. DNA repair in which a missing or damaged base on a single strand is recognized, excised, and replaced by synthesizing a sequence complementary to the remaining strand.

BF Blood flow.

BH domains Features of proapoptotic proteins (BH1-4) are essential for homo- and heterocomplex formation, as well as to induce cell death. Proapoptotic homologs can be subdivided into two major subtypes, the multidomain Bax subfamily (e.g., Bax and Bak), which possesses BH1-3 domains, and the BH3-only subfamily (e.g., Bad and Bid).

Bid A proapoptotic Bcl-2 related protein, which links the extrinsic and intrinsic apoptotic pathways.

Bilayer Arrangement of phospholipids in biological membranes.

Biolistic transformation Method of introducing DNA into cells by use of highly accelerated DNA-coated metal particles.

BMP Bone morphogenetic protein, a class of ligands that bind specific membrane-bound receptors involved in signaling events in early cardiomyocyte differentiation.

BNP Brain natriuretic peptide; hemodynamic marker of neurohumoral and vascular stress.

bp Base pairs.

Blunt end Results from the breaking of double-stranded DNA into two complementary strands of equal length.

BMDC Bone-marrow–derived cells.

BRDU Bromodeoxyuridine; a DNA synthesis inhibitor.

BrS Brugada syndrome. A form of idiopathic ventricular fibrillation, which, in children, can be inherited as an autosomal-dominant trait with variable penetrance and can lead to sudden death in healthy young individuals; a subset of cases have *SCN5A* mutations.

BSS Bernard–Soulier syndrome, a rare autosomal-recessive disorder caused by mutations in various polypeptides in the GpIb/IX/V complex, the principal platelet receptor for vWF.

CA Catalase; antioxidant enzyme.

CABG Coronary artery bypass grafting.

CAD Coronary artery disease (See ischemic heart disease).

Calcineurin Intracellular Ca^{++}-regulated phosphatase implicated as a mediator of cardiac hypertrophy.

CaM Calmodulin; an intracellular Ca^{++} sensor that selectively activates downstream signaling pathways in response to local changes in Ca^{++}.

CaMK Ca^{++}/CaM-dependent protein kinase.

Chameleons Fluorescent indicators for Ca^{++}, which can be targeted to specific intracellular locations.

cAMP Cyclic AMP; second messenger used in extensively in cell signaling. Product of adenylyl cyclase (AC).

CAR Coxsackievirus-adenovirus receptor; cell-surface receptor specifically interacts with both coxsackieviruses of group B (CVB) and adenoviruses.

Cardiomyocyte A single cell of a heart muscle.

Cardiolipin Anionic phospholipid located primarily in mitochondrial inner membrane.

Carnitine Carrier molecule involved in the transport of long-chain fatty acids into the mitochondria for β-FAO.

Caspases Intracellular cysteine proteases activated during apoptosis that cleave substrates at their aspartic acid residues.

Caveolae Vesicular organelles, which are specialized subdomains of the plasma membrane particularly abundant in cardiovascular cells, that function both in protein trafficking and signal transduction.

CCCP Carbonyl cyanide m-chlorophenyl hydrazone. A potent uncoupler.

CCS Cardiac conduction system, a heterogenous complex of cells responsible for establishing and maintaining the rhythmic excitation of the mature heart.

cDNA Complementary DNA. DNA fragment that is synthesized from the mRNA strand by reverse transcriptase. This DNA copy of a mature RNA lacks the introns that are present in the genomic DNA.

cDNA library Collection of cDNAs synthesized from the mRNA of an organism cloned into a vector.

Cell cycle The period between the release of a cell as one of the progeny of a mitotic division and its own subsequent division into two daughter cells.

Cell fusion Fusion of two somatic cells creating a hybrid cell.

Central dogma Flow of genetic information going from DNA to RNA to protein

CETP Cholesteryl ester transfer protein plays a role in reverse cholesterol transport with transfer of cholesteryl ester-rich HDL to triglyceride-rich lipoproteins (VLDL).

CGRP Calcitonin gene-related peptide; cardioprotective agent.

Chagas' disease Infectious disease that often results in myocarditis and/or dilated cardiomyopathy endemic to South and Central America caused by *Trypanosoma cruzi. (T. cruzi)*, a flagellated protozoan parasite.

Chaperone Protein that assists in the proper folding and assembly into larger complexes of unfolded or misfolded proteins.

Char syndrome Autosomal-dominant syndrome with patent ductus arteriosus (PDA), facial dysmorphology, and fifth-finger malformation caused by mutations in *TFAP2B*, the gene encoding a transcription factor AP2β, which is highly expressed in neural crest cells.

CHD Congenital heart defect.

Cholesterol 7-hydroxylase deficiency Deficiency in the first enzyme in the pathway for cholesterol catabolism and bile synthesis leads to high levels of LDL cholesterol and hypercholesterolemia.

Chromatin The complex of DNA and histone and nonhistone proteins found in the nucleus of a eukaryotic cell that constitutes the chromosomes.

Chromosome A very long, continuous piece of DNA, which holds together many genes, regulatory elements, and other intervening nucleotide sequences.

Chromosome walking A technique used to identify and sequence long segments of a DNA strand (e.g., a chromosome).

Chylomicron Large triglyceride-rich particles containing apoB48 packaged from dietary lipids in the intestinal enterocyte that transport exogenous lipids to the liver.

***Cis*-acting elements** Regulatory DNA sequences that affect the expression of genes only on the molecule of DNA where they reside; not protein encoding.

CK Creatine kinase. Both mitochondrial and cytosolic isoforms of this enzyme catalyze the reversible phosphorylation of creatine by ATP to form the high-energy compound phosphocreatine.

CL Cardiolipin, anionic phospholipid present in mitochondrial membranes. Deficiency in Barth syndrome. Involved in stabilization of ETC complexes and in apoptosis.

Codon A 3-nucleotide sequence in mRNA specifying a unique amino acid.

Complementary DNA A DNA copy of a mature mRNA template.

Complex I NADH-ubiquinone oxidoreductase.

Complex II Succinate CoQ oxidoreductase.

Complex III CoQ-cytochrome *c* oxidoreductase.

Complex IV Cytochrome *c* oxidase.

Complex V Oligomycin-sensitive ATP synthase. Also termed F0-F1 ATPase.

Congenic rats Strains produced by repeated backcrosses to an inbred strain, with selection of a particular marker from the donor strain.

Connexins A group of transmembrane proteins that form gap junctions between cells.

Consomic rats Strains produced by repeated backcrossing of a whole chromosome onto an inbred strain.

CoQ Coenzyme Q (also ubiquinone). Electron carrier and antioxidant.

Cosmid vector A type of plasmid constructed by the insertion of *cos* sequences from Phage lambda).

COT curve Expresses the ratio of the concentration of denatured single-stranded DNA to the initial concentration of DNA (Co) times time or Cot; used in DNA-DNA renaturation to assess sequence homology and complexity.

COX Cytochrome *c* oxidase (complex IV).

CP Cardioprotection.

CpG islands GC-rich regions of DNA often found in promoter regions.

CPT-I Carnitine palmitoyltransferase I.

CPT-II Carnitine palmitoyltransferase II.

CPVT Catecholaminergic polymorphic ventricular tachycardia; characterized by syncope, seizures, or SD, in response to exercise or emotional stress, and affecting mainly young children with morphologically normal hearts. Mutations in *RyR2* (autosomal dominant) or in *CASQ2* encoding calsequestrin 2 (autosomal recessive) have been found leading to CPVT.

Cristae Folding of inner mitochondrial membrane to enlarge the surface area.

CR Caloric restriction, a restricted dietary regimen that has been shown to increase life span in a number of organisms, including mammals, and may have antiaging effects in the heart.

CRP C-reactive protein; a significant marker of inflammation and atherosclerotic progression. Serum CRP levels are predictive of future cardiovascular events.

CsA Cyclosporin A. An inhibitor of PT pore opening.

CT-1 Cardiotrophin, an interleukin 6-related cytokine, has been shown to promote both the survival and proliferation of cultured neonatal cardiomyocytes.

CTnI Cardiac troponin I; widely used marker of myocardial ischemia and necrosis.

CTnT Cardiac troponin T; widely used marker of myocardial ischemia and necrosis.

CVB Coxsackievirus B viruses; commonly identified infectious agents (particularly CVB3) causing human viral myocarditis.

CVD Cardiovascular disease.

CyP-D Cyclophilin D. CsA-binding mitochondrial matrix protein component of the PT pore.

Cytochrome A family of proteins that contain heme as a prosthetic group involved in electron transfer and identifiable by their absorption spectra.

Cytochrome *c* A mitochondrial protein involve in ETC at complex IV. Its release from the mitochondrial into the cytosol is a trigger of caspase activation and early myocardial apoptosis.

DAG Diacylglycerol; second messenger produced by phospholipase C activation (see PLC).

Danon disease Glycogen storage disease with hypertrophic cardiomyopathy.

DCFH-DA 2′,7′-Dichlorodihydrofluorescein diacetate, membrane-permeable fluorometric indicator of ROS levels in the cytosol.

DCA Dichloroacetate. By inhibiting PDH kinase, DCA stimulates PDH, promoting aerobic oxidation and reducing lactic acidosis.

DCM Dilated cardiomyopathy.

Delayed pre-conditioning Often referred to as a second window of protection, this pathway appears about 12–24 h after the pre-conditioning event and lasts several days.

DES Drug-eluting stent.

Desmosome Highly organized intercellular junctions that provide mechanical integrity to tissues by anchoring intermediate filaments to sites of strong adhesion.

Desmoglein A transmembrane desmosomal protein; mutations lead to ARVD.

Desmoplakin Desmosomal protein; mutations lead to either dominant or recessive ARVD.

Dexrazoxane Antioxidant that prevents site Fe-based oxidative damage by chelating free iron; provides clinical cardioprotection against doxorubicin-induced oxidative damage.

DGGE Denaturing gradient gel electrophoresis, mutation screening technique.

DGS DiGeorge or velocardiofacial syndrome, characterized by an assortment of craniofacial defects, conotruncal heart abnormalities (PTA, TOF, and VSD), aortic arch defects, and hypoplasia most commonly caused by large-scale deletions at chromosome 22 q11.

DHF 2′,7′-Dichlorofluorescin, membrane-permeable fluorometric indicator of ROS levels (particularly in mitochondrial matrix).

DHT Dihydrotestosterone.

2-Dimensional electrophoresis Technique for separating proteins on the basis of their size and charge differences.

DIC Disseminated intravascular coagulation.

Differential display Technique used to identify genes that are differentially expressed; RNA from the samples being compared is reverse transcribed, and the cDNA is further amplified by use f random primers. Genes that are differentially expressed in the chosen samples can be identified by electrophoresis.

DISC Death-inducing signaling complex, a multiprotein complex involved in the extrinsic apoptotic pathway triggered by the binding of specific ligands to the death receptor.

DNA footprinting A technique that detects DNA–protein interaction knowing that a protein bound to DNA may protect that DNA from enzymatic cleavage.

DNA primary structure The specific nucleotide sequence from the beginning to the end of the molecule.

DORV Double outlet right ventricle; a rare group of cardiac anomalies characterized by both great arteries (pulmonary and aorta) arising primarily from the right ventricle. Patients with DORV have been reported with mutations in connexin (C×43) or with deletions in chromosome 22q11.

Double helix A molecular model of DNA made of two complementary antiparallel strands of the bases guanine, adenine, thymine, and cytosine, covalently linked through phosphodiester bonds. Each strand forms a helix, and the two helices are held together through hydrogen bonds, ionic forces, hydrophobic interactions, and van der Waals forces.

Doxorubicin Also called Adriamycin. Used to treat leukemia but also causes extensive mitochondrial defects, myocardial apoptosis, and induces cardiomyopathy.

DQAsomes Liposome-like vesicles formed in aqueous medium with a dicationic amphiphile dequalinium used as a mitochondrial-specific delivery system for gene therapy.

D-loop Noncoding regulatory region of mtDNA involved in controlling its replication and transcription.

EAM Experimental autoimmune myocarditis.

Early pre-conditioning Also called acute or classic pre-conditioning, results from brief periods of ischemia applied 1–2 h before the index ischemia, occurs within a few minutes after the initial stimulus, and lasts for 2–3 h.

EB Embryoid bodies, aggregations of embryonic stem cells that can differentiate spontaneously *in vitro* to a variety of cell types, including cardiomyocytes.

EC Endocardial cushions.

ECM Extracellular matrix.

ED-FRAP Enzyme-dependent fluorescence recovery after photobleaching used to measure the dehydrogenase activities associated with mitochondrial NADH generation.

Ees End-systolic elastance, a major determinant of systolic function and ventricular-arterial interaction.

Electroporation Method to transfect cells with either exogenous genes or proteins by use of an electrical field.

EMT Endothelial-mesenchymal transdifferentiation.

Endonucleases Enzymes that cleave DNA or RNA at specific sites within the polynucleotide chain.

Enhancer A regulatory, short region of DNA that can dramatically enhance specific gene expression, often lying outside the promoter.

ENOS Endothelial nitric oxide synthase.

EPC Endothelial progenitor cells.

Epigenetic Acquired and reversible modification of genetic material (e.g., methylation).

Episomes Genetic elements with extrachromosomal location (e.g., plasmids).

Epitope Part of a foreign organism or its proteins that is recognized by the immune system and targeted by antibodies, cytotoxic T cells, or both.

ER Endoplasmic reticulum. A membrane-bound cytosolic compartment where lipids and membrane-bound proteins are synthesized.

ErbB proteins Family of receptor tyrosine kinases, which mediate cell proliferation, migration, differentiation, adhesion, and apoptosis in numerous cell types, binding to a wide variety of ligands (e.g., EGF, neuregulins, TGF and HB-EGF) and contribute to regulating endocardial cushion remodeling and valve formation.

ERE Estrogen response element, specific regulatory sites within the DNA of target nuclear genes, which bound by ERs, produce long-term genomic effects of testosterone.

ERK Extracellular regulated kinase.

ER Estrogen receptor.

ES Embryonic stem cell.

Essential hypertension The most common form of hypertension for which there is no recognized primary cause.

EST Expressed sequence tags. A sequence fragment of a transcribed protein-coding or non-protein-coding DNA.

ET Endothelin, signaling peptides (ET-1, ET-2, and ET-3) modulate contractile function and growth stimulation of cardiomyocytes by binding specific G-protein–coupled receptors and triggering downstream signaling (e.g., DAG, IP3).

ETC Electron transport chain. A series of complexes in the mitochondrial inner membrane to conduct electrons from the oxidation of NADH and succinate to oxygen.

Euchromatin Less condensed chromosomal regions containing more active genes.

Exon Segment of a gene that remains after the splicing of the primary RNA transcript and contains the coding sequences as well as 5′ and 3 untranslated regions.

Exonucleases Enzymes that cleave nucleotides one at a time from an end of a polynucleotide chain.

Expression vector A vector that contains elements necessary for high-level and accurate transcription and translation of an inserted cDNA in a particular host or tissue.

FA Friedreich ataxia. An autosomal-dominant neuromuscular disorder with frequent HCM.

FACS Fluorescence-activated cell sorting, which can evaluate and isolate specific cell types and cell-cycle stages.

FAD Flavin adenine dinucleotide. Common coenzyme of dehydrogenases. In the ETC, FAD is covalently linked to SDH.

FADD Fas-associated via death domain; adaptor protein recruiting procaspase into the apoptotic-promoting complex DISC.

FADH$_2$ Flavin adenine dinucleotide (reduced form).

Familial ligand-defective apoB A common monogenic-dominant hypercholesterolemic disorder resulting from mutation and deficiency of apolipoprotein B, the major protein of LDL, which reduces its affinity and binding to LDLR.

FAO Fatty acid oxidation.

FasL Fas ligand; death ligand in extrinsic apoptotic pathway.

FCHL Familial combined hyperlipidemia; characterized by elevated levels of plasma triglycerides, LDL, and VLDL-cholesterol, or both is the most common discrete hyperlipidemia and a common cause of premature atherosclerosis.

FDA Fluorescein diacetate; a fluorochrome used to discriminate between necrosis and apoptosis in intact cardiomyocytes.

FDCM Familial dilated cardiomyopathy.

FDG-PET Fluoro-deoxy-glucose-positron-emission tomography; this glucose analogue is used in the evaluation of levels of myocardial bioenergetic substrates such as glucose by imaging with a PET scanner.

FGF Fibroblast growth factor.

FH Familial hypercholesterolemia is an autosomal-dominant disorder characterized by elevated cholesterol and premature CAD and is the result of mutations that affect the LDL receptor (LDLR).

FHCM Familial hypertrophic cardiomyopathy.

Fish eye disease A rare autosomal-dominant disorder affecting HDL levels because of a deficiency in lecithin: cholesterol acyltransferase (LCAT).

Frameshift mutation Changes in the reading frame resulting from either an insertion or deletion of nucleotides.

FRDA Gene for frataxin; a mitochondrial-localized protein. Mutations in FRDA are responsible for Friedreich ataxia (FA).

FRTA Free radical theory of aging.

FRET Fluorescence resonance energy transfer; technique useful in monitoring the fluctuations (and localization) of molecules (e.g., cyclic AMP) in living cells by use of fluorophores.

Functional genomics A branch of molecular biology that makes use of the enormous amount of data produced by genome sequencing to delineate genome function.

GATA Family of zinc finger–containing transcription factors that contribute to the activation of the cardiac-specific gene program involved in cardiac cell differentiation.

GC Guanylyl cyclase.

Gel shift Differential mobility during gel electrophoresis is used to gauge whether a specific DNA fragment (containing the regulatory motif of interest) is bound by an extract of nuclear proteins. The specific protein–DNA interaction is detectable by retardation in the mobility of the DNA fragment that is successfully bound.

Gene product The protein, tRNA, or rRNA encoded by a gene.

Gene transfection Introduction of DNA into eukaryotic cells.

Genetic code Correspondence between nucleotide triplets (codon) and specific amino acids in proteins.

Genome Total genetic information carried by a cell or an organism.

Genomic library Collection of DNA fragments (each inserted into a vector molecule) representative of the entire genome.

Genotype Genetic constitution of a cell or an organism.

GFP Green fluorescent protein. Useful marker for imaging localized proteins.

GH Growth hormone.

GIK Glucose, insulin, and potassium. Applied as a metabolic "cocktail" to provide beneficial preconditioning effects to injured myocardium.

GK rat Goto–Kakizaki rat, experimental model of type 2 diabetes.

GLP-1 Glucagon-like peptide 1, cardioprotective agent.

GLUT Glucose transporter.

Glycolysis Cytosolic-located metabolic pathway present in all cells catalyzing the anaerobic conversion of glucose to pyruvate.

Glycosylation An enzyme-directed site-specific process, resulting in the addition of carbohydrate residues to proteins and lipids.

Gordon's syndrome Also pseudohypoaldosteronism type II (PHA type II), a rare monogenic Mendelian trait characterized by familial hypertension with increased renal salt reabsorption, impaired K^+ and H^+ excretion, and low renin activity found to be due to mutations in the WNK gene.

GPCR G-protein–coupled receptors.

G-protein A heterotrimeric membrane-associated GTP-binding protein involved in cell signaling pathways; activated by specific hormone or ligand binding to a 7-helix transmembrane receptor protein.

GPx Glutathione peroxidase. An antioxidant enzyme with both mitochondrial and cytosolic isoforms.

GRK G-protein–regulated kinases.

GSH Reduced glutathione.

GSK-3B Glycogen synthase kinase 3B; negative regulator of cardiac hypertrophy and of both normal and pathological stress-induced growth.

GSSG Glutathione disulfide.

GTP Guanosine triphosphate.

HA Hyaluronic acid; a glycosaminoglycan composed of alternating glucuronic acid and N-acetylglucosamine residues, present in the ECM, to expand the extracellular space, regulate ligand availability, and direct remodeling events in the cardiac jelly.

HAGR Human Aging Genomic Resources.

HAND proteins Basic helix-loop-helix transcription factors (e.g., eHAND, dHAND) that play critical roles in early cardiac development.

HCM Hypertrophic cardiomyopathy.

HDL High-density lipoprotein.

Helicase Enzymes that separate the strands of DNA using energy derived from ATP hydrolysis.

Heterochromatin Condensed regions of chromosomes containing less active genes.

Heteroplasmy Presence of more than one genotype in a cell.

H-FABP Heart-type fatty acid–binding protein, an intracellular binding protein, with potential clinical use as an indicator of cardiac ischemia and necrosis.

5-HD 5-Hydroxydecanoic acid; selective mitoK$_{ATP}$ channel blocker.

HF Heart failure.

HGP Human Genome Project.

HiCM Histiocytoid cardiomyopathy.

HIF Hypoxia-inducible factor.

Histones Chief proteins of chromatin acting as spools around which nuclear DNA winds. They play a role in regulation of gene expression.

HLHS Hypoplastic left heart syndrome; a heterogeneous group of developmental abnormalities in which there is a small or absent left ventricle with hypoplastic mitral and aortic valves rendering the left ventricle nonfunctional with systemic outflow obstruction; seems to have a genetic component in its etiology.

HNE 4-Hydroxynonenal. A major product of endogenous lipid peroxidation.

H$_2$O$_2$ Hydrogen peroxide; a form of ROS and marker of oxidative stress.

HO-1 Heme oxygenase; antioxidant enzyme with cardioprotective function.

Homocysteine A reactive amino acid intermediate in methionine metabolism whose adverse effects include endothelial dysfunction with associated platelet activation and thrombus formation and accumulation of vascular atherosclerotic lesions.

Homoplasmy Presence of a single genotype in a cell.

HOP Homeodomain only protein; a small divergent protein that lacks certain conserved residues required for DNA binding, initiates gene expression early in cardiogenesis, and is involved in control of cardiac growth during embryogenesis and early prenatal development, and acts by transcription factor recruitment and chromatin remodeling.

HOS Holt–Oram syndrome; a rare inherited disease characterized mainly by upper limb and CHD resulting from mutations and haploinsufficiency of *TBX5*.

HRE Hypoxia response element.

HRT Hormone replacement therapy.

HSC Hematopoietic stem cells.

HSP Heat-shock protein. A family of chaperones involved in protein folding.

2-Hybrid system Method to detect proteins that interact with each other using yeast gene expression.

Hybridization Binding of nucleic acid sequences through complementary base pairing. The hybridization rate is influenced by temperature, G-C composition, extent of homology, and length of the sequences involved.

Hydrophobic Lipophilic. Insoluble in water.

IAA Interrupted aortic arch; an extremely rare CHD defined as the loss of luminal continuity between the ascending and descending aorta caused by a variety of genetic defects.

ICD Implantable cardioverter-defibrillator, implantation effective for treatment of short QT.

IFM Interfibrillar mitochondria.

IGF-1 Insulin-like growth factor; stimulates proliferative cardiomyocyte pathways and cell growth.

IMA Ischemia-modified albumin; indicator of early myocardial ischemia and ACS

IMAC Anion channel of the inner mitochondrial membrane.

Infective endocarditis Microbial infection of the endocardial surface of the heart, which commonly involves the heart valves.

***In situ* hybridization** Technique using DNA probes to localize specific transcripts within the cell in conjunction with microscopy.

Integral membrane protein Protein with at least one transmembrane segment requiring detergent for solubilization.

Integrins Class of transmembrane, cell-surface receptor molecules that constitute part of the link between the extracellular matrix and the cardiomyocyte cytoskeleton and that act as signaling molecules and transducers of mechanical force.

Intermembrane space Space between inner and outer membranes.

Intron A segment of a nuclear gene that is transcribed into the primary RNA transcript but is excised during RNA splicing and not present in the mature transcript.

Ion channels Multisubunit transmembrane protein complexes that perform the task of mediating selective flow of millions of ions per second across cell membranes and are the fundamental functional units of biological excitability.

Ionophore Small hydrophobic molecule that promotes the transfer of specific ions through the membrane bilayer.

IP3 Inositol trisphosphate; intracellular second messenger produced by phospholipase C (see PLC).

IPC Ischemic pre-conditioning.

Iron-sulfur center Nonheme iron ions complexed with cysteine chains and inorganic sulfide atoms making a protein capable of conducting electrons in electron transport or redox reactions.

Iroquois homeobox genes Family of homeodomain-containing transcription factor genes, implicated in cardiac chamber–specific gene expression.

I/R Ischemia/reperfusion.

Ischemic heart disease Also called coronary artery disease (CAD) and coronary heart disease, this condition is caused by narrowing of the coronary arteries, thereby causing decreased blood supply to the heart.

Isoforms Related form of the same protein generated by alternative splicing, transcriptional starts or encoded by entirely different genes.

Isoschizomers Pairs of restriction enzymes specific to the same DNA recognition sequence.

JC-1 Fluorometric dye used for measuring/imaging mitochondrial membrane potential.

Karyotype A snapshot of the number of chromosomes in the normal diploid cell, as well as their size distribution.

Kawasaki disease An acute self-limited vasculitis of infancy and early childhood has undetermined etiology and is the leading cause of acquired pediatric heart disease in the United States and Japan.

KCOs Potassium channel openers (e.g., nicorandil, diazoxide, and pinacidil); can mediate cardioprotection.

kDa Kilodalton; measure of mass with proteins.

KIR Potassium inward rectifier.

KLOTHO A single-pass transmembrane protein that functions in signaling pathways that suppress aging and that has beta-glucuronidase activity.

Knock-out mutation A null mutation in a gene, abolishing its function (usually in transgenic mouse); allows evaluation of its phenotypic role.

Krebs cycle Central metabolic pathway of aerobic respiration occurring in the mitochondrial matrix; involves oxidation of acetyl groups derived from pyruvate to CO_2, NADH, and H2O. The NADH from this cycle is a central substrate in the OXPHOS pathway. Also termed TCA or citric acid cycle.

KSS Kearns–Sayre syndrome. A mitochondrial neuropathy characterized by ptosis, ophthalmoplegia, and retinopathy with frequent cardiac conduction defects and cardiomyopathy.

LA Lipoic acid, a potent thiol antioxidant and mitochondrial metabolite, seems to increase low molecular weight antioxidants, decreasing age-associated oxidative damage.

LAD Left anterior descending.

LBBB Left bundle branch block.

LCAD Long-chain acyl CoA dehydrogenase involved in FAO.

LCFA Long-chain fatty acid.

LCHAD Long-chain 3-hydroxylacyl-CoA dehydrogenase.

LDL Low-density lipoprotein, a cholesteryl ester–rich particle (containing only apoB100) whose plasma levels are elevated in several monogenic disorders of lipoprotein metabolism and lead to atherosclerosis.

LDLR Low-density lipoprotein receptor, cell-surface receptor in liver or peripheral tissues responsible for LDL removal from blood; defective LDLR results in FH.

LHON Leber hereditary optical neuropathy.

Liddle's syndrome An autosomal-dominant monogenic form of hypertension with both hypokalemia and increased sodium reabsorption because of specific defects in either the β- or γ-subunit of the epithelial sodium channel (*ENaC*) causing gain-of-function of channel activation.

Ligand Any molecule that binds to a specific site on a protein or a receptor molecule.

Ligase Enzymes that join together two molecules in an energy-dependent process; involved in DNA replication and repair. Extensively used in recombinant DNA.

Liposomes Lipid spheres with a fraction of aqueous fluid in the center used as vectors for gene transfection with plasmid DNA or oligonucleotides.

LQT Long QT syndrome; prolongation of the Q-T interval is a significant cause of syncope and SCD in children; delayed or prolonged repolarization of the cardiac myocyte can be acquired (e.g., drugs) or congenital (e.g., mutations in specific ion channels).

LS LEOPARD syndrome; an autosomal-dominant syndrome characterized by multiple lentigines and cafe-au-lait spots, cardiac conduction abnormalities, obstructive cardiomyopathy, pulmonary stenosis, retardation of growth, and deafness, allelic to NS and caused by mutations in *PTPN11*, a gene encoding tyrosine phosphatase SHP-2.

LTA Lymphotoxin-alpha; an inflammation-mediating cytokine implicated in coronary artery plaque formation; polymorphic gene variants associated with MI.

Luciferase ATP-dependent photoprotein luciferase used to fluorometrically assess the specific organelle ATP levels.

LVAD Left-ventricular assist devices.

LVH Left-ventricular hypertrophy.

MAPCs Multipotent adult precursor cells.

MAPK Mitogen-activated protein kinases. A family of conserved serine/threonine protein kinases activated as a result of a wide range of signals involved in cell proliferation and differentiation; includes JNK and ERK.

Marfan syndrome This autosomal-dominant connective tissue disorder presents with skeletal, ocular, and cardiovascular abnormalities, including high neonatal mortality caused by polyvalvular involvement, aortic root, fatal aortic dissection, or aortic insufficiency and severe CHF.

Matrix Space enclosed by the mitochondrial inner membrane.

MBs Molecular beacons. Hairpin-forming oligonucleotides labeled at one end with a quencher and at the other end with a fluorescent reporter dye.

MCAD Medium-chain acyl-CoA dehydrogenase, FAO enzyme.

MCD Malonyl-CoA decarboxylase, an enzyme involved in regulation of malonyl CoA turnover.

MCM Mitochondrial cardiomyopathy.

MELAS Mitochondrial encephalomyopathy with lactic acidosis and strokelike episodes.

Membrane potential Combination of proton and ion gradients across the inner membrane making the inside negative relative to the outside.

MERRF Mitochondrial cytopathy including myotonus, epilepsy, and ragged-red fibers.

MetSyn Metabolic syndrome.

MHC Myosin heavy chain.

MI Myocardial infarction.

Microarray A range of oligonucleotides immobilized onto a surface (chip) that can be hybridized to determine quantitative transcript expression or mutation detection.

Mineralocorticoid-induced hypertension A monogenic autosomal-dominant form of an early-onset hypertension, markedly exacerbated during pregnancy caused by mutations in the *MR* hormone–binding domain.

Minisatellites Repetitive and variable DNA sequences, generally GC-rich, ranging in length from 10 to more than 100 bp.

Missense mutation Mutation that causes substitution of one amino acid for another.

MitoK$_{ATP}$ channel Activation of the ATP-sensitive inner membrane mitoKATP channel has been implicated as a central signaling event (both as trigger and end effector) in IPC and other cardioprotection pathways.

MitoQ Synthetic ubiquinone analog that can be selectively targeted to mitochondria used to provide antioxidant cardioprotection.

Mito VitE Synthetic analog of vitamin E that can reduce mitochondrial lipid peroxidation and protein damage and accumulate after oral administration at therapeutic concentrations within the cardiac tissue.

MLA Monophosphoryl lipid A, a nontoxic derivative of the endotoxin pharmacophore lipid A, cardioprotective agent.

MLC Myosin light chain.

MLP Muscle LIM protein, localized in the cardiomyocyte cytoskeleton, a positive regulator of myogenic differentiation.

MMPs Metalloproteinases; enzymes involved in extracellular matrix remodeling.

MOMP Mitochondrial outer-membrane permeabilization; an apoptotic event in part mediated by binding of pro-apoptotic proteins (e.g., Bad, Bax, Bid) to mitochondria.

Mobile carrier Small molecule shuttling electrons between complexes in the ETC.

Modifier gene A gene that modifies a trait encoded by another gene.

Motif homology searching Search for patterns between proteins, which can prove highly informative about the structural and functional properties of the encoded protein.

MPO Myeloperoxidase, indictor of pathological inflammation and for risk of ACS.

MR Mineralocorticoid receptor.

mRNA Messenger RNA. Specifies the amino acid sequence of a protein; translated into protein on ribosomes. Transcripts of RNA polymerase II.

MPG N-2-mercaptopropionylglycine, a free radical scavenger.

MSC Mesenchymal stem cells.

MT Metallothionein. An inducible antioxidant metal-binding protein with cardioprotective properties.

mtDNA Mitochondrial DNA.

mtTFA Mitochondrial transcription factor A (also called *TFAM*).

MTOR Mammalian target of rapamycin.

MTP Mitochondrial trifunctional protein, part of mitochondrial FAO.

Mutation Change occurring in the genetic material (usually DNA or RNA).

MVH Midventricular hypertrophy. An unusual pattern of hypertrophy in which papillary muscle hypertrophy leads to obstruction of the midventricular cavity.

MVP Mitral valve prolapse.

NADH Nicotinamide adenine dinucleotide (reduced form).

NARP Neuropathy, ataxia, retinitis pigmentosa.

ND1 One of seven ND subunits in mtDNA encoding complex I.

NER Nucleotide excision repair.

Neuregulin Family of endocardial-expressed peptide growth factors acting as potent ligands for tyrosine kinase receptors (ErbBs) and shown to promote growth and differentiation of embryonic cardiomyocytes, epithelial, and glial cells.

NFAT Nuclear factor of activated T cells, a family of transcription factors controlled by the Ca^{++}-regulated phosphatase, calcineurin.

NFATc A member of the NFAT family, exclusively expressed on the endocardium that plays a prominent role in the morphogenesis of the semilunar valves and septa.

NF-κB Nuclear factor-kappa B. Family of transcription factors involved in the control of a number of normal cellular and organismal processes, including immune and inflammatory responses, developmental processes, cellular growth, and apoptosis.

NO Nitric oxide; vasodilator.

Nonmendelian inheritance Cytoplasmic inheritance caused by genes located in mitochondria.

Nonsense codon Any one of three triplets (UAG, UGA, UAA) that cause termination of protein synthesis (same as stop codon).

Nonsense mutation Change in DNA specifying replacement of an amino acid codon by a nonsense codon.

Northern blot Molecular technique by which RNA separated by electrophoresis is transferred and immobilized for the detection of specific transcripts by hybridization with labeled probe.

NOS Nitric oxide synthase.

Notch A receptor family mediating an evolutionarily conserved signaling pathway involved in cell fate specification. Mutations in Notch signaling regulator have been implicated in aortic valve disease.

NRF-1 and NRF-2 Nuclear respiratory factors. Transcription factors that modulate expression of nuclear DNA encoded mitochondrial proteins.

NS Noonan syndrome. An autosomal-dominant disorder characterized by craniofacial dysmorphia, short stature, and cardiovascular defects, including pulmonary valvular stenosis, HCM, and ASD, allelic to LS and caused by mutations in *PTPN11*, a gene encoding tyrosine phosphatase SHP-2.

nt Nucleotide; the basic unit of DNA composed of a purine or pyrimidine base, a sugar (deoxyribose), and a phosphate group.

NT-proBNP N-terminal pro-brain natriuretic peptide; indicator of hemodynamic stress and for risk of congestive HF.

NTG Nitroglycerin.

Nucleases Enzymes that catalyze the degradation of DNA (DNAse) or RNA (RNAse); specific nucleases have been identified that target either the 5' or 3' ends of DNA (exonuclease) or that can digest nucleic acids from internal sites (endonucleases).

Null mutation Ablation or knock-out of a gene.

NKX2.5 An NK-class homeodomain-containing transcription factor homologous to *Drosophila melanogaster tinman* gene is crucial in cardiac differentiation processes, including the establishment or maintenance of a ventricular gene expression program. Although not essential for the specification of heart cell lineage, or for heart tube formation, it is required for completion of the looping of the heart.

OH Origin of replication for mtDNA, heavy strand.

OL Origin of replication for mtDNA, light strand.

Oligomycin Specific inhibitor of mitochondrial ATP synthase and OXPHOS.

Oligonucleotide Short polymer of DNA or RNA that is usually synthetic in origin.

ORF Open reading frame sequence that can be translated into a protein. It contains a start codon and a stop codon.

ORI Origin of replication. Unique DNA sequence at which DNA replication is initiated; from this point replication may proceed either bidirectionally or unidirectionally.

OS Oxidative stress.

oxLDL Oxidized LDL, a primary substrate for macrophage activation and involved in atherosclerosis progression.

OXPHOS Oxidative phosphorylation. A process in mitochondria in which ATP formation is driven by electron transfer from NADH and FADH2 to molecular oxygen and by the generation of a pH gradient and chemiosmotic coupling.

PAAs Pharyngeal arch arteries.

PAGE Polyacrylamide gel electrophoresis.

PAI-1 Plasminogen activator inhibitor-1; a principal regulator of fibrinolysis.

PAR Protease-activated receptors; these G-coupled transmembrane receptors are activated by extracellular proteolytic cleavage by serine proteases such as thrombin and trypsin.

Paraoxonase Antioxidant enzyme.

PARP Poly (ADP-ribose) polymerase.

PCR Polymerase chain reaction. An amplification of DNA fragments using a thermostabile DNA polymerase and paired oligonucleotide primers subjected to repeated reactions with thermal cycling.

PDA Patent ductus arteriosus; a relatively common CHD that results when the ductus arteriosus, a muscular artery connecting the pulmonary artery to the descending aorta, fails to remodel and close after birth, resulting in a left-to-right shunt.

PDGF Platelet-derived growth factor.

PDH Pyruvate dehydrogenase (Also **PDC**).

PDK-1 Phosphoinositide-dependent kinase 1; enzyme downstream of PIP3 production in the PI3K pathway that becomes activated in part by its translocation to the plasma membrane and proximity to its substrates, which include Akt (PKB).

Penetrance The proportion of individuals with a specific genotype expressing the related phenotype.

Peptide Short polymer of amino acids that can be produced synthetically.

Peripheral membrane protein Protein associated with membrane by means of protein–protein interactions; solubilized by changes in pH or salt.

Peroxisome Small membrane-bounded organelle that uses oxygen to oxidize organic molecules, including fatty acids, and contains enzymes that generate and degrade hydrogen peroxide (H_2O_2) (e.g., catalase).

pFOX inhibitor Partial fatty acid oxidation inhibitors; by partially blocking FAO and increasing glucose utilization, these agents ameliorate hemodynamic and LV function in HF and myocardial ischemia.

PGC-1α Peroxisome proliferator-activated receptor γ co-activator. Transcriptional regulator of mitochondrial bioenergetic, and biogenesis operative during physiological transitions.

Phage display Technique involving fusing proteins with a bacteriophage coat protein resulting in the display of the fused protein on the exterior surface of the phage, whereas the DNA encoding the fusion protein resides within; this permits the selection of proteins (and their physically attached DNAs) for specific binding characteristics (e.g., antibody, DNA, or specific ligand-binding) or functional features (e.g., enzymatic assay) by an *in vitro* screening process.

Pharmaceutical pre-conditioning A large variety of drugs, including the targeted use of volatile anesthetics, potassium channel openers, nitric oxide donors, and modulators of downstream pathways including erythropoietin, statins, insulin, and pyruvate, have been shown to mimic ischemic pre-conditioning and provide cardioprotection when either substituted for the pre-conditioning period or applied at reperfusion.

Pharmacogenetics Study of the role of inheritance in interindividual variation in drug response.

Pharmacogenomics A branch of pharmaceutics dealing with the influence of genetic changes on drug response by correlating gene expression or SNPs with the drug's effect.

Phenotype Observable physical characteristics of a cell or organism resulting from the interaction of its genetic constitution (genotype) with its environment.

Phospholamban Negative regulator of SERCA.

Phosphorothioate-modified antisense oligonucleotides Substitution of sulfur for one of the oxygens in the phosphate backbone renders the oligonucleotide more stable to nuclease degradation.

PIP3 Phosphatidylinositol 3,4,5-triphosphate; product of PI3K activity.

PI3K Phosphatidylinositol 3-kinase.

PKA Protein kinase A. Activated by cAMP.

PKB Protein kinase B; also called *Akt*.

PKC Protein kinase C.

Plakoglobin Cytoplasmic protein of catenin family, a key component of desmosomes and adherens junctions. Mutant forms can promote ARVD and Naxos disease.

Plakophilin Desmosomal protein with mutational variants causing ARVD.

Plasmid A relatively autonomous replicating nonchromosomal DNA molecule primarily found in bacteria, which can be used as a vector for transferring recombinant genes to cells or tissues.

PLC Phospholipase C; a potent effector enzyme catalyzing the hydrolysis of inositol phospholipids and production of second messengers such as IP3 and DAG in response to activation by agonists (e.g., acetylcholine) binding to membrane-bound receptors in concert with G-proteins. This signaling promotes a downstream increase in intracellular Ca^{++} levels and PKC activation, which modulate myocardial contraction

Pleiotropic mutation A single mutation with multiple (often unrelated) effects on organism.

PNA Peptide nucleic acids; an alternative delivery system for nucleic acids to mitochondria.

POBA Plain old balloon angioplasty.

Polyadenylation Addition of a sequence of polyadenylic acid (poly A residues) to the 3′ end of most messenger RNAs after its translation.

Polygenic A large number of genes each contributing a small amount to the phenotype.

Polγ Nuclear-encoded catalytic subunit of mtDNA polymerase γ.

Porin Pore-forming protein in the outer mitochondrial membrane (see VDAC).

Positional cloning A technique used to identify polymorphisms that flank the mapped allele, as well as to map and to clone mutant alleles that are not tagged for rapid cloning.

Post-conditioning A series of brief interruptions of reperfusion applied at the very onset of reperfusion can reduce infarct size and apoptosis and provide cardioprotection.

Post-translational modification Post-synthetic modification of proteins by glycosylation, phosphorylation, proteolytic cleavage, or other covalent changes involving side chains or termini.

PPARs Peroxisome proliferator-activated receptors. Nuclear receptor transcription factors that function as transcriptional regulators in a variety of tissues, including the heart.

PPH Primary pulmonary hypertension, also called pulmonary arterial hypertension, (PAH). A rare autosomal-dominant disease with incomplete penetrance characterized by distinctive changes in pulmonary arterioles that lead to increased pulmonary artery pressures, right ventricular failure, and death that can be acquired or congenital arising most commonly from mutations in BMPR2, ALK1, or ENG (encoding endoglin).

Primer Short nucleotide sequence that is paired with one strand of DNA and provides a free 3′-OH end at which a DNA polymerase starts the synthesis of a nascent chain.

Promoter Noncoding regulatory region of DNA sequence upstream of the gene coding sequences involved in the binding of RNA polymerase to initiate transcription.

Protein kinase Enzyme that transfers the terminal phosphate group of ATP to a specific amino acid of a target protein.

Proteome Entire complement of proteins contained within the eukaryotic cell.

Pseudogenes Nonfunctional genes that are likely relics of evolution with strong homology to functional gene.

p70S6K 70-kDa Ribosomal protein S6 kinase. It plays a key role in translational control of cell proliferation in response to growth factors in mammalian cells.

PTA Persistent truncus arteriosus.

PT pore Permeability transition pore. A nonspecific megachannel in the mitochondrial inner membrane.

PUFA Polyunsaturated fatty acids.

QTL Quantitative trait loci for blood-pressure regulation have been identified in rat studies by genome-wide scanning and linkage studies in the rat.

RA Retinoic acid; plays role in cardiomyocyte differentiation.

RACKs Receptors for activated C kinase.

Ras A small G-protein (see small G-proteins).

Ras-GTP Activated Ras.

RBBB Right bundle-branch block.

RCM Restrictive cardiomyopathy.

Reading frame Contiguous and nonoverlapping set of three nucleotide codons in DNA or RNA used to predict amino acid sequence.

Real-Time PCR Quantitative PCR technique uses simultaneous DNA amplification and quantification often using fluorescent dyes that intercalate with double-stranded DNA and modified DNA oligonucleotide probes) that fluoresce when hybridized with a complementary DNA.

Recombinant DNA An artificial DNA sequence resulting from the combination of two DNA sequences in a plasmid.

Redox reactions Oxidation-reduction reactions in which there is a transfer of electrons from an electron donor (the reducing agent) to an electron acceptor (oxidizing agent).

Remote conditioning Pre-conditioning, which is not confined to one organ but also limits infarct size in remote, non-pre-conditioned organs.

Reporter gene A gene that is attached to another gene or regulatory element to be identified in cell culture, animals or plants.

Restenosis The re-closing or re-narrowing of an artery after an interventional procedure such as angioplasty or stent placement.

Restriction endonucleases Endonucleases that recognize a specific sequence in a DNA molecule (usually palindromic) and cleave the DNA at or near that site.

RFLP Restriction fragment-length polymorphism. A variation in the length of restriction fragments because of presence or absence of a restriction site.

RGS proteins Regulators of G-protein–signaling proteins; a family of proteins that accelerate intrinsic GTP hydrolysis on α-subunits of trimeric G proteins and play crucial roles in the physiological regulation of G-protein–mediated cell signaling.

Rhodamine 123 A fluorescent dye used to stain mitochondria in living cells.

Rhod-2AM Fluorescent calcium indicator used to assess Ca^{++} uptake, localization, and levels in cardiomyocytes.

Ribosome A factory-like organelle that builds proteins from a set of genetic instructions. Composed of rRNA and ribosomal proteins, it translates mRNA into a polypeptide chain.

Ribozyme RNA molecule with endonucleolytic activity, which can be used to selectively target specific gene expression.

RNAi RNA interference; use of a specific double-stranded RNA (dsRNA) construct to post-transcriptionally silence specific gene expression.

RNA polymerase Enzyme responsible for transcribing DNA as template into RNA.

RNS Reactive nitrogen species.

ROS Reactive oxygen species, including superoxide, hydroxyl radicals, and hydrogen peroxide.

Rotenone Specific inhibitor of complex I activity.

RRF Ragged red fiber.

rRNA Ribosomal RNA. A central component of the ribosome.

RTK Receptor tyrosine kinase; this large family of proteins includes receptors for many growth factors and insulin; ligand binding results in dimerization and phosphorylation of downstream signaling targets, as well as autophosphorylation.

RT-PCR Reverse transcription (RT) of RNA to DNA with the enzyme reverse transcriptase can be combined with traditional PCR to allow the amplification and determination of the abundance of specific RNA.

RXR Retinoid X receptor. On binding 9-cis retinoic acid, RXR acts as a heterodimer and as a repressor or activator of specific gene transcription, playing a key role in cardiac development and physiological gene expression.

Ryanodine receptor Major SR Ca^{++} release channel in cardiac muscle; mutations in the cardiac isoform encoded by RyR2 result in ARVD and CVPT.

SAGE Serial analysis of gene expression. Quantitative analysis of RNA transcripts by use of short sequence tags to generate a characteristic expression profile.

SAME Syndrome of apparent mineralocorticoid excess. A rare, autosomal-recessive form of hypertension presenting in infancy or childhood as an acquired syndrome resulting from the inhibition of 11 β-hydroxysteroid dehydrogenase, an enzyme responsible for converting cortisol to cortisone or from mutations in the renal-specific isoform gene for 11 β-hydroxysteroid dehydrogenase (11-β *HSD*).

SA node The sinoatrial node is a group of specialized cells in the top of the right atrium that produces electrical impulses (a relatively simple action potential) that travel down to eventually reach the ventricular muscle causing the heart to contract and serving as the "natural" pacemaker of the heart.

SCAD Short-chain acyl-CoA dehydrogenase, enzyme involved in FAO.

SCD Sudden cardiac death; death resulting from cardiac arrest.

SD Sudden death.

SDH Succinate dehydrogenase. A TCA cycle enzyme associated with respiratory complex II.

SDDS Sudden dysrhythmia death syndrome.

SDS Sodium dodecyl sulfate. An ionic detergent used for the solubilization, denaturation of proteins, and their size separation in PAGE.

Septal defect A hole in the wall of the heart separating the atria (ASD) or in the wall separating the ventricles (VSD).

SERCA Sarcoplasmic reticulum Ca^{++}-ATPase. There are three major isoforms that are variably expressed in different muscle types.

SERM Selective estrogen receptor modulators.

SHR Spontaneously hypertensive rats.

Sick sinus syndrome The failure of the sinus node to regulate the heart's rhythm.

Signal sequence Amino acid sequence for targeting proteins into specific organelles (e.g., mitochondria, nucleus) often, but not invariably, located at N-terminus.

Silent mutation Mutation that alters a particular codon, but not the amino acid and does not affect protein or phenotype.

SINES Single interspersed elements. Family of repeated sequence present in the human genome.

siRNA Small interfering RNA. Sometimes known as short interfering RNA, they are a class of 20–25 nucleotide-long RNA molecules that interfere with the expression of genes.

Sitosterolemia A rare autosomal-recessive hypercholesterolemic disorder caused by mutations in either sterolin-1 or -2, members of the ABC transporter family.

Smac/Diablo Mitochondrial intermembrane protein released into the cytosol during early apoptosis stimulating caspase activation.

Smads A group of cytoplasmic signaling proteins that upon phosphorylation in response to BMP, TGF-β, and BMPR signaling, translocate to the nucleus and directly regulate gene transcription.

Small G-proteins Superfamily of guanine nucleotide–binding proteins, including Ras, Rho, Rab, Ran, and ADP ribosylation factor(s) that act as molecular switches to regulate cardiac myocyte hypertrophy and survival associated with cell growth and division, cytoskeletal events, vesicular transport, and myofibrillar apparatus. As with heterotrimeric G-proteins, they are activated by exchange of GDP to GTP and inactivated by return to a GDP-bound state but not mediated by agonist-occupied receptors, rather primarily mediated by activation of guanine nucleotide exchange factors (GEFs).

SMC Smooth muscle cell.

Snail/Slug factors A group of zinc-finger transcription factors that primarily act as transcriptional repressors.

S-nitrosylation A ubiquitous post-translational modification involving the covalent attachment of NO to cysteine thiol moieties on targeted proteins.

SNP Single nucleotide polymorphism.

SOD Superoxide dismutase. An antioxidant ROS-scavenging enzyme with both mitochondrial (MnSOD) and cytosolic (CuSOD) isoforms.

Southern blot Detection of separated restriction fragments after size separation on agarose gels, transfer to membranes, and hybridization with labeled gene probes.

SP cells Side population cells; rare groups of multipotent progenitor cells capable of proliferation and differentiation.

SPECT Single-photon emission computed tomography used to assess myocardial metabolism and screen for CAD.

SPI Serine protease inhibitors, termed serpins, are key regulators of numerous cardiovascular pathways that initiate inflammation, coagulation, angiogenesis, apoptosis, extracellular matrix composition, and complement activation responses.

Splicing Reaction in the nucleus in which introns are removed from primary nuclear RNA and exons joined to generate mRNA.

SR Sarcoplasmic reticulum. A network of internal membranes in muscle-cell cytosol that contains high Ca^{++} concentration, which is released on excitation.

SRF Serum response factor, a transcription factor required for the formation of vertebrate mesoderm leading to the origin of the cardiovascular system.

SSCP Single strand conformation polymorphism; technique of mutation detection.

SSM Subsarcolemmal mitochondria.

Statins HMG-CoA reductase inhibitors used to treat patients with elevated plasma LDL.

STEMI ST-segment elevation myocardial infarction.

Sticky ends Results from restriction endonuclease enzymes making a staggered cut across the two strands of DNA with one of the two molecules containing a short single-stranded unique overhang; useful in recombinant DNA engineering.

Stop codon See nonsense codon.

Supravalvular aortic stenosis Discrete narrowing of the ascending aorta resulting from mutations in the gene encoding a component of the extracellular matrix (i.e., elastin).

SUR Sulfonylurea receptor.

SVT Supraventricular tachycardia.

Syntenic homology The presence of two or more genetic loci on the same chromosome. Also refer to the similarity in content and organization between chromosomes of different species for example.

TAG Intracellular triacylglycerols.

Taq polymerase Thermostable DNA polymerase isolated from the bacterium *Thermus aquaticus* used extensively in PCR.

T box genes Family of highly conserved transcription factors involved in early embryonic cell fate decisions, regulatory development of extraembryonic structures, embryonic patterning, and many aspects of organogenesis.

TCA cycle Tricarboxylic acid cycle (see Krebs cycle).

TCE Trichloroethylene; a halogenated hydrocarbon associated with increased CHD incidence in animal studies and in pregnant women exposed to TCE-contaminated water.

TD Tangier disease; a rare monogenic autosomal-co-dominant atherosclerotic disease characterized by the absence of HDL and very low plasma levels of apoA. It is caused by mutations in the ATP binding cassette transporter gene (*ABCA1*).

TdP *Torsade de pointes*; a polymorphic ventricular tachycardia that can be followed by syncope and SD; this can be acquired by exercise (swimming) or be congenital (any of the LQTs).

Telomere Special structure containing tandem repeats of a short G-rich sequence present at the end of a chromosome.

Telomerase An enzyme that recognizes the G-rich strand and elongates it using an RNA template that is a component of the enzyme itself.

TERC Telomerase RNA component.

TERT Telomerase reverse transcriptase catalytic subunit.

TF Tissue factor; protein present in endothelium, platelets, and leukocytes is crucial for thrombin formation.

TGA Transposition of the great arteries, the most common cyanotic neonatal lesion has shown in several cases a monogenic inheritance pattern.

TGCE Temperature gradient capillary electrophoresis; a sensitive technique coupling heteroduplex analysis with capillary electrophoresis used to efficiently scan an entire coding region to identify a wide spectrum of mutations.

TGF Transforming growth factor.

TH Thyroid hormone (also thyroxin); a stimulus of cardiac hypertrophic growth and myocardial mitochondrial biogenesis.

Thrombospondins Family of extracellular matrix glycoproteins with a role in platelet adhesion, modulation of vascular injury, coagulation, and angiogenesis and MI.

Titin Large polypeptide, anchored in the Z-disc spanning the sarcomere contributes to sarcomere organization, myofibrillar elasticity, and myofibrillar cell signaling.

TLR Toll-like receptors involved in the innate immunity signaling response of the macrophage, including pattern recognition of pathogens and oxidized LDL, leukocyte recruitment, and production of local inflammation and downstream signaling in atherosclerotic progression.

TNF-α Tumor necrosis factor-α.

TIM Protein complex in mitochondrial inner membrane required for protein import.

TIMP Tissue inhibitor of the metalloproteinase.

TOF Tetralogy of Fallot; most common form of complex CHD featuring VSD, obstructed right ventricular outflow, aortic dextroposition and right ventricular hypertrophy.

TOM Protein complex in mitochondrial outer membrane required for protein import.

Topoisomerases Enzymes that change the supercoiling of DNA.

T3 Triiodothyronine, active form of TH.

TPA Tissue-type plasminogen activator, a primary regulator of fibrinolysis.

TR Thyroid hormone receptor; mediates both nuclear genomic effects of TH (largely as a transcription factor) and nongenomic effects of TH.

***Trans*-acting elements** Regulatory elements that mediate specific gene expression. They are not located within or near the gene (e.g., proteins that bind and regulate specific promoters).

Transgenesis Introduction an exogenous gene—called a transgene—into a living organism so that the organism will exhibit a new property and transmit that property to its offspring.

Transgenic animal Animal that has stably incorporated one or more genes from another cell or organism.

Transcript Specific RNA product of DNA transcription.

Transcription factor Protein required for the initiation of transcription by RNA polymerase at specific sites and functioning as a regulatory factor in gene expression.

Transcriptome Comprehensive transcript analysis for expression profiling.

Translation Synthesis of protein from the mRNA template at the ribosome.

Transposons Sequences of DNA that can move/transpose around to different locations within the genome of a cell.

Triplet Repeat syndromes Inherited neuromuscular disorders caused by expanded repeats of trinucleotide sequences within specific genes including Friedreich ataxia (FA) and myotonic muscular dystrophy (MMD).

TRF2 Telomere repeat-binding factor; telomere-associated protein critical for the control of telomere structure and function.

tRNA Transfer RNA. A small RNA molecule used in protein synthesis as an adaptor between mRNA and amino acids.

TS Timothy syndrome (also termed LQT8); a multiorgan disease that includes lethal cardiac dysrhythmias, webbing of fingers and toes, congenital heart disease, immune deficiency, intermittent hypoglycemia, cognitive abnormalities, and autism. It is due to a missense mutation (G406R) in the *CACNA1c* gene, encoding the L-type Ca^{++} α-subunit Cav1.2.

TTP Thrombotic thrombocytopenic purpura; an autosomal-recessive relapsing form of severe thrombotic microangiopathy characterized by marked thrombocytopenia, systemic platelet aggregation, erythrocyte fragmentation, and organ ischemia and caused by mutations in the *ADAMTS13* gene.

UCP Uncoupling protein.

Uncoupler Protein or other molecule capable of uncoupling electron transport from oxidative phosphorylation.

UPA Urokinase-type plasminogen activator.

Vasculogenesis The *de novo* formation of the first primitive vascular plexus and post-natal vascularization.

VCAM-1 Vascular cell adhesion molecule.

VDAC Voltage-dependent anion channel in mitochondrial outer membrane (see porin).

VDR Vitamin D receptor; involved in signaling cardiac morphogenesis.

VEGF Vascular endothelial growth factor.

Versican An ECM-localized chondroitin sulfate proteoglycan that binds HA is expressed in the pathways of neural crest cell migration and in pre-chondrogenic regions and has been associated with valvulogenesis in the developing heart.

VGCC Voltage-gated calcium channels.

VLCAD Very long-chain acyl CoA-dehydrogenase; enzyme involved in mitochondrial β-oxidation of fatty acids.

VLDL Very low-density lipoprotein; a triglyceride-rich lipoprotein containing apoB100 that progressively becomes enriched in cholesteryl ester (CE) as a result of CE transfer from HDL and is converted by lipolysis to LDL and/or taken up as VLDL remnants by the liver.

von Willebrand disease Most common human congenital bleeding disorder, in which both quantitative and qualitative abnormalities of the vWF glycoprotein have been implicated.

vWF von Willebrand factor; blood glycoprotein involved in coagulation.

Western blot Immunochemical detection of proteins immobilized on a filter after size separation by PAGE.

Wild type The common genotype or phenotype of a given organism occurring in nature.

Williams syndrome A rare autosomal-dominant disorder characterized by supravalvular aortic stenosis and stenoses of systemic and/or pulmonary arteries; it has been associated with a large-scale deletion in chromosome 7 (including the elastin gene).

WPW Wolff–Parkinson–White syndrome presents with hypertrophic cardiomyopathy, ventricular preexcitation, conduction defects, and accumulation of cardiac glycogen.

X-linked inheritance pattern The presence of the gene of interest on the X-chromosome.

XO Xanthine oxidase, cytosolic enzyme involved in purine metabolism, involved in myocardial ROS production (e.g., superoxide radicals) particularly after I/R injury.

YAC Yeast artificial chromosomes; shuttle vectors designed to contain yeast chromosomal elements (i.e., centromeric and telomeric sequences) that segregate as chromosomes and allow the incorporation of very large inserts of heterologous DNA.

Z-discs Cardiomyocyte component positioned at the junction between the cytoskeleton and the myofilaments, providing a physical connection between the sarcomere, nucleus, membrane, and sarcoplasmic reticulum (SR) with a role in cardiac contraction and signaling.

Index

A

A. *See* Adenine
AAV. *See* Adenovirus-associated virus
Abiotrophia, 425t
AC. *See* Adenylyl cyclase
ACC. *See* Acetyl-CoA carboxylase
ACE. *See* Angiotensin-converting enzyme
Acetylases, 92, 103
Acetylcholine, 283t
Acetyl-CoA carboxylase (ACC), 475, 476
Acetyl-l-carnitine (ALCAR), 605
Acquired cardiac diseases, 188–189
Acquired dysrhythmias. *See* Cardiac dysrhythmias, acquired
ACS. *See* Acute coronary syndromes
ACS cells. *See* Adult cardiac stem cells
ACTC. *See* Cardiac actin, gene
Action potential (AP), 514, 543
    ion currents and ventricular, 514–515
        ventricular myocyte's, 515f
            five phases in, 514–515
Acute coronary syndromes (ACS), 262t, 263, 267, 268t, 269
    biomarkers in myocardial ischemia and, 619–621, 620t
Acute myocardial infarction (AMI), 281
    negative vascular modulators and, 327, 332
Acyl-CoA synthetase, 474, 475f
ADAMs, 453
Adenine (A), 3
Adenosine, 283t
Adenosine triphosphate (ATP), 287
    glucose metabolism and, 473, 477
    IPC and levels of, 287, 288
Adenoviral vectors, 28, 29t
Adenovirus-associated virus (AAV), 28, 29t
Adenoviruses, 415, 416, 417, 420, 421
Adenylyl cyclase (AC), 83–84
    isoform of, 83
Adrenergic receptors, 77–79
    $\beta 1$-, 77, 78, 103
        stimulation of, 77, 78, 78f, 103
    $\beta 2$-, 78, 103
    cardiac aging and, 583
Adrenoreceptors (ARs), 77–79. *See also* G-protein-coupled receptors
    $\alpha 1$-, 78, 103
        seven transmembrane-spanning domain of, 78, 104
    $\alpha 2$-, 78–79, 103
    GPCRs and, 103
Adult cardiac stem (ACS) cells, 62
    markers of, 63t
    transplantation of, 62–63
        advantages/limitations in, 62–63, 65t
AF. *See* Atrial fibrillation
Affinity tag, 14
Age, MI and, 262f
Aging. *See also* Cardiac aging
    cardiac aging and, 605–606
Aging heart. *See* Cardiac aging
AGT. *See* Angiotensinogen
AKAPs (specific PKA anchoring proteins), 84, 85
Akt, 85–87
    isoforms of, 85
    signaling, 86f
    stimuli activating myocardial, 86t
Alagille syndrome, 167, 167t, 168
ALCAR. *See* Acetyl-l-carnitine
Alcohol/substance abuse
    alcoholic cardiomyopathy and, 396
    cardiac development and, 147
    cocaine and, 147, 420t
    fetal alcohol syndrome in, 147
    hypertension and, 341, 353
Alleles, 9
    codominant, 9
    dominant, 9
    polymorphic, 9
    recessive, 9
Allele-specific oligonucleotides (ASO), 21
ALOX5AP. *See* 5-lipoxygenase activating protein
AMI. *See* Acute myocardial infarction
Amino-terminus, 7
AMISTAD clinical trial, 302, 626
AMP-activated protein kinase (AMPK)
    ACC activity and, 476
    fatty acid transporters and, 476
    glycogen turnover and, 479, 480
    signaling, 474
Amphetamines, 420
AMPK. *See* AMP-activated protein kinase
Andersen syndrome, 523, 529
Androgens, 558–559
    CVDs and, 558–559
    genomic/non-genomic effects of, 558–559
Androstenedione, 556f

Anesthetics
    cardioprotection and, 293–294
    pre-conditioning signaling events and, 293t
ANF. *See* Atrial natriuretic factor
Ang II. *See* Angiotensin, II
Ang-1. *See* Angiopoietin-1
Angioblasts, 316, 318, 322f, 332
    bFGF, VEGF-R2 and, 332
Angiogenesis. *See also* Cardiac neovascularization
    activators of, 319t
    arteriogenesis *v.* vasculogenesis *v.*, 322f
    definition of, 315
    dysregulators of, 327–328
    gene therapy for, 36
    genes/gene products, and modulatory factors in, 316, 318–322
    inhibitors of, 326–327, 326t
    pathway of, 317f
        key molecular factors in, 317f
    therapeutic, 331
    vascular remodeling during, 325–326
Angiopoietin-1 (Ang-1), 316, 318, 319t, 321, 332
    VEGF, Notch pathway, TGF and, vii
Angiotensin, 77, 79
    II, 79, 103
Angiotensin-converting enzyme (ACE), 79
    CAD and, 216t
    MI and, 268t, 269, 270
Angiotensinogen (AGT), 261, 268t, 273
Animal models
    cardiac neovascularization and transgenic mouse, 328–329
    CVDs and, 40t
    CVDs and gender in, 560–562, 561t
    DCM and transgenic, 385, 386t
    dysrhythmias/conduction defects and, 536, 537t, 538t
    endocarditis and, 429
    fatty acid/glucose metabolism defects and transgenic mice, 485–487
    genetic HCM and, 372–373
    HF and, 442–443
    HF and transgenic, 446–447, 446t, 447t, 461
    hypertension and, 346–347, 354
    L-R identity and, 175t
    MI and, 272
    myocarditis and, 421–423
    transgenic expression system and, 15t, 20f
    transgenic technologies and, 165, 168, 169, 170, 179, 192
    VEGF and transgenic overexpression in, 315, 316, 318, 319t, 321, 328, 332
Ankyrins, 516, 522t, 523, 524
Anterior field, 123, 148
Anthracyclines, 420
Antimicrobial therapy, endocarditis and, 430–431, 430t
Antiphospholipid antibodies (APLA), 240, 241–242
α2-antiplasmin (α2-AP), 233
α2-AP. *See* α2-antiplasmin
Antisense oligonucleotide approach, 29–30
Antisense strand, 4, 6
Antisense strategies, 29
Antithrombotic pathways, 234f

Aortic arch, defects in, 172–173. *See also* Interrupted aortic arch
AP. *See* Action potential
APLA. *See* Antiphospholipid antibodies
Apolipoprotein
    A-1 (ApoA-1), 216t
    A-V (ApoA5), 216t
    B (ApoB, FLB), 212, 216t
    E (ApoE), 216t
        MI and, 261, 266
Apoptosis
    cell-cycle signaling and, 97–99, 98t
    extrinsic/intrinsic pathways of, 449–452, 450f
    imaging of cardiomyocyte, 54
    mitochondrial pathway of, 290f
    mitochondria's role in, 104
α1-ARs. *See* Adrenoreceptors
α2-ARs. *See* Adrenoreceptors
ARH. *See* Autosomal-recessive hypercholesterolemia
Aromatase, 556f, 559, 560, 565
Arrhythmogenic right ventricular dysplasia/cardiomyopathy (ARVD/ARVC), 181, 183, 363. *See also* Cardiomyopathies
    chromosomal loci in, 389
    classification of cardiomyopathies and, 364t
    clinical manifestations of, 388–389
    desmosomal genes in, 390–391
    genes associated with familial, 390t
    genetics of, 389–391
    pathophysiological mechanisms in, 391
    RYR2 in, 389–390
    TGF-β3 in, 391
ARs. *See* Adrenoreceptors
Arsenic, myocarditis and, 420t
Arterial aging, 591
    cardiac aging and, 591–592
Arteriogenesis, 322–324. *See also* Cardiac neovascularization
    angiogenesis *v.* vasculogenesis *v.*, 322f
    definition of, 315
    flow chart of, 323f
        molecular factors in, 323f
    stimulation of, 332
ARVC. *See* Arrhythmogenic right ventricular dysplasia/cardiomyopathy
ARVD/ARVC. *See* Arrhythmogenic right ventricular dysplasia/cardiomyopathy
ASO. *See* Allele-specific oligonucleotides
Aspergillus, 416t
Atherosclerosis, 211
    cell-based inflammatory/immune responses in, 217, 218, 218f
    DNA testing for, 220–221
    future therapies of, 242
    gene profiling in, 221
    gene therapy and, 242
    inflammation and, 214–218, 243
    knock-out mouse models and, 219–220, 219t
        genes in, 219t
    molecular mechanisms of, 211
Atherosclerotic lesion, MI and rupture of, 219

Atherothrombosis, 223
ATM. *See* Atomic force microscopy
Atomic force microscopy (ATM), 600
   cardiac aging and, 600
ATP. *See* Adenosine triphosphate
Atrial fibrillation (AF), 529. *See also* Familial atrial fibrillation
   cardiac structural defects of, 532–534
      genotype and phenotype in, 532, 533, 533t
Atrial natriuretic factor (ANF), 56t
   chamber specificity and, 119, 148
Atrial septal defect (ASD), 168–169, 168t
   conduction defects and, 169, 192
   gene loci and genesis of, 168, 169, 192
Atrioventricular (AV)
   cushions, 130–131
   junction, 130–131
Atrioventricular canal (AVC), 168t, 169–170
   developmental formation of, 169, 170, 192
Atrioventricular septal defects (AVSD), 169, 170, 192
   CHARGE syndrome and, 176, 193
   CRELD1 and, 170, 192
Atrioventricular (AV) valves, 118
Autophagy
   cardiac aging and, 589–590, 607
   cell death/renewal and, 449
Autosomal inheritance pattern, 9
Autosomal-recessive hypercholesterolemia (ARH), 212, 212t, 243
AV valves. *See* Atrioventricular valves
AVC. *See* Atrioventricular canal
AVSD. *See* Atrioventricular septal defects
5-aza-2-deoxycytidine, 58t
Azide, 420t

B

BAC vectors. *See* Bacterial artificial chromosome vectors
Bacteria
   endocarditis and, 425t
   myocarditis and, 416t
Bacteria expression system, 15t
Bacterial artificial chromosome (BAC) vectors, 14t
Bacteriophages, 11
Baculovirus expression system, 15t
Bad, 85, 86f, 96, 98f, 100, 104. *See also* Proapoptotic proteins
Barth syndrome, 183
Bartonella henselae, 425t
Bases, 3
Basic fibroblast growth factor-2, 58t
Basic helix-loop-helix (bHLH) transcription factors, 122, 148
   migration of cardiac precursors and, 122–123
Bax, 97, 98f, 99, 104. *See also* Proapoptotic proteins
Behçet's disease, 232
Bernard-Soulier syndrome (BSS), 239
bFGF. *See* Fibroblast growth factor, basic
bHLH transcription factors. *See* Basic helix-loop-helix transcription factors
Bid, 97, 98f, 99, 104. *See also* Proapoptotic proteins
Biomarkers, cardiac disease, 619–621
   myocardial ischemia/ACS and, 619–621, 620t

Biopanning, 15
2′,7′-bis(carboxyethyl)-5,6-carboxyfluorescein, 52, 52t
Bites (snake, spider), 420t
Blood coagulation. *See* Coagulation, blood
Blood pressure (BP), 341
   hypertension and, 341, 342, 354
Blunt ends, 12
BM cells. *See* Bone marrow cells
BMP. *See* Bone morphogenetic protein
BMPRs. *See* Bone morphogenetic protein, receptors
BNP. *See* Brain natriuretic peptide
Bone marrow (BM) cells, 61
   transplantation of, 61–62
      advantages/concerns in, 61–62, 65t
      clinical studies of, 61, 66
      markers in, 63t
Bone morphogenetic protein (BMP), vii, 57
   -2, 58t
   pathway, 131, 133f
      EMT and, 133f
   receptors (BMPRs), 104
      mutations in, 349, 350f, 351, 354
   signaling inhibition, 58t
Bone resorption, estrogen and, 556, 557f
BOOST clinical trial, 61
Borrelia, 416t
BP. *See* Blood pressure
Brachydactyly, 343t, 344
Bradykinin, 283t
Brain natriuretic peptide (BNP), 620t
BRDU. *See* Bromodeoxyuridine
Brenner, Sydney, 625
Bromodeoxyuridine (BRDU), 55
BrS. *See* Brugada syndrome
Brucella, 425t
Brugada syndrome (BrS), 524–527, 543
   molecular and cellular characterization of, 524–527
   ST-segment abnormalities in leads V1-V3 in, 524, 525, 525t
   types 1-3 in, 525t
BSS. *See* Bernard-Soulier syndrome
C. *See* Citrate; Cytosine

C

Ca^{++} CaM regulated kinase, 89, 90f
Ca^{++} regulation, 77, 103
Ca^{++}/CaM-dependent protein kinase (CaMK), 89, 90f
   isoforms of, 89
CAD. *See* Coronary artery disease
CADILLAC trial, 568
Calcein-AM, 52t
Calcineurin, 89, 90f
Calcium imaging, 52–53
Calcium, LVH and, 374
Calcium/ion flux, 52–53, 66
Calmodulin (CaM), 89
Caloric restriction, cardiac aging and, 603–604, 607
Calsequestrin (CASQ2), 527, 527t, 528, 543
CaM. *See* Calmodulin

Ca^{++}-mediated kinase signaling, 89–91, 90f
    calcineurin/calmodulin in, 89, 90f
    GRKs in, 89–91
    MAP kinases in, 91
CaMK. See Ca^{++}/CaM-dependent protein kinase
cAMP. See Cyclic AMP
Campylobacter, 416t
Candida, 416t
Carbamate. See Organophosphate
Carbon monoxide, myocarditis and, 420t
Carboxyl-terminus, 7
Cardiac actin, 368–369
    gene (ACTC), 377
Cardiac aging
    adrenergic receptors in, 583
    AFM and, 600
    arterial aging in, 591–592
    autophagy and, 589–590
    caloric restriction for, 603–604, 607
    cardiomyopathy of, 592–593
    cardiovascular changes and, 579–580, 606
    cell-based therapy for, 603
    cellular damage/cell loss in, 586, 606
    comparative genomics and, 598–599
    CP in, 602
    drug therapy/metabolic interventions for, 604–605
    endothelial remodeling in, 592–593
    energy metabolism in, 593–594
    epigenetic/environmental factors' influence on, 590–591
    extracellular remodeling in, 592
    gender differences in, 590
    gene expression in, 599–600
    genetic contribution to, 598
    genomic therapies for, 602
    GH and IGF-1 in, 584–585
    GPCRs in, 583
    inflammatory mechanisms/signaling in, 582–583
    mechanisms/pathways of, 580–591, 606
    mitochondrial function in, 595–598
        approaches to rescue, 605
        mtDNA damage in, 579, 595–598, 596f, 607
    muscarinic receptors in, 583
    myocardial ischemia/reperfusion injury and, 595
    new approaches in assessment of, 600–602
    pharmacogenomics and, 602–603
    post-genomic appraisal of, 579–618
    proteomics and, 600
    recombinant DNA and gene expression techniques for, 600–602
    ROS's role on oxidative stress and signaling of, 580–582, 581f
    SERCA and thyroid hormone in, 584
    signaling events and, 102
    skeletal muscle changes and, 594–595
    SNPs/transcription analysis in, 599
    susceptibility genes and genetic makeup in, 590
    telomeres and telomere-related proteins in, 586–588, 587f, 588f, 589f
    therapies for, 602–605
    thrombosis in, 592

Cardiac chambers
    dorsoventral specification and formation of, 128
    early specification of, vii
    growth/maturation of, 129
        nuclear regulators of, 130
    looped tubular heart with, 123f
    septation of, 130
Cardiac conduction defects (CCD)
    animal models of dysrhythmias and, 536, 537t, 538t
    gene defects in, 531, 531t
    mouse models of dysrhythmias and, 536, 537t
Cardiac conduction system (CCS), 142–145
    connexin genes and, 142, 143, 149
    ErbB and, 145, 149
    HOP and, 143, 144, 149
    molecular markers of, 149
    Nkx2.5 and, 142, 143, 149
    research on origins of, vii
    Tbx5 and, 142, 143, 149
Cardiac development, 117
    aortic/pulmonic valve formation in, 140–142
    AV cushions in, 130–131
        AV junction and, 130–131
    AV valve formation in, 131–140
        ErbB receptors and, 137–140, 148
        factors implicated in, 132t
    cardiac precursors in, 121–123
    CCS in, 142–145
    cell differentiation and mesoderm development in, 118–121
    chamber formation in, 128
    chamber growth/maturation in, 129
        nuclear regulators of, 130
    chamber septation in, 130
    dorsoventral specification in, 128
    environmental factors' influence on, 145–148, 149
        cocaine in, 147
        drugs in, 145, 146, 149
        fetal alcohol syndrome in, 147
        hypoxia in, 147–148
        maternal diseases in, 147
        toxins in, 146
        vitamins/micronutrients in, 146–147
    GATA-4 and, 102, 104
    gene expression's role in, 118
    L-R identity in, 128–129
    molecular and genetic analysis of, 117–163
    overview of, 117–118, 118f
    PAAs development in, 145
    semilunar valve formation in, 140–142
    signaling in, 102, 104
    tube looping and segmentation in, 117–118, 123–128, 123f
Cardiac dysrhythmias
    acquired, 532–536, 543
    animal models for conduction defects and, 536, 537t, 538t, 543
    causality of, 513–514, 543
    CCD and, 530–531, 531t
    drug-related, 535
        pharmacogenetics and, 535
    fatty acid/glucose metabolism defects and, 490–491
    gender differences in outcome of, 564–565

Cardiac dysrhythmias (*Continued*)
   gene/cell-based therapies for, 37–38, 541, 544
   genetic screening and risk stratification of, 537–540, 539f, 543
      molecular/genetic data in, 539f
   HGP and, 513, 541, 542
   infant/children, 180–181
      mutations in, 180t, 181, 193
   inherited, 520–532
      channelopathies and, 513, 522
   lethal, 523, 534, 535, 537, 543
   mitochondrial function/FAO defects and, 535–536
   molecular and cellular characterization of, 513, 520–532
   mouse models of conduction defects and, 536, 537t
   pharmacogenomics/pharmacogenetics and other therapeutic options in, 540–541, 543
   post-genomic approaches to, 536–541
   PVCs in, 534
   SD and, 513–554
   SDDS in, 534
   SNPs and, 513, 536, 537, 541
   ventricular, 534
   ventricular AP and ion currents in, 514–515
Cardiac neovascularization, 315–340
   angiogenesis in, 315–322
      dysregulators of, 327–328
      inhibitors of, 326–327, 326t
      vascular remodeling during, 325–326
   arteriogenesis in, 322–324
   cell transplant and, 330–331
   gene therapy and, 329
   gene transfer/overexpression and, 330
   key features in, 322f
   mechanisms of, 315–328
   PDGF-CC and, 319, 330, 332
   restenosis and, 328
   siRNA/Antisense RNA and, 329–330
   therapy and, 329–331
   transgenic mouse models and, 328–329
Cardiac outflow tract. *See* Outflow tract
Cardiac precursor cells, 121
   differentiation of, 121–122
   migration of, 122–123
Cardiac progenitor cells, 123
   anterior field and, 123, 148
Cardiac remodeling, in HF, 441, 447–453, 448f, 461
Cardiac stem cells. *See* Adult cardiac stem cells
Cardiac valves
   defects in, 170
   developmental formation of, 170, 192
   mouse genes implicated in formation of, 171t
Cardiology, post-genomic, 3
   biomarkers and, 619–621, 620t
      microarray analysis in, 623
   cardiac cell engineering and, vii, viii, 628–629
   diagnostic challenges in, 627–628
   future of, 619–636
   gene engineering and, 628
   genetic counseling in, 627–628
   introduction to, 3–10
   mitochondrial medicine in, 621, 623–625
   new/future therapeutic options in, 628–630
   pre-clinical cell studies *v.* clinical medicine in, 626–627
   proteomics and, 621–623
   systems biology and, viii, 621, 625–626
   vessel engineering and, 629–630
Cardiomyocytes, 3, 8
   apoptosis in, 449–452, 450f
      imaging of, 54
   cardiac aging and loss of, 586, 606
   cell cultures of, 55
   contractile elements of, 449
   differentiation, 57
      ES cell-mediated, 57–58, 58t
      regulatory elements in ES cell, 58t
      signaling pathways in, 58t
   fatty acids and, 473
   fluorescent dyes used in, 52t
   genetic defects in molecular targets within, 186, 187f
   glucose and, 476, 477
   growth factors in proliferation of, 55–56, 56t
      gene transfer manipulation in, 56t
   hypertrophy of, 449
   metabolic imaging of, 51–52, 66
   molecular basis in proliferation of, vii, 191
   regeneration of, 451–452
   in vitro culturing of, 54–56, 66
      cell cycle and cell sorting techniques in, 54–55
Cardiomyopathies, 363–414. *See also* Dilated cardiomyopathy; Hypertrophic cardiomyopathy; Mitochondrial cardiomyopathy; Restrictive cardiomyopathy
   acquired, 396–397
   of aging: remodeling, 592–593
   alcoholic, 396
   Chagas' disease in, 396–397
   classification of, 364t
   diabetic, 395–396
      defective genes in, 396t
   fatty acid metabolism defects in HF and, 183–184, 183t
      genetic loci in, 183t
   fatty acid oxidation and, 183–185
   fatty acid/glucose metabolism defects and, 488–490
   gene therapy for, 38–39
   inherited
      gene therapy and, 39, 41
   metabolic, 391–393
   MtDNA mutations in, 392t
   nuclear mutations and mitochondrial, 394t
      clinical evidence of, 393–394
   pediatric, 181–183
      genetic defects in, 182t
      population-based study of, 183, 193
Cardioprotection
   cardiac aging and, 602
Cardioprotection (CP), 191, 194, 281–314. *See also* Ischemic pre-conditioning; Pharmaceutical pre-conditioning; Post-conditioning
   anesthetics and, 293–294
   cellular pathways and molecular events in, 281–314
   clinical application of, 301–302
      clinical trials and, 302

Cardioprotection (*Continued*)
    human studies and, 301–302
    diabetes and, 296–297
    future perspective of, 302
    interventions/drugs in, 283t
    IPC and, 282–293
    pharmaceutical pre-conditioning and, 293–298
    post-conditioning and, 298–300
    remote conditioning and, 300–301
Cardiovascular cells. *See* Cells, cardiovascular
Cardiovascular diseases (CVDs), 15
    animal gender in, 560–562
        differences on cardiovascular phenotypes by, 560–562, 561t
    cardiac aging and, 579, 606
    gender and, 555–578
        animal and *in vitro* studies in, 560–562, 561t
        cardiovascular phenotype development in, 562–563
        gonadal hormones' role in, 559–560
        human studies in, 562–563
        sex steroid hormones in, 555–559
        therapy in, 566–568
    gender differences in outcome of, 564–566
        CAD and, 565–566
        cardiac hypertrophy and, 565
        dysrhythmias and, 564–565
        HF and, 564
        hypertension and, 565
    gene profiling in, 33t
    gene therapy for, 35–43
        animal models and, 40t
    HF and, 441
    mechanisms/pathophysiology of, vii
    pediatric
        chromosome map of inherited, 166f
    targeted gene expression in, 29, 30t
Cardiovascular gene expression, 27–50
Cardiovascular signaling pathways, 77–113
    Ca++-mediated kinase signaling and, 89–91, 90f
    cardiac aging and, 102
    cardiac development and, 102, 104
    cell-cycle signaling and apoptosis in, 97–99, 98t
    components of, 77
    ECM in, 101, 102
    effectors in, 83–84
    ER and, 97, 104
    extracellular signals, 101–102
    G-proteins in, 82–83
    infancy of, 77, 102
    kinases and phosphatases in, 84–89
    metabolic signals in, 101
    mitochondrial kinases in, 96–97, 96t
    mitochondrial retrograde signaling in, 97
    mitochondrial signaling in, 93–96, 104
    mitochondrial translocation in, 97
    nuclear receptor transcription factors in, 91–92, 92t, 103
    physiological growth and, 77
    plasma membrane signaling in, 93
    receptors in, 77–82
    stress signals in, 99, 100–101, 104
    survival/growth signals in, 99, 100, 104
    translation control in, 92–93
Carnitine, 474, 475
    deficiency, 184, 488t, 489, 490
Carnitine palmitoyltransferase
    I (CPT-I), 475, 475f, 476
        PPAR and, 480, 481t
    II (CPT-II), 474, 475, 475f
Carvedilol, 283t
Cascades, 103
CASQ2. *See* Calsequestrin
CAT. *See* Chloramphenicol acetyltransferase
Catalase, 94f, 95, 104, 580, 581f
Catecholaminergic polymorphic ventricular tachycardia (CPVT), 527–528
    cardiac channelopathies of, 527t
    molecular and cellular characterization of, 527–528
Catecholamines (norepinephrine, epinephrine), 420t
Caveolae, 84, 104
CCD. *See* Cardiac conduction defects
CCS. *See* Cardiac conduction system
CDK4, 56
cDNA, 14
    expression, 14–15
    libraries, 15
Cell cycle, 7–8, 8f
Cell engineering, cardiac, vii, viii, 628–629
Cell-based therapy
    cardiac aging and, 603
    cardiac dysrhythmias and, 37–38, 541, 544
Cell-cycle signaling, apoptosis and, 97–99, 98t
Cells, cardiovascular. *See also* Cardiomyocytes
    growth of, 54–55
    HF and death/renewal of, 449–452, 450f
    imaging techniques for, 51–54
CellTracker Blue CMAC, 52, 52t
Cellular biology. *See* Molecular cellular biology
Cellular damage/cell loss, cardiac aging and, 586, 606
Cellular imaging techniques, 51–54
    calcium/ion flux and, 52–53, 66
    electrophysiological, 52–53
    metabolic, 51–52, 66
    signaling and real-time measurements in, 53–54
Cellular techniques, 51–75
Cellular transplantation, 56–64, 66
    HF and, 458–459, 461
Central dogma, 4
Ceramide, 283t
CETP. *See* Cholesteryl ester transfer protein
cGMP. *See* Cyclic GMP
Chagas' disease, 396–397. *See also* Cardiomyopathies
Chlamydiae, 416t, 425t
Channelopathies, cardiac, 513, 522. *See also* Cardiac dysrhythmias, inherited
    CPVT, 527t
    LQTS, 522, 522t, 523, 524
    SIDS and, 531, 532t
Char syndrome, 165, 173, 192
    TFAP2b and, 165

CHARGE syndrome, 176, 193
   AVSD and, 176, 193
   TOF and, 176, 193
CHD. *See* Congenital heart disease
Chelerythrine, 283t
Chloramphenicol acetyltransferase (CAT), 20t, 21
Chloromethyl-X-rosamine (CMX-ros), 52t
Cholesteryl ester transfer protein (CETP), 216t
   MI and, 261, 267, 268t
Chromosome 22, PTA and deletion of, 174, 192
Chromosome walking, 14
Chromosomes, 7
   pediatric cardiovascular disorders and map of, 166f
Chromatin, 7
Cis-acting elements, 6
Citrate (C), 476, 478f, 479
CK-MB. *See* Creatine kinase MB
Cloning, therapeutic, 64, 66
Clotting, blood. *See* Coagulation, blood
Clozapine, 420t
CMX-ros. *See* Chloromethyl-X-rosamine
CNG ion channels. *See* Cyclic nucleotide gated ion channels
Coagulation, blood
   common pathway factors in, 227–233
      factor XIIIa in, 232–233
      fibrinogen in, 229–230, 229t
      FV and, 227–228
      FX in, 228–229
      protein C pathway and, 231
      PT in, 230–231
      TAFI in, 232
      thrombin in, 231
      thrombomodulin in, 231–232
   intrinsic pathway of, 222f, 225–227
      F8 in, 227
      F12 in, 226
      factor IX in, 226–227
      FXI in, 225–226
   pathway of, 221–233, 222f
      F7 in, 224–225
      TF and, 222–223
   thrombosis and, 221
CoA-Sh, 474
Cocaine
   cardiac development and, 147
   myocarditis and, 420t
Codominant alleles, 9
Coenzyme Q, 283t
Colony hybridization, 15
Comparative genomics. *See* Genomics
Complementary base pairing, 3, 4f
Complementary DNA. *See* cDNA
Congenital heart disease (CHD)
   cardiac malformations in, 167–176, 168t
   chromosomal map of inherited pediatric CVDs and, 166f
   expression profiling and, 176–177, 193
   future research for, 190–192, 194
   gene defects in, 168t
      frequency of, 168t
   gene therapy for, 39, 41, 165

   genetic diagnostic approaches to, 165, 189–190, 193–194
   genetic/molecular factors linked to, 165, 166f
   inborn errors causing, 165, 167t
      cardiac phenotype/syndrome in, 167t
      genes affected in, 167t
   molecular genetic analysis of, 165–176
   prenatal diagnosis of, 190
   single gene mutations and, 165, 167, 167t, 192
   in utero detection of, 190, 194
Congenital vascular defects, 177–179
Connexin genes, 132t, 136
   CCS and, 142, 143, 149
Contractile elements, HF and, 449
Contrast-enhanced magnetic resonance imaging, 51
Copper, 420t
Coronary artery disease (CAD), 211
   ACE and, 216t
   candidate genes associated with, 216t
   DNA testing and, 220–221
   ethnicity/race, MI and, 270–271
   gender differences in outcome of, 565–566
   gene variation and, 220
   genetic/environmental risk factors for, 212t
   HF and, 445
   ischemia and, 281
   molecular markers of, 221
   single gene defects in, 212t
Coronary thrombosis, genes involved in MI and, 263–265, 264t
Corynebacterium diphtheriae, 416t
Cosmid vectors, 14t
Coxiella burnetii, 425t
Coxsackieviruses, 415, 416, 417, 421, 422, 423
CP. *See* Cardioprotection
CPT. *See* Carnitine palmitoyltransferase
CPVT. *See* Catecholaminergic polymorphic ventricular tachycardia
C-reactive protein (CRP), 220, 240, 241
   atherogenesis and, 241
   thrombosis and, 240, 243
Creatine kinase MB (CK-MB), 620t
CRELD1 (cysteine rich with EGF domains), 170, 192
Cromakalim, 283t
CRP. *See* C-reactive protein
Cryptococcus, 416t
CT-1, 56t
C-terminus, 7
cTnI. *See* Troponins, I
cTnT. *See* Troponins, T
CuSOD, 94f, 104, 581f
CVDs. *See* Cardiovascular diseases
Cyclic AMP (cAMP), 53
   real-time measurements of, 53–54, 66
Cyclic GMP (cGMP), 103
   PKG and, 88, 89, 103
Cyclic nucleotide gated (CNG) ion channels, 54
Cyclin B1, 55, 56t
Cyclin B1-CDC2 (cell division cycle 2 kinase), 55, 56t
Cyclin D1, 55, 56t
Cyclin D2, 55, 56t

Cyclopentanone prostaglandins, 481
Cyclophosphamide. *See* Organophosphate
Cyclosporin A, 283t
Cysteine rich with EGF domains. *See* CRELD1
cyt*c*. *See* Cytochrome c
Cytochrome c (cyt*c*), 478f
  release, 487
Cytochrome p450, 95, 104
  synthesis of steroid hormones and, 556f, 568
Cytokines
  cascade, 188, 193
  MI and, 265, 266, 267, 271, 273
Cytosine (C), 3
Cytotoxic lymphocyte response, 178, 193

D

DCFH-DA. *See* 2′,7″-Dichlorofluorescin diacetate
DCM. *See* Dilated cardiomyopathy
Deacetylases, 103
Decoys, 31f
"Defective force generation" hypothesis, 385–386
"Defective force transmission" hypothesis, 385
Delayed pre-conditioning, 291–293
  cellular events occurring in, 292f
Delivery methods, gene therapy, 35f
Denaturing gradient gel electrophoresis (DGGE), 21
Denaturing high-performance liquid chromatography (DHPLC), 190, 193
Deoxyribose, 3
Dequalinium, 624
Desmin gene, DCM and, 379–380
DGGE. *See* Denaturing gradient gel electrophoresis
dHAND. *See* HAND proteins
DHF. *See* Dihydrofluorescein diacetate
DHPLC. *See* Denaturing high-performance liquid chromatography
di-4-ANEPPS, 52t, 53
Diabetes mellitus
  cardioprotection and, 296–297
  fatty acid/glucose metabolism defects and, 491–492
  MI and, 262f
  polymorphisms associated with MetSyn, insulin resistance and, 493t
  type-1
    laterality defect and, 175, 192
  type-2
    atherogenesis and, 211, 214, 215, 223, 241, 243
Diabetic cardiomyopathy. *See* Cardiomyopathies, diabetic
4,5-Diaminofluorescein-2/diacetate, 52, 55
Diazoxide, 283t
Dicer, 31
2′,7″-Dichlorofluorescin diacetate (DCFH-DA), 51, 52t
Dideoxyribonucleotides, 11
Diet
  cardiac aging and restrictions in, 603–604
  MI and, 262f
DiGeorge syndrome, 174
Dihydroethidium, 52t
Dihydrofluorescein diacetate (DHF), 51, 52t
Dihydrorhodamine 123, 51, 52t

Dihydroxyvitamin D3, 56t
Dilated cardiomyopathy (DCM). *See also* Cardiomyopathies
  ACTC gene in, 377
  cardiac troponin T in, 379
  chromosomal loci in, 376–377
  classification of cardiomyopathies and, 364t
  clinical manifestations of, 375–376
  desmin gene in, 379–380
  dystrophin gene in, 382–383
    mutations in, 384t
  emerin in, 381, 382
  genes implicated in, 377–385, 377t, 398
  genes involved in, 16, 17t
  genetic defects in, 181, 182, 182t, 183
  genetics of familial, 376
  human chromosome map of genetic loci involved in, 378f
  LMNA gene in, 381–382
  MYBP3 gene in, 379
  myocarditis as precursor to, 415, 416, 417, 423, 431
  myosin heavy chain genes in, 378–379
  pathophysiological mechanisms of, 385–387
    altered gene expression's role in, 386–387
    "defective force generation" hypothesis in, 385–386
    "defective force transmission" hypothesis in, 385
    nuclear lamina proteins' role in, 386
  SGCD in, 380
  TAZ in, 383–384
  titin gene in, 379
  transgenic animals and gene alterations in, 386t
  transgenic models of, 385
  tropomyosin genes in, 379
  VCL in, 380–381
Diphtheroids, 426t
Diploid human cells, 8
DMD. *See* Duchenne muscular disease; Dystrophin gene
DMPK. *See* Myotonin protein kinase
DNA, 3
  CAD and testing of, 220–221
  double-strand structure of, 3–4, 4f
  mitochondria, 8, 9f
  polymorphisms, 513, 536, 540, 543
  primary structure, 3
  replication, 4, 5f
  secondary structure, 3
  strands of, 3
DNA analysis, 11–15
DNA chip analysis. *See* Microarray analysis
DNA footprinting, 21
DNA hybridization, 17
DNA polymerase, 7–8
DNA-RNA hybridization, 19–21
  mutation detection and, 21
DNP, 283t, 287
Dominant alleles, 9
Dopamine, 420t
Dorsoventral specification, 128
DORV. *See* Double outlet right ventricle
Double helix, 3
Double outlet right ventricle (DORV), 168t, 174–175
  mouse mutations and phenotype of, 174, 192

Double-stranded RNA (dsRNA), 30
Down syndrome, 168, 170
DQAsomes, 624
Drugs
    cardiac aging, metabolic interventions and, 604–605
    cardiac development influenced by, 145, 146, 149
    CP, interventions and, 283t
    dysrhythmias associated with, 535
    IPC mimicking with, 283t, 293–298
dsRNA. *See* Double-stranded RNA
Duchenne muscular disease (DMD), 10f
    inheritance pattern of, 10f
Dyslipidemia, 211, 213, 220, 223, 243
Dysrhythmias. *See* Cardiac dysrhythmias
Dystrophin gene (DMD), 382–383
Dystrophin protein, 382, 383
    domains of, 383f
    E2F2, 56t

E

Early growth response factor-1 (Egr-1), 328, 332
Ebselen, 605
ECM. *See* Extracellular matrix
EcoR1, 12f
ECs. *See* Endocardial cushions
ED-FRAP. *See* Enzyme-dependent fluorescence recovery after photobleaching
Effectors, 77, 83–84
    AC as, 83–84
    caveolae, 84
    PLC as, 84
EGF. *See* Epidermal growth factor
Egr-1. *See* Early growth response factor-1
eHAND. *See* HAND proteins
Electron transfer flavoprotein (ETF), 478f
    dehydrogenase, 478f
Electron transport chain (ETC), 51, 58t, 443, 444f, 478f
Electrophoretic mobility shift assay, 21
Electrophysiological imaging, 52–53
ELN gene, 178, 193
Embryonic stem (ES) cells, 57–59
    cardiomyocyte differentiation in, 57–58
        regulatory elements in, 58t
        signaling pathways in, 58t
    development of, 57f
    transplantation of, 56–60
        advantages/limits in, 59–60, 65t
        markers in, 63t
EMT. *See* Endothelial-mesenchymal transdifferentiation
Endocardial cushions (ECs), 118
    formation of, 131, 149
Endocarditis, 415, 424–431
    animal models of, 429
    bacteria of, 425t
    diagnosis of, 429–430
    infectious agents associated with, 426t
    infectious agents (blood culture-negative) of, 425t
    microbial pathogens in, 424–426
    pathogenesis of, 426–429, 427f
    primary risk factors in, 424t
    treatment of, 430–431
        antimicrobial therapy in, 430–431, 430t
        surgical intervention in, 431
Endomyocardial biopsy, 622
Endoplasmic reticulum (ER), 97
    signaling, 97, 104
Endothelial cell(s)
    angiogenesis pathway and, 315, 316, 317f, 318, 332
    molecular profiling of, 329
Endothelial cell activation, 188, 193
Endothelial function
    MI and, 262f
Endothelial nitric oxide synthase (eNOS), 216t
    gene transfer, 320, 321, 332
    MI and, 261, 268t, 269
Endothelial progenitor cells (EPCs), 61
    angiogenic properties of early, 330, 331
    gene therapy and, 329
    telomeres and, 586
    vasculogenesis and, 322f, 325, 332
Endothelial remodeling, cardiac aging and, 592–593
Endothelial-mesenchymal transdifferentiation (EMT), 131
    pathway of TGFβ/BMP/Notch signaling in, 133f
Endothelin (ET), 77, 79, 283t
    signaling peptides, 79, 103
        ET-1, ET-2, ET-3 as, 79
Energy metabolism, cardiac aging and, 593–594
Energy metabolism profiling, HF and, 455
Enhancers, 21
eNOS. *See* Endothelial nitric oxide synthase
Enterococci spp., 426t
    antimicrobial therapy for, 430t
Enteroviruses, myocarditis and, 415, 416–417
Environmental factors
    cardiac aging and epigenetic, 590–591
    cardiac development influenced by, 145–148, 149
    MI influenced by, 262f
Environmental toxins. *See* Toxins, environmental
Enzyme-dependent fluorescence recovery after photobleaching (ED-FRAP), 51
Enzymes, 3. *See also* Polymerases
    cell cycle and, 7–8
    reverse transcriptase, 4
    telomerase, 8
EPCs. *See* Endothelial progenitor cells
Epicardium, 129, 148
    chamber growth/maturation and, 129
Epidermal growth factor (EGF), 56t
    ERK1/2 and, 79, 80
Epigenetic factors
    cardiac aging and environmental, 590–591
Epinephrine. *See* Catecholamines
Episomal plasmids, 13
EPO. *See* Erythropoietin
Epstein-Barr virus, 416t
ER. *See* Endoplasmic reticulum
ErbB receptors, 137–140, 148
    CCS and, 145, 149
    signaling of, 138f
ERE. *See* Estrogen response elements

ERK1/2. *See* Extracellular signal-regulated kinases 1/2
ERR. *See* Estrogen-related receptor
ERs. *See* Estrogen receptors
Erythromelalgia, 239
Erythropoietin (EPO)
    factor (rHuEpo), 56t
    pharmaceutical pre-conditioning and, 294–295
ES cells. *See* Embryonic stem cells
EST. *See* Expressed sequence tags
Estradiol, 556, 556f, 557, 559, 560
Estrogen(s), 270, 555
    bone resorption and, 556, 557f
    CVDs and, 555–558
    ERs and, 555–558
        overview of interactions between, 557f
    genomic effects of, 557–558
    non-genomic effects of, 556–557
Estrogen receptors (ERs), 555–558
    overview of interactions between estrogens and, 557f
Estrogen response elements (ERE), 557, 569
Estrogen-related receptor (ERR), 483, 484, 484f
ET. *See* Endothelin
ET-1 signaling peptides, 79
ET-2 signaling peptides, 79
ET-3 signaling peptides, 79
ETC. *See* Electron transport chain
ETF. *See* Electron transfer flavoprotein
Ethanol, 420t
Ethnicity, MI, race and, 270–271
Eucaryotic cell cycle, 7–8, 8f
Euchromatin, 7, 8f
Exercise
    cardiac aging and, 605
    MI and, 262f
Exons, 6
Expressed sequence tags (EST), 16
    HGP and, 16
Expression profiling, 176
    CHD and, 176–177, 193
    HLHS and, 193
Expression systems, comparison of, 15t
Extracellular matrix (ECM), 101, 102
    AVC and, 169
    cardiac remodeling and, 452–453
        MMPs' role in, 452, 453
Extracellular remodeling, cardiac aging and, 592
Extracellular signal-regulated kinases 1/2 (ERK1/2), 80–81
Extracellular signals, 101–102
Extrinsic pathway, 449
    apoptosis, intrinsic pathway and, 449–452, 450f

F

F. *See* Fumarate
F7. *See* Factor VII
F8. *See* Factor VIII
F12. *See* Factor XII
FA. *See* Friedreich ataxia
FABP. *See* Fatty acid binding protein
FABPpm. *See* Plasma membrane-associated fatty acid binding proteins

Factor IX, 226–227
Factor V (FV), 227–228
    MI and, 261, 263, 264t, 267, 273
Factor VII (F7), 224–225
    MI and, 261, 263, 264t, 273
Factor VIII (F8), 227
Factor X (FX), 228–229
Factor XI (FXI), 225–226
Factor XII (F12), 226
Factor XIIIa, 232–233
FAF. *See* Familial atrial fibrillation
FAK. *See* Focal adhesion kinase
Familial atrial fibrillation (FAF), 529
    genes involved in, 529, 529t, 530
Familial combined hyperlipidemia (FCHL), 212t, 213
Familial defective apolipoprotein B, 212–214, 212 t
Familial dilated cardiomyopathy (FDCM). *See* Dilated cardiomyopathy
Familial hypercholesterolemia (FH), 211–212
Familial hypertrophic cardiomyopathy (FHCM). *See* Hypertrophic cardiomyopathy
FAO. *See* Fatty acid oxidation
FAT/CD36. *See* Fatty acid translocase
FATP. *See* Fatty acid transport protein
Fatty acid and glucose metabolism
    cardiomyocytes and, 473
    cardiomyopathy and defects in, 488–490
    cellular perspective of, 473–485
    diabetes/hyperglycemia and defects in, 491–492
    diagnostic advances for defects in, 494
    disorders
        cardiomyopathy and, 183t
        HF and, 183t
    effects of dysfunctional or perturbations in, 485
        cardiac structure/function changes from, 485–494
        cellular and animal models in, 485–487, 486t
    fatty acid transporters in, 473–476
    glucose carriers and, 476–477
    HF and defects in, 488
    import and oxidation, 473–476, 475f
    metabolic dysfunction and cardiovascular disease phenotypes in, 488–494
    metabolic therapies for defects in, 494–495
    MetSyn, insulin resistance, and, 492–494
        polymorphisms associated with, 493t
    myocardial apoptosis and metabolism in, 487–488
        molecular mechanism in, 487, 488
Fatty acid binding protein (FABP), 475, 475f, 476
    PPAR and, 481t
    transgenic mice and, 486t
Fatty acid β-oxidation
    four reaction steps of, 477
    OXPHOS, TCA cycle and, 473, 477–480, 478f
Fatty acid oxidation (FAO), 477
    cardiomyopathy and, 183–185
    cellular location of glucose oxidation and, 480
    dysrhythmias associated with defects in mitochondrial function and, 535–536
    MTP and, 489
    PGC-1α and, 482–485

Fatty acid oxidation (FAO), 477 (*Continued*)
  PPARs and, 480–482
  regulatory elements in, 480–485
Fatty acid translocase (FAT/CD36), 474, 475, 475f, 488t, 490
Fatty acid transport protein (FATP), 474
  -1, 474
Fatty acids, role of, 473
FBN1 gene. *See* Fibrillin gene
FCCP, 283t
FCHL. *See* Familial combined hyperlipidemia
FDCM. *See* Dilated cardiomyopathy
FDG-PET. *See* Fluorodeoxyglucose positron emission tomography
FFAU. *See* Unbound free fatty acids
FGF. *See* Fibroblast growth factor
FHCM. *See* Familial hypertrophic cardiomyopathy
Fibrillin *(FBN1)* gene, 177, 178, 193
Fibrinogen, 229–230, 620t
  MI and, 261, 263, 264t, 265, 273
  polymorphisms of, 229t
  thrombosis and, 240
Fibrinolytic system, 233–237
  α2-AP and, 236–237
  clot dissolution and, 234f
  PAIs and, 233, 235–236
  plasminogen and, 233–234
  tPA and, 234–235
  uPA and, 235
Fibrinopeptide A, 620t
Fibroblast growth factor (FGF), 55
  -2, 56t
  basic (bFGF), 316, 324, 326t, 332
  ERK1/2 and, 79, 80
  L-R signaling and, 148
  receptor, 56t
Fish-eye disease, 212t, 213, 243
5-LO. *See* 5-Lipoxygenase
FLB. *See* Apolipoprotein, B
Fluo-3, 52, 52t, 53
Fluorescence resonance energy transfer (FRET), 53, 54
Fluorescence-activated cell sorting (FACS), 55
Fluorescent dyes, in cardiomyocytes, 52t
Fluorescent imaging, 53, 53f
Fluorodeoxyglucose positron emission tomography (FDG-PET), 52
Fluorouracil, 420t
Focal adhesion kinase (FAK), 320, 321, 332
Forkhead box O (FOXO) transcription factors, 320, 332
FOXO transcription factors. *See* Forkhead box O transcription factors
Frameshift, 7
Framingham Heart Study, 562, 585
Frataxin protein [FRDA], 178, 193
  deficiency of, 193
FRDA. *See* Frataxin protein
Free radical theory of aging (FRTA), 580
FRET. *See* Fluorescence resonance energy transfer
Friedreich ataxia (FA), 178–179, 193
  molecular pathology in, 179f
FRTA. *See* Free radical theory of aging

Fumarate (F), 478f
Fungi, myocarditis and, 416t
Fura-2, 52, 52t, 53
Fusion proteins, 14
FV. *See* Factor V
FX. *See* Factor X
FXI. *See* Factor XI

G

G. *See* Guanine
GATA family of transcription factors, 102
  cardiac cell differentiation and, 118, 120f, 148
  migration of cardiac precursors and, 122, 148
GATA-4, vii, 57, 58, 58t, 59, 62, 62t, 165
  acetylation, 58t
  cardiac development and, 102, 104
  cardiac septal defects and, 165, 167t
  linear heart tube and, 118f
Gel shift, 21
Gender. *See also* Cardiovascular diseases
  cardiac aging and differences in, 590
  CVD and, 555–578
  MI and, 262f
Gene(s)
  ARVD and, 390t
  atherosclerosis, mouse models, and, 219t
  CAD and candidate, 216t
  CAD and defects in single, 212t
  cardiac dysrhythmias and mutations in, 180t
  cardiac valve formation and mouse, 171t
  cardiomyocyte proliferation and overexpressed, 56t
  CHD, inborn errors and, 167t
  DCM and, 16, 17t
  definition of, 3
  desmosomal, 390–391
  diabetic cardiomyopathy and defective, 396t
  expression of, 4–7
    DNA-RNA hybridization and, 19–21
  FAF and, 529, 529t, 530
  FDCM and implicated, 377–385, 377t, 398
  FHCM and implicated, 365–370, 365t, 398
  β-globin, 6f
  HCM and, 16, 17t, 365t
  HCM and non-sarcomeric, 369–370
  hypertension and candidate, 346t
  LQTS and, 16, 17t
  L-R positioning, animal studies and, 175t
  Marfan syndrome and, 16, 17t
  MI and inflammatory response, 265t
  MI and other cardiovascular, 268t
  MI and stroke, 267, 268t
  MI and thrombotic, 264t
  MI, gender and, 270
  modifier, 10
  nuclear environment of, 7
  nuclear receptor transcription factors and cardiac metabolic, 92t
  pediatric cardiomyopathy and defective, 182t
  PPH and candidate, 351t
  reporter, 20t, 21

Gene(s) (*Continued*)
  SERCA, 32, 33
  structure of, 3–7
  therapeutic, 35f
Gene blockades, 31f
Gene chip technology, 190, 194
Gene engineering, 628
Gene expression profiling, HF and, 454
Gene libraries, 15
Gene mapping, 165
  inherited pediatric CVDs and, 166f
Gene mutations, single, CHD pathogenesis and, 165, 167, 167t
Gene profiling, 19, 33–34
  atherosclerosis and, 221
  CVD and, 33t
  hypertension and, 347–349
  limitations in, 33–34
  MI, ischemia and, 271–272
  PPH and, 352, 352t, 353
Gene therapy, 35–43
  angiogenesis and, 36
  atherosclerosis and, 242
  cardiac dysrhythmias and, 37–38
  cardiac neovascularization and, 329
  cardiomyopathy and, 38–39
  clinical studies of, 41
  congenital heart disease and, 39, 41
  CVDs, animal models and, 40t
  delivery methods in, 35f
  DNA delivery in, 41–42
  elements of, 35f
  humans and, 35, 43
  hypertension and, 36–37
  inherited cardiomyopathies and, 39, 41
  myocardial protection and, 39
  myocardial transplantation and, 39
  myocarditis, inflammation, and, 38
  restenosis and, 36
  ribozymes and, 35
  targets for, 34f, 36–42
  therapeutic genes and, 35f
  vectors in, 35f
Gene transfection, vectors and, 27–28
Gene transfer manipulation, of cardiomyocyte proliferation, 56t
Genetic code, 7
Genetic counseling, 627–628
Genetic diagnostic approaches, CHD and, 165, 189–190, 193–194
Genetic information, flow of, 4, 5f, 6–7
Genetic screening, viii
  cardiac dysrhythmias, risk stratification and, 537–540, 539f, 543
Genetics, 3
Genome. *See* Human genome
Genomic
  DNA, 14
  imprinting, 172, 192
  library, 15
  therapies and cardiac aging, 602

Genomics, 16–21
  cardiac aging and comparative, 598–599
  human genome and comparative, 598, 599, 607
Genotype marking, 63
Genotypes, 9
GFPs. *See* Green fluorescent proteins
GH. *See* Growth hormone
Giant cell myocarditis. *See* Myocarditis, giant cell
GIK. *See* Glucose-insulin-potassium
Glanzmann's thrombasthenia, 239
Glibenclamide, 283t
Glitazones, 481
β-globin gene, 6f
Glucagon-like peptide, 283t
Glucocorticoid-remediable aldosteronism (GRA), 342, 343t
Glucose, 56t
Glucose carriers
  fatty acid metabolism and, 476–477
Glucose metabolism. *See* Fatty acid and glucose metabolism
Glucose oxidation, cellular location of FAO and, 480
Glucose-insulin-potassium (GIK), 283t, 295–296
GLUT family of transporters, 474, 476, 477, 481, 496
  transgenic mice and, 485, 486t
Glutathione peroxidase (GPx), 580, 581f, 588f, 597, 605, 606, 607
Glycogen phosphorylase, 479, 480
Glycogen turnover, 479
Gonadal hormones, cardiac pathology and, 559–560
Gordon's syndrome (PHAII), 341, 343t, 344, 353
GPCRs. *See* G-protein-coupled receptors
G-protein-coupled receptors (GPCRs), 77, 82, 83, 83t, 103. *See also* Adrenoreceptors
  ARs and, 103
  cardiac aging and, 583
G-protein-regulated kinases (GRKs), 89–91
G-proteins, 82–83
GPx. *See* Glutathione peroxidase
GRA. *See* Glucocorticoid-remediable aldosteronism
Gram-negative bacilli, 426t
Green fluorescent proteins (GFPs), 20t, 21, 53
  cell identity and, 63
  signaling and real-time measurements with, 53–54, 66
GRKs. *See* G-protein-regulated kinases
Growth hormone (GH), 584
  cardiac aging, IGF-1 and, 584–585
Growth/survival signals, 100, 104
Guanine (G), 3

H

H_2O_2. *See* Hydrogen peroxide
HACEK, 425, 425t, 426t
  antimicrobial therapy for, 430t
HAND proteins, 126
  cardiac development and, 126–128, 126f, 148
  dHAND, vii, 118f, 126, 126f, 165
  eHAND, 118f, 126, 126f
  gene expression of, 126–127, 126f
  MEF2c and, 126–128, 148
  right ventricular hypoplasia and, 165

Haploid human cells, 8
Haplotype markers, 513, 529, 532, 536, 537, 541, 542, 543
HCM. *See* Hypertrophic cardiomyopathy
HDLs. *See* High-density lipoproteins
HDOA, 22
Heart and Estrogen/Progestin Replacement Study (HERS), 555, 566
Heart failure (HF), 281
   alterations in myocardial energetics in, 443, 444f
   animal models and, 442–443
   apoptosis in, 449–452
      intrinsic and extrinsic pathways of, 450–452, 450f
   autologous skeletal myoblast transfer and, 459
   CAD and, 445
   cardiac remodeling in, 441, 447–453, 461
      ECM in, 452–453, 461
      progression of, 448–449, 448f
   cell death and renewal in, 449–452
   cell transplantation and, 458–459, 461
   cellular hypertrophy in, 449
   contractile elements in, 449
   drug metabolism and, 459–460
   energy metabolism profiling in, 455
   fatty acid metabolism disorders and, 183t
   fatty acid/glucose metabolism defects and, 488
   gender differences in outcome of, 564
   gene expression profiling in, 454, 461
      transcriptome and proteome in, 454
   gene therapy for, 38–39, 457–458, 461
   genetic influences on drug targets in, 460
   genetics of, 455–456
   global analysis of gene function in, 454–455
   global and specific analysis of expression in, 454–455, 461
   insights from patients with LVADs in, 453
   intracellular Ca++ cycling profiling in, 455
   markers for diagnosis, prognostication, disease monitoring, and clinical decision-making in, 456–457
   mitochondrial bioenergetic pathways in, 443, 444f
   molecular analysis of, 441–472
   molecular triggers and cellular pathways leading to, 442f
   molecular/cellular basis of, 442–446
   myocarditis and, 415, 416, 417, 431
   neurohormonal/cytokine activation in, 448–449, 448f
   oxidative stress and, 443–446
   pediatric, 187
   PGC-1 and its metabolic pathway in, 443, 445f
   pharmacogenetics/pharmacogenomics in, 459, 460t
   pre-clinical models of, 442–443
   stem cell therapy for, 458–459
   therapeutic options for, 457–460, 461
   transgenic models, 446–447, 461
      fatty acid defects in, 446–447, 447t
      metabolopathies in, 446–447, 446t
   triggers of, 441, 461
   underlying disorders in, 441
Heart fatty acid binding protein (H-FABP), 620t
Heart, formation of. *See* Cardiac development
Heavy metals, myocarditis and, 420t
Helix, 3
   double, 3

Hematopoietic stem cells (HSCs), 61
Hemostasis, platelets and, 221, 237, 238
Heparin-induced thrombocytopenia (HIT), 239
Heparin-sulfate proteoglycan perlecan, 170, 192
Hepatic lipase, 216t
Hepatitis C virus, 416t
HERG blockers, 535, 543
HERS. *See* Heart and Estrogen/Progestin Replacement Study
Heterochromatin, 7, 8f
Heterotaxy. *See* Laterality defect
Heterozygous organisms, 9
Hexokinase (HK), 477, 478, 480
HF. *See* Heart failure
H-FABP. *See* Heart fatty acid binding protein
HGP. *See* Human Genome Project
hhLIM, 57, 58t
HIF-1. *See* Hypoxia-inducible factor-1
High sensitivity C-reactive protein (hsCRP), 620t
High-density lipoproteins (HDLs), 243
   CAD and levels of, 212–214, 243
High-throughput capillary array electrophoresis, 190, 193
HindII, 12f
HindIII, 12f
Histones, 7
HIT. *See* Heparin-induced thrombocytopenia
HIV disease. *See* Human immunodeficiency virus disease
HK. *See* Hexokinase
HLHS. *See* Hypoplastic left heart syndrome
Holter monitoring, 534, 543
Holt-Oram syndrome, 165, 167t, 169, 170
Homeodomain only protein (HOP), 130, 143, 149
   CCS and, 143, 144, 149
Homocysteine, 240–241
Homozygous organisms, 9
HOP. *See* Homeodomain only protein
Hormesis, 604
Hormone replacement therapy (HRT), 555, 565, 566, 567, 569
Hormones. *See* Androgens; Estrogens; Gonadal hormones; Sex steroid hormones, CVDs and; Testosterone
HRT. *See* Hormone replacement therapy
hsCRP. *See* High sensitivity C-reactive protein
HSCs. *See* Hematopoietic stem cells
HSPG2 gene, 170, 171, 192
Human cardiovascular diseases. *See* Cardiovascular diseases
Human genome, 3
   comparative genomics and, 598, 599, 607
   SNPs in, 10
Human Genome Project (HGP), vii
   cardiac aging and, 579, 598
   cardiac dysrhythmias and, 513, 541, 542
   EST approach and, 16
   MI and, 262
   post-genomic cardiology and, 621
Human immunodeficiency virus (HIV) disease, 418
   myocarditis and, 418
HUPO Plasma Proteome Project pilot phase, 622
Hybridization
   DNA, 17
   DNA-RNA, 19–21

Hybridization (*Continued*)
  gene structure and, 17
  northern blot and, 18f
  southern blot and, 18f
Hydrogen bonding, 3, 4f
Hydrogen peroxide (H_2O_2), 322, 332
Hypercholesterolemia, 212
Hyperglycemia, fatty acid/glucose metabolism defects and, 491–492
Hyperhomocysteinemia, 240, 243
Hyperlipoproteinemia (type III), 212t, 213
Hypertension
  animal models of, 346–347
  BP and, 341, 342, 354
  candidate genes in primary, 346, 346t
  elderly and, 341
  future directions for, 353
  gender differences in outcome of, 565
  gene expression in, 347–349
  gene profiling in, 347–349
  gene therapy for, 36–37
  genetic contributions to, 342, 343f
  genomic analysis of essential, 344–346
  genomic loci associated with essential, 345t
  mendelian forms of, 342, 343t, 344
  MI and, 262f
  pharmacogenetics and, 353
  primary pulmonary (PPH), 349–353, 354
    BMPR2 and ALK-1 mutations in, 349, 350f, 351, 354
    candidate genes for, 351t
    gene profiling in sporadic/familial, 352, 352t, 353
  primary risk genes in, 346
  primary risks in, 341–342
  systemic and pulmonary, 341–362
  targeted gene deletion and mice BP in, 348t
Hypertriglyceridemia, 212t, 213, 214
Hypertrophic cardiomyopathy (HCM), 9, 10. *See also* Cardiomyopathies
  animal models of genetic, 370–375
    cardiac troponin I gene in, 372–373
    cTnT gene mutations in, 371–372
    gene alterations in, 371t
    MLC in, 373
    other non-sarcomeric models in, 373–374
    sarcomeric mutations in, 370–371
    α-tropomyosin in, 373
  cardiac actin in, 368–369
  cardiac troponins in, 367–368
  classification of cardiomyopathies and, 364t
  clinical manifestations of, 363–364
  genes implicated in familial, 365–370, 365t, 398
  genes involved in, 16, 17t
  genetic defects in, 181, 182, 182t, 183
  genetics of, 364–370
  human chromosome map of genetic loci involved in, 366f
  LVH *v.*, 364
  MLC and, 368
  MYBPC3 in, 367
  β-myosin heavy chain in, 365–367
  non-sarcomeric genes in, 369–370

  PRKAG2 mutations leading to WPW and, 369t
  triggers/effectors of LVH in, 374–375
  Z disc proteins and, 369
Hypertrophy, 447
  cellular, 449
  eccentric, 447
  gender differences in outcome of, 565
Hypoplastic left heart syndrome (HLHS), 168t, 171–172, 192
  expression profiling and, 193
Hypoxia
  angiogenesis and, 315, 316, 318, 322, 332
  cardiac development and, 147–148
Hypoxia-inducible factor-1 (HIF-1), 315, 316, 317, 318, 332

I

IAA. *See* Interrupted aortic arch
IC. *See* Isocitrate
Icariin, 58t
ICD. *See* Implantable cardioverter-defibrillator
Idebenone, 283t
IL-10. *See* Interleukin-10
IMA. *See* Ischemia-modified albumin
Imaging techniques. *See* Cellular imaging techniques
Implantable cardioverter-defibrillator (ICD), 528
in vitro fertilization (IVF), 628
Index ischemia, 282
Inducible nitric oxide synthase (iNOS), 582
Infectious agents
  Endocarditis and blood culture-negative, 425t
  Endocarditis with, 426t
Inflammation
  atherosclerosis and, 214–218, 243
  genes in MI and, 265–267, 265t
  MI and, 262f
Inflammatory mechanisms/signaling, cardiac aging and, 582–583
Inhalants, myocarditis and, 420t
Inheritance patterns, 9
  autosomal, 9
  Duchenne muscular disease and, 10f
  maternal, 9
  X-linked, 9, 10f
Inherited dysrhythmias. *See* Cardiac dysrhythmias, inherited
iNOS. *See* Inducible nitric oxide synthase
In-stent restenosis (ISR), 623
Insulin, pharmaceutical pre-conditioning and, 295, 296
Insulin receptor substrate (IRS), 474
  -1, 474, 476, 492, 493t
  -2, 474
Insulin/insulin-like growth factor (IGF), 55, 56t
  -1, 53, 55, 56, 57, 58t
    GH and, 584–585
  receptor, 56t
Interleukin-2, 420t
Interleukin-6 (IL-6), 620t
Interleukin-10 (IL-10), 620t
Interphase, 7, 8f
Interrupted aortic arch (IAA), 168t, 173–174
  phenotypes of PTA and, 173, 192
  22q11 deletion and, 192

Intracellular calcium cycling profiling, HF and, 455
Intrinsic pathway, apoptosis, extrinsic pathway and, 450–452, 451f
Introns, 6
INVLM. *See* Isolated non-compaction of left ventricular myocardium
Ion channels, cardiac, 515–520
   subunits of, 520, 521t
      cardiac currents and, 515, 516, 517t
      genes in, 521t
      nomenclature of, 521t
      proteins in, 521t
      type of, 521t
   voltage-gated Ca++, 516, 517, 518f
   voltage-gated K+, 517–520, 519f
      structure and function of, 517–520, 519f
   voltage-gated Na+, 516, 518f
Ion currents, 514
   ion channel subunits of major cardiac, 515, 516, 517t
   pharmacological and physiological modulators of specific, 515, 516t
      activators as, 516t
      blockers/inhibitors as, 516t
   ventricular AP and, 514–515
IONA trial, 302
IPC. *See* Ischemic pre-conditioning
Iron, 420t
IRS. *See* Insulin receptor substrate
Irx4, 125, 148
Ischemia
   biomarkers in ACS and myocardial, 619–621, 620t
   cardiac aging and reduced tolerance of myocardial, 595
   gene profiling in MI and, 271–272
   MI and, 261–280
Ischemia-modified albumin (IMA), 620t
Ischemic pre-conditioning (IPC), 282. *See also* Pharmaceutical pre-conditioning
   ATP levels in, 287, 288
   cellular/molecular events in, 282–293
      schematic representation of, 285f
   delayed, 291–293
      cellular events occurring in, 292f
   drugs and mimicking of, 283t, 293–298
   end effectors of, 288–289
   K_{ATP} channels and, 286–287
      mitoK_{ATP} and KCOs in, 286, 287
   MAPKs in, 286
   mediators of early, 284
   mitochondrial ETC activity in, 287–288
   mitochondrial functional plasticity and, 288t
   nonmitochondrial targets in, 289–290
   PI3K pathway and, 284–285
   PKC and, 285–286
   post-conditioning *v.*, 299f, 300t
   ROS generation and, 288
   transcriptional/nuclear events in, 291
   triggering of early, 283–284
   tyrosine in, 286
Isocitrate (IC), 478f
Isoforms, 6

AC, 83
Akt, 85
CaMK, 89
PKC, 87, 88
PPARs, 91
Isolated non-compaction of left ventricular myocardium (INLVM), 183
Isoschizomers, 11
ISR. *See* In-stent restenosis
IVF. *See* in vitro fertilization

J
JC-1, 51, 52t
Jervell Lange Nielsen syndrome (JLNS), 520, 522t
JLNS. *See* Jervell Lange Nielsen syndrome
JNKs. *See* c-Jun N-terminal kinases
c-Jun N-terminal kinases (JNKs), 79, 81, 91, 94f, 96

K
K_{ATP} channels, 93
   IPC and, 286–287
Karyotype, 7
Kawasaki disease, 188, 193
KCOs. *See* Potassium channel openers
Ketoglutarate (KG), 478f
KG. *See* Ketoglutarate
Kidney disease, 341, 353
Kinases
   ERK1/2, 79–80
   JNK as, 79, 81, 91, 94f, 96
   MAP, 91
   mitochondrial, 96–97, 96t
   p38 MAP, 56t
   phosphatases and, 84–89
   PI3K as, 85–87
   PKA as, 84–85, 103
   PKB as, 85–87, 103
   PKC as, 87–88, 103
   PKG as, 88–89, 103
   serine/threonine, 77, 87, 90, 103
Klotho protein, 585

L
LA. *See* Lupus anticoagulants
Lac Z (β-galactosidase), 20t, 21
Lamin A/C (LMNA)
   DCM and proteins of, 386
   gene, 381–382
Lamin protein, 381, 382
   structure of, 382f
Lasofoxifene, 555, 567
Laterality defect (heterotaxy), 175–176, 192
   diabetes mellitus type-1 and, 175, 192
   RA and, 175, 192
LCAD. *See* Long-chain acyl CoA dehydrogenase
LCAT. *See* Lecithin/cholesterol acyltransferase
LCFAs. *See* Long-chain fatty acids
LCHAD. *See* Long-chain 3-hydroxylacyl-CoA dehydrogenase
LDL. *See* Low-density lipoprotein

LDL receptor (LDLR), 212, 212t, 213
LDLR. *see* LDL receptor
Lead, 420t
Lecithin/cholesterol acyltransferase (LCAT), 213, 214
Left ventricle outflow tract obstruction (LVOTO), 171
Left ventricular (LV) dysfunction, 447, 451, 453, 461
Left ventricular hypertrophy (LVH), 364, 367, 370.
    See also Hypertrophic cardiomyopathy
  calcium and, 374
  effectors of, 374–375
  gender differences in outcome of, 565
  HCM *v.*, 364
  myocardial energetics and, 374
  triggers of, 374
Left-right (L-R) identity, 128–129, 148
  animal studies and genes implicated in, 175t
Left-ventricular assist devices (LVADs), 453
  insights from patients with, 453
Legionella spp., 425t
Lenegre-Lev disease, 530
Lentivirus, 28, 29t
Leopard syndrome, 167, 170
Liddle's syndrome, 341, 343t, 344, 353
Ligands
  SRs, TLRs, and, 217, 217t
LIM, 57, 58t
Lipid metabolism, MI and, 262f
Lipid oxidation, 214, 243
Lipoprotein lipase (LPL), 212t, 214
  PPAR and, 481, 481t
Liposomes, 28, 29t
5-Lipoxygenase (5-LO), 216t
5-Lipoxygenase activating protein (ALOX5AP), 216t
Lithium, 420t
Liver x receptor (LXR), 216t
LMNA. *See* Lamin A/C
Long QT syndrome (LQTS), 543
  cardiac dysrhythmias and, 180
  channelopathies of, 522, 522t, 523, 524
  diagnostic criteria for, 524, 525t
    electrocardiogram findings in, 525t
  genes involved in, 16, 17t
  molecular and cellular characterization of, 522–524
  physical activities and, 524, 543
  SIDS and, 531
Long-chain 3-hydroxylacyl-CoA dehydrogenase (LCHAD), 478f
  defects in, 488t, 489
Long-chain acyl CoA dehydrogenase (LCAD), 475f, 477, 478f
  PPAR-α and, 481t
Long-chain fatty acids (LCFAs), 473, 474
Looped heart tube, 123f
Low-density lipoprotein (LDL), 214
  metabolism of, 214, 215f
  oxidation of, 214, 243
LPL. *See* Lipoprotein lipase
LQTS. *See* Long QT syndrome
L-R identity. *See* Left-right identity
LTA. *See* Lymphotoxin-α

Lucifer yellow, 52t
Luciferase, 20t, 21
Lupus anticoagulants (LA), 241, 242, 243
LV dysfunction. *See* Left ventricular dysfunction
LVADs. *See* Left-ventricular assist devices
LVOTO. *See* Left ventricle outflow tract obstruction
LXXLL motif, 482f, 483
Lymphotoxin-α (LTA), 216t
  MI and, 261, 265t, 266, 273

M

M. *See* Malate; Mitotic stage
mABs. *See* Monoclonal antibodies
mAChR. *See* Muscarinic acetylcholine receptor
Macrophage activation
  MI and, 262f, 266
Malate (M), 478f
Malonyl-CoA, 475, 475f, 476
Malonyl-CoA decarboxylase (MCD), 476
Mammalian cell expression system, 15t
MAP kinases (MAPKs), 91
  IPC, tyrosine and, 286
MAPCs. *See* Multipotent adult precursor cells
MAPKs. *See* Map kinases
Marfan syndrome, 177–178, 193
  genes involved in, 16, 17t
Mass spectroscopy, 23
Maternal diseases, cardiac development and, 147
Maternal inheritance pattern, 9
Maternal lupus, cardiac development and, 147
Matrix metalloproteinases (MMPs), 265, 620t
  ECM remodeling and, 452–453
  MI and, 265–266
MBs. *See* Molecular beacons
MCAD. *See* Medium-chain acyl CoA dehydrogenase
MCD. *See* Malonyl-CoA decarboxylase
MCM. *See* Mitochondrial cardiomyopathy
MCP-1. *See* Monocyte chemoattractant protein
Medium-chain acyl CoA dehydrogenase (MCAD), 475t, 477, 478f, 480
  defects in, 488, 488t, 489, 490
  PPAR-α activation and, 481, 481t
MEF2c. *See* Myocyte enhancer factor 2c
Meiosis, 7
Meningococcus, 416t
Mesenchymal stem cells (MSCs), 61
Mesoderm specification, Wnt pathway and, 122, 136
Mesp(1,2), 122
  migration of cardiac precursors and, 118f, 122, 148
Metabolic conditioning, glucose, insulin and, 295–296
Metabolic imaging, 51–52, 66
Metabolic signals, 101
Metabolic syndrome (MetSyn), 214, 218, 241, 243
  insulin resistance and, 492–494
    polymorphisms associated with diabetes and, 493t
Metaphase, 7
Metavinculin gene (VCL), 380–381
Methylenetetrahydrofolate reductase (MTHFR), 216t, 240
MetSyn. *See* Metabolic syndrome
MI. *See* Myocardial infarction

Microarray (DNA chip) analysis, 19, 33, 165, 176, 177, 188
  cardiac aging, drug therapy and, 604–605
  post-genomic cardiology, biomarkers and, 623
Microbial pathogens, endocarditis and, 424–426
Micronutrients, cardiac development and, 146–147
Mineralocorticoid receptor (MR), 342
  mutations in, 343t
Minisatellite sequences, 17
Mismatch detection/cleavage, 21
Missense mutations, 9
MitoChip, 22
Mitochondria, 8
  apoptosis and, 104, 290f
  cardiac aging and function of, 595–598, 596f, 605
  dysrhythmias associated with defects in FAO and, 535–536
  imaging techniques and, 51–53
  IPC and functional plasticity of, 288t
  signaling at, 93–95
    ROS generation and, 93, 94f, 104
Mitochondrial cardiomyopathy (MCM), 185–186
  pathogenic mitochondrial DNA mutations and, 185f
Mitochondrial K_{ATP} channel (mitoK_{ATP}), 95
  IPC and, 286, 287
Mitochondrial kinases, 96–97, 96t
Mitochondrial medicine, 621, 623–625
Mitochondrial retrograde signaling, 97
Mitochondrial translocation, 97
Mitochondrial trifunctional protein (MTP; TFP), 184, 477, 478f
  defects in, 488, 488t
  FAO and, 489
  transgenic mice and, 485
mitoK_{ATP}. See Mitochondrial K_{ATP} channel
mitoQ, 624
Mitosis, 7
Mitotic stage (M), 7, 8f
Mitotracker green, 52t
Mitotracker red, 52t
Mitral valve prolapse (MVP), 170
MLA. See Monophosphoryl lipid A
MLC. See Myosin light chain
MLCK. See Myosin light chain kinase
MMD. See Myotonic muscular dystrophy
MMP-3. See Stromelysin-1
MMPs. See Matrix metalloproteinases
MnSOD, 581f
  overexpression, 322, 332
Modifier genes, 10
Modifier loci, HF progression and, 455, 456, 461
Modifiers, MI, 262f
Molecular beacons (MBs), 22
Molecular cellular biology. See also Cardiology, post-genomic
  central dogma of, 4
  clinical cardiology and, 3
Molecular diagnostic approaches. See Genetic diagnostic approaches
Molecular genetic analysis, 165
  cardiac development and, 117–163
  CHD and, 165–176

  pediatric CVD pathogenesis and, 165
Molecular profiling, endothelial cells and, 329
Monoclonal antibodies (mABs), 15
  CVD and, 16t
Monocyte chemoattractant protein (MCP-1), 620t
Monocyte/macrophages recruitment, 188, 193
Monogenic traits, 9
Monophosphoryl lipid A (MLA), 283t, 298
MORE (Multiple Outcomes of Raloxifene) trial, 567
Mosaic trisomy, 172
Motif homology searching, 11
Mouse genes, cardiac valve formation and, 171t
MPO. See Myeloperoxidase
MR. See Mineralocorticoid receptor
mRNA, 4
MSCs. See Mesenchymal stem cells
mtDNA, 8, 9f
  aging heart and damage to, 579, 595–598, 596f, 607
  cardiomyopathies and mutations of, 392t
  proteins in, 8, 9f
MTHFR. See Methylenetetrahydrofolate reductase
MTP. See Mitochondrial trifunctional protein
Multipotent adult precursor cells (MAPCs), 61
Muscarinic acetylcholine receptor (mAChR), 78f, 79
  cardiac aging, GPCRs and, 583
Muscarinic receptors, 78f, 79, 103
  cardiac aging and, 583
Mutation detection, DNA-RNA hybridization and, 21
Mutations, 9–10. See also Gene mutations, single
  cardiac dysrhythmias and gene, 180t
MVP. See Mitral valve prolapse
Mycobacteria, myocarditis and, 416t
Mycoplasma, 416t
Mycoplasma spp., 425t
Myeloperoxidase (MPO), 620t
Myoblasts, skeletal, 57, 60
  transplantation of, 60–61
    advantages/limitations in, 60–61, 65t
Myocardial energetics
  HF and alterations in, 443, 444f
  LVH and, 374
Myocardial infarction (MI), 219, 261–280
  ACE and, 268t, 269, 270
  AGT and, 261, 268t, 273
  animal studies and, 272
  ApoE and, 261, 266
  candidate gene approach to, 262–270
  CETP and, 261, 267, 268t
  chromosomal loci linked to, 262t
  cytokines and, 265, 266, 267, 271, 273
  eNOS and, 261, 268t, 269
  ethnicity/race in, 270–271
  F7 and, 261, 263, 264t, 273
  fibrinogen and, 261, 263, 264t, 265, 273
  FV and, 261, 263, 264t, 267, 273
  gene profiling in ischemia and, 271–272
  gene variants and, 267, 268t, 269–270
  gene variants in stroke and, 267, 268t
  genes involved in coronary thrombosis and, 263–265, 264t
  genes involved in inflammatory response in, 265–267, 265t

Myocardial infarction (MI) (*Continued*)
  genes/gender and, 270
  genesis of, 261, 262f, 272
    environmental factors in, 261, 262f, 272
    modifiers in, 261, 262f, 272
    susceptibility genes in, 262f
  genome-wide approach to, 261–262
  HGP and, 262
  ischemia and, 261–280
  linkage studies and, 261–262, 262t, 272
  LTA and, 261, 265t, 266, 273
  macrophage activation and, 262f, 266
  MMP-3 and, 261, 265, 265t, 270, 273
  MMPs and, 265–266
  PAI-1 and, 261, 263, 264t, 270, 273
  PGIIIa and, 261, 263, 264t, 273
  TSP genes and, 263, 264, 264t, 273
Myocardial protection, gene therapy and, 39
Myocardial transplantation, gene therapy and, 39
Myocardin, 118f, 119, 148
Myocarditis, 188, 193, 415–424
  animal models and, 421–423
  DCM and, 415, 416, 417, 423, 431
  diagnosis of, 420–421
  etiological agents of infectious, 416t
  gene therapy for, 38
  giant cell, 416
  HF and, 415, 416, 417, 431
  HIV and, 418
  nonviral infective, 418–419
  parasites in, 419–420
  three stages of, 416
  timecourse of events in mice viral, 422f
  toxic, 420
    toxic agents and, 420t
  treatment of, 423–424
  viral, 416–418
    enteroviruses in, 415, 416–417
    other viruses in, 417–418
  viral infection and, 188–189, 193
  viral-mediated mechanisms in pathogenesis of, 421t
Myocyte enhancer factor 2c (MEF2c), 126
  HAND proteins and, 126–128, 148
Myoglobin, 620t
Myosin binding protein C (MYBP3), 367
  DCM and, 379
  HCM and, 367
Myosin heavy chain genes, DCM and, 378–379
Myosin light chain (MLC), 148, 620t
  animal models of genetic HCM and, 373
  HCM and, 368
Myosin light chain kinase (MLCK), 89, 90f
β-MHC. *See* β-Myosin heavy chain
β-Myosin heavy chain (β-MHC)
  HCM in, 365–367
    benign and malignant mutations of, 367t
Myotonic muscular dystrophy (MMD), 178, 193
Myotonin protein kinase [DMPK], 178, 193
MYBPC. *See* Myosin binding protein C

N

NADH. *See* Nicotinamide adenine dinucleotide
NAD(P)H oxidase, 94f, 95, 104
NADPH oxidase, 94f, 581f
Neovascularization, cardiac. *See* Cardiac neovascularization
Neural crest cells, 140, 145
  CHARGE syndrome and, 176, 193
  PAAs and, 145
Neuregulins (NRG)
  1, 56t, 82
  2, 82
Neurohormonal inhibition, 441
NFAT family, 134
  model for NFATc1 of, 135f
NF-κB, 91, 103
Nicorandil, 283t
Nicotinamide adenine dinucleotide (NADH), 51
  FAO and, 477, 478f, 479, 485
Nitric Oxide (NO), 55, 58t
  ROS and, 582
Nitroglycerin (NTG), 283t, 298
Nkx2.5, vii, 102, 104, 165
  cardiac differentiation processes and, 118, 118f, 119, 120, 121, 148
  CCS and, 142, 143, 149
  conduction defects and, 165
NO. *See* Nitric Oxide (NO)
Nodal gene, 129, 148
Nonsense mutations, 9
Nonviral infective myocarditis. *See* Myocarditis, nonviral infective
Noonan syndrome, 167, 170
Norepinephrine. *See* Catecholamines
Northern blots, 19
  hybridization and, 18f
Northwick Park Heart study, 229
Not1, 12f
Notch pathway, 138
  signaling, 133
    EMT and, 133f
  VEGF, angiopoietin, TGF and, vii
NRF-1/2. *See* Nuclear respiratory factor-1,-2
NRG. *See* Neuregulins
N-terminal pro-brain natriuretic peptide (NT-proBNP), 620t
N-terminus, 7
NTG. *See* Nitroglycerin
NT-proBNP. *See* N-terminal pro-brain natriuretic peptide
Nuclear receptor transcription factors. *See* Transcription factors, nuclear receptor
Nuclear regulators, of chamber growth/maturation, 130
Nuclear respiratory factor-1,-2 (NRF-1/2), 484f
Nuclease protection, 19
Nucleosomes, 7
Nucleotides, 3
  types of, 3

O

OAA. *See* Oxaloacetate
Obesity, MI and, 262
Ocular migraine, 239
OECs. *See* Outgrowth endothelial cells

OFT. *See* Outflow tract
Oligonucleotide-ligation assay, 22
1-DE. *See* One-dimensional gel electrophoresis
One-dimensional gel electrophoresis (1-DE), 22
Opioids, 283t
Organophosphate (carbamate, cyclophosphamide), 420t
OS. *See* Oxidative stress
Outflow tract (OFT), 118, 120, 124, 139, 149
   defects in, 172–173
Outgrowth endothelial cells (OECs), 330, 331
Overexpression, 28, 31f
Oxaloacetate (OAA), 478f
Oxidative phosphorylation (OXPHOS), 483, 484, 487, 488, 495
   fatty acid β-oxidation, TCA cycle and, 477–480, 478f
   metabolism, 165, 186, 193
Oxidative stress (OS)
   HF and, 443–446
   signaling of aging heart, ROS and, 580–582, 581f
   theory of aging, 580, 581, 582
OXPHOS. *See* Oxidative phosphorylation
Oxytocin, 58t

P

p38 MAP kinase, 56t
PAAs. *See* Pharyngeal arch arteries
PAIs. *See* Plasminogen activator inhibitors
Palindromes, 11
Paramyxovirus, 416t
Paraoxonase-1 (PON-1), 216t
Parasites
   myocarditis and, 416t, 419–420
PARs. *See* Protease-activated receptors
Parvovirus, 416t
Patent ductus arteriosus (PDA), 168t, 173
   Char syndrome and, 165, 173, 192
Pax3, 118f, 146
   PAAs and, 145, 149
PC. *See* Protein C
PCR amplification. *See* Polymerase chain reaction amplification
PCSK9, 212t, 213
PDA. *See* Patent ductus arteriosus
PDC. *See* Pyruvate dehydrogenase complex
PDE4D. *See* Phosphodiesterase 4D
PDGF. *See* Platelet-derived growth factor
Pediatric heart failure. *See* Heart failure
Penetrance, 9
Peptide-conjugated nucleic acids (PNA), 23, 624
Permeability transition (PT) pore, 96, 104, 581f
Peroxisome proliferator-activated receptor gamma
      coactivator-1 (PGC-1), 92, 92t
   -α, 482–485
      functional domains of, 482, 482f, 483
      metabolic pathway of
         FAO and, 483–485, 484f
         HF and, 443, 445f
Peroxisome proliferator-activated receptors (PPARs), 91–92
   -α regulated genes
      cardiac metabolism and, 481, 481t
      activation
         MCAD and, 481, 481t

   CAD and, 216t
   cofactors in, 91, 92
   FAO and, 480–482
   isoforms of, 91
Persistent truncus arteriosus (PTA), 174
   chromosome 22 deletion and, 174, 192
   phenotypes of IAA and, 173, 192
PFK-1. *See* Phosphofructokinase
PGC-1. *See* Peroxisome proliferator-activated receptor gamma
      coactivator-1
PGIIIa. *See* Platelet glycoprotein IIIa
PHA type II. *See* Pseudohypoaldosteronism type II
Phage display, 15
Pharmaceutical pre-conditioning, 293–298
Pharmacogenetics/pharmacogenomics, 191, 221
   cardiac aging and, 602–603
   cardiac dysrhythmias and, 540–541
      drug-related, 535
   HF and, 459, 460t
   hypertension and, 353
Pharmacogenomics
   HF and, 459, 460t
Pharyngeal arch arteries (PAAs), 145
   development of, 145
   neural crest cells in, 145
   Pax3 and, 145, 149
   semaphorin and, 145, 149
Phenothiazines, 420t
Phenotypes, 9
Phosphatases, kinases and, 84–89
Phosphate groups, 3
Phosphodiesterase 4D (PDE4D), 216t
Phosphofructokinase (PFK-1), 478, 479
Phosphoinositide 3 kinases (PI3K), 85–87
   pathway of, 284, 484
Phospholipase C (PLC), 84
Phosphorylation, 103
PI3K. *See* Phosphoinositide 3 kinases
Pinacidil, 283t
Pitx2, 118f, 129, 148
   L-R identity and, 129, 148
PKA. *See* Protein kinase A
PKB. *See* Protein kinase B
PKC. *See* Protein kinase C
PKG. *See* Protein kinase G
Plasma membrane
   signaling at, 93
      sarcolemmal K atp channel and, 93
Plasma membrane-associated fatty acid binding proteins
      (FABPpm), 474
Plasmid vectors, 11, 13f, 29t
   gene transfection and, 27
Plasmids, 7, 13f
Plasminogen, 233–234
Plasminogen activator inhibitors (PAIs), 216t
   -1, 620t
   fibrinolytic system and, 233, 235–236
   MI and, 261, 263, 264t, 270, 273
Platelet(s), 237–240
   adhesion and aggregation of, 237, 238f

Platelet(s) (*Continued*)
  hemostasis regulation and, 221, 237, 238
  membrane receptors of, 238t
Platelet glycoprotein IIIa (PGIIIa), 261, 263, 264t, 273
Platelet-derived growth factor (PDGF), 55, 56t
  cardiac neovascularization and, 330
  -CC, 319, 330, 332, 333
  ERK1/2 and, 79, 80
PLC. *See* Phospholipase C
Pleiotropic cardiac malformations, 165, 192
PNA. *See* Peptide-conjugated nucleic acids
Point mutations, 9
Polygenic traits, 9
Polymerase chain reaction (PCR) amplification, 11, 17–19, 189
  cycles in, 19f
  myocarditis and, 188
  real-time, 18–19
  reverse transcriptase and, 18
Polymerases
  DNA, 7–8
  RNA, 4
Polymorphic alleles, 9
Polypeptide chain, 3
PON-1. *See* Paraoxonase-1
Positional cloning, 14
Positron emission tomography (PET), 51, 52, 625
Post-conditioning, 298. *See also* Ischemic pre-conditioning
  cardioprotection and, 298–300
  IPC *v.*, 300t
  pre-conditioning *v.*, 299f
Post-genomic cardiology. *See* Cardiology, post-genomic
Potassium channel openers (KCOs), 286, 287
PPARs. *See* Peroxisome proliferator-activated receptors
PPH. *See* Hypertension, primary pulmonary
Pre-conditioning. *See* Ischemic pre-conditioning (IPC); Pharmaceutical pre-conditioning
Pregnenolone, 556f
Premature ventricular contractions (PVCs), 534. *See also* Cardiac dysrhythmias
Primary dysrhythmias. *See* Cardiac dysrhythmias, inherited
Primary nuclear transcript, 6
Primary pulmonary hypertension (PPH). *See* Hypertension, primary pulmonary
Primary structure, DNA, 3
PRIME. *See* Prospective Epidemiological Study of Myocardial Infarction
Primer extension assays, 19
Prions, 4
PRKAG2 gene
  HCM, WPW and mutations of, 369t
Proapoptotic proteins, 85, 98f, 100, 104
  Bad as, 85, 86f, 96, 98f, 100, 104
  Bax as, 97, 98f, 99, 104
  Bid as, 97, 98f, 99, 104
PROCAM study. *See* Prospective Cardiovascular Munster study
Progesterone, 555, 556f, 559, 560, 566, 567, 568, 569
Promoter regions, 4, 5f
Prospective Cardiovascular Munster (PROCAM) study, 229
Prospective Epidemiological Study of Myocardial Infarction (PRIME), 229

Protease
  MI and, 262f
Protease-activated receptors (PARs), 80–81, 80f, 103
Protein(s), 3
  modification of, 7
  MtDNA and, 8, 9f
  transcription and, 4, 5f, 6
  translation and, 4, 5f, 6–7
Protein analysis, 21–23
  1-DE, 22
  2-DE, 22
  yeast 2-hybrid, 22
Protein C (PC), 231
  pathway, 231
Protein kinase A (PKA), 84–85, 103
  AKAPs and, 84, 85
Protein kinase B (PKB), 85–87, 103. *See also* Akt
Protein kinase C (PKC), 87–88, 103
  IPC and, 285–286
  isoforms of, 87, 88
Protein kinase G (PKG), 88–89
  cGMP and, 88, 89, 103
Proteome, 454
Proteomic analysis, 165, 177, 190, 193, 194
  cardiac aging and, 600
  post-genomic cardiology, new markers and, 621–623
Prothrombin (PT), 230–231
Protozoa, myocarditis and, 416t
P-selectin, 620t
Pseudogenes, 17
Pseudohypoaldosteronism type II (PHA type II), 344. *See also* Gordon's syndrome (PHAII)
PT. *See* Prothrombin
PT pore. *See* Permeability transition pore
PTA. *See* Persistent truncus arteriosus
PTEN gene, 321, 322
Pulmonary hypertension. *See* Hypertension, primary pulmonary
PVCs. *See* Premature ventricular contractions
Pyruvate, 281, 283t, 296
Pyruvate dehydrogenase complex (PDC), 479, 480

Q

QTLs. *See* Quantitative trait loci
Quantitative trait loci (QTLs)
  animal models of hypertension and, 346–347, 354
Quercetin, 283t

R

RA. *See* Retinoic acid
Race/ethnicity, MI and, 270–271
RACKs. *See* Receptors for activated C-kinase
R-α-lipoic acid, 605
Raloxifene, 555, 559, 566, 567
Rapamycin, 56t
Raynaud's phenomenon, 239
RCM. *See* Restrictive cardiomyopathy
Reactive oxygen species (ROS), 93, 443, 444, 445
  cell signaling and role of, 95
  generation, 93, 94f, 104
    IPC and, 288

Reactive oxygen species (ROS), 93, 443, 444, 445 (*Continued*)
  negative effects of, 93–95, 104
  signaling, 93, 94f, 104
    mitochondria and, 93
    signaling of aging heart, oxidative stress and, 580–582, 581f
Reading frame, 7
Real-time PCR, 18–19
Rearrangements, 9
Receptor tyrosine kinases (RTKs), 81–82, 81f, 103
Receptors, 77–82, 103
  adrenergic, 77–79, 103
  growth factors in, 79–80
  mAChR, 78f, 79, 583
  muscarinic, 78f, 79, 103
Receptors for activated C-kinase (RACKs), 88
  cardiac aging, adrenergic receptors and, 583
Recessive alleles, 9
Recombinant DNA, 11–14
  cardiac aging, gene expression techniques and, 600–602
Recombinant human vascular endothelial growth factor protein (rhVEGF), 324, 331, 333
Recombinant proteins, CVD and, 16t
Recombinant vectors. *See* Vectors
Regulatable promoters, 14
Remote conditioning, 300–301. *See also* Ischemic pre-conditioning
Renin-angiotensin, 346t
  pathway, 342, 346, 353
Reperfusion injury, cardiac aging and, 595
Replication, DNA, 4, 5f
Reporter genes, 20t, 21
Restenosis
  cardiac neovascularization and, 328
  gene therapy for, 36
Restriction endonuclease digestion, 12f
Restriction endonuclease enzymes, 11, 12f
Restriction fragment length polymorphism (RFLP) analysis, 21, 189
Restriction maps, 11
Restrictive cardiomyopathy (RCM). *See also* Cardiomyopathies
  chromosomal loci and disease genes in, 387–388
  classification of cardiomyopathies and, 364t
  clinical manifestations of, 387
Retinoic acid (RA), 58t, 124, 148
  laterality defect and, 175, 192
  TGA and, 171, 192
Retinoid X receptor (RXR), 91, 92, 92t, 480, 483, 484f, 487
Retroviruses, 4, 29t
  vectors and, 28
Reverse transcriptase, 4
  PCR and, 18
Reverse transcription, 4, 5f
RFLP analysis. *See* Restriction fragment length polymorphism analysis
Rheumatic cardiac disease, 188
Rhod-2, 52, 52t, 53, 53f
Rhodamine 123, 52t
rHuEpo. *See* Erythropoietin; Erythropoietin, factor
rhVEGF. *See* Recombinant human vascular endothelial growth factor protein
Rhythm disorders. *See* Cardiac dysrhythmias
Ribose, 4
Ribosomes, 4
Ribozymes, 32f
  gene therapy and, 35
  trans-cleaving, 32f
Rickettsiae, myocarditis and, 416t
RISC. *See* RNA-induced silencing ribonucleoprotein complexes
RNA interference (RNAi), 30–33
RNA polymerase, 4
RNA transcript, processing of, 4, 5f, 6, 6f
RNAi. *See* RNA interference
RNAi-mediated gene knock-down, 33
RNA-induced silencing ribonucleoprotein complexes (RISC), 31
RNase A mismatch cleavage, 21
Romano-Ward syndrome (RWS), 520, 522t
ROS. *See* Reactive oxygen species
RTKs. *See* Receptor tyrosine kinases
RWS. *See* Romano-Ward syndrome
Ryanodine receptor type 2 (RYR2)
  ARVD/ARVC and, 389–390
  CPVT and, 527, 527t, 528, 543
RYR2. *See* Ryanodine receptor type 2

S

S. *See* Succinate
S phase, 7, 8f
SAGE. *See* Serial analysis of gene expression
Salmonella, 416t
SAME. *See* Syndrome of apparent mineralocorticoid excess
Sanger technique, 11
δ-Sarcoglycan gene (SGCD), 380
Sarcolemmal K_{ATP} channel, 93
Sarcoplasmic reticulum (SR), 53
Sarcoplasmic reticulum Ca++ ATPase (SERCA)
  cardiac aging, thyroid hormone and, 584
  gene, 32, 33
SCAD. *See* Short-chain acyl CoA dehydrogenase
Scavenger receptors (SRs), 217
  ligands for TLRs and, 217, 217t
sCD40L. *See* Soluble CD40 ligand
S-CoA. *See* Succinyl CoA
SD. *See* Sudden death
SDDS. *See* Sudden dysrhythmia death syndrome
Second window of protection (SWOP), 291–293, 292f. *See also* Ischemic pre-conditioning, delayed
Secondary dysrhythmias. *See* Cardiac dysrhythmias, acquired
Secondary structure, DNA, 3
Selectable markers, 12, 13t
Selective estrogen receptor modulators (SERMs), 555, 559, 566, 567, 568, 569
Selective receptor modulators (SRMs), 566, 567
Semaphorin, 145
  PAAs and, 145, 149
Semilunar valves, formation of, 140–142
Seminaphthorhodafluor-1, 52t
Sengers syndrome, 186
SERCA. *See* Sarcoplasmic reticulum Ca++ ATPase
Serial analysis of gene expression (SAGE), 19, 33

Serine protease inhibitors (SPI), 327, 332
Serine/threnonine kinases, 77, 87, 90, 103
SERMs. *See* Selective estrogen receptor modulators
Serpins, 327, 332
Serum response factor (SRF), 119
    lateral plate mesoderm differentiation and, 118f, 148
Sex. *See* Gender
Sex steroid hormones, CVDs and, 555–559
    androgens in, 558–559
        genomic/non-genomic effects of, 558–559
    estrogen and ERs in, 555–558, 557f
        genomic effects of, 557–558
        non-genomic effects of, 556–557
    flowchart depicting synthesis of, 556f
SGCD. *See* δ-Sarcoglycan gene
Shigella, 416t
Short hairpin RNAs (shRNAs), 31
Short QT syndrome, 528–529, 543
    molecular and cellular characterization of, 528–529
    molecular genetic data in, 528, 528t, 529
Short-chain acyl CoA dehydrogenase (SCAD), 474, 475, 475f, 477
    transgenic mice and, 485
shRNAs. *See* Short hairpin RNAs
Shuttle vectors, 13
Side population (SP) cells, 61
SIDS. *See* Sudden infant death syndrome
Signaling pathways. *See* Cardiovascular signaling pathways
Sildenafil, 283t, 298
Silent mutations, 513, 528, 535
SINES. *See* Single interspersed elements
Single interspersed elements (SINES), 17
Single nucleotide polymorphisms (SNPs), 10
    cardiac aging, transcription analysis and, 599
    cardiac dysrhythmias and, 513, 536, 537, 541
        HF progression and, 456, 461
        human genome and, 10
Single ventricle (SV), 172
Single-photon emission computed tomography (SPECT), 51
Single-strand conformation polymorphism (SSCP) analysis, 21, 189
siRNA. *See* Small interfering RNA
Sitosterolemia, 212t, 213, 243
Situs ambiguus, 175, 176, 193
Situs inversus, 175, 176
Situs solitus, 175, 176
Skeletal muscle changes, cardiac aging and, 594–595
Skeletal myoblasts. *See* Myoblasts, skeletal
SLE. *See* Systemic lupus erythematosus
Smad proteins, 132
Small interfering RNA (siRNA), 31f
SMCs. *See* Smooth muscle cells
SMILE. *See* Study of Myocardial Infarctions Leiden
Smoking, MI and, 262f
Smooth muscle cells (SMCs), 315
    angiogenesis pathway and, 315, 316, 317f, 318, 319
SNPs. *See* Single nucleotide polymorphisms
SOD. *See* Superoxide dismutase
Soluble CD40 ligand (sCD40L), 620t
Soluble fiber, 620t
Southern blots, 17
    hybridization and, 18f
SP cells. *See* Side population cells

Specific PKA anchoring proteins. *See* AKAPs
SPECT. *See* Single-photon emission computed tomography
Sphingosine, 283t
SPI. *See* Serine protease inhibitors
SR. *See* Sarcoplasmic reticulum
SRF. *See* Serum response factor
SRMS. *See* Selective receptor modulators
SRs. *See* Scavenger receptors
SSCP analysis. *See* Single-strand conformation polymorphism analysis
Staphylococcus spp., 426t
    antimicrobial therapy for, 430t
Statins, 283t, 297–298
Stem cells, 57. *See also* Adult cardiac stem cells; Bone marrow; Embryonic stem cells
    diagnosis and, 165, 191, 194
    engineering of, 64–65
    identity of, 63–64
Steroid hormones. *See* Sex steroid hormones, CVDs and
Sticky ends, 11
Stings (wasp/bee), 420t
Stop codons, 7
Strands, DNA, 3
Streptococcus spp., 426t
    antimicrobial therapy for, 430t
Stress, MI and, 262f
Stress signals, 100–101, 104
Stroke
    gene variants in MI and, 267, 268t
    hypertension and, 341, 342, 353
Stromelysin-1 (MMP-3), 261, 265, 265t, 270, 273
Study of Myocardial Infarctions Leiden (SMILE), 232
Substance abuse. *See* Alcohol/substance abuse
Succinate (S), 478f
Succinyl CoA (S-CoA), 478f
Sudden death (SD), 415. *See also* Cardiac dysrhythmias, SD and
    myocarditis and, 415, 431
Sudden dysrhythmia death syndrome (SDDS), 534. *See also* Cardiac dysrhythmias
Sudden infant death syndrome (SIDS), 181, 531
    cardiac channelopathies and, 531, 532t
    LQTS and, 531
Sugar-phosphate backbone, 3, 4f
Sugars, 3
    deoxyribose, 3
    ribose, 4
Superoxide dismutase (SOD), 580
Survival/growth signals, 100, 104
Susceptibility genes
    cardiac aging and, 590
    MI and, 262f
SWOP. *See* Second window of protection
Syndrome of apparent mineralocorticoid excess (SAME), 342, 343t
Systemic hypertension. *See* Hypertension
Systemic lupus erythematosus (SLE), 241
Systems biology, viii, 621, 625–626

T

T. *See* Thymine
Taffazin gene (TAZ), 383–384
TAFI. *See* Thrombin-activatable fibrinolysis inhibitor

Reactive oxygen species (ROS), 93, 443, 444, 445 (*Continued*)
  negative effects of, 93–95, 104
  signaling, 93, 94f, 104
    mitochondria and, 93
    signaling of aging heart, oxidative stress and, 580–582, 581f
Reading frame, 7
Real-time PCR, 18–19
Rearrangements, 9
Receptor tyrosine kinases (RTKs), 81–82, 81f, 103
Receptors, 77–82, 103
  adrenergic, 77–79, 103
  growth factors in, 79–80
  mAChR, 78f, 79, 583
  muscarinic, 78f, 79, 103
Receptors for activated C-kinase (RACKs), 88
  cardiac aging, adrenergic receptors and, 583
Recessive alleles, 9
Recombinant DNA, 11–14
  cardiac aging, gene expression techniques and, 600–602
Recombinant human vascular endothelial growth factor protein (rhVEGF), 324, 331, 333
Recombinant proteins, CVD and, 16t
Recombinant vectors. *See* Vectors
Regulatable promoters, 14
Remote conditioning, 300–301. *See also* Ischemic pre-conditioning
Renin-angiotensin, 346t
  pathway, 342, 346, 353
Reperfusion injury, cardiac aging and, 595
Replication, DNA, 4, 5f
Reporter genes, 20t, 21
Restenosis
  cardiac neovascularization and, 328
  gene therapy for, 36
Restriction endonuclease digestion, 12f
Restriction endonuclease enzymes, 11, 12f
Restriction fragment length polymorphism (RFLP) analysis, 21, 189
Restriction maps, 11
Restrictive cardiomyopathy (RCM). *See also* Cardiomyopathies
  chromosomal loci and disease genes in, 387–388
  classification of cardiomyopathies and, 364t
  clinical manifestations of, 387
Retinoic acid (RA), 58t, 124, 148
  laterality defect and, 175, 192
  TGA and, 171, 192
Retinoid X receptor (RXR), 91, 92, 92t, 480, 483, 484f, 487
Retroviruses, 4, 29t
  vectors and, 28
Reverse transcriptase, 4
  PCR and, 18
Reverse transcription, 4, 5f
RFLP analysis. *See* Restriction fragment length polymorphism analysis
Rheumatic cardiac disease, 188
Rhod-2, 52, 52t, 53, 53f
Rhodamine 123, 52t
rHuEpo. *See* Erythropoietin; Erythropoietin, factor
rhVEGF. *See* Recombinant human vascular endothelial growth factor protein

Rhythm disorders. *See* Cardiac dysrhythmias
Ribose, 4
Ribosomes, 4
Ribozymes, 32f
  gene therapy and, 35
  trans-cleaving, 32f
Rickettsiae, myocarditis and, 416t
RISC. *See* RNA-induced silencing ribonucleoprotein complexes
RNA interference (RNAi), 30–33
RNA polymerase, 4
RNA transcript, processing of, 4, 5f, 6, 6f
RNAi. *See* RNA interference
RNAi-mediated gene knock-down, 33
RNA-induced silencing ribonucleoprotein complexes (RISC), 31
RNase A mismatch cleavage, 21
Romano-Ward syndrome (RWS), 520, 522t
ROS. *See* Reactive oxygen species
RTKs. *See* Receptor tyrosine kinases
RWS. *See* Romano-Ward syndrome
Ryanodine receptor type 2 (RYR2)
  ARVD/ARVC and, 389–390
  CPVT and, 527, 527t, 528, 543
RYR2. *See* Ryanodine receptor type 2

S

S. *See* Succinate
S phase, 7, 8f
SAGE. *See* Serial analysis of gene expression
Salmonella, 416t
SAME. *See* Syndrome of apparent mineralocorticoid excess
Sanger technique, 11
δ-Sarcoglycan gene (SGCD), 380
Sarcolemmal K_{ATP} channel, 93
Sarcoplasmic reticulum (SR), 53
Sarcoplasmic reticulum Ca++ ATPase (SERCA)
  cardiac aging, thyroid hormone and, 584
  gene, 32, 33
SCAD. *See* Short-chain acyl CoA dehydrogenase
Scavenger receptors (SRs), 217
  ligands for TLRs and, 217, 217t
sCD40L. *See* Soluble CD40 ligand
S-CoA. *See* Succinyl CoA
SD. *See* Sudden death
SDDS. *See* Sudden dysrhythmia death syndrome
Second window of protection (SWOP), 291–293, 292f. *See also* Ischemic pre-conditioning, delayed
Secondary dysrhythmias. *See* Cardiac dysrhythmias, acquired
Secondary structure, DNA, 3
Selectable markers, 12, 13t
Selective estrogen receptor modulators (SERMs), 555, 559, 566, 567, 568, 569
Selective receptor modulators (SRMs), 566, 567
Semaphorin, 145
  PAAs and, 145, 149
Semilunar valves, formation of, 140–142
Seminaphthorhodafluor-1, 52t
Sengers syndrome, 186
SERCA. *See* Sarcoplasmic reticulum Ca++ ATPase
Serial analysis of gene expression (SAGE), 19, 33

Serine protease inhibitors (SPI), 327, 332
Serine/threnonine kinases, 77, 87, 90, 103
SERMs. *See* Selective estrogen receptor modulators
Serpins, 327, 332
Serum response factor (SRF), 119
    lateral plate mesoderm differentiation and, 118f, 148
Sex. *See* Gender
Sex steroid hormones, CVDs and, 555–559
    androgens in, 558–559
        genomic/non-genomic effects of, 558–559
    estrogen and ERs in, 555–558, 557f
        genomic effects of, 557–558
        non-genomic effects of, 556–557
    flowchart depicting synthesis of, 556f
SGCD. *See* δ-Sarcoglycan gene
Shigella, 416t
Short hairpin RNAs (shRNAs), 31
Short QT syndrome, 528–529, 543
    molecular and cellular characterization of, 528–529
    molecular genetic data in, 528, 528t, 529
Short-chain acyl CoA dehydrogenase (SCAD), 474, 475, 475f, 477
    transgenic mice and, 485
shRNAs. *See* Short hairpin RNAs
Shuttle vectors, 13
Side population (SP) cells, 61
SIDS. *See* Sudden infant death syndrome
Signaling pathways. *See* Cardiovascular signaling pathways
Sildenafil, 283t, 298
Silent mutations, 513, 528, 535
SINES. *See* Single interspersed elements
Single interspersed elements (SINES), 17
Single nucleotide polymorphisms (SNPs), 10
    cardiac aging, transcription analysis and, 599
    cardiac dysrhythmias and, 513, 536, 537, 541
    HF progression and, 456, 461
    human genome and, 10
Single ventricle (SV), 172
Single-photon emission computed tomography (SPECT), 51
Single-strand conformation polymorphism (SSCP) analysis, 21, 189
siRNA. *See* Small interfering RNA
Sitosterolemia, 212t, 213, 243
Situs ambiguus, 175, 176, 193
Situs inversus, 175, 176
Situs solitus, 175, 176
Skeletal muscle changes, cardiac aging and, 594–595
Skeletal myoblasts. *See* Myoblasts, skeletal
SLE. *See* Systemic lupus erythematosus
Smad proteins, 132
Small interfering RNA (siRNA), 31f
SMCs. *See* Smooth muscle cells
SMILE. *See* Study of Myocardial Infarctions Leiden
Smoking, MI and, 262f
Smooth muscle cells (SMCs), 315
    angiogenesis pathway and, 315, 316, 317f, 318, 319
SNPs. *See* Single nucleotide polymorphisms
SOD. *See* Superoxide dismutase
Soluble CD40 ligand (sCD40L), 620t
Soluble fiber, 620t
Southern blots, 17
    hybridization and, 18f
SP cells. *See* Side population cells

Specific PKA anchoring proteins. *See* AKAPs
SPECT. *See* Single-photon emission computed tomography
Sphingosine, 283t
SPI. *See* Serine protease inhibitors
SR. *See* Sarcoplasmic reticulum
SRF. *See* Serum response factor
SRMS. *See* Selective receptor modulators
SRs. *See* Scavenger receptors
SSCP analysis. *See* Single-strand conformation polymorphism analysis
Staphylococcus spp., 426t
    antimicrobial therapy for, 430t
Statins, 283t, 297–298
Stem cells, 57. *See also* Adult cardiac stem cells; Bone marrow; Embryonic stem cells
    diagnosis and, 165, 191, 194
    engineering of, 64–65
    identity of, 63–64
Steroid hormones. *See* Sex steroid hormones, CVDs and
Sticky ends, 11
Stings (wasp/bee), 420t
Stop codons, 7
Strands, DNA, 3
Streptococcus spp., 426t
    antimicrobial therapy for, 430t
Stress, MI and, 262f
Stress signals, 100–101, 104
Stroke
    gene variants in MI and, 267, 268t
    hypertension and, 341, 342, 353
Stromelysin-1 (MMP-3), 261, 265, 265t, 270, 273
Study of Myocardial Infarctions Leiden (SMILE), 232
Substance abuse. *See* Alcohol/substance abuse
Succinate (S), 478f
Succinyl CoA (S-CoA), 478f
Sudden death (SD), 415. *See also* Cardiac dysrhythmias, SD and
    myocarditis and, 415, 431
Sudden dysrhythmia death syndrome (SDDS), 534. *See also* Cardiac dysrhythmias
Sudden infant death syndrome (SIDS), 181, 531
    cardiac channelopathies and, 531, 532t
    LQTS and, 531
Sugar-phosphate backbone, 3, 4f
Sugars, 3
    deoxyribose, 3
    ribose, 4
Superoxide dismutase (SOD), 580
Survival/growth signals, 100, 104
Susceptibility genes
    cardiac aging and, 590
    MI and, 262f
SWOP. *See* Second window of protection
Syndrome of apparent mineralocorticoid excess (SAME), 342, 343t
Systemic hypertension. *See* Hypertension
Systemic lupus erythematosus (SLE), 241
Systems biology, viii, 621, 625–626

T

T. *See* Thymine
Taffazin gene (TAZ), 383–384
TAFI. *See* Thrombin-activatable fibrinolysis inhibitor

TAGs. *See* Triacylglycerols
Tamoxifen, 555, 559, 561, 566, 567
Tangier disease (TD), 212t, 213, 243
Taq 1, 12
Taq polymerase, 18
Targeted gene expression, CVD and, 29, 30t
TAZ. *See* Taffazin gene
T-box transcription factors, 124, 125–126, 148
Tbx5, vii, 118f, 124, 125, 126, 127, 165
   CCS and, 142, 143, 149
   Holt-Oram syndrome and, 165, 167t, 169, 170
TD. *See* Tangier disease
TdP. *See* torsade de pointes
Telomerase, 8
   CVD, telomeres and, 589f
Telomerase reverse transcriptase (TERT), 586, 587f, 601, 602
Telomere repeat-binding factor (TRF2), 586
Telomeres, 8
   cardiac aging, telomere-related proteins and, 586–588, 587f, 588f, 589f
   CVD, telomerase and, 589f
   EPCs and, 586
Temperature gradient capillary electrophoresis (TGCE), 22
TERT. *See* Telomerase reverse transcriptase
Testosterone, 555, 556f, 558, 559, 569
Tetralogy of Fallot (TOF), 167, 168, 168t, 169
   CHARGE syndrome and, 176, 193
Tetramethylrhodamine methyl ester, 52t
TF. *See* Tissue factor
TFAP2B, 165, 173, 192
TFP. *See* Mitochondrial trifunctional protein
TFPI genes, 223
   TF and, 223–224
TGA. *See* Transposition of the great arteries
TGCE. *See* Temperature gradient capillary electrophoresis
TGF. *See* Transforming growth factor
Therapeutic angiogenesis, 331
Therapeutic cloning, 64, 66
Therapeutic genes
   gene therapy and, 35f
Thiazolidines (TZDs), 481
Thrombin, 237
Thrombin-activatable fibrinolysis inhibitor (TAFI), 232
Thrombomodulin, 231–232
Thrombosis, 211, 221
   cardiac aging and, 592
   coagulation and, 221
   coagulation pathways in, 221–233, 222f
   future therapies of, 242
   MI and, 262f
   TF's role in, 223
Thrombospondin (TSP) genes, 261
   -1 (TSP-1), 263, 264, 264t, 273
   -2 (TSP-2), 263, 264, 264t, 273
   -4 (TSP-4), 263, 264, 264t, 273
Thrombospondins (TSPs), 261
Thymine (T), 3
Thyroid hormone
   cardiac aging, SERCA and, 584
Timothy syndrome, 522t, 523
TIMPs. *See* Tissue inhibitor of metalloproteinases

Tinman gene, 118. *See also* Nkx2.5
Tissue factor (TF), 221, 222–223
   coagulation clotting cascade and, 222, 222t
   MI and, 264
   TFPI genes and, 223–224
   thrombosis and, 223
Tissue inhibitor of metalloproteinases (TIMPs)
   4 (TIMP-4), 326t, 330, 333
   ECM, MMPs and, 452, 453
Tissue-type plasminogen activator (tPA), 233, 234–235
Titin gene
   DCM and, 379
   HCM and, 369
TLR. *See* Toll-like receptors
TOF. *See* Tetralogy of Fallot
Toll-like receptors (TLR), 80, 103
   ligands for SRs and, 217, 217t
Toll-R. *See* Toll-like receptors
torsade de pointes (TdP), 524, 535, 543
Toxic agents, myocarditis and, 420, 420t
Toxic myocarditis. *See* Myocarditis, toxic
Toxins, environmental, 145
Toxoplasma gondii, 416t
tPA. *See* Tissue-type plasminogen activator
Traits, 9
Trans-acting factors, 6
Trans-cleaving ribozyme, 32f
Transcription, 4, 5f, 6
   reverse, 4, 5f
Transcription analysis, 599
   cardiac aging, SNPs and, 599
Transcription factors, nuclear receptor, 91–92, 92t
   co-activators and, 92t
   NF-κB, 91, 103
   PPAR, 91–92
      PGC-1 in, 92, 92t
      RXR in, 91, 92, 92t
Transcriptome, 454
Transcriptome analysis. *See* Gene profiling
Transforming growth factor (TGF), 55
   -β, 56t, 58t
      EMT and, 133f
      L-R identity and, 129, 148
   -β3, 391
   VEGF, angiopoietin, Notch pathway and, vii
Transgene expression, 28–29
Transgenic animal expression system, 15t, 20f
Transgenic technologies, 165, 168, 169, 170, 179, 192
Translation, 4, 5f, 6–7
Translation control, 92–93
Transposition of the great arteries (TGA), 168f, 170–171
   HSPG2 gene and, 170, 171, 192
   RA and, 171, 192
Transposon elements, 27
Trastuzumab, 420t
TRF2. *See* Telomere repeat-binding factor
Triacylglycerols (TAGs), 474
Tricarboxylic acid (TCA) cycle, 483, 496
   fatty acid β-oxidation, OXPHOS and, 477–480, 478f
Trichinella spinalis, 416t
Tricyclic antidepressants, 420t

Triglycerides, 243
  elevated, 212t
Triplet Repeat syndromes, 178, 193
tRNA molecules, 6, 7
α-Tropomyosin, HCM and, 373
Troponins
  animal models of genetic HCM and, 372–373
  DCM and, 379
  HCM and, 367–368
  I (cTnI), 372–373, 620, 620t
  mutations of T, 367t
  T (cTnT), 371, 619, 620t
    HCM and mutations in, 371–372
Trypanosome cruzi
  Chagas' disease and, 396, 397, 419
  Myocarditis and, 416t, 419, 420
TSPs. *See* Thrombospondins
Tuberculosis, myocarditis and, 416t
Tubular heart
  cardiac chambers in looped, 123f
  looping/segmentation of, 123–128
22q11 deletion syndrome, 172, 192
  IAA and, 192
2-DE. *See* Two-dimensional electrophoresis
Two-dimensional electrophoresis (2-DE), 22
Tyrosine, IPC, MAPKs and, 286
TZDs. *See* Thiazolidines

U

Unbound free fatty acids (FFAU), 620t
Uniparental disomy (UPD), 172
Univentricular heart, 172, 192
uPA. *See* Urokinase-type plasminogen activator
UPD. *See* Uniparental disomy
Upstream transcription factor-1 (USF-1), 216t
Uracil (U), 4
Urokinase-type plasminogen activator (uPA), 233, 235
USF-1. *See* Upstream transcription factor-1

V

Valves, cardiac. *See* Cardiac valves
Varicella-zoster, 416t
Vascular endothelial growth factor (VEGF), 65
  Notch pathway, angiopoietin, TGF and, vii
  receptor 2 (VEGF-R2), 316, 319t, 321, 328, 332
  transgenic overexpression of, 315, 316, 318, 319t, 321, 328, 332
Vascular remodeling
  angiogenesis and, 325–326
Vasculogenesis. *See also* Cardiac neovascularization
  angiogenesis *v.* arteriogenesis *v.*, 322f
  definition of, 315, 332
VCL. *See* Metavinculin gene
Vectors
  comparison of, 14t
    gene transfer/expression and, 29t
    gene therapy and, 35f
    gene transfection and, 27–28
VEGF. *See* Vascular endothelial growth factor
Ventricle septal defect (VSD), 167–168, 168t
  animal gene mutations and, 169

Ventricular dysrhythmias. *See* Cardiac dysrhythmias, ventricular
Very long-chain acyl CoA dehydrogenase (VLCAD), 477, 482t
  defects in, 488t, 489, 494
  transgenic mice and, 485, 486t
Very low-density lipoprotein (VLDL), 212t, 213, 217f, 220
Vessel engineering, 629–630
Viral infection, 193
  myocarditis and, 188–189, 193
Viral myocarditis. *See* Myocarditis, viral
Viruses. *See also* Enteroviruses, myocarditis and; Human immunodeficiency virus disease
  myocarditis and, 416t, 417
Vitamins, cardiac development and, 146–147
VIVA (Vascular endothelial growth factor in Ischemia for Vascular Angiogenesis), 324
VLCAD. *See* Very long-chain acyl CoA dehydrogenase
VLDL. *See* Very low-density lipoprotein
voltage-gated
  Ca^{++}, 516, 517, 518f
  K^+, 517–520, 519f
    structure and function of, 517–520, 519f
  Na^+, 516, 518f
von Willebrand disease, 239
von Willebrand factor (vWF), 263, 268t, 620t
vWF. *See* von Willebrand factor

W

Western immunoblot analysis, 22
WHI studies. *See* Women's Health Initiative studies
Whipple's disease bacillus, 425t
William's syndrome (WS), 178, 193
Wilson's disease, 186
WISE study. *See* Women's Ischemia Syndrome Evaluation Study
WNK gene, 343t, 344, 353
Wnt 11, vii, 57, 58t
Wnt pathway, mesoderm specification and, 122, 136
Wolff-Parkinson-White (WPW) syndrome, 181, 531
  inherited and sporadic, 531–532
  molecular and genetic characterization of, 531–532
  PRKAG2 mutations in HCM and, 369t
Women's Health Initiative (WHI) studies, 565, 566, 568
Women's Ischemia Syndrome Evaluation (WISE) Study, 567
WPW syndrome. *See* Wolff-Parkinson-White syndrome
WS. *See* William's syndrome

X

Xanthine oxidase (XO), 94f, 95, 104, 581f
X-linked inheritance pattern, 9, 10f
XO. *See* Xanthine oxidase

Y

YAC vectors. *See* Yeast artificial chromosome vectors
Yeast 2-hybrid, 22
Yeast artificial chromosome (YAC) vectors, 14t
Yeast expression system, 15t

Z

Z disc proteins, 369
ZIC3 gene, 176, 193

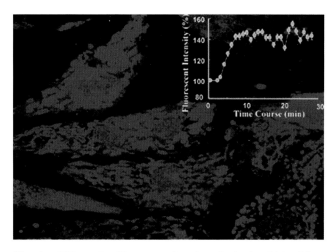

Chapter 4 Figure 1 Fluorescent imaging of mitochondrial calcium in cultured neonatal rat cardiomyocytes. Cells were loaded with Rhod-2, a fluorescent indicator of mitochondrial calcium level at 4° for 60 min followed by incubation at 37° for 30 min. The graph inside shows the time-lapse increases in mitochondrial-localized Rhod-2 fluorescence in cardiomyocytes subjected to H_2O_2 treatment. (Reproduced with permission from Springer Science and Business Media, New York.)

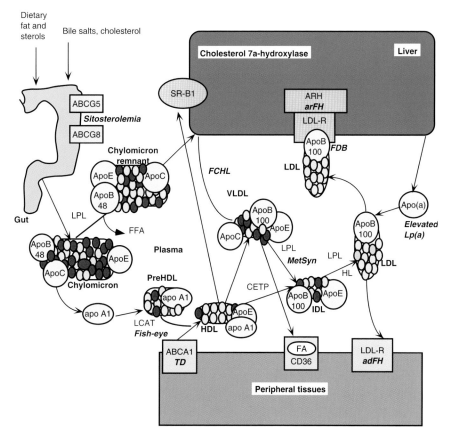

Chapter 8 Figure 1 Overview of LDL metabolism in humans. Dietary cholesterol and triglycerides are packaged with apolipoproteins in enterocytes of the small intestine, secreted into the lymphatic system as chylomicrons. As chylomicrons circulate, the core triglycerides are hydrolyzed by lipoprotein lipase (LPL), resulting in the formation of chylomicron remnants that are rapidly removed by the liver. In liver, dietary cholesterol has several possible fates: it can be esterified and stored, packaged into VLDL particles and secreted into plasma or into bile; or converted into bile acids and secreted into bile. VLDL particles secreted into the plasma undergo lipolysis to form VLDL remnants, including IDL. VLDL remnants are also removed by the liver by the LDLR, and the remainder mature into LDL, the major cholesterol transport particle in the blood. ABCG5 and ABCG8 are located predominantly in the enterocytes of the duodenum and jejunum, the sites of uptake of dietary sterols. Mutations in either transporter cause an increase in delivery of dietary sterols to the liver and a decrease in secretion into the bile.

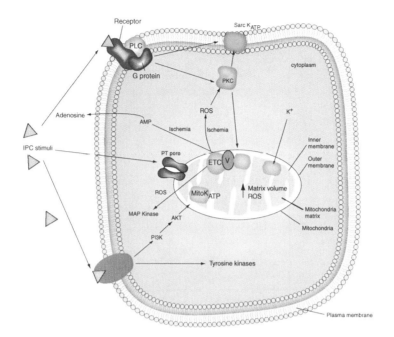

Chapter 10 Figure 1 Schematic representation depicting cellular events occurring in early ischemic preconditioning. Agonists of G-protein–coupled receptors, including adenosine and bradykinin, can stimulate parallel signaling pathways by stimulating either phosphoinositide 3-kinase (PI3K) or tyrosine kinase-dependent cascades, or a PKC-dependent cascade involving activation of phospholipase C (PLC) can confer cardioprotection. Multiple mitochondrial events integral to cardioprotection include the increased modulation of mitoK$_{ATP}$ channels (and to a lesser extent, sarcolemmal $_{KATP}$ channels), mitochondrial reactive oxygen species (ROS) generation, modulation of mitochondrial calcium intake, permeability transition (PT) pore opening, and electron transport chain (ETC) activity. MitoK$_{ATP}$ channel opening also provides positive feedback by altering components such as ROS, mitogen-activated protein kinase (MAPK), or protein kinase C (PKC) activation. PI3K activation results in further signaling by downstream targets such as Akt, eNOS, GSK-3β, PKC, mTOR, and p70S6K. PI3K activation also plays a role in the activation of PKC and perhaps other signaling events (such as the activation of mitoK$_{ATP}$).

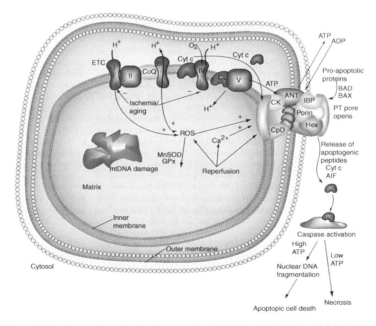

Chapter 10 Figure 2 The mitochondrial pathway of apoptosis. An array of cellular signals, including ROS, trigger the apoptotic pathway regulated by proapoptotic proteins (e.g., Bax, Bid, and Bad) binding to the outer mitochondrial membrane and leading to mitochondrial outer-membrane permeabilization (MOMP) and PT pore opening. Elevated levels of mitochondrial Ca^{++}, as well as ETC-generated ROS, also promote PT pore opening. This is followed by the release of cytochrome c (Cyt c), Smac, endonuclease G (Endo G), and apoptosis-inducing factor (AIF) from the mitochondrial intermembrane space to the cytosol and apoptosome formation leading to caspase and endonuclease activation, DNA fragmentation, and cell death. Association of Bax with mitochondria is prevented by antiapoptotic proteins (e.g., Bcl-2). Also shown are proteins making up the PT pore, including hexokinase (Hex), adenine nucleotide translocator (ANT), creatine kinase (CK), cyclophilin D (CyP-D), and porin (VDAC), as well as the inner membrane phospholipid cardiolipin (CL).

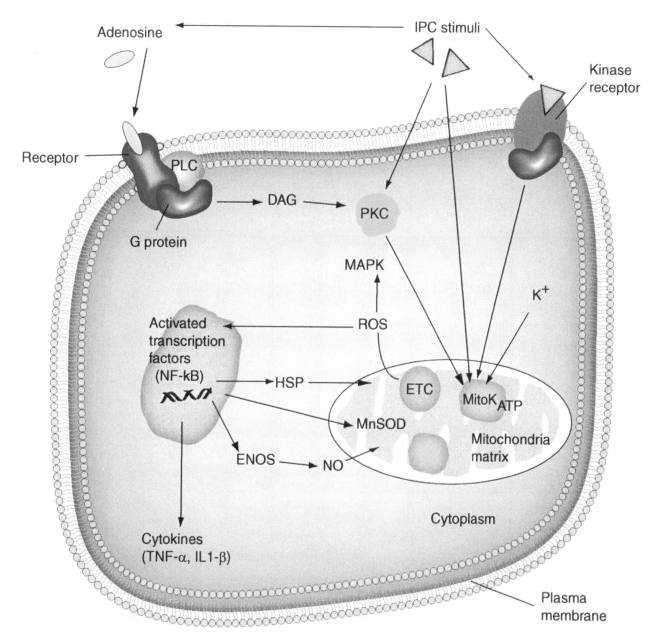

Chapter 10 Figure 3 Schematic representation depicting cellular events occurring in delayed IPC. In delayed IPC, preconditioning stimuli impact sarcolemmal receptors by use of a variety of agonists and triggering a series of parallel signal transduction pathways involving G-coupled proteins and protein kinases, including PLC, PKC, and production of diacylglycerol (DAG). In addition, modulation of mitoK$_{ATP}$ channel opening and extensive ROS involvement occur as noted with the acute preconditioning model. Also involved is a protein kinase–mediated activation of specific nuclear transcription factors (e.g., NF-κB), leading to increased gene expression and synthesis of protective proteins to enable sustained cardioprotection. These include mitochondrial Mn-superoxide dismutase (MnSOD), eNOS, and several stress-activated, heat shock proteins (HSPs), ETC proteins, and the activation of cytokines (including TNF-α and IL-1β). Both eNOS and iNOS produce increased NO that provides cardioprotective stimulation of mitochondrial metabolism, These proteins provide a variety of levels of cytoprotection by acting at a variety of cellular sites including mitochondria.

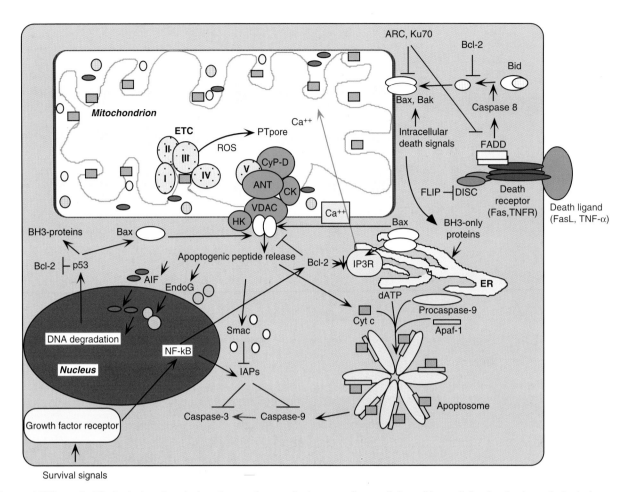

Chapter 15 Figure 5 The intrinsic and extrinsic pathways of apoptosis. An array of extracellular and intracellular signals trigger the intrinsic apoptotic pathway regulated by proapoptotic proteins (e.g., Bax, Bid, and Bak) binding to outer mitochondrial membrane leading to mitochondrial outer-membrane permeabilization and PT pore opening. Elevated levels of mitochondrial Ca^{++}, as well as ETC-generated ROS, also promote PT pore opening. This is followed by the release of cytochrome c (Cytc), Smac, endonuclease G (Endo G), and apoptosis-inducing factor (AIF) from the mitochondrial intermembrane space to the cytosol. This leads to apoptosome formation (with Cytc) leading to caspase 9 activation, DNA fragmentation stimulated by the nuclear translocation of AIF and EndoG, whereas inhibition of IAP (by Smac) further promotes the activation of caspases 9 and 3. Bax and Bid mediate mitochondrial membrane permeabilization, and antiapoptogenic proteins (e.g., Bcl-2) prevent apoptogen release. Also shown are major proteins making up the PT pore, including hexokinase (Hx), adenine nucleotide translocator (ANT), creatine kinase (CK), cyclophilin D (CyP-D), and porin (VDAC). The extrinsic pathway is initiated by ligand binding to death receptors, leading to recruitment of FADD and DISC, which stimulate the activation of caspase 8, resulting in caspase 3 activation and Bid cleavage (a C-terminal fragment of Bid targets mitochondria). FLIP, ARC, and Ku70 can stem this pathway's progression at specific points. Intracellular stimuli trigger ER release of Ca^{++} through both Bax and BH3-protein interactions. Also shown is the survival pathway triggered by survival stimuli, mediated by growth factor receptors, transcription factor activation (e.g., NF-κB), and enhanced expression of IAPs and Bcl-2.